REMOTE SENSING HANDBOOK
VOLUME II

LAND RESOURCES MONITORING, MODELING, AND MAPPING WITH REMOTE SENSING

Remote Sensing Handbook

Remotely Sensed Data Characterization, Classification, and Accuracies

Land Resources Monitoring, Modeling, and Mapping with Remote Sensing

Remote Sensing of Water Resources, Disasters, and Urban Studies

REMOTE SENSING HANDBOOK
VOLUME II

LAND RESOURCES MONITORING, MODELING, AND MAPPING WITH REMOTE SENSING

Edited by

Prasad S. Thenkabail, PhD
United States Geological Survey (USGS)

CRC Press
Taylor & Francis Group
Boca Raton London New York

CRC Press is an imprint of the
Taylor & Francis Group, an **informa** business

CRC Press
Taylor & Francis Group
6000 Broken Sound Parkway NW, Suite 300
Boca Raton, FL 33487-2742

First issued in paperback 2019

© 2016 by Taylor & Francis Group, LLC
CRC Press is an imprint of Taylor & Francis Group, an Informa business

No claim to original U.S. Government works

ISBN-13: 978-1-4822-1795-7 (hbk)
ISBN-13: 978-0-367-86897-0 (pbk)

Library of Congress Cataloging-in-Publication Data

Land resources monitoring, modeling, and mapping with remote sensing / editor, Prasad S. Thenkabail.
   pages cm
  Includes bibliographical references and index.
  ISBN 978-1-4822-1795-7 (alk. paper)
  1. Remote sensing. 2. Natural resources--Remote sensing. I. Thenkabail, Prasad Srinivasa, 1958- editor of compilation.

G70.4.L33 2015
621.36'78--dc23

2015001323

Visit the Taylor & Francis Web site at
http://www.taylorandfrancis.com

and the CRC Press Web site at
http://www.crcpress.com

I dedicate this work to my revered parents whose sacrifices gave me an education, as well as to all those teachers from whom I learned remote sensing over the years.

Contents

SECTION I   Vegetation and Biomass

SECTION II   Agricultural Croplands

SECTION III Rangelands

SECTION IV Phenology and Food Security

SECTION V Forests

SECTION VI Biodiversity

SECTION VII Ecology

SECTION VIII Land Use/Land Cover

SECTION IX Carbon

SECTION X Soils

SECTION XI Summary

Foreword: Satellite Remote Sensing Beyond 2015

Satellite remote sensing has progressed tremendously since Landsat 1 was launched on June 23, 1972. Since the 1970s, satellite remote sensing and associated airborne and in situ measurements have resulted in vital and indispensible observations for understanding our planet through time. These observations have also led to dramatic improvements in numerical simulation models of the coupled atmosphere–land–ocean systems at increasing accuracies and predictive capabilities. The same observations document the Earth's climate and are driving the consensus that *Homo sapiens* is changing our climate through greenhouse gas (GHG) emissions.

These accomplishments are the combined work of many scientists from many countries and a dedicated cadre of engineers who build the instruments and satellites that collect Earth observation (EO) data from satellites, all working toward the goal of improving our understanding of the Earth. This edition of the Remote Sensing Handbook (*Remotely Sensed Data Characterization, Classification, and Accuracies*; *Land Resources Monitoring, Modeling, and Mapping with Remote Sensing*; and *Remote Sensing of Water Resources, Disasters, and Urban Studies*) is a compendium of information for many research areas of our planet that have contributed to our substantial progress since the 1970s. The remote sensing community is now using multiple sources of satellite and in situ data to advance our studies, whatever they may be. In the following paragraphs, I will illustrate how valuable and pivotal satellite remote sensing has been in climate system study over the last five decades. The chapters in the handbook provide many other specific studies on land, water, and other applications using EO data of the last five decades.

The Landsat system of Earth-observing satellites has led the way in pioneering sustained observations of our planet. From 1972 to the present, at least one and sometimes two Landsat satellites have been in operation (Irons et al. 2012). Starting with the launch of the first NOAA–NASA Polar Orbiting Environmental Satellites NOAA-6 in 1978, improved imaging of land, clouds, and oceans and atmospheric soundings of temperature was accomplished. The NOAA system of polar-orbiting meteorological satellites has continued uninterrupted since that time, providing vital observations for numerical weather prediction. These same satellites are also responsible for the remarkable records of sea surface temperature and land vegetation index from the advanced very-high-resolution radiometers (AVHRRs) that now span more than 33 years, although no one anticipated these valuable climate records from this instrument before the launch of NOAA-7 in 1981 (Cracknell 1997).

The success of data from the AVHRR led to the design of the moderate-resolution imaging spectroradiometer (MODIS) instruments on NASA's Earth-Observing System (EOS) of satellite platforms that improved substantially upon the AVHRR. The first of the EOS platforms, Terra, was launched in 2000; and the second of these platforms, Aqua, was launched in 2002. Both of these platforms are nearing their operational life, and many of the climate data records from MODIS will be continued with the visible infrared imaging radiometer suite (VIIRS) instrument on the polar orbiting meteorological satellites of NOAA. The first of these missions, the NPOES Preparation Project (NPP), was launched in 2012 with the first VIIRS instrument that is operating currently among several other instruments on this satellite. Continuity of observations is crucial for advancing our understanding of the Earth's climate system. Many scientists feel that the crucial climate observations provided by remote sensing satellites are among the most important satellite measurements because they contribute to documenting the current state of our climate and how it is evolving. These key satellite observations of our climate are second in importance only to the polar orbiting and geostationary satellites needed for numerical weather prediction.

The current state of the art for remote sensing is to combine different satellite observations in a complementary fashion for what is being studied. Let us review climate change as an excellent example of using disparate observations from multiple satellite and in situ sources to observe climate change, verify that it is occurring, and understand the various component processes:

1. *Warming of the planet, quantified by radar altimetry from space*: Remotely sensed climate observations provide the data to understand our planet and what forces our climate. The primary climate observation comes from radar altimetry that started in late 1992 with Topex/Poseidon and has been continued by Jason-1 and Jason-2 to provide an uninterrupted record of global sea level. Changes in global sea level provide unequivocal evidence if our planet is warming,

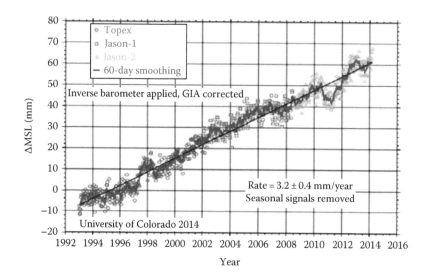

FIGURE F.1 Warming of the planet quantified by radar altimetry from space. Sea level determined from three radar altimeters from late 1992 to the present shows global sea level increases of ~3 mm/year. Sea level is the unequivocal indicator of the Earth's climate—when sea level rises, the planet is warming; when sea level falls, the planet is cooling. (From Gregory, J.M. et al., *J. Climate*, 26(13), 4476, 2013.)

cooling, or staying at the same temperature. Radar altimetry from 1992 to date has shown global sea level increases of ~3 mm/year, and hence, our planet is warming (Figure F.1). Sea level rise has two components, thermal expansion and ice melting in the ice sheets of Greenland and Antarctica, and to a much lesser extent, in glaciers.

2. The Sun is not to blame for global warming, based on solar irradiance data from satellites. Next, we consider two very different satellite observations and one in situ observing system that enable us to understand the causes of sea level variations: total solar irradiance, variations in the Earth's gravity field, and the Argo floats that record ocean temperature and salinity with depth, respectively.

Observations of total solar irradiance have been made from satellites since 1979 and show total solar irradiance has varied only ±1 part in 500 over the past 35 years, establishing that our Sun is not to blame for global warming

(Figure F.2). Thus, we must look to other remotely sensed climate observations to explain and confirm sea level rise.

3. Sea level rise of 60% is explained by a mass balance of melting of ice measured by GRACE satellites. Since 2002, we have measured gravity anomalies from the Gravity Recovery and Climate Experiment Satellite (GRACE) dual satellite system. GRACE data quantify ice mass changes from the Antarctic and Greenland ice sheets (AIS and GIS) and concentrations of glaciers, such as in the Gulf of Alaska (GOA) (Luthcke et al. 2013). GRACE data are truly remarkable—their retrieval of variations in the Earth's gravity field is quantitatively and directly linked to mass variations. With GRACE data, we are able to determine for the first time the mass balance with time of the AIS and GIS and concentrations of glaciers on land. GRACE data show sea level rise of 60% explained by ice loss from

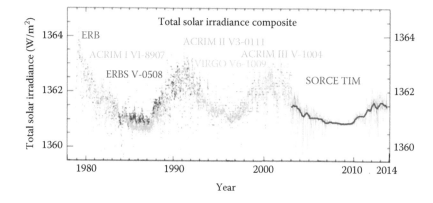

FIGURE F.2 The Sun is not to blame for global warming, based on solar irradiance data from satellites. Total solar irradiance reconstructed from multiple instruments dates back to 1979. The luminosity of our Sun varies only 0.1% over the course of the 11-year solar cycle. (From Froehlich, C., *Space Sci. Rev.*, 176(1–4), 237, 2013.)

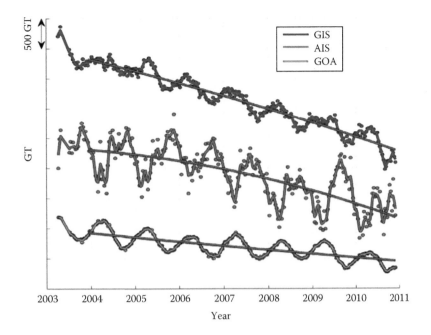

FIGURE F.3 Sea level rise of 60% explained by mass balance of melting of ice measured by GRACE satellites. Ice mass variations from 2003 to 2010 for the Antarctic ice sheets (AIS), Greenland ice sheets (GIS) and the Gulf of Alaska (GOA) glaciers using GRACE gravity data. (From Luthcke, S.B. et al., *J. Glaciol.*, 59(216), 613, 2013.)

land (Figure F.3). GRACE data have many other uses, such as indicating changes in groundwater storage, and readers are directed to the GRACE project's website if interested (http://www.csr.utexas.edu/grace/).

4. Sea level rise of 40% is explained by thermal expansion in the planet's oceans measured by in situ ~3700 drifting floats. The other contributor to sea level rise is thermal expansion in the planet's oceans. This necessitates using diving and drifting floats in the Argo network to record temperature with depth (Roemmich et al. 2009 and Figure F.4). Argo floats are deployed from ships; they then submerge and descend slowly to 1000 m depth, recording temperature, pressure, and salinity as they descend. At 1000 m depth, they drift for 10 days continuing their measurements of temperature and salinity. After 10 days, they slowly descend to 3000 m and then ascend to the surface, all the time recording their measurements. At the surface, each float transmits all the data collected on the most recent excursion to a geostationary satellite and then descends again to repeat this process.

Argo temperature data show that 40% of sea level rise results from the warming and thermal expansion of our oceans. Combining radar altimeter data, GRACE data, and Argo data provides a confirmation of sea level rise and shows what is

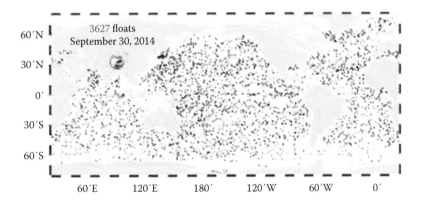

FIGURE F.4 Sea level rise of 40% explained by thermal expansion in the planet's oceans measured by in situ ~3700 drifting floats. This is the latest picture of the 3627 Argo floats that were in operation on September 30, 2014. These floats provide the data needed to document thermal expansion of the oceans. (From http://www.argo.ucsd.edu/.)

responsible for it and in what proportions. With total solar irradiance being near constant, what is driving global warming can be determined. The analysis of surface in situ air temperature coupled with lower tropospheric air temperature and stratospheric temperature data from remote sensing infrared and microwave sounders shows that the surface and near surface are warming while the stratosphere is cooling. This is an unequivocal confirmation that greenhouse gases are warming the planet.

Many scientists are actively working to study the Earth's carbon cycle, and there are several chapters in the handbook that deal with the components of this undertaking. Much like simultaneous observations of sea level, total solar irradiance, the gravity field, ocean temperature, surface temperature, and atmospheric temperatures were required to determine if the Earth is warming and what is responsible; the carbon cycle (Figure F.5) will require several complementary satellite and in situ observations (Cias et al. 2014).

Carbon cycles through reservoirs on the Earth's surface in plants and soils exist in the atmosphere as gases, such as carbon dioxide (CO_2) and methane (CH_4), and in ocean water in phytoplankton and marine sediments. CO_2 and CH_4 are released into the atmosphere by the combustion of fossil fuels, land cover changes on the Earth's surface, respiration of green plants, and decomposition of carbon in dead vegetation and in soils, including carbon in permafrost. The atmospheric concentrations of CO_2 and CH_4 control atmospheric and oceanic temperatures through their absorption of outgoing long-wave radiation and thus also indirectly control sea level via the regulation of planetary ice volumes.

Satellite-borne sensors provide simultaneous global carbon cycle observations needed for quantifying carbon cycle processes, that is, to measure atmospheric CO_2 concentrations and emission sources, to measure land and ocean photosynthesis, to measure the reservoir of carbon in plants on land and its change, to measure the extent of biomass burning of vegetation on land, and to measure soil respiration and decomposition, including decomposing carbon in permafrost. In addition to the required satellite observations, in situ observations are needed to confirm satellite-measured CO_2 concentrations and determine soil and vegetation carbon quantities. Understanding the carbon cycle requires a full court press of satellite and in situ observations because all of these observations must be made at the same time. Many of these measurements have been made over the past 30–40 years, but new measurements are needed to quantify carbon storage in vegetation, atmospheric measurements are needed to quantify CH_4 and CO_2 sources and sinks, better measurements are needed to quantify land respiration, and more explicit numerical carbon models need to be developed.

Similar work needs to be performed for the role of clouds and aerosols in climate because these are fundamental to understanding our radiation budget. We also need to improve our understanding of the global hydrological cycle.

The remote sensing community has made tremendous progress over the last five decades as discussed in this edition of the handbook. Chapters on aerosols in climate, because these are fundamental, provide comprehensive understanding of land and water studies through detailed methods, approaches, algorithms, synthesis, and key references. Every type of remote

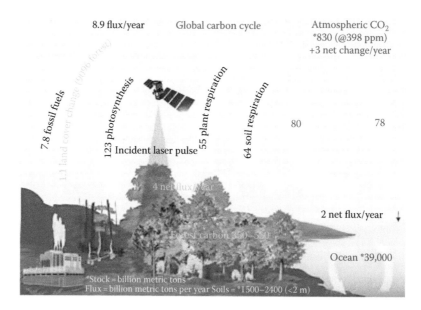

FIGURE F.5 Global carbon cycle measurements from a multitude of satellite sensors. A representation of the global carbon cycle showing our best estimates of carbon fluxes and carbon reservoirs. A series of satellite observations are needed simultaneously to understand the carbon cycle and its role in the Earth's climate system. (From Cias, P. et al., *Biogeosciences*, 11(13), 3547, 2014.)

sensing data obtained from systems such as optical, radar, light detection and ranging (LiDAR), hyperspectral, and hyperspatial is presented and discussed in different chapters. *Remotely Sensed Data Characterization, Classification, and Accuracies* sets the stage with chapters in this book addressing remote sensing data characteristics, within and between sensor calibrations, classification methods, and accuracies taking a wide array of remote sensing data from a wide array of platforms over the last five decades. *Remotely Sensed Data Characterization, Classification, and Accuracies* also brings in technologies closely linked with remote sensing such as global positioning system (GPS), global navigation satellite system (GNSS), crowdsourcing, cloud computing, and remote sensing law. In all, the 82 chapters in the 3 volumes of the handbook are written by leading and well-accomplished remote sensing scientists of the world and competently edited by Dr. Prasad S. Thenkabail, Research Geographer-15, at the United States Geological Survey (USGS).

We can look forward in the next 10–20 years to improving our quantitative understanding of the global carbon cycle, understanding the interaction of clouds and aerosols in our radiation budget, and understanding the global hydrological cycle. There is much work to do. Existing key climate observations must be continued and new satellite observations will be needed (e.g., the recently launched NASA's Orbiting Carbon Observatory-2 for atmospheric CO_2 measurements), and we have many well-trained scientists to undertake this work and continue the legacy of the past five decades.

References

Ciais, P. et al. 2014. Current systematic carbon-cycle observations and the need for implementing a policy-relevant carbon observing system. *Biogeosciences* 11(13): 3547–3602.

Cracknell, A. P. 1997. *The Advanced Very High Resolution Radiometer (AVHRR)*. Taylor & Francis, U.K., 534pp.

Froehlich, C. 2013. Total solar irradiance: What have we learned from the last three cycles and the recent minimum? *Space Science Reviews* 176(1–4): 237–252.

Gregory, J. M. et al. 2013. Twentieth-century global-mean sea level rise: Is the whole greater than the sum of the parts? *Journal of Climate* 26(13): 4476–4499.

Irons, J. R., Dwyer, J. L., and Barsi, J. A. 2012. The next Landsat satellite: The Landsat data continuity mission. *Remote Sensing of Environment* 122: 11–21. doi:10.1016/j.rse.2011.08.026.

Luthcke, S. B., Sabaka, T. J., Loomis, B. D. Arendt, A. A., McCarthy, J. J., and Camp, J. 2013. Antarctica, Greenland, and Gulf of Alaska land-ice evolution from an iterated GRACE global mascon solution. *Journal of Glaciology* 59(216): 613–631.

Roemmich, D. and the Argo Steering Team, 2009. Argo—The challenge of continuing 10 years of progress. *Oceanography* 22(3): 46–55.

Compton J. Tucker
Earth Science Division
Goddard Space Flight Center
National Aeronautics and Space Administration
Greenbelt, Maryland

Preface: Remote Sensing Advances of the Last 50 Years and a Vision for the Future

The overarching goal of the Remote Sensing Handbook (*Remotely Sensed Data Characterization, Classification, and Accuracies*; *Land Resources Monitoring, Modeling,* and *Mapping with Remote Sensing*; and *Remote Sensing of Water Resources, Disasters, and Urban Studies*), with 82 chapters and about 2500 pages, was to capture and provide the most comprehensive state of the art of remote sensing science and technology development and advancement in the last 50 years, by clearly demonstrating the (1) scientific advances, (2) methodological advances, and (3) societal benefits achieved during this period, as well as to provide a vision of what is to come in the years ahead. The three books are, to date and to the best of my knowledge, the most comprehensive documentation of the scientific and methodological advances that have taken place in understanding remote sensing data, methods, and a wide array of land and water applications. Written by 300+ leading global experts in the area, each chapter (1) focuses on a specific topic (e.g., data, methods, and applications), (2) reviews the existing state-of-the-art knowledge, (3) highlights the advances made, and (4) provides guidance for areas requiring future development. Chapters in the books cover a wide array of subject matter of remote sensing applications. The Remote Sensing Handbook is planned as a reference material for remote sensing scientists, land and water resource practitioners, natural and environmental practitioners, professors, students, and decision makers. The special features of the Remote Sensing Handbook include the following:

1. Participation of an outstanding group of remote sensing experts, an unparalleled team of writers for such a book project
2. Exhaustive coverage of a wide array of remote sensing science: data, methods, and applications
3. Each chapter being led by a luminary and most chapters written by teams who further enriched the chapters
4. Broadening the scope of the book to make it ideal for expert practitioners as well as students
5. Global team of writers, global geographic coverage of study areas, and a wide array of satellites and sensors
6. Plenty of color illustrations

Chapters in the books cover the following aspects of remote sensing:

State of the art
Methods and techniques
Wide array of land and water applications
Scientific achievements and advancements over the last 50 years
Societal benefits
Knowledge gaps
Future possibilities in the twenty-first century

Great advances have taken place over the last 50 years using remote sensing in the study of the planet Earth, especially using data gathered from a multitude of Earth observation (EO) satellites launched by various governments as well as private entities. A large part of the initial remote sensing technology was developed and tested during the two world wars. In the 1950s, remote sensing slowly began its foray into civilian applications. During the years of the Cold War, remote sensing applications, both civilian and military, increased swiftly. But it was also an age when remote sensing was the domain of a very few top experts and major national institutes, having multiple skills in engineering, science, and computer technology. From the 1960s onward, there have been many governmental agencies that have initiated civilian remote sensing. The National Aeronautics and Space Administration (NASA) and the United States Geological Survey (USGS) have been in the forefront of many of these efforts. Others who have provided leadership in civilian remote sensing include, but are not limited to, the European Space Agency (ESA) of the European Union, the Indian Space Research Organization (ISRO), the Centre National d'Études Spatiales (CNES) of France, the Canadian Space Agency (CSA), the Japan Aerospace Exploration Agency (JAXA), the German Aerospace Center (DLR), the China National Space Administration (CNSA), the United Kingdom Space Agency (UKSA), and the Instituto Nacional de Pesquisas Espaciais (INPE) of Brazil. Many private entities have launched and operated satellites. These government and private agencies and enterprises launched and operated a wide array of satellites and sensors that captured the data of the planet Earth in various regions of the electromagnetic spectrum and in various spatial, radiometric, and temporal resolutions, routinely and repeatedly. However, the real thrust for remote sensing advancement came during the last decade of the twentieth century and the beginning of the twenty-first century. These initiatives included a launch of a series of new-generation EO satellites to gather data more frequently and routinely, release of pathfinder datasets, web enabling the data for free by many agencies (e.g., USGS release of the entire Landsat archives as well as real-time acquisitions of the world for free dissemination by web-enabling), and providing processed data ready to users (e.g., surface reflectance products of moderate-resolution imaging spectroradiometer [MODIS]). Other efforts like Google Earth made remote sensing more popular and brought in a new platform for easy visualization and navigation of remote sensing data. Advances in computer hardware and software made it possible to handle Big Data. Crowdsourcing, web access, cloud computing, and mobile platforms added a new dimension to how remote sensing data are used. Integration with global positioning systems (GPS) and global navigation satellite systems (GNSS) and inclusion of digital secondary data (e.g., digital elevation, precipitation, temperature) in analysis have made remote sensing much more powerful. Collectively, these initiatives provided a new vision in making remote sensing data more popular, widely understood, and increasingly used for diverse applications, hitherto considered difficult. The free availability of archival data when combined with more recent acquisitions has also enabled quantitative studies of change over space and time. The Remote Sensing Handbook is targeted to capture these vast advances in data, methods, and applications, so a remote sensing student, scientist, or a professional practitioner will have the most comprehensive, all-encompassing reference material in one place.

Modern-day remote sensing technology, science, and applications are growing exponentially. This growth is a result of a combination of factors that include (1) advances and innovations in data capture, access, and delivery (e.g., web enabling, cloud computing, crowdsourcing); (2) an increasing number of satellites and sensors gathering data of the planet, repeatedly and routinely, in various portions of the electromagnetic spectrum as well as in an array of spatial, radiometric, and temporal resolutions; (3) efforts at integrating data from multiple satellites and sensors (e.g., sentinels with Landsat); (4) advances in data normalization, standardization, and harmonization (e.g., delivery of data in surface reflectance, intersensor calibration); (5) methods and techniques for handling very large data volumes (e.g., global mosaics); (6) quantum leap in computer hardware and software capabilities (e.g., ability to process several terabytes of data); (7) innovation in methods, approaches, and techniques leading to sophisticated algorithms (e.g., spectral matching techniques, and automated cropland classification algorithms); and (8) development of new spectral indices to quantify and study specific land and water parameters (e.g., hyperspectral vegetation indices or HVIs). As a result of these all-around developments, remote sensing science is today very mature and is widely used in virtually every discipline of the earth sciences for quantifying, mapping, modeling, and monitoring our planet Earth. Such rapid advances are captured in a number of remote sensing and earth science journals. However, students, scientists, and practitioners of remote sensing science and applications have significant difficulty gathering a complete understanding of the various developments and advances that have taken place as a result of their vastness spread across the last 50 years. Therefore, the chapters in the Remote Sensing Handbook are designed to give a whole picture of scientific and technological advances of the last 50 years.

Today, the science, art, and technology of remote sensing are truly ubiquitous and increasingly part of everyone's everyday life, often without the user knowing it. Whether looking at your own home or farm (e.g., see the following figure), helping you navigate when you drive, visualizing a phenomenon occurring in a distant part of the world (e.g., see the following figure), monitoring events such as droughts and floods, reporting weather, detecting and monitoring troop movements or nuclear sites, studying deforestation, assessing biomass carbon, addressing disasters such as earthquakes or tsunamis, and a host of other applications (e.g., precision farming, crop productivity, water productivity, deforestation, desertification, water resources management), remote sensing plays a pivotal role. Already, many new innovations are taking place. Companies such as the Planet Labs and Skybox are planning to capture very-high-spatial-resolution imagery (typically, sub-meter to 5 meters), even videos from space using a large number of microsatellite constellations. There are others planning to launch a constellation of hyperspectral or other sensors. Just as the smartphone and social media connected the world, remote sensing is making the world our backyard. No place goes unobserved and no event gets reported without a satellite or other kinds of remote sensing images or their derivatives. This is how true liberation for any technology and science occurs.

Google Earth can be used to seamlessly navigate and precisely locate any place on Earth, often with very-high-spatial-resolution data (VHRI; submeters to 5 m) from satellites such as IKONOS, QuickBird, and GeoEye (Note: the image below is from one of the VHRI). Here, the editor-in-chief (EiC) of this handbook located his village home (Thenkabail) and surroundings that have land covers such as secondary rainforests, lowland paddy farms, areca nut plantations, coconut plantations, minor roads, walking routes, open grazing lands, and minor streams (typically, first and second order) (note: land cover detailed is based on the ground knowledge of the EiC). The first primary school attended by him is located precisely. Precise coordinates (13 degree 45 minutes 39.22 seconds northern latitude, 75 degrees 06 minutes 56.03 seconds eastern longitude) of Thenkabail's village house on the planet and the date

of image acquisition (March 1, 2014) are noted. Google Earth images are used for visualization as well as for numerous science applications such as accuracy assessment, reconnaissance, determining land cover, and establishing land use for various ground surveys. It is widely used by lay people who often have no idea on how it all comes together but understand the information provided intuitively. This is already happening. These developments make it clear that we not only need to understand the state of the art but also have a vision of where the future of remote sensing is headed. Therefore, in a nutshell, the goal of this handbook is to cover the developments and advancement of six distinct eras in terms of data characterization and processing as well as myriad land and water applications:

1. *Pre–civilian remote sensing era of the pre-1950s*: World War I and II when remote sensing was a military tool

2. *Technology demonstration era of the 1950s and 1960s*: Sputnik-I and NOAA AVHRR era of the 1950s and 1960s

3. *Landsat era of the 1970s*: when the first truly operational land remote sensing satellite (Earth Resources Technology Satellite or ERTS, later renamed Landsat) was launched and operated in the 1970s and early 1980s by United States

4. *Earth observation era of the 1980s and 1990s*: when a number of space agencies began launching and operating satellites (e.g., Landsat 4,5 by the United States; SPOT-1,2 by France; IRS-1a, 1b by India) from the middle to late 1980s onward till the middle of 1990s

5. *Earth observation and the first decade of the New Millennium era of the 2000s*: when data dissemination to users became as important as launching, operating, and capturing data (e.g., MODIS Terra\Aqua, Landsat-8, Resourcesat) in the late 1990 and the first decade of the 2000s

6. *Second decade of the New Millennium era starting in the 2010s*: when new-generation micro-\nanosatellites (e.g., PlanetLabs, Skybox) are added to the increasing constellation of multiagency sensors (e.g., Sentinels, and the next generation of satellites such as SMAP, hyperspectral satellites like NASA's HyspIRI and others from private industry)

Motivation for the Remote Sensing Handbook started with a simple conversation with Irma Shagla-Britton, acquisitior editor for remote sensing and GIS books of Taylor & Francis Group/CRC Press, way back in early 2013. Irma was informally getting my advice about "doing a new and unique book" on remote sensing. Neither the specific subject nor the editor was identified. What was clear to me though was that I certainly did not want to lead the effort. I was nearing the end of my third year of recovery from colon cancer, and the last thing I wanted to do was to take any book project, forget a multivolume remote sensing magnum opus, as it ultimately turned out. However, mostly out of courtesy for Irma, I did some preliminary research. I tried to identify a specific topic within remote sensing where there was a sufficient need for a full-fledged book. My research showed that there was not a single book that would provide a complete and comprehensive coverage of the entire subject of remote sensing starting from data capture, to data preprocessing, to data analysis, to myriad land and water applications. There are, of course, numerous excellent books on remote sensing, each covering a specific subject matter. However, if a student, scientist, or practitioner of remote sensing wanted a standard reference on the subject, he or she would have to look for numerous books or journal articles and often a coherence of these topics would still be left uncovered or difficult to comprehend for students and even for many experts with less experience. Guidance on how to approach the study of remote sensing and capture its state of the art and advances remained hazy and often required referring to a multitude of references that may or may not be immediately available, and if available, how to go about it was still hazy to most. During this process, I asked myself, several times, what remote sensing book will be most interesting, productive, and useful to a broad audience? The answer, each time, was very clear: "A complete and comprehensive coverage of the state-of-the-art remote sensing, capturing the advances that have taken place over the last 50 years, which will set the stage for a vision for the future." When this became clear, I started putting together the needed topics to achieve such a goal. Soon I realized that the only way to achieve this goal was through a multivolume book on remote sensing. Because the number of chapters was more than 80, this appeared to be too daunting, too overwhelming, and too big a project to accomplish. Yet I sent the initial idea to Irma, who I thought would say "forget it" and ask me to focus on a single-volume book. But to my surprise, Irma not only encouraged the idea but also had a number of useful suggestions. So what started as intellectual curiosity turned into this full-fledged multivolume Remote Sensing Handbook.

However, what worried me greatly was the virtual impossibility (my thought at that time) of gathering the best authors. What was also crystal clear to me was that unless the very best were attracted to the book project, it was simply not worth the effort. I had made up my mind to give up the book project, unless I got the full support of a large number of the finest practitioners of remote sensing from around the world. So, I spent a few weeks researching the best authors to lead each chapter and wrote to them to participate in the Remote Sensing Handbook project. What really surprised me was that almost all the authors I contacted agreed to lead and write a chapter. This was truly surreal. These are extremely busy people of great scientific reputation and achievements. For them to spend the time, intellect, and energy to write an in-depth and insightful book chapter spread across a year or more is truly amazing. Most also agreed to put together a writing team, as I had requested, to ensure greater perspective for each chapter. In the end, we had 300+ authors writing 82 chapters.

At this stage, I was somewhat drawn into the project as if by destiny and felt compelled to go ahead. One of the authors who agreed to lead the chapter mentioned "…..whether it was even possible." This is exactly what I felt, too. But I had reached the stage of no return, and I took on the book project with all the seriousness it deserved. It required some real changes to my lifestyle: professional and personal. Travel was reduced to bare minimum during most of the book project. Most weekends were spent editing, writing, and organizing, and other social activities were reduced. Accomplishing such complex work requires the highest levels of discipline, planning, and strategy. But, above all, I felt blessed with good health. By the time the book is published, I will have completed about 5 years from my colon cancer surgery and chemotherapy. So I am as happy to see this book released as I am with the miracle of cancer cure (I feel confident to say so).

But it is the chapter authors who made it all feasible. They amazed me throughout the book project. First, the quality and content of each of the chapters were of the highest standards. Second, with very few exceptions, chapters were delivered on time. Third, edited chapters were revised thoroughly and returned on time. Fourth, all my requests on various formatting and quality enhancements were addressed. This is what made the three-volume Remote Sensing Handbook possible and if I may say so, a true *magnum opus* on the subject. My heartfelt gratitude to these great authors for their dedication. It has been my great honor to work with these dedicated legends. Indeed, I call them my *heroes* in a true sense.

Overall, the preparation of the Remote Sensing Handbook took two and a half years, from the time book chapters and authors were being identified to its final publication. The three books are designed in such a way that a reader can have all three books as a standard reference or have individual books to study specific subject areas. The three books of Remote Sensing Handbook are

Remotely Sensed Data Characterization, Classification, and Accuracies: 31 Chapters
Land Resources Monitoring, Modeling, and Mapping with Remote Sensing: 28 Chapters
Remote Sensing of Water Resources, Disasters, and Urban Studies: 27 Chapters

There are about 2500 pages in the 3 volumes.

The wide array of topics covered is very comprehensive. The topics covered in *Remotely Sensed Data Characterization, Classification, and Accuracies* include (1) satellites and sensors; (2) remote sensing fundamentals; (3) data normalization, harmonization, and standardization; (4) vegetation indices and their within- and across-sensor calibration; (5) image classification methods and approaches; (6) change detection; (7) integrating remote sensing with other spatial data; (8) GNSS; (9) crowdsourcing; (10) cloud computing; (11) Google Earth remote sensing; (12) accuracy assessments; and (13) remote sensing law.

The topics covered in *Land Resources Monitoring, Modeling, and Mapping with Remote Sensing* include (1) vegetation and biomass, (2) agricultural croplands, (3) rangelands, (4) phenology and food security, (5) forests, (6) biodiversity, (7) ecology, (8) land use/land cover, (9) carbon, and (10) soils.

The topics covered in *Remote Sensing of Water Resources, Disasters, and Urban Studies* include (1) hydrology and water resources; (2) water use and water productivity; (3) floods; (4) wetlands; (5) snow and ice; (6) glaciers, permafrost, and ice; (7) geomorphology; (8) droughts and drylands; (9) disasters; (10) volcanoes; (11) fire; (12) urban areas; and (13) nightlights.

There are many ways to use the Remote Sensing Handbook. A lot of thought went into organizing the books and chapters. So you will see a *flow* from chapter to chapter and book to book. As you read through the chapters, you will see how they are interconnected and how reading all of them provides you with greater in-depth understanding. Some of you may be more interested in a particular volume. Often, having all three books as reference material is ideal for most remote sensing experts, practitioners, or students; however, you can also refer to individual books based on your interest. We have also made attempts to ensure the chapters are self-contained. That way you can focus on a chapter and read it through, without having to be overly dependent on other chapters. Taking this perspective, there is a slight (~5%–10%) material that may be repeated in some of the chapters. This is done deliberately. For example, when you are reading a chapter on LiDAR or radar, you don't want to go all the way back to another chapter (e.g., Chapter 1, *Remotely Sensed Data Characterization, Classification, and Accuracies*) to understand the characteristics of these sensors.

Similarly, certain indices (e.g., vegetation condition index [VCI], temperature condition index [TCI]) that are defined in one chapter (e.g., on drought) may be repeated in another chapter (also on drought). Such minor overlaps are helpful to the reader to avoid going back to another chapter to understand a phenomenon or an index or a characteristic of a sensor. However, if you want a lot of details on these sensors or indices or phenomena or if you are someone who has yet to gain sufficient expertise in the field of remote sensing, then you will have to read the appropriate chapter where there is in-depth coverage of the topic.

Each book has a summary chapter (the last chapter of each book). The summary chapter can be read two ways: (1) either as a last chapter to recapture the main points of each of the previous chapters or (2) as an initial overview to get a feeling for what is in the book. I suggest the readers do it both ways: Read it first before going into the details and then read it at the end to recollect what was said in the chapters.

It has been a great honor as well as a humbling experience to edit the Remote Sensing Handbook (*Remotely Sensed Data Characterization, Classification, and Accuracies; Land Resources Monitoring, Modeling, and Mapping with Remote Sensing; and Remote Sensing of Water Resources, Disasters, and Urban Studies*). I truly enjoyed the effort. What an honor to work with luminaries in this field of expertise. I learned a lot from them and am very grateful for their support, encouragement, and deep insights. Also, it has been a pleasure working with outstanding professionals of Taylor & Francis Group/CRC Press. There is no joy greater than being immersed in pursuit of excellence, knowledge gain, and knowledge capture. At the same time, I am happy it is over. The biggest lesson I learned during this project was that if you set yourself to a task with dedication, sincerity, persistence, and belief, you will have the job accomplished, no matter how daunting.

I expect the books to be standard references of immense value to any student, scientist, professional, and practical practitioner of remote sensing.

Prasad S. Thenkabail, PhD
Editor-in-Chief

Acknowledgments

The Remote Sensing Handbook (*Remotely Sensed Data Characterization, Classification, and Accuracies*; *Land Resources Monitoring, Modeling, and Mapping with Remote Sensing*; and *Remote Sensing of Water Resources, Disasters, and Urban Studies*) brought together a galaxy of remote sensing legends. The lead authors and coauthors of each chapter are internationally recognized experts of the highest merit on the subject about which they have written. The lead authors were chosen carefully by me after much thought and discussions, who then chose their coauthors. The overwhelming numbers of chapters were written over a period of one year. All chapters were edited and revised over the subsequent year and a half.

Gathering such a galaxy of authors was the biggest challenge. These are all extremely busy people, and committing to a book project that requires a substantial work load is never easy. However, almost all those whom I asked agreed to write the chapter, and only had to convince a few. The quality of the chapters should convince readers why these authors are such highly rated professionals and why they are so successful and accomplished in their field of expertise. They not only wrote very high quality chapters but delivered on time, addressed any editorial comments timely without complaints, and were extremely humble and helpful. What was also most impressive was the commitment of these authors for quality science. Three lead authors had serious health issues and yet they delivered very high quality chapters in the end, and there were few others who had unexpected situations (e.g., family health issues) and yet delivered the chapters on time. Even when I offered them the option to drop out, almost all of them wanted to stay. They only asked for a few extra weeks or months but in the end honored their commitment. I am truly honored to have worked with such great professionals.

In the following list are the names of everyone who contributed and made possible the Remote Sensing Handbook. In the end, we had 82 chapters, a little over 2500 pages, and a little over 300 authors.

My gratitude to the following authors of chapters in *Remotely Sensed Data Characterization, Classification, and Accuracies*. The authors are listed in chapter order starting with the lead author.

- Chapter 1, Drs. Sudhanshu S. Panda, Mahesh Rao, Prasad S. Thenkabail, and James P. Fitzerald
- Chapter 2, Natascha Oppelt, Rolf Scheiber, Peter Gege, Martin Wegmann, Hannes Taubenboeck, and Michael Berger
- Chapter 3, Philippe M. Teillet
- Chapter 4, Philippe M. Teillet and Gyanesh Chander
- Chapter 5, Rudiger Gens and Jordi Cristóbal Rosselló
- Chapter 6, Dongdong Wang
- Chapter 7, Tomoaki Miura, Kenta Obata, Javzandulam T. Azuma, Alfredo Huete, and Hiroki Yoshioka
- Chapter 8, Michael D. Steven, Timothy Malthus, and Frédéric Baret
- Chapter 9, Sunil Narumalani and Paul Merani
- Chapter 10, Soe W. Myint, Victor Mesev, Dale Quattrochi, and Elizabeth A. Wentz
- Chapter 11, Mutlu Ozdogan
- Chapter 12, Jun Li and Antonio Plaza
- Chapter 13, Claudia Kuenzer, Jianzhong Zhang, and Stefan Dech
- Chapter 14, Thomas Blaschke, Maggi Kelly, and Helena Merschdorf
- Chapter 15, Stefan Lang and Dirk Tiede
- Chapter 16, James C. Tilton, Selim Aksoy, and Yuliya Tarabalka
- Chapter 17, Shih-Hong Chio, Tzu-Yi Chuang, Pai-Hui Hsu, Jen-Jer Jaw, Shih-Yuan Lin, Yu-Ching Lin, Tee-Ann Teo, Fuan Tsai, Yi-Hsing Tseng, Cheng-Kai Wang, Chi-Kuei Wang, Miao Wang, and Ming-Der Yang
- Chapter 18, Daniela Anjos, Dengsheng Lu, Luciano Dutra, and Sidnei Sant'Anna
- Chapter 19, Jason A. Tullis, Jackson D. Cothren, David P. Lanter, Xuan Shi, W. Fredrick Limp, Rachel F. Linck, Sean G. Young, and Tareefa S. Alsumaiti
- Chapter 20, Gaurav Sinha, Barry J. Kronenfeld, and Jeffrey C. Brunskill
- Chapter 21, May Yuan
- Chapter 22, Stefan Lang, Stefan Kienberger, Michael Hagenlocher, and Lena Pernkopf
- Chapter 23, Mohinder S. Grewal
- Chapter 24, Kegen Yu, Chris Rizos, and Andrew Dempster
- Chapter 25, D. Myszor, O. Antemijczuk, M. Grygierek, M. Wierzchanowski, and K.A. Cyran
- Chapter 26, Fabio Dell'Acqua
- Chapter 27, Ramanathan Sugumaran, James W. Hegeman, Vivek B. Sardeshmukh, and Marc P. Armstrong
- Chapter 28, John Bailey
- Chapter 29, Russell G. Congalton

- Chapter 30, P.J. Blount
- Chapter 31, Prasad S. Thenkabail

My gratitude to the following authors of chapters in *Land Resources Monitoring, Modeling, and Mapping with Remote Sensing*. The authors are listed in chapter order starting with the lead author.

- Chapter 1, Alfredo Huete, Guillermo Ponce-Campos, Yongguang Zhang, Natalia Restrepo-Coupe, Xuanlong Ma, and Mary-Susan Moran
- Chapter 2, Frédéric Baret
- Chapter 3, Wenge Ni-Meister
- Chapter 4, Clement Atzberger, Francesco Vuolo, Anja Klisch, Felix Rembold, Michele Meroni, Marcio Pupin Mello, and Antonio Formaggio
- Chapter 5, Agnès Bégué, Damien Arvor, Camille Lelong, Elodie Vintrou, and Margareth Simoes
- Chapter 6, Pardhasaradhi Teluguntla, Prasad S. Thenkabail, Jun Xiong, Murali Krishna Gumma, Chandra Giri, Cristina Milesi, Mutlu Ozdogan, Russell G. Congalton, James Tilton, Temuulen Tsagaan Sankey, Richard Massey, Aparna Phalke, and Kamini Yadav
- Chapter 7, David J. Mulla, and Yuxin Miao
- Chapter 8, Baojuan Zheng, James B. Campbell, Guy Serbin, Craig S.T. Daughtry, Heather McNairn, and Anna Pacheco
- Chapter 9, Prasad S. Thenkabail, Pardhasaradhi Teluguntla, Murali Krishna Gumma, and Venkateswarlu Dheeravath
- Chapter 10, Matthew Clark Reeves, Robert A. Washington-Allen, Jay Angerer, E. Raymond Hunt, Jr., Ranjani Wasantha Kulawardhana, Lalit Kumar, Tatiana Loboda, Thomas Loveland, Graciela Metternicht, and R. Douglas Ramsey
- Chapter 11, E. Raymond Hunt, Jr., Cuizhen Wang, D. Terrance Booth, Samuel E. Cox, Lalit Kumar, and Matthew C. Reeves
- Chapter 12, Lalit Kumar, Priyakant Sinha, Jesslyn F. Brown, R. Douglas Ramsey, Matthew Rigge, Carson A. Stam, Alexander J. Hernandez, E. Raymond Hunt, Jr., and Matt Reeves
- Chapter 13, Molly E. Brown, Kirsten M. de Beurs, and Kathryn Grace
- Chapter 14, E.H. Helmer, Nicholas R. Goodwin, Valéry Gond, Carlos M. Souza, Jr., and Gregory P. Asner
- Chapter 15, Juha Hyyppä, Mika Karjalainen, Xinlian Liang, Anttoni Jaakkola, Xiaowei Yu, Mike Wulder, Markus Hollaus, Joanne C. White, Mikko Vastaranta, Kirsi Karila, Harri Kaartinen, Matti Vaaja, Ville Kankare, Antero Kukko, Markus Holopainen, Hannu Hyyppä, and Masato Katoh
- Chapter 16, Gregory P. Asner, Susan L. Ustin, Philip A. Townsend, Roberta E. Martin, and K. Dana Chadwick
- Chapter 17, Sylvie Durrieu, Cédric Véga, Marc Bouvier, Frédéric Gosselin, and Jean-Pierre Renaud Laurent Saint-André

- Chapter 18, Thomas W. Gillespie, Andrew Fricker, Chelsea Robinson, and Duccio Rocchini
- Chapter 19, Stefan Lang, Christina Corbane, Palma Blonda, Kyle Pipkins, and Michael Förster
- Chapter 20, Conghe Song, Jing Ming Chen, Taehee Hwang, Alemu Gonsamo, Holly Croft, Quanfa Zhang, Matthew Dannenberg, Yulong Zhang, Christopher Hakkenberg, Juxiang Li
- Chapter 21, John Rogan and Nathan Mietkiewicz
- Chapter 22, Zhixin Qi, Anthony Gar-On Yeh, and Xia Li
- Chapter 23, Richard A. Houghton
- Chapter 24, José A.M. Demattê, Cristine L.S. Morgan, Sabine Chabrillat, Rodnei Rizzo, Marston H.D. Franceschini, Fabrício da S. Terra, Gustavo M. Vasques, and Johanna Wetterlind
- Chapter 25, E. Ben-Dor and José A.M. Demattê
- Chapter 26, Prasad S. Thenkabail

My gratitude to the following authors of chapters in *Remote Sensing of Water Resources, Disasters, and Urban Studies*. The authors are listed in chapter order starting with the lead author.

- Chapter 1, Sadiq I. Khan, Ni-Bin Chang, Yang Hong, Xianwu Xue, and Yu Zhang
- Chapter 2, Santhosh Kumar Seelan
- Chapter 3, Trent W. Biggs, George P. Petropoulos, Naga Manohar Velpuri, Michael Marshall, Edward P. Glenn, Pamela Nagler, and Alex Messina
- Chapter 4, Antônio de C. Teixeira, Fernando B. T. Hernandez, Morris Scherer-Warren, Ricardo G. Andrade, Janice F. Leivas, Daniel C. Victoria, Edson L. Bolfe, Prasad S. Thenkabail, and Renato A. M. Franco
- Chapter 5, Allan S. Arnesen, Frederico T. Genofre, Marcelo P. Curtarelli, and Matheus Z. Francisco
- Chapter 6, Sandro Martinis, Claudia Kuenzer, and André Twele
- Chapter 7, Chandra Giri
- Chapter 8, D. R. Mishra, Shuvankar Ghosh, C. Hladik, Jessica L. O'Connell, and H. J. Cho
- Chapter 9, Murali Krishna Gumma, Prasad S. Thenkabail, Irshad A. Mohammed, Pardhasaradhi Teluguntla, and Venkateswarlu Dheeravath
- Chapter 10, Hongjie Xie, Tiangang Liang, Xianwei Wang, and Guoqing Zhang
- Chapter 11, Qingling Zhang, Noam Levin, Christos Chalkias, and Husi Letu
- Chapter 12, James B. Campbell and Lynn M. Resler
- Chapter 13, Felix Kogan and Wei Guo
- Chapter 14, Felix Rembold, Michele Meroni, Oscar Rojas, Clement Atzberger, Frederic Ham, and Erwann Fillol
- Chapter 15, Brian Wardlow, Martha Anderson, Tsegaye Tadesse, Chris Hain, Wade T. Crow, and Matt Rodell
- Chapter 16, Jinyoung Rhee, Jungho Im, and Seonyoung Park
- Chapter 17, Marion Stellmes, Ruth Sonnenschein, Achim Röder, Thomas Udelhoven, Stefan Sommer, and Joachim Hill

- Chapter 18, Norman Kerle
- Chapter 19, Stefan Lang, Petra Füreder, Olaf Kranz, Brittany Card, Shadrock Roberts, and Andreas Papp
- Chapter 20, Robert Wright
- Chapter 21, Krishna Prasad Vadrevu and Kristofer Lasko
- Chapter 22, Anupma Prakash and Claudia Kuenzer
- Chapter 23, Hasi Bagan and Yoshiki Yamagata
- Chapter 24, Yoshiki Yamagata, Daisuke Murakami, and Hajime Seya
- Chapter 25, Prasad S. Thenkabail

These authors are "who is who" in remote sensing and come from premier institutions of the world. For author affiliations, please see "Contributors" list provided a few pages after this. My deepest apologies if I have missed any name. But, I am sure those names are properly credited and acknowledged in individual chapters.

The authors not only delivered excellent chapters, they provided valuable insights and inputs for me in many ways throughout the book project.

I was delighted when Dr. Compton J. Tucker, senior Earth scientist, Earth Sciences Division, Science and Exploration Directorate, NASA Goddard Space Flight Center (GSFC), agreed to write the foreword for the book. For anyone practicing remote sensing, Dr. Tucker needs no introduction. He has been a *godfather* of remote sensing and has inspired a generation of scientists. I have been a student of his without ever really being one. I mean, I have not been his student in a classroom but have followed his legendary work throughout my career. I remember reading his highly cited paper (now with citations nearing 4000!):

- Tucker, C.J. (1979) Red and photographic infrared linear combinations for monitoring vegetation, *Remote Sensing of Environment*, **8(2)**,127–150.

That was in 1986 when I had just joined the National Remote Sensing Agency (NRSA; now NRSC), Indian Space Research Organization (ISRO). After earning his PhD from the Colorado State University in 1975, Dr. Tucker joined NASA GSFC as a postdoctoral fellow and became a full-time NASA employee in 1977. Since then, he has conducted path-finding research. He has used NOAA AVHRR, MODIS, SPOT Vegetation, and Landsat satellite data for studying deforestation, habitat fragmentation, desert boundary determination, ecologically coupled diseases, terrestrial primary production, glacier extent, and how climate affects global vegetation. He has authored or coauthored more than 170 journal articles that have been cited more than 20,000 times, is an adjunct professor at the University of Maryland, is a consulting scholar at the University of Pennsylvania's Museum of Archaeology and Anthropology, and has appeared on more than twenty radio and TV programs. He is a fellow of the American Geophysical Union and has been awarded several medals and honors, including NASA's Exceptional Scientific Achievement Medal, the Pecora Award from the U.S. Geological Survey (USGS), the National Air and Space Museum Trophy, the Henry Shaw Medal from the Missouri Botanical Garden, the Galathea Medal from the Royal Danish Geographical Society, and the Vega Medal from the Swedish Society of Anthropology and Geography. He was the NASA representative to the U.S. Global Change Research Program from 2006 to 2009. He was instrumental in releasing the AVHRR 32-year (1982–2013) Global Inventory Monitoring and Modeling Studies (GIMMS) data. I strongly recommend that everyone read his excellent foreword before reading the book. In the foreword, Dr. Tucker demonstrates the importance of data from EO sensors from orbiting satellites to maintaining a reliable and consistent climate record. Dr. Tucker further highlights the importance of continued measurements of these variables of our planet in the new millennium through new, improved, and innovative EO sensors from Sun-synchronous and/or geostationary satellites.

I am very thankful to my USGS colleagues for their encouragement and support. In particular, I mention Edwin Pfeifer, Dr. Susan Benjamin, Dr. Dennis Dye, Larry Gaffney, Miguel Velasco, Dr. Chandra Giri, Dr. Terrance Slonecker, Dr. Jonathan Smith, and Dr. Thomas Loveland. There are many other colleagues who made my job at USGS that much easier. My thanks to them all.

I am very thankful to Irma Shagla-Britton, acquisition editor for remote sensing and GIS books at Taylor & Francis Group/CRC Press. Without her initial nudge, this book would never have even been completed. Thank you, Irma. You are doing a great job.

I am very grateful to my wife (Sharmila Prasad) and daughter (Spandana Thenkabail) for their usual unconditional love, understanding, and support. They are always the pillars of my life. I learned the values of hard work and dedication from my revered parents. This work wouldn't have come about without their sacrifices to educate their children and their silent blessings. I am ever grateful to my former professors at The Ohio State University, Columbus, Ohio, United States: Prof. John G. Lyon, Dr. Andrew D. Ward, Prof. (Late) Carolyn Merry, Dr. Duane Marble, and Dr. Michael Demers. They have taught, encouraged, inspired, and given me opportunities at the right time. The opportunity to work for six years at the Center for Earth Observation of Yale University (YCEO) was incredibly important. I am thankful to Prof. Ronald G. Smith, director of YCEO, for his kindness. At YCEO, I learned and advanced myself as a remote sensing scientist. The opportunities I got from working for the International Institute of Tropical Agriculture (IITA), Africa and International Water Management Institute (IWMI) that had a global mandate for water were very important, especially from the point of view of understanding the real issues on the ground. I learned my basics of remote sensing mainly working with Dr. Thiruvengadachari of the National Remote Sensing Agency/Center (NRSA/NRSC), Indian Space Research Organization (ISRO), India, where I started my remote sensing career as a young scientist. I was just 25 years old then and had joined NRSA after earning my masters of engineering (hydraulics and water resources) and bachelors of engineering (civil engineering). During my first day in the office, Dr. Thiruvengadachari asked me how much remote sensing did I know. I said, "zero" and instantly thought that I would be thrown out of the room. But he said "very good" and gave me a manual on remote sensing from the Laboratory for Applications of Remote Sensing (LARS), Purdue. Those were the days where there was no formal training in remote sensing in any Indian universities. So my remote sensing lessons began working practically on projects and one of our first projects was "drought monitoring for India using NOAA AVHRR data." This was an intense period of learning remote sensing by actually practicing it on a daily basis. Data came on 9 mm tapes; data were read on massive computing systems; image processing was done, mostly working on night shifts by booking time on centralized computing; field work was conducted using false color composite outputs and topographic maps (not the days of global positioning systems); geographic information system was in its infancy; and a lot of calculations were done using calculators. So when I decided to resign my NRSA job and go to the United States to do my PhD, Dr. Thiruvengadachari told me, "Prasad, I am losing my right hand, but you can't miss opportunity." Those initial wonderful days of learning from Dr. Thiruvengadachari will remain etched in my memory. Prof. G. Ranganna of the Karnataka Regional Engineering College (KREC; now National Institute of Technology), Karnataka, India, was/is one of my most revered gurus. I have learned a lot observing him, professionally and personally, and he has always been an inspiration. Prof. E.J. James, former director of the Center for Water Resources Development and Management (CWRDM), was another original guru from whom I have learned the values of a true professional. I am also thankful to my good old friend Shri C. J. Jagadeesha, who is still working for ISRO as a senior scientist. He was my colleague at NRSA/NRSC, ISRO, and encouraged me to grow as a scientist. This Remote Sensing Handbook is a blessing from the most special ones dear to me. Of course, there are many, many others to thank especially many of my dedicated students over the years, but they are too many to mention here. I thank the truly outstanding editing work performed by Arunkumar Aranganathan and his team at SPi Global.

It has been my deep honor and great privilege to have edited the Remote Sensing Handbook. I am sure that I won't be taking on any such huge endeavors in the future. I will need time for myself, to look inside, understand, and grow. So thank you all, for making this possible.

Prasad S. Thenkabail, PhD
Editor-in-Chief

Editor

Prasad S. Thenkabail, PhD, is currently working as a research geographer-15 with the U.S. Geological Survey (USGS), United States. Currently, at USGS, Prasad leads a multi-institutional NASA MEaSUREs (Making Earth System Data Records for Use in Research Environments) project, funded through NASA ROSES solicitation. The project is entitled Global Food Security-Support Analysis Data at 30 m (GFSAD30) (http://geography.wr.usgs.gov/science/croplands/index.html also see https://www.croplands.org/). He is also an adjunct professor at three U.S. universities: (1) Department of Soil, Water, and Environmental Science (SWES), University of Arizona (UoA); (2) Department of Space Studies, University of North Dakota (UND); and (3) School of Earth Sciences and Environmental Sustainability (SESES), Northern Arizona University (NAU), Flagstaff, Arizona.

Dr. Thenkabail has conducted pioneering scientific research work in two major areas:

1. Hyperspectral remote sensing of vegetation
2. Global irrigated and rainfed cropland mapping using spaceborne remote sensing

His research papers on these topics are widely quoted. His hyperspectral work also led to his working on the scientific advisory board of Rapideye (2001), a German private industry satellite. Prasad was consulted on the design of spectral wavebands.

In hyperspectral research, Prasad pioneered in the following:

1. The design of optimal hyperspectral narrowbands (HNBs) and hyperspectral vegetation indices (HVIs) for agriculture and vegetation studies.
2. Certain hyperspectral data mining and data reduction techniques such as now widely used concepts of lambda by lambda plots.
3. Certain hyperspectral data classification methods. This included the use of a series of methods (e.g., discriminant model, Wilk's lambda, Pillai trace) that demonstrate significant increases in classification accuracies of land cover and vegetation classes as determined using HNBs as opposed to multispectral broadbands.

In global croplands, Prasad conducted seminal research that led to the first global map of irrigated and rainfed cropland areas using multitemporal, multisensor remote sensing, one book, and a series of more than ten novel peer-reviewed papers.

In 2008, for one of these papers, Prasad (lead author) and coauthors (Pardhasaradhi Teluguntala, Trent Biggs, Murali Krishna Gumma, and Hugh Turral) were the second-place recipients of the 2008 John I. Davidson American Society of Photogrammetry and Remote Sensing (ASPRS) President's Award for practical papers. The paper proposed a novel spectral matching technique (SMT) for cropland classification. Earlier, Prasad (lead author) and coauthors (Andy Ward, John Lyon, and Carolyn Merry), won the 1994 Autometric Award for outstanding paper on remote sensing of agriculture from ASPRS. Recently, Prasad (seccond author) with Michael Marshall (lead author), won the ASPRS ERDAS award for best scientific paper on remote sensing for their hyperspectral remote sensing work.

Earlier to this **path-breaking Remote Sensing Handbook**, Prasad has published two seminal books (both published by Taylor & Francis Group/CRC Press) related to hyperspectral remote sensing and global croplands:

- Thenkabail, P.S., Lyon, G.J., and Huete, A. 2011. *Hyperspectral Remote Sensing of Vegetation*. CRC Press/Taylor & Francis Group, Boca Raton, FL, 781pp.

Reviews of this book:

- http://www.crcpress.com/product/isbn/9781439845370.
- Thenkabail, P., Lyon, G.J., Turral, H., and Biradar, C.M. 2009. *Remote Sensing of Global Croplands for Food Security*. CRC Press/Taylor & Francis Group, Boca Raton, FL, 556pp (48 pages in color).

Reviews of this book:

- http://www.crcpress.com/product/isbn/9781420090093.
- http://gfmt.blogspot.com/2011/05/review-remote-sensing-of-global.html.

He has guest edited two special issues for the American Society of Photogrammetry and Remote Sensing (PE&RS):

- Thenkabail, P.S. 2014. Guest editor of special issue on "Hyperspectral remote sensing of vegetation and agricultural crops." *Photogrammetric Engineering and Remote Sensing* 80(4).
- Thenkabail, P.S. 2012. Guest editor for Global croplands special issue. *Photogrammetric Engineering and Remote Sensing* 78(8).

He has also guest edited a special issue on global croplands for the *Remote Sensing Open Access Journal* (ISSN 2072-4292):

- Thenkabail, P.S. 2010. Guest editor: Special issue on "Global croplands" for the MDPI remote sensing open access journal. Total: 22 papers. http://www.mdpi.com/journal/remotesensing/special_issues/croplands/.

Prasad is, currently editor-in-chief, *Remote Sensing Open Access Journal,* an on-line journal, published by MDPI; editorial board member, *Remote Sensing of Environment;* editorial advisory board member, *ISPRS Journal of Photogrammetry and Remote Sensing.*

Prior to joining USGS in October 2008, Dr. Thenkabail was a leader of the remote sensing programs of leading institutes International Water Management Institute (IWMI), 2003–2008; International Center for Integrated Mountain Development (ICIMOD), 1995–1997; International Institute of Tropical Agriculture (IITA), 1992–1995.

He also worked as a key remote sensing scientist for Yale Center for Earth Observation (YCEO), 1997–2003; Ohio State University (OSU), 1988–1992; National Remote Sensing Agency (NRSA) (now NRSC), Indian Space Research organization (ISRO), 1986–1988.

Over the years, he has been a principal investigator (PI) of NASA, USGS, IEEE, and other funded projects such as inland valley wetland mapping of African nations, characterization of eco-regions of Africa (CERA), which involved both African savannas and rainforests, global cropland water use for food security in the twenty-first century, automated cropland classification algorithm (ACCA) within WaterSMART (Sustain and Manage America's Resources for Tomorrow) project, water productivity mapping in the irrigated croplands of California and Uzbekistan using multisensor remote sensing, IEEE Water for the World Project, and drought monitoring in India, Pakistan, and Afghanistan.

The USGS and NASA selected Dr. Thenkabail to be on the Landsat Science Team (2007–2011) for a period of five years (http://landsat.gsfc.nasa.gov/news/news-archive/pol_0005.html; http://ldcm.usgs.gov/intro.php). In June 2007, his team was recognized by the Environmental System Research Institute (ESRI) for "special achievement in GIS" (SAG award) for their tsunami-related work (tsdc.iwmi.org) and for their innovative spatial data portals (http://waterdata.iwmi.org/dtView-Common.php; earlier http://www.iwmidsp.org). Currently, he is also a global coordinator for the Agriculture Societal Beneficial Area (SBA) of the Committee for Earth Observation

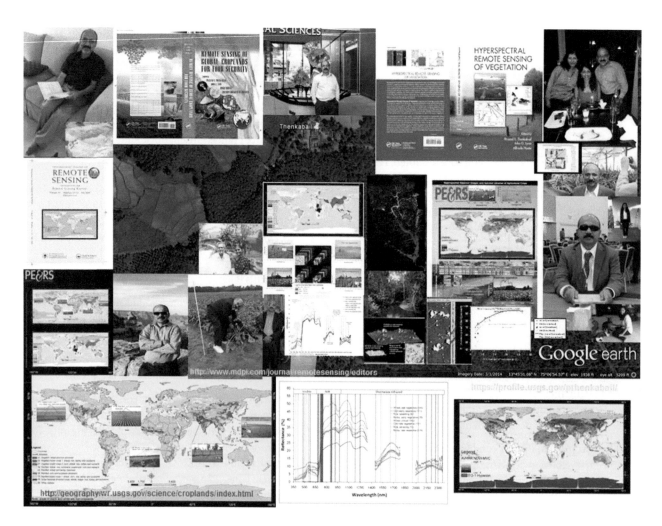

Systems (CEOS). He is active in the Group on Earth Observation (GEO) agriculture and water efforts through Earth observation. He was a co-lead of the Water for the World Project (IEEE effort). He is the current chair of the International Society of Photogrammetry and Remote Sensing (ISPRS) Working Group WG VIII/7: "Land Cover and Its Dynamics, including Agricultural & Urban Land Use" for the period 2013–2016. Thenkabail earned his PhD from The Ohio State University (1992). His master's degree in hydraulics and water resources engineering (1984) and bachelor's degree in civil engineering (1981) were from India. He began his professional career as a lecturer in hydrology, water resources, hydraulics, and open channel in India. He has 100+ publications, mostly peer-reviewed research papers in major international remote sensing journals: http://scholar.google.com/citations?user=9IO5Y7 YAAAAJ&hl=en. Prasad has about 30 years' experience working as a well-recognized international expert in remote sensing and geographic information systems (RS/GIS) and their application to agriculture, wetlands, natural resource management, water resources, forests, sustainable development, and environmental studies. His work experience spans over 25+ countries spread across West and Central Africa, Southern Africa, South Asia, Southeast Asia, the Middle East, East Asia, Central Asia, North America, South America, and the Pacific.

Contributors

Jay Angerer
Blackland Research and Extension
  Center
Texas A&M AgriLife Research
Texas A&M University
Temple, Texas

Damien Arvor
Joint Research Unit "Littoral,
  Environment, Remote Sensing and
  Geomatics"—COSTEL laboratory
French National Centre for Scientific
  Research
Rennes, France

Gregory P. Asner
Department of Global Ecology
Carnegie Institution for Science
Stanford, California

Clement Atzberger
University of Natural Resources and Life
  Sciences (BOKU)
Wien, Austria

Frédéric Baret
L'Unité Mixte de Recherche
Institut National de la Recherche
  Agronomique
Avignon, France

Agnès Bégué
Joint Research Unit "Spatial Information
  and Analysis for Territories and
  Ecosystems"
French Agricultural Research Centre for
  International Development
Montpellier, France

E. Ben-Dor
Department of Geography
Tel Aviv University
Tel Aviv, Israel

Palma Blonda
Consiglio Nazionale delle Ricerche
  (CNR)
Istituto di Studi sui Sistemi Intelligenti
  per l'Automazione (ISSIA)
Bari, Italy

D. Terrance Booth
Rangeland Resources Research Unit
High Plains Grasslands Research Station
Agricultural Research Service
United States Department of Agriculture
Cheyenne, Wyoming

Marc Bouvier
UMR TETIS Territoires, Environnement,
  Télédétection et Information Spatiale
Centre de Montpellier
Montpellier, France

Jesslyn F. Brown
Earth Resources Observation and
  Science (EROS) Center
U.S. Geological Survey
Sioux Falls, South Dakota

Molly E. Brown
Department of Geographical Sciences
University of Maryland
College Park, Maryland

James B. Campbell
Virginia Tech
Blacksburg, Virginia

Sabine Chabrillat
Remote Sensing Section
GFZ German Research Centre for
  Geosciences
Helmholtz Centre Potsdam
Potsdam, Germany

K. Dana Chadwick
Carnegie Institution for Science
Stanford, California

Jing Ming Chen
Department of Geography and Program
  in Planning
University of Toronto
Toronto, Ontario, Canada

Russell G. Congalton
Department of Geography
University of New Hampshire
Durham, New Hampshire

Christina Corbane
Institut national de recherche en sciences
  et technologies pour l'environnement
  et l'agriculture (IRSTEA)—UMR
  TETIS—Maison de la Télédétection
Montpellier, France

Samuel E. Cox
Wyoming State Office
Bureau of Land Management
Cheyenne, Wyoming

Holly Croft
Department of Geography and Program
  in Planning
University of Toronto
Toronto, Ontario, Canada

Matthew Dannenberg
Department of Geography
University of North Carolina at Chapel
  Hill
Chapel Hill, North Carolina

Craig S.T. Daughtry
USDA-ARS Hydrology and Remote
  Sensing Laboratory
Beltsville, Maryland

Kirsten M. de Beurs
Department of Geography
University of Oklahoma
Norman, Oklahoma

José A.M. Demattê
Department of Soil Science
"Luiz de Queiroz" College of Agriculture
University of São Paulo
São Paulo, Brazil

Venkateswarlu Dheeravath
United Nations World Food Program
Erbil, Iraq

Sylvie Durrieu
UMR TETIS Territoires, Environnement,
Télédétection et Information Spatiale
Centre de Montpellier
Montpellier, France

Antonio Formaggio
National Institute for Space Research
São José dos Campos, Brazil

Michael Förster
Institute for Landscape Architecture and
Environmental Planning
Geoinformation in Environmental
Planning Lab
Technische Universität Berlin
Berlin, Germany

Marston H.D. Franceschini
Department of Soil Science
"Luiz de Queiroz" College of Agriculture
University of São Paulo
São Paulo, Brazil

Andrew Fricker
Department of Geography
University of California
Los Angeles, California

Thomas W. Gillespie
Department of Geography
University of California
Los Angeles, California

Chandra Giri
Earth Resources Observation and
Science (EROS) Center
U.S. Geological Survey
Sioux Falls, South Dakota

Valéry Gond
Agricultural Research for Development
Forest Ecosystems Goods and Services
Montpellier, France

Alemu Gonsamo
Department of Geography and Program
in Planning
University of Toronto
Toronto, Ontario, Canada

Nicholas R. Goodwin
Department of Science, Information
Technology, Innovation and the Arts
Remote Sensing Centre
Ecosciences Precinct
Brisbane, Queensland, Australia

Frédéric Gosselin
UR EFNO Écosystèmes Forestiers
Centre de Nogent-sur-Vernisson
Nogent-sur-Vernisson, France

Kathryn Grace
Geography Department
University of Utah
Salt Lake City, Utah

Murali Krishna Gumma
Remote Sensing and GIS Division
International Crops Research Institute
for the Semi-Arid Tropics
Hyderabad, India

Christopher Hakkenberg
Curriculum for the Environment and
Ecology
University of North Carolina at Chapel
Hill
Chapel Hill, North Carolina

E.H. Helmer
International Institute of Tropical
Forestry
United States Department of Agriculture
Forest Service
Río Piedras, Puerto Rico

Alexander J. Hernandez
Department of Wildland Resources
Utah State University
Logan, Utah

Markus Hollaus
Department of Geodesy and
Geoinformation
Vienna University of Technology
Vienna, Austria

Markus Holopainen
Centre of Excellence in Laser Scanning
Research
University of Helsinki
Helsinki, Finland

Richard A. Houghton
Woods Hole Research Center
Falmouth, Massachusetts

Alfredo Huete
Plant Functional Biology and Climate
Change Cluster
University of Technology
Sydney, New South Wales, Australia

E. Raymond Hunt, Jr.
Hydrology and Remote Sensing
Laboratory
Beltsville Agricultural Research Center
Agricultural Research Service
United States Department of Agriculture
Beltsville, Maryland

Taehee Hwang
Department of Geography
Indiana University
Bloomington, Indiana

Hannu Hyyppä
Centre of Excellence in Laser Scanning
Research
Aalto University
Helsinki, Finland

Juha Hyyppä
Centre of Excellence in Laser Scanning
Research
Finnish Geodetic Institute
Masala, Finland

Anttoni Jaakkola
Centre of Excellence in Laser Scanning
Research
Finnish Geodetic Institute
Masala, Finland

Harri Kaartinen
Centre of Excellence in Laser Scanning
    Research
Finnish Geodetic Institute
Masala, Finland

Ville Kankare
Centre of Excellence in Laser Scanning
    Research
University of Helsinki
Helsinki, Finland

Kirsi Karila
Centre of Excellence in Laser Scanning
    Research
Finnish Geodetic Institute
Masala, Finland

Mika Karjalainen
Centre of Excellence in Laser Scanning
    Research
Finnish Geodetic Institute
Masala, Finland

Masato Katoh
Institute of Mountain Science
Shinshu University
Matsumoto, Japan

Anja Klisch
University of Natural Resources and Life
    Sciences (BOKU)
Wien, Austria

Antero Kukko
Centre of Excellence in Laser Scanning
    Research
Finnish Geodetic Institute
Masala, Finland

Ranjani Wasantha Kulawardhana
Spatial Sciences Laboratory
Department of Ecosystem Science and
    Management
Texas A&M University
College Station, Texas

Lalit Kumar
School of Environmental and Rural
    Science
University of New England
Armidale, New South Wales, Australia

Stefan Lang
Department of Geoinformatics - Z_GIS
University of Salzburg
Salzburg, Austria

Camille Lelong
Joint Research Unit "Spatial Information
    and Analysis for Territories and
    Ecosystems"
French Agricultural Research Centre for
    International Development
Montpellier, France

Juxiang Li
College of Ecology and Environmental
    Sciences
East China Normal University
Shanghai, People's Republic of China

Xia Li
School of Geography and Planning
Sun Yat-sen University
Guangzhou, Guangdong, People's
    Republic of China

Xinlian Liang
Centre of Excellence in Laser Scanning
    Research
and
Finnish Geodetic Institute
Masala, Finland

Tatiana Loboda
Department of Geographical Sciences
University of Maryland
College Park, Maryland

Thomas Loveland
Earth Resources Observation and
    Science (EROS) Center
U.S. Geological Survey
Sioux Falls, South Dakota

Xuanlong Ma
Plant Functional Biology and Climate
    Change Cluster
University of Technology
Sydney, New South Wales, Australia

Roberta E. Martin
Department of Global Ecology
Carnegie Institution for Science
Stanford, California

Richard Massey
School of Earth Sciences and
    Environmental Sustainability
Northern Arizona University
Flagstaff, Arizona

Heather McNairn
Agriculture and Agri-Food Canada
Ottawa, Ontario, Canada

Marcio Pupin Mello
Lettere E Filosofia Department
Boeing Research & Technology
São José dos Campos, Brazil

Michele Meroni
Joint Research Centre
European Commission
Ispra (VA), Italia

Graciela Metternicht
Institute of Environmental Studies
University of New South Wales
Sydney, New South Wales, Australia

Yuxin Miao
College of Resource and Environmental
    Sciences
China Agricultural University
Beijing, People's Republic of China

Nathan Mietkiewicz
Graduate School of Geography
Clark University
Worcester, Massachusetts

Cristina Milesi
Ames Research Center
National Aeronautics and Space
    Administration
Moffett Field, California

Mary Susan Moran
Southwest Watershed Research Center
Agricultural Research Service
United States Department of Agriculture
Tucson, Arizona

Cristine L.S. Morgan
Department of Soil and Crop Sciences
Texas A&M University
College Station, Texas

David J. Mulla
Department of Soil, Water and Climate
University of Minnesota
St. Paul, Minnesota

Wenge Ni-Meister
Department of Geography
Hunter College of the City University
    of New York
New York, New York

Mutlu Ozdogan
Department of Forest and Wildlife Ecology
and
Nelson Institute for Environmental
    Studies
University of Wisconsin
Madison, Wisconsin

Anna Pacheco
Agriculture and Agri-Food Canada
Ottawa, Ontario, Canada

Aparna Phalke
Department of Forest and Wildlife Ecology
and
Nelson Institute for Environmental
    Studies
University of Wisconsin
Madison, Wisconsin

Kyle Pipkins
Institute for Landscape Architecture and
    Environmental Planning
Geoinformation in Environmental
    Planning Lab
Technische Universität Berlin
Berlin, Germany

Guillermo Ponce-Campos
Southwest Watershed Research Center
Agricultural Research Service
United States Department of Agriculture
Tucson, Arizona

Zhixin Qi
Department of Urban Planning and
    Design
The University of Hong Kong
Hong Kong SAR, People's Republic of
    China

R. Douglas Ramsey
Department of Wildland Resources
Utah State University
Logan, Utah

Matthew C. Reeves
Rocky Mountain Research Station
United States Department of Agriculture
    Forest Service
Missoula, Montana

Felix Rembold
Environmental Science, Agronomy
Joint Research Centre
European Commission
Ispra (VA), Italia

Jean-Pierre Renaud
Office National des Forêts
Nancy, France

Natalia Restrepo-Coupe
Plant Functional Biology and Climate
    Change Cluster
University of Technology
Sydney, New South Wales, Australia

Matthew Rigge
Earth Resources Observation and
    Science (EROS) Center
U.S. Geological Survey
Sioux Falls, South Dakota

Rodnei Rizzo
Department of Soil Science
University of São Paulo
São Paulo, Brazil

Chelsea Robinson
Department of Geography
University of California
Los Angeles, California

Duccio Rocchini
Department of Biodiversity and
    Molecular Ecology
Fondazione Edmund Mach
Trento, Italy

John Rogan
Graduate School of Geography
Clark University
Worcester, Massachusetts

Laurent Saint-André
UR BEF, Biogéochimie des Ecosystèmes
    Forestiers
INRA
Champenoux, France

and

UMR Eco&Sols, Ecologie Fonctionnelle
    et Biogéochimie des Sols et
    Agro-écosystèmes
CIRAD
Montpellier, France

Temuulen Tsagaan Sankey
School of Earth Sciences and
    Environmental Sustainability
Northern Arizona University
Flagstaff, Arizona

Guy Serbin
Soil Science and Remote Sensing
Teagasc Food Research Centre
Dublin, Ireland

Margareth Simoes
Department of Computer Engineering
Rio de Janeiro State University
Embrapa—Brazilian Agriculture
    Research Corporation
Rio de Janeiro, Brazil

Priyakant Sinha
School of Environmental and Rural
    Science
University of New England
Armidale, New South Wales, Australia

Conghe Song
Department of Geography
University of North Carolina at Chapel
    Hill
Chapel Hill, North Carolina

Carlos M. Souza, Jr.
Amazon Institute of People and the
    Environment
Instituto do Homen e Meio Ambiente da
    Amazônia
Belém, Brazil

Carson A. Stam
Department of Wildland Resources
Utah State University
Logan, Utah

Pardhasaradhi Teluguntla
U.S. Geological Survey
Flagstaff, Arizona

and

Bay Area Environmental Research
    Institute
Sonoma, California

Fabrício da S. Terra
Institute of Agricultural Sciences
Federal University of Jequitinhonha and
    Mucuri Valleys
Minas Gerais, Brazil

Prasad S. Thenkabail
U.S. Geological Survey
Flagstaff, Arizona

James Tilton
Goddard Space Flight Center
National Aeronautics and Space
    Administration
Greenbelt, Maryland

Philip A. Townsend
Forest and Wildlife Ecology
University of Wisconsin
Wisconsin, Madison

Susan L. Ustin
The Center for Spatial Technologies and
    Remote Sensing (CSTARS)
University of California
Oakland, California

Matti Vaaja
Centre of Excellence in Laser Scanning
    Research
Aalto University
Helsinki, Finland

Gustavo M. Vasques
Embrapa Soils
Rio de Janeiro, Brazil

Mikko Vastaranta
Department of Forest Sciences
University of Helsinki
Helsinki, Finland

Cédric Véga
Laboratoire de l'Inventaire Forestier
Institut National de l'Information
    Geographique et Forestière
Nancy, France

and

Institut Français de Pondichéry
Pondicherry, India

Elodie Vintrou
Joint Research Unit "Spatial Information
    and Analysis for Territories and
    Ecosystems"
French Agricultural Research Centre for
    International Development
Montpellier, France

Francesco Vuolo
University of Natural Resources and Life
    Sciences (BOKU)
Wien, Austria

Cuizhen Wang
Department of Geography
University of South Carolina
Columbia, South Carolina

Robert A. Washington-Allen
Environmental Tomography Laboratory
Department of Geography
University of Tennessee
Knoxville, Tennessee

Johanna Wetterlind
Department of Soil and Environment
Swedish University of Agricultural
    Sciences
Skara, Sweden

Joanne C. White
Natural Resources Canada
Ottawa, Ontario, Canada

Michael Wulder
Natural Resources Canada
Ottawa, Ontario, Canada

Jun Xiong
School of Earth Sciences and
    Environmental Sustainability
Northern Arizona University
Flagstaff, Arizona

Kamini Yadav
Department of Geography
University of New Hampshire
Durham, New Hampshire

Anthony Gar-On Yeh
Department of Urban Planning and
    Design
The University of Hong Kong
Hong Kong SAR, People's Republic of
    China

Xiaowei Yu
Finnish Geodetic Institute
and
Centre of Excellence in Laser Scanning
    Research
Masala, Finland

Quanfa Zhang
Key Laboratory of Aquatic Botany and
    Watershed Ecology
Wuhan Botanical Garden
Chinese Academy of Sciences
Wuhan, Hubei, People's Republic of
    China

Yongguang Zhang
Institute for Space Sciences
Freie Universität Berlin
Berlin, Germany

Yulong Zhang
Department of Geography
University of North Carolina at Chapel
    Hill
Chapel Hill, North Carolina

and

Key Laboratory of Aquatic Botany and
    Watershed Ecology
Wuhan Botanical Garden
Chinese Academy of Sciences
Wuhan, Hubei, People's Republic of
    China

Baojuan Zheng
School of Geographical Sciences and
    Urban Planning
Arizona State University
Tempe, Arizona

Vegetation and Biomass

I

Monitoring Photosynthesis from Space

Alfredo Huete
University of Technology

Guillermo
Ponce-Campos
*USDA/ARS Southwest
Watershed Research Center*

Yongguang Zhang
Freie Universität Berlin

Natalia Restrepo-Coupe
University of Technology

Xuanlong Ma
University of Technology

Mary Susan Moran
*USDA/ARS Southwest
Watershed Research Center*

Acronyms and Definitions

ANPP	Aboveground net primary production
APAR	Absorb photosynthetically active radiation
BIOME-BGC	BioGeochemical Cycles Model
BPLUT	Biome-properties look-up table
DOE	Department of Energy
EC	Eddy covariance
eLUE	Ecosystem LUE
EVI	Enhanced vegetation index
FAO	Food and Agriculture Organization
FLEX	Fluorescence explorer satellite
fAPAR	Fraction of APAR
fPAR	Fraction of PAR
G-R	Greenness and radiation
GOME-2	Global Ozone Monitoring Experiment-2
GOSAT	Greenhouse Gases Observing Satellite
GPP	Gross primary productivity
HyspIRI	Hyperspectral infrared imager
LAI	Leaf area index
Lidar	Light detection and ranging
LST	Land surface temperature
LSWI	Land surface water index
LUE	Light-use efficiency
LUT	Look up table
MTCI	MERIS terrestrial chlorophyll index
MERIS	Medium-resolution imaging spectrometer
MODIS	Moderate-resolution imaging spectroradiometer
NASS	National Agricultural Statistics Service
NCAR	National Centre for Atmospheric Research
NCEP	National Centres for Environmental Prediction
NDVI	Normalized difference vegetation index
NEE	Net ecosystem exchange
NIR	Near infrared
NPP	Net primary production
OCO-2	Orbiting Carbon Observatory-2
PAR	Photosynthetically active radiation
PVI	Perpendicular vegetation index
SIF	Solar-induced chlorophyll fluorescence
SPOT-4	Satellite Pour l'Observation de la Terre-4
SW	Short-wave downward solar radiation
T-G	Temperature and Greenness
Ta	Air temperature
TC-GVI	Tasselled cap green vegetation index
TROPOMI	Tropospheric Monitoring Instrument satellite
USDA	U.S. Department of Agriculture

VI   Vegetation index
VPD   Vapor pressure deficit
VPM   Vegetation photosynthesis model
VPRM   Vegetation photosynthesis and respiration model
WDRVI   Wide dynamic range vegetation index

1.1 Introduction

Vegetation productivity is defined as the process by which plants use sunlight to produce organic matter from carbon dioxide through photosynthesis. Gross primary productivity (GPP), or photosynthesis, is the rate of carbon fixation or total plant organic matter produced per unit of time and over a defined area, whereas the amount of carbon fixed by plants and accumulated as biomass is known as terrestrial net primary production (Cramer et al. 1999; Zhao and Running 2010). Productivity forms the basis of terrestrial biosphere functioning and carbon, energy, and water budgets. Accurate estimates of plant productivity across space and time are thus necessary for quantifying carbon balances at regional to global scales (Lieth 1975; Schimel 1998). Vegetation productivity is generally limited by the availability of spatially and temporally varying plant resources (e.g., nutrients, light, water, and temperature) (Field et al. 1995; Churkina and Running 1998; Nemani et al. 2003) (Figure 1.1). Improved knowledge of the main drivers and resource constraints of plant productivity is thus needed for predictable assessments of climate change.

1.1.1 Measures of Productivity

Measures of productivity are essential in global change studies; yet despite their importance, they are quite challenging to obtain or sample (Baldocchi et al. 2001). The assessment of plant production is carried out in various ways, from plot measurements and plant harvests, micrometeorological fluxes, remote sensing, and through empirical and process-based models that may involve remote sensing data inputs. In situ measures include methods that vary with biome type, for example, tree inventories, litter traps, grassland forage estimates, and agricultural harvests and market statistics. Plot-level methods measure aboveground net primary production (ANPP) that often involves destructive sampling during peak biomass periods. Established long-term experimental plots enable cross-site production comparisons; however, they are also amenable to many uncertainties due to differences in site-based procedures, and in some cases, inconsistent sampling methods over time at a given site (Sala et al. 1988; Biondini et al. 1991; Moran et al. 2014). GPP has traditionally been estimated from plot level ANPP measurements by correcting for respiratory losses (Field et al. 1995). Agricultural yield statistics (USDA NASS) combined with maps of cropland areas provide large-scale NPP estimates from local to national level census statistics (Monfreda et al. 2008; Guanter et al. 2014).

A global network of micrometeorological tower sites, known as FLUXNET, now provide continuous measurements of carbon, water, and energy exchanges between ecosystems and the atmosphere (Running et al. 1999). This yields information on seasonal dynamics and interannual variations of net ecosystem exchange (NEE) of carbon dioxide between the land surface and the atmosphere (Baldocchi et al. 2001; Verma et al. 2005). This has yielded quite valuable in situ data to independently evaluate and assess uncertainties in carbon models and satellite carbon products, as they are applied to global change studies.

Satellite imaging sensors offer synoptic-scale observations of ecosystem states and landscape dynamics, and are seen as invaluable tools to help fill the large spatial gaps of in situ measurements, and constrain and improve the accuracies of models. Remote sensing complements the restrictive coverage afforded by experimental plots and eddy covariance (EC) tower flux

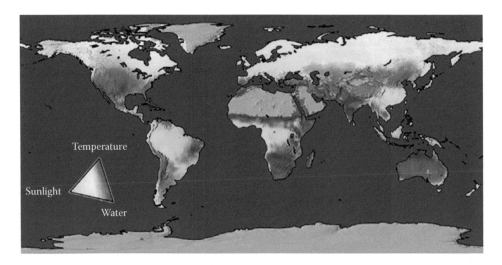

FIGURE 1.1 Potential limits to vegetation net primary production based on fundamental physiological limits of solar radiation, water balance, and temperature. Greener colors depict biomes increasingly limited by radiation, while red colors are water-limited and blue colors temperature-limited. Many regions are limited by more than one factor. (Adapted from Nemani, R.R. et al., *Science*, 300(5625), 1560, 2003.)

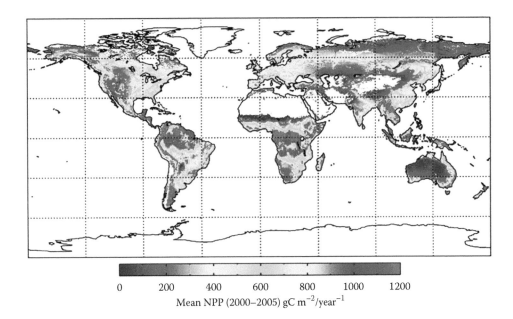

FIGURE 1.2 MODIS net primary production satellite product (MOD17). Example showing the mean NPP across years 2000–2005 for the global terrestrial surface. The highest production is seen across the equatorial zone encompassing southeast Asia, the Amazon basin, and equatorial Africa. The least productive regions appear in Australia and the Sahelian region. (Courtesy of Numerical Terradynamic Simulation Group, University of Montana, Missoula, MT.)

measurements, facilitating observations of broad-scale patterns of ecosystem functioning. This renders remote sensing a powerful tool for studying vegetation productivity at local, regional, and global scales (Gitelson et al. 2006).

The integration of independently derived tower measured carbon fluxes with satellite data is the focus of many investigations across many ecosystems from sparse shrublands to mesic grasslands, and tropical to temperate forests. Estimates of daily GPP and annual NPP are now routinely produced operationally over the global terrestrial surface at 1 km spatial resolution through production efficiency models with near real-time satellite data inputs from the moderate-resolution imaging spectroradiometer (MODIS) (Turner et al. 2006) (Figure 1.2).

Finally, there are many empirical, diagnostic, and process-based models that have been developed over the past decades to monitor and assess vegetation productivity, with many of these methods employing remote sensing data in conjunction with micrometeorological carbon flux measurements to varying extents.

1.1.2 Lidar

Traditionally, national-scale carbon monitoring has been accomplished with networks of field inventory plots (FAO 2007), which provide direct carbon measurements of only very small areas of forest, and are further difficult to install, monitor, and maintain over time (Chambers et al. 2009). Airborne laser technology called light detection and ranging (lidar) offers much potential for terrestrial carbon assessments. Lidar measures the physical structure of woody vegetation, from sparse shrublands to dense forests, and can serve as a

reliable replacement for inventory plots in areas lacking field data (Lefsky et al. 2002; Zolkos et al. 2013). Thus, lidar integration with field inventory plots can provide calibrated lidar estimates of aboveground carbon stocks, which can then be scaled up using satellite data on vegetation cover, topography, and rainfall from satellite data to model carbon stocks (Asner et al. 2013). Opportunities to fuse temporally dynamic vegetation optical measurements with lidar have promising potential for better assessments of not only standing wood biomass, but also forest disturbance, biomass loss, and carbon accumulation through forest regrowth (Lefsky et al. 2002; Asner et al. 2010).

1.2 Remote Sensing and Net Primary Production

1.2.1 NDVI–fAPAR Relationships

Remote sensing approaches to estimate productivity generally employ spectral measures of vegetation, which are used for estimating their capacity to absorb photosynthetically active radiation (APAR). Vegetation productivity is directly related to the interaction of solar radiation with the plant canopy, based on the original logic of Monteith (1972), who suggested that productivity of stress-free annual crops was linearly related to vegetation absorbed PAR. Spectral vegetation indices (VIs) such as the normalized difference vegetation index (NDVI) (Tucker 1979), the perpendicular vegetation index (PVI) (Richardson and Wiegand 1977), and the tasselled cap green vegetation index (TC-GVI) (Kauth and Thomas 1976) were consequently developed over croplands and grasslands.

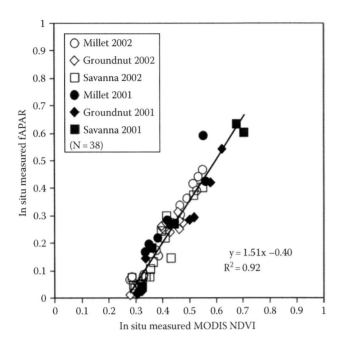

FIGURE 1.3   Linear relationship between in situ NDVI and field measured fAPAR across multiple cropland and biome sites in Africa. (From Fensholt, R. et al., *Remote Sens. Environ.*, 91(3–4), 490, 2004.)

The NDVI is written as follows:

$$NDVI = \frac{(\rho_{NIR} - \rho_{red})}{(\rho_{NIR} + \rho_{red})}, \qquad (1.1)$$

where ρ_{NIR} and ρ_{red} are spectral reflectance values (unitless) that exploit the chlorophyll-absorbing red band relative to the non-absorbing and high scattering near-infrared (NIR) band. Asrar et al. (1984) showed the NDVI was linearly related with vegetation absorption of light energy (APAR) or fraction of APAR (fAPAR), and thereby related to productivity through the potential capacity of vegetation to absorb light for photosynthesis (Figure 1.3). The linear relationship between NDVI and fAPAR has been documented through field measurements (Ruimy et al. 1994; Fensholt et al. 2004) and theoretical analyses (Sellers 1985; Goward and Huemmrich 1992; Myneni and Williams 1994).

1.2.2 Annual Integrated Estimates of Productivity

Several studies suggest that annual vegetation productivity status can be captured with the annual NDVI integral, used as surrogate measures of fAPAR. Goward et al. (1985) used integrated NDVI values derived from the advanced-very-high-resolution radiometer (AVHRR) and found good relationships between NPP and integrated NDVI over annual growing periods of North American biomes (Figure 1.4). Wang et al. (2004) found that the NDVI integral over the early growing season was strongly correlated to in situ forest measurements of diameter increase and tree ring width in the U.S. central Great Plains. They also found

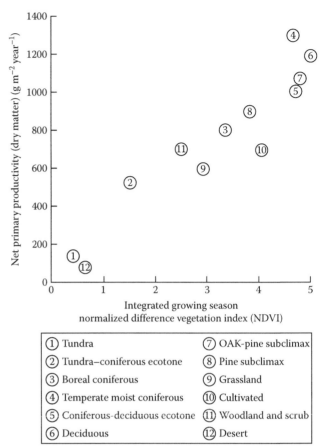

① Tundra	⑦ OAK-pine subclimax
② Tundra–coniferous ecotone	⑧ Pine subclimax
③ Boreal coniferous	⑨ Grassland
④ Temperate moist coniferous	⑩ Cultivated
⑤ Coniferous-deciduous ecotone	⑪ Woodland and scrub
⑥ Deciduous	⑫ Desert

FIGURE 1.4   Relationship between biome averaged integrated NDVI from NOAA-AVHRR sensors and net primary productivity rates. (From Goward, S.N. et al., *Vegetatio*, 64(1), 3, 1985.)

the previous year integrated NDVI was well correlated with current year increases in tree height growth.

The annual integrated VI offers a robust approximation of vegetation productivity, because, in general, VIs provide both a measure of the capacity to absorb light energy, as well as reflect recent environmental stress acting on the canopy, with stress forcings showing up as reductions in NDVI expressed as either less chlorophyll and/or less foliage (Running et al. 2004). Photosynthesis or primary production is essentially integrator of resource availability, and according to the resource optimization theory (Field et al. 1995), ecological processes tend to adjust plant characteristics over time periods of weeks or months to match the capacity of the environment to support photosynthesis and maximize growth.

Ponce-Campos et al. (2013) compiled in situ field measures of ANPP across 10 sites in the United States, ranging from arid grassland to forest and directly compared annual integrated values of the MODIS enhanced vegetation index (EVI, or iEVI) (Figure 1.5). Using a log–log relation to account for the uneven distribution of ANPP estimates over time, the iEVI was found to be an effective surrogate to estimate ANPP and quantify vegetation dynamics:

$$ANPP = 51.42 \times iEVI^{1.15}, \qquad (1.2)$$

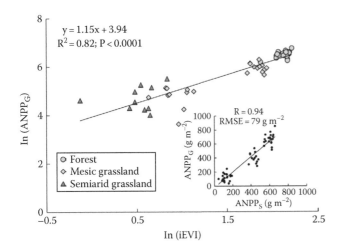

FIGURE 1.5   Relationship between in situ estimates of aboveground net primary production (ANPPg) and annual integrated EVI (iEVI) derived from MODIS data (2000–2009) for 10 sites across several biomes. The solid line represents the linear regression used to estimate ANPP from iEVI (ANPPs). (From Ponce-Campos, G.E. et al., *Nature*, 494(7437), 349, 2013.)

$$EVI = G \cdot \frac{\rho_{NIR} - \rho_{red}}{\rho_{NIR} + L + C_1 \cdot \rho_{red} - C_2 \cdot \rho_{blue}} \qquad (1.3)$$

where

ρ_{NIR}, ρ_{red}, and ρ_{blue} are atmospherically corrected spectral reflectances

G is a gain factor

C_1 and C_2 are aerosol resistance coefficients

L functions as the soil-adjustment factor, with all terms dimensionless (Huete et al. 2002)

In the MODIS EVI product, G = 2.5, L = 1, and C_1 and C_2 are 6.0 and 7.5, respectively.

Moran et al. (2014) found plot-scale measurements of ANPP at arid and mesic grassland sites were significantly related to MODIS iEVI over a decadal time period in a log–log relation ($r^2 = 0.71$, $P < 0.01$). Zhang et al. (2013) studied the ecological impacts of rainfall intensification on vegetation productivity through the use of iEVI as a surrogate measure of ANPP. They found extreme precipitation patterns, associated with heavy rainfall events followed by longer dry periods, caused higher water stress conditions that resulted in strong negative influences on ANPP across biomes and reduced rainfall use efficiencies (20% on average) (Figure 1.6).

1.2.3 Growing Season Phenology Relationships

The annual life cycle of plant species and vegetation canopies have large effects on rates of photosynthesis and annual productivity. Phenological factors such as leaf age and life

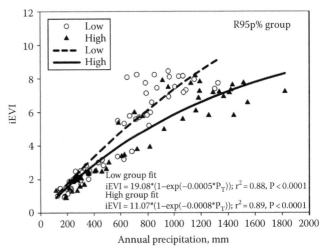

FIGURE 1.6   Relationship of growing season integrated MODIS EVI, as a surrogate of annual primary production, and annual precipitation for low vs. high rainfall variability. This demonstrates the negative influence of precipitation intensification for a wide range of biome types. (Zhang, Y. et al., *J. Geophys. Res. Biogeosci.*, 118(1), 148, 2013.)

expectancy play important roles on productivity (Wilson et al. 2001) with some production models explicitly incorporating phenophase periods, such as bud burst to full leaf expansion, and full expansion to dormancy (Xiao et al. 2004). LST satellite data and/or meteorological air temperature data (Ta) are also used to identify biologic inactive seasonal periods, for example, masking cold temperature time intervals from the EVI or NDVI integrals.

Often, there is also a need to synchronize the satellite data with scheduled or variable destructive sampling dates. Generally, in situ measures of productivity are made at discrete times within the growing season or may be associated with variable sampling times with uncertain estimates of the dates of peak greenness. In such cases, remote sensing data provides better temporal stability and opportunities to reduce productivity uncertainties. For example, Moran et al. (2014) found significant improvements in productivity–iEVI relationships across a range of grassland sites, when the EVI was only partially integrated from the beginning to the peak of the growing season period (rather than the full season). This was due to the synchronization of time periods to peak biomass periods when grassland ANPP destructive sampling are typically conducted.

Numerous efforts have been made to improve upon the characterization of the plant growing season at regional scales using satellite-based phenology models. Software packages such as Timesat (Jönsson and Eklundh 2004) can be used to quantitatively model the growing season and facilitate the temporal synchronization of in situ production measures with satellite data. A summary of the various remote sensing methods that have been used in estimating net primary productivity is shown in Table 1.1.

TABLE 1.1   Examples of Remote Sensing Methods of Deriving Net Primary Productivity with Some References

Net Primary Productivity Measurement	Biome/Location	Satellite Products Used	Method/Approach	Equation	R^2	Reference
Annual NPP	Across different North American biomes from tundra to forest to crops and deserts	Integrated growing season NDVI from NOAA/AVHRR	Linear regression between NPP and integrated NDVI	NA	0.89 / 0.94 (excl. crops)	Goward et al. (1985)
Growing season NPPgs, Early growing season NPPegs, Tree ring width, stem growth, and litterfall	Natural and plantation forests in Central Great Plains, North America	NOAA/AVHRR NDVI integrated across (1) growing season (late April–October); (2) early growing season (May–June); and (3) annual year.	Linear regression between NPP and integrated NDVI for growing season, early growing season, and annual year	NA	Growing season, 0.86 / Early growing season, 0.83	Wang et al. (2004)
Annual above ground net primary productivity (ANPP)	Ten sites ranging from forests to mesic and semiarid grasslands to forest (USA)	Annual integrated values of MODIS enhanced vegetation index, iEVI	Log-log relation between ANPP and iEVI	$\text{ANPP} = 51.42 \times \text{iEVI}$ 1.15 ANPP $(g\ m^{-2})$	0.82	Ponce-Campos et al. (2013)
Annual above ground net primary productivity (ANPP)	Arid to mesic grasslands	Annual integrated values of MODIS enhanced vegetation index, iEVI	Log-log relation between ANPP and iEVI	NA	0.71	Moran et al. (2014)
Aboveground carbon density (ACD)	Mangroves, dry, moist and wet forests	LiDAR top-of-canopy height	Exponential relation between ACD and top-of-canopy height (H)	$\text{ACD} = 0.359 \times H^{1.7676}$	0.86, calibration plots / 0.92, validation plots	Asner et al. (2013)

1.3 Remotely Sensed Production Efficiency Models

Remote sensing estimates of GPP and net primary production (NPP) have been implemented at global scales, based on the light-use efficiency (LUE) equation that defines the amount of carbon fixed through photosynthesis as proportional to the solar energy absorbed by green vegetation multiplied by the efficiency with which the absorbed light is used in carbon fixation (Monteith 1972; Monteith and Unsworth 1990):

$$GPP = \varepsilon \times APAR = \varepsilon \times fAPAR \times PAR \quad (1.4)$$

where

ε is the efficiency of conversion of absorbed light into aboveground biomass, or light-use efficiency

APAR is integrated over a time period

fAPAR is derived through spectral VI relationships (Asrar et al. 1984; Sellers 1985; Goward and Huemmrich 1992; Ruimy et al. 1994).

The LUE concept has been widely adopted by the remote sensing community to assess and extrapolate carbon processes through knowledge of two conversion coefficients: the fAPAR and ε. Although fAPAR is readily estimated using remotely sensed "greenness" measures, ε is very difficult to measure as it dynamically varies with plant functional type, vegetation phenophase, and different environmental stress conditions (Ruimy et al. 1995; Turner et al. 2003; Sims et al. 2006; Jenkins et al. 2007). As a result, there are scarce measurements of ε available,

particularly at the landscape scale, and potential or maximum LUE values have only been specified for a limited set of biome types, with these values downregulated by environmental stress scalars derived from meteorological inputs (Zhao et al. 2005; Heinsch et al. 2006).

1.3.1 BIOME-BGC Model

The BIOME-BGC (BioGeochemical Cycles) model calculates daily GPP as a function of incoming solar radiation, conversion coefficients, and environmental stresses (Running et al. 2004). This was implemented as the first operational standard satellite product for MODIS (MOD17), providing global estimates of global GPP (Figure 1.7), expressed as follows:

$$GPP = \varepsilon_{max} \times 0.45 \times SW_{rad} \times fPAR \times f(VPD) \times f(T_{min}), \quad (1.5)$$

where

ε_{max} is the maximum light-use efficiency (g C MJ^{-1}) obtained from a biome-properties look-up table (BPLUT)

SW_{rad} is short-wave downward solar radiation (MJ^{-1} day^{-1}), of which 45% is assumed to be PAR

$f(VPD)$ and $f(T_{min})$ are vapor pressure deficit and air temperature reduction scalars for the biome specific ε_{max} values

fAPAR is directly input from the MODIS FPAR (MOD15) product (Running et al. 2004; Zhao et al. 2005)

MODIS FPAR retrievals are physically based and use biome-specific look-up tables (LUTs) generated using a three-dimensional radiative transfer model (Myneni et al. 2002).

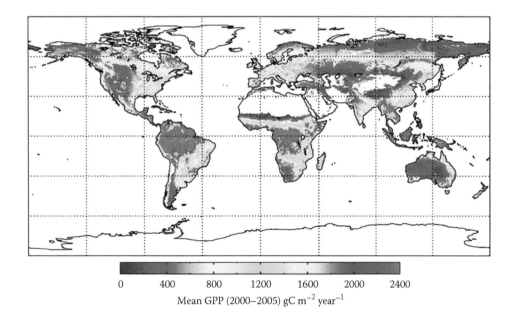

Mean GPP (2000–2005) gC m^{-2} year^{-1}

FIGURE 1.7   MODIS gross primary production satellite product (MOD17). Example showing the mean GPP across years 2000–2005 for the global terrestrial surface. The highest rates of photosynthesis are seen in the tropical forests of southeast Asia, the Amazon basin, and equatorial Africa. The lowest rates of photosynthesis are seen in Australia, South Africa, western North America, the Sahel, and Atacama desert. (Courtesy of Numerical Terradynamic Simulation Group, University of Montana, Montana, MT.)

The reduction scalars encompass LUE variability resulting from water stress (high daily VPD) and low temperatures (low daily minimum temperature T_{min}) (Running et al. 2004). The MODIS GPP product is directly linked to remote sensing and weather forecast products and can provide near real-time information on productivity and the influence of anomalies such as droughts. A consistent forcing meteorology is based upon the NCEP/NCAR (National Centres for Environmental Prediction/National Centre for Atmospheric Research) Reanalysis II datasets (Kanamitsu et al. 2002) (Figure 1.7).

Using these satellite products, Zhao and Running (2010) found that global NPP declined slightly by 0.55 petagram carbon (Pg C, with Pg = 10^{15} g = 1 billion metric tonnes) due to drought from 2000 to 2009. Ichii et al. (2007) used the BIOME-BGC model to simulate seasonal variations in GPP for different rooting depths, from 1 to 10 m, over Amazon forests and determine which rooting depths best estimated GPP consistent with satellite-based EVI, and thereby were able to map rooting depths at regional scales and improve the assessments of carbon, water, and energy cycles in tropical forests.

The utility and accuracy of MODIS GPP/NPP products have been validated in various FLUXNET studies, which have also demonstrated the value of independent tower flux measures to better understand the satellite-based GPP/NPP products (Kang et al. 2005; Leuning et al. 2005; Zhao et al. 2005, 2006; Turner et al. 2006). These studies highlight the capabilities of MODIS GPP to correctly predict observed fluxes at tower sites, but also draw attention to some of the uncertainties associated with use of coarse resolution and interpolated meteorology inputs, uncertainties with the LUT-based values, noise and uncertainties in the satellite fAPAR inputs, and difficulties in constraining the light-use efficiency term (Zhao et al. 2005; Heinsch et al. 2006; Yuan et al. 2010; Sjöström et al. 2013). Since meteorological inputs are often not available at sufficiently detailed temporal and spatial scales, they can introduce substantial errors into the carbon exchange estimates.

Turner et al. (2006) concluded that although the MODIS NPP/GPP products are generally responsive to spatial–temporal trends associated with climate, land cover, and land use, they tend to overestimate GPP at low productivity sites and underestimate GPP at high productivity sites. Similarly, Sjostrom et al. (2013) found that although MODIS-GPP described seasonality at 12 African flux tower sites quite well, it tended to underestimate tower GPP at the dry sites in the Sahel region due to uncertainties in the meteorological and fAPAR input data and the underestimation of ε_{max}. Jin et al. (2013) reported the MODIS GPP product to substantially underestimate tower GPP during the green-up phase at a woodland savanna site in Botswana, while overestimating tower-GPP during the brown-down phase.

Some studies have found that when properly parameterized with site-level meteorological measurements, MODIS GPP becomes more closely aligned with flux tower derived GPP (Turner et al. 2003; Kanniah et al. 2009; Sjostrom et al. 2013). Kanniah et al. (2011), however, found that utilizing site-based meteorology could only improve GPP estimates during the wet season over northern Australian savannas, and suggested the MODIS GPP product has a systematic limitation in the estimation of savanna GPP in arid and semiarid areas due to the lack of the representation of soil moisture. Sjöström et al. (2013) also found soil moisture information to be quite important for accurate GPP estimates in drier African savannas.

1.3.2 Vegetation Index: Tower GPP Relationships

There have also been many attempts to estimate GPP based solely on remote sensing inputs, thereby minimizing or eliminating the need for meteorological and LUE information. Spectral VIs have been directly related to EC tower carbon flux measurements (Rahman et al. 2005; Gitelson et al. 2006; Sims et al. 2006; Sjöström et al. 2011). Monteith and Unsworth (1990) noted that VIs can legitimately be used to estimate the rate of processes that depend on absorbed light, such as photosynthesis and transpiration.

Wylie et al. (2003) reported a strong relationship between biweekly aggregated NDVI and daytime CO_2 flux in a sagebrush-steppe ecosystem, while Rahman et al. (2005) found that EVI can provide reasonably accurate estimates of GPP across a wide range of North American ecosystems, including dense forests. However, the strength of the linear relationships between EVI and tower GPP in temperate forests was greater in seasonally contrasting deciduous forests compared with evergreen forests (Rahman et al. 2005; Sims et al. 2006). Sims et al. (2006) further noted that when data from the winter period of inactive photosynthesis were excluded, the EVI—tower GPP relationship was better than that between tower GPP and MODIS GPP (Figure 1.8). Olofsson et al. (2008) reported strong correlations between EVI and GPP across Northern Europe, while NDVI showed problems with saturation in such areas of high biomass. NDVI saturation is attributed to the strong weighing of the red band, which is primarily absorbed by the uppermost leaf layer of a dense crop or forest canopy while the nonabsorbing NIR band is able to penetrate 5–7 leaf layers. Thus, the more NIR-sensitive indices, such as EVI, PVI, TC-GVI, and linear mixture models are less prone to saturate (Huete et al. 2002, 2006).

Sjöström et al. (2011) found EVI was able to track the seasonal dynamics of tower GPP closely across African tropical savanna ecosystems. Ma et al. (2013) similarly observed good convergences between MODIS EVI and tower GPP across northern Australian mesic and xeric tropical savannas, confirming the potentials to link these two independent data sources for accurate estimation of savanna GPP. Strongly linear and consistent relationships between EVI and tower GPP were also shown in dry to humid tropical forest sites in Southeast Asia and the Amazon (Xiao et al. 2005; Huete et al. 2006, 2008).

These relationships have shown the EVI to estimate GPP with relatively high accuracy, thus greatly simplifying carbon balance models and potentially offering opportunities for

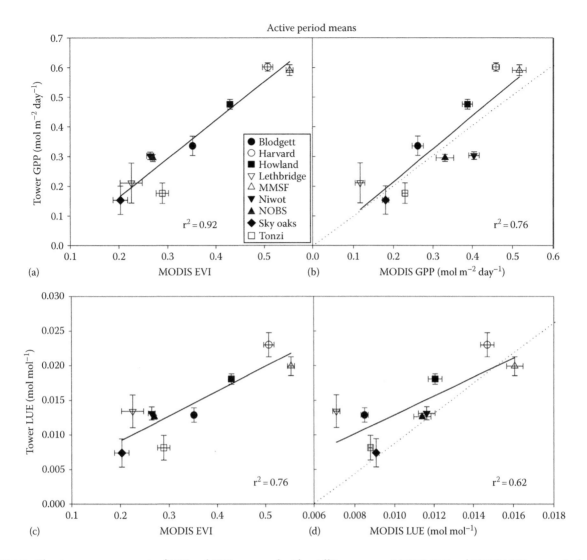

FIGURE 1.8 Flux tower measurements of GPP and LUE compared with satellite measures, MODIS GPP and MODIS EVI, respectively, over a range of North American biome types. (From Sims, D.A. et al., *J. Geophys. Res.*, 111, 2006.)

region-wide scaling of carbon fluxes. The relationships between EVI and tower GPP are partly a result of fairly good correlations between LUE and EVI that make an independent estimate of LUE less necessary. Sims et al. (2006) reported that LUE derived from nine flux towers in North America was well correlated with EVI ($R^2 = 0.76$; Figure 1.8), while Wu et al. (2011) reported moderate correlation between EVI and tower LUE in temperate and boreal forest ecosystems in North America. Further, the 16-day averaging period removes much of the influences of short-term fluctuations in solar radiation and other environmental parameters, thereby minimizing the need for climatic drivers. On the other hand, such relationships were weaker in evergreen forests relative to deciduous ones and one study in an evergreen oak forest showed no correlation between EVI and LUE (Goerner et al. 2009); thus, correlations between EVI and LUE may be a result of covariations between fAPAR and LUE.

1.3.3 Temperature and Greenness Model (T-G)

The simple VI "greenness" model, defined as the straightforward relationship between VIs and GPP, although potentially useful in certain cases, exhibits various limitations due to its inability to always recognize between growth and inactive growth periods, in which spectral "greenness" may show little change. These inactive periods are associated with evergreen vegetation in winter months with low temperatures as well as evergreen vegetation growing in Mediterranean climates in which high temperature, vapor pressure deficit, and soil drought limit growth (Sims et al. 2008; Vickers et al. 2012).

For these reasons, Sims et al. (2008) introduced the temperature and greenness (T-G) model, using combined daytime LST (Wan 2008) and EVI products from MODIS. They found the T-G model substantially improved the correlation between predicted and measured GPP at 11 EC flux tower sites across

North American biomes compared with the MODIS GPP product or MODIS EVI alone, while keeping the model based entirely on remotely sensed variables without any ground-based meteorological inputs (Sims et al. 2008). The T-G model may be described as follows:

$$GPP = (EVI_{scaled} \times LST_{scaled}) \times m \qquad (1.6)$$

$$LST_{scaled} = min\left[\left(\frac{LST}{30}\right); \ (2.5 - (0.05 \times LST))\right] \qquad (1.7)$$

$$EVI_{scaled} = EVI - 0.10 \qquad (1.8)$$

where

  LST$_{scaled}$ sets GPP to zero when LST is less than zero, and defines the inactive winter period

  EVI$_{scaled}$ adjusts EVI values to a zero baseline value in which GPP is known to be zero

  m is a scalar that varies between deciduous and evergreen sites, with units of mol C m^{-2} day^{-1}

LST$_{scaled}$ also accounts for low temperature limitations to photosynthesis when LST is between 0°C and 30°C, and accounts for high temperature and high VPD stress in sites that exceed LST values of 30°C (Sims et al. 2008) (Figure 1.9).

LST is closely related to VPD and thus can provide a measure of drought stress (Hashimoto et al. 2008), consistent with the BIOME-BGC model, where temperature and VPD are used as scalars directly modifying LUE (Running et al. 2004). LST is a useful measure of physiological activity of the upper canopy leaves, provided that leaf cover is great enough that LST is not significantly affected by soil surface temperature. Thus, the T-G model has been found less useful in sparsely vegetated

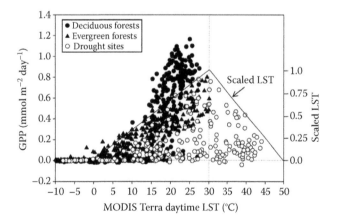

FIGURE 1.9 GPP measured at the EC flux towers as a function of daytime LST measured by the MODIS satellite. Solid line represents scaled LST from T-G model. GPP is enhanced by increasing temperatures, but only to approximately 30°C before being negatively influenced. (From Sims, D.A. et al., *Remote Sens. Environ.*, 112(4), 1633, 2008.)

ecosystems (e.g., shrubs) where soil surface temperatures significantly influence derived LST values, rendering them less useful as indicators of plant physiology. As an example, Ma et al. (2014) found coupling EVI with LST showed no improvements in predicting savanna GPP compared with using EVI alone over the relatively open tropical savannas in northern Australia, with appreciable soil exposure. This may also be due to temperature not being a limiting factor or significant driver of photosynthesis in tropical savannas (Leuning et al. 2005; Cleverly et al. 2013; Kanniah et al. 2013b).

1.3.4 Greenness and Radiation (G-R) Model

Chlorophyll-related spectral indices have also been coupled with measures of light energy, PAR, to provide robust estimates of GPP:

$$GPP = VI_{chl} \times PAR_{toc} \qquad (1.9)$$

where

  PAR$_{toc}$ is the top-of-canopy measured PAR (MJ m^{-2} day^{-1})

  VI$_{chl}$ is a chlorophyll-related spectral index

Peng et al. (2013) described two types of chlorophyll spectral indices, (1) commonly used VIs, such as EVI and the wide dynamic range vegetation index (WDRVI), which indirectly indicate total chlorophyll content through "greenness" estimates and (2) chlorophyll indices, such as the MERIS terrestrial chlorophyll index (MTCI), which directly represent the leaf chlorophyll content. The WDRVI equation is,

$$WDRVI = \frac{\left(a^* \ \rho_{NIR} - \rho_{red}\right)}{\left(a^* \ \rho_{NIR} + \rho_{red}\right)} \qquad (1.10)$$

where a is a weighing coefficient with value between 0.1 and 0.2 (Gitelson 2004; Gitelson et al. 2006).

MTCI is the ratio of the difference in reflectance between an NIR and red edge band and the difference in reflectance between red edge and red band as

$$MTCI = \frac{\left(\rho_{753.75} - \rho_{708.75}\right)}{\left(\rho_{708.75} - \rho_{681.25}\right)}, \qquad (1.11)$$

where $\rho_{753.75}$, $\rho_{708.75}$, and $\rho_{681.25}$ are reflectances in the center wavelengths of the MERIS narrow-band channel settings (Dash and Curran 2004).

Canopy level chlorophyll represents a community property that is most relevant in quantifying the amount of absorbed radiation used for productivity (Whittaker and Marks 1975; Dawson et al. 2003). Long- or medium-term changes (weeks to months) in canopy chlorophyll are related to canopy stress, phenology, and photosynthetic capacity of the vegetation (Ustin et al. 1998; Zarco-Tejada et al. 2002). Ciganda et al. (2008) showed that for the same LAI amount, the chlorophyll content during the green-up stage might be more than two times higher than the chlorophyll content in leaves in the reproductive and senescence stages. In the G-R model, both fAPAR and LUE are driven by total chlorophyll content with strong correlations between GPP/PAR and canopy chlorophyll content (Gitelson et al. 2006; Peng et al. 2011).

Ma et al. (2014) found significant improvements in the use of G-R models, relative to EVI alone, for predicting tower GPP, demonstrating the importance of this quantity as a critical driver of savanna vegetation productivity (Whitley et al. 2011; Kanniah et al. 2013a). The R-G model has been successfully applied in estimating GPP in natural ecosystems (Sjöström et al. 2011; Wu et al. 2011, 2014) and croplands, including maize, soybeans, and wheat (Wu et al. 2010; Peng et al. 2011; Peng and Gitelson 2012).

Site-based PAR_{toc} measurements, however, may exhibit uncertainties associated with high-frequency fluctuations that are difficult to extrapolate beyond the tower sensor footprint and, moreover, scale regionally. Therefore, other measures of PAR that have been used include "potential" PAR, or maximal clear-sky PAR ($PAR_{potential}$) (Gitelson et al. 2012; Peng et al. 2013; Rossini et al. 2014) and top-of-atmosphere PAR (PAR_{toa}). $PAR_{potential}$ can be calibrated from long-term PAR_{toc} measurements or modeled using an atmosphere radiative transfer code (Kotchenova and Vermote 2007).

Gitelson et al. (2012) found an improved performance of $PAR_{potential}$ relative to PAR_{toc} noting that decreases in PAR_{toc} during the day do not always imply a decrease in GPP. Further, Kanniah et al. (2013a) showed that the negative forcings of wet season cloud cover on Australian tropical savannas were partly compensated by enhanced LUE resulting from a greater proportion of diffuse radiation. Ma et al. (2014) found that coupling of EVI with PAR_{toa} better predicted GPP than coupling EVI with PAR_{toc} and attributed this to tower sensor-based measurement uncertainties of PAR_{toc}, as well as better approximations of meteorological controls on GPP by PAR_{toa}.

Two definitions of LUE become apparent in G-R models, with this term either defined as the ratio of GPP to APAR or defined as the ratio of GPP to PAR (Gower et al. 1999), with the latter sometimes referred to as ecosystem-LUE or eLUE:

$$\varepsilon = \frac{GPP}{APAR} = \frac{GPP}{fAPAR \times PAR}, \tag{1.12}$$

$$eLUE = \frac{GPP}{PAR} = fAPAR \times \varepsilon \tag{1.13}$$

An advantage of using chlorophyll-based VIs in G-R models is that the biological drivers of photosynthesis, fAPAR and light-use efficiency (ε) resulting from environmental stress and leaf age phenology, are combined into eLUE, thereby simplifying remote sensing–based productivity estimates.

1.3.5 Vegetation Photosynthesis Model (VPM)

Xiao et al. (2004) developed a mostly satellite-based vegetation photosynthesis model (VPM) that estimates GPP using satellite inputs of EVI and the land surface water index (LSWI):

$$GPP = \varepsilon \times fAPAR_{chl} \times PAR_{toc} \tag{1.14}$$

$$\varepsilon = \varepsilon_{max} \times T_{scalar} \times W_{scalar} \times P_{scalar} \tag{1.15}$$

where
  $fAPAR_{chl}$ is estimated as a linear function of EVI
  PAR_{toc} is measured at the site
  T_{scalar}, W_{scalar}, P_{scalar} are scalars for the effects of temperature, water, and leaf phenology on vegetation, respectively (Figure 1.10)

T_{scalar} is based on air temperature and uses minimum, maximum, and optimum temperature for photosynthesis at each time step; W_{scalar} is based on satellite-derived LSWI that accounts for the effect of water stress on photosynthesis:

$$W_{scalar} = \frac{\left(1 + LSWI\right)}{\left(1 + LSWI_{max}\right)}, \tag{1.16}$$

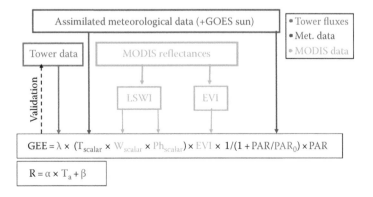

FIGURE 1.10 Schematic diagram of the VPRM utilizing EVI, LSWI, and scalars for temperature, leaf phenology, and canopy water content, T_{scalar}, P_{scalar}, and W_{scalar}, respectively. The VPM model uses primarily remote sensing data along with air temperatures, while the VPRM model additionally assimilates tower flux and meteorological information. (From Mahadevan, P. et al., *Global Biogeochem. Cycles*, 22(2), GB2005, 2008.)

$$LSWI = \frac{\left(\rho_{nir} - \rho_{swir}\right)}{\left(\rho_{nir} + \rho_{swir}\right)}, \qquad (1.17)$$

where

ρ_{swir} is the reflectance in a broadband shortwave infrared band (e.g., MODIS, 1580–1750 nm)

$LSWI_{max}$ is the maximum value for the growing season

P_{scalar} accounts for the effect of leaf age on photosynthesis and is dependent on the growing season life expectancy of the leaves (Wilson et al. 2001). P_{scalar} is calculated over two phenophases as

$$P_{scalar} = \frac{\left(1 + LSWI\right)}{2} \qquad (1.18)$$

from bud burst to full leaf expansion, and $P_{scalar} = 1$, after full expansion (Xiao et al. 2004).

The VPM model has been applied to both MODIS and SPOT-4 VEGETATION sensor data to produce tower-calibrated estimates of GPP across a wide range of biomes, including evergreen and deciduous forests, grasslands, and shrub sites in temperate North America and in seasonally moist tropical evergreen forests in the Amazon (Xiao et al. 2005; Mahadevan et al. 2008; Jin et al. 2013).

Mahadevan et al. (2008) further developed the vegetation photosynthesis and respiration model (VPRM), a satellite-based assimilation scheme that estimates hourly values of NEE using EVI, LSWI, and high-resolution meteorology observations of sunlight and air temperature (Figure 1.10). NEE represents the difference between uptake (photosynthesis) and loss (respiration) processes that vary over a wide range of timescales (Goulden et al. 1996; Katul et al. 2001). The VPRM model provides fine-grained fields of surface CO_2 fluxes for application in inverse models at continental and smaller scales (Mahadevan et al. 2008). This capability is presently limited by the number of vegetation classes for which NEE can be constrained using EC tower flux data. A summary of the various remote sensing–based estimates of GPP is shown in Table 1.2.

1.3.6 Photochemical Reflectance Index (PRI)

There is also much interest in reducing the uncertainties in GPP models through direct remote sensing assessments of LUE. The photochemical reflectance index (PRI) is a hyperspectral index that provides a scaled LUE measure, or photosynthetic efficiency, based on light absorption processes by carotenoids (Gamon et al. 1992; Middleton et al. 2011),

$$PRI = \frac{\left(\rho_{531\,nm} - \rho_{570\,nm}\right)}{\left(\rho_{531\,nm} + \rho_{570\,nm}\right)} \qquad (1.19)$$

Spectral variations at 531 nm are closely associated with the dissipation of excess light energy by xanthophyll pigments (a major carotenoid group of yellow pigments) in order to protect the photosynthetic leaf apparatus (Ripullone et al. 2011). Carotenoids function in processes of light absorption in plants as well as protecting plants from the harmful effects of high light conditions; hence, lower carotenoid/chlorophyll ratios indicate lower physiological stress (Peñuelas et al. 1995; Guo and Trotter 2004).

Several studies have shown the linear relationship between PRI and LUE over different vegetation types (e.g., Nichol et al. 2000, 2002). Rahman et al. (2004) produced a "continuous field" retrieval of LUE from satellite data, using the PRI as a proxy of LUE, without the need of LUTs or predetermined biome-specific LUE values. They suggested that the variations found in the continuous LUE fields must be taken into account to accurately estimate CO_2 fluxes of terrestrial vegetation. However, Barton and North (2001) showed that PRI was most sensitive to changes in leaf area index (LAI), and Gitelson et al (2006) noted that in order to use PRI to predict LUE, one would need an independent estimate of LAI.

The upcoming potential launches of new hyperspectral missions, such as hyperspectral infrared imager (HyspIRI), will provide future data fusion opportunities for the scaling and extension of leaf physiologic processes and phenology from species and ecosystem to regional and global scales.

1.4 Spaceborne Fluorescence Measures

Sunlight absorbed by chlorophyll in photosynthetic organisms is mostly used to drive photosynthesis, but some radiation can also be dissipated as heat or reradiated at longer wavelengths (650–850 nm). This NIR light re-emitted from illuminated plants, as a by-product of photosynthesis, is termed as solar-induced chlorophyll fluorescence (SIF), and it has been found to strongly correlate with GPP (Baker 2008; Meroni et al. 2009). Chlorophyll fluorescence may be conceptualized as

$$SIF = \varepsilon_f \times PAR \times fAPAR \qquad (1.20)$$

where ε_f is the yield of fluorescence photons (i.e., the fraction of absorbed PAR photons that are re-emitted from the canopy as SIF photons). This expression can be combined with the GPP-based LUE equation to remove the parallel dependence of both processes on APAR to yield

$$GPP = \frac{SIF \times \varepsilon_p}{\varepsilon_f} \qquad (1.21)$$

Empirical studies at the leaf and canopy scale indicate that the two LUE terms tend to covary under the conditions of the satellite measurement (Flexas et al. 2002). SIF data provides information on both the light absorbed and the efficiency with which it is being used for photosynthesis. It is an independent measurement, linked to chlorophyll absorption, providing unique information on photosynthesis relative to VIs. Further, SIF is more dynamic than greenness, and will respond much more quickly to environmental stress, through both change in stress-induced light-use efficiency and canopy light absorption (Porcar-Castell et al. 2014).

TABLE 1.2 Examples of Remote Sensing Methods of Deriving Gross Primary Productivity with Some References

Gross Primary Productivity Measurement	Biome/Location	Satellite Products Used	Other Non-Satellite Drivers	Method/Approach	Equation	R^2	Reference
BIOME-BGC (BioGeochemical Cycles) MODIS GPP/NPP, where GPP is Gross Primary Productivity and NPP is Net Primary Productivity	Continental	MODIS FPAR (MOD15); photosynthetic active radiation (PAR) as $0.45 \times$ SWrad (shortwave downward solar radiation)	Maximum light use efficiency (ε_{max}) from a biome-properties look-up table and maximum daily vapour pressure deficit (VPD) and minimum daily air temperature (T_{min}) from forcing meteorology.	See equation	$GPP = \varepsilon_{max} \times 0.45 \times SWrad \times fPAR \times f(VPD) \times f(Ta)$	NA	Running et al. (2004)
GPP and light use efficiency (LUE)	North American ecosystems from evergreen needleleaf and deciduous forest to grassland to savanna	MODIS EVI	NA	Linear regression	$GPP = m \times EVI + b$; $LUE = m \times EVI + b$	0.76 MODIS GPP-GPP; 0.92 EVI–GPP; 0.76 EVI–LUE; 0.62 MODIS LUE–LUE	Sims et al. (2006)
GPP	Tropical forests and converted pastures at the Amazon basin	MODIS EVI	NA	Linear regression	$GPP = m \times EVI + b$	0.5	Huete et al. (2006)
GPP and maximum Net Ecosystem Exchange (NEE_{max})	Northern Europe ecosystems from evergreen needleleaf and deciduous forest to grasslands	MODIS EVI	NA	Linear regression	$GPP = m \times EVI + b$; $NEE_{max} = m \times EVI + b$	GPP–EVI: 0.81 deciduous, 0.69, coniferous forests NEE_{max}–EVI:0.83 deciduous. 0.72, coniferous forests	Olofsson et al. (2008)
GPP	Dry to humid tropical forest sites in Southeast Asia	MODIS EVI	NA	Linear regression	$GPP = 8282 \times EVI + 2118$, GPP (kgC ha⁻¹ month⁻¹)	0.74	Huete et al. (2008)
GPP	African tropical savanna ecosystems including shrubland, woodlands, crops and grasslands	MODIS EVI	NA	Linear regression	$GPP = m \times EVI + b$	NA	Sjöström et al. (2011)
GPP	Northern Australian mesic and xeric tropical savannas	MODIS EVI MODIS GPP product	Eddy covariance measured water availability index (EF) and PAR	Linear regression	$GPP = m \times EVI + b$; $GPP = b + m(MODIS\ GPP)$; $GPP = b + m(EVI \times PAR)$; $GPP = b + m(EVI \times PAR \times EF)$	Linear regression EVI–GPP ranges from 0.89 (woodland) to 0.52 (wooded grassland)	Ma et al. (2013)
Temperature and Greenness Model (T–G)	North American ecosystems from evergreen needleleaf and deciduous forest to grassland to savanna	MODIS daytime land surface temperature (LST) and EVI	NA	See equation	$GPP = (EVI_{scaled} \times LST_{scaled}) \times m$ $LST_{scaled} = \min(LST/30)$; $(2.5 - (0.05 \times LST))$ $EVI_{scaled} = EVI - 0.10$	NA	Sims et al. (2008)
Greenness and Radiation (G–R) model	Crops, including soybean and maize–soybean rotation	MODIS NDVI and a chlorophyll-related spectral index (VI_{chl}): EVI or wide dynamic range vegetation index (WDRVI)	PAR_{toc} is the top of canopy measured PAR	See equation	$GPP = VI_{chl} \times PAR_{toc}$; $GPP = NDVI \times PAR_{toc}$	0.84 $GPP = EVI2 \times PAR_{toc}$ 0.87 GPP = Red edge $NDVI \times PAR_{toc}$ 0.9–0.9 $GPP = EVI \times PAR_{toc}$	Peng and Gitelson (2012)

(Continued)

TABLE 1.2 (Continued) Examples of Remote Sensing Methods of Deriving Gross Primary Productivity with Some References

Gross Primary Productivity Measurement	Biome/Location	Satellite Products Used	Other Non-Satellite Drivers	Method/Approach	Equation	R^2	Reference
Greenness and Radiation (G–R) model	Northern Australian mesic and xeric tropical savannas	MODIS EVI	PAR_{toa} is the top of atmosphere PAR	See equation	$GPP = EVI \times PAR_{toa}$	NA	Ma et al. (2014)
Temperature and Greenness Model (T–G) and Greenness and Radiation (G–R) model	Temperate and boreal forest ecosystems in North America	MODIS EVI	NA	Linear regression	$GPP = m \times EVI + b$; $GPP = EVI_{Scaled} \times LST_{Scaled}$	T–G model GPP:0.27 to 0.91 at non-forests ~0.9 at deciduous forests 0.28–0.91 evergreen forests	Wu et al. (2011)
Vegetation Photosynthesis Model (VPM)	Single temperate deciduous broadleaf ecosystem forest	MODIS EVI, NDVI, LSWI, water (W_{scalar}), leaf phenology (P_{scalar})	Temperature (air) and leaf phenology information T_{scalar}, P_{scalar} respectively	See equation	$GPP = \varepsilon \times fAPARchl \times PARtoc$ $\varepsilon = \varepsilon_{max} \times T_{scalar} \times W_{scalar} \times P_{scalar}$ $GPP = -74.4 + 179.4 \times NDVI$ $GPP = -68.3 + 299.7 \times EVI$, GPP (gC m^{-2} 10–day)	GPP-NDVI, 0.64 GPP-EVI, 0.84 GPP-VPM GPP, 0.92	Xiao et al. (2004)
Vegetation Photosynthesis and Respiration Model (VPRM) Net Ecosystem Exchange, NEE = GPP–ecosystem respiration	Nine vegetation classes, including evergreen and deciduous forests, grasslands, and shrub sites in North America	MODIS EVI, and LSWI	Incoming solar radiation and air temperature	Model	NA	Monthly NEE – VPRM NEE ranges from 0.96 (deciduous temperate forest) to >1 at grasses and agricultural areas	Mahadevan et al. (2008)
Light use efficiency, LUE	Crops: sunflower	Photochemical reflectance index (PRI)	NA	Linear regression	$LUE = m \times PRI + b$	NA	Gamon et al. (1992)
LUE	Crops: corn	PRI	NA	Linear regression	$LUE = 1.37 \times PRI_{570} - 0.04$, LUE (mol C mol^{-1} APAR)	0.66	Middleton et al. (2011)
GPP	Cropland and grassland ecosystems	Solar-induced chlorophyll fluorescence (SIF)	NA	Linear regression	US croplands: $GPP = -0.88 + 3.55 \times SIF$ Europe grasslands: $GPP = 0.35 + 3.71 \times SIF$ All sites: $GPP = -0.17 + 3.48 \times SIF$, GPP (gC m^{-2} day^{-1})	0.92, US croplands 0.79, Europe grasslands 0.87, All sites	Guanter et al. (2014)

Global space-based estimates of SIF have recently become available through the Japanese Greenhouse Gases Observing Satellite (GOSAT) using solar absorption, where Fraunhofer lines are used to derive fluorescence estimates. Subsequently, global SIF data with better spatial and temporal sampling are now produced from the Global Ozone Monitoring Experiment-2 (GOME-2) instrument onboard the Metop-A platform (Joiner et al. 2013) and the Orbiting Carbon Observatory-2 (OCO-2) launched in July 2014 (Frankenberg et al. 2014). Preparatory studies are also underway for the future European fluorescence explorer (FLEX) satellite mission (Meroni et al. 2009). Whereas OCO-2 and GOME-2 were not designed specifically to measure fluorescence and estimate only a single-wavelength SIF, the FLEX mission will provide measurements characterizing the spectral shape of fluorescence emission and enable estimates of photosynthesis rates under different vegetation stress conditions. In addition, the future Sentinel-5 Precursor Tropospheric Monitoring Instrument (TROPOMI) (Veefkind et al. 2012) satellite mission will provide advance spectrometer and fluorescence data with significantly finer spatial resolution.

Chlorophyll fluorescence provides estimates of actual photosynthetic rates, as opposed to estimates of potential photosynthesis that are typically derived using spectral VIs, fAPAR and LAI products. Satellite-based SIF retrievals have thus been shown to be highly correlated with GPP estimates derived at global and seasonal scales (Frankenberg et al. 2011; Guanter et al. 2012). Guanter et al. (2014) showed satellite SIF retrievals provided direct measures of GPP of cropland and grassland ecosystems and a more direct link with photosynthesis than found with vegetation greenness measures, such as VIs. Their SIF-based GPP estimates were similar to flux-tower comparisons and found to be significantly more accurate than empirical and process-based productivity models, which underestimated GPP by as much as 50%–75%. This study, along with Zhang et al. (2014), has shown the potential of SIF data to improve carbon cycle models and provide more accurate projections of ecosystem and agricultural productivity and climate impacts on production.

1.5 Discussion

The simple LUE-based productivity equation introduced by Monteith (1972) comprises a great deal of biological and biophysical complexity, resulting in numerous productivity modeling approaches that attempt to deal with this complexity in different ways. GPP is proportional to the incident shortwave radiation, the fractional absorption of that energy (fAPAR), and the efficiency with which the absorbed radiation is converted to fixed carbon, ε. The different modeling approaches tend to emphasize one term or the other of the LUE equation, with remote sensing–based algorithms focusing on the fAPAR term, or more recently, the $fAPAR_{chlorophyll}$ component. Others have focused on the LUE term as the primary determining factor of productivity either focusing on the biome level versus species specificity of LUE variability (e.g., Ahl et al. 2004) or focusing on the meteorologic scalars of LUE with potential incorporation of soil moisture as

an LUE regulator. Kanniah et al. (2009) noted that strong seasonal variations in LUE at tropical mesic savanna sites were primarily driven by the dynamics in understorey grasses. There is also much attention on the role of the PAR term in explaining seasonal and year-to-year growth variability of plant productivity, including the incorporation of light quality (direct and diffuse) to complement data of radiation quantities.

With the increasing use of satellite data for large-scale productivity assessments, it has become appealing to calibrate such data with in situ productivity measures, such as from EC tower sites. Glenn et al. (2008) suggested that remote sensing is very suitable as a scaling tool of productivity when ground data are available. Remote sensing can greatly simplify the upscaling of ecosystem processes, such as photosynthesis, from an expansive network of flux towers to larger landscape units and to regional scales, as the measurement footprint of flux towers at least partially overlaps the pixel size of daily-return satellites (e.g., 250 m for MODIS). Further, as top-of-canopy measurements, flux towers do not require knowledge of LAI or details of canopy architecture to estimate fluxes facilitating their comparisons with remote sensing measures that similarly involve community properties resulting from integrative, top-of-canopy radiation interactions. However, tower data of fluxes potentially offer much more than simply validating and/or calibrating remote sensing products and models. An understanding of why satellite–flux tower relationships hold, or do not hold, will greatly advance and contribute to our comprehension of the carbon cycle mechanisms and scaling factors at play.

The validation of satellite-based productivity products remains challenging due to a variety of spatial and temporal scaling issues (Morisette et al. 2002; Turner et al. 2004). These include the matching of large satellite pixels (~1 km) with field plot scale measurements in both time and space. Li et al. (2008) demonstrated limitations associated with disparate footprints between satellite and tower flux measurements and the need for Landsat spatial resolutions for flux footprint matching, particularly in nonforested canopies. Plot-level ANPP measurements are commonly made at scales from 1 m² to 0.01 km², while the matching MODIS footprint may range from 62,500 m² to 1 km².

From a temporal perspective, plants respond to the dynamics of environmental variables through stomatal closure and other diurnal adjustments that cannot be easily sensed by satellite sensors (e.g., MODIS). Variation in LUE is likely to be significant over shorter, daily time frames when water or temperature stress develops. However, at moderate to longer (e.g., weekly to monthly) time scales, plants tend to increase leaf area under favorable environments as an investment of resources into their photosynthetic apparatus, and reduce LAI under stress when leaves are expensive to produce and maintain. Thus, at longer time scales, there would be a convergence of satellite greenness signals with biologic and structural canopy properties. SIF, however, is seen as one way to increase the effective temporal remote sensing of vegetation photosynthesis, essentially to near real-time.

Despite these challenges, continuing advances made in global weather-forecasting accuracies and the development of new satellite sensor technologies, including fluorescence, hyperspectral, thermal, and lidar, now enable a more thorough coupling of the environmental conditions that plants experience with improved characterization of their biophysical states, and with better monitoring capabilities to track plant responses to environmental changes. These advances are providing a better understanding of the dynamics of terrestrial productivity and the use of satellite data to drive productivity models of the land surface.

References

Ahl, D.E., Gower, S.T., Mackay, D.S., Burrows, S.N., Norman, J.M., and Diak, G.R. 2004. Heterogeneity of light use efficiency in a northern Wisconsin forest: Implications for modeling net primary production with remote sensing. *Remote Sensing of Environment*, 93(1/2), 168–178.

Asner, G., Mascaro, J., Anderson, C., Knapp, D., Martin, R., Kennedy-Bowdoin, T., van Breugel, M. et al. 2013. High-fidelity national carbon mapping for resource management and REDD+. *Carbon Balance and Management*, 8(1), 7.

Asner, G.P., Powell, G.V.N., Mascaro, J., Knapp, D.E., Clark, J.K., Jacobson, J., Kennedy-Bowdoin, T. et al. 2010. High-resolution forest carbon stocks and emissions in the Amazon. *Proceedings of the National Academy of Sciences of the United States*, 107(38), 16738–16742.

Asrar, G., Fuchs, M., Kanemasu, E.T., and Hatfield, J.L. 1984. Estimating absorbed photosynthetic radiation and leaf area index from spectral reflectance in wheat. *Agronomy Journal*, 76, 2, 300–306.

Baker, N.R. 2008. Chlorophyll fluorescence: A probe of photosynthesis in vivo. *Annual Review of Plant Biology*, 59(1), 89–113.

Baldocchi, D., Falge, E., Lianhong, G., Olson, R., Hollinger, D., Running, S., Anthoni, P. et al. 2001. FLUXNET: A new tool to study the temporal and spatial variability of ecosystem-scale carbon dioxide, water vapor, and energy flux densities. *Bulletin of the American Meteorological Society*, 82(11), 2415.

Barton, C.V.M. and North, P.R.J. 2001. Remote sensing of canopy light use efficiency using the photochemical reflectance index: Model and sensitivity analysis. *Remote Sensing of Environment*, 78(3), 264–273.

Biondini, M.E., Lauenroth, W.K., and Sala, O.E. 1991. Correcting estimates of net primary production: Are we overestimating plant production in rangelands? *Journal of Range Management*, 44(3), 194–198.

Chambers, J.Q., Negrón-Juárez, R.I., Hurtt, G.C., Marra, D.M., and Higuchi, N. 2009. Lack of intermediate-scale disturbance data prevents robust extrapolation of plot-level tree mortality rates for old-growth tropical forests. *Ecology Letters*, 12(12), E22–E25.

Churkina, G. and Running, S.W. 1998. Contrasting climatic controls on the estimated productivity of global terrestrial biomes. *Ecosystems*, 1(2), 206–215.

Ciganda, V., Gitelson, A., and Schepers, J. 2008. Vertical profile and temporal variation of chlorophyll in maize canopy: Quantitative "Crop Vigor" indicator by means of reflectance-based techniques. *Agronomy Journal*, 100(5), 1409–1417.

Cleverly, J., Boulain, N., Villalobos-Vega, R., Grant, N., Faux, R., Wood, C., Cook, P.G., Yu, Q., Leigh, A., and Eamus, D. 2013. Dynamics of component carbon fluxes in a semi-arid Acacia woodland, central Australia. *Journal of Geophysical Research: Biogeosciences*, 118(3), 1168–1185.

Cramer, W., Kicklighter, D., Bondeau, A., Moore III, B., Churkina, G., Nemry, B., Ruimy, A., Schloss, A., Intercomparison, T., and Model, P.O.T.P.N. 1999. Comparing global models of terrestrial net primary productivity (NPP): Overview and key results. *Global Change Biology*, 5(S1), 1–15.

Dash, J. and Curran, P.J. 2004. The MERIS terrestrial chlorophyll index. *International Journal of Remote Sensing*, 25(23), 5403–5413.

Dawson, T.P., North, P.R.J., Plummer, S.E., and Curran, P.J. 2003. Forest ecosystem chlorophyll content: Implications for remotely sensed estimates of net primary productivity. *International Journal of Remote Sensing*, 24(3), 611–617.

FAO. 2007. *State of the World's Forests 2007*. FAO, Rome, Italy.

Fensholt, R., Sandholt, I., and Rasmussen, M.S. 2004. Evaluation of MODIS LAI, fAPAR and the relation between fAPAR and NDVI in a semi-arid environment using in situ measurements. *Remote Sensing of Environment*, 91(3–4), 490–507.

Field, C.B., Randerson, J.T., and Malmström, C.M. 1995. Global net primary production: Combining ecology and remote sensing. *Remote Sensing of Environment*, 51(1), 74–88.

Flexas, J., Bota, J., Escalona, J.M., Sampol, B., and Medrano, H. 2002. Effects of drought on photosynthesis in grapevines under field conditions: An evaluation of stomatal and mesophyll limitations. *Functional Plant Biology*, 29(4), 461–471.

Frankenberg, C., Butz, A., and Toon, G.C. 2011. Disentangling chlorophyll fluorescence from atmospheric scattering effects in O2 A-band spectra of reflected sun-light. *Geophysical Research Letters*, 38(3), L03801.

Frankenberg, C., O'Dell, C., Berry, J., Guanter, L., Joiner, J., Köhler, P., Pollock, R., and Taylor, T.E. 2014. Prospects for chlorophyll fluorescence remote sensing from the orbiting carbon observatory-2. *Remote Sensing of Environment*, 147(0), 1–12.

Gamon, J.A., Peñuelas, J., and Field, C.B. 1992. A narrow-waveband spectral index that tracks diurnal changes in photosynthetic efficiency. *Remote Sensing of Environment*, 41(1), 35–44.

Gitelson, A.A. 2004. Wide dynamic range vegetation index for remote quantification of biophysical characteristics of vegetation. *Journal of Plant Physiology*, 161(2), 165–173.

Gitelson, A.A., Keydan, G.P., and Merzlyak, M.N. 2006. Three-band model for noninvasive estimation of chlorophyll, carotenoids, and anthocyanin contents in higher plant leaves. *Geophysical Research Letters*, 33(11), L11402.

Gitelson, A.A., Peng, Y., Masek, J.G., Rundquist, D.C., Verma, S., Suyker, A., Baker, J.M., Hatfield, J.L., and Meyers, T. 2012. Remote estimation of crop gross primary production with Landsat data. *Remote Sensing of Environment*, 121(0), 404–414.

Glenn, E.P., Huete, A.R., Nagler, P.L., and Nelson, S.G. 2008. Relationship between remotely-sensed vegetation indices, canopy attributes and plant physiological processes: What vegetation indices can and cannot tell us about the landscape. *Sensors*, 8(4), 2136–2160.

Goerner, A., Reichstein, M., and Rambal, S. 2009. Tracking seasonal drought effects on ecosystem light use efficiency with satellite-based PRI in a Mediterranean forest. *Remote Sensing of Environment*, 113(5), 1101–1111.

Goulden, M.L., Munger, J.W., Fan, S.-M., Daube, B.C., and Wofsy, S.C. 1996. Measurements of carbon sequestration by long-term eddy covariance: Methods and a critical evaluation of accuracy. *Global Change Biology*, 2(3), 169–182.

Goward, S.N. and Huemmrich, K.F. 1992. Vegetation canopy PAR absorptance and the normalized difference vegetation index: An assessment using the SAIL model. *Remote Sensing of Environment*, 39(2), 119–140.

Goward, S.N., Tucker, C., and Dye, D. 1985. North American vegetation patterns observed with the NOAA-7 advanced very high resolution radiometer. *Vegetatio*, 64(1), 3–14.

Gower, S.T., Kucharik, C.J., and Norman, J.M. 1999. Direct and indirect estimation of leaf area index, fAPAR, and net primary production of terrestrial ecosystems. *Remote Sensing of Environment*, 70(1), 29–51.

Guanter, L., Frankenberg, C., Dudhia, A., Lewis, P.E., Gómez-Dans, J., Kuze, A., Suto, H., and Grainger, R.G. 2012. Retrieval and global assessment of terrestrial chlorophyll fluorescence from GOSAT space measurements. *Remote Sensing of Environment*, 121(0), 236–251.

Guanter, L., Zhang, Y., Jung, M., Joiner, J., Voigt, M., Berry, J.A., Frankenberg, C. et al. 2014. Global and time-resolved monitoring of crop photosynthesis with chlorophyll fluorescence. *Proceedings of the National Academy of Sciences of the United States*, 111(14), E1327–E1333.

Guo, J. and Trotter, C.M. 2004. Estimating photosynthetic light-use efficiency using the photochemical reflectance index: Variations among species. *Functional Plant Biology*, 31(3), 255–265.

Hashimoto, H., Dungan, J.L., White, M.A., Yang, F., Michaelis, A.R., Running, S.W., and Nemani, R.R. 2008. Satellite-based estimation of surface vapor pressure deficits using MODIS land surface temperature data. *Remote Sensing of Environment*, 112(1), 142–155.

Heinsch, F.A., Maosheng, Z., Running, S.W., Kimball, J.S., Nemani, R.R., Davis, K.J., Bolstad, P.V. et al. 2006. Evaluation of remote sensing based terrestrial productivity from MODIS using regional tower eddy flux network observations. *IEEE Transactions on Geoscience and Remote Sensing*, 44(7), 1908–1925.

Huete, A., Didan, K., Miura, T., Rodriguez, E.P., Gao, X., and Ferreira, L.G. 2002. Overview of the radiometric and biophysical performance of the MODIS vegetation indices. *Remote Sensing of Environment*, 83(1–2), 195–213.

Huete, A.R., Restrepo-Coupe, N., Ratana, P., Didan, K., Saleska, S.R., Ichii, K., Panuthai, S., and Gamo, M. 2008. Multiple site tower flux and remote sensing comparisons of tropical forest dynamics in Monsoon Asia. *Agricultural and Forest Meteorology*, 148(5), 748–760.

Ichii, K., Hashimoto, H., White, M.A., Potter, C., Hutyra, L.R., Huete, A.R., Myneni, R.B., and Nemani, R.R. 2007. Constraining rooting depths in tropical rainforests using satellite data and ecosystem modeling for accurate simulation of gross primary production seasonality. *Global Change Biology*, 13(1), 67–77.

Jenkins, J.P., Richardson, A.D., Braswell, B.H., Ollinger, S.V., Hollinger, D.Y., and Smith, M.L. 2007. Refining light-use efficiency calculations for a deciduous forest canopy using simultaneous tower-based carbon flux and radiometric measurements. *Agricultural and Forest Meteorology*, 143(1–2), 64–79.

Jin, C., Xiao, X., Merbold, L., Arneth, A., Veenendaal, E., and Kutsch, W.L. 2013. Phenology and gross primary production of two dominant savanna woodland ecosystems in Southern Africa. *Remote Sensing of Environment*, 135, 0, 189–201.

Joiner, J., Guanter, L., Lindstrot, R., Voigt, M., Vasilkov, A.P., Middleton, E.M., Huemmrich, K.F., Yoshida, Y., and Frankenberg, C. 2013. Global monitoring of terrestrial chlorophyll fluorescence from moderate spectral resolution near-infrared satellite measurements: Methodology, simulations, and application to GOME-2. *Atmos Meas Tech*, 6(10), 2803–2823.

Jönsson, P. and Eklundh, L. 2004. TIMESAT—A program for analyzing time-series of satellite sensor data. *Computers & Geosciences*, 30(8), 833–845.

Kanamitsu, M., Ebisuzaki, W., Woollen, J., Yang, S.-K., Hnilo, J.J., Fiorino, M., and Potter, G.L. 2002. NCEP–DOE AMIP-II Reanalysis (R-2). *Bulletin of the American Meteorological Society*, 83(11), 1631–1643.

Kang, S., Running, S.W., Zhao, M., Kimball, J.S., and Glassy, J. 2005. Improving continuity of MODIS terrestrial photosynthesis products using an interpolation scheme for cloudy pixels. *International Journal of Remote Sensing*, 26(8), 1659–1676.

Kanniah, K.D., Beringer, J., and Hutley, L. 2013a. Exploring the link between clouds, radiation, and canopy productivity of tropical savannas. *Agricultural and Forest Meteorology*, 182–183(0), 304–313.

Kanniah, K.D., Beringer, J., and Hutley, L.B. 2011. Environmental controls on the spatial variability of savanna productivity in the Northern Territory, Australia. *Agricultural and Forest Meteorology*, 151(11), 1429–1439.

Kanniah, K.D., Beringer, J., and Hutley, L.B. 2013b. Response of savanna gross primary productivity to interannual variability in rainfall: Results of a remote sensing based light use efficiency model. *Progress in Physical Geography*, 37(5), 642–663.

Kanniah, K.D., Beringer, J., Hutley, L.B., Tapper, N.J., and Zhu, X. 2009. Evaluation of collections 4 and 5 of the MODIS gross primary productivity product and algorithm improvement at a tropical savanna site in northern Australia. *Remote Sensing of Environment*, 113(9), 1808–1822.

Katul, G.G., Leuning, R., Kim, J., Denmead, O.T., Miyata, A., and Harazono, Y. 2001. Estimating CO_2 source/sink distributions within a rice canopy using higher-order closure model. *Boundary-Layer Meteorology*, 98(1), 103–125.

Kauth, R.J. and Thomas, G.S. 1976. The tasseled cap. A graphic description of the spectral-temporal development of agricultural crops as seen by Landsat, *2nd International Symposium on Machine Processing of Remotely Sensed Data*, IEEE Cat. 76, CH-1103-1, Purdue University, West Lafayette, IN and Proc. of LACIE Symp., Houston, TX.

Kauth, R.J. and Thomas, G. 1976. The tasselled cap—A graphic description of the spectral-temporal development of agricultural crops as seen by Landsat. *LARS Symposia*, Paper 159.

Kotchenova, S.Y. and Vermote, E.F. 2007. Validation of a vector version of the 6S radiative transfer code for atmospheric correction of satellite data. Part II. Homogeneous Lambertian and anisotropic surfaces. *Applied Optics*, 46, 20, 4455–4464.

Lefsky, M.A., Cohen, W.B., Harding, D.J., Parker, G.G., Acker, S.A., and Gower, S.T. 2002. Lidar remote sensing of aboveground biomass in three biomes. *Global Ecology and Biogeography*, 11(5), 393–399.

Leuning, R., Cleugh, H.A., Zegelin, S.J., and Hughes, D. 2005. Carbon and water fluxes over a temperate Eucalyptus forest and a tropical wet/dry savanna in Australia: Measurements and comparison with MODIS remote sensing estimates. *Agricultural and Forest Meteorology*, 129(3–4), 151–173.

Li, F., Kustas, W.P., Anderson, M.C., Prueger, J.H., and Scott, R.L. 2008. Effect of remote sensing spatial resolution on interpreting tower-based flux observations. *Remote Sensing of Environment*, 112, 2, 337–349.

Lieth, H. 1975. Primary production of the major vegetation units of the world, in H. Lieth, and R. Whittaker (eds.), *Primary Productivity of the Biosphere*. Springer, Berlin, Germany, Vol. 14, pp. 203–215.

Ma, X., Huete, A., Yu, Q., Coupe, N.R., Davies, K., Broich, M., Ratana, P. et al. 2013. Spatial patterns and temporal dynamics in savanna vegetation phenology across the North Australian Tropical Transect. *Remote Sensing of Environment*, 139(0), 97–115.

Ma, X., Huete, A., Yu, Q., Restrepo-Coupe, N., Beringer, J., Hutley, L.B., Kanniah, K.D., Cleverly, J., and Eamus, D. 2014. Parameterization of an ecosystem light-use-efficiency model for predicting savanna GPP using MODIS EVI. *Remote Sensing of Environment*, 154(0), 253–271.

Mahadevan, P., Wofsy, S.C., Matross, D.M., Xiao, X., Dunn, A.L., Lin, J.C., Gerbig, C., Munger, J.W., Chow, V.Y., and Gottlieb, E.W. 2008. A satellite-based biosphere parameterization for net ecosystem CO2 exchange: Vegetation photosynthesis and respiration model (VPRM). *Global Biogeochemical Cycles*, 22(2), GB2005.

Meroni, M., Rossini, M., Guanter, L., Alonso, L., Rascher, U., Colombo, R., and Moreno, J. 2009. Remote sensing of solar-induced chlorophyll fluorescence: Review of methods and applications. *Remote Sensing of Environment*, 113(10), 2037–2051.

Middleton, E., Huemmrich, K., Cheng, Y., and Margolis, H. 2011. Spectral bioindicators of photosynthetic efficiency and vegetation stress, in *Hyperspectral Remote Sensing of Vegetation*, Thenkabail, P., Lyon, J, and Huete, A. (eds.), CRC Press, pp. 265–288.

Monfreda, C., Ramankutty, N., and Foley, J.A. 2008. Farming the planet: 2. Geographic distribution of crop areas, yields, physiological types, and net primary production in the year 2000. *Global Biogeochemical Cycles*, 22(1), GB1022.

Monteith, J.L. 1972. Solar radiation and productivity in tropical ecosystems. *Journal of Applied Ecology*, 9(3), 747.

Monteith, J.L. and Unsworth, M.H. 1990. *Principles of Environmental Physics*, 2nd edn. Antony Rowe Ltd., Eastbourne, U.K.

Moran, M.S., Ponce-Campos, G.E., Huete, A., McClaran, M.P., Zhang, Y., Hamerlynck, E.P., Augustine, D.J. et al. 2014. Functional response of U.S. grasslands to the early 21st-century drought. *Ecology*, 95(8), 2121–2133.

Morisette, J.T., Privette, J.L., and Justice, C.O. 2002. A framework for the validation of MODIS land products. *Remote Sensing of Environment*, 83(1–2), 77–96.

Myneni, R.B., Hoffman, S., Knyazikhin, Y., Privette, J.L., Glassy, J., Tian, Y., Wang, Y. et al. 2002. Global products of vegetation leaf area and fraction absorbed PAR from year one of MODIS data. *Remote Sensing of Environment*, 83(1–2), 214–231.

Myneni, R.B. and Williams, D.L. 1994. On the relationship between FAPAR and NDVI. *Remote Sensing of Environment*, 49(3), 200–211.

Nemani, R.R., Keeling, C.D., Hashimoto, H., Jolly, W.M., Piper, S.C., Tucker, C.J., Myneni, R.B., and Running, S.W. 2003. Climate-driven increases in global terrestrial net primary production from 1982 to 1999. *Science* (New York), 300(5625), 1560–1563.

Nichol, C.J., Huemmrich, K.F., Black, T.A., Jarvis, P.G., Walthall, C.L., Grace, J., and Hall, F.G. 2000. Remote sensing of photosynthetic-light-use efficiency of boreal forest. *Agricultural and Forest Meteorology*, 101(2–3), 131–142.

Nichol, C.J., Lloyd, J.O.N., Shibistova, O., Arneth, A., RÖSer, C., Knohl, A., Matsubara, S., and Grace, J. 2002. Remote sensing of photosynthetic-light-use efficiency of a Siberian boreal forest. *Tellus B*, 54(5), 677–687.

Olofsson, P., Lagergren, F., Lindroth, A., Lindström, J., Klemedtsson, L., Kutsch, W., and Eklundh, L. 2008. Towards operational remote sensing of forest carbon balance across Northern Europe. *Biogeosciences*, 5(3), 817–832.

Peng, Y. and Gitelson, A.A. 2012. Remote estimation of gross primary productivity in soybean and maize based on total crop chlorophyll content. *Remote Sensing of Environment*, 117(0), 440–448.

Peng, Y., Gitelson, A.A., Keydan, G., Rundquist, D.C., and Moses, W. 2011. Remote estimation of gross primary production in maize and support for a new paradigm based on total crop chlorophyll content. *Remote Sensing of Environment*, 115(4), 978–989.

Peng, Y., Gitelson, A.A., and Sakamoto, T. 2013. Remote estimation of gross primary productivity in crops using MODIS 250 m data. *Remote Sensing of Environment*, 128(0), 186–196.

Peñuelas, J., Filella, I., and Gamon, J.A. 1995. Assessment of photosynthetic radiation-use efficiency with spectral reflectance. *New Phytologist*, 131(3), 291–296.

Ponce-Campos, G.E., Moran, M.S., Huete, A., Zhang, Y., Bresloff, C., Huxman, T.E., Eamus, D. et al. 2013. Ecosystem resilience despite large-scale altered hydroclimatic conditions. *Nature*, 494(7437), 349–352.

Porcar-Castell, A., Tyystjärvi, E., Atherton, J., van der Tol, C., Flexas, J., Pfündel, E.E., Moreno, J., Frankenberg, C., and Berry, J.A. 2014. Linking chlorophyll a fluorescence to photosynthesis for remote sensing applications: Mechanisms and challenges. *Journal of Experimental Botany*, 65(15), 4065–4095.

Rahman, A.F., Cordova, V.D., Gamon, J.A., Schmid, H.P., and Sims, D.A. 2004. Potential of MODIS ocean bands for estimating CO2 flux from terrestrial vegetation: A novel approach. *Geophysical Research Letters*, 31(10), L10503.

Rahman, A.F., Sims, D.A., Cordova, V.D., and El-Masri, B.Z. 2005. Potential of MODIS EVI and surface temperature for directly estimating per-pixel ecosystem C fluxes. *Geophysical Research Letters*, 32(19), L19404.

Richardson, A.J. and Wiegand, C.L. 1977. Distinguishing vegetation from soil background information. *Photogrammetric Engineering and Remote Sensing*, 43(12), 1541–1552.

Ripullone, F., Rivelli, A.R., Baraldi, R., Guarini, R., Guerrieri, R., Magnani, F., Peñuelas, J., Raddi, S., and Borghetti, M. 2011. Effectiveness of the photochemical reflectance index to track photosynthetic activity over a range of forest tree species and plant water statuses. *Functional Plant Biology*, 38(3), 177–186.

Rossini, M., Migliavacca, M., Galvagno, M., Meroni, M., Cogliati, S., Cremonese, E., Fava, F. et al. 2014. Remote estimation of grassland gross primary production during extreme meteorological seasons. *International Journal of Applied Earth Observation and Geoinformation*, 29(0), 1–10.

Ruimy, A., Jarvis, P., Baldocchi, D., and Saugier, B. 1995. CO$_2$ fluxes over plant canopies and solar radiation: A review. *Advances in Ecological Research*, 26, 1–68.

Ruimy, A., Saugier, B., and Dedieu, G. 1994. Methodology for the estimation of terrestrial net primary production from remotely sensed data. *Journal of Geophysical Research: Atmospheres*, 99(D3), 5263–5283.

Running, S.W., Baldocchi, D.D., Turner, D.P., Gower, S.T., Bakwin, P.S., and Hibbard, K.A. 1999. A global terrestrial monitoring network integrating tower fluxes, flask sampling, ecosystem modeling and EOS satellite data. *Remote Sensing of Environment*, 70(1), 108–127.

Running, S.W., Heinsch, F.A., Zhao, M., Reeves, M., Hashimoto, H., and Nemani, R.R. 2004. A continuous satellite-derived measure of global terrestrial primary production. *BioScience*, 54(6), 547–560.

Sala, O.E., Parton, W.J., Joyce, L.A., and Lauenroth, W.K. 1988. Primary production of the central grassland region of the United States. *Ecology*, 69(1), 40–45.

Schimel, D.S. 1998. Climate change: The carbon equation. *Nature*, 393(6682), 208–209.

Sellers, P.J. 1985. Canopy reflectance, photosynthesis and transpiration. *International Journal of Remote Sensing*, 6(8), 1335–1372.

Sims, D.A., Rahman, A.F., Cordova, V.D., El-Masri, B.Z., and Baldocchi, D.D. 2006. On the use of MODIS EVI to assess gross primary productivity of North American ecosystems. *Journal of Geophysical Research*, 111.

Sims, D.A., Rahman, A.F., Cordova, V.D., El-Masri, B.Z., Baldocchi, D.D., Bolstad, P.V., Flanagan, L.B. et al. 2008. A new model of gross primary productivity for North American ecosystems based solely on the enhanced vegetation index and land surface temperature from MODIS. *Remote Sensing of Environment*, 112(4), 1633–1646.

Sjöström, M., Ardö, J., Arneth, A., Boulain, N., Cappelaere, B., Eklundh, L., de Grandcourt, A. et al. 2011. Exploring the potential of MODIS EVI for modeling gross primary production across African ecosystems. *Remote Sensing of Environment*, 115(4) 1081–1089.

Sjöström, M., Zhao, M., Archibald, S., Arneth, A., Cappelaere, B., Falk, U., de Grandcourt, A. 2013. Evaluation of MODIS gross primary productivity for Africa using eddy covariance data. *Remote Sensing of Environment*, 131(0), 275–286.

Tucker, C.J. 1979. Red and photographic infrared linear combinations for monitoring vegetation. *Remote Sensing of Environment*, 8, 2, 127–150.

Turner, D.P., Ollinger, S.V., and Kimball, J.S. 2004. Integrating remote sensing and ecosystem process models for landscape- to regional-scale analysis of the carbon cycle. *BioScience*, 54(6), 573–584.

Turner, D.P., Ritts, W.D., Cohen, W.B., Gower, S.T., Running, S.W., Zhao, M., Costa, M.H. et al. 2006. Evaluation of MODIS NPP and GPP products across multiple biomes. *Remote Sensing of Environment*, 102(3–4), 282–292.

Turner, D.P., Urbanski, S., Bremer, D., Wofsy, S.C., Meyers, T., Gower, S.T., and Gregory, M. 2003. A cross-biome comparison of daily light use efficiency for gross primary production. *Global Change Biology*, 9(3), 383–395.

USDA-NASS. USDA National Agricultural Statistics Service Cropland Data Layer. Published crop-specific data layer. Technical report, USDA-NASS, Online, 37, 2010

Ustin, S.L., Roberts, D.A., Pinzón, J., Jacquemoud, S., Gardner, M., Scheer, G., Castañeda, C.M., and Palacios-Orueta, A. 1998. Estimating canopy water content of chaparral shrubs using optical methods. *Remote Sensing of Environment*, 65(3), 280–291.

Veefkind, J.P., Aben, I., McMullan, K., Förster, H., de Vries, J., Otter, G., Claas, J. et al. 2012. TROPOMI on the ESA Sentinel-5 Precursor: A GMES mission for global observations of the atmospheric composition for climate, air quality and ozone layer applications. *Remote Sensing of Environment*, 120(0), 70–83.

Verma, S.B., Dobermann, A., Cassman, K.G., Walters, D.T., Knops, J.M., Arkebauer, T.J., Suyker, A.E. et al. 2005. Annual carbon dioxide exchange in irrigated and rainfed maize-based agroecosystems. *Agricultural and Forest Meteorology*, 131(1–2), 77–96.

Vickers, D., Thomas, C.K., Pettijohn, C., Martin, J.G., and Law, B.E. 2012. Five years of carbon fluxes and inherent water-use efficiency at two semi-arid pine forests with different disturbance histories. *Tellus B*, 64, 17159.

Wan, Z. 2008. New refinements and validation of the MODIS land-surface temperature/emissivity products. *Remote Sensing of Environment*, 112(1), 59–74.

Wang, J., Rich, P.M., Price, K.P., and Kettle, W.D. 2004. Relations between NDVI and tree productivity in the central Great Plains. *International Journal of Remote Sensing*, 25(16), 3127–3138.

Whitley, R.J., Macinnis-Ng, C.M.O., Hutley, L.B., Beringer, J., Zeppel, M., Williams, M., Taylor, D., and Eamus, D. 2011. Is productivity of mesic savannas light limited or water limited? Results of a simulation study. *Global Change Biology*, 17(10), 3130–3149.

Whittaker, R. and Marks, P. 1975. Methods of assessing terrestrial productivity, in H. Lieth, and R. Whittaker (eds.), *Primary Productivity of the Biosphere*, Springer, Berlin, Germany, Vol. 14, pp. 55–118.

Wilson, K.B., Baldocchi, D.D., and Hanson, P.J. 2001. Leaf age affects the seasonal pattern of photosynthetic capacity and net ecosystem exchange of carbon in a deciduous forest. *Plant, Cell & Environment*, 24(6), 571–583.

Wu, C., Chen, J.M., and Huang, N. 2011. Predicting gross primary production from the enhanced vegetation index and photosynthetically active radiation: Evaluation and calibration. *Remote Sensing of Environment*, 115(12), 3424–3435.

Wu, C., Gonsamo, A., Zhang, F., and Chen, J.M. 2014. The potential of the greenness and radiation (GR) model to interpret 8-day gross primary production of vegetation. *ISPRS Journal of Photogrammetry and Remote Sensing*, 88(0), 69–79.

Wu, C., Niu, Z., and Gao, S. 2010. Gross primary production estimation from MODIS data with vegetation index and photosynthetically active radiation in maize. *Journal of Geophysical Research: Atmospheres*, 115, 12.

Wylie, B.K., Johnson, D.A., Laca, E., Saliendra, N.Z., Gilmanov, T.G., Reed, B.C., Tieszen, L.L., and Worstell, B.B. 2003. Calibration of remotely sensed, coarse resolution NDVI to CO2 fluxes in a sagebrush–steppe ecosystem. *Remote Sensing of Environment*, 85(2), 243–255.

Xiao, X., Zhang, Q., Braswell, B., Urbanski, S., Boles, S., Wofsy, S., Moore III, B., and Ojima, D. 2004. Modeling gross primary production of temperate deciduous broadleaf forest using satellite images and climate data. *Remote Sensing of Environment*, 91(2), 256–270.

Xiao, X., Zhang, Q., Saleska, S., Hutyra, L., De Camargo, P., Wofsy, S., Frolking, S., Boles, S., Keller, M., and Moore III, B. 2005. Satellite-based modeling of gross primary production in a seasonally moist tropical evergreen forest. *Remote Sensing of Environment*, 94(1), 105–122.

Yuan, W., Liu, S., Yu, G., Bonnefond, J.-M., Chen, J., Davis, K., Desai, A.R. et al. 2010. Global estimates of evapotranspiration and gross primary production based on MODIS and global meteorology data. *Remote Sensing of Environment*, 114(7), 1416–1431.

Zarco-Tejada, P.J., Miller, J.R., Mohammed, G.H., Noland, T.L., and Sampson, P.H. 2002. Vegetation stress detection through chlorophyll a + b estimation and fluorescence effects on hyperspectral imagery. *Journal of Environmental Quality*, 31(5), 1433–1441.

Zhang, Y., Guanter, L., Berry, J.A., Joiner, J., van der Tol, C., Huete, A., Gitelson, A., Voigt, M., and Köhler, P. 2014. Estimation of vegetation photosynthetic capacity from space-based measurements of chlorophyll fluorescence for terrestrial biosphere models. *Global Change Biology*, 20(12), 3727–3742.

Zhang, Y., Moran, Moran, M.S., Nearing, M.A., Ponce Campos, G.E., Huete, A.R., Buda, A.R., Bosch, D.D. et al. 2013. Extreme precipitation patterns and reductions of terrestrial ecosystem production across biomes. *Journal of Geophysical Research: Biogeosciences*, 118(1), 148–157.

Zhao, M., Heinsch, F.A., Nemani, R.R., and Running, S.W. 2005. Improvements of the MODIS terrestrial gross and net primary production global data set. *Remote Sensing of Environment*, 95(2), 164–176.

Zhao, M. and Running, S.W. 2010. Drought-induced reduction in global terrestrial net primary production from 2000 through 2009. *Science*, 329(5994), 940–943.

Zhao, M., Running, S.W., and Nemani, R.R. 2006. Sensitivity of moderate resolution imaging spectroradiometer (MODIS) terrestrial primary production to the accuracy of meteorological reanalyses. *Journal of Geophysical Research: Biogeosciences*, 111(G1), G01002.

Zolkos, S.G., Goetz, S.J., and Dubayah, R. 2013. A meta-analysis of terrestrial aboveground biomass estimation using lidar remote sensing. *Remote Sensing of Environment*, 128(0), 289–298.

Canopy Biophysical Variables Retrieval from the Inversion of Reflectance Models

Frédéric Baret
INRA UMR-EMMAH/UMT-CAPTE

Acronyms and Definitions

J	Cost function
W	Covariance matrix of measurement and model uncertainties
GAI_{eff}	Effective GAI
$\hat{V}$	Estimated values of canopy biophysical variables
V_p	Prior values of canopy biophysical variables
R	Vector of observed BRF values
$\hat{R}$	Vector of simulated BRF values
$FAPAR_{bs}$	Black sky FAPAR
GAI_{true}	True GAI
$FAPAR_{ws}$	White sky FAPAR
C	Covariance matrix of the prior distribution of variables
N	Number of configurations
Po	Gap fraction
σ^2	Variance associated to measurement and model uncertainties
AVHRR	Advanced very-high-resolution radiometer

BRF	Bidirectional reflectance factor
CHRIS	Compact high-resolution imaging spectrometer
DMC	Disaster Monitoring Constellation
ECV	Essential climate variable
FAPAR	Fraction of absorbed photosynthetically active radiation
FIPAR	Fraction of intercepted photosynthetically active radiation
FORMOSAT	FORMOse SATellite
GAI	Green area index
GEOV1	Geoland2, version 1
GF	Green fraction
GIMMS	Global Inventory Modeling and Mapping Studies
GLAI	Green leaf area index
GPR	Gaussian process regression
LAI	Leaf area index
LANDSAT	LAND SATellite
LUT	Look-up table
MERIS	Medium-resolution imaging spectrometer
MGVI	MERIS global vegetation index

MODIS   Moderate-resolution imaging spectrometer
NDVI   Normalized difference vegetation index
NNT   Neural network
PAI   Plant area index
POLDER   Polarization and directionality of Earth's reflectance
RT   Radiative transfer
SLC   Soil leaf Canopy
SVM   Support vector Machine
VI   Vegetation index

2.1 Introduction

Estimates of canopy biophysical characteristics are required for a wide range of agricultural, ecological, hydrological, and meteorological applications. These should cover exhaustively large spatial domains at several scales: from the very local one corresponding to precision agriculture where cultural practices are adapted to the within-field variability, through resources and environmental management generally approached at the landscape scale, up to biogeochemical cycling and vegetation dynamics investigated at national, continental, and global scales. Remote sensing observations answer these requirements with spatial resolution spanning from kilometric down to decametric resolution observations according to the nomenclature proposed by Morisette (2010). Further, remote sensing from satellites brings the unique capacity to monitor the dynamics required to access the functioning of the vegetation.

Few biophysical variables have been recognized as essential climate variables (ECVs) for their key role in the main vegetation canopy processes such as photosynthesis and evapotranspiration (GCOS 2011). These ECVs include the leaf area index (LAI) and the fraction of absorbed photosynthetically active radiation (FAPAR). Since the 1980s, considerable improvement in the quality of terrestrial estimates of LAI and FAPAR derived from satellite or airborne systems has been achieved due to the advances of measurement capability of satellite instruments and to our understanding of the radiation regime within vegetation canopies (Liang 2004). However, remote sensing observations sample the radiation field reflected or emitted by the surface, and thus do not provide directly LAI or FAPAR estimates. It is therefore necessary to transform the radiance values recorded by the sensor into LAI or FAPAR values. The retrieval algorithms used should ideally be accurate, precise, and computationally efficient. Most importantly, they should require minimal calibration since they are supposed to be applied over diverse locations, seasons, and conditions (Walthall et al. 2004).

Many methods have been proposed to retrieve land surface characteristics from remote sensing observations (Baret and Buis 2007; Goel 1989; Houborg and Boegh 2008; Kimes et al. 2000; Laurent et al. 2013; Myneni et al. 1988; Pinty and Verstraete 1991a; Verger et al. 2011a). They include empirical methods with calibration over experimental datasets. These simple methods are limited by the size and diversity of the calibration dataset as well as by the uncertainties attached to the ground

measurements. More complex ones based on the use of radiative transfer models have been proposed where no in situ calibration dataset is required. Radiative transfer models describe the physical processes involved in the photon transport within vegetation canopies. They simulate the radiation field reflected by the surface for a given observational configuration, once the vegetation and the background are known. Retrieving canopy characteristics from the radiation field as sampled by the sensor aboard satellite needs to "invert" the radiative transfer model, that is, to estimate some input variables from the measurement of the outputs of the model.

This chapter aims at reviewing how canopy biophysical variables may be derived both from kilometric and decametric resolution remote sensing observations. It will be illustrated by LAI and FAPAR variables that will first be defined before describing the principles of the radiative transfer model inversion used to retrieve them. Then, the theoretical performances of LAI and FAPAR will be investigated. Several techniques that improve the retrievals will be discussed in detail. Finally, the possible combinations of methods, products, and sensors will be presented. A conclusion will highlight the main issues to tackle, suggesting future research avenues.

2.2 Several Definitions of LAI and FAPAR

2.2.1 Leaf Area Index: LAI, GLAI, PAI, GAI, Effective and Apparent Values

LAI is defined as half of the total developed area of green vegetation elements per unit ground area ($m^2 \ m^{-2}$) (Chen and Black 1992; Stenberg 2006). It is a structural variable, which describes the size of the interface for exchange of energy and mass between the canopy and the atmosphere. It governs photosynthesis, transpiration, and rain interception processes. For photosynthesis and transpiration, the LAI definition should be restricted to the green active area leading to the green leaf area index (GLAI) definition. Further, the area of other organs such as stems, branches, or fruits should be accounted for if they are green, leading to the green area index (GAI) definition. LAI, GLAI, and GAI may be measured using destructive techniques. However, this is tedious and time consuming and indirect methods based on canopy gap fraction (Po) measurements have been developed (Jonckheere et al. 2004; Weiss et al. 2004). Since no distinction is made by these devices between green and nongreen elements, neither between leaves and the other elements, the actual quantity measured is plant area index (PAI). However, directional photos taken from the top of the canopy may be also used to compute the green fraction (GF) defined as the fraction of green area seen in the considered direction. Assuming that the green leaves are mostly at the top of the canopy, which is generally the case, such technique provides an estimate of the GAI (Baret et al. 2010). Similarly, remote sensing observations are mainly sensitive to the green elements of the canopy and, thus, are mostly related to the GAI (Duveiller et al. 2011; Raymaekers et al. 2014).

TABLE 2.1　Definitions of LAI, GLAI, GAI, and PAI and the Associated Indirect Measurement Methods

		Only Green	Green + Non-Green	Only Leaves	All Elements	Indirect Measurement Method
LAI	Leaf area index		✓	✓		Only destructive methods
GLAI	Green leaf area index	✓		✓		Only destructive methods
GAI	Green area index	✓			✓	GF from the top, remote sensing
PAI	Plant area index		✓		✓	Po measurements

Note: All quantities are expressed in $m^2 \cdot m^{-2}$.

Table 2.1 clearly shows that indirect methods are mainly accessing GAI and PAI depending on the capacity to distinguish green from nongreen elements.

The derivation of PAI or GAI from indirect measurements requires some assumptions on canopy architecture. The turbid medium assumption is the most commonly used, considering that leaves have infinitesimal size and are randomly distributed in the canopy volume. However, this simple assumption is not always verified by actual canopies, leaves having a finite dimension and being clumped at several scales, including the shoot (leaves grouped into shoots), plant (shoots grouped into plants), stand (plants grouped into stands) to landscape (stands distributed in the landscape). This creates artifacts in the estimation of the corresponding PAI (Walter et al. 2003) from gap fraction measurements or GAI from reflectance measurements (Chen et al. 2005). Therefore, "effective" and "apparent" quantities need to be introduced to complement the actual "true" PAI or GAI definitions. The effective PAI or GAI is the quantity that can be derived from the directional gap fraction or GF based on Miller's formula (Miller 1967) that assumes leaves randomly distributed in the canopy volume (Ryu et al. 2010). However, the application of Miller's formula requires the measurement of Po or GF in all the directions of the hemisphere, which is rarely possible. We therefore estimate an "apparent" PAI or GAI value that depends on the directional sampling used. Similarly, estimates of GAI from remote sensing are "apparent" values (Martonchik 1994) that will depend on the observational configuration used, and the inverse technique employed including the assumptions on canopy architecture embedded in the radiative transfer model considered as we will see in the following sections.

2.2.2 FAPAR: Illumination Conditions and Green/Nongreen Elements

FAPAR is defined as the fraction of the photosynthetically active radiation (PAR, solar radiation in the 400–700 nm spectral domain) absorbed by a vegetation canopy (Mõttus et al. 2011). FAPAR is widely used as input in a number of primary productivity models (McCallum et al. 2009). It is therefore necessary to consider only the green photosynthetically active elements, that is, the green parts of the canopy. Similar to what was presented for the LAI definition, FAPAR measurements can be computed from the radiation balance in the 400–700 nm PAR spectral domain (Mõttus et al. 2011). The FAPAR value can be also approximated by the fraction of intercepted radiation, FIPAR, that is,

the complement to unity of the gap faction (Asrar 1989; Begué et al. 1991; Gobron et al. 2006; Russel et al. 1989). However, it is not possible to distinguish the absorption or interception of the light by the green elements from that of the nongreen elements using these measurement techniques. Conversely, measurements of the GF from the top of the canopy in the illumination direction provide a direct estimate of the FIPAR.

FAPAR and FIPAR variables are not intrinsic properties of the vegetation, but result from the interaction of the light with the canopy. FAPAR and FIPAR will thus depend on the illumination conditions. Similarly to albedo (Martonchik 1994), the illumination conditions could be described by a component coming only from the sun's direction, the black sky FAPAR or FIPAR, and a diffuse component coming from the sky hemisphere, the white sky FAPAR or FIPAR. The black sky FAPAR or FIPAR values depend on the sun's direction. Most FAPAR products are defined as the black sky values corresponding to the sun's position at the time of the satellite overpass (Weiss et al. 2014), that is, around 10:30 solar time. Note that the black sky FAPAR or FIPAR values at 10:00 local solar time have been demonstrated to be a good estimation of the daily integrated value of FAPAR or FIPAR (Baret et al. 2004).

2.3 Radiative Transfer Model Inversion Methods

The light reflected by the canopy results from the radiative transfer processes within the vegetation. It depends on canopy state variables as well as on the illumination conditions and the observational configuration that defines the sampling of the reflectance field: wavebands, view direction(s), frequency of observations, and spatial resolution. State variables characterizing the canopy structure and the optical properties of the vegetation elements include therefore some of the variables of interest for applications such as LAI (Figure 2.1). Other variables such as FAPAR can also be computed from the knowledge of the canopy state variables and the illumination configuration considered using the same radiative transfer model.

The causal relationship between the variables of interest and remote sensing data corresponds to the forward (or direct) problem. They could be either described through empirical relationships calibrated over experiments or using radiative transfer models based on a more or less close approximation of the actual physical processes, canopy architecture, and optical properties of the elements including the background. Conversely, retrieving

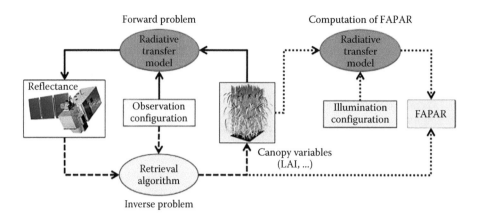

FIGURE 2.1 Forward (solid lines) and inverse (dashed lines) problems in remote sensing. The computation of FAPAR (Fraction of Absorbed Photosynthetically Active Radiation) from the retrieved canopy variables is also illustrated.

the variables of interest from remote sensing measurements corresponds to the inverse problem, that is, developing algorithms to estimate the variables of interest from remote sensing data as observed in a given configuration. Prior information on the type of surface and on the distribution of the variables of interest can also be included in the retrieval process to improve the performance as we will see later. Note that the estimation of FAPAR is achieved in two steps: first, the canopy state variables are retrieved by inverting a radiative transfer model. Then, the FAPAR is computed under specific illumination conditions using the same radiative transfer model and the estimates of canopy state variables.

The retrieval techniques can be split into two main approaches depending on whether the emphasis is on the inputs (the canopy

biophysical variables–driven approach) or the outputs (radiometric data–driven approach) of the radiative transfer model (Figure 2.2).

2.3.1 Radiometric Data–Driven Approach: Minimizing the Distance between Observed and Simulated Reflectance

The radiometric data–driven approach focuses on the outputs of the radiative transfer model: it aims to find the best match between the measured reflectance values and those simulated by a radiative transfer model (Figure 2.2, right). The misfit is quantified by a cost function (J) that should account for measurements and model uncertainties. It can be theoretically derived

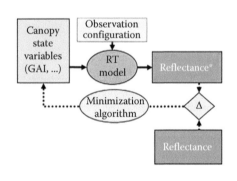

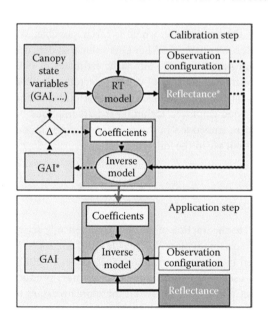

FIGURE 2.2 The two main approaches used to estimate canopy characteristics from remote sensing data for GAI estimation. On the left side, the approach focuses on radiometric data showing the solution search process leading to the estimated *LAI* value, *GAI**. On the right side, the approach focuses on the biophysical variables showing the calibration of the inverse model (top) and the application using the inverse model with its calibrated coefficients (bottom). "Δ" represents the cost function to be minimized over the biophysical variables (right) or over the radiometric data (left).

from the maximum likelihood (Tarentola 1987) assuming that uncertainties associated to each configuration used are independent and Gaussian distributed:

$$J = (R - \hat{R})^t \cdot W^{-1} \cdot (R - \hat{R}) \qquad (2.1)$$

where

R and $\hat{R}$ are, respectively, the vectors of observed and estimated reflectances

W is the covariance matrix of uncertainties

One main limitation in applying this formalism is the difficulty in obtaining the covariance matrix W. In most cases, just the diagonal terms corresponding to the variance associated to the uncertainties (σ^2) are known. In these conditions, Equation 2.1 simplifies into the normalized Euclidean distance:

$$J = \sum_{n=1}^{N} \frac{(R_n - \hat{R}_n)^2}{\sigma_n^2} \qquad (2.2)$$

where N is the number of configurations used (bands, directions, etc.). More sophisticated cost functions have been proposed to include a regularization term that prevents the solution to be too far away from its prior expectation. This will be reviewed later in a separate section.

Several techniques have been used to get the solution corresponding to the minimum of the cost function: iterative minimization including the simplex algorithm (Nelder and Mead 1965), gradient descent–based algorithms (Bacour et al. 2002a; Bicheron and Leroy 1999; Combal et al. 2000, 2002; Goel and Deering 1985; Goel et al. 1984; Goel and Thompson 1984; Jacquemoud et al. 1995; Kuusk 1991a,b; Lauvernet et al. 2008; Pinty et al. 1990; Privette et al. 1996; Voßbeck et al. 2010), Monte Carlo Markov chains (Zhang et al. 2005), simulated annealing (Bacour 2001), and genetic algorithms (Fang et al. 2003; Renders and Flasse 1996). One of the major difficulties associated with these techniques is the possibility to get suboptimal solutions that correspond to a local minimum of the cost function. This can be mitigated by using several initial guesses spread over the space of canopy realization as well as allowing some flexibility and randomness along the search path toward the solution. This is unfortunately achieved at the expense of additional computation time. However, the process can be speeded up by using an analytical expression of the gradient of the cost function, that is, the adjoint model (Lauvernet et al. 2008). Further, to increase the computation speed, the actual radiative transfer model could be emulated into a metamodel that additionally eases the derivation of the adjoint model (Jamet et al. 2005). Note that the metamodel can be considered as an interpolation between a set of simulated cases that can be used to populate a look-up table (LUT) as described in the following. To limit the problem of possible local minimum when iteratively minimizing the cost function, a regularization term could be added based on the knowledge of the prior distribution of the input variables

(Tarentola 1987) or integrating some constraints. This will be more detailed in a following sections.

LUTs working on precomputed simulations containing the input canopy variables and the corresponding simulated reflectance values have also been used directly without interpolation (Darvishzadeh et al. 2008; Ganguly et al. 2012; Knyazikhin et al. 1998; Vohland et al. 2010; Weiss et al. 2000). This technique is more tractable in terms of computation requirements and limits the possibility to get trapped in a local minimum of the cost function, as this cost function is evaluated systematically over each case of the LUT. To populate the LUT, the space of canopy realization has to be sampled to represent the surface response, that is, with better sampling where the sensitivity of reflectance to canopy characteristics is the higher (Combal et al. 2002; Weiss et al. 2000). This is different from the sampling of the training database required in canopy biophysical variable–driven approaches as explained earlier. The cases in the LUT are sorted according to the cost function value (J). Then, the solution may be considered as the one corresponding to the best match obtained with the minimum value of J, similarly to what is done with the iterative minimization techniques. It can be also defined as a fraction of the initial population of cases such as in Combal et al. (2002) and Weiss et al. (2000) or using a threshold defined by measurements and model uncertainties (Knyazikhin et al. 1998). A more rigorous way of exploiting the solutions would be to weigh each case according to its likelihood as done in the GLUE method (Beven and Binsley 1992; Makowski et al. 2002).

2.3.2 Canopy Biophysical Variable–Driven Approach: Machine Learning

This approach belongs to the machine learning type of algorithm that requires first to calibrate an inverse parametric model (Figure 2.2, right). The calibration mainly consists in adjusting the coefficients of the inverse model to minimize the distance between the estimated variable of interest (GAI in this example) and the ones populating the calibration dataset. For FAPAR, the RT model is used a second time to simulate the corresponding FAPAR values for a given illumination condition. The inverse model is then calibrated by minimizing the distance between the estimated FAPAR value and the one simulated in the calibration dataset. Once calibrated, the parametric inverse model can be used in the forward mode to compute the variables of interest from the observed reflectance values. The learning dataset can be generated either using simulations of radiative transfer models, or based on concurrent experimental measurements of the variables of interest and reflectance data.

The inverse model may be calibrated both over experimental or synthetic datasets (Asrar et al. 1984; Chen et al. 2002; Deng et al. 2006; Huete 1988; Richardson et al. 1992; Verrelst et al. 2012; Wiegand et al. 1990, 1992). However, the use of experimental datasets may be limited by its representativeness regarding the possible conditions encountered over the targeted surfaces, that is, combinations of geometrical configurations, type of vegetation and state, including variability in development stage, stress level and type, and background (bare soil, understory) and its state (roughness, moisture). Measurement errors associated both to the variables of interest and to the reflectance values may also propagate to uncertainties and biases in the algorithm and should be explicitly accounted for (Fernandes and Leblanc 2005; Huang et al. 2006). Further, since ground measurements have a footprint ranging from few meters to few decameters, specific sampling designs should be developed to represent the sensor pixel (Weiss et al. 2007). This task is obviously more difficult for medium and coarse resolution sensors (Camacho et al. 2013; Morisette et al. 2006; Weiss et al. 2007). Radiative transfer models could be used efficiently to generate a calibration dataset covering a wide range of situations and configurations (Bacour et al. 2006; Banari et al. 1996; Baret and Guyot 1991; Baret et al. 2007; Ganguly et al. 2012; Gobron et al. 2000; Huete et al. 1997; Myneni et al., 2002; Leprieur et al. 1994; Rondeaux et al. 1996; Sellers 1985; Verstraete and Pinty 1996).

2.3.2.1 Vegetation Index (VI)-Based Approaches

The simplest methods are based on the calibration of linear or polynomial multiple regression functions where the dependent variable is the biophysical variable of interest. The independent variables are either the top of canopy reflectance in few bands, or a transform and/or a combination of these reflectances resulting into a vegetation index (VI). VIs are designed to minimize the influence of confounding factors such as soil reflectance (Baret and Guyot 1991; Richardson and Wiegand 1977) or atmospheric effects (Huete and Lui 1994). The strong nonlinearity between reflectances and canopy variables is reduced using these reflectance transforms or band combination allowing using linear statistical models. Based on these principles, operational algorithms developed for medium-resolution sensors are currently used: MGVI for MERIS further extended to other sensors (Gobron et al. 2008), MODIS back-up algorithm based on NDVI (Myneni et al. 2002), and POLDER algorithm based on DVI computed from bidirectional reflectance factor (BRF) (Roujean and Lacaze 2002). Nevertheless, although quite often effective, VIs are intrinsically limited by the empiricism of their design and the small number of bands concurrently used (generally 2–3).

2.3.2.2 Machine Learning Approaches

Alternatively, more sophisticated machine learning methods have been proposed since the beginning of the 1990s. Neural networks (NNTs) have been used intensively (Abuelgasim et al. 1998; Atkinson and Tatnall 1997; Danson et al. 2003; Gong et al. 1999; Kimes et al. 1998; Smith 1993). Baret et al. (1995) and Verger et al. (2011a) demonstrated that NNTs used with individual bands were performing better than classical approaches based on VIs. Fang and Liang (2005) found that NNTs were performing similarly as the projection pursuit multiple regression. It was applied over MERIS (Bacour et al. 2006) and VEGETATION (Baret et al. 2007) kilometer spatial resolution data. The principles have been also applied at decametric resolution over airborne POLDER (Weiss et al. 2002a), LANDSAT (Fang and Liang 2003), CHRIS (Verger et al. 2011a), and FORMOSAT (Claverie

et al. 2013) sensors. Although NNTs are becoming very popular, Verrelst et al. (2012) investigated alternative machine learning methods including support vector regression (SVM) and Gaussian process regression (GPR). They demonstrated the interest of GPR when the training was achieved over experimental datasets. However, when applied to a large number of simulated cases, GPR is limited by the computation capacity (Mackay 2003). Further, one advantage of the GPR is the possibility to get an estimation of the associated uncertainties when applied to experimental data. In the case of model simulations, the uncertainties attached to the reflectance measurements need to be specified, which is not an easy task.

The training dataset is obviously a major component of the machine learning methods. It should represent the distributions and codistributions of the input canopy biophysical variables. This is where the prior information is mainly embedded in machine learning methods that can be considered as a Bayesian approach. The density of cases that populate the space of canopy realization may rapidly decrease as a function of its dimensionality defined by the number of required canopy variables. Experimental plans may be conveniently used to limit local sparseness of the training dataset (Bacour et al. 2002b). Machine learning systems can be also considered as smoothers. They thus mainly "interpolate" between cases in the training dataset. Extrapolation outside the definition domain (corresponding to the convex hull of the input reflectance of the training dataset) is likely to provide unrealistic estimates. Further, cases that are simulated but never observed may be discarded to get a more compact training dataset and efficient learning process (Baret and Buis 2007). However, it requires compiling a large database of reflectance measurements that should be representative of all the possible situations available.

2.3.3 Pros and Cons Associated to the Retrieval Approaches

The pros and cons of the several approaches just briefly reviewed are as follows:

- *Computation requirements*: Machine learning approaches, once calibrated, are obviously very little demanding in terms of computation. The inverse model is generally relatively simple and could be run very quickly. However, the calibration (or learning or training) process could require large computer resources, particularly for complex parametric model with a significant number of coefficients to be tuned and when the training dataset is large. The implementation of a LUT technique in algorithmic operational chains is very efficient, because the radiative transfer model is run offline. Conversely, iterative minimization methods require large computer resources because of its iterative nature. Improvements are possible using a metamodel. Further, automatic segmentation or discretization of the reflectance space (Pinty et al. 2011) will also reduce the number of inversions to be completed over a whole set of images.

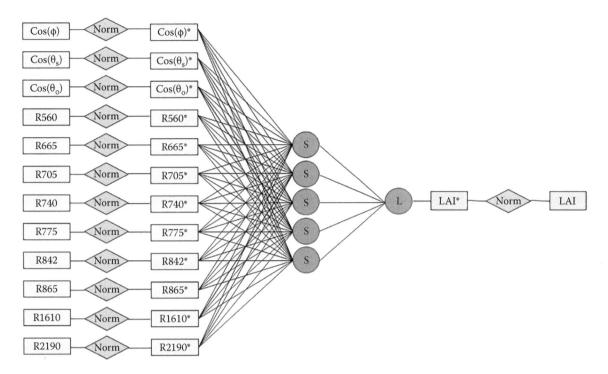

FIGURE 2.3 An example of an NNT used to estimate LAI from Sentinel2 top of canopy reflectance. "Norm" represents the normalization of the inputs or output (LAI). "S" and "L" represent tangent-sigmoid and linear transfer functions associated to each neuron. ϕ and θ_s, θ_o represent, respectively, the relative azimuth between sun and view directions, sun and view zenith angles. "R560" to "R2190" represent the top of canopy reflectance in the several Sentinel 2 bands. (From Baret, F. et al., *S2PAD—Sentinel-2 MSI—Level 2B Products Algorithm Theoretical Basis Document*, Vega, GmbH, Avignon, France, 2009.)

- *Flexibility of the observational configuration*: Iterative optimization methods allow retrieving canopy characteristics from several observational configurations. It is even possible to invert radiative transfer models concurrently over several pixels. This opens great potentials for exploiting additional temporal or spatial constraints as we will see later. LUT could theoretically cope with variable configurations at the expense of the dimensionality and thus the size of the tables, making them more difficult to manipulate. Conversely, machine learning methods require a fixed number of inputs. The characteristics of the configuration need thus to be used as inputs of the inverse parametric model as illustrated in Figure 2.3 where the illumination and view directions are explicitly used. However, this increases the dimensionality of the system, making the calibration step more demanding and more difficult. One alternative is to calibrate several parametric models for each individual configuration and then select the proper calibrated inverse model.

- *Integration of prior information*: The radiometric data–driven approaches integrate the prior information directly in the cost function within the regularization term (see Equation 2.3). However, in the case of LUTs, it is also possible to restrict the simulations to the range of situations to be encountered as is done within the MODIS LAI and FAPAR algorithm that depends on the

biome type considered (Shabanov et al. 2005). For the machine learning approaches, prior information is introduced through the distributions and codistributions of the inputs of the radiative transfer model: when LAI and FAPAR have to be estimated under situations where the type of canopies and their stage of development are known, it is more efficient to calibrate a specific inverse model for each individual situation. Note that Qu et al. (2008) proposed to use Bayesian networks where model simulations could be exploited along with a description of the distribution of the variables that may depend on growth stages or canopy types.

- *Associated uncertainties*: The radiometric data–driven approaches allow getting some estimates of the uncertainties associated to the solution by propagating the uncertainties associated to the measurements and to the model using the partial derivatives of the cost function with regard to the measurements (Lauvernet 2005). When using LUTs, uncertainties could be estimated by Monte Carlo methods or approximated by the standard deviation of the ensemble of solutions defined by the uncertainties in the measurements (Knyazikhin et al. 1998). For machine learning methods, the error on the measurements may be assessed in different ways as proposed by Aires et al. (2004), which may also include the errors associated to the retrieval process itself. A more simple alternative solution is also proposed by Baret et al. (2013) based on the training dataset. Although the estimation of uncertainties of the retrievals is possible, it is generally limited by the poor knowledge on the input uncertainties associated to the reflectance measurements and radiative transfer models used. Knyazikhin et al. (1998) used a 20% relative uncertainty applied to MODIS top of canopy reflectance for LAI and FAPAR retrieval. Baret et al. (2007) proposes to use an additive uncertainty around 0.05 and a multiplicative uncertainty around 3%. This example shows that the uncertainties attached to each band are poorly known. Further, the structure of the uncertainties may also play an important role and is unfortunately very difficult to describe.

- *Robustness of the retrieval and quality assessment*: A quality index needs to be associated to the retrieved values to inform about the status of the inversion process. For iterative optimization techniques, it could be the criteria used to the stop the iterations (Gilbert 2002). As a matter of fact, the algorithm may sometimes encounter numerical problems occurring generally with very small values of *J*. Conversely, no numerical problems are expected for LUT and machine learning approaches, and the quality index should mainly indicate whether the input reflectances were inside the definition domain and if the output solution is in the expected range of variation (Baret et al. 2013). The performances of the approach will both depend on the minimization algorithm itself and on the level of ill-posedness of the inverse problem as a function of measurement configuration and model and measurement uncertainties.

2.4 Theoretical Performances of Biophysical Variables Estimation

Several biophysical variables are potentially accessible as reviewed previously. However, depending on the assumptions on canopy structure and the observational configuration considered, the "apparent" values retrieved from remote sensing observations will be associated with contrasted performances. Further, several possible definitions for GAI and FAPAR need also to be discussed in terms of the associated uncertainties. The theoretical estimation performances were thus investigated using a simple numerical experiment. The SLC radiative transfer model (Verhoef and Bach 2007) coupled with the PROSPECT model (Jacquemoud and Baret 1990) was used to simulate the canopy reflectance in the Sentinel 2 (Malenovský et al. 2012) bands for a large set of combination of canopy characteristics (Figure 2.5) covering the expected range of variation of each of the canopy, leaf, and soil input variables. The seven bands considered (560, 670, 705, 740, 865, 1610, and 2190 nm) were chosen to sample the main absorption features of chlorophyll and water. The SLC model allows simulating leaf clumping at the plant scale: plants are randomly sown and are represented by ellipsoidal envelopes filled with randomly distributed leaves. The leaf clumping is mainly driven here by the crown fraction, that is, the fraction of ground area casted by the crowns in the vertical direction. Therefore, SLC allows also simulating turbid medium canopies when the crowns cover fully the background (crown cover = 1.0). Three typical sun positions and five view directions were considered. The black sky FAPAR ($FAPAR_{bs}$) and white sky FAPAR ($FAPAR_{ws}$), the green fraction (*GF*) and the effective GAI, GAI_{eff} were simulated in addition to the input GAI, GAI_{true}. The simulated dataset was used as a LUT to retrieve the five variables of interest: $FAPAR_{bs}$, $FAPAR_{ws}$, GF, GAI_{eff}, and GAI_{true}. A subsample of the simulated cases was used as the test dataset. The corresponding reflectances were contaminated with realistic measurement uncertainties. The solution is finally selected as the case in the LUT that corresponds to the minimum of the cost function presented in Equation 2.2, where σ^2 is the variance of the reflectance of the test dataset computed from the measurement uncertainties introduced. Note that no constraints or prior information was used in the cost function. The retrieval was achieved over turbid medium or clumped test cases using LUT based either on turbid medium or clumped canopy structure assumption. More details can be found in Kandasamy et al. (2010).

Results presented in Figure 2.4 show that GF and $FAPAR_{ws}$ are the best estimated variables. Further, the good performances are relatively independent from the assumptions on canopy structure. The black sky, $FAPAR_{bs}$, is well estimated, with however a significant degradation of the retrieval performances when the test cases correspond to clumped canopies. GAI values are much more difficult to estimate, particularly the actual GAI_{true} value for the clumped test cases. Conversely, the effective GAI, GAI_{eff}, provides relatively stable performances independent from the assumptions on canopy structure. Note that the turbid medium test cases retrieved with a LUT made of clumped canopies provide poorer estimates as compared to those derived from the turbid medium LUT. Although turbid medium cases are included in the LUT made with clumped canopies, the degradation of performances is explained by the smaller number of turbid medium cases contained in the LUT populated with clumped canopies (less cases with crown cover close to 1.0). Further, possible ambiguities between turbid medium and clumped cases providing very similar reflectance values may be encountered. The variability of retrieval performances depending on the observation configuration is larger for the $FAPAR_{bs}$ and GAI_{eff} and more particularly for GAI_{true}, in agreement with the overall performances associated to the retrieval of these variables.

This simple numerical experiment demonstrates that the retrieval of the true GAI from monodirectional reflectance measurements is likely to be relatively inaccurate, particularly in the case of clumped canopies. The use of a clumped canopy model in the inversion process does not help the retrieval: more constraints or prior information is needed to compensate for the additional unknown variables required to describe canopy clumping as discussed in Section 2.5.

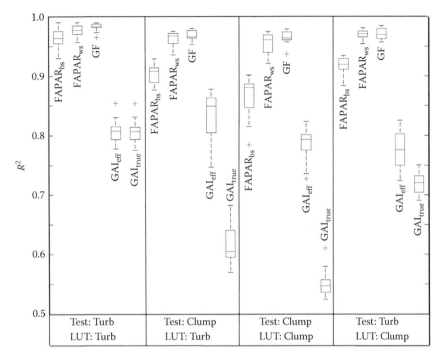

FIGURE 2.4 Theoretical performances (R^2 of the regression between the reference variable and the estimated value) of [FAPAR$_{bs}$, FAPAR$_{ws}$, GF, GAI$_{eff}$, and GAI$_{true}$] estimation depend on the test cases considered (turbid or clumped) and the assumptions on canopy structure used in the LUT (turbid or clumped). Box plot representation of the results for the 15 observational configurations (three sun position and five view directions). The median value is the red line in the box plot that contains 50% of the data. Whiskers extend to the extreme values except if they are considered as outliers that are represented by a red "+."

2.5 Mitigating the Underdetermined and Ill-Posed Nature of the Inverse Problem

2.5.1 Underdetermination and Ill-Posedness of the Inverse Problem

Canopy reflectance models will depend on a set of input variables characterizing the several components: soil, leaf, and canopy structure (Figure 2.5). Several models have been proposed to describe the soil reflectance. They are either physically based ones mostly focusing on the bidirectional variability of the reflectance (Cierniewski et al. 2002; Hapke 1981; Jacquemoud et al. 1992; Liang and Townshend 1996) or more empirical ones describing the spectral variability (Bach and Mauser 1994; Liu et al. 2002; Price 1990). At least six parameters are required to describe both the directional and spectral variation of soil properties. Leaf reflectance and transmittance may be simulated from the knowledge of its composition in the main absorbers (chlorophyll, water, and dry matter), the mesophyll structure, and the surface features (Dawson et al. 1998; Jacquemoud and Baret 1990; Jacquemoud et al. 2009). At least four parameters

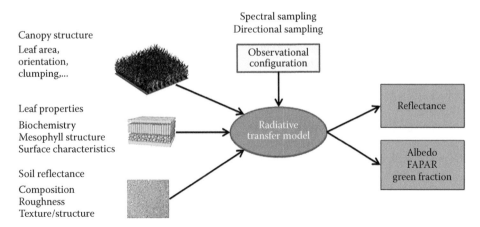

FIGURE 2.5 The radiative transfer models used to simulate canopy reflectance and additional canopy properties as a function of the leaf, soil, and canopy characteristics.

are required here. The simplest description of canopy structure could be achieved with two parameters: GAI and the orientation of the leaves. Therefore, the whole spectral and directional reflectance field of the canopy could be simulated with at least 12 parameters that are mainly unknown and should be estimated through radiative transfer model inversion. This needs to be solved with at least the same amount of independent reflectance measurements provided by the observational configuration, that is, combination of bands and view or illumination conditions.

The actual dimensionality of remote sensing measurements has been evaluated in different ways, generally by considering independently the spectral and directional dimensions. Several studies report that the bidirectional reflectance distribution function could be decomposed using empirical or semiempirical orthogonal functions with generally 2–4 kernels (Bréon et al. 2002; Lucht 1998; Weiss et al. 2002b; Zhang et al. 2002a,b). Other studies report a high level redundancy between bands (Green and Boardman 2001; Liu et al. 2002; Price 1990, 1994; Thenkabail et al. 2004) with a dimensionality varying between 5 and 60 depending on the data considered and the method used to quantify the dimensionality. More recently, Laurent et al. (2011) found a dimensionality of 3–4 using a singular decomposition method applied to CHRIS images having a high spectral resolution and several view directions. This finding confirmed those of Settle (2004) and Simic and Chen (2008) showing a high degree of redundancy between bands and directions. It is therefore clear that in most situations, the radiative transfer model inversion is an underdetermined problem, since the number of unknown variables to be estimated is larger than the actual dimensionality of the observations.

Because of its under-determination and uncertainties attached to models and measurements, the inverse problem is generally ill-posed: the solution is not unique and does not depend continuously on the observations (Garabedian 1964). In these conditions, very similar reflectance spectra simulated by a radiative transfer model (Figure 2.2, left) may correspond to a wide range of solutions. This may be due to two main factors:

- *Lack of sensitivity of canopy reflectance to a given variable*: This is the case for large GAI values because of the well-known saturation problem: a small variation in the measurements may correspond to a very large variation in the retrieved GAI value. Under the same high GAI conditions, the retrieved soil reflectance will be very poor, since the measured reflectance will be no more sensitive to soil background reflectance.
- *Compensation between variables*: This is obviously the case when some variables appear combined together always the same way in the model, such as in the form of a product: it is thus impossible to estimate separately each variable in this situation. However, this is also observed for other variables that are not formally appearing as products in the model as reported by several authors (Baret and Buis 2007; Baret et al. 1999; Shoshany 1991; Teillet et al. 1997; Weiss et al. 2000).

The ill-posedness of radiative transfer model inversion should be mitigated by exploiting additional information (Baret et al. 2000; Combal et al. 2001; Myneni et al., 2002). This could be achieved both by using prior information on the distribution of the variables, and by exploiting some constraints on the variables. Further, reducing the model uncertainties when possible by a proper selection of the radiative transfer model will also improve the accuracy of the retrieval. These issues will be investigated separately in the following sections.

2.5.2 Reducing Model Uncertainties

The realism of the radiative transfer model impacts largely the retrieval performances. The model should be physically sound and the embedded assumptions on canopy architecture and leaf and soil optical properties should be consistent with the actual canopy considered. The soil is relatively well described mainly by empirical models as reviewed in a previous section. The leaf optical properties are also quite well described by the PROPSECT (Jacquemoud et al. 2009) or LIBERTY (Dawson et al. 1998; Moorthy et al. 2008) models, at least if the directional effects are not considered (Comar et al. 2014). The canopy architecture is therefore recognized as the main limiting factor in the modeling of vegetation reflectance. To account for particular architectural features of a given canopy, prior knowledge on the type of vegetation viewed is therefore mandatory. Depending on the spatial resolution of the observation and the heterogeneity of the scene, this information is not always accessible. Observations at kilometric spatial resolution are often corresponding to a mix of different vegetation types, making the use of specific radiative transfer models challenging. Conversely, at decametric spatial resolution, pixels are more likely to be "pure" and the type of vegetation may be more easily identified. In these conditions, the inversion using a radiative transfer model for which the architecture is described in a more realistic way will reduce the error associated to the radiative transfer model and contribute to improve the retrieval performances. Lopez-Lozano (2008) compared the inversion of a turbid medium reflectance model where leaves are assumed randomly distributed within the canopy volume and of infinitesimal size to that of a 3D model adapted to maize and vineyard crops (Figure 2.6). The results showed clearly that GAI estimation improved a lot with a 3D description of canopy architecture for the vineyard case, where the turbid medium assumption is very far from reality as compared to the maize case. The estimated GAI using a turbid medium shows a systematic underestimation due to the leaf clumping: when the assumptions on canopy architecture are not verified by the canopy observed, the retrieved GAI value will be termed "apparent." This apparent GAI value is the one, which is accessible from the remote sensing measurements and the interpretation pipeline. It will thus depend on the inverse technique and on the radiative transfer model used. In addition, the apparent value may also strongly depend on the observational configuration used, as in the case of the vineyard canopy where the row orientation has to be accounted for. Using more realistic canopy

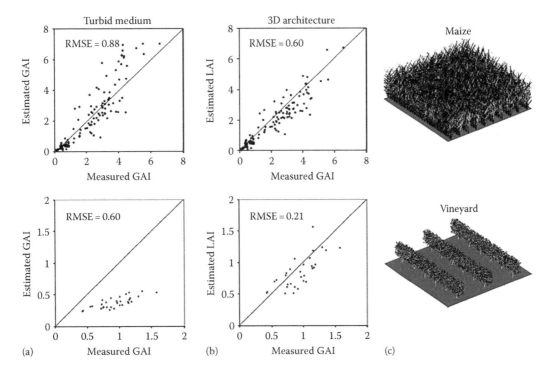

FIGURE 2.6 Comparison of GAI retrieval performances when using a turbid medium radiative transfer model (a) and 3D realistic canopy architecture (b). Examples over maize (c, top) and vineyard (c, bottom). RMSE is the root mean square error. (From Lopez-Lozano, R., Tecnologias de informacion geografica en la cartografia de parametros biofisicos de parcelas de maiz y vina para agricultura de precision, in *Geografia y ordenacion del territorio*, Universidad de Zaragoza, Zaragoza, Spain, 2008, p. 211.)

architecture implies more variables to describe the vertical and horizontal distributions of the green area density. The gain in realism obtained at the expense of additional unknown canopy variables should be counterbalanced by prior information on the distribution of these additional canopy structure variables.

2.5.3 Using Prior Information

The prior information characterizes the knowledge available on the distribution and codistribution of the input variables of the radiative transfer models. It is used directly in machine learning approaches to generate a calibration dataset that reflects this knowledge. For LUTs and iterative optimization methods, prior information is introduced in the cost function through a regularization term:

$$J = \underbrace{(R - \hat{R})^t \cdot W^{-1} \cdot (R - \hat{R})}_{\text{Radiometric information}} + \underbrace{(\hat{V} - V_p)^t \cdot C^{-1} \cdot (\hat{V} - V_p)}_{\text{Prior information}} \quad (2.3)$$

where
  $\hat{V}$ and V_p are, respectively, the vectors of the estimated and prior values of the input biophysical variables
  C is the covariance matrix characterizing the prior information

Note that the first part of this equation corresponds to Equation 2.1. The second part of Equation 2.3 corresponds to the distance between the values of the estimated variables and those of the prior information. The theory behind this equation derives from Bayes' theorem (Bayes and Price 1763) that was extensively used in parameter estimation (Tarantola 2005). However, if the theory is well known, it is not yet largely used in the community (Combal et al. 2002; Lauvernet et al. 2008; Lewis et al. 2012; Pinty et al. 2011).

Implementing the cost function as expressed by Equation 2.3 requires some reasonable estimates of covariance matrices W and C as well as of the prior values V_p. The terms of W should reflect both measurement and radiative transfer model uncertainties. While some rough estimates of the measurement uncertainties could be derived from the sensor specification, model uncertainties are far more difficult to estimate. Further, they may depend significantly on the situation considered, such as low or high vegetation amount and the discrepancy between the canopy structure description embedded in the radiative transfer model and that of the observed canopy. It is even more difficult to estimate the covariance terms in W: measurement and model uncertainties may have important structure that translates into high covariance terms that are however very poorly known. When using simultaneously a large number of configurations as in the case of hyperspectral observations, these covariance terms will allow weighing properly the several configurations used. It thus accounts for the large redundancy exhibited between spectral bands. The difficulty to estimate the covariance terms in W explains why a small number of configurations are often selected when a larger number is available as in the case of hyperspectral and/or multidirectional observations. Thus, the reduction of

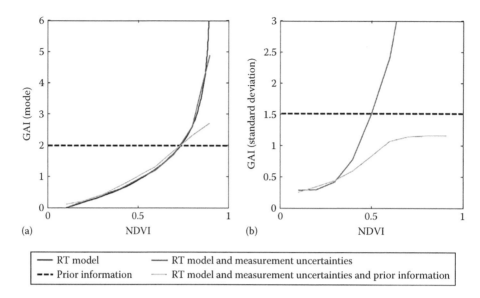

FIGURE 2.7 (a) Mode of the distribution of the solution (GAI) of the inverse problem as a function of the measured value. In this case, GAI is estimated using a simple empirical RT model (NDVI = RT(GAI)). The mode corresponds to the maximum PDF value, that is, the maximum likelihood. Four estimates are displayed: using only prior information (i.e., no measurements are used) (prior information); using RT model (GAI = RT^{-1}(NDVI)) assumed to be perfect with perfect measurements (no uncertainties accounted for); using RT model and measurements with their associated uncertainties; using RT model and measurements with their associated uncertainties and prior information. (b) The standard deviation of the distribution of the solution is also displayed for several cases. The case with perfect RT model and measurements is not displayed here, because its standard deviation is null by definition. (From Baret, F. and Buis, S., Estimating canopy characteristics from remote sensing observations. Review of methods and associated problems, in Liang, S., ed., *Advances in Land Remote Sensing: System, Modeling, Inversion and Application*, Springer, New York, 2007, pp. 171–200.)

the dimensionality of the input observations is highly desired in most retrieval problems (Tenenbaum et al. 2000). For machine learning methods, a reduced dimensionality is also beneficial, since the number of coefficients of the inverse parametric model will grow with the number of observations used as inputs, making the calibration process more difficult and instable.

Introducing prior information in the inversion process improves the precision by reducing the variability of the posterior distribution of the estimated variables. However, this is achieved at the expense of accuracy: the solution is biased toward the prior information value as observed in Figure 2.7.

2.5.4 Using Additional Constraints

2.5.4.1 Temporal Constraints

The dynamics of canopy structure and leaf optical properties results from incremental processes under the control of climate, soil, and the genetic characteristics of the plants. Very brutal and chaotic time courses are therefore not expected, at the exception of accidents such as fire, flooding, harvesting, or lodging. The smooth character of the dynamics of canopy variables may be exploited as additional constraints in the retrieval process as proposed by Lewis et al. (2012). The use of models describing the time course of some of the variables was proposed by Kötz et al. (2005) to improve remote sensing estimates of GAI in maize crops: results show a significant improvement of estimates, particularly for the larger GAI values where saturation of reflectance is known to be a problem. The semi-empirical nature of the

model with parameters having some biological meaning, allows to accumulate prior information on them for efficient exploitation. However, the results show that the improvement in GAI retrieval is mainly coming from the "smoothing" effect of the model: fitting the GAI dynamics model over the instantaneous estimates corresponding to each individual date of observation provides similar performances (Kötz et al. 2005). This explains why compositing techniques applied to kilometric resolution observations are very popular: very little prior information is available on the dynamics of the surface except the expected smoothness of the temporal profiles (Atkinson et al. 2012; Chen et al. 2004; Kandasamy et al. 2013; Lewis et al. 2012; Refslund et al. 2013; Verbesselt et al. 2010b; Zhu et al. 2013). The usual sigmoidal shape of vegetation growth and senescence curves (Jönsson and Eklundh 2004; Zhang et al. 2003), and the possible use of the climatology, (Samain et al. 2007; Verger et al. 2012a) have been also exploited.

2.5.4.2 Spatial Constraints

Most of the algorithms are currently applied to independent pixels, neglecting the possible use of spatial structures as observed on most images. However, some authors attempted to exploit these very obvious patterns at high spatial resolution. The "object retrieval" approach proposed by Atzberger (2004) is based on the use of covariance between variables as observed over a limited cluster of pixels representing the same class of object such as an agricultural field. Results show significant improvement of the retrieval performances for GAI, chlorophyll and water contents, presumably because of a better handling of compensations between GAI and leaf inclination in the retrieval process as suggested by Atzberger (2004). The principles were further extended using simple heuristics that could apply at the field scale for agriculture applications (Atzberger and Richter 2012). This implies that objects, sometimes called "patches," are first identified, which is now becoming a very common approach in remote sensing image segmentation techniques (Blaschke 2010; Peña-Barragán et al. 2011; Vieira et al. 2012). The objects need then to be classified to exploit some features shared by the pixels of a single patch.

2.5.4.3 Holistic Retrieval over Coupled Models: From Inversion to Assimilation

Retrieval of characteristics of some element of the system without solving the whole system at once will be suboptimal: each element of the system imposes constraints on other elements through the radiative transfer physical processes, temporal or spatial constraints as seen previously. This is clearly demonstrated in the case of the radiative coupling between the leaves and the canopy: when estimating structural canopy characteristics from bottom of the atmosphere reflectance measurements in several bands and directions, the inversion process may be split into several parallel and independent inversions for each band. The leaf characteristics, that is, reflectance and transmittance, sometimes grouped into the single scattering albedo, need to be estimated (Pinty and Verstraete 1991b) for each of the bands considered. This may lead to inconsistent estimates of the structure characteristics derived from the inversion applied independently on each band. Further, it may lead to spectrally inconsistent leaf optical properties estimates, since no spectral constraints coming from a leaf optical properties model are imposed. Solving the whole system at once using coupled leaf and canopy radiative transfer models will therefore improve the consistency of the estimates by imposing the spectral constraints coming from the leaf radiative transfer model. The interest of such holistic approach was recently highlighted by Laurent et al. (2011) when using coupled canopy and atmosphere radiative transfer models.

Lauvernet et al. (2008) proposed a "multitemporal patch" inversion scheme to account both for spatial and temporal constraints. Reflectance data are here considered observed from the top of the atmosphere. Atmosphere/canopy/leaf/soil radiative transfer models are thus coupled to simulate top of the atmosphere reflectance from the set of input variables of each submodel. Spatial and temporal constraints are based on the assumption that the atmosphere is stable over a limited area (typically few kilometers) but varies from date to date, and that surface characteristics vary only marginally over a limited temporal window (typically ± 7 days) but may strongly change from pixel to pixel (Hagolle et al. 2008). This has obviously important consequences on the underdetermined nature of the inverse problem, since atmospheric characteristics will be shared between the pixels of a patch, while vegetation characteristics will be shared during a limited time period. Results

demonstrate the interest of the approach for the estimation of most of the variables, particularly for the aerosol characteristics and for canopy characteristics such as GAI.

However, the improvement of retrieval performances based on such holistic approach is gained at the expense of additional complexity in terms of the number of unknown variables to be estimated and of the computational resources required to run the coupled models. Machine learning approaches may reach their limits in such conditions. Iterative optimization efficiently implemented using the adjoint model (Lauvernet et al. 2008; Lewis et al. 2012; Voßbeck et al. 2010) provides a convenient solution. This could be used ultimately to couple the radiative transfer model to a functional-structural plant model as proposed by Weiss et al. (2001). However, considerable efforts are still needed to describe the dynamics of the canopy structure consistently with both the radiative transfer modeling and with the canopy functioning.

2.6 Combination of Methods and Sensors to Improve the Retrievals

2.6.1 Hybrid Methods and Ensemble Products

Verger et al. (2008) demonstrated that NNTs could be used efficiently to replace the actual MODIS algorithm (Shabanov et al. 2005) that is based on a LUT method: NNTs were calibrated over an empirical training dataset containing the MODIS top of canopy BRF values and the corresponding MODIS LAI products. This approach is therefore different from calibrating a machine learning algorithm directly on radiative transfer model simulations as done by Bacour et al. (2006) or Baret et al. (2007). It is termed hybrid, because a canopy biophysical-driven method is calibrated over the outputs of a radiometric data–driven approach. This principle was later used to relate the long time series of AVHRR NDVI VI (Tucker et al. 2005) to LAI and FAPAR MODIS products during an overlapping period between both sensors (2000–2009) (Zhu et al. 2013).

With the compilation of results derived from several initiatives dedicated to the validation of global remote sensing products, the performances of products started to be quantified in a more representative way (Garrigues et al. 2008). This allowed to select the more consistent available products and to eventually combine them and propose a new "ensemble" product that capitalizes over past development efforts (Figure 2.8): a training dataset is first built that contains a globally representative sample of MODIS and CYCLOPES products along with reflectance as measured by a sensor from which the "ensemble" product is generated (Baret et al. 2013; Verger et al. 2014; Xiao et al. 2014). The original biophysical products in the training database need to share the same spatial and temporal support to be consistently combined. This is achieved by applying interpolation methods that will further smooth possible spatial or temporal discrepancies. A weighed average of the original products is then computed to get the fused products (Figure 2.8). The weights are derived from the results of the validation of the original products

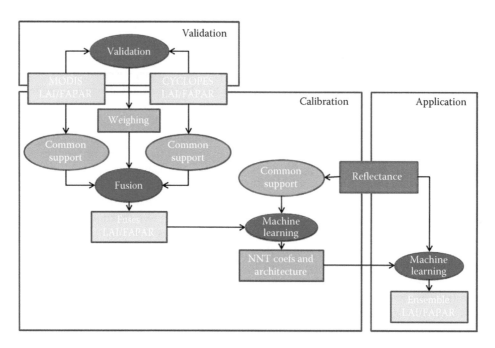

FIGURE 2.8 Principle of the GEOV1 (Baret et al. 2013), GEOV2 (Verger et al. 2014), and GLASS (Xiao et al. 2014) algorithms to generate "ensemble" products.

using either the associated uncertainties (Xiao et al. 2014) or heuristic arguments (Baret et al. 2013). The fused products and the corresponding reflectance values are then used to calibrate a machine learning algorithm. The calibrated machine learning algorithm is finally used to transform the reflectance values into the corresponding "ensemble" product (Figure 2.8). In the case of GEOV1 (Baret et al. 2013), the transformation is applied using a backpropagation NNT over each individual observation to get instantaneous fused LAI and FAPAR values. A smoothing and gap-filling algorithm is then applied over the time series of fused products (Verger et al. 2011b). In the case of GLASS products (Xiao et al. 2014), a whole year of reflectance observations in the red and near-infrared is used to get the corresponding yearly time series of LAI products using a generalized regression NNT (Specht 1991). Results show that these ensemble products are generally overperforming the original products (Camacho et al. 2013; Fang et al. 2013; Xiao et al. 2014).

2.6.2 Combining Sensors to Build Long, Dense, and Consistent Time Series

Monitoring the dynamics at the seasonal or multiannual scale allows to better characterize the canopy functioning including the phenology (Ganguly et al. 2010; Jönsson and Eklundh 2004) and detect anomalies (Bessemoulin et al. 2004; Ciais et al. 2005), breaks (Verbesselt et al. 2010a) or trends (Alcaraz-Segura et al. 2010; de Jong et al. 2012; Fensholt et al. 2012; Herrmann et al. 2005) across long time series of consistent observations. The revisit frequency, consistency, and length of the period when observations are accumulated are the main limiting factors when exploiting the time series. For global scale applications,

observations are currently provided by kilometric spatial resolution sensors on polar orbit (Figure 2.10). They have a relatively large swath allowing to map the whole Earth within 1 day. However, this potential daily observation frequency is reduced because of the cloud occurrence. The combination of observations by different sensors will provide only marginal gain in terms of the number of cloud-free dates of observations because of the strong spatiotemporal correlation of the distribution of clouds: Yang et al. (2006) reported no improvement when combining MODIS products derived from AQUA and TERRA. However, Hagolle et al. (2005) reported an improvement of both the completeness and the precision of top of canopy reflectance when compositing the two VEGETATION instruments as compared to the use of a single one. Similarly, Verger et al. (2011b) fused MODIS and VEGETATION data, which resulted both in a significant reduction of the fraction of missing products as well as an improvement of the accuracy and precision of LAI estimates. These contrasting results are explained by the very different compositing algorithms used in these studies, highlighting the importance of the compositing process that mainly consists in smoothing and eventually gap-filling the time series (Kandasamy et al. 2013).

The interest of fusing the data coming from several sensors is obvious when considering the decametric spatial resolution observations for which several days are needed to map the whole Earth. However, except in the case of the RapidEye and DMC constellation of satellites (Röser et al. 2005), very little attention has been carried out on the fusion between different decametric resolution satellites. Although the satellites currently orbiting provide great potentials for seasonal monitoring of the vegetation at decametric spatial resolution, this has not yet been exploited

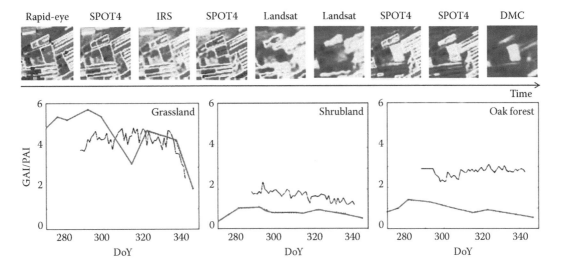

FIGURE 2.9 Exploitation of a heterogeneous constellation of satellites to derive seasonal variation of GAI. On top, the GAI images derived from each individual sensor along the growing season. On bottom, the seasonal variation over three sites. The red line corresponds to GAI estimates. The black line corresponds to reference ground measurements of PAI. Unpublished results obtained on the Crau site in South-East France (43.5° latitude, 4.9° longitude). The algorithm used for all the sensors is similar to that described by Verger et al. (2011a) with no compositing applied to the data. The difference between satellite estimates and ground measurements is mainly explained by the difference in the definition of the variable accessed from remote sensing (GAI) and that measured on the ground (PAI).

because of the difficulty and cost associated to the images of these sensors that are used commercially. However, the development of the fusion between different decametric satellites does not pose great technical difficulties as illustrated by Figure 2.9: a very good temporal consistency of estimates derived from different sensors using the same algorithm is generally observed. This confirms the results of Verger et al. (2008) and Gobron et al. (2008) who demonstrated that applying a single algorithm to different sensors provides generally consistent products if the differences in observational configurations are carefully accounted for.

The fusion between decametric resolution images and daily kilometric resolution data is very appealing, because it potentially provides daily decametric products. However, this combination has been rarely investigated for deriving decametric dynamics of biophysical variables. It has mainly been applied for classification (Karkee et al. 2009), for reflectances (Faivre and Fischer 1997), including pan-sharpening (Fasbender et al. 2008), and for vegetation indices (Cardot et al. 2008; Gao et al. 2006). More studies should be directed toward the development of the fusion between biophysical variables obtained from decametric and kilometric spatial resolution sensors.

The succession of several kilometric sensors allows building long time series of global observations since 1981 (Figure 2.10). However, the consistency between the several sensors used to build the time series has to be very high in order to identify possible trends that may be very small (Beck et al. 2011). This is currently achieved by applying a single algorithm to the succession of sensors available. Zhu et al. (2013) transformed the long time series of NDVI derived from the several AVHRR sensors (Figure 2.10) into LAI and FAPAR by calibrating an NNT on MODIS products during and overlapping period between

AVHRR and MODIS. The consistency and the compositing are here achieved at the NDVI level, based on the GIMMS products (Tucker et al. 2005). Verger et al. (2012b) built also a long time series of observations based on AVHRR up to 2000, and then, using VEGETATION data. The input reflectance values were carefully processed according to Nagol et al. (2009). Then, NNTs were calibrated over an overlapping period among AVHRR, VEGETATION, and MODIS. The LAI and FAPAR products from MODIS and VEGETATION were fused to be used as target products. Finally, a compositing algorithm was applied to eliminate outliers, smooth out the resulting data and fill possible gaps (Verger et al. 2012a).

2.7 Conclusion

This review of retrieval techniques for canopy biophysical variables shows that great advancement in the maturity of the algorithms has been achieved in recent years. Several products were released to the wider community, mainly derived from kilometric resolution sensors as illustrated by Table 2.2. The multiplicity of products allows building enough confidence from the consistency observed between some of them as well as in ground measurements. The validation exercise is therefore mandatory to identify possible problems, improve the products, and finally quantify the associated uncertainties. The root mean square error (RMSE) values associated to FAPAR are in the order of 0.10–0.15 in absolute value (Weiss et al. 2014), while LAI is estimated within an RMSE slightly smaller than 1.0. However, the currently limited number of available ground measurements at the kilometric resolution limits the evaluation of the accuracy of remote sensing products (Garrigues et al. 2008).

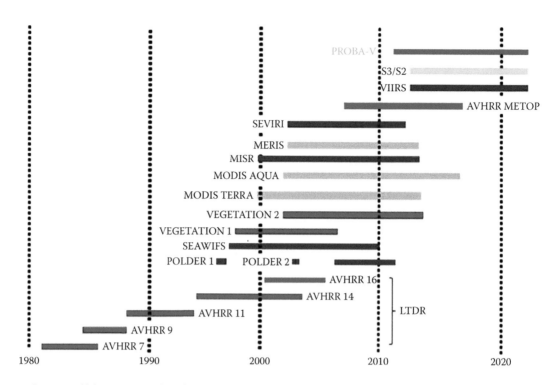

FIGURE 2.10 The series of kilometric spatial resolution sensors available from 1980 up to now.

TABLE 2.2 Several LAI and FAPAR Global Products Currently Available

Products	Sensors	LAI	FAPAR	Spatial Resolution	Time Sampling (days)	Time Period	Reference
MODIS C5	MODIS	✓	✓	1 km	8	2000	Myneni et al. (2002)
CYCLOPES V3	VEGETATION	✓	✓	0.009°	10	1999–2007	Baret et al. (2007)
GLOBCARBON	VEGETATION	✓	✓	0.009°	30	1999–2007	Deng et al. (2006)
JRC-FAPAR	SEAWIFS		✓	2 km	1	1997–2006	Gobron et al. (2006)
JRC-TIP	MODIS	✓	✓	0.01°	16	2000	Pinty et al. (2010)
GIMMS_3g	AVHRR	✓	✓	8 km	30	1981–2013	Ganguly et al. (2010)
GLASS	MODIS/AVHRR	✓	✓	1 km	10	1981–2014	Xiao et al. (2012)
GEOV1_VEG	VEGETATION	✓	✓	0.009°	10	1999	Baret et al. (2012)
GEOV2_VEG	VEGETATION	✓	✓	0.009°	10	1999	Verger et al. (2014)

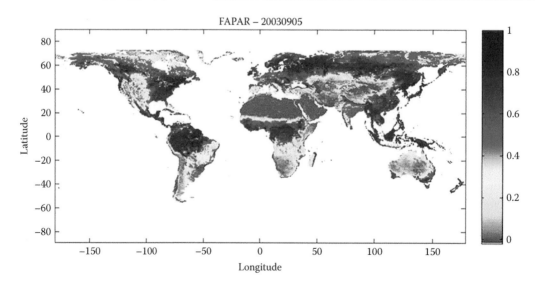

FIGURE 2.11 A global map of GEOV1 FAPAR product for September 5, 2003.

When very few information or constraints are available as it is the case for the kilometric resolution observations, the variables that are the better estimated are the GF and the black and white sky FAPAR or FIPAR ones. Conversely, the "apparent" LAI derived from the reflectance observations is more closely linked with the effective GAI, GAI_{eff}, while the true LAI is poorly estimated with uncertainties that are dependent on the observational configuration. This finding should be much better reflected to the users of current LAI products derived from remote sensing, although some attempts were proposed to correct for this effect (Xiao et al. 2012). However, focusing on the GF variable will allow reaching more easily a better consistency with the canopy functioning models that have their own specific description of the canopy architecture.

Two main types of canopy biophysical variables retrieval approaches were identified. Canopy biophysical variable–driven approaches when trained over empirical datasets with ground measurements of the canopy variables defined in a consistent way with the variables accessible from remote sensing observations would be ideal: they implicitly integrate the measurement uncertainties, while no model uncertainties have to be included, since no radiative transfer model is used. Further, machine learning approaches are very computer efficient once trained, allowing easy implementation within operational processing chains. However, because of the difficulty of getting a representative sampling of cases to populate the training dataset, training over a database made of radiative transfer model simulations is often preferred. The radiative transfer models need to be well adapted to the type of canopy they target. Unfortunately, using a more realistic description of the canopy architecture requiring more variables may create problems in the inversion process if no additional prior information or constraints are exploited. Radiometric data–driven approaches such as iterative minimization appear to be very appealing to handle a wide range of constraints and prior information that may be available at the decametric resolution. This may ultimately lead to the assimilation of calibrated radiances into structural–functional vegetation models coupled with atmospheric models that is currently in the infancy stage of development. The expected increasing accessibility of frequent decametric observations will certainly push investigations in such a direction, exploiting explicitly the whole set of available information and knowledge on physical and biological processes.

References

Abuelgasim, A.A., Gopal, S., and Strahler, A.H. (1998). Forward and inverse modelling of canopy directional reflectance using a neural network. *International Journal of Remote Sensing, 19*, 453–471.

Aires, F., Prigent, C., and Rossow, W.B. (2004). Neural network uncertainty assessment using Bayesian statistics: A remote sensing application. *Neural Computation, 16*, 2415–2458.

Alcaraz-Segura, D., Liras, E., Tabik, S., Paruelo, J., and Cabello, J. (2010). Evaluating the consistency of the 1982–1999 NDVI trends in the Iberian Peninsula across four time-series derived from the AVHRR sensor: LTDR, GIMMS, FASIR, and PAL-II. *Sensors, 10*, 1291–1314.

Asrar, G. (1989). *Theory and Applications of Optical Remote Sensing*. New York: Wiley.

Asrar, G., Fuchs, M., Kanemasu, E.T., and Hatfield, J.L. (1984). Estimating absorbed photosynthetic radiation and leaf area index from spectral reflectance in wheat. *Agronomy Journal, 76*, 300, 306.

Atkinson, P.M., Jeganathan, C., Dash, J., and Atzberger, C. (2012). Inter-comparison of four models for smoothing satellite sensor time-series data to estimate vegetation phenology. *Remote Sensing of Environment, 123*, 400–417.

Atkinson, P.M. and Tatnall, A.R.L. (1997). Neural network in remote sensing. *International Journal of Remote Sensing, 18*, 699–709.

Atzberger, C. (2004). Object-based retrieval of biophysical canopy variables using artificial neural nets and radiative transfer models. *Remote Sensing of Environment, 93*, 53–67.

Atzberger, C. and Richter, K. (2012). Spatially constrained inversion of radiative transfer models for improved LAI mapping from future Sentinel-2 imagery. *Remote Sensing of Environment, 120*, 208–218.

Bach, H. and Mauser, W. (1994). Modelling and model verification of the spectral reflectance of soils under varying moisture conditions. In *International Geoscience and Remote Sensing Symposium* (pp. 2354–2356). Pasadena, CA: IEEE.

Bacour, C. (2001). Contribtion à la détermination des paramètres biophysiques des couverts végétaux par inversion de modèles de réflectance: Analyse de sensibilité et configurations optimales. In *Méthodes physiqes en télédétection* (p. 206). Paris, France: Université Paris 7—Denis Diderot.

Bacour, C., Baret, F., Béal, D., Weiss, M., and Pavageau, K. (2006). Neural network estimation of LAI, fAPAR, fCover and LAIxCab, from top of canopy MERIS reflectance data: Principles and validation. *Remote Sensing of Environment, 105*, 313–325.

Bacour, C., Jacquemoud, S., Leroy, M., Hautecoeur, O., Weiss, M., Prévot, L., Bruguier, N., and Chauki, H. (2002a). Reliability of the estimation of vegetation characteristics by inversion of three canopy reflectance models on airborne POLDER data. *Agronomie, 22*, 555–566.

Bacour, C., Jacquemoud, S., Tourbier, Y., Dechambre, M., and Frangi, J.P. (2002b). Design and Analysis of numerical experiments to compare four canopy reflectance models. *Remote Sensing of Environment, 79*, 72–83.

Banari, A., Huete, A.R., Morin, D., and Zagolski, F. (1996). Effets de la couleur et de la brillance du sol sur les indices de végétation. *International Journal of Remote Sensing, 17*, 1885–1906.

Baret, F., Bacour, C., Weiss, M., Pavageau, K., Béal, D., Bruniquel, V., Regner, P., Moreno, J., Gonzalez, C., and Chen, J. (2004). Canopy biphysical variables estimation from MERIS observations based on neural networks and radiative transfer modelling: Principles and validation. In ESA (Ed.), *ENVISAT Conference*. Salzburg, Austria: ESA.

Baret, F. and Buis, S. (2007). Estimating canopy characteristics from remote sensing observations. Review of methods and associated problems. In S. Liang (Ed.), *Advances in Land Remote Sensing: System, Modeling, Inversion and Application* (pp. 171–200). New York: Springer.

Baret, F., Clevers, J.G.P.W., and Steven, M.D. (1995). The robustness of canopy gap fraction estimates from red and near infrared reflectances: A comparison of approaches. *Remote Sensing of Environment, 54*, 141–151.

Baret, F., De Solan, B., Lopez-Lozano, R., Ma, K., and Weiss, M. (2010). GAI estimates of row crops from downward looking digital photos taken perpendicular to rows at 57.5° zenith angle. Theoretical considerations based on 3D architecture models and application to wheat crops. *Agricultural and Forest Meteorology, 150*, 1393–1401.

Baret, F. and Guyot, G. (1991). Potentials and limits of vegetation indices for LAI and APAR assessment. *Remote Sensing of Environment, 35*, 161–173.

Baret, F., Hagolle, O., Geiger, B., Bicheron, P., Miras, B., Huc, M., Berthelot, B. et al. (2007). LAI, fAPAR and fCover CYCLOPES global products derived from VEGETATION: Part 1: Principles of the algorithm. *Remote Sensing of Environment, 110*, 275–286.

Baret, F., Knyazikhin, J., Weiss, M., Myneni, R., and Pragnère, A. (1999). Overview of canopy biophysical variables retrieval techniques. In *ALPS'99*, Meribel, France.

Baret, F., Weiss, M., Bicheron, P., and Bethelot, B. (2009). *S2PAD—Sentinel-2 MSI—Level 2B Products Algorithm Theoretical Basis Document*. Avignon, France: Vega, GmbH.

Baret, F., Weiss, M., Lacaze, R., Camacho, F., Makhmara, H., Pacholcyzk, P., and Smets, B. (2013). GEOV1: LAI and FAPAR essential climate variables and FCOVER global time series capitalizing over existing products. Part1: Principles of development and production. *Remote Sensing of Environment, 137*, 299–309.

Baret, F., Weiss, M., Troufleau, D., Prévot, L., and Combal, B. (2000). Maximum information exploitation for canopy characterisation by remote sensing. *Remote Sensing in Agriculture* (pp. 71–82). Cirencester, U.K.: Association of Applied Biologists.

Bayes, T. and Price, R. (1763). An essay towards solving a problem in the Doctrine of Chance. (By the late Rev. Mr. Bayes, communicated by Mr. Price, in a letter to John Canton, A. M. F. R. S.) *Philosophical Transactions of the Royal Society of London, 53*, 370–418.

Beck, H.E., McVicar, T.R., van Dijk, A.I., Schellekens, J., de Jeu, R.A., and Bruijnzeel, L.A. (2011). Global evaluation of four AVHRR–NDVI data sets: Intercomparison and assessment against Landsat imagery. *Remote Sensing of Environment, 115*, 2547–2563.

Begué, A., Desprat, J.F., Imbernon, J., and Baret, F. (1991). Radiation use efficiency of pearl millet in the Sahelian zone. *Agricultural and Forest Meteorology, 56*, 93–110.

Bessemoulin, P., Bourdette, N., Courtier, P., and Manach, J. (2004). La canicule d'aout 2003 en France et en Europe. *La météorologie, 46*, 25–33.

Beven, K.J. and Binsley, A.M. (1992). The future of distributed models: Model calibration and uncertainty predictions. *Hydrological Processes, 6*, 279–298.

Bicheron, P. and Leroy, M. (1999). A method of biophysical parameter retrieval at global scale by inversion of a vegetation reflectance model. *Remote Sensing of Environment, 67*, 251–266.

Blaschke, T. (2010). Object based image analysis for remote sensing. *ISPRS Journal of Photogrammetry and Remote Sensing, 65*, 2–16.

Bréon, F.M., Maignan, F., Leroy, M., and Grant, I. (2002). Analysis of hot spot directional signatures measured from space. *Journal of Geophysical Research, 107*, AAC 1 1–AAC 1 15.

Camacho, F., Cernicharo, J., Lacaze, R., Baret, F., and Weiss, M. (2013). GEOV1: LAI, FAPAR essential climate variables and FCOVER global time series capitalizing over existing products. Part 2: Validation and intercomparison with reference products. *Remote Sensing of Environment, 137*, 310–329.

Cardot, H., Maisongrande, P., and Faivre, R. (2008). Varying-time random effects models for longitudinal data: unmixing and temporal interpolation of remote sensing data. *Journal of Applied Statistics, 35*, 827–846.

Chen, J., Jönsson, P., Tamura, M., Gu, Z., Matsushita, B., and Eklundh, L. (2004). A simple method for reconstructing a high quality NDVI time series data set based on the Savitzky-Golay filter. *Remote Sensing of Enviroment, 91*, 332–344.

Chen, J.M. and Black, T.A. (1992). Defining leaf area index for non-flat leaves. *Plant, Cell and Environment, 15*, 421–429.

Chen, J.M., Menges, C.H., and Leblanc, S.G. (2005). Global mapping of foliage clumping index using multi-angular satellite data. *Remote Sensing of Environment, 97*, 447–457.

Chen, J.M., Pavlic, G., Brown, L., Cihlar, J., Leblanc, S., White, H.P., Hall, R.J. et al. (2002). Derivation and validation of Canada wide coarse resolution leaf area index maps using high resolution satellite imagery and ground masurements. *Remote Sensing of Environment, 80*, 165–184.

Ciais, P., Reichstein, M., Viovy, N., Granier, A., Ogée, J., Allard, V., Aubinet, M. et al. (2005). Europe wide reduction in primary productivity caused by the heat and drought in 2003. *Nature, 437*, 529–533.

Cierniewski, J., Verbrugghe, M., and Marlewski, A. (2002). Effects of farming works on soil surface bidirectional reflectance measurements and modelling. *International Journal of Remote Sensing, 23*, 1075–1094.

Claverie, M., Vermote, E., Weiss, M., Baret, F., Hagolle, O., and Demarez, V. (2013). Validation of coarse spatial resolution LAI and FAPAR time series over cropland in southwest France. *Remote Sensing of Environment, 139*, 216–230.

Comar, A., Baret, F., Obein, G., Simonot, L., Meneveaux, D., Vienot, F., and de Solan, B. (2014). ACT: A leaf BRDF model taking into account the azimuthal anisotropy of monocotyledonous leaf surface. *Remote Sensing of Environment, 143*, 112–121.

Combal, B., Baret, F., and Weiss, M. (2001). Improving canopy variables estimation from remote sensing data by exploiting ancillary information. Case study on sugar beet canopies. *Agronomie, 22*, 2–15.

Combal, B., Baret, F., Weiss, M., Trubuil, A., Macé, D., Pragnère, A., Myneni, R., Knyazikhin, Y., and Wang, L. (2002). Retrieval of canopy biophysical variables from bi-directional reflectance data. Using prior information to solve the ill-posed inverse problem. *Remote Sensing of Environment, 84*, 1–15.

Combal, B., Ochshepkov, S.L., Sinyuk, A., and Isaka, H. (2000). Statistical framework of the inverse problem in the retrieval of vegetation parameters. *Agronomie, 20*, 65–77.

Danson, F.M., Rowland, C.S., and Baret, F. (2003). Training a neural network with a canopy reflectance model to estimate crop leaf area index. *International Journal of Remote Sensing, 24*, 4891–4905.

Darvishzadeh, R., Skidmore, A., Schlerf, M., and Atzberger, C. (2008). Inversion of a radiative transfer model for estimating vegetation LAI and chlorophyll in a heterogeneous grassland. *Remote Sensing of Environment, 112*, 2592.

Dawson, T.P., Curran, P.J., and Plummer, S.E. (1998). LIBERTY-Modeling the effects of leaf biochemical concentration on reflectance spectra. *Remote Sensing of Environment, 65*, 50–60.

de Jong, R., Verbesselt, J., Schaepman, M., and De Bruin, S. (2012). Trend changes in global greening and browning: contribution of short-term trends to longer-term change. *Global Change Biology, 18*, 642–655.

Deng, F., Chen, J.M., Chen, M., and Pisek, J. (2006). Algorithm for global leaf area index retrieval using satellite imagery. *IEEE Transactions on Geoscience and Remote Sensing, 44*, 2219–2229.

Duveiller, G., Weiss, M., Baret, F., and Defourny, P. (2011). Retrieving wheat Green Area Index during the growing season from optical time series measurements based on neural network radiative transfer inversion. *Remote Sensing of Environment, 115*, 887.

Faivre, R. and Fischer, A. (1997). Predicting crop reflectances using satellite data observing mixed pixels. *Journal of Agricultural, Biological and Environmental Statistics, 2*, 87–107.

Fang, H., Jiang, C., Li, W., Wei, S., Baret, F., Chen, J.M., Garcia-Haro, J. et al. (2013). Characterization and intercomparison of global moderate resolution leaf area index (LAI) products: Analysis of climatologies and theoretical uncertainties. *Journal of Geophysical Research: Biogeosciences, 118*, 529–548.

Fang, H. and Liang, S. (2003). Retrieving leaf area index with a neural network method: simulation and validation. *IEEE Transactions on Geoscience and Remote Sensing, 41*, 2052–2062.

Fang, H. and Liang, S. (2005). A hybrid inversion method for mapping leaf area index from MODIS data: Experiments and application to broadleaf and needleleaf canopies. *Remote Sensing of Enviroment, 94*, 405–424.

Fang, H., Liang, S., and Kuusk, A. (2003). Retrieving leaf area index using a genetic algorithm with a canopy radiative transfer model. *Remote Sensing of Enviroment, 85*, 257–270.

Fasbender, D., Radoux, J., and Bogaert, P. (2008). Bayesian data fusion for adaptable image pansharpening. *IEEE Transactions on Geoscience and Remote Sensing, 46*, 1847–1857.

Fensholt, R., Langanke, T., Rasmussen, K., Reenberg, A., Prince, S.D., Tucker, C., Scholes, R.J. et al. (2012). Greenness in semi-arid areas across the globe 1981–2007—An Earth Observing Satellite based analysis of trends and drivers. *Remote Sensing of Environment, 212*, 144–158.

Fernandes, R. and Leblanc, S.G. (2005). Parametric (modified least squares) and non parametric (Theil-Sen) linear regressions for predicting biophysical parameters in the presence of measurements errors. *Remote Sensing of Enviroment, 95*, 303–316.

Ganguly, S., Friedl, M.A., Tan, B., Zhang, X., and Verma, M. (2010). Land surface phenology from MODIS: Characterization of the Collection 5 global land cover dynamics product. *Remote Sensing of Environment, 114*, 1805.

Ganguly, S., Nemani, R.R., Zhang, G., Hashimoto, H., Milesi, C., Michaelis, A., Wang, W. et al. (2012). Generating global Leaf Area Index from Landsat: Algorithm formulation and demonstration. *Remote Sensing of Environment, 122*, 185–202.

Gao, F., Masek, J.G., Schwaller, M., and Forrest, H. (2006). On the blending of the Landsat and MODIS surface reflectance. *IEEE Transactions on Geoscience and Remote Sensing, 44*, 2207–2219.

Garabedian, P. (1964). *Partial Differential Equations*. John Willey and Sons, New York.

Garrigues, S., Lacaze, R., Baret, F., Morisette, J., Weiss, M., Nickeson, J., Fernandes, R. et al. (2008). Validation and intercomparison of global leaf area index products derived from remote sensing data. *Journal of Geophysical Research, 113*.

GCOS (2011). Global climate observing system—Systematic observation requirements for satellite-based products for climate—2011 update, Supplemental details to the satellite based component of the implementation plan for the global observing system for climate in support of the UNFCCC (2010 Update) (p. 138). Geneva, Switzerland: World Meteorological Organization.

Gilbert, J.C. (2002). *Optimisation Différentiable: Théorie et Algorithmes*. INRIA Rocquencourt.

Gobron, N., Pinty, B., Aussedat, O., Chen, J.M., Cohen, W.B., Fensholt, R., Gond, V. et al. (2006). Evaluation of fraction of absorbed photosynthetically active radiation products for different canopy radiation transfer regimes: Methodology and results using Joint Research Center products derived from SeaWiFS against ground-based estimations. *Journal of Geophyisical Research, 111*, doi:10.1029/2005JD006511.

Gobron, N., Pinty, B., Aussedat, O., Taberner, M., Faber, O., Melin, F., Lavergne, T., Robustelli, M., and Snoeij, P. (2008). Uncertainty estimates for the FAPAR operational products derived from MERIS -Impact of top-of-atmosphere radiance uncertainties and validation with field data. *Remote Sensing of Environment, 112,* 1871–1883.

Gobron, N., Pinty, B., Verstraete, M., and Widlowski, J.L. (2000). Advanced vegetation indices optimized for up-coming sensors: design, performances and applications. *IEEE Transactions on Geoscience and Remote Sensing, 38,* 2489–2505.

Goel, N.S. (1989). Inversion of canopy reflectance models for estimation of biophysical parameters from reflectance data. In G. Asrar (Ed.), *Theory and Applications of Optical Remote Sensing* (pp. 205–251). Wiley Interscience, New York, pp. 205–251.

Goel, N.S. and Deering, D.W. (1985). Evaluation of a canopy reflectance model for LAI estimation through its inversion. *IEEE Transactions on Geoscience and Remote Sensing, GE-23,* 674–684.

Goel, N.S., Strebel, D.E., and Thompson, R.L. (1984). Inversion of vegetation canopy reflectance models for estimating agronomic variables. II. Use of angle transforms and error analysis as illustrated by the SUIT model. *Remote Sensing of Environment, 14,* 77–111.

Goel, N.S. and Thompson, R.L. (1984). Inversion of vegetation canopy reflectance models for estimating agronomic variables. IV. Total inversion of the SAIL model. *Remote Sensing of Environment, 15,* 237–253.

Gong, P., Wang, D.X., and Liang, S. (1999). Inverting a canopy reflectance model using a neural network. *International Journal of Remote Sensing, 20,* 111–122.

Green, R.O. and Boardman, J. (2001). Exploration of the relationship between information content and signal to noise ratio and spatial resolution in AVIRIS spectral data. In R.O. Green (Ed.), *AVIRIS Workshop,* Pasadena, CA.

Hagolle, O., Dedieu, G., Mougenot, B., Debaecker, V., Duchemin, B., and Meygret, A. (2008). Correction of aerosol effects on multi-temporal images acquired with constant viewing angles: Application to Formosat-2 images. *Remote Sensing of Environment, 112,* 1689.

Hagolle, O., Lobo, A., Maisongrande, P., Cabot, F., Duchemin, B., and De Pereyra, A. (2005). Quality assessment and improvement of temporally composited products of remotely sensed imagery by combination of VEGETATION 1 and 2 images. *Remote Sensing of Environment, 94,* 172–186.

Hapke, B. (1981). Bidirectional reflectance spectroscopy. 1. Theory. *Journal of Geophysical Research, 86,* 3039–3054.

Herrmann, S.M., Anyamba, A., and Tucker, C.J. (2005). Recent trends in vegetation dynamics in the African Sahel and their relationship to climate. *Global Environmental Change, 15,* 394–404.

Houborg, R. and Boegh, E. (2008). Mapping leaf chlorophyll and leaf area index using inverse and forward canopy reflectance modeling and SPOT reflectance data. *Remote Sensing of Environment, 112,* 186.

Huang, D., Yang, W., Tan, B., Rautiainen, M., Zhang, P., Hu, J., Shabanov, N., Linder, S., Knyazikhin, Y., and Myneni, R. (2006). The importance of measurement error for deriving accurate reference leaf area index maps and validation of the MODIS LAI products. *IEEE Transactions on Geoscience and Remote Sensing, 44,* 1866–1871.

Huete, A.R. (1988). A soil adjusted vegetation index (SAVI). *Remote Sensing of Environment, 25,* 295–309.

Huete, A.R. and Lui, H.Q. (1994). An error and sensitivity analysis of the atmospheric and soil correcting variants of the NDVI for the MODIS-EOS. *IEEE Transactions on Geoscience and Remote Sensing, 32,* 897–905.

Huete, A.R., Liu, H.Q., Batchily, K., and van Leeuwen, W. (1997). A comparison of vegetation indices over a global set of TM images for EOS-MODIS. *Remote Sensing of Environment, 59,* 440–451.

Jacquemoud, S. and Baret, F. (1990). PROSPECT: A model of leaf optical properties spectra. *Remote Sensing of Environment, 34,* 75–91.

Jacquemoud, S., Baret, F., Andrieu, B., Danson, M., and Jaggard, K. (1995). Extraction of vegetation biophysical parameters by inversion of the PROSPECT+SAIL model on sugar beet canopy reflectance data. Application to TM and AVIRIS sensors. *Remote Sensing of Environment, 52,* 163–172.

Jacquemoud, S., Baret, F., and Hanocq, J.F. (1992). Modeling spectral and directional soil reflectance. *Remote Sensing of Environment, 41,* 123–132.

Jacquemoud, S., Verhoef, W., Baret, F., Bacour, C., Zarco-Tejada, P.J., Asner, G.P., François, C., and Ustin, S.L. (2009). PROSPECT + SAIL models: A review of use for vegetation characterization. *Remote Sensing of Environment, 113,* S56–S66.

Jamet, C., Thiria, S., Moulin, C., and Crepon, M. (2005). Use of a neurovariational inversion for retrieving oceanic and atmospheric constituents from ocean color imagery: A feasibility study. *Journal of Atmospheric and Oceanic Technology, 22,* 460–475.

Jonckheere, I., Fleck, S., Nackaerts, K., Muys, B., Coppin, P., Weiss, M., and Baret, F. (2004). Review of methods for in situ leaf area index determination. Part I: Theories, sensors and hemispherical photography. *Agricultural and Forest Meteorology, 121,* 19–35.

Jönsson, P. and Eklundh, L. (2004). TIMESAT—A program for analyzing time series of satellite sensor data. *Computers and Geosciences, 30,* 833–845.

Kandasamy, S., Baret, F., Verger, A., Neveux, P., and Weiss, M. (2013). A comparison of methods for smoothing and gap filling time series of remote sensing observations: application to MODIS LAI products. *Biogeosciences, 10,* 4055–4071.

Kandasamy, S., Lopez-Lozano, R., Baret, F., and Rochdi, N. (2010). The effective nature of LAI as measured from remote sensing observations. In *Proceedings of the 2010 IEEE International Geoscience and Remote Sensing Symposium,* Honolulu, HI (pp. 789–792).

Karkee, M., Steward, B.L., Tang, L., and Aziz, S.A. (2009). Quantifying sub-pixel signature of paddy rice field using an artificial neural network. *Computers and Electronics in Agriculture, 65*, 65–76.

Kimes, D.S., Knyazikhin, Y., Privette, J.L., Abuelgasim, A.A., and Gao, F. (2000). Inversion methods for physically-based models. *Remote Sensing Reviews, 18*, 381–439.

Kimes, D.S., Nelson, R.F., Manry, M.T., and Fung, A.K. (1998). Attributes of neural networks for extracting continuous vegetation variables from optical and radar measurements. *International Journal of Remote Sensing, 19*, 2639–2663.

Knyazikhin, Y., Martonchik, J.V., Myneni, R.B., Diner, D.J., and Running, S.W. (1998). Synergistic algorithm for estimating vegetation canopy leaf area index and fraction of absorbed photosynthetically active radiation from MODIS and MISR data. *Journal of Geophysical Research, 103*, 32257–32275.

Kötz, B., Baret, F., Poilvé, H., and Hill, J. (2005). Use of coupled canopy structure dynamic and radiative transfer models to estimate biophysical canopy characteristics. *Remote Sensing of Environment, 95*, 115–124.

Kuusk, A. (1991a). Determination of vegetation canopy parameters from optical measurements. *Remote Sensing of Environment, 37*, 207–218.

Kuusk, A. (1991b). The inversion of the Nilson-Kuusk canopy reflectance model, a test case. In *International Geoscience and Remote Sensience Symposium 1991*, Helsinki, Finland, pp. 1547–1550.

Laurent, V.C., Verhoef, W., Damm, A., Schaepman, M.E., and Clevers, J.G. (2013). A Bayesian object-based approach for estimating vegetation biophysical and biochemical variables from APEX at-sensor radiance data. *Remote Sensing of Environment, 139*, 6–17.

Laurent, V.C.E., Verhoef, W., Clevers, J.G.P.W., and Schaepman, M.E. (2011). Inversion of a coupled canopy-atmosphere model using multi-angular top-of-atmosphere radiance data: A forest case study. *Remote Sensing of Environment, 115*, 2603–2612.

Lauvernet, C. (2005). Assimilation variationnelle d'observations de télédétection dans les modèles de fonctionnement de la végétation: utilisation du modèle adjoint et prise en compte de contraintes spatiales. In *Mathématiques appliquées* (p. 205). Grenoble, France: Université Joseph Fourier.

Lauvernet, C., Baret, F., Hascoët, L., Buis, S., and Le Dimet, F.X. (2008). Multitemporal-patch ensemble inversion of coupled surface-atmosphere radiative transfer models for land surface characterization. *Remote Sensing of the Enviroment, 112*, 851–861.

Leprieur, C., Verstraete, M.M., and Pinty, B. (1994). Evaluation of the performance of various vegetation indices to retrieve vegetation cover from AVHRR data. *Remote Sensing Reviews, 10*, 265–284.

Lewis, P., Gómez-Dans, J., Kaminski, T., Settle, J., Quaife, T., Gobron, N., Styles, J., and Berger, M. (2012). An earth observation land data assimilation system (EO-LDAS). *Remote Sensing of Environment, 120*, 219–235.

Liang, S. (2004). *Quantitative Remote Sensing of Land Surfaces.* Hoboken, NJ: Wiley-Interscience.

Liang, S. and Townshend, J.R.G. (1996). A parametric soil BRDF model: A four stream approximation for multiple scattering. *International Journal of Remote Sensing, 17*, 1303–1315.

Liu, W., Baret, F., Gu, X., Tong, Q., Zheng, L., and Zhang, B. (2002). Relating soil surface moisture to reflectance. *Remote Sensing of Environment, 81*, 238–246.

Lopez-Lozano, R. (2008). Tecnologias de informacion geografica en la cartografia de parametros biofisicos de parcelas de maiz y vina para agricultura de precision. In *Geografia y ordenacion del territorio* (p. 211). Zaragoza, Spain: Universidad de Zaragoza.

Lucht, W. (1998). Expected retrieval accuracies of bidirectional reflectance and albedo from EOS-MODIS and MISR angular sampling. *Journal of Geophysical Research, 103*, 8763–8778.

Mackay, D.J.C. (Ed.) (2003). *Information Theory, Inference and Learning Algorithms.* Cambridge, U.K.: Cambridge University Press.

Makowski, D., Wallach, D., and Tremblay, M. (2002). Using Bayesian approach to parameter estimation; comparison of the GLUE and MCMC methods. *Agronomie, 22*, 191–203.

Malenovský, Z., Rott, H., Cihlar, J., Schaepman, M.E., García-Santos, G., Fernandes, R., and Berger, M. (2012). Sentinels for science: Potential of sentinel −1, −2, and −3 missions for scientific observations of ocean, cryosphere, and land. *Remote Sensing of Environment, 120*, 91–101.

Martonchik, J.V. (1994). Retrieval of surface directional reflectance properties using ground level multiangle measurements. *Remote Sensing of Environment, 50*, 303–316.

McCallum, I., Wagner, W., Schmullius, C., Shvidenko, A., Obersteiner, M., Fritz, S., and Nilsson, S. (2009). Satellite-based terrestrial production efficiency modeling. *Carbon Balance and Management, 4*, 8.

Miller, J.B. (1967). A formula for average foliage density. *Australian Journal of Botany, 15*, 141–144.

Moorthy, I., Miller, J.R., and Noland, T.L. (2008). Estimating chlorophyll concentration in conifer needles with hyperspectral data: An assessment at the needle and canopy level. *Remote Sensing of Environment, 112*, 2824–2838.

Morisette, J.T. (2010). Toward a standard nomenclature for imagery spatial resolution. *International Journal of Remote Sensing, 31*, 2347–2349.

Morisette, J., Baret, F., Privette, J.L., Myneni, R.B., Nickeson, J., Garrigues, S., Shabanov, N. et al. (2006). Validation of global moderate resolution LAI Products: a framework proposed within the CEOS Land Product Validation subgroup. *IEEE Transactions on Geoscience and Remote Sensing, 44*, 1804–1817.

Mõttus, M., Sulev, M., Baret, F., Reinart, A., and Lopez, R. (2011). Photosynthetically active radiation: Measurement and modeling. In R. Meyers (Ed.), *Encyclopedia of Sustainability Science and Technology* (pp. 7902–7932). Springer, New York.

Myneni, R.B., Gutschick, V.P., Asrar, G., and Kanemasu, E.T. (1988). Photon transport in vegetation canopies with anisotropic scattering part II. Discrete-ordinates/exact-kernel technique for one-angle photon transport in slab geometry. *Agricultural and Forest Meteorology, 42,* 17–40.

Myneni, R.B., Hoffman, S., Knyazikhin, Y., Privette, J.L., Glassy, J., Tian, Y., Wang, Y. et al. (2002). Global products of vegetation leaf area and absorbed PAR from year one of MODIS data. *Remote Sensing of Environment, 83,* 214–231.

Myneni, R.B., Hoffman, S., Knyazikhin, Y., Privette, J.L., Glassy, J., Tian, Y., Wang, Y., Song, X., Zhang, Y., Smith, G.R., Lotsch, A., Friedl, M., Morisette, J.T., Votava, P., Nemani, R.R., S.W., (2002). Running Global products of vegetation leaf area and fraction absorbed PAR from year one of MODIS data *Remote Sensing of Environment, 83* pp. 214–231.

Nagol, J.R., Vermote, E.F., and Prince, S.D. (2009). Effects of atmospheric variation on AVHRR NDVI data. *Remote Sensing of Environment, 113,* 392–397.

Nelder, J.A. and Mead, R.A. (1965). A simplex method for function optimization. *Computer Journal, 7,* 308–313.

Peña-Barragán, J.M., Ngugi, M.K., Plant, R.E., and Six, J. (2011). Object-based crop identification using multiple vegetation indices, textural features and crop phenology. *Remote Sensing of Environment, 115,* 1301–1316.

Pinty, B., Jung, M., Kaminski, T., Lavergne, T., Mund, M., Plummer, S., Thomas, E., and Widlowski, J.L. (2011). Evaluation of the JRC-TIP 0.01° products over a mid-latitude deciduous forest site. *Remote Sensing of Environment, 115,* 3567–3581.

Pinty, B. and Verstraete, M.M. (1991a). Extracting information on surface properties from bidirectional reflectance measurements. *Journal of Geophysical Research, 96,* 2865–2874.

Pinty, B. and Verstraete, M.M. (1991b). Extracting information on surface properties from bidirectional reflectance measurements. *Journal of Geophysical Research, 96,* 2865–2874.

Pinty, B., Verstraete, M.M., and Dickinson, R.E. (1990). A physical model of the bidirectional reflectance of vegetation canopies. 2. Inversion and validation. *Journal of Geophysical Research, 95,* 11767–11775.

Price, J. (1994). How unique are spectral signatures? *Remote Sensing of Environment, 49,* 181–186.

Price, J.C. (1990). On the information content of soil reflectance spectra. *Remote Sensing of Environment, 33,* 113–121.

Privette, J.L., Emery, W.J., and Schimel, D.S. (1996). Inversion of a vegetation reflectance model with NOAA AVHRR data. *Remote Sensing of Environment, 58,* 187–200.

Qu, Y., Wang, J., Wan, H., Li, X., and Zhou, G. (2008). A Bayesian network algorithm for retrieving the characterization of land surface vegetation. *Remote Sensing of Environment, 112,* 613–622.

Raymaekers, D., Garcia, A., Di Bella, C., Beget, M.E., Llavallol, C., Oricchio, P., Straschnoy, J., Weiss, M., and Baret, F. (2014). SPOT-VEGETATION GEOV1 biophysical parameters in semi-arid agro-ecosystems. *International Journal of Remote Sensing, 35,* 2534–2547.

Refslund, J., Dellwik, E., Hahmann, A., Barlage, M., and Boegh, E. (2013). Development of satellite green vegetation fraction time series for use in mesoscale modeling: Application to the European heat wave 2006. *Theoretical and Applied Climatology, 117,* 1–16.

Renders, J.-M. and Flasse, S.P. (1996). Hybrid methods using genetic algorithms for global optimization. *IEEE Transactions on Systems, Man, and Cybernetics, 26,* 243–258.

Richardson, A.J. and Wiegand, C.L. (1977). Distinguishing vegetation from soil background information. *Photogrammetric Engineering and Remote Sensing, 43,* 1541–1552.

Richardson, A.J., Wiegand, C.L., Wanjura, D.F., Dusek, D., and Steiner, J.L. (1992). Multisite analyses of spectral-biophysical data for sorghum. *Remote Sensing of Environment, 41,* 71–82.

Rondeaux, G., Steven, M.D., and Baret, F. (1996). Optimization of soil adjusted vegetation indices. *Remote Sensing of Environment, 55,* 95–107.

Röser, H.-P., Sandau, R., and Valenzuela, A. (2005). Small satellites for earth observation: Selected proceedings of the 5th international symposium of the international academy of astronautics, Berlin, April 4–8, 2005, 459pp.

Roujean, J.L. and Lacaze, R. (2002). Global mapping of vegetation parameters from POLDER multiangular measurements for studies of surface-atmosphere interactions: A pragmatic method and validation. *Journal of Geophysical Research, 107,* ACL 6 1–ACL 6 14.

Russel, G., Jarvis, P.G., and Monteith, J.L. (1989). Absorption of radiation by canopies and stand growth. In G. Russel, B. Marshall, and P.G. Jarvis (Eds.), *Plant Canopies: Their Growth, Form and Function* (pp. 21–39). New York: Cambridge University Press.

Ryu, Y., Nilson, T., Kobayashi, H., Sonnentag, O., Law, B.E., and Baldocchi, D.D. (2010). On the correct estimation of effective leaf area index: Does it reveal information on clumping effects? *Agricultural and Forest Meteorology, 150,* 463–472.

Samain, O., Roujean, J.L., and Geiger, B. (2007). Use of Kalman filter for the retrieval of surface BRDF coefficients with time-evolving model based on ECOCLIMAP land cover classification. *Remote Sensing of Enviroment, 112*(4), 1337–1346.

Sellers, P.J. (1985). Canopy reflectance photosynthesis and transpiration. *International Journal of Remote Sensing, 3,* 1335–1372.

Settle, J. (2004). On the dimensionality of multi-view hyperspectral measurements of vegetation. *Remote Sensing of Enviroment, 90,* 235–242.

Shabanov, N.V., Huang, D., Yang, W., Tan, B., Knyazikhin, Y., Myneni, R.B., Ahl, D.E., Gower, S.T., and Huete, A.R. (2005). Analysis and optimization of the MODIS leaf area index algorithm retrievals over broadleaf forests. *IEEE Transactions on Geoscience and Remote Sensing, 43,* 1855–1865.

Shoshany, M. (1991). The equifinality of bidirectional distribution functions of various microstructures. *International Journal of Remote Sensing, 12,* 2267–2281.

Simic, A. and Chen, J.M. (2008). Refining a hyperspectral and multiangle measurement concept for vegetation structure assessment. *Canadian Journal of Remote Sensing, 34,* 174–171.

Smith, J.A. (1993). LAI inversion using backpropagation neural network trained with multiple scattering model. *IEEE Transactions on Geoscience and Remote Sensing, 31,* 1102–1106.

Specht, D.F. (1991). A general regression neural network. *IEEE Transactions on Neural Networks, 2,* 568–576.

Stenberg, P. (2006). A note on the G-function for needle leaf canopies. *Agricultural and Forest Meteorology, 136,* 76.

Tarantola, A. (2005). *Inverse Problem Theory Problem and Methods for Model Parameter Estimation.* Philadelphia, PA: Society for Industrial and Applied Mathematics.

Tarentola, A. (1987). *Inverse Problem Theory. Methods for Data Fitting and Model Parameter Estimation.* New York: Elsevier Science Publisher B.V.

Teillet, P.M., Gauthier, R.P., Staenz, K., and Fournier, R.A. (1997). BRDF equifinality studies in the context of forest canopies. In G. Guyot and T. Phulpin (Eds.), *Physical Measurements and Signatures in Remote Sensing* (pp. 163–170). Courchevel, France/Rotterdam, the Netherlands: Balkema.

Tenenbaum, J.B., de Silva, V., and Langford, J.C. (2000). A global geometric framework for nonlinear dimensionality reduction. *Science, 290,* 2319–2323.

Thenkabail, P.S., Enclona, E.A., Ashton, M.S., and Van Der Meer, B. (2004). Accuracy assessments of hyperspectral waveband performance for vegetation analysis applications. *Remote Sensing of Enviroment, 91,* 354–376.

Tucker, C.J., Pinzón, J.E., Brown, M.E., Slayback, D.A., Pak, E.W., Mahoney, R., Vermote, E., and El Saleous, N. (2005). An extended AVHRR 8-km NDVI dataset compatible with MODIS and SPOT vegetation NDVI data. *International Journal of Remote Sensing, 26,* 4485–4498.

Verbesselt, J., Hyndman, R., Newnham, G., and Culvenor, D. (2010a) Detecting trend and seasonal changes in satellite image time series. *Remote Sensing of Environment, 114,* 106.

Verbesselt, J., Hyndman, R., Zeileis, A., and Culvenor, D. (2010b). Phenological change detection while accounting for abrupt and gradual trends in satellite image time series. *Remote Sensing of Environment, 114,* 2970.

Verger, A., Baret, F., and Camacho de Coca, F. (2011a). Optimal modalities for radiative transfer-neural network estimation of canopy biophysical characteristics: evaluation over an agricultural area with CHRIS/PROBA observations. *Remote Sensing of Environment, 115,* 415–426.

Verger, A., Baret, F., and Weiss, M. (2008). Performances of neural networks for deriving LAI estimates from existing CYCLOPES and MODIS products. *Remote Sensing of Environment, 112,* 2789–2803.

Verger, A., Baret, F., and Weiss, M. (2011b). A multisensor fusion approach to improve LAI time series. *Remote Sensing of Environment, 115,* 2460–2470.

Verger, A., Baret, F., and Weiss, M. (2014). Near real-time vegetation monitoring at global scale. *IEEE Journal of Selected Topics in Applied Earth Observations and Remote Sensing, 99,* 1–9.

Verger, A., Baret, F., Weiss, M., Kandasamy, S., and Vermote, E. (2012a). The CACAO method for smoothing, gap filling and characterizing anomalies in satellite time series. *IEEE Transactions on Geoscience and Remote Sensing, 51,* 1963–1972.

Verger, A., Baret, F., Weiss, M., Lacaze, R., Makhmara, H., and vermote, E. (2012b). Long term consistent global GEOV1 AVHRR biophysical products. In *First EARSeL Workshop on Temporal Analysis of Satellite Images,* Mykonos, Greece, pp. 1–6.

Verhoef, W. and Bach, H. (2007). Coupled soil-leaf-canopy and atmosphere radiative transfer modeling to simulate hyperspectral multi-angular surface reflectance and TOA radiance data. *Remote Sensing of Enviroment, 109,* 166–182.

Verrelst, J., Muñoz, J., Alonso, L., Delegido, J., Rivera, J.P., Camps-Valls, G., and Moreno, J. (2012). Machine learning regression algorithms for biophysical parameter retrieval: Opportunities for sentinel −2 and −3. *Remote Sensing of Environment, 118,* 127–139.

Verstraete, M. and Pinty, B. (1996). Designing optimal spectral indexes for remote sensing applications. *IEEE Transactions on Geoscience and Remote Sensing, 34,* 1254–1265.

Vieira, M.A., Formaggio, A.R., Rennó, C.D., Atzberger, C., Aguiar, D.A., and Mello, M.P. (2012). Object based image analysis and data mining applied to a remotely sensed Landsat time-series to map sugarcane over large areas. *Remote Sensing of Environment, 123,* 553–562.

Voßbeck, M., Clerici, M., Kaminski, T., Lavergne, T., Pinty, B., and Giering, R. (2010). An inverse radiative transfer model of the vegetation canopy based on automatic differentiation. *Inverse Problems, 26,* 15.

Vohland, M., Mader, S., and Dorigo, W. (2010). Applying different inversion techniques to retrieve stand variables of summer barley with PROSPECT + SAIL. *International Journal of Applied Earth Observation and Geoinformation, 12,* 71–80.

Walter, J.M.N., Fournier, R.A., Soudani, K., and Meyer, E. (2003). Integrating clumping effects in forest canopy structure: an assessment through hemispherical photographs. *Canadian Journal of Remote Sensing, 29,* 388–410.

Walthall, C.L., Dulaney, W.P., Anderson, M.C., Norman, J.M., Fang, H., and Liang, S. (2004). A comparison of empirical and neural network approaches for estimating corn and soybean leaf area index from Landsat ETM+ imagery. *Remote Sensing of Enviroment, 92,* 465–474.

Weiss, M., Baret, F., Block, T., Koetz, B., Burini, A., Scholze, B., Lecharpentier, P. et al. (2014). On line validation exercise (OLIVE): A web based service for the validation of medium resolution land products. Application to FAPAR products. *Remote Sensing, 6,* 4190–4216.

Weiss, M., Baret, F., Garrigues, S., Lacaze, R., and Bicheron, P. (2007). LAI, fAPAR and fCover CYCLOPES global products derived from VEGETATION. part 2: Validation and comparison with MODIS Collection 4 products. *Remote Sensing of Environment, 110,* 317–331.

Weiss, M., Baret, F., Leroy, M., Hautecoeur, O., Bacour, C., Prévot, L., and Bruguier, N. (2002a). Validation of neural net techniques to estimate canopy biophysical variables from remote sensing data. *Agronomie, 22,* 547–554.

Weiss, M., Baret, F., Myneni, R., Pragnère, A., and Knyazikhin, Y. (2000). Investigation of a model inversion technique for the estimation of crop charcteristics from spectral and directional reflectance data. *Agronomie, 20,* 3–22.

Weiss, M., Baret, F., Smith, G.J., Jonckheered, I., and Coppin, P. (2004). Review of methods for in situ leaf area index determination, part II: Estimation of LAI, errors and sampling. *Agricultural and Forest Meteorology, 121,* 37–53.

Weiss, M., Jacob, F., Baret, F., Pragnère, A., Bruchou, C., Leroy, M., Hautecoeur, O., Prévot, L., and Bruguier, N. (2002b). Evaluation of kernel-driven BRDF models for the normalization of Alpilles/ReSeDA POLDER data. *Agronomie, 22,* 531–536.

Weiss, M., Troufleau, D., Baret, F., Chauki, H., Prévot, L., Olioso, A., Bruguier, N., and Brisson, N. (2001). Coupling canopy functioning and canopy radiative transfer models for remote sensing data assimilation. *Agricultural and Forest Meteorology, 108,* 113–128.

Wiegand, C.L., Gerberman, A.H., Gallo, K.P., Blad, B.L., and Dusek, D. (1990). Multisite analyses of spectral-biophysical data for corn. *Remote Sensing of Environment, 33,* 1–16.

Wiegand, C.L., Maas, S.J., Aase, J.K., Hatfield, J.L., Pinter, P.J.J., Jackson, R.D., Kanemasu, E.T., and Lapitan, R.L. (1992). Multisite analysis of spectral biophysical data for wheat. *Remote Sensing of Environment, 42,* 1–22.

Xiao, Z., Liang, S., Wang, J., Chen, P., Yin, X., Zhang, L., and Song, J. (2014). Use of general regression neural networks for generating the GLASS leaf area index product from time-series MODIS surface reflectance. *IEEE Transactions on Geoscience and Remote Sensing, 52,* 209–223.

Xiao, Z., Liang, S., Wang, J., Yin, X., Xiang, Y., Song, J., and Ma, H. (2012). GLASS leaf area index product derived from MODIS time series remote sensing data. In *IGARSS,* IEEE, Munich, Germany.

Yang, W., Shabanov, N.V., Huang, D., Wang, W., Dickinson, R.E., Nemani, R.R., Knyazikhin, Y., and Myneni, R.B. (2006). Analysis of leaf area index products from combination of MODIS Terra and Aqua data. *Remote Sensing of Environment, 104,* 297–312.

Zhang, Q., Xiao, X., Braswell, B., Linder, E., Baret, F., and Moore Iii, B. (2005). Estimating light absorption by chlorophyll, leaf and canopy in a deciduous broadleaf forest using MODIS data and a radiative transfer model. *Remote Sensing of Environment, 99,* 357.

Zhang, X., Friedl, M.A., Schaaf, C.B., Strahler, A.H., Hodges, J.C.F., Gao, F., Reed, B.C., and Huete, A.R. (2003). Monitoring vegetation phenology using MODIS. *Remote Sensing of Environment, 84,* 471–475.

Zhang, Y., Shabanov, N., Knyazikhin, Y., and Myneni, R.B. (2002a). Assessing the information content of multiangle satellite data for mapping biomes I. Theory. *Remote Sensing of Environment, 80,* 435–446.

Zhang, Y., Tian, Y., Myneni, R.B., Knyazikhin, Y., and Woodcock, C.E. (2002b). Assessing the information content of multiangle satellite data for mapping biomes I. Statistical analysis. *Remote Sensing of Environment, 80,* 418–434.

Zhu, Z., Bi, J., Pan, Y., Ganguly, S., Anav, A., Xu, L., Samanta, A., Piao, S., Nemani, R.R., and Myneni, R.B. (2013). Global data sets of vegetation leaf area index (LAI)3g and fraction of photosynthetically active radiation (FPAR)3g derived from global inventory modeling and mapping studies (GIMMS) normalized difference vegetation index (NDVI3g) for the period 1981 to 2011. *Remote Sensing, 5,* 927–948.

3

Aboveground Terrestrial Biomass and Carbon Stock Estimations from Multisensor Remote Sensing

Wenge Ni-Meister
Hunter College of the City University of New York

Acronyms and Definitions

AGB	Aboveground biomass
ASTER	Advanced spaceborne thermal emission and reflection radiometer
AVHRR	Advanced very-high-resolution radiometer
AVIRIS	NASA's airborne visible/infrared imaging spectrometer
BEF	Bartlett experimental forest (BEF) in central new Hampshire (USA)
CC	Canopy cover
DBH	Crown diameter at breast height
DGVI	First/second derivative of red edge normalized to 626–795 nm baseline
DWEL	Dual-wavelength Echidna® LiDAR
ETM	Enhanced thematic mapper
EVI	Echidna® validation instrument
GLAS/ICESat	Geoscience laser altimeter system (GLAS), on board the ice, cloud, and land elevation satellite (ICESat)
LVIS	Land vegetation and ice sensor
MCH	Mean canopy vertical height profiles, the distance from ground (digital terrain models) to the approximate centroid of the tree crowns
MODIS	Moderate-resolution imaging spectroradiometer
NDVI	Normalized difference of vegetation index
NDWI	Normalized different of water index
NIR	Near infrared
PALSAR	Phased array-type L-band synthetic aperture radar
QSCAT	Quick scatterometer
RH25	Relative height (RH) to the ground elevation at which 25% of the accumulated full-waveform energy occurs
RH50	Relative height (RH) to the ground elevation at which 50% of the accumulated full-waveform energy occurs
RH75	Relative height (RH) to the ground elevation at which 75% of the accumulated full-waveform energy occurs

RH100	Relative height (RH) to the ground elevation at which 100% of the accumulated full-waveform energy occurs
SIR-C/X-SAR	Spaceborne imaging radar-C/X-band synthetic aperture radar
SLICER	Scanning LiDAR imager of canopies by echo recovery
SPOT HRV	Le Système Pour l'Observation de la Terre high resolution visible
SRTM	Shuttle radar topography mission
TIR	Thermal infrared

3.1 Introduction

Recent global observation systems provide measurements of horizontal and vertical vegetation structures of ecosystems, which will be critical for estimating global carbon storage and assessing ecosystem response to climate change and natural and anthropogenic disturbances. Remote sensing overcomes the limitations associated with sparse field surveys; it has been used extensively as a basis for inferring forest structure and aboveground biomass (AGB) over large areas. This chapter summarizes recent progress on the AGB estimate using remote sensing technology including strength and weakness of using optical passive, radar, and LiDAR remote sensing and fusion of multisensor for the AGB estimates. It lays out the potential of remote sensing in the AGB and carbon storage estimates at large scales for meeting the requirements under the United Nations Framework Convention on Climate Change (UNFCCC) for measuring, reporting, and verification. The purpose of this chapter is to review recent progress on the AGB estimates using remote sensing data.

3.1.1 Importance of the Terrestrial Ecosystem Carbon and Carbon Change Estimates

Vegetation biomass is a crucial ecological variable for understanding the evolution and potential future changes of the climate system. Global carbon stored in vegetation is comparable in size to atmospheric carbon and plays an important role in the global carbon cycle (Houghton 2005). Changes of forest biomass in time can be used as an essential climate variable (ECV), because it is a direct measure of sequestration or release of carbon between terrestrial ecosystems and the atmosphere. During productive seasons, forests take up carbon dioxide (CO_2) from the atmosphere and store it as plant biomass, while they release CO_2 to the atmosphere during deforestation, decomposition, and biomass burning. Changes in the amount of vegetation biomass due to deforestation significantly affect the global atmosphere by acting as a net source of carbon. The Global Climate Observing System (GCOS) recognizes the AGB and associated carbon stocks of the world's forests as an ECV (Hollman et al. 2013).

However, the terrestrial carbon cycle is the most uncertain component of the global carbon cycle (Heimann and Reichstein 2008). Large uncertainties in terrestrial carbon cycle arise from inadequate data on the current state of the land surface vegetation

structure and the carbon density of forests. Consequently, there is an urgent need for improved datasets that characterize the global distribution of AGB, especially in the tropics. Therefore, a global assessment of biomass and its dynamics is an essential input to climate change prognostic models and mitigation and adaptation strategies.

3.1.2 Importance of Tropical Rain Forests in Carbon Storage

Tropical forests are disappearing rapidly due to land conversion, selective cutting, and fires. The single biggest direct cause of tropical forest loss is due to conversion of forests to cropland and pasture. Humans harvest timber for construction and fuel, and wildfires pose a big threat to Amazon forests. The tropics exhibit a rising trend of forest loss, increasing by 2101 km²/year, half of which occurred in South American rain forests (Hansen et al. 2013), with the recent report of reduced rate of forest loss from high of over 40,000 km²/year in 2003–2004 and a low of under 20,000 km²/year in 2010–2011 in central America. However, this decreasing rate of loss was counterbalanced by increased forest loss from other tropical forest regions (Figure 3.1).

Land use, land-use change, and forestry sector is the second-largest source of anthropogenic greenhouse gas (GHG) emissions, dominated by tropical deforestation (Canadell et al. 2007). The loss of Amazon forest releases huge carbon to the atmosphere. Tropical deforestation contributes about one-eighth to one-fifth of total anthropogenic CO_2 emissions to the atmosphere (Houghton 2005, 2007, Houghton et al. 2012). However, the magnitude of these emissions has remained poorly constrained. Emissions from land-use change remains as one of the most uncertain components of the global carbon cycle. Global carbon emission estimates using different approaches and the uncertainties associated with each approach range from 10% to 34% (Houghton et al. 2012). A recent estimate of gross carbon emissions across tropical regions between 2000 and 2005 was 0.81 petagram of carbon per year (PgC/year) (Harris et al. 2012), which was only 25%–50% of recently published estimates (FAO 2010, Pan et al. 2011, Baccini et al. 2012). Huge discrepancy exists in the carbon emission estimates.

The lack of reliable estimates of forest carbon storage and rates of deforestation and forestation result in the uncertainties of terrestrial carbon emission estimates (Houghton 2005, Houghton et al. 2009, 2012). Estimates of the biomass storage disagree with biomass obtained from large-scale wood-volume inventories (Houghton et al. 2001). Large uncertainties in the carbon stock estimates contribute to the broad range of possible emissions of carbon from tropical deforestation and degradation (Houghton 2005).

Reducing emissions from deforestation and forest degradation (REDD) in developing countries launched by the UNFCCC provides positive incentives to individuals, communities, projects, and government agencies, in developing countries to reduce GHG emissions from forests through monetary compensation. REDD was extended as REDD+ to include conservation, sustainable management of forests, and the enhancement of forest

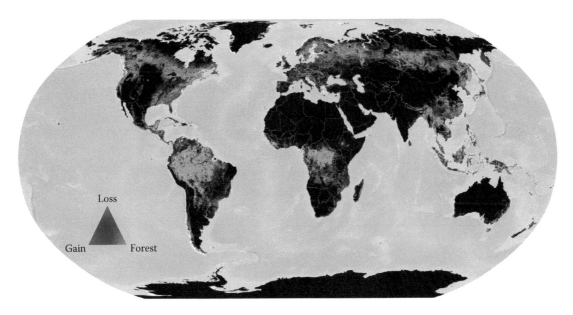

FIGURE 3.1   Global distribution of forest cover change, ca. 1990–2000. The false-color composite was aggregated from 30 m to 5 km grid cells. Forest loss is represented in red, forest gain in blue, and persistent forest in green. Colors are stretched in the proportion of 1 (forest): 4 (gain): 4 (loss). (For interpretation of the references to color in this figure legend, the reader is referred to the web version of this chapter.) (From Kim, D.K. et al., *Remote Sens. Environ.*, ISSN 0034–4257. Available online September 26, 2014, http://dx.doi.org/10.1016/j.rse.2014.08.017.)

carbon stocks. As a mechanism under the multilateral climate change agreement, REDD+ is a vehicle to financially reward developing countries for their verified efforts to reduce emissions and enhance removals of GHGs through a variety of forest management options.

Efforts to mitigate climate change through REDD depend on mapping and monitoring of tropical forest carbon stocks and emissions over large geographic areas. There are many challenges to making REDD work, and mapping forest carbon stocks and emissions at the high resolution demanded by investors and monitoring agencies remains a technical barrier. Foremost among the challenges is quantifying nations' carbon emissions from deforestation and forest degradation, which requires information on forest clearing and carbon storage.

3.1.3 Summary of Methods Used to Estimate Terrestrial Biomass and Carbon Stocks

Vegetation AGB is defined as the mass per unit area (Mg/ha) of live or dead plant organic matter. Forest biomass consists of AGB and below-ground biomass. AGB represents all living biomass above the soil including stem, stump, branches, bark, seeds, and foliage, while below-ground biomass consists of all living roots excluding fine roots (less than 2 mm in diameter) (FAO 2010). Because AGB is relatively easy to measure and it accounts for the majority of the total accumulated biomass in forest ecosystem, AGB is usually estimated in many studies as to forest biomass. At the level of individual plants and forest stand levels, aboveground and belowground biomass are different, but share strikingly similar scaling exponents (Figure 3.2) (Cheng and Niklas 2007). Below-ground biomass is often estimated based on AGB. This review mainly focuses on AGB estimates.

Biomass estimate methods range from simple to more complex methods. The biome-averaged method is to estimate the biome-averaged AGB first, and the spatial distribution of biomass is mapped based on biome type. A more complex method is to develop species- and site-specific allometric models depending on bole diameter at breast height (DBH; cm) or diameter and tree height. The plot estimates of national forest inventories are commonly aggregated to represent forest biomass at national or regional scales (Brown et al. 1989, Jenkins 2003, Gibbs et al. 2007, Goetz et al. 2009).

3.1.4 Role of Remote Sensing in Terrestrial Ecosystem Carbon Estimates

Recent global observation systems provide measurements of horizontal and vertical vegetation structure of ecosystems, which will be critical for estimating global carbon storage and assessing ecosystem response to climate change and natural and anthropogenic disturbances. Remote sensing overcomes the limitations associated with sparse field surveys; it has been used extensively as a basis for inferring forest structure and AGB over large areas. Although no sensor has been developed that is capable of providing direct measures of vegetation biomass, the radiometry is sensitive to vegetation structure (crown size, tree density height), texture, and shadow, which are correlated with AGB. Three types of remote sensing data are often used, which are

1. Optical remote sensing
2. Radar (radio detection and ranging, microwave) data
3. LiDAR (light detection and ranging) data

Optical spectral reflectances are sensitive to vegetation structure (leaf area index (LAI), crown size, and tree density),

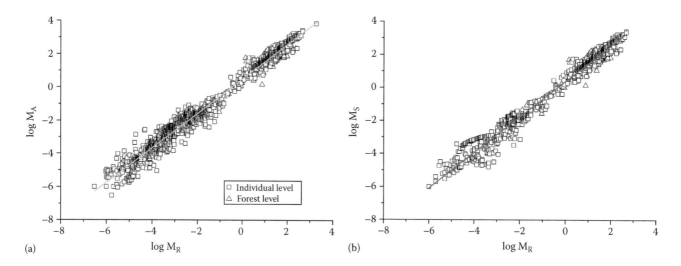

FIGURE 3.2 Log–log bivariate plots of above- vs. belowground (root) biomass (M_A vs. M_R) (a) and stem vs. root biomass (M_S vs. M_R) (b) at the level of individual plants (n = 1406) and Chinese forest samples (n = 1534). (From Cheng, D.L. and Niklas, K.J., *Ann. Botany*, 99, 95, 2007.)

TABLE 3.1 Strengths and Limitations of Conventional Methods to Estimate Aboveground Biomass and Forest Carbon Stocks

Methods	Descriptions	Input Parameters	Strengths	Limitations	References
Direct measurement	• Harvest all trees • Dry them • Weigh the biomass	n/a	• Very accurate	• Very small areas	Brown et al. (1989)
Biome average	• Estimate average forest carbon stocks for each biome based on inventory data • Map carbon stocks based on land cover types	• Land cover types • Averaged biomass for each biome	• Easy and quick • Globally consistent • Low cost	• Low accuracy • Lost local variations	FAO (2010)
Species-based allometric method	• Use allometric relationships to estimate AGB based on DBH	• DBH • Species	• Easy to implement	• Low accuracy if the allometric relationship is not local	Jenkins et al. (2004)
Woody volume and woody density based allometric method	• Use generalized allometric relationships for all species stratified by broad forest types or ecological zones	• DBH • Tree height • Wood density • Forest types (dry or wet forest)	• Quite accurate • Effective for tropical forests	• Need extra wood density and tree height measurement	Brown (2002) and Chave et al. (2005, 2014)

texture, and shadow, which are correlated with AGB. Radar data are directly related to AGB through measuring dielectric and geometrical properties of forests (Le Toan et al. 2011). LiDAR remote sensing is promising in characterizing vegetation vertical structure and height, which are then associated to AGB (Drake et al. 2002a,b, Lefsky et al. 2005a,b). Vegetation structure characteristics measured from satellite data are linked to field-based AGB estimates, and their relationships are used to map large-scale AGB from satellite data. Recently, remote sensing has been extensively used as a robust tool in delivering forest structure and AGB because it provides a practical means of acquiring spatially distributed forest biomass from local to continental areas (Houghton 2005, Lu 2006, Zhang and Kondragunta 2006, Goetz and Dubayah 2011).

3.1.5 Specific Topics Covered in This Chapter

Significant progress has been made in recent years regarding the large-area application of spaceborne remote sensing for the mapping of terrestrial ecosystem carbon stocks, which manifested in the release of several regional- to continental-scale maps of AGB. This paper reviews recent progress of terrestrial AGB and carbon stock estimations from remote sensing. It focuses on not only the current state of remote sensing of biomass using one particular sensor, but also recent progress on biomass mapping through fusion of multisensors. First, we brief the traditional method of AGB estimates, and then summarize what types of remote sensing data being used for biomass estimates followed by a summary of research methods. Later sections provide recent progresses on biomass estimates using optical, radar, and LiDAR sensors and fusion of multisensors. Finally, we discuss the strengths and potential improvement of remote sensing approaches for mapping terrestrial ecosystem biomass and carbon stocks and point out future research directions.

3.2 Conventional Methods of Carbon Stock Estimates

Table 3.1 lists the strengths and limitations of conventional methods for carbon stock estimates. The direct method is to harvest trees, dry, and weigh the biomass. It is the most accurate method; however, the most labor intensive. For a small area, the most direct way to measure the carbon stored in aboveground living forest biomass is to harvest all trees, dry them, and weigh the biomass. The dry biomass can be converted to carbon content by taking half of the biomass weight (carbon content ≈50% of biomass). This method is destructive, expensive, extremely time consuming, and impractical for any large regions.

No methodology can directly measure forest carbon stocks across a landscape. Different methods are used to approximate large-scale carbon stocks ranging from simple empirical to more complex physically based methods. At the national level, the Intergovernmental Panel on Climate Change (IPCC) proposed different tiers of carbon stocks quality, ranging from Tier 1 (simplest to use; globally available data) up to Tier 3 (high-resolution methods specific for each country and repeated through time) (Gibbs et al. 2007).

3.2.1 Biome-Average Methods

The simplest one is to use the biomass average for each biome to approximate a nation's carbon stocks (IPCC's Tier 1) (Houghton et al. 2001, Gibbs et al. 2007). Biome averages are compiled based on tree harvesting measurements and analysis of forest inventory data archived by the United Nations Food and Agricultural Organization (FAO) (Gibbs et al. 2007).

This method has both strengths and limitations. Biomes account for major bioclimatic gradients such as temperature, precipitation, and geologic substrate; it is a quick and easy way to estimate forest carbon stocks based on biomes. Besides, biome averages are free and easily accessible to map global forest carbon systematically. It provides a starting point for a country to access their carbon emission from disturbance. However, biome averages were generally focused on mature stands and were based on a few plots that may not adequately represent the biome or region. Further, forest carbon stocks vary significantly with slope, elevation, drainage class, soil type, and land-use history within each biome; therefore, an average value cannot adequately represent the variation for an entire forest category or country. Finally, the carbon stock estimates over disturbed areas could also be biased as the carbon stocks for the new growth systematically differ from the biome-average values (Houghton et al. 2001).

3.2.2 Allometric Biomass Methods

Another commonly used approach is the allometric-based biomass and carbon stock estimates (IPCC's Tier 2 or 3). It depends on forest inventory measurements to develop allometric relationships between tree diameters at breast height (DBH) alone or in combination with tree height with AGB. Ground-based DBH and height measurements in large areas are converted to forest carbon stocks using allometric relationships. Many allometric equations for estimating AGB have been published in the past (Brown et al. 1989, West et al. 1999, Brown 2002, Jenkins 2003, Jenkins et al. 2004, Chave et al. 2005, 2014). Two allometric-based approaches are commonly used to estimate AGB.

3.2.2.1 Species-Based Allometric Method

The first one is a species-based approach to estimate biomass based on a given tree DBH. It requires measuring the diameter for each individual tree and allometric equations for each individual tree species. For example, Jenkins et al. (2004) developed a set of generalized allometric regression models to predict AGB in tree components for all tree species in the United States. It is used by the USDA Forest Service, Forest Inventory and Analysis program to estimate the U.S. national forest carbon estimates. This approach provides a nationally consistent method for the estimation of biomass and C stocks at large scales and requires only a single field-based variable—tree DBH (1.37 m)—as input.

3.2.2.2 Woody Volume—and Woody Density—Based Allometric Method

The second approach is a more generalized one, using woody volume and wood density to calculate biomass (Brown 2002, Chave et al. 2005, 2014). Developing allometric equations for each individual species can be very difficult. However, grouping all species together and using generalized allometric relationships, stratified by broad forest types or ecological zones, is highly effective, particularly for the tropics because DBH alone explains more than 95% of the variation in aboveground tropical forest carbon stocks, even in highly diverse regions (Brown 2002).

Chave et al. (2005) developed generalized allometric equations for the pan-tropics based on an exceptionally large dataset of 2410 trees across a wide range of forest types. They included wood density and tree height within their models and proposed a global forest classification system that contains three climatic categories (dry, moist, and wet) to account for climatic constraints determining the AGB variation. Very recently, Chave et al. (2014) updated their allometric equations and developed a single model using trunk diameter, total tree height, and wood-specific gravity across tropical vegetation types, with no detectable effect of region or environmental factors. The new allometric models should contribute to improving the accuracy of biomass assessment protocols in tropical vegetation types and to improving accuracy of carbon stock estimates for tropical forests.

Studies show that the most important parameters in estimating biomass (in decreasing order of importance) were diameter, wood density, tree height, and forest type (classified as dry, moist, or wet forest). Including tree height reduced the standard error of biomass estimates from 19.5% to 12.5% (Chave et al. 2005). Tree biomass estimation was significantly improved by including wood density (Brown et al. 1989) and tree height (Brown 1997, Nogueira et al. 2008) in the allometric models in addition to tree diameter. However, measuring height (H) and wood density (q) requires additional work, increasing project time and costs. This approach is not often used as it required additional height measurements for each individual tree.

Despite the difficulty, more and more studies demonstrated the importance of these parameters for biomass estimates. For example, studies by Feldpausch et al. (2011, 2012) demonstrate that incorporating height in biomass estimates for the pantropical region improves biomass estimates by lowering it. For tropical forests, carbon storage can be overestimated by 35 PgC if height is ignored. The study by Domke et al. (2012) for the United States also demonstrates similar results. Domke et al. (2012) compared estimates of carbon stocks using Jenkin's and a tree height–based approach—the component ratio method (Woodall et al. 2011)—for the 20 most abundant tree species in the 48 states of the United States and found the method incorporating height decreased national carbon stock estimates by an average of 16% for the species. These results implicate that tree height, an important allometric factor, needs to be included in future forest biomass estimates to reduce error in the estimates of tropical carbon stocks and emissions due to deforestation and to improve accuracy of national and global forest carbon.

3.3 Remote Sensing Data

A variety of remote sensing systems have been used to estimate AGB estimates: passive optical remote sensing, radar, and LiDAR. Table 3.2 summarizes the characteristics of satellite sensors used to estimate AGB and carbon storage.

3.3.1 Passive Optical Remote Sensing Data

Passive remote sensors measure different wavelengths of reflected solar radiation, providing two-dimensional information that can be indirectly linked to biophysical properties of vegetation and AGB and carbon stocks. Several optical satellite instruments are available for mapping AGB and carbon stocks at different spatial scales. The spatial resolutions of these satellite data range from meter to kilometer scales. Those data span from 1970s to current, and some recent satellite data are collected on a daily scale. The most popular optical remote sensing satellite data being used to map AGB are multispectral satellite data at various spatial resolutions.

NOAA's advanced very-high-resolution radiometer (AVHRR) and NASA's moderate-resolution imaging spectroradiometer (MODIS) data are promising in producing biomass at continental and global scales (Dong et al. 2003, Baccini et al. 2004, 2008, Zhang and Kondragunta 2006, Blackard et al. 2008). AVHRR

provides global observations at 1 km scale every one or two days since 1979. MODIS aboard the Aqua and Terra satellites has imaged the entire globe approximately every two days at resolutions of 250–500 m, dating as far back as 2000. These datasets are used alone or fused with other remote sensing data to provide AGB and carbon stock estimates at large scales.

Landsat Thematic Mapper (TM), Enhanced Thematic Mapper Plus (ETM+), and Advanced Spaceborne Thermal Emission and Reflection Radiometer (ASTER) provide biomass estimates at local and regional scales at high spatial resolution (Muukkonen and Heiskanen 2005, Zhang and Kondragunta 2006, Pflugmacher et al. 2014). Landsat provides four decades of imagery of the entire globe at 30 m spatial resolution, the longest continuous record of space-based moderate-resolution land remote sensing data freely available to the public. With the advantages of being free and long-term data records, methods of using spectral information or more complicated methods using both spectral and temporal information or fusion with other remote sensing data have been developed to estimate AGB estimates. Landsat images are invaluable data sources to AGB and carbon stock estimates. ASTER, an imaging instrument on board Terra launched in December 1999, images the earth at 15 m resolution in visible to near-infrared spectrum, which is the most sensitive to vegetation properties. Other passive optical systems such as multiangular data from MISR on board Terra and airborne/spaceborne hyperspectral data from AVIRIS and EO1 sensors are also used for biomass estimates (Anderson et al. 2008, Chopping et al. 2009).

3.3.2 Radar Data

Radar data physically measure biomass through the interaction of the radar waves with tree scattering elements. The widely used active radar data for biomass estimates are from spaceborne synthetic aperture radar (SAR) sensors, such as the L-band Advanced Land Observing Satellite (ALOS), Phased Array-Type L-band Synthetic Aperture Radar (PALSAR), the C-band European remote sensing satellite (ERS)/SAR, RADARSAT/SAR or Environmental Satellite (ENVISAT)/Advanced Synthetic Aperture Radar, and the X-band TerraSAR-X instrument, which transmit microwave energy at wavelengths from 3.0 (X-band) to 23.6 cm (L-band).

ERS and ENVISAT operated by European Space Agency (ESA) collect C-band SAR data since 1991. Canadian RADARSAT has collected C-band data since 1995. German TerraSAR-X was in space since 2010. Those data have been used to estimate AGB with low density. The L-band PALSAR was launched by Japan Aerospace Exploration Agency. ALOS/PALSAR was operated in orbit from January 2006 until April 2011. It shows a great potential for forestry applications in the boreal regions due to high signal/noise ratio, high resolution (~20 m), provision of cross-polarized data, and because data are being systematically collected across the Northern Hemisphere. ALOS2 was launched in 2014, and PALSAR-2 has updated features of PALSAR. A spaceborne P-band SAR, which would be less affected by saturation at

Aboveground Terrestrial Biomass and Carbon Stock Estimations from Multisensor Remote Sensing

TABLE 3.2 Characteristics of Satellite Sensors Used to Estimate Aboveground Biomass

	Sensor Characteristics		Sensor	Spectral Range	Spatial Resolution	Spatial Coverage	Temporal Resolution	Temporal Coverage
Active	LiDAR	Ground LiDAR	• EVI, DEWL	1064, 1548 nm	Site level	Site level	Discontinuous	2000s
		Small-footprint LiDAR	• Optech ALTM 3100C • Leica ALS50-II • Riegl LMS-Q140i-60	1064 nm	Foot-meter scale	Local	Discontinuous	1988–now
		Medium-footprint LiDAR	• LVIS	1064 nm	15–25 m	Regional	Discontinuous	1999–now
		Large-footprint LiDAR	• GLAS	1064 nm	60–90 m	Global	Discontinuous	2003–2009
	Radar	P-band	• Biomass	200 m	<50 m	Semi-Global No Europe/USA	25–45 days	Scheduled launch in 2020
		L-band	• ALOS-PALSAR • ALOS-PALSAR(2)	15–30 cm	7–89 m	Global	46 days	2006–2011 (PALSAR) 2014–present (PALSAR2)
		X/C band	• ERS • ENVISAT • RADARSAT • TerraSAR-X	2.5–7.5 cm	• ERS:30 m • ENVISAT:30–90 m • RADARSAT:1–100 m • TerraSAR-X: 1–16 m	Global	3, 35, and 336 days	• 1995–present • ERS:1991–2011 • ENVISAT:2002–2012 • RADARSAT:1995–present • TerraSAR-X: 2007–present
Passive	Multispectral/hyperspatial		• IKONOS • QuickBird • Orbit view	VIS-NIR	1–5 m	Global	No regular repeat cycle	2000–present
	Multispectral high spatial		• Landsat • SPOT HRV • ASTER	VIS-TIR	30 m	Global	16 days	1972–present
	Multispectral coarse resolution		• MODIS • AVHRR	VIS-TIR	1 km	Global	Daily	2000–present
	Multispectral and multiangular		• MISR	VIS-NIR	1 km	Global	Daily	1999–present
	Hyperspectral		• AVRIS • Hyperion	VIS-IR	4–20, 30 m	Global	Discontinuous	2000–present

ASTER, advanced spaceborne thermal emission and reflection radiometer; AVHRR, advanced very-high-resolution radiometer; AVIRIS, airborne visible/infrared imaging spectrometer; DWEL, dual-wavelength Echidna® LiDAR; EVI, Echidna® validation instrument; GLAS, geoscience laser altimeter system, on board the ice, cloud, and land elevation satellite (ICESat); LVIS, land vegetation and ice sensor; MODIS, moderate resolution imaging spectroradiometer; NIR, near-infrared; PALSAR, phased array-type L-band synthetic aperture radar; SPOT HRV, Le Syst`eme Pour l'Observation de la Terre High Resolution Visible; TIR, Thermal infrared.

higher biomass levels, is planned to launch in the coming years in the frame of the Earth Explorer Program of the ESA. Many airborne L-band and P-band data were also collected for biomass estimates. The major advantage of all SAR systems is their weather and daylight independency.

3.3.3 LiDAR Data

LiDAR is an active remote sensing system based on laser ranging, which measures the distance between a sensor and the target surface. Vegetation LiDAR systems typically emit at wavelengths between 900 and 1064 nm and record the time during which the emitted laser pulse is reflected off an object and returns to the sensor. The time-return interval is used to calculate the range (distance) between the sensor and the object. LiDAR provides direct and indirect measurements of vegetation structure, which can be used to estimate global carbon storage. Recent advances in LiDAR technology have made LiDAR data widely available to study vegetation structure characteristics and forest biomass.

LiDAR systems are classified as small-footprint LiDAR (laser footprint less than 1 m scale) and large-footprint LiDAR (laser footprint 10 m or greater) based on the size of laser footprint or discrete-return and full-waveform recording based on how laser energy is recorded (Dubayah and Drake 2000, Wulder et al. 2012). Discrete-return systems record single or multiple returns from a given laser pulse. As the laser signal is reflected back to the sensor, large peaks, (i.e., bright returns) represent discrete objects in the path of the laser beam and are recorded as discrete points. Most small-footprint LiDAR system record discrete energy returns. In contrast, full-waveform-recording LiDAR systems digitize the entire reflected energy from a return, resulting in complete sub-meter vertical vegetation profiles. The waveform is a function of canopy height and vertical distribution of foliage, as it is made up of the reflected energy from the surface area of canopy components such as foliage, trunks, twigs, and branches, at varying heights within the large footprint. The total waveform is therefore a measure of both the vertical distribution of vegetation surface area and the distribution of the underlying ground height. Waveform-recording instruments are mainly large-footprint LiDAR systems; however, recent advances made full-waveform instruments with increasingly smaller footprint sizes available.

Small-footprint multiple return LiDAR data have been collected in many regions of the globe, and more recently small-footprint scanning waveform systems have become operational. Such small-footprint airborne LiDAR systems are available on a commercial basis and are now used at the operational level in forest resource inventories (Næsset and Gobakken 2008). At standard level, ground-based LiDAR data, such as EVI, were collected and used for AGB estimates (Strahler et al. 2008, Ni-Meister et al. 2010). With many ground LiDAR system, complex and detailed vegetation structure data have been recorded over various study sites.

The Geoscience Laser Altimeter System (GLAS) was a large-footprint spaceborne full-waveform profiling LiDAR carried on the Ice, Cloud, and land Elevation Satellite (ICESat) for 2003–2009. GLAS was the first spaceborne LiDAR, and global measurement of canopy height was one of the science objectives of the ICESat mission (Zwally et al. 2002). The size and shape of the GLAS footprints vary from 50 to 65 m in diameter and from elliptical to circular, depending on the date of the acquisition. The pulses are spaced approximately 172 m apart.

Airborne data have also been collected using a Scanning LiDAR Imager of Canopies by Echo Recovery (SLICER) with a 15 m footprint and the Laser Vegetation and Ice Sensor (LVIS) with a 20 m/25 m footprint over several large areas for improved vegetation structure characterization since 1998 (Blair et al. 1999). This large-footprint LiDAR system records full-waveform laser energy returns. These global, regional, and local LiDAR data can provide the detailed vegetation structure and biomass maps necessary for carbon models and ecosystem process studies.

3.4 Research Approaches/Methods

Many methods are adopted to convert field-measured AGB at local scale to large scale based on remote sensing measurements or extrapolating from small-scale LiDAR and field measurements to large-scale maps of AGB. Common methods include linear statistical models, support vector machines, nearest neighbor-based methods, random forest, and Gaussian processes (e.g., Figure 3.3). The most common approach is line statistical regression (Fassnacht et al. 2014), then nonparametric nearest neighbor, machine learning (Zhao et al. 2011, Carreiras et al. 2012), random forest (Baccini et al. 2012), and Gaussian

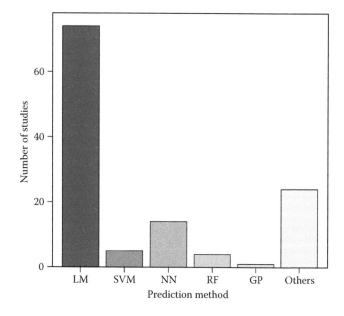

FIGURE 3.3 Frequency distribution of the prediction methods used for aboveground biomass estimates. LM, linear regression model; SVM, support vector machines; NN, nearest neighbor-based methods; RF, random forest; GP, Gaussian processes. (From Fassnacht, F.E. et al., *Remote Sens. Environ.*, 154, 102, 2014.)

processes (Zhao et al. 2011) (see Figure 3.3 for a summary by Fassnacht et al. 2014). Some physically based or semiempirical models have also been used (Saatchi et al. 2007).

3.4.1 Nonparametric Methods

With recent advancement in geospatial statistical methods and ongoing technology improvement in performing expensive statistical computations, the nonparametric method appears more prevalent in more recent studies (Baccini et al. 2004, 2008). These methods perform recursive partitioning of datasets, make no assumptions regarding the distribution and correlation of the input data, effectively solve complex nonlinear relationships between the response and predictor variables, and show great advantages for nonlinear problems and often perform better than standard linear regression models.

3.4.1.1 Tree-Based Models

Tree-based models (Breiman et al. 1984), a nonparametric approach, are a fundamental tool in data mining. They perform recursive partitioning of datasets to capture nonlinear relationships between the response and predictor variables for predicting a categorical (classification tree) or continuous (regression tree) outcome. This method has been previously used in remote sensing field to predict for classification and continuous variables (Baccini et al. 2004, 2008). Tree-based models are known for their simplicity and efficiency when dealing with domains with large number of variables and cases. However, it can also lead to poor decision in lower levels of the tree due to the unreliability of estimates based on small samples of cases.

A commonly used tree-based model in AGB estimate is random forest (Breiman 2001, Breiman et al. 1984). Random forest constructs a multitude of decision trees at training time in which different bootstrap samples of the data are used to estimate each tree and outputting the class corresponding to the individual trees. The resulting model is more accurate and less sensitive to noise in input data relative to conventional tree-based modeling algorithms.

The use of random forest for biomass estimate demonstrates the advantages of the nonparametric statistical method. For example, Baccini et al. (2008) compared the performance between random forest with more traditional multiple regression analysis, and they found that the traditional regression–explained variance is 71% compared to 82% from random forest of their AGB in their study region. LiDAR data, in combination with a random forest algorithm and a large number of reference sample units on the ground, often yield the lowest error for biomass predictions and become very popular in most research efforts on biomass estimates. There are, however, limitations to the random forest model in the prediction phase. The model tends to overpredict low biomass values and underpredict high biomass values. This trend is intrinsic of regression tree-based models whose predictions are the average of the values within the terminal node. Different methods are used in different remote sensing field for biomass estimates.

3.5 Remote Sensing–Based Aboveground Biomass Estimates

Different remote sensing datasets were used to estimate AGB. Table 3.3 listed strengths and limitations of using different remote sensing data to estimate AGB and forest carbon stocks. Details are discussed in the following text.

TABLE 3.3 Strengths and Limitations of Using Different Remote Sensing Data to Estimate Aboveground Biomass and Forest Carbon Stocks

Remote Sensing Data Types	Measured Forest Structure Parameters Inputs	Methods	Strengths	Limitations	References
Passive optical remote sensing	• Reflectances • Spectral indices • Tree shadows • Height for sparse canopy • Stand age • Land cover types	• Linear regression • Nonparametric method	• High spatial imaging capability • Consistent at all scales • Free for most imageries except for very high spatial data	• Saturation at high biomass	• Baccini et al. (2004, 2008) • Dong et al. (2003) • Chopping et al. (2009) • Zhang and Kondragunta (2006)
RADAR	• Radar signals • Woody volume • Crown center height	• Linear regression • Nonparametric method • Physical models	• Accurate at low biomass • High spatial imaging capability • Free data	• Saturation at high biomass • Impact from underneath topography/roughness and soil wetness	• Le Toan et al. (1992, 2004, 2011) • Askne and Santoro (2005) • Carreiras et al. (2012) • Cartus et al. (2012) • Chowdhury et al. (2014)
LiDAR	• Tree height • Height metrics • Foliage profiles • Crown sizes	• Linear regression • Nonparametric method	• Most accurate • Free data	• Sparse samplings for spaceborne LiDAR data • Small regions for small-footprint LiDAR data	• Asner et al. (2012, 2014) • Blair et al. (1999) • Drake et al. (2002a,b) • Dubayah and Drake (2000) • Garcia et al. (2010) • Lefsky et al. (2005a,b)

3.5.1 Optical Remote Sensing

Optical remote sensing data have been extensively used to map AGB. One simple method to map AGB is to use remotely sensed land cover classification maps where each class is assigned an average value of biomass density based on literature estimates or forest inventories. The IPCC Tier 1 approach was applied to the study area using their prescribed forest carbon density values combined with land cover data generated from the globally available land cover dataset, Global Land Cover 2000. Land cover data were reclassified as forest or nonforest, using all forest classes of GLC2000. Aboveground carbon densities were assigned to each land cover class using IPCC values.

The other more commonly used method is the determination of relationships between in situ biomass density and remote sensing characteristics/signals that can be consistently mapped over large regions (Saatchi et al. 2007). This approach has the advantage of providing spatially consistent and continuous values of the amount of biomass present at any given location. The suite of freely available optical satellite sensors, such as Landsat, AVHRR, and MODIS, has been used extensively to map AGB based on statistical relationships between ground-based measurements and satellite-observed surface reflectance, or vegetation indices or tree canopy attributes are derived from optical satellite data (Lu 2006). Spectral reflectances of optical remote sensing are the simplest variables in biomass estimates. Vegetation indices are particularly useful in biomass observations because it enhances green vegetation signals and minimizes the impacts from surface and atmospheric effects. Alternatively, tree canopy attributes such as LAI, tree cover, crown size, density, and tree shadow fraction derived from optical satellite data are considered to be effective proxies of AGB. Tree shadow fraction is an indicator of vertical vegetation structure, which can be an indicator of biomass.

At continental and global scale biomass mapping, the coarse spatial resolution optical sensors, such as the NOAA AVHRR (1.1 km) and MODIS (250 m to 1 km), have been useful for forest biomass estimates due to the good trade-off between spatial resolution, image coverage, and frequency in data acquisition (Lu 2006). Dong et al. (2003) used the normalized difference vegetation index (NDVI) estimate provided by the AVHRR sensor to estimate forest biomass at continental scale. A regression model was developed to relate AGB to latitude and the inverse of the AVHRR NDVI. Their results were encouraging for a study at this scale, but were ultimately unreliable for small-area, high-accuracy forest inventories required by small property owners seeking to quantify their forests.

Recent studies using MODIS data using random forest (Baccini et al. 2004, 2008) found that the shortwave infrared (SWIR) bands (MODIS bands 6 [1628–1652 nm] and 7 [2105–2155 nm]) are particularly sensitive to forest structural parameters (crown size and tree density), texture, and shadow, which are correlated with AGB. They have found a negative relationship between AGB and SWIR reflectance. They argue that SWIR signal is a strong indicator of tree shadows, which is related to stand age structure. Generally, the structure of young forests is often characterized by a single canopy layer, high density, relatively few canopy gaps, and trees of roughly the same size. Conversely, older forests are characterized by a mix of tree ages and sizes, canopy gaps, and multiple canopy layers resulting in increases in the shadow component, thus decreases in SWIR reflectance. Baccini et al. (2008) report a high accuracy, with the map explaining 82% of the variance in AGB for 10% of field plots held back for validation, with a root mean square error (RMSE) of 50.5 Mg/ha. However, many other studies using MODIS data have various successes (Blackard et al. 2008, Anaya et al. 2009). The main limitation is that MODIS signals are not very sensitive to high biomass values (Lu 2006, Zheng et al. 2007, Anaya et al. 2009).

For quantifying biomass at local to regional scales, data provided by finer spatial resolution instruments, such as Landsat TM (Lu et al. 2005) and ASTER (Muukkonen and Heiskanen 2005, 2007), are required. Typically, finer spatial resolution satellite data have been used as an intermediate step when relating ground reference data with coarser spatial resolution data, usually by regression techniques. For example, Muukkonen and Heiskanen (2005, 2007) used stand-wise forest inventory data and moderate-resolution ASTER data to estimate biomass with coarse-resolution MODIS data for a large area with good accuracy. The demonstrated approach can be used as a cost-effective tool to produce preliminary biomass estimates for large areas where more accurate national or large-scale forest inventories do not exist.

The Landsat series of satellites has proven to be a successful venture, providing decades of free-access moderate-resolution multispectral imagery. To estimate forest biomass, many of the studies used band combinations of the Landsat data and vegetation indices in a regression with a variety of standard field variables including mean height, Lorey's mean height (mean stand height weighted by basal area per tree), maximum height, crown width, and others. These efforts met with varying degrees of success. Cartus et al. (2014) reported a great success of using Landsat to map biomass than radar data. Landsat data, in the form of a canopy density product, was an important predictor for the AGB of forests in Africa (Avitabile et al. 2012) and in the Amazon (Saatchi et al. 2007). Canopy density metrics works well on open canopies (i.e., primarily during early successional stages of forest development). Biomass differences between forests with closed canopies are not captured. Foody et al. (2003) employed a feed-forward neural network to model forest biomass and was successful in extracting forest biomass with high levels of accuracy. Foody et al. (2003) also note a key issue in remote sensing of biomass: the inability of models to transfer from study site to study site. Empirical models built from satellite imagery rarely transfer from one study area to another, even if the study sites are composed of similar forest species and climatic conditions. Small forest plots are not represented well by image pixels larger than their spatial extent (Lu 2006), and complex biophysical environments are not well represented at the scale of Landsat data.

Recent advances of mapping disturbance using Landsat data lead to a new approach to map AGB dynamics using forest disturbance and recover history maps derived from Landsat (Powell et al. 2010, Pflugmacher et al. 2012, 2014, Main-Knorn et al. 2013). With recently developed algorithms that characterize trends in disturbance (e.g., year of onset, duration, and magnitude) and post-disturbance regrowth, the new method improved Landsat-based mapping of current biomass across large regions. The new approach includes information on vegetation trends prior to the date to enhance Landsat's spectral relationships with biomass. The method was tested in various forests in Oregon (USA), Arizona, Minnesota, Montana, and Europe using Landsat-based disturbance and recovery metrics. They found that the new method substantially improved predictions of AGB compared to models based on only single-date reflectance. Conversely, they also found that their method performed significantly better in estimating AGB dead than LiDAR models, and single-date Landsat data failed completely.

Chopping et al. (2008) investigated the usability of Multiangle Imaging SpectroRadiometer (MISR) on board the Terra satellite to measure woody biomass and other forest parameters for large parts of Arizona and New Mexico. The advantage of MISR over active or other passive sensors is timely and extensive estimates of forest biomass and other parameters at low cost.

Gonzalez et al. (2010) used QuickBird's panchromatic band to automatically detect tree crowns and then used regression techniques to estimate biomass from the diameter of each tree crown. They found that the QuickBird imagery resulted in higher error and lower total biomass estimates than the LiDAR data due to the shadowing that interfered with the crown detection algorithm. The cost of acquiring the images from these sensors is prohibitive for most research purposes. While the spatial resolution offered by these sensors is excellent for crown delineation, care must be taken with shadowing and other effects of sun angle and tree height, further reducing the utility of these data for small-area forest quantification (Gleason and Im 2011).

3.5.2 Radar

Radar signals are sensitive to dielectric and geometric properties of forests and are thus directly related to measurements of AGB. The ability of radar sensors to measure biomass mainly depends on how deep the radar signals can penetrate into the canopy. The longer the wavelength is, the deeper the penetration is. The L- and P-band backscatter, particularly HV- and HH-polarized backscatter, is strongly dependent on biomass amount (Le Toan et al. 1992, Ranson and Sun 1994, Imhoff 1995, Saatchi 2007, Saatchi et al. 2012). P-band backscatter shows stronger dependence on biomass than L-band backscatter. The radar backscatter increases approximately linearly with increasing biomass until it is saturated at a certain biomass level that varies with the radar wavelength (Imhoff 1995). The biomass level for backscatter saturation is about 200 tons/ha at P-band, 100 tons/ha at L-band, and 30–50 tons/ha at X- and C-bands (Le Toan et al. 2011).

The observed relationship between radar backscatter and biomass can be physically illustrated using electromagnetic scattering models (Sun and Ranson 1995). HV backscatter is dominated by volume scattering from the woody elements in the trees, so that HV is strongly related to AGB. For the HH and VV polarizations, ground conditions can affect the biomass–backscatter relationship, because HH backscatter comes mainly from trunk-ground scattering, while VV backscatter results from both volume and ground scattering.

Application of the radar biomass estimation at continental or globe scale is best at 1.0 ha scale (100 m × 100 m pixel size). At this scale, the distribution of AGB over the landscape is both stationary and normal, and the radar resolution is large enough to reduce the speckle noise and the geolocation error between radar pixel and the plot location. Errors associated with the biomass estimation from radar backscatter or height measurements at this scale can be reduced to acceptable levels (10%–20%) for mapping the AGB globally (Saatchi et al. 2011, 2012).

SAR sensors on board several satellites (ERS-1, JERS-1, ENVISAT, and RADARSAT) with C- and X-bands were used to quantify forest carbon stocks in relatively homogeneous or young forests, but the signal tends to saturate at fairly low biomass levels (~50–100 tons C/ha) (Le Toan et al. 2004). Mountainous or hilly conditions also increase errors. Several studies have used the phased array-type L-band SAR (PALSAR) on board the Japanese ALOS launched in 2005 to estimate biomass and carbon stocks in sparse canopies from African savanna woodlands to boreal forests (Carreiras et al. 2012, Cartus et al. 2012, Peregon and Yamagata 2013, Mermoz et al. 2014). Those studies found that ALOS/PALSAR data can successfully map AGB in sparse canopies when aggregating the ALOS biomass maps at large scale (county scale or hectare scale). Synergistic use of L- and X-band SAR can provide large-scale AGB (Englhart et al. 2011). They combined multitemporal TerraSAR-X x-band and ALOS PALSAR L-band to estimate large-scale biomass for tropical forests with $r^2 = 0.53$ with an RMSE of 79 tons/ha.

Many studies have demonstrated that radar backscattering works best only to estimate biomass for sparse canopy. As an alternative to SAR backscatter intensity, recent advancement in interferometric radar analysis techniques such as polarimetric and interferometric radar (PolInSAR) has shown great potential to predict biomass (Askne and Santoro 2005). These interferometric techniques allow for a characterization of the vertical forest structure and thus a more immediate estimation of forest biophysical attributes. Coherence saturation levels are generally higher than those reported for backscatter intensity. Under favorable conditions, correlations exist for values of up to 250–300 tons/ha (Santoro et al. 2007, Chowdhury et al. 2014). The backscattering intensity for C- and X-bands is not very good for forest biomass estimation. But the InSAR coherence and the phase center height of X-band InSAR can be used for the purpose. However, the potential to implement such experimental techniques across large areas depends on suitable configurations of future spaceborne SAR missions. With the advancement in interferometric radar analysis techniques, radar data have a great potential for global biomass estimates due to its independence from clouds and therefore the possibility to obtain continuous global coverage.

3.5.3 LiDAR

Use of LiDAR to estimate forest biomass has accelerated rapidly in recent years. Observation from both discrete and full-return LiDAR can be translated into various forest structure metrics such as maximum canopy height and multistrata heights aboveground as well as characteristic height at which different proportions of the total reflected energy are returned to the sensor. The various derived metrics can be related to AGB, typically via correlative model with associated field measurements (Goetz and Dubayah 2011, Wulder et al. 2012).

Many studies have demonstrated the strong relationship between AGB and LiDAR-measured height metrics, ranging from boreal conifers to equatorial rain forests. LiDAR has been widely used to map AGB using different LiDAR system. LiDAR is recognized as the state-of-the-art remote sensing technology for mapping AGB because it is much less sensitive to the saturation problem, compared to conventional remote sensing optical and radar data. We summarize recent progress on LiDAR-based biomass mapping activities from the following two perspectives:

3.5.3.1 Small-Footprint Discrete-Return LiDAR

AGB has been estimated successfully with remote sensing, especially using small-footprint discrete LiDAR data (Nelson 1988, Nelson et al. 2004, Næsset and Nelson 2007, Næsset and Gobakken 2008, García et al. 2010). Nelson et al. (2004) demonstrated that tree height obtained from airborne LiDAR is a good predictor of biomass for large area averages. Næsset and Goabakken (2008) found that LiDAR tree height and forest density were able to explain 88% and 85% of the variability in aboveground and belowground biomass, respectively, for 1395 sample plots in the coniferous boreal zone of Norway. These studies often use LiDAR data alone or in combination with passive optical or radar data.

Most studies were conducted based on regression equations relating vegetation biomass to LiDAR-derived variables across different scales from individual tree to plot and stand scales. The plot-based approach commonly involves field-measured biomass regressed against derived statistics from plot-level LiDAR data. The LiDAR statistics can be from the individual returns or from the height of canopy (also called canopy height model [CHM]). This approach adopts distributional metrics such as the mean canopy height and the standard deviation of the canopy height derived from the CHM or the raw returns. These metrics are then used in conjunction with regression equations to predict forest properties (Nelson 1988, 2004, Garcia et al. 2010). However, many recent studies used LiDAR return intensities rather than height metrics to estimate biomass. Garcia et al. (2010) found that several biomass estimation models based on LiDAR intensity or height combined with intensity data provide better biomass estimate than using height metrics alone.

3.5.3.2 Large-Footprint Full-Waveform LiDAR

Large-footprint full-waveform systems have been shown to provide accurate estimates of AGB in tropical and temperate deciduous, conifer, and mixed forests over a wide range of conditions. Over the past decade, several airborne Land Vegetation Ice System (LVIS) and SLICER LiDAR systems have demonstrated the ability to retrieve AGB over various biomes ranging from boreal conifers to equatorial rain forests (Drake et al. 2002a,b, Lefsky et al. 2005b, Anderson et al. 2006, 2008, Dubayah et al. 2010). Most studies adapted stepwise multiple regressions to predict ground-based measures of stand structure from both conventional canopy structure indices include mean and maximum canopy surface height, canopy cover, and indices derived from the canopy height profile (CHP), vegetation height metrics: RH100, RH75, RH50, and RH25 defined as the relative height (RH), relative to the ground elevation, at which 100%, 75%, 50%, and 25%, respectively, of the accumulated full-waveform energy occurs (Blair et al. 1999).

The GLAS, on board the ICESat, is a full-waveform digitizing LiDAR system with a nominal footprint size of ~65 m that acquires information on topography and the vertical structure of the vegetation (Zwally et al. 2002, Carabajal and Harding 2005, Harding and Carabajal 2005). A series of studies using GLAS data have successfully demonstrated the capabilities of GLAS data for estimating forest biomass on ground plots in tropical, temperate, and conifer forests (Lefsky et al. 2005a, Boudreau et al. 2008, Nelson et al. 2009, Baccini et al. 2012).

One major limitation of current spaceborne LiDAR systems (i.e., ICESat GLAS) is the lack of imaging capabilities and the fact that it provides sparse sampling information on the forest structure. To overcome this problem, it has been fused with other data to map large-scale AGB. Boudreau et al. (2008) and Nelson et al. (2009) used a multiphase sampling approach to relate GLAS waveforms to airborne profiling LiDAR measurements and profiling LiDAR measurement to field estimates of total aboveground dry biomass in Québec, Canada, and Siberia, USSR. Some combines with optical remote sensing images with GLAS data to map biomass at large scales (Baccini et al. 2008, 2012). Another issue of ICESat data is that the LiDAR waveform mixes LiDAR energy returns from both vegetation and underneath topography. To mitigate this problem, researchers have limited their analyses to area with <10 DEG slope (Nelson et al. 2009). Lefsky et al. (2005a,b) uses waveform shapes to remove the impact of underneath topography on waveform. Yang et al. (2011) developed a physical approach to remove the underneath topography effect. It is important to evaluate the accuracy, precision, and sources of uncertainty involved in using GLAS for large-scale biomass estimation in different regions of the world.

Full-waveform instruments such as GLAS (and LVIS and SLICER) must use high pulse energies in order to penetrate dense canopy and detect the ground surface. As a result of the high pulse energies, the pulse rate must be low, which limits the spatial sampling and resolution of these instruments.

Furthermore, the width of the pulse "acts as a low-pass filter, thereby smoothing the waveform and limiting the vertical resolution of the canopy features". This also broadens the return from the ground and reduces its amplitude thus making its detection more difficult.

3.5.4 Multisensor Fusion

The use of LiDAR data, particularly spaceborne data, is limited by its sparse spatial sampling. Both radar and passive optical remote sensing provide large scale of imaging capability. However, both optical and SAR estimates of AGB are limited by a loss of sensitivity with increasing biomass, commonly known as "saturation." A promising development is to combine radar/passive optical data with LiDAR to develop models that improve biomass estimates by exploiting the strengths of each sensor. The fusion of metrics from multiple sensors has

produced biomass models with high accuracy. While results have been variable, multisensor fusion can produce models with accuracy levels similar to or better than those of LiDAR alone (see Table 3.4 for a summary).

Many studies investigate if additional hyperspectral signature from hyperspectral data or radar and optical imaging capability besides LiDAR measurements improve biomass estimates (Anderson et al. 2008, Gonzalez et al. 2010, Sun et al. 2011, Swatantran et al. 2011). The results vary. But most studies found that LiDAR provides the best biomass estimates, and additional optical passive or radar data do not improve biomass estimates.

However, another series of multisensor fusion study for AGB is fusion of airborne LiDAR, spaceborne radar, Landsat, and field data to map AGB at large scales through two stages of upscaling: scaling from field measurements to airborne LiDAR scale, then from airborne LiDAR scale to spaceborne radar

TABLE 3.4 Capabilities to Estimate Aboveground Biomass and Forest Carbon Stocks through Multisensor Fusion

Multisensors	Study Area	Biomass Parameters	Method	Resolution	Accuracy	References
GLAS/ICESat, MODIS, SRTM, and QSCAT	• Tropical Forests: Latin America Sub-Saharan Africa SE Asia	Lorey's height	Maximum entropy	1, 10, 100 km	• 1 km scale: ±6%–53% • 10 km scale: ±5% • 100 km scale: ±1%	Saatchi et al. (2011)
GLAS/ICESat, MODIS, and SRTM	Pantropical forest	• Waveform metrics • Surface reflectance • Temperature • Topography	Random forest	500 m	• Tropical America: ±8.4/117.7 = 7% • Africa: ±8.4/64.5 = 13% • Asia: ±3.0/46.5 = 6%	Baccini et al. (2012)
Airborne LiDAR, GLAS/ICESat, and MODIS	Colombia and Peru	• MCH • Surface reflectance • Temperature • Topography	Random forest	1.1 km	RMSE Colombia: ±15.7 Mg C/ha Peru: ±17.6 Mg C/ha	Barccini and Asner (2013)
GLAS, MODIS	South-central Siberia	• GLAS waveform metrics • MODIS land cover	Neural network	500 m	• <10^0 slope: ±11.8/163.4 = 7% • >10^0 slope: ±12.4/171.9 = 7%	Nelson et al. (2009)
Airborne LiDAR GLAS/ICESat Landsat ETM+ SRTM	Quebec, Canada 1.3 M km²	• GLAS waveform metrics • Land cover	Regression	30 m	• Carbon density: ±2.2/39 = 6% • Total carbon: ±0.3/4.9 = 6% • R^2 = 0.56–0.65	Boudreau et al. (2008)
GLAS/ICESat Landsat	CA	• Tree height • LAI	Regression	30 m	• RMSE: 40–150 Mg C/ha • Relative error: ±40%	Zhang et al. (2014)
Airborne LiDAR Landsat	Peruvian Amazon 43 Mha	• MCH • Forest cover	Regression	0.1 and 5 ha	• At 0.1 and 5 ha: RMSE = 23 and 5 Mg C/ha	Asner et al. (2010)
LVIS and AVIRIS	Bartlett forest	• RH50 • AVIRIS MNF	Stepwise regression	20 m	• RMSE improved from 0.55 to 0.51 when combined Mg/ha • Adjusted R^2 from 027,0.3 to 0.39 • Fusion reduced error by 5%–8%	Anderson et al. (2008)
LVIS and AVIRIS	Sierra Nevada, CA	• RH100, RH75, RH50, RH25 • NDVI, NDWI, DGVI, CC	Regression	20 m Species based	• r^2 = 0.84, RMSE = 58.78 Mg/ha • No significant improvement fusing AVIRIS and LiDAR comparing to LiDAR alone	Swatantran et al. (2011)

AVIRIS MNF, AVIRIS minimum noise fraction transform (MNF) rotation; CC, Canopy cover; DGVI, First/second derivative of red edge normalized to 626–795 nm baseline; Lorey's height, basal area weighted height of all trees >10 cm in diameter; MCH, mean canopy vertical height profiles, the distance from ground (digital terrain models) to the approximate centroid of the tree crowns; NDVI, Normalized difference of vegetation index; NDWI, Normalized different of water index; RH100, Relative height (RH) to the ground elevation at which 100% of the accumulated full-waveform energy occurs; RH75, Relative height (RH) to the ground elevation at which 75% of the accumulated full-waveform energy occurs; RH50, Relative height (RH) to the ground elevation at which 50% of the accumulated full-waveform energy occurs; RH25, Relative height (RH) to the ground elevation at which 25% of the accumulated full-waveform energy occurs; SRTM, Shuttle radar topography mission.

scale (Asner 2009, Nelson et al. 2009, Asner et al. 2010, 2012, Nelson 2010, Asner and Mascaro 2014). Baccini et al. (2008) generated AGB estimates of tropical Africa from MODIS data using GLAS height metrics (average height and height of median energy or HOME metrics). Asner et al. (2010, 2012) use airborne LiDAR and Landsat data together with field data to map AGB and carbon at high spatial scale in Amazon. Nelson et al. (2009) and Nelson (2010) combine field data, airborne LiDAR, and spaceborne GLAS data to map AGB at large scales in boreal forests.

Most recent development on biomass and carbon estimates using remote sensing data is large regional mapping of AGB through multisensor fusion. Those activities include fusion LiDAR and multispectral data (Asner 2009, Asner et al. 2012, Baccini et al. 2012) together with radar data (Saatchi et al. 2011). Two independent studies have produced pantropical maps of AGB at 500 and 1 m spatial resolutions (Saatchi et al. 2011, Baccini et al. 2012). These two maps have been widely used by subnational- and national-level activities in relation to REDD+.

Both maps use similar input data layers and are driven by the same spaceborne LiDAR dataset providing systematic forest height and canopy structure estimates, but use different ground datasets for calibration and different spatial modeling methodologies. Field data were upscaled to GLAS footprint level (70 m) over a broad range of conditions in tropical Africa, America, and Asia based on the statistical relationships between LiDAR metrics and filed AGB, then GLAS footprint biomass was scaled to 500 m wall-to-wall biomass map through a random forest machine learning using MODIS BRDF, surface temperature, and SRTM digital elevation data (Baccini et al. 2012).

Saatchi et al. (2011) calibrated ICESat/GLAS Lorey's height (basal area–weighted height of all trees >10 cm in diameter) to AGB using field data collected from 4079 in situ inventory plots across three tropical continents. These AGB estimates were extrapolated from inventory plots (0.25 ha) to the entire landscape at 1 km scale based on spatial imagery from multiple sensors (MODIS, shuttle radar topography mission [SRTM], and quick scatterometer—[QSCAT]) using a data fusion model based on the maximum entropy (MaxEnt) approach. This benchmark map of biomass carbon stocks over 2.5 billion ha of forests on three continents, encompassing all tropical forests, for the early 2000s (see Figure 3.4).

A recent study compared these two maps and found significant difference in their AGB estimates over a wide variety of forest cover types and scales; however, at country level, there is general agreement, with much of the country-level difference explained by the choice of different allometric equations (Mitchard et al. 2013). These two maps were also compared to a high-resolution, locally calibrated map (Asner 2009). A further limitation present in both studies is the lack of local wood density or diameter–height calibration. Both are known to vary considerably across the landscape but using constant wood density or/and diameter-height relationship smooth out the variations of AGB estimates. This has an important implication for REDD+—it appears we have the algorithms and tools to estimate biomass stocks with some certainty.

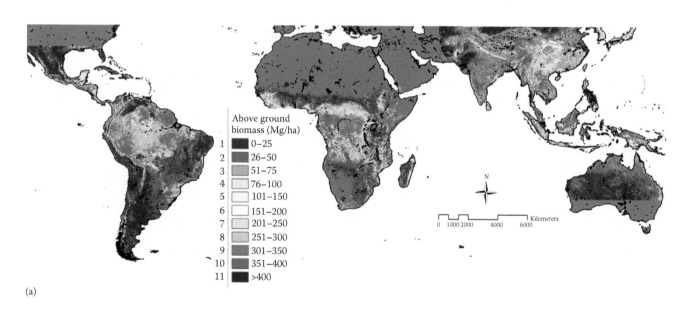

(a)

FIGURE 3.4   Distribution of forest aboveground biomass (Saatchi et al. 2011). (a) Forest aboveground biomass is mapped at 1 km spatial resolution. The study region was bounded at 30° north latitude and 40° south latitude to cover forests of Latin America and sub-Saharan Africa and from 60° to 155° east and west longitude. The map was colored on the basis of 25–50 Mg/ha AGB classes to clearly show the overall spatial patterns of forest biomass in tropical regions. Histogram distributions of forest area (at 10% tree cover) for each biomass class were calculated by summing the pixels over Latin America in.   *(Continued)*

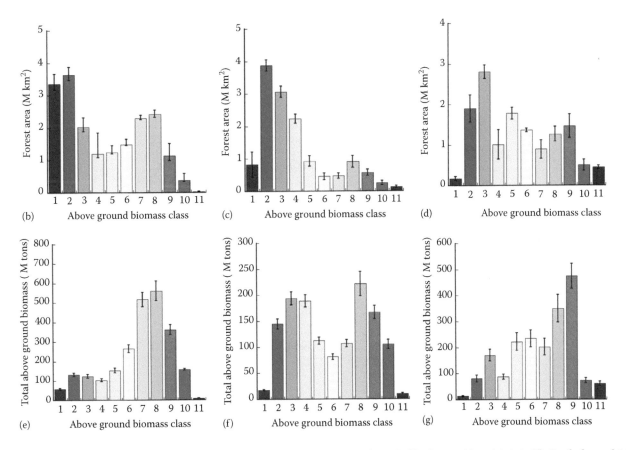

FIGURE 3.4 (*Continued*) Distribution of forest aboveground biomass (Saatchi et al. 2011). (b) Africa in (c), and Asia in (d). Similarly, total AGB for each class was computed by summing the values in each region with distributions provided for Latin America in (e), Africa in (f), and Asia in (g). All error bars were computed by using the prediction errors from spatial modeling.

3.6 Summary

A variety of remote sensing data types including optical, LiDAR, and RADAR (mostly SAR) are used to estimate biomass. The most frequently applied sensors were discrete-return airborne LiDAR, spaceborne multispectral, and airborne or spaceborne RADAR systems (Figure 3.5) (Fassnacht et al. 2014).

Several studies were conducted for an analysis of reported biomass accuracy estimates using different remote sensing platforms (airborne and spaceborne) and sensor types (optical, radar, and LiDAR) (Zolkos et al. 2013) (Goetz and Dubayah 2011). These studies reported that LiDAR is significantly better at estimating biomass than passive optical or radar sensors used alone (Figure 3.6). AGB models developed from airborne LiDAR metrics are significantly more accurate than those using radar or passive optical data. The LiDAR model error is positively correlated with the magnitude of AGB and varies at higher biomass and decreases with plot size (Figure 3.7). Fusion of LiDAR and other sensors does not always improve biomass estimates. The spatial extent of airborne LiDAR is typically restricted to relatively small areas (tens of km²) and is also often integrated with imaging sensors for larger area mapping. Airborne LiDAR metrics–produced AGB models were significantly more accurate than those based on the spaceborne GLAS instrument due to its

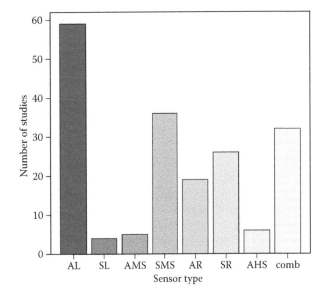

FIGURE 3.5 Frequency distribution of the data sources (sensors) for aboveground biomass estimates. AL, airborne LiDAR; SL, spaceborne LiDAR; AMS, airborne multispectral; SMS, spaceborne multispectral; AR, airborne RADAR; SR, spaceborne RADAR; AHS, airborne hyperspectral; comb, studies using data from at least two sensors. (From Fassnacht, F.E. et al., *Remote Sens. Environ.*, 154, 102, 2014.)

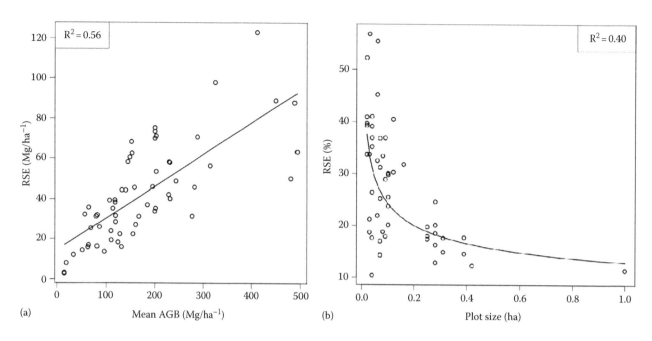

FIGURE 3.6 LiDAR model RSE vs. mean field-estimated AGB for (a) 51 LiDAR-only studies and (b) RSE (%) variability with plot size for 48 studies. (From Zolkos, S.G. et al., *Remote Sens. Environ.*, 128, 289, 2013.)

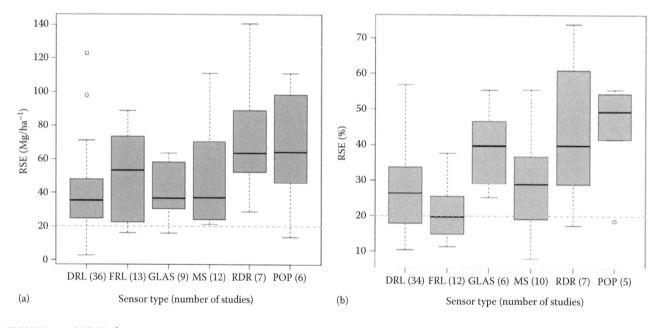

FIGURE 3.7 (a) RSE of remote sensing—AGB regression models, with dotted horizontal line at RSE = 20 Mg/ha, and (b) RSE (%) (RSE standardized by mean AGB from field measurements) categorized by sensor type, with dotted horizontal line at RSE = 20% of mean AGB. The number of studies for each type is indicated in parentheses. Not all studies reported both mean AGB and RSE values; hence, the slight disparity in sample sizes between mean RSE and AGB by sensor type. DRL, discrete-return LiDAR; FRL, full-return LiDAR; MS, multisensor; RDR, radar, POP, passive optical. (From Zolkos, S.G. et al., *Remote Sens. Environ.*, 128, 289, 2013.)

sparse samplings. The sparse sampling density of GLAS requires fusion with image data for any AGB mapping application, with associated losses in model accuracy.

Previous studies have reported that the error varies with forest types, with higher accuracies for biomass estimates in coniferous stands compared with hardwood stands (Nelson et al. 2004, Ni-Meister et al. 2010). Studies also reported that model errors tend to decrease with increasing plot size (Frazer et al. 2011). Large plot size lowers between-plot variance and has greater spatial overlap and is more resilient to GPS positional errors (Frazer et al. 2011).

Zolkos et al. (2013) reported that the error from LiDAR and multisensor models, but not radar or passive optical alone, may satisfy measurement, reporting, and verification (MRV) guidelines, particularly in tropical forests. The best LiDAR model and fusion of LiDAR and imaging satellite data at large spatial extents have demonstrated accuracies that may be suitable for carbon accounting purposes at the project level.

3.7 Conclusions and Future Directions

Tremendous progress has been made to estimate AGB in the last decade or so. With the development of new LiDAR technology, a number of investigators have developed innovative approaches to fuse passive optical imagery or radar imaging data with spatially extend point-based estimates of biophysical parameters derived from LiDAR to develop high-quality wall-to-wall AGB maps with unprecedented accuracy and spatial resolution. Particularly, the synergy of spaceborne large-footprint LiDAR (ICESAT GLAS) and medium-resolution optical data, primarily from MODIS, has been exploited to map canopy height and biomass at regional to continental scales. Combining with high-quality forest loss maps, these high-quality carbon stock maps are being used to estimate carbon emission due to forest cover change at regional and continental scales (Baccini et al. 2012, Harris et al. 2012). Those large-scale maps of biomass, carbon, and carbon emission can be extremely useful for REDD and global carbon monitoring program and have the potential to substantially reduce uncertainty in global carbon exchanges and net carbon budgets.

Developing accurate and consistent biomass maps is still challenging. One issue is a lack of large-scale densely sampled LiDAR data at continental and global scales. We call an urgent need for a LiDAR mission to quantify forest carbon store and carbon change at global scale. Currently a laser-based instrument called the Global Ecosystem Dynamics Investigation LiDAR is being developed for the International Space Station, which will provide a unique 3D view of the earth's forest structure, as the valuable information for global carbon estimates. Combining high vertical and spatial resolution, photon-counting systems might overcome the limitations of full-waveform low-detector sensitivity and restricted vertical and spatial resolution. However, how much vegetation structure properties can be retrieved from future spaceborne LiDAR missions, ICESat-II, scheduled to launch in 2017 with a 10 kHz, 532 nm micropulse photon counting laser altimeter still needs further investigation. With recent advancement in polarimetric and interferometric radar (PolInSAR), fusion LiDAR and PolInSAR may have a great potential to provide accurate AGB estimates at centennial and global scales. However, implementation of such experimental techniques across large areas heavily depends on suitable configurations of future spaceborne SAR missions.

With increasing use of remote sensing technology to map AGB and carbon stocks at large scales, their calibration will still rely on the accuracy of ground-based carbon storage estimation. Accuracy of aboveground estimates using remote sensing data depends heavily on the accuracy of allometric equations chosen. Different allometric equations used to calibrate the remote sensing data resulted in different carbon estimates. Recent studies suggested that regional variation allometric equations were

an important source of variation in tree AGB (Feldpausch et al. 2012, Goodman 2014). With the recently updated allometric equations for tropical forests (Chave et al. 2014), remote sensing products could be improved.

There is an urgent need for improved datasets that characterize the global distribution of AGB, especially in the tropics. For the UN Framework Convention on Climate Change to implement the Reduced Emissions from Deforestation and Degradation (REDD) scheme, more accurate and precise country-based carbon inventories are needed. With recent progress made on biomass and carbon store estimates at continental scales and recently published global high-resolution (30 m) forest cover change maps (Hansen et al. 2013), accurate estimates of global carbon store and carbon emission estimates are possible in the next future.

Acknowledgment

This review was partially supported by NASA under the contract number NNX10AG28G.

References

Anaya, J. A., E. Chuvieco, and A. Palacios-Orueta (2009) Aboveground biomass assessment in Colombia: A remote sensing approach. *Forest Ecology and Management*, 257, 1237–1246.

Anderson, J., M. E. Martin, M. L. Smith, R. O. Dubayah, M. A. Hofton, P. Hyde, B. E. Peterson, J. B. Blair, and R. G. Knox (2006) The use of waveform lidar to measure northern temperate mixed conifer and deciduous forest structure in New Hampshire. *Remote Sensing of Environment*, 105, 248–261.

Anderson, J., L. Plourde, M. Martin, B. Braswell, M. Smith, R. Dubayah, M. Hofton, and J. Blair (2008) Integrating waveform lidar with hyperspectral imagery for inventory of a northern temperate forest. *Remote Sensing of Environment*, 112, 1856–1870.

Askne, J. and M. Santoro (2005) Multitemporal repeat pass SAR interferometry of boreal forests. *IEEE Transactions on Geoscience Remote Sensing*, 43, 10.

Asner, G. P. (2009) Tropical forest carbon assessment: Integrating satellite and airborne mapping approaches. *Environmental Research Letters*, 4, 034009.

Asner, G. P., J. K. Clark, J. Mascaro, G. A. Galindo García, K. D. Chadwick, D. A. Navarrete Encinales, G. Paez-Acosta et al. (2012) High-resolution mapping of forest carbon stocks in the Colombian Amazon. *Biogeosciences*, 9, 2683–2696.

Asner, G.P., R.E. Martin, R. Tupayachi, C.B. Anderson, F. Sinca, L. Carranza-Jimenez, et al. (2014). Amazonian functional diversity from forest canopy chemical assembly. *Proceedings of the National Academy of Sciences*, 111, 5604–5609.

Asner, G. P. and J. Mascaro (2014) Mapping tropical forest carbon: Calibrating plot estimates to a simple LiDAR metric. *Remote Sensing of Environment*, 140, 614–624.

Asner, G. P., G. V. N. Powellb, J. Mascaroa, D. E. Knappa, J. K. Clarka, J. Jacobsona, T. Kennedy-Bowdoina et al. (2010) High-resolution forest carbon stocks and emissions in the Amazon. *Proceedings of the National Academy of Sciences of the United States of America*, 107, 16738–16742.

Avitabile, V., A. Baccini, M. A. Friedl, and C. Schmullius (2012) Capabilities and limitations of Landsat and land cover data for aboveground woody biomass estimation of Uganda. *Remote Sensing of Environment*, 117, 366–380.

Baccini, A. and G.P. Asner. (2013) Improving pantropical forest carbon maps with airborne LiDAR sampling. *Carbon Management*, 4, 591–600.

Baccini, A., M. A. Friedl, C. E. Woodcock, and R. Warbington2 (2004) Forest biomass estimation over regional scales using multisource data. *Geophysical Research Letters*, 31, L10501.

Baccini, A., S. J. Goetz, W. S. Walker, N. T. Laporte, M. Sun, D. Sulla-Menashe, J. Hackler et al. (2012) Estimated carbon dioxide emissions from tropical deforestation improved by carbon-density maps. *Nature Climate Change*, 2, 182–185.

Baccini, A., N. Laporte, S. J. Goetz, M. Sun, and H. Dong (2008) A first map of tropical Africa's above-ground biomass derived from satellite imagery. *Environmental Research Letters*, 3, 045011.

Blackard, J., M. Finco, E. Helmer, G. Holden, M. Hoppus, D. Jacobs, A. Lister, G. Moisen, M. Nelson, and R. Riemann (2008) Mapping U.S. forest biomass using nationwide forest inventory data and moderate resolution information. *Remote Sensing of Environment*, 112, 1658–1677.

Blair, J. B., D. L. Rabine, and M. A. Hofton (1999) The Laser Vegetation Imaging Sensor (LVIS): A medium-altitude, digitization-only, airborne laser altimeter for mapping vegetation and topography. *ISPRS Journal of Photogrammetry and Remote Sensing*, 54, 8.

Boudreau, J., R. Nelson, H. Margolis, A. Beaudoin, L. Guindon, and D. Kimes (2008) Regional aboveground forest biomass using airborne and spaceborne LiDAR in Québec. *Remote Sensing of Environment*, 112, 3876–3890.

Breiman, L. (2001) Random forests. *Machine Learning*, 45, 5–32.

Breiman, L., J. H. Friedman, R. A. Olshen, and C. J. Stone (1984) *Classification and Regression Trees*. Wadsworth, Belmont, CA.

Brown, S. (1997) Estimating biomass and biomass change of tropical forests. A primer. FAO Forest Resources Assessment Publication No.134. Roma. p. 55.

Brown, S. (2002) Measuring carbon in forests: Current status and future challenges. *Environmental Pollution*, 116, 9.

Brown, S., A. J. Gillespie, and A. E. Lugo (1989) Biomass estimation methods for tropical forests with applications to forest inventory data. *Forest Science*, 35, 21.

Canadell, J. G., C. Le Quere, M. R. Raupach, C. B. Field, E. T. Buitenhuis, P. Ciais, T. J. Conway, N. P. Gillett, R. A. Houghton, and G. Marland (2007) Contributions to accelerating atmospheric CO_2 growth from economic activity, carbon intensity, and efficiency of natural sinks. *Proceedings of the National Academy of Sciences of the United States of America*, 104, 18866–18870.

Carabajal, C. C. and D. J. Harding (2005) ICESat validation of SRTM C-band digital elevation models. *Geophysical Research Letters*, 32, L22S01.

Carreiras, J. M. B., M. J. Vasconcelos, and R. M. Lucas (2012) Understanding the relationship between aboveground biomass and ALOS PALSAR data in the forests of Guinea-Bissau (West Africa). *Remote Sensing of Environment*, 121, 426–442.

Cartus, O., J. Kellndorfer, W. Walker, C. Franco, J. Bishop, L. Santos, J. M. M. Fuentes (2014) A national, detailed map of forest aboveground carbon stocks in Mexico. *Remote Sensing*, 6(6), 5559–5588.

Cartus, O., M. Santoro, and J. Kellndorfer (2012) Mapping forest aboveground biomass in the Northeastern United States with ALOS PALSAR dual-polarization L-band. *Remote Sensing of Environment*, 124, 466–478.

Chave, J., C. Andalo, and S. Brown (2005) Tree allometry and improved estimation of carbon stocks and balance in tropical forests. *Oecologia*, 145, 87–99.

Chave, J., M. Rejou-Mechain, A. Burquez, E. Chidumayo, M. S. Colgan, W. B. Delitti, A. Duque et al. (2014) Improved allometric models to estimate the aboveground biomass of tropical trees. Global Change Biology, 20, 3177–3190.

Cheng, D. L. and K. J. Niklas (2007) Above- and below-ground biomass relationships across 1534 forested communities. *Annals of Botany*, 99, 95–102.

Chopping, M., G. Moisen, L. Su, A. Laliberte, A. Rango, J.V. Martonchik, and D.P.C. Peters (2008) Large area mapping of southwestern forest crown cover, canopy height, and biomass using MISR, *Remote Sensing of Environment*, 112, 2051–2063.

Chopping, M., A. Nolin, G. G. Moisen, J. V. Martonchik, and M. Bull (2009) Forest canopy height from the Multiangle Imaging SpectroRadiometer (MISR) assessed with high resolution discrete return lidar. *Remote Sensing of Environment*, 113, 2172–2185.

Chowdhury, T. A., C. Thiel, and C. Schmullius (2014) Growing stock volume estimation from L-band ALOS PALSAR polarimetric coherence in Siberian forest. *Remote Sensing of Environment*, 155, 129, e4.

Domke, G. M., C. W. Woodall, J. E. Smith, J. A. Westfall, and R. E. McRoberts (2012) Consequences of alternative tree-level biomass estimation procedures on U.S. forest carbon stock estimates. *Forest Ecology and Management*, 270, 108–116.

Dong, J., R. K. Kaufmann, R. B. Myneni, C. J. Tucker, P. E. Kaupp, J. Liski, W. Buermann, V. Alexeyev, and M. K. Hughes (2003) Remote sensing estimates of boreal and temperate forest woody biomass: Carbon pools, sources, and sinks. *Remote Sensing of Environment*, 84, 8.

Drake, J. B., R. Dubayah, R. G. Knox, D. B. David, B. Clark, and J. B. Blaire (2002a) Sensitivity of large-footprint lidar to canopy structure and biomass in a neotropical rainforest. *Remote Sensing of Environment*, 81, 15.

Drake, J. B., R. O. Dubayah, D. B. Clark, R. G. Knox, J. B. Blair, M. A. Hofton, R. L. Chazdon, J. F. Weishampel, and S. D. Prince (2002b) Estimation of tropical forest structural characteristics using large-footprint lidar. *Remote Sensing of Environment*, 79, 15.

Dubayah, R. and J. Drake (2000) Lidar remote sensing for forestry. *Journal of Forestry*, 98, 12.

Dubayah, R. O., S. L. Sheldon, D. B. Clark, M. A. Hofton, J. B. Blair, G. C. Hurtt, and R. L. Chazdon (2010) Estimation of tropical forest height and biomass dynamics using lidar remote sensing at La Selva, Costa Rica. *Journal of Geophysical Research-Biogeosciences*, 115, 1–17.

Englhart, S., V. Keuck, and F. Siegert (2011) Aboveground biomass retrieval in tropical forests—The potential of combined X- and L-band SAR data use. *Remote Sensing of Environment*, 115, 1260–1271.

FAO (2010) Food and agriculture organization of the United Nations global forest resources assessment 2010. FAO Forestry Paper 163, Rome, Italy.

Fassnacht, F. E., F. Hartig, H. Latifi, C. Berger, J. Hernández, P. Corvalán, and B. Koch (2014) Importance of sample size, data type and prediction method for remote sensing-based estimations of aboveground forest biomass. *Remote Sensing of Environment*, 154, 102–114.

Feldpausch, T. R., L. Banin, O. L. Phillips, T. R. Baker, S. L. Lewis, C. A. Quesada, K. Affum-Baffoe et al. (2011) Height-diameter allometry of tropical forest trees. *Biogeosciences*, 8, 1081–1106.

Feldpausch, T. R., J. Lloyd, S. L. Lewis, R. J. W. Brienen, M. Gloor, A. Monteagudo Mendoza, G. Lopez-Gonzalez et al. (2012) Tree height integrated into pantropical forest biomass estimates. *Biogeosciences*, 9, 3381–3403.

Foody, G. M., D. S. Boyd, M. E. J. Cutler (2003) Predictive relations of tropical forest biomass from Landsat TM data and their transferability between regions, *Remote Sensing of Environment*, 85(4), 463–474.

Frazer, G. W., S. Magnussen, M. A. Wulder, and K. O. Niemann (2011) Simulated impact of sample plot size and co-registration error on the accuracy and uncertainty of LiDAR-derived estimates of forest stand biomass. *Remote Sensing of Environment*, 115, 636–649.

García, M., D. Riaño, E. Chuvieco, and F. M. Danson (2010) Estimating biomass carbon stocks for a Mediterranean forest in central Spain using LiDAR height and intensity data. *Remote Sensing of Environment*, 114, 816–830.

Gibbs, H. K., S. Brown, J. O. Niles, and J. A. Foley (2007) Monitoring and estimating tropical forest carbon stocks: Making REDD a reality. *Environmental Research Letters*, 2, 045023.

Gleason, C. J. and J. Im (2011) A review of remote sensing of forest biomass and biofuel: Options for small-area applications. *Giscience & Remote Sensing*, 48, 141–170.

Goetz, S. J., A. Baccini, N. T. Laporte, T. Johns, W. Walker, J. Kellndorfer, R. A. Houghton, and M. Sun (2009) Mapping and monitoring carbon stocks with satellite observations: A comparison of methods. *Carbon Balance Management*, 4, 2.

Goetz, S. J. and R. Dubayah (2011) Advances in remote sensing technology and implications for measuring and monitoring forest carbon stocks and change. *Carbon Management*, 2, 14.

Gonzalez, P., G. P. Asner, J. J. Battles, M. A. Lefsky, K. M. Waring, and M. Palace (2010) Forest carbon densities and uncertainties from Lidar, QuickBird, and field measurements in California. *Remote Sensing of Environment*, 114, 1561–1575.

Goodman, R. C. (2014) The importance of crown dimensions to improve tropical tree biomass estimates. *Ecological Applications*, 24, 9.

Hansen, M. C., P. V. Potapov, R. Moore, M. Hancher, S. A. Turubanova, A. Tyukavina, D. Thau et al. (2013) High-resolution global maps of 21st-century forest cover change. *Science*, 342, 850–853.

Harding, D. J. and C. C. Carabajal (2005) ICESat waveform measurements of within-footprint topographic relief and vegetation vertical structure. *Geophysical Research Letters*, 32, L21S10.

Harris, N. L., S. Brown, S. C. Hagen, S. S. Saatchi, S. Petrova, W. Salas, M. C. Hansen, P. V. Potapov, and A. Lotsch (2012) Baseline map of carbon emissions from deforestation in tropical regions. *Science*, 336, 1573–1576.

Heimann, M. and M. Reichstein (2008) Terrestrial ecosystem carbon dynamics and climate feedbacks. *Nature*, 451, 289–292.

Hollman, R., C. J. Merchant, R. Saunders, C. Downy, M. Buchwitz, A. Cazenave, E. Chuvieco et al. (2013) The ESA climate change initiative—Satellite data records for essential climate variables. *Bulletin of American Meteorology Society*, 10, 12.

Houghton, R., K. T. Lawrance, J. L. Hackler, and S. Brown (2001) The spatial distribution of forest biomass in Brazilian Amazon: A comparison of estimates. *Global Climate Biology*, 4, 16.

Houghton, R. A. (2005) Aboveground forest biomass and the global carbon balance. *Global Change Biology*, 11, 945–958.

Houghton, R. A. (2007) Balancing the global carbon budget. *Annual Review of Earth and Planetary Sciences*, 35, 313–347.

Houghton, R. A., F. Hall, and S. J. Goetz (2009) Importance of biomass in the global carbon cycle. *Journal of Geophysical Research*, 114, 1–13.

Houghton, R. A., J. I. House, J. Pongratz, G. R. van der Werf, R. S. DeFries, M. C. Hansen, C. L. Quere, and N. Ramankutty (2012) Carbon emissions from land use and land-cover change. *Biogeosciences*, 9, 18.

Imhoff, M. (1995) Radar Backscatter and biomass saturation: Ramifications for global biomass inventory. *IEEE Transactions on Geoscience and Remote Sensing*, 33, 9.

Jenkins, J. C. (2003) National-scale biomass estimators for United States tree species. *Forest Science*, 49, 24.

Jenkins, J. C., D. C. Chojnacky, L. S. Heath, and R. A. Birdsey (2004) Comprehensive database of diameter-based biomass regressions for North American tree species. United States Department of Agriculture Forest Service, General Technical Report NE-319, Newtown Square, PA.

Kim, D. K., J. O. Sexton, P. Noojipady, C. Huang, A. Anand, S. Channan, M. Feng, and J. R. Townshend (2014) Global, landsat-based forest-cover change from 1990 to 2000. *Remote Sensing of Environment*. ISSN 0034-4257. Available online September 26, 2014, http://dx.doi.org/10.1016/j.rse.2014.08.017.

Le Toan T., A. Beaudoin, J. Riom, and D. Guyon (1992) Relating forest biomass to SAR data. *IEEE Transactions on Geoscience and Remote Sensing*, 30, 8.

Le Toan T., S. Quegan, M. Davidson, H. Balzter, P. Paillou, K. Papathanassiou, S. Plummer, S. Saatchi, H. Shugart, and L. Ulander (2011) The BIOMASS mission: Mapping global forest biomass to better understand the terrestrial carbon cycle. *Remote Sensing of Environment*, 115, 11.

Le Toan T., S. Quegan, I. Woodward, M. Lomas, N. Delbart, and G. Picard (2004) Relating radar remote sensing of biomass to modelling of forest carbon budgets. *Journal of Climatic Change*, 67, 4.

Lefsky, M. A., D. J. Harding, M. Keller, W. B. Cohen, C. C. Carabajal, F. D. Espirito-Santo, M. O. Hunter, and R. de Oliveira (2005a) Estimates of forest canopy height and aboveground biomass using ICESat. *Geophysical Research Letters*, 32, L22S02.

Lefsky, M. A., A. T. Hudak, W. B. Cohen, and S. A. Acker (2005b) Geographic variability in lidar predictions of forest stand structure in the Pacific Northwest. *Remote Sensing of Environment*, 95, 532–548.

Lu, D. S. (2006) The potential and challenge of remote sensing-based biomass estimation. *International Journal of Remote Sensing*, 27, 1297–1328.

Lu, D. S., M. Mateus Batistella, and E. Moran (2005) Satellite estimation of aboveground biomass and impacts of forest stand structure. *Photogrammetric Engineering & Remote Sensing*, 71, 967–974.

Main-Knorn, M., W. B. Cohen, R. E. Kennedy, W. Grodzki, D. Pflugmacher, P. Griffiths, and P. Hostert (2013) Monitoring coniferous forest biomass change using a Landsat trajectory-based approach. *Remote Sensing of Environment*, 139, 277–290.

Mermoz, S., T. Le Toan, L. Villard, M. Réjou-Méchain, and J. Seifert-Granzin (2014) Biomass assessment in the Cameroon savanna using ALOS PALSAR data. *Remote Sensing of Environment*, 155, 109–119.

Mitchard, E. T., S. S. Saatchi, A. Baccini, G. P. Asner, S. J. Goetz, N. L. Harris, and S. Brown (2013) Uncertainty in the spatial distribution of tropical forest biomass: A comparison of pan-tropical maps. *Carbon Balance Management*, 8, 10.

Muukkonen, P. and J. Heiskanen (2005) Estimating biomass for boreal forests using ASTER satellite data combined with standwise forest inventory data. *Remote Sensing of Environment*, 99, 434–447.

Muukkonen, P. and J. Heiskanen (2007) Biomass estimation over a large area based on standwise forest inventory data and ASTER and MODIS satellite data: A possibility to verify carbon inventories. *Remote Sensing of Environment*, 107, 617–624.

Næsset, E. and T. Gobakken (2008) Estimation of above- and below-ground biomass across regions of the boreal forest zone using airborne laser. *Remote Sensing of Environment*, 112, 3079–3090.

Næsset, E. and R. Nelson (2007) Using airborne laser scanning to monitor tree migration in the boreal–alpine transition zone. *Remote Sensing of Environment*, 110, 357–369.

Nelson, R. (1988) Estimating forest biomass and volume using airborne laser data. *Remote Sensing of Environment*, 24, 21.

Nelson, R. (2010) Model effects on GLAS-based regional estimates of forest biomass and carbon. *International Journal of Remote Sensing*, 31, 1359–1372.

Nelson, R., K. J. Ranson, G. Sun, D. S. Kimes, V. Kharuk, and P. Montesano (2009) Estimating Siberian timber volume using MODIS and ICESat/GLAS. *Remote Sensing of Environment*, 113, 691–701.

Nelson, R., A. Short, and M. Valenti (2004) Measuring biomass and carbon in delaware using an airborne profiling LIDAR. *Scandinavian Journal of Forest Research*, 19, 12.

Ni-Meister, W., S. Y. Lee, A. H. Strahler, C. E. Woodcock, C. Schaaf, T. A. Yao, K. J. Ranson, G. Q. Sun, and J. B. Blair (2010) Assessing general relationships between aboveground biomass and vegetation structure parameters for improved carbon estimate from lidar remote sensing. *Journal of Geophysical Research-Biogeosciences*, 115, 1–12.

Nogueira, E. M., P. M. Fearnside, B. W. Nelson, R. I. Barbosa, and E. W. H. Keizer (2008) Estimates of forest biomass in the Brazilian Amazon: New allometric equations and adjustments to biomass from wood-volume inventories. *Forest Ecology and Management*, 256, 1853–1867.

Pan, Y., R. A. Birdsey, J. Fang, R. Houghton, P. E. Kauppi, W. A. Kurz, O. L. Phillips et al. (2011) A large and persistent carbon sink in the world's forests. *Science*, 333, 988–993.

Peregon, A. and Y. Yamagata (2013) The use of ALOS/PALSAR backscatter to estimate above-ground forest biomass: A case study in Western Siberia. *Remote Sensing of Environment*, 137, 139–146.

Pflugmacher, D., W. B. Cohen, and R. E. Kennedy (2012) Using landsat-derived disturbance history (1972–2010) to predict current forest structure. *Remote Sensing of Environment*, 122, 146–165.

Pflugmacher, D., W. B. Cohen, R. E. Kennedy, and Z. Yang (2014) Using landsat-derived disturbance and recovery history and lidar to map forest biomass dynamics. *Remote Sensing of Environment*, 151, 124–137.

Powell, S. L., W. B. Cohen, S. P. Healey, R. E. Kennedy, G. G. Moisen, K. B. Pierce, and J. L. Ohmann (2010) Quantification of live aboveground forest biomass dynamics with Landsat time-series and field inventory data: A comparison of empirical modeling approaches. *Remote Sensing of Environment*, 114, 1053–1068.

Ranson, K. J. and G. Sun (1994) Mapping biomass in a northern forest using multifrequency SAR data *IEEE Transactions on Geoscience and Remote Sensing*, 32, 8.

Saatchi, S., L. Ulander, M. Williams, S. Quegan, T. LeToan, H. Shugart, and J. Chave (2012) Forest biomass and the science of inventory from space. *Nature Climate Change*, 2, 826–827.

Saatchi, S. S., N. L. Harris, S. Brown, M. Lefsky, E. T. A. Mitchard, W. Salas, B. R. Zutta et al. (2011) Benchmark map of forest carbon stocks in tropical regions across three continents. *Proceedings of the National Academy of Sciences of the United States of America*, 108, 9899–9904.

Saatchi, S. S., R. A. Houghton, R. C. Dos Santos AlvalÁ, J. V. Soares, and Y. Yu (2007) Distribution of aboveground live biomass in the Amazon basin. *Global Change Biology*, 13, 816–837.

Santoro, M., A. Shvidenko, I. McCallum, J. Askne, and C. Schmullius (2007) Properties of ERS-1/2 coherence in the Siberian boreal forest and implications for stem volume retrieval. *Remote Sensing of Environment*, 106, 154–172.

Strahler, A. H., D. L. B. Jupp, C. E. Woodcock, C. B. Schaaf, T. Yao, F. Zhao, X. Yang et al. (2008) Retrieval of forest structural parameters using a ground-based lidar instrument (Echidna®). *Canadian Journal of Remote Sensing*, 34, 14.

Sun, G. and K. J. Ranson (1995) A three dimensional radar backscattering model for forest canopies. *IEEE Transactions on Geoscience and Remote Sensing*, 33, 11.

Sun, G., K. J. Ranson, Z. Guo, Z. Zhang, P. Montesano, and D. Kimes (2011) Forest biomass mapping from lidar and radar synergies. *Remote Sensing of Environment*, 115, 2906–2916.

Swatantran, A., R. Dubayah, D. Roberts, M. Hofton, and J. B. Blair (2011) Mapping biomass and stress in the Sierra Nevada using lidar and hyperspectral data fusion. *Remote Sensing of Environment*, 115, 2917–2930.

West, G. B., J. H. Brown, and B. J. Enquist (1999) A general model for the structure and allometry of plant vascular systems. *Nature*, 400, 4.

Woodall, C. W., L. S. Heath, G. M. Domke, and M. C. Nichols (2011) Methods and equations for estimating aboveground volume, biomass, and carbon for trees in the U.S. Forest Inventory, 2010. United States Department of Agriculture, Forest Service, General Technical Report NRS-88, Newtown Square, PA.

Wulder, M. A., J. C. White, R. F. Nelson, E. Naesset, H. O. Orka, N. C. Coops, T. Hilker, C. W. Bater, and T. Gobakken (2012) Lidar sampling for large-area forest characterization: A review. *Remote Sensing of Environment*, 121, 196–209.

Yang, W. Z., W. Ni-Meister, and S. Lee (2011) Assessment of the impacts of surface topography, off-nadir pointing and vegetation structure on vegetation lidar waveforms using an extended geometric optical and radiative transfer model. *Remote Sensing of Environment*, 115, 2810–2822.

Zhang, G., S. Ganguly, R. R. Nemani, M. A. White, C. Milesi, H. Hashimoto, W. Wang, S. Saatchi, Y. Yu, and R. B. Myneni (2014) Estimation of forest aboveground biomass in California using canopy height and leaf area index estimated from satellite data, *Remote Sensing of Environment*, 151, 44–56.

Zhang, X. and S. Kondragunta (2006) Estimating forest biomass in the USA using generalized allometric models and MODIS land products. *Geophysical Research Letters*, 33, L09402.

Zhao, K., S. Popescu, X. Meng, Y. Pang, and M. Agca (2011) Characterizing forest canopy structure with lidar composite metrics and machine learning. *Remote Sensing of Environment*, 115, 1978–1996.

Zheng, D., L. S. Heath, and M. J. Ducey (2007) Forest biomass estimated from MODIS and FIA data in the Lake States: MN, WI and MI, USA. *Forestry*, 80, 265–278.

Zolkos, S. G., S. J. Goetz, and R. Dubayah (2013) A meta-analysis of terrestrial aboveground biomass estimation using lidar remote sensing. *Remote Sensing of Environment*, 128, 289–298.

Zwally, H. J., B. Schutz, W. Abdalati, J. Abshire, C. Bentley, A. Brenner, J. Bufton et al. (2002) ICESat's laser measurements of polar ice, atmosphere, ocean, and land. *Journal of Geodynamics*, 34, 11.

II

Agricultural Croplands

4

Agriculture

Clement Atzberger
*University of Natural Resources
and Life Sciences (BOKU)*

Francesco Vuolo
*University of Natural Resources
and Life Sciences (BOKU)*

Anja Klisch
*University of Natural Resources
and Life Sciences (BOKU)*

Felix Rembold
*Joint Research Centre (JRC),
European Commission*

Michele Meroni
*Joint Research Centre (JRC),
European Commission*

Marcio Pupin Mello
Boeing Research & Technology

Antonio Formaggio
*National Institute for
Space Research*

Acronyms and Definitions

AFI	Area fraction image
AgRISTARS	Agriculture and Resource Inventory Surveys through Aerospace Remote Sensing
ASTER	Advanced Spaceborne Thermal Emission and Reflection Radiometer
AVHRR	Advanced very-high-resolution radiometer
BN	Bayesian network
BOKU	University of Natural Resources and Life Science, Vienna
CERES	Name of crop growth model
CGM	Crop growth model
CNDVI	Crop-specific NDVI
CORINE	Coordination of Information on the Environment Programme
CROPSYST	Name of crop growth model
CWSB	Crop water satisfaction boundary model
DM	Data mining
DSS	Decision support system
DT	Decision tree
ECMWF	European Centre for Medium-Range Weather Forecasts
ERTS	Earth Resources Technology Satellite
ETa	Actual evapotranspiration
ETM+	Enhanced thematic mapper
EU	European Union
FAO	Food and Agriculture Organization of the United Nations
fAPAR	Fraction of absorbed photosynthetically active radiation
FAS	Foreign Agricultural Service
FEWS-NET	Famine Early Warning System
GDP	Gross domestic product
GEO	Group of earth observations
GEOSS	Global Earth Observation System of Systems
GeoWiki	Project name
GHG	Greenhouse gas
GLAM	Global agricultural monitoring
GRAMI	Name of crop growth model

IFPRI	International Food Policy Research Institute	RMSE	Root mean square error
INPE	National Institute for Space Research of Brazil	SAM	Spectral angle mapping
IPCC	Intergovernmental Panel on Climate Change	Sentinel	Name of satellite
IT	Information technology	SOS	Start of season
JECAM	Joint Experiment for Crop Assessment and Monitoring	SPOT	Satellite Pour l'Observation de la Terre
JRC	Joint Research Center of the European Commission	STICS	Name of crop growth model
		Suomi-NPP	Suomi National Polar-Orbiting Partnership Satellite
LACIE	Large area crop inventory experiment		
LAI	Leaf area index	SVAT	Soil vegetation atmosphere transfer modeling
LULC	Land use/land cover	TM	Thematic mapper
MARS	Monitoring agriculture by remote sensing	U.S.	United States
MLC	Maximum likelihood classifier	USAID	United States Agency for International Development
MODIS	Moderate-resolution imaging spectroradiometer		
		USDA	United States Department of Agriculture
MVC	Maximum value composit(ing)	VCI	Vegetation condition index
NASA	National Aeronautics and Space Administration	VGT	Name of sensor (vegetation) onboard SPOT
		VHR	Very high resolution
NDVI	Normalized difference vegetation index	VI	Vegetation index
NLCD	National Land Cover Database	VITO	Flemish Institute for Technological Research
NN	Neural network	WOFOST	Name of crop growth model
NOAA	National Oceanic and Atmospheric Administration		

4.1 Introduction

The development of satellite-based remote sensing technologies was, for a long time, driven by agricultural information needs (Becker-Reshef et al., 2010a). In the United States, for example, preliminary research and development of civil satellite monitoring is reported having started in the early 1970s (launch of Earth Resources Technology Satellite later renamed Landsat-1) following unanticipated severe wheat shortages in Russia (Figure 4.1) (Pinter et al., 2003).

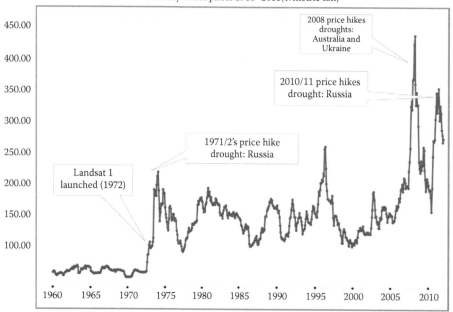

FIGURE 4.1   Monthly wheat prices from 1960 to 2011 (in USD per metric ton). (From http://wmp.gsfc.nasa.gov/uploads/science/slides/Justice_ASP-WR_2012-09-06.pdf.)

TABLE 4.1 Specific Challenges of Agricultural Activities Favoring the Use of Remote Sensing Data Compared to Other Data Sources

Agricultural production depends on physical landscape (e.g., soil type), as well as climatic driving variables and agricultural management practices, all these factors being highly variable in space and time
Agricultural production follows strong seasonal patterns related to the biological life cycle of crops
Productivity can be quickly affected by unfavorable growing conditions, pests, and diseases
Many agricultural items are perishable
Agricultural trade and prices are globally linked and therefore affecting the actions of various stakeholders ranging from farmer to traders and governments
Agricultural commodities are subjected to excessive market speculation, resulting in price spikes often affecting the poorest people most strongly

Although the focus of remote sensing has broadened over the years, agriculture is still important as shown by the large—and increasing—number of publications dealing with remote sensing and agriculture. Not surprisingly, most remote sensing scientific conferences have at least one session dealing with agriculture.

The importance of remote sensing in agriculture stems from the fact that agricultural activities face specific challenges not common to other economic sectors (Table 4.1). As a result, agricultural activities have to be monitored from local to global scales at high temporal frequency.

In recent years, we observed an increased use of remote sensing data and related technologies in agricultural production systems. First, remote sensing data have found their entrance in precision farming aiming to increase agricultural efficiency (Moran et al., 1997; Seelan et al., 2003; Mulla, 2013). Second, remote sensing is also a very valuable tool for monitoring agricultural expansion (e.g., following deforestation) (Galford et al., 2008; Gibbs et al., 2010). Finally, by providing timely, comprehensive, objective, transparent, accurate, and unbiased data, remotely derived information can eventually prevent excessive market speculation and resulting price spikes (Naylor, 2011).

High—and volatile—food prices repeatedly restrict food access in the most vulnerable parts of the world (Figure 4.2). For example, between 2006 and 2008, average world prices for rice, wheat, corn, and soybeans rose between 107% and 217%. This demonstrates that remote sensing has more to offer than just an ecological and economic component in monitoring systems. Indeed, the example

shows the social component of remote sensing, as the poorest people are usually the most affected by rising food prices.

One can expect that the impact of remote sensing data in agriculture and agronomy will continue to increase in the future, as the agricultural sector itself is under high pressure. A number of external drivers require a quick and widespread adaptation of agriculture practices:

- Agriculture must strongly increase its production for feeding the nine-billion people predicted by mid-century (Foley et al., 2011).
- Agricultural production and productivity must be increased while minimizing the environmental impact of agriculture (Zaks and Kucharik, 2011).
- Agriculture must cope with climate change (Olesen and Bindi, 2002).
- Agriculture must deal with land users not involved in food production (e.g., use of agricultural land for biofuel production, and urban expansion) (Demirbas and Balat, 2006).

To avoid information gaps, the progress of these necessary adaptations has to be monitored through appropriate agricultural monitoring systems. For example, policy makers and stakeholders should be informed about the state of the agricultural sector and the pathway that led to the current situation. Information is also critical for delivering feedbacks to decision makers regarding the actual impact of their policies and investments. Reliable

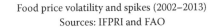

Food price volatility and spikes (2002–2013)
Sources: IFPRI and FAO

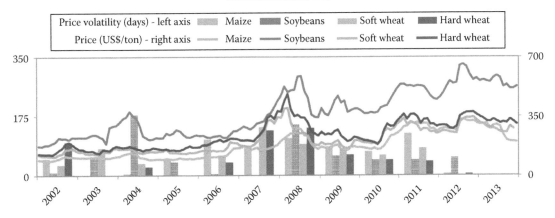

FIGURE 4.2 Food price volatility (left axis) and price (right axis) between 2002 and 2013. Price spikes refer to a steep rise in prices over a short period, whereas volatility is defined as high dispersion of prices around the average market price. (From http://www.ifpri.org/sites/default/files/publications/2020resilienceconfbr16.pdf.)

TABLE 4.2 Strengths and Applications of Remote Sensing in the Fields of Agronomy and Agriculture

Biomass and yield estimates

Crop acreage information

Objective and unbiased assessment of crop conditions over large (agricultural) areas with high revisit frequency

Mapping of disturbances and stresses

Assessment of disastrous climatic events on agricultural production

Identification of cropping patterns and agricultural production systems

Provision of baseline information for index-based (agricultural) insurances

Information for helping understanding possible effects of climate change

Identification of areas with yield gaps

Mapping of crop phenological development

Mapping of irrigated areas and water requirements

Increased productivity efficiency through precision farming

Monitoring of agricultural expansion/farmland abandonment

information also facilitates risk reduction and would lead to optimized statistical analyses at a range of scales, enabling a timely and accurate national to regional agricultural statistical reporting.

To cope with these conditions and information needs, two important requirements have to be met:

1. Information has to be provided globally at a reasonably detailed spatial scale and with a frequent updating frequency (Bruinsma, 2003).
2. Information is needed in due time—information is worth little, if it becomes available too late (FAO, 2011).

Remote sensing can significantly contribute to provide a timely and accurate picture of the agricultural sector. Remote sensing is probably also the most cost-efficient means for gathering timely, detailed, and reliable information over large areas with high revisit frequency (Table 4.2).

Remote sensing techniques are particularly well suited for assessing the two components of crop production (GEO, 2013):

1. Yield (e.g., Doraiswamy et al., 2005; Zhang et al., 2005; Bernardes et al. 2012; Duveiller et al., 2013; Meroni et al., 2013a; Mulianga et al., 2013; Rembold et al., 2013)
2. Acreage (e.g., Gallego, 2004; Fritz et al., 2008; Galford et al., 2008; Pittman et al., 2010; Boryan et al., 2011; Mello et al., 2013ac)

Moreover, the remotely retrieved information permits decision makers to better anticipate the effects of (disastrous) climatic events (predicted to increase in strength and frequency) and to get an objective and unbiased spatial picture over large areas (for risk assessment). Remotely sensed data can also be used as baseline information to provide cost-efficient (index-based) insurance schemes stimulating investments of smallholder farmers (De Leeuw et al., 2014). By putting the current situation in a historical context, an agricultural monitoring system permits better understanding of the possible effects of climate change (for preparedness and mitigation) and identification of areas with the highest yield potential—a prerequisite for closing the huge yield gaps in many parts of the world. In addition, crop phenological

information (Sakamoto et al., 2005; Shen et al., 2013), stress situations (Gu et al., 2007; Rembold et al., 2013), and disturbances (Zhan et al., 2002; Verbesselt et al., 2010) can be detected. Finally, remote sensing is also well suited for documenting the state of the land surface, and existing image archives provide ample material to study how agriculture changed over the past decades (Cousins, 2001; Dramstad et al., 2002).

4.2 Agricultural Challenges

4.2.1 Limiting the Environmental Impacts of Agriculture

Agriculture and natural resources are both under strong pressure. The main drivers are population growth, increasing consumption of calorie- and meat-intensive diets, and an increasing use of cropland for bioenergy production (Hill et al., 2006; FAO, 2009; Pelletier and Tyedmers, 2010; Foley et al., 2011).

The resulting negative impacts of current crop production are manifold and can be related to agricultural expansion and intensification (Foley et al., 2011; Tilman et al., 2011):

- Biodiversity is threatened by land clearing and habitat fragmentation (Dirzo and Raven, 2003).
- Greenhouse gas (GHG) emissions from land clearing, crop production, and fertilization contribute already to one-third of global GHG emissions (Burney et al., 2010).
- Global nitrogen and phosphorus cycles have been disrupted, with impacts on water quality, aquatic ecosystems, and marine fisheries (Vitousek et al., 1997; Canfield et al., 2010).
- Freshwater resources are depleted, as nearly 80% of freshwater currently used by humans is for irrigation (Postel et al., 1996; Thenkabail et al., 2009; Thenkabail, 2010).

4.2.2 Coping with Increasing Global Food Demand

As demonstrated by Tilman et al. (2011), on a global scale, per capita food demand is closely related to per capita gross domestic product (GDP). For example, people in the richest countries (group A—the United States, for instance) consume roughly 8000 kcal day^{-1} compared to an average consumption of 4000 kcal day^{-1} for people in groups C and D (Brazil and Indonesia, respectively).

Assuming that the GDP and global population will continue to increase in the future, the past trend of strongly increasing food demand is expected to last for three to four decades. Tilman et al. (2011), for example, project that per capita demand for crops will double between 2005 and 2050. Following these assumptions, the strongest increases (in absolute values) are predicted within economic groups C–E (Figure 4.3).

Based on this and other forecasts, most agronomists and international food organizations, such as the Food and Agriculture Organization of the United Nations (FAO), agree that food production must grow substantially for meeting the world's future food security and sustainability needs. At the same time,

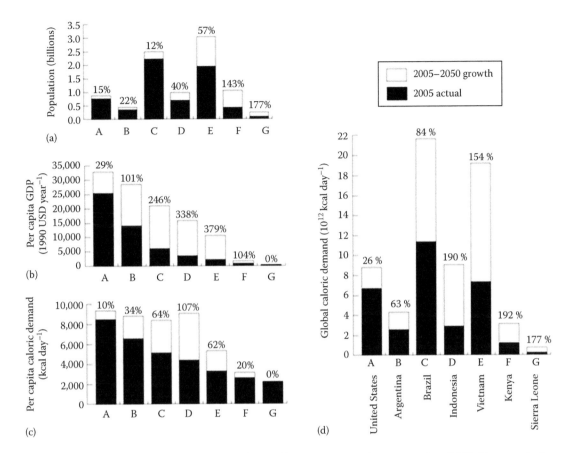

FIGURE 4.3 Increasing global food demand: (a) Global population, (b) per capita gross domestic product (GDP), (c) per capita demand for crop calories, and (d) global demand for crop calories. Indicators for 2005 in black and projected 2050 increases (white; percent increases above bars). Nations were assigned to economic groups A–G based on their rankings per capita GDP (average for 2000–2007). Group A had the highest and group G had the lowest per capita GDP. The predictions are based on forecasts of per capita real income in 2050 (b) and by a projected 2.3 billion person increase in global population (a). The predictions within each economic group also take into account shifts in food demand and quality with increased per capita real GDP and assume that GDP increases by 2.5% annually between 2000 and 2050. (Modified from Tilman, D. et al., *Proc. Natl. Acad. Sci. USA*, 108, 20260, 2011.)

agriculture's environmental footprint must shrink dramatically (The Royal Society, 2005; Godfray et al., 2010; Foley et al., 2011).

Hence, in the coming decades, a crucial challenge for humanity will be meeting future food demands without undermining further the integrity of the earth's environmental systems (Mueller et al., 2012). The necessary transformation will have to take place in times of climate change, adding supplementary difficulties (Jones and Thornton, 2003; Trnka et al., 2014). For example, it is expected that temperature and precipitation patterns will change in the next decades, with more frequent extreme meteorological conditions (IPCC, 2007, 2013; Godfray et al., 2010). The necessary agricultural transition phase should be monitored at various temporal and spatial scales.

4.2.3 Pathways for Increasing Agricultural Production

The environmental impacts of an increased global crop production will depend on how this increase is pursued (Foley et al., 2011; Tilman et al., 2011). Production could be increased by

agricultural extensification or intensification. Extensification implies clearing or adapting additional land for crop production. Intensification, on the other hand, achieves higher yields through increased inputs, improved agronomic practices (e.g., drop irrigation), improved crop varieties, and other innovations.

According to Tilman et al. (2011), the "land sparing trajectory" (i.e., intensification) to an increased global production is the preferred solution, as closing the yield gap would minimize both land clearing and GHG emissions, compared to a continuation of current practices of extensification in the poorer countries ("past trend trajectory"). The yield gap is here defined as the difference between realized productivity and the best that can be achieved using current plant material. On a global scale, huge differences in yield gap exist, exemplified in Figure 4.4 for cereals.

This view on intensification is also shared by Foley et al. (2011). Their analysis showed how many calories could be produced by closing existing yield gaps (Figure 4.5 top). In some countries, additional calories could be produced by allocating a higher fraction of the cropland to growing food crops (crops that are directly

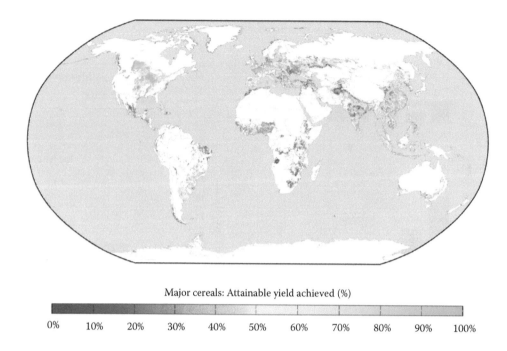

Major cereals: Attainable yield achieved (%)

0%   10%   20%   30%   40%   50%   60%   70%   80%   90%   100%

FIGURE 4.4   Average yield gaps for major cereal crops: corn, wheat, and rice. The yield gap is the differences between the potential yield and the realized yield at a given location. (From Mueller, N.D. et al., *Nature*, 490, 254, 2012.)

consumed by people) instead of using this land for animal feed, bioenergy crops, fibers, etc. (Figure 4.5, center and bottom).

If adopted, the proposed land sparing trajectory could meet the 2050 projected global crop demand, while clearing *only* 0.2 billion ha of land globally (compared to 1.0 billion ha from "past trend trajectory") and producing global GHG emissions of *just* 1 Gt·year⁻¹ (instead of 3 Gt·year⁻¹). In particular, Foley et al. (2011) suggested that tremendous progress could be made by simultaneously adopting the following five strategies:

1. Halting agricultural expansion
2. Closing yield gaps on underperforming lands
3. Increasing cropping efficiency
4. Shifting diets to less meat demanding ones
5. Reducing waste within the agricultural production chain

Together, these five strategies could double food production, while greatly reducing the environmental impacts of agriculture. Similar conclusions are drawn by Godfray et al. (2010), promoting a "multifaceted and linked global strategy" to ensure sustainable and equitable food security.

4.3 Remote Sensing for Assessing Yield and Biomass

Agricultural vegetation develops from sowing to harvest as a function of meteorological driving variables (e.g., temperature, sunlight, and precipitation). Plant growth is further modified by soil and plant characteristics (genetics) as well as farming practices. As changes in crop vigor, density, health, and productivity affect canopy optical properties, crop development and growth can be monitored remotely (Jones and Vaughan, 2010).

The relationship between the spectral properties of crops and their biomass/yield has been recognized since the very first spectrometric field experiments. For example, Tucker and coworkers showed already in the early 1980s that an agricultural crop can be monitored through its spectral reflectance properties (Tucker, 1979; Tucker et al., 1980). The use of spectral data was studied extensively by using satellite imagery after the launch of the first civil earth observation satellite (Landsat-1). However, only since the early 1980s, with the growing availability of low-spatial-resolution images from the advanced very-high-resolution radiometer (AVHRR) sensor on board of meteorological satellite series known as National Oceanic and Atmospheric Administration (NOAA), similar analyses have been extended to large areas, including many countries in arid and semiarid climates (Johnson et al., 1987; Hutchinson, 1991). Thanks to their large swath width, low-resolution systems have a much better synoptic view and temporal revisit frequency compared to high-spatial-resolution sensors. The intrinsic drawback of these sensors is, of course, related to their low spatial resolution, with pixel sizes of about 1 km², that is, far above typical field sizes. As a consequence, recorded spectral radiances are mostly composed by mixed information from several surface types. This seriously complicates the interpretation (and validation) of the signal, as well as the reliability of the derived information products. Several approaches for deriving sub-pixel information exist, but reveal serious limitations (Foody and Cox, 1994; Atkinson et al., 1997; Busetto et al., 2008; Atzberger and Rembold, 2013).

Grassland productivity for large areas, such as the Sahel region, was investigated by using AVHHR images by Tucker et al. (1983) and Prince (1991a,b). Other studies were made to move directly to the prediction of grain yield instead of total biomass

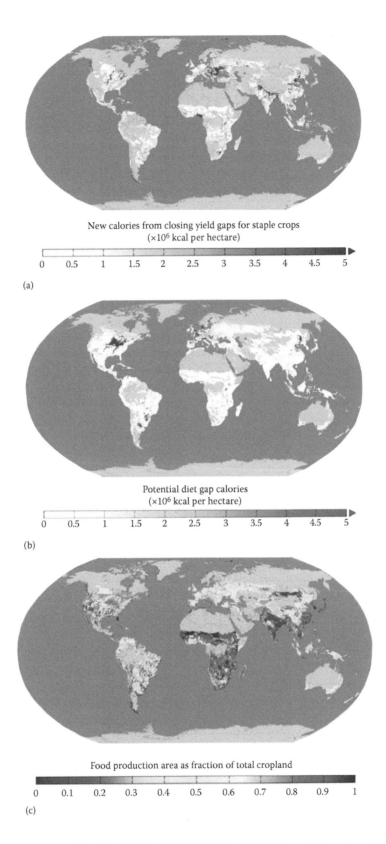

FIGURE 4.5 Pathways for increasing agricultural production: (a) Additional calories that could be produced by closing current yield gaps of crops, (b) increased food supply (in calories) by shifting crops to 100% human food and away from current mix of uses, and (c) fraction of cropland that is allocated in 2000 to growing food crops (crops that are directly consumed by people) versus all other crop uses, including animal feed and bioenergy crops. (From Foley, J.A. et al., *Nature,* 478, 337, 2011.)

by using field measured radiances (Tucker et al., 1981), Landsat images (Pinter et al., 1981; Barnett and Thompson, 1983), and finally NOAA AVHRR normalized difference vegetation index (NDVI) (Quarmby et al., 1993). With the increasing popularity of low-resolution satellite images for monitoring large geographic areas, an early warning of water stress as indicator for lowered final productivity became a well-established practice (Henricksen and Durkin, 1986; Johnson et al., 1987; Maselli et al., 1993). Both at national and regional levels, experimental crop monitoring systems were put in place starting in the late 1970s in the United States with the Large Area Crop Inventory Experiment (LACIE) and continuing in the 1980s in the EU with the Monitoring Agriculture with Remote Sensing (MARS) project. In many cases, these systems led to operational services that are still in existence today.

Nowadays, a much larger range of satellite sensors regularly provides data covering a wide spectral range (from optical through microwave) and using both active and passive devices (Belward and Skøien, 2014). Data are acquired from various orbits and in different spatial and temporal resolutions. For analyzing the recorded images, and for deriving the sought information, a large number of analysis tools have been developed (Macdonald and Hall, 1980; Verstraete et al., 1996; Justice et al., 2002). Besides analyzing the recorded spectral and temporal signatures (e.g., Badhwar et al., 1982; Lobell and Asner, 2004; Wardlow et al., 2007; Udelhoven et al., 2009; Vuolo and Atzberger, 2012; Mello et al., 2013b), one can also analyze the directional reflectance properties of vegetation (e.g., Clevers et al., 1994; Barnsley et al., 1997; Gobron et al., 2002; Vuolo et al., 2008; Koukal and Atzberger, 2012; Schlerf and Atzberger, 2012). Further useful information can be retrieved from the spatial arrangement of the pixels, that is, the texture of the image (Vintrou et al., 2012) as well as object size and association (Blaschke, 2010).

For the remainder of this chapter, we distinguish five main groups of techniques for mapping crop biomass and yield estimation. The five groups also summarize the evolution from purely qualitative to more quantitative and process-based approaches and hence—in some way—the history of agricultural remote sensing:

1. Qualitative crop monitoring
2. Regression modeling
3. Application of Monteith's efficiency equation
4. Assimilation of remote sensing data into (mechanistic and dynamic) crop growth models (CGMs)
5. Data mining (DM) approaches

Not surprisingly, some techniques can be seen as partially belonging to two different groups, while other methods may not strictly fit into any of these major subdivisions. However, the adopted simplification is believed to help the reader distinguish the main broad approaches that can be found in this field.

4.3.1 Qualitative Crop Monitoring

Crop monitoring methods that are based on the qualitative (or comparative) interpretation of remote sensing–derived indicators are in the following summarized under the term "qualitative crop monitoring." In general, these methods are based on the comparison of the actual crop status to previous seasons or to what can be assumed to be the average or "normal" situation. Detected divergences (or "anomalies") are then used to draw conclusions on possible yield limitations.

For qualitative crop growth monitoring, a large number of remotely sensed vegetation indices or biophysical products have been used. Most studies, however, used the NDVI for studying agricultural (and natural) vegetation. The usefulness of vegetation characterization by using arithmetic combinations of vegetation reflectances in different spectral bands (so called "vegetation indices") was established in the early 1980s by Tucker, Deering, and coworkers (Deering, 1978; Tucker, 1979; Tucker et al., 1980) that proposed the NDVI, using the red and near-infrared reflectances. The NDVI became subsequently the most popular indicator for studying vegetation health and crop production using qualitative approaches. The NDVI has been later demonstrated to hold a close relation to the canopy leaf area index (LAI) and fraction of absorbed photosynthetically active radiation (fAPAR) (Baret and Guyot, 1991; Prince, 1991a). Due to its almost linear relation with fAPAR, the NDVI can therefore also be seen as an indirect measure of primary productivity.

Low-resolution satellites are best suited for regional to continental monitoring of vegetation using this technique as they offer a high temporal revisit frequency with an extended geographical coverage at low data costs per unit area.

Crop monitoring systems making use of "anomaly maps" are particularly useful in arid and semiarid countries, where temporal and geographic rainfall variability leads to high interannual fluctuations in primary production and to a large risk of famines (Hutchinson, 1991). These environmental situations, along with the wide extent of the areas to monitor and the generally poor availability of efficient agricultural data collection systems, represent a scenario where qualitative monitoring can produce valid information for releasing early warnings about possible crop stress. Such systems are typically used in many food-insecure countries by FAO, FEWS-NET (Famine Early Warning System) of United States Agency for International Development (USAID), and the MARS project of the European Commission.

However, qualitative crop monitoring is not necessarily linked to an early warning context in arid areas but can also be very useful to get a quick overview of vegetation stress for large areas in temperate climatic zones. An example is given in Figure 4.6, which depicts vegetation condition index (VCI) anomalies (May–July) for 2009, 2010, and 2011 in Central Europe. VCI (Kogan, 1995) scales the NDVI value of a given 10-day period (dekad) within its min–max range as derived from the historical archive of observations for that dekad. France and Germany reported low cereal yields in 2011 and good yields in 2009 (with intermediate values in 2010). This is well reflected in the 3-monthly VCI values (May–July) for these two countries.

Vegetation performance anomaly detection with low-resolution images continues to be a fundamental component of agricultural (and drought) monitoring systems at the regional scale.

NDVI anomalies (May–July) over Central Europe
(2009–2011)

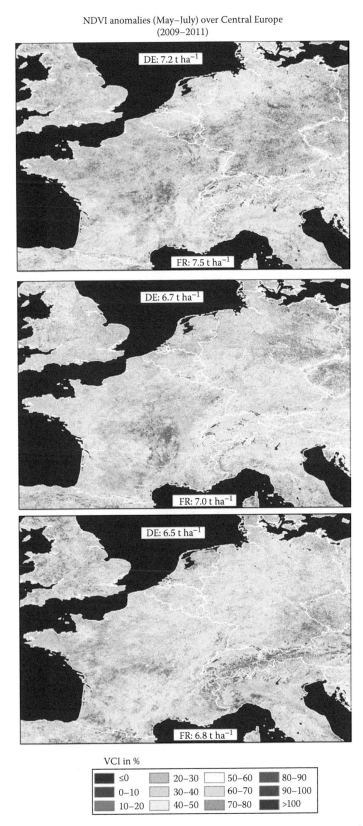

VCI in %

■ ≤0	■ 20–30	□ 50–60	■ 80–90
■ 0–10	■ 30–40	■ 60–70	■ 90–100
■ 10–20	□ 40–50	■ 70–80	■ >100

FIGURE 4.6 NDVI anomalies (3-monthly VCI) from 2009 (a), 2010 (b), and 2011 (c) over central Europe. The displayed VCI values are from filtered and gap-filled moderate-resolution imaging spectroradiometer (MODIS) data and always refer to the period from May to end of July (12 weeks). Cereal yields (in t/ha) according to World bank are reported for France (FR) and Germany (DE). (From data.worldbank.org/indicator/AG.YLD.CREL.KG and own data.)

For applications at more detailed scales, the limitations created by the mixed nature of low-resolution pixels are being progressively reduced by the higher resolution offered by new sensors (Belward and Skøien, 2014). However, the continuity of existing systems remains crucial for ensuring the availability of long time series as needed by the majority of the yield prediction methods used today (Rembold et al., 2013).

4.3.2 Crop Yield Predictions Using Regression Analysis

In the previous section, approaches have been described using (low-resolution) satellite imagery for providing qualitative indications of crop growth (e.g., crop growth worse/better than average). In this section, two methods will be described that quantify the expected yield (e.g., in t/ha) using regression models. In contrast to the qualitative approaches, the regression approaches must necessarily be calibrated using appropriate reference information. In most cases, agricultural statistics and, specifically, crop yield are used as reference information. Of course, this prerequisite limits its applicability in many regions of the world. We will distinguish purely remote sensing–based approaches

and mixed approaches where additional bioclimatic predictor variables are used.

The already mentioned relationship between vegetation indices/fAPAR and biomass enables the early estimation of crop yield, since yield of many crops is mainly determined by the photosynthetic activity of agricultural plants in certain periods prior to harvest (Baret et al., 1989; Benedetti and Rossini, 1993). As fAPAR and NDVI are linearly linked, NDVI is often used as an independent variable in empirical regression models to estimate final crop grain yield (the dependent variable).

The basic assumption of this method is that sufficiently long and consistent time series of both remote sensing images and agricultural statistics are available. The latter are normally aggregated at the level of subnational administrative units, for which average NDVI values can be extracted. At the aggregation stage, it can be decided if pixels are weighted or not according to crop coverage. Examples of NDVI/yield regressions for cereals at national level are shown in Figure 4.7.

Many studies reported useful statistical relationships using NDVI values at the peak of the growing season and final crop yield. The different empirical techniques appear to be relatively accurate for crops with low final production because biomass

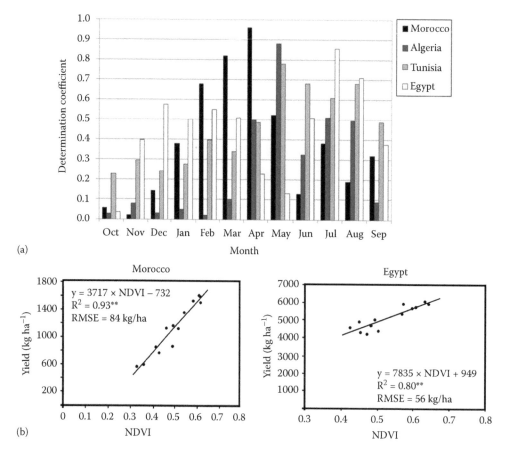

FIGURE 4.7 NDVI/yield linear regressions for cereals in North Africa. (a) Evolution of the coefficient of determination (R^2) between NDVI and yield over time. (b) Scatter plots between NDVI and cereal yield for Morocco (left) and Egypt (right). Each dot corresponds to the annual yield for agricultural areas at national level and to the monthly NDVI best correlated to yield. (Modified from Maselli, F. and Rembold, F., *Photogramm. Eng. Remote Sens.*, 67, 593, 2001.)

is the limiting factor to yield, and the relationship between LAI and the vegetation response (NDVI) is below the range of saturation (Delécolle et al., 1992). Empirical relationships also appear to be relatively accurate for grass crops, where dry matter (DM) is the harvestable yield.

Linear regression models relating NDVI to crop yield have, for example, been developed by Rasmussen (1992) and Groten (1993) for Burkina Faso and by Maselli et al. (1993) for Niger. The same and other investigations showed that yield forecasting can be obtained by the use of NDVI data of specific periods, which depend on the eco-climatic conditions of the areas and the types of crop grown (Hayes and Decker, 1996; Lewis et al., 1998; Maselli et al., 2000).

It has to be noted that the correlation between crop yield and spectral measurements varies during the growing season, and regression coefficients show strong temporal variations (Rudorff and Batista, 1990a,b). Established relationships are therefore, to some degree, "good fortune" and usually time and site specific (Baret et al. 1989). In cases where the aboveground biomass is not the harvestable yield, one has also to consider that the relation between crop yield and spectral data is only indirect (Hayes and Decker, 1996). Besides classical (multiple) linear regression, other statistical techniques such as partial least square regression or principle component regression may be more appropriate to model the relation between the sought variable(s) and the spectral reflectances (Hansen and Schjoerring, 2003; Nguyen et al., 2006; Atzberger et al., 2010).

Various authors postulated that accumulated radiometric data are more closely related to crop production than instantaneous measurements. Several choices of temporal NDVI integration can be found, reaching from the simple selection of the maximum NDVI value of the season, to the average of the peak values (plateau) to the sum of the total NDVI values of the total crop cycle. A recent example for winter wheat yield estimation at national level is provided by Meroni et al. (2013a) for Tunisia. Instead of using a fixed integration period, the integral is computed between the start of the growing period and the beginning of the descending phase. The two dates are computed for each pixel and each crop season separately.

Pinter et al. (1981) argued that the accumulation of radiometric data was similar to a measure of the duration of green leaf area. They consequently related yield of wheat and barley to an accumulated NDVI index and obtained satisfactory results. However, their results reveal that the performance of the integration is only optimum if it starts at a specific phenological event (i.e., at heading stage). When the optimum data could not be specified accurately, predictions were less accurate.

For the area of North America, Goward and Dye (1987) showed that an integrated NDVI from NOAA AVHRR gave a good description of the produced biomass. Tucker et al. (1983) found a strong correlation between the integrated NOAA-7 NDVI data and end-of-season aboveground dry biomass for ground samples collected over a 3-year period in the Sahel region. The correlation was higher than the one obtained from instantaneous NDVI values.

A less used technique involves the concept of aging or senescence, first developed by Idso et al. (1980). Idso and coworkers found that yield of wheat could be estimated by an evaluation of the rate of senescence as measured by a ratio index following heading. The lower the rate of senescence, the larger the yield as stressed plants begins to senesce sooner. The same technique was later applied by Baret and Guyot (1986). They confirmed that final yield production in winter wheat was correlated with the senescence rate. More recently, Koaudio et al. (2012) used LAI trajectories during the senescence phase to estimate wheat yield in the European Union. Examples of regression-based yield prediction studies are summarized in Table 4.3.

One important limitation of the regression approach is (as for any other empirical approach) that most of the mentioned studies are linked to the environmental characteristics of specific geographic areas or are limited by the availability of large and homogeneous data sets of low-resolution data. A common problem in crop monitoring and yield forecasting in many countries of the world is the difficulty in extending locally calibrated forecasting methods to other areas or to other scales.

One should also note that where the crop area is not known, the NDVI/yield relationship does not provide information on final crop production, which is what many users of crop monitoring information are ultimately interested in.

In many cases, the predictive power of remotely sensed indicators can be improved by adding independent meteorological (or bioclimatic) variables into the regression models. Several bioclimatic variables have proven to be highly correlated with yield for certain crops in specific areas (Lewis et al., 1998; Rasmussen, 1998; Reynolds et al., 2000). These variables can be measured either directly (like rainfall coming from synoptic weather stations) or by satellites (like rainfall estimates) or can be the result of other models as it is normally the case of agro-meteorological variables like ETa (actual evapotranspiration) or soil moisture.

Potdar et al. (1999) observed that the spatiotemporal rainfall distribution can be successfully incorporated into crop yield models (in addition to vegetation indices), to predict crop yield of different cereal crops grown in rain-fed conditions. Such hybrid models often show higher correlation and predictive capability compared to models using solely remote sensing indicators (Manjunath et al., 2002; Balaghi et al., 2008) as the input variables complement each other. The bioclimatic variables introduce information about the environmental drivers of vegetation growth (e.g., solar radiation, temperature, air humidity, and soil water availability), whereas the spectral component introduces information about the actual growth outcome of such drivers, thus indirectly taking into account crop management, varieties, and other stresses not directly considered by the agro-meteorological models (Rudorff and Batista, 1990a). However, it must be noted that many bioclimatic indicators, especially if they are derived from satellites as well, are not really independent from vegetation indices. The interrelation of the different input variables should be considered and corrected when integrating bioclimatic and spectral indicators into multiple regression models.

TABLE 4.3   Examples of Regression-Based Yield Prediction Studies

Target Crop	Yield Data Aggregation	Geographic Location	Predictor	Sensor	Crop Mask Used	Specific Processing	Regression Type	R^2	Reference
Sugarcane	Field-level data	Brazil	NIR/RED ratio	MSS	Not applicable	Single image before harvest	Linear	0.50–0.69[a]	Rudorff et al. (1990a)
Millet	Field-level data	Burkina Faso	Dekadal NDVI	NOAA-AVHRR	Not applicable	NDVI integration during a fixed reproductive phase	Linear	0.93[b]	Rasmussen (1992)
Millet and sorghum	FAO stats (subdistrict level)	Niger	Dekadal NDVI	NOAA-AVHRR	No	Single and fixed date standardized NDVI, standardization of yield	Linear	0.28–0.72[b]	Maselli et al. (1993)
Millet	Official stats (provincial level)	Burkina Faso	Dekadal NDVI	NOAA-AVHRR	No	Single-date NDVI, integration, multi-dekad multiple regression	Linear and quadratic	Up to 0.87[b]	Groten (1993)
Wheat	Official stats (sub-provincial level)	Italy	Dekadal NDVI	NOAA-AVHRR	No	NDVI integration during a fixed grain filling period	Linear	0.52[b]	Benedetti and Rossini (1993)
Maize	Official stats (production district level)	United States	Weekly NDVI	NOAA-AVHRR		Transformation into VCI (vegetation condition index)	Quadratic	0.54[b]	Hayes and Decker (1996)
Maize	Official stats (district level)	Kenya	Dekadal NDVI	NOAA-AVHRR	No	Annual maximum NDVI	Linear	0.56[a]	Lewis et al. (1998)
Millet	Field-level data	Senegal	9-Day NDVI (plus environmental and climatic data)	NOAA-AVHRR	Not applicable	NDVI integration during a fixed grain filling period	Linear (and multi-linear)	0.76[b] (0.88[b])	Rasmussen (1998)
Millet and sorghum	FAO stats (subdistrict level)	Niger	Dekadal NDVI, corrected for background effect	NOAA-AVHRR	Yield-masking approach (see Section 3.6)	Standardization of NDVI and yield, selection of the contiguous 3 dekads having maximum correlation with yield	Multi-linear	0.62[b]	Maselli et al. (2000)
Main cereals	FAO stats (country level)	North Africa	Monthly NDVI composite	NOAA-AVHRR	Yield-masking approach (see Section 3.6)	Selection of the monthly NDVI composite having maximum correlation with yield	Linear	0.65–0.93[a]	Maselli and Rembold (2001)
Maize	Official stats (province-level production)	Kenya	Deakadal NDVI (plus modeled meteo data, and water balance model output)	SPOT-VGT	Yes	Area fraction cover weighting of NDVI, NDVI integration during a fixed growing season	Multi-linear	0.81[a], 0.83[b]	Rojas (2007)
Wheat	Official stats (province level)	Marocco	Dekadal NDVI (plus rainfall and air temperature)	NOAA-AVHRR	Yes	Selection of best yield predictor combination	Stepwise regression	0.97[a]	Balaghi et al. (2008)
Wheat	Official stats (province level)	Belgium and northern France	GAI (Green area index)	MODIS	Yes	Maximum GAI value and phenology metric	Multi-linear	0.70–0.72[a]	Koaudio et al. (2012)
Wheat	Official stats (district level)	Tunisia	Dekadal FAPAR	SPOT-VGT	Yes	Cumulative FAPAR value during growing season (as from phenology retrieval)	Panel regression model	0.77[a]	Meroni et al. (2013a)

[a] Cross-validated.

[b] Fitting.

Rasmussen (1998) used multiple regression models by introducing environmental information such as grazing pressure, density, and percentage of cultivated land, and arrived to explain 88% of the millet grain yield variance. Rojas (2007) used the actual evapotranspiration (ETa) calculated by the FAO CWSB model and the CNDVI as independent variables in a regression analysis in order to estimate maize yield in Kenya during the first cropping season. CNDVI and ETa combined explained 83% of the maize crop yield variance with a root mean square error (RMSE) of 0.33 t/ha (coefficient of variation of 21%). The optimal prediction capability of the independent variables was 20 and 30 days for the short and long maize crop cycles, respectively. If validated over long time series, such models are expected to be utilized in an operational way.

Although linear regression modeling is likely the most common method to produce yield predictions by using remote sensing–derived indicators together with bioclimatic information, this is not the only one. Numerous other methods have been developed that include, for instance, similarity analysis and (nonlinear) neural networks (NN; Stathakis et al., 2006).

4.3.3 Use of Monteith's Efficiency Equation

Remotely sensed images were first proposed in the 1980s for assessing and mapping the crop's assimilation potential. One of the first steps in this direction was the introduction of the Monteith's light-use efficiency equation (Monteith, 1972, 1977). In this approach, it is assumed that the biomass production can be described as the simple multiplication of three variables: the incident photosynthetically active radiation (PAR, 400–700 nm); the PAR fraction, which is actually absorbed by the vegetation layer (fAPAR), and finally ε_b, the energy to DM conversion factor.

The approach has a sound physiological basis as the biomass production of a crop is linearly related to the amount of photosynthetically active solar radiation (PAR) absorbed (Tucker and Sellers, 1986). Other important climatic and ecological factors (e.g., temperature conditions and water/nutrient availability) controlling actual photosynthesis can be used to modulate the ε_b.

The amount of radiation available to the photosynthetic process is the absorbed solar radiation (APAR) and is a function of the incoming PAR and the crop's PAR interception capacity, fAPAR:

$$fAPAR = \frac{APAR}{PAR} \qquad (4.1)$$

fAPAR depends mainly (but not solely) on the leaf area of the canopy (Baret et al., 1989). Generally, an exponential relation between LAI and fAPAR is admitted:

$$fAPAR = fAPARmax \, (1 - \exp(-k \times LAI)) \qquad (4.2)$$

with fAPARmax between 0.93 and 0.97 and extinction coefficient k between 0.6 and 2.2 (Baret et al., 1989). The close link between fAPAR and LAI also explains why so many studies attempt mapping leaf area (Guérif and Delécolle, 1993).

Similarly, a close link between NDVI and fAPAR has been confirmed from both theoretical considerations and experimental field studies (Myneni and Williams, 1994). The studies agree that a quasi-linear relation between NDVI and fAPAR can be assumed:

$$fAPAR = a + b \times NDVI \qquad (4.3)$$

Most studies reviewed by Atzberger (1997) found a slope (b) between 1.2 and 1.4 and an intercept (a) between −0.2 and −0.4. The negative intercept reflects the fact that the NDVI of bare soils (i.e., fAPAR = 0) is often between 0.2 and 0.4.

The relation between fAPAR and NDVI is not surprising because PAR interception and canopy reflectance/NDVI are functionally interdependent as they both depend on the same factors (Baret, 1988; Baret et al., 1989). The main factors determining PAR interception and canopy reflectance/NDVI are (in the order of decreasing importance) (1) LAI, (2) leaf optical properties (especially leaf pigment concentration), (3) leaf angle distribution, (4) soil optical properties, and (5) the sun–target–sensor geometry.

The mechanism by which the incident PAR is transformed into DM can be written as (Steinmetz et al., 1990)

$$\Delta DM = PAR \times fAPAR \times \varepsilon_b \qquad (4.4)$$

with

ΔDM is the net primary production (NPP) (g·m^{-2}·day^{-1})

PAR is the incident photosynthetically active radiation (MJ·m^{-2}·day^{-1})

fAPAR is the fraction of incident PAR that is intercepted and absorbed by the canopy (dimensionless)

ε_b is the light-use efficiency of absorbed photosynthetically active radiation (g·MJ^{-1}).

When calculated over the entire growth cycle—and in the absence of growth stresses—the light-use efficiency (ε_b) is relatively constant for crops like winter wheat (with a value of about 2.0 g·MJ^{-1}) (Baret et al., 1989). However, the light-use efficiency is not constant when calculated over small periods of the growth cycle (Steinmetz et al., 1990; Leblon et al., 1991). The short-term variability of the light-use efficiency is a result of temperature, nutrient, and water conditions that eventually can lead to plant stress.

Remotely sensed data can be well used in Monteith's efficiency equation (Equation 4.4) if one manages to map the seasonal cycle of fAPAR (i.e., if enough images are available so that the full temporal fAPAR profile can be reconstructed). Incident PAR must be also known (e.g., from meteorological stations) or estimated (e.g., using general circulation model as done by ECMWF, using meteorological satellite observations as in Roerink et al., 2012). As explained, at the same time, the light-use efficiency (ε_b) must either be relatively constant/known or should be assessed using other remote sensing inputs (e.g., from thermal data revealing plant stress).

Provided that enough images are available, the seasonal integration of radiometric measurements theoretically improves the

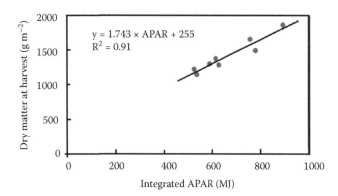

FIGURE 4.8 Linear relation between the seasonally integrated absorbed PAR (from sowing to harvest) and dry matter at harvest (g·m⁻²). Each point corresponds to one commercial winter wheat plot (n = 9). (From Atzberger, C., *Estimates of Winter Wheat Production through Remote Sensing and Crop Growth Modelling: A Case Study on the Camargue Region,* Verlag für Wissenschaft und Forschung, Berlin, Germany, 1997.)

capability of estimating biomass compared to one-time measurements, since the approach is based on sound physical and biological theory, whereas the relationship between instantaneous measurements of canopy reflectance and biomass is mainly empirical (Baret et al., 1989). For example, Figure 4.8 shows the close correspondence between seasonally integrated absorbed PAR (fAPAR × PAR) and the DM at harvest for nine commercial winter wheat plots in the Camargue region of France (Atzberger, 1997).

Nowadays, fAPAR is routinely assessed using various approaches and algorithms (Verstraete et al., 1996; Gobron et al., 2002; Baret et al., 2013) and applied to different sensors (VGT, MODIS, AVHRR, and others). Likewise, operational NPP products based on Monteith's formula are available.

Monteith's efficiency equation has been further extended to include, for example, temperature dependency of photosynthesis and respiration. For example, VITO (Eerens et al., 2004) uses the following formula for the NPP calculation:

$$\Delta DM = PAR \times fAPAR \times \varepsilon_b \times p(T) \times CO_2 fert \times (1 - r(T)) \quad (4.5)$$

where

ΔDM is the increase in DM or NPP (g·m⁻²·day⁻¹)

PAR is incident photosynthetically active solar radiation (MJ·m⁻²·day⁻¹)

fAPAR is fraction of intercepted and absorbed PAR calculated by means of a linear equation from NDVI (dimensionless)

ε_b is photosynthetic efficiency (g·MJ⁻¹)

p(T) is normalized temperature dependency factor (dimensionless)

CO_2fert is normalized CO_2 fertilization factor [85] (dimensionless)

r(T) is fraction of assimilated photosynthesis consumed by autotrophic respiration; r is modeled as a simple linear function of daily mean air temperature

Hence, compared to Equation 4.4, ε_b is reduced/increased as a function of temperature and CO_2 content to mimic the earlier-mentioned plant reactions to changing growth conditions. Similar approaches are often used in NPP approaches (Goward and Dye, 1987).

To calculate final yield (Y) from Equation 4.4 or Equation 4.5, it has to be assumed that a portion of the cumulated biomass at the end of the growing season (the harvest index, HI) is the harvestable yield, that is,

$$Y = HI \times \sum_{sowing}^{harvest} \Delta DM \quad (4.6)$$

The HI may be obtained by traditional regression analysis between primary production and statistical crop yields.

A number of studies found that the use of cumulated DM over the crop growing period gives more reliable results compared with NDVI for crop yield forecasting in many Mediterranean and Central Asian countries. For corn, Gallo et al. (1985), for example, found that the cumulated daily absorbed PAR explained 73% of the variance in the observed grain yield. The absorbed PAR was computed from the daily incident PAR and fAPAR predicted from NDVI. Only 56% and 58% of variance were accounted by the cumulated LAI and cumulated NDVI, respectively. Similarly, Meroni et al. (2013a) found that the cumulative value of APAR during the growing season explained 80% of the wheat yield variability in Tunisia. Several other studies using this technique are summarized in Table 4.4.

The main disadvantage of models based on Monteith's efficiency equation relates to their need for complete series of fAPAR information from sowing to harvest. Such information is currently provided (at the necessary temporal frequency) only at coarse spatial resolution. Of course, these data are often too coarse to resolve, for example, individual fields.

4.3.4 Remote Sensing Data Assimilation into Dynamic Crop Growth Models

The approaches described in the previous sections aimed either to qualitatively assess vegetation vigor (by comparing observed vegetation greenness against the "normal" situation) or to quantitatively estimate the crop yield using semiempirical regression techniques.

In this section, we will introduce a group of techniques involving modeling of crop physiology including feedback mechanisms. Approaches in this group of techniques are also known as crop growth modeling, soil–vegetation–atmosphere transfer (SVAT) modeling, or agro-meteorological modeling.

As defined by Delécolle et al. (1992), crop growth modeling involves the use of mathematical simulation models formalizing the analytical knowledge previously gained by plant physiologists. The models describe the primary physiological mechanisms of crop growth (e.g., phenological development,

TABLE 4.4   Examples of Light-Use Efficiency-Based Yield Estimation Studies

Target Crop	Yield Data Aggregation	Geographic Location	Variables and Sources	Crop Mask Used	LUE Specification	R^2	Reference
Maize	Field level data	United States	VIs from field spectroscopy, PAR from field measurements	N.A.	Integration during the whole season, constant ε_b	0.73[a]	Gallo et al. (1985)
Rice (two cultivars)	Field-level data	France	VIs from field spectroscopy, PAR from meteorological station	N.A.	Integration during the growing season with variable ε_b	~0.80[a]	Leblon et al. (1991)
Wheat	Field-level data	France	fAPAR from SPOT-HRV, PAR from meteorological station	N.A.	Integration during the whole season, constant and fixed ε_b and HI	0.76[a]	Atzberger (1997)
Wheat, rice, cotton, and sugarcane	Official stats (district level)	Pakistan	AVHRR NDVI, linearly scaled to FAPAR, AVHRR surface temperature	Yes	Integration during the whole season, ε_b modulated using temperature and water constraints	Relative RMSE[b]: 26%–49%	Bastiananssen and Ali (2003)
Wheat	Official stats (province level)	Italy	AVHRR NDVI, linearly scaled to FAPAR	Yes	Integration during the whole season, constant ε_b, HI derived from NDVI	0.73–0.77[b]	Moriondo et al. (2007)
Wheat	Official stats (district level)	Tunisia	Dekadal SPOT-VGT FAPAR, PAR from ECMWF model	Yes	Integration during the growing phase of the season, constant ε_b	0.80[b]	Meroni et al. (2013a)
Maize and soybeans	Official stats (county level)	United States	MODIS GPP estimates	Yes	Integration during the growing season, fixed ε_b and HI	0.66–0.77[b]	Xin et al. (2013)

[a] Fitting.
[b] Cross-validated.

photosynthesis, DM portioning, and organogenesis), as well as their interactions with the underlying environmental driving variables (e.g., air temperature, soil moisture, and nutrient availability) using mechanistic equations (Delécolle et al., 1992). Importantly, state variables (such as phenological development stage, biomass, LAI, and soil water content) are updated in a computational loop that is usually performed daily (Guérif and Delécolle, 1993) (Figure 4.9).

In the computational loop (Figure 4.9), model state variables such as development stage, organ dry mass, and LAI are linked to environmental driving variables such as temperature and precipitation, which are usually provided with a daily time step (Delécolle et al., 1992). Soil and plant parameters are used to mimic the plant's reaction to these driving variables. Whereas model state variables are constantly updated within the computational loop, model parameters remain unchanged during the simulation run (e.g., soil texture information). All state variables have to be initialized at the beginning of the simulation run.

It worth noting that Monteith efficiency equation (described in the previous section) lacks the computational loop and feedbacks included in CGMs and is therefore not a dynamic model, albeit it represents a physical description of the growth process.

CGMs are excellent analytical tools because they exhibit three distinct characteristics that distinguish them from the previously described approaches (Delécolle et al., 1992):

1. They are dynamic in that they operate on a time step for ordering input data and updating state variables.

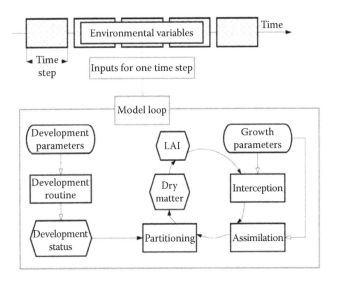

FIGURE 4.9   Simplified scheme of a crop process model. Model state variables such as development phase, organ dry mass, or leaf area index are linked to input variables, including weather, geographic, and management variables. (Modified from Delécolle, R. et al., *ISPRS J. Photogramm.*, 47, 145, 1992.)

2. They contain parameters that allow a general scheme of equations to be adopted to the specific growth behavior of different crop species.

3. They include a strategy for describing phenological development of a crop to order organ appearance and portioning/division of photosynthetic products.

TABLE 4.5   Techniques Used for Assimilating Remote Sensing Data in Dynamic Crop Growth Models

Technique	Example
Recalibration or re-parameterization of CGM	Maas (1988a,b, 1992), Bouman (1992, 1995), Clevers and van Leeuwen (1996), Launay and Guérif (2005), Clevers et al. (1994), Guérif and Duke (1998)
Reinitialization of CGM	Bach and Mauser (2003), Nouvellon et al. (2001), Guérif and Duke (1998), Doraiswamy et al. (2003)
Forcing of CGM	Maas (1988a,b), Bouman (1995), Clevers et al. (2002)
Updating of CGM	Bach and Mauser (2003), Pellenq and Boulet (2004)

Note that the provided examples sometimes combine two techniques. The "forcing" technique has been added to the table to complete the list, albeit it is not strictly speaking an assimilation technique.

The first CGMs were developed by the end of World War II (Sinclair and Seligman, 1996). In subsequent decades, they became both more complex and potentially more useful (Boote et al., 1996). Deterministic CGMs have been validated for cereals, as well as for potato, sugar beet, oilseed, rice, canola, and sunflower. Most of these models include water and energy balance modules and run on a daily time basis over the whole life cycle of a crop. Prominent models are, for example, CERES (Jones and Kiniry, 1986), WOFOST (Supit et al., 1994), OILCROPSUN (Villalobos et al., 1996), CROPSYST (Stöckle et al., 2003), and STICS (Brisson et al., 1998). Some simpler models (without water and energy balance) such as SAFY (Duchemin et al., 2008) and GRAMI (Maas, 1992) also exist. More sophisticated models attempt to integrate numerous factors that affect crop growth and development, such as plant available soil water, temperature, wind, genetics, management choices, and pest infestations. Currently, attempts are made to permit the integration and combination of various submodels from different model developers describing a specific plant behavior (e.g., phenology) (Donatelli et al., 2010).

The strength of CGMs as research tools resides in their ability to capture the soil–environment–plant interactions, but their initialization and parameterization generally require a number of physiological and pedological parameters that are not easily available. In addition, given the high model parameterization, careful validation strategies have to be employed for obtaining the required predictive power (Bellochi et al., 2010).

CGMs are covered here in some detail because CGM and remote sensing nicely complement each other: CGMs provide a continuous estimate of crop growth over time, while remote sensing provides temporally discontinuous but spatially detailed pictures of crop actual status (e.g., LAI) within a given area (Clevers and van Leeuwen, 1996; Guérif and Duke, 2000; Doraiswamy et al., 2003, 2004; Padilla et al., 2012). The complementary nature of remote sensing and crop growth modeling was first recognized by S. Maas from USDA, who described routines for using satellite-derived information in mechanistic crop models (Maas, 1988ab).

Remotely sensed images are particularly useful in spatially distributed modeling frameworks (Moulin et al., 1998; Weiss et al., 2001; Running and Nemani, 1988). In this case, all model inputs and parameters have to be provided in spatialized form. As remote sensing provides spatial status maps, the use of remotely sensed information makes the CGM more robust (Moulin et al., 1998; Guérif and Duke, 2000; Doraiswamy et al., 2003).

Spatialized information is readily available concerning many meteorological driving variables (e.g., from global circulation models like ECMWF). However, other parameters and initial conditions required by GCMs may not be available spatially, for instance, (1) soil, plant, and management parameters and (2) initial values of all crop state variables (Doraiswamy et al., 2003). In the following, we present different approaches for using remote sensing data to fill this gap in spatially distributed crop growth modeling (Table 4.5). All ideas are extracted from the outstanding paper of Delécolle et al. (1992). Useful overviews are also given in Moulin et al. (1998), Bach and Mauser (2003), and Dorigo et al. (2007).

In the most straightforward way, remote sensing may be used to parameterize and/or initialize CGMs. Hereafter, the term "parameterization" refers to the provision of model parameters required by crop growth and agro-meteorological models, for example, soil texture information, photosynthetic pathway information, crop type, and sowing date. The term "initialization" refers to the provision of model state variables at the start of the simulation (e.g., the soil water content at sowing).

For the purpose of parameterization or initialization, satellite imagery covering different wavelength ranges (i.e., optical to microwave) may be combined. In the simplest case, remotely sensed data are used to provide information about crop type. With known crop type, plant-specific parameter settings can be assigned (therefore the term parameterization). Optical imagery of bare soil conditions may be used to map soil organic matter content, soil texture, and soil albedo (Ben-Dor, 2002; Viscarra Rossel et al., 2009). These three model parameters are often used in CGMs as they influence nutrient release, water capacity, and radiation budget (Ungaro et al., 2005). Other imagery (e.g., microwave) may be used to provide an estimate of soil water content at the beginning of the simulation run, that is, at sowing (Wagner et al., 2007). This will result in a model initialization, as the state variable "soil water content" has been attributed a value for the start of the simulation.

In the "recalibration" or "re-parameterization" approach, one assumes that some parameters of the CGM are inaccurately calibrated, although the model as a whole is formally adequate (Delécolle et al., 1992). By providing "reference" observations of some key vegetation properties (e.g., remotely derived LAI), some crop model parameters can be calibrated (Figure 4.10), provided that such parameters do have an effect on the vegetation property as described by the model. This is usually achieved by (iteratively) adjusting the model parameters until measured and simulated temporal profiles of the selected variable (here LAI values) match

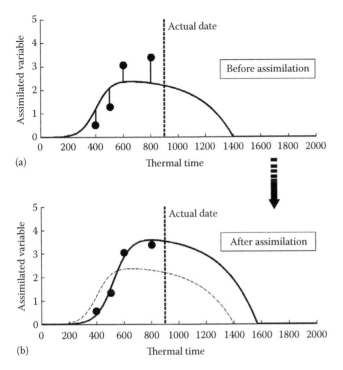

(a)

(b)

FIGURE 4.10 Schematic description of the recalibration method using remotely sensed state variables as inputs (here the assimilated variable corresponds to LAI). The crop growth model simulates the leaf development (LAI) over time. (a) Without assimilation, the simulated LAI is far from the four LAI observations. (b) After assimilation (e.g., after the nonlinear minimization procedure), new model coefficients are assigned to the crop growth model such that the residues between observed and simulated LAI are minimized. (Modified from Houles, V., Mise au point d'un outile de modulation intra-parcellaire de la fertilization azotee du ble d'hiver base sur la teledetection et un modele de culture, Thèse de Doctorat, Institut National Agronomique Paris-Grignon, France, 2004.)

each other (Doraiswamy et al., 2003). In spatially distributed modeling, this recalibration has of course to be done pixel by pixel.

The "reinitialization" of CGMs works in a very similar way; however, instead of adjusting model parameters, one simply tunes the initial values of state variables until a good match between observed and simulated state variables is obtained. In both cases, the remote sensing–derived variables are considered as an absolute reference for the model simulation. The exact timing of the remotely sensed observations is of minor importance. Already as few as one reference observations are useful (Atzberger, 1997; Launay and Guerif, 2005; Baret et al., 2007). However, the more satellite observations are available and the better they are distributed across the growing season, the more/better model parameters can be calibrated and/or initialized (Doraiswamy et al., 2003).

Alternatively, one may also choose to infer important state variables from remotely sensed data for each time step of the model simulation (e.g., LAI) for direct ingestion into the model, thus "forcing" the model to follow the remotely sensed information (Figure 4.11, left). Such a simplification makes CGMs very similar to the Monteith efficiency equation (Equation 4.4), as one breaks the computational loop in the model shown in Figure 4.10. As the model does no longer determine the values of that variable by itself, inconsistent model states may result (Delécolle et al., 1992).

In a very similar way, remotely sensed data are used in the "updating" of CGMs (Figure 4.11, right). One simply replaces simulated values of crop state variables by remotely sensed values each time these are available (not necessarily at each time step). The computations then continue with these updated values until new (remote sensing) inputs are provided. As for the "forcing" method, the replacement of simulated by observed state variables may result in inconsistent model states as one does not

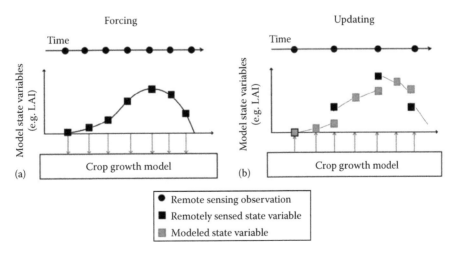

(a)

(b)

- ● Remote sensing observation
- ■ Remotely sensed state variable
- ▪ Modeled state variable

FIGURE 4.11 Schematic description of "forcing" and "updating" methods. (a) In the forcing approach, the complete time profile of a crop state variable (here: LAI) is reconstructed from remote sensing data and introduced (e.g., "forced") into the dynamic crop growth model at each time step in the simulation. (b) In the updating approach, the crop growth model is run (with standard parameter settings) until a remotely sensed state variable is available (in black). When new observations are available, the simulated state variable at this point is replaced by the remotely observed state variable and the crop growth simulation is continued without changing the parameter setting until another (if any) new observation becomes available. (Modified from Dorigo, W.A. et al., *Int. J. Appl. Earth Observ. Geoinform.*, 9(2), 165, 2007.)

correct for apparent errors in the model calibration, which are causing the differences between simulated and observed state variables.

4.3.5 Yield-Correlation Masking

One obstacle to successful modeling and prediction of crop yields using remotely sensed imagery is the identification of suitable image masks (Kastens et al., 2005). Problems and possible solution will be described hereafter with major ideas extracted from the outstanding paper of Kastens et al. (2005). According to this paper, image masking involves restricting the analysis to a subset of a region's pixels, rather than using all of the pixels in the scene. Cropland masking, where all sufficiently cropped pixels are included in the mask regardless of crop type, has been shown to generally improve crop yield forecasting ability. Doraiswamy and Cook (1995), for example, used 3 years of AVHRR NDVI imagery to assess spring wheat yields in North and South Dakota in the United States. They concluded that the most promising way to improve the use of AVHRR NDVI for estimating crop yields at regional scales would be to use better crop masks. This was also confirmed by Lee et al. (2000). They used a 10-year, biweekly AVHRR data set to forecast corn yields in the U.S. state of Iowa. They found that the most accurate forecasts of crop yield were made using accumulated NDVI and a cropland mask. Similarly, Maselli and Rembold (2001) found that application of cropland masks improved relationships between NDVI and final yield in four Mediterranean countries. For simplicity, in the remaining of this section, we will refer to NDVI although any other remote sensing indicator of vegetation biomass and vigor can be used (e.g., fAPAR, LAI, other VIs).

For crop yield forecasting, the ideal approach would be to use crop-specific masks. This would allow one to consider only the remote sensing information pertaining to the crop of interest. However, when such masking is applied to multiple years of imagery, several difficulties arise (Becker-Reshef et al., 2010b). A major problem relates to the widespread practice of crop rotation. In areas with crop rotation, a single and static crop-specific mask would not be appropriate. Instead, year-specific masks would be needed. For retrospective analysis, this implies the production of a crop mask for each year of interest, a challenging but still feasible task using the observed NDVI temporal profiles. However, in operational yield forecasting, this task presents even greater difficulties, as only incomplete growing season NDVI information is available. This is especially true early in the season, when the crop has low biomass and does not produce a large NDVI response.

A more feasible alternative to crop-specific masking is cropland masking, which refers to using pixels dominated by "arable land." The studies discussed earlier used this approach. Cropland masks usually are derived from existing land use/land cover (LU/LC) maps. If the area of interest was not subject to major land use changes with regard to cropland during the period of interest, a single mask can be applied. Albeit simpler to realize compared to crop-specific masking (i.e., one mask

per crop type and year), it has to be considered that all agricultural crops are now lumped in the general class of "cropland." Thus, crop-specific growth patterns are neglected.

To overcome the shortcomings related to cropland masking and crop-specific masking, Kastens et al. (2005) proposed a new masking technique, called yield-correlation masking. The main idea behind this concept is that all vegetated pixel in a region (i.e., crops and natural vegetation) integrate the season's cumulative growing conditions in some fashion. Hence, in the yield-masking approach, all pixels are considered for use in crop yield prediction. In practical terms, yield-correlation masking generates a unique mask for each NDVI variable (e.g., each time step at which NDVI is available) and each combined pair of crop x region. The technique is initiated by correlating each of the historical, pixel-level NDVI variable values with the region's yield history. The highest correlating pixels are retained for further processing and evaluation of the (NDVI) variable at hand. A diagram outlining this process for a single NDVI variable is shown in Figure 4.12.

Though computationally intensive, the yield-correlation masking technique overcomes the major problems afflicting crop-specific masking and cropland masking. Unlike these approaches, yield-correlation masking readily can be applied to low-producing regions and regions possessing sparse crop distribution. Also, since yield-correlation masks are not constrained to include pixels dominated by cropland, they are not necessarily hindered by the weak and insensitive NDVI responses exhibited by crops early in their respective growing seasons. Furthermore, once the issue of identifying optimal mask size (i.e., determining how many pixels should be included in the masks) is addressed, the entire modeling procedure becomes completely automated.

The most important appeal of yield-correlation masking is that no land cover map is required to implement the procedure, while the procedure results in forecasts of comparable accuracy to those obtained when using cropland masking or crop-specific masks (Kastens et al., 2005). The procedure requires only an adequate time series of imagery and a corresponding record of the region's crop yields. Problems regarding this approach can be expected when the land cover/land use of the selected yield proxies changes. In addition, the procedure used to select the pixels to be retained in the mask increases the parameterization of the final yield forecast model, so that its predictive power must be carefully scrutinized. The combined use of data sets from different sensors remains difficult, given the observed large discrepancies described, for example, in Meroni et al. (2013b).

A recent application of the yield-masking approach is presented in Mello et al. (2014), and described hereafter. The study is focused on the estimation of sugarcane yield at the municipal level in Brazil. For each municipality, yearly sugarcane yield data from 2003 to 2012 are available as reference information (IBGE, 2013).

To model yield, weekly smoothed and gap-filled MODIS NDVI time series (MOD13) from BOKU University (Austria) was used (Atzberger, 2015). A rectangular area was chosen so that a buffer of >60 km around the centroid of each of the five municipalities is covered.

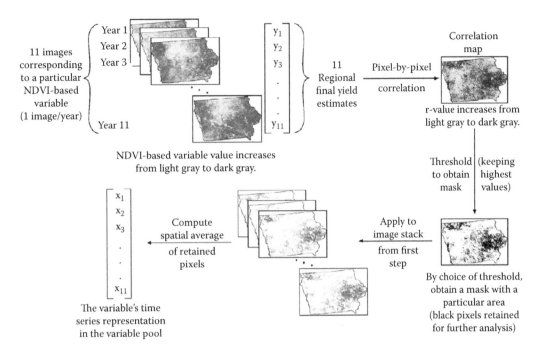

NDVI-based variable value increases
from light gray to dark gray.

The variable's time
series representation
in the variable pool

FIGURE 4.12 Illustration of the yield-masking approach involving a data set of 11 years. (From Kastens, J.H. et al., *Remote Sens. Environ.*, 99, 341, 2005.)

To simulate within-seasonal yield forecasts in near-real time for 2012, the official yield data from 2003 to 2011 were used to select, within the area covered by the MODIS time series, the 100 pixels where the NDVI time series best match the yield time series. These 100 pixels are called proxies. The quality of the match between official yield and remote sensing time series was based on the RMSE between the two variables. Before calculating the RMSE, both variables were standardized to zero mean and unit standard deviation.

Proxy selection was done in weekly intervals from January to June 2012, always considering the average of the last 10 weeks (starting from the week of interest) of standardized NDVI. From the selected proxies of each week, yield was estimated using the median of the (standardized) NDVI values. The official yield data for 2012 (not used to select proxies) were then used to assess the weekly estimates over the year 2012.

The weekly differences between official and modeled yield are shown (in % difference) in Figure 4.13 for each of the five municipalities as well as for all municipalities together; values above zero indicate overestimated yield.

The thick gray line, representing the average of the difference for the five municipalities tested, revealed that the remote sensing approach proposed showed an increasing overestimation of the sugarcane yield between February and June 2012. The development of sugarcane reaches its maximum in March (which is also the end of the rainy season), when sugar accumulation period usually ends. As sugarcane yield is strongly influenced by the sugar contents, this period is critical to define yield (Rudorff and Batista, 1990a). In fact, Figure 4.13 showed that the period from January to March presented the best estimates for remote sensing–based yield (when the differences

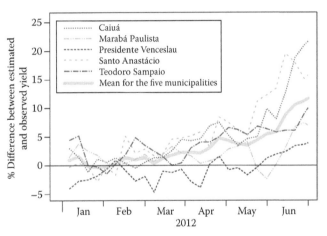

FIGURE 4.13 Difference (given in %) between the sugarcane yield estimated based on the yield-masking approach and the official yield published by IBGE (2013). (From Mello, M.P. et al., Near real time yield estimation for sugarcane in Brazil combining remote sensing and official statistical data, in *Proceedings of the 34rd IEEE International Geoscience and Remote Sensing Symposium (IGARSS 2014)*, IEEE, Montreal, Quebec, Canada (in press), 2014.)

between remote sensing–based and official yield were close to zero). From April to June, the remote sensing approach tended to overestimate sugarcane yield for all municipalities evaluated. The reduced performances of the model are justified by the fact that this period represents the start of the harvest season in São Paulo, which ends in December (Rudorff et al., 2010). Although remote sensing–based estimates in January were found to be already valuable to predict sugarcane yield with good accuracy for this particular year, in operational yield forecasting, it would be important to monitor yield throughout several months, especially until April, to spot possible reduction of the forecasted yield, since heavy rains as well as frost in some growing areas may affect the sugar concentration and, consequently, yield (Monteiro and Sentelhas, 2014).

4.4 Crop Acreage Estimation

Cropland areas are often characterized by a diverse mosaic of LULC types that change over various spatial and temporal scales in response to different management practices and agricultural policies (e.g., Galford et al., 2008; Epiphanio et al., 2010; Hostert et al., 2011; Kuemmerle et al., 2011; Atzberger and Rembold, 2013). As a result, detailed regional-scale cropping patterns need to be mapped on a repetitive basis (Wardlow et al., 2007; Atzberger and Rembold, 2012; Vieira et al., 2012). As described earlier, in many approaches, crop masks are also necessary for building yield models and for obtaining signals related to the class of interest (e.g., the actual crop being investigated). Some important applications requiring information about cropped surfaces and crop type area are listed in Table 4.6.

For example, information about crop extent (often referred to as "acreage") is necessary to better understand the role and response of regional cropping practices in relation to various environmental issues (e.g., climate change, groundwater depletion, soil erosion) that potentially threaten the long-term sustainability of major agricultural producing areas (Galford et al., 2008). Monitoring the time and location of land cover changes is important for establishing links between policy decisions, regulatory actions, and subsequent land use activities, as outlined by Galford et al. (2008). Determining the physical and temporal patterns of agricultural extensification or expansion and intensification is the first step in understanding their implications, for example, for long-term crop production and environmental, agricultural, and economic sustainability (Galford et al., 2008). The acreage of the different crops must also be known for each growing season for accurate production estimates (Gallego, 2005;

TABLE 4.6 Main Applications of Agriculture-Related LULC Maps Derived from Remote Sensing

Application	Example
Estimation of agricultural production	Bolton and Friedl (2013), Becker-Reshef et al. (2009), Becker-Reshef et al. (2010b), Pittman et al. (2010)
Monitoring of agricultural management practices	Brown and Pervez (2014), Han et al. (2012)
Monitoring of the effects of climate change on agriculture	Brink and Eva (2009), Romo-Leon et al. (2014)
Monitoring of agricultural policies	De Beurs and Henebry (2004), Bryan et al. (2009)
Assessment of the impact of agriculture on natural resources	Cardille and Foley (2003), Dale and Polasky (2007), Duro et al. (2007)
Crop masking	Becker-Reshef et al. (2010a), Kogan et al. (2013)
Unmixing of coarse-resolution satellite data	Oleson et al. (1995), Maselli et al. (1998), Busetto et al. (2008)

Baruth et al., 2008). The recent review paper of Olofsson et al. (2014) offers a guidance for accurate estimation and change in land use, whereas the excellent work of Wardlow et al. (2007) provides a valuable discussion of different approaches focused on crop mapping. The following is a summary of the latter work.

4.4.1 Crop Mapping Using High-Resolution Satellite Data

Remotely sensed data from satellite-based sensors have proven useful for large-area LULC characterization due to their synoptic and repeat coverage. Considerable progress has been made classifying LULC patterns using multispectral, high-resolution Landsat TM data as a primary input (Vogelmann et al., 2001).

In most cases, crop maps are generated by supervised classification (Congalton et al., 1998; Beltran and Belmonte, 2001). EO images for classification are generally acquired at key phenological stages for optimizing class separability. These approaches are labor and cost intensive, and require amounts of cloud-free high–spatial resolution imagery. This impedes an operational implementation over large areas and in multiple years (Lobell and Asner, 2004). Data availability—particularly if specific crop stages need to be imaged—is often insufficient (Annoni and Perdiago, 1997). Although generally feasible, the problems mentioned have limited the possibility of automatically updating land cover maps over large areas at regular (annual) intervals (Chang et al., 2007). For example, for the United States, it is expected that the Landsat-based National Land Cover Database (NLCD) will result in a 6-year delay between data collection and product availability (Lunetta et al., 2006). The NLCD of 2011, for example, became available at the time of writing (2014).

Conventional pixel-based procedures of digital classification occasionally reveal difficulties regarding the automatic pattern recognition, mainly because of the phenological variability of crops, different cropping systems, and nonuniform measurement conditions (e.g., atmospheric disturbances) (Vieira et al., 2012). This is particularly true in cases using only single-date imagery. In such a context, object-based image analysis (OBIA) using multi-temporal (satellite) imagery appears promising.

The most common approach used to generate image objects is image segmentation (Pal and Pal, 1993; Benz et al., 2004). The segmentation process is the subdivision of an image into homogeneous regions through the grouping of pixels in accordance with determined criteria of homogeneity and heterogeneity (Haralick and Shapiro, 1985; Comaniciu and Meer, 2002). For each object created in a segmentation process, spectral, textural, morphic, and contextual attributes can be generated and may be employed in image analysis (Blaschke, 2010). In very-high-resolution (VHR) images, textural and shape information is particularly important.

After the process of outlining objects in an image, the next step is to assign them to a certain class, by comparing objects identified in the image with patterns previously defined, thus performing the classification of image objects considering them thematically homogeneous. This is what is called object-oriented classification (Whiteside and Ahmad, 2005).

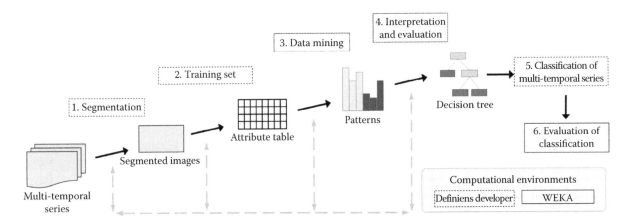

FIGURE 4.14 Flowchart illustrating the main stages that are part of the OBIA+ DM approach proposed to classify sugarcane areas (RH) from Landsat time-series images. Hachures illustrate the different computational environments used in each methodological stage. Broken-lined arrows indicate iteration possibilities. (From Vieira, M.A. et al., *Remote Sens. Environ.*, 123, 553, 2012.)

Having a sometimes huge set of attributes for automatic classification available, it may be very difficult for a human to identify the optimum descriptive attributes of the objects for a successful classification (Witten and Frank, 2005). In such cases, DM techniques can be employed. DM techniques enable the automatic generation of a structure of knowledge (Silva et al., 2008). Among the many available DM techniques, decision trees (DTs)—and ensemble of DTs called "random forest" (Breiman, 2001)—are particular easy and appealing. An important advantage of classification trees is that they are structurally explicit, allowing for clear interpretation of the links between the dependent variable of class membership and the independent variables of remote sensing and/or ancillary data (Lawrence and Wright, 2001). An operational example of the application of DT algorithm is the annual national-level coverage product of the United States (Cropland Data Layer; Boryan et al., 2011).

DTs have been preferred to statistical classifiers such as maximum likelihood classifier because they do not make implicit assumptions about normal distributions in the input data (Friedl and Brodley, 1997). As stated by Brown de Colstoun et al. (2003), DT classifiers can also accept a wide variety of input data, including non–remotely sensed ancillary data and in the form of both continuous and/or categorical variables. Further advantages of DTs include an ability to handle data measured on different scales, lack of any assumptions concerning the frequency distributions of the data in each of the classes, flexibility, and ability to handle nonlinear relationships between features and classes (Friedly and Brodley, 1997).

Peña-Barragán et al. (2011) successfully combined OBIA and DT methodology for identification of 13 major crops cultivated in the agricultural area of Yolo County (California, USA). They explored the use of several vegetation indices and textural features derived from visible, near-infrared, and short-wave infrared bands of ASTER satellite scenes collected during three distinct growing season periods. Their multi-seasonal assessment of a large number of crop types and field status reported an overall accuracy of 79%.

Brown de Colstoun et al. (2003) used multi-temporal ETM+/ Landsat-7 data and a DT classifier to map 11 types of land cover classes, acquiring a final land cover map with an overall accuracy of 82%. Grouping the 11 land cover classes in forest vs. non-forest classes, this same accuracy was 99.5%.

Vieira et al. (2012) combined OBIA and DT to map harvest-ready sugarcane in Brazil. To derive the binary map indicating the area of harvest-ready sugarcane, four Landsat images (TM-5 and ETM+) acquired between September 2000 and March 2001 were used. An overview of the methodology and processing steps is outlined in Figure 4.14. For image segmentation, eCognition software was used and resulted in a quite precise delineation of the different field boundaries (Figure 4.15). A large number of

FIGURE 4.15 Example of the segmentation result used by Vieira et al. (2012) for mapping harvest-ready sugarcane. The underlying RGB composite consists of a TM image taken in the month of February 2011, with composition R(4) G(5) B(3). (From Vieira, M.A. et al., *Remote Sens. Environ.*, 123, 553, 2012.)

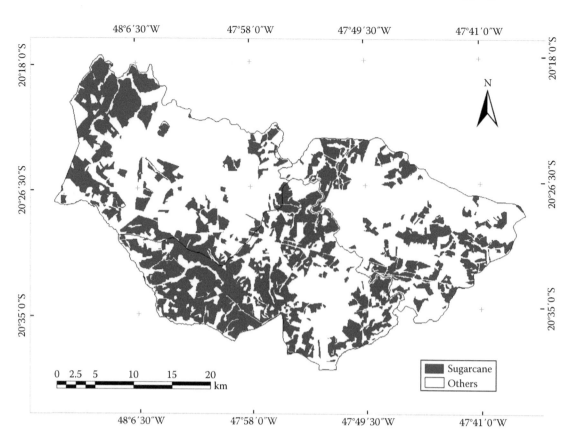

FIGURE 4.16 Map of harvest-ready sugarcane areas using and integrated OBIA and DT approach. (From Vieira, M.A. et al., *Remote Sens. Environ.,* 123, 553, 2012.)

attributes were afterward extracted for each polygon (object). The attributes included spectral, spatial, and textural features as described in Blaschke (2010).

The map shown in Figure 4.16, depicting the location of harvest-ready sugarcane fields, was derived by application of the trained DT. Validation using an independent set of 500 reference points not used during DT training yielded an overall accuracy of 94% (Kappa 0.84).

Interestingly, only a small set of features was selected by the DT for obtaining these good results (i.e., NDVI, spectral signatures, and one textural feature). As expected, multi-temporal information was necessary to differentiate between harvest-ready sugarcane and the other land uses. Textural attributes were relevant where and when areas with high-biomass sugarcane were confounded with other high-biomass areas (e.g., forests). Spatial attributes (e.g., shape, dimension) were not selected in this study area since most agricultural fields, both sugarcane and the other crops, had similar geometric characteristics.

4.4.2 Crop Mapping Using Medium- to Coarse-Resolution Satellite Data

At national to global scales, advances have been made in LULC classification using multi-temporal, coarse-resolution data such as SPOT-VGT or NOAA-AVHRR (Loveland et al., 2000; Defries

et al., 1998). The high temporal resolution of satellite time series data allows land cover types to be discriminated based on their unique phenological (seasonal) characteristics (de Fries et al., 1998; Vuolo and Atzberger, 2012, 2014). However, few of these mapping efforts have classified detailed, crop-related LULC patterns, particularly at the annual time step required to reflect common agricultural LULC changes (Wardlow et al., 2007). This is mainly due to the mixed nature of coarse-resolution pixels. For similar reasons, it is not surprising that existing LULC maps often reveal strong differences, making harmonization attempts necessary (Vancutsem et al., 2013). Indeed, most globally available land cover products reveal significant differences, even if maps are recoded into a few (broad) classes. This is exemplified in Figure 4.17 for Europe.

The observed differences constitute a real limitation for using remote sensing data. This is unfortunate, as coarse-resolution satellite data will constitute for a number of years the main input for regional-scale crop mapping and monitoring protocols. The situation will probably change significantly only with the launch of Sentinel-2 (scheduled for mid-2015) offering five-daily global revisit time at 10 m spatial resolution. Today, only the mentioned coarse-resolution data sets have wide geographic coverage and high temporal resolution. This is achieved at the expenses of their spatial resolution that, compared to the granularity of the landscape (i.e., typical field size), is often inadequate. Remotely

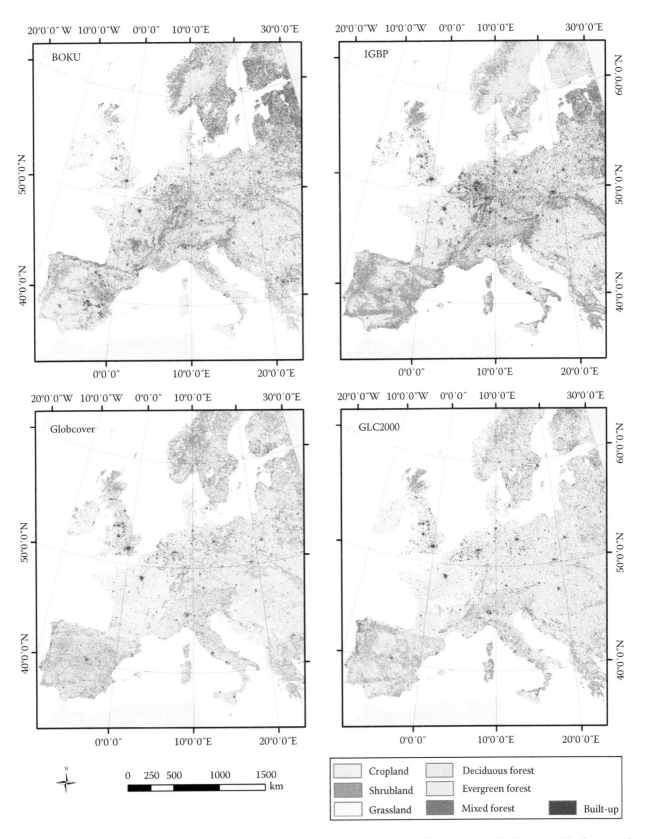

FIGURE 4.17 Different LULC maps of Europe demonstrating the large uncertainty in available information. The legends of the four maps have been harmonized to ease interpretation. (From Vuolo, F. and Atzberger, C., *Remote Sens., 4,* 3143, 2014.)

sensed data from traditional sources, such as Landsat, fulfill the spatial resolution needs, but are limited for such a protocol, due to their temporal resolution, availability, and/or cost.

For areas with relatively large field sizes, MODIS provides interesting data at 250 m spatial resolution in particular compared to (1–8 km) AVHRR and 1 km SPOT-VGT data. This offers an opportunity for a more detailed, large-area LULC characterization by providing global coverage with daily revisit frequency and intermediate spatial resolution (Justice et al., 2002). The data set is available at no cost, including 16-day composites of NDVI and enhanced vegetation index updated every 8 days. Several studies have already successfully demonstrated the potential of these data for detailed LULC characterization in an agricultural setting (Lobell and Asner, 2004; Wardlow et al., 2007; Lunetta et al., 2010).

Wardlow et al. (2007), for example, found that MODIS time series at 250 m ground resolution had sufficient temporal and radiometric resolution to discriminate major crop types and crop-related land use practices in Kansas, United States. For each crop, a unique multi-temporal VI profile consistent with the known crop phenology was detected. Most crop classes were separable at some point during the growing season based on their phenology-driven differences expressed in the VI temporal trajectory. Even regional intra-class variations were detected, reflecting the climate and planting date gradient in the study area. They also found that MODIS's 250 m spatial resolution was an appropriate scale at which to map the general cropping patterns of the U.S. Central Great Plains.

Lunetta et al. (2010) used MODIS 16-day NDVI composite data to successfully develop annual cropland and crop-specific map products (corn, soybeans, and wheat) for the Laurentian Great Lakes Basin. The crop area distributions and changes in crop rotations were characterized by comparing annual crop map products for 2005, 2006, and 2007.

Other studies also confirmed that crop area estimations were significantly improved since the introduction of the MODIS sensor with 250 m ground resolution (Lunetta et al., 2006; Chang et al., 2007; Fritz et al., 2008; Wardlow and Egbert, 2008). Not surprisingly, the best results were obtained for agricultural areas such as the central plains of the United States and the Don River basin in Russia, where typical field sizes are large. Of course, for other parts of the world with (much) smaller field sizes, the resolution of MODIS can be much less adequate.

4.4.3 Fractional Abundances from Medium- to Coarse-Resolution Satellite Data

The use of coarse-resolution data is effective because it offers numerous advantages: global coverage and low cost, high temporal frequency, easy processing at a regional to continental scale, availability of long-term records (e.g., from 1980 thanks to AVHRR instruments onboard of NOAA satellites), and finally continuity of data provision as ensured by several current (MODIS, Suomi-NPP, SPOT-VGT, Proba-V) and planned (Sentinel-3) missions. However, because of sub-pixel heterogeneity, the application of traditional hard classification approaches faces intrinsic methodological limitations and may result in significant errors in the estimated crop areas (Defries et al., 1995; Chang et al., 2007).

To address sub-pixel heterogeneity common for many areas of the world with fragmented landscapes, Quarmby (1992) used linear mixture model techniques applied to coarse-resolution data. Hansen et al. (2002) used the continuous field algorithm for mapping vegetative traits, such as tree cover, using MODIS data. In the continuous field approach, each coarse-resolution pixel is characterized as 0%–100% cover of a vegetation class, ameliorating the primary limitation of coarse spatial resolution data (Chang et al., 2007).

Several authors combined high-resolution images with NOAA AVHRR 1 km imagery to improve sub-pixel crop monitoring capabilities (Maselli et al., 1998; Doraiswamy et al., 2004). However, insufficient contrast between endmembers often leads to unstable solutions, resulting in inaccurate fraction images (Lobell and Asner, 2004). On the other hand, too few endmembers will fail to correctly represent the input signature.

A probabilistic linear unmixing approach with MODIS spectral/temporal data was developed and tested by Lobell and Asner (2004). The approach estimates sub-pixel fractions of crop area based on the temporal reflectance signatures throughout the growing season. In this approach, endmember sets are constructed using Landsat data to identify pure pixels, mainly located within large fields. Rather than defining endmembers with a single spectrum, endmembers are defined as a set of spectra that represent the full range of potential variability. The uncertainty in endmember fractions arising from endmember variability can then be quantified using Monte Carlo techniques. The performance of the proposed approach was assessed over Mexico and the Southern Great Plains and varied depending on the scale of comparison. Coefficients of determination ranged from greater 80% for crop cover within areas over 10 km^2 to roughly 50% for estimating crop area within individual MODIS pixels.

Several studies used spectral angle mapping (SAM) for measuring interannual crop area changes based on NDVI time series from NOAA-AVHRR (Rembold and Maselli, 2004, 2006). The studies found that it was feasible to derive relatively accurate interannual winter crop acreage changes for the region of Tuscany, Italy. However, good results could be obtained only by estimating the crop acreage changes of single years from the average of a high number (seven) of reference years. The results were significantly worse by using less or single years of reference data.

Regression tree analysis was used by Chang et al. (2007) for the percentage of the corn and soybean area mapping using 500 m MODIS time series data set. The strength of the regression tree is its use as a DM tool. Numerous phenological measures and data transformations may be input to such a model to identify which ones are the most useful for crop-type discrimination.

Verbeiren et al. (2008) used NNs and monthly maximum value compositing of SPOT-VGT (between March and October) to model the area fraction images (AFI) of eight classes in 2003 for Belgium. Relatively good results were obtained, especially if the initial (pixel-based) results were aggregated to coarser regional levels. The portability across growing seasons was investigated in an accompanying paper on the same data set by Bossyns et al. (2007). The NNs were trained on data of one growing season and then applied to SPOT-VGT of the training year plus three additional seasons. High and stable accuracies of the estimated AFIs were obtained for the training data. For example, at regional level, the R^2 for winter wheat of the training years was ~0.8 (0.67–0.87). On average, however, this value decreased by 0.45 units when the networks were applied to different seasons, probably because of a too high interannual variability of the endmembers.

To better cope with the natural year-to-year variability of NDVI profiles of vegetated surfaces, Atzberger and Rembold (2013) trained networks with AVHRR time series. The target variable represented the sub-pixel winter crop fractional coverage. To permit the net distinguishing for various proportions of non-arable land within the mixed pixels (e.g., forested areas and urban land), CORINE land cover information was used as additional input. A positive impact was demonstrated regarding the concurrent use of ancillary information. In-season predictions improved compared to the mentioned work of Rembold and Maselli (2004, 2006) using the same data set and linear prediction models. On average (median), 79% of the spatial variability of the (sub-pixel) winter crop abundances was explained by the NN approach (Figure 4.18).

For the individual years, the cross-validated R^2 ranged between 0.70 (1988) and 0.82 (2000). The cross-validated RMSE values were around ~10% (relative to the winter crop area). For the year 2000, Figure 4.19 shows the relation between the winter wheat area fractions independently derived from Landsat imagery and the modeled results.

The same approach was tested by Atzberger and Rembold (2012) for its portability across years and its usefulness to derive regional statistics. Data from 3 years between 1988 and 2001 were used to train the NN (Figure 4.20). The trained net was then applied to the period 2002–2009.

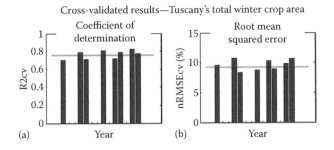

FIGURE 4.18 Spatialization of winter crop acreages (AFIw) using neural networks: cross-validated results showing (a) coefficient of determination, and (b) relative root mean squared error between predicted and Landsat TM/ETM+-derived reference information. Each bar corresponds to 1 year of the 8-year dataset. The horizontal red line corresponds to the median of the 8 years. (From Atzberger, C. and Rembold, F., *Remote Sens.*, 5(3), 1335, 2013.)

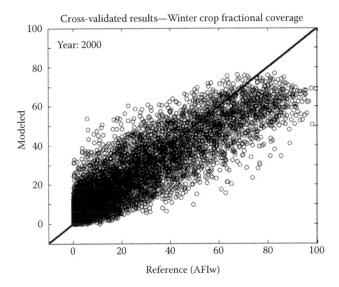

FIGURE 4.19 Scatterplot between reference winter crop abundances (AFIw) of Tuscany at 1 km scale (x-axis) and modeled results (y-axis). The modeled results were obtained in a leave-one-out approach. The reference information was obtained from Landsat images. (From Atzberger, C. and Rembold, F., *Remote Sens.*, 5(3), 1335, 2013.)

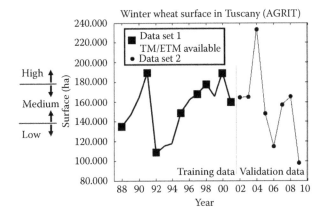

FIGURE 4.20 Reference winter wheat surface data for the region of Tuscany, Italy. Large squares indicate the 8 years for which high-resolution TM/ETM+ imagery is available (data set 1). The second data set (data set 2; shown as small dots) was used for assessing the portability of networks across growing seasons. For stratified sampling, data set 1 was split into three categories: low (1988, 1992), medium (1995, 1997, 1998, 2001), and high winter crop acreages (1991, 2000). (From Atzberger, C. and Rembold, F., *Eur. J. Remote Sens.*, 45, 371, 2012.)

Despite the fact that 2 years of the validation data set had (extreme) conditions not previously seen by the NN (e.g., with exceptionally high and low winter wheat areas, respectively), the NN performed remarkably well (Figure 4.21).

Other studies also used successfully NNs. For example, Atkinson et al. (1997) showed how NN can be used for unmixing single-date (five wavebands) AVHRR imagery to map sub-pixel proportional land cover. The use of NN for estimating sub-pixel land cover from temporal signatures was investigated by Karkee et al. (2009). Braswell et al. (2003) demonstrated that

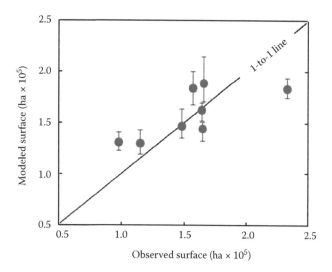

FIGURE 4.21 Estimated total winter crop acreage for Tuscany (Italy) obtained with sixteen neural nets trained with 3 years of available reference data (data set 1 of Figure 4.20) and applied to the time series (2002–2009) (data set 2). The red points and whiskers indicate the simulated averages and standard deviations across the sixteen individual nets. The 1-to-1 line is also drawn. (From Atzberger, C. and Rembold, F., *Eur. J. Remote Sens.*, 45, 371, 2012.)

network-based nonlinear regression offers significant improvement relative to linear unmixing for the estimation of sub-pixel land cover fractions in the heterogeneous disturbed areas of Brazilian Amazonia. The improvement was related to the fact that linear unmixing assumes the existence of pure sub-pixel classes (endmembers) with fixed reflectance signatures. The NN approach proposed by Braswell et al. (2003) estimates nonlinear relationships between each land cover fraction and spectral-directional reflectances, without making assumptions about the physics of sub-pixel mixing.

4.4.4 Combined Use of Satellite Data and Ancillary Information

The combined use of remotely sensed data and ancillary information was presented by Mello et al. (2013c) for the case of soybean mapping in Mato Grosso State, Brazil. The approach is based on a computer-aided Bayesian networks (BNs). These networks are able to incorporate experts' knowledge in complex classification tasks and therefore help to characterize phenomena through plausible reasoning inferences based on evidence. Mato Grosso State (total size of about 900,000 km²) was selected by Mello et al. (2013c) as a study region because it is the largest Brazilian soybean producer (about 30% of the total domestic production) and an important global hub for tropical agricultural production.

For Mato Grosso, tabulated agricultural statistics at municipality level exist, which are however released only with a delay of about 2 years after harvest. The absence of timely and spatial data restricts investigations related to crop monitoring and forecast. It also hinders the monitoring of the possible spread of soybean cropping into new, sometimes environmentally sensitive,

areas. As such, there is demand for the use of remote sensing images as an accurate, efficient, timely, and cost-effective way to monitor agricultural crops (Rudorff et al., 2010).

Bayes's theorem, which is used in BN, updates the knowledge (prior probability) of a specific event in light of additional evidence (conditional probabilities), allowing one to have a plausible reasoning based on a degree of belief (posteriori probability) (McGrayne, 2011). Therefore, observations made upon variables that are related to a particular phenomenon may be used to develop plausible reasoning about the phenomenon, its causes, and consequences (Jaynes, 2003). When the number of variables increases or even when the complexity of the interactions among the variables involved in a specific phenomenon rises, the BN is a representation suited to model and handle such tasks (Pearl, 1988; Jensen and Nielsen, 2007).

The joint probability of any instantiation (sometimes called realization) of all the variables in a BN can be computed as the product of only n probabilities. Thus, one can determine any probability of the form

$$P(V_1 \,|\, V_2, \ldots, V_n) \qquad (4.7)$$

where V_i are sets of variables with known values. This ability to compute posterior probabilities given some evidence is called inference. In the case of using Equation 4.7 for inferences about a phenomenon, Mello et al. (2013c) named "target variable" the variable that represents the phenomenon, and "context variables" the variables that are somehow related to the phenomenon.

Besides remotely sensed spectral and temporal information, several other context variables are closely related with soybean occurrence in a given field (e.g., soil type and distance to roads and other infrastructure facilities) (Garrett et al., 2013). In the mentioned study, this information was combined within a BN structure to optimize soybean identification and mapping.

The selected context variables used in the study of Mello et al. (2013c) to compose the model are listed in Table 4.7. From expert knowledge, it is known that each context variable influences soybean occurrence.

The resulting probability image (PI) is shown in Figure 4.22. The PI shows the spatial distribution of (the probability of) soybean crops throughout Mato Grosso territory in crop year 2005/2006. Green-colored pixels represent areas with higher probability of soybean presence based on observation of the context variables. Mello et al. (2013c) found a high agreement of the

TABLE 4.7 Summary of the Six Context Variables Used in the Soybean Mapping Case Study of Mello et al. (2013c)

Variable	Description
C	CEI[a] value in the current crop year (2005/2006)
L	CEI[a] value in the last crop year (2004/2005)
A	Soil aptitude
T	Terrain slope (given in %)
W	Distance to the nearest water body (given in km)
R	Distance to the nearest road (given in km)

[a] Crop Enhancement Index.

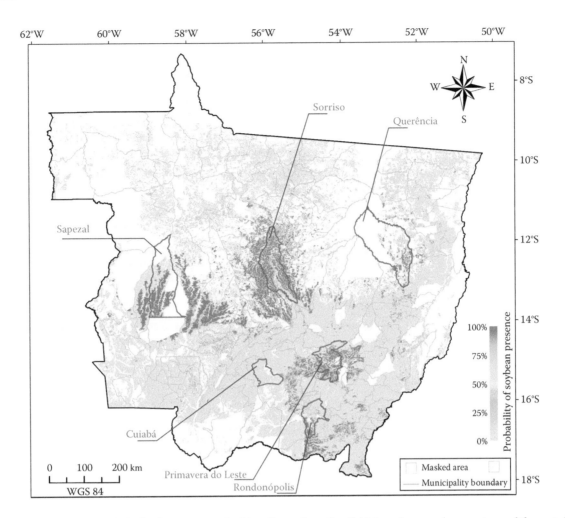

FIGURE 4.22 Probability image (PI) of soybean presence for Mato Grosso State, Brazil. Main soybean producer centers and the capital, Cuiabá, are highlighted. The color indicates the calculated probability of soybean presence in 2005/2006 given the observations made for the context variables. (From Mello, M.P. et al., Spatial statistic to assess remote sensing acreage estimates: An analysis of sugarcane in São Paulo State, Brazil, in *Proceedings of the 33rd IEEE International Geoscience and Remote Sensing Symposium (IGARSS 2013)*, IEEE, Melbourne, Victoria, Australia, 2013a, pp. 4233–4236; Mello, M.P. et al., *IEEE Trans. Geosci. Remote Sens.*, 51(4), 1897, 2013b; Mello, M.P. et al., *Remote Sens.*, 5(11), 5999, 2013c.)

mapped soybean acreage with (independent) official statistics. Moreover, the BN approach proposed by the authors quantified the influence of each context variable on soybean mapping, stating that remote sensing data were the most important variables used to infer about soybean occurrence.

4.4.5 Accuracy Considerations

As Olofsson et al. (2014) pointed out, "a key strength of remote sensing is that it enables spatially exhaustive, wall-to-wall coverage of the area of interest. However, as might be expected with any mapping process, the results are rarely perfect. Placing spatially and categorically continuous conditions into discrete classes may result in confusion at the categorical transitions. Error can also result from the mapping process, the data used, and analyst biases (Foody, 2010)." Map users on the other hand are acutely interested in understanding the quality of the provided maps (Olofsson et al., 2014).

The mentioned paper of Olofsson et al. (2014) provides excellent guidance on how to assess map accuracy in a consistent and transparent manner. An example of good practice is also provided. We therefore invite interested readers to consult this work as well as work published by Foody (2002), Strahler et al. (2006), Foody (2010), and Gallego (2012).

4.5 Crop Development and Phenology

The phenological dynamics of terrestrial ecosystems—both natural vegetation and agricultural crops—reflect the response of the earth's biosphere to inter- and intra-annual dynamics of the earth's climate and hydrologic regimes (Zhang et al., 2003). Example NDVI images of four different months during 2005 are shown in Figure 4.23. The maps that cover large parts of Europe are derived from MODIS data at 250 m ground resolution (Atzberger and Klisch, 2014). Also shown are selected temporal profiles extracted from the same data set.

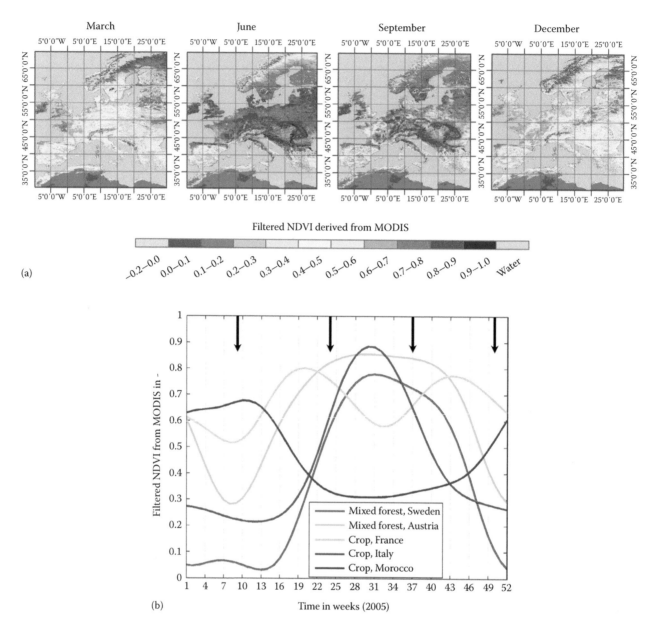

FIGURE 4.23 Temporal variability of vegetation biomass (NDVI) over Europe/Maghreb. (a) Observed spatial NDVI pattern throughout the year 2005 derived from MODIS. From left to right are shown: March (week 12), June (week 24), September (week 37), and December (week 50). (b) Example temporal NDVI profiles at weekly temporal resolution from 2005 for five randomly selected pixels. Black arrows indicate the timing of the NDVI values relative to the aforementioned displayed maps. (From Atzberger, C. and Klisch, A., 2014.)

Once again, moderate to coarse spatial resolution data possess significant potential for monitoring vegetation/crop dynamics for several reasons:

- Synoptic global coverage
- Frequent temporal sampling (e.g., daily)
- Short leap time (usually less than 3 days)
- Free and easy access

Using such (NDVI) time series, it is, for example, possible to extract (Figure 4.24) and monitor simple "phenological" events, such as the start (Figure 4.25) and peak of vegetation growth (Figure 4.26). For both land surface phenological events, a sometimes huge interannual variability can be observed (Figure 4.26, bottom).

Mapping of a crop's phenological development is important as the phenology is sometimes closely related to biomass production and crop yield (Meroni et al., 2014a). For example, cool summers may result in delayed heading and thus decreased yields. Besides, the temporal signature of vegetated surfaces is also useful for distinguishing land cover types and for mapping land use change (Badhwar et al., 1982; Wardlow et al., 2007; Galford et al., 2008; Vuolo and Atzberger, 2012). Several applications for land surface phenology (LSP) products are listed in Table 4.8.

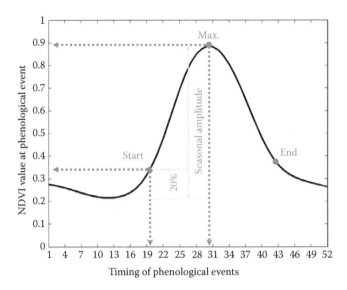

FIGURE 4.24 Basic phenological metrics that can be extracted from NDVI time series: start of season, maximum/peak of season, and end of season. For extracting the start of season (SOS), the relative threshold approach (20% of seasonal amplitude) is used and illustrated with gray lines. (From Atzberger, C. and Klisch, A., 2014.)

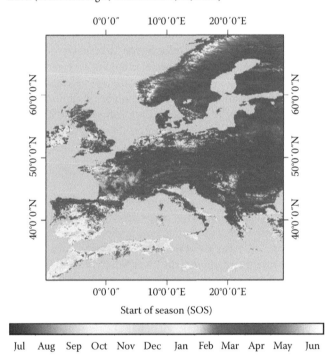

FIGURE 4.25 Spatial distribution of start of season (SOS) derived from MODIS time series in 2007 over Europe/Maghreb. Land pixels without vegetation and water surfaces are masked out (in gray). (From Atzberger, C. et al., 2014.)

This "multiple use" makes phenological metrics very interesting within agricultural monitoring systems. Croplands, for example, present a more complex phenology than natural land cover, due to their many peaks resulting from multiple crops planted sequentially within a growing season (Galford et al., 2008; Arvor et al.,

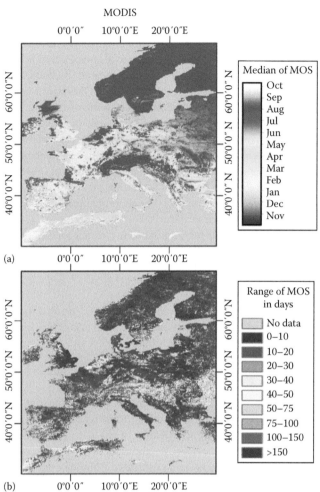

FIGURE 4.26 Spatial distribution of maximum of season (MOS) derived from MODIS time series. (a) averaged MOS (median) over all years (2003–2011) and (b) interannual range. Land pixels without vegetation and water surfaces are masked out (in gray). (From Atzberger, C. et al., 2014.)

TABLE 4.8 Applications of LSP Products Related to Agriculture

Application	Example
Mapping of crop development and conditions	Sakamoto et al. (2011), Kawamura et al. (2005)
Use in data assimilation approaches within crop growth models	Moulin et al. (1998), Boschetti et al. (2009)
Mapping of crop type and crop rotation	Moro and Manjunath (2013), Galford et al. (2008), Peña-Barragán et al. (2011)
Mapping of land cover and land use change	Galford et al. (2010)
Use as predictor in yield models	Xin et al. (2002), Meroni et al. (2013a), Bolton and Friedl (2013)

2011; Atkinson et al., 2012) (Figure 4.23, bottom). Additionally, the uniform cover of green leaves in an agricultural field creates very high observed greenness, especially as compared to the bare soils left after harvest. Consequently, several studies have identified land cover based on specific properties of the observed green leaf phenology, such as start of season, dry season minimums, and amplitude of maximums (e.g., Badhwar et al., 1982; Guérif et al., 1996; Zhang et al., 2003; Bradley et al., 2007; Galford et al., 2008). Other studies analyze the detected phenological indicators with respect to climate variables (Udelhoven et al., 2008) and/or run trend analysis (Jong et al., 2011). As CGMs have to order organ appearance and assimilation portioning/distribution, the phenological stage of a crop has to be simulated (mostly as a simple function of accumulated growing degree days since planting). Hence, externally provided information about crop emergence, etc., can be assimilated into such models (Moulin et al., 1998).

To determine the timing of vegetation green-up and senescence from remotely sensed VI time series, a number of different approaches have been developed. Following Beck et al. (2006), the different methods can be grouped in two categories:

1. Methods estimating the timing of single phenological events (Reed et al., 1994; White et al., 1997; Badeck et al., 2004)
2. Methods modeling the entire time series using a mathematical function (Jönsson and Eklundh, 2002; Stöckli and Vidale, 2004; Beck et al., 2006)

Approaches belonging to the two groups are summarized in Table 4.9. Relevant references are also given.

Modeling VI time series as such has the advantage of conserving a maximum amount of information in the VI data, while reducing the dimensionality of the data (Jönsson and Eklundh, 2004). Therefore, in addition to the phenological dates, other parameters can be estimated from the models' output (Beck et al., 2006). However, such methods are difficult to apply for large regions and generally do not apply well for ecosystems characterized by multiple growth cycles (e.g., double- or triple-cropping systems and semiarid systems with multiple

TABLE 4.9 Methods for Determining Land Surface Phenological Events from EO Data Such As Green-Up and Senescence

Methods timing single phenological events	
Use of specific (NDVI) thresholds	White et al. (1997), Lloyd (1990), Atzberger et al. (2014)
Detection of the largest (NDVI) increase	Kaduk and Heinmann (1996)
Use of backward-looking moving averages	Reed et al. (1994)
Methods modeling the entire time series	
Use of principal component analysis	Hirosawa et al. (1996)
Use of Fourier and harmonic analysis	Atkinson et al. (2012), Azzali and Menenti (2000), Jakubauskas et al. (2001)
Use of wavelet decomposition	Anyamba and Eastman (1996), Sakamoto et al. (2005)
Curve fitting (global)	Zhang et al. (2003), Beck et al. (2006), Jönsson and Eklundh (2002), Meroni et al. (2014b)
Curve fitting (local)	Zhang et al. (2003)

rainy seasons). This was demonstrated, for example, recently by Atkinson et al. (2012) over India.

The traditional Fourier transform, for example, expects periodicity in the data not always given (e.g., in the case of land use change). Additionally, application of Fourier transforms often reveals spurious oscillations (Hermance, 2007). This happens frequently when many harmonics have to be combined for fitting nontrivial temporal patterns (e.g., related to double/triple cropping).

Nonstationary data with irregular temporal shapes is better handled by the wavelet transform (Galford et al., 2008). In an agricultural application, wavelet-smoothed time series were successfully used to identify the start of the growing season and the time of harvest with relatively low errors (±2 weeks) (Sakamoto et al., 2005). Wavelet analysis is capable of handling the range of agricultural patterns that occur through time, as well as the spatial heterogeneity of fields that result from precipitation and management decisions, because the transform is localized in time and frequency.

Curve fitting using predefined functions (e.g., double logistic) is another approach modeling the entire time series (Badhwar et al., 1982; Guérif et al., 1992; Beck et al., 2006; Meroni et al., 2014b). A fitted curve simplifies the parameterization necessary for identification of metrics, such as start of season. In addition, data gaps are easily handled. A drawback of curve-fitting approaches is that *a priori* information is necessary to inform the algorithm about the number of cropping seasons within a 12-month period and the probable location of vegetation peaks (Jönsson and Eklundh, 2004). A large number of additional temporal features can be extracted using software like TimeStats (Udelhoven, 2011).

4.6 Existing Operational Agricultural Monitoring Systems

Agriculture monitoring is not a new concern. In fact, the basics of geometry and land surveying were developed in ancient Egypt (Luiz et al., 2011). The aim was assessing cultivated areas affected by water-level fluctuations of the River Nile, with the purposes of taxation and famine prevention.

Today, probably more urgently as ever before, a regional to global agricultural intelligence is needed to respond to various societal needs. For example, national and international agricultural policies, global agricultural trade, and organizations dealing with food security issues heavily depend on reliable and timely crop production information (Becker-Reshef et al., 2010a).

Agricultural monitoring systems should be able to provide timely information on crop production, status, and yield in a standardized and regular manner at the (sub)regional to the national level. Estimates should be provided as early as possible during the growing season(s) and updated periodically through the season until harvest. Based on the information provided, stakeholders are enabled to take early decisions and identify geographically the areas with large variation in production and productivity. The system should provide homogeneous and interchangeable data sets with statistically valid precision and accuracy. Probably, only (satellite) remote sensing—combined with sophisticated modeling tools—can provide such information in a timely manner, over large areas, in sufficient spatial detail and with reasonable costs (Macdonald and Hall, 1980).

The first agricultural monitoring system based on remote sensing data was developed in the United States in the 1970s (Pinter et al., 2003; Becker-Reshef et al., 2010a). In 1974, the USDA, together with NASA and NOAA, initiated LACIE (Bauer, 1979). The goal of this experiment was to improve domestic and international crop forecasting methods (Pinter et al., 2003). With enhancements that became available from the NOAA AVHRR sensor, the Agriculture and Resource Inventory Surveys Through Aerospace Remote Sensing program was initiated in the early 1980s (Pinter et al., 2003; Becker-Reshef et al., 2010a). At this stage, the NOAA AVHRR sensor allowed for the first time a daily global monitoring. Through the research conducted in these NASA–USDA joint programs, the considerable potential for use of remotely sensed information for monitoring and management of agricultural lands was identified.

One of the most recent efforts that NASA and the USDA Foreign Agricultural Service (FAS) have initiated is the Global Agricultural Monitoring (GLAM) Project (Becker-Reshef et al., 2009, 2010a). The GLAM project focuses on applying data from NASA's MODIS instrument to feed FAS decision support system needs (pecad.fas.usda.gov/).

Besides the GLAM system, there are currently several other regional to global operational agricultural monitoring systems providing critical agricultural information at a range of scales (Pinter et al., 2003; Becker-Reshef et al., 2010a; GEO, 2013; Rembold et al., 2013) (Table 4.10).

However, the USDA FAS with its GLAM system is currently the only provider of regular, timely, objective crop production

forecasts at a *global* scale. This unique capability is in part afforded by the USDA's partnership with NASA, providing global coverage of the earth observation data, as well as analysis tools for crop condition monitoring and production assessment at the global scale (Becker-Reshef et al., 2010a).

The GLAM project is also playing a leadership role in the Group on Earth Observations (GEO) agricultural monitoring component AG-07-03. GEO itself is part of Global Earth Observation System of Systems (GEOSS), providing decision-support tools to a wide variety of users. Recently, the GEOGLAM initiative was created integrating GLAM into GEOSS (Soares et al., 2011). The group defined observation requirements that remote sensing data should meet, ranging from meteorological conditions to area and yield estimates (Figure 4.27).

The graph exemplifies that different approaches (and different sensors) will be needed to access the requested information encompassing local to global scales. Many of the requested qualitative to quantitative information needs were covered within this chapter.

4.7 Conclusions and Recommendations

The chapter demonstrated the strong role remote sensing plays within the agricultural sector. The number of applications is huge. However, the most important applications focus on yield and area estimation. Such information is regularly needed for various decision makers. The information need is expected to increase in the future, as the agricultural sector is in a very dynamic phase (e.g., for meeting food requirements and environmental restrictions). Remotely sensed information can help to identify yield gaps and to monitor related agricultural practices. In parallel, environmentally sensitive areas can be identified for protective purposes. With appropriate preprocessing of time series (e.g., gap filling and smoothing), phenological indicators, such as start of the growing season, can probably be estimated with acceptable accuracy (e.g., 7–10 days). Vegetation anomalies related to local meteorological conditions (e.g., droughts) can be readily detected from space and combined with other data sources to identify stress-affected regions. This information not only is important for organizations dealing with food security, but can also help to identify a region's vulnerability to (drought) stress. Finally, the detection and monitoring of (permanent) land cover changes is best achieved using remotely sensed data. This is, for example, important for establishing links between policy decisions, regulatory actions, and subsequent land use activities.

Although we mainly described approaches using globally available (moderate to coarse resolution) data sets (plus some examples using Landsat-type data), it is clear that additional information can be derived from (very) high spatial resolution data (plus ground sensors). Thus, besides investments in the agricultural sector, the related monitoring components should be strengthened. Elements of the necessary monitoring component exist, but should be further integrated and consolidated.

TABLE 4.10 Major Global Agricultural Monitoring Systems Making Strong Use of Remotely Sensed Inputs

Name	Monitoring System	Access Points
GLAM	USDA (FAS) Global agricultural monitoring system	glam1.gsfc.nasa.gov/
FEWS-NET	USAID Famine Early Warning System	fews.net/
GIEWS	UN Food and Agriculture Organization (FAO) Global Information and Early Warning System	fao.org/giews/
MARS	JRC's Monitoring Agricultural ResourceS action of the European Commission	mars.jrc.ec.europa.eu/
GMFS	European Union Global Monitoring of Food Security program	gmfs.info/
CropWatch	Crop Watch Program at the Institute of Remote Sensing Applications of the Chinese Academy of Sciences (CAS)	cropwatch.com.cn/en/

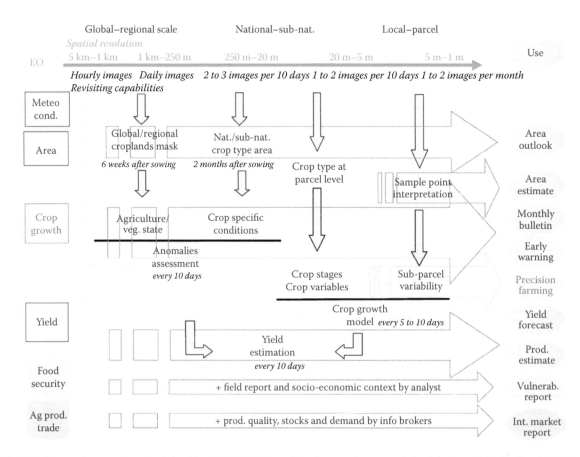

FIGURE 4.27　Observation requirements defined by GEOGLAM. (From http://wmp.gsfc.nasa.gov/uploads/science/slides/Justice_ASP-WR_2012-09-06.pdf.)

Similar to the objectives of the GEOGLAM initiative (Soares et al., 2011), the following recommendations are drawn (Atzberger, 2013):

- Agriculture depends strongly on the timeliness of the provided information. Information is worth little if it comes (too) late (FAO, 2011). Thus, the issue of timeliness should be dealt with in all developments.
- Product developers have only limited access to ground truth information to evaluate their products under various environmental settings. International efforts are needed to establish such networks of validation sites (Justice et al., 2000; Baret et al., 2006; Morisette et al., 2006; Olofsson et al., 2012; 2014; Stehman et al., 2012). This also requires substantial funding by space agencies and/or environmental institutions.
- More use should be made of crowd-sourced information (Fritz et al., 2009, 2012; Foody and Boyd, 2013; Foody et al., 2013). Interesting attempts are, for example, GeoWiki (Fritz et al., 2009), JECAM (jecam.org/), and USAID (Silversmith and Tulchin, 2013).
- Space agencies and sensor developers spend huge amounts of money for precise radiometric calibration of the deployed instruments. However, these efforts have little positive effect unless the much stronger radiometric distortions introduced by the atmosphere are removed. Operational

implementations of precise atmospheric correction algorithms are mandatory. Instead of relying on (aerosol) climatologies, the algorithms should be fed with local atmospheric properties (probably also derived from satellites).

- In the future, multisensor studies will become frequent. Thus, sensor intercalibration studies are urgently needed (Meroni et al., 2013b).
- Access of data and derived products is sometimes still too complicated. Efforts are necessary to permit users to visualize (and possibly download) information products in a very simple way (such as realized in Google Earth).
- Approaches are still very scattered and not always implemented in operational processing chains. Funding organizations should facilitate international cooperation, while limiting administrative burdens. With the new generation of (Sentinel) images, IT solutions are to be developed sustaining the processing of huge data sets as well as the cooperative development of algorithms, etc. (Wagner et al., 2014). For example, with Sentinel-2, a global coverage at 10 m spatial resolution and three-to-five daily updating frequency will be achieved. This amount of data cannot be handled any longer using traditional approaches.
- For potential users, the wide variety of products can be confusing. Efforts are necessary to clearly explain the purpose (and limits) of a given product.

References

Annoni, A., Perdigao, V. *Technical and Methodological Guide for Updating CORINE Land Cover Data Base.* EUR 17288EN. Ispra, Italy: European Commission, 1997.

Anyamba, A., Eastman, J.R. Interannual variability of NDVI over Africa and its relation to El Niño/Southern Oscillation. *Int. J. Remote Sens.* 1996, *17*, 2533–2548.

Arvor, D., Jonathan, M., Simoes Penello Meirelles, M., Durbreuil, V., Durieux, L. Classification of MODIS EVI time series for crop mapping in the state of Mato Grosso, Brazil. *Int. J. Remote Sens.* 2011, *32*, 7847–7871.

Atkinson, P.M., Cutler, M.E.J., Lewis, H. Mapping sub-pixel proportional land cover with AVHRR imagery. *Int. J. Remote Sens.* 1997, *18*, 917–935.

Atkinson, P.M., Jeganathan, C., Dash, J., Atzberger, C. Inter-comparison of four models for smoothing satellite sensor time-series data to estimate vegetation phenology. *Remote Sens. Environ.* 2012, *123*, 400–417.

Atzberger, C. Advances in remote sensing of agriculture: Context description, existing operational monitoring systems and major information needs. *Remote Sens.* 2013, *5*, 949–981.

Atzberger, C. Satellite data processing at BOKU, Vienna, 2015. http://ivfl-info.boku.ac.at/index.php/eo-data-processing. Accessed April 8, 2015.

Atzberger, C. *Estimates of Winter Wheat Production through Remote Sensing and Crop Growth Modelling: A Case Study on the Camargue Region.* Berlin, Germany: Verlag für Wissenschaft und Forschung, 1997.

Atzberger, C., Formaggio, A.R., Shimabukuro, Y.E. et al. Obtaining crop-specific time profiles of NDVI: The use of unmixing approaches for serving the continuity between SPOT-VGT and PROBA-V time series. *Int. J. Remote Sens.* 2014, *35*(7), 2615–2638.

Atzberger, C., Guerif, M., Baret, F., Werner, W. Comparative analysis of three chemometric techniques for the spectroradiometric assessment of canopy chlorophyll content in winter wheat. *Comput. Electron. Agric.* 2010, *73*, 165–173.

Atzberger, C., Klisch, A., Mattiuzzi, M. et al. Phenological metrics derived over the european continent from NDVI3g data and MODIS time series. *Remote Sens.* 2014, *6*(1), 257–284.

Atzberger, C., Rembold, F. Portability of neural nets modelling regional winter crop acreages using AVHRR time series. *Eu. J. Remote Sens.* 2012, *45*, 371–392.

Atzberger, C., Rembold, F. Mapping the spatial distribution of winter crops at sub-pixel level using AVHRR NDVI time series and neural nets. *Remote Sens.* 2013, *5*(3), 1335–1354.

Azzali, S., Menenti, M. Mapping vegetation-soil-climate complexes in southern Africa using temporal Fourier analysis of NOAA-AVHRR NDVI data. *Int. J. Remote Sens.* 2000, *21*, 973–996.

Bach, H., Mauser, W. Methods and examples for remote sensing data assimilation in land surface process modeling. *IEEE Trans. Geosci. Remote Sens.* 41(7): 1629–1637, 2003.

Badeck, F.-W., Bondeau, A., Böttcher, K., Doktor, D., Lucht, W., Schaber, J., Sitch, S. Responses of spring phenology to climate change. *New Phytol.* 2004, *162*, 295–309.

Badhwar, G.D., Austin, W.W., Carnes, J.G. A semi-automatic technique for multitemporal classification of a given crop within a Landsat scene. *Pattern Recog.* 1982, *15*, 217–230.

Balaghi, R., Tychon, B., Eerens, H., Jlibene, M. Empirical regression models using NDVI, rainfall and temperature data for the early prediction of wheat grain yields in Morocco. *Int. J. Appl. Earth Obs. Geoinf.* 2008, *10*, 438–452.

Baret, F. Un modele simplifie de reflectance et d'absorptance d'un couvert vegetal. In *Proceedings of the Fourth International Colloquium on Spectral Signatures of Objects in Remote Sensing ESA SP-287*, Aussois, France, 1988, pp. 113–120.

Baret, F., Guyot, G. Monitoring of the ripening period of wheat canopies using visible and near infra red radiometry. *Agronomie* 1986, *6*, 509–516.

Baret, F., Guyot, G. Potentials and limits of vegetation indices for LAI and APAR assessment. *Remote Sens. Environ.* 1991, *35*, 161–173.

Baret, F., Guyot, G., Major, D.J. Crop biomass evaluation using radiometric measurements. *Photogrammetria* 1989, *43*, 241–256.

Baret, F., Houlès, V., Guérif, M. Quantification of plant stress using remote sensing observations and crop models: The case of nitrogen management. *J. Exp. Bot.* 2007, *58*(4): 869–880.

Baret, F., Morissette, J.T., Fernandes, R.A. et al. Evaluation of the representativeness of networks of sites for the global validation and intercomparison of land biophysical products: Proposition of the CEOS-BELMANIP. *IEEE Trans. Geosci. Remote Sens.* 2006, *44*(7), 1794–1803.

Baret, F., Weiss, M., Lacaze, R., Camacho, F., Makhmara, H., Pacholcyzk, P., Smets, B. GEOV1: LAI, FAPAR essential climate variables and FCOVER global time series capitalizing over existing products. Part 1: Principles of development and production. *Remote Sens. Environ.* 2013, *137*, 299–309.

Barnett, T.L., Thompson, D.R. Large-area relation of landsat MSS and NOAA-6 AVHRR spectral data to wheat yields. *Remote Sens. Environ.* 1983, *13*, 277–290.

Barnsley, M.J., Allison, D., Lewis, P. On the information content of multiple view angle (MVA) images. *Int. J. Remote Sens.* 1997, *18*, 1937–1960.

Baruth, B., Royer, A., Genovese, G., Klisch, A. The use of remote sensing within the MARS crop yield monitoring system of the European Commission. *Int. Arch. Photogramm. Remote Sens. Spat. Inf. Sci.* 2008, *36*, 935–941.

Bastiananssen, W.G.M., Ali, S. A new crop yield forecasting model based on satellite measurements applied across the Indus Basin, Pakistan. *Agri. Ecosyst. Environ.* 2003, *94*, 321–340.

Bauer, M.E. LACIE: An experiment in global crop forecasting. *Crops Soils Mag.* 1979, *31*, 5–7.

Beck, P.S.A., Atzberger, C., Høgda, K.A., Johansen, B., Skidmore, A.K. Improved monitoring of vegetation dynamics at very high latitudes: A new method using MODIS NDVI. *Remote Sens. Environ.* 2006, *100*, 321–334.

Becker-Reshef, I., Justice, C., Doorn, B., Reynolds, C., Anyamba, A., Tucker, C.J., Korontzi, S. NASA's contribution to the Group on Earth Observation (GEO) global agricultural monitoring system of systems. *Earth Obs.* 2009, *21*, 24–29.

Becker-Reshef, I., Justice, C.O., Sullivan, M. et al. Monitoring global croplands with coarse resolution Earth observation: The Global Agriculture Monitoring (GLAM) project. *Remote Sens.* 2010a, *2*, 1589–1609.

Becker-Reshef, I., Vermote, E., Lindeman, M., Justice, C. A generalized regression-based model for forecasting winter wheat yields in Kansas and Ukraine using MODIS data. *Remote Sens. Environ.* 2010b, *114*(6): 1312–1323.

Bellocchi, G., Rivington, M., Donatelli, M., Matthews, K. Validation of biophysical models: Issues and methodologies. A review. *Agron. Sustain. Dev.* 2010, *30*, 109–130.

Beltran, C.M., Belmonte, A.C. Irrigated crop area estimation using Landsat TM imagery in La Mancha, Spain. *Photogramm. Eng. Remote Sens.* 2001, *67*(10), 1177–1184.

Belward, A.S., Skøien, J.O. Who launched what, when and why: Trends in global land-cover observation capacity from civilian earth observation satellites. *ISPRS J. Photogramm. Remote Sens.* 2014. http://www.sciencedirect.com/science/article/pii/S0924271614000720.

Ben-Dor, E. Quantitative remote sensing of soil properties. *Adv. Agron.* 2002, *75*, 173–243.

Benedetti, R., Rossini, P. On the use of NDVI profiles as a tool for agricultural statistics: The case study of wheat yield estimate and forecast in Emilia Romagna. *Remote Sens. Environ.* 1993, *45*, 311–326.

Benz, U.C., Hofmann, P., Willhauck, G., Lingenfelder, I., Heynen, M. Multi-resolution, object-oriented fuzzy analysis of remote sensing data for GIS-ready information. *ISPRS J. Photogramm. Remote Sens.* 2004, *58*(3–4), 239–258.

Bernardes, T., Moreira, M.A., Adami, M., Giarolla, A., Rudorff, B.F.T. Monitoring biennial bearing effect on coffee yield using MODIS remote sensing imagery. *Remote Sens.* 2012, *4*, 2492–2509.

Blaschke, T. Object based image analysis for remote sensing. *ISPRS J. Photogramm. Remote Sens.* 2010, *65*(1), 2–16.

Bolton, D.K., Friedl, M.A. Forecasting crop yield using remotely sensed vegetation indices and crop phenology metrics. *Agri. Forest Meteorol.* 2013, *173*, 74–84.

Boote, K.J., Jones, J.W., Pickering, N.B. Potential uses and limitations of crop models. *Agron. J.* 1996, *88*, 704–716.

Boryan, C., Yang, Z., Mueller, R., Craig, M. Monitoring US agriculture: The US department of agriculture, national agricultural statistics service, cropland data layer program. *Geocarto Int.* 2011, *26*(5), 341–358.

Boschetti, M., Stroppiana, D., Brivio, P.A. Multi-year monitoring of rice crop phenology through time series analysis of MODIS images. *Int. J. Remote Sens.* 2009, *30*(18), 4643–4662.

Bossyns, B., Eerens, H., van Orshoven, J. Crop area assessment using sub-pixel classification with a neural network trained for a reference year. In *Proceedings of the Fourth International Workshop on the Analysis of Multi-Temporal Remote Sensing Imagery*, Leuven, Belgium, July 18–20, 2007, pp. 1–8.

Bouman, B.A.M. Linking physical remote-sensing models with crop growth simulation-models, applied for sugar-beet. *Int. J. Remote Sens.* 1992, *13*(14), 2565–2581.

Bouman, B.A.M. Crop modeling and remote-sensing for yield prediction. *Netherlands J. Agri. Sci.* 1995, *43*(2), 143–161.

Bradley, B.A., Jacob, R.W., Hermance, J.F., Mustard, J.F. A curve fitting procedure to derive inter-annual phenologies from time series of noisy satellite NDVI data. *Remote Sens. Environ.* 2007, *106*, 137–145.

Braswell, B.H., Hagen, S.C., Frolking, S.E., Salas, W.A. A multivariable approach for mapping sub-pixel land cover distributions using MISR and MODIS: Application in the Brazilian Amazon region. *Remote Sens. Environ.* 2003, *87*, 243–256.

Breiman, L. Random forests. *Mach. Learn.*, 2001, *45*(1), 5–32.

Brink, A.B., Eva, H.D. Monitoring 25 years of land cover change dynamics in Africa: A sample based remote sensing approach. *Appl. Geogr.* 2009, *29*, 501–512.

Brisson, N., Mary, B., Ripoche, D. et al. STICS: A generic model for the simulation of crops and their water and nitrogen balances. I. Theory and parameterization applied to wheat and corn. *Agronomie* 1998, *18*, 311–346.

Brown, J.F., Pervez, M.S. Merging remote sensing data and national agricultural statistics to model change in irrigated agriculture. *Agri. Syst.* 2014, *127*, 28–40.

Brown de Colstoun, E.C., Story, M.H., Thompson, C., Commisso, K., Smith, T.G., Irons, J.I. National Park vegetation mapping using multitemporal Landsat 7 data and a decision tree classifier. *Remote Sens. Environ.* 2003, *85*(03), 316–327.

Bruinsma, J. *World agriculture: Towards 2015/2030: An FAO perspective*. London, U.K.: Earthscan Publication Ltd, 2003.

Bryan, B.A., Hajkowicz, S., Marvanek, S., Young, M.D. Mapping economic returns to agriculture for informing environmental policy in the Murray-Darling Basin, Australia. *Environ. Model. Assess.* 2009, *14*(3), 375–390.

Burney, J.A., Davis, S.J., Lobell, D.B. Greenhouse gas mitigation by agricultural intensification. *Proc. Natl. Acad. Sci. U. S. A.* 2010, *107*, 12052–12057.

Busetto, L., Meroni, M., Colombo, R. Combining medium and coarse spatial resolution satellite data to improve the estimation of sub-pixel NDVI time series. *Remote Sens. Environ.* 2008, *112*, 118–131.

Canfield, D.E., Glazer, A.N., Falkowski, P.G. The evolution and future of Earth's nitrogen cycle. *Science* 2010, *330*, 192–196.

Cardille, J.A., Foley, J.A. Agricultural land-use change in Brazilian Amazônia between 1980 and 1995: Evidence from integrated satellite and census data. *Remote Sens. Environ.* 2003, *87*, 551–562.

Chang, J., Hansen, M.C., Pittman, K., Carroll, M., DiMiceli, C. Corn and soybean mapping in the United States using MODIS time-series data sets. *Agron. J.* 2007, *99*, 1654–1664.

Clevers, J.G.P.W., Büker, C., van Leeuwen, H.J.C., Bouman, B.A.M. A framework for monitoring crop growth by combining directional and spectral remote sensing information. *Remote Sens. Environ.* 1994, *50*, 161–170.

Clevers, J.G.P.W., van Leeuwen, H.J.C. Combined use of optical and microwave remote sensing data for crop growth monitoring. *Remote Sens. Environ.* 1996, *56*, 42–51.

Clevers, J.G.P.W., Vonder, O.W., Jongschaap, R.E.E., Desprats, J.F., King, C., Pre´vot, L., Bruguier, N. Using SPOT data for calibrating a wheat growth model under mediterranean conditions. *Agronomie* 2002, *22*, 687–694.

Comaniciu, D., Meer, P. Mean shift: A robust approach toward feature space analysis. *IEEE Trans. Pattern Anal. Mach. Intell.* 2002, *24*(5), 603–619.

Congalton, R.G., Balogh, M., Bell, C., Green, K. Mapping and monitoring agricultural crops and other land cover in the Lower Colorado River Basin. *Photogramm. Eng. Remote Sens.* 1998, *64*(11), 1107–1113.

Cousins, S.A.O. Analysis of land-cover transitions based on 17th and 18th century cadastral maps and aerial photographs. *Landscape Ecol.* 2001, *16*(1), 41–54.

Dale, V.H., Polasky, S. Measures of the effects of agricultural practices on ecosystem services. *Ecol. Econ.* 2007, *64*(2), 286–296.

De Beurs, K., Henebry, G.H. Land surface phenology, climatic variation, and institutional change: Analyzing agricultural land cover change in Kazakhstan. *Remote Sens. Environ.* 2004, *89*, 497–509.

De Leeuw, J., Keah, H., Hadgu, K.M., Shee, A., Vrieling, A., Biradar, C., Atzberger, C., Turvey, C. The Potential and Uptake of Remote Sensing in Insurance: A Review, *Remote Sens.* 2014, *6*(11), 10888–10912.

Deering, D.W. Rangeland reflectance characteristics measured by aircraft and spacecraft sensors. PhD Thesis, Texas A&M University, College Station, TX, 1978.

Defries, R.S., Field, C.B., Fung, I. et al. Mapping the land-surface for global atmosphere-biosphere models—Toward continuous distributions of vegetations functional properties. *J. Geophys. Res.* 1995, *100*, 20867–20882.

Defries, R.S., Hansen, M., Townshend, J.R.G., Sohlberg, R. Global land cover classification at 8 km spatial resolution: The use of training data derived from Landsat imagery in decision tree classifiers. *Int. J. Remote Sens.* 1998, *19*, 3141–3168.

Delécolle, R., Maas, S.J., Guérif, M., Baret, F. Remote sensing and crop production models: Present trends. *ISPRS J. Photogramm.* 1992, *47*, 145–161.

Demirbas, M.F., Balat, M. Recent advances on the production and utilization trends of bio-fuels: A global perspective. *Energy Convers. Manage.* 2006, *47*(15–16), 2371–2381.

Dirzo, R., Raven, P.H. Global state of biodiversity and loss. *Annu. Rev. Environ. Resour.* 2003, *28*, 137–167.

Donatelli, M., Russell, G., Rizzoli, A.E. et al. A component-based framework for simulating agricultural production and externalities. In *Environmental and Agricultural Modelling* (Brouwer, F.M., Ittersum, M.K., eds.). Dordrecht, The Netherlands: Springer, 2010, pp. 63–108.

Doraiswamy, P.C., Cook, P.W. Spring wheat yield assessment using NOAA AVHRR data. *Can. J. Remote Sens.* 1995, *21*, 41–51.

Doraiswamy, P.C., Hatfield, J.L., Jackson, T.J., Akhmedov, B., Prueger, J., Stern, A. Crop condition and yield simulations using Landsat and MODIS. *Remote Sens. Environ.* 2004, *92*, 548–559.

Doraiswamy, P.C., Moulin, S., Cook, P.W., Stern, A. Crop yield assessment from remote sensing. *Photogramm. Eng. Remote Sens.* 2003, *69*(6), 665–674.

Doraiswamy, P.C., Sinclair, T.R., Hollinger, S., Akhmedov, B., Stern, A., Prueger, J. Application of MODIS derived parameters for regional crop yield assessment. *Remote Sens. Environ.* 2005, *97*, 192–202.

Dorigo, W.A., Zurita-Milla, R., de Wit, A.J.W., Braziled, J., Singh, R., Schaepman, M.E. A review on reflective remote sensing and data assimilation techniques for enhanced agroecosystem modeling. *Int. J. Appl. Earth Observ. Geoinform.* 2007, *9*(2), 165–193.

Dramstad, W.E., Fjellstad, W.J., Strand, G.H. et al. Development and implementation of the Norwegian monitoring programme for agricultural landscapes. *J. Environ. Manage.* 2002, *64*, 49–63.

Duchemin, B., Maisongrande, P., Boulet, G., Benhadj, I. A simple algorithm for yield estimates: Evaluation for semi-arid irrigated winter wheat monitored with green leaf area index. *Environ. Model. Soft.* 2008, *23*, 876–892.

Duro, D.C., Coops, N.C., Wulder, M.A., Han, T. Development of a large area biodiversity monitoring system driven by remote sensing. *Progr. Phys. Geogr.* 2007, *31*(3), 235–260.

Duveiller, G., Lopez, R., Baruth, B. Enhanced processing of low spatial resolution fAPAR time series for sugarcane yield forecasting and monitoring. *Remote Sens.* 2013, *5*(3), 1091–1116.

Eerens, H., Piccard, I., Royer, A., Orlandi, S. *Methodology of the MARS Crop Yield Forecasting System. Vol. 3: Remote Sensing Information, Data Processing and Analysis.* Ispra, Italy: Joint Research Centre European Commission, 2004, p. 76.

Epiphanio, R.D.V., Formaggio, A.R., Rudorff, B.F.T., Maeda, E.E., Luiz, A.J.B. Estimating soybean crop areas using spectral-temporal surfaces derived from MODIS images in Mato Grosso, Brazil. *Pesquisa Agropecuária Brasileira* 2010, *45*, 72–80.

FAO. *The State of Food Insecurity in the World 2009: Economic Crises—Impacts and Lessons Learned.* Rome, Italy: Food and Agriculture Organization of the United Nations (FAO), 2009.

FAO. *Global Strategy to Improve Agricultural and Rural Statistics,* Report No. 56719-GB. Rome, Italy: FAO, 2011.

Foley, J.A., Ramankutty, N., Brauman, K.A. et al. Solutions of a cultivated planet. *Nature* 2011, *478*, 337–342.

Foody, G.M. Assessing the accuracy of land cover change with imperfect ground reference data. *Remote Sens. Environ.* 2010, 114, 2271–2285.

Foody, G.M. Status of land cover classification accuracy assessment. *Remote Sens. Environ.* 2002, *80*, 185–201.

Foody, G.M., Boyd, D.S. Using volunteered data in land cover map validation: Mapping west African forests. *IEEE J. Select. Topics Appl. Earth Observ. Remote Sens.* 2013, *6*(3), 1305–1312.

Foody, G.M., Cox, D.P. Sub-pixel land cover composition estimation using a linear mixture model and fuzzy membership functions. *Int. J. Remote Sens.* 1994, *15*, 619–631.

Foody, G.M., See, L., Fritz, S., Van der Velde, M., Perger, C., Schill, C., Boyd, D.S. Assessing the accuracy of volunteered geographic information arising from multiple contributors to an internet based collaborative project. *Trans. GIS*, 2013, *17*(6), 847–860.

Friedl, M.A., Brodley, C.E. Decision tree classification of land cover from remotely sensed data. *Remote Sens. Environ.*, 1997, *61*(3), 399–408.

Fritz, S., Massart, M., Savin, I., Gallego, J., Rembold, F. The use of MODIS data to derive acreage estimations for larger fields: A case study in the south-western Rostov region of Russia. *Int. J. Appl. Earth Obs. Geoinf.* 2008, *10*, 453–466.

Fritz, S., McCallum, I., Schill, C., Perger, C., Grillmayer, R., Achard, F., Kraxner, F., Obersteiner, M. Geo-Wiki.og: The use of crowdsourcing to improve global land cover. *Remote Sens.* 2009, *1*, 345–354.

Fritz, S., Purgathofer, P., Kayali, F. et al. Landspotting: Social gaming to collect vast amounts of data for satellite validation. *Geophys. Res. Abstr.* 2012, *14*, EGU2012-13173.

Galford, G.L., Melillo, J., Mustard, J.F., Cerri, C.E.P., Cerri, C.C. The Amazon frontier of land-use change: Croplands and consequences for greenhouse gas emissions. *Earth Interact.* 2010, *14*(15), 1–24.

Galford, G.L., Mustard, J.F., Melillo, J., Gendrin, A., Cerri, C.C., Cerri, C.E.P. Wavelet analysis of MODIS time series to detect expansion and intensification of row-crop agriculture in Brazil. *Remote Sens. Environ.* 2008, *112*, 576–587.

Gallego, F.J. Remote sensing and land cover estimation. *Int. J. Remote Sens.* 2004, *25*, 3019–3047.

Gallego, F.J. Stratified sampling of satellite images with a systematic grid of points. *ISPRS J. Photogramm.* 2005, *59*, 369–376.

Gallego, F.J. The efficiency of sampling very high resolution images for area estimation in the European Union. *Int. J. Remote Sens.* 2012, *33*(6), 1868–1880.

Gallo, K.P., Daughtry, C.S.T., Bauer, M.E. Spectral estimation of absorbed photosynthetically active radiation in corn canopies. *Remote Sens. Environ.* 1985, *17*, 221–232.

Gibbs, H.K., Ruesch, A.S., Achard, F., Clayton, M.K., Holmgren, P., Ramankutty, N., Foley, J.A. Tropical forests were the primary sources of new agricultural land in the 1980s and 1990s. *Proc. Natl. Acad. Sci. U. S. A.* 2010, *107*(38), 16732–16737.

Gobron, N., Pinty, B., Verstraete, M.M., Widlowski, J.-L., Diner, D.J. Uniqueness of multiangular measurements— Part II: Joint retrieval of vegetation structure and photosynthetic activity from MISR. *IEEE Trans. Geosci. Remote Sens.* 2002, *40*, 1574–1592.

Godfray, H.C.J., Beddington, J.R., Crute, I.R. et al. Food security: The challenge of feeding 9 billion people. *Science* 2010, *327*, 812–818.

Goward, S.N., Dye, D.G. Evaluating North American net primary productivity with satellite observations. *Adv. Space Res.* 1987, *7*, 165–174.

Groten, S.M.E. NDVI-crop monitoring and early yield assessment of Burkina Faso. *Int. J. Remote Sens.* 1993, *14*, 1495–1515.

Group on Earth Observations (GEO). *Global Agricultural Monitoring System of Systems*, Task AG-07–03a. Available at: http://www.earthobservations.org/cop_ag_gams.shtml (accessed on February 18, 2013).

Gu, Y., Brown, J.F., Verdin, J.P., Wardlow, B. A five-year analysis of MODIS NDVI and NDWI for grassland drought assessment over the central Great Plains of the United States. *Geophys. Res. Lett.* 2007, *34*, L06407.

Guérif, M., Blöser, B., Atzberger, C., Clastre, P., Guinot, J.P., Delecolle, R. Identification de parcelles agricoles à partir de la forme de leur évolution radiométrique au cours de la saison de culture. *Photo Interpret.* 1996, *1*, 12–22.

Guérif, M., Delécolle, R. Introducing remotely sensed estimates of canopy structure into plant models. In *Canopy Structure and Light Microclimate. Characterization and Applications* (Varlet-Grancher, C., Bonhomme, R., Sinoquet, H., eds.). Paris, France: INRA, 1993, pp. 479–490.

Guérif, M., Duke, C.L. Calibration of the SUCROS emergence and early growth module for sugarbeet using optical remote sensing data assimilation. *Eur. J. Agron.* 1998, *9*, 127–136.

Guérif, M., Duke, C.L. Adjustment procedures of a crop model to the site specific characteristics of soil and crop using remote sensing data assimilation. *Agr. Ecosyst. Environ.* 2000, *81*, 57–69.

Han, W., Yang, Z., Di, L., Mueller, R. CropScape: A Web service based application for exploring and disseminating US conterminous geospatial cropland data products for decision support. *Comp. Elec. Agri.* 2012, *84*, 111–123.

Hansen, M.C., DeFries, R.S., Townshend, J.R.G., Sohlberg, R., Dimiceli, C., Carroll, M. Towards an operational MODIS continuous field of percent tree cover algorithm: Examples using AVHRR and MODIS data. *Remote Sens. Environ.* 2002, *83*, 303–319.

Hansen, P.M., Schjoerring, J.K. Reflectance measurement of canopy biomass and nitrogen status in wheat crops using normalized difference vegetation indices and partial least squares regression. *Remote Sens. Environ.* 2003, *86*(4), 542–553.

Haralick, R., Shapiro, L., 1985. Image segmentation techniques. *Comp. Vision Graph. Image Proc.* 29(1), 100–132.

Hayes, M.J., Decker, W.L. Using NOAA AVHRR data to estimate maize production in the United States Corn Belt. *Int. J. Remote Sens.* 1996, *17*, 3189–3200.

Henricksen, B.L., Durkin, J.W. Growing period and drought early warning in Africa using satellite data. *Int. J. Remote Sens.* 1986, *7*, 1583–1608.

Hermance, J.F. Stabilizing high-order, non-classical harmonic analysis of NDVI data for average annual models by damping model roughness. *Int. J. Remote Sens.* 2007, *28*, 2801–2819.

Hill, J., Nelson, E., Tilman, D., Polasky, S., Tiffany, D. Environmental, economic, and energetic costs and benefits of biodiesel and ethanol biofuels. *Proc. Natl. Acad. Sci. U. S. A.* 2006, *103*, 1120–11210.

Hirosawa, Y., Marsh, S.E., Kliman, D.H. Application of standardized principal component analysis of land-cover characterization using multitemporal AVHRR data. *Remote Sens. Environ.* 1996, *58*, 267–281.

Hostert, P., Kuemmerle, T., Prishchepov, A., Sieber, A., Lambin, E.F., Radeloff, V.C. Rapid land use change after socioeconomic disturbances: The collapse of the Soviet Union *versus* Chernobyl. *Environ. Res. Lett.* 2011, *6*, 04520.

Houles, V. Mise au point d'un outile de modulation intra-parcellaire de la fertilization azotee du ble d'hiver base sur la teledetection et un modele de culture. Thèse de Doctorat, Institut National Agronomique Paris-Grignon, France, 2004.

Hutchinson, C.F. Uses of satellite data for famine early warning in sub-Saharan Africa. *Int. J. Remote Sens.* 1991, *12*, 1405–1421.

IBGE. Sistema IBGE de Recuperação Automática (SIDRA)—Produção Agrícola Municipal (PAM). Rio de Janeiro, RJ, Brazil, 2013. Available at: http://www.sidra.ibge.gov.br/ (accessed on December 7, 2013).

Idso, S.B., Pinter, P.J., Jr., Jackson, R.D., Reginato, R.J. Estimation of grain yields by remote sensing of crop senescence rates. *Remote Sens. Environ.* 1980, *9*, 87–91.

IPCC (Intergovernmental Panel on Climate Change). *Climate Change 2007: Synthesis Report. Contribution of Working Groups I, II and III to the Fourth Assessment Report of the Intergovernmental Panel on Climate Change.* Geneva, Switzerland: IPCC, 2007.

IPCC (Intergovernmental Panel on Climate Change). *Climate Change 2013: The Physical Science Basis. Contribution of Working Group I to the Fifth Assessment Report of the Intergovernmental Panel on Climate Change* (Stocker, T.F., Qin, D., Plattner, G.-K., Tignor, M., Allen, S.K., Boschung, J., Nauels, A., Xia, Y., Bex, V., Midgley, P.M., eds.). Cambridge, U.K.: Cambridge University Press, 2013, 1535 pp.

Jakubauskas, M.E., Legates, D.R., Kastens, J.H. Harmonic analysis of time-series AVHRR NDVI data. *Photogramm. Eng. Remote Sens.* 2001, *67*, 461–470.

Jaynes, E.T. *Probability Theory: The Logic of Science.* Cambridge, UK: Cambridge University Press, 2003, p. 727.

Jensen, F.V., Nielsen, T.D. *Bayesian Networks and Decision Graphs*, 2nd edn. New York: Springer, 2007, p. 447.

Johnson, G.E., van Dijk, A., Sakamoto, C.M. The use of AVHRR data in operational agricultural assessment in Africa. *Geocarto. Int.* 1987, *2*, 41–60.

Jones, C.A., Kiniry, J.R. *Ceres-Maize: A Simulation Model of Maize Growth and Development*, 1st edn. College Station, TX: Texas A&M University, 1986.

Jones, H.G., Vaughan, R.A. *Remote Sensing of Vegetation: Principles, Techniques, and Applications.* Oxford Press, Oxford, U.K., 2010, 384p.

Jones, P.G., Thornton, P.K. The potential impacts of climate change on maize production in Africa and Latin America in 2055. *Global Environ. Change* 2003, *13*, 51–59.

Jong, R., de Bruin, S., de Wit, A.J.W., de Schaepman, M.E., Dent, D.L. Analysis of monotonic greening and browning trends from global NDVI time-series. *Remote Sens. Environ.* 2011, *115*, 692–702.

Jönsson, P., Eklundh, L. Seasonality extraction by function fitting to time-series of satellite sensor data. *IEEE Trans. Geosci. Remote Sens.* 2002, *40*, 1824–1832.

Jönsson, P., Eklundh, L. TIMESAT—A program for analyzing time-series of satellite sensor data. *Comput. Geosci.* 2004, *30*, 833–845.

Justice, C., Belward, A., Morisette, J., Lewis, P., Privette, J., Baret, F. Developments in the validation of satellite sensor products for the study of the land surface. *Int. J. Remote Sens.* 2000, *21*(17): 3383–3390.

Justice, C.O., Townshend, J.R.G., Vermote, E.F., Masuoka, E., Wolfe, R.E., Saleous, N., Roy, D.P., Morisette, J.T. An overview of MODIS Land data processing and product status. *Remote Sens. Environ.* 2002, *83*, 3–15.

Kaduk, J., Heinmann, M. A prognostic phenology scheme for global terrestrial carbon cycle models. *Clim. Res.* 1996, *6*, 1–19.

Karkee, M., Steward, B.L., Tang, L., Aziz, S.A. Quantifying subpixel signature of paddy rice field using an artificial neural network. *Comput. Electron. Agric.* 2009, *65*, 65–76.

Kastens, J.H., Kastens, T.L., Kastens, D.L.A., Price, K.P., Martinko, E.A., Lee, R.-Y. Image masking for crop yield forecasting using AVHRR NDVI time series imagery. *Remote Sens. Environ.* 2005, *99*, 341–356.

Kawamura, K., Akiyama, T., Yokota, H., Tsutsumi, M., Yasuda, T., Watanabe, O., Wang, G., Wang, S. Monitoring of forage conditions with MODIS imagery in the Xilingol steppe, Inner Mongolia. *Int. J. Remote Sens.* 2005, *26*(7), 1423–1436.

Koaudio, L., Duveiller, G., Djaby, B., El Jarroudi, M., Defourny, P., Tychon, B. Estimating regional wheat yield from the shape of decreasing curves of green area index temporal profiles retrieved from MODIS data. *Int. J. Appl. Earth Observ. Geoinform.* 2012, *18*, 111–118.

Kogan, F.N. Droughts of the late 1980s in the United-States as derived from Noaa polar-orbiting satellite data. *Bulletin of the American Meteorological Society.* 1995, *76*(5), 655–668.

Kogan, F., Kussul, N., Adamenko, T. et al. Winter wheat yield forecasting in Ukraine based on Earth observation, meteorological data and biophysical models. *Int. J. Appl. Earth Observ. Geoinform.* 2013, *23*, 192–203.

Koukal, T., Atzberger, C. Potential of multi-angular data derived from a digital aerial frame camera for forest classification. *IEEE J. Sel. Top. Appl. Earth Obs. Remote Sens.* 2012, *5*, 30–43.

Kuemmerle, T., Olofsson, P., Chaskovskyy, O., Baumann, M., Ostapowicz, K., Woodcock C., Houghton, R.A., Hostert, P., Keeton, W.S., Radeloff, V.C. Post-Soviet farmland abandonment, forest recovery, and carbon sequestration in western Ukraine. *Global Change Biol.* 2011, *17*, 1335–1349.

Launay, M., Guerif, M. Assimilating remote sensing data into a crop model to improve predictive performance for spatial applications. *Agri. Ecosyst. Environ.* 2005, *111*, 321–339.

Lawrence, R.L., Wright, A. Rule-based classification systems using classification and regression tree (CART) analysis. *Photogramm. Eng. Remote Sens.* 2001, *67*(10), 1137–1142.

Leblon, B., Guerif, M., Baret, F. The use of remotely sensed data in estimation of PAR use efficiency and biomass production of flooded rice. *Remote Sens. Environ.* 1991, *38*, 147–158.

Lee, R., Kastens, D.L., Price, K.P., Martinko, E.A. Forecasting corn yield in Iowa using remotely sensed data and vegetation phenology information. In *Proceedings of the Second International Conference on Geospatial Information in Agriculture and Forestry*, Lake Buena Vista, FL, January 10–12, 2000, Volume II, pp. 460–467.

Lewis, J.E., Rowland, J., Nadeau, A. Estimating maize production in Kenya using NDVI: Some statistical considerations. *Int. J. Remote Sens.* 1998, *19*, 2609–2617.

Lloyd, D. A phenological classification of terrestrial vegetation cover using shortwave vegetation index imagery. *Int. J. Remote Sens.* 1990, *11*, 2269–2279.

Lobell, D.B., Asner, G.P. Cropland distributions from temporal unmixing of MODIS data. *Remote Sens. Environ.* 2004, *93*, 412–422.

Loveland, T.R., Reed, B.C., Brown, J.F., Ohlen, D.O., Zhu, Z., Yang, L., Merchant, J.W. Development of a global land cover characteristics database and IGBP DISCover from 1 km AVHRR data. *Int. J. Remote Sens.* 2000, *21*, 1303–1330.

Luiz, A.J.B., Formaggio, A.R., Epiphanio, J.C.N. Objective sampling estimation of crop area based on remote sensing images. In *Computational Methods for Agricultural Research. Advances and Applications* (Prado, H.A., Luiz, A.J.B., Chaib Filho, H., eds.) Chapter 4. Hershey, PA: IGI-Global-Global, 2011, pp. 73–95.

Lunetta, R.S., Knight, J.F., Ediriwickrema, J., Lyon, J.G., Worthy, L.D. Land-cover change detection using multi-temporal MODIS NDVI data. *Remote Sens. Environ.* 2006, *105*, 142–154.

Lunetta, R.S., Shao, Y., Ediriwickrema, J., Lyon, J.G. Monitoring agricultural cropping patterns across the Laurentian Great Lakes Basin using MODIS-NDVI data. *Int. J. Appl. Earth Obs. Geoinf.* 2010, *12*, 81–88.

Maas, S.J. *GRAMI: A Crop Growth Model That Can Use Remotely Sensed Information ARS 91.* Washington, DC: U.S. Department of Agriculture, 1992, p. 78.

Maas, S.J. Use of remotely-sensed information in agricultural crop growth models. *Ecol. Model.* 1988a, *41*, 247–268.

Maas, S.J. Using satellite data to improve model estimates of crop yield. *Agron. J.* 1988b, *80*(4), 655–662.

Macdonald, R.B., Hall, F.G. Global crop forecasting. *Science* 1980, *208*, 670–679.

Manjunath, K.R., Potdar, M.B., Purohit, N.L. Large area operational wheat yield model development and validation based on spectral and meteorological data. *Int. J. Remote Sens.* 2002, *23*, 3023–3038.

Maselli, F., Conese, C., Petkov, L., Gilabert, M.A. Environmental monitoring and crop forecasting in the Sahel through the use of NOAA NDVI data. A case study: Niger 1986–89. *Int. J. Remote Sens.* 1993, *14*, 3471–3487.

Maselli, F., Gilabert, M.A., Conese, C. Integration of high and low resolution NDVI data for monitoring vegetation in Mediterranean environments. *Remote Sens. Environ.* 1998, *63*, 208–218.

Maselli, F., Rembold, F. Analysis of GAC NDVI data for cropland identification and yield forecasting in Mediterranean African countries. *Photogramm. Eng. Remote Sens.* 2001, *67*, 593–602.

Maselli, F., Romanelli, S., Bottai, L., Maracchi, G. Processing of GAC NDVI data for yield forecasting in the Sahelian region. *Int. J. Remote Sens.* 2000, *21*, 3509–3523.

McGrayne, S.B. *The Theory that would not Die: How Bayes' Rule Cracked the Enigma Code, Hunted Down Russian Submarines, and Emerged Triumphant from Two Centuries of Controversy.* New Haven, CT: Yale University Press, 2011, p. 336.

Mello, M.P., Aguiar, D.A., Rudorff, B.F.T., Pebesma, E., Jones, J., Santos, N.C.P. Spatial statistic to assess remote sensing acreage estimates: An analysis of sugarcane in São Paulo State, Brazil. In *Proceedings of the 33rd IEEE International Geoscience and Remote Sensing Symposium (IGARSS 2013)*, IEEE, Melbourne, Australia, 2013a, p. 4233–4236.

Mello, M.P., Atzberger, C., Formaggio, A.R. Near real time yield estimation for sugarcane in Brazil combining remote sensing and official statistical data. In *Proceedings of the 34rd IEEE International Geoscience and Remote Sensing Symposium (IGARSS 2014)*, IEEE, Quebec, Canada, 2014, 5064–5067.

Mello, M.P., Risso, J., Atzberger, C., Aplin, P., Pebesma, E., Vieira, C.A.O., Rudorff, B.F.T. Bayesian Networks for Raster Data (BayNeRD): Plausible reasoning from observations. *Remote Sens.* 2013c, *5*(11), 5999–6025.

Mello, M.P., Vieira, C.A.O., Rudorff, B.F.T., Aplin, P., Santos, R.D.C., Aguiar, D.A. STARS: A new method for multitemporal remote sensing. *IEEE Trans. Geosci. Remote Sens.* 2013b, *51*(4), 1897–1913.

Meroni, M., Atzberger, C., Vancutsem, C., Gobron, N., Baret, F., Lacaze, R., Eerens, H., Leo, O. Evaluation of agreement between space remote sensing SPOT-VGT fAPAR time series. *IEEE Trans. Geosci. Remote Sens.* 2013b, *99*, 1–12.

Meroni, M., Marinho, E., Sghaier, N., Verstrate, M.M., Leo, O. Remote sensing based yield estimation in a stochastic framework—Case study of durum wheat in Tunisia. *Remote Sens.* 2013a, *5*, 539–557.

Meroni, M., Rembold, F., Verstraete, M., Gommes, R., Schucknecht, A., Beye, G. Investigating the relationship between the inter-annual variability of satellite-derived vegetation phenology and a proxy of biomass production in the Sahel. *Remote Sens.* 2014a, *6*(6), 5868–5884.

Meroni, M., Verstraete, M.M., Rembold, F., Urbano, F., Kayitakire, F. A phenology-based method to derive biomass production anomaly for food security monitoring in the Horn of Africa. *Int. J. Remote Sens.* 2014b, *37*(7), 2472–2492.

Monteiro, L.A., Sentelhas, P.C. Potential and actual sugarcane yields in Southern Brazil as a function of climate conditions and crop management. *Sugar Tech.* 2014, 16(3), 264–276.

Monteith, J.L. Solar radiation and productivity in tropical ecosystems. *J. Appl. Ecol.* 1972, 9, 747–766.

Monteith, J.L. Climate and the efficiency of crop production in Britain. *Phil. Trans. R. Soc. Lond.* 1977, *281*, 277–294.

Moran, M.S., Inoue, Y., Barnes, E.M. Opportunities and limitations for image-based remote sensing in precision crop management. *Remote Sens. Environ.* 1997, 61(3), 319–346.

More, R., Manjunath, K.R. Deducing rice crop dynamics and cultural types of Bangladesh using geospatial techniques. *J. Ind. Soc. Remote Sens.* 2013, 41(3), 597–607.

Moriondo, M., Maselli, F., Bindi, M. A simple model of regional wheat yield based on NDVI data. *Eur. J. Agronomy* 2007, *26*, 266–274.

Morisette, J.T., Baret, F., Privette, J.L. et al. Validation of global moderate-resolution LAI products: A framework proposed within the CEOS land product validation subgroup. *IEEE Trans. Geosci. Remote Sens.* 2006, 44(7), 1804–1814.

Moulin, S., Bondeau, A., Delécolle, R. Combining agricultural crop models and satellite observations: From field to regional scales. *Int. J. Remote Sens.* 1998, 19(6), 1021–1036.

Mueller, N.D., Gerber, J.S., Johnston, M., Ray, D.K., Ramankutty, N., Foley, J.A. Closing yield gaps through nutrient and water management. *Nature* 2012, *490*, 254–257.

Mulianga, B., Begue, A., Simoes, M., Todoroff, P. Forecasting regional sugarcane yield based on time integral and spatial aggregation of MODIS NDVI. *Remote Sens.* 2013, 5(5), 2184–2199.

Mulla, D.J. Twenty five years of remote sensing in precision agriculture: Key advances and remaining knowledge gaps. *Biosyst. Eng.* 2013, *114*(4), 358–371.

Myneni, R.B., Williams, D.L. On the relationship between FAPAR and NDVI. *Remote Sens. Environ.* 1994, *49*, 200–211.

Naylor, R. Expanding the boundaries of agricultural development. *Food Security* 2011, *3*, 233–251.

Nguyen, H.T., Kim, J.H., Nguyen, A.T., Nguyen, L.T., Shin, J.C., Lee, B.W. Using canopy reflectance and partial least squares regression to calculate within-field statistical variation in crop growth and nitrogen status of rice. *Precis. Agri.* 2006, *7*, 249–264.

Nouvellon, Y., Moran, M.S., Lo Seen, D., Bryant, R., Rambal, S., Ni, W., Begue, A., Chehbouni, A., Emmerich, W.E., Heilman, P., Qi, J. Coupling a grassland ecosystem model with Landsat imagery for a 10-year simulation of carbon and water budgets. *Remote Sens. Environ.* 2001, *78*(1–2), 131–149.

Olesen, J.E., Bindi, M. Consequences of climate change for European agricultural productivity, land use and policy. *Eur. J. Agron.* 2002, *16*(4), 239–262.

Oleson, K.W., Sarlin, S., Garrison, J., Smith, S., Privette, J.L., Emery, W.J. Unmixing multiple land-cover type reflectances from coarse spatial resolution satellite data. *Remote Sens. Environ.* 1995, *54*, 98–112.

Olofsson, P., Foody, G.M., Herold, M., Stehman, S.V., Woodcock, C.E., Wulder, M.A. Good practices for estimating area and assessing accuracy of land change. *Remote Sens. Environ.* 2014, *148*, 42–57.

Olofsson, P., Stehman, S.V., Woodcock, C.E., Sulla-Menashe, D., Sibley, A.M., Newell, J.D., Friedl, M.A., Herold, M. A global land-cover validation data set, part I: Fundamental design principles. *Int. J. Remote Sens.* 2012, *33*(18), 5768–5788.

Padilla, F.M., Maas, S., Gonzales-Dugo, M.P., Raja, N., Mansilla, F., Gavilan, P., Dominguez, J. Wheat yield monitoring in Southern Spain using the GRAMI model and a series of satellite images. *Field Crop. Res.* 2012, *130*, 145–154.

Pal, N.R., Pal, S.K. A review on image segmentation techniques. *Pattern Recog.* 1993, *26*(9), 1277–1299.

Pearl, J. *Probabilistic Reasoning in Intelligent Systems: Networks of Plausible Inference*, 1st edn. San Francisco, CA: Morgan Kaufmann, 1988, p. 552.

Pellenq, J., Boulet, G. A methodology to test the pertinence of remote-sensing data assimilation into vegetation models for water and energy exchange at the land surface. *Agronomie* 2004, *24*(4), 197–204.

Pelletier, N., Tyedmers, P. Forecasting potential global environmental costs of livestock production 2000–2050. *Proc. Natl. Acad. Sci. U. S. A.* 2010, *107*, 18371–18374.

Peña-Barragán, J.M., Ngugi, M.K., Plant, R.E., Six, J. Object-based crop identification using multiple vegetation indices, textural features and crop phenology. *Remote Sens. Environ.* 2011, *115*(6), 1301–1316.

Pinter, P.J., Jr., Jackson, R.D., Idso, S.B., Reginato, R.J. Multidate spectral reflectance as predictors of yield in water stressed wheat and barley. *Int. J. Remote Sens.* 1981, *2*, 43–48.

Pinter, P.J., Jr., Ritchie, J.C., Hatfield, J.L., Hart, G.F. The agricultural research service's remote sensing program: An example of interagency collaboration. *Photogramm. Eng. Remote Sens.* 2003, *69*, 615–618.

Pittman, K., Hansen, M.C., Becker-Reshef, I., Potapov, P.V., Justice, C.O. Estimating global cropland extent with multi-year MODIS data. *Remote Sens.* 2010, *2*(7), 1844–1863.

Postel, S., Daily, G.C., Ehrlich, P.R. Human appropriation of renewable fresh water. *Science* 1996, *271*, 785–788.

Potdar, M.B., Manjunath, K.R., Purohit, N.L. Multi-season atmospheric normalization of NOAA AVHRR derived NDVI for crop yield modeling. *Geocarto. Int.* 1999, *14*, 51–56.

Prince, S.D. A model of regional primary production for use with coarse resolution satellite data. *Int. J. Remote Sens.* 1991a, *12*, 1313–1330.

Prince, S.D. Satellite remote sensing of primary production: Comparison of results for Sahelian grasslands 1981–1988. *Int. J. Remote Sens.* 1991b, *12*, 1301–1311.

Quarmby, N.A. Linear mixture modelling applied to AVHRR data for crop area estimation. *Int. J. Remote Sens.* 1992, *13*, 415–425.

Quarmby, N.A., Milnes, M., Hindle, T.L., Silleos, N. The use of multi-temporal NDVI measurements from AVHRR data for crop yield estimation and prediction. *Int. J. Remote Sens.* 1993, *14*, 199–210.

Rasmussen, M.S. Assessment of millet yields and production in northern Burkina Faso using integrated NDVI from the AVHRR. *Int. J. Remote Sens.* 1992, *13*, 3431–3442.

Rasmussen, M.S. Developing simple, operational, consistent NDVI-vegetation models by applying environmental and climatic information. Part II: Crop yield assessment. *Int. J. Remote Sens.* 1998, *19*, 119–139.

Reed, B.C., Brown, J.F., Vanderzee, D., Loveland, T.R., Merchant, J.W., Ohlen, D.O. Measuring phenological variability from satellite imagery. *J. Veg. Sci.* 1994, *5*, 703–714.

Rembold, F., Atzberger, C., Rojas, O., Savin, I. Using low resolution satellite imagery for yield prediction and yield anomaly detection. *Remote Sens.* 2013, *5*(4), 1704–1733.

Rembold, F., Maselli, F. Estimating inter-annual crop area variation using multi-resolution satellite sensor images. *Int. J. Remote Sens.* 2004, *25*, 2641–2647.

Rembold, F., Maselli, F. Estimation of inter-annual crop area variation by the application of spectral angle mapping to low resolution multitemporal NDVI images. *Photogram. Eng. Remote Sens.* 2006, *72*, 55–62.

Reynolds, C.A., Yitayew, M., Slack, D.C., Hutchinson, C.F., Huetes, A., Petersen, M.S. Estimating crop yields and production by integrating the FAO Crop Specific Water Balance model with real-time satellite data and ground-based ancillary data. *Int. J. Remote Sens.* 2000, *21*, 3487–3508.

Roerink, G.J., Bojanowski, J.S., de Wit, A.J.W., Eerens, H., Supit, I., Leo, O., Boogaard, H.L. Evaluation of MSG-derived global radiation estimates for application in a regional crop model. *Agri. Forest Meteorol.* 2012, *160*, 36–47.

Rojas, O. Operational maize yield model development and validation based on remote sensing and agro-meteorological data in Kenya. *Int. J. Remote Sens.* 2007, 28, 3775–3793.

Romo-Leon, J.R., van Leeuwen, W.J.D., Castellanos-Villegas, A. Using remote sensing tools to assess land use transitions in unsustainable arid agro-ecosystems. *J. Arid Environ.* 2014, *106*, 27–35.

Rudorff, B.F.T., Aguiar, D.A., Silva, W.F., Sugawara, L.M., Adami, M., Moreira, M.A. Studies on the rapid expansion of sugarcane for ethanol production in São Paulo State (Brazil) using Landsat data. *Remote Sens.* 2010, *2*(4), 1057–1076.

Rudorff, B.F.T., Batista, G.T. Yield estimation of sugarcane based on agrometeorological-spectral models. *Remote Sens. Environ.* 1990a, *33*(3), 183–192.

Rudorff, B.F.T., Batista, G.T. Spectral response of wheat and its relationship to agronomic variables in the tropical region. *Remote Sens. Environ.* 1990b, *31*, 53–63.

Running, S.W., Nemani, R.R. Relating seasonal patterns of the AVHRR vegetation index to simulated photosynthesis and transpiration of forests in different climates. *Remote Sens. Environ.* 1988, *24*, 347–367.

Sakamoto, T., Wardlow, B.D., Gitelson, A.A. Detecting spatio-temporal changes of corn developmental stages in the U.S. corn belt using MODIS WDRVI data. *IEEE Trans. Geosci. Remote Sens.* 2011, *49*(6 part 1), art. no. 5682035, 1926–1936.

Sakamoto, T., Yokozawa, M., Toritani, H., Shibayama, M., Ishitsuka, N., Ohno, H. A crop phenology detection method using time-series MODIS data. *Remote Sens. Environ.* 2005, *96*, 366–374.

Schlerf, M., Atzberger, C. Vegetation structure retrieval in Beech and Spruce forests using spectrodirectional satellite data. *IEEE J. Sel. Top. Appl. Earth Obs. Remote Sens.* 2012, *5*, 8–17.

Seelan, S.K., Laguette, S. Casady, G.M., Seielstad, G.A. Remote sensing applications for precision agriculture: A learning community approach. *Remote Sens. Environ.* 2003, *88*(1–2), 157–169.

Shen, Y., Di, L., Wu, L., Yu, G., Tang, H., Yu, G., Shao, Y. Hidden Markov models for real-time estimation of corn progress stages using MODIS and meteorological data. *Remote Sens.* 2013, *5*(4), 1734–1753.

Silva, M.P.S., Câmara, G., Escada, M.I.S., Souza, R.C.M. Remote-sensing image mining: Detecting agents of land-use change in tropical forest areas. *Int. J. Remote Sens.* 2008, *29*(16), 4803–4822.

Silversmith, A., Tulchin, D. Crowdsourcing applications for agricultural development in Africa. 2013, USAID briefing paper.

Sinclair, T.R., Seligman, N.G. Crop modeling: From infancy to maturity. *Agron. J.* 1996, *88*, 698–704.

Soares, J., Williams, M., Jarvis, I. et al. *The G20 Global Agricultural Monitoring Initiative (GEO-GLAM)*, Technical report, 2011, p. 16.

Stathakis, D., Savin, I.Y., Nègre, T. Neuro-fuzzy modeling for crop yield prediction. In *Proceedings of the ISPRS Commission VII Symposium "Remote Sensing: From Pixels to Processes"*, Enschede, The Netherlands, May 8–11, 2006.

Stehman, S.V., Olofsson, P., Woodcock, C.E., Herold, M., Friedl, M.A. A global land-cover validation data set, II: Augmenting a stratified sampling design to estimate accuracy by region and land-cover class. *Int. J. Remote Sens.* 2012, *33*(22), 6975–6993.

Steinmetz, S., Guerif, M., Delecolle, R., Baret, F. Spectral estimates of the absorbed photosynthetically active radiation and light-use efficiency of a winter wheat crop subjected to nitrogen and water deficiencies. *Int. J. Remote Sens.* 1990, *11*, 1797–1808.

Stöckle, C.O., Donatelli, M., Nelson, R. CropSyst, a cropping systems simulation model. *Eur. J. Agron.* 2003, *18*, 289–307.

Stöckli, R., Vidale, P.L. European plant phenology and climate as seen in a 20-year AVHRR land-surface parameter dataset. *Int. J. Remote Sens.* 2004, *25*, 3303–3330.

Strahler, A.H., Boschetti, L., Foody, G.M. et al. *Global Land Cover Validation: Recommendations for Evaluation and Accuracy Assessment of Global Land Cover Maps.* EUR 22156 EN-DG. Luxembourg: Office for Official Publications of the European Communities, 2006, 48pp.

Supit, I., Hooijer, A.A., Diepen, C.A. Van *System Description of the WOFOST 6.0 Crop Simulation Model Implemented in CGMS.* Ispra, Italy: European Commission, 1994.

The Royal Society. Reaping the benefits: Science and the sustainable intensification of global agriculture, 2005. Available at: http://royalsociety.org/Reapingthebenefits (accessed on February 18, 2013).

Thenkabail, P.S. Global croplands and their importance for water and food security in the twenty-first century: Towards an ever green revolution that combines a second green revolution with a blue revolution. *Remote Sens.* 2010, *2*, 2305–2231.

Thenkabail, P.S., Biradar, C.M., Noojipady, P. et al. Global irrigated area map (GIAM), derived from remote sensing, for the end of the last millennium. *Int. J. Remote Sens.* 2009, *30*, 3679–3733.

Tilman, D., Balzer, C., Hill, J., Befort, B.L. Global food demand and the sustainable intensification of agriculture. *Proc. Natl. Acad. Sci. U. S. A.* 2011, *108*, 20260–20264.

Trnka, M., Rotter, R.P., Ruiz-Ramos, M., Kersebaum, K.C., Olesen, J. E., Zalud, Z., Semenov, M.A. Adverse weather conditions for European wheat production will become more frequent with climate change. *Nature Clim. Change*, 2014. doi:10.1038/nclimate2242.

Tucker, C.J. Red and photographic infrared linear combinations for monitoring vegetation. *Remote Sens. Environ.* 1979, *8*, 127–150.

Tucker, C.J., Holben, B.N., Elgin, J.H., Jr., McMurtrey, J.E., III. Relationship of spectral data to grain yield variation. *Photogramm. Eng. Remote Sens.* 1980, *46*, 657–666.

Tucker, C.J., Holben, B.N., Elgin, J.H., Jr., McMurtrey, J.E., III. Remote sensing of total dry-matter accumulation in winter wheat. *Remote Sens. Environ.* 1981, *11*, 171–189.

Tucker, C.J., Sellers, P.J. Satellite remote sensing of primary production. *Int. J. Remote Sens.* 1986, *7*, 1395–1416.

Tucker, C.J., Vanpraet, C., Boerwinkel, E., Gaston, A. Satellite remote sensing of total dry matter production in the Senegalese Sahel. *Remote Sens. Environ.* 1983, *13*, 461–474.

Udelhoven, T. TimeStats: A software tool for the retrieval of temporal patterns from global satellite archives. *IEEE J. Sel. Top. Appl. Earth Obs. Remote Sens.* 2011, *4*, 310–317.

Udelhoven, T., Stellmes, M., Del Barrio, G., Hill, J. Modelling the NDVI–rainfall relationship in Spain (1989–1999) using distributed lag models. *Int. J. Remote Sens.* 2008, *30*, 1961–1976.

Udelhoven, T., van der Linden, S., Waske, B. et al. Hypertemporal classification of large areas using decision fusion, *IEEE Geoscience and Remote Sensing Letters.* 2009, *6*(3), 592–596.

Ungaro, F., Calzolari, C., Busoni, E. Development of pedotransfer functions using a group method of data handling for the soil of the PianuraPadano-Veneta region of North Italy: Water retention properties. *Geoderma* 2005, *124*, 293–317.

Vancutsem, C., Marinho, E., Francois, K., See, L., Fritz, S. Harmonizing and combining existing land cover/land use datasets for cropland area monitoring at the African continental scale. *Remote Sens.* 2013, *5*, 19–41.

Verbeiren, S., Eerens, H., Picard, I., Bauwens, I., van Orshoven, J. Sub-pixel classification of SPOT-VGT time series for the assessment of regional crop areas in Belgium. *Int. J. Appl. Earth Obs. Geoinf.* 2008, *10*, 486–497.

Verbesselt, J., Hyndman, R., Newnham, G., Culvenor, D. Detecting trend and seasonal changes in satellite image time series. *Remote Sens. Environ.* 2010, *114*, 106–115.

Verstraete, M.M., Pinty, B., Myneni, R.B. Potential and limitations of information extraction on the terrestrial biosphere from satellite remote sensing. *Remote Sens. Environ.* 1996, *58*, 201–214.

Vieira, M.A., Formaggio, A.R., Rennó, C.D., Atzberger, C., Aguiar, D.A., Mello, M.P. Object Based Image Analysis and Data Mining applied to a remotely sensed Landsat time-series to map sugarcane over large areas. *Remote Sens. Environ.* 2012, *123*, 553–562.

Villalobos, F.J., Hall, A.J., Ritchie, J.T., Orgaz, F. OILCROP-SUN: A development, growth, and yield model of the sunflower crop. *Agron. J.* 1996, *88*, 403–415.

Vintrou, E., Sourmare, M., Bernard, S., Begue, A., Baron, C., Lo Seen, D. Mapping fragmented agricultural systems in the Sudano-Sahelian environments of Africa using Random Forest and ensemble metrics of coarse resolution MODIS imagery. *Photogramm. Eng. Remote Sens.* 2012, *78*, 839–848.

Viscarra Rossel, R.A., Cattle, S.R., Ortega, A., Fouad, Y. In situ measurements of soil colour, mineral composition and clay content by vis-NIR spectroscopy. *Geoderma* 2009, *150*, 253–266.

Vitousek, P.M., Aber, J.D., Howarth, R.W., Likens, G.E., Matson, P.A., Schindler, D.W., Schlesinger, W.H., Tilman, D.G. Human alteration of the global nitrogen cycle: Sources and consequences. *Ecol. Appl.* 1997, *7*, 737–750.

Vogelmann, J.E., Howard, S.M., Yang, L., Larson, C.R., Wylie, B.K., van Driel, N. Completion of the 1990s National Land Cover Data set for the conterminous United States from Landsat Thematic Mapper data and ancillary data sources. *Photogramm. Eng. Remote Sens.* 2001, *67*, 650–662.

Vuolo, F., Atzberger, C. Exploiting the classification performance of support vector machines with multi-temporal moderate-resolution imaging spectroradiometer (MODIS) data in areas of agreement and disagreement of existing land cover products. *Remote Sens.* 2012, *4*, 3143–3167.

Vuolo, F., Atzberger, C. Improving land cover maps in areas of disagreement of exist-ing products using NDVI time series of MODIS—Example for Europe. *Photogramm. Fernerkund. Geoinform.* 2014, *2014*(5), 393–407.

Vuolo, F., Dini, L., D'Urso, G. Retrieval of Leaf Area Index from CHRIS/PROBA data: An analysis of the directional and spectral information content. *Int. J. Remote Sens.* 2008, *29*, 5063–5072.

Wagner, W., Fröhlich, J., Wotawa, G., Stowasser, R., Staudinger, M., Hoffmann, C., Walli, A., Federspiel, C., Aspetsberger, M., Atzberger, C., Briese, C., Notarnicola, C., Zebisch, M., Boresch, A., Enenkel, M., Kidd, R., von Beringe, A., Hasenauer, S., Naeimi, V., Mücke, W., Addressing Grand Challenges in Earth Observation Science: The Earth Observation Data Centre for Water Resources Monitoring. *Proceedings of the ISPRS Technical Commission VII Symposium 2014*, Sunar, F., Altan, O., Taberner M., (ed.); ISPRS Annals of the Photogrammetry, Remote Sensing and Spatial Information Sciences, Volume II-7, 2014, 81–88.

Wagner, W., Naeimi, V., Scipal, K., Jeu, R., Martínez-Fernández, J. Soil moisture from operational meteorological satellites. *Hydrogeol. J.* 2007, *15*, 121–131.

Wardlow, B.D., Egbert, S. Large-area crop mapping using time-series MODIS 250 m NDVI data: An assessment for the U.S. Central Great Plains. *Remote Sens. Environ.* 2008, *112*, 1096–1116.

Wardlow, B.D., Egbert, S.L., Kastens, J.H. Analysis of time-series MODIS 250 m vegetation index data for crop classification in the US Central Great Plains. *Remote Sens. Environ.* 2007, *108*, 290–310.

Weiss, M., Troufleau, D., Baret, F., Chauki, H., Prévot, L., Olioso, A., Bruguier, N., Brisson, N. Coupling canopy functioning and radiative transfer models for remote sensing data assimilation. *Agric. Forest Meteorol.* 2001, *108*, 113–128.

White, M.A., Thornton, P.E., Running, S.W. A continental phenology model for monitoring vegetation responses to interannual climatic variability. *Global Biogeochem. Cycles* 1997, *11*, 217–234.

Whiteside, T., Ahmad, W. A comparison of object-oriented and pixel based classification methods for mapping land cover in North Australia. In: *Proceedings of Spatial Science Institute Biennial Conference SSC2005. Spatial Intelligence, Innovation and praxis: The national biennial Conference of the Spatial Science Institute.* Spatial Sciences Institute, Melbourne, Australia, 2005, pp. 1225–1231.

Witten, I., Frank, E. *Data Mining: Practical Machine Learning Tools and Techniques*, 2nd edn. San Francisco, CA: Morgan Kaufmann Publishers, 2005, 524p.

Xin, J., Yu, Z., van Leeuwen, L., Driessen, P.M. Mapping crop key phenological stages in the North China Plain using NOAA time series images. *Int. J. Appl. Earth Observ. Geoinform.* 2002, *4*(2), 109–117.

Xin, Q., Gong, P., Yu, C., Broich, M., Suyker, A.E., Myneni, R.B. A production efficiency model-based method for satellite estimates of corn and soybean yields in the midwestern US. *Remote Sens.* 2013, *5*, 5926–5943.

Zaks, D.P.M., Kucharik, C.J. Data and monitoring needs for a more ecological agriculture. *Environ. Res. Lett.* 2011, *6*, 1–10.

Zhan, X., Sohlberg, R.A., Townshend, J.R.G., DiMiceli, C., Carroll, M.L., Eastman, J.C., Hansen, M.C., DeFries, R.S. Detection of land cover changes using MODIS 250 m data. *Remote Sens. Environ.* 2002, *83*, 336–350.

Zhang, P., Anderson, B., Tan, B., Huang, D., Myneni, R. Potential monitoring of crop production using a satellite-based Climate-Variability Impact Index. *Agr. Forest Meteorol.* 2005, *132*, 344–358.

Zhang, X., Friedl, M.A., Schaaf, C.B., Strahler, A.H., Hodges, J.C.F., Gao, F., Reed, B.C., Huete, A. Monitoring vegetation phenology using MODIS. *Remote Sens. Environ.* 2003, *84*, 471–475.

5

Agricultural Systems Studies Using Remote Sensing

Agnès Bégué
CIRAD, UMR TETIS

Damien Arvor
CNRS, UMR 6554
LETG-Rennes-COSTEL

Camille Lelong
CIRAD, UMR TETIS

Elodie Vintrou
CIRAD, UMR TETIS

Margareth Simoes
Rio de Janeiro State University

Acronyms and Definitions

ABC	Brazilian Low Carbon Agriculture
ADI	Area Diversity Index
CLUI	Cultivated Land Utilization Index
DEM	Digital Elevation Model
EVI	Enhanced Vegetation Index
FAO	Food and Agriculture Organization
FEWS-NET	Famine Early Warning System-Network
GAP	Good Agricultural Practices
GEOBIA	GEOgraphic Object-Based Image Analysis
LULC	Land Use/Land Cover
MCI	Multiple Cropping Index
MODIS	MODerate resolution Imaging Spectroradiometer
NDVI	Normalized Difference Vegetation Index
SPAM	Spatial Allocation Model
USAID	United State Agency for International Development

5.1 Introduction

The world population is expected to reach 9.3 billion in 2050 (UN, 2010). To feed this population, the Food and Agriculture Organization's last global projection exercise forecasted that the world's agricultural production will need to increase by approximately 70% by 2050, compared with the 2005 production levels (FAO, 2011). Approximately 80% of the increased agricultural production will need to come from yield increases, and higher cropping intensities such as increased multiple cropping and/or shortening of fallow periods.

Such evolutions must cope with climate change (characterized by changing rainfall patterns and an increasing number of extreme weather events) and its consequences (changing distributions of plant and vector-borne diseases, and increased crop yield variability), more competition for land (increased competition between food and bioenergy production), and the associated increased environmental pressures (e.g., overexploitation of ground water resources, water quality degradation, and soil degradation). As a consequence, in addition to the need to increase crop production, another major agricultural challenge is the task of improving the management of natural resources, especially through the adoption of more environmental-friendly practices, such as ecological intensification or conservation agriculture. Major agricultural powers such as Europe and Brazil have launched ambitious programs, for example, the Good Agricultural Practices (GAP) guidelines and the ABC Program (Brazilian Low Carbon Agriculture Program), respectively. These programs give a special role to multifunctional landscapes to establish sustainable agriculture. Landscapes must be considered a whole land use system at the heart of human–nature relationships that need to be efficiently managed to preserve and restore ecosystem services (DeFries and Rosenzweig, 2010), and to contribute to sustainable solutions, especially regarding food security challenges (Verburg et al., 2013). In view of these global challenges, there is an urgent need to better characterize agricultural systems at the regional and global scales, with a particular emphasis on the various pathways toward agricultural intensification. Those systems are the key to understanding land use sustainability in agricultural territories.

Although everyone agrees on the need to qualify agricultural systems at the regional scale, few examples exist in the literature. Leenhardt et al. (2010) reviewed cropping system descriptions and locations at the regional scale, and concluded that both remain highly unclear for most world regions. The FAO continental farming system maps (Dixon et al., 2001) and the U.S. Agency for International Development (USAID) Famine Early Warning Systems Network (FEWS NET) national livelihood maps for Africa (USAID, 2009) are produced at very broad scales. More detailed, regional maps of rice areas in southeast Asia (Bridhikitti and Overcamp, 2012) or sugarcane areas in Brazil (Adami et al., 2012) have recently been produced using remote sensing data only. But these simple approaches, based on the dominant crop type with limited consideration of land management, are insufficient to draw a complete picture of coupled human–environment systems (Verburg et al., 2009).

So, evolving from traditional remote sensing land cover mapping to land use system mapping is not straightforward and requires processing new data, implementing new methods, and, above all, an enhanced integration between land science research disciplines (Verburg et al., 2009; Koschke et al., 2013). Vaclavik et al. (2013) derived a global representation of land use systems using land use intensity datasets, environmental conditions, and socioeconomic indicators. Land use intensity was derived from satellite-based land cover maps and subnational statistics. The authors noted that the scope of the study was limited, because the quality of the statistical datasets they used was geographically distributed unevenly worldwide. Kuemmerle et al. (2013) proposed a review of the current

input (crop type, cropping frequency, capital, labor intensity, etc.) and output (yields and carbon stock, etc.) land intensity metrics that could be provided directly or indirectly by satellite remote sensing. They concluded that satellite-based approaches are still experimental in that domain and cannot readily be applied across large areas. Despite these issues, new opportunities are arising.

The objective of the present study is to give an overview of remote sensing–based approaches for regional mapping of agricultural systems and to illustrate the diversity of these approaches through case studies. To do this, we propose and introduce a general framework, including satellite data and land mapping approaches, to characterize agricultural systems at different scales. These approaches are illustrated by three case studies representing a wide diversity of agricultural systems across the tropical world. Based on these case studies and a literature review, the opportunities and challenges for agricultural systems mapping at regional and global scales are discussed, and further research is proposed.

5.2 Roles of Remote Sensing in the Assessment of Agricultural Systems

5.2.1 Diversity of the Agricultural Systems in the World

To our knowledge, the most complete global agricultural map is the map produced by the FAO and the World Bank (Dixon et al., 2001), which covers the six main regions of the developing world. This map represents 72 farming systems (Figure 5.1a) that were defined according to (1) the available natural resource base (water, land, climate, altitude, etc.), (2) the dominant pattern of

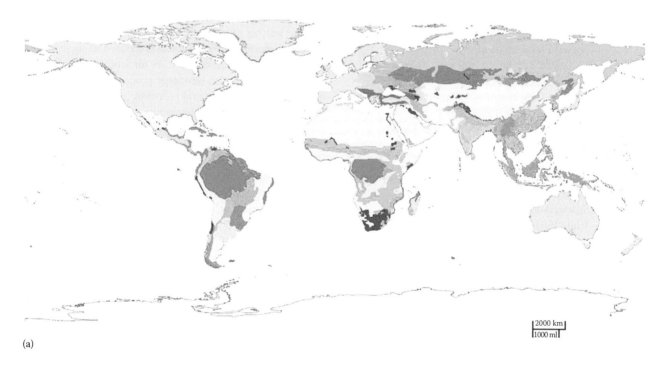

(a)

FIGURE 5.1 Farming system maps of the developing regions of the world (Dixon et al., 2001): (a) the original FAO 72-class map (see Dixon et al., 2001 for legend).

(Continued)

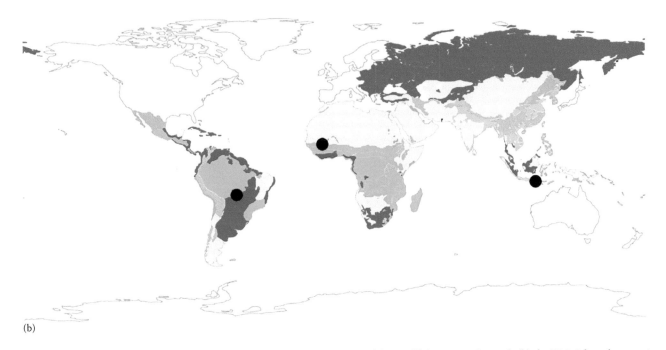

(b)

FIGURE 5.1 (*Continued*)   Farming system maps of the developing regions of the world (Dixon et al., 2001): (b) the FAO 8-broad categories (see Table 5.1 for legend). Black dots in (b) correspond to the location of the three case studies.

TABLE 5.1   Broad category of farming systems (Dixon et al., 2001)

	Farming System Name	Characteristics
⬤	Irrigated farming systems	Dominated by smallholder producers
⬤	Wetland rice based	Dominated by smallholder producers, dependent upon seasonal rains supplemented by irrigation
⬤	Rainfed farming systems in humid (and subhumid) areas	Dominated by smallholder producers, characterized by specific dominant crops or mixed crop-livestock systems
⬤	Rainfed farming systems in steep and highland areas	Dominated by smallholder producers, often mixed crop-livestock systems
⬤	Rainfed farming systems in dry or cold areas	Dominated by smallholder producers, with mixed crop-livestock and pastoral systems merging into systems with very low current productivity
⬤	Mixed large commercial and small holder	Dualistic, across a variety of ecologies and with diverse production patterns
⬤	Coastal artisanal fishing mixed	Dominated by smallholder producers, incorporates mixed farming elements
⬤	Urban based	Dominated by smallholder producers, typically focused on horticultural and livestock production

farm activities and household livelihoods, including relationship to markets, and (3) the intensity of production activities. These detailed farming systems are grouped into eight broad categories (Figure 5.1b; Table 5.1). It is interesting to note that seven out of the eight broad farming systems categories are based on smallholder producers (less than 2 ha land, according to FAO).

5.2.2 A Conceptual Framework Based on Land Mapping Issues

Remote sensing–based information can play different roles in the assessment of agricultural systems. Figure 5.2 illustrates how satellite images can help derive "land maps" (land cover, land use, and land use system maps; ① in Figure 5.2) using

various processing approaches (② in Figure 5.2). In the case of agriculture-dominated landscapes, these "land maps" can be interpreted as "agricultural system" maps (cropland, cropping system, and farming system; ③ in Figure 5.2).

Based on this framework, monitoring and mapping agricultural systems using remote sensing require clearly defined concepts and objects, that is, which "land maps" to monitor which "agricultural systems"? In the proposed conceptual framework (Figure 5.2), we tried to build bridges between the land maps (land cover, land use, and land use system), that can be obtained with the contribution of remote sensing data, and the agricultural systems (cropland, cropping system, and farming system, respectively) that are addressed in this chapter. These bridges are based on a set of definitions and hypotheses that are presented hereafter.

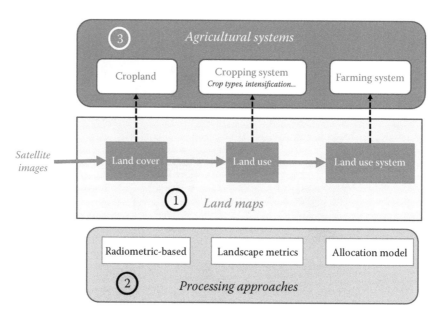

FIGURE 5.2   Conceptual framework used in this study.

- Land cover addresses the description of the land surface in terms of soil and vegetation layers, including natural vegetation, crops, and human structures (Burley, 1961). Land use refers to the purpose for which humans exploit the land cover (Lambin et al., 2006), including land management techniques (Verburg et al., 2009). In remote sensing-derived maps, mixed land use/land cover (LULC) legends are often used, because concepts concerning land cover and land use activities are closely related and, in many cases, can be used interchangeably (Anderson et al., 1976). Cropping systems are defined, at least, by the dominant crop type. Crop types, or at least crop groups (e.g., winter and summer crops; Atzberger and Rembold, 2013), are often represented in these satellite-derived LULC maps. More recently, information on the intensification mode, such as the use of irrigation (e.g., Thenkabail et al., 2010) or the adoption of multiple cropping (e.g., Arvor et al., 2011), appears in the LULC maps, improving the characterization of the cropping systems using remote sensing data.

- Land use system can be defined as a coupled human–environment system. It describes how land, as an essential resource, is being used and managed. Remote sensing data do not record human activities and thus cannot be directly used for land use system mapping. Photointerpreters historically used patterns, tones, textures, shapes, and site associations to derive initial land cover information into land use information (Anderson et al., 1976). This approach is consistent with Verburg et al. (2009) who proposed obtaining land use system maps from land cover maps supplemented by observations, inferred from landscape structures. Farming systems, defined by most experts as a combination of biophysical, socioeconomic, and human elements of a farm, can be seen as the land use system version for agriculture.

To conclude, LULC mapping can be obtained by classifying satellite images, while land use system mapping needs a larger view and must be approached on a larger scale (landscape scale).

5.2.3 Processing Approaches

A large panel of methods and tools to produce agricultural system maps from remote sensing data are described in the literature. The methods can be grouped into three types: radiometric-based method, landscape approach, and allocation models.

5.2.3.1 Radiometric-Based Methods

Radiometric-based methods are largely used for cropland and crop type mapping. Most of the publications report pixel or object-based classifications, and photointerpretation methods. Examples are discussed in Chapter 4, and this topic will not be further discussed in this chapter.

Beyond crop type, many examples concerning remote sensing and cropping practices are found in the literature. Most of the methods are based on statistical relationships between surface variables and image variables (reflectance, spectral index, texture index, etc.), while others use signal-processing techniques. The examples listed in Table 5.2 show that there is a strong link between the type of cropping practice and the sensor. High-resolution image primarily identifies intercropping and mixed-cropping, and agroforestry composition and structure. High image acquisition frequency usually helps to identify double cropping practices, crop types or groups of crop types, and sowing/harvest dates, while spectral richness is used to distinguish cultivars. Irrigation, crop residues, and tillage practices are mainly obtained through multispectral image analyses conducted at different scales depending on the structure of the fields.

TABLE 5.2   Literature Examples of Use of Remote Sensing for Mapping Cropping Practices

Cropping Practice	Crop (Sensor)	Example of Studies
Crop variety	Sugarcane (Hyperion)	Galvao et al. (2005)
	Sugarcane (Landsat)	Fortes and Dematte (2006)
Double cropping	Soybean and others (MODIS)	Arvor et al. (2011)
	Cereals (MODIS)	Qiu et al. (2014)
Harvest date	Sugarcane (SPOT)	Lebourgeois et al. (2007)
	Sugarcane (SPOT)	El Hajj et al. (2009)
Sowing date	Soybean (MODIS)	Maatoug et al. (2012)
Harvest mode	Sugarcane (Landsat, DMC)	Aguiar et al. (2011)
	Sugarcane (Landsat, CBERS)	Goltz et al. (2009)
Irrigation	Various crops (MODIS)	Gumma et al. (2011)
	Wheat (FORMOSAT, ASAR)	Hadria et al. (2009)
	Review	**Ozdogan et al. (2010)**
Crop residue	Various crops (Landsat)	Pacheco et al. (2006)
	Review	**Zhang et al. (2011)**
Tillage	Wheat (FORMOSAT, ASAR)	Hadria et al. (2009)
	Various crops (Landsat)	Sullivan et al. (2008)
Row orientation and width	Vineyard (aerial photos)	Delenne et al. (2008)
	Olive groves (QuickBird)	Amoruso et al. (2009)
	Orchards (Ikonos)	Aksoy et al. (2012)
	Vineyard, cereals (aerial photos)	Lefebvre et al. (2011)

Note: References in bold are review papers.

A detailed analysis of the publications on cropping practices and remote sensing shows that, even if the proportion of publications addressing this issue is increasing (4% of the total remote sensing and agriculture publications in the 1990s, and 9% currently), these publications primarily concern only one cropping practice at a time, and the analyses are generally conducted at local scale. Literature on the cropping system itself is still limited in terms of the number of publications (2% of the total published remote sensing and agriculture papers), and does not progress significantly.

5.2.3.2 Landscape Approach

Cropland and crop type maps can be viewed as a mosaic of patches, where the patches are the landscape elements. In that case, landscape metrics can be used to characterize the agricultural system. The term "landscape metrics" refers to indices developed for categorical map patterns (McGarigal, 2014). Landscape metrics exist at the patch, class (patch type), and landscape levels. At the class and landscape levels, some of the metrics quantify the landscape composition (e.g., the relative abundance of crop patch types), while others quantify the landscape configuration (e.g., the position, connectivity, or the edge-to-area ratios of the cropland).

Although very few articles use landscape metrics to characterize agrosystems compared to ecosystems (see review by Uuemaa et al., 2013), some of them use crop class metrics as an input for ecological studies (e.g., Pocas et al., 2011), and a few use landscape research for agricultural perspectives. The aim of these latter is generally to evaluate different policies on agricultural landscapes or to assess the sustainability of the agricultural

systems. For example, Plexida et al. (2014) discussed the role of modern cultivation methods in the simplification of landscape patterns in central Greece. They showed that the landscape in the agricultural lowlands was characterized by connectedness (high values of patch cohesion index) and simple geometries (low values of fractal dimension index), whereas the landscape pattern of the pastoral uplands was found to be highly diverse (high Shannon diversity index). Panigrahy et al. (2005) and Panigrahy et al. (2011) used landscape composition metrics to assess and evaluate the efficiency and sustainability of the agricultural systems in India. They proposed and calculated three indices, namely, the multiple cropping index (MCI), area diversity index (ADI), and cultivated land utilization index (CLUI), using three satellite-derived seasonal land cover maps. The MCI measures the cropping intensity as the number of crops grown temporally in a particular area over a period of 1 year, the ADI measures the multiplicity of crops or farm products planted in a single year, and the CLUI measures how efficient the available land area has been used over the year (see Panigrahy et al., 2005 for formula). The indices were categorized as high, medium, and low to evaluate the cropping system performance in each of the districts.

An example of landscape metrics based on the spatial configuration of the classes is given in Colson et al. (2011). They used eight landscape metrics to quantify and investigate the spatial patterns of cattle pasture and cropland throughout the states of Pará, Mato Grosso, Rondônia, and Amazonas, and concluded that these metrics showed evidence of a possible measure for discerning the patterns of agriculture attached to a certain state.

5.2.3.3 Spatial Allocation Modeling

Global cropping system maps (crop type and irrigation) are emerging at coarse resolution (see Anderson et al. (2014) for the description and comparison of these products). They are based on statistical data downscaled at the administrative level into grid-cell specific values. An illustrative example of spatial allocation is the spatial allocation model (SPAM), developed at the mesoscale by You and Wood (2006) and You et al. (2009), to spatially disaggregate crop production data (acreage and yield) within geopolitical units (e.g., countries or subnational provinces and districts), using a cross-entropy approach. The pixel-scale allocations are performed by compiling and merging relevant spatially explicit data, including production statistics, satellite-derived land cover data, biophysical crop suitability assessments, and population density. In such models, remote sensing is mainly used to locate cropland at regional scales as an input for the allocation models (to spatially disaggregate statistics data for instance), while the crop-determining factors are generally established by expertise or statistical analyses (Leenhardt et al., 2010). Recent examples showed that satellite images can also be used to understand and model the environmental drivers of cropping systems. For example, Jasinski et al. (2005) used a multiple logistic regression to model the role of environmental variables (vegetation type, soil type, altitude, slope, and rainfall) in the southeastern Amazonian cropland dynamics previously assessed using remote sensing data. More recently, Arvor et al. (2014) showed that the adoption of intensive double cropping practices was related to the spatial variability of rainfall regimes and favored by a high annual rainfall, a long rainy season, and a low variability of the onset date.

However, a major drawback of the spatial allocation models approach is that it is not always possible to obtain deterministic relations between easily accessible factors (climate, soil, etc.) and cropping system elements, especially in "intensive systems" compared to "traditional systems," which are more dependent on environmental factors (Figure 5.3). According to Jouve (2006), in

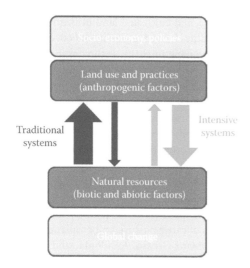

FIGURE 5.3 Relative weights of the determining factors in the traditional and intensive agricultural systems.

southern countries where traditional systems are important and make little use of modern means of production (mechanization, fertilization), the farmer's capacity to artificialize their environment and get rid of the environmental constraints is limited. In those cases, the relationship between the cropping systems and environmental conditions is strong, and the spatial distribution of the cropping systems reflects more the environmental differences than the farming differences. Additionally, the relationship can be identified at the rural community scale. Inversely, in intensive systems, the determining factors approach is more difficult to set up and the spatial allocation models can be more difficult to implement.

5.3 Examples of Agricultural System Studies Using Remote Sensing

Three case studies—agroforestry in Bali (Indonesia), double cropping in the southeastern Amazon (Brazil), and traditional rain-fed agriculture in Mali—were selected to illustrate the use of remote sensing for mapping agricultural systems. Two of them, Bali and Mali, are characterized by smallholder agriculture, while the Brazilian case is characterized by commercial agriculture (Figure 5.1b). These case studies are far from representing all of the possible uses of remote sensing, but they illustrate a diversity of technical and scientific approaches, while addressing some worldwide agricultural issues (geographic certification, agricultural system sustainability, food security, etc.).

5.3.1 Presentation of the Case Studies

5.3.1.1 Agro-Forestry in Bali

In tropical regions, small stakeholders' agroforestry is the most common traditional cropping system. It associates different crops inside a single plot, with multifunctional trees to produce fruits, cash-crops, wood, medicines, shading, or to conserve biodiversity in various proportions and organizations. This system allows a relative sustainability in food diversification, but not in incomes, which depends on the trading market fluctuations. Agroforestry is promoted by agronomists for environmental and livelihood quality, and is questioned by socioeconomists because of the cash-crop vulnerability. This emphasizes the need for evaluating the actual environmental, social, and economic benefits of such cropping systems. Remote sensing studies now propose new tools to objectively characterize the agroforestry systems at the intraplot scale (Peña-Barragán et al., 2004; Mougel and Lelong, 2008; Aksoy et al., 2012; Ursani et al., 2012; Coltri et al., 2013; Guillen-Climent et al., 2014), at the farm level (distribution among neighbors), and to replace it in the landscape matrix (Lei et al., 2012; Wästfelt et al., 2012). This allows associating different environmental, agricultural, and socioeconomic conditions in integrated analyses to understand the drivers of agricultural choices and resilience (Fox et al., 1994; Gobin et al., 2001; Kunwar, 2010), and the level of productivity and quality of the production.

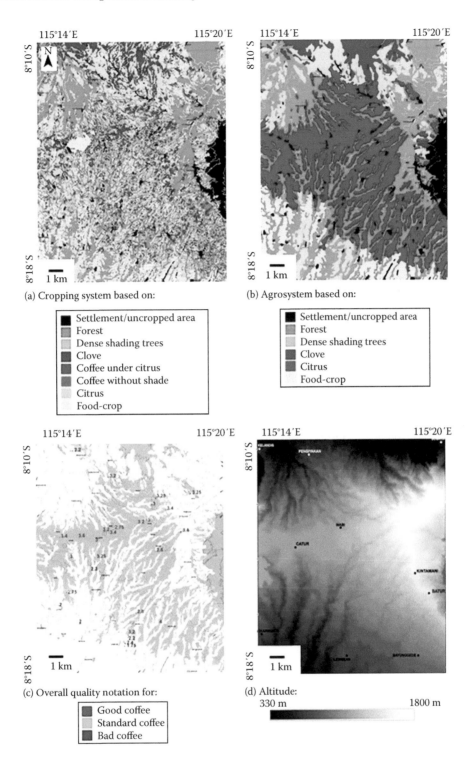

FIGURE 5.4   Map products in Central Bali: (a) main cropping systems map derived from QuickBird image visual interpretation, (b) agrosystems map derived from spatial analysis of the cropping system map, (c) location map of 40 sampled coffees and quality notation rate for each type of aromatic value, and (d) digital elevation model derived from topographic maps.

The case study presented in this chapter is situated in Bali, an active volcanic island of Indonesia. Coffee is cropped almost everywhere in the central highlands. The study focused on a 220 km² area located in Kintamani county, which is famous for its coffee crops. The landscape is shaped by the local topography, which ranges from 300 to 1800 m (Figure 5.4). This work aims at producing a cropping system map in order to understand coffee quality drivers, and helps in delimitating an area labeled by the distinction of the protected geographical indication on Arabica coffee.

5.3.1.2 Double Cropping in Southeastern Amazon

For nearly 40 years, the Brazilian southeastern Amazon experiences severe agricultural dynamics. Cropland expanded dramatically to support commercial cultivation of important commodities such as soybean, maize, and cotton. The severity of the agricultural dynamics explains the abundance of large-scale monitoring studies using remote sensing. To date, most remote sensing–based studies were carried out with MODIS data for three reasons: (1) monitoring such a large area requires a huge number of high remote sensing data to be processed, (2) high cloud cover rates during the rainy season prevent the acquisition of good-quality, high-resolution images during the cropping period, and (3) the mean field area is about 180 ha so that even 250 m medium-resolution images are valid for crop type mapping. Consequently, most MODIS-based approaches to date were based on the interpretation of vegetation index (NDVI or EVI) time series. Such time series have long been successfully used to estimate cropland areas, thus evidencing the rapid agricultural expansion during the 2000s (Anderson et al., 2003; Morton et al., 2006).

In Mato Grosso state, Arvor et al. (2012) estimated that net cropped areas increased by 43% between 2000 and 2007, reaching an area of 55,988 km². In the same time, farmers adopted new agricultural management practices to intensify the production process. The cultivation of two successive crops, such as soybean and cotton, benefits from a long rainy season (Arvor et al., 2014) and regular rainfall from mid-September to late May. In this context, the Mato Grosso case study aims at producing a cropping system map showing the main crop type and the intensification practices in relation to the rainfall, and a land use system map to analyze the agricultural transition in Mato Grosso.

5.3.1.3 Rain-Fed Agriculture in Mali

In the Sudano–Sahelian region, farming is the main source of income for many people, where millet and sorghum are the main food crops. The vast majority of the population (80%) consists of subsistence farmers. A few larger farms produce crops for sale (cash crops), mainly cotton and peanuts. In the Sudano–Sahelian

zone, the strong dependence on rain-fed agriculture implies exposure to climate variability in addition to the impacts that population growth has on food security. Key deliverables of food security systems for crop monitoring consist of early estimates of cultivated area and crop-type distribution, cropping practices, detection of growth anomalies, and crop yield estimates. Unfortunately, the national statistics can be deficient in insecure countries, and remote sensing has an important role to play in delivering information for crop monitoring (e.g., Hutchinson, 1991; Thenkabail et al., 2009). Remote sensing techniques face numerous challenges for crop mapping in regions where the cropland is fragmented, made of small, highly heterogeneous fields covered with many trees. In Mali, Vintrou et al. (2011) showed that 20%–40% of cropland classification errors using MODIS is inherent to the structure of the landscape.

Southern Mali case study aims at producing farming system map (food-producing, intensive, and mixed agricultures) in support to food security analyses (USAID, 2009). Because local factors, such as climate, soil, water availability, access to markets, and fertilizers, influence the agricultural systems, mapping these systems can help to determine which region and which population may be vulnerable to different hazards. Additionally, the cropping system map can be used for spatialized agrometeorological modeling and forecasting at regional scales (see example in Vintrou et al., 2014).

5.3.2 Remote Sensing Data and Methods

The data (remote sensing images, ancillary data) and methods used to produce agricultural maps are presented in Table 5.3 for the three case studies.

In Bali, a multispectral QuickBird image at 0.6 m resolution was photointerpreted to delineate the field limits and identify six cropping systems based on the field survey: citrus monocrop, coffee monocrop without shade, coffee associated with light shadow (citrus), coffee under dense shadow (erythrina, albizias, leucaenas, etc.), clove crops associated or not with coffee, and food-crops. An agrosystem map was then obtained by applying a majority filter (1 ha square corresponding to a dozen of

TABLE 5.3   Typology, Data, and Methods Used to Produce Agricultural System Maps for the Three Case Studies

Case Study (Area)	Agriculture Type	Satellite Data (Acquisition Year)	Other Data	Method	Map Products
Bali island (220 km²)	Smallholder agriculture	QuickBird bundle (2003)	DEM 760 ground survey points	Photointerpretation Spatial analysis (majority filter; 1 ha window)	Cropping system Farming system (agrosystem)
Mato Grosso (906,000 km²)	Commercial agriculture	MOD13Q1 EVI product (2005–2008)		Pixel-based supervised classification Landscape analysis (land cover and land use classes metrics; 770 km² window)	Crop type Cropping system Farming system
Southern Mali (165 790 km²)	Smallholder agriculture	MOD13Q1 NDVI product (2007) MCD12Q2 phenology product (2007)	100 villages field survey (2001–2004) Cropland map at 250 m resolution. Climate type, DEM and population 4000 villages location	Texture analysis (MODIS NDVI) Landscape analysis (land cover classes metrics; 100 km² window) Random forest classification	Farming system

crop plots) on the cropping system map, and was defined by its upper vegetation layer in four classes: citrus, clove, dense shading trees, and food crops. The term *agrosystem* is preferred here to the term *farming system* whose definition goes beyond what is studied in this case.

In Mato Grosso, MOD13Q1 EVI products acquired during 2005–2008 period were used to produce a cropping system map showing the main crop types (soybean, corn, and cotton), and their intensification practices (monocropping and double cropping). Arvor et al. (2013a) used a landscape approach to better characterize the land use system across the state. The strategy consisted of applying a regular grid where each cell represented an approximation of a district territory (a district was considered as an administrative sublevel, below the municipality level). There were 1,175 districts in Mato Grosso, a total of 906,000 km², and the grid cell was fixed at 27.75×27.75 km², approximating an area of 770 km². A set of landscape indices was then computed for each cell based on MODIS-based land use classifications and deforestation maps. Those indices referred to the proportion of wilderness areas, the proportion of cropped areas in deforested areas, and the proportion of intensive practices observed in cropped areas. Some thresholds were applied to identify different land use systems, such as presettlement area, noncropland occupation, cropland occupation, noncropland consolidation, cropland consolidation, noncropland intensifying, cropland intensifying, and intensive cropland.

In Mali, the field size and MODIS spatial resolution prevent from producing a crop type map. We then mapped directly the farming system map using a 3-class typology. This typology was defined at the village scale, and based on a field survey carried out in 100 villages in southern Mali (Soumare, 2008). The typology was created using expert knowledge, and considering the main crop types cultivated in the village and the intensification of production (use of fertilizers, equipment, livestock, etc.): the "food-producing agriculture" class groups the millet- and sorghum-based agricultural systems, the "intensive agriculture" class includes farms with maize and cotton, and the "mixed agriculture" class encompasses farms where both coarse grain (sorghum) and a cash crop (cotton) are found (Vintrou et al., 2012). A random forest algorithm (Breiman, 2001) was trained on the 100-village dataset, and on a set of 30 variables composed of 4 spectral metrics (annual maximum, annual mean, annual amplitude, and seasonal mean from May through November; MOD13Q1 product), 12 texture indices (maximum and mean of the variance and skewness indices, calculated with a pattern size of 7 MODIS pixels for March, June, and September; MOD13Q1 product), 7 phenology metrics (MCD12Q2 product), 3 spatial metrics (the fraction of cropped area, number of cultivated patches, and the mean cultivated patch size inside a 10×10 km² area centered on the village; MCD12Q1 product), 3 environmental indices (climate type, maximum, and mean of elevation), and 1 population index. All of the indices were extracted for cropland only.

The random forest model trained on the 100-village ground survey was applied to the 4000 villages in south Mali.

5.3.3 Results

5.3.3.1 Agroforestry in Bali

In Bali, the cropping system map is presented in Figure 5.4a. Photointerpretation performed on the ground-truth plots showed that confusion between citrus and coffee under citrus is less than 10%, whereas other class errors lie below 2%. The analysis of the distribution of each cropping system showed that the most frequent are the citrus-based crops (18%) and those shaded by large trees (15%), followed by the food-crops (12%), and the associated coffee and citrus crops (10%). The mean size of a plot is approximately 0.7 ha, but the clove plots are generally bigger (1.2 ha) and the food-crops are smaller (0.3 ha).

The agrosystem map is presented in Figure 5.4b. The citrus-based agrosystem is largely dominant. Coffee, as being cropped below the dominant trees, does not appear in the map legend.

At first glance, the cropping system and agrosystem spatial distribution looks complex because of a number of factors, such as a north/south contrast, altitude, and local geographic characteristics, such as river network density, slope, exposition to wind, and the presence of lava-flows and forests. The cropping and agrosystem maps were then used to analyze the distribution of each agricultural system, in relation to altimetry because of the strong relationship between coffee quality and altitude (Florinsky, 1998; Wintgens, 2004; Montagnon, 2006). The area covered by all of the different cropping systems is plotted for each 100 m-altitude bin, between 1000 and 1800 m in Figure 5.5a, while Figure 5.5b represents the altitude distribution for the area covered by the coffee-based cropping systems alone. The two principal coffee-based cropping systems were found to be those dominated by citrus or dense shading trees. The former is most common at high altitudes (64% from 1200 to 1400 m), while the latter dominates coffee crops at lower altitudes (68% below 1100 m). The third coffee-based cropping system, dominated by clove shading, covers a small acreage and is spatially restricted. It is present at the lowest altitudes, mainly below 1100 m (68%) and 1200 m (28%). The unshaded coffee monoculture is not typical in this territory.

The coffee samples location and sensorial quality rates were plotted in both the cropping and agrosystem maps to understand the spatial distribution of the coffee characteristics at the two scales (Figure 5.4c). A landscape analysis provided spatial and topographic distribution information about the three coffee quality classes, and helped to identify the relationships between quality of coffee beans and the local and regional environments. This integrated analysis suggests that good coffee is only found in the citrus-dominated agrosystem, even if it is not cultivated in association with citrus at the plot level, and cropped above 1200 m. This area was validated by both the coffee farmers and the traders, and accepted by the Indonesian government as the official limits of the labeled territory.

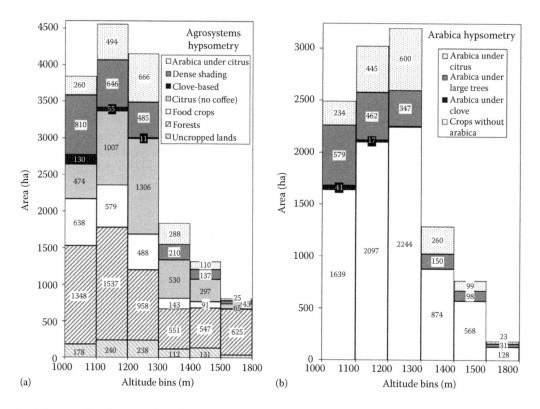

FIGURE 5.5 Areal altimetry distribution per bin of 100 m between 1000 and 1800 m in Kintamani territory in Bali: (a) per cropping system class, and (b) per coffee-based cropping system class (Arabica monocrop is not displayed, because it covers less than 2 ha).

5.3.3.2 Double Cropping in Southeastern Amazon

Time series of vegetation indices were used to detect crop types and cropping practices using an analysis of agricultural calendars. The producers undertake two successive harvests per rainy season: they cultivate soybean from late September to early February, and then cultivate maize or cotton until June or July. The double cropping systems show very different patterns in their vegetation index time series and can be easily discriminated (Arvor et al., 2011). The user's and producer's accuracies of the cropland were higher than 95%. Main crop types were also correctly detected (Figure 5.6) with good kappa index (0.68) and overall accuracy (74%). Once the double cropping classes are grouped (i.e., the "soybean + corn" and "soybean + cotton" classes; Figure 5.7a), the user's and producer's accuracies increased up to 95% and 86%, respectively. The main uncertainties to be considered in these maps refer to sorghum or millet that is sometimes sown after the soybean harvest (to prevent soil erosion from intense rainfall) and can thus be confused with maize. Such issue highlights a main limitation of EVI time series–based classification (different crops with similar agricultural calendars may be confused) that could be overcome with a better spatial and radiometric resolution (since only blue, red, and near-infrared bands are used to compute the EVI used in that work).

Beyond such limitations, those results are in agreement with results obtained by different authors (Galford et al., 2008; Arvor et al., 2011; Brown et al., 2013) who successfully mapped double cropping systems in Mato Grosso and confirmed the

generalization of such intensive practices. Arvor et al. (2012) estimated that the proportion of croplands permanently covered by double cropping vegetation during the rainy season increased from 35% to 62% between 2000 and 2007. This trend raises a major issue regarding the sustainability of cropland systems in Mato Grosso. Fu et al. (2013) proved that the length of the rainy season is decreasing in the southern Amazon, which leads to the question of whether the adoption of double cropping practices would still be viable in the changing climate. Even if intensive practices are a relevant strategy to contain deforestation, it raises new issues regarding agricultural sustainability in that region.

The land use system map shows a good overview of the soybean agricultural frontier in the southeastern Amazon (Figure 5.7b). It demonstrates the efficiency of public policies to simultaneously contain deforestation (through the creation of protected areas) and encourage crop expansion (through the construction of important infrastructures, such as the Trans-Amazonian roads).

5.3.3.3 Rain-Fed Agriculture in Mali

The random forest model classified the agricultural systems with an estimated overall accuracy of 60% calculated from out-of-bag observations (Figure 5.8). The "food-producing agriculture" class was dominant in the Sudano–Sahelian part of the area. Sorghum and millet are well adapted to this zone, because they are resistant, and have a short growth cycle of about 90 days. In the traditional cotton basin, the dominant system is agroforestry/pastoral agriculture mainly with rain-fed crops.

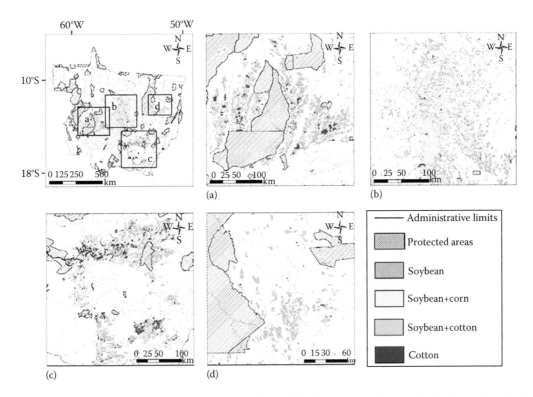

FIGURE 5.6 Maps of the three main crop types (soybean, corn, and cotton) for the 2006–2007 harvest for the four main agricultural regions in Mato Grosso: (a) Parecis plateau, (b) along the BR163 highway, (c) southeastern region, (d) eastern region (Arvor et al., 2011). Maps were obtained through supervised classification of MODIS vegetation index (EVI) time series.

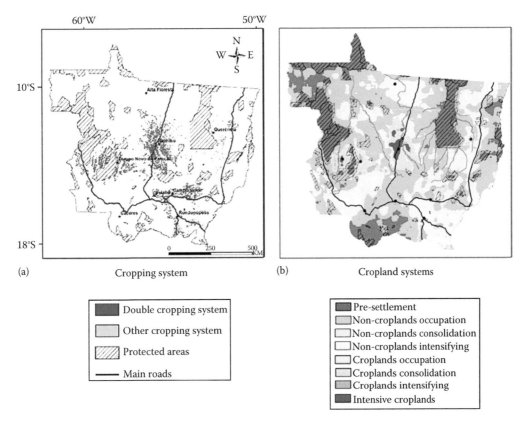

FIGURE 5.7 Maps of (a) cropping systems and (b) land use systems obtained from MODIS vegetation index time series and landscape analysis for the 2006–2007 harvest. (From Arvor, D. et al., *Appl. Geog.*, 32, 702, 2012; Arvor, D. et al., *GeoJournal*, 78, 833, 2013a.)

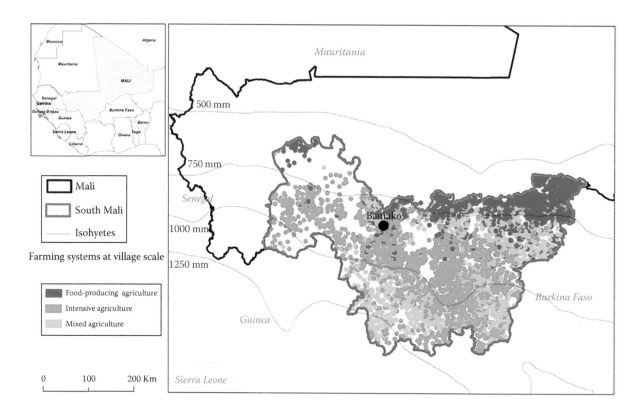

FIGURE 5.8 Village-based farming systems in South Mali predicted by the random forest model (Vintrou et al., 2012). The model was based on 100 village samples, and 30 MODIS-derived and socioenvironmental metrics calculated on agricultural areas.

Agriculture is focused on cotton, the main cash crop, and corresponds to the class "intensive agriculture". The Sudanian zone part of the area is also a cotton-based system zone, but is more diversified, with the simultaneous presence of "intensive agriculture" and "mixed agriculture" systems. The length of the rainy season in this region makes it possible to grow a wide range of species. Farmers usually cultivate different species and varieties to ensure a certain degree of production stability.

Class errors ranged from 30% to 50%. Globally, producer's and user's accuracies were reasonably balanced for each class (less than 10% difference): the village agricultural systems were estimated correctly. Misclassifications can be explained by three main factors: (1) the small size of the crop patches compared to the 250 m spatial resolution of MODIS sensor, and the natural and crop vegetation seasonal synchronization due to a short rainy season, (2) the size of the training dataset (100 villages), and (3) the definition of the classes (a rough proportion of different crop types, and crop intensification variables) that is expert dependent and includes variables that cannot be directly related to landscape features.

The analysis of the contribution of the different metrics (Figure 5.9) shows the role of the texture of the MODIS images in the classification of the cropland, even if the fields are not visible

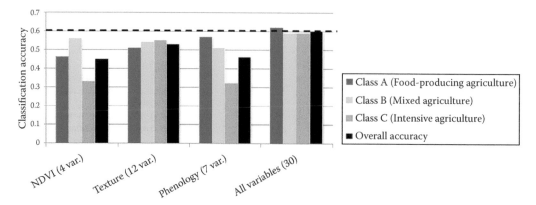

FIGURE 5.9 Accuracy of class and overall classification of random forest run with different sets of metrics (NDVI, texture, and phenology metrics). The dotted line corresponds to the overall accuracy obtained with all the metrics.

at the MODIS resolution. The field crop information is hidden in these broad images, but can be identified with landscape metrics, such as image texture indices. This indirect analysis was confirmed by Bisquert et al. (2015) who showed that the texture of broad-scale images is an important variable for land stratification in relation to land cover, even if the land cover units are not detectable.

5.4 Discussion

While remote sensing approaches have proven to be efficient for cropland (land cover) mapping, they still remain ill-suited for cropping system (land use) monitoring at the regional and global scales because of their inability to distinguish crop types and the associated practices (Monfreda et al., 2008). In this section, we consider the main present limitations of remote sensing studies for regional mapping of cropping systems, and introduce some emerging research areas to overcome such limitations. We then discuss the opportunity to work on an extended landscape agronomy approach.

5.4.1 Difficulties of Mapping the Cropping Systems at Regional Scales

Remote sensing–based land use maps suffer from uncertainties related to the spatial and temporal resolutions of the observing system, and to the landscape structure.

The spatial resolution issue is particularly true for smallholders agriculture (Figure 5.1b), for which remote sensing data are unable to resolve individual fields (Ozdogan, 2010). Rather than a sensor resolution issue, it should actually be considered as a scale issue to be addressed through the concept of H-resolution and L-resolution (Strahler et al., 1986; Blaschke et al., 2014). H- and L-resolution terms are different from high and low spatial resolution images as generally mentioned in remote sensing studies. In the latter, the resolution refers to the sensor spatial resolution independently of the geographic objects concerned. H-resolution model is valid when scene objects are much larger than the image spatial resolution; thus, several pixels may represent a single object (a field, a tree, etc.). Meanwhile, L-resolution model is when objects are much smaller than the image spatial resolution. An image may contain both H- and L-resolution information (Hay et al., 2001). Marceau et al. (1994) place the limit between H- and L- when the dimension of the resolution cells is ½–¾ the size of the objects of interest in the scene. This threshold should be a guide for assessing whether the analysis should be performed at H- or L-resolution.

- For an H-resolution situation—agricultural fields in Mato Grosso using MODIS sensor—a cropping system can be assessed directly by characterizing crop types and their associated cropping practices using inner field information (derived from relatively pure pixels).
- For an L-resolution situation—cropped trees in Bali using QuickBird sensor or cropped fields in Mali using MODIS sensor—pixels correspond to a mixture of different crop (or trees) types and other landscape elements (natural vegetation, water bodies, buildings, roads, etc.).

The temporal resolution issue in crop mapping is highly dependent on the environmental and agronomic conditions. For example, in tropical dry areas where rainfall is the main driver of vegetation growth (e.g., the Sahelian part of Mali), natural and cultivated vegetation are difficult to separate using phenology. In equatorial areas (e.g., Bali) characterized by a low seasonality, it is difficult to discriminate crops due to fluctuating crop calendars. However, even in regions with contrasted seasons (e.g., Mato-Grosso), different cropping systems with similar agricultural calendars cannot be separated using MODIS EVI time series. A better temporal resolution (less than 16 days) would surely improve crop discrimination in most of the agricultural systems.

The quality of the land maps produced by image pixel–based classification is usually evaluated using a set of indices (producer's accuracy, user's accuracy, overall accuracy, and kappa index), which are commonly calculated from an error matrix (or confusion matrix; see Congalton and Green, 1999). While such accuracy metrics have been widely accepted by the scientific community for a long time, they have also been regularly criticized (Pontius and Millones, 2011). These metrics tell nothing about the source of error that can be linked to the performance of the classification algorithm, or to the resolution of the remotely sensed data (Boschetti et al., 2004). For instance, Vintrou et al. (2011) using the Pareto boundary method showed that in Mali, 20%–40% of cropland classification errors using MODIS data is inherent to the landscape structure. In this context, new processing and evaluation approaches are required to better consider landscape properties in order to overcome these limitations and allow an efficient monitoring of farming systems at regional scale.

5.4.2 Emerging Remote Sensing Research

There was a challenge in land cover mapping in the 2000s, and today, there is a challenge in land use system mapping. It is an emerging area for the remote sensing community that needs to focus on land use and land function (Verburg et al., 2009). It requires developing new data, methods, and a further integration of the disciplines involved in land science research. These developments are presented hereafter according to the resolution situation (the direct and indirect cases).

When the landscape elements are larger than the pixel size (H-resolution situation), many examples in the literature showed that cropping practices can be directly assessed (Table 5.2). Except for rare examples of mapping crop type and cropping intensity in regions where the plot size is compatible with broad scale sensors (Mato Grosso case study), the research was mainly developed at local scale, and for one practice at a time. To further characterize regional scale cropping practices, research needs to focus on developing automatic or semiautomatic crop type classification procedures, and on the combination of different sensors to catch different practices in the same area. Another way to work at broader scales is to properly translate local findings to larger regions by using case study results from specific land functions (Verburg et al., 2009). This approach needs to define the spatial extent and function for the local studies representative of a region. Land stratification into homogeneous landscape units could be a way to reach this objective. Bisquert et al. (2015) showed that processing broad-scale remote sensing data with spectral and textural segmentation techniques permits to delineate radiometrically homogeneous landscapes that were consistent in terms of land cover.

When the landscape elements are smaller than the pixel size (L-resolution situation), research needs to focus on the role of landscape as an indirect mean to characterize the cropping systems. Research on landscape metrics for agricultural systems characterization must be pursued and enhanced. Furthermore, given the multidimensional nature of agricultural systems, focusing on multiple metrics within a system perspective is needed (Kuemmerle et al., 2013). As the current approaches based on the remote sensing data are not sufficient to develop a comprehensive understanding of situational changes for multiple land functions, remote sensing–based metrics should be completed by other types of metrics, such as socioeconomic descriptors (demography, ethnic spatialized data, etc.). To merge heterogeneous information, new data-processing tools, such as fuzzy logic and data-mining tools (Korting et al., 2013; Vintrou et al., 2013), must be tested to characterize and map agricultural systems and processes.

To implement both approaches (direct and indirect), the scientific community should benefit from recent promising advances in remote sensing such as geographic object-based image analysis (GEOBIA) and ontologies. GEOBIA is based on the hypothesis that partitioning an image into objects is related to the way humans conceptually organize the landscape to comprehend it (Hay and Castilla, 2008). It is actually based on two main components. First, a segmentation delineates regions (objects) of the image that have common attributes. Second, the approach incorporates the user (expert) knowledge in the image-processing operation to produce reliable maps. However, to date, GEOBIA is still limited by important issues related to product evaluation and knowledge management. Indeed, it is still unclear how to assess a segmentation quality (actually considered as an ill-posed problem), even if Clinton et al. (2010) proposed interesting metrics to assess GEOBIA segmentation goodness through vector-based measures. Although the integration of knowledge expertise in the image interpretation process is a main strength of GEOBIA, it can also be considered as a main limitation as long as two experts do not share a consensual knowledge (Belgiu et al., 2014). In such a context, it is likely that knowledge representation techniques such as ontologies can play a pivotal role (Arvor et al., 2013b). This point is especially meaningful in the case of agricultural system mapping where expert knowledge is crucial and often difficult to formalize. In case of land cover products, Comber et al. (2005) investigated the semantic and ontological meanings of land cover classes and concluded that current paradigms for reporting data quality do not adequately communicate the producer's knowledge. In case of land use and land use system products, the ontological meaning of the classes is even more difficult to formalize. For example, agricultural practices such as double cropping or no-tillage have been studied in various regions of the

world although they might correspond to different practices on the ground (different types of crop, different levels of soil management). In conclusion, ontologies might play an important role to allow the comparison of complex and heterogeneous land maps.

5.4.3 Toward an Extended Landscape Agronomy Approach

Landscape and agronomy have long been considered as closely associated. The first references on the relationship between agricultural landscapes and field management appeared in the 1990s (e.g., Baudry, 1993; Deffontaines et al., 1995) and addressed how farming activities produce agricultural landscapes, that is, explain the spatial distribution of patches (fields and associated boundaries). Since then, very few studies were published on the relationship between agricultural practices and landscape properties (e.g., Herzog et al., 2006; Galli et al., 2010). Most of the research focused on the characterization and understanding of landscape patterns to relate them to ecological issues (e.g., Baudry, 1993; Herzog et al., 2006). Benoit et al. (2012) argued why and how agronomy can contribute to landscape research with a conceptual model. He suggested a new perspective on farming practices as a crucial driver in the landscape pattern–agricultural process relationship. He proposed to develop a new research area called landscape agronomy (see also Rizzo et al., 2013) defined as "the relations among farming practices, natural resources and landscape patterns, which are involved in the dynamics of agricultural landscapes."

We previously mentioned that few landscape studies related to agricultural issues use remote sensing. Although it is now widely understood that cropping practices adopted in agricultural systems shape rural landscapes, we believe it is time to use landscape agronomy and quantitative remote sensing sciences. Applying concepts of landscape ecology to agricultural systems monitoring and mapping is a major idea. The case studies from Bali and Mato Grosso illustrate this new trend in landscape agronomy research and show that, thanks to its ability to identify spatial land cover patterns at local (Bali) and regional (Mato-Grosso) scales, remote sensing has become an essential source of information to identify agricultural systems.

However, landscape agronomy research will have to face the same limitations as landscape research. These limitations concern the numerous sources of error or uncertainty with producing land cover/land use maps from remote sensing imagery, and on the choice of the landscape metrics, which need to show a close association with the processes to be detected (Newton et al., 2009; Hurni et al., 2013). Another source of limitation is the simplistic approach of thematic mapping and the derivation of two-dimensional pattern metrics in landscape ecology (Newton et al., 2009), while remote sensing data have the potential to provide a three-dimensional characterization of landscapes and their component parts (as seen in Bali study case) and quantitative surface variables (as seen in Mali study case) that could be directly integrated in the landscape analysis. We showed through the Mali study case that the agricultural landscapes could be indirectly characterized by using a set of satellite-derived metrics (spectral, textural, and temporal metrics) without going through a thematic map of the crop types. This approach is essential when the ratio between the field size and the sensor spatial resolution is low (L-resolution)—land use maps cannot be produced, but it can also be used in H-resolution situation.

5.5 Conclusions

It is widely recognized that accurate, updated, and spatially explicit information on cropping systems (and thus cropping intensity) is urgently needed at the global and regional scales to provide insight into the direction and magnitude of world agricultural production in terms of crop type acreage and yield (Lobell and Field, 2007), and in terms of agricultural impacts on natural environments (Galford et al., 2008) and water resources (Thenkabail et al., 2010). Additionally, information is needed locally to monitor resources, preserve cultural landscapes, and for land certification (Jouve, 2006). This information is not yet included in the regional land cover datasets, and remote sensing entirely overlooks the actual practice of agriculture (what is grown, how it is grown, and what inputs are used) at this scale (Monfreda et al., 2008).

In this chapter, we showed how the current generation of Earth observation systems can contribute to the characterization of agricultural systems locally and regionally, through bibliographic studies and three case studies. We showed that remote sensing's ability to describe cropping systems is mainly related to the ratio between the spatial resolution of the sensor and the size of the landscape elements. This ratio determines if the fields (or the trees) can be identified by the observation system, or if the remote sensing data offers only a view of the cropland in its environment. This latter case leads to the development of new tools and methods to indirectly connect the spatial patterns of the agricultural landscape to the cropping management practices over large territories.

This bibliographic overview shows that the research community is now at a turning point where landscape research is not devoted to ecological issues only, but has started to embrace agricultural matters also. We believe that landscape agronomy is on the right track, and that the current and future Earth observing systems (such as Landsat8 and Sentinel-2) will have an important role to play in this new research area.

References

Adami, M., Mello, M.P., Aguiar, D.A., Rudorff, B.F.T., Souza, A.F., 2012. A web platform development to perform thematic accuracy assessment of sugarcane mapping in south-central Brazil. *Remote Sensing*, 4, 3201–3214.

Aguiar, D.A., Rudorff, B.F.T., Silva, W.F., Adami, M., Mello, M.P., 2011. Remote sensing images in support of environmental protocol: Monitoring the sugarcane harvest in Sao Paulo State, Brazil. *Remote Sensing*, 3, 2682–2703.

Aksoy, S., Yalniz, I.Z., Tasdemir, K., 2012. Automatic detection and segmentation of orchards using very high resolution imagery. *IEEE Transactions on Geoscience and Remote Sensing*, 50, 3117–3131.

Amoruso, N., Baraldi, A., Tarantino, C., Blonda, P., 2009. Spectral rules and geostatistic features for characterizing olive groves in Quickbird images. *IEEE International Geoscience and Remote Sensing Symposium*. IEEE, Cape Town, South Africa, vol 4, pp. 228–231.

Anderson, J.R., Hardy, E.E., Roach, J.T., Witmer, R.E., 1976. A land use and land cover classification system for use with remote sensor data. United States Government Printing Office, Washington, DC (US1), 41pp.

Anderson, L.O., Rojas, E., Shimabukuro, Y., 2003. *Avanço da soja sobre os ecossistemas cerrado e floresta no Estado do Mato Grosso*. XI Simpósio Brasileiro de Sensoriamento Remoto, Belo Horizonte, Brésil, pp. 19–25.

Anderson, W., You, L.Z., Wood, S., Wood-Sichra, U., Wu, W., 2014. A comparative analysis of global cropping systems models and maps. IFPRI, Washington, DC, 33pp.

Arvor, D., Dubreuil, V., Ronchail, J., Simoes, M., Funatsu, B.M., 2014. Spatial patterns of rainfall regimes related to levels of double cropping agriculture systems in Mato Grosso (Brazil). *International Journal of Climatology*, 34, 2622–2633.

Arvor, D., Dubreuil, V., Simoes, M., Bégué, A., 2013a. Mapping and spatial analysis of the soybean agricultural frontier in Mato Grosso, Brazil, using remote sensing data. *GeoJournal*, 78, 833–860.

Arvor, D., Durieux, L., Andres, S., Laporte, M.A., 2013b. Advances in geographic object-based image analysis with ontologies: A review of main contributions and limitations from a remote sensing perspective. *ISPRS Journal of Photogrammetry and Remote Sensing*, 82, 125–137.

Arvor, D., Jonathan, M., Meirelles, M.S.P., Dubreuil, V., Durieux, L., 2011. Classification of MODIS EVI time series for crop mapping in the state of Mato Grosso, Brazil. *International Journal of Remote Sensing*, 32, 7847–7871.

Arvor, D., Meirelles, M., Dubreuil, V., Begue, A., Shimabukuro, Y.E., 2012. Analyzing the agricultural transition in Mato Grosso, Brazil, using satellite-derived indices. *Applied Geography*, 32, 702–713.

Atzberger, C., Rembold, F., 2013. Mapping the spatial distribution of winter crops at sub-pixel level using AVHRR NDVI time series and neural nets. *Remote Sensing*, 5, 1335–1354.

Baudry, J., 1993. Landscape dynamics and farming systems—Problems of relating patterns and predicting ecological changes. In: Bunce, R.G.H., Ryszkowski, L., Paoletti, M.G. (Eds.), Landscape Ecology and Agroecosystems. Lewis Publishers, Boca Raton, FL, pp. 21–40.

Belgiu, M., Hofer, B., Hofmann, P., 2014. Coupling formalized knowledge bases with object-based image analysis. *Remote Sensing Letters*, 5, 530–538.

Benoit, M., Rizzo, D., Marraccini, E., Moonen, A.C., Galli, M., Lardon, S., Rapey, H., Thenail, C., Bonari, E., 2012. Landscape agronomy: A new field for addressing agricultural landscape dynamics. *Landscape Ecology*, 27, 1385–1394.

Bisquert, M., Bégué, A., Deshayes, M., 2015. A methodology for delineating landscapes patches at the regional scale using OBIA techniques applied to MODIS time series of vegetation and texture indices. *International Journal of Applied Earth Observation and Geoinformation*, 37, 72–82.

Blaschke, T., Hay, G.J., Kelly, M., Lang, S., Hofmann, P., Addink, E., Queiroz Feitosa, R., van der Meer, F., van der Werff, H., van Coillie, F., Tiede, D., 2014. Geographic object-based image analysis—Towards a new paradigm. *ISPRS Journal of Photogrammetry and Remote Sensing*, 87, 180–191.

Boschetti, L., Flasse, S.P., Brivio, P.A., 2004. Analysis of the conflict between omission and commission in low spatial resolution dichotomic thematic products: The Pareto Boundary. *Remote Sensing of Environment*, 91, 280–292.

Breiman, L., 2001. Random forests. *Machine Learning*, 45, 5–32.

Bridhikitti, A., Overcamp, T.J., 2012. Estimation of Southeast Asian rice paddy areas with different ecosystems from moderate-resolution satellite imagery. *Agriculture Ecosystems & Environment*, 146, 113–120.

Brown, J.C., Kastens, J.H., Coutinho, A.C., Victoria, D.D., Bishop, C.R., 2013. Classifying multiyear agricultural land use data from Mato Grosso using time-series MODIS vegetation index data. *Remote Sensing of Environment*, 130, 39–50.

Burley, T.M., 1961. Land use or land utilization? *The Professionnal Geographer*, 13, 18–20.

Clinton, N., Holt, A., Scraborough, J., Yan, L., Gong, P., 2010. Accuracy assessment measures for object-based image segmentation goodness. *Photogrammetric Engineering and Remote Sensing*, 76, 289–299.

Colson, F., Bogaert, J., Ceulemans, R., 2011. Fragmentation in the Legal Amazon, Brazil: Can landscape metrics indicate agricultural policy differences? *Ecological Indicators*, 11, 1467–1471.

Coltri, P.P., Zullo, J., Goncalves, R.R.D., Romani, L.A.S., Pinto, H.S., 2013. Coffee crop's biomass and carbon stock estimation with usage of high resolution satellites images. *IEEE Journal of Selected Topics in Applied Earth Observations and Remote Sensing*, 6, 1786–1795.

Comber, A.J., Fisher, P.F., Wadsworth, R.A., 2005. You know what land cover is but does anyone else?…an investigation into semantic and ontological confusion. *International Journal of Remote Sensing*, 26, 223–228.

Congalton, R.G., Green, K., 1999. *Assessing the Accuracy of Remotely Sensed Data: Principles and Practices*. Boca Raton, FL: Lewis Publishers Inc., 137pp.

Deffontaines, J.P., Thenail, C., Baudry, J., 1995. Agricultural systems and landscape patterns—How can we build a relationship? *Landscape and Urban Planning*, 31, 3–10.

DeFries, R., Rosenzweig, C., 2010. Toward a whole-landscape approach for sustainable land use in the tropics. *Proceedings of the National Academy of Sciences of the United States of America*, 107, 19627–19632.

Delenne, C., Durrieu, S., Rabatel, G., Deshayes, M., Bailly, J.S., Lelong, C., Couteron, P., 2008. Textural approaches for vineyard detection and characterization using very high spatial resolution remote sensing data. *International Journal of Remote Sensing*, 29, 1153–1167.

Dixon, J., Gulliver, A., Gibbon, D., 2001. *Farming Systems and Poverty: Improving Farmers' Livelihoods in a Changing World*. FAO, Rome, Italy and World Bank, Washington, DC.

El Hajj, M., Bégué, A., Guillaume, S., Martiné, J.-F., 2009. Integrating SPOT-5 time series, crop growth modeling and expert knowledge for monitoring agricultural practices—The case of sugarcane harvest on Reunion Island. *Remote Sensing of Environment*, 113, 2052–2061.

FAO, 2011. *Looking Ahead in World Food and Agriculture: Perspectives to 2050*. FAO, Rome, Italy, 539pp.

Florinsky, I.V., 1998. Combined analysis of digital terrain models and remotely sensed data in landscape investigations. *Progress in Physical Geography*, 22, 33–60.

Fortes, C., Dematte, J.A.M., 2006. Discrimination of sugarcane varieties using Landsat 7 ETM+ spectral data. *International Journal of Remote Sensing*, 27, 1395–1412.

Fox, J., Kanter, R., Yarnasarn, S., Ekasingh, M., Jones, R., 1994. Farmer decision-making and spatial variables in northern Thailand. *Environmental Management*, 18, 391–399.

Fu, R., Yin, L., Li, W.H., Arias, P.A., Dickinson, R.E., Huang, L., Chakraborty, S., Fernandes, K., Liebmann, B., Fisher, R., Myneni, R.B., 2013. Increased dry-season length over southern Amazonia in recent decades and its implication for future climate projection. *Proceedings of the National Academy of Sciences of the United States of America*, 110, 18110–18115.

Galford, G.L., Mustard, J.F., Melillo, J., Gendrin, A., Cerri, C.C., Cerri, C.E.P., 2008. Wavelet analysis of MODIS time series to detect expansion and intensification of row-crop agriculture in Brazil. *Remote Sensing of Environment*, 112, 576–587.

Galli, M., Bonari, E., Marraccini, E., Debolini, M., 2010. Characterisation of agri-landscape systems at a regional level: A case study in northern tuscany. *Italian Journal of Agronomy*, 5, 285–294.

Galvao, L.S., Formaggio, A.R., Tisot, D.A., 2005. Discrimination of sugarcane varieties in southeastern Brazil with EO-1 hyperion data. *Remote Sensing of Environment*, 94, 523–534.

Gobin, A., Campling, P., Deckers, J., Feyen, J., 2001. Integrated land resources analysis with an application to Ikem (southeastern Nigeria). *Landscape and Urban Planning*, 52, 95–109.

Goltz, E., Arcoverde, G.F.B., de Aguiar, D.A., Rudorff, B.F.T., Maeda, E.E., 2009. Data mining by decision tree for object oriented classification of the sugar cane cut kinds. *IEEE International Geoscience and Remote Sensing Symposium*. IEEE, New York, pp. 405–408.

Guillen-Climent, M.L., Zarco-Tejada, P.J., Villalobos, F.J., 2014. Estimating radiation interception in heterogeneous orchards using high spatial resolution airborne imagery. *IEEE Geoscience and Remote Sensing Letters*, 11, 579–583.

Gumma, M.K., Thenkabail, P.S., Nelson, A., 2011. Mapping irrigated areas using MODIS 250 meter time-series data: A study on krishna river basin (India). *Water*, 3, 113–131.

Hadria, R., Duchemin, B., Baup, F., Le Toan, T., Bouvet, A., Dedieu, G., Le Page, M., 2009. Combined use of optical and radar satellite data for the detection of tillage and irrigation operations: Case study in Central Morocco. *Agricultural Water Management*, 96, 1120–1127.

Hay, G.J., Castilla, G., 2008. Geographic object-based image analysis (GEOBIA): A new name for a new discipline. In: Blaschke, T., Lang, S., Hay, G.J. (eds.), *Object-Based Image Analysis*. Springer, Berlin, Germany, pp. 75–89.

Hay, G.J., Marceau, D., Dube, P., Bouchard, A., 2001. A multiscale framework for landscape analysis: Object-specific analysis and upscaling. *Landscape Ecology*, 16, 471–490.

Herzog, F., Steiner, B., Bailey, D., Baudry, J., Billeter, R., Bukácek, R., De Blust, G. et al., 2006. Assessing the intensity of temperate European agriculture at the landscape scale. *European Journal of Agronomy*, 24, 165–181.

Hurni, K., Hett, C., Epprecht, M., Messerli, P., Heinimann, A., 2013. A texture-based land cover classification for the delineation of a shifting cultivation landscape in the Lao PDR using landscape metrics. *Remote Sensing*, 5, 3377–3396.

Hutchinson, C.F., 1991. Uses of satellite data for famine early warning in sub-Saharan Africa. *International Journal of Remote Sensing*, 12, 1405–1421.

Jasinski, E., Morton, D., DeFries, R., Shimabukuro, Y., Anderson, L., Hansen, M., 2005. Physical landscape correlates of the expansion of mechanized agriculture in Mato Grosso, Brazil. *Earth Interactions*, 9, 1–18.

Jouve, P., 2006. Cropping systems and farming land space organisation: A comparison between temperate and tropical farming systems. *Cahiers Agricultures*, 15, 255–260.

Korting, T.S., Fonseca, L.M.G., Camara, G., 2013. GeoDMA-geographic data mining analyst. *Computers & Geosciences*, 57, 133–145.

Koschke, L., Fürst, C., Lorenz, M., Witt, A., Frank, S., Makeschin, F., 2013. The integration of crop rotation and tillage practices in the assessment of ecosystem services provision at the regional scale. *Ecological Indicators*, 32, 157–171.

Kuemmerle, T., Erb, K., Meyfroidt, P., Muller, D., Verburg, P.H., Estel, S., Haberl, H. et al., 2013. Challenges and opportunities in mapping land use intensity globally. *Current Opinion in Environmental Sustainability*, 5, 484–493.

Kunwar, P., Kachhwaha, T.S., Kumar, A., Agrawal, A.K., Singh, A.N., Mendiratta, N., 2010. Use of high-resolution IKONOS data and GIS technique for transformation of landuse/landcover for sustainable development. *Current Science*, 98, 204–212.

Lambin, E.F., Geist, H.J., Rindfass, R.R., 2006. Introduction: Local processes with global impacts. In: Lambin, E.F., Geist, H.J. (eds.), *Land-Use and Land-Cover Change: Local Processes and Global Impacts*. Springer-Verlag, Berlin, Germany, pp. 1–8.

Lebourgeois, V., Bégué, A., Degenne, P., Bappel, E., 2007. Improving sugarcane harvest and planting monitoring for smallholders with geospatial technology: The Reunion Island experience. *International Sugar Journal*, 109, 109–117.

Leenhardt, D., Angevin, F., Biarnes, A., Colbach, N., Mignolet, C., 2010. Describing and locating cropping systems on a regional scale. A review. *Agronomy for Sustainable Development*, 30, 131–138.

Lefebvre, A., Corpetti, T., Moy, L.H., 2011. Estimation of the orientation of textured patterns via wavelet analysis. Pattern Recognition Letters, 32, 190–196.

Lei, Z., Bingfang, W., Liang, Z., Peng, W., 2012. Patterns and driving forces of cropland changes in the Three Gorges Aarea, China. *Regional Environmental Change*, 12, 765–776.

Lobell, D.B., Field, C.B., 2007. Global scale climate—Crop yield relationships and the impacts of recent warming. *Environmental Research Letters*, 2: 014002, 7pp.

Maatoug, L., Arvor, D., Simoes, M., Bégué, A., 2012. Monitoring crop phenology in Mato Grosso (Brazil) using remote sensing data. XV Sociedad Especialistas Latino-americana en PErcepción Remota (SELPER) Symposium, Cayenne, Guyane (FR), 8 pp.

Marceau, D.J., Howarth, P.J., Gratton, D.J., 1994. Remote-sening and the measurement of geographical entities in a forested environment. 1. The scale and spatial aggregation problem. *Remote Sensing of Environment*, 49, 93–104.

McGarigal, K., 2014. FRAGSTATS documentation. 182pp. Available at the following web site: http://www.umass.edu/landeco/research/fragstats/documents/fragstats.help.4.2.pdf (Accessed on April 2015).

Monfreda, C., Ramankutty, N., Foley, J.A., 2008. Farming the planet: 2. Geographic distribution of crop areas, yields, physiological types, and net primary production in the year 2000. *Global Biogeochemical Cycles*, 22 (1), GB1022, 19pp.

Montagnon, C., 2006. *Coffee: Terroirs and Quality*. Cemagref/INRA/CIRAD, Versailles, France.

Morton, D., DeFries, R., Shimabukuro, Y., Anderson, L., Arai, E., del Bon Espirito-Santo, F., Freitas, R., Morisette, J., 2006. Cropland expansion changes deforestation dynamics in the southern Brazilian Amazon. *Proceedings of the National Academy of Sciences of the United States of America*, 103, 14637–14641.

Mougel, B., Lelong, C., 2008. Classification and information extraction in very high resolution satellite images for tree crops monitoring. *28th EARSel Symposium Remote Sensing for a Changing Europe*, Istanbul, Turkey, 13pp.

Newton, A.C., Hill, R.A., Echeverria, C., Golicher, D., Benayas, J.M.R., Cayuela, L., Hinsley, S.A., 2009. Remote sensing and the future of landscape ecology. *Progress in Physical Geography*, 33, 528–546.

Ozdogan, M., 2010. The spatial distribution of crop types from MODIS data: Temporal unmixing using independent component analysis. *Remote Sensing of Environment*, 114, 1190–1204.

Pacheco, A., McNairn, H., Smith, A.M., 2006. Multispectral indices and advanced classification techniques to detect percent residue cover over agricultural crops using Landsat data. *Proceedings of SPIE, Remote Sensing and Modeling of Ecosystems for Sustainability III*, 62981C, San Diego, California, USA.

Panigrahy, S., Manjunath, K.R., Ray, S.S., 2005. Deriving cropping system performance indices using remote sensing data and GIS. *International Journal of Remote Sensing*, 26, 2595–2606.

Panigrahy, S., Ray, S.S., Manjunath, K.R., Pandey, P.S., Sharma, S.K., Sood, A., Yadav, M., Gupta, P.C., Kundu, N., Parihar, J.S., 2011. A spatial database of cropping system and its characteristics to aid climate change impact assessment studies. *Journal of the Indian Society of Remote Sensing*, 39, 355–364.

Peña-Barragán, J.M., Jurado-Expósito, M., López-Granados, F., Atenciano, S., Sánchez-de la Orden, M., Garcia-Ferrer, A., Garcia-Torres, L., 2004. Assessing land-use in olive groves from aerial photographs. *Agriculture, Ecosystems and Environment*, 103, 117–122.

Plexida, S.G., Sfougaris, A.I., Ispikoudis, I.P., Papanastasis, V.P., 2014. Selecting landscape metrics as indicators of spatial heterogeneity—A comparison among Greek landscapes. *International Journal of Applied Earth Observation and Geoinformation*, 26, 26–35.

Pocas, I., Cunha, M., Pereira, L.S., 2011. Remote sensing based indicators of changes in a mountain rural landscape of Northeast Portugal. *Applied Geography*, 31, 871–880.

Pontius, R.G., Millones, M., 2011. Death to Kappa: Birth of quantity disagreement and allocation disagreement for accuracy assessment. *International Journal of Remote Sensing*, 32, 4407–4429.

Qiu, B.W., Zhong, M., Tang, Z.H., Wang, C.Y., 2014. A new methodology to map double-cropping croplands based on continuous wavelet transform. *International Journal of Applied Earth Observation and Geoinformation*, 26, 97–104.

Rizzo, D., Marraccini, E., Lardon, S., Rapey, H., Debolini, M., Benoit, M., Thenail, C., 2013. Farming systems designing landscapes: Land management units at the interface between agronomy and geography. *Geografisk Tidsskrift: Danish Journal of Geography*, 113, 71–86.

Sullivan, D.G., Strickland, T.C., Masters, M.H., 2008. Satellite mapping of conservation tillage adoption in the Little River experimental watershed, Georgia. *Journal of Soil and Water Conservation*, 63, 112–119.

Soumare, M., 2008. Dynamique et Durabilité des Systemes Agraires à Base de Coton au Mali. Université de Paris X Nanterre (FR), Paris, France, 373pp.

Strahler, A.H., Woodcock, C.E., Smith, J.A., 1986. On the nature of models in remote sensing. *Remote Sensing of Environment*, 20, 121–139.

Thenkabail, P.S., Hanjra, M.A., Dheeravath, V., Gumma, M., 2010. A holistic view of global croplands and their water use for ensuring global food security in the 21st century through advanced remote sensing and non-remote sensing approaches. *Remote Sensing*, 2, 211–261.

Thenkabail, P., Lyon, G.J., Turral, H., Biradar, C.M., 2009. *Remote Sensing of Global Croplands for Food Security*. CRC Press, Taylor & Francis Group, Boca Raton, FL.

UN, 2010. *World Population Prospects, the 2010 Revision*. United Nations, Department of Economic and Social Affairs, Population Division, New York, ST/ESA/SER.A/307; Sales No. E.11.XIII.6.

Ursani, A.A., Kpalma, K., Lelong, C.C.D., Ronsin, J., 2012. Fusion of textural and spectral information for tree crop and other agricultural cover mapping with very-high resolution satellite images. *IEEE Journal of Selected Topics in Applied Earth Observations and Remote Sensing*, 5, 225–235.

USAID, 2009. Application of the livelihood zone maps and profiles for food security analysis and early warning. Guidance for Famine Early Warning Systems Network (FEWS NET) Representatives and Partners. USAID, Washington, DC, pp. 23.

Uuemaa, E., Mander, Ü., Marja, R., 2013. Trends in the use of landscape spatial metrics as landscape indicators: A review. *Ecological Indicators*, 28, 100–106.

Vaclavik, T., Lautenbach, S., Kuemmerle, T., Seppelt, R., 2013. Mapping global land system archetypes. *Global Environmental Change*, 23, 1637–1647.

Verburg, P.H., Mertz, O., Erb, K.-H., Haberl, H., Wu, W., 2013. Land system change and food security: Towards multi-scale land system solutions. *Current Opinion in Environmental Sustainability*, 5, 494–502.

Verburg, P.H., van de Steeg, J., Veldkamp, A., Willemen, L., 2009. From land cover change to land function dynamics: A major challenge to improve land characterization. *Journal of Environmental Management*, 90, 1327–1335.

Vintrou, E., Begue, A., Baron, C., Saad, A., Lo Seen, D., Traore, S.B., 2014. A comparative study on satellite- and model-based crop phenology in West Africa. *Remote Sensing*, 6, 1367–1389.

Vintrou, E., Desbrosse, A., Begue, A., Traore, S., Baron, C., Lo Seen, D., 2011. Crop area mapping in West Africa using landscape stratification of MODIS time series and comparison with existing global land products. *International Journal of Applied Earth Observation and Geoinformation*, 14, 83–93.

Vintrou, E., Ienco, D., Begue, A., Teisseire, M., 2013. Data mining, a promising tool for large-area cropland mapping. *IEEE Journal of Selected Topics in Applied Earth Observations and Remote Sensing*, 6, 2132–2138.

Vintrou, E., Soumare, M., Bernard, S., Begue, A., Baron, C., Lo Seen, D., 2012. Mapping fragmented agricultural systems in the Sudano-Sahelian environments of Africa using random forest and ensemble metrics of coarse resolution MODIS imagery. *Photogrammetric Engineering and Remote Sensing*, 78, 839–848.

Wästfelt, A., Tegenu, T., Nielsen, M.M., Malmberg, B., 2012. Qualitative satellite image analysis: Mapping spatial distribution of farming types in Ethiopia. *Applied Geography*, 32, 465–476.

Wintgens, J.-N., 2004. Factors influencing the quality of green coffee. In: Wintgens, J.-N. (ed.), *Coffee: Growing, Processing, Sustainable Production: A Guidebook for Growers, Processors, Traders, and Researchers*, Wiley-VCH, Weihnheim, Germany, pp. 789–809.

You, L.Z., Wood, S., 2006. An entropy approach to spatial disaggregation of agricultural production. *Agricultural Systems*, 90, 329–347.

You, L.Z., Wood, S., Wood-Sichra, U., 2009. Generating plausible crop distribution maps for Sub-Saharan Africa using a spatially disaggregated data fusion and optimization approach. *Agricultural Systems*, 99, 126–140.

Zhang, M., Li, Q.Z., Meng, J.H., Wu, B.F., 2011. Review of crop residue fractional cover monitoring with remote sensing. *Spectroscopy and Spectral Analysis*, 31, 3200–3205.

6

Global Food Security Support Analysis Data at Nominal 1 km (GFSAD1km) Derived from Remote Sensing in Support of Food Security in the Twenty-First Century: Current Achievements and Future Possibilities

Pardhasaradhi Teluguntla
U.S. Geological Survey
and
Bay Area Environmental
Research Institute

Prasad S. Thenkabail
U.S. Geological Survey

Jun Xiong
U.S. Geological Survey
and
Northern Arizona University

Murali Krishna Gumma
International Crops Research
Institute for the Semi Arid Tropics

Chandra Giri
U.S. Geological Survey,
(EROS) Center

Cristina Milesi
NASA Ames Research Center

Mutlu Ozdogan
University of Wisconsin

Russell G. Congalton
University of New Hampshire

James Tilton
NASA Goddard Space Flight Center

Temuulen Tsagaan
Sankey
Northern Arizona University

Richard Massey
Northern Arizona University

Aparna Phalke
University of Wisconsin

Kamini Yadav
University of New Hampshire

Acronyms and Definitions

ACCA	Automated cropland classification algorithm
ASTER	Advanced spaceborne thermal emission and reflection radiometer
AVHRR	Advanced very-high-resolution radiometer
AWiFS	Advanced wide field sensor
CDL	The Cropland Data Layer (CDL) was created by the USDA, National Agricultural Statistics Service
CEOS	Committee on Earth Observing Satellites (CEOS)
EDS	Euclidean distance similarity
FPA	Full pixel areas
GCAD	Global cropland area database
GCE	Global cropland extent
GCE V1.0	Global cropland extent version 1.0
GDEM	ASTER-derived digital elevation data
GEO	Group on Earth Observations
GEOSS	Global Earth Observation System of Systems
GFSAD	Global food security support analysis data
GIMMS	Global Inventory Modeling and Mapping Studies
JERS SAR	Japanese Earth Resources Satellite-1 (JERS-1)
ISDB IA	Ideal Spectra Data Bank on Irrigated Areas
LEDAPS	Landsat Ecosystem Disturbance Adaptive Processing System
MFDC	Mega File Data Cube
MODIS	Moderate-resolution imaging spectroradiometer
MSAS	Modified spectral angle similarity
NASS	National Agricultural Statistics Service of USDA
NDVI	Normalized difference vegetation index
NOAA	National Oceanic and Atmospheric Administration
SAR	Synthetic aperture radar
SCS	Spectral correlation similarity
SIT	Strategic Implementation Team
SMT	Spectral matching techniques
SPA	Subpixel areas
SPOT	Système Pour l'Observation de la Terre
SSV	Spectral similarity value
USDA	United States Department of Agriculture
USGS	United States Geological Survey
VGT	Vegetation sensor of SPOT satellite
VHRI	Very-high-resolution imagery
VHRR	Very-high-resolution radiometer

6.1 Introduction

The precise estimation of the global agricultural cropland—extents, areas, geographic locations, crop types, cropping intensities, and their watering methods (irrigated or rain-fed; type of irrigation)—provides a critical scientific basis for the development of water and food security policies (Thenkabail et al., 2010, 2011, 2012, Turral et al., 2009). By year 2100, the global human population is expected to grow to 10.4 billion under median fertility variants or higher under constant or higher fertility variants (Table 6.1) with over three-quarters living in developing countries and in regions that already lack the capacity to produce enough food. With current agricultural practices, the increased demand for food and nutrition would require about 2 billion hectares of additional cropland, about twice the equivalent to the land area of the United States, and lead to significant increases in greenhouse gas emissions (GHG) associated with agricultural practices and activities (Tillman et al., 2011). For example, during 1960–2010, world population more than doubled from 3 to 7 billion. The nutritional demand of the population also grew swiftly during this period from an average of about 2000 calories per day per person in 1960 to nearly 3000 calories per day per person in 2010. The food demand of increased population along with increased nutritional demand during this period was met by the "green revolution," which more than tripled the food production, even though croplands decreased from about 0.43 ha per capita to 0.26 ha per capita (FAO, 2009; Funk and Brown, 2009). The increase in food production during the green revolution was the result of factors such as: (1) expansion of irrigated croplands, which had increased in 2000 from 130 Mha in the 1960s to between 278 Mha (Siebert et al., 2006) and 467 Mha (Thenkabail et al., 2009a,b,c), with the larger estimate due to consideration of cropping intensity; (2) increase in yield and per capita production of food (e.g., cereal production from 280 to 380 kg/person and meat from 22 to 34 kg/person (McIntyre, 2008); (3) new cultivar types (e.g., hybrid varieties of wheat and rice, biotechnology); and (4) modern agronomic and crop management practices (e.g., fertilizers, herbicide, pesticide applications).

Although modern agriculture met the challenge to increase food production last century, lessons learned from the twentieth century "green revolution" and our current circumstances impact the likelihood of another such revolution. The intensive

TABLE 6.1 World Population (Thousands) Under All Variants, 1950–2100

Year	Medium Fertility Variant	High Fertility Variant	Low Fertility Variant	Constant Fertility Variant
1950	2,529,346	2,529,346	2,529,346	2,529,346
1955	2,763,453	2,763,453	2,763,453	2,763,453
1960	3,023,358	3,023,358	3,023,358	3,023,358
1965	3,331,670	3,331,670	3,331,670	3,331,670
1970	3,685,777	3,685,777	3,685,777	3,685,777
1975	4,061,317	4,061,317	4,061,317	4,061,317
1980	4,437,609	4,437,609	4,437,609	4,437,609
1985	4,846,247	4,846,247	4,846,247	4,846,247
1990	5,290,452	5,290,452	5,290,452	5,290,452
1995	5,713,073	5,713,073	5,713,073	5,713,073
2000	6,115,367	6,115,367	6,115,367	6,115,367
2005	6,512,276	6,512,276	6,512,276	6,512,276
2010	6,916,183	6,916,183	6,916,183	6,916,183
2015	7,324,782	7,392,233	7,256,925	7,353,522
2020	7,716,749	7,893,904	7,539,163	7,809,497
2025	8,083,413	8,398,226	7,768,450	8,273,410
2030	8,424,937	8,881,519	7,969,407	8,750,296
2035	8,743,447	9,359,400	8,135,087	9,255,828
2040	9,038,687	9,847,909	8,255,351	9,806,383
2045	9,308,438	10,352,435	8,323,978	10,413,537
2050	9,550,945	10,868,444	8,341,706	11,089,178
2055	9,766,475	11,388,551	8,314,597	11,852,474
2060	9,957,399	11,911,465	8,248,967	12,729,809
2065	10,127,007	12,442,757	8,149,085	13,752,494
2070	10,277,339	12,989,484	8,016,514	14,953,882
2075	10,305,146	13,101,094	7,986,122	15,218,723
2080	10,332,223	13,213,515	7,954,481	15,492,520
2085	10,358,578	13,326,745	7,921,618	15,775,624
2090	10,384,216	13,440,773	7,887,560	16,068,398
2095	10,409,149	13,555,593	7,852,342	16,371,225
2100	10,433,385	13,671,202	7,815,996	16,684,501

Source: UNDP, *Human Development Report 2012: Overcoming Barriers: Human Mobility and Development*, New York, United Nations, 2012.

use of chemicals has adversely impacted the environment in many regions, leading to salinization and decreasing water quality and degrading croplands. From 1960 to 2000, worldwide phosphorous use doubled from 10 million tons (MT) to 20 MT, pesticide use tripled from near zero to 3 MT, and nitrogen use as fertilizer increased to a staggering 80 MT from just 10 MT (Foley et al., 2007; Khan and Hanjra, 2008). Diversion of croplands to biofuels is taking water away from food production (Bindraban et al., 2009), even as the economic, carbon sequestration, environmental, and food security impacts of biofuel production are proving to be a net negative (Gibbs et al., 2008; Lal and Pimentel, 2009; Searchinger et al., 2008). Climate models predict that the hottest seasons on record will become the norm by the end of the century in most regions of the world—a prediction that bodes ill

for feeding the world (Kumar and Singh, 2005). Increasing per capita meat consumption is increasing agricultural demands on land and water (Vinnari and Tapio, 2009). Cropland areas are decreasing in many parts of the world due to urbanization, industrialization, and salinization (Khan and Hanjra, 2008). Ecological and environmental imperatives, such as biodiversity conservation and atmospheric carbon sequestration, have put a cap on the possible expansion of cropland areas to other lands such as forests and rangelands (Gordon et al., 2009). Crop yield increases of the green revolution era have now stagnated (Hossain et al., 2005). Given these factors and limitations, further increase in food production through increase in cropland areas and/or increased allocations of water for croplands is widely considered unsustainable or simply infeasible.

Clearly, our continued ability to sustain adequate global food production and achieve future food security in the twenty-first century is challenged. So, how does the world continue to meet its food and nutrition needs? Solutions may come from biotechnology and precision farming. However, developments in these fields are not currently moving at rates that will ensure global food security over the next few decades (Foley et al., 2011). Further, there is a need for careful consideration of possible adverse effects of biotechnology. We should not be looking back 30–50 years from now with regrets, like we are looking back now at many mistakes made during the green revolution. During the green revolution, the focus was only on getting more yield per unit area. Little thought was given to the serious damage done to our natural environments, water resources, and human health as a result of detrimental factors such as uncontrolled use of herbicides, pesticides, and nutrients, drastic groundwater mining, and salinization of fertile soils due to overirrigation. Currently, there are discussions of a "second green revolution" or even an "evergreen revolution," but definitions of what these terms actually mean are still debated and are evolving (e.g., Monfreda et al., 2008). One of the biggest issues that has not been given adequate focus is the use of large quantities of water for food production. Indeed, an overwhelming proportion (60%–90%) of all human water use in the World, for example, goes for producing their food (Falkenmark and Rockström, 2006). But such intensive water use for food production is no longer sustainable due to increasing competition for water in alternative uses (EPW, 2008), such as urbanization, industrialization, environmental flows, biofuels, and recreation. This has brought into sharp focus the need to grow more food per drop of water (or crop water productivity or crop per drop) leading to the need for a "blue revolution" in agriculture (Pennisi, 2008).

A significant part of the solution lies in determining how global croplands are currently used and how they might be better managed to optimize the use of resources in food production. This will require development of an advanced global cropland area database (GCAD) with an ability to map global croplands and their attributes routinely, rapidly, consistently, and with sufficient accuracies. This in turn requires the creation of a framework of best practices for cropland mapping and an advanced global geospatial information system on global croplands. Such a system would need to be consistent across nations and regions by providing information on issues such as the composition and location of cropping, cropping intensities (e.g., single, double crop), rotations, crop health/vigor, and irrigation status. Opportunities to establish such a global system can be achieved by fusing advanced remote sensing data from multiple platforms and agencies (e.g., http://eros.usgs.gov/ceos/satellites_midres1.shtml; http://www.ceos-cove.org/index.php) in combination with national statistics, secondary data (e.g., elevation, slope, soils, temperature, and precipitation), and the systematic collection of field level observations. An example of such a system on a regional scale is USDA, NASS Cropland Data Layer (CDL), which is a raster, georeferenced, crop-specific land cover data layer with a ground resolution of 30 m (Johnson and Mueller, 2010). The GCAD will be a major contribution to Group on Earth Observations (GEO) Global Agricultural Monitoring Initiative (GLAM), to the overarching vision of GEO Agriculture and Water Societal Beneficial Areas (GEO Ag. SBAs), G20 Agriculture Ministers initiatives, and ultimately to the Global Earth Observation System of Systems (GEOSS). These initiatives are also supported by the Committee on Earth Observing Satellites (CEOS) Strategic Implementation Team (SIT).

Within the context of the above facts, the overarching goal of this chapter is to provide a comprehensive overview of the state-of-art of global cropland mapping procedures using remote sensing as characterized and envisioned by the "Global Food Security Support Analysis Data @ 30 m (GFSAD30)" project working group team. First, the chapter will provide an overview of *existing cropland maps* and their characteristics along with establishing the gaps in knowledge related to global cropland mapping. Second, *definitions* of cropland mapping along with key parameters involved in cropland mapping based on their importance in food security analysis, and cropland naming conventions for standardized cropland mapping using remote sensing will be presented. Third, *existing methods and approaches* for cropland mapping will be discussed. This will include the type of remote sensing data used in cropland mapping and their characteristics along with discussions on the secondary data, field-plot data, and cropland mapping algorithms. Fourth, currently *existing global cropland products* derived using remote sensing will be presented and discussed. Fifth, a *synthesis* of all existing products leading to a composite global cropland extent version 1.0 (GCE V1.0) is presented and discussed. Sixth, a *way forward* for advanced global cropland mapping is visualized.

6.2 Global Distribution of Croplands and Other Land Use and Land Cover: Baseline for the Year 2000

The first comprehensive global map of croplands was created by Ramankutty et al. in 1998. A more current version for the year 2000 shows the spatial distribution of global croplands along with other land use and land cover classes (Figure 6.1). This provides a first view of where global croplands are concentrated and helps us to focus on the appropriate geographic locations for detailed cropland studies. Water and snow (Class 8 and 9, respectively) have zero croplands and occupy 44% of the total terrestrial land surface. Further, forests (Class 6) occupy 17% of the terrestrial area and deserts (Class 7) an additional 12%. In these two classes, <5% of the total croplands exist. Therefore, in order to study croplands systematically and intensively, one must prioritize mapping in the areas of Classes 1–5 (26% of the terrestrial area) where >95% of all global croplands exist, with the first 3 classes (Class 1, 2, and 3) having ~75% and the next 2 ~20%. In the future, it is likely some of the noncroplands may be converted to croplands (e.g., especially in Africa where large farmlands are introduced in recent years in otherwise overwhelmingly small-holder dominant farming) or vice versa, highlighting the need for repeated and systematic global mapping of croplands. Segmenting the world into cropland versus noncropland areas routinely will help us understand and study these change dynamics better.

6.2.1 Existing Global Cropland Maps: Remote Sensing and Non–Remote Sensing Approaches

There are currently six major global cropland maps: (1) Thenkabail et al. (2009a,b), (2) Ramankutty and Foley (1998), (3) Goldewijk et al. (2011), (4) Portmann et al. (2010), (5) Pittman et al. (2010), and (6) Yu et al. (2013). These studies estimated the total global cropland area to be around 1.5 to 1.7 billion hectares for the year 2000 as a baseline. However, there are two significant differences in these products: (1) spatial disagreement on where the actual croplands are, and (2) irrigated to rain-fed cropland proportions and their precise spatial locations. Globally, cropland areas have increased from around 265 Mha in year 1700 to around 1471 Mha in year 1990, while the area of pasture has increased approximately sixfold from 524 to 3451 Mha (Foley et al., 2011). Ramankutty and Foley (1998) estimated the cropland and pasture to represent about 36% of the world's terrestrial surface (148,940,000 km^2), of which, according to different studies, roughly 12% is croplands and 24% pasture. Multiple studies (Goldewijk et al., 2011; Portmann et al., 2010; Ramankutty et al., 2008) integrated agricultural statistics and census data from the national systems with spatial mapping technologies involving geographic information systems (GIS) to derive global cropland maps.

Thenkabail and others (2009a,b, 2011) produced the first remote sensing–based global irrigated and rain-fed cropland maps and statistics through multisensor remote sensing data fusion along with secondary data and in situ data. They further used five dominant crop types (wheat, rice, corn, barley, and soybeans) using parcel-based inventory data (Monfreda et al., 2005, 2008; Portmann et al., 2010; Ramankutty et al., 2008) to produce a classification of global croplands with crop dominance (Thenkabail et al., 2012). The five crops account for about 60% of the total global cropland

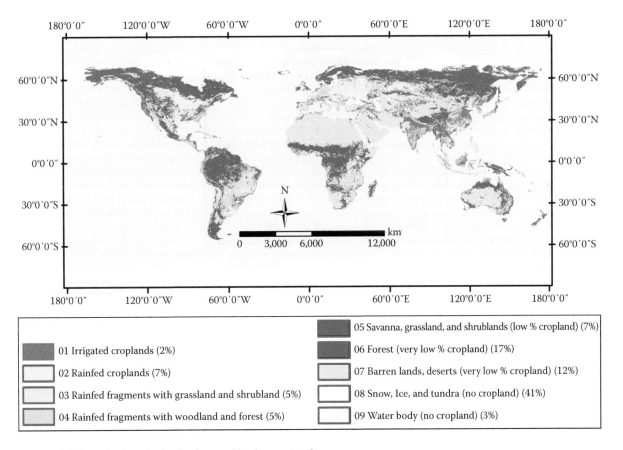

FIGURE 6.1 Global croplands and other land use and land cover: Baseline.

areas. The precise spatial location of these crops is only an approximation due to the coarse resolution (approximately 1 km²) and fractional representation (1%–100% crop in a pixel) of the crop data in each grid cell of all the maps from which this composite map is produced (Thenkabail et al., 2012). The existing global cropland datasets also differ from each other due to inherent uncertainties in establishing the precise location of croplands, the watering methods (rain-fed versus irrigated), cropping intensities, crop types and/or dominance, and crop characteristics (e.g., crop or water productivity measures such as biomass, yield, and water use). Improved knowledge of the uncertainties (Congalton and Green, 2009) in these estimates will lead to a suite of highly accurate spatial data products (Goodchild and Gopal, 1989) in support of crop modeling, food security analysis, and decision support.

6.3 Key Remote Sensing–Derived Cropland Products: Global Food Security

The production of a repeatable global cropland product requires a standard set of metrics and attributes that can be derived consistently across the diverse cropland regions of the world. Four key cropland information systems attributes that have been identified for global food security analysis and that can be readily derived from remote sensing include (Figure 6.2): (1) cropland extent/areas,

(2) watering methods (e.g., irrigated, supplemental irrigated, and rain-fed), (3) crop types, and (4) cropping intensities (e.g., single crop, double crop, and continuous crop). Although not the focus of this chapter, many other parameters are also derived in local regions, such as: (5) precise location of crops, (6) cropping calendar, (7) crop health/vigor, (8) flood and drought information, (9) water use assessments, and (10) yield or productivity (expressed per unit of land and/or unit of water). Remote sensing is specifically suited to derive the four key products over large areas using fusion of advanced remote sensing (e.g., Landsat, Resourcesat, MODIS) in combination with national statistics, ancillary data (e.g., elevation, precipitation), and field-plot data. Such a system, at the global level, will be complex in data handling and processing and requires coordination between multiple agencies leading to development of a seamless, scalable, transparent, and repeatable methodology. As a result, it is important to have a systematic class labeling convention as illustrated in Figure 6.3. A standardized class identifying and labeling process (Figure 6.3) will enable consistent and systematic labeling of classes, irrespective of analysts. First, the area is separated into cropland versus noncropland. Then, within the cropland class, labeling will involve (Figure 6.3): (1) cropland extent (cropland versus noncropland), (2) watering source (e.g., irrigated versus rain-fed), (3) irrigation source (e.g., surface water, ground water), (4) crop type or dominance, (5) scale (e.g., large or contiguous, small or fragmented), and (6) cropping intensity (e.g., single crop, double crop). The detail at which one maps at each stage and each

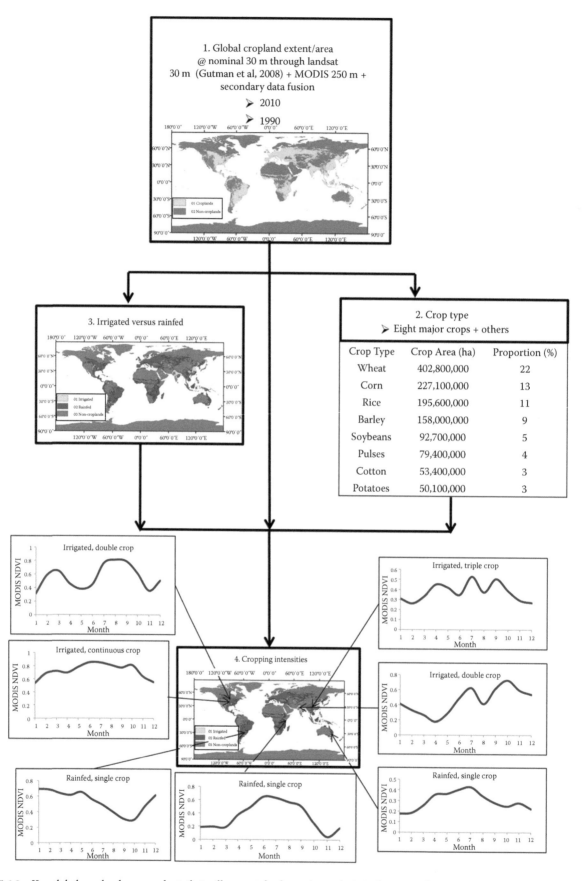

FIGURE 6.2   Key global cropland area products that will support food security analysis in the twenty-first century.

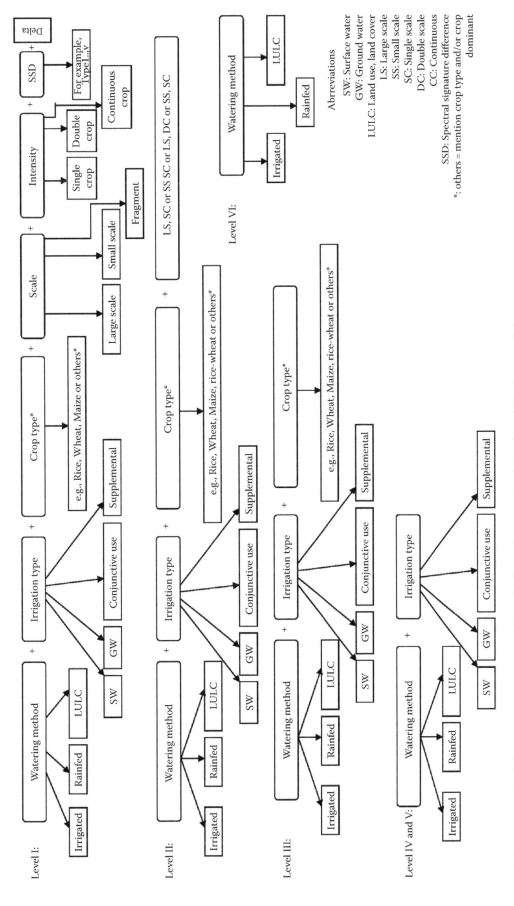

FIGURE 6.3 Cropland class naming convention at different levels. Level I is most detailed and Level IV is least detailed.

parameter would depend on many factors such as resolution of the imagery, available ground data, and expert knowledge. For example, if there is no sufficient knowledge on whether the irrigation is by surface water or ground water, but it is clear that the area is irrigated, one could just map it as irrigated without mapping greater details on the type of irrigation. But, for every cropland class, one has the potential to map the details as shown in Figure 6.3.

6.4 Definition of Remote Sensing–Based Cropland Mapping Products

Key to effective mapping is a precise and clear definition of what will be mapped. It is the first and primary step, with different definitions leading to different products. For example, irrigated areas are defined and understood differently in different applications and contexts. One can define them as areas that receive irrigation at least once during their crop growing period. Alternatively, they can be defined as areas that receive irrigation to meet at least half their crop water requirements during the growing season. One other definition can be that these are areas that are irrigated throughout the growing season. In each of these cases, the extent of irrigated area mapped will vary. Similarly, croplands can be defined as all agricultural areas irrespective of the types of crops grown or they may be limited to food crops (and not the fodder crops or plantation crops). So, it is obvious that having a clear understanding of the definitions of what we map is extremely important for the integrity of the products developed. We defined cropland products as follows:

- *Minimum mapping unit*: The minimum mapping unit of a particular crop is an area of 3 by 3 (0.81 ha) Landsat pixels identified as having the same crop type.
- *Cropland extent*: All cultivated plants harvested for food, feed, and fiber, *including* plantations (e.g., orchards, vineyards, coffee, tea, rubber).
- *What is a cropland pixel?*: sub-pixel composition is used to calculate area. This involves multiplying full pixel area (FPA) with cropland area fraction (CAF). CAF provides what % of pixel is cropped. So, sub-pixel area/actual area = FPA*CAF
- *Irrigated areas*: Irrigation is defined as artificial application of any amount of water to overcome crop water stress. Irrigated areas are those areas that are irrigated one or more times during crop growing season.
- *Rain-fed areas*: Areas that have no irrigation whatsoever and are precipitation dependent.
- *Cropping intensity*: Number of cropping cycles within a 12-month period.
- *Crop type*: Eight crops (wheat, corn, rice, barley, soybeans, pulses, cotton, and potatoes), that occupy approx. 70% global cropland areas are considered. The rest of the crops are under "others". However, in particular continents where other crops like sugarcane or cassava etc. are important, they will be mapped as well.

6.5 Data: Remote Sensing and Other Data for Global Cropland Mapping

Cropland mapping using remote sensing involves multiple types of data: satellite data with a consistent and useful global repeat cycle, secondary data, statistical data, and field plot data. When these data are used in an integrated fashion, the output products achieve highest possible accuracies (Thenkabail et al., 2009b,c).

6.5.1 Primary Satellite Sensor Data

Cropland mapping will require satellite sensor data across spatial, spectral, radiometric, and temporal resolutions from a wide array of satellite/sensor platforms (Table 6.2) throughout the growing season. These satellite sensors are "representative" of hyperspectral, multispectral, and hyperspatial data. The data points per hectare (Table 6.2, last column) will indicate the spatial detail of agricultural information gathered. In addition to satellite-based sensors, it is always valuable to gather ground-based hand-held spectroradiometer data from hyperspectral sensors (Thenkabail et al., 2013), and/or imaging spectroscopy from ground-based, airborne, or space borne sensors for validation and calibration purposes (Thenkabail et al., 2011). Much greater details of a wide array of sensors available to gather data are presented in Chapters 1 and 2 of *Remotely Sensed Data Characterization, Classification, and Accuracies*.

6.5.2 Secondary Data

There is a wide array of secondary or ancillary data such as the ASTER-derived digital elevation data (GDEM), long (50–100 years) records of precipitation and temperature (CRU), digital maps of soil types, and administrative boundaries. Many secondary data are known to improve crop classification accuracies (Thenkabail et al., 2009a,b). The secondary data will also form core data for the spatial decision support system and final visualization tool in many systems.

6.5.3 Field-Plot Data

Field-plot data (e.g., Figure 6.4) will be used for purposes such as: (1) class identification and labeling; (2) determining irrigated area fractions (AFs), and (3) establishing accuracies, errors, and uncertainties. At each field point (e.g., Figure 6.3), data such as cropland or noncropland, watering method (irrigated or rain-fed), crop type, and cropping intensities are recorded along with GPS locations, digital photographs, and other information (e.g., yield, soil type) as needed. Field plot data will also help in gathering an ideal spectral data bank of croplands. One could use the precise locations and the crop characteristics and generate coincident remote sensing data characteristics (e.g., MODIS time-series monthly NDVI).

TABLE 6.2   Characteristics of Some of the Key Satellite Sensor Data Currently Used in Cropland Mapping

Satellite Sensor	Wavelength Range (μm)	Spatial Resolution (m)	Spectral Bands (#)	Temporal (days)	Radiometric (bits)	Data Points (per ha)
A. Hyperspectral						
EO-1 Hyperion			196	16	16	11.1 points for 30 m pixel
VNIR	0.43–0.93	30				(0.09 ha per pixel)
SWIR	0.93–2.40	30				
B. Advanced multispectral						
Landsat TM			7/8	16	8	
Multispectral						
Band 1	0.45–0.52	30				44.4 points for 15 m pixel
Band 2	0.53–0.61	30				11.1 points for 30 m pixel
Band 3	0.63–0.69	30				2.77 points for 60 m pixel
Band 4	0.78–0.90	30				0.69 points for 120 m pixel
Band 5	1.55–1.75	30				
Band 6	10.40–12.50	120/60				
Band 7	2.09–2.35	30				
Panchromatic	0.52–0.90	15				
EO-1 ALI			10	16	16	
Multispectral						
Band 1	0.43–0.45	30				
Band 2	0.45–0.52	30				
Band 3	0.52–0.61	30				
Band 4	0.63–0.69	30				
Band 5	0.78–0.81	30				
Band 6	0.85–0.89	30				
Band 7	1.20–1.30	30				
Band 8	1.55–1.75	30				
Band 9	2.08–2.35	30				
Panchromatic	0.48–0.69	10				
ASTER			14	16	8	
VNIR		15				
Band 1	0.52–0.60					
Band 2	0.63–0.69					
Band 3N/3B	0.76–0.86					
SWIR		30				
Band 4	1.600–1.700					
Band 5	2.145–2.185					
Band 6	2.185–2.225					
Band 7	2.235–2.285					
Band 8	2.295–2.365					
Band 9	2.360–2.430					
TIR		90				1.23 points for 90 m
Band 10	8.125–8.475					
Band 11	8.475–8.825					
Band 12	8.925–9.275					
Band 13	10.25–10.95					
Band 14	10.95–11.65					
MODIS						
MOD09Q1		250	2	1	12	0.16 points for 250 m
Band 1	0.62–0.67					
Band 2	0.84–0.876					

(*Continued*)

TABLE 6.2 (*Continued*)   Characteristics of Some of the Key Satellite Sensor Data Currently Used in Cropland Mapping

Satellite Sensor	Wavelength Range (μm)	Spatial Resolution (m)	Spectral Bands (#)	Temporal (days)	Radiometric (bits)	Data Points (per ha)
MOD09A1		500	7[a]/36	1	12	0.04 points for 500 m
Band 1	0.62–0.67					
Band 2	0.84–0.876					
Band 3	0.459–0.479					
Band 4	0.545–0.565					
Band 5	1.23–1.25					
Band 6	1.63–1.65					
Band 7	2.11–2.16					
C. Hyperspatial						
GeoEye-1						
Multispectral		1.65	5	<3	11	
Band 1	0.45–0.52					59,488 points for 0.41 m
Band 2	0.52–0.60					26,874 points for 0.61 m
Band 3	0.63–0.70					10,000 points for 1 m
Band 4	0.76–0.90					3673 points for 1.65 m
Panchromatic	0.45–0.90	0.41				1679 points for 2.44 m
IKONOS			5	3	11	
Multispectral		4				
Band 1	0.45–0.52					625 points for 4 m
Band 2	0.51–0.60					400 points for 5 m
Band 3	0.63–0.70					236 points for 6.5 m
Band 4	0.76–0.85					100 points for 10 m
Panchromatic	0.53–0.93	1				44.4 points for 15 m
QuickBird			5	1–6	11	
Multispectral		2.44				
Band 1	0.45–0.52					
Band 2	0.52–0.60					
Band 3	0.63–0.69					
Band 4	0.76–0.90					
Panchromatic	0.45–0.90	0.61				
RapidEye		5–6.5	5	1–6	16	
Band 1	0.44–0.51					
Band 2	0.52–0.59					
Band 3	0.63–0.68					
Band 4	0.69–0.73					
Band 5	0.76–0.85					

[a] MODIS has 36 bands, but we considered only the first 7 bands (Mod09A1).

6.5.4 Very-High-Resolution Imagery Data

Very-high-resolution (submeter to 5 m) imagery (VHRI; see hyperspatial data characteristics in Table 6.2) is widely available these days from numerous sources. These data can be used as ground samples in localized areas to classify as well as verify classification results of the coarser resolution imagery. For example, in Figure 6.5, VHRI tiles identify uncertainties existing in cropland classification of coarser resolution imagery. VHRI is specifically useful for identifying croplands versus noncroplands (Figure 6.5). They can also be used for identifying irrigation based on associated features such as canals and tanks.

6.5.5 Data Composition: Mega File Data Cube (MFDC) Concept

Data preprocessing requires that all the acquired imagery is harmonized and standardized in known time intervals (e.g., monthly, biweekly). For this, the imagery data is either acquired or converted to at-sensor reflectance (see Chander et al., 2009; Thenkabail et al., 2004) and then converted to surface reflectance using Landsat Ecosystem Disturbance Adaptive Processing System (LEDAPS) codes for Landsat (Masek et al., 2006) or similar codes for other sensors. All data are processed and mosaicked to required geographic levels (e.g., global, continental). One method to organize these disparate but colocated datasets is through the

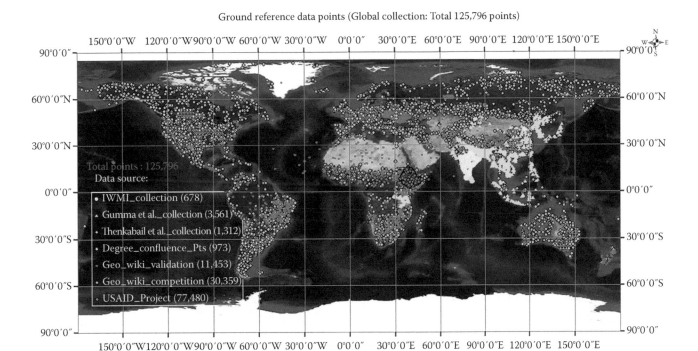

FIGURE 6.4 Field plot data for cropland studies collected over the globe.

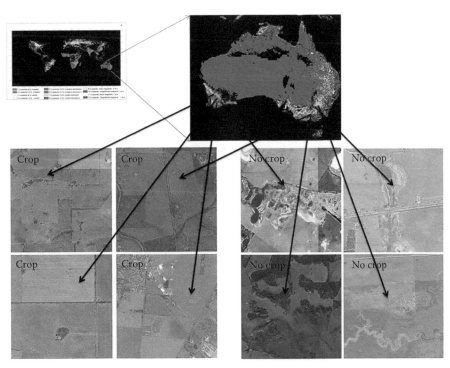

FIGURE 6.5 Very-high-resolution imagery used to resolve uncertainties in cropland mapping of Australia.

use of a MFDC. Numerous secondary datasets are combined in an MFDC, which is then stratified using image segmentation into distinct precipitation-elevation-temperature-vegetation zones. Data within the MFDC can include ASTER-derived refined digital elevation from SRTM (GDEM), monthly long-term precipitation, monthly thermal skin temperature, and forest cover and density. This segmentation allows cropland mapping to be focused; creating distinctive segments of MFDCs and analyzing them separately for croplands will enhance accuracy. For example, the likelihood of croplands in a temperature zone of <280°K is very low. Similarly, croplands in elevation above 1500 m will be of distinctive characteristics (e.g., patchy, on hilly terrain most likely plantations of coffee or tea). Every layer of data is geolinked (having precisely same projection and datum and are georeferenced to one another).

The purpose of MFDC (MFDC; see Thenkabail et al., 2009b for details) is to ensure numerous remote sensing and secondary data layers are all stacked one over the other to form a data cube akin to hyperspectral data cube. This approach has been used by X to map croplands in Y (reference). The MFDC allows us to have the entire data stack for any geographic location (global to local) as a single file available for analysis. For example, one can classify 10s or 100s or even 1000s of data layers (e.g., monthly MODIS NDVI time series data for a geographic area for an entire decade along with secondary data of the same area) stacked together in a single file and classify the image. The classes coming out of such a MFDC inform us about the phenology along with other characteristics of the crop.

6.6 Cropland Mapping Methods

6.6.1 Remote Sensing–Based Cropland Mapping Methods for Global, Regional, and Local Scales

There is a growing literature on cropland mapping across resolutions for both irrigated and rain-fed crops (Friedl et al., 2002; Gumma et al., 2011; Hansen et al., 2002; Kurz and Seelan, 2007; Loveland et al., 2000; Olofsson et al., 2011; Ozdogan and Woodcock, 2006; Thenkabail et al., 2009a,c; Wardlow and Egbert, 2008; Wardlow et al., 2006, 2007). Based on these studies, an ensemble of methods that is considered most efficient include: (1) spectral matching techniques (SMTs) (Thenkabail et al., 2007a, 2009a,c); (2) decision tree algorithms (DeFries et al., 1998); (3) Tassel cap brightness-greenness-wetness (Cohen and Goward, 2004; Crist and Cicone, 1984; Masek et al., 2008); (4) space-time spiral curves and change vector analysis (Thenkabail et al., 2005); (5) phenology (Loveland et al., 2000; Wardlow et al., 2006); and (6) climate data fusion with MODIS time-series spectral indices using decision tree algorithms and subpixel classification (Ozdogan and Gutman, 2008). More recently, cropland mapping algorithms that analyze end-member spectra have been used for global mapping by Thenkabail et al. (2009a, 2011).

6.6.2 Spectral Matching Techniques (SMTs) Algorithms

SMTs (Thenkabail et al., 2007a, 2009a, 2011) are innovative methods of identifying and labeling classes (see illustration in Figures 6.6 and 6.7a). For each derived class, this method identifies its characteristics over time using MODIS time-series data (e.g., Figure 6.6). NDVI time-series or other metrics (Biggs et al., 2006; Dheeravath et al., 2010; Thenkabail et al., 2005, 2007a) are analogous to spectra, where time is substituted for wavelength. The principle in SMT is to match the shape, or the magnitude or both to an ideal or target spectrum (pure class or "end-member"). The spectra at each pixel to be classified is compared to the end-member spectra and the fit is quantified using the following SMTs (Thenkabail et al., 2007a): (1) spectral correlation similarity (SCS)—a shape measure; (2) spectral similarity value (SSV)—a shape and magnitude measure; (3) Euclidian distance similarity (EDS)—a distance measure; and (4) modified spectral angle similarity (MSAS)—a hyperangle measure.

6.6.2.1 Generating Class Spectra

The MFDC (Section 6.4.5) of each of segment (Figures 6.6 and 6.7a) is processed using ISOCLASS K-means classification to produce a large number of class spectra with a unsupervised classification technique that are then interpreted and labeled. In more localized applications, it is common to undertake a field-plot data collection to identify and label class spectra. However, at the global scale, this is not possible due to the enormous resources required to cover vast areas to identify and label classes. Therefore, SMTs (Thenkabail et al., 2007a) to match similar classes or to match class spectra from the unsupervised classification with a library of ideal or target spectra (e.g., Figure 6.6a) will be used to identify and label the classes.

6.6.2.2 Creating Ideal Spectra Data Bank (ISDB)

The term "ideal or target" spectra refers to time-series spectral reflectivity or NDVI generated for classes for which we have precise location-specific ground knowledge. From these locations, signatures are extracted using MFDC, synthesized, and aggregated to generate a few hundred signatures that will constitute an ISDB (e.g., Figures 6.6 and 6.7a).

6.6.2.3 Matching Class Spectra with Ideal Spectra Using Spectral Matching Techniques (SMTs)

Once the class spectra are generated, they are compared with ideal spectra to match, identify, and label classes. Often quantitative spectral matching techniques like spectral correlation similarity R-square (SCS R-square) and spectral similarity value (SSV) are used (Thenkabail et al., 2007a).

6.7 Automated Cropland Classification Algorithm

The first part of the automated cropland classification algorithm (ACCA) method involves knowledge capture to understand and map agricultural cropland dynamics by: (1) identifying croplands versus noncroplands and crop type/dominance based on SMTs, decision trees tassel cap bispectral plots, and very-high-resolution imagery; (2) determining watering method (e.g., irrigated or rainfed) based on temporal characteristics (e.g., NDVI), crop water requirement (water use by crops), secondary data (elevation, precipitation, temperature), and irrigation structure (e.g., canals and wells); (3) establishing croplands that are large scale (i.e., contiguous) versus small scale (i.e., fragmented); (4) characterizing cropping intensities (single, double, triple, and continuous cropping); (5) interpreting MODIS NDVI temporal bispectral plots to identify and label classes; and (6) using in situ data from very-high-resolution imagery, field-plot data, and national statistics (see Figure 6.7b for details). The second part of the method establishes accuracy of the knowledge-captured agricultural map (Congalton, 1991 and 2009) and statistics by comparison with national statistics, field-plot data, and very-high-resolution imagery. The third part of the method makes use of the captured knowledge to code and map cropland

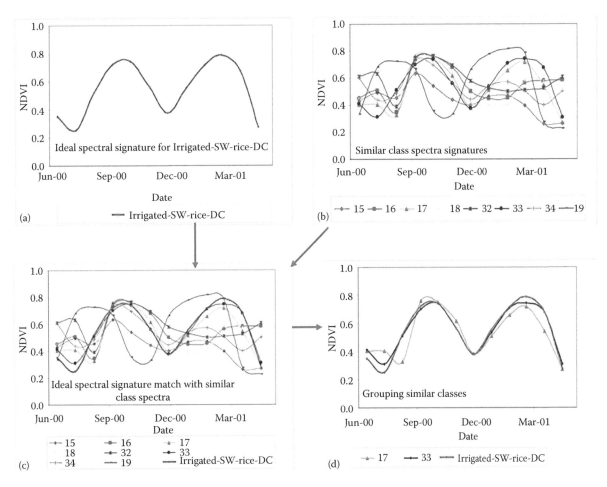

FIGURE 6.6 SMT. In SMTs, the class temporal profile (NDVI curves) are matched with the ideal temporal profile (quantitatively based on temporal profile similarity values) in order to group and identify classes as illustrated for a rice class in this figure. (a) Ideal temporal profile illustrated for "irrigated- surface-water-rice-double crop"; (b) some of the class temporal profile signatures that are similar; (c) ideal temporal profile signature (Figure 6.6a) matched with class temporal profiles (Figure 6.6b); and (d) the ideal temporal profile (Figure 6.6a, in deep green) matches with class temporal profiles of Classes 17 and 33 perfectly. Then one can label Classes 17 and 33 to be same as the ideal temporal profile ("irrigated-surface-water-rice-double crop"). This is a qualitative illustration of SMTs. For quantitative methods, refer to Thenkabail et al. (2007a).

dynamics through an automated algorithm. The fourth part of the method compares the agricultural cropland map derived using an automated algorithm (classified data) with that derived based on knowledge capture (reference map). The fifth part of the method applies the tested algorithm on an independent dataset of the same area to automatically classify and identify agricultural cropland classes. The sixth part of the method assesses accuracy and validates the classes derived from independent dataset using an automated algorithm (Thenkabail et al., 2012; Wu et al., 2014a,b).

6.8 Remote Sensing–Based Global Cropland Products: Current State-of-the-Art Maps, Their Strengths, and Limitations

Remote sensing offers the best opportunity to map and characterize global croplands most accurately, consistently, and repeatedly. Currently, there are three global cropland maps that have

been developed using remote sensing techniques. In addition, we also considered a recent MODIS global land cover and land use map where croplands are included. We examined these maps to identify their strengths and weaknesses, to see how well they compare with each other, and to understand the knowledge gaps that need to be addressed. These maps were produced by:

1. Thenkabail et al. (2009b, 2011; Biradar et al., 2009)
2. Pittman et al. (2010)
3. Yu et al. (2013)
4. Friedl et al. (2010)

Thenkabail et al. (2009b, 2011; Figure 6.8; Table 6.3) used a combination of AVHRR, SPOT VGT, and numerous secondary (e.g., precipitation, temperature, and elevation) data to produce a global irrigated area map (Thenkabail et al., 2009b, 2011) and a global map of rain-fed cropland areas (Biradar et al., 2009; Thenkabail et al., 2011; Figure 6.8; Table 6.3). Pittman et al. (2010; Figure 6.9; Table 6.4) used MODIS 250 m data to map global cropland extent. More recently, Yu et al. (2013; Figure 6.10; Table 6.5) produced a

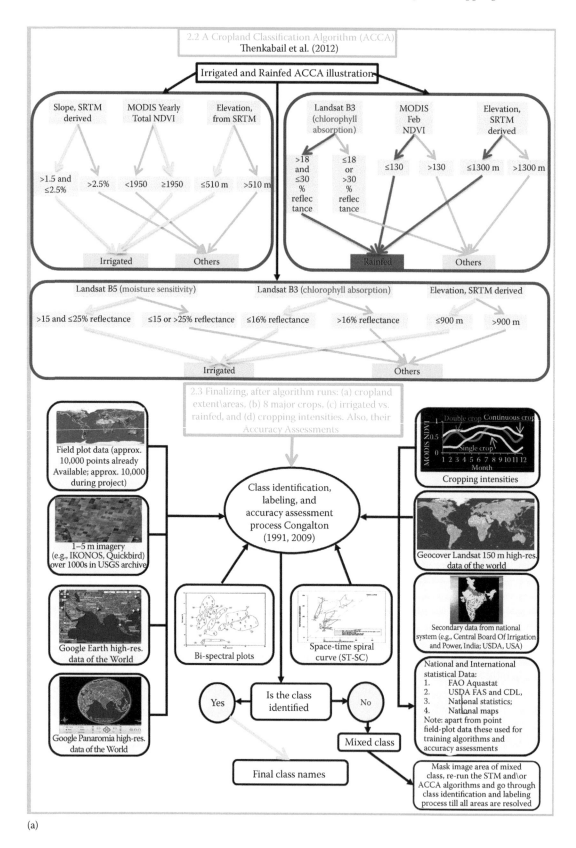

(a)

FIGURE 6.7 (a) Cropland mapping method illustrated here for a global scale (see Thenkabail et al., 2009b, 2011). The flowchart demonstrates comprehensive global cropland mapping methods using multisensor, multidate remote sensing, secondary, field plot, and very-high-resolution imagery data.

(Continued)

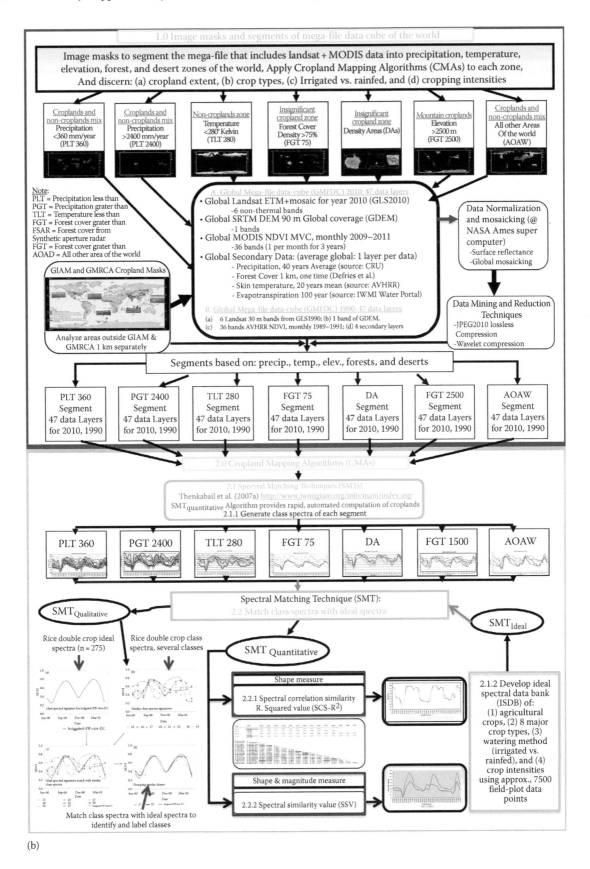

FIGURE 6.7 (*Continued*)   (b) Cropland mapping methods illustrated for a global scale. Top half shows ACCA (see Thenkabail and Wu, 2012; Wu et al., 2014a) and bottom half shows class identification and labeling process.

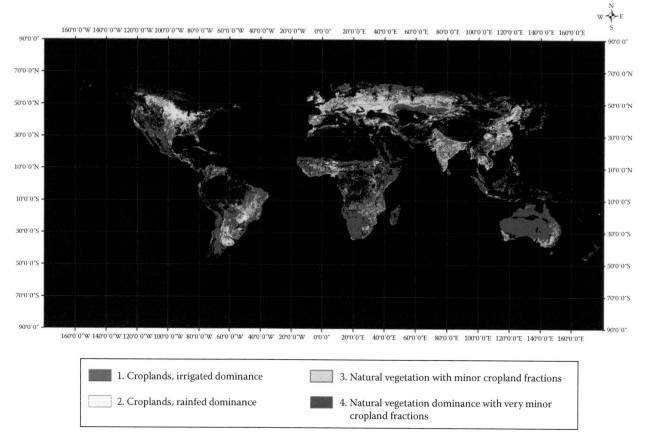

FIGURE 6.8   Global cropland product by Thenkabail et al. (2011, 2009b) using the method illustrated in Figure 6.7 and described in Section 6.1.1 (details in Thenkabail et al., 2011, 2009b). This includes irrigated and rain-fed areas of the world. The product is derived using remotely sensed data fusion (e.g., NOAA AVHRR, SPOT VGT, JERS SAR), secondary data (e.g., elevation, temperature, and precipitation), and in situ data. Total area of croplands is 2.3 billion hectares.

nominal 30 m resolution cropland extent of the world. These three global cropland extent maps are the best available current state-of-the-art products. Friedl et al. (2010; Figure 6.11; Table 6.6) used 500 m MODIS data in their global land cover and land use product (MCD12Q1) where croplands were one of the land cover classes. The methods, approaches, data, and definitions used in each of

TABLE 6.3   Global Cropland Extent at Nominal 1-km Based on Thenkabail et al. (2009b, 2011)[a]

Class #	Class Description (Names)	Pixels (1 km)	Percent (%)
1	Croplands, irrigated dominance	9,359,647	40
2	Croplands, rain-fed dominance	14,273,248	60
3	Natural vegetation with minor cropland fractions	5,504,037	
4	Natural vegetation dominance with very minor cropland fractions	44,170,083	
		23,632,895	100

[a] Total of approximately 2.3 billion hectares; Note that these are FPAs. Actual area is SPA. The SPA is not estimated here. See Thenkabail et al. (2007b) for the methods for calculating SPAs; % calculated based on Class 1 and 2. Class 3 and 4 are very small cropland fragments.

these products differ extensively. As a result, the cropland extents mapped by these products also vary significantly. The areas in Tables 6.3 through 6.6 only show the full pixel areas (FPAs) and not subpixel areas (SPAs). SPAs are actual areas, which can be estimated by reprojecting these maps to appropriate projections and calculating the areas. For the purpose of this chapter, we did not estimate SPAs. However, a comparison of the FPAs of the four maps (Figures 6.8 through 6.11) shows significant differences in the cropland areas (Tables 6.3 through 6.6) as well as significant differences in the precise locations of the croplands (Figures 6.8 through 6.11), the reasons for which are discussed in the next section.

6.8.1 Global Cropland Extent at Nominal 1 km Resolution

We synthesized the four global cropland products discussed and produced a unified global cropland extent map GCE V1.0 at nominal 1 km (Table 6.7a; Figure 6.12a). The process involved resampling each global cropland product to a common resolution of 1 km and then performing GIS data overlays to determine where the cropland extents matched and where they differed.

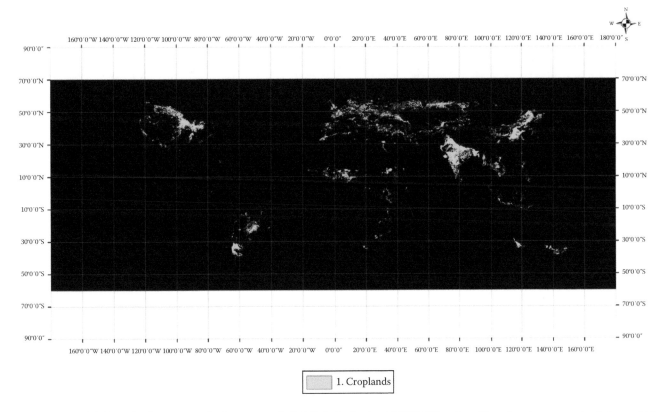

1. Croplands

FIGURE 6.9   Global cropland extent map by Pittman et al. (2010) derived using MODIS 250 m data. There is only one cropland class, which includes irrigated and rain-fed areas of the world. There is no discrimination between rain-fed and irrigated areas. Total area of croplands is 0.9 billion hectares.

TABLE 6.4   Global Cropland Extent at Nominal 250 m Based on Pittman et al. (2010)[a]

Class #	Class Description (Names)	Pixels (1 km)	Percent (%)
1	Croplands	8,948,507	100

[a] Total of approximately 0.9 billion hectares. Note that these are FPAs. Actual area is SPA. SPA is not estimated here. See Thenkabail et al. (2007b) for the methods for calculating SPAs; % calculated based on Class 1.

Figure 6.12a shows the aggregated global cropland extent map with its statistics in Table 6.7a. Class 1 in Figure 6.12a and Table 6.7a provides the global cropland extent included in all four maps. Actual area of this extent is not calculated yet, but it includes approximately 2.3 billion hectares FPAs (Table 6.7a). The spatial distribution of these 2.3 billion hectares is demonstrated as Class 1 in Figure 12a. Classes 2 and 3 are areas with minor or very minor cropland fractions. Class 2 and Class 3 are classes with large areas of natural vegetation and/or desert lands and other lands.

Figure 6.12b and Table 6.7b demonstrate where and by how much the four products match with one another. For example, 2,802,397 pixels (Class 1, Table 6.7b; Figure 6.12b) are croplands that are irrigated. Some of the products do not separately classify irrigated versus rain-fed croplands, although all four products show where croplands are. We first identified where all four

products match as croplands and then added irrigation status or other indicators (e.g., irrigation dominance, rain-fed; Table 6.7b) from the product by Thenkabail et al. (2009b, 2011).

Table 6.7b and Figure 6.12b show 12 classes of which Classes 1 and 2 are croplands with irrigated agriculture, Classes 3 and 4 are croplands with rain-fed agriculture, Classes 5 and 6 are croplands where irrigated agriculture dominates, Classes 7 and 8 are croplands where rain-fed agriculture dominates, and Classes 9–12 are areas with minor or very minor cropland fractions. Classes 9–12 are those with large areas of natural vegetation and\or desert lands and other lands.

Interestingly, and surprisingly as well, only 20% (Class 1 and 3; Table 6.7b; Figure 6.12b) of the total cropland extent are matched precisely in all four products. Further, 49% (Class 1, 2, 3, 4, and 7; Table 6.7b; Figure 6.12b) of the total cropland areas match in at least three of the four products. This implies that all the four products have considerable uncertainties in determining the precise location of the croplands. The great degree of uncertainty in the cropland products can be attributed to factors including

1. Coarse resolution of the imagery used in the study
2. Definition of mapping products of interest
3. Methods and approaches adopted
4. Limitations of the data

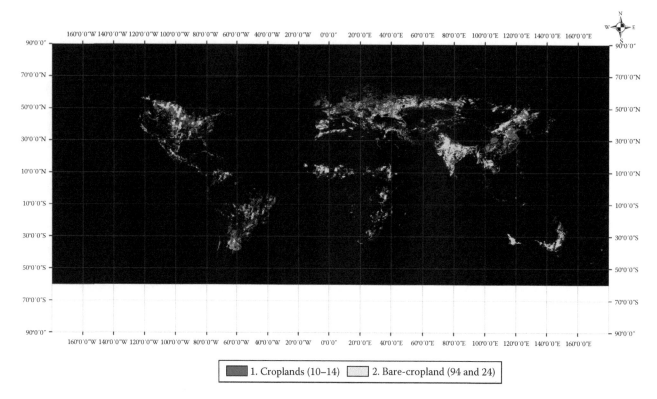

FIGURE 6.10   Global cropland extent map by Yu et al. (2013) derived at nominal 30 m data. Total area of croplands is 2.2 billion hectares. There is no discrimination between rain-fed and irrigated areas.

TABLE 6.5   Global Cropland Extent at Nominal 30 m Based on Yu et al. (2013)[a]

Class #	Class Description (Names)	Pixels (1 km)	Percent
1	Croplands (Classes 10–14)	7,750,467	35
2	Bare-cropland (Classes 94 and 24)	14,531,323	65
		22,281,790	100

[a] Total of approximately 2.2 billion hectares. Note that these are FPAs. Actual area is SPA. SPA is not estimated here. See Thenkabail et al. (2007b) for the methods for calculating SPAs; % calculated based on Class 1 and 2.

Table 6.7c and Figure 6.12c show five classes of which Classes 1 and 2 are croplands with irrigated agriculture, Class 3 is cropland with rain-fed agriculture, Classes 4 and 5 have ONLY minor or very minor cropland fractions. We recommend the use of this aggregated five class global cropland map (Figure 12c and Table 6.7c) produced based on the four major cropland mapping efforts [i.e., Thenkabail et al. (2009a, 2011), Pittman et al. (2010), Yu et al. (2013), and Friedl et al. (2010)] using remote sensing. This map (Figure 6.12c; Table 6.7c) provides clear consensus view on of four major studies on global:

- Cropland extent location
- Cropland watering method (irrigation versus rain-fed)

The product (Figure 6.12c; Table 6.7c) does not show where the crop types are or even the crop dominance. However, cropping intensity can be gathered using multitemporal remote sensing over these cropland areas.

6.9   Change Analysis

Once the croplands are mapped (Figure 6.13), we can use the time-series historical data such as continuous global coverage of remote sensing data from NOAA very-high-resolution radiometer (VHRR) and advanced VHRR (AVHRR), Global Inventory Modeling and Mapping Studies (GIMMS; 1982–2000), and MODIS time-series (2001–present) to help build an inventory of historical agricultural development (e.g., Figures 6.13 and 6.14). Such an inventory will provide information including identifying areas that have switched from rain-fed to irrigated production (full or supplemental), and noncropped to cropped (and vice versa). A complete history will require systematic analysis of remotely sensed data as well as a systematic compilation of all routinely populated cropland databases from the agricultural departments of all countries throughout the world. The differences in pixel sizes in AVHRR versus MODIS will: (1) influence class

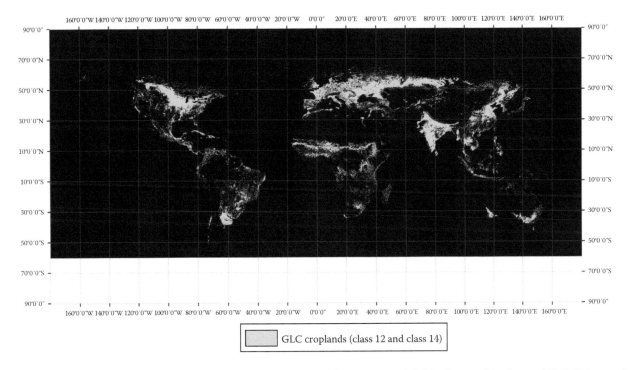

GLC croplands (class 12 and class 14)

FIGURE 6.11 Global cropland classes (Class 12 and Class 14) extracted from MODIS Global land use and land cover (GLC) 500 m product MCD12Q2 by Friedl et al. (2010). Total area of croplands is 2.7 billion hectares. There is no discrimination between rain-fed and irrigated cropland areas.

TABLE 6.6 Global Cropland Extent at Nominal 500 m Based on Friedl et al. (2010)[1]

Class #	Class Description (Names)	Pixels (1 km)	Percent
1	Global croplands (Class 12 and 14)	27,046,084	100

[a] Approximately, total 2.7 billion hectares based on Class 12 and 14. Note that these are FPAs. Actual area is SPA. SPA is not estimated here. See Thenkabail et al. (2007b) for the methods for calculating SPAs.

identification and labeling, and (2) cause different levels of uncertainties. We will address these issues by determining SPAs and uncertainties involved in class accuracies and uncertainties in areas at various spatial resolutions using methods detailed in recent work of this team (Ozdogan and Woodcock, 2006; Thenkabail et al., 2007b; Velpuri et al., 2009). Change analyses (Tomlinson, 2003) are conducted in order to investigate both the spatial and temporal changes in croplands (e.g., Figures 6.13 and 6.14) that will help establish: (1) change in total cropland areas, (2) change in spatial location of cropland areas, (3) expansion on croplands into natural vegetation, (4) expansion of irrigation, (5) change from croplands to biofuels, and (6) change from croplands to urban. Massive reductions in cropland areas in certain parts of the world will be detected, including cropland lost as a result of reductions in available ground water supply due to overdraft (Jiang, 2009; Rodell et al., 2009; Wada et al., 2012).

6.10 Uncertainties of Existing Cropland Products

Currently, the main causes of uncertainties in areas reported in various studies (Ramankutty et al., 2008; Thenkabail et al., 2009a,c) can be attributed to, but not limited to: (1) reluctance of national and state agencies to furnish the census data on irrigated area and concerns of their institutional interests in sharing of water and water data; (2) reporting of large volumes of census data with inadequate statistical analysis; (3) subjectivity involved in the observation-based data collection process; (4) inadequate accounting of irrigated areas, especially minor irrigation from groundwater, in national statistics; (5) definitional issues involved in mapping using remote sensing as well as national statistics; (6) difficulties in arriving at precise estimates of AFs using remote sensing; (7) difficulties in separating irrigated from rain-fed croplands; and (8) imagery resolution in remote sensing. Other limitations include (Thenkabail et al., 2009a, 2011)

1. Absence of precise spatial location of the cropland areas for training and validation
2. Uncertainties in differentiating irrigated areas from rain-fed areas
3. Absence of crop types and cropping intensities
4. Inability to generate cropland maps and statistics, routinely
5. Absence of dedicated web\data portal for dissemination cropland products

TABLE 6.7   Global Cropland Extent at Nominal 1-km Based on Four Major Studies: Thenkabail et al. (2009b, 2011), Pittman et al. (2010), Yu et al. (2013), and Friedl et al. (2010).

Class #	Class Description (Names)	Pixels (1 km)	Percent (%)
(a) *Three class map*[a]			
1	Croplands	23,493,936	100
2	Cropland minor fractions	13,700,176	
3	Cropland very minor fractions	44,662,570	
(b) *Twelve class map*[b]			
1	Croplands all 4, irrigated	2,802,397	12
2	Croplands 3 of 4, irrigated	289,591	1
3	Croplands all 4, rain-fed	1,942,333	8
4	Croplands 3 of 4, rain-fed	427,731	2
5	Croplands, 2 of 4, irrigation dominance	3,220,330	14
6	Croplands, 2 of 4, irrigation dominance	1,590,539	7
7	Croplands, 3 of 4, rain-fed dominance	6,206,419	26
8	Croplands, 2 of 4, rain-fed dominance	3,156,561	13
9	Croplands, minor fragments, 2 of 4	3,858,035	17
10	Croplands, very minor fragments, 2 of 4	6,825,290	
11	Croplands, minor fragments, 1 of 4	6,874,886	
12	Croplands, very minor fragments, 1 of 4	44,662,570	
	Class 1–9 total	23,493,936	100
(c) *Five class map*[c]			
1	Croplands, irrigation major	3,091,988	13
2	Croplands, irrigation minor	4,810,869	21
3	Croplands, rain-fed	11,733,044	50
4	Croplands, rain-fed minor fragments	3,858,035	16
5	Croplands, rain-fed very minor fragments	13,700,176	
	Classes 1–4 total	23,493,936	100.0%

[a] Approximately 2.3 billion hectares (Class 1) of cropland is estimated. But this is full pixel area (FPA). Actual area is sub pixel area (SPA). SPA is not estimated here. See Thenkabail et al. (2007b) for the methods for calculating SPAs; % calculated based on Class 1; Class 2 and 3 are minor/very minor cropland fragments.

[b] Approximately 2.3 billion hectares (Class 1–9) of cropland is estimated. But this is FPA. Actual area is SPA. SPA is not estimated here. See Thenkabail et al. (2007b) for the methods for calculating SPAs; % calculated based on Class 1–9; Classes 10, 11, and 12 are minor cropland fragments; All 4 means, all 4 studies agreed.

[c] Approximately 2.3 billion hectares (Class 1–4) of cropland is estimated. But this is FPA. Actual area is SPA. SPA is not estimated here. See Thenkabail et al. (2007b) for the methods for calculating SPAs; % calculated based on Class 1–4; Class 5 is very minor cropland fragments.

These limitations are a major hindrance in accurate/reliable global, regional, and country-by-country water use assessments that in turn support crop productivity (productivity per unit of land, kg/m^2) studies, water productivity (productivity per unit of water, kg/m^3) studies, and food security analyses. The higher degrees of uncertainty in coarser resolution data are a result of an inability to capture fragmented, smaller patches of croplands accurately, and the homogenization of both crop and noncrop land within areas of patchy land cover distribution. In either case, there is a strong need for finer spatial resolution to resolve the confusion.

6.11 Way Forward

Given the aforementioned issues with existing maps of global croplands, the way forward will be to produce global cropland maps at finer spatial resolution and applying a suite of advanced analysis methods. Previous research has shown that at finer spatial resolution, the accuracy of irrigated and rain-fed area class delineations improves, because at finer spatial resolution, more fragmented and smaller patches of irrigated and rain-fed croplands can be delineated (Ozdogan and Woodcock, 2006; Velpuri et al., 2009). Further, greater details of crop characteristics such as crop types (e.g., Figure 6.15) can be determined at finer spatial resolutions. Crop type mapping will involve the use of advanced methods of analysis such as data fusion of higher spatial resolution images from sensors such as Resourcesat\Landsat and AWiFS\MODIS (e.g., Table 6.2) supported by extensive ground surveys and ideal spectral data bank (ISDB) (Thenkabail et al., 2007a). Harmonic analysis is often adopted to identify crop types (Sakamoto et al., 2005) using methods such as the conventional Fourier analysis and adopting a Fourier filtered cycle similarity (FFCS) method. Mixed classes are resolved using hierarchical crop mapping

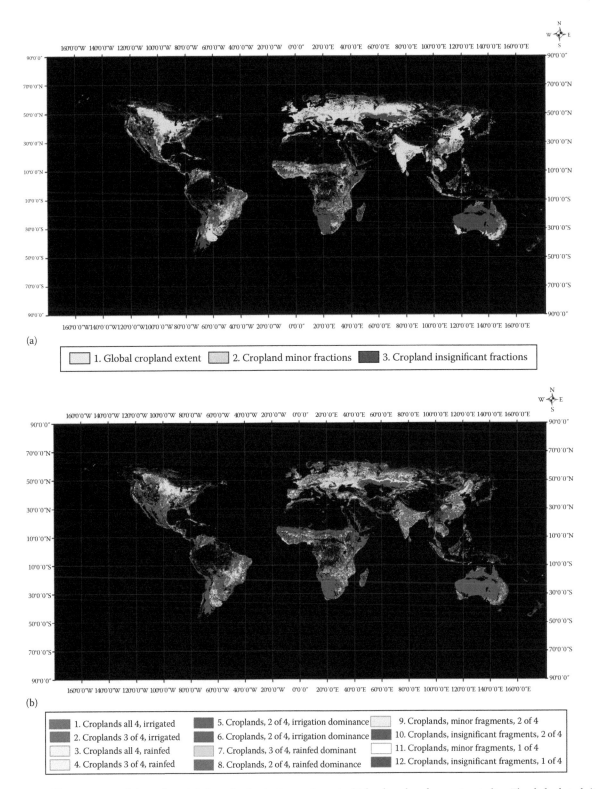

(a)

□ 1. Global cropland extent	▨ 2. Cropland minor fractions	■ 3. Cropland insignificant fractions

(b)

■ 1. Croplands all 4, irrigated	■ 5. Croplands, 2 of 4, irrigation dominance	□ 9. Croplands, minor fragments, 2 of 4
■ 2. Croplands 3 of 4, irrigated	■ 6. Croplands, 2 of 4, irrigation dominance	■ 10. Croplands, insignificant fragments, 2 of 4
□ 3. Croplands all 4, rainfed	▨ 7. Croplands, 3 of 4, rainfed dominant	□ 11. Croplands, minor fragments, 1 of 4
▨ 4. Croplands 3 of 4, rainfed	■ 8. Croplands, 2 of 4, rainfed dominance	■ 12. Croplands, insignificant fragments, 1 of 4

FIGURE 6.12 (a) An aggregated three class global cropland extent map at nominal 1 km based on four major studies: Thenkabail et al. (2009a, 2011), Pittman et al. (2010), Yu et al. (2013), and Friedl et al. (2010). Class 1 is total cropland extent; total cropland extent is 2.3 billion hectares (FPAs). Class 2 and Class 3 have ONLY minor fractions of croplands. Refer to Table 6.7a for cropland statistics of this map. (b) A disaggregated twelve class global cropland extent map derived at nominal 1-km based on four major studies: Thenkabail et al. (2009a, 2011), Pittman et al. (2010), Yu et al. (2013), and Friedl et al. (2010). Classes 1–9 are cropland classes that are dominated by irrigated and rain-fed agriculture. Classes 10–12 have ONLY minor or very minor fractions of croplands. Refer to Table 6.7b for cropland statistics of this map.   *(Continued)*

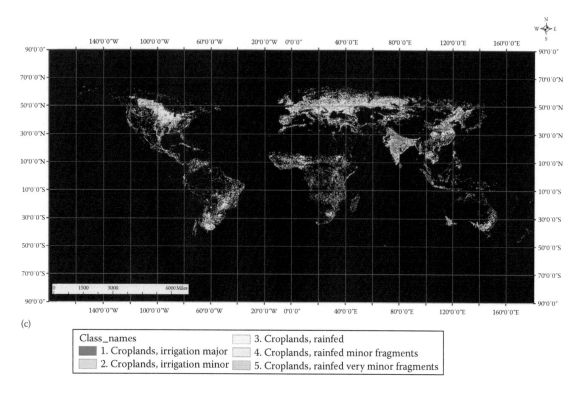

(c)

Class_names
1. Croplands, irrigation major
2. Croplands, irrigation minor
3. Croplands, rainfed
4. Croplands, rainfed minor fragments
5. Croplands, rainfed very minor fragments

FIGURE 6.12 (*Continued*) (c) A disaggregated five class global cropland extent map derived at nominal 1-km based on four major studies: Thenkabail et al. (2009a, 2011), Pittman et al. (2010), Yu et al. (2013), and Friedl et al. (2010). Classes 1–5 are cropland classes, that are dominated by irrigated and rain-fed agriculture. However, Class 4 and Class 5 have ONLY minor or very minor fractions of croplands. Refer to Table 6.7c for cropland statistics of this map. Note: *Irrigation major*: areas irrigated by large reservoirs created by large and medium dams, barrages, and even large ground water pumping. *Irrigation minor*: areas irrigated by small reservoirs, irrigation tanks, open wells, and other minor irrigation. However, it is very hard to draw a strict boundary between major and minor irrigations and in places, there can be significant mixing. Major irrigated areas such as the Ganges basin, California's central valley, Nile basin, etc., are clearly distinguishable as major irrigation, and in other areas major and minor irrigation may be intermixed.

protocol based on decision tree algorithm (Wardlow and Egbert, 2008). Irrigated versus rain-fed croplands will be distinguished using spectral libraries (Thenkabail et al., 2007b) and ideal spectral data banks (Thenkabail et al., 2007a, 2009a). Similar classes will be grouped by matching class spectra with ideal spectra based on SMTs (SMTs; Thenkabail et al., 2007a). Details such as crop types are crucial for determining crop water use, crop productivity, and water productivity, leading to providing crucial information needed for food security studies. However, the high spatial resolution must be fused with high temporal resolution data in order to obtain time-series spectra that are crucial for monitoring crop growth dynamics and cropping intensity (e.g., single crop,

double crop, and continuous year round crop). Numerous other methods and approaches exist. But, the ultimate goal using multisensor remote sensing is to produce croplands products such as

1. Cropland extent\area
2. Crop types (initially focused on eight crops that occupy 70% of global croplands)
3. Irrigated versus rain-fed croplands
4. Cropping intensities\phenology (single, double, triple, and continuous cropping)
5. Cropped area computation
6. Cropland change over space and time

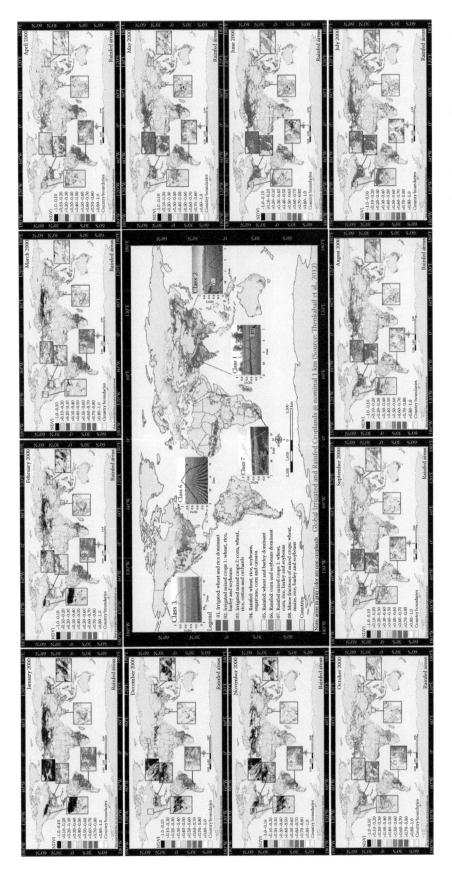

FIGURE 6.13 Center image of global cropland (irrigated and rainfed) areas @ 1 km for year 2000 produced by overlying the remote sensing derived product of the International Water Management Institute (IWMI; Thenkabail et al., 2012, 2011, 2009a,b; http://www.iwmigiam.org) over five dominant crops (wheat, rice, maize, barley, and soybeans) of the world produced by Ramankutty et al. (2008). The five crops constitute about 60% of all global cropland areas. The IWMI remote sensing product is derived using remotely sensed data fusion (e.g., NOAA AVHRR, SPOT VGT, and JERS SAR), secondary data (e.g., elevation, temperature, and precipitation), and *in situ* data. Total area of croplands is 1.53 billion hectares, of which 399 million hectares is total area available for irrigation (without considering cropping intensity) and 467 million hectares is annualized irrigated areas (considering cropping intensity). Surrounding NDVI images of irrigated areas: From January to December irrigated area NDVI dynamics is produced using NOAA AVHRR NDVI. The irrigated areas were determined by Thenkabail et al. (2011, 2009a,b).

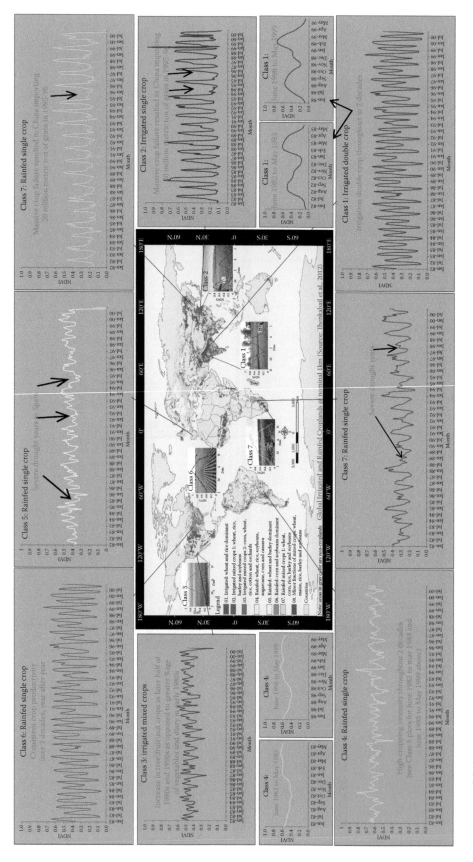

FIGURE 6.14　Global agricultural dynamics over two decades illustrated here for some of the most significant agricultural areas of the World. Once we establish GCAD2010 and GCAD1990 at nominal 30 m resolution for the entire world, we will use AVHRR-MODIS monthly MVC NDVI time-series from 1982 to 2017 to provide a continuous time history of global irrigated and rain-fed croplands, establish their spatial and temporal changes, and highlight the hot spots of change. The GCAD2010, GCAD1990, and GCAD four decade's data will be made available on USGS global cropland data portal (currently under construction): http://powellcenter.usgs.gov/current_projects.php#GlobalCroplandsAbstract. Further, the need to map accurately specific cropland characteristics such as crop types and watering methods (e.g., irrigated versus rain-fed) is crucial in food security analysis. For example, the importance of irrigation to global food security is highlighted in a recent study by Siebert and Döll (2010) who show that without irrigation, there would be a decrease in production of various foods including dates (60%), rice (39%), cotton (38%), citrus (32%), and sugarcane (31%) from their current levels. Globally, without irrigation, cereal production would decrease by a massive 43%, with overall cereal production, from irrigated and rain-fed croplands, decreasing by 20%.

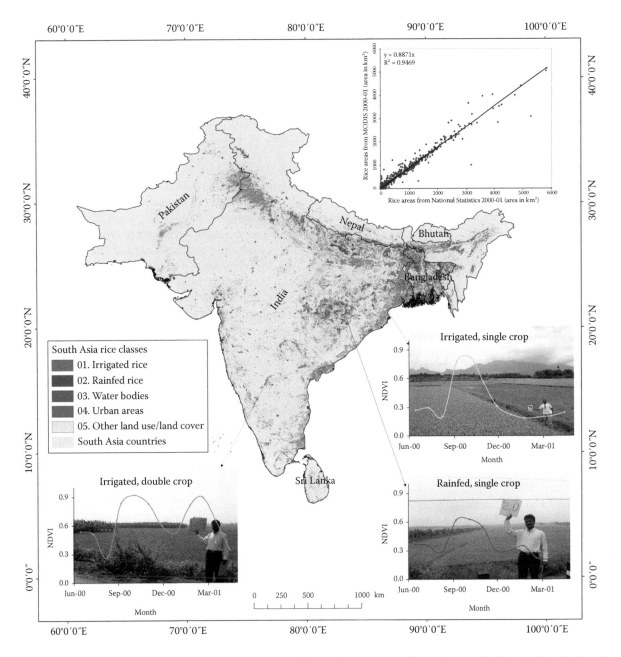

FIGURE 6.15 Rice map of south Asia produced using the method illustrated in Figure 6.6. (From Gumma, M. K. et al., *J. Appl. Rem. Sens.*, 5, 053547, September 1, 2011, doi:10.1117/1.3619838, 2011.)

6.12 Conclusions

This chapter provides an overview of the importance of global cropland products in food security analysis. It is obvious that only remote sensing from Earth-observing (EO) satellites provides consistent, repeated, high-quality data for characterizing and mapping key cropland parameters for global food security analysis. Importance of definitions and class naming conventions in cropland mapping has been reiterated. Typical EO systems and their spectral, spatial, temporal, and radiometric characteristics useful for cropland mapping have been highlighted. The chapter provides a review of various cropland mapping methods

used at global, regional, and local levels. Some of the remote sensing methods for global cropland mapping have been illustrated. The current state-of-the-art provides four-key global cropland products (-e.g., Figure 6.12) derived from remote sensing, based on the work conducted by four major studies (Thenkabail et al. (2009a, 2011, Pittman et al. 2010, Yu et al. 2013, and Friedl et al. 2010). These studies were conducted using: (1) time-series of multisensor data and secondary data, (2) 250 m MODIS time-series data, (3) 30 m Landsat data, and (4) a MODIS 500 m time-series derived cropland classes from a land use\land cover product has been used. These four studies help synthesized, at nominal 1 km, to obtain a consensus cropland mask of the world (global cropland extent version 1.0 or GCE V1.0). It was demonstrated from these products that the uncertainty in location of croplands in any one given product is quite high and no single product maps croplands particularly well. Therefore, a synthesis identifies where some or all of these products agree and where they disagree. This provides a starting point for the next level of more detailed cropland mapping at 250 and 30 m (see ongoing efforts at: http://geography.wr.usgs.gov/science/croplands/ and https://www.croplands.org/). The five key cropland products identified to be derived from remote sensing are: (1) cropland extent\areas, (2) cropping intensities, (3) watering method (irrigated versus rain-fed), (4) crop type, and (5) cropland change over time and space. From these primary products, one can derive crop productivity and water productivity. Such products have great importance and relevance in global food security analysis.

Authors recommend the use of composite global cropland map (see Figure 6.12c; Table 6.7c) that provides clear consensus view on of four major cropland studies on global

- Cropland extent location
- Cropland watering method (irrigation versus rain-fed)

The nominal 1 km product (Figure 6.12c and Table 6.7c) does not show where the crop types are or even where the crop dominance occur. However, cropping intensity can be generated using multitemporal remote sensing for every pixel over these cropland areas.

Acknowledgments

The authors thank NASA Making Earth Science Data Records for Use in Research Environments (MEaSUREs) solicitation for funding this research. Support by USGS Powell Center for a working group on global croplands is much appreciated. We thank the global food security support analysis data @ 30 m (GFSAD30) project team for inputs. Figures 6.1 and 6.2 were produced by Dr. Zhuoting Wu, USGS Mendenhall Post Doctoral Researcher, we thank her for it.

References

Biggs, T., Thenkabail, P. S., Krishna, M., GangadharaRao, P., and Turral, H. (2006). Vegetation phenology and irrigated area mapping using combined MODIS time-series, ground surveys, and agricultural census data in Krishna River Basin, India. *International Journal of Remote Sensing*, 27 (19), 4245–4266.

Bindraban, P. S., Bulte, E. H., and Conijn, S. G. (2009). Can large-scale biofuel production be sustained by 2020? *Agricultural Systems*, 101, 197–199.

Biradar, C. M., Thenkabail, P. S., Noojipady, P., Li, Y., Dheeravath, V., Turral, H., Velpuri, M. et al. (2009). A global map of rainfed cropland areas (GMRCA) at the end of last millennium using remote sensing. *International Journal of Applied Earth Observation and Geoinformation*, 11 (2), 114–129. doi:10.1016/j.jag.2008.11.002.

Chander, G., Markham, B. L., and Helder, D. L. (2009). Summary of current radiometric calibration coefficients for Landsat MSS, TM, ETM+, and EO-1 ALI sensors. *Remote Sensing of Environment*, 13 (5), 893–903. ISSN 0034-4257, http://dx.doi.org/10.1016/j.rse.2009.01.007. http://www.sciencedirect.com/science/article/pii/S0034425709000169. Last accessed March 15th, 2014.

Cohen, W. B. and Goward, S. N. (2004). Landsat's role in ecological applications of remote sensing. *BioScience*, 54, 535–545.

Congalton, R. (1991). A review of assessing the accuracy of classifications of remotely sensed data. *Remote Sensing of Environment*, 37, 35–46.

Congalton, R. (2009). Accuracy and error analysis of global and local maps: Lessons learned and future considerations. In: *Remote Sensing of Global Croplands for Food Security*, P. Thenkabail, J. Lyon, H. Turral, and C. Biradar (eds). CRC/Taylor & Francis, Boca Raton, FL, pp. 441–458.

Congalton, R. and Green, K. (2009). *Assessing the Accuracy of Remotely Sensed Data: Principles and Practices*, 2nd edn. CRC/Taylor & Francis, Boca Raton, FL, 183pp.

Crist, E. P. and Cicone, R. C. (1984). Application of the tasseled cap concept to simulated Thematic Mapper data. *Photogrammetric Engineering and Remote Sensing*, 50, 343–352.

DeFries, R., Hansen, M., Townsend, J. G. R., and Sohlberg, R. (1998). Global land cover classifications at 8 km resolution: The use of training data derived from Landsat imagery in decision tree classifiers. *International Journal of Remote Sensing*, 19, 3141–3168.

Dheeravath, V., Thenkabail, P. S., Chandrakantha, G, Noojipady, P., Biradar, C. B., Turral, H., Gumma, M. I., Reddy, G. P. O., and Velpuri, M. (2010). Irrigated areas of India derived using MODIS 500m data for years 2001–2003. *ISPRS Journal of Photogrammetry and Remote Sensing*, 65 (1), 42–59. http://dx.doi.org/10.1016/j.isprsjprs.2009.08.004.

EPW (2008). Food Security Endangered: Structural changes in global grain markets threaten India's food security. *Economic and Political Weekly*, 43, 5.

FAO (2009). *Food Outlook: Outlook Global Market Analysis*. Rome: FAO.

Foley, J. A., DeFries, R., Asner, G. P., Barford, C., Bonan, G., Carpenter, S. R., Chapin, F. S. et al. (2011). Solutions for a cultivated planet. *Nature*, 478 (7369), 337–342.

Foley, J. A., Monfreda, C., Ramankutty, N., and Zaks, D. (2007). Our share of the planetary pie. *PNAS*, 104, 12585–12586.

Falkenmark, M. and Rockström, J. (2006). The new blue and green water paradigm: Breaking new ground for water resources planning and management. *Journal of Water Resource Planning and Management*, 132, 1–15.

Friedl, M. A., McIver, D. K., Hodges, J. C. F., Zhang, X. Y., Muchoney, D., and Strahler, A. H. (2002). Global land cover mapping from MODIS: Algorithms and early results. *Remote Sensing of Environment*, 83, 287–302.

Friedl, M. A., Sulla-Menashe, D., Tan, B., Schneider, A., Ramankutty, N. S., and Huang, X. M. (2010). MODIS Collection 5 global land cover: Algorithm refinements and characterization of new datasets. *Remote Sensing of Environment*, 114 (1), 168–182.

Funk, C. and Brown, M. (2009). Declining global per capita agricultural production and warming oceans threaten food security. *Food Security*, 1 (3), 271–289. doi:10.1007/s12571-009-0026-y.

Gibbs, H. K., Johnston, M., Foley, J. A., Holloway, T., Monfreda, C., Ramankutty, N., and Zaks, D. (2008). Carbon payback times for crop-based biofuel expansion in the tropics: The effects of changing yield and technology. *Environment Research Letters*, 3(2008), 034001, 10pp. doi:10.1088/1748–9326/3/3/034001.

Goldewijk, K., Beusen, A., de Vos, M., and van Drecht, G. (2011). The HYDE 3.1 spatially explicit database of human induced land use change over the past 12,000 years. *Global Ecology and Biogeography*, 20 (1), 73–86. doi:10.1111/j.1466–8238.2010.00587.x.

Goodchild, M. and Gopal, S. (eds.) (1989). *The Accuracy of Spatial Databases*. Taylor & Francis, New York, 290pp.

Gordon, L. J., Finlayson, C. M., and Falkenmark, M. (2009). Managing water in agriculture for food production and other ecosystem services. *Agricultural Water Management*, 97, 512–519. doi: 10.1016/j.agwat.2009.03.017.

Gumma, M. K., Nelson, A., Thenkabail, P. S., and Singh, A. N. (2011). Mapping rice areas of South Asia using MODIS multi temporal data. *Journal of Applied Remote Sensing*, 5, 053547 (September 1, 2011). doi:10.1117/1.3619838.

Gutman, G., Byrnes, R., Masek, J., Covington, S., Justice, C., Franks, S., and Headley, R. (2008). Towards monitoring Land-cover and land-use changes at a global scale: The global land survey 2005. *Photogrammetric Engineering and Remote Sensing*, 74 (1): 6–10.

Hansen, M. C., DeFries, R. S., Townshend, J. R. G., Sohlberg, R., Dimiceli, C., and Carroll, M. (2002). Towards an operational MODIS continuous field of percent tree cover algorithm: Examples using AVHRR and MODIS data. *Remote Sensing of Environment*, 83, 303–319.

Hossain, M., Janaiah, A., and Otsuka, K. (2005). Is the productivity impact of the Green Revolution in rice vanishing? *Economic and Political Weekly* 5595–9600.

Jiang, Y. (2009). China's water scarcity. *Journal of Environmental Management*, 90, 3185–3196.

Johnson, D. M. and Mueller, R. (2010). The 2009 cropland data layer. *Photogrammetric Engineering and Remote Sensing*, 76 (11), 1201–1205.

Khan, S. and Hanjra, M. A. (2008). Sustainable land and water management policies and practices: A pathway to environmental sustainability in large irrigation systems. *Land Degradation and Development*, 19, 469–487.

Kumar, M. D. and Singh, O. P. (2005). Virtual water in global food and water policy making: Is there a need for rethinking? *Water Resources Management*, 19, 759–789.

Kurz, B. and Seelan, S. K. Use of remote sensing to map irrigated agriculture in areas overlying the Ogallala Aquifer, U.S. In Remote Sensing of Global Croplands for Food Security; Thenkabail, P. S., Turral, H., Lyon, J. G., Biradar, C., Eds.; CRC Press: Boca Raton, FL, 2009.

Lal, R. and Pimentel, D. (2009). Biofuels from crop residues. *Soil and Tillage Research*, 93, 237–238.

Loveland, T. R., Reed, B. C., Brown, J. F., Ohlen, D. O., Zhu, Z., and Yang, L. (2000). Development of global land cover characteristics database and IGBP DISCover from 1 km AVHRR data. *International Journal of Remote Sensing*, 21, 1303–1330.

McIntyre, B. D. (2008). International assessment of agricultural knowledge, science and technology for development (IAASTD): Global report. Includes Bibliographical References and Index. ISBN 978-1-59726-538-6, Oxford, U.K.

Masek, J. G., Huang, C., Wolfe, R., Cohen, W., Hall, F., Kutler, J., and Nelson, P. (2008). North American forest disturbance mapped from a decadal Landsat record. *Remote Sensing of Environment*, 112, 2914–2926.

Masek, J. G., Vermote, E. F., Saleous, N., Wolfe, R., Hall, F. G., Huemmrich, Gao, F., Kutler, J., and Lim, T. K. (2006). A Landsat surface reflectance data set for North America, 1990–2000. *Geoscience and Remote Sensing Letters*, 3, 68–72.

Monfreda, C., Patz, J. A., Prentice, I. C., Ramankutty, N., and Snyder, P. K. (2005). Global consequences of land use. *Science* July 22; 309 (5734), 570–574.

Monfreda, C., Ramankutty, N., and Foley, J. A. (2008). Farming the planet: 2. Geographic distribution of crop areas, yields, physiological types, and net primary production in the year 2000. *Global Biogeochemical Cycles* 22 (1), GB1022.

Olofsson, P., Stehman, S. V., Woodcock, C. E., Sulla-Menashe, D., Sibley, A.M., Newell, J.D. et al. (2012). A global land cover validation dataset, I: Fundamental design principles. *International Journal of Remote Sensing*, 33, 5768–5788.

Ozdogan, M. and Gutman, G. (2008). A new methodology to map irrigated areas using multi-temporal MODIS and ancillary data: An application example in the continental US. *Remote Sensing of Environment*, 112 (9), 3520–3537.

Ozdogan, M. and Woodcock, C. E. (2006). Resolution dependent errors in remote sensing of cultivated areas. *Remote Sensing of Environment*, 103, 203–217.

Pennisi, E. (2008). The Blue revolution, drop by drop, gene by gene. *Science*, 320 (5873): 171–173. doi:10.1126/science.320.5873.171

Pittman, K., Hansen, M. C., Becker-Reshef, I., Potapov, P. V., and Justice, C. O. (2010). Estimating global cropland extent with multi-year MODIS data. *Remote Sensing* 2 (7), 1844–1863.

Portmann, F., Siebert, S., and Döll, P. (2010). MIRCA2000—Global monthly irrigated and rainfed crop areas around the year 2000: A new high-resolution data set for agricultural and hydrological modelling. *Global Biogeochemical Cycles*, 24 GB0003435. 24pp.

Ramankutty, N., Evan, A. T., Monfreda, C., and Foley, J. A. (2008). Farming the planet: 1. Geographic distribution of global agricultural lands in the year 2000. *Global Biogeochemical Cycles*, 22. 19pp. doi:10.1029/2007GB002952.

Ramankutty, N. and Foley, J. A. (1998). Characterizing patterns of global land use: An analysis of global croplands data. *Global Biogeochemical Cycles* 12 (4), 667–685.

Rodell, M., Velicogna, I., and Famiglietti, J. S. (2009). Satellit-based estimates of groundwater depletion in India. *Nature*, 460, 999–1002. doi:10.1038/nature08238.

Sakamoto, T., Yokozawa, M., Toritani, H., Shibayama, M., Ishitsuka, N., Ohno, H. (2005), A crop phenology detection method using time-series MODIS data. *Remote Sensing of Environment*, 96, 366–374.

Searchinger, T., Heimlich, R., Houghton, R. A., Dong, F., Elobeid, A., Fabiosa, J., Tokgoz, S., Hayes, D., and Yu, T. H. (2008). Use of U.S. croplands for biofuels increases greenhouse gases through emissions from land-use change. *Science*, 319, 1238–1240.

Siebert, S. and Döll, P. (2010). Quantifying blue and green virtual water contents in global crop production as well as potential production losses without irrigation. *Journal of Hydrology*. 384(3–4), 198–217. doi:10.1016/j.jhydrol.2009.07.031.

Siebert, S., Hoogeveen, J., and Frenken, K. (2006). *Irrigation in Africa, Europe and Latin America—Update of the digital global map of irrigation areas to version 4*, Frankfurt Hydrology Paper 05, 134 pp. Institute of Physical Geography, University of Frankfurt, Frankfurt am Main, Germany and Rome, Italy.

Thenkabail, P. S., Biradar, C. M., Noojipady, P., Cai, X. L., Dheeravath, V., Li, Y. J., Velpuri, M., Gumma, M., and Pandey, S. (2007b). Sub-pixel irrigated area calculation methods. *Sensors Journal* (special issue: Remote Sensing of Natural Resources and the Environment (Remote Sensing Sensors Edited by Assefa M. Melesse). 7: 2519–2538. http://www.mdpi.org/sensors/papers/s7112519.pdf.)

Thenkabail, P. S., Biradar, C. M., Noojipady, P., Dheeravath, V., Li, Y. J., Velpuri, M., Gumma, M. et al. (2009b). Global irrigated area map (GIAM), derived from remote sensing, for the end of the last millennium. *International Journal of Remote Sensing*, 30 (14), 3679–3733. July, 20, 2009.

Thenkabail, P. S., Enclona, E. A., Ashton, M. S., Legg, C., Jean De Dieu, M. (2004). Hyperion, IKONOS, ALI, and ETM+ sensors in the study of African rainforests. *Remote Sensing of Environment*, 90, 23–43.

Thenkabail, P. S., GangadharaRao, P., Biggs, T., Krishna, M., and Turral, H., (2007a). Spectral matching techniques to determine historical land use/land cover (LULC) and irrigated areas using time-series AVHRR pathfinder datasets in the Krishna river Basin, India. *Photogrammetric Engineering and Remote Sensing*, 73 (9), 1029–1040. (Second Place Recipients of the 2008 John I. Davidson ASPRS President's Award for Practical papers).

Thenkabail, P. S., Hanjra, M. A., Dheeravath, V., Gumma, M. (2011). Global croplands and their water use remote sensing and non-remote sensing perspectives. In: Weng, Q. (ed.), *Advances in Environmental Remote Sensing: Sensors, Algorithms, and Applications*, Taylor & Francis, pp. 383–419.

Thenkabail, P. S., Hanjra, M. A., Dheeravath, V., Gumma, M. A. (2010). A holistic view of global croplands and their water use for ensuring global food security in the 21st century through advanced remote sensing and non-remote sensing approaches. *Journal Remote Sensing*, 2 (1), 211–261. doi:10.3390/rs2010211. http://www.mdpi.com/2072-4292/2/1/211.

Thenkabail, P. S., Knox, J. W., Ozdogan, M., Gumma, M. K., Congalton, R. G., Wu, Z., Milesi, C. et al. (2012). Assessing future risks to agricultural productivity, water resources and food security: How can remote sensing help? *Photogrammetric Engineering and Remote Sensing*, August 2012 Special Issue on Global Croplands: Highlight Article. 78(8), 773–782.

Thenkabail, P. S., Lyon, G. J., Turral, H., and Biradar, C. M. (2009a). *Remote Sensing of Global Croplands for Food Security*. CRC Press-Taylor & Francis group, Boca Raton, FL, London, New York. pp. 556 (48 pages in color). Published in June, 2009.

Thenkabail, P. S., Lyon, G. J., Turral, H., and Biradar, C. M. (2009c). *Remote Sensing of Global Croplands for Food Security*. Boca Raton, FL, London, New York: CRC Press/Taylor & Francis Group, Published in June.

Thenkabail, P. S., Mariotto, I., Gumma, M. K., Middleton, E. M., Landis, and D. R., Huemmrich, F. K. (2013). Selection of hyperspectral narrowbands (HNBs) and composition of hyperspectral twoband vegetation indices (HVIs) for biophysical characterization and discrimination of crop types using field reflectance and Hyperion/EO-1 data. *IEEE Journal of Selected Topics in Applied Earth Observations and Remote Sensing* 6 (2), 427–439, April 2013. doi:10.1109/JSTARS.2013.2252601.

Thenkabail, P. S., Schull, M., and Turral, H. 2005. Ganges and Indus River Basin Land Use/Land Cover (LULC) and irrigated area mapping using continuous streams of MODIS data. *Remote Sensing of Environment*, 95 (3), 317–341.

Thenkabail, P. S. and Wu, Z. (2012). An automated cropland classification algorithm (ACCA) for Tajikistan by combining landsat, MODIS, and secondary data. *Remote Sensing*, 4 (10), 2890–2918.

Tillman, D., C. Balzer, J. Hill, and B. L. Befort, 2011. Global food demand and the sustainable intensification of agriculture, *Proceedings of the National Academy of Sciences of the United States of America*, November 21. doi:10.1073/pnas.1116437108. Tomlinson, R. 2003. Thinking about Geographic Information Systems Planning for Managers, 283 ESRI Press.

Turral, H., Svendsen, M., and Faures, J. (2009). Investing in irrigation: Reviewing the past and looking to the future. *Agricultural Water Management*. 97(2010), 551–560. doi:101.1016/j.agwt.2009.07.012.

UNDP (2012). *Human Development Report 2012: Overcoming Barriers: Human Mobility and Development*. New York: United Nations.

Velpuri, M., Thenkabail, P. S., Gumma, M. K., Biradar, C. B., Dheeravath, V., Noojipady, P., and Yuanjie, L. (2009). Influence of resolution or scale in irrigated area mapping and area estimations. *Photogrammetric Engineering and Remote Sensing* (PE&RS). 75 (12), December 2009, 1383–1395.

Vinnari, M. and Tapio, P. (2009). Future images of meat consumption in 2030. *Futures*, 41 (5), 269–278. doi:10.1016/j.futures.2008.11.014.

Wada, Y., van Beek, L. P. H., and Bierkens, M. F. P. (2012). Nonsustainable groundwater sustaining irrigation: A global assessment. *Water Resources Research*, 48, W00L06. doi:10.1029/2011WR010562.

Wardlow, B. D. and Egbert, S. L. (2008). Large-area crop mapping using time-series MODIS 250 m NDVI data: An assessment for the U.S. Central Great Plains. *Remote Sensing of Environment*, 112, 1096–1116.

Wardlow, B. D., Egbert, S. L., and Kastens, J. H. (2007). Analysis of time-series MODIS 250 m vegetation index data for crop classification in the U.S. Central Great Plains. *Remote Sensing of Environment*, 108, 290–310.

Wardlow, B. D., Kastens, J. H., and Egbert, S. L. (2006). Using USDA crop progress data for the evaluation of greenup onset date calculated from MODIS 250-meter data. *Photogrammetric Engineering and Remote Sensing*, 72, 1225–1234.

Wu, Z., Thenkabail, P. S., and Verdin, J. (2014b). An automated cropland classification algorithm (ACCA) using Landsat and MODIS data combination for California. *Photogrammetric Engineering and Remote Sensing*, 80 (1): 81–90.

Wu, Z., Thenkabail, P. S., Zakzeski, A., Mueller, R., Melton, F., Rosevelt, C., Dwyer, J., Johnson, J., and Verdin, J. P. (2014a). Seasonal cultivated and fallow cropland mapping using modis-based automated cropland classification algorithm. *Journal of Applied Remote Sensing* 0001;8 (1), 083685. doi:10.1117/1.JRS.8.083685.

Yu, L., Wang, J., Clinton, N., Xin, Q., Zhong, L., Chen, Y., and Gong, P. (2013). FROM-GC: 30 m global cropland extent derived through multisource data integration. *International Journal of Digital Earth*. 12pp. doi:10.1080/17538947.2013.822574.

7

Precision Farming

David J. Mulla
University of Minnesota

Yuxin Miao
China Agricultural University

Acronyms and Definitions

AI	Aphid index
B	Blue wave band
CCCI	Canopy chlorophyll content index
CIred edge	Red-edge chlorophyll index
COA	Certificates of authorization
CWSI	Crop water stress index
DHT	Double Hough transformation
DSSI	Damage sensitive spectral index
FAA	Federal Aviation Administration
FLDA	Fisher linear discriminant analysis
F_m	Maximal fluorescence
FNIR	Far–near infrared
F_o	Minimal fluorescence level
F_v	Variable fluorescence
G	Green wave band
GIS	Geographic information system
GNDVI	Green normalized difference vegetative index
GPS	Global positioning system
HSI	Hue, saturation, and intensity
LAI	Leaf area index
MRESAVI	Modified RESAVI
MTCI	MERIS terrestrial chlorophyll index

MZ	Management zone
NDRE	Normalized difference red edge
NDVI	Normalized difference vegetative index
NIR	Near infrared
NNI	Nitrogen nutrition index
OMNBR	Optimal multiple narrow band reflectance index
OSAVI	Optimized soil-adjusted vegetation index
PRI	Photochemical reflectance index
PSII	Photosystem II
R	Red wave band
REDVI	Red-edge difference vegetation index
REIP	Red-edge inflation point
RERDVI	Red-edge renormalized difference vegetation index
RERVI	Red-edge ratio vegetation index
RESAVI	Red edge soil-adjusted vegetation index
RVI	Ratio vegetation index
SAVI	Soil-adjusted vegetation index
SWIR	Shortwave infrared
TCARI	Transformed chlorophyll absorption reflection index
TIR	Thermal infrared
UAV	Unmanned aerial vehicle
UV	Ultraviolet
VIS	Visible
VRT	Variable rate technology

7.1 Introduction

The world population is increasing rapidly, and by 2050, it is estimated that there will be nearly nine billion people to feed (Cohen, 2003). Agricultural production to feed this large population will be severely constrained by a lack of additional arable land combined with a diminishing supply of water and increasing pressure to protect the quality of water resources beyond the edge of agricultural fields. These constraints mean that it will be increasingly imperative to prevent losses in crop productivity due to water stress, nutrient deficiencies, weeds, insects, and crop diseases. These losses in productivity often occur at specific locations within fields and at critical growth stages. They are not typically uniform in severity across locations within a field. Thus, farmers must take measures to identify where crop stress occurs in a timely fashion, they must identify what is causing crop stress, and they must try to use management practices that overcome crop stress at specific locations and times.

This chapter provides an overview of remote sensing techniques used in precision farming to efficiently identify locations affected by crop stress. Crop stresses discussed include water stress, nutrient deficiencies, insect damage, disease infestations, and weed pressure. Crop stresses are typically identified by professional scouts who walk through fields looking for characteristic symptoms on crop leaves and stems. Remote sensing offers the potential to improve the efficiency of locating areas of crop stress and identifying which type of stress is present. For each stress, the key wavelengths and spectral indices that can be used to identify crop stress are reviewed. The relative advantages and disadvantages of satellite, airplane, unmanned aerial vehicles (UAVs), and proximal sensing platforms are discussed. The chapter concludes by identifying key knowledge gaps that must be overcome in order to accelerate the adoption of remote sensing in precision agriculture.

7.2 Precision Farming

Precision farming is one of the top 10 revolutions in agriculture (Crookston, 2006), ranking below conservation tillage, fertilizer and herbicide management, and improved crop genetics. It can be generally defined as doing the right management practices at the right location, in the right rate, and at the right time. Management practices commonly used in precision farming include variable rate fertilizer (Diacono et al., 2013) or pesticide application, variable rate seeding or tillage, and variable rate irrigation. Precision farming offers several benefits, including improved efficiency of farm management inputs, increases in crop productivity or quality, and reduced transport of fertilizers and pesticides beyond the edge of field (Mulla et al., 1996).

Precision farming is also known as precision agriculture or site-specific crop management. Precision farming as it is practiced today had its beginnings in the mid-1980s with two contrasting philosophies, namely, farming by soil (Larson and Robert, 1991) versus grid soil sampling for delineation of management zones (MZs) (Bhatti et al., 1991; Mulla, 1991, 1993).

Precision farming aims to improve site-specific agricultural decision-making through collection and analysis of data, formulation of site-specific management recommendations, and implementation of management practices to correct the factors that limit crop growth, productivity, and quality (Mulla and Schepers, 1997).

Precision farming has always relied on technology for data collection and analysis at specific locations and times across agricultural fields. The earliest technology was geographic information system (GIS), followed by variable rate spreaders, yield monitors, global positioning system (GPS), and remote sensing. As technology has improved, the scale at which management actions are implemented has become finer spatially and temporally. Ultimately, technology will lead to the ability to manage individual plants within an agricultural field in real time (Freeman et al., 2007; Shanahan et al., 2008).

Adoption rates of technology in precision agriculture vary widely (Whipker and Akridge, 2006). GPS (including autosteer) and yield monitors are widely used. Variable rate spreaders are moderately popular. Remote sensing has not yet been widely adopted for use in precision agriculture (Moran et al., 1997; Mulla, 2013). The main reasons include the difficulty in interpreting spectral signatures, the slow processing time for data, the high expense, and the need to collect confirmatory data from ground surveys in order to diagnose causative factors for anomalous spectral reflectance data. Clearly, there is a significant scope for improving the interpretation and utility of remote sensing data for precision agriculture.

Remote sensing in precision farming started with Landsat Thematic Mapper (TM) imagery for improved mapping of soil fertility patterns across complex agricultural landscapes (Bhatti et al., 1991). Proximal sensing of soil organic matter content or weeds was also developed for early application in precision farming, and this approach now includes detection of crop nutrient deficiencies. Commercial satellite imagery was first provided to agricultural users at the beginning of the twenty-first century with IKONOS and QuickBird. Spatial and spectral resolution and return frequencies of satellite remote sensing platforms have improved rapidly since then with the advent of RapidEye, GeoEye, and WorldView imagery. Satellite imagery is typically unavailable on days with significant cloud cover.

Interest in remote sensing from airplanes and UAVs has recently been very intense (Berni et al., 2009; Zhang and Kovacs, 2012; Huang et al., 2013). One of the most active emerging areas of research in precision agriculture uses cameras mounted on UAVs. The UAVs are relatively inexpensive, can be deployed rapidly at low altitudes when crop stress is starting to appear, and have the flexibility to be flown during windy or partially cloudy conditions. Their limitations include a ban on their use for commercial purposes; difficulty in obtaining certificates of authorization (COA) from the Federal Aviation Administration (FAA); inability to carry heavy cameras, mounts, and GPS units; and short battery life. UAVs also have other advantages and disadvantages, which are described more fully in Section 7.12.

Several companies offer precision farming services that rely on remote sensing. These include companies that are based primarily on satellite imagery, including DigitalGlobe, Satellite Imaging Corp., Geosys SST/GeoVantage, and Winfield Solutions. Companies that offer equipment for proximal sensing of crop nutrient deficiencies include Trimble's GreenSeeker (Solie et al., 1996), AgLeader's OptRx (Holland et al., 2012), Topcon's CropSpec (Reusch et al., 2010), and Yara's N-sensor (Link and Reusch, 2006). Trimble also offers equipment for proximal sensing of weeds (WeedSeeker; Hanks and Beck, 1998). Numerous companies offer aerial remote sensing services with panchromatic imagery, broadband multispectral imagery or hyperspectral imagery. One example is InTime Corp., which operates a fleet of airplanes that collect remote sensing imagery for cotton, vegetable, rice, and tree crops. This imagery is used for crop scouting and prescription maps for variable rate growth regulator applications on cotton and variable rate herbicide, insecticide, or fertilizer applications.

Commercial applications of remote sensing for precision farming have not always been successful. John Deere's AgriServices division partnered with GeoVantage in 2006 to provide the OptiGro precision remote sensing service to farmers. This service proved to be unprofitable for John Deere, and they sold it to GeoVantage in 2008.

7.3 Management Zones

Conventional agriculture involves uniform management of fields. In contrast, precision agriculture involves customized management in areas that are much smaller than fields (e.g., a 1 ha farm can be divided into 10,000 pixels of 1 m^2 and one can monitor each of these 10,000 pixels or any combination of them as a unique MZ as described in the following text). MZs (Mulla, 1991, 1993) are used in precision farming to divide field regions that differ in their requirements for fertilizer, pesticide, irrigation, seed, or tillage. MZs are relatively homogeneous units within the field that differ from one another in their response to fertilizer, irrigation, or pesticides. They can be delineated based on differences in crop yield, soil type, topography, or soil properties (fertility, moisture content, pH, organic matter, etc.). Remote sensing has been used to delineate MZs based on variations in soil organic matter content (Mulla, 1997; Fleming et al., 2004; Christy, 2008). Boydell and McBratney (2002) used 11 years of Landsat TM imagery for a cotton field to identify MZs based on yield stability.

7.4 Irrigation Management

Water stress is one of the major causes for loss of crop productivity (Moran et al., 2004). Irrigation is widely used to overcome crop water stress but, when applied uniformly, can lead to drawdown of water supply and environmental pollution. In precision irrigation, also known as variable rate irrigation (Sadler et al., 2005), sprinkler heads deliver water at rates that are varied using either microprocessors (Stark et al., 1993) or solenoids connected to manifolds (Omary et al., 1997). Nozzle spray rates are varied depending on spatial patterns in soil moisture (Hedley and Yule, 2009), crop stress (Bastiaanssen and Bos, 1999), or soil or landscape patterns, including rock outcroppings (Sadler et al., 2005). Variable rate irrigation uses water more efficiently than uniform irrigation, leading to better water conservation and improved environmental quality, without affecting crop yield.

Remote sensing can be used in variable rate irrigation applications to detect crop water stress through thermal infrared (TIR) (Moran et al., 2004; Rud et al., 2014) or microwave (Vereecken et al., 2012) sensing. TIR sensing can be used to measure canopy temperature and crop water stress, and this measurement, when combined with reflectance measurements in the red and near-infrared (NIR) regions, can be used to construct reflectance index-temperature space graphs that lead to identification of field locations where nutrient and/or water stress occurs (Lamb et al., 2014). TIR sensing can also be used to infer crop water stress by measuring a crop water stress index (CWSI) that is proportional to the difference between canopy and air temperatures (Moran et al., 2004) but also depends on the atmospheric vapor pressure deficit. CWSI values are estimated relative to the canopy and air temperatures for a nonstressed (well-watered) crop. This method works well for full crop canopies in close proximity to a well-watered section of the crop. Meron et al. (2010) developed a simplified approach for estimating CWSI that involves TIR measurements of canopy temperature relative to the temperature of a nearby artificial reference surface consisting of a wet, white fabric covering polystyrene floating in a container of water. Care must be taken to segment thermal images in fields with partial canopy cover in order to eliminate errors due to high soil temperatures. Meron et al. (2010) and Rud et al. (2014) showed that TIR measurements of CWSI based on the artificial reference surface approach could be used to develop maps showing spatial patterns in crop water stress with an 82% accuracy relative to leaf water potential measurements. These maps were useful for guiding the application of variable rates of irrigation.

7.5 Crop Scouting

Crop scouting is used for timely detection of crop stressors that pose an economic risk to production (Linker et al., 1999; Fishel et al., 2001; Mueller and Pope, 2009). If detected at an early stage, management actions can be taken to control crop water stress and nutrient deficiencies, kill weeds or insects, and eradicate crop diseases. Crop scouting traditionally involves having a trained professional walk in a predetermined pattern through an agricultural field in order to conduct a limited and somewhat random sampling to detect and identify crop stress. This approach is time-consuming and labor intensive, and it does not guarantee that the sampling strategy covered the right spatial locations or occurred at the right time. Remote sensing offers the potential for improved crop scouting, with better spatial and temporal coverage than would be possible with a trained professional walking through fields. While remote sensing can accurately identify locations where crop stress is occurring, remote sensing alone is often unable to distinguish between crop stress caused by nutrient deficiencies, weed or insect pressure, or crop diseases. This inability has slowed the adoption of remote sensing in precision farming.

7.6 Wavelengths and Band Ratios of Interest in Precision Farming

Remote sensing in precision farming has focused on reflectance in the visible (VIS) and NIR, emission of radiation in the TIR, and fluorescence in the VIS spectrum. Remote sensing of soil is responsive to spatial patterns in soil moisture and organic matter content, as well as soil carbonate and iron oxide content. Remote sensing of crop canopies in the VIS spectrum responds to plant pigments such as chlorophyll *a* and *b*, anthocyanins, and carotenoids (Pinter et al., 2003; Blackburn, 2007; Hatfield et al., 2008). Plant pigments absorb radiation in narrow wavelength bands centered around 430 nm (blue or B) and 650 nm (red or R) for chlorophyll *a* and 450 nm (B) and 650 nm (R) for chlorophyll *b*. Wavelengths with low absorption characteristics conversely have high reflectance, particularly in the green (550 nm) wavelength. Remote sensing of crops in the NIR spectrum (particularly at 780, 800, and 880 nm) responds to crop canopy biomass and leaf area index (LAI), leaf orientation, and leaf size and geometry. Plant pigments and crop canopy architecture in turn respond to many crop stresses, including water stress (Bastiaanssen et al., 2000), nutrient deficiencies (Samborski et al., 2009), crop diseases (West et al., 2003), and infestations of insects (Seelan et al., 2003) or weeds (Lamb and Brown, 2001; Thorp and Tian, 2004). As a result, remote sensing has often proved useful at indirectly detecting crop stresses for applications in precision farming.

In contrast to broadband multispectral reflectance imagery collected with older satellite platforms such as Landsat, QuickBird, and IKONOS, recent attention in remote sensing has turned to analysis of narrow bands (10 nm wide) collected using hyperspectral imagery (Miao et al., 2009; Thenkabail et al., 2010; Yao et al., 2010). The hyperspectral data cube can be used to represent crop reflectance over large areas at each of these narrow bands (Figure 7.1; Nigon et al., 2014), illustrating the large amount of spatial and spectral information collected with hyperspectral imaging. In theory, hyperspectral imaging offers the capability of sensing a wide variety of soil and crop characteristics simultaneously, including moisture status, organic matter, nutrients, chlorophyll, carotenoids, cellulose, LAI, and crop biomass (Haboudane et al., 2002, 2004; Goel et al., 2003). Thenkabail et al. (2000) showed that hyperspectral data can be used to construct three general categories of predictive spectral indices, including (1) optimal multiple narrowband reflectance indices (OMNBR), (2) narrowband normalized difference vegetative indices (NDVIs), and (3) soil-adjusted vegetation indices (SAVIs). Only two to four narrow bands were needed to describe plant characteristics with OMNBR. The greatest information about plant characteristics in OMNBR includes the longer red wavelengths (650–700 nm), shorter green wavelengths

FIGURE 7.1 Hyperspectral data cube for an irrigated Minnesota potato field showing the spatial and spectral resolution available with hyperspectral imaging. The circular slices in front represent a combination of reflectance values at red, green, and blue wavelengths, whereas the cubical slices in the back represent narrowband reflectance across a broad range of VIS and NIR wavelengths.

(500–550 nm), red edge (720 nm), and two NIR (900–940 and 982 nm) spectral bands. The information in these bands is only available in narrow increments of 10–20 nm and is easily obscured in broad multispectral bands that are available with older satellite imaging systems. The best combination of two narrow bands in NDVI-like indices was centered in the red (682 nm) and NIR (920 nm) wavelengths but varied depending on the type of crop (corn, soybean, cotton, or potato) as well as the plant characteristic of interest (LAI, biomass, etc.). Analysis of hyperspectral imagery can potentially involve advanced chemometric methods that are not possible with broadband multispectral imagery, including (1) lambda–lambda plots, (2) spectral derivatives, (3) discriminant analysis, and (4) partial least squares analysis (Jain et al., 2007; Alchanatis and Cohen, 2010, Li et al., 2014b, Yuan et al., 2014).

The sharp contrast in reflectance behavior between the red and NIR portions of the spectrum is the motivation for development of spectral indices that are based on ratios of reflectance values in the VIS and NIR regions (Sripada et al., 2008). Commonly used spectral reflectance indices (Table 7.1) include NDVI (NDVI = (NIR − red)/(NIR + red)), green NDVI, and ratio vegetation index (RVI = NIR/R). These indices, along with indices that are based on reflectance in the red-edge spectrum region (700–740 nm), have been found to be very sensitive to crop canopy chlorophyll and nitrogen status due to the rapid change in leaf reflectance caused by the strong absorption by pigments in the red spectrum and leaf scattering in the NIR spectrum (Hatfield et al., 2008; Nguy-Robertson et al., 2012).

TABLE 7.1 Multispectral Broadband Vegetation Indices or Commercial Sensor Midpoint Wavelengths Available for Use in Precision Agriculture

Index	Definition	References
GNDVI	$(NIR - G)/(NIR + G)$	Gitelson et al. (1996)
MSAVI2	$0.5 \times [2 \times (NIR + 1) - SQRT((2 \times NIR + 1)^2 - 8 \times (NIR - (R))]$	Qi et al. (1994)
NDVI	$(NIR - R)/(NIR + R)$	Rouse et al. (1973)
OSAVI	$(NIR - R)/(NIR + R + 0.16)$	Rondeaux et al. (1996)
REIP	$R/(NIR + R + G)$	Sripada et al. (2005)
RVI	NIR/R	Jordan (1969)
SAVI	$1.5 \times [(NIR - R)/(NIR + R + 0.5)]$	Huete (1988)
Crop Circle ACS 430	$R_{670}, R_{730}, R_{780}$	Holland et al. (2012)
CropSpec	R_{730}, R_{805}	Reusch et al. (2010)
GreenSeeker	R_{650}, R_{770}	Solie et al. (1996)
Yara N sensor ALS	$R_{730}, R_{760}, R_{900}, R_{970}$	Link and Reusch (2006)

G refers to green reflectance, NIR to near infrared, and R to red reflectance. For commercial sensors, Rx refers to the center wavelength x of the reflectance band used by the sensor.

Several red-edge-based vegetation indices such as transformed chlorophyll absorption reflection index (TCARI) have been identified from hyperspectral imagery (Haboudane et al., 2002) for estimating crop nitrogen status (Table 7.2). For example, red-edge inflation point (REIP; Guyot et al., 1988) uses a red band (670 nm), two red-edge bands (700 and 740 nm), and an NIR band (780 nm). It accurately estimated nitrogen supply to the plant, plant nitrogen concentration and uptake, and the nitrogen nutrition index (NNI) and was not affected significantly by interfering factors (e.g., zenith angle of the sun, cloud cover, and soil color) (Heege et al., 2008; Mistele and Schmidhalter, 2008). The canopy chlorophyll content index (CCCI) is an integrated index based on the theory of 2D planar domain illustrated by Clarke et al. (2001) using three bands (red, red-edge, and NIR). It uses NDVI as a surrogate for ground cover to separate soil signal from plant signal and the normalized difference red-edge (NDRE) index as a measure of canopy nitrogen status (Fitzgerald et al., 2010). It is not significantly affected by ground cover (Fitzgerald et al., 2010) and worked well for estimating plant nitrogen status in the early growing season of maize (Li et al., 2014a). Other red-edge indices include red-edge chlorophyll index (CIred edge) (Gitelson et al., 2005), red-edge ratio index (Erdle et al., 2011), DATT index (Datt, 1999), medium-resolution imaging spectrometer terrestrial chlorophyll index (MTCI) (Shiratsuchi et al., 2011), red-edge soil-adjusted vegetation index (RESAVI), modified RESAVI (MRESAVI), red-edge difference vegetation index (REDVI), and red-edge renormalized difference vegetation index (RERDVI) (Cao et al., 2013).

The ultraviolet (UV), violet, and blue spectral regions have also been found to be important for estimating plant nitrogen concentration (Li et al., 2010). Wang et al. (2012) developed a new three-band vegetation index using NIR, red-edge, and blue bands $[(R_{924} - R_{703} + 2 \times R_{423})/(R_{924} + R_{703} - 2 \times R_{423})]$, which was found to be closely related to wheat and rice leaf nitrogen concentration. Far NIR (FNIR) and shortwave infrared (SWIR) bands were found to be important for estimating plant aboveground biomass (Thenkabail et al., 2004; Gnyp et al., 2014). These bands are currently missing from the commercial active canopy sensors commonly used in precision farming.

TABLE 7.2 Hyperspectral Narrowband Vegetation Indices Available for Use in Precision Agriculture

Index	Definition	References
Aphid index (AI)	$(R_{576} - R_{908})/(R_{756} - R_{716})$	Mirik et al. (2007)
CIred edge	$(R_{753}/R_{709}) - 1$	Gitelson et al. (2005)
DATT index	$(R_{850} - R_{710})/(R_{850} - R_{680})$	Datt (1999)
Damage sensitive spectral index (DSSI)	$(R_{576} - R_{868} - R_{508} - R_{540})/[(R_{716} - R_{868}) + (R_{508} - R_{540})]$	Mirik et al. (2007)
Leafhopper index (LHI)	$(R_{761} - R_{691})/(R_{550} - R_{715})$	Prabhakar et al. (2011)
MTCI	$(R_{754} - R_{709})/(R_{709} - R_{681})$	Dash and Curran (2004)
NDRE	$(R_{790} - R_{720})/(R_{790} + R_{720})$	Barnes et al. (2000)
REIP	$700 + 40 \times \{[(R_{667} - R_{782})/2 - R_{702}]/(R_{738} + R_{702})\}$	Guyot et al. (1988)
Red edge ratio index	(R_{760}/R_{730})	Erdle et al. (2011)
PK index	$(R_{1645} - R_{1715})/(R_{1645} - R_{1715})$	Pimstein et al. (2011)
PRI	$(R_{531} - R_{570})/(R_{531} + R_{570})$	Gamon et al. (1992)
S index	$(R_{1260} - R_{660})/(R_{1260} + R_{660})$	Mahajan et al. (2014)
TCARI	$3 \times [(R_{700} - R_{670}) - 0.2 \times (R_{700} - R_{550})(R_{700}/R_{670})]$	Haboudane et al. (2002)

R refers to reflectance at the wavelength (nm) in subscript. NIR refers to near-infrared reflectance.

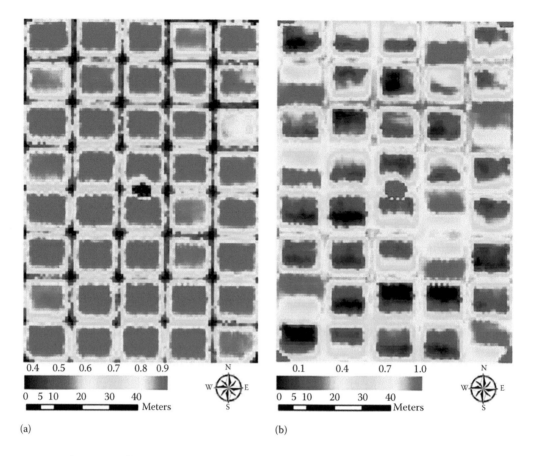

(a)

(b)

FIGURE 7.2 Hyperspectral estimates of (a) NDVI values and (b) TCARI-OSAVI values for small plots in a Minnesota potato field with two crop varieties receiving a wide range of nitrogen fertilizer application rates and timings. NDVI values exhibit a small range of values due to saturation. In contrast, TCARI-OSAVI values exhibit a large range of values and are better suited for identifying differences in nitrogen stress for each variety.

The commonly used NDVI can easily become saturated at moderate to high canopy coverage conditions (Figure 7.2; Nigon et al., 2014). One reason is due to the normalization effect embedded in the calculation formula of this index (Nguy-Robertson et al., 2012; Gnyp et al., 2014), and another reason is due to the different transmittance of red and NIR radiation through the crop canopy leaves. The saturation effect of NDVI can be partially addressed by using RVI or wavelengths having similar penetration into the canopy (Van Niel and McVicar, 2004; Gnyp et al., 2014; Li et al., 2014a).

It should be noted that the sensitive spectral reflectance bands for precision farming change at different crop growth stages in response to crop growth and development (Li et al., 2010; Gnyp et al., 2014). Different vegetation indices are needed for different crops, with different crop growth parameters at different growth stages (Hatfield and Prueger, 2010).

Fluorescence of leaf chlorophyll is an emerging research area in precision farming (Tremblay et al., 2012). When leaves that have been in the dark are exposed to UV or blue light, chlorophyll *a* in photosystem II (PSII) is excited to the first singlet state (Sayed, 2003), and upon decay to the ground energy state, these molecules are capable of fluorescence. Leaf fluorescence is affected by many factors including the wavelength and intensity of incident light, temperature, canopy structure, and leaf

chlorophyll content, which may be affected by crop stresses from water, nitrogen, and salinity (Sayed, 2003; Tremblay et al., 2012). On first exposure to light, quinine acceptors in PSII are maximally oxidized (Baker and Rosenqvist, 2004), leading to a minimal fluorescence level (F_o). After further exposure to light, maximal fluorescence (F_m) may be attained, indicating that all electron acceptors are reduced (Baker and Rosenqvist, 2004). Interpretation of plant stress levels is often based on combinations or ratios of these two parameters (Sayed, 2003; Baker and Rosenqvist, 2004; Tremblay et al., 2012). Variable fluorescence (F_v) is defined as $F_m - F_o$, and F_v/F_m represents the photochemical efficiency of PSII (Tremblay et al., 2012). High values of F_o indicate plant stress (Tremblay et al., 2012), whereas low values of F_v/F_m indicate nitrogen stress (Baker and Rosenqvist, 2004). Diagnosis of specific types of crop stress may be facilitated by combining fluorescence spectroscopy with hyperspectral or multispectral imaging (Moshou et al., 2012).

7.7 Nutrient Deficiencies

Crop nutrient deficiencies are a major cause of crop stress and reductions in crop yield or quality. Nutrient deficiencies may be caused by macronutrients such as nitrogen, phosphorus, or potassium, or by micronutrients such as sulfur, calcium, magnesium, or zinc. Nutrient deficiencies often cause changes in leaf pigment concentrations, particularly for chlorophyll *a* and *b*. Changes in chlorophyll *a* or *b* content can be detected using remote sensing in the green (550 nm) and red-edge (710 nm) wavelengths. Nutrient deficiencies from either macro- or micronutrients cause spectral reflectance of crop leaves to increase in the green portion of the spectrum. Reflectance spectra of deficient leaves alone are insufficient in many cases to determine which nutrient is responsible for the deficiency and what rate or formulation of fertilizer is needed to correct the deficiency. Crop deficiencies also cause changes in crop biomass that can be detected using NIR reflectance.

Crop scout professionals have learned to distinguish and identify nutrient deficiencies based on coloration, pattern, location, and timing of the deficiency. Several examples for corn illustrate the approach used by crop scouts (Mueller and Pope, 2009). Nitrogen deficiency in corn appears as a yellowing of leaf color, starting with lower leaves. Deficiencies first appear at leaf tips and progress toward the base of the leaf in a v-shaped pattern. Phosphorus deficiency appears as red to purple leaf tips in the older leaves of young corn plants that appear to have stunted growth. Newly emerged leaves do not show phosphorus deficiencies, and the distinctive coloration associated with phosphorus deficiencies disappears when the crop grows to a meter or more in height. Potassium deficiency appears in corn as a yellowing along the edges of leaves at growth stage V6. It is often associated with conditions that lead to poor rooting depth. Remote sensing offers the potential to identify characteristic colors, patterns, and locations on a plant affected by nutrient deficiencies if the spatial resolution of imagery is on the order of a few centimeters.

Nutrient deficiencies that are detected and diagnosed in a timely fashion can be corrected using variable rate technology (VRT). VRT involves applying the right rate of fertilizer, at the right blend, in the right location, and at the right time. There is a long history of VRT in precision farming, with a primary focus on correcting nutrient deficiencies caused by phosphorus or nitrogen. In the earliest application of remote sensing for precision farming, Landsat TM images were used along with auxiliary data from soil sampling to develop maps showing spatial variability in phosphorus fertilizer recommendations for a wheat farm in Washington State (Bhatti et al., 1991). Landsat imagery was used to estimate spatial patterns in soil organic matter content, which were indirectly correlated with spatial patterns in soil phosphorus.

Proximal sensing of crops is currently the primary tool used to detect nutrient deficiencies for variable rate application of fertilizer. This is based on research that showed nitrogen deficiencies could be detected using spectral reflectance in the green, red, red edge, and NIR portions of the spectrum. Commercial sensors used in precision farming to detect crop nitrogen deficiencies (Figure 7.3; Table 7.1) are mainly active crop canopy sensors with their own light sources to avoid the influence of different environmental light conditions, including the GreenSeeker, Crop Circle, CropSpec, and Yara N-sensor (Barker and Sawyer, 2010; Kitchen et al., 2010; Shaver et al., 2011). GreenSeeker operates

FIGURE 7.3 Active crop canopy sensors commonly used in precision farming in the United States. (GreenSeeker, left; Crop Circle ACS 430, middle; Crop Circle ACS 470, right.)

in the red (650 nm) and NIR (770 nm). Crop Circle ACS 210 operates in the green (590 nm) and NIR (880 nm), while Crop Circle ACS 430 has red (670 nm), red edge (730 nm), and NIR (780 nm) bands. Crop Circle ACS 470 sensor also has three bands but is user-configurable with a choice of six spectral bands covering blue (450 nm), green (550 nm), red (650, 670 nm), red edge (730 nm), and NIR (>760 nm) regions (Cao et al., 2013). CropSpec operates in the red edge (730 nm) and NIR (805 nm). Yara's traditional N-sensor operates at 730 (red) and 760 (NIR) nm. A newer version of the Yara N sensor allows the operator to select four reflectance bands between 730 and 970 nm.

One limitation of the GreenSeeker, Yara N, CropSpec, and Crop Circle sensors is that they cannot directly estimate the amount of N fertilizer needed to overcome crop N stress (Samborski et al., 2009). Instead, sensor readings have to be compared to readings in reference strips receiving sufficient N fertilizer (Blackmer and Schepers, 1995; Raun et al., 2002; Sripada et al., 2008; Kitchen et al., 2010). These comparisons are the basis for N fertilizer response functions that relate sensor readings to the amount of N fertilizer needed to overcome crop N stress (Scharf et al., 2011). Clay et al. (2012) have shown that for wheat, when both water and nitrogen stress occur simultaneously, N fertilizer recommendations based on NDVI values are more accurate when reference strips have both sufficient nitrogen and insufficient moisture (water stress) in comparison with reference strips with both sufficient nitrogen and sufficient moisture (no water stress). Kitchen et al. (2010) found that use of Crop Circle sensors was able to accurately identify N stress in corn 50% of the time in 22 field studies conducted over 4 years across a wide range of soil types in Missouri.

Phosphorus deficiencies typically appear as changes in reflectance in the NIR and blue portions of the spectrum. There has been little research on remote sensing methods to distinguish nitrogen, phosphorus, and potassium deficiencies in crops (Pimstein et al., 2011; Mahajan et al., 2014). Spectral signatures for nitrogen, phosphorus, and potassium deficiency show responses at different wavelengths (Pimstein et al., 2011). NDVI values (such as those estimated using GreenSeeker technology) are often not able to distinguish between N and P deficiencies (Grove and Navarro, 2013). To distinguish nitrogen, phosphorus, and potassium deficiencies in wheat, Pimstein et al. (2011) proposed new spectral indices that require collecting reflectance data in the SWIR region (1450, 1645, and 1715 nm). These new indices were able to predict P or K deficiency with an accuracy ranging from 78% to 80%, but accuracy levels decreased as variability in crop biomass increased. Mahajan et al. (2014) found that distinguishing between sulfur and nitrogen deficiency in wheat required the collection of SWIR data. They proposed a sulfur deficiency index that involves an NDVI-like ratio of reflectances at 1260 and 660 nm (Mahajan et al., 2014). The performance of the sulfur index was nominally better than other standard vegetative indices, including NDVI and SAVI.

In order to distinguish between different types of nutrient deficiencies, remote sensing must rely on more than changes in reflectance at key wavelengths. A diagnosis with remote sensing must also be able to detect where on the plant (upper vs. lower leaves, leaf tips or edges, etc.) symptoms of deficiency occur and in what pattern. These patterns change over time, and early detection is important. High-resolution imagery at the scale of centimeter-size pixels is needed for early detection; otherwise it will be difficult to identify whether or not symptoms of deficiency are in upper or lower leaves, at leaf tips or basal regions, or along the edges or in interveinal regions of the leaf. For deficiencies that tend to occur in young plants, remote sensing must be able to compensate for reflectance from bare soil; hence, spectral indices such as SAVI (Huete, 1988), modified SAVI (MSAVI; Qi et al., 1994), or optimized SAVI (OSAVI; Rondeaux et al., 1996) may be useful.

7.8 Insect Detection

Insects cause crop damage by sucking plant sap, eating plant tissue, or damaging crop roots. Examples include European corn borer and Russian wheat aphid. These damages usually result in decreased crop biomass and deformed or stripped leaves. Because decreased biomass also occurs in response to other crop stressors, identifying insect damage via remote sensing has proved challenging.

Insect growth and development is more strongly linked with temperature and growing degree days than crop phenology (Hicks and Naeve, 1998; MacRae, 1998). Insects can first appear in a variety of locations, including along edges of fields, on undersides of leaves, or in the soil. It is difficult to detect insects in soil or on the undersides of leaves with remote sensing. Remote sensing often detects crop damage caused by insects, rather than the insects themselves. Harmful insects should be detected and identified before they can cause significant damage to crops. Proper identification is important because control methods vary by insect species.

Remote sensing is not widely used in precision farming for detecting insect infestations. Franke and Menz (2007) used hyperspectral imaging from an airplane in Iowa corn plots inoculated with European corn borer. Spectral indices were largely ineffective at differentiating inoculated plots from control plots during the first generation of insect growth. NDVI was consistently able to identify inoculated plots during the second generation of corn borer growth. These results show that it is difficult to use remote sensing for early detection of European corn borer. Mirik et al. (2007) used a handheld hyperspectral radiometer to measure reflectance in the VIS and NIR wavelengths for Texas, Colorado, and Oklahoma winter wheat plots with and without significant Russian wheat aphid infestations. Their results showed that aphid damage resulted in changes in biomass that reduced NIR reflectance in infested plants relative to undamaged plants. They also showed increased reflectance in the green portion of the spectrum due to changes in chlorophyll content of leaves for infested plants relative to uninfested plants. They proposed using an aphid index (AI) and a damage sensitive spectral index (DSSI) to detect Russian wheat aphid damage (Table 7.2). AI is estimated based on $(R_{576} - R_{908})/(R_{756} - R_{716})$, where R is reflectance and the subscript denotes the wavelength (nm) of interest. DSSI is more complicated and is estimated using $(R_{716} - R_{868} - R_{508} - R_{540})/[(R_{716} - R_{868}) + (R_{508} - R_{540})]$. Because the field of view for the handheld spectrometer was narrow, there was little mixing of pixels from infested and uninfested leaves, something that would be a significant impediment if reflectance measurements were obtained using satellites. Aphid damage was identified in four fields at different times of the year with an accuracy ranging from 46% to 80% using the AI.

Prabhakar et al. (2011) used hyperspectral imaging to detect leafhopper damage in cotton. They found that leafhopper damage was associated with decreases in the content of chlorophyll *a* and *b* pigments in leaves. The best spectral indices for identifying leafhopper damage were based on changes in leaf reflectance in the VIS (376, 496, and 691 nm) and NIR (761, 1124, and 1457 nm) portions of the spectrum. A leafhopper index defined as $(R_{761} - R_{691})/(R_{550} - R_{715})$ could explain from 46% to 82% of the variability in leafhopper damage across three fields. A number of other spectral indices also performed relatively well, including NDVI, OSAVI, AI, and DSSI (Table 7.2).

7.9 Disease Detection

Diseases are caused by infestations of virus, fungi, or bacteria. They can affect any part of the plant, including leaves, stalks, roots, or grain. Damage to leaves often occurs as lesions or pustules that may lead to white, tan, brown, or orange leaf colors (Mueller and Pope, 2009). Lesions can occur in shapes as varied as spots, rectangles, or strips that vary in size and area. Each disease has a specific location where infection tends to occur and each is associated with different shapes and colors of infected areas. Infected plants may eventually become stunted and have chlorotic or necrotic leaves (Mirik et al., 2011). Early detection of disease is essential to limit economic damage (Sankaran et al., 2010).

Spectral characteristics of crops are often affected by disease, as described by West et al. (2003). Disease propagules often influence reflectance in the VIS spectrum. Necrotic or chlorotic damage affects chlorophyll content and reflectance in the green and red-edge regions. Senescence affects reflectance in the red to NIR region. Stunting and reduced leaf area influences NIR reflectance. Impacts of disease on photosynthesis affect fluorescence in the spectral region between 450–550 and 690–740 nm (West et al., 2003). Crop disease also affects transpiration rates and water contents of leaves; these effects can be detected in the shortwave and TIR regions.

Remote sensing is not widely used to detect crop disease in precision farming; however, research has shown that remote sensing has the potential to be used for such purposes (Table 7.2). Remote sensing has been used to detect fungal and viral infections in soybean (Das et al., 2013) and wheat (Muhammed, 2005; Huang et al., 2007; Mewes et al., 2011; Mirik et al., 2011). Yellow rust infections of wheat in China were detected with a 91%–97% accuracy over 2 years using aerial hyperspectral remote sensing and a photochemical reflectance index (PRI) (Huang et al., 2007). Values of PRI were estimated using reflectance values at 531 and 570 nm. Fluorescence at 550 and 690 nm was also useful for distinguishing wheat leaves infected with yellow rust from uninfected leaves (Bravo et al., 2004). Wheat infected with septoria leaf blotch in France was accurately distinguished from uninfected wheat using a combination of NDVI and TIR measurements (Nicolas, 2004). Infestations of powdery mildew and leaf rust on wheat in Germany were difficult to detect at early stages of infection with QuickBird-like NDVI values (Franke and Menz, 2007), with an accuracy of only 57%. This is because at early stages of infection, reflectance in the red portion of the spectrum is affected, but NIR reflectance is not (Lorenzen and Jensen, 1989). At more advanced stages of infection, plant canopy structure and biomass are affected, causing changes in NIR reflectance that result in large decreases in NDVI values and higher accuracy (89%) in detecting infection.

Yuan et al. (2014) used hyperspectral imaging to simultaneously detect and distinguish damage to wheat leaves caused by yellow rust and powdery mildew diseases and Russian wheat aphids. Reflectance in leaves damaged by disease and insect generally increased relative to undamaged leaves at wavelengths between 500 and 690 nm. Distinguishing between disease and insect damage required analysis of reflectance in the NIR portion of the spectrum between 750 and 1300 nm. Powdery mildew and aphid damage caused reflectance in this region to decrease, whereas reflectance in this region increased for yellow rust damage. Partial least squares regression of reflectance in these regions, along with spectral derivative parameters and conventional spectral indices such as AI (Table 7.3), could explain 73% of the variability in intensity of wheat damage by the three stressors studied. Distinguishing damage from yellow rust versus powdery mildew versus aphids with hyperspectral imaging and Fisher linear discriminant analysis was more challenging, however, especially at low intensities of infestation. Further work is needed to extend the research of Yuan et al. (2014) to entire crop canopies.

TABLE 7.3 Spectral Indices or Commercial Sensors Available for Diagnosis of Nutrient Deficiencies, Crop Disease, and Insect or Weed Infestations in Precision Agriculture

Index	N, P, or K	Disease	Insects	Weeds
Aphid index (AI)			X	
CIred edge	X			
DATT index	X			
Damage sensitive spectral index (DSSI)			X	
Fluorescence	X	X		
Leafhopper index (LHI)			X	
MERIS TCI	X			
NDRE	X			
NDVI	X	X	X	X
REIP	X			
Red edge ratio index	X			
RVI	X			X
PK index	X			
PRI		X		
SAVI (or related)	X		X	X
S index	X			
TCARI	X			
Crop Circle ACS 430	X			
CropSpec	X			
GreenSeeker	X			
Yara N sensor	X			
WeedSeeker				X

7.10 Weed Detection

Weeds compete with crops for light, water, and nutrients. Above critical weed density thresholds, crop yields and quality will decline substantially. In most fields, weed infestations are not uniform; rather, weeds tend to occur in patches or clusters, leaving up to 80% of the field free of weeds (Wiles et al., 1992; Lamb and Brown, 2001). Because of this, there has been quite a bit of interest in precision farming (variable rate herbicide application) to control weeds that occur in patches while avoiding herbicide application in areas without weeds (Stafford and Miller, 1993; Mulla et al., 1996; Hanks and Beck, 1998; Khakural et al., 1999). Variable rate herbicide application is especially of interest in Europe, where genetically modified crops (such as Roundup Ready soybean) are not allowed.

Weeds can be identified using remote sensing based on their spectral signatures, leaf shape, and organization of the weedy plant. Detecting and identifying weeds in a bare soil that is crop-free is easier than detecting and identifying weeds in an actively growing crop (Thorp and Tian, 2004; López-Granados, 2011). Detecting weeds that occur in large, dense clusters is easier with aerial remote sensing than identifying small, isolated weeds.

Remote sensing with satellites or airplanes is adequate for detecting weeds that occur in large, dense clusters within a crop or in bare crop-free soil (Lamb and Brown, 2001). Ground-based proximal sensing is more suited than aerial remote sensing to detect and identify small, isolated weeds in a growing crop (Thorp and Tian, 2004). Proximal sensing has been used for real-time monitoring and spraying of weeds from a field herbicide applicator (López-Granados, 2011). A commercial example of this technology is WeedSeeker (Hanks and Beck, 1998), which uses gallium arsenide photoelectric emitters to detect weeds growing in bare soil or in a crop canopy (Sui et al., 2008). This technology is best suited to detecting weeds at intermediate growth stages that are growing between crop rows. It is not well suited to detecting recently emerged weeds (Thorp and Tian, 2004).

Zwiggelaar (1998) reviewed remote sensing methods for distinguishing weeds from soils or crops. Remote sensing is only useful if weeds have a spectral signature that is uniquely different from surrounding bare soil or crops and if the spatial resolution of images is fine enough to detect individual weeds or patches of weeds (Lamb and Brown, 2001). Distinguishing weeds from soil is often based on graphing reflectance in the red portion of the spectrum versus reflectance in the NIR portion of the spectrum. A graph of these two reflectance bands for bare soil gives the soil line (Wiegand et al., 1991). For fields with mixtures of bare soil and weeds, the presence of weeds increases with vertical distance above the soil line along the NIR axis. Graphs of red versus NIR reflectance are commonly referred to as tasseled cap transformations. Band ratios have also been used to distinguish weeds from bare soil. The most common approach for detection of weeds in bare soil is to use the NDVI ratio (Table 7.3). This ratio has the advantage of canceling out effects of shadows produced by weeds. Reflectance from bare soil can also be diminished through use of SAVI (OSAVI, MSAVI, etc).

Spectral reflectance patterns of weeds and crops are in general very similar when bare soil is absent (Zwiggelaar, 1998; Lamb and Brown, 2001). When bare soil is present, reflectance values at two wavelengths (e.g., 758 and 658 nm) can be used along with discriminant analysis to distinguish crops from weeds from soil (Borregaard et al., 2000). RVI (= NIR/R) and NDVI have also often been used to discriminate between weeds and crops (Table 7.3), especially when crops occur in systematic rows and weeds occur as patches between rows. Detection of weeds at early growth stages is very challenging (López-Granados, 2011), especially if they occur in recently germinated crops with similar physiology (e.g., grassy weeds in cereal crops or broad leaf weeds in dicotyledonous crops). Detection is easier at later growth stages, when spectral differences between weeds and crops are greatest (López-Granados, 2011). Accuracies at discriminating weeds from bare soil range from 75% to 92%, while accuracies in distinguishing one weed species from another often range between 61% and 88% (Thorp and Tian, 2004; López-Granados, 2011).

7.11 Machine Vision for Weed Discrimination

Discrimination between weeds and crops requires high spatial resolution of imagery (Zwiggelaar, 1998). Remote sensing images with a spatial resolution of tens of meters will not be sufficient for discrimination of weeds and crops. Images at a spatial resolution of tens of centimeters to a meter are needed to distinguish plants from weeds (Lamb and Brown, 2001; Rasmussen et al., 2013). However, even spectral indices at this fine scale of resolution are often by themselves not sufficient because crops and weeds often have similar reflectance signatures. Crops and weeds are more easily distinguished based on differences in their canopy or leaf shapes, heights, and structures. These features can be described and distinguished from one another using machine vision analysis of color images or video imagery (Gée et al., 2008; Burgos-Artizzu et al., 2011).

Discrimination of one weed species from another is more challenging than discriminating weeds from crops. Gibson et al. (2004) used supervised classification of weeds in soybean based on aerial remote sensing in the yellow, green, red, and NIR bands. While weedy areas could be distinguished from soybeans or bare soil with accuracies of greater than 90%, distinguishing giant foxtail from velvetleaf had accuracy levels ranging from 41% to 83%.

Machine vision is commonly used for precision farming applications of discriminating weeds from bare soil or crops (Thorp and Tian, 2004). There are two basic steps in discriminating weeds (Gée et al., 2008; Burgos-Artizzu et al., 2011). The first is distinguishing regions with vegetation from regions with bare soil (segmentation). The second is distinguishing weeds from crops (discrimination). As an example of this two-step process, Gée et al. (2008) used a red–green–blue color image in various row crops to estimate an excess green index (Gée et al., 2008), which was then reclassified into black (soil) and white (vegetation) components. The reclassified image was then subjected to a double Hough transformation (DHT) to identify the position of the linear crop rows. Blobs of white (vegetation) that were offset from rows were assumed to be weeds. Burgos-Artizzu et al. (2011) use real-time analysis of video imagery to perform these same two steps and were able to accurately identify 85% of the weeds in a field of maize.

Examples of machine vision for precision weed management are numerous. The University of Tokyo developed an autonomous vehicle for mechanical weeding and variable rate application of chemicals (Torii, 2000). This vehicle is guided along crop rows based on a hue, saturation, and intensity transformation. Tillet et al. (2008) used real-time machine vision in conjunction with a mechanical weeder to reduce weed populations in cabbage by 62%–87%. Blasco et al. (2002) used machine vision with a robotic weeder that produced an electrical discharge of 15,000 V. These studies both show that it is possible to use precision farming techniques to avoid using herbicides to control weeds.

7.12 Remote Sensing Platforms

Remote sensing imagery for precision farming can be obtained using satellites, airplanes, UAVs, ground robots, or agricultural machinery (Moran et al., 1997; Zhang and Kovacs, 2012; Mulla, 2013). Remote sensing imagery from satellites has improved in spatial resolution, spectral resolution, and the frequency of return visits since the launch of Landsat in the 1970s. Spatial

TABLE 7.4   Characteristics of Data Gathered from Satellite Sensors of Different Eras Suitable for Precision Farming

Satellite/Sensor	Spatial Resolution (m)	Spectral Bands (Number of Bands)	Data Points or Pixels per Hectare
MODIS-Terra	250–1000 m	36	0.16, 0.01
Terra EOS ASTER	15, 30, 90 m (VIS, SWIR, TIR)	4, 6, 5	44.4, 11.1, 1.26
Landsat-7 TM	15 m (P), 30 m (M)	7	44.4, 11.1
ALI	10 m (P), 30 m (M)	1, 9	100, 11.1
Hyperion	30	220 (400–2500 nm)	11.1
IRS-1C LISS	5 m (P), 23.5 m (M)	3	400, 18.1
IRS-1D LISS	5 m (P), 23.5 m (M)	3	400, 18.1
SPOT-1,2,3,4 HRV	10 m (P), 20 m (P)	4	100, 25
Landsat-4,5 TM	30 m (M)	7	11.1
Landsat-1,2,3 MSS	56 × 79	4	2.26

M, multispectral; P, panchromatic; VIS, visible; SWIR, shortwave infrared; TIR, thermal infrared.

TABLE 7.5   Characteristics of Data Gathered from Very-High-Spatial-Resolution Satellites/Sensors Suitable for Precision Farming

Satellite/Sensor	Spatial Resolution (m)	Spectral Resolution (Number of Bands)	Data Points or Pixels per Hectares
IKONOS 2	0.82 m (P), 4 m (M)	4	14,872, 625
QuickBird	0.61 m (P), 4 m (M)	4	26,874, 625
EROS A	1.82 m (P)	1	3,020
RapidEye	5 (M)	4 + red-edge	400
GeoEye-1	1.65 (M)	4	3,673
WorldView-3	1.24 (M), 3.7 (SWIR)	8 (M), 8 (SWIR)	6,504, 730
AISA Eagle	1 (H)	63	10,000
Tetracam Mini-MCA6	0.066 (M)	5 + red-edge	2,295,684

M, multispectral; P, panchromatic; H, hyperspectral.

resolution has improved from 30 m with Landsat 4 to 1.24 m with WorldView-3 for multispectral satellite imagery (Tables 7.4 and 7.5). Spectral resolution (number of bands) has improved from four broad bands in the blue, green, red, and NIR regions to multiple narrowband imagery in the purple, blue, green, yellow, red, red edge, and NIR wavelengths. Return frequencies have improved from several weeks to a day or 2. Despite these improvements, satellite imagery in the VIS and NIR regions still suffers from an inability to penetrate cloud cover. Furthermore, there are continuing issues with satellite providers who are unable to reliably provide agricultural imagery at desired time intervals.

Aerial remote sensing imagery offers excellent capabilities for precision farming applications. Spatial resolution is typically a meter or better, and spectral resolution ranges from broadband blue, green, red, and NIR to hyperspectral imaging (e.g., with the AISA Eagle camera; Table 7.5). Aerial imaging can typically be obtained when and where it is needed with high reliability. Cloud cover is a continuing challenge for remote sensing from airplanes. Even though airplanes can fly below cloud cover, shadows from clouds cause difficulties in interpreting imagery.

Remote sensing imagery obtained by proximal sensing from agricultural equipment is very popular in precision farming. Examples include on-the-go sensing from fertilizer spreaders for variable rate application of nitrogen fertilizer and on-the-go sensing from herbicide sprayers for variable rate application of herbicides. Sensors used for proximal sensing are typically limited to two or three narrow bands of reflectance, thereby limiting the number of spectral indices that can be used to diagnose causes of stress. This is particularly limiting in mature crops with LAI values greater than three for sensors that calculate NDVI values. The NDVI values are less sensitive to spatial variations in chlorophyll content of leaves in mature crop canopies than at earlier growth stages.

Researchers are beginning to explore the use of UAVs for acquisition of remote sensing imagery (Figure 7.4). UAVs typically include fixed-wing aircraft or helicopters that fly at altitudes of roughly 100 m (Zhang and Kovacs, 2012). Because of the low altitude, many images are typically acquired, and these must be tiled or mosaicked together to produce a continuous image of the field or farm of interest (Gómez-Candón et al., 2014). Fixed-wing aircraft generally have longer flight time (greater power supply) and payload capacity than helicopters. Aircraft have faster flight speeds than helicopters, and this may result in blurring of images due to the low altitude. Helicopter UAVs have the advantages of flexibility and less space restriction by allowing vertical takeoff and the ability to land vertically, hover, and fly forward, backward, and laterally as compared with fixed-wing UAVs, allowing them to inspect isolated small fields closer to obstructions, which may be difficult for fixed-wing UAVs (Huang et al., 2013). Helicopters are generally more stable than aircraft, resulting in fewer problems with variations in viewing angle from one image to another. Remote sensing imagery from UAVs has

FIGURE 7.4 Different types of UAVs used in precision farming: (a) fixed-wing aircraft, (b) helicopter, (c) quadrocopter, and (d) octocopter.

very high spatial resolution, typically on the order of 7–50 cm (Table 7.5; Tetracam Mini-MCA6). This allows individual plants to be studied. However, it also requires special care in correcting geometric distortion. Cameras used on UAVs range from inexpensive digital cameras that provide panchromatic images to expensive multispectral cameras that provide narrowband reflectance in the blue, yellow, green, red, red edge, and NIR regions of the spectrum (Table 7.5; Tetracam Mini-MCA6). Promising results have been obtained using UAV-based remote sensing for estimating crop LAI, biomass, plant height, nitrogen status, water stress, weed infestation, yield, and grain protein content (Berni et al., 2009; Swain et al., 2010; Samseemoung et al., 2012; Bendig et al., 2013). It is expected to become a major remote sensing platform for precision farming in the future.

7.13 Knowledge Gaps

Remote sensing applications in precision farming have increased dramatically over the last 25 years (Mulla, 2013). This increased adoption is associated with investments in precision farming research, coupled with improvements in the spatial and spectral resolution and return frequency of aerial remote sensing imagery, and the development of proximal sensors. Aerial and proximal remote sensing are primarily used for variable rate application of irrigation water and nitrogen fertilizer or for detection of weeds. Remote sensing is not widely used for detection of crop stresses by insects or plant diseases and is rarely used for detection of nutrient deficiencies other than nitrogen.

There is a pressing need for broader use of proximal and remote sensing in precision farming. Current applications of remote sensing are rarely able to simultaneously identify locations of a field afflicted with crop stress and distinguish between stresses caused by water, nutrients, weeds, insects, and disease. Furthermore, remote sensing is rarely able to distinguish between stresses caused by different types of nutrients, different types of diseases, or different types of insects. The main reason for this failure is that remote sensing applications typically rely only on spectral signatures at a few important wavelengths (green, red, red edge, and NIR) or combinations of these wavelengths where different types of crop stress have similar influences on chlorophyll content of leaves and adverse effects on crop biomass or canopy structure (Table 7.3). Distinguishing between stresses caused by water, nutrients, weeds, insects, and disease will require fusion of remote sensing information (e.g., hyperspectral and fluorescence spectroscopy) that are sensitive to these influences and effects, combined with machine vision to identify the locations on a plant (stems or leaves, leaf tips or leaf edges, and upper leaves or lower leaves), colors of stress (yellow, purple, red, brown, white, etc.), and the shapes associated with stresses (e.g., monocotyledonous vs. dicotyledonous weeds, spots vs. stripes).

Further development of remote sensing applications in precision farming will require multidisciplinary efforts by experts in crop water, nutrient, weed, insect, and disease stresses working collaboratively with experts in remote sensing and engineering. At present, these types of multidisciplinary team efforts are rare. Further development of remote sensing applications in precision farming will require use of high-resolution (centimeter scale) aerial imagery at key wavelengths to identify locations affected by crop stress, coupled with proximal sensing and machine vision to differentiate between different types of crop stress in order to diagnose the problem. Platforms to collect remote sensing imagery must be capable of deployment at intervals of at least every week during the growth of the crop, and these platforms must be capable of distinguishing between stresses caused by water, nutrients, weeds, diseases, and insects. UAVs and proximal sensors offer significant potential to address these capabilities, and further research with these platforms and sensors is encouraged.

7.14 Conclusions

Precision farming is one of the top 10 revolutions in agriculture (Crookston, 2006). It can be generally defined as doing the right management practices at the right location, in the right rate, and at the right time. Precision farming offers several benefits, including improved efficiency of farm management inputs, increases in crop productivity or quality, and reduced transport of fertilizers and pesticides beyond the edge of field.

Losses in crop productivity often occur nonuniformly at specific locations within fields and at critical growth stages. Crop stress must be detected in a timely fashion, the type of stress causing it must be identified, and management practices must be implemented at the right locations and times to overcome crop stress.

Research applications of remote sensing in precision farming are numerous and include techniques for detecting water stress, nitrogen stress, weed infestations, fungal disease, and insect damage. Remote sensing has shown the ability to identify locations experiencing stress, with accuracies ranging from 50% to 80% for nutrient stress, 46% to 82% for insect damage, 57% to 97% for crop disease, and 75% to 92% for weeds. Accuracy depends on the growth stage of crop, the level of crop stress, the spectral index used for assessment of stress, and the spatial and spectral resolution of remote sensing.

Significant advances have been made in identifying key wavelengths and spectral indices at which these stresses influence the reflectance or fluorescence properties of plant pigments and crop canopy architecture. However, little research has been conducted on detecting locations affected by crop stress and simultaneously distinguishing between different types of crop stress. A basic problem is that remote sensing does not typically respond directly to water, nutrient, weed, insect, or disease stresses; rather it responds indirectly to the changes in chlorophyll or crop canopy architecture caused by these crop stresses. For this reason, remote sensing has not yet been widely adopted by farmers for routine use in precision agriculture. The main reasons include the difficulty in interpreting spectral signatures, the slow processing time for data, the high expense, and the need to collect confirmatory data from ground surveys in order to diagnose causative factors for anomalous spectral reflectance data. Clearly, there is a significant scope for improving the interpretation and utility of remote sensing data for precision agriculture.

Researchers have focused significant effort on identifying key wavelengths at which areas with crop stress can be distinguished from areas without crop stress. These wavelengths, and spectral indices based on them, typically occur in the green, red, red edge, and NIR bands. Significant progress has been made in identifying spectral indices that respond to changes in leaf pigmentation or canopy biomass and architecture, or indices that are capable of eliminating interference from shadows and soil background effects. As the spatial resolution of remote sensing imagery used in precision farming has improved (from 30 m to submeter resolution), techniques for discriminating crops, soils, and weeds have also improved. As spectral bandwidth has decreased (from broadband blue, green, red, and NIR to narrowband hyperspectral and fluorescence spectroscopy), researchers have discovered that crop stress is more easily detected with narrow bands (10–20 nm wide) rather than broad bands (50–100 nm wide) at these key wavelengths. Narrowband hyperspectral imagery is amenable to image analysis with advanced chemometric techniques that allow for better diagnosis of crop stress, including lambda–lambda plots, derivative analysis, and partial least squares analysis.

Less progress has been made in the use of remote sensing coupled with computer vision for differentiating between specific types of crop stress based on the location within the plant where stress occurs and the shape or color of the stressor. Advances in computer vision are needed that required collaborative research by multidisciplinary teams of agronomists, engineers, and remote sensing experts working with high-resolution hyperspectral and video imagery that is capable of viewing individual plants. High-resolution imagery is increasingly possible because of improvements in camera technology and proximal sensors deployed on UAVs or ground vehicles that collect imagery at short distances from the growing crop.

References

Alchanatis, V. and Y. Cohen. 2010. Spectral and spatial methods of hyperspectral image analysis for estimation of biophysical and biochemical properties of agricultural crops. Chapter 13. In: Thenkabail, P. S., J. G. Lyon, and A. Huete (eds.), *Hyperspectral Remote Sensing of Vegetation*. CRC Press, Boca Raton, FL, 705pp.

Baker, N. R. and E. Rosenqvist. 2004. Applications of chlorophyll fluorescence can improve crop production strategies: An examination of future possibilities. *J. Exp. Botany* 55:1607–1621.

Barker, D. W. and J. E. Sawyer. 2010. Using active canopy sensors to quantify corn nitrogen stress and nitrogen application rate. *Agron. J.* 102:964–971.

Barnes, E. M., T. R. Clarke, S. E. Richards, P. D. Colaizzi, J. Haberland, M. Kostrzewski, and P. Waller. 2000. Coincident detection of crop water stress, nitrogen status and canopy density using ground based multispectral data. In: Robert, P. C., R. H. Rust, and W. E. Larson (eds.), *Proceedings of the Fifth International Conference on Precision Agriculture*, Bloomington, MN, pp. 16–19.

Bastiaanssen, W. G. M. and M. G. Bos. 1999. Irrigation performance indicators based on remotely sensed data: A review of literature. *Irrigation Drainage Syst.* 13:291–311.

Bastiaanssen, W. G. M., D. J. Molden, and I. W. Makin. 2000. Remote sensing for irrigated agriculture: Examples from research and possible applications. *Agric. Water Manage.* 46:137–155.

Bendig, J., A. Bolten, and G. Bareth. 2013. UAV-based imaging for multi-temporal, very high resolution crop surface models to monitor crop growth variability. *PFG* 2013(6): 0551–0562.

Berni, J. A. J., P. J. Zarco-Tejada, L. Suárez, and E. Fereres. 2009. Thermal and narrowband multispectral remote sensing for vegetation monitoring from an unmanned aerial vehicle. *IEEE Trans. Geosci. Remote Sens.* 47(3):722–738.

Bhatti, A. U., D. J. Mulla, and B. E. Frazier. 1991. Estimation of soil properties and wheat yields on complex eroded hills using geostatistics and thematic mapper images. *Remote Sens. Environ.* 37:181–191.

Blackburn, G. A. 2007. Hyperspectral remote sensing of plant pigments. *J. Exp. Bot.* 58:855–867.

Blackmer, T. M. and J. S. Schepers. 1995. Use of a chlorophyll meter to monitor nitrogen status and schedule fertigation for corn. *J. Prod. Agric.* 8:56–60.

Blasco, J., N. Aleixos, J. M. Roger, G. Rabatel, and E. Molto. 2002. Robotic weed control using machine vision. *Biosys. Eng.* 83(2):149–157.

Borregaard, T., H. Nielsen, L. Norgaard, and H. Have. 2000. Crop-weed discrimination by line imaging spectroscopy. *J. Agric. Eng. Res.* 75:389–400.

Boydell, B. and A. McBratney. 2002. Identifying potential within-field management zones from cotton-yield estimates. *Precis. Agric.* 3(1):9–23.

Bravo, C., D. Moshou, R. Oberti, J. West, A. McCartney, L. Bodria, and H. Ramon. 2004. Foliar disease detection in the field using optical sensor fusion. *Agric. Eng. Int. CIGR J. Sci. Res. Dev.* Manuscript FP 04 008. VI.

Burgos-Artizzu, X. P., A. Ribeiro, M. Guijarro, and G. Pajares. 2011. Real-time image processing for crop/weed discrimination in maize fields. *Comp. Electron. Agric.* 75(2): 337–346.

Cao, Q., Y. Miao, H. Wang, S. Huang, S. Cheng, R. Khosla, and R. Jiang. 2013. Non-destructive estimation of rice plant nitrogen status with Crop Circle multispectral active canopy sensor. *Field Crops Res.* 154:133–144.

Christy, C. D. 2008. Real-time measurement of soil attributes using on-the-go near infrared reflectance spectroscopy. *Comp. Electron. Agric.* 61:10–19.

Clarke, T. R., M. S. Moran, E. M. Barnes, P. J. Pinter, and J. Qi. 2001. Planar domain indices: A method for measuring a quality of a single component in two-component pixels. In: *Proceedings of the IEEE International Geoscience and Remote Sensing Symposium* [CD ROM], Sydney, New South Wales, Australia, July 9–13, 2001, pp. 1279–1281.

Clay, D. E., T. P. Kharel, C. Reese, D. Beck, C. G. Carlson, S. A. Clay, and G. Reicks. 2012. Winter wheat crop reflectance and nitrogen sufficiency index values are influenced by nitrogen and water stress. *Agron. J.* 104:1612–1617.

Cohen, J. E. 2003. Human population: The next half century. *Science* 302:1172–1175.

Crookston, K. 2006. A top 10 list of developments and issues impacting crop management and ecology during the past 50 years. *Crop Sci.* 46:2253–2262.

Das, D. K., S. Pradhan, V. K. Sehgal, R. N. Sahoo, V. K. Gupta, and R. Singh. 2013. Spectral reflectance characteristics of healthy and yellow mosaic virus infected soybean (*Glycine max* L.) leaves in a semiarid environment. *J. Agrometeor.* 15:37–39.

Dash, J. and P. J. Curran. 2004. The MERIS terrestrial chlorophyll index. *Int. J. Remote Sens.* 25:5403–5413.

Datt, B. 1999. A new reflectance index for remote sensing of chlorophyll content in higher plants: Tests using eucalyptus leaves. *J. Plant Physiol.* 154:30–36.

Diacono, M., P. Rubino, and F. Montemurro. 2013. Precision nitrogen management of wheat: A review. *Agron. Sustain. Dev.* 33:219–241.

Erdle, K., B. Mistele, and U. Schmidhalter. 2011. Comparison of active and passive spectral sensors in discriminating biomass parameters and nitrogen status in wheat cultivars. *Field Crops Res.* 124:74–84.

Fishel, F. M., W. C. Bailey, M. Boyd, W. G. Johnson, M. O'Day, L. E. Sweets, and W. J. Wiebold. 2001. *Introduction to Crop Scouting.* IPM Manual 1006. University of Missouri, Columbia, MO.

Fitzgerald, G., D. Rodriguez, and G. O'Leary. 2010. Measuring and predicting canopy nitrogen nutrition in wheat using a spectral index—The canopy chlorophyll content index (CCCI). *Field Crops Res.* 116:318–324.

Fleming, K. L., D. F. Heermann, and D. G. Westfall. 2004. Evaluating soil color with farmer input and apparent soil electrical conductivity for management zone delineation. *Agron. J.* 96:1581–1587.

Franke, J. and G. Menz. 2007. Multi-temporal wheat disease detection by multi-spectral remote sensing. *Precis. Agric.* 8:161–172.

Freeman, K. W., K. Girma, D. B. Arnall, R. W. Mullen, K. L. Martin, R. K. Teal, and W. R. Raun. 2007. By-plant prediction of corn forage biomass and nitrogen uptake at various growth stages using remote sensing and plant height. *Agron. J.* 99:530–536.

Gamon, J. A., J. Penuelas, and C. B. Field. 1992. A narrow-waveband spectral index that tracks diurnal changes in photosynthetic efficiency. *Remote Sens. Environ.* 41:35–44.

Gée, C., J. Bossu, G. Jones, and F. Truchetet. 2008. Crop/weed discrimination in perspective agronomic images. *Comp. Electron. Agric.* 60:49–59.

Gibson, K. D., R. Dirks, C. R. Medlin, and L. Johnston. 2004. Detection of weed species in soybean using multispectral digital images. *Weed Technol.* 18:742–749.

Gitelson, A. A., A. Viña, V. Ciganda, D. C. Rundquist, and T. J. Arkebauer. 2005. Remote estimation of canopy chlorophyll content in crops. *Geophys. Res. Lett.* 32:L08403.1–L08403.4.

Gitelson, A. A., Y. J. Kaufmann, and M. N. Merzlyak. 1996. Use of a green channel in remote sensing of global vegetation from EOS-MODIS. *Remote Sens. Environ.* 58:289–298.

Gnyp, M. L., Y. Miao, F. Yuan, S. L. Ustin, K. Yu, Y. Yao, S. Huang, and G. Bareth. 2014. Hyperspectral canopy sensing of paddy rice aboveground biomass at different growth stages. *Field Crops Res.* 155:42–55.

Goel, P. K., S. O. Prasher, J. A. Landry, R. M. Patel, R. B. Bonnell, A. A. Viau, and J. R. Miller. 2003. Potential of airborne hyperspectral remote sensing to detect nitrogen deficiency and weed infestation in corn. *Comp. Electron. Agric.* 38:99–124.

Gómez-Candón, D., A. I. De Castro, and F. López-Granados. 2014. Assessing the accuracy of mosaics from unmanned aerial vehicle (UAV) imagery for precision agriculture purposes in wheat. *Precis. Agric.* 15:44–56.

Grove, J. H. and M. M. Navarro. 2013. The problem is not N deficiency: Active canopy sensors and chlorophyll meters detect P stress in corn and soybean. In: Stafford, J. V. (ed.), *Precision Agriculture '13*. Wageningen Academic Publishers, Wageningen, the Netherlands, pp. 137–144.

Guyot, G., F. Baret, and D. J. Major. 1988. High spectral resolution: Determination of spectral shifts between the red and infrared. *Intl. Arch. Photogram. Remote Sens.* 11:750–760.

Haboudane, D., J. R. Miller, E. Pattey, P. J. Zarco-Tejada, and I. B. Strachan. 2004. Hyperspectral vegetation indices and novel algorithms for predicting green LAI of crop canopies: Modeling and validation in the context of precision agriculture. *Remote Sens. Environ.* 90:337–352.

Haboudane, D., J. R. Miller, N. Tremblay, P. J. Zarco-Tejada, and L. Dextraze. 2002. Integrated narrow-band vegetation indices for prediction of crop chlorophyll content for application to precision agriculture. *Remote Sens. Environ.* 81:416–426.

Hanks, J. E. and J. L. Beck. 1998. Sensor-controlled hooded sprayer for row crops. *Weed Technol.* 12:308–314.

Hatfield, J. L., A. A. Gitelson, S. Schepers, and C. L. Walthall. 2008. Application of spectral remote sensing for agronomic decisions. *Agron. J.* 100:117–131.

Hatfield, J. L. and J. H. Prueger. 2010. Value of using different vegetative indices to quantify agricultural crop characteristics at different growth stages under varying management practices. *Remote Sens.* 2:562–578.

Hedley, C. B. and I. J. Yule. 2009. Soil water status mapping and two variable-rate irrigation scenarios. *Precis. Agric.* 10:342–355.

Heege, H. J., S. Reusch, and E. Thiessen. 2008. Prospects and results for optical systems for site-specific on-the-go control of nitrogen-top-dressing in Germany. *Precis. Agric.* 9:115–131.

Hicks, D. R. and S. L. Naeve. 1998. *The Minnesota Soybean Field Book*. University of Minnesota Extension Service, St. Paul, MN.

Holland, K. H., D. W. Lamb, and J. S. Schepers. 2012. Radiometry of proximal active optical sensors (AOS) for agricultural sensing. *IEEE J. Sel. Topics Appl. Earth Observ. Remote Sens.* 5:1793–1802.

Huang, W., D. W. Lamb, Z. Niu, L. Liu, and J. Wang. 2007. Identification of yellow rust in wheat by in situ and airborne spectrum data. *Precis. Agric.* 8(4–5):187–197.

Huang, Y., S. J. Thomson, W. C. Hoffmann, Y. Lan, and B. K. Fritz. 2013. Development and prospect of unmanned aerial vehicle technologies for agricultural production management. *Int. J. Agric. Biol. Eng.* 6(3):1–10.

Huete, A. 1988. A soil adjusted vegetation index (SAVI). *Remote Sens. Environ.* 25:295–309.

Jain, N., S. S. Ray, J. P. Singh, and S. Panigrahy. 2007. Use of hyperspectral data to assess the effects of different nitrogen applications on a potato crop. *Precis. Agric.* 8:225–239.

Jordan, C. F. 1969. Derivation of leaf area index from quality of light on the forest floor. *Ecology* 50:663–666.

Khakural, B. R., P. C. Robert, and D. R. Huggins. 1999. Variability of corn/soybean yield and soil/landscape properties across a southwestern Minnesota landscape. In: Robert, P. C. et al. (eds.), *Proceedings of the Fourth International Conference on Precision Agricultural*, St. Paul, MN, July 19–22, 1998, ASA, CSSA, SSSA, Madison, WI, pp. 573–579.

Kitchen, N. R., K. A. Sudduth, S. T. Drummond, P. C. Scharf, H. L. Palm, D. F. Roberts, and E. D. Vories. 2010. Ground-based canopy reflectance sensing for variable-rate nitrogen corn fertilization. *Agron. J.* 102:71–84.

Lamb, D. W. and R. B. Brown. 2001. Remote-sensing and mapping of weeds in crops. *J. Agric. Eng. Res.* 78:117–125.

Lamb, D. W., D. A. Schneider, and J. N. Stanley. 2014. Combination active optical and passive thermal infrared sensor for low-level airborne crop sensing. *Precis. Agric.* 15:523–531. doi: 10.1007/s11119-014-9350-0.

Larson, W. E. and P. C. Robert. 1991. Farming by soil. In: Lal, R. and F. J. Pierce (eds.), *Soil Management for Sustainability*. Soil and Water Conservation Society, Ankeny, IA, pp. 103–112.

Li, F., Y. Miao, G. Feng, F. Yuan, S. Yue, X. Gao, Y. Liu, B. Liu, S. L. Ustin, and X. Chen. 2014a. Improving estimation of summer maize nitrogen status with red edge-based spectral vegetation indices. *Field Crops Res.* 157:111–123.

Li, F., Y. Miao, S. D. Hennig, M. L. Gnyp, X. Chen, L. Jia, and G. Bareth. 2010. Evaluating hyperspectral vegetation indices for estimating nitrogen concentration of winter wheat at different growth stages. *Precis. Agric.* 11:335–357.

Li, F., B. Mistele, Y. Hu, X. Chen, and U. Schmidhalter. 2014b. Reflectance estimation of canopy nitrogen content in winter wheat using optimised hyperspectral spectral indices and partial least squares regression. *Eur. J. Agron.* 52:198–209.

Link, A. and S. Reusch. 2006. Implementation of site-specific nitrogen application—Status and development of the YARA N-Sensor. In: *NJF Seminar 390, Precision Technology in Crop Production Implementation and Benefits*, Norsk Jernbaneforbund, Stockholm, Sweden, pp. 37–41.

Linker, H. M., J. S. Bacheler, H. D. Coble, E. J. Dunphy, S. R. Koenning, and J. W. Van Duyn. 1999. Integrated pest management soybean scouting manual. North Carolina Cooperative Extension Service Publication No. AG-385. North Carolina State University. Raleigh, NC.

López-Granados, F. 2011. Weed detection for site-specific weed management: Mapping and real time approaches. *Weed Res.* 51:1–11.

Lorenzen, B. and A. Jensen. 1989. Changes in leaf spectral properties induced in barley by cereal powdery mildew. *Remote Sens. Environ.* 27:201–209.

MacRae, I. 1998. Scouting for insects in wheat, alfalfa and soybeans. University of Minnesota Extension Service, St. Paul, MN.

Mahajan, G. R., R. N. Sahoo, R. N. Pandey, V. K. Gupta, and D. Kumar. 2014. Using hyperspectral remote sensing techniques to monitor nitrogen, phosphorus, sulphur and potassium in wheat (*Triticum aestivum* L.). *Precis. Agric.* 15(5):499–522. doi: 10.1007/s11119-014-9348-7.

Meron, M., J. Tsipris, V. Orlov, V. Alchanatis, and Y. Cohen. 2010. Crop water stress mapping for site-specific irrigation by thermal imagery and artificial reference surfaces. *Precis. Agric.* 11:148–162.

Mewes, T., J. Franke, and F. Menz. 2011. Spectral requirements on airborne hyperspectral remote sensing data for wheat disease detection. *Precis. Agric.* 12:795–812.

Miao, Y., D. J. Mulla, G. Randall, J. Vetsch, and R. Vintila. 2009. Combining chlorophyll meter readings and high spatial resolution remote sensing images for in-season site-specific nitrogen management of corn. *Precis. Agric.* 10:45–62.

Mirik, M., G. J. Michels, S. Kassymzhanova-Mirik, and N. C. Elliott. 2007. Reflectance characteristics of Russian wheat aphid (Hemiptera: Aphididae) stress and abundance in winter wheat. *Comp. Electron. Agric.* 57:123–134.

Mirik, M., Y. Aysan, and F. Sahin. 2011. Characterization of *Pseudomonas cichorii* isolated from different hosts in Turkey. *Int. J. Agric. Biol.* 13:203–209.

Mistele, B. and U. Schmidhalter. 2008. Estimating the nitrogen nutrition index using spectral canopy reflectance measurements. *Eur. J. Agron.* 29:184–190.

Moran, M. S., Y. Inoue, and E. M. Barnes. 1997. Opportunities and limitations for image-based remote sensing in precision crop management. *Remote Sens. Environ.* 61:319–346.

Moran, M. S., C. D. Peters-Lidard, J. M. Watts, and S. McElroy. 2004. Estimating soil moisture at the watershed scale with satellite-based radar and land surface models. *Can. J. Remote Sens.* 30:805–826.

Moshou, D., I. Gravalos, D. K. C. Bravo, R. Oberti, J. S. West, and H. Ramon. 2012. Multisensor fusion of remote sensing data for crop disease detection. In: Thakur, J.K., Singh, S.K., Ramanathan, A., Prasad, M.B.K., Gossel, W. (Eds.), *Geospatial Techniques for Managing Environmental Resources*. Springer, Dordrecht, the Netherlands, pp. 201–219.

Mueller, D. and R. Pope. 2009. *Corn Field Guide: A Reference for Identifying Diseases, Insect Pests and Disorders of Corn*. Iowa State University, University Extension, Ames, AI.

Muhammed, H. H. 2005. Hyperspectral crop reflectance data for characterising and estimating fungal disease severity in wheat. *Biosys. Eng.* 91(1):9–20.

Mulla, D. J. 1991. Using geostatistics and GIS to manage spatial patterns in soil fertility. In: Kranzler, G. (ed.), *Automated Agriculture for the 21st Century*. ASAE, St. Joseph, MI, pp. 336–345.

Mulla, D. J. 1993. Mapping and managing spatial patterns in soil fertility and crop yield. In: Robert, P., W. Larson, and R. Rust (eds.), *Soil Specific Crop Management*. ASA, Madison, WI, pp. 15–26.

Mulla, D. J. 1997. Geostatistics, remote sensing and precision farming. In: Stein, A. and J. Bouma (eds.), *Precision Agriculture: Spatial and Temporal Variability of Environmental Quality*. Ciba Foundation Symposium 210. Wiley, Chichester, U.K., pp. 100–119.

Mulla, D. J. 2013. Twenty five years of remote sensing in precision agriculture: Key advances and remaining knowledge gaps. *Biosys. Eng.* 114:358–371.

Mulla, D. J., C. A. Perillo, and C. G. Cogger. 1996. A site-specific farm-scale GIS approach for reducing groundwater contamination by pesticides. *J. Environ. Qual.* 25:419–425.

Mulla, D. J. and J. S. Schepers. 1997. Key processes and properties for site-specific soil and crop management. In: Pierce, F. J. and E. J. Sadler (eds.), *The State of Site Specific Management for Agriculture*. ASA/CSSA/SSSA, Madison, WI, pp. 1–18.

Nguy-Robertson, A., A. Gitelson, Y. Peng, A. Viña, T. Arkebauer, and D. Rundquist. 2012. Green leaf area index estimation in maize and soybean: Combining vegetation indices to achieve maximal sensitivity. *Agron. J.* 104:1336–1347.

Nicolas, H. 2004. Using remote sensing to determine of the date of a fungicide application on winter wheat. *Crop Prot.* 23:853–863.

Nigon, T. J., D. J. Mulla, C. J. Rosen, Y. Cohen, V. Alchanatis, and R. Rud. 2014. Evaluation of the nitrogen sufficiency index for use with high resolution, broadband aerial imagery in a commercial potato field. *Precis. Agric.* 15:202–226.

Omary, M., C. R. Camp, and E. J. Sadler. 1997. Center pivot irrigation system modification to provide variable water application depths. *Appl. Eng. Agric.* 13(2):235–239.

Pimstein, A., A. Karnieli, S. K. Bansal, and D. J. Bonfil. 2011. Exploring remotely sensed technologies for monitoring wheat potassium and phosphorus using field spectroscopy. *Field Crops Res.* 121:125–135.

Pinter, Jr., P. J., J. L. Hatfield, J. S. Schepers, E. M. Barnes, M. S. Moran, C. S. T. Daughtry, and D. R. Upchurch. 2003. Remote sensing for crop management. *Photogr. Eng. Remote Sens.* 69:647–664.

Prabhakar, M., Y. G. Prasad, M. Thirupathi, G. Sreedevi, B. Dharajothi, and B. Venkateswarlu. 2011. Use of ground based hyperspectral remote sensing for detection of stress in cotton caused by leafhopper (Hemiptera: Cicadellidae). *Comp. Electron. Agric.* 79:189–198.

Qi, J., A. Chehbouni, A. R. Huete, Y. H. Keer, and S. Sorooshian. 1994. A modified soil vegetation adjusted index. *Remote Sens. Environ.* 48:119–126.

Rasmussen, J., J. Nielsen, F. Garcia-Ruiz, S. Christensen, and J. C. Streibig. 2013. Potential uses of small unmanned aircraft systems (UAS) in weed research. *Weed Res.* 53:242–248.

Raun, W. R., J. B. Solie, G. V. Johnson, M. L. Stone, R. W. Mullen, K. W. Freeman, W. E. Thomason, and E. V. Lukina. 2002. Improving nitrogen use efficiency in cereal grain production with optical sensing and variable rate application. *Agron. J.* 94:815–820.

Reusch, S., J. Jasper, and A. Link. 2010. Estimating crop biomass and nitrogen uptake using CropSpec™, a newly developed active crop-canopy reflectance sensor. In: Khosla, R. (ed.), *Proceedings of the 10th International Conference on Precision Agriculture*, Denver, CO.

Rondeaux, G., M. Steven, and F. Baret. 1996. Optimization of soil-adjusted vegetation indices. *Remote Sens. Environ.* 55:95–107.

Rouse, J. W. Jr., R. H. Hass, J. A. Schell, and D. W. Deering. 1973. Monitoring vegetation systems in the great plains with ERTS. In: *Proceedings of the Third Earth Resources Technology Satellite (ERTS) Symposium*, Vol. 1, NASA SP-351, NASA, Washington, DC, pp. 309–317.

Rud, R., Y. Cohen, V. Alchanatis, A. Cohen, A. Levi, R. Brikman, C. Shenderey et al. 2014. Crop water stress index derived from multi-year ground and aerial thermal images as an indicator of potato water status. *Precis. Agric.* 15:273–289. doi: 10.1007/s11119-014-9351-z.

Sadler, E. J., R. G. Evans, K. C. Stone, and C. R. Camp. 2005. Opportunities for conservation with precision irrigation. *J. Soil Water Cons.* 60(6):371–379.

Samborski, S. M., N. Tremblay, and E. Fallon. 2009. Strategies to make use of plant sensors-based diagnostic information for nitrogen recommendations. *Agron. J.* 101:800–816.

Samseemoung, G., P. Soni, H. P. W. Jayasuriya, and V. M. Salokhe. 2012. Application of low altitude remote sensing (LARS) platform for monitoring crop growth and weed infestation in a soybean plantation. *Precis. Agric.* 13:611–627.

Sankaran, S., A. Mishra, R. Ehsani, and C. Davis. 2010. A review of advanced techniques for detecting plant diseases. *Comp. Electron. Agric.* 72(1):1–13.

Sayed, O. H. 2003. Chlorophyll fluorescence as a tool in cereal crop research. *Photosynthetica* 41:321–330.

Scharf, P. C., D. K. Shannon, H. L. Palm, K. A. Sudduth, S. T. Drummond, N. R. Kitchen, L. J. Mueller, V. C. Hubbard, and L. F. Oliveira. 2011. Sensor-based nitrogen applications out-performed producer-chosen rates for corn in on-farm demonstrations. *Agron. J.* 103:1683–1691.

Seelan, S. K., S. Laguette, G. M. Casady, and G. A. Seielstad. 2003. Remote sensing applications for precision agriculture: A learning community approach. *Remote Sens. Environ.* 88:157–169.

Shanahan, J. F., N. R. Kitchen, W. R. Raun, and J. S. Schepers. 2008. Responsive in-season nitrogen management for cereals. *Comp. Electron. Agric.* 61:51–62.

Shaver, T. M., R. Khosla, and D. G. Westfall. 2011. Evaluation of two crop canopy sensors for nitrogen variability determination in irrigated maize. *Precis. Agric.* 12:892–904.

Shiratsuchi, L., R. Ferguson, J. Shanahan, V. Adamchuk, D. Rundquist, D. Marx, and G. Slater. 2011. Water and nitrogen effects on active canopy sensor vegetation indices. *Agron. J.* 103:1815–1826.

Solie, J. B., W. R. Raun, R. W. Whitney, M. L. Stone, and J. D. Ringer. 1996. Optical sensor based field element size and sensing strategy for nitrogen. *Trans. ASAE* 39:1983–1992.

Sripada, R. P., R. W. Heiniger, J. G. White, and R. Weisz. 2005. Aerial color infrared photography for determining late-season nitrogen requirements in corn. *Agron. J.* 97:1443–1451.

Sripada, R. P., J. P. Schmidt, A. E. Dellinger, and D. B. Beegle. 2008. Evaluating Multiple indices from a canopy reflectance sensor to estimate corn N requirements. *Agron. J.* 100:1553–1561.

Stafford, J. V. and P. C. H. Miller. 1993. Spatially selective application of herbicide to cereal crops. *Comp. Electron. Agric.* 9:217–229.

Stark, J. C., I. R. McCann, B. A. King, and D. T. Westermann. 1993. A two-dimensional irrigation control system for site specific application of water and chemicals. *Agron. Abs.* 85:329.

Sui, R., J. A. Thomasson, and J. Hanks. 2008. Ground-based sensing system for weed mapping in cotton. *Comp. Electron. Agric.* 60(1):31–38.

Swain, K. C., S. J. Thomson, and H. P. W. Jayasuriya. 2010. Adoption of an unmanned helicopter for low-altitude remote sensing to estimate yield and total biomass of a rice crop. *Trans. ASABE* 53(1):21–27.

Thenkabail, P. S., E. A. Enclona, M. S. Ashton, and B. Van Der Meer. 2004. Accuracy assessments of hyperspectral waveband performance for vegetation analysis applications. *Remote Sens. Environ.* 91:354–376.

Thenkabail, P. S., J. G. Lyon, and A. Huete. 2010. Hyperspectral remote sensing of vegetation and agricultural crops: Knowledge gain and knowledge gap after 40 years of research. Chapter 28. In: Thenkabail, P. S., J. G. Lyon, and A. Huete (eds.), *Hyperspectral Remote Sensing of Vegetation*. CRC Press, Boca Raton, FL, 705pp.

Thenkabail, P. S., R. B. Smith, and E. De Pauw. 2000. Hyperspectral vegetation indices and their relationships with agricultural crop characteristics. *Remote Sens. Environ.* 71:158–182.

Thorp, K. R. and L. Tian. 2004. A review on remote sensing of weeds in agriculture. *Precis. Agric.* 5(5):477–508.

Tillet, N. D., T. Hague, A. C. Grundy, and A. P. Dedousis. 2008. Mechanical within-row weed control for transplanting crops using computer vision. *Biosys. Eng.* 99(2):171–178.

Torii, T. 2000. Research in autonomous agriculture vehicles in Japan. *Comp. Electron. Agric.* 25(1–2):133–153.

Tremblay, N., Z. Wang, and Z. G. Cerovic. 2012. Sensing crop nitrogen status with fluorescence indicators. A review. *Agron. Sustain. Dev.* 32:451–464.

Van Niel, T. G. and T. R. McVicar. 2004. Current and potential uses of optical remote sensing in rice-based irrigation systems: A review. *Aust. J. Agric. Res.* 55(2):155–185.

Vereecken, H., L. Weihermüller, F. Jonard, and C. Montzka. 2012. Characterization of crop canopies and water stress related phenomena using microwave remote sensing methods: A review. *Vadose Zone J.* 11:1–23.

Wang, W., X. Yao, X. Yao, Y. Tian, X. Liu, J. Ni, W. Cao, and Y. Zhu. 2012. Estimating leaf nitrogen concentration with three-band vegetation indices in rice and wheat. *Field Crops Res.* 129:90–98.

West, J. S., C. Bravo, R. Oberti, D. Lemaire, D. Moshou, and H. A. McCartney. 2003. The potential of optical canopy measurement for targeted control of field crop disease. *Annu. Rev. Phytopathol.* 41:593–614.

Whipker, L. D. and J. D. Akridge, 2006. Precision agricultural services dealership survey results. Staff paper, Department of Agricultural Economics, Purdue University, West Lafayette, IN.

Wiegand, C. L., A. J. Richardson, and D. E. Escobar. 1991. Vegetation indices in crop assessment. *Remote Sens. Environ.* 35:105–119.

Wiles, L. J., G. G. Wilkerson, H. J. Gold, and H. D. Coble. 1992. Modeling weed distribution for improved postemergence control decisions. *Weed Sci.* 40:546–553.

Yao, H. L. Tang., L. Tian, R. L. Brown, D. Bhatnagar, and T. E. Cleveland. 2010. Using hyperspectral data in precision farming applications. Chapter 25. In: Thenkabail, P. S., J. G. Lyon, and A. Huete (eds.), *Hyperspectral Remote Sensing of Vegetation.* CRC Press, Boca Raton, FL, 705pp.

Yuan, L., Y. Huang, R. W. Loraamm, C. Nie, J. Wang, and J. Zhang. 2014. Spectral analysis of winter wheat leaves for detection and differentiation of diseases and insects. *Field Crops Res.* 156:199–207.

Zhang, C. and J. M. Kovacs. 2012. The application of small unmanned aerial systems for precision agriculture: A review. *Precis. Agric.* 13:693–712.

Zwiggelaar, R. 1998. A review of spectral properties of plants and their potential use for crop/weed discrimination in row-crops. *Crop Prot.* 17:189–206.

8

Remote Sensing of Tillage Status

Baojuan Zheng
Arizona State University

James B. Campbell
Virginia Tech

Guy Serbin
Teagasc Food Research Centre

Craig S.T. Daughtry
*USDA-ARS Hydrology and
Remote Sensing Laboratory*

Heather McNairn
Agriculture and Agri-Food Canada

Anna Pacheco
Agriculture and Agri-Food Canada

Acronyms and Definitions

ALI	Advanced Land Imager
ALOS PALSAR	Advanced Land Observation Satellite Phased Array type L-band Synthetic Aperture Radar
ASAR	Advanced Synthetic Aperture Radar
ASTER	Advanced Spaceborne Thermal Emission and Reflection Radiometer
AVIRIS	Airborne Visible/Infrared Imaging Spectrometer
CAI	Cellulose absorption index
CCD	Coherent change detection
COSMO-SkyMed	Constellation of Small **S**atellites for the Mediterranean Basin Observation
CP	Compact polarization
CTIC	Conservation Technology Information Center
DP	Dual polarization
EnMAP	Environmental Mapping and Analysis Program
ERS	European Remote Sensing
ESA	European Space Agency
Hyperion	First spaceborne hyperspectral sensor onboard Earth Observing-1(EO-1)
HyspIRI	Hyperspectral Infrared Imager
IEM	Integral Equation Model
Landsat-4, 5 TM	Thematic Mapper
Landsat-7 ETM+	Enhanced Thematic Mapper Plus
Landsat 8 OLI	Operational Land Imager
LCA	Lignin cellulose absorption
LUT	Lookup table
minNDTI	Minimum NDTI
MODIS	Moderate Imaging Spectroradiometer
NDI	Normalized difference index
NDTI	Normalized difference tillage index
NDVI	Normalized difference vegetation index
NIR	Near infrared
NPV	Nonphotosynthetic vegetation
PolInSAR	Polarimetric interferometric
PPD	Copolarized phase difference
QP	Quadrature polarization
RISAT	Radar Imaging Satellite
RMS	Root mean square
SAOCOM	SAtélite Argentino de Observación COn Microondas; Argentine Microwaves Observation Satellite
SARs	Synthetic aperture radars
SINDRI	Shortwave infrared normalized difference residue index
SMAP	Soil Moisture Active Passive
SP	Single polarization
SPOT	Satellites Pour l'Observation de la Terre or Earth-observing satellites
STARFM	Spatial and temporal adaptive reflectance fusion model

STI	Simple tillage index
STIR	Soil Tillage Intensity Rating
SWIR	Shortwave infrared
TanDEM-X	TerraSAR-X add-on for Digital Elevation Measurement
VNIR	Visible, near-infrared
USDA	U.S. Department of Agriculture
USGS	U.S. Geological Survey

8.1 Introduction

Tillage prepares the seedbed by mechanical disturbance to loosen and smooth the soil surface, often mixing topsoil with surface organic debris to aerate soil, assist in weed suppression, control insects and pests, and, in midlatitudes, promote springtime warming and drying. Tillage has been practiced, in varied forms, throughout the world since antiquity. During the 1700s and 1800s, innovations in designs of plowshares greatly increased tillage effectiveness by increasing depth of the disturbed soil and by turning the surface soil to more completely mix surface crop residue (also referred to as plant litter, senescent vegetation, or nonphotosynthetic vegetation [NPV]) with disturbed soil.

For millennia, mechanical disturbance of the soil was accomplished using hand tools and animal power. By the mid-nineteenth century, steam-powered tractors (later replaced by internal combustion engines) greatly increased tillage efficiency and speed and expanded tillage into a wider range of slopes, topography, and ecosystems. Notable impacts of mechanization are the expansion of tillage into formerly uncultivated environments, especially prairies and steppes of several continents that have since become some of the most productive agricultural systems, but also some of the most susceptible to drought and erosion. Mechanization also led to further innovations in designs of specialized tillage implements and to increases in tillage operations, which often created the context for soil and water erosion.

Detrimental impacts of tillage include increased wind and water erosion; increased soil compaction, especially in the context of mechanization; decreased soil organic matter; reduced water infiltration; and increased amounts of nutrients reaching streams and rivers. By the 1940s, increased awareness of detrimental aspects of tillage (Faulkner 1943), combined with availability of herbicides, led to alternative practices to minimize adverse aspects of tillage. Such practices include increased use of tillage instruments that minimize soil disturbance and leave crop residue on the soil surface.

Recognized environmental benefits of conservation tillage systems include reduced soil erosion from wind and water, carbon emission reductions, and improvements of air, soil, and water quality (Wander and Drinkwater 2000). Long-term adoption of conservation tillage practices can increase soil organic matter content and, hence, can potentially sequester atmospheric carbon into soils (Lal 2004). Conservation tillage practices increase soil water infiltration, improve nutrient cycling, and, in general, improve water quality because of improved retention of soil nutrients (Karlen et al. 1994). Soil quality is improved because

accumulation of surface organic matter increases aggregate stability and higher levels of crop residues provide shelter and food for wildlife. As for economic perspective, conservation tillage practices decrease labor and fuel costs because of reduced tillage operations and reduced fertilizer requirements as a result of improved soil quality (West and Marland 2002). Conservation tillage, especially no-till, requires fewer field operations and reduces the number of field days needed to plant a crop. As a result, it reduces the risk of delayed planting due to unfavorable weather conditions and also provides possibilities to practice double-cropping.

As alternative tillage practices gained acceptance and were implemented, conservationists needed objective data to gauge the extent and benefits of their use. The Soil Tillage Intensity Rating (STIR), developed by USDA-Natural Resource Conservation Services (NRCS), provides a physically based evaluation of tillage systems across the spectrum from true no-till to conventional plow systems (USDA-NRCS 2014). STIR requires information on (1) each tillage implement used, (2) the operating speed of the implement, (3) the depth of tillage, and (4) the fraction of the total soil surface disturbed by the tillage implement. STIR provides robust evaluations of complex tillage systems and crop rotations for conservation planning. However, STIR is impractical for surveys over many fields and large regions.

Tillage intensity can also be characterized by the fraction of the soil surface covered by crop residue. The Conservation Technology Information Center (CTIC) defined the following categories of tillage based on the crop residue cover on the soil surface shortly after planting: conventional tillage has <15% residue cover, reduced tillage has 15%–30% residue cover, and conservation tillage has >30% residue cover (CTIC 2014). This less robust definition of tillage intensity has a few caveats that must be considered, for example, fields where crop residues were harvested for feed or biofuel may have low crop residue cover without soil-disturbing tillage.

Over time, varied efforts to collect information on tillage intensity have included visual assessment, field measurements, agricultural censuses, and remote-sensing techniques. Such information is required by a number of agroecosystem models and is important for assessing the impacts of tillage practices on soil erosion, soil carbon sequestration, and water quality. Field measurements and agricultural surveys to acquire tillage information are time-consuming and difficult. Moreover, it is unrealistic to survey every single field using these methods over large regions and over time. Therefore, it is of great interest to develop techniques that can routinely and systematically map tillage practices. Synoptic remote-sensing imagery offers opportunities to provide spatial–temporal information on tillage practices efficiently at low costs. The first investigation on the potential of using remote-sensing imagery to map crop residues can be traced back to Gausman et al. (1975). Thereafter, both aerial and satellite imagery were tested to differentiate different tillage practices and estimate crop residue cover. For instance, Airborne Visible/Infrared Imaging Spectrometer (AVIRIS) data were found to be useful for crop residue cover estimation (Daughtry et al. 2005).

Although aerial imagery, properly timed and collected at suitable resolutions, offers the capability to assess soil tillage status, the broadscale surveys require the areal coverage, revisit capabilities, and spectral channels that are, as a practical matter, available only through satellite observation systems. Here we discuss the two main classes of satellite systems with the potential for routine broadscale tillage assessment: (1) optical remote sensing (visible, near-infrared [NIR], and midinfrared imaging sensors) and (2) microwave remote sensing (synthetic aperture radar [SAR]).

8.2 Field Assessment of Crop Residue Cover

Methods appropriate for assessing crop residue cover in fields can be grouped into intercept and photographic techniques (Morrison et al. 1993). Intercept methods use a system of grid points, crosshairs, or points along a line where the presence or absence of residue is determined. The standard technique used by USDA-NRCS is the line–point transect method where a 15–30 m line with 100 evenly spaced markers along the line is stretched diagonally across the crop rows in the field and markers intersecting crop residue are counted. Accuracy of the line–point transect method depends on the length of the line, the number of points per line, and the skill of the observer. At least 500 points must be observed to determine corn residue cover to within 15% of the mean (Laflen et al. 1981). Significant modifications to the line–point transect method include the use of measuring tapes, meter sticks, and wheels with pointers (Corak et al. 1993; Morrison et al. 1993). However, the line–point transect is impractical for monitoring crop residue cover in many fields over broad areas in a timely manner.

For the photographic method, a color or color infrared digital camera is used to take multiple vertical photographs within a sampling area where residue conditions appear visually homogeneous. A grid or crosshairs is superimposed on the digital images and the points intersecting residue are counted. Software programs, such as SamplePoint, can randomly select sample points within each image for the user to identify and can tabulate the proportion of each class (Booth et al. 2006). Alternatively, the image may be classified into soil and residue classes using objective image analysis procedures. Classification errors occur when the spectral differences between soil and residues classes are not sufficiently large for discrimination. Shortly after harvest, crop residues are often much brighter than soils, but as the residues decompose, the residues may be brighter or darker than the soil. The best time to acquire information of tillage practices in the field is shortly after sowing and before crop emergence, which is also the optimal time window to acquire images to map tillage practices.

The CTIC, established at Purdue University in 1983 as clearinghouse for tillage and conservation information, has conducted field surveys to assess tillage status in the United States (http://www.ctic.purdue.edu). For the CTIC surveys, trained observers visually assessed tillage status in fields at regular intervals along selected routes through participating counties. The survey provided county-level estimates of overall tillage practices. The roadside assessment task is subject to various degrees of error and uncertainty because it mainly relies on visual interpretation.

TABLE 8.1 Tillage Types and Their Corresponding Crop Residue Cover

Tillage Category	Tillage Types	Description	Crop Residue Cover (%)
Conservation	No-till/strip-till	Minimal soil disturbance (<25%)	>30 (likely >70)
	Ridge till	Residue left on the surface between ridges	>30
	Mulch till	100% Soil surface disturbance	>30
Nonconservation	Reduced till	15%–30%	15–30
	Conventional till or intensive till	<15%	<15

The quality of the data has also varied from time to time and from county to county due to a variety of reasons, such as unfavorable weather conditions at the time of survey and inconsistent levels of experience among the observers. Finally, some counties have stopped acquiring tillage data after the national survey program was discontinued in 2004 (CTIC 2014).

Limited soil tillage information is available for other countries. Canada conducts tillage inventory as part of its 5-year census of agriculture. Tillage practices are reported by province in three categories: (1) tillage incorporating most of the crop residue into the soil, (2) tillage retaining most of the crop residue on the surface, and (3) no-till seeding or zero-till seeding. Thus, it is difficult and impractical to evaluate tillage practices over time, and by nation, because of wide variations in field data collection, survey responses, and agricultural censuses (Zheng et al. 2014). The tillage categories defined by Canada are less precise than the CTIC definitions. Definitions of tillage categories may slightly differ from one country to another and even differ from organization to organization. To evaluate tillage practices for a particular field using visual assessment or remote-sensing methodologies, we have to link the ground surface status observed from the ground, air, or space to types of tillage practices. Although soil texture and smoothness can be one of the indicators for different tillage status, the amount of crop residues left on the ground after planting are often considered as the most reliable indicator. Here, we list types of tillage practices and their expected crop residue cover according to CTIC and NRCS's definitions in Table 8.1. Globally, a systematic monitoring of soil tillage is needed to manage the finite soil resources as demand for food, feed, fiber, and fuel intensifies.

8.3 Monitoring with Optical Remote Sensing

Optical remote-sensing imagery is valuable for monitoring biophysical properties of various objects on the Earth. Crop residue, although spectrally similar to soils, has a unique absorption feature near 2100 nm. The absorption depth becomes deeper as the amount of crop residue increases. Thus, optical remote-sensing imagery provides a better capability for estimating crop residue cover than does radar data. This section firstly describes spectral properties of soils, green vegetation, and NPV, following with Section 8.3.2 on tillage spectral indices based on spectral differences among soils, green vegetations, and NPV. Section 8.3.3 reviews tillage assessment using different remote-sensing platforms, followed by Section 8.3.4, which discusses current challenges and future possibilities.

8.3.1 Spectral Properties of Soils, Green Vegetation, and Nonphotosynthetic Vegetation

Soil tillage intensity is defined by the proportion of the soil surface covered by crop residue shortly after planting. Green vegetation may also be present in the field as the planted crop or as weeds. This section focuses on the spectral properties of soils, green vegetation, and crop residues.

8.3.1.1 Spectral Properties of Soils

Soil reflectance typically increases monotonically with increasing wavelength (Figure 8.1). Major contributors to the reflectance spectra of soils include moisture content, iron oxide content, organic matter content, particle-size distribution, mineralogy, and soil structure (Baumgardner et al. 1986; Ben-Dor 2002). Stoner and Baumgardner (1981) measured the spectral reflectance of 485 soil samples representing 10 soil taxonomic orders and identified 5 distinct soil reflectance curve forms. Soil organic matter content and iron oxide content were the primary factors determining shape of the reflectance spectra.

In general, soil reflectance decreases as soil moisture content, organic matter content, and iron oxide content increase. Spectral reflectance is strongly correlated with soil organic matter among soils from the same parent materials (Henderson et al. 1992). Reflectance spectra of soils may also have absorption features near 2210 nm that are associated with Al-OH in phyllosilicate clays (Figures 8.1 and 8.2) (Serbin et al. 2009b). However, mineral absorption features evident in the reflectance spectra of dry soils are often obscured by the strong absorption of water in the reflectance spectra of wet soils (Stoner et al. 1980; Daughtry et al. 2004).

Soil tillage roughens the soil surface and often decreases soil reflectance, but the effect is short-lived and soil reflectance increases as the soil surface is smoothed by precipitation or additional tillage operations (Irons et al. 1989). As water wets the soil surface and fills pore spaces, soil reflectance decreases.

8.3.1.2 Spectral Properties of Green Vegetation

Reflectance of solar radiation from a dense canopy of actively growing green plants is characterized by three distinct regions: visible, NIR, and shortwave infrared (SWIR) (Figure 8.1). In the visible wavelength region (400–700 nm), chlorophyll and other leaf pigments strongly absorb blue and red wavelengths,

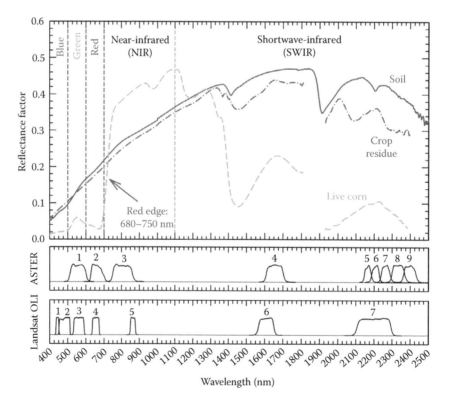

FIGURE 8.1 Spectra of a soil, corn residue, and live corn canopy for the visible through SWIR and relative spectral response (RSR) for ASTER and Landsat OLI bands. Note that reflectance values vary from sample to sample. (Adapted from Daughtry, C.S.T. et al., *Agron. J.,* 97(3), 864, 2005, doi: 10.2134/agronj2003.0291.)

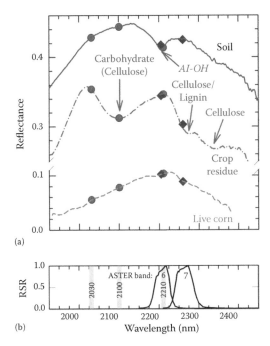

FIGURE 8.2 (a) Spectra of a soil, corn residue, and live corn canopy in the SWIR and (b) RSR for ASTER bands 6 and 7 and 11 nm wide CAI bands centered at 2030, 2100, and 2210 nm. (Adapted from Serbin, G. et al., *Remote Sens. Lett.*, 4(6), 552, 2013, doi: 10.1080/2150704x.2013.76747p.)

which largely determines the reflectance and transmittance spectra (Thomas and Gausman 1977). In the NIR wavelength region (700–1200 nm), there is very little absorption, and spectral reflectance and transmittance are largely determined by leaf mesophyll structure and cell wall–air interfaces (Slaton et al. 2001). Reflectance and transmittance in the SWIR wavelength region (1200–2500 nm) are affected primarily by the amount of water in the leaves (Hunt 1989; Yilmaz et al. 2008). Thus, a distinguishing spectral characteristic of green vegetation is the steplike transition from low reflectance and low transmittance in the visible region to high reflectance and high transmittance in the NIR (Figure 8.1). Soils and NPV lack this spectral feature. Spectral vegetation indices that exploit this fundamental spectral feature are particularly sensitive to green vegetation, for example, the normalized difference vegetation index (NDVI) (Rouse et al. 1973; Asrar et al. 1989).

8.3.1.3 Spectral Properties of Nonphotosynthetic Vegetation

NPV broadly refers to any senesced vegetation and includes crop residues, which are the portions of a cultivated crop remaining in the field after harvest. Initially, crop residues may completely cover the soil surface, but when the soil is tilled or the crop residues are harvested for feed or biofuel, crop residue cover decreases. Crop residues on the soil surface decrease soil erosion, increase soil organic matter, and improve soil quality (Lal et al. 1998). Quantification of crop residue cover is required to assess the effectiveness and extent of conservation tillage practices.

The reflectance spectra of both soils and crop residues lack the unique spectral signature of green vegetation (Figure 8.1). Crop residues and soils are spectrally similar and differ only in amplitude in the 400–1100 nm wavelength region, which makes quantification of crop residue cover by spectral reflectance challenging (Streck et al. 2002). Crop residues may be brighter than the soil shortly after harvest, but as residues weather and decompose, they may become either brighter or darker than the soil (Nagler et al. 2000; Daughtry et al. 2010). Residue water content also has impacts on its spectral properties. The presence of water in crop residues decreases reflectance across all wavelengths (Daughtry 2001). Thus, assessing crop residue cover with broadband multispectral data can be challenging and may require extensive local calibration data.

An alternative approach for discriminating crop residues from soils is based on detecting absorption features in the 2100–2350 nm wavelength regions that are associated with cellulose and lignin in crop residues (Workman and Weyer 2008). High residue water content can obscure the absorption feature at 2100 nm (Daughtry 2001). Increases in soil moisture content also decrease our ability to separate crop residue from soils (Daughtry 2001). Thus, it becomes more difficult to discriminate crop residue from soils as residue and soil water content increases. As illustrated in Figure 8.2, these absorption features are not shared by common soil minerals but are obscured by the strong absorption of water often present in soils, crop residues, and green vegetation, which can significantly attenuate the cellulose and lignin absorption features (Daughtry and Hunt 2008; Serbin et al. 2009a).

8.3.2 Spectral Indices for Assessing Crop Residue Cover

Spectral vegetation indices designed for assessing green vegetation, such as NDVI, cannot distinguish soil and crop residues. Numerous tillage or residue indices use various combinations of visible, NIR, and shortwave multispectral bands to discriminate crop residues from soils. The index best suited for crop residue cover estimation from single scenes is the cellulose absorption index (CAI), which specifically targets this feature. It has the distinct advantage that crop residues always have CAI > 0, live vegetation $\approx$ 0, and soils $\leq$ 0 (Figure 8.3). The CAI is defined as the relative intensity of the absorption feature at 2100 nm, which is attributed to an O–H stretching and C–O bending combination in cellulose and other carbohydrates in crop residues. CAI is measured using three relatively narrow (10–30 nm spectral resolution depending on the sensors) spectral bands—two on the shoulders and one near the center of the absorption feature at 2100 nm (Nagler et al. 2000) (Table 8.2). CAI is effective in discriminating crop residues from soils for dry to moderately moist mixtures of crop residues and soils but less effective for mixtures of wet crop residues and soils (Daugthtry 2001).

Additional spectral indices that also target the cellulose and lignin absorption features of crop residues have used the relatively narrow (30–90 nm) SWIR bands of the Advanced

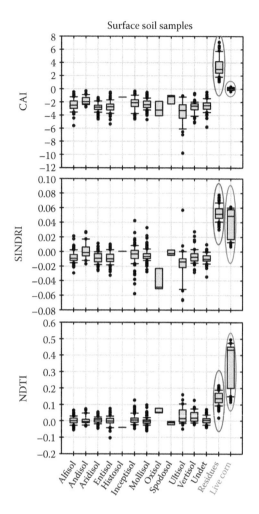

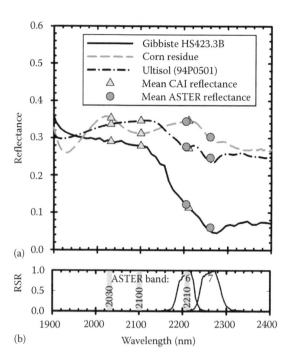

FIGURE 8.4 (a) Spectra of Gibbsite HS423.3B (Clark et al. 2007), a gibbsitic Ultisol (Brown et al. 2006), and corn residue with convolved spectral band values. (b) Relative spectral response functions (RSR) for 11-nm wide bands centered at 2030, 2100, and 2210 nm (CAI), and ASTER bands 6 and 7 (SINDRI). (Adapted from Serbin, G. et al., *Remote Sens. Lett.*, 4(6), 552, 2013, doi: 10.1080/2150704x.2013.767479.)

FIGURE 8.3 Spectral index values for surface soils, crop residues, and live corn canopy. (Adapted from Serbin, G. et al., *Soil Sci. Soc. Am. J.*, 73(5), 1545, 2009a, doi: 10.2136/sssaj2008.0311; Serbin, G. et al., *Remote Sens. Lett.*, 4(6), 552, 2013, doi: 10.1080/2150704x.2013.767479.)

Spaceborne Thermal Emission and Reflection Radiometer (ASTER) on the NASA Terra satellite, that is, the lignin cellulose absorption (LCA) and the shortwave infrared normalized difference residue index (SINDRI) (Daughtry et al. 2005; Serbin et al. 2009c). For two-band normalized difference indices (NDIs), the ASTER-based SINDRI performs well and targets a decrease in reflectance associated with cellulose and lignin features between ASTER SWIR bands 6 and 7 (Serbin et al. 2009c; Table 8.2). However, SINDRI is sensitive to green vegetation (Figures 8.2 and 8.3) and certain soil minerals (Figure 8.4), which also

experience reflectance decreases between these bands, such that it may not work well for a limited number of soils or where emerged crops may be present (Serbin et al. 2013).

While Landsat Thematic Mapper (TM)/Enhanced Thematic Mapper (ETM) bands 5 and 7 and Landsat 8 Operational Land Imager (OLI) bands 6 and 7 are too wide and not properly placed to capture the cellulose absorption feature at 2100 nm, they can be used for tillage estimation via normalized difference tillage index (NDTI) (van Deventer et al. 1997; Table 8.2). In addition to NDTI, NDI (McNairn and Protz 1993) and simple tillage index (STI) (van Deventer et al. 1997) are Landsat-based tillage indices. Serbin et al. (2009a) showed that NDTI performed the best of several Landsat-based tillage indices but underperformed in comparison to CAI and the ASTER-based LCA. Furthermore, NDTI was found to lack adequate contrast for a number of soils with high content of kaolinite or smectite and had a much stronger signal for live vegetation than either crop residues or soil minerals (Figure 8.3). In Figure 8.3, the median values of NDTI for crop residues are

TABLE 8.2 Selected Tillage Indices and Their Calculation

Sensor	Tillage Indices	Formula	Description	References
Landsat TM and ETM+	NDTI	(B5 − B7)/(B5 + B7)	B5, B7: Landsat bands 5 and 7.	Van Deventer et al. (1997)
AVIRIS Hyperion	CAI	$100 \times [0.5(R_{2030} + R_{2210}) - R_{2100}]$	R_{2030} and R_{2210} are the reflectances of the shoulders at 2030 and 2210 nm; R_{2100} is at the center of the absorption.	Daughtry et al. (2005) Daughtry et al. (2006)
ASTER	LCA SINDRI	$100(2 \times B6 - B5 - B8)$ (B6 − B7)/(B6 + B7)	B5, B6, B7, B8: ASTER shortwave infrared bands 5, 6, 7, and 8.	Daughtry et al. (2005) Serbin et al. (2009a)

consistently higher than the median values of surface soils. However, discrimination of some combinations of soils and crop residues may be difficult without adequate quantities of local data for calibration and validation. For example, the NDTI values of most crop residues may not differ significantly from NDTI values of soils with high content of kaolinite or smectite (Serbin et al. 2009a). As the fraction of green vegetation in a scene increases, NDTI also increases, which alters the estimation of crop residue cover. One approach is to exclude pixels with green vegetation using an NDVI threshold (Thoma et al. 2004; Daughtry et al. 2005). Another robust approach to reduce effects of soil and green vegetation on estimates of crop residue cover is to identify the minimum NDTI (minNDTI) values from multitemporal NDTI data, because the minNDTI values were found to be well correlated with crop residue cover (Zheng et al. 2012, 2013a). This method was found to be similar in accuracy to single collects using SINDRI or CAI (Figure 8.5) (Zheng et al. 2013a). However, as we can see in Figure 8.5 that minNDTI results in higher root mean squared errors (RMSE), NDTI is more subject to the negative influences of soil moisture and soil organic carbon than SINDRI and CAI (Zheng et al. 2013a).

8.3.3 Tillage Assessment Using Airborne and Satellite Imagery

Until recently, most assessments of crop residue cover and tillage intensity were snapshots of conditions using single dates of multispectral imagery. For example, various spectral indices using Landsat TM bands 5 and 7 successfully differentiated conventional tillage from conservation tillage using logistic regression (van Deventer et al. 1997; Gowda et al. 2001). Other classification methods (e.g., minimum distance, Mahalanobis

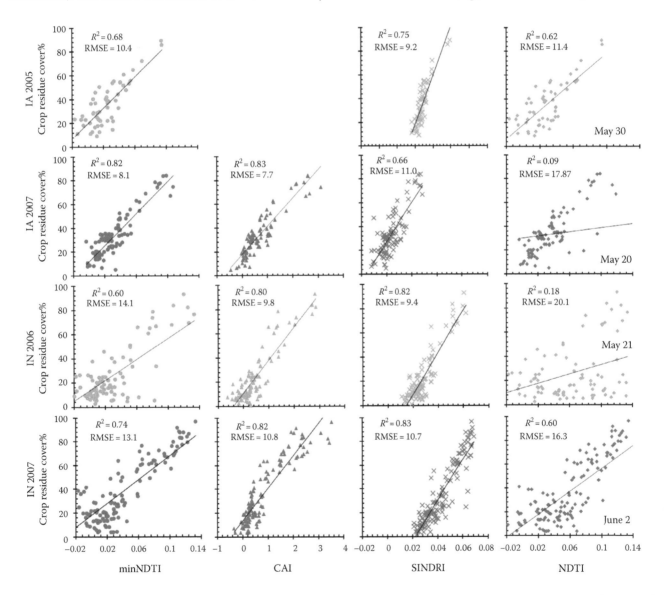

FIGURE 8.5 Comparison of minNDTI and single-scene CAI, SINDRI, and NDTI in two study areas: Ames, IA (year 2005 and 2007), and Fulton, IN (year 2006 and 2007). (Adapted from Zheng, B. et al., *J. Soil Water Conserv.*, 68(2), 120, 2013a.)

distance, maximum likelihood, spectral angle mapping, and cosine of the angle concept) and data mining approaches (e.g., random forest classifier and support vector machine) have been examined for identifying two broad tillage categories (South et al. 2004; Bricklemyer et al. 2006; Watts et al. 2008; Sudheer et al. 2010; Samui et al. 2012). These studies demonstrated the capability of Landsat TM imagery to discriminate between two broad tillage categories (i.e., conventional and conservation tillage) (van Deventer et al. 1997; Gowda et al. 2001) but fell short of achieving the reliability and consistency required for operational applications. Based on previous studies, it remains unclear which classification approach performs the best in classifying tillage categories. Research also has been conducted to test the feasibility of estimating crop residue cover using Landsat data (McNairn and Protz 1993; Thoma et al. 2004; Daughtry et al. 2006). These studies used single-date multispectral images and yielded mixed results. The inconsistent results of these studies may be related to the spectral resolution of Landsat TM data, different image pre-processing strategies to correct for atmospheric transmittance, spatial and temporal variations in soils, and green vegetation.

Tillage indices developed using hyperspectral and advanced multispectral (e.g., ASTER) data have provided consistent assessments of crop residue cover across years and study sites (Table 8.3; Figure 8.5). These tillage indices (e.g., CAI, SINDRI) detect absorption features associated with cellulose and lignin and are robust for discriminating crop residues from soils and green vegetation. However, the sensor systems with the appropriate spectral bands have very limited spatial and temporal coverage, which limits their usefulness for monitoring crop residue cover and tillage intensity over large areas. Finally, the SWIR bands of ASTER needed to characterize residue cover are no longer available due to detector failure in April 2008 (NASA/JPL 2008). Spaceborne multispectral imagery, however, is favorable due to its ability to provide extended repetitive coverage of the Earth. Landsat TM/ETM+ imagery, thus, is extremely attractive for monitoring tillage practices and crop residue cover over large areas because it is freely available and provides a long-term synoptic view of the Earth with a 16-day revisit frequency.

Timing of image acquisition is very important for monitoring agricultural resources because agricultural land surfaces change rapidly as growers prepare soils for planting and as crops emerge from soils, mature, and are harvested. It is well recognized that soil and residue status change rapidly during the planting season and vary in space and time (McNairn et al. 2001; Watts et al. 2008), but tillage and crop residue mapping have been long treated as a one-time mapping effort using only one image at a time, until Watts et al. (2011) incorporated temporal dimensions into tillage mapping. Zheng et al. (2012) emphasized the need to consider varied timings of tillage and planting in tillage mapping and significantly improved mapping accuracy using multitemporal Landsat imagery (Table 8.3). Minimum NDTI values were extracted from a time-series Landsat image that included images from 1 to 2 months before expected planting date to 1–2 months after planting date (Zheng et al. 2012). The method was designated as minNDTI and forms an effective way to minimize confounding effects of green vegetation (Zheng et al. 2012). Figure 8.6 shows a tillage map and its corresponding NDTI values of Champaign County, Illinois. The left image in Figure 8.6 is the minNDTI values extracted from a time-series NDTI image. Agricultural fields managed with conservation tillage are relatively brighter because higher levels of crop residue cover result in higher NDTI values. The multitemporal approach requires the use of surface reflectance Landsat data products, which are available from EarthExplorer (http://earthexplorer.usgs.gov/) and

TABLE 8.3 Summary of Studies in Crop Residue Estimation Using Remote-Sensing Imagery

Sensor	n[a]	Image Dates	Indices or Methods	R²	References
Landsat TM	266	4/18/1990	NDI	0.74	McNairn and Protz (1993)
Landsat ETM+	468	03/28/2000	NDI	0.38	Thoma et al. (2004)
		06/03/2001	STI	0.47	
		11/10/2001	NDTI	0.48	
Landsat TM	54	06/12/2004	NDI	0.14	Daughtry et al. (2006)
			NDTI	0.11	
SPOT	Varied	Varied	Spectral unmixing	0.58–0.78	Pacheco and McNairn (2010)
Landsat TM	39	05/28/2008		0.69	
Hyperion	54	05/03/2004	CAI	0.85	Daughtry et al. (2006)
Landsat TM[b]	Varied	Varied	NDTI	0.004–0.64	Serbin et al. (2009c)
ASTER[c]	Varied	Varied	LCA	0.39–0.86	
			SINDRI	0.61–0.87	
Airborne hyperspectral data	Varied	Varied	CAI	0.72–0.89	
Landsat TM and ETM+	31	Multitemporal	minNDTI	0.89	Zheng et al. (2012)
Landsat TM and ETM+	Varied	Multitemporal	minNDTI	0.66–0.89	Zheng et al. (2013a)

[a] n denotes number of samples.

[b] Data were simulated using ASTER data when Landsat TM imagery was unavailable.

[c] Data were simulated using airborne hyperspectral data when ASTER imagery was unavailable.

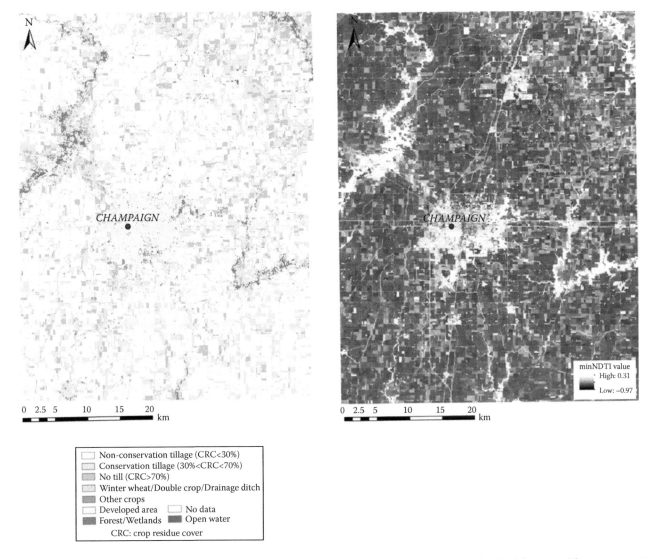

FIGURE 8.6 2006 Tillage map of Champaign County (left), Illinois, and its corresponding minNDTI values (right) extracted from a time-series NDTI image. Agricultural fields with brighter tones indicate higher levels of crop residue cover, which corresponds to the conservation tillage category.

USGS EROS Science Processing Architecture (ESPA) ordering interface (https://espa.cr.usgs.gov). The minNDTI approach was also applied to six additional datasets collected in different regions of the United States and the technique was comparable to CAI and SINDRI in achieving similar classification accuracy of three tillage categories (Zheng et al. 2013a). Zheng et al. (2013a) reported 68%–86% overall accuracies for three tillage categories—a significant improvement compared to 42%–56% accuracies reported by Thoma et al. (2004). However, the minNDTI approach cannot address the effects of surface soil variability as its performance was degraded when applied to a larger geographical area. Nevertheless, a multitemporal approach has shown a substantial potential to track changes of tillage practices over time and space using freely available Landsat and Landsat-like data (Watts et al. 2011; Zheng et al. 2012, 2013a).

8.3.4 Summary

8.3.4.1 Challenges

The primary challenges for operational tillage mapping using optical remote-sensing imagery include the following: (1) Revisit rates of moderate-spatial-resolution imagery are not frequent enough to capture the rapid changes in agricultural land surfaces during planting season, (2) there is limited spatial coverage of satellite hyperspectral imagery, (3) there are confounding effects of soil background and green vegetation, and (4) there is a lack of transferability of locally developed models.

Landsat is currently the best satellite system to provide the capabilities for long-term and broadscale tillage assessment. Although the minNDTI technique showed promises in tillage mapping at large scales, the 8-day revisit rate of combined Landsat 8 OLI and 7 ETM+ cannot guarantee adequate numbers of cloud-free observations to capture the *recently tilled surface.* In tropical regions or other areas that have persistent cloud cover, one may be lucky to obtain two or three cloud-free images per year. The data gap issues of Landsat 7 ETM+ imagery also prevent rapid application of the minNDTI technique because additional image preprocessing skills are required to fill the missing data. Zheng et al. (2013b) have presented an easy way to fill the missing data for broadscale tillage mapping using the multiscale segmentation method. Landsat images with partial cloud cover can be incorporated into the time series; however, estimation of tillage status for the cloud-contaminated pixels could be less accurate, and a quality assessment map should be provided to inform users about locations of cloud and cloud shadow pixels (Zheng et al. 2014).

The spatial and temporal adaptive reflectance fusion model (STARFM) (Gao et al. 2006), which produces cloud-free synthetic Landsat images with 30 m spatial resolution at Moderate Imaging Spectroradiometer (MODIS) temporal frequency, could be an alternative option to enhance temporal resolution for tillage mapping. The enhanced STARFM (Zhu et al. 2010), future improvement of data fusion techniques, and the higher quality of Landsat 8 and the European Space Agency (ESA) Sentinel-2 data could open possibilities to provide data optimized in both temporal and spatial resolutions for tillage assessment. However, the potential to incorporate data fusion techniques into minNDTI technique to improve our ability to map tillage practices currently remains unknown and required for future investigation (Zheng et al. 2014).

Locally developed empirical models often show degraded performance when applied to the same location over time or to a broader region. Variations in weather, soil, and terrain conditions across landscape are the main reasons for the degraded performance when a model is extrapolated to new situations. Zheng et al. (2013a) reported superior performance of local models than a *universal* model and highlighted negative impacts of local variation in terrain, moisture, and soil color upon crop residue estimation. Thus, estimation of crop residue cover with broadband multispectral may require extensive local calibration data. Alternatively, the effects of soil variation can be reduced or minimized using local soil-adjusted tillage indices (Biard and Baret 1997) or the spectral unmixing approach (Pacheco and McNairn 2010). The spectral unmixing approach has the potential to map crop residue cover over large geographic regions as the approach is insensitive to variations in soil and residue when end-members are retrieved directly from the image (Pacheco and McNairn 2010). However, future work is required to examine how well the unmixing approach performs in the presence of green vegetation.

Much of the research to apply remote sensing to tillage assessment has been developed in the context of midlatitude agriculture, characterized by distinct seasonal cycles, large field sizes, common use of monoculture, or reduced crop diversity, over large regions. In other regions of the world, or in irrigated regions, there may be a much larger range of crops, with a variety of planting and harvesting dates, not synchronized with

each other, and smaller field sizes—in such situations, the tillage assessment task requires different strategies than may be effective in midlatitude regions.

8.3.4.2 Future Capabilities

At the time of this writing, due to the limited availability of hyperspectral data, the minNDTI approach is probably the most effective method to map tillage practices at broadscale using optical remote-sensing imagery. The minNDTI can be applied to Landsat 7 ETM+ and Landsat 8 OLI, which together provides an 8-day observation cycle. The OLI imagery has potential to enhance our ability to accurately estimate crop residue cover with its narrower spectral bands and 12-bit dynamic range, as indicated by Galloza et al. (2013), who found that the Advanced Land Imager (ALI) has better capability to discriminate crop residues from soils than Landsat TM data.

The upcoming launch of ESA Sentinel-2 satellite will provide enhanced Landsat-type data with <5-day revisit time. Sentinel-2 is particularly useful for monitoring the rapid changes of agricultural lands. Operational tillage assessment is likely to involve multisensor multidate image fusion and could be implemented using Landsat and Sentinel-2 data together. The planned hyperspectral satellite missions, including ESA Environmental Mapping and Analysis Program (EnMAP) and NASA Hyperspectral Infrared Imager (HyspIRI), will also make contribution to large-scale tillage assessment. These hyperspectral data can be used to calculate CAI. Fusion of hyperspectral and multispectral images could estimate crop residue cover at the multispectral spatial extent with improved accuracy (Galloza et al. 2013). The WorldView-3 satellite launched in August 2014 includes SWIR bands equivalent to ASTER SWIR sensor (DigitalGlobe, 2014), which can be used to derive SINDRI for crop residue estimation. The very high spatial resolution (3.7 m) of WorldView-3 SWIR data will permit fine-scale assessment of crop residue cover, soil texture, and soil roughness.

8.4 Monitoring with SAR

8.4.1 Introduction

SARs are considered active remote-sensing sensors as they generate pulses of energy that are propagated toward a target. SARs then record the energy scattered by the target, back toward the radar antenna. The strength (intensity) of the received or *backscattered* signal is measured as sigma naught (σ^0), expressed in decibels (dB). Since these sensors provide their own source of energy, SARs are able to collect data day or night. SARs generate energy at microwave frequencies (0.2–300 GHz), with Earth-observing SAR satellites typically operating at X-band (2.40–3.75 cm; 8.0–12.5 GHz), C-band (3.75–7.5 cm; 4.0–8.0 GHz), and L-band (15–30 cm; 1.0–2.0 GHz) (Lewis and Henderson 1998) (Table 8.4). These lower frequencies are unaffected by the presence of cloud and haze. Given this context and the sensitivity of microwaves to soil conditions, SARs are an important data source for mapping and monitoring tillage and residue.

TABLE 8.4 Selected Civilian Spaceborne Radar Sensors

Frequency (in GHz)		Sensor	Polarization[a]	Incidence Angle (°)	Resolution (m)	Swath (km)	Dates of Operation
X	8.600	COSMO-SkyMed 1	SP, DP	25–50	1–100	10–200	2007–
		COSMO-SkyMed 2	SP, DP				2007–
		COSMO-SkyMed 3	SP, DP				2008–
		COSMO-SkyMed 4	SP, DP				2010–
	8.650	TerraSAR-X	SP, DP, QP	15–60	0.25–40	4–270	2007–
	8.650	TanDEM-X	SP, DP, QP	15–60	0.25–40	4–270	2010–
C	5.300	RADARSAT-1	SP (HH)	10–60	8–100	45–500	1995–2013
	5.300	ERS-2	SP (VV)	20–26	30	100	1995–2011
	5.331	Envisat ASAR	SP, DP	15–45	10–1000	5–405	2002–2012
	5.350	RISAT-1	SP, DP, QP, CP	12–55	1–50	25–223	2012–
	5.405	RADARSAT-2	SP, DP, QP	10–60	3–100	18–500	2007–
	5.405	RADARSAT Constellation	SP, DP, QP, CP	10–60	1–500	5–500	2018
	5.405	Sentinel 1A	SP, DP	20–45	5–40	80–400	2014–
		Sentinel 1B	SP, DP	20–45	5–40	80–400	2016
L	1.200	ALOS/PALSAR-1	SP, DP, QP	8–60	10–100	20–350	2006–2011
	1.200	ALOS/PALSAR-2	SP, DP, QP, CP	8–60	1–100	25–490	2014–
	1.260	SMAP	SP, VV/HH/HV[b]	40	1–3 (km)	1000	2015–
	1.275	SAOCOM 1A	SP, DP, QP, CP	17–51	10–100	20–350	2015
		SAOCOM 1B					2016

[a] In the polarization column, SP = single polarization, DP = dual polarization, QP = quadrature polarization, and CP = compact polarization.

[b] SMAP has now been launched. Thus this should say "SMAP acquires radar imagery simultaneously in VV, HH, and HV.

8.4.2 Critical Variables for Tillage Assessment

The interaction of microwaves with a target and the characteristics of the scatter that results from this interaction are a function of the condition of the target as well as the SAR sensor specifications. SAR response is driven by the dielectric permittivity, roughness, and structural properties of the target. In the context of tillage monitoring, SARs are sensitive to small-scale roughness and large macrostructures produced by farming implements, as well as volumetric soil moisture. In addition to their spatial resolution, SARs are characterized by their frequency, incidence angle, and polarization—configurations that also affect the target interaction.

8.4.2.1 Sensitivity of SAR to Soil Characteristics

8.4.2.1.1 Surface Roughness

Random and periodic roughness determines the angular scattering pattern with diffuse scattering increasing as roughness increases. For agricultural fields, roughness is created by land management activities (principally tillage and seedbed preparation) modified over time by water and wind erosion. Roughness is defined by two parameters: the root mean square (RMS) variance and surface correlation length (l). RMS describes the surface's random vertical statistical variability relative to a reference surface; while correlation length is an autocorrelation function that measures the statistical independence of surface heights at two points (Ulaby et al. 1986). For very smooth surfaces, as expected from no-till fields, the random roughness (RMS) is small and the height of every point is correlated with

the height of every other point (hence l is large). In this case, most microwave energy is forward scattered and backscatter is low. Inversely randomly rough surfaces, created by tillage, result in more diffuse scattering with a greater proportion of the incident energy scattered back to the sensor. These surfaces have higher RMS, short correlation lengths, and higher backscatter.

8.4.2.1.2 Dielectric Permittivity

The intensity of backscatter from soils is largely determined by the soil permittivity (dielectric constant), while the angular pattern of microwave scattering is governed by the surface roughness. The permittivity ε is a frequency-dependent complex quantity [$\varepsilon(f) = \varepsilon'(f) - j\varepsilon''(f)$], where the real component ε' describes the polarizability of a material when an electric field is applied and the imaginary component ε'' energy losses (Hasted 1973). Dielectric losses are due to relaxation ε''_{ref} and direct current electrical conductivity σ in S/m: $\varepsilon''(f) = \varepsilon''_{ref}(f) + \sigma/2\pi f\varepsilon_0$, where ε_0 is the permittivity of free space ($8.854 \cdot 10^{-12}$ F/m). On agricultural fields (without vegetation cover), scattering occurs at the air/soil boundary as a dielectric discontinuity exists at this interface. The majority of dry soils have ε' of 3–8, and bulk soil permittivity increases with water content. This is due to the much greater, albeit frequency-dependent, permittivity of water, which at 1.4 GHz ranges from 84.1 – j10.7 at 5°C to 74.5 – j4.1 at 35°C, 69.0 – j32.1 ~ 71.4 – j14.6 at 5.3 GHz, and 49.2 – j39.7 ~ 65.1 – j23.7 at 9.6 GHz for pure water where $s = 0$ S/m for pure water where $\sigma = 0$ S/m. Increases in either part of the permittivity will increase soil reflectivity. Electromagnetic wavelength is an inverse function of ε'; thus the wavelength becomes shorter within the soil as it becomes wetter. As backscatter intensity is a function of permittivity, a strong linear relationship exists between soil moisture and backscatter. The depth of sensitivity within the soil volume is dependent upon three parameters: the SAR configuration, soil moisture, and bulk soil ε''. Penetration depth is an inverse function of bulk soil permittivity and, thus, soil moisture and conductivity. Consequently, SARs respond to moisture over deeper volumes as soils dry. Regardless, sensitivity is still near surface with this depth approximately equivalent to the microwave wavelength (Boisvert et al. 1995).

8.4.2.1.3 Residue

If vegetation (green or senesced vegetation or postharvest residue) is present, SAR response will be affected if water is present in the vegetation. Residue is considered "dead" vegetation, and thus its effect on backscatter is often assumed insignificant, effectively transparent to the incident microwaves. This assumption has proven invalid in circumstances where residue retains water. The impact of residue on backscatter varies depending upon the volume of water held, a function of the amount and type of residue (McNairn et al. 2001). Jackson and O'Neill (1991) reported that residue can retain significant moisture with McNairn et al. (2001) measuring up to 60% and 40%–50% moisture in corn and barley residue, respectively, following rain events.

8.4.2.1.4 Row Direction

Land management practices (planting, harvesting, and tillage) can create row effects and row direction relative to the radar look direction impacts SAR response. When row direction is perpendicular to the look direction, SAR response is stronger when compared to a look direction parallel to rows (Beaudoin et al. 1990; McNairn et al. 1996). Producers follow a rectangular pattern operating parallel to the long and short axes of fields. This practice creates a "bow-tie" effect visible on SAR imagery where, within a single field, backscatter is significantly higher for the axis of the field oriented perpendicular to the sensor.

8.4.2.2 Impact of SAR Configuration

SAR sensors are defined by three configurations—frequency (GHz, or cm, if characterized as free-space wavelength), incidence angle (degrees), and polarization. These configurations affect how microwaves interact with the target in terms of backscatter intensity and scattering characteristics. SAR configurations can be selected to maximize sensitivity to the target property of interest (soil moisture, surface roughness, or residue). Alternatively, as these properties are confounded in the microwave signal, multiple configurations can be used together to resolve individual contributions.

8.4.2.2.1 Frequency

As well as affecting penetration depth, SAR frequency determines sensitivity to surface roughness. Thus, surface roughness must be considered relative to frequency. Surfaces are defined as *rough* or *smooth* according to the Rayleigh criterion. Surfaces are smooth if $h < \lambda/25 \sin\tau$ and rough if $h > \lambda/4.4 \sin\tau$ where

h is the RMS, λ is the wavelength, and τ is the depression angle (Sabins 1986). Assuming flat terrain, τ is the complement of the incidence angle ($\theta = 90-\tau$). In practice, this means that a field will appear rougher (higher backscatter) at shorter wavelengths (i.e., X-band) than at longer wavelengths (i.e., L-band). With this strong dependency, the choice of wavelength is especially important when monitoring tillage. Short-wavelength (high-frequency) SARs will see many fields as *rough* and thus may not differentiate among tillage classes at the upper ranges of roughness. Several studies (Pacheco et al. 2010; Aubert et al. 2011; Panciera et al. 2013) reported that X-band data from TerraSAR-X were not well suited for roughness mapping when RMS was high. Panciera et al. (2013) found that TerraSAR-X backscatter was sensitive to roughness (RMS), which fell between 0.5 and 1.5 cm, but that the signal saturated beyond 2 cm. Conversely, large-wavelength (low-frequency) SARs may view even tilled fields as *smooth*. Nevertheless, numerous studies have reported sensitivity of C- and L-band responses to roughness and residue (McNairn et al. 2001, 2002; Baghdadi et al. 2008). Baghdadi et al. (2008) compared three frequencies (X-, C-, and L-band) demonstrating that sensitivity to roughness increased with wavelength.

8.4.2.2.2 Incidence Angle

Regardless of the target, backscatter decreases with increasing incidence angle, which is defined as the angle between the radar beam and a line perpendicular to the surface. The rate of decrease is target dependent, with backscatter decreasing with angle at a higher rate when soils are smooth. This differential rate of decrease can be used to separate smooth from rough fields, if fields are imaged at contrasting incidence angles (McNairn et al. 1996). As simultaneous multiangle data are typically unavailable from spaceborne SARs, a simpler approach is to select an incidence angle that maximizes sensitivity to surface roughness. Steeper (smaller) angles minimize roughness contributions to backscatter and are thus more suited to estimate soil moisture, while shallower (larger) angles maximize roughness effects on backscatter (McNairn et al. 1996). Similarly, larger angles are more sensitive to residue as soil moisture contributions are minimized, and more microwave interaction occurs with residue at these angles (McNairn et al. 2001). Although these larger angles are more suited to roughness and residue applications, contributions from soil moisture are not completely eliminated. Aubert et al. (2011) noted that the range in X-HH backscatter due to surface roughness increased as incidence angle increased, with backscatter varying 3.5 and 1.9 dB at angles of 50° and 25°, respectively. Baghdadi et al. (2008) reported a slightly larger range in X-band backscatter (5.5 dB at 50°–52° and 4 dB at 26°–28°).

8.4.2.2.3 Polarization

Polarization is defined by the orientation of the electric field vectors of the transmitted and received electromagnetic wave. Polarization should be considered relative to the target structure and response interpreted according to the characteristics of scattering from the target, including the sources of scattering and the randomness of the scatter. Scattering is categorized as single bounce (surface), multiple (volume), or double bounce. Targets usually produce more than one type of scattering although typically one source dominates. For smooth soils devoid of residue, surface single-bounce scattering dominates. Rough soils result in multiple scattering of microwaves. Residue (depending on the amount and water content) also causes multiple scattering and, if residue is vertically oriented, double-bounce events may also contribute.

Most SAR sensors transmit and receive microwaves in the horizontal (H) and/or vertical (V) linear polarizations (Table 8.4). Early satellites transmitted and received microwaves in a single linear polarization (European Remote Sensing [ERS]-1 and 2 [VV], Japanese Earth Resources Satellite [JERS]-1 [HH], and RADARSAT-1 [HH]). Next-generation sensors (i.e., Advanced Synthetic Aperture Radar [Envisat ASAR]) transmitted and/or received in both linear polarizations, which permitted acquisition of like (HH and/or VV) and cross (HV or VH) polarizations. When targets are physically oriented parallel to the polarization of the incident wave, greater microwave interaction occurs. This is most obvious for targets like crops where their vertical structure aligns well with vertical transmitted waves. Consequently, a VV configuration provides more information on crops than HH. For soils without residue, horizontal or vertical orientation is absent and thus HH and VV backscatter is correlated. A linear cross polarization response (HV or VH) results when the transmitted wave (i.e., H) is repolarized to its orthogonal polarization (i.e., V). Repolarization of H to V (or V to H) occurs as a result of multiple scattering (at least two bounces), and thus a target must be able to cause more than a single scatter event to elicit an HV or VH response. Smooth soils, devoid of structure, are dominated by single-bounce forward scattering and produce very low cross-polarized backscatter. For soils with random roughness or residue (assuming moisture in the residue), incident waves experience multiple scattering and higher cross polarization response is observed. McNairn et al. (2001) reported that, of all the linear polarizations, the cross polarization was most sensitive to the amount of residue. The cross polarization has the advantage of being insensitive to planting, harvesting, or tillage row direction (McNairn and Brisco 2004). This is important considering that Brisco et al. (1991) established that row direction from tillage significantly impacted like-polarized backscatter.

8.4.2.2.4 Polarimetry

Some satellites (i.e., ALOS PALSAR, RADARSAT-2, and TerraSAR-X) are polarimetric capable. Polarimetric sensors capture the complete characterization of the scattering field meaning that they record all four mutually coherent channels (HH, VV, HV, and VH), with phase information between orthogonal polarizations retained and processed. Any linear, elliptical, or circular polarization can be synthesized from polarimetric data. Circular polarizations are described by their handedness (direction of rotation) relative to the observer. Right-handed circular waves (R) rotate clockwise (relative to observer), while left-handed waves (L) rotate counterclockwise. The application of circular polarizations for agriculture has received limited attention although for soils, circular and linear backscatter is highly correlated (Sokol et al. 2004). As with linear polarizations, multiple scattering must occur to change the handedness of the transmitted circular polarization. Roughness or residue can cause two or more bounces, changing the handedness and resulting in a higher circular copolarization (RR or LL) response (recall rotation is defined relative to the observer). Indeed, de Matthaeis et al. (1992) observed high circular cross-polarized backscatter (LR) returns for surfaces with dominant surface scattering. Circular copolarized (RR) backscatter increases when the mechanisms producing volume scattering dominate (McNairn et al. 2002).

Polarimetric data can be processed to extract additional parameters, which characterize scattering and thus tillage and residue conditions. SARs transmit completely polarized waves but with multiple scattering, microwaves become completely or partially depolarized. The degree of depolarization (or proportion of unpolarized energy) is indicative of the randomness of scattering within the target. Smooth soils create little depolarization (Evans and Smith 1991). The degree of depolarization increases with roughness and residue cover as the phase becomes unpredictable from point to point within the target. The degree of depolarization can be measured by pedestal height with height increasing as roughness increases or in the presence of residue (van Zyl 1989; de Matthaeis et al. 1991; McNairn et al. 2002; Adams et al. 2013a). Adams et al. (2013a) also reported that the dynamic range of the degree of polarization (Δ_{POL}) was sensitive to roughness and residue. Δ_{POL} is the difference between the maximum and minimum degree of polarization and reflects the heterogeneity of scattering mechanisms within the target (Touzi et al. 1992).

Absolute phase (φ) of a scattered wave is a function of distance from the target and carries no target scattering information (Langman and Inggs 1994). However, the difference in the phase between two orthogonal polarizations (i.e., H and V) is of interest for tillage monitoring. Shifts in the phase (characterized by the copolarized phase difference [PPD] [$\varphi_{VV} - \varphi_{HH}$]) occur due to double-bounce or multiple scattering. For smooth soils with minimal contributions from multiple scattering, HH and VV are in phase and mean PPD is close to zero (Evans et al. 1988). A vertical structure can cause a double bounce and here PPD values approach 180° (de Matthaeis et al. 1991). Large phase differences are typically associated with cropped fields although high PPD values have been observed for standing senesced crops (McNairn et al. 2002). Ulaby et al. (1987) reported that plowed and disked fields, as well as those with corn and soybean residue, had a mean PPD close to zero. However, the standard deviation of the phase difference among the disked, plowed, residue and standing crops was very different. These results were confirmed by McNairn et al. (2002) where multiple scattering in residue caused a highly varying PPD with a noise-like distribution for these fields. The copolarized complex correlation coefficient (ρ_{HH-VV}) measures the decorrelation of the phase and some sensitivity to residue has also been reported (Adams et al. 2013a).

Methods that decompose the SAR signal have drawn considerable interest with the Cloude–Pottier (Cloude and Pottier 1997) and Freeman–Durden (Freeman and Durden 1998) decompositions showing sensitivity to tillage and residue. Cloude–Pottier decomposes the signal into a set of eigenvectors (which characterize the scattering mechanism) and eigenvalues (which estimate the intensity of each mechanism) (Alberga et al. 2008). From the eigenvalues, entropy (H) and anisotropy (A) are calculated. H measures the degree of randomness of the scattering (from 0 to 1); values near zero are characteristic of single scatter targets (i.e., smooth soils). Rough soils and those with residue have larger contributions from multiple scattering. This increase in randomness of scattering is measured as an increase in H. Anisotropy estimates the relative importance of the dominant scattering mechanism and the contribution from secondary and tertiary scattering mechanisms. Zero A identifies two mechanisms of approximately equal proportions, while values approaching 1 indicate that the second mechanism dominates the third (Lee and Pottier 2009). The Cloude–Pottier decomposition also calculates the average alpha ($\bar{\alpha}$) angle (0°–90°), which identifies the dominant scattering source (Alberga et al. 2008). Smooth soils with single-bounce scattering have angles close to 0°, volume scatterers close to 45°, and double bounce nearing 90°. Adams et al. (2013a) reported that H and $\bar{\alpha}$ were significantly correlated with roughness and percent crop residue. The Freeman–Durden decomposition separates the total power of every SAR resolution cell into contributions from three scattering mechanisms—volume (multiple), double-bounce, and single-bounce (surface) scattering. Adams et al. (2013b) demonstrated that H, $\bar{\alpha}$, and the Freeman–Durden multiple scattering could statistically separate fields with different harvesting, tillage, and residue conditions, particularly at higher incidence angles (49°). In addition, the best separability was found between unharvested or fields not tilled and conventionally tilled fields; fields under conservation tillage were confused with other tillage classes (Adams et al. 2013b).

8.4.3 Methods

8.4.3.1 Change Detection and Classification

Change detection identifies and measures differences between two (or more) images, indicated by a change in SAR response or in derived surface properties (roughness, residue). Several SAR metrics can be used to capture change and include (1) incoherent SAR backscatter (HH, VV, HV, and VH), (2) degree of polarization, (3) copolarized phase parameters, (4) decomposition parameters, and (5) coherent change. When change is measured directly from SAR response, consideration must be given to the confounding effects of target parameters, SAR configuration, and sensor calibration. To isolate change in SAR response due to roughness (or residue), soil moisture must not vary and thus the period between acquisitions should be minimized. Since frequency, incidence angle, and polarization affect target interaction, images must have the exact same SAR configuration. For spaceborne SARs, this means using exact repeat orbits. Constellations of satellites (such as the planned Canadian

RADARSAT Constellation) will be of interest for change detection since repeat acquisitions in the same SAR configuration will be possible within a short period of time. Finally, SARs must be well calibrated; scene to scene calibration of spaceborne sensors is typically well below 1 dB. If changes in derived properties (roughness, residue) are used, errors in methods or model performance will be carried forward in the change detection process. Whatever metric is adopted, interpretation of the change is required. This means that a threshold must be determined, above which change is considered significant. In addition, change must be linked to information meaningful for tillage monitoring (type of implement used, tillage or residue class, change in residue amount).

McNairn et al. (1998) applied a simple change detection approach to a pair of RADARSAT-1 (HH) images acquired one week apart. The incidence angle difference between the Standard Mode 2 and 3 images was limited to 6° and was considered of secondary importance. In the one week separating the first from second acquisition, C-HH backscatter remained stable (average difference of 0.7 dB) for fields not tilled. No rain fell during the week, and the small difference was attributed to the 6° difference in angles. For fields that were tilled, the average change (increase) in backscatter was 5.6 dB. This technique (Figure 8.7) enabled the identification of broad conservation tillage classes (no-till, intermediate, and tilled) and flagged fields where harvesting and tillage had occurred. Hadria et al. (2008) combined SAR (Envisat ASAR) and optical data to classify broad categories of tillage. The authors used a combination of image thresholding and decision tree classification. Envisat ASAR was especially helpful at differentiating smooth surfaces (no-till) from other rougher (tilled) surfaces.

Coherent change detection (CCD) exploits the coherence between two polarimetric complex images acquired at different times but in the same imaging geometry (Milisavljević et al. 2010). A pixel-by-pixel correlation of the coherence between the images reveals changes in the target; if no change has taken place, the pixels remain correlated. This technique requires that the target is coherent, allowing changes in coherence from image to image to be measured. Random phase characterizes most distributed natural targets like forests and crops. These targets typically have low coherence and are not ideal candidates for this method. As well, external effects like wind can cause these targets to temporally decorrelate. Polarimetric interferometric (PolInSAR) may be useful in optimizing coherence for detecting change in distributed targets like crops (Li et al. 2014). Although CCD for tillage change detection has not been explored, this approach may be capable of observing changes from tillage activities.

8.4.3.2 Semiempirical and Physical Models

Physical scattering models estimate backscatter using the soil's physical properties and sensor configurations. Soil properties include the dielectric constant, RMS, and correlation length. The small perturbation and Kirchhoff models (geometrical optics and physical optics models) are two common physical models. However, these models are not suited to targets with

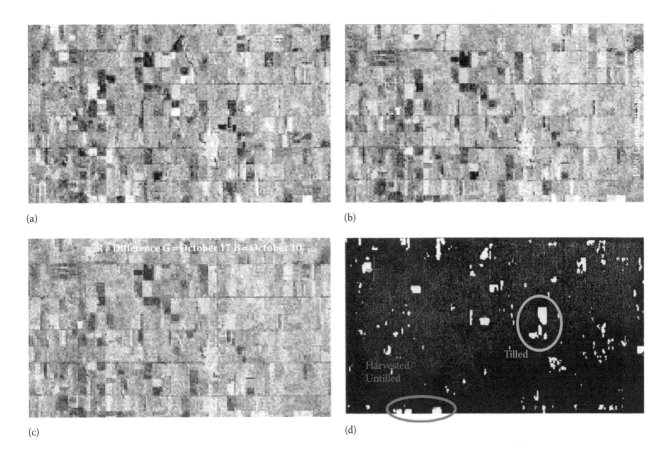

(a)

(b)

R = Difference G = October 17 B = October 10

Tilled

Harvested/
Untilled

(c)

(d)

FIGURE 8.7   Detection of tillage and harvesting activities using RADARSAT-1 over an agricultural site in Canada (Altona, Manitoba). Standard beam mode images were collected on October 10 (a) and October 17 (b) in 1996. A difference image (c) and a change detection product (d) were produced from the backscatter. (Adapted from McNairn, H. et al., *Can. J. Remote Sens.*, 24, 28, 1998.)

multiple sources of scattering and large ranges of roughness, as expected from agricultural fields. The Integral Equation Model (IEM) (Fung et al. 1992) integrates these two models and is better adapted for targets with surface and multiple scattering and with roughness ranging from smooth to rough.

The goodness of fit between backscatter predicted by the IEM and that observed by SARs has varied depending on the roughness, frequency, and incidence angle. Speculation has been that in many cases, the error in IEM-simulated backscatter is due to inaccurate representation of the correlation length (l), a parameter difficult to adequately measure in the field (Merzouki et al. 2010). As a solution, Baghdadi et al. (2004) proposed a calibrated version of the IEM, introducing an optimum correlation length (l_{opt}). The optimum correlation length is derived from a set of equations that relates correlation length (l) to RMS, as a function of polarization and incidence angle (Baghdadi et al. 2006). Simulated backscatter from the calibrated IEM has more closely matched backscatter from the C-band SAR backscatter (Merzouki et al. 2010). Figure 8.8a is an example of a surface roughness (RMS height [hRMS]) map derived from this study (Merzouki et al. 2010). Rahman et al. (2008) also derived surface roughness over sparsely vegetated fields using Envisat ASAR and the IEM in a multiangle approach. The image-derived RMS (2.19 cm) overestimated the field-derived RMS (0.79 cm)

(Figure 8.8b). The subsurface rock fragments may have caused multiple bounce interactions, thus increasing response and generating a larger *radar-perceived* roughness (Rahman et al. 2008).

Inversion of the IEM or calibrated IEM is difficult due to the complexity of the model. As well, multiple unknowns in the IEM (dielectric constant, RMS, and l) and the calibrated IEM (dielectric constant and RMS) require multiple sources of SAR information. In this case, a lookup table (LUT) approach can be used to estimate roughness or dielectric from SAR response (Merzouki et al. 2011). Forward runs of the model are used to create the LUTs with incremental steps in dielectric, RMS, l, and incidence angle and their modeled backscatter (in HH and VV). Direct search functions are used to find the LUT entry that minimizes the difference between the measured (from SAR sensor) and modeled (from IEM) backscatter. This LUT entry provides the model estimate of soil dielectric and surface roughness. With multiple unknowns, multiple SAR configurations are needed to solve the IEM (three unknowns) or calibrated IEM (two unknowns). Typically, SAR data acquired at two polarizations (i.e., HH and VV) are used with the calibrated IEM. With a third unknown (l), an additional source of backscatter is needed to implement the original IEM. One approach is to use SAR data acquired at two polarizations (HH and VV) and two contrasting incidence angles.

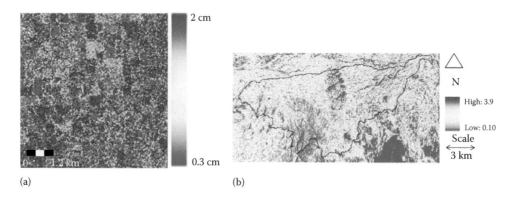

FIGURE 8.8   Surface roughness maps derived from radar images and the IEM over two agricultural sites: an area within the Red River Watershed in Southern Manitoba in Canada (a) and Walnut Gulch Experimental Watershed in Arizona in the United States (b). The surface roughness map for the Red River Watershed is expressed in hRMS in centimeters. (Adapted from Merzouki et al. 2010.) The Walnut Gulch Experimental Watershed surface roughness map is defined by the hRMS variation of the surface at centimeter scale. The solid line represents the boundary of the Watershed. (Adapted from Rahman, M.M., et al., *Remote Sens. Environ.*, 112(2), 391, 2008, doi:10.1016/j.rse.2006.10.026.)

The Oh (Oh et al. 1992; Oh 2004) and Dubois (Dubois et al. 1995) models are semiempirical models created from the collection of large experimental datasets and subsequently empirically relating soil dielectric (directly or via the Fresnel reflectivity), RMS, the wavelength (through the wave number), and the incidence angle to SAR backscatter. Oh modeled backscatter from all three linear polarizations (HH, VV, and HV) and for three frequencies (X, C, and L). In contrast, the Dubois model uses only the copolarized backscatter (HH and VV) and was developed using data collected only at L-band. As with the IEM, the Oh model can be inverted using a LUT. The Dubois model is easily inverted by solving the model's two backscatter equations. Because these models were created with experimental data, application of these models to target conditions or SAR configurations beyond those of the experimental data used to create them may yield uncertain results. Indeed, Merzouki et al. (2010) found that these models tended to overestimate backscatter when modeled backscatter was compared to that measured by RADARSAT-2, which would lead to an overestimation of RMS. The Oh model resulted in larger errors between modeled and measured backscatter on smoother fields (<2 cm). Conversely, errors were greater on rougher fields (>1.5 cm) for the Dubois model.

Hajnsek et al. (2003) developed a model to invert surface roughness by coupling a Bragg scattering term and a roughness variable derived from the scattering entropy, anisotropy and alpha angle. This model was validated against airborne polarimetric L-band (E-SAR) data and yielded low RMSE (19%). Figure 8.9 shows a roughness map created using this approach.

8.4.4 Linking Radar Products to Tillage Information

SAR sensors can provide information on roughness (RMS) and residue, as well as changes in these conditions. However, to be meaningful, roughness and residue must be linked to information of interest such as tillage implement or tillage class. This linkage is required for applications such as watershed management, soil erosion risk assessment or estimation of carbon sequestration. Establishing this linkage is not a simple task given the complexity and dynamics of tillage activities. Producers use a combination of tillage implements and tillage occurs periodically and at a range of soil depths and directions. Tillage-induced roughness also varies depending upon soil texture and moisture and is modified over time by erosion events. Winter crops and weeds present on fields also complicate tillage mapping. How to link SAR-derived products and tillage information will vary depending on the approach used to create these products. For example, if models are used to estimate roughness (RMS), an association between RMS and tillage operation could be established. Such an approach was proposed by Jackson et al. (1997). However, the roughness (RMS) created by each tillage implement, and sequences of tillage applications, is likely to vary field to field due to soil conditions, erosion, and characteristics of the implement itself. Consequently, a much larger database of roughness responses to tillage is required, and these data must be acquired over regions with varying tillage systems. For example, Pacheco et al. (2010) found that in eastern Canada, some conservation tilled fields (chiseled plowed) had greater roughness (RMS) than conventional tilled fields (moldboard plowed). As well, RMS varied greatly within the chisel class, creating confusion when attempting to use backscatter to identify classes. Classifications or change detection approaches typically identify broad tillage classes (untilled, conservation, and conventional). While these classes may be useful for some mapping applications (identifying adoption of no-till for carbon sequestration), they may not be adequate for others (erosion modeling).

8.4.5 Summary

Given the dynamics of tillage activities during the preseeding and postharvest seasons, SAR sensors can be a valuable data source for time-critical applications (McNairn et al. 1998). With longer wavelengths, SAR data acquisition is unaffected by atmospheric

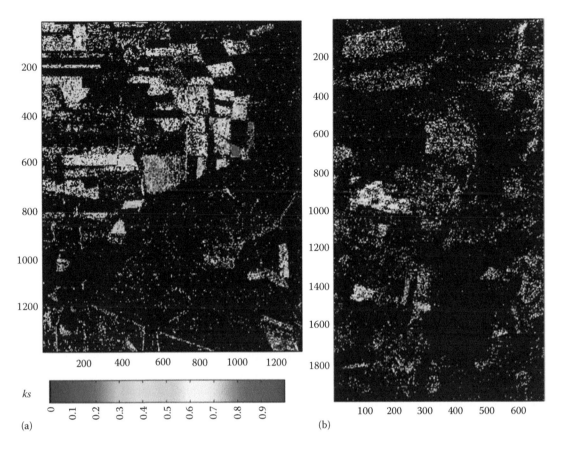

FIGURE 8.9 Estimated surface roughness, ranging from 0 to 1, over two study sites in Germany: Elbe-Auen (a) and Weiherbach (b). Areas in black represent data gaps. (Adapted from Hajnsek, I. et al., *IEEE Trans. Geosci. Remote Sens.*, 41(4), 727, 2003, doi: 10.1109/tgrs.2003.810702.)

conditions such as cloud cover and haze. The number of SAR satellites in orbit continues to increase and the engineering behind these satellites has led to a greater diversity in SAR configurations. This means that users now have choices in incidence angle and polarization and, in some cases, access to polarimetric data (Table 8.4). Research has demonstrated that success in this application will be best achieved when data can be accessed at more than one frequency and polarization. The choice of incidence angle and polarization is clear with researchers agreeing that shallower angles and cross polarizations are best for roughness and residue mapping. The availability of polarimetric-capable sensors is relatively recent, and thus more research is needed to develop methods to exploit these complex data. The primary challenge is the coupling of roughness, soil moisture, and residue in the SAR response. This coupling complicates the extraction of tillage information from the signal but can be accomplished by exploiting SAR data acquired at multiple configurations (frequency, angle, or polarization). Planned, and recently launched, satellites include the C-band RADARSAT Constellation (Canada), C-band Sentinel-1A and B (ESA), and L-band SAOCOM-1A and B (Comisión Nacional de Actividades Espaciales (CONAE)) (Table 8.4). These satellites will provide frequent data at a range of angles and polarizations and promise to provide an important source of data for monitoring tillage.

8.5 Review and Outlook

This chapter has summarized recent progress to advance applications of remote-sensing technologies to broadscale assessment of tillage status. Nonetheless, important challenges remain. Here we recap some of the key elements of current research to apply remote-sensing technologies to broadscale, site-specific tillage assessment and then highlight some of the principal challenges this effort faces as further research progresses (see also Zheng et al. 2014).

Optical remote sensing and SAR data provide different capabilities for tillage assessment. Whereas optical remote-sensing imagery provides the spectral basis for detection of crop residues on soil surface, SAR data provide information on soil physical properties, such as roughness and texture, which can reveal the nature of tillage practices. With the presence of green vegetation, both SAR and optical remote-sensing data have difficulties to discriminate different tillage categories. Remotely sensed imagery sensitive to radiation near 2100 nm cellulose absorption bands provides the best opportunity to estimate crop residue cover and to map tillage practices. In this context, the best three tillage indices are CAI, SINDRI, and NDTI. Because current satellite hyperspectral systems cannot provide systematic spatial coverage, at present, multispectral imagery now forms the preferred candidate for a broadscale tillage assessment. Multitemporal imagery is required to provide accurate assessment on tillage practices for regions with diverse crop calendars—a range of dates for soil preparation and planting schedules. The upcoming launch of several new satellite systems with optical sensors will offer solid opportunities to enhance our ability to monitor rapid changes of agricultural lands, providing timely, and low-cost, information for monitoring site-specific tillage assessment.

8.5.1 Challenges—Optical Systems

As noted previously, optical systems provide capabilities for monitoring tillage in a systematic manner. Yet they are subject to disruptive influences of soil moisture variations and uneven terrain. Possible solutions include (1) development of terrain and soil data layers that can guide interpretations of image data in such areas and (2) development of specialized indices or other strategies to detect or adjust for spectral variations caused by these effects. Further, although current systems can provide revisit intervals adequate in key agricultural regions, these capabilities may not be adequate in other regions, where higher cloud cover may require more frequent revisit capabilities to acquire cloud-free coverage necessary for the temporal sequences required for the minNDTI strategy.

From evaluation of the SINDRI and CAI tillage indices, we know that carefully, and narrowly, defined spectral channels are effective in tillage assessment. However, it seems unlikely that future satellite systems are likely to incur the costs of designing and operating new bands to support a single application mission. As a result, future opportunities for optical tillage assessment seem likely to be based on the NDTI model (which relies upon broadly defined spectral channels, but ones that support a range of application missions), relying upon the sequential imagery to apply strategies, such as the minNDTI.

8.5.2 Challenges—SAR Systems

Although specific strategies for application of SAR for monitoring tillage status are still under development, it has great potential for systematic tillage assessment, in part because of its ability to acquire data in the presence of cloud cover and the potential to extract a suit of terrain measurements as part of a tillage assessment mission. As reported here, current research has been successful in applying radar fundamentals to the tillage assessment task, although the multiplicity of system variables that interact with each other and with the landscape offers challenges in isolating tillage information. The SAR tillage effort has yet to scale current findings to examine larger regions, allowing identification of unexpected effects of local terrain, interactions between agricultural practices, and the geometries of varied SAR satellite systems.

8.5.3 Challenges—Sequential Observations

Monitoring tillage status by remote sensing by its nature requires broadscale observation of very large regions. Within such broad regions, weather, terrain, and local practices vary, necessarily dispersing tillage and planting data operations over intervals of several weeks. Because the tillage event is ephemeral, soon concealed by the foliage of the emerging crop, it must be assessed as it occurs, not at a later date. As a result, a single *snapshot* satellite image can capture only a partial record of a region's tillage pattern. This effect is significant regardless of the sensor system or tillage assessment strategy—sequential imagery of the entire planting season is necessary to observe the correct tillage status of a landscape. Otherwise, the inventory will record only a portion of the tillage operations within the area. In this context, both SAR and optical satellite systems are challenged to provide reliable coverage in the sense that current revisit intervals of optical systems are subject to disruption by cloud cover, and current SAR systems are challenged to simultaneously provide the spatial detail, broadscale coverage, and revisit intervals necessary to observe the full planting season.

8.5.4 Challenges—Global Tillage Monitoring

Current research to apply remote sensing to tillage assessment has been developed largely in midlatitudes, in regions characterized by large fields, simple crop calendars that apply for very large areas, limited numbers of crops, known crop rotation sequences, and availability of supporting data. These conditions may apply in many of the other major grain-producing regions (e.g., Brazil, China, Argentina, Ukraine, and Mexico), where current tillage assessment strategies may transfer. Many of the world's other agricultural regions present much different conditions that do not favor their direct transfer. For irrigated crops, there may be several planting cycles. Many tropical regions are characterized by smaller fields and complicated crop calendars, so investigators may require mastery of detailed knowledge of a diversity of cropping systems and irrigation practices, which may all vary within short distances. Such agricultural systems may exhibit levels of spatial and temporal variability that will greatly complicate applications of remote-sensing strategies that have been successful in the context of midlatitude agricultural systems.

8.5.5 Challenges—Field and Validation Data

Further advances in tillage assessment will require development of additional strategies for collection of field data for preparation of assessment model and for valuation of survey findings. Field data collection campaigns following established and co-coordinated protocols have a role in broadscale survey, especially when it is feasible to mobilize a network or experienced volunteers to support campaigns. However, such efforts inevitably encounter logistical problems, especially when unfavorable weather creates uncertainties or prevents acquisition of viable imagery. Work to investigate alternative strategies, including the feasibility of using commercial satellite imagery to collect tillage observations to support model development and validation of project findings, deserves attention.

References

Adams, J. R., A. A. Berg, H. McNairn, and A. Merzouki. 2013a. Sensitivity of C-band SAR polarimetric variables to unvegetated agricultural fields. *Can. J. Remote Sens.* 39 (1): 1–16. doi: 10.5589/m13-003.

Adams, J. R., T. L. Rowlandson, S. J. McKeown, A. A. Berg, H. McNairn, and S. J. Sweeney. 2013b. Evaluating the Cloude–Pottier and Freeman–Durden scattering decompositions for distinguishing between unharvested and post-harvest agricultural fields. *Can. J. Remote Sens.* 39 (4): 318–327. doi: 10.5589/m13-040.

Alberga, V., G. Satalino, and D. K. Staykova. 2008. Comparison of polarimetric SAR observables in terms of classification performance. *Int. J. Remote Sens.* 29 (14): 4129–4150. doi: 10.1080/01431160701840182.

Asrar, G., R. B. Myneni, and E. T. Kanemasu. 1989. Estimation of plant-canopy attributes from spectral reflectance measurements. In *Theory and Applications of Optical Remote Sensing*, G. Asrar (ed.), pp. 252–296. New York: John Wiley.

Aubert, M., N. Baghdadi, M. Zribi, A. Douaoui, C. Loumagne, F. Baup, M. El Hajj, and S. Garrigues. 2011. Analysis of TerraSAR-X data sensitivity to bare soil moisture, roughness, composition and soil crust. *Remote Sens. Environ.* 115 (8): 1801–1810. doi: http://dx.doi.org/10.1016/j.rse.2011.02.021.

Baghdadi, N., I. Gherboudj, M. Zribi, M. Sahebi, C. King, and F. Bonn. 2004. Semi-empirical calibration of the IEM backscattering model using radar images and moisture and roughness field measurements. *Int. J. Remote Sens.* 25 (18): 3593–3623. doi: 10.1080/01431160310001654392.

Baghdadi, N., N. Holah, and M. Zribi. 2006. Calibration of the integral equation model for SAR data in C-band and HH and VV polarizations. *Int. J. Remote Sens.* 27 (4): 805–816. doi: 10.1080/01431160500212278.

Baghdadi, N., M. Zribi, C. Loumagne, P. Ansart, and T. P. Anguela. 2008. Analysis of TerraSAR-X data and their sensitivity to soil surface parameters over bare agricultural fields. *Remote Sens. Environ.* 112 (12): 4370–4378. doi: 10.1016/j.rse.2008.08.004.

Baumgardner, M. F., L. F. Silva, L. L. Biehl, and E. R. Stoner. 1986. Reflectance properties of soils. In *Advances in Agronomy*, N. C. Brady (ed.), pp. 1–44. Academic Press, Orlando, FL.

Beaudoin, A., T. Le Toan, and Q. H. J. Gwyn. 1990. SAR observations and modeling of the C-band backscatter variability due to multiscale geometry and soil moisture. *IEEE Trans. Geosci. Remote Sens.* 28 (5): 886–895. doi: 10.1109/36.58978.

Ben-Dor, E. 2002. Quantitative remote sensing of soil properties. *Adv. Agron.*, 75:173–243.

Biard, F. and F. Baret. 1997. Crop residue estimation using multiband reflectance. *Remote Sens. Environ.* 59 (3): 530–536.

Boisvert, J. B., Q. H. J. Gwyn, B. Brisco, D. Major, and R. J. Brown. 1995. Evaluation of soil moisture techniques and microwave penetration depth for radar applications. *Can. J. Remote Sens.* 21: 110–123.

Booth, D. T., S. E. Cox, and R. D. Berryman. 2006. Point sampling digital imagery with 'SamplePoint'. *Environ. Monit. Assess.* 123: 97–108.

Bricklemyer, R. S., R. L. Lawrence, P. R. Miller, and N. Battogtokh. 2006. Predicting tillage practices and agricultural soil disturbance in north central Montana with Landsat imagery. *Agric. Ecosyst. Environ.* 114 (2–4): 210–216. doi: 10.1016/j.agee.2005.10.005.

Brisco, B., R. J. Brown, B. Snider, G. J. Sofko, J. A. Koehler, and A. G. Wacker. 1991. Tillage effects on the radar backscattering coefficient of grain stubble fields. *Int. J. Remote Sens.* 12 (11): 2283–2298. doi: 10.1080/01431169108955258.

Brown, D. J., K. D. Shepherd, M. G. Walsh, M. Dewayne Mays, and T. G. Reinsch. 2006. Global soil characterization with VNIR diffuse reflectance spectroscopy. *Geoderma* 132 (3–4): 273–290. doi: 10.1016/j.geoderma.2005.04.025.

Clark, R. N., G. A. Swayze, R. Wise, K. E. Livo, T. M. Hoefen, R. F. Kokaly, and S. J. Sutley. 2007. USGS digital spectral library splib06a. Available online at: http://speclab.cr.usgs.gov/spectral.lib06/ (Accessed January 11, 2013).

Cloude, S. R. and E. Pottier. 1997. An entropy based classification scheme for land applications of polarimetric SAR. *IEEE Trans. Geosci. Remote Sens.* 35 (1): 68–78. doi: 10.1109/36.551935.

Corak, S. J., T. C. Kaspar, and D. W. Meek. 1993. Evaluating methods for measuring residue cover. *J. Soil Water Conserv.* 48: 700–704.

CTIC. 2014. National crop residue management survey. http://www.ctic.purdue.edu/CRM/ (accessed January 21, 2014).

DigitalGlobe, 2014. Content collection. Available online at: https://www.digitalglobe.com/about-us/content-collection#satellites&worldview-3. (Accessed April 8, 2015).

Daughtry, C. S. T. 2001. Discriminating crop residues from soil by shortwave infrared reflectance. *Agron. J.* 93 (1): 125–131.

Daughtry, C. S. T., P. C. Doraiswamy, E. R. Hunt, Jr., A. J. Stern, J. E. McMurtrey, III, and J. H. Prueger. 2006. Remote sensing of crop residue cover and soil tillage intensity. *Soil Tillage Res.* 91 (1–2): 101–108.

Daughtry, C. S. T. and E. R. Hunt, Jr. 2008. Mitigating the effects of soil and residue water contents on remotely sensed estimates of crop residue cover. *Remote Sens. Environ.* 112 (4): 1647–1657. doi: 10.1016/j.rse.2007.08.006.

Daughtry, C. S. T., E. R. Hunt, Jr., P. C. Doraiswamy, and J. E. McMurtrey, III. 2005. Remote sensing the spatial distribution of crop residues. *Agron. J.* 97 (3): 864–871. doi: 10.2134/agronj2003.0291.

Daughtry, C. S. T., E. R. Hunt, Jr., and J. E. McMurtrey, III. 2004. Assessing crop residue cover using shortwave infrared reflectance. *Remote Sens. Environ.* 90 (1): 126–134. doi: 10.1016/j.rse.2003.10.023.

Daughtry, C. S. T., G. Serbin, J. B. Reeves, III, P. C. Doraiswamy, and E. R. Hunt, Jr. 2010. Spectral reflectance of wheat residue during decomposition and remotely sensed estimates of residue cover. *Remote Sens.* 2 (2): 416–431.

de Matthaeis, P., P. Ferrazzoli, G. Schiavon, and D. Solimini. 1991. Radar response to vegetation parameters: Comparison between theory and MAESTRO-1 results. Paper presented at the *Proceedings of the International Geoscience and Remote Sensing Symposium (IGARSS 1991)*, Espoo, Finland.

de Matthaeis, P., P. Ferrazzoli, G. Schiavon, and D. Solimini. 1992. Agriscatt and MAESTRO: Multifrequency radar experiments for vegetation remote sensing. Paper presented at the *MAESTRO-1/AGRISCATT: Radar Techniques for Forestry and Agricultural Applications*, Final Workshop, Noordwijk, the Netherlands.

Dubois, P. C., J. Van Zyl, and T. Engman. 1995. Measuring soil moisture with imaging radars. *IEEE Trans. Geosci. Remote Sens.* 33 (4): 915–926. doi: 10.1109/36.406677.

Evans, D. L., T. G. Farr, J. J. van Zyl, and H. A. Zebker. 1988. Radar polarimetry: Analysis tools and applications. *IEEE Trans. Geosci. Remote Sens.* 26 (6): 774–788. doi: 10.1109/36.7708.

Evans, D. L. and M. O. Smith. 1991. Separation of vegetation and rock signatures in thematic mapper and polarimetric SAR images. *Remote Sens. Environ.* 37 (1): 63–75. doi: 10.1016/0034-4257(91)90051-7.

Faulkner, E. H. 1943. *Plowman's Folly*. New York: Grosset & Dunlap.

Feng, G., Masek, J., Schwaller, M., and Hall, F. 2006. On the blending of the Landsat and MODIS surface reflectance: Predicting daily Landsat surface reflectance. *Geoscience and Remote Sensing, IEEE Transactions on*, 44 (8), 2207–2218.

Freeman, A. and S. L. Durden. 1998. A three-component scattering model for polarimetric SAR data. *IEEE Trans. Geosci. Remote Sens.* 36 (3): 963–973. doi: 10.1109/36.673687.

Fung, A. K., Z. Li, and K. S. Chen. 1992. Backscattering from a randomly rough dielectric surface. *IEEE Trans. Geosci. Remote Sens.* 30 (2): 356–368. doi: 10.1109/36.134085.

Galloza, M. S., M. M. Crawford, and G. C. Heathman. 2013. Crop residue modeling and mapping using landsat, ALI, hyperion and airborne remote sensing data. *IEEE J. Sel. Topics Appl. Earth Observ.* 6 (2): 446–456. doi: 10.1109/jstars.2012.2222355.

Gao, F., J. Masek, M. Schwaller, and F. Hall. 2006. On the Blending of the Landsat and MODIS Surface reflectance: Predict daily landsat surface reflectance, *IEEE Transactions on Geoscience and Remote Sensing*, 44 (8): 2207–2218.

Gausman, H. W., Gerbermann, A. H., Wiegand, C. L., Leamer, R. W., Rodriguez, R. R., and Noriega, J. R. (1975). Reflectance differences between crop residues and bare soils. *Soil Sci. Soc. Am. J.* 39 (4): 752–755.

Gowda, P. H., B. J. Dalzell, D. J. Mulla, and F. Kollman. 2001. Mapping tillage practices with landstat thematic mapper based logistic regression models. *J. Soil Water Conserv.* 56 (2): 91–96.

Hadria, R., B. Duchemin, F. Baup, T. Le Toan, A. Bouvet, G. Dedieu, and M. Le Page. 2009. Combined use of optical and radar satellite data for the detection of tillage and irrigation operations: Case study in Central Morocco. *Agric. Water Manage.* 96 (7): 1120–1127. doi: 10.1016/j.agwat.2009.02.010.

Hajnsek, I., E. Pottier, and S. R. Cloude. 2003. Inversion of surface parameters from polarimetric SAR. *IEEE Trans. Geosci. Remote Sens.* 41 (4): 727–744. doi: 10.1109/tgrs.2003.810702.

Hasted, J. B. 1973. *Aqueous Dielectrics*. London, U.K.: Chapman & Hall.

Henderson, T. L., M. F. Baumgardner, D. P. Franzmeier, D. E. Stott, and D. C. Coster. 1992. High dimensional reflectance analysis of soil organic matter. *Soil Sci. Soc. Am. J.* 56 (3): 865–872. doi: 10.2136/sssaj1992.03615995005600030031x.

Hunt, Jr., E. R. 1989. Detection of changes in leaf water content using near- and middle-infrared reflectances. *Remote Sens. Environ.* 30 (1): 43–54. doi: 10.1016/0034-4257(89)90046-1.

Irons, J. R., R. A. Weismiller, and G. W. Petersen. 1989. Soil reflectance. In *Theory and Applications of Optical Remote Sensing*, G. Asrar (ed.). New York: John Wiley & Sons, pp. 66–106.

Jackson, T. J., H. McNairn, M. A. Weltz, B. Brisco, and R. Brown. 1997. First order surface roughness correction of active microwave observations for estimating soil moisture. *IEEE Trans. Geosci. Remote Sens.* 35 (4): 1065–1068. doi: 10.1109/36.602548.

Jackson, T. J. and P. E. O'Neill. 1991. Microwave emission and crop residues. *Remote Sens. Environ.* 36 (2): 129–136. doi: 10.1016/0034-4257(91)90035-5.

Karlen, D. L., N. C. Wollenhaupt, D. C. Erbach, E. C. Berry, J. B. Swan, N. S. Eash, and J. L. Jordahl. 1994. Long-term tillage effects on soil quality. *Soil Tillage Res.* 32 (4): 313–327. doi:10.1016/0167-1987(94)00427-G.

Laflen, J. M., M. Amemiya, and E. A. Hintz. 1981. Measuring crop residue cover. *J. Soil Water Conserv.* 36 (6): 341–343.

Lal, R. 2004. Soil carbon sequestration to mitigate climate change. *Geoderma* 123 (1–2): 1–22. doi:10.1016/j.geoderma.2004.01.032.

Lal, R., J. M. Kimble, R. F. Follett, and C. V. Cole. 1998. *The Potential of U.S. Cropland to Sequester Carbon and Mitigate the Greenhouse Effect*. Boca Raton, FL: Lewis Publishers.

Langman, A. and M. R. Inggs. 1994. The use of polarimetry in subsurface radar. Paper presented at the *Proceedings in International Geoscience and Remote Sensing Symposium, 1994 (IGARSS '94). Surface and Atmospheric Remote Sensing: Technologies, Data Analysis and Interpretation*, August 8–12, 1994.

Lee, J. S. and E. Pottier. 2009. *Polarimetric Radar Imaging: From Basics to Applications*. New York: CRC Press.

Lewis, A. J. and F. M. Henderson. 1998. Radar fundamentals: The geoscience perspective. In *Principles and Applications of Imaging Radar. Manual of Remote Sensing*, F. M. Henderson and A. J. Lewis (eds.). New York: John Wiley & Sons, pp. 131–181.

Li, Y., T. Liu, G. Lampropoulos, H. McNairn, J. Shang, and R. Touzi. 2014. RADARSAT-2 POLInSAR coherence optimization for agriculture crop change detection. Paper presented at the *Proceedings in International Geoscience and Remote Sensing Symposium (IGARSS 2014)*, Quebec City, Quebec, Canada.

McNairn, H., J. B. Boisvert, D. J. Major, Q. H. J. Gwyn, R. J. Brown, and A. M. Smith. 1996. Identification of agricultural tillage practices from C-band radar backscatter. *Can. J. Remote Sens.* 22: 154–162.

McNairn, H. and B. Brisco. 2004. The application of C-band polarimetric SAR for agriculture: A review. *Can. J. Remote Sens.* 30 (3): 525–542. doi: 10.5589/m03-068.

McNairn, H., C. Duguay, J. Boisvert, E. Huffman, and B. Brisco. 2001. Defining the sensitivity of multi-frequency and multi-polarized radar backscatter to post-harvest crop residue. *Can. J. Remote Sens.* 27 (3): 247–263.

McNairn, H., C. Duguay, B. Brisco, and T. J. Pultz. 2002. The effect of soil and crop residue characteristics on polarimetric radar response. *Remote Sens. Environ.* 80 (2): 308–320. doi: 10.1016/s0034-4257(01)00312-1.

McNairn, H. and R. Protz. 1993. Mapping corn residue cover on agricultural fields in Oxford County, Ontario, using Thematic Mapper. *Can. J. Remote Sens.* 19: 152–158.

McNairn, H., D. Wood, Q. H. J. Gwyn, and R. J. Brown. 1998. Mapping tillage and crop residue management practices with RADARSAT. *Can. J. Remote Sens.* 24: 28–35.

Merzouki, A., H. McNairn, and A. Pacheco. 2010. Evaluation of the Dubois, Oh, and IEM radar backscatter models over agricultural fields using C-band RADARSAT-2 SAR image data. *Can. J. Remote Sens.* 36 (S2): S274–S286. doi: 10.5589/m10-055.

Merzouki, A., H. McNairn, and A. Pacheco. 2011. Mapping soil moisture using RADARSAT-2 data and local autocorrelation statistics. *IEEE J. Sel. Topics Appl. Earth Observ.* 4 (1): 1–10.

Milisavljević, N., D. Closson, and I. Bloch. 2010. Detecting human-induced scene changes using coherent change detection in SAR images. Paper presented at the *ISPRS TC VII Symposium—100 Years ISPRS*, Vienna, Austria, July 5–7.

Morrison, J. E., C.-H. Huang, D. T. Lightle, and C. S. T. Daughtry. 1993. Residue measurement techniques. *J. Soil Water Conserv.* 48 (6): 478–483.

Nagler, P. L., C. S. T. Daughtry, and S. N. Goward. 2000. Plant litter and soil reflectance. *Remote Sens. Environ.* 71 (2): 207–215. doi:10.1016/S0034-4257(99)00082-6.

NASA/JPL. 2008. ASTER user advisory. http://asterweb.jpl.nasa.gov/swir-alert.asp (accessed January 15, 2014).

Oh, Y. 2004. Quantitative retrieval of soil moisture content and surface roughness from multipolarized radar observations of bare soil surfaces. *IEEE Trans. Geosci. Remote Sens.* 42 (3): 596–601. doi: 10.1109/tgrs.2003.821065.

Oh, Y., K. Sarabandi, and F. T. Ulaby. 1992. An empirical model and an inversion technique for radar scattering from bare soil surfaces. *IEEE Trans. Geosci. Remote Sens.* 30 (2): 370–381. doi: 10.1109/36.134086.

Pacheco, A. and H. McNairn. 2010. Evaluating multispectral remote sensing and spectral unmixing analysis for crop residue mapping. *Remote Sens. Environ.* 114 (10): 2219–2228. doi:10.1016/j.rse.2010.04.024.

Pacheco, A. M., H. McNairn, and A. Merzouki. 2010. Evaluating TerraSAR-X for the identification of tillage occurrence over an agricultural area in Canada. Proc. SPIE 7824, Remote Sensing for Agriculture, Ecosystems, and Hydrology XII, 78240P. doi: 10.1117/12.868218.

Panciera, R., F. MacGill, M. Tanase, K. Lowell, and J. Walker. 2013. Sensitivity of TerraSAR-X-band data to surface parameters in bare agricultural areas. Paper presented at the *IEEE 2013 International Geoscience and Remote Sensing Symposium* (*IGARSS*), July 21–26, 2013.

Rahman, M. M., M. S. Moran, D. P. Thoma, R. Bryant, C. D. Holifield Collins, T. Jackson, B. J. Orr, and M. Tischler. 2008. Mapping surface roughness and soil moisture using multi-angle radar imagery without ancillary data. *Remote Sens. Environ.* 112 (2): 391–402. doi:10.1016/j.rse.2006.10.026.

Rouse, J. W., Jr., R. H. Haas, J. A. Schell, and D. W. Deering. 1973. Monitoring the vernal advancement and retrogradation (green wave effect) of natural vegetation. Prog. Rep. RSC 1978-1, Remote Sensing Center, Texas A&M Univ., College Station, 93p. (NTIS No. E73-106393).

Sabins, F. F. 1986. *Remote Sensing: Principles and Interpretation.* San Francisco, CA: Freeman Publishers.

Samui, P., P. H. Gowda, T. Oommen, T. A. Howell, T. H. Marek, and D. O. Porter. 2012. Statistical learning algorithms for identifying contrasting tillage practices with Landsat Thematic Mapper data. *Int. J. Remote Sens.* 33 (18): 5732–5745. doi: 10.1080/01431161.2012.671555.

Serbin, G., C. S. T. Daughtry, E. R. Hunt, Jr., D. J. Brown, and G. W. McCarty. 2009a. Effect of soil spectral properties on remote sensing of crop residue cover. *Soil Sci. Soc. Am. J.* 73 (5): 1545–1558. doi: 10.2136/sssaj2008.0311.

Serbin, G., C. S. T. Daughtry, E. R. Hunt, Jr., J. B. Reeves III, and D. J. Brown. 2009b. Effects of soil composition and mineralogy on remote sensing of crop residue cover. *Remote Sens. Environ.* 113 (1): 224–238.

Serbin, G., E. R. Hunt, Jr., C. S. T. Daughtry, G. W. McCarty, and P. C. Doraiswamy. 2009c. An improved ASTER index for remote sensing of crop residue. *Remote Sens.* 1 (4): 971–991.

Serbin, G., E. R. Hunt, Jr., C. S. T. Daughtry, and G. W. McCarty. 2013. Assessment of spectral indices for cover estimation of senescent vegetation. *Remote Sens. Lett.* 4 (6): 552–560. doi: 10.1080/2150704x.2013.767479.

Shelton, D. P., R. Kanable, and P. L. Jasa. 1993. Estimating percent residue cover using the line-transect method. http://www.ianrpubs.unl.edu/live/g1931/build/g1931.pdf (assessed February 5, 2014).

Slaton, M. R., E. R. Hunt, Jr., and W. K. Smith. 2001. Estimating near-infrared leaf reflectance from leaf structural characteristics. *Am. J. Bot.* 88 (2): 278–284. doi: 10.2307/2657018.

Sokol, J., H. NcNairn, and T. J. Pultz. 2004. Case studies demonstrating the hydrological applications of C-band multipolarized and polarimetric SAR. *Can. J. Remote Sens.* 30 (3): 470–483. doi: 10.5589/m03-073.

South, S., J. Qi, and D. P. Lusch. 2004. Optimal classification methods for mapping agricultural tillage practices. *Remote Sens. Environ.* 91 (1): 90–97. doi: 10.1016/j.rse.2004.03.001.

Stoner, E. R. and M. F. Baumgardner. 1981. Characteristic variations in reflectance of surface soils. *Soil Sci. Soc. Am. J.* 45 (6): 1161–1165. doi: 10.2136/sssaj1981.03615995004500060031x.

Stoner, E. R., M. F. Baumgardner, R. A. Weismiller, L. L. Biehl, and B. F. Robinson. 1980. Extension of laboratory-measured soil spectra to field conditions. *Soil Sci. Soc. Am. J.* 44 (3): 572–574. doi: 10.2136/sssaj1980.03615995004400030028x.

Streck, N. A., D. Rundquist, and J. Connot. 2002. Estimating residueal wheat dry matter from remote sensing measurements. *Photogramm. Eng. Remote Sens.* 68: 1193–1201.

Sudheer, K. P., P. Gowda, I. Chaubey, and T. Howell. 2010. Artificial neural network approach for mapping contrasting tillage practices. *Remote Sens.* 2 (2): 579–590.

Thoma, D. P., S. C. Gupta, and M. E. Bauer. 2004. Evaluation of optical remote sensing models for crop residue cover assessment. *J. Soil Water Conserv.* 59 (5): 224–233.

Thomas, J. R. and H. W. Gausman. 1977. Leaf reflectance vs. leaf chlorophyll and carotenoid concentrations for eight crops. *Agron. J.* 69 (5): 799–802. doi: 10.2134/agronj1977.00021962006900050017x.

Touzi, R., S. Goze, T. Le Toan, A. Lopes, and E. Mougin. 1992. Polarimetric discriminators for SAR images. *IEEE Trans. Geosci. Remote Sens.* 30 (5): 973–980. doi: 10.1109/36.175332.

Ulaby, F. T., D. Held, M. C. Donson, K. C. McDonald, and T. B. A. Senior. 1987. Relating polaization phase difference of SAR signals to scene properties. *IEEE Trans. Geosci. Remote Sens.* GE-25 (1): 83–92. doi: 10.1109/tgrs.1987.289784.

Ulaby, F. T., R. K. Moore, and A. K. Fung. 1986. *Microwave Remote Sensing: Active and Passive*, Vol. III. Reading, MA: Addison-Wesley.

USDA-NRCS. 2014 The soil tillage intensity rating (STIR). Available on-line at: http://www.nrcs.usda.gov/Internet/FSE_DOCUMENTS/stelprdb1119754.pdf (accessed February 28, 2014).

van Deventer, A. P., A. D. Ward, P. H. Gowda, and J. G. Lyon. 1997. Using thematic mapper data to identify contrasting soil plains and tillage practices. *Photogramm. Eng. Remote Sens.* 63 (1): 87–93.

van Zyl, J. J. 1989. Unsupervised classification of scattering behavior using radar polarimetry data. *IEEE Trans. Geosci. Remote Sens.* 27 (1): 36–45. doi: 10.1109/36.20273.

Wander, M. M. and L. E. Drinkwater. 2000. Fostering soil stewardship through soil quality assessment. *Appl. Soil Ecol.* 15 (1): 61–73. doi: 10.1016/S0929-1393(00)00072-X.

Watts, J. D., R. L. Lawrence, P. R. Miller, and C. Montagne. 2009. Monitoring of cropland practices for carbon sequestration purposes in north central Montana by Landsat remote sensing. *Remote Sens. Environ.* 113 (9): 1843–1852. doi: 10.1016/j.rse.2009.04.015.

Watts, J. D., S. L. Powell, R. L. Lawrence, and T. Hilker. 2011. Improved classification of conservation tillage adoption using high temporal and synthetic satellite imagery. *Remote Sens. Environ.* 115 (1): 66–75. doi: 10.1016/j.rse.2010.08.005.

West, T. O. and G. Marland. 2002. A synthesis of carbon sequestration, carbon emissions, and net carbon flux in agriculture: Comparing tillage practices in the United States. *Agric. Ecosyst. Environ.* 91 (1–3): 217–232. doi: 10.1016/S0167-8809(01)00233-X.

Workman, J. J. and L. Weyer. 2008. *Practical Guide to Interpretive Near-Infrared Spectroscopy*. Boca Raton, FL: Taylor & Francis Group.

Yilmaz, M. T., E. R. Hunt, Jr., and T. J. Jackson. 2008. Remote sensing of vegetation water content from equivalent water thickness using satellite imagery. *Remote Sens. Environ.* 112 (5): 2514–2522. doi: 10.1016/j.rse.2007.11.014.

Zheng, B., J. B. Campbell, and K. M. de Beurs. 2012. Remote sensing of crop residue cover using multi-temporal Landsat imagery. *Remote Sens. Environ.* 117 (0): 177–183. doi: 10.1016/j.rse.2011.08.016.

Zheng, B., J. B. Campbell, G. Serbin, and C. S. T. Daughtry. 2013a. Multi-temporal remote sensing of crop residue cover and tillage practices: A validation of the minNDTI strategy in the United States. *J. Soil Water Conserv.* 68 (2): 120–131.

Zheng, B., J. B. Campbell, G. Serbin, and J. M. Galbraith. 2014. Remote sensing of crop residue and tillage practices: Present capabilities and future prospects. *Soil Tillage Res.* 138 (0): 26–34. doi: 10.1016/j.still.2013.12.008.

Zheng, B., J. B. Campbell, Y. Shao, and R. H. Wynne. 2013b. Broad-scale monitoring of tillage practices using sequential landsat imagery. *Soil Sci. Soc. Am. J.* 77 (5): 1755–1764. doi: 10.2136/sssaj2013.03.0108.

Zhu, X., J. Chen, F. Gao, X. Chen, and J. G. Masek. 2010. An enhanced spatial and temporal adaptive reflectance fusion model for complex heterogeneous regions. *Remote Sens. Environ.* 114 (11): 2610–2623. doi: 10.1016/j.rse.2010.05.032.

Hyperspectral Remote Sensing for Terrestrial Applications

Prasad S. Thenkabail
U.S. Geological Survey

Pardhasaradhi Teluguntla
U.S. Geological Survey
and
Bay Area Environmental
Research Institute

Murali Krishna Gumma
International Crops Research
Institute for the Semi Arid Tropics

Venkateswarlu
Dheeravath
United Nations World
Food Programme

Acronyms and Definitions

ASD	Analytical spectral devices
AISA	Airborne imaging spectrometer for applications
AVIRIS	Airborne visible/infrared imaging spectrometer sensor
CHRIS PROBA	Compact High Resolution Imaging Spectrometer Project for On-Board Autonomy, Belgian Satellite
DHVIs	Derivative hyperspectral vegetation indices (DHVIs)
DNs	Digital numbers
EnMAP	Environmental Mapping and Analysis Program, Genrman's hyperspectral satellite mission
EO-1	Earth Observing-1 satellite of NASA
GnyLi	A hyperspectral vegetation index involving 5 hyperspectral narrow bands developed by Martin Gnyp Leon, Fei Li, and Georg Bareth et al.
HICO	Hyperspectral Imager for Coastal Oceans sensor, NASA's Hyperspectral Imager for the Coastal Ocean (HREP-HICO)
HBSIs	Hyperspectral biomass and structural indices
HNBs	Hyperspectral narrow bands
HVIs	Hyperspectral vegetation indices (HVIs)
HyspIRI	Hyperspectral infrared imager, next-generation hyperspectral sensor by NASA
MBHVI	Multiple band hyperspectral vegetation indices
MNF	Minimum noise fraction
NASA	National Atmospheric and Space Administration

OHNBs	Optimum hyperspectral narrow bands
OMI	*Ozone Monitoring Instrument onboard Aura satellite*
PCA	Principal component analysis
PRISMA	*Hyperspectral* Precursor and Application Mission or PRecursore IperSpettrale della Missione Applicativa of Italy
SCIAMACHY	Scanning Imaging Absorption spectroMeter for Atmospheric CartograpHY, hyperspectral sensor onboard European Space Agencies (ESA's) ENVISAT
SMA	Spectral mixture analysis
SMT	Spectral matching techniques
SVM	Support vector machines
TBHVIs	Two-band hyperspectral vegetation indices
VNIR	Visible and nearinfrared (VNIR)
WSA	Whole spectral analysis

9.1 Introduction

Remote sensing data are considered hyperspectral when the data are gathered from numerous wavebands, contiguously over an entire range of the spectrum (e.g., 400–2500 nm). Goetz (1992) defines hyperspectral remote sensing as "The acquisition of images in hundreds of registered, contiguous spectral bands such that for each picture element of an image it is possible to derive a complete reflectance spectrum." However, Jensen (2004) defines hyperspectral remote sensing as "The simultaneous acquisition of images in many relatively narrow, contiguous and/or non contiguous spectral bands throughout the ultraviolet, visible, and infrared portions of the electromagnetic spectrum."

Overall, the three key factors in considering data to be hyperspectral are the following:

1. *Contiguity in data collection*: Data are collected contiguously over a spectral range (e.g., wavebands spread across 400–2500 nm).
2. *Number of wavebands*: The number of wavebands by itself does not make the data hyperspectral. For example, if there are numerous narrowbands in 400–700 nm wavelengths, but have only a few broadbands in 701–2500 nm, the data cannot be considered hyperspectral. However, even relatively broad bands of width, say, for example, 30 nm bandwidths spread equally across 400–2500 nm, for a total of ~70 bands, are considered hyperspectral due to contiguity.
3. *Bandwidths*: Often, hyperspectral data are collected in very narrow bandwidths of ~1 to ~10 nm, contiguously over the entire spectral range (e.g., 400–2500 nm). Such narrow bandwidths are required to get hyperspectral signatures. But one can have a combination of narrowbands and broadbands spread across the spectrum and meet the criterion for hyperspectral remote sensing.

In summary

Remote sensing data are called hyperspectral when the data are collected contiguously over a spectral range, preferably in narrow bandwidths and in reasonably high number of bands.

Such a definition will meet many requirements and expectations of hyperspectral data.

Hyperspectral remote sensing is also referred to as imaging spectroscopy since data for each pixel are acquired in numerous contiguous wavebands resulting in (1) 3d image cube and (2) hyperspectral signatures. The various forms and characteristics of hyperspectral data (imaging spectroscopy) are illustrated in Figures 9.1 through 9.7. The distinction between hyperspectral and multispectral is based on the narrowness and contiguous nature of the measurements, not the "number of bands" (Qi et al., 2012).

The overarching goal of this chapter is to provide an introduction to hyperspectral remote sensing, its characteristics, data mining approaches, and methods of analysis for terrestrial application. First, hyperspectral sensors from various platforms are

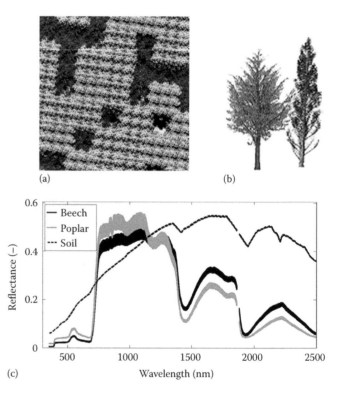

FIGURE 9.1 Tree spectra. Analytical Spectral Devices (ASD) FieldSpec JR spectroradiometer. Hyperspectral shape-based unmixing to improve intra- and interclass variabilities for forest and agro-ecosystem monitoring. A detail of a 30-by-30 m image pixel of the virtual forest consisting of two species with a different structure, with 10% of the trees removed to include gaps in the canopy (a). An example of a virtual tree for the two species, used to build up the forest, is shown in (b), while the spectral variability of the two species and the soil is given as well (c). (From Tits, L. et al., *ISPRS J. Photogramm. Remote Sens.*, 74, 163, 2012.)

noted. Second, data mining to overcome data redundancy is enumerated. Third, concept of Hughes's phenomenon and the need to overcome it are highlighted. Fourth, hyperspectral data analysis methods are presented and discussed. Methods section includes approaches to optimal band selection, deriving hyperspectral vegetation indices (HVIs) and various classification methods.

9.2 Hyperspectral Sensors

Hyperspectral data (or imaging spectroscopy) are gathered from various sensors. These are briefly discussed in the following text.

9.2.1 Spectroradiometers

The most common and widely used over last 50 years is hand-held or platform-mounted spectroradiometers. Typically, spectroradiometers gather hyperspectral data ~1 nm wide narrowbands over the entire spectral range (e.g., 400–13,500 nm). For example, Figure 9.1 illustrates the hyperspectral data gathered for Beech versus Poplar forests (Thomas, 2012; Tits et al., 2012; Zhang, 2012; Tanner, 2013) based on FieldSpec Pro FR spectroradiometer manufactured by Analytical Spectral Devices (ASD). Data are acquired over 400–2,500 nm at every 1 nm bandwidth. Gathering spectra at any given location involved optimizing the integration time (typically set at 17 ms), providing foreoptic information, recording dark current, collecting white reference reflectance, and then obtaining target reflectance at set field of view such as 18° (Thenkabail et al., 2004a). Data are either in radiance (W m^{-2} sr^{-1} μm^{-1}) or reflectance factor as shown in Figure 9.1 or in percentage.

9.2.2 Airborne Hyperspectral Remote Sensing

Airborne hyperspectral remote sensing platform is the next most common hyperspectral data, which has a history of over 30 years. The most common is the airborne visible/infrared

imaging spectrometer (AVIRIS) by NASA's Jet Propulsion Laboratory (JPL). As an imaging spectrometer, AVIRIS gathers data in 614-pixel swath, in 224 bands, over 400–2500 nm wavelength. The data can be constituted as image cube (e.g., Figure 9.2; [Guo et al., 2013]). Figure 9.2 shows hyperspectral imaging data gathered by AVIRIS over an agricultural area. The hyperspectral signatures of tilled versus untilled lands of corn and soybean farms as well as few other crops are illustrated by Guo et al. 2013 (Figure 9.2). Spectral reflectivity of notill corn fields is highest in the red (around 680 nm). In contrast, grass/pasture and woods are highest around 680 nm, and reflectivity is highest for these land covers in the near-infrared (NIR; 760–900 nm). The healthy grass/pasture and woods also absorb heavily around 960–970 nm range. There are many other unique features that can even be observed qualitatively by someone trained in imaging spectroscopy.

Another frequently used airborne hyperspectral imager is the Australian HyMap. It has 126 wavebands over 400–2500 nm. The data captured by HyMap are illustrated in Figure 9.3 (Andrew and Ustin, 2008). Typical characteristics of healthy vegetation for certain species is obvious as described earlier for wavelengths centered in red and NIR. In contrast, the soil and the litter have comparable spectra, with litter having higher reflectivity than soil in NIR and SWIR bands. Water absorbs heavily in NIR and SWIR, and hence the reflectivities are very low or zero (Figure 9.3).

9.2.3 Spaceborne Hyperspectral Data

In the year 2000, NASA launched the first civilian spaceborne hyperspectral imager called Hyperion onboard Earth Observing-1 (EO-1) satellite. Hyperion gathers data in 242 bands spread across 400–2500 nm. Each band is 10 nm wide. Of the original 242 Hyperion bands, 196 are unique and calibrated: bands 8 (427.55 nm) to 57 (925.85 nm) from the visible and near-infrared (VNIR) sensors, and bands 79 (932.72 nm) to 224

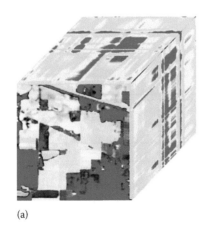
(a)

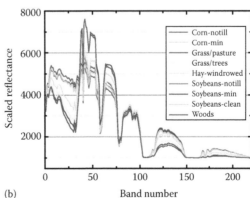
(b)

FIGURE 9.2 Corn-till. AVIRIS Indian Pines data set: (a) 3D hyperspectral cube and (b) the scaled reflectance plot. (From Guo, X. et al., *ISPRS J. Photogramm. Remote Sens.*, 83, 50, 2013.)

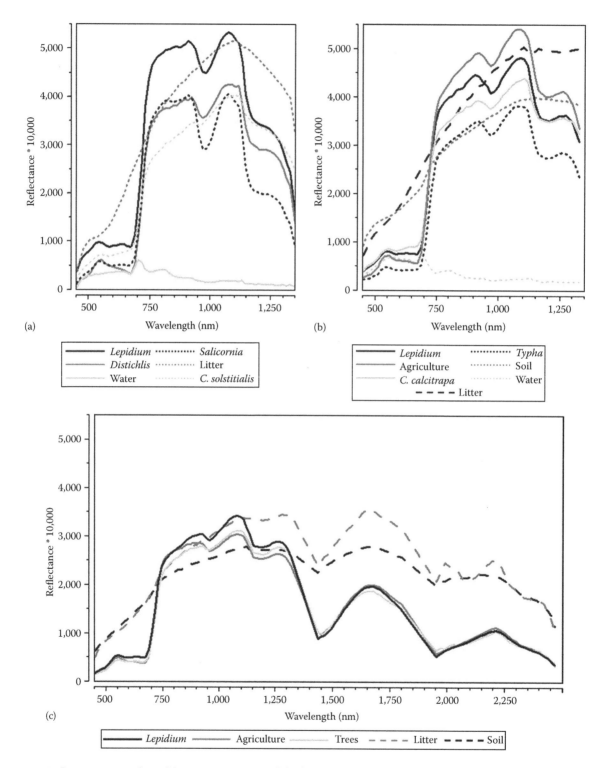

FIGURE 9.3 Reflectance spectra derived from HyMap imagery of the dominant species at (a) Rush Ranch, (b) Jepson Prairie, and (c) Consumes River Preserve. These spectra were used as training end members for the mixture-tuned matched filtering (MTMF). (From Andrew, M.E. and Ustin, S.L., *Remote Sens. Environ.,* 112, 4301, 2008.)

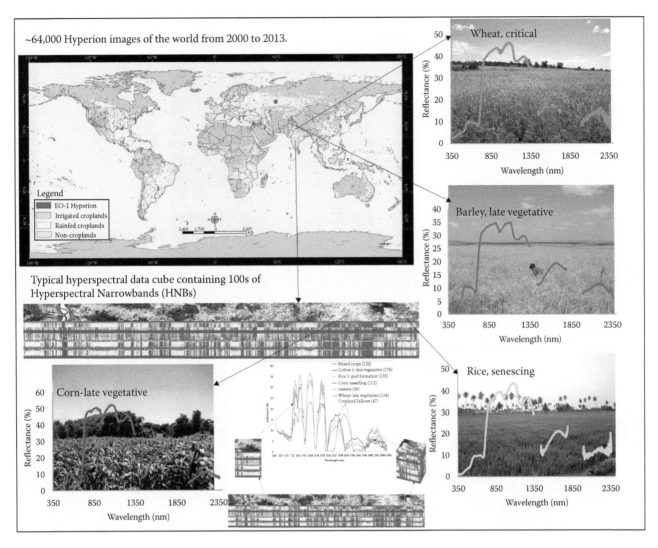

FIGURE 9.4 EO-1 Hyperion is the first spaceborne civilian hyperspectral sensor that was launched in year 2000 and has so far acquired ~64,000 images of the world (see the area covered by Hyperion images marked in red on global image). Each image is 7.5 km by 185 km, has 242 bands over 400–2500 nm. A single such image data cube is shown in the center with spectral signatures derived from the Hyperion sensor shown for few land cover themes. Typical ASD spectroradiometer gathered hyperspectral data of crops are shown in photos. The gaps in ASD hyperspectral data are in areas of atmospheric windows where data is too noisy and hence deleted. (Plotted using Data available from http://earthexplorer.usgs.gov/; http://eo1.gsfc.nasa.gov/.)

(2395.53 nm) from the SWIR sensors (Thenkabail et al., 2004b). The redundant and uncalibrated bands are in the spectral range: 357–417, 936–1068, and 852–923 nm. The 196 bands are further reduced to 157 bands after dropping bands in atmospheric windows: 1306–1437, 1790–1992, and 2365–2396 nm ranges, which show high noise level (Thenkabail et al., 2004b).

From year 2000 to 2014, Hyperion has acquired ~64,000 images spread across the world (Figure 9.4) that are now freely available from the U.S. Geological Survey's (USGS) EarthExplorer and Glovis portals. Each image is 7.5 km by 185 km with a pixel resolution of 30 m. The data cubes composed from these images allow us to derive hyperspectral signature banks of various land cover or cropland themes (e.g., Figure 9.4). Figure 9.5a illustrates two Hyperion images acquired over California as well as a number of hyperspectral signatures of major crops gathered using ASD field spectroradiometer.

9.2.4 Unmanned Aerial Vehicles

Hyperspectral sensors are increasingly carried onboard unmanned aerial vehicles (UAVs; Colomina and Molina, 2014). The UAVs are fast evolving as widely used remote sensing platform. A wide array of UAVs (e.g., Figure 9.5b) are currently used to carry hyperspectral sensors as well as many different types of sensors.

9.2.5 Multispectral versus Hyperspectral

Whereas multispectral broadband data-acquired from sensors such as the Landsat ETM+ only offer few possibilities, in contrast Hyperion offers many possibilities for visualizations and quantification of terrestrial earth features (e.g., Figure 9.6). In Figure 9.6, depiction of different false color composites (FCCs) of Hyperion (e.g., RGB: 843, 680, 547 nm; or RGB: 680, 547, 486 nm, and so on)

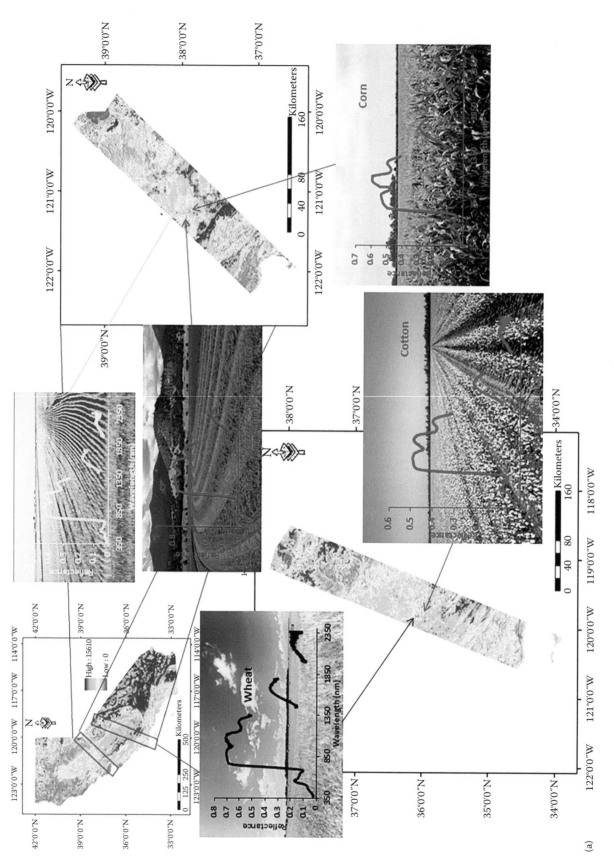

FIGURE 9.5 Hyperspectral spectral signatures of some of the major crops of California. The depicted spectral signatures are representative of the particular crops measured using ASD spectroradiometer. Two Hyperion images (each of 7.5 km-by-185 km) are also illustrated. (a) Microdrone MD4-1000 flying over the experimental crop. (From Torres-Sánchez, J. et al., *Comput. Electron. Agric.*, 103, 104, 2014.) *(Continued)*

(b)

FIGURE 9.5 (*Continued*) Hyperspectral spectral signatures of some of the major crops of California. The depicted spectral signatures are representative of the particular crops measured using ASD spectroradiometer. Two Hyperion images (each of 7.5 km-by-185 km) are also illustrated. (a) Microdrone MD4–1000 flying over the experimental crop.

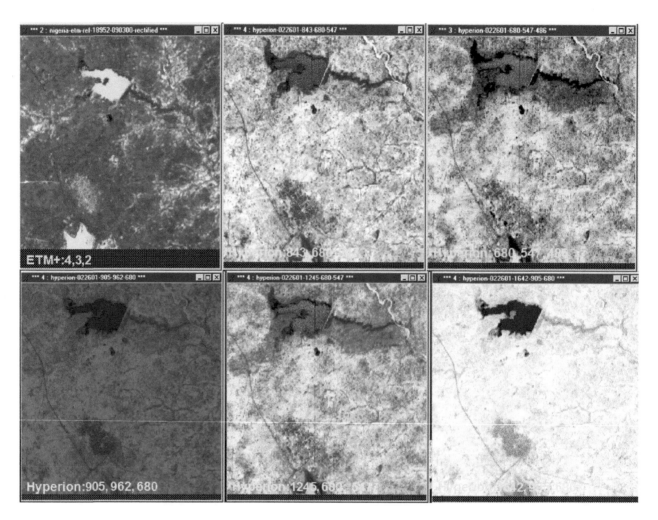

FIGURE 9.6 Hyperion images displayed in a number of different combinations of false color composites (FCCs) (e.g., wavebands centered at 843, 680, 547 nm, which are NIR, red, green as RGB FCC) and compared with classic RGB 4, 3, 2 (NIR, red, green) FCC combination of Landsat ETM+ data on top left. Unlike multispectral data, hyperspectral data offer numerous different opportunities to depict, quantify, and study the Planet Earth.

and comparison with FCC of Landsat ETM+ bands 4, 3, 2 clearly demonstrate, even by visual observation, the many possibilities that exist with Hyperion. For example, a seven-band Landsat will provide 21 unique indices ($7 \times 7 = 49$ indices – 7 indices on the diagonal of the matrix divided by 2 since the values above and below the matrix are transpose of each other). In contrast, 157-band clean Hyperion data (after reduced from original 242 bands by eliminating bands in atmospheric windows and uncalibrated bands) allow for 12,246 unique indices ($157 \times 157 = 24,640$ indices—157 indices on the diagonal of the matrix divided by 2 since the values above and below the matrix are the transpose of each other).

9.2.6 Hyperspectral Data: 3D Data Cube Visualization and Spectral Data Characterization

One quick way to visualize the hyperspectral data is to create 3D cubes as illustrated by an EO-1 Hyperion data in Figure 9.7. The 3D cube basically is a data layer stack of 242 bands over

400–2500 nm. Looking through this stack, when there is same color along the bands 1–242, it indicates less diversity in data. The spectral regions with significant diversity are in different color (e.g., red versus cyan in Figure 9.7). Hyperion digital numbers (DNs) are 16-bit radiances and are stored as 16-bit signed integer, which are then converted to radiances using a scaling factor provided in the header file, then to at-sensor reflectance, and finally to ground reflectance (see Thenkabail et al., 2004b). So, a click on any pixel will give reflectances in 242 bands, which is then plotted as hyperspectral signature (e.g., Figure 9.6) and analyzed quantitatively.

9.2.7 Past, Present, and Near-Future Spaceborne Hyperspectral Sensors

Hyperspectral sensors are of increasing interest to the remote sensing community given its their natural inherent advantages over multispectral sensors (Qi et al., 2012; Thenkabail et al., 2012a). As a result, we are seeing a number of spaceborne

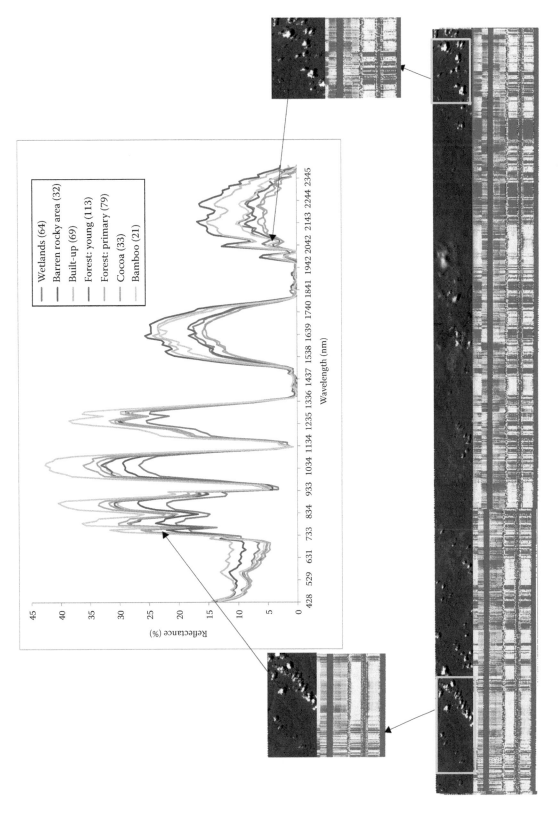

FIGURE 9.7 Hyperspectral signatures derived from Hyperion data cube for certain land cover themes. The numbers within brackets show sample sizes.

hyperspectral imagers for Ocean, Atmosphere, and Land (Table 9.1). These include (Table 9.1) NASA's Hyperion, HyspIRI, OMI, HICO, German's EnMap, Italy's PRISMA, ESA's SCIAMACHY, and CHRIS PROBA (Miura and Yoshioka, 2012; Ortenberg, 2012; Qi et al., 2012). There are also current initiatives from private industry in the commercial sector, like that from Boeing to launch hyperspectral sensors. The spatial, spectral, radiometric, and temporal characteristics of some of the key ocean, atmospheric, and land observation spaceborne hyperspectral data are provided in Table 9.1.

9.2.8 Data Normalization Hyperspectral Data

We illustrate the hyperspectral data normalization taking the case of Hyperion data. The DNs of the Hyperion level 1 products are 16-bit radiances and are stored as 16-bit signed integers. The DNs were converted to radiances (W m^{-2} sr^{-1} μm^{-1}) using an appropriate scaling (e.g., for a Hyperion image dated March 21, 2002, factor: 40 for visible and VNIR, and 80 for SWIR). However, users should check the header file of the image they work with to determine the exact scaling factor for their image.

Radiance (W m^{-2} sr^{-1} μm^{-1}) for VNIR bands = DN/40
Radiance (W m^{-2} sr^{-1} μm^{-1}) for SWIR bands = DN/80
Radiance to at-sensor top of atmosphere reflectance is then calculated using

$$Reflectance\ (\%) = n\frac{\pi L_\lambda d^2}{ESUN_\lambda \cos\theta_S}$$

where,

TOA reflectance (at-satellite exoatmospheric reflectance)
L_λ is the radiance (W m^{-2} sr^{-1} μm^{-1})
d is the earth-to-sun distance in astronomic units at the acquisition date (see Markham and Barker, 1987)
$ESUN_\lambda$ is the irradiance (W m^{-2} sr^{-1} μm^{-1}) or solar flux (Neckel and Labs, 1984)
θ_s is the solar Zenith angle

Note: θ_s is solar Zenith angle in degrees (i.e., 90° minus the sun elevation or sun angle when the scene was recorded as given in the image header file).

Atmospheric correction methods include (1) dark object subtraction technique (Chavez, 1988), (2) improved dark object subtraction technique (Chavez, 1989), (3) radiometric normalization technique: Bright and dark object regression (Elvidge et al., 1995), and (4) 6S model (Vermote et al. 2002). Readers with further interest in this topic are referred to Chapters 4 through 8 in *Remotely Sensed Data Characterization, Classification, and Accuracies* and Chander et al. (2009).

9.3 Data Mining and Data Redundancy of Hyperspectral Data

Data mining is one of the critical first steps in hyperspectral data analysis. The primary goal of data mining is to eliminate redundant data and retain only the useful data. Data volumes are reduced through data mining methods such as feature selection (e.g., principal component analysis (PCA), derivative analysis, and wavelets), lambda-by-lambda correlation plots (Thenkabail et al., 2000), minimum noise fraction (MNF) (Green et al., 1988; Boardman and Kruse, 1994), and HVIs (e.g., Thenkabail et al., 2014). Data mining methods lead to (Thenkabail et al., 2012b) (1) reduction in data dimensionality, (2) reduction in data redundancy, and (3) extraction of unique information.

It is a well-known fact that wavebands adjacent to one another (e.g., 680 nm versus 690 nm or 550 nm versus 560 nm) are often highly correlated for a given application. In various research papers, Thenkabail et al. (2000, 2004a,b, 2010, 2012b, 2014), Numata (2012), and Thenkabail and Wu (2012) showed that in a large stack of 242 bands in a Hyperion data, typically ~10% of the wavebands (~20 bands) are very useful in agricultural cropland or vegetation studies. It means for any one given application (e.g., agriculture), a large number of bands are likely to be redundant. So, the goal of the data mining is to identify and eliminate redundant bands. This will help eliminate unnecessary processing of redundant data, at the same time retaining the optimal power of hyperspectral data. This process is of great importance at a time when "big data" are the norm of the times.

However, eliminating redundant bands needs to be done with considerable care and expertise. What is redundant for one application (e.g., agriculture; [Yao et al., 2011]) may be critical for another application (e.g., geology).

Data mining requires merging of different disciplines such as digital imagery, pattern recognition, database, artificial intelligence, machine learning, algorithms, and statistics. There are various models of data mining. The generic concept of data mining is illustrated in Figure 9.8 (Lausch et al., 2014). Figure 9.9 (Lausch et al., 2014) shows data mining model applications for studies in soil clay content and soil organic content.

9.4 Hughes' Phenomenon and the Need for Data Mining

If the number of bands remains high, the number of observations required to train a classifier increases exponentially to maintain classification accuracies, which is called Hughes's phenomenon (Thenkabail et al. 2012a, b). For example, Thenkabail et al. (2004a,b) used 20 Hyperion bands to classify five crop types and achieve an accuracy of 90%. Relative to this, the seven-band Landsat data provided only an accuracy of 60% in classifying the same five crops. However, the number of observation points (e.g., ground data) to train and test the algorithms will be exponentially higher for the Hyperion data relative to Landsat data because larger numbers of bands are involved with Hyperion. So, one needs to weigh the higher classification accuracies achieved using greater number of bands versus the resources required to gather exponentially

TABLE 9.1 Characteristics of Spaceborne Hyperspectral Sensors (Either in Orbit or Planned for Launch) for Ocean, Atmosphere, Land, and Water Applications Compared with ASD Spectroradiometer[a]

Sensor, Satellite[c]	Spatial (m)	Spectral (#)	Swath (km)	Band Range (μm)	Band Widths (μm)	Irradiance (W m−Wsr−r μm−μ)	Data Points (# per ha)	Launch (Date)
1. Coastal hyperspectral spaceborne imagers								
a. HICO, ISS USA	90	128	42	353–1,080	5.7	See data in Neckel and Labs (1984). Plot it	0.81	2009–present
2. Atmosphere\ozone hyperspectral spaceborne imagers								
a. OMI, Aura USA	13,000 × 12,000	740	145	270–500	0.45–1	See data in Neckel and Labs (1984). Plot it	1/16,900	2004–present
b. SCIAMACHY, ENVISAT ESA	30,000 × 60,000	~2,000	960	212–2,384	0.2–1.5	See data in Neckel and Labs (1984). Plot it	1/180,000	2002–present
3. Land and water hyperspectral spaceborne imagers								
a. Hyperion, EO-1 USA	30	220 (196[b])	7.5	196 effective Calibrated bands VNIR (band 8–57) 427.55–925.85 nm SWIR (band 79–224) 932.72–2,395.53 nm	10 nm wide (approx.) for all 196 bands	See data in Neckel and Labs (1984). Plot it and obtain values for Hyperion bands	11.1	2000–present
b. CHRIS, PROBA ESA	25	19	17.5	200–1,050	1.25–11	Same as above	16	2001–present
c. HyspIRI VSWIR USA	60	210	145	210 bands in 380–2,500 nm	10 nm wide (approx.) for all 210 bands	See data in Neckel and Labs (1984). Plot it	2.77	2020+
d. HyspIRI TIR USA	60	8	145	7 bands in 7,500–12,000 nm and 1 band in 3,000–5,000 nm (3,980 nm center)	7 bands in 7,500–12,000 nm	See data in Neckel and Labs (1984). Plot it	2.77	2020+
e. EnMAP Germany	30	92 / 108	30	420–1,030 / 950–2,450	5–10 / 10–20	Same as above	11.1	2015+
f. PRISMA Italy	30	250	30	400–2,500	<10	Same as above	11.1	2014+
4. Land and Water Hand-held spectroradiometer								
a. ASD spectroradiometer	1,134 cm² @ 1.2 m Nadir view 18° Field of view	2,100 bands 1 nm width between 400 and 2,500 nm	N/A	2,100 effective bands	1 nm wide (approx.) in 400–2,500 nm	See data in Neckel and Labs (1984). Plot it and obtain values for Hyperion bands	88,183	Last 30+ years

Sources: Thenkabail, P.S. et al., 2012; Thenkabail, P.S. et al., *Photogramm. Eng. Remote Sens.*, 80, 697, 2014; Qi, J. et al., Hyperspectral sensor systems and data characteristics in global change studies, Chapter 3, in Thenkabail, P.S., Lyon, G.J., and Huete, A. 2012, *Hyperspectral Remote Sensing of Vegetation*, CRC Press/Taylor & Francis Group, Boca Raton, FL, London, FL, New York, 2011, pp. 69–92.

[a] Information for the table modified and adopted from Thenkabail et al., 2011, Thenkabail et al., 2014, and Qi et al., 2014.

[b] Of the 242 bands, 196 are unique and calibrated. These are: (1) Band 8 (427.55 nm) to band 57 (925.85 nm) that are acquired by visible and near-infrared (VNIR) sensor; and (2) Band 79 (932.72 nm) to band 224 (2395.85 nm) that are acquired by short wave infrared (SWIR) sensor

[c] HICO, **Hyperspectral** Imager for the Coastal Ocean onboard International Space Station; OMI, Ozone Monitoring Instrument onboard AURA of NASA; SCIAMACHY, Scanning Imaging Absorption Spectrometer for Atmospheric CHartographY of ESA; Hyperion EO-1, hyperspectral sensor onboard; EO-1, Earth observing 1; CHRIS PROBA, Compact High Resolution Imaging Spectrometer Project for On Board Autonomy satellite of ESA; HyspIRI VSWIR, Hyperspectral Infrared Imager Visible to Short Wavelength InfraRed of NASA; HyspIRI TIR, Hyperspectral Infrared Imager thermal infrared of NASA; Environmental Mapping and Analysis Program of Germany; PRISMA, PRecursore IperSpettrale della Missione Applicativa of Italy.

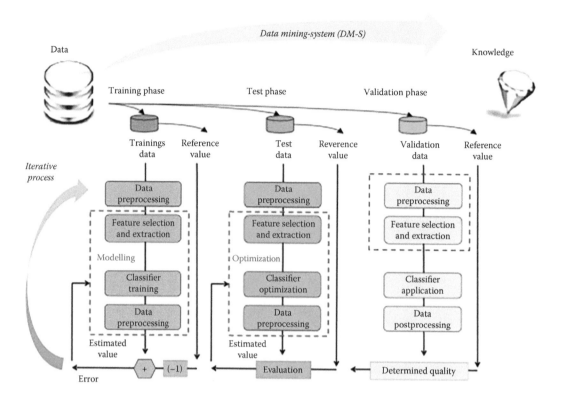

FIGURE 9.8 Data mining 1. Data mining and linked open data—New perspectives for data analysis in environmental research. Data mining process with the data mining system (DM-S) in the phases: (1) training phase, (2) test phase, and (3) validation phase. The data mining process works in a comparable way in all types of data mining types like text mining or web mining (changed according to Fayyad et al., (1996) and Tanner (2013). (From Lausch, A. et al., *Ecol. Model.*, 2014.)

higher number of observation (e.g., ground data) required to train and test the algorithms. So, higher accuracy by as much as 30% using 20 hyperspectral narrowbands (HNBs) when compared with seven-band Landsat will justify the greater number of ground data required. However, beyond 20 bands, increase in accuracy per increase in wavebands becomes asymptotic (e.g., Thenkabail et al., 2004a,b, 2012b). These studies, for example, show that when 40 Hyperion bands were used, the classification accuracies increased only by another 5% (from 90% with 20 bands to 95% with 40 bands). Here using 20 additional Hyperion bands (from 20 to 40) cannot be justified since the ground observation needed to train and test the algorithm will also increase exponential for 40 bands relative to 20. So, the key aim is to balance the higher classification accuracies with an optimal number of bands such as 20 instead too few or too many (e.g., 7 or 40). By doing so, we achieve a number of goals:

1. Increased classification accuracies with optimal number of bands.
2. Significantly reduced data redundancies with optimal number of bands.
3. Overcoming Hughes's phenomenon by using optimal number of bands (e.g., 20) in which observation data (ground data) to train and test the algorithms will be kept to reasonable levels.

9.5 Methods of Hyperspectral Data Analysis

Hyperspectral data analysis methods are broadly grouped under two categories (Bajwa and Kulkarni, 2012):

1. Feature extraction methods
2. Information extraction methods

Under each of the earlier two categories, specific unsupervised and supervised classification approaches exist (Figure 9.10) (Bajwa and Kulkarni, 2012; Plaza et al., 2012). Methods of classifying vegetation classes or crop types or vegetation species using HNBs are discussed extensively in this chapter and include unsupervised classification, supervised approaches, spectral angle mapper (SAM), artificial neural networks, and support vector machines (SVMs), multivariate or partial least square regressions (PLSR), and discriminant analysis (Thenkabail et al., 2012a).

Fundamental philosophies of hyperspectral data analysis involve two approaches:

1. Optimal hyperspectral narrowbands (OHNBs) where only a selective number of nonredundant bands are used (e.g., ~20 off Hyperion OHNBs are used).
2. Whole spectral analysis (WSA) where all the bands in the continuum (e.g., all 242 Hyperion bands in 400–2500 nm) are used.

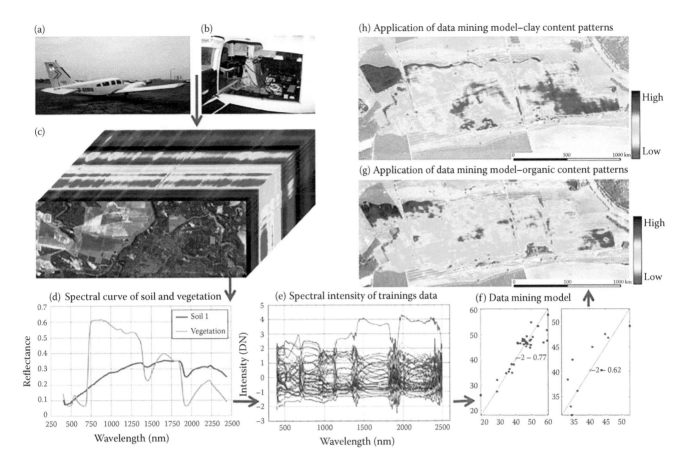

FIGURE 9.9 Data mining 2. (a, b) Data mining and linked open data—New perspectives for data analysis in environmental research. Airborne hyperspectral AISA-Eagle/HAWK remote sensor mounted on Piper, (c) CIR-image from hyperspectral sensors of the AISA-EAGLE/HAWK (AISA-DUAL) 400–2500 nm with data cube, 367 spectral bands with 2 m recorded ground resolution, date of recording Mai 2012 with a Piper, Region Schäfertal—Bode Catchment, (d) Spectral curve of ground truth sampling points for soil and vegetation in the test site. (e) Spectral intensity curves of imaging hyperspectral data, (f) data mining model, (g) application of the best data mining model on airborne hyperspectral image data for quantification and recognition of organic content patterns, and (h) pattern of clay content. (From Lausch, A. et al., *Ecol. Model.*, 2014.)

9.6 Optimal Hyperspectral Narrowbands

Determining wavebands that are optimal for different studies requires a thorough study of these subjects. For example, the importance of the wavebands for different studies such as vegetation, geology, and water are all different. So, determining optimal OHNBs requires subject knowledge and considerable experience working with hyperspectral data. Based on the synthesis of the extensive studies conducted by Thenkabail et al. (2000, 2002, 2004a,b, 2012, 2013, 2014), the OHNBs for agriculture and vegetation studies are established and presented in Table 9.2. Each of these HNBs is identified for their importance in studying one or more of vegetation and crop biophysical and biochemical characteristics. Most of these bands are also very distinct from one another; so none of them are redundant. Using some combination of these bands will help better quantify the biophysical and biochemical characteristics of vegetation and agricultural crops (Alchanatis and Cohen, 2012; Pu, 2012). In the following sections and subsections, we will demonstrate how these HNBs are used in classifying, modeling, and mapping agricultural croplands and other vegetation.

Table 9.2 shows that over 400–2500 nm range of the spectrum, there are 28 bands (e.g., ~12% of the 242 Hyperion bands in 400–2500 nm range) that are optimal in the study of agriculture and vegetation. However, the redundant bands here (i.e., agriculture and vegetation applications) may be very useful in other applications such as geology (Ben-Dor, 2012). For example, the critical absorption bands for studying minerals like biotite, kaolinite, hematite, and others are shown in Table 9.3. Unlike the vegetation and cropland bands, the HNBs required for mineralogy are quite different (Vaughan et al., 2011; Slonecker, 2012).

The earlier fact clearly establishes the need to determine OHNBs that are application specific.

9.7 Hyperspectral Vegetation Indices

One of the most common, powerful, and useful form of feature selection methods for hyperspectral data is based on the calculation of HVIs (Clark, 2012; Colombo et al., 2012; Galvão, 2012;

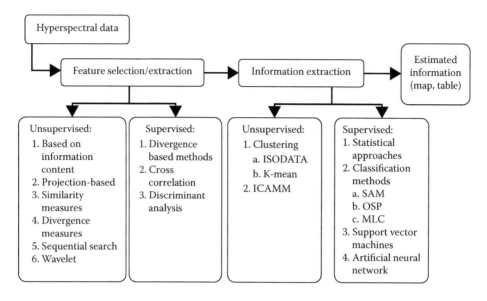

FIGURE 9.10 Hyperspectral data analysis methods. (From Bajwa, S. and Kulkarni, S.S., Hyperspectral data mining, Chapter 4, in Thenkabail, P.S., Lyon, G.J., and Huete, A., *Hyperspectral Remote Sensing of Vegetation*, CRC Press/Taylor & Francis Group, Boca Raton, FL/London, U.K./New York, 2012, pp. 93–120.)

Gitelson, 2012a,b; Roberts, 2012). The HVIs achieve two important goals of hyperspectral data analysis:

1. Compute many specific targeted HVIs to help model biophysical and biochemical quantities.
2. Reduce the data volume (mine the data) to eliminate all redundant bands for a given application.

There are several approaches to deriving HVIs. These are briefly presented and discussed.

9.7.1 Two-Band Hyperspectral Vegetation Indices

The two-band hyperspectral vegetation indices (TBHVIs) are defined as follows (Thenkabail et al., 2000):

$$TBHVI_{ij} = \frac{(R_j - R_i)}{(R_j + R_i)} \qquad (9.1)$$

where, i, j = 1 … N, with N = number of narrowbands. Hyperion 242 bands offer the possibility of 29,161 unique indices (242 * 242 = 58,564 − 242 = 58,322 divided by 2 resulting in C_{242}^2= 29,161; −242 because the values on the diagonal of matrix of 242 * 242 are unity, divided by 2 because the values above the diagonal of the matrix and below the diagonal of matrix are transpose of one another). However, as defined in Section 9.2.3, only 157 of the 242 Hyperion bands are useful after removing the wavebands in the atmospheric windows and those that are uncalibrated. This will still leave C_{157}^2= 12,246 unique TBHVIs.

Any one of the crop biophysical or biochemical quantity (e.g., biomass, leaf area index, nitrogen) is correlated with each one of the 12,246 TBHVIs (Stroppiana et al., 2012; Zhu et al., 2012). This will result for each crop variable (e.g., biomass) a total of 12,246 unique models, each providing an R-square. Figure 9.11 shows the contour plot of 12,246 R-square values plotted for (1) rice crop wet biomass with TBHVIs (Figure 9.11; above the diagonal) and (2) barley crop wet biomass with TBHVIs (Figure 9.11, below the diagonal). The areas with "bull's-eye" are regions of rich information having high R-square values, whereas the areas in gray are redundant bands with low R-square values. Based on these lambda (λ_1) versus lambda (λ_2) plots (Figure 9.11), the optimal waveband centers (λ) and widths ($\Delta\lambda$) are determined (Table 9.2). Table 9.2 shows the optimal wavebands (λ), wavebands centers (λ), and widths ($\Delta\lambda$) based on numerous studies (Thenkabail et al., 2000, 2002, 2004a,b, 2012, 2013, 2014), and a meta-analysis of these studies.

9.7.1.1 Refinement of Two-Band HVIs

Further refinement of each of the two-band HVIs (TBHVIs) is possible by computing (1) soil-adjusted versions of TBHVIs and (2) atmospheric corrected versions of TBHVIs. Interested readers can read more on this topic at Thenkabail et al. (2000).

9.7.2 Multi-Band Hyperspectral Vegetation Indices

The multi-band hyperspectral vegetation indices (MBHVIs) are computed as follows (Thenkabail et al., 2000; Li et al., 2012):

$$MBHVI_i = \sum_{j=1}^{N} a_{ij} R_j \qquad (9.2)$$

where

$MBHVI_i$ is the crop variable i

R is the reflectance in bands j (j = 1 − N with N = 242 for Hyperion)

a is the coefficient for reflectance in band j for ith variable

TABLE 9.2 Optimal (Nonredundant) Hyperspectral Narrowbands to Study Vegetation and Agricultural Crops[a, b, c]

Waveband Number (#)	Waveband Range (λ)	Waveband Center (λ)	Waveband Width ($\Delta\lambda$)	Importance and Physical Significance of Waveband in Vegetation and Cropland Studies
A. Ultraviolet				
1	373–377	375	5	fPAR, leaf water: fraction of photosynthetically active radiation (fPAR), leaf water content
B. Blue bands				
2	403–407	405	5	Nitrogen, Senescing: sensitivity to changes in leaf nitrogen reflectance changes due to pigments is moderate to low. Sensitive to senescing (yellow and yellow green leaves).
3	491–500	495	10	Carotenoid, Light use efficiency (LUE), Stress in vegetation: Sensitive to senescing and loss of chlorophyll\browning, ripening, crop yield, and soil background effects
C. Green bands				
4	513–517	515	5	Pigments (Carotenoid, Chlorophyll, anthocyanins), Nitrogen, Vigor: positive change in reflectance per unit change in wavelength of this visible spectrum is maximum around this green waveband
5	530.5–531.5	531	1	Light use efficiency (LUE), Xanophyll cycle, Stress in vegetation, pest and disease: Senescing and loss of chlorophyll\browning, ripening, crop yield, and soil background effects
6	546–555	550	10	Chlorophyll: Total chlorophyll; Chlorophyll/carotenoid ratio, vegetation nutritional and fertility level; vegetation discrimination; vegetation classification
7	566–575	570	10	Pigments (Anthocyanins, Chlorophyll), Nitrogen: negative change in reflectance per unit change in wavelength is maximum as a result of sensitivity to vegetation vigor, pigment, and N.
D. Red bands				
8	676–685	680	10	Biophysical quantities and yield: leaf area index, wet and dry biomass, plant height, grain yield, crop type, crop discrimination
E. Red-edge bands				
9	703–707	705	5	Stress and chlorophyll: Nitrogen stress, crop stress, crop growth stage studies
10	718–722	720	5	Stress and chlorophyll: Nitrogen stress, crop stress, crop growth stage studies
11	700–740	700–740	700–740	Chlorophyll, senescing, stress, drought: first-order derivative index over 700–740 nm has applications in vegetation studies (e.g., blue-shift during stress and red-shift during healthy growth)
F. Near infrared (NIR) bands				
12	841–860	850	20	Biophysical quantities and yield: LAI, wet and dry biomass, plant height, grain yield, crop type, crop discrimination, total chlorophyll
13	886–915	900	20	Biophysical quantities, Yield, Moisture index: peak NIR reflectance. Useful for computing crop moisture sensitivity index, NDVI; biomass, LAI, Yield.
G. Near infrared (NIR) bands				
14	961–980	970	20	Plant moisture content Center of moisture sensitive "trough"; water band index, leaf water, biomass;
H. Far near infrared (FNIR) bands				
15	1073–1077	1075	5	Biophysical and biochemical quantities: leaf area index, wet and dry biomass, plant height, grain yield, crop type, crop discrimination, total chlorophyll, anthocyanin, carotenoids
16	1178–1182	1080	5	Water absorption band
17	1243–1247	1245	5	Water sensitivity: water band index, leaf water, biomass. Reflectance peak in 1050–1300 nm
I. Early short-wave infrared (ESWIR) bands				
18	1448–1532	1450	5	Vegetation classification and discrimination: ecotype classification; plant moisture sensitivity. Moisture absorption trough in early short wave infrared (ESWIR)
19	1516–1520	1518	5	Moisture and biomass: A point of most rapid rise in spectra with unit change in wavelength in SWIR. Sensitive to plant moisture.
20	1648–1652	1650	5	Heavy metal stress, Moisture sensitivity: Heavy metal stress due to reduction in Chlorophyll Sensitivity to plant moisture fluctuations in ESWIR. Use as an index with 1548 or 1620 or 1690 nm.
21	1723–1727	1725	5	Lignin, biomass, starch, moisture: sensitive to lignin, biomass, starch Discriminating crops and vegetation.
J. Far short-wave infrared (FSWIR) bands				
22	1948–1952	1950	5	Water absorption band: highest moisture absorption trough in FSWIR. Use as an index with any one of 2025, 2133, and 2213 nm Affected by noise at times.
23	2019–2027	2023	8	Litter (plant litter), lignin, cellulose: litter-soil differentiation: moderate to low moisture absorption trough in FSWIR. Use as an index with any one of 2025, 2133, and 2213 nm

(Continued)

TABLE 9.2 (*Continued*)   Optimal (Nonredundant) Hyperspectral Narrowbands to Study Vegetation and Agricultural Crops[a, b, c]

Waveband Number (#)	Waveband Range (λ)	Waveband Center (λ)	Waveband Width (Δλ)	Importance and Physical Significance of Waveband in Vegetation and Cropland Studies
24	2131–2135	2133	5	Litter (plant litter), lignin, cellulose: typically highest reflectivity in FSWIR for vegetation. Litter- soil differentiation
25	2203–2207	2205	5	Litter, lignin, cellulose, sugar, starch, protein; Heavy metal stress: typically, second highest reflectivity in FSWIR for vegetation. Heavy metal stress due to reduction in Chlorophyll
26	2258–2266	2262	8	Moisture and biomass: moisture absorption trough in far short-wave infrared (FSWIR). A point of most rapid change in slope of spectra based on land cover, vegetation type, and vigor.
27	2293–2297	2295	5	Stress: sensitive to soil background and plant stress
28	2357–2361	2359	5	Cellulose, protein, nitrogen: sensitive to crop stress, lignin, and starch

Sources: Modified and adopted from Thenkabail, P.S. et al., *Remote Sens. Environ.*, 71, 158, 2000; Thenkabail, P.S. et al. (2002); Thenkabail, P.S. et al., *Remote Sens. Environ.*, 90, 23, 2004a; Thenkabail, P.S. et al., *Remote Sens. Environ.*, 91, 354, 2004b; Thenkabail et al. (2012, 2013); Thenkabail, P.S. et al., *Photogramm. Eng. Remote Sens.*, 80, 697, 2014.

[a] Most hyperspectral narrowbands (HNBs) that adjoin one another are highly correlated for a given application. Hence from a large number of HNBs, these non-redundant (optimal) bands are selected.

[b] These optimal HNBs are for studying vegetation and agricultural crops. When we use some or all of these wavebands, we can attain highest possible classification accuracies in classifying vegetation categories or crop types.

[c] Wavebands selected here are based on careful evaluation of large number of studies.

TABLE 9.3   Subpixel Mineral Mapping of a Porphyry Copper Belt Using EO-1 Hyperion Data

Hyperion Band (#)	Wavelength (nm)	Feature	Minerals	Mineral Characteristic
210, 217	2254, 2324	Absorption	Biotite	Potassic-biotitic alteration zone
205	2203	Absorption	Muscovite and illite	Al–OH vibration in minerals with muscovite deeper absorption than illite
201, 205	2163, 2203	Absorption	Kaolinite	Al–OH vibration
14, 79, 205	487, 932, 2203	Absorption	Goethite	
14, 53, 205	487, 884, 2203	Absorption	Hematite	
79211205	932, 2264, 2203	Absorption	Jarosite	
201	2163	Absorption	Pyrophyllite	Al–OH and Mg–OH
218	2335	Absorption	Chlorite	Al–OH and Mg–OH

Source: Adopted and modified from information in manuscript by Hosseinjani Zadeh, M. et al., *Adv. Space Res.*, 53, 440, 2014.

The process of modeling involves running stepwise linear regression models (e.g., using MAXR algorithm in Statistical Analysis System (SAS, 2009) with any one biophysical or biochemical variable (e.g., biomass) as dependent variable and the numerous HNBs as independent variables (e.g., 157 of the 242 useful bands of Hyperion). In this modeling approach, we will get the best one-band, two-band, three-band, and so on to best n-band model. The best one-band model is the one in which the biomass (taken as example) has highest R-square value with a single band out of the total 157 Hyperion HNBs. Then, we obtain the best two-band model, in which two HNBs provide a best R-square value with biomass. Similarly, the best three-band, best four-band, and best n-band (e.g., all 157 Hyperion bands) models are obtained, even though, theoretically, all 157 bands can be involved in providing a 157-band biomass model that is usually meaningless due to overfitting of data. However, a plot of R-square values (y-axis) versus the number of bands (x-axis) will show us when an increase in R-square values with the addition of wavebands becomes asymptotic. Alternatively, we can also consider additional bands, when there is at least an increase of 0.03 or higher in R-square value when additional bands are added. So, the approach we can use is to look at one-band model and see its R-square. Then, when two-band model increases R-square value by at least 0.03 (a threshold we can set), then consider the two-band model; otherwise, retain

the one-band model as final. At some stage, we will notice that addition of a band does not increase R-square value by more than 0.03. Typically, we have noticed that anywhere between 3 and 10 HNBs explain optimal variability in most agricultural crop and vegetation variables. Beyond these 3–10 bands, the increase in R-square per increase in band is insignificant or asymptotic. However, which 3–10 bands within 400–2500 nm will, often, vary is based on the type of crop variable.

Through MBHVIs, we can establish the following:

1. How many HNBs are required to achieve an optimal R-square for any biophysical or biochemical quantity?
2. Which HNBs are involved in providing optimal R-square?
3. Through this process, we can determine which are important HNBs and which are redundant. However, the best approach to achieve this is by a study conducted for many crops, involving several crop variables, and based on data from multiple sites and years. Table 9.2 provides one such summary.

These MBHVIs take advantage of the key absorption and reflective portions of the spectrum (e.g., Figure 9.12; [Gnyp et al., 2014]). Taking advantage of four HNBs, two reflective (900 and 1050 nm) and two absorptive (955 and 1220 nm), Gnyp et al. constitute an MBHVI (Equation 9.1). In their paper, Gnyp et al. (2014) clearly

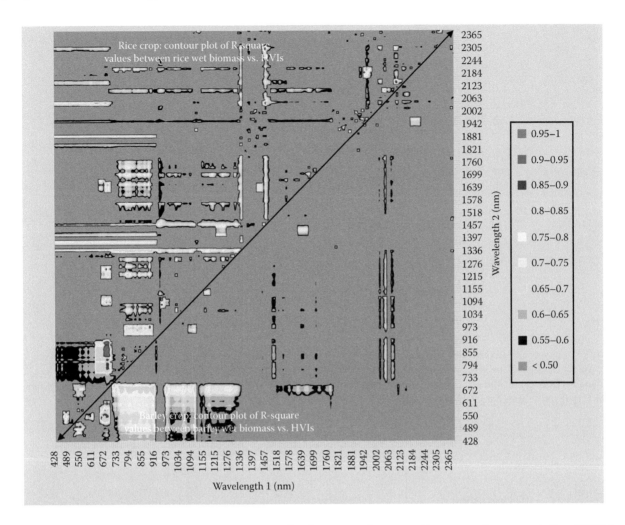

FIGURE 9.11 Lambda (λ) versus Lambda (λ) plot of R-square values between wet biomass and hyperspectral vegetation indices (HVIs) for the rice crop (above the diagonal) and barley crop (below the diagonal).

demonstrate the significantly higher R-square values provided by such a multiband HVIs when compared with other two-band HVIs (e.g., in Figure 9.13, GnyLi has a much higher R-square value relative to other indices). Interesting and maybe noteworthy that while the typical saturation effect (lack of sensitivity) at higher biomass amounts is still present, it is evidently somewhat less severe with GnyLi than the others (except REP but it has lower r2). Also, research by Thenkabail et al. (2004a, b), Mariotto et al. (2013), and Marshall and Thenkabail (2014) has demonstrated that anywhere between 3 and 10 HNBs involved in multiband HVIs explain greatest variability in modeling various biophysical and biochemical quantities for various agricultural crops.

However, it needs to be noted that the specific band centers and band widths are not as definitive as shown in Figure 9.12 or/ and Equation 9.1. This is because, with crop type and crop growing conditions, the specific reflective maxima (900 and 1050 nm) and reflective minima (955 and 1220 nm) shown in Figure 9.12 and Equation 9.1 can vary. For example, the moisture absorption maxima can be at 750, 760, 770, or 780 nm (Thenkabail et al., 2012, 2013) or can be at 755 nm as shown in Figure 9.12 and Equation 9.1. As a result, we performed meta-analysis of a

number of papers to come with the recommendations of HNB centers and HNB widths (Table 9.2) that are optimal for use in HVI computations across crops and vegetation.

$$\text{GnyLi} = \frac{R_{900} \times R_{1050} - R_{955} \times R_{1220}}{R_{900} \times R_{1050} + R_{955} \times R_{1220}} \quad (9.3)$$

9.8 The Best Hyperspectral Vegetation Indices and Their Categories

Based on extensive research over the last decade (Thenkabail et al., 2000, 2002, 2004a,b, 2012, 2013, 2014), six distinct categories of two-band TBHVIs (Table 9.4) are considered most significant and important in order to study specific biophysical and biochemical quantities of agriculture and vegetation. Author recommends that in future, researchers use these HVIs, derived using HNBs, for their studies to quantify and model biophysical and biochemical quantities of various agricultural crops and vegetation of different types. The values of two such indices are illustrated. These are (1) hyperspectral biomass and structural index 1 (HBSI1; Thenkabail et al., 2014), derived using the Hyperion bands centered around

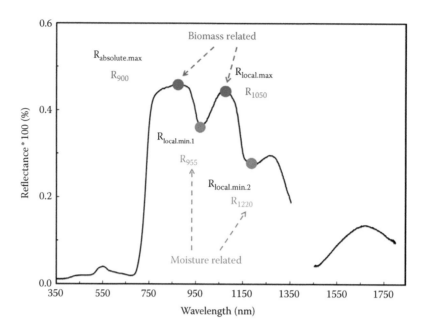

FIGURE 9.12 New index. Development and implementation of a multiscale biomass model using hyperspectral vegetation indices for winter wheat in the North China Plain. Reflectance of winter wheat and its characteristic peaks and troughs with the reflectance maxima and minima in the NIR and SWIR domains. These peaks were used to compute the VI GnyLi. R is the reflectance value (%) at a specific wavelength (nm). (From Gnyp, M.L. et al., *Int. J. Appl. Earth Obs. Geoinf.*, 33, 232, 2014.)

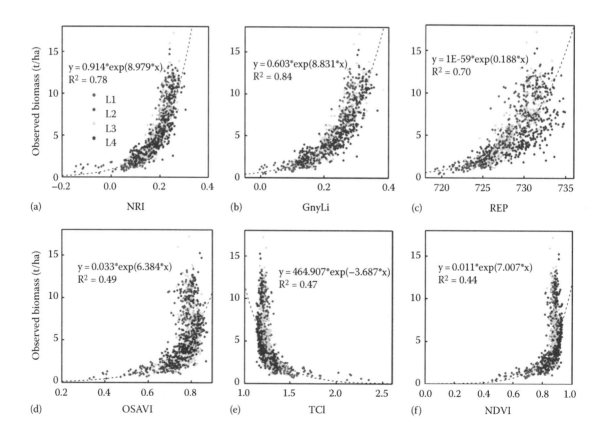

FIGURE 9.13 Observed biomass versus various HVIs and also GnyLi (see Figure 9.20). R is the reflectance value (%) at a specific wavelength (nm). (From Gnyp, M.L. et al., *Int. J. Appl. Earth Obs. Geoinf.*, 33, 232, 2014.)

TABLE 9.4 Hyperspectral Vegetation Indices or HVIs

Band Number (#)	Hyperspectral Narrowband (λ_1)	Bandwidth ($\Delta\lambda_1$)	Hyperspectral Narrowband (λ_2)	Bandwidth ($\Delta\lambda_2$)	Hyperspectral Vegetation Index (HVI)	Best Index Under Each Category
1. Hyperspectral biomass and structural indices (HBSIs) (to best study biomass, LAI, plant height, and grain yield)						
HBSI1	855	20	682	5	(855 − 682)/(855 + 682)	HBSI: Hyperspectral biomass and structural index
HBSI2	910	20	682	5	(910 − 682)/(910 + 682)	
HBSI3	550	5	682	5	(550 − 682)/(550 + 682)	
2. Hyperspectral biochemical indices (HBCIs) (pigments like carotenoids, anthocyanins as well as Nitrogen, chlorophyll)						
HBCI8	550	5	515	5	(550 − 515)/(550 + 515)	HBCI: Hyperspectral biochemical index
HBCI9	550	5	490	5	(550 − 490)/(550 + 490)	
3. Hyperspectral Red-edge indices (HREIs) (to best study plant stress, drought)						
HREI14	700 − 740	40	First-order derivative integrated over red-edge.			HREI: Hyperspectral red-edge index
HREI15	855	5	720	5	(855 − 720)/(855 + 720)	
4. Hyperspectral water and moisture indices (HWMIs) (to best study plant water and moisture)						
HWMI17	855	20	970	10	(855 − 970)/(855 + 970)	HWMI: Hyperspectral water and moisture index
HWMI18	1075	5	970	10	(1075 − 970)/(1075 + 970)	
HWMI19	1075	5	1180	5	(1075 − 1180)/(1075 + 1180)	
HWMI20	1245	5	1180	5	(1245 − 1180)/(1245 + 1180)	
5. Hyperspectral light-use efficiency index (HLEI) (to best study light use efficiency or LUE)						
HLUE24	570	5	531	1	(570 − 531)/(570 + 531)	HLEI: Hyperspectral light-use efficiency index
6. Hyperspectral lignin cellulose index (HLCI) (to best study plant lignin, cellulose, and plant residue)						
HLCI25	2205	5	2025	1	(2205 − 2025)/(2205 + 2025)	HLCI: Hyperspectral lignin cellulose index

Sources: Modified and adopted from Thenkabail, P.S. et al., *Photogramm. Eng. Remote Sens.*, 80, 697, 2014.
Note: Also see wavebands in Table 9.2 used to derive these indices.

855 and 682 nm (each with 10 nm width), is applied to an agricultural area to determine biomass (Figure 9.14); and (2) photochemical reflectance index (PRI) for stress detection (e.g., Figure 9.15; Middleton et al., 2012). The importance of wavebands in computing the indices for various biophysical and biochemical is illustrated in Figure 9.16. Reader is encouraged to compare Figure 9.15 with Table 9.4 and Table 9.2 for better understanding of HNBs (Table 9.2), HVIs (Table 9.4), and their importance (Figure 9.16) in studies pertaining to crops and vegetation.

9.9 Whole Spectral Analysis

A number of chapters discuss the usefulness and utility of using whole spectra (e.g., continuous and entire spectra over 400–2500 nm) for analysis using such methods as PLSR, wavelet analysis, continuum removal, SAM, and spectral matching techniques (SMTs) (Thenkabail et al., 2012).

9.9.1 Spectral Matching Techniques

SMTs (Thenkabail et al., 2007) involves the following:

1. *Ideal or target spectral library creation*: Collecting ideal or target spectra (e.g., specific crops, specific species, specific mineral) and creating a spectral library.
2. Class spectra collection.
3. Matching class spectra with ideal spectra to identify and label classes.

The principal approach in SMT is to match the shape or the magnitude or (preferably) both to an ideal or target spectrum (pure class or "end member"). Thenkabail et al. (2007) proposed and implemented SMT for multitemporal data illustrated later (Figure 9.17). The qualitative pheno-SMT approach concept remains the same for hyperspectral data (replace the number of bands of temporal data with the number of hyperspectral bands).

The quantitative SMTs consist of (Thenkabail et al., 2007) (1) spectral correlation similarity—a shape measure; (2) spectral similarity value—a shape and magnitude measure; (3) Euclidian distance similarity—a distance measure; and (4) modified spectral angle similarity—a hyper angle measure.

9.9.2 Continuum Removal through Derivative Hyperspectral Vegetation Indices

The derivative hyperspectral vegetation indices (DHVIs) are computed by integrating index over a certain wavelength (e.g., 600–700 nm or 700–760 nm). The equation is

$$\text{DHVI} = \sum \frac{\lambda_n (\rho'(\lambda_i) - (\rho'(\lambda j))}{\lambda_1 \Delta\lambda_I} \tag{9.4}$$

where
i and j are band numbers
λ is the center of wavelength

The process of obtaining DHVI value for 600–700 nm is as follows: (1) DHVI1 = lambda 1 (e.g., λ_1 = 600 nm) versus lambda 2 (e.g.,

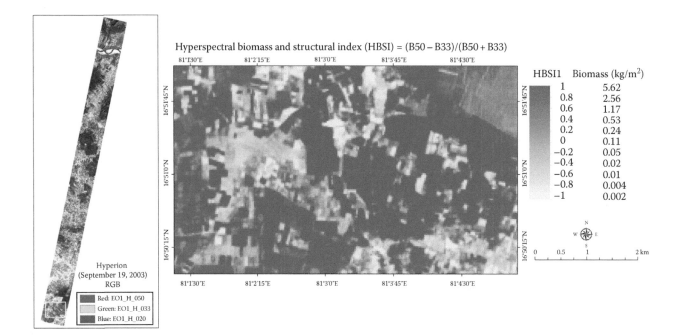

FIGURE 9.14 Spatial depiction of a hyperspectral biomass and structural index 1 (HBSI1) as applied to an agricultural area. One of the HVIs (HBSI1) in mapping wet biomass for a study area using Hyperion hyperspectral data. The red area in the z-scale can be stretched further to show better biomass variability with change in HBSI1. For example, HBSI1 0.4 = 0.53 and HBSI 0.6 = 1.16, HBSI1 0.8 = 2.56, and HBSI1 5.62. The current stretch does not adequately show these differences (as much of the higher end is in red). However, if we stretch between HBSI1 from 0.4 to 1.0, then the biomass differences in this HBSI1 range, which is 0.53–5.62, will show up in better contrast. The relationship between HBSI1 and biomass is nonlinear due to saturation of indices at the higher end of the biomass. However, this saturation is much lower for hyperspectral index like HBSI1 when compared to broadband NDVI.

$\lambda_2 = 610$ nm). The difference in the reflectivity of these two bands is then divided by their bandwidth ($\Delta\lambda_I = 10$ nm) and (2) DHVI2 = the process is repeated for lambda 1 (e.g., $\lambda_1 = 610$ nm) versus lambda 2 (e.g., $\lambda_2 = 620$ nm). The difference in reflectivity of these two bands is then divided by their bandwidth ($\Delta\lambda_I = 10$ nm) and (3) DHVIn = so on to lambda 1 (e.g., $\lambda_1 = 690$ nm) versus lambda 2 (e.g., $\lambda_2 = 700$ nm). The difference in reflectivity of these two bands is then divided by their bandwidth ($\Delta\lambda_I = 10$ nm). Finally, add DHVI1, DHVI2, and so on to DHVIn to get single an integrated DHVI value over the entire 600–700 nm range.

The DHVIs can be derived over various wavelengths such as 400–2500 nm, 500–600 nm, 600–800 nm, and any other wavelength you find useful for the particular application. There are opportunities to further investigate the significance of DHVIs over different wavelengths for a wide array of applications.

9.10 Principal Component Analysis

Another common, powerful, and useful feature selection method for hyperspectral data analysis is PCA. The PCA performs following functions:

1. *Reduces data volumes*: This happens since the PCA generates numerous principal components (PCs) (as many as the number of wavebands), but the first few PCs explain almost all the variability of data. The first PC (PC1) explains the highest, followed by the other. Since each PC is constituted based on the information from all the bands (e.g., PC1 = factor loading for band 1 * band 1 reflectivity + ⋯ + factor loading for band n * band n reflectivity), the PCs have the power of hyperspectral bands, but does not have all the redundancy of the same.

2. *Provides a new single band of information* (e.g., PC1, PC2), each of which (e.g., PC1) actually has the information derived from all the HNBs. These new bands of information (e.g., PC1) can then be used to classify an area (e.g., to establish crop types) or used to model crop biophysical or biochemical quantities.

3. *The power of PCs* can be used to discriminate crop types, or land cover themes, or species (e.g., Figure 9.18).

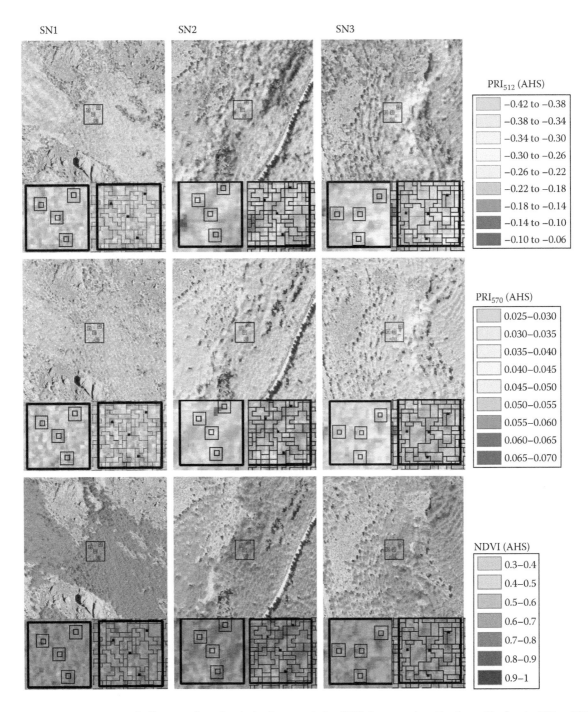

FIGURE 9.15 Assessing structural effects on photochemical reflectance index (PRI) for stress detection in conifer forests. PRI_{512}, PRI_{570}, and NDVI obtained from the AHS airborne sensor from three study areas of *Pinus nigra* with different levels of stress: SN1, SN2, and SN3. At the bottom of each image, two zoom images of a central plot, one pixel based displaying 1 × 1 and 3 × 3 resolutions and the other at object level. *Note*: PRI_{512} is a normalized index involving a waveband centered at 512 and 531 nm, whereas PRI_{570} is a normalized index involving a waveband centered at 570 and 531 nm. Airborne hyperspectral scanner (AHS) (Sensytech Inc., currently Argon St. Inc., Ann Arbor, MI) acquiring 2 m spatial resolution imagery in 38 bands in the 0.43–12.5 μm spectral range. (From Hernández-Clemente, R. et al., *Adv. Space Res.*, 53, 440, 2011.)

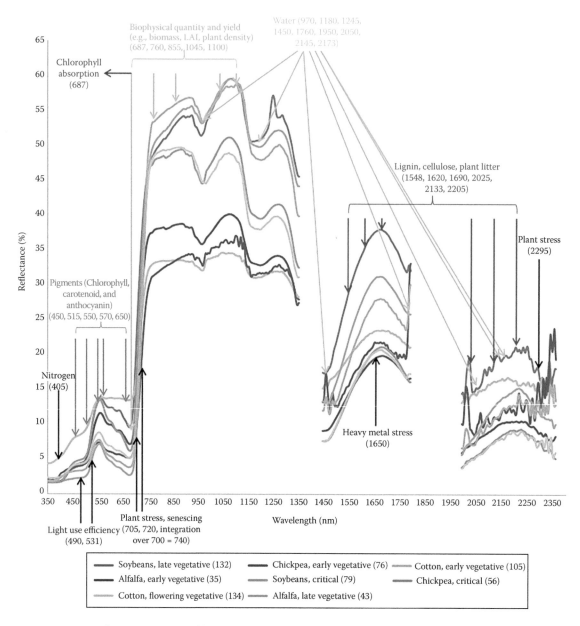

FIGURE 9.16 Importance of various portions of hyperspectral data in characterizing biophysical and biochemical quantities of crops and vegetation.

9.11 Spectral Mixture Analysis of Hyperspectral Data

Hyperspectral data have great ability to distinguish specific objects based on their unique signatures. For example, wheat versus barley crops are distinguished based on the spectral reflectivity in two HNBs, each of 10 nm wide, and centered at 687 and 855 nm (e.g., Figure 9.19). However, often, we find multiple objects or classes within a pixel. In situations like that, we will need to perform spectral mixture analysis (SMA) and an independent component analysis, in order to unmix the spectral signatures within each pixel.

The reference spectra for SMA are derived from "end members" (e.g., Figure 9.20). Once all the materials in the image are identified, then it is possible to use linear or nonlinear spectral unmixing to find out how much of each material is in each pixel.

The concept of unmixing hyperspectral data is illustrated by showing Hyperion unmixing of (1) vegetation fractional cover in Figure 9.21 and (2) minerals in Figure 9.22. Subpixel mineral mapping of a porphyry copper belt using EO-1 Hyperion data in Figure 9.23 involved mineral spectra extracted from Hyperion compared to convolved spectra from field samples and reference library spectra (Figures 9.20 and 9.21; Hosseinjani Zadeh et al., 2014). Extensive discussions on linear and nonlinear SMAs can be found in Plaza et al. (2012).

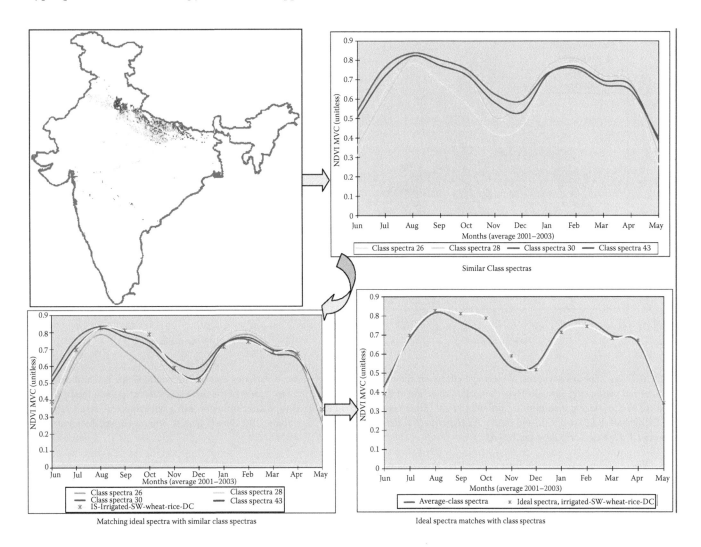

FIGURE 9.17 Pheno-spectral matching technique (SMTs). In SMTs, the class temporal profiles (NDVI curves) are matched with the ideal temporal profile (quantitatively based on temporal profile similarity values) in order to group and identify classes as illustrated for a rice class in this figure. Illustration of double-crop (DC) irrigation. The NDVI spectra of the four classes (C-26, C-28, C-30, and C-43) of DC irrigation are "matched" with ideal spectra (shaded in yellow) for the same. This is a qualitative illustration of SMTs. For quantitative methods, refer to Thenkabail et al. 2007.

9.12 Support Vector Machines

SVMs are a machine learning supervised classification approach. Unlike the feature selection approach, data dimensionality is not an issue here. Any number of bands can be used. The process involves supervised training of classes, based on sufficient and accurate knowledge of the class (e.g., ground data), where one can use all or some of the hyperspectral bands to train the algorithm. Once the algorithm is sufficiently trained, it can be run on rest of the data to gather the same class occurring in other areas. Figure 9.24a shows the classification performed using all 272 AISA hyperspectral bands based on SVM algorithm. In Figure 9.24b, the same classification is performed using only 51 of the most important AISA hyperspectral bands. Results of the 51-band classification output (Figure 9.24b) are comparable to 272-band classification output (Figure 9.24a) in most areas; there is

significant uncertainty in the northern portion of the image. Studies have shown that by using only 1% of training pixels per class, almost 90% overall classification accuracies are obtained using SVM methods (Bajwa and Kulkarni, 2012; Ramsey III and Rangoonwala, 2012).

9.13 Random Forest and Adaboost Tree-Based Ensemble Classification and Spectral Band Selection

Random forest and Adaboost are two tree-based ensemble classifiers. These classifiers serve two purposes:

1. Help select hyperspectral bands that are important as well as those that are redundant.
2. Classify hyperspectral data through decision tree-based classifiers.

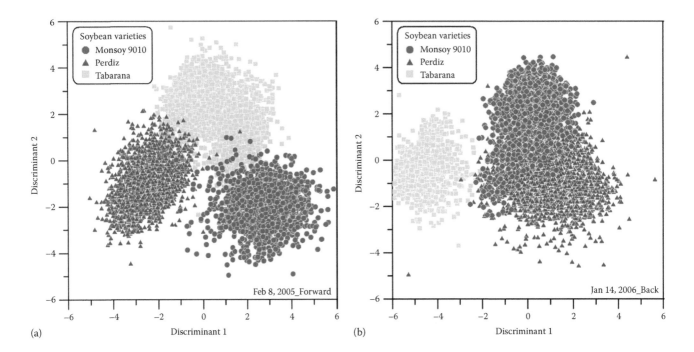

FIGURE 9.18 Species soybeans. View angle effects on the discrimination of soybean varieties and on the relationships between vegetation indices and yield using off-nadir Hyperion data. Projection of the Hyperion discriminant scores of the three soybean varieties in the (a) forward and (b) backscattering directions for different years. (From Galvao, L.S. et al., Crop type discrimination using hyperspectral data, Chapter 17, in Thenkabail, P.S., Lyon, G.J., and Huete, A., *Hyperspectral Remote Sensing of Vegetation*, CRC Press/Taylor & Francis Group, Boca Raton, FL/London, U.K./New York, 2009, pp. 397–422.)

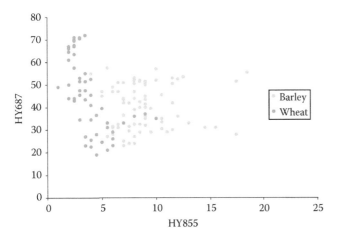

FIGURE 9.19 Differentiating corn from soybeans using two hyperspectral narrowbands.

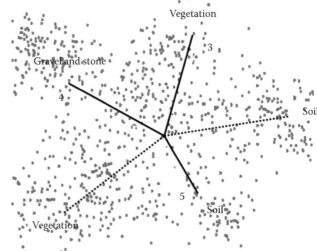

FIGURE 9.20 End member. Arid land characterization with EO-1 Hyperion hyperspectral data. End member extraction in n-dimension visualizer using bands 3, 4, and 5 of the minimum noise fraction (MNF) transform Hyperion image. (From Jafari, R. and Lewis, M.M., *Int. J. Appl. Earth Obs. Geoinf.*, 19, 298, 2012.)

This approach has been discussed in great detail by Chan and Paelinckx (2008) for thorough classification of detailed ecotopes using hyperspectral data (Figures 9.25 and 9.26). They gathered extensive hyperspectral data for (Figure 9.25) (1) 6 grassland classes and (2) 10 tree classes. In terms of accuracy performance, random forest and Adaboost are almost the same, and both have outperformed a neural network classifier (Chan and Paelinckx, 2008). Both feature selection routines, the best-first search and the out-of-bag ranking index under random forest, are successful in identifying substantially smaller band subsets that attained almost the same accuracy as all the bands (e.g., Figure 9.24; Chan and Paelinckx, 2008). There are many approaches to selecting the spectral wavebands for obtaining best classification results. For agriculture

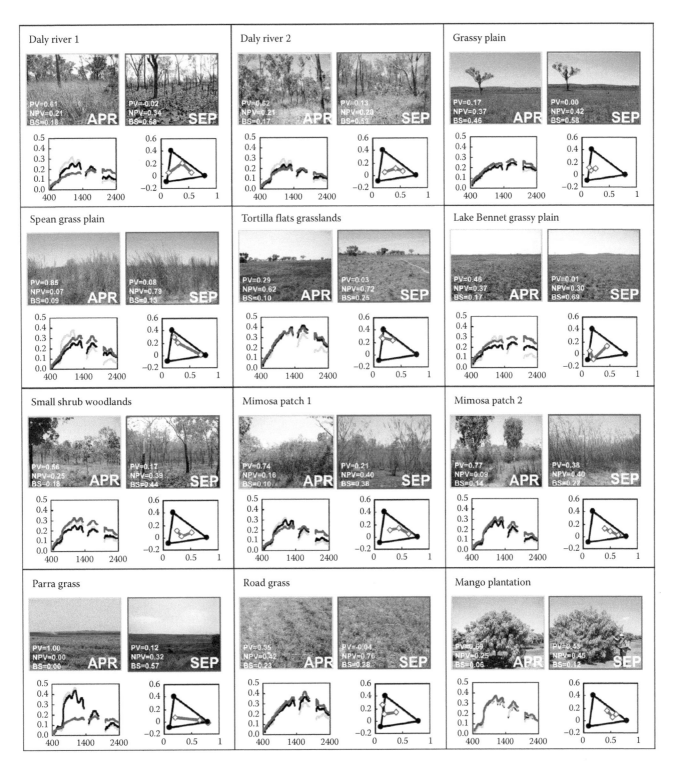

FIGURE 9.21 Unmixing qualitative assessment of Hyperion unmixing of vegetation fractional cover. Qualitative validation of the fractional cover estimated with Hyperion imagery. Each set of pictures and graphs corresponds to one of 12 sites visited from May 16 to 19 and August 29 to 31, 2005. The left graphs show the reflectance spectra derived from Hyperion images for the April (green curve), July (black curve), and September (blue curve) images from 400 to 2400 nm. The right graphs show the position of each spectrum in the normalized difference vegetation index or NDVI (x-axis) (detecting live, green vegetation) and cellulose absorption index or CAI (y-axis) (detecting non-photosynthetic vegetation) space from April to September (red dots and line) and the position of the end members (black lines). The derived photosynthetic vegetation (f_{PV}), non-photosynthetic vegetation (f_{NPV}), and bare soil (f_{BS}) are shown in each picture and are critical for natural resource management and for modeling carbon dynamics. (From Guerschman, J.P. et al., *Remote Sens. Environ.*, 113, 928, 2009.)

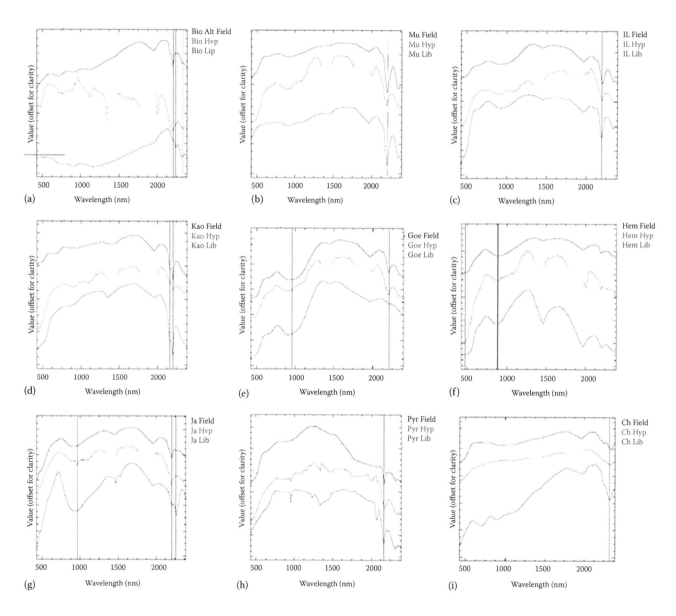

FIGURE 9.22 Mineral mapping. Subpixel mineral mapping of a porphyry copper belt using EO-1 Hyperion data. Mineral spectra extracted from Hyperion comparing to convolved spectra from field samples and reference library spectra. (a) Biotite (Bio), (b) Muscovite (Mu), (c) Illite (Il), (d) Kaolinite (Kao), (e) Goethite (Goe), (f) Hem (Hem), (g) Jarosite (Ja), (h) pyrophyllite (Pyr), and (i) Chlorite (Ch). Hyp and Lib are abbreviations of Hyperion and Library, respectively. The red vertical lines indicate locations of diagnostic absorption features. (From Hosseinjani Zadeh, M. et al., *Adv. Space Res.*, 53, 440, 2014.)

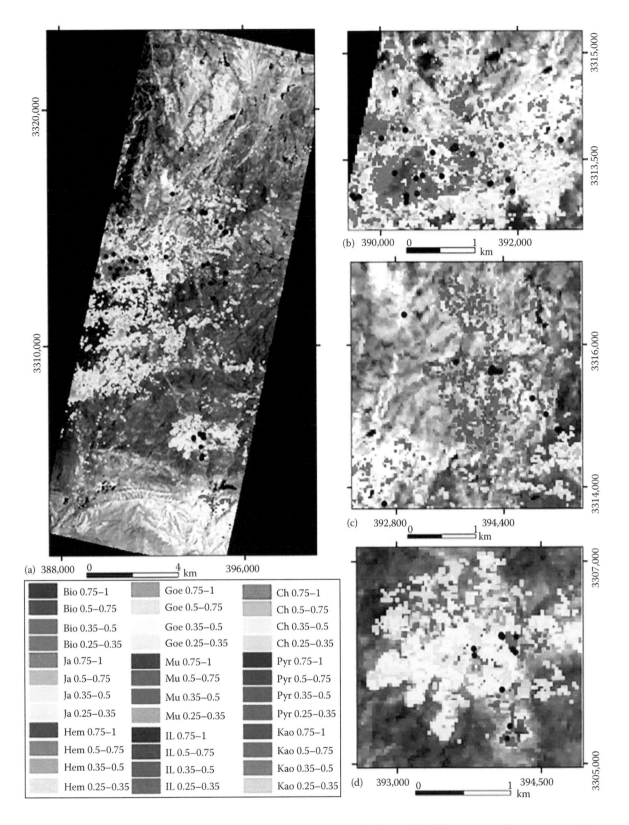

FIGURE 9.23 Mineral mapping. Subpixel mineral mapping of a porphyry copper belt using EO-1 Hyperion data. Thematic mineral maps using subpixel mixture tuned matched filtering (MTMF) method. (a) Final classification image map of alteration minerals derived from MTMF algorithm. (b) Sarcheshmeh mine, (c) Sereidun, and (d) Darrehzar. Bio, Mu, Il, Kao, Goe, Hem, Ja, Pyr, and Ch indicate Biotite, Muscovite, Illite, Kaolinite, Goethite, Hematite, Jarosite, pyrophyllite, and Chlorite, respectively. These values indicate percentages of each mineral at the pixel. For instance, value of 0.25 shows that 25% of pixel contains the selected mineral. (From Hosseinjani Zadeh, M. et al., *Adv. Space Res.*, 53, 440, 2014.)

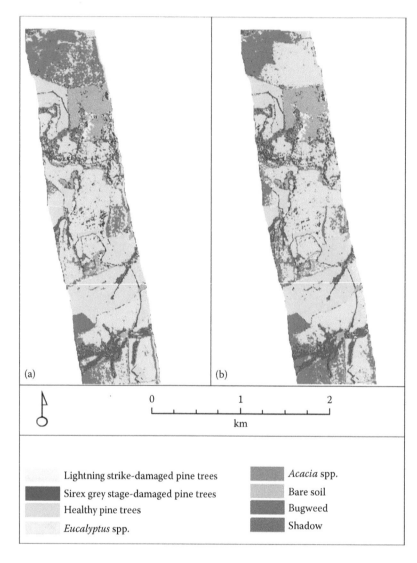

FIGURE 9.24 SVM. Detecting Sirex noctilio gray-attacked and lightning-struck pine trees using airborne hyperspectral data, random forest, and support vector machine classifiers. Classification maps obtained using support vector machine (SVM) classification algorithm, all (a) and the 51 most important (b) Airborne Imaging System for different Applications (AISA) Eagle spectral bands. The AISA image spatial resolution was about 2 m, and there were 272 spectral bands ranging from 393.23 to 994.09 nm (VNIR: visible near-infrared) with bandwidths between 2 and 4 nm. (From Abdel-Rahman, E.M. et al., *ISPRS J. Photogramm. Remote Sens.*, 88, 48, 2014.)

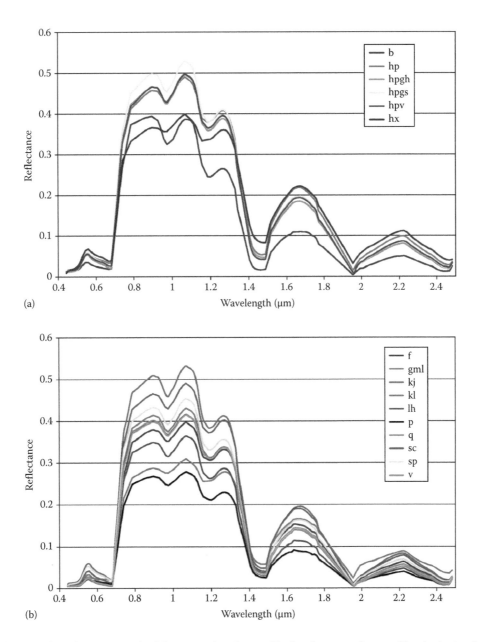

FIGURE 9.25 Evaluation of Random Forest and Adaboost tree-based ensemble classification and spectral band selection for ecotope mapping using airborne hyperspectral imagery. Mean spectrum of the (a) 6 grassland classes and (b) 10 tree classes. *Notes*: b, grassland, arable land; hp, grassland, species poor improved grassland (normally more homogenous for the whole parcel); hpgh, grassland, semi-natural grassland; hpgs, grassland, species rich improved grassland (between hpgh and hp); hpv, grassland, grassland with patches hp and either patches hpgs or hpgh; hx, grassland, grass monocultures (equal to arable land sown with grasses of one or more years); f, tree/tall_veg, deciduous forest<comma> dominated by beech (*Fagus* sp.); gml, tree/tall_veg, plantation of deciduous tree species other than beech, oak, alder, and poplar; kj, tree/tall_veg, tall tree orchard; kl, tree/tall_veg, low tree orchard; lh, tree/tall_veg, poplar plantation; p, tree/tall_veg, conifer plantation; q, tree/tall_veg, deciduous forest<comma> dominated by oak trees (*Quercus* sp.); sp, tree/tall_veg sc, scrubs of clearings and scrubs on abandoned land; sp, tree/tall_veg, thorn ticket; v, tree/tall_veg,Woodland of alluvial soil<comma> fens and bogs (mostly dominated by alder, *Alnus* sp.). (From Chan, J.C.-W. and Paelinckx, D., *Remote Sens. Environ.*, 112, 2999, 2008.)

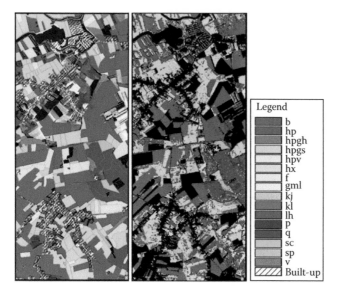

Legend
- b
- hp
- hpgh
- hpgs
- hpv
- hx
- f
- gml
- kj
- kl
- lh
- p
- q
- sc
- sp
- v
- Built-up

FIGURE 9.26 Evaluation of random forest and Adaboost tree-based ensemble classification and spectral band selection for ecotope mapping using airborne hyperspectral imagery. On the left is the ground truth image of the biological valuation map. On the right is the biological valuation map classification based on airborne hyperspectral data using 99 trials of Adaboost with 21 bands selected by the best-first search method. The black areas represent unclassified land covers that have been masked out. (From Chan, J.C.-W. and Paelinckx, D., *Remote Sens. Environ.*, 112, 2999, 2008.)

TABLE 9.5 Best Hyperspectral Narrowband (HNB) Combinations Based on Number of Bands Available to Classify Crops or Vegetation

Best 4 bands	550, 680, 850, 970
Best 6 bands	550, 680, 850, 970, 1075, 1450
Best 8 bands	550, 680, 850, 970, 1075, 1180, 1450, 2205
Best 10 bands	550, 680, 720, 850, 970, 1075, 1180, 1245, 1450, 2205
Best 12 bands	550, 680, 720, 850, 910, 970, 1075, 1180, 1245, 1450, 1650, 2205
Best 16 bands	490, 515, 550, 570, 680, 720, 850, 900, 970, 1075, 1180, 1245, 1450, 1650, 1950, 2205
Best 20 bands	490, 515, 531, 550, 570, 680, 720, 850, 900, 970, 1075, 1180, 1245, 1450, 1650, 1725, 1950, 2205, 2262, 2359

Sources: Adapted and modified from Thenkabail, P.S. et al., 2012, 2013; Thenkabail, P.S. et al., *Photogramm. Eng. Remote Sens.*, 80, 697, 2014.

and vegetation studies, one could use various combination of band selection (e.g., Table 9.5) depending on the number of bands one decided to use, classification accuracies desired, and the need to overcome Hughes's phenomenon.

9.14 Conclusions

This chapter provides an overview of hyperspectral remote sensing for terrestrial applications. First, the chapter defines hyperspectral remote sensing or imaging spectroscopy. Second, characteristics of hyperspectral data acquired from three distinct platforms are discussed: (1) ground-based or handheld or truck-mounted spectroradiometers, (2) airborne, and (3) spaceborne. Third, the needs for data mining to eliminate redundant bands are discussed. Various data mining methods are presented. Fourth, the importance of understanding Hughes's phenomenon and approaches to overcome the same are highlighted. Fifth, methods of hyperspectral analysis are presented and discussed. These methods include feature extraction methods and information extraction methods. OHNBs best suited for agricultural and vegetation studies are determined from meta-analysis. HVIs, two-band and multi-band versions, best suited for agricultural and vegetation studies are also determined from meta-analysis. The WSA was performed through SMTs and continuum removal derivative HVIs. Hyperspectral image classification for land cover and species types was performed using such methods like SMA, SVMs, and tree-based ensemble classifiers such as random forest and Adaboost.

References

Abdel-Rahman, E.M., Mutanga, O., Adam, E., Ismail, R., 2014. Detecting Sirex noctilio grey-attacked and lightning-struck pine trees using airborne hyperspectral data, random forest and support vector machines classifiers. *ISPRS Journal of Photogrammetry and Remote Sensing*, 88, 48–59.

Alchanatis, V., Cohen, Y., 2012. Spectral and spatial methods for hyperspectral image analysis for estimation of biophysical and biochemical properties of agricultural crops. Chapter 13, in Thenkabail, P.S., Lyon, G.J., and Huete, A. *Hyperspectral Remote Sensing of Vegetation*. CRC Press/Taylor & Francis Group, Boca Raton, FL/London, U.K./New York. pp. 239–308.

Andrew, M.E., Ustin, S.L., 2008. The role of environmental context in mapping invasive plants with hyperspectral image data. *Remote Sensing of Environment*, 112, 4301–4317.

Bajwa, S., Kulkarni, S.S., 2012. Hyperspectral data mining. Chapter 4, in Thenkabail, P.S., Lyon, G.J., and Huete, A. *Hyperspectral Remote Sensing of Vegetation*. CRC Press/Taylor & Francis Group, Boca Raton, FL/London, U.K./New York. pp. 93–120.

Ben-Dor, E., 2012. Characterization soil properties using reflectance spectroscopy. Chapter 22, in Thenkabail, P.S., Lyon, G.J., and Huete, A. *Hyperspectral Remote Sensing of Vegetation*. CRC Press/Taylor & Francis Group, Boca Raton, FL/London, U.K./New York. pp. 513–560.

Boardman, J. W., Kruse, F. A., 1994. Automated spectral analysis: A geological example using AVIRIS data, north Grapevine Mountains, Nevada. In *Proceedings of ERIM 10th Thematic Conference on Geologic Remote Sensing*, Environmental Research Institute of Michigan, Ann Arbor, MI. pp. I-407–I-418.

Chan, J.C.-W., Paelinckx, D., 2008. Evaluation of Random Forest and Adaboost tree-based ensemble classification and spectral band selection for ecotope mapping using airborne hyperspectral imagery. *Remote Sensing of Environment*, 112, 2999–3011.

Chander, G., Markham, B.L., Helder, D.L. 2009. Summary of current radiometric calibration coefficients for Landsat MSS, TM, ETM+, and EO-1 ALI sensors. *Remote Sensing of Environment*, 113, 893–903.

Chavez, P.S., 1988. An improved dark-object subtraction technique for atmospheric scattering correction of multispectral data. *Remote Sensing of Environment*, 24, 459–479.

Chavez, P.S., 1989. Radiometric calibration of Landsat thematic mapper multispectral images. *Photogrammetric Engineering and Remote Sensing*, 55, 1285–1294.

Clark, M.L., 2012. Identification of canopy species in tropical forests using hyperspectral data. Detecting and mapping invasive species using hyperspectral data. Chapter 18, in Thenkabail, P.S., Lyon, G.J., and Huete, A. *Hyperspectral Remote Sensing of Vegetation*. CRC Press/Taylor & Francis Group, Boca Raton, FL/London, U.K./New York. pp. 423–446.

Colombo, R., Lorenzo, B., Michele, M., Micol, R., Cinzia, P., 2012. Optical remote sensing of vegetation water content. Chapter 10, in Thenkabail, P.S., Lyon, G.J., and Huete, A. *Hyperspectral Remote Sensing of Vegetation*. CRC Press/Taylor & Francis Group, Boca Raton, FL/London, U.K./New York. p. 227.

Colomina, I., Molina, P. 2014. Unmanned aerial systems for photogrammetry and remote sensing: A review. *ISPRS Journal of Photogrammetry and Remote Sensing*, 92, 79–97, ISSN 0924–2716, http://dx.doi.org/10.1016/j.isprsjprs.2014.02.013.

Elvidge, C.D., Yuan, D., Weerackoon, R.D., Lunetta, R.S. 1995. Relative radiometric normalization of landsat multispectral scanner (MSS) data using an automatic scattergram controlled regression. *Photogrammetric Engineering and Remote Sensing*, 61, 1255–1260.

Fayyad, U., Piatetsky-Shapiro, G., Smyth, P., 1996. The KDD process for extracting useful knowledge from volumes of data. *Communications of the ACM*, 39, 27–34.

Galvão, L.S., 2012. Crop type discrimination using hyperspectral data. Chapter 17, in Thenkabail, P.S., Lyon, G.J., and Huete, A. *Hyperspectral Remote Sensing of Vegetation*. CRC Press/Taylor & Francis Group, Boca Raton, FL/London, U.K./New York. pp. 397–422.

Galvão, L.S., Roberts, D.A., Formaggio, A.R., Numata, I., and Breunig, F.M., 2009. View angle effects on the discrimination of soybean varieties and on the relationships between vegetation indices and yield using off-nadir Hyperion data. *Remote Sensing of Environment*, 113, 846–856, DOI:10.1016/j.rse.2008.12.010.

Gitelson, A., 2012a. Non-destructive estimation of foliar pigment (chlorophylls, carotenoids, and anthocyanins) contents: Evaluating a semi-analytical three-band model. Chapter 6, in Thenkabail, P.S., Lyon, G.J., and Huete, A. *Hyperspectral Remote Sensing of Vegetation*. CRC Press/Taylor & Francis Group/Boca Raton, FL/London, U.K./New York. pp. 141–166.

Gitelson, A., 2012b. Remote estimation of crop biophysical characteristics at various scales. Chapter 15, in Thenkabail, P.S., Lyon, G.J., and Huete, A. *Hyperspectral Remote Sensing of Vegetation*. CRC Press/Taylor & Francis Group, Boca Raton, FL/London, U.K./New York. pp. 329–360.

Gnyp, M.L., Bareth, G., Li, F., Lenz-Wiedemann, V.I., Koppe, W., Miao, Y., Hennig, S.D., Jia, L., Laudien, R., Chen, X., 2014. Development and implementation of a multiscale biomass model using hyperspectral vegetation indices for winter wheat in the North China Plain. *International Journal of Applied Earth Observation and Geoinformation*, 33, 232–242.

Goetz, A.F.H., 1992. Imaging Spectrometry for Earth Observations. *Episodes*, 15, 7–14.

Green, A. A., Berman, M., Switzer, P., Craig, M.D., 1988. A transformation for ordering multispectral data in terms of image quality with implications for noise removal. *IEEE Transactions on Geosciences and Remote Sensing*, 26(1), 65–74.

Guerschman, J.P., Hill, M.J., Renzullo, L.J., Barrett, D.J., Marks, A.S., Botha, E.J., 2009. Estimating fractional cover of photosynthetic vegetation, non-photosynthetic vegetation and bare soil in the Australian tropical savanna region upscaling the EO-1 Hyperion and MODIS sensors. *Remote Sensing of Environment*, 113, 928–945.

Guo, X., Huang, X., Zhang, L., Zhang, L., 2013. Hyperspectral image noise reduction based on rank-1 tensor decomposition. *ISPRS Journal of Photogrammetry and Remote Sensing*, 83, 50–63.

Hernández-Clemente, R., Navarro-Cerrillo, R.M., Suárez, L., Morales, F., Zarco-Tejada, P.J., 2011. Assessing structural effects on PRI for stress detection in conifer forests. *Remote Sensing of Environment*, 115, 2360–2375.

Hosseinjani Zadeh, M., Tangestani, M.H., Roldan, F.V., Yusta, I., 2014. Sub-pixel mineral mapping of a porphyry copper belt using EO-1 Hyperion data. *Advances in Space Research*, 53, 440–451.

Jafari, R., Lewis, M.M., 2012. Arid land characterisation with EO-1 Hyperion hyperspectral data. *International Journal of Applied Earth Observation and Geoinformation*, 19, 298–307.

Jensen, J.R. 2004. *Introductory Digital Image Processing: A Remote Sensing Perspective* (3rd edn.). Prentice-Hall, Upper Saddle River, New Jersey.

Lausch, A., Schmidt, A., Tischendorf, L., 2014. Data mining and linked open data–New perspectives for data analysis in environmental research. *Ecological Modelling*, 295, 5–17.

Li, J., Li, C., Xhao, D., Gang, C., 2012. Hyperspectral narrowbands and their indices on assessing nitrogen contents of cotton crop applications. Chapter 24, in Thenkabail, P.S., Lyon, G.J., and Huete, A. *Hyperspectral Remote Sensing of Vegetation*. CRC Press/Taylor & Francis Group, Boca Raton, FL/London, U.K./New York. pp. 579–590.

Mariotto, I., Thenkabail, P.S., Huete, A., Slonecker, E.T., Platonov, A., 2013. Hyperspectral versus multispectral crop-productivity modeling and type discrimination for the HyspIRI mission. *Remote Sensing of Environment*, 139, 291–305.

Markham, B.L. and Barker, J.L., 1987. Radiometric properties of U.S. processed Landsat MSS data. *Remote Sensing of Environment*, 22, 39–71.

Marshall, M., Thenkabail, P., 2014. Biomass modeling of four leading world crops using hyperspectral narrowbands in support of HyspIRI mission. *Photogrammetric Engineering and Remote Sensing*, 80, 757–772.

Middleton, E., Huemmrich, K.F., Cheng, Y., B, Margolis, H., 2012. Spectral bio-indicators of photosynthetic efficiency and vegetation stress. Chapter 12, in Thenkabail, P.S., Lyon, G.J., and Huete, A. *Hyperspectral Remote Sensing of Vegetation*. CRC Press/Taylor & Francis Group, Boca Raton, FL/ London, U.K./New York. pp. 265–288.

Miura, T., Yoshioka, H., 2012. Hyperspectral data in long-term cross-sensor vegetation index continuity for global change studies. Chapter 26, in Thenkabail, P.S., Lyon, G.J., and Huete, A. 2011. *Hyperspectral Remote Sensing of Vegetation*. CRC Press/Taylor & Francis Group, Boca Raton, FL/ London, U.K./New York. pp. 611–636.

Neckel, H. and Labs, D., 1984. The solar radiation between 3300 and 12500 A. *Solar Physics* 90, 205–258. DOI: 10.1007/ BF00173953.

Numata, I., 2012. Characterization on pastures using field and imaging spectrometers. Chapter 9, in Thenkabail, P.S., Lyon, G.J., and Huete, A. 2012. *Hyperspectral Remote Sensing of Vegetation*. CRC Press/Taylor & Francis Group, Boca Raton, FL/London, U.K./New York. pp. 207–226.

Ortenberg, F., 2012. Hyperspectral sensor characteristics airborne, spaceborne, hand-held, and truck-mounted and integration of hyperspectral data with LiDAR. Chapter 2, in Thenkabail, P.S., Lyon, G.J., and Huete, A. 2012. *Hyperspectral Remote Sensing of Vegetation*. CRC Press/ Taylor & Francis Group, Boca Raton, FL/London, U.K./ New York. pp. 39–68.

Plaza, A., Plaza, J., Martin, G., S, S., 2012. Hyperspectral data processing algorithms. Chapter 5, in Thenkabail, P.S., Lyon, G.J., and Huete, A. 2012. *Hyperspectral Remote Sensing of Vegetation*. CRC Press/Taylor & Francis Group, Boca Raton, FL/London, U.K./New York. pp. 121–140.

Pu, R., 2012. Detecting and mapping invasive plant species by using hyperspectral data. Chapter 19, in Thenkabail, P.S., Lyon, G.J., and Huete, A. 2012. *Hyperspectral Remote Sensing of Vegetation*. CRC Press/Taylor & Francis Group, Boca Raton, FL/London, U.K./New York. pp. 447–468.

Qi, J., Inoue, Y., Wiangwang, N., 2012. Hyperspectral sensor systems and data characteristics in global change studies. Chapter 3, in Thenkabail, P.S., Lyon, G.J., and Huete, A. 2012. *Hyperspectral Remote Sensing of Vegetation*. CRC Press/Taylor & Francis Group, Boca Raton, FL, London, FL, New York. pp. 69–92.

Ramsey III, E.W., Rangoonwala, A., 2012. Hyperspectral remote sensing of wetlands. Chapter 21, in Thenkabail, P.S., Lyon, G.J., and Huete, A. 2012. *Hyperspectral Remote Sensing of Vegetation*. CRC Press/Taylor & Francis Group, Boca Raton, FL/London, U.K./New York. pp. 487–512.

Roberts, D.A., 2012.. Hyperspectral vegetation indices. Chapter 14, in Thenkabail, P.S., Lyon, G.J., and Huete, A. 2012. *Hyperspectral Remote Sensing of Vegetation*. CRC Press/ Taylor & Francis Group, Boca Raton, FL, London, U.K./ New York. pp. 309–328.

SAS, 2009. SAS Institute. SAS/STAT user's guide and software release, version 9.2 Ed. SAS Institute, Cary, NC.

Slonecker, T., 2012. Hyperspectral analysis of the effects of heavy metals on vegetation reflectance. Chapter 23, in Thenkabail, P.S., Lyon, G.J., and Huete, A. 2012. *Hyperspectral Remote Sensing of Vegetation*. CRC Press/ Taylor & Francis Group, Boca Raton, FL/London, U.K./ New York. pp. 561–578.

Stroppiana, D., Fava, F., Boschetti, M., Brivio, P.A., 2012. Estimation of nitrogen content in crops and pastures using hyperspectral vegetation indices. Chapter 11, in Thenkabail, P.S., Lyon, G.J., and Huete, A. 2012. *Hyperspectral Remote Sensing of Vegetation*. CRC Press/Taylor & Francis Group, Boca Raton, FL/London, U.K./New York. pp. 245–264.

Tanner, R., 2013. Data mining–das etwas andere eldorado. *Technologie IT-Methoden*, 8, 37–42.

Thenkabail, P., GangadharaRao, P., Biggs, T., Krishna, M., Turral, H., 2007. Spectral matching techniques to determine historical land-use/land-cover (LULC) and irrigated areas using timeseries 0.1-degree AVHRR Pathfinder datasets. *Photogrammetric Engineering and Remote Sensing* 73, 1029–1040.

Thenkabail, P.S., Enclona, E.A., Ashton, M.S., Legg, C., De Dieu, M.J., 2004a. Hyperion, IKONOS, ALI, and ETM+ sensors in the study of African rainforests. *Remote Sensing of Environment* 90, 23–43.

Thenkabail, P.S., Enclona, E.A., Ashton, M.S., Van Der Meer, B., 2004b. Accuracy assessments of hyperspectral waveband performance for vegetation analysis applications. *Remote Sensing of Environment* 91, 354–376.

Thenkabail, P.S., Gumma, M.K., Teluguntla, P., Mohammed, I.A., 2014. Hyperspectral remote sensing of vegetation and agricultural crops. *Photogrammetric Engineering and Remote Sensing*, 80, 697–709.

Thenkabail, P.S., Hanjra, M.A., Dheeravath, V., Gumma, M., 2010. A holistic view of global croplands and their water use for ensuring global food security in the 21st century through advanced remote sensing and non-remote sensing approaches. *Remote Sensing*, 2, 211–261.

Thenkabail, P.S., Lyon, G.J., Huete, A., 2012a. Advances in hyperspectral remote sensing of vegetation and agricultural crops. Chapter 1, in Thenkabail, P.S., Lyon, G.J., and Huete, A. 2012. *Hyperspectral Remote Sensing of Vegetation*. CRC Press/Taylor & Francis Group, Boca Raton, FL/London, U.K./New York. pp. 3–29.

Thenkabail, P.S., Lyon, J.G., Huete, A., 2012b. Synthesis on hyperspectral remote sensing of vegetation. Chapter 28, in Thenkabail, P.S., Lyon, G.J., and Huete, A. 2012. *Hyperspectral Remote Sensing of Vegetation: Current Status and Future Possibilities*. CRC Press/Taylor & Francis Group, Boca Raton, FL/London, U.K./New York. pp. 663–668.

Thenkabail, P.S., Mariotto, I., Gumma, M.K., Middleton, E.M., Landis, D.R., and Huemmrich, F.K. 2013. Selection of hyperspectral narrowbands (HNBs) and composition of hyperspectral twoband vegetation indices (HVIs) for biophysical characterization and discrimination of crop types using field reflectance and Hyperion/EO-1 data. *IEEE Journal of Selected Topics in Applied Earth Observations And Remote Sensing*, 6(2), 427–439, April 2013. DOI: http://dx.doi.org/10.1109/JSTARS.2013.2252601 10.1109/JSTARS.2013.2252601.

Thenkabail, P.S., Smith, R.B., and De-Pauw, E. 2002. Evaluation of narrowband and broadband vegetation indices for determining optimal hyperspectral wavebands for agricultural crop characterization. *Photogrammetric Engineering and Remote Sensing*, 68(6), 607–621.

Thenkabail, P.S., Smith, R.B., De Pauw, E., 2000. Hyperspectral vegetation indices and their relationships with agricultural crop characteristics. *Remote Sensing of Environment*, 71, 158–182.

Thenkabail, P.S., Wu, Z., 2012. An automated cropland classification algorithm (ACCA) for Tajikistan by combining Landsat, MODIS, and secondary data. *Remote Sensing*, 4, 2890–2918.

Thomas, V., 2012. Hyperspectral remote sensing for forest management. Chapter 20, in Thenkabail, P.S., Lyon, G.J., and Huete, A. 2012. *Hyperspectral Remote Sensing of Vegetation*. CRC Press/Taylor & Francis Group, Boca Raton, FL/London, U.K./New York. pp. 469–486.

Tits, L., De Keersmaecker, W., Somers, B., Asner, G.P., Farifteh, J., Coppin, P., 2012. Hyperspectral shape-based unmixing to improve intra-and interclass variability for forest and agroecosystem monitoring. *ISPRS Journal of Photogrammetry and Remote Sensing*, 74, 163–174.

Torres-Sánchez, J., Peña, J.M., de Castro, A.I., López-Granados, F. 2014. Multi-temporal mapping of the vegetation fraction in early-season wheat fields using images from UAV. *Computers and Electronics in Agriculture*, 103, 104–113, ISSN 0168–1699, http://dx.doi.org/10.1016/j. compag.2014.02.009. (http://www.sciencedirect.com/science/article/pii/S0168169914000568).

Vaughan, R.G., Titus, T.N., Johnson, J.R., Hagerty, J., Gaddis, L., Soderblom, L.A., Geissler, P., 2011. Hyperspectral analysis of rocky surfaces on the earth and other planetary systems. Chapter 27, in Thenkabail, P.S., Lyon, G.J., and Huete, A. 2012. *Hyperspectral Remote Sensing of Vegetation*. CRC Press/Taylor & Francis Group, Boca Raton, FL/London, U.K./New York. pp. 637–662.

Vermote, E.F., El Saleous, N.Z., Justice, C.O., 2002. Atmospheric correction of MODIS data in the visible to middle infrared: First results. *Remote Sensing of Environment*, 83(1–2), 97–111.

Yao, H., Tang, L., Tian, L., Brown, R.L., Bhatnagar, D., Cleveland, T.E., 2011. Using hyperspectral data in precision farming applications. Chapter 25, in Thenkabail, P.S., Lyon, G.J., and Huete, A. 2012. *Hyperspectral Remote Sensing of Vegetation*. CRC Press/Taylor & Francis Group, Boca Raton, FL/London, U.K./New York. pp. 591–610.

Zhang, Y., 2012. Forest leaf chlorophyll content study using hyperspectral remote sensing. Chapter 7, in Thenkabail, P.S., Lyon, G.J., and Huete, A. 2012. *Hyperspectral Remote Sensing of Vegetation*. CRC Press/Taylor & Francis Group, Boca Raton, FL/London, U.K./New York. pp. 167–186.

Zhu, Y., Wang, W., Yao, X., 2012. Estimating leaf nitrogen concentration (LNC) of cereal crop with hyperspectral data. Chapter 8, in Thenkabail, P.S., Lyon, G.J., and Huete, A. 2012. *Hyperspectral Remote Sensing of Vegetation*. CRC Press/Taylor & Francis Group, Boca Raton, FL/London, U.K./New York. pp. 187–206.

III

Rangelands

10

Global View of Remote Sensing of Rangelands: Evolution, Applications, Future Pathways

Matthew C. Reeves
U.S.D.A Forest Service

Robert A. Washington-Allen
University of Tennessee

Jay Angerer
Texas A&M University

E. Raymond Hunt, Jr.
USDA-ARS Beltsville Agricultural Research Center

Ranjani Wasantha Kulawardhana
Texas A&M University

Lalit Kumar
University of New England

Tatiana Loboda
University of Maryland

Thomas Loveland
U.S. Geological Survey EROS Center

Graciela Metternicht
University of New South Wales

R. Douglas Ramsey
Utah State University

Acronyms and Definitions

AATSR	Advanced Along-Track Scanning Radiometer
AI	Aridity Index
ANPP	Aboveground net primary productivity
ASSOD	Assessment of the Status of Human-induced Soil Degradation in South and Southeast Asia
ATSR	Along-Track Scanning Radiometer
AUM	Animal unit month
AVHRR	Advanced Very High Resolution Radiometer
BRDF	Bidirectional reflectance distribution function
BT	Brightness temperature
CBI	Composite burn index
CSIRO	Commonwealth Scientific and Industrial Research Organization

DISCover	International Geosphere-Biosphere Programme Data and Information System, Global Land Cover Classification	MEA	Millennium Ecosystem Assessment
		MERIS	Medium Resolution Imaging Spectrometer
		MODIS	Moderate Resolution Imaging Spectroradiometer
DLDD	Desertification, land degradation, and drought	MSI	Moisture Stress Index
dNBR	Differenced normalized burn ratio	MSS	Multispectral Scanner
EDR	Environmental Data Record	MTBS	Monitoring Trends in Burn Severity
Eionet	European Environment Information and Observation Network	MWIR	Mid-wave Infrared
		NASA	National Aeronautics and Space Administration
EM	Electromagnetic	NDBR	Normalized difference burn ratio
ENVISAT	Environmental Satellite	NDII	Normalized Difference Infrared Index
EOS	Earth Observing System	NDVI	Normalized Difference Vegetation Index
ERS-2	European Remote Sensing (satellite)	NDWI	Normalized Difference Water Index
ERTS	Earth Resources Technology Satellite	NIR	Near infrared (0.7–1.0 μm)
ESA	European Space Agency	NLCD	National Land Cover Database
ETM+	Enhanced Thematic Mapper Plus	NOAA	National Oceanic and Atmospheric Administration
EVI	Enhanced vegetation index	NPoP	National Polar-orbiting Partnership
FAO	Food and Agricultural Organization of the United Nations	NPP	Net primary productivity
		NWCG	National Wildfire Coordinating Group
fPAR	Fraction of Photosynthetically Active Radiation	OLI	Optical Land Imager
FRE	Fire radiative energy (in Joules)	P	Precipitation
FROM-GLC	Fine Resolution Observation and Monitoring of Global Land Cover	PET	Potential Evapotranspiration
		PHYGROW	Phytomass Growth Simulation Model
FRP	Fire radiative power (in Watts)	PSNnet	Moderate Resolution Imaging Spectroradiometer net photosynthesis product
GAC	Global Area Coverage		
GDAS	Global Data Assimilation System	RdNBR	Relativized differenced normalized burn ratio
GEF	Global Environmental Facility	Rio+20	United Nations Conference on Sustainable Development
GEO-5	Fifth Global Environment Outlook		
GEO BON	Group on Earth Observations Biodiversity Observation Network	RUE	Rain use efficiency
		SAVI	Soil Adjusted Vegetation Index
GIMMS	Global Inventory Modeling and Mapping Studies	SDGs	Sustainable development goals
GIS	Geographic information system	SSI	Soil Stability Index
GLADA	Global Assessment of Land Degradation and Improvement	SST	Sea Surface Temperature
		SOVEUR	Mapping of Soil and Terrain Vulnerability in Central and Eastern Europe
GLADIS	Global Land Degradation Information System		
GLASOD	Global assessment of human-induced soil degradation	SPOT	Satellite Pour l'Observation de la Terre (French)
		SWIR	Shortwave infrared (1.1–2.4 μm)
GLC2000	Global Land Cover 2000	$SWIR_{2.2}$	Shortwave infrared (2.08–2.35 μm)
GPP	Gross primary production	Tg	teragrams
GVMI	Global Vegetation Moisture Index	TIROS-N	Television Infrared Observation Satellite-Next Generation
HRVIR	Haute Résolution dans le Visible et l'Infra-Rouge (French)		
		TIRS	Thermal Infrared Sensor
IDRISI	a geographic information system and remote sensing software produced by Clark University	TM	Thematic Mapper (Landsat)
		TNDVI	Transformed Normalized Difference Vegetation Index
IGBP	International Geosphere–Biosphere Programme		
IRS	Indian Remote Sensing	TVI	Transformed Vegetation Index
J	Joules	UMD	University of Maryland
JPSS	Joint Polar Satellite System	UNCCD	United Nations Convention to Combat Desertification
LADA	Land Degradation Assessment in Drylands		
LAI	Leaf area index	UNEP	United Nations Environment Program
LCCS	Land cover classification system	USFWS	United States Fish and Wildlife Service
LEWS	Livestock Early Warning System	USGS	United States Geological Survey
LNS	Local net primary productivity scaling	VASClimO	Variability Analyses of Surface Climate Observations
LUS	Land use system		
LWCI	Leaf Water Content Index	VGT	VEGETATION sensor onboard SPOT satellite
LWIR	Long Wave Infrared	VI	Vegetation Index

10.1 Introduction

The term "rangeland" is rather nebulous, and there is no single definition of rangeland that is universally accepted by land managers, scientists, or international bodies (Lund, 2007; Reeves and Mitchell, 2011). Dozens and possibly hundreds (Lund, 2007) of definitions and ideologies exist because various stakeholders often have unique objectives requiring different information. For the purpose of describing the role of remote sensing in a global context, it is, however, necessary to provide definitions to orient the reader. The Food and Agricultural Organization (FAO) of the United Nations convened a conference in 2002 and again in 2013 to begin addressing the issue of harmonizing definitions of forest-related activities. Based on this concept, here rangelands are considered lands usually dominated by nonforest vegetation. The Society for Range Management defines rangelands as (SRM, 1998)

> Land on which the indigenous vegetation (climax or natural potential) is predominantly grasses, grass-like plants, forbs, or shrubs and is managed as a natural ecosystem. If plants are introduced, they are managed similarly. Rangelands include natural grasslands, savannas, shrublands, many deserts, tundra, alpine communities, marshes, and wet meadows.

Rangelands occupy a wide diversity of habitats and are found on every continent except Antarctica. Excluding Antarctica and barren lands, rangelands occupy 52% of the Earth's surface based on the land cover analysis presented in Figure 10.1.

Figure 10.1 is based on the 2005 Moderate Resolution Imaging Spectroradiometer (MODIS) Collection 4.5, 1 km² land cover (the University of Maryland [UMD] classification), and suggested rangeland classes for this dataset are closed shrubland, open shrubland, woody savanna, savanna, and grassland. Using these classes, Russia, Australia, and Canada are the top three countries with the most rangelands (Table 10.1) representing 18%, 10%, and 8% of the global extent, respectively. The large areal extent of rangelands, high cost of field data collection, and quest for societal well-being have, for decades, provided rich opportunity for remote sensing to aid in answering pressing questions.

10.2 History and Evolution of Global Remote Sensing

The application of digital remote sensing to rangelands is as long as the history of digital remote sensing itself. Before the launch of the Earth Resources Technology Satellite (ERTS)—later renamed Landsat—scientists were evaluating the use of multispectral aerial imagery to map soils and range vegetation (Yost and Wenderoth, 1969). During the late 1960s, the promise of ERTS, designed to drastically improve our ability to update maps and study Earth resources, particularly in developing countries, was eagerly anticipated by a number of government agencies (Carter, 1969). With the ERTS launch on July 23, 1972, a flurry of research activity aimed at the application of this new data source to map Earth resources began. Practitioners who pioneered the use of satellite-based digital remote sensing found the new data source a significant value for rangeland assessments (e.g., Rouse et al., 1973, 1974; Bauer, 1976). This early work established many of the basic techniques still in use today to assess and monitor global rangelands. The following subsections discuss the evolution of remote-sensing data, methods, and approaches in various decades.

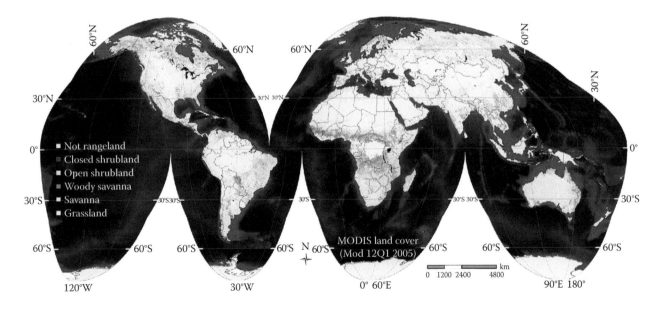

FIGURE 10.1 Global distribution of land cover types (MODIS MOD12Q1, 2005; University of Maryland Classification), considered rangelands for this chapter.

TABLE 10.1   Global Area of Rangeland Vegetation Types Estimated Using MODIS Land Cover Data (Mod12Q1) for the Top 12 Countries with the Most Rangeland.

Country	Area (km²)	CSL	Grassland	OSL	Savanna	Woody Savanna	Rangeland Area	Rangeland Proportion (%)
Russia	16,851,940	5,461	795,938	8,174,738	170,456	1,223,381	10,369,974	62
Australia	7,706,142	13,543	182,983	4,690,912	505,136	620,265	6,012,839	78
Canada	9,904,700	1,187	271,855	3,901,991	54,738	509,117	4,738,888	48
United States	9,450,720	78,929	1,777,542	2,077,055	95,380	673,199	4,702,105	50
China	9,338,902	42,548	1,745,760	1,002,771	73,717	399,032	3,263,828	35
Brazil	8,507,128	15,879	278,859	136,105	1,852,468	541,479	2,824,790	33
Kazakhstan	2,715,976	512	1,793,967	171,930	1,859	14,538	1,982,806	73
Argentina	2,781,013	88,877	363,509	1,094,845	121,035	94,377	1,762,643	63
Mexico	1,962,939	64,011	217,212	556,928	85,889	194,310	1,118,350	57
Sudan	2,490,409	8,210	278,848	205,781	404,276	163,169	1,060,284	43

CSL is closed shrubland, OSL is open shrubland, and rangeland proportion is the rangeland area column divided by the area column multiplied by 100.

10.2.1 Beginning of Landsat MSS Era, 1970s

In this first decade of satellite-based digital remote sensing, rangeland scientists quickly assessed the capabilities of this new tool across the globe (Rouse et al., 1973; Graetz et al., 1976). Work by Rouse et al. (1973), in what would later become the Normalized Difference Vegetation Index (NDVI) (Rouse et al., 1974), applied multitemporal ERTS (Landsat 1) at 79 m² spatial resolution data to the grasslands of the central Great Plains of the United States and documented that the normalized ratio of the multispectral scanner (MSS) near-infrared (NIR) (band 7) and red band (band 5) was sensitive to vegetation dry biomass, percent green, and moisture content (Figure 10.2). They also determined that within uniform grasslands, field-based estimates of moisture content and percent green cover accounted for 99% of the variation in their "Transformed Vegetation Index" (TVI). The TVI was later renamed to the Transformed Normalized Difference Vegetation Index (TNDVI) (Deering et al., 1975) and is calculated as the square root of the NDVI plus an arbitrary constant (0.5 in their case). This transformation of the NDVI was done to avoid negative values.

The NDVI is, to date, one of the most widely used vegetation index on a global basis. Figure 10.2 shows the graphic published by Rouse et al. (1973) identifying the tight relationship between field-derived green biomass and the TVI. The significance of Figure 10.2 is the demonstration of potential to track vegetation growth across time, thus documenting the ability for remote-sensing instruments to monitor vegetation dynamics and the importance of systematic and uninterrupted collection of remotely sensed imagery.

Another significant development during this first decade of satellite-based remote sensing was the "tasseled cap transformation" (Kauth and Thomas, 1976). The tasseled cap (or "Kauth–Thomas transformation" to some) employed principal component analysis to understand the covariate nature of the four MSS spectral bands and extract from those data the primary ground features, or components, influencing the spectral signature. The tasseled cap and its eventual successor—the brightness, greenness, wetness transform (Crist and Cicone, 1984) applied to the Landsat Thematic Mapper (TM) sensor—has been a widely used tool for many land resource applications (Hacker, 1980; Graetz et al., 1986; Todd et al., 1998).

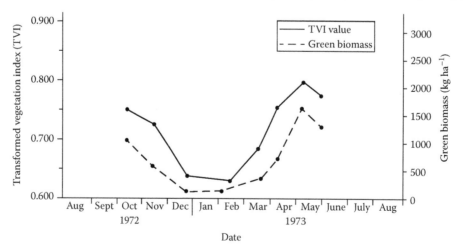

FIGURE 10.2   ERTS-1 TVI values versus green biomass. (Original from Rouse, J.W. et al., Monitoring vegetation systems in the Great Plains with ERTS, in *Proceedings of the Third ERTS Symposium*, Washington, DC, 1973, pp. 309–317.)

The NDVI and the tasseled cap provided the ability to convert reflectance values collected across multiple spectral bands into biophysically focused data layers, thus giving range managers and ecologists a tool by which to directly assess and monitor vegetation growth.

10.2.2 Multiple Sensor Era, 1980s

With the development of the NDVI and the launch of the Television Infrared Observation Satellite-Next Generation (TIROS-N) satellite carrying the Advanced Very High Resolution Radiometer (AVHRR) in October of 1978, remote-sensing practitioners now had the means to monitor temporal vegetation dynamics across very large areas (Tucker, 1979). The 1 km^2 resolution of the AVHRR was ideal for continental-scale monitoring, which was not possible with Landsat images given the computing power and data storage capacities of that era. Further, a 1-day global repeat cycle provided the ability to track phenological changes in vegetation growth within and between years—a feature also not possible with the 18- and 16-day repeat cycles of the Landsat platforms. Gray and McCrary (1981) showed the utility of the AVHRR for vegetation mapping and noted that vegetation indices derived from this sensor could be related to plant growth stress due to water deficits. This relationship, coupled with the high temporal repeat interval of the TIROS-N, led to the use of the NDVI to monitor the impact of drought on grasslands across the Sahel region of Africa (Tucker et al., 1983) and by direct inference predict the impact of drought to local human populations (Prince and Tucker, 1986).

The application of the NDVI to semiarid landscapes was somewhat problematic due to generally low vegetation canopy cover in these environments and the fact that background soil brightness tended to influence the resulting NDVI values (Elvidge and Lyon, 1985). The soil-adjusted vegetation index (SAVI) (Huete, 1988) was developed as a simple modification to the NDVI to account for the influence of soil on the reflectance properties of green vegetation. The SAVI has been used widely within semiarid environments where vegetation cover is low. The 1980s also saw great strides in satellite-based terrestrial remote sensing with the launch of Landsat 4 in July of 1982 and Landsat 5 in March of 1985, as well as the launch of the French Satellite Pour l'Observation de la Terre (SPOT) in 1986. Each platform carried sensors with slightly different capabilities, but each focused their spectral resolution on the red and NIR portions of the electromagnetic spectrum, save one. The Landsat TM was a significant improvement over its predecessor, the MSS. Not only were the spatial and radiometric resolutions improved, but also the TM supported two additional spectral bands calibrated to the shortwave infrared portion of the electromagnetic (EM) spectrum. This significant addition provided the ability to monitor leaf moisture (Tucker, 1980, Hunt and Rock, 1989) as well as identify and map recent wildfires (Chuvieco and Congalton, 1988, Key and Benson, 1999a,b).

While the work with AVHRR in Africa expanded and new sensors were becoming readily available, researchers in Australia were evaluating the applicability of Landsat images to monitoring and assessment of rangelands. Work by Dean Graetz, now retired from the Commonwealth Scientific and Industrial Organisation (CSIRO) of Australia, was instrumental in fostering use of satellite remote sensing to monitor rangelands (Graetz et al., 1983, 1986, 1988; Pech et al., 1986; Graetz, 1987). This work, coupled with other CSIRO scientists such as Geoff Pickup (Pickup and Nelson, 1984; Pickup and Foran, 1987; Pickup and Chewings, 1988), firmly established Australia as a leader in the use of remote sensing for rangeland monitoring and assessment.

Researchers in Australia had similar problems applying digital imagery to semiarid rangelands as did the United States and Africa teams. The difficulty in applying imagery collected by the Landsat sensors to rangeland assessment is documented by Tueller et al. (1978) and McGraw and Tueller (1983), who found that the spectral differences among semiarid range plant communities were so small that they approached the noise level of the imagery. Even with these limitations, Robinove et al. (1981) and Frank (1984) developed methodologies for using albedo to measure soil erosion on rangelands. Pickup and Nelson (1984) developed the soil stability index (SSI) by using the ratio of the MSS green band divided by the NIR, plotted against the ratio of the red divided by the NIR. This comparison between the two ratios provided a quantitative measure of soil stability. Further, a temporal sequence of SSI images could be used as a monitoring tool to identify changes in landscape state (Pickup and Chewings, 1988). As research progressed in the use of imagery on rangelands through the 1980s, the US civilian remote-sensing program began a transition to private sector management of the Landsat program. Issues of data cost and data licensing arose placing financial and legal limitations on research and data sharing. Still, research and application continued into the 1990s with an increased demand by federal land managers for landscape-level information.

10.2.3 Advanced Multisensor Era, 1990s

In 1989 and throughout the 1990s, the US Fish and Wildlife Service (USFWS) and the US Geological Survey (USGS) embarked on a number of large-scale land cover mapping efforts across the United States. The Gap Analysis Program initiated by the USFWS and later absorbed into the USGS was designed as a spatial database to identify landscapes of high biological diversity and evaluate their management status (Scott et al., 1993). The Gap Analysis was built around the linkage between wildlife habitat relationship (WHR) models and a detailed land cover map. This linkage allowed the WHR database to be spatially visualized by relating habitat parameters to land cover. The significance of this effort to remote sensing is that at the time, no one had attempted to map vegetation across landscapes requiring multiple frames of radiometrically normalized satellite imagery. The first digitally produced land cover map derived from a statistical classification of a 14-image mosaic of radiometrically normalized Landsat TM imagery was completed for the state of Utah in 1995 by Utah State University (Homer et al., 1997). Programs like the Gap Analysis, coupled with the advent of the publicly available Internet in 1991, provided the impetus for a new brand of remote sensing centering on large data and improved data access and product delivery. During the late 1980s, the National Aeronautics and Space Administration (NASA) was envisioning the need to provide rapid data access to users. At the time, image acquisition and delivery to the end user required a minimum of a few weeks. There was a need for time critical imagery by users and to meet that demand; NASA set a goal of data delivery to within 24 h of acquisition. Even with the advent of data transfer through the Internet, a 24-h lag between acquisition and delivery is a relatively new phenomenon of the mid-2000s.

10.2.4 New Millennium Era, 2000s

In this era, noteworthy changes to the remote-sensing community, including dramatic improvements in data availability, spatial and spectral resolution, and temporal frequency (Figure 10.3), were made. Commonly used high-spatial-resolution sensors launched during this time including IKONOS, QuickBird, GeoEye-1, and WorldView-2 exhibit spatial resolutions in the multispectral domain of 4, 2.4, 1.65, and 2 m², respectively.

These sensors have enabled improvements in species discrimination (e.g., Everitt et al., 2008; Mansour et al., 2012) and stand-level attributes such as canopy cover (e.g., Sant et al., 2014). Use of QuickBird for identifying giant reed (*Arundo donax*) improved both user's and producer's accuracy by an average of 12% over use of SPOT 5 alone (Everitt et al., 2008). Similarly, Sant et al. (2014) used IKONOS imagery to quantify percent vegetation cover and explained 5% more variation than using Landsat (r^2 of 0.79 versus 0.84) alone. Hyperspectral data emanating from this era also enable greater discrimination of many biophysical features than multispectral sensors alone especially in the realm of invasive species mapping. Parker and Hunt (2004) distinguished leafy spurge with the Airborne Visible/Infrared Imaging Spectrometer (AVIRIS) data with overall accuracy of 95%, while Oldeland et al. (2010) detected bush encroachment by *Acacia* spp. ($r^2 = 0.53$). These improved capabilities emanate not only from improved sensor characteristics in the 2000s, but also greatly improved data availability.

In 1999, the launch of Landsat 7, coupled with new sensors from a host of other countries as well as commercial, high-spatial-resolution sensors, ushered a new era of global assessment and monitoring of natural and human landscapes. With the end of private sector management of the Landsat program in 1999, imagery was again placed in the public domain, and costs for Landsat

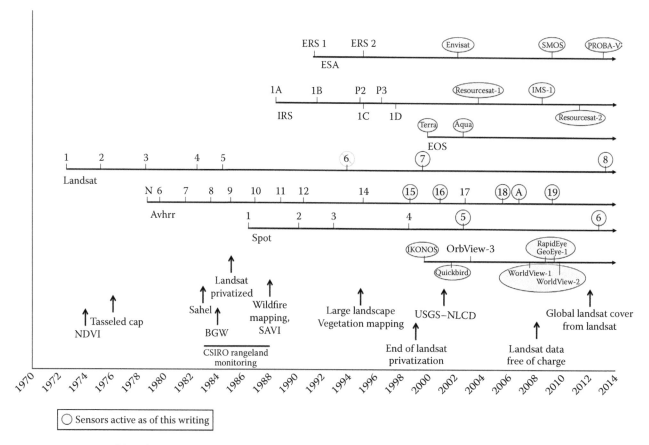

FIGURE 10.3 History of digital remote sensing sensors used in research and monitoring of rangelands since the advent of the technology in the early 1970's. Specific research milestones and policy changes are also noted.

imagery were reduced to $600 per scene (previously set at $4400 per scene) for Landsat 7 Enhanced Thematic Mapper Plus (ETM+) imagery and $450 per scene for Landsat 5 TM. This reduction in cost, coupled with the free exchange of data between collaborators, boosted research and application of satellite remote sensing. Further, the replacement of the AVHRR as the primary global sensor with the much-advanced MODIS with 36 spectral bands spanning the 405–14,384 nm range provided the ability for scientists to model, map, and monitor not only land cover but also net primary productivity (NPP) among other metrics. The now 15-year history of the MODIS sensor aboard two platforms (Terra and Aqua) has provided an unprecedented source of global land cover dynamics data freely available to land managers and scientists. In 2008, the USGS made all Landsat data accessed through the Internet free of charge. With this policy change, scene requests at the USGS Earth Resources Observation and Science Center jumped from 53 images per day to about 5800 images per day. This increase in data demand and delivery has arguably resulted in research in the 2000s centered on the copious use of imagery across multiple temporal and landscape scales. Commercial satellites such as the IKONOS, launched in 1999, QuickBird in 2001, and the WorldView and GeoEye satellites launched between 2007 and 2009 has provided on-demand access to high spatial resolution (submeter to a few meters) that allows data integration between a wide array of platforms and spatial scales (Sant et al., 2014).

10.3 State of the Art

Millions of people depend on rangelands for their livelihood. This dependence raises numerous concerns about the health, maintenance, and management of rangelands from local to global perspectives. Discerning and describing how rangelands are changing at multiple spatial and temporal scales requires the integration of sensors that possess specific characteristics. The current suite of government-sponsored and commercial sensors suitable for regional to global analysis span the spatial range of submeter to 1 km^2, a temporal range of daily to bimonthly (temporal resolution is inversely proportional to spatial resolution), and all have the capacity to image landscapes in the visible and NIR (Figure 10.4). The most commonly used sensors for global applications, however, have spatial resolutions of between 250 and 1000 m^2 (e.g., MODIS, AVHRR, and Visible Infrared Imaging Radiometer Suite [VIIRS]) and exhibit high temporal frequency, numerous spectral bands, but relatively low spatial resolution. Sensors best suited for regional to local applications (e.g., Landsat, SPOT, WorldView, and GeoEye) have higher spatial resolutions (submeter to 30 m^2) and lower temporal repeat cycles.

The present role of remote sensing for characterizing five globally significant phenomena are discussed hereafter, including land degradation, fire, food security, land cover, and vegetation response to global change (Table 10.2). These factors

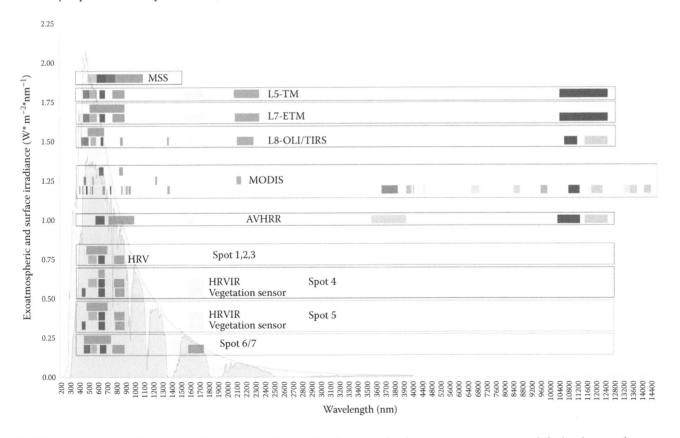

FIGURE 10.4 Exoatmosphreic and surface irradiance for wavelengths across the electromagnetic spectrum and the bandpasses of sensors (colored squares within each sensor box) commonly used for rangeland studies and monitoring.

TABLE 10.2   Four Most Common Sensors for Regional and Global Applications, Their Characteristics, and Example Applications

Satellite (Sensors)	Characteristics (a Is Spatial Resolution, b Is Launch Date, c Is Swath Width, and d Is Revisit Time)	Rangeland Application Examples	References
Landsat (5, 7, 8) (Thematic Mapper, Enhanced Thematic Mapper Plus [ETM+], Optical Land Imager [OLI])	(a) 15 (panchromatic), 30 (multispectral), 100 (thermal), (b) 1999 (ETM+) and 2013 (OLI), (c) 185 km × 170 km, and (d) 16 days	Fire (often dNBR, NBR, LWCI)	
		Burn severity (dNBR, RdNBR, tasseled cap brightness)	Key and Benson (2006), Miller and Thode (2007), and Loboda et al. (2013)
		Burned area mapping (Eidenshenk et al., 2007)	Eidenshink et al. (2007)
		Fuel moisture (variety of indices such as NDVI, NDII, and LWCI)	Chuvieco et al. (2002)
		Vegetation attributes	
		Land cover (varied methods)	Gong et al. (2013), Fry et al. (2011), and Rollins (2009)
		Leaf area index (LAI)/Fraction of Photosynthetically Active Radiation (fPAR) absorbed by vegetation (radiative transfer and vegetation indices)	Shen et al. (2014)
		Net primary production (NPP) (multisensor fusion and process modeling)	Li et al. (2012)
		Degradation (change detection and residual trend analysis)	Jabbar and Zhou (2013)
SPOT (VEGETATION)	(a) 1000, (b) 1998, (c) 2250 km, and (d) 1–2 days	Fire	
		Burned area mapping dNBR (NDVI, NDWI)	Silva et al. (2005) and Tansey et al. (2004)
		Fuel moisture (primarily NDVI, NDWI)	Verbesselt et al. (2007)
		Vegetation attributes	
		Land cover (GLC2000)	Bartholomé and Belward (2005)
		NPP/abundance (NDVI, process modeling)	Telesca and Lasaponara (2006), Geerken et al. (2005), and Jarlan et al. (2008)
		Degradation (trend analysis)	Fang and Ping (2010)
Aqua and Terra (Moderate Resolution Imaging Spectroradiometer)	(a) 250 (red, NIR), 500 (multispectral), 1000 (multispectral); (b) 2000 (Terra), 2002 (Aqua); (c) 2230 km; and (d) 1–2 days	Fire (often dNBR, NBR, LWCI)	
		Active fire detection (thermal anomalies and fire radiative potential)	Giglio et al. (2003, 2009)
		Burned area evaluation (SWIR VI and change detection)	Roy et al. (2008)
		Burn severity (time-integrated dNBR)	Veraverbeke et al. (2011)
		Fuel moisture (empirical relations and radiative transfer modeling; many vegetation indices [GVMI,NDWI, MSI, etc.])	Yebra et al. (2008) and Sow et al. (2013)
		Vegetation attributes	
		Land cover (varied methods)	Friedl et al. (2010)
		LAI/fPAR absorbed by vegetation (radiative transfer modeling)	Myneni et al. (2002) and Wenze et al. (2006)
		NPP (process modeling)	Running et al. (2004), Reeves et al. (2006), Zhao et al. (2011)
		Degradation (rain use efficiency, local NPP scaling, trend and condition analysis)	Bai et al. (2008), Prince et al. (2009), and Reeves and Bagget (2014)
		Livestock Early Warning System (time series analysis of NDVI, and biomass)	Angerer (2012) and Yu et al. (2011)

(Continued)

TABLE 10.2 (*Continued*) Four Most Common Sensors for Regional and Global Applications, Their Characteristics, and Example Applications

Satellite (Sensors)	Characteristics (a Is Spatial Resolution, b Is Launch Date, c Is Swath Width, and d Is Revisit Time)	Rangeland Application Examples	References
National Oceanic and Atmospheric Administration (Advanced Very High Resolution Radiometer)	(a) 1000 m, (b) NOAA-15 (1998), NOAA-16 (2000), NOAA-18 (2005), NOAA-19 (2009) satellite series (1980 to present). The approximate scene size is 2400 km × 6400 km	**Fire**	
		Active fire detection (thermal anomalies and NDVI)	Pu et al. (2004), Flasse and Ceccato (1996), and Dwyer et al. (2000)
		Burned area evaluation (multitemporal multithreshold approach)	Barbosa et al. (1999)
		Fuel moisture (NDVI)	Paltridge and Barber (1988) and Eidenshink et al. (2007)
		Vegetation attributes	
		Land cover (unsupervised and supervised time series analysis)	Loveland et al. (2000) and Hansen et al. (2000)
		LAI/fPAR absorbed by vegetation (radiative transfer modeling, feedforward neural network)	Myneni et al., (2002), Ganguly (2008), and Zhu and Southworth (2013)
		NPP (time-integrated NDVI)	An et al. (2013)
		Degradation (NDVI and rainfall use efficiency)	Wessels et al. (2004) and Bai et al. (2008)

Many sensors that may have use for evaluating rangeland are not included. Svoray et al. (2013) provide a larger number of example applications in rangeland environments, but this table focuses largely on globally applicable sensors and global applications.

are not mutually exclusive and often exhibit significant interaction. Using remote sensing at global scales provides insight to what may be anticipated in the future and indicates regions where ecological thresholds have been crossed, beyond which decreased goods and services from rangelands can be expected.

10.3.1 Rangeland Degradation

Land and soil degradation are accelerating, and drought is escalating worldwide. At the UN Conference on Sustainable Development (Rio+20), world leaders acknowledged that desertification, land degradation, and drought (DLDD) are challenges of a global dimension affecting the sustainable development of all countries, especially developing countries. Drylands are often identified and classified according to the aridity index (AI), which is defined as P/PET where P is the annual precipitation and PET is the potential evapotranspiration. Drylands yield AI values ≤ 0.65. Despite decades of research, standards to measure progression of land degradation (e.g., global mapping and monitoring systems) remain elusive, but remote sensing plays a significant role.

10.3.1.1 Soil and Land Degradation and Desertification: What Is the Difference?

Land degradation and desertification have been sometimes used synonymously. Land degradation refers to any reduction or loss in the biological or economic productive capacity of the land (UNCCD, 1994) caused by human activities, exacerbated by natural processes, and often magnified by the impacts of climate change and biodiversity loss. In contrast, desertification only occurs in drylands and is considered as the last stage of land degradation (Safriel, 2009).

10.3.1.2 Role of Remote Sensing for Monitoring Rangeland Degradation

Much research conducted over the last decade has been on remotely sensed biophysical indicators of land degradation processes (e.g., soil salinization, soil erosion, waterlogging, and flooding), without integration of socioeconomic indicators (Metternicht and Zinck 2003, 2009; Allbed and Kumar 2013). Studies from the 1970s onward have related soil erosion severity to variations in spectral response. Good reviews of spectrally based mapping of land degradation are found in Metternicht and Zinck (2003), Bai et al. (2008), Marini and Talbi (2009), and Shoshanya et al. (2013). Moreover, research work from the 1990s and 2000s (Metternicht. 1996; Vlek et al. 2010; Le et al., 2012; Shoshanya et al., 2013) reports the benefits of a synergistic use of satellite- and/or airborne remote sensing with ground-based observations to provide consistent, repeatable, cost-effective information for land degradation studies at regional and global scales. Hereafter follows a brief description of some of the most frequent applications of remote sensing applied in "global or subglobal assessments" of land degradation. These remotely sensed products include biomass and vegetation health modeling via NDVI and NPP, rain use efficiency (RUE), and local NPP scaling.

10.3.1.3 Biomass and Vegetation Health Modeling as an Indicator of Degradation

The biomass produced by soil and other natural resources can be a proxy for land health (Nkonya et al., 2013). In this vein, Bai et al. (2008) framed land degradation in the context of the Land Degradation Assessment in Drylands (LADA) program as long-term loss of ecosystem function and productivity and used trends in 8 km² NDVI from the Global Inventory Modeling and

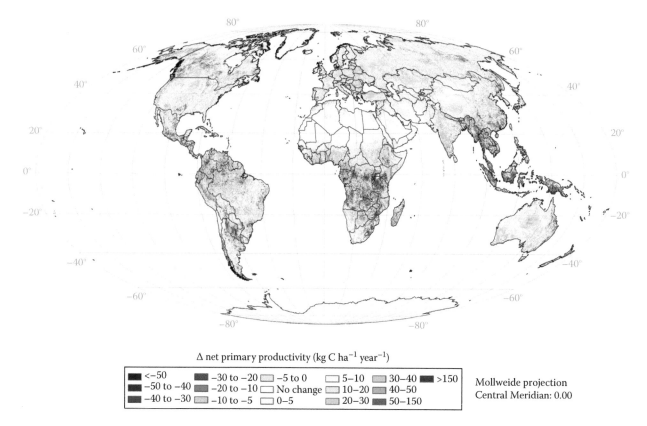

Δ net primary productivity (kg C ha⁻¹ year⁻¹)

■ <−50	■ −30 to −20	□ −5 to 0	□ 5–10	▨ 30–40	■ >150
■ −50 to −40	▨ −20 to −10	□ No change	□ 10–20	▨ 40–50	
▨ −40 to −30	▨ −10 to −5	□ 0–5	▨ 20–30	▨ 50–150	

Mollweide projection
Central Meridian: 0.00

FIGURE 10.5 Global change in NDVI scaled in terms of NPP using MODIS 1 km² 8-day composite net photosynthesis data (1981–2003). NDVI is a proxy indicator of changes in NPP. (From Bai, Z.G. et al., *Soil Use Manage.*, 24, 223, 2008.)

Mapping Studies (GIMMS) as a "proxy indicator" of changes in NPP. Figure 10.5 represents changes in NPP from 1981 to 2003 resulting from fusion of GIMMS NDVI and MODIS 1 km² NPP (Bai et al., 2008). The NDVI is related to variables such as leaf area index (LAI) (Myneni et al., 1997), the fraction of photosynthetically active radiation (fPAR) absorbed by vegetation, and NPP. This explains why many NPP estimates derived from remote-sensing approaches are based on LAI, and fPAR commonly from the AVHRR onboard the National Oceanic and Atmospheric Administration (NOAA) satellite, and the MODIS on the Terra and Aqua satellites (Ito, 2011). One caveat to remotely sensed estimates of NPP for degradation analyses is the need for comparison with ground-measured biophysical parameters such as NPP, LAI, or soil erosion (or salinization) for accuracy assessment (Bai et al., 2008; Le et al., 2012).

10.3.1.4 Rain Use Efficiency

RUE (ratio of NPP to rainfall) can be used to distinguish between the relatively low NPP of drylands associated with inherent moisture deficit and the additional decline in primary production due to land degradation (Le Houérou, 1984; Le Houérou et al., 1988; Pickup, 1996). In the context of the LADA project, Bai et al. (2008) estimated RUE from the ratio of the annual sum of NDVI (derived from MODIS and NOAA AVHRR) to annual rainfall and used it to identify and isolate areas where declining productivity was a function of drought (Figure 10.6).

Figure 10.6 was produced using the same GIMMS NDVI data as Figure 10.5 in concert with Variability Analyses of Surface Climate Observations (VASClimO)-gridded precipitation data at 0.5° resolution. This recalibration process was thought to yield a proxy index for land degradation, assuming that a decline in vegetation for any other reason than rainfall (and temperature) differences would be an expression of some form of degradation.

Statistical analysis showed 2% of the land area exhibited a negative trend at the 99% confidence level, 5% at the 95% confidence level, and 7.5% at the 90% confidence level (Bai et al., 2008). A drawback of this mapping approach is that an area of land degradation much smaller than 8 km² (pixel size of the GIMMS AVHRR) must be severe to significantly change the signal from a much larger surrounding area. In addition, the application of RUE to identify degraded landscapes has been somewhat controversial and misinterpreted as an indicator of degradation (Prince et al., 2007) since the RUE is highly variable (Fensholt and Rasmussen, 2011). In addition, errors in gridded precipitation data can add significant uncertainty, and noise to a degradation analysis suggesting analyses based solely on remotely sensed data may be beneficial (Reeves and Baggett, 2014).

10.3.1.5 Local NPP Scaling

Prince (2002) developed the local net primary productivity scaling (LNS) approach. Though the LNS approach can be applied to data of any resolution, derived from a host of sensors

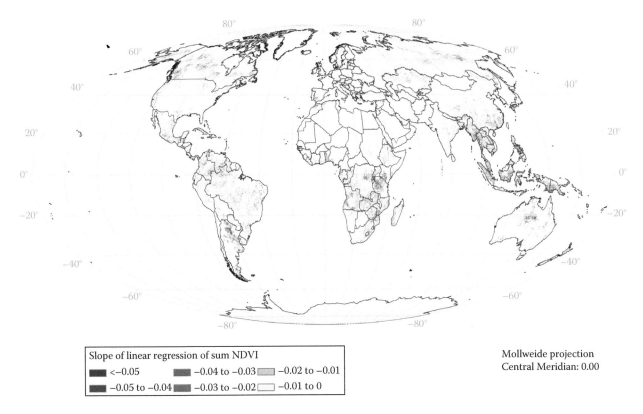

Slope of linear regression of sum NDVI

■ <−0.05	■ −0.04 to −0.03	▨ −0.02 to −0.01
■ −0.05 to −0.04	■ −0.03 to −0.02	☐ −0.01 to 0

Mollweide projection
Central Meridian: 0.00

FIGURE 10.6   Average RUE-adjusted NDVI, from GIMMS-AVHRR 8 km² and VASClimO at 0.5° spatial resolution. (From Bai, Z.G. et al., *Soil Use Manage.*, 24, 223, 2008.)

yielding visible and infrared bandpasses, AVHRR and Terra MODIS are commonly used. The LNS approach compares seasonally summed NDVI (ΣNDVI) of a single pixel to that of highest pixel value (or, commonly, the 90th percentile) observed in homogeneous biophysical land units (e.g., similar soils, climate, and landforms). The highest ΣNDVI value is assumed as a proxy for the potential aboveground NPP (ANPP) for each unit, and the other ΣNDVI values are rescaled accordingly. Prince et al. (2009) applied the LNS approach at national scales in Zimbabwe using MODIS 250 m² NDVI and concluded that 17.6 Tg C year⁻¹ were lost due to degradation. Similarly, Wessels et al. (2007) used 1 km² time-integrated NDVI in northeastern South Africa. More recently, Fava et al. (2012) used annual summations of MODIS 250 m² NDVI resolution in an LNS study for assessing pasture conditions in the Mediterranean resulting in a mean agreement of 65% with field-based classes of degradation. In a variant of the LNS approach, Reeves and Baggett (2014) used the mean 250 m² MODIS NDVI response of like-kind sites compared with reference conditions using a time series analysis to identify degradation on the northern and southern Great Plains, United States. With this approach, 11.5% of the region was estimated to be degraded.

10.3.1.6 Global Assessment of Land Degradation: The Evolution of Remote Sensing Use

The use of remote-sensing data in global programs of land degradation assessment is related to the history of the global assessment of human-induced soil degradation (GLASOD), the global

LADA (LADA-Global Assessment of Land Degradation and Improvement [GLADA]), and the Global Land Degradation Information System (GLADIS) programs, funded by the global organizations such as United Nations Environment Program (UNEP), the UN FAO, and the Global Environmental Facility (GEF). Table 10.3 summarizes the objectives, methods, and main outputs derived from these programs, including the use of remote-sensing technologies in their implementation.

The GLASOD, an expert-opinion-based study (Table 10.3), and Oldeman et al. (1991) had two follow-up assessments, namely, the regional assessments of soil degradation status in South and Southeast Asia (Assessment of the Status of Human-induced Soil Degradation in South and Southeast Asia [ASSOD]) and Central and Eastern Europe (Soil and Terrain Vulnerability in Central and Eastern Europe [SOVEUR]) and the global LADA project, under UNEP/FAO. The LADA had the objectives of developing and testing effective methodological frameworks land degradation assessment, at global, national, and subnational scales. The global component of LADA (i.e., GLADA) provided a baseline assessment of global trends in land degradation using a range of indicators collected by processing satellite data and existing global databases (NPP, RUE, AI, rainfall variability, and erosion risk) as described in Bai et al. (2008). The GLADA was implemented between 2006 and 2009, based on 22 years (1981–2003) of fortnightly NDVI data collection and processing (Table 10.3). The project developed and validated a harmonized set of methodologies for the assessment of land use, land degradation, and

TABLE 10.3   Cursory Comparison between Major Global Rangeland Degradation Efforts

Program	Objective	Methodology—Remote Sensing Usage
Global assessment of human-induced soil degradation (GLASOD) (UNEP) (1987–1990)	Produce a world map of human-induced "soil degradation," on the basis of incomplete knowledge, in the shortest possible time	No remote sensing; expert-based approach; distinguishes "types" of soil degradation, based on perceptions; it is "not a measure" of land degradation

Land Degradation Assessment in Drylands (LADA)-GLADA—global project, under UNEP/FAO (2006–2009)	Assess (quantitative, qualitative, and georeferenced) land degradation at global, national, and subnational levels to identify status, driving forces and impacts and trends of land degradation in drylands; identify "hot" (degradation) and "bright" (improvement) spots	The global LADA was based on 22 years (1981–2003) of fortnightly NDVI data, derived from GIMMS and MODIS-related NPP (MOD 17) Method
		
		    Identify degrading areas (negative trend in sum of NDVI)
		    Eliminate false alarms of productivity decline by masking out urban areas, areas with a positive correlation between rainfall and NDVI and a positive NDVI-RUE
		    Produce RUE-adjusted NDVI map
		    Calculate NDVI trends for remaining areas

LADA-Global Land Degradation Information System (GLADIS) FAO-UNEP-GEF (2006–2010)	Focus on land degradation as a process resulting from pressures on a given status of the ecosystem resources	Remote sensing is used for biomass status and trends, based on a correction factor to the GLADA-RUE-adjusted NDVI, to present trends in NDVI (1981–2006) translated in greenness losses and gains distinguished by climatic and human-induced (e.g., deforestation from FAO-FRA dataset) causes. Outputs are a series of global maps on the "status and trends" of the main ecosystem services considered and radar graphs

Sources: Oldeman (1996); Bai, Z.G. et al., *Soil Use Manage.*, 24, 223, 2008; Nachtergaele et al., (2010).
Prepared by Metternicht, G.

land management practices at global, national, subnational, and local levels (Ponce-Hernandez and Koohafkan, 2004).

The GLADIS was developed by FAO, UNEP, and the GEF using preexisting data and newly developed global databases to inform decision makers on all aspects of land degradation. The GLADIS developed a global land use system (LUS) classification and mapping using a set of pressures and threat indicators at the global level, allowing access to information at country, LUS, and pixel (5 arc-minute resolution) levels. It accounts for socioeconomic factors of land degradation, using a variety of ancillary data to this end. Lastly, Zika and Erb (2009) produced a global estimate of NPP losses caused by human-induced dryland degradation using existing datasets from GLASOD and other sources. Table 10.3 shows an evolution in the use of remote-sensing technology from the first global assessments (GLASOD), expert based, with no use of remote-sensing imagery, to the latest LADA-GLADIS, heavily reliant on remote-sensing derived data coupled with an ecosystem approach. The GLASOD estimated that 20% of drylands ("excluding" hyperarid areas) was affected by soil degradation. A study commissioned by the Millennium Assessment based on regional datasets ("including" hyperarid drylands) derived from literature reviews, erosion models, field assessments, and remote sensing found lower levels of land degradation in drylands, to be around 11% (although coverage was not complete) (Lepers et al., 2005). The LADA project reported that over the period of 1981–2005, 23.5% of the global land area was being degraded. On the other hand, Zika and Erb (2009) report that approximately 2% of the global terrestrial NPP is lost each year due to dryland degradation, or between 4% and 10% of the potential NPP in drylands. Figure 10.7 is a compilation of the global extent of drylands and human-induced dryland degradation, produced for the fifth Global Environment Outlook

(GEO-5) based on research of Zika and Erb (2009) who express dryland degradation in croplands and grasslands as a function of NPP losses.

The three dryland area zones (top of the figure) are derived on basis of the AI. Only dryland areas (arid, semiarid, and dry subhumid), characterized by an AI between 0.05 and 0.65, are considered. Degradation is assessed by calculating the difference of the potential NPP (NPP_0) and current NPP (NPP_{act}). NPP losses due to human-induced degradation amount to 965 Tg C year^{-1}, giving evidence that about 4%–10% of the potential production in drylands is lost every year due to human-induced soil degradation. The largest losses are occurring in the Sahelian and Chinese arid and semiarid regions, followed by the Iranian and Middle Eastern drylands and to a lesser extent the Australian and Southern African regions (UNEP, 2012) (Table 10.4) (Figure 10.5).

A loss of NPP in the range of 20%–30% means reductions of potential productivity in that range; in most pixels of Figure 10.7, productivity losses range between 0% and 5% of their NPP_0. The results presented in Figures 10.5 and 10.7 illustrate the scope and patterns of degradation but must only be considered as rough estimates (Zika and Erb, 2009). Major uncertainties related to the results arise from three assumptions: (a) estimates of degradation extent, (b) assumptions on NPP losses due to degradation processes, and (c) potential NPP as a proxy for production potential.

In recognition of the scope of degradation globally, the UN Conference on Sustainable Development (Rio+20) prompted the international community to develop universal sustainable development goals providing a timely opportunity to respond to the threat of soil and land degradation (Koch et al., 2013). Despite over 30 years of applied research in this area, however, the need to provide a baseline and method from which to measure degradation still remains (Gilbert, 2011).

TABLE 10.4 Estimates of NPP Losses due to Dryland Degradation, Regional Breakdown

Region	Degraded Dryland[a]		NPP Loss[b]	
	1000 km²	%	Tg C year⁻¹	%
Central Asia and Russian Federation	1,432	19.5	250	26
Eastern and Southeastern Europe	391	55.5	73	8
Eastern Asia	1,887	45.3	50	5
Latin America and the Caribbean	1,206	18.8	98	10
Northern Africa and Western Asia	1,207	33.8	70	7
Northern America	607	11.3	51	5
Oceania and Australia	866	13.2	24	2
Southeastern Asia	45	40.4	10	1
Southern Asia	1,437	30.9	106	11
Sub-Saharan Africa	2,597	22.8	215	22
Western Europe	128	24.7	18	2
Total	11,802	23.2	965	100

Source: Zika, M.E. and Erb, K.H., *Ecol. Econ.*, 69, 310, 2009.

[a] Percentage of dryland area.

[b] Estimated NPP losses associated with dryland degradation (see Zika and Erb (2009) for more detail).

For regional refinements to degradation analyses, radar satellite-based aboveground biomass estimations by Carreiras et al. (2012), or regional vegetation cover (Dong et al., 2014), could aid degradation analyses since cloud issues faced by LADA-GLADA and GLADIS could be mitigated. Additionally, Blanco et al. (2014) propose ecological site classification of semiarid rangelands enabling more refined spatial units across which remote sensing can be conducted. Finally, engaging citizens in knowledge production (including field verification of remotely sensed derived information), as fostered by current global (UNEPLive, Future Earth, Group on Earth Observations Biodiversity Observation Network) and subglobal initiatives (Eionet of the European Environmental Agency), could address the significant lack of ground truthing of previous global land degradation studies.

10.3.2 Fire in Global Rangeland Ecosystems

The extremely wide range of rangeland environments makes it virtually impossible to develop generalized statements about global fire regimes. However, the general composition of fuel and fuel characteristics defines some specifics of fire occurrence common for these ecosystems. Vegetation of rangelands

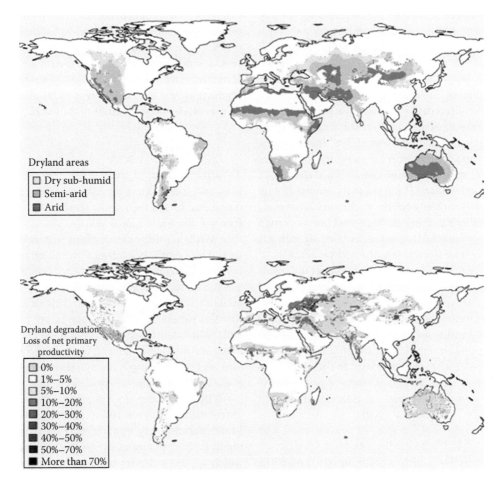

FIGURE 10.7 Global extent of drylands and human-induced dryland degradation. (From UNEP, 2012. Redrawn from Zika, M.E. and Erb, K.H., *Ecol. Econom.*, 69, 310, 2009; We thank UNEP and the GEO-5 process for use of the figure.)

is characterized by fast growth and slow decomposition rates (Vogl, 1979) leading to considerable buildup of surface litter. The majority of fuels in these ecosystems, with the possible exception of chaparral systems, are flash and fine fuels (<0.25 in diameter), which dry out rapidly (i.e., 1-h time lag fuels) and burn readily (National Wildfire Coordinating Group, 2012). Therefore, it is not unusual for these ecosystems to transition from low-fire danger state to extreme-fire danger state over a comparatively short period. Contiguity and loading of fuel in these ecosystems is highly variable both spatially and temporally: interannual variation in fuel loading often exceeds 110% (Ludwig, 1987). While fire is currently a common and widespread disturbance agent globally in rangelands, its prominence is expected to rise under projected climate change. Past and ongoing satellite monitoring and mapping of rangeland fire extent provide a much needed baseline for assessment of potential future change in fire occurrence and its impact on ecosystem functioning.

10.3.2.1 Satellite Monitoring of Ongoing Burning

The hotspot detections from the nighttime top of atmosphere radiance data from the Along-Track Scanning Radiometer (ATSR-2) and Advanced ATSR (AATSR) were used to build the first World Fire Atlas (Jenkins et al., 1997). Neither of the source instruments was designed to support fire detection specifically, and therefore, the algorithms were based on suboptimal ranges of electromagnetic radiation (at brightness temperature [BT] centered on 3.7 and 11.8 μm) using a suite of simple thresholds (Arino et al., 2012). The MODIS was, however, designed with a specific goal to enhance fire-mapping capabilities (Kaufman et al., 1998). MODIS collects daily global observations from Terra ~11:30 a.m. and 11:30 p.m. and Aqua at ~1:30 a.m. and 1:30 p.m. equatorial crossing time. In addition, several "fire" channels were included in the instrument to support fire monitoring: two 4 μm channels (channel 21 with 500 K saturation level and channel 22 with 331 K saturation level) and 11 μm channel (channel 31 with 400 K saturation level) at 1 km^2 nominal resolution (Giglio et al., 2003). The flexibility of switching the high- and low-saturation 4 μm channels in the contextual active fire detection algorithm is particularly important for tropical savanna environments.

The MODIS active fire product is the first product to include fire characterization metrics in addition to the binary "fire/no fire" masks. Fire radiative power (FRP), expressed in watts (W) is an instantaneous measurement of power released by ongoing burning during the satellite overpass (Kaufman et al., 1996a,b) and are estimated using an empirical relationship established in Kaufman et al. (1998). FRP is directly related to the intensity of biomass burning and, when integrated overtime to fire radiative energy (FRE) expressed in joules (J), is linearly related to biomass consumption (Wooster et al., 2005).

10.3.2.2 Satellite Estimates of Burned Area

Unlike active fire detection, which is primarily based on BT in mid- and long-infrared spectrum, burned area estimates are most frequently based on changes in surface reflectance due to burning observable within the visible (0.4–0.6 μm), NIR (0.7–1.0 μm), and shortwave infrared (SWIR 1.1–2.4 μm) spectrum. The relatively short wavelength of radiation in this range determines that burned area mapping relies on clear-surface observations and is strongly limited by considerable aerosol contamination from smoke during the burning process and high cloud cover in high northern latitudes.

The first multiyear global burned area products were developed from data acquired by VEGETATION (VGT) (onboard SPOT), ATSR-2 (onboard ERS-2), Medium Resolution Imaging Spectrometer (MERIS), and AATSR (onboard Environmental Satellite [ENVISAT]) instruments (Plummer et al., 2006) within the GLOBCARBON initiative. The suite of fire products developed from the MODIS 500 m^2 data includes two global burned area algorithms. The MCD45 algorithm (Roy et al., 2008) is based on detection of rapid changes in surface reflectance within a MODIS 500 m^2 pixel (Figure 10.8).

The MCD64 algorithm (Giglio et al., 2009) relies on detection of persistent changes in vegetation state and subsequent attribution of the change to burning by comparison to active fire occurrence within a specified spatiotemporal window. A detailed study in Central Asia (Loboda et al., 2012) has shown that MODIS-based products deliver spatially accurate estimates of burned area in Central Asia. However, MCD45 on average underestimates the total amount of burned area by ~30%, whereas MCD64 estimates are considerably closer to Landsat-based assessments (~18% underestimation). The independent accuracy assessment results within drylands of Central Asia are similar to those in North America (Giglio et al., 2009). This makes MODIS-based products appear to deliver a reasonable estimate of fire impact on grasslands and shrublands of the world.

10.3.2.3 Remote-Sensing Methods for Fire Impact Characterization

The large footprint of savanna fires, remote locations of tundra fires, and overall short longevity of scars of grass- and shrub-dominated fires make remote sensing the only viable source of data for consistent global postfire characterization of burned area. While a healthy debate about what constitutes burn severity and how much the ecological definition ranges across ecosystems is still ongoing in the fire science community (French et al., 2008), the Monitoring Trends in Burn Severity (MTBS) program established the baseline definition. This includes the assumption that this parameter can be mapped from remotely sensed data and is ultimately based on a combination of "visible changes in living and nonliving biomass, fire byproducts (scorch, char, and ash), and soil exposure" among other components (Eidenshink et al., 2007). The same ranges of electromagnetic spectrum (visible–NIR–SWIR), therefore, constitute the basis for the strongest differentiation between soil, vegetation, char, and ash components characterizing burn severity as those used most commonly for burned area mapping. It is not surprising that the first widely applied index for mapping and quantifying burn severity is based on the normalized difference of NIR and SWIR in 2.2 μm range (SWIR$_{2.2}$) originally developed by Lopez-Garcia and Caselles (1991) for burned area mapping. The Normalized

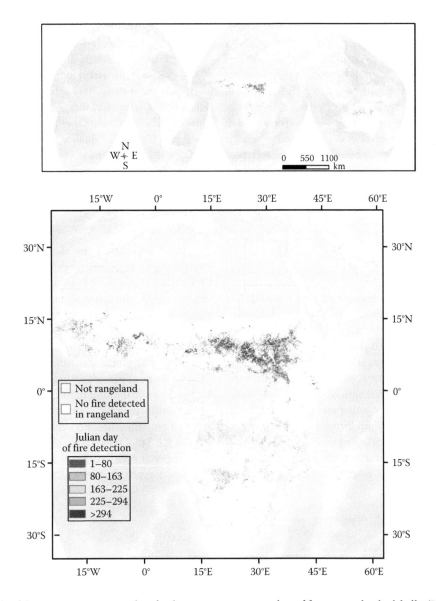

FIGURE 10.8 Example of the MCD45 MODIS product for depicting approximate date of fire in rangelands globally. (Prepared by Matt Reeves. MCD45 MODIS data for 2013.)

Difference Burn index (NDBR), as it was subsequently named by Key and Benson (1999a,b), is calculated as follows:

$$NBR = \frac{NIR - SWIR_{2.2}}{NIR + SWIR_{2.2}}$$

where

NIR refers to the TM band 4 (0.76–0.90 μm)

$SWIR_{2.2}$ refers to band 7 (2.08–2.35 μm)

Key and Benson (1999a,b) aimed to capture the fire-induced changes to the proportions of soil, char, ash, and vegetation through differencing the preburn and postburn NDBR measurement within a fire perimeter. This approach (differenced normalized burn ratio [dNBR], calculated as dNBR = $NBR_{pre-burn}$ − $NBR_{post-burn}$) has become the most widely applied

metric of burn severity across all ecosystems in the United States (Eidenshink et al., 2007).

Compared to forest cover, where the original assessment of dNBR were closely related to field measurements of burn severity expressed through a composite burn index (CBI) (Key and Benson, 2006, Allen and Sorbel, 2008), these grass- and shrub-dominated ecosystems have a low amount of aboveground biomass and are spatially highly heterogeneous. Thus, the magnitude of change between preburn and postburn surface conditions is considerably more muted and uneven. To account for the initial lower fuel loading in these ecosystems, an adjustment to dNBR, named relativized dNBR (RdNBR), was developed by Miller and Thode (2007). This index is calculated as follows:

$$RdNBR = \frac{dNBR}{\sqrt{\left|NBR_{pre-burn}/1100\right|}}$$

Although RdNBR versus CBI assessments show that RdNBR is more robust in assessing burn severity compared to dNBR in grass- and shrub-dominated ecosystems (Miller and Thode, 2007; Loboda et al., 2013), it does not overcome a major limitation of spectral signature change due to fire in NIR/SWIR spectral space within these ecosystems.

It is likely that the success rate of any one spectral index in mapping and quantifying burn severity depends strongly on the specific proportions of grass, woody biomass, exposed soil, geographic location (as related to frequency of observation allowing for a wider range of mapping days and different sun-sensor geometries), moisture status during image acquisition, and the timing of mapping.

10.3.3 Food Security: Role of Remote Sensing in Forage Assessment

On rangelands, quantifying the amount of forage available to livestock on a near real-time basis using traditional methods (e.g., clipping vegetation along transects) can be costly, time consuming, and logistically challenging. A lack of information for making livestock management decisions at critical times could lead to loss of livestock due to lack of forage, or lead to vegetation overuse, which, in turn, could result in rangeland degradation (Weber et al., 2000). Therefore, having an objective means of setting stocking rates on rangelands based on productivity will allow rangeland managers to better adapt to changing weather conditions.

Because of the large areal cover that remote-sensing products provide, in addition to the greater temporal frequencies of collection compared to traditional field sampling over large areas, the use of remote-sensing imagery is attractive for assessing vegetation production on rangelands. Multiple satellite platforms exist that are useful for rangeland forage assessments and early warning systems. Two approaches have generally been used for assessing rangeland forage conditions using remote-sensing imagery. These include (1) empirical approaches that estimate the forage biomass or quality based on a statistical relationship between the spectral bands (or some combination of bands) in the imagery and field-collected vegetation data and (2) process models that use remote-sensing data as inputs for predicting vegetation biomass or quality.

10.3.3.1 Empirical Approaches

Empirical approaches for assessing rangeland forage conditions using remote-sensing products generally involve the use of a statistical relationship between the remote-sensing spectral response or product variable and data collected from field measurements (Dungan, 1998). Using the empirical approach example in Figure 10.9, a MODIS 250 m^2 maximum value composite and NDVI value of 7500 correspond to approximately 3414 kg ha^{-1} of annual production, after accounting for unavailability ($\phi = 0.15$) and suggested utilization ($\upsilon = 0.5$) results in stocking rate of 5.3 animal unit month's (AUM) ha^{-1}.

In a similar manner, Tucker et al. (1983) used both a linear and logarithmic regression between the ground-collected biomass data in the Sahel region and AVHRR NDVI to predict biomass on a regional scale. Al-Bakri and Taylor (2003) used a linear regression approach to predict shrub biomass production for

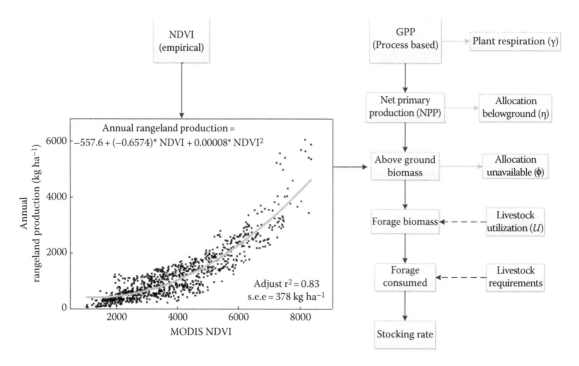

FIGURE 10.9 Process for estimating stocking rate from remote-sensing data, either empirically or using a process model. GPP is determined from land cover type, spectral vegetation indices, incident photosynthetically active radiation, and climate-dependent radiation use efficiencies or empirically. Solid arrows represent reductions based on physiology, while dashed arrows represent critical management decisions are determined.

rangelands in Jordan using 7.6 km² AVHRR NDVI. Both these studies reported accounting for >60% of the variation in herbaceous biomass with AVHRR NDVI alone using linear regression against biomass. In the Xilingol steppe of Inner Mongolia, Kawamura et al. (2005) used 500 m² MODIS enhanced vegetation index (EVI) to predict live biomass and total biomass of livestock forage with linear regression models, which accounted for 80% of the variation in live biomass and 77% of the variation in total biomass. In the Tibetan Autonomous Prefecture of Golog, Qinghai, China, Yu et al. (2011) used the 250 m² resolution MODIS NDVI to estimate aboveground green biomass using regression relationships between the NDVI and field-collected biomass data (r² of 0.51) from sites across the region.

As with forage biomass, empirical approaches can be used for forage quality assessments generally involving examining statistical relationships between forage quality variables such as crude protein or energy and spectral information from remote-sensing imagery. For example, Thoma et al. (2002) used simple linear regression with AVHRR NDVI as the independent variable to predict forage quality and quantity on rangelands in Montana, United States. Their analysis indicated reasonable relationships between NDVI and live biomass (r² = 0.68) and nitrogen in standing biomass (r² = 0.66). Similarly, Kawamura et al. (2005) used regression relationships between field-collected data and MODIS EVI to predict live and dead biomass and crude protein in standing biomass. They found good predictability between standing live biomass and total biomass (live + dead) (r² = 0.77–0.80), but correlations with crude protein were poor (r² = 0.11).

Remote-sensing imagery provides a dense and exhaustive dataset that can serve as a secondary variable for geostatistical interpolation given that a correlation exists (both direct and spatial) between the primary and secondary variable (Dungan, 1998). Use of MODIS NDVI in the cokriging analysis of forage crude protein provides reasonable during the dry season (r² = 0.69) but less so during the wet season (r² = 0.51) (Awuma et al., 2007) likely because the amount of unpalatable shrub cover increased the greenness signal in the NDVI in some of the sampling areas that did not contribute to the available forage.

10.3.3.2 Process Models Using Remote-Sensing Inputs

One problem that has been noted for regression models that use remote-sensing variables is that they violate the regression assumption of no autocorrelation in the predictor variable(s) (Dungan, 1998; Foody, 2003). Since most remote-sensing data are inherently autocorrelated, violation of this assumption may reduce the effectiveness of the regression model (Dungan, 1998). One way of overcoming the autocorrelation problems is to use process models that are driven by remotely sensed input variables on a pixel-by-pixel basis. Reeves et al. (2001) describe such an approach for predicting rangeland biomass using remote-sensing products from the MODIS system and a light use efficiency model for plant growth. Hunt and Miyake (2006) used a similar light use efficiency model approach for estimating stocking rates for livestock at 1 km² resolution in Wyoming, United

States (Figure 10.9). Using the approach of Hunt and Miyake (2006), the stocking rate is estimated as gross primary production (GPP) $(1 - \chi)(1 - \eta)(1 - \phi) \upsilon$ (AUM/273 kg month^{-1}). From Hunt and Miyake (2006), the parameters for grasslands are approximately $\chi = 0.48$, $\eta = 0.79$, $\phi = 0.15$, and $\upsilon = 0.5$ where χ is autotrophic respiration, η is belowground carbon allocation, ϕ is carbon allocation to nonpalatable stems and other vegetation, and υ is an estimated accepted level of utilization. Therefore, a monthly GPP of 11,000 kg ha^{-1} month^{-1} is about 1.7 AUM's ha^{-1}, but this is just one method of using process models parameterized with remote-sensing inputs.

An example of a process-based modeling approach for forage quantity assessment at the regional level is the Livestock Early Warning Systems (LEWS) in East Africa (Stuth et al., 2003a, 2005) and Mongolia (Angerer, 2012) (Figure 10.10).

Figure 10.10 presents results of the LEWS applied in Mongolia in 2013. Note the significant decline of forage in southwestern Mongolia in 2013. The LEWS was developed to provide near real-time estimates of forage biomass and deviation from average conditions (anomalies) to provide pastoralists, policy makers, and other stakeholders with information on emerging forage conditions to improve risk management decision making. The LEWS combines MODIS 250 m² NDVI, field data collection from a series of monitoring sites, simulation model outputs, and statistical forecasting, to produce regional maps of current and forecast forage conditions and anomalies. The system uses the Phytomass Growth Simulation model (PHYGROW) (Stuth et al., 2003b), parameterized with the MODIS 250 m² NDVI, as the primary tool for estimating available forage. Model verification indicates the model performs well in estimating forage biomass (Stuth et al., 2005). For example, model verification across monitoring sites in Mongolia indicated a good correspondence between the PHYGROW predicted biomass and observed field data (r² = 0.76) with forage biomass ranging from 3 to 1230 kg ha^{-1}. PHYGROW had a tendency to underestimate forage biomass across sites by 14% with an overall mean bias error of −18 kg ha^{-1} (Angerer, 2008).

10.3.4 Rangeland Vegetation Response to Global Change: The Role of Remote Sensing

Monitoring global change is an increasingly important endeavor (Running et al., 1999) since ecosystem goods and services, essential to human survival, are directly linked to the health of the biosphere (Fox et al., 2009). The Earth is a dynamic system with many interacting components that are complex and highly variable in space and time. Though change has always been present, human activities have influenced rates and extent of change beyond historical ranges (Vitousek, 1992; Levitus et al., 2000; Foley et al., 2005). Global change involves terrestrial, aquatic, oceanic, and atmospheric systems and cycles and is not limited to climate change alone (Beatriz and Valladares, 2008). Other factors such as invasive species, habitat change, overexploitation,

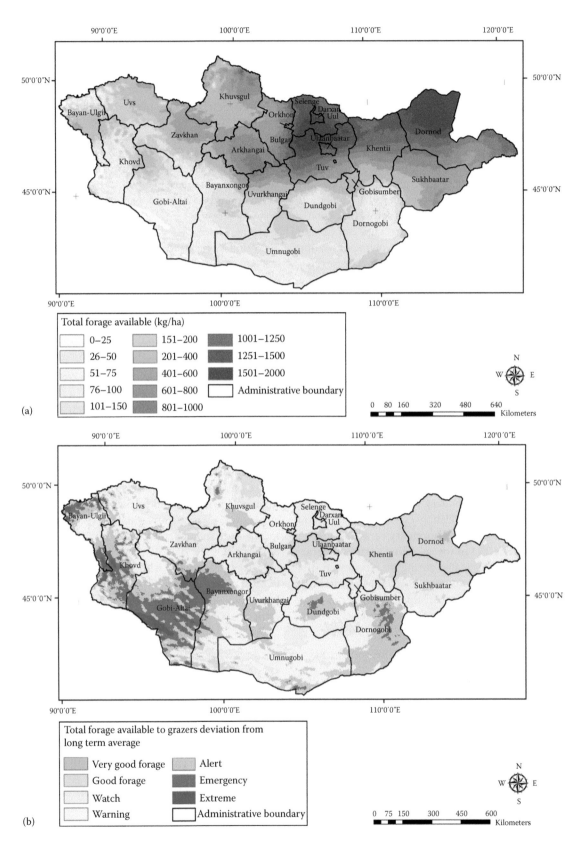

FIGURE 10.10 Panel (a) represents total forage available (kg ha⁻¹) during August 2013 for the Mongolia LEWS. Panel (b) represents a map of forage deviation from long-term average (i.e., forage anomaly) for August 2013. Note areas in southwestern Mongolia experiencing emergency to extreme drought conditions.

and pollution are equally or even more important to the Earth's future (Millennium Ecosystem Assessment, 2005). Thus, the goal of global monitoring is aimed at characterizing "human habitability" through evaluation of vegetation that provides food, fiber, and fuel (Running et al., 1999) to a rapidly growing population. In the burgeoning field of global change monitoring, satellite remote sensing is increasingly more important. Only remote sensing offers a truly synoptic perspective of our surroundings and is therefore a critical tool for describing the type, rate, and extent of change unfolding across the globe. This is especially true for rangeland ecosystems that experienced losses of about 700 million ha by 1983 due to agriculture. In the United States alone, an estimated 75 million ha of former rangelands have been converted to agricultural land use since Euro-American settlement (Reeves and Mitchell, 2011) (Figure 10.11). The impacts of global change, such as climate impacts and land conversion, are often quantified through evaluation of vegetation cover and NPP in the context of the global carbon budget (Running et al., 1999).

10.3.4.1 Vegetation Productivity

Given the lack of field-referenced data available for determining productivity for rangelands globally, ecosystem modeling, remote sensing (Hunt and Miyake, 2006; Fensholt et al., 2006; Reeves et al., 2006), or a combination of both (Jinguo et al., 2006; Wylie et al., 2007; Xiao et al., 2008) can be used to estimate spatial and temporal trends across large areas. Many studies have evaluated the growth, total production, and health of rangeland vegetation, but two general approaches are normally applied that are very similar to the procedures outlined in the food security section. The first approach involves directly sensing, via radiometric measurement, the amount of growth that has occurred over a given time period.

Direct quantification of biomass across rangeland vegetation types requires a set of spatially explicit field samples describing

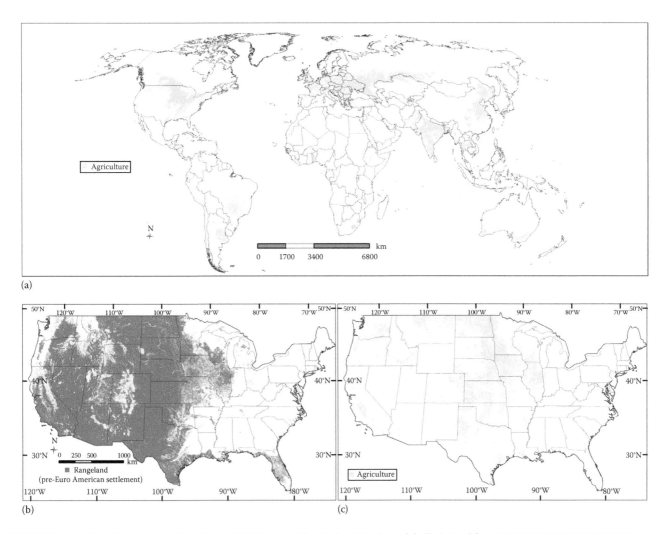

(a)

(b)

(c)

FIGURE 10.11 Panel (a) represents the estimated distribution of agricultural land use globally derived from MODIS MOD12Q1, 2006; University of Maryland Classification. Also shown is the hypothesized pre-Euro-American extent of rangeland (From Reeves, M.C. and Mitchell, J.E., *Rangeland Ecol. Manage.*, 64, 1, 2011.) as is shown in Panel (b), while Panel (c) demonstrates areas of former rangeland now in agricultural production (estimated using the Biophysical Settings data product from the Landfire Project; Rollins, 2009).

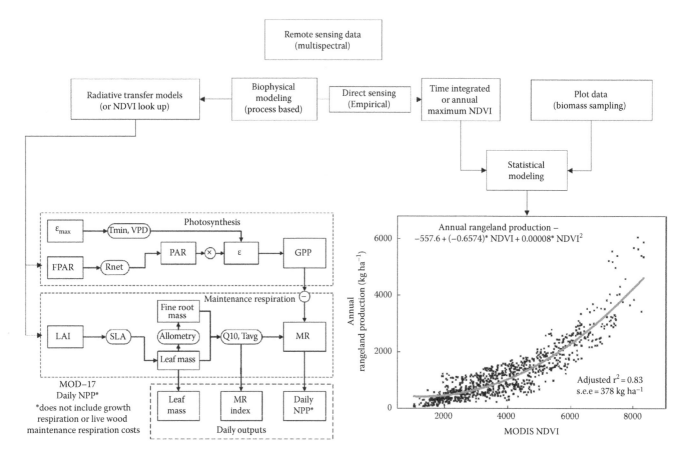

FIGURE 10.12 Direct sensing and biophysical modeling (process modeling) are two methods for estimating productivity of rangeland landscapes.

the amount of peak biomass or annual production. Once field data are collected and properly scaled, statistical models can be developed to describe the relationship between NDVI and biomass (Figure 10.12) that can, in turn, be used to monitor the response of vegetation through time. If peak biomass is estimates are sought, the annual maximum NDVI value should work reasonably well, but if annual production estimates are desired, a time integration of NDVI is usually employed (e.g., Paruelo et al., 1997).

Though NDVI has been widely used for monitoring global vegetation conditions, it exhibits well-known saturation characteristics at relatively higher levels of biomass. The EVI can be used, with some success to overcome the saturation limitations inherent in NDVI. The saturation component of the NDVI signal, however, does not render it less useful for most applications. The reason for this is that across the range of productivity levels expected in most rangeland environments, the response is linear (Skidmore and Ferwerda, 2008).

The second approach for monitoring growth, total production, and health of vegetation involves use of remote sensing for quantifying canopy parameters, such as LAI, and fPAR, which, in turn, become part of a vegetation modeling system (Figure 10.12). Such a system is exemplified by the MODIS NPP algorithm (MOD17), which provides gross and NPP products at 1 km² resolution for the entire globe.

This approach is more sophisticated than direct sensing of biomass but enables carbon accounting for the global extent of rangelands. The modeling approach also requires a good deal more information including biome specific physiological parameters (Running et al., 2004). In addition, since this type of modeling approach requires meteorological and land cover information, it is directly informed by land cover/land use changes associated with global change. The NPP of rangeland vegetation from 2000 to 2012 is depicted in Figure 10.13, which demonstrates the type of ecosystem analysis possible with the MODIS NPP product.

Figure 10.13 was created using a time series analysis from 2000 to 2012 of the MODIS-derived annual NPP and Collection 4.5 land cover products. From this analysis, significant overlap and similarities between the savanna and woody savanna land cover classes are evident. These similarities suggest similar biophysical and bioclimatic conditions are present in these two classes or confusion exists between the classes. The close relationship between woody savanna and savanna could also be related to spatial commingling of the two types, which could be alleviated using higher-resolution imagery. Multisensor fusion between MODIS (high temporal resolution) and Landsat (e.g., ETM+—high spatial resolution) can be used to explore why woody savannas and savannas are performing very similarly.

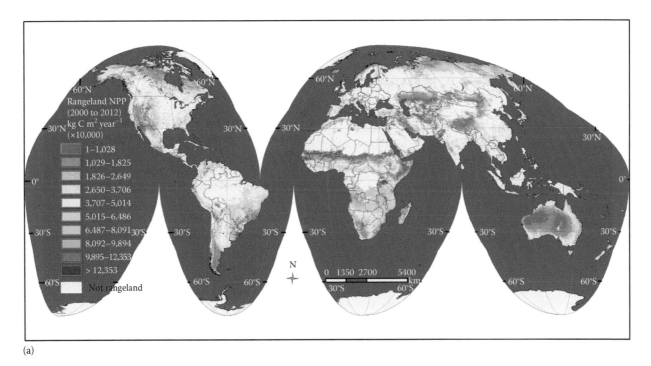

(a)

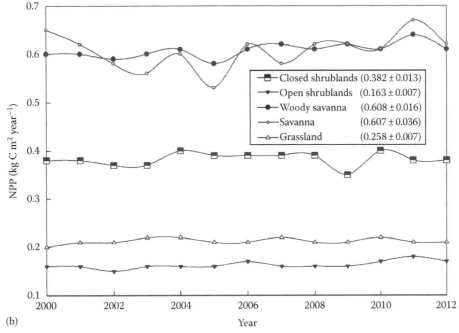

(b)

FIGURE 10.13 Panel (a) represents the mean (2000–2012) global distribution of rangeland NPP from the MODIS NPP (MOD17) product. Panel (b) represents the time series (2000–2012) of global rangeland NPP from the MOD17 product.

Roy et al. (2008) used MODIS 500 m² bidirectional reflectance distribution function spectral model parameters and the sun-sensor geometry to estimate ETM surface reflectance to fill temporal gaps between suitable ETM+ overpasses. This process resulted in prediction errors in the NIR dataspace of about 12% overall. Directly incorporating effects from changing climate, land cover, and associated vegetation responses simultaneously enables improved analysis of global change effects on rangeland

environments. One major goal of satellite remote sensing is observation of vegetation over large areas and for long periods of time. The appropriate length of observation depends on the behavior of the phenomena to be studied. Developing long-term observations requires much effort to ensure continuity across new sensors with varying bandpasses and associated target-atmospheric effects, drifts in calibration, and filter degradation (Huete et al., 2002).

10.3.4.2 Extending Remote Sensing Time Series Using Cross-Sensor Calibration

Recent ecological research has shown that declines in dryland productivity (often estimated measured using trends in NDVI and/or NPP), and increases in soil loss are due to the synergistic effects of extreme climatic events and land management practices. In particular, livestock grazing and El Niño and La Niña events have 3- to 7-year return intervals (Holmgren and Scheffer, 2001; Holmgren et al., 2006; Washington-Allen et al., 2006) indicating that 10–20 years of continuous data is required to replicate, monitor, and assess the influence of land use practices and these extreme events (Washington-Allen et al., 2006).

Sensors have finite life spans, and developing long-term observations often requires using multiple sources of data to develop a continuous, compatible dataset. The extension of time series is challenging due to drifts in calibration, filter degradation, and band locations (Miura et al., 2006). These characteristics create errors and uncertainties that vary with the landscape and sensors being evaluated. As examples, red and NIR spectral channels from AVHRR are relatively broad occupying the spectral space between 580–680 and 730–1000 nm, respectively. In contrast, MODIS provides more narrow bands in the red and NIR space at 620–670 and 841–876 nm, respectively. The broader AVHRR red channel incorporates a portion of the green reflectance region (500–600 nm) (Figure 10.4) inevitably yielding a different spectral response of vegetation than MODIS.

The approaches for extending a satellite data time series via sensor (or product) cross-calibration involve remote-sensing data fusion that accounts for multisensory, multitemporal, multiresolution, and multifrequency image data from operational satellites (Pohl and Van Gederen, 1998; Zhang et al., 2010). Extension of satellite data records to produce time series of NDVI or NPP data typically involve

1. Development of equations to simulate the spectral responses of individual channels (e.g., Suits et al., 1988)
2. Development of calibration equations to simulate the vegetation indices derived from other sensors (e.g., Steven et al., 2003; Tucker et al., 2005)
3. Cross-calibration of NDVI (e.g., from AVHRR) and NPP data products (e.g., from the MODIS sensor) to back cast the NPP record

These techniques have been explored in a good number of studies and indicate suitable relations between sensors, but results are often inconclusive (Fensholt et al., 2009). Suits et al. (1988) determined that multiple regression analysis compared to principle component analysis was the best approach for spectral response substitution between Landsat and AVHRR sensors. Steven et al. (2003) found that vegetation indices from Landsat, SPOT, AVHRR, and MODIS were strongly linearly related, which allowed them to develop a table of conversion coefficients that allowed simulation of NDVI and SAVI across these sensors within a 1%–2% margin of error. With the exception of AVHRR, which was designed for other purposes, most high temporal resolution sensors have

similar sensitivity to green vegetation. In addition, vegetation indices from many global platforms can be calibrated to within approximately ± 0.02 units if surface reflectance (as opposed to top of atmosphere) is used (Steven et al., 2003). Fensholt and Proud (2012) compared the GIMMS 3g 8 km² NDVI archive with MODIS 1 km² NDVI and showed that global trends exhibit similar tendencies but significant local and regional differences were present, especially in more xeric environments. A comprehensive analysis of four long-term AVHRR-based NDVI datasets with MODIS and SPOT NDVI datasets for the common period (from 2001 to 2008) clearly demonstrated lower correlations in more xeric regions such as the southwest and Great Basin of the United States (Scheftic et al., 2014). Similarly, Gallo et al. (2005) reported that 90% of the variation between 1 km² MODIS and AVHRR NDVI can be explained by a simple linear relationship, while Miura et al. (2006) developed translation equations to emulate MODIS NDVI from AVHRR resulting in an r² of 0.97. Despite these successes, trend analyses from AVHRR can differ strongly from those estimated with MODIS and SPOT-VGT (Steven et al., 2003) and lead to spurious conclusions. Unlike MODIS, AVHRR does not provide additional necessary channels permitting analysis of atmospheric composition for suitable atmospheric correction (Yin et al., 2012). Therefore, cross-sensor calibration must be carefully planned and should leverage the strengths of previous efforts. Most efforts aimed for extending time series to improve trend analyses involve spectral calibration, either of individual band passes or indices. For monitoring global change and ecosystem performance, however, it is useful to quantify NPP trends given its link with the global carbon cycle and paramount importance to maintaining goods and services. Bai et al. (2008, 2009) developed a 23-year time series of global NPP data from 1982 to 2003 using the overlap period (2000–2003) between 1 km² MODIS NPP and the mean annual sum of 8 km² AVHRR GIMMS, for LADA program of FAO. Next, linear regression was applied to 4-year mean, global, annual sum of NDVI from the GIMMS dataset and MODIS NPP to generate a single empirical equation between these two datasets. The resulting equation was then used to produce an 8 km² NPP time series from 1982 to 2003. Wessels (2009) critiqued the approach of Bai et al. (2008) arguing that spatial variability was reduced and unaccounted for by using a single mean equation rather than a pixel-by-pixel approach. As a result, the following case study used a pixel-wise regression approach for establishing relationships between 8 km² GIMMS NDVI and 1 km² MODIS NPP. The goal of this case study was to produce a continuous, compatible dataset describing annual NPP from 1982 to 2009 using both 8 km² AVHRR GIMMS from Tucker et al. (2005) and 1 km² MODIS net photosynthesis. A more recent version of GIMMS AVHRR NDVI (GIMMS 3g) data is available from 1981 to 2011 at 1/12th° spatial resolution.

10.3.4.2.1 Case Study

The strategy suggested by Steven et al. (2003) and Wessels (2009) was followed for calibrating 8 km² pixel resolution GIMMS annual ΣNDVI from 1982 to 2006 to MODIS NPP data aggregated from 1 to 8 km² using the 2000 to 2006 overlap period between these two

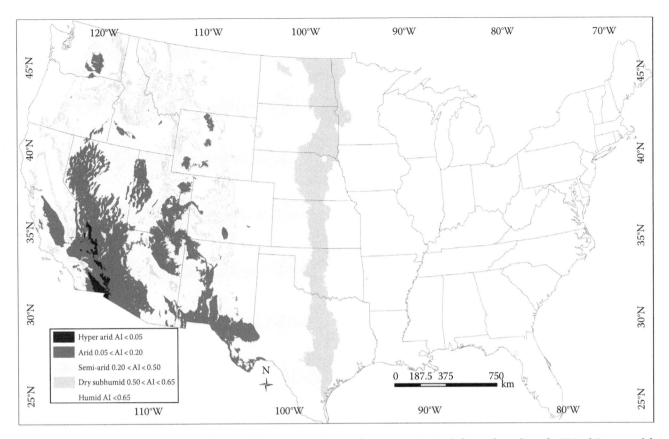

FIGURE 10.14 Distribution of aridity index (annual precipitation/potential evapotranspiration) classes throughout the United States used for aggregating NPP estimates for coterminous US rangelands.

time series. Collection 5 annual estimates of MODIS NPP from 2000 to 2006 and GIMMS ΣNDVI time series were subset to the rangeland portion of the contiguous United States and classified according to varying levels of aridity using AI (Figure 10.14). The AI of drylands (AI $\leq$ 0.65) is partitioned into four classes including the hyperarid, arid, semiarid, and dry subhumid classes.

10.3.4.2.1.1 Application and Validation of Linear Regression Approach The Taiga Earth Trend Modeler from IDRISI was used to conduct a simple linear regression on a pixel-by-pixel basis between the two time series using the years 2000, 2002, 2004, and 2006. This was done so that a holdout dataset could be retained for comparing predicted and observed NPP. Across all pixels in the rangeland domain, the mean NDVI was 0.03 and mean NPP was 281.6 g C m^{-2} year^{-1}. The mean equation across all pixels was

$$Y = 0.03 * X + (-31.7)$$

where

X is the annual GIMMS ΣNDVI

Y is the predicted 8 km^2 MODIS NPP and r^2 = 0.41 (Figure 10.15)

Panels A, B, and C in Figure 10.15 represent the estimated slope, intercept, and r^2 of a linear regression for each pixel in the study area between GIMMS NDVI and MODIS NPP for the years 2000, 2002, 2004, and 2006. Predicted MODIS NPP was subsequently

compared to the observed MODIS NPP (Table 10.5). Figure 10.16 indicates a strong relationship between monthly integrated 8 km^2 GIMMS NDVI and monthly integrated 1 km^2 MODIS net photosynthesis (PSNnet) over the domain of coterminous US rangelands.

The net photosynthesis is a major component of the annual NPP product. To derive the final model to extend the NPP time series, the pixel-level regressions developed were applied to the annual GIMMS ΣNDVI from 1982 to 1999. To these data, the MODIS NPP time series from 2000 to 2009 were added, thus extending the final time series from 1982 to 2009.

Using the final time series, temporal and spatial variations in NPP response can be quantified. The mean NPP for each class from 1982 to 2009 was 95 $\pm$ 28 for hyperarid, 115 $\pm$ 47 for arid, 218 $\pm$ 114 for semiarid, and 370 $\pm$ 117 (g C m^{-2} year^{-1}) for the dry subhumid class. In addition, the temporal trend (not accounting for temporal autocorrelation) of NPP within each AI class was as follows: hyperarid (r^2 = 0.08, p = 0.08), arid (r^2 = 0.01, p = 0.37), semiarid (r^2 = 0.25, p = 0.004), and dry subhumid (r^2 = 0.22, p = 0.006) (Figure 10.17).

Using this approach, significant carbon gains were detected for both semiarid and arid systems. In addition, the positive response in arid and semiarid systems agrees with conclusions by Reeves and Baggett (2014) that significant increasing trends have been observed from 2000 to 2012 across much of the US rangeland domain, owed mostly to increased precipitation.

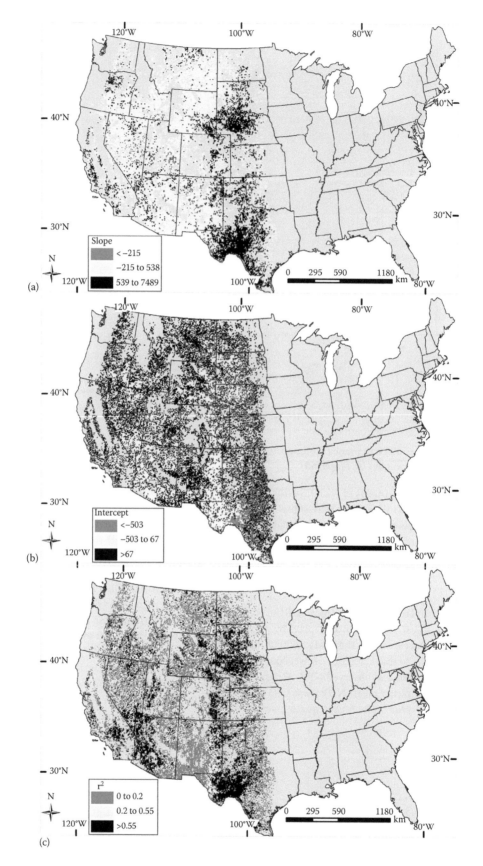

FIGURE 10.15   The resulting pixel-to-pixel linear regression models that were developed to calibrate GIMMS annual ΣNDVI to the MODIS NPP time series for the years 2000, 2002, 2004, and 2006. Panels (a), (b), and (c) represent the slope, intercept, and r^2 values for each pixel.

TABLE 10.5   Comparison of Predicted and Observed Values across the Extent of Rangelands in the Coterminous U.S. (g C m² year⁻¹)

Year	g C m² year⁻¹			
	Minimum	Median	Mean	SD
2001	16.5	179	211	123
	0.1	**186**	**220**	**137**
2003	15.3	189	218	131
	2.4	**184**	**217**	**131**
2005	19.8	236	126	126
	0.1	**210**	**157**	**157**

Bold numbers are predicted values based on the pixel level regression equations depicted in Figure 10.16.

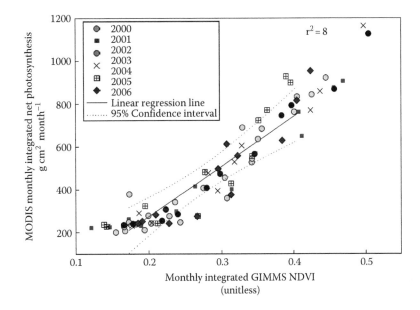

FIGURE 10.16   Relationship between monthly integrated MODIS-derived 1 km² net photosynthesis and GIMMS km² NDVI aggregated across all coterminous U.S. rangelands.

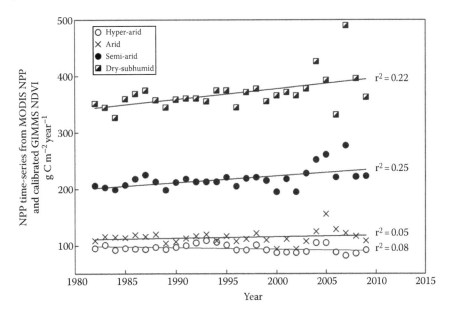

FIGURE 10.17   Rangeland mean NPP from 1982 to 2009 across four zones of AI including hyperarid, arid, semiarid, and dry subhumid.

The results portrayed in Figure 10.17 demonstrate improved chances for successfully interpreting vegetation response to global change through increasing the time series of satellite observation.

10.3.5 Remote Sensing of Global Land Cover

Global land cover data are essential to most global change research objectives, including the assessment of current global environmental conditions and the simulation of future environmental scenarios that ultimately lead to public policy development. In addition, land cover data are applied in national- and subcontinental-scale operational environmental and land management applications (e.g., weather forecasting, fire danger assessments, resource development planning, and the establishment of air quality standards). Land cover characteristics are integral to many Earth system processes (Hansen et al., 2000),

in addition to providing information for carbon exchange and general circulation models. A common and important application of global land cover information is inference of biophysical parameters, such as LAI and fPAR, which influence global-scale climate and ecosystem process models. Use of these models and monitoring the state of the Earth's rangelands is needed for global change research, especially given the influence of growing anthropogenic disturbances (Lambin et al., 2001; Jung et al., 2006; Xie et al., 2008).

One of the remote-sensing community's grand challenges is to provide globally consistent but locally relevant land cover information (Estes et al., 1999). Evaluations of remote-sensing-based global land cover datasets have shown general agreement of patterns and total area of different land covers at the global level but have more limited agreement in spatial patterns at local to regional levels (McCullum et al., 2006) (Figure 10.18). Figure 10.18 demonstrates the difficulty in deriving rangeland

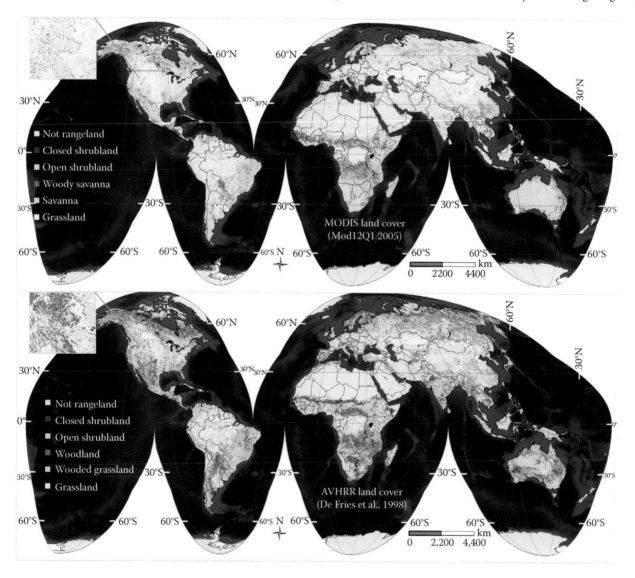

FIGURE 10.18   Comparison between 1 km² MODIS land cover data (Mod12Q1; UMD classification) and AVHRR-derived land cover (DeFries et al., 1998) using the simple biosphere model legend (Table 10.6).

area estimates using data from AVHRR (DeFries et al., 1998) and the MODIS Mod12Q1 (2005).

Both datasets have global coverage at 1 km² resolution but have different legends and classification techniques. Global mapping presents special challenges since the geographic variability of both land cover and remote-sensing inputs add complexity that can lead to inconsistent results. The evolution of global land cover datasets over the past 30 years has attempted to meet the grand challenge while adhering to general remote-sensing land cover–mapping standards dealing with accuracy, consistency, and repeatability.

The earliest contemporary efforts to provide global land cover data did not rely on remote-sensing inputs but instead was based on the developer's expertise and the quality of information from best available sources (Matthews, 1983; Olson, 1983; Wilson and Henderson-Sellers, 1985). These maps were coarse (i.e., 1° × 1°) in resolution but thematically detailed. Global land cover mapping based on remote sensing advanced rapidly in the 1990s when NOAA polar-orbiting data from the AHVRR were compiled into global coverage. Initially, 4 km² AVHRR Global Area Coverage Pathfinder data aggregated to 1° × 1° (DeFries and Townshend, 1994) and later to 8 km² resolution (DeFries et al., 1998) were inputs to the first remote-sensing–based global land cover products.

The International Geosphere-Biosphere Programme (IGBP) served as the catalyst for a worldwide effort led by the USGS to generate a 1992–1993 set of 1 km² resolution AVHRR global 11-day maximum NDVI composites (Eidenshink and Faundeen, 1994). Also under IGBP auspices, these data were used to produce the first 1 km² resolution global land cover dataset using the 17-class International Geosphere-Biosphere Programme Global Land Cover Classification (IGBP DISCover) legend (Loveland et al., 2000) (Figure 10.18). Hansen et al. (2000) followed with the completion of a 1 km², 12-class land cover dataset (UMD land cover map). These two maps served as the foundation for future global-mapping initiatives since their development experiences and map strength and weaknesses provided valuable lessons for the next generation of maps.

The NASA Earth Observing System's ambitious global land product program based on multiresolution MODIS data established a new state of the art in global land cover mapping. MODIS global land cover based on 500 m resolution imagery and the 17-class IGBP DISCover legend started in the 2001 and since then has been updated annually (Friedl et al., 2002). This ongoing activity represents the only sustained global land cover initiative. In the 2000s, European global land cover projects contributed significantly to advancing global land cover understanding. The Global Land Cover 2000 (GLC2000) project used SPOT vegetation instrument data to produce a 22-class 1 km² resolution land cover dataset (Bartholomé and Belward, 2005). In a follow-on effort, the European Space Agency sponsored a follow-on project, GlobCover, that used ENVISAT MERIS imagery to generate the highest-resolution (300 m) global land dataset ever. The MERIS-based map contained 22 land cover classes based on the United Nations-sponsored international standard—land cover classification system (LCCS).

The most recent global land cover dataset is the unprecedented China-led Fine Resolution Observation and Monitoring of Global Land Cover (FROM-GLC) dataset that is based on Landsat 5 and 7 TM/ETM+ and other high-resolution Earth observation data spanning the first decade of the twenty-first century (Gong et al., 2013). The FROM-GLC dataset with 29 land cover classes establishes new standards for high-resolution land cover mapping and monitoring.

In addition to the thematic land cover mapping efforts described earlier (Table 10.6), global "continuous fields" products provide quantitative estimates of the percent tree cover within each grid cell. DeFries et al. (1999) developed global percent tree cover data using 1 km² AVHRR imagery, and Hansen et al. (2003) created similar products using MODIS.

10.3.5.1 Comparative Investigations of Global Land Cover Datasets

With a relatively large number of global land cover datasets available, users face a challenge in understanding which one is best suited for their application. The differences in spatial resolution, temporal properties, land cover legend, and quality complicate the selection. Land cover legend and quality are particularly significant factors. Accuracy assessments that provide insights into data quality are available for some of the global products. For example, both the IGBP DISCover and GLC2000 datasets were evaluated using an independent accuracy assessment. DISCover accuracy was measured at 66.9% (Scepan, 1999). Mayaux et al. (2006) determined that the overall GLC2000 product accuracy was 68.6%. The MODIS land cover dataset accuracy was assessed based on a comparison with training data, with the results showing 78.3% agreement (Friedl et al., 2002). The more recent GlobCover land cover dataset's (Table 10.6) independent accuracy was measured to be 73.0%. Finally, the China-led fine-resolution global land cover product was determined to have an overall accuracy of 71.5% (Gong et al., 2013). Accuracy assessments were not produced for UMD global land cover datasets.

The overall accuracies mask the significant variations in per class accuracies (e.g., Scepan, 1999 estimates that the DISCover individual class accuracies varied from 40% to 110%). The class accuracy variations, as well as variations in land cover legends and class definitions, make cover-specific applications problematic. As a response to this problem, a number of global dataset comparison studies have been undertaken, which focus on determining dataset strengths and weaknesses. Some have used independent datasets to look at regions or continents, such as Tchuenté et al.'s (2011) evaluation of GLC2000, GlobCover, and MODIS land cover for Africa and Frey and Smith's (2007) evaluation of IGBP DISCover and MODIS land cover over western Siberia. Other comparisons have looked at agreement between datasets across the globe. For example, Hansen and Reed (2000) compared UMD and IGBP DISCover products; Giri et al. (2005) compared MODIS and GLC2000; McCullum et al. (2006) compared IGBP, UMD, GLC2000, and MODIS products; and Fritz and See (2007) compared MODIS, GLC2000, and GlobCover.

TABLE 10.6   Summary of Characteristics of the Major Remote Sensing Global Land Cover Datasets

Database	Source	Vintage	Resolution	Land Cover Content (Suggested Rangeland Classes)	Strengths	Weaknesses
Global AVHRR NDVI land cover (De Fries and Townshend, 1994)	AVHRR	1987	1.0°² latitude	11 (3) land cover classes—based on simple biosphere model	First remote-sensing-based depiction of global land cover	Coarse resolution, applications limited to global circulation model applications.
Global AVHRR land cover (De Fries et al., 1998)	AVHRR Global Area Coverage Pathfinder	1987	8 km²	14 (5) land cover classes—based on the simple biosphere model	Improved spatial resolution provided more realistic view of global land cover	Land cover classes were general and specific to one application requirement
IGBP DISCover (Loveland et al., 2000)	AVHRR local area coverage	1992–1993	1 km²	17 (5) IGBP DISCover land cover classes and other land cover legends	Highest-resolution global land cover to date, validated based on statistical design	Variable image quality contributed to unevenness of land cover accuracy
UMD global land cover (Hansen et al., 2000)	AVHRR local area coverage	1992–1993	1 km²	12 (5) land cover classes	Based on an automated analysis strategy	Not validated, affected by variable image quality
MODIS global land cover (Friedl et al., 2002)	MODIS	2001–present, produced annually	500 m²	17 (5) IGBP DISCover land cover	Uses highest-quality remotely sensed inputs available, based on rigorous automated methods	Unknown accuracy due to the lack of a design-based map validation
GLC2000 (Bartholome and Belward, 2005)	SPOT 4 VEGETATION	2000	1 km²	22 (5) land cover classes	Based on standardized land cover legend, validated results	Affected by variable image quality
GlobCover (Arino et al., 2007)	ENVISAT MERIS	2005–2006	300 m²	22 (4) land cover classes, UN Land Cover Classification System	Based on standardized land cover legend, validated results, and highest-resolution imagery to date	Regional variability in image quality increased uncertainty of results in some parts of the world
Fine resolution global land cover (Gong et al., 2013)	Landsat 5 and 7	Nominally 2005–2006	30 m²	29 (6) land cover classes	Highest-resolution dataset ever produced	Limited temporal inputs resulted in regional inconsistencies

McCullum et al. (2006) concluded that while there is general agreement at the global level in total area and general land cover patterns; there is limited agreement when looking at specific spatial distributions.

Perhaps the most definitive effort to understand the difference in global datasets comes from Herold et al. (2008). In this study, the IGBP DISCover, UMD, GLC2000, and MODIS land cover datasets were harmonized by crosswalking the different land cover classes to a common classification standard—the UN LCCS (Di Gregorio, 2005). Thirteen classes were defined, and the original accuracy assessment samples associated with the various products were used to determine per class and overall accuracy for each harmonized product. Cover types with large homogeneous extents, such as barren, cultivated, and managed, shrublands, and snow and ice, are more consistently represented in global products than smaller, discontinuous classes. All products show a limited ability to consistently represent mixed classes.

As the quality and resolution of remotely sensed data used for global land cover mapping improves, the logical expectation is that overall and individual class accuracies will also improve. Fritz et al. (2011) emphasize the continued uncertainty in global land cover products, especially in land cover classes associated with agriculture and some forest groups. They suggest that increased use of in situ data is the key to improving global land cover datasets.

10.4 Future Pathways of Global Sensing in Rangeland Environments

Remote sensing has created unprecedented capacity to study the Earth by providing repeated measurements of biological phenomena at global scales. Since the first regional applications of NDVI (one of the earliest regional applications found is Rouse et al., 1973) (Section 10.2), the study of the global rangeland situation has benefitted greatly from advancements made in a relatively short period of time. Though future uses of remote-sensing data will be used in unexpected ways, obvious areas of enhancement and progress are anticipated. These future pathways can be expressed in distinct areas including data availability, processing improvement, and biophysical product improvement.

The design and intended application of spaceborne sensors will continue to evolve, and a wider variety of satellite systems including radar and lidar could be quite beneficial in the future. If the past provides a glimpse into the future, new sensors with improved capabilities will be developed, but it is unclear, however, whether improved spatial, spectral, and temporal resolution of satellite remote sensing will provide the greatest advancements in the evaluations of rangelands on a global scale. The ability to extract surface features and quantify biophysical properties will still be limited by the same factors presently hindering remote sensing of rangelands. Characteristics such as soil background, leaf anatomy and physiology, and relatively low biomass conspire to hinder remote sensing of rangelands. Very little can be done to change these situations, and as a result, future pathways should include a focus on data continuity, increased data availability, better computer processing systems, and global campaigns for collecting field-referenced data.

Remote-sensing data continuity is important to monitoring global rangelands, and loss of this critical aspect will significantly weaken our ability to understand what the biosphere is indicating. The need for continuity is recognized in the Land Remote Sensing Policy Act of 1992, which states

The continuous collection and utilization of land remote sensing data from space are of major benefit in studying and understanding human impacts on the global environment, in managing the Earth's resources, in carrying out national security functions, and in planning and conducting many other activities of scientific, economic, and social importance.

Since the first civilian spaceborne missions (e.g., Landsat 1), the global monitoring community and government agencies have been reasonably successful in providing the needed continuity. The Landsat program is a good example of the flow and continuity with incremental improvements with each successive launch generally maintaining a 30 m^2 resolution benchmark. If archive data from Landsat 4 (deployed in 1982) are included, 32 years of 30 m^2 spatial resolution from the TM sensor in visible and NIR (at the minimum) are available. Landsat 8, launched on February 11, 2013, is the most recent addition to the suite of Landsat satellite launches and provides an example of maintaining continuity with previous missions while improving capability. Landsat 8 contains the Operational Land Imager (OLI) and the Thermal Infrared Sensor (TIRS), which provide global coverage at varying resolutions. The OLI provides two new spectral bands for detecting cirrus clouds and the other for coastal zone observations. Now that the entire archive of Landsat data has been made freely and publically available, usage has increased exponentially. The unprecedented data availability has and will continue to lead to new algorithmic and ecological discoveries.

Increased data usage may signal greater interest in remote sensing but certainly tracks the increased microprocessor speed over the last decade (Figure 10.19). As processing speed and memory have increased so has the level of algorithmic sophistication and spatial domain for analysis. Indeed, the global remote-sensing community is poised for improved characterization capabilities, due to new data policies and concurrent advances in computing (Hansen and Loveland, 2011).

Even a decade ago, it would have been unthinkable to regularly process and store a global time series of satellite imagery with a pixel resolution of less than about 250 m^2. Although it is certainly possible to monitor rangelands globally at 30 m^2, it will be a monumental task. Each TM path/row contains 0.534 GB in the seven multispectral and thermal channels and approximately 0.234 GB for the panchromatic band. Since roughly 16,396

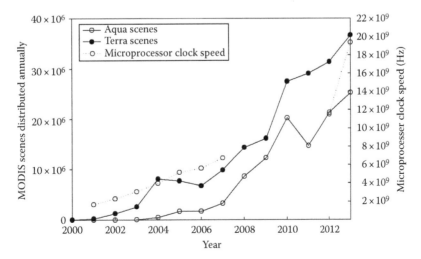

FIGURE 10.19 Microprocessor speed and MODIS data usage (microprocessor speed data courtesy of https://www.raptureready.com/accessed April 1, 2014; MODIS usage data courtesy of B. Ramachandran NASA Earth Observing System, LP DAAC.)

TABLE 10.7   Spectral Channels and Suggested Usefulness

		Band No.	Driving EDR(s)	Spectral Range (μm)	Horiz. Sample Interval (km) (Track × Scan)	
					Nadir	End of Scan
Reflective bands	VisNIR	M1	Ocean color aerosol	0.402–0.422	0.742 × 0.259	1.60 × 1.58
		M2	Ocean color aerosol	0.436–0.454	0.742 × 0.259	1.60 × 1.58
		M3	Ocean color aerosol	0.478–0.498	0.742 × 0.259	1.60 × 1.58
		M4	Ocean color aerosol	0.545–0.565	0.742 × 0.259	1.60 × 1.58
		I1	Imagery EDR	0.600–0.680	0.371 × 0.387	0.80 × 0.789
		M5	Ocean color aerosol	0.662–0.682	0.742 × 0.259	1.60 × 1.58
		M6	Atmospheric correction	0.739–0.754	0.742 × 0.776	1.60 × 1.58
		I2	NDVI	0.846–0.885	0.371 × 0.387	0.80 × 0.789
		M7	Ocean color aerosol	0.846–0.885	0.742 × 0.259	1.60 × 1.58
	S/WMIR	M8	Cloud particle size	1.230–1.250	0.742 × 0.776	1.60 × 1.58
		M9	Cirrus/cloud cover	1.371–1.386	0.742 × 0.776	1.60 × 1.58
		I3	Binary snow map	1.580–1.640	0.371 × 0.387	0.80 × 0.789
		M10	Snow fraction	1.580–1.640	0.742 × 0.776	1.60 × 1.58
		M11	Clouds	2.225–2.275	0.742 × 0.776	1.60 × 1.58
Emissive Bands		I4	Imagery clouds	3.550–3.930	0.371 × 0.387	0.80 × 0.789
		M12	SST	3.660–3.840	0.742 × 0.776	1.60 × 1.58
		M13	SST fires	3.973–4.128	0.742 × 0.259	1.60 × 1.58
	LWIR	M14	Cloud top properties	8.400–8.700	0.742 × 0.776	1.60 × 1.58
		M15	SST	10.263–11.263	0.742 × 0.776	1.60 × 1.58
		I5	Cloud imagery	10.500–12.400	0.371 × 0.387	0.80 × 0.789
		M16	SST	11.538–12.488	0.742 × 0.776	1.60 × 1.58

The LWIR are long-wave infrared bands while the S/MWIR are short- to mid-wave infrared bands.
Source: Adapted from Schueler et al. (2003).

scenes are required for global coverage (including oceans), that is an estimated 12.3 TB of data for a single 16-day period. The repeat frequency or revisit cycle is 16 days (~22 periods per year), so the total amount of data since 1999 is near 4208 TB. Based on an online storage price of $0.08 per month per GB (https://cloud.google.com/products/cloud-storage/), the storage cost is tantamount to roughly 4 million dollars per year. While this represents a significant amount of data and resources, a growing number of global applications at 30 m² spatial resolution can be expected. Indeed, this past year has seen the production of a Landsat-based global database of tree cover at 30 m² resolution (Sexton et al., 2013), and work is underway to develop long-term (Landsat period of record) land cover dynamics on a global scale (Sexton et al., 2013).

Presently, numerous efforts aimed at global remote sensing of rangelands are based on MODIS sensors aboard the Terra and Aqua satellites. Since 2006, the number of scenes annually distributed from MODIS data from both Aqua and Terra has increased by 7.6 million per year (about 181 TB year⁻¹) (Figure 10.19). This use is a testament to the breadth of vetted science data products offered globally. Continuity between MODIS and future global Earth-observing satellites is provided by the Suomi National Polar-orbiting Partnership (NPoP) satellite. Suomi NPoP was launched in 2011 with five key instruments, but the instrument with greatest application, to rangelands globally, and similarity with the AVHRR and MODIS predecessors is the VIIRS. The VIIRS instrument observes the Earth and

atmosphere at 22 visible and infrared wavelengths (Table 10.7). Suomi NPP is the bridge between the current NASA research Earth-observing satellites and future NOAA missions, specifically the Joint Polar Satellite System (JPSS) (Lee et al., 2006). The JPSS is a joint program between NOAA, NASA, and the Defense Weather Satellite System, tasked with developing the next-generation requirements for environmental research, weather forecasting, and climate monitoring (npp.gsfc.nasa.gov/viirs.html). The JPSS provides operational continuity of satellite-based observations and products through a series of advanced spacecraft of which Suomi NPoP is a member. The next two satellites to be launched include JPSS 1 and JPSS 2, both of which will contain, among others, the VIIRS instrument. The JPSS 1 platform is scheduled to be launched in 2017, while JPSS 2 is scheduled for launch in 2021.

The continuity of land remote-sensing instruments is well established and provides a critical component to researchers involved with global change research in rangeland environments. Most future global issues will emulate present concerns. In other words, the problems, or area of focus, today (e.g., vegetation trends, land degradation, and fire processes) will continue and perhaps intensify in the future.

Regardless of the increasingly important roles remote sensing will play, georeferenced field data will play an equally critical aspect of biospheric monitoring (Baccini et al., 2007). Fritz and See (2011) suggest that increased use of in situ data is the key to improving global datasets. The collection, maintenance, analysis, and distribution of georeferenced field data, however, are a time-consuming and resource-intensive exercise, especially over regional or global domains. In this vein, the citizen scientist is an underutilized concept that can be cheaply and effectively employed to globally collect biospheric observations. Citizen science can be defined as

> the systematic collection and analysis of data; development of technology; testing of natural phenomena; and the dissemination of these activities by researchers on a primarily avocational basis.

OpenScience (2011)

These open networks promote interactions between scientists, society, and policymakers leading to decision making by scientific research conducted by amateur or nonprofessional scientists (Socientize, 2013). Advancements in communication and technology are credited with aiding the growth of citizen scientists (Silverton, 2009). Collectively, citizen science efforts from around the globe could possibly provide powerful venues for validating and calibrating future remote-sensing efforts.

10.5 Conclusions

Rangelands are found extensively throughout the world covering about 50% of the global land mass. The remoteness, harsh conditions, and high interannual variation in productivity make remote sensing the most cost-effective and efficacious tool for evaluating the status and health of rangelands globally. Global remote sensing has unique constraints from a remote-sensing perspective and spatial resolution is often sacrificed in place of temporal resolution. A broad suite of sensors possessing various spectral channels, revisit times, and spatial resolutions are available for regional to global rangeland applications. However, most global applications, especially those sponsored for national or international applications (e.g., LADA, IGBP), use AVHRR, SPOT-VGT, MODIS, and to a lesser degree TM. Additionally, a large number of biophysical phenomena can be investigated with the myriad of sensors, but as discussed in this chapter, we focused on the globally relevant issues of degradation, fire, land cover, food security, and global change. In this chapter, we demonstrate sensors, data, algorithms, strengths, and limitations of various methods to address these globally significant issues.

Though estimates vary, the proportion of degraded rangelands is around 23% globally (Table 10.4). The use and interpretation of RUE for evaluating degradation patterns is controversial (Prince et al., 2007), but alternative techniques are subject to similar issues and assumptions. Thus, when considering degradation, especially in a global context, a model ensemble approach (e.g., combine local NPP scaling, rainfall use efficiency, and NPP trend analysis) may be most useful to indicate trends and identify where action is needed to lessen detrimental effects on goods and services.

Most global land cover efforts have limited thematic resolution of rangeland classes (average number of rangeland classes is 4.75; Table 10.6). However, computational resources and algorithmic complexity is sufficient to produce higher spatial and thematic resolution land cover maps as inaugurated by studies such as Gong et al. (2013) and Hansen et al. (2013). Land cover and land use will continue to evolve in response to broadscale disturbance and global change. As a result, monitoring global change and extent and severity of fire has been the focus of many algorithms, national programs, and sensors. As an example, the MODIS sensor aboard both the Terra and Aqua platforms was designed with fire monitoring in mind with channels 21, 22, 31, and 33. Burn severity evaluation is a relatively new capability since the AVHRR and SPOT-VGT sensors lack the spectral channels necessary for contemporary algorithms. Likewise, the advent of the MODIS—derived NPP product (Running et al., 2004)—has spawned numerous studies aimed at evaluating NPP patterns globally. In this chapter, we demonstrate rather unchanged NPP trajectories in the rangeland domain but also identify cases where higher spatial and thematic resolution products are needed to further understand patterns. Despite these relatively unchanged temporal trajectories globally, drought and degradation are detrimental on a regional basis and regularly threaten the security of food derived from rangelands. The LEWS, driven by MODIS-derived 250 m² NDVI, is a useful program to provide guidance local governments and international aid organizations. As world population continues to grow, it is likely that the programs like LEWS will become increasingly important. These issues emphasize the critical importance of mission and spectral continuity. The recent launch of Landsat 8 and Suomi NPoP is a critical stepping stone to future efforts, but compared to their predecessors, they possess a distinctive lack of present use, given their recent recentness.

References

Al-Bakri, J. T. and J. C. Taylor. 2003. Application of NOAA AVHRR for monitoring vegetation conditions and biomass in Jordan. *Journal of Arid Environments*, 54, 579–593.

Allbed, A. and L. Kumar. 2013. Soil salinity mapping and monitoring in arid and semi-arid regions using remote sensing technology: A review. *Advances in Remote Sensing*, 2, 373–385.

Allen, J. L. and B. Sorbel. (2008). Assessing the differenced Normalized Burn Ratio's ability to map burn severity in the boreal forest and tundra ecosystems of Alaska's national parks. *International Journal of Wildland Fire*, 17, 463–475.

An, N., K. P. Price, and J. M. Blair. (2013). Estimating aboveground net primary productivity of the tallgrass prairie ecosystem of the Central Great Plains using AVHRR NDVI. *International Journal of Remote Sensing*, 34, 3717–3735.

Angerer, J. P. (2008). Examination of high resolution rainfall products and satellite greenness indices for estimating patch and landscape forage biomass. *PhD Dissertation*. Texas A&M University, College Station, Texas.

Angerer, J. 2012. Gobi forage livestock early warning system. In *Conducting National Feed Assessments*, eds. M. B. Coughenour and H. P. S. Makkar, pp. 115–130. Rome, Italy: Food and Agriculture Organization.

Arino, O., D. Gross, F. Ranera, L. Bourg, M. Leroy, P. Bicheron, and R. Witt. (2007). *GlobCover: ESA service for global land cover from MERIS.* International Geoscience and Remote Sensing Symposium, Barcelona, Spain. pp. 2412–2415.

Arino, O., S. Casadio, and D. Serpe. 2012. Global night-time fire season timing and fire count trends using the ATSR instrument series. *Remote Sensing of Environment*, 116, 226–238.

Awuma, K. S., J. W. Stuth, R. Kaitho, and J. Angerer. (2007). Application of Normalized Differential Vegetation Index and geostatistical techniques in cattle diet quality mapping in Ghana. *Outlook on Agriculture*, 36, 205–213.

Baccini, A., M. A. Friedl, C. E. Woodcock, and Z. Zhu. 2007. Scaling field data to calibrate and validate moderate spatial resolution remote sensing models. *Photogrammetric Engineering and Remote Sensing*, 73, 945–954.

Bai, E., T. W. Boutton, X. B. Wu, F. Liu, and S. R. Archer. 2009. Landscape-scale vegetation dynamics inferred from spatial patterns of soil δ13C in a subtropical savanna parkland. *Journal of Geophysical Research*, 114, G01019.

Bai, Z. G., D. L. Dent, L. Olsson, and M. E. Schaepman. 2008. Proxy global assessment of land degradation. *Soil Use and Management*, 24, 223–234.

Barbosa, P. M., J. M. Grégoire, and J. M. C. Pereira. 1999. An algorithm for extracting burned areas from time series of AVHRR GAC data applied at a continental scale. *Remote Sensing of Environment*, 69, 253–263.

Bartholomé, E. and A. S. Belward. 2005. GLC2000: A new approach to global land cover mapping from Earth observation data. *International Journal of Remote Sensing*, 26, 1959–1977.

Bauer, M. E. 1976. Technological basis and applications of remote sensing of the Earth's resources. *IEEE Transactions on Geoscience Electronics*, 14, 3–9.

Beatriz, A. and F. Valladares. 2008. International efforts on global change research. In *Earth Observation of Global Change: The Role of Satellite Remote Sensing in Monitoring the Global Environment*, ed. E. Chuvieco, pp. 1–21. Dordrecht, the Netherlands: Springer.

Blanco, P., H. d. Valle, P. Bouza, G. Metternicht, and L. Hardtke. 2014. Ecological site classification of semiarid rangelands: Synergistic use of Landsat and Hyperion imagery. *International Journal of Applied Earth Observation and Geoinformation*, 29, 11–21.

Carreiras, A., M. Vasconcelos, and R. Lucas. 2012. Understanding the relationship between aboveground biomass and ALOS PALSAR data in the forests of Guinea-Bissau (West Africa). *Remote Sensing of Environment*, 121, 426–442.

Carter, L. J. 1969. Earth resources satellite: Finally off the ground? *Science*, 163, 796–798.

Chuvieco, E. and R. G. Congalton. 1988. Mapping and inventory of forest fires from digital processing of TM data. *Geocarto International*, 3, 41–53.

Chuvieco, E., D. Riano, I. Aguado, and D. Cocero. (2002). Estimation of fuel moisture content from multitemporal analysis of Landsat Thematic Mapper reflectance data: applications in fire danger assessment. *International Journal of Remote Sensing*, 23, 2145–2162.

Crist, E. P. and R. C. Cicone. 1984. Application of the tasseled cap concept to simulated Thematic Mapper data. *Photogrammetric Engineering and Remote Sensing*, 50, 343–352.

Deering, D. W., J. W. Rouse, R. H. Haas, and J. A. Schell. 1975. Measuring forage production of grazing units from Landsat MSS data. In *Tenth International Symposium on Remote Sensing of Environment*, pp. 1169–1178. Ann Arbor, MI: University of Michigan.

DeFries, R. S., M. Hansen, J. R. G. Townshend, and R. Solberg. 1998. Global land cover classifications at 8 km spatial resolution: The use of training data derived from Landsat imagery in decision tree classifiers. *International Journal of Remote Sensing*, 19, 3141–3168.

DeFries, R. S. and J. R. G. Townshend. 1994. NDVI-derived land cover classifications at a global scale. *International Journal of Remote Sensing*, 15, 3567–3586.

DeFries, R. S., J. R. G. Townshend, and M. C. Hansen. 1999. Continuous fields of vegetation characteristics at the global scale at 1-km resolution. *Journal of Geophysical Research: Atmospheres*, 104, 16911–16923.

Di Gregorio, A. 2005. *Land Cover Classification System: Classification Concepts and User Manual: LCCS*. Rome, Italy: Food & Agriculture Organization.

Dong, J., X. Xiao, S. Sheldon, C. Biradar, G. Zhang, N. D. Duong, M. Hazarika, K. Wikantika et al. 2014. A 50-m forest cover map in southeast Asia from ALOS/PALSAR and its application on forest fragmentation assessment. *PLoS ONE*, 9, e8580. doi:10.1371/journal.pone.0085801.

Dungan, J. 1998. Spatial prediction of vegetation quantities using ground and image data. *Remote Sensing*, 19, 267–285.

Dwyer, E., S. Pinnock, J. M. Gregorie, and J. M. C. Pereira. 2000. Global spatial and temporal distribution of vegetation fire as determined from satellite observations. *International Journal of Remote Sensing*, 21, 1289–1302.

Eidenshink, J., B. Schwind, K. Brewer, Z. Zhu, B. Quayle, and S. Howard. 2007. A project for monitoring trends in burn severity. *Fire Ecology*, 3, 3–19.

Eidenshink, J. C. and J. L. Faundeen. 1994. The 1 km AVHRR global land data set: First stages in implementation. *International Journal of Remote Sensing*, 104, 3443–3462.

Elvidge, C. D. and R. J. P. Lyon. 1985. Influence of rock soil spectral variation on the assessment of green biomass. *Remote Sensing of Environment*, 17, 265–279.

Estes, J., A. Belward, T. Loveland, J. Scepan, A. Strahler, J. Townshend, and C. Justice. 1999. The way forward. *Photogrammetric Engineering and Remote Sensing*, 65, 1089–1093.

Everitt, J. H., C. Yang, R. Fletcher, and C. J. Deloach. 2008. Comparison of QuickBird and SPOT 5 satellite imagery for mapping giant reed. *Journal of Aquatic Plant Management*, 46, 77–82.

Fang, H. and W. Ping. 2010. Vegetation change of ecotone in west of northeast china plain using time-series remote sensing data. *Chinese Geographical Society*, 20, 167–175.

Fava, F., R. Colombo, S. Bocchi, and C. Zucca. 2012. Assessment of Mediterranean pasture condition using MODIS normalized difference vegetation index time series. *Journal of Applied Remote Sensing*, 6, 063530-1-063530-12.

Fensholt, R., K. Rasmussen, T. T. Nielsen, and C. Mbow. (2009). Evaluation of earth observation based long term vegetation trends—Intercomparing NDVI time series trend analysis consistency of Sahel from AVHRR GIMMS, Terra MODIS and SPOT VGT data. *Remote Sensing of Environment*, 113, 1886–1898.

Fensholt, R. and S. R. Proud. 2012. Evaluation of Earth observation based global long term vegetation trends—Comparing GIMMS and MODIS global NDVI time series. *Remote Sensing of Environment*, 119, 131–147.

Fensholt, R. and K. Rasmussen. 2011. Analysis of trends in the Sahelian 'rain-use efficiency' using GIMMS NDVI, RFE and GPCP rainfall data. *Remote Sensing of Environment*, 115, 438–451.

Fensholt, R. I., M. S. Sandholt, S. Rasmussen, S. Stisen, and A. Diou. 2006. Evaluation of satellite based primary production modeling in the semi-arid Sahel. *Remote Sensing of Environment*, 105, 173–188.

Flasse, S. and P. Ceccato. 1996. A contextual algorithm for AVHRR fire detection. *International Journal of Remote Sensing*, 17, 419–424.

Foley, J., R. DeFries, and G. P. Asner. 2005. Global consequences of land use. *Science*, 309, 570–574.

Foody, G. M. 2003. Geographical weighting as a further refinement to regression modelling: An example focused on the NDVI–rainfall relationship. *Remote Sensing of Environment*, 88, 283–293.

Fox, W. E., D. W. McCollum, J. E. Mitchell, L. E. Swanson, U. P. Kreuter, J. A. Tanaka, G. R. Evans, and H. T. Heintz. (2009). An Integrated Social, Economic, and Ecologic Conceptual (ISEEC) Framework for Considering Rangeland Sustainability. *Society & Natural Resources*, 22, 593–606.

Frank, T. D. 1984. Assessing change in the surficial character of a semiarid environment with Landsat residual images. *Photogrammetric Engineering and Remote Sensing*, 50, 471–480.

French, N. H. F., J. L. Allen, R. J. Hall, E. E. Hoy, E. S. Kasischke, K. A. Murphy, and D. L. Verbyla. 2008. Using Landsat data to assess fire and burn severity in the North American boreal forest region: An overview. *International Journal of Wildland Fire*, 17, 443–462.

Frey, K. E. and L. C. Smith. 2007. How well do we know northern land cover? Comparison of four global vegetation and wetland products with a new ground-truth database for West Siberia. *Global Biogeochemical Cycles*, 21, 1–15.

Friedl, M. A., D. K. McIver, J. C. F. Hodges, X. Y. Zhang, D. Muchoney, A. H. Strahler, C. E. Woodcock, S. Gopal et al. 2002. Global land cover mapping from MODIS: Algorithms and early results. *Remote Sensing of Environment*, 83, 287–302.

Friedl, M. A., D. Sulla-Menashe, B. Tan, A. Schneider, N. Ramankutty, A. Sibley, and X. M. Huang. (2010). MODIS Collection 5 global land cover: Algorithm refinements and characterization of new datasets. *Remote Sensing of Environment*, 114, 168–182.

Fritz, S. and L. M. See. 2007. Identifying and quantifying uncertainty and spatial disagreement in the comparison of Global Land Cover for different applications. *Global Change Biology*, 14, 1057–1075.

Fritz, S., L. See, I. McCallum, C. Schill, M. Obersteiner, M. van der Velde, H. Boettcher, P. Havlík et al. 2011. Highlighting continued uncertainty in global land cover maps for the user community. *Environmental Research Letters*, 6, 1–7.

Fry, J., G. Xian, S. Jin, J. Dewitz, C. Homer, L. Yang, C. Barnes, N. Herold et al. 2011. Completion of the 2006 national land cover database for the conterminous United States. *Photogrammetric Engineering and Remote Sensing*, 77, 858–864.

Gallo, K., L. Li, B. Reed, J. Eidenshink, and J. Dwyer. (2005). Multi-platform comparisons of MODIS and AVHRR normalized difference vegetation index data. *Remote Sensing of Environment*, 99, 221–231.

Ganguly, S., A. Samanta, M. A. Schull, N. V. Shabanov, C. Milesi, R. R. Nemani, Y. Knyazikhin, and R. B. Myneni. 2008. Generating vegetation leaf area index Earth system data record from multiple sensors. Part 2: Implementation, analysis and validation. *Remote Sensing of Environment*, 112, 4318–4332.

Geerken, R., N. Batikha, D. Celisj, and E. Depauw. 2005. Differentiation of rangeland vegetation and assessment of its status: Field investigations and MODIS and SPOT VEGETATION data analyses. *International Journal of Remote Sensing*, 26, 4499–4526.

Giglio, L., J. Descloitres, C. O. Justice, and Y. J. Kaufman. 2003. An enhanced contextual fire detection algorithm for MODIS. *Remote Sensing of Environment*, 87, 272–282.

Giglio, L., T. Loboda, D. P. Roy, B. Quayle, and C. O. Justice. 2009. An active-fire based burned area mapping algorithm for the MODIS sensor. *Remote Sensing of Environment*, 113, 408–420.

Gilbert, N. 2011. Science enters desert debate. *Nature*, 447, 262.

Giri, G., Z. L. Zhu, and B. Reed. 2005. A comparative analysis of the global land cover 2000 and MODIS land cover data sets. *Remote Sensing of Environment*, 94, 123–132.

Gong, P., J. Wang, L. Yu, Y. Zhao, Y. Zhao, L. Liang, Z. Niu, X. Huang et al. 2013. Finer resolution observation and monitoring of global land cover: First mapping results with Landsat TM and ETM+ data. *International Journal of Remote Sensing*, 34, 2607–2654.

Graetz, R. D. 1987. Satellite remote-sensing of Australian range-lands. *Remote Sensing of Environment*, 23, 313–331.

Graetz, R. D., D. M. Carneggie, R. Hacker, C. Lendon, and D. G. Wilcox. 1976. A qualitative evaluation of LANDSAT imagery of Australia rangelands. *Australian Rangeland Journal*, 1, 53–59.

Graetz, R. D., M. R. Gentle, R. P. Pech, J. F. O'Callaghan, and G. Drewien. 1983. The application of Landsat image data to rangeland assessment and monitoring: An example from South Australia. *Australian Rangeland Journal*, 5, 63–73.

Graetz, R. D., R. P. Pech, and A. W. Davis. 1988. The assessment and monitoring of sparsely vegetated rangelands using calibrated Landsat data. *International Journal of Remote Sensing*, 9, 1201–1222.

Graetz, R. D., R. P. Pech, M. R. Gentle, and J. F. Ocallaghan. 1986. The application of Landsat image data to rangeland assessment and monitoring—The development and demonstration of a land image-based resource information-system (LIBRIS). *Journal of Arid Environments*, 10, 53–80.

Gray, T. I. and D. G. McCrary, 1981. The environmental vegetation index, a tool potentially useful for arid land management. *AgRISTAR Report No. EW-N-1-04076*, Johnson Space Center, Houston, Texas 17132. 7 pp.

Hacker, R. 1980. Prospects for satellite applications in Australian rangelands. *Tropical Grasslands*, 14, 289.

Hansen, M. C., R. S. DeFries, J. R. G. Townshend, M. Carroll, C. Dimiceli, and R. A. Dimiceli. 2003. Global percent tree cover at a spatial resolution of 500 meters: First results of the MODIS vegetation continuous fields algorithm. *Earth Interactions*, 7, 1–15.

Hansen, M. C., R. S. DeFries, J. R. G. Townshend, and R. Sohlberg. 2000. Global land cover classification at the 1 km spatial resolution using a classification tree approach. *International Journal of Remote Sensing*, 21, 1331–1364.

Hansen, M. C. and T. R. Loveland. 2011. A review of large area monitoring of land cover change using Landsat data. *Remote Sensing of Environment*, 122, 66–74.

Hansen, M. C., P. V. Potapov, R. Moore, M. Hancher, S. A. Turubanova, A. Tyukavina, D. Thau, S. V. Stehman et al. 2013. High-Resolution global maps of 21st-century forest cover change. *Science*, 342, 850–853.

Hansen, M., and B. Reed, (2000). A comparison of the IGBP DISCover and University of Maryland 1 km global land cover products. *International Journal of Remote Sensing*, 21, 1365–1373.

Herold, M., P. Mayaux, C. E. Woodcock, A. Baccini, and C. Schmullius. 2008. Some challenges in global land cover mapping: An assessment of agreement and accuracy in existing 1 km datasets. *Remote Sensing of the Environment*, 112, 2538–2556.

Holmgren, M., and M. Scheffer. (2001). El Niño as a Window of Opportunity for the Restoration of Degraded Arid Ecosystems. *Ecosystems*, 4, 151–159.

Holmgren, M., P. Stapp, C.R. Dickman, C. Gracia, S. Graham, J. R. Gutiérrez, C. Hice, F. Jaksic, D. A. Kelt, M. Letnic, M. Lima, B. C. López, P. L. Meserve, W. B. Milstead, G. A. Polis, M. A. Previtali, M. Richter, S. Sabaté, and F. A. Squeo, (2006). Extreme climatic events shape arid and semiarid ecosystems. *Frontiers in Ecology and the Environment*, 4, 87–95.

Homer, C. G., R. D. Ramsey, T. C. Edwards, and A. Falconer. 1997. Landscape cover-type modeling using a multi-scene thematic mapper mosaic. *Photogrammetric Engineering and Remote Sensing*, 63, 59–67.

Huete, A. R. 1988. A soil-adjusted vegetation index (SAVI). *Remote Sensing of Environment*, 25, 295–309.

Huete, A., K. Didan, T. Miura, E. P. Rodriguez, X. Gao, and L. G. Ferreira. 2002. Overview of the radiometric and biophysical performance of the MODIS vegetation indices. *Remote Sensing of Environment*, 83, 195–213.

Hunt, E. R. and B. A. Miyake. 2006. Comparison of stocking rates from remote sensing and geospatial data. *Rangeland Ecology & Management*, 59, 11–18.

Hunt, E. R. and B. N. Rock. 1989. Detection of changes in leaf water content using near- and middle-infrared reflectances. *Remote Sensing of Environment*, 30, 43–54.

Ito, A. 2011. A historical meta-analysis of global terrestrial net primary productivity: Are estimates converging? *Global Change Biology*, 17, 3161–3175.

Jabbar, M. T. and J. Zhou. 2013. Environmental degradation assessment in arid areas: A case study from Basra Province, southern Iraq. *Environmental Earth Sciences*, 70, 2203–2214.

Jarlan, L., S. Mangiarotti, E. Mougin, P. Mazzega, P. Hiernaux, and V. L. Dantec. 2008. Assimilation of SPOT/VEGETATION NDVI data into a Sahelian vegetation dynamics model. *Remote Sensing of Environment*, 112, 1381–1394.

Jenkins, G., K. Mohr, V. Morris, and O. Arino. 1997. The role of convective processes over the Zaire- Congo basin to the southern hemispheric ozone maximum. *Journal of Geophysical Research*, 102, 18963–18980.

Jinguo, Y., N. Zheng, and C. Wang. 2006. Vegetation NPP distribution based on MODIS data and CASA model—A case study of northern Hebei Province. *Chinese Geographical Science*, 16, 334–341.

Jung, M., K. Henkel, M. Herold, and G. Churkina, (2006). Exploiting synergies of global land cover products for carbon cycle modeling. *Remote Sensing of Environment*, 101, 534–553.

Kaufman, Y. J., P. V. Hobbs, V. Kirchhoff, P. Artaxo, L. A. Remer, B. N. Holben, M. D. King, D. E. Ward, E. M. Prins, K. M. Longo, L. F. Mattos, C. A. Nobre, J. D. Spinhirne, Q. Ji, A. M. Thompson, J. F. Gleason, S. A. Christopher, and S. C. Tsay, (1998). Smoke, Clouds, and Radiation-Brazil (SCAR-B) experiment. *Journal of Geophysical Research-Atmospheres*, 103, 31783–31808.

Kaufman, Y. J., C. O. Justice, L. P. Flynn, J. D. Kendall, E. M. Prins, L. Giglio, D. E. Ward, W. P. Menzel et al. 1996a. Potential global fire monitoring from EOSMODIS. *Journal of Geophysical Research: Atmospheres*, 103, 32215–32238.

Kaufman, Y. J., L. A. Remer, R. D. Ottmar, D. E. Ward, R. R. Li, R. Kleidman, R. S. Fraser, L. Flynn et al. 1996b. Relationship between remotely sensed fire intensity and rate of emission of smoke: SCAR-C experiment. In *Global Biomass Burning*, ed. J. Levin, pp. 685–696. Cambridge, MA: The MIT Press.

Kauth, R. J. and G. S. Thomas. 1976. The tasseled cap—A graphic description of the spectral-temporal development of agricultural crops as seen by LANDSAT. In *LARS Symposia*, Paper 159, pp. 4B-41–4B-51. West Lafayette, IN: Purdue e-Pubs, Purdue University.

Kawamura, K., T. Akiyama, H. Yokota, and M. Tsutsumi. 2005. Monitoring of forage conditions with MODIS imagery in the Xilingol steppe, Inner Mongolia. *International Journal of Remote Sensing*, 26, 1423–1436.

Key, C. H. and N. C. Benson. 1999a. The normalized burn ratio (NBR): A Landsat TM radiometric measure of burn severity. http://nrmsc.usgs.gov/research/ndbr.htm (accessed on March 31, 2014).

Key, C. H. and N. C. Benson. 1999b. Measuring and remote sensing of burn severity. In *Proceedings of the Joint Fire Science Conference*, eds. L. F. Neuenschwander, K. C. Ryan, and G. E. Goldberg, Vol. II, p. 284. Boise, ID: University of Idaho and the International Association of Wildland Fire.

Key, C. H. and N. C. Benson. 2006. Landscape assessment: Ground measure of severity, the Composite Burn Index, and remote sensing of severity, the Normalized Burn Index. In *FIREMON: Fire Effects Monitoring and Inventory System, For. Serv. Gen. Tech. Rep. RMRS-GTR-164*, eds. D. C. Lutes et al., pp. CD: LA1–LA51. Ogden, UT: USDA For. Serv. Rocky Mt. Res. Stn.

Koch, A., A. McBratney, M. Adams, D. Field, R. Hill, J. Crawford, B. Minasny, R. Lal et al. 2013. Soil security: Solving the global soil crisis. *Global Policy*, 4, 434–444.

Lambin, E. F., B. L. Turner, and J. Helmut. 2001. The causes of land-use and land-cover change: Moving beyond the myths. *Global Environmental Change*, 11, 261–269.

Le, Q. B., L. Tamene, and P. L. G. Vlek. 2012. Multi-pronged assessment of land degradation in West Africa to assess the importance of atmospheric fertilization in masking the processes involved. *Global Planetary Change*, 92, 71–81.

Le Houérou, H. N. 1984. Rain-use efficiency: A unifying concept in arid-land ecology. *Journal of Arid Environments*, 7, 213–247.

Le Houérou, H. N., R. L. Bingham, and W. Skerbek. 1988. Relationship between the variability of primary production and the variability of annual precipitation in world arid lands. *Journal of Arid Environments*, 15, 1–18.

Lee, T. E., S. Miller, F. J. Turk, C. Schueler, R. Julian, S. Deyo, P. Dills, and S. Wang. 2006. The NPOESS VIIRS day/night visible sensor. *Bulletin of the American Meteorological Society*, 87, 191–199.

Lepers, E., E. F. Lambin, A. C. Janetos, R. DeFries, F. Achard, N. Ramankutty, and R. J. Scholes. 2005. A synthesis of information on rapid land-cover change for the period 1981–2000. *BioScience*, 55, 115–124.

Levitus, S., J. I. Antonov, T. P. Boyer, and C. Stephens. 2000. Warming of the world ocean. *Science*, 287, 2225–2229.

Li, S., C. Potter, and C. Hiatt, (2012). Monitoring of Net Primary Production in California Rangelands Using Landsat and MODIS Satellite Remote Sensing. *Natural Resources, Vol.03No.02*, 10.

Loboda, T. V., N. H. F. French, C. Hight-Harf, L. Jenkins, and M. E. Miller. 2013. Mapping fire extent and burn severity in Alaskan tussock tundra: An analysis of the spectral response of tundra vegetation to wildland fire. *Remote Sensing of Environment*, 134, 194–209.

Loboda, T. V., L. Giglio, L. Boschetti, and C. O. Justice. 2012. Regional fire monitoring and characterization using global NASA MODIS fire products in dry lands of Central Asia. *Frontiers in Earth Science*, 6, 196–205.

Lopez-Garcia, M. J. and V. Caselles. 1991. Mapping burns and natural reforestation using Thematic Mapper data. *Geocarto International*, 6, 31–37.

Loveland, T. R., B. C. Reed, J. F. Brown, D. O. Ohlen, L. Yang, and J. W. Merchant. 2000. Development of a global land cover characteristics database and IGBP DISCover from 1 km AVHRR data. *International Journal of Remote Sensing*, 29, 3–10.

Ludwig, J. 1987. Primary productivity in arid lands: Myths and realities. *Journal of Arid Environments*, 13, 1–7.

Lund, G. H. 2007. Accounting for the worlds rangelands. *Rangelands*, 29, 3–10.

Mansour, K., O. Mutango, T. Everson, and E. Adam. 2012. Discriminating indicator grass species for rangeland degradation assessment using hyperspectral data resampled to AISA Eagle resolution. *ISPRS Journal of Photogrammetry and Remote Sensing*, 70, 56–65.

Marini, A. and M. Talbi. 2009. *Desertification and Risk Analysis Using High and Medium Resolution Satellite Data*, 271 pp. Dordrecht, the Netherlands: Springer.

Matthews, E. 1983. Global vegetation and land use: New high-resolution data bases for climate studies. *Journal of Climatology and Applied Meteorology*, 22, 474–487.

Mayaux, P., H. Eva, J. Gallego, A. H. Strahler, M. Herold, S. Agrawal, S. Naumov, E. E. D. Miranda et al. 2006. Validation of the global land cover 2000 map. *IEEE Transactions on Geoscience and Remote Sensing*, 44, 1728–1738.

McCullum, I., M. Obersteiner, S. Nilsson, and A. Shvidenko. 2006. A spatial comparison of four satellite derived 1 km global land cover datasets. *International Journal of Applied Earth Observation and Geoinformation*, 8, 246–255.

McGraw, J. F. and P. T. Tueller. 1983. Landsat computer-aided analysis techniques for range vegetation mapping. *Journal of Range Management*, 36, 627–631.

Metternicht, G. 1996. *Detecting and Monitoring Land Degradation Features and Processes in the Cochabamba Valleys, Bolivia.* Enschede, the Netherlands: International Institute for Geo-Information Science and Earth Observation, ITC Publication No.36.

Metternicht, G. and J. Zinck. 2003. Remote sensing of soil salinity: Potentials and constraints. *Remote Sensing of Environment,* 85, 1–20.

Metternicht, G. and J. Zinck. 2009. *Remote Sensing of Soil Salinization—Impact on Land Management.* London, U.K.: CRC Press.

Millennium Ecosystem Assessment. 2005. *Ecosystems and Human Well-Being: Desertification Synthesis.* Washington, DC: Island Press.

Miller, J. D. and A. E. Thode. 2007. Quantifying burn severity in a heterogeneous landscape with a relative version of the delta Normalized Burn Ratio (dNBR). *Remote Sensing of Environment,* 109, 66–80.

Miura, T., A. Huete, and H. Yoshioka. (2006). An empirical investigation of cross-sensor relationships of NDVI and red/near-infrared reflectance using EO-1 Hyperion data. *Remote Sensing of Environment,* 100, 223–236.

Myneni, R.B., S. Hoffman, Y. Knyazikhin, J.L. Privette, J. Glassy, Y. Tian, Y. Wang, X. Song, Y. Zhang, G. R. Smith, A. Lotsch, M. Friedl, J. T. Morisette, P. Votava, R. R. Nemani, and S. W. Running, (2002). Global products of vegetation leaf area and fraction absorbed PAR from year one of MODIS data. *Remote Sensing of Environment,* 83, 214–231.

Myneni, R. B., R. R. Nemani, and S. Running. 1997. Estimation of global leaf area index and absorbed PAR using radiative transfer models. *IEEE Transactions on Geoscience and Remote Sensing,* 35, 1380–1393.

Nachtergaele, F., M. Petri, R. Biancalani, G. Van Lynden, and H. Van Velthuizen, (2010). Global Land Degradation Information System (GLADIS). Beta Version. An Information Database for Land Degradation Assessment at Global Level. Land Degradation Assessment in Drylands Technical Report, (17).

National Wildfire Coordinating Group (NWCG). 2012. Glossary of Wildland fire terminology. http://www.nwcg.gov/pms/pubs/glossary/ (accessed March 31, 2014).

Nkonya, E., J. von Braun, A. Mirzabaev, Q. B. Le, H. Y. Kwon, and O. Kurui. 2013. Economics of land degradation initiative: Methods and approach for global and national assessments. ZEF-discussion papers on development policy No. 183, Bonn, Germany, pp. 40.

Oldeland, J., W. Dorigo, D. Wesuls, and N. Jürgens. 2010. Mapping bush encroaching species by seasonal differences in hyperspectral imagery. *Remote Sensing,* 2, 1416–1438.

Oldeman, L. R. 1998. Soil degradation: A threat to food security? Report 98/01. International Soil Reference and Information Centre, Wageningen.

Oldeman, L., R. Hakkeling, and W. Sombroek. 1991. *World Map of the Status of Human-Induced Soil Degradation: An Explanatory Note* (2nd edn.). Wageningen, the Netherlands: International Soil Reference and Information centre.

Olson, J. S. 1983. *Carbon in Live Vegetation of Major World Ecosystems,* ORNL-5862. Environmental Sciences Division Publication No. 1997. Oak Ridge, TN: Oak Ridge National Laboratory.

OpenScientist. 2011. Finalizing a Definition of "Citizen Science" and Citizen Scientists." http://www.openscientist.org/2011/09/finalizing-definition-of-citizen.html (accessed March 31, 2014).

Paltridge, G. W. and I. Barber. 1988. Monitoring grassland dryness and fire potential in Australia with NOAA/AVHRR data. *Remote Sensing of Environment,* 25, 381–394.

Parker, A. E. and E. R. Hunt. 2004. Accuracy assessment for detection of leafy spurge with hyperspectral imagery. *Journal of Range Management,* 57, 106–112.

Paruelo, J. M., H. E. Epstein, W. K. Lauenroth, and I. C. Burke. 1997. ANPP estimates from NDVI for the central grassland region of the United States. *Ecology,* 78, 953–958.

Pech, R. P., R. D. Graetz, and A. W. Davis. 1986. Reflectance modeling and the derivation of vegetation indexes for an Australian semiarid shrubland. *International Journal of Remote Sensing,* 7, 389–403.

Pickup, G. 1996. Estimating the effects of land degradation and rainfall variation on productivity in rangelands, an approach using remote sensing and models of grazing and herbage dynamics. *Journal of Applied Ecology,* 33, 819–832.

Pickup, G. and V. H. Chewings. 1988. Forecasting patterns of soil-erosion in arid lands from Landsat MSS data. *International Journal of Remote Sensing,* 9, 69–84.

Pickup, G. and B. D. Foran. 1987. The use of spectral and spatial variability to monitor cover change on inert landscapes. *Remote Sensing of Environment,* 23, 351–363.

Pickup, G. and D. J. Nelson. 1984. Use of Landsat radiance parameters to distinguish soil-erosion, stability, and deposition in arid central Australia. *Remote Sensing of Environment,* 16, 195–209.

Plummer, S., O. Arino, M. Simon, and W. Steffen. 2006. Establishing an earth observation product service for the terrestrial carbon community: The GLOBCARBON initiative. *Mitigation and Adaptation Strategies for Global Change,* 11, 97–111.

Pohl, C. and J. L. Van Genderen, (1998). Review article Multisensor image fusion in remote sensing: Concepts, methods and applications. *International Journal of Remote Sensing,* 19, 823–854.

Ponce-Hernandez, R. and P. Koohafkan. 2004. Methodological framework for land degradation assessment in drylands (LADA) (simplified version). ftp://ftp.fao.org/agl/agll/lada/LADA-Methframwk-simple.pdf (accessed March 31, 2014).

Prince, S. D. 2002. Spatial and temporal scale for detection of desertification. In *Do Humans Create Deserts*, eds. J. F. Reynolds and M. Stafford-Smith, pp. 23–40. Berlin, Germany: Dahlem University Press.

Prince, S. D. and C. J. Tucker, (1986). Satellite Remote-Sensing of Rangelands in Botswana .2. Noaa Avhrr and Herbaceous Vegetation. *International Journal of Remote Sensing*, 7, 1555–1570.

Prince, S. D., I. Becker-Reshef, and K. Rishmawi. 2009. Detection and mapping of long-term land degradation using local net production scaling: Application to Zimbabwe. *Remote Sensing of Environment*, 113, 1046–1057.

Prince, S. D., K. J. Wessels, C. J. Tucker, and S. E. Nicholson. 2007. Desertification in the Sahel: A reinterpretation of a reinterpretation. *Global Change Biology*, 13, 1308–1313.

Pu, R., P. Gong, Z. Li, and J. Scarborough. 2004. A dynamic algorithm for wildfire mapping with NOAA/AVHRR data. *International Journal of Wildland Fire*, 13, 275–285.

Reeves, M. and L. S. Bagget. 2014. A remote sensing protocol for identifying rangelands with degraded productive capacity. *Ecological Indicators*, 43, 172–182.

Reeves, M. C. and J. E. Mitchell. 2011. Extent of coterminous US rangelands: Quantifying implications of differing agency perspectives. *Rangeland Ecology and Management*, 64, 1–12.

Reeves, M. C., J. C. Winslow, and S. W. Running. 2001. Mapping weekly rangeland vegetation productivity using MODIS algorithms. *Journal of Range Management*, 54, 90–105.

Reeves, M. C., M. Zhao, and S. W. Running. 2006. Applying improved estimates of MODIS productivity to characterize grassland vegetation dynamics. *Journal of Rangeland Ecology and Management*, 59, 1–10.

Robinove, C., J. P. S. Chavez, D. Gehring, and R. Holmgren. 1981. Arid land monitoring using Landsat albedo difference images. *Remote Sensing of Environment*, 11, 133–156.

Rollins, Matthew G. "LANDFIRE: A nationally consistent vegetation, wildland fire, and fuel assessment." *International Journal of Wildland Fire* 18.3 (2009): 235–249.

Rouse, J. W., R. H. Haas, J. A. Schell, and D. W. Deering. 1973. Monitoring vegetation systems in the Great Plains with ERTS. In *Proceedings of the Third ERTS Symposium*, Washington, DC, pp. 309–317.

Rouse Jr., J. W., R. H. Hass, J. A. Schell, D. Deering, and J. C. Harlan. 1974. *Monitoring the Vernal Advancement and Retrogradation (Greenwave Effect) I of Natural Vegetation*, 390. College Station, TX: Remote Sensing Center, Texas A&M University.

Roy, D. P., L. Boschetti, C. O. Justice, and J. Ju. 2008. The collection 5 MODIS burned area product—Global evaluation by comparison with the MODIS active fire product. *Remote Sensing of Environment*, 112, 3960–3707.

Running, S. W., D. D. Baldocchi, D. P. Turner, S. T. Gower, P. S. Bakwin, and K. A. Hibbard. 1999. A global terrestrial monitoring network integrating tower fluxes, flask sampling, ecosystem modeling and EOS satellite data. *Remote Sensing of Environment*, 70, 108–127.

Running, S. W., R. R. Nemani, F. A. Heinsch, M. Zhao, M. Reeves, and H. Hashimoto. 2004. A continuous satellite-derived measure of global terrestrial primary production. *Bioscience*, 54, 547–560.

Safriel, U. 2009. Deserts and desertification: Challenges but also opportunities. *Land Degradation and Development*, 20, 353–366.

Sant, E. D., G. E. Simonds, R. D. Ramsey, and R. T. Larsen, (2014). Assessment of sagebrush cover using remote sensing at multiple spatial and temporal scales. *Ecological Indicators*, 43, 297–305.

Scepan, J. 1999. Thematic validation of high-resolution global land-cover data sets. *Photogrammetric Engineering and Remote Sensing*, 65, 1051–1060.

Scheftic, W., X. Zeng, P. Broxton, and M. Brunke, (2014). Intercomparison of seven NDVI products over the United States and Mexico. *Remote Sensing*, 6, 1057–1084.

Schueler, C., J. E. Clement, L. Darnton, F. DeLuccia, T. Scalione, and H. Swenson. 2003. VIIRS sensor performance. International Geoscience And Remote Sensing Symposium Proceedings, Volume 1. Toulouse, France. p: 369–372.

Scott, J. M., F. Davis, B. Csuti, R. Noss, B. Butterfield, C. Groves, H. Anderson, S. Caicco et al. 1993. Gap analysis—A geographic approach to protection of biological diversity. *Wildlife Monographs*, 123, 1–41.

Sexton, J. O., X. P. Song, M. Feng, P. Noojipady, A. Anand, C. Huang, D. Kim, K. M. Collins et al. 2013. Global, 30-m resolution continuous fields of tree cover: Landsat-based rescaling of MODIS vegetation continuous fields with lidar-based estimates of error. *International Journal of Digital Earth*, 6(5), 427–448.

Shen, L., Z. Li, and X. Guo. 2014. Remote sensing of leaf area index (LAI) and a spatiotemporally parameterized model for mixed grasslands. *International Journal of Applied Science and Technology*, 4, 46–61.

Shoshanya, M., N. Goldshleger, and A. Chudnovsky. 2013. Monitoring of agricultural soil degradation by remote-sensing methods: A review. *International Journal of Remote Sensing*, 34, 6152–6181.

Silva, J. M. N., A. C. L. Sá, and J. M. C. Pereira. 2005. Comparison of burned area estimate derived from SPOT-VEGETATION and Landsat ETM+ data on Africa: Influence of spatial pattern and vegetation type. *Remote Sensing of Environment*, 2, 188–201.

Silverton, J. 2009. A new dawn for citizen science. *Trends in Ecological Evolution*, 24, 467–201.

Skidmore, A. K. and J. G. Ferwerda. 2008. Resource distribution and dynamics. In *Resource Ecology: Spatial and Temporal Dynamics of Foraging*, eds. H. H. T. Prins and F. van Langevelde. Wageningen, the Netherlands: Springer.

Socientize Project (2013-12-01). 2013. Green Paper on Citizen Science: Citizen Science for Europe: Towards a better society of empowered citizens and enhanced research. Socientize consortium.

Society for Range Management [SRM], Glossary Update Task Group. 1998. *Glossary of Terms Used in Range Management* (4th edn.), 32 pp. Denver, CO: Society for Range Management.

Sow, M., C. Mbow, C. Hély, R. Fensholt, and B. Sambou, (2013). Estimation of herbaceous fuel moisture content using vegetation indices and land surface temperature from MODIS data. *Remote Sensing*, 5, 2617–2638.

Steven, M. D., T. J. Malthus, F. Baret, H. Xu, and M. J. Chopping. 2003. Intercalibration of vegetation indices from different sensor systems. *Remote Sensing of Environment*, 88, 412–422.

Stuth, J. W., J. Angerer, R. Kaitho, K. Zander, A. Jama, C. Heath, J. Bucher, W. Hamilton, R. Conner, D. Inbody. (2003a) The Livestock Early Warning System (LEWS): Blending technology and the human dimension to support grazing decisions. *Arid Lands Newsletter*, University of Arizona. Available from http://cals.arizona.edu/OALS/ALN/aln53/stuth.html (accessed 4/17/2015).

Stuth, J. W., D. Schmitt, R. C. Rowan, J. P. Angerer, K. Zander, (2003b) PHYGROW Users Guide and Technical Documentation. Texas A&M University. Available from http://cnrit.tamu.edu/physite/PHYGROW_userguide.pdf (accessed April 15, 2015).

Stuth, J.W., J. Angerer, R. Kaitho, A. Jama, R. Marambii, (2005a) Livestock early warning system for Africa rangelands. In Boken V. K., Cracknell A. P. and Heathcote R. L. (eds), *Monitoring and Predicting Agricultural Drought: A global Study*. Oxford University Press, New York, pp. 283–294.

Suits, G., W. Malila, and T. Weller. 1988. Procedures for using signals from one sensor as substitutes for signals of another. *Remote Sensing of Environment*, 25, 395–408.

Svoray, T., A. Perevolotsky, and P. M. Atkinson, (2013). Ecological sustainability in rangelands: The contribution of remote sensing. *International Journal of Remote Sensing*, 34, 6216–6242.

Tansey, K., J. M. Gregoire, E. Binaghi, L. Boschetti, P. A. Brivio, D. Ershov, S. Flasse, R. Fraser et al. 2004. A global inventory of burned area at 1 km resolution for the year 2000 derived from SPOT Vegetation data. *Climatic Change*, 67, 345–377.

Tchuenté, K., A. Thibaut, J. L. Roujean, and S. M. D. Jong. 2011. Comparison and relative quality assessment of the GLC2000, GLOBCOVER, MODIS and ECOCLIMAP land cover data sets at the African continental scale. *International Journal of Applied Earth Observation and Geoinformation*, 13, 207–219.

Telesca, L. and R. Lasaponara. 2006. Quantifying intra-annual persistent behaviour in SPOT-VEGETATION NDVI data for Mediterranean ecosystems of southern Italy. *Remote Sensing of Environment*, 101, 95–103.

Thoma, D. P., D. W. Bailey, D. S. Long, G. A. Nielsen, M. P. Henry, M. C. Breneman, and C. Montagne. 2002. Short-term monitoring of rangeland forage conditions with AVHRR imagery. *Journal of Range Management*, 55, 383–389.

Todd, S. W., R. M. Hoffer, and D. G. Milchunas. 1998. Biomass estimation on grazed and ungrazed rangelands using spectral indices. *International Journal of Remote Sensing*, 19, 427–438.

Tucker, C. J. 1979. Red and photographic infrared linear combinations for monitoring vegetation. *Remote Sensing of Environment*, 8, 127–150.

Tucker, C. J. 1980. Remote sensing of leaf water content in the near infrared. *Remote Sensing of Environment*, 10, 23–32.

Tucker, C. J., J. E. Pinzon, M. E. Born, D. A. Slayback, E. W. Pak, R. Mahoney, E. F. Vermote, and N. E. Saleous. 2005. An extended AVHRR 8-km NDVI dataset compatible with MODIS and SPOT vegetation NDVI data. *International Journal of Remote Sensing*, 26, 4485–4498.

Tucker, C. J., C. Vanpraet, E. Boerwinkel, and A. Gaston. 1983. Satellite remote sensing of total dry matter production in the Senegalese Sahel. *Remote Sensing of the Environment*, 13, 461–474.

Tueller, P. T., F. R. Honey, and I. J. Tapley. 1978. Landsat and photographic remote sensing for arid land applications in Australia. In *International Symposium on Remote Sensing of Environment, Proceedings*, Ann Arbor, Michigan. pp. 2177–2191.

United Nations Convention to Combat Desertification (UNCCD). 1994. Article 2 of the Text of the United Nations Convention to Combat Desertification. http://www.unccd.int/Lists/SiteDocumentLibrary/conventionText/conv-eng.pdf (accessed March 31, 2014).

UNEP. 2012. Global Environment Outlook: GEO-5. Nairobi, Kenya: United Nations Environment Programme. 551 p.

Veraverbeke, S., S. Lhermitte, W. W. Verstraeten, and R. Goossens. 2011. Time-integrated MODIS burn severity assessment using the multi-temporal differenced normalized burn ratio (dNBR(MT)). *International Journal of Applied Earth Observation and Geoinformation*, 13, 52–58.

Verbesselt, J., B. Somers, S. Lhermitte, I. Jonckheere, J. van Aardt, and P. Coppin. 2007. Monitoring herbaceous fuel moisture content with SPOT VEGETATION time-series for fire risk prediction in savanna ecosystems. *Remote Sensing of Environment*, 108, 357–368.

Vitousek, P. M. 1992. Global environmental change: An introduction. *Annual Review of Ecology, Evolution, and Systematics*, 23, 1–14.

Vlek, P., Q. B. Le, and L. Tamene. 2010. Assessment of land degradation, its possible causes and threat to food security in Sub-Saharan Africa. In *Food Security and Soil Quality*, eds. R. Lal and B. A. Stewart, pp. 57–86. Boca Raton, FL: CRC Press.

Vogl, R. 1979. Some basic principles of grassland fire management. *Environmental Management*, 3, 51–57.

Washington-Allen, R. A., N. E. West, R. D. Ramsey, and R. A. Efroymson. 2006. A protocol for retrospective remote sensing-based ecological monitoring of rangelands. *Rangeland Ecology Management*, 59, 19–29.

Weber, G. E., K. Moloney, and F. Jeltsch. 2000. Simulated long-term vegetation response to alternative stocking strategies in savanna rangelands. *Plant Ecology*, 150, 77–96.

Wenze, Y., B. Tan, D. Huang, M. Rautiainen, N. V. Shabanov, Y. Wang, J. L. Privette, K. F. Huemmrich et al. 2006. MODIS leaf area index products: From validation to algorithm improvement. *IEEE Transactions on Geoscience and Remote Sensing*, 44, 1885–1898.

Wessels, K. J. (2009). Letter to the Editor-Comments on 'Proxy global assessment of land degradation' by Bai et al. (2008). *Soil Use and Management*, 25, 91–92.

Wessels, K. J., S. D. Prince, P. E. Frost, and D. van Zyl. 2004. Assessing the effects of human-induced land degradation in the former homeland of northern South Africa with a 1 km AVHRR NDVI time-series. *Remote Sensing of Environment*, 91, 47–67.

Wessels, K. J., S. D. Prince, J. Malherbe, J. Small, P. E. Frost, and D. van Zyl. 2007. Can human-induced land degradation be distinguished from the effects of rainfall variability? A case study in South Africa. *Journal of Arid Environments*, 68, 271–297.

Wilson, M. F. and A. Henderson-Sellers. 1985. A global archive of land cover and soils data for use in general circulation climate models. *International Journal of Climatology*, 5, 119–143.

Wooster, M. J., G. Roberts, G. Perry, and Y. J. Kaufman. 2005. Retrieval of biomass combustion rates and totals from fire radiative power observations: Calibration relationships between biomass consumption and fire radiative energy release. *Journal of Geophysical Research: Atmospheres*, 110, 1–24.

Wylie, B. K., E. A. Fosnight, T. G. Gilmanov, A. B. Frank, J. A. Morgan, M. R. Haferkamp, and T. P. Meyers. 2007. Adaptive data-driven models for estimating carbon fluxes in the northern Great Plains. *Remote Sensing of Environment*, 106, 399–413.

Xiao, J., Q. Zhuang, D. D. Baldocchi, B. E. Law, A. D. Richardson, J. Chen, R. Oren, G. Starr et al. 2008. Estimation of net ecosystem carbon exchange for the conterminous United States by combining MODIS and AmeriFlux data. *Agricultural and Forest Meteorology*, 148, 1827–1847.

Xie, Y., Z. Sha, and M. Yu, (2008). Remote sensing imagery in vegetation mapping: A review. *Journal of Plant Ecology*, 1, 9–23.

Yebra, M., E. Chuvieco, and D. Riaño, (2008). Estimation of live fuel moisture content from MODIS images for fire risk assessment. *Agricultural and Forest Meteorology*, 148, 523–536.

Yin, H., T. Udelhoven, R. Fensholt, D. Pflugmacher, and P. Hostert, (2012). How Normalized Difference Vegetation Index (NDVI) Trends from Advanced Very High Resolution Radiometer (AVHRR) and Systeme Probatoire d'Observation de la Terre VEGETATION (SPOT VGT) Time Series Differ in Agricultural Areas: An Inner Mongolian Case Study. *Remote Sensing*, 4, 3364–3389.

Yost, E. and S. Wenderoth. 1969. Ecological applications of multispectral colour aerial photography. In *Remote Sensing in Ecology*, ed. P. L. Johnson, pp. 46–62. Athens, GA: University of Georgia Press.

Yu, L., L. Zhou, W. Liu, and H. K. Zhou. 2011. Using remote sensing and GIS technologies to estimate grass yield and livestock carrying capacity of alpine grasslands in Golog Prefecture China. *Pedosphere*, 17, 419–424.

Zhang, L., D. J. Jacob, X. Liu, J. A. Logan, K. Chance, A. Eldering, and B. R. Bojkov, (2010). Intercomparison methods for satellite measurements of atmospheric composition: Application to tropospheric ozone from TES and OMI. *Atmos. Chem. Phys.*, 10, 4725–4739.

Zhao, M., S. Running, F. Heinsch, and R. Nemani,(2011). MODIS-Derived Terrestrial Primary Production. In B. Ramachandran, C.O. Justice, & M.J. Abrams (Eds.), *Land Remote Sensing and Global Environmental Change* (pp. 635–660): Springer New York.

Zhao, M. and S. W. Running. 2010. Drought induced reduction in global terrestrial net primary production from 2000 to 2009. *Science*, 329, 940–943

Zhu, L. and J. Southworth. 2013. Disentangling the relationships between net primary production and precipitation in Southern Africa savannas using satellite observations from 1982 to 2010. *Remote Sensing*, 5, 3803–3825.

Zika, M. E. and K. H. Erb. 2009. The global loss of net primary production resulting from human-induced soil degradation in drylands. *Ecological Economics*, 69, 310–388.

11

Remote Sensing of Rangeland Biodiversity

E. Raymond Hunt, Jr.
USDA ARS Beltsville
Agricultural Research Center

Cuizhen Wang
University of South Carolina

D. Terrance Booth
USDA ARS High Plains
Grasslands Research Station

Samuel E. Cox
Bureau of Land Management

Lalit Kumar
University of New England

Matthew C. Reeves
USDA Forest Service

Acronyms and Definitions

ACRIS	Australian Collaborative Rangelands Information System
ASTER	Advanced spaceborne thermal emission and reflection radiometer on NASA Terra
ATREM	Atmospheric removal program
AVHRR	Advanced very-high-resolution radiometer
AVIRIS	Airborne visible infrared imaging spectrometer, NASA JPL hyperspectral sensor
AWiFS	Advanced wide-field sensor, on India's ResourceSat
DOY	Day of year
ENVI	Environment for Visualizing Images, Exelis Visual Information Solutions (Boulder, CO)
EOS	End of growing season
ETM+	Enhanced thematic mapper plus sensor on Landsat 7
EVI	Enhanced vegetation index
fPAR	Fraction of absorbed photosynthetically active radiation calculated from spectral vegetation indices
GIS	Geographic Information System
HSR	High spatial resolution
Hysp	Hyperspectral (high spectral resolution)
HyspIRI	Hyperspectral infrared imager, mission planned by NASA
ISODATA	Iterative self-organizing data analysis technique algorithm
LiDAR	Light detection and ranging
LOS	Length of growing season
LRR	Land resource region
LRU	Land resource unit
LSA	Light sport aircraft
MERIS	Medium-resolution imaging spectrometer
MLRA	Major land resource area
MNF	Minimum noise fraction
MODIS	Moderate-resolution imaging spectroradiometer on NASA Terra and Aqua satellites
MTMF	Mixture tuned matched filter
MVC	Maximum value composite
NASA	National Aeronautics and Space Administration, United States
NASS	National Agricultural Statistics Service, United States Department of Agriculture

NDVI	Normalized difference vegetation index
NIR	Near infrared (0.725–1.2 µm wavelength)
NOAA	National Oceanic and Atmospheric Administration, United States
NRC	National Research Council, United States
NRCS	National Resource Conservation Service, United States Department of Agriculture
OLI	Operational Land Imager, sensor on Landsat 8
PC	Principal components
PFT	Plant functional type
PROSAIL	Combined PROSPECT and Scattering by Arbitrarily Inclined Leaves models
RSAC	Remote Sensing Applications Center, USDA Forest Service
SAM	Spectral angle mapper
SAVI	Soil adjusted vegetation index
SCM	Spectral correlation measure
SID	Spectral information divergence
SOS	Start of growing season
SVM	Support vector machine
SWIR	Shortwave infrared (1.2–2.5 µm wavelength)
TM	Thematic mapper, sensor on Landsat's 4 and 5
UAS	Unmanned aircraft systems, also known as drones or unmanned aerial vehicles
USDA	United States Department of Agriculture
USGS	United States Geological Survey
VIIRS	Visible infrared imaging radiometer Suite

11.1 Introduction

Rangelands are a type of land cover dominated by grasses, grass-like plants, broadleaf herbaceous plants (forbs), shrubs, and isolated trees, usually in which large herbivores evolved as part of the ecosystem. In many rangelands, the large herbivores were replaced by livestock, and thus, livestock grazing represents a major land use for production of food and fiber. Sustainability is maintained by species diversity (Hooper et al., 2005; Tilman et al., 2006, 2012; Zavaleta et al., 2010; Reich et al., 2012), and reduction of biodiversity is expected to be one of the major consequences of global climatic change (Soussanna and Lüscher, 2007; Janetos et al., 2008; McKeon et al., 2009; Pereira et al., 2010, 2012; Belgacem and Louhaichi, 2013; Joyce et al., 2013; Polley et al., 2013). Along with climatic change, invasions of nonnative species are threatening rangeland sustainability by decreasing native species diversity (Ricciardi, 2007; Bradley et al., 2009; Lavergne et al., 2010; Ziska et al., 2011).

Rangelands cover large sparsely populated areas; thus, remote sensing is becoming much more important for rangeland monitoring, whether the objectives are sustainable livestock foraging or maintenance of other ecosystem goods and services with climatic change. Different stakeholders at the national, state/province, and landscape scales have different needs, which may require remotely sensed data at different spatial, temporal, or spectral resolutions. Reeves et al. (Chapter 10) examined the relationship between rangeland productivity and climate, which highlights some of the direct mechanisms on how rangelands may respond to increased atmospheric CO_2, global warming, and changes in precipitation regimes. Methods and techniques of rangeland characterization using remote sensing are covered by Kumar et al. (Chapter 12).

Monitoring the biodiversity of rangelands is critical for maintaining sustainability; therefore, one of the major challenges for remote sensing is how to use imagery to estimate biodiversity (Ludwig et al., 2004; Gillespie et al., 2008; John et al., 2008; Huang and Asner, 2009; Ward and Kutt, 2009). Diversity of plant functional types (PFTs) may be a good indicator of biodiversity, and determining the diversity of PFTs by remote sensing may be easier than determining plant species richness (Ustin and Gamon, 2010). However, land-cover and land-use maps based on global-scale PFTs (Running et al., 1995; Friedl et al., 2010) do not provide sufficient information for conserving and managing rangelands.

This chapter examines data types and methods for remote sensing of biodiversity at different resolutions: spectral, temporal, and spatial. Medium-resolution sensors, such as the Landsat 8 Operational Land Imager (OLI), generally have resolutions on the order of 10–60 m, 10–20 days, and 4–10 bands for spatial, temporal, and spectral resolutions, respectively. High spectral resolution (hyperspectral) sensors generally have 100 or more contiguous bands, which are used to determine a reflectance spectrum of the land surface. High temporal resolution is usually provided by satellites, such as the moderate-resolution imaging spectroradiometer (MODIS), that have a broad swath (>400 km) in order to cover the Earth frequently. Commercial vendors provide satellite data with panchromatic bands of about 0.5 m and multispectral bands of about 2–3 m spatial resolution, but for this chapter, high spatial resolution (HSR) data have pixel sizes of 10 cm or less, and are acquired by aircraft- or ground-based imaging.

To be useful for rangeland managers, remote sensing data must be able to estimate biodiversity at a landscape scale, and the information must be compatible with rangeland management systems based on ecological sites, rangeland state-and-transition models, and rangeland health. With over 40 years of data acquired from the Landsat satellites, and extensive programs of research, Landsat imagery has not been used routinely to provide information necessary to affect managers' decisions on which land areas are used for a given purpose at a given time. Remote sensing data providers and analysts must adapt to the needs of rangeland managers, and the needs are increasingly being driven by issues of sustainability, based on maintaining biodiversity. Therefore, we briefly describe rangeland management concepts and then describe remote sensing data and methods related to rangeland biodiversity. From this perspective, we conclude that HSR data acquired from aircraft provide the necessary data and will have the highest impact on management decisions. However, HSR data have their own unique challenges.

11.2 Biodiversity and Rangeland Management

A plant community is a set of interacting species co-occurring at a given site, with dominant species identified by mass and typical longevity of an individual. Management of U.S. rangelands was built on Clements' (1916) theory of plant succession where the dominant plant species changed over time to a set of species in equilibrium with climate, which is defined as the climax plant community (Brown, 2010). These communities were assumed to have the highest sustainable productivity and greatest resistance to invasive species. Overgrazing was assumed to reverse the plant community to an earlier stage of succession (Sampson, 1919). For monitoring and management, rangeland sites were defined by assuming the climax plant community was the dominant plant community at the onset of European immigration and settlement into the Western United States.

However, it was recognized that with natural disturbances such as fire, drought, and grazing, plant succession was much more dynamic and variable. Furthermore, after severe overgrazing and soil erosion, it was unlikely that the climax community could re-establish itself without costly interventions. In response to the deficiencies of Clementsian succession, two ideas emerged in parallel (Briske et al., 2005): state-and-transition models (Brown, 1994, 2010) and rangeland health (NRC, 1994). To better guide range management based on these two ideas, the concept of rangeland sites was replaced with the concept of ecological sites

in which the various plant communities are in a dynamic equilibrium determined by the natural disturbance regime.

11.2.1 Ecological Sites and State-and-Transition Models

An ecological site is a conceptual division of the landscape, defined as a distinctive kind of land based on recurring soil, landform, geological, and climate characteristics that differs from other kinds of land in its ability to produce distinctive kinds and amounts of vegetation and in its ability to respond similarly to management actions and natural disturbances (Caudle et al., 2013).

In the United States, the top levels of a hierarchical classification system based on soil, landform, geology, and climate are the land resource region (LRR) and major land resource area (MLRA), defined by the USDA NRCS (2006). Subdividing MLRAs based on finer scale differences in climate, geomorphology, and soils, the next level down in the hierarchy is the land resource unit (LRU), LLRs, MLRAs, LRUs, and ecological sites are provisional and may be revised with more information (Moseley et al., 2010; Caudle et al., 2013). In an MLRA or LRU, ecological sites are identified by a reference state (State 1) and reference plant community (Community Phase 1.1), based on the dominant vegetation thought to be present at the time of European immigration and settlement (Figure 11.1). Ecological sites are divided into a series of alternative states (states 2, 3,... to some number *N*),

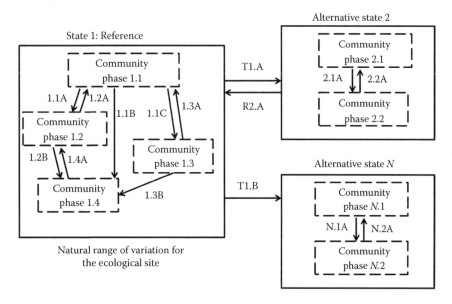

FIGURE 11.1 State-and-transition model for an ecological site in the USA. Outer boxes (solid lines) represent stable ecological states (*A, B,..., N*) within which changes in community phases (dashed lines) result from natural disturbances and succession (arrows). State 1 is the reference state and community phase 1 represents the historic climax plant community. Other community phases represent the range of natural variation and may be identified by dominant plant species. Transitions from one state to another stable state occur when a threshold is crossed (bold arrows), which is usually irreversible without intensive inputs.

which recognize new sets of stable communities that become self perpetuating (Figure 11.1). Land areas in most alternative states are not barren wastes; with appropriate management, these areas may be used sustainably by preventing further degradation.

The definition of an ecological site raises an important drawback for remote sensing—the reference plant community is usually inferred from a wide variety of sources, and mapped using geographic information data. An ecological site is not defined by the plant community that is currently occupying it; therefore, remotely sensed land cover, at any resolution, cannot be used to define an ecological site. Whereas ecological sites are not defined by the plant communities, ecological states may be mapped using characteristic plant communities determined from high-resolution aerial photographs (Steele et al., 2012).

Within an ecological state (e.g., Reference State 1, Figure 11.1), various plant community phases result from natural disturbances (including grazing by livestock) and subsequent plant succession leads back to the reference community phase (Figure 11.1). Together, all of the community phases in Reference State 1 represent the natural range of variation found over time (Caudle et al., 2013). Land areas in Community Phases 1.2, 1.3, and 1.4 are not considered degraded simply, because the plant community is not in Phase 1.1, even though there may be fewer species and less biomass. Change detection must distinguish between changes within a state compared to a change of state. For example, there is less foliar mass in grasslands either after a drought (a frequent occurrence within the range of natural variability) or after severe overgrazing by livestock (which frequently entails a state change; Stafford Smith et al., 2007). Detecting a change using remote sensing is not evidence that a state change has occurred (Bastin et al., 2012; Bradley, 2013). State-and-transition models are provisional hypotheses that describe the ecological processes leading to transitions (T1.A and T1.B, Figure 11.1) from one state to another (Caudle et al., 2013).

For example, invasion of downy brome (*Bromus tectorum* L.) into Great Basin sagebrush ecosystems increased the frequency of fire, which enhanced the dominance of downy brome in a feedback loop (Balch et al., 2013).

Thresholds are the conceptual boundaries dividing alternative states, which are crossed during transitions (Briske et al., 2005, 2006; Bestelmeyer, 2006). Ecological resilience describes how much disturbance an ecological site can withstand without crossing a threshold into an alternative state; operationally, resilience may be defined as multiple species having the same ecological function (Elmqvist et al., 2003; Folke et al., 2004; Allen et al., 2005; Briske et al., 2008).

For most stakeholders, the goal for monitoring rangelands is to determine if an ecological site is in the process of transitioning to alternative state. These transitions are not distributed evenly over an area or over time, so monitoring plant communities requires a larger and longer perspective (Bestelmeyer et al., 2011; Williamson et al., 2012). Determining the amount of species diversity, or at least an array of PFTs, is a primary objective for developing science-based, cost-effective tools for rangeland monitoring based on remote sensing.

11.2.2 Biodiversity Metrics for Managing Australian Rangelands

Rangelands, including tropical savannas, woodlands, shrublands, and grasslands, make up 75% of Australia, with 55% of these rangelands being grazed by livestock. The Australian Collaborative Rangelands Information System (ACRIS) was set up in 2002 to support the Commonwealth and state governments in better managing the rangelands (Bastin et al., 2009; Eyre et al., 2011a; Oliver et al., 2014). In Australia, each state has its own regulations and methods of assessing and managing rangelands, and ACRIS is the overarching body that supports this by collating and synthesizing the monitoring data and making these data available to interested parties. This information assists Natural Resource Management organizations, state governments, and the Commonwealth in planning and reporting obligations, and evaluating the effectiveness of investments.

For determining the health, condition, and biodiversity of rangelands, extensive surveys on a repeated basis are essential. Surveys of species presence and abundance are time consuming and expensive if undertaken at broad scales. For rapid monitoring, three states developed multimetrics for biodiversity to help in the assessment of site conditions:

1. BioCondition (in Queensland)
2. Habitat Hectares (in Victoria)
3. BioMetric (in New South Wales)

BioCondition is a vegetation and biodiversity assessment framework developed by the Queensland Department of Resource Management to provide onsite guidance to beginners and document the assessment process for future revision and comparison (Eyre et al., 2011b). Various surrogates are used to represent the health and condition of the environment being assessed. The BioCondition Assessment Tool uses cameras on mobile devices to take visual evidence of flora and uses an interactive means to identify the flora. Keeping a visual record and associating this with a description can be used by experts at a later stage for validation purposes. Repeated measurements and comparisons of the BioCondition index provide a measure of how well a terrestrial ecosystem is functioning and this can then be linked to biodiversity values. It should be noted that BioCondition is more geared to be used at the local or property scale.

In Victoria, the Department of Sustainability and Environment has developed the Habitat Hectares method for estimating the quality of an area of vegetation (Parkes et al., 2003). It is mainly geared toward native vegetation and includes site-based measures of quality and quantity of vegetation and condition within the landscape context. Habitat Hectares provides a step-by-step approach to habitat and landscape assessment in the field and includes useful tips for ensuring consistency of application. Vegetation condition is determined by utilizing variables such as presence and amounts of weeds, amounts of log and leaf litter, cover and diversity of understory, and canopy cover and presence of older trees. Repeat measurements and comparisons against predetermined benchmarks allow for the calculation of native vegetation losses and gains.

BioMetric is a terrestrial biodiversity assessment tool used in New South Wales (Gibbons et al., 2008, 2009). It is mainly applicable for assessment at the paddock or property scale and assesses losses of biodiversity from proposed activities, gains in biodiversity from proposed offsets, or gains in biodiversity as a result of management actions. In BioMetric, the vegetation is assessed against benchmarks, which are quantitative measures of the range of variability in the condition compared to pre-European settlement. Vegetation condition benchmarks are available by vegetation class, and BioMetric compares the current or predicted future condition against this benchmark to denote scores that are then converted to a metric.

Currently, no satellite or aircraft data products are used to determine BioCondition, Habitat Hectares, or BioMetric, although research is being conducted in this area. To conclude, monitoring rangeland biodiversity is the basis of management in Australia, so the question is how biodiversity could be measured more efficiently and accurately by remote sensing.

11.2.3 Assessing Rangeland Health by Remote Sensing

Rangeland health is the degree to which soils and vegetation are maintained, which would sustain the kinds and amounts of vegetation that would typically occur for that site (NRC, 1994;

Pyke et al., 2002). A series of 17 qualitative indicators (with 2 optional indicators, Table 11.1) are intended to help people with some training to determine rangeland health on the ground in a consistent manner. An overall rating of rangeland health is determined by a preponderance of evidence (Pellant et al., 2005).

Multiple indicators of rangeland health (Table 11.1) are related to the ecological processes leading to transitions between alternative states at an ecological site (Caudle et al., 2013). These indicators may show that a site is either at risk for a transition or has crossed the threshold to an alternative state (Figure 11.1) when monitored on the ground (Herrick et al., 2005a,b). Probably, not all indicators need to be determined (MacKinnon et al., 2011); the three most important in Table 11.1 are

1. Bare ground cover (Indicator 4)
2. Vegetation composition (Indicator 12)
3. Presence of invasive species (Indicator 16)

Gaps of bare ground may be the single most important indicator for rangeland health (Booth and Tueller, 2003).

Table 11.1 (third column) lists the potential data sources for 15 of the 19 indicators. The data resolution, which provides the information about an ecological process, is given instead of the sensor name, because the spectral, temporal, and spatial resolutions overlap among different sensors. For detecting gaps of bare ground (Indicator 4), HSR provides direct measurements of cover, whereas hyperspectral (Hysp) and medium-resolution

TABLE 11.1   Indicators of Rangeland Health

Rangeland Health Indicator	Assessment	Potential Data Type
1. Rills in soil	Active soil erosion by water	HSR
2. Water flow patterns	Water infiltration/runoff	HSR
3. Pedestals/terracettes	Active soil erosion by wind or water	
4. Bare ground cover/gap sizes	Potential soil erosion by wind or water	HSR, Hysp, medium resolution
5. Gullies	Active soil erosion by water	HSR, LiDAR
6. Wind-scoured/deposition areas	Active soil erosion by wind	HSR
7. Litter movement	Soil erosion by wind or water	HSR
8. Soil surface resistance to erosion	Soil quality	
9. Soil surface loss/degradation	Soil quality	
10. Community composition and distribution	Water infiltration/runoff	HSR
11. Compaction layer	Water infiltration/runoff	
12. Vegetation composition/functional groups	Biogeochemical cycles	HSR, Hysp
13. Plant mortality/decadence	Population dynamics	HSR
14. Litter amount	Soil quality	Hysp
15. Net primary production/green leaves	Biogeochemical cycles	fPAR
16. Invasive plants	Population dynamics and biogeochemical cycles	HSR, Hysp, fPAR
17. Perennial plant reproduction	Population dynamics	HSR
18. Biological crusts on soil (optional)	Soil quality	Hysp
19. Vertical vegetation structure (optional)	Animal communities	HSR, LiDAR

Source: Pellant, M. et al., Interpreting indicators of rangeland health, Version 4, 122pp., Technical Reference 1734-6, United States Department of Interior, Bureau of Land Management, Denver, CO, 2005.

HSR, high spatial resolution; Hysp, high spectral resolution (hyperspectral); LiDAR, light detection and ranging; fPAR, high temporal resolution estimating the fraction of absorbed photosynthetically active radiation.

By examining several qualitative indicators to establish a "preponderance of evidence," an overall assessment of soil and site stability, hydrologic function, and biotic integrity may be made. In the third column, we suggest the potential remote sensing methods suitable for monitoring the indicator.

data could be used either for direct estimates using spectral unmixing or indirect estimates using spectral indices.

Image classification for vegetation composition (Indicator 12) is one of the primary applications and research areas in remote sensing (Lu and Weng, 2007; Franklin, 2010). In general, the two methods for classification are supervised and unsupervised (Kumar et al., 2015 [Chapter 12]). Supervised classification uses known areas on the ground to create rules (training) for assigning a pixel to a specific category. Unsupervised classification groups similar pixels, which are assigned to different categories afterward. Independent areas on the ground are then used for assessing the accuracy of the classification (Congalton and Green, 2008; Stehman and Foody, 2009; Olofsson et al., 2013). Based on the PROSAIL model (Jacquemoud et al., 2009), the major interacting variables affecting spectral reflectance from a plant canopy are the following: (1) leaf area index; (2) plant structure and leaf angle distribution; (3) soil background reflectance; (4) positions of the sun, target, and sensor; and (5) leaf spectral reflectance and transmittance. A plant community or functional type will have similar structural and spectral properties, and to the extent, they have constant leaf area index and plant density on similar landscapes, the community, or functional type will comprise a single class on an image.

Remote sensing of invasive species (Indicator 16) is based on some detectable difference between the invasive and native species (Hunt et al., 2003; Underwood et al., 2003; Asner, 2004; Madden, 2004; Franklin, 2010; He et al., 2011; Pu, 2012; Bradley, 2013), either spectrally (Section 11.4.2), temporally (Section 11.5.3), or spatially (Section 11.6.3). Most invasive plant species need to be detected at the initial stages of infestation for control, perhaps limiting the usefulness of medium-resolution and high temporal resolution sensors. In the Western United States alone, there are more than 300 species of invasive weeds. According to DiTomaso (2000), the five most problematic are

1. Downy brome (*B. tectorum* also called cheatgrass)
2. Yellow starthistle (*Centaurea solstitialis* L.)
3. Spotted knapweed (*Centaurea stoebe* L., the synonym *Centaurea maculosa* is more common in the literature)
4. Diffuse knapweed (*Centaurea diffusa* Lam.)
5. Leafy spurge (*Euphorbia esula* L.)

Added to this list is tamarisk (*Tamarix* spp., also called saltcedar), a shrub spreading along rivers and streams in the Western United States (Nagler et al., 2011).

11.2.4 Remote Sensing for Animal Biodiversity

The use of remote sensing to estimate animal diversity is increasing. Leyequien et al. (2007) listed five variables that relate to animal needs for food and shelter (Table 11.2). Habitat suitability uses land-cover class determined with medium-resolution satellite data, particularly bird species (Gottschalk et al., 2005). Productivity and phenology (Variables 2 and 3, Table 11.2) are important variables that are determined from high temporal resolution data, which are related to animal biodiversity (Pettorelli

TABLE 11.2 Variables for the Remote Sensing of Animal Biodiversity

Variable	Potential Data Type
1. Habitat suitability	Medium resolution
2. Photosynthetic productivity	fPAR
3. Multitemporal patterns	fPAR
4. Habitat structure	Medium resolution, LiDAR, HSR
5. Forage quality	Hysp

Abbreviations are as defined in Table 11.1.

Source: Leyequien, et al., *International Journal of Applied Earth Observation and Geoinformation*, 9, 1, 2007.

et al., 2011). Correlations between animal species diversity and plant production were established before satellite data were available (Gaston, 2000), although the underlying causes for the correlations are being debated. Habitat structure (Variable 4, Table 11.2) combines several attributes in a given area: variation in the amount of bare soil and vegetation, variation in the occurrence of PFTs, and variation of shadows related to the vertical structure of vegetation. Estimates of image heterogeneity are strongly related to habitat structure and biodiversity (Section 11.3.2). Vertical habitat structure is measured directly at HSR using either light detection and ranging (LiDAR) or stereo-HSR. Forage quality (Variable 5, Table 11.2) depends on the protein and fiber consumed by herbivores, and may be detectable with hyperspectral remote sensing (Section 11.4.2) or with commercial satellite sensors that have bands at the red edge of the chlorophyll absorption spectrum.

11.3 Medium-Resolution Remote Sensing

The Landsat series of satellites are the archetype of medium-resolution remote sensing and have provided data globally for over 40 years. Landsat's 4 and 5 carried the thematic mapper (TM) sensor (Figure 11.2), Landsat 7 carries the enhanced thematic mapper plus (ETM+), and the recently launched Landsat 8 carries the operational land imager (OLI) with two new bands (Figure 11.2).

Red and near-infrared (NIR) spectral indices have been a standard method in analysis of multispectral data, since the Landsat 1 was launched to enhance differences between soil and vegetation (Figure 11.2) and to reduce effects of atmospheric transmittance and solar irradiance from either time of year or topography. The spectral indices used most frequently in remote sensing are the normalized difference vegetation index (NDVI; Rouse et al., 1974; Tucker, 1979), the soil adjusted vegetation index (SAVI; Huete, 1988), and the enhanced vegetation index (EVI; Huete et al., 2002).

Seasonal and annual precipitation totals in rangelands have high variability, so drought is relatively frequent. Vegetation index differences among images acquired on the same day of the year for different years may be from: (1) drought, (2) recent grazing, (3) fire, or (4) a state change of the ecological site. Long-term monitoring is the only way to distinguish among these possibilities to account for the natural range of variation at a

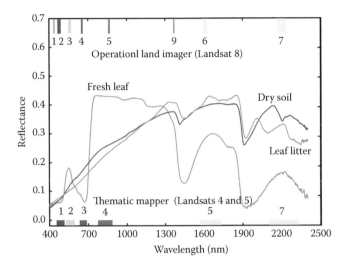

FIGURE 11.2 Spectral reflectances of a healthy green leaf (green line), old leaf litter (gray), and dry soil (brown line). Along the bottom are the wavebands for the Landsats 4 and 5 thematic mapper and along the top are the wavebands for the Landsat 8 operational line imager. The width of each bar shows the spectral width of each waveband.

single ecological site (Washington-Allen et al., 2006; Bastin et al., 2014). Comparisons of vegetation indices for a specific area in relation to the average for all areas of that ecological site will highlight areas that have changed or are in the process of changing to another ecological state (Maynard et al., 2007; Williamson et al., 2012).

Wildlife and livestock grazing patterns in rangelands are not random; they selectively graze areas with high forage quality (Ramoelo et al., 2012; Zengeya et al., 2013). There are several management changes that affect livestock grazing patterns at the landscape scale, such as fences and locations for water (Washington-Allen et al., 2004; Bastin et al., 2012). In contrast, wildlife feed in areas with sufficient cover from predators. In the future, determining the nonrandom grazing patterns at a landscape scale may be important information to manage biodiversity.

11.3.1 Spectral Unmixing

There is a large spectral difference between vegetation and soils, primarily at red and NIR wavelengths (Figure 11.2). Differences in the short-wave infrared region are small if the vegetation and soil are either dry or moist, and are large if the leaves are moist and the soils are dry (Figure 11.2). Linear spectral mixture models are generally thought of as a method for analyzing hyperspectral data (Roberts et al., 1993); however, Adams et al. (1986) developed this method using multispectral data. There are several important assumptions for linear spectral unmixing:

1. The fractional covers are nonnegative and sum to one
2. No multiple scattering among the spectral components (endmembers)
3. All of the spectral components are known

Distinct patches of vegetation, litter, and bare soil meet these assumptions; whereas within a patch of vegetation, multiple scattering between individual plants and soil creates nonlinear mixing.

The simplest case of a linear spectral mixture model has two spectral components (S_1 and S_2):

$$S_\lambda = f S_1 + (1 - f) S_2 \qquad (11.1)$$

where

S_λ is the sensor measurement
f is the percentage of the first component
$(1 - f)$ is the percentage of the second component

If the two components are vegetation and soils (Figure 11.2), Equation 11.1 may be rearranged and solved for f:

$$f = \frac{(S_\lambda - S_2)}{(S_1 - S_2)} \qquad (11.2)$$

Thus, the percentage of bare soil may be calculated directly from the sensor data, and an indicator of rangeland health becomes unambiguously measured.

From Equation 11.2, vegetation indices based on NIR and red wavebands are nonlinearly related to the fractional cover of vegetation and soil (Jiang et al., 2006; Montandon and Small, 2008). Empirical relationships between fractional cover and vegetation indices have problems, because there is a large amount of variability in soil spectra, particularly in rangelands where plant cover is low. Equation 11.2 is also used to normalize NDVI for a site (S_λ) based on maximum NDVI for the year (S_1) and minimum NDVI for the year (S_2), where f is called the vegetation condition index (Kogan et al., 2003).

Equation 11.1 may be extended to include other spectral components such as plant litter, other nonphotosynthetic vegetation, and crop residue (Kuemmerle et al., 2006; de Asis and Omasa, 2007; Numata et al., 2007; Davidson et al., 2008; Guerschman et al., 2009). In some areas, broad-band spectral indices may be used to calculate plant litter cover based on statistical regressions (McNairn and Protz, 1993; van Deventer et al., 1997; Marsett et al., 2006). However, it is often difficult to distinguish bare soil from plant litter with medium-resolution data, particularly when the soil and litter are moist (Daughtry and Hunt, 2008).

11.3.2 Habitat Heterogeneity and Structure

In general, landscapes with a large diversity of cover types and vertical structure also have a high amount of species diversity (Fuhlendorf and Engle, 2001; Tews et al., 2004; Gillespie et al., 2008). Image texture is a general term for the amount of heterogeneity in gray-scale values surrounding a pixel, and there are different statistical formulae (randomness, variance, skewness, entropy, correlation, and more) used for calculating texture (Haralick et al., 1973; Franklin and Wulder, 2002; Wood et al., 2012). It is recognized that heterogeneity is a function of scale

(Haralick et al., 1973; Franklin and Wulder, 2002), so image texture needs to be evaluated with both medium-resolution data and higher-resolution commercial satellite data (Johansen and Phinn, 2006).

Plant species richness is strongly related to remotely sensed texture for different regions (Gould, 2000; Dobrowski et al., 2008). Commercial satellite sensors such as DigitalGlobe's WorldView-2 have a panchromatic band with about 0.5-m pixel resolution and multispectral bands with about 2-m pixel resolution. Plant species richness can be determined with these data using various methods of analysis (Hall et al., 2012; Mansour and Mutanga, 2012; Adelabu et al., 2013; Dalmayne et al., 2013; Müllerová et al., 2013). When pixels are smaller than image features, object-based classification creates clusters from adjacent pixels based on spectral information and texture (Yu et al., 2006; Dobrowski et al., 2008; Blaschke, 2010).

Remotely sensed image texture may have more information when flow direction of water and other resource flows are included. Ludwig et al. (2000, 2002) developed a spatial index (leakiness) to examine the connections among patches of bare soil, which is related to water flow patterns and the potential to lose soil and nutrients. Later, Ludwig et al. (2006, 2007), using ideas based on spectral unmixing, extended the leakiness index to landscapes using Landsat thematic mapper data.

Habitat suitability for a given bird species is usually known, so land-cover data from medium-resolution satellites have been used to predict species distribution (Gottshalk et al., 2005). Small-scale disturbances create mosaics of different vegetation types and the resulting heterogeneity was associated with greater species richness (Fuhlendorf et al., 2006). Image texture is sensitive to small variations of both vertical structure and vegetation mosaics for the determination of bird species richness in a variety of habitats (St. Louis et al., 2006, 2009; Bellis et al., 2008; Culbert et al., 2012; Wood et al., 2013).

11.3.3 Assessment of Medium Resolution

During most of the operational life of Landsats 4 and 5, the data were licensed commercially and thus had limited availability for rangeland management. The economic value per area of rangeland is low, but the total value is high because of the large areas of rangelands on Earth. The U.S. Geological Survey has made the entire archive of Landsat data available free of charge (http://landsat.usgs.gov), and analysis of this long-term archive will provide important insights into the patterns and processes of ecological states (Washington-Allen et al., 2006; Hernandez and Ramsey, 2013; Bastin et al., 2014).

We included the new commercial satellites (IKONOS, GeoEye, and WorldView) with pixel sizes between 2 and 5 m for the multispectral bands in this section, because the commercial sensors will be analyzed in large part with the same techniques that are currently used to analyze medium-resolution data. Specifically, commercial satellite data will be used to estimate vegetation biomass using spectral vegetation indices, classify habitat suitability from land cover, and estimate vegetation structure with image texture. Some commercial satellite sensors include a band at the red edge of the chlorophyll absorption spectrum, which provides better information on forage quality compared to current medium-resolution sensors. But the main disadvantage of commercial satellites is that these are pointed to specific areas, thereby missing adjacent areas on that satellite orbit.

11.4 High Spectral Resolution

Acquisition and analysis of high spectral resolution data, properly called imaging spectrometer data, but ubiquitously called hyperspectral remote sensing, obtains radiance data in numerous, narrow, contiguous bands over a target (Green et al., 1998). With atmospheric correction, the reflectance spectrum of the target is calculated (Gao et al., 1993). From the reflectance spectrum, the identity of the target is determined based on chemical composition.

Linear spectral unmixing may be able to distinguish one spectral component (endmembers) more than the number of bands (Equation 11.1), if the components are known. However, the reflectances at one waveband are highly correlated with those nearby, so the effective number of spectral components is much less than the number of bands (Thorp et al., 2013). Usually, the number of spectral components is not known, or there is multiple, nonlinear scattering among them, which increases the complexity of the spectral unmixing.

Spectral matching compares a pixel's spectrum with some reference spectrum acquired from: a spectral library, field-acquired spectra, or the image itself. The advantage of spectral matching is that the total number of component spectra in the mixture does not need to be known. It must be decided *a priori* the amount of similarity for a match; a typical value for a match is 5.7° or 0.1 rad. The common spectral matching algorithms are: the spectral angle mapper (SAM; Kruse et al., 1993), mixture tuned matched filter (MTMF; Boardman and Kruse, 2011), spectral information divergence (SID; Chang, 2000), spectral correlation measure (SCM; van der Meer, 2006), and Tetracorder (Clark et al., 2003).

SAM calculates a vector angle (Θ) between a reference spectrum $\mathbf{R} = (R\lambda_1, R\lambda_2,..., R\lambda_n)$ and a target spectrum $\mathbf{T} = (T\lambda_1, T\lambda_2,..., T\lambda_n)$, where λ_1 to λ_n are the spectral wavelengths:

$$\Theta = \arccos\left[\frac{(\mathbf{R} \bullet \mathbf{T})}{(\|\mathbf{R}\| \cdot \|\mathbf{T}\|)}\right] \qquad (11.3)$$

where

$\|\mathbf{R}\|$ and $\|\mathbf{T}\|$ are vector normalizations
$\mathbf{R \cdot T}$ is the vector dot product

The spectral angle between the leaf and soil spectra in Figure 11.2 is 28° or 0.49 rad. A major advantage (or disadvantage in some instances) of SAM is that differences in brightness are removed by the vector normalization, so large values of Θ are from spectral differences only.

11.4.1 Spectral Separability of Plant Species

The challenge using SAM and other spectral matching algorithms is determining the threshold value for determining a match, so that there are not large numbers of false positives and false negatives. A large threshold value will increase the number of false negatives and a small threshold value will increase the number of false positives. The value also depends on the variability of the target and reference spectra; therefore, how variable are the spectra from different species? To minimize the variation within a species, 10 leaves were acquired from four species growing in a common garden. Spectral reflectances were measured using a field portable spectrometer collecting light from an integrating sphere (Hunt et al., 2004).

The median spectral angles within a species were less than the median angles compared to the other species (Figure 11.3). The largest separations of spectral angles between within-species and among-species were for *Tripsacum dactyloides* (L.) and *Sorghastrum nutans* (L.) Nash. Using the default angle for classification (5.7° or 0.1 rad), only five leaves of three species were not spectrally similar to others.

These results were expected based on previous results (Carter et al., 2005; Clark et al., 2005; Irisarri et al., 2009; Cho et al., 2010; Martin et al., 2011). Furthermore, the PROSPECT leaf optics model (Jacquemoud et al., 1996, 2009; Feret et al., 2008) accurately predicts leaf reflectance and transmittance with only four parameters: the chlorophyll content, the liquid water content, the dry matter content, and a leaf structure parameter

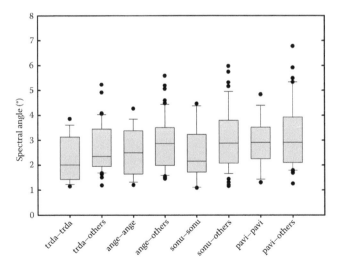

FIGURE 11.3 Spectral angles from leaf spectral reflectances (400–2400 nm) for four grass species: *T. dactyloides* (trda, gamagrass); *Andropogon gerardii* (ange, big bluestem); *S. nutans* (sonu, yellow Indian grass); and *Panicum virgatum* (pavi, switch grass). The mean spectrum was calculated for each species and used as a reference spectrum. Θ (Equation 11.3) was calculated for each leaf of the same species (e.g., trda-trda) and each leaf of the other three species (e.g., trda-others). The center line is the median, the boxes show the range of the 25th to 75th percentiles, the error bars show the range of the 10th to 90th percentiles, and outliers are shown as single points.

equivalent to the number of parallel plates reflecting and transmitting radiation. Therefore, there may not be sufficient degrees of freedom to differentiate species using SAM or other methods (Mansour et al., 2012). Spectral absorption of radiation is based on the types and amounts of chemical bonds (Shenk et al., 2008), so while there may be unique organic compounds in different plant species, there will be few differences in the types and amounts of the chemical bonds. Differences in leaf structure are also highly variable and depend on the conditions during leaf development. Leaf water and chlorophyll contents are affected by current environmental conditions; so much of the variation in leaf reflectance spectra is not related to species.

11.4.2 Plant Chemical Composition

High spectral resolution data have potential for determining forage quality (Variable 5, Table 11.2), which is important for estimating animal biodiversity and livestock distribution (Leyequien et al., 2007; Skidmore et al., 2010; Knox et al., 2011). Near-infrared (NIR) spectroscopy is a standard laboratory method for analysis of dried plant materials for protein and fiber contents (Shenk et al., 2008), and hyperspectral sensors provide coverage of the same wavelength regions. One problem is determination of forage quality when the vegetation has high water content, because the spectral absorption of water dominates the shortwave infrared (SWIR) reflectance spectrum (Ramoelo et al., 2011; Ustin et al., 2012). As the spectral absorption coefficients for liquid water are reasonably well known (Ustin et al., 2012), it is possible to remove the spectral features caused by water to estimate proteins and dry matter (Ramoelo et al., 2011; Wang et al., 2011b). Extensive studies with field spectrometers show strong potential for determination of forage quality (Starks et al., 2004; Knox et al., 2011), but the algorithms require extensive calibration and may not be appropriate for other species at other locations.

The areal cover of bare soil and litter are important indicators of rangeland health (Indicators 4 and 14, respectively, Table 11.1) and are readily detectable using high spectral resolution data (Figure 11.1). Cellulose and lignin are synthesized by plants and these substances are not found in soil, so narrow-band indices emphasizing spectral features in the SWIR are linearly related to the cover of litter over bare soil (Daughtry, 2001; Nagler et al., 2003; Serbin et al., 2009, 2013). SWIR absorption features of bare soil (Figure 11.1) exposed from erosion are also detectable by the advanced spaceborne thermal emission and reflection radiometer (ASTER) on NASA's Terra satellite (Vrieling et al., 2007; Gill and Phinn, 2008; Serbin et al., 2009).

Biological soil crusts (Indicator 18, Table 11.1) are mixtures of cyanobacteria, fungi, and bacteria that form on the soil surface in arid and semiarid ecosystems. Chlorophyll a in the cyanobacteria creates a small absorption feature at 680 nm wavelength distinct from soil (Karnieli et al., 2003; Weber et al., 2008; Ustin et al., 2009). The problem is that the chlorophyll-a feature is also found in algae, lichens, moss, and plants; reflectances in the SWIR allow the separation of biological soil crusts from small amounts of plants and plant-like classes (Ustin et al., 2009).

11.4.3 Detection of Invasive Plant Species

Whereas leaves of most rangeland species may not have much spectral diversity, distinctive leaves and flowers of invasive plants may result in characteristic spectral signatures that may be detected with high spectral resolution data (Table 11.3). Leafy spurge is a noxious invasive weed infesting large areas of the U.S. Great Plains. It is clonal, forming dense stands of genetically identical aboveground shoots spreading from a single root system. Flowers are clustered at or near the top of the shoot (umbels), and the flower cover is frequently over 20% in a single clone (Hunt et al., 2007). The flower bracts' yellow-green color is from a small amount of chlorophyll and a large ratio of carotenoids to chlorophylls (Hunt et al., 2004). When leafy spurge

is found, one management option is the release of insects that specifically feed on the roots (larvae) and shoots (adult) for biological control (Anderson et al., 2003; Lym, 2005; Samuel et al., 2008; Lesica and Hanna, 2009).

Hyperspectral sensors are successful in detecting leafy spurge based on the yellow-green bracts (O'Neill et al., 2000; Parker Williams and Hunt, 2002, 2004; Dudek et al., 2004; Glenn et al., 2005; Lawrence et al., 2006). Furthermore, medium-resolution satellite sensors generally were not very successful (Hunt and Parker Williams, 2006; Mladinich et al., 2006; Stitt et al., 2006; Hunt et al., 2007). The flower bracts of leafy spurge are readily detectable with aerial photography and videography (Everitt et al., 1995; Anderson et al., 1996).

TABLE 11.3   Invasive Plant Species in Rangelands Identified with High Spectral Resolution Data

Study	Species	Methods
O'Neill et al. (2000), Parker Williams and Hunt (2002, 2004), Dudek et al. (2004), Glenn et al. (2005), Hunt et al. (2007), Lawrence et al. (2006), Mitchell and Glenn (2009)	Leafy spurge (*E. esula* L.)	MNF, MTMF, SAM, supervised classification
Lass et al. (2002)	Spotted knapweed (*C. stoebe* L., *C. maculosa* is a commonly used synonym)	SAM
Underwood et al. (2003, 2007)	Ice plant (*Carpobrotus edulis* [L. N. E. Br.]), Jubata grass/Pampas grass (*Cortaderia jubata* [Lemoine ex Carrière] Stapf), Blue gum (*Eucalyptus globulus* Labill.)	MNF, continuum removal, spectral indices, supervised classification
Anderson et al. (2005)	Tamarisk (*Tamarix chinensis* Lour.; *Tamarix gallica* L.; and *Tamarix ramosissima* Ledeb.)	PC, spectral indices
Lass et al. (2005)	Yellow starthistle (*C. solstitialis* L.) and Babysbreath (*Gypsophila paniculata* L.)	SAM
Mundt et al. (2005)	Hoary cress (*Cardaria draba* L.)	MNF, MTMF, SAM
Andrew and Ustin (2006, 2008)	Perennial pepperweed (*Lepidium latifolium* L.)	MNF, MTMF
Ge et al. (2006, 2007)	Yellow starthistle	Spectral indices, SAM
Miao et al. (2006, 2007)	Yellow starthistle	Band selection, feature extraction
Mirik et al. (2006, 2013)	Musk thistle (*Carduus nutans* L.)	Regression, SVM
Narumalani et al. (2006, 2009)	Tamerisk, Musk thistle, Canada thistle (*Cirsium arvense* L.), Reed canary grass (*Phalaris arundinacea* L.), Russian olive (*Elaeagnus angustifolia* L.)	ISODATA, MNF, SAM
Cheng et al. (2007)	Kudzu (*Pueraria montana* [Lour.] Merr.)	MNF, MTMF, SAM
Hamada et al. (2007)	Tamarisk	MNF, spectral indices, MTMF, stepwise discriminant analysis, hierarchical clustering, root sum squared differential area
Hestir et al. (2008)	Perennial pepperweed, Water hyacinth (*Eichhornia crassipes* [Mart.] Solms), Brazilian waterweed (*Egeria densa* Planch.)	MNF, band indices, spectral mixture analysis, SAM, continuum removal
Noujdina and Ustin (2008)	Downy brome (*B. tectorum* L.)	MNF, MTMF
Pu (2008), Pu et al. (2008)	Tamarisk	PC, spectral indices
Wang et al. (2008)	Sericea lespedeza (*Lespedeza cuneata* [Dum. Cours.] G. Don)	ISODATA, spectral derivatives
Carter et al. (2009)	Tamarisk	MNF, supervised classification
Yang et al. (2009), Yang and Everitt (2010a,b)	Ashe juniper (*Juniperus ashei* J. Buchholz), Broom snakeweed (*G. sarothrae* [Pursh.] Britt. and Rusby), water hyacinth	MNF, SAM, supervised classification
Wang et al. (2010a), Bentivegna et al. (2012)	Cut-leaved teasel (*Dipsacus laciniatus* L.)	SAM, supervised classification, SID
Fletcher et al. (2011), Yang et al. (2013)	Tamarisk	SVM
Miao et al. (2011)	Tamarisk	MNF, PC, supervised classification, SAM
Olsson and Morisette (2014)	Buffelgrass (*Pennisetum ciliare* [L.] Link)	MNF, MTMF, random forest

MNF, minimum noise fraction; MTMF, mixture tuned matched filtering; SAM, spectral angle mapper; ISODATA, interactive self-organizing data analysis technique algorithm; PC, principal components; and SVM, support vector machine.

Parker Williams and Hunt (2002, 2004) analyzed two airborne visible infrared imaging spectrometer (AVIRIS) scenes acquired on July 6, 1999, over Devils Tower National Monument and surrounding areas in Crook County, Wyoming. Two sets of plots were acquired: one with the cover of leafy spurge estimated ($N = 66$) and one with a simple classification of presence or absence ($N = 146$). Two scenes were atmospherically corrected to land-surface reflectance using the ATREM version 3.1 program (Gao et al., 1993) and processed using the "Spectral Hourglass" approach (Boardman and Kruse, 2011). Dimensionality was reduced using the minimum noise fraction (MNF) in the environment for visualizing images (ENVI version 3.2, Research Systems Inc.), and spectral signatures for leafy spurge were picked from the AVIRIS images using the pixel purity index and *n*-dimensional visualizer (Boardman and Kruse, 2011).

MTMF (Boardman and Kruse, 2011) was used to identify pixels with possible leafy spurge. The MTMF score was related to leafy spurge cover with an R^2 of 0.66 (Parker Williams and Hunt, 2002), and the classification accuracy was 87% (Parker Williams and Hunt, 2004). However, accuracies of leafy spurge detection with MTMF appear to be dependent on the number of other spectral signatures in a scene (Dudek et al., 2004; Glenn et al., 2005; Mitchell and Glenn, 2009). Using the data from Parker Williams and Hunt (2002, 2004), 11 AVIRIS scenes were combined into a single image for analysis; the overall classification accuracy using MTMF was no better than chance (Hunt et al., 2007). The type of spectral mixing needs to be considered, differences in plant density may change the amount of nonlinear mixing (Mitchell and Glenn, 2009).

There were many studies using high spectral resolution data to detect invasive species; two species commonly studied were tamarisk and yellow starthistle (Table 11.3). The detection of these invasive species may seem to be at odds with Figure 11.3 and related studies. In Table 11.3, the reflectance spectra of invasive species were in context with native species, soils, and subtle variations in spatial arrangement, leaf area index, or leaf angle distribution. The leaf spectra in Figure 11.3 were isolated from any context. Successful use of hyperspectral data for detection of invasive species was based on knowing the invasive was present and having areas for training of supervised classification algorithms (Underwood et al., 2007).

11.4.4 Assessment of High Spectral Resolution

There are only two indicators of rangeland health (Table 11.1) and one variable for animal biodiversity (Table 11.2) for which the preferred method of remote sensing is high spectral resolution. The large spectral difference between vegetation and soils allows accurate assessments of relative cover; so one of the three most important indicators is reliably determined with hyperspectral sensors. Compared to other methods, the suitability of hyperspectral remote sensing for rangeland monitoring depends on detection of the other two critical indicators: invasive species and composition of PFTs. The extensive discussion in Section 11.4.3 shows that more research is required, perhaps

using simulated hyperspectral infrared imager (HyspIRI) mission data produced with high-altitude AVIRIS imagery (http://hyspiri.jpl.nasa.gov; last checked June 12, 2014).

Compared to other sensors, it is more expensive to acquire an airborne hyperspectral sensor, and it takes more time and expertise to analyze hyperspectral imagery, so mapping the distribution of an invasive species over a landscape may not be cost effective. Geospatial species-distribution models classify suitable habitat over larger areas based on combinations of climate, vegetation, soils, topography, and other variables within a geographic information system. There are too many combinations of factors to establish field plots for complete model testing. Classified hyperspectral imagery for small areas should be used to test species geospatial niche models, and then the geospatial models should be used to classify suitable habitat over larger areas (Underwood et al., 2004; Andrew and Ustin, 2009; Hunt et al., 2010).

Global biodiversity is one of the important science questions for the NASA HyspIRI mission (Roberts et al., 2012). The total number of species in a small area (alpha diversity) and the total number of species in a large area (gamma diversity) are more encompassing assessments of biodiversity. Alpha and gamma diversities are interrelated by beta diversity, which is a measure of the proportion of shared species between two points along an environmental gradient. Rocchini et al. (2010a,b) hypothesized spectral diversity is an important method for estimating beta diversity. If this hypothesis is validated, then high spectral resolution data will become invaluable for estimating biodiversity, even though invasive species or PFTs may not be detected.

11.5 High Temporal Resolution

Repeated satellite acquisitions reveal seasonal patterns and growth of vegetation; however, cloud cover may obscure vegetation during critical time periods of active plant growth. High temporal resolution data acquire data every few days, which are composited to form weekly to monthly cloud-free time series. Even in rangelands, "where the skies are not cloudy all day" (Higley and Kelly, 1873), medium-resolution imagery is frequently unusable because of cloud occurrence during the onset of the plant growing season. Meteorological satellites, such as NOAA's advanced very-high resolution radiometer (AVHRR) and NASA's MODIS, use the change in NDVI over the year to determine phenology and the fraction of absorbed photosynthetically active radiation (fPAR). Gross and net primary production are calculated from the total absorbed photosynthetically active radiation and light use efficiency (Hunt et al., 2003; Reeves et al. [Chapter 10]; Kumar et al. [Chapter 12]). Productivity is an indicator of rangeland health (Indicator 15, Table 11.1) and a variable for predicting animal biodiversity (Variable 2, Table 11.2).

Over a small region, interannual variation in phenology is related to climatic variability (Reed et al., 1994; Schwartz et al., 2002), whereas consistent spatial variation in phenology is related to PFT (Loveland et al., 1991; Kremer and Running, 1993; Peters et al., 1997; Aragón and Oesterheld, 2008; Gu et al., 2010).

Grasses (Poaceae) have two functional types based on photosynthetic pathway: cool-season grasses with C_3 photosynthesis, and warm-season grasses with C_4 photosynthesis. Physiologically, C_3 photosynthesis is greater at cool temperatures and C_4 photosynthesis is higher at warm temperatures, but temperature affects many different processes, not just photosynthesis (Sage and Kubien, 2007). With global climatic change, warmer temperature may favor C_4 grasses, whereas elevated atmospheric CO_2 may promote C_3 plants (Morgan et al., 2007). Therefore, C_3/C_4 relative abundance is a complex response of climate change, and will affect ecosystem biogeochemical cycles at regional and global scales.

Temperate grasslands in North America, South America, Australia, Africa, and Asia have mixtures of C_3 and C_4 grasses with the ratio depending on the amount of rainfall during the warm and cool seasons (Winslow et al., 2003). Determining the ratios of C_3 and C_4 grasses is a challenge, because the reflectance spectra may not be separable (Adjorlolo et al., 2012a,b). High temporal resolution AVHRR or medium-resolution imaging spectrometer (MERIS) data was used to separate C_3 and C_4 grasses (Goodin and Henebry, 1997; Tieszen et al., 1997; Ricotta et al., 2003; Foody and Dash, 2007, 2010). Medium-resolution imagery acquired two or three times per year also showed differences based on temperature (Davidson and Csillag, 2001; Peterson et al., 2002; Guo et al., 2003). High temporal resolution data may add more information by subdividing both C_3 and C_4 functional types into tall and short functional types (Wang et al., 2013).

11.5.1 Phenology Metrics

The first step for determining phenological metrics from high temporal resolution data is applying a low-pass filter to the time series of NDVI to remove the effects of subpixel clouds, atmospheric conditions, and view angles. Several smoothing algorithms have been used: a nonlinear median filter (Reed et al., 1994), an upper-envelope three-point filter (Gu et al., 2006), and a second-order polynomial filter (Savitzky and Golay, 1964).

Most studies define a series of metrics from the smoothed NDVI time series (Table 11.4). These metrics may be determined by a sequence of logical statements; tools have been developed

TABLE 11.4 Common Phenology Metrics for Determining PFTs

Metric	Definition
Start of season (SOS)	Before Peak NDVI, DOY when NDVI > threshold
End of season (EOS)	After peak NDVI, DOY when NDVI < threshold
Length of season (LOS)	EOS − SOS
Peak NDVI	Maximum NDVI in time series
Peak date	DOY of Peak NDVI
Cumulative NDVI (Σ − NDVI)	Sum NDVI when SOS < NDVI < EOS

These metrics have to be determined with a complete time series of NDVI. DOY, Day of the Year.

to automate calculation of the various metrics: fitting the NDVI time series into piecewise logistic functions (Zhang et al., 2003), fitting the time series using a quadratic function (de Beurs and Henebry, 2004), and in the open-source TIMESAT program (Jönsson and Eklundh, 2004). In rangelands, these mathematical approaches may produce errors, because grasses often have prolonged growing season and asymmetric trajectories (de Beurs and Henebry, 2010). For this reason, some studies directly use the polynomial-smoothed time series to extract phenological metrics directly (Wang et al., 2010b, 2011a, 2013).

11.5.2 Grass Functional Types

Dividing the U.S. Great Plains at 100° West Longitude, east of this meridian is primarily subhumid tallgrass prairie with 500–750 mm of annual precipitation (Figure 11.4a). West of this meridian are the southern shortgrass steppe and the northern mixed-grass prairie (Figure 11.4a), which get about 250–500 mm of precipitation annually. Precipitation strongly controls grassland productivity resulting in the designations, tallgrass and shortgrass. The latitudinal C_3/C_4 variations and longitudinal tall/short differences are combined into tallgrass C_3, tallgrass C_4, shortgrass C_3, and shortgrass C_4. The latter three PFTs correspond to naturally occurring ecological regions; however, tallgrass C_3 may be the result of seeding cool-season species (e.g., tall fescue) into pastures for higher forage production (Wang et al., 2013).

Grassland areas were determined from the USGS National Land-Cover Database (Homer et al., 2004); the grassland areas were not divided into functional types. Time series of the 500-m, 8-day Terra MODIS surface reflectance products (MOD09A1) for the years 2000–2009 show phenological variations of C_3 and C_4 grasses (Figure 11.5). Tallgrass C_3 and C_4 have higher NDVI values than shortgrass C_3 and C_4. Summer dormancy and fall growth for tallgrass C_3 was seen in some years as a second NDVI peak in the autumn (Figure 11.5), but this phenological feature was much less important than an earlier start of growing season (*SOS*), later end of growing season (EOS), and therefore longer length of growing season (LOS). Shortgrass C_3 have much earlier *SOS* than shortgrass C_4, and both have shorter LOS than the tallgrass PFTs (Figure 11.5).

The four grass PFTs in the U.S. Great Plains were mapped over the 10-year period (2000–2009) with a decision tree approach (Figure 11.4b). Because of interannual variability in temperature and precipitation, a given pixel would be classified as one PFT for some years and would be classified as another PFT for other years. For example, during drought, tallgrass C_3 and shortgrass C_3 *SOS* may be delayed and EOS may be earlier, so that the phenological features become similar to those of tallgrass C_4 and shortgrass C_4, respectively.

Therefore, delineation of grass C_3/C_4 PFTs is problematic using data from only a single year. A long time series of data were required for reliable separation (Winslow et al., 2003; Wang et al., 2013). The final classification in Figure 11.4b displays the distributions where one PFT was selected for at least 6 of the 10 years (Wang et al., 2013). The four PFTs followed the

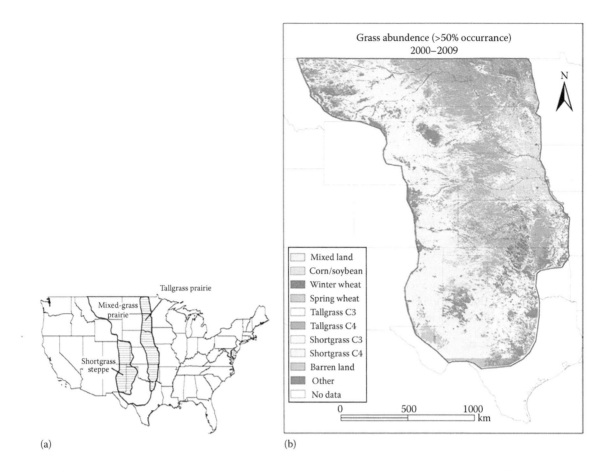

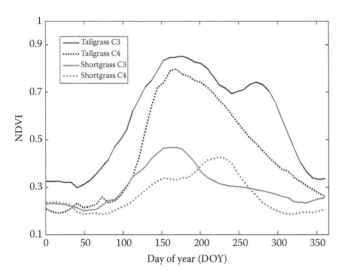

(a)

(b)

FIGURE 11.4 (a) Ecological regions of the tallgrass prairie, shortgrass steppe, and northern mixed-grass prairie. (b) Distributions of four grass functional types in the U.S. Great Plains: tallgrass C_3, tallgrass C_4, shortgrass C_3, and shortgrass C_3. Tallgrass C_3 is not a naturally occurring ecological region in the U.S. Great Plains and may be the result of seeding C_3 species into pastures. Interannual variability of precipitation and temperature affect the NDVI time series, so overall assignment of a pixel to one of the four grassland functional types was determined when a single functional type was the classification result in 6 of the 10 years.

FIGURE 11.5 Examples of NDVI time series of the four grass functional types: tallgrass C_3, tallgrass C_4, shortgrass C_3, and shortgrass C_3. The NDVI values of pure pixels are extracted from the 46 MODIS scenes acquired in 2007.

longitudinal transition of ecological regions based on precipitation. However, latitudinal trends caused by temperature were not present, because east of 100° West, tallgrass C_3 areas were south of tallgrass C_4 areas. Splitting grasses into tallgrass and shortgrass highlighted the occurrence of a new ecological region; the areas classified as tallgrass C_3 are located in areas generally considered to be pastures (Wang et al., 2013). Because more of the variation in temporal profiles was now explained, the resulting classification had somewhat increased accuracy.

11.5.3 Invasive Plant Species

Few invasive species are detected from the large pixels of high temporal resolution data (e.g., AVHRR and MODIS). One species is downy brome in the Western United States; Bradley and Mustard (2005, 2006) showed that the temporal variation between downy brome and co-occurring species was large enough to classify downy brome invasion using AVHRR. Singh and Glenn (2009) and Clinton et al. (2010) replicated this study using a time series of Landsat and MODIS data, respectively.

Two other species that may be detected using high temporal resolution data are broom snakeweed (*Gutierrezia sarothrae* [Pursh] Britton and Rusby; Peters et al., 1992) and Lehmann lovegrass (*Eragrostis lehmanniana* Nees; Huang and Geiger, 2008; Huang et al., 2009). Broom snakeweed is a C_3 shrub surrounded by C_4 grasses, so it was detected based on its earlier *SOS*. Lehmann lovegrass is a C_4 grass that crowds out native C_4 grasses; during senescence, it produces a litter layer that is dense, bright, and yellow, which is detectable even with large pixel sizes (Huang and Geiger, 2008; Huang et al., 2009).

Two important questions about invasive species are addressed using ecological niche or habitat suitability models: (1) what areas on Earth may be susceptible to an invasive species, and (2) how will the areas expand or contract with anthropogenic global warming (Peterson, 2003; Thuiller et al., 2005; Andrew and Ustin, 2009; Stohlgren et al., 2010). MODIS vegetation index data are important for predicting the potential distribution of tamarisk (Morisette et al., 2006) and purple loosestrife (*Lythrum salicaria* L.; Anderson et al., 2006). In these studies, time series of NDVI are used as a surrogate for meteorological data, because even with 1-km pixel size, the data are much more densely distributed compared to available weather station data.

11.5.4 Assessment of High Temporal Resolution

The specifications for NASA MODIS and the recently launched Suomi National Polar-orbiting Partnership Visible Infrared Imaging Radiometer Suite (VIIRS) were developed in part based on the long history of AVHRR for global remote sensing of interannual climatic variability. Most land-cover classifications using AVHRR data were designed to be used by global ecosystem models, so fewer PFTs were defined. Because of limited computing power, there were limitations placed on the spatial resolution in order to provide high-quality data at high temporal resolution (Townshend and Justice, 1988). Therefore, variations in phenology related to taxonomy, PFTs, and other considerations at smaller scales were rarely explored because of the large pixel sizes.

Two satellite systems have broad swaths and high temporal resolution with much smaller pixel sizes: (1) the Indian Space Research Organization's advanced wide field sensor (AWiFS) and (2) the international Disaster Monitoring Constellation (Wang et al., 2010c). However, data from these two sensors are only available from commercial venders and are not freely available from public archives (Goward et al., 2012).

With more powerful computers available, trade-offs between high temporal resolution and HSR are no longer required. For example, the U.S. Department of Agriculture (USDA) National Agricultural Statistics Service (NASS) previously used AWiFS data (56–70 m pixels) and currently uses Deimos-1 data (22 m pixels) to produce annual cropland data layers for the USA (Boryan et al., 2011). The potential of high temporal resolution

data for monitoring rangeland biodiversity is simply not known. Making AWiFS, Deimos, and similar multispectral sensor data available worldwide is an important investment for monitoring global biodiversity (Wang et al., 2010c).

11.6 High Spatial Resolution

While ground and aerial photographs have been used for natural resource management since the last century (Cooper, 1924), advances in digital sensors and computer processing have made HSR the newest dimension in remote sensing. The term "HSR" is subjective, changes with time, and can mean pixel sizes from 0.25 mm to 2 m with the current state of the art. Smaller pixel sizes must be acquired from aircraft flying at lower altitudes, usually resulting in smaller areas covered per image (<1 ha compared to 27,000 ha for WorldView-2 data). HSR images are also referred to as "very-large scale" in reference to the map scale (e.g., 1:500 compared to 1:24,000 for a 7.5′ × 7.5′ topographic map).

HSR imagery is useful in most aspects of rangeland resource management including assessments of ground cover, riparian condition, wildlife habitat, woodlands, and other resource concerns. One of the characteristics of HSR images is very high pixel-to-pixel variability, but each pixel is spectrally pure (Figure 11.6). With larger pixel sizes, each pixel is a heterogeneous mixture of different spectral components (Figure 11.6). At the scale of 1 m, spectral unmixing or spectral indices are required to measure the amount of bare ground, so the advantages of HSR satellite data compared to medium-resolution satellite data are not clear. On the other hand, at the scale of 0.4-mm HSR aircraft data, the established techniques of remote sensing are inadequate and new techniques are being developed and tested. Alternatively, established techniques of visual photointerpretation can often be updated for use with digital imagery. As shown by Figure 11.6, the resolution of HSR data needs to be from 0.25 to 10 mm in order to visually interpret the data effectively.

11.6.1 Ground Imaging

Most ground assessments continue to be interpretations and judgments based on monitoring a limited number of nonrandomly selected sites. It simply has not been practical to do otherwise. However, ground and aerial digital photography are replacing or augmenting time-consuming field methods, and are increasing the objectivity of rangeland monitoring (Booth et al., 2008; Moffet, 2009; Sadler et al., 2010; Booth and Cox, 2011). Digital photography and subsequent photo interpretation can increase sampling rates over traditional field sampling methods (Luscier et al., 2006; Cagney et al., 2011). Reduced monitoring time and associated labor costs can increase monitoring precision by allowing increased sample numbers.

Methods for obtaining nonaerial, nadir-looking digital images include a free-hand method (Cagney et al., 2011) and

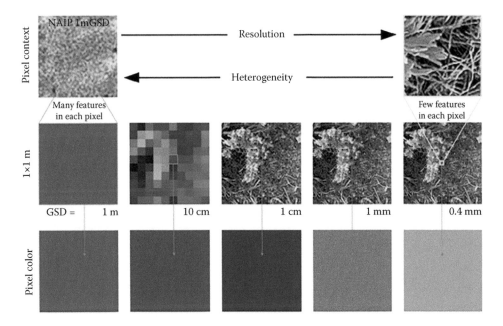

FIGURE 11.6 Effect of spatial resolution (also called ground sample distance) on pixel color and visual interpretation. The top left panel shows a 100 × 100-m area of rangeland in southeast Idaho at 1-m resolution to illustrate the heterogeneity of the imaged area. The middle row shows a 1 × 1-m patch of rangeland in all panels with the lowest spatial resolution on the left (1-m pixel size) and highest resolution on the right (0.4-mm pixel size). The top right panel shows an enlargement of the 0.4-mm resolution image; the red dot in the center (shown by the blue arrow) is one 0.4-mm pixel situated on a sagebrush leaf. The lower row shows pixel color change from increasing resolution from 1 m to 0.4 mm. (Figure adapted from Weber, K.T. et al., *Rangeland Ecol. Manag.*, 66, 82, 2013, used with permission.)

the use of staffs, tripods, stands (Booth et al., 2004; Booth and Cox, 2011), booms, and gantries (Louhaichi et al., 2010), vehicle mounts and other aids to reduce camera motion. Furthermore, these tools allow for a range of camera positions above ground level and a range of image resolutions. For a permanent record, the digital images are archived at much lower cost compared to film negatives or prints. Even where sampling will be done primarily from aircraft, acquisition of ground images is an important part of monitoring, since it provides the opportunity to capture nadir images with less motion blur than can be obtained from the air.

11.6.2 Aerial Imaging

Some image users require boundary-to-boundary coverage, and HSR mosaics may be made from airborne data at considerable expense. However, many users do not require mosaics, and can manage effectively at lower cost by capturing single-image samples that are systematically distributed across watersheds, allotments, and other landscape-scale management units (Figure 11.7). Geographic Information Systems (GIS) should be used to draft sampling plans that define intended locations for ground or aerial image acquisition to acquire a statistically adequate sample that represents the natural range of variation in the areas monitored (Blumenthal et al., 2007; Booth and Cox, 2008).

Platforms used for HSR aerial-image acquisition include manned conventional and light sport airplanes (LSAs), manned

and unmanned helicopters, and unmanned fixed-wing aircraft (Booth and Cox, 2011). Surprisingly, contracting for a piloted LSA was the least expensive, because fewer personnel were required to be on site for safety. Furthermore, only LSAs and helicopters were able to fly slow enough to avoid excessive motion blur with image resolutions of 1-mm resolution or less (Figure 11.8). Contract costs for manned helicopters are usually more expensive than for LSAs, but helicopters (manned or unmanned) can fly during windy conditions when LSAs cannot safely operate. To mitigate risk, LSAs need to be equipped with rocket-deployed, whole-plane parachutes in the event of mechanical failure. Frequently in windy areas, flights are made early in the day before strong winds and atmospheric turbulence develop. A potential problem with early morning flights is the presence of shadows that may affect estimates of bare ground cover.

Manned helicopters and LSAs can carry multiple cameras allowing for nested, simultaneously acquired, multiresolution images: for example, 1-, 10-, and 20-mm pixel sizes. This capability allowed users to acquire wide fields of view *and* high resolution, a capability that proved important for the study of fire intervals in shrub ecosystems (Moffet, 2009), and monitoring invasive species (Blumenthal et al., 2007, 2012; Booth et al., 2010).

Aerial surveys acquired imagery with 1-mm pixels that overcame the need to depend on subjective selection of "representative" study areas. Summarizing, aerial surveys acquiring

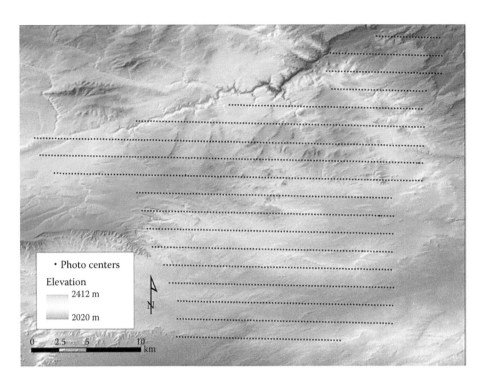

FIGURE 11.7 Flight plan in Wyoming's Red Desert (center: 43.37°N, 108.40°W) showing 1457 image acquisition points that were 200 m apart along east–west flight lines spaced 1600 m apart. Three images were collected at each acquisition point: (1) an area of 3 × 4 m with 0.9-mm pixels; (2) an area of 24 × 36 m, with 7-mm pixels; and (3) an area of 48 × 72 m with 14-mm pixels.

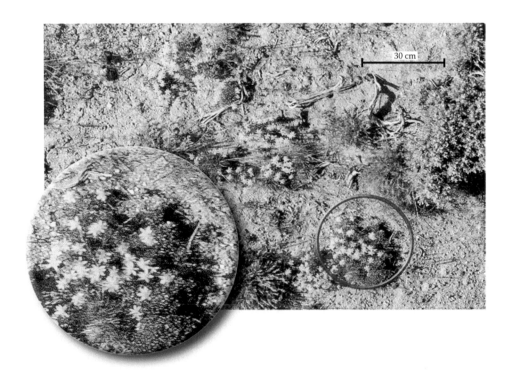

FIGURE 11.8 Portion of a 3 × 4-m aerial image (0.9-mm pixel size) acquired from a LSA traveling 84 km/h at an altitude of 103 m above ground level in the Red Desert of central Wyoming. The main image shows bare soil, litter, shadow, grasses, sagebrush (*Artemisia tridentata* Nutt.), and yellow rabbitbrush (*Chrysothamnus viscidiflorus* [Hook.] Nutt.), while the circular inset (circled area on main image enlarged twofold) shows goldenweed (*Stenotus acaulis* [Nutt.] Nutt.).

multiresolution imagery that includes 1-mm pixel data have repeatedly demonstrated

1. Lower cost than extensive conventional ground sampling for areas greater than about 200 ha
2. Practical acquisition of large sample numbers
3. Reduced sample-collection time
4. Collection of many samples within short phenological windows
5. Creation of permanent records that may be examined any time in the future
6. Capability of capturing submeter details for detecting ecologically important changes

Measurement of bare ground from 1-mm pixel images is just one example of the different indicators of rangeland health that may be monitored.

11.6.3 Visual Analysis

For measuring ground cover, image resolution should be at least 1-mm resolution (Booth and Cox, 2009). Attempts to accurately measure ground-cover from 10-, 20-, and 50-mm resolution imagery were unsuccessful (Booth and Cox, 2009; Duniway et al., 2012; Weber et al., 2013). Figure 11.8 shows that 1-mm resolution images are essential to capturing the detail needed for identifying species. Current methods for analysis of this type of imagery include visual-analysis-facilitating software programs (Booth et al., 2006a,b). These programs are accurate when used

with appropriate image resolutions (you cannot measure what you cannot see), and by people with adequate field experience.

SamplePoint is free software (available online at www.SamplePoint.org; last accessed May 30, 2014) that facilitates point sampling of digital images. Because the sample point is always a single pixel of the image (Figure 11.9a), where the spatial resolution is equal or less than 1 mm, the analysis has a potential accuracy of 92%. To use the program, images are loaded from a computer directory, and a database is created to store the data entered. The number and pattern of sample points (grid or random) are determined by the user (Figure 11.9b). The predetermined classes are associated with buttons along the bottom of the image. *SamplePoint* automatically moves from point to point when the class information is entered by clicking one of the buttons. The software allows image magnification (zoom), and it will support up to three monitors, each of which displays the image with different levels of magnification.

Images need not be orthorectified or georeferenced for use in SamplePoint. Thus, users with no GIS experience can use the software effectively. This is an important point, because other point-sampling tools like Image Interpreter Tool (Duniway et al., 2012) and Digital Mylar (Clark et al., 2004; RSAC, 2014) do require georeferenced imagery. That adds a workflow step that makes the software more difficult and time-consuming to use for those who have GIS experience, and virtually impossible for those who do not. The cited tools function very well with orthoimagery, as intended. Nevertheless, for ground- or aerial-based photo plot monitoring, a georeferenced image prerequisite

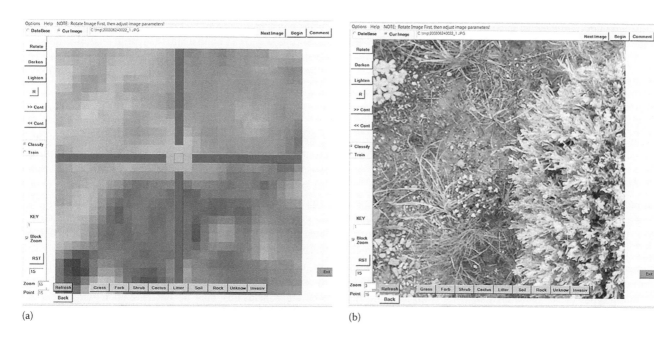

(a) (b)

FIGURE 11.9 SamplePoint analysis of a ground photograph in Wyoming's Red Desert associated with aerial images (Figure 11.8) collected using the flight plan in Figure 11.7. Classification is performed on a single pixel within a 9-pixel array framed by the red crosshair by clicking one of the user-defined buttons along the bottom of the screen (a). The same sampling point is shown zoomed out (b) to show the context of the point that is hitting bare ground in between spiny phlox (*Phlox hoodii* Richardson) and sagebrush (*A. tridentata* Nutt.). Note the systematic grid sampling pattern of the classification crosshairs.

restricts usage of these tools because (1) it is impractical to geo-reference hundreds of disconnected plot-based images and (2) commercial submeter orthoimagery is too expensive to acquire frequently. To illustrate, over the last 6 years, the Bureau of Land Management contracted only one submeter orthoimage collection in the state of Wyoming: 30-cm resolution at $18/km². Collection of the 2–3-cm orthoimagery that Duniway et al. (2012) reported as yielding reliable estimates for many cover types will cost substantially more than $18/km², limiting usage among agencies where even contracting for 30-cm orthoimagery is usually cost-prohibitive. As internal collection capabilities improve, cost may become less of a factor.

First-time users of 1-mm resolution imagery are often surprised at the number of species they can identify in an image. This leads to an initial presumption that they can use *SamplePoint* to measure cover by species. While this may be true for appropriately spaced woody species, measuring cover by PFT (grasses, grass-like plants, and forbs) is usually more practical except where and when plant species are distinctive. *SamplePoint* does *not* correct for user biases that may occur due to personal interpretations of protocol (e.g., what is litter), or correct for conditions such as age that can influence color perception. In fact, the variation among *SamplePoint* users was found to be about equal with that of users of the line-point intercept (Moffet, 2009). Additionally, multiple people used *SamplePoint* to analyze the same set of images showing there are user biases in class selection. When biases are found, the image data set can be re-analyzed. Thus, data verifiability and the capability to significantly increase sampling are key advantages of using *SamplePoint* with image-based monitoring.

Two other programs are also freely available at http://www.SamplePoint.org, *SampleFreq* and *ImageMeasurement*. *SampleFreq* functions similarly to *SamplePoint*, but it facilitates nested frequency sampling. *ImageMeasurement* is used to measure areas and lengths from nadir imagery of known scale. Originally, this program was used to measure stream and channel widths with other ecological indicators for riparian areas. *ImageMeasurement* is unique among similar programs by incorporating exact image resolution for every image in a multi-image data set.

11.6.4 Unmanned Aircraft Systems

Unmanned aircraft systems (UAS) may also be effective platforms for remote sensing (Hardin and Jackson, 2005; Rango et al., 2006, 2009; Hardin and Hardin, 2010). Some of the first UAS that were used in agricultural and natural resource management were made from radio-control model aircraft (Tomlins and Lee, 1983; Quilter and Anderson, 2000, 2001). Now, there are many types of advanced civilian UAS, from high-altitude long-endurance to low-altitude short-endurance aircraft (Watts et al., 2012), so few generalizations may be made about current UAS flight capabilities. LSAs may fly in steady wind conditions of 32 km/h when there are no wind gusts. Two particular UAS (UX5 and Gatewing X100, Trimble Navigation Limited, Westminster, CO) are claimed to be able to fly in 65 km/h winds (Trimble Navigation, 2013), because the automatic pilot updates

aircraft controls at 100 Hz (Southard, 2013). The ability of UAS to fly in moderate to high winds may facilitate timely collection of aerial data for biodiversity.

Like manned aircraft, UAS may be configured with medium-resolution, high spectral resolution, or HSR sensors. Furthermore, LiDAR, thermal infrared, radar, and other sensors are available for UAS platforms. Most of the research uses small low-flying UAS with digital cameras to measure ground cover (Hardin et al., 2007; Laliberte et al., 2007, 2010; Breckenridge and Dakins, 2011; Breckenridge et al., 2011). Hardin et al. (2007) using small UAS to detect squarrose knapweed found that overall detection accuracy was low but the rate of false positives was miniscule. Breckenridge et al. (2011) also attempted to determine presence and sex of sage grouse (*Centrocercus urophasianus*) using decoys to represent grouse during spring mating season. At a height of 73 m, 100% of the male and 80% of the female decoys could be detected by a skilled observer, whereas at 305 m height, 90% of the male and only 10% of the female decoys could be detected by the skilled observer.

With small UAS and digital cameras, it is easy to acquire many more images than would be possible to analyze using aids such as *SamplePoint*; so automated processing or preprocessing is required. Many people using these systems have backgrounds in image processing, in which the typical workflow would be production of orthorectified image mosaics. Two general methods used in orthorectification are the scale invariant feature transform (Lowe, 2004) and structure from motion (Turner et al., 2012; Westoby et al., 2012). Classification of land-cover types from these image mosaics is not very accurate when using methods developed for multispectral satellite data, because the spectral properties of a small pixel have little connection with its cover type (Figure 11.6). Instead, the spatial relationships among pixels are used in an object-based classification (Luscier et al., 2006; Laliberte et al., 2007, 2010; Laliberte and Rango, 2011).

11.6.5 Assessment of High Spatial Resolution

Remote sensing with HSR sensors can be used for monitoring numerous indicators of rangeland health (Table 11.1) in order to get a preponderance of evidence for evaluation. Aircraft are necessary for HSR data acquisition; image analysis currently requires human inputs. The typical image-processing workflow for aerial photography is to georectify and orthorectify the large number of photographs to produce a single image with boundary-to-boundary coverage. We suggest that this time-consuming effort is not necessary for rangeland monitoring. By assuming each photograph is one plot in a statistical sample, accurate conclusions may be determined regarding the land area as a whole. Furthermore, from the geographic coordinates of each photograph, the plot-scale results may be combined for landscape-scale information by kriging or cokriging.

Unsupervised or supervised classification of remote sensed imagery requires areas of a known category for training, which is a problem for early detection of invasive species. Computer-facilitated interpretation of HSR data may be used to detect plant species new to an area, because photographs from other areas would be used as examples. Furthermore, with low-cost data storage, HSR acquisitions may be archived for re-examination.

Most of the airborne HSR data used for method development were acquired using a manned light sport aircraft (Booth and Cox, 2009, 2011), and the costs for LSA remote sensing were less than other methods. The decade-long LSA effort included a strong emphasis on safety, and there were no incidents during the thousands of LSA flight-hours. Even though the availability of high-quality components has made UAS much more reliable than a few years ago, there will be flight and control failures. The expanse of rangelands is large, so the number of flight hours required for monitoring will be large; using either LSA or UAS as low-altitude platforms should facilitate the acquisition of HSR imagery for rangeland monitoring.

11.7 Conclusions

Rangeland state-and-transition models account for differences in plant communities occurring at an ecological site, and these models are being adopted in different countries for rangeland management. Based on the natural range of variation, recently, disturbed ecological sites in the reference state may have lower diversity and biomass compared to an alternative state. Therefore, after accounting for interannual variation in precipitation, detecting lower biomass with spectral vegetation indices is not necessarily equal to detecting rangeland degradation. Instead, degraded rangelands crossed a transition into an alternative ecological state such that the area will not recover over time to its former diversity, biomass, and structure. Operational monitoring of rangelands by remote sensing needs to be based on the information required for management in order to prevent degradation.

As discussed in this chapter, there are many examples showing how sensors with different combinations of high spectral resolution, high temporal resolution, and HSR may be used to measure biodiversity, usually by detecting various indicators of rangeland health. We attempted to make the best possible case for rangeland monitoring with each data type. *At the current time, we conclude that facilitated image interpretation of HSR data is the best method for rangeland monitoring for management based on state-and-transition models.* This conclusion is based on one narrowly focused objective, providing information required for modifying management practices in order to protect rangeland diversity and prevent rangeland degradation. Other objectives, such as estimating biomass forage production, require other sources of remotely sensed data.

Medium-resolution satellite data have been available for decades, and the only rangeland health indicator that may be reliably estimated is the amount of bare ground (Table 11.1). Transitions to alternative states may be recognized afterward, but perhaps too late to prevent further degradation. High temporal resolution data from operational meteorological satellites are necessary for monitoring vegetation phenology, functional types, and effects of drought. Phenology using higher spatial resolution data from satellites currently in orbit is an extremely

important new source of information, and there should be an international effort to archive these data for worldwide availability. Multiple years worth of data need to have been acquired and available in order to account for interannual variability in temperature and precipitation.

High spectral resolution data may be used to detect and measure bare soil, many different invasive species, and most PFTs, the three most important indicators of rangeland health (MacKinnon et al., 2011). Combined hyperspectral sensors and LiDAR in aircraft (Asner et al., 2007; Asner and Martin, 2009; Kampe et al., 2010) are promising, because it is much easier to identify pure spectral components. Airborne sensors have pixel sizes from 5 to 30 m, so image texture would help to identify areas of spatial heterogeneity for animal habitat. Finally, litter cover, biological crusts, and forage quality may be determined by chemical composition.

Hyperspectral remote sensing does well when either the analyst has some prior information on where the target is located (such as invasive species) or the spectral signature is invariant (cellulose and lignin for litter cover). Soil minerals have spectral signatures, whereas soil types are classified based on pedogenesis. Plant species do not have signatures in the way that soil minerals and biological compounds do, but this conclusion is being re-evaluated using more advanced airborne sensors (Asner and Martin, 2009, 2011; Asner et al., 2011). Furthermore, Rocchini et al. (2010a,b) hypothesized spectral diversity *per se* is important for estimating biodiversity, which is parallel to the claim that spatial diversity from image texture is important for estimating biodiversity. Object-based classification combines both spectral and spatial diversity and may become an effective method.

Rangeland monitoring with HSR data provides information directly relevant for management based on rangeland health, to avoid transitions to more degraded states. There may be a large difference between the scales at which information is useful compared to the scales at which the data is acquired to produce the information. The scale for information required by rangeland managers is the landscape, but acquisition of HSR data may be the most cost-effective method of obtaining the required information.

Disclaimer

Mention of product names is for information only, does not imply government endorsement, and other products may be equally suitable.

References

Adams, J. B., Smith, M. O., and Johnson, P. E. 1986. Spectral mixture modeling: A new analysis of rock and soil types at the Viking Lander 1 site. *Journal of Geophysical Research*, 91, 8098–8112.

Adelabu, S., Mutanga, O., Adam, E., and Cho, M. A. 2013. Exploiting machine learning algorithms for tree species classification in a semiarid woodland using RapidEye image. *Journal of Applied Remote Sensing*, 7, 073480-1.

Adjorlolo, C., Cho, M. A., Mutanga, O., and Ismail, R. 2012a. Optimizing spectral resolutions for the classification of C$_3$ and C$_4$ grass species, using wavelengths of known absorption features. *Journal of Applied Remote Sensing*, 6, 063560.

Adjorlolo, C., Mutanga, O., Cho, M. A., and Ismail R. 2012b. Challenges and opportunities in the use of remote sensing for C$_3$ and C$_4$ grass species discrimination and mapping. *African Journal of Range and Forage Science*, 29, 47–61.

Allen, C. R., Gunderson, L., and Johnson, A. R. 2005. The use of discontinuities and functional groups to assess relative resilience in complex systems. *Ecosystems*, 8, 958–966.

Anderson, G. L., Everitt, J. H., Escobar, D. E., Spencer, N. R., and Andrascik, R. J. 1996. Mapping leafy spurge (*Euphorbia esula*) infestations using aerial photography and geographic information systems. *Geocarto International*, 11, 81–89.

Anderson, G. L., Prosser, C. W., Wendel, L. E., Delfosse, E. S., and Faust R. M. 2003. The ecological areawide management (TEAM) of leafy spurge program of the United States Department of Agriculture—Agricultural Research Service. *Pest Management Science*, 59, 609–613.

Anderson, G. L., Carruthers, R. I., Ge, S., and Gong, P. 2005. Monitoring of invasive *Tamarix* distribution and effects of biological control with airborne hyperspectral remote sensing. *International Journal of Remote Sensing*, 26, 2487–2489.

Anderson, R. P., Peterson, A. T., and Egbert S. L. 2006. Vegetation-index models predict areas vulnerable to purple loosestrife (*Lythrum salicaria*) invasion in Kansas. *Southwestern Naturalist*, 51, 471–480.

Andrew, M. E. and Ustin, S. L. 2006. Spectral and physiological uniqueness of perennial pepperweed (*Lepidium latifolium*). *Weed Science*, 54, 1051–1062.

Andrew, M. E. and Ustin, S. L. 2008. The role of environmental context in mapping invasive plants with hyperspectral image data. *Remote Sensing of Environment*, 112, 4301–4317.

Andrew, M. E. and Ustin, S. L. 2009. Habitat suitability modeling of an invasive plant with advanced remote sensing data. *Diversity and Distributions*, 15, 627–640.

Aragón, R. and Oesterheld, M. 2008. Linking vegetation heterogeneity and functional attributes of temperate grasslands through remote sensing. *Applied Vegetation Science*, 11, 117–130.

de Asis, A. M. and Omasa, K. 2007. Estimation of vegetation parameter for modeling soil erosion using linear spectral mixture analysis of Landsat ETM+ data. *ISPRS Journal of Photogrammetry and Remote Sensing*, 62, 309–324.

Asner, G. P. 2004. Biophysical remote sensing signatures of arid and semiarid ecosystems. In *Remote Sensing for Natural Resource Management and Environmental Monitoring. Manual of Remote Sensing*, 3rd Edition, Volume 4, pp. 53–109 (S. L. Ustin, ed.). Hoboken, NJ: John Wiley and Sons.

Asner, G. P. and Martin, R. E. 2009. Airborne spectranomics: Mapping canopy chemical and taxonomic diversity in tropical forests. *Frontiers in Ecology and Environment*, 7, 269–276.

Asner, G. P. and Martin, R. E. 2011. Canopy phylogenetic, chemical and spectral assembly in a lowland Amazonian Forest. *New Phytologist*, 189, 999–1012.

Asner, G. P., Knapp, D. E., Kennedy-Bowdoin, T., Jones, M. O., Martin, R. E., Boardman, J., and Field, C. B. 2007. Carnegie airborne observatory: In-flight fusion of hyperspectral imaging and waveform light detection and ranging (wLiDAR) for three-dimensional studies of ecosystems. *Journal of Applied Remote Sensing*, 1, 013536.

Asner, G. P., Martin, R. E., Knapp, D. E., Tupayachi, R., Anderson, C., Carranza, L., Martinez, P., Houcheime, M., Sinca, F., and Weiss, P. 2011. Spectroscopy of canopy chemicals in humid tropical forests. *Remote Sensing of Environment*, 115, 3587–3598.

Balch, J. K., Bradley, B. A., D'Antonio, C. M., and Gómez-Dans, J. 2013. Introduced annual grass increases regional fire activity across the arid western USA. *Global Change Biology*, 19, 173–183.

Bastin, G., Scarth, P., Chewings, V., Sparrow, A., Denham, R., Schmidt, M., O'Reagain, P., Shepard, R., and Abbot, B. 2012. Separating grazing and rainfall effects at regional scale using remote sensing imagery: A dynamic reference-cover method. *Remote Sensing of Environment*, 121, 443–457.

Bastin, G. N., Stafford Smith, D. M., Watson, I. W., and Fisher, A. 2009. The Australian collaborative rangelands information system: Preparing for a climate of change. *Rangeland Journal*, 31, 111–125.

Bastin, G. N., Denham, R., Scarth, P., Sparrow, A., and Chewings, V. 2014. Remotely-sensed analysis of ground-cover change in Queensland's rangelands, 1988–2005. *Rangeland Journal*, 36, 191–204.

Belgacem, A. O. and Louhaichi, M. 2013. The vulnerability of native rangeland plant species to global climate change in the West Asia and North African regions. *Climatic Change*, 119, 451–463.

Bellis, L. M., Pidgeon, A. M., Radeloff, V. C., St-Louis, V., Navarro, J. L., and Martella, M. B. 2008. Modeling habitat suitability for greater rheas based on satellite image texture. *Ecological Applications*, 18, 1956–1966.

Bentivegna, D. J., Smeda, R. J., and Wang, C. 2012. Detecting cutleaf teasel (*Dipsacus laciniatus*) along a Missouri highway with hyperspectral imagery. *Invasive Plant Science and Management*, 5, 155–163.

Bestelmeyer, B. T. 2006. Threshold concepts and their use in rangeland management and restoration: The good, the bad, and the insidious. *Restoration Ecology*, 14, 325–329.

Bestelmeyer, B. T., Goolsby, D. P., and Archer, S. R. 2011. Spatial perspectives in state-and-transition models: A missing link to land management. *Journal of Applied Ecology*, 48, 746–757.

de Beurs, K. M. and Henebry, G. M. 2004. Land surface phenology, climatic variation, and institutional change: Analyzing agricultural land cover change in Kazakhstan. *Remote Sensing of Environment*, 89, 497–509.

de Beurs, K. M. and Henebry, G. M. 2010. Spatio-temporal statistical methods for modeling land surface phenology. In *Phenological Research*, pp. 177–208 (I. L. Hudson and M. R. Keatley, eds). New York: Springer Science and Business Media.

Blaschke, T. 2010. Object-based image analysis for remote sensing. *ISPRS Journal of Photogrammetry and Remote Sensing*, 65, 2–16.

Blumenthal, D., Booth, D. T., Cox, S. E., and Ferrier, C. E. 2007. Large-scale aerial images capture details of invasive plant populations. *Rangeland Ecology & Management*, 60, 523–528.

Blumenthal, D. M., Norton, A. P., Cox, S. E., Hardy, E. M., Liston, G. E., Kennaway, L., Booth, D. T., and Derner, J. D. 2012. *Linaria dalmatica* invades south-facing slopes and less grazed areas in grazing-tolerant mixed-grass prairie. *Biological Invasions*, 14, 395–404.

Boardman, J. W. and Kruse, F. A. 2011. Analysis of imaging spectrometer data using N-dimensional geometry and a mixture-tuned matched filter approach. *IEEE Transactions on Geoscience and Remote Sensing*, 49, 4138–4152.

Booth, D. T. and Cox, S. E. 2008. Image-based monitoring to measure ecological change in rangeland. *Frontiers of Ecology and Management*, 6, 185–190.

Booth, D. T. and Cox, S. E. 2009. Dual-camera, high-resolution aerial assessment of pipeline revegetation. *Environmental Monitoring and Assessment*, 158, 23–33.

Booth, D. T. and Cox, S. E. 2011. Art to science: Tools for greater objectivity in resource monitoring. *Rangelands*, 33(4), 27–34.

Booth, D. T. and Tueller, P. T. 2003. Rangeland management using remote sensing. *Arid Land Research and Management*, 17, 455–467.

Booth, D. T., Cox, S. E., and Berryman, R. D. 2006a. Point sampling digital imagery with 'Samplepoint'. *Environmental Monitoring and Assessment*, 123, 97–108.

Booth, D. T., Cox, S. E., and Berryman, R. D. 2006b. Precision measurements from very-large scale aerial digital imagery. *Environmental Monitoring and Assessment*, 112, 293–307.

Booth, D. T., Cox, S. E., Louhaichi, M., and Johnson, D. E. 2004. Lightweight camera stand for close-to-earth remote sensing. *Journal of Range Management*, 57, 675–678.

Booth, D. T., Cox, S. E., Meilke, T., and Zuuring, H. R. 2008. Ground-cover measurements: Assessing correlation among aerial and ground-based measurements. *Environmental Management*, 42, 1091–1100.

Booth, D. T., Cox, S. E., and Teel, D. 2010. Aerial assessment of leafy spurge (*Euphorbia esula* L.) on Idaho's Deep Fire Burn. *Native Plants Journal*, 11, 327–339.

Boryan, C., Yang, Z., Mueller, R., and Craig, M. 2011. Monitoring US agriculture: The US Department of Agriculture, National Agricultural Statistics Service, Cropland Data Layer Program. *Geocarto International*, 26, 341–358.

Bradley, B. A. 2013. Remote detection of invasive plants: A review of spectral, textural and phenological approaches. *Biological Invasions*, 16, 1411–1425. DOI 10.1007/s10530-013-0578-9.

Bradley, B. A. and Mustard, J. F. 2005. Identifying land cover variability distinct from land cover change: Cheatgrass in the Great Basin. *Remote Sensing of Environment*, 94, 204–213.

Bradley, B. A. and Mustard, J. F. 2006. Characterizing the landscape dynamics of an invasive plant and risk of invasion using remote sensing. *Ecological Applications*, 16, 1132–1147.

Bradley, B. A., Blumenthal, D. M., Wilcove, D. S., and Ziska, L. H. 2009. Predicting plant invasions in an era of global change. *Trends in Ecology and Evolution*, 25, 310–318.

Breckenridge, R. P. and Dakins, M. 2011. Evaluation of bare ground on rangelands using unmanned aerial vehicles: A case study. *GIScience & Remote Sensing*, 48, 74–85.

Breckenridge, R. P., Dakins, M., Bunting, S., Harbour, J. L., and White, S. 2011. Comparison of unmanned aerial vehicle platforms for assessing vegetation cover in sagebrush steppe ecosystems. *Rangeland Ecology & Management*, 64, 521–532.

Briske, D. D., Fuhlendorf, S. D., and Smeins, F. E. 2005. State-and-transition models, thresholds, and rangeland health: A synthesis of ecological concepts and perspectives. *Rangeland Ecology & Management*, 58, 1–10.

Briske, D. D., Fuhlendorf, S. D., and Smeins, F. E. 2006. A unified framework for assessment and application of ecological thresholds. *Rangeland Ecology & Management*, 59, 225–236.

Briske, D. D., Bestelmeyer, B. T., Stringham, T. K., and Shaver P. L. 2008. Recommendations for development of resilience-based state-and-transition models. *Rangeland Ecology & Management*, 61, 359–367.

Brown, J. R. 1994. State and transition models for rangelands. 2. Ecology as a basis for rangeland management: Performance criteria for testing models. *Tropical Grasslands*, 28, 206–213.

Brown, J. R. 2010. Ecological sites: Their history, status and future. *Rangelands*, 32(6), 5–8.

Cagney, J., Cox, S. E., and Booth, D. T. 2011. Comparison of point intercept and image analysis for monitoring rangeland transects. *Rangeland Ecology & Management*, 64, 309–315.

Carter, G. A., Knapp, A. K., Anderson, J. E., Hoch, G. A., and Smith, M. D. 2005. Indicators of plant species richness in AVIRIS spectra of a mesic grassland. *Remote Sensing of Environment*, 98, 304–316.

Carter, G. A., Lucas, K. L., Blossom, G. A., Lassitter, C. L., Holiday, D. M., Mooneyhan, D. S., Fastring, D. R., Holcombe, T. R., and Griffith, J. A. 2009. Remote sensing and mapping tamarisk along the Colorado River, USA: A comparative use of summer-acquired Hyperion, Thematic Mapper, and Quickbird data. *Remote Sensing*, 1, 318–329.

Caudle, D., Sanchez, H., DiBenedetto, J., Talbot, C., and Karl, M. 2013. *Interagency Ecological Site Handbook for Rangelands*. Washington, DC: US DOI Bureau of Land Management, USDA Forest Service, and USDA Natural Resource

Conservation Service (http://http://www.ars.usda.gov/Research/docs.htm?docid = 18502, last accessed February 28, 2014).

Chang, C.-I. 2000. An information-theoretic approach to spectral variability, similarity, and discrimination for hyperspectral image analysis. *IEEE Transactions on Information Theory*, 46, 1927–1932.

Cheng, Y. B., Tom, E., and Ustin, S. L. 2007. Mapping an invasive species, kudzu (*Pueraria montana*), using hyperspectral imagery in western Georgia. *Journal of Applied Remote Sensing*, 1, 013514.

Cho, M. A., Debba, P., Mathieu, R., Naidoo, L., van Aardt, J., and Asner, G. P. 2010. Improving discrimination of savanna tree species through a multiple-endmember spectral angle mapper approach: Canopy level analysis. *IEEE Transactions on Geoscience and Remote Sensing*, 48, 4133–4142.

Clark, J., Finco, M., Warbington, R., and Schwind, B. 2004. Digital Mylar: A tool to attribute vegetation polygon features over high-resolution imagery. In *Proceedings of the Tenth Forest Service Remote Sensing Applications Conference*, April 5–9, Salt Lake City, UT (http://www.fs.fed.us/r5/rsl/publications/, last accessed March 9, 2014).

Clark, M. L., Roberts, D. A., and Clark, D. B. 2005. Hyperspectral discrimination of tropical rainforest tree species at leaf to crown scales. *Remote Sensing of Environment*, 96, 375–398.

Clark, R. N., Swayze, G. A., Livo, K. E., Kokaly, R. F., Sutley, S. J., Dalton, J. B., McDougal, R. R., and Gent, C. A. 2003. Imaging spectroscopy: Earth and planetary remote sensing with the USGS Tetracorder and expert systems. *Journal of Geophysical Research*, 108, E12, 5131.

Clements, F. E. 1916. *Plant Succession: An Analysis of the Development of Vegetation*, Volume 242. Washington, DC: Carnegie Institute of Washington.

Clinton, N. E., Potter, C., Crabtree, B., Genovese, V., Gross, P., and Gong, P. 2010. Remote sensing-based time-series analysis of cheatgrass (*Bromus tectorum* L.) phenology. *Journal of Environmental Quality*, 39, 955–963.

Congalton, R. G. and Green, K. 2008. *Assessing the Accuracy of Remotely Sensed Data: Principles and Practices*. Boca Raton, FL: CRC Press, 183pp.

Cooper, W. S. 1924. An apparatus for photographic recording of quadrats. *Journal of Ecology*, 12, 317–321.

Culbert, P. D., Radeloff, V. C., St-Louis, V., Flather, C. H., Rittenhouse, C. D., Albright, T. P., and Pidgeon, A. M. 2012. Modeling broad-scale patterns of avian species richness across the Midwestern United States with measures of satellite image texture. *Remote Sensing of Environment*, 118, 140–150.

Dalmayne, J., Möckel, T., Prentice, H. C., Schmid, B. C., and Hall, K. 2013. Assessment of fine-scale plant species beta diversity using WorldView-2 satellite spectral dissimilarity. *Ecological Informatics*, 18, 1–9.

Daughtry, C. S. T. 2001. Discriminating crop residues from soil by shortwave infrared reflectance. *Agronomy Journal*, 93, 125–131.

Daughtry, C. S. T. and Hunt Jr., E. R. 2008. Mitigating the effects of soil and residue water contents on remotely sensed estimates of crop residue cover. *Remote Sensing of Environment*, 112, 1647–1657.

Davidson, A. and Csillag, F. 2001. The influence of vegetation index and spatial resolution on a two-date remote sensing-derived relation to C4 species coverage. *Remote Sensing of Environment*, 75, 138–151.

Davidson, E. A., Asner, G. P., Stone, T. A., Neill, C., and Figueiredo, R. O. 2008. Objective indicators of pasture degradation from spectral mixture analysis of Landsat imagery. *Journal of Geophysical Research*, 113, G00B03.

van Deventer, A. P., Ward, A. D., Gowda, P. H., and Lyon, J. G. 1997. Using Thematic Mapper data to identify contrasting soil plains to tillage practices. *Photogrammetric Engineering & Remote Sensing*, 63, 87–93.

DiTomaso, J. M. 2000. Invasive weeds in rangelands: Species, impacts, and management. *Weed Science*, 48, 255–265.

Dobrowski, S. Z., Safford, H. D., Cheng, Y. B., and Ustin, S. L. 2008. Mapping mountain vegetation using species distribution modeling, image-based texture analysis, and object-based classification. *Applied Vegetation Science*, 11, 499–508.

Dudek, K. B., Root, R. R., Kokaly, R. F., and Anderson, G. L. 2004. Increased spatial and temporal consistency of leafy spurge maps from multidate AVIRIS imagery: A modified, hybrid linear spectral mixture analysis/mixture-tuned matched filtering approach. In *Proceedings of the 13th JPL Airborne Earth Science Workshop* (R. O. Green, ed.). Pasadena, CA: Jet Propulsion Laboratory.

Duniway, M. C., Karl, J. W., Schrader, S., Baquera, N., and Herrick J. E. 2012. Range and pasture monitoring using high resolution aerial imagery: A repeatable image interpretation approach. *Environmental Monitoring and Assessment*, 184, 789–804.

Elmqvist, T., Folke, C., Nyström, M., Peterson, G., Bengtsson, J., Walker, B., and Norberg, J. 2003. Response diversity, ecosystem change, and resilience. *Frontiers of Ecology and Management*, 1, 488–494.

Everitt, J. H., Anderson, G. L., Escobar, D. E., Davis, M. R., Spencer, N. R., and Andrascik, R. J. 1995. Use of remote sensing for detecting and mapping leafy spurge (*Euphorbia esula*). *Weed Technology*, 9, 599–609.

Eyre, T. J., Fisher, A., Hunt, L. P., and Kutt, A. S. 2011a. Measure it to better manage it: A biodiversity monitoring framework for the Australian rangelands. *Rangeland Journal*, 33, 239–253.

Eyre, T. J., Kelly, A. L., Neldner, V. J., Wilson, B. A., Ferguson, D. J., Laidlaw, M. J., and Franks A. J. 2011b. *BioCondition: A Condition Assessment Framework for Terrestrial Biodiversity in Queensland*. Assessment Manual. Version 2.1. Brisbane, Queensland, Australia: Department of Environment and Resource Management (DERM), Biodiversity and Ecosystem Sciences.

Feret, J. B., François, C., Asner, G. P., Gitelson, A. A., Martin, R. E., Bidel, L. P. R., Ustin, S. L., le Maire, G., and Jaquemoud, S. 2008. PROSPECT-4 and 5: Advances in the leaf optical properties model separating photosynthetic pigments. *Remote Sensing of Environment*, 112, 3030–3043.

Fletcher, R. S., Everitt, J. H., and Yang, C. 2011. Identifying saltcedar with hyperspectral data and support vector machines. *Geocarto International*, 26, 195–209.

Folke, C., Carpenter, S., Walker, B., Scheffer, M., Elmqvist, T., Gunderson, L., and Holling, C. S. 2004. Regime shifts, resilience, and biodiversity in ecosystem management. *Annual Review of Ecology, Evolution, and Systematics*, 35, 557–581.

Foody, G. M. and Dash, J. 2007. Discriminating and mapping the C3 and C4 composition of grasslands in the northern Great Plains, USA. *Ecological Informatics*, 2, 89–93.

Foody, G. M. and Dash, J. 2010. Estimating the relative abundance of C_3 and C_4 grasses in the Great Plains from multi-temporal MTCI data: Issues of compositing period and spatial generalizability. *International Journal of Remote Sensing*, 31, 351–362.

Franklin, S. E. 2010. *Remote Sensing for Biodiversity and Wildlife Management*. New York: McGraw-Hill, 346pp.

Franklin, S. E. and Wulder, M. A. 2002. Remote sensing methods in medium spatial resolution satellite data land cover classification of large areas. *Progress in Physical Geography*, 26, 173–205.

Friedl, M. A., Sulla-Menashe, D., Tan, B., Schnieder, A., Ramankutty, N., Sibley, A., and Huang, X. 2010. MODIS collection 5 global land cover: Algorithm refinements and characterization of new datasets. *Remote Sensing of Environment*, 114, 168–182.

Fuhlendorf, S. D. and Engle, D. M. 2001. Restoring heterogeneity on rangelands: Ecosystem management based on evolutionary grazing patterns. *Bioscience*, 51, 625–632.

Fuhlendorf, S. D., Harrell, W. C., Engle, D. M., Hamilton, R. G., Davis, C. A., and Leslie Jr., D. M. 2006. Should heterogeneity be the basis for conservation? Grassland bird response to fire and grazing. *Ecological Applications*, 16, 1706–1716.

Gao, B. C., Heidebrecht, K. B., and Goetz, A. F. H. 1993. Derivation of scaled surface reflectances from AVIRIS data. *Remote Sensing of Environment*, 44, 145–163.

Gaston, K. J. 2000. Global patterns in biodiversity. *Nature*, 405, 220–227.

Ge, S., Everitt, J., Carruthers, R., Gong, P., and Anderson, G. 2006. Hyperspectral characteristics of canopy components and structure for phonological assessment of an invasive weed. *Environmental Monitoring and Assessment*, 120, 109–126.

Ge, S., Xu, M., Anderson, G. L., and Carruthers, R. I. 2007. Estimating yellow starthistle (*Centaurea solstitialis*) leaf area index and aboveground biomass with the use of hyperspectral data. *Weed Science*, 55, 671–678.

Gibbons, P., Briggs, S. V., Ayers, D. A., Doyle, S., Seddon, J., McElhinny, C., Jones, N., Sims, R., and Doody J. S. 2008. Rapidly quantifying reference conditions in modified landscapes. *Biological Conservation*, 141, 2483–2493.

Gibbons, P., Briggs, S. V., Ayers, D. A., Seddon, J., Doyle, S., Cosier, P., McElhinny, C., Pelly, V., and Roberts, K. 2009. An operational method to assess impacts of land clearing on terrestrial biodiversity. *Ecological Indicators*, 9, 26–40.

Gill, T. K. and Phinn, S. R. 2008. Estimates of bare ground and vegetation cover from Advanced Spaceborne Thermal Emission and Reflection Radiometer (ASTER) short-wave-infrared reflectance imagery. *Journal of Applied Remote Sensing*, 2, 023511.

Gillespie, T. W., Foody, G. M., Rocchini, D., Giorgi, A. P., and Saatchi, S. 2008. Measuring and modeling biodiversity from space. *Progress in Physical Geography*, 32, 203–221.

Glenn, N. F., Mundt, J. T., Weber, K. T., Prather, T. S., Lass, L. W., and Pettingill, J. 2005. Hyperspectral data processing for repeat detection of small infestations of leafy spurge. *Remote Sensing of Environment*, 95, 399–412.

Goodin, D. G. and Henebry, G. M. 1997. A technique for monitoring ecological disturbance in tallgrass prairie using seasonal NDVI trajectories and a discriminant function mixture model. *Remote Sensing of Environment*, 61, 270–278.

Gottschalk, T. K., Huettmann, F., and Ehlers, M. 2005. Thirty years of analyzing and modeling avian habitat relationships using satellite imagery data: A review. *International Journal of Remote Sensing*, 26, 2631–2656.

Gould, W. 2000. Remote sensing of vegetation, plant species richness, and regional biodiversity hotspots. *Ecological Applications*, 10, 1861–1870.

Goward, S. N., Chander, G., Pagnutti, M., Marx, A., Ryan, R., Thomas, N., and Tetrault, R. 2012. Complementarity of ResourceSat-1 AWiFS and Landsat TM/ETM+ sensors. *Remote Sensing of Environment*, 123, 41–56.

Green, R. O., Eastwood, M. L., Sarture, C. M., Chrien, T. G., Aronsson, M., Chippendale, B. J., Faust, J. A. et al. 1998. *Imaging Spectroscopy Environment*, 65, 227–248.

Gu, Y., B´elair, S., Mahfouf, J., and Deblonde, G. 2006. Optional interpolation analysis of leaf area index using MODIS data. *Remote Sensing of Environment*, 104, 283–296.

Gu, Y., Brown, J. F., Miura, T., van Leeuwen, W. J. D., and Reed, B. C. 2010. Phenological classification of the United States: A geographic framework for extending multi-sensor time-series data. *Remote Sensing*, 2, 526–544.

Guerschman, J. P., Hill, M. J., Renzullo, L. J., Barrett, D. J., Marks, A. S., and Botha, E. J. 2009. Estimating fractional cover of photosynthetic vegetation, non-photosynthetic vegetation and bare soil in the Australian tropical savanna region upscaling the EO-1 Hyperion and MODIS sensors. *Remote Sensing of Environment*, 113, 928–945.

Guo, X., Price, K. P., and Stiles, J. 2003. Grasslands discriminant analysis using Landsat TM single and multitemporal data. *Photogrammetric Engineering & Remote Sensing*, 69, 1255–1262.

Hall, K., Reitalu, T., Sykes, M. T., and Prentice, H. C. 2012. Spectral heterogeneity of QuickBird satellite data is related to fine-scale plant species spatial turnover in semi-natural grasslands. *Applied Vegetation Science*, 15, 145–157.

Hamada, Y., Stow, D. A., Coulter, L. L., Jafolla, J. C., and Hendricks, L.W. 2007. Detecting Tamarisk species (*Tamarix* spp.) in riparian habitats of Southern California using high spatial resolution hyperspectral imagery. *Remote Sensing of Environment*, 109, 237–248.

Haralick, R. M., Shanmugam, K., and Dinstein, I. 1973. Textural features for image classification. *IEEE Transactions on Systems, Man and Cybernetics*, 3, 610–621.

Hardin, P. J. and Hardin, T. J. 2010. Small-scale remotely piloted vehicles in environmental research. *Geography Compass*, 4/9, 1297–1311.

Hardin, P. J. and Jackson, M. W. 2005. An unmanned aerial vehicle for rangeland photography. *Rangeland Ecology & Management*, 58, 439–442.

Hardin, P. J., Jackson, M. W., Anderson, V. J., and Johnson, R. 2007. Detecting squarrose knapweed (*Centaurea virgata* Lam. ssp. *squarrosa* Gugl.) using a remotely piloted vehicle: A Utah case study. *GIScience & Remote Sensing*, 44, 203–219.

He, K. S., Rocchini, D., Neteler, M., and Nagendra, H. 2011. Benefits of hyperspectral remote sensing for tracking plant invasions. *Diversity and Distributions*, 17, 381–392.

Hernandez, A. J. and Ramsey, R. D. 2013. A landscape similarity index: Multitemporal remote sensing to track changes in big sagebrush ecological sites. *Rangeland Ecology & Management*, 66, 71–81.

Herrick, J. E., Van Zee, J. W., Havstad, K. M., Burkett, L. M., and Whitford, W. G. 2005a. *Monitoring Manual for Grassland, Shrubland, and Savanna Ecosystems, Volume I: Quick Start*, 36pp. Las Cruces, NM: USDA-ARS Jornada Experimental Range.

Herrick, J. E., Van Zee, J. W., Havstad, K. M., Burkett, L. M., and Whitford, W. G. 2005b. *Monitoring Manual for Grassland, Shrubland, and Savanna Ecosystems, Volume II: Design, Supplementary Methods and Interpretation*, 200pp. Las Cruces, NM: USDA-ARS Jornada Experimental Range.

Hestir, E. L., Khanna, S., Andrew, M. E., Santos, M. J., Viers, J. H., Greenberg, J. A., Rajapakse, S. S., and Ustin S. L. 2008. Identification of invasive vegetation using hyperspectral remote sensing in the California Delta ecosystem. *Remote Sensing of Environment*, 112, 4034–4047.

Higley, B. M. and Kelly, D. E. 1873. Home on the range. Smith County Pioneer, Kansas.

Homer, C., Huang, C., Yang, L., Wylie, B., and Coan, M. 2004. Development of a 2001 National Landcover Database for the United States. *Photogrammetric Engineering & Remote Sensing*, 70, 829–840.

Hooper, D. U., Chapin III, F. S., Ewel, J. J., Hector, A., Inchausti, P., Lavorel, S., Lawton, J. H. et al. 2005. Effects of biodiversity on ecosystem functioning: A consensus of current knowledge. *Ecological Monographs*, 75, 3–35.

Huang, C. and Asner, G. P. 2009. Applications of remote sensing to alien invasive plant studies. *Sensors*, 9, 4869–4889.

Huang, C. and Geiger, E. L. 2008. Climate anomalies provide opportunities for large-scale mapping of non-native plant abundance in desert grasslands. *Diversity and Distributions*, 14, 875–884.

Huang, C., Geiger, E. L., van Leeuven, W. J. D., and Marsh, S. E. 2009. Discrimination of invaded and native species sites in a semi-desert grassland using MODIS multi-temporal data. *International Journal of Remote Sensing*, 30, 897–917.

Huete, A., Didan, K., Miura, T., Rodriguez, E. P., Gao, X., and Ferreira, L. G. 2002. Overview of the radiometric and biophysical performance of the MODIS vegetation indices. *Remote Sensing of Environment*, 83, 195–213.

Huete, A. R. 1988. A soil-adjusted vegetation index (SAVI). *Remote Sensing of Environment*, 25, 295–309.

Hunt Jr., E. R. and Parker Williams, A. E. 2006. Detection of flowering leafy spurge with satellite multispectral imagery. *Rangeland Ecology & Management*, 59, 494–499.

Hunt Jr., E. R., Everitt, J. H., Ritchie, J. C., Moran, M. S., Booth, D. T., Anderson, G. L., Clark, P. E., and Seyfried M. S. 2003. Applications and research using remote sensing for rangeland management. *Photogrammetric Engineering & Remote Sensing*, 69, 675–693.

Hunt Jr., E. R., McMurtrey III, J. E., Parker Williams, A. E., and Corp, L. A. 2004. Spectral characteristics of leafy spurge (*Euphorbia esula*) leaves and flower bracts. *Weed Science*, 52, 492–497.

Hunt Jr., E. R., Daughtry, C. S. T., Kim, M. S., and Parker Williams, A. E. 2007. Using canopy reflectance models and spectral angles to assess potential of remote sensing to detect invasive weeds. *Journal of Applied Remote Sensing*, 1, 013506.

Hunt Jr., E. R., Gillham, J. H., and Daughtry, C. S. T. 2010. Improving potential geographic distribution models for invasive plants by remote sensing. *Rangeland Ecology & Management*, 63, 505–513.

Irisarri, J. G. N., Oesterheld, M., Vernón, S. R., and Paruelo, J. M. 2009. Grass species differentiation through canopy hyperspectral reflectance. *International Journal of Remote Sensing*, 30, 5959–5975.

Jacquemoud, S., Ustin, S. L., Verdebout, J., Schmuck, G., Andreoli, G., and Hosgood, B. 1996. Estimating leaf biochemistry using the PROSPECT leaf optical properties model. *Remote Sensing of Environment*, 56, 194–202.

Jacquemoud, S., Verhoef, W., Baret, F., Bacour, C., Zarco-Tejada, P. J., Asner, G. P., François, C., and Ustin S. L. 2009. PROSPECT+SAIL models: A review of use for vegetation characterization. *Remote Sensing of Environment*, 113(suppl), S56–S66.

Janetos, A., Hansen, L., Inouye, D., Kelly, B. P., Meyerson, L., Peterson, B., and Shaw R. 2008. Biodiversity. In M. Walsh, ed, *The Effects of Climate Change on Agriculture, Land Resources, Water Resources, and Biodiversity in the United States*, Chapter 5, pp. 151–182, Synthesis and Assessment Product 4.3. Washington, DC: U.S. Climate Change Science Program and the Subcommittee on Global Change Research.

Jiang, Z., Huete, A. R., Chen, J., Chen, Y., Li, J., Yan, G., and Zhang, X. 2006. Analysis of NDVI and scaled difference vegetation index retrievals of vegetation fraction. *Remote Sensing of Environment*, 101, 366–378.

Johansen, K. and Phinn, S. 2006. Mapping structure parameters and species composition of riparian vegetation using IKONOS and Landsat ETM+ data in Australian tropical savannahs. *Photogrammetric Engineering & Remote Sensing*, 72, 71–80.

John, R., Chen, J., Lu, N., Guo, K., Liang, C., Wei, Y., Noormets, A., Ma, K., and Han, X. 2008. Predicting plant diversity based on remote sensing products in the semi-arid region of Inner Mongolia. *Remote Sensing of Environment*, 112, 2018–2032.

Jönsson, P. and Eklundh, L. 2004. TIMESAT—A program for analyzing time-series of satellite sensor data. *Computers & Geoscience*, 30, 833–845.

Joyce, L. A., Briske, D. D., Brown, J. R., Polley, H. W., McCarl, B. A., and Bailey, D. W. 2013. Climate change and North American Rangelands: Assessment of mitigation and adaptation strategies. *Rangeland Ecology & Management*, 66, 512–528.

Kampe, T. U., Johnson, B. R., Kuester, M., and Keller, M. 2010. NEON: The first continental-scale observatory with airborne remote sensing of vegetation biochemistry and structure. *Journal of Applied Remote Sensing*, 4, 043510.

Karnieli, A., Kokaly, R. F., West, N. E., and Clark, R. N. 2003. Remote sensing of biological soil crusts. In *Biological Soil Crusts: Structure, Function, and Management*, Ecological Studies Volume 150, pp. 431–455 (J. Belnap and O. L. Lange, eds.). Berlin, Germany: Springer-Verlag.

Knox, N. M., Skidmore, A. K., Prins, H. H. T., Asner, G. P., van der Werff, H., de Boer, W. F., van der Waal, C., de Knegt, H. J., Kohi, E.,M., Slotow, R., and Grant R. C. 2011. Dry season mapping of savanna forage quality, using the hyperspectral Carnegie Airborne Observatory sensor. *Remote Sensing of Environment*, 115, 1478–1488.

Kogan, F., Gitelson, A., Zakarin, E., Spivak, L., and Lebed, L. 2003. AVHRR-based spectral vegetation index for quantitative assessment of vegetation state and productivity: Calibration and validation. *Photogrammetric Engineering & Remote Sensing*, 69, 899–906.

Kremer, R. G. and Running, S. W. 1993. Community type differentiation using NOAA/AVHRR data within a sagebrush-steppe ecosystem. *Remote Sensing of Environment*, 46, 311–318.

Kruse, F. A., Lefkoff, A. B., Boardman, J. W., Heidebrecht, K. B., Shapiro, A. T., Barloon, P. J., and Goetz, A. F. H. 1993. The Spectral Image Processing System (SIPS)—Interactive visualization and analysis of imaging spectrometer data. *Remote Sensing of Environment*, 44, 145–163.

Kuemmerle, T., Röder, A., and Hill, J. 2006. Separating grassland and shrub vegetation by multidate pixel-adaptive spectral mixture analysis. *International Journal of Remote Sensing*, 27, 3251–3271.

Kumar, L., Sinha, P., Brown, J. F., Ramsey, R. D., Rigge, M., Stam, C. A., Hernandez, A. J., Hunt E. R., and Reeves, M. C. Rangeland, grassland, shrublands monitoring: Methods and approaches. In *Remote Sensing Handbook. Volume II: Land Resources: Monitoring, Modeling, and Mapping*, Chapter 12, in press (P.S. Thenkabali, ed.). Boca Raton, FL: CRC Press.

Laliberte, A., Herrick, J. E., Rango, A., and Winters, C. 2010. Acquisition, orthorectification, and object-based classification of unmanned aerial vehicle (UAV) imagery for rangeland monitoring. *Photogrammetric Engineering & Remote Sensing*, 76, 661–672.

Laliberte, A. S. and Rango, A. 2011. Image processing and classification procedures for analysis of sub-decimeter imagery acquired with an unmanned aircraft over arid rangelands. *GIScience & Remote Sensing*, 48, 4–23.

Laliberte, A. S., Rango, A., Herrick, J. E., Fredrickson, E. L., and Burkett, L. 2007. An object-based image analysis approach for determining fractional cover of senescent and green vegetation with digital plot photography. *Journal of Arid Environments*, 69, 1–14.

Lass, L. W., Thill, D. C., Shafii, B., and Prather, T. S. 2002. Detecting spotted knapweed (*Centaurea maculosa*) with hyperspectral remote sensing technology. *Weed Technology*, 16, 426–432.

Lass, L. W., Prather, T. S., Glenn, N. F., Weber, K. T., Mundt, J. T., and Pettingill, J. 2005. A review of remote sensing of invasive weeds and example of the early detection of spotted knapweed (*Centaurea maculosa*) and babysbreath (*Gypsophila paniculata*) with a hyperspectral sensor. *Weed Science*, 53, 242–251.

Lavergne, S., Mouquet, N., Thuiller, W., and Ronce, O. 2010. Biodiversity and climate change: Integrating evolutionary and ecological responses of species and communities. *Annual Review of Ecology, Evolution, and Systematics*, 41, 321–350.

Lawrence, R. L., Wood, S. D., and Sheley, R. L. 2006. Mapping invasive plants using hyperspectral imagery and Breiman Cutler classifications (Random Forest). *Remote Sensing of Environment*, 100, 356–362.

Lesica, P. and Hanna, D. 2009. Effect of biological control on leafy spurge (*Euphorbia esula*) and diversity of associated grasslands over 14 years. *Invasive Plant Science and Management*, 2, 151–157.

Leyequien, E., Verrelst, J., Slot, M., Schaepman-Strub, G., Heitkönig, I. M. A., and Skidmore, A. 2007. Capturing the fugitive: Applying remote sensing to terrestrial animal distribution and diversity. *International Journal of Applied Earth Observation and Geoinformation*, 9, 1–20.

Louhaichi, M., Johnson, M. D., Woerz, A. L., Jasra, A. W., and Johnson, D. E. 2010. Digital charting technique for monitoring rangeland vegetation cover at local scale. *International Journal of Agriculture and Biology*, 12, 406–410.

Loveland, T. R., Merchant, J. W., Ohlen, D. O., and Brown, J. F. 1991. Development of a land-cover characteristics database for the conterminous U.S. *Photogrammetric Engineering & Remote Sensing*, 57, 1453–1463.

Lowe, D. G. 2004. Distinctive image features from scale-invariant keypoints. *International Journal of Computer Vision*, 60, 91–110.

Lu, D. and Weng, Q. 2007. A survey of image classification methods and techniques for improving classification performance. *International Journal of Remote Sensing*, 28, 823–870.

Ludwig, J. A., Wiens, J. A., and Tongway, D. J. 2000. A scaling rule for landscape patches and how it applies to conserving soil resources in savannas. *Ecosystems*, 3, 84–97.

Ludwig, J. A., Eager, R. W., Bastin, G. N., Chewings, V. H., and Liedloff, A. C. 2002. A leakiness index for assessing landscape function using remote sensing. *Landscape Ecology*, 17, 157–171.

Ludwig, J. A., Tongway, D. J., Bastin, G. N., and James, C. D. 2004. Monitoring ecological indicators of rangeland functional integrity and their relation to biodiversity at local and regional scales. *Austral Ecology*, 29, 108–120.

Ludwig, J. A., Eager, R. W., Liedloff, A. C., Bastin, G. N., and Chewings, V. H. 2006. A new landscape leakiness index based on remotely sensed ground-cover data. *Ecological Indicators*, 6, 327–336.

Ludwig, J. A., Bastin, G. N., Chewings, V. H., Eager, R. W., and Liedloff, A. C. 2007. Leakiness: A new index for monitoring the health of arid and semiarid landscapes using remotely sensed vegetation cover and elevation data. *Ecological Indicators*, 7, 442–454.

Luscier, J. D., Thompson, W. L., Wilson, J. M., Gorham, B. E., and Dragut, L. D. 2006. Using digital photographs and object-based image analysis to estimate percent ground cover in vegetation plots. *Frontiers of Ecology and Management*, 4, 408–413.

Lym, R. G. 2005. Integration of biological control agents with other weed management technologies: Success from the leafy spurge (*Euphorbia esula*) IPM program. *Biological Control*, 35, 366–375.

MacKinnon, W. C., Karl, J. W., Toevs, G. R., Taylor, J. J., Karl, M., Spurrier, C. S., and Herrick, J. E. 2011. BLM core terrestrial indicators and methods. Tech Note 440. Denver, CO: U.S. Department of the Interior, Bureau of Land Management, National Operations Center.

Madden, M. 2004. Remote sensing and geographic information system operations for vegetation mapping of invasive exotics. *Weed Technology*, 18, 1457–1463.

Mansour, K. and Mutanga, O. 2012. Classifying increaser species as an indicator of different levels of rangeland degradation using WorldView-2 imagery. *Journal of Applied Remote Sensing*, 6, 063558-1.

Mansour, K., Mutanga, O., and Everson, T. 2012. Remote sensing based indicators of vegetation species for assessing rangeland degradation: Opportunities and challenges. *African Journal of Agricultural Research*, 7, 3261–3270.

Marsett, R. C., Qi, J., Heilman, P., Biedenbender, S. H., Watson, M. C., Amer, S., Weltz, M., Goodrich, D., and Marset, R. 2006. Remote sensing for grassland management in the arid southwest. *Rangeland Ecology & Management*, 59, 530–540.

Martin, M. P., Barreto, L., Riaño, D., Fernandez-Quintanilla, C., and Vaughan, P. 2011. Assessing the potential of hyperspectral remote sensing for the discrimination of grass weeds in winter cereal crops. *International Journal of Remote Sensing*, 31, 49–67.

Maynard, C. L., Lawrence, R. L., Nielson, G. A., and Decker, G. 2007. Ecological site descriptions and remotely sensed imagery as a tool for rangeland evaluation. *Canadian Journal of Remote Sensing*, 33, 109–115.

McKeon, G. M., Stone, G. S., Syktus, J. I., Carter, J. O., Flood, N. R., Ahrens, D. G., Bruget, D. N. et al. 2009. Climate change impacts on northern Australian rangeland livestock carrying capacity: A review of issues. *Rangeland Journal*, 31, 1–29.

McNairn, H. and Protz, R. 1993. Mapping corn residue cover on agricultural fields in Oxford County, Ontario, using Thematic Mapper. *Canadian Journal of Remote Sensing*, 19, 152–159.

van der Meer, F. 2006. The effectiveness of spectral similarity measures for the analysis of hyperspectral imagery. *International Journal of Applied Earth Observation and Geoinformation*, 8, 3–17.

Miao, X., Gong, P., Swope, S., Pu, R., Carruthers, R., Anderson, G. L., Heaton, J. S., and Tracy, C. R. 2006. Estimation of yellow starthistle abundance through CASI-2 hyperspectral imagery using linear spectral mixture models. *Remote Sensing of Environment*, 101, 329–341.

Miao, X., Gong, P., Swope, S., Pu, R., Carruthers, R., and Anderson, G. L. 2007. Detection of yellow starthistle through band selection and feature extraction from hyperspectral imagery. *Photogrammetric Engineering & Remote Sensing*, 73, 1005–1015.

Miao, X., Patil, R., Heaton, J. S., and Tracy, R. C. 2011. Detection and classification of invasive saltcedar through high spatial resolution airborne hyperspectral imagery. *International Journal of Remote Sensing*, 32, 2131–2150.

Mirik, M., Steddom, K., and Michels Jr., G. J. 2006. Estimating biophysical characteristics of musk thistle (*Carduus nutans*) with three remote sensing instruments. *Rangeland Ecology & Management*, 59, 44–54.

Mirik, M., Ansley, R. J., Steddom, K., Jones, D. C., Rush, C. M., Michels Jr., G. J., and Elliot, N. C. 2013. Remote distinction of a noxious weed (musk thistle: *Carduus nutans*) using airborne hyperspectral imagery and the support vector machine classifier. *Remote Sensing*, 5, 612–630.

Mitchell, J. J. and Glenn, N. F. 2009. Subpixel abundance estimates in mixture-tuned matched filtering classifications of leafy spurge (*Euphorbia esula* L.). *International Journal of Remote Sensing*, 30, 6099–6119.

Mladinich, C. S., Ruiz Bustos, M., Stitt, S., Root, R., Brown, K., Anderson, G. L., and Hager, S. 2006. The use of Landsat 7 Enhanced Thematic Mapper Plus for mapping leafy spurge. *Rangeland Ecology & Management*, 59, 500–506.

Moffet, C. A. 2009. Agreement between measurements of shrub cover using ground-based methods and very large scale aerial imagery. *Rangeland Ecology & Management*, 62, 268–277.

Montandon, L. M. and Small, E. E. 2008. The impact of soil reflectance on the quantification of the green vegetation fraction from NDVI. *Remote Sensing of Environment*, 112, 1835–1845.

Morgan, J. A., Milchunas, D. G., LeCain, D. R., West, M., and Mosier, A. R. 2007. Carbon dioxide enrichment alters plant community structure and accelerates shrub growth in the shortgrass steppe. *Proceedings of the National Academy of Sciences*, 104, 14724–14729.

Morisette, J. T., Jarnevich, C. S., Ullah, A., Cai, W., Pedelty, J. A., Gentle, J. E., Stohlgren, T. J., and Schnase, J. L. 2006. A tamarisk habitat suitability map for the continental United States. *Frontiers of Ecology and Management*, 4, 11–17.

Moseley, K., Shaver, P. L., Sanchez, H., and Bestelmeyer, B. T. 2010. Ecological site development: A gentle introduction. *Rangelands* 32(6), 16–22.

Müllerová, J., Pergl, J., and Pyšek, P. 2013. Remote sensing as a tool for monitoring plant invasions: Testing the effects of data resolution and image classification approach on the detection of a model plant species *Heracleum mantegazzianum* (giant hogweed). *International Journal of Applied Earth Observation and Geoinformation*, 25, 55–65.

Mundt, J. T., Glenn, N. F., Weber, K. T., Prather, T. S., Lass, L. W., and Pettingill, J. 2005. Discrimination of hoary cress and determination of its detection limits via hyperspectral image processing and accuracy assessment techniques. *Remote Sensing of Environment*, 96, 509–517.

Nagler, P. L., Inoue, Y., Glenn, E. P., Russ, A. L., and Daughtry, C. S. T. 2003. Cellulose absorption index (CAI) to quantify mixed soil-plant litter scenes. *Remote Sensing of Environment*, 87, 310–325.

Nagler, P. L., Glenn, E. P., Jarnevich, C. S., and Shafroth, P. B. 2011. Distribution and abundance of saltcedar and Russian olive in the Western United States. *Critical Reviews of Plant Science*, 30, 508–523.

Narumalani, S., Mishra, D. R., Burkholder, J., Merani, P. B. T., and Wilson, G. 2006. A comparative evaluation of ISODATA and Spectral Angle Mapping for the detection of saltcedar using airborne hyperspectral imagery. *Geocarto International*, 21, 59–66.

Narumalani, S., Mishra, D. R., Wilson, R., Reece, P., and Kohler, A. 2009. Detecting and mapping four invasive species along the floodplain of North Platte River, Nebraska. *Weed Technology*, 23, 99–107.

National Research Council (NRC). 1994. *Rangeland Health, New Methods to Classify, Inventory, and Monitor Rangelands*, 180pp. Washington, DC: National Academy Press.

Noujdina, N. V. and Ustin, S. L. 2008. Mapping downy brome (*Bromus tectorum*) using multidate AVIRIS data. *Weed Science*, 56, 173–179.

Numata, I., Roberts, D. A., Chadwick, O. A., Schimel, J., Sampson, F. R., Leonidas, F. C., and Soares, J. V. 2007. Characterization of pasture physical biophysical properties and the impact of grazing intensity using remotely sensed data. *Remote Sensing of Environment*, 109, 314–327.

Oliver, I., Eldridge, D. J., Nadolny, C., and Martin, W. K. 2014. What do site condition multi-metrics tell us about species biodiversity? *Ecological Indicators*, 38, 262–271.

Olofsson, P., Foody, G. M., Stehman, S. V., and Woodcock, C. E. 2013. Making better use of accuracy data in land change studies: Estimating accuracy and area and quantifying uncertainty using stratified estimation. *Remote Sensing of Environment*, 129, 122–131.

Olsson, A. D. and Morisette, J. T. 2014. Comparison of simulated HyspIRI with two multispectral sensors for invasive species mapping. *Photogrammetric Engineering & Remote Sensing*, 80, 217–227.

O'Neill, M., Ustin, S. L., Hager, S., and Root, R. 2000. Mapping the distribution of leafy spurge at Theodore Roosevelt National Park using AVIRIS. In *Proceedings of the 10th AVIRIS Airborne Geoscience Workshop* (R. O. Green ed.). Pasadena, CA: Jet Propulsion Laboratory.

Parker Williams, A. E. and Hunt Jr., E. R. 2002. Estimation of leafy spurge cover from hyperspectral imagery using mixture tuned matched filtering. *Remote Sensing of Environment*, 82, 446–456.

Parker Williams, A. E. and Hunt Jr., E. R. 2004. Accuracy assessment for detection of leafy spurge with hyperspectral imagery. *Journal of Range Management*, 57, 106–112.

Parkes, D., Newell, G., and Cheal, D. 2003. Assessing the quality of native vegetation: The 'habitat hectares' approach. *Ecological Management & Restoration*, 4(suppl), S29–S38.

Pellant, M., Shaver, P., Pyke, D. A., and Herrick, J. E. 2005. *Interpreting Indicators of Rangeland Health*, Version 4, 122pp. Technical Reference 1734-6. Denver, CO: United States Department of Interior, Bureau of Land Management.

Pereira, H. M., Leadley, P. W., Proença, V., Alkemade, R., Scharlemann, J. P. W., Fernandez-Manjarrés, J. F., Araújo, M. G. et al. 2010. Scenarios for global biodiversity in the 21st Century. *Science* 330, 1496–1501.

Pereira, H. M., Navarro, L. M., and Martins, I. S. 2012. Global biodiversity change: The bad, the good, and the unknown. *Annual Review of Environment and Resources*, 37, 25–50.

Peters, A. J., Reed, B. C., Eve, M. D., and McDaniel, K. C. 1992. Remote sensing of broom snakeweed (*Gutierrezia sarothrae*) with NOAA-10 spectral image processing. *Weed Technology*, 6, 1015–1020.

Peters, A. J., Eve, M. D., Holt, E. H., and Whitford, W. G. 1997. Analysis of desert plant community growth patterns with high temporal resolution satellite spectra. *Journal of Applied Ecology*, 34, 418–432.

Peterson, A. T. 2003. Predicting the geography of species' invasions via ecological niche modeling. *Quarterly Review of Biology*, 78, 419–433.

Peterson, D. L., Price, K. P., and Martinko, E. A. 2002. Discriminating between cool season and warm season grassland cover types in northeastern Kansas. *International Journal of Remote Sensing*, 23, 5015–5030.

Pettorelli, N., Ryan, S., Mueller, T., Bunnefeld, N., Jędrejewska, B., Lima, M., and Kausrud, K. 2011. The normalized difference vegetation index (NDVI): Unforeseen successes in animal ecology. *Climate Research*, 46, 15–27.

Polley, H. W., Briske, D. D., Morgan, J. A., Wolter, K., Bailey, D. W., and Brown, J. R. 2013. Climate change and North American Rangelands: Trends, projections, and implications. *Rangeland Ecology & Management*, 66, 493–511.

Pu, R. 2008. Invasive species change detection using artificial neural networks and CASI hyperspectral imagery. *Environmental Monitoring and Assessment*, 140, 15–32.

Pu, R. 2012. Detecting and mapping invasive plant species by using hyperspectral data. In *Hyperspectral Remote Sensing of Vegetation*, pp. 447–465 (P. S. Thenkabali, J. G. Lyon, and A. Huete, eds.). Boca Raton, FL: CRC Press-Taylor & Francis Group.

Pu, R., Gong, P., Tian, Y., Miao, X., Carruthers, R. I., and Anderson, G. L. 2008. Using classification and NDVI differences for monitoring sparse vegetation coverage: A case study of saltcedar in Nevada, USA. *International Journal of Remote Sensing*, 29, 3987–4011.

Pyke, D. A., Herrick, J. E., Shaver, P., and Pellant, M. 2002. Rangeland health attributes and indicators for qualitative assessment. *Journal of Range Management*, 55, 584–597.

Quilter, M. C. and Anderson, V. J. 2000. Low altitude/large scale aerial photographs: A tool for range and resource managers. *Rangelands*, 22(2), 13–17.

Quilter, M. C. and Anderson, V. J. 2001. A proposed method for determining shrub utilization using (LA/LS) imagery. *Journal of Range Management*, 54, 378–381.

Ramoelo, A., Skidmore, A. K., Schlerf, M., Mathieu, R., and Heitkönig, I. 2011. Water-removed spectra increase the retrieval accuracy when estimating savanna grass nitrogen and phosphorus concentrations. *ISPRS Journal of Photogrammetry and Remote Sensing*, 66, 408–417.

Ramoelo, A., Skidmore, A. K., Cho, M. A., Schlerf, M., Mathieu, R., and Heitkönig, I. M. A. 2012. Regional estimation of Savanna grass nitrogen using the red-edge band of the spaceborne RapidEye sensor. *International Journal of Applied Earth Observation and Geoinformation*, 19, 151–162.

Rango, A., Laliberte, A., Steele, C., Herrick, J. E., Bestelmeyer, B., Schmugge, T., Roanhorse, A., and Jenkins, V. 2006. Using unmanned aerial vehicles for rangelands: Current applications and future potentials. *Environmental Practice*, 8, 159–168.

Rango, A., Laliberte A., Herrick, J. E., Winters, C., Havstad, K., Steele, C., and Browning, D. 2009. Unmanned aerial vehicle-based remote sensing for rangeland assessment, monitoring and management. *Journal of Applied Remote Sensing*, 3, 033542.

Reed, B. C., Brown, J. F., VanderZee, D., Loveland, T. R., Merchant, J. W., and Ohlen, D. O. 1994. Measuring phenological variability from satellite imagery. *Journal of Vegetation Science*, 5, 703–714.

Reeves, M. C., Angerer, J., Hunt, E. R., Kulawardhana, W., Kumar, L., Loboda, T., Loveland, T., Metternicht, G., Ramsey, R. D., and Washington-Allen, R. A. A global view of remote sensing of rangelands: Evolution, applications, and future pathways. *Remote Sensing Handbook. Volume II: Land Resources: Monitoring, Modeling, and Mapping*, Chapter 10, in press (P.S. Thenkabali, ed.). Boca Raton, FL: CRC Press.

Reich, P. B., Tilman, D., Isbell, F., Mueller, K., Hobbie, S. E., Flynn, D. F. B., and Eisenhauer, N. 2012. Impacts of biodiversity loss escalate through time as redundancy fades. *Science*, 336, 587–592.

Ricciardi, A. 2007. Are modern biological invasions an unprecedented form of global change? *Conservation Biology*, 21, 329–336.

Ricotta, C., Reed, B. C., and Tieszen, L. T. 2003. The role of C_3 and C_4 grasses to interannual variability in remotely sensed ecosystem performance over the US Great Plains. *International Journal of Remote Sensing*, 24, 4421–4431.

Roberts, D. A., Smith, M. O., and Adams, J. B. 1993. Green vegetation, nonphotosynthetic vegetation, and soils in AVIRIS data. *Remote Sensing of Environment*, 44, 255–269.

Roberts, D. A., Quattrochi, D. A., Hulley, G. C., Hook, S. J., and Green, R. O. 2012. Synergies between VSWIR and TIR data for the urban environment: An evaluation of the potential for the Hyperspectral Infrared Imager (HyspIRI) Decadal Survey mission. *Remote Sensing of Environment*, 117, 83–101.

Rocchini, D., Balkenhol, N., Carter, G. A., Foody, G. M., Gillespie, T. W., He, K. S., Kark, S. et al. 2010a. Remotely sensed spectral heterogeneity as a proxy of species diversity: Recent advances and open challenges. *Ecological Informatics*, 5, 318–329.

Rocchini, D., He, K. S., Oldeland, J., Wesuls, D., and Neteler, M. 2010b. Spectral variation versus species β-diversity at different spatial scales: A test in African highland savannas. *Journal of Environmental Monitoring*, 12, 825–831.

Rouse, J. W., Haas, R. H., Schell, J. A., and Deering, D. W. 1974. Monitoring vegetation systems in the Great Plains with ERTS. In *Third Earth Resources Technology Satellite-1 Symposium, Volume 1: Technical Presentations*, pp. 309–317, NASA SP-351 (S. C. Freden, E. P. Mercanti, and M. Becker, eds.). Washington, DC: National Aeronautics and Space Administration.

RSAC (Remote Sensing Applications Center). 2014. Digital Mylar. Available at http://www.fs.fed.us/eng/rsac/digitalmylar/, accessed June 9, 2014.

Running, S. W., Loveland, T. R., Pierce, L. L., Nemani, R., and Hunt Jr., E. R. 1995. A remote sensing based vegetation classification logic for global land cover analysis. *Remote Sensing of Environment*, 51, 39–48.

Sadler, R. J., Hazelton, M., Boer, M., and Grierson, P. F. 2010. Deriving state-and-transition models from an image series of grassland pattern dynamics. *Ecological Modelling*, 221, 433–444.

Sage, R. F. and Kubien, D. S. 2007. The temperature response of C_3 and C_4 photosynthesis. *Plant, Cell & Environment*, 30, 1086–1106.

Sampson, A. W. 1919. Plant succession relation to range management, 76 p. USDA Technical Bulletin No. 791. Washington, DC: United States Department of Agriculture.

Samuel, L. W., Kirby, D. R., Norland, J. E., and Anderson, G. L. 2008. Leafy spurge suppression by flea beetles in the Little Missouri Drainage Basin, USA. *Rangeland Ecology & Management*, 61, 437–443.

Savitzky, A. and Golay, M. J. E. 1964. Smoothing and differentiation of data by simplified least squares procedures. *Analytical Chemistry*, 36, 1627–1639.

Schwartz, M. D., Reed, B. C., and White, M. A. 2002. Assessing satellite-derived start-of-season measures in the Conterminous USA. *International Journal of Climate*, 22, 1793–1805.

Serbin, G., Hunt Jr., E. R., Daughtry, C. S. T., McCarty, G. W., and Doraiswamy, P. C. 2009. An improved ASTER index for remote sensing of crop residue. *Remote Sensing*, 1, 971–991.

Serbin, G., Hunt Jr., E. R., Daughtry, C. S. T., and McCarty, G. W. 2013. Assessment of spectral indices for cover estimation of senescent vegetation. *Remote Sensing Letters*, 4, 552–560.

Shenk, J. S., Workman Jr., J. J., and Westerhaus, M. O. 2008. Application of NIR spectroscopy to agricultural products. In *Handbook of Near-Infrared Analysis*, 3rd Edition, pp. 347–386 (D. A. Burns and E. W. Ciurczak, eds.). Boca Raton, FL: CRC Press.

Singh, N. and Glenn, N. F. 2009. Multitemporal spectral analysis for cheatgrass (*Bromus tectorum*) classification. *International Journal of Remote Sensing*, 30, 3441–3462.

Skidmore, A. K., Ferwerda, J. G., Mutanga, O., Van Wieren, S. E., Peel, M., Grant, R. C., Prins, H. H. T., Balcik, F. B., and Venus, V. 2010. Forage quality of savannas – Simultaneously mapping foliar protein and polyphenols for trees and grass using hyperspectral imagery. *Remote Sensing of Environment*, 114, 64–72.

Soussanna, J. –F. and Lüscher, A. 2007. Temperate grasslands and global atmospheric change: A review. *Grass and Forage Science*, 62, 127–134.

Southard, G. 2013. A little UAV grows up. In *American Society of Photogrammetry and Remote Sensing 2013 Annual Conference*, March 27, Baltimore, MD. http://www.asprs.org/a/publications/proceedings/Baltimore2013/southard.pdf (last accessed April 7, 2015).

Stafford Smith, D. M., McKeon, G. M., Watson, I. W., Henry, B. K., Stone, G. S., Hall, W. B., and Howden, S. M. 2007. Learning from episodes of degradation and recovery in variable Australian rangelands. *Proceedings of the National Academy of Sciences*, 104, 20690–20695.

Starks, P. J., Coleman, S. W., and Phillips, W. A. 2004. Determination of forage chemical composition using remote sensing. *Rangeland Ecology & Management*, 57, 635–640.

Steele, C. M., Bestelmeyer, B. T., Burkett, L. M., Smith, P. L., and Yanoff, S. 2012. Spatially explicit representation of state-and-transition models. *Rangeland Ecology & Management*, 65, 213–222.

Stehman, S. V., Giles M., and Foody, G. M. 2009. Accuracy assessment. In *The SAGE Handbook of Remote Sensing*, pp. 297–309 (T. A. Warner, G. M. Foody, and M. D. Nellis, eds.). London, U.K.: Sage Publications.

Stitt, S., Root, R., Brown, K., Hager, S., Mladinich, C., Anderson, G. L., Dudek, K., Ruiz Bustos, M., and Kokaly, R. 2006. Classification of leafy spurge with Earth Observing-1 Advanced Land Imager. *Rangeland Ecology & Management*, 59, 507–511.

St-Louis, V., Pidgeon, A. M., Radeloff, V. C., Hawbaker, T. J., and Clayton, M. K. 2006. High-resolution image texture as a predictor of bird species richness. *Remote Sensing of Environment*, 105, 299–312.

St-Louis, V., Pidgeon, A. M., Clayton, M. K., Locke, B. A., Bash, D., and Radeloff, V. C. 2009. Satellite image texture and a vegetation index predict avian biodiversity in the Chihuahuan desert of New Mexico. *Ecography*, 32, 468–480.

Stohlgren, T. J., Ma, P., Kumar, S., Rocca, M., Morisette, J. T., Jarnevich, C. S., and Benson, N. 2010. Ensemble habitat mapping of invasive plant species. *Risk Analysis*, 30, 224–235.

Tews, J., Brose, U., Grimm, V., Tielbörger, K., Wichmann, M. C., Schwager, M., and Jeltsch, F. 2004. Animal species diversity driven by habitat heterogeneity/diversity: The importance of keystone structures. *Journal of Biogeography*, 31, 79–92.

Thorp, K. R., French, A. N., and Rango, A. 2013. Effect of image spatial and spectral characteristics on mapping semi-arid rangeland vegetation using multiple endmember spectral mixture analysis (MESMA). *Remote Sensing of Environment*, 132, 120–130.

Thuiller, W., Richardson, D. M., Pyšek, P., Midgley, G. F., Hughes, G. O., and Rouget, M. 2005. Niche-based modeling as a tool for predicting the risk of alien plant invasions at a global scale. *Global Change Biology*, 11, 2234–2250.

Tieszen, L. L., Reed, B. C., Bliss, N. B., Wylie, B. K., and DeJong, D. D. 1997. NDVI, C_3 and C_4 production and distributions in Great Plains grassland land cover classes. *Ecological Applications*, 7, 59–78.

Tilman, D., Reich, P. B., and Knops, J. M. H. 2006. Biodiversity and ecosystem stability in a decade-long grassland experiment. *Nature*, 441, 629–632.

Tilman, D., Reich, P. B., and Isbell, F. 2012. Biodiversity impacts ecosystem productivity as much as resources, disturbance, or herbivory. *Proceedings of the National Academy of Sciences*, 109, 10394–10397.

Tomlins, G. F. and Lee, Y. J. 1983. Remotely piloted aircraft – An inexpensive option for large-scale aerial photography in forestry applications. *Canadian Journal of Remote Sensing*, 9, 76–85.

Townshend, J. R. G. and Justice, C. O. 1988. Selecting the spatial resolution of satellite sensors required for global monitoring of land transformations. *International Journal of Remote Sensing*, 9, 187–236.

Trimble Navigation Limited. 2013. *Trimble Unmanned Aircraft Systems for Surveying and Mapping*. Westminster, CO: Trimble Navigation Limited (http://www.trimble.com/Survey/unmanned-aircraft-systems.aspx, last accessed June 9, 2014).

Tucker, C. J. 1979. Red and photographic infrared linear combinations for monitoring vegetation. *Remote Sensing of Environment*, 8, 127–150.

Turner, D., Lucieer, A., and Watson, C. 2012. An automated technique for generating georectified mosaics from ultra-high resolution unmanned aerial vehicle (UAV) imagery, based on structure from motion (StM) point clouds. *Remote Sensing*, 4, 1392–1410.

Underwood, E., Ustin, S., and DiPietro, D. 2003. Mapping nonnative plants using hyperspectral imagery. *Remote Sensing of Environment*, 86, 150–161.

Underwood, E. C., Klinger, R., and Moore, P. E. 2004. Predicting patterns of non-native plant invasions in Yosemite National Park, California, USA. *Diversity and Distributions*, 10, 447–459.

Underwood, E. C., Ustin, S. L., and Ramirez, C. M. 2007. A comparison of spatial and spectral image resolution for mapping invasive plants in coastal California. *Environmental Management*, 39, 63–83.

USDA NRCS. 2006. United States Department of Agriculture, Natural Resource Conservation Service. *Land Resource Regions and Major Land Resource Areas of the United States, the Caribbean, and the Pacific Basin.* Handbook 206. Washington, DC: United States Department of Agriculture (http://www.nrcs.usda.gov/wps/portal/nrcs/detail/soils/survey/?cid = nrcs142p2_053624, last accessed February 28, 2014).

Ustin, S. L. and Gamon, J. A. 2010. Remote sensing of plant functional types. *New Phytologist*, 186, 795–816.

Ustin, S. L., Valko, P. G., Kefauver, S. C., Santos, M. J., Zimpfer, J. F., and Smith, S. D. 2009. Remote sensing of biological soil crust under simulated climate change manipulations in the Mojave Desert. *Remote Sensing of Environment*, 113, 317–328.

Ustin, S. L., Riaño, D., and Hunt Jr., E. R. 2012. Estimating canopy water content from spectroscopy. *Israel Journal of Plant Science*, 60, 9–23.

Vrieling, A., Rodrigues, S. C., Bartholomeus, H., and Sterk, G. 2007. Automatic identification of erosion gullies with ASTER imagery in the Brazilian Cerrados. *International Journal of Remote Sensing*, 28, 2723–2738.

Wang, C., Zhou, B., and Palm, H. L. 2008. Detecting invasive sericea lespedeza (*Lespedeza cuneata*) in Mid-Missouri pastureland using hyperspectral imagery. *Environmental Management*, 41, 853–862.

Wang, C., Bentivegna, D. J., Smeda, R. J., and Swanigan, R. E. 2010a. Comparing classification approaches for mapping cut-leaved teasel in highway environments. *Photogrammetric Engineering & Remote Sensing*, 76, 567–575.

Wang, C., Jamison, B. E., and Spicci, A. A. 2010b. Trajectory-based warm season grassland mapping in Missouri prairies with multi-temporal ASTER imagery. *Remote Sensing of Environment*, 114, 531–539.

Wang, C., Fritschi, F. B., Stacey, G., and Yang, Z. 2011a. Phenology-based assessment of perennial energy crops in North American Tallgrass Prairie. *Annals of the Association of American Geographers*, 101, 741–751.

Wang, C., Hunt Jr., E. R., Zhang, L., and Guo, H. 2013. Spatial distributions of C_3 and C_4 grass functional types in the U.S. Great Plains and their dependency on inter-annual climate variability. *Remote Sensing of Environment*, 128, 90–101.

Wang, K., Franklin, S. E., Guo, X., and Cattet, M. 2010c. Remote sensing of ecology, biodiversity and conservation: A review from the perspective of remote sensing specialists. *Sensors*, 10, 9647–9667.

Wang, L., Hunt Jr., E. R., Qu, J., Hao, X., and Daughtry, C. S. T. 2011b. Estimating dry matter content of fresh leaves from the residuals between leaf and water reflectance. *Remote Sensing Letters*, 2, 137–145.

Ward, D. P. and Kutt, A. S. 2009. Rangeland biodiversity assessment using fine scale on-ground survey, time series of remotely sensed ground cover and climate data: An Australian savanna case study. *Landscape Ecology*, 24, 495–507.

Washington-Allen, R. A., Van Niel, T. G., Ramsey, R. D., and West, N. E. 2004. Remote sensing-based piosphere analysis. *GIScience & Remote Sensing*, 41, 136–154.

Washington-Allen, R. A., West, N. E., Ramsey, R. D., and Efromson, R. A. 2006. A protocol for retrospective remote sensing-based ecological monitoring of rangelands. *Rangeland Ecology & Management*, 59, 19–29.

Watts, A. C., Ambrosia, V. G., and Hinckley, E. A. 2012. Unmanned aircraft systems in remote sensing and scientific research: Classification and considerations of use. *Remote Sensing*, 4, 1671–1692.

Weber, B., Olehowski, C., Knerr, T., Hill, J., Deutschewitz, K., Wessels, D. C. J., Eitel, B., and Büdel, B. 2008. A new approach for mapping of biological soil crusts in semidesert areas with hyperspectral imagery. *Remote Sensing of Environment*, 112, 2187–2201.

Weber, K. T., Chen, F., Booth, D. T., Raza, M., Serr, K., and Gokhale, B. 2013. Comparing two ground-cover measurement methodologies for semiarid rangelands. *Rangeland Ecology & Management*, 66, 82–87.

Westoby, M. J., Brasington, J., Glasser, N. F., Hambrey, M. J., and Reynolds, J. M. 2012. 'Structure-from-Motion' photogrammetry: A low-cost, effective tool for geosciences applications. *Geomorphology*, 179, 300–314.

Williamson, J. C., Bestelmeyer, B. T., and Peters, D. P. C. 2012. Spatiotemporal patterns of production can be used to detect state change across an arid landscape. *Ecosystems*, 15, 34–47.

Winslow, J. C., Hunt Jr., E. R., and Piper, S. C. 2003. The influence of seasonal water availability on global C_3 vs C_4 grassland biomass and its implications for climate change research. *Ecological Modelling*, 163, 153–173.

Wood, E. M., Pidgeon, A. M., Radeloff, V. C., and Keuler, N. S. 2012. Image texture as a remotely sensed measure of vegetation structure. *Remote Sensing of Environment*, 121, 516–526.

Wood, E. M., Pidgeon, A. M., Radeloff, V. C., and Keuler, N. S. 2013. Image texture predicts avian density and species richness. *PLoS-ONE*, 8, e63211.

Yang, C. and Everitt, J. H. 2010a. Comparison of hyperspectral imagery with aerial photography and multispectral imagery for mapping broom snakeweed. *International Journal of Remote Sensing*, 31, 5423–5438.

Yang, C. and Everitt, J. H. 2010b. Mapping three invasive weeds using airborne hyperspectral imagery. *Ecological Informatics*, 5, 429–439.

Yang, C., Everitt, J. H., and Johnson, H. B. 2009. Applying image transformation and classification techniques to airborne hyperspectral imagery for mapping Ashe juniper infestations. *International Journal of Remote Sensing*, 30, 2741–2758.

Yang, C., Everitt, J. H., and Fletcher, R. S. 2013. Evaluating airborne hyperspectral imagery for mapping saltcedar infestations in west Texas. *Journal of Applied Remote Sensing*, 7, 073556-1.

Yu, Q., Gong, P., Clinton, N., Biging, G., Kelly, M., and Schirokauer, D. 2006. Object-based detailed vegetation classification with airborne high spatial resolution remote sensing imagery. *Photogrammetric Engineering & Remote Sensing*, 72, 799–811.

Zavaleta, E. S., Pasari, J. R., Hulvey, K. B., and Tilman, G. D. 2010. Sustaining multiple ecosystem functions in grassland communities requires higher biodiversity. *Proceedings of the National Academy of Sciences*, 107, 1443–1446.

Zengeya, F. M., Mutanga, O., and Murwira, A. 2013. Linking remotely sensed forage quality estimates from WorldView-2 multispectral data with cattle distribution in a savanna landscape. *International Journal of Applied Earth Observation and Geoinformation*, 21, 513–524.

Zhang, X., Friedl, M. A., Schaaf, C. B., Strahler, A. H., Hodges, J. C. F., Gao, F., Reed, B. C., and Huete, A. 2003. Monitoring vegetation phenology using MODIS. *Remote Sensing of Environment*, 84, 471–475.

Ziska, L. H., Blumenthal, D. M., Runion, G. B., Hunt Jr., E. R., and Diaz-Soltero, H. 2011. Invasive species and climate change: An agronomic perspective. *Climatic Change*, 105, 13–42.

12

Characterization, Mapping, and Monitoring of Rangelands: Methods and Approaches

Lalit Kumar
University of New England

Priyakant Sinha
University of New England

Jesslyn F. Brown
U.S. Geological Survey

R. Douglas Ramsey
Utah State University

Matthew Rigge
InuTec, Contractor to U.S. Geological Survey

Carson A. Stam
Utah State University

Alexander J. Hernandez
Utah State University

E. Raymond Hunt, Jr.
USDA ARS Beltsville Agricultural Research Center

Matthew C. Reeves
USDA Forest Service

Acronyms and Definitions

ANN	Artificial neural network
ANPP	Annual net primary production
ASTER	Advanced spaceborne thermal emission and reflection radiometer
AVHRR	Advanced very high resolution radiometer
AVIRIS	Airborne visible infrared imaging spectrometer
BPS	Biophysical settings
CASI	Compact airborne spectrographic imager
CFFDRS	Canadian Forest Fire Danger Rating System
CRLRS	Cottonwood Range and Livestock Research Station
DEM	Digital elevation model
DT	Decision tree
ES	Ecological sites
ETM	Enhanced thematic mapper
EVI	Enhanced vegetation index
FA	Factor analysis
FAPAR	Fraction of absorbed photosynthetically active radiation
FBFM	Fire behavior fuel models
FRAGSTAT	Spatial Pattern Analysis Program for Categorical Maps
GIS	Geographic Information Systems
GLOVIS	U.S. Geological Survey Global Visualization Viewer
GPP	Gross primary productivity
GS	Gramm–Schmidt
GVI	Green vegetation index
HSI	Hue saturation intensity
HMM	Hidden Markov model
LAI	Leaf area index
LiDAR	Light detection and ranging
LSWI	Land surface water index
MAE	Mean absolute error

MESMA	Multiple endmember spectral mixture analysis
MDM	Minimum distance to mean
MLC	Maximum likelihood classification
MODIS	Moderate resolution imaging spectroradiometer
NAIP	National Agricultural Imagery Project
NDSVI	Normalized difference soil-adjusted vegetation index
NDVI	Normalized difference vegetation index
NDWI	Normalized difference water index
NIR	Near-infrared
NPP	Net primary productivity
PAR	Photosynthetically active radiation
PCA	Principal component analysis
PCC	Pearson's correlation coefficients
PG	Gross photosynthesis
PRSIM	Parameter-elevation regressions on independent slopes model
PVI	Perpendicular vegetation index
RED	Red reflected radiant flux
RF	Random forest
RMSE	Root mean square error
RVI	Ratio vegetation index
SAC	Spectral angle classifier
SAVI	Soil-adjusted vegetation index
SMA	Spectral mixture analysis
SMU	Soil map units
sNDVI	Scaled normalized difference vegetation index
SPOT	Satellite Pour l'Observation de la Terre
SSAC	Supervised spectral angle classifier
SSURGO	Soil Survey Geographic Database
SWRGAP	South west regional gap
TC	Tasselled cap
TM	Thematic mapper
TNC	The Nature Conservancy
TRMI	Topographic relative moisture index
UDWR	Utah Division of Wildlife Resources
USAC	Unsupervised spectral angle classifiers
USGS	U.S. Geological Survey
UTM	Universal transverse mercator
VCF	Vegetation continuous fields
VI	Vegetation index
WGS	World Geodetic System

12.1 Introduction

While there are many definitions of rangeland, the central theme of all these is that it is land on which the dominating vegetation is mainly grasses, grass-like plants, forbs, shrubs, and isolated trees. Rangelands include shrublands, natural grasslands, woodlands, savannahs, tundra, and many desert regions. A distinguishing factor of rangelands from pasture lands is that they grow primarily native vegetation, rather than plants established by humans. Rangelands are also managed mainly through extensive practices such as managed livestock grazing and prescribed fire rather than more intensive agricultural practices and the use of fertilizers. Rangelands worldwide are known to provide a wide range of desirable goods and services, including but not limited to livestock forage, wildlife habitat, wood products, mineral resources, water, and recreation space. Large populations depend on rangelands for their livelihoods, hence effective monitoring and management is crucial for sustainable production, health, and biodiversity of these systems.

Effective monitoring of rangelands has proven logistically and statistically difficult using field-based monitoring methods alone due to the sheer size, range, and complexity of rangelands. Mapping and monitoring of rangelands, especially those in a disturbed state or under rapid change, requires data that are extensive, accurate, timely, and with regular repeat coverage. All this makes remote sensing an ideal platform for rangeland monitoring as recent developments in sensor capabilities means we have repeat coverage from multiple satellites, spectral resolutions sufficient to distinguish many rangeland vegetation species and communities, spatial resolutions allowing monitoring and management at micro-scales and costs that are a small fraction of a few decades ago. Remote sensing data are more easily available and the systems cover almost the entire world, and certainly all the regions where rangelands occur.

There is a plethora of literature available that describe various uses of remote sensing data for rangeland monitoring, ranging from mapping species distribution, biomass, degradation, woody cover, net primary production, biodiversity, change detection, fuel loads, fire extents and frequency, invasive species encroachment, livestock foraging, etc. New methods of image classification and interpretation are regularly published, as well as novel techniques of incorporating satellite data with ancillary data to better understand rangeland dynamics. Some of these applications have been discussed in the previous two chapters, with Reeves et al. (Chapter 10) looking at the relationship between rangeland productivity and climate, food security, fire, and rangeland degradation, and Hunt et al. (Chapter 11) exploring rangeland biodiversity using different spectral, spatial, and temporal satellite data.

The focus of this chapter is to present and discuss methods and approaches used in the mapping and monitoring of rangelands. We do this by presenting characterization, mapping, and monitoring of rangelands in specific applications such as (1) phenology and productivity studies; (2) fuel analysis; (3) biodiversity and gap analysis; (4) vegetation continuous fields; and (5) change detection analysis. We summarize some oft-used traditional means of mapping and monitoring rangelands but concentrate on newer developments in this field.

12.2 Rangeland Monitoring Methods Using Vegetation Indices

12.2.1 Rangeland Phenology

Rangeland phenology for vegetation types (shrublands, grasslands, steppes, deserts, and woodlands) is affected by environmental drivers (temperature, precipitation, and sunshine) and factors such as topography (elevation, slope, and aspect), edaphic conditions (variations in soil type, texture, and nutrients), and latitude. Rangeland vegetation growth is the result of overall influence of all environmental drivers and factors and their interaction, and responds more rapidly to the environmental variations as compared to other kinds of vegetation (Reed et al., 1994). Since rangelands are mainly located in dry areas characterized by low and variable annual rainfall (Grice and Hodgkinson, 2002; Tussie, 2004), precipitation regime has a much more significant influence on rangeland vegetation among all the environmental drivers and factors (Reed et al., 1994). They usually respond to precipitation in a pulsed way where their phenology is dependent on discrete rainfall events in terms of productivity, density, and abundance (Rauzi and Dobrenz, 1970). Temperature has also been observed to have direct influence on phenological phases and a large number of studies have been conducted to determine the effects of temperature on the phonological timings of plants (Badeck et al., 2004; Sparks et al., 2000). Livestock grazing is another notable factor that influences rangeland phenology and has impact on rangeland vegetation (Desalew et al., 2010). The grazing-induced vegetation change is dependent on the type of livestock and the composition of vegetation types and hence an important consideration in monitoring or predicting rangeland plant phenology.

The advent of remote sensing technology induced great changes in vegetation phenology studies by providing temporal data at regular intervals. Time series data have been used to predict phenophase in terms of onsets and offsets of the vegetation growing season as well as budburst, flowering, or leaf color–changing dates. A simple way to predict the onset and/or offset of the growing season is to analyze the time series of vegetation indices (VIs). There have been many methods to determine the onset/offset dates, including thresholds, maximum rate of change, or a certain percentage of the greatest VI increase. For example, Rigge et al. (2013) evaluated the productivity and phenology of western South Dakota mixed-grass prairie in the period from 2000 to 2008 using the normalized difference vegetation index (NDVI) derived from moderate resolution imaging spectroradiometer (MODIS) data. They used growing season NDVI images on a weekly basis to produce time-integrated NDVI, a proxy of total annual biomass production, and also integrated seasonally to represent annual production by cool and warm season species (C_3 and C_4, respectively). Heumann et al. (2007) studied phenological change in the Sahel and Soudan, Africa, from 1982 to 2005. They used TIMESAT software to estimate phenological parameters from the advanced very high resolution radiometer (AVHRR) NDVI data set and have found significant positive trends for the length of the growing and end of the growing season for the Soudan and Guinean regions. Kumar et al. (2002) showed that soil type has a significant impact on the early season growth variation of annual vegetation on sandy and clay soils, a fact that is utilized in the movement patterns of graziers in the Sahel.

12.2.2 Vegetation Indices in Rangeland Monitoring

12.2.2.1 What to Measure?

To be effective estimators of biomass, leaf area index (LAI), or percentage cover, spectral indices must be able to differentiate vegetation features from soil features (Todd et al., 1998). For such differentiation, it is a requirement that the soils and vegetation have different reflectance patterns and the spectral index should be sensitive enough to detect the differences. Green vegetation has characteristically low reflectance in the visible portion of the spectrum (lowest in red portion of the spectrum) with a sharp increase in reflectance in the near-infrared portion (Figure 12.1). Most of the commonly used VIs exploit this expected difference in near-infrared and red reflectance for vegetation discrimination and in separating them from nonvegetated areas. For example, the NDVI, computed as NDVI = NIR − Red/NIR + Red, has been used in a wide range of practical remote sensing applications (e.g., Tucker et al., 1985). The NDVI values range between −1 and +1, with dense vegetation having a high NDVI while soil values are low but positive, and water is negative due to its strong absorption of NIR. Tucker (1979) tested various combinations of the red, NIR, and green bands to predict biomass, water and chlorophyll content of grass plots. A strong correlation was observed between NDVI values and chlorophyll content and crop characteristics such as green biomass and leaf water content. Sellers (1985) used a canopy radiative transfer model to show that NDVI is near-linearly related to area-averaged net carbon assimilation and plant transpiration, even at different values of fractional vegetation cover (fc) and LAI over an area of interest.

The TM Tasseled Cap green vegetation index (GVI) is a linear combination of the six reflecting wavebands of Landsat TM (Crist, 1983). The GVI coefficients with the highest values are for the red (negatively loaded) and the near-infrared (positively loaded) wavebands.

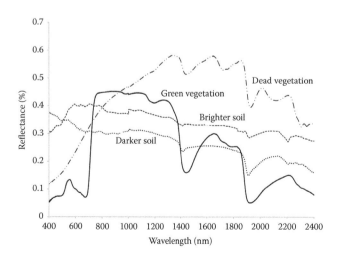

FIGURE 12.1 Idealized representation of spectral reflectance curves for green and dead vegetation, and light and dark soils.

12.2.2.2 Soil Reflectance Variation

As compared to vegetation, soil reflectance patterns are usually quite different and generally increase linearly with increasing wavelength: from visible to near-infrared to mid-infrared (Figure 12.1). Unlike vegetation, soil usually has high reflectance in the visible wavelengths and low reflectance in the near-infrared wavelengths. Figure 12.1 illustrates the contrast in reflectance patterns between dark (low reflecting) and light (high reflecting) soils. The difference between reflectance of soil and vegetation, in case of high-reflecting soils, can be small in the near-infrared wavelengths.

Soil reflectance properties vary considerably with soil type, texture, moisture content, organic matter content, color, and the presence of iron oxide (Hoffer, 1978). Low-reflecting soils are usually dark, high in organic matter or moisture, containing iron oxides, and/or coarse textured. Dry soils show high reflectance values in visible, near-infrared, and mid-infrared regions of the spectrum. In contrast, wet soils show low-reflectance values for these regions (Bowers and Hanks, 1965; Hoffer and Johannes, 1969).

12.2.2.3 Soil Background Impacts on Spectral Indices

In rangelands and semiarid regions, background soil effects lead to soil–vegetation spectral mixing, a major concern in vegetation identification from spectral indices using remote sensing data. Soil, plant, and shadow reflectance components mix interactively to produce composite reflectance (Richardson and Wiegand, 1990). The knowledge on soil reflectance variations for different soil types and conditions and their interaction with vegetation reflectance is essential to differentiate the soil and vegetation reflectance patterns. This helps in assessing the potential effectiveness of remote sensing techniques to map spatial distribution of plant species or communities and estimate biomass.

Generally, the values of ratio-based indices, such as the normalized vegetation index, tend to increase with dark or low-reflecting soil backgrounds (Elvidge and Lyon, 1985; Huete et al., 1985; Todd and Hoffer, 1998). However, a few studies have shown that the Tasseled Cap GVI decreases with low-reflecting soil backgrounds (Huete et al., 1985; Huete and Jackson, 1987). The influence of soil type and moisture content on vegetation index was found less for Landsat TM derived GVI as compared to NDVI when estimating percentage green vegetation cover, based on a two component soil and vegetation model (Todd and Hoffer, 1998). Where soil effects on NDVI are a problem, alternative VIs such as the soil-adjusted vegetation index (SAVI) (e.g., Huete, 1988) or the scaled difference vegetation index (Jiang et al., 2006) can be used.

12.2.2.4 Vegetation Reflectance Variation

The relationship between biomass and spectral indices can also be affected by vegetation condition, distribution, and structure (Todd et al., 1998). The loss in chlorophyll content due to drying of vegetation alters spectral reflectance characteristics in the visible and infrared regions. This phenomenon is a very common occurrence in semiarid rangelands such as the shortgrass steppe (Todd et al., 1998). In case of drying vegetation, the reflectances in both the visible and in the mid-infrared regions of the spectrum increase significantly. The reflectance patterns of dead or dry plant material are more similar to soil than to healthy green vegetation (Hoffer, 1978). Both dry vegetation and most soils have high reflectance in the visible and mid-infrared regions (Hoffer and Johannsen, 1969) and hence the spectral similarities introduce difficulties in remote sensing applications in rangelands.

For a given region, if dry or senescent biomass forms a significant portion of total vegetation present, the spectral distinction between vegetation and soil background is altered (Todd et al., 1998). For bright soil, spectral bands such as Red or indices such as Tasseled Cap brightness index, which are responsive to scene brightness, should help in vegetation discrimination in the presence of dry and/or senescent vegetation that appear less bright than the soil background.

Graetz et al. (1988) found that the RED index, which responds to surface brightness, estimated vegetation cover on high-reflecting soils more accurately than VIs because dry vegetation was less bright than the soil background.

The grazing pattern and intensity also change the relative proportions of the standing dead and green biomass (Sims and Singh, 1978) and cover characteristics (Milchunas et al., 1989) of plant communities within the rangelands. With increasing grazing intensity, the amount of standing dead plant material as well as litter decreases; however, in the absence of grazing, there will be considerable amount of both green and dry vegetation present. The presence of dead and/or drying plant material causes spectral confusion between this dead/drying plant material and the soil background and hence creates difficulties in estimating biomass using remote sensing techniques. Therefore, comparisons between grazed and ungrazed sites can provide insights into the impacts of senescence on biomass detection. In addition, grazing intensity varies considerably with topography (Milchunas et al., 1989), therefore information on the effectiveness of various remote sensing indices in detecting biomass under different grazing intensities could be useful in developing models across landscapes with heterogeneous grazing patterns. The field methods used to sample vegetation may also affect possible relationships between remote sensing indices and biomass.

12.2.2.5 Overview of VIs

VIs combine reflectance measurements from different portions of the electromagnetic spectrum to provide information about vegetation cover on the ground (Campbell, 1996). Healthy green vegetation has distinctive reflectance in the visible and near-infrared regions of the spectrum. In the visible region, and in particular red wavelengths, plant pigments strongly absorb the energy for photosynthesis, whereas in the near-infrared region the energy is strongly reflected by the internal leaf structures. This strong contrast between red and near-infrared reflectance has formed the basis of many different VIs. When applied to multispectral remote sensing images, these indices involve numeric combinations of the sensor bands that record land surface reflectance at various wavelengths.

One of the most promising applications of satellite data is the estimation of net primary productivity over time and space. The use of satellite-derived VIs has been useful for estimating net productivity (Cihlar et al., 1991). VIs are not a direct measure of biomass or primary productivity, but establish empirical relationships with a range of vegetation/climatological parameters (Weiser et al., 1986) such as (1) fraction of absorbed photosynthetically active radiation (FAPAR), (2) LAI, and (3) net primary productivity (NPP). They are simple to understand and their implementation is fast as most of them use spectral band values in a mathematical formulation (e.g., ratio, difference, etc.) While maintaining sensitivity to vegetation temporal characteristics and seasonality, most of these indices reduce sensitivity to topographic effects, soil background, view/sun angle and atmosphere to some extent. While many VIs have

been developed over the years, Table 12.1 lists a few of these and provides references for further investigation.

12.2.2.6 VIs as Proxies for Other Canopy Attributes

12.2.2.6.1 VIs and LAI

The LAI is an estimate of the total leaf area per unit area (Glenn et al., 2008). LAI links VIs to photosynthesis through the absorbed photosynthetically active radiation (PAR) (Tucker and Sellers, 1986). When light travels through a series of leaves, the transmission and therefore reflectance in near infrared of the electromagnetic radiation decreases (Hofer, 1978). On the other hand, the uppermost leaf layers of a green vegetation canopy strongly absorb the red portion of the spectrum and therefore reduce their transmittance to successive leaf layers. The relation between LAI and VIs varies among vegetation types (Peterson et al., 1987). In the case of saturated LAI, the ability to estimate

TABLE 12.1 Vegetation Indices Commonly Used in Rangeland Studies

	Vegetation Index	Formula
1.	SR—simple ratio	$SR = \dfrac{\rho_{NIR}}{\rho_R}$
2.	EVI—environmental vegetation index (Birth and McVey, 1968)	$EVI = \dfrac{\rho_{IR}}{\rho_R}$
3.	NDVI—normalized difference vegetation index (Rouse et al., 1974)	$NDVI = \dfrac{\rho_{IR} - \rho_R}{\rho_{IR} + \rho_R}$
4.	PVI—perpendicular vegetation index (Richardson and Wiegand, 1977)	$PVI = \sqrt{\left(\rho_{R_{soil}} - \rho_{R_{veg}}\right)^2 + \left(\rho_{IR_{soil}} - \rho_{IR_{veg}}\right)^2}$
5.	SAVI—soil-adjusted vegetation index (Huete, 1988)	$SAVI = (1 + L)\dfrac{\left(\rho_{IR} - \rho_R\right)}{\left(\rho_{IR} + \rho_R + L\right)}$
		L is a correction factor whose values range from 0 (high vegetation cover) to 1 (low vegetation).
6.	TSAVI—transformed SAVI (Baret et al., 1989)	$TSAVI = \dfrac{a^*\left(\rho_R - a^*\rho_{IR} - b\right)}{\left(\rho_R + a^*\rho_{IR} - a^*b\right)}$
7.	Kauth–Thomas transformation (Tasseled cap, K–T) (Kauth and Thomas, 1976)	$GVI = -0'2728(TM1) - 0'2174(TM2) - 0'5508(TM3) + 0'722(TM4) + 0'0733(TM5) - 0'1648(TM7)$
8.	GNDVI—green NDVI (Gitelson et al., 1996)	$NDVI_g = \dfrac{\left(\rho_{NIR} - \rho_{green}\right)}{\left(\rho_{NIR} + \rho_{green}\right)}$
9.	GEMI—normalized difference vegetation index (Pinty and Verstraete, 1992)	$GEMI = \eta\left(1 - 0.25\eta\right)\dfrac{\left(\rho_R - 0.125\right)}{\left(1 - \rho_R\right)}$
10.	EVI—enhanced vegetation index (MODIS) Huete et al., 2002)	$EVI = G\dfrac{\rho_{NIR} - \rho_{red}}{\rho_{NIR} + C_1 * \rho_{red} - C_2 * \rho_{blue} + L}$
		L is the canopy background adjustment that addresses nonlinear, differential NIR and red radiant transfer through a canopy, and C_1, C_2 are the coefficients of the aerosol resistance term, which uses the blue band to correct for aerosol influences in the red band.
11.	II—infrared index (Hardisky et al., 1983)	$II = \dfrac{\left(TM4 - TM5\right)}{\left(TM4 + TM5\right)}$
12.	MSI—moisture stress index (Rock et al., 1985)	$MSI = \dfrac{TM5}{TM4}$

biomass using VIs is constrained; however, in rangelands, such as the shortgrass steppe, LAI saturation is improbable.

LAI is derived mathematically and has no direct relationship to fPAR or processes that depend on fPAR (Monteith and Unsworth, 1990). LAI is related to light interception by a canopy (Ri) by

$$Ri = Rs[1 - \exp(-k \, LAI)]$$

where k is a factor that accounts for leaf angles and other factors that affect absorption of Rs within a canopy (Monteith and Unsworth, 1990). Vertical leaved (erectophiles) plants typically absorb less light per unit leaf area than plants with relatively horizontal leaves (planophiles). Within a stand, the coefficient k also varies with respect to plant arrangements. Single-stand plants receive light from all sides of their canopy as compared to dense stands of plants that receive light from the top of the canopy. The fraction of light absorbed by the canopy (fPAR) depends not only on Ri but also on the spectral properties of the leaves. The reflective surfaces present in some leaves reduce the heat gain while other surfaces absorb nearly all of the incident radiation between the visible bands (400 and 700 nm). Therefore, a correlation can easily be established between LAI and NDVI for single-plant species grown under uniform conditions (Asrar et al., 1985). However, the same is not the case with mixed canopies, often the case in remote sensing data, in which a single pixel contains several landscape units.

12.2.2.6.2 VIs and Fractional Cover (fc)

Models use fractional vegetation cover to divide the landscape into areas of vegetation and bare soil (Anderson, 1997; Glenn et al., 2007; Kustas and Norman, 1996; Timmerman et al., 2007). Different methods are used in these models to estimate carbon and moisture fluxes from vegetation and bare soil and the fraction of the landscape that is vegetated (Glenn et al., 2008). Typically, a landscape unit is partitioned into bare soil and vegetation through the use of VIs. For example, NDVI values derived from satellite images and ranging from −1 to +1 are rescaled between 0 and 1 to represent bare soil at values near 0% and 100% vegetation cover at values near to 1 to get fc for a given pixel or area of interest in the scene. Depending on the models used, the scaling is done linearly or in a nonlinear way to represent the vegetation type of interest. Some models require both LAI and fc often estimated by VIs (Anderson, 1997). In a few cases, ground-based information relating to vegetation species and canopy characteristics are also included to improve the estimates. Sometimes, average leaf angles for a particular type of landscape are used to predict both fc and LAI from VIs (Anderson, 1997).

For partially vegetated scenes with LAI in the range of 1–3, Carlson and Ripley (1997) found VIs to be much more closely related to fc than to LAI in case of clumped vegetation, and the relationships between NDVI and fc to be nonlinear. They showed that, for partially vegetated scenes of uniform vegetation type, VIs were a good measure of fc. Other studies on a variety of landscape types have also reported strong linear (Ormsby et al., 1987)

or nonlinear (Li et al., 2005) relationships between VIs and fc. However, at 100% cover, different plant species may have different VIs due to differences in chlorophyll content and canopy structure, and thus create a potential practical problem in using VIs to estimate fc over mixed scenes. Amiri and Shariff (2010), in their study of vegetation cover assessment in semiarid rangelands of Iran, used 26 different VIs to determine suitable indices for vegetation cover and production assessment, and found a significant relationship between NDVI derived from ASTER data with the vegetation cover. In another study, Ajorlo and Abdullah (2007) examined four VIs (NDVI, SAVI, perpendicular vegetation index [PVI]), and ratio vegetation index [RVI]) to assess rangeland degradation in semiarid parts of the Qazvin province, Iran. The results showed that NDVI was a more powerful index for assessing the rangeland degradation as compared to other indices. Jianlong et al. (1998) used NDVI and RVI in grassland study and found good correlation between fresh herbage yields and RVI and NDVI (P < 0.01) in four grassland types with correlation coefficient (r) > 0.679. Fresh herbage yields correlated better with RVI than with NDVI for lowland meadow, hill desert steppe, and mountain meadow, but not for plains desert steppe. Guo et al. (2005) monitored grassland health with remote sensing approaches and assessed the effectiveness of remote sensing in grassland monitoring. They found that it was challenging to use remotely sensed data in mixed grasslands because the large proportion of dead material complicated analysis for indices that were not developed for heterogeneous landscapes, especially in conservation areas. They investigated the relationship between remote sensing data and grassland biophysical measurement, including aboveground biomass and plant moisture content, in the native mixed prairie ecosystem with its high litter component. Their results indicated that the NDVI was not suitable for biomass estimation although a moderate relationship was found between NDVI and plant moisture content. Compared to NDVI, LAI provided promising results on both biomass and plant moisture content estimation. John et al. (2008) evaluated the utility of MODIS-based productivity (GPP and EVI) and surface water content (NDSVI and LSWI) in predicting species richness in the semiarid region of Inner Mongolia, China. They found that these metrics correlated well with plant species richness and could be used in biome- and life form–specific models.

Although different VIs are used in assessing the rangeland degradation, there are still challenges facing the classification of vegetation species in degraded areas where the reflectance is strongly affected by the soil background as a result of relatively sparse vegetation and atmospheric conditions.

12.2.3 Case Study: Rangeland Phenology and Productivity in the Northern Mixed-Grass Prairie, North America

12.2.3.1 Introduction

Rangelands across the northern plains of North America are predominantly mixed-grass prairie communities and can be dominated by either C_3 (cool season) or C_4 (warm season) grass species. These mixed-grass communities exhibit different growth dynamics or phenological patterns that can be detected from satellite remote sensing (Rigge et al., 2013; Tieszen et al., 1997). Although remote sensing cannot detect traditional phenological events such as budding and flowering in individual plants (Tieszen et al., 1997; White et al., 2009), it can detect important landscape-level phenological measures such as the onset of the growing season, the end of the growing season, rate of green-up, and peak vegetation vigor based on satellite image time series over a growing season (Kovalskyy and Henebry, 2012; Reed et al., 1994; Tao et al., 2008; van Leeuwen et al., 2010). In this case study, conducted in the Bad River watershed in western South Dakota, USA, the spatial and temporal dynamics of rangelands as measured by remote sensing indicators or "phenological metrics" varied related to climate, management, and plant photosynthetic pathway (Rigge et al., 2013).

Land management, livestock grazing, invasive species, and prolonged droughts have the potential to change rangeland plant community structure and therefore contribute to altered phenological patterns (Foody and Dash, 2007; Tieszen et al., 1997). These factors also have the potential to modify ecosystem goods and services (Bradley and Mustard, 2008). One benefit in monitoring rangelands is to detect and mitigate any damaging trends caused by land management (Paruelo and Lauenroth, 1998).

Biomass production in rangelands by C_3 and C_4 plants displays significantly different phenological timing that is detectable utilizing satellite remote sensing. Information on rangeland phenological dynamics has been applied to assessing rangeland health monitoring due to inferences that can be made about community composition and presence of invasive species (Boyte et al., 2014; Reed et al., 1994; Rigge et al., 2013; Tateishi and Ebata, 2004). For example, rangeland plant communities dominated by either C_3 or C_4 species can be identified through their unique and asynchronous phenological time series signals (Foody and Dash, 2007). Grasses with C_3 pathways are most active during the cooler spring and fall seasons, while many C_4 grasses are adapted to the hot and dry summer months (Foody and Dash, 2007; Tieszen et al., 1997; Wang et al., 2010). State and transition models indicate that cool season (C_3) grasses tend to dominate historic plant climax communities in many ecological sites (ES) of the northern mixed prairie, while shortgrass (C_4) species generally increase under disturbance such as heavy, continuous, season-long grazing (U.S. Department of Agriculture, 2008). Furthermore, phenological differences are useful for identifying vegetation types. For example, in western South Dakota, riparian vegetation tends to be dominated by C_4 species while upland vegetation communities consist primarily of C_3 plants. In late summer, this difference allows for the detection of a clear riparian vegetation signal (Kamp et al., 2013).

In the northern mixed-grass prairie, the majority of C_3 production occurs in spring and fall, and most C_4 production occurs in summer, although there is both spatial and temporal overlap in production (Ode et al., 1980). For example, both C_3 and C_4 plants actively produce biomass in mid- to late June with no clear separation between the timing of their production (Ode et al.,

1980; USDA 2008). In midsummer (July 1–August 31), production is typically dominated by C_4 grasses in the study area and was therefore used to define the warm season period (Ode et al., 1980; Wang et al., 2010; White, 1983).

12.2.3.2 Methods

12.2.3.2.1 Study Site

The Bad River watershed of western South Dakota (~lat 45°N, long 101°W) is dominated by the Clayey ecological site description in the Major Land Resource Area classification (U.S. Department of Agriculture, 2008). The topography of this region is generally typified by long, smooth slopes, with steeper slopes along well-defined waterways. Bedrock throughout the watershed is Pierre Shale, resulting in soils with a high clay content and low permeability. The climate is semiarid, receiving an average of 398 mm precipitation annually over the 2000–2008 period of which 80% occurred during the growing season of April to September. Annual precipitation is highly variable, with drought and insufficient moisture common. The daily mean temperature ranges from 32°C in July to –14°C in January, with a yearly mean of 8°C (Smart et al., 2007). Analysis was constrained to pixels classified in the National Land Cover Database 2006 as herbaceous cover (Fry et al., 2011).

The study area vegetation is mixed-grass rangeland dominated by C_3 grasses including western wheatgrass (*Pascopyrum smithii Rybd.*) and green needlegrass (*Stipa viridula Trin. & Rupr.*; Smart et al., 2007). Shortgrasses (C_4) include buffalograss (*Bouteloua dactyloides Nutt.*) and blue grama (*Bouteloua gracilis H.B.K.*). Midgrasses are typically C_4 species such as little bluestem (*Schizachyrium scoparium [Michx.] Nash*) and sideoats grama (*Bouteloua curtipendula [Michx.] Torr.*). Little forb and succulent cover exists (Sims and Singh, 1978; USDA, 2008).

12.2.3.2.2 Input Data

MODIS satellite time series data provided an ideal balance between spatial resolution and temporal repeat, and are therefore well-suited for phenology studies across relatively large regions. This study is based on 9 years (2000–2008) of eMODIS weekly composite Terra MODIS imagery at 250 m resolution (Jenkerson, 2010). NDVI values were calculated and rescaled to a range of 0–200 to simplify calculations by eliminating negative values. Hereafter, rescaled NDVI will be referred to as sNDVI (scaled NDVI).

Annual variability in rangeland phenology and productivity as shown in MODIS sNDVI time series plots (Figure 12.2a) is fairly common (Tieszen et al., 1997), where complex herbaceous communities respond to highly variable inter- and intraannual precipitation (Lauenroth and Sala, 1992; Smart et al., 2007).

12.2.3.2.3 Phenological Analysis

Several measures calculated on a pixel by pixel basis were utilized in this study to describe the phenology of rangeland communities (Table 12.2). The start of the season was calculated for each pixel as the first point in time each year when sNDVI reached

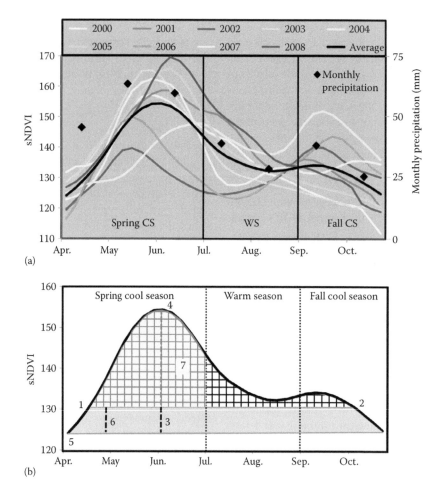

FIGURE 12.2 (a) Phenological profiles for Bad River watershed, South Dakota, USA. from 2000 to 2008 where years are colored according to growing season precipitation (shown on the secondary vertical axis). Colors range from red (for dryer growing seasons to blue (for relatively wetter growing seasons). The left third of the graph shows the spring cool season (CS), the middle shows the warm season (WS), and the right thirds shows the fall CS. (b) Illustration of idealized phenological profile and main phenological indicators used in the study. The entire gridded area corresponds to the growing season time integrated NDVI (TIN) and the blue gridded area corresponds to the spring CS integrated NDVI. (Based on Figures 1 and 3 in Rigge, M. et al., *Rangel. Ecol. Manage.*, 66, 579, 2013.)

TABLE 12.2   Phenological Measures Calculated from Input MODIS Time Series Profiles

Remote Sensing Phenological Metric	Description/Notes
1. SOS—start of season	Time of first occurrence of sNDVI $\geq$20% of growing season amplitude (within a calendar year)
2. EOS—end of season	Time of last occurrence of sNDVI $\geq$20% growing season amplitude (within a calendar year)
3. AMP—amplitude	Difference between the growing season minimum sNDVI and the growing season peak NDVI
4. sNDVI$_{peak}$	Value of peak growing season NDVI
5. sNDVI$_{min}$	Value of minimum growing season NDVI
6. Baseline	Values of $\leq$20% of growing season amplitude (within a calendar year) that are eliminated
7. TIN$_{GS}$	Summation of sNDVI from April 1–October 1 where sNDVI is $\geq$20% of growing season amplitude

20% of the total growing season sNDVI amplitude of that year (vanLeeuwen et al., 2010). The total growing season amplitude was calculated as the difference between the peak sNDVI value (during the April 1–October 31 growing season) and the minimum sNDVI value during this period. Similarly, the end of the season was the time at which the sNDVI value dropped below 20% of the total growing season sNDVI amplitude. This approach better approximated the total seasonal biomass production than simply averaging NDVI across the entire growing season. TIN served as a proxy for growing season total biomass production. Because TIN is influenced by the magnitude and duration of mNDVI values, the saturation effects that occur at larger LAI values (Huete et al., 2002) are minimized. The TIN of the spring cool season (start of season to June 30) was also used in this study (Figure 12.2b).

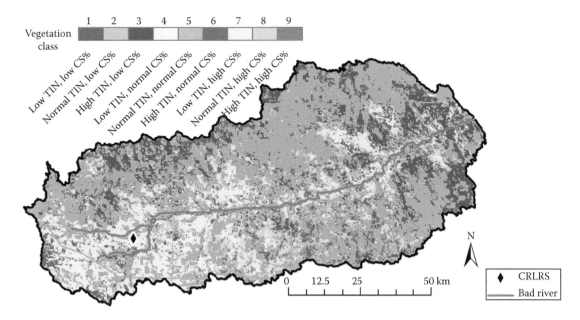

FIGURE 12.3 Nine vegetation classes derived from growing season time-integrated NDVI (TIN) and average cool season percentage (CS%) in the Bad River watershed, South Dakota, USA, shaded grey in the inset map. Light gray areas denote nonrangeland vegetation types that were excluded from this analysis. The Cottonwood Range and Livestock Research Station (CRLRS) is located in the southwest. (Based on Figure 6 in Rigge, M. et al., *Rangel. Ecol. Manage.*, 66, 579, 2013).

Since phenological profiles can reveal community species composition and vegetation communities (Reed et al., 1996; Tateishi and Ebata, 2004) and phenology can strongly influence biomass production (Smart et al., 2007), we combined geospatial data on cool season grasses and TIN to create a watershed-level vegetation map (Figure 12.3). The per-pixel 2000–2008 average TIN was grouped into three classes: (1) within one standard deviation of the study area mean, (2) below one standard deviation of the mean, and (3) above one standard deviation of the mean. A similar approach was used to group the 2000–2008 average cool season percentages into three classes. The TIN and cool season percentage classes were integrated to form nine vegetation classes, representing the average of the 2000–2008 conditions.

Table 12.3 provides vegetation class numbers, generally indicative of the plant community state, with higher numbers closely approaching the western wheatgrass/green needlegrass community (historic climax plant community of the Clayey

ecological site) and lower numbers generally indicative of the blue grama/buffalograss plant community (Smart et al., 2007; U.S. Department of Agriculture, 2008). Classes marked with "+" indicated TIN or CS% over one standard deviation higher than the study area average, "N" indicated TIN or CS% within one standard deviation of the average, and "–" indicated TIN or CS% below one standard deviation of the average.

Spatial patterns of vegetation classes are useful for describing topographic, edaphic, and land management influences on plant communities (Figure 12.3). For example, areas with both low cool season percentage and low TIN (class 1) were likely dominated by buffalograss and blue grama, both low-producing C_4 species. This community was reported to result from strong grazing pressure (Smart et al., 2007; U.S. Department of Agriculture, 2008) in the Clayey ecological site, but might stem from lower than normal warm season moisture in Pierre Shale/badland outcrops.

12.2.3.2.4 Validation Methods

Validation is a critical, yet challenging, component of phenological studies based on remote sensing. Methods were performed based on field plot data and carbon flux tower data. The field plot data were collected for two distinct plant communities (mixed-grass and midgrass-dominated) in fields that were retained under long-term grazing management practices (Dunn et al., 2010). Field-measured vegetation data estimated species composition by weight in late June (peak cool season biomass) and early August (peak warm season biomass). A nonrandom sampling scheme consisted of 14 plots per field represented local variation in soil type, slope, and aspect. The remotely sensed yearly average TIN and cool season percentage

TABLE 12.3 Bad River Watershed Rangeland Vegetation Classes

Vegetation Class	TIN	CS%	Area (%)	Mean TIN	Mean CS%
1	–	–	4.7	268.2	74.5
2	N	–	8.8	329.0	74.7
3	+	–	19.1	389.4	73.7
4	–	N	10.2	279.2	80.3
5	N	N	12.8	327.3	80.0
6	+	N	10.3	376.7	79.8
7	–	+	18.1	270.1	85.7
8	N	+	11.5	325.0	85.0
9	+	+	4.5	370.7	84.7

for each field were regressed against the corresponding field-measured average biomass production and cool season percentage data. The comparison of remotely sensed growing season TIN and phenology to field vegetation data collected at the Cottonwood Range and Livestock Research Station (CRLRS) in the Bad River watershed suggested that the remote observations were successful in describing actual conditions (Rigge et al., 2013). Field-measured cool season biomass production and cool season TIN were greater on the midgrass-dominated pastures than the mixed-grass pastures, following the expected pattern (Sims and Singh, 1978; Smart et al., 2007). Overall, the relationship between field-measured annual biomass production and TIN by pasture was strong ($R^2 = 0.69$, $P < 0.01$, $n = 24$).

A carbon flux tower located on a mixed-grass plant community pasture at the CRLRS was used for validation at that site. Thirty-minute flux tower quality-controlled ecosystem data for 2007 and 2008 were modeled using nonlinear light response curves driven by PAR, soil temperature, and vapor pressure deficit to partition carbon fluxes associated with PAR (i.e., gross primary productivity) from those associated with total ecosystem respiration (Gilmanov et al., 2010). Gross photosynthesis (PG) data were calculated as the sum of day CO_2 flux and respiration minus the rate of change in atmospheric CO_2 storage below the tower. Phenological metrics from the flux tower data were generated using the same methods employed for MODIS data, making the results generated from both data sets more directly comparable. Flux tower PG data were strongly related to sNDVI values ($R^2 = 0.67$, $P < 0.01$) at the overlapping pixel. Similarly, the accumulation of TIN and PG throughout both growing seasons were related ($R^2 > 0.90$ in both years).

12.2.3.3 Conclusions

Information on the timing and magnitude of biomass production can be a useful tool for assessment of rangeland health. Species diversity and variable weather across the northern mixed-grass prairie make modeling native rangelands problematic. However, the MODIS-based phenological indicators were successful in capturing important characteristics of plant community phenology and productivity. The results of this study clarify the spatial and temporal dynamics (inter- and intraannual) of phenology and biomass production in response to precipitation in this region.

This approach could be useful to land managers in adjusting the stocking rate and season of grazing to (1) maximize rangeland productivity and profitability (Dunn et al., 2010) and (2) achieve conservation objectives, through improving understanding of management impacts to rangeland phenology and production. Maps such as these might also be useful to identify areas where land degradation may be occurring or is likely to occur if current management practices are continued. Further, if the phenological metrics of a patch (e.g., pasture) of land is dramatically different than the surrounding landscape or contrasts

with the general landscape gradient, it can be presumed that land management practices on that parcel, not climate, soils, or topography, are primarily responsible. These specific parcels can then be the focus of best management practices, field examination, and data collection.

12.2.4 Rangeland Fuel Load Assessment

Knowledge of the spatial distribution of fuels is critical when characterizing susceptibility of landscapes to wildfire and for estimating expected fire behavior (Arroyo et al., 2008). In the simplest terms, the susceptibility of a rangeland landscape to wildfire is a function of expected fire weather, topography, probability of ignition, juxtaposition of the landscape, and fuelbed characteristics. Many rangelands are dominated by small diameter fuel components, which respond very quickly to environmental conditions such as heating, relative humidity, and precipitation (Cheney and Sullivan, 1997). In addition, the high surface area to volume ratio, high packing ratios and small sizes make rangeland fuels characteristically prone to ignition and very high rates of spread. For example, under extreme fire weather conditions, grassland fires can exceed 28 km h^{-1} making rangeland fires unpredictable. The temporal and spatial variability of fuels add to this uncertainty.

Productivity of rangeland varies much more on an interannual, proportional basis than that of forests (Briggs and Knapp, 1995; Le Houerou and Hoste, 1977; Teague et al., 2008; Zhang et al., 2010). As a result, the fuelbed properties associated fire behavior characteristics and potential risk can change commensurately. In addition, rangelands are the most extensive kind of land cover, occupying nearly 33% of ice-free land globally (Ellis and Ramankutty, 2008). In the coterminous United States alone, there is an estimated 268 million ha that often occur across large, open, remote expanses prone to high and gusty winds. The high variability and large areas inherent in rangeland systems suggest that regular monitoring is needed for properly addressing the fire danger situation.

Regular monitoring of fuels is needed for both strategic (Kaloudis et al., 2005) and tactical purposes (Reeves et al., 2009). Strategic assessment involves planning and resource allocation modeling. In contrast, tactical uses of wildland fuel data involve specific fire behavior projections and assessment for determining things such as arrival time, risks to resources, and suppression tactics.

A complete evaluation of wildland fuels requires accounting for all fuelbed characteristics including such components as fuelbed depth, vegetation structure and species composition, litter, fuel moisture, fuel quantity of various size classes (Anderson, 1982; Ottmar et al., 2007; Sandberg et al., 2001; Scott and Burgan, 2005). The diameters of the fuel particles can be classified based on their time lag. Larger diameter fuels have longer time lags because they respond more slowly to changes in environmental conditions. The time lag categories most often used for fire behavior and danger assessment include 1, 10,

100, and 1000 h corresponding to diameter ranges of 0–0.635, 0.635–2.54, 2.54–7.62, and 7.62–20.32 cm, respectively. These fuelbed components are most strongly influenced by the kinds and amounts of vegetation. The pressing need for timely information, dynamic nature of rangeland vegetation and fuels, and paucity of plot data suggest that remote sensing can play a significant role in the evaluation of rangeland fuels.

Remote sensing has long been used to evaluate vegetation conditions at spatial scales and spatial resolutions varying from plot level evaluation (Blumenthal et al., 2007) to global assessment at 8 km². Mere vegetation classification, however, is generally not sufficient for quantifying fuel conditions. As a result, many approaches have been derived for using remote sensing to evaluate wildland fuels (Arroyo et al., 2008; Reeves et al., 2009; Riaño et al., 2002, 2003, 2007; Rollins, 2009; van Wagtendonk and Root, 2003). Most methods for quantifying fuels across the landscape involve a combination of remote sensing techniques (e.g., spectral analysis for species composition, structure and production; LiDAR for vegetation structure) and modeling or expert systems (Keane and Reeves, 2011; Reeves et al., 2009). Use of satellite remote sensing for capturing temporal dynamics of fuelbed properties is usually most successful for characterizing the herbaceous biomass response. These biomass estimates can subsequently be converted to fuel loadings (1-h time lag category). With woody vegetation, the situation is more complicated and quantifying fuels can be accomplished by first determining vegetation structure (height and cover) (e.g., Chopping et al., 2006; Riaño et al., 2007; Rollins et al., 2009; Vierling et al., 2012). Yet another method is to first determine stand structure (stand cover and height) and species composition and then using allometric relationships to estimate individual fuel components (Means et al., 1996). Once the appropriate fuelbed attributes have been estimated, it is often necessary to invoke expert systems to crosswalk stand level fuel attributes into a fuel model depending on the intended use of the data. This is a critical step because describing all fuel characteristics in an area is exceedingly difficult given the extreme spatial and temporal variation of fuelbed components (Arroyo et al., 2008; Keane, 2013). As a result, description of fuel properties relevant to fire behavior or effects is normally based on classification schemes, which summarize large groups of fuel characteristics. These classes are expressed as "an identifiable association of fuel elements of distinctive species, form, size arrangement, and continuity that will exhibit characteristic fire behavior under defined burning conditions" (Merrill and Alexander, 1987).

Commonly used fuel classification systems include surface fire behavior fuel models (Anderson, 1982; Scott and Burgan, 2005) and fuel loading models (Lutes et al., 2009). The reason for this is that most fire behavior or fire affects processors such as Farsite (Finney, 2005), FlamMap (Finney, 2005), Behave (Andrews and Bevins, 2003), and Promethius (Tymstra et al., 2010) require stylized fuel models or standardized classifications of fuel components. In the United States, many decision support systems and tactical evaluations utilize FarSite and Flammap that generally require stylized fuel models from either Anderson (1982) or Scott and Burgan (2005). Prometheus, on the other hand, is widely used in Canada and components of it have been used in the United States, Spain, Portugal, Sweden, Argentina, Mexico, Fiji, Indonesia, and Malaysia (http://www.nrcan.gc.ca/forests/fires/14470) to estimate fire spread and uses fuel models from the Canadian Forest Fire Danger Rating System (CFFDRS) (FCRDG, 1992; Wotton, 2009).

12.2.5 Case Study: Using Remote Sensing to Aid Rangeland Fuel Analysis

This case study focuses on the rangelands of the coterminous United States where, in places, annual aboveground rangeland production can vary by more than 150% in extreme years. Prominent fire decision support systems in the United States presently require the surface fire behavior fuel models (FBFM) (Anderson, 1982; Scott and Burgan, 2005). As a result, here we present a case study using remote sensing to interannually quantify 1-h time lag fuels for informing mapping processes designed to predict surface FBFM's for all coterminous United States (Figure 12.4).

The fine fuels of rangelands are driven primarily by grasses and forbs. Capturing these interannual fuel dynamics requires a relatively high repeat cycle and reasonably good spatial resolution, both of which are inherent in the 250 m² 16-day composite MODIS NDVI from the MODIS. From the 23 periods in each year from 2000 to 2012, the annual maximum NDVI was chosen to correlate with ground observations of aboveground productivity. Since spatially explicit, consistent, comprehensive ground data do not exist, production estimates from the Soil Survey Geographic database (SSURGO; http://www.nrcs.usda.gov/wps/portal/nrcs/detail/soils/survey/?cid = nrcs142p2_053627) were used as a surrogate for plot-based ground data. For each soil type in the SSURGO database, the expected annual production based on normal (mean), drought (low), and above average (high) growing conditions is available. Thus, for each soil site, represented as a soil map unit (SMU), there are three data points representing expected annual production. Annual maximum NDVI, gridded annual precipitation from the PRISM project, and the biophysical settings (BPS) data layer were used to develop a model for predicting annual production and, ultimately, 1-h time lag fuels through time. For precipitation, the maximum, minimum, and mean annual total precipitation from 2000 to 2012 was selected to correspond with the high, mean, and low productivity for each BPS type. Likewise, the maximum, minimum, and mean annual maximum from 2000 to 2012 was also selected to correspond with the rangeland productivity from the SSURGO soil types. For each BPS type, the range of annual production, range of annual precipitation, and annual maximum NDVI was spatially averaged. For example, average mean annual maximum NDVI, annual summation of precipitation and rangeland productivity were averaged for all shortgrass prairie BPS. The annual

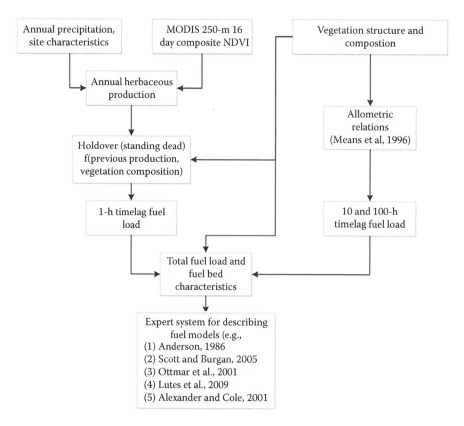

FIGURE 12.4   Fuel flowchart.

productivity model resulted in a simple combination of these predictors and their interactions as

$$
\begin{aligned}
\text{AnnualProduction} = &\ (Y_{int}) + (\text{Precip}_{annsum}*0.141) \\
&+ (\text{NDVI}_{annmax}*3.0056) + \text{BPS} \\
&+ (\text{Precip}_{annsum}*\text{BPS}* - 0.1138) \\
&- (\text{NDVI}_{annmax}*\text{BPS}*-1.2961)
\end{aligned}
$$

where

AnnualProduction is the estimated annual production of rangeland biomass

Precip_{annsum} is the annual sum of precipitation for a BPS unit

NDVI_{annmax} is the average annual maximum NDVI

This model resulted in an R^2 value of 0.94, a bias estimate of 1.43, and a mean absolute error (MAE) of 164 kg ha^{-1}.

This relationship between NDVI, precipitation, and site-specific (BPS) coefficients enabled estimates of rangeland annual production from 2000 to 2012.

After quantifying annual production, estimates of standing dead residue ("holdover"), which add considerably to the total fuel load on a rangeland site, were estimated. This was accomplished using a simple decay function represented in Figure 12.4. As a result, for a given year, the total 1-h fuel load can be described as function of the following form f (annual production +

holdover$_{t-1}$ + holdover$_{t-2}$, holdover$_{t-3}$). These fine fuel loads were combined with estimates of 10 and 100 h time lag fuels of the woody components (e.g., shrubs) of each stand. This was accomplished by combining the cover, height, and species composition from the LANDFIRE project, and quantifying woody biomass and 10 and 100 h time lag fuels using allometric equations from Means et al. (1996) (Figure 12.4). Once the full suite of fuel characteristics were known, surface FBFM's from the Scott and Burgan (2005) suite were estimated for average fuel conditions and then modulated through time based on the interannual change in 1-h time lag fuel loads (Figure 12.5). Figure 12.5 demonstrates how remote sensing is used with climate and site characteristics to estimate 1-h time lag fuels every year. In addition in Figure 12.5, the amount of deviation (as a percent of the 12 year average) in 2000, 2005, and 2011 are shown for the southwestern Ecological Province (Bailey and Hogg, 1986). Finally, the resulting change in the distribution of surface FBFM's can be seen in panel C for all 3 years presented. Note the large increase in 1-h time lag fuels in 2005 in the southwestern ecoregion. This corresponds to a large increase in rainfall and a strong NDVI signal resulting in very high biomass and subsequent changes to surface FBFM's suggesting much greater flame lengths and spread rates (the GR1 is reduced and GR2 is substantially increased). As an example, under the identical environmental parameters with an

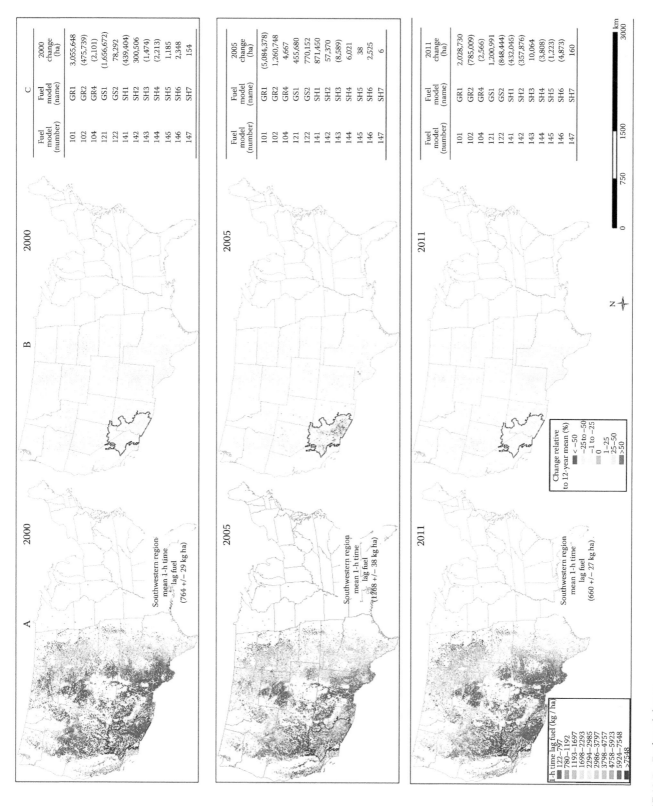

FIGURE 12.5 Fuel modulation.

open wind speed of 24 km h⁻¹, a GR2 surface FBFM results in a spread rate of 9 times greater than a GR1.

This brief case study demonstrates a method for using high temporal resolution satellite remote sensing for quantifying fuels in rangeland landscapes. The important characteristics of a sensor for characterizing fine fuels across large landscapes in a timely manner suggest satellites with relatively high repeat frequencies and appropriate spectral channels for modeling changes in aboveground biomass. Application of the methods outlined here are limited to regions where sufficient numbers of spatially explicit measures (or estimates as in the present work) of biomass or fuels are available.

12.3 Rangeland Vegetation Characterization

12.3.1 Rangeland Biodiversity and Gap Analysis

Rangeland biological diversity (biodiversity) refers to the variety and variability among living organisms and the environments in which they occur and is recognized at species, ecosystem, and landscape level. The goal of biodiversity conservation is to reverse the processes of biotic impoverishment at each of these levels of organization. Gap analysis provides a quick overview of the distribution and conservation status of several components of biodiversity (Scott et al., 1993). Gap analysis is the process by which the distribution of species and vegetation types are compared with the distribution of different land management and land ownership classifications. It seeks to identify gaps (vegetation types and species that are not represented in the network of biodiversity management) that may be filled through establishment of changes in land management practices (Scott et al., 1993). The goal here is to ensure that all ecosystems and areas rich in species diversity are represented adequately in biodiversity management areas. Gap analysis uses vegetation types and vertebrates as indicators of biodiversity. Maps of existing vegetation are prepared from satellite imagery and other sources and entered into a geographic information system (GIS). Because the mapping is carried on a regional scale, the smallest area identified on vegetation maps is generally 100 ha. Vegetation maps are verified through field checks and aerial photographs. Predicted species distributions are based on existing range maps and other distributional data, combined with information on the habitat of each species. Distribution maps for individual species are overlaid in the GIS to produce maps of species richness. An additional GIS layer of land ownership and management status allows identification of gaps in the representation of vegetation types and centers of species richness in biodiversity management areas through a comparison of the vegetation and species richness maps with ownership and management status maps. Gap analysis is a powerful and efficient first step toward setting land management priorities. It provides focus, direction, and accountability for conservation efforts. Areas identified as

important through gap analysis can then be examined more closely for their biological qualities and management needs (Scott et al., 1993).

12.3.2 Case Study: Vegetation Continuous Fields with Regression Trees

12.3.2.1 Introduction

The sagebrush ecosystem is an important and distinguishing natural system of the U.S. Intermountain West. However, an estimated 40% of its pre-European settlement distribution has been reduced due to conversion to agriculture, urban/suburban growth, energy development, invasion by exotic plants, and encroachment by woodlands among others (Wisdom et al., 2005). The sagebrush ecosystem is an example of many other globally distributed natural systems that have been impacted by similar disturbances. Monitoring land-cover change, across vast landscapes, in an efficient manner is touted by many to be a significant strength of digital remote sensing. The holistic view of remotely sensed data allows us to evaluate landscapes in their entirety. Traditional, field-based, monitoring systems are limited to specific locations whose characteristics are then extrapolated across the entire landscape. The number of field-based samples required to characterize landscapes with statistical certainty is often cost prohibitive and in some cases impossible to acquire due to access restrictions. The temporally systematic, landscape-level monitoring that remote sensing offers coupled with modern statistical modeling approaches can provide land managers with much of the information necessary for effective planning and management.

Traditional image interpretation techniques that convert digital remote sensing data into discrete land-cover maps have been used to monitor landscapes (Jin et al., 2013; Vogelmann et al., 2012). These techniques rely on the ability to accurately classify and map vegetation types at multiple time intervals to determine how change has occurred at the vegetation community level. A limitation of this technique is the assumption that adjacent vegetation community types have discrete boundaries (sharp ecotones) and that variation within the community is typically ignored. The reality is that sharp ecotones between community types, while they exist, are typically not the norm. Further, the spatial variation of vegetation cover within a mapped community type is often an important diagnostic of community condition. In traditional classification of imagery, this information is lost.

This case study describes the use of digital remote sensing data coupled with field-based measurements and advanced statistical modeling techniques to map vegetation continuous fields (VCF) in the sagebrush ecosystem. VCF is a relatively new concept that attempts to model percent canopy cover of specific vegetation types using remotely sensed imagery as its primary input. This technique has been used to estimate canopy cover of woody and herbaceous vegetation, as well as bare ground, on a global basis (Defries et al., 2000; Hansen et al., 2003b; Sexton et al., 2013).

A series of VCFs consist of several continuous response surfaces (one for each cover type) in which every pixel value corresponds to a percent canopy cover estimate predicted through a regression model. The VCF offers an advantage over traditional discrete classifications because areas of heterogeneity within vegetation community types are better represented (Hansen et al., 2002).

The objectives of the case study were to (1) test the effectiveness of a regression tree algorithm (random forest [RF]) to model estimates of percent cover and (2) develop a series of VCF models for shrubs, woodland, herbaceous vegetation, and bare ground for a semiarid shrub-steppe landscape.

12.3.2.2 Methods

12.3.2.2.1 Study Area

Our research was conducted in the northwest corner of Box Elder County, Utah (114°2′31. 2′–112°43′40. 8′ west longitude and 41°6′27. 36′–41°59′59. 64′ north latitude) depicted in Figures 12.6 and 12.7. The area covers 1,742,860 ha, with approximately 60% of the county occupied by the Great Salt Lake and barren playa bottoms. Of the remaining area, salt desert scrub occupies about one-fifth of the area while big sagebrush shrubland and steppe covers nearly the same amount (19%). Pinyon-juniper ecosystems are an important part of the landscape making up 12% of the area.

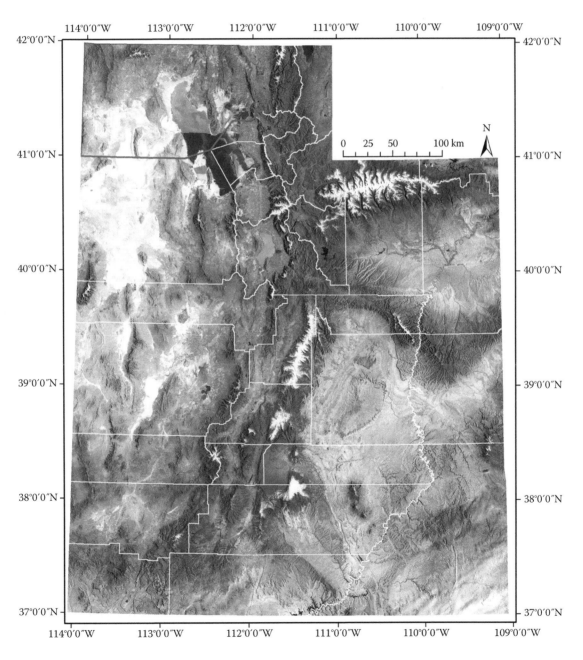

FIGURE 12.6   Utah image mosaic of 2013 Landsat 8 OLI imagery with Box Elder County highlighted in red.

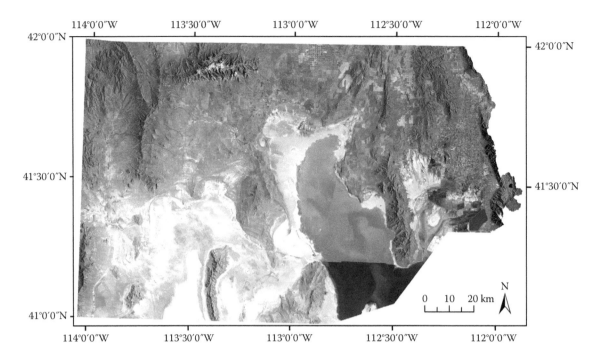

FIGURE 12.7   Landsat 8 OLI image mosaic (2013) of Box Elder County, Utah.

The remainder consists of greasewood flats, montane sagebrush steppe, xeric mixed sagebrush shrubland, and invasive annual grasslands (Program, 2004). The elevation ranges from 1278 m in the lowlands close to the Great Salt Lake to 3027 m at the peak of the Raft River range. The mean elevation is 1520 m.

12.3.2.2.2 Field Data

Field-based estimates of percent canopy cover for shrubs, trees, herbaceous (grasses and forbs) vegetation, and bare ground were used to develop the VCF models. These data were prepared as geo-referenced field points and obtained from different sources: (1) 482 points collected by the South West Regional GAP (SWREGAP) project during 2001 (Lowry et al., 2007), (2) points collected by The Nature Conservancy (TNC) for the Northwest Utah Landscape modeling project in 2007 (Conservancy, 2009), and (3) field points that we collected during a field season in 2007. In total, 135 field observations were available for the year 2007. A fourth data set was available from the Utah Division of Wildlife Resources (UDWR) (Resources, 2010). Figure 12.8 contains the spatial distribution of the different data sets across the study area.

With the exception of the UDWR data set, which are permanent sample plots and follow a standardized and quantitative data gathering method, the rest of the field points were visually assessed (qualitative assessment) in terms of percent canopy cover for shrubs, trees, herbaceous vegetation, and bare ground on an area that resembled a 3 × 3 Landsat TM pixel (approximately 90 × 90 m). Cover estimates were taken independently at four cardinal directions from the center of the plot in the NE, SE, SW, and NW directions. These estimates were averaged to represent the entire plot. The percent cover estimates were recorded using 5% increments for each life form and bare ground (i.e., 0%, 5%, 10%, etc.). The sum of the percent cover for shrubs, herbaceous vegetation, trees, and bare ground totaled 100% at each point. The sampling scheme consisted of locating sites along an elevation range that included Wyoming big sagebrush (*Artemisia tridentata* ssp. wyomingensis), basin big sagebrush (*Artemisia tridentata* ssp. tridentata), and mountain big sagebrush (*Artemisia tridentata* ssp. vaseyana) communities.

12.3.2.2.3 Explanatory Variables: Remote Sensing and Topography

Remotely sensed images and topographic data sets were used as explanatory variables for this modeling. With regard to the remotely sensed data, scenes from the Thematic mapper (TM) sensor of the Landsat 5 satellite, Path 39 Row 31 were obtained. Due to the underlying differences in phenology that most vegetation types exhibit in semiarid landscapes (Bradley and Mustard, 2008), imagery from multiple dates during the summer months of 2001 were acquired. Within year seasonal imagery allowed us to capture major phenological variations that occur during the growing season. Landsat TM imagery collection was concentrated during late spring, midsummer and early fall. An effort to obtain only imagery with the best quality (i.e., minimum cloud cover) was made. Three images were selected for this study. These were collected on April 28, July 01, and October 05, 2001.

Where necessary, imagery was rectified and resampled to a common map projection UTM Zone 12 WGS 1984. Standardization of the imagery was performed by converting the raw digital numbers to exo-atmospheric reflectance values using an image-based atmospheric correction procedure (Chavez, 1996) with the most up-to-date calibration coefficients for the Landsat TM sensor (Chander et al., 2009).

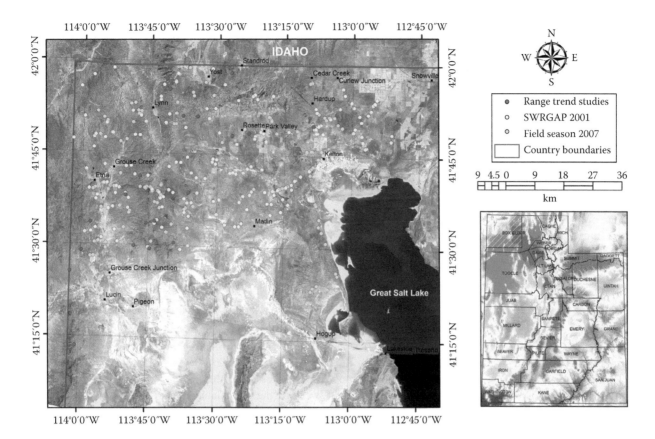

FIGURE 12.8 Distribution of field observations to model multitemporal CVF.

The SAVI was computed for each date. SAVI may be calculated as follows:

$$SAVI = \frac{NIR - RED}{(NIR + RED + L)} * (1 + L) \qquad (12.1)$$

where

NIR is the near-infrared reflected radiant flux
RED is red reflected radiant flux
L is adjustment factor (typically a value of 0.5 is used).

SAVI has been reported to work well in semiarid ecosystems because it minimizes soil background effects that are known to affect other indices such as the NDVI (Huete, 1988; Jensen, 2007). It has been widely reported that a vegetation index such as SAVI may be used to follow the phenological trajectory or seasonal and interannual change in vegetation growth and activity (Jensen 2007).

A new variable was created from the multi-temporal SAVI and named as NDSAVI or the normalized difference SAVI. The NDSAVI takes advantage of the contrast between the spring and the summer SAVI and may be used to improve our understanding of the phenological dynamics of grasses on the landscape. Higher values in the NDSAVI would correspond with higher greenness during the early spring relative to summer whereas low values of NDSAVI would relate to areas that become green later in the growing season. This new variable conveys a multi-temporal signature of greenness variation that may be used to discriminate among different land-cover types and particularly focus on nonnative grasses such as cheatgrass that follows this phenological pattern. Within this environment, this index allows us to identify areas where cheatgrass, a common and significant invasive annual grass, is a major component of the plant community.

NDSAVI was estimated as follows:

$$\frac{SAVIspr - SAVIsum}{SAVIspr + SAVIsum} \qquad (12.2)$$

The Normalized Difference Water Index NDWI (Gao, 1996) was also calculated for each image. NDWI takes advantage of the contrast found between the near and middle infrared bands to provide information about water content. Forest disturbances have been successfully detected using the NDWI (Jin and Sader, 2005), and thus it was appropriate to test its performance in our regression models. The brightness, greenness, and wetness (BGW) transformation (Crist and Kauth, 1986) was also derived for each image. This transformation has been used extensively to monitor condition and changes in soil brightness, vegetation, and moisture content respectively (Jensen, 2007; Lowry et al., 2007).

In addition to the remotely sensed information (Landsat TM spectral bands, SAVI, NDWI, and BGW), derivatives from a 30 m digital elevation model (DEM) including slope, aspect, and landform were obtained. Two transformations of aspect, namely southness and westness indices (Chang et al., 2004) and a modification to the original topographic relative moisture index TRMI (Parker, 1982) were generated. An existing land-cover map from the SWRGAP project (Lowry et al., 2007) was included as an explanatory variable. The inclusion of this type of ancillary information has been shown to greatly improve classification and regression modeling in rangelands (Peterson, 2005).

12.3.2.2.4 *Regression with Random Forests*

12.3.2.2.4.1 *Background*

Random Forests (RF) is a relatively new statistical method that emerged from the machine learning literature, and is based on the same philosophy as CARTs. In RF, multiple bootstrapped regression trees without pruning are created. In a typical bootstrap sample, approximately 63% of the original observations occur at least once (Cutler et al., 2007). The data that are not used in the training set are termed "out-of-bag" observations and are customarily used to provide estimates of errors (Prasad et al., 2006). Out-of-bag samples are also used to calculate variable importance (Cutler et al., 2007). In RF, each tree is grown with a randomized subset of predictors, which equal the square root of the number of variables. In general, 500–2000 trees are grown and averaging aggregates the results. The method is very effective in reducing variance and error in multi-dimensional data sets. One of the strengths of RF is that because it grows a large number of trees, the method tends not to over-fit the data, and because the selection of predictors is random, the bias can be kept low (Prasad et al., 2006). More comprehensive descriptions of the method may be found in Cutler et al. (2007), Lawler et al. (2006), Prasad et al. (2006), and Sutton (2005). The application of RF was done in two phases. First, the best subset of variables to model each response variable: shrubs, trees, herbaceous vegetation, and bare ground were identified. Second, the best subset of variables identified for each response variable was used to model the VCF for that variable.

12.3.2.2.4.2 *Variable Importance and Parsimony*

The underlying principle that the phenological pattern of a given vegetation type should dictate which remotely sensed data sets to use (Bradley and Mustard, 2008) was followed. For example, it is sensible to use only one scene (midsummer for instance) to model bare ground percent cover due to its relatively constant spectral response throughout the year. On the other hand, it makes sense to utilize two to three images (i.e., midsummer and early fall) to model herbaceous vegetation due to its conspicuous phenological signature which peaks during the summer and then significantly decreases during the fall.

In order to develop a simple yet effective model the concept of variable importance (Cutler et al., 2007) was used. This is based on the mean decrease in accuracy concept, and is assessed based on how much poorer the predictions would be if the data for that predictor were permuted arbitrarily. This provides a measure of the impact that a specific variable has in decreasing the precision of prediction. This is a somewhat subjective approach to choosing the most important variables since thresholds of decreasing precision tend to be arbitrary. Once the most important variables are chosen, the correlation coefficients between each variable are evaluated and further used to eliminate highly correlated variables (choosing the variable that makes most ecological/spectral sense) to arrive at a parsimonious set of predictor variables.

12.3.2.2.5 *Regression*

The R package RandomForest (Liaw and Wiener, 2002) was used to develop regression trees to calculate the VCF. Regressions were run separately for each of the four response variables (i.e., shrubs, trees, etc.) using the selected subset of variables determined to be most important. The R package YaImpute (Crookston and Finley, 2008) was used to extract the model for each VCF run, and then applied a predict function to generate a continuous geospatial response surface for the entire study area.

12.3.2.3 Validation and Comparison Metrics

For independent validation purposes, 20% of the field observations were withheld during model development. Pearson's correlation coefficients were calculated for each of the VCF predictions using this withheld set of data and two metrics were further calculated: MAE and root mean square error (RMSE). MAE is the average absolute difference of the predicted value from the field-observed estimate, while RMSE is the square root of the mean squared error (Prasad et al., 2006; Walton, 2008).

12.3.2.4 Results and Discussion

Table 12.4 contains the Pearson's Correlation Coefficients (PCC), MAE, and root mean square error (RMSE). A global average of the PCC was 0.65. Our highest PCC was for herbaceous cover at 0.77 and the lowest was for trees at 0.52. Figures 12.9 and 12.10 illustrate samples of the VCF spatial layers generated using RF.

The use of regression trees to depict sub-pixel heterogeneity has been widely reported in the literature. In rangeland environments, work has been conducted to model woody vegetation cover (Danaher et al., 2004), bare ground cover (Weber et al., 2009), and shrub cover and encroachment (Laliberte et al., 2004). With the exception of the MODIS global continuous vegetation maps (Hansen et al., 2003a), the examples above dealt with only one response variable. In this work there were four response variables. Since a pixel is in essence an integrated

TABLE 12.4   Validation Metrics between MRTS and RF

VCF	Pearson's Correlation	MAE[a]	RMSE[b]
Shrubs	0.72	7.81	10.00
Trees	0.52	12.56	16.69
Herbaceous	0.77	9.39	12.02
Bare ground	0.62	8.15	11.05
Average	0.65		

[a] Mean absolute error.

[b] Root mean square error.

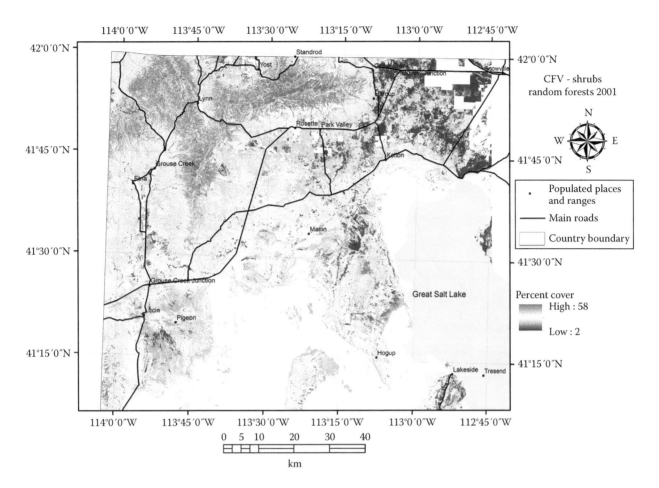

FIGURE 12.9   Percent canopy cover of shrubs as modeled using random forests.

multi-dimensional spectral response of vegetation, bare ground and other features, it makes sense to attempt to decompose that response to understand the land-cover dynamics of a given pixel in relation to the surrounding landscape.

With regards to the ease of understanding, RF has been frequently described as a "black box" (Prasad et al., 2006) because the individual trees cannot be examined separately due to the sheer number of trees that may be generated. This can limit the ability of an analyst to understand the underlying dynamics of the resulting model. Random Forests does provide metrics to aid in interpretation. One metric is variable importance, which can be used to compare relative importance among predictor variables. Such a feature is not available in other regression tree tools and therefore the importance of variables must be determined with a careful data mining process.

12.3.2.5 Implications

The development of a multi-temporal collection of VCF may be used to update information about the status or condition of a particular ecological site as well as characterizing the states and transitions for that site. For instance, a specific spatial unit of an ecological site may be characterized in terms of its occupancy by shrubs, grasses, trees, and bare ground using modeled VCF. Knowledge about the relative dominance of these life forms

in a particular unit may shed light about its current condition relative to a reference condition. The VCF process may provide knowledge about usage of the ground by major life forms and bare ground and in this way pinpoint areas that are diverging from a desired condition.

12.4 Rangeland Change Detection Analysis

12.4.1 Rangeland indicators

Numerous landscape metrics have been developed to characterize the patterns and configurations of different land-cover types in a landscape. Categorical maps generated from remote sensing data can then be used for landscape characterization to better understand spatial arrangements between different cover classes, particularly forest fragments (Read and Lam, 2002). These spatial arrangements are expressed numerically in the form of landscape indices or pattern metrics and have been used in many studies to assess land-cover change and its impact, ecosystem health, or as variables for models that support environmental assessment and planning efforts (e.g., Botequilha Leitão and Ahern, 2002; Fuller, 2001; Gergel, 2005; Griffith et al., 2000; Liu and Cameron, 2001).

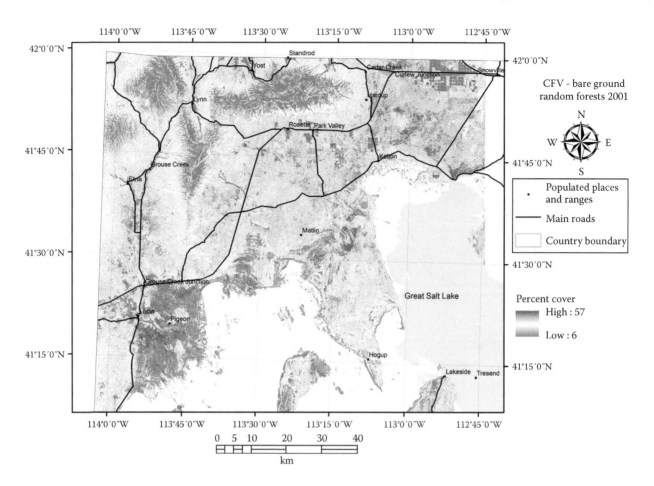

FIGURE 12.10   Percent canopy cover of bare ground as modeled using random forests.

More than 100 pattern metrics can be computed using free-ware such as FRAGSTAT (McGarigal and Marks, 1995), Patch Metrics (Rempel et al., 1999), and others (Crews-Meyer, 2002; Cumming and Vervier, 2002; Stanfield et al., 2002). However, many pattern metrics are highly correlated (Cain et al., 1997; Riitters et al., 1995). Efforts have been made to identify a minimum set of pattern metrics that describe landscape patterns adequately. Multivariate data analysis using principal component analysis (PCA) and factor analysis (FA) are the most commonly used methods to reduce pattern metrics data (Griffith et al., 2000; Honnay et al., 2003; McAlpine and Eyre, 2002; Stanfield et al., 2002). These methods identify a small number of components, which are then interpreted in terms of their dominant characteristics and underlying causes (Griffith et al., 2000). Multivariate data analysis requires large datasets and several landscape units to be statistically consistent (e.g., Cumming and Vervier, 2002; Schmitz et al., 2003). A few empirical landscape studies apply pattern analysis to only one landscape (e.g., Griffith et al., 2000) and are useful because they tackle problems at relevant scales, but the validity of making statistical inferences with such an approach is seriously compromised (Li and Wu, 2004). Gustafson (1998) overcame this by generating artificial landscapes (called neutral models), but the technique was difficult

to relate to pattern metrics in real landscapes (Li and Wu, 2004). Therefore, the behavior of pattern metrics in real landscapes over time needs further investigation (Griffith et al., 2003). Irrespective of the landscape unit used, pattern metrics require rigorous validation in order to be interpreted and applied with confidence (McAlpine and Eyre, 2002).

12.4.2 Pattern Metrics to Measure Landscape Attributes

12.4.2.1 Area

Area distribution pattern metrics include basic attributes of the landscape such as the number of patches (NP) and the total area (CA) of all class patches, expressed in hectares. Mean patch size (AREA_MN in ha) is an intuitive index for measuring aggregation and is particularly suitable for categorical maps (He et al., 2000). Other measures are the standard deviation of the mean patch size (AREA_SD in ha) and the coefficient of variation (AREA_CV in %), determined by Equation 12.3:

$$AREA\_CV = \frac{AREA\_SD}{AREA\_MN} * 100 \qquad (12.3)$$

AREA_CV indicates patch area distribution by determining the difference among patches within one landscape class. Lower values indicate a more uniform class distribution (Batistella et al., 2003; McAlpine and Eyre, 2002), that is if a landscape class is dominated by big patches both AREA_SD and AREA_CV values are large. All of these pattern metrics assign equal weights to each patch. For measuring landscape resistance to fragmentation, large forest patches are most important (Batistella et al., 2003); therefore, metrics such as largest patch index (LPI) are useful. LPI is equal to the percent of the total landscape made up by the largest patch (McAlpine and Eyre, 2002).

12.4.2.2 Edge

Total edge (TE) refers to the length of edge that exists at the interface between land classes (McGarigal and Marks, 1995) while mean patch edge (MPE) indicates the amount of edge per patch (Equation 12.4).

$$MPE = \frac{TE}{NP} \qquad (12.4)$$

12.4.2.3 Patch Shape Complexity

Mean perimeter–area ratio (PAR_MN) is a simple index of patch shape complexity, computed as the mean ratio between patch perimeter and area; it describes the amount of edge per unit area of landscape unit (such as forest) (McGarigal and Marks, 1995). Since the simple ratio is usually affected by patch size, a modified perimeter–area ratio called landscape index (LSI), implied as shape of landscape, is computed using Equation 12.5:

$$LSI = \frac{Perimeter}{2\sqrt{(area^*\pi)}} \qquad (12.5)$$

Higher LSI values indicate higher complexity (McAlpine and Eyre, 2002). Because LSI is influenced both by shape complexity and number of patches (NP), preference is sometimes given to the mean shape index (SHAPE_MN) or the area-weighted mean shape index (SHAPE_AM), which weights larger patches more heavily than smaller patches (Batistella et al., 2003).

Fractal dimension (FRAC) has equally been used to describe patch shape complexity across a range of spatial scales (patch sizes). This overcomes one of the major limitations of the perimeter–area ratio as a measure of shape complexity. The value of FRAC ranges between 1 and 2 and is computed as

$$FRAC = \frac{2\log P}{\log A} \qquad (12.6)$$

where
    P is the perimeter
    A is the area of the patch

Fractal dimension approaches 1 for shapes with very simple perimeters such as squares, and approaches 2 for shapes with highly complex perimeters (Read and Lam, 2002). Mean patch fractal dimension (FRAC_MN) and the area-weighted mean patch fractal dimension (FRAC_AW) are derived measures of shape complexity, interpreted similarly to FRAC with the addition of an individual patch area weighting applied to each patch in the case of FRAC_AW (McGarigal and Marks, 1995).

12.4.2.4 Interior-to-Edge (Core Area)

Core area (CA) represents the area of patch greater than a specified depth-of-edge distance from the perimeter. It is determined by defining a vanishing distance as the distance from a patch boundary inward, to a point where edge effects are eliminated (Baskent and Jordan, 1995). The higher the ratio between the CA and the total area, the lesser the degree of fragmentation (Batistella et al., 2003; Tang et al., 2005). The vanishing distance is arbitrarily set. The number of CAs (NCA) is the number of distinct CAs contained within a patch boundary. Total core area (TCA) is the same as CA except that CA is aggregated (summed) over all patches, approaching CA as patch shapes are simplified. CA distribution in the landscape is measured using the parameters mean CA per patch (ha) (CORE_MN), patch CA standard deviation (CORE_SD) and patch CA coefficient of variation (CORE_CV).

12.4.2.5 Isolation

The distance of patch to its nearest neighbor measures the degree of isolation of that patch. Averaging this distance for all individual land-cover classes or patches in a landscape gives the mean nearest neighbor distance (ENN_MN). Mean proximity index (PROX_MN) is also computed (Equation 12.7) and is a measure of isolation and fragmentation:

$$PROX\_MN = \sum \frac{\sum_i^N \sum_j^{m'} \frac{a_{ij}}{h_{ij}}}{N} \qquad (12.7)$$

where
    a_{ij} is the area of patch ij in the neighborhood of patch i
    h_{ij} the distance between patch i and patch ij
    N is the number of patches
    m' is the number of patches in the neighborhood of patch i

This index allows comparison among landscapes of any size, as long as the search buffer around each patch is the same (Gustafson and Parker, 1992).

12.4.2.6 Contagion/Interspersion

Contagion (CONTAG) provides an effective summary of overall clumpiness of categorical data where lower values indicate fragmentation of larger patches into smaller patches, that is, the patch types are maximally disaggregated and interspersed (equal proportions of all pairwise adjacencies) (Equation 12.8):

$$\text{CONTAG} = \left[1 + \frac{\sum_{i=1}^{m}\sum_{k=1}^{m}\left[(P_i)\left(\frac{g_{ik}}{\sum_{k=1}^{m} g_{ik}}\right)\right]*\left[\ln(P_i)\left(\frac{g_{ik}}{\sum_{k=1}^{m} g_{ik}}\right)\right]}{2\ln(m)} \right] (100)$$

$$(12.8)$$

where

P_i is the proportion of the landscape occupied by path type (class) i

g_{ik} is the number of adjacencies (joins) between pixels of patch types (classes) i and k based on the double-count method

m is the number of patch types (classes) present in the landscape, including the landscape border if present

Aggregation index (AI) is calculated from an adjacency matrix at the class level, AI equals 0 when the patch types are maximally disaggregated. SPLIT and MESH are two subdivision metrics computed as isolation measures (McGarigal and Marks, 1995). The interspersion and juxtaposition index (IJI), a configuration metric, is a measure of relative interspersion of each class at class level and each patch at landscape level (Crews-Meyer, 2002; McAlpine and Eyre, 2002). IJI approaches 100 when all patch types are equally adjacent to all other patch types, and as the distribution of class types depart from evenness, the IJI approaches 0 (Equation 12.9):

$$\text{IJI} = \frac{-\sum_{i=1}^{m}\sum_{k=i+1}^{m}\left[\left(\frac{e_{ik}}{E}\right)*\ln\left(\frac{e_{ik}}{E}\right)\right]}{\ln(0.5m(m-1))}(100) \quad (12.9)$$

where

e_{ik} is the total length (m) of edge in the landscape between patch types (classes) i and k

E is the total length (m) of edge in the landscape, excluding background

m is the number of patch types (classes) present in the landscape, including the landscape border, if present

12.4.2.7 Spatial Heterogeneity

Spatial diversity representing the extent to which all patches are equally adjacent to each other, number of different land-cover classes, and interspersion metrics determine the spatial arrangements of land-cover classes (Trani and Giles, 1999). Patch richness (PR) is a simple metric equal to the number of different patch types present within the landscape boundary. Shannon's diversity index (SHDI) assesses the diversity of patches, and the extent to which one or a few patch types dominate the landscape (McAlpine and Eyre, 2002). SHDI starts at zero and increases without limit and is more sensitive

to the number of patch types than evenness. SHDI is given by Equation 12.10 (Read and Lam, 2002):

$$\text{SHDI} = -\sum_{i=1}^{m}(P_i * \ln P_i) \quad (12.10)$$

where

P_i is the proportion of landscape occupied by patch type (class) i

m is the number of patch types (classes) present in the landscape, excluding the landscape border, if present

The Shannon's evenness index (SHEI) is expressed such that an even distribution of area among patch types results in maximum evenness (Equation 12.11):

$$\text{SHEI} = -\frac{\sum_{i=1}^{m}(P_i * \ln P_i)}{\ln m} \quad (12.11)$$

SHEI varies between 0 and 1, and the highest value is reached when the distribution among patch types is perfectly even (Wickham and Rhtters, 1995). The modified Simpson's diversity (MSIDI) is similar to the SHEI (Equation 12.12):

$$\text{MSIDI} = -\ln\sum_{i=1}^{m}\left(P_i^2\right) \quad (12.12)$$

where P_i is the proportion of the landscape occupied by patch type (class) i. Diversity metrics represent landscape composition in terms of the relative proportions of each patch types (evenness), together with the number of patch types (richness). Detailed information on these indices and their computation can be found in McGarigal and Marks (1995).

Ludwig et al. (2004) monitored ecological indicators of rangeland functional integrity and their relation to biodiversity at local to regional scales. Functional integrity is the intactness of soil and native vegetation patterns and the processes that maintain these patterns. The integrity of these patterns and processes has been modified by clearing, grazing, and fire. Intuitively, biodiversity should be strongly related to functional integrity. Ludwig et al. (2004), based on published work on Australian rangelands, identified several indicators of landscape functional integrity at finer patch and hillslope scales. These indicators, based on the quantity and quality of vegetation patches and inter patch zones, are related to biodiversity. These vegetation-cover and bare-soil patches can be measured using remote sensing data at various resolutions depending on specific rangeland areas. Bastin and James (2000) used Landsat MSS for prevailing dry periods before the major rainfall events and about 6 weeks after them when vegetation growth had peaked for the assessment of biodiversity condition in a rangelands in central Australia. They computed contagion and interspersion scores for land systems and vegetation types. The contagion and interspersion values, as indices of the spatial arrangement of patches in the landscape, indicated the extent to which both water development and land degradation associated with pastoralism had fragmented habitat types. Nondegraded and water-remote patches of pastorally more productive land systems or vegetation types were more dispersed and more fragmented relative to their original state, than for patches of pastorally less productive land systems or vegetation types.

12.4.2.8 Resilience

Ecological resilience is defined as the degree, manner, and pace of the restoration of vegetation attributes after a disturbance (Westman, 1985). Ecological resilience and its characteristics have been quantified statistically in ecosystem simulations (O'Neill, 1976; Westman and O'Leary, 1986), plant community field studies (Westman and O'Leary, 1986), and regional analyses of land degradation (Wessels et al., 2004, 2007). O'Neill (1976) simulated energy flow through a three-compartment (autotrophs, heterotrophs, and detrivores) model for six different biomes to test their ability to recover from a 10% reduction in plant biomass over a 25-year period. A recovery index, that is, a measure of the amount of deviation from a reference equilibrium value or initial state after a disturbance (a measure of malleability), was computed for each year over the 25-year period. The mean malleability of each biome for this period was interpreted to indicate which biomes were least and most resistant to disturbance (O'Neill, 1976). Westman and O'Leary (1986) developed four measures of ecological resilience: elasticity, that is, the rate of recovery from a disturbance; amplitude, that is, the threshold beyond which recovery to a previous reference state no longer occurs; damping, that is, the extent and duration of an ecosystem parameter following disturbance, and malleability. These were used to estimate the responses of various plant functional types within a coastal sage scrub plant community at 5–6 years after fire using field data and for 200 years after fire using a simulation model. Wessels et al. (2004, 2007) used an interannual time series of AVHRR- and MODIS-derived NDVI (or ANPP) from 1985 to 2005 to examine the resilience of a landscape in northeastern South Africa that had been subject to both overpopulation and apparent overgrazing by the forced settlement of pastoralists into "homelands" during the apartheid era from 1910 to 1994. A spatially aggregated approach was used in which the mean annual ΣNDVI for the period of 1985–2005 was compared between nondegraded benchmark areas and degraded areas within the homelands as

$$\text{RDI or PD} = \frac{\text{Non-degraded} \sum \text{NDVI-degraded} \sum \text{NDVI}}{\text{Non-degraded} \sum \text{NDVI}} \times 100$$

(12.13)

where RDI is the relative degradation index or the percent difference (PD) between the mean NDVIs of nondegraded and degraded sites (Wessels et al., 2004, 2007). Wessels et al. (2004, 2007) found that these paired sites were significantly different from each other and that degraded sites had significantly lower mean annual NDVI than did nondegraded sites. Also, there was no indication of recovery (malleability) of degraded sites toward the mean conditions of nondegraded sites from the end of the apartheid era in 1994–2005 (Wessels et al., 2004, 2007). This provided evidence in support of the hypothesis that the observed migrations of former rural homeland populations to jobs in major urban centers, such as Johannesburg and Durban, were due in part to environmental degradation of the homelands. However, at the coarse resolution of MODIS (0.25 km²) and AVHRR (1 km²), it was observed that finer-scale phenomena at the community and species levels were masked or averaged out (Wessels et al., 2007). A number of scaling studies in drylands suggest characteristic length scales for vegetation and bare soil patches of approximately 1–100 m; these scales are probably not amenable to analysis using spatially coarse-resolution sensors such as MODIS or AVHRR (Hudak and Wessman, 1998; Rietkerk et al., 2004). To account for this disparity, Wessels et al. (2007) suggested a staged remote sensing approach to monitoring at regional and national scales in which both coarse-resolution sensors such as MODIS are used in conjunction with relatively finer-resolution data sets such as Landsat.

12.4.3 Change Detection Methods

It is important to monitor and understand change in rangelands so that effective action can be taken to maintain ecological, economic, and social values. Both seasonal conditions and grazing management play a role in vegetation dynamics on pastoral areas. For example, Sinha et al. (2012a) determined the effect of seasonal spectral variability on vegetation and land-cover classification of Landsat TM by comparing accuracies in different seasons in a mid-latitudinal (29°30′–31°0′S) region with summer and winter rainfall, a broad altitudinal range, a temperate to subtropical climate and diverse land uses (e.g., summer and winter crops, and nature conservation). By comparing the observed changes with those expected under the prevailing seasonal conditions and by investigating the response of species known to be adversely or positively affected by livestock grazing, it was possible to conclude that at least some of the positive changes observed could be attributed to the type of grazing management, rather than seasonal conditions alone. In a similar study, Sinha et al. (2012b) used three-date composite land-cover maps through a process called referential refinement and aggregation for understanding the level of land exploitation (extensive vs. intensive) activities carried out in the region. Landscape-scale monitoring is important for providing regional to national-scale intelligence on habitat quality and trends in threats to or drivers of biodiversity, with data obtained using systematic ground-based and remote methods. Landscape-scale monitoring is typically based on remotely obtained or extrapolated data that can be mapped, and use temporal scales appropriate to the indicator (Ludwig et al., 2007). The advantage of landscape-scale monitoring is that it is often relatively cheap to undertake over space and time and provides a broader context for national reporting.

There are a variety of opinions and suggestions about the selection of the most effective and appropriate techniques for change detection studies (Lu et al., 2004). Therefore, it is difficult to specify which change detection algorithm will suit a specific problem. A review of different change detection techniques used in previous research would be of great help in selecting methods to aid in producing good quality change detection results. For any change detection project, the following conditions must be satisfied (Lu et al., 2004): (1) accurate and precise registration between multitemporal images; (2) precise radiometric and atmospheric normalization between multitemporal images; (3) similar phenological or seasonal conditions between multitemporal images; and (4) selection of the same spatial and spectral resolution images if possible.

The aim of change detection studies is to compare the spatial representation of two points in time by controlling all variance caused by differences in variables that are not of interest and to measure changes caused by differences in the variables of interest (Green et al., 1994). Since there are many change detection techniques that can be used in one study, the selection of the most suitable method for a given project is not easy since different approaches, with the same environmental conditions, produce different results (Coppin et al., 2004).

12.4.3.1 Classification-Based Change Detection

Broadly, image classification is a process of drawing meaningful information by differentiation and extraction of different classes or types (e.g., land-use types, vegetation species) from remote sensing data through a number of image processing procedures including image preprocessing. Image is classified either using traditional methods or improved or modified techniques.

The classification group of change detection techniques includes supervised and unsupervised classification followed by postclassification comparison of results of change/no-change identification. The aim of these methods is to produce high-quality, accurate classification results from remote sensing data through the use of adequate numbers of accurate training sample data, to produce accurate change results after comparison. However, the selection of sufficient numbers of high-quality training samples to truly represent each land-use/land-cover class is laborious and time consuming and requires a thorough knowledge of the study area in order to obtain high-quality classification results. The situation is more difficult in the classification of historical data when there is no or insufficient ground or area information, making accurate classification a challenge and often leading to unsatisfactory change detection results (Lu et al., 2004). The change detection results are often represented in the form of matrix showing land-use/land-cover dynamics of pixel changes from one class to another, providing a detailed description of changes over a specified period of time and minimizing the effect of atmospheric and environment difference between the multidate images. A detailed review of quality assessment of different image classification algorithms for land-cover mapping

and accuracy assessment has been summarized by Smits et al. (1999). These techniques have been used by many researches in different types of land-cover change detection analysis with good results (e.g., Li and Zhou, 2009; Petit and Lambin, 2002; Wang et al., 2009; Xiuwan, 2002).

Among traditional methods of classifying remote sensing data, unsupervised and supervised techniques are the most commonly used algorithms. ISODATA clustering and K-means methods are mainly used in unsupervised techniques while the maximum likelihood classification (MLC) and minimum distance to mean (MDM) are commonly applied techniques in supervised methods since the beginning of digital image feature extraction using statistical software. Unsupervised approaches, such as ISODATA clustering and K-means algorithms, are easy to apply for thematic mapping of land-cover or vegetation classification using statistical software packages (Langley et al., 2001) based on iterative learning of remote sensing data defined by the user. In this classification approach, an arbitrary initial cluster vector is assigned first, based on which each pixel is classified as closest to that cluster value. Finally, based on all the pixels present in one cluster, the new cluster mean vector is calculated and the process is repeated until the gap between the iterations becomes smaller than the threshold value (Xie et al., 2008). This method does not require any prior knowledge of the area or theme being studied (Cihlar, 2000) and classifies the image based on natural grouping or spectrally homogenous thematic classes using spectral values of each pixel. The analyst then assigns the spectral classes into thematic information classes of interest (Jensen, 2005). It has the benefit of automatically converting the raw image data into useful information so long as higher accuracy is achieved (Tso and Olsen, 2005). However, since each pixel is treated as spatially independent, a traditional unsupervised classification based on spectral data alone results in poor accuracy. To overcome this, Tso and Olsen (2005) introduced hidden Markov model (HMM) as a fundamental framework to incorporate both the spectral and contextual information for unsupervised classification and achieved higher classification accuracies.

In supervised classification, a prior knowledge about some cover types are obtained in advance through a combination of field work, aerial photo interpretation, existing maps, and other sources. Based on this, the analyst attempts to locate specific sites (class representation) for these cover classes in remote sensing data. These sites are called training sites and used to train the classification algorithm based on spectral characteristics of each site to classify the rest of the image. Statistical parameters such as mean, standard deviation, covariance matrices, correlation matrices, etc., are computed for each training site, based on these statistics, every pixel is then evaluated and assigned to a class of which it has the highest likelihood of being a member of (Jensen, 2005). Various supervised classification algorithms such as parallelepiped (PAR), minimum distance (MID), nearest neighborhood (NN), and maximum likelihood (MLC) may be used to assign an unknown pixel to one of the possible classes; therefore the choice of a particular classifier or decision rule depends on the nature of input data and desired output

(Jensen, 2005). MLC is considered to be classic and most widely used supervised classification technique, applied by many researchers in different studies for satellite image classification (Abdulaziz et al., 2009; Kamusoko and Aniya, 2009; Laba et al., 1997; Langford and Bell, 1997; Munoz-Villers and Blanco, 2008; Rogan et al., 2002; Xiuwan, 2002). In rangeland studies, Wu et al. (2008) evaluated the potential of multispectral Landsat images (MSS and TM) and applied MLC to classify vegetation cover in the degraded land of MuUs sandy land in China. Torahi (2012), in monitoring rangeland dynamics between 1990 and 2006, used MLC for classifying TM and ASTER data and generated detailed rangeland maps and also separated grazing intensity levels in rangelands.

However, in complex areas, the assumption of MLC that data follow Gaussian distribution is not always applicable, resulting in less satisfactory results. Bruin and Gorte (2000) and Strahler (1980) demonstrated the effective use of modified prior probabilities into MLC to improve image classification. In addition to MLC alone, there are studies conducted by researchers using a combination of other supervised classification algorithms, for example, Miller et al. (1998) applied parallelepiped technique to classify water, open land, clouds and cloud shadows, assuming them as discrete classes in spectral space, while other forest classes were classified using MLC in land-cover change study in Northern Forest of New England. Xiuwan (2002) compared several classification methods such as PAR, MID, Mahalanobis distance (MAD), MLC, ISODATA, and a method combining an unsupervised algorithm and training data (CUT) to develop a new classification method and suitable change detection technique for analysis of land-cover change and its regional impacts on sustainable development in Korea. Im et al. (2008) used nearest-neighborhood classification techniques for land-use and land-cover classification and compared the accuracy with other techniques. Munyati et al. (2011) used the hybrid image classification approach to monitor savanna rangeland deterioration in Mokopane, South Africa. For classification, initial clustering was undertaken by unsupervised classification using the ISODATA algorithm. On the resulting cluster image, the field data sites were then located and the spectral signature clusters were assigned names. The named spectral signatures were then used for a final supervised classification.

12.4.3.1.1 *Improved/Modified Classifiers*

Depending upon different geographic conditions and other factors, there is a possibility that same ground covers or vegetation types show different spectral response or different covers show similar spectral response on remote sensing imagery due to spectral mixing, therefore an accurate classification result is very difficult to obtain from traditional supervised or unsupervised classification techniques. Hence, there is always a need for improvement in the classification methods for better classification results. Some of the new methods have been based on traditional supervised or unsupervised approaches through a process of combination, extension or as an expansion to provide better classification results from remote sensing data (e.g., Pal, 2008).

Sohn and Rebello (2002) developed the new spectral angle classifier (SAC), a combination of supervised spectral angle classifier (SSAC) and unsupervised spectral angle classifiers (USAC), based on the fact that the spectra of same types of surface objects are approximately linearly scaled variations of one another due to atmospheric and topographic effects. SAC allows the spectral angle to be used as a metric for measuring angular distances in feature space for classification and clustering of multispectral satellite data and was successfully applied in biotic and land-cover classification (Sohn and Qi, 2005). Collado et al. (2002) found use of spectral mixture analysis (SMA) to be valuable in monitoring desertification processes in the crop-rangeland boundary of Argentina. Savanna rangeland degradation in Namibia was classified by Vogel and Strohbach (2009), who used Landsat TM and ETM+ data. The decision tree classifier was also used. Their results showed that savanna degradation could be classified into the following six classes: vegetation densification, vegetation decrease, complete vegetation loss, long-term vegetation patterns, the recovery of vegetation on formerly bare soils, and no change with an overall accuracy of 73.4%, with class pair accuracies ranging from 80% to 100% for producer and user accuracies. Okin et al. (2001) assessed the utility of AVIRIS satellite imagery for accurately discriminating among vegetation types in the Mojave Desert, USA. Multiple endmember spectral mixture analysis (MESMA) and SMA were performed to estimate the proportion of each ground pixels area that fitted with different cover types. They concluded that AVIRIS showed low potential for classifying vegetation types, with an overall accuracy of only 30% due to low vegetation cover. This is a common problem in rangelands, where the overall classification accuracies are generally lower than for forest or well vegetated areas.

James et al. (2003) suggested that ecological condition of rangelands is a major factor in their environmental quality, their overall performance as watersheds and in wildlife and livestock production. They pointed out that to maintain the quality of rangelands, they must be monitored over time and space and also must take into account topography, climate, soils, plant communities, and animal population. Several studies have shown that extensive field knowledge and support data improves classification accuracy, for example, Wang et al. (2009) used stratified classification approaches based on segmentation of an image into focused area and categories based on existing GIS land-cover data in order to improve classification accuracy. Franklin and Wilson (1992) developed a three-stage classifier that incorporated a quadtree-based segmentation operator, a Gaussian minimum-distance to mean test and final test involving ancillary geomorphometric data and a spectral curve measure and attained significant increase in classification accuracy in less time and with minimum field training data as compared to MLC. Jianlong et al. (1998) used green herbage yield data, environment, and remote sensing data recorded in different grassland types in Fukang County, Xinjiang, from 1991 to 1996. They explored the methods of processing images, analyzing information, and linking of remote sensing data with ground grassland data. Tagestad and Downs (2007) studied landscape measures of rangeland condition using texture methods, highlighting the apparent roughness in the visible surface due to drastic changes in brightness between adjacent pixels. The texture model reduces the color signal in the image and maximizes the texture signal. The field-measured shrub canopy cover in each plot was compared to the corresponding texture ratio values for pixels representing that plot to develop a simple linear regression relationship between shrub canopy cover and image texture. Franklin et al. (2001) used spatial co-occurrence texture measures and MLC to generate higher forest species composition classification accuracies in New Brunswick forest stand than the use of spectral patterns alone. Gong and Howarth (1992) developed and evaluated a contextual method of land-use classification using SPOT data involving two steps: gray-level vector reduction and frequency-based classification and found that the frequency-based classification method was comparatively fast, efficient, and could improve land-use classification accuracies over MLC and was effective in identification of spatially heterogeneous land-use classes. Pinˉeiro et al. (2006) estimated seasonal variation in aboveground production and radiation-use efficiency of temperate rangelands using remote sensing. They evaluated, at a seasonal scale, the relationship between ANPP and the NDVI and estimated the seasonal variations in the coefficient of conversion of absorbed radiation into aboveground biomass and also identified the environmental controls on such temporal changes. Their results indicated that NDVI produced good, direct estimates of ANPP only if NDVI, PAR, and aboveground biomass were correlated throughout the seasons and hence suggested seasonal variations of aboveground biomass associated with temperature and precipitation to be taken into account to generate seasonal ANPP estimates with acceptable accuracy.

Fuzzy methods in the classification of remote sensing images have become popular because of their ability to deal with situations where the geographical boundary is inherently fuzzy or heterogeneous due to the presence of mixed pixels and traditional methods of classification are often incapable of performing satisfactorily (Zang and Foody, 1998). For example, Zhang and Foody (1998) investigated the fuzzy approach for the classification of sub-urban land cover from remote sensing imagery and reported to have advantages over both conventional hard method and partially fuzzy approach. Other similar applications have been by Okeke and Karnieli (2006) for vegetation change study, and Sha et al. (2008) for grassland classification. Discriminating and mapping vegetation degradation at Fowlers Gap Arid Zone Research Station in Western New South Wales, Australia, Lewis (2000) used random forest method to classify perennial vegetation, chenopod shrubs and trees using hyperspectral imaging (CASI). An area of less than 25% was discriminated and mapped. Lewis (2000) concluded that high-spectral resolution imagery had potential for the discrimination of vegetation cover in arid regions.

Recently, decision tree (DT) classifiers have been used for land cover and vegetation classification from remote sensing data based on the concept of splitting a complex decision into several simpler decisions that may be easier to interpret. DT is based on multistage or hierarchal decision scheme or a tree-like structure composed of a root node (containing all data), a set of internal nodes (splits) and a set of terminal nodes (leaves) and processes from top to bottom by moving down the tree where each node of the decision tree structure makes a binary decision that separates either one class or some of the classes from the remaining classes (Chen and Rao, 2008; Xu et al., 2005). DT is relatively simple, explicit, computationally fast, makes no statistical assumptions and can handle data on different measurement scales (Friedl and Brodley, 1997; Pal and Mather, 2003). Chen and Rao (2008) determined the rate and status of grassland degradation and soil salinization based on DT and field investigation with overall classification accuracy of more than 85%. Xu et al. (2005) employed a decision tree regression approach to determine class proportions within a pixel so as to produce a soft classification and the accuracy achieved by DT regression was found to be significantly higher as compared to MLC applied in soft mode and supervised version of fuzzy-c-soft classification, especially when data contained a large proportion of mixed pixels.

Pal and Mather (2003) assessed the utility of DT classifier for land-cover classification using multispectral and hyperspectral data and compared the performance of univariate and multivariate DT with that of ANN and MLC. They concluded that DT performance was always affected by training data size, univariate DT was more systematic than multivariate DT for common training and test data sets and, in the case of univariate DT, a minimum of 300 training pixels per class were needed to the achieve most suitable classification accuracy. They further concluded that DT classifiers were not recommended for high-dimensional data sets. Friedl and Brodley (1997) tested univariate DT, multivariate DT, and a hybrid DT on three different remote sensing data sets and compared the classification results from each DT algorithm with those of MLC and linear discriminant function classifier, and reported that DT, hybrid DT in particular, consistently outperformed MLC and linear function discriminant classifier in regard to classification accuracy. Rogan et al. (2002) compared MKT, MSMA, MLC, and DT to accurately identify changes in vegetation cover in south California and showed that DT classification approach outperformed MLC by nearly 10% regardless of enhancement technique used and using DT classification, MSMA change fractions outperformed MKT change features by nearly 5%. Borak and Strahler (1999) developed a tree-based model for land-cover identification from satellite data for a semiarid region in Cochise County, Arizona, and compared the results with other classifiers such as fuzzy ARTMAP, MLC, and reported that DT could reduce a high-dimension data set to a manageable set of inputs that retained most of the information of the original database. However, fuzzy ARTMAP achieved the highest accuracy in comparison to MLC or DT classifier. Muchoney et al. (2000) compared the classification results obtained from MODIS data for vegetation and land-cover mapping using DT, Gaussian ART, and Fuzzy ART ANN algorithms in Central America and attained high accuracies from DT (88%), Gaussian ART (83%), and Fuzzy ART NN (79%).

Jarman et al. (2011) studied rangeland conditions in the regions of Kvemo Kartli and Samtskhe-Javakheti in southern Georgia, USA, using remote sensing data. They used (NDVI) as the indicator of condition and used object-based image analysis. The Landsat scenes and ancillary data sets were collated within eCognition following a two tiered (multilevel) hierarchical approach. The first level brought together the nonrangeland data sets and the second level classified the rangeland area into relative states of rangeland condition (good, moderate, poor). A rule-based classification was undertaken within eCognition to map the classes. Laliberte et al. (2011) applied IHS transformation on remote sensing data followed by object-based segmentation and classification for structure and species-level rangeland mapping. They developed specific rules to define threshold for broader classes and nearest-neighborhood classification for finer species-level classification. They reported that classification accuracies were highly dependent on the level of detail, number of classes, size of area, and specific mapping objectives.

12.4.3.2 Image Differencing and Image Ratioing

The algebraic technique of change detection analysis is based on the selection of a threshold to determine the changed areas (Lu et al., 2004). In this group of change detection techniques, many researchers have applied image differencing and image ratioing with different combinations of spectral bands as their first choice for identifying change, and derived satisfactory results. Conclusions and recommendations vary about whether image differencing and regression, vegetation index differencing or image ratioing is the best change detection technique. Since each method has been applied to different areas, with different data sets and under different environmental conditions, the decision on the selection of a suitable method is not an easy task (Coppin et al., 2004). For example, Sinha and Kumar (2013a) tested 11 different binary change detection methods and compared their capability in detecting land-cover change/no-change information in different seasons using multidate TM data. They proposed a relatively new approach for optimal threshold value determination for separation of change/no-change areas and found improved results with this as compared to traditional thresholding (Sinha and Kumar, 2013b).

12.4.3.2.1 Transformation

Various linear data transformation techniques such as PCA, Tasselled Cap (TC), Gramm–Schmidt (GS), and chi-square transformations have been applied to multitemporal remote sensing data to identify changes in various land-use/land-cover classes. In PCA, only two bands of the multidate image are used as input instead of all bands (Mas, 1999; Richards, 1986), and hence reducing the data volume and redundancy between bands. After transformation using two bands, the derived components contain information about change and no-change areas as the first and second components, respectively, based on information common or unique in the two input bands (Chavez and Kwarteng, 1989).

PCA is based on three steps: calculation of a variance–covariance matrix, computation of eigenvectors, and linear transformation of data sets (Richards, 1986). Two types of PCA, such as standardized PCA (uses correlation matrix) and nonstandardized PCA (uses covariance matrix), have been used for change detection (Singh and Harrison, 1985). PCA has certain disadvantages in not providing detailed change matrices and difficulty in interpretation, identification, and labeling of changed areas.

TC transformation is carried out by assigning tasselled cap coefficients to spectral bands of two dates, a positive coefficient to the first date and a negative coefficient to the second date, as explained by Crist and Cicone (1984). This is followed by Gramm–Schmidt transformation to make the derived vectors orthogonal to each other. The three transformed images thus obtained contain information about differences in greenness, brightness, and wetness, with highest classification accuracy in the greenness change image. Maynard et al. (2007) classified Landsat 7 ETM+ to identify spectrally anomalous locations on satellite data and their correlation with corresponding ground locations for rangeland evaluation. Their classification was carried out using TC brightness, greenness, and wetness components stratified by ecological site descriptions. PCA and TC have been the most used approaches for detecting change/no-change information. An additional advantage of TC transformation is that the TC coefficients are scene independent compared to PCA coefficients, which are scene dependent.

12.4.4 Case Study: Spectral-Spatial Characteristics of Selected Ecological Sites

12.4.4.1 Introduction

ES characterize land of specific biophysical and plant community properties. Spatially distinct land areas of the same ecological site respond similarly to management actions and natural disturbances (U.S. Department of Agriculture, NRCS, 2012). Therefore, the identification and mapping of ES across large landscapes can be an important tool for rangeland managers. ES, as they are applied by the United States Department of Agriculture Natural Resources Conservation Service (NRCS), are defined on the basis of soils, geomorphology, hydrology, and the plant species composition that occur on those soils. The NRCS spatially ties ESs to soil components mapped within SMU contained in their spatial digital soil surveys.

Bestelmeyer et al. (2009) formulated an approach to identify ESs from Landsat (or similar platforms) interpreted into land-cover maps to identify vegetation distribution. These mapped vegetation areas are used to infer possible ES. In addition to this effort, Maynard et al. (2007) found that there was a high correlation between field measures of productivity and exposed soil when compared to the tasseled cap brightness component extracted from Landsat TM imagery. Differences in brightness have been shown to discriminate between deciduous shrubs (or harvested forest stands) and closed canopy forests (Dymond et al., 2002).

Gamon et al. (1995) discussed the usefulness of the NDVI as an indicator of photosynthetic activity as well as canopy structure and plant nitrogen content. Jensen (2000) showed that NDVI was sensitive to canopy variations including soil visible through canopy openings. While the sensitivity to soil background has typically been seen as a disadvantage of NDVI for vegetation assessment, it could prove useful for studying ESs because areas of the same ES may have a similar amount and type of bare soil. The NDVI values within a polygon, such as an SMU, and the variation in the NDVI within that polygon has also been used to distinguish between cover types (Pickup and Foran, 1987).

Accurately classifying and identifying the spatial extent of ESs on a landscape level is a time-consuming process involving extensive field work to characterize soils. While remotely sensed data cannot yet be used to obtain detailed data about soils, it can be used to identify the unique vegetation components of ESs. Being able to accurately identify the vegetation component of ESs should provide a means by which soil field sample locations can be identified more efficiently. Based on past research, the use of satellite-derived NDVI and brightness, coupled with biophysical geospatial data (elevation, slope, and aspect), should allow areas of the same ES vegetation components to be mapped.

12.4.4.2 Methods

12.4.4.2.1 Study Area

This research was conducted in Rich County, Utah, USA (Figure 12.11), located in the northeastern corner of the state (long 111°30′38.5″–long 111°2′42.2″ west and lat 42°0′0″–lat 42°08′24.3″ north). The western portion of the study area is characterized by high elevations with vegetation consisting of aspen forests, subalpine conifer forests, and scattered mountain sagebrush steppe. Moving east, elevation decreases, and the mountain sagebrush steppe becomes dominant. Central and eastern Rich County is made up of relatively lower elevations with vegetation consisting of basin big sagebrush steppe and shrubland, subalpine grasslands, and agriculture (Figure 12.12). The average elevation is 2093 m. The highest point is Bridger Peak at 2821 m and the lowest point is about 1800 m. The climate is variable and is affected by the changing topography of the county.

12.4.4.2.2 Biophysical Geospatial Data Sets

A series of Landsat 5 TM images (Path 38/Row 31) for each year between 1984 and 2011 with Julian date as close to 207 (July 26th) as possible (given acceptable cloud cover) were collected from the U.S. Geological Survey Global Visualization Viewer (GLOVIS). The Julian date of 207 was chosen by averaging the date for each year that displayed the greatest variance in NDVI between different land-cover types. The dates were obtained by examining line graphs of mean NDVI values collected by the MODIS of evergreen forests, shrubs, and deciduous forests. Figure 12.13 is an example of one of these graphs from 2009.

Of the 28 years' images, 18 were within 20 days of 207, 5 more were within 30 days of 207, and 3 more were within 40 days of 207. The cloud-free scene closest to Julian date 207 from 1987 had a Julian date of 153 and was 54 days off. The year 2001 was the only year that an image was not available due to cloud cover.

Raw pixel values were converted to reflectance using an image-based atmospheric correction (Chavez, 1996) using appropriate calibration coefficients for Landsat 5 TM (Chander et al., 2009). The NDVI was calculated for each image (Rouse et al., 1974) and for each NDVI product we calculated the spatial standard deviation of the NDVI using a 5 × 5 pixel focal window. The brightness component for each year was calculated using the published transformation coefficients for Landsat 5 TM (Crist and Cicone, 1984). Images from every year were utilized, based on literature indicating that longer time series of remotely sensed data were necessary to adequately characterize different ecological states due to inherent year-to-year variance (Hernandez, 2011).

A 30 m DEM produced by the USGS National Elevation Dataset program was used to extract slope and aspect. Elevation, slope, and aspect have been shown to drive microclimatic variation and therefore the spatial distribution and patterns of vegetation (Jin et al., 2008).

12.4.4.2.3 Ecological Sites

To test the process, land-cover types representing five ESs were selected. These were Wyoming big sagebrush (*Artemisia tridentata* ssp. *wyomingensis*) steppe, mountain big sagebrush (*Artemisia tridentata* ssp. *vaseyana*), Utah juniper (*Juniperus osteosperma*), Douglas-fir (*Pseudotsuga menziesii*), and aspen (*Populus tremuloides*). With the exception of Utah juniper, these vegetation components were selected because of their prevalence in the county. Wyoming big sagebrush accounts for much of the foothill vegetation in the study area, and aspen is prevalent in the mountainous section, with Douglas-fir, and mountain big sagebrush as secondary and tertiary types. Utah juniper is not prevalent within the study area; however, it is an important vegetation component due to its potential to encroach into sagebrush steppe communities (Miller and Rose, 1999). Together, these vegetation components represent approximately 71% of the county by area.

Twenty polygons were digitized for each of the five ESs using 2009 National Agricultural Imagery Project (NAIP) 1 m resolution aerial orthoimagery. In total, 100 polygons were created (20 for each ES vegetation component).

Polygons were intersected with the topographic data layers, yearly NDVI imagery, and yearly brightness component images. For each polygon, the mean values of topographic and brightness variables were extracted along with the mean and standard deviation of each NDVI image for each year. From these data, a 28-year mean of the average brightness as well as the mean of the average NDVI and the mean NDVI standard deviation (sdNDVI) were produced. This was done to minimize the effects of interannual climate variability and clouds and provide long-term average values for each polygon of each land-cover type.

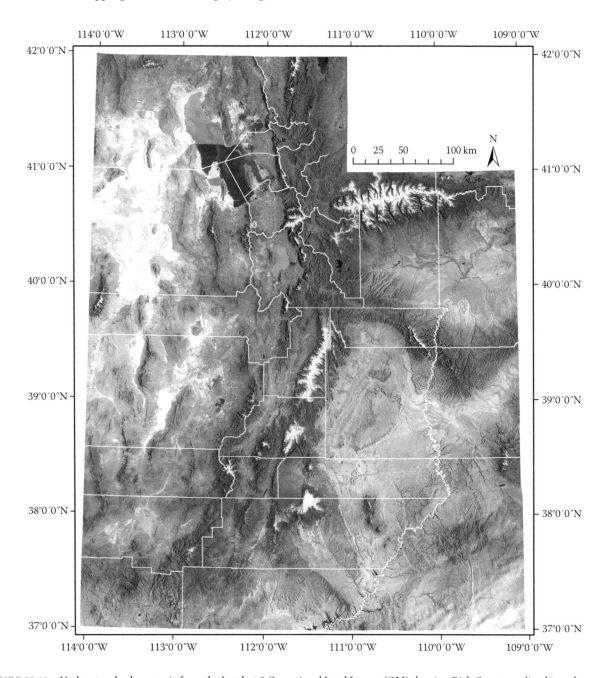

FIGURE 12.11   Utah natural color mosaic from the Landsat 8 Operational Land Imager (OLI) showing Rich County outlined in red.

Interannual climate variability has been shown to affect plant species productivity (Arain et al., 2002; Goulden et al., 1996) and ecological processes (Westerling and Swetnam, 2003). The resulting data matrix was therefore composed of the ES vegetation component name followed by three columns for the DEM derivatives, and three columns for the NDVI, sdNDVI, and brightness.

12.4.4.3 Results

Figure 12.14 shows the 28-year mean of the average NDVI value for each polygon plotted against the 28-year mean of the sdNDVI for each polygon showing that the five ESs occupied unique NDVI mean and spatial variance "niches." Some overlap occurred between Wyoming big sagebrush and Utah juniper and between Douglas-fir and aspen ESs. Figure 12.15 shows the same data points with 28-year mean of the average NDVI value plotted against the 28-year mean of the average brightness. The brightness component was able to cleanly separate Aspen polygons from the Douglas-fir polygons. However, brightness provided little separation between Utah juniper and Wyoming big sagebrush.

Each polygon was plotted against elevation and slope (Figure 12.16) and also against elevation and the cosine of aspect (Figure 12.17). Topographic variables alone were able

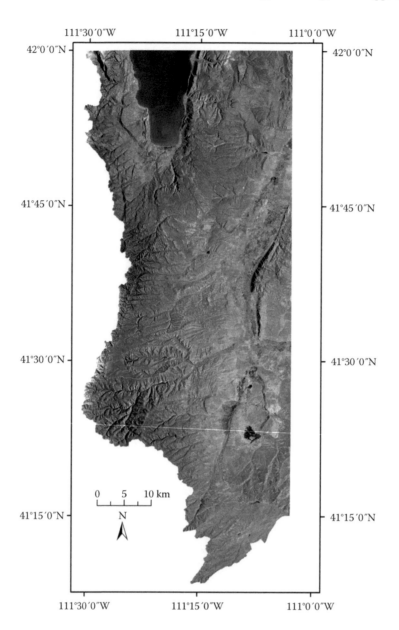

FIGURE 12.12   Rich County, Utah, natural color image from Landsat 8 OLI.

to somewhat separate vegetation components along an eleva-tion gradient (as expected). Slope seemed to be a good variable to separate Utah Juniper from Wyoming big sagebrush and Douglas-fir from Aspen. Aspect was not useful for distinguish-ing between any vegetation types.

The Wyoming big sagebrush polygons collectively had low NDVI and low spatial variation in NDVI. Utah juniper sites had similarly low-average NDVI, but due to high contrast between green juniper trees and a relatively larger amount of bare ground, these sites had higher spatial variation in NDVI. Mountain big sagebrush had higher average NDVI values. This was expected since mountain big sagebrush occurs at higher elevations that receive more precipitation than either Wyoming big sagebrush or Utah juniper and therefore is associated with higher plant production. Aspen polygons tended to have higher NDVI values

compared to Douglas-fir polygons with both ESs having a simi-lar, relatively large distribution of spatial variance. Where the aspen sites are concerned, there was a slight but distinct trend of decreasing composite mean NDVI with increasing mean stan-dard deviation (sdNDVI). Indeed considering the four aspen sites with the highest sdNDVI, three sites consisted of lower aspen canopy cover and the site with the highest sdNDVI con-tained a mix of immature aspen trees and shrubs. The remaining 16 aspen sites were all similar in canopy cover.

12.4.4.4 Discussion

This case study has shown that using variables derived from remotely sensed images as well as biophysical geospatial data, selected ESs can be discriminated on a per-pixel basis. Prediction of the spatial distribution of ESs on a pixel basis

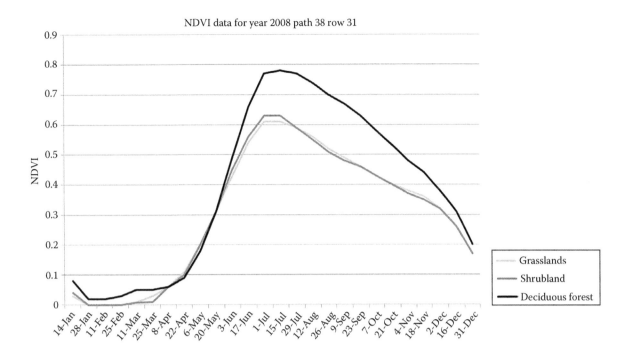

FIGURE 12.13 Line graph of annual fluctuations in NDVI for evergreen forests, shrublands, and deciduous forests. The largest differences in NDVI can be seen in midsummer.

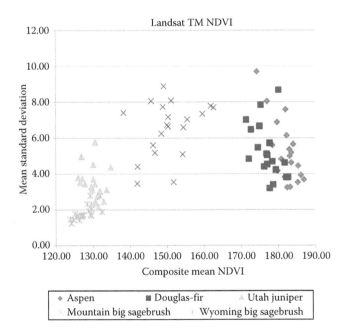

FIGURE 12.14 Scatter plot showing the distribution of each ecological site vegetation component in our study with average NDVI value on the x-axis and the average standard deviation in NDVI on the y-axis. Both of these variables provide separation between vegetation classes.

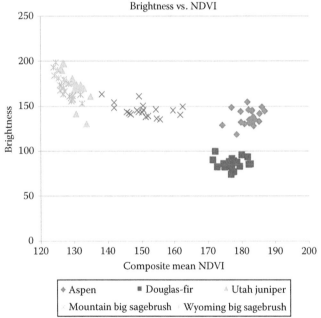

FIGURE 12.15 Scatter plot showing the distribution of each ecological site vegetation component in our study with average NDVI value on the x-axis and the average brightness component (obtained from the tasseled cap transformation) value on the y-axis.

has been suggested as the next step in remote sensing applications to rangeland conservation (Hernandez, 2011). This research has shown that selected ESs can be somewhat cleanly discriminated by utilizing spectral (NDVI and brightness) and spectral-spatial (sdNDVI) variables and therefore could

be mapped by utilizing remotely sensed imagery. This is, of course, not surprising since the remote sensing community has clearly shown the ability to map land cover. What is important here is that by identifying the spatio/temporal spectral nature of selected land-cover types, land-cover change can be tracked.

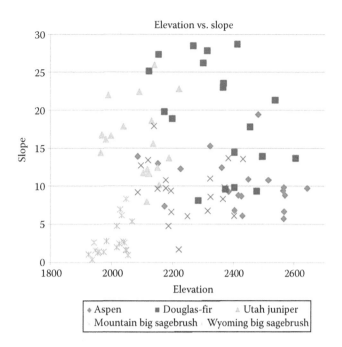

FIGURE 12.16 Scatter plot showing the distribution of each ecological site vegetation component in our study with elevation on the x-axis and slope on the y-axis. Most vegetation classes overlap one another. However, slope does help with separating Wyoming big sagebrush from Utah juniper.

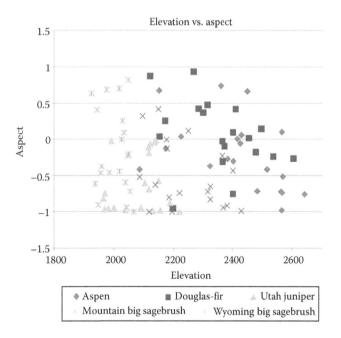

FIGURE 12.17 Scatter plot showing the distribution of each ecological site vegetation component in our study with elevation on the x-axis and aspect on the y-axis. Aspect does not appear to separate any vegetation classes.

ES are used by the USDA-NRCS as a benchmark condition to develop state and transition models (Bestelmeyer et al., 2009; Briske et al., 2005). ES linked to SMU provide a means to compare current land-cover condition to historic conditions (as defined by the ES). By establishing the spectral nature of ESs,

the spectral/spatial signature of soil map polygons can be compared with an expected ES for that polygon. Deviations from the expected response could indicate type and directionality of change, thus providing managers with a powerful land management tool.

12.5 Conclusion

Rangelands are dynamic environments that probably exhibit change more frequently and profoundly than many other systems. Land management, livestock grazing, invasive species, prolonged droughts and fire have the potential to alter rangeland community structure at both the micro and macro levels. Vegetation communities can be dominated by either C_3 or C_4 grass species or have permanent shrublands; thus, productivity is very phenology dependent. Biomass production in rangelands by C_3 and C_4 plants displays significantly different phenological timing that is detectable using satellite remote sensing. The phenological differences are also useful for identifying vegetation types. Such information about rangeland productivity, vegetation species, and timing are useful for rangeland managers in adjusting stocking rate, broad-scale movement of stock, land degradation monitoring, and long-term trends in the health and condition of the systems. This chapter has highlighted numerous methods of achieving these by using remotely sensed imagery.

Accurately classifying and identifying the spatial extent of ES on a landscape scale is a laborious task involving extensive fieldwork to characterize soils. Case studies presented in this chapter show that variables extracted from remote sensing images, together with biophysical geospatial data, can be used to discriminate ES on a per-pixel basis with a relatively high degree of confidence. ES are widely used as a benchmark condition to develop state and transition models and provide a means to compare current land-cover condition with historic conditions, thus indicating type and directionality of change and helping in rangeland management.

Monitoring of fuel in rangelands is also important and this chapter has provided examples of how high temporal resolution satellite remote sensing can be used for quantifying fuel loads in rangeland landscapes. We have also summarized a range of change detection techniques relevant to rangeland monitoring and discussed, through a case study, the effectiveness of a regression tree algorithm (random forest) in modeling estimates of percent cover and developing VCF for a range of vegetation communities in a rangeland environment.

Remote sensing has undergone major transformations since its mainstream application began about 50 years ago. Spatial resolutions of images have shrunk from a few kilometers range to the submeter range, spectral resolutions have moved to the hyperspectral domain and temporal resolutions enable images to be available almost on a daily basis. In conjunction with this development, image classification techniques have made major progress and continue to be developed. New techniques include neural networks, random forest and object-based classifications, and research cited earlier show that they have led to a marked increase in classification accuracies. All these developments have greatly enhanced our ability to map and monitor rangelands to a degree not possible in the past and no doubt will continue to be improved into the future as new techniques and sensor capabilities become mainstream. This chapter has covered some of these developments and highlighted the importance and usefulness of remote sensing for rangeland mapping and monitoring.

References

Abdulaziz, A. M., Hurtado, J. M., and Al-Douri, R. 2009. Application of multitemporal Landsat data to monitor land cover changes in Eastern Nile Delta region, Egypt. *International Journal of Remote Sensing*, 30, 2977–2996.

Ajorlo, M. and Abdullah, B. R. 2007. Develop an appropriate vegetation index for assessing rangeland degradation in semi-arid areas. In *Proceedings of 28th Asian Conference on Remote Sensing*, November 12–16, Kuala Lumpur, Malaysia.

Amiri, F. and Shariff, A. R. B. M. 2010. Using remote sensing data for vegetation cover assessment in semi-arid rangeland of Center Province of Iran. *World Applied Sciences Journal*, 11(12), 1537–1546.

Anderson, H. 1982. Aids to determining fuel models for estimating fire behavior. General Technical Report GTR INT-122. USDA Forest Service, Intermountain Forest and Range Experiment Station, Ogden, UT. p. 22.

Anderson, M. A. 1997. Two-source time-integrated model for estimating surface fluxes using thermal infrared remote sensing. *Remote Sensing of Environment*, 60, 195–216.

Andrews, P. L. and Bevins, C. D. 2003. BehavePlus fire modeling system, version 2.0: overview. In *Proceedings of the Second International Wildland Fire Ecology and Fire Management Congress and Fifth Symposium on Fire and Forest Meteorology*, November 16–20, 2003, American Meteorological Society, Orlando, FL. p. P5.11.

Arain, M. A., Black, T. A., Barr, A. G., Jarvis, P. G., Massheder, J. M., Verseghy, D. L., and Nesic, Z. 2002. Effects of seasonal and interannual climate variability on net ecosystem productivity of boreal deciduous and conifer forests. *Canadian Journal of Forest Research*, 32, 878–891.

Arroyo, L. A., Pascual, C., and Manzanera, J. A. 2008. Fire models and methods to map fuel types: The role of remote sensing. *Forest Ecology and Management*, 256, 1239–1252.

Asrar, G., Kanemasu, E., and Yoshida, M. 1985. Estimates of leaf area index from spectral reflectance of wheat under different cultural practices and solar angles. *Remote Sensing of Environment*, 17, 1–11.

Badeck, F.-W., Bondeau, A., Bottcher, K., Doktor, D., Lucht, W., Schaber, J., and Sitch, S. 2004. Responses of spring phenology to climate change. *New Phytologist*, 162(2), 295–309.

Bailey, R. G. and Hogg, H. C. 1986. A world ecoregions map for resource reporting. *Environmental Conservation*, 13, 195–202. doi:10.1017/S0376892900036237.

Baret, F., Guyot, G., and Major, D. J. 1989. TSAVI: A vegetation index which minimizes soil brightness effects on LAI and APAR estimation. *12th Canadian Symposium on Remote Sensing and IEEE International Geoscience and Remote Sensing Symposium 1990*, July 10–14, 1989. Vancouver, British Columbia, Canada.

Baskent, E. Z. and Jordan, G. A. 1995. Characterizing spatial structure of forest landscapes. *Canadian Journal of Forest Research*, 25, 1830–1849.

Bastin, G. and Jones, C. 2000. Assessment of biodiversity condition in a rangelands environment using remote sensing. CSIRO Sustainable Ecosystems Centre for Arid Zone Research. CSIRO Report, Alice Springs, Northern Territory, Australia.

Batistella, M., Robeson, S., and Moran, E. F. 2003. Settlement design, forest fragmentation, and landscape change in Rondônia, Amazônia. *Photogrammetric Engineering and Remote Sensing*, 69, 805–811.

Bestelmeyer, B. T., Tugel, A. J., Peacock, G. L., Robinett, D. G., Shaver, P. L., Brown, J. R., Herrick, J. E., Sanchez, H., and Havstad, K. M. 2009. State-and-transition models for heterogeneous landscapes: A strategy for development and application. *Rangeland Ecology & Management*, 52, 1–15.

Birth, G. S. and McVey, G. 1968. Measuring the color of growing turf with a reflectance spectrophotometer. *Agronomy Journal*, 60, 640–643.

Blumenthal, D., Booth, D. T., Cox, S. E., and Ferrier, C. E. 2007. Large-scale aerial images capture details of invasive plant populations. *Rangeland Ecological Management*, 60, 523–528.

Borak, J. S. and Strahler, A. H. 1999. Feature selection and land cover classification of MODIS-like data set for a semiarid environment. *International Journal of Remote Sensing*, 20, 919–938.

Bowers, S. A. and Hanks, R. J. 1965. Reaction of radiant energy from soil. *Soil Science*, 100, 130–138.

Boyte, S. P., Wylie, B. K., Major, D. J., and Brown, J. F. 2015. The integration of geophysical and enhanced Moderate Resolution Imaging Spectroradiometer Normalized Difference Vegetation Index data into a rule-based, piecewise regression-tree model to estimate cheatgrass beginning of spring growth. *International Journal of Digital Earth*, 8(2), 116–130.

Bradley, B. A. and Mustard, J. F. 2008. Comparison of phenology trends by land cover class: A case study in the Great Basin, USA. *Global Change Biology*, 14, 334–346.

Briggs, J. M. and Knapp, Z. A. K. 1995. Interannual variability in primary production in tallgrass prairie: Climate, soil moisture, topographic position, and fire as determinants of aboveground biomass. *American Journal of Botany*, 82, 1024–1030.

Briske, D. D., Fuhlendorf, S. D., and Smeins F. E. 2005. State-and-transition models, thresholds, and rangeland health: A synthesis of ecological concepts and perspectives. *Rangeland Ecology & Management*, 58, 1–10.

Bruin, S.D., and Gorte, B.H.G., 2000. Probabilistic image classification using geological map units applied to land-cover change detection. *International Journal of Remote Sensing*, 21(12), 2389–2402.

Cain, D. H., Riitters, K., and Orvis, K. 1997. A multi-scale analysis of landscape statistics. *Landscape Ecology*, 12, 199–212.

Campbell, J. B. 1996. Introduction to Remote Sensing. Taylor & Francis, London.

Carlson, T. and Ripley, D. 1997. On the relationship between fractional vegetation cover, leaf area index, and NDVI. *Remote Sensing of Environment*, 62, 241–252.

Chander, G., Markham, B. L., and Helder, D. L. 2009. Summary of current radiometric calibration coefficients for Landsat MSS, TM, ETM+, and EO-1 ALI sensors. *Remote Sensing of Environment*, 113, 893–903.

Chang, C. R., Lee, P. F., Bai, M. L., and Lin, T. T. 2004. Predicting the geographical distribution of plant communities in complex terrain—A case study in Fushian Experimental Forest, northeastern Taiwan. *Ecography*, 27, 577–588.

Chavez, P. S. 1996. Image-based atmospheric corrections revisited and improved. *Photogrammetric Engineering and Remote Sensing*, 62, 1025–1036.

Chavez, P. S. Jr. and Kwarteng, A. Y. 1989. Extracting spectral contrast in Landsat Thematic Mapper image data using selective principal component analysis. *Photogrammetric Engineering and Remote Sensing*, 55, 339–348.

Chen, S. and Rao, P. 2008. Land degradation monitoring using-multi-temporal Landsat TM/ETM data in a transition zone between grassland and cropland of northeast China. *International Journal of Remote Sensing*, 29, 2055–2073.

Cheney, P. and Sullivan, A. 1997. *Grassfires: Fuel, Weather and Fire Behaviour*, CSIRO Publishing, Collingwood, Victoria, Australia.

Chopping, M., Lihong, S., Rango, A., Martonchik, J. V., Peters, D. P. C., and Laliberte, A. 2006. Remote sensing of woody shrub cover in desert grasslands using MISR with a geometric-optical canopy reflectance model. *Remote Sensing of Environment*, 112, 19–34.

Cihlar, J. 2000. Land cover mapping of large areas from satellites: status and research priorities. *International Journal of Remote Sensing*, 21(67), 1093–1114.

Cihlar, J., St-Laurent, L., and Dyer, J. A. 1991. Relation between the normalized difference vegetation index and ecological variables. *Remote Sensing of Environment*, 35, 279–298.

Collado, A. D., Chuvieco, E., and Camarasa, A. 2002. Satellite remote sensing analysis to monitor desertification processes in the crop-rangeland boundary of Argentina. *Journal of Arid Environment*, 52, 121–133.

Conservancy, T. N. 2009. *Spatial Modeling of the Cumulative Effects of Land Management Actions on Ecological Systems of the Grouse Creek Mountains–Raft River Mountains Region*, Salt Lake City, UT. p. 243.

Coppin, P., Joncheere, I., Nackaerts, K., and Muys, B. 2004. Digital change detection methods in ecosystem monitoring: A review. *International Journal of Remote Sensing*, 25, 1565–1596.

Crews-Meyer, K. A. 2002. Characterizing landscape dynamism using pane-pattern metrics. *Photogrammetric Engineering and Remote Sensing*, 68, 1031–1040.

Crist, E. P. and Kauth, R. J. 1986. The tasseled cap de-mystified. *Photogrammetric Engineering and Remote Sensing*, 52, 81–86.

Crist, E. P. and Cicone, R. C. 1984. A physically-based transformation of Thematic Mapper data—The TM tasseled cap. *IEEE Transactions on Geosciences and Remote Sensing*, 22, 256–263.

Crist, E. P. 1983. The thematic mapper tasseled capped a preliminary formulation. In *Proceedings of 1983 Machine Processing of Remotely Sensed Data Symposium*, Environmental Research Institute of Michigan, Ann Arbor, MI, pp. 357–363.

Crist, E. P. and Cicone, R. C. 1984. Application of the tasseled cap concept to simulated Thematic Mapper data. *Photogrammetric Engineering and Remote Sensing*, 50, 343–352.

Crookston, N. L. and Finley, A. O. 2008. yaImpute: An R package for kNN imputation. *Journal of Statistical Software*, 23, 1–16.

Cumming, S. and Vervier, P. 2002. Statistical models of landscape pattern metrics, with applications to regional scale dynamic forest simulations. *Landscape Ecology*, 17, 433–444.

Cutler, D. R., Edwards, T. C., Beard, K. H., Cutler, A., and Hess, K. T. 2007. Random forests for classification in ecology. *Ecology*, 88, 2783–2792.

Danaher, T., Armston, J., and Collett, L. 2004. A regression model approach for mapping woody foliage projective cover using landsat imagery in Queensland, Australia. In *IEEE International Proceedings of Geoscience and Remote Sensing Symposiuim*, Anchorage, AK. pp. 523–527.

Defries, R. S., Hansen, M. C., and Towsend, J. R. G. 2000. Global continuous fields of vegetation characteristics: A linear mixture model applied to multi-year 8 km AVHRR data. *International Journal of Remote Sensing*, 21, 1389–1414.

Desalew, T., Tegegne, A., Nigatu, L., and Teka, W. 2010. Rangeland condition and feed resources in Metema District, North Gondar Zone, Amhara Region, Ethiopia, IPMS (Improving Productivity and Market Success) of Ethiopian Farmers Project Working Paper 25, Nairobi, Kenya.

Dunn, B. H., Smart, A. J., Gates, R. N., Johnson, P. S., Beutler, M. K., Diersen, M. A., and Janssen, L. L. 2010. Long-term production and profitability from grazing cattle in the northern mixed grass prairie. *Rangeland Ecology and Management*, 63, 233–242.

Dymond, C. C., Mladenoff, D. J., and Radeloff, V. C. 2002. Phenological differences in tasseled cap indices improve deciduous forest classification. *Remotes Sensing of Environment*, 80, 460–472.

Ellis, E. C. and Ramankutty, N. 2008. Putting people in the map: anthropogenic biomes of the world. *Frontier in Ecology and the Environment*, 6, 439–447.

Elvidge, C. D. and Lyon, R. J. P. 1985. Influence of rock-soil spectral variation on the assessment of green biomass. *Remote Sensing of Environment*, 17, 37–53.

Finney, M. A. 2004. FARSITE: Fire Area Simulator—Model development and evaluation. RP-RMRS-004. U.S. Department of Agriculture, Forest Service, Rocky Mountain Research Station, Ogden, UT. pp. 47.

Foody, G. M. and Dash, J. 2007. Discriminating and mapping the C3 and C4 composition of grasslands in the northern Great Plains, USA. *Ecological Informatics*, 2, 89–93.

Franklin, S. E. and Wilson, B. A. 1992. A three-stage classifier for remote sensing of mountain environments. *Photogrammetric Engineering and Remote Sensing*, 58, 449–454.

Franklin, S. E., Maudie, A. J., and Lavigne, M. B. 2001. Using spatial co-occurrence texture to increase forest structure and species composition classification accuracy. *Photogrammetric Engineering and Remote Sensing*, 67, 849–855.

Friedl, M. A. and Brodley, C. E. 1997. Decision tree classification of land cover from remotely sensed data. *Remote Sensing of Environment*, 61, 399–409.

Fry, J. A., Xian, G., Jin, S., Dewitz, J. A., Homer, C. G., Yang, L., Barnes, C. A., Herold, N. D., and Wickham, J. D. 2011. Completion of the 2006 national land cover database for the conterminous united states. *Photogrammetric Engineering and Remote Sensing*, 77, 858–864.

Fuller, D. O. 2001. Forest fragmentation in Loudoun County, Virginia, USA evaluated with multitemporal Landsat imagery. *Landscape Ecology*, 16, 627–642.

Gamon, J. A., Field, C. B., Goulden, M. L., Griffin, K. L., and Hartley, A. E. 1995. Relationships between NDVI, canopy structure, and photosynthesis in three Californian vegetation types. *Ecological Applications*, 5, 28–41.

Gao, B. C. 1996. NDWI—A normalized difference water index for remote sensing of vegetation liquid water from space. *Remote Sensing of Environment*, 58, 257–266.

Gergel, S. E. 2005. Spatial and non-spatial factors: When do they affect landscape indicators of watershed loading? *Landscape Ecology*, 20, 177–189.

Gilmanov, T. G., Aires, L., Barcza, Z., Baron, V. S., Belelli, L., Beringer, J., Billesbach, D. et al. 2010. Productivity, respiration, and light-response parameters of world grassland and agroecosystems derived from flux-tower measurements. *Rangeland Ecology and Management*, 63, 16–39.

Gitelson, A. A., Kaufman, Y., and Merzlyak, M. N. 1996. Use of green channel in remote sensing of global vegetation from EOS-MODIS. *Remote Sensing of Environment*, 58, 289–298.

Glenn, E. P., Huete, A. R., Nagler, P. L., and Nelson, S. G. 2008. Relationship between remotely-sensed vegetation indices, canopy attributes and plant physiological processes: What vegetation indices can and cannot tell us about the landscape. *Sensors*, 8, 2136–2160.

Glenn, E., Huete, A., Nagler, P., Hirschboeck, K., and Brown, P. 2007. Integrating remote sensing and ground methods to estimate evapotranspiration. *Critical Reviews in Plant Sciences*, 26, 139–168.

Gong, P. and Howarth, P. J. 1992. Frequency based contextual classification and gray-level vector reduction for land use identification. *Photogrammetric Engineering and Remote Sensing*, 58, 423–437.

Goulden, M. L., Munger, J. W., Fan, S., Daube, B. C., and Wofsy, S. C. 1996. Exchange of carbon dioxide by a deciduous forest: Response to interannual climate variability. *Science*, 271, 1576–1578.

Graetz, R. D., Pech, R. P., and Davis, A. W. 1988. The assessment and monitoring of sparsely vegetated rangelands using calibrated Landsat data. *International Journal of Remote Sensing*, 9, 1201–1222.

Green, K., Kempka, D., and Lackey, L. 1994. Using remote sensing to detect and monitor land-cover and land-use change. *Photogrammetric Engineering and Remote Sensing*, 60, 331–337.

Grice, A. C. and Hodgkinson, K. C. 2002. *Global Rangelands: Progress and Prospects*. CABI Publishing, New York.

Griffith, J., Stehman, S., Sohl, T., and Loveland, T. 2003. Detecting trends in landscape pattern metrics over a 20-year period using a sampling-based monitoring programme. *International Journal of Remote Sensing*, 24, 175–181.

Griffith, J. A., Martinko, E. A., and Price, K. P. 2000. Landscape structure analysis of Kansas at three scales. *Landscape and Urban Planning*, 52, 45–61.

Guo, X., Zhang, C., Wilmshurst, J. F., and Sissons, R. 2005. Monitoring grassland health with remote sensing approaches. *Prairie Perspectives*, 8, 11–22.

Gustafson, E. J. 1998. Quantifying landscape spatial pattern: What is the state of the art? *Ecosystems*, 1, 143–156.

Gustafson, E. J. and Parker, G. R. 1992. Relationships between landcover proportion and indices of landscape spatial pattern. *Landscape Ecology*, 7, 101–110.

Hansen, M. C., DeFries, R. S., Townshend, J. R. G., Carroll, M., Dimiceli, C., and Sohlberg, R. A. 2003a. Development of 500 meter vegetation continuous field maps using MODIS data. *IEEE International Proceedings of Geoscience and Remote Sensing Symposium*, Toulouse, France. pp. 264–266.

Hansen, M. C., DeFries, R. S., Townshend, J. R. G., Carroll, M., Dimiceli, C., and Sohlberg, R. A. 2003b. Global percent tree cover at a spatial resolution of 500 meters: First results of the MODIS vegetation continuous fields algorithm. *American Meteorological Society*, 7, 1–15.

Hansen, M. C., DeFries, R. S., Townshend, J. R. G., Sohlberg, R., Dimiceli, C., and Carroll, M. 2002. Towards an operational MODIS continuous field of percent tree cover algorithm: Examples using AVHRR and MODIS data. *Remote Sensing of Environment*, 83, 303–319.

Hardisky, M. A., Klemas, V., and Smart, R. M. 1983. The influence of soil salinity, growth form, and leaf moisture on the spectral radiance of *Spartina alterniflora* canopies. *Photogrammetric Engineering and Remote Sensing*, 49, 77–83.

He, H. S., DeZonia, B. E., and Mladenoff, D. J. 2000. An aggregation index AI to quantify spatial patterns of landscapes. *Landscape Ecology*, 15, 591–601.

Hernandez, A. 2011. Spatiotemporal modeling of threats to big sagebrush ecological sites in northern Utah, Dissertation. Utah State University, Logan, UT. 163pp.

Heumann B. W., Seaquis, J. W., Eklundh, L., and Jönsson, P. 2007. AVHRR derived phenological change in the Sahel and Soudan, Africa, 1982–2005. *Remote Sensing of Environment*, 108, 385–392.

Hoffer, R. M. 1978. Biological and physical considerations in applying computer-aided analysis techniques to remote sensor data. Chapter 5 in Remote Sensing: The Quantitative Approach, McGraw-Hill, New York.

Hoffer, R. M. and Johannsen, C. J. 1969. Ecological potentials in spectral signature analysis. Chapter 1, In *Remote Sensing in Ecology*, P. C. Johnson (ed.), University of Georgia Press, Athens, Greece. pp. 1–6.

Honnay, O., Piessens, K., Van Landuyt, W., Hermy, M., and Gulinck, H. 2003. Satellite based land use and landscape complexity indices as predictors for regional plant species diversity. *Landscape and Urban Planning*, 63, 241–250.

Hudak, A. T. and Wessman, C. A. 1998. Textural analysis of historical aerial photography to characterize woody plant encroachment in South African savanna. *Remote Sensing of Environment*, 66(3), 317–330.

Huete, A. R. 1988. A soil-adjusted vegetation index (Savi). *Remote Sensing of Environment*, 25, 295–309.

Huete, A. R. and Jackson, R. D. 1987. Suitability of spectral indices for evaluating vegetation characteristics on arid rangelands. *Remote Sensing of Environment*, 25, 89–105.

Huete, A. R., Jackson, R. D., and Post, D. F. 1985. Spectral response of a plant canopy with different soil backgrounds. *Remote Sensing of Environment*, 17, 37–53.

Huete, A., Didan, K., Miura, T., Rodriguez, E. P., Gao, X., and Ferreira, L. G. 2002. Overview of the radiometric and biophysical performance of the MODIS vegetation indices. *Remote Sensing of Environment*, 83, 195–213.

Hunt Jr., E. R., Wang, C., Booth, D. T., Cox, S. E., Kumar, L. and Reeves, M. C. 2015. Remote Sensing of Rangeland Biodiversity. *Remote Sensing Handbook. Volume II: Land Resources: Monitoring, Modeling, and Mapping*, Chapter 11, (P.S. Thenkabail, ed.). Boca Raton, FL: CRC Press.

Im, J., Jensen, J. R., and Tylliss, J. A. 2008. Object based change detection using correlation image analysis and image segmentation. *International Journal of Remote Sensing*, 29, 399–423.

James, L. F., Young, J. A., and Sanders, K. 2003. A new approach to monitoring rangelands. *Arid Land Research and Management*, 17, 319–328.

Jarman, M., Ties, S., Finch, C., and Keyworth, S. 2011. Remote Sensing Study into Rangeland Condition in Kvemo Kartli and Samtskhe-Javakheti Regions—Final Report. Mercy Corps Alliances. (online:https://www.eda.admin.ch/content/dam/countries/countries-content/georgia/en/resource_en_219265.pdf, Accessed November 2013).

Jenkerson, C., Maiersperger, T., and Schmidt, G. 2010. eMODIS: A user-friendly data source. U.S. Geological Survey, Reston, VA.

Jensen, J. R. 2000. *Remote Sensing of the Environment: An Earth Resources Perspective*. Prentice-Hall, Upper Saddle River, NJ. 592pp.

Jensen, J. R. 2007. *Remote Sensing of the Environment: An Earth Resource Perspective*. Pearson Prentice Hall, Upper Saddle River, NJ.

Jensen, J. R. 2005. *Introductory Digital Image Processing: A Remote Sensing Perspective*. Pearson Prentice Hall, Upper Saddle River, NJ.

Jiang, Z., Huete, A. R., Chen, J., Chen, Y., Li, J., and Yan, G. 2006. Analysis of NDVI and scaled difference vegetation index retrievals of vegetation fraction. *Remote Sensing of Environment*, 101, 366–378.

Jianlong, L., Tiangang, L., and Quangong, C. 1998. Estimating grassland yields using remote sensing and GIS technologies in China. *New Zealand Journal of Agricultural Research*, 41, 31–380028–8233/98/4101–0031 $7.00/0© The Royal Society of New Zealand 1998.

Jin, S. M. and Sader, S. A. 2005. Comparison of time series tasseled cap wetness and the normalized difference moisture index in detecting forest disturbances. *Remote Sensing of Environment*, 94, 364–372.

Jin, S., Yang, L., Danielson, P., Homer, C., Fry, J., and Xian, G. 2013. A comprehensive change detection method for updating the National Land Cover Database to circa 2011. *Remote Sensing of Environment*, 132, 159–175.

Jin, X. M., Zhang, Y. K., Schaepman, M. E., Clevers, J. G., and Su, Z. 2008. Impact of elevation and aspect on the spatial distribution of vegetation in the Qilian mountain area with remote sensing data. *The International Archives of the Photogrammetry, Remote Sensing and Spatial Information Sciences*, 37, 1385–1390.

John, R., Chen, J., Lu, N., Guo, K., Liang, C., Wei, Y., Noormets, A., Ma, K., and Han, X. 2008. Predicting plant diversity based on remote sensing products in the semi-arid region of Inner Mongolia. *Remote Sensing of Environment*, 112, 2018–2032.

Kaloudis, S. T., Lorentzos, N. A., Sideridis, A. B., and Yialouris, C. P. 2005. A decision support system for forest fire management. *Operational Research*, 5(1), 141–152.

Kamp, K. V., Rigge, M., Troelstrup Jr, N. H., Smart, A. J., and Wylie, B. 2013. Detecting channel riparian vegetation response to best-management-practices implementation in ephemeral streams with the use of spot high-resolution visible imagery. *Rangeland Ecology & Management*, 66, 63–70.

Kamusoko, C. and Aniya, M. 2009. Hybrid classification of Landsat data and GIS for land use/cover change analysis of Bindura district, Zimbabwe. *International Journal of Remote Sensing*, 30, 97–115.

Kauth, R. J. and Thomas, G. S. 1976. The Tasseled Cap: A graphic description of the spectratemporal development of agricultural crops as seen by Landsat. In *Proceedings of the Symposium on Machine Processing of Remotely Sensed Data*, Purdue University, West Lafayette, IN. pp. 4B41–4B51.

Keane, R. E. 2013. Describing wildland surface fuel loading for fire management: a review of approaches, methods and systems. *International Journal of Wildland Fire*, 22, 51–62.

Keane, R. E. and Reeves, M. 2012. Use of expert knowledge to develop fuel maps for wildland fire management. In *Expert Knowledge and Its Application in Landscape Ecology*, A. H. Perera, A. Drew, and C. J. Johnson (eds.), Springer Science+Business Media, New York. pp. 211–228.

Kovalskyy, V. and Henebry, G. M. 2012. A new concept for simulation of vegetated land surface dynamics-Part 1: The event driven phenology model. *Biogeosciences*, 9, 141–159.

Kumar, L., Rietkerk, M., van Langevelde, F., van de Koppel, J., van Andel, J., Hearne, J., de Ridder, N., Stroosnijder, L., Prins, H. H. T., and Skidmore, A. K. 2002. Relationship between vegetation recovery rates and soil type in the Sahel of Burkina Faso: Implications for resource utilization at large scales. *Ecological Modelling*, 149, 143–152.

Kustas, W. and Norman, J. 1996. Use of remote sensing for evapotranspiration monitoring over land surfaces. *Hydrological Sciences Journal*, 41, 495–516.

Laba, M., Smith, S. D., and Degloria, S. D. 1997. Landsat based land cover mapping in the lower Yuna river watershed in the Dominican Republic. *International Journal of Remote Sensing*, 18, 3011–3025.

Laliberte, A. S., Rango, A., Havstad, K. M., Paris, J. F., Beck, R. F., McNeely, R., and Gonzalez, A. L. 2004. Object-oriented image analysis for mapping shrub encroachment from 1937 to 2003 in southern New Mexico. *Remote Sensing of Environment*, 93, 198–210.

Laliberte, A. S., Winters, C., and Rango, A. 2011. UAS remote sensing missions for rangeland applications. *Geocarto International*, 26(2), 141–156.

Langford, M. and Bell, W. 1997. Land cover mapping in a tropical hillsides environment: A case study in the Cauca region of Colombia. *International Journal of Remote Sensing*, 18, 1289–1306.

Langley, S. K., Cheshire, H. M., and Humes, K. S. 2001. A comparison of single date and multitemporal satellite image classifications in a semi-arid grassland. *Journal of Arid Environment*, 49, 401–411.

Lauenroth, W. K. and Sala, O. E. 1992. Long-term forage production of North American shortgrass steppe. *Ecological Applications*, 2, 397–403.

Lawler, J. J., White, D., Neilson, R. P., and Blaustein, A. R. 2006. Predicting climate-induced range shifts: model differences and model reliability. *Global Change Biology*, 12, 1568–1584.

Le Houerou, H. N. and Hoste, C. H. 1977. Rangeland production and annual rainfall relations in the Mediterranean Basin and in the African Sahelo-Sudanian Zone. *Journal of Range Management*, 30, 181–189.

Leitão, B. and Ahern, J. 2002. Applying landscape ecological concepts and metrics in sustainable landscape planning. *Landscape and Urban Planning*, 59, 65–93.

Lewis, M. 2000. Discrimination of arid vegetation composition with high resolution CASI imagery. *The Rangeland Journal*, 22(1), 141–167.

Li, B. and Zhou, Q. 2009. Accuracy assessment on multi-temporal land-cover change detection using a trajectory error matrix. *International Journal of Remote Sensing*, 30, 1283–1296.

Li, F., Kustas, W., Preuger, J., Neale, C., and Jackson, T. 2005. Utility of remote sensing-based two-source balance model under low- and high-vegetation cover conditions. *Journal of Hydrometeorology*, 6, 878–891.

Li, H. and Wu, J. 2004. Use and misuse of landscape indices. *Landscape Ecology*, 19, 389–399.

Liaw, A. and Wiener, M. 2002. Classification and regression by randomForest. *R News*, 2, 5.

Liu, A. J. and Cameron, G. N. 2001. Analysis of landscape patterns in coastal wetlands of Galveston Bay, Texas (USA). *Landscape Ecology*, 16, 581–595.

Lowry, J., Ramsey, R. D., Thomas, K., Schrupp, D., Sajwaj, T., Kirby, J., Waller, E. et al. 2007. Mapping moderate-scale land-cover over very large geographic areas within a collaborative framework: A case study of the Southwest Regional Gap Analysis Project (SWReGAP). *Remote Sensing of Environment*, 108, 59–73.

Lu, D., Mausel, P., Brondizio, E., and Moran, E. 2004. Change detection techniques. *International Journal of Remote Sensing*, 25, 2365–2407.

Ludwig, J. A., Tongway, D. J., Bastin, G. N., and James, C. D. 2004. Monitoring ecological indicators of rangeland functional integrity and their relation to biodiversity at local to regional scales. *Austral Ecology*, 29, 108–120.

Lutes D. C., Keane, R. E., and Caratti, J. F. 2009. A surface fuels classification for estimating fire effects. *International Journal of Wildland Fire*, 18, 802–814.

Mas, J. F. 1999. Monitoring land-cover changes: a comparison of change detection techniques. *International Journal of Remote Sensing*, 20, 139–152.

Maynard, C. L., Lawrence, R. L., Nielsen, G. A., and Decker, G. 2007. Ecological site descriptions and remotely sensed imagery as a tool for rangeland evaluation. *Canadian Journal of Remote Sensing*, 33(2), 109–115.

Maynard, C. L., Lawrence, R. L., Nielsen, G. D., and Decker, G. 2007. Ecological site descriptions and remotely sensed imagery as a tool for rangeland evaluation. *Canadian Journal of Remote Sensing*, 33, 109–115.

McAlpine, C. A. and Eyre, T. J. 2002. Testing landscape metrics as indicators of habitat loss and fragmentation in continuous eucalypt forests (Queensland, Australia). *Landscape Ecology*, 17, 711–728.

McGarigal, K. and Marks, B. J. 1995. Spatial pattern analysis program for quantifying landscape structure. General Technical Report PNW-GTR-351. US Department of Agriculture, Forest Service, Pacific Northwest Research Station, Portland, OR.

Means, J. E., Krankina, O. N., Jiang, H., and Li, H. 1996. *Estimating Live Fuels for Shrubs and Herbs With BIOPAK*. Gen. Tech. Rep. PNW-GTR-372. U.S. Department of Agriculture, Forest Service, Pacific Northwest Research Station, Portland, OR. p. 28.

Merrill, D. F. and Alexander, M. E. 1987. *Glossary of Forest Fire Management Terms*. 4th edn. National Research Council of Canada, Canadian Committee on Forest Fire Management, Ottawa, ON.

Milchunas, D. G., Laurenroth, W. K., Chapman, P. L., and Kazempour, M. K. 1989. Effects of grazing, topography, and precipitation on the structure of a semiarid grassland. *Vegetation*, 80, 11–23.

Miller, A. B., Bryant, E. S., and Birnie, R. W. 1998. An analysis of land cover changes in the northern forest of New England using multi temporal Landsat MSS data. *International Journal of Remote Sensing*, 19, 245–265.

Miller, R. F. and Rose, J. A. 1999. Fire history and western juniper encroachment in sagebrush steppe. *Journal of Range Management*, 52, 550–559.

Monteith, J. and Unsworth, M. 1990. *Principles of Environmental Physics*, 2nd edn. Edward Arnold, London, U.K.

Muchoney, D. M., Borak, J., Chi, H., Friedl, M., Gopal, S., Hodges, J., Morrow, N., and Strahler, A. 2000. Application of the MODIS global supervised classification model to vegetation and land cover mapping of Central America. *International Journal of Remote Sensing*, 21, 1115–1138.

Munoz-Villers, L. E. and Lopez-Blanco, J. 2008. Land use/cover changes using Landsat TM/ETM images in a tropical and biodiverse mountainous area of central-eastern Mexico. *International Journal of Remote Sensing*, 29, 71–93.

Munyati, C., Shaker, P., and Phasha, M. G. 2011. Using remotely sensed imagery to monitor savanna rangeland deterioration through woody plant proliferation: A case study from communal and biodiversity conservation rangeland sites in Mokopane, South Africa. *Environmental Monitoring Assessment*, 176, 293–311.

O'Neill, R. V. 1976. Ecosystem persistence and heterotrophic regulation. *Ecology*, 57(6), 1244–1253.

Ode, D. J., Tieszen, L. L., Lerman, J. C. 1980. The seasonal contribution of C3 and C4 plant species to primary production in a mixed prairie. *Ecology*, 61, 1304–1311.

Okeke, F. and Karnieli, A. 2006. Methods for fuzzy classification and accuracy assessment of historical aerial photographs for vegetation change analyses. Part I: algorithm development. *International Journal of Remote Sensing*, 27, 153–76.

Okin, G. S., Roberts, D. A., Murray, B., and Okin, W. J. 2001. Practical limits on hyperspectral vegetation discrimination in arid and semiarid environments. *Remote Sensing of Environment*, 77(2), 212–225.

Ormsby, J., Choudry, B., and Owe, M. 1987. Vegetation spatial variability and its effect on vegetation indexes. *International Journal of Remote Sensing*, 8, 1301–1306.

Ottmar, R. D., Sandberg, D. V., Riccardi, C. L., and Prichard, S. 2007. An overview of the fuel characteristic classification system—Quantifying, classifying, and creating fuelbeds for resource planners. *Canadian Journal of Forest Research*, 37, 2381–2382.

Pal, M. 2008. Artificial immune-based classifier for land-cover classification. *International Journal of Remote Sensing*, 29, 2273–2291.

Pal, M. and Mather, P. M. 2003. An assessment of the effectiveness of decision tree methods for landcover classification. *Remote Sensing of Environment*, 86, 554–565.

Parker, A. J. 1982. The topographic relative moisture index: An approach to soil- moisture assessment in mountain terrain. *Physical Geography*, 3, 160–168.

Paruelo, J. M. and Lauenroth, W. K. 1998. Interannual variability of NDVI and its relationship to climate for North American shrublands and grasslands. *Journal of Biogeography*, 25, 721–733.

Peterson, D. L., Spanner, M. A., Running, S. W., and Teuber, K. B. 1987. Relationships of Thematic Mapper simulator data to leaf area index of temperate coniferous forests. *Remote Sensing of Environment*, 22, 323–341.

Peterson, E. B. 2005. Estimating cover of an invasive grass (*Bromus tectorum*) using tobit regression and phenology derived from two dates of Landsat ETM plus data. *International Journal of Remote Sensing*, 26, 2491–2507.

Petit, C. C. and Lambin, E. F. 2002. Impact of data integration technique on historical land-use/landcover change: Comparing historical maps with remote sensing data in the Belgian Ardennes. *Landscape Ecology*, 17, 117–132.

Pickup, G. and Foran, B. D. 1987. The use of spectral and spatial variability to monitor cover change on inert landscapes. *Remote Sensing of Environment*, 23, 351–363.

Piñeiro, G., Oesterheld, M., and Paruelo, J. M. 2006. Seasonal variation in aboveground production and radiation-use efficiency of temperate rangelands estimated through *remote sensing*. *Ecosystems*, 9, 357–373, DOI: 10.1007/s10021-005-0013-x.

Pinty, B. and Verstraete, M. M. 1992. GEMI: A non-linear index to monitor global vegetation from satellites. *Vegetation*, 101, 15–20.

Prasad, A. M., Iverson, L. R., and Liaw, A. 2006. Newer classification and regression tree techniques: Bagging and random forests for ecological prediction. *Ecosystems*, 9, 181–199.

Program, U. N. G. A. 2004. Provisional digital land cover map for the Southwestern United States. Version 1.0, RS/GIS Laboratory, College of Natural Resources, Utah State University, Logan, U.T.

Rauzi, F. and Dobrenz, A. K. 1970. Seasonal variation of chlorophyll in Western wheatgrass and blue grama. *Journal of Range Management*, 23, 372–373.

Read, J. and Lam, N. S. 2002. Spatial methods for characterising land cover and detecting land-cover changes for the tropics. *International Journal of Remote Sensing*, 23, 2457–2474.

Reed, B. C., Brown, J. F., VanderZee, D., Loveland, T. R., Merchant, J. W., and Ohlen, D. O. 1994. Measuring phenological variability from satellite imagery. *Journal of Vegetation Science*, 5, 703–714.

Reed, B. C., Loveland, T. R., and Tieszen, L. L. 1996. An approach for using AVHRR data to monitor U.S. Great Plains grasslands. *Geocarto International*, 11, 13–22.

Reeves, M., Ryan, K. C., Rollins, M. G., and Thompson, T. 2009. Spatial fuel data products of the LANDFIRE Project. *International Journal of Wildland Fire*, 18, 250–267.

Reeves, M. C., Angerer, J., Hunt, E. R., Kulawardhana, W., Kumar, L., Loboda, T., Loveland, T., Metternicht, G., Ramsey, R. D., and Washington-Allen, R. A. 2015. A global view of remote sensing of rangelands: Evolution, applications, and future pathways. *Remote Sensing Handbook. Volume II: Land Resources: Monitoring, Modeling, and Mapping*, Chapter 10, (P.S. Thenkabail, ed.). Boca Raton, FL: CRC Press.

Rempel, R. S., Carr, A., and Elkie, P. C. 1999. *Patch Analyst User's Manual: A Tool for Quantifying Landscape Structure*, Ontario Ministry of Natural Resources, Boreal Science, Northwest Science & Technology, Thunder Bay, Ontario, Canada.

Resources, U. D. W. 2010. Utah Big Game Range Trend Studies.Available at: http://wildlife.utah.gov/range/statewide%20management%20units.htm. Accessed October 05, 2010.

Riaño, D., Chuvieco, E., Salas, J., Palacios-Orueta, A., and Bastarrika, A. 2002. Generation of fuel type maps from Landsat TM images and ancillary data in Mediterranean ecosystems. *Canadian Journal of Forest Research*, 32(8), 1301–1315.

Riaño, D., Chuvieco, E., Ustin, Salas, S., Rodríguez-Pérez, J., Ribeiro, L., Viegas, D., Moreno, J., and Fernández, H. 2007. Estimation of shrub height for fuel-type mapping combining airborne LiDAR and simultaneous color infrared ortho imaging. *International Journal of Wildland Fire*, 16(3), 341–348.

Riaño, D., Meier, E., Allgöwer, B., Chuvieco, E., and Ustin, S. L. 2003. Modeling airborne laser scanning data for the spatial generation of critical forest parameters in fire behavior modeling. *Remote Sensing of Environment*, 86(2), 177–186.

Richards, J. A. 1986. Thematic mapping from multitemporal image data using the principal components transformation. *Remote Sensing of Environment*, 16, 35–46.

Richardson, A. J. and Wiegand, C. L. 1990. Comparison of two models for simulating the soil-vegetation composite reflectance of a developing cotton canopy. *International Journal of Remote Sensing*, 11, 447–459.

Richardson, A. J. and Wiegand, C. L. 1977. Distinguishing vegetation from soil background information. *Photogrammetric Engineering and Remote Sensing*, 43, 1541–1552.

Rietkerk, M., Dekker, S. C., de Ruiter, P. C., and van de Koppel, J. 2004. Self-organized patchiness and catastrophic shifts in ecosystems. *Science*, 305(5692), 1926–1929.

Rigge, M., Smart, A., Wylie, B., Gilmanov, T., and Johnson, P. 2013. Linking phenology and biomass productivity in South Dakota Mixed-Grass Prairie. *Rangeland Ecology & Management*, 66, 579–587.

Riitters, K. H., O'neill, R., Hunsaker, C., Wickham, J. D., Yankee, D., Timmins, S., Jones, K., and Jackson, B. 1995. A factor analysis of landscape pattern and structure metrics. *Landscape Ecology*, 10, 23–39.

Rock, B. N., Williams, D. L., and Vogelmann, J. E. 1985. Field and airborne spectral characterization of suspected acid deposition damage in red spruce (*Picea rubens*) from Vermont. In *Proceedings of Symposium on Machine Processing of Remotely Sensed Data*, Purdue University, West Lafayette, IN. pp. 71–81.

Rogan, J., Franklin, J., and Roberts, D. A. 2002. A comparison of methods for monitoring multitemporal vegetation change using Thematic Mapper imagery. *Remote Sensing of Environment*, 80, 143–156.

Rollins, M. 2009. LANDFIRE: A nationally consistent vegetation, wildland fire and fuel assessment. *International Journal of Wildland Fire*, 18, 235–249.

Rouse, J. W., Haas, R. H., Schell, J. A., and Deering, D. W. 1974. Monitoring vegetation systems in the Great Plains with ERTS. In *Proceedings of Third ERTS-1 Symposium*, NASA Goddard, NASA SP-351, Washington, DC. pp. 309–317.

Sandberg, D. V., Ottmar, R. D., and Cushon, G. H. 2001. Characterizing fuels in the 21st century. *International Journal of Wildland Fire*, 10, 381–387.

Schmitz, M., De Aranzabal, I., Aguilera, P., Rescia, A., and Pineda, F. 2003. Relationship between landscape typology and socioeconomic structure: Scenarios of change in Spanish cultural landscapes. *Ecological Modelling*, 168, 343–356.

Scott, J. H. and Burgan, R. E. 2005. Standard fire behavior fuel models: A comprehensive set for use with Rothermel's surface fire spread model. General Technical Report RMRS-GTR-153. U.S. Department of Agriculture, Forest Service, Rocky Mountain Research Station, Fort Collins, CO. p. 72.

Scott, J. M., Davis, F., Csuti, B., Noss, R., Butterfield, B., and Groves, C. 1993. Gap analysis: A geographic approach to protection of biological diversity. *Wildlife Monographs*, 123, 3–41.

Sellers, P. 1985. Canopy reflectance, photosynthesis and transpiration. *International Journal of Remote Sensing*, 6, 1335–1372.

Sexton, J. O., Song, X. P., Feng, M., Noojipady, P., Anand, A., Huang, C., Kim, D. H. et al. 2013. Global, 30-m resolution continuous fields of tree cover: Landsat-based rescaling of MODIS vegetation continuous fields with lidar-based estimates of error. *International Journal of Digital Earth*, 6, 427–448.

Sha, Z., Baiy, Y., and Xie, Y. 2008. Using a hybrid fuzzy classifier (HFC) to map typical grassland vegetation in Xilinhe River Basin, Inner Mongolia, China. *International Journal of Remote Sensing*, 29(8), 2317–2337.

Sims, P. L. and Singh, J. S. 1978. Primary production of the central grassland region of the United States. *Ecology*, 69, 40–45.

Sing, A. and Harrison, A. 1985. Standardized principal components. *International Journal of Remote Sensing*, 6, 883–896.

Sinha, P. and Kumar, L. 2013a. Binary images in seasonal land-cover change identification: A comparative study in parts of NSW, Australia. *International Journal of Remote Sensing*, 34(6), 2162–2182.

Sinha, P. and Kumar, L. 2013b. Independent two-step thresholding of binary images in inter-annual land cover change/no-change identification. *ISPRS Journal of Photogrammtery and Remote Sensing*, 81(7), 31–43.

Sinha, P., Kumar, L., and Reid, N. 2012a. Seasonal variation in landcover classification accuracy in diverse region, *Photogrammetric Engineering and Remote Sensing*, 78(3), 770–781.

Sinha, P., Kumar, L., and Reid, N. 2012b. Three-date Landsat TM composite in seasonal land-cover change identification in a mid-latitudinal region of diverse climate and land-use. *Journal of Applied Remote Sensing*, 6(1), 063595, doi: 10.1117/1.JRS.6.063595.

Smart, A. J., Dunn, B. H., Johnson, P. S., Xu, L., and Gates, R. N. 2007. Using weather data to explain herbage yield on three Great Plains plant communities. *Rangeland Ecology Management*, 60, 146–153.

Smits, P. C., Dellepiane, S. G., and Schowengerdt, R. A. 1999. Quality assessment of image classification algorithms for land-cover mapping: A review and a proposal for a cost-based approach. *International Journal of Remote Sensing*, 20, 1461–1486.

Sohn, Y. and Qi, J. 2005. Mapping detailed biotic communities in the upper San Pedro Valley of southeastern Arizona using Landsat 7 ETM + data and supervised spectral angle classifier. *Photogrammetric Engineering and Remote Sensing*, 71, 709–718.

Sohn, Y. and Rebello, N. S. 2002. Supervised and unsupervised spectral angle classifiers. *Photogrammetric Engineering and Remote Sensing*, 68, 1271–1280.

Sparks, T. H., Jeffree, E. P., and Jeffree, C. E. 2000. An examination of the relationship between flowering times and temperature at the national scale using long-term phenological records from the UK. *International Journal of Biometeorology*, 44(2), 82–87.

Stanfield, B. J., Bliss, J. C., and Spies, T. A. 2002. Land ownership and landscape structure: A spatial analysis of sixty-six Oregon (USA) Coast Range watersheds. *Landscape Ecology*, 17, 685–697.

Strahler, A. H. 1980. The use of prior probabilities in maximum likelihood classification of remotely sensed data. *Remote Sensing of Environment*, 10, 1135–1163.

Sutton, C. D. 2005. Classification and regression trees, bagging, and boosting. In *Handbook of Statistics: Data Mining and Data Visualization*, C. R. Rao, E. J. Wegman, and J. L. Solka (eds.), Elsevier Amsterdam, The Netherlands. pp. 303–329.

Tagestad, J. D. and Downs, J. L. 2007. Landscape Measures of Rangeland Condition in the BLM Owyhee Pilot Project: Shrub Canopy Mapping, Vegetation Classification, and Detection of Anomalous Land Areas. Report prepared for U.S. Department of Interior Bureau of Land Management under a Related Services Agreement with the U.S. Department of Energy Contract DE-AC05–76RL01830 Pacific Northwest National Laboratory Richland, Washington, DC.

Tang, J., Corresponding, L. W., and Zhang, S. 2005. Investigating landscape pattern and its dynamics in Daqing, China. *International Journal of Remote Sensing*, 26, 2259–2280.

Tao, F., Yokozawa, M., Zhang, Z., Hayashi, Y., and Ishigooka, Y. 2008. Land surface phenology dynamics and climate variations in the North East China Transect (NECT), 1982–2000. *International Journal of Remote Sensing*, 29, 5461–5478.

Tateishi, R. and Ebata, M. 2004. Analysis of phenological change patterns using 1982–2000 advanced very high resolution radiometer (AVHRR) data. *International Journal of Remote Sensing*, 25, 2287–2300.

Teague, W. R., Ansley, J. R., Pinchak, W. E., Dowhower, S. L., Gerrard, S. A., and Waggoner, J. A. 2008. Interannual herbaceous biomass response to increasing honey mesquite cover on two soils. *Rangeland Ecological Management*, 61, 496–508.

Tieszen, L. L., Reed, B. C., Bliss, N. B., Wylie, B. K., and DeJong, D. D. 1997. NDVI, C3 AND C4 production, and distributions in Great Plains grassland land cover classes. *Ecological Applications*, 7, 59–58.

Timmerman, W., Kustas, W., Anderson, M., and French, A. 2007. An intercomparison of the surface energy balance algorithm for land (SEBAL) and the two-source energy balance (TSEB) modeling schemes. *Remote Sensing of Environment*, 108, 369–384.

Todd, S. W. and Hoffer, R. M. 1998. Responses of spectral indices to variations in vegetation cover and soil background. *Photogrammetric Engineering and Remote Sensing*, 64(9), 915–921.

Todd, S. W., Hoffer, R. M., and Milchunas, D. G. 1998. Biomass estimation on grazed and ungrazed rangelands using spectral indices. *International Journal of Remote Sensing*, 19(3), 427–438.

Torahi, A. A. 2012. Rangeland dynamics monitoring using remotely-sensed data, in Dehdez Area, Iran. *International Conference on Applied Life Sciences*, Konya, Turkey, September 10–12, 2012.

Trani, M. K. and Giles, R. H. 1999. An analysis of deforestation: Metrics used to describe pattern change. *Forest Ecology and Management*, 114, 459–470.

Tso, B. and Olsen, R. C. 2005. Combining spectral and spatial information into hidden Markov models for unsupervised image classification. *International Journal of Remote Sensing*, 26, 2113–2133.

Tucker, C. 1979. Red and photographic infrared linear combinations for monitoring vegetation. *Remote Sensing of Environment*, 8, 127–150.

Tucker, C. J. and Sellers, P. J. 1986. Satellite remote sensing of primary production. *International Journal of Remote Sensing*, 7, 1395–1416.

Tucker, C., Townshend, J., and Goff, T. 1985. African land cover classification using satellite data. *Science*, 227, 229–235.

Tussie, G. D. 2004. Vegetation ecology, rangeland condition and forage resources evaluation in the Borana Lowlands, Southern Oromia, Ethiopia, Georg-August University Goettingen. PhD thesis. Cuvillier Verlag Goettingen, Germany.

Tymstra, C., Bryce, R. W., Wotton, B. M., Taylor, S. W., and Armitage, O. B. 2010. Development and structure of Prometheus: The Canadian wildland fire growth simulation model. Natural Resources Canada, Canadian Forest Service, Northern Forestry Centre, Edmonton, AB. Information Report NOR-X-417. p. 88. Available online (http://cfs. nrcan.gc.ca/bookstore_pdfs/31775.pdf) Verified March 16, 2011.

U.S. Department of Agriculture, National Resources Conservation Service. 2008. Field Office Technical Guide, Major Land Resource Area 063A: Northern Rolling Pierre Shale Plains, Ecological Site Description: Clayey.

U.S. Department of Agriculture, NRCS. 2012. Ecological sites. Available at: http://www.nrcs.usda.gov/wps/portal/nrcs/detail/national/landuse/rangepasture/?cid= STELPRDB1043235. Accessed October 17, 2011.

van Leeuwen, W. J. D., Davison, J. E., Casady, G. M., and Marsh, S. E. 2010. Phenological characterization of desert sky island vegetation communities with remotely sensed and climate time series data. *Remote Sensing*, 2, 388–415.

van Wagtendonk, J. W. and Root, R. R. 2003. The use of multi-temporal Landsat Normalized Difference Vegetation Index (NDVI) data for mapping fuel models in Yosemite National Park, USA. *International Journal of Remote Sensing*, 24(8), 1639–1651.

Vierling, L. A., Xu, Y., Eitel, J. U. H., and Oldow, J. S. 2012. Shrub characterization using terrestrial laser scanning and implications for airborne LiDAR assessment. *Canadian Journal of Remote Sensing*, 38(6), 709–722.

Vogel, M. and Strohbach, M. 2009. Monitoring of savanna degradation in Namibia using Landsat TM/ETM+ data. *IEEE International Geoscience and Remote Sensing Symposium*, Cape Town, South Africa, pp. III-931–III-934.

Vogelmann, J. E., Xian, G., Homer, C., and Tolk, B. 2012. Monitoring gradual ecosystem change using Landsat time series analyses: Case studies in selected forest and rangeland ecosystems. *Remote Sensing of Environment*, 122, 92–105.

Walton, J. T. 2008. Subpixel urban land cover estimation: Comparing Cubist, Random Forests, and support vector regression. *Photogrammetric Engineering & Remote Sensing*, 74, 1213–1222.

Wang, C., Jamison, B. E., and Spicci, A. A. 2010. Trajectory-based warm season grassland mapping in Missouri prairies with multi-temporal ASTER imagery. *Remote Sensing of Environment*, 114, 531–539.

Wang. Y., Mitchell, B. R., Nugranad-Marzilli, J., Bonynge, G., Zhou, Y., and Shriver, G. 2009. Remote sensing of land-cover change and landscape context of National Parks: A case study of northeast temperate network. *Remote Sensing of Environment*, 113, 1453–1461.

Weber, K. T., Alados, C. L., Bueno, C. G., Gokhale, B., Komac, B., and Pueyo, Y. 2009. Modeling bare ground with classification trees in Northern Spain. *Rangeland Ecology & Management*, 62, 452–459.

Weiser, R. L., Asrar, G., Miller, G. P., and Kanemasu, E. T. 1986 Assessing grassland biophysical characteristics from spectral measurements. *Remote Sensing of Environment*, 20, 141–152.

Wessels, K. J., Prince, S. D., Carroll, M., and Malherbe, J. 2007. Relevance of rangeland degradation in semiarid northeastern South Africa to the nonequilbrium theory. *Ecological Applications*, 17(3), 815–827.

Wessels, K. J., Prince, S. D., Frost, P. E., and Van Zyl, D. 2004. Assessing the effects of human-induced land degradation in the former homelands of northern South Africa with a 1 km AVHRR NDVI time-series. *Remote Sensing of Environment*, 91, 47–67.

Westerling, A. L. and Swetnam, T. W. 2003. Interannual to decadal drought and wildfire in the western United States. *Transactions American Geophysical Union*, 84, 545–560.

Westman, W. E. 1985. *Ecology, Impact Assessment, and Environmental Planning*, Wiley, New York.

Westman, W. E. and O'Leary, J. F. 1986. Measures of resilience: The response of coastal sage scrub to fire. *Vegetation*, 65(3), 179–189.

White, L.M. 1983. Seasonal changes in yield, digestibility, and crude protein of vegetative and floral tillers of two grasses. *Journal of Range Management*, 36, 402–405.

White, M. A., de Beurs, K. M., Didan, K., Inouye, D. W., Richardson, A. D., Jensen, O. P., O'Keefe, J. et al. 2009. Intercomparison, interpretation, and assessment of spring phenology in North America estimated from remote sensing for 1982–2006. *Global Change Biology*, 15, 2335–2359.

Wickham, J. and Rhtters, K., 1995. Sensitivity of landscape metrics to pixel size. *International Journal of Remote Sensing*, 16, 3585–3594.

Wisdom, M., Rowland, M., Suring, L., Shueck, L., Meinke, C. and Knick S. 2005. Evaluating species of conservation concern at regional scales. In *Habitat Threats in the Sagebrush Ecosystem: Methods of Regional Assessment and Applications in the Great Basin*, M. Wisdom, M. Rowland, and L. Suring (eds.), Alliance Communications Group, Lawrence, KS. pp. 5–74.

Wotton, B. M. 2009. Interpreting and using outputs from the Canadian Forest Fire Danger Rating System in research applications. *Environmental and Ecological Statistics*, 16(2), 107–131.

Wu, W., De Pauw, E., and Zucca, C. 2008. Land degradation monitoring in the west Muus, China. The International Archives of the Photogrammetry, Remote Sensing and Spatial Information Sciences XXXVII, Part B 8847–858.

Xie, Y., Sha, Z. and Yu, M. 2008. Remote sensing imagery in vegetation mapping: A review. *Journal of Plant Ecology*, 1, 9–23.

Xiuwan, C. 2002. Using remote sensing and GIS to analyse land cover change and its impacts on regional sustainable development. *International Journal of Remote Sensing*, 23, 107–124.

Xu, M., Watanachaturaporn, P., Varshney, P. K. and Arora, M. K. 2005. Decision tree regression for soft classification of remote sensing data. *Remote Sensing of Environment*, 97, 322–336.

Zhang, J. and, Foody, G. M. 1998. A fuzzy classification of suburban landcover from remotely sensed imagery. *International Journal of Remote Sensing*, 19, 2721–2738.

Zhang, L., Wylie, B. K., Ji, L., Gilmanov, T. G. and Tieszen, L. L. 2010. Climate-driven interannual variability in net ecosystems exchange in the northern Great Plains grasslands. *Rangeland Ecology and Management*, 63, 40–50.

Phenology and Food Security

Global Land Surface Phenology and Implications for Food Security

Molly E. Brown
University of Maryland

Kirsten M. de Beurs
University of Oklahoma

Kathryn Grace
University of Utah

Acronyms and Definitions

AVHRR   Advanced very-high-resolution radiometer
CDL   Cropland data layer
LSP   Land surface phenology
MODIS   Moderate-Resolution Imaging Spectroradiometer

13.1 Introduction

Phenology is the scientific study of periodic biological phenomena in relation to climate conditions and habitat factors. Phenology varies by species and is influenced by many factors, such as soil temperature, solar illumination, day length, and soil moisture. Land surface phenology (LSP) is the study of the spatiotemporal patterns in the vegetated land surface as observed by satellite sensors. Agriculture and food production are linked inextricably to the seasonal effects of rainfall and temperature changes. LSP can be used to estimate agriculturally important changes in the start, length, and strength of the growing season, which controls how much food is produced in rainfed agricultural systems (Bolton and Friedl 2013; Koetse and Rietveld 2009). Since the supply of food in many countries is strongly affected by how much food is grown locally, understanding LSP is a critical part of assessing food availability.

Food security is defined as the ability of all people to acquire enough culturally relevant food for an active and healthy life (FAO 2012). Roughly 850 million people, most of them live in the developing world, suffer from undernourishment, an outcome of food insecurity (FAO 2012). To help explain the reasons underlying undernourishment, food security is commonly examined by focusing on four underlying pillars:

- *Availability*—the availability of sufficient quantities of food of appropriate quality, supplied through domestic production or imports (including food aid).
- *Access*—access by individuals to adequate resources for acquiring appropriate foods for a nutritious diet.
- *Utilization*—utilization of food through adequate diet, clean water, sanitation, and health care to reach a state of nutritional well-being where all physiological needs are met.
- *Stability*—to be food secure, a population, household, or individual must have access to adequate food at all times. They should not risk losing access to food as a consequence of sudden shocks or cyclical events (FAO 2008; Godfray et al. 2010; Schmidhuber and Tubiello 2007).

Vegetative variability, as measured with LSP, can impact each of these elements. Extreme events can affect food production directly (Vrieling et al. 2011), affect distribution of food and thus the price of food in regions where supply is low and demand is high (de Beurs and Brown 2013), affect utilization through increased spread of disease (Myers and Patz 2009), and impact stability through extreme events that reduce the ability of farmers to predict weather conditions from 1 year to the next. Farming is becoming even more risky because of heat stress, lack of water, pests, and diseases that interact with ongoing pressures on natural resources. The lack of predictability in the start and length of the growing season affects the ability of farmers to invest in appropriate fertilizer levels or improved, high-yielding varieties (Zaal et al. 2004).

We can use LSP models to better predict interannual variability of food production. LSP models rely on remote sensing observations of vegetation, such as datasets derived from the advanced very-high-resolution radiometer (AVHRR) and the newer Moderate-Resolution Imaging Spectroradiometer (MODIS) sensors on Aqua and Terra satellites. Vegetation and rainfall data can assess variables such as the start of season (SOS), growing season length, and overall growing season productivity (Brown and de Beurs 2008; de Beurs and Henebry 2004, 2010). These metrics are common inputs to crop models that estimate the impact of weather on agricultural area and yield (Bolton and Friedl 2013; Funk and Budde 2009). LSP metrics have a strong relationship with regional food production, particularly those with sufficiently long records to capture local variability. This chapter will focus on reviewing how LSP analysis is done, how it can be used to monitor agriculture and food production, and the links between these observations and food security (FCPN 2007; Zaal et al. 2004).

13.2 Characterizing Land Surface Phenology

The international biological program defined phenology as "the study of the timing of recurrent biological events, the causes of their timing with regard to biotic and abiotic forces, and the interrelation among phases of the same or different species (Lieth 1974)." The importance of phenological observations to understanding the impact of global environmental change has been increasingly recognized (Mu et al. 2013). The impact of the weather on plant phenology can serve as a biological indicator of the impacts of climate change on terrestrial ecosystems, including agroecosystems that support the production of food (Bradley et al. 2011; Schwartz 1992).

The vigor and development of vegetation depends on available moisture and nutrients for plant development. The health of crops can be studied by looking at their phenological characteristics including germination, leaf emergence, and start of senescence (Vrieling et al. 2011). LSP is defined as the spatiotemporal development of the vegetated land source as observed by synoptic satellite sensors (de Beurs and Henebry 2004). Datasets from satellite remote sensing of vegetation can approximate phenological stages and thus characterize the general vegetation behavior (Justice et al. 1985; Reed et al. 1994). A derived metric of particular interest is the seasonally cumulated vegetation index as it is related to net primary productivity (Awaya et al. 2004).

Phenology models produce annual metrics that describe the growing season, including the SOS, length of season, maximum normalized difference vegetation index (maxNDVI) value, and cumulated NDVI over the season (cumNDVI) (Figure 13.1) (Brown et al. 2010; de Beurs and Henebry 2010). Satellite remote sensing can be used to study the spatiotemporal development of the vegetated land surface. When using LSP in assessing food

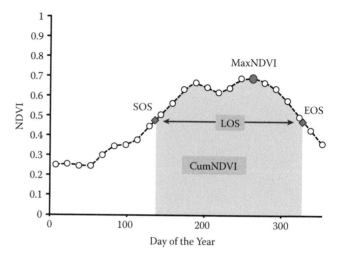

FIGURE 13.1 Start of season, length of season, end of season, maximum NDVI, and cumulative NDVI for a sample curve. (From Brown, M.E. et al., *Remote Sens. Environ.*, 114, 2286, 2010.)

production, the measurement needs to be focused on the agriculturally productive areas of the land surface, including row crops, pasture, and gardens to maximize the representativeness of the satellite-derived assessment of the SOS (White et al. 2009).

In the context of food security, LSP is used to provide a remote estimate of the timing of the start of the agricultural growing season. The SOS metric is a critical parameter for food security assessment and is monitored remotely by many food security and agriculture organizations (Brown 2008). Staple cereal crops in semiarid agricultural zones such as millet and sorghum are often photoperiod sensitive, and thus a sowing delay can translate into a reduction in yield (Brown and de Beurs 2008; Buerkert et al. 2001). Changes in the SOS also may reduce the overall length of the growing season, further reducing the yields obtained in marginal semiarid agroecosystems.

Figure 13.2 provides the growing season length for West Africa for the years 2009 (average), 2011 (very dry), and 2012 (relatively wet after a dry winter). The growing season length is calculated with the threshold percentage method (White et al. 1997) based on MODIS bidirectional reflectance function adjusted reflectance–derived NDVI data (MCD43C4) at the global climate modeling grid (0.05°). The maps show a north–south gradient in the length of the growing season with much shorter growing seasons further north. In addition, 2011 reveals much shorter growing seasons, especially compared to 2012. As an example, we calculated the average length of the growing season for the agricultural regions (cropland percentage according to the MODIS land cover classification > 0) in Mali. We found that the average growing season length ranged from 101 days in the driest year of 2011 to 125 days in 2012. The year 2009 had an average growing season length of 110 days. Figure 13.3 also reveals that the length of the growing season was very short in 2011, more than 48% of the cropped pixels had a growing season of less than 100 days and the median growing season length was 101 days. The short growing seasons in 2001 resulted in failure of most harvests in Mali and

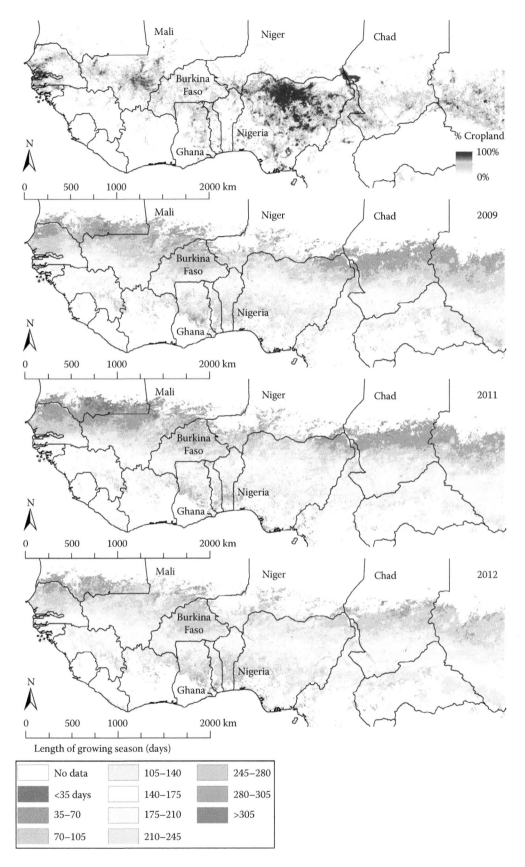

FIGURE 13.2   (Top) Percentage of cropland. (Bottom) Three figures: length of the growing season in West Africa for 2009 (average), 2011 (dry), and 2012 (relatively wet).

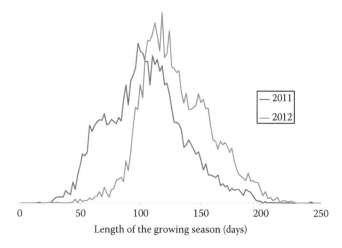

FIGURE 13.3　Histogram of the length of the growing season for 2011 and 2012 for the cropped areas in Mali.

Henebry and de Beurs (2013) provide a basic overview of LSP and the different satellite sensors that are used in studying LSP going back to Landsat 1 in 1972. Indeed, one of the very first LSP studies was performed based on 80 m spatial resolution Landsat data. However, most LSP studies have used observations from the AVHRR, which has provided time series of observations starting in 1982. The AVHRRs have truly been the workhorse platform for the study of LSP (Henebry and de Beurs 2013, and citations within). Since the launch of the MODIS sensors on the Terra and Aqua satellites in 2000 and 2002, MODIS data are also regularly applied in LSP studies as a result of their increased spatial resolution and general accuracy as compared with AVHRR data. While there has been a tremendous increase in the number of published studies that use the term LSP, few of these studies specifically use higher-spatial-resolution (30 m) Landsat data (Fisher et al. 2006; Fisher and Mustard 2007). However, some recent efforts have focused on the use of fused data products (e.g., fused data based on MODIS and Landsat) for LSP studies as well as web-enabled Landsat data (Kovalskyy et al. 2012).

While White et al. (2009) compared a host of different LSP methods for one dataset (AVHRR) to determine which method resulted in data closest to a variety of field observations, we are aware of just one study that compares LSP metrics based on different datasets (Brown et al. 2008). This study applied one methodology, quadratic regression models to AVHRR, SPOT Vegetation, and several MODIS products to determine the start of the growing season. They then compare the results with fields' observed SOS. The results showed that 8 km MODIS data at 8-day temporal resolution resulted in SOS measurements closest to field observations. It is important to note that the field observations were specifically designed for large-scale satellite validation and thus were better matched with 8 km data as opposed to data at finer spatial resolutions.

other countries in West Africa. In comparison, the growing season for 2012 was much longer. Only 16.5% of the pixels showed a growing season with fewer than 100 days and the median growing season was 123 days. Unfortunately, there was civil unrest in Mali in 2012, which still resulted in reduced food insecurity.

White et al. (2009) explored the methodology that is used to estimate phenology metrics with remote sensing data. They found that care must be taken with the modeling approach used and matching the method to the region where the agricultural assessment is done. In the context of food security, tropical semiarid regions dominate the countries where food security is monitored (Figure 13.4); thus, the quadratic and multiple-model fit approaches are necessary to correctly assess variations in phenology relevant to agriculture (White et al. 2009).

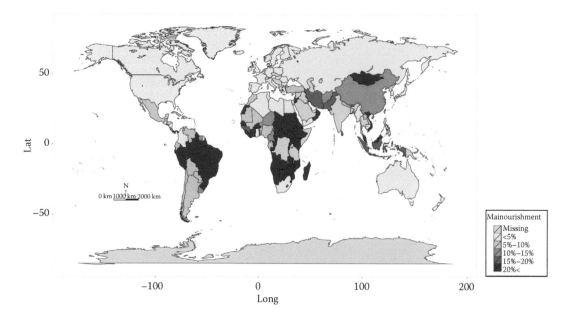

FIGURE 13.4　Map of the percent of the population who are undernourished, 2011–2013, World Food Program data.

13.3 Agriculture and Phenology Metrics

LSP metrics have been used to better connect observed variability in the weather to impacts on food production. Based on agronomic research on the development and response of different cereal varieties to moisture and temperature stress at various stages of development, LSP assessment can provide a quantitative link between remote sensing observations and yield outcomes (Nafziger 2009).

Funk and Budde (2009) used phenology to adjust the SOS across multiple years to increase the reliability of vegetation-derived metrics to assess yield. They noted that Rasmussen (1992) found that early season vegetation data bore no significant relationship to millet yields in Burkina Faso ($r^2 = 0.1$), while values from 30 days after the midseason maxima until the end of season explained 93% of the variation in yields (Funk and Budde 2009; Rasmussen 1992). They found that when removing nonagriculture regions and normalizing the SOS across multiple years, they were able to predict 90% of the variability in yields from 2001 to 2008. This period included the significant change in management of agriculture in the area, which caused a massive decline in productivity outside of weather-induced changes (Funk and Budde 2009).

Bolton and Friedl (2013) used the Funk and Budde (2009) method to derive yield estimates from three different vegetation products and phenology estimates using 500 m MODIS. Their results showed that remotely sensed information related to crop phenology is useful for agricultural monitoring. The best times to predict crop yields were 65–75 days after green-up for maize and 80 days after green-up for soybeans (Bolton and Friedl 2013). For relatively homogenous pixels, the timing of maize and soybeans SOS derived from satellite data appeared to be both detectable and separable (Bolton and Friedl 2013). However, because MODIS acquires data at relatively coarse spatial resolution (250 m), most pixels will include mixtures of crops. Thus, separation of crops based on satellite-derived phenological information is likely to be challenging in many areas (Ozdogan 2010). Thus, most LSP information is used as a measure of moisture availability instead of crop type.

Most satellite-derived agricultural yield estimates using phenology data require a base map of land cover and land use that is used to distinguish land in cultivation from nonagricultural land cover. Bolton and Friedl (2013) tested two data sources for this purpose: (1) the MODIS land cover type product, which provides a 500 m spatial resolution representation of land cover, and (2) the cropland data layer (CDL) created by the USDA, which provides a much higher-resolution representation but is only available for the United States. They found that in regions with high-intensity, industrial agriculture such as in the United States, the MODIS land cover and the CDL were equally effective. In regions with small field size, this will not be the case, however, and the ability of the remote sensing data to identify farm management and moisture-related differences in yield in small fields will be severely limited.

Jain et al. (2013) focused on evaluating the impact of high-resolution satellite data to estimate cropping intensity, or the number of crops planted annually, in regions with small fields. Subsistence agricultural systems are often characterized by small, irregular fields, and these areas often are poorly characterized by methods based on coarse resolution due to the mixing of natural and cropped vegetation in the same pixel (Rasmussen 1992). Using a multiscalar method where 30 m resolution Landsat data are combined with 250 m daily MODIS imagery, Jain et al. (2013) show that using a hierarchical training method that uses the high-resolution spatial information together with the lower spatial but higher temporal information from MODIS allows for an accurate mapping of crop intensity over these small fragmented agricultural systems (Jain et al. 2013). Other authors have also shown that merging high-resolution imagery with lower-resolution imagery in quantitative fusion approaches can improve the results of agricultural assessment (Jin et al. 2010).

Outside of smallholder agricultural regions in the tropics, there is significant food insecurity in the lesser-developed regions of central and eastern Asia. Land abandonment resulting in reduced agricultural production, large changes in economic and governmental institutions, and a crumbling physical infrastructure has resulted in increased vulnerability to moisture conditions.

de Beurs and Henebry (2004) used LSP measurements at 8 km spatial resolution to demonstrate the effect of the collapse of the Soviet Union on agriculture in Kazakhstan. de Beurs and Ioffe (2014) used LSP measurements to determine the number of times agricultural fields were cropped over a 10-year period in European Russia, which is an indicator of potential future cropland abandonment. Prishchepov et al. (2012) found the highest land abandonment was found in Latvia, where 42% of its agricultural land was abandoned from 1990 to 2000, followed by Russia (31%), Lithuania (28%), Poland (14%), and Belarus (13%) (Prishchepov et al. 2012). When abandoned areas are reforested or native grasslands are restored, water quality improves and carbon can be sequestered, but unless economic growth in other sectors occurs, food security of the area could suffer (Gellrich and Zimmermann 2007).

13.4 Food Security and Phenology

Seasonality in food production, where food is produced in one primary growing season, is a common characteristic across many climates. The impact of seasonality in food availability can translate to seasonality in food prices and food security in poor and food-insecure countries with insufficient storage and poorly functioning markets (Alderman et al. 1997; Chen 1991; Crews and Silva 1998; Haddad et al. 1997; Handa and Mlay 2006; Hillbruner and Egan 2008). As Devereux (2012, p. 111) states, "not only does seasonality generate short-term hunger and seasonal food crises, it is also responsible for various 'poverty ratchets' that can have irreversible long-term consequences for household well-being and for productive capacity in rural areas." Coping strategies that have been developed in response to regular reductions in availability and affordability of food involve transfers of assets from poorer households to richer ones at less than full value (Cekan 1992; Devereux 2012).

Understanding the impact of local declines in food production due to the weather relies on understanding the livelihood approach of the people living in the area affected. Livelihood is defined as "the capabilities, assets (including both material and social resources) and activities required for a means of living" (Chambers and Conway 1991). LSP analysis can be used to determine when agricultural production is likely to fall below the needs for farmers in the region. Its relevance to food security depends on the ability of a community to access food if a production decline occurs.

Access to food requires that all individuals have the income sufficient to purchase food that is personally and culturally acceptable (Sen 1981). The concept of access focuses on the ability of households to maintain food consumption in the face of a wide variety of shocks that increase the gap between available income and entitlements and the amount of food that can be purchased with that income at a particular time and place.

Shocks to food security can come from many sources. Droughts can greatly reduce the supply of food, increasing the local food prices (Brown et al. 2012). Personal shocks, such as death of a member of the family, poor health, and loss of employment, can reduce household income and access to food (Gazdar and Mallah 2013). Economic shocks, such as inflation, changes in government policies, changes in public safety nets, and international commodity prices, can reduce the ability of households to attain enough food, despite the fact that the amount of food they produce or their household income has not changed (Brown 2014). Issues of access are described most clearly in social science concepts of individual and household well-being, which capture stress and coping strategies at a variety of scales (Barrett 2010).

13.5 Approaches to Measuring Food Insecurity

There are two comprehensive approaches to measuring food insecurity. One approach involves measurement of the anthropometrics such as body weight, height, and age of representative samples of the population (e.g., demographic and health surveys) (Brown et al. 2014). The other involves measurements of aggregate household and/or individual consumption of food per day of representative samples of the population (e.g., the World Bank's living standard measurement surveys). However, both these methods require extensive surveys of large numbers of people and households. This data collection strategy is costly in terms of time and money, and as a result, these surveys are rarely conducted more than a few times in a decade for a given country. Additionally, a nontrivial limitation of using anthropometry in assessing child nutritional status is the lack of specificity. In other words, changes in body measurements are sensitive to many unobserved factors including intake of essential nutrients, infection, altitude, stress, and genetic background (Barrett 2010).

Thus, for measuring and monitoring food security, many organizations use food prices in both urban and rural areas as a proxy for food access, since large, rapid increases in food prices can result in widespread reduction in food consumption, resulting from widespread declines in food availability and thus food supply in a market (Brown et al. 2009). Many rural dwellers buy food and sell food so they are sensitive to price changes in nearby markets (Brown 2014). Food prices are collected across small towns, in regional capitals, and in the capital city on a monthly basis for this use. Because food prices can be influenced by a number of international and domestic policies or events that are unrelated to local production (Brown 2014), the links between phenology assessment and food security outcomes can quickly become complex. Including LSP assessments with food prices at multiple scales of analysis can help to identify potential "hot spots" of food insecurity.

13.6 Use of LSP in Food Security Assessment in Niger

The use of LSP data can be important in food security assessment, particularly in poor agropastoral regions that are remote and frequently food insecure as illustrated here for Niger. Niger consistently ranks in the bottom five poorest countries in the world for gross domestic product per capita (World Bank 2013). Over half of its 13 million residents are engaged in the agriculture sector, despite the country being semiarid with a short annual growing season and that only 12% of its land is arable (CIA 2012).

In 2013, Ouallam, a region north of Niamey in the Department of Tillaberi, and other agropastoral areas of Niger had 1.2 million people that had difficulty getting enough to eat with the high price of cereal and limited livelihood strategies (Famine Early Warning Systems Network [FEWS NET] 2013a). Niger's Ouallam Department in the Tillaberi Region has an economy that is based mainly on agriculture (harvests of winter millet, sorghum, cowpeas, groundnuts, and peas) and on raising cattle, sheep, and goats. Annual rainfall in this area ranges from 400 to 600 mm. Local crop production normally covers over 40% of household food needs and accounts for 18% of the income of very poor and poor households. Migration and sales of bush products (wood and straw) are also important sources of cash income for local households. The types of foods normally consumed by households in this area are furnished by on-farm production or purchased with income from livestock sales, farm labor, and sales of wood and straw (Senahoun et al. 2011). Most household spending is on cereal purchases by poor and cereal-short households and on school fees for their children's education. An average to high local demand is helping to generate normal to above-normal levels of income from the farm activities (FEWS NET 2013b).

The Ouallam Department of Niger experienced a delay in the start of its growing season, restricting the total length of the season and resulting in a production deficit of 68,000 metric tons (FEWS NET 2013b). Although the 2012 cropping season was above average, the poor and very poor households, who make up 61% of the population of the region, depleted their food stocks from the previous year before the start of the 2013 growing season and thus are vulnerable to food insecurity due to the low productivity (FEWS NET 2013b).

Given this situation, food security analysts were looking at the remote sensing data products available to determine the likely impact of the reported late start of the growing season. Here, we will show the remotely sensed vegetation data that the

FEWS NET uses operationally to identify and quantify the likely impact of variations in the SOS. FEWS NET uses the eMODIS data product derived from daily MODIS reflectance data and recomposited into 10-day observations for ease of comparison with rainfall and other datasets (Jenkerson et al. 2010; Ji et al. 2010). Figure 13.5 shows the Ouallam Region on an NDVI anomaly image calculated from a 10-year mean (from 2001 to 2010) used by FEW NET in its Early Warning Explorer tool for the first 10 days of July. Figure 13.5a and b shows the time

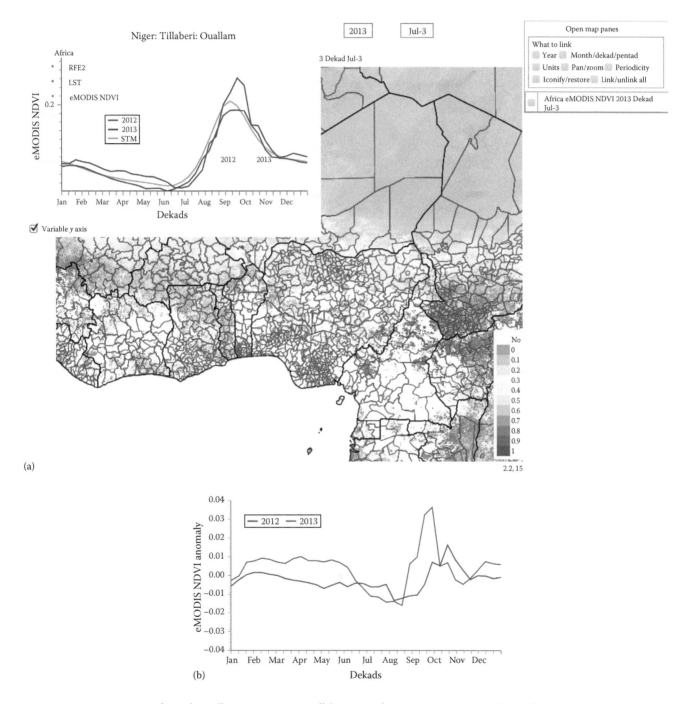

FIGURE 13.5 Comparison of Niger's Ouallam Department in Tillaberi Region's MODIS NDVI: (a) actual NDVI from 2012 and 2013, along with the short-term mean and (b) NDVI anomaly from 2012 and 2013 using the short-term mean.

series of the NDVI as it compared to the previous season and the anomaly of the NDVI. Although the anomaly image showed very little impact of the late start, the time series does show the late start and then a subsequent robust response to significant rains in August.

The late start in July 2013 resulted in below average yields, despite the later robust response to rainfall in August. A shorter growing season due to a late start has long been recognized as a risk factor for crop development (Brown and de Beurs 2008). Because crops in the Sahel are often photoperiod sensitive, a late start cannot be made up later in the season and typically translates into a yield reduction (Buerkert et al. 2001). Thus, the food security analysts using the information from vegetation were able to quantitatively link the late start to identifying regions that were more likely to experience food insecurity as a result in yield declines.

13.7 Using LSP to Contextualize the Relationship between Maize Price and Health in Kenya

Kenya's malnutrition rate has remained high despite a number of improvements in other socioeconomic and health indicators (KNBS 2010). Additionally, Kenya faces important shifts in climate and weather patterns that may be linked to food insecurity (Grace et al. 2013). Recent research has sought to examine the effect of price changes (a measure of food access) on household-level food insecurity outcomes, specifically the birth weight of babies (Grace et al. 2014). A key component of this type of approach to food insecurity and food access is food availability.

Kenya has no annually updated measures of local, household-relevant food production. Instead, to measure local production, the authors calculated the maximum NDVI in a small area (10 km radius) around each community with household-level health data. Given the long time series of available MODIS NDVI data, the authors were able to match births to the relevant growing season's NDVI. Using the MODIS NDVI value as a proxy for local production, the authors identified the importance of local prices on infant birth weights (Grace et al. 2014). The results further suggested that price impacts were dependent on NDVI. In general, when NDVI was high and food prices were low, relative to the areas under study, households were less likely to experience food insecurity, but when MODIS NDVI was low and prices were low, the likelihood of food insecurity increased. The results suggest that food prices alone cannot predict a household's risk for food insecurity; rather, the combination of local production and food prices should be evaluated (Brown 2014).

Because the majority of impoverished, food-insecure countries are not able to collect and disseminate fine-resolution estimates of food production, the potential for NDVI and other measures of LSP to support analyses similar to the Kenya analyses just described is high and very relevant. Ultimately, the use of this type of data can provide an improved understanding of community- and household-level response to food price volatility.

13.8 Discussion

Improving scientific understanding of LSP through the use of remotely sensed data and climate models is necessary for anticipating areas of potential food shortage. LSP models can be combined with data on food prices, health outcomes, and household economics at micro- and macrolevels to more fully examine the links between the physical environment and the people who are most likely impacted by weather changes. This chapter highlighted the work that has been done by physical scientists to advance LSP modeling using remote sensing and climate models. We also highlighted the impacts of food insecurity and undernutrition on human health outcomes. Interdisciplinary research that includes LSP and human health is an important next step in applying the physical science to relevant issues related to development and health.

There are some drawbacks to phenology as a measure of the progress of the growing season. Most LSP models are most effective when most of the growing season has happened, as the curve-fitting algorithms do best when the beginning, middle, and end of the season are present. When only the start is known, estimating the yield impacts from the beginning of the season is very challenging due to uncertainties in the model. Thus, although LSP tools derived from remote sensing of vegetation are quite effective as a retrospective analysis, continued reliance on rainfall-based estimation of the start and peak of the season through crop models is likely for famine early warning organizations interested in the current month's growing conditions.

Another challenge to LSP and other new metrics of remote sensing such as soil moisture from new sensors such as soil moisture active passive and evapotranspiration derived from satellite-derived temperature is the challenge of communicating the value of such new tools to the widespread and diverse food security and humanitarian communities (Brown and Brickley 2012). FEWS NET, for example, is comprised of a central office, four organizations that are experts on remote sensing and biophysical modeling, and 23 offices in food-insecure countries. This makes the organization extremely susceptible to center–periphery problems where new ways of looking at drought and its impact on crop yield move only extremely slowly from the research centers, located mostly in the United States or Europe, to where food security analysis is actually conducted, in the food-insecure countries themselves. Thus, although LSP analysis could be extremely helpful to identify and respond to food insecurity, moving the analysis approach from research to the operational context will take a lot of investment and effort (Brown and Brickley 2012).

13.9 Conclusions

LSP is the study of the changes in start, peak, and end of the growing season on the land surface as observed by satellite sensors. LSP can be used to estimate agriculturally important weather-related impacts on food production, distribution, and cost that may result in changes in food security in vulnerable communities. Food security assessment uses satellite remote sensing to determine how agriculture is changing, the impact of weather on food production, and how these changes affect food availability and access. Research to improve our understanding of environmental and weather drivers of production can provide new and valuable information for the food security community.

References

Alderman, H., Bouis, H., and Haddad, L. (1997). Aggregation, flexible forms, and estimation of food consumption parameters: Comment. *American Journal of Agricultural Economics*, 79, 267.

Awaya, Y., Kodani, E., Tanaka, K., Liu, J., Zhuang, D., and Meng, Y. (2004). Estimation of the global net primary productivity using NOAA images and meteorological data: Changes between 1988 and 1993. *International Journal of Remote Sensing*, 25, 1597–1613.

Barrett, C. (2010). Measuring food insecurity. *Science*, 327, 825–828.

Bolton, D.K. and Friedl, M.A. (2013). Forecasting crop yield using remotely sensed vegetation indices and crop phenology metrics. *Agricultural and Forest Meteorology*, 173, 74–84.

Bradley, A.V., Gerard, G.G., Barbier, N., Weedon, G.P., Anderson, L.O., Huntingford, C., Aragao, L.E.O.C., Zelazowski, P., and Arai, E. (2011). Relationships between phenology, radiation and precipitation in the Amazon region. *Global Change Biology*, 17, 2245–2260.

Brown, M.E. (2008). *Famine Early Warning Systems and Remote Sensing Data*. Heidelberg, Germany: Springer Verlag.

Brown, M.E. (2014). *Food Security, Food Prices and Climate Variability*. London, U.K.: Routledge Press.

Brown, M.E. and Brickley, E.B. (2012). Evaluating the use of remote sensing data in the USAID Famine Early Warning Systems Network. *Journal of Applied Remote Sensing*, 6, 063511.

Brown, M.E. and de Beurs, K. (2008). Evaluation of multi-sensor semi-arid crop season parameters based on NDVI and rainfall. *Remote Sensing of Environment*, 112 (5), 2261–2271.

Brown, M.E., de Beurs, K., and Vrieling, A. (2010). The response of African land surface phenology to large scale climate oscillations. *Remote Sensing of Environment*, 114, 2286–2296.

Brown, M.E., Grace, K., Shively, G., Johnson, K., and Carroll, M. (2014). Using satellite remote sensing and household survey data to assess human health and nutrition response to environmental change. *Population and Environment*, 36, 48–72.

Brown, M.E., Hintermann, B., and Higgins, N. (2009). Markets, climate change and food security in West Africa. *Environmental Science and Technology*, 43, 8016–8020.

Brown, M.E., Tondel, F., Essam, T., Thorne, J.A., Mann, B.F., Leonard, K., Stabler, B., and Eilerts, G. (2012). Country and regional staple food price indices for improved identification of food insecurity. *Global Environmental Change*, 22, 784–794.

Buerkert, A., Moser, M., Kumar, A.K., Furst, P., and Becker, K. (2001). Variation in grain quality of pearl millet from Sahelian West Africa. *Field Crops Research*, 69, 1–11.

Cekan, J. (1992). Seasonal coping strategies in central Mali: Five villages during the 'soudure'. *Disasters*, 16, 66–73.

Chambers, R. and Conway, G. (1991). Sustainable rural livelihoods: Practical concepts for the 21st century. In: *IDS Discussion Paper 296*. London, U.K.: Institute of Development Studies.

Chen, M.A. (1991). *Coping with Seasonality and Drought*. London, U.K.: Sage Publications.

CIA (2012). *Central Intelligence Agency World Factbook*. Washington, DC: United States Government.

Crews, D.E. and Silva, H.P. (1998). Seasonality and human adaptation: Current reviews and trends. *Reviews in Anthropology*, 27, 1–15.

de Beurs, K.M. and Brown, M.E. (2013). The effect of agricultural growing season change on market prices in Africa. In: A. Tarhule (ed.), *Climate Variability—Regional and Thematic Patterns*, pp. 189–203.

de Beurs, K.M. and Henebry, G.M. (2004). Land surface phenology, climatic variation, and institutional change: Analyzing agricultural land cover change in Kazakhstan. *Remote Sensing of Environment*, 89, 497–509.

de Beurs, K.M. and Henebry, G.M. (2010). Spatio-temporal statistical methods for modeling land surface phenology. In: I.L. Hudson and M.R. Keatley (eds.), *Phenological Research: Methods for Environmental and Climate Change Analysis*. (pp. 177–208) London and New York: Springer.

de Beurs, K.M. and Loffe, G. (2014). Use of Landsat and MODIS data to remotely estimate Russia's sown area. *Journal of Land Use Science*, 9(4).

Devereux, S. (2012). Seasonal food crises and social protection. In: B. Harriss-White and J. Heyer (eds.), *The Comparative Political Economy of Development: Africa and South Asia* (p. 358). Oxon, U.K.: Routledge Press.

FAO (2008). *Assessment of the World Food Security and Nutrition Situation*. Rome, Italy: United Nations Food and Agriculture Organization.

FAO (2012). *The State of Food Insecurity in the World*. Rome, Italy: United Nations Food and Agriculture Organization.

FCPN (2007). *Food Situation in the Sahel and West Africa: Should Satisfactory Agricultural and Food Prospects Be Expected?* Food Crises Prevention Network (FCPN). Paris, France.

FEWS NET (2013a). *Niger Food Security Outlook* (p. 8). Washington, DC: US Agency for International Development.

FEWS NET (2013b). *Niger Food Security Outlook: Auspicious Cropping Season Conditions through August Deteriorate in September* (p. 8). Washington, DC: US Agency for International Development Famine Early Warning Systems Network.

Fisher, J., Mustard, J., and Vadeboncoeur, M. (2006). Green leaf phenology at Landsat resolution: Scaling from the field to the satellite. *Remote Sensing of Environment*, 100, 265–279.

Fisher, J. I. and Mustard, J. F. (2007). Cross-scalar satellite phenology from ground, Landsat, and MODIS data. *Remote Sensing of Environment*, 109(3), 261–273.

Funk, C.C. and Budde, M.E. (2009). Phenologically-tuned MODIS NDVI-based production anomaly estimates for Zimbabwe. *Remote Sensing of Environment, 113*, 115–125.

Gazdar, H. and Mallah, H.B. (2013). Inflation and food security in Pakistan: Impact and coping strategies. *IDS Bulletin, 44*, 31–37.

Gellrich, M. and Zimmermann, N.E. (2007). Investigating the regional-scale pattern of agricultural land abandonment in the Swiss mountains: A spatial statistical modeling approach. *Landscape and Urban Planning, 79*, 65–76.

Godfray, H.C.J., Beddington, J.R., Crute, I.R., Haddad, L., Lawrence, D., Muir, J.F., Pretty, J., Robinson, S., Thomas, S.M., and Toulmin, C. (2010). Food security: The challenge of feeding 9 billion people. *Science, 327*, 812–818.

Grace, K., Brown, M.E., and McNally, A. (2014). Examining the link between food price and food insecurity: A multi-level analysis of maize price and birthweight in Kenya. *Food Policy, 46*, 56–65.

Grace, K., Davenport, F., Funk, C., and Lerner, A. (2013). Child malnutrition and climate conditions in Kenya. *Applied Geography, 11*, 164–177.

Haddad, L., Hoddinott, J., and Alderman, H. (1997). *Intrahousehold Resource Allocation in Developing Countries: Models, Methods and Policies*. Washington, DC: IFPRI/JHU Press.

Handa, S. and Mlay, G. (2006). Food consumption patterns, seasonality and market access in Mozambique. *Development Southern Africa, 23*, 541–560.

Henebry, G.M. and de Beurs, K.M. (2013). Remote sensing of land surface phenology: A prospectus. *Phenology: An Integrative Environmental Science*. Amsterdam, the Netherlands, Springer, pp. 385–411.

Hillbruner, C. and Egan, R. (2008). Seasonality, household food security, and nutritional status in Dinajpur, Bangladesh. *Food and Nutrition Bulletin, 29*, 221–231.

Jain, M., Mondal, P., DeFries, R.S., Small, C., and Galford, G.L. (2013). Mapping cropping intensity of smallholder farms: A comparison of methods using multiple sensors. *Remote Sensing of Environment, 134*, 210–223.

Jenkerson, C., Maiersperger, T., and Schmidt, G. (2010). eMODIS: A user-friendly data source. In: *Open-File Report 2010–1055* (p. 10). Sioux Falls, SD: U.S. Geological Survey.

Ji, L., Wylie, B., Ramachandran, B., and Jenkerson, C. (2010). A comparative analysis of three different MODIS NDVI datasets for Alaska and adjacent Canada. *Canadian Journal of Remote Sensing, 36*, S149–S167.

Jin, H., Wang, J., Bo, Y., Chen, G., and Xue, H. (2010). Data assimilation of MODIS and TM observations into CERES-Maize model to estimate regional maize yield. In: *Remote Sensing and Modeling of Ecosystems for Sustainability VII*. San Diego, CA: SPIE-Int Soc Optical Engineering.

Justice, C.O., Townshend, J.R.G., Holben, B.N., and Tucker, C.J. (1985). Analysis of the phenology of global vegetation using meteorological satellite data. *International Journal of Remote Sensing, 6*, 1271–1318.

KNBS (2010). *Kenya Demographic and Health Survey 2008–09*. Calverton, MD: Kenya National Bureau of Statistics (KNBS) and ICF Macro.

Koetse, M.J. and Rietveld, P. (2009). The impact of climate change and weather on transport: An overview of empirical findings. *Transportation Research Part D: Transport and Environment, 14*, 205–221.

Kovalskyy, V. et al. (2011). The suitability of multi-temporal web-enabled Landsat data NDVI for phenological monitoring—A comparison with flux tower and MODIS NDVI. *Remote Sensing Journal, 3*(4), 325–334.

Lieth, H. (1974). *Phenology and Seasonality Modeling* (444 pp.). Berlin, Germany: Springer Verlag.

Mu, J.E., McCarl, B.A., and Wein, A. (2013). Adaptation to climate change: Changes in farmland use and stocking rate in the U.S. *Mitigation and adaptation strategies for Global Change, 18*, 713–730.

Myers, S. and Patz, J. (2009). Emerging threats to human health from global environmental change. *Annual Review of Environmental Resources, 34*, 223–252.

Nafziger, E. (2009). Corn. In: *Illinois Agronomy Handbook* (pp. 13–26). Urbana, IL: University of Illinois.

Ozdogan, M. (2010). The spatial distribution of crop types from MODIS data: Temporal unmixing using independent component analysis. *Remote Sensing of Environment, 114*, 1190–1204.

Prishchepov, A.V., Radeloff, V.C., Baumann, M., Kuemmerle, T., and Müller, D. (2012). Effects of institutional changes on land use: Agricultural land abandonment during the transition from state-command to market-driven economies in post-Soviet Eastern Europe. *Environmental Research Letters, 7*, 024021.

Rasmussen, M.S. (1992). Assessment of millet yields and production in northern Burkina Faso using integrated NDVI from the AVHRR. *International Journal of Remote Sensing, 13*, 3431–3442.

Reed, B.C., Brown, J.F., Vanderzee, D., Loveland, T.R., Merchant, J.W., and Ohlen, D.O. (1994). Measuring phenological variability from satellite imagery. *Journal of Vegetation Science, 5*, 703–714.

Schmidhuber, J. and Tubiello, F.N. (2007). Global food security under climate change. *Proceedings of the National Academy of Sciences, 104*, 19703–19708.

Schwartz, M.D. (1992). Phenology and springtime surface-layer change. *Monthly Weather Review, 120*, 2570–2578.

Sen, A.K. (1981). *Poverty and Famines: An Essay on Entitlements and Deprivation*. Oxford, U.K.: Clarendon Press.

Senahoun, J., Ndiaye, C.I., Haido, A.M., Saidou, O., Tahirou, L., Akakpo, K., and Kountche, B.I. (2011). *Special Report: Inter-Agency Crop and Food Security Assessment Mission to Niger*. Rome, Italy: The World Food Program and the Food and Agriculture Organization.

Vrieling, A., de Beurs, K.M., and Brown, M.E. (2011). Variability of African farming systems from phenological analysis of NDVI time series. *Climatic change, 109*, 455–477.

Walker, J.J., de Beurs, K.M., and Wynne, R.H. (2014). Dryland vegetation phenology across an elevation gradient in Arizona, USA, investigated with fused MODIS and Landsat data. *Remote Sensing of Environment*, 144, 85–97.

White, M.A., Thornton, P.E., and Running, S.W. (1997). A continental phenology model for monitoring vegetation responses to interannual climatic variability. *Global Biogeochemical Cycles*, 11, 217–234.

White, M.A., de Beurs, K., Didan, K., Inouye, D., Richardson, A., Jensen, O., O'Keefe, J. et al. (2009). Intercomparison, interpretation and assessment of spring phenology in North America estimated from remote sensing for 1982 to 2006. *Global Change Biology*, 15, 2335–2359.

World Bank (2013). *Data: The World Bank*. Washington, DC: The World Bank.

Zaal, F., Dietz, T., Brons, J., Van der Geest, K., and Ofori-Sarpong, E. (2004). Sahelian livelihoods on the rebound: A critical analysis of rainfall, drought index and yields in Sahelian Agriculture. In: A.J. Dietz, R. Ruben, and A. Verhagen (eds.), *The Impact of Climate Change on Drylands: With a Focus on West Africa* (pp. 61–77). Dordrecht, the Netherlands: Kluwer Academic Publishers.

V

Forests

14

Characterizing Tropical Forests with Multispectral Imagery

E.H. Helmer
International Institute of Tropical Forestry, USDA Forest Service

Nicholas R. Goodwin
Ecosciences Precinct

Valéry Gond
Forest Ecosystems Goods and Services

Carlos M. Souza, Jr.
Instituto do Homen e Meio Ambiente da Amazônia

Gregory P. Asner
Carnegie Institution for Science

Acronyms and Definitions

AB-C	Aboveground live Biomass in units of Mg C ha^{-1}
ACCA	Automated Cloud Cover Assessment
AGLB	Aboveground Live Forest Biomass in Mg dry weight ha^{-1}
ASTER	Advanced Spaceborne Thermal Emission and Reflection Radiometer
AVHRR	Advanced Very High Resolution Radiometer
AWiFS	Advanced Wide Field Sensor
BB-C	Belowground live Biomass in units of Mg C ha^{-1}
BRDF	Bidirectional Reflectance Distribution Function
CBERS	China–Brazil Earth Resources Satellite
CDM	Clean Development Mechanism
DEM	Digital Elevation Model
DW-C	Dead Wood biomass in units of Mg C ha^{-1}
ESTARFM	Enhanced Spatial and Temporal Adaptive Reflectance Fusion Model
ETM+	Enhanced Thematic Mapper Plus
Fmask	Function of Mask
GHG	Greenhouse Gas
GLAS	Geoscience Laser Altimeter System
GV	Green Vegetation (unitless fraction, range 0–1)
HRG	High-Resolution Geometric
HRV	High-Resolution Visible
HRVIR	High-Resolution Visible and Infrared
HRS	High-Resolution Stereoscopic
INPE	Instituto Nacional de Pesquisas Espaciais
IRMSS	Infrared Multispectral Camera
IRS	Indian Resources Satellite
LI-C	Carbon content of forest floor litter in Mg C ha^{-1}
LISS	Linear Imaging Self-Scanner
MAIAC	Multi-Angle Implementation of Atmospheric Correction for MODIS
MERIS	Medium-Resolution Imaging Spectrometer
Mg	Megagram = 1×10^6 g = 1 metric ton
MISR	Multi-angle Imaging SpectroRadiometer
MODIS	Moderate Resolution Imaging Spectroradiometer

MSS   Multispectral Scanner
MVC   Maximum-Value Compositing
NDFI   Normalized Difference Fraction Index (unitless, range −1 to 1)
NDMI   Normalized Difference Moisture Index
NPV   Non-Photosynthetic Vegetation (unitless fraction, range 0–1)
NDVI   Normalized Difference Vegetation Index
NIR   Near-Infrared
SMA   Spectral Mixture Analysis
STARFM   Spatial and Temporal Adaptive Reflectance Fusion Model
REDD+   Reducing Emissions from Deforestation and Degradation, conservation of forest carbon stocks, sustainable management of forests, or enhancement of forest carbon stocks in developing countries
SO-C   Soil organic carbon in Mg C ha^{-1}
SPOT   Satellite Pour l'Observation de la Terre
SWIR   Shortwave Infrared
TM   Thematic Mapper
UNFCCC   United Nations Framework Convention on Climate Change
WiFS   Wide Field Sensor

14.1 Introduction

Tropical forests abound with regional and local endemic species and house at least half of the species on earth, while covering less than 7% of its land (Gentry, 1988; Wilson, 1988; as cited in Skole and Tucker, 1993). Their clearing, burning, draining, and harvesting can make slopes dangerously unstable, degrade water resources, change local climate, or release to the atmosphere the organic carbon (C) that they store in their biomass and soils as greenhouse gases (GHGs). These forest disturbances accounted for 19% or more of annual human-caused emissions of CO_2 to the atmosphere from the years 2000 to 2010, and that level is more than the global transportation sector, which accounted for 14% of these emissions. Forest regrowth from disturbances removes about half of the CO_2 emissions coming from the forest disturbances (Houghton, 2013; IPCC 2014). Another GHG of concern when considering tropical forests is N_2O released from forest fires.

Tropical forests (including subtropical forests) occur where hard frosts are absent at sea level (Holdridge, 1967), which means low latitudes, and where the dominant plants are trees, including palm trees, tall woody bamboos, and tree ferns. They include former agricultural or other lands that are now undergoing forest succession (Faber-Langendoen et al., 2012). They receive from <1000 mm year^{-1} of precipitation to more than 10 times that much as rainfall or fog condensation. Whether dry or humid, tropical forests have far more species diversity than temperate or boreal forests, and their role in earth's atmospheric GHG budgets is large.

Multispectral satellite imagery, that is, remotely sensed imagery with discrete bands ranging from visible to shortwave

infrared (SWIR) wavelengths, is the timeliest and most accessible remotely sensed data for monitoring these forests. Given this relevance, we summarize here how multispectral imagery can help characterize tropical forest attributes of widespread interest, particularly attributes that are relevant to GHG emission inventories and other forest C accounting: forest type, age, structure, and disturbance type or intensity; the storage, degradation, and accumulation of C in aboveground live tree biomass (AGLB, in Mg dry weight ha^{-1}); the feedbacks between tropical forest degradation and climate; and cloud screening and gap filling in imagery. In this chapter, the term *biomass* without further specification is referring to AGLB.

14.2 Multispectral Imagery and REDD+

14.2.1 Greenhouse Gas Inventories and Forest Carbon Offsets

Multispectral satellite imagery can provide crucial data to inventories of forest GHG sinks and sources. Inventories of GHGs that have forest components include national inventories for negotiations related to the United Nations Framework Convention on Climate Change (UNFCCC). The UNFCCC now includes a vision of compensating countries for reducing greenhouse gas emissions to the atmosphere from deforestation, degradation, sustainable management of forests, or conservation or enhancement of forest C stocks in developing countries (known as REDD+). Inventories of GHG emissions for the UNFCCC Clean Development Mechanism (CDM) may also include forests, and there are other forest carbon offset programs.

Programs like REDD+ could help moderate earth's climate. They could also help conserve tropical forests and raise local incomes, as long as countries make these latter goals a priority in REDD+ planning. Compensation in REDD+ is for organic carbon (C) stored in forest AGLB, dead wood, belowground live biomass, soil organic matter, or litter, as long as the stored C is "produced" by avoided GHG emissions, such as avoided deforestation or avoided degradation of forest C stores.

In forest C offsets, avoided emissions are estimated as the difference between net GHG emissions that would have occurred without implementing change (the *baseline case* or *business-as-usual scenario*) and actual net emissions that are reduced from what they would have been without the management change (the *project case*). Logging, burning, and fragmentation are examples of disturbances that degrade forest C stores. Replacing conventional logging with reduced impact logging reduces associated C emissions and is an example of avoided C emissions. For subnational projects such as those developed under voluntary carbon markets or the CDM, *leakage* must also be subtracted. *Leakage* refers net emissions that a carbon offset project displaces from its location to elsewhere. Examples are deforestation or removals of roundwood or fuelwood in a forest not far from the forest where such activities have ceased for forest C credits.

Many countries and organizations have officially proposed that forest C stored by enrichment planting, or by forest growth or regrowth on lands that were not forest before 1990, should also be explicitly eligible for REDD+ compensation (Parker et al., 2009). These latter activities, afforestation and reforestation, already dominate forest projects developed under the CDM.

14.2.2 Roles of Multispectral Imagery

The United Nations Intergovernmental Panel on Climate Change (IPCC) provides guidelines for GHG emission inventories, including for forest land (IPCC, 2006). Expanded methods based on these guidelines include those from the Verified Carbon Standard program (http://www.v-c-s.org/methodologies/redd-methodology-framework-redd-mf-v15, Avoided Deforestation Partners, 2015; Pearson et al., 2011). Summaries of these guidelines for communities seeking to certify carbon credits for voluntary carbon markets are also available (e.g., Vickers et al., 2012). For each stratum of each land use considered, changes in C stocks are estimated on an annual basis as the net of changes in the C pools as follows (in Mg C year^{-1}) (Equation 14.1, IPCC, 2006):

$$\Delta C_{LU} = \Delta C_{AB-C} + \Delta C_{BB-C} + \Delta C_{DW-C} + \Delta C_{LI-C} + \Delta C_{SO-C} + \Delta C_{HW-C}$$
(14.1)

where

> ΔC_{LU} is the carbon stock changes for a land-use stratum, for example, a forest stratum, in Mg C year^{-1}
> $\Delta C_{SUBSCRIPT}$ represents carbon stock changes for a given pool
> Subscripts denote the following carbon pools in units of Mg C year^{-1}:
> AB-C is the aboveground live biomass carbon
> BG-C is the belowground biomass carbon
> DW-C is the dead wood carbon
> LI-C is the litter carbon
> SO-C is the soil organic carbon
> HW-C is the harvested wood carbon

For forest GHG inventories for REDD+ and other programs, multispectral satellite imagery can be used to estimate some of the key variables for Equation 14.1:

1. Areas of forest strata (e.g., forest types, disturbance/degradation classes, or management)
2. Baseline and ongoing rates of change in the areas of forest strata
3. The AGLB and rates of C accumulation in young forests
4. Point estimates of forest C pools in AGLB with fine-resolution imagery to supplement ground plot data
5. Potentially, forest AGLB if shown to be accurate for a given landscape
6. Potentially, GHG emission factors for forest disturbances if spectral indices of disturbance intensity can be calibrated to correlate well with associated GHG emissions and remaining C pools

Monitoring forest extent over large scales is also crucial to this forest C accounting, and multispectral satellite imagery is the best data for this purpose, but this topic is covered in other chapters of this book (Chapters 15, 17 through 19). Other chapters also cover multispectral image fusion with radar to map forest AGLB (e.g., Saatchi et al., 2011) or estimation of tropical forest biomass with airborne lidar (e.g., Asner et al., 2012). Multi-angular image data can also improve forest age mapping (Braswell et al., 2003).

When using the "stock-difference" method (IPCC, 2006) to quantify the parameters in Equation 14.1, the total C pool for each time period is estimated by multiplying the spatial density of C by the area (in hectares) of the forest stratum. The change in the C pool is estimated as the difference in C pools between two time periods divided by the elapsed time in years (IPCC, 2006). In addition, in Equation 14.1, belowground biomass is usually estimated as a fraction of aboveground biomass with default values by ecological zone, region, or country. Also, when the type of land use is forest, litter can often be ignored.

The average spatial density of carbon in live biomass, in Mg C ha^{-1}, is estimated from the average spatial density of the dry weight of live biomass (in Mg ha^{-1}) multiplied by the C fraction of dry weight biomass. Typically, this C fraction is about 50% of dry weight mass. IPCC (2006) has published default values for average C fraction of dry weight wood biomass by ecological zone. Dry weight is estimated with equations that relate the size of the trees growing in a forest to their dry weight, mainly as gauged by tree stem diameter and height. Then, the estimated dry weights of all trees in a known area are summed. Species-specific or regional equations are sometimes available.

14.3 Characteristics of Multispectral Image Types

Multispectral satellite imagery is available at spatial resolutions ranging from high (<5 m) to medium (5–100 m), to coarse (>100 m) (e.g., Table 14.1). The data usually include reflective bands covering the visible (blue, green, and red) and near-infrared (NIR) wavelengths of the electromagnetic spectrum. Several other sensors include SWIR bands (e.g., Landsat Thematic Mapper [TM] and subsequent Landsat sensors); the sensors aboard the fourth and fifth missions of Satellite Pour l'Observation de la Terre (SPOT 4 high-resolution visible and infrared [HRVIR], SPOT 5 high-resolution geometric [HRG], and the SPOT 4 and 5 Vegetation instruments); the Moderate Resolution Imaging Spectroradiometer (MODIS), the Advanced Wide Field Sensor (AWiFS), and the Infrared Multispectral Scanner Camera aboard the China–Brazil Earth Resources Satellite series [CBERS].

Satellite launches in the years 1998–1999 greatly increased the amount of imagery available for monitoring tropical forests. These launches brought (1) the first public source of high-spatial-resolution imagery (IKONOS, with <5-m pixels); (2) the first medium-resolution imagery (5–100 m pixels) with some degree of consistent global data collection (Landsat 7); (3) the first medium-resolution imagery with fine-resolution panchromatic bands of 10 m (SPOT 4 and Landsat 7, respectively); and

TABLE 14.1   Multispectral Satellite Imagery Most Commonly Used to Characterize Tropical Forests

SatelliteRepeat/Revisit[a] Cycle, Scene Size/Swath Width Quantization	Band	Wavelength (μm)	Distributed Spatial Resolution (m)	Approximate Active Dates
High resolution (<5 m)				
IKONOS	Panchromatic	0.45–0.90	1	September 24, 1999 to present
3- to 5-day revisit	1-Blue	0.445–0.516	4	
11 × 11 km scenes	2-Green	0.506–0.595	4	
11 bits	3-Red	0.632–0.698	4	
	4-Near-infrared	0.757–0.853	4	
QuickBird	Panchromatic	0.45–0.90	0.6	October 18, 2001 to present
2- to 6-day revisit	1-Blue	0.45–0.52	2.4	
18 × 18 km Scenes	2-Green	0.52–0.60		
11 bits	3-Red	0.63–0.69		
	4-Near-infrared	0.76–0.90		
Medium resolution (5–100 m) with high-resolution panchromatic				
SPOT 4 HRVIR; SPOT 5 HRG	Panchromatic	0.51–0.73	2.5	SPOT 4: March 24, 1998 to July 2013
2–3 days Revisit	Panchromatic	0.51–0.73	5	SPOT 5: May 04, 2002 to present
60 × 60 km	Green	0.50–0.59	10	
8 bits	Red	0.61–0.68	10	
	Near-infrared	0.78–0.89	10	
	Shortwave infrared	1.58–1.75	20	
SPOT 1, 2, 3 HRV	Panchromatic	0.51–0.73	10	SPOT 1: February 22, 1986 to September 1990
1- to 3-day revisit	Green	0.50–0.59	20	SPOT 2: January 22, 1990 to July 16, 2009—
60 km × 60 km	Red	0.61–0.68	20	SPOT 3: September 26, 1993 to November 14, 1996
8 bits	Near-infrared	0.78–0.89	20	
Medium resolution (5–100 m)				
Landsat MSS 1,2,3 (4,5)	4 (1)-Blue–green	0.5–0.6	60[b]	Landsat 1: July 23, 1972 to January 06, 1978
16 days repeat	5 (2)-red	0.6–0.7	60[b]	Landsat 2: January 22, 1975 to February 25, 1982
170 × 185 km	6 (3)-Near-infrared	0.7–0.8	60[b]	Landsat 3: March 05, 1978 to March 31, 1983
4 bits	7 (4)-Near-infrared	0.8–1.1	60[b]	
Landsat 4 TM, 5 TM, 7 ETM+	1-Blue	0.45–0.52	30	Landsat 4: July 17, 1982 to December 14, 1993
16 days Repeat	2-Green	0.52–0.60	30	Landsat 5: March 1, 1984 to January 2013
170 × 183 km	3-Red	0.63–0.69	30	Landsat 7: April 15, 1999
8 bits	4-Near-infrared	0.76–0.90	30	
	5-Shortwave infrared	1.55–1.75	30	
	6-Thermal (2 ETM+ bands)	10.40–12.50	L4,5 120[c] (30) L7 60[c] (30)	
	7-Shortwave infrared	2.08–2.35	30	
	8-Panchromatic (L7 only)	0.52–0.90	15	
EO-1 ALI	MS-1'-Coastal aerosol	0.433–0.453	30	November 21, 2000 to present
16-day repeat	MS-1-Blue	0.45–0.515	30	
37 × 42 km	MS-2-Green	0.525–0.605	30	
12 bits	MS-3-Red	0.63–0.69	30	
	MS-4-Near-infrared	0.775–0.805	30	
	MS-4'-Near-infrared	0.845–0.89	30	
	MS-5'-Shortwave infrared	1.2–1.3	30	
	MS-5	1.55–1.75	30	
	MS-7	2.08–2.35	30	
	Panchromatic	0.48–0.69	10	

(Continued)

TABLE 14.1 (*Continued*)   Multispectral Satellite Imagery Most Commonly Used to Characterize Tropical Forests

SatelliteRepeat/Revisit[a] Cycle, Scene Size/Swath Width Quantization	Band	Wavelength (μm)	Distributed Spatial Resolution (m)	Approximate Active Dates
Landsat 8	1-Coastal aerosol	0.433–0.453	30	February 11, 2013—
16-day repeat	2-Blue	0.450–0.515	30	
170 × 183 km	3-Green	0.525–0.600	30	
12 bits	4-Red	0.630–0.680	30	
	5-Near-infrared	0.845–0.885	30	
	6-SWIR 1	1.560–1.660	30	
	7-SWIR 2	2.100–2.300	30	
	8-Panchromatic	0.500–0.680	15	
	9-Cirrus	1.360–1.390	30	
	10-Thermal infrared 1	10.60–11.19	100[c] (30)	
	11-Thermal infrared 2	11.50–12.51	100[c] (30)	
Coarse resolution (>100 m)				
Terra/Aqua MODIS[d] (7 of 36 bands are shown)	1	0.620–0.670	250	Terra (EOS AM): August 12, 1999 to present
1-day revisit	2	0.841–0.876	250	Aqua (EOS PM): May 04, 2002 to present
2330 km Swath Width	3	0.459–0.479	500	
12 bits	4	0.545–0.565	500	
	5	1.230–1.250	500	
	6	1.628–1.652	500	
	7	2.105–2.155	500	
SPOT 4,5 Vegetation 1, 2[d]	0-Blue	0.43–0.47	1150	Aboard SPOT 4: March 24, 1998 to July 2013
1-day revisit	2-Red	0.61–0.68	1150	Aboard SPOT 5: May 04, 2002 to present
2250 km Swath Width	3-Near-infrared	0.78–0.89	1150	
10 bits	SWIR-Shortwave infrared	1.58–1.75	1150	

[a] Revisit cycles change with latitude.

[b] The original MSS pixel size of 79 × 57 m is now resampled to 60 m.

[c] Thermal infrared Landsat bands are now resampled to 30 m.

[d] For coarse-resolution sensors, resolution given is at nadir.

(4) the first coarse-resolution imagery (>100 m pixels) distributed with higher-level preprocessing like atmospheric correction and cloud-minimized compositing (MODIS and SPOT Vegetation). Before IKONOS, remotely sensed reference data had to come from air photos that in many places were costly to obtain and outdated.

The next big advances in tropical forest monitoring with satellite imagery came in 2005–2008, when (1) Google, Inc. and the producers of high-resolution imagery such as QuickBird and IKONOS made high-resolution data viewable on Google Earth for many sites, making reference data free and accessible for subsets of project areas; and (2) the Brazilian National Institute for Space Research (INPE) and the United States Geological Survey (USGS) began to freely distribute Landsat and other imagery with medium spatial resolution, making long, dense time series of medium-resolution imagery available over large areas.

Other sources of multispectral imagery for monitoring tropical forests over large areas that are not shown in Table 14.1, mainly to highlight them here, include the Japan–U.S. Advanced Spaceborne Thermal Emission and Reflection Radiometer (ASTER) (aboard Terra). In addition to 15 m VNIR bands, it has several SWIR and thermal bands with 30–90 m spatial resolution. Data for Brazil and China and nearby areas are also available from CBERS. The series of CBERS satellites, 1, 2, and 2B, collected panchromatic to SWIR images with medium spatial resolution (20–80 m, 113–120 km swath width), and red and NIR images with coarse spatial resolution (260 m, 890 km swath width) from 1999 to 2010, and missions to collect with medium-resolution multispectral imagery with a 5-day revisit cycle are scheduled. In the Indian Resources Satellite (IRS) series, the Wide Field Sensor (WiFS) has a 740 km swath width, 188 m spatial resolution, and red and NIR bands. More recently, the IRS-P6 satellite carries the AWiFS instrument. AWiFS has 60 m pixels for green through SWIR bands, a 740 km swath width, a 5-day revisit cycle, and a SWIR band, combining advantages of imagery with medium and coarse spatial resolutions. The later of the IRS series sensors include data from Linear Imaging Self-Scanner (LISS) with multispectral imagery with a 23.5 m spatial resolution. Ground stations receiving data from CBERS and the IRS satellite series have not covered all of the tropics. Fortunately, that situation should gradually change.

14.4 Preprocessing Imagery to Address Clouds

14.4.1 Cloud Screening

We begin with cloud and cloud shadow screening, as this step is crucial in the image processing chain for characterizing tropical forests. Clouds and their shadows obscure the ground and contaminate temporal trends in reflectance. Automated systems for processing large archives of satellite imagery are becoming more common for natural resource applications and must screen clouds. Clouds are composed of condensed water vapor that form water droplets and scatter visible to NIR light, reducing direct illumination on the surface below and forming a cloud shadow. In multispectral satellite imagery, clouds are characterized by a high albedo (Choi and Bindschadler, 2004), while their shadows have lower reflectance than surrounding pixels. The easiest solution to cloud contamination is to restrict analyses to cloud-free imagery, which may include only dry season imagery for tropical and coastal environments due to frequent cloud cover. Alternatively, methods to screen cloud- and shadow-contaminated pixels can increase the number of observations available (Figure 14.1). Increasing the number of available observations in a time series may also improve the detection of land surface change and reflectance trends.

Manual and semiautomated approaches to cloud screening are undesirable for processing large numbers of images due to the time-consuming nature of the work, which may depend not only on analyst experience but also on image contrast. Several automated approaches have been developed, but separating cloud

and shadow from the land surface is not necessarily straightforward given the diversity of land surfaces coupled with large variations in cloud and shadow optical properties (Zhu et al., 2012a; Zhu et al., 2012b; Goodwin et al., 2013; Lyapustin et al., 2008). A summary of current approaches to cloud and shadow screening for Landsat TM/ETM+, SPOT, and MODIS sensors follows.

14.4.1.1 Landsat TM Imagery

The Landsat TM/ETM+ archives of countries with receiving stations now contain up to three decades of imagery (1984 to present) with varying levels of cloud and cloud shadow. The U.S. Geological Survey is working with other countries to consolidate these archives through consistent processing and distribution through its website (landsat.usgs.gov). Image preprocessing by the Landsat program has included the Automatic Cloud Cover Assessment (ACCA) algorithms for both Landsat-5 TM and Landsat-7 ETM+ missions, which use optical and thermal (ETM+ only) bands to identify clouds (Irish, 2006). It is designed for reporting the percentage of cloud cover over scenes rather than producing per-pixel masks. Further modifications have also been tested for application to Landsat 8 imagery (Scaramuzza et al., 2012), which includes a new cirrus band (1.360–1.390 μm) that is sensitive to aerosol loadings and should improve cloud detection. ACCA is designed to limit the impacts of cloud and scene variability on thresholding. The ETM+ ACCA incorporates two passes: one to conservatively estimate "certain" cloud at the pixel level with a series of spectral and thermal tests. The result is then used to derive scene-based thermal thresholds for the second pass. The error in scene-averaged cloud amount was

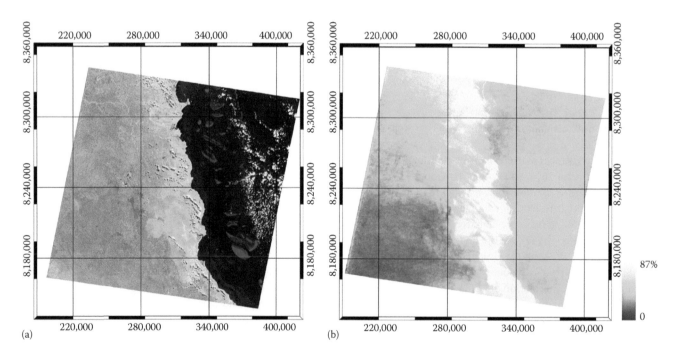

FIGURE 14.1 Illustration of cloud distribution spatially and temporally over tropical forests of north Queensland: (a) Landsat image (RGB: 542, Path/Row: 96/71, and date July 02, 2007) and (b) percentage of observations classified as cloud between 1986 and 2012 (n = 445). Note: High cloud fractions were not included in calculations.

estimated to be around 5% (Irish et al., 2006). Scaramuzza et al. (2012) validated the per-pixel classification of the ETM+ ACCA (pass 1) and found a 79.9% agreement between reference and ACCA at the pixel scale. Using a subset of the same reference set, Oreopoulos et al. (2011) evaluated both per-pixel ACCA masks and a cloud detection algorithm modified from the MODIS Luo–Trishchenko–Khlopenkov algorithm (Luo et al., 2008). Both ACCA and the modified LTK showed greater than 90% agreement with the reference, although like ACCA, the LTK had limited ability to detect thin cirrus clouds. Furthermore, ACCA has been used as the starting point for further cloud masking (Choi and Bindshadler, 2004; Roy et al., 2010; Scaramuzza et al., 2012).

Earlier studies have shown that several approaches work well for classifying clouds and cloud shadows over particular path/rows. One approach is image differencing based on image pairs (Wang, 1999), while other studies have empirically defined thresholds for cloud brightness and coldness in one or more spectral/thermal bands, for example, Landsat TM Bands 1 and 6 (Martinuzzi et al., 2007); Bands 3 and 6 (Huang et al., 2010); Bands 1, 3, 4, and 5 (Oreopoulos et al., 2011); and Bands 1, 4, 5, and 6 (Helmer et al., 2012). The application of these methods to a range of path/rows around the globe, however, remains untested and may encounter issues due to spectral similarities among the wide range of combinations of land surfaces and cloud/cloud shadows.

The automated method that Huang et al. (2010) developed to allow forest change detection in cloud-contaminated imagery considers brightness and temperature thresholds for clouds that are self-calibrated against forest pixels. It requires a digital elevation model to normalize top of atmosphere brightness temperature values and helps to project cloud shadow on the land surface. Published validation data for this method are currently limited to four U.S. images with forest and would benefit from further calibration/validation.

Two additional automated approaches have recently been published: Fmask (Function of mask) (Zhu and Woodcock, 2012) and a time series approach by Goodwin et al. (2013) (Figure 14.2). Fmask integrates existing algorithms and metrics with optical and thermal bands to separate contaminated pixels from land surface pixels. Fmask also considers contextual information for mapping potential cloud shadow using a flood-fill operation applied to the NIR band. Cloud shadows are then identified by linking clouds with their shadow with solar/sensor geometry and cloud height inferred from the thermal Landsat TM Band 6. The results were validated with a global dataset and were a significant improvement to ACCA, with Fmask achieving overall, user's, and producer's accuracies of 96%, 89%, and 92%, respectively compared to 85%, 92%, and 72%, respectively for ACCA.

The time series method uses temporal change to detect cloud and cloud shadow (Goodwin et al., 2013). It smoothes pixel time series of land surface reflectance using minimum and median filters and then locates outliers with multi-temporal image differencing. Seeded region grow is applied to the difference layer using a watershed region grow algorithm to map clusters of change pixels, with clumps smaller than 5 pixels removed to minimize classification speckle. This has the effect of increasing the cloud/shadow detection rate while restricting commission errors; smaller magnitudes of change associated with cloud/cloud shadows are mapped only if they are in the neighborhood of larger changes. Morphological dilation operations were applied to map a larger spatial extent of the cloud and cloud shadow, while shadows were translated along the image plane in the reverse solar azimuth direction to assess the overlap with clouds and confirm the object is a shadow. A comparison with Fmask showed that the time series method could screen more cloud and cloud shadow than Fmask across Queensland, Australia (cloud and cloud shadow producer's accuracies were 8% and 12% higher, respectively).

Several trade-offs exist between these two automated approaches to cloud and shadow screening. The time series method might detect more cloud and cloud shadow, yet Fmask is more computationally efficient and practical for individual images. At present, the time series method is processed using entire time series for each Landsat path/row. For operational systems processing many images, the computational overhead of the time series approach could be worthwhile as it can detect more cloud/shadow contamination. Locations with few cloud-free observations per year and high land-use change are also less desirable for a time series method. In the absence of an atmospheric aerosol correction, pixels contaminated by smoke and haze are more likely to be classified as cloud by the time series method. Neither the Fmask nor the time series method nor previous attempts adequately map high level, semitransparent cirrus cloud (Figure 14.2d–f). New methods for Landsat 8 will likely detect more cloud with the new band sensitive to cirrus clouds. Both Fmask and the time series methods are highly configurable allowing calibration for a localized region or a wider application. Fmask has been calibrated using a global reference set, while the time series approach was calibrated and tested mainly for northeastern Australian conditions.

Although both methods have high accuracy, further improvements could be made particularly to screening cloud shadow. Removing the dependency of a link between cloud and shadow would be a considerable advancement as clouds are often missed or under/overmapped, causing the shadow test to fail. Furthermore, adding thermal information to the time series method has the potential to remove commission errors where bright surfaces such as exposed soil are falsely classified as cloud. Both methods use a series of rules to classify cloud and shadow and have the flexibility to add new algorithms and criteria to improve the detection of contaminated pixels.

14.4.1.2 SPOT Imagery

The spatial and spectral characteristics of SPOT (Satellite Pour l'Observation de la Terre) have similarities to Landsat imagery, with the first satellite launched in 1986 (SPOT 1), and similar methods for screening cloud and cloud shadows should be useful. The main exception is that SPOT lacks a thermal band, which has been useful in discriminating clouds (e.g., ACCA). However, only a limited number of studies have been published

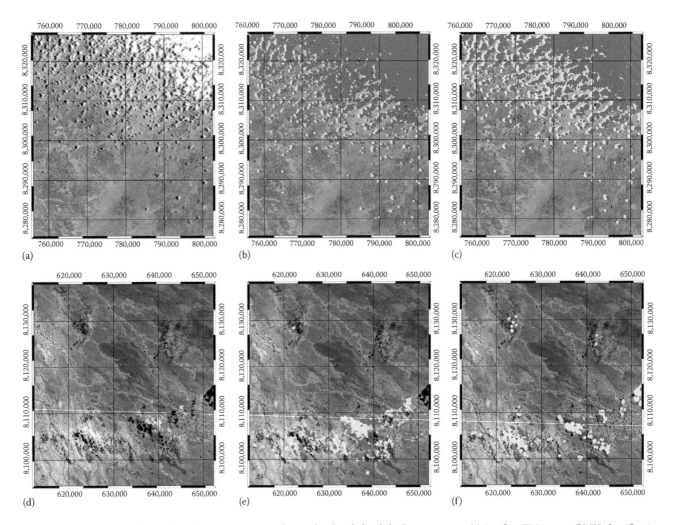

FIGURE 14.2 Examples of Fmask and time series approaches to cloud and cloud shadow screening: (a) Landsat TM image, (b) TS classification, (c) Fmask classification, (d) Landsat TM image, (e) TS classification, and (f) Fmask classification, (a–c) well-detected cumulus cloud and cloud shadow (RGB: 542, Path/Row: 97/71, and date October 10, 1998) and (d–f) a complex example where both methods miss sections of cirrus cloud (RGB: 542, Path/Row: 98/72, and date April 04, 2001).

on screening cloud and cloud shadow from SPOT data. SPOT is a commercially operated sensor, and unlike Landsat TM/ETM+ and MODIS, scenes are typically purchased/tasked with limited cloud cover or would otherwise prove cost prohibitive for many vegetation applications. The New South Wales government of Australia, for example, acquired 1850 images between 2004 and 2012, of which only 313 contain cloud with the maximum cloud cover values <10% (Fisher, 2014).

Le Hégarat-Mascle and André (2009) used a Markov random field framework that assumes that clouds are connected objects, solar/sensor geometry is known, and shadow has a similar shape to its corresponding cloud (excluding the influence of topography). Potential cloud pixels were identified using a relationship between green and SWIR bands; shadows were located using cloud shape, orientation of shadow relative to cloud and SWIR band reflectance, removing objects not part of a cloud–shadow pair. The method was applied to 39 SPOT 4 HRVIR images over West Africa with encouraging results. However, when applying this method, Fisher (2014) found commission errors as bright

surfaces were frequently matched to dark surfaces that were not cloud contaminated. They suggest first masking vegetation and water bodies, then locating marker pixels for clouds and shadows in the green–SWIR space and NIR bands, respectively, then growing objects with the watershed transform. Sensor/solar geometry and object size are also used to match clouds with their shadows.

14.4.1.3 MODIS Imagery

MODIS has a standard cloud product, in contrast to SPOT or until recently Landsat, which includes information on whether a pixel is clear from cloud/shadow contamination. The cloud mask is based on several per-pixel spectral tests and is produced at 250 m and 1 km spatial resolutions (Strabala, 2005). A validation with active ground-based lidar/radar sensors showed an 85% agreement with the MODIS cloud mask (Ackerman et al., 2008).

Recent research has found that time series information can improve cloud detection in MODIS imagery (Lyapustin et al., 2008; Hilker et al., 2012). The cloud-screening method in multi-angle implementation of atmospheric correction, for example, uses a dynamic clear-sky reference image and covariance calculations, in addition to spectral and thermal tests, to locate clouds over land (Lyapustin et al., 2008). In a tropical Amazonian environment, Hilker et al., 2012 demonstrated that this method was better at detecting clouds and increasing the number of usable pixels than the standard product (MYD09GA), which translated into more accurate patterns in NDVI.

14.4.2 Filling Cloud and Scan-Line Gaps

Cloud and cloud shadow screening removes contaminated pixels from analyses but leaves missing data in the imagery and derived products. The scan-line correction error affecting Landsat 7 post-2003 also leaves gaps approximating 20% of affected images (USGS, 2003). Data gaps in maps are aesthetically unappealing, and the derivation of statistics is more difficult. As a result, approaches have been developed to fill data gaps including temporal compositing and fusing imagery from two different sensors.

A range of temporal compositing algorithms have been developed to minimize cloud contamination and noise (Dennison et al., 2007; Flood, 2013). Compositing involves analyzing band/metric values across a date range with an algorithm deciding the pixel value most likely to be cloud/noise free. The choice of algorithm may vary depending on the application and land-cover type. Compositing algorithms have generally been applied to high-temporal-frequency data such as MODIS and AVHRR; however, methods for compositing imagery with a lower temporal resolution have also been developed. For example, the MOD 13 products use the maximum-value compositing algorithm with NDVI as the metric in 16-day and monthly composites of MODIS imagery (Strabala, 2005). Landsat has similarly been composited using a parametric weighting scheme (Griffiths et al., 2013). The result is an image that ideally is free from noise or cloud that can be used as a product itself or the corresponding pixels used to infill data gaps.

The fusion or blending of MODIS and Landsat offers another approach to predict image pixel values within data gaps. These methods integrate medium-spatial-resolution Landsat with temporal trends in reflectance (e.g., seasonality) captured by the higher temporal frequency of MODIS. Roy et al. (2008) integrated the MODIS bidirectional reflectance distribution function (BRDF)/albedo product and Landsat data to model Landsat reflectance. They found that infrared bands were more accurately predicted than visible wavelengths, probably in response to greater atmospheric effects at shorter wavelengths. The spatial and temporal adaptive reflectance fusion model (STARFM) requires a MODIS–Landsat image pair captured on the same day plus a MODIS image on the prediction date and applies spatial weighting to account for reflectance outliers (Gao et al., 2006). Further algorithm development has produced an enhanced STARFM (ESTARFM) method that was found to improve predictions in heterogeneous landscapes (Zhu et al., 2010). However, there are known limitations with blending

or fusing Landsat and MODIS imagery. Solutions involving MODIS will work only post-2000 when imagery was first captured and potentially 2002 onward where stable BRDF predictions are needed (Roy et al., 2008). Furthermore, Emelyanova et al. (2013) found that land-cover type and temporal and spatial variances impact the fusion of MODIS and Landsat as well as the choice of algorithm. Where the temporal variance of MODIS is considerably less than the spatial variance of Landsat, blending may not improve predictions.

Gap filling using Landsat imagery alone has also been performed. Helmer and Ruefenacht (2005) developed a method for predicting Landsat values using two Landsat images for change detection. This method develops a relationship between uncontaminated pixels in an image pair with regression tree models, and it then applies these models to predict the values in areas with missing data in the target image. Additional images are used in the same way to predict pixels in remaining cloud gaps. Langner et al. (2014) segment such pairwise predictive models according to forest type. Approaches using geostatistics have also been developed. Pringle et al. (2009) use an image before and after the target image in geostatistical interpolation to predict values in Landsat 7 SLC-off imagery. Based on their results, they recommend images captured within weeks, rather than months, of each other to limit temporal variance in a tropical savanna environment. Zhu et al. (2012a) also use geostatistics with encouraging results to predict missing Landsat 7 SLC-off data based on the Geostatistical Neighborhood Similar Pixel Interpolator.

A potential limitation with gap filling is the introduction of image noise or artifacts. This is because of differences in vegetation phenology, illumination, and atmospheric effects as gap-filled imagery contains data from multiple dates and/or sensors. These effects can be minimized by atmospheric and illumination corrections as well as methods that seek to balance the distribution of pixel values such as histogram matching, linear regression, or regression trees (Helmer and Ruefenacht, 2007).

14.5 Forest Biomass, Degradation, and Regrowth Rates from Multispectral Imagery

Studies have used multispectral imagery to map or estimate some key inputs to the variables in Equation 14.1 (Section 14.2.2) for forests: forest AGLB (in Mg dry weight ha[-1]), rates of C accumulation in reforesting lands (in Mg dry weight ha[-1] year[-1]), and area or intensity of forest degradation or disturbance (in ha). In addition, multispectral imagery is the most common satellite imagery for mapping tropical forest types, which we discuss in Section 14.6, and AGLB estimates are often more precise and accurate if stratified by forest type.

In this section, we first review work that uses the spectral and textural information in multispectral imagery of high spatial resolution to estimate tropical forest AGLB. We then discuss how the spectral information inherent to multiyear image time series has high sensitivity to the height, AGLB, and age of forests that have established since about 10 years before the start of an image sequence (so as early as 10 years before 1972 for Landsat data), which we refer to here as *young forests*, allowing estimates of biomass and C accumulation rates in reforested lands. Next, we discuss how multispectral imagery from a single epoch of medium- to coarse-spatial-resolution imagery has limited sensitivity to tropical forest age or biomass. Section 14.5.3 focuses on detecting tropical forest degradation at pixel and subpixel scales.

14.5.1 Tropical Forest Biomass from High-Resolution Multispectral Imagery

When considering forest structure mapping, multispectral imagery of high spatial resolution, with pixels ≤5 m, is distinct from imagery with medium spatial resolution because the spatial patterns of dominant and codominant tree crowns are visible. The possibility of detecting tree crown size suggests a way to estimate AGLB by allometry between stem diameters, used to estimate AGLB, and crown size (Asner et al., 2002; Couteron et al., 2005; Palace et al., 2008). Automated crown delineation in these images is more accurate than manual means, but both methods overestimate the area of large crowns and underestimate the frequency of understory and codominant trees (Asner et al., 2002; Palace et al., 2008), such that biomass estimates from crown delineation alone require adjustments.

A new technique, however, predicts the biomass of high-biomass tropical forests with stand-level spatial patterns of tree crowns in images with ~1 m or finer pixels. The new method first applies two-dimensional Fourier transforms to subsets (*samples*) of high-resolution panchromatic images, from which it produces a dataset with a row for each sample of imagery and columns that bin the outputs from the transform so that the columns in each row together form a proxy for the distribution of crown sizes discerned or "apparent" in each image sample. Principal components transformation of this matrix yields axes that serve as predictors in regression models of stand structural parameters, like basal area, AGLB, or "apparent" dominant crown size (calculated by inversion) (Couteron et al., 2005; Barbier et al., 2010; Ploton et al., 2011). Ploton et al. (2011) predicted forest biomass ranging from ~100 to over 600 Mg ha⁻¹ in Western Ghats, India, with IKONOS image extracts downloaded from Google Earth Pro (0.6–0.7 m resolution). Their model explained 75% of the variability in forest biomass. They estimated that the relative uncertainty in AGLB estimates that was due to the remote sensing technique, of <15%, was similar to uncertainties associated with estimating forest AGLB with lidar. With this new technique, AGLB estimates from high-resolution imagery on Google Earth could supplement ground- or lidar-based surveys. The resulting increase in the number and density of AGLB estimates for forests should better characterize the landscape-scale spatial variability in AGLB and increase the precision of forest C-pool estimates.

Related to the earlier work on AGLB are studies that have characterized how gradients in the spatial patterns of tropical

forest canopies correspond with climate. These gradients are apparent in high-resolution imagery, and future changes in these patterns could reflect and help monitor effects of global climate change (Malhi and Román-Cuesta, 2008; Palace et al., 2008; Barbier et al., 2010). Barbier et al. (2010), for example, showed how dominant crown size and canopy size heterogeneity change with climate and substrate across Amazonia.

14.5.2 Biomass, Age, and Rates of Biomass Accumulation in Forest Regrowth

With a long time series of medium-resolution multispectral images such as Landsat, key variables for GHG inventories (and forest C accounting for REDD+) can be mapped and estimated for young tropical forests, including area, age, height, AGLB, and rates of biomass accumulation. Where an image time series spans the age range of young forests, its spectral data can precisely estimate age, which is needed to estimate biomass accumulation rates and can also help estimate the height or AGLB of these forests. Helmer et al. (2009) estimated a landscape-level rate of AGLB accumulation in Amazonian secondary forest by regressing forest biomass estimates from the Geoscience Laser Altimeter System (Figure 14.3) against remotely sensed forest age (R-square = 0.60). The estimated landscape-level biomass accumulation rate of 8.4 Mg ha⁻¹ year⁻¹ agreed well with ground-based studies. Forest age was mapped with an algorithm that automatically processed a time series of Landsat MSS and TM imagery (1975–2003) with self-calibrated thresholds that detect

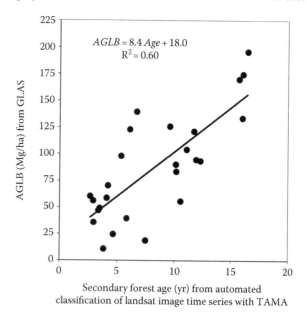

FIGURE 14.3 The average age of secondary forest pixels, as estimated from automatic processing of a time series of Landsat MSS, TM, and ETM+ imagery, in the 150 m window surrounding GLAS waveform centers explained 60% of the variance in GLAS-estimated canopy height and biomass (aboveground live biomass, AGLB, in Mg ha⁻¹ year⁻¹ dry weight). The standard error of the slope and intercept are 1.4 and 13.2, respectively, for 26 observations.

when secondary forests established on previously cleared land. The technique mapped the extent of old-growth forest and age of secondary forest with an overall accuracy of 88%. With the time series, tropical secondary forest >28 years old was accurately distinguished from old-growth forest, even though it was spectrally indistinct in the most recent Landsat scenes. This older secondary forest clearly stored less C than the old-growth forest, being shorter and having much smaller average canopy diameters than nearby old growth.

Forest height and AGLB are strongly related, and the height or AGLB of young forests can be mapped with long time series of Landsat images in tropical (Helmer et al., 2010) and temperate (Li et al., 2011; Plugmacher et al., 2012; Ahmed et al., 2014) regions. With a regression tree model based on the spectral data from all of the images in a time series of cloud-gap-filled Landsat imagery (1984–2005 with 1- to 5-year intervals), Helmer et al. (2010) mapped the height (RMSE = 0.9 m, R-square = 0.84, range 0.6–7 m) and foliage height profiles of tropical semievergreen forest (Figure 14.4). In contrast with mapping the height of old forests, local-scale spatial variability in young forest structure was mapped, because within-patch differences in disturbance intensity and type, and subsequent forest recovery rate, were reflected in the spectral data from the multiyear image stack. This study also mapped forest disturbance type, age, and wetland forest type, with an overall accuracy of 88%, with a decision tree model of the entire time series of cloud-minimized

composite images to better understand avian habitat. As a result, the classification distinguished different agents of forest disturbance, including classes of cleared forests and forests affected by escaped fire, and allowed estimation of rates of forest regrowth. Forest age, vertical structure, and disturbance type explained differences in woody species composition, including abundance of forage species for an endangered Neotropical migrant bird, Kirtland's warbler *Dendroica kirtlandii*.

14.5.3 Limitations to Mapping Forest Biomass or Age with One Multispectral Image Epoch

14.5.3.1 Tropical Forest Biomass with One Image Epoch

Forest biomass mapping with multispectral imagery empirically predicts the AGLB of forested pixels with models that relate forest AGLB or height, from ground plots or lidar, to spectral bands, spectral indices, or spectral texture variables. It remains a challenge (Song, 2013). Forest AGLB is usually estimated in units of Mg dry weight ha^{-1} (see Section 14.2). As more data on stand species composition and species-specific wood densities become available, maps of C storage in forest biomass, as in Asner et al. (2013) and Michard et al. (2014), rather than forest biomass itself, may become more common.

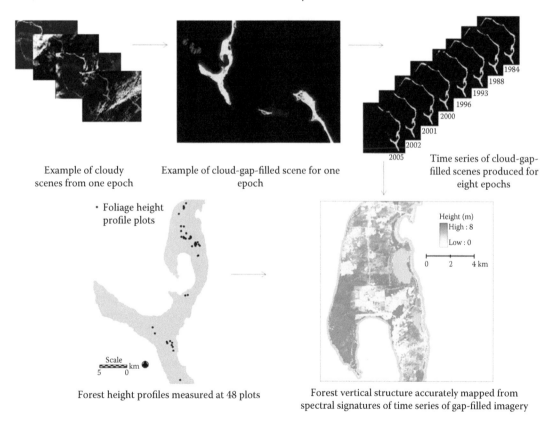

FIGURE 14.4 Tropical dry forest height and foliage height profiles were mapped from a time series of gap-filled Landsat and ALI imagery on the island of Eleuthera, The Bahamas, substituting time for vertical canopy space. The time series was also used to map forest disturbance type and age.

Medium- to coarse-spatial-resolution imagery from one epoch is not that sensitive to small changes in the AGLB or C storage in aboveground biomass of dense tropical forests. (By *epoch*, we mean imagery from one date, one gap-filled or composite image composed of imagery from one to several years, or multiseason imagery from 1 year.) This limited sensitivity appears in biomass mapping models as high per pixel uncertainty that can manifest itself in several ways:

1. Mapping models may explain a minority of variance in reference data (i.e., regressions of predicted vs. observed values have low coefficients of determination or R-squared values of less than 0.50) (e.g., Oza et al., 1996 for volume of Indian deciduous forest; Steininger, 2000 for Bolivian sites; Wijaya et al., 2010 in Indonesia).
2. Mapping models may both underestimate AGLB at high-biomass sites and overestimate AGLB where biomass is low (e.g., Baccini et al., 2008 for tropical Africa; Blackard et al., 2008 for the United States including Puerto Rico; Wijaya et al., 2010).
3. Spectral responses to AGLB may saturate at relatively low levels of around 175 Mg C ha^{-1}. For example, studies indicate that stand-level multispectral responses saturate at 150–170 Mg ha^{-1} for study sites in Brazilian Amazonia (Steininger, 2000; Lu, 2005), ~180 Mg C ha^{-1} in Panama (Asner et al., 2013), and 175 Mg ha^{-1} across Uganda (Avitabile et al., 2012). These saturation levels may be half or less of the biomass of the most structurally complex or old-growth tropical forests in humid lowlands. In many landscapes, the relationship between multispectral data and tropical forest AGLB may saturate at even lower levels.
4. Continental- to global-scale mapping models may not capture gradients in AGLB and C pools that stem from differences in forest allometry and average wood density (Mitchard et al. (2014).

Despite per-pixel uncertainties, estimates of the total forest biomass may be accurate when pixels are summed over large areas that have a wide range of AGLB. This result could happen when the average biomass of pixels covering a large area approaches the mean of the ground or lidar data used to estimate the mapping model. Estimates of total forest AGLB across tropical landscapes can also be accurate if the landscapes has few forest patches with AGLB exceeds the levels where spectral response becomes saturated (e.g., Avitabile et al. 2012).

Texture variables from SPOT 5 imagery may improve mapping models of AGLB, because SPOT 5 imagery has finer spatial resolutions of 10–20 m compared with many other image sources with medium spatial resolution (Table 14.1), but results may still have relative errors of around 20% (Castillo-Santiago et al., 2010). Exceptions may include Asian bamboo forests (Xu et al., 2011) or low-biomass tropical forests.

Mapping models of tropical forest AGLB or height that rely on multispectral imagery benefit from added predictors. Example predictors that may improve models include topography, forest type, climate, soils, geology, or indicators of disturbance like tree canopy cover (Helmer and Lefsky, 2006; Saatchi et al., 2007; Blackard et al., 2008; Asner et al., 2009a; Lefsky, 2010; Wijaya et al., 2010). After including these predictors in mapping models, the variability in the biomass mapped for undisturbed forests may reflect more of the variability in AGLB that stems from regional- to landscape-scale environmental gradients in attributes like rainfall and human-caused disturbance. Maps of these spatial patterns may be useful, but they may not reveal much local-scale AGLB variation.

14.5.3.2 Tropical Forest Age with One Image Epoch

As with AGLB, multispectral imagery from a single image epoch has limited sensitivity to increasing forest age. Many studies show that spectral indices that contrast the mid-infrared bands with the near-infrared or visible bands are the most sensitive indices to tropical forest age, height, and AGLB (e.g., Boyd et al., 1996; Helmer et al., 2000; Steininger, 2000; Thenkabail et al., 2003; Helmer et al., 2010). For example, with Landsat TM or ETM+ data, these indices include the NIR/SWIR ratio, the tasseled cap wetness index (Crist and Cicone, 1984; Huang et al., 2002), the wetness brightness difference index (WBDI) (Helmer et al., 2009), and the normalized difference moisture index (NDMI) (also referred to as the normalized difference structure index and the normalized difference infrared index). The WBDI and NDMI are calculated as

$$WBDI = TC\ Wetness - TC\ Brightness \qquad (14.2)$$

$$NDMI = \frac{(NIR_{b4} - SWIR_{b5})}{(NIR_{b4} + SWIR_{b5})} \qquad (14.3)$$

However, lowland humid tropical forests recovering from previous clearing may become spectrally indistinct from mature forests within 15–20 years (Boyd et al., 1996; Steininger, 2000), though slower-growing tropical forests, like montane or dry forests, can remain spectrally distinct longer (Helmer et al., 2000; Viera et al., 2003). Only a handful of forest age classes can be reliably distinguished in single-date multispectral imagery. Age differences are blurred by differences in disturbance type and intensity that affect regrowth rates and related spectral responses during forest succession (Foody and Hill, 1996; Nelson et al., 2000; Thenkabail et al., 2004; Arroyo-Mora et al., 2005), although age explains more variability in rates of forest regrowth than does disturbance type (Helmer et al., 2010; Omeja et al., 2012).

Recently logged forest has less biomass than old-growth forest, but it may become spectrally indistinct from mature forest within a year or two (Asner et al., 2004a), which is another case in which the forest canopy recovers faster than forest AGLB. In a study in Sabah, Malaysia, conventional logging reduced forest biomass by 67%, but reduced impact logging reduced it by 44% (Pinard and Putz, 1996). In moist forests of Amazonia, AGLB decreased by only 11%–15% after reduced impact logging (Miller et al., 2011).

The youngest regenerating forest patches in landscapes usually do not dominate pixels as large as those of coarse-spatial-resolution imagery like MODIS. The outcome is that maps from such coarse resolution imagery have high error rates for secondary tropical forest. When modeling pixel fractional cover of one or more young forest classes vs. nonforest vs. old forest with MODIS, for example, secondary forest is modeled with the most bias and the least precision (Braswell et al., 2003; Tottrup et al,. 2007). In Amazonia, the model R-square values for the fraction of secondary forest cover were 0.35 for MODIS data alone and 0.61 for MODIS plus MISR data. At the spatial resolution of 1.1 km, corresponding to most of the MISR bands, resulting maps overestimated secondary forest area by 26%. Converting fractional secondary forest cover to discrete classes underestimated secondary forest area by 43% (Braswell et al., 2003). Similarly, Carreiras et al. (2006) concluded that the errors for decision tree classification of secondary forest with SPOT 4 Vegetation across Amazonia were unacceptably high.

14.5.4 Detecting Tropical Forest Degradation with Multispectral Imagery

Tropical forests suffer anthropogenic pressures that perturb their structure and ecological functioning (Vitousek 1994). Human activities that disturb them range from plant collection and human habitation to total deforestation. Many of these forest disturbances can occur at fine spatial scales of less than five to tens of meters, including forest fire (Aragão and Shimabukuro, 2010), recent logging (de Wasseige and Defourny, 2004; Asner et al., 2005; Sist and Ferreira, 2007), road networks (Laporte et al., 2007; Laurance et al., 2009), mining (Peterson and Heemskerk, 2001), and expanding agricultural frontiers (Dubreuil et al., 2012). These human impacts appear like small isolated objects within an ocean of greenness (Souza et al., 2003). They appear as points (logging gaps), lines (roads, trails), both points and lines (logging decks plus skid trails), and with mining areas, both bare soil and pooled water are present.

Although these disturbances can be small, medium-resolution remote sensing techniques can detect and quantify them within homogeneous forest cover (Gond et al., 2004). Compared with fine-scale imagery, images with pixels of 5–30 m have lower or no cost while more frequently covering larger areas of tropical forest. Consequently, medium-resolution imagery constitutes an excellent tool for assessing logging activities in tropical forests across large scales (Asner et al., 2005). Much work to detect finely scaled disturbances of tropical forests uses pixel-level spectra (Section 14.5.3.1). Other work models subpixel spectra to derive continuous variables for monitoring fine-scale disturbances, focusing on the degradation of forest C storage for REDD+ programs and ecosystem models (Section 14.5.3.2).

14.5.4.1 Detecting Fine-Scale Forest Degradation at the Pixel Level

Detecting small canopy gaps and skid trails that have been open for less than 6 months is possible in French Guiana with SPOT 5 HRG images (Gond and Guitet, 2009). The technique developed is based on the local contrast between a photosynthetically active surface (the forest) and one with no or little photosynthetic activity (the gap itself). Using the three main channels dedicated to vegetation identification (red [0.61–0.68 μm], near-infrared [0.79–0.89 μm] and SWIR [1.58–1.75 μm] wavelengths), the contrast between forests and gap is increased enough to be accurately depicted. The detection of an undisturbed forest pixel is made by multiple thresholds on the different reflectances. The advantage of standard remotely sensed data like SPOT 4/5 or Landsat 5/7/8 is the possibility to detect the focused object automatically (Pithon et al., 2013). The automatic processing makes the system operational for tropical forest management and depends only on image availability.

14.5.4.1.1 Road and Trail Detection

Road and trail detection is also a challenge for tropical forest management. Opening, active, and abandoned road and trail networks are a permanent landmark of tropical forest openness and degradation (Laurance et al., 2009). Documenting this dynamic is possible with the 30 years of medium-resolution radiometer archives (Landsat and SPOT). In 2007, Laporte et al. (2007) photo-interpreted Landsat imagery to map the road and trail network across the forests of Central Africa to show which forest areas are endangered by logging activity. When displaying red, NIR, and SWIR channels in red, green, and blue, active roads and trails are "brown"; abandoned roads and trails are "green," and intact tropical forests are "dark green" (de et al., 2004). To automatically process the archives for large areas, Bourbier et al. (2013) proposed a method for using Landsat archive to allow tropical forest managers to visualize the road and trail network dynamism at local (concession) or national scales.

14.5.4.1.2 Mining Detection

Detecting mining activity is slightly different. In general, detecting legal mining is not a real challenge because bare surfaces are prominent and easily mapped. When mining is illegal in tropical forests, however, the bare surface is much smaller and difficult to detect (Almeida-Filho and Shimabukuro, 2002). The additional difficulty comes from the mobility of the illegal miners. A recent abandoned mining site is detectable, but the miners have left. Detecting active mining sites where miners are illegally working is most critical to managers. To map active mining sites in French Guiana, an automatic system using SPOT 5 imagery from a local reception station has been operational since 2008 (Gond et al., unpublished). The system is based on detecting turbid waters resulting from debris washing. Again, the object "turbid water" sharply contrasts with its environment, as with tropical forest vs. bare soil. Using red, NIR, and SWIR channels, turbid water is detected by multiple thresholds on reflectances. So far, the operational system has processed over 1230 SPOT 5 images to ensure regular coverage in space and time of illegal mining activity in French Guiana (Joubert et al., 2012).

14.5.4.2 Detecting Forest Degradation at the Subpixel Level with Spectral Mixture Analysis

Forest degradation in the context of REDD+ can be defined as a persistent reduction in carbon stocks or canopy cover caused by sustained or high-impact disturbance. As a result, forest degradation is often expressed as a complex, three-dimensional change in forest structure related to the introduction of areas of bare soil, piles of dead vegetation created by the residues and collateral damage of removed trees and other plants, and areas with standing dead or damaged tree trunks associated with partial tree fall. Burned forests also leave surface fire scars, indicated by patches of charred vegetation and bare ground (Cochrane et al., 1999; Alencar et al., 2011). Much of tropical forest degradation occurring around the world is driven by selective logging and fires that escape into forests from neighboring clearings. At the multispectral sensor resolution of Landsat, SPOT, and MODIS, it is expected that forest degradation will be expressed in varying combinations of green vegetation (GV), soil, non-photosynthetic vegetation (NPV), and shade within image pixels.

Spectral mixture analysis (SMA) models can be used to decompose the mixture of GV, NPV, soil, and shade reflectances into component fractions known as endmembers (Adams et al., 1995). The SMA has been extensively used throughout the world's tropical forests to detect and map forest degradation (Asner et al., 2009a). For example, subpixel fractional cover of soils derived from the SMA was used to detect and map logging infrastructure including log landings and logging roads (Souza and Barreto, 2000), while the NPV fraction improved the detection of burned forests and of logging damage areas (Cochrane and Souza, 1998; Cochrane et al., 1999). GV and shade enhance the detection of canopy gaps created by tree fall (Asner et al., 2004b) and forest fires (Morton et al., 2011).

SMA models usually assume that the image spectra are formed by a linear combination of n pure spectra, or endmembers (Adams et al., 1995), such that

$$R_b = \sum_{i=1}^{n} F_i R_{i,b} + \varepsilon_b \tag{14.4}$$

for

$$\sum_{i=1}^{n} F_i = 1 \tag{14.5}$$

where
- R_b is the reflectance in band b
- $R_{i,b}$ is the reflectance for endmember i, in band b
- F_i is the fraction of endmember i
- ε_b is the residual error for each band

The SMA model error is estimated for each image pixel by computing the RMS error, given by

$$RMS = \left[n^{-1} \sum_{b=1}^{n} \varepsilon_b \right]^{1/2} \tag{14.6}$$

As mentioned, in the case of degraded forests, the expected endmembers are GV, NPV, soil, and shade fractions. Including a cloud endmember is also possible, which improves the detection and masking of clouds when mapping forest degradation over large areas with long time series of imagery in the Amazon region (Souza et al., 2013). To calibrate the model, the endmembers can be obtained directly from the images (Small, 2004) or from reflectance spectra acquired in the field with a handheld spectrometer (Roberts et al., 2002). The advantage of obtaining endmembers directly from images is that spatial and radiometric calibration between field and sensor observations is not required. The SMA can be automated to make this technique useful for mapping and monitoring large tropical forest regions. A Monte Carlo unmixing technique using reference endmember bundles was proposed for that purpose (Bateson et al., 2000), as well as generic endmember spectral libraries (Souza et al., 2013).

14.5.4.3 Interpreting and Combining Subpixel Endmember Fractions and Derived Indices

The SMA fractions can be combined into indices to further accentuate areas of forest degradation. For example, the normalized difference fraction index (NDFI) was developed to enhance the detection of forest degradation by combining the detection capability of individual fractions (Souza et al., 2005). The NDFI values range from −1 to 1. For intact forests, NDFI values are expected to be high (i.e., about 1) due to the combination of high GV_{shade} (i.e., high GV and canopy shade) and low NPV and soil values. As forest becomes degraded, the NPV and soil fractions are expected to increase, lowering NDFI values relative to intact forest. Bare soil areas will produce NDFI value of −1 because of the absence of GV.

Another approach to SMA allows for uncertainty in the endmember reflectance spectra used for decomposing each pixel into constituent cover types. Referred to as endmember bundles (Bateson et al., 2000), SMA with spectral endmember variability provides a means to estimate GV, NPV, soil, and shade fractions with quantified uncertainty in each image pixel. Using a Monte Carlo approach, Asner and Heidebrecht (2002) developed automated SMA procedures that have subsequently been used to map forest degradation due to logging or understory fire in a wide variety of tropical regions (e.g., Alencar et al., 2011; Carlson et al., 2012; Allnutt et al., 2013; Bryan et al., 2013).

Several mapping algorithms based on spatial and contextual classifiers, decision trees, and change detection have also been applied to SMA results to better map forest degradation using Landsat, SPOT, and MODIS imagery. These techniques are discussed elsewhere (Asner et al., 2009b; Souza and Siqueira, 2013). Additionally, large area mapping and estimates of forest degradation in the Amazon region have also been conducted using these techniques (Asner et al., 2005; Souza et al., 2013).

14.6 Mapping Tropical Forest Types with Multispectral Imagery

14.6.1 Forest Types as Strata for REDD+ and Other C Accounting

Maps of forest type are critical to tropical forest management, including for REDD+ and other GHG inventories. When estimating tropical forest AGLB and other C stores with existing inventory ground plots or lidar data, the estimates are generally stratified by forest type (Asner, 2009; Helmer et al., 2009; Salimon et al., 2011). When designing forest inventories or lidar surveys, stratifying sample locations by forest type improves the efficiency of the sample design (Wertz-Kanounnikoff, 2008), including stratification with types defined by disturbance history (Salk et al., 2013). Stratification by topography or geology may also be important (Ferry et al., 2010; Laumonier et al., 2010) if forest type does not inherently account for related variability in AGLB. An informative review and synthesis of lidar sample design as it relates to forest parameter estimation over large forest areas is available in Wulder et al. (2012). Another important role of forest-type maps based on multispectral satellite imagery is that they are often used to account for the distributions of species and habitats when planning representative reserve systems. For this reason, forest-type maps are also useful to identify where deforestation or wood harvesting is "leaking" to forests that are critical to conserve, but that store less C than forest areas being targeted in REDD+ or carbon offset projects.

Most satellite image–based maps of tropical forest types map classes of forest *formations*. Vegetation formations are defined by growth form and physiognomy. At the simplest level, forest formations may distinguish among closed, open, and wetland forests. More detailed formations may distinguish among forests with different leaf forms or phenology (e.g., deciduous vs. evergreen, broad-leaved vs. needle-leaved, or descriptors that imply a suite of physiognomic characteristics, such as "dry," "montane," or "cloud" forests). More detailed than forest formations are forest *associations*, which distinguish among tree species assemblages. For example, in Figure 14.5, which we discuss in Section 14.6.4, the upper-level headings for forests are forest formations. The subheadings under each forest formation are forest associations.

14.6.2 High-Resolution Multispectral Imagery for Mapping Finely Scaled Habitats

High-resolution imagery makes excellent reference data for calibrating classification and mapping models based on imagery with coarser spatial resolution, but using it as the primary basis

for mapping forest types has several disadvantages. In high-resolution imagery, the within-stand spectral variability of forest types can be large, varying within tree crowns, for example, such that digital classifications at the pixel scale cannot distinguish many forest types. Also these images cover relatively small areas, making them inefficient for mapping forest types over large areas (Nagendra and Rocchini, 2008). Existing archives of high-resolution imagery also lack SWIR bands, which are important in vegetation mapping. Because Landsat ETM+ data have SWIR bands, for example, Thenkabail et al. (2003) found that three floristic tropical forest classes were more distinct in ETM+ data than in IKONOS imagery. Worldview 3, however, will have eight SWIR bands collected at a spatial resolution of 3.7 m.

Yet satellite imagery with high spatial resolution can aid in mapping finely scaled habitats or habitat characteristics. Example habitats are edges or linear features: riparian areas (Nagendra and Rocchini, 2008), roadsides or other corridors, or strands of vegetation types along coastlines. Habitats with high mechanical, chemical, or moisture stress can also be finely scaled. Example stresses are fast-draining substrates where microtopography strongly affects vegetation, like substrates of limestone (Martinuzzi et al., 2008) or sand, or substrates that are also semi-toxic like serpentines. High winds, or drier climate as in savanna ecotones, also lead to finely scale habitats.

Savanna ecosystems, for example, range in tree cover from grassland to forest, which is why we mention them here. Tree cover may change over meters, and high-resolution imagery may be most effective for habitat mapping. Boggs (2010) applied object-oriented classification to 4 m multispectral IKONOS imagery to map tree cover patterns in Mozambique savanna.

In Namibia, tree clusters and grass patches are distinguishable with object-oriented or pixel-level classifications of pan-sharpened QuickBird imagery (0.6 m pixels). In contrast, 10 m multispectral SPOT-5 pixels, though pan-sharpened to 2.5 m, required object-oriented classification (Gibbes et al., 2010).

Object-oriented classification of medium-resolution imagery can indeed sometimes substitute for high-resolution imagery when it can discern finer-scale features of interest that are missed with pixel-level classifications. In Jamaica, Newman et al. (2011) found that object-oriented classification of medium-resolution imagery led to better characterization of roads and forest fragmentation metrics than pixel-level classification did. Object-oriented classification of ASTER data can map savanna habitats in northwest Australia, and it was also more accurate than pixel-level classification (Whiteside et al., 2011). Longer-wave infrared bands were resampled to the 15 m resolution of the visible and NIR bands.

14.6.3 Remote Tree Species Identification and Forest-Type Mapping

Many tropical tree species can be identified by photo interpretation of high-resolution satellite imagery or air photos. With tree crowns in tropical forest often reaching >10 m in diameter,

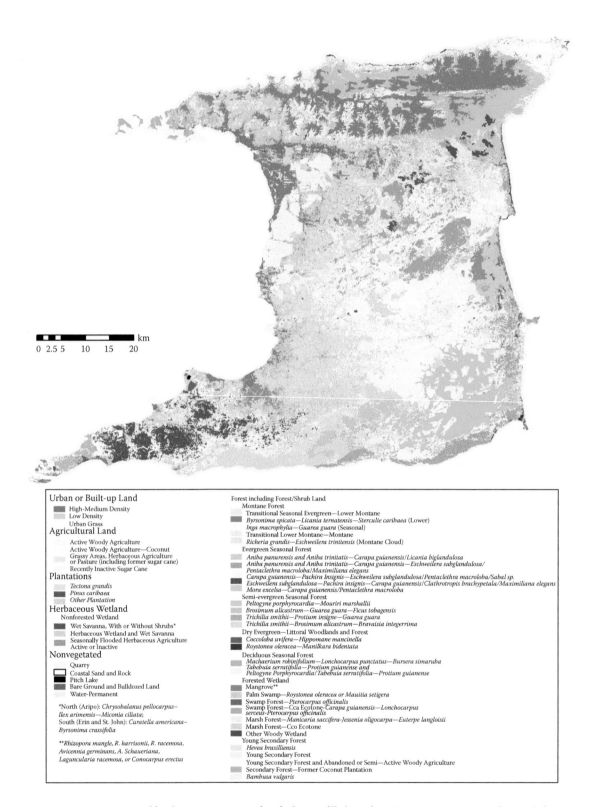

FIGURE 14.5 Forest associations and land cover were mapped with the gap-filled Landsat ETM+ imagery, centered around the year 2007, plus synthetic multiseason imagery developed from three gap-filled TM images from the 1980s that were from the mid to late dry season including from severe drought.

subcrown features are visible. In subtropical to warm-temperate forests of east central Queensland, Australia, Tickle et al. (2006) correctly identified dominant tree species in most of 150 air photo plots with stereo color air photos of scale 1:4000 (~2 m resolution). With these data, they categorized the air photo plots into five genus groups.

In moist forests of Panama, Garzón-López et al. (2013) found that visual analysis of high-resolution color air photos (0.13 m pixels) can reveal spatial distributions of some tropical forest canopy trees. Of 50 common canopy species on a 50 ha plot, 22% had crowns that were distinct in the photos. Of four species tested, interpreters found 40% of the stems that were recorded in field surveys; the resulting maps accurately showed spatial patterns of the species. Sánchez-Azofeifa et al. (2011) concluded that 2.4 m multispectral QuickBird imagery can reveal the spatial distribution and clusters of a species that is conspicuous when flowering, though immature or nonflowering individuals are often missed.

In French Guiana, Trichon and Julien (2006) found that 12 of the 15 most common canopy species or species groups were identifiable, with an accuracy of 87%, in color air photos with scale 1:3700. In the photos, 20%–25% of trees with dbh ≥10 cm, and all trees with dbh ≥20 cm, were visible. For 10 taxa from old-growth Ecuadorian Amazon forest representing a range of crown structures, González-Orozco et al. (2010) found that photo interpretation of large-scale air photos with a dichoto-mous key correctly identified individuals at a rate of >70% for three of the taxa and >50% for two of them.

That photo interpreters can identify many of the dominant species in tropical forest canopies in high-resolution imagery suggests that, given field-based knowledge of the composition and distribution of tree floristic classes (i.e., tree species asso-ciations), which are defined by dominant tree species, floris-tic types of tropical forest can be identified in high-resolution multispectral imagery. Consequently, reference data from photo-interpreting high-resolution multispectral imagery can supplement field data as a source of training and validation data for mapping tropical tree communities with satellite imagery (Helmer et al., 2012).

14.6.4 Mapping Tropical Forest Types with Medium-Resolution Imagery

In mapping tropical forest types with multispectral imagery, spectral similarity among forest classes is a major challenge. Disturbance, differences in topographic illumination, artifacts from filling cloud and other data gaps or from scene mosaicing, all increase class signature variability and consequently increase signature overlap among classes. Secondary forest in a humid montane zone, for example, may be spectrally similar to shade coffee or old-growth forest on highly illuminated slopes. When on a shaded slope, that same secondary montane forest is spec-trally similar to old-growth forest in a less productive zone at higher altitudes (Helmer et al., 2000). Yet digital classifications of multispectral imagery can map many different forest types

with some additions: (1) ancillary geographic data, (2) multisea-son or multiyear imagery or derived phenology, and (3) pixels for training classification models that represent the variability in environmental and image conditions.

Digital maps of environmental data like topography, climate, or geology help distinguish spectrally similar forest types. With Landsat TM/ETM+, linear discriminant function classifications have incorporated ancillary data via post-classification rules based on topography to map eucalyptus forest types (Skidmore, 1989); adding topographic bands to spectral bands to map land-cover and forest physiognomic types (Elomnuh and Shrestha 2000; Helmer et al., 2002; Gottlicher et al., 2009) or distinguish among tree floristic classes (Foody and Cutler, 2003; Salovaara et al., 2005); and classifying imagery by geoclimatic zone (Helmer et al., 2002). Image smoothing or segmentation can improve these classifications by reducing within-class spectral variation (Tottrup, 2004; Thessler et al., 2008).

Tree associations or other floristic classes can be separable with multispectral imagery within an ecological zone, particularly if topographic bands are included. With TM/ETM+ and 18–127 plots, studies have separated three to nine floristic classes within lowland evergreen forest in central Africa, Amazonia, Borneo, or Costa Rica (Foody and Cutler, 2003; Thenkabail et al., 2003; Salovaara et al., 2005; Thessler et al., 2008; Sesnie et al., 2010). Chust et al. (2006) mapped nine floristic subclasses with ETM+ data, elevation, and geographic position over a broad environ-ment across central Panama. With Landsat TM data, Wittmann et al. (2002) mapped three structural classes of Amazonian várzea forests that corresponded to four associations: early suc-cessional low várzea, late secondary and climax low várzea (two associations), and climax high várzea. These studies use spectral data from a single image date and consider only forest; cloudy areas were mapped as such.

When mapping many classes, machine learning classifica-tions more effectively incorporate ancillary environmental data including date bands for gap-filled images. They also do not assume that class spectral distributions are para-metric, and they typically outperform linear classifications. Combining ancillary data and machine learning classifica-tion permits classifications that distinguish many forest and land-cover types, even with noisy, cloud-gap-filled imagery. Examples with TM/ETM+ include decision tree classifications of one or two seasons of cloud-gap-filled Landsat plus ancillary data to map tropical forest physiognomic types and land cover (Kennaway and Helmer, 2007; Helmer et al., 2008; Kennaway et al., 2008). Sesnie et al. (2008) mapped land cover, agriculture type, floristic classes of lowland old-growth forest and three higher-elevation classes based on a map of life zones (*sensu* Holdridge, 1967) with a relatively cloud-free image for each of two scenes. To map tree floristic classes of lowland through montane tropical forest types and land cover in Trinidad and Tobago, Helmer et al. (2012) applied decision tree classifica-tion to recent cloud-gap-filled Landsat imagery stacked with decades-old, gap-filled synthetic multiseason imagery from droughts (Figure 14.5).

Mapping many physiognomic or floristic classes of tropical forest as in the earlier studies requires (1) thousands of training and testing pixels representing the environmental and spectral ranges of each class, including the range of pixel dates where gap-filled imagery was used (Helmer and Ruefenacht, 2007); (2) a band that represents the date of the source image for each pixel in the composite image (a *date band*); and (3) a machine learning classification model. The extensive training data needed are rarely available from field plots. But analysts can learn to identify many physiognomic and floristic classes in remotely sensed imagery given field-based knowledge of general distributions, particularly given free viewing of high-resolution imagery online and Landsat image archives, allowing almost unlimited reference data collection.

Helmer et al. (2012) found that all mono- and bidominant tree floristic classes and many other tree communities in Trinidad and Tobago could be distinguished in reference imagery from nearby associations by (1) unique canopy structure in high-resolution imagery or (2) distinct or unique phenology on specific dates of either high- or medium-resolution reference imagery. For example, distinct canopy structure at high resolution distinguished *Mora excelsa* forests, littoral associations (frequent palms in one; prostrate stems in the other); *Pterocarpus officinalis* swamps, palm swamps, mangroves, and stands of bamboo (*Bambusa vulgaris*), abandoned coconut (*Cocos nucifera*), teak (*Tectona grandis*), pine (*Pinus caribaea*), and Brazilian rubber (*Hevea brasiliensis*). Phenology, including characteristics like flowering, deciduousness, leaf flushes, or inundation, helped to distinguish seven forest associations in high-resolution reference imagery and four associations in phenologically unique Landsat reference scenes. With this knowledge and reference imagery, thousands of training data pixels could be collected.

Including multiseason imagery in classification models of coarse-resolution imagery also improves spectral distinction among tropical forest types (Bohlman et al., 1998, Tottrup, 2004). What is exciting is that we can now think beyond multiseason imagery to multiyear imagery that captures climate or weather extremes or disturbance history. Helmer et al. (2012) found that adding bands from cloud-gap-filled TM imagery from a severe drought that occurred 20 years earlier than the most recent imagery used in the stack of data for classification contributed to the largest increases in accuracy when mapping forest associations in Trinidad. Mapping accuracy of seasonal associations benefited the most. Accuracy improved by 14%–21% for deciduous, 7%–36% for semievergreen, and 3%–11% for seasonal evergreen associations, and by 5%–8% for secondary forest and woody agriculture. Multiyear multispectral imagery that displays different flood stages helps distinguish between upland and periodically flooded tropical forests (Helmer et al., 2009) and among tropical forested wetland types (and can reflect differences in secondary forest species composition by mapping disturbance type as mentioned) (Helmer et al., 2010). In Amazonia, de Carvalho et al. (2013) determined the life cycle length of native bamboo patches with multiyear TM/ETM+ data.

14.6.5 Species Richness and Multispectral Imagery

The tree species richness of tropical forests increases with some of the same variables that influence forest reflectance in multispectral satellite imagery. Richness increases with forest height (among lowland forests with strong edaphic differences), soil fertility (after accounting for rainfall), canopy turnover, and time since catastrophic disturbance; richness decreases with dry season length, latitude, and altitude (Givnish, 1999). We know from forest ground plots that tree species richness also increases with secondary forest age (Whittmann et al., 2002; Chazdon et al., 2007; Helmer et al., 2008). Consequently, over gradients that span from dry to humid, multispectral bands and indices related to vegetation greenness, structure, or disturbance may correlate with species richness. And in fact studies have documented such relationships with single-date Landsat TM or ETM+ imagery (Foody and Cutler, 2006; Nagendra et al., 2010; Hernández-Stefanoni et al., 2011). Single-date multispectral data are unlikely, however, to be sensitive to differences in species richness along short environmental gradients such as among humid evergreen tropical forests. Moreover, an important consideration in biodiversity conservation is that species richness alone does not define conservation value: representation across as many native ecosystems and species as possible is just as important if not more so. Many less productive tropical forest types with less tree species richness, like cloud forests, or forests on harsh or drying soils like those on ultramafic or limestone substrates or ombrotrophic sands, have the most endemic species richness.

14.6.6 Tropical Forest-Type Mapping at Coarse Spatial Scale

In tropical regions extending over large areas, multiseason data from monthly, annual, or multiyear composites of imagery with coarse spatial resolution have supported large-area mapping of tropical forest formations with even linear classification methods (Joshi et al., 2006; Gond et al., 2011, 2013; Pennec et al., 2011; Verheggen et al., 2012). For example, Gond et al. (2011) mapped five classes of forest canopy openness across French Guiana with an unsupervised classification of an annual composite image of SPOT 4 Vegetation data. Across Central Africa, Gond et al. (2013) mapped 14 forest formations with 1 year of 8- and 16-day MODIS image composites. The forest formations were based on leaf phenology and canopy openness. With 1 year of NDVI composite images from the Indian Resource Satellite (IRS 1C) WiFS across India, Joshi et al. (2006) mapped 14 forest formations. The formations were labeled by phenology and climatic class (e.g., *Tropical dry deciduous forest*, *Tropical moist deciduous forest*, and so on). Verheggen et al. (2012) applied unsupervised classification to seasonal and annual composites of MEdium-Resolution Imaging Spectrometer (MERIS) and SPOT 4 Vegetation data for the Congo basin, producing a map with six forest classes that were based on leaf phenology, canopy openness, and elevation class. Producers' and users' accuracies for forest classes in the latter two studies were mostly between 80% and 100%.

Combining ancillary data, monthly image composites of imagery with coarse spatial resolution but high temporal resolution, and decision tree classification has permitted forest classifications at subcontinental to global scales or has distinguished many more forest formations. Decision tree classification of monthly composites of imagery with coarse spatial resolution, and mosaics of such composites, is also used to map tropical forests over large areas. Examples of such large-area maps based on MODIS image composites are of tropical forest ecoregion (Muchoney et al., 2000), biome (Friedl et al., 2002), or forest formation (Carreiras et al., 2006). With decision tree classification of dry season MODIS image composites, Portillo-Quintero and Sánchez-Azofeifa (2010) mapped the extent of two classes of tropical dry forests (*Tropical dry forest and Forests in tropical grasslands, savannas, and shrublands*), for the mainland Neotropics plus the Greater Antilles. Overall accuracy was 82%. The importance of this latter work is that global land-cover maps often misclassify dry tropical forests as some other land cover.

14.6.7 Tropical Forest-Type Mapping and Image Spatial Resolution

Without question, multiseason data greatly improve the number of different physiognomic or floristic classes of tropical forest that can be mapped with multispectral satellite imagery. Monthly image composites or derived phenology metrics, as are possible with coarse-resolution imagery, are optimal. Joshi et al. (2006) qualitatively compared their WiFS-based map of forest types across India with a forest map of the country based on LISS data, which has a pixel resolution of 23.5 m but a 24-day repeat cycle. They concluded that the 5-day revisit cycle of WiFS, which allowed them to incorporate 12 monthly image composites, yielded better information on forest types and other vegetation and land-cover classes, even though WiFS has a spatial resolution of 188 m.

However, tropical forest types can change greatly over small areas, and spatial resolutions coarser than 100–200 m are too coarse to distinguish important differences in forest types in many places. In tropical islands, for example, forest floristic and physiognomic types that are critical to distinguish for conservation planning would be poorly delineated. Medium-resolution imagery with a shorter revisit cycle would greatly improve prospects for mapping tropical forest types with multispectral imagery. This could be more easily accomplished, for example, if AWiFS data, with its 56 m spatial resolution and 5-day revisit cycle, were available for all of the tropics, or if the Landsat program had a constellation of at least four satellites.

In addition, past disturbances affect forest physiognomy and species composition, and some forest classes may become spectrally distinct only during periodic drought and flooding. Consequently, forest-type mapping can also benefit when older satellite imagery or long image time series are incorporated into forest-type mapping, as in Helmer et al. (2010, 2012).

Finally, to distinguish tropical forest types on small mountains, small islands, along coastlines, rivers, and other linear features, or in other finely scaled landscapes, high-resolution imagery will be needed.

14.7 Monitoring Effects of Global Change on Tropical Forests

14.7.1 Progress in Monitoring Tropical Forests at Subcontinental to Global Scales

Tropical forest mapping with coarse-resolution imagery in optical remote sensing is very constrained by cloud cover. Helpfully, its high temporal frequency of acquisition balances the handicaps of cloud-contaminated pixels (McCallum et al., 2006). Historical long time series from NOAA-AVHRR paved the way for this research (Tucker et al., 1985; Townshend et al., 1991). Indeed, the spectral capacities from visible to SWIR of these sensors motivated many applications and technological developments. The identification of tropical forest patterns has improved over time (Holben, 1986; Mayaux et al., 1998; DeFries et al., 2000) and benefits from a large panel of vegetation indices for evaluating photosynthetic activity (Rouse et al., 1974; Huete, 1988; Pinty and Verstraete, 1992; Qi et al., 1994; Gao, 1996).

At the end of the 1990s, the experiences gained from these applications led to new sensors adapted to land surface observation, including SPOT Vegetation (March 1998) and TERRA-MODIS (December 1999) (Friedl et al., 2010). Spatial resolutions were improved from 1.1 km (NOAA-AVHRR) to 1.0 km (Vegetation), 0.3 (MERIS), and 0.5/0.25 km (MODIS). Geo-location was improved. Specific spectral bands dedicated to vegetation were implemented. New sensor technology was developed such as the push-broom system on Vegetation, which avoids large swath distortions. After 15 years of feedback, we may now measure the added value of these sensors.

Research to characterize tropical forests at subcontinental to global scales has become more accurate and precise (Mayaux et al., 2004; Vancutsem et al., 2009) by taking phenology into account (Xiao et al., 2006; Myneni et al., 2007; Doughty and Goulden, 2008; Park, 2009; Brando et al., 2010). Repetitive observation and long temporal archives make possible land-surface observation on 8-, 10-, or 16-day time periods and allow phenology studies to take advantage of both high spectral quality and high observation frequency (Verheggen et al., 2012 for MERIS and Vegetation; Gond et al., 2013 for MODIS). In addition, there are more forest attributes being characterized, including forest edges (to delimit forest patches and more accurately estimate forest areas) (Verheggen et al., 2012; Mayaux et al., 2013), aboveground biomass (Malhi et al., 2006; Saatchi et al., 2007; Baccini et al., 2008), deforestation and forest degradation (Achard et al., 2002; Duveiller et al., 2008; Hansen et al., 2008; Baccini et al., 2012; Desclée et al., 2013), and climate change impacts (Phillips et al., 2009; Lewis et al., 2011; Samanta et al., 2011).

Sensor capabilities and computer capacities now allow the production of global-scale land-cover maps (Bartholomé and Belward, 2005, for Vegetation; Friedl et al., 2002; Hansen et al., 2008, for MODIS), which have greatly improved our knowledge of land surface cover in comparison with previous views obtained from NOAA-AVHRR (DeFries and Townshend, 1994; Loveland and Belward, 1997).

Tropical forest characterizations with multispectral imagery have now begun to address a real challenge: that of monitoring and understanding climate change impacts on the biosphere (Gibson et al., 2011). Tropical forests are particularly threatened by global temperature increases and the possibility of modified rainfall regimes (Zelazowski et al., 2011). These changes will influence vegetation spatial distribution (Parmesan and Yohe, 2003), forest functioning (Nemani et al., 2003), and carbon storage capacity (Stephens et al., 2007), which may in turn affect climate. In this context, monitoring tropical forests with coarse-resolution satellite imagery is of prime importance to understanding biological processes and managing forest resilience. Zhao and Running (2010), for example, showed that large-scale droughts have decreased net primary productivity in the Southern Hemisphere, including tropical Asia and South America. As we discuss later, however, some critical remote sensing problems still need to be addressed before we can effectively monitor some important effects of droughts on tropical forests.

14.7.2 Feedbacks between Tropical Forest Disturbance and Drought

Multispectral imagery can help characterize the positive feedback among tropical forest disturbance, fire, and climate. First, tropical forest clearing dries nearby forest, and multispectral imagery can detect forest clearing. In Amazonia, for example, Briant et al. (2010) delineated forest boundaries with MODIS multispectral bands and found that as the forest becomes more fragmented, drops in MODIS-based indices related to canopy moisture extend further into intact forest, and that the old forest in more fragmented landscapes has lower canopy moisture to begin with. Second, forest cover data also reveal that forests desiccated by fragmentation and other disturbance are more susceptible to fire. Armenteras et al. (2013) used forest fragmentation indices from forest cover maps, along with active fire data from MODIS, which uses MODIS thermal bands, to show that forest fires increase in extent and frequency with fragmentation. Logging also increases forest vulnerability to fire (Uhl and Buschbacher, 1985; Woods, 1989), and as outlined earlier, logging can be detected with medium-resolution multispectral imagery.

A third aspect of the disturbance–fire–climate feedback is that drought magnifies the association between disturbance and fire (Siegert et al., 2001; Alencar et al., 2006). In Amazonia, fire scars mapped with Landsat occurred mostly within 1 km of clearings during normal dry seasons but extended to 4 km from clearings during drought years (Alencar et al., 2006). Some of these studies relied on Landsat imagery to quantify forest fragmentation,

because of its finer spatial resolution, or radar imagery to map fire scars, to avoid clouds.

Amazonian droughts are likely to become more common and severe with climate change. During droughts, reduced forest growth and increased tree mortality cause intact forests to shift from a net sink to a net source of CO_2 to the atmosphere (Lewis et al., 2011). However, monitoring drought effects that are spectrally subtle, like increased tree mortality or changes in phenology, remains a challenge because of residual cloud and aerosol contamination in coarse-resolution multispectral imagery. For example, studies have found that vegetation greenness may increase, decrease, or show no change during drought. The increases could stem from decreased cloud cover, leaf flushes related to increased sunlight, decreased canopy shadow from increased mortality of the tallest trees, or all three of these factors, and despite observation frequency, cloud and smoke contamination in pixels still obscures trends in vegetation greenness (Anderson et al., 2010, Asner and Alencar, 2010; Samanta et al., 2010; Morton et al., 2014). Asner et al. (2004c) suggest that metrics from hyperspectral imagery may be better suited to resolve drought effects on tropical forests because they are sensitive to canopy leaf water content and light-use efficiency. A challenge, then, is to develop a system that, despite cloud and smoke contamination, integrates these different sensors to continuously monitor the feedback between forest fragmentation, logging, fire, and climate.

14.8 Summary and Conclusions

Across spatial scales, increased image access, and data usability are the main factors driving an explosion of progress in characterizing tropical forests with multispectral satellite imagery. The menu of preprocessed image products of the second generation of high-frequency earth observation satellite sensors, MODIS and SPOT Vegetation, along with their improved spatial and spectral resolution, led to a wider group of users applying multispectral imagery across larger areas and in more diverse ways. Products like cloud-screened composites of earth surface reflectance, vegetation indices, quality flags, fire flags, and land cover have enabled efforts to map tropical forest productivity, type, phenology, moisture status, and biomass, and to study the effects of climate change on tropical forests, particularly feedback among drought, fire, and deforestation.

At the scale of medium-resolution imagery, free access to Landsat, and in some cases free access to SPOT imagery, has spawned many new applications that rely on dozens, hundreds, or thousands of scenes, including scenes with scan-line gaps or scenes previously considered too cloudy to bother with. Cloud- and gap-filled Landsat imagery and image time series are now used to automatically detect forest clearing, partial disturbance, or regrowth; quantify degradation of tropical forest C storage; map the age, structure, biomass, height, and disturbance type of secondary tropical forests; automatically and more precisely mask clouds and cloud shadows in imagery; and create detailed maps of forest types in these often cloudy landscapes.

Characterizing tropical forest phenology at medium resolution will now be possible for many places, which will be easier given recent additions to Landsat image preprocessing. Many of these automated applications build on the experiences gained from the high-frequency, coarse-spatial-resolution imagery, and all of them are relevant to REDD+ monitoring, reporting, and verification.

At fine spatial scales, free viewing and low-cost printing of georeferenced high-resolution imagery via online tools like Google Earth and Bing supplement field data for training and testing the earlier products that are based on medium- and coarse-resolution imagery. In addition, scientists have used image products from Google Earth to estimate tropical forest biomass directly. New commercial sensors that produce multispectral satellite imagery with spatial resolutions ≤0.5 m should also allow more disturbance types and tropical tree communities to be remotely identifiable.

Acknowledgments

Thanks to John Armston and Ariel Lugo for their invaluable comments on this text. This research was conducted in cooperation with the University of Puerto Rico and the USDA Forest Service Rocky Mountain Research Station.

References

Achard, F., Eva, H. D., Stibig, H.-J., Mayaux, P., Gallego, J., Richards, T., and Malingreau, J.-P. 2002. Determination of deforestation rates of the world's humid tropical forests. *Science*, 297, 999–1002.

Ackerman, S., Holz, R., Frey, R., Eloranta, E., Maddux, B., and McGill, M. 2008. Cloud detection with MODIS. Part II: Validation. *Journal of Atmospheric and Oceanic Technology*, 25, 1073–1086.

Adams, J. B., Sabol, D. E., Kapos, V., Almeida Filho, R., Roberts, D. A., Smith, M. O., and Gillespie, A. R. 1995. Classification of multispectral images based on fractions of endmembers: Application to land-cover change in the Brazilian Amazon. *Remote Sensing of Environment*, 52, 137–154.

Ahmed, O. S., Franklin, S. E., and Wulder, M. A. 2014. Interpretation of forest disturbance using a time series of Landsat imagery and canopy structure from airborne lidar. *Canadian Journal of Remote Sensing*, 39, 521–542.

Alencar, A., Nepstad, D., and Diaz, M. C. V. 2006. Forest understory fire in the Brazilian Amazon in ENSO and non-ENSO years: Area burned and committed carbon emissions. *Earth Interactions*, 10, 1–17.

Alencar, A., Asner, G. P., Knapp, D., and Zarin, D. 2011. Temporal variability of forest fires in eastern Amazonia. *Ecological Applications*, 21, 2397–2412.

Allnutt, T. F., Asner, G. P., Golden, C. D., and Powell, G. V. 2013. Mapping recent deforestation and forest disturbance in northeastern Madagascar. *Tropical Conservation Science*, 6, 1–15.

Almeida-Filho, R. and Shimabukuro, Y. E. 2002. Digital processing of a Landsat-TM time series for mapping and monitoring degraded areas caused by independent gold miners, Roraima State, Brazilian Amazon. *Remote Sensing of Environment*, 79, 42–50.

Anderson, L. O., Malhi, Y., Aragão, L. E., Ladle, R., Arai, E., Barbier, N., and Phillips, O. 2010. Remote sensing detection of droughts in Amazonian forest canopies. *New Phytologist*, 187, 733–750.

Aragão, L. E. and Shimabukuro, Y. E. 2010. The incidence of fire in Amazonian forests with implications for REDD. *Science*, 328, 1275–1278.

Armenteras, D., González, T. M., and Retana, J. 2013. Forest fragmentation and edge influence on fire occurrence and intensity under different management types in Amazon forests. *Biological Conservation*, 159, 73–79.

Arroyo-Mora, J. P., Sánchez-Azofeifa, G. A., Kalacska, M. E., Rivard, B., Calvo-Alvarado, J. C., and Janzen, D. H. 2005. Secondary forest detection in a Neotropical dry forest landscape using Landsat 7 ETM+ and IKONOS Imagery1. *Biotropica*, 37, 497–507.

Asner, G. P. and Heidebrecht, K. B. 2002. Spectral unmixing of vegetation, soil and dry carbon cover in arid regions: Comparing multispectral and hyperspectral observations. *International Journal of Remote Sensing*, 23, 3939–3958.

Asner, G. P., Palace, M., Keller, M., Pereira, R., Silva, J. N., and Zweede, J. C. 2002. Estimating canopy structure in an Amazon forest from laser range finder and IKONOS satellite observations 1. *Biotropica*, 34, 483–492.

Asner, G. P., Keller, M., Pereira, R., Zweede, J. C., and Silva, J. N. 2004a. Canopy damage and recovery after selective logging in Amazonia: Field and satellite studies. *Ecological Applications*, 14, 280–298.

Asner, G. P., Keller, M., and Silva, J. N. 2004b. Spatial and temporal dynamics of forest canopy gaps following selective logging in the eastern Amazon. *Global Change Biology*, 10, 765–783.

Asner, G. P., Nepstad, D., Cardinot, G., and Ray, D. 2004c. Drought stress and carbon uptake in an Amazon forest measured with spaceborne imaging spectroscopy. *Proceedings of the National Academy of Sciences of the United States of America*, 101, 6039–6044.

Asner, G. P., Knapp, D. E., Broadbent, E. N., Oliveira, P. J., Keller, M., and Silva, J. N. 2005. Selective logging in the Brazilian Amazon. *Science*, 310, 480–482.

Asner, G. P. 2009. Tropical forest carbon assessment: Integrating satellite and airborne mapping approaches. *Environmental Research Letters*, 4, 034009.

Asner, G. P., Knapp, D. E., Balaji, A., and Páez-Acosta, G. 2009a. Automated mapping of tropical deforestation and forest degradation: CLASlite. *Journal of Applied Remote Sensing*, 3, 033543.

Asner, G. P., Rudel, T. K., Aide, T. M., Defries, R., and Emerson, R. 2009b. A contemporary assessment of change in humid tropical forests. *Conservation Biology*, 23, 1386–1395.

Asner, G. P. and Alencar, A. 2010. Drought impacts on the Amazon forest: The remote sensing perspective. *New Phytologist*, 187, 569–578.

Asner, G. P., Mascaro, J., Muller-Landau, H. C., Vieilledent, G., Vaudry, R., Rasamoelina, M., Hall, J. S. et al. 2012. A universal airborne LiDAR approach for tropical forest carbon mapping. *Oecologia*, 168, 1147–1160.

Asner, G. P., Mascaro, J., Anderson, C., Knapp, D. E., Martin, R. E., Kennedy-Bowdoin, T., van Breugel, M. et al. 2013. High-fidelity national carbon mapping for resource management and REDD+. *Carbon Balance and Management*, 8, 1–14.

Avitabile, V., Baccini, A., Friedl, M. A., and Schmullius, C. 2012. Capabilities and limitations of Landsat and land cover data for aboveground woody biomass estimation of Uganda. *Remote Sensing of Environment*, 117, 366–380.

Avoided Deforestation Partners. 2015. VCS Methodology VM0007 REDD+ Methodology Framework (REDD-MF) Version 1.5, Sectoral Scope 14. 42 pp. http://www.v-c-s.org/methodologies/redd-methodology-framework-redd-mf-v15

Baccini, A., Laporte, N., Goetz, S., Sun, M., and Dong, H. 2008. A first map of tropical Africa's above-ground biomass derived from satellite imagery. *Environmental Research Letters*, 3, 045011.

Baccini, A., Goetz, S., Walker, W., Laporte, N., Sun, M., Sulla-Menashe, D., Hackler, J. et al. 2012. Estimated carbon dioxide emissions from tropical deforestation improved by carbon-density maps. *Nature Climate Change*, 2, 182–185.

Barbier, N., Couteron, P., Proisy, C., Malhi, Y., and Gastellu-Etchegorry, J. P. 2010. The variation of apparent crown size and canopy heterogeneity across lowland Amazonian forests. *Global Ecology and Biogeography*, 19, 72–84.

Bartholomé, E. and Belward, A. 2005. GLC2000: A new approach to global land cover mapping from Earth observation data. *International Journal of Remote Sensing*, 26, 1959–1977.

Bateson, C. A., Asner, G. P., and Wessman, C. A. 2000. Endmember bundles: A new approach to incorporating endmember variability into spectral mixture analysis. *IEEE Transactions on Geoscience and Remote Sensing*, 38, 1083–1094.

Blackard, J., Finco, M., Helmer, E., Holden, G., Hoppus, M., Jacobs, D., Lister, A. et al. 2008. Mapping US forest biomass using nationwide forest inventory data and moderate resolution information. *Remote Sensing of Environment*, 112, 1658–1677.

Boggs, G. 2010. Assessment of SPOT 5 and QuickBird remotely sensed imagery for mapping tree cover in savannas. *International Journal of Applied Earth Observation and Geoinformation*, 12, 217–224.

Bohlman, S. A., Adams, J. B., Smith, M. O., and Peterson, D. L. 1998. Seasonal foliage changes in the eastern Amazon basin detected from Landsat Thematic Mapper satellite images. *Biotropica*, 30, 13–19.

Bourbier, L., Cornu, G., Pennec, A., Brognoli, C., and Gond, V. 2013. Large-scale estimation of forest canopy opening using remote sensing in Central Africa. *Bios et Forets des Tropiques*, 315, 3–9.

Boyd, D. S., Foody, G. M., Curran, P., Lucas, R., and Honzak, M. 1996. An assessment of radiance in Landsat TM middle and thermal infrared wavebands for the detection of tropical forest regeneration. *International Journal of Remote Sensing*, 17, 249–261.

Brando, P. M., Goetz, S. J., Baccini, A., Nepstad, D. C., Beck, P. S., and Christman, M. C. 2010. Seasonal and interannual variability of climate and vegetation indices across the Amazon. *Proceedings of the National Academy of Sciences*, 107, 14685–14690.

Braswell, B., Hagen, S., Frolking, S., and Salas, W. 2003. A multivariable approach for mapping sub-pixel land cover distributions using MISR and MODIS: Application in the Brazilian Amazon region. *Remote Sensing of Environment*, 87, 243–256.

Briant, G., Gond, V., and Laurance, S. G. 2010. Habitat fragmentation and the desiccation of forest canopies: A case study from eastern Amazonia. *Biological Conservation*, 143, 2763–2769.

Bryan, J. E., Shearman, P. L., Asner, G. P., Knapp, D. E., Aoro, G., and Lokes, B. 2013. Extreme differences in forest degradation in Borneo: Comparing practices in Sarawak, Sabah, and Brunei. *PloS One*, 8, e69679.

Carlson, K. M., Curran, L. M., Ratnasari, D., Pittman, A. M., Soares-Filho, B. S., Asner, G. P., Trigg, S. N. et al. 2012. Committed carbon emissions, deforestation, and community land conversion from oil palm plantation expansion in West Kalimantan, Indonesia. *Proceedings of the National Academy of Sciences of the United States of America*, 109, 7559–7564.

Carreiras, J., Pereira, J., Campagnolo, M. L., and Shimabukuro, Y. E. 2006. Assessing the extent of agriculture/pasture and secondary succession forest in the Brazilian Legal Amazon using SPOT VEGETATION data. *Remote Sensing of Environment*, 101, 283–298.

Castillo-Santiago, M. A., Ricker, M., and de Jong, B. H. 2010. Estimation of tropical forest structure from SPOT-5 satellite images. *International Journal of Remote Sensing*, 31, 2767–2782.

Chazdon, R. L., Letcher, S. G., Van Breugel, M., Martínez-Ramos, M., Bongers, F., and Finegan, B. 2007. Rates of change in tree communities of secondary Neotropical forests following major disturbances. *Philosophical Transactions of the Royal Society B: Biological Sciences*, 362, 273–289.

Choi, H. and Bindschadler, R. 2004. Cloud detection in Landsat imagery of ice sheets using shadow matching technique and automatic normalized difference snow index threshold value decision. *Remote Sensing of Environment*, 91, 237–242.

Chust, G., Chave, J., Condit, R., Aguilar, S., Lao, S., and Pérez, R. 2006. Determinants and spatial modeling of tree β-diversity in a tropical forest landscape in Panama. *Journal of Vegetation Science*, 17, 83–92.

Cochrane, M. and Souza Jr., C. 1998. Linear mixture model classification of burned forests in the eastern Amazon. *International Journal of Remote Sensing*, 19, 3433–3440.

Cochrane, M. A., Alencar, A., Schulze, M. D., Souza, C. M., Nepstad, D. C., Lefebvre, P., and Davidson, E. A. 1999. Positive feedbacks in the fire dynamic of closed canopy tropical forests. *Science*, 284, 1832–1835.

Couteron, P., Pelissier, R., Nicolini, E. A., and Paget, D. 2005. Predicting tropical forest stand structure parameters from Fourier transform of very high-resolution remotely sensed canopy images. *Journal of Applied Ecology*, 42, 1121–1128.

Crist, E. P. and Cicone, R. C. 1984. A physically-based transformation of Thematic Mapper data—The TM Tasseled Cap. *IEEE Transactions on Geoscience and Remote Sensing*, 22, 256–263.

de Carvalho, A. L., Nelson, B. W., Bianchini, M. C., Plagnol, D., Kuplich, T. M., and Daly, D. C. 2013. Bamboo-dominated forests of the southwest Amazon: Detection, spatial extent, life cycle length and flowering waves. *PLoS One*, 8, e54852.

de Wasseige, C. and Defourny, P. 2004. Remote sensing of selective logging impact for tropical forest management. *Forest Ecology and Management*, 188, 161–173.

DeFries, R. and Townshend, J. 1994. NDVI-derived land cover classifications at a global scale. *International Journal of Remote Sensing*, 15, 3567–3586.

DeFries, R., Hansen, M., Townshend, J., Janetos, A., and Loveland, T. 2000. A new global 1-km dataset of percentage tree cover derived from remote sensing. *Global Change Biology*, 6, 247–254.

Dennison, P. E., Roberts, D. A., and Peterson, S. H. 2007. Spectral shape-based temporal compositing algorithms for MODIS surface reflectance data. *Remote Sensing of Environment*, 109, 510–522.

Desclee, B., Simonetti, D., Mayaux, P., and Achard, F. 2013. Multi-sensor monitoring system for forest cover change assessment in Central Africa. *IEEE Journal on Selected Topics in Applied Earth Observations and Remote Sensing*, 6, 110–120.

Doughty, C. E. and Goulden, M. L. 2008. Seasonal patterns of tropical forest leaf area index and CO2 exchange. *Journal of Geophysical Research: Biogeosciences* (2005–2012), 113, G1.

Dubreuil, V., Debortoli, N., Funatsu, B., Nédélec, V., and Durieux, L. 2012. Impact of land-cover change in the Southern Amazonia climate: A case study for the region of Alta Floresta, Mato Grosso, Brazil. *Environmental Monitoring and Assessment*, 184, 877–891.

Duveiller, G., Defourny, P., Desclée, B., and Mayaux, P. 2008. Deforestation in Central Africa: Estimates at regional, national and landscape levels by advanced processing of systematically-distributed Landsat extracts. *Remote Sensing of Environment*, 112, 1969–1981.

Elumnoh, A. and Shrestha, R. P. 2000. Application of DEM data to Landsat image classification: Evaluation in a tropical wet-dry landscape of Thailand. *Photogrammetric Engineering and Remote Sensing*, 66, 297–304.

Emelyanova, I. V., McVicar, T. R., Van Niel, T. G., Li, L. T., and van Dijk, A. I. 2013. Assessing the accuracy of blending Landsat–MODIS surface reflectances in two landscapes with

contrasting spatial and temporal dynamics: A framework for algorithm selection. *Remote Sensing of Environment*, 133, 193–209.

Faber-Langendoen, D., Keeler-Wolf, T., Meidinger, D., Josse, C., Weakley, A., Tart, D., Navarro, G. et al. 2012. *Classification and Description of World Formation Types*: Hierarchy Revisions Working Group, Federal Geographic Data Committee, FGDC Secretariat, US Geological Survey. Reston, VA, and NatureServe, Arlington, VA.

Ferry, B., Morneau, F., Bontemps, J. D., Blanc, L., and Freycon, V. 2010. Higher treefall rates on slopes and waterlogged soils result in lower stand biomass and productivity in a tropical rain forest. *Journal of Ecology*, 98, 106–116.

Fisher, A. 2014. Cloud and cloud-shadow detection in SPOT5 HRG satellite imagery with automated morphological feature extraction. *Remote Sensing*, 6, 776–800.

Flood, N. 2013. Seasonal composite Landsat TM/ETM+ images using the Medoid (a Multi-Dimensional Median). *Remote Sensing*, 5, 6481–6500.

Foody, G. M. and Hill, R. 1996. Classification of tropical forest classes from Landsat TM data. *International Journal of Remote Sensing*, 17, 2353–2367.

Foody, G. M., Palubinskas, G., Lucas, R. M., Curran, P. J., and Honzak, M. 1996. Identifying terrestrial carbon sinks: Classification of successional stages in regenerating tropical forest from Landsat TM data. *Remote Sensing of Environment*, 55, 205–216.

Foody, G. M. and Cutler, M. E. 2003. Tree biodiversity in protected and logged Bornean tropical rain forests and its measurement by satellite remote sensing. *Journal of Biogeography*, 30, 1053–1066.

Foody, G. M. and Cutler, M. E. 2006. Mapping the species richness and composition of tropical forests from remotely sensed data with neural networks. *Ecological Modelling*, 195, 37–42.

Friedl, M. A., McIver, D. K., Hodges, J. C., Zhang, X., Muchoney, D., Strahler, A. H., Woodcock, C. E. et al. 2002. Global land cover mapping from MODIS: Algorithms and early results. *Remote Sensing of Environment*, 83, 287–302.

Friedl, M. A., Sulla-Menashe, D., Tan, B., Schneider, A., Ramankutty, N., Sibley, A., and Huang, X. 2010. MODIS Collection 5 global land cover: Algorithm refinements and characterization of new datasets. *Remote Sensing of Environment*, 114, 168–182.

Gao, B.-C. 1996. NDWI—A normalized difference water index for remote sensing of vegetation liquid water from space. *Remote Sensing of Environment*, 58, 257–266.

Gao, F., Masek, J. G., Schwaller, M., and Hall, F. 2006. On the blending of the Landsat and MODIS surface reflectance: Predicting daily Landsat. *IEEE Transactions on Geoscience and Remote Sensing*, 44, 2207–2208.

Garzon-Lopez, C. X., Bohlman, S. A., Olff, H., and Jansen, P. A. 2013. Mapping tropical forest trees using High-resolution aerial digital photographs. *Biotropica*, 45, 308–316.

Gentry, A. H. 1988. Tree species richness of upper Amazonian forests. *Proceedings of the National Academy of Sciences*, 85, 156–159.

Gibbes, C., Adhikari, S., Rostant, L., Southworth, J., and Qiu, Y. 2010. Application of object based classification and high resolution satellite imagery for savanna ecosystem analysis. *Remote Sensing*, 2, 2748–2772.

Gibson, L., Lee, T. M., Koh, L. P., Brook, B. W., Gardner, T. A., Barlow, J., Peres, C. A. et al. 2011. Primary forests are irreplaceable for sustaining tropical biodiversity. *Nature*, 478, 378–381.

Givnish, T. J. 1999. On the causes of gradients in tropical tree diversity. *Journal of Ecology*, 87, 193–210.

Gond, V., Bartholomé, E., Ouattara, F., Nonguierma, A., and Bado, L. 2004. Surveillance et cartographie des plans d'eau et des zones humides et inondables en régions arides avec l'instrument VEGETATION embarqué sur SPOT-4. *International Journal of Remote Sensing*, 25, 987–1004.

Gond, V. and Guitet, S. 2009. Elaboration d'un diagnostic post-exploitation par télédétection spatiale pour la gestion des forêts de Guyane. *Bois et Forêts des Tropiques*, 299, 5–13.

Gond, V., Freycon, V., Molino, J.-F., Brunaux, O., Ingrassia, F., Joubert, P., Pekel, J.-F. et al. 2011. Broad-scale spatial pattern of forest landscape types in the Guiana Shield. *International Journal of Applied Earth Observation and Geoinformation*, 13, 357–367.

Gond, V., Fayolle, A., Pennec, A., Cornu, G., Mayaux, P., Camberlin, P., Doumenge, C. et al. 2013. Vegetation structure and greenness in Central Africa from Modis multi-temporal data. *Philosophical Transactions of the Royal Society B: Biological Sciences*, 368, 1625.

González-Orozco, C. E., Mulligan, M., Trichon, V., and Jarvis, A. 2010. Taxonomic identification of Amazonian tree crowns from aerial photography. *Applied Vegetation Science*, 13, 510–519.

Goodwin, N. R., Collett, L. J., Denham, R. J., Flood, N., and Tindall, D. 2013. Cloud and cloud shadow screening across Queensland, Australia: An automated method for Landsat TM/ETM+ time series. *Remote Sensing of Environment*, 134, 50–65.

Göttlicher, D., Obregón, A., Homeier, J., Rollenbeck, R., Nauss, T., and Bendix, J. 2009. Land-cover classification in the Andes of southern Ecuador using Landsat ETM+ data as a basis for SVAT modelling. *International Journal of Remote Sensing*, 30, 1867–1886.

Griffiths, P., Kuemmerle, T., Baumann, M., Radeloff, V. C., Abrudan, I. V., Lieskovsky, J., Munteanu, C. et al. 2013. Forest disturbances, forest recovery, and changes in forest types across the Carpathian ecoregion from 1985 to 2010 based on Landsat image composites. *Remote Sensing of Environment*, 151, 72–88.

Hansen, M. C., Roy, D. P., Lindquist, E., Adusei, B., Justice, C. O., and Altstatt, A. 2008. A method for integrating MODIS and Landsat data for systematic monitoring of forest cover and change in the Congo Basin. *Remote Sensing of Environment*, 112, 2495–2513.

Helmer, E. and Ruefenacht, B. 2005. Cloud-free satellite image mosaics with regression trees and histogram matching. *Photogrammetric Engineering and Remote Sensing*, 71, 1079.

Helmer, E. and Lefsky, M. 2006. Forest canopy heights in Amazon River basin forests as estimated with the Geoscience Laser Altimeter System (GLAS), in *Monitoring Science and Technology Symposium: Unifying Knowledge for Sustainability in the Western Hemisphere*, September 21–25, 2004, C. Aguirre-Bravo, Pellicane, P. J., Burns, D. P., and Draggan, S. (eds.) Denver, CO: Proceedings RMRS-P-37CD/Ogden, UT: U.S. Department of Agriculture, Forest Service, Rocky Mountain Research Station, CD-ROM, pp. 802–808.

Helmer, E. and Ruefenacht, B. 2007. A comparison of radiometric normalization methods when filling cloud gaps in Landsat imagery. *Canadian Journal of Remote Sensing*, 33, 325–340.

Helmer, E., Ruzycki, T. S., Wunderle, J. M., Vogesser, S., Ruefenacht, B., Kwit, C., Brandeis, T. J. et al. 2010. Mapping tropical dry forest height, foliage height profiles and disturbance type and age with a time series of cloud-cleared Landsat and ALI image mosaics to characterize avian habitat. *Remote Sensing of Environment*, 114, 2457–2473.

Helmer, E. H., Brown, S., and Cohen, W. 2000. Mapping montane tropical forest successional stage and land use with multi-date Landsat imagery. *International Journal of Remote Sensing*, 21, 2163–2183.

Helmer, E. H., Ramos, O., López, T. del. M., Quiñones, M., and Diaz, W. 2002. Mapping forest type and land cover of Puerto Rico, a component of the Caribbean biodiversity hotspot. *Caribbean Journal of Science*, 38, 165–183.

Helmer, E. H., Kennaway, T. A., Pedreros, D. H., Clark, M. L., Marcano-Vega, H., Tieszen, L. L., Ruzycki, T. R. et al. 2008. Land cover and forest formation distributions for St. Kitts, Nevis, St. Eustatius, Grenada and Barbados from decision tree classification of cloud-cleared satellite imagery. *Caribbean Journal of Science*, 44, 175–198.

Helmer, E. H., Lefsky, M. A., and Roberts, D. A. 2009. Biomass accumulation rates of Amazonian secondary forest and biomass of old-growth forests from Landsat time series and the Geoscience Laser Altimeter System. *Journal of Applied Remote Sensing*, 3, 033505.

Helmer, E. H., Ruzycki, T. S., Benner, J., Voggesser, S. M., Scobie, B. P., Park, C., Fanning, D. W. et al. 2012. Detailed maps of tropical forest types are within reach: Forest tree communities for Trinidad and Tobago mapped with multiseason Landsat and multiseason fine-resolution imagery. *Forest Ecology and Management*, 279, 147–166.

Hernández-Stefanoni, J. L., Alberto Gallardo-Cruz, J., Meave, J. A., and Dupuy, J. M. 2011. Combining geostatistical models and remotely sensed data to improve tropical tree richness mapping. *Ecological Indicators*, 11, 1046–1056.

Hilker, T., Lyapustin, A. I., Tucker, C. J., Sellers, P. J., Hall, F. G., and Wang, Y. 2012. Remote sensing of tropical ecosystems: Atmospheric correction and cloud masking matter. *Remote Sensing of Environment*, 127, 370–384.

Holben, B. N. 1986. Characteristics of maximum-value composite images from temporal AVHRR data. *International Journal of Remote Sensing*, 7, 1417–1434.

Holdridge, L. R. 1967. *Life Zone Ecology*. San José, CA: Tropical Science Center.

Houghton, R. A. 2013. The emissions of carbon from deforestation and degradation in the tropics: Past trends and future potential. *Carbon Management*, 4, 539–546.

Huang, C., Wylie, B., Homer, C., Yang, L., and Zylstra, G. 2002. Derivation of a Tasseled Cap transformation based on Landsat 7 at-satellite reflectance. *International Journal of Remote Sensing*, 23, 1741–1748.

Huang, C., Thomas, N., Goward, S. N., Masek, J. G., Zhu, Z., Townshend, J. R., and Vogelmann, J. E. 2010. Automated masking of cloud and cloud shadow for forest change analysis using Landsat images. *International Journal of Remote Sensing*, 31, 5449–5464.

Huete, A. R. 1988. A soil-adjusted vegetation index (SAVI). *Remote Sensing of Environment*, 25, 295–309.

Intergovernmental Panel on Climate Change (IPCC). 2006. IPCC Guidelines for National Greenhouse Gas Inventories Volume 4: Agriculture, Forestry and Other Land Use. Chapter 2, Generic methodologies applicable to multiple land-use categories. Institute for Global Environmental Strategies, Hayama Japan. http://www.ipcc-nggip.iges.or.jp/public/2006gl/vol4.html.

IPCC. 2014. *Climate Change 2014: Mitigation of Climate Change, Contribution of Working Group III to the Fifth Assessment Report of the Intergovernmental Panel on Climate Change*, O. Edenhofer, Pichs-Madruga, R., Sokona, Y., Farahani, E., Kadner, S., Seyboth, K., Adler, A. et al. (eds.). Cambridge, U.K./New York: Cambridge University Press.

Irish, R. R., Barker, J. L., Goward, S. N., and Arvidson, T. 2006. Characterization of the Landsat-7 ETM+ automated cloud-cover assessment (ACCA) algorithm. *Photogrammetric Engineering and Remote Sensing*, 72, 1179.

Joshi, P. K. K., Roy, P. S., Singh, S., Agrawal, S., and Yadav, D. 2006. Vegetation cover mapping in India using multi-temporal IRS Wide Field Sensor (WiFS) data. *Remote Sensing of Environment*, 103, 190–202.

Joubert, P., Bourgeois, U., Linarés, S., Gond, V., Verger, G., Allo, S., and Coppel, A. 2012. L'observatoire de l'activité minière, un outil adapté à la surveillance de l'environnement, in *XV° symposium de la Société Savante Latino-Américaine de Télédétection et des Systèmes d'Informations Spatiales (SELPER)*. Cayenne, France.

Kennaway, T. and Helmer, E. 2007. The forest types and ages cleared for land development in Puerto Rico. *GIScience & Remote Sensing*, 44, 356–382.

Kennaway, T. A., Helmer, E. H., Lefsky, M. A., Brandeis, T. A., and Sherrill, K. R. 2008. Mapping land cover and estimating forest structure using satellite imagery and coarse resolution lidar in the Virgin Islands. *Journal of Applied Remote Sensing*, 2, 033521.

Langner, A., Hirata, Y., Saito, H., Sokh, H., Leng, C., Pak, C., and Raši, R. 2014. Spectral normalization of SPOT 4 data to adjust for changing leaf phenology within seasonal forests in Cambodia. *Remote Sensing of Environment*, 143, 122–130.

Laporte, N. T., Stabach, J. A., Grosch, R., Lin, T. S., and Goetz, S. J. 2007. Expansion of industrial logging in Central Africa. *Science*, 316, 1451–1451.

Laumonier, Y., Edin, A., Kanninen, M., and Munandar, A. W. 2010. Landscape-scale variation in the structure and biomass of the hill dipterocarp forest of Sumatra: Implications for carbon stock assessments. *Forest Ecology and Management*, 259, 505–513.

Laurance, W. F., Goosem, M., and Laurance, S. G. 2009. Impacts of roads and linear clearings on tropical forests. *Trends in Ecology & Evolution*, 24, 659–669.

Le Hégarat-Mascle, S. and André, C. 2009. Use of Markov random fields for automatic cloud/shadow detection on high resolution optical images. *ISPRS Journal of Photogrammetry and Remote Sensing*, 64, 351–366.

Lefsky, M. A. 2010. A global forest canopy height map from the moderate resolution imaging spectroradiometer and the geoscience laser altimeter system. *Geophysical Research Letters*, 37, 15.

Lewis, S. L., Brando, P. M., Phillips, O. L., van der Heijden, G. M., and Nepstad, D. 2011. The 2010 amazon drought. *Science*, 331, 554–554.

Li, A., Huang, C., Sun, G., Shi, H., Toney, C., Zhu, Z., Rollins, M. G. et al. 2011. Modeling the height of young forests regenerating from recent disturbances in Mississippi using Landsat and ICESat data. *Remote Sensing of Environment*, 115, 1837–1849.

Loveland, T. and Belward, A. 1997. The IGBP-DIS global 1km land cover data set, DISCover: First results. *International Journal of Remote Sensing*, 18, 3289–3295.

Lu, D. 2005. Aboveground biomass estimation using Landsat TM data in the Brazilian Amazon. *International Journal of Remote Sensing*, 26, 2509–2525.

Luo, Y., Trishchenko, A. P., and Khlopenkov, K. V. 2008. Developing clear-sky, cloud and cloud shadow mask for producing clear-sky composites at 250-meter spatial resolution for the seven MODIS land bands over Canada and North America. *Remote Sensing of Environment*, 112, 4167–4185.

Lyapustin, A., Wang, Y., and Frey, R. 2008. An automatic cloud mask algorithm based on time series of MODIS measurements. *Journal of Geophysical Research: Atmospheres (1984–2012)*, 113, D16.

Malhi, Y., Wood, D., Baker, T. R., Wright, J., Phillips, O. L., Cochrane, T., Meir, P. et al. 2006. The regional variation of aboveground live biomass in old-growth Amazonian forests. *Global Change Biology*, 12, 1107–1138.

Malhi, Y. and Román-Cuesta, R. M. 2008. Analysis of lacunarity and scales of spatial homogeneity in IKONOS images of Amazonian tropical forest canopies. *Remote Sensing of Environment*, 112, 2074–2087.

Martinuzzi, S., Gould, W. A., and González, O. M. R. 2007. Creating cloud-free Landsat ETM+ data sets in tropical landscapes: Cloud and cloud-shadow removal. F. S. US Department of Agriculture (ed.), San Juan, PR: International Institute of Tropical Forestry.

Martinuzzi, S., Gould, W. A., Ramos González, O. M., Martínez Robles, A., Calle Maldonado, P., Pérez-Buitrago, N., and Fumero Caban, J. J. 2008. Mapping tropical dry forest habitats integrating Landsat NDVI, Ikonos imagery, and topographic information in the Caribbean Island of Mona. *Revista de Biologia Tropical*, 56, 625–639.

Mayaux, P., Achard, F., and Malingreau, J.-P. 1998. Global tropical forest area measurements derived from coarse resolution satellite imagery: A comparison with other approaches. *Environmental Conservation*, 25, 37–52.

Mayaux, P., Bartholomé, E., Fritz, S., and Belward, A. 2004. A new land-cover map of Africa for the year 2000. *Journal of Biogeography*, 31, 861–877.

Mayaux, P., Pekel, J.-F., Desclée, B., Donnay, F., Lupi, A., Achard, F., Clerici, M. et al. 2013. State and evolution of the African rainforests between 1990 and 2010. *Philosophical Transactions of the Royal Society B: Biological Sciences*, 368, 1625.

McCallum, I., Obersteiner, M., Nilsson, S., and Shvidenko, A. 2006. A spatial comparison of four satellite derived 1 km global land cover datasets. *International Journal of Applied Earth Observation and Geoinformation*, 8, 246–255.

Miller, S. D., Goulden, M. L., Hutyra, L. R., Keller, M., Saleska, S. R., Wofsy, S. C., Figueira, A. M. S. et al. 2011. Reduced impact logging minimally alters tropical rainforest carbon and energy exchange. *Proceedings of the National Academy of Sciences of the United States of America*, 108, 19431–19435.

Mitchard, E. T. A., Feldpausch, T. R., Brienen, R. J. W., Lopez-Gonzalez, G., Monteagudo, A., Baker, T. R., Lewis, S. L. et al. 2014. Markedly divergent estimates of Amazon forest carbon density from ground plots and satellites. *Global Ecology and Biogeography*, 23, 935–946.

Morton, D. C., DeFries, R. S., Nagol, J., Souza Jr., C. M., Kasischke, E. S., Hurtt, G. C., and Dubayah, R. 2011. Mapping canopy damage from understory fires in Amazon forests using annual time series of Landsat and MODIS data. *Remote Sensing of Environment*, 115, 1706–1720.

Morton, D. C., Nagol, J., Carabajal, C. C., Rosette, J., Palace, M., Cook, B. D., Vermote, E. F. et al. 2014. Amazon forests maintain consistent canopy structure and greenness during the dry season. *Nature*, 506, 221–224.

Muchoney, D., Borak, J., Chi, H., Friedl, M., Gopal, S., Hodges, J., Morrow, N., and Strahler, A. 2000. Application of the MODIS global supervised classification model to vegetation and land cover mapping of Central America. *International Journal of Remote Sensing*, 21, 1115–1138.

Myneni, R. B., Yang, W., Nemani, R. R., Huete, A. R., Dickinson, R. E., Knyazikhin, Y., Didan, K. et al. 2007. Large seasonal swings in leaf area of Amazon rainforests. *Proceedings of the National Academy of Sciences of the United States of America*, 104, 4820–4823.

Nagendra, H. and Rocchini, D. 2008. High resolution satellite imagery for tropical biodiversity studies: The devil is in the detail. *Biodiversity and Conservation*, 17, 3431–3442.

Nagendra, H., Rocchini, D., Ghate, R., Sharma, B., and Pareeth, S. 2010. Assessing plant diversity in a dry tropical forest: Comparing the utility of Landsat and IKONOS satellite images. *Remote Sensing*, 2, 478–496.

Nelson, R. F., Kimes, D. S., Salas, W. A., and Routhier, M. 2000. Secondary forest age and tropical forest biomass estimation using Thematic Mapper imagery: Single-year tropical forest age classes, a surrogate for standing biomass, cannot be reliably identified using single-date tm imagery. *Bioscience*, 50, 419–431.

Nemani, R. R., Keeling, C. D., Hashimoto, H., Jolly, W. M., Piper, S. C., Tucker, C. J., Myneni, R. B. et al. 2003. Climate-driven increases in global terrestrial net primary production from 1982 to 1999. *Science*, 300, 1560–1563.

Newman, M. E., McLaren, K. P., and Wilson, B. S. 2011. Comparing the effects of classification techniques on landscape-level assessments: Pixel-based versus object-based classification. *International Journal of Remote Sensing*, 32, 4055–4073.

Omeja, P. A., Obua, J., Rwetsiba, A., and Chapman, C. A. 2012. Biomass accumulation in tropical lands with different disturbance histories: Contrasts within one landscape and across regions. *Forest Ecology and Management*, 269, 293–300.

Oreopoulos, L., Wilson, M. J., and Várnai, T. 2011. Implementation on Landsat data of a simple cloud-mask algorithm developed for MODIS land bands. *IEEE Geoscience and Remote Sensing Letters*, 8, 597–601.

Oza, M., Srivastava, V., and Devaiah, P. 1996. Estimating tree volume in tropical dry deciduous forest from Landsat TM data. *Geocarto International*, 11, 33–39.

Palace, M., Keller, M., Asner, G. P., Hagen, S., and Braswell, B. 2008. Amazon forest structure from IKONOS satellite data and the automated characterization of forest canopy properties. *Biotropica*, 40, 141–150.

Park, S. 2009. Synchronicity between satellite-measured leaf phenology and rainfall regimes in tropical forests. *Photogrammetric Engineering and Remote Sensing*, 75, 1231–1237.

Parker, C., Mitchell, A., Trivedi, M., Mardas, N., and Sosis, K. 2009. The little REDD+ book. *Global Canopy Foundation*, 132, 139 p.

Parmesan, C. and Yohe, G. 2003. A globally coherent fingerprint of climate change impacts across natural systems. *Nature*, 421, 37–42.

Pearson, T. R. H., Brown, S., and Walker, S. 2011. Guidance Document: Avoided Deforestation Partners VCS REDD Methodology Modules. Published by Climate Focus, LLP.

Pennec, A., Gond, V., and Sabatier, D. 2011. Tropical forest phenology in French Guiana from MODIS time series. *Remote Sensing Letters*, 2, 337–345.

Peterson, G. D. and Heemskerk, M. 2001. Deforestation and forest regeneration following small-scale gold mining in the Amazon: The case of Suriname. *Environmental Conservation*, 28, 117–126.

Pflugmacher, D., Cohen, W. B., and E Kennedy, R. 2012. Using Landsat-derived disturbance history (1972–2010) to predict current forest structure. *Remote Sensing of Environment*, 122, 146–165.

Phillips, O. L., Aragão, L. E., Lewis, S. L., Fisher, J. B., Lloyd, J., López-González, G., Malhi, Y. et al. 2009. Drought sensitivity of the Amazon rainforest. *Science*, 323, 1344–1347.

Pinard, M. A. and Putz, F. E. 1996. Retaining forest biomass by reducing logging damage. *Biotropica*, 28(3), 278–295.

Pinty, B. and Verstraete, M. 1992. GEMI: A non-linear index to monitor global vegetation from satellites. *Vegetatio*, 101, 15–20.

Pithon, S., Jubelin, G., Guitet, S., and Gond, V. 2013. A statistical method for detecting logging-related canopy gaps using high-resolution optical remote sensing. *International Journal of Remote Sensing*, 34, 700–711.

Ploton, P., Pélissier, R., Proisy, C., Flavenot, T., Barbier, N., Rai, S. N., and Couteron, P. 2011. Assessing aboveground tropical forest biomass using Google Earth canopy images. *Ecological Applications*, 22, 993–1003.

Portillo-Quintero, C. A. and Sánchez-Azofeifa, G. A. 2010. Extent and conservation of tropical dry forests in the Americas. *Biological Conservation*, 143, 144–155.

Pringle, M., Schmidt, M., and Muir, J. 2009. Geostatistical interpolation of SLC-off Landsat ETM+ images. *ISPRS Journal of Photogrammetry and Remote Sensing*, 64, 654–664.

Qi, J., Chehbouni, A., Huete, A., Kerr, Y., and Sorooshian, S. 1994. A modified soil adjusted vegetation index. *Remote Sensing of Environment*, 48, 119–126.

Roberts, D., Numata, I., Holmes, K., Batista, G., Krug, T., Monteiro, A., Powell, B. et al. 2002. Large area mapping of land-cover change in Rondônia using multitemporal spectral mixture analysis and decision tree classifiers. *Journal of Geophysical Research*, 107, 8073.

Rouse, J., Haas, R., Schell, J., Deering, D., and Harlan, J. 1974. *Monitoring the Vernal Advancement and Retrogradation (Greenwave Effect) of Natural Vegetation*. College Station, TX: Texas A & M University, Remote Sensing Center.

Roy, D. P., Ju, J., Lewis, P., Schaaf, C., Gao, F., Hansen, M., and Lindquist, E. 2008. Multi-temporal MODIS-Landsat data fusion for relative radiometric normalization, gap filling, and prediction of Landsat data. *Remote Sensing of Environment*, 112, 3112–3130.

Roy, D. P., Ju, J., Kline, K., Scaramuzza, P. L., Kovalskyy, V., Hansen, M., Loveland, T. R. et al. 2010. Web-enabled Landsat Data (WELD): Landsat ETM+ composited mosaics of the conterminous United States. *Remote Sensing of Environment*, 114, 35–49.

Saatchi, S., Houghton, R., Dos Santos Alvala, R., Soares, J., and Yu, Y. 2007. Distribution of aboveground live biomass in the Amazon basin. *Global Change Biology*, 13, 816–837.

Saatchi, S. S., Harris, N. L., Brown, S., Lefsky, M., Mitchard, E. T., Salas, W., Zutta, B. R. et al. 2011. Benchmark map of forest carbon stocks in tropical regions across three continents. *Proceedings of the National Academy of Sciences of the United States of America*, 108, 9899–9904.

Salimon, C. I., Putz, F. E., Menezes-Filho, L., Anderson, A., Silveira, M., Brown, I. F., and Oliveira, L. 2011. Estimating state-wide biomass carbon stocks for a REDD plan in Acre, Brazil. *Forest Ecology and Management*, 262, 555–560.

Salk, C. F., Chazdon, R., and Andersson, K. 2013. Detecting landscape-level changes in tree biomass and biodiversity: Methodological constraints and challenges of plot-based approaches. *Canadian Journal of Forest Research*, 43, 799–808.

Salovaara, K. J., Thessler, S., Malik, R. N., and Tuomisto, H. 2005. Classification of Amazonian primary rain forest vegetation using Landsat ETM+ satellite imagery. *Remote Sensing of Environment*, 97, 39–51.

Samanta, A., Ganguly, S., Hashimoto, H., Devadiga, S., Vermote, E., Knyazikhin, Y., Nemani, R. R. et al. 2010. Amazon forests did not green-up during the 2005 drought. *Geophysical Research Letters*, 37, 5.

Sánchez-Azofeifa, A., Rivard, B., Wright, J., Feng, J.-L., Li, P., Chong, M. M., and Bohlman, S. A. 2011. Estimation of the distribution of Tabebuia guayacan (Bignoniaceae) using high-resolution remote sensing imagery. *Sensors*, 11, 3831–3851.

Scaramuzza, P. L., Bouchard, M. A., and Dwyer, J. L. 2012. Development of the Landsat data continuity mission cloud-cover assessment algorithms. *IEEE Transactions on Geoscience and Remote Sensing*, 50, 1140–1154.

Sesnie, S. E., Gessler, P. E., Finegan, B., and Thessler, S. 2008. Integrating Landsat TM and SRTM-DEM derived variables with decision trees for habitat classification and change detection in complex neotropical environments. *Remote Sensing of Environment*, 112, 2145–2159.

Sesnie, S. E., Finegan, B., Gessler, P. E., Thessler, S., Bendana, Z. R., and Smith, A. M. 2010. The multispectral separability of Costa Rican rainforest types with support vector machines and Random Forest decision trees. *International Journal of Remote Sensing*, 31, 2885–2909.

Siegert, F., Ruecker, G., Hinrichs, A., and Hoffmann, A. 2001. Increased damage from fires in logged forests during droughts caused by El Nino. *Nature*, 414, 437–440.

Sist, P. and Ferreira, F. N. 2007. Sustainability of reduced-impact logging in the Eastern Amazon. *Forest Ecology and Management*, 243, 199–209.

Skidmore, A. K. 1989. An expert system classifies eucalypt forest types using thematic mapper data and a digital terrain model. *Photogrammetric Engineering and Remote Sensing*, 55, 1449–1464.

Skole, D. and Tucker, C. 1993. Tropical deforestation and habitat fragmentation in the Amazon: Satellite data from 1978 to 1988. *Science*, 260, 1905–1910.

Small, C. 2004. The Landsat ETM+ spectral mixing space. *Remote Sensing of Environment*, 93, 1–17.

Song, C. 2013. Optical remote sensing of forest leaf area index and biomass. *Progress in Physical Geography*, 37, 98–113.

Souza Jr., C. and Barreto, P. 2000. An alternative approach for detecting and monitoring selectively logged forests in the Amazon. *International Journal of Remote Sensing*, 21, 173–179.

Souza Jr., C., Firestone, L., Silva, L. M., and Roberts, D. 2003. Mapping forest degradation in the Eastern Amazon from SPOT 4 through spectral mixture models. *Remote Sensing of Environment*, 87, 494–506.

Souza Jr., C. M., Roberts, D. A., and Cochrane, M. A. 2005. Combining spectral and spatial information to map canopy damage from selective logging and forest fires. *Remote Sensing of Environment*, 98, 329–343.

Souza Jr., C. M., Siqueira, J. V., Sales, M. H., Fonseca, A. V., Ribeiro, J. G., Numata, I., Cochrane, M. A. et al. 2013. Ten-year Landsat classification of deforestation and forest degradation in the Brazilian Amazon. *Remote Sensing*, 5, 5493–5513.

Souza Jr., C. M. and Siqueira, J. V. N. 2013. ImgTools: a software for optical remotely sensed data analysis. pp. 1571–1578 in *Anais XVI Simpósio Brasileiro de Sensoriamento Remoto*. Foz do Iguaçu, PR: Instituto Nacional de Pesquisas Espaciais (INPE).

Steininger, M. 2000. Satellite estimation of tropical secondary forest above-ground biomass: Data from Brazil and Bolivia. *International Journal of Remote Sensing*, 21, 1139–1157.

Stephens, B. B., Gurney, K. R., Tans, P. P., Sweeney, C., Peters, W., Bruhwiler, L., Ciais, P. et al. 2007. Weak northern and strong tropical land carbon uptake from vertical profiles of atmospheric CO2. *Science*, 316, 1732–1735.

Strabala, K. I. 2005. MODIS cloud mask user's guide. Madison, Wisconsin: Cooperative Institute for Meteorological Satellite Studies, University of Wisconsin, Madison, 32 pp.

Thenkabail, P. S., Hall, J., Lin, T., Ashton, M. S., Harris, D., and Enclona, E. A. 2003. Detecting floristic structure and pattern across topographic and moisture gradients in a mixed species Central African forest using IKONOS and Landsat-7 ETM+ images. *International Journal of Applied Earth Observation and Geoinformation*, 4, 255–270.

Thenkabail, P. S., Enclona, E. A., Ashton, M. S., Legg, C., and De Dieu, M. J. 2004. Hyperion, IKONOS, ALI, and ETM+ sensors in the study of African rainforests. *Remote Sensing of Environment*, 90, 23–43.

Thessler, S., Sesnie, S., Ramos Bendaña, Z. S., Ruokolainen, K., Tomppo, E., and Finegan, B. 2008. Using k-nn and discriminant analyses to classify rain forest types in a Landsat TM image over northern Costa Rica. *Remote Sensing of Environment*, 112, 2485–2494.

Tickle, P., Lee, A., Lucas, R. M., Austin, J., and Witte, C. 2006. Quantifying Australian forest floristics and structure using small footprint LiDAR and large scale aerial photography. *Forest Ecology and Management*, 223, 379–394.

Tottrup, C. 2004. Improving tropical forest mapping using multi-date Landsat TM data and pre-classification image smoothing. *International Journal of Remote Sensing*, 25, 717–730.

Tottrup, C., Rasmussen, M., Samek, J., and Skole, D. 2007. Towards a generic approach for characterizing and mapping tropical secondary forests in the highlands of mainland Southeast Asia. *International Journal of Remote Sensing*, 28, 1263–1284.

Townshend, J., Justice, C., Li, W., Gurney, C., and McManus, J. 1991. Global land cover classification by remote sensing: Present capabilities and future possibilities. *Remote Sensing of Environment*, 35, 243–255.

Trichon, V. and Julien, M.-P. 2006. Tree species identification on large-scale aerial photographs in a tropical rain forest, French Guiana-application for management and conservation. *Forest Ecology and Management*, 225, 51–61.

Tucker, C. J., Goff, T., and Townshend, J. 1985. African land-cover classification using satellite data. *Science*, 227, 369–375.

Uhl, C. and Buschbacher, R. 1985. A disturbing synergism between cattle ranch burning practices and selective tree harvesting in the eastern Amazon. *Biotropica*, 17(4), 265–268.

USGS. 2003. *Preliminary Assessment of Landsat 7 ETM+ Data Following Scan Line Corrector Malfunction USGS*. Sioux Falls, SD: United States Geological Survey.

Vancutsem, C., Pekel, J.-F., Evrard, C., Malaisse, F., and Defourny, P. 2009. Mapping and characterizing the vegetation types of the Democratic Republic of Congo using SPOT VEGETATION time series. *International Journal of Applied Earth Observation and Geoinformation*, 11, 62–76.

Verhegghen, A., Mayaux, P., De Wasseige, C., and Defourny, P. 2012. Mapping Congo Basin vegetation types from 300 m and 1 km multi-sensor time series for carbon stocks and forest areas estimation. *Biogeosciences*, 9, 5061–5079.

Vickers, B., Trines, E., and Pohnan, E. 2012. Community guidelines for accessing forestry voluntare carbon markets. Bangkok: Regional office for Asia and the Pacific, Food and Agriculture Organization of the United Nations, 196 pp.

Vieira, I. C. G., de Almeida, A. S., Davidson, E. A., Stone, T. A., Reis de Carvalho, C. J., and Guerrero, J. B. 2003. Classifying successional forests using Landsat spectral properties and ecological characteristics in eastern Amazonia. *Remote Sensing of Environment*, 87, 470–481.

Vitousek, P. M. 1994. Beyond global warming: Ecology and global change. *Ecology*, 75, 1861–1876.

Wang, B. 1999. Automated detection and removal of clouds and their shadows from Landstat TM images. *IEICE Transactions on Information and Systems*, 82, 453–460.

Wertz-Kanounnikoff, S. 2008. *Monitoring Forest Emissions, A Review of Methods*. Bogor, Indonesia: CIFOR.

Whiteside, T. G., Boggs, G. S., and Maier, S. W. 2011. Comparing object-based and pixel-based classifications for mapping savannas. *International Journal of Applied Earth Observation and Geoinformation*, 13, 884–893.

Wijaya, A., Liesenberg, V., and Gloaguen, R. 2010. Retrieval of forest attributes in complex successional forests of Central Indonesia: Modeling and estimation of bitemporal data. *Forest Ecology and Management*, 259, 2315–2326.

Wilson, E. O. 1988. The current state of biological diversity, in *Biodiversity*, E. O. Wilson and Peters, F. M. (eds.) Washington, DC: National Academy Press, pp. 3–18.

Wittmann, F., Anhuf, D., and Funk, W. J. 2002. Tree species distribution and community structure of central Amazonian várzea forests by remote-sensing techniques. *Journal of Tropical Ecology*, 18, 805–820.

Woods, P. 1989. Effects of logging, drought, and fire on structure and composition of tropical forests in Sabah, Malaysia. *Biotropica*, 21(4), 290–298.

Wulder, M. A., White, J. C., Nelson, R. F., Næsset, E., Ørka, H. O., Coops, N. C., Hilker, T. et al. 2012. Lidar sampling for large-area forest characterization: A review. *Remote Sensing of Environment*, 121, 196–209.

Xiao, X., Hagen, S., Zhang, Q., Keller, M., and Moore III, B. 2006. Detecting leaf phenology of seasonally moist tropical forests in South America with multi-temporal MODIS images. *Remote Sensing of Environment*, 103, 465–473.

Xu, X., Du, H., Zhou, G., Ge, H., Shi, Y., Zhou, Y., Fan, W. et al. 2011. Estimation of aboveground carbon stock of Moso bamboo (*Phyllostachys heterocycla* var. pubescens) forest with a Landsat Thematic Mapper image. *International Journal of Remote Sensing*, 32, 1431–1448.

Zelazowski, P., Malhi, Y., Huntingford, C., Sitch, S., and Fisher, J. B. 2011. Changes in the potential distribution of humid tropical forests on a warmer planet. *Philosophical Transactions of the Royal Society A: Mathematical, Physical and Engineering Sciences*, 369, 137–160.

Zhao, M. and Running, S. W. 2010. Drought-induced reduction in global terrestrial net primary production from 2000 through 2009. *Science*, 329, 940–943.

Zhu, X., Chen, J., Gao, F., Chen, X., and Masek, J. G. 2010. An enhanced spatial and temporal adaptive reflectance fusion model for complex heterogeneous regions. *Remote Sensing of Environment*, 114, 2610–2623.

Zhu, X., Gao, F., Liu, D., and Chen, J. 2012a. A modified neighborhood similar pixel interpolator approach for removing thick clouds in Landsat images. *IEEE Geoscience and Remote Sensing Letters*, 9, 521–525.

Zhu, Z. and Woodcock, C. E. 2012b. Object-based cloud and cloud shadow detection in Landsat imagery. *Remote Sensing of Environment*, 118, 83–94.

Zhu, Z., Woodcock, C. E., and Olofsson, P. 2012. Continuous monitoring of forest disturbance using all available Landsat imagery. *Remote Sensing of Environment*, 122, 75–91.

15

Remote Sensing of Forests from Lidar and Radar

Juha Hyyppä
Finnish Geospatial Research Institute and Centre of Excellence in Laser Scanning Research

Mika Karjalainen
Finnish Geospatial Research Institute and Centre of Excellence in Laser Scanning Research

Xinlian Liang
Finnish Geospatial Research Institute and Centre of Excellence in Laser Scanning Research

Anttoni Jaakkola
Finnish Geospatial Research Institute and Centre of Excellence in Laser Scanning Research

Xiaowei Yu
Finnish Geospatial Research Institute and Centre of Excellence in Laser Scanning Research

Michael Wulder
Pacific Forestry Center Natural Resources Canada

Markus Hollaus
Vienna University of Technology

Joanne C. White
Pacific Forestry Center Natural Resources Canada

Mikko Vastaranta
University of Helsinki and Centre of Excellence in Laser Scanning Research

Kirsi Karila
Finnish Geospatial Research Institute and Centre of Excellence in Laser Scanning Research

Harri Kaartinen
*Finnish Geospatial Research
Institute and Centre of Excellence
in Laser Scanning Research*

Matti Vaaja
*Centre of Excellence in
Laser Scanning Research*

Ville Kankare
*University of Helsinki and
Centre of Excellence in
Laser Scanning Research*

Antero Kukko
*Finnish Geospatial Research
Institute and Centre of Excellence
in Laser Scanning Research*

Markus Holopainen
*University of Helsinki and
Centre of Excellence in
Laser Scanning Research*

Hannu Hyyppä
*Centre of Excellence in
Laser Scanning Research*

Masato Katoh
Shinshu University

Acronyms and Definitions

2.5D	2.5-Dimensional model
3D	Three dimensional
ALS	Airborne laser scanning/scanner
CHM	Canopy height model
dbh	Diameter at breast height
DEM	Digital elevation model
DSM	Digital surface model
DTM	Digital terrain model
GLAS	Geoscience Laser Altimeter System
ICESat	NASA's Ice, Cloud and Land Elevation Satellite
InSAR	SAR interferometry
ITD	Individual tree detection
LIDAR	Light Detection and Ranging
LS	Laser scanning/scanner
MLS	Mobile laser scanning/scanner
MSN	Most similar neighbor
nDSM	Normalized digital surface model, canopy height model
NDVI	Normalized difference vegetation index
NFI	National Forest Inventory
NN	Nearest Neighbor
RF	Random Forest
RMSE	Root mean squared error
SAR	Synthetic Aperture radar
SWFI	Standwise field inventory
TLS	Terrestrial laser scanning/scanner

15.1 Introduction

This chapter is about collecting three-dimensional (3D) information from lidar and radar and turning that information into valuable forest informatics. For the first time, we present that the processing of all these data, whether lidar or radar, should go into the same pipeline.

Today, it can be seen that many of the future remote sensing processes for forestry will be based on point cloud processing or on elevation models (3D techniques). These required forestry data can be provided not only by both the lidar and the radar but also by photogrammetry. For example, analogous to photogrammetric spatial intersection, a stereo pair of SAR images with different off-nadir angles can be used to calculate the 3D coordinates for corresponding points on the image pair producing point clouds from radar imagery. Also, in SAR interferometry (InSAR), pixel-by-pixel phase difference between two complex SAR images acquired from slightly different perspectives can be converted into elevation differences of the terrain/object. Thus, both lidar and radar can provide data that can be processed in a similar way either using original points or using surface models in a raster form. From the point clouds, you can calculate digital terrain model (DTM), digital surface model (DSM), and canopy height model, normalized digital surface model (CHM/nDSM). The idea is to provide surface model (DSM) and subtract the ground elevation (DTM) from it in order to get a canopy height. Intensity, coherence (in interferometry SAR) and texture can be used to improve the estimates in 3D-based inventory.

In general, there is high synergy between lidar and radar since they are based on the same measuring principle even though they are using different frequencies/wavelengths.

1. Laser intensity calibration has stemmed from the corresponding work with radar backscatter coefficient determination. In the late 1980s, Finnish and Swedish researchers were developing early versions of scanning lidar/radars, and already at that time, the radar return versus time was automatically corrected in the hardware (Hallikainen et al. 1993).
2. The lidar-derived terrain model is the basic information needed in future radar-based forest inventories (FIs). In future, radar-based FI processing will be mainly done using laser scanning (LS) point cloud processing techniques. In large-area FI in near future, satellite-based radar can cover large areas with relatively high repetition rate, and LS can provide important field reference for satellite data calibration.
3. Both lidar and radar are active remote sensing techniques. The major advantages of active remote sensing systems include better penetration of atmosphere; coverage can be obtained at user-specified times, even at night, since it produces its own illumination to the target; images/echoes can be produced from different types of polarized energy; systems may operate simultaneously in several wavelengths/frequencies, and, thus, have multiwavelength/frequency potential.

There are many chapters and books that complement this chapter. Starting from the past literature, the reader is referred to know basics of, for example, lidar/LS from, for example, Hyyppä et al. (2008), radar from Henderson and Lewis (1998), and Jensen (2000). Forestry applications and processing based on active sensors are also covered in Hyyppä et al. (2008) on using LS data in forestry applications and related algorithms; Holopainen et al. (2014) on the estimation of forest stock and yield with lidar; Nelson (2013) on how did we get there and an early history of forestry lidar; Koch (2010) on the status and future using new LS, synthetic aperture radar (SAR), and hyperspectral remote sensing data for forest biomass assessment; and Mallet and Bretar (2009) on full-waveform topographic lidar. In this handbook, there are many complementary chapters, for example, those dealing with lidar processing, tropical rainforests, remote sensing of tree height, terrestrial carbon modeling, and global biomass modeling.

We see that radar and lidar are currently changing the way how operative FI is performed. Remote sensing of forest from lidar and radar is the future of FI at local, regional, or country level. Currently, airborne lidar is operationally applied in the Nordic countries when carrying out standwise FIs. We believe, terrestrial and mobile lidar or LS will be used in collecting detailed field data. Radar data seem to be promising for large-area monitoring applications. Therefore, we planned the content of the chapter in the following way:

- Section 2: Conventional practices to acquire forest resource information.
- Section 3: Basics of lidar and radar.

- Section 4: Obtaining 3D data from lidar and radar for forestry.
- Section 5: Common processing chain for lidar and radar into useful forest informatics.
- Section 6: Finally, we go into new areas, applications, and large-area FI, which lidar and radar techniques are making possible in the near future.

In this chapter, we aim to demonstrate that lidar and radar point clouds are in future processed in a very similar processing chain, which has been originally developed for airborne lidar for stand-level FI in boreal forest area. Additional, we will show large number of future techniques, which will further challenge current operative inventory systems—currently not based on these data.

15.2 Conventional Practices for Acquisition of Forest Resource Information

According to data from the Global Forest Resources Assessment (FRA 2010), the total global forest area is slightly over 4 billion hectares (31% of the total land surface). The five most forest-rich countries are the Russian Federation, Brazil, Canada, the United States, and China. In terms of the ratio of forest cover to total land area, Finland (73% of the land area), Japan, and Sweden (both 69% of the land area) are the world's most extensively forested countries among the industrialized and temperate countries (Table 15.1). Forest monitoring via remote sensing plays a crucial role in the assessment, planning, field data collection, image processing, analysis, and modeling for sustainable forest

TABLE 15.1 Forest Area and Coverage in Selected Developed Countries

Country	Forest Area		Land Area	Country Area
	Million ha	Percentage of Land Area	Million ha	Million ha
Russian Federation	809.1	49	1638.1	1709.8
Brazil	519.5	62	832.5	851.5
Canada	310.1	34	909.4	998.5
United States	304.0	33	916.2	963.2
China	206.9	22	942.5	960.0
Australia	149.3	19	768.2	774.1
Sweden	28.2	69	41.0	45.0
Japan	25.0	69	36.5	37.8
Finland	22.2	73	30.4	33.8
Spain	18.2	36	49.9	50.5
France	16.0	29	55.0	55.2
Germany	11.1	32	34.9	35.7
Norway	10.1	33	30.4	32.4
Italy	9.1	31	29.4	30.1
United Kingdom	2.9	12	24.3	24.4
World	4,033.1	31	13,010.5	13,434.2

Source: FRA, Global forest resources assessment 2010 Available at http://www.fao.org/forestry/fra/fra2010/en/, 2010.

management (SFM). Geographic information system (GIS) is widely employed to manage forest information obtained with remote sensing on a stand polygon basis. The addition of stems by species to each stand polygon could become an important tool for practical forest management operations, such as precise thinning, selective cutting, and harvesting.

15.2.1 Forest Inventory

FI is carried out to support decision-making by the forest owner. Forest resource information is, thus, needed for large-scale strategic planning, operative forest management, and preharvest planning (Table 15.2). National forest inventories (NFIs) are examples of inventories undertaken for large-scale strategic planning for gathering information about nationwide forest resources, such as growing stock volume, forest cover, growth and yield, biomass, carbon balance, and large-scale wood procurement potential. In NFIs, it is important to have unbiased estimates and to obtain information also from small strata. The most conventional strategy for NFI is to use sampling and measure sampling plots at the field. Thus, NFIs does not provide mapping information that is required for operative forest management and preharvest planning. In many countries, information for these purposes has been collected using standwise field inventories (SWFIs). There are many different methods with varying accuracy to carry out SWFI. In general, visual interpretation of aerial images is combined with field sampling and some rapid measurements. For example, in Finland and Sweden, stands are delineated from aerial photographs, and then every stand is visited by a forester. At the stand, basal area is measured using relascope (angle-count method) from various locations. Then, basal-area-weighted mean tree diameter and height are measured by caliper and clinometer. Finally, inventory attributes are generalized to stand. Forest management operations are determined using calculated inventory attributes combined with ocular assessments at the site.

During the last 100 years, FI has transferred from the determination of the volume of logs, trees, and stands, and a calculation of the increment and yield toward more into multiple uses of forests (wildlife, recreation, and watershed management).

Currently, multiple use of forest is increasingly considered as different ecosystem services, and from an operational forestry standpoint, the information required to support "multiple use" can also satisfy characterization of "ecosystems services." However, a major focus of forest assessment is still in obtaining accurate information on the volume and growth of trees in forest plots, stands, and large areas. The forest bound carbon is also an important issue globally. One of the biggest challenges currently in FI research is how to measure and monitor forest biomass and its changes effectively and accurately. Radar and lidar provide tools for that.

15.2.2 Forest Measurements

An FI could be, in principle, based on measuring every tree in a given area, but this is usually not realistic in forestry. As such, the acquisition of forest resource information is typically based on sampling. The most common sample units are a tree and a plot. To obtain usable information-related forest resources, various attributes have to be measured at the tree level. Individual tree measures are summed at the plot level, and then plots are used to obtain forest resource information representative of any given area, such as a stand, a woodlot, a county, or a country. Due to this hierarchy, it is highly important to measure individual tree attributes accurately. It should also be pointed out that some tree attributes are modeled instead of measured, for example, stem volume is usually modeled based on diameter at breast height (dbh) and height of the tree (h), noting that measurements made at a tree or plot level are often selected by some sampling criteria. Table 15.3 summarizes the main tree attributes from the point of view of forest mensuration. Upper diameter is typically taken from the height of 6 m. Height of the crown base is the height from the ground to the lowest green branch or to the lowest complete living whorl of branches. Basal area is the cross-sectional area defined by the dbh.

Some of them can be directly measured or calculated from these direct measurements, while others need to be predicted

TABLE 15.2　Aims and Methods to Collect Forest Resource Information

Aim	Method	Provides a Map	Description
Large-scale strategic planning	National forest inventory (NFI)	No	Nationwide statistics are calculated based on systematic sample of field plots. Field plots are measured tree by tree.
Operational forest management	Standwise field inventory (SWFI)	Yes	Several field plots are measured from every stand. Sample trees are measured from plots.
Preharvest planning	SWFI with additional measurements	Yes	SWFI information is double-checked by measuring additional plots.

TABLE 15.3　Attributes of Trees to Be Measured

Attribute	Unit	Typically Expected Accuracy for Measurement[a]
Height	m	0.5–2 m
Diameter at breast height (dbh)	mm	5–10 mm
Upper diameter (e.g., at height 6 m)	mm	5–10 mm
Height of crown base	m	0.2–0.4 m
Species		
Age	years	5 years
Location	m	0.5–2 m
Basal area	m^2	See diameter accuracy
Volume	m^3	10%–20%
Biomass	kg/m^3	10%–20%
Growth	mm (increment borer)	1 mm

[a] Depends strongly on the use of the data.

through statistical or physical modeling. Traditionally, the following individual tree attributes such as

- Height (and height growth)
- Diameters at different height along the stem (and diameter growth)
- Crown diameter
- Tree species

are measured or determined in the field. Diameter is convenient to measure and is one of the directly measurable dimensions from which tree cross-sectional area, surface area, and volume can be computed. Various instruments and methods have been developed for measuring the tree dimension in the field (Husch et al. 1982; Päivinen et al. 1992; Clark et al. 2000; Gill et al. 2000; Korhonen et al. 2006), such as the following:

- Caliper, diameter tape, and optical devices for diameter measurements
- Level rod, pole, and hypsometers for tree height measurements
- Increment borer for diameter growth measurements

The method used in obtaining the measurements is largely determined by the accuracy required. Past growth of diameter can be obtained from increment borings or cross-section cut. For some species, past height growth may be determined by measuring internodal lengths. Sometimes, it is even necessary to fell the tree to obtain more accurate measurements, for example, to measure the stem volume accurately requires destructive sampling of a tree. Accordingly, direct and indirect methods have been developed for the estimation of such forest attributes. Practically speaking, tree volume is estimated from dbh and possibly together with height and upper diameter for each tree species. The models for stem volume, especially based on diameter information, exist for many commercially important tree species. For example, in Finland, there are volume models v for main tree species based on different inputs (diameter d, height h, diameter at the height of 6 m, d_6):

- $v = f(d)$
- $v = f(d,h)$
- $v = f(d,h,d_6)$

The most accurate and nondestructive way to determine the stem volume of the tree is to use stem curve, that is, the stem diameter as a function of the tree height. Relascope is often used to measure basal area of the sample plots if dbhs of all the trees within the plot are not measured.

15.3 General Features of LS/Lidar and Radar

Active microwave imagery is obtained using instruments and principles that are different from those acquired in the visible, near-, mid-, and thermal infrared portions of the spectrum using passive remote sensing techniques. Therefore, it is necessary to understand the basics of the active microwave systems, such as lidar and radars.

The LS is a *surveying technique* used for mapping topography, vegetation, urban areas, ice, infrastructure, and other targets of interest. The LS is many times referred to as lidar because of the central role of the lidar in the system. Also LS is more commonly used in Europe whereas lidar in the United States. The basic principle of lidar is to use a laser beam to illuminate an object and a photodiode to register the backscatter radiation and to measure the range. More precisely, airborne laser scanning (ALS) is a method based on light detection and ranging measurements from an aircraft, where the precise position and orientation of the sensor is known, and therefore the position (x, y, and z) of the reflecting objects can be determined. In addition to ALS, there is an increasing interest in terrestrial laser scanning (TLS), where the laser scanner is mounted on a tripod or even on a moving platform, that is, mobile laser scanning (MLS). The output of the laser scanner is then a georeferenced point cloud of lidar measurements, including the intensity and possibly waveform information of the returned light. A typical ALS system consists of (1) a laser ranging unit (i.e., lidar), (2) an opto-mechanical scanner, (3) a position and orientation unit, and (4) a control, processing, and storage unit. The laser ranging unit can be subdivided into a transmitter, a receiver, and the optics for both units. These components also apply to other types of LS systems, such as MLS. The receiver optics collects the backscattered light and focuses it onto the detector converting the photons to electrical impulses. The opto-mechanical scanning unit is responsible for the deflection of the transmitted laser beams across the flight track. The type of the applied deflection unit (e.g., oscillating mirror/zigzag scanning, rotating mirror/line scanning, push broom/fiber scanning, and Palmer/conical scanning) defines the scan pattern on the ground. A differential Global Navigation Satellite System (GNSS) receiver provides the position of the laser ranging unit. Its orientation is determined by the pitch, roll, and heading of the aircraft, which are measured by an inertial navigation/measurement system (Hyyppä et al. 2008).

Radar is a similar object-detection system using radio waves to determine the range, altitude, direction, or speed of objects. In remote sensing, especially the backscatter strength is used to object recognition/classification. The transmitted energy illuminates an area on the ground. The radar cross section of the object is defined as the ratio of the backscattered power versus isotropically reflecting object. The radar backscatter coefficient is defined as the radar cross section divided by the illuminated area. The radar backscatter coefficient is used to classify targets. With radar operating wavelengths, surface roughness, moisture, and biomass and vegetation structure are major parameters affecting the backscatter. In addition to backscatter, range (e.g., Hallikainen et al. 1993), polarization response, stereoscopy, various incidence angles, and interferometry have been applied for remote sensing of forests.

Table 15.4 gives a short comparison of lidar and radar in remote sensing of forests.

TABLE 15.4   Comparison of Lidar and Radar in Remote Sensing of Forests

Characteristics	Lidar	Radar
Cloud penetration capacity	No.	Yes.
Coverage can be obtained at user-specified times, even at night	Yes.	Yes.
May penetrate vegetation, sand, and surface layers of snow	No.	Yes.
Penetration to forests	Using canopy gaps and small beams.	May penetrate through leaves and needles.
Images can be produced from different types of polarized energy (HH, HV, VH, VH)	Yes, but not applied.	Usually applied.
May operate simultaneously in several wavelengths (frequencies) and thus has multifrequency potential	Yes, applied currently in bathymetry and hyperspectral lidars.	Yes, common approach.
Can produce overlapping images suitable for stereoscopic viewing and radargrammetry	Not applied.	Yes.
Supports interferometric	Not applied.	Yes.
Applied wavelengths	Typical wavelengths are between 500 and 1550 nm.	The abbreviations associated with the radar frequencies and wavelengths include (K: 18–26.5 GHz, 1,67–1,19 cm; X: 8–12.5 GHz, 3.8–2.4 cm; C: 4–8 GHz, 7.5–3.9 cm; S: 2–4 GHz, 15–7.5 cm; L: 1–2 GHz, 30–15 cm; P: 0.3 –1 GHz, 100–30 cm).
Speckle	Since backscatter is not strongly used, the effect of speckle is not well studied. However, the speckle effect is smaller in magnitude as with radars.	Strong speckle effect with microwave wavelengths, to remove the speckle, the image is usually processed using several looks; thus, an averaging process takes place.
Foreshortening, layover, and shadowing	Do not exist with lidar data, since point clouds always preferred before backscatter information.	Geometric distortions exist in all radar imagery when backscatter is the main data output.
Object roughness	Most of the object are considered as rough with lidar wavelengths.	Incidence angle affects strongly to the radar backscatter with non-rough surfaces.
Biomass measurements	Typically based on point cloud metrics data.	Previously based on backscatter data. Radar backscatter increases approximately linearly with increasing biomass until it saturates at a biomass level that depends on the radar frequency. The lower the frequency, the better the penetration.
Soil moisture	Affects intensity.	L-band radar penetrates to a maximal depth of approximately 10 cm. Shorter wavelengths penetrate to only 1–3 cm. Multi-temporal radar data can be used to measure soil moisture variation.

15.4 Obtaining 3D Data from Forestry

The focus in this chapter is on techniques capable to provide 2.5D/3D (2.5D stands for 3D surface models) data for FI processing to be depicted in Section 15.5. Short state of the art of the lidar/radar techniques providing 3D data covers

- Space-borne lidar
- Space-borne SAR
- InSAR
- SAR radargrammetry
- ALS
- TLS
- MLS

SAR tomography is not covered, since it is still far from practical usability. Since there are many forest attributes to be measured, more focus is on biomass and stem volume, to which all attributes basically correlate. In Section 15.5, there

is additional discussion to measure each attribute separately. Table 15.5 gives main characteristics of each of these systems at the general level.

15.4.1 Space-Borne Lidar

The Geoscience Laser Altimeter System (GLAS) on board NASA's Ice, Cloud and Land Elevation Satellite (ICESat) was the first space-borne lidar mission providing lidar data at a global scale (Zwally et al. 2002; Schutz et al. 2005). ICESat/GLAS was launched in January 2003 and acquired lidar waveform data until October 2009. Several studies also demonstrated the potential of GLAS data to characterize forest structure (e.g., Lefsky et al. 2005, 2007; Rosette et al. 2008; Sun et al. 2008; Ballhorn et al. 2011). Space-borne lidar to provide systematic and widely dispersed measures of vegetation characteristics is required for robust global biomass estimates, among other information needs (Wulder et al. 2013b).

TABLE 15.5 Example Characteristics of the 3D Remote Sensing Systems

Space-borne Lidar	
Potential	Global biomass with remarkably better accuracy than with today's techniques
Possible beam size	Few tens of meters, ranging accuracy of 1–3 m (for canopy height)
Challenges	To obtain high power from space, lifetime of the system
System providers	Currently, NASA with former ICESat, with GEDI[a] recently (2014) funded to place a lidar on the International Space Station
Space-borne SAR	
Resolution	Resolution up to 1 m from space, typically to 10 km–by–10 km imagery
Frequency	C- or X-band
Revisit time	Few days
Interferometry	Repeat-pass interferometry/single-pass interferometry
Feasible	Large-area inventories
Characteristics	Ranging accuracy of 3–5 m (for canopy height), cost of the data is typically high compared to data quality
System providers	DLR (TerraSAR X), ASI (Cosmo), ESA (Sentinel-1), NASDA
Airborne laser scanning	
Point density	Point density 0.5–40 pts/m^2
Elevation accuracy	5–30 cm
Planimetric accuracy	20–80 cm
Operating range	Few hundred meters to several kilometers
Feasible	Cost effective for areas larger than 50 km^2
Characteristics	Homogenous point clouds
System providers	Optech, Leica, Riegl
Mobile laser scanning	
Point density	Point density in the range of 100 to several thousand pts/m^2
Accuracy	Point accuracy of few centimeters (egg) when collected with good GNSS coverage
Operating range	Applicable range of few tens of meters
Feasible	Collecting large data sets for road environment
Characteristics	Relatively high variation (density) in the range data
System providers	StreetMapper, Optech, Riegl, Trimble, Nokia Here, Topcon, IGI, MDL
Terrestrial laser scanning	
Point density	Point density in the range of 10,000 pts/m^2 at the 10 m
Accuracy	Distance accuracy of few mm to 1–2 cm
Operating range	Applicable range of few tens of meters
	Operational scanning range from one to several hundred meters
Feasible	Feasible for small areas less than few tens of meters distance
Characteristics	Processing time challenging: image processing techniques applied; small variation in data, e.g., distance variation low, thus surface normal can be calculated
System providers	Faro, Leica, Riegl, Topcon, Trimble, Zoller & Fröhlich

[a] http://www.nasa.gov/press/2014/july/nasa-selects-instruments-to-track-climate-impact-on-vegetation/#.VAXwT_ldX1Z.

15.4.2 Space-Borne Synthetic Aperture Radar

In the past decades, remarkable amount of research using SAR data has been conducted in the field of FI concentrating usually on stand- or plot-level mean stock volume and/or above-ground biomass (AGB) estimation (e.g., Le Toan et al. 1992; Fransson and Israelsson 1999; Wagner et al. 2003; Rauste 2005; Tokola et al. 2007; Holopainen et al. 2010; Solberg et al. 2010). Results have been promising, but merely not accurate enough for operative FIs. Recently, SAR satellite images have rapidly improved thanks to the latest very-high-resolution SAR satellites (TerraSAR-X/TanDEM-X, COSMO-SkyMed, and Radarsat-2) (Krieger et al. 2007; Torres et al. 2012). Therefore, improvements in the field of SAR-based forest mapping are anticipated as well. In general, SAR images contain the following information at the pixel level:

(1) radar backscattering intensity, (2) phase of the backscattered signal, and (3) range to target pixel.

1. Radar intensity corresponds to the strength of the backscattered signal compared to the strength of the transmitted signal, and it is a function of the SAR system parameters (such as the wavelength and the polarization of the used electromagnetic radiation) and target parameters (such as the target area roughness compared to the used radar wavelength and dielectric properties).

2. The phase information in the single-channel SAR data is quasi-random and, therefore, useless for target interpretation. However, phase information is an essential part of multipolarized data analysis (SAR polarimetry) and InSAR.

3. The range measurement is based on the time-of-flight information of the radar pulse and has typically neglected in biomass estimation tasks.

The use of intensity and backscattering coefficient information in forest resources mapping has been widely studied over the past few decades. In general, the longer radar wavelengths (L-band or for airborne sensors also P- and VHF-band) are more suitable for stem volume estimation than the shorter wavelengths of C-band or X-band (Le Toan et al. 1992; Fransson 1999). The reason for this is that the interaction between radar waves and forest structures in the L-band and P-band occurs on the trunks of trees. On the other hand, in the X-band and C-band, the scattering takes place at the top of the forest canopy, on branches and foliage, contributing apparently less to the information related to stem volume. Even though the relationship between radar intensity and stem volume has been well studied, there still remain practical challenges to be overcome due to topography and seasonal variations in weather. Rauste (2005) was able to obtain a correlation coefficient of 0.85 between stem volume and radiometrically normalized L-band JERS-1 data in Finland. However, the L-band intensity appears to saturate at some level of stem volume. Typically, stem volume levels beyond 100–200 m^3/ha cannot be observed (Fransson and Israelsson 1999; Rauste 2005) due to saturation. In addition to saturation at higher biomass levels, there are other issues, such as speckle and mixture of surface roughness, moisture, and biomass affecting the output of the radar.

The radar pulse illuminates a given surface area that consists of several scattering points. Thus, the returned echo comprises a coherent combination of individual echoes from a large number of points (see Elachi 1987). The result is a single vector representing the amplitude V and phase f ($I–V^2$) of the total echo, which is a vector sum of the individual echoes. This variation is called fading or speckle. Thus, an image of a homogeneous surface with constant reflectivity will result in intensity variation from one resolution element to the next. Speckle gives images recorded with radar a grainy texture.

The radar cross section is defined as the equivalent of a perfectly reflecting area that reflects isotropically (spherically). The backscatter coefficient is defined as the radar cross section divided by the area illuminated. The radar backscatter coefficient is mainly used to classify target characteristics. Surface roughness, moisture, and biomass and vegetation structure are major environmental parameters within the resolution cell that are responsible for backscattering the incident energy. Surface roughness is the terrain property that strongly influences the strength of the radar backscatter. Co-polarization backscatter toward the sensor results from single reflections from canopy components such as the leaves, stems, branches, and trunk, and these returns are generally very strong (called canopy surface scattering). If the energy is scattered multiple times within a distributed volume such as a stand of pine trees, this is often called volume scattering. Radar backscatter increases approximately linearly. With higher biomass levels, it is hard to separate soil moisture and vegetation backscatter contributions.

The phase information can be used to calculate 2D interferometric coherence maps and this way to deriving stem volume and other forest parameters (e.g., Askne et al. 2003; Wagner et al. 2003; Santoro et al. 2007). However, the coherence signal appears to saturate at some point of the biomass hampering the estimation of high biomass values. Even though this is a very promising technology as demonstrated with the ERS-1 and ERS-2 SAR tandem mission, it is not supported by the currently operational SAR satellite missions. Moreover, promising results have been achieved by combining the methods of interferometry and polarimetry, that is, PolInSAR (Papathanassiou and Cloude 2001) or SAR tomography (Reigber 2002). However, the methods of PolInSAR and SAR tomography are still under scientific research, and appropriate SAR satellite data have not been available before TanDEM-X satellite. Also several data fusion studies of SAR and lidar data (scanning or profiling) in forest mapping have been performed (e.g., Hyde et al. 2007; Nelson et al. 2007; Goodenough et al. 2008; Sun et al. 2008; Kellndorfer et al. 2010; Banskota et al. 2011). Overview of using ALS, SAR, and hyperspectral remote sensing data for AGB assessment can be found in Koch (2010). The estimation of AGB solely on the basis of SAR backscatter intensity has proven to be challenging (e.g., Fransson and Israelsson 1999; Rauste 2005; Holopainen et al. 2010).

The most promising approach to determine forest biomass by radar imaging from space is likely to be via canopy height information (i.e., 3D techniques) similarly to LS. Recent studies have shown that elevation information extracted from SAR has potential in the estimation of forest canopy height even close to ALS data (e.g., Solberg et al. 2010; Perko et al. 2011; Karjalainen et al. 2012). Basically, there are two approaches to extract elevation information from the SAR images: (1) InSAR (Massonnet and Feigl 1998; Rosen et al. 2000) and (2) radargrammetry (Leberl 1979; Toutin and Gray 2000).

15.4.2.1 SAR Interferometry

InSAR is a technique in which the pixel-by-pixel phase difference between two complex SAR images acquired from slightly different perspectives can be converted into elevation differences of the terrain (Massonnet and Feigl 1998; Rosen et al. 2000). When the X- or C-band of the radar is considered, the scattering takes place near the top of the forest canopy (Le Toan et al. 1992). Therefore, if the elevation of the ground surface is known (e.g., a DTM is available), then the X- or C-band's interferometric height compared to the ground surface elevation is related to the forest canopy height and accordingly to the stem volume. For forestry applications, simultaneous acquisition of the SAR data used for interferometry is especially advantageous. An example of the use of interferometric data for forest canopy height estimation has been provided by Kellndorfer et al. (2004), who used the C-band's interferometric heights from the shuttle radar topography mission (SRTM) to estimate the forest canopy height. Similar results using the SRTM X-band data were presented by Solberg et al. (2010), who also estimated the AGB based on SRTM elevation values. The tandem-X mission (TDM), launched in 2010, consists of two satellites flying in close formation enabling bistatic acquisition of X-band SAR data. Recent studies (Askne et al. 2013; Solberg et al. 2013) have demonstrated the use of TDM data retrieval of forest biomass of boreal forests.

15.4.2.2 SAR Radargrammetry

SAR radargrammetry is an alternative way to InSAR to extract elevation data from radar data. This is based on stereoscopic measurement of SAR images. Analogous to photogrammetric spatial intersection, a stereo pair of SAR images with different off-nadir angles can be used to calculate the 3D coordinates for corresponding points on the image pair. However, contrary to interferometry, radargrammetry is based on the intensity and range values of SAR data and not on the phase information. The foundations for the stereo-viewing capabilities of radar images were recognized already in the 1960s (see, e.g., La Prade 1963). An example of research looking into the mathematical foundations for calculating 3D coordinates and their expected accuracies is the work by Leberl (1979). When the trajectory of a SAR antenna (position and velocity as functions of time) is known accurately enough in relation to the object coordinate system and when a point target can be clearly identified on two SAR images with different off-nadir angles, the 3D coordinates of the point target can be calculated based on the range information. Typically, the so-called range-Doppler equation system is used as a sensor model, which describes accurately enough the propagation of electromagnetic radiation from the SAR image pixel to the point target and vice versa (Leberl 1979). The Canadian satellite, Radarsat-1, was one of the first SAR satellites to provide images with variable off-nadir angles suitable for radargrammetric processing (Toutin and Gray 2000). The ERS-1 and ERS-2 satellites of the European Space Agency have also provided suitable stereo pairs, but with limited stereo overlap areas (Li et al. 2006). However, only a few studies related to the extraction of forest information from radargrammetry have been published, for example, by Chen et al. (2007). Recent studies by Perko et al. (2011) and Karjalainen et al. (2012) have revealed the potential of radargrammetric 3D data in forest biomass estimation and change detection.

15.4.3 Airborne Laser Scanning, Airborne Lidar

ALS is a method based on lidar measurements from an aircraft, where the precise position and orientation of the sensor is known, and therefore the point cloud (x, y, z) of the reflecting objects can be determined. The first studies of ALS for forestry purposes included standwise mean height and volume estimation (e.g., Næsset 1997a,b), individual-tree-based height determination and volume estimation (e.g., Hyyppä and Inkinen 1999; Hyyppä et al. 2001a), tree-species classification (e.g., Hyyppä et al. 2001; Brandtberg et al. 2003; Holmgren and Persson 2004), and measurement of forest growth and detection of harvested trees (e.g., Hyyppä et al. 2003; Yu et al. 2004, 2006). Today, ALS is becoming a standard technique in the mapping and monitoring of forest resources. By using ALS-based inventory, 5%–20% error in main forest stand attributes at stand level has been obtained. For

overviews on using ALS in FI, see Hyyppä et al. (2008). ALS is a promising technique also for efficient and accurate AGB retrieval because of its capability of direct measurement of vegetation 3D structure. AGB correlates strongly with canopy height. Popescu et al. (2003), Popescu (2007), van Aardt et al. (2008), and Zhao et al. (2009) showed that AGB can be estimated similarly to other forest attributes by means of ALS metrics. The leaf area index (LAI) has also been used as a predictor of AGB (Koch 2010). Interaction of ALS pulse and forest canopy has also been studied using ALS intensity (Korpela et al. 2010).

15.4.4 Terrestrial Laser Scanning

TLS, also known as ground-based lidar, has been shown to be a promising technique for forest field inventories at tree and plot levels. The major advantage of using TLS in forest field inventories lies in its capacity to document the forest in detail. The first commercial TLS system was built by Cyra Technologies (acquired by Leica in 2001) in 1998, and the first papers related to plot-level tree attribute estimation were reported in early 2000s. Currently, TLS has shown to be feasible for collecting basic tree attributes at tree and plot levels, such as dbh and tree position (Maas et al. 2008; Brolly and Kiraly 2009; Murphy et al. 2010; Lovell et al. 2011; Liang et al. 2012a). By reconstructing tree stem, it is possible to derive high-quality stem volume and biomass estimates comparable in accuracy with the best national allometric models (Liang et al. 2014a). TLS data also permit time series analyses because the entire plot can be documented consecutively over time (Liang et al. 2012b). It is expected that TLS will be operationally used in plot-level FIs as soon as the appropriate software becomes available, best practices become known, and general knowledge of these findings is more widely spread.

15.4.5 Mobile Laser Scanning

MLS is based on lidar measurements from a moving platform, where the precise position and orientation of the sensor is known using a navigation system, similar to ALS, and therefore the position of the reflecting objects can be determined either from pulse travel time or from phase information. An MLS system consists of one or several laser scanners. Navigation system consists of various sensors for positioning and determining the rotation angles of the system, while GNSS and inertial measurement unit being the most important parts of the system. Also other mapping sensors, such as cameras, thermal imagers, and spectrometers, can be incorporated into the MLS systems. In principle, MLS is similar to ALS, whereas the platform is not aircraft. The application of MLS in forestry is being recently studied (Lin et al. 2010; Holopainen et al. 2013; Liang et al. 2014c). In the near future, MLS can be seen as a practical means to produce tree maps or inventories in urban forest environments, but in future possibly also in boreal forests and managed forests. However, MLS is still far from a widely used practical application in forestry, but the situation may change due to the rapid development of automatic MLS and TLS data processing.

15.5 Processing 3D Data into Forest Information

It is anticipated that many of the future remote sensing processes for forestry will be based on 3D point cloud or elevation data processing, especially those based on lidar and radar. This is perhaps the first time that lidar-based point cloud processing is proposed as an optimal solution also for radar data.

From the 3D data, you can calculate DTM, DSM, and CHM/nDSM. Today, most 3D techniques require a good DTM, which typically comes from LS, since radar data with available frequency bands do not provide penetration into ground floor due to too large footprint size and no ranging capacity of the applied radars. In optical wavelength, the penetration of signal through vegetation is lower, but since lidar is based on narrow beams finding canopy gaps, the terrain model can be calculated accurately from the lidar data. Lower microwave wavelength penetrates the vegetation layer better, but then also the penetration into the ground increases based on dielectric properties of the ground. From that point of view, lidar data are optimal for topographic mapping. The idea is to provide surface model (DSM) and subtract the ground elevation (DTM) from it in order to get a canopy height. Intensity, coherence (in interferometry SAR), and texture can be used to improve the estimates in 3D-based inventory techniques as well as other lidar metrics.

We see that there are two kinds of processing needed: forest attribute estimation based on single-time point cloud and use of bi-temporal point clouds for change detection (see Figure 15.1). Additionally, direct measurement of individual tree attributes is the third kind of data processing methodology used, but it can be combined with the first two approaches.

Single-point cloud processing for forest attribute data collection includes the following steps. DTM is obtained from ALS data or known beforehand (for SAR). CHM is calculated to get tree heights and tree height metrics. Features (point cloud metrics) are calculated from the data, and nonparametric estimation is applied. Nonparametric estimation requires field plots, which are used for the teaching of the classifier. In addition to

the point cloud metrics, other features, such as individual tree information, texture, waveform lidar features, image processing applied to, for example, DSMs, image-based features (including NDVI), and other channel information and ratios, can be added to improve the prediction. Tree species are predicted also in this phase, and therefore, the system should include features capable to discriminate species. The optimum output, requested by forest companies, of the process is species-specific height distribution of the trees.

The process includes

1. DTM generation
2. DSM and CHM generation
3. Derivation of point cloud metric
4. Prediction of forest attributes using nonparametric estimation

Typical process for change detection is to subtract two DSMs from bi-temporal data after they have been shown to match with elevation. After thresholding, which are used to see real changes, and after filtering/smoothing, segmentation of the changes, areas can be delineated. The changes can then be compared with the real change of the reference. Prediction of the change can also be done using nonparametric estimation.

Direct measurement of individual tree attributes can be based on the detection of individual trees, measurement of the tree heights, diameters, and stem curves directly from the point cloud data.

In the following, these three types of processing are further discussed.

15.5.1 DTM Processing

Removal of low points is an important preprocessing part of DTM and is usually done before ground classification. Low points seem to come below the ground surface, and their origin in forested area may be multiple reflections from trees or the ringing effect (too high return signal entering the receiver). Low points exist also with airborne 3D radar data (Hyyppä 1993).

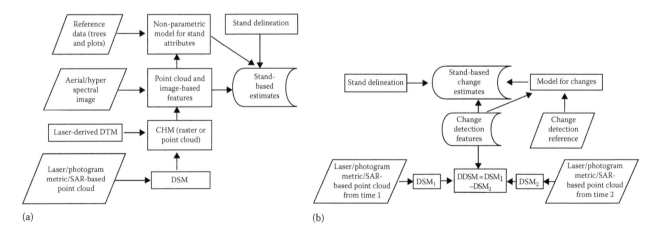

(a)   (b)

FIGURE 15.1   (a) Forest attribute estimation based on single-time point cloud and (b) use of bi-temporal point clouds for change detection.

Forested areas may also include buildings, and multiple reflections from windows and ground can cause low points. A single or group of points may cause an anomaly in the correct ground surface if it is not removed from the point cloud in the first step. A point can be defined as being low if all of its neighboring points in a search window are more than a predefined value higher than the point. One or several low-point removal filtering processes should be done before the DTM filtering (classification of DTM points).

The second step in DTM calculation is the initial DTM point cloud selection. This step is needed to detect building regions. For example, by selecting the lowest points with an 80 m-by-80 m window will remove all buildings less than 80 m in size. This step can also be done by data pyramids. The first and second steps are not always used, but the user should be careful in such cases.

For the final ground point classification, there are several DTM filtering techniques developed. Mathematic morphology is one applied technology. Using operators such as erosion, dilation, closing, and opening can produce DTM and DSM. Vosselman (2000) applied a maximum admissible height difference function for a defined distance. Points within the maximum height difference function were included as ground points. Progressive densification strategy starts with step 2 (initial DTM point cloud selection) and then iteratively increase the amount of accepted terrain points. Axelsson (2000) developed a progressive TIN densification method, which is implemented into TerraScan software. In TerraScan, laser point clouds are first classified to separate ground points from all other points. The program selects local low points on the ground and makes an initial triangulated model. New laser points are then added to the model iteratively, and the actual ground surface is then described more and more precisely. Maximum building size, iteration angle, and distance parameters determine which points are accepted. Kraus and Pfeifer (1998) developed a DTM algorithm for which laser points between terrain points and non-terrain points were distinguished using an iterative prediction of the DTM and weights attached to each laser point, depending on the vertical distance between the expected DTM level and the corresponding laser point. The method officially goes to category surface-based filtering, in which the starting point is that all given points belong to the terrain class and then iteratively remove the points that do not fit to surface model. In the beginning, the method did not use initial DTM, but in order to overcome the limitation in large building areas and in order to speed up the process, initial DTM and data pyramids are applied in hierarchical framework. The method is implemented in SCOP++ (Kraus and Otepka 2005). Additionally, the filtering can be based on segments, that is, segment-based filtering. Either object- or feature-based segmentation can be done, and then filtering is performed to each segment separately. Additionally, waveform and intensity can be used to assist in ground filtering.

A comparison of the filtering techniques used for DTM extraction can be found in a report on International Society for Photogrammetry and Remote Sensing (ISPRS) comparison of filters (Sithole and Vosselman 2004). Selection of the filtering strategy is not a simple process. In practice, the amount of interactive work determines the final quality of the product, but in FI, fully automated DTM calculation is preferred. Examples of commercial software that include DTM generation are REALM, TerraScan, and SCOP++. DTM quality indicators can also be directly calculated from the point cloud and waveform data. Such indicators include, for example, point density, point spacing, terrain slope, echo width, and estimate of the AGB estimated with lidar data.

15.5.2 DSM Processing and Canopy Height Model

Ideally, the first echoes over a forest region come from the surface model of the canopy, whereas the last echoes from the terrain model. The most frequently used method for the creation of a DSM is, therefore, to take the highest echo within a given neighborhood and interpolate the missing heights. Following the creation of the digital terrain (or elevation) model (DTM or digital elevation model (DEM)), a CHM can be calculated by subtracting the height of the ground from the DSM and presented in a raster or TIN height data format.

Airborne lidar measurements tend to underestimate tree height (Nelson et al. 1988; Hyyppä and Inkinen 1999; Lefsky et al. 2002; Rönnholm et al. 2004), and the same happens with 3D radars (Hyyppä 1993). The first echo return comes more often from the shoulder of the tree instead of the tree top. Although a laser pulse hits the top, the tree top may not be wide enough to reflect a recordable return signal. On the other hand, dense undervegetation causes overestimation in the DTM. For these reasons, the CHM is typically underestimated. Other factors affecting tree height measurement accuracy are scanning parameters, such as flying height, pulse density, pulse footprint, applied modeling algorithms, and scanner properties (e.g., sensitivity, field of view, zenith scan angle, and beam divergence), and structure and density of the tree crown (Holmgren 2003; Hopkinson et al. 2006). With deciduous forests, seasonal aspects have to be recognized.

15.5.3 Point Cloud Metrics

The prediction of stand variables is typically based mainly on point height metrics calculated from the ALS data. Nelson et al. (1988) divided features related to the height and density, which is the foundation of the area-based technology. Features such as percentiles calculated from a normalized point height distribution, mean point height, densities of the relative heights or percentiles, standard deviation, and coefficient of variation are generally used (Hyyppä and Hyyppä 1999; Næsset 2002). The percentiles are down to the top heights calculated from the vertical distribution of the point heights, that is, the percentile describes the height at which a certain number of cumulative point heights occur. Density-related features are calculated from the proportion of vegetation hits compared with all hits. A hit

TABLE 15.6 Typical Point Height Metrics Used in Forest Attribute Derivation

No.	Feature	Explanation
	Point Height Metrics	
1	meanH	Mean canopy height calculated as the arithmetic mean of the heights from the point cloud
2	stdH	Standard deviations of heights from the point cloud
3	P	Penetration calculated as a proportion of ground returns to total returns
4	COV	Coefficient of variation
5	H10	10th percentile of canopy height distribution
6	H20	20th percentile of canopy height distribution
7	H30	30th percentile of canopy height distribution
8	H40	40th percentile of canopy height distribution
9	H50	50th percentile of canopy height distribution
10	H60	60th percentile of canopy height distribution
11	H70	70th percentile of canopy height distribution
12	H80	80th percentile of canopy height distribution
13	H90	90th percentile of canopy height distribution
14	maxH	Maximum height
15	D10	10th canopy cover percentile computed as the proportion of returns below 10% of the total height
16	D20	20th canopy cover percentile computed as the proportion of returns below 20% of the total height
17	D30	30th canopy cover percentile computed as the proportion of returns below 30% of the total height
18	D40	40th canopy cover percentile computed as the proportion of returns below 40% of the total height
19	D50	50th canopy cover percentile computed as the proportion of returns below 50% of the total height
20	D60	60th canopy cover percentile computed as the proportion of returns below 60% of the total height
21	D70	70th canopy cover percentile computed as the proportion of returns below 70% of the total height
22	D80	80th canopy cover percentile computed as the proportion of returns below 80% of the total height
23	D90	90th canopy cover percentile computed as the proportion of returns below 90% of the total height

is seen as a vegetation hit from trees or bushes if it has been reflected from over some threshold limit above ground level. All the features are calculated separately for every echo type. The reason for this is that the sampling between echo types is somewhat different (Korpela et al. 2010). Table 15.6 gives a list of typical point height metrics used.

15.5.4 Approaches for Obtaining Forest Data from Point Clouds

Approaches aimed at obtaining forest and forestry data from point cloud data have been divided into two groups: (1) area-based approaches (ABAs) and (2) individual/single-tree detection approaches (ITDs).

ABA prediction of forest variables is based on the statistical dependency between the variables measured in the field and the predictor features derived from the ALS data. The sample unit in the ABA is most often a grid cell, the size of which depends on the size of the field-measured training plot. Stand-level FI results are aggregated by summing and weighting the grid-level predictions inside the stand. When using ITD techniques, individual trees are detected, and tree-level variables, such as height and volume, are measured or predicted from the ALS data, that is, the basic unit is an individual tree. Then, the stand-level FI results are aggregated by summing up the treewise data. The ABA does not make use of the neighborhood data of laser returns. On the other hand, ABAs are based on the height and density data acquired

by the ALS, which are highly correlated with the forest variables. Currently, the ABA is operationally applied in the Nordic countries when carrying out standwise FIs. Some 3 million hectare of Finnish forests is inventoried every year by applying ABA. White et al. (2013a) report on best practices for using the ABA in a forest management context.

ABA is based on accurate training plot–level data, which should represent the whole population and cover the variations in it as much as possible. The efficient selection of the training plot locations requires pre-knowledge of the inventory area. The statistical relation between the predictors and dependent variables to be defined is modeled using training data. The dependent variables are then predicted for all (other) grid cells without training data typically using nonparametric estimation techniques. If stand-level variables are needed, they are calculated by weighting the grid-level predictions inside the stand to the known stand delineation map.

One of the first tests with ABA in Finland was Suvanto et al. (2005). Regression models were developed using laser height metrics for diameter, height, stem number, basal area, and stem volume of 472 reference plots. The predicted accuracies were 9.5%, 5.3%, 18.1%, 8.3%, and 9.8%, respectively, at stand level. Current forest management planning inventories in Scandinavia require species-specific information for growth projections and simulated bucking. Tree species composition has also a major effect on forest value. Maltamo et al. (2006) added predictor features from aerial photographs and existing stand registers to ALS height metrics resulting in plot-level volume estimation accuracy from 13% to 16% depending on the predictors used. Similarly, Packalen and Maltamo (2007) used the k-MSN method to impute species-specific stand variables using ALS metrics and aerial photographs to the same data set as in Suvanto et al. (2005), and the species-specific volume estimates at the stand level were 62.3%, 28.1%, and 32.6% for deciduous Scots pine (*Pinus sylvestris* L.) and Norway spruce (*Piceaabies* L.), to were 62.3%, 28.1%, and 32.6% for deciduous, Scots pine (*Pinus sylvestris* L.) and Norway spruce (*Piceaabies* L.). Thus, there are limitations with current technology, and especially species-specific tree size (height and diameter) distribution information is needed. One possible way is to use more detailed data and use ITD-type processing.

In addition to NN and k-MSN methods, random forest (RF) classifier has also been applied in ABA (Yu et al. 2011). The RF is a nonparametric regression method in which the prediction is obtained by aggregating regression trees, each constructed using a different random sample of the training data and choosing splits of the trees from among the subsets of the available features, randomly chosen at each node. The samples that are not used in training are called "out-of-bag" observations. They can be used to estimate the feature's importance by randomly permutating out-of-bag data across one feature at a time and then estimating the increment in error due to this permutation. The greater the increment, the more important the feature.

Similar ABA approach can also be used to process point clouds provided by radar imagery. According to the first results obtained, the use of stereo SAR data in the prediction of plot-level forest variables is promising. Karjalainen et al. (2012) obtained a relative error (RMSE%) of 34.0% for stem volume prediction. For the other forest variables, that is, the mean basal area, mean diameter at breast height, and mean forest canopy height, the accuracies were 29.0%, 19.7%, and 14.0%, respectively, using RF as nonparametric estimation technique. Typically, such a high level of prediction accuracy cannot be obtained using satellite-borne remote sensing at the plot-level data in the boreal forest zone.

Since there are limitations in the ABA and user's need to get better species-specific tree size distribution data, individual tree approaches have been developed. The basic idea is to derive more detailed information of standing trees that are then used in the prediction of the forest attributes. Thus, area-based prediction can be done also using individual tree-based features, as originally proposed in Hyyppä (1999). That was demonstrated in Hyyppä et al. (2012) in which both individual tree-based and point height metrics were used as the inputs for the RF classifier. Individual tree-based features improved the ABA's accuracy significantly since they had very high correlation, for example, with the reference stem volume. When calculating the importance of the features, most of the individual tree-based features were among the best features confirming that individual tree-based features are applicable in ABA or stand-level inventory in general. When estimating plot-level mean height, the best laser-derived feature was the mean height derived by using the individual tree technique. When estimating dbh, the best laser-derived features were (1) mean canopy height, (2) penetration

to the ground, (3) mean tree height (derived from the extracted individual trees), and (4) mid percentiles. For the estimation of stem volume, the best laser-derived feature was the stem volume derived from extracted individual trees, followed by the basal area derived from extracted individual trees. It is possible to easily derive further laser point height metrics and individual tree-based features.

15.5.5 Tree Locating with ALS

Most of the current approaches for tree detection are based on finding trees from the CHM, which is calculated as a maximum of canopy height values within each raster cell. Thus, the CHM corresponds to the maximum canopy height of the first pulse data. Recently, other approaches have been proposed (Hyyppä et al. 2012) utilizing the canopy penetration capability of the last pulse returns with overlapping trees and correcting past information in this area (Hyyppä et al. 2008). When trees overlap, the surface model corresponding to the first pulse stays high, whereas with last pulse, even a small gap results in a drop in elevation, that is, the trees can be more readily discriminated. The first pulse works, when the whole laser beam penetrates the gap between the crowns, so that the drop is detectable after filtering. With last pulse, the drop in elevation is substantially larger, and the drop can be detected even with overlapping trees since the last pulse is more sensitive to lower canopy levels. The methodology and the applied automatic accuracy assessment are further demonstrated in Figure 15.2, which shows two raster models. The Fmax surface (first pulse surface) model corresponds to the commonly accepted way of finding trees. When comparing Lmin (last pulse surface) and Fmax models, it seems to be easier to discriminate trees from Lmin rather than from Fmax, even though visual processing of laser data in the forest is inferior to the best automated techniques.

Figure 15.3 depicts the percentage of correctly matched trees using four different raster models for tree location. The use of last pulse data gave a higher degree of discrimination between the trees than the use of first pulse data. The use of the raster corresponding to the minimum of last returns resulted in the highest discrimination between trees. An improvement of 6% in ITD is better than that obtained by increasing the pulse density from two to eight pulses per m² reported in Kaartinen and Hyyppä (2008). The improvement in tree detection increased when the density of the forest stand decreases. With the dbh class 5–10 cm, the last pulse resulted in 10% better detection of trees. The results confirm that there is also substantial information for tree detection in last pulse data. Currently, in raster-based processing, this information has been largely neglected. The obtained results would even suggest the use of last pulse data for detection, but we assume that a hybrid model utilizing both the first and last pulse data should be developed, even when processing is done at raster level. The advantages of first pulse data obviously include the lower number of commission errors and the high quality of tree separation when the crowns are not overlapping, whereas the advantage of last pulse is in the separation

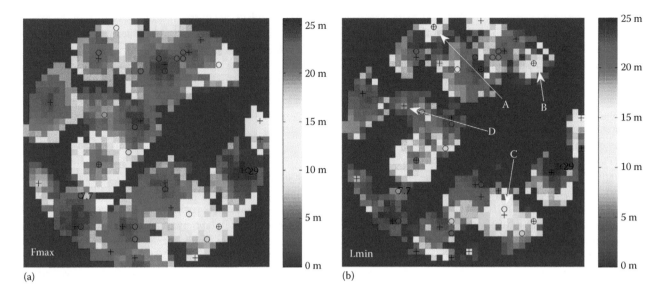

(a)   (b)

FIGURE 15.2 Using last pulse data in tree finding. Tree detection from first return data (a) and last return data (b) circle indicates the location of field measurements and plus lidar-detected location. (From Hyyppä, J. et al., *Remote Sens.*, 4, 1190, 2012.)

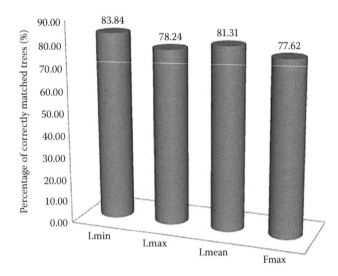

FIGURE 15.3 The percentage of correctly matched trees when tested with 5532 trees surveyed in the field using last pulses (Lmin) versus first pulses (Fmax). (From Hyyppä, J. et al., *Remote Sens.*, 4, 1190, 2012.)

of trees whose crowns overlap. A hybrid model, utilizing the advantages of both pulse types should be developed.

15.5.6 Individual Tree Height Derivation

In order to test the individual tree extraction methods using the same remote sensing data sets, the European Spatial Data Research Organization (EuroSDR) and the ISPRS initiated the Tree Extraction Project in 2005 to evaluate the quality, accuracy, and feasibility of automated tree extraction methods based on high-density laser scanner data and digital aerial images. The project was hosted by the Finnish

Geodetic Institute (FGI). Twelve partners from the United States, Canada, Norway, Sweden, Finland, Germany, Austria, Switzerland, Italy, Poland, and Taiwan participated in the test included in the Tree Extraction Project. The partners were requested to extract trees using the given test data sets. Another objective of the study was to find out how the pulse density impacts on individual tree extraction. The results were published in the project final report (Kaartinen and Hyyppä 2008) and in Kaartinen et al. (2012).

Tree height quality analysis showed that the variability of point density was negligible when compared to variability between the methods (Figure 15.4). With the best models, an RMSE of 60–80 cm was obtained for tree height. High-quality tree height estimates were obtained when using the models FOI, Metla, Texas, and FGI_VWS on the trees these methods had been detected. The results with the best automated models were significantly better than those attained when using the manual process. Both underestimation of tree height and standard deviation were decreased in general as the point density increased. The overestimation produced by the Model Norway in regard to tree height was due to the correction applied to the tree height in the preprocessing phase. The methods capable of finding more trees in the lower classes are obviously suffering; the uncertainty regarding the heights of the extracted tree in the lower levels is greater.

The measurement of tree height using TLS at the plot level has not been thoroughly studied because the visibility of treetops with TLS techniques can be questioned. However, there are past results with TLS showing that tree height is typically underestimated and that the magnitude of estimation error is typically of several meters. In Huang et al. (2011), a −0.26 m bias and a 0.76 m RMSE were reported for one plot (212 stems/ha, sparse stand) using the multi-scan approach. In Brolly and Kiraly (2009), a −0.27 m bias with a 1.82 m RMSE and a −2.37 m

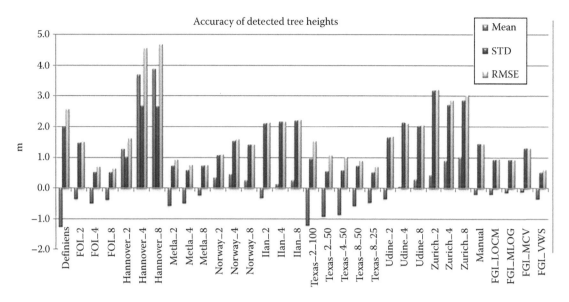

FIGURE 15.4 Tree height comparison. (From Kaartinen, H. et al., *Remote Sens.*, 4, 950, 2012.)

bias with a 3.25 m RMSE were reported for one more dense plot (753 stems/ha) using the single-scan approach. In Hopkinson et al. (2004), an approximate 1.5 m underestimation of tree heights was reported for two medium-density plots (465 and 661 stems/ha) using the multi-scan approach. Maas et al. (2008) depicted a −0.64 m bias and a 4.55 m RMSE for nine trees locating on four plots (212–410 stems/ha) using the single- and multi-scan approaches. Fleck et al. (2011) concluded a 2.41 m RMSE for 45 selected trees on one plot (392 stems/ha) using multi-scan data. The observation of tree tops from the TLS data is possible on sparse sample plots using many scans, as reported in Huang et al. (2011) and Fleck et al. (2011), but not in dense sample plots. Tree tops are most likely shadowed by other parts of the crown in the point cloud, but the use of well-visible trees in sparse plots could be actually used to calibrate ALS-based tree heights. In the ISPRS/EuroSDR Tree extraction test (Kaartinen and Hyyppä 2008), TLS was able to collect tree height information with the level of 10 cm, but very-high-density scanning was made (both as point cloud density and in the number of scans applied).

One of the advantages of using MLS for plot-level inventories lies in the fact that MLS can see many of the invisible tree tops for TLS. Since MLS platform moves all the time during the data acquisition, the gaps to tree tops are more likely visible with MLS than with TLS. This preliminary conclusion still requires detailed scientific studies to be confirmed.

15.5.7 Diameter Derivation

Since diameter cannot be directly measured from ALS, TLS, and MLS have been studied to provide accurate diameters at tree and plot levels. This kind of field data collection could also be used as a substitute for field-measured plot-level data in the ABA prediction of forest attributes.

The most popular processing method today for locating trees and determining diameter is to cut a slice of data from the original point cloud and to identify and model tree stems from this layer by point clustering or circle finding (Simonse et al. 2003; Aschoff and Spiecker 2004; Thies et al. 2004; Watt and Donoghue 2005; Maas et al. 2008; Brolly and Kiraly 2009; Tansey et al. 2009; Huang et al. 2011). This assumes that all trees present a clear stem at the same height at which the slice goes through the point cloud. This assumption is typically not valid in most mixed forests having branches at different heights, and nearby branches may be overlapped in the layer. A study of a mixed deciduous stand showed difficulties even in the manual stem detection in a TLS data layer (Hopkinson et al. 2004). Results from studies of different types of forests are highly variable, indicating the need for more research on these topics. In Liang et al. (2012b), the dbh estimation results are reported at the tree level from five plots having bias of 0.16 cm and the RMSE of 1.29 cm. In Lindberg (2012), the bias and RMSE of tree-level dbh estimates from six plots were 0.16 cm and 3.8 cm, respectively. Table 15.7

TABLE 15.7 Summary of the Plotwise dbh Estimation from the Single-Scan Methods

	Plot			Result	
	Number	Size	Density (stems/ha)	Bias dbh (cm)	RMSE dbh (cm)
Maas et al. (2008)	3	15 m radius	212–410	−0.67 to 1.58	1.80 to 3.25
Brolly and Kiraly (2009)[a]	1	30 m radius	753	−1.6 to 0.5	3.4 to 7.0
Liang and Hyyppä (2013)	5	10 m radius	605–1210	−0.18 to 0.76	0.74 to 2.41

[a] Three detection methods were discussed.

summarizes the accuracy of the plotwise dbh measurements using the single-scan approach. In practice, single-scan, multi-scan, and multi-single-scan techniques can be applied each having their own pros and cons.

In Liang et al. (2014c), an MLS system was tested, and its implications for FIs were discussed. The stem mapping accuracy was 87.5%; the root mean square errors of the dbh estimates and the location were 2.36 cm and 0.28 m, respectively. These results indicate that the MLS system has the potential to accurately map large forest plots, and further research on mapping accuracy and cost–benefit analyses is needed.

15.5.8 Stem Curve Derivation

The tree stem curve, or stem taper, depicts the tapering of the stem as a function of the height. The tree stem curve holds a significant position in forestry, as it is the key input needed in the harvest operation and used in various ecological studies. If the most important part of the stem curve can be determined, it is also possible to derive biomass and stem volume estimates from the trees without knowledge of conventional allometric models relating, for example, diameter and height information to volume and biomass. From that point of view, it is surprising that the noninvasive measurement of stem curves using TLS has not been intensively studied.

Pioneering work of TLS for stem curve includes nine pine trees studied in Henning and Radtke (2006), a spruce tree in Maas et al. (2008), and two trees, one pine and one spruce, in Liang et al. (2011). The RMSE of the stem curve measurements was 4.7 cm in Maas et al. (2008) utilizing single-scan data. In Liang et al. (2011), the RMSE of the stem curve estimation of the pine tree was 1.3 and 1.8 cm with the multi- and single-scan data, respectively, and the RMSE of the curve measurement of the spruce tree was 0.6 cm and 0.6 cm using the multi- and single-scan data, respectively.

The first detailed study on the plot-level automatic measurement of the stem curves of different species and different growth stages using TLS was reported in 2014 (Liang et al. 2014a). From 9 sample plots, 28 trees, 16 pines and 12 spruces, were selected. The plots were scanned utilizing the multi-scan approach. The trees were felled, and the stem curves were manually measured in the field. For comparison, the stem curve was also manually measured from the point cloud data. The stem curves were automatically measured with a mean bias of 0.15 cm and mean RMSE of 1.13 cm at the tree level. The highest diameters measured were between 50.6% and 74.5% of the total tree height, with a mean of 65.8% for pine trees and 61.0% for spruce trees. These results showed that TLS data and automated processing have the capability of accurately measuring stem curves of different species and different growth stages. Surprisingly, the automated processing gave clearly more diameter measurements at the upper part of the stems than with manual measurements from the same data. The difficulty of the manual measurement from point cloud data is that the stem edges are difficult to locate when the stem is partly blocked in the data by other branches.

15.6 Future Challenges

Remote sensing is today changing rapidly forestry inventory. For 100 years, FI was carried out by foresters; today, this task is strongly moving toward the direction which FIs are performed by remote sensing experts and forestry experts capable to handle remote sensing, 3D data, point cloud and image processing. In the following, we try to anticipate the cases where lidar and radar remote sensing steps more deeply into forestry practices.

15.6.1 Concept and Utility of the Lidar Plots

Ground-plot-like measures from airborne scanning lidar in many ways resemble field measures and are termed lidar plots. Ground plots remain invaluable for robust forest characterizations, enabling consistent and reliable measurement of attributes to support FI, mapping, monitoring, modeling, and science. National inventories in many jurisdictions are primarily based upon careful and systematic measurement of ground plots (Kangas and Maltamo 2006). Furthermore, applications that use remotely sensed data to produce forest attribute maps often require ground plot data for building models and validating outcomes. In many jurisdictions, ground plots remain costly to install and, as a result, are often limited in number and extent. For example, remote locations are difficult to access, such as some northern regions of Canada, further precluding the establishment of ground plots. Another example is locations with noncommercial forests that do not have the requisite economic drivers for maintaining up-to-date plot or inventory data sets. This dearth of ground measurements precludes the development of robust large-area FI, mapping, and monitoring applications. As an alternative, Wulder et al. (2012a) have proposed the concept of the lidar plot. Airborne scanning lidar data have been shown to offer attribute characterizations (especially height-related attributes) that are similar and in some cases better than ground measurements (Næsset 2007; Hyyppä et al. 2008).

The concept of the lidar plot is comparable to that of a ground plot with a fixed area. Lidar plots are an area-based summary of a lidar point cloud, whereby descriptive statistics or metrics are generated from the point cloud (e.g., percentiles, mean, and standard deviation), and these are used, with a sample of colocated ground measurements, to model FI attributes of interest such as mean height, dominant height, basal area, volume, and biomass (Næsset 2002). Thus, although the lidar plot concept requires some amount of traditional ground plots, it enables efficient propagation of the ground plot information over large areas, via airborne measurements. The stability of the empirical relationships between metrics and inventory attributes across many different forest types and structures can be attributed to the nature of the lidar data itself: in essence, the lidar metrics represent a detailed measurement of all surfaces within a canopy (foliage, branches, and stems). The lidar point clouds are generalized on a grid as well as vertically, creating a voxel of information that can be simplified. This voxel-based generalization can be implemented in freely available software (such as LASTools and FUSION) that provides unique metrics that in turn are used for model development. Thus, even when lidar data are collected at a lower hit density (i.e., 1 hit/m²) (Jakubowski et al. 2013), or when the vertical structure of the forest is complex (i.e., composed of multiple canopy strata, with a significant understory component) (Vastaranta et al. 2013), meaningful relationships to plot-level forest attributes may still be generated.

Lidar plots are typically square in shape, and their size is determined in concert with the size of the aforementioned ground plots used for model development. Typical plots sizes are 400 m² or 20 m-by-20 m (White et al. 2013a), and the plots must be large enough to contain sufficient lidar hits, to have a more uniform hit density (Næsset 2002), and to enable reasonable attribute estimates (McGaughey 2013). The full swath of the lidar transect (depending on instrument scan angle) may be tessellated into these fixed-area plots, followed by the generation of metrics and estimation of inventory attributes. Figure 15.5 provides a schematic of the lidar plot concept.

Note that opportunistically located ground plots (measured for lidar plot attribute modeling) and lidar transects can be used to improve large-area mapping and monitoring (e.g., Chen et al. 2012a; Magnussen and Wulder 2012; Mora et al. 2013) but may not be appropriate for statistically driven designs for large-area FIs (Wulder et al. 2012a). Chen et al. (2012a) used samples of lidar plots generated from a national collection of lidar transects (Wulder et al. 2012b) as calibration and validation data to support geometric-optical modeling of mean, dominant, and Lorey's height using Landsat imagery. Estimates of vertical forest structure are critical for FI and reporting. Heights were

modeled over the area of a single Landsat scene (185 km-by-185 km) at a 25 m resolution with average estimation errors (RMSE) of 4.9, 4.1, and 4.7 m for mean height, dominant height, and Lorey's height respectively. In this study, the lidar plot data were useful for model development, identification of spectral endmembers required for the geometric-optical model, and parameterization of the model's tree variables. Using a different modeling approach and different lidar transects in combination with higher resolution optical imagery (QuickBird), Chen et al. (2012b) obtained an average error (RMSE) of 3.3 m in the estimation of plot-level mean heights. Mora et al. (2013a) used data from lidar plots (Wulder et al. 2012b) with samples of very high spatial resolution (VHSR; <1 m) imagery to achieve estimation errors (RMSE) of 2.3 m for mean stand height.

Obviously, there is promise in the use of lidar plot data to enable modeling of forest structural attributes across large areas. What is emerging from the literature is that only modest gains in error reduction are possible when using lidar plots with increasingly higher resolution optical imagery (Mora et al. 2013b). These modest gains in accuracy are offset by the increased level of effort and cost associated with using the higher-resolution imagery. Landsat data are free and readily accessible; with each scene covering a markedly larger area (185 km-by-185 km) relative to what is typical for VHSR data (e.g., 10 km-by-10 km). Thus, synergies between lidar plots and Landsat data offer particular advantages for nations such as Canada, with more that 600 million ha of forested ecosystems that require monitoring and reporting information to be collected in a systematic and transparent manner. Lidar plots have demonstrated a unique

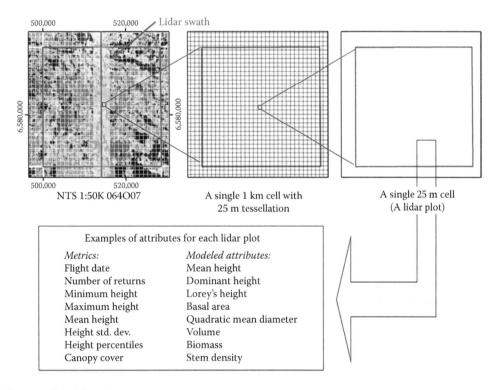

FIGURE 15.5 Schematic of the lidar plot concept.

and valuable role in supporting large-area mapping, monitoring, and modeling for boreal forest ecosystems. The utility has been demonstrated for science- and management-related information needs across broad range of applications. Further transect installation and application over different forest ecosystems (such as tropical) and addressing different management or science questions remain to be undertaken.

15.6.2 Improving Large-Area Mapping of Forest Attributes Using Country-Wide Laser Scanning

ALS/lidar-based FI is now operational/commercial in Scandinavia, Baltic countries, Spain, Switzerland, the United States, Canada, Australia, and New Zealand. An increasing number of countries are applying ALS for national/statewide elevation modeling (e.g., The Netherlands, Switzerland, Finland, Sweden, Austria, Germany, and the United States).

Both in Finland and Sweden, the countrywide lidar data are applied into NFI concept by SLU in Sweden and by remote sensing companies in Finland. The national data consist of 0.5–1.0 return/m^2 and has a maximum scan angle of 20°. The challenges when adopting this technology to national context are the following:

- The scannings have been made during different seasons.
- The scannings have been made with different laser scanners.
- There is incidence angle effect in returns when applying.
- The scannings have been made in different years.

In Sweden, there exist about 30,000 National Forestry Inventory (NFI) plots that can be used to train ALS data sets. Half of Sweden is productive forest land. In Finland, NFI is done by Finnish Forest Research Institute, but their sample plots are not available for industry who is implementing new NFI based on countrywide ALS data. Due to the earlier-mentioned challenges, and due to the fact that NFI aims at low systematic error at large area level, there are limitations where the new NFI developments go. In Finland, the need to have improved NFI stems from industry wanting to know more accurately where the wood resources are.

15.6.3 National Forest Inventory Based on Multitemporal ALS Data

For integrating ALS data into operational FIs and thus for monitoring applications, a regular basis of data acquisition is imperative. Mainly due to economic reasons, one of the most important open questions in the community is, therefore, if ALS data will be available in a regular basis in the future. In order to minimize the costs of ALS data for an individual user or per km^2, several approaches have been tested during the last years. This includes on the one hand the increased technical capabilities of ALS sensors leading to higher pulse repetition frequencies, increased flying heights, and swath widths resulting in lower flying costs. On the other hand, there is also a trend

of ALS campaigns financed by a pool of different users. One of these later examples can be observed at the federal district of Vorarlberg in Austria where after 6 years, a complete ALS reacquisition took place in 2011 for the entire area (~3300 km^2). Multi-epoch ALS allows also new methods for analyzing the development of the forest (e.g., change detection) (e.g., Hyyppä et al. 2003; Yu et al. 2004, 2006; Næsset and Nelson 2007; Hopkinson et al. 2008).

For large area, multi-epoch ALS data acquisition normally varies in terms of ALS sensor on various platforms, acquisition times, and flying properties (e.g., flying height, scan angles). These facts increase the requirements on the applied algorithms for assessing forest parameters.

In general, it is important that for the calculation of the topographic models (i.e., DTM, DSM) and ALS metrics from multi-epoch ALS data, comparable methods will be applied. Therefore, it is strongly recommended that the original 3D ALS point clouds are used as input for current and future analyses.

For the study area Vorarlberg, the DSM was calculated based on a land cover–dependent approach (Hollaus et al. 2010), which is robust against varying point densities, acquisition times, and tree species. This approach uses the strengths of different algorithms for generating the final DSM by using surface roughness information to combine two DSMs, which are calculated based on (1) the highest echo within a raster cell and (2) moving least squares interpolation with a plane as functional model (i.e., a tilted regression plane is fitted through the k-nearest neighbors).

One of the most important error sources are originating from an insufficient geocoding. The experiences in Austria have shown that height differences of stable objects (e.g., roof planes, streets) between the multi-epoch surface models (i.e., DTMs, DSMs) originating from errors in the georeferencing of the individual ALS data sets have to be minimized to ensure that the height differences can be connected to changes of tree heights and consequently to growing stock or biomass changes. To minimize these height differences, least square matching of the DSMs from both acquisition times was applied successfully, whereas a 3D shift was sufficient. It could be demonstrated that for streets, roofs, and bare soils, the mean height differences of 0.17 m could be reduced to 0.07 m (Hollaus et al. 2013).

For analyzing multi-epoch ALS data, the height differences are the most important measure for forest monitoring applications. Apart from the described minimization of height differences from stable objects, one can use one reference DTM due to the assumption that the DTM changes within the forest will be negligible. For the study area in Austria, the DTM derived from the second ALS data set characterized with higher point density compared to the first one was used as reference for detecting forest height changes and biomass changes respectively.

For the district-wide growing stock/biomass estimation, the rasterized nDSM (= DSM-DTM) was used as input for the semiempirical regression model from Hollaus et al. (2009). This method assumes a linear relationship between the growing stock and the ALS-derived canopy volume, stratified according to four canopy height classes to account for height-dependent differences in canopy structure. Each data set was calibrated with the corresponding FI data, and the derived growing stock maps were compared. Finally, the changes of the growing stock maps are detected, whereas the changes are split into exploitation and forest growth. To consider small differences in tree crown representation within each different ALS data set, morphologic operations (i.e., open/close) and a minimum mapping area of 10 m² are applied. Finally, the growing stock change map can be limited to the determined forest area, for example, fulfilling the criterions of the Austrian forest definition (Eysn et al. 2012).

The validation of the growing stock change has shown high agreement between the changes calculated from the sample plot–based FI data (+43.0 m³/ha) and those derived from the ALS-derived growing stock change map (42.5 m³/ha) and shows the high potential of ALS data for integrating them into operational FIs.

15.6.4 Improving Large-Area Mapping of Forest Attributes Using Radargrammetry and Interferometry

Recent experiences with ALS (Næsset 2002; Hyyppä et al. 2008; Vastaranta 2012; Vastaranta et al. 2012) and digital stereo-photogrammetry (Nurminen et al. 2013; White et al. 2013a; Vastaranta et al. 2014) have shown that precise forest biomass estimations can be achieved using these airborne techniques. The foundation of the high biomass estimation accuracy is their ability to measure the forest canopy height and density. On the other hand, majority of satellite data-based forest mapping techniques, until now, have used only the intensity information (reflectance values or SAR backscattering coefficient), that is, 2D information, in estimating forest biomass. Even though in some experiments a good agreement with 2D information and forest biomass has been obtained, the way forward in satellite-based forest mapping appears to be the use of 3D techniques. This chapter enlightens the recent advances in 3D techniques related to large-area forest resources mapping and SAR satellite data. In the future, these techniques can be used in generating forest attribute maps from ground plots measured for NFIs.

In the past years, commercial SAR satellite data have undergone remarkable progress in terms of spatial resolution, geolocation accuracy, and data availability—mainly thanks to the X-band SAR satellite systems of TerraSAR-X, TanDEM-X, and COSMO-SkyMed. Consequently, there has been a growing interest of using aforementioned satellites for 3D forest mapping also. In principle, there are two main techniques to extract 3D/elevation data from satellite SAR images: (1) InSAR and (2) radargrammetry.

InSAR is based on the phase differences of two or more SAR data acquisitions with an appropriate geometric baseline. When the temporal baseline of data acquisitions is close to zero, InSAR data are particularly good for the creation of DEMs, even in the forested areas. In the case of intensity information, X-band is typically considered useless for forest resources mapping

(Holopainen et al. 2010). But in the case of X-band-derived DSMs, extracted elevation values are known to be close to the top of the forest canopy. Therefore, if an accurate DTM exists, derived by other means, it is straightforward to create CHMs that contain forest biomass-related information. The earliest work in 3D techniques was carried out with NASA/DLR SRTM data. The capability to map forest resources from SRTM DSMs was demonstrated, for example, by Kellndorfer et al. (2004) and Solberg et al. (2010). The SRTM mission lasted only 11 days, and it had limited global coverage; therefore, suitable data for InSAR techniques were not available before the TDM by the DLR. The TDM consists of two identical SAR satellites flying in a formation to create InSAR data with temporal baseline virtually close to zero. TanDEM-X data are very interesting from the forest resources mapping point of view. First scientific results have already been published (Askne et al. 2013; Solberg et al. 2013), and there is ongoing EU project Advanced_SAR in this field (Figure 15.6).

Radargrammetry is an alternative approach to extract elevation data from SAR data. Compared to InSAR, radargrammetry does not use phase information, but stereoscopic measurement similar to photogrammetric forward intersection. The main advantages of the radargrammetry method compared to InSAR are that (1) elevation values derived using radargrammetry are absolute values, that is, measured coordinates are directly in the 3D ground coordinate system; (2) more versatile SAR data sources can be used, and there is no restriction of simultaneous image acquisition as in the InSAR case. On the other hand, the main disadvantage of radargrammetry is that elevation models typically contain no-data values and possibly some gross errors, in areas where automatic image matching fails. Nevertheless, DSMs created using X-band radargrammetry are similar to X-band InSAR DSMs. The use of radargrammetric 3D models in forest resources mapping is a fairly new concept, and earliest studies were carried out using Radarsat data by Chen et al. (2007). Thanks to TerraSAR-X SAR data more detailed radargrammetric surface models can be produced compared to earlier SAR satellite systems (Perko et al. 2011). Recently, the use of TerraSAR-X radargrammetric DSMs in forest biomass mapping has been successfully demonstrated in Finland and Norway (Karjalainen et al. 2012; Vastaranta et al. 2012; Solberg et al. 2014; Vastaranta et al. 2014) (Figure 15.7).

Based on the recent scientific studies in 3D SAR techniques and forest resources mapping, there appears to be potential in these techniques for deriving detailed forest attribute maps over large areas with good temporal resolution. Even though ALS and aerial stereo-photogrammetry provide more accurate estimates for forest attributes compared to 3D SAR techniques, there might be a demand for SAR data especially for large-area mapping and especially in the monitoring of changes in the forest structure. The advantage of 3D SAR techniques compared to 2D estimation techniques is that the data derived using 3D techniques can be easily integrated to existing forest attribute inventory processes (see Section 15.5).

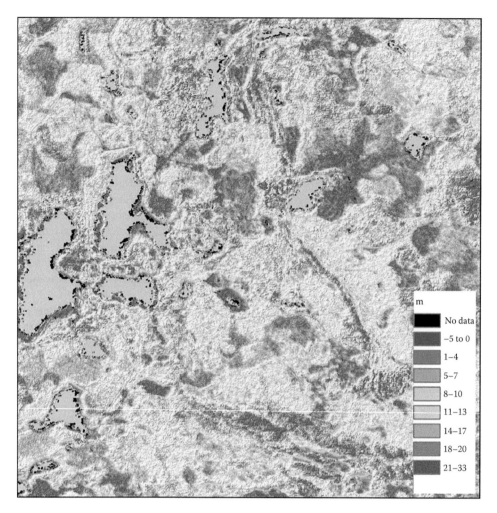

m	
▇	No data
▇	−5 to 0
▇	1–4
▇	5–7
▇	8–10
▇	11–13
▇	14–17
▇	18–20
▇	21–33

FIGURE 15.6 Forest canopy height model (CHM) extracted from TanDEM-X InSAR data and digital terrain model from ALS. (Original data © DLR, 2013 and NLS, processed by Kirsi Karila/FGI.)

FIGURE 15.7 Radargrammetric elevation model in Southern Finland. (Original data © DLR, processed by Mika Karjalainen/FGI.)

15.6.5 Explanatory Power of 3D Remote Sensing Techniques in Forest Attribute Estimation

The currently applied techniques in GMES/Copernicus are based on 2D techniques that are not able to provide reliable information related to biomass and wood volume. Scientific papers have shown that canopy height information derived from remote sensing data is the best feature for retrieval of stem volume and basal areas (Hyyppä and Hyyppä 1999; Hyyppä et al. 2000)—and also to AGB, which is highly correlated with standing volume. In Hyyppä et al. (2000), the accuracy of different satellite data for FI was studied, providing insights into the explanatory power and information content of several remote sensing data sources on the retrieval of stem volume, basal area, and mean tree height. The results showed relatively low accuracy of all 2D techniques, such as Finnish NFI data (based on Landsat

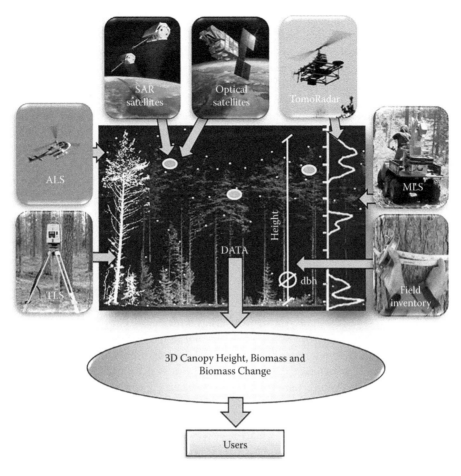

FIGURE 15.8 Comparison and synergies of 3D remote sensing techniques for forestry.

TM data), SAR intensity images, for forest attribute retrieval. Superior performance of LS giving the CHM over all optical 2D techniques was demonstrated in Hyyppä and Hyyppä (1999). By using single laser-based predictor, namely, mean tree height, provided 13.5% error in standwise stem volume estimation, whereas by using imaging spectrometer, a corresponding error of 27% was obtained using six predictors in the model.

Currently, FGI and Helsinki University are repeating the work of Hyyppä et al. (2000) from the 3D perspective: major source for 3D forest data will be compared in terms of accuracy. Also synergies between the data sets are studied. One of the instruments in the comparison is a new profiling ranging radar Tomoradar of the FGI, which is equivalent to Hallikainen et al. (1993). That can be used to calibrate the penetration of space-borne SAR to forests and for comparison with ALS full-waveform technologies (Figure 15.8).

15.6.6 Biomass Mission

Biomass is ESA Explorer 7 satellite aiming to take measurements of forest biomass to assess terrestrial carbon stocks and fluxes. The satellite employs a P-band synthetic aperture

polarimetric radar operating at 435 MHz. Maps of forest biomass and canopy height are aimed to obtain at a resolution of 200 m. The mission will also have an experimental tomographic phase to provide 3D views of forests. The original launch is scheduled for 2020. Sun-synchronous orbit at an altitude of 660 km is planned. Mission life of 5 years is expected, and the tomographic phase is 3 months. Fully polarimetric P-band response is obtained.

Currently, the use of Biomass in European and North American forests is not straightforward. Data collection in these areas is perhaps impossible because of conflicting frequency transmissions by U.S. military radars. The frequencies between 420 and 450 MHz are used for ballistic missile warning and space surveillance network.

How does P-band response differ from those more common L-, C-, and X-band missions? For example, the shorter wavelengths at X-band are attenuated by surface scattering at the top of the canopy by foliage and small branches, and in boreal forests, the ground response is obtained only from canopy gaps. The C-band response is caused by surface scattering at the top of the canopy as well as some volume scattering from the canopy. A small fraction of the energy reaches the ground in addition to energy coming from the canopy gaps. In boreal forest zone, the amount of canopy caps is substantial. L-band microwave energy penetrates farther into the canopy, where volume scattering among the leaves, stems, branches, and trunk causes also depolarization. Large number of pulses may reach the ground, where surface scattering from the soil-vegetation boundary layer may take place. P-band SAR would afford the greatest penetration through the vegetation, and the signal mainly reflects from larger stems and the soil surface (Jensen 2000). Hence, the difference of the X- and P-band-derived elevation model correlates with canopy height, but it is not that accurate as that obtainable from X-band and LS DTM. Advantage of the Biomass mission includes that P-band backscatter correlates with biomass, and the saturation level is much higher than with C- and X-bands. Use of X-, C-, L-, and lidar DTM results in better canopy probing that ever in all forest regions.

15.6.7 Crowdsourcing in Forest Inventory: National Forest Inventories by Commons

The mapping task has been for centuries performed by state organization, and the mapping has been, therefore, extremely centralized. This work has been done by trained staff typically having background in the field of surveying. Since early 2000, it has been possible to map the surrounding by ordinary, non-skilled citizens having GNSS receivers, cameras, and smartphones. The collection of geospatial user-created information is today called by many different terms such as crowdsourcing, collaboratively contributed geographic information, web-based public participation GIS, collaborative mapping, web mapping 2.0, neogeography, WikiMapping, and volunteered geographic information. More commonly, crowdsourcing is understood as geospatial data collection of voluntary citizens who are untrained in the disciplines of geography, cartography, or related fields. Short review of crowdsourcing can be found in Heipke (2010) and Fritz et al. (2009).

In the field of forestry, crowdsourcing has been used to assess the condition of city trees. For example, PhillyTreeMap is a web-based application that allows citizen to input tree information of city forests; today, over 144,000 Philadelphian trees are stored in the database.

Field reference data are conventionally collected at the sample plot level by means of manual measurements, which are both labor intensive and time consuming. Because of the high costs and laboriousness, the number of tree attributes collected is limited. In practice, some of the most important tree attributes are even not measured or sampled. Automated and more cost-effective techniques are needed to provide plotwise field inventory data. Recently, TLS has been shown to be a promising solution for forest-related studies. There is huge lack of open data in the field of plot-level or tree-level data to calibrate the ALS estimates. Today's challenge is the creation of large-area forest resource maps with small costs and as high accuracy as possible for various purposes. Since plotwise field data do not openly exist, and

TLS processing at country level produces significant costs, alternative solution are also developed. Use of crowdsourcing to get field reference for large-area FI has not seriously studied earlier.

In Hyyppä et al. (2014), a new, low-cost inventory schema based on open-access ALS data, individual-tree-based feature extraction, and crowdsourcing as field data collection technique was presented. It demonstrated the advantage of using locally collected field data in improving the quality of the estimates (RMSE) using crowdsourcing and 3D game engine to interactively calibrate ALS tree maps and locate persons for field surveys. The crowdsourcing concept was based on ITD techniques, where individual trees are detected, and tree-level variables, such as height and volume, are measured or predicted from the ALS data, that is, the basic unit is an individual tree. Advantage of the ITD approach over ABA in crowdsourcing concept includes that smaller amount of reference trees is needed compared to that in ABA to get reasonable accuracy, and it is easier of the landowners to measure physically well-established parameters, that is, diameter of the tree (cm) instead of basal area (m²/ha). The disadvantage of the ITD concept, in which suppressed trees are not found, can be overcome by the use of crowdsourcing measuring diameters for the missing trees. Thus, there is synergy with the use of ITD and crowdsourcing. The proposed crowdsourcing concept of Hyyppä et al. (2014) was the following. There are ¾ million forest owners alone in Finland (14% of the population), many of them visit their forests regularly or live close to their forests, and, thus, personal technologies that could be used to collect objectively non-biased reference information could be applicable to collect necessary field data for ALS inventories. For example, if 0.1% of the forest owners in Finland would measure 20 trees of their own forests that would lead to 15,000 of measured trees, which may result on accurate, non-biased countrywide biomass/volume map of unique accuracy. Research has shown that random error of the estimates can be easily solved by having several hundreds of reference plots, but in Hyyppä et al. (2014), local reference individual tree data were used to calibrate ALS data into non-biased estimates. Since GNSS accuracy in forest is not enough for plot- or individual tree-level inventory, a GNSS-assisted, 3D game engine–based approach was developed. First, the GPS location provided by the mobile phone was applied, and then the user uses the CHM generated with ALS data to visually locate themselves. The output of the CHM is visualized in the 3D game engine working in smartphone, and user can select the tree groups or individual trees from CHM to which they like to give field reference. If the CHM and ITD process indicated one tree that corresponds to several trees in the field, the landowner records all trees and their attributes, especially the diameter breast heights and tree species. For suitable plots, he preferably records corresponding information of all trees. Also, if there are omissions in the ITD process that are not part of the any tree group is recorded. Consequently, the landowner finds and locates trees with the help of the CHM and 3D game engine interface, and measures diameters of trees or stem curve of the tree close to ground with the tape. This information is used as local field data together with existing plot- and tree-level data. ITD approach is applied. Nonparametric RF classifier is applied to the link between the field reference and ALS features.

15.6.8 Toward Personal Laser Scanning in Forestry

Personal laser scanning (PLS) is an emerging technology, Hyyppä et al. (2013). The idea first appeared as a backpack-type MLS system, where the scanning and positioning systems were on the operator's back rather than on a vehicle platform. The first system prototype weighted about 30 kg (Kukko et al. 2012). The weight and size of the system limited its usability, for example, in open area for geomorphological terrain modeling (Wang et al. 2013). Last 5 years have witnessed rapid progresses in sensor miniaturization. A new system has been developed and built at the FGI in 2013, with new scanner and multi-constellation navigation systems. The new system weights approximately 10 kg.

The main advantage of PLS lies in its easy movement in various terrain conditions and the fast data collection. By providing professional quality scanning and navigation systems, the collected data can document objects in detail and high precision. PLS has the potential to improve the mapping efficiency compared with conventional field measurement and compensate the limitations of other LS techniques, for example, the TLS is not easy to transfer from site to site, and the MLS has to be utilized in areas where the terrain conditions are easy for the vehicle movement. These characters are very attractive for forest mapping and ecosystem services.

In Liang et al. (2014b), a professional-quality PLS for collecting tree attributes was demonstrated for the first time. The applied 10 kg FGI system consists of a multi-constellation navigation system and an ultra-high-speed phase-shift laser scanner mounted on a rigid baseplate as for a single sensor block. In the data acquisition, the system was tilted by 20° to record the vertical breaklines of tree stems. The forest area utilized in this pilot test of PLS system was 0.2 ha in size. The tree stem detection accuracy was 82.6%; the root mean square errors of the estimations of the tree diameter at breast height and the tree location were 5.06 cm and 0.38 m, respectively. The results showed the first potential of the PLS system in mapping large forest plots. Further research on mapping accuracy and cost–benefit analyses is needed. In addition to collecting tree- and plot-level data for FI, other possible applications in forest ecosystem services include the use of the system in the mapping of canopy gaps, LAI measurements from large areas, documentation and visualization of forest routes feasible for recreation, hiking, berry, and mushroom picking, and for harvester operations.

15.6.9 Toward Mobile-Phone-Embedded Laser Scanning

In Jaakkola et al. (2014), we demonstrated the feasibility of using the Microsoft Kinect depth sensor for tree stem measurements and reconstruction for the purpose of FI. Field reference data in FI are conventionally collected at tree and sample plot. A Kinect sensor capable of capturing several million data points per second from distances up to 4 m is a powerful new tool, and this chapter provides the first example of how it can be used in forestry. The Kinect depth sensor measures the distance to objects in the surrounding environment by transmitting near-infrared light and using structured light for point cloud generation. In the present research, about 100 reference stem diameter measurements were made with tape and caliper. Color (i.e., RGB) and range images acquired by a Kinect system were processed and used to extract tree diameter measurements for the reference tree stems. Kinect-derived tree diameters agreed with the caliper measurements to 2.50 cm (RMSE) and 10% (RMSE%) and with tape measurements to 1.90 cm and 7.3%. The stem curve from the ground to the diameter breast height agreed with a bias of 0.7 cm and random error of 0.8 cm with respect to the reference trunk. As a highly portable and inexpensive system, Kinect provides an easy way to collect tree stem diameter and stem curve information vital to FI. We see Kinect or similar inexpensive instruments (e.g., less than 100€) as a competing technology to TLS and conventional fieldwork using calipers. Measurements made using Kinect could also be acquired using a crowdsourcing approach as a complement to National ALS-based FI. However, further work to automate data processing of Kinect depth sensor data is needed. Google, among others, is already developing Kinect-type sensor to mobile phones (Google 2014).

15.6.10 Toward UAV-Based Laser Scanning

Gordon Petrie made his review (2013) in Geoinformatics about UAV and LS, and he summarized it by saying "Until now, airborne laser scanners from the mainstream system suppliers have not been operated from lightweight UAVs on a commercial basis, mainly due to the size and weight of the scanner systems and the limited payload of these very lightweight aircraft. However, currently, considerable efforts are being made to develop laser scanners that are suitable for use on these aircraft. These are utilizing the technologies that are already being employed in mobile mapping systems, terrestrial (ground-based) laser scanning and autonomous (driverless) vehicles." General review of very small UAV technology for photogrammetry and remote sensing can be found from Colomina and Molina (2014), but in the field of mini-UAV LS, the reference of Petrie (2013) is recommended reading.

Mini-UAV-based ALS data collection has been possible since Jaakkola et al. (2010). Mini-UAVs (<20 kg) have been previously used for mapping purposes using, for example, aerial images. Zhao et al. (2006) depicted a remote-controlled helicopter supplied with navigation sensors, namely, GPS, and a laser range finder. In Jaakkola et al. (2010), the first mini-UAV (FGI Sensei) including the laser scanner, intensity recording, spectrometer, thermal camera, and conventional digital camera was depicted. Laser scanners in the Sensei included an Ibeo Lux and a Sick LMS151 profile lasers. The Ibeo Lux measures points (and echo width) from four different layers simultaneously and is theoretically able to measure up to 38,000 points/s. Sick LMS151 is also able to measure two returning signals and the intensity from each laser pulse. The optical sensors were CCD camera (AVT Pike F-421C having a 2048 × 2048 pixel), Specim V10H spectrometer (spatial res. of 0.067° and a spectral res. of 8.5 nm) at 397–1086 nm, and thermal camera of Flir Photon 320 with a 324 × 256 pixel resolution and a 50° × 38° field of view.

It should be pointed out that this kind of system would be feasible for integrating several mapping concepts using simultaneous measurements done with a single system. Thus, for research purposes, in addition to point cloud data giving the geometry of the objects, simultaneously taken image data including overlapping images, intensity of laser backscatter, spectrometer, and thermal camera would provide even better data source for new algorithms and concept development, especially when recorded in a multitemporal way. That is an area where mini-UAV-based LS is extremely feasible; the use of conventional ALS results in high costs of data and lack of frequently enough collected data.

Mini-UAV LS is also feasible for small-area surveys, for example, collecting airborne data from areas of several hectares. In future, corridor-type applications done by UAV LS can provide cost-effective solution and new markets (BBC 2014).

Mini-UAVs taking images are already in operation, for example, in part of North America. They are especially useful when highly detailed imagery is required over a relatively small area, and in applications such as pre- and post-harvest surveys, compliance monitoring, habitat surveys, establishment, and regeneration surveys. Mini-UAVs are helpful in fire management by monitoring in-progress fires, mapping green-tree retention post-fire, and identifying hotspots with an infrared camera (Launchbury 2014).

15.6.11 Precision Forestry, the Future of Forest Inventory

Currently, the FI is about 2–3B€ annual market, which is mainly based on fieldwork. Inventory is increasingly shifting into using ALS, as has already occurred in many countries, Scandinavia, Austria, and Canada leading this development. Boreal and mountainous forests are easier to be measured due to smaller biodiversity (smaller number of species) and more sparse forest structure. The ALS part can also move to the use of 3D techniques of space-borne imagery or photogrammetry (beyond the scope of this chapter). There is also a need to get tree-level information, in which Japan is one of the leading countries to show example. Japan has a long history of approximately 10 million hectares of conifer plantations under SFM. Timber has been used in important cultural heritage sites since the Asuka and Nara Periods (AD 607–793), including in the oldest existing wooden buildings, castles, temples,

and shrines, and forms an important part of the history of the Japanese Imperial Family, which is one of the longest in the world. Traditional Japanese culture is based on wood and the country's rich natural heritage. The forest survey system in Japan has been taking aerial photographs and recording airborne data every 5 years since 1947. High-resolution optical images and GIS are widely used for practical forestry. However, forest officers use only remote sensing data overlaid with GIS forest polygons and assess boundaries, forest condition, cutting area, and damage area by image interpretation, because the forest resource information extracted from image analysis alone is insufficient. Today, forest officers and landowners require more precise information at the individual tree level on the amount of plantation resource harvested for timber, biomass, and clean energy (Katoh and Gougeon 2012). At present, the coverage of ALS is expanding. Further application of the ALS approach is needed to estimate the number of large trees of each species available for use in traditional wooden buildings (Figure 15.9).

Detailed and up-to-date information is a necessity for implementing sustainable forest resource management practices. To acquire this information, forest companies and governmental organizations are using point cloud–based, forest-inventory methodologies.

When applying the ideas of precision forestry concept, we can reach several economic and ecological benefits (Holopainen et al. 2010):

1. Better integration of forest planning and forest operation planning—with the accurate information from forests, the operation planning can be done with FI data-saving expenses.
2. Measurement of the quality of wood. The amount and size of branches measured with TLS and hopefully with MLS on a logging machine allow more detailed wood quality measurements. Also, the valuable information on the length of non-branched stem can be determined.
3. When FI data are more accurate and updated, the storages of forest industry can optimally move from real storages to forests. Wood acquisition can be planned in higher detail not possible before.
4. When FI gives accurate data to forest operation planning, the bucking of trees can be predicted and performed in a more optimal way significantly affecting the economy of the forest owners.
5. Precision forestry allows easily certification of wood origin, since the location of every stem can be recorded and stored to maps.
6. The value of forest can be more reliably determined.
7. The logistics chain from forests to value-added end-products can be planned in higher detail.

Laser scanners and radar have the potential of revolutionized measurement of vegetation canopy structure.

FIGURE 15.9 Increased need for ALS at the individual tree level in SFM. Left: Akasawa Forest Reserve, old-growth cypress (*Chamaecyparis obtusa*) forest over 350 years under SFM for construction timber for Ise Jingu Shrine in Nagano, Japan. Center: Cutting ceremony for replacing the timber in Ise Jingu Shrine; note use of the traditional axe. Right: Ise Jingu Shinto Shrine in Mie, Japan, which uses 10,000 large cypress in 14 wooden buildings, and the timber of which has been replaced every 20 years since AD 690.

15.7 Summary

In this chapter, we presented some highlights of 3D information collection and processing from lidar and radar and turning that information into valuable forest informatics. We really see that the processing of 3D data will be more and more based on similar tools that are currently available in the lidar community. Some of the most exciting developments anticipated are summarized as follows:

The concept of the lidar plot is comparable to that of a ground plot with a fixed area. Lidar plots are an area-based summary of a lidar point cloud, whereby descriptive statistics or metrics are generated from the point cloud (e.g., percentiles, mean, and standard deviation), and these are used, with a sample of colocated ground measurements, to model FI attributes of interest such as mean height, dominant height, basal area, volume, and biomass.

Both in Finland and in Sweden, the countrywide lidar data are applied into NFI concept. The national data consist of 0.5–1.0 return/m² and have a maximum scan angle of 20°. The challenges of adopting this technology to national context are that the scans have been made during different seasons, the scans have been made with different laser scanners, there is incidence angle effect in returns when applying, and the scannings have been made in different years.

For large-area multi-epoch lidar data acquisition, normally different lidar sensor on various platforms, acquisition times, and flying properties (e.g., flying height, scan angles) are used. These facts increase the requirements on the applied algorithms for assessing forest parameters. Therefore, it is strongly recommended that the original 3D lidar point clouds are used as input for current and future analyses.

Based on the recent scientific studies in 3D SAR techniques and forest resources mapping, there appears to be potential in these techniques for deriving detailed forest attribute maps over large areas with good temporal resolution. Even though ALS and aerial stereo-photogrammetry provide more accurate estimates for forest attributes compared to 3D SAR techniques, there might be a demand for 3D SAR data especially for large-area mapping and especially in the monitoring of changes in the forest structure. The advantage of 3D SAR techniques compared to 2D estimation techniques is that the data derived using 3D techniques can be easily integrated to existing forest attribute inventory processes.

Biomass is ESA Explorer 7 satellite aiming to take measurements of forest biomass to assess terrestrial carbon stocks and fluxes. The satellite employs a P-band synthetic aperture polarimetric radar operating at 435 MHz. Maps of forest biomass and canopy height are aimed to obtain at a resolution of 200 m.

Low-cost inventory schema based on open-access ALS data, individual-tree-based feature extraction, and crowdsourcing as field data collection technique can be performed using locally collected field data in improving the quality of the estimates (RMSE) using crowdsourcing and 3D game engine to interactively calibrate ALS tree maps and locate persons for field surveys. The use of Microsoft Kinect depth-type sensor for tree stem measurements and reconstruction is also feasible.

In addition to these new, exciting developments presented, there are too many new areas, which should have been discussed, such as future space-borne lidar, TomoSAR, UAV-based low-cost SAR, and lidar as well as future airborne SAR development. Also full-waveform lidar is of high possibilities. In the previous chapter, we favored techniques with high potential to practical forestry with shorter time frame.

Acknowledgments

Support from the Academy of Finland (Centre of Excellence in Laser Scanning Research, project 272195) and EC Advanced_SAR are gratefully acknowledged.

References and Related Literature

Aschoff, T. and Spiecker, H. 2004. Algorithms for the automatic detection of trees in laser scanner data. *International Archives of Photogrammetry, Remote Sensing and Spatial Information Sciences* 36: W2.

Askne, J., Santoro, M., Smith, G., and Fransson, J.E.S. 2003. Multitemporal repeat-pass SAR interferometry of boreal forests. *IEEE Transactions on Geoscience and Remote Sensing* 43(6): 1219–1228.

Askne, J.I., Fransson, J.E., Santoro, M., Soja, M.J., and Ulander, L.M. 2013. Model-based biomass estimation of a hemiboreal forest from multitemporal TanDEM-X acquisitions. *Remote Sensing* 5(11): 5574–5597.

Axelsson, P. 2000. DEM generation from laser scanner data using adaptive TIN models. *International Archives of Photogrammetry and Remote Sensing* 33(Part B4): 110–117.

Ballhorn, U., Jubanski, J., and Siegert, F. 2011. ICESat/GLAS data as a measurement tool for peatland topography and peat swamp forest biomass in Kalimantan, Indonesia. *Remote Sensing* 3(9): 1957–1982.

Banskota, A., Wynne, R., Johnson, P., and Emessiene, B. 2011. Synergistic use of very high-frequency radar and discrete-return lidar for estimating biomass in temperate hardwood and mixed forests. *Annals of Forest Science* 68(2): 347–356.

BBC. 2014. Tree-mapping drone start-up has sky-high ambitions. By Mark Bosworth BBC World Service, Helsinki. Available at http://www.bbc.com/news/technology-27485418 (Accessed July 1, 2014).

Brandtberg, T., Warner, T., Landenberger, R., and McGraw, J. 2003. Detection and analysis of individual leaf-off tree crowns in small footprint, high sampling density lidar data from the eastern deciduous forest in North America. *Remote Sensing of Environment* 85: 290–303.

Brolly, G. and Kiraly, G. 2009. Algorithms for stem mapping by means of terrestrial laser scanning. *Acta Silvatica et Lignaria Hungarica* 5: 119–130.

Chen, G., Hay, G.J., and St-Onge, B.A. 2012b. GEOBIA framework to estimate forest parameters from lidar transects, Quickbird imagery and machine learning: A case study in Quebec, Canada. *International Journal of Applied Earth Observation* 15: 28–37.

Chen, G., Wulder, M.A., White, J.C., Hilker, T., and Coops, N.C. 2012a. Lidar calibration and validation for geometric-optical modeling with Landsat imagery. *Remote Sensing of Environment* 124: 384–393.

Chen, Y., Shi, P., Deng, L., and Li, J. 2007. Generation of a top-of-canopy Digital Elevation Model (DEM) in tropical rain forest regions using radargrammetry. *International Journal of Remote Sensing* 28(19): 4345–4349.

Clark, N.A., Wynne, R.H., and Schmoldt, D.L. 2000. A review of past research on dendrometers. *Forest Science* 46: 570–576.

Colomina, I. and Molina, P. 2014. Unmanned aerial systems for photogrammetry and remote sensing: A review. *ISPRS Journal of Photogrammetry and Remote Sensing* 92: 79–97.

Elachi, C. 1987. *Spaceborne Radar Remote Sensing*. IEEE Press, New York, 255 p.

Eysn, L., Hollaus, M., Schadauer, K., and Pfeifer, N. 2012. Forest delineation based on airborne lidar data. *Remote Sending* 4(3): 762–783.

Fleck, S., Mölder, I., Jacob, M., Gebauer, T., Jungkunst, H.F., and Leuschner, C. 2011. Comparison of conventional eight-point crown projections with LIDAR-based virtual crown projections in a temperate old-growth forest. *Annals of Forest Science* 68: 1173–1185.

FRA. 2010. Global forest resources assessment 2010. Available at http://www.fao.org/forestry/fra/fra2010/en/ (Accesssed April 16, 2015.)

Fransson, J. 1999. Analysis of synthetic aperture radar images for forestry applications, Doctoral Thesis. *Acta Universitatis Agriculturae Sueciae, Silvestria*, Vol. 100. Swedish University of Agricultural Sciences, Umeå, Sweden.

Fransson, J.E.S. and Israelsson, H. 1999. Estimation of stem volume in boreal forests using ERS-1 C- and JERS-1 L-band SAR data. *International Journal of Remote Sensing* 20(1): 123–137.

Fritz, S., McCallum, I., Schill, C., Perger, C., Grillmayer, R., Achard, F., Kraxner, F., and Obersteiner, M. 2009. Geo-Wiki.Org: The use of crowdsourcing to improve global land cover. *Remote Sensing* 1: 345–354.

Gill, S.J., Biging, G.S., and Murphy, E.C. 2000. Modeling conifer tree crown radius and estimating canopy cover. *Forest Ecology and Management* 126(3): 405–416.

Goodenough, D., Chen, H., Dyk, A., Hobart, G., and Richardson, A. 2008. Data fusion study between polarimetric SAR, hyperspectral and lidar data for forest information. *IEEE International Symposium on Geoscience and Remote Sensing—IGARSS 2008*, Boston, MA, pp. 281–284.

Google. 2014. Available at: http://www.google.com/atap/projecttango/

Hallikainen, M., Hyyppä, J., Haapanen, J., Tares, T., Ahola, P., Pulliainen, J., and Toikka, M. 1993. A helicopter-borne eight-channel ranging scatterometer for remote sensing, Part I: System description. *IEEE Transactions on Geoscience and Remote Sensing* 31(1): 161–169.

Heipke, C. 2010. Crowdsourcing geospatial data. *ISPRS Journal of Photogrammetry and Remote Sensing* 65: 550–557.

Henderson, F. and Lewis, A. 1998. *Principles and Application of Imaging Radar*: *Manual of Remote Sensing*, Vol. 2, 3rd edn. John Wiley & Sons, New York, 896 p.

Henning, J.G. and Radtke, P.J. 2006. Detailed stem measurements of standing trees from ground-based scanning lidar. *Forest Science* 52(1): 67–80.

Hollaus, M., Eysn, L., Karel, W., and Pfeifer, N. 2013. Growing stock change estimation using Airborne Laser Scanning data. Silvilaser 2013, Peking. Paper ID SL2013–060, 8 p.

Hollaus, M., Mandlburger, G., Pfeifer, N., and Mücke, W. 2010. Land cover dependent derivation of digital surface models from airborne laser scanning data. *International Archives of Photogrammetry, Remote Sensing and the Spatial Information Sciences*. PCV 2010, Paris, France. 39(3): 6.

Hollaus, M., Wagner, W., Schadauer, K., Maier, B., and Gabler, K. 2009. Growing stock estimation for alpine forests in Austria: A robust lidar-based approach. *Canadian Journal of Forest Research* 39(7): 1387–1400.

Holmgren, J. 2003. Estimation of forest variables using airborne laser scanning, PhD Thesis. *Acta Universitatis Agriculturae Sueciae, Silvestria* 278. Swedish University of Agricultural Sciences, Umeå, Sweden.

Holmgren, J. and Persson, Å. 2004. Identifying species of individual trees using airborne laser scanning. *Remote Sensing of Environment* 90: 415–423.

Holopainen, M., Haapanen, E., Karjalainen, M., Vastaranta, M., Hyyppä, J., Yu, X., Tuominen, S., and Hyyppä, H. 2010. Comparing accuracy of airborne laser scanning and TerraSAR-X radar images in the estimation of plot-level forest variables. *Remote Sensing* 2(2): 432–445.

Holopainen, M., Kankare, V., Vastaranta, M., Liang, X., Lin, Y., Vaaja, M., Yu, X. et al. 2013. Tree mapping using airborne, terrestrial and mobile laser scanning—A case study in a heterogeneous urban forest. *Urban Forestry and Urban Greening* 12(2013): 546–553.

Holopainen, M., Vastaranta, M., Liang, X., Hyyppä, J., Jaakkola, A., and Kankare, V. 2014. Estimation of forest stock and yield using Lidar data. In G. Wang and Q. Weng (eds.), *Remote Sensing of Natural Resources*. CRC Press, Boca Raton, FL, pp. 259–290.

Hopkinson, C., Chasmer, L., and Hall, R.J. 2008. The uncertainty in conifer plantation growth prediction from multitemporal lidar datasets. *Remote Sensing of Environment* 112(3): 1168–1180.

Hopkinson, C., Chasmer, L., Lim, K., Treitz, P., and Creed, I. 2006. Towards a universal lidar canopy height indicator. *Canadian Journal of Remote Sensing* 32(2): 139–152.

Hopkinson, C., Chasmer, L., Young-Pow, C., and Treitz, P. 2004. Assessing forest metrics with a ground-based scanning lidar. *Canadian Journal of Forest Research* 34: 573–583.

Huang, H., Li, Z., Gong, P., Cheng, X., Clinton, N., Cao, C., Ni, W., and Wang, L. 2011. Automated methods for measuring dbh and tree heights with a commercial scanning lidar. *Photogrammetric Engineering and Remote Sensing* 77: 219–227.

Husch, B., Miller, C.I., and Beers, T.W. 1982. *Forest Mensuration*, 2nd edn. John Wiley & Sons, New York.

Hyde, P., Nelson, R., Kimes, D., and Levine, E. 2007. Exploring lidar-RaDAR synergy—Predicting aboveground biomass in a Southwestern ponderosa pine forest using lidar, SAR, and InSAR. *Remote Sensing of Environment* 106(1): 28–38.

Hyyppä, J. 1993. Development and feasibility of airborne ranging radar for forest assessment. Doctor of Technology thesis, Helsinki University of Technology, Laboratory of Space Technology, Report, December 1993, 112 p.

Hyyppä, J. 1999. Method for determination of stand attributes and a computer program to perform the method. Finnish patent 117490.

Hyyppä, H. and Hyyppä, J. 1999. Comparing the accuracy of laser scanner with other optical remote sensing data sources for stand attribute retrieval. *Photogrammetric Journal of Finland* 16(2): 5–15.

Hyyppä, J. and Inkinen, M. 1999. Detecting and estimating attributes for single trees using laser scanner. *Photogrammetric Journal of Finland* 16(2): 27–42.

Hyyppä, J., Hyyppä, H., Inkinen, M., Engdahl, M., Linko, S., and Zhu, Y-H. 2000. Accuracy comparison of various remote sensing data sources in the retrieval of forest stand attributes. *Forest Ecology and Management* 128: 109–120.

Hyyppä, J., Hyyppä, H., and Kukko, A. 2001a. Inventorying single trees with laser scanning. *Fourth Airborne Remote Sensing Conference*, San Francisco, CA (Cancelled), 1 p. Abstract.

Hyyppä, J., Hyyppä, H., Leckie, D., Gougeon, F., Yu, X., and Maltamo, M. 2008. Review of methods of small-footprint airborne laser scanning for extracting forest inventory data in boreal forests. *International Journal of Remote Sensing* 29(5): 1339–1366.

Hyyppä, J., Jaakkola, A., Chen, Y., Kukko, A., Kaartinen, H., Zhu, L., Alho, P., and Hyyppä, H. 2013. Unconventional LIDAR mapping from air, terrestrial and mobile. In D. Fritsch (ed.), *Photogrammetric Week 2013*. Wichmann, Berlin, Germany. Available at http://www.ifp.uni-stuttgart.de/publications/phowo13/180Hyyppae.pdf

Hyyppä, J., Kelle, O., Lehikoinen, M., and Inkinen, M. 2001b. A segmentation-based method to retrieve stem volume estimates from 3-dimensional tree height models produced by laser scanner. *IEEE Transactions of Geoscience and Remote Sensing* 39: 969–975.

Hyyppä, J., Yu, X., Hyyppä, H., Vastaranta, M., Holopainen, M., Kukko, A., Kaartinen, H. et al. 2012. Advances in forest inventory using airborne laser scanning. *Remote Sensing* 4: 1190–1207.

Hyyppä, J., Yu, X., Rönnholm, P., Kaartinen, H., and Hyyppä, H. 2003. Factors affecting laser-derived object-oriented forest height growth estimation. *Photogrammetric Journal of Finland* 18(2): 16–31.

Jaakkola, A., Hyyppä, J., Kukko, A., Yu, X., Kaartinen, M., Lehtomäki, M., and Lin, Y. 2010. A low-cost multi-sensoral mobile mapping system and its feasibility for tree measurements. *ISPRS Journal of Photogrammetry and Remote Sensing* 65(6): 514–522.

Jaakkola, A., Hyyppä, J., and Yu, X. 2014. Measuring tree stems with kinect sensor in a crowdsourcing context. Submitted.

Jaakkola, A., Hyyppä, J., Yu, X., Turppa, T., Zhu, L., and Hyyppä, H. 2015. Crowdsourcing in forest inventory—National forest inventories by commons. Submitted.

Jakubowski, M.K., Guo, Q., and Kelly, M. 2013. Trade-offs between lidar pulse density and forest measurement accuracy. *Remote Sensing of Environment* 130: 245–253.

Jensen, J.R. 2000. *Remote Sensing of the Environment—An Earth Resource Perspective*. Prentice Hall, Upper Saddle River, NJ, 544 p.

Kaartinen, H. and Hyyppä, J. 2008. EuroSDR/ISPRS Project, Commission II, "Tree Extraction". Final Report, EuroSDR, European Spatial Data Research, Official Publication No. 53.

Kaartinen, H., Hyyppä, J., Yu, X., Vastaranta, M., Hyyppä, H., Kukko, A., Holopainen, M. et al. 2012. An international comparison of individual tree detection and extraction using airborne laser scanning. *Remote Sensing* 4(4): 950–974.

Kangas, A. and Maltamo, M. (eds.). 2006. *Managing Forest Ecosystems: Forest Inventory: Methodology and Applications*. Springer, Dordrecht, the Netherlands.

Karjalainen, M., Kankare, V., Vastaranta, M., Holopainen, M., and Hyyppä, J. 2012. Prediction of plot-level forest variables using TerraSAR-X Stereo SAR data. *Remote Sensing of Environment* 17(2): 338–347.

Katoh, M. and Gougeon, F.A. 2012. Improving the precision of tree counting by combining tree detection with crown delineation and classification on homogeneity guided smoothed high resolution (50 cm) multispectral airborne digital data. *Remote Sensing* 4(5): 1411–1424.

Kellndorfer, J., Walker, W., Pierce, L., Dobson, C., Fites, J.A., Hunsaker, C., Vonad, J., and Clutter, M. 2004. Vegetation height estimation from shuttle radar topography mission and national elevation datasets. *Remote Sensing of Environment* 93(3): 339–358.

Kellndorfer, J.M., Walker, W.S., LaPoint, E., Kirsch, K., Bishop, J., and Fiske, G. 2010. Statistical fusion of lidar, InSAR, and optical remote sensing data for forest stand height characterization: A regional-scale method based on LVIS, SRTM, Landsat ETM+, and ancillary data sets. *Journal of Geophysical Research* 115: G00E08.

Koch, B. 2010. Status and future of laser scanning, synthetic aperture radar and hyperspectral remote sensing data for forest biomass assessment. *ISPRS Journal of Photogrammetry and Remote Sensing* 65(6): 581–590.

Korhonen, L., Korhonen, K.T., Rautiainen, M., and Stenberg, P. 2006. Estimation of forest canopy cover: A comparison of field measurement techniques. *Silva Fennica* 40(4): 577–588.

Korpela, I., Orka, H., Hyyppa, J., Heikkinen, V., and Tokola, T. 2010. Range and AGC normalization in airborne discrete-return lidar intensity data for forest canopies. *ISPRS Journal of Photogrammetry and Remote Sensing* 65(4): 369–379.

Kraus, K. and Otepka, J. 2005. DTM modeling and visualization—The SCOP approach. In D. Fritsch (ed.), *Photogrammetric Week 2005*. Herbert Wichmann Verlag, Heidelberg, Germany, pp. 241–252.

Kraus, K. and Pfeifer, N. 1998. Determination of terrain models in wooded areas with airborne laser scanner data. *ISPRS Journal of Photogrammetry and Remote Sensing* 53: 193–203.

Krieger, G., Moreira, A., Fiedler, H., Hajnsek, I., Werner, M., Younis, M., and Zink, M. 2007. TanDEM-X: A satellite formation for high-resolution SAR interferometry. *IEEE Transactions on Geoscience and Remote Sensing* 45(11): 3317–3341.

Kukko, A., Kaartinen, H., Hyyppä., J., and Chen, Y. 2012. Multiplatform mobile laser scanning: Usability and performance. *Sensors* 12(9): 11712–11733.

La Prade, G. 1963. An analytical and experimental study of stereo for radar. *Photogrammetric Engineering* 29(2): 294–300.

Launchbury, R. 2014. Unmanned aerial vehicles in forestry. *The Forestry Chronicle* 90(4): 418–420.

Leberl, F. 1979. Accuracy analysis of stereo side-looking radar. *Photogrammetric Engineering and Remote Sensing* 45(8): 1083–1096.

Lefsky, M., Cohen, W., Parker, G., and Harding, D. 2002. Lidar remote sensing for ecosystem studies. *Bioscience* 52: 19–30.

Lefsky, M.A., Harding, D.J., Keller, M., Cohen, W.B., Carabajal, C.C., Del Bom Espirito-Santo, F., Hunter, M.O., and de Oliveira, Jr., R. 2005. Estimates of forest canopy height and aboveground biomass using ICESat. *Geophysical Research Letters* 32: L22S02.

Lefsky, M.A., Keller, M., Pang, Y., De Camargo, P.B., and Hunter, M.O. 2007. Revised method for forest canopy height estimation from Geoscience Laser Altimeter System waveforms. *Journal of Applied Remote Sensing* 1(1): 013537-013537-18.

Le Toan, T., Beaudoin, A., Riom, J., and Guyon, D. 1992. Relating forest biomass to SAR data. *IEEE Transactions on Geoscience and Remote Sensing* 30(2): 403–411.

Li, Z.L., Liu, G.X., and Ding, X.L. 2006. Exploring the generation of digital elevation models from same-side ERS SAR images: Topographic and temporal effects. *Photogrammetric Record* 21(114): 124–140.

Liang, X., Hyyppä, J., Kaartinen, J., Holopainen, M., and Melkas, T. 2012b. Detecting changes in forest structure over time with bi-temporal terrestrial laser scanning data. *ISPRS International Journal of Geo-Information* 1(3): 242–255.

Liang, X., Hyyppä, J., Kankare, V., and Holopainen, M. 2011. Stem curve measurement using terrestrial laser scanning. *Proceedings of Silvilaser 2011*, University of Tasmania, Australia, 6 p.

Liang, X. and Hyyppä, J. 2013. Automatic stem mapping by merging several terrestrial laser scans at the feature and decision levels. *Sensors* 13(2): 1614–1634.

Liang, X., Hyyppä, J., Kukko, A., Kaartinen, H., Jaakkola, A., and Yu, X. 2014c. The use of a mobile laser scanning for mapping large forest plots. *IEEE Geoscience and Remote Sensing Letters* 11(9): 1504–1508.

Liang, X., Kankare, V., Yu, X., and Hyyppä, J. 2014a. Automated stem curve measurement using terrestrial laser scanning. *IEEE Transactions on Geoscience and Remote Sensing* 52(3): 1739–1748.

Liang, X., Kukko, A., Kaartinen, H., Hyyppä, J., Yu, X., Jaakkola, A., and Wang, Y. 2014b. Possibilities of a personal laser scanning system for forest mapping and ecosystem services. *Sensors* 14(1): 1228–1248.

Liang, X., Litkey, P., Hyyppä, J., Kaartinen, H., Vastaranta, M., and Holopainen, M. 2012a. Automatic stem mapping using single-scan terrestrial laser scanning. *IEEE Transactions on Geoscience and Remote Sensing* 50(2): 661–670.

Lin, Y., Jaakkola, A., Hyyppä, J., and Kaartinen, H. 2010. From TLS to VLS: Biomass estimation at individual tree level. *Remote Sensing* 2(8): 1864–1879.

Lindberg, E. 2012. Estimation of canopy structure and individual trees from laser scanning data [WWW Document]. Available at http://pub.epsilon.slu.se/8888/ (Accessed March 21, 2013).

Lovell, J.L., Jupp, D.L.B., Newnham, G.J., and Culvenor, D.S. 2011. Measuring tree stem diameters using intensity profiles from ground-based scanning lidar from a fixed viewpoint. *ISPRS Journal of Photogrammetry and Remote Sensing* 66: 46–55.

Maas, H.G., Bienert, A., Scheller, S., and Keane, E. 2008. Automatic forest inventory parameter determination from terrestrial laser scanner data. *International Journal of Remote Sensing* 29(5): 1579–1593.

Magnussen, S. and Wulder, M.A. 2012. Post-fire canopy height recovery in Canada's boreal forests using airborne laser scanner (ALS). *Remote Sensing* 4: 1600–1616.

Mallet, C. and Bretar, F. 2009. Full-waveform topographic lidar: State-of-the-art. *ISPRS Journal of Photogrammetry and Remote Sensing* 64(1): 1–16.

Maltamo, M., Malinen, J., Packalen, P., Suvanto, A., and Kangas, J. 2006. Nonparametric estimation of stem volume using airborne laser scanning, aerial photography, and stand-register data. *Canadian Journal of Forest Research* 36: 426–436.

Massonnet, D. and Feigl, K.L. 1998. Radar interferometry and its application to changes in the earth's surface. *Reviews of Geophysics* 36(4): 441–500.

McGaughey, R.J. 2013. FUSION/LDV: Software for lidar data analysis and visualization. Version 3.30. U.S. Department of Agriculture Forest Service, Pacific Northwest Research Station, University of Washington, Seattle, WA. Available at http://forsys.cfr.washington.edu/fusion/FUSION_manual.pdf (Accessed May 2013).

Mora, B., Wulder, M.A., Hobart, G.W., White, J.C., Bater, C.W., Gougeon, F.A., Varhola, A., and Coops, N.C. 2013a. Forest inventory stand height estimates from very high spatial resolution satellite imagery calibrated with lidar plots. *International Journal of Remote Sensing* 34(12): 4406–4424.

Mora, B., Wulder, M.A., White, J.C., and Hobart, G. 2013b. Modeling stand height, volume, and biomass from very high spatial resolution satellite imagery and samples of airborne lidar. *Remote Sensing* 5: 2308–2326.

Murphy, G.E., Acuna, M.A., and Dumbrell, I. 2010. Tree value and log product yield determination in radiata pine (Pinus radiata) plantations in Australia: Comparisons of terrestrial laser scanning with a forest inventory system and manual measurements. *Canadian Journal of Forest Research* 40: 2223–2233.

Næsset, E. 1997a. Determination of mean tree height of forest stands using airborne laser scanner data. *ISPRS Journal of Photogrammetry and Remote Sensing* 52: 49–56.

Næsset, E. 1997b. Estimating timber volume of forest stands using airborne laser scanner data. *Remote Sensing of Environment* 61(2): 246–253.

Næsset, E. 2002. Predicting forest stand characteristics with airborne scanning laser using a practical two-stage procedure and field data. *Remote Sensing of Environment* 80: 88–99.

Næsset, E. 2007. Airborne laser scanning as a method in operational forest inventory: Status of accuracy assessments accomplished in Scandinavia. *Scandinavian Journal of Forest Research* 22: 433–442.

Næsset, E. and Nelson, R. 2007. Using airborne laser scanning to monitor tree migration in the boreal–alpine transition zone. *Remote Sensing of Environment* 110(4): 357–369.

Nelson, R. 2013. How did we get there? An early history of forestry lidar. *Canadian Journal of Remote Sensing* 39(1): 6–17.

Nelson, R., Krabill, W., and Tonelli, J. 1988. Estimating forest biomass and volume using airborne laser data. *Remote Sensing of Environment* 24: 247–267.

Nelson, R.F., Hyde, P., Johnson, P., Emessiene, B., Imhoff, M.L., Campbell, R., and Edwards, W. 2007. Investigating RaDAR-lidar synergy in a North Carolina pine forest. *Remote Sensing of Environment* 110(1): 98–108.

Nurminen, K., Karjalainen, M., Yu, X., Hyyppa, J., and Honkavaara, E. 2013. Performance of dense digital surface models based on image matching in the estimation of plot-level forest variables. *ISPRS Journal of Photogrammetry and Remote Sensing* 83: 104–115.

Packalen, P. and Maltamo, M. 2007. The k-MSN method in the prediction of species specific stand attributes using airborne laser scanning and aerial photographs. *Remote Sensing of Environment* 109: 328–341.

Päivinen, R., Nousiainen, M., and Korhonen, K. 1992. Puutunnusten mittaamisen luotettavuus (Accuracy of certain tree measurements). *Folia Forestalia* 787: 18.

Papathanassiou, K.P. and Cloude, S.R. 2001. Single-baseline polarimetric SAR interferometry. *IEEE Transactions on Geoscience and Remote Sensing* 39(11): 2352–2363.

Perko, R., Raggam, H., Deutscher, J., Gutjahr, K., and Schardt, M. 2011. Forest assessment using high resolution SAR data in X-band. *Remote Sensing* 3(4): 792–815.

Petrie, G. 2013. Current developments of laser scanner suitable for use on lightweight UAVs. *Geoinformatics* 8: 16–22.

Popescu, S., Wynne, R., and Nelson, R. 2003. Measuring individual tree crown diameter with lidar and assessing its influence on estimating forest volume and biomass. *Canadian Journal of Remote Sensing* 29(5): 564–577.

Popescu, S.C. 2007. Estimating biomass of individual pine trees using airborne lidar. *Biomass and Bioenergy* 31(9): 646–655.

Rauste, Y. 2005. Multi-temporal JERS SAR data in boreal forest biomass mapping. *Remote Sensing of Environment* 97(2): 263–275.

Reigber, A. 2002. *Airborne Polarimetric SAR Tomography.* Deutsches Zentrum fuer Luft- und Raumfahrt, Forschungsberichte.

Rönnholm, P., Hyyppä, J., Hyyppä, H., Haggrén, H., Yu, X., Pyysalo, U., Pöntinen, P., and Kaartinen, H. 2004. Calibration of laser-derived tree height estimates by means of photogrammetric techniques. *Scandinavian Journal of Forest Research* 19(6): 524–528.

Rosen, P.A., Hensley, S., Joughin, I.R., Li, F.K., Madsen, S.N., Rodriguez, E., and Goldstein, R.M. 2000. Synthetic aperture radar interferometry. *Proceedings of the IEEE,* 88(3): 333–382.

Rosette, J.A.B., North, P.R.J., and Suarez, J.C. 2008. Vegetation height estimates for a mixed temperate forest using satellite laser altimetry. *International Journal of Remote Sensing* 29(5): 1475–1493.

Santoro, M., Shvidenko, A., McCallum, I., Askne, J., and Schmullius C. 2007. Properties of ERS-1/2 coherence in the Siberian boreal forest and implications for stem volume retrieval. *Remote Sensing of Environment* 106(2): 154–172.

Schutz, B.E., Zwally, H.J., Shuman, C.A., Hancock, D., and DiMarzio, J.P. 2005. Overview of the ICESat mission. *Geophysical Research Letters* 32(21).

Simonse, M., Aschoff, T., Spiecker, H., and Thies, M. 2003. Automatic determination of forest inventory parameters using terrestrial laserscanning. *Proceedings of the ScandLaser Scientific Workshop on Airborne Laser Scanning of Forests,* Umeå, Sweden, pp. 252–258.

Sithole, G. and Vosselman, G. 2004. Experimental comparison of filter algorithms for bare-Earth extraction from airborne laser scanning point clouds. *ISPRS Journal of Photogrammetry and Remote Sensing* 59: 85–101.

Solberg, S., Astrup, R., Bollandsas, O.M., Næsset, E., and Weydahl, D.J. 2010. Deriving forest monitoring variables from X-band InSAR SRTM height. *Canadian Journal of Remote Sensing* 36: 68–79.

Solberg, S., Astrup, R., Breidenbach, J., Nilsen, B., and Weydahl, D. 2013. Monitoring spruce volume and biomass with InSAR data from TanDEM-X. *Remote Sensing of Environment* 139: 60–67.

Solberg, S., Riegler, G., and Nonin, P. 2014. Estimating forest biomass from TerraSAR-X stripmap radargrammetry. *IEEE Transactions on Geoscience and Remote Sensing* 53(1): 154–161.

Sun, G., Ranson, K.J., Kimes, D.S., Blair, J.B., and Kovacs, K. 2008. Forest vertical structure from GLAS: An evaluation using LVIS and SRTM data. *Remote Sensing of Environment* 112(1): 107–117.

Suvanto, A., Maltamo, M., Packalen, P., and Kangas, J. 2005. Kuviokohtaisten puustotunnustenennustaminen laserkeilauksella. *Metsätieteen aikakauskirja* 4/2005: 413–428.

Tansey, K., Selmes, N., Anstee, A., Tate, N.J., and Denniss, A. 2009. Estimating tree and stand variables in a Corsican pine woodland from terrestrial laser scanner data. *International Journal of Remote Sensing* 30(19): 5195–5209.

Thies, M., Pfeifer, N., Winterhalder, D., and Gorte, B.G.H. 2004. Three-dimensional reconstruction of stems for assessment of taper, sweep and lean based on laser scanning of standing trees. *Scandinavian Journal of Forest Research* 19: 571–581.

Tokola, T., Letoan, T., Poncet, F.V., Tuominen, S., and Holopainen, M. 2007. Forest reconnaissance surveys: Comparison of estimates based on simulated TerraSar, and optical data. *Photogrammetric Journal of Finland* 20: 64–79.

Torres, R., Snoeij, P., Geudtner, D., Bibby, D., Davidson, M., Attema, E., Potin, P. et al. 2012. GMES Sentinel-1 mission. *Remote Sensing of Environment* 120(S1): 9–24.

Toutin, T. and Gray, L. 2000. State-of-the-art of elevation extraction from satellite SAR data. *ISPRS Journal of Photogrammetry and Remote Sensing* 55(1): 13–33.

van Aardt, J., Wynne, R., and Scrivani, J. 2008. Lidar-based mapping of forest volume and biomass by taxonomic group using structurally homogenous segments. *Photogrammetric Engineering and Remote Sensing* 74(8): 1033–1044.

Vastaranta, M. 2012. Forest mapping and monitoring using active 3D remote sensing. *Dissertationes Forestales* 144: 45 p.

Vastaranta, M., Holopainen, M., Karjalainen, M., Kankare, M., Hyyppä, J., and Kaasalainen, S. 2014. TerraSAR-X stereo radargrammetry and airborne scanning lidar height metrics in imputation of forest aboveground biomass and stem volume. *IEEE Transactions on Geoscience and Remote Sensing* 52: 1197–1204.

Vastaranta, M., Holopainen, M., Karjalainen, M., Kankare, V., Hyyppä, J., Kaasalainen, S., and Hyyppä, H. 2012a. SAR radargrammetry and scanning lidar in predicting forest canopy height. *Proceedings of IEEE International Geoscience and Remote Sensing Symposium*, Munich, Germany, pp. 6515–6518.

Vastaranta, M., Kankare, V., Holopainen, M., Yu, X., Hyyppä, J., and Hyyppä, H. 2012b. Combination of individual tree detection and area-based approach in imputation of forest variables using airborne laser data. *ISPRS Photogrammetry and Remote Sensing* 67: 73–79.

Vastaranta, M., Wulder, M.A., White, J., Pekkarinen, A., Tuominen, S., Ginzler, C., Kankare, V., Holopainen, M., Hyyppä, J., and Hyyppä, H. 2013. Airborne laser scanning and digital stereo imagery measures of forest structure: Comparative results and implications to forest mapping and inventory update. *Canadian Journal of Remote Sensing* 39(5): 382–395.

Vosselman, G. 2000. Slope based filtering of laser altimetry data. *International Archives of Photogrammetry and Remote Sensing* 33(B3/2): 935–942.

Wagner, W., Luckman, A., Vietmeier, J., Tansey, K., Balzter, H., Schmullius, C., Davidson, M. et al. 2003. Large-scale mapping of boreal forest in Siberia using ERS tandem coherence and JERS backscatter data. *Remote Sensing of Environment* 85(2): 125–144.

Wang, Y., Liang, X., Flener, C., Kukko, A., Kaartinen, H., Kurkela, M., Vaaja, M., Hyyppä, H., and Alho, P. 2013. 3D Modeling of coarse fluvial sediments based on mobile laser scanning data. *Remote Sensing* 5(9): 4571–4592.

Watt, P.J. and Donoghue, D.N.M. 2005. Measuring forest structure with terrestrial laser scanning. *International Journal of Remote Sensing* 26: 1437–1446.

White, J.C., Wulder, M.A., Varhola, A., Vastaranta, M., Coops, N.C., Cook, B.D., Pitt, D, and Woods, M. 2013a. A best practices guide for generating forest inventory attributes from airborne laser scanning data using the area-based approach. Information Report FI-X-10, Canadian Forest Service, Canadian Wood Fibre Centre, Pacific Forestry Centre, Victoria, BC, 50 p.

White, J.C., Wulder, M.A., Vastaranta, M., Coops, N.C., Pitt, D., and Woods, M. 2013b. The utility of image-based point clouds for forest inventory: A comparison with airborne laser scanning. *Forests* 4(3): 518–536.

Wulder, M.A, White, J.C, Bater, C.W, Coops, N.C, Hopkinson, C., and Chen, G. 2012b. Lidar plots a new large-area data collection option: Context, concepts, and case study. *Canadian Journal of Remote Sensing* 38(5): 600–618.

Wulder, M.A., White, J.C., Nelson, R.F., Næsset, E., Ørka, H.O., Coops, N.C., Hilker, T., Bater, C.W., and Gobakken, T. 2012a. Lidar sampling for large-area forest characterization: A review. *Remote Sensing of Environment* 121: 196–209.

Yu, X., Hyyppä, J., Kaartinen, H., and Maltamo, M. 2004. Automatic detection of harvested trees and determination of forest growth using airborne laser scanning. *Remote Sensing of Environment* 90: 451–462.

Yu, X., Hyyppä, J., Kukko, A., Maltamo, M., and Kaartinen, H. 2006. Change detection techniques for canopy height growth measurements using airborne laser scanner data. *Photogrammetric Engineering & Remote Sensing* 72(12): 1339–1348.

Yu, X., Hyyppä, J., Vastaranta, M., Holopainen, M., and Viitala, R. 2011. Predicting individual tree attributes from airborne laser point clouds based on random forest technique. *ISPRS Journal of Photogrammetry and Remote Sensing* 66: 28–37.

Zhao, X., Liu, J., and Tan, M. 2006. A remote aerial robot for topographic survey. *Proceedings of the International Conference on Intelligent Robots and Systems*, IEEE/RJS, Beijing, China, October 9–15, 2006, pp. 3143–3148.

Zhao, K., Popescu, S., and Nelson, R. 2009. Lidar remote sensing of forest biomass: A scale-invariant estimation approach using airborne lasers, *Remote Sensing of Environment* 113(1): 182–196.

Zwally, H.J., Schutz, B., Abdalati, W., Abshire, J., Bentley, C., Brenner, A., Bufton, J. et al. 2002. ICESAT's laser measurements of polar ice, atmosphere, ocean, and land. *Journal of Geodynamics* 34: 405–445.

16

Forest Biophysical and Biochemical Properties from Hyperspectral and LiDAR Remote Sensing

Gregory P. Asner
Carnegie Institution for Science

Susan L. Ustin
University of California

Philip A. Townsend
University of Wisconsin

Roberta E. Martin
Carnegie Institution for Science

K. Dana Chadwick
Carnegie Institution for Science

Acronyms and Definitions

ACD	Aboveground carbon density
AIS	Airborne imaging spectrometer
ALI	Advanced Land Imager
A_{max}	Maximum rate of photosynthesis
APAR	Absorbed photosynthetically active radiation
ATLAS	Advanced Topographic Laser Altimeter System
AVIRIS	Airborne visible/infrared imaging spectrometer
CAO	Carnegie Airborne Observatory
CASI	Compact Airborne Spectrographic Imager
CHRIS	Compact High-Resolution Imaging Spectrometer
DASF	Directional area scattering factor
EnMAP	Environmental Mapping and Analysis Program
EO-1	Earth Observing-1 (Hyperion)
ESA	European Space Agency
ETM+	Enhanced Thematic Mapper Plus
ETR	Electron transport rate
EWT	Equivalent water thickness
GLAS	Geoscience Laser Altimeter System
HSI	Hyperspectral imaging
HyspIRI	Hyperspectral and Infrared Imager
ICESat	Ice, Cloud, and land Elevation Satellite
J_{max}	Maximum electric transport rate
LAD	Leaf angle distribution
LAI	Leaf area index
LiDAR	Light detection and ranging
LMA	Leaf mass per area
LUE	Light use efficiency
LVIS	Land, Vegetation, and Ice Sensor
MCH	Mean canopy profile height
N	Nitrogen
NCALM	National Center for Airborne Laser Mapping
NDVI	Normalized difference vegetation index
NEON	National Ecological Observatory Network
NIR	Near infrared
NPP	Net primary productivity
NPQ	Nonphotochemical quenching
NPV	Nonphotosynthetic vegetation
NSF	National Science Foundation
PAR	Photosynthetically active radiation
PLSR	Partial least squares regression
PRI	Photochemical Reflectance Index
SIF	Solar-induced fluorescence
SLA	Specific leaf area
SNR	Signal-to-noise ratio

SWIR   Shortwave infrared
UAV   Unmanned aerial vehicle
V_{cmax} Maximum rate of carboxylation
VNIR   Visible to near infrared
VSWIR   Visible to shortwave infrared

16.1 Introduction

Forests store about three-quarters of all carbon stocks in vegetation in the terrestrial biosphere and harbor an array of organisms that comprise most of this carbon (IPCC 2000). The distribution of carbon and biodiversity in forests is spatially and temporally heterogeneous. The complex, 3D arrangement of plant species and their tissues has always challenged field-based studies of forests. Remote sensing has long endeavored to address these challenges by mapping the cover, structure, composition, and functional attributes of forests, and new approaches are continually being developed to increase the breadth and accuracy of remote measurements.

Over the past few decades, two technologies—hyperspectral imaging (HSI) and light detection and ranging (LiDAR)—have rapidly advanced from use in testbed-type research to applications ranging from ecology to land management. HSI, also known as imaging spectroscopy, involves the measurement of reflected solar radiance in narrow, contiguous spectral bands that form a spectrum for each image pixel. LiDAR uses emitted laser pulses in a scanning pattern to determine the distance between objects such as canopy foliage and ground surfaces. Individually, HSI and LiDAR are advancing the study of forests at landscape to global scales, uncovering new spatial and temporal patterns of forest biophysical and biochemical properties, as well as physiological processes. When combined, HSI and LiDAR can provide ecological detail at spatial scales unachievable in the field. This chapter discusses HSI and LiDAR data sources, techniques, applications, and challenges in the context of forest ecological research.

16.2 HSI and LiDAR Data

16.2.1 HSI Data Sources

The availability of HSI for ecological applications is growing as the utility of these data has increasingly been recognized. HSI can be collected either with airborne sensors that have a limited spatial coverage but high-spatial resolution or with spaceborne sensors capable of capturing data globally, but generally with coarser spatial resolution. There are an expanding number of government, private, and commercial airborne HSI sensors. In addition, one spaceborne HSI sensor—Earth Observing-1 (EO-1) Hyperion—has been in operation as a technology demonstration since November 2000. Other orbital sensors are in the planning or development stages in hopes of further extending the spatial coverage of available imaging spectroscopy (Table 16.1).

Airborne HSI sensors have been operating since the 1980s. An early system was NASA's airborne imaging spectrometer (AIS), followed later by the airborne visible/infrared imaging spectrometer (AVIRIS), which is still in operation and provides data to NASA-supported investigators. Newer instruments including the Carnegie Airborne Observatory (CAO) visible-to-shortwave-infrared (VSWIR) imaging spectrometer provide increased spectral resolution and performance (e.g., signal-to-noise ratio [SNR]) over previous technology (Table 16.1). The U.S. National Science Foundation's (NSF) National Ecological Observatory Network (NEON) has created three copies of the CAO VSWIR, which will provide annual collection of HSI data for each of its core research sites across the United States.

Beyond government and privately funded instruments for research, a number of HSI sensors have been built for commercial applications. For example, the Compact Airborne Spectrographic Imager (CASI, CASI-2, CASI-1500) and HyMap provide high-performance visible-to-near-infrared (VNIR) (365–1052 nm) and VSWIR (440–2500 nm) measurements, respectively (Table 16.1).

TABLE 16.1   Examples of Current and Planned Airborne and Spaceborne HSI

Sensor	Spectral Range (nm)	Spectral Bands	Spectral Resolution (nm)	Spatial Resolution (m)	Reference
Airborne					
AVIRIS	400–2450	224	10	2.0+	Green et al. (1998)
AVIS-2	400–900	64	9	2.0+	Oppelt and Mauser (2007)
CAO VSWIR	380–2510	428	5	0.5+	Asner et al. (2012)
HYDICE	400–2500	206	8–15	1.0+	Basedow et al. (1995)
NEON VSWIR	380–2510	428	5	0.5+	www.neoninc.org
AISA	380–2500	275	3.5–12	1+	www.specim.fi
CASI	365–1052	288	2–10	0.25+	www.itres.com
HyMap	440–2500	100–200	10–20	2.0+	Cocks et al. (1998)
Spaceborne					
EO-1 Hyperion	400–2500	220	10	30	Folkman et al. (2001)
Proba-1 CHRIS	415–1050	18–62	1.3–12	18, 36	Barnsley et al. (2004)
EnMAP (*planned*)	420–2450	98–130	6.5–10	30	Stuffler et al. (2007)
HyspIRI (*planned*)	380–2500	210	10	60	hyspiri.jpl.nasa.gov

In comparison to airborne systems, there are fewer spaceborne sensors collecting hyperspectral data (Table 16.1). NASA's EO-1 Hyperion has far exceeded its intended 1-year life span, performing for over a decade (Riebeek 2010). Thenkabail et al. (2004) showed that Hyperion data, when compared to data from even the most advanced broadband sensors (Enhanced Thematic Mapper Plus [ETM+], IKONOS, and Advanced Land Imager [ALI]) in orbit at that time, yielded models that explained 36%–83% more of the variability in rainforest biomass and produced land use/land cover classifications with 45%–52% higher accuracies. The European Space Agency (ESA) also has a hyperspectral sensor (Compact High-Resolution Imaging Spectrometer [CHRIS]) on board the Proba-1 satellite, which observes in the visible and near-infrared (NIR) portion of the spectrum, though at higher spatial resolutions than Hyperion it is only able to record in 18 bands in this range (Barnsley et al. 2004). In addition, Germany is planning the launch of a hyperspectral sensor Environmental Mapping and Analysis Program (EnMAP) in 2017, and NASA is planning a mission called Hyperspectral and Infrared Imager (HyspIRI) for sometime near the year 2020. The addition of these spaceborne sensors will greatly contribute to the spatial and temporal coverage of hyperspectral data for forest research.

16.2.2 LiDAR Data Sources

LiDAR data sources are both numerous and variable, a reflection of the demand for airborne LiDAR in a wide variety of scientific and engineering applications. Recent and upcoming spaceborne LiDAR systems, described in this section, offer new data for forest monitoring. While the amount of LiDAR data being collected is increasing, there is a great deal of variability in the quality, type (discrete return vs. waveform), and spatial resolution of the resulting data.

LiDAR datasets for the United States are publicly available from a variety of sources. The National Center for Airborne Laser Mapping (NCALM; www.ncalm.cive.uh.edu) uses commercially sourced LiDAR sensors to collect high-resolution data (>2 laser spots m⁻²) for NSF-funded projects or for other select projects. These data are currently made available to the public within 2 years of collection through the NSF-supported OpenTopography program (www.opentopography.org), which provides a platform to access these data, along with other LiDAR datasets contributed by researchers. NASA's Land, Vegetation, and Ice Sensor (LVIS), which has been operating in North America since the late 1990s, provides waveform data at coarser resolution of 10–25 m diameter laser spots in support of NASA studies (Blair et al. 1999). In addition, due to the increasing availability of commercial LiDAR acquisition services, many state and local governments have commissioned datasets. In the United States, the National Oceanic and Atmospheric Administration provides an inventory of these data (http://www.csc.noaa.gov/inventory/). There are no standard characteristics of these datasets, as they all vary with sensor parameters, elevation of data collection, and the density of returns collected.

These heterogeneous data collection conditions hinder general assessments of the quality of these data.

In addition to airborne LiDAR data, NASA's Geoscience Laser Altimeter System (GLAS) Instrument, on board the Ice, Cloud, and land Elevation Satellite (ICESat), was the first spaceborne LiDAR instrument (Abshire et al. 2005). GLAS collected waveform data with 70 m spot diameter and 170 m spot intervals. The GLAS instrument was in operation from 2003 to 2009, and the data are publically available (icesat.gsfc.nasa.gov). The ICESat-2 is expected to launch in 2016, carrying the Advanced Topographic Laser Altimeter System (ATLAS).

16.2.3 Data Quality

The vast majority of HSI and LiDAR instruments have been deployed on aircraft, so the geographic coverage, ground sampling distance (spatial resolution and/or laser spot spacing), flying altitudes, and atmospheric conditions have varied enormously, making comparisons of instrument performances difficult to achieve. Nonetheless, comparative use of these instruments often reveals that sensor performance is paramount to achieving quality estimates of vegetation biophysical and biochemical properties.

Three sensor qualities have proven particularly important in the effort to achieve high-fidelity data output. These include detector uniformity, instrument stability, and SNR performance of the measurement (Green 1998). From the HSI perspective, each of these metrics of quality is important. Uniformity refers to the detailed way in which spectra are collected in the cross track and spectral directions on the instrument detector. Many HSI instruments fail to meet the often-cited 95%–98% absolute uniformity standard. One of the most insidious errors in uniformity occurs in the spectral direction. Most area-array HSI sensors fail to keep the spectral measurement aligned "down spectrum" from the VNIR (e.g., 400–1100 nm) and throughout the shortwave infrared (SWIR) (e.g., 1100–2500 nm), leading to a mismatch in different parts of the spectrum projected onto the Earth's surface. Another HSI performance issue is stability, which refers to the repeatability of the measurement across the imaging detector and/or over time. Much of the stability issue rests in the performance of the electronics and temperature stabilization subsystems. Finally, SNR is a quality that reports the strength and accuracy of the measurement signal relative to noise generated by the electronics and optics. SNR varies widely from instrument to instrument and also with environmental conditions such as temperature and humidity. Readers should be cautious when reviewing potential sources of HSI data, as providers may report SNR on either a bright target (e.g., white reference) or with enlarged camera apertures and/or inappropriately long integration times (equivalent to shutter speed). This will greatly inflate reported SNR values. For vegetation applications, SNR performances should be reported on dark targets in the 5%–8% reflectance range, typical for plants in the visible spectrum (350–700 nm), and with integration times that are appropriate for airborne or spaceborne ground speeds (usually 10 ms).

TABLE 16.2 Forest Biochemical and Physiological Properties Estimated from HSI, along with a Summary of Example Methods (Spectral Indices), Relevant Spectral Bands, Maturity, and References

Vegetation Property	Estimation Method(s)	Relevant Bands (nm)	Maturity Level	Example References
Foliar nitrogen	Normalized difference nitrogen index; band depth analysis; PLSR; RT model inversion	1510, 1680; 400–2500	✓✓	Kokaly (2001), Serrano et al. (2002), Smith et al. (2003), Asner and Vitousek (2005), and Dahlin et al. (2013)
LUE	PRI	531, 570	✓✓	Gamon et al. (1992, 1997), Gamon and Surfus (1999), Stylinksi et al. (2000), Guo and Trotter (2004), Hilker et al. (2008), Filella et al. (2009), Garbulsky et al. (2011), and Ripullone et al. (2011)
Foliar carotenoids	Various narrowband spectral indices	510, 550, 700; 445, 680, 800	✓✓	Gitelson et al. (2002) and Peñuelas et al. (1995)
Foliar anthocyanin	Various narrowband spectral indices	400–700	✓	Gamon and Surfus (1999), Gitelson et al. (2001, 2006), and Van den Berg and Perkins (2005)
APAR	Simple ratio, NDVI	400–700	✓✓✓	Jordan (1969) and Rouse et al. (1974)
LAI	Various narrowband spectral indices; RT model inversion	700–1300	✓✓✓	Rouse et al. (1974), Huete (1988), Gao et al. (1995), Rondeaux et al. (1996), Haboudane et al. (2002), Gitelson (2004), and Lim et al. (2004)
LMA	PLSR	400–2500	✓	Asner et al. (2011)
Foliar chlorophylls	Various narrowband spectral indices; RT model inversion	550, 670, 700; 800–1300; 690–725	✓✓✓	Kim (1994), Daughtry et al. (2000), Zarco-Tejada et al. (2001), Gitelson et al. (2006), and Zhang et al. (2008)
Foliar water	Various narrowband spectral indices	820, 1600; 860, 1240; 900, 970	✓✓	Hunt and Rock (1989), Peñuelas et al. (1997), and Dahlin et al. (2013)
Canopy water	EWT; RT model inversion	800–2500	✓✓✓	Hunt and Rock (1989), Gao and Goetz (1990), Gao (1996), Peñuelas et al. (1997), and Roberts et al. (2004)
Foliar lignin and cellulose	Cellulose absorption index; normalized difference lignin index	2015, 2106, 2195; 1680, 1754	✓✓	Daughtry (2001) and Serrano et al. (2002)
Foliar carbon	PLSR	1500–2500	✓	Dahlin et al. (2013)

Note: Maturity is a metric of relative accuracy as depicted in the literature, with one checkmark indicating low maturity and three checkmarks indicating high maturity. RT, radiative transfer; PLSR, partial least squares regression.

LiDAR measurements also have SNR, uniformity, and stability challenges. The shape, noisiness, and strength of the outbound laser pulses largely affect LiDAR SNR. Commercial LiDARs come in a wide range of SNR performance levels. For forest science, strong pulse strength (e.g., high-wattage laser diodes) is necessary to overcome absorption by the vegetation canopy. In addition, uniformity tends to be overlooked by scientists prior to data source selection; it is highly advisable to select LiDAR instruments that deliver a uniform scan pattern across the swath of the data set. Without strict control over this factor, the user will end up with high data density in the middle of the scan and low-data density at the edges of the scan. Finally, stability is a key issue with LiDAR instrumentation. Many commercial LiDARs exhibit instability as they change temperature, pressure, and humidity, resulting in variability in the quality of the laser data throughout the course of a mapping flight or research campaign.

16.3 HSI Remote Sensing of Forests

Forests, as fundamental components of the Earth's biosphere, have been a major focus of study from the beginning of HSI data collection. HSI provides a quantitative measure of the sunlight reflected from the forest canopy and the properties therein. The extended range and high-fidelity narrowband resolution of HSI offers enhanced capability for mapping forest biochemical and biophysical constituents along with physiological processes that contribute to the shape of the reflectance spectrum (Table 16.2). HSI data are used in a number of ways to assess leaf and canopy properties, namely, semiempirical methods utilizing narrowband spectral indices, regression modeling, and radiative transfer model inversion. As the HSI data quality improves, so do the results derived from these methods. Most recently, HSI combined with improved analytical methods has dramatically advanced species mapping and land cover classification.

16.3.1 Biophysical Properties

HSI data can uncover biophysical properties of ecological significance at both the leaf and canopy scales. Properties related to forest composition and leaf area index (LAI) are perhaps best retrieved from HSI data, whereas some properties like canopy-gap distribution and leaf angle distribution (LAD) are more readily determined from LiDAR. LAI (leaf area per unit ground area, $m^2 \, m^{-2}$) is one of the most important canopy properties because it is directly related to productivity and water use, but variation in LAI can also indicate stress resistance and competition for light (see Waring 1983; Asner et al. 2004a). Field data and models show that LAI and LAD are primary controls on canopy reflectance in dense vegetation (Gong et al. 1992; Asner 1998).

While LAI is detectable from broadband sensors, studies show that HSI data and analysis methods optimized for HSI are more accurate (e.g., Spanner et al. 1994; Gong et al. 1995). Lee et al. (2004) examined four structurally different land cover types and showed that HSI red-edge and SWIR bands produced the best estimates of LAI. Equivalent water thickness (EWT, mm) produces better estimates of LAI than do pigment-based indices such as the normalized difference vegetation index (NDVI) (Roberts et al. 1998), with LAI values (up to nine) that far exceed the sensitivity range of NDVI and other indices (Roberts et al. 2004). Water indices derived from HSI have also been used to quantify loss of LAI from pest-related defoliation and other factors (e.g., White et al. 2007).

At the leaf level, leaf mass per area (LMA, g m^{-2} and its reciprocal; specific leaf area [SLA], m^2 g^{-1}) is a key foliar property that is highly correlated with light harvesting and potential plant productivity (Niinemets 1999; Westoby et al. 2002). LMA can be defined for foliage throughout the canopy or in any given canopy layer, depending upon the ecological question. While there is enormous range in LMA within a given plant functional type and among coexisting species, LMA is broadly correlated with temperature and precipitation at the global level (Wright et al. 2004). Higher temperatures, drier conditions, and higher irradiance are associated with higher values of LMA. Leaves with higher LMA are built for defense and longer life spans, creating higher resource use efficiency per nutrient acquired (Poorter et al. 2009). Conversely, lower LMA values are found in fast-growing species, often with higher nutrient concentrations and photosynthetic rates (Wright et al. 2004). In addition, there is a strong degree of taxonomic organization to LMA within forest communities (Asner et al. 2014). Because LMA is a function of leaf thickness and is correlated with total carbon and nitrogen, it is uniquely detectable in HSI data and has been estimated from inversion of radiative transfer models such as the PROSPECT model (Jacquemoud et al. 2009), chemometric analytical methods (Asner et al. 2011), and HSI-optimized SWIR indices (le Maire et al. 2008). The results from these studies conform to field measurements.

16.3.2 Biochemical Properties

The foremost motivation for biochemical detection is to better assess the spatiotemporal status and trends of forest canopy functioning, especially those related to fluxes of water, carbon, and nutrients. The list of plant biochemicals that have been identified and quantified using HSI data is extensive (Table 16.2) and has received several detailed reviews (Blackburn 2007; Kokaly et al. 2009; Ustin et al. 2009; Homolová et al. 2013). Many studies have found strong correlations between remotely sensed foliar nitrogen content and photosynthetic capacity or net primary production (Kokaly et al. 2009; Townsend et al. 2013), despite the small fraction of biomass comprised nitrogen. Most of these studies have been based on partial least squares regression (PLSR) analysis (Ollinger et al. 2002; Smith et al. 2002; Martin et al. 2008) of the full spectrum or spectral matching and continuum removal techniques (Kokaly 2001). Feilhauer et al. (2011) and Homolová

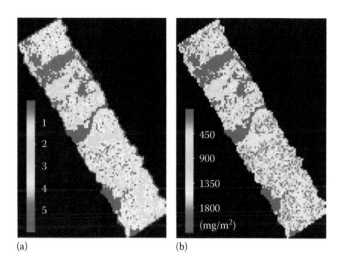

(a) (b)

FIGURE 16.1   (a) LAI image of a black spruce forest (53.2% conifer, 16.1% deciduous species, and 21.1% grass) near Sudbury, Ontario, Canada. The image is derived based on a relationship between the simple ratio (near infrared/R) and LAI (r^2 = 0.88). (b). Chlorophyll a + b content distribution per unit ground area. The image combines the retrieved leaf chlorophyll a + b content for the three cover types (r^2 = 0.47) times the LAI. The chlorophyll data were analyzed using the 4-Scale geometrical–optical model to characterize the effect of structure on above canopy reflectance and inversion of the PROSPECT leaf model to estimate pigment concentration. Data from 72-band Compact Airborne Spectrographic Imager (HSI) averaged from 2 m pixel resolution to 20 m. (Reprinted from Zhang, Y. et al., *Remote Sens. Environ.*, 112, 3234, 2008.)

et al. (2013) show that multiple wavelengths throughout the 400–2500 nm range have enabled nitrogen detection, indicating that nitrogen-related spectral features may vary by site, species, or phenological state.

Vegetation indices (Zarco-Tejada et al. 1999, 2001), semiempirical indices (e.g., Gitelson et al. 2003, 2006), and radiative transfer models (Zarco-Tejada et al. 2001, 2004; Féret et al. 2008, 2011) have been used to characterize growth-related foliar chemicals (e.g., nitrogen and chlorophyll pigments), yet other studies demonstrate that remote sensing of canopy structure also aids quantitative retrieval of biochemical properties (e.g., Zhang et al. 2008; Hernández-Clemente et al. 2012; Knyazikhin et al. 2013a,b,c; Ollinger et al. 2013; Townsend et al. 2013) (Figure 16.1). Asner and Warner (2003) conclude that quantitative information on gap fraction and tree structure is needed to validate or constrain remote sensing models to accurately estimate chemistry and energy exchange. Possible ways to account for structure in the retrieval of foliar chemistry include canopy radiative transfer models, LiDAR, and other methods that account for intra- and intercanopy gaps, self-shading, and stand structure (see Section 16.5.1). Many proposed methods remain untested, including the directional area scattering factor (DASF), which is a function based on three wavelength invariant parameters: canopy interceptance, probability of recollision, and directional gap density (Lewis and Disney 2007; Schull et al. 2007, 2011; Knyazikhin et al. 2013a). Still other researchers have argued that the canopy architecture of a species is an integrated component of its strategy for resource capture and therefore should covary with chemistry (Ollinger et al. 2013; Townsend et al. 2013).

Foliar and canopy water content has also received a significant amount of attention due to its relationship with transpiration and plant water stress (Ustin et al. 2012; Hunt et al. 2013). The water absorption signal has a large effect on plant spectra, from small absorptions in the NIR at 970 and 1240 nm, accessible through HSI data, to a large broad absorption across the entire SWIR (1300–2500 nm). Gao and Goetz (1995) developed one of the first narrowband indices for the quantification of EWT of vegetation. The values derived for EWT from AVIRIS data were tested against field data from the Harvard Forest, Massachusetts. HSI also offers the unique ability to differentiate between different phases of water (atmospheric water vapor and the moisture content of vegetation), for which the absorption maxima are offset by about 40–50 nm (Gao and Geotz 1990). This ability to quantify atmospheric water aids in the statistical modeling of the atmosphere such that water vapor signals can be removed, permitting proper estimation of the underlying liquid water stored in vegetation (Green et al. 1989). Recently, Cheng et al. (2013b) showed that it is possible to monitor small diurnal changes in water content from optimized indices and wavelet analysis that provide information on plant water status and whether root uptake can support full transpiration demand.

Nonpigment materials in the forest canopy range from foliar carbon constituents, such as lignin and cellulose, to dead leaves, stems, or remaining reproductive structures of flowers and fruits. The detection and quantification of these materials, sometimes referred to as dry matter or nonphotosynthetic vegetation (NPV), is often used as an indicator of canopy stress and may be important for quantifying the contribution of plant litter to forest carbon pools. Particularly after foliage has lost pigments and water, the cellulose–lignin absorptions become easily detectable with HSI data through narrowband methods such as the cellulose absorption index (Daughtry 2001; Daughtry et al. 2005), spectral mixture analysis (Asner and Lobell 2000; Roberts et al. 2003a), chemometric approaches like PLSR (Asner et al. 2011), or radiative transfer models (Riaño et al. 2004; Jacquemoud et al. 2009). Kokaly et al. (2007, 2009) used continuum removal combined with a spectral library to reveal a 2–3 nm shift in the cellulose–lignin absorption feature when the concentration of lignin increases, demonstrating the utility of HSI in quantifying subtle variations in canopy carbon. Numerous examples of forest NPV quantification also exist in the HSI literature (e.g., Ustin and Trabucco 2000; Roberts et al. 2004; Guerschman et al. 2009). Dry matter signatures in the HSI spectrum have been used to assess whether canopies were subjected to insect defoliation, drought stress (White et al. 2007; Fassnacht et al. 2014), or root pathogen damage (Santos et al. 2010).

HSI data have significant potential for mapping forest composition at species and community levels, based largely on their biochemical attributes (Figure 16.2). Many examples have been published using various analytical approaches with airborne HSI images (e.g., Martin et al. 1998; Clark et al. 2005; Bunting and Lucas 2006; Bunting et al. 2010), EO-1 Hyperion satellite data (Townsend and Foster 2002), time series of Hyperion data (Kalacska et al. 2007; Somers and Asner 2013), and combinations of airborne HSI imagers and LiDAR (Dalponte et al. 2007; Jones et al. 2010; Colgan et al. 2012a; Naidoo et al. 2012; Baldeck et al. 2014). In recent years, the ability to map species and detailed land cover has significantly

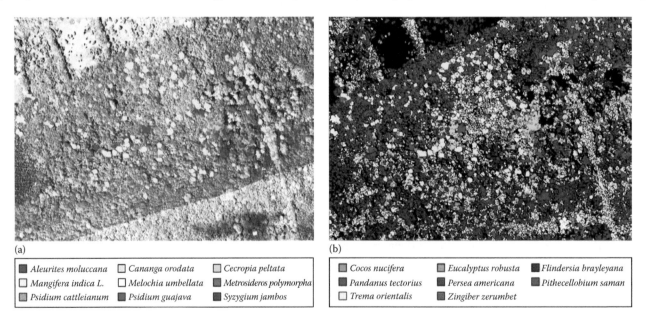

(a)

■ *Aleurites moluccana*	□ *Cananga orodata*	□ *Cecropia peltata*
□ *Mangifera indica L.*	□ *Melochia umbellata*	■ *Metrosideros polymorpha*
■ *Psidium cattleianum*	■ *Psidium guajava*	□ *Syzygium jambos*

(b)

□ *Cocos nucifera*	■ *Eucalyptus robusta*	■ *Flindersia brayleyana*
■ *Pandanus tectorius*	■ *Persea americana*	■ *Pithecellobium saman*
□ *Trema orientalis*	■ *Zingiber zerumbet*	

FIGURE 16.2	(a) A false color composite image of Nanawale Forest Reserve, Hawaii Island (R = 646 nm; G = 560.7 nm; B = 447 nm), with colored polygons showing locations of species data from a field survey. (b) Classification of 17 canopy species based on regularized discriminant analysis (n = 50 samples/species) using CAO VNIR imaging spectrometer data. (Reprinted from Féret, J.-B. and Asner, G.P., *Remote Sens. Environ.*, 115, 2415, 2013.)

improved (Asner 2013). It is likely that this is a consequence of improved instrument performance, especially for high-fidelity HSI data and for the adoption of a wide variety of new analytical methods including radiative transfer models, segmentation and object delineation, and numerous statistical methods such as ensemble classifiers, discriminate analysis, support vector machines, and combined approaches. No one method has yet been shown to work universally across global land cover types with complex environment and terrain interactions. However, several general conclusions can be inferred from these and other studies: (1) the addition of SWIR bands along with VNIR bands often significantly increases the accuracy of mapping forest species; (2) species mapping is further enhanced if HSI data encompass multidate periods that capture phenological patterns, as is consistent with improvements reported for multidate multispectral data (e.g., Wolter et al. 2008); and (3) combining information on tree structure from LiDAR, such as canopy height, diameter, and volume, with HSI data improves results (Féret and Asner 2013).

16.3.3 Canopy Physiology

Imaging spectroscopy can be used to characterize three key physiological processes responsible for carbon uptake in forests: photochemistry, nonphotochemical quenching (NPQ), and fluorescence. Solar radiation, and photosynthetically active radiation (PAR; 400–700 nm) in particular, supplies the energy that drives carbon uptake in forests. The first process, photochemistry, refers directly to the process by which the enzyme ribulose-1,5 bisphosphate carboxylase–oxygenase (RuBisCO) catalyzes RuBP to fix carbon from carbon dioxide. Within the Calvin cycle of C_3 plants (which includes trees), photochemistry is driven by the energy supplied from light harvesting by pigment complexes. The second process, NPQ, relates directly to plant interactions with light. Plants downregulate photosynthesis through a range of processes related to pigment concentrations to either make use of light energy or dissipate it (Demmig-Adams and Adams 2006). Photochemistry and NPQ processes can be characterized through estimation of pigment concentrations or through inference based on changes in leaf pigment pools associated with plant responses to excess light or stresses that prevent them from fully utilizing ambient light energy (Demmig-Adams and Adams 1996). Finally, all plants dissipate light energy through solar-induced fluorescence (SIF), which only occurs as a consequence of photosynthesis and has been found to scale directly to rates of photosynthetic activity (Baker 2008).

Quantifying foliar nitrogen, the key element in RuBisCO and a trait whose concentration within proteins in foliage scales directly with photosynthetic capacity (Field and Mooney 1986; Evans 1989; Reich et al. 1997), provides a measure of the functioning of forest canopies (as described earlier). This functioning includes the capacity for carbon uptake, but photosynthetic downregulation limits carbon uptake under adverse environmental conditions. The most widely used models of

photosynthesis employ the Farquhar model (Farquhar et al. 1980; Farquhar and von Caemmerer 1982), in which the potential photosynthetic performance of a leaf is characterized using two parameters: the maximum rate of carboxylation (V_{cmax}) governed by RuBisCO activity and the maximum electron transport rate (ETR) (J_{max} is the maximum rate of ETR; Farquar and von Caemmerer 1982). Together, these limit the maximum rate of photosynthesis (A_{max}). V_{cmax} is strongly related to N concentration and LMA, that is, the investment by a plant in light harvesting relative to construction and maintenance (Poorter et al. 2009). ETR and J_{max} are more closely related to the processes set in motion by light harvesting in PSI and PSII (PS = photosystem), necessary for the synthesis of adenosine triphosphate (ATP) to drive cellular reactions. Because the Calvin cycle depends on ATP availability to sustain the regeneration of RuBP (which in turn permits carboxylation), photosynthetic capacity is limited by J_{max}. Therefore, the optical properties of foliage related to light harvesting may also facilitate mapping J_{max} from HSI. It should be noted that all photosynthetic parameters of vegetation are sensitive to temperature and moisture, so any remotely sensed estimate of such parameters will be specific to the ambient conditions at the time of measurement (Serbin et al. 2012). HSI has also been used as part of multisensor approaches to characterize net ecosystem photosynthesis (e.g., Rahman et al. 2001; Asner et al. 2004a; Thomas et al. 2006, 2009).

Doughty et al. (2011) successfully related leaf-level spectroscopic measurements to A_{max} but had less success with the other parameters. Variations in V_{cmax} and J_{max} related to temperature were measured in cultivated aspen and cottonwood leaves and accurately predicted similar relationships in plantation trees (Serbin et al. 2012). The ability to map V_{cmax} and J_{max} from imaging spectroscopy is most likely a consequence of the ability to infer these properties from traits that are directly detectable based on known or hypothesized absorption features (e.g., N, LMA, and water; see Kattge et al. 2009 and Cho et al. 2010) and the coordination of these traits with canopy structure (Ollinger et al. 2013). These studies show promise for developing remote sensing methods to map the properties used by modelers to characterize forest physiological function.

Efforts to map parameters directly associated with photochemistry are an area of continuing development in imaging spectroscopy. The discipline of physiological remote sensing using HSI has its roots in efforts to characterize NPQ and how NPQ relates to photosynthetic rates and capacity. This work stems from the development of the Photochemical Reflectance Index (PRI) (Gamon et al. 1992; Peñuelas et al. 1995). While typically associated with the de-epoxidation of xanthophylls for photosynthetic downregulation during NPQ (Bilger and Björkman 1990; Demmig-Adams and Adams 1996), the PRI more generally correlates with total pigment pools and their variation with environmental context (Gamon and Bond 2013). As such, the PRI has been shown to be an indicator of photosynthetic rates and light use efficiency (LUE) (Gamon and Surfus 1999).

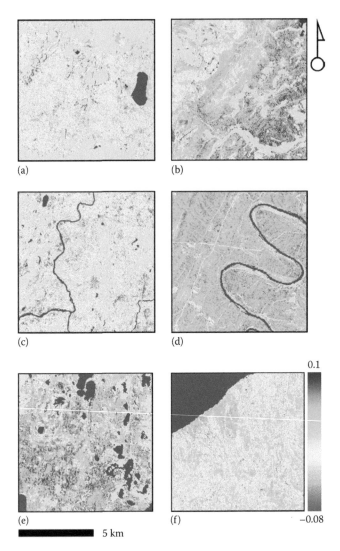

(a)   (b)

(c)   (d)

0.1

(e)   (f)   −0.08

5 km

FIGURE 16.3 Midsummer PRI images derived from 2009 AVIRIS of (a) oak/pine forests in Baraboo/Devil's Lake, Wisconsin; (b) oak and tulip poplar forests in Fernow Experimental Forest, West Virginia; (c) northern hardwood and conifer forests in Flambeau River State Forest, Wisconsin; (d) xeric oak forests in Green Ridge State Forest, Maryland; (e) northern hardwood and subboreal conifers in Ottawa National Forest, Michigan; and (f) hemlock, white pine, and deciduous hardwoods in the Porcupine Mountains, Michigan. Lower values indicate areas of greater vegetation stress. These images illustrate significant variability in forest physiological status across landscapes.

Accounting for species composition, environmental variability, and seasonal responses, the PRI is often correlated with the carotenoid to chlorophyll ratio ($r^2 = 0.50$–0.80), a property linked to photosynthesis and light harvesting (Garbulsky et al. 2011) (Figure 16.3). In addition, Stylinksi et al. (2000) also showed close relationships between the PRI and xanthophyll cycle pigments and modeled electron transport capacity (J_{max}) in leaves of pubescent oak (*Quercus pubescens*). Kefauver et al. (2013) showed

strong relationships between PRI and physiological damage to forests by ozone. A limitation of the PRI has been its species-level sensitivity, that is, relationships between the PRI and photosynthesis are species dependent (Guo and Trotter 2004; Filella et al. 2009; Ripullone et al. 2011). However, Hilker et al. (2008) have shown that PRI data may facilitate retrieval of plant photosynthetic efficiency independent of species composition.

The key physiological processes responsible for productivity of forests can also be addressed by remote sensing through the estimation of light absorption by canopies and its presumed linkage to light harvesting and use in photosynthesis. Under nonstressed conditions, net primary production is linearly related to the absorbed photosynthetically active radiation (APAR; Montieth 1977). This relationship is modulated by LUE. Traditionally, APAR has been successfully calculated from vegetation indices derived from spectral sensors of many varieties (e.g., Field et al. 1995; Sellers et al. 1996). The detection of forest LUE using PRI, and thus potential carbon uptake, has been demonstrated in numerous systems including boreal (Nichol et al. 2000) and conifer forests (Middleton et al. 2009; Atherton et al. 2013), but the utilization of remotely estimated APAR by the canopy for photosynthesis remains a more difficult task. The most common approach to assessing LUE using HSI has been through narrowband indices such as the PRI, which uses the reflectance at 570 and 531 nm (i.e., Gamon et al. 1997), but future developments in retrieving the Farquhar parameters (V_{cmax}, J_{max}) and SIF are likely to provide more robust estimates of key drivers of physiological processes. Ultimately, linkages across methods, for example, estimating LUE using derivations biochemistry (%N) and LAI, may provide a hybrid approach to best map factors important to net primary productivity (NPP) (Green et al. 2003).

Chlorophyll fluorescence provides another means of estimating photosynthetic performance and LUE from HSI data (Meroni et al. 2009). Numerous studies since the early 2000s have demonstrated the capacity of measurements of SIF to accurately characterize seasonal patterns of carbon uptake (Guanter et al. 2007; Frankenburg et al. 2011; Joiner et al. 2011). Under natural conditions, fluorescence and photosynthesis are positively correlated. Energy absorbed in the photosystems is reradiated at longer wavelengths than those absorbed, adding a subtle signal to reflected solar radiation, most notably with peaks around 685 and 740 nm. Measurements of SIF require narrowband data at specific wavelengths in the NIR in which the vegetation fluorescence signal in retrieved reflectance (about 2%) can be distinguished from NIR albedo (>40%) (Berry et al. 2013). Most efforts to date have focused on retrievals of SIF in narrow wavebands (preferably <0.3 nm) ± 20 nm around the solar Fraunhofer lines (wavelengths where there is no incoming solar energy, ~739 nm) or O_2-A band at 760 nm. Generally correlated with the PRI (Zarco-Tejada et al. 2009; Cheng et al. 2013a), fluorescence has also been measured at field sites differing in soil salinity and estimated spatially from airborne HSI data using the PRI index (531 and 570 nm) (Naumann et al. 2008). Zarco-Tejada et al. (2009) estimated fluorescence from infilling of the O_2-A bands at 757.5 and 760.5 nm measured in 1 nm wavelength bands, which minimized confounding effects from variance in chlorophyll and LAI. More recently, Zarco-Tejada et al. (2013) used narrowband spectral indices and fluorescence infilling at 750, 762, and 780 nm, revealing that seasonal spectroscopic trends tracked changes in carbon fluxes. HSI observations continue to pave additional avenues to insight on plant physiological processes.

16.4 LiDAR Remote Sensing of Forests

Whereas HSI provides estimates of the chemical, physiological, and plant compositional properties of forests, LiDAR probes the structural and architectural traits of vegetation as well as the terrain below the canopy (Table 16.3). A large number of synthesis papers have been written on the use of LiDAR for studies of ecosystem structure (e.g., Dubayah and Drake 2000; Lefsky et al. 2002; Lim et al. 2003; Vierling et al. 2008; Wulder et al. 2012), including in other chapters of this book (e.g., Chapter 17). Here, we only briefly highlight the various uses of LiDAR in the context of forest structure, architecture, and biomass; the reader should also read Chapter 17 for further details.

16.4.1 Canopy Structure and Biomass

The height of a forest canopy is a fundamental characteristic that both discrete and waveform LiDAR sensors are capable of describing (Figure 16.4). Even discrete-return datasets that contain only the first and last return from the laser pulse will allow for the calculation of this parameter, after a ground elevation model has been generated from LiDAR data (Lim et al. 2003). While canopy height alone does not provide extensive information on forest structure, it is a parameter related to tree diameters (Feldpausch et al. 2012), and thus to aboveground biomass.

LiDAR can also be used to determine the vertical profile of canopy tissues including foliar and some woody structures (Figure 16.5). Waveform LiDAR instruments collect the full shape of the returning laser pulse, allowing for detailed information on the structure of the canopy (Blair and Hofton 1999; Dubayah and Drake 2000; Ni-Meister et al. 2001). If detailed canopy structure is of interest, but only discrete-return LiDAR data are available, it is possible to use these data to generate a pseudowaveform. This method aggregates discrete returns into bins over spatial extents that incorporate multiple laser spots in order to gain an aggregated understanding of the vertical vegetation profile in the absence of waveform data for each laser pulse (Muss et al. 2011). Vertical profiles are indicative of canopy density, vertical distribution, and the presence of undergrowth, all of which can provide information on the 3D structure and habitat of forests (Parker 1995; Lefsky et al. 1999; Clark and Clark 2000; Weishampel et al. 2000; Drake et al. 2002; Asner et al. 2008; Vierling et al. 2008).

One of the most widespread uses for LiDAR-derived canopy information is in the estimation of aboveground biomass, also known as aboveground carbon density (ACD). Such approaches have been applied in numerous studies of conifer, broadleaf temperate, and tropical forest ecosystems (Nelson 1988; Lefsky et al. 1999, 2002, 2005; Popescu et al. 2003; Næsset and Gobakken 2008; van Aardt et al. 2008; Asner et al. 2012c; Wulder et al. 2012). The mean canopy profile height (MCH) has been used as the canopy structural metric, which relates the LiDAR vertical structure data to ACD (Lefsky et al. 2002; Asner et al. 2009). However, recent studies have indicated that variations in sensor characteristics and settings can cause significant differences in the MCH metric between data acquisitions (Næsset 2009), strongly indicating that top-of-canopy height is a more reliable method for estimating ACD of tropical forests (Asner and Mascaro 2014). The use of LiDAR data to produce estimates of ACD that closely match plot-level estimates allows for the mapping and monitoring of aboveground carbon stocks at landscape scales and, with the further development of spaceborne LiDAR, potentially regional/biome scales.

TABLE 16.3 Forest Structural Properties Estimated from LiDAR, along with an Estimate of Scientific Maturity and Example References

Vegetation Property	Maturity Level	Example References
Total canopy height	✓✓✓	Dubayah and Drake (2000), Ni-Meister et al. (2001), Drake et al. (2002), and Lim et al. (2003)
Mean canopy profile height	✓✓✓	Lefsky et al. (1999, 2002, 2005)
Aboveground biomass	✓✓	Nelson (1988), Lefsky et al. (1999, 2002, 2005), Popescu et al. (2003), Næsset and Gobakken (2008), van Aardt et al. (2008), Asner et al. (2012c), and Wulder et al. (2012)
Leaf area density	✓	Sun and Ranson (2000), Lovell et al. (2003), Riaño et al. (2004), Morsdorf et al. (2006), Richardson et al. (2009), Soldberg et al. (2009), and Vaughn et al. (2013)
Understory presence	✓✓	Zimble et al. (2003) and Asner et al. (2008)

Note: Maturity is a metric of relative accuracy as depicted in the literature, with one checkmark indicating low maturity and three checkmarks indicating high maturity.

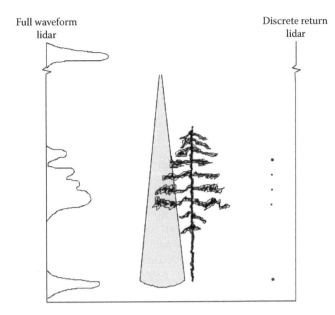

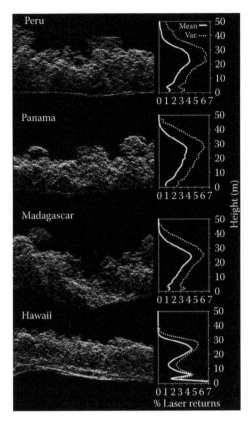

FIGURE 16.4 Illustration of waveform and discrete-return measurements of a tree. While both provide information on the vertical structure of canopies, discrete-return sampling records the returning laser pulse at specified peaks (e.g., first and last pulse) of the return wave, whereas waveform sampling collects the full shape of the returning pulse. (Reprinted from Lim, K. et al., *Prog. Phys. Geogr.*, 27, 88, 2003).

16.4.2 Light Penetration

Canopy gaps, or openings in forest canopies, influence population dynamics of forest trees by affecting forest structure, regeneration dynamics, and species composition (Brokaw 1985; Denslow 1987). Canopy gaps occur at scales ranging from single branches to multiple treefalls and result from disturbances caused by natural tree life cycles (Asner 2013), human processes such as logging (e.g., Nepstad et al. 1999; Asner et al. 2004b; Curran et al. 2004), and environmental factors such large-scale blowdowns (Chambers et al. 2013). Recently, airborne LiDAR data from a number of tropical forests have enabled the measurement of millions of canopy gaps over large spatial scales, both as single measurements and with repeat collections, improving the understanding of static and dynamics gaps, respectively (e.g., Magnussen et al. 2002; Kellner et al. 2009; Udayalakshmi et al. 2011; Armston et al. 2013).

Static canopy-gap size-frequency distributions (known as λ) are strikingly similar across a wide range of tropical forest types on differing geologic substrates and within differing disturbance regimes. This collective evidence suggests consistent turnover rates and similar mechanisms of gap formation across tropical forests (Kellner and Asner 2009; Asner et al. 2013, 2014). Deviations from this stable range of observed λ values potentially provide another metric for detecting and mapping disturbance. In a recent study, repeat LiDAR collections permitted the quantification of positive height changes in a forest canopy in

FIGURE 16.5 LiDAR cross-sectional views of four mature tropical forests in the Peruvian Amazon, Panamanian Neotropics, southeastern Madagascar, and Hawaii depict 3D forest structure along a 100 m long × 20 m wide transect. Right-hand panels show mean and spatial variance of LiDAR vertical canopy profiles for all returns in a 1 km^2 area centered on each cross section. Vertical canopy profiles are generally consistent across the four study sites, yet the Hawaiian forest contains the most pronounced groundcover, understory, and canopy layers. (Reprinted from Asner, G.P. et al., *Oecologia*, 168, 1147, 2012b).

Hawaii and illustrated how size and the proximity to other canopies influenced the outcome of competition for space within this forest (Kellner and Asner 2014).

16.5 Integrating HSI and LiDAR

In recent years, HSI and LiDAR observations have been integrated using two approaches. One method involves the acquisition of HSI and LiDAR data from separate platforms, such as from different aircraft, followed by modeling and analysis steps to fuse the resulting datasets (e.g., Mundt et al. 2006; Anderson et al. 2008; Jones et al. 2010). This is currently the most common approach, and following acquisition, the data must be digitally coaligned using techniques such as image pixel–based coregistration. These efforts usually yield an integrated data "cube" with an average misalignment of one pixel or so, although the scanning and/or array patterns of the HSI and LiDAR data may yield much higher coalignment errors.

A second, rapidly growing approach to HSI and LiDAR data integration involves the comounting of instruments on the same platform, whether on board aircraft or an unmanned aerial vehicle (UAV) (Asner et al. 2007). Integration steps range from colocating the instruments on the same mounting plate on board the aircraft or UAV, to precise time registration of each measurement, to final data fusion using ray tracing models for each instrument (Asner et al. 2012a). Each of these steps is key to producing a highly integrated dataset, reducing coalignment issues such that the data can be treated as one information vector per unit ground sample (e.g., one pixel). The onboard and postflight fusion of HSI and LiDAR data developed and deployed by the CAO (http://cao.carnegiescience.edu) has been replicated and is currently being used by the U.S. NEON's Airborne Operational Platform program (http://www.neoninc.org/science/aop).

16.5.1 Benefits of Data Fusion

The benefits of HSI and LiDAR data fusion include increased data dimensionality, constraints on the interpretation of one portion of the dataset using another portion, and filtering of data to specific observation conditions or specifications. The dimensionality of, or degrees of freedom within, a fused dataset increases with the integration of complementary or orthogonal observations such as chemical or physiological metrics from HSI and structural or architectural measures from LiDAR. A highly demonstrative example can be taken from two integrated HSI–LiDAR datasets collected with the CAO Airborne Taxonomic Mapping System (Figure 16.6). One dataset was collected over a portion of Stanford University in 2011, and the other taken over a remote Amazonian rainforest in the same year. In the Stanford case, the LiDAR data alone contain about 25 degrees of freedom for a 200 ha area comprised buildings with varying architecture, vegetation ranging from grasses to trees, roads and pathways, and other built surfaces. Here, degrees of freedom are quantitatively assessed using principal component analysis, so each degree is orthogonal to or unique from the others (Asner et al. 2012a). A 72-band VNIR image of the same Stanford scene, taken from the same aircraft, contains about 50 degrees of freedom. Combined, the VNIR HSI and LiDAR provide about 100 degrees of freedom. A VSWIR imaging spectrometer on board the same aircraft provides about 260 degrees of freedom in the Stanford case. In conjunction, the LiDAR and VNIR and VSWIR HSI offer more than 330 orthogonally aligned sources of information. In the Amazon forest case, data fusion yields similar increases in data dimensionality, more than doubling the information content by sensor fusion over that which can be achieved by any one sensor.

A second powerful use of combined HSI and LiDAR data involves constraint of interpretation and/or filtering of one data stream relative to the other. Looking down upon a forest canopy, one observes strong variation in bright and dark portions of the canopy, as well as gaps and spectrally inconsistent

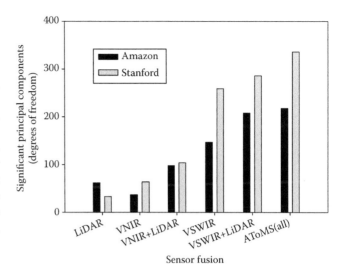

FIGURE 16.6 Integration of HSI and LiDAR sensor hardware, and data streams, provides a uniquely powerful way to greatly increase the inherent dimensionality of the data collected over forested areas and other ecosystems. For two sample 200 ha areas (Stanford University and a lowland Amazonian forest), individual LiDAR and HSI sensors provide highly dimensional data as assessed with principal components analysis. The dimensionality of the data increased when data are analyzed simultaneously. Here, VNIR is a visible-to-near infrared HSI and VSWIR is a visible-to-shortwave infrared HSI. All sensors combined are referred to as the Airborne Taxonomic Mapping Systems on board the CAO. (Reprinted from Asner, G.P., *Remote Sens. Environ.*, 124, 454, 2012a).

observation conditions (Figure 16.7). As a result, reflectance analysis of forests is often an underdetermined problem involving variation in 3D architecture, leaf layering (LAI), and foliar biochemical constituents. This variation in illumination conditions occurs between pixels in high-resolution HSI data and within pixels in lower resolution HSI data. One of many possible ways to constrain observation conditions for improved HSI-based analysis of forest canopy traits is to use the LiDAR (Asner and Martin 2008; Dalponte et al. 2008; Colgan et al. 2012b). For example, LiDAR maps of top-of-canopy structure can be used to precisely model sun and viewing geometry on the canopy surface in each pixel. Combined with simple filtering of the HSI data based on the NDVI or other narrowband indices, an HSI image can be partitioned into regions most suitable for a particular type of analysis. Biochemical analyses are particularly sensitive to this filtering process, and much higher performances in biochemical retrievals can be achieved based on combined HSI–LiDAR filtering (Asner and Martin 2008). Still other approaches to integrate HSI and LiDAR data have yet to be explored, such as in the full 3D analysis and modeling of canopy structural and functional traits. These approaches will become more common with the rise of integrated data fusion systems.

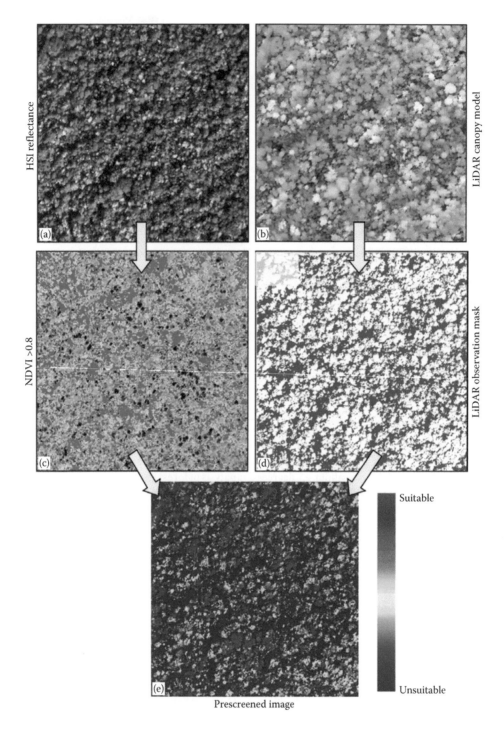

FIGURE 16.7 Prescreening of (a) HSI data using fused (b) LiDAR data. This can be accomplished in various ways, and an example is shown here. (c) A minimum NDVI threshold of 0.8 ensures sufficient foliar cover in the analysis. (d) Combining LiDAR and solar-viewing geometry, a filtering mask is generated to remove pixels in shade or of ground and water surfaces. (e) The resulting suitability image provides an indication of pixels that can be used for biophysical, biochemical, and/or physiological analysis.

16.6 Conclusions

HSI and LiDAR mapping provides independent and highly complementary data on forest canopies and whole ecosystems. Here, we summarized sources of HSI and LiDAR data, their general uses in determining forest structural and functional properties, and the potential value of collecting and analyzing HSI and LiDAR together via hardware integration and data fusion. Much of the science of HSI and LiDAR analysis of forests will remain in the airborne domain until orbital instrumentation is deployed and made available to the scientific research and application communities. In light of the myriad studies found throughout the remote sensing, forest science, and conservation research literature, it is clear that the time is right for a rapid expansion of HSI and LiDAR data collection and sharing efforts worldwide.

Acknowledgments

We thank E. Tasar for providing editing and management support. The CAO is made possible by the Gordon and Betty Moore Foundation, the Grantham Foundation for the Protection of the Environment, the John D. and Catherine T. MacArthur Foundation, the Avatar Alliance Foundation, the W. M. Keck Foundation, the Margaret A. Cargill Foundation, Mary Anne Nyburg Baker and G. Leonard Baker Jr., and William R. Hearst III.

References

Abshire, J. B., X. Sun, H. Riris, J. M. Sirota, J. F. McGarry, S. Palm, D. Yi et al. 2005. Geoscience laser altimeter system (GLAS) on the ICESat Mission: On-orbit measurement performance. *Geophys Res Lett* 32:L21S02.

Anderson, J. E., L. C. Plourde, M. E. Martin, B. H. Braswell, M-L. Smith, R. O. Dubayah, M. A. Hofton et al. 2008. Integrating waveform LiDAR with hyperspectral imagery for inventory of a northern temperate forest. *Remote Sens Environ* 112:1856–1870.

Armston, J., M. Disney, P. Lewis, P. Scarth, S. Phinn, R. Lucas, P. Bunting et al. 2013. Direct retrieval of canopy gap probability using airborne waveform LiDAR. *Remote Sens Environ* 134:24–38.

Asner, G. P. 1998. Biophysical and biochemical sources of variability in canopy reflectance. *Remote Sens Environ* 64:234–253.

Asner, G. P. 2013. Geography of forest disturbance. *Proc Nat/Acad Sci* 110:3711–3712.

Asner, G. P., C. Anderson, R. E. Martin, D. E. Knapp, R. Tupayachi, F. Sinca, and Y. Malhi. 2014. Landscape-scale changes in forest structure and functional traits along an andes-to-amazon elevation gradient. *Biogeosciences* 11:843–856.

Asner, G. P., J. K. Clark, J. Mascaro, G. A. Galindo García, K. Chadwick, D. A. Navarrete Encinales, G. Paez-Acosta et al. 2012c. High-resolution mapping of forest carbon stocks in the Colombian amazon. *Biogeosciences* 9:2683–2696.

Asner, G. P., R. F. Hughes, T. A. Varga, D. E. Knapp, and T. Kennedy-Bowdoin. 2009. Environmental and biotic controls over aboveground biomass throughout a tropical rain forest. *Ecosystems* 12:261–278.

Asner, G. P., R. F. Hughes, P. M. Vitousek, D. E. Knapp, T. Kennedy-Bowdoin, J. Boardman, R. E. Martin, M. Eastwood, and R. O. Green. 2008. Invasive plants alter 3-D structure of rainforests. *Proc Nat/Acad Sci* 105:4519–4523.

Asner, G. P., M. Keller, R. Pereira, J. C. Zweede, and J. N. Silva 2004b. Canopy damage and recovery after selective logging in Amazonia: Field and satellite studies. *Ecol Appl* 14:S280–S298.

Asner, G. P., J. R. Kellner, T. Kennedy-Bowdoin, D. E. Knapp, C. Anderson, and R. E. Martin. 2013. Forest canopy gap distributions in the southern Peruvian Amazon. *PloS One* 8:e60875.

Asner, G. P., D. E. Knapp, J. Boardman, R. O. Green, T. Kennedy-Bowdoin, M. Eastwood, R. E. Martin et al. B. 2012a. Carnegie airborne observatory-2: Increasing science data dimensionality via high-fidelity multi-sensor fusion. *Remote Sens Environ* 124:454–465.

Asner, G. P., D. E. Knapp, T. Kennedy-Bowdoin, M. O. Jones, R. E. Martin, J. Boardman and C. B. Field. 2007. Carnegie Airborne Observatory: In flight fusion of hyperspectral imaging and waveform light detection and ranging (LiDAR) for three-dimensional studies of ecosystems. *J Appl Remote Sens* 1:013536–36–21.

Asner, G. P. and D. B. Lobell. 2000. A biogeophysical approach for automated swir unmixing of soils and vegetation. *Remote Sens Environ* 74:99–112.

Asner, G. P. and R. E. Martin. 2008. Spectral and chemical analysis of tropical forests: Scaling from leaf to canopy levels. *Remote Sens Environ* 112:3958–3970.

Asner, G. P., R. E. Martin, R. Tupayachi, R. Emerson, P. Martinez, F. Sinca, G. V. N. Powell et al. 2011. Taxonomy and remote sensing of leaf mass per area (LMA) in humid tropical forests. *Ecol Appl* 21:85–98.

Asner, G. P. and J. Mascaro. 2014. Mapping tropical forest carbon: Calibrating plot estimates to a simple LiDAR metric. *Remote Sens Environ* 140:614–624.

Asner, G. P., J. Mascaro, H. C. Muller-Landau, G. Vieilledent, R. Vaudry, M. Rasamoelina, J. S. Hall et al. 2012b. A universal airborne LiDAR approach for tropical forest carbon mapping. *Oecologia* 168:1147–1160.

Asner, G. P., D. Nepstad, G. Cardinot, and D. Ray. 2004a. Drought stress and carbon uptake in an Amazon forest measured with spaceborne imaging spectroscopy. *Proc Nat/Acad Sci* 101:6039–6044.

Asner, G. P. and P. M. Vitousek. 2005. Remote analysis of biological invasion and biogeochemical change. *Proc Nat/Acad Sci* 102:4383–4386.

Asner, G. P. and A. S. Warner. 2003. Canopy shadow in IKONOS satellite observations of tropical forests and savannas. *Remote Sens Environ* 87:521–533.

Atherton, J. M., C. J. Nichol, M. Mencuccini, and K. Simpson. 2013. The utility of optical remote sensing for characterizing changes in the photosynthetic efficiency of Norway maple saplings following transplantation. *Int J Remote Sens* 34:655–667.

Baker, N. R. 2008. Chlorophyll fluorescence: A probe of photosynthesis in vivo. *Annu Rev Plant Biol* 59:89–113.

Baldeck, C. A., M. S. Colgan, J.-B. Féret, S. R. Levick, R. E. Martin, and G. P. Asner. 2014. Landscape-scale variation in plant community composition of an African savanna from airborne species mapping. *Ecol Appl* 24:84–93.

Barnsley, M. J., J. J. Settle, M. A. Cutter, D. R. Lobb, and F. Teston. 2004. The PROBA/CHRIS mission: A low-cost small-sat for hyperspectral multiangle observations of the earth surface and atmosphere. *IEEE Trans Geosci Remote Sens* 42:1512–1520.

Berry, J. A., C. Frankenberg, and P. Wennberg. 2013. New methods for measurements of photosynthesis from space. *Keck Institute for Space Studies* 1:72.

Bilger, W. and O. Björkman. 1990. Role of the xanthophyll cycle in photoprotection elucidated by measurements of light-induced absorbance changes, fluorescence and photosynthesis in leaves of *Hedera canariensis*. *Photosynth Res* 25:173–185.

Blackburn, G. A. 2007. Hyperspectral remote sensing of plant pigments. *J Exp Botany* 58:855–867.

Blair, J. B. and M. A. Hofton. 1999. Modeling laser altimeter return waveforms over complex vegetation using high-resolution elevation data. *Geophys Res Lett* 26:2509–2512.

Blair, J. B., D. L. Rabine, and M. A. Hofton. 1999. The laser vegetation imaging sensor: A medium-altitude, digitisation-only, airborne laser altimeter for mapping vegetation and topography. *ISPRS J Photogramm Remote Sens* 54:115–122.

Brokaw, N. V. L. 1985. Treefalls, regrowth, and community structure in tropical forests. In *The Ecology of Natural Disturbance and Patch Dynamics*, edited by S. T. A. Pickett, P. S. White (pp. 53–69). New York: Academic Press.

Bunting, P. and R. Lucas. 2006. The delineation of tree crowns in Australian mixed species forests using hyperspectral compact airborne spectrographic imager (CASI) data. *Remote Sens Environ* 101:230–248.

Bunting, P., R. Lucas, K. Jones, and A. R. Bean. 2010. Characterisation and mapping of forest communities by clustering individual tree crowns. *Remote Sens Environ* 114:231–245.

Chambers, J. Q., R. I. Negron-Juarez, D. M. Marra, A. Di Vittorio, J. Tews, D. Roberts, G. H. P. M. Ribeiro, S. E. Trumbore, and N. Higuchi. 2013. The steady-state mosaic of disturbance and succession across an old growth central amazon forest landscape. *Proc Nat/Acad Sci* 110:3949–3954.

Cheng, T., D. Riaño, A. Koltunov, M. L. Whiting, S. L. Ustin, and J. Rodriguez. 2013b. Detection of diurnal variation in orchard canopy water content using MODIS/ASTER airborne simulator (MASTER) data. *Remote Sens Environ* 132:1–12.

Cheng, Y.-B., E. M. Middleton, Q. Zhang, K. F. Huemmrich, P. K. E. Campbell, L. A. Corp, B. D. Cook et al. 2013a. Integrating solar induced fluorescence and the photochemical reflectance index for estimating gross primary production in a cornfield. *Remote Sens* 5:6857–6879.

Cho, M. A., J. A. van Aardt, R. Main, and B. Majeke. 2010. Evaluating variations of physiology-based hyperspectral features along a soil water gradient in a eucalyptus grandis plantation. *Int J Remote Sens* 31:3143–3159.

Clark, D. B. and D. A. Clark. 2000. Landscape-scale variation in forest structure and biomass in a tropical rain forest. *Forest Ecol Manag* 137:185–198.

Clark, M. L., D. A. Roberts, and D. B. Clark. 2005. Hyperspectral discrimination of tropical rain forest tree species at leaf to crown scales. *Remote Sens Environ* 96:375–398.

Colgan, M., C. Baldeck, J.-B. Féret, and G. Asner. 2012a. Mapping savanna tree species at ecosystem scales using support vector machine classification and BRDF correction on airborne hyperspectral and LiDAR data. *Remote Sens* 4:3462–3480.

Colgan, M. S., G. P. Asner, S. R. Levick, R. E. Martin, and O. A. Chadwick. 2012b. Topo-edaphic controls over woody plant biomass in South African savannas. *Biogeosciences* 9:1809–1821.

Curran, L. M., S. N. Trigg, A. K. McDonald, D. Astiani, Y. M. Hardiono, P. Siregar, I. Caniago, E. Kasischke. 2004. Lowland forest loss in protected areas of Indonesian Borneo. *Science* 303:1000–1003.

Dahlin, K. M., G. P. Asner, and C. B. Field. 2013. Environmental and community controls on plant canopy chemistry in a Mediterranean-type ecosystem. *Proc Nat/Acad Sci* 110:6895–6900.

Dalponte, M., L. Bruzzone, and D. Gianelle. 2007. Tree species classification in the Southern Alps based on the fusion of very high geometrical resolution multispectral/hyperspectral images and LiDAR data. *Remote Sens Environ* 123:258–270.

Dalponte, M., L. Bruzzone, and D. Gianelle. 2008. Fusion of hyperspectral and LiDAR remote sensing data for classification of complex forest areas. *IEEE Trans Geosci Remote Sens* 46:1416–1427.

Daughtry, C. S. T. 2001. Discriminating crop residues from soil by shortwave infrared reflectance. *Agronomy J* 93:125–131.

Daughtry, C. S. T., E. R. Hunt, P. C. Doraiswamy, and J. E. McMurtrey III. 2005. Remote sensing the spatial distribution of crop residues. *Agronomy J* 97:864–871.

Daughtry, C. S. T., C. L. Walthall, M. S. Kim, E. Brown de Colstoun, and J. E. McMurtrey III. 2000. Estimating corn leaf chlorophyll concentration from leaf and canopy reflectance. *Remote Sens Environ* 74:229–239.

Demmig-Adams, B. and W. W. Adams. 1996. The role of the xanthophyll cycle carotenoids in the protection of photosynthesis. *Trends Plant Sci* 1:21–26.

Demmig-Adams, B. and W. W. Adams. 2006. Photoprotection in an ecological context: The remarkable complexity of thermal energy dissipation. *New Phytol* 172:11–21.

Denslow, J. S. 1987. Tropical rainforest gaps and tree species diversity. *Annu Rev Ecol Evol Syst* 18:431–451.

Doughty, C. E., G. P. Asner, and R. E. Martin. 2011. Predicting tropical plant physiology from leaf and canopy spectroscopy. *Oecologia* 165:289–299.

Drake, J. B., R. O. Dubayah, D. B. Clark, R. G. Knox, J. B. Blair, M. A. Hofton, R. L. Chazdon et al. 2002. Estimation of tropical forest structural characteristics using large-footprint LiDAR. *Remote Sens Environ* 79:305–319.

Dubayah, R. and J. Drake. 2000. LiDAR remote sensing for forestry. *J For* 98:44–46.

Evans, J. R. 1989. Photosynthesis and nitrogen relationships in leaves of c3 plants. *Oecologia* 78:9–19.

Farquhar, G. D. and S. von Caemmerer. 1982. Modelling of photosynthetic response to environmental conditions. In *Physiological Plant Ecology II. Water Relations and Carbon Assimilation*, edited by O. L. Lange, P. S. Nobel, C. B. Osmond, and H. Ziegler (pp. 549–587). Berlin, Germany: Springer-Verlag.

Farquhar, G. D., S. von Caemmerer, and J. A. Berry. 1980. A biochemical model of photosynthetic CO_2 assimilation in leaves of c3 species. *Planta* 149:78–90.

Fassnacht, F. E., H. Latifi, A. Ghosh, P. K. Joshi, and B. Koch. 2014. Assessing the potential of hyperspectral imagery to map bark beetle-induced tree mortality. *Remote Sens Environ* 140:533–548.

Feilhauer, H., U. Faude, and S. Schmidtlein. 2011. Combining isomap ordination and imaging spectroscopy to map continuous floristic gradients in a heterogeneous landscape. *Remote Sens Environ* 115:2513–2524.

Feldpausch, T. R., J. Lloyd, S. L. Lewis, R. J. W. Brienen, M. Gloor, A. Monteagudo Mendoza, G. Lopez-Gonzalez et al. 2012. Tree height integrated into pantropical forest biomass estimates. *Biogeosciences* 9:3381–3403.

Féret, J.-B. and G. P. Asner. 2013. Spectroscopic classification of tropical forest species using radiative transfer modeling. *Remote Sens Environ* 115:2415–2422.

Féret, J.-B., C. François, G. P. Asner, A. A. Gitelson, R. E. Martin, L. P. R. Bidel, S. L. Ustin et al. 2008. Prospect −4 and −5: Advances in the leaf optical properties model separating photosynthetic pigments. *Remote Sens Environ* 112:3030–3043.

Féret, J.-B., C. François, A. Gitelson, G. P. Asner, K. M. Barry, C. Panigada, A. D. Richardson et al. 2011. Optimizing spectral indices and chemometric analysis of leaf chemical properties using radiative transfer modeling. *Remote Sens Environ* 115:2742–2750.

Field, C. and H. A. Mooney. 1986. The photosynthesis–nitrogen relationship in wild plants. In *On the Economy of Plant Form and Function*, edited by T. J. Givnish (pp. 25–55). Cambridge, U.K.: Cambridge University Press.

Field, C. B., J. T. Randerson, and C. M. Malmström. 1995. Global net primary production: Combining ecology and remote sensing. *Remote Sens Environ* 51:74–88.

Filella, I., A. Porcar-Castell, S. Munné-Bosch, J. Bäck, M. F. Garbulsky, and J. Peñuelas. 2009. PRI assessment of long-term changes in carotenoids/chlorophyll ratio and short-term changes in de-epoxidation state of the xanthophyll cycle. *Int J Remote Sens* 30:4443–4455.

Frankenburg, C., J. B. Fisher, J. Worden, G. Badgley, S. S. Saatchi, J-E. Lee, G. C. Toon et al. 2011. New global observations of the terrestrial carbon cycle from gosat: Patterns of plant fluorescence with gross primary productivity. *Geophys Res Lett* 38:L17706.

Gamon, G. A. and B. Bond. 2013. Effects of irradiance and photosynthetic downregulation on the photochemical reflectance index in Douglas-fir and ponderosa pine. *Remote Sens Environ* 135:141–149.

Gamon, J. A., J. Peñuelas, and C. B. Field. 1992. A narrow-waveband spectral index that tracks diurnal changes in photosynthetic efficiency. *Remote Sens Environ* 41:35–44.

Gamon, J. A., L. Serrano, and J. S. Surfus. 1997. The photochemical reflectance index: An optical indicator of photosynthetic radiation use efficiency across species, functional types, and nutrient levels. *Oecologia* 112:492–501.

Gamon, J. A. and J. S. Surfus. 1999. Assessing leaf pigment content and activity with a reflectometer. *New Phytol* 143:105–117.

Gao, B. C. 1996. NDWI: A normalized difference water index for remote sensing of vegetation liquid water from space. *Remote Sens Environ* 58:257–266.

Gao, B. C. and A. F. H. Goetz. 1990. Column atmospheric water-vapor and vegetation liquid water retrievals from airborne imaging spectrometer data. *J Geophys Res Atmos* 95:3549–3564.

Gao, B. C. and A. F. H. Goetz. 1995. Retrieval of equivalent water thickness and information related to biochemical-components of vegetation canopies from AVIRIS data. *Remote Sens Environ* 52:155–162.

Garbulsky, M. F., J. Penuelas, J. Gamon, Y. Inoue, and I. Filella. 2011. The photochemical reflectance index (PRI) and the remote sensing of leaf, canopy and ecosystem radiation use efficiencies; a review and metaanalysis. *Remote Sens Environ* 115:281–297.

Gitelson, A. A. 2004. Wide dynamic range vegetation index for remote quantification of biophysical characteristics of vegetation. *J Plant Physiol* 161:165–173.

Gitelson, A. A., U. Gritz, and M. N. Merzlyak. 2003. Relationships between leaf chlorophyll content and spectral reflectance and algorithms for non-destructive chlorophyll assessment in higher plant leaves. *J Plant Physiol* 160:271–282.

Gitelson, A. A., Y. J. Kaufman, R. Stark, and D. Rundquist. 2002. Novel algorithms for remote estimation of vegetation fraction. *Remote Sens Environ* 80:76–87.

Gitelson, A. A., G. P. Keydan, and M. N. Merzlyak. 2006. Three-band model for noninvasive estimation of chlorophyll, carotenoids, and anthocyanin contents in higher plant leaves. *Geophys Res Lett* 33:L11402.

Gitelson, A. A., M. N. Merzlyak, and O. B. Chivkunova. 2001. Optical properties and non-destructive estimation of anthocyanin content in plant leaves. *J Photochem Photobiol* 74:38–45.

Gong, P., R. Pu, and I. R. Miller. 1992. Correlating leaf area index of ponderosa pine with hyperspectral CASI data. *Can J Remote Sens* 78:275–282.

Gong, P., R. Pu, and J. R. Miller. 1995. Coniferous forest leaf area index along the Oregon transect using compact airborne spectrographic imager data. *Photogramm Eng Remote Sensing* 61:107–117.

Green, R. O. 1998. Spectral calibration requirement for earth-looking imaging spectrometers in the solar-reflected spectrum. *Appl Optics* 37:683–690.

Green, R. O., V. Carrère, and J. E. Conel. 1989. Measurement of atmospheric water vapor using the airborne visible infrared imaging spectrometer. P. 6 in *Workshop Imaging Processing*, : Reno, NV: American Society of Photogrammetry and Remote Sensing.

Green, D. S., J. E. Erickson, and E. L. Kruger. 2003. Foliar morphology and canopy nitrogen as predictors of light-use efficiency in terrestrial vegetation. *Agric For Meteorol* 115:163–171.

Guanter, L., L. Alonso, L. Gómez-Chova, J. Amorós-López, J. Vila, and J. Moreno. 2007. Estimation of solar-induced vegetation fluorescence from space measurements. *Geophys Res Lett* 34:L08401.

Guerschman, J. P., M. J. Hill, L. J. Renzullo, D. J. Barrett, A. S. Marks, and E. J. Botha. 2009. Estimating fractional cover of photosynthetic vegetation, non-photosynthetic vegetation and bare soil in the Australian tropical savanna region upscaling the EO-1 hyperion and MODIS sensors. *Remote Sens Environ* 113:928–945.

Guo, J. and C. M. Trotter. 2004. Estimating photosynthetic light-use efficiency using the photochemical reflectance index: Variations among species. *Functional Plant Biol* 31:255–265.

Haboudane, D., J. R. Miller, N. Tremblay, P. J. Zarco-Tejada. 2002. Integrated narrow-band vegetation indices for prediction of crop chlorophyll content for application to precision agriculture. *Remote Sens Environ* 81:416–426.

Hernández-Clemente, R., R. M. Navarro-Cerrillo, and P. J. Zarco-Tejada. 2012. Carotenoid content estimation in a heterogeneous conifer forest using narrow-band indices and prospect + dart simulations. *Remote Sens Environ* 127:298–315.

Hilker, T., N. C. Coops, F. G. Hall, A. Black, M. A. Wulder, Z. Nesic, and P. Krishnan. 2008. Separating physiologically and directionally induced changes in PRI using BRDF models. *Remote Sens Environ* 112:2777–2788.

Homolová, L., Z. Malenovský, J. Clevers, G. García-Santos, and M. E. Schaepman. 2013. Review of optical-based remote sensing for plant trait mapping. *Ecol Complex* 15:1–16.

Huete, A. R. 1988. A soil-adjusted vegetation index (SAVI). *Remote Sens Environ* 25:295–309.

Hunt, E. R. Jr. and B. N. Rock. 1989. Detection of changes in leaf water content using near- and middle-infrared reflectances. *Remote Sens Environ* 30:43–54.

Hunt, E. R. Jr., S. L. Ustin, and D. Riaño. 2013. Remote sensing of leaf, canopy and vegetation water contents for satellite environmental data records. In *Satellite-based Applications of Climate Change*, edited by J. J. Qu, A. M. Powell, and M. V. K. Sivakumar (pp. 335–358). New York: Springer.

IPCC. 2000. Special report on land use, land-use change, and forestry. Edited by R. T. Watson, I. R. Noble, B. Bolin, N. H. Ravindranath, D. J. Verardo, and D. J. Dokeen, Cambridge, U.K.: Cambridge University Press.

Jacquemoud, S., W. Verhoef, F. Baret, C. Bacour, P. J. Zarco-Tejada, G. P. Asner, C. François, and S. L. Ustin. 2009. Prospect + sail: A review of use for vegetation characterization. *Remote Sens Environ* 113:S56–S66.

Joiner, J., Yoshida, Y., Vasilkov, A. P., and Middleton, E. M. 2011. First observations of global and seasonal terrestrial chlorophyll fluorescence from space. *Biogeosciences*, 8(3):637–651.

Jones, T. G., N. C. Coops, and T. Sharma. 2010. Assessing the utility of airborne hyperspectral and LiDAR data for species distribution mapping in the coastal Pacific Northwest, Canada. *Remote Sens Environ* 114:2841–2852.

Jordan, C. F. 1969. Leaf-area index from quality of light on the forest floor. *Ecology* 50:663–666.

Kalacska, M., S. Bohlman, G. A. Sanchez-Azofeifa, K. Castro-Esau, T. Caelli. 2007. Hyperspectral discrimination of tropical dry forest lianas and trees: Comparative data reduction approaches at the leaf and canopy levels. *Remote Sens Environ* 109:406–415.

Kattge, J., W. Knorr, T. Raddatz, and C. Wirth. 2009. Quantifying photosynthetic capacity and its relationship to leaf nitrogen content for global-scale terrestrial biosphere models. *Global Change Biol* 15:976–991.

Kefauver, S. C., J. Peñuelas, and S. Ustin. 2013. Using topographic and remotely sensed variables to assess ozone injury to conifers in the Sierra Nevada (USA) and Catalonia (Spain). *Remote Sens Environ* 139:138–148.

Kellner, J. R. and G. P. Asner. 2009. Convergent structural responses of tropical forests to diverse disturbance regimes. *Ecol Lett* 9:887–897.

Kellner, J. R. and G. P. Asner. 2014. Winners and losers in the competition for space in tropical forest canopies. *Ecol Lett.* doi:10.1111/ele.12256.

Kellner, J. R., D. B. Clark, S. P. Hubbell. 2009. Pervasive canopy dynamics produce short-term stability in a tropical rain forest landscape. *Ecol Lett* 12:155–164.

Kim, M. S. 1994. The use of narrow spectral bands for improving remote sensing estimation of fractionally absorbed photosynthetically active radiation (FAPAR). Masters Thesis, Department of Geography, University of Maryland, College Park, MD.

Knyazikhin, Y., P. Lewis, M. I. Disney, P. Stenberg, M. Mõttus, M. Rautiainen, P. Stenberg et al. 2013b. Reply to Ollinger et al.: Remote sensing of leaf nitrogen and emergent ecosystem properties. *Proc Nat/Acad Sci* 110:E2438–E2438.

Knyazikhin, Y., P. Lewis, M. I. Disney, P. Stenberg, M. Mõttus, M. Rautianinen, R. K. Kaufmann et al. 2013c. Reply to Townsend et al.: Decoupling contributions from canopy structure and leaf optics is critical for remote sensing leaf biochemistry. *Proc Nat/Acad Sci* 110:E1075–E1075.

Knyazikhin, Y., M. A. Schull, P. Stenberg, M. Mõttus, M. Rautiainen, Y. Yang, A. Marshak et al. 2013a. Hyperspectral remote sensing of foliar nitrogen content. *Proc Nat/Acad Sci* 110:E185–E192.

Kokaly, R. F. 2001. Investigating a physical basis for spectroscopic estimates of leaf nitrogen concentration. *Remote Sens Environ* 75:153–161.

Kokaly, R. F., G. P. Asner, S. V. Ollinger, M. E. Martin, and C. A. Wessman. 2009. Characterizing canopy biochemistry from imaging spectroscopy and its application to ecosystem studies. *Remote Sens Environ* 113:S78–S91.

Kokaly, R. F., B. W. Rockwell, S. L. Haire, and T. V. V. King. 2007. Characterization of post-fire surface cover, soils, and burn severity at the Cerro Grande fire, New Mexico, using hyperspectral and multispectral remote sensing. *Remote Sens Environ* 106:305–325.

le Maire, G., C. François, K. Soudani, D. Berveiller, J-Y. Pontailler, N. Bréda, H. Genet et al. 2008. Calibration and validation of hyperspectral indices for the estimation of broadleaved forest leaf chlorophyll content, leaf mass per area, leaf area index and leaf canopy biomass. *Remote Sens Environ* 112:3846–3864.

Lee, K.-S., W. B. Cohen, R. E. Kennedy, T. K. Maiersperger, and S. T. Gower. 2004. Hyperspectral versus multispectral data for estimating leaf area index in four different biomes. *Remote Sens Environ* 91:508–520.

Lefsky, M. A., W. B. Cohen, S. A. Acker, G. G. Parker, T. A. Spies, and D. Harding. 1999. LiDAR remote sensing of the canopy structure and biophysical properties of Douglas-fir western hemlock forests. *Remote Sens Environ* 70:339–361.

Lefsky, M. A., W. B. Cohen, D. J. Harding, G. G. Parker, S. A. Acker, and S. T. Gower. 2002. LiDAR remote sensing of above-ground biomass in three biomes. *Global Ecol Biogeogr.* 11:393–399.

Lefsky, M. A., D. J. Harding, M. Keller, W. B. Cohen, C. C. Carabajal, F. D. B. Espirito-Santo, M. O. Hunter et al. 2005. Estimates of forest canopy height and aboveground biomass using ICESat. *Geophys Res Lett* 35:L22S02.

Lewis, P. and M. Disney. 2007. Spectral invariants and scattering across multiple scales from within-leaf to canopy. *Remote Sens Environ* 109:196–206.

Lim, K., P. Treitz, K. Baldwin, I. Morrison, and J. Green. 2003. LiDAR remote sensing of forest structure. *Prog Phys Geog.* 27:88–106.

Lovell, J. L., D. L. B. Jupp, D. S. Culvenor, and N. C. Coops. 2003. Using airborne and ground-based LiDAR to measure canopy structure in Australian forests. *Can J Remote Sens* 29:607–622.

Magnussen, S., M. Wulder, and D. Seemann. 2002. Stand canopy closure estimated by line sampling. In *Continuous Cover Forestry*, edited by K. von Gadow, J. Nagel, and J. Saborowski (pp. 1–12). Dordrecht, the Netherlands: Springer.

Martin, M. E., S. D. Newman, J. D. Aber, and R. G. Congalton. 1998. Determining forest species composition using high spectral resolution remote sensing data. *Remote Sens Environ* 65:249–254.

Martin, M. E., L. C. Plourde, S. V. Ollinger, M.-L. Smith, B. E. McNeil. 2008. A generalizable method for remote sensing of canopy nitrogen across a wide range of forest ecosystems. *Remote Sens Environ* 112:3511–3519.

Meroni, M., M. Rossini, L. Guanter, L. Alonso, U. Rascher, R. Colombo, and J. Moreno. 2009. Remote sensing of solar-induced chlorophyll fluorescence: Review of methods and applications. *Remote Sens Environ* 113:2037–2051.

Middleton, E. M., Y.-B. Cheng, T. Hilker, T. A. Black, P. Krishnan, N. C. Coops, and K. F. Huemmrich. 2009. Linking foliage spectral responses to canopy-level ecosystem photosynthetic light-use efficiency at a Douglas-fir forest in Canada. *Can J Remote Sens* 35:166–188.

Montieth, J. L. 1977. Climate and the efficiency of crop production in Britain. *Philos Trans R Soc B* 281:277–294.

Morsdorf, F., B. Kötz, K. I. Itten, and B. Allgöwer. 2006. Estimation of LAI and fractional cover from small footprint airborne laser scanning data based on gap fraction. *Remote Sens Environ* 29:607–622.

Mundt, J. T., D. R. Streutker, and N. F. Glenn. 2006. Mapping sagebrush distribution using fusion of hyperspectral and LiDAR classifications. *Photogramm Eng Remote Sens* 72:47–54.

Muss, J. D., D. J. Mladenoff, and P. A. Townsend. 2011. A pseudo-waveform technique to assess forest structure using discrete LiDAR data. *Remote Sens Environ* 115:824–835.

Næsset, E. 2009. Effects of different sensors, flying altitudes, and pulse repetition frequencies on forest canopy metrics and biophysical stand properties derived from small-footprint airborne laser data. *Remote Sens Environ* 113:148–159.

Næsset, E. and T. Gobakken. 2008. Estimation of above- and below-ground biomass across regions of the boreal forest zone using airborne laser. *Remote Sens Environ* 112: 3079–3090.

Naidoo, L., M. A. Cho, R. Mathieu, and G. P. Asner. 2012. Classification of savanna tree species, in the greater Kruger national park region, by integrating hyperspectral and LiDAR data in a random forest data mining environment. *ISPRS J Photogramm Remote Sens* 69:167–179.

Naumann, J. C., J. E. Anderson, and D. R. Young. 2008. Linking physiological responses, chlorophyll fluorescence and hyperspectral imagery to detect salinity stress using the physiological reflectance index in the coastal shrub, *Myrica cerifera*. *Remote Sens Environ* 112:3865–3875.

Nelson, R. 1988. Estimating forest biomass and volume using airborne laser data. *Remote Sens Environ* 24:247–267.

Nepstad, D. C., A. Verissimo, A. Alencar, C. Nobre, E. Lima, P. Lefebvre, P. Schlesinger et al. 1999. Large-scale impoverishment of Amazonian forests by logging and fire. *Nature* 398:505–508.

Nichol, C. J., K. F. Huemmrich, T. A. Black, P. G. Jarvis, C. L. Walthall, J. Grace, and F. G. Hall. 2000. Remote sensing of photosynthetic-light-use efficiency of boreal forest. *Agric For Meteorol* 101:131:142.

Niinemets, U. 1999. Research review. Components of leaf dry mass per area—thickness and density—Alter leaf photosynthetic capacity in reverse directions in woody plants. *New Phytol* 144:35–47.

Ni-Meister, W., D. L. B. Jupp, and R. Dubayah. 2001. Modeling LiDAR waveforms in heterogeneous and discrete canopies. *IEEE Trans Geosci Remote Sens* 39:1943–1958.

Ollinger, S. V., P. B. Reich, S. Frolking, L. C. Lepine, D. Y. Hollinger, and A. D. Richardson. 2013. Nitrogen cycling, forest canopy reflectance, and emergent properties of ecosystems. *Proc Nat/Acad Sci* 110:E2437.

Ollinger, S. V., M. L. Smith, M. E. Martin, R. A. Hallett, C. L. Goodale, and J. D. Aber 2002. Regional variation in foliar chemistry and n cycling among forests of diverse history and composition. *Ecology* 83:339–355.

Parker, G. G. 1995. Structure and microclimate of forest canopies. In *Forest Canopies*, edited by M. D. Lowman and N. M. Nadkarni (pp. 73–106). New York: Academic Press.

Peñuelas, J., F. Baret, and I. Filella. 1995. Semiempirical indexes to assess carotenoids chlorophyll-a ratio from leaf spectral reflectance. *Photosynthetica* 31:221–230.

Peñuelas, J. F., J. Pinol, R. Ogaya, and I. Lilella. 1997. Estimation of plant water concentration by the reflectance water index WI (r900/r970). *Int J Remote Sens* 18:2869–2875.

Poorter, H., U. Niinemets, L. Poorter, I. J. Wright, and R. Villar. 2009. Causes and consequences of variation in leaf mass per area (LMA): A meta-analysis. *New Phytol* 182:565–588.

Popescu, S. C., R. H. Wynne, and R. F. Nelson. 2003. Measuring individual tree crown diameter with LiDAR and assessing its influence on estimating forest volume and biomass. *Can J Remote Sens* 29:564–577.

Rahman, A. F., J. A. Gamon, D. A. Fuentes, D. A. Roberts, and D. Prentiss. 2001. Modeling spatially distributed ecosystem flux of boreal forest using hyperspectral indices from AVIRIS imagery. *J Geophys Res* 106:33579–33591.

Reich, P. B., M. B. Walters, and D. S. Ellsworth. 1997. From tropics to tundra: Global convergence in plant functioning. *Proc Nat/Acad Sci* 94:13730–13734.

Riaño, D. P., F. Valladares, S. Condés, and E. Chuvieco. 2004b. Estimation of leaf area index and covered ground from airborne laser scanner (LiDAR) in two contrasting forests. *Agric For Meteorol* 124:269–275.

Riaño, D. P., P. Vaughan, E. Chuvieco, P. J. Zarco-Tejada, and S. L. Ustin. 2004a. Estimation of fuel moisture content by inversion of radiative transfer models to simulate equivalent water thickness and dry matter content. Analysis at leaf and canopy level. *IEEE Trans Geosci Remote Sens* 43:819–826.

Richardson, J. J., L. M. Moskal, and S.-H. Kim. 2009. Modeling approaches to estimate effective leaf area index from aerial discrete-return LiDAR. *Agric For Meteorol* 149:1152–1160. http://earthobservatory.nasa.gov/Features/EO1Tenth/

Riebeek, H. 2010. Earth observing-1: Ten years of innovation. In *NASA Earth Observatory: Features*.

Ripullone, F., A. R. Rivelli, R. Baraldi, R. Guarini, R. Guerrieri, F. Magnani, J. Peñuelas et al. 2011. Effectiveness of the photochemical reflectance index to track photosynthetic activity over a range of forest tree species and plant water statuses. *Funct Plant Biol* 38:177–186.

Roberts, D. A., P. Dennison, M. Gardner, Y. Hetzel, S. L. Ustin, and C. Lee. 2003. Evaluation of the potential of hyperion for fire danger assessment by comparison to the airborne visible infrared imaging spectrometer. *IEEE Trans Geosci Remote Sens* 41:1297–1310.

Roberts, D. A., M. Gardner, R. Church, S. L. Ustin, G. Scheer, and R. O. Green. 1998. Mapping chaparral in the Santa Monica mountains using multiple endmember spectral mixture models. *Remote Sens Environ* 65:267–279.

Roberts, D. A., S. L. Ustin, S. Ogunjemiyo, J. Greenberg, S. Z. Dobrowski, J. Chen, and T. M. Hinckley. 2004. Spectral and structural measures of northwest forest vegetation at leaf to landscape scale. *Ecosystems* 7:545–562.

Rondeaux, G., M. Steven, and F. Baret. 1996. Optimization of soil-adjusted vegetation indices. *Remote Sens Environ* 55:95–107.

Rouse, J. W., R. H. Haas, J. A. Schell, and D. W. Deering. 1974. Monitoring vegetation systems in the great plains with ERTS. In *Third ERTS Symposium NASA SP-351*, edited by S. C. Fraden, E. P. Marcanti, and M. A. Becker (pp. 309–317). Washington, DC: NASA.

Santos, M. J., J. A. Greenberg, and S. L. Ustin. 2010. Detecting and quantifying southeastern pine senescence effects to red-cockaded woodpecker (*Picoides borealis*) habitat using hyperspectral remote sensing. *Remote Sens Environ* 114:1242–1250.

Schull, M. A., S. Ganguly, A. Samanta, D. Huang, N. V. Shabanov, J. P. Jenkins, J. C. Chiu et al. 2007. Physical interpretation of the correlation between multi-angle spectral data and canopy height. *Geophys Res Lett* 34:L18405.

Schull, M. A., Y. Knyazikhin, L. Xu, A. Samanta, P. L. Carmona, L. Lepine, J. P. Jenkins et al. 2011. Canopy spectral invariants, part 2: Application to classification of forest types from hyperspectral data. *J Quant Spectrosc Radiat Transf* 112:736–750.

Sellers, P. J., S. O. Los, C. J. Tucker, C. O. Justice, D. A. Dazlich, G. J. Collatz, and D. A. Randall. 1996. A revised land surface parameterization (SiB2) for atmospheric GCMS. Part II: The generation of global fields of terrestrial biophysical parameters from satellite data. *J Climate* 9:706–737.

Serbin, S. P., D. N. Dillaway, E. L. Kruger, and P. A. Townsend. 2012. Leaf optical properties reflect variation in photosynthetic metabolism and its sensitivity to temperature. *J Exp Biol* 63:489–502.

Serrano, L., J. F. Peñuelas, and S. L. Ustin. 2002. Remote sensing of nitrogen and lignin in Mediterranean vegetation from AVIRIS data: Decomposing biochemical from structural signals. *Remote Sens Environ* 81:355–364.

Smith, M. L., M. E. Martin, L. Plourde, and S. V. Ollinger. 2003. Analysis of hyperspectral data for estimation of temperate forest canopy nitrogen concentration: Comparison between an airborne (AVIRIS) and a spaceborne (hyperion) sensor. *IEEE Trans Geosci Remote Sens* 41:1332–1337.

Smith, M. L., S. V. Ollinger, M. E. Martin, J. D. Aber, R. A. Hallett, and C. L. Goodale. 2002. Direct estimation of aboveground forest productivity through hyperspectral remote sensing of canopy nitrogen. *Ecol Appl* 12:1286–1302.

Soldberg, S., A. Brunner, K. H. Hanssen, H. Lange, E. Næsset, M. Rautiainen, and P. Stenberg. 2009. Mapping LAI in a Norway spruce forest using airborne laser scanning. *Remote Sens Environ* 113:2317–2327.

Somers, B. and G. P. Asner. 2013. Multi-temporal hyperspectral mixture analysis and feature selection for invasive species mapping in rainforests. *Remote Sens Environ* 136:14–27.

Spanner, M., L. Johnson, J. Miller, R. McCreight, J. Freemantle, J. Runyon, and P. Gong. 1994. Remote sensing of seasonal leaf area index across the Oregon transect. *Ecol Appl* 4:258–271.

Stylinksi, C. D., W. C. Oechel, J. A. Gamon, D. T. Tissue, F. Miglietta, and A. Raschi. 2000. Effects of lifelong [Co₂] enrichment on carboxylation and light utilization of Quercus pubescens Willd. Examined with gas exchange, biochemistry and optical techniques. *Plant Cell Environ* 23:1353–1362.

Sun, G. and K. J. Ranson. 2000. Modeling lidar returns from forest canopies. *IEEE Trans Geosci Remote Sens* 38:2617–2626.

Thenkabail, P. S, E. A. Enclona, M. S. Ashton, C. Legg, and M. J. De Dieu. 2004. Hyperion, IKONOS, ALI, and ETM+ sensors in the study of African rainforests. *Remote Sens Environ* 90:23–43.

Thomas, V., D. A. Flinch, J. H. McCaughey, T. Noland, L. Rich, and P. Treitz. 2006. Spatial modelling of the fraction of photosynthetically active radiation absorbed by a boreal mixedwood forest using a lidar–hyperspectral approach. *Agric For Meteorol* 140:287–307.

Thomas, V., J. H. McCaughey, P. Treitz, D. A. Finch, T. Noland, and L. Rich. 2009. Spatial modelling of photosynthesis for a boreal mixedwood forest by integrating micrometeorological, lidar, and hyperspectral remote sensing data. *Agric For Meteorol* 149:639–654.

Townsend, P. A. and J. R. Foster. 2002. Comparison of EO-1 hyperion to AVIRIS for mapping forest composition in the Appalachian mountains. *IGARSS* 2:793–795.

Townsend, P. A., S. P. Serbin, E. L. Kruger, and J. A. Gamon. 2013. Disentangling the contribution of biological and physical properties of leaves and canopies in imaging spectroscopy data. *Proc Nat/Acad Sci* 110:E1074.

Udayalakshmi, V., B. St-Onge, and D. Kneeshaw. 2011. Response of a boreal forest to canopy opening: Assessing vertical and lateral tree growth with multi-temporal LiDAR data. *Ecol Appl* 21:99–121.

Ustin, S. L., A. A. Gitelson, S. Jacquemoud, M. Schaepman, G. P. Asner, J. A. Gamon, and P. Zarco-Tejada. 2009. Retrieval of foliar information about plant pigment systems from high resolution spectroscopy. *Remote Sens Environ* 113: S67–S77.

Ustin, S. L., D. Riaño, and E. R. Hunt Jr. 2012. Estimating canopy water content from spectroscopy. *Isr J Plant Sci* 60:9–23.

Ustin, S. L. and A. Trabucco. 2000. Analysis of AVIRIS hyperspectral data to assess forest structure and composition. *J For* 98:47–49.

van Aardt, J. A. N., R. H. Wynne, and J. A. Scrivani. 2008. LiDAR-based mapping of forest volume and biomass by taxonomic group using structurally homogenous segments. *Photogramm Eng Remote Sens* 74:1033–1044.

Van den Berg, A. K. and T. D. Perkins. 2005. Nondestructive estimation of anthocyanin content in autumn sugar maple leaves. *HortScience* 40:685–686.

Vaughn, N. R., G. P. Asner, and C. P. Giardina. 2013. Polar grid fraction as an estimator of forest canopy structure using airborne lidar. *Int J Remote Sens* 34:7464–7473.

Vierling, K. T., L. A. Vierling, W. Gould, S. Martinuzzi, and R. Clawges. 2008. Lidar: Shedding new light on habitat characterization and modeling. *Front Ecol Environ* 6:90–98.

Waring, R. H. 1983. Estimating forest growth and efficiency in relation to canopy leaf area. *Adv Ecol Res* 13:327–354.

Weishampel, J. F., J. B. Blair, R. G. Knox, R. Dubayah, and D. B. Clark. 2000. Volumetric LiDAR return patterns from an old-growth tropical rainforest canopy. *Int J Remote Sens* 21:409–415.

Westoby, M., D. S. Falster, A. T. Moles, P. A. Vesk, and I. J. Wright. 2002. Plant ecological strategies: Some leading dimensions of variation between species. *Annu Rev Ecol Syst* 33:125–159.

White, J. C., N. C. Coops, T. Hilker, M. A. Wulder, and A. L. Carroll. 2007. Detecting mountain pine beetle red attack damage with EO-1 hyperion moisture indices. *Int J Remote Sens* 28:2111–2121.

Wolter, P. T., P. A. Townsend, B. R. Sturtevant, and C. C. Kingdon. 2008. Remote sensing of the distribution and abundance of host species for spruce budworm in Northern Minnesota and Ontario. *Remote Sens Environ* 112:3971–3982.

Wright, I. J., P. B. Reich, M. Westoby, D. D. Ackerly, Z. Baruch, F. Bongers, J. Cavender-Bares et al. 2004. The worldwide leaf economics spectrum. *Nature* 428:821–827.

Wulder, M. A., J. C. White, R. F. Nelson, E. Næsset, H. O. Ørka, N. C. Coops, T. Hilker et al. 2012. LiDAR sampling for large-area forest characterization: A review. *Remote Sens Environ* 121:196–209.

Zarco-Tejada, P. J., J. A. J. Berni, L. Suárez, G. Sepulcre-Cantó, F. Morales, and J. R. Miller. 2009. Imaging chlorophyll fluorescence from an airborne narrow-band multispectral camera for vegetation stress detection. *Remote Sens Environ* 113:1262–1275.

Zarco-Tejada, P. J., J. R. Miller, G. H. Mohammed, T. L. Noland, and P. H. Sampson 1999. Optical indices as bioindicators of forest condition from hyperspectral CASI data. *Proceedings of the 19th Symposium of the European Association of Remote Sensing Laboratories (EARSeL)*, Valladolid, Spain.

Zarco-Tejada, P. J., J. R. Miller, G. H. Mohammed, T. L. Noland, and P. H. Sampson. 2001. Scaling-up and model inversion methods with narrow-band optical indices for chlorophyll content estimation in closed forest canopies with hyperspectral data. *IEEE Trans Geosci Remote Sens* 39:1491–1507.

Zarco-Tejada, P. J., J. R. Miller, A. Morales, A. Berjón, and J. Agüera. 2004. Hyperspectral indices and model simulation for chlorophyll estimation in open-canopy tree crops. *Remote Sens Environ* 90:463–476.

Zarco-Tejada, P. J., A. Morales, L. Testi, and F. J. Villalobos. 2013. Spatio-temporal patterns of chlorophyll fluorescence and physiological and structural indices acquired from hyperspectral imagery as compared with carbon fluxes measured with eddy covariance. *Remote Sens Environ* 133:102–115.

Zhang, Y., J. M. Chen, J. R. Miller, and T. L. Noland. 2008. Leaf chlorophyll content retrieval from airborne hyperspectral remote sensing imagery. *Remote Sens Environ* 112:3234–3247.

Zimble, D. A., D. L. Evans, G. C. Carlson, R. C. Parker, S. C. Grado, and P. D. Gerard. 2003. Characterizing vertical forest structure using small-footprint airborne LiDAR. *Remote Sens Environ* 87:171–182.

17

Optical Remote Sensing of Tree and Stand Heights

Sylvie Durrieu
Irstea, UMR TETIS

Cédric Véga
*Institut National de l'Information
Geographique et Forestière and
Institut Français de Pondichéry*

Marc Bouvier
Irstea, UMR TETIS

Frédéric Gosselin
Irstea, UR EFNO

Jean-Pierre Renaud
Office National des Forêts

Laurent Saint-André
*INRA, Unité Biogéochimie des
Ecosystèmes Forestiers (UR 1138)
CIRAD, UMR Eco&Sols*

Acronyms and Definitions

AGB	Aboveground biomass
ALS	Airborne laser scanning
CHM	Canopy Height Model (= DSM–DTM)
DBH	Diameter at breast height
DP	Digital photogrammetry
DSM	Digital surface model
DTM	Digital terrain model
GLAS	Geoscience Laser Altimeter System
GVM	Global vegetation models
IDW	Inverse distance weighted
InSAR	Interferometric synthetic aperture radar
LiDAR	Light Detection And Ranging
LM	Local maxima
NEP	Net productivity
NFI	National Forest Inventory
NIR	Near infrared
POLinSAR	Technique based on the synergy between SAR interferometry and SAR polarimetry
R^2	Determination coefficient
R^2_{adj}	Adjusted determination coefficient
Reco	Total ecosystem respiration
REDD	Reducing emissions from deforestation and forest degradation
RMSE	Root-mean-square error
SAR	Synthetic Aperture Radar
SLICER	Scanning Lidar Imager of Canopies by Echo Recover
SRTM	Shuttle Radar Topography Mission, by extension refers to the resulting digital elevation model (DEM) based on SRTM data
TIN	Triangular irregular network
UAV	Unmanned aerial vehicles
UNFCCC	United Nations Framework Convention on Climate Change
UV	Ultraviolet

17.1 Introduction

Forests cover 30% of continental surfaces and play a key role in climate change regulation, in raw material and renewable energy supply, and in biodiversity conservation. Successfully maintaining all the functions of forest ecosystems through their sustainable management is thus crucial for the future of mankind. However, developing appropriate policies and management practices requires an in-depth knowledge on forest ecosystems within a fast-evolving context, as well as appropriate models to forecast how the way they will respond to management practices and global change.

Remote-sensing data supported by ground observations are considered as key to obtaining quantitative and timely information on forest ecosystems at a variety of scales in space and time, thereby allowing effective monitoring (DeFries et al., 2007; Fuller, 2006; Kleinn, 2002; Simonett, 1969), enhanced ecosystem modeling (Cabello et al., 2012; le Maire et al., 2011; Marsden et al., 2013; Wang et al., 2010), and appropriate management of forest resources (Le Goff et al., 2010; Liu and Han, 2009; Thürig and Kaufmann, 2010).

In addition to forest composition, that is, abundance and distribution of species, forest structure is a key descriptor of forest ecosystems (Wynne, 2006) that can, to some extent, be retrieved from remote-sensing data. Structure refers to the 3D arrangement and characteristics of vegetation compartments, including trunks, branches, twigs, and leaves. A given structure is both a result and a driver of ecosystem functions (Shugart et al., 2010). Forest structure is also directly related to the main biogeochemical (water, nutrient, and carbon) cycles (determining stocks and driving fluxes). It can affect local abiotic factors and is also essential in providing habitats, therefore possibly impacting biodiversity (Couteron et al., 2005). Accurate measurements of forest structure based on remote-sensing data would thus represent a major step toward an in-depth knowledge on forest ecosystems. Several remote-sensing technologies can provide high valuable information for sustainable forest managements. Some of them are particularly promising to provide information on forest structure. The aim of this book chapter is to present two remote-sensing approaches: airborne laser scanning (ALS) and digital photogrammetry (DP) that have both proved their efficiency for characterizing forest structure based on vegetation height measurements, either at the individual tree level or at the stand level. The first section explains why the knowledge of height structure is of major importance for both forest management and ecosystem modeling. In this section, our focus on the two aforementioned optical remote-sensing approaches as promising solutions to extend height measurements in space and time is also justified. In the second section, the concepts and history of both technologies are briefly described, and the resulting 3D data are compared in the framework of forest applications. Sections 17.3 and 17.4 detail how forest height structures can be measured from these 3D data at the tree and stand level, respectively. After a brief conclusion, the last section presents some promising prospects offered by the possibility to monitor changes in vegetation height using ALS or DP 3D data. In this last section, we will also discuss some issues related to spaceborne systems. These systems offer the opportunity to overcome coverage limitations encountered with systems operated from air platforms, thus enabling to extend forest ecosystems monitoring at global scale. Existing spaceborne imaging systems can be used from now to provide DP 3D data from space. For Light Detection and Ranging (LiDAR), ICESat experiment (2003–2009) has demonstrated the possibility to derive global height (Simard et al., 2011) and biomass maps (Baccini et al., 2012;

Saatchi et al., 2011) by combining LiDAR sampling measurements with other satellite data and global products. However, a mission embedding a LiDAR system primarily designed for vegetation monitoring has yet to be developed.

17.2 Why Measure Tree Heights?

17.2.1 Determinants of Heights

Height growth, also called primary tree growth, is a complex process that combines the production of the internodes by the apical bud and their elongation. Every year, one or several growth units can be produced. Apical growth is one of the three main processes that govern tree architecture along with branching and reiteration (Barthelemy and Caraglio, 2007; Guédon et al., 2007). Tree height growth and tree architecture result from an equilibrium between endogenous features (cells arrangements, cell properties) and exogenous factors (competition for light, water, and nutrients). Growth phase patterns and drivers (e.g., climate, between tree competition, forest management) can be disentangled using statistical models (e.g., Markov chains (Chaubert-Pereira et al., 2009), nonlinear models (Saint-André et al., 2008), or process-based models such as MAESPA, a model of forest canopy radiation absorption, photosynthesis and water balance combining the MAESTRA (Multi-Array Evaporation Stand Tree Radiation A) and SPA (Soil-Plant-Atmosphere) ecosystem models (Duursma and Medlyn, 2012)). In natural forest ecosystems, stand height structure is mainly dependent on the tree species and their traits (e.g., shade tolerant, light demander, among others), while it is greatly determined by forester strategies in managed forests leading to stand structures such as even-aged high forest with either a single or mixed species, coppices, coppices with standards, or selection forests. The secondary growth, which results from the cambium activity, increases tree size and biomass. Primary growth and secondary growth are correlated, and the proportions between height and diameter, or between biomass and height and diameter, follow rules that are the same for all trees of a given species growing under the same conditions (climate, forest management). These allometries are widely used to calculate volume and biomass at tree and plot scale (Picard et al., 2012b), and the research domain is very active (Henry et al., 2013). Tree height at a given time is then integrating all previous growth phases and can be used as indicator of the current forest status, for example, biomass or carbon stocks, or future growth.

17.2.2 Importance of Tree Height Distribution for Forest Management and Ecology

Among forest attributes, tree height plays a central role in forest inventories, where it is critical for calculating volumes, siting quality indexes, and assessing needs for silvicultural treatments (Bontemps and Bouriaud, 2014; Pardé, 1956). An evocative example is given by thinning operations in young stands, which are normally planned based on threshold crossings in canopy height. In more mature stands, height indicators such as top, dominant, or Lorey's heights are frequently used for volume or growth estimations. This underlines the importance of height indicators for forest managers.

Regarding ecological modeling, growth and yield models are mainly based on height growth. These models generally apply to even-aged forests that are characterized by a relative tree-population homogeneity: same age (or age range in case of natural regeneration) and a dominant tree species. The growth of these populations has been widely studied (De Perthuis, 1788 in Batho and Garcia, 2006), which gave rise to following generic principles (Assmann, 1970; Dhôte, 1991; García, 2011; Pretzsch, 2009; Skovsgaard and Vanclay, 2008). It is customary to distinguish the population as a whole, then the tree in the stand. This distinction is useful to separate the different factors involved in tree growth into three main components: (1) site fertility in a broad sense, including the ability of the soil to feed the trees (nutrients and water availability) and the climate characteristics of the area, (2) the overall pressure within the population that is appreciated by different indices of density, and finally, (3) the social status of each individual tree, which will define its ability to mobilize resources in its immediate environment.

For these three components, height is of primary importance since dominant height is widely used to define site fertility (or site index, for which preliminary principles were given by De Perthuis in 1788 (Batho and Garcia, 2006) and concepts formalized by Eichhorn (1904) and later discussed by Assmann (1970), and more recently by Bontemps and Bouriaud (2014)). It can also be used to define some stand density/global pressure indicators (ex Hart–Becking factor), even if to a less significant extent than stand density or density indicators based on tree diameters. Finally, height can be used to define the social status of the trees, for example, tree height related to the stand dominant height.

Owing to the importance of forest in climate mitigation, through its capacity to accumulate carbon, improved knowledge of carbon stocks and fluxes in forest ecosystems is needed to better understand carbon cycle and to develop improved climate models. This knowledge is also needed for carbon accounting. After the Kyoto Protocol, the Cancún Agreements provide strong backing for a reducing emissions from deforestation and forest degradation mechanism under the United Nations Framework Convention on Climate Change whereby developed countries would provide positive benefits to developing ones for reducing deforestation, forest degradation, enhancement of forest carbon stocks, and forest conservation. However, in order to implement these mechanisms, forest services in most countries must make more accurate assessment of the forest carbon stocks and carbon stock changes. Changes in forest carbon stocks through time are best appraised by a combination of remote sensing and field-based measurement where height is an essential variable. Most of the current methods used to assess carbon emissions from deforestation and forest degradation are indeed based on the measurement of changes in surface area of the main forest types and on the assessment of a mean biomass value for each type (De Sy et al., 2012; ESA, 2008). Therefore, improving accuracy of both surface areas of forest types and biomass estimations would lead

to improved carbon flux predictions. For biomass estimation, a consensus exists stating that, for a biomass map with a 1 ha resolution, estimation accuracy should not exceed 20 t·ha^{-1} or 20% of field estimations without exceeding 50 t·ha^{-1} (Hall et al., 2011b; Le Toan et al., 2011; Zolkos et al., 2013). However, reaching such accuracy all over the world is highly challenging (Angelsen, 2008; Chave et al., 2003; Pelletier et al., 2011).

Another approach to study carbon stocks and fluxes relies on the use of forest ecosystems functioning models. Up to now, two kinds of model have coexisted (Bellassen et al., 2011). The first one consists of models adapted to stand scale, like growth models previously discussed, which are process-based models that can include information on silvicultural practices. Provided an important local calibration is performed, these models can output reliable simulations of local carbon stock evolutions. The second type consists of global vegetation models (GVMs), which can provide carbon stocks and fluxes at regional scales but with lower accuracy levels. There are two main causes for inaccuracy. The first one lies in pedoclimatic data, which are required to drive the models but are too coarse to reflect local variations. The second cause for inaccuracy is the nonintegration of management impacts, which hampers a reliable modeling of age-dependent variables like biomass (Bellassen et al., 2011). This is why GVMs are currently evolving to better manage intracell variability within coarse grid models and to take into account management impacts. ORCHIDEE-FM, coupling a forest management module (FM) to the Organising Carbon and Hydrology in Dynamic Ecosystems (ORCHIDEE) model (Bellassen et al., 2010) and the second version of the Ecosystem Demography model (Medvigy and Moorcroft, 2012; Medvigy et al., 2009) are examples of this new model generation that can assimilate height or biomass information from National Forest Inventory (NFI) field plots or remote-sensing data. This assimilation led to significant improvement in aboveground biomass dynamic characterization and productivity assessment (Antonarakis et al., 2011; Bellassen et al., 2011), for example, a decrease in error rate of 30% and 50% for total ecosystem respiration and net productivity, respectively, in the study by Bellassen et al. (2011).

Therefore, whatever the approach used and due to its tie link with biomass, vegetation height is an essential parameter to develop scientific knowledge on carbon cycle and to address carbon-accounting issues.

Another main aspect of forest structure is its major role in ecology (Jaskierniak, 2011; Vepakomma et al., 2008). Stand structure affects microclimate, habitat quality, and therefore biodiversity potential. In particular, the gradient in gap sizes is known to influence many parts of biodiversity. Gaps are either part of the natural forest cycle, or result from silvicultural treatments, or are caused by accidental disturbances such as fire, storm, or plant health problems. They have a short-term impact on biodiversity through different mechanisms. The most obvious one is an increased irradiance that can benefit heliophilous species and also has an impact on the microclimate in the patch, with a higher temperature variance in gaps than in forests. Through the removal of trees, gaps also release some soil resources, such as water and nutrients, which can positively impact vascular plant development, for example. Soil disturbances associated with gaps (e.g., pits and mounds) were also found to impact biodiversity. Bouget (2005) reported that the gap effect was on the whole favorable to biodiversity, with the notable exception of shade-preferring groups. And, while small gaps generally have a weaker effect on biodiversity (see Bouget, 2005), surprisingly, benefits were observed for forest vascular plants, for which small openings might be very positive (Duguid and Ashton, 2013). Gaps not only have an effect in the area they cover as some species shun closed forests in the vicinity of the gaps. These species are generally called forest-interior species and have particularly been found among birds (Germaine et al., 1997). For all the reasons, biodiversity indicators were shown to be strongly correlated to the 3D spatial patterns of vegetation (Williams et al., 1994) and wildlife richness to be related to the 3D features of the canopy (Magnussen et al., 2012). Temporal effects have also been reported. Through the succession that occurs after gap formation, gap impacts can evolve and even be reversed. For vascular plants in particular, silvicultures based on large cuttings have been found to have an adverse effect on floristic biodiversity, which occurs decades after gap formation (Duguid and Ashton, 2013). This may be why some authors promote the imitation of the natural disturbance regimes, especially the natural gap dynamics, in managed stands (Angelstam, 1998; Næsset, 2002b).

17.2.3 Limitations of Field Measurements and How Remote Sensing Can Help Meet Information Requirements

From the previous subsections, we can see how important it is to measure tree heights and to monitor the dynamic of both canopy height and gaps at several scales in time and space in order to improve both forest ecosystem modeling and to contribute to their sustainable management.

Total height of a tree is either defined as "the distance between the top and the base of the tree, measured along a perpendicular, dropped from the top" (van Laar and Akça, 2007) or as the stem length. Several methods are used to measure tree height in the field (Larjavaara and Muller-Landau, 2013; Williams et al., 1994). Graduated poles can be used, but without climbing the tree, their use is limited to relative small trees (Larjavaara and Muller-Landau, 2013), usually under 15 m. For bigger trees, height measurements are usually derived from angle and distance measurements (Figure 17.1). Angles are measured using a clinometer. Distances are measured with either a measuring tape, an ultrasonic measuring system, or a laser range finder (Larjavaara and Muller-Landau, 2013). With the latter, which can be used to measure the distance between the operator and the tree top, the sine method can be applied (Figure 17.1). This method is very effective in dense forests as it can be used to carry out measurements from a distance close from the trunk, even by shooting directly up the tree, whereas the tangent method is highly inaccurate in these situations.

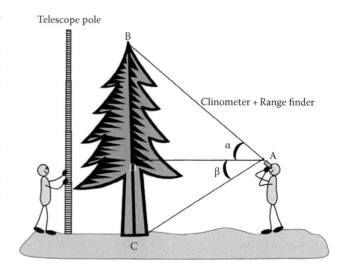

FIGURE 17.1 Height measurement of a vertical tree using a graduated pole (left side of the figure) or a hypsometer measuring angles and distances (right side of the figure). If distance AD is measured, using, for example, an ultrasonic system, tree height can be retrieve using the tangent method ($H_{tree} = BD + DC = AD·tg\alpha + AD·tg\beta$); if distance AB and AC are measured, using, for example, a laser system, sine method can be used ($H_{tree} = BD + DC = AB·sin\alpha + AC·sin\beta$). To account for slope or leaning trees, several variants depending on the measurement system and the targeted accuracy (e.g., DC is sometimes approximated from the height of the surveyor) were developed based on this basic principle.

Mean absolute errors were found to be in the order of 1 m in temperate forests (e.g., from 0.90 to 1.3 m, according to the measurement method used in Williams et al., 1994) and higher for tall trees. For example, in the study by Williams et al. (1994), for one of the two classes with the tallest trees, that is, trees higher than 24 m, errors ranged from 1.6.5 to 3.45 m according to the measurement method used. Measurements are more difficult to achieve and more error prone for broad-leaved trees than for coniferous species, especially during the leaf-on period, and also for leaning trees. In a moist tropical semideciduous forest and during the leaf-on period, which can be considered among one of the most complex measurement conditions, the residual mean-squared error was found to be on average between 5.05 and 6.85 m, using sine and tangent methods, respectively (Larjavaara and Muller-Landau, 2013). Part of the error can be attributed to a bias, and errors depend not only on the measurement method and the stand characteristics but also on the surveyor's expertise (Kitahara et al., 2010; Larjavaara and Muller-Landau, 2013).

Within the inventory process, height is certainly among the most costly data to collect. This is why different height indicators have been developed to minimize the measurement effort (Van Laar and Akça, 2007). Such estimators rely on the relationship between tree size and stand density and depend on other indicators used to assess the biophysical characteristics of the stand (Vanclay, 2009). For example, when using the quadratic mean diameter, which is favored over the arithmetic-mean diameter due to its higher correlation with stand volume and biomass, surveyors will only measure and average the heights of a sample of trees whose diameters at breast height (DBH) are closest to the plot quadratic mean diameter. In even-aged stands, the dominant height or top height is most commonly used. It can be defined as the arithmetic mean of the 100 largest trees per hectare. It is favored over mean height when assessing site quality and modeling growth because it is less affected by stand density and silvicultural practices (Tran-Ha et al., 2011). However, dominant height definitions may vary among users (Jaskierniak, 2011), and more importantly, this measurement can be biased as it depends upon the sampling area (García, 1998). Another measured height indicator is Lorey's mean height. Here, the contribution of each tree to the stand height is weighted by its basal area. Lorey's mean height is thus calculated by multiplying the tree height (h) by the tree basal area (g) and by dividing the sum of these weighted heights by the total stand basal area. Lorey's mean height is less impacted than the aforementioned height indicators by both the mortality and harvesting of smaller trees, and is also well correlated with stand volume (Tran-Ha et al., 2011).

Field measurements are essential but remain labor intensive and costly, and their scope must therefore be limited in space and time. Therefore, a technology accounting for extensive areas and for tree heights at individual or stand scale, as well as providing gap measurements at regular time scale and at lower costs, would represent a real quantum leap in the development of forest ecosystem modeling. It would also help managers who are calling out for more precise inventories with fine spatial and temporal resolutions in order to deal with the growing societal and economic pressure exerted on forests.

The potential of remote-sensing-based technologies to extend forest structure measurements in space and time is widely acknowledge (see, e.g., De Leeuw et al., 2010; De Sy et al., 2012; DeFries et al., 2007; Ostendorf, 2011; Wang et al., 2010). Table 17.1 summarizes forest information that can be derived from main remote-sensing data types. Forest information was classified into three main categories: forest maps, forest structural and biophysical properties, and information on forest changes. The remote-sensing technologies that can be used to characterize forest structure can be identified from Table 17.1 along with the level of data availability.

Forest structure can be indirectly assessed by analyzing, for example, texture in very-high-resolution imagery (Couteron et al., 2005; Proisy et al., 2007). But several remote-sensing techniques, based on LiDAR, optical imagery, or radar technologies, allow direct measurements of the forest 3D structure.

The utility of LiDAR, which is an active optical remote-sensing technology, has been widely demonstrated with respect to forest structure measurements and biomass estimation (Hyyppä et al., 2008; Lim et al., 2003; Naesset, 2007; Wulder et al., 2012). LiDAR remains efficient in closed-canopy tropical areas supporting high-biomass forests greater than 200 t·ha^{-1} (Kellner et al., 2009; Lefsky et al., 2005), where optical vegetation indices and volumetric radar measurements typically saturate (Castro et al., 2003). Photogrammetry has long been used for forest applications, particularly forest inventories, since accurate

3D measurements were render possible thanks to analytic stereoplotters and comparators that became commonly available from the mid-1970s (Vastaranta et al., 2013). This technology has been neglected for many years, probably partly due to the great potential of LiDAR to measure the 3D structure of vegetation and the fast development of ALS systems, systems combining a LiDAR with a scanning device. But photogrammetry has recently received renewed interest for forest applications thanks to developments in DP that enables automatic reconstruction of 3D canopy models based on very-high-resolution optical images. Using small wavelength polarimetric radar systems (e.g., band × (~2.5 to 3.75 cm) radar), accurate digital surface models (DSMs) of the top of the canopy can also be produced using interferometry, also referred as the POLinSAR technique. Indeed, the signal barely penetrates the vegetation at these wavelengths and is then mainly backscattered by the elements of the upper canopy (Soja and Ulander, 2013). Polarimetric radar tomography applied to data acquired with large-wavelength systems with a signal that penetrates deeper into the vegetation, for example, band P (>1 m), has also been used recently to decompose the signal into a few vertical strata producing low-resolution vertical signal profiles in forests (Ho Tong Minh et al., 2014).

However, unlike for radar, the high technological readiness of ALS systems and of both airborne and spaceborne optical imagery systems, as well as the great versatility of the produced data, means that both these technologies are particularly suited to providing information for operational applications or for research in the field of ecosystem modeling. Indeed, the high availability of data has favored the development of processing approaches, some of which are currently used on an operational level.

17.3 Two Promising Optical Remote-Sensing Techniques for Tree Height Measurements: LiDAR and Digital Photogrammetry

17.3.1 LiDAR

17.3.1.1 Principle and Brief History

LiDAR is an active remote-sensing technology based on emission reception of a laser beam. Several kinds of LiDAR systems exist, differential absorption LiDAR, Doppler, and range finder, but most of the systems dedicated to continental surface observation belongs to the range finder class. They assess the distance between the sensor and a target, by measuring the round-trip time for a short laser pulse (in general, near infrared [NIR] or green wavelengths) to travel between the sensor and a target. By combining these range measurements with information on both sensor position and attitude, obtained thanks to a differential GPS and an inertial measurement unit onboard the platform, the position of the target on the Earth's surface can be accurately computed. For semitransmitting mediums, such as forests, the incident pulse might be partly backscattered by the top of the

TABLE 17.1　Overview of the Main Remote-Sensing Data and of Their Use for Forest Ecosystem Monitoring

	Remote Sensing Data — Multispectral optical imagery — Very high resolution data Analogous (airborne): Scale: ~1:5,000 to <1:80,000 Digital: resolution: sub-metric to few meters	Remote Sensing Data — Multispectral optical imagery — High to low resolution data Resolution: tens of meters to few hundreds of meters	2D Imagery — Hyperspectral optical imagery — High resolution Resolution: 1–30 m, depending on flight height and system	2D Imagery — Hyperspectral optical imagery — Low resolution Resolution: 250–1200 m	Synthetic Aperture Radar (SAR) — Airborne Cloud penetration capability	Synthetic Aperture Radar (SAR) — Spaceborne Cloud penetration capability
Data Availability (From a Technological Perspective)	Airborne and spaceborne (IKONOS, QuickBird, Pléiade, WorldView, SPOT…) High but expensive Restricted by cloud cover Operational	High (Landsat, Aster, RapidEye, IRS, MODIS…) Restricted by cloud cover Several free sources	Low (few experimental airborne systems) Restricted by cloud cover	Low Spaceborne (Hyperion) Restricted by cloud cover	Low (few commercial systems)	High
Scale and Coverage	From local to national scale Mapping scale greater than 1:50,000	From regional to global scale according to spatial resolution	From local to national scale	From regional to global scale according to system	Local to regional	Regional to global
Forest Maps — Forest/Non forest	***	***	***	***	*** Using multitemporal datasets	
Coniferous/Deciduous/Mixed	***	***	***	***	** Using polarization information	
Additional Information on Forest Types	** Structural information by texture analysis [4]	* Information on species by time series analysis	*** Higher species classification accuracy than with multispectral data (e.g., +27% ==> 88% final accuracy) [7]	* Hampered by the coarse resolution	** Using multifrequency/ multipolarization data	
Tree Level	* Generalization and computing issues [5]	—	** If sufficient resolution Technically difficult	—	—	
Forest Structural and Biophysical Properties — Cover Rate (%)	**	*	***	*	** Using multifrequency/ multipolarization data	
Density (nb of Trees/ha)	*	—	** If sufficient resolution	*	—	
Height (m)	*	—	*	*	*	
Basal Area (m²)	*	—	*	*	** Using polarization information	
LAI/PAI (m² m⁻²)	* Saturation at low to medium LAI level	*	** If sufficient resolution	* Using appropriate vegetation indices	** Using polarization information	
Wood Volume (m³ ha⁻¹) AGB (Mg ha⁻¹)	* Saturation at low biomass level	*	** Using appropriate vegetation indices, i.e., based on a spectral band sensitive to vegetation water content	** Using appropriate vegetation indices	** [9, 10]	** Using polarization information Saturation for AGB > 200 t.ha⁻¹ (for L or P band, lower saturation levels for X and C bands) [8, 11, 12]
Forest Change Monitoring — Land Cover Change (km²)	***	**	***	**	** Using polarization information	***
Aforestation/Reforestation/ Deforestation (km²)	***	**	***	**	**	**
Degradation	—	—	** If sufficient resolution	* If sufficient resolution	*	**
Disaster (km²)	*** For fire** For storm and phytosanitary problems	**	***	**	**	**
Growth	—	—	—	*	*	*
Gap Dynamics	**	* Depends on Resolution /gap size	** Depends on data availability	*	—	—

(Continued)

TABLE 17.1 (Continued) Overview of the Main Remote-Sensing Data and of Their Use for Forest Ecosystem Monitoring

Remote Sensing Data

2.5D imagery Digital Surface or Terrain Model: Digital photogrammetry, Radar interferometry

3D information: Radar tomography, Lidar

Digital photogrammetry	Radar interferometry	Radar tomography	Lidar — Airborne commercial systems (ALS) [33, 34, 35]	Lidar — Spaceborne ICESat not optimized for forest applications
Surface geometry retrieved from two or multi point of view images [13, 14, 15, 16, 17]	Difference in the phase information between two SAR images provides an ambiguous measurement of the relative terrain altitude due to the periodic nature of the signal. After the phase unwrapping step, aiming at solving the ambiguity, an accurate DSM is obtained. [14, 19, 20]	Raw vegetation profiles (several tens of centimeters from airborne data to several meters expected from BIOMASS mission) are computed using multiple-baseline images. Long wavelengths sensitive to the whole vegetation layer must be used (P and L bands) [20, 22, 23]	Part of the 3D information can be used to provide—DSM, DTM and CHM. Additional information on vegetation vertical structure [30, 31, 32]	
Very High resolution DSM Accuracy: 3 cm to 14 m. Acquisition parameters affecting elevation accuracy: flying height, image resolution, B/H ratio. Aerial photographs or VHR imagers (<4 m). High to Low resolution DSM with depointing imagers Accuracy: 5–40 m	Spaceborne X, C, or L bands (e.g., ERS, J-ERS, ENVISAT, ALOS, TerraSAR, TanDEM-X, SRTM) Z accuracy: better than 2 m up to 15 m (for SRTM). Airborne (e.g., commercial intermap products) Z accuracy: 0.5 m	Airborne	Airborne commercial systems (ALS) [33, 34, 35]	Spaceborne ICESat not optimized for forest applications
High. Available globally (e.g., from ASTER) but with low accuracy for global products. High but expensive. Restricted by cloud cover. Operational. Low. No cloud cover restriction	High. No cloud cover restriction. No P band InSAR data yet available from space; but one panned mission: the ESA BIOMASS mission	Low	High but expensive	No current spaceborn system ICESat 2003–2009 ICESat 2 planned to be launched in 2017
From local to regional or national. Regional to global	Local to regional. Regional (high accurate z data) to global (medium accurate z data)	Local to regional	Local to regional or national non wall to wall acquisitions can be used	Global non wall to wall coverage
*** Using image spectral information when available	** Using the associated amplitude images	** Using the associated amplitude images	** Based on vegetation height	*
*** Using image spectral information when available	** Using the associated amplitude images	** Using the associated amplitude images	—	—
** Direct: using the associated images; Indirect: information on topographic environment can help to identify species	* Indirect: information on topographic environment can help to identify species	** Using the associated amplitude images	** information on forest structure	** Information on forest structure; * Map quality depends on other RS data used for structural information extrapolation
** Improved results when using both DSM and images	—	—	** Generalization and computing issues	—
* Using external DTM or associated images	** Using the associated amplitude images	** Using the associated amplitude images	***	** ICESat2 more suitable but still limitations for dense covers (results from signal modeling) [25, 26]
—	—	—	**	*
** Using external DTM Height information quality depends on elevation accuracy	* Using external DTM Modeled surface depends on the used frequency ==> Direct or indirect height assessment	** Raw vertical profiles	*** for dominant strata; ** for understorey	*** ICESat2 more suitable but still limitations for dense covers results from signal modeling [25,26,27]
* Using external DTM Highly accurate DSM possible (few cm)	** Using the associated amplitude images	** Using the associated amplitude images	**	**
—	** Using the associated amplitude images	** Using the associated amplitude images	***	*** Using ICESat data [28]

(Continued)

TABLE 17.1 (Continued)   Overview of the Main Remote-Sensing Data and of Their Use for Forest Ecosystem Monitoring

	Remote Sensing Data				
	2.5D imagery Digital Surface or Terrain Model		**3D Information**		
	Digital photogrammetry	Radar interferometry	Radar tomography	Lidar	
	Surface geometry retrieved from two or multi point of view images [13, 14, 15, 16, 17]	Difference in the phase information between two SAR images provides an ambiguous measurement of the relative terrain altitude due to the periodic nature of the signal. After the phase unwrapping step, aiming at solving the ambiguity, an accurate DSM is obtained. [14, 19, 20]	Raw vegetation profiles (several tens of centimers from BIOMASS mission) are computed using multiple-baseline images Long wavelengths sensitive to the whole vegetation layer must be used (P and L bands) [20, 22, 23]	Part of the 3D information can be used to provide—DSM, DTM and CHM Additional information on vegetation vertical structure [30, 31, 32]	
	Very High resolution DSM Accuracy: 3 cm to 14 m Acquisition parameters affecting elevation accuracy: flying height, image resolution, B/H ratio Aerial photographs or VHR imagers (<4 m)	Spaceborne X, C, or L bands (e.g., ERS, J-ERS, ENVISAT, ALOS, TerraSAR, TanDEM-X, SRTM) Z accuracy: better than 2 m up to 15 m (for SRTM) / Airborne (e.g., commercial intermap products) Z accuracy: 0.5 m / High to Low resolution DSM with depointing imagers Accuracy: 5–40 m	Airborne	Airborne commercial systems (ALS) [33, 34, 35]	Spaceborne ICESat not optimized for forest applications
**	** Using external DTM Height information quality depends on elevation accuracy	** Using polarization information and height information saturation for AGB > 300 t·ha^{-1} (P band) [20, 21]	** Using vertical profiles	***	** Biomass maps with high discrepancies at the local level [29]
*** Using image spectral information when available	*** Using image spectral information when available	*** Using the associated amplitude images	*** Using the associated amplitude images	—	—
*** Using image spectral information when available	*** Using image spectral information when available	** Using the associated amplitude images and height information from interferometry	** Using the associated amplitude images and height information from interferometry	** Based on vegetation height	—
—	—	** Using the associated amplitude images, height and biomass information from interferometry	** Using the associated amplitude images, height and biomass information from tomography	** The most promising technology but yet little studied topic; e.g., [24]	—
*** Using image spectral information when available	*** Using image spectral information when available	**	**	* Through changes in structure Kind of disaster difficult to identify	**
**	** [18]	—	—	*** Depends on data availability (expensive)	—
**	*	—	—	*** Depends on data availability (expensive)	—

Remote-sensing technology capabilities are evaluated through the quality level of forest information or parameters derived from each data source. Capabilities are classified into four classes: * contribution but poor product quality, ** average product quality, *** good product quality, and —limited contribution to no technical capabilities or no relevant published study identified. Change monitoring capability is a consequence of land cover–type discrimination capability. The operationally effective level results from a combination of data availability, scale coverage, and information quality level. The potential synergy between technologies or between remote-sensing data and ancillary information is not presented in this table

References quoted in the table: [1] Boyd and Danson (2005); [2] De Sy et al. (2012); [3] Turner at al. (2003); [4] Couteron et al. (2005); [5] Ke and Quackenbush (2011); [6] Buckingham and Staenz (2008); [7] Goodenough et al. 2004); [8] Kasischke et al. (1997); [9] Li et al. (2009); [10] Manninen et al. (2005); [11] Castro et al. (2003); [12] Mitchard et al. (2009); [13] Baltasavia (1996c); [14] Gao (2007); [15] Korpela (2004); [16] Naesset (2002a); [17] St-Onge et al. (2004); [18] Véga and St-Onge (2008); [19] Ferretti et al. (2007); [20] Le Toan et al. (2011); [21] Garestier et al. (2009); [22] Ho Tong Minh et al. (2014); [23] Tebaldini and Rocca (2012); [24] Weishampel et al. (2012); [25] Durrieu and Nelson (2013); [26] Rosette et al. (2013); [27] Bolton et al. (2013); [28] Luo et al. (2013); [29] Mitchard et al. (2013); [30] Lim et al. (2003); [31] Wulder et al. (2012); [32] Zolkos et al. (2013); [33] Naesset (2007); [34] Evans et al. (2009); [35] Holmgren et al. (2003).

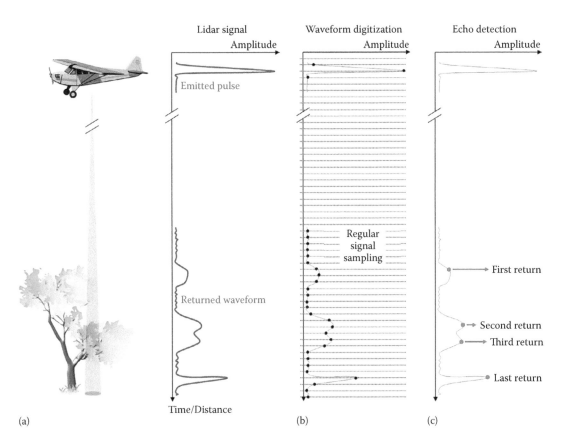

FIGURE 17.2 Principle of LiDAR measurement: photons are backscattered toward the sensor every time the laser beam is partially or totally intercepted by an obstacle. The resulting signal is a waveform (a) that is either digitized at high frequency, for example, every 1 ns, with full-waveform systems (b) or processed in real time by multiecho systems that only record a few returns (c). Depending on the system used, time information might be complemented with additional information such as signal intensity and echo width.

canopy, the understory vegetation, and the ground. The backscattered signal (waveform) thus embeds information on the 3D structure of vegetation covers. It is collected by a telescope and recorded using photodiodes (Figure 17.2).

The idea of measuring distance using light can be traced back to the late 1930s, that is, before the development of lasers. Barthélémy (1946) reported the development, in 1938, of a device that could be used to measure cloud height using high-power flashes of light produced by a spark gap and lasting no more than a few microseconds. In favorable situations, cloud heights up to 7000 m could be measured with this system (Barthélémy, 1946).

Attempts to build the first operational laser (light amplification by stimulated emission of radiation) started in the late 1950s and were successful in 1960 (Nelson, 2013).

In his paper tracing the history of the use of LiDAR in forest applications, Nelson (2013) reported that range finder lasers were initially used over continental surface for topographic purposes thanks to their ability to penetrate forest canopies. Tree measurements became the primary objective rather than just a source of noise in 1976 (Nelson, 2013). The development of digital recording devices and the development of positioning and scanning systems were key steps leading to the current systems, which can be classified depending on (Baltsavias, 1999b; Dubayah and Drake, 2001; Wulder et al., 2007) (1) whether they

fully digitize the return signal (full-waveform systems) or they record multiple echoes (multiecho systems), that is, the range of a finite number of returns, ranging from 1 (first or last) up to 8 returns for current commercial airborne systems, (2) whether they are small footprint (typically in the order of one to a few decimeters) or large footprint systems (tens of meters), and (3) their sampling rate and scanning pattern. Another major feature of LiDAR systems is the laser wavelength. Excluding x-ray and free-electron lasers, lasers exist in a wide spectrum, that is, 50–30,000 nm (Baltsavias, 1999c). However, most of the range finder LiDAR developed for the Earth surface studies operate in the NIR domain (900, 1040–1060, or 1550 nm) (Balstsavias, 1999c). NIR is particularly well fitted because of high atmospheric transmission in this spectral range (Baltsavias, 1999a) and because most targets at the Earth's surface, except water, have a high reflectivity in the NIR. In particular, their reflectivity is higher than in visible wavelengths. In addition, higher powers are allowed with NIR wavelengths than with visible ones while respecting eye safety. This is especially the case for 1550 nm lasers that stay eye safe at higher power than 1064 nm ones. Signal-to-noise ratio is thus increased by using NIR laser. NIR LiDAR instruments are also currently capable of significantly higher pulse rates than green LiDAR (Faux et al., 2009). The latter are mainly used for bathymetry due to the capacity of green wavelength to penetrate into the water. However, after ICESat, which acquired data in the NIR wavelength from 2003 to 2009, ICESat2, the second space LiDAR mission dedicated to Earth surface monitoring and scheduled for launch in 2017, will use a green laser. ICESat2 includes among its secondary scientific objectives measurement of vegetation canopy height. The choice to use a low-energy high-frequency, that is, photon counting, instrument constrained the choice of the wavelength. Indeed, this technology is not mature enough and not yet space qualified in the NIR wavelength. Airborne UV (Allouis et al., 2011) and dual wavelength LiDAR (Hancock et al., 2012) have also been successfully used for forest applications. The later systems were found well suited for separating the vegetation from the ground in complex topography (Hancock et al., 2012) as well as improving land cover classification (Wang et al., 2014).

The fast development of airborne LiDAR systems (most of which being multiecho sensors) (Figure 17.2) has been driven by commercial opportunities presented by environmental issues. Advances in LiDAR technology (e.g., increased pulse rates, increased number of echo digitization or full-waveform recording, scanning device) and advances in data geolocation have rapidly turned LiDAR into a fully functional and operational technology with a steady increase in availability of ALS systems operated by data providers. Full-waveform LiDAR systems were first designed as experimental systems (e.g., Laser Vegetation Imaging Sensor) and developed by NASA as an improved version of a former experimental system called Scanning Lidar Imager of Canopies by Echo Recover, developed in 1994 (Blair et al., 1999; Harding et al., 2001). Geoscience Laser Altimeter System, on ICESat, an ice-centric designed mission that collected data from 2003 to late 2009, was the first spaceborne LiDAR system to measure terrestrial surfaces (Zwally et al., 2002). However, the first commercial full-waveform airborne LiDAR system only became available in 2004 (LiteMapper-5600 LiDAR system based on the Riegl LMS-Q560 laser scanner) (Hug et al., 2004).

17.3.1.2 Measuring Vegetation Height from 3D ALS Data

The increasing use of ALS data for forest applications demonstrates the suitability of ALS technology for forest survey (Wulder et al., 2007) (see also Table 17.1). However, the assessment of forest parameters derived from ALS data is faced with a number of issues related to both the horizontal and vertical sampling characteristics of the measurements and to the nature of the interactions between the signal and the vegetation. The main factors that influence vegetation height accuracy at the footprint, tree, and stand levels are reported in Table 17.2.

By modifying the point distributional properties, the acquisition setup, which includes the laser instrument characteristics (wavelength, energy, pulse frequency, beam divergence, scanning pattern) and the flight parameters (flying altitude and speed, overlap of flight lines), was found to impact the quality of height estimates (Hopkinson, 2007; Næsset, 2009b). Additionally, because the laser signal interacts with vegetation components on

its path to the ground, both the structure and the optical properties of the vegetation also influence the vertical sampling of the vegetation layer (Disney et al., 2010). Consequently, the quality of the information on tree height distribution derived from ALS data depends on the capacity of the data to describe the top of the canopy and the underlying ground, but also on vegetation layering.

When a laser pulse interacts with a tree apex, the total tree height is accurate on condition that the amount of vegetation material in a narrow elevation range is sufficient to backscatter, in a very short laps of time, more energy than required to trigger a return (see, e.g., Wagner et al., 2006 for the theoretical background). According to the nature of the target, that is, its shape, leaf density, and reflectance properties, canopy height is underestimated to varying degrees (Disney et al., 2006; Nelson, 1997). Disney et al. (2010) using a simulation approach reported underestimations of canopy height of approximately 4% and 16% for broad leaves and conifers, respectively.

Also, point positions, and hence height estimates, may depend on the analog detection method used to identify an echo in the backscattered waveform (Disney et al., 2010; Wagner et al., 2004).

The impacts of the ALS system and flight setting on height retrieval have been investigated in numerous studies. Increasing the flight altitude leads to an increase in the footprint size at the Earth surface level and a simultaneous reduction in the pulse energy per unit area. Therefore, the emitted pulse has to penetrate deeper within the canopy before sufficient energy is backscattered to trigger a return, thus leading to higher underestimations of canopy height (Hopkinson, 2007; Lovell et al., 2005; Persson et al., 2002). For instance, Anderson et al. (2006b) reported mean errors of −0.76 m (±0.43 m) and −1.12 m (±0.56) with footprint sizes of 0.33 m and 0.8 m, respectively. However, the increase in footprint size that occurs with higher flight altitude increases the probability of sampling tree tops (Hirata, 2004; Hopkinson, 2007; Næsset, 2009b). Hirata (2004) showed that when increasing the footprint size from 0.3 to 1.2 m by increasing flying altitude from 300 to 1,200 m, a canopy height increase of 0.9 m was obtained for a mountainous stand of Japanese cedar (*Cryptomeria japonica* L.f.).

Another important factor is the point density (Nelson, 1997; Reutebuch et al., 2003; Véga et al., 2012). Point density is closely linked to flight altitude and speed, pulse and scan frequencies, as well as scan angle. It also depends on the overlap between flight lines. Because the likelihood of sampling tree apices increases with point density, lower densities were found to generate higher height underestimations (Disney et al., 2010). However, Hopkinson (2007) stressed that it remains difficult to distinguish the impact of each component. When working at tree level, height is directly assessed using tree top elevation (see Section 17.5), and some authors recommend a point density above 5 pts·m^{-2} to maximize both tree crown detection and the probability of sampling tree apices (Falkowski et al., 2009; Hirata, 2004). On the contrary, assessments of height at plot level, which are mainly performed using models and no longer

TABLE 17.2   Main Factors Influencing Accuracy of Vegetation Height Measurements from LiDAR Data at Three Level of Analysis: Individual Footprint (Pulse), Tree, and Stand Levels

Level of Analysis	Factors Influencing Height Accuracy	Comments	References (Nonexhaustive)
Footprint	Signal triggering method in multi-echo systems or echo detection method in fullwave from data	Impacts both number and position of detected points	Disney et al. (2010), Holmgren et al. (2003), Wagner et al. (2004, 2006)
	Vegetation structure and spectral properties	Height underestimation ranges from ~4% to 7% and is more important within conifers	
	Emitted energy	The higher the energy, the lower the time for triggering a return	
	Footprint size	Underestimation increases with footprint size due to a decrease in irradiance (power/unit area)	
		But the probability to sample a tree apex increases with footprint size	
	Scan angle	No significant impact on measurements for angle <10°	
Individual tree	Pulse density (function of ALS system and flight parameters)	The probability to sample a tree apex increases with pulse density	Hirata (2004), Kaartinen et al. (2012), Véga et al. (2014), Véga and Durrieu (2011)
		The minimum required density depends on crown size: 2–10 pulses m^{-2} for mature stands and saplings	
		Tree structure and terrain are better characterized when density increases	
	Scan angle	Impact of scan angle on point distribution is greater for elongated crowns, for example, conifer crowns	
		Flight line overlapping reduces occlusions thus improving both crown shape and height descriptors	
	Vegetation structure and composition	Quality of height estimates depends on crown shape and radiometric properties	
		Tree detection algorithms perform better in homogenous stands	
		Dominated trees are more difficult to detect and measure	
	Topography	Accuracy of height assessment decreases with slope	
		Height of slanting trees might be biased	
		Crown structure can be distorted by slope normalization	
Plot/stand	Pulse density (function of ALS system and flight parameters)	Height parameters estimated through models; results are less sensitive to pulse density than for tree level analysis	Disney et al. (2010), Evans et al. (2009), Hodgson and Bresnahan (2004), Hopkinson (2007), Næsset (2009b), Véga et al. (2014)
	Scan angle	Scan angles <15° off-nadir to be preferred	
	Vegetation structure and composition	Accuracy is higher in simple structure (e.g., even age single layer stands)	
		Accuracy is higher in coniferous stands	
		Local calibration is required.	
	Topography	DTM quality impact the accuracy of height parameters	

by direct measurements (see Section 17.4), were found to be unaffected by point density (Jakubowski et al., 2013; Lim et al. 2008; Treitz et al., 2012) (Table 17.2).

Besides flight parameters, and because vegetation height is typically computed by subtracting a digital terrain model (DTM) from the elevation of either the nonground points or the DSM of the outer canopy layer, the quality of vegetation height estimates is closely correlated with DTM quality. Thanks to the ability of a light signal to penetrate through vegetation openings, ALS acquisitions can sample the ground. Dedicated algorithms have been developed to classified points into ground and nonground categories and to produce a DTM (Kraus and Pfeifer, 1998; Meng et al., 2010; Sithole and Vosselman, 2004). Overall, ground elevation errors in

LiDAR DTMs are usually less than 30 cm under forest covers (Chauve et al., 2008; Chen, 2010; Hodgson and Bresnahan, 2004; Reutebuch et al., 2003). But significant variations were found depending on both vegetation structure and density, which impacts the way the ground is sampled (Hodgson and Bresnahan, 2004). Lower sampling densities result in coarser terrain modeling. Ackermann (1999) reported penetration rates around 20%–40% for coniferous and deciduous forest types in Europe, but this rate can be locally inferior to a few percent, that is, 2% or 3%, in very dense tropical forests. Under a conifer forest, Reutebuch et al. (2003) reported mean DTM errors of 0.22 m (±0.24 m SD) with errors increasing with canopy densities and ranging from 0.16 m (±0.23 m) within clearcuts to 0.31 m (±0.29 m) within uncut areas. In their study,

Clark et al. (2004) obtained the highest root-mean-square error (RMSE) (1.95 m) within dense, multilayered evergreen canopy in old-growth forests, for which ground elevation was overestimated. In addition, Leckie et al. (2003) noticed that variation of the ground surface at the base of the tree could easily reach ±50 cm. For these authors, the local microtopography, which is hard to model, partly explained the observed 1.3 m (±1.0 SD) tree height underestimation.

Slope can also affect the quality of DTM under vegetation, and elevation errors were found to increase with slope (Hodgson and Bresnahan, 2004; Hodgson et al., 2005). For example, under a multilayered tropical forest, Clark et al. (2004) reported a 0.67 m elevation RMSE increase within steep slopes. This can be explained by the fact that point classification algorithms, which are used to identify ground points before producing the DTM, are mainly based on geometric properties. Due to more similar geometric characteristic between ground and low-vegetation point clouds in the presence of slope, they perform less efficiently. The difficulties involved in classifying ground points in relief and forested environments led to the development of many algorithms (see Meng et al. (2010) for a recent review). Among the several classification algorithms, the triangular irregular network iterative approach (Axelsson, 2000) is still one of the most robust, and it produces good results in a wide range of environments (Véga et al., 2012). Methods developed to process full-waveform data acquired by some recent commercial small footprint systems led to improved geometric information (more echoes extracted and higher target localization accuracy) and also provide additional features such as echo intensity and width that are linked to target properties (Chauve et al., 2009; Reitberger et al., 2006; Wagner et al., 2004). Despite the difficulty involved in decorrelating the influence of geometric and radiometric characteristics of the targets on these features (Ducic et al., 2006), they proved to be very useful in some studies (Chehata et al., 2009) when attempting to improve ground point classification when used in addition to echo locations.

DTM errors in sloping areas clearly impact height assessments. Over a complex terrain in a mountainous area, Véga and Durrieu (2011) found that tree height errors increased as a function of slope. Heights were underestimated for low slopes (i.e., below 25%), while an overestimation trend was found for steeper slopes (i.e., above 25%). However, errors in height were not attributed solely to DTM inaccuracy but were also due to the way heights are derived from ALS data (Véga and Durrieu, 2011). Indeed, when a tree top is identified, height is assessed as the difference between the tree top elevation and the ground elevation and does not always represent the actual tree height. When trees slant, the crown and its associated local maxima (LM) move toward the slope and thus lead to an overestimated height (Figure 17.3). Overall, ALS was found to underestimate vegetation height with a magnitude of underestimation that varied depending on the sensor used, the flight parameters, and the characteristics of the vegetation (Table 17.2).

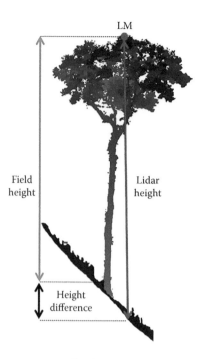

FIGURE 17.3 Error in tree height measurement associated with terrain slope. (Adapted from Véga, C. and Durrieu, S., *Int. J. Appl. Earth Observ. Geoinformat.*, 13(4), 646, 2011.)

17.3.2 Three-Dimensional Modeling of the Canopy by Digital Photogrammetry

17.3.2.1 Principle and Brief History

Since the post–World War I period, large-scale aerial photographs have been extensively used in both forest inventory and monitoring (Spurr, 1960) for many purposes: locating forest areas, mapping forest types, inventorying forest conditions, assessing wood production, monitoring damage due to insects, diseases, and fires, etc. Information on stand structure and composition has been intensively used as the basis of stratification to improve the efficiency of field data collection and the accuracy of results in multistage sampling–designed forest inventory (Korpela, 2004). Stereophotogrammetry, introduced during the same period (Andrews, 1936), was used to assess qualitative and quantitative forest stand structural characteristics and was further developed in the 1940s (Korpela, 2004). The stereophotogrammetry process is analogous to our own perception of depth with normal binocular vision. It is based upon the principle of parallax, which is the apparent displacement of a stationary object due to changes in the observer position (White et al., 2013). It thus requires two images, which were taken from two different viewpoints, and the 3D measurements are deduced from the analysis of the parallaxes that change according to the distance between the objects and the sensor (Figure 17.4).

Stereoscopes and parallax bars have been long used as low-cost viewing and measurement instruments (Korpela, 2004). Analytic stereoplotters and comparators, which have been commonly available since the mid-1970s, enable accurate 3D measurements. The emergence of DP dates back to the 1990s

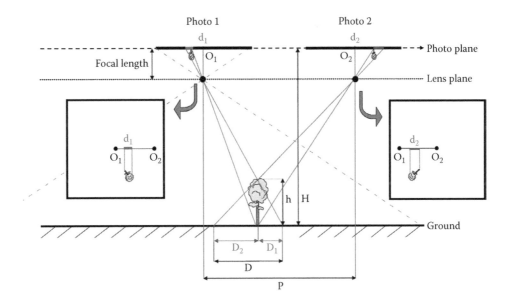

FIGURE 17.4 Principle of stereoscopic measurement, with O_1 and O_2, the centers of the photographs; H: height of the camera above the ground level; h: height of the tree; P: absolute parallax of the tree base; and D: parallax difference of the tree top with reference to the base plane (D_1-D_2). Tree height (h) is computed as the ratio (H–D)/(P–D).

(Maltamo et al., 2009) and was made possible by the development in computer technology that provided significant computing power required by 3D image analysis algorithms (Morgan and Gergel, 2013). DP has benefited from the research works conducted in the graphic, vision, and photogrammetry communities (Remondino and El-hakim, 2006).

To retrieve the three dimensions of surfaces from airborne or spaceborne images, one must fully characterize the geometry of the acquisition system in order to trace the geometric path of the sunrays, from the target to its images, reflected by the observed target. The workflow consists in solving the internal orientation, which defines the geometry of the imaging system, and the exterior orientation, which gives the position and orientation of the sensor at the acquisition time and is often divided into two steps, that is, solving the relative and absolute orientations (Heipke, 1997). Relative orientation is generally solved using coplanarity condition and requires matching points between images to generate tie points and absolute orientation is solved by using control points in which coordinates are known in both the image and mapping reference frames. Once the orientation of the photogrammetric model is known, parallax differences can be used to compute the elevation for each pixel of one image, provided a conjugate point can be identified in another image. This latter stage is also referred as dense matching. Point matching, also known as the correspondence problem, is thus crucial in the photogrammetric workflow. And many algorithms have been developed to solve it, including surface-based or object-based approaches (Barnard and Fischler, 1982; Brown et al., 2003) or a combination of both (Baltsavias et al., 2008; Lisein et al., 2013).

With the current state-of-the-art computers and computational methods, DP does indeed offer several advantages. The complex calculation process named aerial triangulation can be used to process image blocks and not only stereo pairs, thus leading to a reduced number of ground control points required to process an area covered by several images. Furthermore, a multiray matching strategy, made possible by multiview acquisitions, can significantly improve results of both the aerotriangulation and the object reconstruction steps (Thurgood et al., 2004).

The development of DP was also favored by the emergence of novel large-format digital aerial cameras producing significantly improved image quality as they combine very high spatial-resolution and high radiometric sensitivity, for example, radiometric information coded on more than 8 bits (Leberl et al., 2010). The sensors thus overcame some issues linked to the complex interdependency between the radiometric range and the pixel size encountered in analog film imagery, that is, grain noise (Leberl et al., 2010). Image texture is also enhanced, which is an important advantage for image-matching algorithms (White et al., 2013). The processing workflow is also simplified by the elimination of some tricky and time-consuming tasks such as the scanning of analog photographs. State-of-the-art digital aerial cameras, for example, Z/I Imaging Digital Mapping Camera (DMC) series from Intergraph, airborne digital sensor series from Leica, or UltraCam series from Microsoft (formerly Vexcel), can capture large-format multispectral images typically in the red, green, blue, and NIR wavebands (Petrie and Walker, 2007). Some may also have the capability to record simultaneously very large-format panchromatic images (Z/I Imaging DMC [Hinz and Heier, 2000] and UltraCam [http://www.microsoft.com/en-us/ultracam/]).

Digital images can nowadays be acquired with increased within and between flight-line overlaps. While traditional acquisition mainly used around 60% and 20% overlap within and between flight lines, respectively, state-of-the-art technology currently allows for up to 90% and 60% overlap with an add-on cost only linked to the additional airtime when between flight-line overlap is increased (Leberl et al., 2010; Thurgood et al., 2004).

The more recently developed algorithms can also be used to process data acquired by nonmetric cameras, thereby extending the processing capacity beyond standard photogrammetric geometry and products. Data acquired with low-cost light acquisition systems, like consumer-grade cameras embedded on unmanned aerial vehicles (UAV), can now be processed. These cameras have high distortion levels and low geometric stability. Therefore, images that are also characterized by high rotational and angular variations between successive images and significant perspective distortions (Lisein et al., 2013) cannot be processed using conventional photogrammetric software. To address such image constraints, new solutions have been developed. For example, Lisein et al. (2013) used Multi Image Matches for Auto Correlation Methods (MICMAC), an open-source software, combining photogrammetric approach and newly developed computer vision algorithms referred to as Structure from Motion to retrieve canopy structure from very-high-resolution images acquired using an UAV.

The combined and recent developments in sensor technology, positioning systems (triangulation), and processing algorithms offer numerous benefits including reduced occlusion, higher level of automation, limited manual editing, and increased geometry accuracy (Leberl et al., 2010).

17.3.2.2 Measuring Tree Height Using Digital Photogrammetry and Resulting 3D Models

Even if some authors consider that image-based methods are now better at creating 3D points cloud than ALS survey (Leberl et al., 2010), height information retrieval remains challenging in forest environment, and several issues must be addressed.

First, a major drawback of photogrammetry is its inability to provide ground elevation under dense forests, thus preventing vegetation height assessment. Second, only the trees whose crowns are at least partially in direct sunlight are detectable on aerial photographs. And despite progress in sensor technology, suppressed or small shaded trees are still undetectable (Korpela, 2004). Finally, image matching in forest environment remains challenging and an error-prone task due to occlusion, repetitive texture, multilayered objects, or moving objects, that is, change in position of tree tops in windy conditions (Lisein et al., 2013).

With visual or semiautomatic stereo interpretation, the accuracy of tree height measurement has been widely studied (Gong et al., 2002; Korpela, 2004). Accuracy depends on factors related to both images, for example, scale and quality, and targets, for example, crown geometry, radiometric properties, stand structure and topography, and leaf-on versus leaf-off conditions. Some authors announced 10 cm accuracy rates (Gagnon et al., 1993) and others mean absolute errors of 1.8 m (Gong et al., 2002). Korpela (2004), who compiled results from several studies, reported that heights are underestimated due to the inability to measure the very highest top shoots, and this bias tends to increase as photograph scale decreases. With DP, the whole

canopy surface can be automatically reconstructed as a 3D point cloud or a DSM.

But tree height measurement accuracy is also highly dependent on tree base elevation assessment capacities. Ground elevation has been measured within open areas in the neighborhood of the trees (Gong et al., 2002) or, in closed canopies, on targets positioned on the ground (Kovats, 1997). It was then assimilated to the tree base elevation. More conveniently, a range of DTMs has been used in combination with photogrammetric measurements or DSMs to derive forest parameters at the tree or stand level. DTMs extracted from field surveys (Fujita et al., 2003) or from a triangulation of manual photogrammetric measurement in open areas (Næsset, 2002a) were sometimes used. However, such solutions are limited to small areas. St-Onge et al. (2004) first tested the coregistration of a photogrammetric model with an ALS DTM to estimate individual tree height based on manual measurements. Using scanned photographs with a 11.3 cm pixel size, they reported an average underestimation in height of 0.59 m in a white cedar (*Thuja occidentalis*) stand. Korpela (2004) developed a semiautomatic, single-scale template-matching approach to position tree tops on stereo images and then to estimate tree height and crown size using an ALS DTM. The spatial registration of both photo-derived DSM and ALS-derived DTM proved to be a critical step toward canopy height estimation (St-Onge et al., 2004), and various approaches have been proposed to coregister these models (Huang et al., 2009; Lisein et al., 2013; St-Onge et al., 2008).

Despite the potential of DP to accurately measure elevations, most of the image-matching algorithms, originally developed for DTM extraction, proved incapable of consistently generating accurate DSMs over forested areas. St-Onge et al. (2008) reported that, when using precise ALS DTMs, photo-ALS Canopy Height Model (CHM = DSM-DTM) could reconstruct general height patterns but was unable to provide details about both individual tree crowns and small gaps. Similarly, using images at a 1:15,000 scale digitized at 2,000 dpi and processed using an automatic image-matching algorithm (Match-T), Naesset (2002a) reported significant underestimation of mean plot height. The author also indicated that the algorithm was not flexible enough to reconstruct abrupt changes in elevation such as those that characterize canopies. Occlusions, due to the shape of the trees and to the complex 3D structure of the forest canopy, hinder image matching (Lisein et al., 2013). Compared to single stereo models, image block approaches allowing multiview processing were found to improve 3D canopy modeling and tree height estimation due to the reduction of occluded areas (Hirschmugl et al., 2007; Magnani et al., 2000). The quality of the results also depends on the parameters defining the matching strategy (Figure 17.5). For example, Lisein et al. (2013) who tested several matching strategies explained that some omitted isolated trees and that those that were optimized for deciduous canopy reconstruction did not performed very well when used for coniferous crown reconstruction.

Due to the high sensitivity of the canopy model quality to the matching strategy, and whose optimization depends on the cover

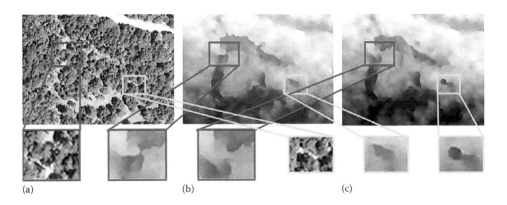

FIGURE 17.5 Illustration of the impact of changes in matching strategy on forest DSM using the MICMAC software. (a) Orthorectified image of the area. The two DSMs (b and c) were both computed using a 0.28 m spatial resolution and a 9 × 9 pixels correlation window. The regularization coefficients (RS) and the correlation thresholds (CT) were different: RS = 0.010 and CT = 0.2 for (b) and RS = 0.005 and CT = 0.0 for (c). Changes in matching parameters impact crown shape (red area) and gap shape (yellow area) that were better reconstructed with the first set of parameters (b).

type, the production of accurate CHM over large areas is a tricky task. To tackle this problem, Baltsavias et al. (2008) proposed a complex method combining both area- and feature-based matching, the self-tuning of matching parameters, the generation of redundant matches, an automatic blunder detection, and a coarse to fine hierarchical matching strategy. The method was able to achieve better results than when using an ALS dataset for canopy height assessment, thus suggesting that methods developed for processing ALS CHM could also be efficient to process high-quality photo-ALS CHM over forests.

17.3.3 Comparison of ALS and Photogrammetric Products

Few studies have compared photogrammetric and ALS products (Baltsavias, 1999c; Leberl et al., 2010; White et al., 2013). As a complement to Table 17.1, Table 17.3 compares main strengths and limitations of both photogrammetric and ALS technologies and derived products.

For acquisition purposes, imagery is considered to be the most advantageous (Table 17.3). Imaging systems, which have a greater field of view, can be operated at both a higher flying speed and altitude than ALS systems for which flight height is dependent on available laser power. As a result, survey planning is easier and less costly with an imager (White et al., 2013). For example, assuming an equal flying altitude and speed, and an equal sidelap, the same area is covered in about one-third of the time with an imager, considering a typical 75° field of view, than with an ALS with a 30° scan angle (Baltsavias, 1999c). Comparing typical acquisitions under optimal flying height and speed, Leberl et al. (2010) found that the flying time would be 13 times longer with an ALS system when compared to a digital camera system to obtain comparable point clouds and elevation results. In addition, imaging continuously benefits from the development of satellite stereo imagery. Despite a lower geometric resolution leading to a decrease in CHM quality (St-Onge et al., 2008), satellite solutions can be used to cover very large areas at low cost (Neigh et al., 2014) and is a viable alternative to

ALS for small-scale forest monitoring when high-quality DTMs already exist. Besides these considerations, another advantage of ALS is that, as an active sensor, it can be operated at any time of the day while the quality of aerial photographs is highly influenced by solar illumination. Indeed, as shadows hamper image matching, flying hours must be carefully chosen to minimize shadowing in the forest canopy (White et al., 2013).

As regards the production of georeferenced point clouds, a 3D point cloud can be obtained more quickly when using ALS systems as coordinates can be, under ideal conditions, automatically computed (Baltsavias, 1999c). However, as DP workflows become both increasingly efficient and automated, this advantage is gradually diminishing (White et al., 2013). Baltsavias (1999c) gives detailed information on the comparative geometric quality of point clouds generated from photogrammetry and ALS. At the same flying height, the geometric resolution of a laser measurement, given by the footprint size and depending on the laser beam divergence, typically 1 mrad, is coarser than the pixel resolution obtained by a digital camera (e.g., with a 15 μm pixel). Concerning relative planimetric/altimetric accuracies, planimetry is typically 1/3 more accurate than elevation with photogrammetry, while it is 2–6 times less accurate with ALS data. These higher planimetric errors will also significantly influence elevation accuracy on sloped terrain (Baltsavias, 1999c). Comparing the accuracy values for identical flying height in the 400–1000 m range, shared by both technologies, the photogrammetric accuracy is, on average, slightly better than with ALS. However, in practice, airborne LiDAR and imagers are not operated in the same conditions, thus making it difficult to compare their real performance levels.

Height point clouds, obtained by subtracting ground elevation from the elevation of ALS and photo-derived 3D points, were also compared in term of height distribution, by comparing percentiles on 400 m² forest plots (Lisein et al., 2013; Vastaranta et al., 2013). Low correlations between lower height percentile values of both point clouds (Lisein et al., 2013; Vastaranta et al., 2013) may be explained by the presence of ALS points

TABLE 17.3 Comparative Summary of Main Strengths and Limitations of Both ALS and DP Technologies and of Resulting 3D Point Clouds and Forest Products (Baltsavias, 1999c; Leberl et al., 2010; Lisein et al., 2013; St-Onge et al., 2008; Vastaranta et al., 2013; White et al., 2013)

		Best Rated	ALS	DP
Data acquisition	System lifetime	DP	Determined by the number of pulses that can be emitted by the laser; equivalent to ~10,000 operating hours. Rapid deterioration may occur with a drastic decrease in output power.	Decades for robust aerial cameras.
	Mission planning	DP	Flight height limited by eye safety (min height) and laser power (max height); typically 500–1000 m. Higher flight height possible (up to 4000 m) but with reduced pulse frequency to avoid signal mixture between successive backscattered signals. Typical scan angle: 20°–40° → mission planning difficult in mountainous areas.	Flight height: from 500 to 3500 m according to plane type (up to 12 km with high-altitude aircraft). Typical effective FOV: 75° → Large areas covered with less flight lines and flying time up to 13 times shorter compared to ALS. → Multiview stereo and optimized B/H might provide higher-quality forest canopy reconstruction.
	Flying conditions	ALS	Few illumination constraints (LiDAR can be operated day and night, winter and summer). System can be operated in both leaf-on and leaf-off conditions.	Acquisitions constrained by solar illumination (impacts radiometric quality), and view angles due to sensitivity of image matching to occlusions and shadows. Leaf-on conditions only to provide forest height products.
Data processing	Production of geolocated 3D point clouds	ALS	-3D coordinates automatically computed by combining information recorded by the LiDAR, the IMU, and the DGPS → reduced processing time.	With increased efficiency and automation level of DP workflows, advantages of ALS have gradually diminished. Final quality of geolocation is less dependent on DGPS and IMU measurements quality than for ALS.
Product quality	3D products	ALS	Planimetry 2–6 times less accurate than altimetry. Sampling pattern and pulse density depend on flight parameters. Information on both the vegetation surface and the ground, allowing to extract a DTM and to further assess vegetation height and structure.	Planimetry 1/3 more accurate than altimetry. Sampling: in theory regular (once to twice the image resolution); in practice depends on image-matching results. At a given cost, higher point density achievable than with ALS. Information only for object surfaces visible in at least two images.
	Forest height products	ALS	Height products accuracy depends on sensor specifications, flight parameters, resulting point density. Small gaps and tree tops better described than on DP products. Accuracy of both ALS and DP height products is a function of vegetation type and structure. In both cases, it is higher within coniferous and even-aged stands. In-depth evaluations of forest height products are still required for accuracy comparisons.	External DTM needed to provide vegetation height. Height products accuracy depends on image scale, B/H, image quality (radiometry and texture), shadow patterns, image-matching algorithm, number of images used during matching, external DTM quality.
	Forest inventory and monitoring	ALS and DP	ALS is the leading technology for acquiring DTMs over forested areas. But the potential of ALS for updating forest information limited due to cost considerations. Higher potential than DP for the assessment of biophysical parameters linked to structural parameters. ALS and DP are complementary for forest mapping and inventory → necessity to optimize alternated acquisitions under cost constraints.	Large-area wall-to-wall coverage at low cost. Multiple use of data: stand delineation, species identification, monitoring of deforestation and disasters. Large archive datasets exist in some countries for retrospective mapping.

within the canopy when DP CHM only describes the top of the canopy. However, the very high correlations observed for higher percentiles, for example, Vastaranta et al. (2013) reported mean differences below 0.2 m for the 70th height percentile values and beyond for 500 circular plots, reveal that most of the points in the ALS dataset were located at the same level as photo-derived points. They mainly describe the outer canopy shape, and there is very little information remaining to describe both the understory and the ground due to occlusion effects.

When further assessing heights at tree level or assessing dominant heights at plot level using height distribution metrics (see Section 17.4.2), estimation accuracy was found to be slightly better with ALS data compared to photo-derived data (Lisein et al., 2013; Vastaranta et al., 2013; White et al., 2013). For instance, Lisein et al. (2013) obtained an adjusted R^2 and a RMSE (%) of 0.94 and 3.7% and, respectively, 0.91 and 4.7% for height estimation at tree level and a R^2 and RMSE (%) of 0.86 and 7.4% and 0.82 and 8.4% for dominant height estimation at stand level (Lisein et al., 2013). Vastaranta (2013) also found a lower RMSE (%) for mean height prediction at plot level with ALS data (7.8%) compared to photo-derived products (11.2%). However, White et al. (2013) also pointed out that there is no rigorous comparison of the relative accuracy of canopy heights derived from ALS and image-based point clouds over a range of forest types and terrain complexities.

The product that is common to both ALS and DP technologies is the 3D model of the Earth surface. Focusing on the structure of the outer canopy surface approximated by computing a raster CHM, several studies reported that tree crowns were wider and less defined in photo-derived CHM (Barbier et al., 2010; Lisein et al., 2013; Vastaranta et al., 2013). Moreover, small gaps and tree tops as well as fine-scale peaks and gaps in the outer canopy were not perfectly modeled on the photo-derived CHM that tend to behave as a smoothed version of ALS CHM (Lisein et al., 2013; St-Onge et al., 2008).

The ability of ALS to provide information on subcanopy forest structure as well as on ground topography, even below closed canopies, means this technology is very suitable for forest inventory. Despite its sensitivity to solar illumination, optical imagery can provide consistent spectral information that can be used to report, for example, on species composition and to assist forest inventory in a way that cannot be done by ALS (Baltsavias, 1999c).

Finally, due to their respective advantages and to their equivalent potential for top vegetation height measurements, ALS and photogrammetry are more complementary than mutually exclusive, and the issue should be to how best optimize alternated acquisitions under cost constraints. For example, Vastaranta et al. (2013) suggest that, for forest mapping and monitoring purpose, ALS data could be acquired every 10 or 20 years, depending on forest and management considerations, and digital stereo imagery could be used to update forest information in the intervening period. However, for an initial inventory, both ALS and imagery, but not necessarily stereo imagery in this case, are acknowledged to be very useful.

17.4 Assessing Height Characteristics at Stand Level

Early LiDAR studies showed an underestimation bias when measuring tree height from LiDAR data, which is due to several factors (see Section 17.3.1.2). Models based on empirical relationships between LiDAR data and forest attributes measured in the field at plot level (Figure 17.6) were thus used to correct this bias and produce stand level estimations and maps of stand height characteristics. These approaches, which have been extended to predict other stand characteristics, are widely used in forest applications and are often referred as area-based approaches.

17.4.1 General Presentation of Area-Based Approaches

The development of models at plot level is usually achieved in two or three stages. The first two stages, which are first the establishment of a predictive model to assess a forest parameter from 3D data, and then its application to the area covered by the 3D

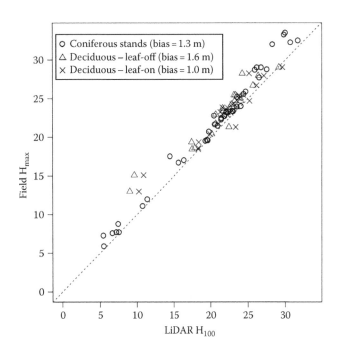

FIGURE 17.6 Maximum LiDAR height (H_{100}) is compared to the height of the tallest tree measured in the field for 93 forest plots. Regardless of the stand type (here coniferous [39 plots] and deciduous [twice 27 plots]) and of acquisition conditions (leaf-on and leaf-off for the deciduous stand), maximum height was underestimated by LiDAR by at least 1 m on average (biases ranging from 1 to 1.6 m). The coniferous stands are pine plantations and trees have a relative flat crown compared to other coniferous species. In addition, LiDAR datasets used for the comparison of maximum heights have point densities high enough to limit the probability to miss tree tops (8, 20, and 18 pulses m^{-2} for coniferous, deciduous leaf-on, and deciduous leaf-off, respectively, and with a footprint size of about 27 cm in all cases).

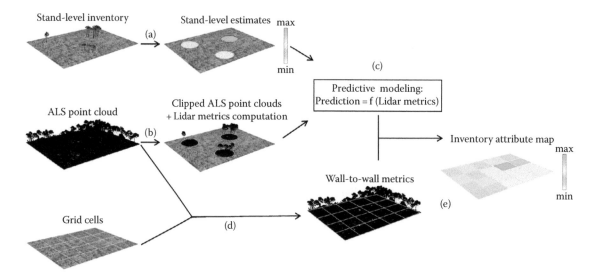

FIGURE 17.7 Principle and main steps of area based approaches used to predict forest parameters from 3D data. (a) Field inventory at plot level; (b) extraction of subpoint clouds for all the inventoried plots; (c) establishment of a predictive model linking forest characteristics to 3D ALS data characteristics and validation of the model; (d) segmentation of the whole area into grid cells with a size similar to the one of the inventory plots; (e) computation of the explanative ALS metrics for each grid cell and application of the model, cell by cell, to obtain a map of forest parameters.

dataset, are described in many studies (e.g., White et al., 2013; Wulder et al., 2012). When wall-to-wall 3D data are available, these stages lead directly to maps. When only 3D measurements samples are available, local forest parameter assessments have to be further extrapolated, as part of a third stage, to produce maps over large areas using other remote-sensing data, for example, optical or radar imageries and other map products available at national or global scale (e.g., SRTM, vegetation maps). The two first stages are briefly presented in the succeeding text and are illustrated in Figure 17.7 for 3D ALS point clouds. However, the principle remains the same when using either DP point clouds or raster CHM. It can also be extended to large footprint LiDAR data processing, even if in this case the metrics used are different from derived from point clouds.

First, a model is developed at plot level to infer forest parameters from metrics derived from either ALS or photogrammetric 3D data (Figure 17.7a through c). The model is calibrated and validated using reference plots measured in the field for which stand characteristics such as dominant height, Lorey's height, basal area, volume, and aerial biomass are inferred from the field measurements (Figure 17.7a). This step requires allometric equations to assess some biophysical parameters, for example, volume and biomass, from structural characteristics such as DBH and height.

For each plot, the corresponding 3D ALS data subset is clipped, and several metrics are derived from either the subpoint clouds or the CHM (Figure 17.7b). Next, using a subset of the available plots as a training set or a cross-validation procedure (e.g., leave-one out), a model is built that predicts stand characteristics measured in the field from the most explanative 3D metrics. Multiple approaches can be used to establish the model. They might be parametric or nonparametric and include maximum likelihood, discriminant analysis, nearest neighbors,

random forests, and various forms of regressions (McRoberts et al., 2010). The selection of the most explanatory variables either is made based on a preliminary statistical analysis or is part of the model construction, for example, when stepwise regression approaches or random forests analysis are used. The model is then validated using either the set of remaining reference ground plots or the plots that were left aside at each repetition when applying a bootstrap approach. This latter approach is to be favored especially when the total number of reference plots is limited.

Once built, the model can be extrapolated to the whole area covered by the ALS 3D dataset. To achieve this, the area is first subdivided into grid cells, the size of which are similar to the size of the reference ground plots (Figure 17.7d), the explanative 3D metrics are then computed for each cell, and the model is used to predict the value of the forest parameter at the cell level for the whole area (Figure 17.7e).

If 3D data do not encompass the whole area, a third stage is required to obtain a map. This consists in building another model that in turn uses the forest parameters assessed using the 3D dataset and links them to new variables extracted from another ancillary dataset that covers the whole area and that is usually made up of optical images. Once built, the model is applied to assess the parameter of interest on the areas that were not covered by the 3D data. Nearest neighbors techniques have been widely used to predict continuous variables based on satellite image data (McRoberts et al., 2010). Other nonparametric approaches, like random forests or neural networks, or parametric approaches based on regression analysis can also be used to achieve this aim (McRoberts et al., 2010).

In Section 17.4.2, we will focus on the first step, that is, the construction of the model linking 3D metrics to forest parameters.

17.4.2 Area-Based Model Implementation

The objective of this section is threefold. First, it aims to illustrate and compare two families of approaches that can be used to build a model to predict stand height characteristics derived from 3D ALS data. Second, it seeks to compare ALS and DP products to assess stand height characteristics through the comparison of results obtained with a model using only the information from the top of the canopy to the models using the whole 3D ALS information. This comparison should illustrate on the added value of having information from the understory, even if it is attenuated by occlusion effects. Third, it presents some issues regarding model accuracy. To achieve the two first objectives, two sites with contrasted stand types were chosen in order to compare the behavior of the three model types by considering several key elements involved in the model construction process. All the models were tested to predict both dominant and Lorey's heights. The first site is a coniferous forest located in the Landes region in the southwest of France (44.40°N, 0.50°W). The site is dominated by monospecific stands of maritime pine (*Pinus pinaster*) in even-aged plantations. The second site is a deciduous forest located in northeastern France (48.53°N, 5.37°E). The site is comprised multilayered broad-leaved stands, dominated by European beech (*Fagus sylvatica*), hornbeams (*Carpinus betulus*), and sycamore maple (*Acer pseudoplatanus*). ALS data were collected over the coniferous and deciduous study sites with a point density of 10 pts m^{-2} and 30 pts m^{-2}, respectively. Field data were collected on 39 15-m radius circular plots within the 2 months that followed the ALS survey for the coniferous site and on 42 circular plots within the year prior to the ALS survey for the broad-leaved forest. Lorey's height (H_L) and dominant height (H_{dom}) were estimated from field measurements for each field plot. H_L was computed as the mean height weighted by basal area. H_{dom} was computed as the mean height of the six largest trees according to their DBH. H_L and H_{dom} were estimated at stand level from ALS metrics using two different approaches that are keeping with the general methodology for area-based approaches described in the previous subsection.

On one hand, we applied a practical process to predict H_L and H_{dom} from ALS data proposed by Næsset (2002b). This approach is typical and widely used to build predictive models of stand characteristics from ALS data. Numerous metrics were derived from the height distributions of first or last LiDAR returns: maximum values (H_{maxf}, H_{maxl}), mean values (H_{meanf}, H_{meanl}), coefficients of variation (H_{cvf}, H_{cvl}), percentiles of the distributions (H_{0f}, H_{10f}...,H_{90f} and H_{0l},...,H_{90l}), and canopy densities (d_{0f}, d_{10f}...,d_{90f} and d_{0l},...,d_{90l}) computed as proportions of ALS hits above a given percentile of the distribution. Stepwise regression was performed in order to select the most explanative LALS metrics that would remain in the final models. No metric with a partial Fisher statistic greater than 0.05 was selected (Næsset, 2002b). A linear relationship among log-transformed variables was applied. Log transformation of stand attributes and ALS metrics was used to accommodate nonlinearity. H_L and H_{dom} were estimated following this methodology for both sites.

We used adjusted R^2 (R^2_{adj}) to account for the number of LALS metrics in the final models. RMSE were also calculated to assess the accuracy of the predictions. This approach is referred to hereafter as the point distribution approach.

On the other hand, we proposed to use a conceptual model to predict H_L and H_{dom} from only four ALS metrics. This model skips the step aiming at selecting the best metrics among a large set of potential ALS metrics. Metrics have been defined to characterize the natural variability of stand structures. In area-based approaches, individual tree heights are not determined. Instead, an average canopy height (μ_{CH}) is easily measured from the 3D LiDAR point cloud and is an important predictive variable (Lefsky et al., 2002). Thereby, μ_{CH} metric was chosen as the first variable and calculated by averaging first return elevations. An indicator of tree height heterogeneity at plot level should be used in addition to μ_{CH}. We used the variance in canopy height (σ^2_{CH}) to characterize tree height heterogeneity as suggested by Magnussen et al. (2012). These two first variables were calculated without taking into account returns that were below a 2 m height threshold, so as to describe the part of the plots that was actually covered by trees. However, the fact that tree attributes measured in the field are related to the whole plot area means that the rate of open areas in each plot must also be evaluated. Gap fraction (P) has been calculated from ALS data as the ratio between the number of first returns below a specified height threshold and the total number of first returns. When calculated in this way, P is related to the penetration of light through the canopy but was found to be well correlated with fractional cover (Hopkinson and Chasmer, 2009). P was thus the third selected metric. The metrics defined earlier only refer to the structural properties of the top of the canopy and can be calculated for either ALS or photo-derived point clouds. To take advantage of the capacity of LiDAR to penetrate into the vegetation and provide information on the vertical crown size and on overtopped trees, we defined an additional metric H_{crown}, as the mean of the crown heights. Crown heights were estimated for each 1 m × 1 m areas included in the plot based on the vertical distribution of ALS points. H_L and H_{dom} were estimated using the four metrics in a log-transformed model. This approach is referred as the mechanistic model.

A third type of model was built in a way similar to the mechanistic model but using only the three metrics that can be computed when using photo-derived point clouds or raster CHMs, that is, μ_{CH}, σ^2_{CH}, and P. This last approach is referred to as the top of canopy mechanistic model.

H_L and H_{dom} have been predicted using the three approaches. A summary of the results is displayed in Table 17.4.

The three approaches satisfactorily estimated height attributes in the coniferous forest. Models provided high R^2_{adj}, all equal to 0.99 despite a higher number of variables for the mechanistic model, and low RMSE. Only one metric remained in the final point distribution model, H_{90f} and H_{maxl}, for H_L and H_{dom}, respectively. RMSE was slightly reduced using the mechanistic model, that is, −0.16 m for both H_L and H_{dom} when compared to the point distribution model. The majority of forest studies have focused

TABLE 17.4   Goodness-of-Fit Statistics for HL and Hdom Predictions Using the Point Distribution, the Mechanistic Model, and the Top of Canopy Mechanistic Models

Approach	Predicted Attribute	Coniferous Forest			Deciduous Forest		
		ALS Metrics	R^2_{adj}	RMSE (m)	ALS Metrics	R^2_{adj}	RMSE (m)
Point distribution model	H_L	H_{90f}	0.99	0.84	H_{100f}, H_{60l}, d_{90f}	0.89	1.97
	H_{dom}	H_{maxl}	0.99	0.88	H_{maxf}, H_{80l}	0.95	1.53
Mechanistic model	H_L	μ_{CH}, σ^2_{CH}, P, H_{crown}	0.99	0.68	μ_{CH}, σ^2_{CH}, P, H_{crown}	0.97	1.11
	H_{dom}	μ_{CH}, σ^2_{CH}, P, H_{crown}	0.99	0.72	μ_{CH}, σ^2_{CH}, P, H_{crown}	0.97	1.08
Top of canopy mechanistic model	H_L	μ_{CH}, σ^2_{CH}, P	0.99	0.68	μ_{CH}, σ^2_{CH}, P	0.96	1.22
	H_{dom}	μ_{CH}, σ^2_{CH}, P	0.99	0.73	μ_{CH}, σ^2_{CH}, P	0.97	1.18

on conifer forests characterized by a quite simple structure (Lim et al., 2003). In such stand types, both stand homogeneity and the absence of understory may explain the very good and similar performances obtained with the three approaches. A single metric describing the top of the canopy (e.g., H_{90f} or H_{maxl}) is sufficient to summarize stand height characteristics. Results were different for the more complex deciduous forest. The first approach provided H_L and H_{dom} predictions with an R^2_{adj} of 0.89 and 0.95, respectively. Three metrics have been selected to predict H_L, while two metrics have been selected to predict H_{dom}. The mechanistic model provided more accurate estimates with an R^2_{adj} of 0.97 and RMSEs reduced by 0.86 and 0.4 m compared to the point distribution model for H_L and H_{dom}, respectively. A more significant improvement was observed for the prediction of H_L, which is a parameter that takes into account overtopped trees and not only the tallest ones. Figure 17.8 illustrates the improvement obtained using the mechanistic model by showing the observed values of H_L against the ones predicted by the three models for the deciduous forest. Applying the mechanistic model across diverse forest area types only required a calibration of the parameters as both the metrics and the model shape were kept from one area to the other one. From these examples, we can see that the mechanistic model proved robust and more efficient than the point distribution model, which is currently the most widely used approach.

Except when considering dominant heights, for which the top of canopy information is sufficient, this example also underlines the value of the information coming from the understory to improve Lorey's height assessments. First, in the point distribution model, one of the selected variables was H_{60l}, which is a parameter that is decorrelated from the variability in height of the canopy surface as seen in Section 17.3.3. Second, when comparing the mechanistic and the top of canopy mechanistic models, the latter performed slightly less well, with a similar R^2_{adj} (0.96 against 0.97) despite using one fewer parameter than the complete mechanistic model and obtained a slightly increased RMSE (+0.11 m).

Even if not as widely used as point distribution approaches, a mechanistic approach appeared to provide a viable alternative solution regarding both robustness and accuracy in order to develop models predicting stand height characteristics from 3D data metrics. Furthermore, even if they performed slightly less well, the models using only the top of canopy information might provide interesting results in a range of stand types, including quite complex ones such as multilayered broad-leaved stands. However, their performance is likely to be lower if the area or the strata on which they are calibrated and applied is characterized by changes in stand structure. In that particular case, a more complete mechanistic model that partly includes the diversity

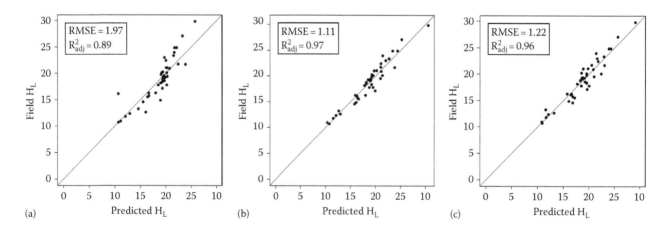

FIGURE 17.8   Observed values of HL against predicted values for the deciduous forest dataset using (a) the point distribution model, (b) the mechanistic model, and (c) the mechanistic top of canopy model.

of structures thanks to an additional parameter, such as H_{crown}, used in our example, is expected to be more robust.

To better cope with the impact of stand characteristics on ALS metrics, a preliminary stratification of forest stands is recommended and would be expected to improve ALS-derived predictions (Næsset, 2002b). Aerial images or very-high-resolution imagery from space (submeter to few meters resolution) can be used to that aim (see Table 17.1). Height distributions derived from ALS could also give some precious information for stratification purposes. But the fact that some metrics derived from 3D data are influenced by acquisitions characteristics (such as laser footprint, ALS sensors, or pulse repetition frequency; see Section 17.3.1.2) also hampers the development of generic models. In general, field calibration plots associated to each new ALS acquisition are required (Gobakken and Næsset, 2008; Næsset, 2009a; Thomas et al., 2006). Trying to identify metrics that are least impacted by the changes in acquisition characteristics would represent an additional step required to provide more robust models. Apart from both stand and acquisition characteristics, other factors such as plot size and coregistration precision were shown to influence models' quality and their derived maps (e.g., Frazer et al., 2011; Maltamo et al., 2009; Strunk et al., 2012). For example, using large plots reduces the probability of edge effects, which may occur when parts of the crowns belonging to trees located outside but close enough to the boundary of a plot are included in the ALS or DP 3D data subsets corresponding to the plot or, conversely, when part of the crowns of trees located inside a plot are excluded when using the plot area to extract the plot-related 3D data subset.

It is also important to bear in mind that LiDAR *sees* all the vegetation within the plot, irrespective of tree social status, while in calibration plots, several trees can be ignored since field measurements may start only at a given minimum diameter. On the contrary, photogrammetric point clouds *see* only the dominant stratum, while part of the overtopped trees is measured in the field (Table 17.3). Field measurement protocols, designed either for NFI or for the purpose of a specific study, can be different from one study to another. Sampling designs can vary, not only in terms of sampling scheme and density but also in terms of plot design (e.g., fixed area, concentric plots with various diameters, fixed number of trees, threshold defining the limit size for measurable trees). Furthermore, tree measurements can also change, for example, measurement of DBH only or both diameters and heights or alternatively all diameters but heights for only a subsample of trees. All these elements are likely to change the field reference estimations and therefore also change the models. Therefore, reflexions and efforts made to harmonize forest inventory procedures (Ferretti, 2010; McRoberts et al., 2009) are likely to increase the generalization level of the models and should contribute to a more widespread use of remote-sensing data in forest applications.

When a model is developed, errors propagate through the entire process. Five main sources of error should be considered in area-based approaches. First, the field measurement errors; they may range from several decimeters to a few meters for height measurements (see Section 17.2.3). Second, for some parameters that cannot be directly measured in the field, such as biomass, another source of error resides in the allometric equations available to estimate the parameters from field measurements. But, unless heights are assessed in the field from DBH measurements, this error will not affect height predictions. A third source of error is the geolocation inaccuracy of both field plots and 3D datasets. A fourth one is the sampling error that is often neglected and mainly linked to the sampling design and the extent of sampling effort. If the whole population is inadequately sampled, the sampling error might be significant. Finally, the models produced using metrics derived from point clouds or CHMs have their own inherent uncertainty. All these sources of error will combine and impact on the quality of the model results.

In the previous part of this section, we focused on models developed to predict the structural and biophysical characteristics of stands. The use of LiDAR in landscape ecology and biodiversity studies is a more recent field of research. However, metrics extracted from LiDAR data or elevation models have already been proposed for the characterization of landscape patterns and structure at several scale (Mücke and Hollaus, 2010; Uuemaa et al., 2009). In addition, the relationships between metrics describing the 3D distribution of the vegetation and the presence or abundance of a given species, or the assemblage of species, have been investigated in several studies. For example, Nelson et al. (2005) used LiDAR data to identify forested sites that might support populations of Delmarva fox squirrels, while Müller and Brandl (2009) and Müller et al. (2014) predicted arthropod diversity and assemblages (Müller et al., 2014; Müller and Brandl, 2009). Bird species abundance and assemblages have also been at the core of several studies (Goetz et al., 2007; Lesak et al., 2011; Müller et al., 2010; Zellweger et al., 2013). All the aforementioned studies deal with fauna biodiversity. And despite the existence of relationships between floristic biodiversity and forests structure (Zilliox and Gosselin, 2013), only a few attempts have been made to study these relationships using either ALS or photo-derived 3D data (Simonson et al., 2012). Modeling the link between biodiversity and environmental conditions remains challenging. Indeed, biodiversity is driven by many processes and taking into account vegetation or landscape structural characteristics exclusively is not sufficient. The ecological context, with regard to abiotic variables, is also of great importance when attempting to explain biodiversity indicators. While several studies do not take into account the ecological context, complementing LiDAR metrics with abiotic variables was shown to improve the predictive power of models (see Zellweger et al., 2014). Furthermore, to tackle the complexity of ecological modeling, multiple regressions, which are widely used to predict forest biophysical parameters, are no longer the most appropriate approach. They were sometimes replaced by statistical approaches that are more suited to the modeling of various parametric distributions and deal with discrete variables while allowing the utilization of nonparametric functions for variable weighting. To this aim, generalized additive models (Goetz et al., 2007), boosted regression trees (Zellweger et al., 2013), or Bayesian models (Zilliox and Gosselin, 2013) have been occasionally used.

17.4.3 Model Extrapolation and Inferences for Large-Area Inventories

For a number of operational inventory applications, many authors (see, e.g., Næsset, 2002b; Naesset, 2007) have demonstrated how wall-to-wall ALS coverage could efficiently be used to improve the accuracy and spatial resolution of field surveys. In Scandinavian countries, LiDAR-derived forest inventories have been made since 1995 (Næsset, 2004). Statistical inferences, sampling design, and statistical properties of LiDAR-derived estimations have only recently received more attention because of the often complex structure of LiDAR surveys (Ben-Arie et al., 2009; Li et al., 2012; Morsdorf et al., 2004; Véga et al., 2014).

Some studies rely directly on existing NFI plots (e.g., in Maltamo et al., 2009), whereas in other studies, specific field campaigns have to be carried out. Sampling design associated with LiDAR surveys is frequently established based on a model-based design (Smits et al., 2012), as opposed to design-based sampling, where systematic or random locations of field calibration plots are being performed (Henry et al., 2013; Naesset, 2007). This aspect is important since model-based estimators are not design unbiased and may reveal potential bias depending on model correctness (Picard et al., 2012a).

Wall-to-wall ALS coverages over extended areas of forests have become more frequent. However, such coverage efforts are still complex and costly to undertake and generate a considerable amount of data to be processed (Uuemaa et al., 2009; Wulder et al., 2012; Zellweger et al., 2013). Furthermore, some systems, such as early LiDAR systems, for example, the Portable Airborne Laser System (PALS), a profiling system (Nelson et al., 2003), or spaceborne systems cannot provide full coverage. ICESat, the first LiDAR mission aimed at measuring terrestrial surfaces, acquired data from 2003 to 2009 and was a profiling system. Due to technological limitations, future space LiDAR missions are likely to embed, at best, multibeam systems. Designed to either reduce survey costs or process data acquired by multibeam or profiling systems, some methods using LiDAR measurement samples were developed to perform extensive forest inventories. These methods also incorporate remote-sensing sources other than LiDAR into existing large-area sample-based forest inventory frameworks (Müller et al., 2010, 2014; Müller and Brandl, 2009; Nelson et al., 2005; Wulder et al., 2012). These approaches still require local calibration of models to link remote-sensing data to forest parameters (Goetz et al., 2007; Zellweger et al., 2014). It is worth noting that all applications do not require map production, and part of the required information can be obtained by analyzing sets of characteristics assessed on samples. Using simulation, Ene et al. (2013) demonstrated that ALS data enhanced forest inventory results and that ALS-aided surveys can be a cost-efficient alternative to traditional field inventories. The latter also provided more accurate results if sampling intensity was optimized (i.e., by optimizing the distance between regular flight lines covering only part of the NFI plots) (Ene et al., 2013).

When models are extrapolated, two points are worth recalling. First, it is important to check if the predictions have been made within the calibration domain. For example, models developed for even-aged stands, or for one specific species, could yield erroneous predictions in multilayer stands or in stands composed of mixed species. Model robustness therefore represents a major issue and should be evaluated across several stand types. It is therefore crucial to be able to characterize the stand types consistently with the calibration domains of the models. However, further difficulties may arise when predictions must be made in edge areas covering several stand types. In some countries, edges between stand types or between forest and nonforest areas can represent a significant share of the NFI plots. In France, for instance, at least 20% of NFI plots are located on an edge. Therefore, forest heterogeneity must not be underestimated when building, validating, and extrapolating models. Finally, when maps are considered as an operational outcome of the models, one must remember, as McRoberts et al. (2010) explain, "the utility of maps is greatly increased when they form an appropriate basis for inferring values of maps parameters describing the populations represented by the maps." Inference means being able to calculate estimates of the mean of the population parameter and of its variance in order to be able to define a $1-\alpha$ confidence interval for this parameter (McRoberts et al., 2010). Both extrapolation of forest structural or biophysical parameters and inference issues are at the core of active research work, and to get a better grasp of these topics, the authors recommend readers to refer to the following papers: Ben-Arie et al. (2009), Ene et al. (2013), Li et al. (2012), McRoberts (2010), and McRoberts et al. (2010).

17.5 Approaches for Individual Tree Height Assessment

Unlike area-based approaches that are poorly sensitive to point density (see Section 17.3.1.2), methods for individual tree height assessment require several height measurements per tree crowns, hence at least a few points/m², and the optimal point density is likely to change according to the stand type and age. Whereas 2 pts·m^{-2} was found to be sufficient in mature stands, at least 10 pts·m^{-2} might be required for saplings (Kaartinen et al., 2012). Early methods used for detecting and characterizing individual trees from ALS data were based on techniques developed to process very-high-resolution optical images (Leckie et al., 2005). To locate individual trees, brightness or color gradients were used in optical imagery (Leckie et al., 2005; Wulder et al., 2000), while ALS CHM-based methods can make use of the geometrical properties of the CHMs, including height, slope, and curvature (Bongers, 2001) for the same purpose. Now that accurate photo-DSMs can be produced with DP, raster-based tree detection approaches can be applied either to ALS-derived or photo-derived CHMs. These approaches are described in Section 17.5.1. Section 17.5.2 presents more recent approaches that directly use 3D point clouds, to improve both tree crown characterization and overtopped tree detection. Finally, Section 17.5.3 presents hybrid approaches based on algorithms that either use both point clouds and raster information or exploit the complementarity between structural and radiometric information.

17.5.1 Raster-Based Approaches

Standard methods for extracting individual trees from a CHM are based on three steps, which are CHM modeling and optimization, detection of tree apices using LM, and development of crown segments around each tree apex (Persson et al., 2002; Popescu et al., 2002; Solberg et al., 2006). When solely focusing on tree height assessment, this last step is not always required in the processing workflow.

17.5.1.1 CHM Modeling and Optimization

The efficiency of raster-based approaches is intimately linked to the quality of the CHM derived from the point clouds. Grid cell size is the first critical parameter. Various studies reported that the optimal size should be of the same order of magnitude as the original point spacing (Vepakomma et al., 2008). Besides pixel size, the point to grid transformation has been widely investigated. Two main approaches are commonly used to compute the initial CHM, from either the first returns or the points classified as nonground points. A common method consists of assigning the maximum Z-value of the points of each grid cell and estimating a value for the empty cells. This can be achieved, for example, by averaging the values of the connected filled cells (Brandtberg et al., 2003; Hyyppä et al., 2001) or by using an inverse distance weighted (IDW) method applied to

a given number of neighboring points among those belonging to the canopy surface (Véga and Durrieu, 2011; Vepakomma et al., 2008). Alternative methods involve interpolating a value at the center of each grid cell from the Z-values of the neighboring points using kriging (Popescu and Wynne, 2004), IDW (Vepakomma et al., 2008), active contour (Persson et al., 2002), or minimum curvature (Solberg et al., 2006) algorithms. When the task is measuring trees, exact interpolation methods might be favored (Kato et al., 2009), but overall, simple interpolation techniques like IDW were found to be sufficiently accurate (Anderson et al., 2006a; Vepakomma et al., 2008). To enhance both tree top detection and crown segmentation algorithms, different procedures can be used to improve the surface described by the CHM prior to further processing. These include point cloud thinning, hole filling, as well as CHM filtering. Thinning procedures are implemented upstream from CHM generation to filter out points within the canopy. For example, an initial outer canopy surface can be defined, for example, by using an active contour algorithm to trace the outer part of the crowns (Persson et al., 2002), and points that are too far below this surface are discarded (Persson et al., 2002; Solberg et al., 2006). Hole-filling algorithms are applied once an initial CHM has been calculated and is aimed at removing irregularities within the canopy surface partly due to crown porosity. Ben-Arie et al. (2009) introduced a 6-step semiautomated pit-filling algorithm based on a Laplacian edge detector to remove pits while preserving edges. Véga and Durrieu (2011) developed an iterative method based on 4–8 connectivity kernels to automatically detect and recalculate local minima (Figure 17.9).

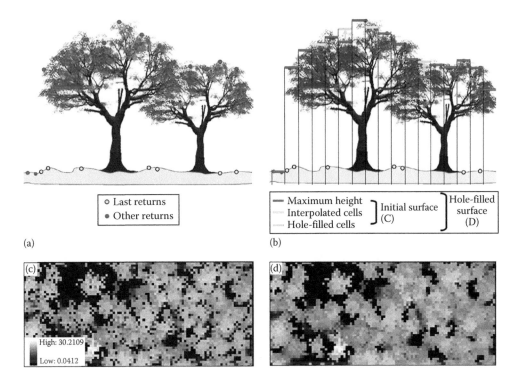

FIGURE 17.9 CHM modeling and optimization: from the point cloud (a) to the CHM (b). (c) and (d), respectively, represent an initial CHM (0.5 m resolution) and its hole-filled version over a coniferous forest (Draix, France).

In general, the workflows include a last smoothing step to reduce commission errors induced by noise within the CHM prior to tree parameter extraction. Gaussian filtering is commonly used (Hyyppä et al., 2001; Morsdorf et al., 2004; Persson et al., 2002; Solberg et al., 2006). But median filtering was also used as it can preserve the original values in the CHM (Popescu et al., 2003). A single smoothing filter can be applied (Popescu et al., 2003) with smoothing intensity, which may be driven by local vegetation height (Koch et al., 2006). Nevertheless, iterative Gaussian filtering is the most widely used approach (Hyyppä et al., 2001; Maltamo et al., 2004; Solberg et al., 2006). In some studies, the results obtained at several filtering levels were combined as optimal filtering intensity level depends on crown size (Persson et al., 2002; Véga and Durrieu, 2011).

17.5.1.2 Detecting Tree Apices

Tree tops are usually considered as LM in the smoothed CHM. An LM can be defined as a pixel containing the highest value in a given neighboring defined by a kernel.

Along with the impact of CHM quality, the efficiency of tree top detection using LM identification mainly depends on the optimization of both kernel size and shape. As explained in various studies, the selection of either an overly small or overly large kernel might lead to commission (i.e., false detection) or omission errors, respectively (Figure 17.10) (Popescu et al., 2002; Reitberger et al., 2009). In addition, while a single kernel size might be sufficient to process CHM acquired over forest plantations with homogenous crowns, approaches using variable kernel sizes had to be developed to improve tree top detection over complex forest structures. Popescu et al. (2002) fixed the window size according to the relationship between tree height and crown diameter they established from field measurements. The method was further improved by introducing a circular window and by using specific relationships for both conifer and hardwood stands (Popescu and Wynne, 2004) (Figure 17.10). Chen et al. (2006) extended the concept of window size dependence on tree height by using a nonlinear power model linking window size to tree height and by then estimating a prediction interval for the optimal window size around the value predicted by the model. This helped to reduce the omissions of trees with crowns smaller than those predicted by the model.

Many other approaches have been proposed to identify trees, such as the second derivative of blob signature (Brandtberg et al., 2003), a multiple morphological opening method to tackle the problem of heterogeneity in crown dimensions (Hu et al., 2014), marked point process models optimized using a multiple births and deaths approach (Zhou et al., 2010), or h-minima transform using distance-transformed images (Chen et al., 2006). The latter was found to be particularly suited for the detection of trees with flat or aggregated crowns.

17.5.1.3 Measuring Tree Crowns

Several methods have also been put forward to reconstructing crown segments and subsequently estimate tree characteristics, such as tree height, crown diameter or area, or crown base height. Most common methods include region-growing approaches (Hyyppä et al., 2001; Persson et al., 2002), watershed analysis (Chen et al., 2006; Kwak et al., 2007; Mei and Durrieu, 2004), morphological analysis (Wang and Glenn, 2008), valley following approach (Leckie et al., 2003), fitting functions (Popescu and Wynne, 2004), ellipse fitting (Véga and Durrieu, 2011), wavelet analysis (Falkowski et al., 2006), or a combination of methods as in Koch et al. (2006) or Hu et al. (2014).

Unlike approaches such as region-growing or watershed approaches, some can be implemented without a tree top detection step, such as the valley following approach (Leckie et al., 2003) or the spatial wavelet analysis proposed by Falkowsky et al. (2006). In few cases, tree top identification and tree delineation are interdependent. For example, in Véga and Durrieu (2011), the initial tree top set evolves as crown contours are identified and refined in a multiscale iterative process. Several methods, most of the region-growing or watershed approaches, tend to produce irregularly shaped crown segments and thus necessitate a shape control process step. Sets of rules have been developed to qualify and constrain the segmentation process based on either height values, areas, distance from the gravity center, or shapes (Hu et al., 2014; Hyyppä et al., 2001; Koch et al., 2006; Solberg et al., 2006; Weinacker et al., 2004). Other methods directly model tree crowns using circular or elliptic shapes (Popescu et al., 2003; Véga and

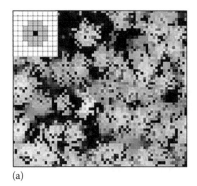 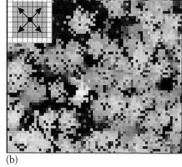 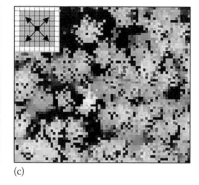

(a)　　　　　　　　　　　　(b)　　　　　　　　　　　　(c)

FIGURE 17.10 Tree top detection in a coniferous area using either a fixed 20 pixel kernel (a) or variable window sizes from using a generic model (b) or a model derived for conifers (c). (From Popescu, S.C. et al., *Comput. Electron. Agric.*, 37, 71, 2002.)

Durrieu, 2011). Here, results might be improved by allowing the tree top to be different from the center of symmetry of the shape used by the model. This is, for example, the case for the approach proposed by Véga and Durrieu (2011) where LM were used to estimate crown radius in various directions and crown elliptic shapes were further adjusted while only using the set of the radius endpoints that were assumed to characterize the crown edge.

Kaartinen et al. (2012) compared several individual tree detection approaches on a given study area and for 3 ALS datasets of different point densities (2, 4, and 8 pts m^{-2}). The comparison was made with regard to the number of correctly detected trees, that is, number of correct matches with field reference, commissions, and omissions. The results were analyzed according to both height classes and tree status. For well-detected trees, tree location accuracy and the accuracy of the assessment of both height and crown dimensions were analyzed. The percentage of detected trees ranged from 25% to 102%. Surprisingly, several automated methods performed better, even in the case of codominant or suppressed trees, than manual detection that identified 70% of the trees (Kaartinen et al., 2012). This point deserves particular attention because, unlike in many other fields, manual detection cannot be used as a reliable reference to assess the performance of automatic tree detection approaches derived from imagery or ALS data (Kaartinen et al., 2012). As expected, higher commission errors were mainly found for the smallest trees. Tree detection was not significantly improved by the increase in point density from 2 to 8 pts m^{-2}. On the contrary, height assessment accuracy was improved with an increase in point density, even if the impact of point density on result accuracy was of less importance compared to the impact of the choice of the tree detection approach. While 75% of the raster-based approached showed an RMSE below 2 m, the best models provided RMSEs ranging from 60 to 80 cm (Kaartinen et al., 2012). In an operational context, the authors emphasized the efficiency of quite simple methods based on LM finding, which achieved a tree detection rate of over 70%, a percentage of matched trees of 60% and of 95% when considering only dominant trees, a commission error of 18%, and an RMSE for height estimates around 80 cm.

17.5.2 Point-Based Approaches

In addition to raster-based approaches, point-based approaches have been developed in order to fully take advantage of the 3D information of ALS point clouds and were expected to detect not only dominant but also overtopped trees. If there is no theoretical reason to not apply point-based approaches to photo-derived point clouds, the utility is limited as these point clouds mainly describe canopy surface, like CHMs. Morsdorf et al. (2004) introduced a voxel-based approach with k-means clustering to detect individual trees and modeled crowns using a paraboloid model. However, as the k-means seeds were based on CHM LM, the method suffers from the same limitations as raster-based approaches. Ferraz et al. (2012) proposed another clustering approach based on a mean-shift algorithm that does not require seed points. The method is used to identify several vegetation

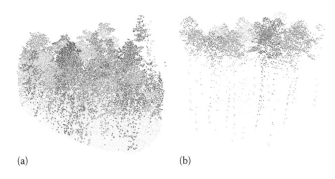

(a)   (b)

FIGURE 17.11 Example of point-based segmentation over a broadleaved forest (566 stem ha^{-1}) (a) and a pine plantation (206 stem ha^{-1}) (b).

layers and then define an adaptive kernel bandwidth parameter optimized for each layer. The method gave very promising results and could detect 98.6% of the dominant trees and approximately 13% of the suppressed ones within stands dominated by eucalyptus and pines. However, the authors stressed that a more sophisticated approach would be required to process more complex forest structures (Ferraz et al., 2010). The normalized cut segmentation introduced by Reitberger et al. (2009) is also not dependent on seed identification. Tested in a mixed mountain forest, the method was used to detect 77%, 32%, and 18% of the trees in the dominant, intermediate, and lower canopy layer, respectively. Li et al. (2012) proposed a distance-based algorithm to sequentially detect trees. The method assumes relative spacing between trees and overcomes the issue of finding LM by using "global maxima," defined during the segmentation process as the highest point not yet associated with a crown. The method was not evaluated in hardwood forests but achieved 86% global detection rate in a mixed conifer forest with a 94% correct detection rate (Li et al., 2012). Recently, Véga et al. (2014) proposed a multiscale dynamic segmentation approach. As in Li et al. (2012), the algorithm is based on global maxima. Each point is considered as a new tree apex or assigned to an existing crown. In the latter case, the algorithm allocates the point to the tree segment, that is, the projection of a given crown in the 2D plane, which is the least changed by the inclusion of this new point considering the change in the surface area of the convex hull of each candidate segment. With this approach, 82% of the trees were correctly detected on average when considering three different forest types (Figure 17.11) (Véga et al., 2014). In the comparative analysis performed by Kaartinen et al. (2012), only one point-based approach was selected to be compared to raster-based approaches. As expected, it enabled better detection of underlayered vegetation than that offered by the raster-based approaches.

17.5.3 Hybrid Approaches

In addition to pure 3D point cloud approaches, other approaches combining raster and point clouds data recently emerged and have proven effective at improving individual tree extraction performance. Using a high-density dataset, Reitberger et al.

(2009) obtained improved results by combining their point-based approach with a watershed segmentation of the CHM and a stem detection applied to the point cloud (86%, 32%, and 17% of correct detection for the dominant, intermediate, and lower canopy layer, respectively, instead of 77%, 32%, and 18% [see Section 17.5.2]). Hu et al. (2014) first segmented a CHM using a multiscale morphological opening approach and flagged uncertain shapes based on shape and height information. Points within the flagged segments were then reprocessed by multi-Gaussian fitting applied to point distribution. The number of fitted Gaussian functions corresponds to the number of trees inside a segment. Splitting and merging operations were then applied to refine the segmentation and derive the final crowns (Hu et al., 2014). In order to tackle issues involved in CHM-based crown extraction, Wang and Glenn (2008) adopted a voxel structure to generate 2D horizontal projection images of the point cloud. For each horizontal layer, a hierarchical morphological algorithm is used to generate crown segments. Tree crowns were then reconstructed using a preorder forest transversal approach to connect the 2D crown segments of the successive layers. The method was able to identify both canopy and overtopped trees but was found to be highly sensitive to the parameters (Wang and Glenn, 2008).

Other studies aimed at exploiting the complementarities between structural information provided by canopy models and radiometric information contained in optical images. This helped to improve individual tree detection and subsequent height measurement accuracy. Popescu and Wynne (2004) used optical data to identify and locate forest types that were then used to adapt the segmentation parameters of their adaptive window size algorithm, thus leading to improved results. Leckie et al. (2003) applied a valley following approach to both multispectral images and ALS CHM in order to extract and characterize individual trees. They reported that crown segmentation was more efficient in dense forests when using spectral images (80%–90% detection) as poor crown outlines were obtained using the ALS CHM. On the contrary, better results were obtained by processing the CHM within open areas where sunlit ground vegetation can cause false detections within images. These results show that combining ALS data with multispectral data might improve crown segmentation, which also holds true for tree top detection (Smits et al., 2012). Suárez et al. (2005) applied an object-based segmentation using both spectral and height data to identify individual trees implemented in ©eCognition software by Definiens. Tested on 345 trees, results showed that ALS underpredicted individual tree heights by 7%–8%. The author also claimed that an ALS and imagery combination could be used to work with lower ALS point densities, thus reducing costs. But, as reported by Kaartinen et al. (2012), methods developed to process ALS data alone are more mature and several approaches based on raster CHM gave better results than hybrid approaches.

A qualitative comparison of the three types of approaches, namely, raster-based, point-based, and hybrid approaches, is provided in Table 17.5.

TABLE 17.5　Summary of Main Strengths and Limitations of Raster-Based, Point-Based, and Hybrid Approaches Developed for Tree Detection and Tree Height Assessment

Approach	Advantages	Limitations
Raster based	• Computational efficiency. • Well established methods. • Widely available image processing tools can be used. • Several methods perform better than manual detection. • Simple methods can achieve tree detection rates >95% for dominated trees, >70% for all trees. • RMSE ranging from 0.6 to 0.8 m for height assessment with the best-performing methods.	• High sensitivity to point density. • Point to raster interpolation can impact height estimation. • CHM smoothing impacts detection of Local Maxima LM and crown boundaries. • Approaches based on LM identification are not well adapted to "flat" crowns. • Crown segments: quality is a function of the method and the chosen parameters (e.g., region growing, watershed, model fitting). • Detection rates highly dependent on the method (from 25% to 102%). • Dominated trees cannot be detected. • Crown parameters such as crown-based height cannot be assessed.
Point based	• No need for height interpolation nor for a smoothing step of a raster CHM. • Improved description of crown shape and attributes; extraction of additional tree parameters such as crown base height. • More adapted to identify small and dominated trees → detection rates >77% considering all trees even in complex stands.	• Sensitive to both pulse sampling pattern and point density. • Lower computational efficiency. • Tree detection and crown segmentation are impacted by the point clustering approach used. • Might be sensitive to sampling patterns.
Hybrid	• Improved efficiency by combining advantages of both raster-based and point-based approaches → optimization of computational efficiency by processing raster data and improvement of results by using additional information on understory provided by 3D point clouds. • When radiometric information is used in addition to 3D point clouds, results are enhanced due to information complementarity (e.g., 9% increase in detection rate of dominant trees in Reitberger et al. (2009)).	• Lower computational efficiency compared to pure raster-based approaches. • Might be affected by the limitations of each source of information depending on the processing workflow.

Compared to area-based approaches, the first major drawback of individual tree approaches is practical in nature. The latter require higher point densities and thus induce an increase in data acquisition costs and in processing time. A second disadvantage is their inconsistent behavior depending on the type of forest on which they are applied. Currently, their performance is insufficient to extract understory trees, whatever the individual tree approach method used. However, the main advantage of individual tree approaches over area-based ones is that they might 1 day provide height distributions in spite of synthetic height indicators. Such distributions would enable improved predictions of other forest parameters such as timber volume or overground biomass (Kaartinen et al., 2012). This is why there is still a lot of commitment to developing powerful and operational approaches in this research field despite the significant challenges involved.

17.6 Conclusion and Perspectives

Tree height and vegetation height distribution are structural characteristics of major importance when attempting to understand and monitor biological and ecological processes at tree, stand, and landscape scales.

Tree height plays a central role in forest inventories, and height growth is a main driver for growth and yield models. Forest structure features, and in particular gaps, also play a major role in forest ecology because of their influence on microclimate and habitat quality and thus on biodiversity potential.

Accurate measurement of tree heights and careful monitoring of the dynamics of both canopy height and gaps at several scales in space and time are known to be crucial to improve both forest ecosystem monitoring and modeling and thus contribute to their sustainable management. But, as field evaluations of tree height and forest structure are labor intensive and costly, they cannot fully meet data requirements. In this chapter, we focused on two remote-sensing technologies, ALS and DP, which are deemed to be mature enough to provide practical and accurate measurements of 3D forest structure. Developing their use to measure and monitor tree height and vegetation structure at several scales, and at regular time scales, is expected to represent a major breakthrough regarding the development of forest ecosystem modeling and to provide precious information to help managers dealing with the increasing societal and economic pressures weighing on forests.

Each technology, although hampered by some major drawbacks, possesses major assets. Today, LiDAR is probably the most promising and mature technology capable of providing direct measurements of 3D forest structure from airborne systems. Despite continuous system improvements, data acquisition remains costly, however, which partly explains the development of inventory methods based on samples of ALS measurements (Ene et al., 2013; Wulder et al., 2012). Furthermore, there is no ongoing or planned spaceborne LiDAR mission having vegetation monitoring as its primary objective. The first ICESat mission was designed for ice monitoring. ICESat2, which is planned to be launched by NASA in 2017, will be the second LiDAR mission

dedicated to Earth surface monitoring. Also primarily designed to characterize and monitor polar ice, vegetation height and biomass measurements are one of its secondary scientific objectives. However, data simulations have demonstrated that, with its current design, ICESat2 will not adequately replace the recently shelved DESDynI vegetation LiDAR mission (Hall et al., 2011a). Part of the scientific community working in the field of remote sensing applied to forest is pushing for the development of an international vegetation LiDAR mission, and several projects of space missions have been submitted and are under evaluation by CNES, JAXA, and NASA, the French, Japanese and American space agencies, respectively (Durrieu et al., 2013; Durrieu and Nelson, 2013; Kobayashi et al., 2013).

Due to recent developments in DP that have provided improved canopy surface reconstruction, this technology has recently benefited from renewed interest for forest applications. Unlike with LiDAR, information is limited to the top of the canopy. However, data access offers several advantages over LiDAR. First, due to the relatively long history of both airborne photography and photogrammetry, large databases of aerial photographs, dating back over 60 years and even beyond, exist in several countries (Véga and St-Onge, 2008). This enables researchers to study forest structure monitoring for long-past time periods. Second, regarding current acquisitions, imagery is considered to offer more advantages than ALS due to easier survey planning and lower costs. Furthermore, the radiometric information provided by imagers is complementary to 3D information and is of high value for forest-type characterization. Finally, digital imagers have long been deemed fit for space applications and several spaceborne systems today provide very-high-resolution multispectral stereoscopic images from which accurate DSM can be obtained (e.g., Aguilar et al., 2014). Overall, due to their respective advantages and to similar potential for top vegetation height measurements, ALS and photogrammetry are more complementary than mutually exclusive, and the objective should be to optimize alternated acquisitions under cost constraints. In particular, combining satellite stereo imagery and ALS DTM may allow the monitoring of vegetation height over large areas at a lower cost than with airborne-based solutions.

Concerning methodological issues, producing accurate surface model in forest environments based on DP remains challenging, and despite considerable improvements made over the last years, further efforts are needed to develop methods that can be used to process different forest types with increased self-tuning functionalities in order to provide user-friendly and powerful tools to foresters. Currently, the methods used to retrieve information on vegetation heights from either ALS or photoderived 3D data can be classified within two families: areabased approaches that assess height characteristics at stand level and approaches for individual tree height assessment. A major advantage of the latter is their capacity to provide height distributions instead of the synthetic height indicators provided by area-based approaches. Such distributions could help improve predictions of other forest parameters such as timber volume or aboveground biomass. However, developing efficient approaches

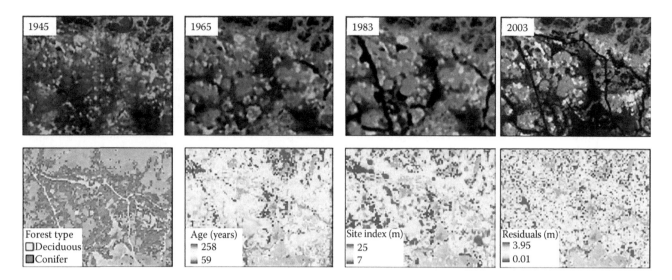

FIGURE 17.12 Time series of DP and ALS CHM (upper line) and derived maps of age and site index along with model residuals at 20 m resolution (lower line). Forest types were computed from a QuickBird image. (Adapted from Véga, C. and St-Onge, B., *Forest Ecol. Manage.*, 257(3), 951, 2009.)

remains very challenging due to the difficulties involved in accurately segmenting trees, in particular those belonging to the understory. In addition, these approaches require high-density point clouds, which increase both acquisition and data management costs as well as processing time. Therefore, despite their interesting features, their operational value is limited. On the contrary, area-based approaches are very efficient and have already been incorporated in NFI in several countries. One way to make the most of both approaches could be the development of methods aimed at assessing height distributions, in a way similar to the attempts to model stem diameter distribution (Gobakken and Næsset, 2004; Thomas et al., 2008).

Finally, another major perspective concerns 3D data time series analysis used to monitor changes in forest structural characteristics. As the development of ALS is quite recent compared to airborne photography, DP should be used once long time series are needed. Time series analysis raises specific methodological issues such as data calibration or the management of data with different qualities within change detection and characterization processes. Monitoring changes in forest structure over time can provide key information on forest growth and disturbances, as well as on forest functioning through the estimation of biomass changes or carbon fluxes. Forest modelers and foresters also need accurate information on height changes to develop models of forest dynamics, estimate annual allowable cuts based on prediction of future yields, or monitor fluctuations of forest carbon stocks under changing climatic and disturbance regime (see, e.g., Coops and Waring, 2001). Véga and St-Onge (2009) demonstrated the potential of height growth monitoring to assess forest site productivity (Figure 17.12). In addition to canopy growth, times series of elevation models also provide insights into disturbances regime such as gaps and gap dynamics (e.g., Tanaka and Nakashizuka, 1997; Vepakomma and Fortin, 2010). While recent studies have focused on the analysis of structural forest changes and gap

dynamics, little attention has been paid to the potential of such information to characterize biodiversity. Indeed, regarding the role of canopy gaps in diversity richness, recent knowledge has mostly been acquired using synchronic approaches. Less work has been devoted to monitoring biodiversity according to gap characteristics or linking past gap dynamics to present-day biodiversity. Yet, this issue warrants further investigation, as this dynamic approach is linked to theoretical constructs about disturbance theory and their forestry counterparts (McCarthy and Burgman, 1995) and, as an integral part of forest biodiversity, could be linked to disturbance dynamics. Documenting gap dynamics could therefore lead to the identification of interesting biodiversity indicators. This type of indicator could also become an interesting feature of forest management and therefore represent a good candidate for the role of sustainable management indicator of forest biodiversity and provide useful information to identify sustainable management practices. Indeed, it has been shown that structurally complex canopy enhanced both biodiversity and productivity (Ishii et al., 2004), and natural disturbance emulation has been proposed as a general approach for ecologically sustainable forests (see, e.g., Kuuluvainen and Grenfell, 2012). To reach these economic and ecological goals will require increased vegetation height monitoring in order to develop and implement silvicultural prescriptions that aim at maintaining forest ecosystem functions and biodiversity (Ishii et al., 2004). And, thanks to their potential for measuring vegetation height characteristics, both optical remote-sensing technologies presented in this chapter are likely to play a major role in reaching these goals.

Acknowledgments

This work is partly based on data and results obtained in the frame of two projects, the Forest Resource Estimation For Energy (FORESEE) project (ANR-2010-BIOE-008) granted by the French National Research Agency (ANR) and the Gestion Forestière, Naturalité et Biodiversité (GNB) project granted by the French ministry in charge of the Ecology through the "Biodiversité, Gestion Forestière et Politiques Publiques" program (convention RESINE CVOJ 000 150, convention MBGD-BGF-1-CVS-092, n°CHORUS 2100 214 651). This work also benefited from data acquired by the Scientific Interest Group "Bassins de Draix, étude de l'érosion en montagne" led by Irstea. The R&D department of Office National des Forêts (ONF) and the Biogeochemical Cycles in Forest Ecosystems research unit of the French National Institute for Agricultural Research (INRA BEF) research unit are supported by the French ANR as part of the "Investissements d'Avenir" program (ANR-11-LABX-0002-01, Lab of Excellence ARBRE). The authors also greatly thank all the people involved in the field measurements from National Radioactive Waste Management Agency (ANDRA), INRA, Irstea (National Research Institute of Science and Technology for Environment and Agriculture), and ONF. Special thanks to Laurent Albrech for his highly valuable work during field campaigns and for field data georeferencement and to Xavier Lucie who provided images of DP DSMs.

References

Ackermann, F., 1999, Airborne laser scanning—Present status and future expectations. *ISPRS Journal of Photogrammetry and Remote Sensing*, 54(2):64–67.

Aguilar, M. A., Del Mar Saldana, M., and Aguilar, F. J., 2014, Generation and quality assessment of stereo-extracted DSM from geoeye-1 and worldview-2 imagery. *IEEE Transactions on Geoscience and Remote Sensing*, 52(2):1259–1271.

Allouis, T., Durrieu, S., Chazette, P., Bailly, J. S., Cuesta, J., Véga, C., Flamant, P., and Couteron, P., 2011, Potential of an ultraviolet, medium-footprint lidar prototype for retrieving forest structure. *ISPRS Journal of Photogrammetry and Remote Sensing*, 66(6 Suppl.):S92–S102.

Anderson, E. S., Thompson, J. A., Crouse, D. A., and Austin, R. E., 2006a, Horizontal resolution and data density effects on remotely sensed LIDAR-based DEM. *Geoderma*, 132(3–4):406–415.

Anderson, J., Martin, M. E., Smith, M. L., Dubayah, R. O., Hofton, M. A., Hyde, P., Peterson, B. E., Blair, J. B., and Knox, R. G., 2006b, The use of waveform lidar to measure northern temperate mixed conifer and deciduous forest structure in New Hampshire. *Remote Sensing of Environment*, 105(3):248–261.

Andrews, G., 1936, Tree-heights from air photographs by simple parallax measurements. *The Forestry Chronicle*, 12(2):152–197.

Angelsen, A., 2008, Moving ahead with REDD: Issues, options and implications, Cifor, Bogor, Indonesia, 156p. http://www.forestday.org/fileadmin/downloads/movingahead2.pdf. (Accessed July 1, 2014.)

Angelstam, P. K., 1998, Maintaining and restoring biodiversity in European boreal forests by developing natural disturbance regimes. *Journal of Vegetation Science*, 9(4):593–602.

Antonarakis, A. S., Saatchi, S. S., Chazdon, R. L., and Moorcroft, P. R., 2011, Using Lidar and Radar measurements to constrain predictions of forest ecosystem structure and function. *Ecological Applications*, 21(4):1120–1137.

Assmann, E., 1970, Principles of forest yield study. *Studies in the Organic Production, Structure, Increment and Yield of Forest Stands*, Pergamon Press, Oxford, NY, 504p.

Axelsson, P., 2000, DEM generation from laser scanner data using adaptive TIN models. *International Archives of Photogrammetry and Remote Sensing*, 33(B4/1; Part 4):111–118.

Baccini, A., Goetz, S. J., Walker, W. S., Laporte, N. T., Sun, M., Sulla-Menashe, D., Hackler et al., 2012, Estimated carbon dioxide emissions from tropical deforestation improved by carbon-density maps. *Nature Climate Change*, 2(3):182–185.

Baltsavias, E., Gruen, A., Eisenbeiss, H., Zhang, L., and Waser, L., 2008, High-quality image matching and automated generation of 3D tree models. *International Journal of Remote Sensing*, 29(5):1243–1259.

Baltsavias, E. P., 1999a, Airborne laser scanning: Basic relations and formulas. *ISPRS Journal of Photogrammetry and Remote Sensing*, 54(2):199–214.

Baltsavias, E. P., 1999b, Airborne laser scanning: Existing systems and firms and other resources. *ISPRS Journal of Photogrammetry and Remote Sensing*, 54(2–3):164–198.

Baltsavias, E. P., 1999c, A comparison between photogrammetry and laser scanning. *ISPRS Journal of Photogrammetry and Remote Sensing*, 54(2–3):83–94.

Barbier, N., Couteron, P., Proisy, C., Malhi, Y., and Gastellu-Etchegorry, J., 2010, The variation of apparent crown size and canopy heterogeneity across lowland Amazonian forests. *Global Ecology and Biogeography*, 19(1):72–84.

Barnard, S. T. and Fischler, M. A., 1982, Computational stereo. *ACM Computing Surveys (CSUR)*, 14(4):553–572.

Barthelemy, D. and Caraglio, Y., 2007, Plant architecture: A dynamic, multilevel and comprehensive approach to plant form, structure and ontogeny. *Annals of Botany*, 99(3):375–407.

Barthélemy, R., 1946, Atmospheric physics—About air soundings—Note from Mr. René Barthélémy, presented by Mr. Camille Guttoni, Session on 11 February 1946. Comptes rendus hebdomadaires des séances de l'Académie de Sciences-Janvier-Juin 1946, Weekly reports of the sessions of the French Academy of sciences- January–June 1946. *Paris*, 220:450–451.

Batho, A. and Garcia, O., 2006, De Perthuis and the origins of site index: A historical note. *FBMIS*, 1:1–10.

Bellassen, V., Delbart, N., Le Maire, G., Luyssaert, S., Ciais, P., and Viovy, N., 2011, Potential knowledge gain in large-scale simulations of forest carbon fluxes from remotely sensed biomass and height. *Forest Ecology and Management*, 261(3):515–530.

Bellassen, V., Le Maire, G., Dhôte, J., Ciais, P., and Viovy, N., 2010, Modelling forest management within a global vegetation model—Part 1: Model structure and general behaviour. *Ecological Modelling*, 221(20):2458–2474.

Ben-Arie, J. R., Hay, G. J., Powers, R. P., Castilla, G., and St-Onge, B., 2009, Development of a pit filling algorithm for LiDAR canopy height models. *Computers & Geosciences*, 35(9):1940–1949.

Blair, J. B., Rabine, D. L., and Hofton, M. A., 1999, The Laser Vegetation Imaging Sensor: A medium-altitude, digitisation-only, airborne laser altimeter for mapping vegetation and topography. *ISPRS Journal of Photogrammetry and Remote Sensing*, 54(2–3):115–122.

Bolton, D. K., Coops, N. C., and Wulder, M. A., 2013, Investigating the agreement between global canopy height maps and airborne Lidar derived height estimates over Canada. *Canadian Journal of Remote Sensing*, 39(Suppl. 1):S139–S151.

Bongers, F., 2001, Methods to assess tropical rain forest canopy structure: An overview. *Plant Ecology*, 153(1–2):263–277.

Bontemps, J.-D. and Bouriaud, O., 2014, Predictive approaches to forest site productivity: Recent trends, challenges and future perspectives. *Forestry*, 87(1):109–128.

Bouget, C., 2005, Short-term effect of windstorm disturbance on saproxylic beetles in broadleaved temperate forests. Part I: Do windthrow changes induce a gap effect ? *Forest Ecology and Management*, 216(1–3):1–14.

Boyd, D. and Danson, F., 2005, Satellite remote sensing of forest resources: Three decades of research development. *Progress in Physical Geography*, 29(1):1–26.

Brandtberg, T., Warner, T. A., Landenberger, R. E., and McGraw, J. B., 2003, Detection and analysis of individual leaf-off tree crowns in small footprint, high sampling density lidar data from the eastern deciduous forest in North America. *Remote Sensing of Environment*, 85(3):290–303.

Brown, M. Z., Burschka, D., and Hager, G. D., 2003, Advances in computational stereo. *IEEE Transactions on Pattern Analysis and Machine Intelligence*, 25(8):993–1008.

Buckingham, R. and Staenz, K., 2008, Review of current and planned civilian space hyperspectral sensors for EO. *Canadian Journal of Remote Sensing*, 34(sup1):S187–S197.

Cabello, J., Fernández, N., Alcaraz-Segura, D., Oyonarte, C., Piñeiro, G., Altesor, A., Delibes, M., and Paruelo, J. M., 2012, The ecosystem functioning dimension in conservation: Insights from remote sensing. *Biodiversity and Conservation*, 21(13):3287–3305.

Castro, K. L., Sanchez-Azofeifa, G. A., and Rivard, B., 2003, Monitoring secondary tropical forests using space-borne data: Implications for Central America. *International Journal of Remote Sensing*, 24(9):1853–1894.

Chaubert-Pereira, F., Caraglio, Y., Lavergne, C., and Guédon, Y., 2009, Identifying ontogenetic, environmental and individual components of forest tree growth. *Annals of Botany*, 104(5):883–896.

Chauve, A., Véga, C., Bretar, F., Allouis, T., Pierrot Deseilligny, M., and Puech, W., 2008, Processing full-waveform lidar data in alpine coniferous forest: Assessing terrain and tree height quality. *International Journal of Remote Sensing*, 30(19):5211–5228.

Chauve, A., Véga, C., Durrieu, S., Bretar, F., Allouis, T., Deseilligny, M. P., and Puech, W., 2009, Advanced full-waveform lidar data echo detection: Assessing quality of derived terrain and tree height models in an alpine coniferous forest. *International Journal of Remote Sensing*, 30(19):5211–5228.

Chave, J., Condit, R., Lao, S., Caspersen, J. P., Foster, R. B., and Hubbell, S. P., 2003, Spatial and temporal variation of biomass in a tropical forest: Results from a large census plot in Panama. *Journal of Ecology*, 91(2):240–252.

Chehata, N., Guo, L., and Mallet, C., 2009, Airborne lidar feature selection for urban classification using random forests. *International Archives of the Photogrammetry, Remote Sensing and Spatial Information Sciences*, 39(Part 3/W8):207–212.

Chen, Q., 2010, Assessment of terrain elevation derived from satellite laser altimetry over mountainous forest areas using airborne lidar data. *ISPRS Journal of Photogrammetry and Remote Sensing*, 65(1):111–122.

Chen, Q., Baldocchi, D., Gong, P., and Kelly, M., 2006, Isolating individual trees in a savanna woodland using small footprint lidar data. *Photogrammetric Engineering & Remote Sensing*, 72(8):923–932.

Clark, M. L., Clark, D. B. and Roberts, D. A., 2004, Small-footprint lidar estimation of sub-canopy elevation and tree height in a tropical rain forest landscape. *Remote Sensing of Environment*, 91(1):68–89.

Coops, N. and Waring, R., 2001, The use of multiscale remote sensing imagery to derive regional estimates of forest growth capacity using 3-PGS. *Remote Sensing of Environment*, 75(3):324–334.

Couteron, P., Pelissier, R., Nicolini, E. A., and Paget, D., 2005, Predicting tropical forest stand structure parameters from Fourier transform of very high-resolution remotely sensed canopy images. *Journal of Applied Ecology*, 42(6):1121–1128.

De Leeuw, J., Georgiadou, Y., Kerle, N., de Gier, A., Inoue, Y., Ferwerda, J., Smies, M., and Narantuya, D., 2010, The function of remote sensing in support of environmental policy. *Remote Sensing*, 2(7):1731–1750.

De Sy, V., Herold, M., Achard, F., Asner, G. P., Held, A., Kellndorfer, J., and Verbesselt, J., 2012, Synergies of multiple remote sensing data sources for REDD+ monitoring. *Current Opinion in Environmental Sustainability*, 4(6):696–706.

DeFries, R., Achard, F., Brown, S., Herold, M., Murdiyarso, D., Schlamadinger, B., and de Souza Jr, C., 2007, Earth observations for estimating greenhouse gas emissions from deforestation in developing countries. *Environmental Science and Policy*, 10(4):385–394.

Dhôte, J., 1991, Modélisation de la croissance des peuplements réguliers de hêtre: Dynamique des hiérarchies sociales et facteurs de production. *Annales des Sciences forestières*, 48(4):389–416.

Disney, M., Lewis, P., and Saich, P., 2006, 3D modelling of forest canopy structure for remote sensing simulations in the optical and microwave domains. *Remote Sensing of Environment*, 100(1):114–132.

Disney, M. I., Kalogirou, V., Lewis, P., Prieto-Blanco, A., Hancock, S., and Pfeifer, M., 2010, Simulating the impact of discrete-return lidar system and survey characteristics over young conifer and broadleaf forests. *Remote Sensing of Environment*, 114(7):1546–1560.

Dubayah, R. O. and Drake, J. B., 2001, Lidar remote sensing for forestry applications. www.geog.umd.edu/vcl/pubs/jof. pdf—January 25, 2001.

Ducic, V., Hollaus, M., Ullrich, A., Wagner, W., and Melzer, T., 2006, 3D Vegetation mapping and classification using full-waveform laser scanning. *Workshop on 3D Remote Sensing in Forestry*, Vienna, Austria, International Archives of Photogrammetry, Remote Sensing and Spatial Information Sciences, 211–217pp.

Duguid, M. C. and Ashton, M. S., 2013, A meta-analysis of the effect of forest management for timber on understory plant species diversity in temperate forests. *Forest Ecology and Management*, 303:81–90.

Durrieu, S., Cherchali, S., Costeraste, J., Mondin, L., Debise, H., Chazette, P., Dauzat, J., Gastellu-Etchegorry, J.-P., N. Baghdadi, N., and Pélissier, R., 2013, Preliminary studies for a vegetation ladar/lidar space mission in France. IGARSS 2013, *IEEE International Geoscience and Remote Sensing Symposium*, International, July 21–26, 2013, Melbourne, Victoria, Australia, IEEE, 4432–4435pp.

Durrieu, S. and Nelson, R. F., 2013, Earth observation from space—The issue of environmental sustainability. *Space Policy*, 29(4):238–250.

Duursma, R. and Medlyn, B., 2012, MAESPA: A model to study interactions between water limitation, environmental drivers and vegetation function at tree and stand levels, with an example application to [CO 2] × drought interactions. *Geoscientific Model Development Discussions*, 5(1):459–513.

Eichhorn, F., 1904, Beziehungen zwischen Bestandshöhe und Bestandsmasse. *Allgemeine Forst-und Jagdzeitung*, 80:45–49.

Ene, L. T., Næsset, E., Gobakken, T., Gregoire, T. G., Ståhl, G., and Holm, S., 2013, A simulation approach for accuracy assessment of two-phase post-stratified estimation in large-area LiDAR biomass surveys. *Remote Sensing of Environment*, 133(0):210–224.

ESA, 2008, Biomass, Mission assessment report, ESA, 132p., http://esamultimedia.esa.int/docs/SP1313-2_BIOMASS. pdf. (Accessed June 15, 2014.)

Evans, J. S., Hudak, A. T., Faux, R., and Smith, A., 2009, Discrete return lidar in natural resources: Recommendations for project planning, data processing, and deliverables. *Remote Sensing*, 1(4):776–794.

Falkowski, M. J., Smith, A. M. S., Hudak, A. T., Gessler, P. E., Vierling, L. A., and Crookston, N. L., 2006, Automated estimation of individual conifer tree height and crown diameter via two-dimensional spatial wavelet analysis of lidar data. *Canadian Journal of Remote Sensing*, 32(2):153–161.

Falkowski, M. J., Wulder, M. A., White, J. C., and Gillis, M. D., 2009, Supporting large-area, sample-based forest inventories with very high spatial resolution satellite imagery. *Progress in Physical Geography*, 33(3):403–423.

Faux, R., Buffington, J. M., Whitley, G., Lanigan, S. , and Roper, B., 2009, Use of airborne near-infrared LiDAR for determining channel cross-section characteristics and monitoring aquatic habitat in Pacific Northwest rivers: A preliminary analysis. *Proceedings of the American Society for Photogrammetry and Remote Sensing*. Pacific Northwest Aquatic Monitoring Partnership, Cook, Washington, DC, pp. 43–60.

Ferraz, A., Bretar, F., Jacquemoud, S., Gonçalves, G., and Pereira, L., 2010, 3D segmentation of forest structure using an adaptive mean shift based procedure. SilviLaser 2010, *Conference on Lidar Applications for assessing Forest Ecosystems*, Freiburg, Germany, 13pp.

Ferretti, A., Monti-Guar, A., Prati, C., Rocca, F., and Massonnet, D., 2007, *InSAR Principles: Guidelines for SAR Interferometry Processing and Interpretation—Part A.* Vol. ESA Publications-TM 19, Noordwijk, the Netherlands, 40pp., http://www.esa.int/esapub/tm/tm19/TM-19_ptA. pdf. (Accessed June 15, 2014.)

Ferretti, M., 2010, Harmonizing forest inventories and forest condition monitoring—The rise or the fall of harmonized forest condition monitoring in Europe? *IForest*, 3(January):1–4.

Frazer, G., Magnussen, S., Wulder, M., and Niemann, K., 2011, Simulated impact of sample plot size and co-registration error on the accuracy and uncertainty of LiDAR-derived estimates of forest stand biomass. *Remote Sensing of Environment*, 115(2):636–649.

Fujita, T., Itaya, A., Miura, M., Manabe, T., and Yamamoto, S.-I., 2003, Long-term canopy dynamics analysed by aerial photographs in a temperate old-growth evergreen broad-leaved forest. *Journal of Ecology*, 91(4):686–693.

Fuller, D. O., 2006, Tropical forest monitoring and remote sensing: A new era of transparency in forest governance? *Singapore Journal of Tropical Geography*, 27(1):15–29.

Gagnon, P., Agnard, J., and Nolette, C., 1993, Evaluation of a soft-copy photogrammetry system for tree-plot measurements. *Canadian Journal of Forest Research*, 23(9):1781–1785.

Gao, J., 2007, Towards accurate determination of surface height using modern geoinformatic methods: Possibilities and limitations. *Progress in Physical Geography*, 31(6):591–605.

García, O., 1998, Estimating top height with variable plot sizes. *Canadian Journal of Forest Research*, 28(10):1509–1517.

García, O., 2011, Dynamical implications of the variability representation in site-index modelling. *European Journal of Forest Research*, 130(4):671–675.

Garestier, F., Dubois-Fernandez, P. C., Guyon, D., and Le Toan, T., 2009, Forest biophysical parameter estimation using L- and P-band polarimetric SAR data. *IEEE Transactions on Geoscience and Remote Sensing*, 47(10):3379–3388.

Germaine, S. S., Vessey, S. H., and Capen, D. E., 1997, Effects of small forest openings on the breeding bird community in a Vermont hardwood forest. *Condor*, 99(3):708–718.

Gobakken, T. and Næsset, E., 2004, Estimation of diameter and basal area distributions in coniferous forest by means of airborne laser scanner data. *Scandinavian Journal of Forest Research*, 19(6):529–542.

Gobakken, T. and Næsset, E., 2008, Assessing effects of laser point density, ground sampling intensity, and field sample plot size on biophysical stand properties derived from airborne laser scanner data. *Canadian Journal of Forest Research*, 38(5):1095–1109.

Goodenough, D. G., Pearlman, J., Chen, H., Dyk, A., Han, T., Li, J., Miller, J., and Niemann, O., 2004, Forest information from hyperspectral sensing. *In Geoscience and Remote Sensing Symposium*, 2004. IGARSS'04. Proceedings. 2004 IEEE International, vol., 4, pp., 2585,2589.

Goetz, S., Steinberg, D., Dubayah, R., and Blair, B., 2007, Laser remote sensing of canopy habitat heterogeneity as a predictor of bird species richness in an eastern temperate forest, USA. *Remote Sensing of Environment*, 108(3):254–263.

Gong, P., Sheng, Y., and Biging, G., 2002, 3D model-based tree measurement from high-resolution aerial imagery. *Photogrammetric Engineering and Remote Sensing*, 68(11):1203–1212.

Guédon, Y., Caraglio, Y., Heuret, P., Lebarbier, E., and Meredieu, C., 2007, Analyzing growth components in trees. *Journal of Theoretical Biology*, 248(3):418–447.

Hall, F., Saatchi, S., and Dubayah, R., 2011a, Preface: DESDynI VEG-3D special issue. *Remote Sensing of Environment*, 115(11):2752.

Hall, F. G., Bergen, K., Blair, J. B., Dubayah, R., Houghton, R., Hurtt, G., Kellndorfer, J. et al., 2011b, Characterizing 3D vegetation structure from space: Mission requirements. *Remote Sensing of Environment*, 115(11):2753–2775.

Hancock, S., Lewis, P., Foster, M., Disney, M., and Muller, J.-P., 2012, Measuring forests with dual wavelength lidar: A simulation study over topography. *Agricultural and Forest Meteorology*, 161:123–133.

Harding, D. J., Lefsky, M. A., Parker, G. G., and Blair, J. B., 2001, Laser altimeter canopy height profiles methods and validation for closed-canopy, broadleaf forests. *Remote Sensing of Environment*, 76(3):283–297.

Heipke, C., 1997, Automation of interior, relative, and absolute orientation. *ISPRS Journal of Photogrammetry and Remote Sensing*, 52(1):1–19.

Henry, M., Bombelli, A., Trotta, C., Alessandrini, A., Birigazzi, L., Sola, G., Vieilledent, G., Santenoise, P., Longuetaud, F., and Valentini, R., 2013, GlobAllomeTree: International platform for tree allometric equations to support volume, biomass and carbon assessment. *iForest—Biogeosciences & Forestry*, 6(6):326.

Hinz, A. and Heier, H., 2000, The Z/I imaging digital camera system. *The Photogrammetric Record*, 16(96):929–936.

Hirata, Y., 2004, The effects of footprint size and sampling density in airborne laser scanning to extract individual trees in mountainous terrain. *International Archives of Photogrammetry, Remote Sensing and Spatial Information Sciences*, 36(Part 8):102–107.

Hirschmugl, M., Ofner, M., Raggam, J., and Schardt, M., 2007, Single tree detection in very high resolution remote sensing data. *Remote Sensing of Environment*, 110(4):533–544.

Ho Tong Minh, D., Le Toan, T., Rocca, F., Tebaldini, S., D'Alessandro, M. M., and Villard, L., 2014, Relating P-band synthetic aperture radar tomography to tropical forest biomass. *IEEE Transactions on Geoscience and Remote Sensing*, 52(2):967–979.

Hodgson, M. E. and Bresnahan, P., 2004, Accuracy of airborne lidar-derived elevation: Empirical assessment and error budget. *Photogrammetric Engineering and Remote Sensing*, 70(3):331–339.

Hodgson, M. E., Jensen, J., Raber, G., Tullis, J., Davis, B. A., Thompson, G., and Schuckman, K., 2005, An evaluation of lidar-derived elevation and terrain slope in leaf-off conditions. *Photogrammetric Engineering and Remote Sensing*, 71(7):817–823.

Holmgren, J., Nilsson, M., and Olsson, H., 2003, Estimation of tree height and stem volume on plots using airborne laser scanning. *Forest Science*, 49(3):419–428.

Hopkinson, C., 2007, The influence of flying altitude, beam divergence, and pulse repetition frequency on laser pulse return intensity and canopy frequency distribution. *Canadian Journal of Remote Sensing*, 33(4):312–324.

Hopkinson, C. and Chasmer, L., 2009, Testing LiDAR models of fractional cover across multiple forest ecozones. *Remote Sensing of Environment*, 113(1):275–288.

Hu, B., Li, J., Jing, L., and Judah, A., 2014, Improving the efficiency and accuracy of individual tree crown delineation from high-density LiDAR data. *International Journal of Applied Earth Observation and Geoinformation*, 26(0):145–155.

Huang, S., Hager, S. A., Halligan, K. Q., Fairweather, I. S., Swanson, A. K., and Crabtree, R. L., 2009, A comparison of individual tree and forest plot height derived from lidar and InSAR. *Photogrammetric Engineering and Remote Sensing*, 75(2):159–167.

Hug, C., Ullrich, A., and Grimm, A., 2004, LiteMapper-5600—A waveform digitizing lidar terrain and vegetation mapping system. *International Conference "Laser-Scanners for Forest and Landscape Assessment"*, Freiburg, Germany, *International Archives of Photogrammetry, Remote Sensing and Spatial Information Sciences*, XXXVI, part 8/W2, 24–29pp.

Hyyppä, J., Hyyppä, H., Leckie, D., Gougeon, F., Yu, X., and Maltamo, M., 2008, Review of methods of small-footprint airborne laser scanning for extracting forest inventory data in boreal forests. *International Journal of Remote Sensing*, 29(5):1339–1366.

Hyyppä, J., Kelle, O., Lehikoinen, M., and Inkinen, M., 2001, A segmentation-based method to retrieve stem volume estimates from 3-D tree height models produced by laser scanners. *IEEE Transactions on Geoscience and Remote Sensing*, 39(5):969–975.

Ishii, H. T., Tanabe, S. I., and Hiura, T., 2004, Exploring the relationships among canopy structure, stand productivity, and biodiversity of temperate forest ecosystems. *Forest Science*, 50(3):342–355.

Jakubowski, M. K., Guo, Q., and Kelly, M., 2013, Tradeoffs between lidar pulse density and forest measurement accuracy. *Remote Sensing of Environment*, 130:245–253.

Jaskierniak, D., 2011, Modelling the effects of forest regeneration on streamflow using forest growth models. PhD dissertation, School of Geography and Environmental Studies, University of Tasmania, Australia, 240p.

Kaartinen, H., Hyyppä, J., Yu, X., Vastaranta, M., Hyyppä, H., Kukko, A., Holopainen, M. et al., 2012, An international comparison of individual tree detection and extraction using airborne laser scanning. *Remote Sensing*, 4(4):950–974.

Kasischke, E. S., Melack, J. M., and Craig Dobson, M., 1997, The use of imaging radars for ecological applications—A review. *Remote Sensing of Environment*, 59(2):141–156.

Kato, A., Moskal, L. M., Schiess, P., Swanson, M. E., Calhoun, D., and Stuetzle, W., 2009, Capturing tree crown formation through implicit surface reconstruction using airborne lidar data. *Remote Sensing of Environment*, 113(6):1148–1162.

Ke, Y. and Quackenbush, L. J., 2011, A review of methods for automatic individual tree-crown detection and delineation from passive remote sensing. *International Journal of Remote Sensing*, 32(17):4725–4747.

Kellner, J. R., Clark, D. B., and Hofton, M. A., 2009, Canopy height and ground elevation in a mixed-land-use lowland Neotropical rain forest landscape. *Ecology*, 90(11):3274–3274.

Kitahara, F., Mizoue, N., and Yoshida, S., 2010, Effects of training for inexperienced surveyors on data quality of tree diameter and height measurements. *Silva Fennica*, 44(4):657–667.

Kleinn, C., 2002, New technologies and methodologies for national forest inventories. *Unasylva*, 53(210):10–15.

Kobayashi, T., Endo, T., Sawada, Y., Endo, S., Hayashi, M., Satoh, Y., Chishiki, Y., and Yamakawa, S., 2013, Waveform simulator and analytical procedure for JAXA's future spaceborne lidar to measure canopy height. SPIE Remote Sensing, Dresden, Germany, September 23, 2013, *International Society for Optics and Photonics*, 8894, 10pp.

Koch, B., Heyder, U., and Welnacker, H., 2006, Detection of individual tree crowns in airborne lidar data. *Photogrammetric Engineering and Remote Sensing*, 72(4):357–363.

Korpela, I., 2004, Individual tree measurements by means of digital aerial photogrammetry. Silva Fennica Monographs 3. 93 p., http://www.scopus.com/inward/record.url?eid=2-s2.0-33746484995, andpartnerID=40&md5=5384dddf6344157bbc40b5a7a5d4c283. (Accessed March 15, 2014.)

Kovats, M., 1997, A large-scale aerial photographic technique for measuring tree heights on long-term forest installations. *Photogrammetric Engineering and Remote Sensing*, 63(6):741–747.

Kraus, K. and Pfeifer, N., 1998, Determination of terrain models in wooded areas with airborne laser scanner data. *Journal of Photogrammetry and Remote Sensing*, 53:193–203.

Kuuluvainen, T. and Grenfell, R., 2012, Natural disturbance emulation in boreal forest ecosystem management—Theories, strategies, and a comparison with conventional even-aged management. *Canadian Journal of Forest Research*, 42(7):1185–1203.

Kwak, D.-A., Lee, W.-K., Lee, J.-H., Biging, G., and Gong, P., 2007, Detection of individual trees and estimation of tree height using LiDAR data. *Journal of Forest Research*, 12(6):425–434.

Larjavaara, M. and Muller-Landau, H. C., 2013, Measuring tree height: A quantitative comparison of two common field methods in a moist tropical forest. *Methods in Ecology and Evolution*, 4(9):793–801.

Le Goff, H., De Grandpré, L., Kneeshaw, D., and Bernier, P., 2010, Sustainable management of old-growth boreal forests: Myths, possible solutions and challenges. *The Forestry Chronicle*, 86(1):63–76.

le Maire, G., Marsden, C., Nouvellon, Y., Grinand, C., Hakamada, R., Stape, J.-L., and Laclau, J.-P., 2011, MODIS NDVI time-series allow the monitoring of *Eucalyptus* plantation biomass. *Remote Sensing of Environment*, 115(10):2613–2625.

Le Toan, T., Quegan, S., Davidson, M. W. J., Balzter, H., Paillou, P., Papathanassiou, K., Plummer, S. et al., 2011, The BIOMASS mission: Mapping global forest biomass to better understand the terrestrial carbon cycle. *Remote Sensing of Environment*, 115(11):2850–2860.

Leberl, F., Irschara, A., Pock, T., Meixner, P., Gruber, M., Scholz, S., and Wiechert, A., 2010, Point clouds: Lidar versus 3D vision. *Photogrammetric Engineering and Remote Sensing*, 76(10):1123–1134.

Leckie, D., Gougeon, F., Hill, D., Quinn, R., Armstrong, L., and Shreenan, R., 2003, Combined high-density lidar and multispectral imagery for individual tree crown analysis. *Canadian Journal of Remote Sensing*, 29(5):633–649.

Leckie, D. G., Gougeon, F. A., Tinis, S., Nelson, T., Burnett, C. N., and Paradine, D., 2005, Automated tree recognition in old growth conifer stands with high resolution digital imagery. *Remote Sensing of Environment*, 94(3):311–326.

Lefsky, M. A., Cohen, W. B., Harding, D. J., Parker, G. G., Acker, S. A., and Gower, S. T., 2002, Lidar remote sensing of above-ground biomass in three biomes. *Global Ecology and Biogeography*, 11(5):393–399.

Lefsky, M. A., Harding, D. J., Keller, M., Cohen, W. B., Carabajal, C. C., Del Bom Espirito-Santo, F., Hunter, M. O., and de Oliveira Jr, R., 2005, Estimates of forest canopy height and aboveground biomass using ICESat. *Geophysical Research Letters*, 32(22):1–4.

Lesak, A. A., Radeloff, V. C., Hawbaker, T. J., Pidgeon, A. M., Gobakken, T., and Contrucci, K., 2011, Modeling forest songbird species richness using LiDAR-derived measures of forest structure. *Remote Sensing of Environment*, 115(11):2823–2835.

Li, K., Shao, Y., Zhang, F., Xu, M., Li, X., and Xia, Z., 2009, Forest parameters estimation using polarimetric SAR data. *Sixth International Symposium on Multispectral Image Processing and Pattern Recognition, International Society for Optics and Photonics*, 7498—article id. 749857, 1–8pp.

Li, W., Guo, Q., Jakubowski, M. K., and Kelly, M., 2012, A new method for segmenting individual trees from the lidar point cloud. *Photogrammetric Engineering & Remote Sensing*, 78(1):75–84.

Lim, K., Hopkinson, C., and Treitz, P., 2008, Examining the effects of sampling point densities on laser canopy height and density metrics. *The Forestry Chronicle*, 84(6):876–885.

Lim, K., Treitz, P., Wulder, M., St-Onge, B., and Flood, M., 2003, LiDAR remote sensing of forest structure. *Progress in Physical Geography*, 27(1):88–106.

Lisein, J., Pierrot-Deseilligny, M., Bonnet, S., and Lejeune, P., 2013, A Photogrammetric workflow for the creation of a forest canopy height model from small unmanned aerial system imagery. *Forests*, 4(4):922–944.

Liu, G. and Han, S., 2009, Long-term forest management and timely transfer of carbon into wood products help reduce atmospheric carbon. *Ecological Modelling*, 220(13–14):1719–1723.

Lovell, J., Jupp, D., Newnham, G., Coops, N., and Culvenor, D., 2005, Simulation study for finding optimal lidar acquisition parameters for forest height retrieval. *Forest Ecology and Management*, 214(1):398–412.

Luo, S., Wang, C., Li, G., and Xi, X., 2013, Retrieving leaf area index using ICESat/GLAS full-waveform data. *Remote Sensing Letters*, 4(8):745–753.

Magnani, F., Mencuccini, M., and Grace, J., 2000, Age-related decline in stand productivity: The role of structural acclimation under hydraulic constraints. *Plant, Cell and Environment*, 23(3):251–263.

Magnussen, S., Næsset, E., Gobakken, T., and Frazer, G., 2012, A fine-scale model for area-based predictions of tree-size-related attributes derived from LiDAR canopy heights. *Scandinavian Journal of Forest Research*, 27(3):312–322.

Maltamo, M., Mustonen, K., Hyyppä, J., Pitkänen, J., and Yu, X., 2004, The accuracy of estimating individual tree variables with airborne laser scanning in a boreal nature reserve. *Canadian Journal of Forest Research*, 34:1791–1801.

Maltamo, M., Packalén, P., Suvanto, A., Korhonen, K., Mehtätalo, L., and Hyvönen, P., 2009, Combining ALS and NFI training data for forest management planning: A case study in Kuortane, Western Finland. *European Journal of Forest Research*, 128(3):305–317.

Manninen, T., Stenberg, P., Rautiainen, M., Voipio, P., and Smolander, H., 2005, Leaf area index estimation of boreal forest using ENVISAT ASAR. *IEEE Transactions on Geoscience and Remote Sensing*, 43(11):2627–2635.

Marsden, C., Nouvellon, Y., Laclau, J.-P., Corbeels, M., McMurtrie, R. E., Stape, J. L., Epron, D., and le Maire, G., 2013, Modifying the G'DAY process-based model to simulate the spatial variability of *Eucalyptus* plantation growth on deep tropical soils. *Forest Ecology and Management*, 301:112–128.

McCarthy, M. A. and Burgman, M. A., 1995, Coping with uncertainty in forest wildlife planning. *Forest Ecology and Management*, 74(1):23–36.

McRoberts, R. E., 2010, Probability-and model-based approaches to inference for proportion forest using satellite imagery as ancillary data. *Remote Sensing of Environment*, 114(5):1017–1025.

McRoberts, R. E., Cohen, W. B., Erik, N., Stehman, S. V., and Tomppo, E. O., 2010, Using remotely sensed data to construct and assess forest attribute maps and related spatial products. *Scandinavian Journal of Forest Research*, 25(4):340–367.

McRoberts, R. E., Tomppo, E., Schadauer, K., Vidal, C., Ståhl, G., Chirici, G., Lanz, A., Cienciala, E., Winter, S. and Smith, W. B., 2009, Harmonizing national forest inventories. *Journal of Forestry*, 107(4):179–187.

Medvigy, D. and Moorcroft, P. R., 2012, Predicting ecosystem dynamics at regional scales: An evaluation of a terrestrial biosphere model for the forests of northeastern North America. *Philosophical Transactions of the Royal Society B: Biological Sciences*, 367(1586):222–235.

Medvigy, D., Wofsy, S., Munger, J., Hollinger, D., and Moorcroft, P., 2009, Mechanistic scaling of ecosystem function and dynamics in space and time: Ecosystem Demography model version 2. *Journal of Geophysical Research: Biogeosciences (2005–2012)*, 114(G1).

Mei, C. and Durrieu, S., 2004, Tree crown delineation from digital elevation models and high resolution imagery. *Proceedings of the ISPRS Working Group Part*, 8(2):3–6.

Meng, X., Currit, N. and Zhao, K., 2010, Ground filtering algorithms for airborne LiDAR data: A review of critical issues. *Remote Sensing*, 2(3):833–860.

Mitchard, E. T., Saatchi, S. S., Baccini, A., Asner, G. P., Goetz, S. J., Harris, N. L., and Brown, S., 2013, Uncertainty in the spatial distribution of tropical forest biomass: A comparison of pan-tropical maps. *Carbon Balance and Management*, 8(10):1–13.

Mitchard, E. T. A., Saatchi, S. S., Woodhouse, I. H., Nangendo, G., Ribeiro, N. S., Williams, M., Ryan, C. M., Lewis, S. L., Feldpausch, T. R., and Meir, P., 2009, Using satellite radar backscatter to predict above-ground woody biomass: A consistent relationship across four different African landscapes. *Geophysical Research Letters*, 36(23):L23401.

Morgan, J. L. and Gergel, S. E., 2013, Automated analysis of aerial photographs and potential for historic forest mapping. *Canadian Journal of Forest Research*, 43(8):699–710.

Morsdorf, F., Meier, E., Kötz, B., Itten, K. I., Dobbertin, M., and Allgöwer, B., 2004, LIDAR-based geometric reconstruction of boreal type forest stands at single tree level for forest and wildland fire management. *Remote Sensing of Environment*, 92(3):353–362.

Mücke, W. and Hollaus, M., 2010, Derivation of 3D landscape metrics from airborne laser scanning data. SilviLaser 2010, *Conference on Lidar Applications for Assessing Forest Ecosytems*, Freiburg, Germany, 11pp.

Müller, J., Bae, S., Röder, J., Chao, A., and Didham, R. K., 2014, Airborne LiDAR reveals context dependence in the effects of canopy architecture on arthropod diversity. *Forest Ecology and Management*, 312:129–137.

Müller, J. and Brandl, R., 2009, Assessing biodiversity by remote sensing in mountainous terrain: The potential of LiDAR to predict forest beetle assemblages. *Journal of Applied Ecology*, 46(4):897–905.

Müller, J., Stadler, J., and Brandl, R., 2010, Composition versus physiognomy of vegetation as predictors of bird assemblages: The role of lidar. *Remote Sensing of Environment*, 114(3):490–495.

Næsset, E., 2002a, Determination of mean tree height of forest stands by digital photogrammetry. *Scandinavian Journal of Forest Research*, 17(5):446–459.

Næsset, E., 2002b, Predicting forest stand characteristics with airborne scanning laser using a practical two-stage procedure and field data. *Remote Sensing of Environment*, 80(1):88–99.

Næsset, E., 2004, Accuracy of forest inventory using airborne laser-scanning: Evaluating the first Nordic full-scale operational project. *Scandinavian Journal of Forest Research*, 19:554–557.

Næsset, E., 2009a, Effects of different sensors, flying altitudes, and pulse repetition frequencies on forest canopy metrics and biophysical stand properties derived from small-footprint airborne laser data. *Remote Sensing of Environment*, 113(1):148–159.

Næsset, E., 2009b, Influence of terrain model smoothing and flight and sensor configurations on detection of small pioneer trees in the boreal–alpine transition zone utilizing height metrics derived from airborne scanning lasers. *Remote Sensing of Environment*, 113(10):2210–2223.

Naesset, E., 2007, Airborne laser scanning as a method in operational forest inventory: Status of accuracy assessments accomplished in Scandinavia. *Scandinavian Journal of Forest Research*, 22(5):433–442.

Neigh, C. S., Masek, J. G., Bourget, P., Cook, B., Huang, C., Rishmawi, K. and Zhao, F., 2014, Deciphering the precision of stereo IKONOS canopy height models for US forests with G-LiHT airborne LiDAR. *Remote Sensing*, 6(3):1762–1782.

Nelson, R., 1997, Modeling forest canopy heights: The effects of canopy shape. *Remote Sensing of Environment*, 60(3):327–334.

Nelson, R., 2013, How did we get here? An early history of forestry lidar. *Canadian Journal of Remote Sensing*, 39(Suppl. 1):S6–S17.

Nelson, R., Keller, C., and Ratnaswamy, M., 2005, Locating and estimating the extent of Delmarva fox squirrel habitat using an airborne LiDAR profiler. *Remote Sensing of Environment*, 96(3):292–301.

Nelson, R., Parker, G., and Hom, M., 2003, A portable airborne laser system for forest inventory. *Photogrammetric Engineering and Remote Sensing*, 69(3):267–273.

Ostendorf, B., 2011, Overview: Spatial information and indicators for sustainable management of natural resources. *Ecological Indicators*, 11(1):97–102.

Pardé, J., 1956, Une notion pleine d'intérêt: La hauteur dominante des peuplements forestiers. *Revue forestière française*, (12):850–856.

Pelletier, J., Ramankutty, N., and Potvin, C., 2011, Diagnosing the uncertainty and detectability of emission reductions for REDD+ under current capabilities: An example for Panama. *Environmental Research Letters*, 6(024005):12.

Persson, A., Holmgren, J., and Söderman, U., 2002, Detecting and measuring individual trees using an airborne laser scanner. *Photogrammetric Engineering & Remote Sensing*, 68(9):925–932.

Petrie, G. and Walker, A. S., 2007, Airborne digital imaging technology: A new overview. *Photogrammetric Record*, 22(119):203–225.

Picard, N., Saint-André, L., and Henry, M., 2012a, Manual for building tree volume and biomass allometric equations: From field measurement to prediction. Food and Agricultural Organization of the United Nations, Rome, and Centre de Coopération Internationale en Recherche Agronomique pour le Développement, Montpellier, 215pp., http://www.fao.org/docrep/018/i3058e/i3058e.pdf.

Picard, N., Saint-André, L., and Henry, M., 2012b, Manual for building tree volume and biomass allometric equations: From field measurement to prediction. Cirad and FAO (Ed.), http://www.fao.org/docrep/018/i3058e/i3058e.pdf. (Accessed January 15, 2015.)

Popescu, S. C. and Wynne, R. H., 2004, Seeing the trees in the forest: Using lidar and multispectral data fusion with local filtering and variable window size for estimating tree height. *Photogrammetric Engineering & Remote Sensing*, 70(5):589–604.

Popescu, S. C., Wynne, R. H., and Nelson, R. F., 2002, Estimating plot-level tree heights with lidar: Local filtering with a canopy-height based variable window size. *Computers and Electronics in Agriculture*, 37:71–95.

Popescu, S. C., Wynne, R. H., and Nelson, R. F., 2003, Measuring individual tree crown diameter with lidar and assessing its influence on estimating forest volume and biomass. *Canadian Journal of Remote Sensing*, 29(5):564–577.

Pretzsch, H., 2009, Forest dynamics, growth, and yield. *From Measurement to Model*, Springer-Verlag, Berlin, Heidelberg, Germany, 664pp.,

Proisy, C., Couteron, P., and Fromard, F., 2007, Predicting and mapping mangrove biomass from canopy grain analysis using Fourier-based textural ordination of IKONOS images. *Remote Sensing of Environment*, 109(3):379–392.

Reitberger, J., Krzystek, P., and Stilla, U., 2006, Analysis of full waveform Lidar data for tree species classification. *Symposium of ISPRS Commission III- Photogrammetric Computer Vision PCV'06*, Bonn, Germany, ISPRS Commission III, XXXVI-part 3, 228–233pp.

Reitberger, J., Schnörr, Cl., Krzystek, P., and Stilla, U., 2009, 3D segmentation of single trees exploiting full waveform LIDAR data. *ISPRS Journal of Photogrammetry and Remote Sensing*, 64:561–574.

Remondino, F. and El-hakim, S., 2006, Image-based 3D modelling: A review. *Photogrammetric Record*, 21(115):269–291.

Reutebuch, S. E., McGaughey, R. J., Andersen, H.-E., and Carson, W. W., 2003, Accuracy of a high-resolution lidar terrain model under a conifer forest canopy. *Canadian Journal of Remote Sensing*, 29(5):527–535.

Rosette, J., North, P. R., Rubio-Gil, J., Cook, B., Los, S., Suarez, J., Sun, G., Ranson, J., and Blair, J., 2013, Evaluating prospects for improved forest parameter retrieval from satellite LiDAR using a physically-based radiative transfer model. *IEEE Journal of Selected Topics in Applied Earth Observations and Remote Sensing*, 6(1):45–53.

Saatchi, S. S., Harris, N. L., Brown, S., Lefsky, M., Mitchard, E. T. A., Salas, W., Zutta, B. R. et al., 2011, Benchmark map of forest carbon stocks in tropical regions across three continents. *Proceedings of the National Academy of Sciences of the United States of America*, 108(24):9899–9904.

Saint-André, L., Laclau, J., Deleporte, P., Gava, J., Gonçalves, J., Mendham, D., Nzila, J., Smith, C., Du Toit, B., and Xu, D., 2008, Slash and litter management effects on Eucalyptus productivity: A synthesis using a growth and yield modelling approach. Site management and productivity in tropical plantation forests. *Proceedings of Workshops in Piracicaba (Brazil)*, November 22–26, 2004 and Bogor (Indonesia), November 6–9, 2006, Center for International Forestry Research (CIFOR), Bogor, Indonesia, 173–189pp.

Shugart, H. H., Saatchi, S., and Hall, F. G., 2010, Importance of structure and its measurement in quantifying function of forest ecosystems. *Journal of Geophysical Research: Biogeosciences (2005–2012)*, 115(D24):1–16.

Simard, M., Pinto, N., Fisher, J. B., and Baccini, A., 2011, Mapping forest canopy height globally with spaceborne lidar. *Journal of Geophysical Research: Biogeosciences* (2005–2012), 116(G4):1–12.

Simonett, D., 1969, Editor's preface. *Remote Sensing of Environment*, 1(1):v.

Simonson, W. D., Allen, H. D., and Coomes, D. A., 2012, Use of an airborne lidar system to model plant species composition and diversity of Mediterranean oak forests. *Conservation Biology*, 26(5):840–850.

Sithole, G. and Vosselman, G., 2004, Experimental comparison of filter algorithms for bare-Earth extraction from airborne laser scanning point clouds. *ISPRS Journal of Photogrammetry and Remote Sensing*, 59(1–2):85–101.

Skovsgaard, J. P. and Vanclay, J. K., 2008, Forest site productivity: A review of the evolution of dendrometric concepts for even-aged stands. *Forestry*, 81(1):13–31.

Smits, I., Prieditis, G., Dagis, S., and Dubrovskis, D., 2012, Individual tree identification using different LIDAR and optical imagery data processing methods. *Biosystems and Information Technology*, 1(1):19–24.

Soja, M. J. and Ulander, L. M. H., 2013, Digital canopy model estimation from TanDEM-X interferometry using high-resolution lidar DEM, In *Proceedings of IEEE Geoscience and Remote Sensing Symposium (IGARSS)*, 21–26 July 2013, pp. 165–168.

Solberg, S., Naesset, E., and Bollandsas, O. M., 2006, Single tree segmentation using airborne laser scanner data in a structurally heterogeneous spruce forest. *Photogrammetric Engineering and Remote Sensing*, 72(12):1369–1378.

Spurr, S. H., 1960, *Photogrammetry and Photo-interpretation with a Section on Applications to Forestry*, Ronald Press Comp., New York, 472p.

St-Onge, B., Jumelet, J., Cobello, M., and Véga, C., 2004, Measuring individual tree height using a combination of stereophotogrammetry and lidar. *Canadian Journal of Forest Research*, 34(10):2122–2130.

St-Onge, B., Véga, C., Fournier, R., and Hu, Y., 2008, Mapping canopy height using a combination of digital stereo-photogrammetry and lidar. *International Journal of Remote Sensing*, 29(11):3343–3364.

Strunk, J., Temesgen, H., Andersen, H.-E., Flewelling, J. P., and Madsen, L., 2012, Effects of lidar pulse density and sample size on a model-assisted approach to estimate forest inventory variables. *Canadian Journal of Remote Sensing*, 38(05):644–654.

Suárez, J. C., Ontiveros, C., Smith, S., and Snape, S., 2005, Use of airborne LiDAR and aerial photography in the estimation of individual tree heights in forestry. *Computers & Geosciences*, 31(2):253–262.

Tanaka, H. and Nakashizuka, T., 1997, Fifteen years of canopy dynamics analyzed by aerial photographs in a temperate deciduous forest, Japan. *Ecology*, 78(2):612–620.

Tebaldini, S. and Rocca, F., 2012, Multibaseline polarimetric SAR tomography of a boreal forest at P-and L-bands. *IEEE Transactions on Geoscience and Remote Sensing*, 50(1):232–246.

Thomas, V., Oliver, R., Lim, K., and Woods, M., 2008, LiDAR and Weibull modeling of diameter and basal area. *The Forestry Chronicle*, 84(6):866–875.

Thomas, V., Treitz, P., McCaughey, J. H., and Morrison, I., 2006, Mapping stand-level forest biophysical variables for a mixedwood boreal forest using lidar: An examination of scanning density. *Canadian Journal of Forest Research*, 36:34–47.

Thurgood, J., Gruber, M., and Karner, K., 2004, Multi-ray matching for automated 3D object modeling. *The International Archives of the Photogrammetry, Remote Sensing and Spatial Information Sciences*, 35:1682–1777.

Thürig, E. and Kaufmann, E., 2010, Increasing carbon sinks through forest management: A model-based comparison for Switzerland with its Eastern Plateau and Eastern Alps. *European Journal of Forest Research*:1–10.

Tran-Ha, M., Cordonnier, T., Vallet, P., and Lombart, T., 2011, Estimation du volume total aérien des peuplements forestiers à partir de la surface terrière et de la hauteur de Lorey. *Revue forestière française*, 63(3):361–378.

Treitz, P., Lim, K., Woods, M., Pitt, D., Nesbitt, D. and Etheridge, D., 2012, LiDAR sampling density for forest resource inventories in Ontario, Canada. *Remote Sensing*, 4(4):830–848.

Turner, W., Spector, S., Gardiner, N., Fladeland, M., Sterling, E. and Steininger, M., 2003, Remote sensing for biodiversity science and conservation. *Trends in Ecology and Evolution*, 18(6):306–314.

Uuemaa, E., Antrop, M., Roosaare, J., Marja, R. and Mander, Ü., 2009, Landscape metrics and indices: An overview of their use in landscape research. *Living Reviews in Landscape Research*, 3(1):1–28.

Van Laar, A. and Akça, A., 2007, *Forest Mensuration*, Managing Forest Ecosystems, Vol. 13, Springer, the Netherlands, 385 p.

Vanclay, J. K., 2009, Tree diameter, height and stocking in even-aged forests. *Annals of Forest Science*, 66(7):1–7.

Vastaranta, M., Wulder, M. A., White, J. C., Pekkarinen, A., Tuominen, S., Ginzler, C., Kankare, V., Holopainen, M., Hyyppä, J., and Hyyppä, H., 2013, Airborne laser scanning and digital stereo imagery measures of forest structure: Comparative results and implications to forest mapping and inventory update. *Canadian Journal of Remote Sensing*, 39(05):382–395.

Véga, C. and Durrieu, S., 2011, Multi-level filtering segmentation to measure individual tree parameters based on Lidar data: Application to a mountainous forest with heterogeneous stands. *International Journal of Applied Earth Observation and Geoinformation*, 13(4):646–656.

Véga, C., Durrieu, S., Morel, J., and Allouis, T., 2012, A sequential iterative dual-filter for Lidar terrain modeling optimized for complex forested environments. *Computers and Geosciences*, 44:31–41.

Véga, C., Hamrouni, A., El Mokhtari, S., Morel, J., Bock, J., Renaud, J.-P., Bouvier, M., and Durrieu, S., 2014, PTrees: A point-based approach to forest tree extraction from lidar data. *International Journal of Applied Earth Observation and Geoinformation*, 33:98–108.

Véga, C. and St-Onge, B., 2008, Height growth reconstruction of a boreal forest canopy over a period of 58 years using a combination of photogrammetric and lidar models. *Remote Sensing of Environment*, 112(4):1784–1794.

Véga, C. and St-Onge, B., 2009, Mapping site index and age by linking a time series of canopy height models with growth curves. *Forest Ecology and Management*, 257(3):951–959.

Vepakomma, U. and Fortin, M.-J., 2010, Scale-specific effects of disturbance and environment on vegetation patterns of a boreal landscape. SilviLaser 2010, *Conference on Lidar Applications for Assessing Forest Ecosytems*, Freiburg, Germany, 18pp.

Vepakomma, U., St-Onge, B., and Kneeshaw, D., 2008, Spatially explicit characterization of boreal forest gap dynamics using multi-temporal lidar data. *Remote Sensing of Environment*, 112(5):2326–2340.

Wagner, W., Ullrich, A., Ducic, V., Melzer, T., and Studnicka, N., 2006, Gaussian decomposition and calibration of a novel small-footprint full-waveform digitising airborne laser scanner. *ISPRS Journal of Photogrammetry and Remote Sensing*, 60(2):100–112.

Wagner, W., Ullrich, A., Melzer, T., Briese, C., and Kraus, K., 2004, From single-pulse to full-waveform airborne laser scanners: Potential and practical challenges. *International Archives of Photogrammetry and Remote Sensing*, 35(B3):201–206.

Wang, C.-K., Tseng, Y.-H., and Chu, H.-J., 2014, Airborne dual-wavelength LiDAR data for classifying land cover. *Remote Sensing*, 6(1):700–715.

Wang, C. and Glenn, N. F., 2008, A linear regression method for tree canopy height estimation using airborne lidar data. *Canadian Journal of Remote Sensing*, 34(2):S217–S227.

Wang, K., Franklin, S. E., Guo, X., and Cattet, M., 2010, Remote sensing of ecology, biodiversity and conservation: A review from the perspective of remote sensing specialists. *Sensors*, 10(11):9647–9667.

Weinacker, H., Koch, B., Heyder, U., and Weinacker, R., 2004, Development of filtering, segmentation and modelling modules for lidar and multispectral data as a fundament of an automatic forest inventory system. *International Archives of Photogrammetry, Remote Sensing and Spatial Information Sciences*, 36(Part 8):50–55.

Weishampel, J. F., Hightower, J. N., Chase, A. F., and Chase, D. Z., 2012, Use of airborne LiDAR to delineate canopy degradation and encroachment along the Guatemala-Belize border. *Tropical Conservation Science*, 5(1):12–24.

White, J. C., Wulder, M. A., Vastaranta, M., Coops, N. C., Pitt, D., and Woods, M., 2013, The utility of image-based point clouds for forest inventory: A comparison with airborne laser scanning. *Forests*, 4(3):518–536.

Williams, M. S., Bechtold, W. A., and LaBau, V., 1994, Five instruments for measuring tree height: An evaluation. *Southern Journal of Applied Forestry*, 18(2):76–82.

Wulder, M., Niemann, K. O., and Goodenough, D. G., 2000, Local maximum filtering for the extraction of tree locations and basal area from high spatial resolution imagery. *Remote Sensing of Environment*, 73(1):103–114.

Wulder, M. A., Bater, C. W., Coops, N. C., Hirata, Y., and Sweda, T., 2007, Advances in laser remote sensing of forests. *Sustainable Development Research Advances*, B. A. Larson, Nova Science Publishers, Inc., Hauppauge, NY, Chapter 8, pp. 223–234.

Wulder, M. A., White, J. C., Nelson, R. F., Næsset, E., Ørka, H. O., Coops, N. C., Hilker, T., Bater, C. W., and Gobakken, T., 2012, Lidar sampling for large-area forest characterization: A review. *Remote Sensing of Environment*, 121:196–209.

Wynne, R. H., 2006, Lidar remote sensing of forest resources at the scale of management. *Photogrammetric Engineering and Remote Sensing*, 72(12):1310–1314.

Zellweger, F., Braunisch, V., Baltensweiler, A., and Bollmann, K., 2013, Remotely sensed forest structural complexity predicts multi species occurrence at the landscape scale. *Forest Ecology and Management*, 307:303–312.

Zellweger, F., Morsdorf, F., Purves, R. S., Braunisch, V., and Bollmann, K., 2014, Improved methods for measuring forest landscape structure: LiDAR complements field-based habitat assessment. *Biodiversity and Conservation*, 23(2):289–307.

Zhou, J., Proisy, C., Descombes, X., Hedhli, I., Barbier, N., Zerubia, J., Gastellu-Etchegorry, J.-P., and Couteron, P., 2010, Tree crown detection in high resolution optical and LiDAR images of tropical forest. *Remote Sensing, International Society for Optics and Photonics*, 78240Q-78240Q-78246pp.

Zilliox, C. and Gosselin, F., 2013, Tree species diversity and abundance as indicators of understory diversity in French mountain forests: Variations of the relationship in geographical and ecological space. *Forest Ecology and Management*, 321:105–116.

Zolkos, S. G., Goetz, S. J., and Dubayah, R., 2013, A meta-analysis of terrestrial aboveground biomass estimation using lidar remote sensing. *Remote Sensing of Environment*, 128:289–298.

Zwally, H. J., Schutz, B., Abdalati, W., Abshire, J., Bentley, C., Brenner, A., Bufton, J. et al., 2002, ICESat's laser measurements of polar ice, atmosphere, ocean, and land. *Journal of Geodynamics*, 34(3–4):405–445.

VI

Biodiversity

Biodiversity of the World: A Study from Space

Thomas W. Gillespie
University of California

Andrew Fricker
University of California

Chelsea Robinson
University of California

Duccio Rocchini
Fondazione Edmund Mach

Acronyms and Definitions

ASAR	Advanced Synthetic Aperture Radar
ASTER	Advanced Spaceborne Thermal Emission and Reflection Radiometer
AVHRR	Advanced Very High Resolution Radiometer
CORINE	Coordination of Information on the Environment
COSMO-SkyMed	Constellation of small Satellites for the Mediterranean basin Observation
DMSP	Defense Meteorological Satellite Program
ENVI	Environment for Visualizing Images
ENVISAT	Environmental Satellite
EOS	Earth Observing System
ESA	European Space Agency
ETM+	Landsat Enhanced Thematic Mapper Plus
GAP	National Gap Analysis Program
GIS	Geographic Information System
GLAS	Geoscience Laser Altimeter System
GPM	Global Precipitation Measurement
ICESat	Ice Cloud and land Elevation Satellite
LAI	Leaf Area Index
JERS	Japanese Earth Resources Satellite
MERIS	Medium Resolution Imaging Spectrometer
MODIS	Moderate Resolution Imaging Spectroradiometer
NASA	National Aeronautics and Space Administration
NDVI	Normalized Difference Vegetation Index
NOAA	National Oceanic and Atmospheric Administration
OLI	Landsat Operational Land Imager
QSCAT	NASA's Quick Scatterometer
RADARSAT	Radar Satellite
SAR-Lupe	Synthetic Aperture Radar-Lupe
SPOT	Satellite Pour l'Observation de la Terre
SRTM	Shuttle Radar Topography Mission
TerraSAR-X	Terra Synthetic Aperture Radar X-band
TIRS	Landsat Thermal Infrared Sensor
TM	Landsat Thematic Mapper
TMI	TRMM (Tropical Rainfall Measuring Mission) Microwave Imager
TRMM	Tropical Rainfall Measuring Mission
UAVs	Unmanned Aerial Vehicles

18.1 Introduction

The Earth is undergoing an accelerated rate of native ecosystem conversion and degradation (Nepstad et al. 1999; Myers et al. 2000; Achard et al. 2002), and there is increased interest in measuring, modeling, and monitoring biodiversity using remote sensing from spaceborne sensors (Nagendra 2001; Kerr and Ostrovsky 2003; Turner et al. 2003; Secades et al. 2014).

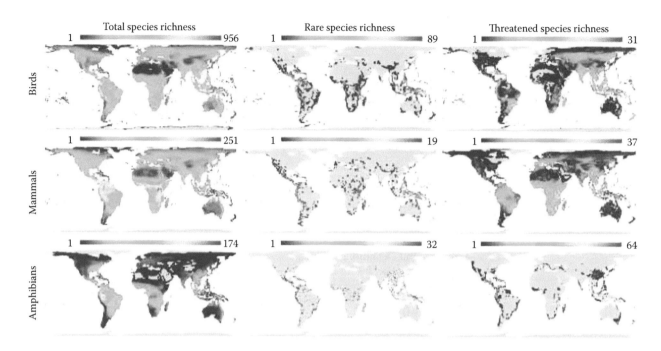

FIGURE 18.1   Total species richness, rare species richness, and threatened species richness for birds, mammals, and amphibians at a global spatial scale. (From Orme, C.D.L. et al., *Nature*, 436, 1016, 2005.)

Biodiversity can be defined as the variation of life forms (genetic, species) within a given ecosystem, region, or the entire Earth. Terrestrial biodiversity, rare and threatened species, tends to be highest near the equator and generally decreases toward the poles because of decreases in temperature and precipitation (Orme et al. 2005; Figure 18.1). However, the distribution of biodiversity is complex and based on a number of environmental and anthropogenic factors over different spatial scales (Whittaker et al. 2001; Field et al. 2009; Jenkins et al. 2013).

Remote sensing has considerable potential as a source of information on biodiversity at a site, landscape, continental, and global spatial scales (Nagendra 2001; Turner et al. 2003). The main attractions of remote sensing as a source of information on biodiversity are that it offers an inexpensive means of deriving complete spatial coverage of environmental information for large areas in a consistent manner that may be updated regularly (Duro et al. 2007; Gillespie et al. 2008). There has been an increase in studies and reviews of biodiversity and remote sensing taking advantage of advances in sensor technology or focusing on broad patterns of variables related to biodiversity (Kerr et al. 2001; Turner et al. 2003; Rocchini 2007a; Pfeifer et al. 2012).

These advances in remote sensing are generally divided into measuring, modeling, and monitoring biodiversity (Nagendra 2001; Turner et al. 2003; Duro et al. 2007). Measuring uses spaceborne sensors to identify either species or individuals, such as the identification of tree species and density, or land cover types associated with species assemblages (such as redwood forest) (Gillespie et al. 2008). Modeling uses spaceborne sensors to create probability models of species distributions and the distributions of biodiversity and associated metrics such as species richness. Monitoring is the use of time series spaceborne data of

measured or modeled biodiversity to study dynamics over time, and this has significant applications for endangered species and ecosystem conservation.

This chapter reviews recent advances in remote sensing that can be used to study biodiversity from space. In particular, this chapter examines ways to measure, model, and monitor biodiversity patterns and processes using spaceborne imagery. First, we examine satellites currently being used to measure biodiversity from space. Second, we examine advances in modeling patterns of species and biodiversity. Third, we examine monitoring applications of remote sensing for the conservation of biodiversity. Finally, we identify spaceborne sensors that can be used to study biodiversity from space.

18.2 Measuring Biodiversity from Space

18.2.1 Mapping Vegetation Types and Invasive Species

There is an increasing desire to identify and map vegetation types associated with biodiversity and native and nonnative species within landscapes from high-resolution spaceborne sensors that have been launched in recent years (Turner et al. 2003; Goodwin et al. 2005; He et al. 2011). High-spatial-resolution imagery has been used to accurately identify some plant species and plant assemblages (Martin et al. 1998; Haara and Haarala 2002; Carleer and Wolff 2004; Foody et al. 2005). Much progress has been made in identifying single species of plants, such as nonnative invasive species, that are of particular interest in natural resource management (He et al. 2011). QuickBird was used to map the invasive nonnative giant reed (*Arundo donax*)

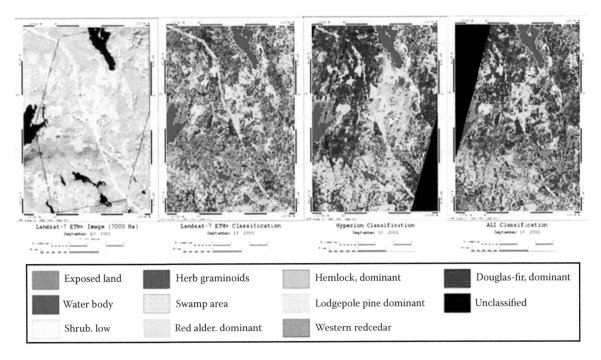

FIGURE 18.2   Classification of tree composition in temperate forests using Landsat ETM+, Hyperion (EO-1), and ALI. (From Goodenough, D.G. et al., *Proc. Int. Geosci. Remote Sens. Symp.*, 2, 882, 2002.)

in southern Texas with 86%–100% accuracy (Everitt et al. 2006). The spaceborne hyperspectral sensor Hyperion has shown potential for identifying the occurrence of select invasive species in southeastern United States like Chinese tallow (*Triadica sebifera*) to within 78% accuracy due to distinct leaf phenology (Ramsey et al. 2005). There has also been significant progress in identifying tree canopies within forest ecosystems. For instance, high-resolution data have been used to identify nonnative invasive species (Fuller 2005; He et al. 2011), native trees (Goodenough et al. 2002; Christian and Krishnayya 2009; Figure 18.2), and mangrove species and mangrove ecosystems at a global spatial scale (Dahdouh-Guebas et al. 2004; Wang et al. 2004; Giri et al. 2011; Heumann 2011).

18.2.2 Mapping Individual Trees

High-spatial-resolution imagery (GeoEye, QuickBird, IKONOS, WorldView) from space has also allowed researchers to address questions that previously were impractical to study from space or on the ground. It is now possible, for instance, for studies to be undertaken at the scale of individual tree crowns over large areas (Hurtt et al. 2003; Clark et al., 2004a). Such data have been used to quantify tree mortality in a tropical rainforest (Clark et al. 2004b) and so may contribute usefully to contentious debates on the issue. Moreover, it may sometimes be possible to achieve high levels of accuracy for some species from satellite as well as airborne sensor data (Carleer and Wolff 2004). There is great potential to manually or digitally identify tree species and canopy attributes from high-resolution imagery. High-resolution imagery is collected primarily from commercial

satellites that are still expensive to acquire ($1000–$4000 for a 10 km²). While the increased pricing of such imagery has put it out of the reach of many ecologists, especially those located in developing countries where the need is perhaps greatest, such cost has decreased with the competition and an increasing number of archived images in the visible spectrum are readily available on Google Earth. Indeed, it is possible to identify species from moderate to high degree of accuracies within many landscapes and forests already, and the accuracies will only increase with increased resolutions (spatial, spectral, and temporal) in the near future.

18.2.3 Mapping Animals from Space

The identification of animals from space is currently difficult because most of the Earth's species are smaller than the largest pixel of current public access satellites (0.5 m) and revisit times are too infrequent for meaningful comparisons. However, high-resolution spaceborne imagery has been used to map large species. Indeed, some groups like whales can be monitored from space with high-resolution WorldView 2 imagery at a 50 cm pixel resolution in the panchromatic (Fretwell et al. 2014; Figure 18.3), and spaceborne remote sensing has been used to survey and discover large colonies of animals like penguin colonies in Antarctica (Schwaller et al. 2013). However, most remote-sensing studies on monitoring fauna have focused on mapping vegetation types or habitat associated with endangered fauna. For instance, giant panda habitat has been monitored over time in China (Jian et al. 2011). Potential great progress in monitoring animals from remote sensing will take off when unmanned aerial vehicles

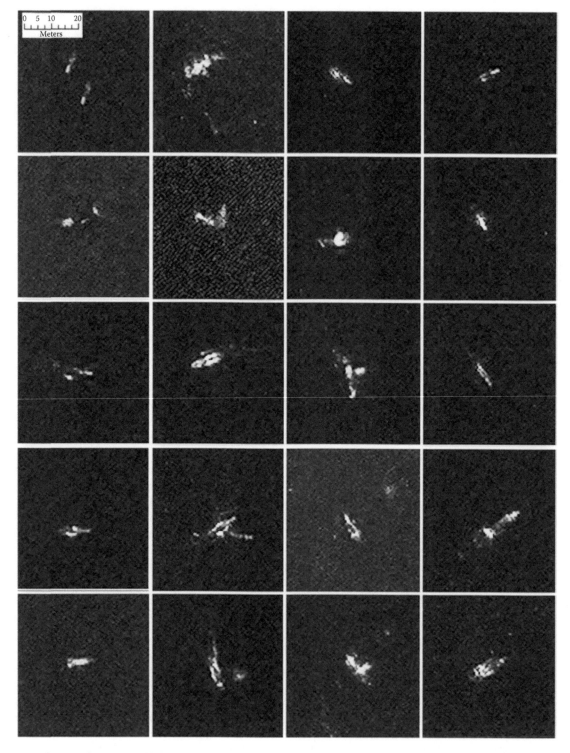

FIGURE 18.3 Selection of 20 comparable false color image chips (bands 1-8-5) from WorldView 2 imagery of whales found along the coast of Argentina. (From Fretwell, P.T. et al., *PLoS One*, 9(2), e88655, 2014.)

(UAVs) start gathering data from a wide array of sensors. UAVs, technology wise, are already quite mature and have been used to monitor elephants in Africa (Schiffman 2014). However, UAVs have several limitations, especially with regard to covering large areas, potential costs, and security concerns.

18.2.4 Mapping Species Assemblages

The production of thematic maps of species assemblages is one of the most common applications of spaceborne remote sensing (Foody 2002). In particular, plant species assemblages or ecosystem distributional patterns within the landscapes, regions, and continents have important applications to natural resource management. In countries with strong and well-funded institutions dedicated to natural resource management, such as the U.S. National Park Service, there is a need to update and standardize landscape dynamic protocols related to vegetation, land cover, and unique resource management needs by region using remote sensing (Fancy et al. 2009). Natural resource agencies in the tropics, where some of the largest decreed protected areas exist (Brooks et al. 2006), do not always have access to the remote-sensing technology or trained individuals to develop and maintain a landscape dynamics change database (Laurance et al. 2012). Numerous large-area, multi-image-based, multiple-sensor land cover mapping programs exist that have resulted in robust and repeatable large-area land cover classifications (Franklin and Wulder 2002; Durio et al. 2007; Gillespie et al. 2008). Franklin and Wulder (2002) undertook an excellent review of large-scale land cover classifications, such as GlobCover, CORINE, and GAP, which generally seek to attain 85% accuracy across all mapping classes using a variety of passive sensors (TM, SPOT, AVHRR, MODIS, ENVI) and to a lesser extent active sensors (RADARSAT, JERS). These land cover classifications provide measurements on the distribution of species assemblages and ecosystems. Recently, there have been a number of advances in methods that can improve the resolution and accuracy of land cover classification. Increased integration of radar data may significantly improve classification accuracy (Saatchi et al. 2001; Boyd and Danson 2005; Li and Chen 2005; Huang et al. 2010).

18.3 Modeling Biodiversity from Space

There are a number of spaceborne metrics that are associated with species and ecosystem distributions that can be used to create probability maps of the distribution of biodiversity (Rocchini et al. 2013) (Table 18.1).

18.3.1 Species Distribution Modeling

Species distribution modeling, also known as ecological niche modeling or spatial modeling, has been growing at a striking rate in the last 20 years (Guisan and Thuiller 2005), providing both estimates of species distributions over space and estimates of bias in the models (Swanson et al. 2012). Species distribution models are based on the presence, absence, or abundance data from museum vouchers or field surveys and environmental predictors to create probability models of species distributions within landscapes, regions, and continents (Guisan and Thuiller 2005). Most environmental predictors used in these species distribution models have been based on geographic information system (GIS) data over different scales (Figure 18.4). There has been an increase in the incorporation of spaceborne remote-sensing data on climate, topography, and land cover that has a great potential to improve the models of species over different spatial scales (Turner et al. 2003). Remote-sensing data on precipitation at 0.1° from NOAA

TABLE 18.1   Remote-Sensing Variables Used for Modeling Biodiversity from Space

Remote-Sensing Variable	Satellite (Sensor)	Pixel Size (m)	Reference
Climate			
Rainfall	TRMM	2, 775	Saatchi et al. (2008)
Rainfall	GPM	250–500	http://pmm.nasa.gov/GPM
Temperature	MODIS	1000	Albright et al. (2011)
Topography			
Elevation and topography	SRTM	30, 90	Elith et al. (2006)
Elevation and topography	ASTER	30	http://asterweb.jpl.nasa.gov/gdem.asp
Land cover			
Vegetation type	Landsat TM, MSS	30, 80	Gottschalk et al. (2005)
NDVI	SPOT, Landsat	20, 30	Leyequien et al. (2007)
LAI	AVHRR, MODIS	1000	Saatchi et al. (2008)
Heterogeneity	SPOT, Landsat	20, 30	Rocchini et al. (2010)
Vegetation structure	QSCAT	1000	Saatchi et al. (2008)
Forest cover/change	Landsat, MODIS	30, 1000	Hansen et al. (2002, 2013)
Old growth	GLAS	1000	Saatchi et al. (2011)
Fire	MODIS	1000	http://modis-fire.umd.edu/
Burned areas	MODIS	500	http://modis-fire.umd.edu/
Energy use	DMSP	1100	http://ngdc.noaa.gov/eog/

satellites (Pearson et al. 2007) and 1000 m from the Tropical Rainfall Measuring Mission (Saatchi et al. 2008) have been used in conjunction with ground-based measurements. This may be superior to traditional GIS estimates of precipitation based on interpolation among widely dispersed climate stations in isolated regions. Topography data have also been a fundamental component of species distribution models (Pearson and Dawson 2003; Elith et al. 2006). Topography data are usually collected from digitized elevation maps, but 90 and 30 m elevation and topography data are available at a near global extent due to the Shuttle Radar Topography Mission (SRTM) and ASTER. These data are increasingly being used in species distribution models (Chaves et al. 2007; Buermann et al. 2008; Saatchi et al. 2008) (Figure 18.5).

18.3.2 Land Cover and Diversity

Land cover classifications collected from spaceborne sensors have long been used to link species distributions with vegetation types and associated habitat preference (Nagendra 2001; Gottschalk et al. 2005; Leyequien et al. 2007). The greatest accuracy was found with nonmobile species such as plants (Pearson et al. 2004). However, vegetation maps as a surrogate for habitat preference have provided insights into the distributions of birds

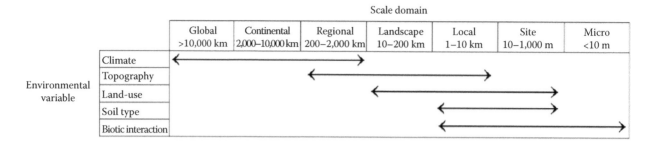

FIGURE 18.4   Environmental predictors used in species distribution models and spatial scale. (From Pearson, R.G. and Dawson, T.P., *Global Ecol. Biogeogr.*, 12, 361, 2003.)

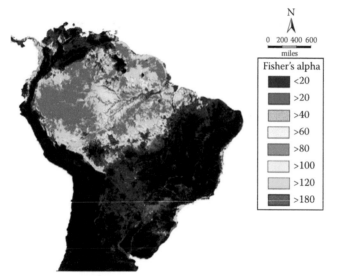

FIGURE 18.5   Tree alpha diversity in South America from Maxent predictions made from remote sensing data and inventory plots (n = 633). (From Saatchi, S. et al., *Remote Sens. Environ.*, 112(5), 2000, 2008.)

(Peterson et al. 2006), herpetofauna (Raxworthy et al. 2003), and insects (Luoto et al. 2002a).

There have been a number of advances in modeling or predicting species richness, alpha diversity, and beta diversity using multisensors that examine relationships over different temporal and spatial scales with increasingly sophisticated methods to improve accuracy. The simplest measure of biodiversity is species richness or the number of species per unit area (i.e., trees per hectare, reptiles per km²). The term diversity is more complex and technically refers to a combination of species richness and weighted abundance or evenness data and is generally quantified as an index (Simpson index, Shannon index, or Fisher's alpha). These indices are used to define alpha diversity, which is the species diversity in one area, community, or ecosystem. Beta diversity refers to the amount of turnover in species composition from one site to another or identifies taxa unique to each area, community, or ecosystem. Beta diversity is more closely related to changes in species similarity or turnover with space. Typically, studies have focused on assessments of species richness with limited attention to other aspects such as species abundance and composition that are difficult to detect from spaceborne sensors (Foody and Cutler

2003; Schmidtlein and Sassin 2004). Information on species richness or diversity may be extracted from remotely sensed data in a variety of ways such as land cover classifications, measures of productivity, and measures of heterogeneity (Nagendra 2001; Kerr and Ostrovsky 2003; Leyequien et al. 2007).

Many studies have related species richness or diversity to information on the land cover mosaic derived from satellite imagery (Nagendra and Gadgil 1999a,b; Gould 2000; Griffiths et al. 2000; Kerr et al. 2001; Oindo et al. 2003; Gottschalk et al. 2005; Leyequien et al. 2007; Gillespie et al. 2008). Through relationships with land cover and habitat suitability, it is possible to assess the diversity of species and assess impacts associated with changes in the habitat mosaic such as fragmentation based on landscape metrics (i.e., area and connectivity) (Kerr et al. 2001; Luoto et al. 2002b, 2004; Cohen and Goward 2004; Fuller et al. 2007; Lassau and Hochuli 2007).

18.3.3 Spectral Indices and Diversity

Most attention has focused on the use of the popular normalized difference vegetation index (NDVI) from passive sensors because it is easy to calculate using the red and near-infrared bands common to almost all passive spaceborne sensors (Oindo and Skidmore 2002; Seto et al. 2004; Gillespie 2005; Lassau and Hochuli 2007; Pettorelli 2013). NDVI has been associated with primary productivity and has been hypothesized to quantify species richness and diversity based on the species–energy theory (Currie 1991; Evans et al. 2005). There have been an increasing number of studies and reviews that have found significant associations between NDVI and diversity (Nagendra 2001; Kerr and Ostrovsky 2003; Leyequien et al. 2007). For plants, many studies have reported significant positive correlations between plant species richness or diversity from plot or region data and NDVI in both temperate (Fairbanks and McGwire 2004; Levin et al. 2007; Rocchini 2007b) and tropical ecosystems (Bawa et al. 2002; Feeley et al. 2005; Gillespie 2005; Cayuela et al. 2006). NDVI can explain between 30% and 87% of the variation in species richness or diversity within a vegetation type, landscape, or region. Results for terrestrial fauna are more complicated given the mobility of faunal species and because NDVI does not directly quantify animal species but species habitats (Leyequien et al. 2007). Similar relationships between NDVI and diversity have been noted for animal taxa such as birds and butterflies within landscapes (Seto et al. 2004; Goetz et al. 2007) and regions (Hulbert and Haskell 2003; Foody 2004b; Ding et al. 2006; Bino et al. 2008). Over the last decade, the NDVI has also proven extremely useful in predicting guild distributions, abundance, and life history traits in space and time (Hurlbert and Haskell 2003; Pettorelli et al. 2011). However, NDVI does not always have a positive relationship with animal species richness and there is no consensus as to which scale results in the greatest accuracy.

Heterogeneity in land cover types, spectral indices, and spectral variability derived from satellite imagery has also been correlated with species richness (Gould 2000; Rocchini 2007b; Rocchini et al. 2010). This is largely based on the hypothesis that heterogeneity in land cover, spectral indices, or spectral variability within an area or landscape is an indicator of habitat heterogeneity, which allows more species to coexist and hence greater species richness (Simpson 1949; Palmer et al. 2002; Carlson et al. 2007; Rocchini et al. 2007, 2010). The variation in land cover types within an area has been associated with species richness for a number of taxa (Gould 2000; Kerr et al. 2001; Leyequien et al. 2007). Variation in spectral indices has been shown to be positively associated with species richness and diversity for a number of taxa in different regions (Gould 2000; Oindo and Skidmore 2002; Fairbanks and McGwire 2004; Levin et al. 2007).

18.3.4 Multiple Sensors and Diversity

Recently, there has been a move toward the use of multiple remote-sensing sensors over different time periods and increasingly sophisticated approaches to modeling diversity over different spatial scales. There are an increasing number of diversity studies that are undertaken using multiple passive sensors (i.e., Landsat, ASTER, QuickBird) (Levin et al. 2007; Rocchini 2007b) or examine relationships with diversity over different time periods (Fairbanks and McGwire 2004; Foody 2005; Levin et al. 2007; Leyequien et al. 2007). These studies are important in the assessment of individual sensors and the effects of seasonality. There has also been an increasing interest in the combination of passive and active sensors to improve species diversity models (Gillespie et al. 2008). Active spaceborne sensors can provide data on the vegetation structure that has been associated with diversity, especially avian diversity, across a number of spatial scales (Imhoff et al. 1997, Bergen et al. 2007; Goetz et al. 2007, Leyequien et al. 2007). Hyperspectral sensors can collect highly detailed information on spectral signatures of species, which has allowed for high-resolution mapping of canopy diversity (Carlson et al. 2007; Asner 2008; Christian and Krishnayya 2009).

18.4 Monitoring Biodiversity from Space

It is well established that biodiversity is greatly threatened by human activity (Myers et al. 2000; Gaston 2005). In particular, land cover changes such as those linked to human-induced habitat loss, fragmentation, and degradation represent the largest current threat to biodiversity (Chapin et al. 2000; Menon et al. 2001; Gaston 2005; Gillespie et al. 2008). Remote sensing can be used

to derive metrics on fragmentation, often in the form of landscape pattern and connectivity indices calculated from a thematic map produced with an image classification analysis (Foody 2001; Gillespie 2005; Lung and Schaab 2006; Kupfer 2012). These metrics can be monitored over time (Table 18.2). Remote sensing may be used to monitor a habitat of interest with a one-class classification approach adopted to focus effort and resources on the class of interest (Foody et al. 2006; Sanchez-Hernandez et al. 2007). This can also reduce problems associated with not satisfying the assumptions of an exhaustively defined set of classes that is commonly made in a standard classification analysis (Sanchez-Azofelfa et al. 2003; Foody 2004a). For instance, Hansen et al. (2013) mapped the spatial extent of all forests at a global spatial scale to 30 m pixel resolution using Landsat imagery (Figure 18.6).

TABLE 18.2 Spaceborne Sensors Commonly Used in Monitoring Biodiversity

Source (Sensor)	Monitoring Use	Strengths	Limitations
High resolution			
Google Earth	Validation, communication	High resolution, free	Temporal gaps, no IR
QuickBird, IKONOS	Area, degradation	High resolution	Cost, coverage
WorldView 3	Area, degradation	Multiband	Cost, coverage
Moderate resolution			
Landsat (TM, ETM+)	Land cover, fragmentation	Long time series, free	Clouds
Landsat (OLI, TIRS)	Temperature, water quality	High resolution	Since 2013
EOS (ASTER)	Land cover, fragmentation	High spectral resolution	On-demand system
SPOT	Land cover, fragmentation	Spectral and spatial resolution	Cost
EOS (Hyperion)	Ecosystem chemistry	Spectral resolution	Signal to noise ratio
Low resolution			
NOAA (AVHRR)	Vegetation indices, thermal	Time series from 1980s	1.1 km pixels
EOS (MODIS)	Vegetation indices, thermal, fire	Time series from 2001	Underutilized
ESA (MERIS)	Vegetation indices, land cover	Highest global land cover map	Not for change detection
Active sensors			
SRTM	Elevation	Bench of canopy height 2001	One off
QSCAT	Canopy structure and moisture	Canopy moisture	2.25 km pixel
TRMM (TMI)	Rainfall	Global comparisons	Underutilized
Envisat (ASAR)	Heterogeneity	High-resolution comparisons	Side angle
ICESat (GLAS)	Biomass	Topography, vegetation height	Sensor failure

FIGURE 18.6   Forest cover and forest cover change from 2000 to 2012 in Banda Aceh, Indonesia. (From Hansen, M.C. et al., *Science*, 342, 850, 2013.)

18.4.1 Remote Sensing of Protected Areas

Protected areas are one of the best ways to conserve biodiversity. Remote sensing has had a major role to play in helping to monitor changes in and around protected areas (Gross et al. 2013; Secades et al. 2014). Remote sensing offers a repeatable, systematic, and spatially exhaustive source of information on key variables such as productivity, disturbance, and land cover that impact biodiversity (Duro et al. 2007; Wright et al. 2007; Gillespie et al. 2008). The provision of data for monitoring large areas is especially attractive in remote and often inaccessible regions (Cayuela et al. 2006; Conchedda et al. 2011). Remote sensing is also often a cost-effective data source (Luoto et al. 2004) and enables rapid biodiversity assessments (Foody 2003; Lassau and Hochuli 2007).

The spatial coverage provided by remote sensing offers the potential to monitor the effectiveness of protected areas, allowing comparisons of changes inside and outside of reserves to be evaluated (Southworth et al. 2006; Wright et al. 2007). For example, even relatively severely logged forest outside of a reserve may represent a significant resource for biodiversity conservation (Cannon et al. 1998; Tang et al. 2010; Figure 18.7). Thus, actions inside and outside of the protected areas are important, supporting the view that biodiversity conservation activities should be undertaken at the level or scale of the landscape (Nagendra et al. 2013). This activity may benefit from remote sensing as its synoptic overview provides information on the entire landscape. Indeed, Crabtree et al. (2009) provide a modeling and spatiotemporal framework for monitoring environmental change using net primary productivity as an ecosystem health indicator.

Remote sensing may be a useful component to general biodiversity assessments, especially in providing data at appropriate spatial

and temporal scales. For example, the biodiversity intactness index was proposed recently as a general indicator of the overall state of biodiversity to aid monitoring and decision-making (Scholes and Biggs 2005). Although there are concerns for its use, notably with the impacts of land degradation, remote sensing may be an important source of data for its derivation (Rouget et al. 2006).

18.5 Spaceborne Sensors and Biodiversity

There has been a dramatic increase in Earth observation satellites and sensors over the last decade, which have been used to measure, model, and monitor biodiversity from space (Table 18.3).

18.5.1 Spectral Sensors and Biodiversity

Passive sensors, which record reflected (visible and infrared wavelengths) and emitted energy (thermal wavelengths), are most frequently used in biodiversity studies. The highest-spatial-resolution data come from commercial satellites, such as GeoEye, WorldView, QuickBird, and IKONOS series, which contain visible and infrared bands used in species and species assemblage mapping. The NASA Landsat series is the most widely used sensor for biodiversity studies due to the ease in which the data can be obtained, long time series, and low cost (Leimgruber et al. 2005). The Landsat series has been used extensively in land cover classifications, diversity models, and conservation studies. The recently launched Landsat 8 satellite will be useful for continuing this time series and biodiversity-related research. Other satellites and sensors such as IRS, SPOT, and ASTER are also used; however, on-demand systems or expensive imagery

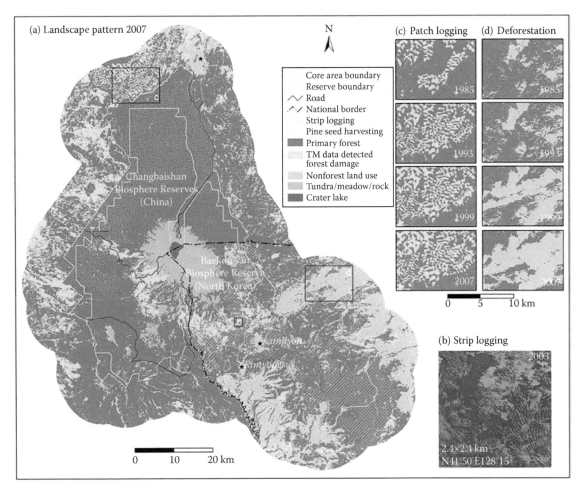

FIGURE 18.7   Landsat landcover and time series of landcover change within and around a protected area in China and North Korea. (From Tang, L. et al., *Biol. Conserv.*, 143, 1295, 2010.)

general results in research that cannot be scaled up to all regions of the globe. The MODIS sensors on EOS satellites have provided extremely useful data for sites, landscape, regional, continental, and global studies of land cover classification and diversity models. These sensors also provide data on temperature, cloud cover, and fire that have been incorporated into biodiversity studies.

18.5.2 Radar Sensors and Biodiversity

Radar is the most common active spaceborne sensor used in biodiversity studies. Radar sensors send and receive a microwave pulse in different wavelengths (i.e., X-, C-, L-bands) to create an image based on radar backscatter, or interferometric radar can be used to provide high-resolution data on elevation and topography. Unlike passive sensors, radar can penetrate cloud cover, providing imagery both day and night regardless of weather conditions. The SRTM provides 30–90 m resolution data on elevation and topography that have been used in species and diversity models. Recent 30 m resolution elevation and topography models have also been created using ASTER's multiangle capability. Radar backscatter from QSCAT and RADARSAT-1 has been used in land cover classification and diversity models.

18.5.3 Lidar Sensors and Biodiversity

Data from the Geoscience Laser Altimeter System (GLAS) aboard the Ice Cloud and land Elevation Satellite (ICESat) offer an unprecedented opportunity for studying biodiversity. The GLAS was the sole instrument on ICESat that had three lasers that emit infrared and visible laser pulses at 1064 and 532 nm wavelengths. The satellite was a profiling sensor with a polar orbit and the tracks have large spatial gaps at low latitudes. GLAS lidar has been used to measure forest structure and terrain in temperate and tropical forests (Pang et al. 2008; Sun et al. 2008; Saatchi et al. 2011). Although GLAS is no longer producing new data, the next generation of spaceborne lidar (ICESat 2) should improve the spatial resolution and point density necessary to study fine-scale processes associated with biodiversity.

18.5.4 Ideal Biodiversity Satellite and Sensor

The ideal biodiversity satellite should include both a high-resolution spectral sensor and a high-resolution lidar sensor. Ideally, 0.5 m pixel resolution is needed for the spectral sensors

TABLE 18.3   Satellites That Can Be Used to Measure, Model, and Monitor Biodiversity from Space

Satellite (Sensor)	Pixel Size (m)	Bands	Revisit Time	Year Operation
Spectral Sensors	Spectral Bands			
High resolution				
QuickBird	0.65, 2.6	5	1–3.5 days	2001
IKONOS	0.8, 3.2	5	3 days	1999
GeoEye 1	0.46, 1.8	5	3 days	2008
WorldView 3	0.3, 1.24, 3.7, 30	29	1 day	2014
Moderate resolution				
Landsat (ETM+)	15, 30, 60, 120	8	16 days	1999
Landsat 8	15, 30, 100	11	16 days	2013
IRS (LISS III)	5, 23, 70	5	5 days	2003
EOS (ASTER)	15, 30, 90	14	On demand	1999
SPOT 5	2.5, 20, 1150	5	26 days	2002
EOS (Hyperion)	30	220	On demand	2000
Low resolution				
NOAA (AVHRR)	1100	5	1 day	1979
EOS (MODIS)	250, 500, 1000	36	1 day	1999
Envisat (MERIS)	300	15	3 days	2002–2012
Active sensors		Bands		
SRTM	30, 90	X, C	None	2000
QSCAT	2500	Ku	4 days	1999–2009
RADARSAT 2	9–100	C	24 days	2007
TRMM (TMI)	5000	X, K, Ka, W	0.5 days	1997
Envisat (ASAR)	30–150	C	35 days	2002–2012
Sentinel-1A	5–100	C	12 days	2014
ICESat 1 (GLAS)	70	1064, 532 nm	8 days	2003–2009

and it should capture the visible, infrared, and thermal bands. The actual wavelengths should correspond to those of Landsat 8 with two bands in the blue wavelength to study water, infrared bands similar to WorldView 3 to study plant species and vegetation, and two bands in the thermal to identify large animals and disturbance like fire. The lidar sensors should have a one meter footprint because this clearly improved estimates of vegetation height and subcanopy topography, both of which are important for modeling and monitoring vegetation structure, species distribution models, and estimates of species richness and biomass. Combining full coverage of the recently launched WorldView 3 and soon to be launched ICESat 2 may come closest to the ideal combination of satellites and sensors needed to monitor biodiversity.

18.6 Conclusions

Spaceborne imagery has made significant contributions to measuring, modeling, and monitoring biodiversity patterns and processes from space. Future research should focus on incorporating recent and new spaceborne sensors; more extensive integration of available field, GIS, and passive and active imagery that can be used across spatial scales; and the collection and dissemination of high-quality field data (Gillespie et al. 2008; Morueta-Holme et al. 2013). The recent developments in satellite and sensor technology will further improve our abilities to measure and model patterns of biodiversity from space. The increase in high-resolution spectral satellites will make it possible to acquire data at enhanced spatial (0.5 m), spectral (visible, infrared, thermal), and radiometric resolutions (11 bit) that can be used to map individual species. Indeed, Google Earth has led the way by providing high-resolution airborne and spaceborne imagery (Loarie et al. 2008). Current radar satellites may be ideal for studying species distributions and diversity patterns, especially in regions with high cloud cover like the tropics. There are multiple satellites (SAR-Lupe, COSMO-SkyMed, TerraSAR-X) that provide elevation and radar backscatter data to 1 m pixel resolution (Gillespie et al. 2007). This will provide valuable multidimensional datasets (vegetation structure, biomass, land cover classifications) that should result in a richer characterization of the environment than conventional passive image datasets (Gillespie et al. 2008). Remote-sensing scientists interested in biodiversity have taken and can take advantage of the different satellite datasets that integrate climate, topography, spectral, lidar, and radar data over a landscape, regional, continental, and global spatial scale. This has increased our understanding of biodiversity, and remote sensing is a useful tool for measuring, modeling, and monitoring biodiversity in near real time and across multispatial scales.

References

Achard, F., H. D. Eva, H. J. Stibig, P. Mayaux, J. Gallego, T. Richards, and J. P. Malingreau. 2002. Determination of deforestation rates of the world's humid tropical forests. *Science* 297:999–1002.

Albright, T. P., A. M. Pidgeon, C. D. Rittenhouse, M. K. Clayton, and C. H. Flather. 2011. Heat waves measured with MODIS land surface temperature data predict changes in avian community structure. *Remote Sensing of Environment* 115:245–254.

Asner, G. P. 2008. Hyperspectral remote sensing of canopy chemistry, physiology, and biodiversity of tropical and subtropical forests. In M. Kalacska, and G. A. Sanchez-Azofeifa (eds.), *Hyperspectral Remote Sensing of Tropical & Subtropical Forests*. Taylor & Francis Group, London.

Bawa, K., J. Rose, K. N. Ganeshaiah, N. Barve, M. C. Kiran, and R. Umashaanker. 2002. Assessing biodiversity from space: An example from the Western Ghats, India. *Conservation Ecology* 6(2):7.

Bergen, K. M., A. M. Gilboy, and D. G. Brown. 2007. Multidimensional vegetation structure in modeling avian habitat. *Ecological Informatics* 2:9–22.

Bino, G., N. Levin, S. Darawshi, N. Van der Hal, A. Reich-Solomon, and S. Kark. 2008. Landsat derived NDVI and spectral unmixing accurately predict bird species richness patterns in an urban landscape. *International Journal of Remote Sensing* 29(13):3675–3700.

Boyd, D. S. and F. M. Danson. 2005. Satellite remote sensing of forest resources: Three decades of research development. *Progress in Physical Geography* 29:1–26.

Brooks, T. M., R. A. Mittermeier, G. A. da Fonseca, J. Gerlach, M. Hoffmann, J. F. Lamoreux, C. G. Mittermeier, J. D. Pilgrim, and A. S. Rodrigues. 2006. Global biodiversity conservation priorities. *Science* 313:58–61.

Buermann, W., S. Saatchi, T. B. Smith, B. R. Zutta, J. A. Chaves, B. Mila, and C. H. Graham. 2008. Predicting species distributions across the Amazonian and Andean regions using remote sensing data. *Journal of Biogeography* 35(7):1160–1176.

Cannon, C. H., D. R. Peart, and M. Leighton. 1998. Tree species diversity in commercially logged Bornean rainforest. *Science* 281:1366–1368.

Carleer, A. and E. Wolff. 2004. Exploitation of very high resolution satellite data for tree species identification. *Photogrammetric Engineering and Remote Sensing* 70:135–140.

Carlson, K. M., G. P. Asner, R. F. Hughes, R. Ostertag, and R. E. Martin. 2007. Hyperspectral remote sensing of canopy biodiversity in Hawaiian lowland rainforests. *Ecosystems* 10:536–549.

Cayuela, L., J. M. Benayas, A. Justel, and J. Salas-Rey. 2006. Modelling tree diversity in a highly fragmented tropical montane landscape. *Global Ecology and Biogeography* 15:602–613.

Chapin, F. S. III, E. S. Zavaleta, V. T. Eviner, R. L. Naylor, P. M. Vitousek, H. L. Reynolds, D. U. Hooper et al. 2000. Consequences of changing biodiversity. *Nature* 405:234–242.

Chaves, J. A., J. P. Pollinger, T. B. Smith, and G. LeBuhn. 2007. The role of geography and ecology in shaping the phylogeography of the speckled hummingbird (*Adelomyia melanogenys*) in Ecuador. *Molecular Phylogenetics and Evolution* 43:795–807.

Christian, B. and N. S. R. Krishnayya. 2009. Classification of tropical trees growing in a sanctuary using Hyperion (EO-1) and SAM algorithm. *Current Science* 96:1601–1607.

Clark, D. B., C. S. Castro, L. D. A. Alvarado, and J. M. Read. 2004b. Quantifying mortality of tropical rain forest trees using high-spatial-resolution satellite data. *Ecology Letters* 7:52–59.

Clark, D. B., J. M. Read, M. L. Clark, A. M. Cruz, M. F. Dotti, and D. A. Clark. 2004a. Application of 1-M and 4-M resolution satellite data to ecological studies of tropical rain forests. *Ecological Applications* 14:61–74.

Cohen, W. B. and S. N. Goward. 2004. Landsat's role in ecological applications of remote sensing. *Bioscience* 54:535–545.

Conchedda, G., E. F. Lambin, and P. Mayaux. 2011. Between land and sea: Livelihoods and environmental changes in mangrove ecosystems of Senegal. *Annals of the Association of American Geographers* 101:1259–1284.

Crabtree, R., C. Potter, R. Mullen, J. Sheldon, S. Huang, J. Harmsen, A. Rodman, and C. Jean. 2009. A modeling and spatio-temporal analysis framework for monitoring environmental change using NPP as an ecosystem indicator. *Remote Sensing of Environment* 113:1486–1496.

Currie, D. J. 1991. Energy and large-scale patterns of animal-and plant-species richness. *American Naturalist* 137:27–49.

Dahdouh-Guebas, F., E. Van Hiel, J. C. W. Chan, L. P. Jayatissa, and N. Koedam. 2004. Qualitative distinction of congeneric and introgressive mangrove species in mixed patchy forest assemblages using high spatial resolution remotely sensed imagery (IKONOS). *Systematics and Biodiversity* 2(2):113–119.

Ding, T., H. Yuan, S. Geng, C. Koh, and P. Lee. 2006. Macro-scale bird species richness patterns of East Asian mainland and islands: Energy, area, and isolation. *Journal of Biogeography* 33:683–693.

Duro, D., N. C. Coops, M. A. Wulder, and T. Han. 2007. Development of a large area biodiversity monitoring system driven by remote sensing. *Progress in Physical Geography* 31:235–260.

Elith, J., C. H. Graham, R. P. Anderson, M. Dudik, S. Ferrier, A. Guisan, R. J. Hijmans et al. 2006. Novel methods improve prediction of species' distributions from occurrence data. *Ecography* 29:129–151.

Evans, K. L., J. J. D. Greenwood, and K. J. Gaston. 2005. Dissecting the species-energy relationship. *Proceedings of the Royal Society B—Biological Sciences* 272:2155–2163.

Everitt, J. H., C. Yang, and C. J. Deloach Jr. 2006. Remote sensing of giant reed with QuickBird satellite imagery. *Journal of Aquatic Plant Management* 43:81–85.

Fairbanks, D. H. K. and K. C. McGwire. 2004. Patterns of floristic richness in vegetation communities of California: Regional scale analysis with multi-temporal NDVI. *Global Ecology and Biogeography* 13:221–235.

Fancy, S. G., J. E. Gross, and S. L. Carter. 2009. Monitoring the condition of natural resources in US national parks. *Environmental Monitoring and Assessment* 151:161–174.

Feeley, K. J., T. W. Gillespie, and J. W. Terborgh. 2005. The utility of spectral indices from Landsat ETM+ for measuring the structure and composition of tropical dry forests. *Biotropica* 37:508–519.

Field, R., B. A. Hawkins, H. V. Cornell, D. J. Currie, J. A. F Diniz-Filho, J.-F. Guégan, D. M. Kaufman et al. 2009. Spatial species-richness gradients across scales: A meta-analysis. *Journal of Biogeography* 36:132–147.

Foody, G. M. 2001. Monitoring the magnitude of land-cover change around the southern limits of the Sahara. *Photogrammetric Engineering and Remote Sensing* 67:841–847.

Foody, G. M. 2002. Status of land cover classification accuracy assessment. *Remote Sensing of Environment* 80:185–201.

Foody, G. M. 2003. Remote sensing of tropical forest environments: Towards the monitoring of environmental resources for sustainable development. *International Journal of Remote Sensing* 24:4035–4046.

Foody, G. M. 2004a. Supervised image classification by MLP and RBF neural networks with and without an exhaustively defined set of classes. *International Journal of Remote Sensing* 25:3091–3104.

Foody, G. M. 2004b. Spatial nonstationarity and scale-dependency in the relationship between species richness and environmental determinants for the sub-Saharan endemic avifauna. *Global Ecology and Biogeography* 13:315–320.

Foody, G. M. 2005. Mapping the richness and composition of British breeding birds from coarse spatial resolution satellite sensor imagery. *International Journal of Remote Sensing* 26:3943–3956.

Foody, G. M., P. M. Atkinson, P. W. Gething, N. A. Ravenhill, and C. K. Kelly. 2005. Identification of specific tree species in ancient semi-natural woodland from digital aerial sensor imagery. *Ecological Applications* 15:1233–1244.

Foody, G. M. and M. E. J. Cutler. 2003. Tree biodiversity in protected and logged Bornean tropical rain forests and its measurement by satellite remote sensing. *Journal of Biogeography* 30:1053–1066.

Foody, G. M., A. Mathur, C. Sanchez-Hernandez, and D. S. Boyd. 2006. Training set size requirements for the classification of a specific class. *Remote Sensing of Environment* 104:1–14.

Franklin, S. E. and M. A. Wulder. 2002. Remote sensing methods in medium spatial resolution satellite data land cover classification of large areas. *Progress in Physical Geography* 26:173–205.

Fretwell, P. T., I. J. Staniland, and J. Forcada. 2014. Whales from space: Counting southern right whales by satellite. *PLoS One* 9(2):e88655.

Fuller, D. O. 2005. Remote detection of invasive Melaleuca trees (*Melaleuca quinquenervia*) in South Florida with multispectral IKONOS imagery. *International Journal of Remote Sensing* 26:1057–1063.

Fuller, R. M., B. J. Devereux, S. Gillings, R. A. Hill, and G. S. Amable. 2007. Bird distributions relative to remotely sensed habitats in Great Britain: Towards a framework for national modeling. *Journal of Environmental Management* 84:586–605.

Gaston, K. J. 2005. Biodiversity and extinction: Species and people. *Progress in Physical Geography* 29:239–247.

Gillespie, T. W. 2005. Predicting woody-plant species richness in tropical dry forests: A case study from south Florida, USA. *Ecological Applications* 15:27–37.

Gillespie, T. W., J. Chu, E. Frankenberg, and D. Thomas. 2007. Assessment and prediction of natural hazards from satellite imagery. *Progress in Physical Geography* 31:459–470.

Gillespie, T. W., G. M. Foody, D. Rocchini, A. P. Giorgi, and S. Saatchi. 2008. Measuring and modelling biodiversity from space. *Progress in Physical Geography* 32:203–221.

Giri, C., E. Ochieng, L. L. Tieszen, Z. Zhu, A. Singh, T. Loveland, J. Masek, and N. Duke. 2011. Status and distribution of mangrove forests of the world using earth observation satellite data. *Global Ecology and Biogeography* 20:154–159.

Goetz, S., D. Steinberg, R. Dubayah, and B. Blair. 2007. Laser remote sensing of canopy habitat heterogeneity as a predictor of bird species richness in an eastern temperate forest, USA. *Remote Sensing of Environment* 108:254–263.

Goodenough, D. G., A. S. Bhogal, A. Dyk, A. Hollinger, Z. Mah, K. O. Niemann, J. Pearlman et al. 2002. Monitoring forests with Hyperion and ALI. *Proceedings of the International Geoscience and Remote Sensing Symposium* 2:882–885.

Goodwin, N., R. Turner, and R. Merton. 2005. Classifying eucalyptus forests with high spatial and spectral resolution imagery: An investigation of individual species and vegetation communities. *Australian Journal of Botany* 53:337–345.

Gottschalk, T. K., F. Huettmann, and M. Ehler. 2005. Thirty years of analyzing and modeling avian habitat relationships using satellite imagery: A review. *International Journal of Remote Sensing* 26:2631–2656.

Gould, W. 2000. Remote Sensing of vegetation, plant species richness, and regional biodiversity hot spots. *Ecological Applications* 10:1861–1870.

Griffiths, G. H., J. Lee, and B. C. Eversham. 2000. Landscape pattern and species richness: Regional scale analysis from remote sensing. *International Journal of Remote Sensing* 21:2685–2704.

Gross, D., G. Dubois, J. F. Pekel, P. Mayaux, M. Holmgren, H. H. T. Prins, C. Rondinini, and L. Boitani. 2013. Monitoring land cover changes in African protected areas in the 21st century. *Ecological Informatics* 14:31–37.

Guisan, A. and W. Thuiller. 2005. Predicting species distribution: Offering more than simple habitat models. *Ecology Letters* 8:993–1009.

Haara, A. and M. Haarala. 2002. Tree species classification using semi-automatic delineation of trees on aerial images. *Scandinavian Journal of Forest Research* 17:556–565.

Hansen, M. C., R. S. DeFries, J. R. G. Townshend, R. Sohlberg, C. Dimiceli, and M. Carroll. 2002. Towards an operational MODIS continuous field of percent tree cover algorithm: Examples using AVHRR and MODIS data. *Remote Sensing of Environment* 83:303–319.

Hansen, M. C., P. V. Potapov, R. Moore, M. Hancher, S. A. Turubanova, A. Tyukavina, D. Thau et al. 2013. High-resolution global maps of 21st-century forest cover change. *Science* 342:850–853.

He, K. S., D. Rocchini, M. Neteler, and H. Nagendra. 2011. Benefits of hyperspectral remote sensing for tracking plant invasions. *Diversity and Distributions* 17:381–392.

Heumann, B. W. 2011. Satellite remote sensing of mangrove forests: Recent advances and future opportunities. *Progress in Physical Geography* 35:87–108.

Huang, S., C. Potter, R. L. Crabtree, S. Hager, and P. Gross, 2010. Fusing optical and radar data to estimate sagebrush, herbaceous, and bare ground cover in Yellowstone. *Remote Sensing of Environment* 114:251–264.

Hurlbert, A. H. and J. P. Haskell. 2003. The effect of energy and seasonality on avian species richness and community composition. *American Naturalist* 161:83–97.

Hurtt, G., X. M. Xiao, M. Keller, M. Palace, G. P. Asner, R. Braswell, E. S. Brondizio et al. 2003. IKONOS imagery for the Large Scale Biosphere–Atmosphere Experiment in Amazonia (LBA). *Remote Sensing of Environment* 88:111–127.

Imhoff, M. L., T. D. Sisk, A. Milne, G. Morgan, and T. Orr. 1997. Remotely sensed indicators of habitat heterogeneity: Use of synthetic aperture radar in mapping vegetation structure and bird habitat. *Remote Sensing of Environment* 60:217–227.

Jenkins, C. N., S. L. Pimm, and L. N. Joppa. 2013. Global patterns of terrestrial vertebrate diversity and conservation. *Proceedings of the National Academy of Sciences* 110:2602–2610.

Jian, J., H. Jiang, G. Zhou, Z. Jiang, S. Yu, S. Peng, S. Liu, and J. Wang. 2011. Mapping the vegetation changes in giant panda habitat using Landsat remotely sensed data. *International Journal of Remote Sensing* 32:1339–1356.

Kerr, J. T. and M. Ostrovsky. 2003. From space to species: Ecological applications for remote sensing. *Trends in Ecology and Evolution* 18:299–305.

Kerr, J. T., T. R. E. Southwood, and J. Cihlar. 2001. Remotely sensed habitat diversity predicts butterfly species richness and community similarity in Canada. *Proceedings of the National Academy of Sciences of the United States of America* 98:11365–11370.

Kupfer, J. A. 2012. Landscape ecology and biogeography: Rethinking landscape metrics in a post-FRAGSTATS landscape. *Progress in Physical Geography* 36:400–420.

Lassau, S. A. and D. F. Hochuli. 2007. Associations between wasp communities and forest structure: Do strong local patterns hold across landscapes? *Austral Ecology* 32:656–662.

Laurance, W. F., D. C. Useche, J. Rendeiro, M. Kalka, C. J. A. Bradshaw, S. P. Sloan, S. G. Laurance et al. 2012. Averting biodiversity collapse in tropical forest protected areas. *Nature* 489:290–294.

Leimgruber, P., C. A. Christen, and A. Laborderie. 2005. The impact of Landsat satellite monitoring on conservation biology. *Environmental Monitoring and Assessment* 106:81–101.

Levin, N., A. Shimida, O. Levanoni, H. Tamari, and S. Kark. 2007. Predicting mountain plant richness and rarity from space using satellite-derived vegetation indices. *Diversity and Distribution* 13:1–12.

Leyequien, E., J. Verrelst, M. Slot, G. Schaepman-Strub, I. M. A. Heitkonig, and A. Skidmore. 2007. Capturing the fugitive: Applying remote sensing to terrestrial animal distribution and diversity. *International Journal of Applied Earth Observation and Geoinformation* 9:1–20.

Li, J. and W. Chen. 2005. A rule-based method for mapping Canada's wetlands using optical, radar, and DEM data. *International Journal of Remote Sensing* 26:5051–5069.

Loarie, S. R., L. N. Joppa, and S. L. Pimm. 2008. Satellites miss environmental priorities. *Trends in Ecology and Evolution* 22(12):630–632.

Lung, T. and G. Schaab. 2006. Assessing fragmentation and disturbance of west Kenyan rainforests by means of remotely sensed time series data and landscape metrics. *African Journal of Ecology* 44:491–506.

Luoto, M., M. Kuussaari, and T. Toivonen. 2002a. Modelling butterfly distribution based on remote sensing data. *Journal of Biogeography* 29:1027–1037.

Luoto, M., T. Toivonen, and R. K. Heikkinen. 2002b. Prediction of total and rare plant species richness in agricultural landscapes from satellite images and topographic data. *Landscape Ecology* 17:195–217.

Luoto, M., R. Virkkala, R. K. Heikkinen, and K. Rainio. 2004. Predicting bird species richness using remote sensing in boreal agricultural-forest mosaics. *Ecological Applications* 14:1946–1962.

Martin, M. E., S. D. Newman, J. D. Aber, and R. G. Congalton. 1998. Determining forest species composition using high spectral resolution remote sensing data. *Remote Sensing of Environment* 65:249–254.

Menon, S., R. G. Pontius, J. Rose, M. L. Khan, and K. S. Bawa. 2001. Identifying conservation-priority areas in the tropics: A land-use change modeling approach. *Conservation Biology* 15:501–512.

Morueta-Holme, N., B. J. Enquist, B. J. McGill, B. Boyle, P. M. Jørgensen, J. E. Ott, R. K. Peet et al. 2013. Habitat area and climate stability determine geographical variation in plant species range sizes. *Ecology Letters* 16:1446–1454.

Myers, N., R. A. Mittermeier, C. G. Mittermeier, G. A. Da Fonseca, and J. Kent. 2000. Biodiversity hotspots for conservation priorities. *Nature* 403:853–858.

Nagendra, H. 2001. Using remote sensing to assess biodiversity. *International Journal of Remote Sensing* 22:2377–2400.

Nagendra, H. and M. Gadgil. 1999a. Biodiversity assessment at multiple scales: Linking remotely sensed data with field information. *Proceedings of the National Academy of Sciences of the United States of America* 96:9154–9158.

Nagendra, H. and M. Gadgil. 1999b. Satellite imagery as a tool for monitoring species diversity: An assessment. *Journal of Applied Ecology* 36:388–397.

Nagendra, H., R. Lucas, J. P. Honrado, R. H. Jongman, C. Tarantino, M. Adamo, and P. Mairota. 2013. Remote sensing for conservation monitoring: Assessing protected areas, habitat extent, habitat condition, species diversity, and threats. *Ecological Indicators* 33:45–59.

Nepstad, D. C., A. Verissimo, A. Alencar, C. Nobre, E. Lima, P. Lefebvre, P. Schlesinger et al. 1999. Large-scale impoverishment of Amazonian forests by logging and fire. *Nature* 398:505–508.

Oindo, B. O. and A. K. Skidmore. 2002. Interannual variability of NDVI and species richness in Kenya. *International Journal of Remote Sensing* 23:285–298.

Oindo, B. O., A. K. Skidmore, and P. De Salvo. 2003. Mapping habitat and biological diversity in the Maasai Mara ecosystem. *International Journal of Remote Sensing* 24:1053–1069.

Orme, C. D. L., R. G. Davies, M. Burgess, F. Eigenbrod, N. Pickup, V. A. Olson, A. J. Webster et al. 2005. Global hotspots of species richness are not congruent with endemism or threat. *Nature* 436:1016–1019.

Palmer, M. W., P. Earls, B. W. Hoagland, P. S. White, and T. Wohlgemuth. 2002. Quantitative tools for perfecting species lists. *Environmetrics* 13:121–137.

Pang, Y., M. A. Lefsky, H. E. Andersen, M. E. Miller, and K. Sherrill. 2008. Validation of the ICESat vegetation product using crown-area-weighted mean height derived using crown delineation with discrete return lidar data. *Canadian Journal of Remote Sensing* 34:471–484.

Pearson, R. G. and T. P. Dawson. 2003. Predicting the impacts of climate change on the distribution of species: Are bioclimate envelope models useful. *Global Ecology and Biogeography* 12:361–371.

Pearson, R. G., T. P. Dawson, and C. Liu. 2004. Modelling species distribution in Britain: A hierarchical integration of climate and land-cover data. *Ecography* 27:285–298.

Pearson, R. G., C. J. Raxworthy, M. Nakamura, and A. T. Peterson. 2007. Predicting species distributions from small numbers of occurrence records: A test case using cryptic geckos in Madagascar. *Journal of Biogeography* 34:102–117.

Peterson, A. T., V. Sánchez-Cordero, E. Martínez-Meyer, and A. G. Navarro-Sigüenza. 2006. Tracking population extirpations via melding ecological niche modeling with land-cover information. *Ecological Modelling* 195:229–236.

Pettorelli, N. 2013. *The Normalized Difference Vegetation Index.* Oxford University Press, London.

Pettorelli, N., S. Ryan, T. Mueller, N. Bunnefeld, B. Jedrzejewska, B. M. Lima, and K. Kausrud. 2011. The Normalized Difference Vegetation Index (NDVI): Unforeseen successes in animal ecology. *Climate Research* 46(1):15–27.

Pfeifer, M., M. Disney, T. Quaife, and R. Marchant. 2012. Terrestrial ecosystems from space: A review of earth observation products for macroecology applications. *Global Ecology and Biogeography* 21:603–624.

Ramsey, E., A. Rangoonwala, G. Nelson, R. Ehrlich, and K. Martella. 2005. Generation and validation of characteristics spectra from EO1 Hyperion image data for detecting the occurrence of the invasive species, Chinese tallow. *International Journal of Remote Sensing* 26:1611–1636.

Raxworthy, C. J., E. Martinez-Meyer, N. Horning, R. A. Nussbaum, G. E. Schneider, M. A. Ortega-Huerta, and A. T. Peterson. 2003. Predicting distributions of known and unknown reptile species in Madagascar. *Science* 426:837–841.

Rocchini, D. 2007a. Distance decay in spectral space in analyzing ecosystem β-diversity. *International Journal of Remote Sensing* 28:2635–2644.

Rocchini, D. 2007b. Effects of spatial and spectral resolution in estimating ecosystem α-diversity by satellite imagery. *Remote Sensing of Environment* 111:423–434.

Rocchini, D., N. Balkenhol, G. A. Carter, G. M. Foody, T. W. Gillespie, K. S. He, S. Kark et al. 2010. Remotely sensed spectral heterogeneity as a proxy of species diversity: Recent advances and open challenges. *Ecological Informatics* 5(5):318–329.

Rocchini, D., G. M. Foody, H. Nagendra, C. Ricotta, M. Anand, K. S. He, V. Amici et al. 2013. Uncertainty in ecosystem mapping by remote sensing. *Computers & Geosciences* 50:128–135.

Rocchini, D., C. Ricotta, and A. Chiarucci. 2007. Using satellite imagery to assess plant species richness: The role of multispectral systems. *Applied Vegetation Science* 10:325–331.

Rouget, M., R. M. Cowling, J. Vlok, M. Thompson, and A. Balmford. 2006. Getting the biodiversity intactness index right: The importance of habitat degradation data. *Global Change Biology* 12:2032–2036.

Saatchi, S., D. Agosti, K. Alger, J. Delabie, and J. Musinsky. 2001. Examining fragmentation and loss of primary forest in Southern Bahian Atlantic forest of Brazil with radar imagery. *Conservation Biology* 15:867–875.

Saatchi, S., W. Buermann, S. Mori, H. ter Steege, and T. Smith. 2008. Modeling distribution of Amazonian tree species and diversity using remote sensing measurements. *Remote Sensing of the Environment* 112(5):2000–2017.

Saatchi, S., N. L. Harris, S. Brown, M. Lefsky, E. T. Mitchard, W. Salas, B. R. Zutta et al. 2011. Benchmark map of forest carbon stocks in tropical regions across three continents. *Proceedings of the National Academy of Sciences* 108:9899–9904.

Sanchez-Azofelfa, G. A., K. L. Castro, B. Rivard, M. R. Kalascka, and R. C. Harriss. 2003. Remote sensing research priorities in tropical dry forest environments. *Biotropica* 35:134–142.

Sanchez-Hernandez, C., D. S. Boyd, and G. M. Foody. 2007. One-class classification for mapping a specific land-cover class: SVDD classification of fenland. *IEEE Transactions on Geoscience and Remote Sensing* 45:1061–1073.

Schifffman, R. 2014. Drones flying high as new tool for field biologists. *Science* 344:459.

Schmidtlein, S. and J. Sassin. 2004. Mapping of continuous floristic gradients in grasslands using hyperspectral imagery. *Remote Sensing of Environment* 92:126–138.

Scholes, R. J. and R. Biggs. 2005. A biodiversity intactness index. *Nature* 434:45–49.

Schwaller, M. R., C. J. Southwell, and L. M. Emmerson. 2013. Continental-scale mapping of Adelie penguin colonies from Landsat imagery. *Remote Sensing of Environment* 139:353–364.

Secades, C., B. O'Connor, C. Brown, and M. Walpole. 2014. Earth observation for biodiversity monitoring: A review of current approaches and future opportunities for tracking progress towards the Aichi Biodiversity Targets. Secretariat of the Convention on Biological Diversity, Montréal, Quebec, Canada. Technical Series No. 72, 183pp.

Seto, K. C., E. Fleishman, J. P. Fay, and C. J. Betrus. 2004. Linking spatial patterns of bird and butterfly species richness with Landsat TM derived NDVI. *International Journal of Remote Sensing* 25:4309–4324.

Simpson, E. H. 1949. Measurement of diversity. *Nature* 163(4148):688.

Southworth, J., H. Nasendra, and D. K. Munroe. 2006. Are parks working? Exploring human-environment tradeoffs in protected area conservation. *Applied Geography* 26:87–95.

Sun, G., K. J. Ranson, D. S. Kimes, J. B. Blair, and K. Kovacs. 2008. Forest vertical structure from GLAS: An evaluation using LVIS and SRTM data. *Remote Sensing of Environment* 112:107–117.

Swanson, A. K., S. Z. Dobrowski, A. O. Finley, J. H. Thorne, and M. K. Schwartz. 2012. Spatial regression methods capture prediction uncertainty in species distribution model projections through time. *Global Ecology and Biogeography* 22:242–251.

Tang, L., G. Shao, Z., Piao, L. Dai, M. A. Jenkins, S. Wang, G. Wu, J. Wu, and J. Zhao. 2010. Forest degradation deepens around and within protected areas in East Asia. *Biological Conservation* 143:1295–1298.

Turner, W., S. Spector, N. Gardiner, M. Fladeland, E. Sterling, and M. Steininger. 2003. Remote sensing for biodiversity science and conservation. *Trends in Ecology and Evolution* 18:306–314.

Wang, L., W. P. Sousab, P. Gong, and G. S. Biging. 2004. Comparison of IKONOS and QuickBird images for mapping mangrove species on the Caribbean coast of Panama. *Remote Sensing of the Environment* 91:432–440.

Whittaker, R. J., K. J. Willis, and R. Field. 2001. Scale and species richness: Towards a general, hierarchical theory of species diversity. *Journal of Biogeography* 28:453–470.

Wright, S. J., G. A. Sanchez-Azofeifa, C. Portillo-Quintero, and D. Davies. 2007. Poverty and corruption compromise tropical forest reserves. *Ecological Applications* 17:1259–1266.

Multiscale Habitat Mapping and Monitoring Using Satellite Data and Advanced Image Analysis Techniques

Stefan Lang
University of Salzburg

Christina Corbane
*Institut national de recherche
en sciences et technologies
pour l'environnement et
l'agriculture (IRSTEA)—UMR
TETIS—Maison de la Télédétection*

Palma Blonda
*Istituto di Studi sui
Sistemi Intelligenti per
l'Automazione (ISSIA)*

Kyle Pipkins
Technische Universität Berlin

Michael Förster
Technische Universität Berlin

Acronyms and Definitions

ALOS	Advanced Land Observing Satellite
AVHRR	Advanced very high resolution radiometer
BIO_SOS	Biodiversity multisource monitoring system [project acronym]
BS	Biodiversity surrogates
CBD	Convention on biological diversity
CHRIS	Compact High Resolution Imaging Spectrometer
CoP	Conference of the parties
DG	Directorate-General
DN	Digital number
DPSIR	Driving forces, pressure, state, impact and response
EC	European Commission
EEA	European Environmental Agency
EMS	Electromagnetic spectrum
EO	Earth observation
EODHaM	Earth observation data for habitat monitoring
ES	Ecosystem services
ESA	European Space Agency
ETC	European Topic Centre
EU	European Union
EUNIS	European Nature Information System
FAO	Food and Agricultural Organization
GEO	Group on Earth Observations
GEOSS	GEO system of systems
GHC	General habitat categories
GMES	Global monitoring for environment and security
HR	High resolution
IPBES	Intergovernmental Panel on Biodiversity and Ecosystem Services
IR	Infrared
LC	Land cover

LCCS	Land Cover Classification System
LiDAR	Light detection and ranging
MAES	Mapping and assessment of ecosystems and their services
MERIS	Medium resolution imaging spectrometer
MIR	Mid-infrared
MODIS	Moderate resolution imaging spectroradiometer
MS.MONINA	Multiscale service for monitoring NATURA 2000 habitats of European community interest [project acronym]
NIR	Near-infrared
nm	Nanometer
NOAA	National Oceanic and Atmospheric Administration
OBIA	Object-based image analysis
PALSAR	Phased Array type L-band Synthetic Aperture Radar
QA4EO	Quality assurance framework for Earth observation
QI	Quality indices
R&D	Research and development
Radar	Radio detection and ranging
SAR	Synthetic aperture radar
SBA	Societal benefit area
SME	Small and medium enterprises
SVM	Support vector machine
SWIR	Shortwave infrared
UAV	Unmanned aerial vehicle
UGV	Unmanned ground vehicles
UN	United Nations
VHR(I)	Very high resolution (image)
VIS	Visible light

19.1 Introduction: The Policy Framework

19.1.1 Monitoring Global Change

"Global change"—a short formula for a multitude of anticipated shifts in societal and environmental domains due to global drivers—calls for spatial monitoring and modeling techniques to better understand the implications and potential dynamics of such changes (Lang et al. 2013a). International initiatives, programs, and visions envisage unified systems based on quality standards for data, products, and services to establish optimized observation capacity to globally monitor land surfaces and oceans, climate, and atmosphere, as well as social systems such as public health, human security, and energy consumption. The global initiative Group on Earth Observations (GEO*) distinguishes nine societal benefit areas of civilian observations systems, among which biodiversity and ecosystems are two. The term Earth observation (EO) comprises all observation

systems that use sensor technologies to capture various kinds of physical parameters. This includes space- or airborne measurement devices (sensors mounted on satellites, aircrafts, or unmanned aerial vehicles), as well as mobile ground devices (e.g., unmanned ground vehicles), and fixed measurement instruments (e.g., ground sensors, buoys, terrestrial laser scanners). All these observation systems taken together form a GEO System of Systems, the GEOSS, which is being implemented over a period of 2005 until 2015 (GEO 2005). More specifically, the GEO biodiversity observation network,[†] for example, evaluates the adequacy of existing biodiversity observation systems to support the Convention on Biological Diversity (CBD) 2020 targets. A list of essential biodiversity variables (Pereira et al. 2013) has been established that should be monitored worldwide in order to follow up the state of biodiversity adequately. Many of these global indicators are relying on remotely sensed imagery.

This GEOSS implementation plan adheres to certain quality criteria, to be assessed by quality indices with respect to data provision, data preprocessing, and the information supplied. According to the quality assurance framework for Earth observation (QA4EO), it needs to be ensured that GEOSS is "implemented in a harmonious and consistent manner throughout all EO communities to the benefit of all stakeholders" (GEO 2010). The QA4EO highlights the role of the user, who should be the "driver for any specific quality requirements" and assess "if any supplied information […] is fit for purpose."

In the context of this chapter, we mainly refer to satellite-borne EO techniques, whose general assets as compared to conventional terrestrial field mapping are summarized in Table 19.1.

The growing need for the civilian use of satellite remote sensing and other EO technologies has born the European program Copernicus,[‡] formerly known as global monitoring for the environment and security (GMES), as a conjoint initiative between the European Commission (EC) and the European Space Agency (ESA). Copernicus is considered the European contribution to GEO fostering the provision of geospatial information and monitoring services in six principal domains, that is, land, water, atmosphere, climate change, emergency response, and human security. It builds on European space infrastructure and the technological capability to turn data into information services. The new Sentinel family of EO satellites, developed by ESA, will provide global coverage with radar and optical data ranging from 10 to 20 m and 60 m "spatial resolution" (in the visible [VIS] and near-infrared [NIR] to shortwave infrared [SWIR] spectral range). Additional data from satellites of the so-called contributing missions will increase both the variety of available data types and the temporal coverage with remotely sensed data. But next to the provision of frequently updated satellite data, we also require the adequate means for an intelligent usage of such data and an efficient analysis of them (Lang 2008). The Copernicus initiative

---

* GEO Secretariat, GEO-Group on Earth observations, www.earthobservations.org/.

[†] Group on Earth Observation (various contributors), GEO BON - GEO Biodiversity Observation Network, http://www.earthobservations.org/geobon.shtml.

[‡] FDC/SpaceTec Partners, Copernicus, www.copernicus.eu/

TABLE 19.1 How General Strengths of Satellite EO-Based Mapping Translates to Habitat Delineation and Characterization

Wide area coverage	• Depending on the spatial resolution, area extents per scene range from a few to several hundreds of square kilometers. • Within the instantaneous view field of a shot, similar atmospheric and illumination conditions apply, so that image statistics are homogenous over a scene and variations within the image can be, by and large, related to changing habitat types or conditions.
Multiscale option	• The technical trade-off between spatial resolution (grain) and area covered (extent) allows for purpose-driven usage. • Habitat mapping implies a multiscale option, assuming fine-scaled observations are required for limited areas ("hot spot") only, and otherwise national or continental investigations cope with lesser detail.
From a distance, no direct contact	• A bird's eye view and the lack of physical contact with the object of interest takes the complex structure of natural habitats, terrain inaccessibility, or disturbance of protected areas.
Multitemporal/ multiseasonal coverage	• A (semi-) permanent installation of space infrastructure enables a repetitive, standardized observation pattern within a given timeframe, whether over several years or seasons. • Repetitive observation under standardized conditions is a crucial prerequisite for monitoring activities prescribed in national and international environmental policies.

has opened new fields of activity to industry (including small and medium enterprises) and research organizations. The financing of so-called core services and GMES initial operations as fundamental information services in all Copernicus domains has led (and will continue to do so) to the stimulation of downstream services in new emerging areas. A key prerequisite for the creation of versatile application domains and related business cases is the provision of the EO data at no cost. The Sentinel program has been designed in such a way that satellite data will be distributed for free and with no limitation for whatsoever (civilian) use.

Biodiversity and habitat monitoring make up such an emerging area. Biodiversity, the variety of life forms, has become a key word for shaping and bundling political will. And biodiversity, if thought of as the information content of life, requires adequate technology to be observable. Satellite EO has started to become a ubiquitous means, to observe the success of policy implementation (Lang et al. 2013a).

The two collaborative projects MS.MONINA* (Multiscale Service for Monitoring Natura 2000 Habitats of Community Interest) and BIO_SOS† (Biodiversity Multi-Source Monitoring

System: from Space to Species), both started 2010, have explored EO data combined with data from ground surveys (Blonda et al. 2012b, Lang et al. 2014). The idea is to set up EO-based (pre-) operational yet economically priced solutions to provide timely information on pressures and impacts, to establish spatial priority for conservation, and to evaluate its effectiveness. MS.MONINA developed advanced data-driven EO-based analysis and modeling tools, specifically tailored to user requirements on all levels of policy implementation. Three (sub-)services were designed, the so-called European Union (EU), state, and site level service, addressing agencies on EU level (e.g., European Topic Centre [ETC] Biodiversity, European Environmental Agency [EEA], and Directorate-General for the environment [DG Env]) providing independent information, national and federal agencies in reporting on the entire territory by utilizing an information layer concept, and local management authorities by advanced mapping methods for status assessment and change maps. BIO_SOS provides cost-effective knowledge-driven EO-based analysis and modeling tools for meeting regulation obligations and for the definition (and effectiveness assessment) of related management strategies and actions. The project developed a preoperational open-source processing system that combines multiseasonal EO satellite data and in situ measurements on the basis of prior expert rules to map land cover (LC) and habitats, their changes, and modifications over time and quantify anthropogenic pressures. Expert rules include prior spectral, spatial, and temporal features characterizing LC classes and habitat classes. The system is cost-effective for mapping large or not accessible areas as in-field reference data are not required for training the system but only for validating the output products (e.g., LC and habitat maps).

In this chapter, we distill the projects' technical outcomes and scientific achievements as reported to the EC and also described in a special issue on *Earth observation for habitat mapping and biodiversity monitoring*, edited by S. Lang and others in 2014. We highlight the great potential of EO data and the achievements of recent technologies but also their challenges and limitations, in support of biodiversity and ecosystem monitoring. In Europe, nature conservation rests upon a strong yet ambitious policy framework with legally binding directives. Also in other parts of the world, the environmental legislation follows ambitious goals that often have to compete with other societal premises such as growth, production, and expansion. Thus, geospatial information products are required at all levels of implementation. With recent advances in EO data availability and the forthcoming of capable analysis tools, we enter a new dimension of satellite-based information services. Recent achievements are showcased and challenges are discussed, using spearheading examples from inside and outside Europe.

19.1.2 Biodiversity and Related Policies

Biodiversity—our "natural capital and life insurance" (European Commission 2011)—is on decline (Isbell 2010, Trochet and Schmeller 2013). This is expected to directly influence the integrity of ecosystem functioning and stability, and thus, ultimately to human well-being (Naeem et al. 2009). In 1992, the United

---

* University of Salzburg, Department for Geoinformatics Z_GIS 2013, Multi-scale service for monitoring Natura 2000 habitats of European Community Interest, www.ms-monina.eu.
† ISSIA CNR 2013, BIO_SOS Homepage, www.biosos.eu.

Global Biodiversity

Natural systems that support economies, lives and livelihoods across the planet
are at risk of rapid degradation and collapse. Natural habitats in most parts of
the world are shrinking, plant and animal populations face constant threat

2010 International Year of Biodiversity

Species under Threat

As of 2009, 36% of 47,677 species are
considered threatend with extinction

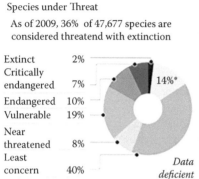

Extinct	2%
Critically endangered	7%
Endangered	10%
Vulnerable	19%
Near threatened	8%
Least concern	40%

14%*

Data deficient

Living Planet Index
Global wild vertebrate species

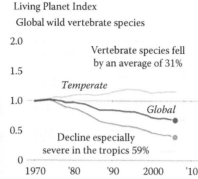

2.0

1.5

1.0

0.5

0

Vertebrate species fell
by an average of 31%

Temperate

Global

Decline especially
severe in the tropics 59%

1970   '80   '90   2000   '10

Protected areas

Nationally desinated in million sq. km

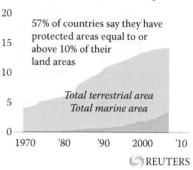

20

15

10

5

0

57% of countries say they have
protected areas equal to or
above 10% of their
land areas

Total terrestrial area
Total marine area

1970   '80   '90   2000   '10

REUTERS

FIGURE 19.1 In the year 2010, the United Nations celebrated the International Year of Biodiversity. In that year, the international community realized that the 1992 target to "halt the loss of biodiversity" has not been reached and more restrictive and better observable measures need to be taken. (From SavingSpecies, Inc., Saving species, www.savingspecies.org.)

Nations joined forces in the CBD, to halt or at least lower the accelerated loss of biodiversity. Next to the challenging nature of this aim as a key global challenge, it remains demanding to monitor and evaluate its success, which requires a concerted, effective use of the latest technology (Lang et al. 2014). As by the end of 2010 (the *International Year of Biodiversity*) the global society became aware that the ambitious goal of "halting biodiversity" has not been reached, the importance of observation techniques became even more important (see Figure 19.1).

The integrity of species and ecosystems is a global phenomenon with continental, regional, and ultimately local implications. This makes biodiversity a "glocalized" phenomenon (Lang et al. 2014). Geographically, this manifests in a hierarchy of scales, from biomes, over (systems of) ecosystems down to communities, populations, and species. Observing and monitoring aspects of biodiversity, at any level and scale, can be approximated by analyzing the composition, variability, and changes of tangible entities (i.e., habitats) and their spatial patterns (Bock et al. 2005). Remote-sensing information complements data obtained through standardized, in situ surveys related to very local aspects of biodiversity, by representing integrated higher level characteristics such as those of ecological neighborhoods (Addicot et al. 1987), defined by the upper (extent, object/scene size) and lower (grain, spatial resolution) limits of data information content and perception (Wiens 1989). The matching of various resolution levels of satellite sensor families (see Section 19.1.2) with the organizational levels of biological systems and organism perception is one aspect—the correspondence with spatial and temporal domains of environmental policies another. Satellite EO has started to become a "democratic tool" to observe what is happening on the different levels of the political framework (Lang et al. 2014).

The EU responded to the recognition that the biodiversity target 2010 would not be met, despite some major successes and the

adoption of a global strategic plan for biodiversity 2011–2020 at the tenth conference of the parties (CoP10) to the CBD, with the *EU biodiversity strategy to 2020*.* The EU biodiversity strategy complements the (general) EU strategy 2020, the EU's growth strategy for the coming decade where six main targets are established: employment, R&D and innovation, climate change and energy, education and poverty, and social exclusion. The strategy's main target is to halt biodiversity loss and the degradation of ecosystem services (ES) in the EU by 2020. To meet this target several subtargets and actions are framed. Projects such as MS.MONINA and BIO_SOS[†] will help to comply with the actions under target 1 "to fully implement the Birds and Habitats Directives." Additionally, the two projects may support action 5 ("mapping and assessment of ecosystems and their services") and action 6a ("set priorities for ecosystem restoration") and 6b ("Development of a Green Infrastructure Strategy 2012") (see Figure 19.2) of the strategy's target 2.

19.1.3 Mapping the State of Ecosystems

A key action of the EU biodiversity strategy is the *Mapping and Assessment of Ecosystems and their Services* in Europe (Maes et al. 2013). Action 5, which aims to improve knowledge on ecosystems and their services, entails EU member states (MS) to map and assess the state of ecosystems and their services in their national territory by 2014, to quantify the economic value of such services,

---

* Our life insurance, our natural capital: an EU biodiversity strategy to 2020—COM(2011) 244 final—http://ec.europa.eu/environment/nature/biodiversity/comm2006/pdf/2020/1_EN_ACT_part1_v7%5B1%5D.pdf.

† See White Paper "Copernicus Biodiversity Monitoring Services" available at http://www.biosos.eu/publ/White_Paper_Biodiversity_Monitoring_BIOSOS_MSMONINA.pdf.

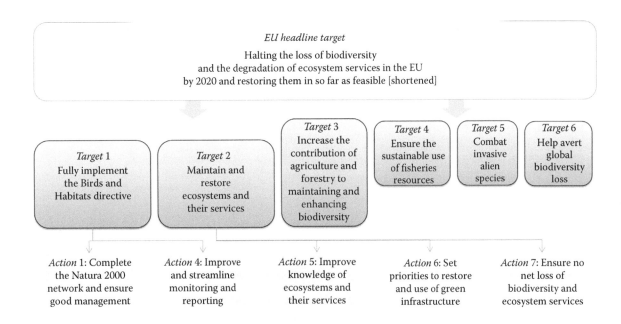

FIGURE 19.2   General and specific objectives of the EU biodiversity strategy 2020 as a strategic policy framework to MS.MONINA and BIO_SOS. (From MS.MONINA user requirement dossier, modified.)

and to promote the integration of these values into accounting and reporting systems at EU and national level by 2020.

The supply of ecosystem services (ES) relates to functions and characteristics of biodiversity aspects including genes, species, and habitats. Next to the collection of biological data such as functional traits of plants, the mapping focuses mainly on ecosystem structure and habitat data. Ideally, all ecosystem types that act as functional units in delivering services should be mapped and evaluated separately. Due to the complexity of the matter, there is no common approach as yet to directly map the "total ecosystem service" for a certain area as cumulative effect of the (sub-)services. Strategies from *vulnerability* or *sensitivity* mapping that likewise integrate a larger set of indicators using the geon approach (cf. Lang et al., "Remotely Sensed Data Characterization, Classification, and Accuracies," Chapter 22) exist and could be adapted. The main challenges with mapping ES are the lack of coverage and resolution in the available habitat maps, a general shortage in time and efforts required, and most importantly the need for coherence and standardization.

Thus, in the two projects which are considered *downstream* to the Copernicus land monitoring service, we followed the EU biodiversity 2010 baseline approach for ecosystem mapping. This implies that LC classes as monitored in Copernicus are translated into habitats and ecosystem types, in the most meaningful way possible to represent broadscale ecosystems, and combined with ecosystem-relevant information. This translation is based on detailed expert analysis of relationships between LC classes (as derived from the Food and Agricultural Organization [FAO] Land Cover Classification System [LCCS] taxonomy; see Section 19.3.1) and habitat classification systems

(i.e., European Nature Information System [EUNIS]) to ensure consistency between these approaches (Tomaselli et al. 2013). Harmonization of the assessment activities of the EU member states is an important ongoing activity but must leave some degrees of freedom to reflect the specific ecological, social, and historical context of each MS. Accordingly, MS are encouraged to use a more detailed habitat typology if available, with the only restriction that the more detailed classes are linkable to the EU-level typology. Habitat maps produced at the local scales usually include detailed data on the associated biodiversity that enable links to ES. Mapping programs have incorporated qualitative descriptors of the mapped habitats under various names (ecosystem state, ecosystem health, ecological integrity, naturalness, vegetation condition, degradation level, etc.). Action 5 of the biodiversity strategy emphasizes the "need to map [...] the state of ecosystems and not simply to map the ecosystems" (see Figures 19.3 and 19.4). In order for maps of habitat conservation status to become a useful input to policies at transnational level, the classification schemes used to evaluate the degradation levels, the habitat categories, and the methods used to assess them should be harmonized (Ichter et al. 2014).

19.1.4 EU Habitats Directive

The EU Habitats Directive (short HabDir, Council Directive 92/43/EEC), essential part of the European endeavor toward the CBD (Trochet and Schmeller 2013), is an ambitious legal instrument to safeguard biodiversity and set aside a network of protected areas (Gruber et al. 2012), called NATURA 2000, currently in completion (Evans 2012). HabDir entails

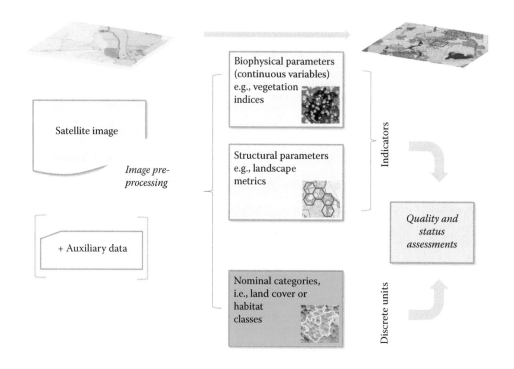

FIGURE 19.3 From imagery via habitat categories to habitat quality assessment.

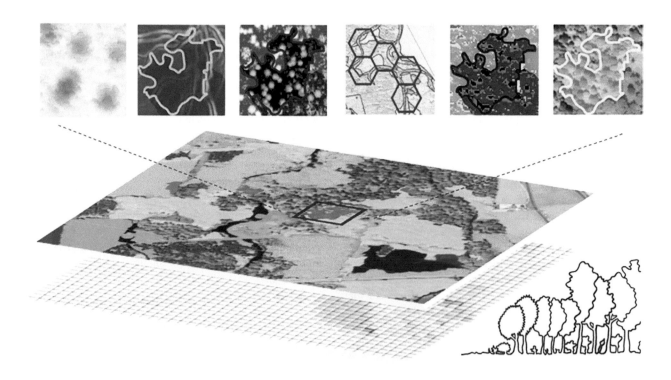

FIGURE 19.4 Habitats are the physical and mappable expression of ecosystems. Satellite Earth observation enables to map and monitor a variety of aspects on habitat distribution, quality, and change in different spatial and temporal scales. (From Lang, S. et al., *Int. J. Appl. Earth Obs. Geoinf.*, (Special Issue), Elsevier, 2014.)

standardized and frequent (every 6 years) standardized monitoring and frequent … and reporting activities with specific responsibilities on all political levels of implementation: (1) the local management authorities for the monitoring of individual protected sites, (2) the EU MS for reporting on the status of the network of protected sites and habitats distribution over the entire territory, and (3) the EU for aggregating this information and the reporting toward the CBD (Figure 19.7; Section 19.3). Updated geospatial information products are required at all three levels, not only to upscale lower level information but also to provide additional independent information on each level. In this framework, EO data and related techniques offer objective yet economically priced solution to (1) provide timely information on pressures and impacts, (2) establish spatial priority for conservation, and (3) collect long-term multiscale baseline information for evaluating the effectiveness of conservation strategies.

19.2 Satellite Sensor Capabilities

Earlier, we have claimed that habitats, as natural systemic areal features, are mappable through satellite remote sensing. With recent advances in EO imagery, we enter a new dimension in habitat and biodiversity mapping. Still, challenges are ahead in terms of data integration, advanced preprocessing and calibration, automated information extraction, ground verification, and product validation, as well as semantic interoperability and exchange (Lang et al. 2014). In the following sections, we take a closer look at the specific requirements habitat mapping poses toward satellite imagery; in other words, what can be achieved in terms of habitat categorization and change qualification. In this section, we shall first look at the capabilities space-based observation technology has developed in recent years to support such demand. Note that the *demand view* and the *supply view* are not distinct realms; in fact, there is interaction and interdependencies among them, (ideally) a mutually fertilizing process that research and development activities, as the projects mentioned earlier, try to catalyze and stimulate.

The key technical characteristics of satellite technology playing a considerable role in habitat mapping are the following:

1. *Spatial resolution*: Habitats are areal features with a certain extent ranging from a few square meters to hundreds of square kilometers. In order to find a commensurate scale of observation, a range of sensor families are at disposal, whereby very-high-resolution (VHR) and high-resolution (HR) sensors play the most important roles, while lower-resolution sensors are more used for differentiating large ecosystems, biomes, etc., on continental scale. Spatial resolution is also a key factor in characterizing within-habitat conditions, such as structure or composition, whereby the resolution needs to be significantly smaller than the extent of the habitat of concern. The number of quantization levels (radiometric resolution) matches spatial resolution. There is no need for having a lot of neighboring pixels with all the same digital number.

There are, for example, 256 levels (8 bits) for Landsat 8 (30/15 m resolution) and 2048 (11 bits) for WorldView-2 (2/0.5 m resolution).

2. *Spectral resolution*: The main feature to be looked at when categorizing and delineating habitats is plant composition. The latter can only be characterized in a satisfying manner, if spectral resolution suffices. The differentiation of vegetation types and plant species require specific wavelengths, such as the so-called red edge, or, by and large, the infrared bands. Multi- and hyperspectral data provide the raster of wavelength that can be used for plant species discrimination. When spectral resolution is not adequate for vegetation discrimination, multitemporal (seasonal) data can be used.

3. *Revisiting time*: The revisiting time of nonflexible sensor units is bound to the orbiting time of the satellite. This is about 16 days for Landsat. Rotating and side-looking sensors increase the revisiting time. Another factor of repeatability can be gained through constellations. These are several, identical satellites that orbit in a given sequence, so to cover each spot on Earth in multiples as compared to a standard revisiting time. Multitemporal, in particular multiseasonal, observations are critical to many, if not most, habitat assessments.

According to Corbane et al. (2014), an obstacle is that, despite the tremendous progress in the applications of remote sensing to habitat mapping, many data types discussed here (hyperspectral, Light detection and ranging [LiDAR], and Radar) might currently be beyond the practical capabilities of the community of practice (Rannow et al. 2014). Furthermore, the costs for imagery and other geospatial data products are still fairly high, while with an overall trend to decline due to market competition. The trend is supported by recent release of free very-high-resolution imagery (VHRI) (e.g., USGS released Orbview-3 data in January 2012) or the upcoming ESA Sentinel series of satellites (Berger et al. 2012). The availability of free geospatial data (e.g., the open street map initiative and open-source image processing and Geographic Information System (GIS) software and tools [e.g., Orfeo Toolbox, Quantum GIS, Geographic Resources Analysis Support System, R]) is also contributing to the democratization of remote sensing and to the decline in the costs of the image processing packages. Still, a challenge for ecologists and conservation biologists is the technical expertise required to handle imagery and other data products (Turner et al. 2003) and thus not recognizing the full potential provided by such datasets and related techniques.

19.2.1 Spatial Resolution: What Detail Can Be Mapped?

Satellite-based EO systems are categorized according to their spatial resolution, while the grouping schema is relative to the highest technical resolution and thereby subject to change. Over the last years, the term "very high resolution" has been used for images with a resolution (i.e., pixel size) around or

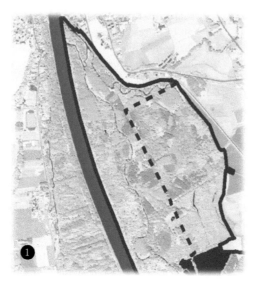

FIGURE 19.5   How WorldView-2 data (left, red outlines indicate boundaries of designated protected areas) translate into habitat classes in a semiautomated image classification process (here EUNIS-3 categories).

smaller than 1 m. We speak of HR images at resolution levels of up to 5 m (see Figure 19.5 and Table 19.2). The highest resolution operationally available is around 0.5 m increasing to 0.3 m with the launch of latest generation VHR satellites, such as WorldView-3.

With habitats being areal features with a certain extent, there is a wide range of scales, from a few square meters, like specific springs or lichen patches, to several square kilometers in the case of the savannah. While the average size of habitats clearly depends on the hierarchical level of habitat categorization (broad habitat categories vs. more specific ones; see below), there are great differences within one scale domain as well. The extent of habitats considered internally homogenous (i.e., residing on the same hierarchical level of organization from an ecological point of view) varies depending on the physical conditions and the species living there. Therefore, it is important to discriminate individual plants such as single trees or distinct features such as shrubs in a specific habitat composition. Table 19.2 provides an overview on the suitability of different sensor groups for distinguishing within broad, physiognomic habitat types.

With highest resolution available, pan-sharpening routines are frequently used to optimize both spatial and spectral resolutions. Still, the fidelity of the pixel information can be too high as well. Aiming, for example, at the extraction of sunlit tree crowns, a resolution of 1–2 m may just be the right resolution to tackle this (Strasser and Lang 2014b). In other words, sensor resolution needs to be commensurate to the observed features, from both directions so to say, not too coarse and not too fine. LiDAR systems, which are described in more detail in the following text, are capable to represent the physical attributes of vegetation canopy structure that are highly correlated with the basic plant community measurements of interest to ecologists (Mücher et al. 2014).

19.2.2 Spectral Resolution: Plant and Plant Feature Discrimination

19.2.2.1 Sun-Source Systems: Optical Sensors

Plant species respond characteristically to light emitted by the sun or an artificial energy source, with specific reflection behavior in the *electromagnetic spectrum* (EMS). Ideally, remotely sensed data of adequate spectral and spatial resolution can be used to distinguish different species, but to identify the appropriate sensor and the appropriate spectral bands can be challenging. Sensors can be grouped into passive and active sensors, depending on the source of energy involved. Passive sensors record the reflectance of sunlight on surfaces, while the spectral resolution corresponds to the number of bands that a sensor is able to acquire from a distinct part of the EMS. Panchromatic sensors, as the name indicates, cover a broad range of the EMS, usually including the VIS and the NIR ranges. Multispectral sensors are sensitive to certain, well-defined portions of the EMS, which are categorized according to their relative and absolute position, such as VIS light (400–700 nm), NIR (750–900 nm), mid-infrared (MIR, 1.55–1.75 μm).

Multispectral sensors have been used since several decades collecting data fairly broad wavebands (typically 4 to recently 8). Multispectral data recorded by spaceborne sensors (e.g., Landsat 7, ASTER, IKONOS, SPOT-5, WorldView-2, QuickBird 2, RapidEye) are useful for LC assessments from local to regional scale (depending on the spatial resolution). Such data are being used to assess habitat conditions for monitoring purposes (Förster et al. 2008, Franke et al. 2012, Spanhove et al. 2012). Advanced studies investigate spectral characteristics of specific sensors for such tasks and address the suitability of multispectral data for an assessment of detailed floristic variation used (Nagendra and Rocchini 2008) in multispectral sensors. This helps assess the floristic composition

TABLE 19.2 Sensor's Suitability for the Characterization of Natural Habitats Based on Corbane et al. (2014) and Ichter et al. (2014).

	Low Spatial Resolution and High Temporal Resolution[a]	Medium to High Spatial/Temporal Resolution[b]	Very High Spatial Resolution[c]	Hyperspectral[d]	Laser Scanning (LiDAR)[e]	Active Microwave Sensors (e.g., SAR)
Forests	± Deciduous/coniferous/ mixed forest, evergreen/deciduous, dense/fragmented	+ Broad types, dominant species using multitemporal imagery	++ Tree species classification, differentiation of structure and age classes, multitemporal.	++	± Assessment of forest parameters (stand density, height, crown width, crown length) species distributions	± Often complementary to the information provided by multispectral imaging
Grasslands	−	++ (multiseasonal imagery): Distinction between marshy grasslands (*Molinia*- or *Juncus*-dominated), unimproved (*Festuca*-dominated), semi-improved and improved	+ (multiseasonal imagery): Grassland types with different levels of agricultural improvement, levels of mowing intensity.	++ Detection of floristic gradients, determination of homogenous cover types	±	+ Distinction between natural grasslands and improved pastures, mowing intensity via swath detection
Heathlands	−	++ (multiseasonal imagery): Distinction between heath types (e.g., *Genista*, *Erica*), four heath types, including ancillary data	++ Seasonal phenological variation can discriminate evergreen *Calluna vulgaris* from deciduous *Vaccinium myrtillus*.	++ Distinction between dry and wet heathland, heathland types (*Calluna, Molinia Deschampsia, Erica*, etc.) and heather age classes	++ Only if types differ in structure or density	−
Wetlands[f]	−	+ Seasonal imagery: mapping extent of seasonally submerged wetlands and some vegetation species, freshwater swamp vegetation, functional wetland types	+ Detection of riparian vegetation species, shallow, submerged vegetation.	++ Distinction between aquatic macrophyte species (*Typha, Phragmites, Scirpus*)	± Surface and terrain models used to better understand characteristics of wetland vegetation species	−

Modified and various remote-sensing techniques (sensors and resolution) are compared for distinguishing between broad physiognomic habitat types. The degree of sensor suitability is indicated as follows:

− = unsuitable, −/+ = partly suitable, + = suitable, ++ = recommended. See detailed references in Corbane et al. (2014).

[a] e.g., NOAA-AVHRR, MODIS.
[b] e.g., Landsat, IRS, SPOT
[c] e.g., IKONOS, QuickBird, GeoEye, WorldView-2, Pléiades.
[d] For example, HyMap, CASI, Hyperion.
[e] To be combined with multispectral/hyperspectral imagery.
[f] Wetlands are not a physiognomic type per se but are various physiognomic types that have adapted to the continuous or temporary presence of water.

within a certain natural habitat (mainly grassland and wet heath and floodplain meadows). Tree species differentiation has been accomplished using 8-band WorldView-2 data for forest management in general (Immitzer et al. 2012) and riparian forest assessment in particular (Strasser and Lang 2014b). Table 19.3 shows a palette of WorldView-2 scenes as used for fine-scaled habitat delineation in MS.MONINA. The recently launched WorldView-3 sensor will provide data with 30 cm maximum spatial resolution.

Narrow spectral bands record in the range of 400–2500 nm. Some sensors cover only parts of this range (e.g., CHRIS/PROBA focuses on VIS/NIR). Due to the large number of wavebands, digital image processing can discriminate biochemical and structural properties of vegetation (Underwood et al. 2003). They demonstrate the potential of hyperspectral data to extract information regarding plant properties (such as leaf pigment, water content, and chemical composition), and thus discriminating tree species in landscapes, and identifying different species. Recent applications of Hyperion hyperspectral imagery include forest biodiversity (Peng et al. 2003), grasslands (Psomas et al. 2011), and invasive species monitoring (Walsh et al. 2008).

19.2.2.2 Active Systems: Radar and LiDAR

Active sensors send an electromagnetic signal and record the travel time of the sent signal and its reflection by a given surface. Active sensors are differentiated by wavelength into microwave (Radar) and laser scanning (LiDAR) systems. Active systems are increasingly used for vegetation mapping, with a number of new satellites (e.g., TerraSAR-X). Gillespie (2005) discusses opportunities for landscape monitoring at finer spatial resolution. The returned signal supplies information about the height and structure of vegetation, especially of woody vegetation, with these relating to forest condition and disturbance regimes (Huang et al. 2013). In particular, X-band radar backscattering has been recommended to differentiate plant species on the basis of canopy architecture (Bouman and van Kasteren 1990). Schuster et al. (2011) proved that distinguishing different grassland swath types for NATURA 2000 monitoring is possible with the TerraSAR-X sensor. The ALOS PALSAR and RADARSAT-2 SAR have shown great potential for mapping wildlife habitat, particularly when combined with optical remote sensing through data fusion (Wang et al. 2009). Moreover, there are recent advances to monitor different shrubland, grassland, and forest habitats with COSMO SkyMed (Ali et al. 2013). While still as a research application, the ability to penetrate the canopies makes microwave instruments a potential tool for measuring biomass and determining vegetation structure.

LiDAR provides highly accurate information on the 3D vegetation structure, derived from pulse characteristics, over a limited area (Puech et al. 2012). For example, data from the airborne laser vegetation imaging sensor enable the mapping of subcanopy topography and canopy heights to within 1 m.

More general, LiDAR is used for the extraction of information on forest structure (vertical information), for example, canopy and tree height, biomass, and volume. According to Turner et al. (2003), the recording of numerous LiDAR return signals (pulses) enables to estimate vegetation density at different heights throughout the canopy and enables 3D profiles of vegetation structure. Besides airborne LiDAR with the limitations of large data volumes, footprint size, and high costs, spaceborne laser technology has been launched by Ice, Cloud, and Land Elevation Satellite/Geoscience Laser Altimeter System, the first laser-ranging instrument for continuous global observations. While LiDAR provides structural attributes of vegetation, little can be derived on the actual species composition suggesting that laser scanning is more a complementary technology (Mücher et al. 2014). In general, optical and LiDAR data acquired simultaneously increase the differentiation of vegetation species. It has also been demonstrated that species distribution models can be improved through airborne LiDAR quantifying vegetation structure within a landscape (Goetz et al. 2007).

19.2.3 Revisiting Time: Phenology

A major asset, and often critical, in using satellite remote sensing for habitat monitoring is that data can be acquired on a regular and repetitive basis, therefore allowing consistent comparisons between image scenes. The frequency of observation by optical spaceborne sensors ranges from several times daily (for coarse spatial resolution sensors, e.g., NOAA-AVHRR, MODIS, and MERIS) to every 16–18 days (e.g., Landsat) although cloud cover, haze, and smoke often limit the number of usable scenes. The new generation of ESA Copernicus satellites with a revisit time of 5 days is expected to be launched around 2015. With a 5-day revisit capacity, the two Sentinel-2 satellites will acquire intra-annual data, thereby allowing the temporal variations in reflectance to be exploited for mapping and monitoring natural habitats. The changes in the seasonality (i.e., phenology) of plants are significant for the differentiation into classes. So far, this information has been widely neglected due to limited access of spatial HR imagery with a high temporal domain (Förster et al. 2012). To include an increasing number of intra-annual images increases the classification accuracy until a certain threshold is reached (Schuster et al. 2015). Images taken over two phenological stages help in discriminating species and habitat classification (Lucas et al. 2011), although some studies conclude that monotemporal data suffice when the SWIR is included (Feilhauer et al. 2013). Moreover, it is possible to analyze the optimal season for acquiring a dataset (Schmidt et al. 2014). Multiannual coverage also helps assessing changes in habitat (such as loss, degradation, and fragmentation) through change detection approaches (see Section 19.4). Monitoring experts appreciate this, because it directs field work on these areas, possibly yielding a significant increase in cost efficiency (Vanden Borre et al. 2011).

TABLE 19.3 Details (Location, Country, Date, Band Combination) and Subsets in Equal Size (2.5 × 2.2 km) and Scale (Original: 1:10,000) of WorldView-2 Satellite Scenes Used for Test and Service Cases in MS.MONINA

Danube riparian forest, National park, Austria [7-6-5]

Doeberitz heathland/Germany [7-6-3]

Salzach riparian forest, Austria [7-6-3]

Lagoons of *Palavas*/France [7-6-3]

Axios delta, Greece [7-6-3]

Rieserferner nature park/Italy [7-6-3]

(Continued)

TABLE 19.3 (Continued) Details (Location, Country, Date, Band Combination) and Subsets in Equal Size (2.5 × 2.2 km) and Scale (Original: 1:10,000) of WorldView-2 Satellite Scenes Used for Test and Service Cases in MS.MONINA

Murnau mire, Germany [7-3-2]

Kalmthoutse heathland, Belgium [7-3-4]

Bierbza National park, Poland [7-6-2]

Eider Treene Sorge lowlands, Germany [7-5-3]

Larzac foothills, France [7-5-3]

Sierra Nevada National park, Spain [7-5-3]

19.2.4 Advanced Image Analysis Techniques

The advancement of imaging technology is a crucial but not the only *ingredient* to cope with the challenges of habitat mapping. As important as high-quality, high-suitable imagery is the utilization of appropriate image analysis techniques (Lang et al. 2014). Figure 19.6 illustrates the importance of analysis steps that follow the actual provision of imagery. Various information types can be extracted from imagery, ranging from biophysical parameters including vegetation indices (Adamczyk and Osberger 2014) and structural parameters (Mairota et al. 2014) to ultimately nominal LC, or more specifically, habitat categories (Adamo et al. 2014) (see Section 19.3 and Figure 19.3). Beyond that, existing or image-derived habitat delineations can be combined with the parameters in order to assess habitat quality, for example, Riedler et al. (2014) (see Section 19.4 and Figure 19.3). Kuenzer et al. (2014) provide

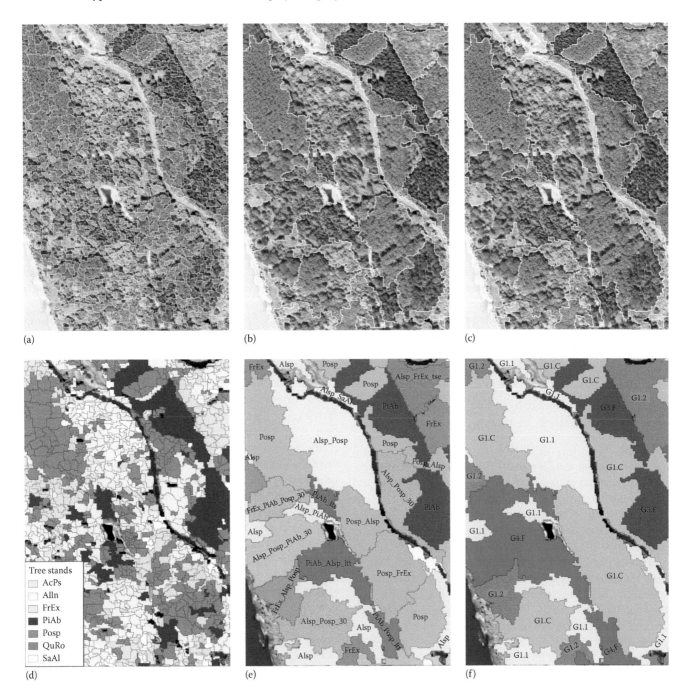

FIGURE 19.6 Hierarchical representation of riparian forest habitats, achieved by multilevel segmentation and class modeling. (a) through (c): Multiscale object delineation (fine to coarse) and (d) through (f): Corresponding classification (tree stands, forest stands, EUNIS classes.) (From Strasser, T. and Lang, S., *Int. J. Appl. Earth Obs. Geoinf.*, 2014b.)

a detailed overview of existing sensor types and related vegetation indices being useful for biodiversity mapping.

Reproducibility, objectivity, transferability, and the increased possibility for quantification have been reported as the main advantages of mapping approaches based on EO data. Semiautomated classification methodologies for EO data provide a more objective, that is, reproducible and transparent outcome as compared to visual interpretation (Lang and Langanke 2006). Over the last years, great advantages have been reported in the use of remote-sensing technology for the mapping and the assessment of habitats in Europe: for an overview, see Vanden Borre et al. (2011). This likewise applies to different broad habitat types (forests, grasslands, wetlands, etc.) and different scales of observations as fine as subhabitat level (Lucas et al. 2011, Strasser and Lang 2014b). Object models can be stored and explicitly called for semiautomated mapping routines (Lucas et al. 2014) or advanced habitat or biotope class models (Tiede et al. 2010).

One strategy to tackle this variability is the use of object-based image analysis (OBIA) (Lang 2008). This is typically done through the combination of spectral behavior and spatial variability, either from the details available through high-spatial-resolution imagery (Johansen et al. 2007, Strasser and Lang 2014b) or through the inclusion of data from active sensors such as LiDAR (Mücher et al. 2014). Object-based class modeling allows for mapping complex, hierarchical habitat systems, such as forest habitats. Forest composition including intermixture of nonnative tree species was modeled in a six-level hierarchical representation in a riparian seminatural forest by Strasser and Lang (2014b). VHRI from WorldView-2 provided the required spatial and spectral details for a multiscale image segmentation and rule-based composition. An image object hierarchy was established to delineate forest stands, stands of homogenous tree species and single trees represented by sunlit tree crowns.

19.3 EO-Based Biodiversity and Habitat Mapping

According to Turner et al. (2003), there are two general approaches to remotely *sensed* biodiversity: (1) direct mapping of individual organisms, species assemblages, or ecological communities using airborne or satellite sensors and (2) indirect sensing of biodiversity-related aspects using environmental parameters as proxies. Many species are confined in their distribution to specific habitats such as woodland, grassland, or sea grass beds that can be directly identified with remote-sensing data. Habitats (see Figure 19.3) as the spatial expression of ecosystems do have a certain extent to be mapped and observed; they function as living space for specific species (both animals and plants) and bear a certain constancy in the 4D space–time physical world.

19.3.1 Land Cover, Habitats, and Indicators

First, we would like to distinguish between LC and habitats, two concepts that frequently cause confusion in the remote-sensing literature (Lang et al. 2013a). According to

EUROSTAT,* LC "corresponds to a physical description of space, the observed (bio) physical cover of the earth's surface. It is that which overlays or currently covers the ground." LC classes representing biophysical categories, such as grassland, woodland, and water bodies, are usually derived from multispectral remote-sensing data by multivariate clustering methods in the feature space. LC data are used at different scales (local, regional, and global) as input variables for biosphere–atmosphere models and terrestrial ecosystem models and respective change assessments as well as proxies of biodiversity distribution (Grillo and Venora 2011). The FAO-LCCS (Di Gregorio and Jansen 2005), as universally applicable classification system, enables a comparison of LC classes regardless of data source, economic sector, or country.

A habitat instead is a "three-dimensional spatial entity that comprises at least one interface between air, water and ground spaces. It includes both the physical environment and the communities of plants and animals that occupy it. It is a fractal entity in that its definition depends on the scale at which it is considered" (Blondel 1979). Habitats are often distinguished into two (or more) stages of naturalness (hemeroby) according to the level of human alteration: "Natural habitats" are considered as the land and water areas where the ecosystem's biological communities are formed largely by native plant and animal species and human activity has not essentially modified the area's primary ecological functions (EEA). "Seminatural" habitats are considered as managed or altered by humans but still "natural" in terms of species diversity and species interrelation complexity (i.e., diversity).

In general, a perfect correspondence of conventional biotope types and spectrally derived vegetation cover is limited (yet from today's point of view, not impossible), due to the practice of manually delineating biotope types from aerial photos and field surveys (Weiers et al. 2004). The EUNIS[†] habitat classification scheme is a hierarchical scheme with six hierarchical levels (Davies et al. 2004). Alternative classification systems such as the general habitat categories (GHC) (Bunce et al. 2008) or the terrestrial ecosystem mapping system (Johansen et al. 2007), using vegetation attributes such as height and leaf phenology, have been proposed in order to more successfully employ EO data in the classification of habitats. A comprehensive mapping system starting from LC classes expressed in FAO-LCCS taxonomy, translated to GHC (Tomaselli et al. 2013), and finally addressing habitat classes according to HabDir has been proposed by Lucas et al. (2014). Such systems require robust and reliable remote sensing–based methods from the start. In this context, Corbane et al. (2014) reviewed the ability of remote sensing to physiognomically distinguish between habitat types at different scales. They report about advances in the use of remote-sensing technology for the mapping and the assessment of habitats in Europe (Vanden Borre et al. 2011). This applies to different broad habitat

---

* European Commission, Europa–eurostat– RAMON, http://ec.europa.eu/eurostat/ramon/nomenclatures.
† European Environment Agency, EUNIS–Welcome to EUNIS Database, http://eunis.eea.europa.eu/.

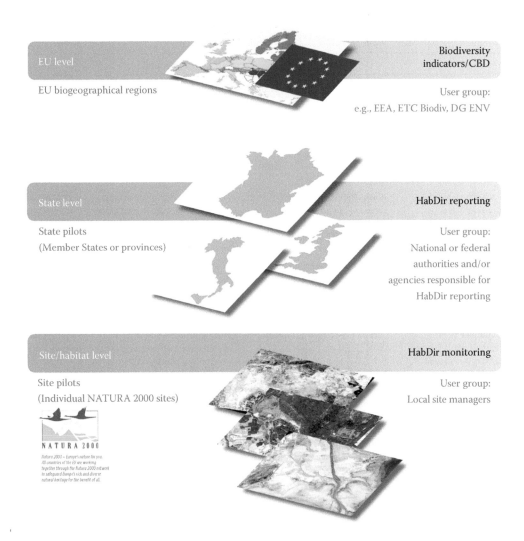

FIGURE 19.7   Three-level information service concept of MS.MONINA. (From Lang, S. et al., Monitoring Natura 2000 habitats at local, regional and European scale, in: Ourevitch, S. (ed.), *Discover What GMES Can Do for European Regions and Cities*. SpaceTec Partners, Brussels, Belgium, 2013b, pp. 58–65.)

types (forests, grasslands, wetlands, etc.) and different scales of observations as fine as subhabitat level. Mapping of broad habitats types using remote sensing is a common practice from the perspective of LC mapping and is generally done at a relatively coarse scale of analysis (Wulder et al. 2004).

19.3.2 Distinguishing Between and Within Broad Habitat Categories

Mapping broad habitats types using remote sensing converges with LC mapping done at a relatively coarse scale of analysis (Wulder et al. 2004). Global LC mapping has been accomplished using the MODIS satellite, at 500 m resolution (Friedl et al. 2010), while country and regional level LC classifications have been accomplished using medium-resolution sensors such as Landsat or SPOT (Fuller et al. 1994, Tiede et al. 2010). More detailed LC boundaries can be obtained using a higher spatial

resolution, multitemporal coverage (Förster et al. 2010) or by including ancillary data (Tiede et al. 2010) or active sensors. In the following, we will discuss how vegetation categories can be distinguished within several broad physiognomic types: forest, grassland, heathland, and wetland (Corbane et al. 2014). Generally, satellite-based habitat mapping can be supported, in terms of plausibility and reliability, through advanced GIS modeling techniques to derive probabilities for the presence of habitats in different biogeographical regions (Förster et al. 2005) and potential habitat ranges under specific assumptions or even changing conditions. The advantages of a high spectral, temporal, or spatial resolution as well as active sensors are summarized and referenced in Table 19.4.

19.3.2.1 Forest Habitats

Low-spatial-resolution data allow rough differentiation of the main forest cover types (deciduous, coniferous, mixed) (Corbane

TABLE 19.4　Influence of Increasing Resolution and the Utilization of Active Sensors on the Possible Detection and Differentiation within Different Habitat Categories

Habitat Type	Influence of Increasing Resolution			Use of Active Sensors
	Spatial	Spectral	Temporal	
Temperate forest	Allows for object-based classification of tree crowns (Immitzer et al. 2012)	Reduces spectral overlap of tree species with similar spectral characteristics (Dalponte et al. 2012).	Enables phenological information (leaf unfolding, coloring, leaf fall) (Wolter et al. 1995).	Improve accuracy via incorporation of canopy height, canopy architecture, and forest structure (Ghosh et al. 2014)
Tropical forest	Reduce the number of mixed pixels between tree species (Nagendra and Rocchini 2008)	Allows for species identification using unique biochemical signatures (Asner et al. 2008b).	Allows an spectral endmember analysis and a detection of tree types (Somers and Asner 2014).	Incorporate forest structural properties to mask nontree gaps (Asner et al. 2008a)
Riparian forest	Allows for object-based classification of tree types (Strasser and Lang 2014a, Suchenwirth et al. 2012)	Allows the estimation of different levels of biomass and subsequent distinction of different types of riparian forest (Filippi et al. 2014).	Possible detection of changes of crown extend and derived health status (Gaertner et al. 2014).	Improve accuracy by means of the structural properties as additional information (Akay et al. 2012)
Grasslands	Allows for object-based classification of homogenous grassland patches (Corbane et al. 2013)	Allows for the detection of floristic gradients (Schmidtlein and Sassin 2004).	Increased accuracy if images are timed to differentiate warm/cool season grasses (Price et al. 2002).	Incorporate information on grassland management practices, such as mowing intensity (Schuster et al. 2011)
Heathland	Allows for the use of object shape complexity in differentiating successional stages (Mac Arthur and Malthus 2008)	Allows for the discrimination of heather age classes (Thoonen et al. 2013).	Multiseasonal imagery allows for distinguishing evergreen and deciduous heath species (Mac Arthur and Malthus 2008).	Can be used to separate shrubs from trees and grassland (Hellesen and Matikainen 2013)
Coastal wetlands	Reduced spectral mixing in heterogeneous species patches (Belluco et al. 2006)	Accuracy is marginally increased, feature selection is necessary (Belluco et al. 2006); spectral libraries may improve classification in properly calibrated images (Schmidt and Skidmore 2003).	Allows for the incorporation of seasonal differences in vegetation communities (Gilmore et al. 2008).	Can be used to separate vegetation height of plant communities (Prisloe et al. 2006)
Inland wetlands	Allows for the distinguishing of small vegetation patches (Everitt et al. 2004)	Hyperspectral imagery has consistently higher accuracy than multispectral (Jollineau and Howarth 2008).	Wetland types may be differentiated by seasonal water regimes and differences in growing season (Davranche et al. 2010).	Can be used to separate vegetation structure to the genus level (Zlinszky et al. 2012)

et al. 2014). The number of differentiated forest classes can be improved when ancillary data, such as terrain (Woodcock et al. 1994), additional geodata (Förster and Kleinschmit 2014), or time series, are used. More detailed analyses can be performed using high-spatial-resolution sensors including image texture analysis, which is indicative for tree species and age classes at the canopy level (Johansen et al. 2007, Immitzer et al. 2012). The use of a time series with high-spatial-resolution data helps distinguish between individual trees, through the use of phenological characteristics such as leaf development and senescence (Key et al. 2001).

The use of hyperspectral imagery allows for an even greater level of detail (Corbane et al. 2014), enabling the distinction of tree types based on reflectance in response to pigment, nutrient, and structural differences between species (Asner et al. 2008a). Still, within-class variability of VHRI (through features such as branches, shadow, and undergrowth) and the between-class spectral mixing of low-resolution imagery (Nagendra and Rocchini 2008) need to be taken into account. Spectral unmixing is possible for lower-resolution sensors but is limited by the

respective sensitivity of the sensor. Forest habitat categories that differ by their understory vegetation, such as *Stellario-Carpinetum* and *Galio-Carpinetum*, which are both oak-hornbeam forest types, can be approached by using additional geodata, such as soil data.

19.3.2.2 Grassland Habitats

In contrast to forests, grassland species are detectable primarily as assemblages and may occur in complex mixtures within habitats (Corbane et al. 2014). Thus, direct remote-sensing approaches are generally limited to the detection of relatively homogenous grassland habitat types, while indirect approaches were found to be successful, such as those that use environmental gradients (Fuller et al. 1994) or usage intensity for mowed seminatural grasslands (Schuster et al. 2011, Buck et al. 2014) (see Figure 19.8). Moderate spatial resolution sensors such as Landsat Thematic Mapper (TM) and Landsat-7 Enhanced Thematic Mapper (ETM+) were used by Lucas et al. (2007) for the classification of grasslands

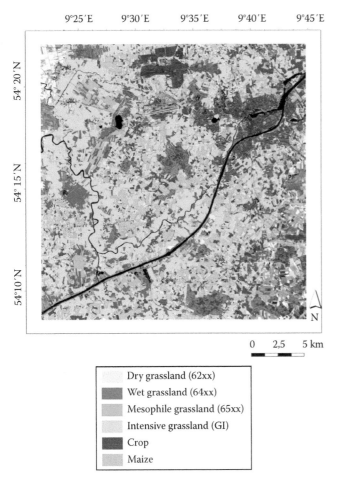

FIGURE 19.8 Grassland classification using SVM classifier of information layers and RapidEye imagery. (From Buck, O. et al., *Int. J. Appl. Earth Obs. Geoinf.*, 2014. Modified.)

allowing only for a broad level of class distinction with respect to grassland improvement levels. Higher-spatial-resolution imagery, using an object-based approach with ancillary data such as elevation and soil type, has proven successful in differentiating between a few dominant grassland species (Laliberte et al. 2007), although other grassland habitats are more difficult to distinguish (Corbane et al. 2013). For relatively homogeneous grasslands, the use of a spectral-temporal library (instead of training areas) with multitemporal HR data has been shown to accurately differentiate grassland types; still, object-based classification methods have been shown to perform better for more heterogeneous grassland types (Förster et al. 2012).

Hyperspectral imagery has been applied to determine floristic gradients, proven to be more useful in habitat identification than single species classifications (Schmidtlein and Sassin 2004). Given this, the necessity of hyperspectral imagery has been debated, as recent research has shown that the spectrum from VIS to SWIR is the most significant for detecting wet and dry grassland floristic gradients (Feilhauer et al. 2013).

19.3.2.3 Heathland Habitats

Generally, heathland habitats are characterized by a mixture of ericaceous dwarf shrub species (e.g., *Calluna vulgaris*), grassland species, and open soil. Since all three components can be reliably spectrally distinguished, these habitats are rather straightforward to detect and to monitor (Corbane et al. 2014). Wet and dry heathland types are more difficult to distinguish with limited spectral separability on moderate-resolution imagery (Diaz Varela et al. 2008). Object-based image analysis has successfully been applied to high-spatial-resolution imagery in identifying dominant heather areas, based on indicator species (Förster et al. 2008) or structural parameters (Langanke et al. 2007). Similarly, more detailed approaches have been done using hyperspectral imagery, where kernel-based reclassification was used to transform the resultant LC classes into heathland habitats (Thoonen et al. 2013). Few specific studies used active sensors in detecting heathland habitats, for example, using LiDAR (Hellesen and Matikainen 2013) data for vertical differentiation of shrub and grassland forms and using HR multichannel SAR data (Bargiel 2013).

19.3.2.4 Wetland Habitats

Wetland vegetation is characterized by high spatial and spectral variability, and is influenced by soil moisture, atmospheric moisture, and the respective hydrological properties of the wetland type (Corbane et al. 2014). This makes traditional vegetation mapping approaches based on the MIR-to-NIR range difficult, due to the relatively higher absorption of this wavelength by water (Adam et al. 2009). Nevertheless, medium-resolution imagery such as Landsat has been used to classify broad wetland habitat types (Mac Alister and Mahaxy 2009). Additionally, the use of ancillary data such as soil type, combined with multispectral imagery, can be used to help differentiate spectrally similar classes (Bock 2003). In terms of mapping dominant species, high-spatial-resolution imagery was successful (Everitt et al. 2004), including submerged vegetation types (Dogan et al. 2009). WorldView-2 data were used by Keramitsoglou et al. (2014) to perform kernel-based classification of river delta vegetation and habitats in Greece. These habitats form a rich yet fine-scaled mosaic of brackish lagoons, saline soils, extensive mudflat, saltwater and freshwater, sand dunes, and also rich vegetation. Mapping remnants of these delta habitats, which are exposed to anthropogenic, mainly agricultural pressure, have proven to be successful to be transferred between similar delta situations.

LiDAR has been used alone to perform genus-level wetland, as well as in combination with VHRI for object-based classification. Using hyperspectral imagery, Schmidt and Skidmore (2003) indicated that it is possible to distinguish between 27 types of salt marsh vegetation using spectral signatures. It remains challenging to distinguish between submerged vegetation types, due to factors such as water turbidity, depth, and bottom reflectance (Jollineau and Howarth 2008).

19.4 Observing Quality, Pressures, and Changes

As the Millennium Ecosystem Assessment stated in 2005,* the Earth's ecosystems have been altered rapidly by human pressure in a short timeframe, of about half a century. In addition to directly reducing existing habitat through land use (LU) conversions, the human impact highly affects the quality of the remaining habitats (Mairota et al. 2014). Indicators derived from remotely sensed data (for an overview, see, e.g., Strand et al. 2007) can help assess and monitor habitat state and conditions. This section describes means and methods how habitat quality can be assessed, pressures on habitat characterized, and changes quantified.

19.4.1 Measuring Habitat Quality

Even more challenging than detecting species and delineating habitat is to obtain robust information about the quality and conservation status of habitat types using remote-sensing data. Biodiversity surrogates (BS) include parameters such as species presence, abundance, probability of site occupancy, aggregate measures such as species richness, diversity, or carrying capacity and can be used to observe the degradation of habitat quality characteristics related to resource availability (e.g., nutrients, refugia), phytomass, vegetation structure, and microclimate (Mairota et al. 2014). Habitat *quality*, according to Lindenmayer et al. (2002), is inherently taxon-specific and scale-dependent with respect to extent and grain, *sensu* Kotliar and Wiens (1990). Since it can be prohibitively expensive to obtain fine-grained habitat quality data at large spatial extents through field surveys alone, remote sensing is used to estimate environmental heterogeneity at differing grains across differing spatial extents (Mairota et al. 2014) and relate it to the variation in species diversity and distribution (Nagendra et al. 2014). VHR data are of particular power (Nagendra et al. 2013) as they enable multiple scale levels to be extracted from one single image (Strasser and Lang 2014b). BS exhibit different relevance to taka and functional groups with varying spectral and textural diversity measurements at different spatial scales and can be predicted with reasonable accuracy using habitat modeling (based on remotely sensed measures of environmental attributes (Mairota et al. 2014).

Approaches to infer habitat quality from remote-sensing data are abundant (Townsend et al. 2009, Rocchini et al. 2010, Costanza et al. 2011). In addition, spatial analysis techniques can be applied in order to quantitatively assess and compare structural indicators related to the actual state (Riedler et al. 2014, Vaz et al.) and conditions of habitats. Related to this is the influence of the complexity of landscape structure (Corbane et al. 2014). Overall, mapping becomes more

challenging when landscapes are more heterogeneous and fine-grained, and the variation between habitats is more continuous (Diaz Varela et al. 2008). Also, the complexity of landscape structure differs between protected areas and their surroundings, and thus different approaches to mapping need to be considered. As landscapes become more heterogeneous and the numbers of potential habitat types increase, modeling approaches of the relationship between species distribution patterns and remotely sensed data gain importance (Schmidtlein and Sassin 2004).

National quality parameters for monitoring the conservation status according to HabDir are defined by different European states, such as Austria (Ellmauer 2005) and Germany (Balzer et al. 2008), as well as Belgium, France, and Denmark. In most cases, the conservation status is assessed by habitat structures (e.g., horizontal and vertical variation, age structure), presence of typical species (mostly flora) in the habitat, abiotic factors (e.g., flooding), and pressures or disturbances of the habitat type (e.g., eutrophication indicators, invasive species). Indicators are usually framed for broad habitat types, as discussed earlier. Within forest habitats, many indicators, such as "percentage of characteristic tree species," can be derived by means of remote sensing (Corbane et al. 2014). Others (e.g., *habitat trees*, very old and degraded living microhabitat-bearing trees with hollows for nesting) remain tricky and require very detailed vertical representations by means of LiDAR data. This also applies to other broad habitat types: Heathlands have been studied extensively in this respect (Delalieux et al. 2012, Spanhove et al. 2012) including degradation stages in bog areas (Langanke et al. 2007). For grasslands, recent advances have been made in monitoring grassland use intensity (Schuster et al. 2011) and shrub encroachment (Lang and Langanke 2006). In addition, *negative* quality criteria can be used, as is the share of invasive species. Many such neophyte species can be detected on remotely sensed imagery based on spectral, phonological, or structural characteristics. This can be supported by modeling potential distribution and susceptibility of specific areas of invasion.

Few approaches exist (as yet), where several indicators are integrated in a quantitative and spatially explicit way in order to receive an aggregated view on habitat quality. Riedler et al. (2014) propose such a strategy for habitat quality assessment in riparian forests. They use a composite indicator (RFI_S) for integrated assessment of habitat quality on patch level and the identification of *hot spots* where management action may be focused. RFI_S is composed of seven indicators (derived from VHRI and LiDAR data) addressing four important attributes of riparian forest quality: (1) tree species composition, (2) vertical forest structure, (3) horizontal forest structure, and (4) water regime. For the aggregation of the RFI_S, expert-based and statistical weighting were applied. Measures of improvement or conservation can be specifically designed through the decomposition of the overall indicator into its underlying components (see Figure 19.9).

---

* Millenium Ecosystem Assessment 2005, Millenium Ecosystem Assessment, http://www.millenniumassessment.org/

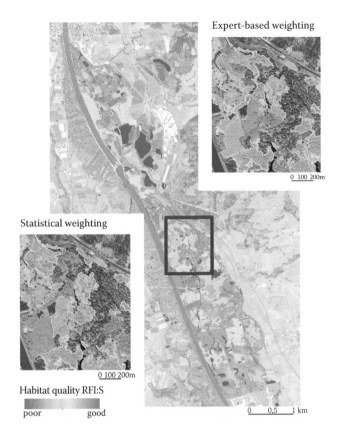

Expert-based weighting

Statistical weighting

0 100 200m

0 100 200m

Habitat quality RFI:S

poor   good

0   0,5   1 km

FIGURE 19.9   Composite indicator for riparian forest habitat quality assessment. (From Riedler, B. et al., *Int. J. Appl. Earth Obs. Geoinf.*, 2014.)

19.4.2 Identifying Pressures and Changes

Both loss of habitat and decline of habitat quality can be linked to anthropogenic pressures, which affect the provisioning of essential ES (Nagendra et al. 2014) for human well-being. International bodies such as the Intergovernmental Panel on Biodiversity and Ecosystem Services* (IPBES) have stressed the need to assess human pressures on biodiversity and ES across all scales. Even protected areas, through direct or indirect anthropogenic impact, continue to experience anthropogenic pressure; thus for effective management response, spatial knowledge of the type and location of pressure is required (Nagendra et al. 2013).

EO data and associated techniques, coupled with landscape pattern analysis and habitat modeling as well as BS estimates, can provide critical information on changes in state and condition of habitats, which in turn can be used to infer evidence of pressures. Nagendra et al. (2014) propose a unified approach to facilitate the provision of value-added products from EO sources for biodiversity conservation purposes. The proposed approach builds on the "driving forces, pressure, state, impact and response" framework (EEA 1995) and is based on the

definition of four broad categories of changes in state, which can be mapped and monitored through EO analyses.

Within the BIO_SOS Earth observation data for habitat monitoring (EODHaM) system (Lucas et al. 2014), six types of change assessments with respect to LC/habitat classes or geometry are distinguished, namely, changes in (1) LCCS classes (or GHCs), (2) specific LCCS component codes, (3) the number of extracted objects belonging to the same category, (4) object size and geometry (splitting or merging), (5) EO-derived measurements (e.g., LiDAR-derived height or vegetation indices), and (6) calculated landscape indicators useful for subsequent biodiversity indicators quantification, for example, species distribution in Ficetola et al. (2014).

In the following,[†] we highlight some specific types of pressure and/or change indicators with respect to broad habitat categories, which often have a dual function in terms of their actual species hosting function and any form of natural resource for anthropogenic use. Among the most generic indicators are habitat extent and habitat fragmentation. Habitat can change in area due to a multitude of driving forces that can each pose different pressures and threats (Nagendra et al. 2014). Changing habitat areas may pinpoint to expanding or shrinking areas of competing LC types like agricultural, industrial, and urban land or might reveal the impact of pollution, climate change, or catastrophes. Forest habitats, for example, degrade as a result of forest plantation that affects their species composition, or artificial fires, a loss of carbon stock and biomass by (over) exploitation, and a decrease of old growth forest. Species composition of grasslands and their layered structures are influenced by human management regime (cutting, grazing, burning) and intensity (fertilization, management frequency) influencing stages of growth and regrowth. Pressures on grasslands most often relate to agricultural practices with potential impact on biodiversity, especially through the use of fertilization, irrigation, and pesticides, with associated threats related to water, air, and soil pollution, drainage. The presence of shrubs and trees might indicate a decrease or even a total abandonment of traditional management practices like hay making or grazing, including active afforestation. Changing indicator values through time may reveal recurrent burning practices, the loss of habitat for species that depend on open grassland on the one hand or grasslands with interspersed trees on the other hand, or even climate change affecting high-elevation grasslands. The presence of open water implies a multitude of potential pressures to biodiversity present in wetland or riverine habitats. Areas with open water can be subject to the extraction of sand, gravel, clay, or other minerals, destructing the habitat of species that entirely or partly rely on open water. The use of open water as a source of gravitational energy or drinking water

___

* IPBES, 2015, Intergovernmental Panel on Biodiversity and Ecosystem Services, http://www.ipbes.net/.

___

† Taken from MS.MONINA Deliverable 5.6 "Framework for identifying threats and pressures on sensitive sites, from remote sensing derived (change) indicators in the site surroundings," available at www.ms-monina.eu.

production can be a threat too, especially when subjected to a management that impedes the development of habitats for particular species (e.g., by recurrent cleaning of basins and water bodies, large and irregular water table fluctuations). The same applies to water bodies used for transport or recreation or marine and freshwater aquaculture. Fishing and harvesting of aquatic resources can be detrimental too when applied in a nonsustainable way. Nonnative aquatic species grown for food by aquaculture can invade neighboring water bodies and even entire catchments when they escape from nurseries. In areas with a high human population density, pollution of open water by all kinds of activities may be ubiquitous, as are changes in the hydrological conditions of the open water or the neighboring areas (e.g., by dredging).

Habitat fragmentation can be evaluated by structural indices (Jaeger 2000) depending on scale and level of spatial explicitness. Fragmentation indices not only consider the absolute shrinking of habitat area but also, and with particular emphasis, the decrease or even loss of functional connectivity. Not only the dynamic of such indices provides information on one habitat type, but also usually habitat fragmentation is coupled with an interplay of changing LU patterns with mutual influence.

19.5 Toward a Biodiversity Monitoring Service

Remote sensing has long been used as a tool for environmental monitoring, especially for vegetation. But while at the global and continental scales, applications using broad LC/habitat categories have been quite successful, success has been harder to achieve at detailed local scales. Indeed, applications in detailed vegetation mapping and monitoring are often demanding in terms of data (requiring both high spectral and high spatial resolution), placing them at the forefront of technological development, with many new approaches still being in a research phase to tackle the widely divergent user needs. In contrast to global mapping initiatives, which have received wide attention and critical evaluations (Bartholome and Belward 2005), local mapping exercises rarely receive any evaluation and validation other than by the user for which they were intended. Hence, it remains unknown whether the chosen method was most appropriate for the situation and problem at hand and whether the method would yield comparable results in a different setting (i.e., the robustness of the method). This impairs the wider application of such methods and adoption by other users with similar problems and their further development toward operationality.

The two European projects, MS.MONINA and BIO_SOS, have addressed these needs, by exploring the potential of EO data in combination with data from ground surveys for supporting management options and reporting of obligations.* The projects

have prepared the ground for establishing services to support a successful implementation of European environmental legislation on all levels. Services, developed in a preoperational mode, underlay four suitability criteria as identified by Vanden Borre et al. (2011): (1) multiscale, that is, addressing multiple scales on all levels of implementation; (2) versatile, with algorithms tailored to the habitat type of interest and different image types; (3) user-friendly, allowing integration of the products into existing workflows; and (4) cost-efficient, providing reliable and reproducible products at an affordable cost, compared to traditional field methods.

Three MS.MONINA (sub-)services were designed, reflecting the different levels of operation, that is, EU, state, and site (see Figure 19.10). This requires a concordant multiuser approach. Each of the service development is tailored to the user and technical requirements that are specific for each level of implementation. User requirements surveys collected all details on existing workflows, data usages, and the responsibilities imposed by HabDir. Based on these requirements, the testing, comparison, and integration of state-of-the-art methodologies are performed. Demonstrators, accompanied by a full-fledged user validation exercise, complete the service evolution plan and the final scoping toward market. MS.MONINA thereby addresses (1) agencies on EU level, that is, ETC biodiversity, the EEA, and DG environment; (2) national and federal agencies in their reporting on sensitive sites and habitats within biogeographical regions on the entire territory; (3) local management authorities by advanced mapping methods for status assessment and change maps of sensitive sites; and (4) all three groups by providing transferable and interoperable monitoring results for an improved information flow between all levels.

Within BIO_SOS, a preoperational knowledge-driven open-source three-stage processing system was developed capable of combining multiseasonal EO data (HR and VHR) and in situ data (including ancillary information and in situ measurements) and for subsequent translation of LC to habitat maps. This system, named EO data for habitat monitoring (Lucas et al. 2014), is based on expert knowledge elicited from botanists, ecologists, remote-sensing experts, and management authorities in order to monitor large and not accessible areas without any ground reference data. Ontologies are used to formally represent the expert knowledge (Arvor et al. 2013). The FAO-LCCS and the GHC taxonomies, from which HabDir Annex I habitats can be defined, were used for describing LC/LU and habitat categories (Tomaselli et al. 2013) and for subsequent translation to habitats. In addition, BIO_SOS focused on the development of a modeling framework for (1) filling the gap between LC/LU and habitat domains (Blonda et al. 2012a) by coupling FAO-LCCS taxonomy with GHCs and EUNIS classification schemes and providing a reliable cost-effective knowledge-driven long-term biodiversity monitoring scheme of protected areas and their surrounds (Tomaselli et al. 2013, Adamo et al. 2014, Kosmidou et al. 2014, Lucas et al. 2014); (2) analyzing appropriate spatial and temporal scales of EO data

---

* See White Paper on "Copernicus Biodiversity Monitoring Services" available at http://www.biosos.eu/publ/White_Paper_Biodiversity_Monitoring_BIOSOS_MSMONINA.pdf.

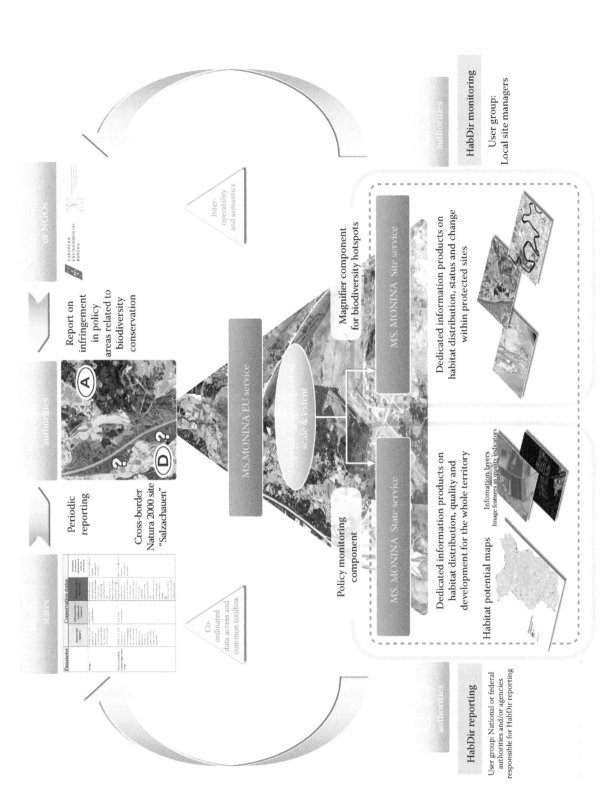

FIGURE 19.10 MS.MONINA 3-level concept of operational service implementation for HabDir-related monitoring requirements (Lang and Pernkopf 2013). The EU service builds on two main components: (1) magnifier component, on-demand provision of habitat distribution and quality indicator maps for biodiversity hotspot sites (e.g., riparian areas, coastal areas), and (2) policy monitoring component, on-demand provision of maps in *rush mode* as a means for external/independent validation of the congruence national biodiversity reports in the case of transboundary protected sites.

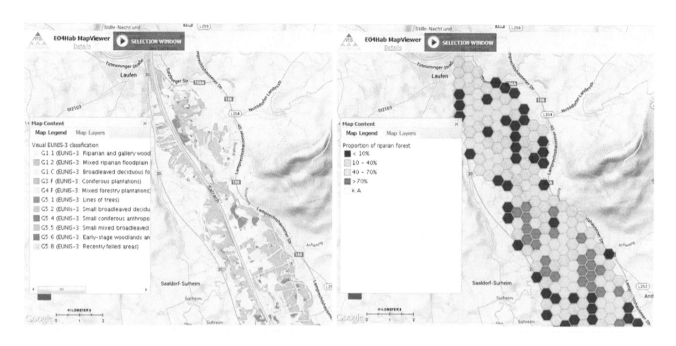

FIGURE 19.11   Transboundary riparian forest site "Salzachauen" (Germany/Austria). (Screenshots taken from MS.MONINA map viewer.)

sources for pressure assessment in the context of existing reference frameworks for pressure assessment and trend extraction (Lang et al. 2014); and (3) handling uncertainty in habitat mapping (Petrou et al. 2014).

Public access web platforms integrate and showcase what is offered by MS.MONINA and BIO_SOS. These mainly include an online service portfolio with specific details on the services to be offered and further information on the respective service cases. Web-GIS and Open Geospatial Consortium-conform metadata geoportals, including an external quality evaluation module (fitness for use and fitness to purpose), contain all geospatial information (see Figure 19.11). A tool repository allows for searching, both thematically and spatially, the available data within the consortium, which are needed by the methodological components and algorithms utilized by the partners for the respective image analysis and geospatial analysis tasks.

References

Adam, E., O. Mutanga, and D. Rugege 2009. Multispectral and hyperspectral remote sensing for identification and mapping of wetland vegetation: A review. *Wetl Ecol Manage*. 18, 281–296.

Adamczyk, J. and A. Osberger 2014. Red-edge vegetation indices for detecting and assessing disturbances in Norway spruce dominated mountain forests. *Int J Appl Earth Obs Geoinf*. 39, 90–99.

Adamo, M. et al. 2014. Expert knowledge for translating land cover/use maps to general habitat categories (GHC). *Landsc Ecol*.

Addicot, J. F., J. M. Aho, M. F. Antolini, D. K. Padilla, J. S. Richardson, and D. A. Soluk 1987. Ecological neighborhoods: Scaling environmental patterns. *Oikos*. 49, 340–346.

Akay, A. E., M. G. Wing, and J. Sessions 2012. Estimating structural properties of riparian forests with airborne lidar data. *Int J Remote Sens*. 7010–7023. *18*, (281–296).

Ali, I., C. Schuster, M. Zebisch, M. Förster, B. Kleinschmit, and C. Notarnicola 2013. First results of monitoring nature conservation sites in alpine region by using very high resolution (VHR) X-band SAR data. *IEEE J Sel Top Appl Earth Observ Remote Sens*. 6, 2265–2274.

Arvor, D., L. Durieux, S. Andrés, and L. Marie-Angélique 2013. Advances in geographic object-based image analysis with ontologies: A review of main contributions and limitations from a remote sensing perspective. *ISPRS J Photogramm*. 82, 125–137.

Asner, G. P., M. O. Jones, R. E. Martin, D. E. Knapp, and R. F. Hughes 2008a. Remote sensing of native and invasive species in Hawaiian forests. *Remote Sens Environ*. *112* (5), 1912–1926.

Asner, G. P., D. E. Knapp, T. Kennedy-Bowdoin, M. O. Jones, R. E. Martin, J. Boardman, and R. F. Hughes 2008b. Invasive species detection in Hawaiian rainforests using airborne imaging spectroscopy and LiDAR. *Remote Sens Environ*. *112*(5), 1942–1955.

Balzer, S., G. Ellwanger, U. Raths, E. Schröder, and A. Ssymank 2008. Verfahren und erste Ergebnisse des nationalen Berichts nach Artikel 17 der FFH-Richtlinie (in German). *Natur und Landschaft*. 83, 111–117.

Bargiel, D. 2013. Capabilities of high resolution satellite radar for the detection of semi-natural habitat structures and grasslands in agricultural landscapes. *Ecol Inform*. 3, 9–16.

Bartholome, E. and A. Belward 2005. GLC2000: A new approach to global land cover mapping from Earth observation data. *Int J Remote Sens*. 26, 1959–1977.

Belluco, E., M. Camuffo, S. Ferrari, L. Modenese, S. Silvestri, A. Marani, and M. Marani 2006. Mapping salt-marsh vegetation by multispectral and hyperspectral remote sensing. *Remote Sens Environ.* 54–67.

Berger, M., J. Moreno, J. A. Johannssen, P. F. Levelt, and R. F. Hansen 2012. ESA's sentinel missions in support of Earth system science. *Remote Sens Environ.* 105 (1), 84–90.

Blonda, P., K. B. Jones, J. Stutte, and P. Dimipoulos 2012a. From space to species. Safeguarding biodiversity in Europe. *Int Innov Environ.* 86–88.

Blonda, P., R. Lucas, and J. P. Honrado 2012b. From space to species: Solutions for biodiversity monitoring. In: S. Ourevitch (ed.) *Window on Copernicus: Discover What GMES Can Do for European Regions and Cities.* Brussels, Belgium: SpaceTec Partners, pp. 66–73.

Blondel, J. 1979. *Biogeógraphie et écologie : Synthèse sur la structure, la dynamique et l'évolution des peuplements de vertébrés terrestres.* Paris, NY: Masson.

Bock, M. 2003. Remote sensing and GIS-based techniques for the classification and monitoring of biotopes: Case examples for a wet grass- and moor land area in Northern Germany. *J Nat Conserv.* 11, 145–155.

Bock, M., G. Rossner, M. Wissen, K. Remm, T. Langanke, S. Lang, and H. Klug 2005. Spatial indicators for nature conservation from European to local scale. *Ecol Ind.* 5, 322–338.

Bouman, B. A. M. and H. W. J. van Kasteren 1990. Ground-based X-band (3-cm wave) radar backscattering of agricultural crops. I. Sugar beet and potato; backscattering and crop growth. *Remote Sens Environ.* 34, 93–105.

Buck, O., V. E. M. Garcia, A. Klink, and K. Pakzad 2014. Using information layers for mapping grassland habitat distribution at local to regional scales. *Int J Appl Earth Obs Geoinf.* 37, 83–89.

Bunce, R. H. G. et al. 2008. A standardized procedure for surveillance and monitoring European habitats and provision of spatial data. *Landsc Ecol.* 1, 11–25.

Corbane, C., S. Alleaume, and M. Deshayes 2013. Mapping natural habitats using remote sensing and sparse partial least square discriminant analysis. *Int J Remote Sens.* 34, 7625–7647.

Corbane, C. et al. 2014. Remote sensing for mapping natural habitats and their conservation status—New opportunities and challenges. *Int J Appl Earth Obs Geoinf.* 37, 7–16.

Costanza, J. K., A. Moody, and R. K. Peet 2011. Multi-scale environmental heterogeneity as a predictor of plant species richness. *Landsc Ecol.* 26, 851–864.

Dalponte, M., Orka, H. O., Gobakken, T., Gianelle, D., and Naesset, E. (2013). Tree species classification in Boreal forests with hyperspectral data. *IEEE Transactions on Geoscience and Remote Sensing,* 51 (5), 2632–2645.

Davies, C. E., M. D. and M. O. Hill 2004. EUNIS habitat classification.http://www.eea.europa.eu/themes/biodiversity/eunis/ eunis-habitat-classification (Accessed September 2014.)

Davranche, A., G. Lefebvre, and B. Poulin 2010. Wetland monitoring using classification trees and SPOT-5 seasonal time series. *Remote Sens Environ.* 114 (3), 552–562.

Delalieux, S., B. Somers, B. Haest, T. Spanhove, J. Vanden Borre, and C. A. Mücher 2012. Heathland conservation status mapping through integration of hyperspectral mixture analysis and decision tree classifiers. *126 Remote Sens Environ.*

Di Gregorio, A. and L. J. M. Jansen 2005. *Land Cover Classification System (LCCS): Classification Concepts and User Manual.* Rome, Italy: Food and Agriculture Organization of the United Nations.

Diaz Varela, R., P. Ramil Rego, S. Calvo Iglesias, and C. Muñoz Sabrino 2008. Automatic habitat classification methods based on satellite images: A practical assessment in the NW Iberia coastal mountains. *Environ Monit Assess.* 144, 229–250.

Dogan, O. K., Z. Akyurek, and M. Beklioglu 2009. Identification and mapping of submerged plants in a shallow lake using quickbird satellite data. *J Environ Manage.* 90, 2138–2143.

EEA, E. E. A. 1995. *Europe's Environment: The Dobris Assessment.* Copenhagen, Denmark: European Environment Agency.

Ellmauer, T. 2005. Umweltbundesamt. European Commission 2005: European Commission GEO 2005, 2010: Group on Earth Observation.

EuropeanCommission 2011. *COM(2011) 244 Final. Our Life Insurance, Our Natural Capital: An EU Biodiversity Strategy To 2020.*

Evans, D. 2012. Building the European Union's Natura 2000 network. *Nat Conserv.* 1, 11–26.

Everitt, J. H., C. S. Yang, R. S. Fletcher, M. R. Davis, and D. L. Drawe 2004. Using aerial color-infrared photography and QuickBird satellite imagery for mapping wetland vegetation. *Geocarto Int.* 19 (4), 15–22.

Feilhauer, H., F. Thonfeld, U. Faude, K. S. He, D. Rocchini, and S. Schmidtlein 2013. Assessing floristic composition with multispectral sensors—A comparison based on monotemporal and multiseasonal field spectra. *Int J Appl Earth Obs Geoinf.* 21, 218–229.

Ficetola, G. F. et al. 2014. Importance of landscape features and Earth observation derived habitat maps for modelling amphibian distribution in the Alta Murgia National. Park *Int J Appl Earth Obs Geoinf.*

Filippi, A. M., I. Gueneralp, and J. Randall 2014. Hyperspectral remote sensing of aboveground biomass on a river meander bend using multivariate adaptive regression splines and stochastic gradient boosting. *Remote Sens Lett.* 5 (5), 432–441.

Förster, M., A. Frick, H. Walentowski, and B. Kleinschmit 2008. Approaches to utilising QuickBird data for the monitoring of NATURA 2000 habitats. *Commun Ecol.* 9, 155–168.

Förster, M. and B. Kleinschmit 2014. Significance analysis of different types of ancillary geodata utilized in a multisource classification process for forest identification in Germany. *IEEE Trans Geosci Remote Sens.* 52 (6), 3453–3463.

Förster, M., B. Kleinschmit, and H. Waltentowski 2005. Comparison of three modelling approaches of potential natural forest habitats in Bavaria, Germany. *Waldökologie Online.* 126–135.

Förster, M., T. Schmidt, C. Schuster, and B. Kleinschmit 2012. Multi-temporal detection of grassland vegetation with RapidEye imagery and a spectral-temporal library. *IEEE Trans Geosci Remote Sens.* 4930–4933.

Förster, M., C. Schuster, and B. Kleinschmit 2010. Significance analysis of multi-temporal RapidEye satellite images in a land cover classification. In: N. J. Tate and P. F. Fisher (eds.) *Accuracy 2010*, Leicester, England, pp. 273–276.

Franke, J., V. Keuck, and F. Siegert 2012. Assessment of grassland use intensity by remote sensing to support conservation schemes. *J Nat Conserv.* 20, 125–134.

Friedl, M. A., D. Sulla-Menashe, B. Tan, A. Schneider, N. Ramankutty, A. Sibley, and X. Huang 2010. MODIS collection 5 global land cover: Algorithm refinements and characterization of new datasets. *Remote Sens Environ.* 114, 168–182.

Fuller, R., G. B. Groom, and A. Jones 1994. The land cover map of Great Britain: An automated classification of Landsat Thematic Mapper data. *Photogramm Eng Remote Sens.* 60, 553–562.

Gaertner, P., M. Foerster, A. Kurban, and B. Kleinschmit 2014. Object based change detection of Central Asian Tugai vegetation with very high spatial resolution satellite imagery. *Int J Appl Earth Obs Geoinf.* 31, 110–121.

GEO 2005. *The Global Earth Observation System of Systems (GEOSS) 10-Year Implementation Plan.* Adopted February 16, 2005.

GEO 2010. *A Quality Assurance Framework for Earth Observation.* Version 4.0.

Ghosh, A., F. E. Fassnacht, P. K. Joshi, and B. Koch 2014. A framework for mapping tree species combining hyperspectral and LiDAR data: Role of selected classifiers and sensor across three spatial scales. *Int J Appl Earth Obs Geoinf.* 26, 49–63.

Gillespie, T. W. 2005. Predicting woody-plant species richness in tropical dry forests: A case study from south Florida, USA. *Ecol Appl.* 15, 27–37.

Gilmore, M. S., E. H. Wilson, N. Barrett, D. L. Civco, S. Prisloe, J. D. Hurd, and C. Chadwick 2008. Integrating multi-temporal spectral and structural information to map wetland vegetation in a lower Connecticut River tidal marsh. *Remote Sens Environ. 112* (11), 4048–4060.

Goetz, S., D. Steinberg, R. Dubayah, and B. Blair 2007. Laser remote sensing of canopy habitat heterogeneity as a predictor of bird species richness in an eastern temperate forest, USA. *Remote Sens Environ.* 108, 254–263.

Grillo, O. and G. Venora 2011. *Biodiversity Loss in a Changing Planet.* Rijeka, Croatia: InTech.

Gruber, B. et al. 2012. "Mind the gap!"—How well does Natura 2000 cover species of European interest? *Nat Conserv.* 3, 45–62.

Hellesen, T. and L. Matikainen 2013. An object-based approach for mapping shrub and tree cover on grassland habitats by use of LiDAR and CIR orthoimages. *Remote Sens.* 5 (2), 558–583.

Huang, W., G. Sun, R. Dubayah, B. Cook, P. Montesano, W. Ni, and Z. Zhang 2013. Mapping biomass change after forest disturbance: Applying LiDAR footprint-derived models at key map scales. *Remote Sens Environ.* 134, 319–332.

Ichter, J., D. Evans, and D. Richard 2014. *Habitat and Vegetation Mapping in Europe—An Overview.* Copenhagen, Denmark: EEA-MNHN.

Immitzer, M., C. Atzberger, and T. Koukal 2012. Tree species classification with random forest using very high spatial resolution 8-Band WorldView-2 satellite data. *Remote Sens.* 4, 2661–2693.

Isbell, F. 2010. Causes and consequences of biodiversity declines. *Nat Edu Knowl. 3* (10), 54.

Jaeger, J. 2000. Landscape division, splitting index, and effective mesh size: New measures of landscape fragmentation. *Landsc Ecol.* 115–130.

Johansen, K., N. C. Coops, S. E. Gergel, and Y. Stange 2007. Application of high spatial resolution satellite imagery for riparian and forest ecosystem classification. *Remote Sens Environ.* 15, 29–44.

Jollineau, M. Y. and P. J. Howarth 2008. Mapping an inland wetland complex using hyperspectral imagery. *Int J Remote Sens. 29* (12), 3609–3631.

Keramitsoglou, I., D. Stratoulias, E. Fitoka, C. Kontoes, and N. Sifakis 2014. A transferability study of the kernel-based reclassification algorithm for habitat delineation. *Int J Appl Earth Obs Geoinf.*

Key, T., T. A. Warner, J. B. McGraw, and M. A. Fajvan 2001. A comparison of multispectral and multitemporal information in high spatial resolution imagery for classification of individual tree species in a temperate hardwood forest. *Remote Sens Environ.* 75, 100–112.

Kosmidou, V. et al. 2014. Harmonization of the Land Cover Classification System (LCCS) with the General Habitat Categories (GHC) classification system: Linkage between remote sensing and ecology. *Ecol Indic.* 36, 290–300.

Kotliar, N. B. and J. Wiens 1990. Multiple scales of patchiness and patch structure: A hierarchical framework for the study of heterogeneity. *Oikos.* 59, 253–260.

Kuenzer, C. et al. 2014. Earth observation satellite sensors for biodiversity monitoring: Potentials and bottlenecks. *Int J Remote Sens. 35* (18), 6599–6647.

Laliberte, A. S., E. L. Frederickson, and A. Rango 2007. Combining decision trees with hierarchical object-oriented image analysis for mapping arid rangelands. *Photogramm Eng Remote Sens.* 73, 197–207.

Lang, S. 2008. Object-based image analysis for remote sensing applications: Modeling reality—Dealing with complexity. In: T. Blaschke, S. Lang, and G. J. Hay (eds.) *Object-Based Image Analysis—Spatial Concepts for Knowledge-Driven Remote Sensing Applications.* Berlin, Germany: Springer, pp. 3–28.

Lang, S., C. Corbane, and L. Pernkopf 2013a. Earth observation for habitat and biodiversity monitoring. In: S. Lang and L. Pernkopf (eds.) *Ecosystem and Biodiversity Monitoring: Best Practice in Europe and Globally* [section title]. Heidelberg, Germany: Wichmann, pp. 478–486.

Lang, S. and T. Langanke 2006. Object-based mapping and object-relationship modelling for land-use classes and habitats. *Photogrammetrie Fernerkundung Geoinformation.* 1, 5–18.

Lang, S. and L. Pernkopf 2013. EO-based monitoring of Europe's most precious habitats inside and outside protected areas. In: K. Bauch (ed.) *5th Symposium for Research in Protected Areas*. Mittersill, Austria: Salzburger Nationalparkfonds, pp. 443–448.

Lang, S., L. Pernkopf, P. Mairota, and E. P. Schioppa 2014. Earth observation for habitat mapping and biodiversity monitoring. *Int J Appl Earth Obs Geoinf.* (Special Issue). Elsevier.

Lang, S., G. Smith, and J. Vanden Borre 2013b. Monitoring Natura 2000 habitats at local, regional and European scale. In: S. Ourevitch (ed.) *Discover What GMES Can Do for European Regions and Cities.* Brussels, Belgium: SpaceTec Partners, pp. 58–65.

Langanke, T., C. Burnett, and S. Lang 2007. Assessing the mire conservation status of a raised bog site in Salzburg using object-based monitoring and structural analysis. *Landscape Urban Plan.* 79, 160–169.

Lindenmayer, D. B., R. B. Cunnigham, C. F. Donnelly, and H. A. Nix 2002. The distribution of birds in a novel landscape context. *Ecol Monogr.* 72, 1–18.

Lucas, R. et al. 2014. The earth observation data for habitat monitoring (EODHAM) system. *Int J Appl Earth Obs Geoinf.* 37, 17–28.

Lucas, R. et al. 2011. Updating the Phase 1 habitat map of Wales, UK, using satellite sensor data. *ISPRS J Photogramm.* 66, 81–102.

Lucas, R., A. Rowlands, A. Brown, S. Keyworth, and P. Bunting 2007. Rule-based classification of multi-temporal satellite imagery for habitat and agricultural land cover mapping. *ISPRS J Photogramm Remote Sens.* 62, 165–185.

Mac Alister, C. and M. Mahaxy 2009. Mapping wetlands in the Lower Mekong Basin for wetland resource and conservation management using Landsat ETM images and field survey data. *J Environ Manage.* 90, 2130–2137.

Mac Arthur, A. A. and J. T. Malthus 2008. An object-based image analysis approach to the classification and mapping of calluna vulgaris canopies. In: *Remote Sensing and Photogrammetry Society Annual Conference,* Falmouth, UK.

Maes, J. et al. 2013. Mapping and assessment of ecosystems and their services. An analytical framework for ecosystem assessments under action 5 of the EU biodiversity strategy to 2020. Discussion paper. Luxembourg, Western Europe: Publications Office of the European Union.

Mairota, P. et al. 2014. Very high resolution Earth observation features for monitoring plant and animal community structure across multiple spatial scales in protected areas. *Int J Appl Earth Obs Geoinf.* 37, 100–105.

Mücher, C. A. et al. 2014. Synergy of airborne LiDAR and Worldview-2 satellite imagery for land cover and habitat mapping: A BIO_SOS-EODHaM case study for the Netherlands. *Int J Appl Earth Obs Geoinf.* 37, 48–55.

Naeem, S., D. E. Bunker, A. Hector, M. Loreau, and C. Perrings 2009. *Biodiversity, Ecosystem Functioning, and Human Wellbeing: An Ecological and Economic Perspective.* Oxford, U.K.: Oxford University Press.

Nagendra, H., R. Lucas, J. P. Honrado, R. H. G. Jongman, C. Tarantino, and M. Adamo 2013. Remote sensing for conservation monitoring: Assessing protected areas, habitat extent, habitat condition, species diversity and threat. *Ecol Indic.* 33, 45–59.

Nagendra, H. and D. Rocchini 2008. High resolution satellite imagery for tropical biodiversity studies: The devil is in the detail. *Biodivers Conserv.* 17 (14), 3431–3442.

Nagendra, H. et al. 2014. Satellite Earth observation data to identify anthropogenic pressures in selected protected areas. *Int J Appl Earth Obs Geoinf.* 37, 124–132.

Peng, G., R. Pu, G. S. Biging, and M. R. Larrieu 2003. Estimation of forest leaf area index using vegetation indices derived from Hyperion hyperspectral data. *IEEE Trans Geosci Remote Sens.* 41, 1355–1362.

Pereira, H. M. et al. 2013. Essential biodiversity variables. *Science.* 339 (6117), 277–278.

Petrou, Z. et al. 2014. A rule-based classification methodology to handle uncertainty in habitat mapping employing evidential reasoning and fuzzy logic. *Pattern Recogn Lett.* 48, 24–33.

Price, K. P., X. Guo, and J. M. Stiles 2002. Optimal landsat TM band combinations and vegetation indices for discrimination of six grassland types in eastern Kansas. *Int J Remote Sens.* 23 (23), 5031–5042.

Prisloe, S., M. Wilson, D. Civco, J. Hurd, and M. Gilmore 2006. Use of Lidar data to aid in discriminating and mapping plant communities in tidal marshes of the lower Connecticut River—Preliminary results. In: *Annual Conference of the American Society for Photogrammetry and Remote Sensing.* Reno, NV: American Society for Photogrammetry and Remote Sensing.

Psomas, A., M. Kneubühler, S. Huber, K. Itten, and N. E. Zimmermann 2011. Hyperspectral remote sensing for estimating aboveground biomass and for exploring species richness patterns of grassland habitats. *Int J Remote Sens.* 32, 9007–9031.

Puech, C., S. Durrieu, and J. S. Bailly 2012. Airborne lidar for natural environments research and applications in France. *Revue Française de Photogrammétrie et de Télédétection.* 54–68.

Rannow, S. et al. 2014. Managing protected areas under climate change: Challenges and priorities. *Environ Manage.* 54 (4), 732–743.

Riedler, B., L. Pernkopf, T. Strasser, G. Smith, and S. Lang 2014. A composite indicator for assessing habitat quality of riparian forests derived from Earth observation data. *Int J Appl Earth Obs Geoinf.* 37, 114–123.

Rocchini, D. et al. 2010. Remotely sensed spectral heterogeneity as a proxy of species diversity: Recent advances and open challenges. *Ecol Inform.* 5, 318–329.

Schmidt, K. S. and A. K. Skidmore 2003. Spectral discrimination of vegetation types in a coastal wetland. *Remote Sens Environ.* 85 (1), 92–108.

Schmidt, T., C. Schuster, B. Kleinschmit, and M. Förster 2014. Evaluating an intra-annual time series for grassland classification—How many acquisitions and what seasonal origin are optimal? *IEEE J Select Top Appl Earth Observ Remote Sens.* 7 (8), 3428–3439.

Schmidtlein, S. and J. Sassin 2004. Mapping of continuous floristic gradients in grasslands using hyperspectral imagery. *Remote Sens Environ.* 92, 126–138.

Schuster, C., I. Ali, P. Lohmann, A. Frick, M. Förster, and B. Kleinschmit 2011. Towards detecting swath events in TerraSAR-X time series to establish Natura 2000 Grassland Habitat Swath Management as monitoring parameter. *Remote Sens.* 3 (12), 1308–1322.

Schuster, C., T. Schmidt, C. Conrad, B. Kleinschmit, and M. Förster 2015. Grassland habitat mapping by intra-annual time series analysis—Comparison of RapidEye and TerraSAR-X satellite data. *Int J Appl Earth Obs Geoinf.* 34, 25–34.

Somers, B. and G. P. Asner 2014. Tree species mapping in tropical forests using multi-temporal imaging spectroscopy: Wavelength adaptive spectral mixture analysis. *Int J Appl Earth Obs Geoinf.* 31, 57–66.

Spanhove, T., J. Vanden Borre, S. Delalieux, B. Haest, and D. Paelinckx 2012. Can remote sensing estimate fine-scale quality indicators of natural habitats? *Ecol Indic.* 18, 403–412.

Strand, H., R. Höft, J. Strittholt, L. Miles, N. Horning, E. Fosnight, and W. Turner 2007. *Sourcebook on Remote Sensing and Biodiversity Indicators.* Montreal, Canada: Secretariat of the Convention on Biological Diversity.

Strasser, T. and S. Lang 2014a. Class modelling of complex riparian forest habitats. *SE Eur J Earth Observ Geom.* 3 (2s), 219–222.

Strasser, T. and S. Lang 2014b. Object-based class modelling for multi-scale riparian forest habitat mapping. *Int J Appl Earth Obs Geoinf.* 37, 19–37.

Suchenwirth, L., M. Foerster, A. Cierjacks, F. Lang, and B. Kleinschmit 2012. Knowledge-based classification of remote sensing data for the estimation of below- and above-ground organic carbon stocks in riparian forests. *Wetl Ecol Manage.* 20 (2), 151–163.

Thoonen, G., T. Spanhove, J. Vanden Borre, and P. Scheunders 2013. Classification of heathland vegetation in a hierarchical contextual framework. *Int J Remote Sens.* 34, 96–111.

Tiede, D., S. Lang, F. Albrecht, and D. Hölbling 2010. Object-based class modeling for cadastre constrained delineation of geo-objects. *Photogramm Eng Remote Sens.* 193–202.

Tomaselli, V. et al. 2013. Translating land cover/land use classifications to habitat taxonomies for landscape monitoring: A Mediterranean assessment. *Landsc Ecol.* 28 (5), 905–930.

Townsend, P. A., T. R. Lookingbill, C. C. Kingdon, and R. H. Gardner 2009. Spatial pattern analysis for monitoring protected areas. *Remote Sens Environ.* 113, 1410–1420.

Trochet, A. and D. Schmeller 2013. Effectiveness of the Natura 2000 network to cover threatened species. *Nat Conserv.* 4, 35–53.

Turner, W., S. Spector, N. Gardiner, M. Fladeland, E. Sterling, and M. Steininger 2003. Remote sensing for biodiversity science and conservation. *Trends Ecol Evol.* 18, 306–314.

Underwood, E. C., S. L. Ustin, and C. M. Ramirez 2003. A comparison of spatial and spectral image resolution for mapping invasive plants in coastal California. *Environ Manage.* 39, 63–83.

Vanden Borre, J., D. Paelinckx, C. A. Mücher, A. Kooistra, B. Haest, G. DeBlust, and A. M. Schmidt 2011. Integrating remote sensing in Natura 2000 habitat monitoring: Prospects on the way forward. *J Nat Conserv.* 19, 116–125.

Vaz, A. S. et al. 2014. Can we predict habitat quality from space? A multi-indicator assessment based on an automated knowledge-driven system. *Int J Appl Earth Obs Geoinf* 37, 106–113.

Walsh, S. J., A. L. McCleary, C. F. Mena, Y. Shao, J. P. Tuttle, and A. A. Gonzales, R. 2008. QuickBird and Hyperion data analysis of an invasive plant species in the Galapagos Islands of Ecuador: Implications for control and land use management. *Remote Sens Environ.* 112, 1927–1941.

Wang, K., S. E. Franklin, X. Guo, Y. He, and G. J. McDermid 2009. Problems in remote sensing of landscapes and habitats. *Prog Phys Geogr.* 33, 747–768.

Weiers, S., M. Bock, M. Wissen, and G. Rossner 2004. Mapping and indicator approaches for the assessment of habitats at different scales using remote sensing and GIS methods. *Landscape Urban Plan.* 67, 43–65.

Wiens, J. 1989. Spatial scaling in ecology. *Funct Ecol.* 3 (4), 385–397.

Wolter, P. T., D. J. Mladenoff, G. E. Host, and T. R. Crow 1995. Improved forest classification in the northern lake states using multi-temporal Landsat imagery. *Photogramm Eng Remote Sens.* 61, 1129–1143.

Woodcock, C. E. et al. 1994. Mapping forest vegetation using Landsat TM imagery and a canopy reflectance model. *Remote Sens Environ.* 50, 240–254.

Wulder, M. A., R. J. Hall, N. C. Coops, and S. E. Franklin 2004. High spatial resolution remotely sensed data for ecosystem characterization. *BioScience.* 54, 511–521.

Zlinszky, A., W. Mücke, H. Lehner, C. Briese, and N. Pfeifer 2012. Categorizing wetland vegetation by airborne laser scanning on Lake Balaton and Kis-Balaton, Hungary. *Remote Sens.* 4 (12), 1617–1650.

VII

Ecology

20

Ecological Characterization of Vegetation Using Multisensor Remote Sensing in the Solar Reflective Spectrum

Conghe Song
University of North Carolina at Chapel Hill

Jing Ming Chen
University of Toronto

Taehee Hwang
Indiana University

Alemu Gonsamo
University of Toronto

Holly Croft
University of Toronto

Quanfa Zhang
Chinese Academy of Sciences

Matthew Dannenberg
University of North Carolina at Chapel Hill

Yulong Zhang
University of North Carolina at Chapel Hill and Chinese Academy of Sciences

Christopher Hakkenberg
University of North Carolina at Chapel Hill

Juxiang Li
East China Normal University

Acronyms and Definitions

AR4	4th Assessment Report
APAR	Absorbed Photosynthetically Active Radiation
ALOS	Advanced Land Observing Satellite
AVHRR	Advanced Very-High-Resolution Radiometer
ARVI	Atmospherically Resistant Vegetation Index
BSMA	Bayesian Spectral Mixtureanalysis
BRDF	Bidirectional Reflectance Distribution Function
ChF	Chlorophyll Fluorescence
DMF	Dimethyl Formamide
DMSO	Dimethyl Sulfoxide
EOS	End of Season
ESA	European Space Agency
FIA	Forest Inventory and Analysis
FLEX	FLuorescence EXplorer

FPAR	Fraction of Absorbed Photosynthetically Active Radiation
GO	Geometric Optical
GORT	Geometric Optical Radiative Transfer
GLAS	Geospatial Laser Altimeter System
GEMI	Global Environmental Monitoring Index
GIMMS	Global Inventory Monitoring and Modeling System
GVI	Global Vegetation Index
GV	Green Vegetation
GPP	Gross Primary Productivity
HRVIR	High-Resolution Visible Infrared
kNN	k-Nearest Neighbor Imputation
LSP	Land Surface Phenology
LAI	Leaf Area Index
LUE	Light Use Efficiency
MIR	Middle Infrared
MODIS	Moderate-Resolution Imaging Spectroradiometer
MNDVI	Modified NDVI
MSR	Modified Simple Ratio
MISR	Multiangle Imaging Spectroradiometer
MSS	Multispectral Scanner Sensors
NBAR	Nadir BRDF-Adjusted Reflectance
NASAEOS	NASA Earth Observing System
NOAA	National Oceanic and Atmospheric Administration
NIR	Near-infrared
NEE	Net Ecosystem Exchange
NPP	Net Primary Production
NAOMI	New AstroSat Optical Modular Instrument
NLI	Nonlinear Index
NPV	Nonphotosynthetic Vegetation
NDMI	Normalized Difference Moisture Index
NDVI	Normalized Difference Vegetation Index
OLI	Operational Land Imager
PNW	Pacific Northwest
PRI	Photochemical Reflectance index
POES	Polar-orbiting Operational Environmental Satellites
RF	Random Forest
RSR	Reduced SR
RDVI	Renormalized Difference Vegetation Index
RBV	Return Beam Vidicon
RMSE	Root Mean Squared Error
SPOT	Satellite Pourl'Observation de la Terre
SWIR	Shortwave-infrared
SR	Simple ratio
SARVI	Soil and Atmosphere-resistant Vegetation Index
SARVI2	Soil and Atmosphere-resistant Vegetation Index 2
SAVI	Soil-adjusted Vegetation Index
SAVI1	Soil-adjusted Vegetation Index 1
SOS	Start of Season
Suomi NPP	Suomi National Polar-orbiting Partnership
TOPS	Terrestrial Observation and Prediction System
TM	Thematic Mapper
TIRS	Thermal Infrared Sensors
NDVI3g	Third-generation GIMMS NDVI
TRAC	Tracing Radiation and Architecture of Canopies
LandTrendr	Trends in Disturbance and Recovery
VI	Vegetation Index
VIIRS	Visible Infrared Imaging Radiometer Suite
WDVI	Weighted Difference Vegetation Index

20.1 Introduction

Vegetation is the primary producer in the terrestrial ecosystem. Vegetation absorbs the energy of electromagnetic radiation from the Sun and converts it to the energy that consumers in the ecosystem can use. As a result, vegetation is the foundation for nearly all the goods and services that terrestrial ecosystems provide to humanity. The advent of optical remote sensing revolutionized our ability to map the characteristics of vegetation wall-to-wall in space and to do so repeatedly, in a cost-efficient manner. Many of these vegetation parameters serve as key inputs to ecological models aiming to understand terrestrial ecosystem functions, at regional to global scales. This chapter summarizes the progress made in characterizing vegetation structure and its ecological functions with optical remote sensing. We first provide a brief review of the development of optical sensors designed primarily for vegetation monitoring. Second, we synthesize the progress made in mapping the physical structure of vegetation with optical sensors, including vegetation cover, vegetation successional stages, biomass, leaf area index (LAI), and its spatial organization, that is, leaf clumping. Third, we review the achievements made in understanding vegetation function with optical remote sensing, particularly vegetation primary productivity and related ecologically important functions. Primary production provides the energy that drives all subsequent ecosystem processes. Optical remote sensing has made it possible to estimate the primary productivity of vegetation over the entire Earth land surface (Running et al. 1994; Zhao et al. 2005).

20.2 Brief History of Key Optical Sensors for Vegetation Mapping

Optical remote sensing is a technique that detects the properties of the Earth's surface from space, using sensors that capture reflected radiation spanning the visible, near-infrared (NIR), and shortwave-infrared (SWIR) wavelengths (~0.4–2.5 μm) (Richards 2013). Different materials absorb and reflect light differently at various wavelengths. Thus, targets can be distinguished by their unique spectral reflectance signatures. Compared to water and bare soil, healthy vegetation generally absorbs more blue and red light in the visible spectrum for photosynthesis but reflects more NIR light (0.7–1.1 μm) to prevent tissue damage (Jones and Vaughan 2010). This unique spectral signature of vegetation is the key to monitoring vegetation structure and function with optical sensors. Since the first man-made

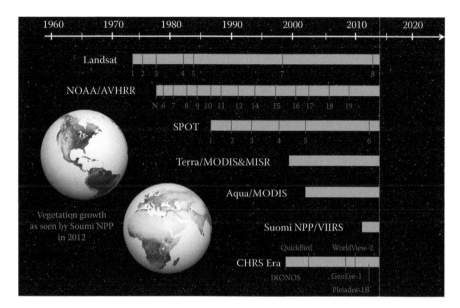

FIGURE 20.1 History of major optical sensors for remote sensing of vegetation reviewed in this chapter. The red line and font indicate the commission date and related satellite, respectively. CHRS is the abbreviation of commercial high-resolution satellite. The two hemisphere images in lower left corner show the global vegetation growth in terms of normalized difference vegetation index derived from the Visible Infrared Imaging Radiometer Suite, instrument aboard the Suomi NPP satellite. (From http://www.nasa.gov/mission_pages/NPP/news/vegetation.html#.Ut3_LRAo7IU.)

satellite (Sputnik 1) was launched in 1957, the development of artificial satellites, which provide the platform for optical sensors, has significantly enhanced the collection of remotely sensed data and offers an efficient platform to obtain vegetation information over large areas (Campbell 2002). Here, we briefly review the history of major optical sensors for the remote sensing of vegetation launched since the first International Symposium on Remote Sensing of Environment held at the University of Michigan in 1962. These programs for optical remote sensing, which follow in order of spatial resolution in this chapter, include National Oceanic and Atmospheric Administration (NOAA)/ Advanced Very-High-Resolution Radiometer (AVHRR), Moderate-Resolution Imaging Spectroradiometer (MODIS)/ Multiangle Imaging Spectroradiometer (MISR), Suomi National Polar-orbiting Partnership (Suomi NPP), Landsat, Satellite Pour l'Observation de la Terre (SPOT), and a series of commercial high-resolution satellites since 1999 (Figure 20.1).

20.2.1 NOAA/AVHRR Program

The AVHRR is a multichannel radiometer carried on the U.S. NOAA family of polar-orbiting operational environmental satellites (POES) (Table 20.1; http://www.ospo.noaa.gov/Operations/POES/index.html). The AVHRR sensor is active on two POES satellites in opposite orbits (ascending and descending), ensuring that every place on Earth can be observed every 6 h. The first AVHRR carried sensors in four spectral channels on TIROS-N (launched in October 1978). This was subsequently improved to a five-channel sensor (AVHRR2) that was initially carried on NOAA-7 (launched in June 1981). The latest sensor is AVHRR3, with six channels, first carried on NOAA-15 (launched May 1998). All AVHRR sensors have the same spatial resolution of 1.09 km at

TABLE 20.1 Summary of NOAA/POES Satellite Family

Satellite	Launch Date	Decommission Date	Sensor
TIROS-N	Oct. 13, 1978	Jan. 30, 1980	AVHRR1
NOAA-6	Jun. 27, 1979	Nov. 16, 1986	AVHRR1
NOAA-7	Jun. 23, 1981	Jun. 07, 1986	AVHRR2
NOAA-8	Mar. 28, 1983	Oct. 31, 1985	AVHRR1
NOAA-9	Dec. 12, 1984	May 11, 1994	AVHRR2
NOAA-10	Sep. 17, 1986	Sep. 17, 1991	AVHRR1
NOAA-11	Sep. 24, 1988	Sep. 13, 1994	AVHRR2
NOAA-12	May 13, 1991	Dec. 15, 1994	AVHRR2
NOAA-14	Dec. 30, 1994	May 23, 2007	AVHRR2
NOAA-15	May 13, 1998	Present	AVHRR3
NOAA-16	Sep. 21, 2000	Present	AVHRR3
NOAA-17	Jun. 24, 2002	Apr. 10, 2013	AVHRR3
NOAA-18	May 20, 2005	Present	AVHRR3
NOAA-19	Feb. 06, 2009	Present	AVHRR3

nadir (Table 20.2). The primary purpose of AVHRR is to monitor clouds and to measure the thermal emission of the Earth. However, the first two bands of AVHRR are sensitive to visible/ NIR radiation, which can be used to detect changes of terrestrial vegetation (Tucker et al. 1985, 2005; Gutman and Ignatov 1998). Based on NOAA/AVHRR, several long-term global vegetation index (GVI) datasets have been established, including NOAA/ NASA Pathfinder normalized difference vegetation index (NDVI) (1981–2000) (http://iridl.ldeo.columbia.edu/SOURCES/.NASA/. GES-DAAC/.PAL/.vegetation/.pal_ndvi.html), Global Inventory Monitoring and Modeling System (GIMMS) NDVI (1981–2011) (http://cliveg.bu.edu/modismisr/lai3g-fpar3g.html), and NOAA's GVI data (1981–2014) (http://www.ospo.noaa.gov/Products/land/gvi/NDVI.html).

TABLE 20.2   Spectral Specifications of NOAA/AVHRR Sensors

Channel Number	Ground Resolution (km)	Spectral Range (µm)	AVHRR1	AVHRR2	AVHRR3
1	1.09	0.58–0.68	√	√	√
2	1.09	0.725–1.00	√	√	√
3A	1.09	1.58–1.64			
3B	1.09	3.55–3.93	√	√	√
4	1.09	10.30–11.30	√	√	√
5	1.09	11.50–12.50		√	√

20.2.2 MODIS and MISR

As the centerpiece of NASA's Earth Science Enterprise, the Earth Observing System consists of a coordinated series of polar-orbiting satellites for continuous observations of the Earth's land, atmosphere, and ocean that offers us a detailed understanding of the biosphere and the dynamics of global change (Justice et al. 2002; Xiong and Barnes 2006). The Terra and Aqua satellites, launched in December 1999 and May 2002, respectively, are two flagships of the Earth Observing System. The MODIS is a key scientific instrument operating on both the Terra and Aqua satellites and is considered to be a major advance over the spectral, spatial, and temporal characteristics of previous sensors (Xiong and Barnes 2006). It has 36 discrete spectral bands ranging from visible through thermal emission bands (wavelengths from 0.4 to 14.4 µm) and 3 ground spatial resolutions (250 m for bands 1 and 2, 500 m for bands 3–7 and 1 km for bands 8–36) (Table 20.3). With complimentary morning (local time 10:30 AM for Terra) and afternoon (local time 1:30 PM for Aqua) observations, the Terra and Aqua sensors can image the entire Earth within 2 days with a swath of 2330 km. The MODIS Characterization Support Team from NASA (http://mcst.gsfc.nasa.gov/) is responsible for converting instrument responses (digital numbers) to the primary calibrated products (radiance and reflectance) (Xiong and Barnes 2006), from which over 50 geophysical science products have been developed by the MODIS Science Team (https://lpdaac.usgs.gov/products/modis_products_table). The MODIS Ecosystem Products include vegetation index (VI) (Huete et al. 2002), LAI (Myneni et al. 2002), vegetation continuous fields (Hansen et al. 2003), gross and net primary productivity (GPP and NPP, respectively) (Zhao and Running 2010), and global evapotranspiration (Mu et al. 2011), among others. These products offer unprecedented perspectives of ecosystem structure and function of the biosphere.

The MISR is another innovative sensor on board the Terra satellite (http://www-misr.jpl.nasa.gov/). It is designed to measure the reflected solar radiation of the Earth's system from nine discrete viewing angles and four visible/NIR spectral bands (Diner et al. 1998). The MISR instrument has nine digital cameras, with one pointing toward nadir and others pointing at forward and backward view angles of 26.1°, 45.6°, 60.0°, and 70.5°. For each direction, the cameras record reflected radiation in four spectral bands (blue, green, red, and NIR). During each orbit, MISR obtains a swath of imagery that is 360 km wide by about 20,000 km long with spatial resolutions of 250 m at nadir and

TABLE 20.3   Spectral and Spatial Resolutions of MODIS Sensors On Board Terra/Aqua

Channel Number	Spectral Range (µm)	Usage	Ground Resolution (m)
1	0.620–0.670	Land cover transformation, vegetation chlorophyll	250
2	0.841–0.876	Cloud amount, vegetation land cover transformation	250
3	0.459–0.479	Soil/vegetation differences	500
4	0.545–0.565	GV	500
5	1.230–1.250	Leaf/canopy differences	500
6	1.628–1.652	Snow/cloud differences	500
7	2.105–2.155	Cloud properties, land properties	500
8	0.405–0.420	Chlorophyll	1000
9	0.438–0.448	Chlorophyll	1000
10	0.483–0.493	Chlorophyll	1000
11	0.526–0.536	Chlorophyll	1000
12	0.546–0.556	Sediments	1000
13h	0.662–0.672	Atmosphere, sediments	1000
13l	0.662–0.672	Atmosphere, sediments	1000
14h	0.673–0.683	ChF	1000
14l	0.673–0.683	ChF	1000
15	0.743–0.753	Aerosol properties	1000
16	0.862–0.877	Aerosol properties, atmospheric properties	1000
17	0.890–0.920	Atmospheric properties, cloud properties	1000
18	0.931–0.941	Atmospheric properties, cloud properties	1000
19	0.915–0.965	Atmospheric properties, cloud properties	1000
20	3.660–3.840	Sea surface temperature	1000
21	3.929–3.989	Forest fires and volcanoes	1000
22	3.929–3.989	Cloud temperature, surface temperature	1000
23	4.020–4.080	Cloud temperature, surface temperature	1000
24	4.433–4.498	Cloud fraction, troposphere temperature	1000
25	4.482–4.549	Cloud fraction, troposphere temperature	1000
26	1.360–1.390	Cloud fraction (thin cirrus), troposphere temperature	1000
27	6.535–6.895	Midtroposphere humidity	1000
28	7.175–7.475	Upper troposphere humidity	1000
29	8.400–8.700	Surface temperature	1000
30	9.580–9.880	Total ozone	1000
31	10.78–11.28	Cloud and surface temperature, forest fires and volcanoes	1000
32	11.77–12.27	Cloud height, forest fires and volcanoes, surface temperature	1000
33	13.19–13.49	Cloud fraction, cloud height	1000
34	13.49–13.79	Cloud fraction, cloud height	1000
35	13.79–14.09	Cloud fraction, cloud height	1000
36	14.09–14.39	Cloud fraction, cloud height	1000

275 m at other angles. The multiangle viewing strategy of MISR provides a unique opportunity to characterize the structure and dynamics of the atmosphere and land surface (Diner et al. 1998). Among other applications, Terra/MISR has been used to retrieve aerosol distribution (Martonchik et al. 2002), measure cloud height (Davies and Molloy 2012), estimate LAI (Hu et al. 2003), extract canopy structure (Chen et al. 2005b), and improve the classification of land cover (Liesenberg et al. 2007).

20.2.3 Suomi NPP/VIIRS

The Suomi NPP, as a major component of NOAA's Joint Polar-Orbiting Satellite System, was designed to provide continuity with NASA's EOS (Justice et al. 2013). The satellite was launched on October 28, 2011, and was named after Verner E. Suomi, a meteorologist at the University of Wisconsin–Madison who is widely recognized as "the father of satellite meteorology" (http://www.nasa.gov/mission_pages/NPP/news/suomi.html). The Visible Infrared Imaging Radiometer Suite (VIIRS) is a key scanning radiometer on board Suomi NPP, which signifies a new era of moderate-resolution imaging capabilities following the legacy of AVHRR and MODIS (Cao et al. 2013). VIIRS is designed to collect imagery of radiometric measurements for the Earth in wavelengths ranging from 0.4 to 12.5 μm (Oudrari et al. 2012). It has 22 spectral bands, including 5 imagery bands (I bands) with 375 m spatial resolution, 1 day–night band (DNB) with 750 m spatial resolution, and 16 moderate-resolution bands (M bands) with 750 m spatial resolution (Table 20.4). VIIRS has

TABLE 20.4   Spectral and Spatial Resolutions of Suomi NPP/VIIRS

Channel Number	Spectral Range (μm)	Description	Ground Resolution (m)
I1	0.6–0.68	Visible/reflective	375
I2	0.85–0.88	NIR	375
I3	1.58–1.64	SWIR	375
I4	3.55–3.93	Medium-wave IR	375
I5	10.5–12.4	Long-wave IR	375
DNB	0.5–0.9	Visible/reflective	750
M1	0.402–0.422	Visible/reflective	750
M2	0.436–0.454		750
M3	0.478–0.488		750
M4	0.545–0.565		750
M5	0.662–0.682		750
M6	0.739–0.754	NIR	750
M7	0.846–0.885		750
M8	1.23–1.25	SWIR	750
M9	1.371–1.386		750
M10	1.58–1.64		750
M11	2.23–2.28		750
M12	3.61–3.79	Medium-wave IR	750
M13	3.97–4.13		750
M14	8.4–8.7	Long-wave IR	750
M15	10.26–11.26		750
M16	11.54–12.49		750

a large ground swath of about 3040 km and provides daily coverage of the entire globe. After about 2 years of calibration and validation, the VIIRS data have achieved provisional maturity (Cao et al. 2013) and are now being used to produce more than 20 land and cryosphere products by NOAA and NASA (Justice et al. 2013), including VI (Vargas et al. 2013), active fire (Csiszar et al. 2014), surface albedo (Wang et al. 2013), and nighttime light distribution (Miller et al. 2012).

20.2.4 Landsat Program

The U.S. Landsat program (http://landsat.usgs.gov) has been collecting images of the Earth's land surface for over four decades, providing the longest continuous archive of the Earth's surface conditions (Markham and Helder 2012). The first Landsat satellite, originally named "Earth Resources Technology Satellite," was launched in 1972. To date, eight Landsat satellites have been launched. All but Landsat 6 successfully reached orbit (Table 20.5). The most recent Landsat satellite, the eighth in the series, was launched in February 2013. The Return Beam Vidicon (RBV), a television camera carried on board Landsat 1 through 3, obtained visible and NIR photographic images, while the Multispectral Scanner (MSS) sensors, which was carried on board Landsat 1 through 5, acquired digital images around the globe nearly continuously from July 1972 to October 1992. Compared with MSS, RBV was rarely used scientifically but considered only for engineering evaluation purposes. It was replaced by the Thematic Mapper (TM) sensor on board Landsat 4 and 5 satellites, which consisted of seven spectral bands with a 16-day repeat cycle and a spatial resolution of 30 m (the thermal infrared band 6 was collected at 120 m spatial resolution). By the time of its decommission in November, 2011. Landsat 5 had orbited the Earth for 28 years—an extraordinary success for NASA—far exceeding its original 3-year design life (https://landsat.usgs.gov/Landsat5Tribute.php).

On Landsat 7, the TM sensor was replaced by the Enhanced Thematic Mapper Plus (ETM+), which included the addition of a panchromatic band 8 at 15 m spatial resolution (Table 20.6) that can be used to *sharpen* the other bands. However, the Scan Line

TABLE 20.5   Landsat Satellites Launched

Satellite	Launch Date	Decommission Date	Orbit Height (km)	Revisit Time (Days)	Sensors
Landsat 1	Jul. 1972	Jan. 1978	917	18	RBV/MSS
Landsat 2	Jan. 1975	Feb. 1982	917	18	RBV/MSS
Landsat 3	Mar. 1978	Mar. 1983	917	18	RBV/MSS
Landsat 4	Jul. 1982	Dec. 1993	705	16	MSS/TM
Landsat 5	Mar. 1984	Dec. 2012	705	16	MSS/TM
Landsat 6	Oct. 1993	Failed	—	—	—
Landsat 7	Apr. 1999	Present	705	16	ETM+
Landsat 8	Feb. 2013	Present	705	16	OLI/TIRS

RBV, return beam vidicon; MSS, multispectral scanner; TM, thematic mapper; ETM+, enhanced thematic mapper plus; OLI, operational land imager; TIRS, thermal infrared sensor.

TABLE 20.6 Spectral and Spatial Resolutions of Landsat Sensors

Sensor	Channel Number	Spectral Range (µm)	Description	Ground Resolution (m)
RBV[a]	1	4.75–5.75	Blue	80
	2	5.80–6.80	Orange–red	80
	3	6.90–8.30	Red–NIR	80
MSS	4	0.5–0.6	Green	57 × 79
	5	0.6–0.7	Red	57 × 79
	6	0.7–0.8	NIR	57 × 79
	7	0.8–1.1	NIR	57 × 79
	8[b]	10.4–12.6	Thermal	57 × 79
TM	1	0.45–0.52	Blue	30
	2	0.52–0.60	Green	30
	3	0.63–0.69	Red	30
	4	0.76–0.90	NIR	30
	5	1.55–1.75	SWIR	30
	6	10.40–12.50	Thermal	120
	7	2.09–2.35	SWIR	30
ETM+	1	0.45–0.52	Blue	30
	2	0.52–0.60	Green	30
	3	0.63–0.69	Red	30
	4	0.77–0.90	NIR	30
	5	1.55–1.75	SWIR	30
	6	10.40–12.50	Thermal	60
	7	2.08–1.35	SWIR	30
	8	0.52–0.90	Pan	15
OLI	1	0.43–0.45	Deep blue	30
	2	0.45–0.51	Blue	30
	3	0.53–0.59	Green	30
	4	0.64–0.67	Red	30
	5	0.85–0.88	NIR	30
	6	1.57–1.65	SWIR1	30
	7	2.11–2.29	SWIR2	30
	8	0.50–0.68	Panchromatic	15
	9	1.36–1.38	Cirrus clouds	30
TIRS	10	10.60–11.19	Thermal	100
	11	11.50–12.51	Thermal	100

Note: See Table 20.5 for sensor abbreviations.

NIR, near infrared; SWIR, shortwave infrared; pan, panchromatic.

[a] Landsat 3 had two RBV cameras with 40 m ground resolution.

[b] Only Landsat 3 had this thermal channel.

Corrector on the satellite failed in May of 2003, causing a permanent loss of about 25% of data toward the scanning edges in all subsequent Landsat 7 images. Fortunately, the successful launch of Landsat 8 in 2013 ensured the continuity of Landsat data. The Operational Land Imager (OLI) sensors on board Landsat 8 include refined versions of the seven TM and ETM+ heritage bands, along with two new bands: a deep blue band for coastal/aerosol studies and a SWIR band for cirrus cloud detection (Table 20.6). Landsat 8 Thermal Infrared Sensors (TIRS) are composed of two thermal bands with a spatial resolution of 100 m (Table 20.6). Both OLI and TIRS sensors provide improved signal-to-noise radiometric

performance quantized over a 12-bit dynamic range compared with the 8-bit instruments for TM and ETM+ sensors.

Conceived in the 1960s, the Landsat program has kept improving its imaging capability and quality while ensuring continuity over the full instrument record (Loveland and Dwyer 2012). To date, it has provided the longest and most geographically comprehensive record of the Earth's surface. Thanks to a data policy change in 2008, all new and archived Landsat images have been made freely available to the public by the U.S. Geological Survey (Woodcock et al. 2008), which has spurred a dramatic increase in scientific applications using Landsat imagery (Wulder et al. 2012).

20.2.5 SPOT Program

The SPOT program is a joint Earth observing satellite family initiated by France in partnership with Belgium and Sweden (http://www.vgt.vito.be/). Since 1986, six SPOT satellites have been successfully launched (Table 20.7). Currently, SPOT 5 and 6 are operational. The High-Resolution Visible (HRV) sensor with one panchromatic (10 m spatial resolution) and three multispectral bands (20 m spatial resolution; green, red, NIR) were carried on board SPOT 1 through 3 (Table 20.8). They have a scene size of 60 × 60 km² and a revisit interval of 1–4 days, depending on the latitude. SPOT 4 featured the High-Resolution Visible Infrared (HRVIR) instrument, which was similar to the HRV but with the addition of a SWIR band and a narrower panchromatic band (Table 20.8). SPOT 5 carries the High-Resolution Geometrical sensor (derived from HRVIR), offering a finer resolution of 2.5–5 m in panchromatic mode and 10 m in multispectral mode (20 m for SWIR) (Table 20.8). SPOT 6 was launched in September 2012, carrying the New AstroSat Optical Modular Instrument (NAOMI). NAOMI is capable of imaging the Earth with a resolution of 1.5 m panchromatic and 6 m multispectral (blue, green, red, NIR) with daily revisit capability, providing the finest level of spatial detail in the history of the SPOT family of satellites (Table 20.8). It is worth noting that the vegetation sensor was carried on board SPOT 4 and 5 (launched in 1998 and 2002, respectively). SPOT/vegetation was designed to provide daily coverage of the entire globe with a spatial resolution of 1.15 km. Unlike many other commercial high-resolution

TABLE 20.7 Summary of SPOT Satellite Family

Satellite	Launch Date	Decommission Date	Orbit Height (km)	Revisit Time (Day)	Sensors
SPOT 1	Feb. 1986	Dec. 1990	832	1–4	HRV
SPOT 2	Jan. 1990	July 2009	832	1–4	HRV
SPOT 3	Sep. 1993	Nov. 1997	832	1–4	HRV
SPOT 4	Mar. 1998	July 2013	832	1–4	Vegetation/HRVIR
SPOT 5	May 2002	Present	832	1–4	Vegetation/HGR
SPOT 6	Sep. 2012	Present	694	1–4	NAOMI

HRV, high-resolution visible; HRVIR, high-resolution visible infrared; NAOMI, new AstroSat optical modular instrument.

TABLE 20.8 Spectral and Spatial Resolutions of Optical Sensors On Board SPOT Satellites

Sensor	Mode	Description	Spectral Range (μm)	Ground Resolution (m)
HRV	Multispectral	Green	0.50–0.59	20
		Red	0.61–0.68	20
		NIR	0.78–0.89	20
	Panchromatic	Pan	0.50–0.73	10
HRVIR	Multispectral	Green	0.50–0.59	20
		Red	0.61–0.68	20
		NIR	0.79–0.89	20
		MIR	1.58–1.75	20
	Panchromatic	Pan	0.61–0.68	10
HGR	Multispectral	Green	0.50–0.59	10
		Red	0.61–0.68	10
		NIR	0.79–0.89	10
		SWIR	1.58–1.75	20
	Panchromatic	Pan	0.51–0.73	5/2.5
NAOMI	Multispectral	Blue	0.45–0.53	6
		Green	0.53–0.59	6
		Red	0.63–0.70	6
		NIR	0.76–0.89	6
	Panchromatic	Pan	0.45–0.75	1.5
Vegetation	Multispectral	Blue	0.43–0.47	1150
		Red	0.61–0.68	1150
		NIR	0.78–0.89	1150
		SWIR	1.58–1.75	1150

Note: Sensor abbreviations seen in Table 20.7.

images, some SPOT/vegetation products are publicly available. The 10-day 1 km global NDVI, for example, is available from May 1998 to the present (http://www.vgt.vito.be/) and has been valuable for studying agriculture, deforestation, and other vegetation changes on a broad scale (Kamthonkiat et al. 2005; Liu et al. 2010).

20.2.6 Commercial High-Resolution Satellite Era

IKONOS, which was launched in 1999, is the first high-resolution commercial Earth observation satellite that collects imagery at submeter (0.82 m for panchromatic band) spatial resolution. This marked the start of a new era of high-resolution Earth observation by commercial satellites, which may revolutionize the future of the entire photogrammetric and remote sensing community (Dial et al. 2003). After IKONOS, a series of commercial civilian satellites with optical sensors were launched to produce panchromatic images with spatial resolutions ranging from less than ½ to 3 m and multispectral images with spatial resolution ranging from 2 to 10 m (Table 20.9). The finer resolution panchromatic bands can be used to sharpen the coarser-resolution multispectral bands, increasing the spatial detail of multispectral images (Zhang and Mishra 2012). Based on the previously launched satellites IKONOS and OrbView-3, the U.S. commercial

company GeoEye, Inc. (merged with DigitalGlobe since January 2013) launched by far the finest-spatial-resolution civilian Earth observation satellite (GeoEye-1) in September 2008. GeoEye-1 provides 0.41 m panchromatic and 1.65 m multispectral (blue, green, red, NIR) imagery and features a revisit time of less than 3 days with a swath of 22.2 km. Based on QuickBird and WorldView-1, the U.S. company DigitalGlobe, Inc. launched the first high-resolution commercial satellite with eight multispectral imaging bands in October 2009. This satellite, known as WorldView-2, is capable of collecting panchromatic imagery at 0.46 m spatial resolution and multispectral (coastal, blue, green, yellow, red, red edge, NIR1 and NIR2) imagery at 1.84 m spatial resolution with an average revisit time of 1.1 days. Compared to the four standard multispectral bands (blue, green, red, NIR), the additional bands increase the spectral information used for vegetation analysis at high spatial resolutions. Depending on budget and usage purposes, other high-resolution commercial satellite images that can be employed include France's Pleiades-1A/B (0.5 m pan, 2 m multispectral), Korea's KOMPSAT-2 (1 m pan, 4 m multispectral), China–Taiwan's FORMOSAT-2 (2 m pan, 4 m multispectral), and Japan's Advanced Land Observing Satellite (ALOS) (2.5 m pan, 10 m multispectral), among others.

20.2.7 Future Direction of Optical Remote Sensing

Optical sensors are poised to acquire increasingly high-quality data across a wide range of spatial, temporal, and spectral resolutions. For instance, the U.S. DigitalGlobe, Inc. is planning to launch its next superspectral, high-resolution commercial satellite named WorldView-3 in 2014. Operating at an expected altitude of 617 km, WorldView-3 will provide 0.31 m panchromatic resolution, 1.24 m multispectral resolution, and 3.7 m SWIR resolution of the Earth with an average revisit time of less than 1 day (http://www.digitalglobe.com). Meanwhile, the European Space Agency (ESA) is carrying out one of the most ambitious Earth observation program to date, called Copernicus. To satisfy the operational needs of Copernicus, up to 30 Sentinel satellites with various sensors will be developed (http://www.esa.int/ESA). The first Sentinel satellite (S1) had been successfully put in orbit in April 2014. Undoubtedly, integrating multiple sources of optical remote sensing will offer a valuable opportunity for the scientific community to investigate and understand the structures and functions of terrestrial ecosystems at different spatial and temporal resolutions (Weng 2011; Richards 2013).

20.3 Optical Remote Sensing of Vegetation Structure

Optical remotely sensed signals originate from the photons in the solar spectrum after interactions with the land surface. Remote sensing signals over vegetated areas are determined by the abundance and spatial organization of vegetation (Li and Strahler 1985; Asrar et al. 1992). Therefore, information about vegetation structure can be derived from optical remotely sensed

TABLE 20.9　Major Commercial High-Resolution Satellites Since 1999

Satellite	Year Launched	Country	Pan Band (µm)/Ground Resolution (m)	Multispectral Bands[b] Ground Resolution (m)	Swath (km)	Revisit Time (Day)
IKONOS	1999	United States	(0.45–0.90) 0.82	(Blue, green, red, NIR) 4	11.3 × 11.3	3–4
QuickBird	2001	United States	(0.405–1.053) 0.61	(Blue, green, red, NIR) 2.44	16.5 × 16.5	1–3.5
OrbView-3	2003	United States	(0.45–0.90) 1	(Blue, green, red, NIR) 4	8 × 8	1–3
FORMOSAT-2	2004	China–Taiwan	(0.45–0.90) 2	(Blue, green, red, NIR) 4	24 × 24	1
CartoSat-1	2005	India	(0.5–0.85) 2.5	—	25 × 25	5
ALOS	2005	Japan	(0.52–0.77) 2.5	(Blue, green, red, NIR) 10	70 × 70	2
EROS-B	2006	Israel	(0.5–0.9) 0.7	—	7 × 7	5–6
KOMPSAT-2	2006	Korean	(0.50–0.90) 1	(Blue, green, red, NIR) 4	15 × 15	1–3
WorldView-1	2007	United States	(0.40–0.90) 0.46[a]	—	17.7 × 17.7	1–5
GeoEye-1	2008	United States	(0.45–0.90) 0.41[a]	(Blue, green, red, NIR) 1.65	22.2 × 22.2	1–3
RapidEye	2008	German	—	(Blue, green, red, NIR) 5	77 × 77	1
WorldView-2	2009	United States	(0.45–0.80) 0.46[a]	(Coastal, blue, green, yellow, red, red edge, NIR1 and NIR2)[c] 1.85	16.4 × 16.4	1–5
Pleiades-1A	2011	France	(0.48–0.83) 0.5	(Blue, green, red, NIR) 2	20 × 20	1
SPOT6	2012	France	(0.450–0.745) 1.5	(Blue, green, red, NIR) 5	60 × 60	1–3
ZY-3	2012	China	(0.50–0.80) 2.1–3.5	(Blue, green, red, NIR) 6	52 × 52	3–5
Pleiades-1B	2012	France	(0.48–0.83) 0.5	(Blue, green, red, NIR) 2	20 × 20	1
GF-1	2013	China	(0.45–0.90) 2	(Blue, green, red, NIR) 8	60 × 60	4

[a] Due to U.S. Government Licensing, the imagery will be made available commercially at ground resolution of 0.5 m.

[b] Different satellites may have slightly different spectral ranges for each of their multispectral bands.

[c] The spectral ranges for WorldView-2 are 0.40–45 µm (coastal), 0.45–0.51 µm (blue), 0.51–0.58 µm (green), 0.585–0.625 µm (yellow), 0.63–0.69 µm (red), 0.705–0.745 µm (red edge), 0.77–0.895 µm (NIR1), and 0.86–1.04 µm (NIR2), respectively.

data. In this section, we review the capabilities of optical remote sensing in deriving information about vegetation cover, forest successional stage, LAI, and biomass, all of which are essential biophysical information to understand terrestrial ecosystem functions.

20.3.1　Vegetation Cover

Vegetation cover is perhaps the simplest measure of vegetation structure that can be derived from remote sensing. The most common approach for mapping vegetation cover from remotely sensed imagery is to assign a single class to each pixel. Vegetation cover can then be estimated as the percentage of pixels classified as vegetation. This approach makes an implicit assumption that

each pixel represents a homogenous cover type. This assumption may be a reasonable one when the pixel size is significantly smaller than the average vegetation patch size. However, this assumption is rarely valid for coarse-resolution remotely sensed imagery because coarse-resolution pixels are generally comprised a mixture of several cover types. Assuming homogeneous land cover composition at the pixel level can lead to substantial errors in estimates of areal abundance (Foody and Cox 1994; Moody and Woodcock 1994).

More accurate estimation of vegetation cover from remotely sensed imagery is usually based on subpixel land cover composition, that is, the fraction of a pixel that is covered by vegetation. The fractional vegetation cover (fc) concept, introduced by Deardorff (1978), is a key component of the current generation of climate models (Zeng et al. 2000). Many methods have been proposed to derive fc from remotely sensed imagery.

20.3.1.1 Regression

A common approach to estimate vegetation fraction cover or percent tree cover is to develop an empirical relationship between ground-based measurements with remotely sensed signals, such as spectral VIs (e.g., NDVI, enhanced VI [EVI]) or suites of other remotely sensed measurements. A variety of model types have been used for this purpose, including ordinary least squares regression (Jiapaer et al. 2011), generalized linear models (Schwarz and Zimmermann 2005), stepwise multiple regression (Cohen et al. 2001), reduced major axis regression (Hayes et al. 2008), and a variety of machine learning methods such as decision trees and neural networks (NNs) (Colditz et al. 2011; Verrelst et al. 2012). At the global scale, the MODIS Vegetation Continuous Fields product (MOD44B) estimates subpixel percentages of tree cover, nontree vegetation cover, and bare ground at 250 m spatial resolution using regression trees and a large suite of metrics calculated from MODIS reflectance data. The algorithm estimates a mean vegetation cover for each node in the regression tree and then uses a linear model fit to the independent variable to fine-tune the tree cover estimation for each node (Hansen et al. 2002, 2003, 2005).

20.3.1.2 Fuzzy Classification

In a typical application of supervised classification of remotely sensed imagery, a single land cover/land use class is assigned to each pixel based on its spectral similarity to training classes (so-called "hard" classifiers). Some of these classifiers can also be modified to predict gradients of class membership—"fuzzy" or "soft" classifications—that provide a relative measure of the similarity of the pixel spectral signature to the class signature. The posterior probabilities from maximum likelihood classifiers, for example, have been used to estimate subpixel land cover fractions, though with limited success (Bastin 1997). Artificial NNs, frequently used in hard classifications, have also been used for deriving subpixel membership functions (Foody 1996; Atkinson et al. 1997).

Some classifiers, such as the fuzzy c-means algorithm, are specifically designed to provide fuzzy membership functions (Foody and Cox 1994). In these approaches, each pixel generally receives a membership value (ranging from 0 to 1) for each class, with the membership values summing to 1. Relatively pure pixels are likely to receive large membership values for a single class, while mixed pixels are more likely to receive intermediate values for multiple classes. The relationship between fuzzy membership values and subpixel land cover fractions can be further improved through a simple regression model based on reference data (Foody and Cox 1994). Fractional land cover obtained using these fuzzy classifiers generally compares favorably with other methods and provides considerable improvement in areal estimates of forest cover over those obtained from hard classifications (Foody and Cox 1994; Atkinson et al. 1997; Bastin 1997).

20.3.1.3 Mixture Models with Spectral Vegetation Indices

Simple two-class mixture models typically assume that pixels in the natural environment are composed of vegetation and soil background. The radiance received at the satellite sensor is therefore assumed to be a mixture of the spectral signatures of vegetation and soil, weighted by their respective fractions. Gutman and Ignatov (1998) proposed a simple linear mixture model based on NDVI to estimate the proportions of vegetation and soil within a pixel:

$$\text{fc} = \frac{\text{NDVI} - \text{NDVI}_s}{\text{NDVI}_v - \text{NDVI}_s} \quad (20.1)$$

where

NDVI is the VI for a given pixel

NDVI_v and NDVI_s are the VIs corresponding to pixels completely covered with dense vegetation and soil, respectively

Other studies have suggested that multiple scattering in vegetation canopies can result in nonlinear relationships between fc and NDVI and have therefore proposed similar alternative models (Carlson and Ripley 1997):

$$\text{fc} = \left(\frac{\text{NDVI} - \text{NDVI}_s}{\text{NDVI}_v - \text{NDVI}_s} \right)^2 \quad (20.2)$$

Equation 20.1 has been applied for global scale estimation of fc (Zeng et al. 2000), with NDVI derived from the maximum 12-month NDVI of each pixel in AVHRR imagery, NDVI_v computed separately for each vegetated land cover class in the IGBP database based on NDVI histograms, and a globally uniform NDVI_s of 0.05 (corresponding to the fifth percentile of the NDVI histogram for the barren or sparsely vegetated category). Results from this model were comparable with, but systematically less than, fc calculated from a more complex global linear mixture model (DeFries et al. 1999, 2000).

20.3.1.4 Spectral Mixture Analysis

The procedure described in the previous section is a special case of a more general technique called spectral mixture analysis (SMA). The generalized formulation of SMA techniques can be represented in matrix notation as

$$\mathbf{x} = \mathbf{M}\mathbf{f} + \mathbf{e}, \quad (20.3)$$

where

$\mathbf{x}$ is a column vector of the observed reflectance (with one element per spectral band)

$\mathbf{M}$ is a matrix of spectral endmembers (with each column representing the spectral signature of pure pixels for each endmember)

$\mathbf{f}$ is a column vector of subpixel proportions for each endmember

$\mathbf{e}$ is a column vector of error residuals

Once **x** and **M** are known, Equation 20.3 can be inverted and solved for the unknown **f** using a variety of techniques (including ordinary least squares), typically with the constraint that the sum of elements in **f** equals unity and each element takes a value within (0,1) (Somers et al. 2011).

The major challenge in the use of SMA is the selection of appropriate endmembers and their spectral signatures. Endmembers can be derived either directly from remotely sensed imagery (image endmembers) (e.g., DeFries et al. 1999; Song 2004) or from field or laboratory measurements (e.g., Adams et al. 1995; Roberts et al. 1998). The number of endmembers that can be used is limited by the dimensionality of the remotely sensed image data. In the case of Landsat imagery, for example, SMA techniques are generally limited to 3–4 endmembers. In many complex landscapes, 3–5 endmembers may be insufficient to represent the spectral and spatial variability within an image. A variety of techniques exist to account for endmember variability (reviewed in Somers et al. 2011), including the multiple endmember SMA technique (Roberts et al. 1998), in which endmember models are selected separately for each pixel in the image from a large library of spectral endmembers to construct numerous candidate models, from which the "best" candidate model is selected for each pixel to perform SMA. Somers et al. (2011) suggest that these types of iterative endmember selection approaches can provide a more effective representation of endmember variability than simple SMA approaches (in which endmember signatures are assumed constant across the entire image). Song (2005) developed the Bayesian spectral mixture analysis (BSMA) to account for endmember signature variation when estimating fc in a pixel. The endmember spectral signature in BSMA is represented by a probability mass function instead of a constant. Deng and Wu (2013) further developed an algorithm that adaptively generates endmember spectral signatures over space to account for endmember spectral signature variations.

20.3.2 Forest Successional Stages

Forest succession can be defined as the change in the 3D architecture and species composition of forest communities through time (Pickett et al. 2013). Successional stage serves as a useful proxy for forest age, as well as competition-mediated demographic and structural development (Peet and Christensen 1988). Successional processes have a profound impact on the provision of ecological goods and services including productivity (Gower et al. 1996), nutrient cycling (Law et al. 2001), and biodiversity (Denslow 1980). Though succession is a continuous process, most models characterize succession as a four-stage process including (1) stand initiation/establishment, (2) stem exclusion/thinning, (3) understory reinitiation/transition, and (4) old growth/steady state (Peet and Christensen 1987; Oliver and Larson 1996). Remote sensing technologies, including physical and empirical-based models, offer an efficient method for monitoring forest succession over large spatial extents.

20.3.2.1 Physical-Based Models

Physical-based models simulate vegetation canopy reflectance based on the physical principles of interaction among incoming solar radiation and canopy structural elements. Forward models like the Li–Strahler model (Li and Strahler 1985) have proven useful for understanding the relationship between vegetation structure and canopy reflectance. The Li–Strahler model is a geometric–optical model that simulates canopy reflectance as viewed by the sensor based on the weighted average of individual scene components within a pixel created by the Sun–tree crown geometry. This model can be inverted to estimate key canopy structure parameters that manifest successional stage, including mean crown size and canopy cover (Franklin and Strahler 1988; Wu and Strahler 1994). Li et al. (1995) improved the Li-Strahler model by representing tree crowns in the Geometric Optical (GO) model as ellipsoids rather than cones and incorporating multiple scattering of photons with a turbid medium Radiative Transfer (RT) model to become the GORT model.

Song et al. (2002) coupled the GORT model with a forest succession model ZELIG (Urban 1990), which simulates stand growth and development, to understand how forest succession changes in the spectral/temporal domain. Using this hybrid model to simulate Landsat TM reflectance of stand succession from open conditions to young, mature, and old-growth stages, they found forest succession produces highly nonlinear temporal trajectories in the tasseled cap brightness/greenness space. The nonlinear spectral/temporal trajectory pattern produced by the GORT–ZELIG simulation compared well with that derived from a time series of Landsat TM images and stand age information from Forest Inventory and Analysis (FIA) stand data collected by the U.S. Forest Service (Song et al. 2007).

20.3.2.2 Empirical-Based Approaches

Empirical-based approaches to the remote sensing of forest succession include (1) indirect space-for-time substitutions using single or multidate imagery and (2) direct monitoring of successional change using multitemporal change detection and time series analysis. While the former is more effective at distinguishing successional stands over large landscapes, the latter is better adapted to capture ongoing successional change in individual stands. Numerous studies show both approaches to produce robust results, though factors of uncertainty remain, such as atmospheric and ground conditions as well as the confounding effects of topography, Sun and view angles, and phenology (Song and Woodcock 2003).

20.3.2.2.1 Space-for-Time Substitution

Given that the short historical record of remotely sensed imagery is often insufficient to capture temporal processes of forest succession that could stretch over centuries, space-for-time substitution uses stands in different successional stages at different locations in space to construct a proxy successional trajectory for a single stand through time. This approach is particularly useful for distinguishing mature and old-growth forests. For example,

Fiorella and Ripple found most raw Landsat TM bands to be inversely correlated with forest age in Pacific Northwest (PNW), with mean TM spectral values tending to be lower for old-growth stands compared with those for mature stands (Fiorella and Ripple 1993a,b). Jakubauskas (1996) found a nonlinear trend in TM spectral reflectance from early to late successional forests in Wyoming resulting from the combined effects of overstory canopy development, increasing canopy shadow, and understory conditions. Spectral indices have likewise proven effective at distinguishing successional stages. For instance, TM 4/5 ratio (Fiorella and Ripple 1993a) and Tasseled Cap wetness have been used to distinguish successional stage in the conifer forests of the PNW (Cohen and Spies 1992; Fiorella and Ripple 1993b; Cohen et al. 1995), while the NDVI/ETM+ band 5 ratio successfully distinguished four successional stages of tropical secondary forests in Brazil (Vieira 2003).

Sabol et al. (2002) mapped structural development and stand age in Washington, United States, using an SMA approach consisting of four spectral endmembers: green vegetation (GV), nonphotosynthetic vegetation (NPV), soil, and shade (topographic shading and canopy shadows). They found successional stage to follow a nonlinear trajectory, characterized by high NPV from slash after clear-cut, to high GV during canopy closure, and finally to higher shade fractions as forests mature and gaps develop. Other techniques using the space-for-time substitution approach include those utilizing spatial predictors. Cohen et al. (1990) identified crown-gap patterning characteristic of different successional stages in Douglas-fir forests by interpreting semivariograms of DN values at different spatial resolutions. They found a pronounced periodicity for the more spatially clumped old-growth crowns compared with the more texturally homogeneous early successional stands. Cohen and Spies (1992) compared spatial versus spectral variables to predict stand structural attributes in Douglas-fir forests, finding the most robust results using textural measures from a 10 m panchromatic SPOT HRV image.

Liu et al. (2008) demonstrated the advantages of multitemporal versus single-date Landsat TM images to distinguish successional groups in the PNW, a result corroborated by others (Song et al. 2007). Multitemporal Landsat imagery has been used to evaluate tropical secondary forest regrowth in Brazil (Steininger 1996) and distinguish secondary forests from agricultural lands and old-growth forests in southern Costa Rica (Helmer et al. 2000). Jiang et al. (2004) used a dense stack of Landsat ETM+ images for successional classification in the PNW, achieving high accuracy for late-seral forests. Bergen and Dronova (2007) used multitemporal Landsat ETM+ data to demonstrate the relationship between ecological land units and the successional pathways of hardwood forests in northern Michigan.

20.3.2.2.2 Multitemporal Change Detection and Time Series Analysis

Multitemporal imagery can be used to capture successional processes by assessing change at the stand/pixel level between two or more dates. While this approach circumvents errors from spatial extrapolation in the space-for-time substitution approach, observation of successional development is limited by the temporal extent of the satellite record. In addition, the success of time series analysis in monitoring subtle successional change over time ultimately hinges on the successful calibration of the image series (Song and Woodcock 2003; Schroeder et al. 2006).

In a classic paper on the subject, Hall et al. (1991) used Landsat MSS images from 1973 and 1983 to infer transition rates in ecological states associated with forest succession in the boreal forests of Minnesota, United States. McDonald et al. (2007) used change vector analysis to validate the prediction of successional models (e.g., Oosting 1942) that predict the transition from pine to hardwood forests in North Carolina using a Landsat time series from 1986 and 2000. Brandt et al. (2012) employed MSS and TM/ETM+ images from 1974 and 2009 to distinguish successional pathways differentially affected by anthropogenic pressure in Yunnan, China.

Provided forest stands were initiated within the satellite record for the area in question, one approach to infer the approximate age for primary (Lawrence and Ripple 1999) and secondary forest (Cohen et al. 2002; Lucas et al. 2002; Schroeder et al. 2007) is to estimate time since the last stand-replacing disturbance. More recently, a number of time series methods have been developed to automate forest disturbance and recovery monitoring in early successional forests by exploiting the relatively long and growing archive available from the Landsat and Landsat-like family of sensors. The central premise of this approach is that changes in vegetation cover such as disturbance and early successional regrowth leave a distinct temporal signal in spectral space that can be identified to derive metrics such as disturbance date and intensity, as well as regeneration rate (Healey et al. 2005; Kennedy et al. 2007). Prominent examples of such automated approaches include the vegetation change tracker (Huang et al. 2010), the Landsat-based detection of Trends in Disturbance and Recovery (LandTrendr) algorithm (Kennedy et al. 2010), and TimeSync (Cohen et al. 2010), a software tool used to aid in image interpretation and validation of time series products. LandTrendr was used to predict current forest structure attributes based on disturbance history, with Landsat-derived predictors performing comparably with, and in some cases better than, LiDAR-based models (Pflugmacher et al. 2012). More recently, forest disturbance detection algorithms have employed multisensor fusion to provide near real-time vegetation change monitoring (Zhu et al. 2012b; Xin et al. 2013).

20.3.3 Leaf Area Index and Clumping Index

Since a leaf surface is a substrate on which major physical and biological processes of plants occur, LAI is arguably the most important vegetation structural parameter and is indispensable for all process-based models for estimating terrestrial fluxes of energy, water, carbon, and other masses. It is therefore of interest not only to the remote sensing community that produces LAI maps but also to ecological, hydrological, and meteorological communities that use LAI products for various modeling purposes (Sellers et al. 1997; Dai et al. 2003; Chen et al. 2005a).

20.3.3.1 Definitions and Ground Measurement Techniques

LAI is defined as one-half the total (all-sided) leaf area per unit ground surface area (Chen and Black 1992; see also review by Jonckheere et al. 2004). It is often indirectly measured using optical instruments that acquire transmitted radiation through a plant canopy, from which the canopy gap fraction is derived. The canopy gap fraction, $P(\theta)$, at zenith angle θ, is related to the plant area index, denoted as L_t, which includes both green leaves and nongreen materials such as stems and branches that intercept radiation. This relation is given by the following equation:

$$P(\theta) = e^{-G(\theta)\Omega L_t/\cos\theta} \qquad (20.4)$$

where

   G(θ) is the projection coefficient, which is determined by the leaf angular distribution (Monsi and Saeki 1953; Campbell 1990)

   Ω is the clumping index, which is related to the leaf spatial distribution pattern (Nilson 1971)

If $P(\theta)$ is measured at one angle and $G(\theta)$ and Ω are known, L_t can be inversely calculated using Equation 20.4. However, both $G(\theta)$ and Ω are generally unknown; therefore, different optical instruments have been developed to measure these unknown parameters.

The Li-Cor LAI 2000 Plant Canopy Analyzer is an optical instrument developed to address the issue of unknown $G(\theta)$ due to nonspherical leaf angle distribution. It measures the diffuse radiation transmission simultaneously in five concentric rings covering the zenith angle range from 0° to 75°, that is, $P(\theta)$ at five angles. These measurements are used to calculate the LAI based on Miller's theorem (Miller 1967):

$$L_e = 2 \int_0^{\pi/2} \ln\frac{1}{P(\theta)} \cos\theta\sin\theta d\theta \qquad (20.5)$$

The original Miller's equation was developed for canopies with random leaf spatial distributions, that is, $\Omega = 1$, and allows the calculation of LAI without the knowledge of $G(\theta)$ when $P(\theta)$ is measured over the full zenith angle range and its azimuthal variation is ignored. LAI and $G(\theta)$ can also be derived simultaneously using multiple angle measurements (Norman and Campbell 1989). For spatially nonrandom canopies, Miller's theorem actually calculates the effective LAI (Chen et al. 1991), expressed as

$$L_e = \Omega L \qquad (20.6)$$

Equation 20.5 can be discretized to calculate L_e using the $P(\theta)$ measurements at five zenith angles by LAI-2000. L_e calculated this way includes all green and nongreen materials above the instrument. With measured L_e, the following equation is proposed to calculate LAI (Chen 1996a):

$$L = \frac{(1-\alpha)L_e}{\Omega} \qquad (20.7)$$

where α is the woody-to-total area ratio. The total area includes both green leaves and nongreen materials such as stems, branches, and attachments (e.g., moss) on branches. The α value is generally in the range of 0.05–0.3 depending mostly on forest age (Chen et al. 2006).

There are also optical techniques for indirect measurement of the clumping index (Chen and Cihlar 1995). These techniques are based on the canopy gap size distribution theory of Miller and Norman (1971). An optical instrument named Tracing Radiation and Architecture of Canopies (TRAC, Chen and Cihlar 1995) was developed to measure the canopy gap size distribution using the solar beam as the probe. In conifer canopies, the gaps between needles within a shoot (a basic collection of needles around the smallest twig) are obscured due to the penumbra effect, and the clumping index derived from TRAC measurements represents the clumping effects at scales larger than the shoot (treated as the foliage element), denoted as Ω_E. According to a random gap size distribution curve based on Miller and Norman's theory, large gaps caused by the nonrandom foliage element distribution, that is, those caused by tree crowns and branches, are identified and removed to reconstruct a random gap size distribution. With this gap removal technique, Ω_E is calculated from the following equation (Chen and Cihlar 1995; Leblanc 2002):

$$\Omega_E(\theta) = \frac{\ln[F_m(0,\theta)]\left[1-F_{mr}(0,\theta)\right]}{\ln[F_{mr}(0,\theta)]\left[1-F_m(0,\theta)\right]} \qquad (20.8)$$

where

   $F_m(0,\theta)$ is the total canopy gap fraction at zenith angle, that is, the accumulated gap fraction from the largest to smallest gaps

   $F_{mr}(0,\theta)$ is the total canopy gap fraction after removing large gaps resulting from the nonrandom foliage element distribution due to canopy structures such as tree crowns and branches

Clumping within individual shoots depends on the density of needles on a shoot. This level of foliage clumping was recognized and estimated in various ways by Oker-Blom (1986), Gower and Norman (1991), Stenberg et al. (1994), Fassnacht et al. (1994), etc. Based on a theoretical development by Chen

(1996a), this clumping is quantified using the needle-to-shoot area ratio (γ_E) as follows:

$$\gamma_E = \frac{A_n}{A_s} \qquad (20.9)$$

where

A_n is half the total needle area (including all sides) in a shoot

A_s is half the shoot area (for a shoot that can be approximated by an ellipsoid, the total shoot area is the ellipsoid surface area, not the projected elliptical area)

To obtain γ_E, shoots need to be sampled from trees of different sizes at different heights, and A_n and A_s need to be measured using laboratory equipment (Chen et al. 1997; Kucharik et al. 1999). For broadleaf forests, the individual leaves are the foliage elements, and therefore $\gamma_E = 1$.

The total clumping of a stand can therefore be written as

$$\Omega = \frac{\Omega_E}{\gamma_E} \qquad (20.10)$$

and the final equation for deriving LAI from indirect measurements is

$$L = \frac{(1-\alpha)L_e\gamma_E}{\Omega_E} \qquad (20.11)$$

Different instruments can be used to measure the different variables in this equation in order to determine LAI.

20.3.3.2 LAI Retrieval Using Remote Sensing Data

Plant leaves intercept solar radiation and selectively absorb part of it for conversion into stored chemical energy by photosynthesis. The unabsorbed radiation is either reflected by the leaf surface or transmitted through the leaves. Healthy plant leaves have distinct reflectance and transmittance spectra relative to soil and other nonliving materials. Optical remote sensing makes use of the contrast between leaf and soil spectral characteristics for retrieving LAI of vegetation. However, vegetation stands have complex 3D canopy architecture, such as tree crowns, branches and shoots in forests, plantation rows in crops, and foliage clumps in shrubs. Remote sensing signals acquired over vegetated area are influenced not only by the amount of leaf area in the canopy but also by the canopy architecture. Seasonal variations of the vegetation background, such as moss/grass cover and snow cover on the forest floor, also greatly influence the total reflectance from a vegetated surface. It has therefore been a challenge to produce consistent and accurate LAI products using satellite measurements. Many remote sensing algorithms have been developed to retrieve LAI with full or partial consideration of the aforementioned factors influencing remote sensing measurements from vegetation. These algorithms are described in the following sections.

20.3.3.2.1 LAI Algorithms Based on Spectral Vegetation Indices

Reflectance spectra of healthy leaves show distinct low values in the red (620–750 nm) wavelengths and high values in NIR (800–1300 nm) wavelengths, and therefore many VIs have been developed using remote sensing measurements in red and NIR bands for estimating LAI and other vegetation parameters (Table 20.10). Liquid water in aboveground living biomass absorbs MIR (1300–2500 nm) radiation, lowering the reflectance in the MIR band. Since foliage biomass interacts most with solar radiation, the MIR reflectance is expected to correlate well with LAI, and some two-band and three-band VIs utilizing the additional information from MIR have been developed for LAI retrieval (Table 20.10).

Not all two-band and three-band VIs are well correlated to LAI. The significance level of the correlation of two-band VIs with LAI varies greatly even though they are constructed using the same two-band reflectance data because these two data are combined in different ways, under different assumptions. An ideal VI for LAI retrieval should preferably have the following properties: (1) it is more or less linearly related to LAI, and (2) it can minimize the impacts of both random and systematic biases that remote sensing errors have on its value. A linear relationship between a VI and LAI is preferred because it is insensitive to the surface heterogeneity within a pixel and induces less error in spatial scaling (Chen 1999). No VIs have so far been found to be linearly related to LAI for all plant functional types. However, some are more linearly related to LAI than others. SR, for example, is more linearly related to LAI than NDVI and SAVI (Chen and Cihlar 1996; Chen et al. 2002). Ideally, VIs would vary with LAI only, or the effects of surface variations other than LAI can be considered by adjusting coefficients or constants in the algorithm. Measured reflectance in different spectral bands is affected by environmental noise, such as subpixel clouds and their shadows, which are not identified in image processing, mixtures of nonvegetative surface features (small water bodies, rock, etc.), fog, smoke, etc. This unwanted noise frequently exists in remote sensing imagery and can dramatically alter the values of VIs. However, the impacts of these types of noise on the reflectances in different spectral bands are often correlated. For example, subpixel clouds would cause red and NIR reflectance to increase simultaneously, while cloud shadows would decrease them in about the same proportion. The same is true for other aforementioned types of noise. Variations in solar and view angle also cause variations of reflectance in various spectral bands in the same direction and in about the same proportions. VIs that are based on ratios of these two bands, such as NDVI and SR, can greatly reduce the impacts of various sources of noise. However, some VIs with sophisticated manipulations of two-band data, such as GEMI, may amplify noise. VIs that cannot be expressed as a function of the ratio of these two-band reflectances, such as SAVI, NLI, and RDVI, will retain the noise. MSR, for example, is developed with the same purpose as RDVI to increase its linearity with LAI, but it is better correlated to LAI than RDVI because it can be expressed as a function of the ratio of NIR and red reflectances, while RDVI cannot. The ability of a VI to minimize unwanted measurement noise is of paramount importance in LAI retrieval because noise in the reflectance measurements can come from many sources and is unavoidable.

TABLE 20.10 Vegetation Indices Useful for LAI Retrieval

Vegetation Index	Definition	References
Normalized difference vegetation index (NDVI)	$NDVI = \dfrac{(\rho_n - \rho_r)}{(\rho_n + \rho_r)}$	Rouse et al. (1974)
Simple ratio (SR)	$SR = \dfrac{\rho_n}{\rho_r}$	Jordan (1969)
Modified simple ratio (MSR)	$MSR = \dfrac{(\rho_n / \rho_r) - 1}{\sqrt{(\rho_n / \rho_r) + 1}}$	Chen (1996b)
Renormalized difference vegetation index (RDVI)	$RDVI = \dfrac{\rho_n - \rho_r}{\sqrt{\rho_n + \rho_r}}$	Roujean and Breon (1995)
Weighted difference vegetation index (WDVI)	$WDVI = \rho_n - \alpha \cdot \rho_r$ where $\alpha = \dfrac{\rho_{n,soil}}{\rho_{r,soil}}$	Clever (1989)
Soil-adjusted vegetation index (SAVI)	$SAVI = \dfrac{(\rho_n - \rho_r)(1 + L)}{(\rho_n + \rho_r + L)}$ $L = 0.5$	Huete (1988)
Soil-adjusted vegetation index 1 (SAVI1)	$SAVI = \dfrac{(\rho_n - \rho_r)(1 + L)}{(\rho_n + \rho_r + L)}$ $L = 1 - 2.12 \cdot NDVI \cdot WDVI$	Qi et al. (1994)
Global environmental monitoring index (GEMI)	$GEMI = \dfrac{\eta(1 - 0.25\eta) - (\rho_r - 0.125)}{(1 - \rho_r)}$ $\eta = \dfrac{2(\rho_n^2 - \rho_r^2) + 1.5\rho_n + 0.5\rho_r}{\rho_n + \rho_r + 0.5}$	Pinty and Verstrate (1992)
Nonlinear index (NLI)	$NLI = \dfrac{(\rho_n^2 - \rho_r)}{(\rho_n^2 + \rho_r)}$	Goel and Qin (1994)
Atmospherically resistant vegetation index (ARVI)	$ARVI = \dfrac{(\rho_n - \rho_{rb})}{(\rho_n + \rho_{rb})}$ $\rho_{rb} = \rho_r - \gamma(\rho_b - \rho_r)$	Kaufman and Tanre (1992)
Soil and atmosphere-resistant vegetation index (SARVI)	$SARVI = \dfrac{(\rho_n - \rho_{rb})(1 + L)}{(\rho_n + \rho_{rb} + L)}$ $L = 0.5$	Huete and Liu (1994)
Soil and atmosphere-resistant vegetation index 2 (SARVI2)	$SARVI2 = \dfrac{2.5(\rho_n - \rho_r)}{(1 + \rho_n + 6\rho_r - 7.5\rho_b)}$	Huete and Liu (1994)
Modified NDVI (MNDVI)	$MNDVI = \dfrac{(\rho_n - \rho_r)}{(\rho_n + \rho_r)}\left(1 - \dfrac{\rho_s - \rho_{smin}}{\rho_{smax} - \rho_{smin}}\right)$	Nemani et al. (1993)
Reduced SR (RSR)	$RSR = \dfrac{\rho_n}{\rho_r}\left(1 - \dfrac{\rho_s - \rho_{smin}}{\rho_{smax} - \rho_{smin}}\right)$	Brown et al. (2000)

Note: See also Chen (1996b).

Three-band VIs have been developed for various purposes. ARVI and SARVI modify NDVI and SAVI, respectively, with the reflectance in the blue band to reduce the atmospheric effect. They are useful when there are insufficient simultaneous atmospheric data to conduct atmospheric correction. MNDVI and RSR introduce a multiplier to NDVI and SR, respectively, based on the reflectance in a MIR band (1600–1800 or 2100–2300 nm). RSR has several advantages over SR for LAI retrieval (Brown et al. 2000): (1) it is more significantly correlated with LAI for different forest types because it is more sensitive to

LAI variation; (2) the differences in the LAI–RSR relationship among different forest types are greatly reduced from those in the LAI–SR relationship, and therefore RSR is particularly useful for mixed cover types; and (3) the influence of the variable background optical properties is much smaller on RSR than on SR because MIR reflectance is highly sensitive to the greenness of the background due to the strong absorption of MIR radiation by grass, moss, and understorey. These advantages of RSR over SR for forest LAI retrieval are confirmed by several independent studies (Eklundh et al. 2003; Stenberg et al. 2004; Wang et al. 2004; Chen et al. 2005c; Tian et al. 2007; Heiskanen et al. 2011). However, RSR is sensitive to soil and vegetation wetness and can increase greatly immediately after rainfall or irrigation, and therefore it is only suitable for forests in LAI retrieval algorithms (Deng et al. 2006).

20.3.3.2.2 LAI Algorithms Based on Radiative Transfer Models

The relationships between LAI and reflectances in individual spectral bands can be simulated using plant canopy radiative transfer models, and LAI algorithms can be developed based on these modeled relationships. Models are useful alternatives to empirical relationships established through correlating VIs or reflectances with LAI measurements because the empirical data are often limited in spatial and temporal coverage and are often location specific. These empirical relationships are also dependent on the quality of ground LAI data, the spectral response functions of remote sensing sensors, the angle of measurements, atmospheric effects, etc. The quality of LAI data can be influenced by the method of LAI measurements, the definition of LAI, and the measurement protocol. Some reported LAI values are actually the effective LAI without considering the clumping effect, and some optical measurements do not include the correction for nongreen materials (see Equation 20.11). Radiative transfer models can theoretically avoid these shortcomings of empirical data, but they need to be calibrated with ground data. In the calibration process, misconceptions and errors in empirical data can also bias the model outcome. For example, some destructive LAI values used for model validation are incorrectly based on the projected area rather than half the total leaf area.

There have been many LAI algorithms developed using radiative transfer models (Table 20.11) for regional and global LAI retrieval. These algorithms are characterized by the radiative transfer modeling method, the ways to consider foliage clumping and background optical properties, and the ways to combine the individual bands. A radiative transfer model, however sophisticated, is an abstract representation of the complex reality, and therefore the modeled relationship between LAI and remote sensing data depends not only on the aforementioned factors but also on how radiative transfer is simulated, such as the ways to consider multiple scattering in the canopy, the assumed leaf angle distributions, and the treatments of diffuse sky radiation. As radiative transfer methods are diverse, it is expected that the simulated relationships between remote sensing data and LAI are quite different among the existing model-based global LAI algorithms. There is a need to calibrate radiative transfer models and LAI algorithms against an accurate ground and remote sensing dataset covering the diverse plant structural types around the globe. The radiation transfer model intercomparison efforts (Pinty et al. 2004; Widlowski et al. 2007) have laid a foundation for further activities to satisfy this need.

20.3.4 Biomass

Biomass is the accumulated net primary production (NPP) in living plants, including both the above- and belowground components. Because of litterfall and mortality, biomass is always less than the sum of annual NPP over the plant's lifetime. It is relatively straightforward to measure the biomass for perennials, but measuring biomass for forests in the real world is extremely laborious. Due to the fact that the majority of the terrestrial biomass is stored in forest ecosystems (Dixon et al. 1994), measuring forest ecosystem biomass has become a major task in global carbon budget studies. In fact, measuring forest biomass defines the state of the art of biomass measurement. Since clearing an extensive area just for measuring biomass would represent an undue disturbance to the ecosystem, measuring forest biomass in the real world usually involves several

TABLE 20.11 Global LAI Products and Their Main Characteristics

	CYCLOPES	ECOCLIMAP	GLOBCARBON	MODIS
Algorithm development	1D turbid media radiative transfer model.	Empirical LAI–NDVI relationships.	Geometric–optical model.	Lookup tables produced using a 3D radiative transfer model.
Clumping consideration	No clumping is considered except consideration of the differences among cover types at the landscape level.	Clumping within shoot and canopy is considered but clumping at the landscape level is not considered.	Clumping is fully considered based on TRAC-measured cover-type specific values.	Clumping is considered through a parameter related to the 3D canopy structure.
Background optical property	Assigned constant values.	Assigned constant values.	Assigned constant values.	Assigned constant values.
Seasonal smoothing	No.	No.	Yes.	No.
References	Baret et al. (2007).	Masson et al. (2003).	Deng et al. (2006).	Knyazikhin et al. (1998), Yang et al. (2006).

Source: Modified after Garrigues, S. et al., *J. Geophys. Res. Biogeosci.*, 113, G02028, 2008, doi:10.1029/2007JG000635.

They include Carbon Cycle and Change in Land Observational Products from an Ensemble of Satellites (CYCLOPES), ECOCLIMAP, Global Biophysical Products for Terrestrial Carbon Studies (GLOBCARBON), and Moderate-Resolution Imaging Spectroradiometer (MODIS).

steps. First, a species-specific allometric relationship is established between biomass and some easy-to-measure structural parameter(s) (typically diameter at breast height and/or occasionally height). This step involves destructive sampling of a number of individuals for each species and is quite expensive. However, these allometric relationships can be reused once they are developed. Such relationships have been documented for the majority of tree species in North America (Grier and Logan 1977; Gholz et al. 1979; Ter-Mikaelian and Korzukhin 1997; Jenkins et al. 2003; Smith et al. 2003). Second, a series of sample plots are made in a region either systematically or randomly (Zhang and Song 2006). Each individual tree species within a sample plot is tallied, and its biomass is calculated using the allometric equations developed. The total biomass for a plot is calculated as the sum of biomass for all individuals. Lastly, the total biomass of a geographic region is estimated based on these sampling plots.

Depending on the rate of growth and length of time accumulating NPP, biomass density is strongly dependent on vegetation successional stage (Song and Woodcock 2002; Pregitzer and Euskirchen 2004). Therefore, an accurate estimation of biomass over a geographic region requires a large number of sampling plots, which are often not practical to make. Optical remote sensing offers a significant advantage over the traditional fieldwork approach in mapping biomass over large areas, as remote sensing–based approaches provide wall-to-wall coverage in space and are much more cost-effective. Numerous approaches using optical remotely sensed data have been developed in the literature and can be summarized into a few categories: (1) nearest neighbor imputation, (2) regression, (3) machine learning algorithms, and (4) biophysical approaches.

20.3.4.1 k-Nearest Neighbor Imputation

k-Nearest neighbor imputation (kNN) takes advantage of spatial autocorrelation of biomass in space. The approach estimates the biomass for a particular location or pixels in a remotely sensed image from the spatial interpolation of biomass in k nearby sampling plots based on distance weighted average (Fazakas and Olsson 1999; Franco-Lopez et al. 2001). Using the field plots from the Swedish National Forest Inventory, Tomppo et al. (2002) used the kNN approach to produce a biomass map with Landsat imagery, then rescaled the biomass map to match that of IRS-1C WiFS imagery, and produced biomass maps over large areas using nonlinear multiple regression. For kNN to be effective, a large number of sampling plots are needed in order to represent the spatial pattern and the whole range of biomass variation in space. The approach was used to develop national biomass maps for Sweden and Finland since 1990 using the national forest inventory sampling plots (Tomppo et al. 2008). Recently, kNN was used with various imputation approaches and remotely sensed data from multiple sensors (Latifi and Koch 2012) and achieved encouraging results for mapping biomass.

20.3.4.2 Regression

Estimating biomass via regression with remotely sensed imagery involves two steps. The first step is the development of an empirical regression model between biomass measured on the ground and remotely sensed data, which can be surface reflectance or a transformation of surface reflectance, such as spectral VIs. Once a robust regression model is established, the second step is to apply the model to the rest of the valid pixels in the image. Many successful applications of this approach have been reported in the literature (Anderson et al. 1993; Roy and Ravan 1996; Fazakas and Olsson 1999; Steininger 2000; Tomppo et al. 2002; Heiskanen 2006; Muukkonen and Heiskanen 2007). Careful examination found that these successful applications were conducted in areas with low biomass density.

Although significant challenges have been encountered using relatively high-spatial-resolution optical imagery to estimate biomass, some success has been achieved using coarse-resolution imagery over large areas. Myneni et al. (2001) and Dong et al. (2003) developed an empirical relationship between cumulative NDVI over the growing season from AVHRR over the forested areas and the woody biomass derived from forest inventory for six countries (Canada, Finland, Norway, Russia, Sweden, and the United States) in 1981–1999 and found a large carbon sink in Eurasian boreal and North American temperate forests. Similarly, Piao et al. (2005) used the GIMMS NDVI (Tucker et al. 2001) and China's forest inventory data to estimate aboveground forest biomass via a nonlinear regression model. The model predicted the aboveground biomass well for the majority of the provinces except for a few outliers, which might be due to errors from the inventory data. Zhang and Kondragunta (2006) used MODIS LAI, land cover types, and vegetation continuous fields to estimate the aboveground biomass for the conterminous United States with a RMSE of 12 t/ha at the state level compared with estimates from U.S. Forest Service FIA data. Le Maire et al. (2011) successfully ($R^2 \approx 0.9$) estimated the forest biomass for young eucalyptus plantations using MODIS time series NDVI images and simple bioclimatic variables. It is counterintuitive that spectral VIs at high spatial resolution performed more poorly in predicting biomass at the plot level than those at coarse spatial resolution that estimate biomass at the continental scale. However, these empirical models provide little insight on the biophysical basis for the strong performance in estimating biomass using coarse-spatial-resolution remotely sensed data over large areas.

20.3.4.3 Machine Learning Algorithms

Machine learning algorithms have several advantages over conventional regression approaches in mapping biomass from remotely sensed data. First, the algorithms do not require normally distributed data. Second, the algorithms do not require the input data layers to be independent from each other. Data layers from multispectral remotely sensed imagery are often correlated. Third, machine learning algorithms are capable of handling nonlinear relationships between biomass and remotely sensed signals. In a Bornean tropical rainforest, Foody et al. (2001) found that artificial NNs using the reflectance of the six optical bands from Landsat TM sensors mapped biomass better than a regression approach using spectral VIs only or the kNN imputation. Foody et al. (2003) further confirmed that NN outperformed the conventional regression approach in mapping biomass in three tropical forest sites in Brazil, Malaysia, and Thailand. Powell et al. (2010) compared reduced major axis regression, kNN, and random forest (RF) algorithms in mapping aboveground biomass in Arizona and Minnesota, United States, using biomass derived from FIA plots and Landsat imagery, and they found all three approaches predict biomass at the pixel level with RMSE well above 50% of the mean biomass. Mutanga et al. (2012) compared RF and multiple regression to estimate biomass for a densely vegetated wetland using narrowband VIs computed from WorldView-2 imagery and found that RF performed better in biomass estimation with RMSE of 0.441 kg/m^2 (12.9% of observed mean biomass). Because it is easy to use data layers from multiple sources with machine learning algorithms, they proved to be effective in integrating remotely sensed data from multiple sensors for mapping aboveground biomass, particularly combining multispectral and LiDAR data. Latifi and Koch (2012) compared kNN imputation with RF to map aboveground biomass using data from airborne scanning LiDAR data and color infrared optical imagery and found that RF produced more accurate results.

In addition to high-resolution imagery, machine learning algorithms have also been used to map biomass with low-resolution imagery, particularly data from MODIS. Combining surface reflectance of the first seven MODIS bands with climate and topographic data, Baccini et al. (2004) mapped aboveground forest biomass for 18 National Forests in California using tree-based regression with reasonable accuracy, but the approach tends to underestimate biomass for stands with high biomass density (>250 t/ha). Houghton et al. (2007) used RF to map aboveground forest biomass for the Russian Federation using MODIS Nadir BRDF-Adjusted Reflectance (NBAR) and forest inventory data. They produced biomass estimates comparable with previous independent estimates. Blackard et al. (2008) mapped the aboveground biomass of the conterminous United States, Alaska, and Puerto Rico, with multiple sources of data, including MODIS imagery, FIA plots, climatic and topographic variables, and other ancillary data. They first divided the conterminous United States into 65 ecological zones and separated the FIA plots into forest and nonforest categories. A separate regression tree model was developed for each ecological zone, Alaska and Puerto Rico. The models tended to overestimate areas with low biomass and underestimate areas with high biomass. Baccini et al. (2008) developed a regression tree model for aboveground biomass using seven-band 1 × 1 km MODIS NBAR data and biomass data derived from NFI for Congo, Cameroon, and Uganda. The model was then applied to the entire tropical Africa. The RMSE was 50.5 t/ha for a biomass range up to 454 t/ha. The estimated aboveground biomass was also highly correlated ($R^2 = 0.90$) with height metrics from ICESat Geospatial Laser Altimeter System

(GLAS). However, the predicted biomass had a positive bias for low biomass and negative bias for high biomass.

A promising recent development in mapping biomass with optical imagery is to combine it with remotely sensed data from LiDAR and/or Radar sensors and produce improved biomass maps. Andersen et al. (2011) integrated Landsat imagery with airborne LiDAR and dual-polarization synthetic aperture radar from ALOS Phased Array L-band Synthetic Aperture Radar (PALSAR) to map forest biomass in the interior Alaska. The ICESat GLAS data were successfully used to map biomass with Landsat (Helmer et al. 2009; Duncanson et al. 2010) and MODIS (Nelson et al. 2009; Baccini et al. 2012) imagery.

20.3.4.4 Biophysical Approaches

Biophysical approaches rely on the physical principles that govern the relationship between vegetation structure and remotely sensed signals. One of the earliest attempts to estimate standing total biomass via optical remote sensing was by Wu and Strahler (1994). The physical principle is the geometric–optical theory for remotely sensed imagery over vegetated landscapes (Li and Strahler 1985). They inverted the remotely sensed data from Landsat TM sensors for tree density and mean tree crown size on a stand basis with a GO model and then used allometry to estimate total standing biomass. However, Wu and Strahler (1994) only tested the model with a limited number of stands. More comprehensive studies by Woodcock et al. (1994, 1997) found that the Li–Strahler model could be used to estimate tree cover effectively, but separation of tree size and stem count was poor. Hall et al. (1995) proved that remotely sensed spectral signals over black spruce forests can be calculated as a linear mixture of sunlit crown, sunlit background, and shadow. Based on the geometric–optical theory, they derived these fractions from stand-level reflectance obtained at nadir by the helicopter-mounted Modular Multiband Radiometer and found that the fraction of shadows was highly correlated to aboveground biomass. Hall et al. (1996) and Peddle et al. (1999) further confirmed the usefulness of fraction of shadows in estimating aboveground biomass. Peddle et al. (1999) found that the fraction of canopy shadows performed 20% better than numerous VIs for estimating aboveground biomass. More recently, Soenen et al. (2010) demonstrated promising results using a geometric–optical canopy reflectance model to estimate tree crown size and stem density and further estimate aboveground biomass with SPOT imagery. Chopping et al. (2008, 2011) developed a similar approach based on the simple geometric–optical model to estimate biomass by taking bidirectional reflectance functions from MODIS and MISR. The approach produced biomass estimation that is comparable with independent data in a low-biomass region, but the approach is computationally intensive and requires some detailed canopy structural parameters that may not be available as a priori knowledge. Hall et al. (2006) developed the BioSTRUCT algorithm to map aboveground biomass and volume in two steps: (1) estimate canopy height and crown closure with Landsat ETM+ imagery and (2) estimate aboveground biomass and stand volume using canopy height and crown closure based on allometric relationships.

20.3.5 Uncertainties, Errors, and Accuracy

Despite the many successes of optical remote sensing in extracting information about vegetation structure, there are varying degrees of uncertainty associated with them. Even for the most simple vegetation estimate, vegetation cover, there is about 10% uncertainty (Hansen et al. 2002; Hayes et al. 2008). Given the same vegetation cover, ecological functions could differ tremendously depending on its successional stages (Law et al. 2001). For example, Liu et al. (2008) found that using Landsat imagery, forest succession in the PNW can only be reliably separated into three broad successional stages (young, mature, and old growth).

LAI is perhaps the most sought-after measure of vegetation structure due to the important role it plays in energy and matter exchange between the land surface and the atmosphere. Several factors prevent accurate estimation of LAI with optical remote sensing. First, remote sensing signals saturate in high LAI (Baret and Guyot 1991; Turner et al. 1999). Second, remnant cloud contamination remains a problem. Lastly, LAI cannot be derived from remotely sensed data analytically. It is an ill-posed mathematical problem because there are too many other factors influencing the remotely sensed signal in addition to LAI (Gobron et al. 1997; Eklundh et al. 2001). As a result, significant uncertainties remain in most of the current LAI products (Song 2013). Numerous studies found that current MODIS LAI products tend to overestimate LAI (Cohen et al. 2003, 2006; Wang et al. 2004; Aragao et al. 2005; Pisek and Chen 2007; Sprintsin et al. 2009).

Numerous studies have found that remotely sensed signals saturate in high-biomass-density areas and spectral VIs poorly predict biomass for forests (Sader et al. 1989; Hall et al. 1995; Peddle et al. 1999). Steininger (2000) found that Landsat TM surface reflectance correlates well with stand structure only when biomass is under 150 t/ha and age is under 15 years in Brazil, and even this is not the case for another study site in Bolivia. Nelson et al. (2000) found a single Landsat TM imagery could not be used to reliably differentiate tropical forest age classes. Lu (2005) found that the spectral signals from Landsat TM imagery can only be used to estimate aboveground biomass for forests with simple structure. Although optical remote sensing has been used to produce numerous key vegetation structure parameters wall-to-wall, improving the accuracy of the estimation remains the major challenge in the foreseeable future.

20.4 Optical Remote Sensing of Vegetation Functions

Vegetation plays a key role in the terrestrial ecosystem that provides vital goods and services upon which the welfare of the humanity depends, such goods as food, fiber, and medicine, and services as soil and water conservation and preservation of biodiversity (Dobson et al. 1997; Salim and Ullsten 1999). Photosynthesis is the entry point of inorganic materials, for example, CO_2, water, and nutrients, into organic forms, such as carbohydrates. The product of photosynthesis provides the matter and energy that drive all subsequence ecosystem processes. Therefore, vegetation primary production is at the core of almost all terrestrial ecosystem goods and services. In this section, we review remote sensing of vegetation functions that are tied to plant photosynthesis, including its seasonal cycles (phenology), the amount of energy used in photosynthesis (fraction of absorbed photosynthetically active radiation [FPAR]), the abundance of photosynthesis apparatus (chlorophyll concentration), and the efficiency of converting the absorbed photosynthetically active radiation (PAR) to carbohydrate (light use efficiency [LUE]).

20.4.1 Vegetation Phenology

Vegetation phenology is the natural rhythm of plant life cycle events, and the timing of these events is largely dependent on climate signals (Körner and Basler 2010). In many temperate forests, winter dormancy must be broken by extended exposure to cold temperatures, and after this chilling requirement has been met, increases in temperature and photoperiod can trigger leaf emergence in spring (Archibold 1995; Zhang et al. 2007). Leaf expansion and shoot growth in some arid and seasonally moist ecosystems can be triggered by the start of the rainy season, while soil moisture depletion may trigger senescence and leaf abscission, though low temperatures may also limit photosynthesis in cooler deserts (Archibold 1995; Jolly and Running 2004; Jolly et al. 2005).

Optical remote sensing offers unprecedented opportunities to observe the synoptic patterns of the timing of plant life cycle events, known as land surface phenology (LSP) (Gonsamo et al. 2012b). While the seasonal patterns of LSP variability are related to plant biological traits, LSP derived from spaceborne optical sensors is distinct from the traditional definition of plant phenology, which aims to understand the timing of recurring biological events, the causes of their timing with regard to biotic and abiotic forces, and the interrelation among phases of the same or different species (Lieth 1974). LSP has strong effects and feedbacks both on climate (Keeling et al. 1996; Peñuelas et al. 2009) and on terrestrial ecosystem functions. The carbon balance of terrestrial ecosystems is highly sensitive to climatic changes in early and late growing seasons (Piao et al. 2007, 2008; Richardson et al. 2010; Wu et al. 2012a,b). Vegetation phenology has been known to be a key and first element in ecosystem response to climate change (Menzel et al. 2006), as well as a major determinant of species distributions (Chuine and Beaubien 2001).

Therefore, changes in LSP events have the potential to broadly impact global carbon fixation, nitrogen cycles, evapotranspiration and ecosystem respiration (Morisette et al. 2009; Richardson et al. 2010), surface meteorology (Schwartz 1992; Bonan 2008a,b; Richardson et al. 2013), interspecific interactions both among plants and between plants and insects, vegetation community structure, and success of invasive species (Willis et al. 2008, 2010; Wolkovich and Cleland 2011; Cleland et al. 2012; Fridley 2012), crop production, frost damage, pollination (Brown and De Beurs 2008), and spreading of diseases (Morisette et al. 2009).

The 4th Assessment Report ("AR4," Parry et al. 2007) of the Intergovernmental Panel on Climate Change—which found that spring onset has been advancing at a rate of between 2.3 and 5.2 days per decade since the 1970s—emphatically concluded that phenology "is perhaps the simplest process in which to track changes in the ecology of species in response to climate change" (Rosenzweig et al. 2007). The spatially integrated nature of LSP—as derived from optical satellite observations of land surface reflectance and their combination in the form of VIs that are associated with the biophysical and biochemical properties of vegetation—has thus received much attention due to its role as a surrogate in detecting the impact of climate change.

20.4.1.1 Vegetation Index for Land Surface Phenology Study

Remote sensing LSP studies use data gathered by satellite sensors that measure wavelengths of visible light as absorbed by leaf pigments, NIR light as reflected by leaf internal structure, and SWIR light as absorbed by leaf in vivo water content. As a plant canopy changes from early spring growth to late-season maturity and senescence, these reflectance and absorptance properties also change. VIs rather than land surface reflectance are especially useful for continental- to global-scale LSP monitoring because it can compensate for changing illumination conditions, surface slope, and viewing angle. Although there are several VIs, the following four are common:

$$NDVI = \frac{NIR - RED}{NIR + RED} \qquad (20.12)$$

$$NDMI = \frac{NIR - SWIR}{NIR + SWIR} \qquad (20.13)$$

$$PI = \begin{cases} 0, \text{if NDVI or NDMI} < 0 \\ (NDVI + NDMI)(NDVI - NDMI) = NDVI^2 - NDMI^2 \\ 0, \text{if PI} < 0 \end{cases}$$

$$(20.14)$$

$$EVI = \frac{G \times (NIR - RED)}{NIR + C1 \times RED - C2 \times BLUE + L} \qquad (20.15)$$

where
  the BLU.E, RED, NIR, and SWIR are surface reflectances in blue, red, NIR, and SWIR spectral bands, respectively
  L is a canopy background adjustment that addresses nonlinear, differential NIR and red radiant transfer through a canopy
  C1 and C2 are the coefficients of the aerosol resistance term, which uses the blue band to correct for aerosol influences in the red band
  G is a gain factor that limits the EVI value to the −1 to +1 range

One of the most widely used VIs is the NDVI. NDVI values range from −1.0 to +1.0. Areas of barren rock, sand, water, ice, and snow usually show very low NDVI values (<0.1), sparse vegetation such as shrubs and grasslands or senescing crops may result in moderate NDVI values, and high NDVI values (>0.6) correspond to dense vegetation such as that found in temperate and tropical forests or crops in their peak growth stage. Numerous studies have shown that NDVI is closely correlated with LAI and fraction of PAR absorbed by vegetation canopy. The NOAA AVHRR archive of NDVI data (Tucker et al. 2005) is generated in the framework of the GIMMS project by careful assembly from different AVHRR sensors, accounting for various deleterious effects, such as calibration loss, orbital drift, and volcanic eruptions. The AVHRR archive is the longest time series for LSP studies, and the results from the analysis of AVHRR-based NDVI revealed significant changes in spring phenology of vegetation during the 1980s and 1990s (Myneni et al. 1997; Eastman et al. 2013). The latest version of the GIMMS NDVI dataset spans the period from July 1981 to December 2011 and is termed NDVI3g (third-generation GIMMS NDVI) from AVHRR sensors (Zhu et al. 2013).

To help discriminate the seasonal dynamics of vegetation phenology from background phenomena such as accumulation and melting of snow, alternative VIs such as normalized difference moisture index (NDMI) have been used (Delbart et al. 2005; Gonsamo et al. 2012a). NDMI is comparable with normalized difference infrared index and normalized difference water index (NDWI). NDMI first decreases with snowmelt and then increases during vegetation greening. NDWI is related to the quantity of water per unit area in the canopy and soil and therefore increases during leaf development and increase in soil moisture content. NDMI time series show that greening-up may start before or after complete snowmelt. If snow did not totally melt before leaf appearance, NDMI first decreases and then increases, displaying a trough at its minimum. If snowmelt is complete before leaf appearance, then NDMI remains stable during a period that may last between a few days and a few weeks before increasing. If greening-up occurs during snowmelt, the NDMI decrease due to snowmelt may mask the NDMI increase due to greening-up, and NDMI may start increasing later than the actual onset of greening-up. If snowmelt and greening-up overlap during a long period, NDMI variations with snowmelt and greening-up compensate for each other, making the NDWI increase start later than the actual onset of greening-up. NDWI time series is sensitive to water intercepted by leaves and to abrupt increases in soil moisture.

VIs are not intrinsic physical quantities, and several attempts have been made to remove the confounding effect of brightness (mainly soil) and wetness (mainly from land surface moisture) from greenness that responds to the development of photosynthetic biomass. For example, NDMI alone cannot capture LSP, since it responds to land surface moisture from both the landscape and vegetation components. NDVI is also affected significantly by both soil moisture and brightness. Therefore, both NDVI and NDMI respond not only to the development of photosynthetic biomass (greenness) but also to soil exposure

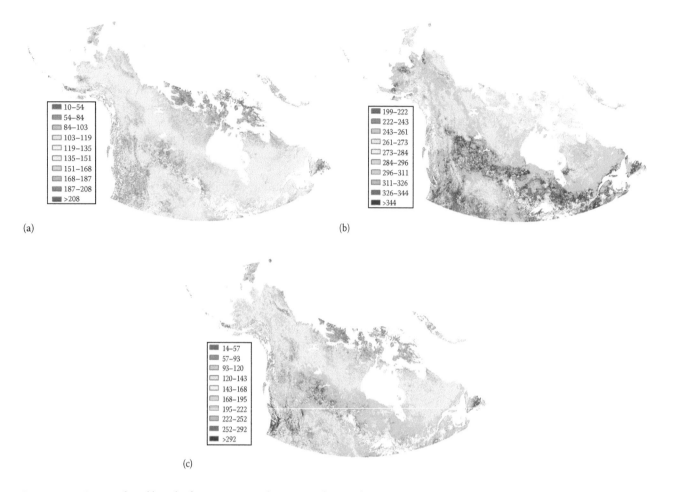

FIGURE 20.2 Start, end, and length of growing season for circumpolar North America (>45°) derived from SPOT VGT sensors using the PI for year 1999. (a) Start of season; (b) end of season; (c) length of season.

(brightness) and snow, soil, and land surface moisture (wetness), suggesting that NDVI or NDMI alone cannot remove the confounding effect of brightness and wetness on an LSP time series. NDVI and NDMI exhibit opposing trends with increasing brightness and wetness and similar trends with increasing greenness. Given these premises, the phenology index (PI) was constructed (Gonsamo et al. 2012a) by combining the merits of NDVI and NDMI (Figure 20.2). PI takes the difference of squared greenness and wetness to remove the soil and snow cover dynamics from key vegetation LSP cycles. The following rationale explains the formulation of PI: (1) NIR reflectance is less than red reflectance for ice, snow, and water resulting in NDVI < 0 for which PI becomes 0; (2) NIR reflectance is less than SWIR for soil and for NPV resulting in NDMI < 0 and NDVI > 0 for which PI becomes 0; (3) if NDMI > NDVI, the GV or land surface is covered by snow for which PI becomes 0; (4) the use of PI instead of NDVI or NDMI masks out the time series of permanently nonvegetated landscape for which NDVI or NDMI may result in a spurious time series due to moisture variations resembling vegetation LSP; and (5) the product of the sum and the difference of NDVI and NDMI gives a pronounced and smooth curve, removes the effect of wetness from the greenness, and avoids the local solution if

we simply consider the use of NDVI once the aforementioned criteria (1, 2, 3, and 4) are met, which may particularly occur in boreal forests due to intermittent loading and unloading of snow. PI is actually the squared greenness minus squared wetness in the growing season and follows the seasonal dynamics of gross (GPP) and net (NPP) primary productivity better than NDVI or NDMI (Gonsamo et al. 2012a,b).

The EVI is developed for use with the MODIS Land Cover Dynamics product (informally called the MODIS Global Vegetation Phenology product). The EVI is a modified NDVI with a soil adjustment factor (L), gain factor (G), and two coefficients (C1 and C2), which describe the use of the blue band in correction of the red band for atmospheric aerosol scattering. The coefficients, G, C1, C2, and L, are empirically determined as 2.5, 6.0, 7.5, and 1.0, respectively. This algorithm has improved sensitivity to high-biomass regions and improved vegetation monitoring through a decoupling of the canopy background signal and a reduction in atmospheric influences (Huete et al. 2002). While NDVI is chlorophyll sensitive, the EVI is more responsive to canopy structural variations, including LAI, canopy type, plant physiognomy, and canopy architecture. EVI is used as a standard VI for LSP study for the NASA MODIS project and has shown not to saturate at high photosynthetic biomass vegetation areas, a great improvement compared to NDVI.

20.4.1.2 Land Surface Phenology Metrics Derivation and Validation

LSP studies pay more attention to critical annual events, such as start of season (SOS), end of season (EOS), and length of season. Earlier studies that used AVHRR NDVI time series focused on a phenologically important threshold value from multitemporal NDVI time series, in which values can be global (like 0.3 NDVI) or locally driven (midvalues of max and min NDVI; White et al. 1997). Since there is no fixed threshold value for SOS and EOS that can be applicable globally, recent LSP studies have focused on curve geometry fitting to extract important dates (e.g., Zhang et al. 2003; Delbart et al. 2005; Gonsamo et al. 2012a,b). Most of these methods involve one or several forms of logistic functions to fit sinusoid models. Jönsson and Eklundh (2004) developed open source LSP extraction software (TIMESAT), which includes double logistic and asymmetric Gaussian functions. These methods to retrieve SOS and EOS estimates from time series of remote sensing data are well summarized in White et al. (2009).

Traditional validation is usually carried out by comparing remote sensing SOS and EOS with specific life cycle events such as budbreak, flowering, or leaf senescence using in situ observations of individual plants or species by network volunteer observations (e.g., PlantWatch Canada Network, USA National Phenology Network) or experts at intensive study sites (e.g., Harvard Forest). Given the lack of in situ data that are comparable with LSP in spatial coverage and landscape representativeness, currently more attention is given to synoptic sensors that have comparable footprints with remote sensing pixels. One of these includes the increasing availability of networks of web cameras (Richardson et al. 2009) that match LSP validation from remote sensing footprints rather than the traditional individual plant phenology networks. However, much work is needed to derive a robust method to extract SOS and EOS from near-surface remote sensing obtained from networked webcams that record repeat LSP in association with the existing eddy covariance flux towers (e.g., the PhenoCam program). Another more objective validation method is the use of GPP estimated based on eddy covariance flux tower measurements (Gonsamo et al. 2012a,b). Several LSP metrics are developed from GPP using curve geometry fitting (Gonsamo et al. 2013), which can be used to validate one or more LSP estimates from remote sensing. The temporal dynamics of GPP as a true photosynthesis phenology provides an objective measures of SOS and EOS. There are also evolving developments in LSP reference measures, such as ground-based spectral and photosynthetic radiation measurements (Richardson et al. 2012), which are expected to help extract the subtle LSP interannual and spatial variability across plant functional types. Remote sensing LSP measures, compared to individual plant or plant organ phenology cycles, integrate the collective effects of atmospheric, environmental, and edaphic conditions as well as interspecific responses to the changing climate.

20.4.2 Fraction of Absorbed Photosynthetically Active Radiation

FPAR is defined as the fraction of incoming PAR (400–700 nm) absorbed by vegetation. It is a key input variable to models that estimate terrestrial ecosystem primary production based on LUE theory driven by remotely sensed data (Potter et al. 1993; Running et al. 1994; Landsberg and Waring 1997). As a result, FPAR has become a much sought-after biophysical product from remote sensing.

20.4.2.1 Estimating FPAR with Empirical Models

FPAR depends on various leaf optical properties (e.g., clumping, leaf angle distribution, and spatial heterogeneity) and is nonlinearly correlated with LAI (e.g., Asrar et al. 1984). FPAR has been empirically related to several VIs from remote sensing data, such as SR, NDVI, greenness, and perpendicular VI (PVI) since the 1980s (Table 20.12). These indices are estimated from various combinations of remotely sensed data from different spectral bands. The SR–FPAR relationship developed in the earlier research was usually done for crops in field using spectroradiometers and was often used to estimate crop production. Kumar and Monteith (1981) developed a linear relationship between SR and FPAR for various crops, and the relationship was further examined by Steven et al. (1983). Asrar et al. (1984) provided the theoretical basis for the NDVI–FPAR relationship, which was further validated with field measurements.

Many studies have found that the intercept (b) of this linear NDVI–FPAR relationship is not zero for nonblack background, but slightly negative (Table 20.13; Figure 20.3). These relations are often dependent on vegetation phenological phases, such as greening-up and senescence (e.g., Hatfield et al. 1984; Wiegand et al. 1991). The slope values (a) in the NDVI–FPAR relations range from 0.95 to 1.386 except for senescence periods. NDVI seems to provide more consistent linear relations with FPAR than SR across different biomes and sensors (Figure 20.3). Although both NDVI and FPAR are not scale invariant, the NDVI–FPAR relationship has been proven to be scale invariant due to its linearity (Myneni and Williams 1994; Friedl et al. 1995; Myneni et al. 1995). For this reason, FPAR was often incorporated as a key scaling measure when fusing multitemporal remote sensing datasets at different scales (Hwang et al. 2011).

20.4.2.2 Estimating FPAR with Biophysical Models

Since a study by Asrar et al. (1984), the NDVI -FPAR relation has been explored for different sensors (Landsat MSS or TM, SPOT, AVHRR, etc.) and biome types at different scales. This linear relationship between FPAR and VI has also been reproduced with several radiative transfer models (Sellers 1985, 1987; Asrar et al. 1992; Myneni and Williams 1994; Knyazikhin et al. 1998). These studies usually found that the NDVI–FPAR relationship is mostly linear regardless of vegetation spatial heterogeneity and leaf optical properties. However, this relationship is sensitive to soil background reflectance. Knyazikhin et al.

TABLE 20.12　Empirical Models for the Fraction of Absorbed PAR with Spectral Vegetation Indices

Index	FPAR Model	Notes	References
SR	$0.34SR - 0.63$	Various crops	Kumar and Monteith (1981)
SR	$0.369\ln(SR) - 0.0353$	Field spectrophotometer/sugar beet	Steven et al. (1983)
SR	$0.0026SR^2 + 0.102SR - 0.006$	Landsat/corn	Gallo et al. (1985)
SR	$0.0294SR + 0.3669$	SPOT/wheat	Steinmetz et al. (1990)
SR	Not specified	Modeling study	Sellers (1987)
NDVI	$1.253NDVI - 0.109$	Landsat/spring wheat	Asrar et al. (1984)
NDVI	$1.200NDVI - 0.184$ (growing) $0.257NDVI + 0.684$ (senescence)	Landsat/wheat	Hatfield et al. (1984)
NDVI	$2.9NDVI^2 - 2.2NDVI + 0.6$ (growing)	Landsat/corn	Gallo et al. (1985)
NDVI	$1.00NDVI - 0.2$	Landsat/coniferous	Peterson et al. (1987)
NDVI	$1.23NDVI - 0.06$	Wheat	Baret and Olioso (1989)
NDVI	$1.33NDVI - 0.31$	Modeling study	Baret and Olioso (1989)
NDVI	$1.240NDVI - 0.228$	Modeling study	Baret et al. (1989)
NDVI	$0.229\exp(1.95NDVI) - 0.344$ (growing) $1.653NDVI - 0.450$ (senescence)	SPOT/cotton and corn	Wiegand et al. (1991)
NDVI	$1.222NDVI - 0.191$[a]	Modeling study	Asrar et al. (1992)
NDVI	$1.254NDVI - 0.205$	Field spectroradiometer/corn and soybean	Daughtry et al. (1992)
NDVI	$1.075NDVI - 0.08$	Modeling study	Goward and Huemmrich (1992)
NDVI	$1.386NDVI - 0.125$	Cereal crop/modeling study	Begue (1993)
NDVI	$1.164NDVI - 0.143$	Modeling study	Myneni and Williams (1994)
NDVI	$1.21NDVI - 0.04$	AVHRR/mixed forests	Goward et al. (1994)
NDVI	$0.95NDVI - 0.02$	Landsat/modeling study	Friedl et al. (1995)
Greenness	Not specified	Landsat/corn	Daughtry et al. (1983)
PVI	$0.036PVI - 0.015$ (growing) $0.037PVI + 0.114$ (senescence)	SPOT and videography/cotton and corn	Wiegand et al. (1991)

SR, simple ratio; NDVI, normalized difference vegetation index, greenness; PVI, perpendicular vegetation index.

[a] Intercept was converted to a negative value based on a scatterplot (Figure 11 in Asrar et al. 1992).

TABLE 20.13　Application of Empirical Models for the Fraction of PAR with the Spectral Vegetation Indices into Global Production Efficiency or Process-Based Biogeochemical Models

FPAR	Model	References
$\min((SR - SR_{min})/(SR_{max} - SR_{min}), 0.95)$[a]	CASA	Potter et al. (1993)
$\{(SR - SR_{min})/(SR_{max} - SR_{min})\}$ $(FPAR_{max} - FPAR_{min})$	SiB2	Sellers et al. (1994)
$1.25NDVI - 0.025$	TURC	Ruimy et al. (1994)
$0.11SR - 0.12$	TURC	Ruimy et al. (1994)
$1.67NDVI - 0.08$	3-PGS	Coops et al. (1998)
$1.67NDVI - 0.08$	Glo-PEM2	Goetz et al. (1999)
$1.21NDVI - 0.04$	3-PGS	Coops (1999)
NDVI	Biome-BGC	Running et al. (2004)
$1.24NDVI - 0.168$	EC-LUE	Rahman et al. (2004)
$0.279SR - 0.294$	NASA-CASA	Potter et al. (2003)
EVI	VPM	Xiao et al. (2004)

[a] SR_{max} and SR_{min} are biome-type dependent.

(1998) mathematically proved the NDVI–FPAR proportionality (FPAR = a·NDVI) if background soil is ideally black.

One of the best known models estimating FPAR is perhaps the MODIS/FPAR algorithm, which simulates bidirectional reflectance factor using a 3D canopy radiative transfer model by biome types (Myneni et al. 2002; Ganguly et al. 2008b). This algorithm assumes biome-specific canopy structure (e.g., leaf orientation distribution; Knyzikhin et al. 1998) and leaf/soil optical properties (Myneni et al. 2002). The algorithm also assumes spectral invariance in canopy transmittance and absorptance at a reference wavelength to those in other wavelengths, which also provides the theoretical basis of the linear NDVI–FPAR relationship (Knyazikhin et al. 1998). In this algorithm, FPAR is calculated by integrating the weighted spectral absorptance over the PAR spectral region (Knyazikhin et al. 1998; Ganguly et al. 2008b). The algorithm was also successfully applied to retrieve FPAR values from AVHRR and Landsat imagery with different spatial and spectral resolutions (Ganguly et al. 2008a, 2012).

20.4.3　Leaf Chlorophyll Content

Leaf pigments such as chlorophyll a and b play a crucial role in plant photosynthesis through the conversion of solar radiation into stored chemical energy, via a series of electron transfers that occur on the thylakoid membranes in chloroplasts. As the amount of solar radiation absorbed by a leaf is primarily a function of the foliar photosynthetic pigments, low concentrations of chlorophyll can limit photosynthetic potential and primary production (Richardson et al. 2002). With a large proportion of nitrogen contained within chlorophyll molecules, leaf

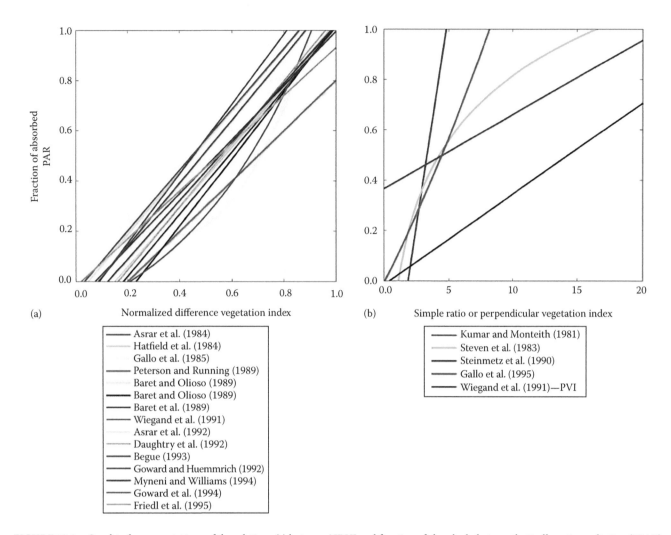

FIGURE 20.3 Graphical representations of the relations (a) between NDVI and fraction of absorbed photosynthetically active radiation (FPAR) and (b) between simple ratio (or perpendicular vegetation index) and FPAR.

chlorophyll is intrinsically linked to carbon and nitrogen cycles. Further, its role in photosynthesis and NPP is important within regional and global carbon models. Decreases in foliar chlorophyll can indicate plant disturbance and stress, for example, from disease, limited water availability, extreme temperature, or pests, and thus act as a bioindicator of plant physiological condition. The importance of chlorophyll to a range of ecological processes has led to a considerable body of research dedicated to deriving chlorophyll content from leaf and canopy reflectance, from laboratory- and field-based studies to airborne and satellite platforms.

Leaf reflectance is controlled by a range of biochemical and physical variables, including chlorophyll, nitrogen, carotenoids, anthocyanins, water, and internal leaf structure, with chlorophyll dominating in the visible wavelengths (400–700 nm). Chlorophyll absorbs strongly in red and blue spectral regions, with maximum absorbance in red wavelengths between 660 and 680 nm and maximum reflectance in green wavelengths (~560 nm) within the visible spectrum. Overlapping absorption from the presence of carotenoids in blue wavelengths often

prevents this region from being useful in chlorophyll estimation (Sims and Gamon 2002). Research has also identified that the absorption feature in red wavelengths readily saturates at relatively low chlorophyll contents, leading to a reduced sensitivity to higher chlorophyll content. This has led to the use of "off-center" wavelengths, with reflectance in wavelengths along the red-edge region (690–750 nm) showing greater sensitivity to subtler changes in chlorophyll content (Curran et al. 1990). This improved sensitivity along the red edge is caused by the increasing chlorophyll content that causes a broadening of the absorption feature centered around 680 nm, shifting the position of the red edge to longer wavelengths. While the relationship between leaf reflectance and chlorophyll content is reasonably well established, particularly for broadleaf species, reflectance sampled from remote platforms is also governed by additional canopy contributions. These include leaf architecture, LAI, clumping, tree density, and nonphotosynthetic canopy elements, along with solar/viewing geometry, ground cover, and understory vegetation, making the relationship between leaf chlorophyll and canopy reflectance complex.

20.4.3.1 Laboratory Extraction of Leaf Chlorophyll Content

Many studies calibrate or validate chlorophyll estimates derived from handheld chlorophyll meters and remotely sensed platforms using laboratory-derived in vitro chlorophyll content. Chlorophyll content is determined by extraction from leaf samples and subsequent spectrophotometric measurements and expressed by weight or, in most cases, by area. A range of organic solvents are typically used to extract chlorophylls and carotenoids from plant tissues, including acetone, methanol, ethanol, dimethyl sulfoxide (DMSO), and *N,N*-dimethylformamide (DMF), which range in optimal extraction time and performance. In a comprehensive study, Minocha et al. (2009) compared the efficiencies of acetone, ethanol, DMSO, and DMF for chlorophyll and carotenoid extraction for 11 species, finding that extraction efficiencies of ethanol and DMF were comparable for analyzing chlorophyll concentrations. DMF was the most efficient solvent for the extraction of carotenoids; however, the toxicity of DMF requires care when using this solvent.

20.4.3.2 Measuring Leaf Chlorophyll Content with SPAD Meter

For rapid, nondestructive leaf chlorophyll measurements taken in the field, a handheld chlorophyll SPAD meter, developed by Minolta corporation, Ltd., is often used, particularly in agricultural applications. The current model (SPAD-502) measures leaf transmittance through a leaf clamped within the meter at two wavelengths 650 and 940 nm. The 650 nm is selected to coincide with the chlorophyll maximum absorbance feature, and 940 nm is used as a reference to compensate for factors such as leaf moisture content and internal structure (Zhu et al. 2012a). The measured SPAD unit value is converted to chlorophyll content using a calibration equation. However, relatively few studies perform their own calibrations with in vitro chlorophyll content, which are likely to vary according to plant species, leaf thickness, and leaf age. Uddling et al. (2007) tested the relationships between chlorophyll content and SPAD values for birch, wheat, and potato. For all three species, the relationships were nonlinear, although for birch and wheat, it was strong ($\sim R^2 = 0.9$), while the potato relationship was weaker ($\sim R^2 = 0.5$). It may therefore be appropriate to develop species-specific calibrations for robust chlorophyll estimation, with consideration for leaf developmental stage.

20.4.3.3 Estimating Leaf Chlorophyll Content with Spectral Vegetation Indices

Empirical spectral VIs are perhaps the most popular and straightforward means of retrieving chlorophyll content from remotely sensed data. There has been a wealth of research devoted to deriving statistical relationships between VIs and biochemical constituents, in order to retrieve chlorophyll content. Spectral indices are usually formulated using ratios of wavelengths that are sensitive to a particular leaf pigment to spectral regions where scattering is mainly driven by leaf

internal structure or canopy structure (i.e., the NIR). Indices including "off-center" wavelengths (690–740 nm) have been shown to be strong indicators of chlorophyll content (Croft et al. 2014) compared to indices containing chlorophyll absorption wavelengths (660–680 nm) due to ready saturation even at low chlorophyll content (Daughtry et al. 2000). Many VIs used in chlorophyll studies are focused along the red edge, including the MERIS Terrestrial Chlorophyll Index (Dash and Curran 2004). Recent research has focused on improving the generality and applicability of spectral indices through testing and modification over a range of species and physiological conditions, using empirical and simulated data. However, many indices have been developed and tested using a few closely related species, at the leaf scale and under controlled laboratory conditions. At the leaf level, surface scattering, internal structural characteristics, and leaf water content affect the relationship between VI and chlorophyll content estimation. Scaling up to a branch or canopy, other factors such as LAI, solar/viewing geometry, and canopy architecture also affect the VI. Background contributions have also been shown to perturb the relationship between chlorophyll and VIs, particularly in sparse or clumped canopies, with low LAI values (Croft et al. 2013). However, these confounding influences are less of a concern in closed broadleaf canopies, which essentially behave as a "big leaf" (Gamon et al. 2010).

20.4.3.4 Estimating Leaf Chlorophyll Content with Radiative Transfer Models

Physically based modeling approaches use radiative transfer models to account for variations in canopy architecture, image acquisition conditions, and background vegetation that may vary in space and time. As radiative transfer models are underpinned by physical laws governing the interaction of radiation at the canopy surface and within the canopy, they provide a direct physical relationship between canopy reflectance and canopy biophysical properties. The most recognized approach for modeling leaf chlorophyll content from remotely sensed data in this manner is through the coupling of a canopy radiative transfer model and a leaf optical model, to firstly retrieve leaf-level reflectance and then derive leaf biochemical constituents from the modeled leaf reflectance (Croft et al. 2013). Several canopy models have been used for this purpose, with the most popular including Scattering by Arbitrarily Inclined Leaves (SAIL) (Verhoef 1984), Discrete Anisotropic Radiative Transfer (DART) (Gastellu-Etchegorry et al. 2004), 4-SCALE (Chen and Leblanc 1997), GeoSAIL (Huemmrich 2001), and FLIGHT (North 1996). The models range from turbid medium models (SAIL), hybrid geometric–optical and radiative transfer models (4-SCALE, GeoSAIL, DART) in which the turbid media are constrained into a geometric form (i.e., a leaf, shoot, branch, and/or crown), and ray-tracing techniques (FLIGHT). At the leaf level, the PROSPECT model (Jacquemoud and Baret 1990) has had widespread validation across a wide range of vegetation species and functional types. In the inverse mode, PROSPECT models chlorophyll and carotenoids from leaf reflectance, along with dry matter, a structural parameter, and equivalent water thickness.

The smaller number of input parameters compared to other leaf-level models, such as LIBERTY (Dawson et al. 1998), means that it is readily inverted. A range of different techniques have been employed to invert the leaf and canopy radiative transfer models, including iterative numerical optimization methods, artificial NNs, vector machine regression, and lookup tables. However, the "ill-posed" problem means that different combinations of the same structural and image acquisition parameters can result in the same canopy reflectance, indicating that some a priori scene information is required to constrain the inversion (Kimes et al. 2000).

20.4.4 Light Use Efficiency

20.4.4.1 Biophysical Basis of Light Use Efficiency

Light provides the necessary energy for plant photosynthesis. LUE measures carbohydrate produced by plants absorbing a unit amount of PAR. Monteith (1972, 1977) initially proposed the theory in estimating crop production. Kumar and Monteith (1981) first applied the LUE theory to estimate crop growth with remote sensing. They successfully estimated crop growth with a constant LUE. This triggered tremendous interest in estimating LUE for different plant communities. Prince (1991) and Ruimy et al. (1994) reviewed LUE values published in the literature and found it varied greatly. In addition to the inherent difference in LUE among different plant communities, numerous external factors also contributed to the apparent variation in LUE values reported in the literature (Prince 1991; Gower et al. 1999), including

- Realized LUE versus maximum LUE
- LUE based on aboveground growth versus total plant growth
- LUE based on PAR absorbed versus intercepted
- LUE based on PAR versus total global radiation
- LUE based on NPP versus GPP
- LUE based on PAR absorbed by GV versus total foliage

In order to reduce confusion, Prince (1991), Gower et al. (1999), and Song et al. (2013) advocated that LUE should be based on absorbed PAR by total foliage in plant canopies, treating the proportion of NPV component as an inherent part of the plant communities.

Use of light for photosynthesis by a single leaf can be characterized by the photosynthesis–light (P–L) response curve (Figure 20.4). The net photosynthesis rate and PAR density is a straight line only when PAR is low. After reaching the light saturation point, the net photosynthesis rate is nonlinearly related to photon flux density. Therefore, leaves in a plant canopy may have very different LUE at any given time of day because leaves could have very different PAR density due to mutual shading and variation in leaf orientation. Plant production of a given plant community is the sum of photosynthesis from all leaves, and the LUE of such production varies with time of day due to changes in incident solar radiation angle and cloudiness. Song et al. (2013) showed that the shorter the time span, the bigger the

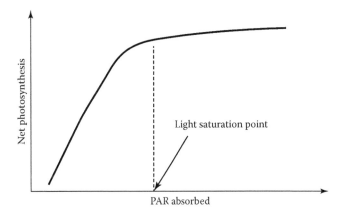

FIGURE 20.4　Photosynthesis–light response curve.

variation of LUE based on simulations with a Farquhar photosynthesis model coupled with a sophisticated canopy radiation transfer model (Song et al. 2009). It takes about a month for LUE to converge to a stable value (Figure 20.5). Despite the highly nonlinear relationship between net photosynthesis rate and PAR density at the instantaneous time scale, the combination of mutual shading of leaves, variation of leaf orientation, and change in solar angle tends to linearize the relationship between absorbed PAR and plant growth over time (Sellers 1985). It is important to note that the LUE estimated by Monteith (1972, 1977) was done over a whole growing season. Goetz and Prince (1999) argued based on plant functional convergence hypothesis (Field 1991) that LUE should converge to a narrow range for GPP among a wide range of plant functional types.

Incident radiation arrives at the top of plant canopies as direct and diffuse radiation. Their relative composition is a function of Sun angle, cloudiness of the atmosphere, and the characteristics of canopy structure (Ni et al. 1997), thus varying with location and canopy structure. Because diffuse radiation can penetrate plant canopies deeper than direct light, plants have a higher LUE for diffuse light. Therefore, LUE should converge to different values for different vegetation biomes, a conclusion that is now generally accepted (Ruimy et al. 1994; Turner et al. 2003; Running et al. 2004). Due to landscape heterogeneity in vegetation biome composition, the biome-dependent LUE creates spatial variation in LUE at the pixel level, particularly with coarse-spatial-resolution imagery, such as that from MODIS (Turner et al. 2002), making direct mapping of LUE from remotely sensed data attractive.

20.4.4.2 Remote Sensing LUE

PAR absorbed by leaf pigments has three pathways within the chloroplast: photochemical quenching (i.e., used for photosynthesis), nonradiative quenching, and photoprotection through which excess energy is dissipated through the xanthophyll cycle and chlorophyll fluorescence (ChF) through which radiation is emitted at longer wavelength than the absorbed light (Coops et al. 2010). Energy directed toward nonradiative quenching and ChF reduces LUE in photosynthesis (Meroni et al. 2009; Coops et al. 2010). Remote sensing of LUE is based on

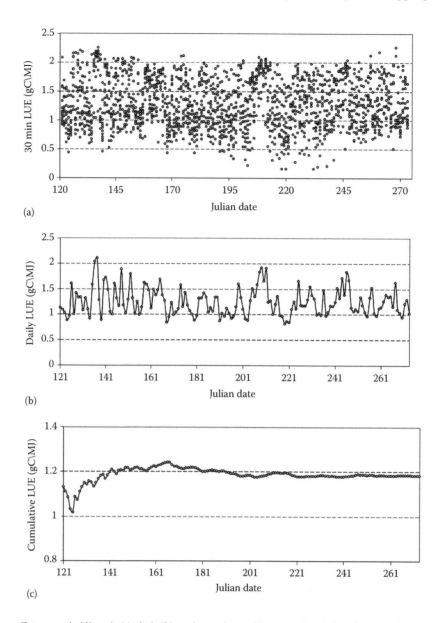

FIGURE 20.5   Light use efficiency at half hourly (a), daily (b), and cumulative (c) temporal scale based on simulation from Song et al. (2009).

the detection of energy that is not used in the photochemical quenching.

One such approach was initially developed by Gamon et al. (1992) based on their discovery that leaf reflectance at 531 nm is related to the xanthophyll cycle (Gamon et al. 1990). When absorbed PAR exceeds photosynthetic capacity, the *excess light* is dissipated through deepoxidation of violaxanthin to zeaxanthin via the intermediate antheraxanthin pigment. The process is reversed when there is insufficient supply of energy (Demmig-Adams 1990). Because zeaxanthin and antheraxanthin have higher absorption coefficients for radiation near 531 nm than violaxanthin, leaf reflectance at 531 nm changes with the xanthophyll cycle. Gamon et al. (1992) developed the photochemical reflectance index (PRI) to measure the leaf reflectance changes at 531 nm

$$PRI = \frac{R_{531} - R_{570}}{R_{531} + R_{570}} \qquad (20.16)$$

where R_{531} and R_{570} are leaf reflectances at 531 and 570 nm, respectively. Here, R_{570} is used as reference reflectance as leaf reflectance changes little at 570 nm with the xanthophyll cycle. Numerous studies confirm that PRI is highly correlated with photosynthetic radiation use efficiency (Peñuelas et al. 1995; Gamon and Surfus 1999; Nichol et al. 2000). Great interest has been generated to map LUE remotely with PRI, particularly with remotely sensed data from MODIS since band 11 is centered at 531 nm. However, MODIS does not have a band centered at 570 nm that can be used as the reference band. Drolet et al. (2005) used bands 11 and 13 to calculate PRI and found that PRI from backscatter images (i.e., near hotspot view) is significantly correlated with LUE estimated from flux tower measurements. Drolet et al. (2008) further demonstrated that PRI calculated with band 14 as the reference band also had a strong relationship with 90 min LUE for a boreal forest in Saskatchewan, Canada. Garbulsky et al. (2008) found that band 12 can also be used as the reference band for characterizing the relationship between PRI and LUE over a Mediterranean forest.

Testing PRI from MODIS with flux tower measurements offers insight on the potential of mapping LUE in space. However, the PRI–LUE relationship can be compromised by the mismatch in scale between the footprint of a flux tower and the corresponding MODIS pixel. Hall et al. (2008) and Hilker et al. (2008, 2010) conducted a series of fieldwork studies that measure PRI on a flux tower with a spectroradiometer in two forests, with one on Vancouver Island, British Columbia, in a Douglas-fir stand and the other in a mature Aspen stand in central Saskatchewan. They found that the downregulation of photosynthesis at the forest scale governs the relationship between PRI and LUE. They also found a stronger relationship between PRI and LUE for sunlit canopies than shaded canopies, providing evidence to explain why only PRI from backscatter MODIS imagery correlated with LUE (Drolet et al. 2005). Hall et al. (2011) proposed a satellite mission, PHOTOSYNSAT, a multiangle along-track satellite to map LUE based on their findings that (1) the first derivative of PRI with respect to the fraction of shadows in the sensor field of view is proportional to LUE and (2) PRI response is independent of vegetation structure and optical properties.

Another approach to estimating LUE remotely is through detecting the energy dissipated in ChF at wavelengths ranging from 650 to 800 nm (Meroni et al. 2009). Remote sensing of ChF is based on three Fraunhofer lines where incident solar radiation is low due to absorption by hydrogen in the solar atmosphere (656.4 nm) and oxygen in the Earth's atmosphere (760.5 and 787.5 nm) (Meroni et al. 2009; Guanter et al. 2010; Liu and Cheng 2010). The ESA's FLuorescence EXplorer (FLEX) mission is aimed at measuring solar-induced ChF in the Fraunhofer lines created by oxygen (Guanter et al. 2010; Mohammed et al. 2012). The feasibility has been demonstrated on the ground with high-spectral-resolution spectroradiometers and fluorometers (Damm et al. 2010; Liu and Cheng 2010; Liu et al. 2013) and in the air and space (Guanter et al. 2007; Zarco-Tejada et al. 2009, 2013; Frankenberg et al. 2011). Due to the weak ChF signal, it is mandatory for FLEX to be successful that the remotely sensed data be collected with high spectral resolution and data have to be corrected with a precise modeling of atmospheric effect in the visible and NIR spectra (Grace et al. 2007; Malenovsky et al. 2009; Guanter et al. 2010).

20.4.5 Gross Primary Productivity/ Net Primary Productivity

GPP measures the net carbohydrate produced during photosynthesis after dark respiration over a unit area in a given time. NPP is the balance of GPP after plant autotrophic respiration. Both GPP and NPP are key carbon fluxes between the terrestrial ecosystem and the atmosphere in the global carbon cycle. However, current eddy covariance technology cannot measure GPP directly, but only measures net ecosystem exchange (NEE), that is, the net difference between GPP and ecosystem respiration, which includes both autotrophic and heterotrophic respirations. Estimation of GPP is accomplished by resorting to models to estimate autotrophic respiration during the daytime. Through careful fieldwork, NPP can be estimated as the total organic matter produced on an annual basis (Clark et al. 2001a,b). However, remote sensing is the only viable option to provide wall-to-wall estimation of primary production.

20.4.5.1 Remote Sensing of Primary Production Based on LUE

Optical remote sensing of terrestrial ecosystem primary production is predominantly based on the LUE theory (Monteith 1972, 1977) because of its sound biophysical basis and simplicity. The general form of LUE models for primary productivity is

$$P = \varepsilon_{max} \times f_{APAR} \times IPAR \times f(E) \qquad (20.17)$$

where P is the primary productivity. The maximum LUE, ε_{max}, is the conversion factor of Absorbed Photosynthetically Active Radiation (APAR) to NPP or GPP under optimal conditions. APAR is generated as the product of the fraction of APAR (f_{APAR}) and incident PAR. The environmental scalar function, f(E), produces a scalar that is between 0 and 1 to reduce the maximum LUE to actual LUE due to environmental stress.

The product of a LUE model can either be NPP or GPP, as listed in Table 20.14. For models with their initial product being GPP, autotrophic respiration is usually simulated to produce NPP, such as the Biome-BGC model that produces the MODIS NPP and GPP products (Zhao et al. 2005).

20.4.5.2 Remote Sensing of Primary Production without LUE

Estimating terrestrial ecosystem primary production does not always rely on LUE. Numerous process-based models, which use remotely sensed land surface biophysical parameter products as key model inputs, but are not based on LUE theory, have been developed. The most commonly used land surface biophysical parameter is perhaps LAI. Running et al. (1989) conducted a pioneering study that integrated remote sensing with ecosystem models to estimate GPP. They used LAI derived from AVHRR NDVI as an input to the Forest-BGC model (Running and Coughlan 1988) and estimated both GPP and transpiration over a 28 × 55 km mountainous region. Nemani et al. (1993) derived an improved LAI product from Landsat TM imagery by correcting the background effects on NDVI using a midinfrared reflectance. The subsequent LAI product was used as an input to the RHESSys model (Band et al. 1993) to estimate evapotranspiration and GPP at the watershed scale. Liu et al. (1997) developed the BEPS model, which estimates NPP using LAI derived from the 10-day composite AVHRR NDVI. Nemani et al. (2009) developed the Terrestrial Observation and Prediction System (TOPS) model to estimate GPP and ET fluxes using the MODIS

TABLE 20.14   Typical LUE Models for Terrestrial Ecosystem Primary Productivity

Model	Productivity Measure	LUE (ε_{max})(gC/MJ)	Environmental Scalars Temperature (T), Vapor Pressure Deficit (D), and Soil Water (θ)	References
Kumar and Monteith	NPP	1.3	N/A	Kumar and Monteith (1981)
BIOMASS	GPP	a	$f(T_{min}) = \begin{cases} 0 & T_{min} < T_{min1} \\ \dfrac{T_{min} - T_1}{T_2 - T_1} & T_{min1} \le T_{min} < T_{min2} \\ 1 & T_{min} \ge T_{min2} \end{cases}$ $f(\theta) = \begin{cases} 0 & \theta < \theta_{min} \\ \dfrac{\theta - \theta_{min}}{\theta_{max} - \theta_{min}} & \theta_{min} \le \theta < \theta_{max} \\ 1 & \theta \ge \theta_{max} \end{cases}$ $f(VPD) = \begin{cases} 0 & D \ge D_2 \\ \dfrac{D_2 - D}{D_2 - D_1} & D_1 \le D < D_2 \\ 1 & D < D_1 \end{cases}$	McMurtrie et al. (1990)
Ruimy et al.	NPP	0.37–2.07 (biome dependent)	N/A	Ruimy et al. (1994)
GLO-PEM	GPP	$\varepsilon_{max} = 55.2a$	$f(T_{min}) = 0$, when $T_{min} < 0$, $f(D) = 1.2 \exp(-0.35D) - 0.2$ $f(\theta) = \dfrac{\alpha_0 - \alpha_d}{\alpha_w - \alpha_d}$	Prince and Goward (1995)
TURC	GPP	1.1	N/A	Ruimy et al. (1996)
3-PGS	GPP	1.8	$f(D) = \exp(-2.5D)$, $f(\theta) = \dfrac{1}{1 + \left\{(1-\theta)/c\right\}^d}$	Coops et al. (1998)
NASA-CASA	NPP	0.389/0.506	$f_1(T) = 0.8 + 0.02 T_{opt} - 0.0005 T_{opt}^2$ $f_2(T) = \dfrac{1.18/(1 + \exp\{0.2(T_{opt1} - 10 - T_a)\})}{1 + \exp\{0.3(-T_{opt1} - 10 - T_a)\}}$ $f(\theta) = 0.5 + \dfrac{AET}{PET}$	Potter et al. (1993), Field et al. (1995), Potter et al. (2003)
MODIS (MOD17)	GPP	0.604–1.259 (biome dependent)	$f(T_{min}) = \begin{cases} 0 & T_{min} < T_1 \\ \dfrac{T_{min} - T_1}{T_2 - T_1} & T_1 \le T_{min} < T_2 \\ 1 & T_{min} \ge T_2 \end{cases}$ $f(D) = \begin{cases} 0 & D \ge D_2 \\ \dfrac{D_2 - D}{D_2 - D_1} & D_1 \le D < D_2 \\ 1 & D < D_1 \end{cases}$	Running et al. (2000), Zhao et al. (2005)
VPM	GPP	Quantum yield	$f(T) = \dfrac{(T_a - T_3)(T_a - T_4)}{\left\{(T_a - T_3)(T_a - T_4)\right\} - (T_a - T_{opt})^2}$ $f(\theta) = \dfrac{1 + LSWI}{1 + LSWI_{max}}$ $LSWI = \dfrac{\rho_{NIR} - \rho_{SWIR}}{\rho_{NIR} + \rho_{SWIR}}$	Xiao et al. (2004a,b)

(Continued)

TABLE 20.14 *(Continued)*　　Typical LUE Models for Terrestrial Ecosystem Primary Productivity

Model	Productivity Measure	LUE (ε_{max})(gC/MJ)	Environmental Scalars Temperature (T), Vapor Pressure Deficit (D), and Soil Water (θ)	References
TOPS (diagnostic version)	GPP	Same with MOD17	Same with MOD17 for T_{min} and VPD $$f(\theta)=\begin{cases} 0 & LWP \geq LWP_2 \\ \dfrac{LWP_2 - LWP}{LWP_2 - LWP_1} & LWP_1 \leq LWP < LWP_2 \\ 1 & LWP < LWP_1 \end{cases}$$	Nemani et al. (2009)
EC-LUE		2.14	$$f(T) = \exp\left[-\left(\frac{T_a - T_{opt}}{T_{opt}}\right)^2\right]$$ $$f(\theta) = \frac{1}{\beta+1} = \frac{LE}{LE+H}$$ Same with MOD17 for T_{min} and VPD f(T) of VPM model used in the old version	Yuan et al. (2007, 2010)

α, quantum yield (moles CO_2 mole $PPFD^{-1}$).

α_0, α_w, and α_d, the slope between NDVI and T_s, and the wet and dry edge slopes (°C $NDVI^{-1}$).

β, Bowen ratio (= H/LE).

$\boldsymbol{\theta}$, calculated soil saturation (%).

θ_{max} and θ_{min}, soil saturation of root zone at field capacity (−0.004 MPa) and permanent wilting point (−1.5 MPa) (%).

c and d, texture-specific parameters for the water stress scalar of 3-PGS model.

T_a, T_s, and T_{min}, air, surface, and daily min temperature (°C).

T_1 and T_2, biome-specific parameters for the T_{min} scalar.

T_3 and T_4, min and max temperature for photosynthetic activity in VPM model.

T_{opt}, optimal temperature for photosynthetic activity (°C) (25 for EC-LUE model, monthly average air temperature when NDVI reaches its maximum in NASA-CASA model.

VPD or D, vapor pressure deficit (kPa).

D_1 and D_2, biome-specific parameters for the VPD scalar.

LWP, predawn leaf water potential (MPa), calculated from θ and soil textural parameters.

LWP_1 and LWP_2, biome-specific parameters for the LWP scalar (min spring LWP and stomatal closure LWP) (MPa).

AET and PET, actual and potential evapotranspiration.

LE and H, latent and sensible heat.

LSWI, land surface water index.

$LSWI_{max}$, max LSWI within the plant growing season for individual pixels.

LAI product as an input. More advanced use of optical remote sensing for primary production estimation takes advantage of multiangular remote sensing to derive more complex canopy biophysical parameters (e.g., the foliage clumping index) that can be integrated into process-based photosynthesis models to provide more detailed partitioning of solar radiation into sunlit and shaded leaves, potentially improving NPP and GPP estimates over the "big-leaf" models based on LAI (Chen et al. 2003, 2012). These types of models closely couple carbon and water fluxes through stomatal conductance, often by linking the Farquhar photosynthesis (Farquhar et al. 1980) and Penman–Monteith (Monteith 1965) evapotranspiration models.

20.4.6 Uncertainties, Errors, and Accuracy

The functional products of vegetation derived from remote sensing directly, such as phenology and LUE, or indirectly, such as GPP and NPP, are theoretically invalidatable because we never have perfect reference data. The remotely sensed phenology parameters, such as the start and the end of growing season, can only be derived through statistical analysis of image time series, typically on a spectral VI (Zhang et al. 2003). LSP is an outcome of synaptic phenomena for all vegetation within the entire pixel, which is usually quite large for the sake of frequent repeat coverage (Myneni et al. 1997; Tucker et al. 2005; Zhu et al. 2013). The synoptic seasonal timing over a sizable area cannot be accurately recorded on the ground as each individual plant may have a different phenology. Although there are now cameras mounted on flux towers, which provide a much better record of the actual synoptic phenology (Richardson et al. 2007), it is nearly impossible to have a perfect spatial match between the area of the camera field of view and the satellite pixel. In addition, the direct derivation of evergreen forest phenology, which demonstrates little seasonal greenness change, remains a challenge using remotely sensed imagery (Dannenberg 2013).

Similarly, the validation of GPP/NPP estimates from remote sensing over a large area is impossible because no reliable reference data are available. On the site basis, carbon fluxes from eddy covariance flux towers are frequently used for model evaluations (Xiao et al. 2004a,b; Yuan et al. 2007, 2014). However, flux towers do not directly measure GPP/NPP, but NEE that is the net difference between the carbon absorbed by plant photosynthesis and that released by ecosystem respiration. Ecosystem respiration has to be subtracted from NEE in order to get GPP. Currently, there is no direct measurement for daily ecosystem respiration when photosynthesis happens, though it can be modeled based on the ecosystem respiration rate at night (Reichstein et al. 2005; Lasslop et al. 2010), adding uncertainty to the flux tower–derived GPP for model evaluation (Schaefer et al. 2012). An alternative approach for model evaluation is comparing independent estimates, such as that from atmospheric inverse modeling and that from ground inventory and analysis (Pacala et al. 2001; Piao et al. 2009). Although this does not validate any model outputs in the strict sense, it does provide more credibility to the model output. Much research and new technologies are needed in this regard in the future.

20.5 Future Directions

Optical remote sensing will continue to play a critical role in monitoring the spatial temporal dynamics of vegetation in the foreseeable future, and it will continue to provide key land surface biophysical parameter data layers to models at regional to global scales that aim to simulate and project the changes of terrestrial ecosystem functions in the future due to global environmental changes. Recently launched remote sensing missions, such as the Landsat 8 and Suomi NPP satellites by the United States and the SPOT 6 by France, will continue to provide pivotal optical remotely sensed data in the immediate future. Planning of remote sensing missions in the United States is guided by the Decadal Survey generated by the National Research Council of the National Academy of Sciences (NRC 2007). The planned mission, HyspIRI, combines a hyperspectral visible SWIR imaging spectrometer with a multispectral thermal infrared spectrometer to map vegetation composition and its health. The ESA's Sentinel mission, Sentinel-2, will continue to provide optical remotely sensed imagery globally as an enhanced continuity of SPOT and Landsat-type data. At the same time, high-spatial-resolution optical remotely sensed data will continue to boom in the private market. Therefore, both the quality and the quantity of optical remotely sensed data will increase in the foreseeable future.

In addition to remotely sensed data in the optical domain, Radar and LiDAR remote sensing missions will increase in the future at the same time. The abundance of remotely sensed data from multiple sensors will lead to synergistic use of remotely sensed data from different sensors to extract vegetation information, particularly by fusing information from LiDAR, Radar, and optical sensors (Lefsky et al. 2005; Gao et al. 2006; Nelson et al. 2009; Gray and Song 2012). Although empirical approaches will continue to be critical to understand the relationship between remotely sensed data and vegetation structure at local to regional scales, more physically based approaches that are applicable globally will continue to advance, providing increasingly higher quality vegetation information that will enable improved understanding of the roles vegetation plays in the terrestrial ecosystem.

Acknowledgment

This work is partly supported by NSF Grant DEB-1313756.

References

Adams, J. B., D. E. Sabol, V. Kapos, R. A. Filho, D. A. Roberts, M. O. Smith, and A. R. Gillespie. 1995. Classification of multispectral images based on fractions of endmembers: Application to land-cover change in the Brazilian Amazon. *Remote Sens Environ* 52:137–154.

Andersen, H. E., J. Strunk, H. Temesgen, D. Atwood, and K. Winterberger. 2011. Using multilevel remote sensing and ground data to estimate forest biomass resources in remote regions: A case study in the boreal forests of interior Alaska. *Canadian J Remote Sens* 37(6):596–611.

Anderson, G. L., J. D. Hanson, and R. H. Haas. 1993. Evaluating Landsat Thematic Mapper derived vegetation indices for estimating above-ground biomass on semiarid rangelands. *Remote Sens Environ* 45:165–175.

Aragao, L. E. O. C., Y. E. Shimabukuro, F. D. B. Espírito-Santo, and M. Williams. 2005. Spatial validation of the Collection 4 MODIS LAI product in Eastern Amazonia. *IEEE Trans Geosci Remote Sens* 43:2526–2534.

Archibold, O. W. 1995. *Ecology of World Vegetation*. London, U.K.: Chapman & Hall.

Asrar, G., M. Fuchs, E. T. Kanemasu, and J. L. Hatfield. 1984. Estimating absorbed photosynthetic radiation and leaf-area index from spectral reflectance in wheat. *Agron J* 76:300–306.

Asrar, G., R. B. Myneni, and B. J. Choudhury. 1992. Spatial heterogeneity in vegetation canopies and remote-sensing of absorbed photosynthetically active radiation—A modeling study. *Remote Sens Environ* 41:85–103.

Atkinson, P. M., M. E. J. Cutler, and H. Lewis. 1997. Mapping sub-pixel proportional land cover with AVHRR imagery. *Int J Remote Sens* 18(4):917–935. doi:10.1080/014311697218836.

Baccini, A., M. A. Friedl, C. E. Woodcock, and A. Warbington. 2004. Forest biomass estimation over regional scales using multisource data. *Geophys Res Lett* 31:L10501.

Baccini, A., S. J. Goetz, W. S. Walker, N. T. Laporte, M. Sun, D. Sulla-Menashe, J. Hackler, P. S. A. Beck, R. Dubayah, and M. A. Friedl. 2012. Estimated carbon dioxide emissions from tropical deforestation improved by carbon density maps. *Nature Clim Chang* 2:182–185.

Baccini, A., N. Laporte, S. J. Goetz, M. Sun, and H. Dong. 2008. A first map of tropical Africa's above-ground biomass derived from satellite imagery. *Environ Res Lett* 3:045011.

Band, L. E., P. Patterson, R. Nemani, and S. W. Running. 1993. Forest ecosystem processes at the watershed scale: Incorporating hillslope hydrology. *Agric Forest Meteorol* 63:93–126.

Baret, F. and G. Guyot. 1991. Potentials and limits of vegetation indexes for LAI and APAR assessment. *Remote Sens Environ* 35:161–173.

Baret, F., G. Guyot, and D. Major. 1989. Crop biomass evaluation using radiometric measurements. *Photogrammetria* 43:241–256.Baret, F., O. Hagolle, B. Geiger, P. Bicheron, B. Miras, M. Huc, B. Berthelot, F. Nino, M. Weiss, and O. Samain. 2007. LAI, fAPAR and fCover CYCLOPES global products derived from VEGETATION: Part 1: Principles of the algorithm. *Remote Sens Environ* 110:275–286.

Baret, F. and A. Olioso. 1989. Estimation à partir de mesures de réflectance spectrale du rayonnement photosynthétiquement actif absorbé par une culture de blé. *Agronomie* 9:885–895.

Bastin, L. 1997. Comparison of fuzzy c-means classification, linear mixture modelling and MLC probabilities as tools for unmixing coarse pixels. *Int J Remote Sens* 18(17):3629–3648. doi:10.1080/014311697216847.

Bégué, A. 1993. Leaf area index, intercepted photosynthetically active radiation, and spectral vegetation indices: A sensitivity analysis for regular-clumped canopies. *Remote Sens Environ* 46:45–59.

Bergen, K. M. and I. Dronova. 2007. Observing succession on aspen-dominated landscapes using a remote sensing-ecosystem approach. *Landscape Ecol* 22(9):1395–1410.

Blackard, J. A., M. V. Finco, E. H. Helmer, G. R. Holden, M. L. Hoppus, D. M. Jacobs, A. J. Lister, G. G. Moisen, M. D. Neison, and R. Riemann. 2008. Mapping US forest biomass using nationwide forest inventory data and moderate resolution information. *Remote Sens Environ* 112:1658–1677.

Bonan, G. B. 2008a. *Ecological Climatology: Concepts and Applications*, 2nd edn. Cambridge, U.K.: Cambridge University Press.

Bonan, G. B. 2008b. Forests and climate change: Forcings, feedbacks, and the climate benefits of forests. *Science* 320:1444–1449.

Brandt, J. S., T. Kuemmerle, H. Li, G. Ren, J. Zhu, and V. C. Radeloff. 2012. Using Landsat imagery to map forest change in southwest China in response to the national logging ban and ecotourism development. *Remote Sens Environ* 121:358–369.

Brown, L. J., J. M. Chen, and S. G. Leblanc. 2000. Short wave infrared correction to the simple ratio: An image and model analysis. *Remote Sens Environ* 71:16–25.

Brown, M. E. and K. M. de Beurs. 2008. Evaluation of multisensor semi-arid crop season parameters based on NDVI and rainfall. *Remote Sens Environ* 112(5):2261–2271.

Campbell, G. S. 1990. Derivation of an angle density function for canopies with ellipsoidal leaf angle distributions. *Agric Forest Meteorol* 49:173–176.

Campbell, J. B. 2002. *Introduction to Remote Sensing*. Taylor & Francis.

Cao, C., J. Xiong, S. Blonski, Q. Liu, S. Uprety, X. Shao, Y. Bai, and F. Weng. 2013. Suomi NPP VIIRS sensor data record verification, validation, and long-term performance monitoring. *J Geophys Res Atmos* 118:11664–11678.

Carlson, T. N. and D. A. Ripley. 1997. On the relation between NDVI, fractional vegetation cover, and leaf area index. *Remote Sens Environ* 62:241–252.

Chen, J. M. 1996a. Optically-based methods for measuring seasonal variation in leaf area index of boreal conifer forests. *Agric Forest Meteorol* 80:135–163.

Chen, J. M. 1996b. Evaluation of vegetation indices and a modified simple ratio for boreal applications. *Can J Remote Sens* 22:229–242.

Chen, J. M. 1999. Spatial scaling of a remotely sensed surface parameter by contexture. *Remote Sens Environ* 69:30–42.

Chen, J. M. and T. A. Black. 1992. Defining leaf area index for non-flat leaves. *Plant Cell Environ* 15:421–429.

Chen, J. M., T. A. Black, and R. S. Adams. 1991. Evaluation of hemispherical photography for determining plant area index and geometry of a forest stand. *Agric Forest Meteorol* 56:129–143.

Chen, J. M., X. Chen, W. Ju, and X. Geng. 2005a. Distributed hydrological model for mapping evapotranspiration using remote sensing inputs. *J Hydrol* 305:15–39.

Chen, J. M. and J. Cihlar. 1995. Plant canopy gap-size analysis theory for improving optical measurements of leaf area index. *Appl Optics* 34(27):6211–6222.

Chen, J. M. and J. Cihlar. 1996. Retrieving leaf area index for boreal conifer forests using Landsat TM images. *Remote Sens Environ* 55:153–162.

Chen, J. M., A. Govind, O. Sonnentag, Y. Zhang, A. Barr, and B. Amiro. 2006. Leaf area index measurements at Fluxnet Canada forest sites. *Agric Forest Meteorol* 140:257–268.

Chen, J. M. and S. G. Leblanc. 1997. A four-scale bidirectional reflectance model based on canopy architecture. *IEEE Trans Geosci Rem Sens* 35:1316–1337.

Chen, J. M., J. Liu, S. G. Leblanc, R. Lacaze, and J.-L. Roujean. 2003. Multi-angular optical remote sensing for assessing vegetation structure and carbon absorption. *Remote Sens Environ* 84(4):516–525.

Chen, J. M., C. Menges, and S. Leblanc. 2005b. Global mapping of foliage clumping index using multi-angular satellite data. *Remote Sens Environ* 97:447–457.

Chen, J. M., G. Mo, J. Pisek, F. Deng, M. Ishozawa, and D. Chan. 2012. Effects of foliage clumping on global terrestrial gross primary productivity. *Global Biogeochem Cycles* 26:GB1019. doi:10.1029/2010GB003996.

Chen, J. M., G. Pavlic, L. Brown, J. Cihlar, S. G. Leblanc, H. P. White, R. J. Hall, D. R. Peddle, D. J. King, and J. A. Trofymow. 2002. Derivation and validation of Canada-wide coarse resolution leaf area index maps using high-resolution satellite imagery and ground measurements. *Remote Sens Environ* 80:165–184.

Chen, J. M., P. M. Rich, S. T. Gower, J. M. Norman, and S. Plummer. 1997. Leaf area index of boreal forests: Theory, techniques, and measurements. *J Geophys Res* 102(D24):29429–29443.

Chen, X., L. Vierling, D. Deering, and A. Conley. 2005c. Monitoring boreal forest leaf area index across a Siberian burn chronosequence: A MODIS validation study. *Int J Remote Sens* 26(24):5433–5451.

Chopping, M., G. G. Moisen, L. Su, A. Laliberte, A. Rango, J. V. Martonchik, and D. P. C. Peters. 2008. Large area mapping of southwestern forest crown cover, canopy height, and biomass using the NASA Multiangle Imaging Spectro-Radiometer. *Remote Sens Environ* 112:2051–2063.

Chopping, M., C. B. Schaaf, F. Zhao, Z. Wang, A. W. Nolin, G. G. Moisen, J. V. Martonchik, and M. Bull. 2011. Forest structure and aboveground biomass in the southwestern United States from MODIS and MISR. *Remote Sens Environ* 115:2943–2953.

Chuine, I. and E. G. Beaubien. 2001. Phenology is a major determinant of tree species range. *Ecol Lett* 4:500–551.

Clark, D. A., S. Brown, D. W. Kicklighter, J. Q. Chambers, J. R. Thomlinson, and J. Ni. 2001a. Measuring net primary production in forests: Concepts and field methods. *Ecol Appl* 11(2):356–370.

Clark, D. A., S. Brown, D. W. Kicklighter, J. Q. Chambers, J. R. Thomlinson, J. Ni, and E. A. Holland. 2001b. Net primary production in tropical forests: An evaluation and synthesis of existing field data. *Ecol Appl* 11(2):371–384.

Cleland, E. E., J. M. Allen, T. M. Crimmins, J. A. Dunne, S. Pau, S. E. Travers, E. S. Zavaleta, and E. M. Wolkovich. 2012. Phenological tracking enables positive species responses to climate change. *Ecology* 93(8):1765–1771.

Clevers, J. G. P. W. 1989. The applications of a weighted infrared vegetation index for estimating leaf area index by correcting for soil moisture. *Remote Sens Environ* 29:25–37.

Cohen, W. B., T. K. Maiersperger, T. A. Spies, and D. R. Oetter. 2001. Modelling forest cover attributes as continuous variables in a regional context with Thematic Mapper data. *Int J Remote Sens* 22(12):2279–2310.

Cohen, W. B., T. K. Maiersperger, D. P. Turner, W. D. Ritts, D. Pflugmacher, R. E. Kennedy, A. Kirschbaum, S. W. Running, M. Costa, and S. T. Gower. 2006. MODIS land cover and LAI Collection 4 product quality across nine sites in the Western Hemisphere. *IEEE Trans Geosci Remote Sens* 44:1843–1857.

Cohen, W. B., T. K. Maiersperger, Z. Yang, S. T. Gower, D. P. Turner, W. D. Ritts, M. Berterretche, and S. W. Running. 2003. Comparisons of land cover and LAI estimates derived from ETM+ and MODIS for four sites in North America: A quality assessment of 2000/2001 provisional MODIS products. *Remote Sens Environ* 88:233–255.

Cohen, W. B. and T. A. Spies. 1992. Estimating structural attributes of Douglas-fir/western hemlock forest stands from Landsat and SPOT imagery. *Remote Sens Environ* 41(1):1–17.

Cohen, W. B., T. A. Spies, R. J. Alig, D. R. Oetter, T. K. Maiersperger, and M. Fiorella. 2002. Characterizing 23 years (1972–95) of stand replacement disturbance in western Oregon forests with Landsat imagery. *Ecosystems* 5:122–137.

Cohen, W. B., T. A. Spies, and G. A. Bradshaw. 1990. Semivariograms of digital imagery for analysis of conifer canopy structure. *Remote Sens Environ* 34(3):167–178.

Cohen, W. B., T. A. Spies, and M. Fiorella. 1995. Estimating the age and structure of forests in a multi-ownership landscape of western Oregon, U.S.A. *Int J Remote Sens* 16(4):721–746.

Cohen, W. B., Z. Yang, and R. Kennedy. 2010. Detecting trends in forest disturbance and recovery using yearly Landsat time series: 2. TimeSync—Tools for calibration and validation. *Remote Sens Environ* 114(12):2911–2924.

Colditz, R. R., M. Schmidt, C. Conrad, M. C. Hansen, and S. Dech. 2011. Land cover classification with coarse spatial resolution data to derive continuous and discrete maps for complex regions. *Remote Sens Environ* 115(12):3264–3275. doi:10.1016/j.rse.2011.07.010.

Coops, N. C. 1999. Improvement in predicting stand growth of Pinus radiata (D. Don) across landscapes using NOAA AVHRR and Landsat MSS imagery combined with a forest growth process model (3-PGS). *Photogramm Eng Remote S* 65:1149–1156.

Coops, N. C., T. Hilker, F. G. Hall, C. J. Nichol, and G. G. Drolet. 2010. Estimation of light-use efficiency of terrestrial ecosystems from space: A status report. *BioScience* 60(10):788–797.

Coops, N. C., R. Waring, and J. Landsberg. 1998. Assessing forest productivity in Australia and New Zealand using a physiologically-based model driven with averaged monthly weather data and satellite-derived estimates of canopy photosynthetic capacity. *Forest Ecol Manag* 104:113–127.

Croft, H., J. M. Chen, and Y. Zhang. 2014. The applicability of empirical vegetation indices for determining leaf chlorophyll content over different leaf and canopy structures. *Ecol Complex* 17:119–130.

Croft, H., J. M. Chen, Y. Zhang, and A. Simic. 2013. Modelling leaf chlorophyll content in broadleaf and needle leaf canopies from ground, CASI, Landsat TM 5 and MERIS reflectance data. *Remote Sens Environ* 133:128–140.

Csiszar, I., W. Schroeder, L. Giglio, E. Ellicott, K. P. Vadrevu, C. O. Justice, and B. Wind. 2014. Active fires from the Suomi NPP Visible Infrared Imaging Radiometer Suite: Product status and first evaluation results. *J Geophys Res Atmos.* 119:803–816.

Curran, P. J., J. L. Dungan, and H. L. Gholz. 1990. Exploring the relationship between reflectance red edge and chlorophyll content in slash pine. *Tree Physiol* 7:33–48.

Dai, Y., X. Zeng, R. E. Dickinson, I. Baker, G. B. Bonan, M. G. Bosilovich, A. S. Denning et al. 2003. The common land model. *Bull Am Meteor Soc.* 84:1013–1023.

Damm, A., J. Elbers, A. Erler, B. Gioli, K. Hamdi, R. Hutjes, M. Kosvancova, M. Meroni, F. Miglietta, and A. Moersch. 2010. Remote sensing of sun-induced fluorescence to improve modeling of diurnal courses of gross primary production (GPP). *Global Change Biol* 16(1):171–186.

Dannenberg, M. P. 2013. ENSO-induced variability in terrestrial vegetation dynamics in the Western United States. M.A. Thesis. University of North Carolina at Chapel Hill, Chapel Hill, NC.

Dash, J. and P. J. Curran. 2004. The MERIS terrestrial chlorophyll index. *Int J Remote Sens* 25(23):5403–5413.

Daughtry, C. S. T., K. Gallo, and M. E. Bauer. 1983. Spectral estimates of solar radiation intercepted by corn canopies. *Agron J* 75:527–531.

Daughtry, C. S. T., K. Gallo, S. Goward, S. Prince, and W. Kustas. 1992. Spectral estimates of absorbed radiation and phytomass production in corn and soybean canopies. *Remote Sens Environ* 39:141–152.

Daughtry, C. S. T., C. L. Walthall, M. S. Kim, E. B. De Colstoun, and J. E. McMurtrey III. 2000. Estimating corn leaf chlorophyll concentration from leaf and canopy reflectance. *Remote Sens Environ* 74:229–239.

Davies, R. and M. Molloy. 2012. Global cloud height fluctuations measured by MISR on Terra from 2000 to 2010. *Geophys Res Lett* 39:L03701. doi:10.1029/2011GL050506.

Dawson, T. P., P. J. Curran, and S. E. Plummer. 1998. LIBERTY—Modeling the effects of leaf biochemical concentration on reflectance spectra. *Remote Sens Environ* 65:50–60.

Deardorff, J. W. 1978. Efficient prediction of ground surface temperature and moisture, with inclusion of a layer of vegetation. *J Geophys Res* 83(C4):1889. doi:10.1029/JC083iC04p01889.

DeFries, R. S., M. C. Hansen, J. R. G. Townshend, A. C. Janetos, and T. R. Loveland. 2000. A new global 1-km dataset of percentage tree cover derived from remote sensing. *Global Change Biol* 6:247–254.

DeFries, R. S., J. R. G. Townshend, and M. C. Hansen. 1999. Continuous fields of vegetation characteristics at the global scale at 1-km resolution. *J Geophys Res* 104(D14):16911. doi:10.1029/1999JD900057.

Delbart, N., L. Kergoat, T. Le Toan, J. Lhermitte, and G. Picard. 2005. Determination of phenological dates in boreal regions using normalized difference water index. *Remote Sens Environ* 97(1):26–38.

Demmig-Adams, B. 1990. Carotenoids and photoprotection in plants: A role for the xanthophyll zeaxanthin. *Biochim Biophys Acta* 1020:1–24.

Deng, C. and C. Wu. 2013. A spatially adaptive spectral mixture analysis for mapping subpixel urban impervious surface distribution. *Remote Sens environ* 133:62–70.

Deng, F., J. M. Chen, S. Plummer, and M. Chen. 2006. Global LAI algorithm integrating the bidirectional information. *IEEE T Geosci Remote* 44:2219–2229.

Denslow, J. 1980. Patterns of plant species diversity during succession. *Oecologia* 46(1):18–21.

Dial, G., H. Bowen, F. Gerlach, J. Grodecki, and R. Oleszczuk. 2003. IKONOS satellite, imagery, and products. *Remote Sens Environ* 88:23–36.

Diner, D. J., J. C. Beckert, T. H. Reilly, C. J. Bruegge, J. E. Conel, R. A. Kahn, J. V. Martonchik, T. P. Ackerman, R. Davies, and S. A. Gerstl. 1998. Multi-angle Imaging SpectroRadiometer (MISR) instrument description and experiment overview. *IEEE T Geosci Remote* 36:1072–1087.

Dixon, R. K., A. M. Solomom, S. Brown, R. A. Houghton, M. C. Trexier, and J. Wisniewski. 1994. Carbon pools and flux of global forest ecosystems. *Science* 263:185–190.

Dobson, A. P., A. D. Bradshaw, and J. M. Baker. 1997. Hopes for the future: Restoration ecology and conservation biology. *Science* 277:515–522.

Dong, J., R. K. Kaufmann, R. B. Myneni, C. J. Tucker, P. E. Kauppi, J. Liski, W. Buermann, V. Alexeyev, and M. K. Hughes. 2003. Remote sensing of boreal and temperate forest woody biomass: Carbon pools, sources and sinks. *Remote Sens Environ* 84:393–410.

Drolet, G. G., K. F. Huemmrich, F. G. Hall, E. M. Middleton, T. A. Black, A. G. Barr, and H. A. Margolis. 2005. A MODIS-derived photochemical reflectance index to detect inter-annual variations in the photosynthetic light-use efficiency of a boreal deciduous forest. *Remote Sens Environ* 98:212–224.

Drolet, G. G., E. M. Middleton, K. F. Huemmrich, F. G. Hall, B. D. Amiro, A. G. Barr, T. A. Black, J. H. McCaughey, and H. A. Margolis. 2008. Regional mapping of gross light-use efficiency using MODIS spectral indices. *Remote Sens Environ* 112:3064–3078.

Duncanson, L. I., K. O. Niemann, and M. A. Wulder. 2010. Integration of GLAS and Landsat TM data for aboveground biomass estimation. *Can J Remote Sens* 36:129–141.

Eastman, J. R., F. Sangermano, E. A. Machado, J. Rogan, and A. Anyamba. 2013. Global trends in seasonality of normalized difference vegetation index (NDVI), 1982–2011. *Remote Sens* 5(10):4799–4818.

Eklundh, L., K. Hall, H. Eriksson, J. Ardö, and P. Pilesjö. 2003. Investigating the use of Landsat thematic mapper data for estimation of forest leaf area index in southern Sweden. *Can J Remote Sens* 29(3):349–362.

Eklundh, L., L. Harrie, and A. Kuusk. 2001. Investigating relationships between Landsat ETMþ sensor data and leaf area index in a boreal conifer forest. *Remote Sens Environ* 78:239–251.

Farquhar, G. D., S. von Caemmerer, and J. A. Berry. 1980. A biochemical model of photosynthetic CO_2 assimilation in leaves of C3 species. *Planta* 149:78–90.

Fassnacht, K., S. T. Gower, J. M. Norman, and R. E. McMurtrie. 1994. A comparison of optical and direct methods for estimating foliage surface area index in forests. *Agric Forest Meteorol* 71:183–207.

Fazakas, Z. and M. N. H. Olsson. 1999. Regional forest biomass and wood volume estimation using satellite and ancillary data. *Agric Forest Meteorol* 98–99:417–425.

Field, C. B. 1991. Ecological scaling of carbon gain to stress and resource availability. In: H. A. Mooney, W. E. Winner, and E. J. Pell (eds.), *Response of Plants to Multiple Stresses*. San Diego, CA: Academic Press, pp. 35–65.

Field, C. B., J. T. Randerson, and C. M. Malmstrom. 1995. Global net primary production: Combining ecology and remote sensing. *Remote Sens Environ* 51:74–88.

Fiorella, M. and W. Ripple. 1993a. Analysis of conifer forest regeneration using Landsat thematic mapper data. *Photogramm Eng Remote Sens* 59(9):1383–1388.

Fiorella, M. and W. Ripple. 1993b. Determining successional stage o temperate coniferous forests with landsat satellite data. *Photogramm Eng Remote Sens* 59(2):239–246.

Foody, G. M. 1996. Relating the land-cover composition of mixed pixels to artificial neural network classification output. *Photogramm Eng Remote Sens* 62(5):491–499.

Foody, G. M., D. S. Boyd, and M. E. J. Cutler. 2003. Predictive relations of tropical forest biomass from Landsat TM data and their transferability between regions. *Remote Sens Environ* 85:463–474.

Foody, G. M. and D. P. Cox. 1994. Sub-pixel land cover composition estimation using a linear mixture model and fuzzy membership functions. *Int J Remote Sens* 15(3):619–631. doi:10.1080/01431169408954100.

Foody, G. M., M. E. Cutler, J. McMorrow, D. Pelz, H. Tangki, D. S. Boyd, and I. Douglas. 2001. Mapping the biomass of Bornean tropical rain forest from remotely sensed data. *Global Ecol Biogeogr* 10:379–387.

Franco-Lopez, H., A. R. Ek, and M. E. Bauer. 2001. Estimating and mapping of forest stand density, volume, and cover type using the k-nearest neighbors method. *Remote Sens Environ* 77:251–274.

Frankenberg, C., J. B. Fisher, J. Worden, G. Badgley, S. S. Saatchi, J.-E. Lee, G. C. Toon, A. Bulz, M. Jung, and A. Kuze. 2011. New global observations of the terrestrial carbon cycle from GOSAT: Patterns of plant fluorescence with gross primary productivity. *Geophys Res Lett* 38:L17706.

Franklin, J. and A. H. Strahler. 1988. Invertible canopy reflectance modeling of vegetation structure in semiarid woodland. *IEEE Trans Geosci Remote* 26(6):809–825.

Friedl, M. A., F. W. Davis, J. Michaelsen, and M. A. Moritz. 1995. Scaling and uncertainty in the relationship between the NDVI and land surface biophysical variables: An analysis using a scene simulation model and data from FIFE. *Remote Sens Environ* 54:233–246.

Fridley, J. D. 2012. Extended leaf phenology and the autumn niche in deciduous forest invasions. *Nature* 485(7398):359–362.

Gallo, K., C. Daughtry, and M. E. Bauer. 1985. Spectral estimation of absorbed photosynthetically active radiation in corn canopies. *Remote Sens Environ* 17:221–232.

Gamon, J. A., C. Coburn, L. B. Flanagan, K. F. Huemmrich, C. Kiddle, G. A. Sanchez-Azofeifa, D. R. Thayer, L. Vescovo, D. Gianelle, and D. A. Sims. 2010. SpecNet revisited: Bridging flux and remote sensing communities. *Can J Remote Sens* 36:376–390.

Gamon, J. A., C. B. Field, W. Bilger, O. Bjorkman, A. L. Fredeen, and J. Penuelas. 1990. Remote sensing of the xanthophyll cycle and chlorophyll fluorescence in sunflower leaves and canopies. *Oecologia* 85:1–7.

Gamon, J. A., J. Peñuelas, and C. B. Field. 1992. A narrow-waveband spectral index that tracks diurnal changes in photosynthetic efficiency. *Remote Sens Environ* 41:35–44.

Gamon, J. A. and J. S. Surfus. 1999. Assessing leaf pigment content and activity with a reflectometer. *New Phytol* 143:105–117.

Ganguly, S., R. R. Nemani, G. Zhang, H. Hashimoto, C. Milesi, A. Michaelis, W. Wang, P. Votava, A. Samanta, and F. Meiton. 2012. Generating global leaf area index from Landsat: Algorithm formulation and demonstration. *Remote Sens Environ* 122:185–202.

Ganguly, S., A. Samanta, M. A. Schull, N. V. Shabanov, C. Milesi, R. R. Nemani, Y. Knyazikhin, and R. B. Myneni. 2008a. Generating vegetation leaf area index Earth system data record from multiple sensors. Part 2: Implementation, analysis and validation. *Remote Sens Environ* 112:4318–4332.

Ganguly, S., M. A. Schull, A. Samanta, N. V. Shabanov, C. Milesi, R. R. Nemani, Y. Knyazikhin, and R. B. Myneni. 2008b. Generating vegetation leaf area index earth system data record from multiple sensors. Part 1: Theory. *Remote Sens Environ* 112:4333–4343.

Gao, F., J. Masek, M. Schwaller, and F. Hall. 2006. On the blending of the Landsat and MODIS surface reflectance: Predicting daily Landsat surface reflectance. *IEEE Trans Geogsci Remote Sens* 44(8):2207–2218.

Garbulsky, M. F., J. Peñuelas, D. Papale, and I. Filella. 2008. Remote estimation of carbon dioxide uptake by a Mediterranean forest. *Global Change Biol* 14(12):2860–2867.

Garrigues, S., R. Lacaze, F. Baret, J. T. Morisette, M. Weiss, J. E. Nickeson, R. Fernandes, S. Plummer, N. V. Shabanov, and R. B. Myneni. 2008. Validation of intercomparison of global leaf area index products derived from remote sensing data. *J Geophys Res Biogeo* 113:G02028. doi:10.1029/2007JG000635.

Gastellu-Etchegorry, J. P., E. Martin, and F. Gascon. 2004. DART: A 3D model for simulating satellite images and studying surface radiation budget. *Int J Remote Sens* 25:73–96.

Gholz, H. L., C. C. Grier, A. G. Campbell, and A. T. Brown. 1979. Equations for estimating biomass and leaf area of plants in the Pacific Northwest. Forest Research Laboratory, School of Forestry, Oregon State University.

Gobron, N., B. Pinty, and M. M. Verstraete. 1997. Theoretical limits to the estimation of the Leaf Area Index on the basis of visible and near-infrared remote sensing data. *IEEE Trans Geosci Remote Sens* 35:1438–1445.

Goel, N. S. and W. Qin. 1994. Influences of canopy architecture on relationships between various vegetation indices and LAI and FPAR: A computer simulation. *Remote Sens Rev* 10:309–347.

Goetz, S. J. and S. D. Prince. 1999. Modeling terrestrial carbon exchange and storage: Evidence and implications of functional convergence in light-use-efficiency. *Adv Ecol Res* 28:57–92.

Goetz, S. J., S. D. Prince, S. N. Goward, M. M. Thawley, and J. Small. 1999. Satellite remote sensing of primary production: An improved production efficiency modeling approach. *Ecol Model* 122:239–255.

Gonsamo, A., J. M. Chen, and P. D'Odorico. 2013. Deriving land surface phenology indicators from CO_2 eddy covariance measurements. *Ecol Indic* 29:203–207.

Gonsamo, A., J. M. Chen, D. T. Price, W. A. Kurz, and C. Wu. 2012a. Land surface phenology from optical satellite measurement and CO_2 eddy covariance technique. *J Geophys Res Biogeosci* 117(G3):1–18.

Gonsamo, A., J. M. Chen, C. Wu, and D. Dragoni. 2012b. Predicting deciduous forest carbon uptake phenology by upscaling FLUXNET measurements using remote sensing data. *Agric Forest Meteorol* 165:127–135.

Goward, S. N. and K. F. Huemmrich. 1992. Vegetation canopy PAR absorptance and the normalized difference vegetation index: An assessment using the SAIL model. *Remote Sens Environ* 39:119–140.

Gower, S. T., C. J. Kucharik, and J. M. Norman. 1999. Direct and indirect estimation of leaf area index, fAPAR, and net primary production of terrestrial ecosystems. *Remote Sens Environ* 70:29–51.

Goward, S. N., R. H. Waring, D. G. Dye, and J. L. Yang. 1994. Ecological remote-sensing at OTTER—Satellite macroscale observations. *Ecol Appl* 4:322–343.

Gower, S. T., R. E. McMurtrie, and D. Murty. 1996. Aboveground net primary production decline with stand age: Potential causes. *Trends Ecol Evol* 11(9):378–382.

Gower, S. T. and J. M. Norman. 1991. Rapid estimation of leaf area index in conifer and broadleaf plantations. *Ecology* 72:1896–1900.

Grace, J., C. Nichol, M. Disney, P. Lewis, T. Quaife, and P. Bowyer. 2007. Can we measure terrestrial photosynthesis from space directly, using spectral reflectance and fluorescence? *Global Change Biol* 13(7):1484–1497.

Gray, J. and C. Song. 2012. Mapping leaf area index using spatial, spectral, and temporal information from multiple sensors. *Remote Sens Environ* 119:173–183.

Grier, C. C. and R. S. Logan. 1977. Old-growth *Pseudotsuga-Menziesii* of a western Oregon watershed: Biomass distribution and production budgets. *Ecol Monogr* 47:373–400.

Guanter, L., L. Alonso, L. Gómez-Chova, J. Amorós-López, J. Vila, and J. Moreno. 2007. Estimation of solar-induced vegetation fluorescence from space measurements. *Geophys Res Lett* 34:L08401.

Guanter, L., L. Alonso, L. Gómez-Chova, M. Meroni, R. Preusker, J. Fischer, and J. Moreno. 2010. Developments for vegetation fluorescence retrieval from spaceborne high-resolution spectrometry in the O_2-A and O_2-B absorption bands. *J Geophys Res* 115:D19303.

Gutman, G. and A. Ignatov. 1998. The derivation of the green vegetation fraction from NOAA/AVHRR data for use in numerical weather prediction models. *Int J Remote Sens* 19(8):1533–1543.

Hall, F. G., D. B. Botkin, D. E. Strebel, K. D. Woods, and S. J. Goetz. 1991. Large-scale patterns of forest succession as determined by remote sensing. *Ecology* 72(2):628–640.

Hall, F. G., T. Hilker, and N. C. Coops. 2011. PHOTOSYNSAT, photosynthesis from space: Theoretical foundations of a satellite concept and validation from tower and spaceborne data. *Remote Sens Environ* 115:1918–1925.

Hall, F. G., T. Hilker, N. C. Coops, A. Lyapustin, K. F. Huemmrich, E. Middleton, H. Margolis, G. Drolet, and T. A. Black. 2008. Multi-angle remote sensing of forest light use efficiency by observing pri variation with canopy shadow fraction. *Remote Sens Environ* 112:3201–3211.

Hall, F. G., D. R. Peddle, and E. F. Ledrew. 1996. Remote sensing of biophysical variables in boreal forest stands of Picea mariana. *Int J Remote Sens* 17(15):3077–3081.

Hall, F. G., J. R. Townshend, and E. T. Engman. 1995. Status of remote sensing algorithms for estimation of land surface parameters. *Remote Sens Environ* 51:135–156.

Hall, R. J., R. S. Skakun, E. J. Arsenault, and B. S. Case. 2006. Modeling forest stand structure attributes using Landsat ETM+ data: Application to mapping of aboveground biomass and stand volume. *Forest Ecol Manage* 225:378–390.

Hansen, M. C., R. S. DeFries, J. R. G. Townshend, M. Carroll, C. Dimiceli, and R. A. Sohlberg. 2003. Global percent tree cover at a spatial resolution of 500 meters: First results of the modis vegetation continuous fields algorithm. *Earth Interact* 7(10):1–15. doi:10.1175/1087-3562(2003)007<0001:GPTCAA>2.0.CO;2.

Hansen, M. C., R. S. DeFries, J. R. G. Townshend, R. Sohlberg, C. Dimiceli, and M. Carroll. 2002. Towards an operational MODIS continuous field of percent tree cover algorithm: Examples using AVHRR and MODIS data. *Remote Sens Environ* 83(1–2):303–319. doi:10.1016/S0034-4257(02)00079-2.

Hansen, M. C., J. R. G. Townshend, R. S. DeFries, and M. Carroll. 2005. Estimation of tree cover using MODIS data at global, continental and regional/local scales. *Int J Remote Sens* 26(19):4359–4380. doi:10.1080/01431160500113435.

Hatfield, J., G. Asrar, and E. Kanemasu. 1984. Intercepted photosynthetically active radiation estimated by spectral reflectance. *Remote Sens Environ* 14:65–75.

Hayes, D. J., W. B. Cohen, S. A. Sader, and D. E. Irwin. 2008. Estimating proportional change in forest cover as a continuous variable from multi-year MODIS data. *Remote Sens Environ* 112(3):735–749.

Healey, S., W. Cohen, Y. Zhiqiang, and O. Krankina. 2005. Comparison of Tasseled Cap-based Landsat data structures for use in forest disturbance detection. *Remote Sens Environ* 97(3):301–310.

Heiskanen, J. 2006. Estimating aboveground tree biomass and leaf area index in a mountain birch forest using ASTER satellite data. *Int J Remote Sens* 27:1135–1158.

Heiskanen, J., M. Rautiainen, L. Korhonen, M. Mõttus, and P. Stenberg. 2011. Retrieval of boreal forest LAI using a forest reflectance model and empirical regressions. *Int J Appl Earth Obs* 13(4):595–606.

Helmer, E. H., S. Brown, and W. B. Cohen. 2000. Mapping montane tropical forest successional stage and land use with multi-date Landsat imagery. *Int J Remote Sens* 21(11):2163–2183.

Helmer, E. H., M. A. Lefsky, and D. A. Roberts. 2009. Biomass accumulation rates of Amazonian secondary forest and biomass of old-growth forests from Landsat time series and the Geoscience Laser Altimeter System. *J Appl Remote Sens* 3:033505.

Hilker, T., N. C. Coops, F. G. Hall, T. A. Black, B. Chen, P. Krishnan, M. A. Wulder, P. S. Sellers, E. M. Middleton, and K. F. Huemmrich. 2008. A modeling approach for upscaling gross ecosystem production to the landscape scale using remote sensing data. *J Geophys Res* 113:G03006.

Hilker, T., F. G. Hall, N. C. Coops, A. Lyapustin, Y. Wang, Z. Nesic, N. Grant, T. A. Black, M. A. Wulder, and N. Kljun. 2010. Remote sensing of photosynthetic light-use efficiency across two forested biomes: Spatial scaling. *Remote Sens Environ* 114:2863–2874.

Houghton, R. A., D. Butman, A. G. Bunn, O. N. Krankina, P. Schlesinger, and T. A. Stone. 2007. Mapping Russian forest biomass with data from satellites and forest inventories. *Environ Res Lett* 2:045032.

Hu, J., B. Tan, N. Shabanov, K. A. Crean, J. V. Martonchik, D. J. Diner, Y. Knyazikhin, and R. B. Myneni. 2003. Performance of the MISR LAI and FPAR algorithm: A case study in Africa. *Remote Sens Environ* 88:324–340.

Huang, C., S. N. Goward, J. G. Masek, N. Thomas, Z. Zhu, and J. E. Vogelmann. 2010. An automated approach for reconstructing recent forest disturbance history using dense Landsat time series stacks. *Remote Sens Environ* 114(1):183–198.

Huemmrich, K. F. 2001. The GeoSail model: A simple addition to the SAIL model to describe discontinuous canopy reflectance. *Remote Sens Environ* 75:423–431.

Huete, A. R. 1988. A soil adjusted vegetation index (SAVI). *Remote Sens Environ* 25:295–309.

Huete, A. R., K. Didan, T. Miura, E. P. Rodriguez, X. Gao, and L. G. Ferreira. 2002. Overview of the radiometric and biophysical performance of the MODIS vegetation indices. *Remote Sens Environ* 83:195–213.

Huete, A. R. and H. Q. Liu. 1994. An error and sensitivity analysis of the atmospheric- and soil-correcting variants of the NDVI for the MODIS-EOS. *IEEE Trans Geosci Remote* 32:897–905.

Hwang, T., C. Song, P. V. Bolstad, and L. E. Band. 2011. Downscaling real-time vegetation dynamics by fusing multi-temporal MODIS and Landsat NDVI in topographically complex terrain. *Remote Sens Environ* 115:2499–2512.

Jacquemoud, S. and F. Baret. 1990. PROSPECT: A model of leaf optical properties spectra. *Remote Sens Environ* 34:75–91.

Jakubauskas, M. 1996. Thematic Mapper characterization of lodgepole pine seral stages in Yellowstone National Park, USA. *Remote Sens Environ* 56:118–132.

Jenkins, J. C., D. C. Chojnacky, L. S. Heath, and R. A. Birdsey. 2003. National-scale biomass estimators for United States tree species. *Forest Sci* 49:12–35.

Jiang, H., J. R. Strittholt, P. A. Frost, and N. C. Slosser. 2004. The classification of late seral forests in the Pacific Northwest, USA using Landsat ETM+ imagery. *Remote Sens Environ* 91:320–331.

Jiapaer, G., X. Chen, and A. Bao. 2011. A comparison of methods for estimating fractional vegetation cover in arid regions. *Agric Forest Meteorol* 151(12):1698–1710. doi:10.1016/j.agrformet.2011.07.004.

Jolly, W. M., R. Nemani, and S. W. Running. 2005. A generalized, bioclimatic index to predict foliar phenology in response to climate. *Global Change Biol* 11(4):619–632.

Jolly, W. M. and S. W. Running. 2004. Effects of precipitation and soil water potential on drought deciduous phenology in the Kalahari. *Global Change Biol* 10:303–308.

Jonckheere, I., S. Fleck, K. Nackaerts, B. Muys, P. Coppin, M. Weiss, and F. Baret. 2004. Methods for leaf area index determination. Part I: Theories, techniques and instruments. *Agric Forest Meteorol* 121:19–35.

Jones, H. G. and R. A. Vaughan. 2010. *Remote Sensing of Vegetation: Principles, Techniques, and Applications*. Oxford, U.K.: Oxford University Press.

Jönsson, P. and L. Eklundh. 2004. TIMESAT—A program for analyzing time-series of satellite sensor data. *Comput Geosci* 30(8):833–845.

Jordan, C. F. 1969. Derivation of leaf area index from quality of light on the forest floor. *Ecology* 50:663–666.

Justice, C. O., M. O. Román, I. Csiszar, E. F. Vermote, R. E. Wolfe, S. J. Hook, M. Friedl, Z. Wang, C. B. Schaaf, and T. Miura. 2013. Land and cryosphere products from Suomi NPP VIIRS: Overview and status. *J Geophys Res Atmos* 118:9753–9765.

Justice, C. O., J. Townshend, E. Vermote, E. Masuoka, R. Wolfe, N. Saleous, D. Roy, and J. Morisette. 2002. An overview of MODIS Land data processing and product status. *Remote Sens Environ* 83:3–22.

Kamthonkiat, D., K. Honda, H. Turral, N. Tripathi, and V. Wuwongse. 2005. Discrimination of irrigated and rainfed rice in a tropical agricultural system using SPOT VEGETATION NDVI and rainfall data. *Int J Remote Sens* 26:2527–2547.

Kaufman, Y. J. and D. Tanre. 1992. Atmospherically resistant vegetation index (ARVI) for EOS-MODIS. *IEEE Trans Geosci Remote* 30:261–270.

Keeling, C. D., J. F. S. Chin, and T. P. Whorf. 1996. Increased activity of northern vegetation inferred from atmospheric CO_2 measurements. *Nature* 382:146–49.

Kennedy, R. E., W. B. Cohen, and T. A. Schroeder. 2007. Trajectory-based change detection for automated characterization of forest disturbance dynamics. *Remote Sens Environ* 110(3):370–386.

Kennedy, R. E., Z. Yang, and W. B. Cohen. 2010. Detecting trends in forest disturbance and recovery using yearly Landsat time series: 1. LandTrendr—Temporal segmentation algorithms. *Remote Sens Environ* 114(12):2897–2910.

Kimes, D. S., Y. Knyazikhin, J. L. Privette, A. A. Abuelgasim, and F. Gao. 2000. Inversion methods for physically-based models. *Remote Sens Rev* 18:381–439.

Knyazikhin, Y., J. V. Martonchik, R. B. Myneni, D. J. Diner, and S. W. Running. 1998. Synergistic algorithm for estimating vegetation canopy leaf area index and fraction of absorbed photosynthetically active radiation from MODIS and MISR data. *J Geophys Res* 103(D24):32257–32275.

Körner, C. and D. Basler. 2010. Phenology under global warming. *Science* 327(5972):1461–1462.

Kucharik, C. J., J. M. Norman, and S. T. Gower. 1999. Characterization of radiation regimes in nonrandom forest canopies: Theory, measurements, and a simplified modeling approach. *Tree Physiol* 19(11):695–706.

Kumar, M. and J. Monteith. 1981. Remote sensing of crop growth. In: Smith, H. Ed. *Plants and the Daylight Spectrum: Proceedings of the First International Symposium of the British Photobiology Society*, Leicester, U.K., January 5–8, 1981, pp. 133–144.

Landsberg, J. J. and R. H. Waring. 1997. A generalised model of forest productivity using simplified concepts of radiation-use efficiency, carbon balance and partitioning. *Forest Ecol Manag* 95(3):209–228.

Lasslop, G., M. Reichstein, D. Papale, A. D. Richardson, A. Arneth, A. Barr, P. Stoy, and G. Wohlfahrt. 2010. Separation of net ecosystem exchange into assimilation and respiration using a light response curve approach: Critical issues and global evaluation. *Global Change Biol.* 16(1):187–208.

Latifi, H. and B. Koch. 2012. Evaluation of most similar neighbour and random forest methods for imputing forest inventory variables using data from target and auxiliary stands. *Int J Remote Sens* 33(21):6668–6694.

Law, B., P. Thornton, J. Irvine, P. Anthoni, and S. Tuyl. 2001. Carbon storage and fluxes in ponderosa pine forests at different developmental stages. *Global Change Biol* 7:755–777.

Lawrence, R. and W. Ripple. 1999. Calculating change curves for multitemporal satellite imagery: Mount St. Helens 1980–1995. *Remote Sens Environ* 67:309–319.

Le Maire, G., C. Marsden, Y. Nouvellon, C. Grinand, R. Hakamada, J.-L. Stape, and J.-P. Laclau. 2011. MODIS NDVI time-series allow the monitoring of *Eucalyptus* plantation biomass. *Remote Sens Environ* 115:2613–2625.

Leblanc, S. G. 2002. Correction to the plant canopy gap-size analysis theory used by the Tracing Radiation and Architecture of Canopies instrument. *Appl Optics* 41(36):7667–7670.

Lefsky, M. A., D. P. Turner, M. Guzy, and W. B. Cohen. 2005. Combining lidar estimates of aboveground biomass and Landsat estimates of stand age for spatially extensive validation of modeled forest productivity. *Remote Sens Environ* 95:549–558.

Li, X. and A. Strahler. 1985. Geometric-optical modeling of a conifer forest canopy. *IEEE Trans Geosci Remote* GE-23(5):705–721.

Li, X. W., A. H. Strahler, and C. E. Woodcock. 1995. A hybrid geometric optical-radiative transfer approach for modeling albedo and directional reflectance of discontinuous canopies. *IEEE Trans Geosci Remote* 33(2):466–480.

Liesenberg, V., L. S. Galvão, and F. J. Ponzoni. 2007. Variations in reflectance with seasonality and viewing geometry: Implications for classification of Brazilian savanna physiognomies with MISR/Terra data. *Remote Sens Environ* 107:276–286.

Lieth, H. (ed.). 1974. *Phenology and Seasonality Modeling*. New York: Springer.

Liu, J., J. M. Chen, J. Cihlar, and W. M. Park. 1997. A process-based boreal ecosystem productivity simulator using remote sensing inputs. *Remote Sens Environ* 62:158–175.

Liu, L. and Z. Cheng. 2010. Detection of vegetation light-use efficiency based on solar-induced chlorophyll fluorescence separated from canopy radiance spectrum. *IEEE J Sel Top Appl* 3(3):306–312.

Liu, L., Y. Zhang, Q. Jiao, and D. Peng. 2013. Assessing photosynthetic light-use efficiency using a solar-induced chlorophyll fluorescence and photochemical reflectance index. *Int J Remote Sens* 34(12):4264–4280.

Liu, S., T. Wang, J. Guo, J. Qu, and P. An. 2010. Vegetation change based on SPOT-VGT data from 1998 to 2007, northern China. *Environ Earth Sci* 60:1459–1466.

Liu, W., C. Song, T. A. Schroeder, and W. B. Cohen. 2008. Predicting forest successional stages using multitemporal Landsat imagery with forest inventory and analysis data. *Int J Remote Sens* 29(13):3855–3872.

Loveland, T. R. and J. L. Dwyer. 2012. Landsat: Building a strong future. *Remote Sens Environ* 122:22–29.

Lu, D. 2005. Aboveground biomass estimation using Landsat TM data in the Brazilian Amazon. *Int J Remote Sens* 26:2509–2525.

Lucas, R. M., M. Honzák, I. Do Amaral, P. J. Curran, and G. M. Foody. 2002. Forest regeneration on abandoned clearances in central Amazonia. *Int J Remote Sens* 23(5):965–988.

Malenovský, Z., K. B. Mishra, F. Zemek, U. Rascher, and L. Nedbal. 2009. Scientific and technical challenges in remote sensing of plant canopy reflectance and fluorescence. *J Exp Bot* 60(11):2987–3004.

Markham, B. L. and D. L. Helder. 2012. Forty-year calibrated record of earth-reflected radiance from Landsat: A review. *Remote Sens Environ* 122:30–40.

Martonchik, J. V., D. J. Diner, K. A. Crean, and M. A. Bull. 2002. Regional aerosol retrieval results from MISR. *IEEE Trans Geosci Remote* 40:1520–1531.

Masson, V., J. L. Champeaux, F. Chauvin, C. Meriguer, and R. Lacaze. 2003. A global database of land surface parameters at 1 km resolution in meteorological and climate models. *J Climate* 16:1261–1282.

McDonald, R. I., P. N. Halpin, and D. L. Urban. 2007. Monitoring succession from space: A case study from the North Carolina Piedmont. *Appl Veg Sci* 10(2):193.

McMurtrie, R. E., D. A. Rook, and F. M. Kelliher. 1990. Modelling the yield of Pinus radiate on a site limited by water and nitrogen. *Forest Ecol Manag* 30:381–413.

Menzel, A., T. H. Sparks, N. Estrella, E. Koch, A. Aasa, R. Ahas, K. Alm-Kübler, P. Bissolli, O. Braslavská, and Briede. 2006. European phenological response to climate change matches the warming pattern. *Global Change Biol* 12(10):1969–1976.

Meroni, M., M. Rossini, L. Guanter, L. Alonso, U. Rascher, R. Colombo, and J. Moreno. 2009. Remote sensing of solar-induced chlorophyll fluorescence: Review of methods and applications. *Remote Sens Environ* 113:2037–2051.

Miller, E. E. and J. M. Norman. 1971. Sunfleck theory for plant canopies 1. Lengths of sunlit segments along a transect. *Agron J* 63(5):735–738.

Miller, J. B. 1967. A formula for average foliage density. *Aust J Bot* 15:141–144.

Miller, S. D., S. P. Mills, C. D. Elvidge, D. T. Lindsey, T. F. Lee, and J. D. Hawkins. 2012. Suomi satellite brings to light a unique frontier of nighttime environmental sensing capabilities. *Proc Natl Acad Sci USA* 109:15706–15711.

Minocha, R., G. Martinez, B. Lyons, and S. Long. 2009. Development of a standardized methodology for quantifying total chlorophyll and carotenoids from foliage of hardwood and conifer tree species. *Can J Forest Res* 39:849–861.

Mohammed, G. H., J. Moreno, Y. Goulas, A. Huth, E. Middleton, F. Miglietta, L. Nedbal, U. Rascher, W. Verhoef, and M. Drusch. 2012. European Space Agency's Fluorescence Explorer mission: Concept and applications. In: *AGU Fall Meeting*, San Francisco, CA, December 3–7, 2012.

Monsi, M. and T. Saeki. 1953. Uber den Lichifktor in den Pflanzengesellschaften und Scine Bedeutung fur die Stoffprodcktion. *Jpn J Bot* 14:22–52.

Monteith, J. L. 1965. Evaporation and the environment. In: *Proceedings of the 19th Symposium of the Society for Experimental Biology*, Cambridge University Press, New York, pp. 205–233.

Monteith, J. L. 1972. Solar radiation and productivity in tropical ecosystems. *J Appl Ecol* 9(3):747–766.

Monteith, J. L. 1977. Climate and the efficiency of crop production in Britain. *Philos Trans R Soc B* 281:277–294.

Moody, A. and C. E. Woodcock. 1994. Scale-dependent errors in the estimation of land-cover proportions: Implications for global land-cover dataset. *Photogramm Eng Remote Sens* 60(5):585–594.

Morisette, J. T., A. D. Richardson, A. K. Knapp, J. I. Fisher, E. A. Graham, J. Abatzoglou, B. E. Wilson, D. D. Breshears, G. M. Henebry, J. M. Hanse, and L. Liang. 2009. Tracking the rhythm of the seasons in the face of global change: Phenological research in the 21st century. *Front Ecol Environ* 7(5):253–260.

Mu, Q., M. Zhao, and S. W. Running. 2011. Improvements to a MODIS global terrestrial evapotranspiration algorithm. *Remote Sens Environ* 115:1781–1800.

Mutanga, O., E. Adam, and M. A. Cho. 2012. High density biomass estimation for wetland vegetation using WorldView-2 imagery and random forest regression algorithm. *Int J Appl Earth Obs Geoinf* 18:399–406.

Muukkonen, P. and J. Heiskanen. 2007. Biomass estimation over a large area based on standwise forest inventory data and ASTER and MODIS satellite data: A possibility to verify carbon inventories. *Remote Sens Environ* 107:617–624.

Myneni, R. B., J. Dong, C. J. Tucker, R. K. Kaufmann, P. E. Kauppi, J. Liski, L. Zhou, V. Alexeyev, and M. K. Hughes, M. K. 2001. A large carbon sink in the woody biomass of Northern forests. *Proc Natl Acad Sci USA* 98(26):14784–14789.

Myneni, R. B., S. Hoffman, Y. Knyazikhin, J. Privette, J. Glassy, Y. Tian, Y. Wang, X. Song, Y. Zhang, and G. R. Smith. 2002. Global products of vegetation leaf area and fraction absorbed PAR from year one of MODIS data. *Remote Sens Environ* 83:214–231.

Myneni, R. B., C. D. Keeling, C. J. Tucker, G. Asrar, and R. R. Nemani. 1997. Increased plant growth in the northern high latitudes from 1981 to 1991. *Nature* 386:698–702.

Myneni, R. B., S. Maggion, J. Iaquinto, J. L. Privette, N. Gobron, B. Pinty, D. S. Kimes, M. M. Verstraete, and D. L. Williams. 1995. Optical remote sensing of vegetation: Modeling, caveats, and algorithms. *Remote Sens Environ* 51:169–188.

Myneni, R. B. and D. L. Williams. 1994. On the Relationship between FAPAR and NDVI. *Remote Sens Environ* 49:200–211.

National Research Council (NRC). 2007. *Earth Science and Application from Space: National Imperatives for the Next Decade and Beyond*. Washington, DC: The National Academies Press.

Nelson, R., K. J. Ranson, G. Sun, D. S. Kimes, V. Kharuk, and P. Montesano. 2009. Estimating Siberian timber volume using MODIS and ICESat/GLAS. *Remote Sens Environ* 113:691–701.

Nelson, R. F., D. S. Kimes, W. A. Salas, and M. Routhier. 2000. Secondary forest age and tropical forest biomass estimation using thematic mapper imagery. *Bioscience* 50:419–431.

Nemani, R., H. Hashimoto, P. Votava, F. Melton, W. Wang, A. Michaelis, L. Mutch, C. Milesi, S. Hiatt, and M. White. 2009. Monitoring and forecasting ecosystem dynamics using the terrestrial observation and prediction system (TOPS). *Remote Sens Environ* 113(7):1497–1509.

Nemani, R., L. Pierce, S. Running, and L. Band. 1993. Forest ecosystem processes at the watershed scale: Sensitivity to remotely-sensed leaf area index estimates. *Int J Remote Sens* 14:2519–2534.

Ni, W., X. Li, C. E. Woodcock, J. Boujean, and R. E. Davis. 1997. Transmission of solar radiation in boreal conifer forests: Measurements and models. *J Geophy Res* 102(D24):29555–29566.

Nichol, C. J., K. F. Huemmrich, T. A. Black, P. G. Jarvis, C. L. Walthall, J. Grace, and F. G. Hall. 2000. Remote sensing of photosynthetic-light-use efficiency of boreal forest. *Agric Forest Meteorol* 101:131–142.

Nilson, T. 1971. A theoretical analysis of the frequency of gaps in plant stands. *Agr Meteorol* 8:25–38.

Norman, J. M. and G. S. Campbell. 1989. Canopy structure. In: R. W. Pearcy, J. Ehlerlnger, H. A. Mooney, and P. W. Rundel (eds.), *Plant Physiological Ecology: Field Methods and Instrumentation*. New York: Chapman & Hall, pp. 301–325.

North, P. 1996. Three-dimensional forest light interaction model using a Monte Carlo method. *IEEE T Geosci Remote* 34:946–956.

Oker-Blom, P. 1986. Photosynthetic radiation regime and canopy structure in modeled forest stands. *Acta Forest Fennica* 197:1–44.

Oliver, C. D. and B. C. Larson. 1996. *Forest Stand Dynamics*. New York: John Wiley & Sons, Inc. ISBN 0-471-13833-9.

Oosting, H. J. 1942. An ecological analysis of the plant communities of Piedmont, North Carolina. *Am Midl Nat* 28:1–126.

Oudrari, H., J. McIntire, D. Moyer, K. Chiang, X. Xiong, and J. Butler. 2012. Preliminary assessment of Suomi-NPP VIIRS on-orbit radiometric performance. SPIE Optical Engineering Applications, International Society for Optics and Photonics, pp. 851011–851024.

Pacala, S. W., G. C. Hurtt, D. Baker, P. Peylin, R. A. Houghton et al. 2001. Consistent land- and atmosphere-based US carbon sink estimates. *Science* 292(5525):2316–2320.

Parry, M. L., O. F. Canziani, J. P. Palutikof, P. J. van der Linden, and C. E. Hanson (eds.). 2007. *Climate Change 2007: Impacts, Adaptation and Vulnerability*. Contribution of

Working Group II to the Fourth Assessment Report of the Intergovernmental Panel on Climate Change. Cambridge University Press, Cambridge, U.K., 976pp.

Peddle, D. R., F. G. Hall, and E. F. LeDrew. 1999. Spectral mixture analysis and geometric-optical reflectance modeling of boreal forest biophysical structure. *Remote Sens Environ* 67:288–297.

Peet, R. K. and N. L. Christensen. 1987. Competition and tree death. *BioScience* 37(8):586–595.

Peet, R. K. and N. L. Christensen. 1988. Changes in species diversity during secondary forest succession on the North Carolina piedmont. In: H. J. During, M. J. A. Werger, and J. H. Willems (eds.), *Diversity and Pattern in Plant Communities*. The Hague, the Netherlands: SPB Academic Publishing, pp. 233–245.

Peñuelas, J., I. Filella, and J. A. Gamon. 1995. Assessment of photosynthetic radiation-use efficiency with spectral reflectance. *New Phytol* 131:291–296.

Peñuelas, J., T. Rutishauser, and I. Filella. 2009. Phenology feedbacks on climate change. *Science* 324(5929):887–888.

Peterson, D. L., M. A. Spanner, S. W. Running, and K. B. Teuber. 1987. Relationship of thematic mapper simulator data to leaf area index of temperate coniferous forests. *Remote Sens Environ* 22:323–341.

Pflugmacher, D., W. B. Cohen, and R. E. Kennedy. 2012. Using Landsat-derived disturbance history (1972–2010) to predict current forest structure. *Remote Sens Environ* 122:146–165.

Piao, S. L., J. Y. Fang, P. Ciais, P. Peylin, Y. Huang, S. Sitch, and T. Wang. 2009. The carbon balance of terrestrial ecosystems in China. *Nature* 458(7241):U1009–U1082.

Piao, S., P. Ciais, P. Friedlingstein, P. Peylin, M. Reichstein, S. Luyssaert, H. Margolis, J. Fang, A. Barr, and A. Chen. 2008. Net carbon dioxide losses of northern ecosystems in response to autumn warming. *Nature* 451:49–52.

Piao, S., P. Friedlingstein, P. Ciais, N. Viovy, and J. Demarty. 2007. Growing season extension and its impact on terrestrial carbon cycle in the Northern Hemisphere over the past 2 decades. *Global Biogeochem Cycles* 21(3):GB3018.

Piao, S. L., J. Y. Fang, B. Zhu, and K. Tan. 2005. Forest biomass carbon stocks in China over the past 2 decades: Estimation based on integrated inventory and satellite data. *J Geophys Res* 110:G01006.

Pickett, S. T. A., M. L. Cadenasso, and S. J. Meiners. 2013. Vegetation dynamics. In: E. Van der Maarel and J. Franklin (eds.), *Vegetation Ecology*, 2nd edn. New York: John Wiley & Sons. ISBN 978-1-4443-3889-8.

Pinty, B., N. Gobron, J.-L. Widlowski, T. Lavergne, and M. M. Verstraete. 2004. Synergy between 1-D and 3-D radiation transfer models to retrieve vegetation canopy properties from remote sensing data. *J Geophys Res* 109:D21205. doi:10.1029/2004JD005214.

Pinty, B. and M. M. Verstrate. 1992. GEMI: A non-linear index to monitor global vegetation from satellites. *Vegetation* 101:15–20.

Pisek, J. and Chen, J. M. 2007. Comparison and validation of MODIS and VEGETATION global LAI products over four BigFoot sites in North America. *Remote Sens Environ* 109:81–94.

Potter, C., S. Klooster, R. Myneni, V. Genovese, P. Tan, and V. Kumar. 2003. Continental-scale comparisons of terrestrial carbon sinks estimated from satellite data and ecosystem modeling 1982–1998. *Global Planet Change* 39:201–213.

Potter, C. S., J. T. Randerson, C. B. Field, P. A. Matson, P. M. Vitousek, H. A. Mooney, and S. A. Klooster. 1993. Terrestrial ecosystem production—A process model-based on global satellite and surface data. *Global Biogeochem Cycles* 7:811–841.

Powell, S. L., W. B. Cohen, S. P. Healey, R. E. Kennedy, G. G. Moisen, K. B. Pierce, and J. L. Ohmann. 2010. Quantification of live aboveground forest biomass dynamics with Landsat time-series and field inventory data: A comparison of empirical modeling approaches. *Remote Sens Environ* 114:1053–1068.

Pregitzer, K. S. and E. S. Euskirchen. 2004. Carbon cycling and storage in world forests: Biome patterns related to forest age. *Global Change Bio* 10(12):2052–2077.

Prince, S. D. 1991. A model of regional primary production for use with coarse resolution satellite data. *Int J Remote Sens* 12(6):1313–1330.

Prince, S. D. and S. N. Goward. 1995. Global primary production: A remote sensing approach. *J Biogeograph* 22(4–5):815–835.

Qi, J., A. Chehbouni, A. R. Huete, Y. H. Kerr, and S. Sorooshian. 1994. A modified soil adjusted vegetation index. *Remote Sens Environ* 48:119–126.

Rahman, A., V. Cordova, J. Gamon, H. Schmid, and D. Sims. 2004. Potential of MODIS ocean bands for estimating CO_2 flux from terrestrial vegetation: A novel approach. *Geophys Res Lett* 31:L10503.

Reichstein, M., E. Falge, D. Baldocchi, D. Papale, M. Aubinet, P. Berbigier, C. Bernhofer, N. Buchmann, T. Gilmanov, and A. Granier. 2005. On the separation of net ecosystem exchange into assimilation and ecosystem respiration: Review and improved algorithm. *Global Change Biol* 11(9):1424–1439.

Richards, J. A. 2013. *Remote Sensing Digital Image Analysis: An Introduction*. Springer.

Richardson, A. D., R. S. Anderson, M. A. Arain, A. G. Barr, G. Bohrer, G. Chen, J. M. Chen, P. Ciais, K. J. Davis, and A. R. Desai. 2012. Terrestrial biosphere models need better representation of vegetation phenology: Results from the North American Carbon Program Site Synthesis. *Global Change Biol* 18(2):566–584.

Richardson, A. D., T. A. Black, P. Ciais, N. Delbart, M. A. Friedl, N. Gobron, D. Y. Hollinger, W. L. Kutsch, B. Longdoz, and S. Luyssaert. 2010. Influence of spring and autumn phenological transitions on forest ecosystem productivity. *Philos Trans R Soc B* 365(1555):3227–3246.

Richardson, A. D., B. H. Braswell, D. Y. Hollinger, J. P. Jenkins, and S. V. Ollinger. 2009. Near-surface remote sensing of spatial and temporal variation in canopy phenology. *Ecol Appl* 19(6):1417–1428.

Richardson, A. D., J. P. Jenkins, B. H. Braswell, BH, D. Y. Hollinger, S. V. Ollinger, and M. L. Smith. 2007. Use of digital webcam images to track spring green-up in a deciduous broadleaf forest. *Oecologia* 152(2):323–334.

Richardson, A. D., S. P. Duigan, and G. P. Berlyn. 2002. An evaluation of noninvasive methods to estimate foliar chlorophyll content. *New Phytol* 153:185–194.

Richardson, A. D., T. F. Keenan, M. Migliavacca, Y. Ryu, O. Sonnentag, and M. Toomey. 2013. Climate change, phenology, and phenological control of vegetation feedbacks to the climate system. *Agric Forest Meteorol* 169:156–173.

Roberts, D. A., M. Gardner, R. Church, S. Ustin, G. Scheer, and G. O. Green. 1998. Mapping chaparral in the Santa Monica Mountains using multiple endmember spectral mixture models. *Remote Sens Environ* 65:267–279.

Rosenzweig, C., G. Casassa, D. J. Karoly, A. Imeson, C. Liu, A. Menzel, S. Rawlins, T. L. Root, B. Seguin, and P. Tryjanowski. 2007. Assessment of observed changes and responses in natural and managed systems. In: M. L. Parry, O. F. Canziani, J. P. Palutikof, P. J. van der Linden, and C. E. Hanson (eds.), *Climate Change 2007: Impacts, Adaptation and Vulnerability*. Contribution of Working Group II to the Fourth Assessment Report of the Intergovernmental Panel on Climate Change. Cambridge, U.K.: Cambridge University Press, pp. 79–131.

Roujean, J. L. and F. M. Breon. 1995. Estimating PAR absorbed by vegetation from bidirectional reflectance measurements. *Remote Sens Environ* 51:375–384.

Rouse, J. W., R. H. Hass, J. A. Shell, and D. W. Deering. 1974. Monitoring vegetation systems in the Great Plains with ERTS-1. In: *Third Earth Resources Technology Satellite Symposium*, Washington, D.C., Vol. 1, pp. 309–317.

Roy, P. S. and S. Ravan. 1996. Biomass estimation using satellite remote sensing data—An investigation on possible approaches for natural forest. *J Biosci* 21:535–561.

Ruimy, A., B. Saugier, and G. Dedieu. 1994. Methodology for the estimation of terrestrial net primary production from remotely sensed data. *J Geophys Res Atmos* 99:5263–5283.

Ruimy, A., G. Dedieu, and B. Saugier. 1996. TURC: A diagnostic model of continental gross primary productivity and net primary production. *Global Biogeochem Cycles* 10(2):269–285.

Running, S. W. and J. C. Coughlan. 1988. A general model of forest ecosystem processes for regional applications I. Hydrologic balance, canopy gas exchange and primary production processes. *Ecol Model* 42:125–154.

Running, S. W., C. O. Justice, V. Salomonson, D. Hall, J. Barker, Y. J. Kaufman, A. H. Strahler, A. R. Huete, J.-P. Muller, and V. Vanderbill. 1994. Terrestrial remote sensing science and algorithms planned for EOS/MODIS. *Int J Remote Sens* 15(17):3587–3620.

Running, S. W., P. E. Thornton, R. Nemani, and J. M. Glassy. 2000. Global terrestrial gross and net primary productivity from the Earth Observing System. In: Sala, O. E., R. B. Jackson, and H. A. Mooney et al. (eds.), *Methods in Ecosystem Science*. Springer: New York, pp. 44–57.

Running, S. W., R. R. Nemani, F. A. Heinsch, M. Zhao, M. Reeves, and H. Hashimoto. 2004. A continuous satellite-derived measure of global terrestrial primary production. *BioScience* 54:547–560.

Running, S. W., R. R. Nemani, D. L. Peterson, L. E. Band, D. F. Potts, L. L. Pierce, and M. A. Spanner. 1989. Mapping regional forest evapotranspiration and photosynthesis by coupling satellite data with ecosystem simulation. *Ecology* 70(4):1090–1101.

Sabol, D. E., A. R. Gillespie, J. B. Adams, M. O. Smith, and C. J. Tucker. 2002. Structural stage in Pacific Northwest forests estimated using simple mixing models of multispectral images. *Remote Sens Environ* 80(1):1–16.

Sader, S. A., R. B. Waide, W. T. Lawrence, and A. T. Joyce. 1989. Tropical forest biomass and successional age class relationships to a vegetation index derived from Landsat TM data. *Remote Sens Environ* 28:143–156.

Salim, E. and O. Ullsten. 1999. *Our Forests Our Future*. Cambridge University Press, Cambridge, U.K.

Schaefer, K., C. R. Schwalm, C. Williams, M. A. Arain, A. Barr, J. M. Chen, K. J. Davis, D. Dimitrov, T. W. Hilton, and D. Y. Hollinger. 2012. A model-data comparison of gross primary productivity: Results from the North American Carbon Program site synthesis. *J Geophys Res Biogeosci* 117(G3).

Schroeder, T. A., W. B. Cohen, C. Song, M. J. Canty, and Z. Yang. 2006. Radiometric correction of multi-temporal Landsat data for characterization of early successional forest patterns in western Oregon. *Remote Sens Environ* 103(1):16–26.

Schroeder, T. A., W. B. Cohen, and Z. Yang. 2007. Patterns of forest regrowth following clearcutting in western Oregon as determined from a Landsat time-series. *Forest Ecol Manag* 243:259–273.

Schwartz, M. D. 1992. Phenology and springtime surface-layer change. *Mon Weather Rev* 120:2570–2578.

Schwarz, M. D. and N. E. Zimmermann. 2005. A new GLM-based method for mapping tree cover continuous fields using regional MODIS reflectance data. *Remote Sens Environ* 95(4):428–443. doi:10.1016/j.rse.2004.12.010.

Sellers, P. J. 1985. Canopy reflectance, photosynthesis and transpiration. *Int J Remote Sens* 6:1335–1372.

Sellers, P. J. 1987. Canopy reflectance, photosynthesis, and transpiration. 2. The role of biophysics in the linearity of their interdependence. *Remote Sens Environ* 21:143–183.

Sellers, P., C. Tucker, G. Collatz, S. Los, C. Justice, D. Dazlich, and D. Randall. 1994. A global 1 by 1 NDVI data set for climate studies. Part 2: The generation of global fields of terrestrial biophysical parameters from the NDVI. *Int J Remote Sens* 15:3519–3545.

Sellers, P. J., R. E. Dickinson, D. A. Randall, A. K. Betts, F. G. Hall, J. A. Berry, G. J. Collatz, A. S. Denning, H. A. Mooney, and C. A. Nobre. 1997. Modeling the exchange of energy, water, and carbon between continents and the atmosphere. *Science* 275:502–509.

Sims, D. A. and J. A. Gamon. 2002. Relationships between leaf pigment content and spectral reflectance across a wide range of species, leaf structures and developmental stages. *Remote Sens Environ* 81:337–354.

Smith, J. E., L. S. Heath, and J. C. Jenkins. 2003. Forest volume-to-biomass models and estimates of mass for live and standing dead trees of US forests. General technical report NE-298. USDA Forest Service, Northeastern Research Station, Delaware, OH.

Soenen, S. A., D. R. Peddle, R. J. Hall, C. A. Coburn, and F. G. Hall. 2010. Estimating aboveground forest biomass from canopy reflectance model inversion in mountainous terrain. *Remote Sens Environ* 114(7):1325–1337.

Somers, B., G. P. Asner, L. Tits, and P. Coppin. 2011. Endmember variability in spectral mixture analysis: A review. *Remote Sens Environ* 115(7):1603–1616. doi:10.1016/j.rse.2011.03.003.

Song, C. 2004. Cross-sensor calibration between Ikonos and Landsat ETM+ for spectral mixture analysis. *IEEE Geosci Remote Sens* 1(4):272–276.

Song, C. 2005. Spectral mixture analysis for subpixel vegetation fractions in the urban environment: How to incorporate endmember variability? *Remote Sens Environ* 95(2):248–263.

Song, C. 2013. Optical remote sensing of forest leaf area index and biomass. *Prog Phys Geog* 37(1):98–113.

Song, C., M. P. Dannenberg, and T. Hwang. 2013. Optical remote sensing of terrestrial ecosystem primary productivity. *Prog Phys Geog* 37(6):834–854.

Song, C., G. Katul, R. Oren, L. E. Band, C. L. Tague, P. C. Stoy, and H. R. McCarthy. 2009. Energy, water, and carbon fluxes in a loblolly pine stand: Results from uniform and gappy canopy models with comparisons to eddy flux data. *J Geophys Res* 114:G04021.

Song, C., T. Schroeder, and W. Cohen. 2007. Predicting temperate conifer forest successional stage distributions with multi-temporal Landsat Thematic Mapper imagery. *Remote Sens Environ* 106(2):228–237.

Song, C. and C. E. Woodcock. 2002. The spatial manifestation of forest succession in optical imagery—The potential of multiresolution imagery. *Remote Sens Environ* 82:271–284.

Song, C. and C. Woodcock. 2003. Monitoring forest succession with multitemporal Landsat images: Factors of uncertainty. *IEEE Trans Geosci Remote* 41(11):2557–2567.

Song, C., C. Woodcock, and X. Li. 2002. The spectral/temporal manifestation of forest succession in optical imagery: The potential of multitemporal imagery. *Remote Sens Environ* 82:285–302.

Sprintsin, M., A. Karnieli, P. Berliner, E. Rotenberg, D. Yakir, and S. Cohen. 2009 Evaluating the performance of the MODIS leaf area index (LAI) product over a Mediterranean dryland planted forest. *Int J Remote Sens* 30:5061–5069.

Steininger, M. K. 1996. Tropical secondary forest regrowth in the Amazon: Age, area and change estimation with Thematic Mapper data. *Int J Remote Sens* 17(1):9–27.

Steininger, M. K. 2000. Satellite estimation of tropical secondary forest above-ground biomass: Data from Brazil and Bolivia. *Int J Remote Sens* 21:1139–1157.

Steinmetz, S., M. Guerif, R. Delecolle, and F. Baret. 1990. Spectral estimates of the absorbed photosynthetically active radiation and light-use efficiency of a winter wheat crop subjected to nitrogen and water deficiencies. *Remote Sens* 11:1797–1808.

Stenberg, P., S. Linder, H. Smolander, and J. Flower-Ellis. 1994. Performance of the LAI-2000 plant canopy analyzer in estimating leaf area index of some Scots pine stands. *Tree Physiol* 14:981–995.

Stenberg, P., M. Rautiainen, T. Manninen, P. Voipio, and H. Smolander. 2004. Reduced simple ratio better than NDVI for estimating LAI in Finnish pine and spruce stands. *Silva Fenn* 38(1):3–14.

Steven, M., P. Biscoe, and K. Jaggard. 1983. Estimation of sugar beet productivity from reflection in the red and infrared spectral bands. *Int J Remote Sens* 4:325–334.

Ter-Mikaelian, M. T. and M. D. Korzukhin. 1997. Biomass equations for sixty-five North America tree species. *Forest Ecol Manage* 97:1–24.

Tian, Q., Z. Luo, J. M. Chen, M. Chen, and F. Hui. 2007. Retrieving leaf area index for coniferous forest in Xingguo County, China with Landsat ETM+ images. *J Environ Manage* 85(3):624–627.

Tomppo, E., M. Nilsson, M. Rosengren, P. P. Aalto, and R. Kennedy. 2002. Simultaneous use of Landsat-TM and IRS-1C WiFS data in estimating large area tree stem volume and aboveground biomass. *Remote Sens Environ* 82:156–171.

Tomppo, E., H. Olsson, G. Stahl, M. Nilsson, O. Hagner, and M. Katila. 2008. Combining national forest inventory field plots and remote sensing data for forest databases. *Remote Sens Environ* 112:1982–1999.

Tucker, C. J., J. E. Pinzon, M. E. Brown, D. A. Slayback, E. W. Pak, R. Mahoney, E. F. Vermote, and N. El Saleous. 2005. An extended AVHRR 8-km NDVI dataset compatible with MODIS and SPOT vegetation NDVI data. *Int J Remote Sens* 26:4485–4498.

Tucker, C. J., D. A. Slayback, J. E. Pinzon, S. O. Los, R. B. Myneni, and M. G. Taylor. 2001. Higher northern latitude normalized difference vegetation index and growing season trends from 1982 to 1999. *Int J Biometeorol* 45:184–190.

Tucker, C. J., J. R. G. Townshend, and T. E. Goff. 1985. African land-cover classification using satellite data. *Science* 227(4685):369–375.

Turner, D. P., W. B. Cohen, R. E. Kennedy, K. S. Fassnacht, and J. M. Briggs. 1999. Relationships between leaf area index and Landsat TM spectral vegetation indices across three temperate zone sites. *Remote Sens Environ* 70:52–68.

Turner, D. P., S. T. Gower, W. B. Cohen, M. Gregory, and T. K. Maiersperger. 2002. Effects of spatial variability in light use efficiency on satellite-based NPP monitoring. *Remote Sens Environ* 80(3):397–405.

Turner, D. P., S. Urbanski, D. Bremer, S. C. Wofsy, T. Meyers, S. T. Gower, and M. Gregory. 2003. A cross-biome comparison of daily light use efficiency for gross primary production. *Global Change Biol* 9:383–396.

Uddling, J., J. Gelang-Alfredsson, K. Piikki, and H. Pleijel. 2007. Evaluating the relationship between leaf chlorophyll concentration and SPAD-502 chlorophyll meter readings. *Photosynth Res* 91:37–46.

Urban, D. L. 1990. *A Versatile Model to Simulate Forest Pattern: A User's Guide to ZELIG Version 1.0.* Charlottesville, VA: University of Virginia.

Vargas, M., T. Miura, N. Shabanov, and A. Kato. 2013. An initial assessment of Suomi NPP VIIRS vegetation index EDR. *J Geophys Res-Atmos* 118:12301–12316.

Verhoef, W. 1984. Light scattering by leaf layers with application to canopy reflectance modeling: The SAIL model. *Remote Sens Environ* 16:125–141.

Verrelst, J., J. Muñoz, L. Alonso, J. Delegido, J. P. Rivera, G. Camps-Valls, and J. Moreno. 2012. Machine learning regression algorithms for biophysical parameter retrieval: Opportunities for Sentinel-2 and -3. *Remote Sens Environ* 118:127–139.

Vieira, I. 2003. Classifying successional forests using Landsat spectral properties and ecological characteristics in eastern Amazônia. *Remote Sens Environ* 87(4):470–481.

Wang, D., S. Liang, T. He, and Y. Yu. 2013. Direct estimation of land surface albedo from VIIRS data: Algorithm improvement and preliminary validation. *J Geophys Res Atmos* 118:12577–12586.

Wang, Y., C. E. Woodcock, W. Buermann, P. Stenberg, P. Voipio, H. Smolander, T. Häme, Y. Tian, J. Hu, and Y. Knyazikhin. 2004. Evaluation of the MODIS LAI algorithm at a coniferous forest site in Finland. *Remote Sens Environ* 91(1):114–127.

Weng, Q. 2011. *Advances in Environmental Remote Sensing: Sensors, Algorithms, and Applications.* CRC Press.

White, M. A., P. E. Thornton, and S. W. Running. 1997. A continental phenology model for monitoring vegetation responses to interannual climatic variability. *Global Biogeochem Cycles* 11(2):217–234.

White, M. A., K. M. de Beurs, K. Didan, D. W. Inouye, A. D. Richardson, O. P. Jensen, J. O'Keefe, G. Zhang, and R. R. Nemani. 2009. Intercomparison, interpretation, and assessment of spring phenology in North America estimated from remote sensing for 1982–2006. *Global Change Bio* 15:2335–2359.

Widlowski, J. L., M. Taberner, B. Pinty, V. Bruniquel-Pinel, M. Disney, R. Fernandes, J.-P. Gastellu-Etchegorry, N. Gobron, A. Kuusk, and T. Lavergne. 2007. Third radiation transfer model intercomparison (RAMI) exercise: Documenting progress in canopy reflectance models. *J Geophys Res* 112:D09111. doi:10.1029/2006JD007821.

Wiegand, C., A. Richardson, D. Escobar, and A. Gerbermann. 1991. Vegetation indices in crop assessments. *Remote Sens Environ* 35:105–119.

Willis, C. G., B. Ruhfel, R. B. Primack, A. J. Miller-Rushing, and C. C. Davis. 2008. Phylogenetic patterns of species loss in thoreau's woods are driven by climate change. *Proc Natl Acad Sci USA* 105(44):17029–17033.

Willis, C. G., B. R. Ruhfel, R. B. Primack, A. J. Miller-Rushing, J. B. Losos, and C. C. Davis. 2010. Favorable climate change response explains non-native species' success in thoreau's woods. *PLoS ONE* 5(1):e8878.

Wolkovich, E. M. and E. E. Cleland. 2011. The phenology of plant invasions: A community ecology perspective. *Front Ecol Environ* 9(5):287–294.

Woodcock, C. E., R. Allen, M. Anderson, A. Belward, R. Bindschadler, W. Cohen, F. Gao, S. N. Goward, D. Helder, and E. Helmer. 2008. Free access to Landsat imagery. *Science* 320:1011.

Woodcock, C. E., J. B. Collins, S. Gopal, V. D. Jakabhazy, X. Li, S. A. Macomber, S. Ryherd, V. J. Harward, J. Levitan, Y. Wu, and R. Warbington. 1994. Mapping forest vegetation using Landsat TM imagery and a canopy reflectance model. *Remote Sens Environ* 50:240–254.

Woodcock, C. E., J. B. Collins, V. D. Jakabhazy, X. Li, S. A. Macomber, and Y. Wu. 1997. Inversion of the Li-Strahler canopy reflectance model for mapping forest structure. *IEEE Trans Geosci Remote Sens* 35:405–414.

Wu, C., J. M. Chen, A. Gonsamo, D. T. Price, T. A. Black, and W. A. Kurz. 2012a. Interannual variability of net carbon exchange is related to the lag between the end-dates of net carbon uptake and photosynthesis: Evidence from long records at two contrasting forest stands. *Agric Forest Meteorol* 164:29–38.

Wu, C., A. Gonsamo, J. M. Chen, W. A. Kurz, D. T. Price, P. M. Lafleur, R. S. Jassal, D. Aragoni, G. Bohrer, and C. M. Gough. 2012b. Interannual and spatial impacts of phenological transitions, growing season length, and spring and autumn temperatures on carbon sequestration: A North America flux data synthesis. *Global Planet Change* 92:179–190.

Wu, Y. and A. Strahler. 1994. Remote estimation of crown size, stand density, and biomass on the Oregon transect. *Ecol Appl* 4(2):299–312.

Wulder, M. A., J. G. Masek, W. B. Cohen, T. R. Loveland, and C. E. Woodcock. 2012. Opening the archive: How free data has enabled the science and monitoring promise of Landsat. *Remote Sens Environ* 122:2–10.

Xiao, X. M., D. Hollinger, J. Aber, M. Goltz, E. A. Davidson, Q. Y. Zhang, and B. Moore. 2004. Satellite-based modeling of gross primary production in an evergreen needleleaf forest. *Remote Sens Environ* 89:519–534.

Xiao, X. M., D. Hollinger, J. Aber, M. Goltz, E. A. Davidson, Q. Y. Zhang, and B. Moore. 2004a. Satellite-based modeling of gross primary production in an evergreen needleleaf forest. *Remote Sens Environ* 89:519–534.

Xiao, X. M., Q. Y. Zhang, B. Braswell, S. Urbanski, S. Boles, S. Wofsy, M. Berrien, and D. Ojima. 2004b. Modeling gross primary production of temperate deciduous broadleaf forest using satellite images and climate data. *Remote Sens Environ* 91(2):256–270.

Xin, Q., P. Olofsson, Z. Zhu, B. Tin, and C. E. Woodcock. 2013. Toward near real-time monitoring of forest disturbance by fusion of MODIS and Landsat data. *Remote Sens Environ* 135:234–247.

Xiong, X. and W. Barnes. 2006. An overview of MODIS radiometric calibration and characterization. *Adv Atmos Sci* 23:69–79.

Yang, W., B. Tan, D. Huang, M. Rautiainen, N. V. Shabanov, Y. Wang, J. L. Privette et al. 2006. MODIS leaf area index products: From validation to algorithm improvement. *IEEE Trans Geosci Remote Sens* 44(7):1885–1898.

Yuan, W., W. Cai, J. Xia, J. Chen, S. Liu, W. Dong, L. Merbold, B. Law, A. Arain, and J. Beringer. 2014. Global comparison of light use efficiency models for simulating terrestrial vegetation gross primary production based on the LaThuile database. *Agr Forest Meteorol* 192:108–120.

Yuan, W., S. Liu, G. Zhou, G. Zhou, L. L. Tieszen, D. Baldocchi, C. Bernhofer, H. Gholz, A. H. Goldstein, and M. L. Goulden. 2007. Deriving a light use efficiency model from eddy covariance flux data for predicting daily gross primary production across biomes. *Agric Forest Meteorol* 143(3):189–207.

Zarco-Tejada, P. J., J. A. J. Berni, L. Suárez, G. Sepulcre-Cantó, F. Morales, and J. R. Miller. 2009. Imaging chlorophyll fluorescence with an airborne narrow-band multispectral camera for vegetation stress detection. *Remote Sens Environ* 113(6):1262–1275.

Zarco-Tejada, P. J., A. Morales, L. Testi, and F. J. Villalobos. 2013. Spatio-temporal patterns of chlorophyll fluorescence and physiological and structural indices acquired from hyperspectral imagery as compared with carbon fluxes measured with eddy covariance. *Remote Sens Environ* 133:102–122.

Zeng, X., R. E. Dickinson, A. Walker, M. Shaikh, R. S. DeFries, and J. Qi. 2000. Derivation and evaluation of global 1-km fractional vegetation cover data for land modeling. *J Appl Meteorol* 39:826–840.

Zhang, X., M. A. Friedl, C. B. Schaaf, A. H. Strahler, J. C. F. Hodges, F. Gao, B. C. Reed, and A. Huete. 2003. Monitoring vegetation phenology using MODIS. *Remote Sens Environ* 84(3):471–475.

Zhang, X., D. Tarpley, and J. T. Sullivan. 2007. Diverse responses of vegetation phenology to a warming climate. *Geophys Res Lett* 34:L19405.

Zhang, X. Y. and S. Kondragunta. 2006. Estimating forest biomass in the USA using generalized allometric models and MODIS land products. *Geophys Res Lett* 33:L09402.

Zhang, Y. and R. K. Mishra. 2012. A review and comparison of commercially available pan-sharpening techniques for high resolution satellite image fusion. In: *IEEE International Geoscience and Remote Sensing Symposium (IGARSS 2012)*, IEEE, pp. 182–185.

Zhang, Y. and C. Song. 2006. Impacts of afforestation, deforestation and reforestation on forest cover in China from 1949 to 2003. *J Forest* 104(7):383–387.

Zhao, M., F. A. Heinsch, R. R. Nemani, and S. W. Running. 2005. Improvements of the MODIS terrestrial gross and net primary production global data set. *Remote Sens Environ* 95(2):164–176.

Zhao, M. and S. W. Running. 2010. Drought-induced reduction in global terrestrial net primary production from 2000 through 2009. *Science* 329:940–943.

Zhu, J., N. Tremblay, and Y. Liang. 2012a. Comparing SPAD and atLEAF values for chlorophyll assessment in crop species. *Can J Soil Sci* 92:645–648.

Zhu, Z., J. Bi, Y. Pan, S. Ganguly, A. Anav, L. Xu, A. Samanta, S. Piao, R. R. Nemani, and R. B. Myneni. 2013. Global data sets of vegetation leaf area index (LAI) 3g and fraction of photosynthetically active radiation (FPAR) 3g derived from global inventory modeling and mapping studies (GIMMS) normalized difference vegetation index (NDVI3g) for the period 1981 to 2011. *Remote Sens* 5(2):927–948.

Zhu, Z., C. E. Woodcock, and P. Olofsson. 2012b. Continuous monitoring of forest disturbance using all available Landsat imagery. *Remote Sens Environ* 122:75–91.

VIII

Land Use/Land Cover

Land Cover Change Detection

John Rogan
Clark University

Nathan Mietkiewicz
Clark University

Acronyms and Definitions

ALI	Advanced Land Imager
ASTER	Advanced Spaceborne Thermal Emission and Reflection Radiometer
AutoMCU	Automated Monte Carlo Unmixing
AVHRR	Advanced Very High Resolution Radiometer
AVHRR NDVI3g	Third-generation GIMMS NDVI from AVHRR sensors
B	Brightness band derived from the Tasseled Cap transformation
BG	Bare ground
B_r	Rescaled brightness band
CLASlite	Carnegie Landsat Analysis System–Lite
CONUS	Contiguous United States
DI	Disturbance index
DI$'$	Disturbance index prime
dNBR	Delta Normalized Burn Ratio
EROS	Earth Resources Observation and Science
EVI	Enhanced vegetation index
FIA	Forest Inventory Analysis
G	Greenness band derived from the Tasseled Cap transformation
GIMMS	Global Inventory Modeling and Mapping Studies
GIS	Geographic information systems
GPS	Global Positioning System
G_r	Rescaled greenness band
LAI	Leaf area index
Landsat ETM+	Landsat Enhanced Thematic Mapper+
Landsat OLI-TIRS	Landsat Operational Land Imager–Thermal Infrared Sensor
Landsat TM	Landsat Thematic Mapper
LandTrendr	Landsat-based Detection of Trends in Disturbance and Recovery
LCMMP	California Land Cover Mapping and Monitoring Program
LST	Land surface temperature
MaFoMP	Massachusetts Forest Monitoring Program
MGDI	MODIS Global Disturbance Index
MODIS	Moderate-Resolution Imaging Spectroradiometer
MTBS	Monitoring Trends in Burn Severity
NDVI	Normalized difference vegetation index
NDWI	Normalized difference wetness index
NPV	Nonphotosynthetic vegetation
NRCS	Natural Resources Conservation Service
PAR	Photosynthetically active radiation
SPOT	Satellite Pour l'Observation de la Terre
STAARCH	Spatial Temporal Adaptive Algorithm for mapping Reflectance Change
STARFM	Spatial and Temporal Adaptive Reflectance Fusion Model

UN-REDD	United Nations Programme on Reducing Emissions from Deforestation and Forest Degradation
USDS-FS	United States Department of Agriculture Forest Service
USGS	United States Geological Survey
USGS-LCCP	United States Geological Survey Land Cover Characterization Program
VCT	Vegetation change tracker
W	Wetness band derived from the Tasseled Cap transformation
WELD	Web-Enabled Landsat Data
W_r	Rescaled wetness band

21.1 Introduction

The purpose of this chapter is to explore the current trends in land cover change detection and to identify those trends that are potentially transformative to our understanding of land change, as well as identify knowledge/information gaps that should require attention in the future. The current level of understanding of the scale and pace of land cover change is inadequate (Frey and Smith 2007; Turner et al. 2007; Hansen et al. 2013). However, it is understood that land cover change is an undisputed component of global environmental change (Kennedy et al. 2014). Land cover changes and their impacts range widely from regional temperature warming to land degradation and biodiversity loss and from diminished food production to the spread of infectious diseases (Vitousek et al. 1997; Farrow and Winograd 2001). Land cover change, manifested as either land cover modification or conversion, can occur at all spatial scales, and changes at local scales can have cumulative impacts at broader scales (Stow 1995).

The long-standing challenge facing scientists and policy makers are the paucity of comprehensive data, at local, regional, and national levels, on the types and rates of land cover changes, and even less systematic evidence on the causes/drivers and consequences of those changes (Walker 1998). Such data can be generated through a dual approach: (1) based on direct or indirect observations, for the regions and time periods for which data exist (Franklin 2002), and (2) based on projections by models (Lambin et al. 1999). A key element for the successful implementation of this dual approach is the monitoring of land cover on a systematic, operational basis (Strahler et al. 1996; Lunetta and Elvidge 1998; Townshend and Justice 2002; Wulder and Coops 2014).

In data-rich locations, such as the United States, federal resource inventory programs, such as the U.S. Forest Service Forest Inventory and Analysis (FIA) program (Gillespie et al. 1999) and the Natural Resources Conservation Service (NCRS 2000), have provided valuable statistical information on land cover dynamics for over 35 years. These agencies provide plot-level information for remote sensing land cover mapping projects (Franklin et al. 2000). However, there is also a need for spatially explicit, thematically comprehensive data products that can be provided by remotely sensed data (Loveland et al. 2002). For example, the U.S. Geological Survey's Land Cover

Characterization Program (USGS-LCCP) is designed to document the rates, causes, and consequences of land cover change from 1973 to present, using Landsat North American Landscape Characterization (NALC) data (Soulard et al. 2014). The program area spans 84 ecoregions of the conterminous United States. Another example of comprehensive large-area land cover assessment is the Canadian Forest Service Earth Observation for Sustainable Development of Forests (EOSD) program (http://www.nrcan.gc.ca/), which monitors Canada's forest cover with Landsat imagery (Wood et al. 2002). Additionally, the European Coordination of Information on the Environment (CORINE) program (http://land.copernicus.eu/pan-european/corine-land-cover) maps land cover and land use (LCLU) (44 categories) using a variety of medium-resolution satellite data from 1990 to present.

In data-poor locations, data derived from remote sensing are often the only source of information available for land cover monitoring (Lambin et al. 1999). This situation places added pressure on remote sensing practitioners to produce accurate change maps using replicable methods, which cannot be verified using the traditional suite of map accuracy tools (Rogan and Chen 2004; Dorais and Cardille 2011). The inclusion of land cover change in international agreements such as the Kyoto Protocol under the United Nations Framework Convention on Climate Change (UNFCC), as well as the growing popularity of the United Nations Programme on Reducing Emissions from Deforestation and Forest Degradation (UN-REDD and REDD+), makes it essential to advance initiatives to monitor land cover change effectively (DeFries and Townsend 1999). Increased Landsat data availability (Wulder and Coops 2014) and the growing trend in automated mapping and change detection algorithms will likely open up the current data bottleneck such that developing countries can create more precise estimates of land change (Zhu and Woodcock 2014).

In addition to the technical advantages of remotely sensed data, the reduced data cost, increased accessibility and availability, and increased understanding of the information derived from these data have facilitated the launch of large-area remote sensing–based monitoring programs/initiatives (Loveland et al. 2002; Eidenshink et al. 2007), as well as global-scale medium spatial resolution change map data sets (Hansen et al. 2013). Therefore, these data, in concert with enabling technologies such as global positioning systems (GPSs) and geographic information systems (GISs), can form the information base upon which sound and cost-effective monitoring decisions can be made (Lunetta 1998).

While a large body of work has accumulated regarding land cover change monitoring using remotely sensed data (e.g., see reviews by Nelson 1983; Singh 1989; Hobbs 1990; Mouat et al. 1993; Stow 1995; Coppin and Bauer 1996; Macleod and Congalton 1998; Ridd and Liu 1998; Mas 1999; Civco et al. 2002; Coppin et al. 2002, 2004; Gong and Xu 2003; Wulder and Franklin 2006), little guidance exists for addressing large-area change mapping, especially in an operational context (Dobson and Bright 1994;

Loveland et al. 2002). Thus, in light of the exciting potential for future operational land cover monitoring programs, and in acknowledgement of the large amount of new, disparate methods currently employed in change detection studies in the literature, this chapter presents a review of the key requirements and chief challenges of land cover change monitoring.

A general classification of the spatial resolution of remote sensing platforms produces three categories (Rogan and Chen 2004): (1) coarse resolution (≥250 m) (e.g., Advanced Very High Resolution Radiometer [AVHRR]); (2) medium resolution (<250 m but ≥20 m) (e.g., Landsat Multispectral Scanner [MSS]); and (3) fine resolution (<20 m) (e.g., WorldView-2).

21.2 Land Cover Change Detection and Monitoring: Theory and Practice

Figure 21.1 presents a conceptual scheme of a forest environment and demonstrates that land cover change can result in alterations (increase or decrease) in the abundance, composition, and condition of remote sensing scene elements over various spatial and temporal resolutions (Stow et al. 1990). Conversion is shown in Figure 21.1b. In contrast, modification (Figure 21.1c and d) involves maintenance of the existing cover type in the face of changes to its scene elements (i.e., change in abundance and condition).

Detection and monitoring land cover change across large areas are two of the most important tasks that remote sensing

data and technology can accomplish (Woodcock et al. 2001). Land cover change detection, one of the most common uses of remotely sensed data, is possible when changes in the surface phenomena of interest result in detectable changes in radiance, emittance (Lunetta and Elvidge 1998), Light Detection and Ranging (LIDAR) return values (Wulder et al. 2007), or microwave backscatter values (Rignot and Vanzyl 1993; Grover et al. 1999), which implicitly involves spatial patterns of change (Crews-Meyer 2002).

Khorram et al. (1999) explored the spatial context of land cover change and stated that spatial entities either (1) become a different category; (2) expand, shrink, or change shape; (3) shift position; or (4) fragment or coalesce. These concepts are well understood by remote sensing practitioners, and especially the resource management community, worldwide, but less so by ecology, sociology, and vulnerability communities.

However, in the last 10 years, a number of important developments have occurred that have helped improve the adoption of land change information by scientific communities that had not done so previously. Land change science (Turner et al. 2007) has emerged as an interdisciplinary field that seeks to understand LCLU dynamics as a coupled human–environment system. This burgeoning theoretical field claims Earth observation data as a crucial component and so has effectively exposed land cover mapping and monitoring practices to a broad audience of anthropologists, economists, and sociologists. Another important development is the opening of the Landsat archive in 2008

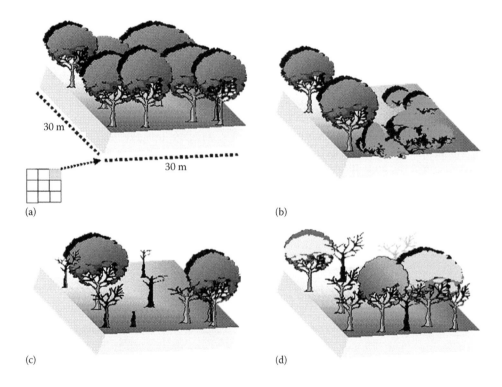

(a) (b)

(c) (d)

FIGURE 21.1 Conceptual scheme representing land-cover changes from Time 1 (represented by (a)) to Time 2 (represented by (b), (c), and/or (d)): (b) change in composition; (c) change in abundance; and (d) change in condition, of vegetation cover, which influence the spectral quantity and quality of solar reflected radiation received by a Landsat sensor (30 m pixel).

(Wulder et al. 2012). The availability of dense time series of moderate spatial resolution Landsat imagery (since 1972 to present) has already had significant impacts on the ecology community (Kennedy et al. 2014) as temporal sequences and trajectories of importance to ecological conservation are now mostly matched by Landsat time stacks. Overall, therefore, we can expect to see, in the near future, remotely sensed data being used to test or verify theories in a much broader array of disciplines than ever before.

Most terrestrial surfaces are comprised complex configurations of land cover attributes (Turner et al. 1999). These range from being mainly *natural* to those that are largely human-dominated (Turner and Dale 1991). Land cover change is viewed in terms of modifications in component attributes within either natural or human-dominated land cover or conversions from natural to human-dominated land cover (Lambin et al. 1999). Despite the recognized importance of land cover modifications (e.g., wind or insect damage), and in contrast to conversions (i.e., forest loss due to agriculture gain), they are not as well documented at operational scales (Lambin et al. 2001). This is partly due to the fact that modifications occur at many different spatial scales and are often too subtle and cryptic to be mapped with a high level of confidence (Ekstrand 1990; Gong and Xu 2003). Therefore, land cover modification analysis requires that a greater level of detail be accommodated in remote sensing analysis.

Macleod and Congalton (1998) listed four aspects of change detection that are important when monitoring land cover using remote sensing data: (1) detecting changes that have occurred (Fung 1990; Lunetta et al. 2002), (2) identifying the nature of the change (Hayes and Sader 2001; Seto et al. 2002), (3) measuring the areal extent of the change (Stow et al. 1990; Rogan et al. 2003), and (4) assessing the spatial pattern of the change (Crews-Meyer 2002; Read 2003). Therefore, change monitoring initiatives/programs (i.e., both current and planned) should try to accommodate these four factors, in addition to appreciating the magnitude, duration, and rate of changes that can occur (Rogan and Chen 2004). Additionally, the burgeoning operational monitoring paradigm represents a shift away from the paradigm of the ubiquitous two-date *end-to-end* change detection approach (i.e., only two dates used in analysis), due to their greater temporal scope (Kasischke et al. 2004).

21.3 Trends in Land Cover Change Detection and Monitoring

21.3.1 Historical Trends: Eight Epochs

The history of land cover change mapping and monitoring has witnessed five distinct periods, determined by the evolution of remote sensor technology, and research needs, related to resource management mandates and various scientific research interests:

1. Early case studies (late 1970s) were exploratory and primarily focused on urban change detection (Todd 1977).
2. Research then shifted to case study applications (early mid-1980s) in natural environments, based on the needs of resource management agencies and the burgeoning interest in carbon sequestration (Singh 1989).
3. Successful applications and experience (mid–late 1990s) led to more widespread applications of remote sensing over large areas and using a wide variety of methods (Lambin and Strahler 1994).
4. Improved sensor technology facilitated the increased interest in less-researched fields, such as urban applications of remote sensing, the cryosphere, and coastal-ocean research (mid-1990s–present) (Rashed et al. 2001), and the new approach adopted by the Moderate-Resolution Imaging Spectroradiometer (MODIS) science team to provide image information products such as global land cover (Friedl et al. 2011). Large-area, high spatial resolution remote sensing became possible in 1994, when the U.S. government allowed civil commercial companies to market high spatial resolution satellite remote sensing data (i.e., 1 and 4 m spatial resolution) (Glackin 1998).
5. Today, a 40-year archive of Landsat imagery, a 22-year archive of AVHRR Global Inventory Modeling and Mapping Studies (GIMMS) normalized difference vegetation index (NDVI) data, and a 15-year archive of MODIS imagery and information products, coupled with an explosion in image time series research and increased automation, have made operational regional–global-scale land change monitoring a reality (Wulder and Coops 2014). Table 21.1 presents a comparison of AVHRR, MODIS, and Landsat data in terms of spatial

TABLE 21.1   Comparison of AVHRR, MODIS, and Landsat in Terms of Spatial and Temporal Resolution

Sensor/Program	Temporal Lineage	Temporal Resolution	Geographic Coverage	Spatial Resolution	Information Content	Information
AVHRR-GIMMS	1982–2012	Biweekly composites	Global	1/12° (8 km at the equator)	NDVI	http://glcf.umd.edu/data/gimms/
MODIS	1999–present	Daily and 8-day composites	Global	250, 500, 1000 m	Multispectral/biophysical products	http://modis.gsfc.nasa.gov/
Landsat	1972–present	16 days	Regional	30 m Global Land Survey global coverage: 1970, 1990, 2000, 2005, 2010	Multispectral	http://landsat.gsfc.nasa.gov/ http://landsat.usgs.gov/science_GLS.php

and temporal resolution. Clearly, the high temporal coverage AVHRR and MODIS data are optimal for regional–global analysis, but they can only provide this coverage at coarse spatial resolution. On the other hand, Landsat data are provided at much finer spatial scales (30 m) but are mostly limited to local–regional coverage. However, the Global Land Survey initiative provides global Landsat coverage for five dates between the early 1970s and 2010. Spatial resolution is a key-limiting factor in the ability of remote sensing imagery to resolve land cover and land cover change classes. This is because spatial scale exerts a strong influence on the ability to extract information from remotely sensed data sets and requires careful specification and analysis. As a result, the question of which remotely sensed data are appropriate for specific land cover change monitoring applications remains an open one. Obviously, the resolvability of land cover change increases with higher spatial resolution. However, high spatial resolution imagery is not typically needed to accurately detect general land cover changes (the goal of large-area monitoring studies) in most environments (Franklin and Wulder 2002). Studying a variety of environments, Townshend and Justice (1988) reported that spatial resolutions coarser than about 200 m undermined the reliable detection of land cover changes. Pax-Lenney and Woodcock (1997) examined the impact of coarsening the spatial resolution on the accuracy of areal estimates of agricultural fields in Egypt (30–120–240–480–960 m). Most of the coarse-resolution estimates were within 10% of the original 30 m estimates. Therefore, medium spatial resolution data remain the optimal choice for most land cover change studies, but more research over time will challenge this assertion in the interest of global-scale estimation and cost reduction, using coarse spatial resolution data, relative to the particular application.

21.3.2 Cause of Land Cover Change

A brief survey of the number of new remote sensing journals shows that 24 journals have been launched since 2007 (an increase of 60% in a 7-year time span). The remarkable proliferation of new journals likely reflects the growing user community and wealth of new remote sensing applications, enabled by a growing time series of free data and also the increased availability of open source software packages (e.g., Quantum GIS). Today, techniques to perform change detection have become numerous as a result of increasing versatility in manipulating digital data and growing computing power (Rogan and Chen 2004). The sheer number of published articles and the importance to resource management indicate both the degree to which remote sensing is used and the proliferation of methods employed. One dimension of this proliferation is progress in developing new and improved ways of detecting change, while another dimension is the wide variety of kinds of changes being monitored (Table 21.2). Table 21.2 presents

the dominant causes of multitemporal land cover change in natural and human-dominated environments and their temporal and physical characteristics. Each change event can result in very different magnitude (i.e., small–large), duration (i.e., days to decades), and temporal rates (i.e., slow–fast) (Aldrich 1975; Gong and Xu 2003). Understanding the magnitude, duration, and rate of land cover disturbances has severe implications for the success of a land cover monitoring study because it permits researchers to determine the most appropriate sensor, derived data set, frequency of acquisition, level of image processing, and reproducible map legend.

It is important to note that not all land change disturbances are equally important in change detection studies, and not all disturbances may be detected as confidently as others (Gong and Xu 2003). For example, land changes of lesser concern to forest managers include those related to interannual variability and growth variation caused by climate variability, whereas, to global change modelers, the last type of change is of chief concern (Turner et al. 1999). A key issue in change detection is understanding how the types of change affect land cover and also how they interact with one another. For example, phenological vegetation change, which varies temporally across scales ranging from years to decades, often interacts with more temporally discrete changes, such as burn scar vegetation depletion and postfire regeneration (Rogan et al. 2002).

21.4 Land Cover Change Detection Approaches

21.4.1 Monotemporal Change Detection: Products for Real Time and Specific Disturbance Types

Numerous land change applications, using only a single image date (i.e., monotemporal change detection) (Coppin and Bauer 1996, p. 217), which focus on a specific change event, have successfully detected a variety of land cover disturbances. These disturbances include water stress (Running and Donner 1987; Running and Pierce 1990), wildfires (Patterson and Yool 1998; Rogan and Franklin 2001), forest thinning (Nilson et al. 2001), forest pest damage (Leckie et al. 1988; Vogelmann and Rock 1988; Joria and Ahearn 1991; Franklin et al. 1994), forest mortality (Ekstrand 1990), and the effects of pollution on vegetation vigor (Pitblado and Amiro 1982; Toutoubalina and Rees 1999).

Monotemporal applications are an effective application of "swapping time for space." Applications of remotely sensed data for disturbance-specific monitoring have considerable advantages, including savings in processing time and reduced costs (Patterson and Yool 1998). Further, end users may require a *quick look* at a particular disturbance for rapid response in the case of mudslide, wildfire, or flood events. A good example of this is the U.S. Forest Service rapid-response wildfire detection project that relies on MODIS active fire detection data (USFS 2004). However, monotemporal approaches rely heavily on assumptions

TABLE 21.2   Causes of Land Cover Change and Their Magnitude, Duration, and Rate

Cause	Magnitude	Duration	Rate	References
Phenology	Small–medium	Days–months	Medium	Goodin et al. (2002), Jakubauskas et al. (2002), Zhang et al (2003)
Regeneration	Small–medium	Days–decades	Slow	Fiorella and Ripple (1993), Lawrence and Ripple (2000)
Drought	Small–medium	Months–years	Slow	Peters et al. (1993), Jacobberger-Jellison (1994)
Flooding	Medium–large	Days–weeks	Medium–fast	Blasco et al. (1992), Michener and Houhoulis (1997), Rogan et al. (2001), Zhan et al. (2002)
Wildfire	Small–large	Days–weeks	Fast	Patterson and Yool (1998), Rogan and Yool (2000)
Disease	Small–large	Days–years	Slow–medium	Wilson et al. (2002), Kelly and Meentemeyer (2002)
Insect attack	Small–large	Days–years	Slow–fast	Muchoney and Haack (1994), Chalifoux et al. (1998), Radeloff et al. (1999)
Ice storm	Small–large	Years	Medium–fast	Dupigny-Giroux et al. (2002), Millward and Kraft (2004), Olthoff et al. (2004)
Mortality	Medium–large	Days–years	Slow–fast	Collins and Woodcock (1996), Allen and Kupfer (2000)
Water/nitrogen stress	Small–medium	Days–years	Slow–fast	Running and Donner (1987), Penuelas et al. (1994)
Pollution	Small–large	Years	Slow	Ekstrand (1994), Rock et al. (1988), Rees and Williams (1997), Diem (2002), Tommervik et al. (2003)
Thinning	Medium–large	Days	Fast	Olsson (1995), Nilson et al. (2001), Peddle et al. (2003a)
Clear-cutting	Large	Days	Fast	Hayes and Sader (2001)
Replanting	Small–medium	Days–decades	Fast	Coppin and Bauer (1996), Levien et al. (1999)
Mining	Large	Days–decades	Medium	Cadac (1998)
Grazing	Small–medium	Days–decades	Slow–medium	Rees et al. (2003)
Wind throw	Large	Days	Medium–fast	Mukai and Hasegawa (2000), Kundu et al. (2001), Lindemann and Baker (2002)
Erosion	Small–medium	Days–weeks	Fast	Dwivedi et al. (1997), Hong and Iisaka (1987), Michalek et al. (1993), Rosin and Hervas (2002)
Environmental quality	Small–large	Months–years	Slow	Fung and Siu (2000)
Fragmentation	Small–large	Days	Fast	Wickham et al. (1999), Millington et al. (2003)
Conversion	Large	Years–decades	Slow–medium	Jha and Unni (1994), Loveland et al. (2002)
Desertification	Small	Years–decades	Slow	Robinove et al. (1981), Pilon et al. (1988)

Source: After Gong and Xu (2003).

about the initial state of land cover in the particular study area (Ekstrand 1994). Indeed, an important factor in the success of these studies is that prechange information (e.g., predisturbance spectral information) and stand and landscape characteristics (e.g., stratification of mixed vegetation canopies, stand-based analysis, slope, and aspect) are controlled to minimize confusion between change and unchanged land cover types (Ekstrand 1990). This implies that prechange, or predisturbance spectral, and/or land cover information are needed to robustly resolve monotemporal disturbances using remotely sensed data (Franklin 2001). For monotemporal (rapid response) applications, coarse spatial resolution data acquired by sensors such as AVHRR, Satellite Pour l'Observation de la Terre (SPOT) Vegetation, and MODIS data are appropriate. Image preprocessing requirements are minimal, but a spectral transformation (e.g., vegetation index) would be useful to separate the disturbance signal (e.g., wildfire or flooding) from the undisturbed background and facilitate simple spectral change thresholding, if required.

Recent advances in real-time disaster response management provide an informative application of monotemporal change detection. The International Charter on Space and Major Disasters (http://www.disasterscharter.org) was founded in 1999, after the catastrophic Hurricane Mitch struck Central America. The Charter aims at providing a unified system of space data acquisition and delivery to locations affected by natural disasters and receives imagery contributions from a group of 15 international participating Earth observation agencies. Additionally, the United Nations Platform for Space-based Information for Disaster Management and Emergency Response (UN-SPIDER program) was established in 2006 to serve as a gateway to space information for disaster management support (http://www.un-spider.org/). These two disaster response programs rely on high spatial resolution data to achieve their goals.

While high spatial resolution sensors cannot conveniently or cost effectively provide wall-to-wall coverage for large-area change mapping applications due to data cost and volume, they are invaluable as a source of ground reference information for medium- and coarse-resolution products/applications and for operational monitoring studies over small spatial extents (Stow et al. 2002). Technological advances in sensor design allow aerial photographic precision and quality in these satellite-based data and permit the investigation of thematic information at the highest order in both natural and urban/suburban landscapes. Though promising, change detection using high spatial resolution data requires further research and development (Rogan and Chen 2004). Data costs, compared to free Landsat data, for example, are very high. Other issues include the impact of off-nadir view angles on change detection and the increasing need for object-based mapping (Stow et al. 2004). Further, geometric distortion is a vexing problem for most airborne data sets (see Franklin and Wulder 2002).

21.4.2 Bitemporal Change Detection: Map Comparison and Disturbance Analysis

In the vast majority of land cover change studies, imagery from one date is compared to another date. Within this paradigm of analyzing images as *endpoints*, there has been a tremendous variety of methods developed and used. This proclivity of bitemporal studies has been caused by several factors: (1) There are fewer data to analyze, (2) studies have been conducted to satisfy burgeoning short-term resource management needs, (3) various researchers have needed a straightforward scenario in order to compare and evaluate a variety of change detection techniques to find an optimal method, (4) most studies have been conducted in regions of limited spatial extent and landscape heterogeneity, and (5) these studies have focused on a single disturbance event (e.g., flooding, fire, logging, or pest infestation) in environmentally (e.g., tropical forests) or politically (e.g., municipalities) important regions. Thus, while bitemporal change detection will continue to serve its purpose for a long time to come, its efficiency and consistency over large, heterogeneous areas has yet to be fully examined (Rogan et al. 2003). However, the potential for moderate spatial resolution analysis in land change monitoring is enormous (Zhu and Woodcock 2013).

21.4.2.1 Bitemporal Change Detection Methods

The selection of an appropriate change detection technique depends on the information requirements, data availability and quality, time and cost constraints, analysis skill, and experience (Johnson and Kasischke 1998). Table 21.3 presents a summary of a variety of land cover change detection methods and their advantages and disadvantages for operational monitoring. Twelve change detection methods are compared according to their status in terms of operational use, as well as their relative strengths and weaknesses. The chief division between the 12 methods occurs between postclassification comparison (i.e., categorical change) and the suite of existing continuous change detection techniques (e.g., image differencing).

The choice of either categorical or continuous comparison must be based on an understanding of the spectral and spatial impact of a given land cover disturbance or range of disturbances. If land cover attributes are expected to change category (e.g., forest to urban), then postclassification comparison is suitable, if not optimal. However, in many ecosystems, complete land cover conversion rarely occurs over short time intervals (i.e., 3–5 years). In effect, modification in condition and abundance is more common than conversion (Coppin and Bauer 1994; Rogan et al. 2002). Therefore, this makes continuous comparison a more suitable choice of change detection approach for monitoring land cover modifications, especially over relatively short time intervals (i.e., 2–5 years). When longer time periods are considered (e.g., 5–10 years), then categorical comparison

may be more suitable, as actual land cover conversion may be more likely to occur. In situations where digital data are not available for earlier time periods (e.g., pre-1972), categorical comparison is the only feasible approach (e.g., a land cover map of 1775 can be compared to a 1990 land cover map) (Petit and Lambin 2002).

21.4.2.2 Map-Updating Approaches

Another interesting trend in bitemporal change mapping is the use of novel map-updating approaches. Postclassification comparison has been implemented in hundreds of land change case studies, but it is problematic in many land change monitoring scenarios (Stow et al. 1980). Over large areas, land change mapping is challenging for some of the following reasons: (1) Data issues such as cost, platform continuity, availability of aerial photographs, or in situ data inhibit comprehensive spatial and temporal coverage and (2) cloud cover, nonstationarity in landscape features, and phenological variability further limit the usability of available imagery. In combination, these challenges make the task of remapping an entire landscape for a second or even third iteration very expensive and possibly unachievable at an acceptable level of map accuracy (Rogan and Chen 2004). Actual land change due to categorical conversions (e.g., forest to urban) or within-category modifications (e.g., timber harvest) usually occupies only a small portion of a pair of 34,000 km^2 Landsat images (e.g., less than 20%) (Rogan et al. 2003) such that independent remapping of a landscape for a new time period is not warranted as long as there are no drastic changes to a land monitoring protocol (e.g., new map legend, change to incompatible new data sources) (Rogan and Chen 2004).

There are two main methods of map updating present in the remote sensing literature: (1) human-interpreted delineation of new changes using multitemporal data and (2) digital change detection of multitemporal imagery to detect a specific type of disturbance, such as urban sprawl, or forest damage. Feranec et al. (2007) implemented a human-interpretation method of change detection with visually interpreted aerial photography to update the CORINE 44 category land cover map for 1990 and 2000. The 2000 land cover map was created by visually and manually editing polygons of change in the original 1990 classification with overall accuracy above 85%. Other studies have used more automated methods of predating and postdating land cover maps to monitor forest change. Wulder et al. (2008) implemented a technique to postdate a 2000 land cover map to 2003 land cover conditions to detect forest clear-cuts using the near-infrared band from Landsat TM/Enhanced Thematic Mapper+ (ETM+), SPOT-4, and Advanced Spaceborne Thermal Emission and Reflection Radiometer (ASTER) data. Forest clear-cuts were detected using an ordinal ranking method that assigns pixels a value based on its reflectance relative to all other pixels. Detected clear-cuts were integrated into the preexisting 2000 EOSD eight-category land cover product. We expect that new innovative approaches to map updating will emerge in the next decade as remote sensing practitioners merge change mapping and resource inventory in a mutually beneficial process.

TABLE 21.3 Summary of a Variety of Land Cover Change Detection Methods and Their Advantages and Disadvantages for Operational Monitoring

Change Detection Method and Status[a]	Advantages	Disadvantages
Postclassification comparison (PCC)	Provides detailed *from–to* information	Only complete class changes are detected
status = I	Can be used with different sensors and with different spatial and spectral resolutions	Heavily dependent on the accuracy of input maps and consistency between mapping methods
	Permits the use of data with interdate phenological differences	Costs often prohibitive over large areas
	Less sensitive to radiometric/geometric errors	
Composite analysis (CA)	Requires only a single classification	Can require a large number of classes and a large calibration data set
Status = I	Can be applied to both raw and enhanced data (e.g., vegetation indices, albedo)	Separation of spectral changes from temporal changes can be difficult
	Makes effective use of prechange (reference) image	
Image differencing (ID)	Can be applied to both raw and enhanced data	Requires optimization of change/no change threshold
status = I	Provides detailed information on "within class change"	Difference image interpretation can be difficult
		Cannot differentiate spectral differences resulting from different original spectral values
		Highly sensitive to radiometric/geometric errors
		Does not provide *from–to* information
Image ratioing (IR)	Can be applied to both raw and enhanced data	Highly sensitive to radiometric/geometric errors
status = I	Can mitigate atmospheric and sun angle effects	Threshold optimization can be difficult, as change is nonlinearly represented
Change vector analysis (CVA)	Can be applied to both raw and enhanced data	Highly sensitive to radiometric/geometric errors
status = F	Provides detailed *from–to* information	Change-direction outputs are difficult to interpret with a large number of input bands
		Change magnitude thresholding is subjective
Multitemporal Kauth Thomas (MKT)	Results are intuitive	Coefficients are sensor dependent
status = I	Produces suites of change, no change, and noise features	Highly sensitive to radiometric/geometric errors
	Standardized coefficients permit application and comparison over time and space	
Multitemporal spectral mixture analysis (MSMA)	Results are intuitive (biophysically)	Sensitive to choice of end-member type
	Can be used to compare fraction estimates across different sensors and platforms	
Principal components analysis (PCA)	Can be applied to both single-date, composite multidate, and composite ID data	Components can be difficult to interpret
status = I	Reduces multispectral data sets into features representing change, no change, and noise	Threshold optimization can be difficult
	In multitemporal analysis, standardized components can minimize atmospheric and sun angle differences	Statistically based, so limited in space and time
		Sensitive to disproportionate amounts of variance in the imagery
Multivariate alteration detection (MAD)	Reduces multispectral data sets into features representing change, no change, and noise	Has not been widely used
status = E	Can be used to compare information from different sensors	
	Insensitive to disproportionate amounts of variance in imagery	
Multitemporal visualization	Simple and intuitive	Qualitative
status = I	Permits inspection of three dates of imagery as RGB	Does not provide *from–to* information
Knowledge-based approaches	Automatic detection of change	Complicated approach to develop
status = F		Have not been widely used
Cross-correlation analysis (CCA)	Allows for direct updating of land cover maps	Has not been widely used
status = F		

[a] Status of the method in an operational context for land cover change monitoring: I, implemented in operational context; F, feasible in an operational context; E, experimental.

21.4.3 Temporal Trend Analysis: Automation and Big Data

Over the last four decades, voluminous amounts of digital data have been gathered from an ever growing number of satellites and sensors continuously monitoring the Earth, atmosphere, and oceans. Fortunately, the massive increase in available data has coincided with a rise in computing power, and since the widespread popularization of online mapping platforms and user-generated geographic information, often linked to the release of Google Earth™ in 2005, a broader user base for the "Geoweb" has developed (Elwood 2011). The most significant change in the practice of land cover change mapping and monitoring has come from this "Big Data" paradigm, also known as "data-intensive science" (Kelling et al. 2009).

21.4.3.1 Hypertemporal Remote Sensing Data in Trend Analysis

Trend, or temporal trajectory, analysis involves the application of data acquired on a large number of observation dates (i.e., hypertemporal) (inter- and intra-annual), traditionally using coarse spatial resolution, spectrally transformed imagery (e.g., NDVI, photosynthetically active radiation, and leaf area index estimates derived from AVHRR and MODIS). This topic is reviewed thoroughly by Henebry and de Beurs (2013). Once assembled, temporal–spectral profiles can be useful for describing high-frequency land cover modifications over coarse spatial scales (Eastman et al. 2009). The study of land surface phenology has witnessed a large increase in remote sensing practitioners and applications as a method for studying the patterns of plant and animal growth cycles, due to the increase in freely available information/data sets. Phenological events are sensitive to climate variation such that phenology data provide timely baseline information for documenting trends in agriculture, irrigation, and forest growth rates and detecting the impacts of climate change on multiple scales (Henebry and de Beurs 2013). The increased complexity that remote sensing practitioners face when working with hypertemporal data sets is now being ameliorated through new software functionality. For example, the Earth Trends Modeler is an integrated suite of tools within IDRISI software for the analysis of image time series data and allows the user to perform and analyze trend analysis results in both graphic and cartographic format (http://www.clarklabs.org/).

Information from trend analysis can provide information on landscape or land surface phenological variability for finer spatial resolution studies so that change related to disturbances can be potentially separated from climate (temperature and precipitation) variability (Borak et al. 2000). High temporal, coarse spatial resolution imagery has also been used effectively to document the prevailing trends in vegetation phenology over large areas to guide the acquisition of medium spatial resolution imagery (i.e., to reduce commission errors caused by uneven intra- and interannual green up) (Rees et al. 2003). As such, changes inherently linked to seasonality can potentially be separated from

other land cover changes (Coppin et al. 2002). However, spatial resolution is often a limiting factor in these studies, especially when examining subtle land cover changes (Rees et al. 2003).

21.4.3.2 Challenges of Trend Analysis

One of the most challenging aspects of trend analysis is that it requires a high level of image preprocessing to account for sensor and platform differences, sensor drift, etc. (Coppin et al. 2004). Trend analysis can be performed using coarse-to-medium spatial resolution data, although coarse-resolution data are more plentiful. Substantial preprocessing is required, given the large volume of data and the need for a high level of geometric and radiometric consistency. While classification is not essential, the use of image transformations to reduce data volume in size is essential. Most large-area programs utilize categorical comparison approaches to detect and monitor land cover change. While this development is noteworthy, and expected to continue, the land change science community requires information on land cover modifications, which conversion-focused programs cannot efficiently or reliably provide. However, there is potential for improvement with increased data availability and accessibility and growing experience with and understanding of sensors and imagery in large-area scenarios (Franklin 2001; Rogan and Chen 2004).

21.4.3.3 Medium-Resolution Data for Trend Analysis

A very promising new development is the advancement of data fusion, which involves the blending of multiple colocated images to produce a hybrid information product that minimizes the limitations of each contributing data set (Walker et al. 2012). A typical fusion combination merges low temporal/high spatial resolution data with high temporal/low spatial resolution data methods to extend the temporal profile of Landsat data using daily or 8-day MODIS reflectance data (Gao et al. 2006).

Medium spatial resolution data sources are considered optimal to obtain sufficient thematic detail for large-area monitoring applications. Fortunately, the last decade has witnessed the growth in availability of medium spatial resolution data sets such as the Web-Enabled Landsat Data (WELD) program (Roy et al. 2010). Since January 2008, the USGS survey has been providing free terrain-corrected and radiometrically calibrated Landsat data via the Internet. The WELD system is being expanded to the global scale to provide monthly and annual Landsat 30 m information for any terrestrial non-Antarctic location for six 3-year epochs spaced every 5 years from 1985 to 2010. The WELD products are developed specifically to provide consistent data that can be used to derive land cover as well as biophysical products for assessment of land surface dynamics (Roy et al. 2010).

21.4.4 Comparison of Several Automated Change Detection Approaches

In recent years, much attention has been focused on automating the detection of land cover change, specifically forest disturbance, across broad landscapes, and using dense image time series stacks.

TABLE 21.4　Comparison of Seven Prominent Change Detection Algorithms according to Ease of Use, Computation Time, Data Type, and Functionality

Algorithms	Ease of Use	Computation Time	Data Type	Cost	Available to Use	Highlights Deforestation	Highlights Degradation	Source
DI	2	NA	L	Free	Y	Y	N	Healey et al. (2005)
DI′	2	NA	L	Free	Y	Y	Y	Hais et al. (2009)
CLASlite	1	1	L,S,A,M	Free	Y—with permission	Y	Y	Asner et al. (2009)
VCT	2	1	L,S,IRS	Free	Y	Y	Y	Huang et al. (2010)
LandTrendr	3	3	L	Free	Y—requires ENVI	Y	Y	Kennedy et al. (2010)
MGDI	1	NA	M	Free	N	Y	N	Mildrexler et al. (2009)
STAARCH	3	NA	L,M	Free	Y	Y	Y	Hilker et al. (2009)

DI, disturbance index; DI′, disturbance index prime; MGDI, MODIS Global Disturbance Index; CLASlite, Carnegie Landsat Analysis System Lite; VCT, Vegetation Change Tracker; LandTrendr, Landsat-based Detection of Trends in Disturbance and Recovery; STAARCH, Spatial Temporal Adaptive Algorithm for mapping Reflectance Change; NA, not available; L, Landsat 4 and 5 Thematic Mapper (TM), Landsat 7 Enhanced Thematic Mapper Plus (ETM+); S, Satellite Pour l'Observation de la Terre 4 and 5 (SPOT); A, Advanced Spaceborne Thermal Emission and Reflection Radiometer (ASTER); Moderate Resolution Imaging Spectrometer (MODIS); IRS, Indian Remote Sensing Satellite; ENVI, Exelis Visual Information Solutions.

Many spectral disturbance indices (DIs) (Healey et al. 2005; Hais et al. 2009; Mildrexler et al. 2009) and software platforms (Asner et al. 2009; Hilker et al. 2009; Huang et al. 2010; Kennedy et al. 2010) have been created to monitor forest disturbance, each with their own relative strengths and weaknesses (Table 21.4).

21.4.4.1 Disturbance Index

Healey et al. (2005) developed a novel combination of the Tasseled Cap features (brightness [B], greenness [G], and wetness [W]) to highlight forest disturbances over single and multidate Landsat image time series, known as the DI. The DI is a linear combination of the B, G, and W features where each feature is rescaled to one standard deviation above or below the mean forest value of the landscape under investigation, resulting in the equation

$$DI = B_r - (G_r + W_r)$$

where r indicates the rescaled features. The DI is most sensitive to stand-replacing, discrete disturbances, which create a strong, stable, and relatively predictable spectral signal across space and time. Alternatively, the DI is less robust in landscapes where rapid postdisturbance succession occurs, such that the disturbance signal is weakened by increased understory vegetation growth and heterogeneity.

21.4.4.2 Disturbance Index′

Hais et al. (2009) refined the DI to account for gradual disturbances across landscapes and forest stands exhibiting rapid succession (i.e., increased greenness) in understory vegetation. The disturbance index′ (DI′) equation is as follows:

$$DI' = W_r - B_r$$

By removing the greenness band from the original DI equation, the DI′ showed a heightened sensitivity to both discrete (i.e., clear-cut,

windthrow, avalanche) and gradual disturbances (i.e., defoliation, insect mortality) across space and time when compared to the DI, G, B, W, and the normalized difference wetness index (NDWI).

21.4.4.2.1 MODIS Global Disturbance Index

The MODIS Global Disturbance Index (MGDI; Mildrexler et al. 2009) is an automated change detection algorithm, which fuses the MODIS Reflectance product, Land Surface Temperature (LST), and MODIS enhanced vegetation index (EVI) data to detect large-area forest disturbances at global, continental, and subcontinental scales. The MGDI uses annual maximum LST composites to detect large changes in land-surface energy and links those changes to the EVI signal, thus detecting discrete disturbances. Due to the scales at which the algorithm is optimized for, disturbances such as wildland fire events, hurricane damage, large-scale windthrow, clear-cuts, and land clearing for agriculture will be the major landscape modifiers captured over the time series.

21.4.4.3 CLASlite

Carnegie Landsat Analysis System–Lite (CLASlite) (V 3.1) is a stand-alone, fully automated software package used to map forest cover, deforestation, and forest degradation over broad spatial extents and long time series by experts and nonexperts alike (Asner et al. 2009). CLASlite boasts a 1 h processing time on a standard Windows PC for a 30 m spatial resolution image across 10,000 km². CLASlite enables users to input raw data from a variety of satellite platforms (Landsat 4, 5, 7, 8; ASTER; Advanced Land Imager [ALI]; SPOT 4, 5; MODIS) where an automation procedure atmospherically corrects, cloud masks, and classifies images across multiple dates with little user input (see Asner et al. 2009 for more details). The CLASlite algorithm utilizes a spectral mixture procedure called Automated Monte Carlo Unmixing (AutoMCU) to classify forest/nonforested areas for one or multiple image dates. Although the spectral libraries used in this procedure are optimized for tropical forests (>300,000 spectral signatures), it has also been shown to classify temperate forests with great success (see case study in the following text).

21.4.4.4 Vegetation Change Tracker

The vegetation change tracker (VCT) (Huang et al. 2010) is an automated algorithm used to delineate forest change across 12 or more Landsat time series stacks with little to no user parameterization for closed or near closed forest canopies. The VCT algorithm will automatically create initial masks (i.e., clouds, cloud shadows, water) and temporally normalize for all scenes, calculate forest features, temporally interpolate masked land areas, and create a composite output image of all locations that experienced a disturbance for each time step. Additionally, the VCT algorithm calculates multiple types of change magnitude measures and tracks postdisturbance vegetation processes (i.e., succession). The VCT disturbance mapping technique is ideal for discrete, land-clearing events but works poorly for nonstand clearing events (i.e., thinning, selective logging, insect outbreak).

21.4.4.5 LandTrendr

The Landsat-based Detection of Trends in Disturbance and Recovery (LandTrendr; Kennedy et al. 2010) is an algorithm that enables the user to systematically analyze a dense Landsat time series stack to produce robust short-term disturbance and long-term vegetation trend maps. Users are able to provide dense Landsat time series stacks into the LandTrendr, which are atmospherically corrected (Cos(t) algorithm), masked (smoke, cloud, cloud shadow, water), and temporally segmented as a means to capture landscape disturbances. Output images and figures provide a wealth of information that quantify landscape dynamics over the time series stack, allowing for a much more detailed assessment than bitemporal change methods can provide.

21.4.4.6 Spatial Temporal Adaptive Algorithm for Mapping Reflectance Change

The Spatial Temporal Adaptive Algorithm for Mapping Reflectance Change (STAARCH; Hikler et al. 2009) blends Landsat and MODIS data to enhance the temporal resolution of Landsat (16 days) to MODIS (8 days). The STAARCH model employs Healey et al.'s (2005) DI to detect landscape changes, where the DI calculation is completely automated. To aid in heterogeneous landscapes, the STAARCH model uses the minimum standard deviation of forest spectral values to increase the sensitivity of the DI to spectral forest change (i.e., disturbance). Additionally, this algorithm is able to create synthetic Landsat images for a given study area/period for each available MODIS scene used. To note, this algorithm builds upon and improves the performance of the Spatial and Temporal Adaptive Reflectance Fusion Model (STARFM) algorithm (Gao et al. 2006).

21.4.4.7 Summary and Comparison of Automated Change Methods

To summarize the aforementioned change detection indices and algorithms, it is necessary to evaluate their purposes accordingly (Tables 21.4 and 21.5). For high spatial and temporal resolution rapid change detection, it would be most advantageous to employ the CLASlite or the VCT algorithm. To evaluate longer-term environmental landscape dynamics, where computational power and time are not limiting, the LandTrendr would be the most appropriate algorithm of choice. The two DIs (DI and DI′) would be most efficiently utilized under the conditions where forest change detection across time would benefit from manual preprocessing steps to accommodate multidate disparities. Additionally, the MGDI would allow for a more sophisticated approximation of landscape disturbances across a very large area. Lastly, the STAARCH algorithm not only allows for a highly accurate downscaling of MODIS to Landsat pixel scale but also accommodates an automated DI calculation; therefore, this would be the algorithm of choice if large spatial extents combined with a need for high spatial and temporal resolution is necessary. It is imperative to assess each change detection algorithm based on their strengths, weakness, and best fit for the research objectives and scales (both spatially and temporally).

TABLE 21.5 Comparison of Seven Prominent Change Detection Algorithms according to the Degree of Automation with respect to Atmospheric Correction, Cloud Masking, Image Calibration, and Mosaicking

Algorithms	Atmospheric Correction	Cloud Mask	Calibration	Mosaic Multiimage
DI	N	N	Y	N
DI′	N	N	Y	N
MGDI	N	N	Y	Y
CLASlite	Y—6S	Y	Y	N
VCT	Y—LEDAPS	Y	Y	N
LandTrendr	Y—Cos(t)	Y	Y	Y
STAARCH	N	Y	Y	N

DI, disturbance index; DI′, disturbance index prime; MGDI, MODIS Global Disturbance Index; CLASlite, Carnegie Landsat Analysis System Lite; VCT, Vegetation Change Tracker; LandTrendr, Landsat-based Detection of Trends in Disturbance and Recovery; STAARCH, Spatial Temporal Adaptive Algorithm for Mapping Reflectance Change; LEDAPS, Landsat Ecosystem Disturbance Adaptive Processing System; Cos(t), cosine of theta.

21.5 Accuracy Assessment: Beyond Statistics

"It is extremely difficult to implement a consistent, comprehensive, quantitative accuracy assessment for large-area change maps" (Loveland et al. 2002, p. 1094). Following the detection and classification/mapping of land cover change, it is preferable that the accuracy of the change maps be assessed. This topic is reviewed in detail by Olofsson et al. (2014). Accuracy assessment serves as a guide to the map quality and to reveal uncertainty and its likely implications to the end user. Accuracy assessment for change detection studies is more challenging than for single-date studies (Congalton 1991; Khorram et al. 1999). This is because change classes usually represent a very small portion of the change image, or thematic map. Additionally, when performing retrospective change detection, acquiring an adequate database of historical reference materials, such as historic aerial photographs, can be very difficult, if not impossible (Biging et al. 1998). The provision of archived imagery by Google Earth provides an important component to addressing the more vexing concerns in land change accuracy assessment (Dorais and Cardille 2011). Unfortunately, the remote sensing community has tended to focus exclusively on the calculation of map accuracy/validation statistics to demonstrate the validity of a method or the worth of a land cover map (Rogan and Chen 2004). While having statistical information about map accuracy is very useful, it ignores many other facets of a change map that are vital to making sure that true change has been captured (Ghimere et al. 2010). These important facets include estimating the potential outcome of the mapping exercise, estimating the areal dominance of categories, and determining the desired shape, location, association, and configuration of mapped categories.

Based on 10 years of experience mapping forest, wetland, and urban change in Massachusetts, the Massachusetts Forest Monitoring Program (MaFoMP) (Rogan et al. 2010) developed the following list of eight steps to pursue when mapping change over a 40-year time period using all available cloud-free Landsat MSS, TM, and ETM+ imagery:

Step 1—Determine optimal data needs, image processing steps based on scene model (Strahler at al. 1986; Phinn et al. 2000), and desired map legend (e.g., Anderson et al. 1976).

Step 2—Determine optimal response design, support size, and sampling design (identify the trade-offs between support size and cost-logistical feasibility) (see Olofsson et al. 2014 for more details).

Step 3—Qualitatively estimate success of mapping project based on previous experience and literature (e.g., expected outcomes—"last time we achieved 80% overall accuracy").

Step 4—Estimate expected category area/dominance using maps from other sources or your knowledge of the study area (e.g., categories A and B should comprise over 70% of the study area, whereas categories C and D should comprise less than 2% of the study area).

Step 5—Estimate expected category shape, location, association, and configuration (e.g., categories F and G will fall only on the coast in long linear strips, associated with ocean water).

Step 6—Quantitatively estimate overall accuracy and per-class accuracy using validation data (should be appropriate support and sampling design). For a general purpose map, all categories should be ranked equal in importance (thus a balance must be struck between omission and commission errors) such that per-class accuracy should be equal. For a phenomenon-specific map (e.g., forest loss), certain categories should be ranked higher in importance than others such that omission errors should be avoided at all allowable costs, whereas certain levels of commission error are permissible (e.g., it is more important not to miss a rare category than it is to falsely map it). Keep in mind that resubstitution accuracy (i.e., using calibration data as validation data) can be a reasonable first-cut measure of your potential mapping success (Rogan et al. 2003).

Step 7—Engage in postclassification editing/filtering to achieve a product that *looks right*. This may make you return to your original training data and redo the work, especially in heterogeneous locations.

Step 8—Evaluate the map such that the end user can employ it wisely for a task that you may not have thought of (e.g., let the map user know your decisions/activities for Steps 1–8 earlier).

21.6 Massachusetts Case Study: CLASlite

This case study explores the application of CLASlite (Asner et al. 2009) mapping and disturbance detection software to map forest and forest change in Massachusetts. CLASlite can operate with a variety of satellite data types, including Landsat, SPOT, ASTER, ALI, and MODIS. Landsat TM, ETM+, and Operational Land Imager–Thermal Infrared Sensor (OLI-TIRS) data were acquired for 9 individual years spanning nearly three decades (Table 21.6) across eastern Massachusetts (Figure 21.2). Four Landsat tiles were downloaded for each respective year and georeferenced using image-to-image registration to an existing orthorectified Landsat image (http://www.landsat.org). All images were registered to an average root-mean-square error of less than one pixel.

Following the manual coregistration procedure, each scene was processed for each of the 9 years using CLASlite (Version 3.1; Asner et al., 2009). CLASlite is an automated change detection and mapping software optimized for tropical forests but was used here to test the feasibility across spatially heterogeneous temperate forested landscapes such as Massachusetts. CLASlite requires limited user interaction in the four main processing steps (image calibration, fraction image creation, forest cover mapping, and deforestation and disturbance delineation), which is optimal for rapid forest cover mapping spanning multiple dates.

TABLE 21.6 Detailed Description of Scene Date, Spatial Location, and Sensor Type Used

| Acquisition Date | Landsat Scene | | Landsat Sensor |
	Path	Row	
August 8, 1985	12	30	TM
August 8, 1985	12	31	TM
September 1, 1985	13	30	TM
September 1, 1985	13	31	TM
August 15, 1993	12	30	TM
August 15, 1993	12	31	TM
July 5, 1993	13	30	TM
July 5, 1993	13	31	TM
August 21, 1995	12	30	TM
August 21, 1995	12	31	TM
July 15, 1999	13	30	TM
July 15, 1999	13	31	TM
July 31, 1999	12	30	ETM+
July 31, 1999	12	31	ETM+
July 23, 2002	12	30	TM
July 23, 2002	12	31	TM
July 10, 2009	12	30	TM
July 10, 2009	12	31	TM
August 18, 2009	13	30	TM
August 18, 2009	13	31	TM
August 30, 2010	12	30	TM
August 30, 2010	12	31	TM
September 6, 2010	13	30	TM
September 6, 2010	13	31	TM
July 17, 2011	12	30	TM
July 17, 2011	12	31	TM
June 16, 2011	13	30	TM
July 7, 2011	13	31	TM
August 6, 2013	12	30	OLI TIRS
August 6, 2013	12	31	OLI TIRS
September 30, 2013	13	30	OLI TIRS
September 30, 2013	13	31	OLI TIRS

First, all scenes were individually imported into CLASlite by specifying the required ancillary and metadata information. During image calibration, CLASlite uses 6S radiative transfer code to atmospherically correct each scene and convert the output images from radiance values to reflectance. Second, CLASlite employs a Monte Carlo (AutoMCU; Asner et al. 2002) spectral decomposition algorithm to partition each scene into proportional fractional cover types of bare ground (B), photosynthetic vegetation (PV), and nonphotosynthetic vegetation (NPV) for every pixel (Figure 21.2). During this stage, the user is able to specify the degree to which clouds and water bodies are masked out of the resulting image. Third, CLASlite delineates forest versus nonforest pixels based on a user-defined threshold based on proportional PV against B and NPV constituents (Figure 21.3). Finally, CLASlite evaluates the fractional and reflectance images to produce disturbance and degradation classifications for each time step. As defined by

Asner et al. (2009), deforestation refers to a diffuse thinning of the forest canopy, while degradation quantifies any spatial or temporal persistence of forest disturbance. In this case study, CLASlite maps the location of deforestation and forest disturbance in eight eras: 1985–1993, 1993–1995, 1995–1999, 1999–2002, 2002–2009, 2009–2010, 2010–2011, and 2011–2013 (Figure 21.4).

CLASlite forest cover maps for each time period were validated using two independent approaches. The first method employed the 30 m resolution MaFoMP land cover maps (Rogan et al. 2010) for the years 1984, 1990, 2000, and 2009 to produce a cross tabulation matrix of quantity agreement and allocation agreement with the associated CLASlite forest cover images. This assessment determined the degree to which pixels of similar land cover type (forest or nonforest) are in agreement with the 30 m MaFoMP maps (MaFoMP 2011; Table 21.7). Errors of omission and commission were reported for each year as a percentage of all pixels in spatial and quantity agreement or disagreement to the MaFoMP map (Table 21.8). Kappa values and the Cramer's V statistic were reported for each year (Table 21.9).

Additionally, CLASlite change maps were validated using a randomly sampled collection of 200 classified pixels that were used to compare the CLASlite delineated pixel values to high spatial resolution Google Earth imagery (Dorais and Cardille 2011; Google, Inc. 2014). The second assessment allowed for an independent evaluation of quantity and allocation pixel agreement to determine the degree to which the CLASlite outputs are correctly classifying forest versus nonforest land cover types. We used available Google Earth imagery that was closest in temporal proximity to the CLASlite-generated forest cover maps. The original fine spatial resolution data were acquired from DigitalGlobe (i.e., WorldView-2 data). Additionally, the deforestation caused by the June 2011 tornado was validated via 50 randomly sampled points using a 2011 Google Earth image captured post tornado.

21.6.1 CLASlite Results

21.6.1.1 Forest Cover Mapping

Forest cover maps produced through an iterative thresholding procedure of the AutoMCU fraction images resulted in a 508 km² net reduction in forest from 1985 to 2009 (Figure 21.3). Comparatively, the MaFoMP maps generated a 566 km² reduction in forest from 1984 to 2009, demonstrating that CLASlite was within a 10% range of similar transitions over a similar time period. The CLASlite-generated forest cover–type maps resulted in an 81% kappa agreement with the MaFoMP maps and an average 85% accuracy when validated with randomly sampled Google Earth imagery.

21.6.1.2 Deforestation and Disturbance Mapping

Between 1985 and 2013, the study area exhibited a net forest change of 2301 km², equating to 19.5% of the study area (Table 21.10). The largest total amount of forest change was

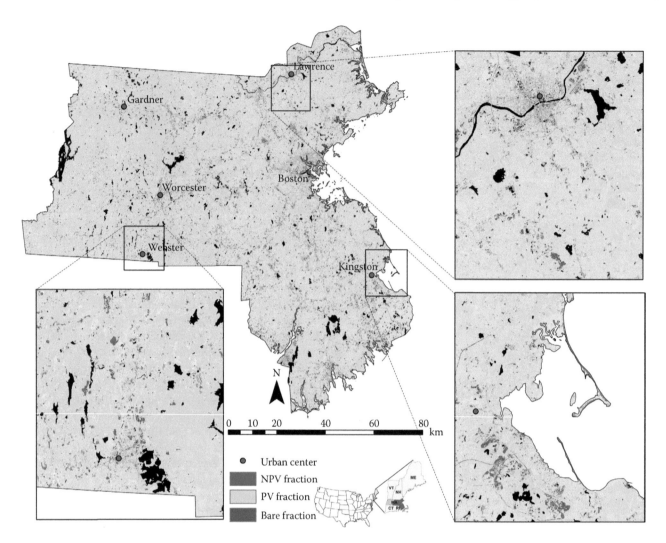

FIGURE 21.2   Study area in Central Massachusetts fraction composite image produced by CLASlite's AutoMCU with examples of rural (Webster), urban (Lawrence), and coastal (Kingston) landscapes.

observed between the interval of 1985–1993, followed by 1995–1999 and 2002–2009, respectively (Figure 21.5), representing 13.5% of the total area that was converted from forest to nonforest (Table 21.10). A visual assessment of the deforestation and disturbance results indicated that forest change was overestimated due to subtle variation in forest phenology, though CLASlite was able to detect most major land-clearing disturbances across one to many years.

21.6.1.3 Gardner, Massachusetts, Forest Change

The case study located in Gardner, Massachusetts (Figure 21.4), illustrated the rural to urban land conversion, a common trend throughout the study area. Forest cover was reduced by 15.2% from 1985 (105 km^2) to 2013 (84 km^2). Across all years, a systematic and continuous shift from forest to nonforest cover types is revealed (Figure 21.4). CLASlite forest cover maps for 1985 report 105 km^2, compared to the MaFoMP maps of 106 km^2. Concomitantly, the 2009 CLASlite output reported 87 km^2 of

forested area remaining in Gardner, MA, compared to 96 km^2 in the MaFoMP product. The area differences between the 2009 classifications were less than 4% of the total case study area of Gardner, Massachusetts. Similar to the eastern Massachusetts deforestation and disturbance mapping, the amount of area affected by forest change in Gardner was overestimated. The total forest change from 1985 to 2013 was reported as being 33 km^2 (23%), where the greatest era of change was 1985–1993, followed by 1995–1999 and 2002–2009.

21.6.1.4 2011 Tornado Disturbance

On June 1, 2011, a 37 km long and 0.8 km wide tornado track touchdowned across southcentral Massachusetts (Figure 21.6). Using the 2010–2011 CLASlite deforestation output, we produced a detailed rendition of the tornado disturbed areas, encompassing 20.3 km^2 over the 60 km track (Figure 21.6). Two years posttornado disturbance, the 2013 forest cover image reported 4.8 km^2 of forest succession along the disturbance

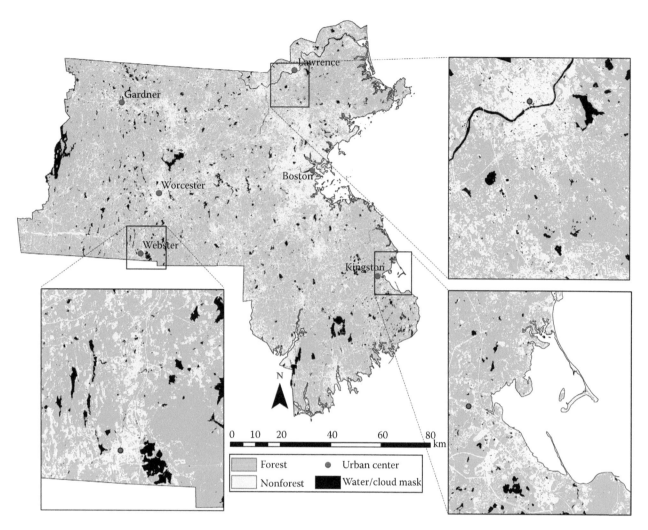

FIGURE 21.3 Statewide automated forest cover image for 2013, with examples of rural (Webster), urban (Lawrence), and coastal (Kingston) landscapes. CLASlite produced cloud/water mask is delineated in black.

edges, while 15.2 km² was still in a disturbed state. Based on 50 randomly sampled points, the agreement was 93% across the tornado track.

21.7 Knowledge Gaps and Future Directions

A remote sensing renaissance has begun. Not since the launch of Earth Resources Technology Satellite 1 in 1972 has the remote sensing community witnessed a more empowering era. Since the mid-1990s, most of the information bottlenecks to operational-style remote sensing research and application have begun to be opened wide for effective and sustainable Earth observation science. The MODIS and Landsat science teams have tenaciously pushed for free, accurate data, and information products, that can be accessed by the rapidly growing global user community. At the same time, high spatial resolution data are available globally from a variety of private companies, most notably (for view only) the Google Earth corporation, at 1–4 m. Importantly, the fields of Landscape

Ecology and Land Change Science have claimed remotely sensed data as an invaluable component of their respective scientific practice. International charters such as the UN-SPIDER initiative rely completely on Earth observation data to draw attention to natural and humanitarian crises. As the content of this chapter highlights, the increased availability of coarse, medium, and high spatial resolution data and the surge in efficient automated methods place remote sensing science in a better place than it has ever been in 40 years. In the next 10 years, remote sensing practitioners can expect to see a multiplier effect with regard to remote sensing applications, as data, methods, and continued advocacy accumulate and expand to new fields and new problems. The following list highlights the current knowledge gaps and future directions for the remote sensing land change community:

1. Ironically, as more and more data become available, more data are needed. Referring to the Landsat program, there will be increasing demand for Landsat MSS data and also TM data that have not yet been catalogued. The collection and processing of these data from various agencies

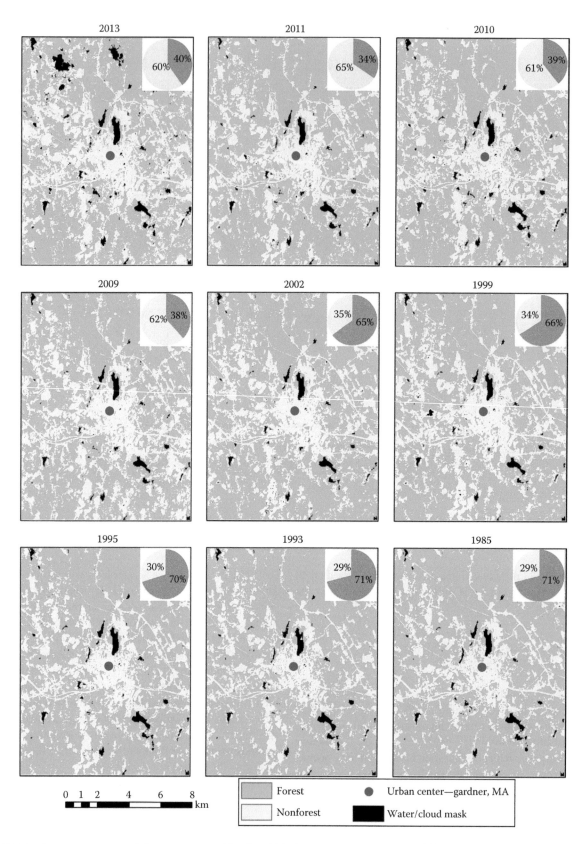

FIGURE 21.4 Forest cover temporal change sequence of Gardner, MA, from 1985 to 2013. Note, each forest cover scene has the percent proportion of pixels for forest and nonforest land cover classes.

TABLE 21.7   Cross Tabulation Assessment Showing Pixel Agreement for Like Years in Terms of Percent of Total Pixel

MAFOMP

	Year	Class	1984 Forest	1984 Nonforest	1990 Forest	1990 Nonforest	2000 Forest	2000 Nonforest	2009 Forest	2009 Nonforest
CLASlite	1985	Forest	0.545	0.078	—	—	—	—	—	—
		Nonforest	0.121	0.256	—	—	—	—	—	—
	1993	Forest	—	—	0.520	0.090	—	—	—	—
		Nonforest	—	—	0.085	0.305	—	—	—	—
	2002	Forest	—	—	—	—	0.510	0.118	—	—
		Nonforest	—	—	—	—	0.072	0.299	—	—
	2009	Forest	—	—	—	—	—	—	0.497	0.078
		Nonforest	—	—	—	—	—	—	0.128	0.297

TABLE 21.8   Kappa (a) and Cramer's V (b) Statistics Showing the Relative Pixel Agreement Accuracy of the CLASlite Forest Cover Classification to MaFoMP Imagery across Four Time Steps

		Kappa			
MAFOMP					
	Year	1984	1990	2000	2009
CLASlite	1985	0.8183	—	—	—
	1993	—	0.8275	—	—
	2002	—	—	0.8179	—
	2009	—	—	—	0.81637
		Cramer's V			
MAFOMP					
	Year	1984	1990	2000	2009
CLASlite	1985	0.7839	—	—	—
	1993	—	0.7864	—	—
	2002	—	—	0.7966	—
	2009	—	—	—	0.781

TABLE 21.9   Random Sample Pixel Percent Agreement of Forest Cover Types of the CLASlite Classification Against High-Resolution Google Earth Imagery

Google Earth™

	Year	Class	1995 Forest	1995 Nonforest	2003 Forest	2003 Nonforest	2008 Forest	2008 Nonforest	2010 Forest	2010 Nonforest	2013 Forest	2013 Nonforest
CLASlite	1995	Forest	0.638	0.064	—	—	—	—	—	—	—	—
		Nonforest	0.037	0.25	—	—	—	—	—	—	—	—
	2002	Forest	—	—	0.613	0.032	—	—	—	—	—	—
		Nonforest	—	—	0.048	0.296	—	—	—	—	—	—
	2009	Forest	—	—	—	—	0.608	0.322	—	—	—	—
		Nonforest	—	—	—	—	0.032	0.317	—	—	—	—
	2010	Forest	—	—	—	—	—	—	0.585	0.032	—	—
		Nonforest	—	—	—	—	—	—	0.037	0.335	—	—
	2013	Forest	—	—	—	—	—	—	—	—	0.5945	0.0594
		Nonforest	—	—	—	—	—	—	—	—	0.0324	0.308

Note: With increasing time there is a direct relationship to decreasing forest and increasing nonforest agreement.

TABLE 21.10   Change Statistics per Era

Era	Deforestation Total	Deforestation (%)	Disturbance Total	Disturbance (%)	Forest Change Total
1985–1993	565.37	4.81	318.21	2.71	883.58
1993–1995	57.1	0.49	29.42	0.25	86.52
1995–1999	250.76	2.13	168.48	1.43	419.24
1999–2002	105.49	0.90	40.57	0.35	146.05
2002–2009	215.35	1.83	119.56	1.02	334.91
2009–2010	81.14	0.69	65.39	0.56	146.54
2010–2011	82.88	0.71	82.52	0.70	165.4
2011–2013	44.97	0.38	74.01	0.63	118.97
Total	1403.06	11.94	898.15	7.65	2301.21

Forest change is the sum of disturbance and deforestation.

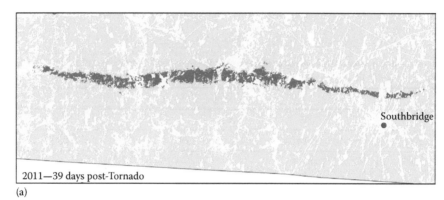

2011—39 days post-Tornado

(a)

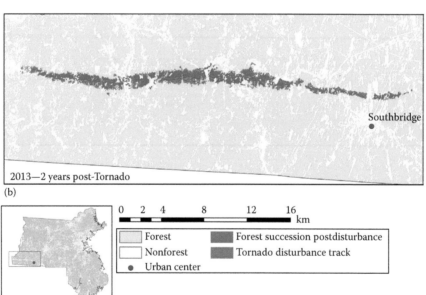

2013—2 years post-Tornado

(b)

FIGURE 21.5   Deforested tornado track as depicted by the CLASlite 2010–2011 deforestation class output (a). Two years post disturbance (b), note that successional infill (blue) has dominated the outer edges of the tornado track, while the interior of the tornado track (red) is still in a deforested state.

throughout the world will greatly extend the reach of the Landsat program, especially to developing countries—the very locations where land change scientists focus their research. Additionally, the cost of high spatial resolution data is problematic. One to four meter data are indispensible for locations where in situ data are unavailable, but these data can currently only be purchased by governments or government-affiliated research initiatives.

2. Given the importance placed currently on land cover modifications by the land change science community, it is important to distinguish their occurrence from land cover conversions. This is a difficult task because both types of

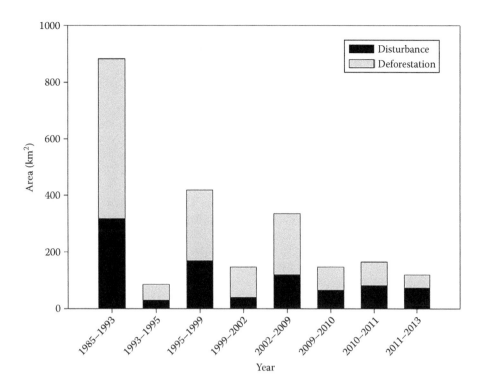

FIGURE 21.6 Area deforested in km² per time step across eastern Massachusetts.

change can result in similar magnitudes of reflectance in a change detection scenario. New methods are needed to ameliorate this problem, especially in developing countries where operational data availability can be scarce.

3. The remote sensing change detection community has laid a strong framework on the back of optical remote sensing imagery. While this paradigm is highly rewarding, optical data are limited in a variety of situations, especially concerning mapping in cloud-prone and data-poor locations. The next decade should hopefully see an expansion in the availability use of large-area radar and LIDAR data collections such that landscape monitoring will be as complete in Cameroon as it is in the United States.

4. All land cover change detection and monitoring relies on the availability of accurate land cover/use information for every location where remotely sensed data are captured. Unfortunately, the process of conducting change detection for a given location is hampered by the paucity of reliable ground reference, wildlife habitat, agricultural land use, and ecological disturbance information. In the next decade, it is hoped that this knowledge gap will be at least partially filled through continued land cover/use mapping efforts, as well as map data sharing.

Acknowledgments

The authors thank David Wilkie and Robert Rose (Wildlife Conservation Society) who inspired the ideas that laid the basis for this chapter. Thanks are also extended to Luisa Young (Clark University), Doug Stow (San Diego State University), and Janet Franklin (Arizona State University) for their contributions to this work.

References

Anderson, J.R., E.E. Hardy, J.T. Roach, and R.E Witmer, 1976, A land use and land-cover classification system for use with remote sensor data. U.S. Geological Survey Professional Paper 964, Washington, DC.

Asner, G.P. and K.B. Heidebrecht, 2002, Spectral unmixing of vegetation, soil and dry carbon cover in arid regions: Comparing multispectral and hyperspectral observations. *International Journal of Remote Sensing,* 23, 3939–3958.

Asner, G.P., D.E. Knapp, A. Balaji, and G. Paez-Acosta, 2009, Automated mapping of tropical deforestation and forest degradation: CLASlite. *Journal of Applied Remote Sensing,* 3, 033543-033543-24.

Biging, G.S., Colby, D.R., and Congalton, R.G., 1998, Sampling systems for change detection accuracy assessment. In *Remote Sensing Change Detection: Environmental Monitoring Methods and Applications* (R.S. Lunetta and C.D. Elvidge, Eds.), Ann Arbor Press, Ann Arbor, MI, pp. 281–308.

Blasco, F., M.F. Bellan, and M.U. Chaudhury, 1992, Estimating the extent of floods in Bangladesh using SPOT data. *Remote Sensing of Environment,* 39,167–178.

Borak, J.S., E.F. Lambin, and A.H. Strahler, 2000, The use of temporal metrics for land-cover change detection at course spatial scales. *International Journal of Remote Sensing,* 21, 1415–1432.

Cadac, J., November 16–20, 1998, Application of change detection algorithms for mine environment monitoring. In *Proceedings of the 19th Asian Conference on Remote Sensing,* Manila, Philippines.

Canadian Forest Service Earth Observation for Sustainable Development of Forests, 2014, ESOD, http://www.nrcan.gc.ca/.

Chalifoux, S., F. Cavayas, and J.T. Gray, 1998, Map-guided approach for automatic detection of Landsat TM images of forest stands damaged by spruce budworm. *Photogrammetric Engineering and Remote Sensing,* 64(6), 629–635.

Cihlar, J., 2000, Landcover mapping of large areas from satellites status and research priorities. *International Journal of Remote Sensing,* 21, 1093–1114.

Cihlar, J. and L.J.M. Jansen, 2001, From land-cover to land use: A methodology for efficient land use mapping over large areas. *Professional Geographer,* 53(2), 275–289.

Civco, D.L., J.D. Hurd, E.H. Wilson, M. Song, and Z. Zhang, April 22–26, 2002, A comparison of land use and land-cover change detection methods. In *Proceedings of ASPRS-ACSM Annual Conference and FIG XXII Congress,* Washington, DC.

Cohen, W.B. and M. Fiorella, 1998, Comparison of methods for detecting conifer forest change with Thematic Mapper imagery. In *Remote Sensing Change Detection Environmental Monitoring Methods and Applications,* (R.S. Lunetta and C.D. Elvidge, Eds.), Ann Arbor Press, Chelsea, MI, 318pp.

Collins, J.B. and C.E. Woodcock, 1994, Change detection using the Gramm-Schmidt transformation applied to mapping forest mortality. *Remote Sensing Environment,* 50, 267–279.

Collins, J.B. and C.E. Woodcock, 1996, An assessment of several linear change detection techniques for mapping forest mortality using multitemporal Landsat TM data. *Remote Sensing Environment,* 56, 66–77.

Congalton, R.G., 1991, A review of assessing the accuracy of classifications of remotely sensed data, *Remote Sensing of Environment,* 37, 35–46.

Coppin, P.R. and M.E. Bauer, 1994, Processing of multitemporal landsat TM imagery to optimize extraction of forest cover change features. *IEEE Transactions on Geoscience and Remote Sensing,* 32, 918–927.

Coppin, P.R. and M.E. Bauer, 1996, Digital change detection in forest ecosystems with remote sensing imagery. *Remote Sensing Reviews,* 13, 207–234.

Coppin, P., I. Jonckheere, B. Muys, and E. Lambin, 2002, Digital change detection methods in natural ecosystem monitoring. In *Proceedings of the First Annual Workshop on the Analysis of Multi-temporal Remote Sensing Images* (L. Bruzzone, and P. Smits, Eds.), World Scientific Publishing Co., Private Ltd., 440p.

Coppin, P., I. Jonckheere, K. Nackaerts, B. Muys, and E. Lambin, 2004, Digital change detection methods in ecosystem monitoring: A review. *International Journal of Remote Sensing,* 25(9), 1565–1596.

Coppin, P., K. Nackaerts, L. Queen, and K. Brewer, 2001, Operational monitoring of green biomass change for forest management. *Photogrammetric Engineering and Remote Sensing,* 67(5), 603–611.

Crews-Meyer, K.A., 2002, Characterizing landscape dynamism using paneled-pattern metrics. *Photogrammetric Engineering and Remote Sensing,* 68(10), 1031–1040.

DeFries, R. and J. Townshend, 1999, Global land-cover characterization from satellite data: From research to operational implementation? *Global Ecology and Biogeography Letters,* 8(5), 367–379.

Diem, J.E., 2002, Remote sensing of forest health in southern Arizona, USA: Evidence for ozone-induced foliar injury. *Environmental Management,* 29(3), 373–384.

Dobson, J. and E. Bright, 1994, Large-area change analysis The Coast-watch Change analysis Project (C-CAP). In *Proceedings of Pecora 12,* August 24–26, 1993, Sioux Falls, South Dakota, American Society for Photogrammetry and Remote Sensing, Bethesda, MD, pp. 73–81.

Dorais, A. and J. Cardille, 2011, Strategies for incorporating high-resolution Google Earth databases to guide and validate classifications: Understanding deforestation in Borneo. *Remote Sensing,* 3(6), 19.

Dupigny-Giroux, L.-A., C.F. Blackwell, S.W. Bristow, and G.M. Olson, 2002, Landscape response to ice disturbance and consecutive moisture extremes. *International Journal of Remote Sensing,* 24(10), 2105–2129.

Dwivedi, R.S., A.B. Kumar, and K.N. Tewari, 1997, The utility of multi-sensor data for mapping eroded lands. *International Journal of Remote Sensing,* 18(11), 2303–2318.

Eastman, J.R., F. Sangermano, B. Ghimire, H. Zhu, H. Chen, N. Neeti, Y. Cai, E.A. Machado, and S.C. Crema, 2009, Seasonal trend analysis of image time series. *International Journal of Remote Sensing,* 30, 2721–2726.

Eidenshink, J., B. Schwind, K. Brewer, Z.-L. Zhu, B. Quayle, and S. Howard, 2007, A project for monitoring trends in burn severity. *Fire Ecology,* 3, 3–21.

Ekstrand, S., 1994, Assessment of forest damage with Landsat TM: Correction for varying forest stand characteristics. *Remote Sensing Environment,* 47, 291–302.

Ekstrand, S.P., 1990, Detection of moderate damage on Norway Spruce using Landsat TM and digital stand data. *IEEE Transactions on Geoscience and Remote Sensing,* 28(4), 685–692.

Elvidge, C.D., T. Miura, W.T. Jansen, D.P. Groeneveld, and J. Ray, 1998, Monitoring trends in wetland vegetation using a Landsat MSS time series. In *Remote Sensing Change Detection Environmental Monitoring Methods and Applications* (R.S. Lunetta and C.D. Elvidge, Eds.), Ann Arbor Press, Chelsea, MI, 318pp.

Elwood, S., 2011, Geographic information science: Visualization, visual methods, and the Geoweb. *Progress in Human Geography,* 35(3):401–408.

European Coordination of Information on the Environment, 2012, Corine Land Cover, http://land.copernicus.eu/pan-european/corine-land-cover/.

Farrow, A. and M. Winograd, 2001, Land use modeling at the regional scale: an input to rural sustainability indicators for Central America. In (A. Veldkamp and E. Lambin, Eds.) Modeling land use/cover change from local to regional scales. *Agricultural Ecosystems Environment*, 85, 249–268.

Feranec, J., G. Hazeu, S. Christensen, and Jaffrain, G., 2007, Corine land cover change detection in Europe (case studies of The Netherlands and Slovakia). *Land Use Policy,* 24(1), 234–247.

Fiorella, M. and W.J. Ripple, 1993, Analysis of conifer forest regeneration using Landsat TM data. *Photogrammetric Engineering and Remote Sensing*, 59, 1383–1387.

Franklin, S.E., 2001, *Remote Sensing for Sustainable Forest Management.* Lewis Publishers, Boca Raton, FL, 407pp.

Franklin, S.E., R.T. Gillespie, B.D. Titus, and D.B. Pike, 1994, Aerial and satellite sensor detection of Kalmia agustifolia at forest regeneration sites in central Newfoundland. *International Journal of Remote Sensing*, 15(13), 2553–2557.

Franklin, S.E. and M.A. Wulder, 2002, Remote sensing methods in medium spatial resolution satellite data land-cover classification of large areas. *Progress in Physical Geography*, 26(2), 173–205.

Frey, K.E. and L.C. Smith, 2007, How well do we know northern land cover? Comparison of four global vegetation and wetland products with a new ground-truth database for West Siberia. *Global Biogeochemical Cycles* 21, GB1016, doi: 10.1029/2006GB002706.

Friedl, M.A., D. Sulla-Menashe, B. Tan, A. Schneider, N. Ramankutty, A. Sibley, and X. Huang, 2010, MODIS collection 5 global land cover: Algorithm Refinements and characterization of datasets. *Remote Sensing of Environment*, 114, 168–182.

Fung, T., 1990, An assessment of TM imagery for land-cover change detection. *IEEE Transactions on Geoscience and Remote Sensing*, 28, 681–684.

Fung, T. and E. LeDrew, 1988, The determination of optimal threshold levels for change detection using various accuracy indices. *Photogrammetric Engineering and Remote Sensing*, 54, 1449–1454.

Fung, T. and W. Siu, 2000, Environmental quality and its changes, an analysis using NDVI. *International Journal of Remote Sensing*, 21(5), 1011–1024.

Gao, F., J. Masek, M. Schwaller, and F. Hall, 2006, On the blending of the Landsat and MODIS surface reflectance: Predicting daily Landsat surface reflectance. *IEEE Transactions on Geoscience and Remote Sensing*, 44, 2207–2218.

Ghimire, B., J. Rogan, and J. Miller, 2010, Contextual land-cover classification: Incorporating spatial dependence in land-cover classification models using random forests and the Getis statistic. *Remote Sensing Letters*, 45–54.

Gillespie, A.J.R., 1999, Rationale for a national annual forest inventory program. *Journal of Forestry*, 97(12), 16–20.

Glackin, D.L., 1998, International space-based remote sensing overview 1980–2007. *Canadian Journal of Remote Sensing*, 24, 307–314.

Gong, P. and B. Xu, 2003, Multi-spectral and multi-temporal image processing approaches: Part 2. Change detection. In M. Wulder and S.E. Franklin, Eds., *Methods and Applications for Remote Sensing of Forests Concepts and Case Studies*, Kluwer Academic Publishers, Dordrecht, the Netherlands, pp. 301–333.

Goodin, D.G., J.A., Harrington, Jr., and B.C. Rundquist, 2002, Land-cover change and associated trends in surface reflectivity and vegetation index in southwest Kansas: 1972–1992. *Geocarto International*, 17(1), 43–50.

Google Inc., 2014, Google Earth™ 7.1.2.2041. Available at: https://www.google.com/earth/.

Graetz, R.D., 1990, Remote sensing of terrestrial ecosystem structure: An ecologist's pragmatic view. In R.J. Hobbs and H.A. Mooney, Eds. *Remote Sensing of Biosphere Functioning*, Springer, New York, pp. 5–30.

Hais, M., M. Jonášová, J. Langhammer, and T. Kučera, 2009, Comparison of two types of forest disturbance using multitemporal Landsat TM/ETM+ imagery and field vegetation data. *Remote Sensing of Environment*, 113(4), 835–845.

Hansen, M.C., P.V. Potapov, R. Moore, M. Hancher, S.A. Turubanova, A. Tyukavina, D. Thau et al., 2013, High-resolution global maps of 21st-century forest cover change. *Science,* 342(6160), 850–853.

Hayes, D.J. and S.A. Sader, 2001, Comparison of change-detection techniques for monitoring tropical forest clearing and vegetation regrowth in a time series. *Photogrammetric Engineering and Remote Sensing*, 67(9), 1067–1075.

Healey, S.P, W.B. Cohen, Y. Zhiqiang, and O.N. Krankina, 2005, Comparison of Tasseled Cap-based Landsat data structures for use in forest disturbance detection. *Remote Sensing of Environment,* 97(3), 301–310.

Henebry, G.M. and K.M. de Beurs, 2013, Remote sensing of land surface phenology: A prospectus. In M.D. Schwartz, Ed., *Phenology: An Integrative Environmental Science*, 2nd edn., Springer. Chapter 21, pp. 385–411.

Hilker, T., M.A. Wulder, N.C. Coops, J. Linke, G. McDermid, J.G. Masek, F. Gao, and J.C. White, 2009, A new data fusion model for high spatial- and temporal-resolution mapping of forest disturbance based on Landsat and MODIS. *Remote Sensing of Environment*, 113, 1613–1627.

Hobbs, R.J., 1990, Remote sensing of spatial and temporal dynamics of vegetation. In R.J. Hobbs and H.A. Mooney Eds., *Remote Sensing of Biosphere Functioning*, Ecological. Studies, 79, Springer-Verlag Inc., New York, pp. 203–222.

Huang, C., S.N. Goward, J.G. Masek, N. Thomas, Z. Zhu, and J.E. Vogelmann, 2010, An automated approach for reconstructing recent forest disturbance history using dense Landsat time series stacks. *Remote Sensing of Environment,* 114(1), 183–198.

Jaccobberger-Jellison, P.A., 1994, Detection of post-drought environmental conditions in the Tombouctou region. *International Journal of Remote Sensing*, 15(16), 3138–3197.

Jakubauskas, M.E., D.R. Legates, and J. Kastens, 2001, Harmonic analysis of time-series AVHRR NDVI data. *Photogrammetric Engineering and Remote Sensing*, 67(4), 461–470.

Jakubauskas, M.E., D.L. Peterson, J.H. Kastens, and D.R. Legates, 2002, Time series remote sensing of landscape-vegetation interactions in the Southern Great Plains. *Photogrammetric Engineering and Remote Sensing*, 68, 1021–1030.

Jha, C.S. and N.V.M. Unni, 1994, Digital change detection of forest conversion of a dry tropical Indian forest region. *International Journal of Remote Sensing*, 15(13), 2543–2552.

Johnson, R.D. and E.S. Kasischke, 1998, Automatic detection and classification of land-cover characteristics using change vector analysis. *International Journal of Remote Sensing*, 19, 411–426.

Joria, P.E. and S.C. Ahearn, 1991, A comparison of the SPOT and Landsat Thematic Mapper Satellite Systems for detecting Gypsy Moth defoliation in Michigan. *Photogrammetric Engineering and Remote Sensing*, 57(12), 1605–1612.

Kasischke, E.S., L.L. Bourgeau-Chavez, and N.H.F. French, 1994, Observations of variations in ERS-1 SAR image intensity associated with forest fires in Alaska. *IEEE Transactions in Geoscience and Remote Sensing*, 32, 206–210.

Kelling, S., W.M. Hochachka, D. Fink, M. Riedewald, R. Caruana, G. Ballard, and G. Hooker, 2009, Data-intensive science: A new paradigm for biodiversity studies. *BioScience*, 59(7), 613–620.

Kelly, M. and R.K. Meentemeyer, 2002, Landscape dynamics of the spread of sudden oak death. *Photogrammetric Engineering and Remote Sensing*, 68(10), 1001–1009.

Kennedy, R.E., S. Andrefouet, W.B. Cohen, C. Gomez, P. Griffiths, M. Hais, S.P. Healey et al., 2014, Bringing an ecological view of change to landsat-based remote sensing. *Frontiers in Ecology and the Environment*, 12(6), 339–346.

Kennedy, R.E., Z. Yang, and W.B. Cohen, 2010, Detecting trends in forest disturbance and recovery using yearly Landsat time series: LandTrendr—Temporal segmentation algorithms. *Remote Sensing of Environment*, 114(12), 2897–2910.

Khorram, S., G. Biging, N. Chrisman, D. Colby, R. Congalton, J. Dobson, R. Ferguson, M. Goodchild, J.R. Jensen, and T. Mace, 1999, *Accuracy Assessment of Land-Cover Change Detection*, ASPRS Monograph Series, American Society for Photogrammetry and Remote Sensing, Bethesda, MD.

Kundu, S.N., A.K. Sahoo, S. Mohapatra, and R.P. Singh, 2001, Change analysis using IRS-P4 OCM data after the Orissa super cyclone. *International Journal of Remote Sensing*, 22(7), 1383–1389.

Lambin, E.F. and A.H. Strahler, 1994a, Change vector analysis in multi-temporal space a tool to detect and categorize land-cover change processes using high temporal resolution satellite data. *Remote Sensing of Environment*, 48, 231–244.

Lambin, E.F. and A.H. Strahler, 1994b, Indicators of land-cover change for change-vector analysis in multitemporal space at coarse spatial scales. *International Journal of Remote Sensing*, 15(10), 2099–2119.

Lambin, E.F., B.L. Turner, H.J. Geist, S.B. Agbola, A. Angelsen, J.W. Bruce, O.T. Coomes et al., 2001, The causes of land-use and land-cover change moving beyond the myths. *Global Environmental Change*, 11, 261–269.

Lambin, E.F., X. Baulies, N. Bockstael, G. Fischer, T. Krug, R. Leemans, E.F. Moran et al., 1999, *Land-Use and Land-Cover Change (LUCC) Implementation Strategy*. IGBP Report 48, IHDP Report 10.

Lawrence, R.L. and W.J. Ripple, 2000, Fifteen years of revegetation of Mount St. Helens: A landscape-scale analysis. *Ecology*, 81, 2742–2752.

Leckie, D.G., P.M. Teillet, G. Fedosejevs, and D.P. Ostaff, 1988, Reflectance characteristics of cumulative defoliation of balsam fir. *Canadian Journal of Forest Research*, 18, 1008–1016.

Levien, L.M., C.S. Fischer, P.D. Roffers, B.A. Maurizi, J. Suero, and X. Huang, 1999, A machine-learning approach to change detection using multi-scale imagery. In *Proceedings of ASPRS Annual Conference*, Portland, OR.

Lindemann, J.D. and W.L. Baker, 2002, Using GIS to analyse a severe forest blowdown in the southern Rocky Mountains. *International Journal of Geographic Information Science*, 16(4), 377–399.

Loveland, T.R, T.L. Sohl, S.V. Stehman, A.L. Gallant, K.L. Sayler, and D.E. Napton, 2002, A strategy for estimating the rates of recent United States land-cover changes. *Photogrammetric Engineering and Remote Sensing*, 68, 1091–1099.

Lu, D., P. Mausel, E. Brondizio, and E. Moran, 2002, Change detection techniques, *International Journal of Remote Sensing*, 25(12), 2365–2401.

Lunetta, R.S., 1998, Project formulation and analysis approaches. In R.S. Lunetta and C.D. Elvidge, Eds., *Remote Sensing Change Detection Environmental Monitoring Methods and Applications*, Ann Arbor Press, Chelsea, MI, 318pp.

Lunetta, R.S. and C.D. Elvidge, Eds., 1998, *Remote Sensing Change Detection Environmental Monitoring Methods and Applications*, Ann Arbor Press, Chelsea, MI, 318pp.

Lunetta, R.S., 2002, Multi-temporal remote sensing analytical approaches for characterizing landscape change. In L. Bruzzone and P. Smits, Eds., *Proceedings of the First Annual Workshop on the Analysis of Multi-Temporal Remote Sensing Images*, World Scientific Publishing Co., Private Ltd., 440p.

Lunetta, R.S., J. Ediriwickrema, D. Johnson, J.G. Lyon, and A. McKerrow, 2002, Impacts of vegetation dynamics on the identification of land-cover change in a biologically complex community in North Carolina, USA. *Remote Sensing of Environment*, 82, 258–270.

Macleod, R.D. and R.G. Congalton, 1998, A quantitative comparison of change detection algorithms for monitoring eelgrass from remotely sensed data. *Photogrammetric Engineering and Remote Sensing*, 64, 207–216.

Macomber, S.A. and C.E. Woodcock, 1994, Mapping and monitoring conifer mortality using remote sensing in the Lake Tahoe Basin. *Remote Sensing of Environment*, 50, 255–266.

MaFoMP, 2011, Massachusetts Forest Monitoring Program (MaFoMP), Human-Environment Regional Observatory of Central Massachusetts (HERO-CM). Available online at: http://www.clarku.edu/department/hero/researcharea/forestchange.cfm

Mas, J.F., 1999, Monitoring land-cover changes: a comparison of change detection techniques. *International Journal of Remote Sensing*, 20, 139–152.

Michener, W.K. and P.F. Houhoulis, 1997, Detection of vegetation changes associated with extensive flooding in a forested ecosystem. *Photogrammetric Engineering and Remote Sensing*, 63(12), 1363–1374.

Mildrexler, D. J., M. Zhao, and S.W Running, 2009, Testing a MODIS Global Disturbance Index across North America. *Remote Sensing of Environment*, 113(10), 2103–2117.

Millington, A.C., X.M. Velez-Liendo, and A.V. Bradley, 2003, Scale dependence in multitemporal mapping of forest fragmentation in Bolivia: Implications for explaining temporal trends in landscape ecology and applications to biodiversity conservation. *ISPRS Journal of Photogrammetry and Remote Sensing*, 57, 289–299.

Millward, A.A. and C.E. Kraft, 2004, Physical influences of landscape on a large-extent ecological disturbance: The northeastern North American ice storm of 1998. *Landscape Ecology*, 19, 99–111.

Mouat, D.A., G.G. Mahin, and J. Lancaster, 1993, Remote sensing techniques in the analysis of change detection. *Geocarto International*, 2, 39–50.

Muchoney, D.M. and B. Haack, 1994, Change detection for monitoring forest defoliation. *Photogrammetric Engineering and Remote Sensing*, 60, 1243–1251.

Mukai, Y. and I. Hasegawa, 2000, Extraction of damaged areas of windfall trees by typhoons using Landsat TM data. *International Journal of Remote Sensing*, 21(4), 647–654.

Nelson, R.F., 1983, Detecting forest canopy change due to insect activity using Landsat MSS. *Photogrammetric Engineering and Remote Sensing*, 49, 1303–1314.

Nilson, T., H. Olsson, J. Anniste, T. Lukk, and J. Praks, 2001, Thinning-caused change in reflectance of ground vegetation in boreal forest. *International Journal of Remote Sensing*, 22(14), 2763–2776.

NRCS (Natural Resources Conservation Service), 2000, *Summary Report: 1997 Natural Resources Inventory* (*Revised*). U.S. Department of Agriculture, Washington, DC, 90p.

Olofsson, P., Foody, G.M., Herold, M., Stehman, S.V., Woodcock, C.E., and Wulder, M.A., 2014, Good practices for assessing accuracy and estimating area of land change. *Remote Sensing of Environment*, (148), 42–57.

Olsson, H., 1995, Reflectance calibration of Thematic Mapper data for forest change detection. *International Journal of Remote Sensing*, 50, 221–230.

Olthof, I., D.J. King, and R.A. Lautenschlager, 2004, Mapping deciduous storm damage using Landsat and environmental data. *Remote Sensing of Environment*, 89, 484–496.

Patterson, M.W. and S.R. Yool, 1998, Mapping fire-induced vegetation mortality using Landsat Thematic Mapper data a comparison of linear transformation techniques. *Remote Sensing of Environment*, 65, 132–142.

Pax Lenney, M. and C.E. Woodcock, 1997, The effect of spatial resolution on monitoring the status of agricultural lands. *Remote Sensing of Environment*, 61(2), 210–220.

Penuelas, J., J.A., Gamon, A.L. Fredeen, and J. Merino, 1994, Reflectance indexes associated with physiological-changes in nitrogen-limited and water-limited sunflower leaves. *Remote Sensing of Environment*, 48, 135–146.

Peters, A.J., B.C. Reed, M.D. Eve, and K.M. Havstad, 1993, Satellite assessment of drought impact on native plant communities of southeastern New Mexico, U.S.A. *Journal of Arid Environments*, 24, 305–319.

Petit, C.C. and E.F. Lambin, 2002, Long-term land-cover changes in the Belgian Ardennes (1775–1929): model-based reconstruction vs. historical maps. *Global Change Biology*, 8(7), 616–630.

Phinn, S.R., C. Menges, G.J.E. Hill, and M. Stanford, 2000, Optimizing remotely sensed solutions for monitoring, modeling, and managing coastal environments. *Remote Sensing of Environment*, 73, 117–132.

Radeloff, V.C., D.J. Mladenoff, and M.S. Boyce, 1999, Detecting jack pine budworm defoliation using spectral mixture analysis: Separating effects from determinants. *Remote Sensing of Environment*, 69, 156–169.

Rashed, T., J.R. Weeks, M.S. Gadalla, and A.G. Hill, 2001, Revealing the anatomy of cities through spectral mixture analysis of multispectral satellite imagery a case study of the greater Cairo region, Egypt. *Geocarto International*, 16, 5–15.

Rees, W. and M. Williams, 1997, Monitoring changes in landcover induced by atmospheric pollution in the Kola peninsula, Russia, using Landsat-MSS data. *International Journal of Remote Sensing*, 18(8), 1703–1723.

Rees, W.G., M. Williams, and P. Vitebsky, 2003, Mapping landcover change in a reindeer herding area of the Russian Arctic using Landsat TM and ETM+ imagery and indigenous knowledge. *Remote Sensing of Environment*, 85, 441–452.

Ridd, M.K. and J. Liu, 1998, A comparison of four algorithms for change detection in an urban environment. *Remote Sensing of Environment*, 63, 95–100.

Rignot, E.J.M. and J.J. Vanzyl, 1993, Change detection techniques for ERS-1 SAR data. *IEEE Transactions on Geoscience and Remote Sensing*, 31(4), 1039–1046.

Rock, B.N., T. Hoshizaki, and J.R. Miller, 1988, Comparison of in situ and airborne spectral measurements of the blue shift associated with forest decline. *Remote Sensing of Environment*, 24, 109–127.

Rogan, J. and D. Chen, 2004, Remote sensing technology for mapping and monitoring land-cover and land use change. *Progress in Planning*, 61, 301–325.

Rogan, J. and J. Franklin, 2001, Mapping wildfire burn severity in southern California forests and shrublands using enhanced Thematic Mapper imagery. *Geocarto International*, 16(4), 89–99.

Rogan, J. and S.R. Yool, 2001, Mapping fire-induced vegetation depletion in the Peloncillo Mountains, Arizona and New Mexico. *International Journal of Remote Sensing*, 22, 3101–3121.

Rogan, J., Bumbarger, N., D. Kulakowski, Z. Christman, D. Runfola, and S.D. Blanchard, 2010, Improving forest type discrimination with mixed lifeform classes using fuzzy classification. *Canadian Journal of Remote Sensing*, 11, 699–708.

Rogan, J., J. Franklin, and D.A. Roberts, 2002, A comparison of methods for monitoring multitemporal vegetation change using Thematic Mapper imagery. *Remote Sensing of Environment*, 80, 143–156.

Rogan, J., J. Miller, D.A. Stow, J. Franklin, L. Levien, and C. Fischer, 2003, Land cover change mapping in California using classification trees with Landsat TM and ancillary data. *Photogrammetric Engineering and Remote Sensing*, 69(7), 793–804.

Roy, D.P., J. Ju, K. Kline, P.L. Scaramuzza, V. Kovalskyy, M.C. Hansen, T.R. Loveland, E.F. Vermote, and C. Zhang, 2010, Web-enabled Landsat Data (WELD): Landsat ETM+ Composited Mosaics of the Conterminous United States, *Remote Sensing of Environment*, 114, 35–49.

Running, S.W. and B.D. Donner, 1987, Water stress response after thinning lodgepole pine stands in Montana. In *Management of Small-Stem Stands of Lodgepole Pine*. U.S. Forest Service Int F.R.E.S, pp. 111–117. General Techical Report, INT-237.

Seto, K.C., C.E. Woodcock, C. Song, X. Huang, J. Lu, and R. Kaufmann, 2002, Monitoring land-use change in the Pearl River Delta using Landsat TM. *International Journal of Remote Sensing*, 23(10), 1985–2004.

Singh, A., 1989, Digital change detection techniques using remotely sensed data. *International Journal of Remote Sensing*, 10, 989–1003.

Skole, D.L. and C. Tucker, 1993, Tropical deforestation and habitat fragmentation in the Amazon: Satellite data from 1978 and 1988. *Science*, 260, 1905–1910.

Sohl, T.L., 1999, Change analysis in the United Arab Emirates: An investigation of techniques. *Photogrammetric Engineering and Remote Sensing*, 65, 475–484.

Soulard, C.E., W. Acevedo, R.F. Auch, T.L. Sohl, M.A. Drummond, B.M. Sleeter, D.G. Sorenson et al., 2014, Land cover trends dataset, 1973–2000: U.S. Geological Survey Data Series 844, 10p., http://dx.doi.org/10.3133/ds844, http://landcover.usgs.gov/.

Stow, D., 1995, Monitoring ecosystem response to global change multitemporal remote sensing analysis. In (J. Moreno and W. Oechel, Eds.), *Anticipated Effects of a Changing Global Environment in Mediterranean Type Ecosystems*, Springer-Verlag, New York, pp. 254–286.

Stow, D., L. Coulter, A. Johnson, and A. Petersen, 2004, Monitoring detailed land-cover changes in shrubland habitat reserves using multi-temporal IKONOS data, *Geocarto International*, 19(2), 95–102.

Stow, D.A., 1999, Reducing misregistration effects for pixel-level analysis of land cover change. *International Journal of Remote Sensing*, 20, 2477–2483.

Stow, D.A., D. Collins, and D. McKinsey, 1990, Land use change detection based on multi-date imagery from different satellite sensor systems. *Geocarto International*, 5(3), 1–12.

Stow, D.A., L. Tinney, and J. Estes, 1980, Deriving land use/land-cover change statistics from Landsat A study of prime agricultural land. In *Proceedings of the 14th International Symposium on Remote Sensing of the Environment*, Ann Arbor, MI, pp. 1227–1237.

Strahler, A.H., C.E. Woodcock, and J.A. Smith, 1986, On the nature of models in remote sensing, *Remote Sensing of Environment*, 20, 121–139.

Todd, W.J., 1977, Urban and regional land use change detected by using Landsat data. *Journal of Research by the United States Geological Survey*, 5, 527–534.

Tommervik, H., K.A. Hogda, and I. Solheim, 2003, Monitoring vegetation changes in Pasvik (Norway) and Pechenga in Kola Peninsula (Russia) using multitemporal Landsat MSS/TM data. *Remote Sensing of Environment*, 85, 370–388.

Toutoubalina, O.V. and G.W. Rees, 1999, Remote sensing of industrial impact of Arctic vegetation around Noril'sk, northern Russia: Preliminary results. *International Journal of Remote Sensing*, 20, 2979–2990.

Townshend, J.R.G. and C.O Justice, 2002, Towards operational monitoring of terrestrial systems by moderate-resolution remote sensing. *Remote Sensing of Environment*, 83, 351–359.

Townshend, J.R.G. and C.O. Justice, 1988, Selecting the spatial resolution of satellite sensors required for global monitoring of land transformations. *International Journal of Remote Sensing*, 9(2), 187–236.

Turner, B. L. II, E. Lambin, and A. Reenberg, 2007, The emergence of land change science for global environmental change and sustainability. *Proceedings, National Academy of Sciences of the United States of America*, 104(52), 20666–20671.

Turner, B.L., II, D. Skole, S. Sanderson, G. Fischer, L.O. Fresco, and R. Leemans, 1999, Land-use and land-cover change science/research plan. IGBP Report No. 35 and HDP Report No. 7. Stockholm International Geosphere-Biosphere Programme.

Turner, B.L., II., 1997, The sustainability principle in global agendas: Justifications for understanding land use/cover change. *Geographical Journal*, 163(12), 133–140.

Veldkamp, A. and E.F. Lambin, 2001, Predicting land-use change. *Agriculture, Ecosystems and Environment*, 85, 1–6.

Vitousek, P.M., H.A. Mooney, J. Lubchenco, and J.M. Melillo, 1997, Human domination of earth's ecosystems. *Science*, 277, 494–499.

Vogelmann, J.E. and B.N. Rock, 1988, Assessing forest damage in high-elevation coniferous forest in Vermont and New Hampshire using Thematic Mapper data. *Remote Sensing of Environment*, 24, 227–246.

Walker, J.J., K. de Beurs, R.H. Wynne, and F. Gao, 2010, An evaluation of data fusion products for the analysis of dryland forest phenology. *Remote Sensing of Environment*, 117, 381–393.

Wang, J., P.M. Rich, and K.P. Price, 2003, Temporal responses of NDVI to precipitation and temperature in the central Great Plains, USA. *International Journal of Remote Sensing*, 24(11), 2345–2364.

Wickham, J.D., K.B. Jones, K.H. Riitters, T.G. Wade, and R.V. O'Neill, 1999, Transitions in forest fragmentation: implications for restoration opportunities at regional scales. *Landscape Ecology*, 14, 137–145.

Wood, J.E., M.D. Gillis, D.G. Goodenough, R.J. Hall, D.G. Leckie, J.L. Luther, and M.A. Wulder, June 24–28, 2002, Earth observation for sustainable development of forests (EOSD): Project Overview. In *Proceedings of the International Geoscience and Remote Sensing Symposium (IGARSS) and 24th Symposium of the Canadian Remote Sensing Society*, Toronto, Ontario, Canada.

Wulder, M.A. and N.C. Coops, 2014, Make Earth observations open access. *Nature*, 513, 30–31.

Wulder, M.A., C.R. Butson, and J.C. White, 2008, Cross-sensor change detection over a forested landscape: Options to enable continuity of medium spatial resolution measures. *Remote Sensing of Environment*, 112, 796–809.

Zhan, X., R.A. Sohlberg, J.R.G. Townshend, C. DiMiceli, M.L. Carroll, J.C. Eastman, M.C. Hansen, and R.C. DeFries, 2002, Detection of land-cover changes using MODIS 250 m data. *Remote Sensing of Environment*, 83, 336–350.

Zhang, X., M.A. Friedl, C.B. Schaaf, A.H. Strahler, J.C.F. Hodges, F. Gao, B.C. Reed, and A. Huete, 2003, Monitoring vegetation phenology using MODIS. *Remote Sensing of the Environment*, 84, 471–475.

Zhu, Z. and C.E. Woodcock, 2014, Continuous change detection and classification of land cover using all available Landsat data, *Remote Sensing of Environment*, 144, 152–171.

22

Land Use and Land Cover Mapping and Change Detection and Monitoring Using Radar Remote Sensing

Zhixin Qi
The University of Hong Kong

Anthony Gar-On Yeh
The University of Hong Kong

Xia Li
Sun Yat-sen University

Acronyms and Definitions

AVHRR	Advanced very-high-resolution radiometer
AVNIR	ALOS visible and nearinfrared
CVA	Change vector analysis
DT-CWT	Dual-tree complex wavelet transform
EM	Expectation maximization
ETM+	Enhanced Thematic Mapper Plus
HMCs	Hidden markov chains
ICM	Iterated conditional mode
InSAR	Interferometric synthetic aperture radar
K&I	Kittler and Illingworth
LJ-EM	Landgrebe and Jackson expectation maximization
LULC	Land use and land cover
MAP	Maximum a posteriori
MCP	Multiscale change profile
ML	Maximum likelihood
MRF	Markov random field
MSS	Multispectral scanner
MuMGDs	Multisensory multivariate gamma distributions
OOIA	Object-oriented image analysis
PAMIR	Phased array multifunctional imaging radar
PCC	Postclassification comparison
PolInSAR	Polarimetric interferometric synthetic aperture radar
PolSAR	Polarimetric synthetic aperture radar
RADAR	Radio detection and ranging
SAR	Synthetic aperture radar
SBA	Split-based approach
SIR	Shuttle imaging Radar
TM	Thematic mapper

22.1 Introduction

Land use and land cover (LULC) mapping and monitoring are one of the most important application areas of remote sensing. Land use refers to the human activities on land, which are directly related to the land (Clawson and Stewart, 1965). It usually emphasizes on the functional role of land in socioeconomic activities, such as agriculture, industry, commerce, transportation, construction, and recreation. These activities are abstract and not always observable from remotely sensed images, and inference based on surrogates often has to be made to identify the land use. Land cover, on the other hand, implies the vegetation and artificial constructions covering the land surface (Burley, 1961). It encompasses natural features such as vegetation, urban areas, water, barren land, or others that are concrete and directly visible on remotely sensed images. Land cover does not describe the use of land, and the use of land may be different for lands with the same cover type. For instance, a land cover type of forest may be used for timber production, wildlife management, or recreation; it might be private land, a protected watershed, or a popular state park.

The importance of mapping, quantifying, and monitoring LULC and its change have been widely recognized in the scientific community as a key element in a variety of applications, such as ecological monitoring, habitat assessment, wildlife management, enforcement, exposure and risk assessment, global change monitoring, environmental impact assessment, state and local planning, hazardous waste remedial action, and regulatory policy development (Lo, 1998). The accurate and timely information on LULC patterns and changes has grown in importance in recent years with our increasing concern over the conflict between economic development and ecological change. The knowledge of the present distribution and area of different LULC types as well as information on their changing proportions are needed by planners, legislators, and governmental officials to determine better land use policy, to project transportation and utility demand, to identify future development pressure points and areas, and to implement effective plans for regional development (Anderson et al., 1976).

Remote-sensing technology has been employed extensively in LULC investigation because of its capability to observe land surface consistently and repetitively and its advantages of cost and time savings for large areas. A LULC classification scheme has been developed for use with remote-sensing data (Anderson et al., 1976). The main characteristics of this scheme are its emphasis on resources rather than people and its capability to provide different levels of classification according to the scale and spatial resolution of the images. The development of such classification scheme has facilitated the mapping, modeling, and measurement of many LULC applications. The scheme includes four classification levels in accordance with the image scale (Table 22.1). However, the general relationship between the classification level and the data source is not intended to restrict uses to particular scales, either in the original data source or in the final map product. For example, Level I LULC information could be not only gathered by a Landsat type of satellite or high-altitude imagery but also interpreted from conventional large-scale aircraft imagery or compiled by ground survey. Similarly, several Level II and III categories have been interpreted from Landsat data. The classification scheme for the first and second levels has been presented by Anderson et al. (1976) (Table 22.2). Levels beyond these two must be designed by users according to their needs.

TABLE 22.1 Classification Level and Corresponding Typical Data Characteristics

Classification Level	Typical Data Characteristics
I	LANDSAT (formerly ERTS) type of data
II	High-altitude data at 40,000 ft (12,400 m) or above (less than 1:80,000 scale)
III	Medium-altitude data taken between 10,000 and 40,000 ft (3,100 and 12,400 m) (1:20,000 to 1:80,000 scale)
IV	Low-altitude data taken below 10,000 ft (3,100 m) (more than 1:20,000 scale)

Source: Anderson, J.R. et al., *A Land Use and Land Cover Classification System for Use with Remote Sensor Data*, United States Government Printing Office, Washington, DC, 1976.

TABLE 22.2 LULC Classification System for Use with Remote-Sensing Data

Level I		Level II	
1	Urban or built-up land	11	Residential
		12	Commercial and services
		13	Industrial
		14	Transportation, communications, and utilities
		15	Industrial and commercial complexes
		16	Mixed urban or built-up land
		17	Other urban or built-up land
2	Agricultural land	21	Cropland and pasture
		22	Orchards, groves, vineyards, nurseries, and ornamental horticultural areas
		23	Confined feeding operations
		24	Other agricultural land
3	Rangeland	31	Herbaceous rangeland
		32	Shrub and brush rangeland
		33	Mixed rangeland
4	Forest land	41	Deciduous forest land
		42	Evergreen forest land
		43	Mixed forest land
5	Water	51	Streams and canals
		52	Lakes
		53	Reservoirs
		54	Bays and estuaries
6	Wetland	61	Forested wetland
		62	Nonforested wetland
7	Barren land	71	Dry salt flats
		72	Beaches
		73	Sandy areas other than beaches
		74	Bare exposed rock
		75	Strip mines quarries, and gravel pits
		76	Transitional areas
		77	Mixed barren land
8	Tundra	81	Shrub and brush tundra
		82	Herbaceous tundra
		83	Bare ground tundra
		84	Wet tundra
		85	Mixed tundra
9	Perennial snow or ice	91	Perennial snowfields
		92	Glaciers

Optical remote-sensing images have been widely applied for a myriad of LULC investigation objectives. The Advanced Very High Resolution Radiometer (AVHRR) embarked on the National Oceanic Atmospheric Administration series of satellites has been predominantly used for global- to continental-scale LULC investigation because of its large swath width and twice-daily global coverage (Lambin and Ehrlich, 1997). Compared with the AVHRR, the Moderate Resolution Imaging Spectroradiometer on the Terra and Aqua satellites has enhanced spatial, radiometric, and spectral capabilities and has also been widely applied to large-scale LULC mapping and monitoring (Lunetta et al., 2006). However, the low resolution (250 m^{-1} km) of these sensors limits their ability to reveal detailed spatial distribution of LULC patterns and changes. Balancing the trade-offs involving spatial detail, areal coverage, and availability of historical data, medium-resolution images (10–90 m) obtained from Landsat 5 Thematic Mapper (TM) and Multispectral Scanner (MSS), Landsat 7 Enhanced Thematic Mapper Plus (ETM+), and Advanced Spaceborne Thermal Emission and Reflection Radiometer are currently the most commonly used datasets for LULC mapping and monitoring. Numerous studies have been carried out on the use of the visible to shortwave infrared bands of these datasets for forestry and agricultural land cover analysis and urban development monitoring (Adams et al., 1995; French et al., 2008). Recently, new sensors with higher spatial resolutions (0.5–6 m), such as SPOT 5 and 6, QuickBird, IKONOS, and WorldView 1 and 2, have emerged and held tremendous promise for investigating LULC with increased spatial detail (Chang et al., 2010; Pu and Landry, 2012; Wang et al., 2004).

However, optical remote sensing is limited by weather conditions. Difficulties are encountered in collecting timely LULC information in tropical and subtropical regions that are characterized by frequent cloud cover. Although some optical sensors such as WorldView 2 can deliver data with high temporal resolution, they cannot guarantee that the images collected at short intervals are unaffected by clouds (DigitalGlobe, 2013). Furthermore, the small coverage and high cost limit routine use of the data obtained by these sensors to investigate LULC information. Because of the shortage of cloud-free images, it might be unfeasible to collect timely LULC information using optical remote sensing. Being capable of transmitting and receiving its own electromagnetic waves with the antenna, radio detection and ranging (RADAR) remote sensing is nearly weather independent and can acquire imagery day and night (Figure 22.1). Furthermore, the development of synthetic aperture radar (SAR) improves the resolution beyond the limitation of physical antenna aperture. Therefore, radar remote sensing has become an effective tool for LULC mapping and monitoring in the perpetually cloud-covered tropical and equatorial regions of the world where many developing countries with the greatest need for LULC data are found. The timely information on LULC and its change is necessary for these countries to develop policies that will enable the maintenance of good balance between land development and environmental protection. In addition, compared with conventional optical remote sensing, radar remote sensing provides a different way to observe the Earth. Radar backscattering from terrain is mainly affected by (1) geometrical factors related to structural attributes of the surface and any overlying vegetation cover relative to the sensor parameters of wavelength and viewing geometry and (2) electrical factors determined by the relative dielectric constants of soil and vegetation at a given wavelength (Dobson et al., 1995). Therefore, radar can provide somewhat complementary information to optical data; and hence, classification can be significantly improved when both suites of sensors are used together.

22.2 Radar System Parameters and Development

Imaging radar can be considered a relatively new remote-sensing system in comparison to optical photography. Radar image tone, which represents the radar return signal strength, is governed by two sets of parameters: (1) target parameters, such as roughness at a variety of scales, dielectric properties, angularity and orientation of the target, target spacing, signal penetration, and signal enhancement, and (2) sensor parameters. The primary radar system parameters influencing the intensity and patterns of radar returns from the observed objects are the frequency, polarization, and incidence angle.

Radar frequency and wavelength are interrelated as seen in Equation 22.1:

$$c = f\lambda \tag{22.1}$$

where
  c is the speed of light (3×10^8 ms^{-1})
  f is the frequency
  λ is the wavelength

In order to calculate λ in centimeters (cm), the value of c is given in terms of cm (3×10^{10} cm s^{-1}) and the value of frequency in terms of hertz (Hz). Radar transmits a microwave signal toward the targets and detects the backscattered portion of the signal. Microwaves are electromagnetic waves with frequencies between 300 MHz (0.3 GHz) and 300 GHz in the electromagnetic spectrum. They are included in radio waves that are electromagnetic waves within the frequencies 30 kHz to 300 GHz. Radar systems can be categorized into different bands according to the variation in frequency (Table 22.3). The definition and nomenclature for these bands were established by the U.S. military during World War II (Waite, 1976). Although other classification systems were established outside of the United States, the system presented in Table 22.3 is the most widely used.

Polarization is a property of waves that describes the orientation of their oscillations. For transverse waves such as many electromagnetic waves, it describes the orientation of the oscillations in the plane perpendicular to the wave's direction of travel. As shown in Figure 22.2, propagating electromagnetic radiation has three vector fields that are mutually orthogonal. The direction of

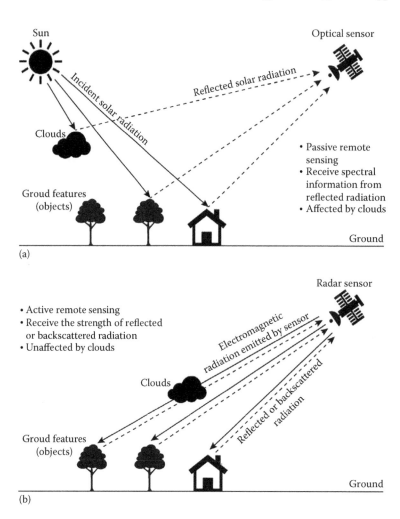

FIGURE 22.1   (a) Optical remote sensing and (b) radar remote sensing.

TABLE 22.3   Radar Bands and Frequencies

Radar Frequency Band	Wavelength (cm)	Frequency Range (MHz)
P	136.00–77.00	220–390
UHF	100.00–30.00	300–1,000
L	30.00–15.00	1,000–2,000
S	15.00–7.50	2,000–4,000
C	7.50–3.75	4,000–8,000
X	3.75–2.40	8,000–12,500
Ku	2.40–1.67	12,500–18,000
K	1.67–1.18	18,000–26,500
Ka	1.18–0.75	26,500–40,000
Millimeter	<0.75	<40,000

Source: Waite, W.P., Historical development of imaging radar, in Lewis, A.J. (ed.), *Geoscience Applications of Imaging Radar Systems, RSEMS (Remote Sensing of the Electro Magnetic Spectrum)*, Association of American Geographers, Omaha, pp. 1–22, 1976.

propagation is one vector, and electric and magnetic fields make up the other two vector fields. Active microwave energy, as well as other frequencies of electromagnetic radiation, has a polarized component defined by the electric field vector of the radiation (Figure 22.2). Most of the radar systems are linear polarized

systems that operate using horizontally (H) or vertically (V) polarized microwave radiation. For these systems, polarization refers to the orientation of the radar beam relative to the Earth's surface (Figure 22.3). If the electric vector field is parallel to the Earth's surface, the wave would be designated horizontally polarized. If it is perpendicular to the Earth's surface, the wave would be designated vertically polarized. In an active system, energy is both transmitted and received. Therefore, the linear polarization can be mixed and matched to provide the four most common linear polarization schemes, namely, HH, HV, VH, and VV (Figure 22.3).

Incidence angle, defined as the angle between the radar line of slight and the local vertical (Figure 22.4) with respect to the geoid, is also a major factor influencing the radar backscatter and the appearance of objects on the imagery, caused by foreshortening or radar layover. In general, reflectivity from distributed scatters decreases with increasing incidence angles (Lewis and Henderson, 1998). Incidence angle incorporating look angle and the curvature of the Earth is shown in Figure 22.4a. This model assumes a level terrain on constant slope angle. In contrast, Figure 22.4b illustrates the *local incidence angle* and takes into account the local slope angle. For example, surface roughness changes as a function of the local incidence angle.

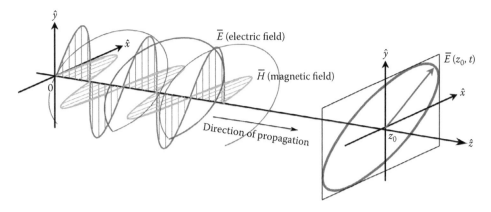

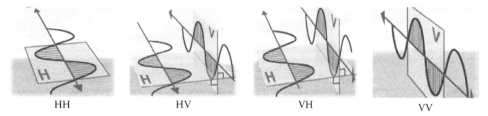

FIGURE 22.2   Components of an electromagnetic wave. The plane of polarization is defined by the electric field. (From European Space Agency, http://earth.eo.esa.int/polsarpro/Manuals/1_What_Is_Polarization.pdf. Accessed April 9, 2015.)

FIGURE 22.3   Linear polarization schemes. (From European Space Agency, Special Features of ASAR, https://earth.esa.int/handbooks/asar/CNTR1-1-5.htm. Accessed April 9, 2015.)

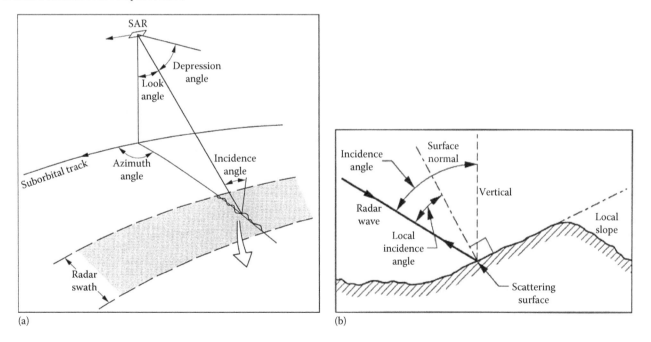

FIGURE 22.4   (a) Schematic diagrams of system and (b) local incidence angle. (From Lewis, A.J. and Henderson, F.M., Radar fundamentals: The geoscience perspective, in Lewis, A.J. and Henderson, F.M., eds., *Manual of Remote Sensing Volume 2—Principles and Applications of Imaging Radar*, John Wiley & Sons, Inc., New York, 1998.)

The development of radar systems has great impact on the application of radar remote sensing in LULC mapping and monitoring. Imaging radar systems could be divided into two categories: spaceborne SAR systems and airborne SAR systems. System parameters for some widely used SAR systems are summarized in Tables 22.4 through 22.6. Airborne imaging SARs are mainly meant for technology development as well as application developments. Early studies on the use of radar remote sensing in LULC investigation were mainly based on airborne radar imagery (Henderson, 1975; Ulaby et al., 1982). The first civilian SAR

TABLE 22.4   Spaceborne SAR Systems 1978–1994

	SEASAT	SIR-A	SIR-B	Kosmos 1870	ALMAZ	ERS-1	JERS-1	SIR-C/X-SAR
Launch date	June 26, 1978	November 12, 1981	October 5, 1984	July 25, 1987	March 31, 1991	July 16, 1991	February 11, 1992	April 9, 1994, September 30, 1994
Country	USA	USA	USA	USSR	USSR	Europe	Japan	USA
Spacecraft	Seasat	Shuttle	Shuttle	Salyut	Salyut	ERS-1	JERS-1	Shuttle
Lifetime	3 months	2.5 days	8 days	2 years	1.5 years	9 years	6 years	10 days
Band	L	L	L	S	S	C	L	L, C; X
Frequency (GHz)	1.275	1.278	1.282	3.0	3.0	5.25	1.275	1.25, 5.3; 9.6
Polarization	HH	HH	HH	HH	HH	VV	HH	L + C: Quad/X: VV
Incident angle (degrees)	23	50	15–64	30–60	30–60	23	39	15–55
Range resolution (m)	25	40	25	30	15–30	26	18	10–30
Azimuth resolution (m)	25	40	58–17	30	15	28	18	30
Swath width (km)	100	50	10–60	20–45	20–45	100	75	15–60
Repeat cycle (days)	17, 3	Nil	Nil	Variable	Nil	3; 35; 176	44	Nil

TABLE 22.5   Spaceborne SAR Systems 1994–2010

	ERS-2	RADARSAT-1	ENVISAT	PALSAR	TerraSAR-X	RADARSAT-2	TanDEM-X
Launch date	April 21, 1995	November 4, 1995	March 1, 2002	January 24, 2006	June 15, 2007	December 14, 2007	June 21, 2010
Country	Europe	Canada	Europe	Japan	Germany	Canada	Germany
Spacecraft	ERS-2	RADARSAT-1	ENVISAT	ALOS	TerraSAR-X	RADARSAT-2	TanDEM-X
Lifetime	11 years	17 years	10 years	5 years	5 years (design)	8 years (design)	5.5 years (design)
Band	C	C	C	L	X	C	X
Frequency (GHz)	5.3	5.3	5.3	1.27	9.65	5.4	9.65
Polarization	VV	HH	Quad	Quad	Quad	Quad	Quad
Incident angle (degrees)	23	10–59	15–45	8–60	20–55	10–60	20–55
Range resolution (m)	26	8–100	30–1000	7–100	1–18.5	3–100	1–18.5
Azimuth resolution (m)	28	8–100	30–1000	7–100	1–18.5	3–100	1–18.5
Swath width (km)	100	50–500	5–400	20–350	10–100	18–500	10–100
Repeat cycle (days)	35	24	35	46	11	24	11

TABLE 22.6   Airborne SAR Systems

	AIRSAR	SAR580	EMISAR	E-SAR	PISAR	STAR-1	UAVSAR
Country	USA	Canada	Denmark	Germany	Japan	Canada	USA
Aircraft	DC-8	CV-580	Gulfstream	DO-228	Gulfstream	Cessna	Gulfstream
Band	C, L, P	X, C	C, L	X, C, L, P	X, L	X	L
Frequency (GHz)	5.3, 1.3, 0.44	9.3, 5.3	5.3, 1.25	9.6, 5.3, 1.3, 0.45	9.55, 1,27	9.6	1.26
Polarization	Quad	Quad	Quad	Single, quad	Quad	HH	Quad
Incident angle	20–60	0–85	30–60	25–60	10–60	45–80	25–60
Range resolution (m)	7.5	6–20	2.4, 8	2–4	3	6, 12	0.5
Azimuth resolution (m)	2	<1 to 10	2.4, 8	2–4	3.2	6	1.5
Swath width (km)	6–20	18–63	12, 24, 48	3	4.3, 19.3, 19.6	40, 60	16

mission in space was United States' Sea Satellite (Seasat), which operated from early July to mid-September in 1978. The design lifetime of Seasat is 2 years. Unfortunately, the spacecraft failed after 3 months of SAR operation due to the problem of power system. However, data provided by Seasat proved to be of high quality and immense interest to the science and application communities. This interest and active research by geoscientists were further augmented in the late 1970s and 1980s by imagery from NASA's Shuttle Imaging Radar (SIR-A and SIR-B) systems along with several systematic airborne SAR projects such as Canada Centre for Remote Sensing (CCRS) SAR-580 campaign. SIR-A

and SIR-B extended the baseline established by Seasat in the dimension of incident angle (Table 22.4). Although these SAR systems proved to be useful for LULC investigation, they were only occasionally launched to collect experimental data within a very short period. The routine investigation of LULC information using radar remote sensing has become practical after some orbital radar systems with SAR, such as ERS-1 and ERS-2, JERS-1, and RADARSAT-1, were made available for regular data collection. However, these orbital SAR systems also have limitations because of only one single frequency available. Some studies indicated that the single-frequency orbital SAR systems can be confused with separating and mapping LULC classes (Li and Yeh, 2004). To overcome the difficulty in single-frequency SAR data, some researchers utilized polarimetric SAR (PolSAR) imagery acquired by SIR-C/X-SAR or airborne SAR systems to investigate LULC information (Lee et al., 2001; Pierce et al., 1994). Their results showed that PolSAR measurements can achieve better classification results than single-polarization SAR. The use of PolSAR data in LULC mapping and monitoring has become an important research topic since PolSAR images have been made available through orbital PolSAR systems, such as Environmental Satellite (ENVISAT), Advanced Synthetic Aperture Radar (ASAR), Advanced Land Observing Satellite (ALOS), Phased Array type L-band Synthetic Aperture Radar (PALSAR), TerraSAR-X, and RADARSAT-2. In addition, constellations such as COnstellation of small Satellites for the Mediterranean basin Observation (COSMO)-SkyMed composed of several satellites equipped with SAR have been available for observing the Earth (Covello et al., 2010). The increasing availability of multidimensional (multifrequency, multipolarized, multitemporal, and multi-incidence angle) digital radar data permits the generation of *true* color radar imagery rather than *colorized* single-channel radar imagery (Figure 22.5).

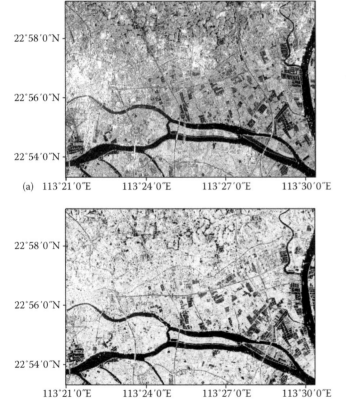

(a) 113°21′0″E 113°24′0″E 113°27′0″E 113°30′0″E

(b) 113°21′0″E 113°24′0″E 113°27′0″E 113°30′0″E

0 2.5 5 km   N

FIGURE 22.5 SAR data of the South Guangzhou area, China. (a) RADARSAT-1 single-polarization data (HH) acquired on December 10, 2005. (b) RADARSAT-2 fully polarimetric data (Pauli RGB composition) acquired on March 21, 2009.

22.3 Radar System Parameter Consideration for LULC Mapping

As introduced previously, the primary radar system parameters that influence the intensity and patterns of radar returns from the observed objects are wavelength, polarization, and incidence angle. Different combinations of these parameters are normally selected to optimize an application. A number of studies have been conducted on the influence of radar frequency and polarization on LULC mapping. Several findings were made for the interaction of the target and polarized signal (Lewis and Henderson, 1998). When the plane of polarization of the transmitted microwave radiation is parallel to the dominant plane of linear features, the like-polarized radar return will be stronger than the transmitted and received signal in the orthogonal plane. For example, it can be expected that VV will have a stronger returned signal than HH if the target, such as a wheat field, has a strong vertical component. The like-polarized image (HH or VV) will have a stronger returned signal than the cross or depolarized image (HV or VH). Only the part of the transmitted signal that is depolarized has the potential of being recorded in the cross-polarized dataset. Depolarization of the transmitted radar signal is primarily results of (1) quasi-specular reflection from corner reflectors, (2) multiple scattering from rough surfaces, and (3) multiple-volume scattering due to inhomogeneities.

Land cover classification was implemented using multifrequency (P, L, C bands) PolSAR images (Chen et al., 1996). The land cover classes included forest, water, bare soil, grass, and eight other types of crops. The radar response of crop types to frequency and polarization states were analyzed for classification based on three configurations: (1) multifrequency and single-polarization images, (2) single-frequency and multipolarization images, and (3) multifrequency and multipolarization images. The classification results showed that using partial information, P-band multipolarization images and multiband HH polarization images, had better classification accuracy, while with a full configuration, namely, multiband and multipolarization, gave the best discrimination capability.

Lee et al. (2001) addressed the land use classification capabilities of fully PolSAR versus dual-polarization and single-polarization SAR for P-, L-, and C-band frequencies. A variety of polarization combinations were investigated for application to crop and tree age classification. They found that L band fully PolSAR data were best for crop classification, but P band was best for forest age classification, because longer wavelength electromagnetic waves provided higher penetration. For dual polarization classification, the HH and VV phase difference was important for crop classification but less important for tree age classification. Also, for crop classification, the L-band complex HH and VV achieved correct classification rates almost as good as for full PolSAR data, and for forest age classification, P-band HH and HV should be used in the absence of fully polarimetric data. In all cases, they indicated that multifrequency fully PolSAR is highly desirable. Similar results were also found by Turkar et al. (2012) in the investigation

TABLE 22.7 Optimal SAR System Parameters for Monitoring Land Surface Characteristics

Application Area	Radar Frequency	Polarization	Incidence Angle[a]	Resolution[b]	Sampling Frequency[c]
Vegetation mapping	Multiple frequency data optimal—as a minimum two frequencies (one high, one low) required	Multipolarization and polarimetric data desired, especially with single frequency systems	Both low and high desired	High Resolutions desirable for mapping smaller sampling units	Low for multiple channel systems, high for single channel systems
Biomass estimation	L- or P-band optimal, as a minimum L- and C-band required	Cross-polarization data most sensitive; multipolarization data improve biomass algorithms	Low	For small forest stands, fine resolution; for larger area studies, low resolution	Low can be used—sampling at proper phenologic stage and under optimum weather conditions important
Monitoring flooded forests	L- and P-band optimal, but some sensitivity at C-band if no leaves present and HH polarization used	HH polarization most sensitive, but VV polarization can be used	Lower required	Higher resolutions may be important for mapping narrow features	High sampling frequencies usually important
Monitoring coastal/low stature wetlands	X or C-band	HH or VV	Low	High or low, depending on ecosystem patch size	High
Monitoring tundra inundation	X or C-band	HH or VV	Low	High or low, depending on ecosystem patch size	High
Monitoring fire-disturbed boreal forests	X or C-band	HH or VV	Low	High or low, depending on ecosystem patch size	High
Detection of frozen/thawed vegetation	Multiple frequencies	Multiple polarizations	Low or high	High or low, depending on ecosystem patch size	High

Source: Kasischke, E.S. et al., *Remote Sens. Environ.*, 59, 141, 1997.

[a] Low incidence angles = 20°–40°, high incidence angles = 40°–60°.

[b] High resolution = 20–40 m, low resolution > 100 m.

[c] High sampling frequency = once every 2 weeks; low sampling frequency = once per year.

of classification capabilities of fully and partially PolSAR data for C- and L-band frequencies. They observed that L-band fully PolSAR data worked better than C band for classification of various land covers. The forest class was well classified with L-band PolSAR data, but it was poorly classified in C band because of dominant scattering from treetops.

There are also many studies on the influence of incidence angle on the intensity and patterns of radar returns from the observed objects. The results show that the choice of optimum incidence angle varies with applications. For example, a geological application generally prefers images acquired with large incidence angles because geometric distortion is minimal and shadowing provides enhancement of topographic relief. On the other hand, a moderate incident angle is required for accurate settlement detection and urban land cover mapping. The incidence angles of less than 20°–23° are of minimum utility for settlement detection and urban analysis, and the amount of information and accuracy of interpretation increases on the image acquired at 41° but decreases on the image acquired at 51° (Henderson, 1995; Henderson and Xia, 1997). Lichtenegger et al. (1991) found that SIR-A imagery was better for land use mapping than the Seasat imagery because of its larger incidence angle. Gauthier et al. (1998) indicated that backscatter coefficients extracted from ERS-1 SAR data over an agricultural area were found to be sensitive to incidence angle. Lang et al. (2008) found that a subtle trend of generally decreasing backscatter with increasing incidence angle increased attenuation of energy from double-bounce

and multipath scattering and possibly increased specular reflectance of the surface layer with increasing incidence angle, and they hypothesized that this decrease was caused by lower transmissivity of the crown layer. Paris (1983) has also indicated that higher incidence angle increases the path length of SAR signal through the vegetation volume, resulting in higher interaction with crop canopy. Ford et al. (1986) stated that as incidence angle increases, there is sensitivity to surface roughness and decreased sensitivity to topography. The incidence angle has also proven to be sensitive to surface roughness and soil moisture (Rahman et al., 2008; Srivastava et al., 2009).

Kasischke et al. (1997) have summarized the optimum SAR parameters for the use of imaging radar systems in several land surface applications (Table 22.7).

22.4 Classification of Radar Imagery

The classification of radar images mainly involves three steps: image preprocessing, feature extraction and selection, and selection of a suitable classifier.

22.4.1 Image Preprocessing

The preprocessing of radar images mainly includes radiometric calibration, geometric calibration, and speckle filtering. Radiometric calibration is a procedure meant to correctly estimate the target reflectance from the measured incoming radiation. Pixel values of radar images are directly related to the radar backscatter of the scene after radiometric calibration. Geometric calibration aims to tie the line/pixel positions in the image coordinates to the geographical latitude/longitude. Speckle is one of the main problems of radar image classification because a homogeneous zone on the ground still has a granular aspect and a statistical distribution with a large standard deviation (Durand et al., 1987). Image filtering is necessary for suppressing the speckle before the classification of radar images. Many adaptive filters for speckle reduction have been proposed, for example, Lee (Lee, 1981), Lee sigma (Lee, 1983), Frost (Frost et al., 1982), Kuan (Kuan et al., 1985), gamma maximum a posteriori (MAP) (Lopes et al., 1993), refined Lee (Lee et al., 1999b), gamma Wilkinson Microwave Anisotropy Probe (Solbo and Eltoft, 2004), and improved Lee sigma (Lee et al., 2009) filters.

Despite their spatial adaptive characteristic, which tends to preserve the signal's high-frequency information, filter applications often give the desired speckle reduction but also an undesired degradation of the geometrical details of the investigated scene. The selection of suitable filter and window size is commonly a heuristic process, in which different filters with different window size are applied to a specific radar image and their performances are compared. Some studies have been conducted on the comparison and selection of appropriate filters. Lee et al. (1999b) stated that refined Lee filter is suitable for PolSAR images because it effectively preserves polarimetric information and retains subtle details while reducing the speckle effect in homogeneous areas. Capstick and Harris (2001) assessed the capability of Lee, Lee sigma, local region, Frost, gamma MAP, and simulated annealing filters to improve classification accuracy in agricultural applications and found the best ones are Lee sigma, gamma MAP, and simulated annealing. Nyoungui et al. (2002) evaluated nine speckle reduction techniques based on their applicability to supervised land cover classification from SAR images. Issues related to suppression of speckle in a uniform area, preservation of edges, and texture preservation were pursued in these filters. The results showed that speckle suppression techniques based on the wavelet transform performs the best, followed by the modified K-nearest neighbors and Lee's local statistic filters.

22.4.2 Feature Extraction and Selection

The primary feature used in radar image classification is the intensity of each pixel that represents the proportion of microwave backscattered from that area on the ground. The pixel intensity values are often converted to a physical quantity called the backscattering coefficient or normalized radar cross section measured in decibel (dB) units with values ranging from +5 dB for very bright objects to −40 dB for very dark surfaces. The backscatter radar intensity depends on a variety of factors: types, sizes, shapes, and orientations of the scatterers in the target area; moisture content of the target area; frequency and polarization of the radar pulses; as well as the incident angles of the radar beam. Therefore, in addition to tonal value of radar intensity images, features that can be extracted from radar

images and have proved to be useful for LULC classification are mainly related to the textural, speckle, polarimetric, interferometric, multifrequency, and multi-incidence angle information.

22.4.2.1 Textural Features

Most of the existing orbital radar systems are single-frequency types and may create confusion during the separation and mapping of LULC classes; this confusion stems from the limited information obtained by single-frequency systems (Ulaby et al., 1986). One way to compensate for the limited information from single-frequency radar data is to derive more features, such as texture, for the classification of radar images in addition to the tonal information of pixels. A variety of textural features have been extracted from radar images and proved to be useful for radar image classification. Miranda et al. (1998) performed classification of JERS-1 SAR data from the rainforest-covered area of the Uaupes River (Brazil) using the semivariogram textural classifier. It was found the semivariogram textural classifier increased the discrimination between upland and flooded vegetation. Kurosu et al. (1999) studied texture statistics in multitemporal JERS-1 SAR single-look imagery and demonstrated a significant improvement in the classification accuracy achieved by using the textural features as additional inputs to the classifier. Fukuda and Hirosawa (1999) derived a wavelet-based texture feature set and successfully applied it to multifrequency PolSAR images of an agricultural area. The classification results indicated that texture was an essential key to the classification of land cover in SAR images. Simard et al. (2000) investigated some texture measures in a study to assess the map-updating capabilities of ERS-1 SAR images in urban areas. The texture measures included histogram measures, wavelet energy, fractal dimension, lacunarity, and semivariograms. The conclusion was that texture improved the classification accuracy, and the measures that performed best were mean intensity, variance, weighted-rank fill ratio, and semivariogram. Rajesh et al. (2001) compared the performance of textural features for characterization and classification of SAR images based on the sensitivity of texture measures for grey-level transformation and multiplicative noise of different speckle levels. Texture features based on grey-level ran length, texture spectrum, power spectrum, fractal dimension, and coocurrence have been considered. They found that fractal, cooccurrence, and texture spectrum-based features performed better, with the maximum classification accuracy for sand texture.

22.4.2.2 Speckle Features

It is well known that the quality of SAR data is degraded by speckle noise, superposing the true radiometric and textural information of the radar image. However, Esch et al. (2011) demonstrated that the information on the local development of speckle can be used for the differentiation of basic land cover types in a high-resolution single-polarized TerraSAR-X strip map images. Combined with local backscatter intensity, the information on the local speckle behavior can be used for the implementation of a straightforward preclassification of single-polarized TerraSAR-X strip map images, showing overall accuracies of 77%–86%. The results show that unsupervised speckle analysis in high-resolution SAR images supplies valuable information for a differentiation of the water, open land, woodland, and urban area.

22.4.2.3 Polarimetric Features

A distinctive characteristic of a PolSAR system is the utilization of polarized waves. The observed polarimetric signatures of the electric field backscattered by the scene depend strongly on the scattering properties of the image objects. In comparison with conventional single-polarization SAR, the inclusion of SAR polarimetry allows for the discrimination of different types of scattering mechanisms that leads to a significant improvement in the quality of classification results (Lee et al., 2001). However, early studies on PolSAR image classification were often conducted on the intensity images of different polarization and their simple combination such as VH/HH and VH/VV (Wu and Sader, 1987). Recently, some polarimetric decomposition techniques have been introduced for the interpretation and classification of PolSAR images. Polarimetric decomposition techniques aim to separate a received signal by the radar as the combination of the scattering responses of simpler objects presenting an easier physical interpretation, which can be used to extract the corresponding target types in PolSAR images. Pauli decomposition is a well-known decomposition method commonly used for PolSAR data (Cloude and Pottier, 1996). In the Pauli decomposition, if the transmit and receive antennas coincide, the backscattering matrix elements can be arranged into a vector: $\mathbf{k} = (a, b, c) / \sqrt{2} = (S_{hh} + S_{vv}, S_{hh} - S_{vv}, 2S_{hv}) / \sqrt{2}$. The polarimetric parameters from the Pauli decomposition are associated for three elementary scattering mechanisms: a stands for single or odd-bounce scattering, b represents double or even-bounce scattering, and c denotes volume scattering. As shown in Figure 22.5b, Pauli RGB composition image can be formed with intensities $|a|^2$(blue), $|b|^2$(red), and $|c|^2$(green), which correspond to clear physical scattering mechanisms. The Pauli RGB composition image has become the standard for PolSAR image display and has often been used for visual interpretation.

With the three elements referred to as the Pauli components of the signal, the 3×3 coherency matrix $\mathbf{T_3}$ is defined as the expected value of $\mathbf{kk}^{*T}$ (Lee and Pottier, 2009):

$$\mathbf{T_3} = \begin{bmatrix} T_{11} & T_{12} & T_{13} \\ T_{12}^* & T_{22} & T_{23} \\ T_{13}^* & T_{23}^* & T_{33} \end{bmatrix}$$

$$= \frac{1}{2} \begin{bmatrix} |S_{hh} + S_{vv}|^2 & (S_{hh} + S_{vv})(S_{hh} - S_{vv})^* & 2(S_{hh} + S_{vv})S_{hv}^* \\ (S_{hh} - S_{vv})(S_{hh} + S_{vv})^* & |S_{hh} - S_{vv}|^2 & 2(S_{hh} - S_{vv})S_{hv}^* \\ 2S_{hv}(S_{hh} + S_{vv})^* & 2S_{hv}(S_{hh} - S_{vv})^* & 4|S_{hv}|^2 \end{bmatrix}$$

$$(22.2)$$

where
    * denotes the conjugate
    | | denotes the module

The covariance matrix $\mathbf{C_3}$ is a close relative of the coherency matrix $\mathbf{T_3}$ (Lee and Pottier, 2009). They contain the same information, but this information comes in different forms. In addition to the Pauli decomposition, many other decomposition methods have been proposed to express the measured backscattering matrix as a combination of the scattering responses of simpler objects or to separate coherency or covariance matrix as the combination of second-order descriptors corresponding to simpler or canonical objects presented as an easier physical interpretation. These widely used decomposition methods are the Huynen (Huynen, 1970), Cloude (Cloude, 1985), Barnes (Barnes, 1988), Holm (Holm and Barnes, 1988), Krogager (Krogager, 1990), Van Zyl (Rignot and Vanzyl, 1993), H/A/Alpha (Cloude and Pottier, 1997), Freeman (Freeman and Durden, 1998), Yamaguchi (Yamaguchi et al., 2005), and Touzi (Touzi, 2007). The RGB composition images that present some of these decompositions are shown in Figure 22.6.

A number of classification methods based on decomposition results have been explored (Cloude and Pottier, 1997; Ferro-Famil et al., 2001; Lee et al., 1999a, 1994; Shimoni et al., 2009). The results of these methods indicated that polarimetric decomposition significantly improved the classification accuracy of PolSAR images. Qi et al. (2012) integrated polarimetric parameters extracted using different polarimetric decomposition methods into the classification of RADARSAT-2 PolSAR image. The results showed that polarimetric parameters were important in identifying different vegetation types and distinguishing between vegetation and urban/built-up areas.

22.4.2.4 Interferometric Features

Most SAR applications make use of the amplitude of the return signal and ignore the phase information. However, SAR interferometry uses the phase of the reflected radiation. It uses two images of the same area taken from the same or slightly different positions and finds the difference in phase between them, producing an image known as an interferogram. The interferogram is measured in radians of phase difference and is recorded as repeating fringes, which each represent a full 2π cycle (Figure 22.7). Coherence is the magnitude of an interferogram's pixels, divided by the product of the magnitudes of the original image's pixels (Figure 22.7). It is usually calculated on a small window of pixels at a time, from the complex interferogram and images. In an interferogram, the coherence serves as a measure of the quality of an interferogram, indicating a tiny and invisible change occurred in the images or the information of the surface type, such as vegetation and rocks. High coherence makes for attractive, not-noisy interferograms, whereas low coherence makes unattractive, noisy interferograms. Substantial improvements in radar image classification can be achieved by integrating interferometric information that is related to the structure and complexity of the observed objects into the classification. Askne et al. (1997) presented a model

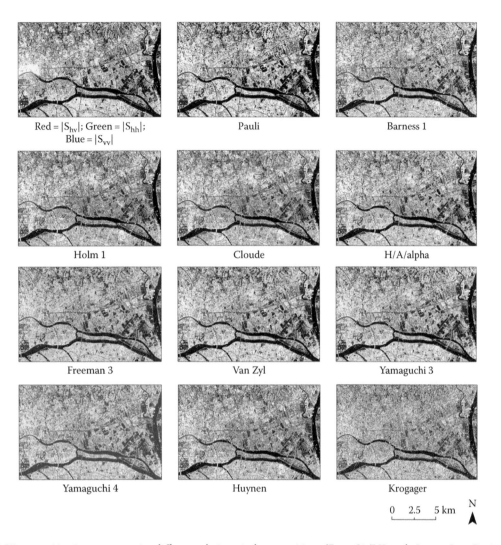

Red = |S$_{hv}$|; Green = |S$_{hh}$|; Blue = |S$_{vv}$|

Pauli

Barness 1

Holm 1

Cloude

H/A/alpha

Freeman 3

Van Zyl

Yamaguchi 3

Yamaguchi 4

Huynen

Krogager

0   2.5   5 km

N

FIGURE 22.6 RGB composition images presenting different polarimetric decompositions. (From Qi, Z.X. et al., *Remote Sens. Environ.*, 118, 21, 2012.)

for interferometric SAR (InSAR) observations of forests. Such observations provide complementary information to the intensity observations and provide new information on coherence and effective interferometric tree height. This study showed the possibility of discriminating forested and nonforested areas using interferometric information. The phase stability of anthropogenic structures between SAR images has led several researchers to propose long timescale phase correlation, or coherence, as a good measure of urban extent, and thus an appropriate tool for mapping urban change (Grey et al., 2003; Qi et al., 2012). The three optimum complex polarimetric interferometric coherences (Papathanassiou and Cloude, 2001) were extracted from two repeat-pass RADARSAT-2 PolSAR images for LULC classification by (Qi et al., 2012). The polarimetric interferometric coherences were found to be very useful in reducing the confusion between urban and nonurban areas (Figure 22.7). As shown in Figure 22.7, there is a strong contrast between urban and nonurban areas in the images of polarimetric interferometric parameters. The repeat cycle of RADARSAT-2 is

24 days, which produces a very strong temporal decorrelation for nonurban areas, such as croplands and natural vegetation. Croplands and natural vegetation are significantly influenced by temporal decorrelation and lose coherence within a few days or weeks as a result of growth, movement of scatterers, and changing moisture conditions. In contrast, within urban/built-up areas, coherence remains high even between image pairs separated by a long time interval.

22.4.2.5 Multifrequency Information

Many studies have been conducted on LULC classification using airborne multifrequency SAR systems or fusing SAR data obtained by different orbital SAR systems with different frequencies (Dobson et al., 1996; Pierce et al., 1994). All these studies indicated that multifrequency SAR data achieves higher accuracy than single-frequency SAR data in the classification. Pierce et al. (1994) performed land cover classification using SIR-C/X-SAR imagery. Using L and C bands alone, the single-scene classification accuracies were quite good, with each image better than 90%. With the addition

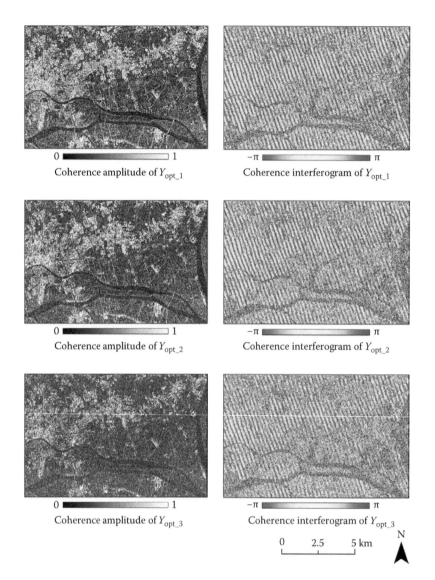

FIGURE 22.7 Polarimetric interferometric parameters extracted using PolSAR interferometry techniques for LULC classification. (From Qi, Z.X. et al., *Remote Sens. Environ.*, 118, 21, 2012.)

of X-band data, the overall accuracies improved to 98%, due to the enhanced ability to distinguish the major tree classes. Dobson et al. (1996) combined C-band ERS- 1 data with L-band JERS-1 data for land cover classification. They found that the results in a classification procedure for the composite image were superior to that obtained from either of the two sensors alone.

22.4.2.6 Multiple Incidence Angle Information

Since radar backscatter from targets is affected by the incidence angle, observing targets with different incidence angle modes can provide more information to LULC classification. Cimino et al. (1986) used multiple incidence angle SIR-B data to discriminate various forest types by their relative brightness versus incidence angle signatures. The results of this study indicated that (1) different forest species, and structures of a single species, may be discriminated using multiple incidence angle radar

imagery and (2) it is essential to consider the variation in backscatter due to incidence angle when analyzing and comparing data collected varying frequencies and polarizations. Grunsky (2002) investigated the potential of the use of multibeam RADARSAT-1 radar imagery in assisting terrain mapping over large areas. Principal components analysis was applied to ascending and descending standard beam modes with incidence angles of 20°–27° (S1) and 45°–49° (S7). The resulting components yielded imagery that highlights geomorphology, geologic structure, variation in vegetation, and an indirect measure of moisture in the study area.

22.4.2.7 Feature Selection

Given that many potential features could be integrated into the classification of radar images, feature selection presents a problem in the classification. Using all available features in classification is improper because computation is intensive and some features may degrade classification performance (Laliberte et al., 2006). Since the recent introduction of polarimetric decomposition theorems, which have brought about abundant polarimetric parameters, the problems of feature selection have become more intractable. Therefore, the selection of suitable features is essential for successfully implementing radar image classification. Many approaches, such as principal component analysis, minimum noise fraction transform, discriminant analysis, decision boundary feature extraction, nonparametric weighted feature extraction, wavelet transform, and spectral mixture analysis, may be used for feature extraction, in order to reduce the data redundancy inherent in remotely sensed data or to extract specific land cover information. Most of these feature selection approaches can be found in the Weka software (Witten et al., 2011).

22.4.3 Selection of Classifiers

After the determination of features used for classification, the next step is to select a suitable classifier to implement the classification based on the selected features. A number of classifiers have been developed for the classification of remote-sensing data, and each classifier has its own strengths and limitations (Lu and Weng, 2007). Many classifiers have been tested with SAR imagery, and these classifiers can be grouped as parametric and nonparametric classifiers.

22.4.3.1 Parametric Classifier

The parametric classifiers assume that the classification dataset follows a specific distribution such as Gaussian distribution and that the statistical parameters (e.g., mean vector and covariance matrix) generated from the training samples are representative. Maximum likelihood (ML) classifier is one of the most widely used parametric classifiers with radar images. Lim et al. (1989) introduced an ML decision rule based on the multivariate complex Gaussian distribution of the elements of the coherent scattering matrix. In order to reduce the effects of speckle in PolSAR images, data are generally processed through incoherent averaging and are represented by coherency matrices. Lee et al. (1994) developed a Wishart classifier by introducing the ML decision rule based on the multivariate complex Wishart distribution for the polarimetric coherency matrix. A k-mean algorithm was applied to iteratively assign the pixels of the PolSAR image to one of the classes using the ML rule. Lee et al. (1999a) proposed an unsupervised classification by combining the decomposition technique with unsupervised classification based on the Wishart classifier. The H/A/Alpha decomposition technique was used to provide an initial guess of the pixel distribution into the classes that produces a better convergence of the unsupervised classification algorithm. This unsupervised classification method has been extended to the classification of multifrequency PolSAR data (Ferro-Famil et al., 2001).

Other commonly used parametric classifiers include MAP, Bayesian, iterated conditional mode (ICM) contextual classifiers, fuzzy c-means clustering, and Markov random field (MRF) models. Rignot et al. (1992) implemented unsupervised

classification of PolSAR data by applying MAP to the covariance matrix. This method used both polarimetric amplitude and phase information, was adapted to the presence of image speckle, and did not require an arbitrary weighing of the different polarimetric channels. Vanzyl and Burnette (1992) classified PolSAR images using a Bayesian classifier in which the classification was done iteratively. The results showed that only a few iterations were necessary to improve the classification accuracy dramatically. A hierarchical Bayesian classifier was developed for the classification of short vegetation using multifrequency PolSAR data (Kouskoulas et al., 2004). It was shown that this classifier outperformed ML classifier with Gaussian assumption. Freitas et al. (2008) applied the ICM classifier to the classification of P-band PolSAR data. The ICM classifier enabled the use of contextual information to improve the classification accuracy. Based on the complex Wishart distribution of the complex covariance matrix, the fuzzy c-means clustering algorithm was used for unsupervised segmentation of multilook PolSAR images (Du and Lee, 1996). Dong et al. (2001) implemented segmentation and classification of PolSAR data using a Gaussian MRF model. The model performed the classification based on image objects, which usually consists of multipixels, providing reliable measurement statistics and texture characteristics. Tison et al. (2004) proposed a classification method that uses mathematical model relying on the Fisher distribution and the logmoment estimation for high-resolution SAR images over urban areas. Their contribution was the choice of an accurate model for high-resolution SAR images over urban areas and its use in a Markovian classification algorithm.

22.4.3.2 Nonparametric Classifiers

Nonparametric classifiers do not assume a particular probability density distribution of the input data, and no statistical parameters are needed to separate images. These classifiers are thus especially suitable for the incorporation of nonintensity data into the classification of SAR images. Among the most commonly used nonparametric classifiers for radar images are neural networks (Bruzzone et al., 2004; Hara et al., 1994; Tzeng and Chen, 1998), support vector machines (Hosseini et al., 2011; Lardeux et al., 2009; Tan et al., 2011), decision trees (Qi et al., 2012; Simard et al., 2000), and knowledge-based classifiers (Dobson et al., 1996; Pierce et al., 1994). There are several advantages of using nonparametric classifiers with SAR images. First, nonparametric classifiers may provide better classification accuracies than parametric classifiers in complex landscapes (Hosseini et al., 2011; Lardeux et al., 2009). Second, nonparametric classifiers are easy to integrate different types of data (e.g., textural, spatial, polarimetric, interferometric, GIS ancillary data) into the classification of radar images because of their nonparametric nature (Hosseini et al., 2011; Simard et al., 2000; Tan et al., 2011). Third, they are easy to be used with multiple (e.g., multitemporal) images (Bruzzone et al., 2004; Qi et al., 2012). Furthermore, each of these classifiers has its own strengths. The neutral networks have arbitrary decision boundary capability and provide fuzzy output values to take into account class mixture and the degree of membership of a pixel (Tzeng and Chen, 1998). The advantage of SVMs for data classification is their ability to be used as an efficient algorithm for nonlinear classification problems, particularly in the case of extracting feature vectors from fully PolSAR data (Hosseini et al., 2011). Decision tree algorithms are efficient in selecting features and implementing classification as well as provide clear classification rules that can be easily interpreted based on the physical meaning of the features used in the classification (Qi et al., 2012). This is very helpful in providing physical insight for the classification of radar images. Knowledge-based classifiers are transportable and robust because they are defined using theoretical understanding, as verified by empirical evidence, of the knowledge of the physics involved in the sensor/scene interaction (Dobson et al., 1996; Pierce et al., 1994). There are also some other nonpolarimetric classifiers, such as sigma-tree structured near-neighbor classifiers (Barnes and Burki, 2006), spectral graph partitioning (Ersahin et al., 2010), and subspace (Bagan et al., 2012). Readers who want to have a detailed description of a specific classification approach should refer to cited references.

22.4.3.3 Object-Oriented Classification

Conventional classification of using remote-sensing images is often based on pixel-based methods. When applied to radar images, pixel-based methods have two disadvantages. First, they are prone to speckle noise and could produce false alarms. Second, pixel-based methods are difficult to use in the extraction of spatial and textural information that is helpful in improving the classification accuracy of remote-sensing images (Myint et al., 2011). Object-oriented image analysis (OOIA) has been increasingly used for the classification of remote-sensing data (Benz et al., 2004). It enables the acquisition of a variety of textural and spatial features for improving the accuracy of remote-sensing classification by delineating objects from remote-sensing images. On the other hand, image objects are much easier to manipulate and utilize than pure pixels. eCognition is so far the most commonly used object-oriented classification package (Benz et al., 2004). Object-oriented radar image classification has also been carried out by many studies (Li et al., 2009; Liu et al., 2008; Qi et al., 2012). All these studies indicated that OOIA provides better results than pixel-based approaches and that textural information in PolSAR images is helpful in enhancing the classification accuracy. Furthermore, OOIA is helpful in improving the accuracy of radar image classification by reducing the speckle effect (Qi et al., 2012).

22.5 Change Detection Methods for Radar Imagery

Many change detection methods have been developed for the use of remote-sensing data (Lu et al., 2004). However, there is still a general lack of the use of radar remote sensing for LULC change detection. Indeed, studies related to radar imagery change detection are fewer and more recent than optical-based ones. Change detection methods for radar imagery can be divided into two categories: (1) unsupervised change detection methods and (2) methods combining unsupervised change detection with postclassification comparison (PCC).

22.5.1 Unsupervised Change Detection Methods

Unsupervised change detection methods for radar images are usually made of three steps: (1) image preprocessing, (2) comparing two images to generate a change magnitude map (e.g., Figure 22.8), and (3) applying threshold methods to the change magnitude map to separate *change* from *no change*. Image ratioing, statistical measure, and image differencing are the most widely utilized unsupervised change detection methods for radar imagery.

22.5.1.1 Image Ratioing

Image ratioing has been widely used in change detection with single-channel (i.e., single frequency and single polarization) radar images because it can effectively reduce multiplicative noise in the images (Rignot and Vanzyl, 1993). The comparison of SAR images can be carried out according to a ratio/logarithmic ratio (log-ratio) operator. An optimal threshold value is then applied to the ratio/log-ratio image (change magnitude image) to identify changes. "Trial-and-error" procedures are typically used to determine optimal threshold value (Dierking and Skriver, 2002; Singh, 1989). However, such manual operations typically turn out to be time consuming; in addition, the quality of their results critically depends on the visual interpretation of the user. To overcome this problem, many automatic thresholding algorithms have been proposed for the analysis of ratio/log-ratio images. Bazi et al. (2005) introduced a generalized Gaussian model for a log-ratio images and applied Kittler and Illingworth (K&I) thresholding algorithm to the log-ratio image to automatically detect changes. The modified K&I criterion was derived under the generalized Gaussian assumption for modeling the distributions of changed and unchanged classes. This parametric model was chosen because it is capable of better fitting the conditional densities of classes in the log-ratio image. However, in this method, changes are assumed to be on one side of the histogram of the log-ratio image, which is not true for all change detection problems. In particular, changes may be present on both sides of the histogram of the log-ratio image. Therefore, the method was further improved by combining generalized Gaussian distributions with a multiple-threshold version of K&I to detect changes on both sides of the histogram of log-ratio images (Bazi et al., 2006). Additionally, Moser and Serpico (2006) developed a K&I minimum-error thresholding algorithm to take into account the non-Gaussian distribution of the amplitude values of SAR images. This method could be applied to images acquired by two distinct sensors with different bands and polarizations. Carincotte et al. (2006) calculated change magnitude images using log-ratio detector and used a fuzzy version of hidden Markov chains to classify the log-ratio images into change and no-change classes. This method took

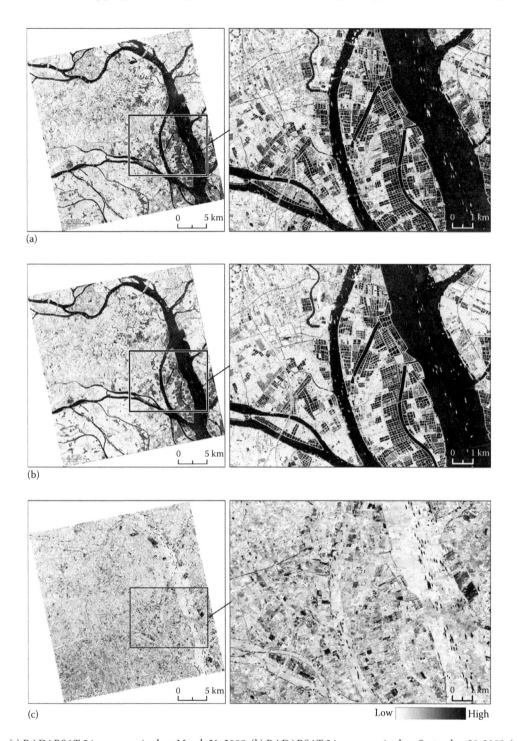

FIGURE 22.8   (a) RADARSAT-2 image acquired on March 21, 2009. (b) RADARSAT-2 image acquired on September 29, 2009. (c) Change magnitude calculated using change vector analysis. (From Qi and Yeh 2013.)

into account the fuzzy aspect of the scene behavior and change detection complexity and reached a satisfactory reliability level in the context of SAR images.

Many advanced methods have also been developed based on image ratioing to address different issues in change detection with radar images. Bujor et al. (2004) developed a method that applies a log-cumulants to the detection of spatiotemporal discontinuities in multitemporal SAR images. The contrast and

the heterogeneity information were extracted by a *multitemporal* application of the ratio of local means and by new 3D texture parameters based on the log-cumulants. After that, the resulting attributes that measure the time variability or the presence of spatial features were merged. An interactive fuzzy fusion approach was proposed to provide end users with a simple and easily understandable tool for tuning the change detection results. This change detection method could enable geophysicists to detect regions that contain spatial features (roads, rivers, etc.) or temporal change (flooded areas, coastline erosion, etc.). Bovolo and Bruzzone (2005) developed an approach that exploits a wavelet-based multiscale decomposition of the log-ratio image (obtained by a comparison of the original multitemporal SAR images) aimed at achieving different scales (levels) of representation of the change signal. Each scale was characterized by a different trade-off between speckle reduction and preservation of geometrical details. For each pixel, a subset of reliable scales was identified on the basis of a local statistic measure applied to scale-dependent log-ratio images. The final change detection result was obtained according to an adaptive scale-driven fusion algorithm. This method could improve the accuracy and geometric fidelity of the change detection map. Gamba et al. (2006) jointly used feature-based and pixel-based techniques to address the problem of change detection from SAR images. Image ratioing was used for deriving the first rough change map at the pixel level, and then linear features were extracted from multiple SAR images and compared to confirm pixel-based changes. This method proved to be effective in dealing with misregistration errors caused by reprojection problems or difference in the sensor's viewing geometry, which are common in multitemporal SAR images. Bovolo and Bruzzone (2007) proposed a split-based approach (SBA) to automatic and unsupervised change detection in large-size multitemporal SAR images. The method consisted of three steps: (1) a split of the computed ratio image into subimages, (2) an adaptive analysis of each subimage, and (3) an automatic split-based threshold-selection procedure. The SBA could detect changes in a consistent and reliable way in images of large size also when the extension of the changed area is small. Thus, it could be suitable for defining a system for damage assessment in multitemporal SAR images.

Single-channel SAR images may result in poor discrimination between changed and unchanged areas because of the limited spectral and polarization information. Compared with single-channel SAR, multichannel (e.g., multipolarization and/or multifrequency) SAR presents a great potential and is expected to provide an increased discrimination capability while maintaining the insensitivity to atmospheric and Sun-illumination conditions. Change detection methods based on image ratioing have also been developed for change detection with multichannel SAR images. Moser et al. (2007) combined the Landgrebe and Jackson expectation–maximization (LJ-EM) algorithm with a SAR-specific version of the Fisher transform to iteratively compute a scalar transformation of the multichannel ratio image that optimally discriminates *change* from *no change*. An MRF approach was also integrated into the method to take advantage of the contextual information, which is crucial to reduce the impact of speckle on the change detection results. This method can be used to generate change maps from multichannel SAR images acquired over the same geographic region in different polarizations or at different frequencies at different times. However, its main drawback is that, even though a convergent behavior is experimentally observed, no theoretical

proof of convergence is available yet. Moser and Serpico (2009) proposed an unsupervised automatic contextual change detection method for multichannel amplitude SAR images based on image ratioing and MRFs. Each channel of the SAR amplitude ratio image was considered as a separate "information source," and an additional source was derived from the spatial context; the multisource data-fusion task was addressed together with the image thresholding task using an MRF-based approach. In order to estimate model parameters, a case-specific novel formulation of LJ-EM was developed and combined with method of log-cumulants. This choice also overcame the convergence drawback of the approach proposed by Moser et al. (2007) because of the robust analytical properties of EM-based estimation procedures.

22.5.1.2 Statistical Measure

Inglada and Mercier (2007) developed a statistical similarity measure for change detection in multitemporal SAR images. This measure was based on the evolution of the local statistics of the image between two dates. The local statistics were estimated by using a cumulant-based series expansion, which approximated probability density functions in the neighborhood of each pixel in the image. The degree of evolution of the local statistics was measured using the Kullback–Leibler divergence. An analytical expression for this detector was given, allowing a simple computation that depends on the four first statistical moments of the pixels inside the analysis window only. This proposed change detector outperformed the classical mean ratio detector, and the fast computation of this detector allowed a multiscale approach in the change detection for operational use. The so-called multiscale change profile was introduced to yield change information on a wide range of scales and to better characterize the appropriate scale.

Chatelain et al. (2008) studied a new family of distributions constructed from multivariate gamma distributions to model the statistical properties of multisensory SAR images. These distributions referred to as multisensory multivariate gamma distributions (MuMGDs) were potentially interesting for detecting changes in SAR images acquired by different sensors having different numbers of looks. This study compared different estimators for the parameters of MuMGDs. These estimators were based on the ML principle, the method of inference function for margins, and the method of moments. The estimated correlation coefficient of MuMGDs showed interesting properties for detecting changes in radar images with different numbers of looks.

When working with multilook fully PolSAR data, an appropriate way of representing the backscattered signal consists of the covariance matrix. For each pixel, this is a 3×3 Hermitian positive definite matrix that follows a complex Wishart distribution. Based on this distribution, Conradsen et al. (2003) proposed a test statistic for equality of two such matrices, and an associated asymptotic probability for obtaining a smaller value of the test statistic was derived and applied successfully to change detection in PolSAR data. If used with HH, VV, or HV data only, the test statistic reduced to the well-known test statistic for equality of the scale parameters in two gamma distributions. The derived test statistic and the associated significance measure could be applied as a line or edge detector in fully PolSAR data. The new polarimetric test statistic was much more sensitive to the differences than test statistics based only on the backscatter coefficients.

22.5.1.3 Image Differencing

Image differencing is one of the most widely used unsupervised change detection methods for the use of optical remote-sensing images. It has also been used for change detection with single-channel SAR images (Cihlar et al., 1992; Villasenor et al., 1993). The procedure of image differencing is similar to that of image ratioing, but the change magnitude in image differencing methods is obtained by subtracting amplitude or intensity images of two SAR images acquired at different times. A widely accepted assumption is that the distribution of pixels in the change and no-change areas in the difference image can be approximated as a mixture of Gaussian distributions (Camps-Valls et al., 2008). Thus, the EM algorithm is commonly used on change magnitude to detect changes because it finds clusters by determining a mixture of Gaussians that fit a given dataset (Bruzzone and Prieto, 2000). In addition to using radar amplitude images, Grey et al (2003) applied image differencing technique to SAR interferometric coherence data to map urban change.

Celik and Ma (2010) proposed an unsupervised change detection algorithm for satellite images by conducting probabilistic Bayesian inferencing to perform unsupervised thresholding over subband difference image generated at the various scales and directional subbands using the dual-tree complex wavelet transform (DT-CWT) for representation. Aside from the intrascale information, the interscale information inherently provided by the DT-CWT was exploited to effectively reduce both the false-alarm and miss-detection rates. Extensive simulation results showed that this algorithm consistently performed quite well on both objective and subjective change detection performance evaluation—under either noise-free or zero-mean Gaussian (or speckle) noise interference cases. Furthermore, the correct- and false-classification rates were almost invariant (insensitive) with respect to noise powers added to the images.

22.5.1.4 Other Methods

Other approaches have also been proposed in the literature to deal with radar change detection, including multitemporal coherence analysis (Rignot and Vanzyl, 1993), integration of segmentation with multilayer perceptron and Kohonen neural networks (White, 1991), case-based reasoning (Li and Yeh, 2004), ML approach (Lombardo and Pellizzeri, 2002), radon transform and Jeffrey divergence (Zheng and You, 2013), and graph-cut and generalized Gaussian model (Zhang et al., 2013).

22.5.2 Combing Unsupervised Change Detection and Postclassification Comparison

Although unsupervised change detection approaches are relatively simple, straightforward, and easy to implement and interpret, they cannot determine types of land cover change. PCC can provide information on both the changed areas and the type of land cover change in these areas because it performs change detection by comparing separate supervised classifications of images acquired at different times (Lu et al., 2004). However, PCC is limited by the accuracies of the classification. PCC results exhibit accuracies similar to the product of the accuracies of each individual classification (Stow et al., 1980). Most orbital SAR systems are single-frequency types, and SAR images suffer from serious speckle noise caused by the SAR system's coherent nature. When applied to SAR images, PCC may yield poor results because of the poor classification accuracies caused by the speckle noise and limited spectral information. Furthermore, change detection and classification of using radar images are often based on pixel-based methods, which are prone to speckle noise and difficult to use in the extraction of spatial and textural information.

A method that integrates change vector analysis (CVA) and PCC with OOIA has been developed to detect land cover changes from two repeat-pass PolSAR images (Qi and Yeh, 2013). OOIA allows for land cover change detection performed on image objects and the incorporation of textural and spatial information of the image objects. It was integrated into the method to suppress the speckle effect and extract textural and spatial information to support PolSAR image classification. In this method, two PolSAR images acquired over the same area at different times were segmented hierarchically to delineate land parcels (image objects) (Figure 22.9). The hierarchical segmentation of PolSAR images not only avoided inconsistencies in the delineation of land parcels but also delineated the changed land parcels. As shown in Figure 22.9, the changed land parcels, such as land parcels a and b_2, were delineated in the hierarchical segmentation. Afterwards, CVA was combined with PCC to detect land cover changes based on OOIA. Parcel-based CVA was performed with the features extracted from the coherency matrices of the PolSAR images to calculate the change magnitude map (Figure 22.8). The EM algorithm was applied to the change magnitude map to identify the changed land parcels. PCC that was based on a parcel-based classification approach, which integrated polarimetric decomposition, decision tree algorithms, and support vector machines, was then used to determine the type of land cover change for the changed land parcels. The combination of CVA and PCC detected different types of land cover change and also reduced the effect of the classification errors on the land cover change detection. The main advantage of this method is that it provides information on both the changed areas and the type of land cover change in these areas. Compared with conventional

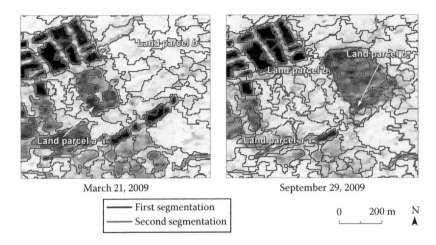

FIGURE 22.9   Hierarchical segmentation of RADARSAT-2 PolSAR images acquired at different times. (From Qi and Yeh 2013.)

PCC that was based on the Wishart-supervised classification, this change detection method significantly reduced overall error and false-alarm rates.

22.6 Applications of Radar Imagery in LULC Mapping and Monitoring

The major applications of radar remote sensing in LULC mapping and monitoring include LULC classification and change detection, forestry inventory and mapping, crop and vegetation identification, application on urban environment, snow and ice mapping, application to wetlands, and shoreline change detection.

22.6.1 LULC Classification and Change Detection

Many studies have reported that radar remotely sensed images can provide valuable information for timely LULC classification and change detection (Table 22.8). Peng et al. (2005) found that stereo RADARSAT-1 SAR images provided valuable data sources for land cover mapping, especially in mountainous areas where cloud cover is a problem for optical data collection and topographical data are not always available. The joint use of InSAR and textural features extracted from SAR images and its application to LULC classification has been assessed by many studies. Strozzi et al. (2000) produced two land use maps and a forest map of three different areas in Europe by using ERS-1/2 SAR interferometry. The three areas represented various geomorphological regions with different cover types. Their classification results showed that land use classification accuracies on the order of 75% are possible with, in the best case, simultaneous forest and nonforest accuracies of around 80%–85%. Qi et al. (2012) implemented land cover classification by using textural, polarimetric, and interferometric information extracted from RADARSAT-2 PolSAR images. The overall accuracy of 86.64% was achieved for classifying land cover types including built-up areas, water, barren land, crop/natural vegetation, lawn, banana, and forests.

If single-frequency SAR systems are used, there is generally a considerable degree of ambiguity between different LULC types (Li et al., 2012). Combining multifrequency SAR scenes has proved to be a valuable tool for distinguishing different LULC classes. Shimoni et al. (2009) investigated the complementarity and fusion of different frequencies (L and P band), PolSAR and polarimetric interferometric (PolInSAR) data obtained by Experimental airborne SAR (E-SAR) for land cover classification. The results showed that the overall accuracy for each of the fused sets was better than the accuracy for the separate feature sets. Moreover, fused features from different SAR frequencies were complementary and adequate for land cover classification and that PolInSAR was complementary to PolSAR information and that both were essential for producing accurate land cover classification. Evans and Costa (2013) used multitemporal L-band ALOS/PALSAR, C-band RADARSAT-2, and Envisat/ASAR data to map ecosystems and create a lake distribution map of the lower Nhecolandia subregion in the Brazilian Pantanal. They provided the first fine spatial-resolution classification showing the spatial distribution of terrestrial and aquatic habitats for the entire subregion of lower Nhecolandia. Holecz et al. (2009) generated land cover maps and changes over large areas by fusing single-date ALOS PALSAR fine/dual beam with multitemporal Envisat ASAR mode/alternating polarization images. The results clearly demonstrated that the synergetic use enabled the reliable identification of key land cover types (in particular cropped areas, bare soil areas, forestry, forest clear cut, forest burnt areas, water bodies) and their evolution over time.

The use of multitemporal SAR data can also increase the number of reliably distinguishable LULC classes. Land cover classification was conducted by using a time series of 14 ERS-1/2 SAR tandem image pairs (Engdahl and Hyyppa, 2003). A total of 14 tandem coherence images and two coherence images with a longer temporal baseline (36 and 246 days) were used in the classification. The overall accuracy for six classes, field/open land, dense forest, sparse forest, mixed urban, dense urban, and water, was found to be 90% with kappa coefficient of 0.86. Huang et al. (2008) found that multitemporal ERS-2 SAR imagery had

TABLE 22.8 Studies on the Application of Radar Imagery in LULC Classification and Change Detection

LULC Types	Data	Accuracy	Strengths	Limitations	Reference
Bare soil, natural forest, pasture, planted forest, sugar-cane plantations, soya plantations, urban or built-up areas, and water	ERS-1 SAR and Landsat-5 TM images	94.85%	Compared with using TM data alone, combining TM with SAR increases the classification accuracy for urban, pasture and forest	Both SAR and TM images are needed	Kuplich et al. (2000)
Urban areas, water, forest, and open land	ERS-1/2 InSAR data	75%	Interferometric information is used in LULC classification	Meteorological condition effect	Strozzi et al. (2000)
Field/open land, dense forest, sparse forest, mixed urban, dense urban, and water	ERS-1/2 InSAR data	90%	Interferometric coherence carries more land cover related information than the backscattered intensity	A large number of tandem pairs are needed	Engdahl and Hyyppa (2003)
Forest, glacial ice, grass, rock, sandy soil, shrub, snow, and water	A stereo pair of RADARSAT-1 images	83%	Suitable for land cover mapping in mountainous areas	Feature extraction and selection needs to be optimized	Peng et al. (2005)
Upland forest, lowland forest, advanced successional vegetation, intermediate successional vegetation, initial successional vegetation, degraded pasture, cultivated pasture, agroforestry, coffee plantation, infrastructure, water, and non-vegetation lowland	Landsat ETM+ and RADARSAT-1 data	72.07%	Incorporation of data fusion and textures increases classification accuracy by approximately 5.8%–6.9% compared to Landsat ETM+ data	Both SAR and ETM+ images are needed	Lu et al. (2007)
Residential, road and dike, cotton field, paddy field, mixed field, orchard, forest land, river and gulf, canal and pond, aqua-farm ponds, tideland and wild land, and nonvegetated saline soil land	Multitemporal ERS-2 SAR imagery	77.34%	Multitemporal SAR imagery has great potential for LULC mapping or resource investigations in coastal zones under rapid development	The best time window for image acquisition should be considered	Huang et al. (2008)
Built-up areas, rural residential areas, bare land, paddy fields, vegetable land, orchards, forest, river, and fishponds	RADARSAT-1 images	75.2%	SAR provides a unique opportunity for detecting LULC changes within short intervals (e.g., monthly) in tropical and sub-tropical regions	Monthly SAR images are needed	Li et al. (2009)
Residences, roads, forests, wheat and corn fields, pastures, abandoned areas, bare soil, and rivers	E-SAR L- and P-band	75.36%	Fused features from different SAR frequencies are complementary and adequate for LULC classification	Results could be further improved by introducing spatial information into the fusion.	Shimoni et al. (2009)
Forest, succession, agro-pasture, water, wetland, and urban	ALOS PALSAR or RADARSAT-2 data	72.2% (L-band); 54.7% (P-band)	L-band data provides much better land cover classification than C-band data	LULC classification with either L- or C-band is a challenge for fine LULC classification system	Li et al. (2012)
Built-up, water, barren land, forest, lawn, banana, and cropland/natural vegetation	RADARSAT-2 data	86.64%	Incorporation of textural, polarimetric, and interferometric features into the classification	PolSAR images segmentation remains a challenge	Qi et al. (2012)
Forests, paddy fields, croplands, lotus fields, grasslands, golf courses, parks, settlements, and water	ALOS AVNIR-2 and PALSAR images	90.34%	Combining AVNIR-2 and PALSAR data produce better accuracy	Both AVNIR-2 and PALSAR data are needed	Bagan et al. (2012)
Forest woodland, open wood savanna, open grass savanna, swampy grassland, agriculture, vazantes, fresh water lake, and brackish lake	ALOS PALSAR, RADARSAT-2, and ENVISAT ASAR data	83%	Combining dual-season, C and L-band, is essential for providing a relatively high overall accuracy of land cover classification	SAR data of different frequencies are needed	Evans and Costa (2013)

great potential for land use mapping or resource investigations in coastal zones under rapid development. They stated that multitemporal SAR data can be regarded as the first choice for monitoring land uses and their dynamic changes in coastal zones that are often affected by heavy cloud or rainy weather.

SAR data and optical data provide complementary information, and their combination often leads to increased classification accuracy. The use of Landsat MSS and Seasat SAR data was evaluated in discriminating suburban and regional cover in the eastern fringe area of the Denver, Colorado, metropolitan area (Toll, 1985). The Seasat SAR data provided a measure of surface geometry that complemented the reflective characteristics of Landsat MSS visible and near-infrared data. The integration of Landsat imagery with SAR data obtained by ERS-1, RADARSAT-1, and ALOS PALSAR for LULC classification have been carried out by many studies (Kuplich et al., 2000; Larranaga et al., 2011; Lu et al., 2007). All these studies reported a significant improvement achieved by the integration in LULC classification. Bagan et al. (2012) evaluated the potential of combined ALOS visible and near-infrared (AVNIR-2) with PALSAR fully PolSAR data for land cover classification. They confirmed that, when the combined optical AVNIR-2, PALSAR, and polarimetric coherency matrix data were used, the classification accuracy of was better than that when other data combinations were used.

SAR data are also useful in the detection of LULC changes. Villasenor et al. (1993) found that the temporal changes between repeat-pass ERS-1 SAR images of the North Slope of Alaska were largely due to changes in soil and vegetation liquid water content induced by freeze/thaw events. This confirmed the viability of radar backscatter intensity comparisons using repeat-pass images as a means of change detection. Orbital SAR data have proved to be a unique opportunity for detecting land use changes within short intervals in tropical and subtropical regions with cloud cover (Li et al., 2009). By using object-oriented analysis with case-based reasoning, Li et al. (2009) successfully detected land use changes at monthly intervals by using multitemporal RADARSAT-1 SAR images.

22.6.2 Forestry Inventory and Mapping

Radar remote sensing can contribute to the inventory and mapping of forest as well as to an understanding of ecosystem process (Table 22.9). Early research to apply imaging radar to forest mapping was conducted using multiple incident angle SIR-B data (Cimino et al., 1986). The research found that different forest species might be discriminated using multiple incidence angle radar imagery and the variation in backscatter due to incidence angle should be considered when analyzing and comparing data collected at varying frequencies and polarizations. After the availability of some orbital radar systems, such as the ERS-1/2, JERS-1, and RADARSAT-2, SAR images obtained by these systems have been widely used in forestry inventory and mapping. The use of ERS-1/JERS-1 SAR composites was shown to be very promising for forest mapping, and the textural information

of ERS-1 and JERS-1 SAR images significantly improved the classification accuracies (Kurvonen and Hallikainen, 1999; Solaiman et al., 1999). Furthermore, interferometric coherence maps derived from ERS-1 and ERS-2 SAR images and from JERS-1 SAR images were found to be an important source of information for biophysical characteristics in regenerating and undisturbed areas of a forest (Luckman et al., 2000).

The capability of PolSAR imagery in forestry inventory and mapping has been investigated by using SIR-C/X-SAR, Envisat ASAR, ALOS PALSAR, and RADARSAT-2. SIR-C imagery was used in mapping land cover types and monitoring deforestation in Amazon rainforest (Saatchi et al., 1997). The SIR-C data delineated five classes including primary forest, secondary forest, pasture crops, quebrada, and disturbed forest with approximately 72% accuracy. The comparison of SIR-C data acquired in April (wet period) and October (dry period) indicated that multitemporal data could be used for monitoring deforestation. Chand and Badarinath (2007) analyzed the capability of Envisat ASAR C-band data in forest parameter retrieval and forest-type classification over deciduous forests. They found a significant correlation between SAR backscatter and biometric parameters, and backscatter values typically increased with increase in basal area, volume, stem density, and dominant height. Santoro et al. (2007) found that the high-coherence difference between forests and bare fields suggested the possibility to use the Envisat coherence for forest/nonforest mapping and estimation of biophysical properties of short vegetation. The ability of PALSAR data in supporting forestry mapping was assessed comprehensively (Longepe et al., 2011; Walker et al., 2010). The assessments confirmed PALSAR data as an accurate source for spatially explicit estimates of forest cover. Liesenberg and Gloaguen (2013) evaluated the backscattering intensity, polarimetric features, interferometric coherence, and texture parameters extracted from PALSAR imagery for forest classification. It was found that forest classes were characterized by low temporal backscattering intensity variability, low coherence, and high entropy and that overall accuracies were affected by precipitation events on the date and prior SAR date acquisition. Polarimetric features extracted from quad-polarization L band increased classification accuracies when compared to single and dual polarization alone. Polychronaki et al. (2013) found that PALSAR could be applied for rapid burned area assessment, especially to areas where cloud cover and fire smoke inhibit accurate mapping of burned areas when optical data are used.

The usefulness of airborne NASA/Jet Propulsion Laboratory (JPL) Airborne Synthetic Aperture Radar (AIRSAR) multifrequency temporal PolSAR data was examined for identifying forest (Ranson and Sun, 1994). With principal component analysis of temporal datasets (winter and late summer), the SAR images were classified into general forest categories such as softwood, hardwood, regeneration, and clearing with better than 80% accuracy. However, classifications from single-date images suffered in accuracy. The winter image had significant confusion of softwoods and hardwoods with a strong tendency to overestimate hardwoods. The increased double-bounce

TABLE 22.9　Studies on the Applications of Radar Imagery in Forestry Inventory and Mapping

Purpose	Data	Accuracy	Strengths	Limitations	Reference
Forest type classification (softwood, hardwood, regeneration, and clearing)	AIR SAR	>80%	Temporal (winter and late summer) SAR images are suitable for forest type classification	Winter images have significant confusion of softwoods and hardwoods with a strong tendency to overestimate hardwoods	Ranson and Sun (1994)
Forest type classification (primary forest, secondary forest, pasture-crops, quebradao, and disturbed forest)	SIR-C/X-SAR	72%	Multitemporal data can be used for monitoring deforestation	Data acquired during the wet season are not suitable for accurate land cover classification	Saatchi et al. (1997)
Forest type classification (coniferous, mixed forest, deciduous, and mire)	ERS-1 and JERS-1 SAR images	66%	The textural information of a multitemporal set of ERS-1 and JERS-1 SAR images has a higher information value for the forest type classification than the SAR image intensity	Weather and seasonal conditions have a significant effect on the textural information of SAR images	Kurvonen and Hallikainen (1999)
Measurements of disturbed tropical forest	ERS-1 and JERS-1 SAR images	79.2%	Coherence from both ERS tandem acquisitions and JERS data is useful for differentiating between forest and nonforest and may include useful information both on the density of regenerating forest and the characteristics of mature forest	Time delay between the ground data campaign and ERS data acquisition may have influenced the result	Luckman et al. (2000)
Forest type classification (xylia dominated, teak mixed, settlements, degraded forest, water bodies, fallow/barren, agriculture, mixed forest, riverine forest); forest parameter (stem density, basal area, and dominant height) retrieval	ENVISAT ASAR	89.2%	Seasonal data of ENVISAT ASAR improves the mapping accuracy; A reasonable correlation of backscatter values derived from ASAR with plot-level biometric parameters	SAR data due to layover and foreshortening effects limits the data utilization	Chand and Badarinath (2007)
Forest mapping (forest and nonforest)	ALOS PALSAR	92.4%	Confirming the ability of modern imaging radar in providing for accurate and timely wall-to-wall mapping and monitoring of forest cover	Fusion of multisensor and multi-temporal radar and optical data is expected to provide better results	Walker et al. (2010)
Forest mapping (natural forests and other land cover types)	ALOS PALSAR	86%	Confirming the high potential of PALSAR sensor for forest monitoring at regional level	The classification accuracy will likely increase if multitemporal PALSAR acquisitions are integrated	Longepe et al. (2011)
Forest type classification (primary forest, riparian forest, advanced secondary forest, intermediate secondary forest, initial secondary forest, water, and pasture)	PALSAR and Landsat TM	85.5%	Forest classes are characterized by low temporal backscattering intensity variability, low coherence and high entropy	Incidence angle and precipitation events on the date and prior data acquisition should be taken into account in mapping	Liesenberg and Gloaguen (2013)

scattering of the radar beam from conifer stands because of lowered dielectric constant of frozen needles and branches was the contributing factor for the misclassifications.

22.6.3 Crop and Vegetation Identification

Research in the use of imaging radar for investigating LULC has also been closely related to crop and vegetation identification (Table 22.10). Early studies indicated that radar backscatter was significantly affected by the effect of soil moisture, surface roughness, and vegetation cover (Wang et al., 1986). Freeman et al. (1994) found that multifrequency, polarimetric radar backscatter signatures extracted from calibrated and noise-corrected NASA/JPL AIRSAR data were useful in classifying several different ground cover types in agricultural areas. JERS-1 SAR data were used for separating basic land cover categories such as savannas, forests, and flooded vegetation by Simard et al. (2000). The textural information extracted from the JERS-1 imagery was found to be particularly useful for refining flooded vegetation classes. McNairn et al. (2009) compared the capability of L-band PALSAR PolSAR, C-band Envisat ASAR, and RADARSAT-1 SAR data for crop classification. Using all L-band linear polarizations, corn, soybeans, cereals, and hay pasture were classified to an overall accuracy of 70%, while a more

temporally rich C-band dataset provided an accuracy of 80%. However, larger biomass crops were well classified using the PALSAR data, whereas C-band data were needed to accurately classify low-biomass crops. With a multifrequency dataset, an overall accuracy of 88.7% was reached, and many individual crops were classified to accuracies better than 90%. These results were competitive with the overall accuracy achieved using three Landsat images (88%).

The existing orbital SARs have only one single band. Although useful, when taken alone, each of these orbital SARs will encounter limitations for crop and vegetation classification because of signal saturation at high levels of biomass (Dobson et al., 1995). One possible solution is to increase the temporal information as compensation. Schotten et al. (1995) assessed the capability of multitemporal ERS-1 SAR data in discriminating between the crop types for land cover inventory purposes. An overall classification accuracy of 80% was achieved for the classification of 12 crop types. Li and Yeh (2004) demonstrated that multitemporal RADARSAT-1 SAR images were suitable for monitoring the rapid changes of cultivation systems in a subtropical region. Park and Chi (2008) used multitemporal RADARSAT-1 data with HH polarization and Envisat ASAR data with VV polarization for the classification of typical five land cover classes in an agricultural area. The results indicated that the use of multiple

TABLE 22.10 Studies on Crop and Vegetation Identification Using Radar Imagery

Crop/Vegetation Types	Data	Accuracy	Strengths	Limitations	Reference
Surfaces, short vegetation, and tall vegetation	ERS-1 and JERS-1 SAR data	90%	The SARs provide information on the structure of the surface and the overlying vegetation cover that is complementary to the greenness information provided by NDVI	The study requires a sequential processor that classifies terrain and selects the appropriate class-specific retrieval algorithms	Dobson et al. (1995)
Low mangrove, urban, swamp, temporarily flooded vegetation, permanently, flooded vegetation, woody savanna, forest, grass savanna, open forest, and raphia	JERS-1 SAR data	84%	Radar backscatter amplitude is important for separating basic land cover categories such as savannas, forests, and flooded vegetation. Texture is useful for refining flooded vegetation classes. Temporal information from SAR images of two different dates is explicitly used to identify swamps and temporarily flooded vegetation	A tradeoff between classification accuracy and spatial resolution must be reached	Simard et al. (2000)
Banana, grass, lotus, water, sugar cane, rice paddy, fishponds, and built-up areas	RADARSAT-1 SAR data	85%	Multitemporal satellite SAR images are suitable for monitoring the rapid changes of cultivation systems in a subtropical region	Multitemporal SAR images are needed	Li and Yeh (2004)
Ambrosia dumosa, Larrea tridentata, encelia farinosa, mixed scrub, olneya lesota, parkinsonia microphylla, and desert pavement	Landsat TM and ERS-1 SAR data	88.89%	Combing ERS-1 SAR imagery and Landsat TM imagery together increases classification accuracy	Both TM and SAR data are needed	Shupe and Marsh (2004)
Corn, soybeans, cereals, and hay-pasture	ALOS PALSAR and RADARSAT-1 SAR data	88.70%	The results reported in this study emphasize the value of polarimetric, as well as multifrequency SAR, data for crop classification	Access to multipolarization data promises to further advance the use of SAR for agricultural applications	McNairn et al. (2009)
Broad-leaved, fine-leaved, no grain (unploughed), grain (ploughed), winter grain, and spring grain	TerraSAR-X data	90%	Multitemporal TerraSAR-X data are suitable for monitoring agricultural land use and its related ecosystems	The study is based on pixel-based MLC method	Bargiel and Herrmann (2011)

polarization SAR data with a proper feature extraction stage would improve classification accuracy. Bargiel and Herrmann (2011) used a stack of l4 spotlight TerraSAR-X images for the classification of agricultural land use in two areas with different population density, agricultural management, as well as geological and geomorphological conditions. Overall accuracy for all classes for the two areas was 61.78% and 39.25%, respectively. Accuracies improved notably for both regions (about 90%) when single vegetation classes were merged into groups of classes. They indicated that SAR imagery could serve as basis for monitoring systems for agricultural land use and its related ecosystems. Yonezawa et al. (2012) used RADARSAT-2 PolSAR images to monitor and classify rice fields and found that multitemporal observation by PolSAR has great potential to be utilized for estimating rice-planted areas and monitoring rice growth.

Some studies were also carried out on the classification of vegetation and agricultural crops by integrating SAR data and Landsat TM imagery (Ban, 2003; Shupe and Marsh, 2004). These studies showed that the synergy of SAR and Landsat TM data could produce much better classification accuracy than that of Landsat TM alone only when careful consideration is given to the temporal compatibility of SAR and visible and infrared data.

22.6.4 Application on Urban Environment

The urban environment is also an important area for radar application (Table 22.11). Cao and Jin (2007) found that urban terrain surfaces could be well classified by fusing Landsat ETM+ and ERS-2 SAR images. Liao et al. (2008) found that urban areas could be detected by jointly using coherence and intensity characteristics of ERS-1/2 SAR imagery based on an unsupervised change detection approach. Ban et al. (2010) fused QuickBird multispectral data and multitemporal RADARSAT-1 fine-beam SAR data for urban land cover mapping and found that decision level fusion of QuickBird classification and RADARSAT SAR classification was able to take advantage of the best classifications of both optical and SAR data, thus significantly improving the classification accuracies of several LULC classes. Vidal and Moreno (2011) applied TerraSAR-X and aerial optical data to the change detection of isolated housing in agricultural areas. They concluded that high-resolution radar images such as TerraSAR-X images are an excellent complement to optical high-resolution images for carrying out isolated housing change detection. Hu and Ban (2012) implemented urban land cover classification using multitemporal RADARSAT-2 ultra-fine beam SAR data, and an accuracy of 81.8% was achieved for the classification. Majd et al. (2012) assessed the potential of a single polarimetric radar image of high spatial resolution, acquired by the airborne Radar Aeroporte Multi-spectral d'Etude des Signatures (RAMSES) SAR sensor of Office national d'études et de recherches aérospatiales (ONERA), for the classification of urban areas. The results highlighted the potential of such data to discriminate urban land cover types, and the overall accuracy reached 84%. However, the results also showed a problematic confusion between roofs and trees. Multitemporal

multi-incidence angle Envisat ASAR and Chinese HJ-1B multispectral were fused for detailed urban land cover mapping (Ban and Jacob, 2013). The best classification result (80%) was achieved using the fusion of eight-date Envisat ASAR and Huan Jing 1B (HJ-1B) data. Niu and Ban (2013) employed multitemporal RADARSAT-2 high-resolution SAR images for urban land cover classification. Six-date PolSAR data in both ascending and descending passes were acquired in a rural–urban fringe area, and major land cover classes included high-density residential areas, low-density residential areas, industrial and commercial areas, construction sites, parks, golf courses, forests, pasture, water, and two types of agricultural crops. The best classification result was achieved using all six-date data (kappa = 0.91), while very good classification results (kappa = 0.86) were achieved using only three-date PolSAR data. The results demonstrated that the combination of both the ascending and the descending PolSAR data with an appropriate temporal span was suitable for urban land cover mapping.

Airborne high-resolution SAR images have also been widely used in investigating urban environment. Gamba et al. (2000) presented a procedure for the extraction and characterization of building structures from the 3D terrain elevation data provided by interferometric InSAR measurements. Dierking and Skriver (2002) addressed the detection of changes in multitemporal polarimetric radar images acquired at C and L band by the airborne EMISAR system, focusing on small objects (such as buildings) and narrow linear features (such as roads). They found that the radar intensities were better suited for change detection than the correlation coefficient and the phase difference between the copolarized channels. Urban height and classification map were retrieved from RAMSES high-resolution InSAR images by Tison et al. (2007). The results obtained on real images were compared to ground truth and indicated a very good accuracy in spite of limited image resolution. Brenner and Roessing (2008) demonstrated the potential of very-high-resolution radar imaging of urban areas by means of SAR and interferometric imaging. The corresponding data were acquired with the X-band phased array multifunctional imaging radar (Figure 22.10). They stated that high-resolution InSAR will be an important basis for upcoming radar-based urban analysis.

22.6.5 Snow and Ice Mapping

The ability of radar remote sensing to image through darkness and cloud cover is a key to its applications in snow and ice mapping in the temperate and polar regions, which are in darkness for much of the year. Albright et al. (1998) used SIR-C/X-SAR images to map snow and glacial ice on the rugged north slope of Mount Everest. SIR-C/X-SAR data were able to identify and map scree/talus, dry snow, dry snow-covered glacier, wet snow-covered glacier, and rock-covered glacier, as corroborated by comparison with existing surface cover maps and other ancillary information. Multitemporal RADARSAT-1 SAR images were also proved to be effective in the classification of ice types by Weber et al. (2003). Zakhvatkina et al. (2013) classified sea ice

TABLE 22.11 Studies on the Applications of Radar Imagery on the Urban Environment

Purpose	Data	Accuracy	Strengths	Limitations	Reference
Urban land use classification (water, grass, building, road, and flat field)	Landsat ETM+ and ERS-2 SAR data	>90%	Fused images from infrared ETM+ and microwave SAR images can yield better classification of complex terrain surfaces	Landsat ETM+ and ERS-2 SAR images are needed simultaneously	Cao and Jin (2007)
Urban land use classification (high-density built-up areas, low-density built-up areas, roads, forest, parks, golf courses, water and several types of agricultural land)	Quickbird MS and RADARSAT SAR images	89.50%	Decision level fusion of RADARSAT SAR and Quickbird classification results are able to take advantage of the best classification of both optical and SAR images	The accuracies of commercial industrial areas and low-density residential areas remain relatively low	Ban et al. (2010)
Change detection of isolated housing	Aerial optical images and TerraSAR-X SAR data	94.83%	High resolution TerraSAR-X images are an excellent complement to optical high-resolution images for carrying out isolated housing change detection	Optical and high resolution SAR images are needed	Vidal and Moreno (2011)
Urban land use classification (high-density built-up area, low-density built-up area, roads, airport, forest, low vegetation, golf course, grass/pasture, bare fields, and water)	ENVISAT ASAR and HJ-1B data	80%	Fusion of SAR and optical images provides complementary information, thus yielding higher classification accuracy than SAR or optical data alone	SAR data and optical data are needed simultaneously	Ban and Jacob (2013)
Urban land use classification (high-density residential areas, low-density residential areas, industrial and commercial areas, construction sites, parks, golf courses, forests, pasture, water, and two types of agricultural crops)	RADARSAT-2 SAR images	90%	Combination of both the ascending and the descending polarimetric SAR data with an appropriate temporal span is suitable for urban land cover mapping	Multitemporal ascending and descending images are needed	Niu and Ban (2013)
Extraction and characterization of building structures	IFSAR data	Footprint error: 1%–37%; Height error: −11 to 4 m	Building footprint, height and position, as well as its description with a simple 3-D model, are recovered from interferometric radar data	Building footprints are largely underestimated; layover/shadowing effects	Gamba et al. (2000)
Joint retrieval of urban height map and classification from high-resolution interferometric SAR images	RAMSES X-band data	Root mean square error is around 2.5 m	An original high-level processing chain is proposed for the computation of a digital surface model (DSM) over urban areas	The major limit of DSM computation remains the initial spatial and altimetric resolutions that need to be made more precise	Tison et al. (2007)
Investigating the potential of very high-resolution radar imaging of urban areas (subdecimeter resolution)	PAMIR X-band data		High-resolution interferometric SAR can overcome the immanent layover situation in urban areas	Coregistration mismatches	Brenner and Roessing (2008)

in the Central Arctic using Envisat ASAR images and found that it was necessary to use textural features in addition to the backscattering coefficients for sea ice classification. The results of the classification showed that the average correspondences with the expert analysis amount to 85%, 83%, and 80% for multiyear ice, deformed first-year ice, and level first-year ice, respectively. Warner et al. (2013) used RADARSAT-2 data for ice detection during summer and found that the physical and electromagnetic properties of the ice surfaces were virtually identical with few differences in the scattering of microwave energy.

22.6.6 Other Applications

Wang and Allen (2008) utilized ALOS PALSAR HH and JERS-1 HH SAR data to delineate estuarine shorelines and to study shoreline changes of North Carolina coast, United States. The results supported further monitoring of shorelines in estuaries using active remote sensing. Evans et al. (2010) used multitemporal C-band RADARSAT-2 and L-band ALOS/PALSAR data to map ecosystems and created spatial–temporal maps of flood dynamics in the Brazilian Pantanal. The cross-sensor,

FIGURE 22.10　Zoomed subset of the high-resolution SAR image obtained by PAMIR. (From Brenner, A.R. and Roessing, L., *IEEE Trans. Geosci. Remote Sens.*, 46, 2971, 2008.)

multitemporal SAR data were found to be useful in mapping both land cover and flood patterns in wetland areas. The generated maps would be a valuable asset for defining habitats required to conserve the Pantanal biodiversity and to mitigate the impacts of human development in the region. Cornforth et al. (2013) contrasted and quantified the impacts of cyclone Sidr and anthropogenic degradation on mangroves using PALSAR imagery. This study illustrated how different threats experienced by mangroves could be detected and mapped using radar-based information, to guide management action.

22.7　Future Developments

Currently operating satellite SAR systems such as TanDEM-X, TerraSAR-X, RADARSAT-2, and COSMO-SkyMed open up the opportunity to carry out research in digital mapping of global or regional scale LULC using fully PolSAR data. Fully polarized SAR data of different frequency will become more widely available to both the scientific and natural resource management communities after the availability of SAR data provided by forthcoming satellite SAR missions, such as RADARSAT Constellation, Sentinel-1 Constellation, ALOS-2 PALSAR, and Tandem-L missions. The RADARSAT Constellation is the evolution of the RADARSAT program with the objective of ensuring C-band data continuity, improved operational use of SAR, and improved system reliability over the next decade (Flett et al., 2009). The mission development has begun in 2005, with satellite launches planned for 2018. The three-satellite configuration will provide daily access to 95% of the world to users. The increase in revisit frequency introduces a range of applications that are based on regular collection of data and creation of composite images that highlight changes over time. Such applications are particularly useful for monitoring changes such as those induced by climate change, land use evolution, coastal change, urban subsidence, and even human impacts

on local environments. As a part of the Copernicus program of European Space Agency, Sentinel-1 constellation consists of two satellites orbiting 180° apart and images the entire Earth every 6 days (Torres et al., 2012). The mission will benefit numerous services, such as Arctic sea-ice extent monitoring; routine sea-ice mapping; surveillance of the marine environment, including oil-spill monitoring and ship detection for maritime security; monitoring land surface for motion risks; mapping for forest; water and soil management; and mapping to support humanitarian aid and crisis situations. ALOS-2 is the follow-on Japan Aerospace Exploration Agency (JAXA); L-SAR (L-band SAR) satellite mission of ALOS-1 approved by the Japanese government in late 2008 (Suzuki et al., 2013). The overall objective is to provide data continuity to be used for cartography, regional observation, disaster monitoring, and environmental monitoring. ALOS-2 will continue the L-SAR observations of the ALOS PALSAR and will expand data utilization by enhancing its performance. ALOS-2 will have a spotlight mode (1–3 m) and a high-resolution mode (3–10 m), while PALSAR has a 10 m resolution. The observation frequency of ALOS-2 will be improved by greatly expanding the observable range of the satellite up to about three times, through an improvement in observable areas (from 870 to 2320 km), as well as giving ALOS-2 a right-and-left looking function. Tandem-L is a proposal for an L-band PolSAR and InSAR mission to monitor Earth's dynamics with unprecedented accuracy and resolution (Moreira et al., 2011). A wide spectrum of scientific mission objectives is covered, including producing a global inventory of forest height and biomass, large-scale measurements of the Earth deformation, systematic observation of glacial motion, soil moisture, and ocean surface currents. Tandem-L foresees the deployment of two spacecraft flying in close formation similarly to the TanDEM-X mission. The instrument will feature many technical innovations, such as the combined use of a reflector antenna and digital beamforming techniques on receive to achieve large swath coverage, high sensitivity and ambiguity rejection at the same time. All these forthcoming SAR missions will further enhance the capability of radar remote sensing in LULC mapping and monitoring. In combination with other sensor types, radar's power in extracting details will be further improved.

Radar imaging with its near-all weather, day/light capability will be an invaluable tool for investigating timely LULC information in the perpetually cloud-covered tropical and equatorial regions of the world where many developing countries with the greatest need for LULC data are found. One of the important applications of radar remote sensing in future is the monitoring of unauthorized land development, which is a growing problem in many developing countries. In China, for example, the rapid urbanization has caused a rapid decline in the supply of arable land (Seto et al., 2000). Although the government has introduced a number of policies aimed at arable land preservation (Lichtenberg and Ding, 2008), losses of arable land are still occurring because of the relentless unauthorized expansion of construction land, especially by local governments illegally leasing land and using farmland for nonagricultural construction (Wang and Scott, 2008). In addition, unauthorized land development has led to many environmental problems, such as urban sprawl, forest degradation, and soil erosion, which have posed a major threat to healthy urban development (Yeh and Li, 1996). Radar remote sensing, which is independent of weather conditions and day light, is a promising tool for monitoring land developments on a regular short-term monthly basis for early detecting and preventing unauthorized ones.

There are still a number of significant challenges and issues that need to be addressed in order for radar remote sensing to achieve its full potential for LULC mapping and monitoring:

1. Developing backscatter models for different land cover types and developing computer algorithms designed specifically for analyzing multifrequency, multipolarization, multi-incidence angle, and multitemporal SAR data. This will lead to improvement in model inversion or the technique of estimation of land cover information from indirect measurements.
2. Quantifying the full range of factors that result in temporally varying signatures on SAR imagery, with the influence of rain and dew, and plant phenology being the principal uncertainties. For example, seasonal agricultural or natural vegetation growth may cause problems in the detection of human-induced land development activities in particular seasons.
3. Developing techniques to account for the effects of topography.
4. The relative utility of polarimetric, multifrequency radar data versus optical data for LULC investigation has to be determined.

Future efforts on the aforementioned issues will further enhance and advance the growing importance of radar remote sensing in LULC mapping and monitoring at local, regional, and global levels in the future.

References

Adams, J.B., Sabol, D.E., Kapos, V., Almeida, R., Roberts, D.A., Smith, M.O., and Gillespie, A.R. 1995. Classification of multispectral images based on fractions of endmembers—Application to land-cover change in the Brazilian Amazon, *Remote Sensing of Environment*, 52, 137–154.

Albright, T.P., Painter, T.H., Roberts, D.A., Shi, J.C., Dozier, J., and Fielding, E. 1998. Classification of surface types using SIR-C/X-SAR, Mount Everest Area, Tibet, *Journal of Geophysical Research-Planets*, 103, 25823–25837.

Anderson, J.R., Hardy, E.E., Roach, J.T., and Witmer, R.E. 1976. *A Land Use and Land Cover Classification System for Use with Remote Sensor Data*. United States Government Printing Office, Washington, DC.

Askne, J.I.H., Dammert, P.B.G., Ulander, L.M.H., and Smith, G. 1997. C-band repeat-pass interferometric SAR observations of the forest, *IEEE Transactions on Geoscience and Remote Sensing*, 35, 25–35.

Bagan, H., Kinoshita, T., and Yamagata, Y. 2012. Combination of AVNIR-2, PALSAR, and polarimetric parameters for land cover classification, *IEEE Transactions on Geoscience and Remote Sensing*, 50, 1318–1328.

Ban, Y.F. 2003. Synergy of multitemporal ERS-1 SAR and Landsat TM data for classification of agricultural crops, *Canadian Journal of Remote Sensing*, 29, 518–526.

Ban, Y.F., Hu, H.T., and Rangel, I.M. 2010. Fusion of Quickbird MS and RADARSAT SAR data for urban land-cover mapping: Object-based and knowledge-based approach, *International Journal of Remote Sensing*, 31, 1391–1410.

Ban, Y.F. and Jacob, A. 2013. Object-based fusion of multitemporal multiangle ENVISAT ASAR and HJ-1B multispectral data for urban land-cover mapping, *IEEE Transactions on Geoscience and Remote Sensing*, 51, 1998–2006.

Bargiel, D. and Herrmann, S. 2011. Multi-temporal land-cover classification of agricultural areas in two European regions with high resolution spotlight TerraSAR-X data, *Remote Sensing*, 3, 859–877.

Barnes, C.F. and Burki, J. 2006. Late-season rural land-cover estimation with polarimetric-SAR intensity pixel blocks and sigma-tree-structured near-neighbor classifiers, *IEEE Transactions on Geoscience and Remote Sensing*, 44, 2384–2392.

Barnes, R.M. 1988. Roll-invariant decompositions for the polarization covariance matrix, *Proceedings of Polarimetry Technology Workshop*, Redstone Arsenal, AL.

Bazi, Y., Bruzzone, L., and Melgani, F. 2005. An unsupervised approach based on the generalized Gaussian model to automatic change detection in multitemporal SAR images, *IEEE Transactions on Geoscience and Remote Sensing*, 43, 874–887.

Bazi, Y., Bruzzone, L., and Melgani, F. 2006. Automatic identification of the number and values of decision thresholds in the log-ratio image for change detection in SAR images, *IEEE Geoscience and Remote Sensing Letters*, 3, 349–353.

Benz, U.C., Hofmann, P., Willhauck, G., Lingenfelder, I., and Heynen, M. 2004. Multi-resolution, object-oriented fuzzy analysis of remote sensing data for GIS-ready information, *ISPRS Journal of Photogrammetry and Remote Sensing*, 58, 239–258.

Bovolo, F. and Bruzzone, L. 2005. A detail-preserving scale-driven approach to change detection in multitemporal SAR images, *IEEE Transactions on Geoscience and Remote Sensing*, 43, 2963–2972.

Bovolo, F. and Bruzzone, L. 2007. A split-based approach to unsupervised change detection in large-size multitemporal images: Application to tsunami-damage assessment, *IEEE Transactions on Geoscience and Remote Sensing*, 45, 1658–1670.

Brenner, A.R. and Roessing, L. 2008. Radar imaging of urban areas by means of very high-resolution SAR and interferometric SAR, *IEEE Transactions on Geoscience and Remote Sensing*, 46, 2971–2982.

Bruzzone, L., Marconcini, M., Wegmuller, U., and Wiesmann, A. 2004. An advanced system for the automatic classification of multitemporal SAR images, *IEEE Transactions on Geoscience and Remote Sensing*, 42, 1321–1334.

Bruzzone, L. and Prieto, D.F. 2000. Automatic analysis of the difference image for unsupervised change detection, *IEEE Transactions on Geoscience and Remote Sensing*, 38, 1171–1182.

Bujor, F., Trouve, E., Valet, L., Nicolas, J.M., and Rudant, J.P. 2004. Application of log-cumulants to the detection of spatiotemporal discontinuities in multitemporal SAR images, *IEEE Transactions on Geoscience and Remote Sensing*, 42, 2073–2084.

Burley, T.M. 1961. Land use or land utilization?, *The Professional Geographer*, 13, 18–20.

Camps-Valls, G., Gomez-Chova, L., Munoz-Mari, J., Rojo-Alvarez, J.L., and Martinez-Ramon, M. 2008. Kernel-based framework for multitemporal and multisource remote sensing data classification and change detection, *IEEE Transactions on Geoscience and Remote Sensing*, 46, 1822–1835.

Cao, G.Z. and Jin, Y.Q. 2007. A hybrid algorithm of the BP-ANN/GA for classification of urban terrain surfaces with fused data of Landsat ETM+ and ERS-2 SAR, *International Journal of Remote Sensing*, 28, 293–305.

Capstick, D. and Harris, R. 2001. The effects of speckle reduction on classification of ERS SAR data, *International Journal of Remote Sensing*, 22, 3627–3641.

Carincotte, C., Derrode, S., and Bourennane, S. 2006. Unsupervised change detection on SAR images using fuzzy hidden Markov chains, *IEEE Transactions on Geoscience and Remote Sensing*, 44, 432–441.

Chand, T.R.K. and Badarinath, K.V.S. 2007. Analysis of ENVISAT ASAR data for forest parameter retrieval and forest type classification—A case study over deciduous forests of central India, *International Journal of Remote Sensing*, 28, 4985–4999.

Chang, N.B., Han, M., Yao, W., Chen, L.C., and Xu, S.G. 2010. Change detection of land use and land cover in an urban region with SPOT-5 images and partial Lanczos extreme learning machine, *Journal of Applied Remote Sensing*, 4, 1–15.

Chatelain, F., Tourneret, J.Y., and Inglada, J. 2008. Change detection in multisensor SAR images using bivariate gamma distributions, *IEEE Transactions on Image Processing*, 17, 249–258.

Chen, K.S., Huang, W.P., Tsay, D.H., and Amar, F. 1996. Classification of multifrequency polarimetric SAR imagery using a dynamic learning neural network, *IEEE Transactions on Geoscience and Remote Sensing*, 34, 814–820.

Cihlar, J., Pultz, T.J., and Gray, A.L. 1992. Change Detection with Synthetic Aperture Radar, *International Journal of Remote Sensing*, 13, 401–414.

Cimino, J., Brandani, A., Casey, D., Rabassa, J., and Wall, S.D. 1986. Multiple incidence angle Sir-B experiment over Argentina—Mapping of forest units, *IEEE Transactions on Geoscience and Remote Sensing*, 24, 498–509.

Clawson, M. and Stewart, C.L., 1965. *Land Use Information, a Critical Survey of U.S. Statistics Including Possibilities for Greater Uniformity*. The John Hopkins Press for Resources for the Future, Inc., Baltimore, MD.

Cloude, S.R. 1985. Target decomposition-theorems in radar scattering, *Electronics Letters*, 21, 22–24.

Cloude, S.R. and Pottier, E. 1996. A review of target decomposition theorems in radar polarimetry, *IEEE Transactions on Geoscience and Remote Sensing*, 34, 498–518.

Cloude, S.R. and Pottier, E. 1997. An entropy based classification scheme for land applications of polarimetric SAR, *IEEE Transactions on Geoscience and Remote Sensing*, 35, 68–78.

Conradsen, K., Nielsen, A.A., Sehou, J., and Skriver, H. 2003. A test statistic in the complex Wishart distribution and its application to change detection in polarimetric SAR data, *IEEE Transactions on Geoscience and Remote Sensing*, 41, 4–19.

Cornforth, W.A., Fatoyinbo, T.E., Freemantle, T.P., and Pettorelli, N. 2013. Advanced Land observing satellite phased array type L-band SAR (ALOS PALSAR) to inform the conservation of mangroves: Sundarbans as a case study, *Remote Sensing*, 5, 224–237.

Covello, F., Battazza, F., Coletta, A., Lopinto, E., Fiorentino, C., Pietranera, L., Valentini, G., and Zoffoli, S. 2010. COSMO-SkyMed an existing opportunity for observing the Earth, *Journal of Geodynamics*, 49, 171–180.

Dierking, W. and Skriver, H. 2002. Change detection for thematic mapping by means of airborne multitemporal polarimetric SAR imagery, *IEEE Transactions on Geoscience and Remote Sensing*, 40, 618–636.

DigitalGlobe, 2013. Tasking the DigitalGlobe constellation. https://www.digitalglobe.com/sites/default/files/DG_SATTASKING_WP_forWeb.pdf (Accessed April 9, 2015.)

Dobson, M.C., Pierce, L.E., and Ulaby, F.T. 1996. Knowledge-based land-cover classification using ERS-1/JERS-1 SAR composites, *IEEE Transactions on Geoscience and Remote Sensing*, 34, 83–99.

Dobson, M.C., Ulaby, F.T., and Pierce, L.E. 1995. Land-cover classification and estimation of terrain attributes using synthetic-aperture radar, *Remote Sensing of Environment*, 51, 199–214.

Dong, Y., Milne, A.K., and Forster, B.C. 2001. Segmentation and classification of vegetated areas using polarimetric SAR image data, *IEEE Transactions on Geoscience and Remote Sensing*, 39, 321–329.

Du, L. and Lee, J.S. 1996. Fuzzy classification of earth terrain covers using complex polarimetric SAR data, *International Journal of Remote Sensing*, 17, 809–826.

Durand, J.M., Gimonet, B.J., and Perbos, J.R. 1987. SAR data filtering for classification, *IEEE Transactions on Geoscience and Remote Sensing*, 25, 629–637.

Engdahl, M.E. and Hyyppa, J.M. 2003. Land-cover classification using multitemporal ERS-1/2 InSAR data, *IEEE Transactions on Geoscience and Remote Sensing*, 41, 1620–1628.

Ersahin, K., Cumming, I.G., and Ward, R.K. 2010. Segmentation and classification of polarimetric SAR data using spectral graph partitioning, *IEEE Transactions on Geoscience and Remote Sensing*, 48, 164–174.

Esch, T., Schenk, A., Ullmann, T., Thiel, M., Roth, A., and Dech, S. 2011. Characterization of land cover types in TerraSAR-X images by combined analysis of speckle statistics and intensity information, *IEEE Transactions on Geoscience and Remote Sensing*, 49, 1911–1925.

European Space Agency, http://earth.eo.esa.int/polsarpro/Manuals/1_What_Is_Polarization.pdf. (Accessed April 9, 2015.)

European Space Agency, Special Features of ASAR, https://earth.esa.int/handbooks/asar/CNTR1-1-5.htm. (Accessed April 9, 2015.)

Evans, T.L. and Costa, M. 2013. Landcover classification of the Lower Nhecolandia subregion of the Brazilian Pantanal Wetlands using ALOS/PALSAR, RADARSAT-2 and ENVISAT/ASAR imagery, *Remote Sensing of Environment*, 128, 118–137.

Evans, T.L., Costa, M., Telmer, K., and Silva, T.S.F. 2010. Using ALOS/PALSAR and RADARSAT-2 to map land cover and seasonal inundation in the Brazilian pantanal, *IEEE Journal of Selected Topics in Applied Earth Observations and Remote Sensing*, 3, 560–575.

Ferro-Famil, L., Pottier, E., and Lee, J.S. 2001. Unsupervised classification of multifrequency and fully polarimetric SAR images based on the H/A/alpha-Wishart classifier, *IEEE Transactions on Geoscience and Remote Sensing*, 39, 2332–2342.

Flett, D., Crevier, Y., and Girard, R. 2009. The radarsat constellation mission: Meeting the Government of Canada's needs and requirements, *2009 IEEE International Geoscience and Remote Sensing Symposium*, 1–5, 1161–1163.

Ford, J.P., Cimino, J.B., Holt, B., and Ruzek, M.R. 1986. *Shuttle Imaging Radar Views of the Earth From Challenger: The SIR-B Experiment*, JPL Publication 86–10, Jet Propulsion Laboratory, Pasadena, CA, p. 135.

Freeman, A. and Durden, S.L. 1998. A three-component scattering model for polarimetric SAR data, *IEEE Transactions on Geoscience and Remote Sensing*, 36, 963–973.

Freeman, A., Villasenor, J., Klein, J.D., Hoogeboom, P., and Groot, J. 1994. On the use of multifrequency and polarimetric radar backscatter features for classification of agricultural crops, *International Journal of Remote Sensing*, 15, 1799–1812.

Freitas, C.D., Soler, L.D., Anna, S.J.S.S., Dutra, L.V., dos Santos, J.R., Mura, J.C., and Correia, A.H. 2008. Land use and land cover mapping in the Brazilian Amazon using polarimetric airborne P-band SAR data, *IEEE Transactions on Geoscience and Remote Sensing*, 46, 2956–2970.

French, A.N., Schmugge, T.J., Ritchie, J.C., Hsu, A., Jacob, F., and Ogawa, K. 2008. Detecting land cover change at the Jornada Experimental Range, New Mexico with ASTER emissivities, *Remote Sensing of Environment*, 112, 1730–1748.

Frost, V.S., Stiles, J.A., Shanmugan, K.S., and Holtzman, J.C. 1982. A model for radar images and its application to adaptive digital filtering of multiplicative noise, *IEEE Transactions on Pattern Analysis and Machine Intelligence*, 4, 157–166.

Fukuda, S. and Hirosawa, H. 1999. A wavelet-based texture feature set applied to classification of multifrequency polarimetric SAR images, *IEEE Transactions on Geoscience and Remote Sensing*, 37, 2282–2286.

Gamba, P., Dell'Acqua, F., and Lisini, G. 2006. Change detection of multitemporal SAR data in urban areas combining feature-based and pixel-based techniques, *IEEE Transactions on Geoscience and Remote Sensing*, 44, 2820–2827.

Gamba, P., Houshmand, B., and Saccani, M. 2000. Detection and extraction of buildings from interferometric SAR data, *IEEE Transactions on Geoscience and Remote Sensing*, 38, 611–618.

Gauthier, Y., Bernier, M., and Fortin, J.P. 1998. Aspect and incidence angle sensitivity in ERS-1 SAR data, *International Journal of Remote Sensing*, 19, 2001–2006.

Grey, W.M.F., Luckman, A.J., and Holland, D. 2003. Mapping urban change in the UK using satellite radar interferometry, *Remote Sensing of Environment*, 87, 16–22.

Grunsky, E.C. 2002. The application of principal components analysis to multi-beam RADARSAT-1 satellite imagery: A tool for land cover and terrain mapping, *Canadian Journal of Remote Sensing*, 28, 758–769.

Hara, Y., Atkins, R.G., Yueh, S.H., Shin, R.T., and Kong, J.A. 1994. Application of neural networks to radar image classification, *IEEE Transactions on Geoscience and Remote Sensing*, 32, 100–109.

Henderson, F.M. 1975. Radar for small-scale land-use mapping, *Photogrammetric Engineering and Remote Sensing*, 41, 307–319.

Henderson, F.M. 1995. An analysis of settlement characterization in central-Europe using Sir-B radar imagery, *Remote Sensing of Environment*, 54, 61–70.

Henderson, F.M. and Xia, Z.G. 1997. SAR applications in human settlement detection, population estimation and urban land use pattern analysis: A status report, *IEEE Transactions on Geoscience and Remote Sensing*, 35, 79–85.

Holecz, F., Barbieri, M., Cantone, A., Pasquali, P., and Monaco, S. 2009. Synergetic use of multi-temporal Alos Palsar and Envisat Asar data for topographic/land cover mapping and monitoring at national scale in Africa, *2009 IEEE International Geoscience and Remote Sensing Symposium*, 1–5, 256–259.

Holm, W.A. and Barnes, R.M. 1988. On radar polarization mixed state decomposition theorems, *Proceedings of the 1988 USA National Radar Conference*, Ann Arbor, MI, pp. 20–21.

Hosseini, R.S., Entezari, I., Homayouni, S., Motagh, M., and Mansouri, B. 2011. Classification of polarimetric SAR images using Support Vector Machines, *Canadian Journal of Remote Sensing*, 37, 220–233.

Hu, H.T. and Ban, Y.F. 2012. Multitemporal RADARSAT-2 ultra-fine beam SAR data for urban land cover classification, *Canadian Journal of Remote Sensing*, 38, 1–11.

Huang, M.X., Shi, Z., and Gong, J.H. 2008. Potential of multitemporal ERS-2 SAR imagery for land use mapping in coastal zone of Shangyu City, China, *Journal of Coastal Research*, 24, 170–176.

Huynen, J.R. 1970. Phenomenological theory of radar targets. PhD dissertation, Drukkerij Bronder-offset N. V., Rotterdam, the Netherlands.

Inglada, J. and Mercier, G. 2007. A new statistical similarity measure for change detection in multitemporal SAR images and its extension to multiscale change analysis, *IEEE Transactions on Geoscience and Remote Sensing*, 45, 1432–1445.

Kasischke, E.S., Melack, J.M., and Dobson, M.C. 1997. The use of imaging radars for ecological applications—A review, *Remote Sensing of Environment*, 59, 141–156.

Kouskoulas, Y., Ulaby, F.T., and Pierce, L.E. 2004. The Bayesian hierarchical classifier (BHC) and its application to short vegetation using multifrequency polarimetric SAR, *IEEE Transactions on Geoscience and Remote Sensing*, 42, 469–477.

Krogager, E. 1990. New decomposition of the radar target scattering matrix, *Electronics Letters*, 26, 1525–1527.

Kuan, D.T., Sawchuk, A.A., Strand, T.C., and Chavel, P. 1985. Adaptive noise smoothing filter for images with signal-dependent noise, *IEEE Transactions on Pattern Analysis and Machine Intelligence*, 7, 165–177.

Kuplich, T.M., Freitas, C.C., and Soares, J.V. 2000. The study of ERS-1 SAR and Landsat TM synergism for land use classification, *International Journal of Remote Sensing*, 21, 2101–2111.

Kurosu, T., Uratsuka, S., Maeno, H., and Kozu, T. 1999. Texture statistics for classification of land use with multitemporal JERS-1 SAR single-look imagery, *IEEE Transactions on Geoscience and Remote Sensing*, 37, 227–235.

Kurvonen, L. and Hallikainen, M.T. 1999. Textural information of multitemporal ERS-1 and JERS-1 SAR images with applications to land and forest type classification in boreal zone, *IEEE Transactions on Geoscience and Remote Sensing*, 37, 680–689.

Laliberte, A.S., Koppa, J., Fredrickson, E.L., and Rango, A. 2006. Comparison of nearest neighbor and rule-based decision tree classification in an object-oriented environment, *2006 IEEE International Geoscience and Remote Sensing Symposium*, 1–8, 3923–3926.

Lambin, E.F. and Ehrlich, D. 1997. Land-cover changes in sub-Saharan Africa (1982–1991): Application of a change index based on remotely sensed surface temperature and vegetation indices at a continental scale, *Remote Sensing of Environment*, 61, 181–200.

Lang, M.W., Townsend, P.A., and Kasischke, E.S. 2008. Influence of incidence angle on detecting flooded forests using C-HH synthetic aperture radar data, *Remote Sensing of Environment*, 112, 3898–3907.

Lardeux, C., Frison, P.L., Tison, C., Souyris, J.C., Stoll, B., Fruneau, B., and Rudant, J.P. 2009. Support vector machine for multifrequency SAR polarimetric data classification, *IEEE Transactions on Geoscience and Remote Sensing*, 47, 4143–4152.

Larranaga, A., Alvarez-Mozos, J., and Albizua, L. 2011. Crop classification in rain-fed and irrigated agricultural areas using Landsat TM and ALOS/PALSAR data, *Canadian Journal of Remote Sensing*, 37, 157–170.

Lee, J.S. 1981. Refined filtering of image noise using local statistics, *Computer Graphics and Image Processing*, 15, 380–389.

Lee, J.S. 1983. Digital image smoothing and the sigma filter, *Computer Vision Graphics and Image Processing*, 24, 255–269.

Lee, J.S., Grunes, M.R., Ainsworth, T.L., Du, L.J., Schuler, D.L., and Cloude, S.R. 1999a. Unsupervised classification using polarimetric decomposition and the complex Wishart classifier, *IEEE Transactions on Geoscience and Remote Sensing*, 37, 2249–2258.

Lee, J.S., Grunes, M.R., and de Grandi, G. 1999b. Polarimetric SAR speckle filtering and its implication for classification, *IEEE Transactions on Geoscience and Remote Sensing*, 37, 2363–2373.

Lee, J.S., Grunes, M.R., and Kwok, R. 1994. Classification of multi-look polarimetric SAR imagery-based on complex Wishart distribution, *International Journal of Remote Sensing*, 15, 2299–2311.

Lee, J.S., Grunes, M.R., and Pottier, E. 2001. Quantitative comparison of classification capability: Fully polarimetric versus dual and single-polarization SAR, *IEEE Transactions on Geoscience and Remote Sensing*, 39, 2343–2351.

Lee, J.S. and Pottier, E., 2009. *Polarimetric Radar Imaging from Basics to Applications*. CRC Press, New York.

Lee, J.S., Wen, J.H., Ainsworth, T.L., Chen, K.S., and Chen, A.J. 2009. Improved sigma filter for speckle filtering of SAR imagery, *IEEE Transactions on Geoscience and Remote Sensing*, 47, 202–213.

Lewis, A.J. and Henderson, F.M. 1998. Radar fundamentals: The geoscience perspective. In: A.J. Lewis and F.M. Henderson (eds.), *Manual of Remote Sensing Volume 2—Principles and Applications of Imaging Radar*. John Wiley & Sons, Inc., New York, pp. 567–629.

Li, G.Y., Lu, D.S., Moran, E., Dutra, L., and Batistella, M. 2012. A comparative analysis of ALOS PALSAR L-band and RADARSAT-2 C-band data for land-cover classification in a tropical moist region, *ISPRS Journal of Photogrammetry and Remote Sensing*, 70, 26–38.

Li, X. and Yeh, A.G. 2004. Multitemporal SAR images for monitoring cultivation systems using case-based reasoning, *Remote Sensing of Environment*, 90, 524–534.

Li, X., Yeh, A.G.O., Qian, J.P., Ai, B., and Qi, Z.X. 2009. A matching algorithm for detecting land use changes using case-based reasoning, *Photogrammetric Engineering and Remote Sensing*, 75, 1319–1332.

Liao, M.S., Jiang, L.M., Lin, H., Huang, B., and Gong, J.Y. 2008. Urban change detection based on coherence and intensity characteristics of SAR imagery, *Photogrammetric Engineering and Remote Sensing*, 74, 999–1006.

Lichtenberg, E. and Ding, C.G. 2008. Assessing farmland protection policy in China, *Land Use Policy*, 25, 59–68.

Lichtenegger, J., Dallemand, J.F., Reichart, P., Rebillard, P., and Buchroithner, M. 1991. Multi-sensor analysis for land use mapping in Tunisia, *Earth Observation Quarterly*, 33, 1–6.

Liesenberg, V. and Gloaguen, R. 2013. Evaluating SAR polarization modes at L-band for forest classification purposes in Eastern Amazon, Brazil, *International Journal of Applied Earth Observation and Geoinformation*, 21, 122–135.

Lim, H.H., Swartz, A.A., Yueh, H.A., Kong, J.A., Shin, R.T., and Vanzyl, J.J. 1989. Classification of Earth terrain using polarimetric synthetic aperture radar images, *Journal of Geophysical Research-Solid Earth and Planets*, 94, 7049–7057.

Liu, J., Pattey, E., and Nolin, M.C. 2008. Object-based classification of high resolution SAR images for within field homogeneous zone delineation, *Photogrammetric Engineering and Remote Sensing*, 74, 1159–1168.

Lo, C.P. 1998. Applications of imaging radar to land use and land cover mapping. In: F.M. Henderson and A.J. Lewis (eds.), *Manual of Remote Sensing—Principles and Applications of Imaging Radar*. John Wiley & Sons, Inc., New York, pp. 705–729.

Lombardo, P. and Pellizzeri, T.M. 2002. Maximum likelihood signal processing techniques to detect a step pattern of change in multitemporal SAR images, *IEEE Transactions on Geoscience and Remote Sensing*, 40, 853–870.

Longepe, N., Rakwatin, P., Isoguchi, O., Shimada, M., Uryu, Y., and Yulianto, K. 2011. Assessment of ALOS PALSAR 50 m orthorectified FBD data for regional land cover classification by support vector machines, *IEEE Transactions on Geoscience and Remote Sensing*, 49, 2135–2150.

Lopes, A., Nezry, E., Touzi, R., and Laur, H. 1993. Structure detection and statistical adaptive speckle filtering in SAR images, *International Journal of Remote Sensing*, 14, 1735–1758.

Lu, D., Batistella, M., and Moran, E. 2007. Land-cover classification in the Brazilian Amazon with the integration of Landsat ETM plus and Radarsat data, *International Journal of Remote Sensing*, 28, 5447–5459.

Lu, D., Mausel, P., Brondizio, E., and Moran, E. 2004. Change detection techniques, *International Journal of Remote Sensing*, 25, 2365–2407.

Lu, D. and Weng, Q. 2007. A survey of image classification methods and techniques for improving classification performance, *International Journal of Remote Sensing*, 28, 823–870.

Luckman, A., Baker, J., and Wegmuller, U. 2000. Repeat-pass interferometric coherence measurements of disturbed tropical forest from JERS and ERS satellites, *Remote Sensing of Environment*, 73, 350–360.

Lunetta, R.S., Knight, J.F., Ediriwickrema, J., Lyon, J.G., and Worthy, L.D. 2006. Land-cover change detection using multi-temporal MODIS NDVI data, *Remote Sensing of Environment*, 105, 142–154.

Majd, M.S., Simonetto, E., and Polidori, L. 2012. Maximum likelihood classification of single high-resolution polarimetric SAR images in urban areas, *Photogrammetrie Fernerkundung Geoinformation*, 2012, 395–407.

McNairn, H., Shang, J.L., Jiao, X.F., and Champagne, C. 2009. The contribution of ALOS PALSAR multipolarization and polarimetric data to crop classification, *IEEE Transactions on Geoscience and Remote Sensing*, 47, 3981–3992.

Miranda, F.P., Fonseca, L.E.N., and Carr, J.R. 1998. Semivariogram textural classification of JERS-1 (Fuyo-1) SAR data obtained over a flooded area of the Amazon rainforest, *International Journal of Remote Sensing*, 19, 549–556.

Moreira, A., Krieger, G., Younis, M., Hajnsek, I., Papathanassiou, K., Eineder, M., and De Zan, F. 2011. Tandem-L: A mission proposal for monitoring dynamic earth processes, *2011 IEEE International Geoscience and Remote Sensing Symposium (IGARSS)*, Vancouver, British Columbia, Canada, July 24–29, 2011, pp. 1385–1388.

Moser, G., Serpico, S., and Vernazza, G. 2007. Unsupervised change detection from multichannel SAR images, *IEEE Geoscience and Remote Sensing Letters*, 4, 278–282.

Moser, G. and Serpico, S.B. 2006. Generalized minimum-error thresholding for unsupervised change detection from SAR amplitude imagery, *IEEE Transactions on Geoscience and Remote Sensing*, 44, 2972–2982.

Moser, G. and Serpico, S.B. 2009. Unsupervised change detection from multichannel SAR data by Markovian data fusion, *IEEE Transactions on Geoscience and Remote Sensing*, 47, 2114–2128.

Myint, S.W., Gober, P., Brazel, A., Grossman-Clarke, S., and Weng, Q.H. 2011. Per-pixel vs. object-based classification of urban land cover extraction using high spatial resolution imagery, *Remote Sensing of Environment*, 115, 1145–1161.

Niu, X. and Ban, Y.F. 2013. Multi-temporal RADARSAT-2 polarimetric SAR data for urban land-cover classification using an object-based support vector machine and a rule-based approach, *International Journal of Remote Sensing*, 34, 1–26.

Nyoungui, A.N., Tonye, E., and Akono, A. 2002. Evaluation of speckle filtering and texture analysis methods for land cover classification from SAR images, *International Journal of Remote Sensing*, 23, 1895–1925.

Papathanassiou, K.P. and Cloude, S.R. 2001. Single-baseline polarimetric SAR interferometry, *IEEE Transactions on Geoscience and Remote Sensing*, 39, 2352–2363.

Paris, J.F. 1983. Radar backscattering properties of corn and soybeans at frequencies of 1.6, 4.75, and 13.3 Ghz, *IEEE Transactions on Geoscience and Remote Sensing*, 21, 392–400.

Park, N.W. and Chi, K.H. 2008. Integration of multitemporal/polarization C-band SAR data sets for land-cover classification, *International Journal of Remote Sensing*, 29, 4667–4688.

Peng, X.L., Wang, J.F., and Zhang, Q.F. 2005. Deriving terrain and textural information from stereo RADARSAT data for mountainous land cover mapping, *International Journal of Remote Sensing*, 26, 5029–5049.

Pierce, L.E., Ulaby, F.T., Sarabandi, K., and Dobson, M.C. 1994. Knowledge-based classification of polarimetric SAR images, *IEEE Transactions on Geoscience and Remote Sensing*, 32, 1081–1086.

Polychronaki, A., Gitas, I.Z., Veraverbeke, S., and Debien, A. 2013. Evaluation of ALOS PALSAR imagery for burned area mapping in Greece using object-based classification, *Remote Sensing*, 5, 5680–5701.

Pu, R.L. and Landry, S. 2012. A comparative analysis of high spatial resolution IKONOS and WorldView-2 imagery for mapping urban tree species, *Remote Sensing of Environment*, 124, 516–533.

Qi, Z. and Yeh, A.G.O. 2013. Integrating change vector analysis, post-classification comparison, and object-oriented image analysis for land use and land cover change detection using RADARSAT-2 polarimetric SAR images. In: S. Timpf and P. Laube (eds.), *Advances in Spatial Data Handling*. Springer, Berlin, Germany, pp. 107–123.

Qi, Z.X., Yeh, A.G.O., Li, X., and Lin, Z. 2012. A novel algorithm for land use and land cover classification using RADARSAT-2 polarimetric SAR data, *Remote Sensing of Environment*, 118, 21–39.

Rahman, M.M., Moran, M.S., Thoma, D.P., Bryant, R., Collins, C.D.H., Jackson, T., Orr, B.J., and Tischler, M. 2008. Mapping surface roughness and soil moisture using multi-angle radar imagery without ancillary data, *Remote Sensing of Environment*, 112, 391–402.

Rajesh, K., Jawahar, C.V., Sengupta, S., and Sinha, S. 2001. Performance analysis of textural features for characterization and classification of SAR images, *International Journal of Remote Sensing*, 22, 1555–1569.

Ranson, K.J. and Sun, G.Q. 1994. Northern forest classification using temporal multifrequency and multipolarimetric SAR images, *Remote Sensing of Environment*, 47, 142–153.

Rignot, E., Chellappa, R., and Dubois, P. 1992. Unsupervised segmentation of polarimetric SAR data using the covariance-matrix, *IEEE Transactions on Geoscience and Remote Sensing*, 30, 697–705.

Rignot, E.J.M. and Vanzyl, J.J. 1993. Change detection techniques for Ers-1 SAR data, *IEEE Transactions on Geoscience and Remote Sensing*, 31, 896–906.

Saatchi, S.S., Soares, J.V., and Alves, D.S. 1997. Mapping deforestation and land use in Amazon rainforest by using SIR-C imagery, *Remote Sensing of Environment*, 59, 191–202.

Santoro, M., Askne, J.I.H., Wegmuller, U., and Werner, C.L. 2007. Observations, modeling, and applications of ERS-ENVISAT coherence over land surfaces, *IEEE Transactions on Geoscience and Remote Sensing*, 45, 2600–2611.

Schotten, C.G.J., Vanrooy, W.W.L., and Janssen, L.L.F. 1995. Assessment of the capabilities of multitemporal Ers-1 SAR data to discriminate between agricultural crops, *International Journal of Remote Sensing*, 16, 2619–2637.

Seto, K.C., Kaufmann, R.K., and Woodcock, C.E. 2000. Landsat reveals China's farmland reserves, but they're vanishing fast, *Nature*, 406, 121–121.

Shimoni, M., Borghys, D., Heremans, R., Perneel, C., and Acheroy, M. 2009. Fusion of PolSAR and PolInSAR data for land cover classification, *International Journal of Applied Earth Observation and Geoinformation*, 11, 169–180.

Shupe, S.M. and Marsh, S.E. 2004. Cover- and density-based vegetation classifications of the sonoran desert using Landsat TM and ERS-1 SAR imagery, *Remote Sensing of Environment*, 93, 131–149.

Simard, M., Saatchi, S.S., and De Grandi, G. 2000. The use of decision tree and multiscale texture for classification of JERS-1 SAR data over tropical forest, *IEEE Transactions on Geoscience and Remote Sensing*, 38, 2310–2321.

Singh, A. 1989. Digital change detection techniques using remotely-sensed data, *International Journal of Remote Sensing*, 10, 989–1003.

Solaiman, B., Pierce, L.E., and Ulaby, F.T. 1999. Multisensor data fusion using fuzzy concepts: Application to land-cover classification using ERS-1/JERS-1 SAR composites, *IEEE Transactions on Geoscience and Remote Sensing*, 37, 1316–1326.

Solbo, S. and Eltoft, T. 2004. Gamma-WMAP: A statistical speckle filter operating in the wavelet domain, *International Journal of Remote Sensing*, 25, 1019–1036.

Srivastava, H.S., Patel, P., Sharma, Y., and Navalgund, R.R. 2009. Large-area soil moisture estimation using multi-incidence-angle RADARSAT-1 SAR data, *IEEE Transactions on Geoscience and Remote Sensing*, 47, 2528–2535.

Stow, D.A., Tinney, L.R., and Estes, J.E., 1980. Deriving land use/land cover change statistics from Landsat: A study of prime agricultural land, *Proceedings of the 14th International Symposium on Remote Sensing of Environment*, Ann Arbor, MI.

Strozzi, T., Dammert, P.B.G., Wegmuller, U., Martinez, J.M., Askne, J.I.H., Beaudoin, A., and Hallikainen, M.T. 2000. Land use mapping with ERS SAR interferometry, *IEEE Transactions on Geoscience and Remote Sensing*, 38, 766–775.

Suzuki, S., Kankaku, Y., and Osawa, Y. 2013. ALOS-2 current status and operation plan, *Sensors, Systems, and Next-Generation Satellites XVII*, 8889.

Tan, C.P., Ewe, H.T., and Chuah, H.T. 2011. Agricultural crop-type classification of multi-polarization SAR images using a hybrid entropy decomposition and support vector machine technique, *International Journal of Remote Sensing*, 32, 7057–7071.

Tison, C., Nicolas, J.M., Tupin, F., and Maitre, H. 2004. A new statistical model for Markovian classification of urban areas in high-resolution SAR images, *IEEE Transactions on Geoscience and Remote Sensing*, 42, 2046–2057.

Tison, C., Tupin, F., and Maitre, H. 2007. A fusion scheme for joint retrieval of urban height map and classification from high-resolution interferometric SAR images, *IEEE Transactions on Geoscience and Remote Sensing*, 45, 496–505.

Toll, D.L. 1985. Analysis of digital LANDSAT MSS and SEASAT SAR data for use in discriminating land cover at the urban fringe of Denver, Colorado, *International Journal of Remote Sensing*, 6, 1209–1229.

Torres, R. et al. 2012. GMES Sentinel-1 mission, *Remote Sensing of Environment*, 120, 9–24.

Touzi, R. 2007. Target scattering decomposition in terms of roll-invariant target parameters, *IEEE Transactions on Geoscience and Remote Sensing*, 45, 73–84.

Turkar, V., Deo, R., Rao, Y.S., Mohan, S., and Das, A. 2012. Classification accuracy of multi-frequency and multi-polarization SAR images for various land covers, *IEEE Journal of Selected Topics in Applied Earth Observations and Remote Sensing*, 5, 936–941.

Tzeng, Y.C. and Chen, K.S. 1998. A fuzzy neural network to SAR image classification, *IEEE Transactions on Geoscience and Remote Sensing*, 36, 301–307.

Ulaby, F.T., Kouyate, F., Brisco, B., and Williams, T.H.L. 1986. Textural information in SAR images, *IEEE Transactions on Geoscience and Remote Sensing*, 24, 235–245.

Ulaby, F.T., Li, R.Y., and Shanmugan, K.S. 1982. Crop classification using airborne radar and Landsat data, *IEEE Transactions on Geoscience and Remote Sensing*, 20, 42–51.

Vanzyl, J.J. and Burnette, C.F. 1992. Bayesian classification of polarimetric SAR images using adaptive a-priori probabilities, *International Journal of Remote Sensing*, 13, 835–840.

Vidal, A. and Moreno, M.R. 2011. Change detection of isolated housing using a new hybrid approach based on object classification with optical and TerraSAR-X data, *International Journal of Remote Sensing*, 32, 9621–9635.

Villasenor, J.D., Fatland, D.R., and Hinzman, L.D. 1993. Change detection on Alaska's North Slope using repeat-pass ERS-1 SAR images, *IEEE Transactions on Geoscience and Remote Sensing*, 31, 227–236.

Waite, W.P. 1976. Historical development of imaging radar. In: A.J. Lewis (ed.), *Geoscience Applications of Imaging Radar Systems, RSEMS (Remote Sensing of the Electro Magnetic Spectrum)*. Association of American Geographers, Omaha, pp. 1–22.

Walker, W.S., Stickler, C.M., Kellndorfer, J.M., Kirsch, K.M., and Nepstad, D.C. 2010. Large-area classification and mapping of forest and land cover in the Brazilian Amazon: A comparative analysis of ALOS/PALSAR and Landsat data sources, *IEEE Journal of Selected Topics in Applied Earth Observations and Remote Sensing*, 3, 594–604.

Wang, J.R., Engman, E.T., Shiue, J.C., Rusek, M., and Steinmeier, C. 1986. The Sir-B observations of microwave backscatter dependence on soil-moisture, surface-roughness, and vegetation covers, *IEEE Transactions on Geoscience and Remote Sensing*, 24, 510–516.

Wang, L., Sousa, W.P., Gong, P., and Biging, G.S. 2004. Comparison of IKONOS and QuickBird images for mapping mangrove species on the Caribbean coast of Panama, *Remote Sensing of Environment*, 91, 432–440.

Wang, Y. and Allen, T.R. 2008. Estuarine shoreline change detection using Japanese ALOS PALSAR HH and JERS-1 L-HH SAR data in the albemarle-pamlico sounds, north Carolina, USA, *International Journal of Remote Sensing*, 29, 4429–4442.

Wang, Y.M. and Scott, S. 2008. Illegal farmland conversion in China's urban periphery: Local regime and national transitions, *Urban Geography*, 29, 327–347.

Warner, K., Iacozza, J., Scharien, R., and Barber, D. 2013. On the classification of melt season first-year and multi-year sea ice in the Beaufort Sea using Radarsat-2 data, *International Journal of Remote Sensing*, 34, 3760–3774.

Weber, F., Nixon, D., and Hurley, J. 2003. Semi-automated classification of river ice types on the Peace River using RADARSAT-1 synthetic aperture radar (SAR) imagery, *Canadian Journal of Civil Engineering*, 30, 11–27.

White, R.G. 1991. Change detection in SAR imagery, *International Journal of Remote Sensing*, 12, 339–360.

Witten, I.H., Frank, E., and Hall, M.A. 2011. *Data Mining: Practical Machine Learning Tools and Techniques*. Morgan Kaufmann, Burlington, MA.

Wu, S.T. and Sader, S.A. 1987. Multipolarization SAR data for surface-feature delineation and forest vegetation characterization, *IEEE Transactions on Geoscience and Remote Sensing*, 25, 67–76.

Yamaguchi, Y., Moriyama, T., Ishido, M., and Yamada, H. 2005. Four-component scattering model for polarimetric SAR image decomposition, *IEEE Transactions on Geoscience and Remote Sensing*, 43, 1699–1706.

Yeh, A.G.O. and Li, X. 1996. Urban growth management in the Pearl River Delta—An integrated remote sensing and GIS approach, *ITC Journal*, 1, 77–86.

Yonezawa, C., Negishi, M., Azuma, K., Watanabe, M., Ishitsuka, N., Ogawa, S., and Saito, G. 2012. Growth monitoring and classification of rice fields using multitemporal RADARSAT-2 full-polarimetric data, *International Journal of Remote Sensing*, 33, 5696–5711.

Zakhvatkina, N.Y., Alexandrov, V.Y., Johannessen, O.M., Sandven, S., and Frolov, I.Y. 2013. Classification of sea ice types in ENVISAT synthetic aperture radar images, *IEEE Transactions on Geoscience and Remote Sensing*, 51, 2587–2600.

Zhang, X.H., Chen, J.W., and Meng, H.Y. 2013. A novel SAR image change detection based on Graph-Cut and generalized Gaussian model, *IEEE Geoscience and Remote Sensing Letters*, 10, 14–18.

Zheng, J. and You, H.J. 2013. A new model-independent method for change detection in multitemporal SAR images based on radon transform and Jeffrey divergence, *IEEE Geoscience and Remote Sensing Letters*, 10, 91–95.

Carbon

23

Global Carbon Budgets and the Role of Remote Sensing

Richard A. Houghton
Woods Hole Research Center

Acronyms and Definitions

ALOS	Advanced land observing satellite
AVHRR	Advanced very high resolution radiometer
ETM+	Enhanced thematic mapper plus
FAO	Food and Agriculture Organization
FAO/JRC	Food and Agriculture Organization/Joint Research Center
FRA	Forest Resources Assessment
GLAS	Geoscience laser altimeter system
GOSAT	Greenhouse gases observing satellite
INPE	National Space Agency of Brazil
IPCC	Intergovernmental Panel on Climate Change
LULCC	Land use and land-cover change
MODIS	Moderate resolution imaging spectrometer
NDVI	Normalized difference vegetation index
NOAA	National Oceanic and Atmospheric Administration
OCO	Orbiting carbon observatory
PALSAR	Phased array type L-band synthetic aperture radar
RED	Reduced Emissions from Deforestation
REDD	Reduced Emissions from Deforestation and forest Degradation
REDD+	Same as REDD but with (1) conservation, (2) sustainable management of forests, and (3) enhancement of forest carbon stocks
UNFCCC	United Nations Framework Convention on Climate Change

23.1 Global Carbon Budget

In its simplest formulation, the global carbon cycle consists of four terms (atmosphere, land, ocean, and fossil fuels) (Table 23.1). The natural fluxes of carbon between land and atmosphere are 120–150 PgC/year (1 petagram carbon equals 10^{15} gC, or 1 billion metric ton C, or 3.67 billion metric ton CO_2), as a result of global photosynthesis and respiration (including fire). Similar fluxes of 90–120 PgC/year occur between ocean and atmosphere as a result of physical, chemical, and biological processes. These are not the fluxes of the global carbon *budget*, however. Instead, the global carbon budget usually refers to the anthropogenic perturbation to the global carbon budget. The fluxes resulting from anthropogenic perturbation are 1–2 orders of magnitude smaller than the natural flows (Table 23.1).

This chapter reviews the global carbon budget and the role of remote sensing, past, present, and future, in helping to define it.

TABLE 23.1   Stocks and Flows of Carbon

Carbon Stocks (PgC)	
Atmosphere	850
Land	2,000
Vegetation	500
Soil	1,500
Ocean	39,000
Surface	700
Deep	38,000
Fossil fuel reserves	5,000
Natural Flows (PgC/year)	
Atmosphere–oceans	90
Atmosphere–land	120
The Global Carbon Budget: Anthropogenic Perturbations (PgC/year averaged over 2000–2009)	
Fossil fuels	7.8 (±0.4)
Land use change	1.0 (±0.5)
Atmospheric increase	4.0 (±0.1)
Oceanic uptake	2.4 (±0.5)
Residual terrestrial sink	2.4 (±0.8)

The focus is on terrestrial ecosystems; very little is written about the role of satellites in measuring either emissions of carbon from fossil fuels or uptake of carbon by the oceans. Within the context of terrestrial ecosystems, the emphasis is on the emissions and uptake of carbon resulting from disturbance and recovery, particularly those disturbances caused by land use and land cover change (LULCC) or management.

Two broad types of explanatory mechanisms account for the loss and accumulation of carbon on land: (1) disturbances and recovery (structural mechanisms) and (2) the differential effects of environmental change (e.g., CO_2, N deposition, climate) on photosynthesis and respiration (metabolic mechanisms). The two types are not clearly distinct, for example, a forest recovering from wood harvest may grow faster because of CO_2 fertilization. Nevertheless, the distinction is useful: it is implicit or explicit in carbon models; the two mechanisms generally operate at different scales; and they are measured with different instruments. From a remote sensing perspective, the first mechanism, structural, involves changes in canopy structure (demography); the second, metabolic, may involve changes in canopy greenness.

23.1.1 Contemporary Carbon Budget

During the first decade of the twenty-first century, the emissions of carbon from combustion of fossil fuels averaged 7.8 (±0.4) PgC/year (Table 23.1). These emissions are determined from (largely economic) data on the production and consumption of coal, oil, and gas (Andres et al., 2012). Another 1.0 (±0.5) PgC/year was released to the atmosphere as a result of LULCC. That source is a net flux that includes both larger emissions and partially offsetting sinks, all attributable to land management, and this is discussed in greater detail in Sections 23.2 and 23.3. The amount of carbon accumulating in the atmosphere each year is based on

measurements, such as those by the National Oceanographic and Atmospheric Administration (NOAA) (Conway and Tans, 2012). Those accumulations averaged 4.0 (±0.1) PgC/year over the period 2000–2009. The amount of carbon taken up by the world's oceans was 2.4 (±0.5) PgC/year, determined by a number of global biogeochemical ocean models (Le Quéré et al., 2013). And the amount of carbon accumulating in terrestrial ecosystems, *not driven by management* (i.e., LULCC), was 2.4 (±0.8) PgC/year. That sink is calculated by difference from the other terms in the global carbon budget. It makes the global carbon budget balance. It is commonly referred to as the residual terrestrial sink. The mechanisms driving that sink are thought to include CO_2 fertilization, N deposition, and changes in climate, but the relative contributions of these mechanisms are uncertain. Indeed, the total sink is greater than 2.4 PgC/year because that value represents a net sink, and there are undoubtedly sources as well, such as enhanced respiration associated with permafrost thaw (Natali et al., 2014).

Keeping track of the global carbon budget annually is crucial, not only to understand how much of the carbon emitted to the atmosphere stays there and how much accumulates on land and in the ocean (Table 23.1), but because changes in the partitioning of emissions among these reservoirs (atmosphere, land, and ocean) may provide the first indication that the global carbon cycle is changing, perhaps in response to climatic change. In particular, the fraction of carbon emissions that remains airborne has been remarkably constant (~50%) for 50 years (despite large interannual variations). As emissions have approximately doubled since the 1960s, so have the sinks on land and ocean, so the fraction remaining airborne has remained the same. And those sinks have kept the atmospheric increase at only half of what it would have been if all of the emissions had remained in the atmosphere. In other words, the sinks have dampened global warming by about half. Whether those fractions will continue as the Earth warms is a question with both scientific and policy implications. Most of the feedbacks known or imaginable between carbon and climate suggest that a warming will lead to more carbon emissions (or less uptake), that is, that positive feedbacks will prevail. But there is surprisingly little evidence of a decline in land and ocean sinks yet (Ballantyne et al., 2012), although the issue is controversial (Canadell et al., 2007; Knorr, 2009; Le Quéré et al., 2009; Gloor et al., 2010; Raupach et al., 2014).

23.1.2 History of Carbon Cycle Research

Global carbon budgets were not possible to construct until after 1957, when Charles David Keeling (1928–2005) began continuous measurement of carbon dioxide concentrations at Mauna Loa, Hawaii, and the South Pole. Those measurements provided a consistent and reliable record of carbon dioxide concentrations that was required to demonstrate, first of all, the rate at which carbon dioxide was increasing in the atmosphere.

At about the same time that Keeling began his measurements, a community of climate scientists began stepping up the construction of global climate models that calculated the changes in climate expected from increased concentrations of greenhouse gases. The models were based on the physics of atmospheric circulation and the physics of radiation. To predict the rate of warming, however, and not just the equilibrium warming expected from a doubling of carbon dioxide, two additional pieces of information are required: the amount of carbon dioxide added to the atmosphere each year and the residence time of the greenhouse gas in the atmosphere. These pieces of information enable the prediction of how rapidly the carbon dioxide concentration in the atmosphere could double and thus how rapidly the Earth's temperature could increase in response to such a doubling (estimated at 1°C–4°C). Actually, the concentration of carbon dioxide does not need to double, because other greenhouse gases contribute as well, but their combined effects can be calculated in carbon dioxide equivalents. Carbon dioxide is the dominant greenhouse gas under human control. Over the last 100 years, it has accounted for more radiative forcing than all the other greenhouse gases combined and is expected to do so in the future.

The annual amount of carbon dioxide emitted globally from burning of fossil fuels is obtained from statistics on oil, coal, and gas production. The residence time of carbon dioxide (and other greenhouse gases) in the atmosphere is not as readily determined. The amount of carbon dioxide emitted from fossil fuel burning that stays in the atmosphere (and for how long) can be determined from an evaluation of the global carbon budget, and that evaluation has been reconstructed annually since about 1960, based on Keeling's initial measurements of carbon dioxide at two locations, now expanded by NOAA and others to about 200 locations over the Earth (Conway and Tans, 2012). The atmosphere is well mixed, and all of the stations show nearly the same annual rate of growth in carbon dioxide concentration, but the spatial and seasonal variability in concentration is useful for sorting out where the emitted carbon is going (land or sea).

The problem with using the atmospheric residence time to project the rate of increase in concentration is that the residence time observed over the last decades, during which the Earth's average temperature has increased by about 0.75°C, is not necessarily the residence time on an Earth that may grow to be 1°C–6°C warmer over the next decades. The processes that control the uptake of carbon dioxide by the world's oceans and by terrestrial ecosystems are affected by climate (e.g., temperature and moisture) and by the concentration of carbon dioxide itself. These processes drive feedbacks in the carbon–climate system.

The observation that surface temperatures seem to have increased at a much slower rate since 1997 may be the result of a bias in the coverage of global temperature measurements. Correcting for the undersampling at high latitudes, where the increases in temperature have been the greatest, shows an average global rate of warming consistent with rate observed before the late 1990s (Cowtan and Way, 2014).

23.1.3 Sources and Sinks of Carbon from Land

In the 1960s and 1970s, there were no independent estimates of the net annual flux of carbon between land and atmosphere,

and the net flux was calculated by difference to make the global carbon budget balance. Using that approach, the net terrestrial uptake for the period 2000–2009 averaged 2.4 (±0.8) PgC/year (Table 23.1). And the global carbon budget is balanced.

In the early 1980s, the first independent estimates of the terrestrial carbon flux were advanced, based on census data (nonspatial) concerning changes in the area of forests (Moore et al., 1981; Houghton et al., 1983). Deforestation was occurring in many tropical countries, and the carbon held in the trees and soils of these forests was released to the atmosphere with deforestation.

23.1.4 Bookkeeping Model

These early analyses developed a "bookkeeping" model, which used annual rates of land cover change and biome-averaged growth and decomposition rates per hectare to calculate annual changes in carbon pools as a result of management (Figure 23.1) (Houghton et al., 1983). For example, conversion of native vegetation to cultivated land causes 25%–30% of the soil organic carbon in the top meter to be lost (Post and Kwon, 2000; Guo and Gifford, 2002; Murty et al., 2002; Don et al., 2011). This tracking approach assigns an average carbon density to the biomass and soils of a small number of ecosystem types (e.g., deciduous forest, grassland). Considerable uncertainty arises because, even within the same ecosystem type, the spatial variability in carbon density is large, partly from variations in soils and microclimate and partly from past disturbances and recovery.

The approach in the early 1980s was not based on remote sensing. Instead, historic changes in croplands and pastures, aggregated at national or continental scales, were obtained from national and international statistics. Not being spatially explicit, the data did not specify the ecosystem type that was converted to new agricultural land. That specification required independent data, such as maps of natural ecosystems and their overlap with the distribution of croplands and pastures. Data on historical changes in land cover were reconstructed from a variety of national and international historical narratives and national land use statistics as well as from population data (Houghton, 2003).

In this highly aggregated approach, the world was divided into 10 major regions, and each region was assigned 2–6 natural ecosystem types (Houghton et al., 1983). Since the original work in 1983, analyses with nonspatial data have been refined over the years, including lands besides forests and changes in land cover besides deforestation. The calculated flux of carbon has been called the deforestation flux, but it is more accurately referred to as the flux from LULCC. It includes increases in forest area (reforestation, afforestation) as well as deforestation; and it includes losses and gains of carbon per hectare within forests as a result of wood harvest and forest growth. Ideally, the net flux would include all changes in terrestrial carbon brought about by management. During the period 2000–2009, the flux from LULCC was a net release of

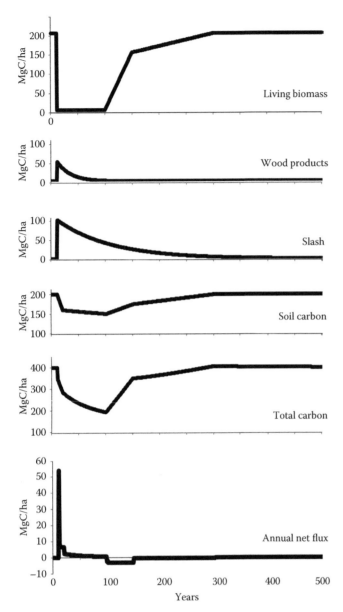

23.1.5 Spatial Analyses

One of the weaknesses of national (nonspatial) data on the areas of cropland and pasture is that the changes through time are net changes in area, not gross changes. Net changes in land cover underestimate gross sources and sinks of carbon that result from simultaneous clearing for, and abandonment of, agricultural lands and thus may underestimate areas of secondary forests and their carbon sinks.

Spatially explicit approaches to historic reconstructions get around this weakness. They were first developed around the year 2000. In one approach, agricultural expansion was distributed spatially on the basis of population density (Klein Goldewijk, 2001). In another, past areas were derived by hind casting of the current distribution of agricultural lands (Ramankutty and Foley, 1999). The data sets of these approaches have been updated and extended to the preindustrial past (Pongratz et al., 2008; Klein Goldewijk et al., 2011). The approaches must make assumptions, just as the nonspatial approach did, about whether agricultural expansion occurs at the expense of grasslands or forests. The distinctions are important because different locations have different carbon stocks, and the carbon flux resulting from LULCC depends on both rates of land cover change and the carbon density of the lands affected. Remote sensing–based information has also been combined with regional tabular statistics to reconstruct spatially explicit land cover changes covering more than the satellite era (Ramankutty and Foley, 1999; Klein Gooldewijk, 2001; Pongratz et al., 2008).

Two spatial data sets, along with the nonspatial data of Houghton (1999, 2003, and updates), have been used in most of the analyses of LULCC: the SAGE data set, including cropland areas from 1700 to 1992 (Ramankutty and Foley, 1999) and the HYDE data set, including both cropland and pasture areas (Klein Goldewijk, 2001). The difference in emissions estimates using these three data sets accounts for about 15% of the difference in flux estimates over the period 1850–1990 (Shevliakova et al., 2009). Other recent data sets, such as the ones compiled by Hurtt et al. (2006) and Pongratz et al. (2008), are based on combinations of SAGE, HYDE, and Houghton data sets, including updates (Houghton, 2010; Houghton et al., 2012).

The results of 13 recent analyses of LULCC (spatial and nonspatial) consistently show a net source of about 1.0 PgC/year to the atmosphere in recent decades (Houghton et al., 2012). Over the longer period, 1850–2012, the annual sources and sinks of carbon from anthropogenic perturbation to the global carbon budget are shown in Figure 23.2. The emissions from fossil fuels have increased steadily through time, now accounting for ~90% of anthropogenic emissions of carbon. But before about 1900, the net emissions from LULCC were higher than fossil fuel emissions. The emissions from LULCC have not varied much from ~1 PgC/year in the last decades.

Figure 23.2 suggests that for budget purposes, there are actually five terms in the global carbon budget. The land appears twice, first as net emissions from LULCC (management) and second

FIGURE 23.1 Idealized response curves of the bookkeeping model. The curves define the annual per hectare changes in carbon pools in response to management. (Updated from Houghton et al., 1983.) The bottom panel shows the annual source (sink) to the atmosphere.

1.0 PgC/year (~11% of total anthropogenic carbon emissions) (Houghton et al., 2012).

Notice that adding another source of carbon to the global carbon budget leaves it unbalanced (Table 23.1). There must be another sink to compensate. Since the oceans and atmosphere are accounted for, the sink must be on land, a sink not attributable to LULCC but, instead, to environmental effects. During the period 2000–2009, this residual terrestrial sink was 2.4 (±0.8) PgC/year. Again, it is determined by difference, although there are a number of global dynamic vegetation models that calculate an annual net uptake of similar magnitude in (unmanaged) terrestrial ecosystems (Le Quéré et al., 2013).

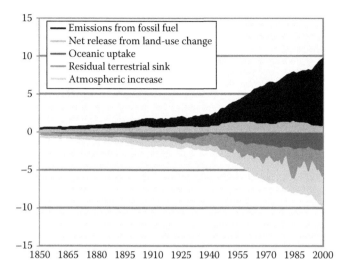

FIGURE 23.2 Annual sources (+) and sinks (−) of carbon in the global carbon budget. (From Le Quéré, C. et al., *Earth Syst. Sci. Data*, 5, 165, 2013.)

as net sinks attributable to processes other than management. These nonmanagement processes are believed to result from natural and indirect anthropogenic effects, such as the effects of elevated CO_2 on plant growth, the effect of greater nitrogen (N) availability (fertilizers, N-fixation through combustion of fossil fuels), and the effects of a changing climate on the growth and respiration of vegetation and on the decay of organic matter in litter and soils. In other words, nature has had an effect on terrestrial carbon storage (and fluxes), at least as great (and in the opposite direction) as the effects of management.

It is important to note that both of these terrestrial fluxes are *net* fluxes. Management is responsible for carbon sinks in growing forests as well as for carbon sources from deforestation. And the residual terrestrial sink is composed of both increased plant growth and increased respiration and decay.

23.2 LULCC, Disturbances, and Recovery

The terrestrial fluxes in Table 23.1 are not evenly distributed over the Earth; forests play a dominant role (Pan et al., 2011). The role of tropical forests in the global carbon cycle is nearly neutral, in part because field data from unmanaged forests suggest an increase carbon storage (Phillips et al., 2008; Gloor et al., 2012), while LULCC results in a loss of carbon of a similar magnitude (Houghton et al., 2009, 2012; Le Quéré et al., 2013) (Table 23.2). The emissions of carbon from LULCC are, themselves, uncertain because of differing estimates of deforestation rates and carbon densities (Baccini et al., 2012; Harris et al., 2012; Houghton et al., 2012).

Outside the tropics, the net carbon balance of forests is reasonably well documented because analyses are based on data from systematic forest inventories. Nevertheless, the question remains: how much of the observed carbon sink is attributable to recovery from a past disturbance, as opposed to an

TABLE 23.2 Role of Forests in the Global Carbon Budget (PgC/year). Carbon Sinks are Negative, Sources Positive

	LULCC	Residual Terrestrial Sink	Total
Tropical forests	1.3	−1.2	0.1
Forests of the temperate and boreal zones	−0.2	−1.0	−1.2
Total	1.1	−2.2	−1.1

Source: Modified from Pan, Y. et al., *Science*, 333, 988, 2011.

environmentally enhanced rate of growth (Williams et al., 2012; Zhang et al., 2012; Fang et al., 2014)?

Table 23.2 also shows that the residual terrestrial sink is roughly the same in tropical forests and in temperate zone and boreal forests (~1.0 PgC/year). The major difference between the regions is the effect of LULCC—high rates of deforestation (and thus emissions) in the tropics versus a small sink outside the tropics where regrowth is slightly greater than emissions. In tropical forests, the sink from environmental effects offsets the emissions from LULCC, for a total balance close to zero. Outside the tropics, both LULCC and environmental effects result in sinks. Overall, the global terrestrial carbon sink is in northern midlatitude forests.

23.2.1 Use of Satellite Data

The use of satellite data to help determine the sources and sinks of carbon from disturbance and recovery is discussed here (Section 23.3). Section 23.4 will address the use of satellite data to help locate and identify other metabolic changes in terrestrial carbon storage.

Satellite data at moderate spatial resolution (30–100 m) have been used to document land cover change since the mid-1980s but only at local or regional scales. Recently, Landsat data have been used to determine global rates of forest loss and gain (Hansen et al., 2013) (Figure 23.3). Furthermore, satellite data have recently begun to be used to document the carbon density of aboveground woody vegetation (Saatchi et al., 2011; Baccini et al., 2012).

23.2.1.1 Rates of Change in Forest Area

A major uncertainty in estimating the net flux of carbon from LULCC has always been, and remains, rates of change in the areas of forests (deforestation and reforestation). Satellites were first recognized as being useful for documentation of these changes in 1984 (Woodwell et al., 1984). One of the earliest studies using Landsat data was of deforestation rates in Amazonia (Skole and Tucker, 1993).

With a time series of satellite data on land cover change, it is possible to estimate the changes. In general, satellite data alleviate the concerns of bias, inconsistency, and subjectivity in country reporting (Grainger, 2008). Depending on the spatial and temporal resolution, satellite data can also distinguish between gross and net losses of forest area. However, increases in forest area are more difficult to define with satellite data than

FIGURE 23.3   Landsat images showing clearings (light blue) within a forested landscape (red) in the state of Rondonia, Brazil. Resolution: 30 m.

deforestation because the growth of trees is a more gradual process. Furthermore, although satellite data are good for measuring losses of forest area, identifying the types of land use that follow deforestation (e.g., croplands, pastures, shifting cultivation) requires repeated looks at the land. Exceptions include the regional studies by Morton et al. (2006) and Galford et al. (2008).

Satellite-based methods for measuring changes in forest area include both high-resolution sample-based methods and wall-to-wall mapping analyses. Sample-based approaches employ systematic or stratified random sampling to quantify gains or losses of forest area at national, regional, and global scales (Achard et al., 2002, 2004; Hansen et al., 2008a, 2010). Systematic sampling provides a framework for forest area monitoring. The United Nations' Food and Agriculture Organization (UN-FAO) forest resource assessment remote sensing survey uses samples at every latitude/longitude intersection to quantify biome and global-scale forest change dynamics from 1990 to 2005 (Food and Agriculture Organization and Joint Research Centre [FAO/ JRC, 2012]). Other sampling approaches stratify by intensity of change, thereby reducing sample intensity. Achard et al. (2002) provided an expert-based stratification of the tropics to quantify forest cover loss from 1990 to 2000 using whole Landsat image pairs. Hansen et al. (2008a, 2010) employed moderate resolution imaging spectrometer (MODIS) data as a change indicator to stratify biomes into regions of homogeneous change for Landsat sampling.

Sampling methods such as described earlier provide regional estimates of forest area and change with uncertainty bounds, but they do not provide a spatially explicit map of forest extent or change. Wall-to-wall mapping does. While coarse-resolution data sets (>4 km) have been calibrated to estimate wall-to-wall changes in area (DeFries et al., 2002), recent availability of moderate spatial resolution data (<100 m), typically Landsat imagery (30 m), allows a more finely resolved approach. Historical

methods rely on photointerpretation of individual images to update forest cover on annual or multiyear bases, such as with the Forest Survey of India (Global Forest Survey of India, 2008) or the Ministry of Forestry Indonesia products (Government of Indonesia/World Bank, 2000). Advances in digital image processing have led to an operational implementation of mapping annual forest cover loss, for example, with the Brazilian PRODES (INPE, 2010) and the Australian national carbon accounting products (Caccetta et al., 2007). These two systems rely on cloud-free data to provide single-image/observation updates on an annual basis. Persistent cloud cover has limited the derivation of products in regions such as the Congo Basin and Insular Southeast Asia (Ju and Roy, 2008). For such areas, Landsat data can be used to generate multiyear estimates of forest cover extent and loss (Hansen et al., 2008b; Broich et al., 2011a). For regions experiencing forest change at an agro-industrial scale, MODIS data provide a capability for integrating Landsat-scale change to annual time steps (Broich et al., 2011b).

In general, moderate spatial resolution imagery is limited in tropical forest areas by data availability. Currently, Landsat is the only source of data at moderate spatial resolution available for tropical monitoring, but to date, an uneven acquisition strategy among bioclimatic regimes limits the application of generic biome-scale methods with Landsat. No other system has the combination of (1) global acquisitions, (2) historical record, (3) free and accessible data, and (4) standard terrain-corrected imagery, along with robust radiometric calibration, which Landsat does. Future improvements in moderate spatial resolution monitoring can be obtained by increasing the frequency of data acquisition.

The primary weakness of satellite data is that they are not available before the satellite era (Landsat began in 1972). Long time series are required for estimating legacy emissions of past land use activity. Although maps, at varying resolutions, exist for many parts of the world, spatial data on LULCC became available at a global level only after 1972, at best. In fact, there are many gaps in the coverage of the Earth's surface before 1999 when the first global acquisition strategy for moderate spatial resolution data was undertaken with the Landsat Enhanced Thematic Mapper Plus (EMT+) sensor (Arvidson et al., 2001). The long-term plan of Landsat ETM+ data includes annual global acquisitions of the land surface, but cloud cover and phenological variability limit the ability to provide annual global updates of forest extent and change. The only other satellite system that can provide global coverage of the land surface at moderate resolution is the advanced land observing satellite (ALOS) phased array type L-band synthetic aperture (PALSAR) radar instrument, which also includes an annual acquisition strategy for the global land surface (Rosenqvist et al., 2007). However, large-area forest change mapping using radar data has not yet been implemented.

23.2.1.2 Uncertainties

Since 1990, forest areas have been reported at 5-year intervals by the UN-FAO. Those estimates (FAO, 2001, 2006, 2010) have been used frequently by analyses calculating the carbon fluxes from LULCC. These FAO data rely on reporting by individual countries. They are more accurate for some countries than for others and are not without inconsistencies and ambiguities (Grainger, 2008). Revisions in the reported rates of deforestation from one 5-year forest resources assessment (FRA) to the next may be substantial due to different methods or data being used.

To estimate the uncertainty in reported rates of deforestation, the FAO began sampling with data from Landsat in the early 1990s to determine the area of forest and changes in that area (FAO, 1996). But the satellite-based estimates are not consistent with the country-based estimates. The FRA 2010 (FAO, 2010), based on data reported by countries, reports a declining rate of tropical deforestation, while the FAO/JRC (2011) study, based on a sampling with Landsat, reports an increasing rate for the years 2000–2005.

As mentioned earlier, the country-based estimates of deforestation from the FAO (2010) are different from estimates determined with Landsat data (FAO/JRC, 2011). There are at least two explanations, besides the explanation generally given, that data from satellites are more consistent and objective. First of all, changes in forest cover as measured from satellites include both natural and anthropogenic disturbances, as caused, for example, by wildfire and clearing for agriculture, respectively. They may also include clear-cut harvests. Census data for agriculture, on the other hand, include only the conversion of forest to cropland or pasture, not natural disturbances, and not harvested forests, which are still defined as forest. These differences in observation and definition raise the important issue of attribution (Section 23.3.8): remote sensing observes all disturbances, not simply those attributable to agricultural conversion and forest management. Everything else being equal, estimates of disturbance observed by remote sensing should be higher than estimates based on management.

Another possible explanation is that census data may include the deforestation of small land parcels, much less than one hectare. Such small clearings may be missed even with Landsat data of 30 m resolution. In the Democratic Republic of Congo, small clearings added 35% to the rate of deforestation obtained by a more traditional analysis of change with Landsat (Tyukavina et al., 2013). Rates of degradation and associated carbon emissions are even more uncertain, as they are not as easily observed with satellite data (Huang and Asner, 2010).

And, finally, one other approach for estimating deforestation rates should be mentioned: satellite detection of forest fires (van der Werf et al., 2010). The approach provides an estimate of forest loss as long as deforestation is accompanied with burning. The approach does not identify LULCC if fire is absent, for example, harvest of wood. Nor does it distinguish between intentional deforestation fires and escaped wildfires. The approach combines estimates of burned area (Giglio et al., 2010) with complementary observations of fire occurrence (Giglio et al., 2003). It makes assumptions about how much fire is for clearing. At province or country level, clearing rates calculated this way capture up to about 80% of the variability and also 80% of the total clearing rates found by other approaches (Hansen et al., 2008a; INPE, 2010). Two advantages of the fire-counting approach are (1) that it allows for an estimate of interannual variability in LULCC emissions and (2) that the emissions of carbon monoxide from burning, routinely monitored by satellites, provide a much larger departure from background conditions than emissions of CO_2 (e.g., van der Werf et al., 2008).

23.2.1.3 Biomass Density

The second type of information required for calculating the emissions of carbon from LULCC is the carbon density of the forests being deforested or harvested.

As mentioned earlier, nonspatial analyses assigned average carbon densities to the vegetation and soils of a small number of natural ecosystems found in each of the 10 major regions. Until recently, data on the distribution of carbon density were not adequate for finer spatial detail. A study of Amazonia, for example, showed that none of the seven different maps of biomass density were in agreement as to the total biomass of the region or even where the largest and smallest densities were to be found (Houghton et al., 2001).

With respect to calculating carbon emissions, the spatial co-occurrence of both forest loss and carbon density is especially important and only available with spatially detailed data. Average carbon densities and average rates of forest loss over large regions may yield accurate estimates of carbon emissions if the disturbances are distributed randomly. But if disturbances, particularly LULCC, affect forests with carbon densities that are systematically different from the mean carbon density, that difference will bias emissions estimates. One way to counter that bias is to colocate changes in area with carbon densities—at the spatial resolution of disturbance.

23.2.1.4 Spatial Analyses

Recently, new satellite techniques have been applied to estimate aboveground carbon densities (Goetz and Dubayah, 2011).

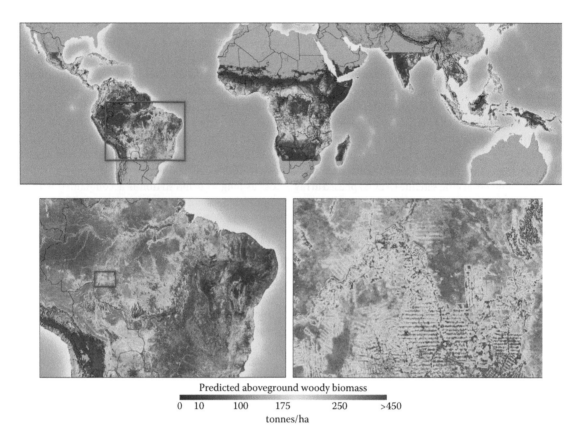

Predicted aboveground woody biomass

0   10   100   175   250   >450

tonnes/ha

FIGURE 23.4 Aboveground carbon density in woody vegetation throughout the tropics. Resolution: 500 m. (From Baccini, A. et al., *Nat. Clim. Change*, 2, 182, doi:10.1038/nclimate1354, 2012.)

Examples of mapping aboveground carbon density over large regions include work with MODIS (Houghton et al., 2007), multiple satellite data (Saatchi et al., 2007, 2011), Radar (Treuhaft et al., 2009), and LiDAR (Baccini et al., 2012) (see Goetz et al., 2009 and Goetz and Dubayah, 2011, for reviews). While the accuracy of fine-scale satellite-based estimates may be lower than site-based inventory measurements (inventory data are generally used to calibrate satellite algorithms), the satellite data are far less intensive to collect, can cover a wide spatial area, and thus can better capture the spatial and temporal variability in aboveground carbon density. By matching carbon density to the forests actually being deforested, this approach has the potential to increase the accuracy of flux estimates, especially in tropical areas where variability of carbon density is high and data availability is poor. Recently published maps of forest biomass are in greater agreement and at finer spatial resolution that previous maps but differences still remain (Mitchard et al., 2013).

One method used to determine the carbon densities of the forests being deforested or those in close proximity to those being deforested is the approach by Baccini et al. (2012). They used a 500 m × 500 m grid of aboveground biomass density determined from MODIS data, calibrated with circa 5.5 million geoscience laser altimeter system (GLAS) shots, which in turn, were calibrated with field measurements at more than

400 locations in the tropics (Figure 23.4) (Baccini et al., 2012). An advantage of the approach is that the GLAS shots and field plots were at similar scales. A second approach might use not the average aboveground carbon density of a MODIS pixel but the sample point data from GLAS estimates to determine the aboveground biomass density in the vicinity of the deforestation or degradation. This approach would not assume an average aboveground carbon density, but it would miss many of the forests being deforested.

Neither approach yields a carbon density at the resolution of forest loss (30–60 m with Landsat), and thus there is still the potential for bias if deforestation or degradation takes place in forest patches systematically different from the mean density of 500 m × 500 m cells. This mismatch in scale is one of the largest sources of uncertainty (bias).

It is interesting to note, however, that the relative error of the first approach is highest in regions with low biomass density and lowest in dense humid forest characterized by high biomass density (Baccini et al., 2012). This observation suggests that errors in biomass density will contribute relatively little to the error in flux estimates calculated for deforestation.

23.2.1.5 Measurement of Changes in Carbon Density

If one can measure aboveground biomass density from satellite, then it should be possible to estimate *changes* in biomass density

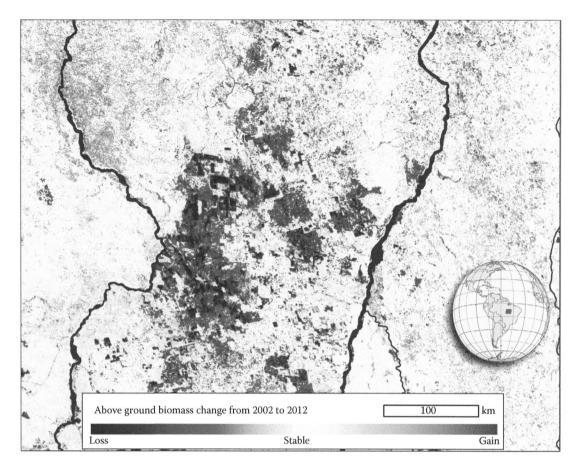

FIGURE 23.5 A 190 km × 215 km region northeast of the Xingu basin in Brazil, showing gains and losses of aboveground live woody biomass density (Mg/ha) between 2002 and 2012.

directly with time series data (Figure 23.5). Some of the changes will result from changes in forest area, as measured by the first approach. But some changes in biomass density may occur without a change in cover type. Such gains and losses in carbon density that exceed the gains and losses from changes in forest area are presumably a measure of growth and degradation.

The traditional approach (changes in area, with average densities assigned) estimates only those density changes related to outright clearing (and recovery). The second, more direct approach estimates changes from both clearing *and* any other factors affecting carbon density. Because the two approaches observe different processes, the difference defines changes in biomass density attributable to degradation (and growth) and/or environmental change (e.g., CO_2, climate, N deposition).

Measuring *changes* in aboveground carbon density is a new approach, but it provides a method for estimating carbon sources and sinks that is more direct than identifying disturbance first, and then assigning a carbon density or change in carbon density (Houghton and Goetz, 2008). The direct measurement of change in density will still require models and ancillary data for full carbon accounting (i.e., changes in soil, slash, and wood products) to yield the total flux of carbon (Section 23.3.6). Furthermore, estimation of change, by itself, does not distinguish between

deliberate LULCC activity and indirect anthropogenic or natural drivers (attribution), because deforestation, as discussed earlier, may result from either management or natural disturbance. Nevertheless, estimation of change in aboveground carbon density has clear potential for improving calculations of sources and sinks of carbon.

Direct measurement of biomass density can provide a continuous range of biomass densities and thus might be expected to yield a more precise estimate of carbon *change* through time than estimates based on changes in area (Houghton and Goetz, 2008). But the change would have to be greater than the uncertainty surrounding any one measurement. Thus, *trends* in MODIS-based biomass density may be more compelling than a single change between two years even if the years are far apart.

Whether measuring changes in forest area (and assigning carbon density) or measuring changes in aboveground carbon density directly, most pixels will probably appear unchanged because rates of growth in mature forests are slow relative to (1) the time interval of observation and (2) the error associated with measurement. However, one might expect two other less common outcomes. First, both approaches might yield a downward trend in carbon density (deforestation and/or degradation) or an upward trend (growth). In these cases, both area changes and density changes would be consistent, although not necessarily equal.

The direct measurement of density change might be expected to indicate greater carbon sinks than the combined approach because global lands, in general, are a net carbon sink despite the fact that managed lands, globally, are a net source (Table 23.1) (Le Quéré et al., 2013). To the extent the terrestrial carbon sink is in aboveground biomass, it should be observable; and the location of changes in carbon *not* attributable to disturbance and recovery would be most instructive for locating the residual terrestrial sink.

In sum, direct measurement of change in aboveground density has the advantage of bypassing the classification step for identifying type of change. But the trade-off is that, without an understanding of LULCC, the observed changes cannot be readily attributed to cause (i.e., anthropogenic or natural, harvest or clearing).

23.3 Policy Realm: Issues Inherent in Estimating the Flux of Carbon from LULCC with an Example Using RED, REDD, and REDD+

RED, REDD, and REDD+ are policy mechanisms proposed for reducing emissions of greenhouse gases under the UN Framework Convention on Climate Change (UNFCCC). *RED* refers to *re*ducing *e*missions from *d*eforestation. If a developing country can demonstrate that it had reduced its emissions of carbon from deforestation, it is eligible for carbon credits. The mechanism was expanded (*REDD*) to include a second "D"—forest degradation. A demonstrated reduction in emissions from forest degradation would also qualify for additional carbon credits. REDD was subsequently expanded to *REDD+*, which adds conservation, the sustainable management of forests, and the enhancement of forest carbon stocks. These three REDD mechanisms are more an application of carbon science to policy than they are a component of the global carbon budget. Nevertheless, the (reduced) emissions of carbon associated with these mechanisms provide a context for discussing issues related to measuring the role of land in the global carbon budget. The paragraphs in the following text discuss these issues.

RED. It is the simplest of the three mechanisms, conceptually and practically. But one still has to

1. Agree on a definition for deforestation
2. Assign a carbon density to the area deforested
3. Decide whether to count committed or actual emissions (see Section 23.3.3)

REDD. When forest degradation (the second "D") is added (REDD), accounting is more difficult.

The same three requirements that apply to RED apply to REDD. And in addition, one must

4. Agree on a definition for degradation (What thresholds for changes in carbon density?)
5. Decide whether to count gross emissions or net emissions

REDD+. For REDD+, emissions and sinks are both counted, emissions from deforestation and degradation and sinks from the enhancement of forest carbon stocks. From the perspective of the atmosphere, reduced emissions are equivalent to increased stocks. Growth as well as degradation is counted. And in addition to deforestation and forest degradation, both reforestation and afforestation are also counted as they represent means for the enhancement of forest carbon stocks.

Nine issues are discussed briefly in the following text.

23.3.1 Definitions

Two recent studies used different estimates of change in forest area to estimate very different (a factor of 3) emissions of carbon from tropical deforestation. Baccini et al. (2012) used FAO rates of deforestation; Harris et al. (2012) used Hansen's rates of forest loss. As discussed earlier, the two estimates of change in forest area are not measures of the same process. *Deforestation*, as defined by the FAO and the Intergovernmental Panel on Climate Change (IPCC), is the permanent conversion of forest cover to another cover. *Forest loss* includes lands deforested, but it also includes temporary losses of forest cover from fires or logging (not defined as deforestation by the IPCC) (see Houghton, 2013a). The difference is important in the UNFCCC intent for accounting for carbon credits and debits. The intent is to reward carbon management, but not to reward natural effects (i.e., attribution). The difference is also important in accounting for differences among estimates of forest loss and estimates of carbon emissions.

23.3.2 Assigning a Carbon Density to the Areas Deforested

This issue is addressed in detail in Section 23.2.1.3.

23.3.3 Committed Versus Actual Emissions (Legacy Effects)

In the process of deforestation and forest degradation, only some of the carbon initially held in the forest is released to the atmosphere in the year of the activity (Figure 23.6). Some of the wood may be removed from the forest and converted to wood products with average lifetimes of a year (fuelwood) to centuries (lumber used in buildings). And some wood may accumulate on the forest floor as woody debris. These pools of woody material will decay over decades and release carbon to the atmosphere well after the actual activity (legacy flux). The same legacy effects pertain to growing forests, only in reverse. That is, harvested forests that are allowed to recover will accumulate carbon for centuries, or until they are harvested again.

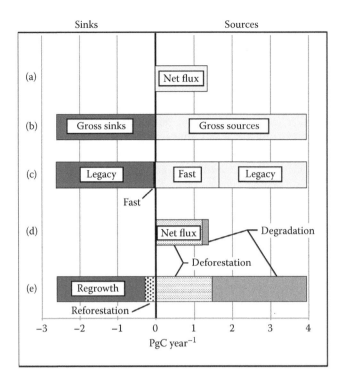

FIGURE 23.6 Annual fluxes of carbon from LULCC. (a) Net emissions, (b) gross sources and sinks, (c) gross fluxes divided into fast (year of disturbance) and legacy (from previous disturbances), (d) net emissions divided into those from deforestation and degradation, (e) gross uptake divided between growth in existing forests and growth of new forests, and gross emissions divided between deforestation and degradation. (From Houghton, R.A. et al., *Biogeosciences*, 9, 5125, 2012.)

Counting the carbon emissions from deforestation and forest degradation can be done in at least two ways. If the total change in carbon density is counted in the year of the activity, the emissions are referred to as committed emissions. If the changes in living and dead biomass, woody debris, harvested products, and soils are tracked through time, then the resulting net emissions reflect actual emissions. Committed emissions are more easily computed. But they cannot be verified by independent measurements of carbon flux, for example, based on forest inventory measurements, eddy covariance measurements, or inverse calculations based on variations of atmospheric CO_2 concentrations. They are easily calculated but not verifiable by independent methods. Remote sensing can help with measuring the sink in growing biomass, but the sources of carbon from dead biomass, woody debris, wood products, and soil are not observable from space.

Two other issues follow from these legacy effects: gross versus net emissions and initial conditions.

23.3.4 Gross and Net Emissions of Carbon from LULCC

Gross emissions refer to the releases of carbon from living biomass, dead biomass, woody debris, wood products, and soils. Gross uptake (or sinks) of carbon refers to the accumulation of carbon in living and dead biomass and soils as a forest grows. Together, gross emissions and gross sinks yield the net flux of carbon (Figure 23.6).

Obviously, gross emissions are greater than net emissions except in the case of deforestation, when they are equal. Gross emissions may be much greater than the net emissions if the gross emissions and gross sinks offset each other in time, that is, if the emissions from decaying wood are balanced by the uptake of carbon in recovering forests.

With deforestation, net and gross emissions are the same (there is no regrowth). But with forest degradation, either net or gross emissions may be counted. Forest degradation is often followed by or offset by forest regrowth, which raises the question: Is regrowth a part of degradation, or should only the gross emissions be counted? If the emphasis is strictly on emissions, then using gross emissions is defensible, using estimates of either committed or actual emissions. But increasing carbon sinks on land is equivalent, in terms of carbon, to reducing emissions and, besides, offers another management opportunity. That is, an emphasis on reducing gross emissions may miss a larger potential for increasing sinks, just as reducing withdrawals from a bank account is only one way to achieve a higher balance. And if reducing emissions and increasing sinks are equivalent, then the emphasis shifts from gross to net emissions, where net emissions are defined as the sum of gross emissions and gross uptake. Net emissions are what affect the atmospheric concentrations of CO_2. Again, remote sensing may help monitor growth of aboveground forest biomass but not the decay of dead wood, wood products, or soil.

23.3.5 Initial Conditions

Legacy effects (growing forests and accumulated pools of decaying wood) have a large effect on calculated emissions and sinks. If one wants to know the emissions in the year 2000, for example, the history of disturbance before 2000 is important. It determines the areas of forest recovering from disturbance as well as the magnitude of carbon pools decaying. Without accounting for that history, one misses the sinks in secondary forests and the sources of carbon from landfills, for example.

There are three ways to handle legacy effects. The simplest way (1) is to count committed emissions. That accounting ignores legacy effects. For actual emissions, however, one must either (2) reconstruct the history of LULCC and disturbance for the years before 2000 (in this example) or (3) determine the age structure of forests in 2000 and the pools of wood in products and woody debris. These two pools (secondary forests and decay pools) determine the current sources and sinks of carbon as well as future sources and sinks. Soils, also, may be either losing or gaining carbon from earlier disturbances. Remote sensing can help with the first and third approaches; the second (reconstruction of history) requires a historic approach.

23.3.6 Full Carbon Accounting

Full carbon accounting refers to the changes in all pools of carbon, not just those in aboveground biomass. The net and gross

2000–2009

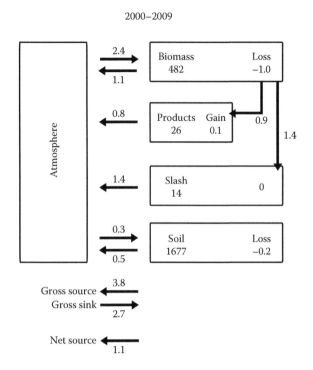

Gross source ← 3.8
Gross sink → 2.7

Net source ← 1.1

FIGURE 23.7 Average annual fluxes of carbon and changes in terrestrial pools of carbon as a result of global LULCC. (From Houghton, R.A., Role of forests and impact of deforestation in the global carbon cycle, in F. Achard and M.C. Hansen, eds., *Global Forest Monitoring from Earth Observation*, CRC Press, Boca Raton, FL, 2013c, pp. 15–38.)

emissions of carbon from disturbance and recovery include, also, the sources of carbon from burning on site; from decay of stumps, roots, and plant material left on site; and from decay of soil organic matter if the soils are cultivated (Figure 23.7). A full global accounting must also consider the decay of wood products removed from the forest. Gross rates of carbon uptake result from the accumulation of carbon in growing forests (vegetation and soils). Remote sensing may help with measuring changes in aboveground living biomass (either gains or losses) but not with the other pools.

The good news is that living biomass is estimated to account for 75%–90% of the total net flux of carbon from LULCC (Houghton, 2003). Soils accounted for most of the rest. Estimates of dead and fallen aboveground biomass density (including coarse wood debris) may be obtained from the literature, documented for both natural forests and changes as a result of disturbance (e.g., wood harvest, fire, shifting cultivation) (e.g., Harmon et al., 1990; Delaney et al., 1998; Chambers et al., 2000; Idol et al., 2001). Distinguishing among types of disturbance helps define the amount of coarse woody debris. For example, harvests and fires remove some wood; storms and insects kill trees but leave the wood. Clearing forests for pastures in Amazonia leaves more debris (dead biomass) than clearing for cultivated agriculture (Morton et al., 2008).

As a first approximation, belowground biomass can be estimated as a fraction of aboveground biomass (e.g., 21%; Houghton et al., 2001). Alternative estimates may be obtained

from documented relationships between forest types or climatic variables and belowground biomass (Cairns et al., 1997; Mokany et al., 2006; Cheng and Niklas, 2007).

Data on harvested wood products and wood removed from deforested sites may be compiled at country level from FAO production yearbooks and FAOstat (2011). Those data can be used together with case studies in the literature covering different types of disturbance and different ecosystems to simulate harvests for the years before satellite data.

A significant fraction of soil organic carbon is lost with cultivation and may accumulate again if croplands are abandoned and forests return. Changes vary with land management and type of disturbance, but cultivation seems to produce a consistent change (a loss of 25%–30% of the upper 1 m of soil) (Post and Kwon, 2000; Guo and Gifford, 2002; McMurty et al., 2002; Don et al., 2011).

23.3.7 Accuracy and Precision

In general, the errors are smallest for deforestation and committed emissions (based on aboveground biomass). These are the observations most directly obtained from satellites. The errors increase for degradation, for actual (delayed) emissions, for net emissions, and for pools of carbon other than aboveground biomass (i.e., belowground carbon, harvested products, decay pools [slash, logging debris]). The errors are largest for distinguishing between managed and natural effects (attribution), especially when considering sinks.

23.3.8 Attribution

Separating sinks that result from natural and indirect processes from sinks that result from management (i.e., attribution) is difficult at any scale, plot level to satellite. And how should emissions or sinks of carbon from natural processes be counted? For example, if unmanaged forests burn, should the emissions count as a carbon debit? It is perhaps better to change the focus from attribution to a focus on those management practices that lead to the greatest net sinks (or lowest net sources), regardless of attribution. With such an approach, all carbon accumulation would be counted as a credit; all emissions would be counted as debits. And the net flux, whatever the cause, defines the credit/debit. This may seem unfair for a country whose forests burn (because of drought), but that country is credited in subsequent years as those forests recover. In choosing the most appropriate policy, it is perhaps more important to bear in mind the net effect on the atmosphere than the causes (anthropogenic or natural) of the emissions.

23.3.9 Uncertainties

The two primary pieces of information for determining changes in terrestrial carbon attributable to LULCC (rates of LULCC and carbon density) contribute about equally to the uncertainty of flux estimates. Before the use of satellite data, the uncertainties contributed by these two variables were each about ±0.3 PgC/year (Houghton et al., 2012). With the use of satellite data for both rates of land cover change and carbon density, these uncertainties should be much lower (Baccini et al., 2012; Harris et al., 2012). The use of satellite data has helped reduce uncertainties substantially.

Nevertheless, there are other issues that contribute to making the terrestrial net emissions considerably more uncertain. Some of these uncertainties may be reduced in the future through new satellite data or new analyses with satellite data.

The issues include more subtle forms of LULCC than clear-cutting and the clearing of forests for croplands and pastures. They involve distinguishing among the land uses following forest clearing (croplands, pastures, shifting cultivation, plantations, etc.). There are also issues about accounting for land degradation, agricultural management, and fire management. Satellite data may help identify some of these more subtle effects.

And there are also processes not yet well accounted for in analyses of global LULCC, for example, losses of carbon from wetlands, especially the draining and burning of peatlands in Southeast Asia; the sources and sinks of carbon associated with settled lands (urban and exurban); woody encroachment (or transitions between woody and herbaceous vegetation types); and lateral transport of carbon resulting from erosion and redeposition (Houghton et al., 2012).

And last, but not least, is the issue of environmental effects. The actual fluxes of carbon from LULCC are affected by variations in temperature and moisture (including longer growing seasons) and long-term trends in CO_2 concentrations and nitrogen loading. These environmental changes affect the annual sources and sinks from managed lands and also the magnitude of the residual terrestrial sink in lands not managed (next section). Not only do the actual fluxes of carbon vary as a result of environmental variation, but also the estimates calculated by different models vary depending on the models. In particular, estimates of the flux of carbon from LULCC vary as much or more from the way environmental effects are modeled than they do from data on LULCC alone (Gasser and Ciais, 2013; Houghton, 2013b; Pongratz et al., 2014).

23.4 Residual Terrestrial Sink

The residual terrestrial sink is the net flux of carbon from the sum of all processes not accounted for in analyses of LULCC. In the ideal case, where all effects of management are included in analyses of LULCC, the residual flux would be the result of natural and indirect effects on terrestrial carbon storage. But as discussed in Section 23.3.9., many of the more subtle forms of management are not included in analyses of LULCC. Furthermore, it is unclear whether some of the processes that affect carbon storage on land are the result of management or environmental change. Is woody encroachment the effect of grazing or fire management, or is it the effect of climate change or of increased CO_2 in the atmosphere? Are the emissions from burning tropical peatlands the result of management or El Niño? The answer is probably that both management and environmental change are involved. And rather than debate the role of intention in these processes, it is more important to try to quantify the net and gross fluxes

for each. Such an approach would yield a graph similar to Figure 23.2 but with a greater number of smaller bands of terrestrial fluxes, not just the two (anthropogenic and natural) that appear in Figure 23.2. Each of the bands appearing now would be broken into individual processes. And as scientists estimate the net flux of carbon for more and more processes, the residual (unexplained) sink (or source) should get smaller and smaller.

The residual terrestrial sink is defined by difference. It is the sink required to balance the global carbon budget. But there may be several ways to quantify it more directly, using remote sensing. One way was discussed in Section 23.2.1.5. The argument there was that changes in aboveground biomass density observed in locations without LULCC might reveal areas of carbon decline or accumulation not caused directly by human activity.

There are at least two additional ways that satellite data might be used to constrain the global carbon budget. One is with repeated measurements of CO_2 in the atmospheric column, and one is with measurement of canopy photosynthesis.

23.4.1 Orbiting Carbon Observatory

Repeated measurements of CO_2 in the column of air over the Earth's surface should add a spatial dimension to the observed seasonal oscillation of CO_2 concentrations, that is, low concentrations in late summer in the northern hemisphere after the growing season in which photosynthesis has exceeded respiration and high concentrations in early spring after the season in which respiration has exceeded photosynthesis (Houghton, 1987). With many more observations of this oscillation, one may be able to deduce sources and sinks of carbon over the surface of the Earth. With frequent enough sampling, day–night changes as well as seasonal changes might be documented.

The OCO-2 satellite, launched in July 2014, is designed to make measurements of total CO_2 in air over the Earth's surface, with repeat coverage every 16 days (Boesch et al., 2011; Table 23.3). The approach for calculating the net flux for any area is similar to the inverse calculations based on ~200 air sampling stations (Conway and Tans, 2012) but with much higher temporal frequency and greater spatial coverage. Greenhouse gases observing satellite (GOSAT) offers the same approach but with coarser spatial resolution.

The global coverage of these satellites means that the data can be used to deduce sources and sinks of carbon, not only from

TABLE 23.3 Orbiting Carbon Observatory (OCO-2)

Science objective: Determine the global geographic distribution of CO_2 sources and sinks.

Measurement approach: The mission will not directly measure CO_2 sources and sinks. Rather, it will use variations in the column-averaged CO_2 mole fraction of air in data assimilation models to infer sources and sinks. The CO_2 data are derived from spectrometers measuring the intensity of sunlight reflected from the presence of CO_2 in a column of air.

Orbit: The OCO-2 will fly in a near-polar, sun-synchronous orbit, viewing the same location on earth once every 16 days.

Launch date: July 2014.

Reference: Boesch et al. (2011).

land but from the ocean as well. And the terrestrial sources and sinks will include all mechanisms, including both management (LULCC) and natural effects (residual terrestrial sink). However, the fossil fuel contribution must be factored out, and the transport or mixing of air must be accounted for.

23.4.2 Satellite Monitoring of Vegetation Activity (Greenness)

Some of the most interesting continental and global measurements of terrestrial metabolism are those of vegetation activity, greenness, or photosynthesis. The most common index is the normalized difference vegetation index (NDVI), as measured with advanced very high resolution radiometer (AVHRR) or MODIS. Such measurements have been correlated with the annual growth of CO_2 in the atmosphere (Myneni et al., 1997).

A major limitation of the approach is that it cannot close the carbon budget; it misses respiration. Thus, the flux is a gross flux, calculated with algorithms to be gross primary production or net primary production. Respiration is indeed related

to photosynthesis, nearly balancing it in most cases. But the imbalance determines whether the net flux is a source or a sink. For example, Myneni et al. (1995) found, surprisingly, that the years with the greatest growth rates of CO_2 coincided with the greenest years. The observation seems contrary to what would be expected if moisture limited photosynthesis. But the finding does not contradict that expectation. Rather, respiration is even more sensitive to moisture than photosynthesis, such that respiration exceeds photosynthesis in moist (green) years, increasing the net emissions from land.

Although seasonal and interannual variations in NDVI fail to capture the respiratory fluxes of carbon, and thus may not be of direct use to carbon budgeting, they are very important for observing physiological responses of photosynthesis to variations in climate. For example, NDVI from AVHRR and MODIS has been used to look for year-to-year variation and trends in greenness (plant productivity) at high latitudes (e.g., Beck and Goetz, 2011) (Figure 23.8). An initial greening of tundra and boreal forests over the 1980s subsequently reversed in many boreal forest areas to declining productivity ("browning")

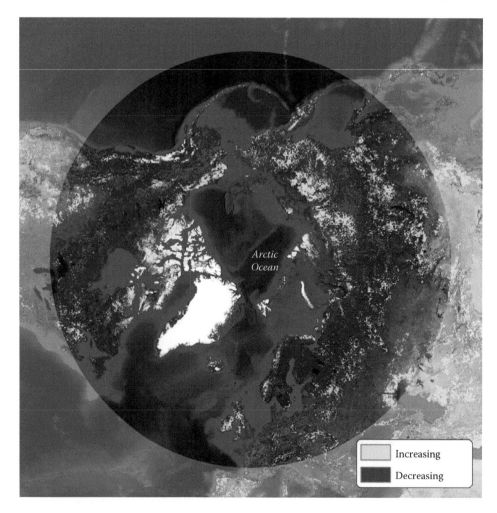

FIGURE 23.8   Circumpolar trends in photosynthetic activity as recorded by NDVI over the period 1982–2005. (Modified from Goetz, S.J. et al., Recent changes in Arctic vegetation: satellite observations and simulation model predictions, in G. Gutman and A. Reissell, eds., *Arctic Land Cover and Land Use in a Changing Climate*, Springer-Verlag, Amsterdam, The Netherlands, pp. 9–36, 2011, doi:10.1007/978-90-481-9118-5_2.)

during the 1990–2000s, while tundra areas continued to systematically increase in greenness. The browning observed in northern forests is believed to result from an increase in summer droughts (high vapor pressure deficits) associated with a warming climate (Goetz et al., 2011).

Other examples of physiological responses to variations in moisture come from the tropics. Some analyses with satellite data suggested that tropical forests in Amazonia were light limited rather than moisture limited because productivity was apparently greater during the dry season (less cloudy) (Nemani et al., 2003). Recent work has shown, however, that the increase in dry season greenness was an artifact of variations in sun–sensor geometry. Correcting for bidirectional reflectance eliminates the seasonal changes in surface reflectance, suggesting that the forests of the region may not be light limited (Morton et al., 2014).

23.5 Conclusions

Over the years, remote sensing has played an ever increasing role in helping to evaluate the global carbon budget, and that trend will continue. This review suggests that data from remote sensing are used in at least five ways to help constrain the terrestrial component of the budget. The primary role of remote sensing before 2010 was in measuring rates of change in the areas of forest. Although such data were spatial, they were used nonspatially in bookkeeping models to calculate sources and sinks of carbon with empirical response curves that assigned carbon densities to vegetation and soil of different types of ecosystems and different land uses. The net emissions from LULCC defined one term in the global carbon budget.

When satellites were also used to measure the aboveground carbon density of forests, the colocation of deforestation rates and carbon densities enabled a more accurate assignment of carbon density to the forests deforested and degraded, thus yielding a more accurate estimate of the emissions of carbon from disturbance and recovery (Baccini et al., 2012; Harris et al., 2012). Uncertainties remain because definitions of deforestation vary among analyses, because different methods are used to estimate emissions, and because other components of an ecosystem besides its aboveground biomass density contribute to carbon dynamics (Houghton, 2013a).

An active area of current research focuses on a third approach: using satellite data to estimate *changes* in aboveground carbon density. The approach is more comprehensive than the first (based on identifying areas disturbed) because some changes in carbon density occur without a change of cover type. Because the direct measurement of change is more sensitive to changes within a cover type (forest degradation and growth), it may yield an estimate closer to the *total net* change in forest carbon, including changes attributable to disturbance and recovery but also changes attributable to environmental change (e.g., enhanced or retarded growth). The magnitude and geographic distribution of the differences between the two approaches may suggest the explanatory mechanisms, including feedbacks to climate change, and where further research should focus.

Satellite data (e.g., NDVI) have been used for decades to identify and measure changes in photosynthetic activity. Although the data do not enable construction of a strict carbon budget, because respiration must be approximated, they have been vital in revealing physiological responses to environmental trends and variability, particularly drought.

Starting in 2014, the OCO will be providing data on spatial and temporal variations in CO_2 concentrations over the Earth's surface. Those data will be used, in a fifth approach, to infer the net sources and sinks of carbon from all processes, including both terrestrial ecosystems and the oceans.

This brief summary of satellites in the global carbon budget suggests a few observations. First, there are crosscutting themes that repeatedly arise in all methods employing satellite data. The most common themes are net versus gross fluxes, attribution, and full carbon accounting. A major justification for attribution is to separate the processes that affect carbon storage into those that can be managed from those that cannot. There's a difference, of course, between understanding the sources and sinks of carbon and being able to manage them. In the end, whatever reduces emissions or increases sinks is helpful for mitigating climate change.

Second, there is a trade-off between comprehensive coverage (full carbon accounting of all processes) and attribution. Those approaches that are most comprehensive, including OCO as well as direct measurement of changes in terrestrial carbon density, yield net fluxes but do not distinguish among specific processes. Measures of greenness are specific to photosynthesis but do not account for changes in all of the other pools of carbon. Measures of disturbance and recovery account for a portion of the terrestrial carbon budget but miss fluxes in undisturbed lands.

Data from satellites require some form of modeling and/or ancillary data to yield estimates of changes in carbon storage or fluxes. For example, changes in terrestrial pools of carbon belowground or on the surface need to be estimated or modeled with data from field measurements. Data from the OCO will require models of atmospheric transport to infer sources and sinks and will need ancillary data to partition those fluxes into fossil fuel, terrestrial, and oceanic components. Thus, satellite data and modeling must work in combination to evaluate some of the terms in the global carbon budget. No single satellite provides the data necessary to evaluate even one term.

Finally, it is important to note that satellite observations are of generally short-term processes, days to months. These observations are sensitive to metabolic and physical processes, but are not necessarily sufficient to predict, or even record, the long-term effects of climate change on land and oceanic carbon storage. Monitoring for the understanding of longer-term effects requires a commitment to long-term data acquisition and data continuity (Tollefson, 2013; Wunsch et al., 2013).

Acknowledgment

Support for this chapter was from NASA's Carbon Monitoring System program, and that support is gratefully acknowledged.

References

Achard, F., H. D. Eva, H.-J. Stibig, P. Mayaux, J. Gallego, T. Richards, and J.-P. Malingreau. 2002. Determination of deforestation rates of the world's humid tropical forests. *Science* 297: 999–1002.

Achard, F., H. D. Eva, P. Mayaux, H.-J. Stibig, and A. Belward. 2004. Improved estimates of net carbon emissions from land cover change in the tropics for the 1990s. *Global Biogeochem. Cycles* 18, GB2008. doi:10.1029/2003GB002142.

Andres, R. J. et al. 2012. A synthesis of carbon dioxide emissions from fossil-fuel combustion. *Biogeosciences* 9: 1845–1871. doi:10.5194/bg-9-1845-2012.

Arvidson, T., J. Gasch, and S. N. Goward. 2001. Landsat 7's long-term acquisition plan—An innovative approach to building a global imagery archive. *Remote Sens. Environ.* 78: 13–26.

Baccini, A., S. J. Goetz, W. S. Walker, N. T. Laporte, M. Sun, D. Sulla-Menashe, J. Hackler et al. 2012. Estimated carbon dioxide emissions from tropical deforestation improved by carbon-density maps. *Nat. Clim. Change* 2: 182–185. doi:10.1038/nclimate1354.

Baker, T. R., O. L. Phillips, Y. Malhi, S. Almeida, L. Arroyo, A. Di Fiore, T. Erwin, N. Higuchi, T. J. Killeen, S. G. Laurance, W. F. Laurance, S. L. Lewis, A. Monteagudo, D. A. Neill, P. N. Vargas, N. C. A. Pitman, J. N. M. Silva, and R.V. Martínez. 2004. Increasing biomass in Amazonian forest plots. *Philosophical Transactions of the Royal Society London* B 359: 353–365.

Ballantyne, A. P., C. B. Alden, J. B. Miller, P. P. Tans, and J. W. C. White. 2012. Increase in observed net carbon dioxide uptake by land and oceans during the past 50 years. *Nature* 488: 70–72. doi:10.1038/nature11299.

Beck, P. S. A. and S. J. Goetz. 2011. Satellite observation of changes in high latitude vegetation productivity changes between 1982 and 2008: Ecological variability and regional differences. *Environ. Res. Lett.* 6: 045501. doi:10.1088/1748–93266/4/045501.

Boesch, H., D. Baker, B. Connor, D. Crisp, and C. Miller. 2011. Global characterization of CO_2 column retrievals from shortwave-infrared satellite observations of the Orbiting Carbon Observatory-2 mission. *Remote Sens.* 3 (2): 270–304. doi:http://dx.doi.org/10.3390/rs3020270.

Broich, M., M. C. Hansen, P. Potapov, B. Adusei, E. Lindquist, and S. V. Stehman. 2011b. Time-series analysis of multi-resolution optical imagery for quantifying forest cover loss in Sumatra and Kalimantan, Indonesia. *Int. J. Appl. Earth Observ. Geoinform.* 13: 277–291.

Broich, M., M. Hansen, F. Stolle, P. Potapov, B. A. Margono, and B. Adusei. 2011a. Remotely sensed forest cover loss shows high spatial and temporal variation across Sumatra and Kalimantan, Indonesia 2000–2008. *Environ. Res. Lett.* 6 (1): 014010.

Caccetta, P. A., S. L. Furby, J. O'Connell, J. F. Wallace, and X. Wu. 2007. Continental monitoring: 34 years of land cover change using Landsat imagery. In: *32nd International Symposium on Remote Sensing of Environment*, June 25–29, 2007, San José, CA.

Cairns, M. A., S. Brown, E. H. Helmer, and G. A. Baumgardner. 1997. Root biomass allocation in the world's upland forests. *Oecologia* 111: 1–11.

Canadell, J. G., C. Le Quéré, M. R. Raupach, C. B. Field, E. T. Buitenhuis, P. Ciais, T. J. Conway, N. P. Gillett, R. A. Houghton, and G. Marland. 2007. Contributions to accelerating atmospheric CO_2 growth from economic activity, carbon intensity, and efficiency of natural sinks. *Proc. Natl. Acad. Sci. U.S.A.* 104: 18866–18870.

Chambers, J. Q., N. Higuchi, J. P. Schimel, L. V. Ferreira, and J. M. Melack. 2000. Decomposition and carbon cycling of dead trees in tropical forests of the central Amazon. *Oecologia* 122: 380–388.

Cheng, D.-L. and K. J. Niklas. 2007. Above- and below-ground biomass relationships across 1534 forested communities. *Ann. Bot.* 99 (1): 95–102.

Conway, T. J. and P. P. Tans. 2012. Trends in atmospheric carbon dioxide. Available at http://www.esrl.noaa.gov/gmd/ccgg/trends

Cowtan, K. and R. G. Way. 2014. Coverage bias in the HadCRUT4 temperature series and its impact on recent temperature trends. *Q.J.R. Meteorol. Soc.* 140: 1935–1944.

DeFries, R. S., R. A. Houghton, M. C. Hansen, C. B. Field, D. Skole, and J. Townshend. 2002. Carbon emissions from tropical deforestation and regrowth based on satellite observations for the 1980s and 90s. *Proc. Natl. Acad. Sci. U.S.A.* 99: 14256–14261.

Delaney, M., S. Brown, A. E. Lugo, A. Torres-Lezama, and N. Bello Quintero. 1998. The quantity and turnover of dead wood in permanent forest plots in six life zones of Venezuela. *Biotropica* 30: 2–11.

Don, A., J. Schumacher, and A. Freibauer. 2011. Impact of tropical land-use change on soil organic carbon stocks—A meta-analysis. *Global Change Biol.* 17: 1658–1670. doi:10.1111/j.1365-2486.2010.02336.x.

Fang, J., T. Kato, Z. Guo, Y. Yang, H. Hu, H. Shen, X. Zhao, A. Kishimodo, Y. Tang, and R. A. Houghton. 2014. Environmental change enhances tree growth: Evidence from Japan's forests. *Proc. Natl. Acad. Sci. U.S.A.* in press.

FAO. 1996. Forest Resources Assessment 1990: Survey of tropical forest cover and study of change processes. FAO Forestry Paper 130, FAO, Rome, Italy.

FAO. 2001. Global Forest Resources Assessment 2000. Main report, FAO Forestry Paper 140, Rome, Italy.

FAO. 2006. Global Forest Resources Assessment 2005. FAO Forestry Paper 147, Rome, Italy.

FAO. 2010. Global Forest Resources Assessment 2010. FAO Forestry Paper 163, Rome, Italy.

FAO/JRC. 2011. Global forest land-use change from 1990 to 2005. Initial results from a global remote sensing survey. http://www.fao.org/forestry/fra/remotesensingsurvey/en/

FAO/JRC. 2012. Global forest land-use change from 1990 to 2005. http://foris.fao.org/static/data/fra2010/RSS_Summary_Report_lowres.pdf, 2012.

FAOSTAT. 2011. http://faostat.fao.org/site/377/default.aspx#ancor

Galford, G. L., J. F. Mustard, J. Melillo, A. Gendrin, C. C. Cerri, and C. E. P. Cerri. 2008. Wavelet analysis of MODIS time series to detect expansion and intensification of row-crop agriculture in Brazil. *Remote Sens. Environ.* 112: 576–587.

Gasser, T. and P. Ciais. 2013. A theoretical framework for the net land-to-atmosphere CO_2 flux and its implications in the definition of "emissions from land-use change. *Earth Syst. Dynam.* 4: 171–186. doi:10.5194/esd-4-171-2013.

Giglio, L., J. Descloitres, C. O. Justice, and Y. J. Kaufman. 2003. An enhanced contextual fire detection algorithm for MODIS. *Remote Sens. Environ.* 87: 273–282. doi:10.1016/S0034-4257(03)00184-6.

Giglio, L., J. T. Randerson, G. R. van der Werf, P. S. Kasibhatla, G. J. Collatz, D. C. Morton, and R. S. DeFries. 2010. Assessing variability and long-term trends in burned area by merging multiple satellite fire products. *Biogeosciences* 7: 1171–1186. doi:10.5194/bg-7-1171-2010.

Global Forest Survey of India. 2008. State of the Forest Report 2005. Forest Survey of India, Ministry of Environment and Forests, Dehradun, India.

Gloor, M., J. L. Sarmiento, and N. Gruber. 2010. What can be learned about carbon cycle climate feedbacks from the CO_2 airborne fraction? *Atmos. Chem. Phys.* 10: 7739–7751.

Gloor, M., L. Gatti, R. Brienen, T. R. Feldpausch, O. L. Phillips, J. Miller, J. P. Ometto et al. 2012. The carbon balance of South America: A review of the status, decadal trends and main determinants. *Biogeosciences* 9: 5407–5430.

Goetz, S. and R. Dubayah. 2011. Advances in remote sensing technology and implications for measuring and monitoring forest carbon stocks and change. *Carbon Manage.* 2 (3): 231–244.

Goetz, S. J., A. Baccini, N. T. Laporte, T. Johns, W. Walker, J. Kellndorfer, R. A. Houghton, and M. Sun. 2009. Mapping and monitoring carbon stocks with satellite observations: A comparison of methods. *Carbon Balance Manage.* 4: 2. doi:10.1186/1750-0680-4-2.

Goetz, S. J., H. E. Epstein, U. S. Bhatt, G. J. Jia, J. O. Kaplan, H. Lischke, Q. Yu et al. 2011. Recent changes in Arctic vegetation: satellite observations and simulation model predictions. In: G. Gutman and A. Reissell (eds.), *Arctic Land Cover and Land Use in a Changing Climate* (pp. 9–36). Amsterdam, The Netherlands: Springer-Verlag. doi:10.1007/978-90-481-9118-5_2.

Government of Indonesia/World Bank. 2000. *Deforestation in Indonesia: A Review of the Situation in 1999.* Jakarta, Indonesia: Government of Indonesia/Work Bank.

Grainger, A. 2008. Difficulties in tracking the long-term trend of tropical forest area. *Proc. Natl. Acad. Sci. U.S.A.* 105 (2): 818–823.

Guo, L. B. and R. M. Gifford. 2002. Soil carbon stocks and land use change: A meta analysis. *Global Change Biol.* 8: 345–360.

Hansen, M. C., P. V. Potapov, R. Moore, M. Hancher, S. A. Turubanova, A. Tyukavina, D. Thau et al. 2013. High-resolution global maps of 21st-century forest cover change. *Science* 342: 850–853.

Hansen, M. C., S. V. Stehman, and P. V. Potapov. 2010. Quantification of global gross forest cover loss. *Proc. Natl. Acad. Sci. U.S.A.* 107: 8650–8655.

Hansen, M. C., S. V. Stehman, P. V. Potapov, T. R. Loveland, J. R. G. Townshend, R. S. DeFries, B. Arunarwati et al. 2008a. Humid tropical forest clearing from 2000 to 2005 quantified using multi-temporal and multi-resolution remotely sensed data. *Proc. Natl. Acad. Sci. U.S.A.* 105: 9439–9444.

Hansen, M. C., D. Roy, E. Lindquist, C. O. Justice, and A. Altstaat. 2008b. A method for integrating MODIS and Landsat data for systematic monitoring of forest cover and change in the Congo Basin. *Remote Sens. Environ.* 112: 2495–2513.

Harmon, M. E., W. K. Ferrell, and J. F. Franklin. 1990. Effects on carbon storage of conversion of old-growth forests to young forests. *Science* 247: 699–702.

Harris, N. L., S. Brown, S. C. Hagen, S. S. Saatchi, S. Petrova, W. Salas, M. C. Hansen, P. V. Potapov, and A. Lotsch. 2012. Baseline map of carbon emissions from deforestation in tropical regions. *Science* 336: 1573–1576.

Houghton, R. A. 1987. Biotic changes consistent with the increased seasonal amplitude of atmospheric CO_2 concentrations. *J. Geophys. Res.* 92: 4223–4230.

Houghton, R. A. 1999. The annual net flux of carbon to the atmosphere from changes in land use 1850–1990. *Tellus* 51B: 298–313.

Houghton, R. A. 2003. Revised estimates of the annual net flux of carbon to the atmosphere from changes in land use and land management 1850–2000. *Tellus* 55B: 378–390.

Houghton, R. A. 2010. How well do we know the flux of CO_2 from land-use change? *Tellus B* 62 (5): 337–351. doi:10.1111/j.1600-0889.2010.00473.x.

Houghton, R. A. 2013a. The emissions of carbon from deforestation and degradation in the tropics: Past trends and future potential. *Carbon Manage.* 4 (5): 539–546.

Houghton, R. A. 2013b. Keeping management effects separate from environmental effects in terrestrial carbon accounting. *Global Change Biol.* 19: 2609–2612.

Houghton, R. A. 2013c. Role of forests and impact of deforestation in the global carbon cycle. In: F. Achard and M. C. Hansen (eds.), *Global Forest Monitoring from Earth Observation* (pp. 15–38). Boca Raton, FL: CRC Press.

Houghton, R. A. and S. J. Goetz. 2008. New satellites help quantify carbon sources and sinks. *Eos* 89 (43): 417–418.

Houghton, R. A., D. Butman, A. G. Bunn, O. N. Krankina, P. Schlesinger, and T. A. Stone. 2007. Mapping Russian forest biomass with data from satellites and forest inventories. *Environ. Res. Lett.* 2: 045032. doi:10.1088/1748-9326/2/4/045032.

Houghton, R. A., J. E. Hobbie, J. M. Melillo, B. Moore, B. J. Peterson, G. R. Shaver, and G. M. Woodwell. 1983. Changes in the carbon content of terrestrial biota and soils between 1860 and 1980: A net release of CO_2 to the atmosphere. *Ecological Monographs* 53: 235–262.

Houghton, R. A., J. I. House, J. Pongratz, G. R. van der Werf, R. S. DeFries, M. C. Hansen, C. Le Quéré, and N. Ramankutty. 2012. Carbon emissions from land use and land-cover change. *Biogeosciences* 9: 5125–5142. doi:10.5194/bg-9-5125-2012.

Houghton, R. A., K. T. Lawrence, J. L. Hackler, and S. Brown. 2001. The spatial distribution of forest biomass in the Brazilian Amazon: A comparison of estimates. *Global Change Biol.* 7: 731–746.

Houghton, R. A., M. Gloor, J. Lloyd, and C. Potter. 2009. The regional carbon budget. In: M. Keller, M. Bustamante, J. Gash, and P. Silva Dias (eds.), *Amazonia and Global Change*, Geophysical Monograph Series 186 (pp. 409–428). Washington, DC: American Geophysical Union.

Huang, M. and G. P. Asner. 2010. Long-term carbon loss and recovery following selective logging in Amazon forests. *Global Biogeochem. Cycles* 24, GB3028. doi:10.1029/2009GB003727.

Hurtt, G. C., S. Frolking, M. G. Fearon, B. Moore, E. Shevliakova, S. Malyshev, S. W. Pacala, and R. A. Houghton. 2006. The underpinnings of land-use history: Three centuries of global gridded land-use transitions, wood harvest activity, and resulting secondary lands. *Global Change Biol.* 12: 1–22.

Idol, T. W., R. A. Filder, P. E. Pope, and F. Ponder. 2001. Characterization of coarse woody debris across a 100 year chronosequence of upland oak-hickory forest. *Forest Ecol. Manage.* 149: 153–161.

Instituto Nacional de Pesquisas Espaciais (INPE). 2010. Deforestation estimates in the Brazilian Amazon. INPE, São José dos Campos, http://www.obt. inpe.br/prodes/

Ju, J. and D. P. Roy. 2008. The availability of cloud-free Landsat ETM+ data over the conterminous United States and globally. *Remote Sens. Environ.* 112: 1196–1211.

Klein Goldewijk, K. 2001. Estimating global land use change over the past 300 years: The HYDE Database. *Global Biogeochem. Cycles* 15: 417–433.

Klein Goldewijk, K., A. Beusen, G. van Drecht, and M. de Vos. 2011. The HYDE 3.1 spatially explicit database of human-induced global land-use change over the past 12,000 years. *Global Ecol. Biogeogr.* 20: 73–86.

Knorr, W. 2009. Is the airborne fraction of anthropogenic CO_2 emissions increasing? *Geophys. Res. Lett.* 36: L21710. doi:10.1029/2009GL040613.

Le Quéré, C., M. R. Raupach, J. G. Canadell, G. Marland, L. Bopp, P. Ciais, T. J. Conway et al. 2009. Trends in the sources and sinks of carbon dioxide. *Nat. GeoSci.* 2: 831–836.

Le Quéré, C., R. J. Andres, T. Boden, T. Conway, R. A. Houghton, J. I. House, G. Marland, G. P. Peters, G. van der Werf, A. Ahlström, R. M. Andrew, L. Bopp, J. G. Canadell, P. Ciais, S. C. Doney, C. Enright, P. Friedlingstein, C. Huntingford, A. K. Jain, C. Jourdain, E. Kato , R. F. Keeling, K. Klein Goldewijk, S. Levis, P. Levy, M. Lomas, B. Poulter, M. R. Raupach, J. Schwinger, S. Sitch, B. D. Stocker, N. Viovy, S. Zaehle, and N. Zeng. 2013. The global carbon budget 1959–2011. *Earth Syst. Sci. Data* 5: 165–185.

Mitchard, E. T. A., S. S. Saatchi, A. Baccini, G. P. Asner, S. J. Goetz, N. L. Harris, and S. Brown. 2013. Uncertainty in the spatial distribution of tropical forest biomass: A comparison of pan-tropical maps. *Carbon Balance Manage.* 8: 10, http://www.cbmjournal.com/content/8/1/10.

Mokany, K., R. J. Raison, and A. S. Prokushkin. 2006. Critical analysis of root:shoot ratios in terrestrial biomes. *Global Change Biol.* 12: 84–96.

Moore, B., R. D. Boone, J. E. Hobbie, R. A. Houghton, J. M. Melillo, B. J. Peterson, G. R. Shaver, C. J. Vorosmarty, and G. M. Woodwell. 1981. A simple model for analysis of the role of terrestrial ecosystems in the global carbon budget. In: B. Bolin (ed.), *Carbon Cycle Modelling*, SCOPE 16 (pp. 365–385). New York: John Wiley & Sons.

Morton, D. C., J. Nagol, C. C. Carabajal, J. Rosette, M. Palace, B. D. Cook, E. F. Vermote, D. J. Harding, and P. R. North. 2014. Amazon forests maintain consistent canopy structure and greenness during the dry season. *Nature* 506 (7487): 221–224. doi:10.1038/nature13006.

Morton, D. C., R. S. DeFries, J. T. Randerson, L. Giglio, W. Schroeder, and G. R. van der Werf. 2008. Agricultural intensification increases deforestation fire activity in Amazonia. *Global Change Biol.* 14 (10): 2262–2276.

Morton, D. C., R. S. DeFries, Y. E. Shimabukuro, L. O. Anderson, E. Arai, F. delBonEspirito-Santo, R. Freitas, and J. Morisette. 2006. Cropland expansion changes deforestation dynamics in the southern Brazilian Amazon. *Proc. Natl. Acad. Sci. U.S.A.* 103 (39): 14637–14641. doi:10.1073/pnas.0606377103.

Murty, D., M. F. Kirschbaum, R. E. McMurtrie, and H. McGilvray. 2002. Does conversion of forest to agricultural land change soil carbon and nitrogen? A review of the literature. *Global Change Biol.* 8: 105–123.

Myneni, R. B., C. D. Keeling, C. J. Tucker, G. Asrar, and R. R. Nemani. 1997. Increased plant growth in the northern high latitudes from 1981–1991. *Nature* 386: 698–701.

Myneni, R. B., S. O. Los, and G. Asrar, G. 1995. Potential gross primary productivity of vegetation from 1982–1990. *Geophys. Res. Lett.* 22: 2617–2620.

Natali, S. M., E. A. G. Schuur, E. E. Webb, C. E. Hicks Pries, and K. G. Crummer. 2014. Permafrost degradation stimulates carbon loss from experimentally warmed tundra. *Ecology* 95 (3): 602–608.

Nemani, R. R., C. D. Keeling, H. Hashimoto, W. M. Jolly, S. C. Piper, C. J. Tucker, R. B. Myneni, and S. W. Running. 2003. Climate-driven increases in global terrestrial net primary production from 1982–1999. *Science* 300: 1560–1563.

Pan, Y., R. A. Birdsey, J. Fang, R. Houghton, P. E. Kauppi, W. A. Kurz, O. L. Phillips et al. 2011. A large and persistent carbon sink in the world's forests. *Science* 333: 988–993.

Phillips, O. L., S. L. Lewis, T. R. Baker, K.-J. Chao, and N. Higuchi. 2008. The changing Amazon forest. *Philos. Trans. R. Soc. B* 363: 1819–1828.

Pongratz, J., C. H. Reick, R. A. Houghton, and J. I. House. 2014. Terminology as a key uncertainty in net land use and land cover change carbon flux estimates. *Earth Syst. Dyn.* 5: 177–195.

Pongratz, J., C. Reick, T. Raddatz, and M. Claussen. 2008. A reconstruction of global agricultural areas and land cover for the last millennium. *Global Biogeochem. Cycles* 22: GB3018. doi:10.1029/2007GB003153.

Post, W. M. and K. C. Kwon. 2000. Soil carbon sequestration and land-use change: Processes and potential. *Global Change Biol.* 6: 317–327.

Ramankutty, N. and J. A. Foley. 1999. Estimating historical changes in global land cover: Croplands from 1700 to 1992. *Global Biogeochem. Cycles* 13: 997–1027.

Raupach, M. R., M. Gloor, J. L. Sarmiento, J. G. Canadell, T. L. Frölicher, T. Gasser, R. A. Houghton, C. Le Quéré, and C. M. Trudinger. 2014. The declining uptake rate of atmospheric CO$_2$ by land and ocean sinks. *Biogeosciences*, in press.

Rosenqvist, A., M. Shimada, N. Ito, and M. Watanabe. 2007. ALOS PALSAR: A pathfinder mission for global-scale monitoring of the environment. *IEEE Trans. Geosci. Remote Sens.* 45: 3307–3316.

Saatchi, S. S., N. L. Harris, S. Brown, M. Lefsky, E. T. A. Mitchard, W. Salas, B. R. Zutta et al. 2011. Benchmark map of forest carbon stocks in tropical regions across three continents. *Proc. Natl. Acad. Sci. U.S.A.* 108: 9899–9904.

Saatchi, S. S., R. A. Houghton, R. C. dos Santos Alvala, J. V. Soares, and Y. Yu. 2007. Distribution of aboveground live biomass in the Amazon basin. *Global Change Biol.* 13: 816–837.

Shevliakova, E., S. Pacala, S. Malyshev, G. Hurtt, P. C. D. Milly, J. Caspersen, L. Sentman, J. Fisk, C. Wirth, and C. Crevoisier. 2009. Carbon cycling under 300 years of land use change: Importance of the secondary vegetation sink. *Global Biogeochem. Cycles* 23: 1–16.

Skole, D. L. and C. J. Tucker. 1993. Tropical deforestation and habitat fragmentation in the Amazon: Satellite data from 1978 to 1988. *Science* 260: 1905–1910.

Tollefson, J. 2013. Budget crunch hits Keeling's curves. *Nature* 503: 321–322.

Treuhaft, R. N., B. D. Chapman, J. R. dos Santos, F. G. Gonçalves, L. V. Dutra, P. M. L. A. Graça, and J. B. Drake. 2009. Vegetation profiles in tropical forests from multibaseline interferometric synthetic aperture radar, field, and lidar measurements. *J. Geophys. Res.* 114, D23110. doi:10.1029/2008JD011674.

Tyukavina, A., S. Stehman, P. Potapov, S. A. Turubanova, A. Baccini, S. J. Goetz, N. T. Laporte, R. A. Houghton, and M. C. Hansen. 2013. National-scale estimation of gross forest aboveground carbon loss: A case study of the Democratic Republic of the Congo. *Environ. Res. Lett.* 8:044039 (14pp); doi:10.1088/1748-9326/8/4/044039.

van der Werf, G. R., J. Dempewolf, S. N. Trigg, J. T. Randerson, P. S. Kasibhatla, L. Giglio, D. Murdiyarso et al. 2008. Climate regulation of fire emissions and deforestation in equatorial Asia. *Proc. Natl. Acad. Sci. U.S.A.* 105: 20350–20355.

van der Werf, G. R., J. T. Randerson, L. Giglio, G. J. Collatz, M. Mu, P. S. Kasibhatla, D. C. Morton, R. S. DeFries, Y. Jin, and T. T. van Leeuwen. 2010. Global fire emissions and the contribution of deforestation, savanna, forest, agricultural, and peat fires (1997–2009). *Atmos. Chem. Phys.* 10: 11707–11735. doi:10.5194/acp-10-11707-2010.

Williams, C. A., G. J. Collatz, J. Masek, and S. Goward. 2012. Carbon consequences of forest disturbance and recovery across the conterminous United States. *Global Biogeochem. Cycles* 26: GB1005. doi:10.1029/2010GB003947.

Woodwell, G. M., J. E. Hobbie, R. A. Houghton, J. M. Melillo, B. Moore, A. Park, B. J. Peterson, and G. R. Shaver. 1984. Measurement of changes in the vegetation of the earth by satellite imagery. In: G. M. Woodwell (ed.), *The Role of Terrestrial Vegetation in the Global Carbon Cycle: Measurement by Remote Sensing*, SCOPE 23 (pp. 221–240). Chichester, U.K.: John Wiley & Sons.

Wunsch, C., R. W. Schmitt, and D. J. Baker. 2013. Climate change as an intergenerational problem. *Proc. Natl. Acad. Sci. U.S.A.* 110: 4435–4436.

Zhang, F., J. M. Chen, Y. Pan, R. A. Birdsey, S. Shen, W. Ju, and L. He. 2012. Attributing carbon changes in conterminous U.S. forests to disturbance and non-disturbance factors from 1901 to 2010. *J. Geophys. Res.: Biogeosci.* 117 (G2): 18

Soils

24

Spectral Sensing from Ground to Space in Soil Science: State of the Art, Applications, Potential, and Perspectives

José A.M. Demattê
University of São Paulo

Cristine L.S. Morgan
Texas A&M University

Sabine Chabrillat
Helmholtz Centre Potsdam

Rodnei Rizzo
University of São Paulo

Marston H.D. Franceschini
University of São Paulo

Fabrício da S. Terra
Federal University of Jequitinhonha and Mucuri Valleys

Gustavo M. Vasques
Embrapa Soils

Johanna Wetterlind
Swedish University of Agricultural Sciences

Acronyms and Definitions

AL	Aluminum	CEC	Cation exchangeable capacity
Bic-K	Plant-available potassium	CK	CoKriging
Ca	Calcium	CR	Continuum removed
Cal. Scale	The intended geographical scale for calibration	CV	Cross-validation
		DN	Digital number
		EC	Electrical conductivity
		Evoi	Expected value of information

F	Farm or field	RS	Remote sensing
FAR	Far infrared	RT	Regression tree
Fe	Iron	SAR	Synthetic aperture radar
FSS	Field spectral sensing	SEP/SECV	Standard error of prediction or cross-validation
G	Global (more than one country)		
GAM	Generalized additive model	Si	Silicon
GIS	Geographic information system	SIC	Soil inorganic carbon
GPS	Global positioning system	SL	Spectral library
GSS	Ground spectral sensing	SM	Soil moisture
H	hyperspectral sensor	SMA	Spectral mixture analysis
Hg	Mercury	SMAP	Soil moisture active passive
IC	Inorganic carbon	SNR	Signal-to-noise ratio
K	Potassium	SOC	Soil organic carbon
KED	Kriging with external drift	SOM	Soil organic matter
LIDAR	Light detection and ranging	Sr	Strontium
L-MEB	L-band microwave emission of the biosphere	SS	Spectral sensing
		SSS	Space—Aerial and Orbital—Spectral Sensing
LOO	Leave-one-out		
LSS	Laboratorial spectral sensing	SVMR	Support vector machine regression
M	Multispectral	SWIR	Shortwave infrared
MAOM	Mineral-associated organic matter	TC	Total carbon
Mg	Magnesium	tbd	to be demonstrated
MIR	Middle infrared (2,500–25,000 nm)	TIR	Thermal infrared
Mn	Manganese	UV	Ultraviolet
N	Number of published figures used in to derive the range and median of R^2 and the errors	UV–VIS–NIR	250–2500 nm (the exact spectral range in the individual studies may deviate but will stay within these ranges)
N. cal.	Number of calibration samples	V	Vanadium
Na	Sodium	VIS	Visible
NDII	Normalized difference index	VIS–NIR	350–2500 nm
NDVI	Normalized difference vegetation index	λ	Wavelength in nm
NIR	Near-infrared (780–2500 nm)	ν and δ CoKriging	Energy levels of fundamental vibration in microscopic interactions
NPV	Nonphotosynthetic vegetation		
NSMI	Normalized soil moisture index	γ	Gamma
OC	Organic carbon, sometimes calculated from soil organic matter by multiplying with 0.58		
P	Phosphorous		
PAWC	Plant-available water capacity		
Pb	Lead		
PLMR2	Polarimetric L-band multibeam radiometer		
PLSR-K	PLSR with kriging interpolation		
POM	Particulate organic matter		
PS	Proximal sensing		
PSR	Penalized-spline signal regression		
PV	Photosynthetic vegetation		
Pred.	Prediction/validation using independent data (samples not used to calibrate the model)		
R/N	Regional or national		
RCGb	Index related to soil weathering		
RID	Reflexion inflexion difference		
RK	Regression-kriging		
RMSEP/RMSECV	Root mean square error of prediction and cross-validation		

Industrial and agricultural activities are developing faster than the public policy on the use of soil resources. The world needs more information about soil for land use planning and interpretative purposes. Spectral sensing (SS) has emerged as a major discipline in remote sensing (RS) science in the past years providing important tools to assist in soil information gathering, mapping, and monitoring. This chapter aims to discuss the role of SS (covering the visible, infrared, thermal, microwave, and gamma ranges of the spectrum) in soil science based on different sensors, scales, and platforms (laboratory, field, aerial, and orbital). We review the state of the art and provide guidance on how to use SS for several purposes, for example, soil classification and mapping, attribute quantification, soil management, conservation, and monitoring. Research has shown that SS has the capability to quantify soil attributes, such as clay, sand, soil organic matter (SOM), soil organic carbon (SOC), cation exchangeable capacity (CEC), Fe_2O_3, carbonates, and mineralogy with reliable and repeatable results. Other soil attributes including pH, Ca, Mg, K, N, P, and heavy metals have also been evaluated with variable outcomes. Laboratory and field-based measurements are more accurate than aerial or space-based measurements as they are conducted under more controlled environments that are less affected by external factors, such as mixing in the field-of-view, vegetation cover, stone cover, water content, and atmospheric conditions. Nevertheless, soil is typically evaluated from space using multispectral sensors on board satellites, which offer many options in terms of temporal and spatial coverage and resolution, and are commonly available free of charge. On the other hand, hyperspectral images are less commonly applied due to their more limited choices of temporal and spatial resolutions and difficulty of processing, despite their great potential to correlate with various soil properties. Other SS techniques, such as passive gamma spectroscopy, provide data for surface and below-surface soil inference, primarily relating to the clay content and types of soil minerals, while microwave (i.e., radar) spectroscopy is mainly used in the study of soil moisture. In soil science, there are promising results and growing interest for visible-near-infrared (VIS-NIR) and middle infrared (MIR) spectroscopy as they allow quick, nondestructive, and cost-effective estimation of soil properties, reducing the need for sample preparation and the use of reagents, minimizing pollution. It has been observed that MIR spectroscopy can quantify properties, such as clay, clay-sized mineralogy, SOC, and inorganic carbon (C), more accurately than VIS-NIR. Both physical (descriptive interpretation of spectral information) and statistical (mathematical approach) methods proved to be useful depending on soil and environmental conditions under study. We observed that the most important limitation of VIS-NIR and MIR spectroscopy for soil classification are their inability to detect soil morphological properties (e.g., soil structure). In the case of VIS-NIR space SS, the limitation is that the radiation only penetrates a few centimeters into the soil surface. On the other hand, satellite-based VIS-NIR data can be used for delineation of soil boundaries supporting soil survey and mapping. SS applicability is also increasing in precision agriculture (PA), coupled with on-the-go sensors that measure soil properties with high sampling density and in real time. Future advances in SS include (1) extraction of moisture effects from intact and field moist spectra, allowing a comparison with laboratory measurements; (2) development of local, regional, or global soil spectral libraries and their appropriate use; and (3) combining multiple sources of sensed data for better soil inference. Country-based soil spectral libraries started in the early 1980s and today we are moving toward a global spectral library (SL) with contribution from as many as 90 countries. Soil spectral libraries, from global to local, will be the future of soil analysis carrying both spatial and hyperspectral data to derive soil information. SS has the advantage of providing quantitative data, and thus reducing the subjectivity of soil spatial information for decision making. SS techniques are powerful when combined with geoprocessing, landscape modeling, geology, and geomorphology. The past and new studies on soil ground SS indicate strong information with a great perspective on all SS platforms, specially for existing hyperspectral aerial and orbital sensors and new ones that are being developed and will be launched soon (2017–2020).

The goal of all SS techniques is to deliver spatially and spectrally accurate, reliable, and transferable information on soil properties. In order to achieve this, SS applications need to properly account for specific advantages and limitations of each sensor, depending on the overall aim. In summary, it is clear that SS can be applied in any field of interest of soil science, depending only on the user's creativity.

24.1 Soils

24.1.1 Definition and Classification

Soil might be defined as the nonconsolidated part of the terrestrial crust or more specifically as "a continuous and three-dimensional natural body, in constant development, formed by organic and mineral constituents, including solid, liquid and gaseous phases organized in specific structure on a certain pedological medium" (IUSS Working Group WRB, 2014). A *soil body*, as defined in pedology, is governed by the interaction of soil factors and formation processes that *sculpt* the soils, and is usually represented by basic units called pedons. As stated by Buol et al. (2011), soil reveals a vertical arrangement of components that change, often gradually, as one traverses the landscape. Our understanding of soil is limited without the use of chemical, mineralogical, biological, and physical quantification techniques, which characterize samples. Besides, soils can be dismembered, sampled, and autopsied, but this analysis will help only if we understand what a soil is and how it functions in the ecosystem. Identifying the horizontal and vertical arrangements of soil attributes, that is, across space and at the topsoil, subsoil, and genetic diagnostic horizons, is essential for soil classification. *Perhaps, no single problem has plagued soil scientists more than the identification of the spatial boundaries of an individual soil on the landscape.* SS may contribute in identifying horizontal and vertical soil boundaries.

Due to the complex interaction that takes place in the pedogenetic processes, the grouping of similar pedons in soil mapping units, by defined boundaries, is normally based on landscape and soil profile descriptions. To recognize the soil as a distinct individual is very efficient from several points of view. First, it allows us to structure our knowledge in the form of individual groups, also known as classes. Second, it facilitates the drawing of thematic maps where classes are spatially delineated (Legros, 2006). For better communication between members of the soil science community, various soil classification systems have been established, such as Soil Taxonomy (Soil Survey Staff, 2014) and World Reference Base for Soil Resources (IUSS Working Group WRB, 2014).

24.1.2 How Does Soil Form?

Soil formation starts with the weathering of the parent material (e.g., the original rock) through physical and chemical processes over time, where climate and organisms have a major role,

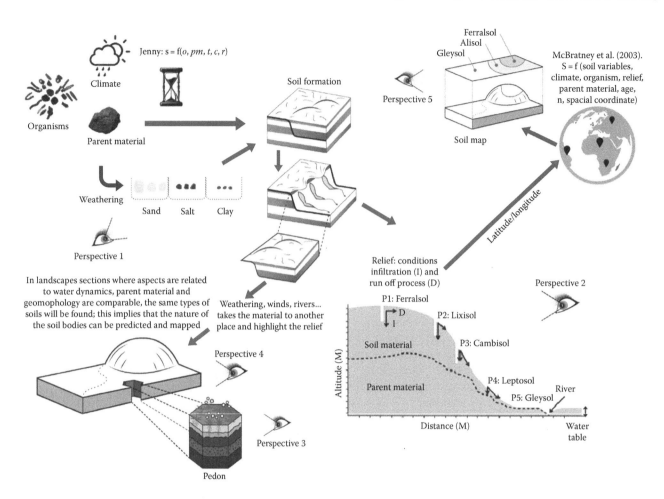

FIGURE 24.1 Conceptual soil and landscape formation. Perspective 1: microperspective, intrinsic soil characteristics, such as mineralogy; Perspective 2: longitudinal vision; Perspective 3: soil profile or site perspective; Perspective 4: related with the soil surface composed by the landscape elements, such as landforms, slopes, and drainage patterns; Perspective 5: spatial vision and distribution of the soil.

and all are influenced by relief (Figure 24.1). The soil-forming factors described here represent soil formation processes that occur under different time and spatial scales and with variable intensities. This concept can be summarized by the conceptual equation proposed by Jenny (1941): S = f (c, o, r, p, t); c (or cl), climate; o, organisms; r, relief; p, parent materials; t, time. With the advent of new methods and technologies for soil evaluation, such as global positioning systems and geographic information systems (GISs), it has been possible to comprehend soils from a different point of view. McBratney et al. (2003) proposed a renewed model, known as "scorpan," where two factors are added to Jenny's equation, namely, the s factor corresponding to soil data available at the beginning of the mapping process, including soil maps, data acquired by means of remote or proximal sensing (PS), and expert pedological knowledge; and the n factor representing the geographic position of the soil. The collection of soil-forming factors (s, c, o, r, p, a, n) represents the underlying landscape characteristics that have allowed the formation of a particular soil class. To completely understand the soil, considering this conceptual soil formation process, and to be able to model this process and map soils, it is important to have different views (Figure 24.1). Thus, we can *look* at soil from

five different perspectives as suggested (Figure 24.1): Perspective 1: micro-perspective, intrinsic soil characteristics, such as mineralogy; Perspective 2: longitudinal vision; Perspective 3: soil profile or site perspective; Perspective 4: related with the soil surface composed by the landscape elements, such as landforms, slopes, and drainage patterns; Perspective 5: spatial vision and distribution of the soil from space. All of these points of view will assist users to understand and visualize the soils as a complete body in which SS can greatly contribute, and we will discuss along this chapter.

24.2 Why Is Soil Important?

The world population is expected to exceed 8 billion people by 2050 (FAO, 2013). Today we have about 867 million chronically undernourished people in the world. At present, more than 1.5 billion ha (about 12%) of the globe's land surface (13.4 billion ha) is or can be used for crop production, 28% (3.7 billion ha) is under forest, and 24%–35% (4.6 billion ha) comprises grasslands and woodland ecosystems (FAO, 2011, 2013). To accomplish the task of feeding almost 9 billion people, one of the most important strategies is closing the yield gap (Godfray et al.,

2010), which requires good land use planning for the existent fields and also the upcoming ones. What has been observed is that farmers who seek higher yields (although with higher costs) promote a better use of natural resources (e.g., fertilizers, water, and soil) and consequently are responsible for a lower impact in the environment. In 2006, the World Conference for Structural Geospatial Data Base indicated the necessity to develop soil maps, which are basic information for crop management planning and consequently has the potential to improve food production.

If on one hand we need to use soils, on the other we are degrading them. Mahmood (1987) calculated by 1986 that around 1100 km³ of sediment had been lost from soils and accumulated in the world's reservoirs, consuming almost one-fifth of the global soil storage capacity. Lal (2003) determined a rate of 201.1 Pg year⁻¹ of soil erosion. Indiscriminate use of herbicides and pesticides all over the world leads to soil and water pollution (Center for Food Safety, 2008; Singh and Ghoshal, 2010). Some herbicides, such as glyphosate and atrazine, might reduce enzyme activity and populations of organisms in soil (Sannino and Gianfreda, 2001). Microbial community structure, often used as an indicator in monitoring soil quality, is affected by various environmental and plant growth factors, such as moisture, temperature, nutrient availability, and management practices (Petersen et al., 2002; Ratcliff et al., 2006).

Soils are relevant not only for food security or environmental quality, but also in climate change issues. In fact, according to Lang (2008), the atmospheric abundance of carbon dioxide (CO_2) has increased by 36%, from 280 ppm in 1750 to 381 ppm in 2006 (Canadell et al., 2007) and in 2014 reached 398.87 ppm (Dlugokencky and Pieter, 2014).

Land use change contributed 158 Pg C, where the deforestation and the attendant biomass burning, and soil tillage along with erosion, contributed with an estimated emission of 78 ± 12 Pg C from world soils (Lal, 1999). Soils can sequester C by increasing stable SOC and soil inorganic carbon (SIC) stocks through judicious land use and recommended management practices. Research has shown a clear link between SOC (stock and change) and the type of soil, which stresses the need for soil maps. Bruinsma (2009) indicates that until 2050, "Arable land would expand by some 70 million ha (or less then 5 percent), the expansion of land in developing countries than about 120 million ha (or 12 percent) being offset by a decline of some 50 million ha (or 8 percent) in the develop countries." In fact, recent work (Spera et al., 2014) indicates agriculture expansion in one of the most important states in Brazil, that is, Mato Grosso. Authors observed 3.3 million ha of mechanized agriculture in 2001, of which 500,000 ha had two commercial crops per growing season (double cropping). By 2011, Mato Grosso had 5.8 million ha of mechanized agriculture, of which 2.9 million ha was double-cropped, an increase of 76%. This is a clear indication of the necessity to have strategies on land use planning to reach a high-quality use of soils.

Studies that address comprehension of soils, soil function, and soil in agriculture and society have several applications such as soil classification, pedological and attribute mapping, soil monitoring, soil conservation, experimental design, spatial allocation for agriculture, food improvement, local sustainability, environment productivity systems, agriculture systems, specific plant and soil interactions, PA, irrigation systems, determination of land use capacity and land aptitude, and many others. Today, to improve agricultural productivity while securing future water quality (and quantity) as well with environment quality, it is necessary to understand soil function in natural and agricultural ecosystems that are tied together over landscapes, watersheds, and larger spatial extents. Nonetheless, soil management occurs at the 1–100 m scale, while soil policy occurs at broader spatial scales (political boundaries). Hence, knowledge of soils that includes detailed spatial information is imperative to understand the impact of management and policy decisions on soils, water, and the environment at large. In the following sections, we show how SS can assist in all these aspects.

24.3 Role of Spectral Sensing in Soil Science

24.3.1 Concepts

RS has copious definitions in the literature. The most simple and effective one refers to the "acquisition of information about an object by detecting its reflected or emitted energy without being in direct physical contact with it" (Colwell, 1997; Jensen, 2006). Therefore, the term RS can be applied for a radiation-detecting sensor installed at any platform or level of acquisition, for instance, laboratory, field, aerial, or orbital (Jensen, 2006). Recently, the expression PS was proposed for soil studies (Viscarra Rossel et al., 2011a). This concept focuses on detecting information about an object in a short distance, primarily in the field, using not only spectral sensors, but also any measurement device. For spectral sensor applications, the concepts of PS and RS, although obvious in some contexts, can be confused. Generally speaking, the term "remote" means that the sensor is "far" from the target, while the term "proximal" means the sensor is "near" the target, but the specific distance that distinguishes one from the other (1 cm, 1 m, or 1 km) is not defined. In fact, the word takes in consideration arbitrary "distance." Commonly (by convention or tradition), RS is applied mostly for orbital and aerial acquisition levels, whereas PS is used for laboratory and field sensors. In fact, as stated by Jensen (2006), all of these concepts are correct in their appropriate context. Thus, in this chapter, we define SS as (1) "spectral": related to the electromagnetic radiation spectrum coming from an object that is dependent on its characteristics and composition and (2) "sensing": referring to the acquisition of this spectral information without directly touching the object. This definition removes the relativity from terms proximal and remote. The SS terminology can be divided into space spectral sensing (SSS), which includes aerial and orbital (also known as RS), and ground spectral sensing (GSS), divided into field spectral sensing (FSS) when spectra are taken in natural conditions in the field, and laboratory spectral sensing (LSS) when spectra are read in the lab using benchtop instruments. GSS is also known as PS. Thus, by definition, SS can be applied using different acquisition distances (scales) and sources, from ground to space. This is our suggestion to better discuss the chapter and to not have the pretension to substitute the traditional terms RS and PS.

24.3.2 How Spectral Sensing Contributes to Soil Science? Why Is It a New Perspective of Science?

The usual way to study soils is by discretizing the knowledge into basic disciplines such as physics, microbiology, pedology, and chemistry. Each discipline has accepted methodologies for analysis, which will allow them to understand and quantify soil properties. The interpretation is related to the results that are commonly obtained from traditional soil laboratory analysis involving chemical, physical, and biological assessments. Spectral readings are physical information that, in many cases, requires models to relate to soil properties. The five perspectives (or points of view) to *look* at soils indicated in Figure 24.1 can

also be observed by an SS point of view. In fact, these perspectives can be related with Figure 24.2. The microscale perspective, related with the soil analysis, can be related with the ground, laboratory, or field sensors. SS can study soils by the longitudinal perspective, where we can *see* the relief from only one perspective, along a toposequence. The soil profile is related to the observation of a single point, which can be detected by ground sensing on surface, inside a pit or a borehole, or a pixel from an image. In the case of the surface landscape perspective, the perception of SS corresponds to the combination of all elements of the relief (shape, slope, and height). Finally, the spatial perspective, which is the visualization of soil with its boundaries, can be detected from space. The pedologist combines different perspectives to analyze soils depending on the objective. Each one of these will add information about the soil.

24.3.3 History and Evolution

We can consider the beginning of SS when people saw and distinguished objects (in our case, soils) by their shape or color (using the eye). The NIR spectral range from 780 to 2500 nm was discovered by Herschel in the year 1800 (Hershel, 1800).

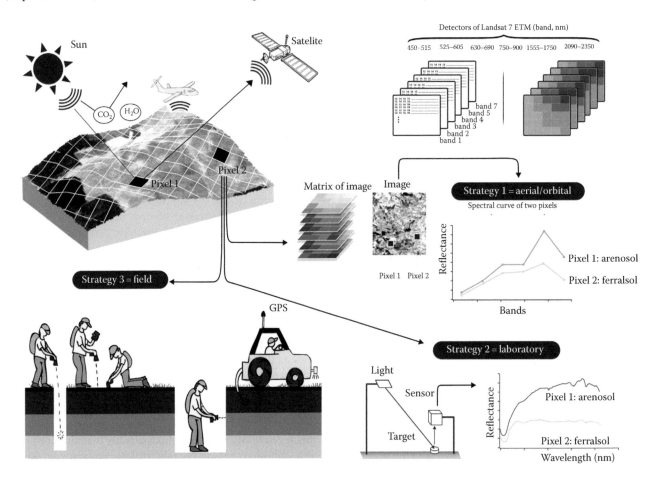

FIGURE 24.2 Alternatives for soil evaluation from ground to space: strategy 1: surface spectral curve from orbital source; strategy 2: ground (laboratory) spectral curve of a soil sample taken from the same pixel (location) of the image; strategy 3: ground spectral data in field (at the topsoil or subsoil measured inside a pit or with a sensor inside the hole to measure underneath).

According to Stark et al. (1986), the NIR was discovered before the MIR region due to the instruments available at the time. But both spectral regions were neglected during years and started their studies only in the 1960s (Stark et al., 1986). Before this period, SS was applied using aerial photographs. In 1826, Joseph Nicéphore Niépce (1765–1833) took the first photography in France (Gernsheim and Gersheim, 1952). In 1890, Arthur Batut published *Aerial Photography by Kites*, which shows the use of aerial photography for agronomy, archaeology, exploration, and military uses (Batut, 1890). Aerial photographs for soil survey and applications were determined by Buringh (1960). Later, Vink (1964) published a very important work, which demonstrated how to use the aerial photographs for a soil survey, with examples given by Goosen (1967) and afterward Hilwig et al. (1974).

In the early 1900s, Bernad Keen and William Haines built the first on-the-go soil sensor (Keen and Haines, 1925). Four decades later, Obukhov and Orlov (1964) selected the 750 nm spectral region to estimate SOC, and Bowers and Hanks (1965) published a paper about the correlation between soil reflectance, soil moisture, and particle size. These pioneering studies were followed by Hunt and Salisbury (1970) and Hunt (1982), which proved that water and minerals in the soil have unique spectral fingerprints that can be identified, quantified, and further used for the assessment of other soil properties. This took soil spectroscopy to another level, and the first scientists to systematically gather soil spectral information and publish it in the form of a soil spectral atlas were Stoner et al. (1980). These works guided spectroscopy for decades. In fact, findings from Vink (1964) and Stoner and Baumbgardner (1981) took Demattê et al. (2001, 2004) to integrate and apply these technologies with the first soil mapping system using spectral information combined with aerial images.

The combination of different sensors and platforms for soil assessment started with Krinov (1947), who combined laboratory data, field spectrographs, and aerial spectrographs adapted to operate from aircraft, in the VIS–NIR region (400–910 nm) for soil research. In fact, maybe the most important event for SS happened in 1957, when the former USSR managed to keep the first satellite in Earth orbit (Sputnik). In the 1960s, a new era began for SS with new sensors onboard satellites capable of taking pictures of Earth (McDonald, 1997). In 1965, the United States (Conterminous, Alaska, and Hawaii) was mapped by the Corona satellite (Clark, 1999a). The first Landsat was launched in 1972, called Landsat-MSS (Later Landsat-1). In 1999, a new generation of multispectral satellite sensors were launched, including Landsat-7 ETM+, Terra Moderate-Resolution Imaging Spectroradiometer (MODIS) and Advanced Spaceborne Thermal Emission and Reflection Radiometer (ASTER), CBERS IRS (Brazil and China), and IKONOS-2. In 2001, QuickBird was launched; in 2002, NASA launched the Aqua satellite carrying a MODIS sensor, and in 2005, Google started using data from satellites creating the Google Earth—a great upgrade on SS for humanity. Almost in parallel, important advances in hyperspectral sensing occurred, starting with an airborne platform—the AVIRIS sensor with 224 bands (400–2500 nm)—operational in

1989 (Vane et al., 1993), followed by the spaceborne Hyperion sensor with 220 bands (400–2500 nm) in the Earth-Observing 1 satellite, launched in 2000 (USGS, 2014).

24.4 Theory behind Soil Spectral Sensing

24.4.1 Visible, Near-Infrared, Shortwave-Infrared, and Mid-Infrared

The spectral ranges most used in soil science include the ultraviolet (UV: 200–380 nm), visible (VIS: 350–700), near-infrared (NIR: 700–1,000 nm), shortwave infrared (SWIR: 1,000–2,500 nm), and mid-infrared (MIR: 2,500–25,000 nm). The region 700–2,500 is generally referred to as NIR, whereas the MIR region at 2,500–25,000 can be divided into thermal infrared (TIR: 8,000–14,000 nm) and far infrared (14,000–25,000 nm). Fundamental studies on soil and SS in these ranges were conducted in the laboratory. Spectral reflectance is extremely complex since it is affected by the concentration, size, and arrangement of components occurring in soil samples. All sample constituents mix and interact to reach the final reflectance information, however, with different aspects of interactions among them (Bowers and Hanks, 1965). The measured soil spectra result from the combination of the intrinsic spectral behavior of different soil constituents including organic materials, mineral materials, and water, which interact differently with the incident light (Clark, 1999b). The percentage of incident light that is reflected by the soil at different wavelengths constitutes the soil's reflectance behavior and is represented by the soil's spectral curve, or simply soil spectrum. Thus, soil spectra are a result of microscopic interactions between the atoms and molecules of the soil and the incident light, which penetrates the first 10–50 µm of soil. These interactions generate specific absorption features at different wavelengths and for different soils, and are mainly affected by soil particle size, porosity for energy dispersion, SOM, and soil water (Jensen, 2006). Therefore, each soil sample has a unique spectral behavior and a unique spectral curve.

Microscopic interactions with light occur in mineral (siloxane, oxyhydroxy, silanol, aluminol, and ferrol surfaces) and organic (carboxyl and phenolic hydroxyl) functional groups of soils at the same place where other soil physical–chemical reactions take place as well (Alleoni et al., 2009). Thus, these interactions are dependent on soil composition with respect to types of atomic and molecular structures, strength of bonds among atoms and molecules, and ionic impurities in soils, such as chemical elements adsorbed on colloidal particles (Hollas, 2004). Spectral absorption features related to microscopic effects will appear only if enough energy is available to promote atomic transitions and/or molecular vibrations (Bernath, 2005). There are two levels of electromagnetic energy absorption that must be considered in soil microscopic interaction: atomic and molecular. Atomic refers to radiation required to produce changes in the energetic level of electrons linked to ions in the soil, causing electronic transitions and rearrangement of charges in atoms

(Sherman and Waite, 1985). Atomic interactions usually take place from VIS to NIR, except for metals (Cu, Fe, Mn, Cr, and Ti) and other elements (Si, Al, K, Ca, and OH-) whose electronic transitions do not produce spectral features at this range.

Light interactions at the molecular level are responsible for the vibrational processes of molecules with its functional groups (e.g., –COOH, –CH$_x$, NH$_x$) in the soil and usually take place after 1100 nm. Molecular vibrations need less energy, but produce more intense and defined absorption features (Stenberg et al., 2010). Molecular vibrations include stretching, bending, and torsion, and are divided into fundamental and nonfundamental vibrations. Nonfundamental vibrations are considered secondary vibrations of lower intensity, such as propagations (or reverberations) of fundamental vibrations, which can be overtones or combination tones. They usually occur in the NIR as a result of interactions of radiation with molecular functional groups, for example, –CO$_x$, –PO$_x$, –CH$_x$, and H$_2$O (Viscarra Rossel et al., 2011b). Fundamental vibrations occur in the MIR and are usually derived from associations of Fe^{3+}, Al^{3+}, Mn^{2+}, and Si^{4+} with O in soil oxides (Viscarra Rossel et al., 2008). The vibrational absorptions in VIS–NIR are overtones and combination frequencies of absorption bands that occur in the MIR (Figure 24.2).

Table 24.1 summarizes the most important wavebands and respective soil attributes. Absorption features have been coming from interaction between electromagnetic energy and soil functional groups in colloidal surfaces. It has been observed a great variety of organic and inorganic functional groups, such as carboxyl, alkyl, methyl, phenolic, aliphatic groups, and others for organic ones, and, mainly, hydroxyl (–OH), silanol (Si]–OH), and aluminol (Al]–OH) for inorganic ones on outer surfaces of clay minerals (phyllosilicates) (Table 24.1). Overlaps of bands and positioning modifications can be observed because molecules of soil constituents are all mixed in a complex system and do not behave harmoniously (Bishop et al., 1994). Vis–NIR–SWIR spectral ranges have shown features related to electronic transition of iron from oxides (hematite) and hydroxides (goethite) in the VIS and NIR, and other features due to nonfundamental vibrations (overtones and combination tones) on molecular structures of 2:1 (vermiculite, smectite, illite, and others) and 1:1 (hematite) clay mineral and Al oxide (gibbsite) have been identified from SWIR. Features due to soil organic compounds have also been identified along VIS–NIR–SWIR spectral ranges (Table 24.1). Due to the stronger microscopic interaction of MIR radiation and soil organic and mineral particles by fundamental vibrations (Janik et al., 1998), a greater number of absorption features can be observed in this range when compared to Vis–NIR–SWIR. This amount of features results in much more information about soil properties in MIR, except for iron contents that have not been identified in this range. Besides the 2:1 and 1:1 clay minerals, features of quartz (from sand particles), phosphates, carbonates, borates, nitrates, and a great variety of organic components have been identified along MIR (Table 24.1). Nonfundamental and fundamental absorptions from Vis–NIR–SWIR and MIR ranges were compiled from Oinuma and Hayashi (1965), Bowers and Hanks (1965), Hunt and Salisbury (1970), White (1971), van der Marel and Beutelspacher (1975), Stoner and Baumgardner (1981), Baumgardner et al. (1985), Sherman and Waite (1985), White and Roth (1986), Nguyen et al. (1991), Salisbury and D'aria (1992, 1994), Srasra et al. (1994), Janik et al. (1998), Fernandes et al. (2004), Madejova and Komadel (2001), Viscarra Rossel et al. (2006, 2008), Stevens et al. (2008), Richter et al. (2009), and Viscarra Rossel and Behrens (2010).

24.4.2 Microwaves

Microwaves constitute the portion of the electromagnetic radiation (spectrum) with wavelengths ranging from 0.1 to 100 cm (frequencies from 300 MHz to 300 GHz) and photon energy from 1.24 meV to 1.24 µeV. The earth's surface emits low levels of microwave radiation, which is modeled by the Rayleigh–Jeans law for black bodies (Jensen, 2006). The radiometric magnitude detected by the sensors in microwave spectral range is called brightness temperature and is expressed by TB = εT, where T is an object's physical temperature and ε is the emissivity. Passive microwave radiometers detect energy in the spectral range between 0.15 and 0.30 cm, with frequencies from 1 to 200 GHz. These sensors have poor spectral and spatial resolutions due to low microwave energy emitted from the earth. Nevertheless, there is growing interest in the measurement of passive microwave radiation to monitor soil moisture conditions (Liu et al., 2012; Mladenova et al., 2014). Active microwave systems, also known as RADAR (radio detection and ranging), differ from passive systems because they provide an energy source. This energy (incident radiation) interacts with the target, and the response is detected by the sensor. Because they provide their own energy, applications are possible during the day and night. Moreover, differently from VIS–NIR–MIR, microwaves can pass through clouds and plant canopy (depending on the spectral range), and thus can be used under bad weather condition (Woodhouse, 2005) and for ground and understory examination.

24.4.3 Thermal Infrared

Any object with temperature greater than absolute zero (0 K, –273.26°C, –459.69°F) emits some amount of radiant energy, which is generally correlated to the object's temperature (Jensen, 2005). Energy emissions occur in the TIR range, but remotely sensed data acquired aiming to quantify radiant flux are generally restricted to ranges near 8,000–9,000 nm or near 10,000–12,000 nm, since these ranges correspond to "atmospheric windows," in which water vapor (H$_2$O), carbon dioxide (CO$_2$), and ozone (O$_3$), among other components, absorb less energy.

Assuming the imaged surface as an ideal black body, TIR radiance measured by the sensor would be proportional to the energy given by Planck's radiation law (Richards, 2013). However, real surfaces are not ideal black bodies, but instead radiate selectively and emit a certain proportion of the energy

TABLE 24.1 Absorption Features (Band Assignments) of Soil Constituents in the Vis–NIR–SWIR (350–2500 nm) and MIR (4000–400 cm^{-1}) Spectral Ranges

Spectral Range	λ (nm)	Functional Groups	Soil Compounds	Microscopic Interactions	Interaction Modes	Types of Propagation
Visible	404	Fe^{3+} ion	Hematite	Atomic level	Electronic transition	
	430		Goethite			
	444		Hematite			
	480		Goethite			
	520		Hematite			
	650		Goethite			
			Hematite			
Near-infrared	751	N–H	Organic compounds (amine)			$4v_7$ overtone
	825	C–H	Organic compounds (aromatic carbon)			$2v_6$ overtone
	850	Fe^{3+} ion	Hematite	Atomic level	Electronic transition	
	853	C–H	Organic compounds (alkyl)			$4v_8$ overtone
	877					$4v_9$ overtone
	940	Fe^{3+} ion	Goethite	Atomic level	Electronic transition	
	1000	N–H	Organic compounds (amine)	Molecular level	Nonfundamental vibration	$3v_7$ overtone
SWIR	1100	C–H	Organic compounds (aromatic carbon)			$2v_6$ overtone
	1135	(H–O–H) + (O–H)	Soil water			$v_1 + v_2 + v_3$ combination tone
	1138	C–H	Organic compounds (alkyl)			$3v_8$ overtone
	1170					$3v_9$ overtone
	1380	(H–O–H) + (O–H)	Soil water			$v_1 + v_3$ combination tone
	1395	2(O–H)	Kaolinite			$2v_{1a}$ overtone
	1414					$2v_{1b}$ overtone
			2:1 vermiculite			$2v_4$ overtone
			Smectite			$2v_4$ overtone
			Mica (illite)			$2v_4$ overtone
	1449	C=O	Organic compound (carboxylic acids)			$4v_{10}$ overtone
	1455	(H–O–H) + (O–H)	Soil water Hygroscopic water			$2v_2 + v_3$ combination tone e overtone
	1500	N–H	Organic compounds (amine)			$2v_7$ overtone
	1524	C=O	Organic compounds (amide)			$4v_{11}$ overtone
	1650	C–H	Organic compounds (aromatic carbon)			$2v_6$ overtone
	1706	C–H	Organic compounds (alkyl)			$2v_8$ overtone
			Organic compounds (aliphatics)			$4v_{12}$ overtone
	1730		Organic compounds (methyl)			$4v_{13}$ overtone
	1754		Organic compounds (alkyl)			$2v_9$ overtone
	1800	Ca, Mg, Fe, Mn, Sr, Ba—BO$_3^{2-}$	Borate			$3v_5$ overtone
		Ca, Mg, Fe, Mn, Sr, Ba—CO$_3^{2-}$	Carbonate			
	1852	C–H	Organic compounds (methyl)			$4v_{13}$ overtone
	1915	(H–O–H) + (O–H)	Soil water Hygroscopic water			$v_2 + v_3$ combination tone
	1930	C=O	Organic compound (carboxylic acids)			$3v_{10}$ overtone
	1961	C–OH	Organic compound (phenolic)			$4v_{14}$ overtone

(Continued)

TABLE 24.1 (*Continued*) Absorption Features (Band Assignments) of Soil Constituents in the Vis–NIR–SWIR (350–2500 nm) and MIR (4000–400 cm^{-1}) Spectral Ranges

Spectral Range	λ (nm)	Functional Groups	Soil Compounds	Microscopic Interactions	Interaction Modes	Types of Propagation
	1980	Ca, Mg, Fe, Mn, Sr, Ba—BO$_3^{2-}$	Borate			$3v_5$ overtone
		Ca, Mg, Fe, Mn, Sr, Ba—CO$_3^{2-}$	Carbonate			
	2033	C=O	Organic compound (amide)			$3v_{11}$ overtone
	2060	N–H	Organic compounds (amine)			$v_7 + \delta_b$ combination tone
	2135	Ca, Mg, Fe, Mn, Sr, Ba—BO$_3^{-2}$	Borate			$3v_5$ overtone
		Ca, Mg, Fe, Mn, Sr, Ba—CO$_3^{2-}$	Carbonate			
	2137	C–O	Organic compound (polysaccharides)			$4v_{15}$ overtone
	2160	(O–H) + (Al–OH)	Kaolinite			$v_{1a} + \delta$ combination tone
	2205					$v_{1b} + \delta$ combination tone
			2:1 vermiculite			
			Smectite			
			Mica (illite)			
	2230	AlFe–OH	Smectite			$v_{1b} + \delta_a$ combination tone
	2260	(O–H) + (Al–OH)	Gibbsite			$v_{1b} + \delta$ combination tone
	2275	C–H	Organic compounds (aliphatics)			$4v_{12}$ overtone
	2316	C–H	Organic compounds (methyl)			$3v_5$ overtone
		[3.(CO$_3^{2-}$)]	Carbonate			
	2307	C–H	Organic compounds (methyl)			$3v_{13}$ overtone
	2336	Ca, Mg, Fe, Mn, Sr, Ba—BO$_3^{2-}$	Borate			$3v_5$ overtone
		Ca, Mg, Fe, Mn, Sr, Ba—CO$_3^{2-}$	Carbonate			
	2350	(O–H) + (Al–OH)	Mica (illite)			$v_{1b} + \delta$ combination tone
	2381	C–O	Organic compounds (carbohydrate)			$4v_{16}$ overtone
	2382					$4v_5$ overtone
	2450	(O–H) + (Al–OH)	Mica (illite)			$v_{1b} + \delta$ combination tone
	2469	C–H	Organic compounds (methyl)			$3v_{13}$ overtone

Spectral Range	Wavenumber (cm^{-1})	Functional Groups	Soil Compounds	Microscopic Interactions	Interaction Modes	Types of Propagation
Mid-infrared	3695 (v_{1a})	(Al—O–H)OH	Kaolinite	Molecular level	Fundamental vibration	Stretching
			Halloysite			
			Smectite			
			Mica (illite)			
	3670		Kaolinite			
	3653		Smectite			
			Mica (illite)			
	3620 (v_{1b})		Kaolinite			
			Halloysite			
			Chlorite Al-rich			
			Smectite			
			Mica (biotite)			

(*Continued*)

TABLE 24.1 (*Continued*)   Absorption Features (Band Assignments) of Soil Constituents in the Vis–NIR–SWIR (350–2500 nm) and MIR (4000–400 cm⁻¹) Spectral Ranges

Spectral Range	Wavenumber (cm⁻¹)	Functional Groups	Soil Compounds	Microscopic Interactions	Interaction Modes	Types of Propagation
			Mica (muscovite)			
			Mica (illite)			
	3575 (v_4)	O–H	Hydroxyl			
	3560	O–Al–OH	Nontronite			
	3550		Mica (biotite)			
			Chlorite Al-rich			
	3529		2:1 vermiculite			
			Kaolinite			
			Gibbsite			
	3484 (v_1)	O–H	Water			
	3448 (v_3)	PO_4^{3-}	Phosphate			
		O–Al–OH	2:1 vermiculite			
			Kaolinite			
			Gibbsite			
	3400		Halloysite			
	3394	(H–O–H) + (O–H)	2:1 vermiculite			
			Kaolinite			
	3340		Chlorite Al-rich			
	3330 (v_7)	N–H	Organic compounds (amine)			
	3278 (v_1)	O–H	Water			
	3330–3030	N–H	Ammonium (nitrate)			
	3030 (v_6)	C–H	Organic compounds (aromatic carbon)			
	2930 (v_8)		Organic compounds (alkyl)			
			Asymmetric–symmetric doublet			
	2924		Organic compounds (aliphatic)			
	2850 (v_9)		Organic compounds (alkyl)			
			Asymmetric–symmetric doublet			
	2843		Organic compounds (aliphatic)			
	2341		CO_2 (breath)			
	2233	(–C Ξ C–H, –C Ξ C–)	Organic compounds (alkyne groups)	Molecular level	Fundamental vibration	Stretching
		Si–O	Quartz			
	2133	(–C Ξ C–H, –C Ξ C–)	Organic compounds (alkyne groups)			
		Si–O	Quartz			
	1975	Si–O	Quartz			
	1867					
	1790					
	1725 (v_{10})	C=O	Organic compound (carboxylic acids)			
	1678	Si–O	Quartz			
	1645 (v_2)	H–O–H	Water			
	1640	O–Al–OH	Halloysite			
			Smectite			
			Nontronite			
			2:1 vermiculite			
	1640 (v_{11})	C=O	Organic compound (amides—protein)			
	1628	O–H	Kaolinite			Deformation
			Smectite			

(*Continued*)

TABLE 24.1 (*Continued*)   Absorption Features (Band Assignments) of Soil Constituents in the Vis–NIR–SWIR (350–2500 nm) and MIR (4000–400 cm^{-1}) Spectral Ranges

Spectral Range	Wavenumber (cm^{-1})	Functional Groups	Soil Compounds	Microscopic Interactions	Interaction Modes	Types of Propagation
			Mica (illite)			
			2:1 vermiculite			
		Si–O	Quartz			Stretching + deformation
	1610 (δ_b)	N–H	Organic compound (amine)			
	1527	Si–O	Quartz			
		COO–	Organic compound (symmetric)			
		N–H + C=N	Organic compound (amide)			
		C=C	Organic compound (aromatic carbon)			
	1497 (ν_4)	Si–O	Quartz			
		COO–	Organic compound (symmetric)			
		N–H + C=N	Organic compound (amide)			
		C=C	Organic compound (aromatic carbon)			
	1490–1410	C–O	Carbonate			
	1485–1390	N–H	Ammonium (nitrate)			
	1465 (ν_{12})	C–H	Organic compound (aliphatics)			
	1445 (ν_{13})		Organic compound (methyl)			
	1435	Ca—CO$_3^{2-}$	Calcite			
	1415 (ν_5)	Ca, Mg, Fe, Mn, Sr, Ba—BO$_3^{2-}$	Borate			
		Ca, Mg, Fe, Mn, Sr, Ba—CO$_3^{2-}$	Carbonate			
	1362 (ν_4)	OH + C–O	Organic compound (phenolic)			
		C–O	CH$_2$ and CH$_3$ groups (methyl)			
		COO– and –CH	Organic compound (aliphatics)			
		Si–O	Quartz			
	1350 (ν_{13})	C–H	Organic compounds (methyl)			
Thermal infrared	1275 (ν_{14})	C–OH	Organic compounds (phenolic)			
	1170 (ν_{15})	C–O	Organic compounds (polysaccharide)			
	1157	C–OH	Organic compounds (aliphatic)			
		O–Al–OH	Smectite			Deformation
			Mica (illite)			
			2:1 vermiculite			
	1111	Si–O–Si	Kaolinite			Stretching
			Halloysite			
			Smectite			
			Mica (illite)			
			Mica (muscovite)			
			2:1 vermiculite			
			Microcline			
			Nontronite			Deformation
		O–Al–OH	Gibbsite			
	1100–1000	P–O	Phosphate			
	1085–1050	C–O	Carbonate			
	1050 (ν_{16})	C–O	Organic compound (carbohydrate)			
	1018 (ν_5)	C–C	Organic compound (aliphatics)			Stretching
		C–O	Organic compounds (polysaccharide)			
			Organic compounds (carbohydrates)			

(*Continued*)

TABLE 24.1 (*Continued*) Absorption Features (Band Assignments) of Soil Constituents in the Vis–NIR–SWIR (350–2500 nm) and MIR (4000–400 cm⁻¹) Spectral Ranges

Spectral Range	Wavenumber (cm⁻¹)	Functional Groups	Soil Compounds	Microscopic Interactions	Interaction Modes	Types of Propagation
		Si–O–Si	Kaolinite			
			Smectite			
			Mica (illite)			
			2:1 vermiculite			
		Si–O	Quartz			
		Al–O	Gibbsite			
	926 (δ)	Al—O–H	Kaolinite			Deformation
		((Al,Al)—O–H)	Smectite			
			Mica (illite)			
			2:1 vermiculite			
			Gibbsite			
	885 (δₐ)	AlFe–OH	Smectite			Stretching
	875–860	C–O	Carbonate			
	877	Ca—CO₃²⁻	Calcite			
	814	Si–O	Quartz			
			Kaolinite			Deformation
		((Al,Fe)—O–H)	Smectite			
		((Al,Mg)—O–H)	Mica (illite)			
			2:1 vermiculite			
	791	Si–O–Si	Kaolinite			Stretching
			Mica (illite)			
			2:1 vermiculite			
	752		Kaolinite			
			Mica (illite)			
			2:1 vermiculite			
Mid-infrared	750–680	C–O	Carbonate			
	712	Ca—CO₃²⁻	Calcite			
	702	Si–O–Si	Kaolinite			
			Smectite			
			Mica (illite)			
			2:1 vermiculite			
	635–5000	P–O	Phosphate			
	517	Si–O–Si	Kaolinite			Deformation
			Smectite			
			Mica (illite)			
			2:1 vermiculite			
	436	Si–O	Quartz			
			Quartz			

λ, wavelength in nm; ν and δ, energy levels of fundamental vibration in microscopic interactions.

that would be emitted by a black body, at the same temperature. The ratio between real radiance and black body radiance, at a given temperature, is the concept of emissivity. This is an intrinsic property of the material, and thus, the radiant flux that leaves a surface in TIR spectroscopy is proportional to the material's emissivity and temperature (Derenne, 2003). In the case of soils, different factors can affect the spectral response, such as soil composition and moisture. Minerals such as quartz present absorption features in the TIR range, called reststrahlen spectral features, which occur from 8,000 to 10,000 nm (Salisbury and D'Aria, 1992). However, an increase in SOM can reduce reststrahlen features, that is, increase emissivity, especially in soils with more than 2% of organic matter (OM; Breunig et al., 2008). In cases of clay coatings on quartz grains and increases in soil water content, the reststrahlen spectral features are attenuated as well (Salisbury and D'Aria, 1992).

24.4.4 Gamma Ray

Gamma ray has initiated studies in rocks, such as performed by Gregory and Horwood (1961), and can be better understood in Grasty (1979). The electromagnetic energy in the γ-ray range is characterized by short wavelengths (on the order of 0.01 nm or less), with high frequency, and high-energy photons (more than 0.04 MeV). Ground or airborne γ-ray spectrometry uses the fact that γ-ray photons have discrete energies, which are characteristic of radioactive isotopes from which they originate (passive system). Consequently, it is possible to determine the source of the radiation by measuring the energies of γ-ray photons (IAEA, 2003).

Radioactive isotopes of elements that emit gamma radiation are called radionuclides. Many naturally occurring elements have radionuclides, but only potassium (^{40}K), cesium (^{137}Cs), and the decay series of uranium (^{238}U and ^{235}U and their daughters) and thorium (^{232}Th and its daughters) are abundant in the environment and produce γ-rays of sufficient energy and intensity to be measured by γ-ray spectrometry (Mahmood et al., 2013).

An energy spectrum is typically measured by γ-ray spectrometers over the 0–3 MeV range, generally in 256 channels or more (Dierke and Werban, 2013). In the conventional approach, γ-ray measurements are used to monitor four spectral regions of interest, corresponding to the energy levels of K (1.460 MeV), U (1.765 MeV), and Th (2.614 MeV), and to the total radioactivity over the 0.4–2.81 MeV range (Viscarra Rossel et al., 2007). The γ-ray measurements taken by spectrometers are counts of the decay rate (intensity) at the specific energy. The measured intensity (counts per second) can be converted into the activity of the nuclides ($Bq\ kg^{-1}$), into the concentrations (% for K, ppm for U and Th), or into the dose rate ($nGy\ h^{-1}$) using a calibration method (IAEA, 2003).

The concentration of γ-ray emitting radionuclides in soils depends on different soil properties, which are the result of physical and chemical composition of parent rock, as well as soil genesis under different climatic conditions (Dierke and Werban, 2013). Mineralogy of source rocks, clay content, and type of clay minerals are properties that directly influence the radionuclide concentration in sediments and soils (Serra, 1984). When measured from the top, approximately 50% of the observed γ-rays originate from the top 0.10 m of dry soil, 90% from the top 0.30 m, and 95% from the upper 0.5 m of the profile (Taylor et al., 2002). Soil water content and bulk density can attenuate γ-rays. For example, radiation attenuation increases by approximately 1% for each 1% increase in volumetric water content, while a dry soil with a bulk density of 1.6 mg m^{-3} causes a decrease in the radiation to half its value at each 10 cm compared to a moist soil (Cook et al., 1996). The radiation decrease caused by air is much smaller, for example, 121 m are needed to reduce the radiation to half its value considering a 2 MeV source; hence, detection of γ-rays is possible from airborne platforms (Viscarra Rossel et al., 2007).

24.5 Strategies for Soil Evaluation by Spectral Sensing

Strategy is the art of applying knowledge and capabilities to reach a goal. In the case of SS, an effective strategy requires knowledge of the limitations and capabilities of the instrument making measurements. The goal is detecting and identifying soil properties for the purpose of making an inference

regarding the capability of the soil for a specific function. Thus, there are several strategies to use on soil evaluation based on SS, but each strategy is different and should depend on the objectives and tools available. The strategies described here are related to research with the missions of creating, determining, and investigating methods for mapping soil attributes. In fact, quantification of SOM, SOC, pH, clay, sand, and other soil constituents for the purpose of converting to other soil properties using pedotransfer functions and ultimately making a soil inference is the goal of SS, and a challenging one. Strategies may focus on GSS or SSS, and the spatial modeling of the soil constituents can be complemented by terrain models (i.e., digital elevation models—DEMs) and other sources of information (Figure 24.2). As we observe in Figure 24.2, there are several strategies to reach soil information. Strategy 1 implies on using aerial or orbital data. Both will get only soil surface information since the energy penetrates a few centimeters into the soil. Orbital information has to be well atmospheric corrected and transformed into reflectance (as will be seen in subsequent sections). On the other hand, when an area with bare soil is properly detected, the spectra of the pixel can have a great relationship with a soil type. A case study indicated in Figure 24.2 illustrates the location of a pixel with an arenosol and another with an ferralsol, reaching spectra completely different. Thus, the importance of this strategy is to detect boundaries and spots of a specific soil type (related only with the surface information) and/or a specific soil attribute (i.e., clay or iron content). If this strategy has the limitation to detect soils in depth, we can change to strategy 2. In this case, soil samples can be collected in field and prepared in laboratory in a controlled condition. As we can see in Figure 24.2, the sample collected in the center of the same pixel (surface sample) has similar spectral intensities. This is the basis of the importance of laboratory spectra information as a pattern to understand aerial or orbital data. Despite this, strategy 2 allows the user to collect soil samples in different depths and determine spectra from horizons for future evaluation. Thus, it is possible to make the relationship between the soil surface information obtained in strategy 1 with the soil surface and subsurface information obtained in strategy 2, thus making a link between these two. Although the laboratory strategy has inconveniences, since we have to go to field and collect samples, in many cases, soil scientists need the information in situ. In this case, we have strategy 3, which allows the user to get soil information *inside* a hole to reach underneath information. Another method is to get the surface information. In this case, we can use two methods: with the fiber and with the contact probe. The first use the natural light of the sun and has as inconvenience the alteration of radiation intensity and water bands. The use of the contact probe does not have these inconveniences, since it has its own source. On the other hand, we have to take more care with the contact probe in relation to contamination since the equipment gets into contact with each sample. Still in strategy 3, to read spectra undersurface of soils, we can use the probe inside a hole, or directly inside a pit. Another important approach is on-the-go sensors in tractors or vehicles, where sensors can be used inside the soil as

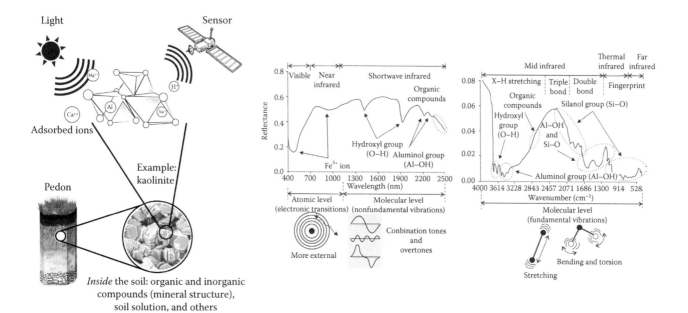

FIGURE 24.3 Illustration of interaction between soil and energy, and soil spectral curve showing fundamental absorption features of selected functional groups in the MIR (2,500 nm–25,000 nm or 4,000 cm^{-1}–400 cm^{-1}), and overtones and combination tones in the visible and near-infrared (400–2,500 nm).

made by implements or sensors that measure the surface of the soils. Thus, Figure 24.3 demonstrates that there are several situations in which sensors can assist soil evaluation, all of them with advantages and limitations and can be used as the necessities of the user. The following sections will show works that use these strategies for several approaches.

24.5.1 Strategies for Soil Sampling

A key set of questions when developing an SS strategy involves deciding where and how to collect soil samples for the calibration of an instrument or validation of a derived map. Regardless of the answers to the where and how questions, it must be recognized that collecting soil samples in the field is expensive and time consuming. Once collected, the samples still must be prepared and analyzed in the laboratory, which also has costs. Hence, most times the key question is this: How can the number of samples reduce? To collect a minimum of representative samples, probably the most used type of sampling is stratified sampling, when landscapes are divided into homogeneous sectors, and each sector is sampled. The usual approach is to collect in each sector a number of samples proportional to the area of the sector. Another common approach is to define the number of samples proportional to the complexity of the sector, with more heterogeneous sectors receiving more samples.

A simple and effective form of stratification is based on the theory of soil change along a toposequence, and thus, a toposequence (meaning positions along a slope) can be used as the spatial stratifier. In general, the concept is summarized as follows: (1) get images from the area, (2) correlate them with an elevation map, and (3) mark the places to collect soils from top to bottom, using the toposequence system. With several toposequences

(transects along slopes), the most different soils in the area can be likely identified. Other types of data can be used as strata, particularly SS data plotted across a landscape.

A more automated stratification system that can use multiple SS and terrain data is the conditioned Latin hypercube (cLHS) algorithm (Minasny and Mcbratney, 2006), which, according to the authors, is a stratified random procedure that provides an efficient way of sampling variables from their multivariate distributions. In fact, Mulder et al. (2013) used the cLHS to assess the variation of soil properties at a regional scale by using the first three principal components of an ASTER satellite image and a DEM. According to them, the sampling approach was successful in representing major soil variation. The cLHS algorithm has been used to map multiple soil properties and included GSS data such as electrical conductivity and γ-ray data (Viscarra Rossel et al., 2007; Adamchuk et al., 2011). In one case, Viscarra Rossel et al. (2008) reduced 1878 samples with MIR information to 213 samples chosen for laboratorial analysis using cLHS. Other sampling methods have been proposed, and they all stratify the area based on auxiliary information. A method by Simbahan and Dobermann (2006) uses many variables (soil series, relative elevation, slope, electrical conductivity, and soil surface reflectance) for sample optimization. A variance quadtree algorithm successively divides the study area into strata, and each stratum has a similar variation (Minasny et al., 2007; Rongjiang et al., 2012).

24.5.2 Strategies for Soil Attribute Prediction and Mapping

Soil analysis is the basic source of data for many applications, notably agriculture and environmental monitoring. The conventional laboratory analysis based on wet chemistry is the most important source of this information with more than 100 years of knowledge. On the other hand, the need for faster information with environmental care has taken research to find other technologies. Since many soil attributes have strong relationship with spectra, their quantification using chemometric techniques constitutes one of the main goals of SS analysis. Since the pioneering work of Schreier (1977), who made quantitative measurements of soil attributes from ground-to-space sensors, this line of research has evolved to more advanced methods including the NIRA method performed by Ben-Dor and Banin (1995a,b). However, SS still relies on wet chemistry analysis as the basic source of data and greatly benefits from existing databanks. Notably in the early part of this decade, soil attribute prediction using SS has been more important, and many papers were published on the quantification of several soil elements by SS, including N, Ca, Mg, K, P, Na, pH, clay, sand, silt, mineralogy, OM, carbon, carbonates, CEC, micronutrients, Fe_2O_3, Al_2O_3, heavy metals, and others, with prediction quality varying for different attributes. History shows how we are going forward on this task: Schreier (1977) used simple regression for each band, which evolved to the use of band depth (Clark et al., 1990)—although with AVIRIS data; then we reached Coleman et al. (1991) using a multivariable analysis with selected bands; afterward using the reflexion inflexion difference (Nanni and Demattê, 2006a); until the use of all spectra with PLS method (Viscarra Rossel et al., 2006). Today, Vohland et al. (2014) raise the discussion about the effects of selected variables on the quantification of soil attributes. Thus, papers indicate that discussion is still needed on this subject to reach the best results and are discussed in this section as summarized in Tables 24.2 (ground data) and 24.3 (aerial and space data).

24.5.2.1 Strategies for Ground Spectral Sensing

Soil properties assessment using spectroscopic approaches in UV–VIS–NIR–SWIR–MIR ranges (from 100 to 25,000 nm) has gained importance in the last 20 years (Ben-Dor, 2011). Soil properties with direct relation to the spectral signature have better prediction performances, especially SOC, calcium carbonate, clay content, clay mineralogy, Fe and Al oxides, and water content (Brown, 2007; Waiser et al, 2007; Morgan et al., 2009; Reeves III, 2010). Other soil chemical attributes, such as, pH, Ca, CEC, N (and all its forms), and Mg, have been predicted with variable accuracy. These properties are not directly related to the soil spectral response, that is, they do not cause absorption features, and their prediction depend on good correlations with spectrally active properties (Kuang et al., 2012).

The correlation between spectra and attributes is generally found using an array of mathematical and statistical procedures. Early in the use of spectroscopy for quantifying soil constituents, linear methods, such as stepwise multiple linear regression (MLR), principal components regression (PCR), and partial least squares regression (PLSR), were used. PCR and PLSR can deal with a large number of predictor variables that are highly collinear (this is the case of spectral curves) (Varmuza and Filzmoser, 2009). The PLSR provides better results than PCR and is by far

the most used method for predicting soil properties; however, non-normal distributions of soil properties, for example, SOC, hinder the use of linear models like stepwise MLR and PLSR (Gobrecht et al., 2014). Nonlinear modeling methods offer an alternative in these cases and include support vector machines (SVMs), regression trees (RTs) and its derivations, and artificial neural networks (ANNs). They are sometimes more complicated to interpret, but can deal with any nonnormal distributions, collinearities, and missing data better than linear methods (Brown et al., 2006; Viscarra Rossel and Behrens, 2010; Gobrecht et al., 2014).

Usually, spectral preprocessing methods are applied to correct nonlinearities, measurement, and sample variations, and for noise reduction in spectra (Stenberg and Viscarra Rossel, 2010). Examples of these approaches are transformation from reflectance (R) to log 1/R (absorbance), and the Kubelka–Munk transformation. Wavelet transformations have been used to reduce the number of variables in a multiple regression model and clarify interpretation of predictors, performing as well as PLSR (Ge et al., 2007). Other pretreatments are suggested to mitigate scattering and path length variation between samples such as multiplicative scatter correction (Geladi et al., 1985), standard normal variate transformation (Barnes et al., 1989), smoothing, and first and second derivatives of spectra (Stenberg and Viscarra Rossel, 2010), including Savitzky–Golay transformation (Savitzky and Golay, 1964).

Because of its importance to the soil condition, hydrologic cycle (infiltration and water retention), and global C cycle, interest in estimating SOC with SS is very high. Bellon-Maurel and McBratney (2011) reviewed the spectroscopic prediction of SOC and its use in carbon stock evaluation; they concluded that MIR had better results than VIS–NIR. The difficulty is that MIR requires laborious sample preparation (soil ground and sieved at a 200 mesh, and sometimes diluted in KBr) and is not practical for field measurements (Reeves III, 2010). Some studies report predictions of SOC using spectroscopic techniques with rather large R^2 (>0.90) (Stenberg et al., 2010; Kuang et al., 2012; McDowell et al., 2012). Gmur et al. (2012) characterized soil attributes by spectral signature measurement with ground sensor using classification and RTs and obtained predictions of N (R^2 0.91), carbonate (R^2 0.95), total carbon (TC; R^2 0.93), and OM (R^2 0.98) in the soil.

Using classification and RT statistical methods, concentrations of N (R^2 0.91), carbonate (R^2 0.95), TC (R^2 0.93), and OM (R^2 0.98) were obtained.

When evaluating the performance of these prediction models, it is important to analyze the geographical extent and the availability and independence of validation and calibration data (Bellon-Maurel and McBratney, 2011). Soriano-Disla et al. (2014) reviewed the performance of LSS and FSS for predicting SOC. They concluded that overall MIR (20 studies) performed better (coefficient of determination—R^2—of 0.93) than NIR (23 studies; R^2 of 0.85). Adding UV, VIS, or multiple combinations of wavelength ranges, that is, VIS–NIR, NIR–MIR, UV–VIS–NIR, or VIS–NIR–MIR, did not necessarily improve

TABLE 24.2 Summary of Revision Related with Quantification of Soil Attributes by Ground SS

Element[a]	Spectral Range[b]	Element Range[c]	N. Cal.	Cal. Scale	R² Range	R² Median	R² N	RMSEP/RMSECV, *SEP/SECV* (%) Range	RMSEP/RMSECV, *SEP/SECV* (%) Median	N	References
A. Lab—dried											
Clay (%)	UV–Vis–NIR	70–90	121–207	R/N	0.61–0.72	0.67	2	12.3, 8.9	12.3, 8.9	2	Islam et al. (2003) and Pirie et al. (2005)
	Vis–NIR	70 to >90	457–4184	G	0.73–0.78	0.77	3	7.5–12	9.5	3	Brown et al. (2006), Ramirez-Lopez et al. (2013), Shepherd and Walsh (2002)
	Vis–NIR	20 to >90	30 to >1000	R/N	0.02–0.91	0.78	22	2–8.5, *4.9–13.7*	5.3, *7.8*	26	Dematté and Garcia (1999), Nanni and Dematté (2006), Bricklemyer and Brown (2010), Chang et al. (2005), Curcio et al. (2013), Genot et al. (2011), Gogé et al. (2012, 2014), Islam et al. (2003), Kinoshita et al. (2012), Knadel et al. (2013), Ramirez-Lopez et al. (2013, 2014), Sankey et al. (2008), Stenberg (2010), Summers et al. (2011), Thomsen et al. (2009), Waiser et al. (2007), Vendrame et al. (2012), Viscarra Rossel and Behrens (2010), Viscarra Rossel et al. (2009), Viscarra Rossel and Lark (2009), Viscarra Rossel and Webster (2012)
	Vis–NIR	<5–80	16 to >100	F	0.39–0.82	0.70	11	0.36–6.3	1.8	11	Debaene et al. (2014), Mahmood et al. (2012), McCarty and Reeves (2006), Ramirez-Lopez et al. (2014), Wetterlind and Stenberg (2010)
	NIR	30–90	35 to >400	R/N	0.50–0.94	0.71	10	3.9–16, *5–10.3*	5.3, *8.6*	11	Ben-Dor and Banin (1995a), Chang et al. (2001), Islam et al. (2003), Malley et al. (2000), Stenberg et al. (2002), Wang et al. (2013b), Waruru et al. (2014)
	NIR	20–50	20–52	F	0.47–0.86	0.63	6	1.9–5.7	3.1	6	Igne et al. (2010), Sudduth et al. (2010), Wetterlind et al. (2008), Viscarra Rossel et al. (2006)
	MIR	30 to >90	60–663	R/N	0.55–0.94	0.86	11	3.7–10.5	6.6	8	D'Acqui et al. (2010), Ge et al. (2014b), Janik et al. (2009), Janik and Skjemstad (1995), Minasny and McBratney (2008), Minasny et al. (2008, 2009), Pirie et al. (2005), Viscarra Rossel and Lark (2009)
	MIR	20	116–209	F	0.67–0.77	0.70	3	1.6 to ~2	1.7	3	Igne et al. (2010), McCarty and Reeves (2006), Viscarra Rossel et al. (2006)
CEC (cmol kg⁻¹)	UV–Vis–NIR	13–97	49–205	R/N	0.47–0.64	0.52	3	1.3–5.8, *4.3*	3.5, *4.3*	3	Islam et al. (2003), Lu et al. (2013), Pirie et al. (2005)
	Vis–NIR	55–165	740–4184	G	0.74–0.88	0.81	2	3.8–6.7	5.3	2	Brown et al. (2006), Shepherd and Walsh (2002)
	Vis–NIR	14–92	94 to >2000	R/N	0.68–0.91	0.81	7	1–7.8, *3.9*	2.1, *3.9*	8	Chang et al. (2001), Dunn et al. (2002), Genú et al. (2011), Gogé et al. (2014), Islam et al. (2003), Kinoshita et al. (2012), Vendrame and Webster (2012)
	Vis–NIR	16	50–299	F	0.73–0.83	0.78	2	1.4	1.4	3	van Groenigen et al. (2003), Nanni and Dematté (2006), Sudduth et al. (2010)
	NIR	29–91	35–1100	R/N	0.64–0.73	0.69	4	9.6, *3.3–8.5*	9.6, *4.1*	4	Ben-Dor and Banin (1995b), Genot et al. (2011), Islam et al. (2003), Waruru et al. (2014)
	NIR	2–7	49	F	0.13	0.13	1	1.04	1.04	1	Viscarra Rossel et al. (2006)
	MIR	36–94	60–662	R/N	0.77–0.92	0.86	5	2–4.8	4.6	3	D'Acqui et al. (2010), Janik et al. (2009), Minasny and McBratney (2008, 2009), Pirie et al. (2005)
	MIR	5–16	50	F	0.34–0.56	0.45	2	0.9–2.1	1.5	2	van Groenigen et al. (2003), Viscarra Rossel et al. (2006)
OC (%)	UV–Vis–NIR	1–5	49–207	R/N	0.76–0.83	0.76	3	0.12–0.5, *0.44*	0.31, *0.44*	3	Islam et al. (2003), Lu et al. (2013), Pirie et al. (2005)

(Continued)

TABLE 24.2 (Continued)   Summary of Revision Related with Quantification of Soil Attributes by Ground SS

Element[a]	Spectral Range[b]	Element Range[c]	N. Cal.	Cal. Scale	R² Range	R² Median	R² N	RMSEP/RMSECV, SEP/SECV (%) Range	Median	N	References
	Vis–NIR	2–33	<50 to >3,000	R/N	0.31–0.96	0.84	29	0.06–0.88, 0.28–2.94	0.35, 0.72	27	Bricklemyer and Brown (2010), Chang and Laird (2002), Conforti et al. (2013), Daniel et al. (2003), Deng et al. (2013), Doetterl et al. (2013), Dunn et al. (2002), Fystro (2002), Gogé et al. (2012, 2014), Kinoshita et al. (2012), Knadel et al. (2013), Liu and Liu (2013), Moron and Cozzolino (2002), Nocita et al. (2011), Ramirez-Lopez et al. (2013), Sankey et al. (2008), Stenberg (2010), Summers et al. (2011), Tian et al. (2013), Wang et al. (2013a), Viscarra Rossel and Behrens (2010), Viscarra Rossel and Lark (2009), Vohland and Emmerling (2011), Vohland et al. (2014)
	Vis–NIR	<1–14	25–287	F	0.12–0.99	0.75	15	0.07–1.6, 0.33	0.3, 0.33	20	Debaene et al. (2014), Fontan et al. (2011), He et al. (2007), Heinze et al. (2013), Kuang and Mouazen (2011), Kuang and Mouazen (2013b), McCarty and Reeves (2006), Wetterlind and Stenberg (2010), Yang et al. (2012)
	NIR	2–60	39 to >2,000	R/N	0.45–0.99	0.83	21	0.14–0.83, 0.16–2.95	0.32, 0.61	21	Ben-Dor and Banin (1995b), Cambule et al. (2012), Chen et al. (2011), Dalal and Henry (1986), Dunn et al. (2002), Fidêncio et al. (2002), Genot et al. (2011), Islam et al. (2003), Malley et al. (2000), McCarty et al. (2002, 2010), Miltz and Don (2012), Nocita et al. (2011), Rabenarivo et al. (2013), Reeves et al. (2006), Stenberg et al. (2002), Todorova et al. (2009), Zornoza et al. (2008)
	NIR	1–10	20–299	F	0.34–0.95	0.93	7	0.1–0.77	0.18	8	Reeves and McCarty (2001), Viscarra Rossel et al. (2006), Guerrero et al. (2014), Sudduth et al. (2010), Wetterlind et al. (2008), Xie et al. (2011)
	MIR	3–8	87–1545	G	0.93–0.95	0.94	2	0.02–0.25	0.14	2	Bornemann et al. (2008), Kamau-Rewe et al. (2011)
	MIR	2–30	31–560	R/N	0.77–0.98	0.93	14	0.11–0.61	0.35	13	D'Acqui et al. (2010), Ge et al. (2014b), Grinand et al. (2012), Janik and Skjemstad (1995), Masserschmidt et al. (1999), McCarty et al. (2002, 2010), Minasny et al. (2009), Pirie et al. (2005), Rabenarivo et al. (2013), Reeves et al. (2006), Viscarra Rossel and Lark (2009), Vohland et al. (2014)
	MIR	1–3	118–217	F	0.73–0.96	0.95	3	0.11–0.15, 0.67	0.14, 0.67	3	McCarty and Reeves (2006), Viscarra Rossel et al. (2006), Xie et al. (2011),
TC (%)	Vis–NIR	2–55	30 to >400	R/N	0.66–0.95	0.88	6	0.02–2.8, 0.3	0.54, 0.3	6	Chang and Laird (2002), Chang et al. (2005), McDowell et al. (2012), Sørensen and Dalsgaard (2005), Thomsen et al. (2009), Vendrame et al. (2012)
	Vis–NIR	1–17	16–429	F	0.01–0.93	0.76	11	0.16–1.1, 0.1	0.19, 0.1	10	Fontan et al. (2011), Kuang and Mouazen (2011), Mahmood et al. (2012), van Groenigen et al. (2003), Wetterlind et al. (2010), Yang et al. (2012)
	NIR	4–5	83–91	G	0.85–0.94	0.90	2	0.35–0.41	0.38	2	Barthes et al. (2006), Brunet et al. (2007)
	NIR	10–28	150 to >700	R/N	0.85–0.87	0.86	3	0.54–0.79	0.71	3	Chang et al. (2001), McCarty et al. (2002), Reeves et al. (2006)
	NIR	1	209	F	~0.65	~0.65	1	~0.16	~0.16	1	Igne et al. (2010)
	MIR	10–50	56–660	R/N	0.79–0.97	0.95	7	0.24–2.28	0.34	7	Janik et al. (2009), Ludwig et al. (2008), McDowell et al. (2012), Minasny et al. (2008, 2009), Reeves et al. (2006), McCarty et al. (2002), Minasny and McBratney (2008)
	MIR	1	50–209	F	0.01 to ~0.85	0.43	2	~0.1	~0.1	1	Igne et al. (2010), van Groenigen et al. (2003)

(Continued)

TABLE 24.2 (Continued) Summary of Revision Related with Quantification of Soil Attributes by Ground SS

Element[a]	Spectral Range[b]	Element Range[c]	N. Cal.	Cal. Scale	R² Range	R² Median	R² N	RMSEP/RMSECV, SEP/SECV (%) Range	Median	N	References
IC (%)	Vis–NIR	13	4184	G	0.83	0.83	1	0.62	0.62	1	Brown et al. (2006)
	Vis–NIR	3–4	76–86	R/N	0.96–0.98	0.97	2	0.15–0.19	0.17	2	Chang and Laird (2002), Chang et al. (2005)
	Vis–NIR	1–2	90–492	F	0.53–0.95	0.74	2	0.08, *0.15*	0.08, *0.15*	2	Fontan et al. (2011), Yang et al. (2012)
	NIR	7	177	R/N	0.87	0.87	1	0.31	0.31	1	McCarty et al. (2002)
	MIR	7–10	60–418	R/N	0.82–0.97	0.96	6	0.02–0.42, *0.28*	0.12, *0.28*	6	D'Acqui et al. (2010), Ge et al. (2014b), Grinand et al. (2012), McCarty et al. (2002), Reeves et al. (2006)
B. Lab—field moist											
Clay (%)	Vis–NIR	40–90	187–2000	R/N	0.76–0.77	0.77	2	5.25	9	2	Chang et al. (2005), Ge et al. (2014a)
	MIR	20	209	F	~0.7	~0.7	1	~2	~2	1	Igne et al. (2010)
OC (%)	Vis–NIR	3–9	75–2000	R/N	0.53–0.94	0.86	6	0.38–0.73, *0.15–0.67*	0.68, *0.41*	6	Chang et al. (2005), Fystro (2002), Ge et al. (2014a), Mouazen et al. (2007, 2010), Terhoeven-Urselmans et al. (2008)
	Vis–NIR	1–7	47–104	F	0.58–0.82	0.72	4	0.2–1.7	1.23	4	Kuang and Mouazen (2012), Kuang and Mouazen (2013b)
TC (%)	Vis–NIR	4	162	R/N	0.85	0.85	1	0.42	0.42	1	Chang et al. (2005)
	MIR	1	209	F	~0.6	~0.6	1	~0.15	~0.15	1	Igne et al. (2010)
C. In field (including at site and on-the-go measurements)											
Clay (%)	Vis–NIR	40–60	311 to >1000	R/N	0.17–0.83	0.78	3	6.1–7.9, *9.0*	7.0, *9.0*	3	Bricklemyer and Brown (2010), Waiser et al. (2007), Viscarra Rossel et al. (2009)
	Vis–NIR	20	209	F	~0.7	~0.7	1	~2	~2	1	Igne et al. (2010)
	Gamma	30–40	13–660	R/N	0.86	0.86	1	4–12	6	1	Petersen et al. (2012), van der Klooster et al. (2011)
	Gamma	<5–40	7–70	F	0.63–0.94	0.72	8	0.81–6.56	3.1	8	Mahmood et al. (2013), Piikki et al. (2013), Priori et al. (2014), Taylor et al. (2010), van der Klooster et al. (2011), van Egmond et al. (2010), Viscarra Rossel et al. (2009)
OC (%)	Vis–NIR	2–11	28–765	R/N	0.0–0.91	0.84	7	0.31–1.16, *0.35*	0.42, *0.35*	6	Gomez et al. (2008b), Gras et al. (2014), Kusumo et al. (2008), Bricklemyer and Brown (2010), Daniel et al. (2003), Denis et al. (2014), Nocita et al. (2011)
	Vis–NIR	1–37	15 to >400	F	0.65–0.90	0.75	3	0.07–7.15, *0.56*	1.74, *0.56*	3	Knadel et al. (2011), Kuang and Mouazen (2013a, b), Reeves et al. (2010), Wijaya et al. (2001)
	NIR	2–3	24–120	R/N	0.53–0.86	0.74	4	0.18–0.52, *0.42–0.53*	0.36, *0.48*	4	Christy (2008), Cozzolino et al. (2013), McCarty et al. (2010), Nocita et al. (2011), Sudduth and Hummel (1993)
	NIR	1–3	11–216	F	0.61–0.95	0.82	4	0.09–0.27	0.18	4	Kweon and Maxton (2013), Reeves et al. (2010), Schirrmann et al. (2012)
TC (%)	Vis–NIR	1–3	38–209	F	~0.6–0.89	0.86	3	0.16–0.19	0.16	3	Igne et al. (2010), Kodaira and Shibusawa (2013), Kusumo et al. (2011)
	NIR	1–2	45–78	F	0.46–0.68	0.57	2	0.15–0.48	0.18	2	Huang et al. (2007), Munoz and Kravchenko (2011)

[a] CEC, cation exchangeable capacity; OC, organic carbon; IC, inorganic carbon; TC, total carbon.

[b] The exact spectral range in the individual studies may deviate to some extent but will stay within ranges as described later.

[c] This is included to give an idea of the variation in the element range (i.e., clay content) in the studies used to calculate the prediction statistics since this has a large impact on the prediction statistics. The individual ranges are presented as percentage units, and "element range" in the table is the range of these for the studies used.

TABLE 24.3   Summary of Revision Related to Quantification of Soil Attributes by Aerial and Orbital Sensors

Soil Properties	Sensor	Nr. Samples (Total/Validation)	Modeling Approach	Validation	Validation/Calibration Results			References
					R^2 (P)/R^2 (CV)	RMSEP/RMSECV	SEP/SECV	
Multispectral								
SOC	ATLAS[a]	31–40	MLR	—	0.63–0.91	0.11%–0.22%	—	Chen et al. (2008)
	SPOT[b]	10–11/17–27	MLR	Pred.	0.55–0.72	4.06–6.57 g kg⁻¹	—	Vaudour et al. (2013)
OM	LANDSAT 5[b]	378/95	MLR	Pred.	0.41	—	—	Fiorio et al. (2010)
	LANDSAT 5[b]	184	MLR	—	0.561	—	—	Nanni and Demattê (2006a)
	LANDSAT 5[b]	164	LR	—	0.79	1.5 g kg⁻¹	—	Dogan and Kılıç (2013)
	LANDSAT 7[b]	110/155	MLR	Pred.	0.27	—	—	Demattê et al. (2007a)
Total carbon	IKONOS[b]	144–222/14–24	MLR	Pred.	0.11–0.61	0.11%–0.24%	—	Sullivan et al. (2005)
	LANDSAT 7[b]	78	MLR	Jackknifing (10-fold)	0.33–0.46	0.27%–0.25%	—	Huang et al. (2007)
Total N	LANDSAT 7[b]	164	LR	—	0.612	—	0.06%	Dogan and Kılıç (2013)
Clay	LANDSAT 5[b]	378/95	MLR	Pred.	0.61	—	—	Fiorio et al. (2010)
	LANDSAT 5[b]	184	MLR	—	0.675	—	—	Nanni and Demattê (2006a)
	LANDSAT 7[b]	110/155	MLR	Pred.	0.63	—	—	Demattê et al. (2007a)
Sand	LANDSAT 5[b]	378/95	MLR	Pred.	0.63	—	—	Fiorio et al. (2010)
	LANDSAT 5[b]	184	MLR	—	0.525	—	—	Nanni and Demattê (2006a)
	LANDSAT 7[b]	110/155	MLR	Pred.	0.67	—	—	Demattê et al. (2007a)
Silt	ASTER[b]	22/5	ER	Pred.	0.63	19.33 g kg⁻¹	—	Breunig et al. (2008)
	LANDSAT 5[b]	378/95	MLR	Pred.	0.54	—	—	Fiorio et al. (2010)
	LANDSAT 5[b]	184	MLR	—	0.508	—	—	Nanni and Demattê (2006a)
	LANDSAT 7[b]	110/155	MLR	Pred.	0.29	—	—	Demattê et al. (2007a)
CEC	LANDSAT 5[b]	378/95	MLR	Pred.	0.45	—	—	Fiorio et al. (2010)
	LANDSAT 5[b]	184	MLR	—	0.551	—	—	Nanni and Demattê (2006a)
	LANDSAT 7[b]	44	MLR	—	0.57	—	—	Ghaemi et al. (2013)
	LANDSAT 7[b]	110/155	MLR	Pred.	0.26	—	—	Demattê et al. (2007a)
	LANDSAT 5[b]	113	MLR-RK/RT/ GAM-RK/KED	Jackknifing (100-fold)	—	6.48–5.53/7.27/ 6.94–6.18/5.41 $cmol_c$ kg⁻¹	—	Bishop and McBratney (2001)

(Continued)

TABLE 24.3 (Continued) Summary of Revision Related to Quantification of Soil Attributes by Aerial and Orbital Sensors

Soil Properties	Sensor	Nr. Samples (Total/Validation)	Modeling Approach	Validation	R² (P)/R² (CV)	RMSEP/RMSECV	SEP/SECV	References
Fe₂O₃ (total)	LANDSAT 5[b]	184	MLR	—	0.725	—	—	Nanni and Dematté (2006a)
SiO₂				—	0.598	—	0.60%	
TiO₂				—	0.727	—	—	
CaCO₃	LANDSAT 7[b]	164	LR	—	0.716	—	—	Dogan and Kılıç (2013)
pH	LANDSAT 7[b]	164	LR	—	0.659	—	0.3	Dogan and Kılıç (2013)
Total P	LANDSAT 7[b]	111	MLR/CK/RK	CV	0.46/—/—	—	356.1/279.2/238.8 mg kg⁻¹	Rivero et al. (2007)
	ASTER[b]	111	MLR/CK/RK	CV	0.39/—/—	—	281.8/238.2/200.1 mg kg⁻¹	
K	LANDSAT 7[b]	110/155	MLR	Pred.	0.11	—	—	Dematté et al. (2007a)
Ca	LANDSAT 7[b]	110/155	MLR	Pred.	0.13	—	—	Dematté et al. (2007a)
Mg	LANDSAT 7[b]	110/155	MLR	Pred.	0.19	—	—	Dematté et al. (2007a)
H + Al	LANDSAT 7[b]	110/155	MLR	Pred.	0.18	—	—	Dematté et al. (2007a)
Basis (CEC saturation)	LANDSAT 5[b]	378/95	MLR	Pred.	0.01	—	—	Fiorio et al. (2010)
Al (CEC saturation)	LANDSAT 7[b]	110/155	MLR	Pred.	0.21	—	—	Dematté et al. (2007a)
	LANDSAT 5[b]	378/95	MLR	Pred.	0.13	—	—	Fiorio et al. (2010)
EC	LANDSAT 7[b]	164	LR	—	0.498	—	131.3	Dogan and Kılıç (2013)
Hyperspectral								
SOC	AHS-160[a]	68/16	PLSR	Pred.	0.53/0.75	2.42/1.18 g kg⁻¹	—	Bartholomeus et al. (2011)
	CHRIS-PROBA[b]	72/24	RK/PLSR/PLSR-K	Pred.	—	1.06–1.16 g kg⁻¹	—	Casa et al. (2013a)
	AHS-160[a]	88–91	PLSR	CV (LOO)	0.93–0.96	3.68–4.9 g kg⁻¹	—	Denis et al. (2014)
	AVNIR[a]	321	MLR	—	0.2692	0.081%	—	De Tar et al. (2008)
	HyMap[a]	67/29	PLSR	Pred.	0.71	—	2.07/1.64 g kg⁻¹	Gerighausen et al. (2012)
	EO-1 Hyperion[b]	72	PLSR	CV (LOO)	0.51	0.73%	—	Gomez et al. (2008b)
	HyMap[a]	95	PLSR	CV (LOO)	0.02	2.6 g kg⁻¹	—	Gomez et al. (2012)
	HyMap[a]	204	PLSR	CV (LOO) and Pred.	0.83	1.10/1.05 g kg⁻¹	—	Hbirkou et al. (2012)
	SpecTIR[a]	269	PLSR	CV (LOO)	0.65	0.0019	—	Hively et al. (2011)
	EO-1 Hyperion[b]	49/14	PLSR/MLR	CV (LOO—PLSR)/CV (LOO—MLR) and Pred. (MLR)	0.63 (PLSR)/0.50 (P—MLR)/0.63 (CV—MLR)	1.6 g kg⁻¹ (PLSR)	—	Lu et al. (2013)
	HyMap[a]	72	PLSR/MLR	CV (LOO)	0.90 (PLSR)/0.86 (MLR)	0.29% (PLSR)/0.22% (MLR)	—	Selige et al. (2006)

(Continued)

TABLE 24.3 (Continued)   Summary of Revision Related to Quantification of Soil Attributes by Aerial and Orbital Sensors

Soil Properties	Sensor	Nr. Samples (Total/Validation)	Modeling Approach	Validation	R² (P)/R² (CV)	RMSEP/RMSECV	SEP/SECV	References
					Validation/Calibration Results			
	CASI[a]	227/57	PLSR	Pred.	0.85	—	5.1/4.8 g kg⁻¹	Stevens et al. (2006)
	AHS-160[a]	197/102	PLSR	Pred.	0.86/0.89	3.56/3.01 g kg⁻¹	—	Stevens et al. (2010)[c]
		188/101	PSR	Pred.	0.88/0.91	3.20/2.54 g kg⁻¹	—	
		201/101	SVMR	Pred.	0.84/0.99	4.20/0.43 g kg⁻¹	—	
	AHS-160[a]	400/126	PLSR/PSR/SVMR	Pred.	0.56–0.73/0.85–0.98	4.74–6.11/1.37–3.45 g kg⁻¹	—	Stevens et al. (2012)[c]
	CASI[a]	47	MLR-PCA/ANN-PCA	CV (10-fold CV)	0.745 (MLR)/0.590 (ANN)	0.49% (MLR)/0.592% (ANN)	—	Uno et al. (2005)
	EO-1 Hyperion[b]	28/08	PLSR	Pred.	0.48/0.56	0.33%/0.43%	—	Zhang et al. (2013)
OM	DAIS-7915[a]	62	MLR	Pred.	0.827	—	0.015%/0.003%	Ben-Dor et al. (2002)
	AVNIR[a]	321	MLR	—	0.4857	0.08%	—	De Tar et al. (2008)
	SpecTIR[a]	269	PLSR	CV (LOO)	0.75	0.4%	—	Hively et al. (2011)
	EO-1 Hyperion[b]	28/08	PLSR	Pred.	0.74/0.72	0.66%/0.72%	—	Zhang et al. (2013)
POM	EO-1 Hyperion[b]	18	PLSR	CV (LOO)	0.67	4.56%	—	Anne et al. (2014)
MAOM					0.74	2.450%	—	
Labile C					0.93	6.72 mg g⁻¹	—	
Stable C					0.71	31.51 mg g⁻¹	—	
N total	CHRIS-PROBA[b]	73/24	RK/PLSR/PLSR-K	Pred.	—	0.144–0.139 g kg⁻¹	—	Casa et al. (2013a)
	HyMap[a]	72	PLSR/MLR	CV (LOO)	0.92/0.87	0.03%/0.02%	—	Selige et al. (2006)
	EO-1 Hyperion[b]	28/08	PLSR	Pred.	0.70/0.63	0.032%/0.033%	—	Zhang et al. (2013)
Labile N	EO-1 Hyperion[b]	18	PLSR	CV (LOO)	0.96	0.34 mg g⁻¹	—	Anne et al. (2014)
Stable N					0.69	2.58 mg g⁻¹	—	
Clay	MIVIS[a]	80/29	PLSR	Pred.	0.48	7.20%	—	Casa et al. (2013b)
	CHRIS-PROBA[b]				0.52	6.87%	—	Casa et al. (2013a)
	CHRIS-PROBA[b]	132/44	RK/PLSR/PLSR-K	Pred.		5.33%–5.82%	—	
	AVNIR[a]	321	MLR		0.6708	3.23%	—	De Tar et al. (2008)
	SIM-GA[a]	40/11	Regression with band depth (2210 nm)	Pred.	0.5994		—	Garfagnoli et al. (2013)
	HyMap[a]	67/29	PLSR	Pred.	0.85		19.41/20.34 g kg⁻¹	Gerighausen et al. (2012)
	HyMap[a]	52	PLSR	CV (LOO)	0.64	49.60 g kg⁻¹	—	Gomez et al. (2008a)
	HyMap[a]	95	PLSR	CV (LOO)	0.67	42.15 g kg⁻¹	—	Gomez et al. (2012)

(Continued)

TABLE 24.3 (Continued) Summary of Revision Related to Quantification of Soil Attributes by Aerial and Orbital Sensors

Soil Properties	Sensor	Nr. Samples (Total/Validation)	Modeling Approach	Validation	R² (P)/R² (CV)	RMSEP/RMSECV	SEP/SECV	References
	SpecTIR[a]	269	PLSR	CV (LOO)	0.66	2.2%	—	Hively et al. (2011)
	HyMap[a]	33	Regression with CR data (2206 nm)	CV (LOO)	0.61	54 g kg⁻¹	—	Lagacherie et al. (2008)
		19			0.60	130 g kg⁻¹	—	
		33 + 19			0.58	82 g kg⁻¹	—	
	AISA-Dual Vis–NIR[a]	152/30	PLSR/RTs	CV (LOO) and Pred.	0.81/0.78/0.78/0.77	87/67/93/69 g kg⁻¹	—	Lagacherie et al. (2013)
	HyMap[a]	72	PLSR/MLR	CV (LOO)	0.71/0.65	4.2%/3.8%	—	Selige et al. (2006)
	EO-1 Hyperion[b]	28/08	PLSR	Pred.	0.51/0.83	5.46%/2.21%	—	Zhang et al. (2013)
Sand	EO-1 Hyperion[b]	18	PLSR	CV (LOO)	0.58	7.02%	—	Anne et al. (2014)
	MIVIS[a]	80/29	PLSR	Pred.	0.64	7.82%	—	Casa et al. (2013b)
	CHRIS-PROBA[b]				0.45	9.32%	—	
	CHRIS-PROBA[b]	132/44	RK/PLSR/PLSR-K	Pred.	—	6.80%–7.40%	—	Casa et al. (2013a)
	AVNIR[a]	321	MLR	—	0.8063	4.83%	—	De Tar et al. (2008)
	HyMap[a]	95	PLSR	CV (LOO)	0.20	90.47 g kg⁻¹	—	Gomez et al. (2012)
	SpecTIR[a]	269	PLSR	CV (LOO)	0.79	7.9%	—	Hively et al. (2011)
	AISA-Dual Vis–NIR[a]	152/30	PLSR/regression trees	CV (LOO) and Pred.	0.75/0.83/ 0.78/0.77	119/97/111/107 g kg⁻¹	—	Lagacherie et al. (2013)
	HyMap[a]	72	PLSR/MLR	CV (LOO)	0.95/0.87	9.7%/12.9%	—	Selige et al. (2006)
Silt	MIVIS[a]	80/29	PLSR	Pred.	0.32	3.28%	—	Casa et al. (2013b)
	CHRIS-PROBA[b]				0.23	3.43%	—	
	CHRIS-PROBA[b]	132/44	RK/PLSR/PLSR-K	Pred.	—	3.12%–3.67%	—	Casa et al. (2013a)
	AVNIR[a]	321	MLR	—	0.7518	3.34%	—	De Tar et al. (2008)
	HyMap[a]	95	PLSR	CV (LOO)	0.17	74.84 g kg⁻¹	—	Gomez et al. (2012)
	SpecTIR[a]	269	PLSR	CV (LOO)	0.79	6.9%	—	Hively et al. (2011)
Silt + clay	EO-1 Hyperion[b]	18	PLSR	CV (LOO)	0.82	1.95%	—	Anne et al. (2014)
Soil moisture (gravimetric)	EO-1 Hyperion[b]	18	PLSR	CV (LOO)	0.82	3.02%	—	Anne et al. (2014)
	DAIS-7915[a]	62	MLR	Pred.	0.645	—	0.14%/0.045%	Ben-Dor et al. (2002)
	HyMap[a]	205	Linear regression with NSMI index	—	0.82	2.30%	—	Haubrock et al. (2008a,b)
	EO-1 Hyperion[b]	28/08	PLSR	Pred.	0.40/0.79	5.12%/3.13%	—	Zhang et al. (2013)

(Continued)

TABLE 24.3 (Continued)   Summary of Revision Related to Quantification of Soil Attributes by Aerial and Orbital Sensors

Soil Properties	Sensor	Nr. Samples (Total/Validation)	Modeling Approach	Validation	Validation/Calibration Results			References
					R^2 (P)/R^2 (CV)	RMSEP/RMSECV	SEP/SECV	
Saturated soil moisture (gravimetric)	DAIS-7915[a]	62	MLR	Pred.	0.759	—	0.021%/0.019%	Ben-Dor et al. (2002)
	AVNIR[a]	321	MLR	—	0.4859	3.07%	—	De Tar et al. (2008)
Available soil water	CHRIS-PROBA[b]	132/44	RK/PLSR/PLSR-K	Pred.	—	2.50%–2.79%	—	Casa et al. (2013a)
Iron oxides	ROSIS[a]	35/16	Redness index	Pred.	—	—	4.97%–10.59%/4.38%–11.39%	Bartholomeus et al. (2007)[d]
			Spectral feature area (550 nm)		—	—	5.73%–9.08%/6.41%–7.26%	
			Standard deviation after CR		—	—	5.79%–7.71%/6.39%–6.43%	
	HyMap[a]	95	PLSR	CV (LOO)	0.78	0.28 g 100g⁻¹	—	Gomez et al. (2012)
Fe_2O_3 (total)	AVIRIS[a]	22	Regression with band depth (1710 nm)	—	0.83	—	—	Galvão et al. (2001)
TiO_2				—	0.74	—	—	
Al_2O_3			Regression with band depth (2200 nm)	—	0.68	—	—	
$CaCO_3$	HyMap[a]	52	PLSR	CV (LOO)	0.77	76.67 g kg⁻¹	—	Gomez et al. (2008a)
	HyMap[a]	95	PLSR	CV (LOO)	0.76	64.32 g kg⁻¹	—	Gomez et al. (2012)
	HyMap[a]	33	Regression with CR data (2341 nm)	CV (LOO)	0.59	113 g kg⁻¹	—	Lagacherie et al. (2008)
		19			0.61	133 g kg⁻¹		
		33 + 19			0.47	132 g kg⁻¹		
RCGb	HyMap[a]	30	Spectral index	—	0.75	—	—	Baptista et al. (2011)
CEC	HyMap[a]	95	PLSR	CV (LOO)	0.62	1.84 cmolc kg⁻¹	—	Gomez et al. (2012)
	AISA-Dual Vis–NIR[a]	147/30	PLSR/RT	CV (LOO) and Pred.	0.71/0.79/0.72/0.79	4.6/3.3/4.6/3.4 Meq 100 g⁻¹	—	Lagacherie et al. (2013)
	EO-1 Hyperion[b]	49	PLSR	CV (LOO)	0.40	1.55 cmolc kg⁻¹	—	Lu et al. (2013)
pH	DAIS-7915[a]	62	MLR	Pred.	0.528	—	0.26/0.146	Ben-Dor et al. (2002)
	AVNIR[a]	321	MLR	—	0.6164	0.076	—	De Tar et al. (2008)
	HyMap[a]	95	PLSR	CV (LOO)	0.31	0.37	—	Gomez et al. (2012)
	SpecTIR[a]	269	PLSR	CV (LOO)	0.51	0.40	—	Hively et al. (2011)

(Continued)

TABLE 24.3 (Continued) Summary of Revision Related to Quantification of Soil Attributes by Aerial and Orbital Sensors

Soil Properties	Sensor	Nr. Samples (Total/Validation)	Modeling Approach	Validation	Validation/Calibration Results			References
					R^2 (P)/R^2 (CV)	RMSEP/RMSECV	SEP/SECV	
Extractable P	EO-1 Hyperion[b]	49/14	PLSR/MLR	CV (LOO—PLSR)/CV (LOO—MLR) and Pred. (MLR)	0.68 (PLSR)/0.65 (P—MLR)/0.83 (CV—MLR)	0.19 (PLSR)	—	Lu et al. (2013)
	AVNIR[a]	321	MLR	—	0.6975	4.32 mg kg[-1]	—	De Tar et al. (2008)
Total P	EO-1 Hyperion[b]	49/14	PLSR/MLR	CV (LOO—PLSR)/CV (LOO—MLR) and Pred. (Indep. data—MLR)	0.62 (PLSR)/ 0.54 (P—MLR)/0.74 (CV—MLR)	0.2 g kg[-1] (PLSR)	—	Lu et al. (2013)
	EO-1 Hyperion[b]	28/08	PLSR	Pred.	0.25/0.44	176.67/141.86 mg kg[-1]	—	Zhang et al. (2013)
K	AVNIR[a]	321	MLR	—	0.6391	45.03 mg kg[-1]	—	De Tar et al. (2008)
	SpecTIR[a]	269	PLSR	CV (LOO)	0.59	89.5 mg kg[-1]	—	Hively et al. (2011)
Ca	AVNIR[a]	321	MLR	—	0.6188	9.51 meq L[-1]	—	De Tar et al. (2008)
	SpecTIR[a]	269	PLSR	CV (LOO)	0.69	166.1 mg kg[-1]	—	Hively et al. (2011)
Mg	AVNIR[a]	321	MLR	—	0.582	4.32 meq L[-1]	—	De Tar et al. (2008)
	SpecTIR[a]	269	PLSR	CV (LOO)	0.69	50.3 mg kg[-1]	—	Hively et al. (2011)
Na	AVNIR[a]	321	MLR	—	0.6224	9.11 meq L[-1]	—	De Tar et al. (2008)
Al	SpecTIR[a]	269	PLSR	CV (LOO)	0.76	104.7 mg kg[-1]	—	Hively et al. (2011)
Mn	SpecTIR[a]	269	PLSR	CV (LOO)	0.62	19.6 mg kg[-1]	—	Hively et al. (2011)
Zn	SpecTIR[a]	269	PLSR	CV (LOO)	0.64	1.4 mg kg[-1]	—	Hively et al. (2011)
Cl	AVNIR[a]	321	MLR	—	0.7376	12.24 meq L[-1]	—	De Tar et al. (2008)
Fe	SpecTIR[a]	269	PLSR	CV (LOO)	0.75	49.3 mg kg[-1]	—	Hively et al. (2011)
EC	DAIS-7915[a]	62	MLR	Pred.	0.665	—	4.58/4.36 ms cm[-1]	Ben-Dor et al. (2002)
	CHRIS-PROBA[b]	74/24	RK/PLSR/ PLSR-K	Pred.	—	123.74–129.6 µS cm[-1]	—	Casa et al. (2013a)
	AVNIR[a]	321	MLR	—	0.6696	1.96 dS m[-1]	—	De Tar et al. (2008)
Bulk density	EO-1 Hyperion[b]	18	PLSR	CV (LOO)	0.82	0.03 g cm[-3]	—	Anne et al. (2014)

CK, CoKriging; CR, continuum removed; KED, kriging with external drift; MAOM, mineral-associated organic matter; NSMI, normalized soil moisture index; PLSR-K, PLSR with kriging interpolation; POM, particulate organic matter; PSR, penalized-spline signal regression; RCGb, Index related to soil weathering; RK, regression-kriging; RT, regression trees; SVMR, support vector machine regression.

[a] Airborne sensor.

[b] Orbital sensor.

[c] These results concern models fitted locally using soil types to stratify the area.

[d] Quantifications were done using linear regression approach, and indices as predictive variables.

prediction accuracy. The UV range has been used for SOM prediction because it allows the detection of any molecule with alternating double and single bonds (Schulthess, 2011).

Nitrogen is part of SOM, relating to SOC, since the absolute majority of this element is in the form of organic N at a C:N rate of about 10:1 in SOM. Organic N has specific absorption features, such as those originated by amide groups, although these features are expected to be very weak due to the very low N content in soil (generally below 1%) (Stenberg et al., 2010). Thus, N may be better predicted based on a correlation with spectrally active soil components, such as SOC. Mean R^2 values for predictions of total N in various soil studies were about 0.9 using MIR and 0.86 using VIS–NIR (Soriano-Disla et al., 2014).

Prediction of soil particle fractions (clay, silt, and sand contents) using spectroscopic techniques is reported as feasible in literature. Soil texture is especially related to the mineralogical composition of soils, mainly of clay minerals and the quartz content, and thus it directly influences the soil spectra. Indeed Araújo et al. (2014a) indicated that when a SL was clustered into smaller datasets based on mineralogy and geology prior to model calibration, the R^2 value increased. Mean values of R^2 for clay and sand predictions, in validations datasets, using MIR (0.80 and 0.83, considering 14 and 11 studies, respectively) were almost equivalent for those obtained using VIS–NIR wavelengths (Soriano-Disla et al., 2014).

Another physical soil property of interest in spectroscopy is soil moisture. Different forms of water in soil (hydration, hygroscopic, and free water) are all active regarding electromagnetic energy absorption (Ben-Dor, 2011). In a given area, water (soil moisture) is related with quartz, clay minerals, and SOM, which are also spectrally active soil properties (Kuang et al., 2012). In general, Soriano-Disla et al. (2014) reported mean R^2 values for volumetric water content predictions of 0.89 for UV–VIS–NIR, 0.86 for NIR and VIS–NIR, and 0.83 for MIR wavelength ranges. The most important wavelengths for predicting soil water are 1350–1450, 1890–1990, and 2220–2280 nm (Zhu et al., 2010). Table 24.1 indicates the relation of these and other wavebands with respective soil attributes.

Iron and aluminum oxides, hydroxides, and oxi-hydroxides (most commonly hematite, goethite, and gibbsite) together with clay minerals (kaolinite, illite, and smectites) originate the main absorption features in the VIS–NIR range (Ben-Dor, 2011). Vendrame et al. (2012) predicted kaolinite, gibbsite, goethite, and hematite contents, measured by the sulfuric acid method, for Brazilian soils, using radiometric data in the NIR, and obtained R^2 of 0.83, 0.78, 0.56, and 0.60, respectively. Sellito et al. (2009) reported R^2 values of 0.46 and 0.80 for goethite and hematite prediction in soils using a simple spectral band depth calculation in the VIS range, as corroborated by a similar method performed by Richter et al. (2009) for iron oxides. Summers et al. (2011) quantified carbonate and iron oxide contents through VIS–NIR and obtained R^2 values of 0.66 and 0.61, respectively.

An important soil property that affects soil fertility and soil management is CEC, which in most cases have good prediction models, with reported mean R^2 values of 0.85, 0.81, and 0.84

for MIR, NIR, and VIS–NIR, considering 13, 9, and 8 studies, respectively (Soriano-Disla et al., 2014). Plant nutrients (Ca, Mg, K, and P) do not present absorption features in UV–VIS–NIR–MIR, except when these elements are present as constituents in molecular groups (Kuang et al., 2012). For predictions of exchangeable Ca content, mean R^2 values of 0.82, 0.75, and 0.80 were reported considering MIR, NIR, and VIS–NIR ranges in 8, 11, and 7 studies, respectively, with similar trends for Mg (Soriano-Disla et al., 2014). On the other hand, highly weathered soils from the tropics usually have very low carbonate and Ca contents, which can explain the low values obtained by Nanni and Demattê (2006a). Predictions of exchangeable K are generally less accurate (Kuang et al., 2012; Soriano-Disla et al., 2014). Inaccurate predictions are reported, in most cases, for available P content in soils, with mean R^2 of 0.35, 0.48, and 0.49 for MIR, NIR, and VIS–NIR, respectively (Soriano-Disla et al., 2014).

Predictions of soil pH depend mainly on correlation with spectrally active attributes, such as SOM and clay content. An important aspect to be considered is that soil acidity can be developed from different sources (organic or inorganic), generating instabilities in models created over large soil variabilities or geographical areas (Stenberg and Viscarra Rossel, 2010). Despite this, relatively accurate results have been reported with mean R^2 values as great as 0.79 (VIS–NIR) considering 18, 21, and 15 studies for MIR, NIR, and VIS–NIR wavelengths, respectively (Soriano-Disla et al., 2014).

Despite the indications of papers in this section, Table 24.2 aggregates more information about the subject. The quantification of clay, CEC, OC, IC, and TC certainly has strong results. In laboratory conditions, we observe average of R^2 between 0.67 and 0.78 depending on the scale of the work and with low error (RMSEP/RMSECV, SEP/SECV). In this exploratory evaluation, we found 25 papers with regional or national focus, and only 3 for global scale. Despite this, the global scale reached 0.77 of R^2 very similar to the regional or national, with less or more number of samples. This shows that number of samples, although important, is not a prerogative, if global or regional, but the representation of the distribution. Anyway, all results indicate the feasibility and repeatable strong results on the quantification of clay in laboratory conditions.

It is important to observe that there are much more works related with VIS–NIR (about 58 in this revision) and only 11 with MIR. Similar differences are observed for the other elements. This is mostly due to the lack of MIR equipments (Stark et al., 1986) and the difficulty on use them (Reeves III, 2010). On the other hand, the results reached with MIR are higher than VIS–NIR, going from an average of 0.71–0.78 of R^2, where MIR reached a maximum of 0.86. For CEC, we go from a maximum of 0.81 in VIS–NIR to 0.86 in MIR. For organic C, the differences are even more important, where we have an R^2 range of 0.75–0.93 and 0.93–0.95 for VIS–NIR and MIR, respectively. Thus, although not so great, in laboratory conditions, MIR presents better results than VIS–NIR. In an important approach, Viscarra Rossel et al. (2010) performed thematic maps of iron oxides and color by spectra information of all Australia.

24.5.2.2 Strategies for Space Spectral Sensing

Evaluating soils from aerial or orbital sensor is certainly a difficult but interesting task. Imagine trying to seek a soil property such as SOM with a sensor about 1–800 km away from the target! There are numerous factors interfering on the signal. Despite the challenge, new methods of atmospheric correction and sensors provide the possibility of improving current results. The key for detecting reliable soil attribute information from a pixel is directly related to the correct detection of areas of bare soil in the image. Opportunities for further research are plentiful; there are still issues to be studied, such as the alterations of soils due to agricultural management and contamination. Despite these, this section describes several works with this approach, and they are summarized in Table 24.3.

A considerable amount of studies have evaluated the use of multispectral data to predict soil properties. Sullivan et al. (2005) evaluated high-spatial-resolution IKONOS multispectral reflectance data for successfully mapping soil properties in regions of Alabama. Chen et al. (2008) used the ATLAS sensor (2 m of spatial resolution and 8 bands in the VIS–NIR–SWIR) data to predict SOC, reporting an R^2 of 0.63 and root mean square error (RMSE) of 0.22%. Löhrer et al. (2013) identified and mapped mineralogical composition of Pleistocene sediments using ASTER and SPOT-5 images associated with spectral laboratory, obtaining good results.

Several studies in the tropics were successful in quantifying soil properties using a technique to detect bare soil (Demattê et al., 2009a—see Section 24.7). An average R^2 of 0.67 was obtained for clay content estimation using MLR (Demattê and Nanni, 2003; Demattê et al., 2005; Nanni and Demattê, 2006a, b; Fiorio et al., 2010). Using the same methodology, Demattê et al. (2007a) studied pixels with two different information of soils with a high variance on texture and reached a value of 0.86 R^2 for clay. Using Landsat, Demattê et al. (2009b) and Fiorio and Demattê (2009) predicted clay, Al_2O_3, Fe_2O_3, and weathering indicators, Ki, SiO_2, and TiO_2, with similar R^2 values of 0.61, 0.68, 0.67, 0.54, 0.65, and 0.72, respectively. To monitor the spatial variation of CEC, Ghaemi et al. (2013) developed a model based on Landsat ETM data in a semiarid area in the Neyshabur Province, Iran, where the model correctly classified the spatial variation of CEC in 45%–65% of the cases. Additionally with Landsat, Masoud (2014) predicted salt abundance in soils of an area in Burg Al-Arab, Egypt, using Landsat-based spectral mixture analysis (SMA) and soil spectroscopy, achieving an R^2 of 0.88. Still with Landsat, Dogan and Kiliç (2013) studied the relationship between soil variables and the Landsat seven bands using digital number (DN) values and found good correlation for pH, OM, $CaCO_3$, and N with band 5.

Using ASTER with eight bands, Vicente and Souza Filho (2011) observed R^2 of 0.65–0.79 between spectral signatures measured from soil samples and ASTER pixels to map kaolinite and iron oxides using a mixture-tuned matched filtering (MTMF) approach. Similarly, Genú et al. (2013a) combined ground and ASTER pixel reflectance with multiple endmember

spectral mixture analysis (MESMA) for mapping SOM, reaching a 60% agreement between them. As to aggregate all information of ASTER, including its 2–14 µm range, Hewson et al. (2012) studied the sensor SL (SWIR and TIR spectral signature) to evaluate the spectral indices for composition and texture of natural soil samples.

Despite the multispectral data, hyperspectral has great importance due to the high number of bands. On the prediction of SOC, Gomez et al. (2008b) reported an R^2 of 0.66, RMSE of 0.61%, and residual prediction deviation (RPD) of 1.69 for laboratory measurements, and an R^2 of 0.51, RMSE of 0.73% and RPD of 1.43 for Hyperion data, while Hbirkou et al. (2012) reached 0.83 of R^2 with HyMap. Stevens et al. (2012) using the AHS-160 hyperspectral sensor obtained high accuracy with R^2 of 0.73 and RMSE of 4.7 g kg^{-1} applying a soil-type calibration strategy and PLSR to derive the prediction model for SOC contents. For N, Selige et al. (2006) reported similar performance for PLSR and MLR models, using HyMap data, to predict total N content, with R^2 of 0.92 and RMSE of 0.03% for PLSR. Afterward, Zhang et al. (2013) also performed total N content quantification but used a Hyperion image and reported an R^2 of 0.70.

The influence of soil particle distribution, for example, clay and sand, on spectra is great, and studies such as Selige et al. (2006) evaluated the potential of ASS (HyMap sensor) to predict these properties. The authors found R^2 values of 0.71 and 0.95 for clay and sand contents, respectively. De Tar et al. (2008) used MLR with AVNIR hyperspectral imagery (60 spectral bands from 429 to 1010 nm) and reported R^2 values of 0.81, 0.75, and 0.67 for sand, silt, and clay contents, respectively. Gomez et al. (2008a, 2012) performed clay content quantification through HyMap sensor data and obtained R^2 of 0.64 and 0.67, with RMSE of 49.6 and 42.2 g kg^{-1} respectively using PLSR. Hively et al. (2011) using HyperSpecTIR sensor data obtained R^2 of 0.79, 0.79, and 0.66 for sand, silt, and clay content predictions, respectively. Casa et al. (2013b) compared different hyperspectral sensors—the airborne MIVIS (VIS–NIR–SWIR) and the spaceborne CHRIS-PROBA (VIS–NIR)—with spatial resolutions of 4.8 and 17 m, respectively, and obtained R^2 values of 0.48 and 0.52, and RMSE of 7.20 and 6.87, for clay prediction with both sensors, respectively.

Moisture is highly related to the soil spectra because water content affects the baseline height (albedo) and causes several spectral absorption features (Lobel and Asner, 2002). Whiting et al. (2004) proposed a spectral technique to estimate SM content, in which an inverted Gaussian function is fitted centered on the assigned fundamental water absorption region at 2800 nm, and the area of the inverted function accurately estimated the water content within an RMSE of 2.7% and R^2 of 0.94, for laboratory spectral data. Ben-Dor et al. (2002) applied MLR to predict soil gravimetric moisture using DAIS-7915 hyperspectral sensor and obtained an R^2 of 0.64 for the calibrated model. Later, Zhang et al. (2012) used the entire spectral range measured by the Hyperion to model soil gravimetric moisture using PLSR. The authors reported an R^2 of 0.79, RMSE of 3.13%, and RPD of 2.22 for the calibration dataset (cross-validation), and R^2 of 0.40, RMSE of 5.12%, and RPD of 1.15 for the validation dataset.

The soil mineralogy has also been studied using SSS. Galvão et al. (2001) evaluated AVIRIS data to quantify TiO_2, Fe_2O_3, and Al_2O_3 contents and reported R^2 values of 0.74, 0.83, and 0.68, respectively. Bartholomeus et al. (2007) observed correlation between spectral bands of ROSIS hyperspectral sensor and iron content determined by the dithionite–citrate method, as high as 0.5, for spectral measurements near 650 nm. Lagacherie et al. (2008) and Gomez et al. (2012) predicted $CaCO_3$ content using band depth measurements (near 2341 nm) from HyMap data and obtained R^2 values between 0.47 and 0.77. Baptista et al. (2011) applied a spectral index (RCGb) to estimate soil weathering degree using HyMap and reported a value of 0.75 R^2 between the proportions of kaolinite and gibbsite in soil samples. Alterations of minerals are also evaluated, for example, Molan et al. (2014) used HyMap to map the distribution of altered clay minerals in Iran.

In terms of the soil chemical attributes predicted by SSS, De Tar et al. (2008) estimated Ca, Mg, K, P, and pH using an MLR approach and reported R^2 values of 0.62, 0.58, 0.64, 0.70, and 0.62, and RMSE values of 9.51 meq L^{-1}, 4.32 meq L^{-1}, 45.03 mg kg^{-1}, 4.32 mg kg^{-1}, and 0.076, respectively, using AVNIR hyperspectral data. Hively et al. (2011) also evaluated predictions of Ca, Mg, K, and pH using HyperSpecTIR sensor data and obtained R^2 values from 0.51 to 0.69. Gomez et al. (2012) report a less optimistic result for pH reaching a maximum R^2 of 0.31. Lu et al. (2013) used Hyperion imagery and PLSR to predict soil total P and pH reporting R^2 of 0.62 and 0.68, respectively. Using hyperspectral images (AISA-Dual VIS–NIR), Lagacherie et al. (2013) predicted soil attributes such as clay, sand, and CEC, in profiles with R^2 between 0.71 and 0.81 for topsoil samples. Gomez et al. (2012) report a less optimistic result reaching a maximum R^2 of 0.31. Lu et al. (2013) used Hyperion imagery to predict soil total P and pH reporting R^2 of 0.62 and 0.68, respectively. Using hyperspectral images (AISA-Dual VIS–NIR), Lagacherie et al. (2013) predicted soil attributes such as clay, sand, and CEC, in profiles with R^2 between 0.55 and 0.81. Gomez et al. (2012) quantified CEC and obtained an RMSE of 1.84 $cmol_c$ kg^{-1} using PLSR with HyMap data. Using Hyperion, Lu et al. (2013) reported cross-validation RMSE of 1.55 $cmol_c$ kg^{-1} for CEC quantification. Figure 24.4a demonstrates the potential of HyMap data to develop a soil map of iron (Richter et al., 2007). The masked and no-iron extractable areas are associated with other components in the image, such as water surfaces or high vegetation cover areas. As another example, Figure 24.4b shows EnMAP sensor-based predictions for clay content in topsoils of exposed surfaces in agricultural areas of southern Spain, where error estimates were about $R^2 \sim 0.5$ (Chabrillat et al., 2014). Another interesting work (van der Meer and de Jong, 2003) takes into account a discussion about hyperspectral sensing and band depth on the quantification of soil attributes. The authors indicate a simple linear interpolation method introduced to estimate absorption band parameters from hyperspectral image data. By applying this hyperspectral data, it has been demonstrated that absorption feature maps correspond favorably with the main alteration phases characterizing the systems studied. Thus, the derived feature maps allow

enhancing the analysis of airborne hyperspectral image data for surface compositional mapping. Despite the quantification of soil attributes, images have great importance to provide information on their spatial distribution as observed by Seid et al. (2013).

The soil property prediction and mapping based on SS data have been done for a while. By far, the most studied soil properties are clay content and SOC (Table 24.3). This is related to the fact that clay and SOC, as well as mineralogy and water content, are spectrally active attributes in soil (Stenberg et al., 2010). Moreover, these attributes may potentially be predicted with high accuracy. Although attributes like soil cations are not spectrally active, these might be predicted correctly using remotely sensed radiometric data. In these cases, Stenberg et al. (2010) state that these attribute variations on an area are related to spectrally active attributes, and consequently satisfactory coefficients of determination (R^2) can be obtained. Considering the R^2 reported for clay and SOC prediction, it is highly variable with a range of 0.01–0.9. This fact proves that although the technique (multi- and hyperspectral) is capable of predicting soil attributes, other conditions need to be accomplished in order to obtain models considered robust. One of these conditions is to correct the radiometric data regarding the influence of atmospheric components, which significantly improves the quantitative potential of soil spectra. Besides, soil moisture contents and nonphotosynthetic vegetation (NPV) mixture with soil in the imaged pixels reduce the predictive potential of algorithms.

Many techniques have been applied to soil property prediction, but those related to multivariate statistics are predominant. When considering multispectral sensors, the MLR is the most applied. On the other hand, the PLSR are recurrent in hyperspectral data modeling (Table 24.3). Furthermore, there is currently a trend in using multispectral images as secondary data in the mapping process, using concomitant other information easily accessible, such as DEM and terrain attributes derived from these. Due to the constrained spectral resolution present in these sensors, using multispectral images as a primary source on soil modeling might be a limiting factor. Moreover, with the availability of new hyperspectral sensors, the diffusion of more robust statistical algorithms, in addition to greater capacity for computational processing, turned the hyperspectral sensing more reliable as a primary source for soil prediction, considering the great quantity of information provided by this kind of data.

From Table 24.3 we also observe the following: (1) R^2 results vary from very low (under 0.3) to high (over 0.8) and very strong (over 0.9). (2) In a general vision, multispectral sensors with low number of bands such as IKONOS show the lowest results. (3) Other multispectral data, the most used being landsat satellite, have variable, but rather good, results with a strong background on processing and understanding over the last 44 years. (4) Hyperspectral information should be the best results, as previously discussed; on the other hand, many of them are similar to landsat data. This can be due to the limited experience with these sensors when related with landsat; thus these will probably be improved in the future with upcoming opportunities, and these techniques have to be encouraged. (5) Interesting to see

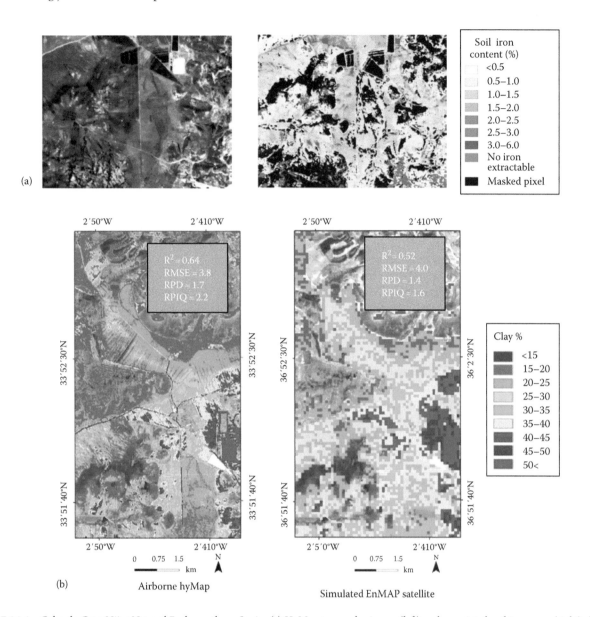

FIGURE 24.4 Cabo de Gata-Nijar Natural Park, southern Spain: (a) HyMap true-color image (left) and associated soil iron map (right). (From Richter et al., 2007.) (b) Potential of upcoming EnMAP hyperspectral satellite for quantitative surface soil mapping. Soil clay content maps (Cabo de Gata natural Park, Spain) based on airborne HyMap (left) and spaceborne simulated EnMAP hyperspectral images (right). (From Chabrillat et al., 2014.) Model error metrics calculated using independent validation data.

that several soil attributes have been achieved since clay, sand, silt, carbon, OM, Ca, Mg, K, P, N, carbonates, iron, and many others. (6) Many airborne hyperspectral sensors have been used and tested such as HyMap and SpecTIR, which indicates that community is going toward these equipments. (7) Considering repeatable results, the most reliable soil attributes to be quantified are clay, OM, OC, carbonates, and Fe_2O_3. (8) There are great results for other elements such as N, Al, Mn, Zn, Cl, and Fe, but these have very low content and low effect directly in spectra, and thus it is suggested that more studies be conducted to prove that the results are reliable.

Finally, it is interesting to observe that, for example, clay soil attribute has great results from aerial or orbital, multi- or hyperspectral sensing, and, in many cases, have similar R^2 when compared with observations when quantified with ground sensing (Table 24.2). The point is that theoretically, results should present the following *sequence* from the best to the worse results: laboratory sensing—field sensing—aerial hyperspectral sensing—orbital hyperspectral—orbital multispectral sensing. On the other hand, the results observed in literature, as indicated in these sections (and some summarized in Tables 24.2 and 24.3), take us to the following points: (1) there are overlapping results between the *sequence*, (2) we can find a multispectral sensor with a better result of a hyperspectral or vice versa, (3) we can have a laboratory result worse than an orbital multispectral data, and (4) we can have very low results for chemical elements such as K with laboratory sensor and a good result from a hyperspectral airborne sensor. This brings us to some conclusions: (1) the quantification of soil attributes is a reality and can be done from any platform; (2) some soil attributes have more literature consistency because they are repeatable and have a theoretical background that explains the results, not relying only on R^2 values; (3) the differences have several factors such as number and homogeneity or heterogeneity of data, quality of data, atmospheric correction factors, date and quality of images, quality of soil analysis, soil moisture effects when analyzing aerial or orbital data, quality of the sensor data, statistical analysis, spectra data processing methods, and others. Thus, it is important to take these into account in future studies looking toward reaching better results.

24.5.2.3 Strategies for Thermal Infrared

Thermal information has been widely used for geology (e.g., van der Meer et al., 2012), and now its importance to soil evaluation is being explored. On the other hand, it has been very poorly explored for soil studies, despite its exciting approach. The determination of soil temperature can be related with soil organisms, OM alterations, soil moisture, culture development, and others. Some works have explored this approach. Zhan et al. (2014) show that thermal SSS can be used to estimate soil temperatures. The result was generated using data from the MODIS satellite and demonstrated that soil temperatures with a spatial resolution of 1 km under snow-free conditions can be generated at any time of a clear-sky day. Comparison between the MODIS and ground-based soil temperatures shows that the accuracy lies between 0.3 and 2.5 K with an average of approximately 1.5 K.

Using another strategy, Zhao et al. (2014) used time-series remotely sensed data, including thermal imagery extracted from MODIS combined with vegetation indices (soil-adjusted vegetation index—SAVI) calculated from Landsat ETM+ to estimate the spatial variation of SOM by land surface diurnal temperature difference in China. They suggested that time-series remotely sensed data can provide tools for mapping SOM. Soliman et al. (2011) described a method by thermal inertia using standardized principal component analysis (PCA) applied to a time series of TIR images. The images were taken from a camera mounted at a height of 17 m above ground level using a mobile hydraulic boom lift, and results were well related with intrinsic soil physical properties, such as soil bulk density and porosity.

24.5.3 Strategies for Soil Classification

Soil classification is a dynamic procedure that requires knowledge on soil science and spatial sciences. The data derived from the soil profile (control section) analysis are the most important information for soil classification of a pedon (smallest, three-dimensional unit at the surface of the earth that is considered as an individual soil). Many classification systems require soil characterization from the laboratory as well as a complete morphological description of the soil horizons. The morphological evaluation of soil structure in the field, for example, requires determination of the type, shapes, and size of soil structural peds, where the soil structure is determined by the activity of

soil biota (macro- and microorganisms), clay content, mineralogy, OM, and soil aggregation. This requires opening soil pits of variable dimensions where the soil is described and classified based on the upper 2 m profile. This is a time-consuming procedure. Thus, the advantage of using SS on soil classification is to infer soil properties using sensors.

The radiometric data can be used to aid soil classification following three main approaches: (1) by quantifying soil properties and then using them as input data for the taxonomic classification; (2) by analyzing the spectral curves of single horizons in search for class-specific spectral patterns, since every spectrum constitutes a unique fingerprint that relates to the soil horizon properties; and (3) by analyzing the spectral curves of all soil horizons simultaneously, combining and interpreting this information to achieve a soil taxonomic classification. The last strategy can be done only using ground-based sensors, directly in field or by bringing sampled horizons into the laboratory. Some methods have been proposed to do this type of analysis, for example, going inside a pit with the spectroradiometer to collect radiometric data (Viscarra Rossel et al., 2009), inserting an optic fiber in a hole and evaluating the spectra at different depths or horizons (Ben-Dor et al., 2008), or collecting samples with an auger or hydraulic core and analyzing them using spectral data measured in the laboratory (Demattê et al., 2004a; Waiser et al., 2007; Morgan et al., 2009)—see Figure 24.2. When using soil surface sensors (orbital, aerial, or ground), the approach is different, since only surface information can be assessed this way (Figure 24.2). Hartemink and Minasny (2014) have recently discussed the importance and future of soil morphometrics, where they describe several soil sensors that can assist in soil classification, including VIS–NIR–SWIR–MIR radiometers, ground-penetrating radar (GPR), electrical resistivity meters, cone penetrometer, hyperspectral core scanner, x-ray fluorescence meter, and others. Despite these possibilities, there is one important limitation: until today, no equipment was able to indicate directly the shapes and sizes of peds, and thus, technologies still rely on other (chemical and physical) qualities of the soil profile to infer soil classification. On the other hand, to reach a soil classification, we also need to understand the behavior of the attributes in different soils, such as made for class texture (Franceschini et al., 2013), clay activity (Demattê et al., 2007b), weathering indexes (Galvão et al., 2008), mineralogy (Madeira Netto, 2001), and electrical conductivity (Ucha et al., 2002). Other interesting techniques have evolved in this area. Schuler et al. (2011) indicate that the gamma spectrometry is a promising potential as a tool to distinguish Reference Group WRB soil profile, field, and the landscape scale, but needs to be verified in other regions of the world.

24.5.3.1 Strategies for Ground Spectral Sensing

One of the first attempts to classify soils using spectra was performed by Condit (1970) and complemented by Stoner and Baumgardner (1981). In general, evaluation is done based on the spectrum of each horizon individually (Galvão et al., 1997). Later on, Demattê et al. (2002) proposed the analysis of the spectral morphology considering the following aspects of the curve: complete shape, reflectance intensity (albedo), and absorption features, which were used to characterize several different soils from the tropics (Nanni et al., 2011, 2012, 2014). This has recently had an upgrade with the development of the morphological interpretation of reflectance spectrum, Demattê et al. (2014), which indicate a detailed system for users to look at spectra shapes looking toward soil classification.

On the other hand, quantitative methods have been used to directly predict soil classes from soil spectra. For example, Vasques et al. (2014) introduced a system where the spectra of different soil layers (depths) are evaluated at the same time and reached 90% accuracy for some classes. Bellinaso et al. (2010) reached an 80% agreement between soils classified by spectra and soils classified in the field following traditional soil survey protocols, as corroborated by Viscarra Rossel and Webster (2011). Thus, it is important to aggregate quantitative and descriptive analysis to reach the best result on soil classification.

24.5.3.2 Strategies for Space Spectral Sensing

Unlike in GSS, data acquired by SSS are usually not enough to classify soils accurately. These aerial or orbital sensors measure only the superficial layer of soil and consequently do not represent the entire profile. Despite that, SSS can still have a great contribution in soil discrimination and mapping by allowing to detect spatial patterns in soil surface variation that relate to the spatial distribution of soil classes. Several studies using Landsat data (Demattê et al., 2004c, 2005, 2007a, 2009a; Nanni et al., 2011, 2012, 2014; Genú et al., 2013b) demonstrated a high potential for soil discrimination. Basically, these studies used images atmospherically corrected and transformed into reflectance, detected bare soil spots, and correlated these pixels with the field soil classification. They found high correlation among the spectral curves of similar soil classes (see example in Figure 24.2, strategy 1). The authors indicated that the soil surface patterns are related to underlying soil variation and dynamics within the soil profile that are specific to each soil class, allowing this correlation. On the other hand, authors emphasized that this correlation is not present in all cases. It is possible that two very different soils (e.g., an arenosol and a lixisol) have a sandy surface that could present similar spectra and thus not be discriminated. In this case, we would have to aggregate a new strategy to outline this problem, such as relief information. We have to underline that this strategy can allow the *separation* of different soils without necessarily giving them *names* (classes). This is helpful, for example, to find similar soils and delineate soil mapping or management units. Their names (the classification itself) can be given in the field using other strategies (Figure 24.2, Strategies 2 and 3).

24.5.4 Strategies for Soil Class Mapping

Soil class (survey) maps, when used at the appropriate cartographic scale, allow the improvement of the land productive capacity preventing environmental degradation. New methodologies and strategies are needed to map soils in unmapped areas efficiently fostering sustainable food production. Soil SS have the potential to assist users in efficient soil mapping. Like other applications, the user has to understand that each SS technique, based on field, laboratory, airborne, or orbital sensors, needs to be explored and adapted to the different objectives presented, considering their advantages and limitations, the desired characteristics of the final products, including the type of product (model, map, or both), scale, and level of accuracy, and the final users of the information.

24.5.4.1 Strategies for Ground Spectral Sensing

When spatial data are available, a soil map can be created by following this suggested sequence: (1) assign the sampling points where soil samples will be collected in the field, based on the methodology described in Section 24.5.1; (2) describe the soil and collect the soil samples at the sampling sites; (3) take samples to the laboratory and prepare them for spectra readings; (4) acquire, process, and interpret spectral data; (5) based on spectra information, choose the most representative one, which will go to traditional wet analysis; and (6) wet versus spectra ones will create models for quantification. The other samples will be quantified by spectra information; (7) relate spectral data to known soil properties and classes; and (8) interpolate the known soil information across the area of interest using spatial interpolation techniques. With portable equipment and appropriate calibration models, the third step can be replaced by spectral measurements made directly in the field. FSS measurements produce less accurate predictions of soil properties than LSS. However, more observations can be made because there is no need for transport or sample preparation. Nonetheless, GSS measurements can be augmented by traditional soil profile evaluation in the laboratory for a more accurate classification of the soil unit, as suggested in Section 24.5.4. Demattê et al. (2001, 2004a) have demonstrated how soil spectra data measured at sampling points can be used to derive soil maps. These authors used ground-based spectroscopy for pedological purposes, collecting spectra (VIS–NIR–SWIR) in soil profiles along landscapes and relating spectra to soil classes using a spectral pedological databank with descriptive and quantitative information associated to the radiometric measurements. These methods were used to create a pedological map that was well correlated (79%) with the results of the traditional mapping approach.

24.5.4.2 Strategies for Space Spectral Sensing

Soil sensing data are valuable for providing either primary or auxiliary information, which can be used to interpolate soil properties assessed using other techniques. Preferably, the spectral imagery must be atmospherically corrected and transformed into reflectance before use. Generally, only pixels with bare soil are considered to study soils using SSS, and techniques to identify these pixels must be applied before soil characterization. Reflectance values at each pixel of designated soil observations can be related to image data bases of surface spectra. Though the use of orbital/aerial SS data does not provide direct measurements beyond topsoil spectra, this information is a first indication of soil spatial variability and can be combined with undersurface soil properties from another data source to create complete soil maps. For example, Demattê et al. (2012b) combined aerial photographs and radar and laboratory spectral information to develop soil maps. In fact, the use of orbital images in bare soil mapping is restricted in areas under haze or clouds, under shades, and under vegetation (crop fields or natural vegetation). Also, in situations where legacy soil data are scarce or unavailable, remotely sensed soil data can be an important source to fill data gaps, for example, by interpolation (McBratney et al., 2003; Mulder et al., 2011).

Boettinger et al. (2008) used Landsat data as environmental covariates for digital soil mapping (DSM) in arid and semiarid regions. In this example, data successfully represented environmental covariates for vegetation (e.g., normalized difference vegetation index—NDVI, fractional vegetation cover) and parent material (e.g., band ratios diagnostic for gypsic and calcareous materials). Browning and Duniway (2011) presented a semiautomated method to map soils with Landsat ETM+ imagery and high-resolution (5 m) terrain (IFSAR) data. Later, Grinand et al. (2008) predicted soil distribution using Landsat ETM imagery, terrain factors (e.g., elevation and slope), land cover, and lithology maps. Hengl et al. (2007) employed multiple covariates including six terrain parameters, MODIS enhanced vegetation index images, and a polygon map of 17 physiographic regions of Iran, to map soil classes. Hansena et al. (2009) used coarse-resolution soil maps, combined with NDVI, normalized difference IR index, SWIR reflectance, slope, and two relative elevation layers, to map soils in Uganda. Although with low spatial resolution (1 km), Hou et al. (2011) used MODIS to identify different soil types, reaching a Kappa index of 0.75.

The importance of images in soil mapping is more than simply aiding in soil classification. They help to detect geographic boundaries among soils to delineate soil mapping units. In fact, Nanni et al. (2014) tested several SS sources, from ground sensors and spectra from images, to evaluate which source could reach better performance for soil mapping. The traditional soil survey map had 53 polygons, while the ground spectra determined 22 polygons, and Landsat gave 35 polygons. Working with aerial photographs, Demattê et al. (2001) defined 12 polygons versus 15 polygons delineated by traditional field soil survey.

Considering the use of hyperspectral imagery, Galvão et al. (2008) described the use of AVIRIS data to map several soil properties related with soil classification such as Fe_2O_3 and Ki (weathering index)—Figure 24.5. They mapped the selected soil properties using hyperspectral images and found good agreement between weathering indexes and elevation. Chabrillat et al. (2002) also used AVIRIS data, but mapped soil expansive clays (Figure 24.6a) considering differences between spectral signatures of mineral endmembers (Figure 24.6b). Chabrillat et al. (2002) examined the potential of optical RS and in particular hyperspectral imagery for the detection and mapping of expansive clays in the Front Range Urban Corridor in Colorado, USA. Here spectral endmembers were extracted from the images

without field knowledge and were implemented with different algorithms for clay mapping. Results showed that spectral discrimination and identification of variable clay mineralogy (smectite, illite/smectite, kaolinite) related to variable swelling potential was possible using different algorithms, in the presence of significant vegetation cover. Field checks have shown that the maps of clay type derived from the imagery and interpretation in terms of swell hazard are accurate. The main limitations for the expansive clay soil detection were in case of a heavy vegetation cover (forest or green grass) and when the reflectance of the soils is approximately 10% or less.

24.5.5 Integrating Strategies

Mapping soils accurately and at detailed enough spatial resolutions requires merging multiple types of spatial data and statistical techniques to maximize soil inference from these datasets. SS data from orbital and airborne platforms can provide information supporting DSM. SS involving field and laboratory passive optical sensors and active microwave instruments have been used at regional and coarser scales to map soil mineralogy, texture, moisture, SOC, salinity, and carbonate content. Ballabio et al. (2012) employed a vegetation-based approach for mapping SOC in alpine grassland. They used a map of the properties of the plant communities created by combining high-resolution multispectral images and light detection and ranging data. Additionally, McBratney et al. (2003) discuss methods to find quantitative relationships between soil properties or classes and their environment. One issue of particular difficulty in DSM is the combination of data collected at different spatial resolutions and with different accuracies. For example, a soil pedon description and subsequent laboratory analysis contain the highest-quality information, but the spatial extent is extremely sparse. GSS data, for example, collected across a soil profile, can be more detailed, but still done on a point support. Moreover, if GSS data are collected without proper soil description, its use is limited for soil mapping. On the other hand, SSS data provide seamless ground coverage to assess soil spatial variation, but their spectral resolution and signal-to-noise ratio (SNR) are usually poorer than laboratory and field sensors. These three types of data acquisition can be combined using a neighborhood-type statistical technique (Zhu et al., 2004). Additionally, filtering techniques can be employed to overlay multiple scales of remotely sensed soil data as suggested by Behrens et al. (2010).

24.5.6 Strategies to Infer Soils from Vegetation

SS data obtained directly from bare soils are not always available to allow soil characterization. Often, soil, photosynthetic vegetation (PV), and NPV occur associated on the landscape, and the fractions they occupy in pixels vary with land use and environmental characteristics of the studied area. For this reason, the use of SS to study soils in densely vegetated areas relies on indirect relations between vegetation and soil properties (Mulder et al., 2011). Vegetation has been used to assess soil characteristics and

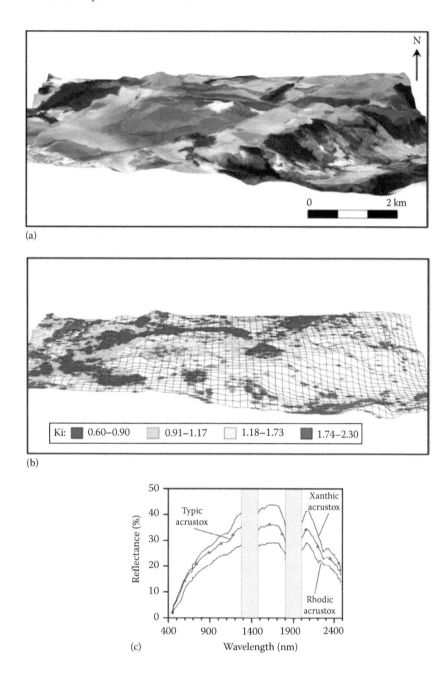

FIGURE 24.5   (a) Digital elevation model, (b) map of weathering index Ki (Al_2O_3/SiO_2) from surface, with higher values indicating more weathered soils, and (c) spectral curves generated by the AVIRIS hyperspectral sensor. Ki: higher values lower weathered soils. (Extracted from Galvão et al., 2008.)

variability in a number of studies (Bartholomeus et al., 2011). For example, Asner et al. (2003) related SOC and N to fractional cover by PV and NPV and were able to show the trends in these soil properties at the ecosystem level. Kooistra et al. (2003, 2004) used vegetation spectral characteristics obtained by a handheld spectrometer in the field, to estimate Zn, Ni, Cd, Cu, and Pb in a floodplain, indicating the potential of using SS techniques for the classification of contaminated areas. Vegetation indices, like NDVI, have been applied as indicators of crop growth and site quality (Sommer et al., 2003). Sumfleth and Duttmann (2008) found, in a study carried out in paddy soil landscapes in

southeastern China, that NDVI values are related to SOC, N, and silt contents, and to some terrain attributes. In specific cases, the combination of several spectral indices can minimize the vegetation influence on estimated soil properties like iron content, as verified by Bartholomeus et al. (2007). Mann et al. (2011) used NDVI, among other covariates, to map the productivity of a citrus grove and stated that the productivity zones could be used successfully to plan soil sampling and characterize soil variation in new fields.

Mulder et al. (2011) indicated two possible approaches to retrieve soil properties from vegetation SS data: through plant

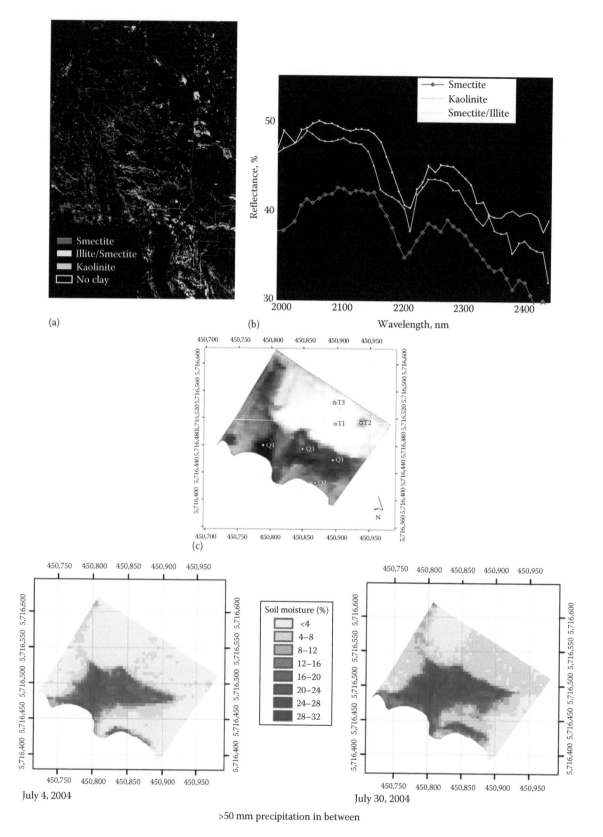

FIGURE 24.6 (a) Map of soil clay type in Colorado, USA, produced from the RGB image of three abundance maps, extracted from AVIRIS high-altitude images covering an area of ~20 × 37 km. (From Chabrillat et al., 2002.) (b) Clay endmembers spectra. (From Chabrillat et al., 2002.) (c) Multitemporal surface soil moisture maps in Welzow Bio-monitoring Recultivation Area, Germany, produced from HyMap airborne images based on NSMI method. (From Haubrock et al., 2008a,b.) Top: HyMAP true-color image and location of different soil features, bottom: Soil moisture maps.

functional types (PFTs) and Ellenberg indicator values. The former states that resource limitations, response to disturbance, biotic factors, and other environmental aspects are related to morphological and physiological adaptations in plants, and the ability to detect functional types with RS relies on the extent to which such relationships are generalized (Ustin and Gamon, 2010). As an example, Schmidtlein (2004) used the spectral characterization of PFT to discriminate different soil units. In the same way, Ellenberg indicator values can be used to retrieve soil properties, like soil moisture, pH, and fertility, and hyperspectral imagery can be used to derive these values (Schmidtlein, 2004; Mulder et al., 2011).

24.5.7 Strategies for Soil Management and Precision Agriculture

The main goal of PA is site-specific management, considering soil and crop factors to maximize the efficiency of the applied resources. These resources include seed (seeding rates and site-specific genetics), fertilizers, soil amendments such as using lime, herbicides, pesticides, and growth regulators. Information needs by PA include high-resolution (<10 m) maps of soil properties, such as plant-available water and nutrient status in the rooting zone. Thus, accurate indication of soil properties with dense spatial coverage is important, and this can be done with SS. For example, VIS–NIR models were used by Wetterlind et al. (2010) to assist field soil characterization using small local calibrations and national libraries, reaching good results. SS is particularly useful to PA because it allows, besides posterior soil characterization, real-time assessment of soil properties and their variation. This is the most successful data source that might relate the well-being of the crop to nutrient needs. Many PA users desire soil sensing strategies that assess nutrient content in the soil linking to plants' needs, although this is a difficult task as already demonstrated in Section 24.5.2.1. Looking at this approach, Tekin et al. (2013) used AgroSpec (Tec5) mobile, fiber type, and VIS–NIR spectrophotometer to measure online soil pH (field measurement). They produce variable-rate lime recommendation maps with results $R^2 = 0.81$, RMSEP = 0.20, and RPD = 2.14.

Today, SS can assist in PA implementation providing data on soil properties such as clay content, SOM, SOC, and CEC, which are related to nutrient supply and retention as well as productivity potential. Nonetheless, laboratory and controlled experiments are important to understand the crop response to the environment, including soil variability, and perhaps can provide a better basis for relating measurable soil properties to nutrient availability and soil water storage, among others. One attempt to combine lab knowledge with spectroradiometer data is described by Demattê et al. (2003a). They identified alterations in soil spectra due to fertilizer application but concluded that it was not possible to identify P in soils. However, lime application in the soil affected soil spectra (Demattê et al., 2003a). This was corroborated by Araújo et al. (2013), who reached a value of 0.90 R^2 predicting liming requirement for a sandy and a clayey oxisol using spectroscopy.

Most SS studies focusing on PA use GSS combined with lab-measured soil properties (clay, SOC, SIC, and others). A review on GSS is provided by Kuang et al. (2012). Mouazen et al. (2007) developed a soil sensing system consisting of an optical probe mounted on a soil penetrometer to measure SOC, SM, pH, and P. Calibration models were developed using laboratory soil data and were validated using spectra from field measurements. Estimation of soil moisture content was satisfactory ($R^2 = 0.89$), whereas the estimates of SOC, pH, and P were not as well matched to the corresponding reference values (R^2 of 0.73, 0.71, and 0.69, respectively). Later, Kodaira and Shibusawa (2013) found a strong correlation between CEC ($R^2 = 0.89$) and on-the-go spectroscopy data.

To actually measure the soil solution, Viscarra Rossel and Walter (2004) built a soil analysis system comprising a batch-type mixing chamber with two inlets for 0.01 M $CaCl_2$ solution and water, respectively. The system was tested in the laboratory using soil solutions of 91 Australian soils and in a 17 ha agricultural field to estimate lime requirements. The system produced an RMSE of 0.2 for pH (R^2 of 0.66). However, the coefficient of determination for pH buffer estimates was not high ($R^2 = 0.49$). Despite this, results indicate a great potential, and research is still required. Due to this, Ballari et al. (2012) propose the use of a network of mobile sensors associated with the expected value of information (Evoi) and mobility restrictions to reduce the costs of monitoring phenomena such as soil and natural radioactivity. Christy (2008) used a field spectrophotometer to provide several soil attributes in real time, obtaining an RMSE of 0.52% and 0.67 R^2 to soil OM.

The success of on-the-go SS, thus, has several tasks. Gras et al. (2014) focused at optimizing the acquisition procedure of topsoil VNIR spectra in the field with the view to predict soil properties. Obtaining good VNIR cross-validation for calcium carbonate, total nitrogen, OM, and exchangeable potassium, RPD reached up to 9.1, 2.9, 2.8, and 3.0, respectively.

24.5.8 Strategies for Soil Conservation

To perform land use planning and promote soil conservation, information about a specific area is needed such as relief and slope data, erosion susceptibility, soil classes, vegetation cover, among management factors. Additionally, it is also necessary to identify areas that have already exhibited problems, so that corrective measures can be taken. SS can be used to identify possible problems in the soil and monitor the effects of management decisions by looking at soil and plant spectral responses. Brodský et al. (2013) mapped the SOC through VNIR spectroscopy with R^2 over 0.7 and RPD over 1.5 in eroded areas at the farm level. King et al. (2005) and Vrieling (2006) presented an interesting review describing erosion mapping techniques integrating inputs derived from SSS and additional data sources into runoff and erosion prediction models. Similarly, Shruthi et al. (2014) detected erosion effects and monitored variations in erosion dynamics and degradation levels using Ikonos-2 and GeoEye-1. Both D'Oleire-Oltmanns et al. (2012) and Peter et al. (2014) used an unmanned aerial vehicle (UAV) for monitoring soil erosion in the Souss Basin (Morocco), where the imagery data were used to quantify gully and badland erosion in 2D and 3D and to analyze the erosion susceptibility of surrounding areas.

Different acquisition levels of SS have been combined for soil conservation studies. Garfagnoli et al. (2013) used a hyperspectral dataset acquired with an airborne Hyper SIM-GA sensor from Selex Galileo simultaneously with ground soil spectral signatures to monitor soil degradation processes. Liberti et al. (2009) assessed how accurately a badland area can be identified from Landsat TM and ETM data. The authors found that the combined use of SS and auxiliary morphological information significantly improved the mapping of badlands over large areas with heterogeneous landscape features. With another approach, Nadal-Romero et al. (2012) assessed badland dynamics using multitemporal Landsat TM and ETM imagery for the period 1984–2006 in Spain, and the results showed that NDVI helped in revealing degraded areas.

Martínez-Casasnovas (2003) presented a method to compute the rate of retreat of gully walls and the associated rate of sediment production caused by erosion by integrating multitemporal aerial photos and multiresolution DEMs in Catalonia, Spain. Ries and Marzolff (2003) designed a hot-air blimp as a platform sensor to obtain large-scale aerial photographs from the Barranco de Las Lenas (Spain) with very high spatial and temporal resolution for monitoring the development and dynamics of erosion. In southern Italy, Conforti et al. (2013), combining GSS in the VIS–NIR spectral range with aerial photo interpretation and geostatistics, predicted and mapped SOM content, relating it back to water erosion processes. Vågen et al. (2013) combined Landsat ETM+ imagery, systematic field methodologies, IR spectroscopy, and ensemble modeling techniques for landscape-level assessments of land degradation risk and soil condition. The Landsat prediction was robust, with R-squared values of 0.86 for pH and 0.79 for SOC, and was used to create maps for these soil properties. Moreover, they developed models for mapping soil erosion and root depth restrictions, with an accuracy of about 80% for both variables.

24.5.9 Strategies for Soil Monitoring

Soil monitoring implies observing the soil over time providing data/information to assure that it stays healthy (chemically, physically, and biologically) and secured against environmental or human degradation. Soil and food security are priorities in today's global agenda, and soil monitoring is an essential activity for achieving sustainable food production and sustainable development. Specifically, soil monitoring studies are also interested in observing the temporal, spatial, and concentration changes in organic and inorganic contaminants in the soil. However, perhaps the most common soil monitoring applications relate to the maintenance of soil fertility over time. In this context, SS can be used to monitor soil condition (i.e., quality) spatially and temporally to support management decisions that enable soil stabilization and improvement.

24.5.9.1 Strategies for Ground Spectral Sensing

Kemper and Stefan (2002) used VIS–NIR reflectance spectroscopy to estimate concentrations of contaminants in soils in Spain with an R^2 of 0.82 for As, 0.96 for Hg, 0.95 for Pb, and 0.87 for S. Vohland et al. (2009) performed a spectroscopy approach to quantify the same previous elements in floodplain soils using VIS–NIR laboratory data and reported R^2 values between 0.60 and 0.71. Jean-Philippe et al. (2012) detected several heavy metals, in particular Hg, with a performance prediction (R^2) of around 0.91. Song et al. (2012) observed relationships between Cr, Cu, and As, and absorption features caused by iron oxides, clay minerals, and SOM, suggesting that they are strongly bounded to these soil constituents. Araújo et al. (2014b) used VIS–NIR–MIR spectroscopy to assess soil contamination with Cr by tannery sludge. They observed strong alterations of absorption features in selected wavelengths (500–600 nm and 2600 wavelength cm^{-1}) and an overall decrease in reflectance intensity across the spectrum promoted by the sludge.

Brunet et al. (2009) monitored chlordecone (a toxic insecticide) used in banana plantations in the French West Indies and determined its content by NIR spectroscopy in andosols, nitisols, and ferralsols. Conventional analyses and spectral predictions were poorly correlated for chlordecone contents higher than 12 mg kg^{-1}. However, 80% of samples were correctly predicted when the dataset was divided into three or four classes of chlordecone content. Chakraborty et al. (2010) quantified total petroleum hydrocarbons (TPHs) in contaminated soils in situ by using VIS–NIR diffuse reflectance spectroscopy and later mapped them in the field using the same spectral range (Chakraborty et al., 2012a,b). To determine TPH's content in soil, PLSR and boosted RT models were used, and the best performance for validation showed an R^2 of 0.64 and an RPD of 1.70. Schwartz et al. (2012) and Reuben and Mouazen (2013) detected diesel-contaminated soils by SS and assessed the relationships between petroleum hydrocarbon concentrations, soil moisture, and clay content. Forrester et al. (2013) also developed calibration models based on PLSR using NIR and MIR data from a diffuse reflectance infrared Fourier-transform (DRIFT) sensor for predicting TPH concentrations in contaminated soils in southeastern Australia. The authors confirmed DRIFT spectroscopy associated with PLSR as capable to provide accurate models for TPH prediction, where the MIR range outperformed NIR deriving high-quality predictions (RPD = 3.7, R^2 = 0.93). A common by-product of sugarcane industry is the vinasse (fermentation residue), which can be used as an important K fertilizer. On the other hand, if used in high quantities, it can pollute soils and make them very saline. Monitoring this by-product, Demattê et al. (2004b) observed differences by VIS–NIR information. Another interesting approach was performed by Shi et al. (2014b), which monitored arsenic in agricultural soils by spectral reflectance of rice plants. Other SS techniques have been used such as that by Radu et al. (2013), which clarified that portable XRF can be used to track and quickly identify possible hot spots of pollution and trends in the elementary distributions.

24.5.9.2 Strategies for Space Spectral Sensing

In a review about image spectroscopy to study soil properties and applications, Ben-Dor et al. (2009) provided some case studies in which different aerial and orbital sensors, with distinct resolutions, were used. The cases addressed by the authors included a variety of soil science applications, such as soil degradation, salinity, erosion and deposition, mapping and classification, pedogenesis, contamination, water content, and swelling. Reschke and Hüttich (2014) used multitemporal Landsat data combined with high-spatial-resolution satellite data to extract sub-pixel information of coastal and inland wetland classes for environmental goals. Aerial or orbital detection of reflected radiation from solids has been widely employed in soil monitoring applications (Schwartz et al., 2013). These authors developed steps to identify hydrocarbons in soil.

Adar et al. (2014) studied the automatic identification of soil changes in different surface soil types by HySpex SS in the VIS–NIR and SWIR spectral ranges. This study identified a gradual change over time in SOM, soil crusting, and compaction, and demonstrated that the SWIR range allowed better change detection than the VIS–NIR. Ghosh et al. (2012) used EO-1 Hyperion data to identify salt-affected soils, correctly identifying highly affected soils in 84.4% of the cases.

Regarding hyperspectral SS monitoring of heavy metals in soils, Shi et al. (2014a) discussed the applicability of SS for mapping soil contamination over large areas. The authors also presented methodologies to estimate heavy metal concentrations using VIS–NIR imaging spectroscopy and reported good results. In a similar research, Choe et al. (2008) used combined data of geochemistry, field spectroscopy, and hyperspectral sensing (HyMAP) to map heavy metal pollution in stream sediments in Rodalquilar mining area (Spain). In fact, when narrowbands are used, they have the opportunity to provide greater focus on targeting specific waveband spots and are likely to provide greater accuracies or R-squares compared to broadbands and their indices, as shown in these papers.

24.5.10 Strategies for Microwave (RADAR) and Gamma Ray

According to a review made by Mulder et al. (2011), the feasibility of determining soil texture, moisture, and salinity by active and passive microwave SS was scaled from medium to high, whereas the feasibility of determining land cover and degradation ranged from low to medium. Microwave SS has been applied to measure the thermal radiation emitted by bare soil, which is mainly affected by its moisture content and temperature (Parrens et al., 2014), where soil temperature is also dependent on soil mineralogy and SOM (van Lier, 2010).

Alternatively, gamma ray spectroscopy is related with the detection of three basic elements: uranium (^{238}U), thorium (^{232}Th), and potassium (^{40}K) (Minty, 1997). One of the advantages of microwave or gamma ray sensors, compared with VIS–NIR–MIR reflectance meters, is that they are less influenced by vegetation cover and climate. One exception, for example, is very dense vegetation cover (as in tropical rainforest), where microwave and gamma rays might be attenuated by the vegetation.

24.5.10.1 Ground-Penetrating Radar

GPR has been used since the 1970s. In fact, Johnson et al. (1979) determined the first information about GPR in soil survey. The GPR emits electromagnetic radiation pulses to the soil, and variations in some soil properties such as moisture, porosity, salinity, SOM, texture, and mineralogical content (iron oxides, high clay activity, and others) affect this radiation (Pozdnyakova, 1999). GPR uses wavelengths with high frequencies in the microwave spectral range (from 10 MHz to 2.5 GHz) (Daniels, 2004; Jol, 2009). Typically, interfaces between horizons (as in soil layering) produce changes in reflectance that are recognizable in GPR images (Stroh et al., 2001; Doolittle and Butnor, 2009). Since the 1980s, studies have demonstrated the applicability of GPR as a tool for the characterization of organic and mineral undersurface horizons regarding their thickness and lateral variability (Collins et al., 1990), and for the identification of lithic contact (Doolittle et al., 1988). GPR is primarily limited to soils with coarser texture, low electrical conductivity (Ucha et al., 2002), and higher (or high enough) moisture contents (Ardekani, 2013).

The use of GPR to identify argillic and cambic B-horizons has been useful for understanding pedogenetic processes (Inman et al., 2002) that are important for soil mapping. Doolittle et al. (2007) produced the "Ground-Penetrating Radar Soil Suitability Map of the Conterminous United States," which limited areas rated as being "unsuited" for GPR to saline and sodic soils, reassessed calcareous and gypsiferous soils, and provided a mineralogy override for soils with low activity clay, where the efficiency of the equipment is restricted to a clay content of 35% (Mahmoudzadeh et al., 2012). The great majority of GPR applications in soil science are concentrated in hydropedology (Doolittle et al., 2012; Zhang and Doolittle, 2014), with studies on the variations in water table depths and groundwater flow patterns (Doolitle et al., 2006), vertical moisture dynamics in a soil profile (Steelman and Endres, 2012), and quantification of soil water content in the vadose zone (Minet et al., 2012; Yochim et al., 2013) at the field scale. Transport of contaminants and agrochemicals in subsurface has also been investigated with GPR (Glaser et al., 2012; McGlashan et al., 2012). In fact, Yoder et al. (2001) identified offsite movement of waterborne agrochemicals using conventional soil survey combined with electromagnetic induction (EMI) and GPR. They concluded that EMI mapping provides rapid identification of areas of high potential for offsite movement of subsurface water, GPR mapping of areas identified by EMI mapping provides a means to identify features that are known to conduct concentrated lateral flow of water, and combining the capabilities of EMI and GPR instrumentation makes possible the surveys of large areas that would otherwise be impossible or unfeasible to characterize.

Other uses of GPR include the assessment of soil porosity (Causse and Sénéchal, 2006), soil compaction (Tosti et al., 2013), and delineation of agriculture management zones (André et al., 2012).

24.5.10.2 Aerial and Orbital Radar

A good review about microwave SSS and soil salinity was written by Metternicht and Zinck (2001). Bell et al. (2001) used an airborne polarimetric synthetic aperture radar (SAR) for mapping soil salinity in the Alligator River Region of the Northern Territory in Australia. According to Hasan et al. (2014), vegetation cover is still a main factor in the attenuation, scattering, and absorption of the microwave emissions from the soil that impacts its brightness. These authors used the airborne Polarimetric L-band Multibeam Radiometer 2 (PLMR2) and the L-band Microwave Emission of the Biosphere (L-MEB) model to simulate microwave emissions from the soil–vegetation layer and to retrieve surface soil moisture in Germany (moisture retrieval with an RMSD of 0.035 $m^3 m^{-3}$ when compared to ground-based measurements). NASA's soil moisture active passive (SMAP) mission will carry in 2014 the first combined spaceborne L-band radiometer and SAR system with the objective of mapping near-surface soil moisture with high (~3 km), low (~36 km), and intermediate resolutions (~9 km). For that, Panciera et al. (2014) conducted three experiments combining field and airborne sources to provide prototype data for the development and validation of soil moisture retrieval algorithms applicable to the SMAP mission.

The sensitivity of spaceborne SAR is well established for soil moisture. However, soil moisture monitoring can be confounded by the effects of vegetation and surface roughness. In this context, Singh and Kathpalia (2007) proposed an approach based on a genetic algorithm with the inclusion of empirical modeling to determine the soil moisture, texture, and roughness with backscattered data from ERS-2 SAR. Kornelsen and Coulibaly (2013) presented a critical review about technical and methodological advances, limitations, and potential of SAR. They concluded that soil moisture estimation can be retrieved with multi-angular SAR without in situ measurements. Fatras et al. (2012) analyzed the potential of the radar altimeter aboard ENVISAT to successfully estimate the surface SM in a semiarid region in Northern Mali, with correlation coefficients (r) higher than 0.8 between SM and the backscattering coefficient, and SM predictions with RMSE <2%.

24.5.10.3 Aerial and Ground Gamma Ray

Passive gamma ray spectrometry is a fast and cost-efficient tool for developing a spatial map of soil properties related to clay content, mineralogy, and soil weathering. It can be used proximally or through an airborne sensor. Initial airborne gamma surveys started with geologists for mineral and lithological exploration (Graham and Bonham-Carter, 1993). Later on, Cook et al. (1996) examined the ability of ground and airborne systems to detect the spatial distribution of soil-forming materials across the landscape and distinguished highly weathered from fresh soil materials, which was also done by Wilford et al. (2011). This usefulness was recently observed by Gooley et al. (2014). They used gamma ray spectrometry from aerial and ground sensing

to develop a DSM of the available water content (QWC) across an irrigated area through map soil properties.

Approximately 50% of the observed gamma rays originate from the top 0.10 m of dry soil, 90% from the top 0.30 m reaching 95% from the upper 0.5 m of the profile (Taylor et al., 2002). Different factors can attenuate gamma rays through the soil, such as moisture and bulk density. Radiation attenuation increases by approximately 1% for each 1% increase in volumetric water content, while a dry soil with a bulk density of 1.6 Mg m^{-3} causes a decrease in the radiation to half its value at each 10 cm (Cook et al., 1996). The radiation decrease caused by air is much smaller, for example, 121 m are needed to reduce the radiation to half its value considering a 2 MeV source, thus making possible the detection of gamma rays from airborne platforms (Viscarra Rossel et al., 2007). Pracilio et al. (2006) demonstrated the use of gamma ray radiometric mapping of clay and plant-available potassium contents at the farm scale, obtaining R^2 up to 0.68 for clay and 0.60 for potassium.

The gamma region of electromagnetic spectrum has been applied successfully for studying the properties of cultivated soils (Medhat, 2012), such as field capacity, porosity, moisture content, and bulk density. There is a rising interest in this spectral range for applications in DSM. Gamma ray spectrometry was found to be an accurate predictor of topsoil clay content in alluvial soils (Piikki et al., 2013). Also van der Klooster et al. (2011) investigated the prediction of soil clay contents in three marine clay districts in the Netherlands with R-squared varied between 0.50 and 0.70.

Vulfson et al. (2013) merged data from microwave and gamma ranges for monitoring soil water content in the root zone and showed strong correlations ($R^2 > 0.9$) between field and laboratory measurements. Evidence suggests that gamma ray spectrometry can be used for assessing SOC (Dierke and Werban, 2013), which is commonly associated with clay content. Dent et al. (2013) obtained gamma radiometric data from an airborne to imply soil proprieties. This investigation confirmed that airborne radiometric has the capability to map different parent material and indirectly can infer on soil texture.

Table 24.2 indicates some results for clay content with an on-the-go sensor, reaching 0.86 of R^2. On the other hand, Table 24.4 indicates more soil attributes so we can have an idea of its utility. Attributes such as clay, silt, sand, fine sand, coarse sand, gravel, plant-available water capacity (PAWC), EC, OC, pH, and K obtained variable R^2 results. The best results were for carbon and clay attributes with R^2 0.89 and 0.83, respectively.

Soderstrom and Eriksson (2013) studied the use of aerial and ground-based gamma radiometry to assess Cd contamination risk in food production. In a field with Cd content in the fluvial sediment, gamma ray measurements allowed to improve mapping of contamination risk in relation to a general soil map. Their results show that geological maps and gamma radiation mapping, calibrated with a few analyses of Cd concentrations in soils and crops, can be used for risk classification of soils at the regional scale. In fact, from aerial gamma, there are several that can be evaluated as is shown in Table 24.4. Results of R^2 are variable since 0.02 for Ca until 0.93 for Sr.

TABLE 24.4 Summary of Revision Related with Quantification of Soil Attributes by Ground and Aerial Gamma Platforms

	R^2			RMSE			References
	Range	Median	N	Range	Median	N	
A. Aerial							
Clay (%)	0.53–0.68	0.59	5	2.40–13.65	3.1	5	Pracilio et al. (2006), Martelet et al. (2013)[a]
Silt (%)	0.34	0.34	1	10.41	10.41	1	Martelet et al. (2013)[a]
Sand (%)	0.56	0.56	1	18.53	18.53	1	Martelet et al. (2013)[a]
Gravel (%)	0.13	0.13	1	6.73	6.73	1	Martelet et al. (2013)[a]
Bic-K (mg kg^{-1})	0.04–0.55	0.53	4	103–145	127	4	Pracilio et al. (2006)
Al (g kg^{-1})	0.50	0.50	1	1.18	1.18	1	Martelet et al. (2013)[a]
Si (g kg^{-1})	0.56	0.56	1	2.18	2.18	1	Martelet et al. (2013)[a]
Ca (g kg^{-1})	0.02	0.02	1	2.98	2.98	1	Martelet et al. (2013)[a]
Fe (g kg^{-1})	0.43	0.43	1	0.74	0.74	1	Martelet et al. (2013)[a]
Mn (g kg^{-1})	0.24	0.24	1	279.80	279.80	1	Martelet et al. (2013)[a]
Mg (g kg^{-1})	0.35	0.35	1	0.16	0.16	1	Martelet et al. (2013)[a]
Na (g kg^{-1})	0.35	0.35	1	0.19	0.19	1	Martelet et al. (2013)[a]
Pb (g kg^{-1})	0.48	0.48	1	7.33	7.33	1	Martelet et al. (2013)[a]
Sr (g kg^{-1})	0.93	0.93	1	4.19	4.19	1	Martelet et al. (2013)[a]
V (g kg^{-1})	0.83	0.83	1	11.13	11.13	1	Martelet et al. (2013)[a]
B. Proximal							
Clay (%)	0.17–0.95	0.83	19	1–8.40	2	18	Viscarra-Rossel et al. (2007), Wong et al. (1999), Priori et al. (2014), Van der Klooster et al. (2011)
Silt (%)	0.40–0.89	0.40	3	2.29–2.46	2.38	2	Viscarra-Rossel et al. (2007), Wong et al. (1999)
Sand (%)	0.65–0.85	0.77	4	6.7–7.9	7.4	1	Priori et al. (2014)
Fine sand (%)	0.05–0.31	0.18	2	3.23–3.96	3.60	2	Viscarra Rossel et al. (2007)
Coarse sand (%)	0.3–0.73	0.55	2	6.73	9.31	2	Priori et al. (2014)
Gravel (%)	0.49–0.58	0.51	4	0.11	0.10–0.11	4	Priori et al. (2014)
PAWC (mm)	0.50	0.50	1	11.4	11.4	1	Wong et al. (2009)
EC (ms m^{-1})	0.30–0.60	0.45	2	27.96–46.55	37.26	2	Viscarra Rossel et al. (2007)
OC (g kg^{-1})	0.89	0.89	1				Wong et al. (1999)
pH$_{Ca}$	0.40–0.63	0.52	2	0.43–0.48	0.46	2	Viscarra Rossel et al. (2007)
Colwell-P (mg kg^{-1})	0.68	0.68	1			1	Wong et al. (1999)
K (mg kg^{-1})	0.61	0.61	1	83.57	83.57	1	Viscarra Rossel et al. (2007)

Bic-K, Plant-available potassium; Al, aluminum; Si, silicon; Ca, calcium; Fe, iron; Mn, manganese; Mg, magnesium; Na, sodium; Pb, lead; Sr, strontium; V, vanadium; PAWC, plant-available water capacity; EC, electrical conductivity in units milliSiemens per meter; OC, organic carbon; pH$_{Ca}$, pH using 0.01 M CaCl$_2$; Colwell-P, phosphorous using the NaHCO$_3$ method; K, bicarbonate-extractable potassium.

[a] Models include morphological variables.

A comparison between ground and airborne information can be extracted from Table 24.4. We observed that clay had R^2 maximum and average of 0.95 and 0.83 respectively, for ground information. On the other hand, aerial presented 0.68 and 0.59 of R^2, lower than field data. This is true since in field, the sensor has a higher spatial resolution and less interference factor. But we rather have to consider that aerial data are good. Another important information given by gamma are the great R^2 for parent material elements such as strontium and vanadium that showed an R^2 of 0.93 and 0.83, which can certainly assist in relation with soils.

24.5.11 Strategies for In Situ Spectral Sensing

Pretreating the soil samples before scanning, that is, drying and sieving, is common and improves the quality and repeatability of the spectral data acquired in the laboratory, reducing the negative influence of variable soil moisture or soil particle size in data acquisition (Tekin et al., 2012). However, the goal is to scan the soils in the field without pretreatment. However, field scans are subject to other sources of variation, including variations in viewing angle, illumination, soil roughness, soil moisture, soil temperature, and soil structure, in the presence of specific features (e.g., redoximorphic features and clay films), among others, all affecting the quality of the measurements.

Applying SS to field studies reduces costs associated with collection, transport, preparation, and analysis of soil samples. As already stated, LSS measurements are made under controlled conditions with standard protocols, which provide a minimum of interference in the acquired radiometric data (Ben-Dor, 2011), whereas for in situ measurements, there are more possibilities for interference due to soil and environmental conditions that cause variations affecting data acquisition. Other in situ factors that affect data quality include noise associated with tractor vibration, sensor-to-soil distance variation (Mouazen et al., 2007), stones, plant roots, and difficulties of matching the position of soil samples collected for validation with corresponding spectra collected from the same position (Kuang et al., 2012).

Water, in the form of soil moisture and soil moisture variability, is a fundamental concern to consider when using SS measurements of unprocessed soil surfaces or soil cores from the field (Waiser et al., 2007; Nduwamungu et al., 2009). Many working in spectroscopy have attempted to address the problem of soil moisture on SS data. However, the solutions usually include linear transformations and or require prior knowledge of the soil moisture, posing difficulties for applications in the field. An effective strategy is to create calibration models using only field-collected spectra (Waiser et al., 2007; Morgan et al., 2009; Bricklemyer and Brown, 2010). However, this approach does not contribute to available soil SLs that require accompanying soil samples. Aiming to reduce the effects of water in the soil spectra without prior knowledge, a technique called "external parameter orthogonalization (EPO)" has been developed and will be discussed in Section 24.9.

As previously discussed, several factors limit the field application of VIS–NIR spectroscopy, such as temperature, luminosity, climatic conditions, sample surface roughness, organic residues, and appropriate equipment to obtain subsurface spectra. Some solutions have been proposed to address these issues, for example, by Ben-Dor et al. (2008) and Ge et al. (2014a). FSS has obtained good results for clay content (Viscarra Rossel et al., 2009) and SOC (Gomez et al., 2008b) prediction in Australian soils, as well as in Texas for clay content (Waiser et al., 2007), and both SOC and SIC (Morgan et al., 2009). Chakraborty et al. (2012a,b) demonstrated the feasibility of VIS–NIR analysis in the field to map TPH contamination in soil in an extensive area. Field measurements can also have issues when working with noncontact equipments (see Figure 24.2, strategy 3) due to atmospheric interferences. For example, the 1400- and 1900-nm bands cannot be detected hindering the interpretation of soil mineralogy, as observed by Fiorio et al. (2014). Thus, it is better to use a contact probe in the field, avoiding atmospheric interferences and sunlight variations.

Soil MIR analysis is not widely used in the field mainly due to (1) the need for sample preparation and (2) strong water absorption in naturally moist soils leading to spectral distortion and total absorption (Ge et al., 2014b). As an alternative, Ge et al. (2014b) used attenuated total reflectance (ATR) as a technique to obtain MIR spectra of neat soil samples. Accordingly, MIR-ATR can be a promising and powerful tool for soil characterization combining the advantages of both VIS–NIR (minimum sample preparation and high analysis throughput) and diffuse reflectance MIR (better model performance) (Ge et al., 2014b).

Table 24.2 indicates quantification of soil attributes in field conditions. The lab-field moist situation indicates high values for all elements going from 0.7 to 0.85 of R^2. Specifically in field conditions (or on-the-go), we have still important results for clay (R^2 0.78), OC with 0.84 and TC with 0.86. Gamma also has interesting results reaching 0.86 of R^2 for clay measurement.

24.5.12 Soil Spectral Libraries

The first soil SL was built by Condit (1970) and complemented by Stoner et al. (1980) as an atlas, published afterward by Stoner and Baumgardner (1981) with soils mainly from the United States and some samples from Paraná state, Brazil. After that, Epiphânio et al. (1992) and Formaggio et al. (1996) constructed an SL comprising 14 soil classes for one Brazilian state, while Bellinaso et al. (2010) reached six Brazilian states with ~8000 soil samples. Even though these and other SLs (e.g., Clark, 1999b) are promising, there are still few examples including a wide diversity of soil classes (Ben-Dor et al., 1999; Chang et al., 2001; Malley et al., 2004). The first publication using an SL with global samples was presented by Brown et al. (2006).

After 2000, several SL initiatives appeared, including the ICRAF-ISRIC world soil SL, composed of 785 soil profiles from 58 countries from Africa, Europe, Asia, and the Americas. Viscarra Rossel and Webster (2011) described a large SL with ~4000 soil profiles covering the Australian continent. An SL covering the United States has been collected under the Rapid Carbon Assessment project (Soil Survey Staff, 2013) with 144,833 VIS–NIR spectral data for 32,084 soil profiles. The European SL called LUCAS (Land Use/Cover Area Frame Survey; http://eusoils.jrc.ec.europa.eu/projects/Lucas) consists of about 20,000 topsoil samples, collected from 23 countries in Europe, and measured for 13 soil properties in a single laboratory (Stevens et al., 2013). Another important example of soil SL was the one built by Baldridge et al. (2009), called ASTER SL. This SL is a compilation of 2400 spectra of soils, rocks, minerals, and other related materials. Soil SL initiatives in other countries include Brazil (Bellinaso et al., 2010), Czech Republic (Brodsky et al., 2011), France (Gogé et al., 2012), Denmark (Knadel et al., 2012), Mocambique (Cambule et al., 2012), and Spain (Bas et al., 2013).

Soil SLs can be applied for many purposes, including (1) modeling of soil attributes; (2) soil survey, classification, and mapping; (3) soil contamination and monitoring, by extracting the baseline electromagnetic properties of soils, which can be compared with any contaminated samples; (4) communication among researchers (soil classification has several systems, but spectra are the same!); and (5) development of field, aerial, and space sensors, among others. To understand the usefulness of soil SL, consider the following example: the interested parties (farmers or researchers) could send their soil samples to a central SL (e.g., a national or global SL) where they would be scanned and the spectral curves stored, or they could send already acquired soil spectral curves that compose their local SLs. Local SL can be explored for personal interests (e.g., soil monitoring) and also feed global SL, growing a global repository. Once having a global SL, spectral curves from a profile of an unknown could be compared with other spectra from the global SL and a preliminary soil classification or the SOC or clay content could be estimated.

The ideal scale (global, continental, regional, local, or farm) for a soil SL application has had much inquiry, and the general result is that the spatial scale of coverage and application depends. This topic was initially raised by Coleman et al. (1993), which concluded that there was evidence that regional/local scale is the most reliable. Later on, Demattê and Garcia (1999) observed better results for soil modeling with a local soil SL than with a regional one. Brown (2007) used a global VIS–NIR SL for local soil characterization and landscape modeling in a second-order Uganda watershed. In brief, a soil attribute may be estimated in a farm using samples collected at the farm to constitute a local SL, or the soil spectra collected at the farm can be compared against other spectra in a global SL to retrieve predictions. One strategy, called "spiking," combines global library with local samples to get better predictions at the local level (Brown et al., 2006; Wetterlind and Stenberg, 2010). Others have tried using pedological knowledge to subsample a geographically more extensive SL (Ge et al., 2011). In the subsampling context, parent material filter improved predictions of clay content; however, for SOC subsampling, the first three principal components coupled with Mahalanobis distance was more effective (Ge et al., 2011). Similar results were found by Araújo et al. (2014a) when analyzing the spectra of 7172 tropical soil samples. They found that separating the global dataset into more mineralogically uniform clusters improved predictive performance of clay content regardless of the geographical origin, showing that probably physically based, soil-related stratification criteria in libraries

offer better results. An interesting study by Ramirez-Lopez et al. (2013) developed the spectrum-based learner, which indicates the best performance of data to reach high quality of quantification when using complex data. Another important discussion is how to use the dataset to reach best results. Debaene et al. (2014) found little significant increase in the prediction capacity of soil attributes with the use of an entire dataset, watching an increase in the R^2 of 0.63–0.72 for SOC and R^2 of 0.71–0.73 for clay. The point if local, regional, or global is certainly an actual discussion. Genot et al. (2011) built a methodological framework for the use of NIR spectroscopy on a local and global scale by spectral treatment and regression methods. In addition, evaluated the ability of NIR spectroscopy to predict total organic carbon (TOC; $R^2 = 0.91$ local and $R^2 = 0.70$ global), TN ($R^2 = 0.73$ local and $R^2 = 0.61$ global), clay ($R^2 = 0.64$ local and $R^2 = 0.61$ global), and CEC ($R^2 = 0.73$ local and $R^2 = 0.43$ global) above several soil conditions.

Despite these discussions, results have suggested that the best scale for an SL is very much application dependent. The application will define the precision needed. Generally, developing a global library does not exclude embracing a local or physically based library, and the user will need to decide which scale to use. Figure 24.7 suggests the sequence on how to construct and use soil SLs.

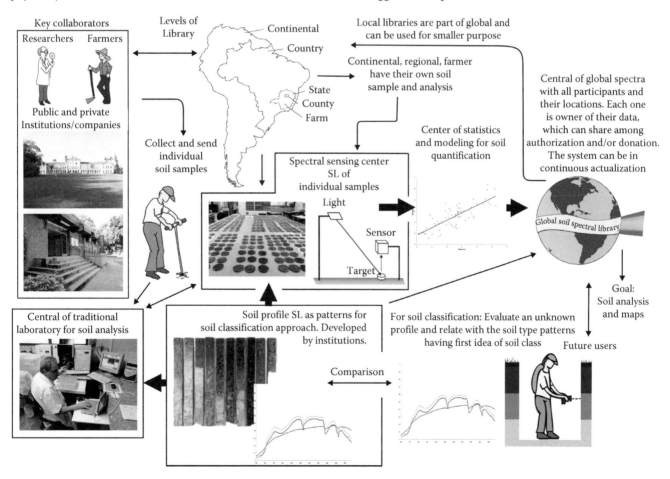

FIGURE 24.7 Illustration on how to contract and use soil spectral libraries. (Photograph of Luiz de Queiroz College of Agriculture, Department of Soil Science, Luis Silva [technician from Department of Soil Science], soil profile pictures from Texas A&M University.)

24.6 Soil Spectral Behavior at Different Acquisition Levels

Soil spectra result from the interaction of many soil attributes with the electromagnetic energy. According to Demattê and Terra (2014), descriptive spectral analyses of soil horizons are important because their shapes have direct relation with mineralogical and organic constituents that, in turn, result from the pedogenetic processes acting in the soil profile. This took them to define the term "spectral pedology," which can be summarized as "a detailed and accurate evaluation of soil spectral behavior obtained by proximal and/or RS, and analyzed by its qualitative (shape, absorption features and reflectance intensity/albedo) and/or quantitative information, where the convergence of evidences guides to a probable soil classification or behavior." This section aims to indicate some important spectral features of soils that relate to their attributes and taxonomic classes.

24.6.1 Spectral Behavior at the Ground Level

Minerals from soils have different and specific spectral shapes (Figure 24.8a), and each one contributes to the total soil spectrum. Features at 1300–1400, 1800–1900, and 2200–2500 nm (hydroxyl groups) are strong indicatives of clay content and type (Zhu et al., 2010). Soil particle size has great influence on spectra. In this case, a coarser texture increases the scatter of energy (reduces reflection), and the apparent absorbance increases as path length increases. In fact, Figure 24.8b presents great differences of energy reflection from clay to sandy soils with different angles (shapes) and intensities of energy with highest peaks occurring after 2000 nm (Demattê, 2002; Franceschini et al., 2013).

The SOM is another important attribute affecting not only specific features in the NIR–MIR region but also the overall spectrum in the VIS region. Udelhoven et al. (2003a) showed the relation between soil brightness and SOC and developed several systems for its analysis (Udelhoven et al., 2003b,c).

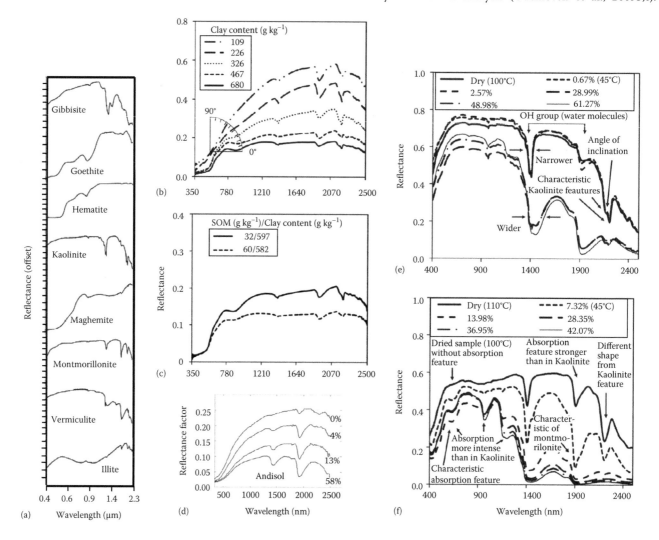

FIGURE 24.8 VIS–NIR spectra of (a) soil minerals. (From Clark et al., 2007.) (b) Soils with different clay contents, indicating angular variations. (From Franceschini et al., 2013; Demattê, 2002.) (c) Soils with variable organic matter contents. (From Demattê et al., 2003a.) (d) Soils with different moisture contents (From Lobel and Asner, 2002.) and pure kaolinite (e) and pure montmorillonite (f) with varying moisture contents. (From Demattê et al., 2006.)

Demattê (2002) indicated that spectra of superficial soil layers had lower reflectance intensities than undersurface ones, which was related to lower SOC contents (Figure 24.8c). Stenberg et al. (2010) stated that, although soil samples tend to get darker colors with increasing SOM, other soil properties, such as texture and moisture, also influence the soil brightness, thus SOM would only be a useful indicator in a specific situation. The same authors have suggested that there are specific features in spectra that would provide a better correlation with SOM, for example, bands around 1100, 1600, 1700–1800, 2000, 2200, and 2400 nm. The spectral features of SOM in soil VIS–NIR spectra are explained by combination and vibration modes of organic functional groups (Chen and Inbar, 1994).

The VIS–NIR–SWIR ranges are strongly influenced by the soil water content. Water in soils can be incorporated to the clay mineral lattice, filling pore spaces as free liquid water or adsorbed on a surface as hygroscopic water. In the first case, the water is related to the mineralogy of the soil samples and directly affects features near 1400 and 1900 nm bands (Hunt et al., 1971). Bishop et al. (1994) also related the vibrations of bound water in the interlayer mineral lattice to the features in 1400 and 1900 nm. In fact, Stenberg et al. (2010) stated that the bands related to vibrations of bound water occurs at shorter wavelengths close to 1400 and 1900 nm, but the hygroscopic water appears as shoulders at 1468 and 1970 nm. Besides, hygroscopic and free pore water are responsible for reducing the albedo of soil spectrum (Figure 24.8d; Lobel and Asner, 2002). Once the surroundings of soil particles are changed from air to water, the refractive index decreases, in other words, changes in the medium surrounding soil particles affect the average degree of forward scattering (Ishida et al., 1991), reducing the reflectance.

Differences in mineral spectra are related to their chemical composition and structure. For example, kaolinite has well-delineated features at 1400 and 2200 nm, but a weak signal in 1900 nm, compared to other minerals (Figure 24.8e). For instance, the spectra of montmorillonite have a strong feature at 1900 nm and a different shape close to 2200 nm (Figure 24.8f; Demattê et al., 2006). Another important mineral in tropical soils is the gibbsite, which has an aluminum octahedral structure with spectral features occurring mainly close to 1400 and 2265 nm (Figure 24.8a, top curve).

Hunt et al. (1971) summarized the physical mechanisms responsible for Fe^{2+} (ferrous) and Fe^{3+} (ferric) spectral activities in the VIS–NIR range and indicated that iron oxides and hydroxides are spectrally active attributes due to the electronic transition of iron cations. Due to their absorption features and overall spectra shapes, the presence of goethite and hematite alters the shape of spectra of soils (e.g., Figures 24.8a through d, and 24.9a). Demattê and Garcia (1999) compared the VIS range from the spectra of oxidic soils and indicated that soils with a predominance of goethite have a narrower feature with higher reflectance between 400 and 570 nm, whereas the opposite occurs with the predominance of hematite (Figures 24.8 and 24.9). In soils derived from carbonate rocks, that is, rocks composed generally by calcite and dolomite minerals, the carbonate

groups ($-CO_3$) are spectrally active causing specific absorption features due to the C–O bonds (Ben-Dor, 2011). A mineral commonly found in temperate soils is montmorillonite, which is a highly enriched Al smectite and consequently presents Al–OH bonds, resulting in a feature at 2160–2170 nm in soil spectra (Figures 24.8f and 24.9c). Other smectites affect the soil spectra in specific regions, so we suggest the referred studies for further information (Stenberg et al., 2010; Ben-Dor, 2011).

As stated by Demattê and Garcia (1999), it is possible to discriminate and identify soils with different weathering conditions. For example, compare the spectra of an oxisol (highly weathered soil) (Figure 24.9a, Typic hapludox) with that of a vertisol (less weathered soil) (Figure 24.9a, aquic hapludert). In fact, the degree of weathering can determine the relative amounts of kaolinite and montmorillonite present in the soil, with important differences in spectral features (Figure 24.9d). Moreover, soils with higher weathered mineralogy have higher iron oxide contents, which add other specific spectral features. Demattê et al. (2003b) found that crystalline hematite and goethite are responsible for the spectral features at 400–850 nm. On the other hand, when these iron forms were extracted from the soil in the laboratory, the concave shapes disappeared from this region (400–850 nm) (Figure 24.9e). For clayey soils, reflectance decreases in intensity around 1200 nm, due to the remaining presence of magnetite. Several wavebands and respective soil attributes can be seen in Table 24.1.

Classification of soils by their spectral behavior must be made by interpretation of all horizons. For example, a ferralsol (oxisol) has only minor differences regarding reflectance intensity and spectral features among spectra collected at different depths, that is, from surface versus subsurface horizons (Figure 24.10a and b). On the other hand, spectra of a lixisol (ultisol) show differences between these horizons, mainly after 2000 nm, due to differences in sand content (quartz mineral). In fact, the absorption features of quartz can be observed in the MIR spectra, mostly between 1111 and 1300 cm^{-1} (in wavenumber) (Figure 24.10d). Observe that there is a great difference in the quartz peak (~1300 cm^{-1}) between the A and B horizons of the lixisol (Figure 24.10d), and no difference in the ferralsol (Figure 24.10b), which is in line with the characteristics of these soils. In fact, ferralsols do not have great differences in clay contents with depth, as opposite to lixisols. These findings allow to relate the shapes and trends in spectra to specific soil classes, aiding in soil classification.

24.6.2 Spectral Behavior at the Space Level: From Aerial to Orbital Platforms

Compared to ground-based sensors, aerial and orbital ones have low SNR due to the larger atmospheric path length, decreased spatial and spectral resolution, geometric distortions, and spectral ambiguity caused by recording multiple signals from adjacent targets. On the other hand, space-based sensors have larger ground coverage (Obade and Lal, 2013). Data from some satellite sensors, such as Landsat ETM, ASTER, and MODIS, may be

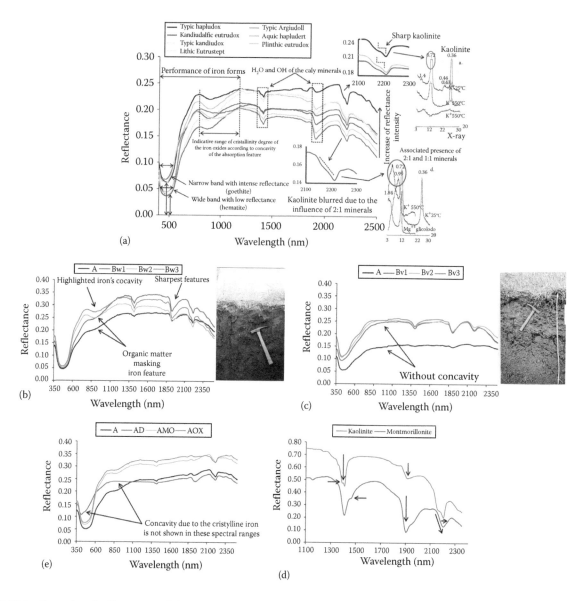

FIGURE 24.9 Examples of soil spectra and their interpretation. (a: From Demattê and Terra, 2014; b, c: From Demattê et al., 2012a; d: From Demattê et al., 2006; e: From Demattê et al., 2003b.) A, raw soil (testimony); AMO, soil without organic matter; AOX, soil without amorphous (oxalate-extracted) iron; AD, soil without any type of iron.

freely downloaded, not requiring a portable radiometer, which is an expensive tool.

The first aspect to be analyzed is the feasibility of ground and space information for soil assessment (attribute prediction, soil classification, and mapping). Figure 24.11 illustrates the study of Demattê et al. (2009a) and indicates the position on the landscape of nitisols, arenosols, and oxisols in a Landsat TM image (RGB = 5, 4, 3). Darker and lighter colors are related with clayey and sandy soils, respectively. These and other areas were evaluated, and the spectra of pixels were compared with ground spectra by correlating 294 laboratory spectra (Figure 24.11c) simulating Landsat TM spectra (Figure 24.11d) with Landsat TM spectra obtained directly from the pixel in the image (Figure 24.11e). They observed similar trends between spectra simulated from the ground sensor and that collected from space. Clayey

and oxidic soils (nitisols) are very different from sandy soils (arenosols) in many aspects. Of course, ground information is more accurate due to the higher spectral resolution. The main differences were related to the reflectance intensities from bands 5–7, where drops in the Landsat TM spectra are related to moisture. Other studies proved the capabilities of Landsat data to differentiate moisture content in soils (Shih and Jordan, 1992; Vicente-Serrano et al., 2004). Satellite images can provide a good discrimination of soil classes and consequently support soil mapping either visually or through DSM. In the given example, the image colors range from a dark blue to magenta corresponding to a sequence of nitisol, rhodic oxisol, and typic oxisol.

As another example, Figure 24.12 presents soil spectra of the same spot taken from ground (2151 bands), Landsat TM (6 bands), and Hyperion (220 bands). From ground to aerial or

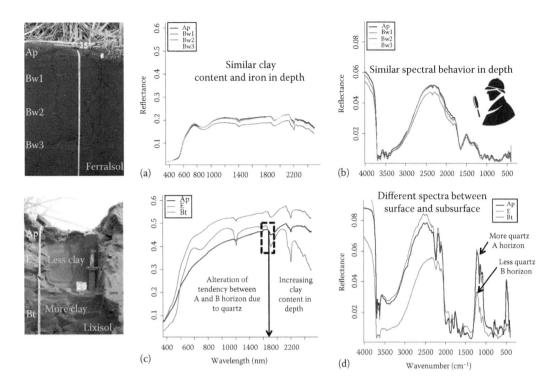

FIGURE 24.10 Methods of soil profile spectral assessment for soil classification: Ferralsol (oxisol) in VIS–NIR (a), and MIR (b); and Lixisol (Ultisol) in VIS–NIR (c), and MIR (d).

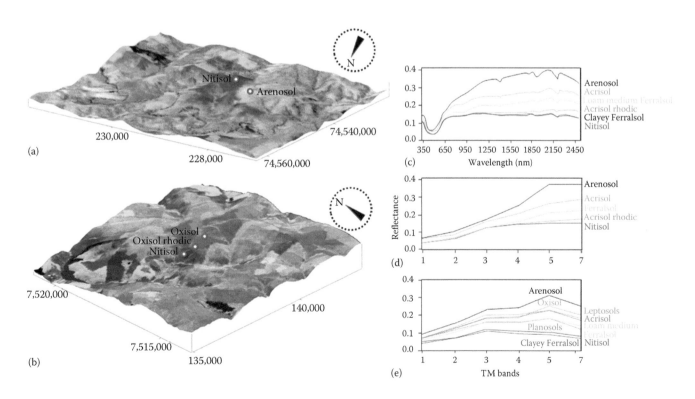

FIGURE 24.11 Landsat TM image compositions (RGB = 5,4,3) with some parts with bare soil indicating areas with clayey and sandy soils (a, b), and soil spectral curves from laboratory (c), laboratory convoluted to Landsat TM (d), and Landsat TM directly (e). (From Demattê et al., 2009a.)

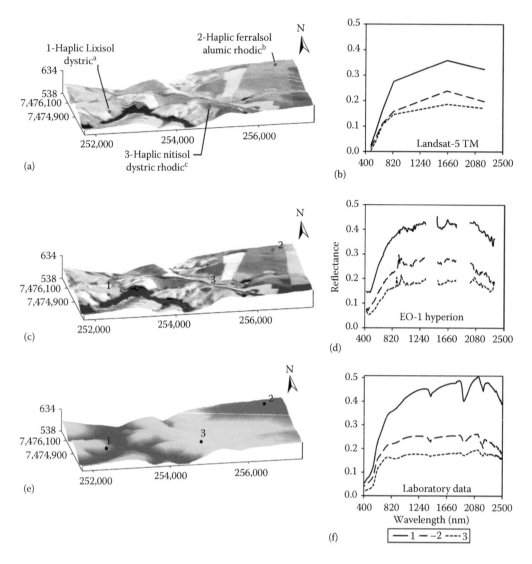

FIGURE 24.12 Landsat-5 TM image composition (R: 1650 nm, G: 830 nm, B: 660 nm) showing the position of soils (a) and their respective Landsat TM spectra (b), Hyperion image composition (R: 1648 nm, G: 823 nm; B: 661 nm) (c), and Hyperion spectra of the same soils (d), digital elevation model (e), and laboratory VIS–NIR spectra of the same soils (f). a—clay content of 138.7 g kg^{-1}; iron oxides content <180 g kg^{-1}; b—clay content of 255.4 g kg^{-1}; iron oxides content <180 g kg^{-1}; c—clay content of 507.1 g kg^{-1}; iron oxides content 180 and 360 g kg^{-1}. Digital elevation model derived from SRTM (Shuttle Radar Topography Mission) data.

orbital sensors, water bands differ among spectra due to atmospheric absorption. Compared to the multispectral sensors, many inferences can be made, although the quality of soil attribute predictions and spectral characterization are limited by the SNR and spectral resolution. Comparing the spectral data acquired from different levels, such as Hyperion, at the field or in the laboratory, the influence of the SNR is clearly observed. Due to the lower spectral resolution, multispectral sensors, such as Landsat TM, do not present important features, and consequently many spectrally active attributes cannot be predicted (e.g., minerals). Despite this, Landsat data have a good temporal resolution, that is, information about soils is continuously generated. In this example, older soils (oxisols and nitisols) presented lower reflectance intensities, while arenosols and inceptisols

(younger soils) presented higher reflectance intensity. Some similarities among spectra from different sensors are observed, for example, the reflectance intensity and convexity, which are greater in the Typic Kandiudult, and lower in the rhodic hapludox. According to Demattê (2002), arenosols and some acrisols present higher reflectance intensities at the IR band (Landsat TM band 5) and increasing albedo from VIS to NIR due to their lower clay content. It is important to notice that, despite the great spectral resolution of Hyperion (Figure 24.12b), spectra are very noisy, which makes its interpretation difficult.

Soil monitoring based on orbital and aerial sensors is another useful method, although interferences due to the soil moisture content in the field, atmospheric conditions, and intensity of illumination in different periods need to be overcome.

24.7 Space Spectral Sensing: Factors to Be Considered for Soil Studies

24.7.1 Data Used for Soil Characterization

An important characteristic of radiometric data acquired by remote sensors is that it is generally recorded and delivered to users as DNs. Since the image formed by each spectral band is a monochrome image, it is usual to refer to DNs as gray levels (Varshney and Arora, 2004). These DNs or gray levels are scaled integer numbers obtained from quantization of the electromagnetic energy that reaches the sensor, and they are not a physical energy measure (Liang, 2004). Generally, DNs have a linear relation with values of radiance at the top of atmosphere, which are a physical energy measure, and they can be converted to these values using available algorithms that take in parameters related to the sensor and to the environment at the time of acquisition (Lillesand et al., 2007).

DN values represent different reflectance intensities (brightness) in distinct images, that is, these values are not comparable between images. However, they can still be used to compare visually different images or to analyze the relative brightness in the same image (Campbell and Wynne, 2011). Although the radiometric data can be statistically analyzed using DNs or radiance at the top of atmosphere, these data can be compared only among sensors and acquisition levels after processing them to surface reflectance, especially when using hyperspectral data (Ustin et al., 2004). Surface reflectance is achieved by correcting the radiometric data regarding the influence of atmospheric

conditions, sunlight, and viewing angle. Radiative transfer models are usually used for this task. It is a necessary procedure if the objective of the study includes the characterization of biophysical properties of the imaged target, since suitable differences in the signal can have great influence in the results obtained, especially in studies of quantitative nature (Liang, 2004; Jensen, 2005; Lillesand et al., 2007). Only after radiometric data correction to surface reflectance, it is possible to compare space-based spectra with spectra measured in the field or laboratory (Ustin et al., 2004). Figure 24.13 shows examples of these different radiometric products. Besides radiometric correction, an important preprocessing step is to apply geometric correction, assuring that the image pixels are correctly georeferenced, that is, they match their true position on the surface of the earth.

To make the image as representative as possible of the scene being recorded it is necessary to rectify and correct the data measured (Richards, 2013). This is necessary because geometric and radiometric distortions hamper an accurate representation of the surface reflectance in the spectral bands measured. Geometric distortions are shape and scale alterations of the obtained pixels, while radiometric distortions concern to the inaccurate transformation of surface reflectance to DNs (Varshney and Arora, 2004). Taken in account that great part of SS data is provided to the user after complete or partial registration and correction for errors caused by sensor malfunction, the radiometric distortions, concerning the brightness values assigned to the pixels, are an important obstacle between the user and the accurate spectral information. Radiative transfer models are generally used to perform radiometric correction of remotely sensed data.

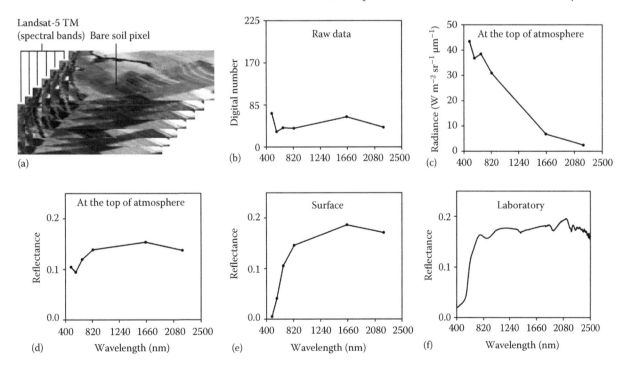

FIGURE 24.13 Landsat-5 TM images from spectral bands 1–5 and 7, showing a pixel predominantly with bare soil (a), with reflectance values plotted as DN (b), radiance at the top of the atmosphere (c), reflectance at the top of the atmosphere, and (d) reflectance at the target (surface reflectance) (e), and corresponding laboratory VIS–NIR spectrum of a soil sample collected in the same pixel (f).

These models consider the scattering and absorption properties of atmospheric components and, in this way, make it possible to transform radiance at the top of atmosphere to surface reflectance (Figure 24.13).

Looking at Figure 24.13, is possible to notice that the radiance data (Figure 24.13b) preserve the general irradiance spectra of the energetic source, that is, the sun, since these data are not corrected for wavelength dependence of radiation that reaches the earth or for atmospheric influences. After transformation to reflectance at the top of the atmosphere (Figure 24.13d), data no more resemble the sun irradiance spectra, but still are influenced by atmospheric components, especially in the VIS wavelengths (from 400 to 700 nm), where absorption and scattering are stronger. Finally, the surface reflectance spectrum (Figure 24.13e) includes correction to the effects caused by atmospheric components and resembles the spectrum taken in the laboratory (Figure 24.13f) for a soil sample collected in the same spot. In fact, Demattê and Nanni (2003) observed great differences between spectra in DN and surface reflectance, arguing that different soil types were distinguishable only when DNs were transformed into surface reflectance.

24.7.2 Temporal and Spatial Variations: Implications to the Spectral Sensing of Soil in Croplands and Natural Areas

Variations in soils and their surrounding landscapes result from natural and anthropic processes occurred in the past and present, which operate at diverse temporal and spatial scales. Thus, the occurrence and variation of lithological formations, soil bodies, natural, or cultivated vegetation can be gradual or abrupt with great complexity in their development in space and time (De Jong and van der Meer, 2004).

Soil characterization through remotely sensed data can be hampered by the landscape complexity as well as by imagery resolution limitations, since the detail level that can be assessed by images is determined by its resolution (Campbell and Wynne, 2011). The components mixture in the pixels has always to be considered when using remote sensors to study soils. Even images with great spatial resolution will generally contain mixture at some degree in its pixels. To illustrate the influence of mixing different materials on soil spectra, in this case, PV, NPV, and lime were added to bare soil in gradual amounts (Figure 24.14). Considering the sequence of soil + PV spectra, it is important to highlight that chlorophyll and other pigments cause absorption in the VIS region of the electromagnetic spectrum, with characteristic features from 350 to 700 nm, and that water in the leafs causes absorption in wavelengths near 970, 1200, 1450, 1950, and 2250 nm (Kokaly et al., 2009; Ustin et al., 2009). In the case of NPV, besides absorption near 1400 and 1900 nm due to water, the main feature occurs at 2100 nm caused by cellulose and other structural components, since sugars and nonstructural components are readily degraded by microorganisms (Nagler et al., 2003).

More specifically, adding PV to soil (Figure 24.14a) alters the albedo and attenuates or suppresses soil spectral features. This happens especially in VIS–NIR (from 400 to 950 nm), where characteristic absorption features of iron oxides, hydroxides, and oxihydroxides are present in soil, and also in wavelengths near 2200 nm, where absorption features of phyllosilicates and gibbsite occur (Stenberg et al., 2010). As already said, strong absorption features caused by pigments influence the VIS reflectance. However, it is also important to emphasize the effect that PV cover has from 680 to 700 nm, a region called "red edge," to longer wavelengths. In the "red edge" region, reflectance increases in the edge between the chlorophyll absorption feature, in the red wavelengths, and the multiple scattering caused by the cell wall–air interface within leafs, in the NIR (Treitz and Howarth, 1999). For wavelengths longer than the "red edge," a pronounced alteration in albedo occurs as the fraction of PV increases in relation to the soil fraction in the field of view.

Soil mixture with NPV (Figure 24.14b) causes soil spectral feature attenuation or suppression together with increases in albedo, the same trend observed in the mixture of soil and PV (Figure 24.14a). However, the influence of this mixture in soil spectra is stronger, since 45% of NPV cover will mask all soil absorption features almost completely; however, the spectral feature near 880 nm, for example, caused mainly by hematite (Stenberg et al., 2010), is still visible, although attenuated, even with high PV. The mixture with NPV influences soil spectra markedly near 2100 nm, where absorption features related to cellulose and other structural components in NPV occur, thus masking the phyllosilicate absorption band near 2200 nm. The example presented in Figure 24.14 follows the same principle of the study made by Nagler et al. (2003), corroborating their results, as well as the conclusions of Serbin et al. (2009) and Hbirkou et al. (2012), which state that the VIS–NIR–SWIR spectral signature of PV is distinct from those obtained for soil and NPV. NPV spectra are very similar to soil spectra and can severely affect soil characterization and soil attribute prediction using spectral data. Along the same lines, we observed different changes of soil spectral behavior between clayey and sandy soils upon adding lime, which will affect aerial/orbital image interpretation as well (Figure 24.14c and d). Thus, accurate ground laboratory measurements are always recommended as a basis to correctly interpret aerial/orbital images.

Chabrillat et al. (2002) observed that clay minerals can be identified and mapped even when the PV fraction in a pixel is between 40% and 50% or the NPV is between 20% and 30%. On the other hand, Bartholomeus et al. (2011) found that a PV fraction greater than 5% can already decrease accuracy, and a fraction greater than 20% can hamper soil property prediction using spectral information. Hbirkou et al. (2012) reported higher performance of SOC prediction by HyMap hyperspectral airborne sensor in recently tilled areas with bare soil ($R^2 = 0.73$), when compared to areas containing 10% of weeds and crop residues ($R^2 = 0.61$) or about 30% of straw residues ($R^2 = 0.34$).

Thus, in studies searching to describe or predict soil properties through remotely sensed data, it is usual to use masks and remove pixels with mixture levels that can hamper an accurate analysis. A methodology developed to isolate pixels containing

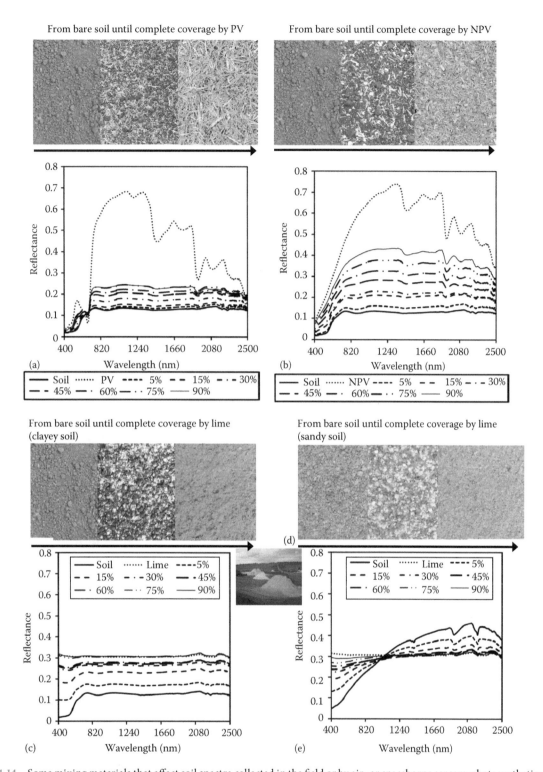

FIGURE 24.14 Some mixing materials that affect soil spectra collected in the field or by air- or spaceborne sensors: photosynthetic vegetation—PV (a), nonphotosynthetic vegetation—NPV (b), lime applied to a clayey soil (560 g kg⁻¹ of clay) (c) and a sandy soil (30 g kg⁻¹ of clay) (d), and (e) fertilizers at the field. (Figure assistence by Marco Antonio Melo Bortolleto.)

mostly exposed soil in imagery obtained by remote sensors is described by Demattê et al. (2009a). In brief, they used a semi-automated approach consisting of the following steps: (1) visually evaluate different spectral band compositions, for example, in the case of Landsat-5 TM, they used bands 3 (660 nm), 2 (560 nm), and 1 (485 nm), and bands 5 (1650 nm), 4 (830 nm), and 3 in RGB compositions, respectively; (2) evaluate the pixels to be analyzed using the SAVI (Huete, 1989); (3) use the soil line concept (Baret et al., 1993) to check if the pixels have characteristic behavior of bare soil areas; and (4) compare pixel spectra with an SL of soils obtained from laboratory and orbital sensors for surface samples (see Figure 24.11 for examples of spectral curves). The pixels that correspond to bare soil in all the steps are classified as so. Another example of bare soil detection methodology was applied by Chabrillat et al. (2011) and Gomez et al. (2012) using HyMap imagery analyzed in the HYperspectral SOil MApper (HYSOMA) software (available free-of-charge at http://www.gfz-potsdam.de/hysoma). The approach is similar to that described by Madeira Netto et al. (2007) in which NDVI (Tucker, 1979) is used to mask out pixels containing a high proportion of PV. For this, a threshold was assigned to the NDVI values (0.3) evaluating known bare soil areas in the image and the coherence of the resulting mask. As suggested by Madeira Netto et al. (2007), Chabrillat et al. (2011), and Gomez et al. (2012) evaluated absorption features near 2100 nm, caused by cellulose and other structural components in NPV, in areas covered by significant amounts of NPV. Chabrillat et al. (2011) used the cellulose absorption index (CAI; Nagler et al., 2003) and a threshold-based

method to select areas of bare soil for soil attribute estimation (Figure 24.15). In the end of this analysis, depending on the environmental conditions in the study sites, variable results are obtained for pixel classification as bare soil or mixed cover, based on the predominance of soil features in the pixel spectral signature.

One of the most commonly used techniques for pixel unmixing is the SMA, introduced by Horwitz et al. (1971). The SMA models consider that each pixel is a mix of different components (endmembers), and thus, the pixel spectral signature can be decomposed as a linear or nonlinear combination of the spectra of the endmembers (e.g., soil, vegetation, and shape), considering that these represent all the individual components present in the scene. To compose the pixel spectral signature, the endmember spectra are weighted according to the respective percent fraction of each endmember in the pixel. Guerschman et al. (2009) described an approach to determine fractional cover of bare soil, PV, and NPV using a linear SMA of NDVI and CAI images adapted for hyperspectral sensors.

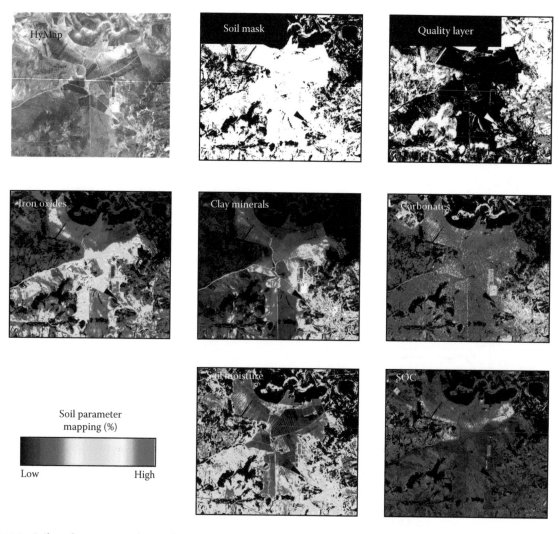

FIGURE 24.15 Soil attribute mapping from airborne HyMap imagery over Cabo de Gata Natural Park, Spain. Map outputs were created using the HYSOMA software. (From Chabrillat et al., 2011.)

Due to the simplicity of linear SMA models, they are applied more frequently in remotely sensed data analysis, although they are suitable especially to mixtures in which the components are segregated spatially. However, when the components to be analyzed are found intimately associated in a pixel, nonlinear models are more indicated. For example, a bare soil pixel could be roughly decomposed into three intimately associated endmembers: (1) the soil matrix composed of solid mineral and organic materials, (2) water in pores and associated with particles of the soil matrix, and (3) air taking the remaining soil volume.

Spectral unmixing is possible only because the electromagnetic energy interacts with each pixel component differently depending on their characteristics as it is multiple scattered (Keshava and Mustard, 2002). The choice between linear and nonlinear SMA models is still controversial, and it depends on the desired level of accuracy of the fraction assessment (Somers et al., 2011), since nonlinear models are more complex and difficult to implement. One of the most important limitations of the conventional linear SMA is that only one spectrum (endmember) is assigned to represent a specific scene (pixel) component, when in reality, each endmember might have variable spectral signatures in an RS image. For example, in the case of vegetation, biochemical composition and physical structure can vary between different species and phenological stages, among other factors, creating a wide variability in the spectral signatures of this component (Varshney and Arora, 2004). To deal with spatial and temporal endmember variabilities, approaches such as MESMA (Roberts et al., 1998) were proposed. In this case, multiple endmembers are considered for each component in an iterative way in the SMA. This methodology was one of the first attempts to manage the endmember variability for each component considered in the SMA. For this, SLs that represent component variability are used. In a review, Somers et al. (2011) indicated other methodologies to deal with spatial and temporal endmember variabilities.

To implement the conventional linear SMA, it is assumed that all the components (endmembers) in the scene have their spectral signatures known and available to be used in the analysis. However, partial unmixing techniques have been developed in order to avoid the necessity to know all endmembers to estimate the abundance for the components of interest. The MTMF (Boardman and Kruse, 2011) is one of the most widespread partial unmixing techniques. An example of the MTMF application is given in Chabrillat et al. (2002), who estimated the clay mineral abundance using hyperspectral imagery. The methodology allowed to quantify the amount of different types of clay with different swelling potentials (2:1 clay minerals—smectite and illite—versus 1:1 clay minerals—kaolinite) exposed at the soil surface in AVIRIS hyperspectral airborne imagery (Figure 24.6a and b).

Sophisticated methodologies to estimate components abundance in pixels have been made available and tested, for example, the approach based on neural networks described by Licciardi and Del Frate (2011) for hyperspectral data. Although bare soil area identification for subsequent processing is a viable alternative, it limits the extent of the analysis and mapping to the identified bare soil pixels, resulting in information loss in other areas. This motivated the use of information about vegetation to describe soil characteristics and variability, as detailed in Section 24.5.6. In addition, Bartholomeus et al. (2011) developed a technique entitled residual spectral unmixing (RSU), which removes the vegetation influence in the pixels' reflectance spectra. The authors stated that RSU can be applied to obtain continuous spectral information in the imaged area. Table 24.5 presents the main differences between airborne and orbital sensors and will be discussed later.

24.8 Comparison between Classical and Spectral Sensing Techniques for Soil Analysis

Chemical analyses are largely used for the evaluation of the fertility of crop soils and environment monitoring. Nevertheless, these analyses demand high costs and long periods to obtain results, which are the main obstacles in producing soil information. It is estimated that in Brazil, the number of chemical analyses of soils reached 1 million in 2001, demanding huge amounts of reagents and generating chemical waste (Raij et al., 2001). If such waste is mishandled or inadequately disposed of, it can result in soil and water contamination. These analyses have the support of strong background and reliable results. For example, Cantarella et al. (2006) found consistent results for chemical analysis in support of fertilizer application. Lime recommendation for a soil with 32% of base saturation reached the target value of 70% $\pm$ 8% in 74% of the cases. In about 90% of the cases, fertilizer recommendations were on or close to the target rates. Sizable deviations of the fertilizer recommendations for P and K that could affect profit occurred in less than 5% of the results reported. In terms of mineralogical properties, Lugassi et al. (2014) studied the potential of reflectance spectroscopy across the VIS–NIR–SWIR spectral region in combination with thermal analysis for soil mineralogy assessment. They concluded that the sensitivity of SS is higher than that of x-ray diffractometry.

SS can overcome the issues related to time, cost, and environmental pollution of classical soil analysis. However, whether these analyses can be definitely substituted by SS still needs to be answered. In general, soil reflectance spectra are directly affected by chemical and physical chromophores, as indicated by Ben-Dor et al. (1999). The spectral response is also a product of the interaction between soil constituents, calling for a precise understanding of all chemical and physical reactions in soils. A review of soil attributes already available from reflectance spectroscopy can be found in Malley et al. (2004) and Viscarra Rossel et al. (2011a,b). Janik et al. (1998) already questioned if spectra could replace soil extractions. This intriguing question raised several papers along years. In fact, Brown et al. (2006) and Demattê and Nanni (2006) made comparisons between wet laboratory and spectra and stated that the first could not be substituted, but optimized, using spectral information. Despite the

TABLE 24.5   Current and Upcoming Sensor Systems Providing Optical Data for Soil Attribute Mapping

| | | | | | Spatial Resolution (m) | Spectral Characteristics | | | |
Platform	Sensor	Origin	Start of Operation	Subsystem		Number of Bands	Wavelength Coverage (μm)	Spectral Resolution (nm)[a]	Spatial Coverage
Spaceborne optical sensors[b]									
Operating/ ready for launch	Landsat 8[m]	United States	2013	VNIR–TIR	30/100	9	0.45–12.50	—	Global
	MODIS[m]	United States	1999	VNIR–TIR	250/1000	36	0.40–14.40	—	Global
	ASTER[m]	Japan/United States	1999	VNIR–TIR	15/30/90	14	0.52–11.65	—	Global
	MERIS[m]	ESA	2002 (until 2012)	VNIR	300/1200	15	0.41–1.05	—	Global
	Hyperion[h]	United States	2000	VNIR–SWIR	30	242	0.36–2.58	10	Regional
	CHRIS[h]	ESA	2001	VNIR	17/34	6/18/37	0.40–1.05	5.6–32.9	Regional
	AVNIR-2[m]	Japan	2006	VNIR	10	4	0.42–0.89	—	Global
	HJ-1A[h]	China	2008	VNIR	100	128	0.45–0.95	5	Regional
	HySI[h]	India	2008	VNIR	506	64	0.40–0.95	~10	Global
	HICO[h]	United States	2009	VNIR	90	102	0.35–1.08	5.7	Regional
	HSE Resurs-P[h]	Russia	2013	VNIR	30	192	0.40–0.96	5–10	Regional
	Sentinel-2[m]	ESA	2015	VNIR–SWIR	10/20/60	13	0.44–2.28	—	Global
Under development	EnMAP[h]	Germany	2017	VNIR–SWIR	30	242	0.42–2.45	6.5/10	Regional
	HISUI[h]	Japan	2017	VNIR–SWIR	30	185	0.40–2.50	10/12.5	Regional
	PRISMA[h]	Italy	2017	VNIR–SWIR	30	237	0.40–2.50	~12	Regional
Planned	HIPXIM-P[h]	France	~2019	VNIR–SWIR	8	>200	0.40–2.50	10	Regional
	HyspIRI[h]	United States	≥2020	VNIR–TIR	60	>200/6	0.38–12.30	10/530	Global
	Shalom[h]	Israel/Italy	TBD	VNIR–SWIR	10	200	0.40–2.50	10	Regional
Airborne hyperspectral sensors									
	AHS-160	Daedalus, Spain	~2004	VNIR–TIR	2.5–10	48	0.45–13	12–550	Local
	aisaEAGLE	Specim, Finland	~2004	VNIR	~0.5–5	~488	0.40–0.97	3.3	Local
	aisaHAWK	Specim, Finland	~2004	SWIR	~0.5–5	254	0.93–2.5	12	Local
	aisaFENIX	Specim, Finland	2013	VNIR–SWIR	~0.5–5	620	0.38–2.5	3.5–12	Local
	aisaOWL	Specim, Finland	2013	TIR	tbd	84	8–12	100	Local
	APEX	VITO, Belgium	2009	VNIR–SWIR	2–10	300	0.38–2.50	5–10	Local
	AVIRIS	JPL, USA	1987	VNIR–SWIR	4–20	224	0.38–2.500	10	Local
	CASI	ITRES, Canada	~1985	VNIR	0.25–1.5	288	0.38–1.05	>3.5	Local
	HyMap	HyVista	1996	VNIR–SWIR	2–10	128	0.45–2.480	13–17	Local
	HySpex-1600	Norsk, Norway	2010	VNIR	0.5–5	160	0.40–1.0	3.7	Local
	HySpex-320	Norsk, Norway	2010	SWIR	2–20	256	1.0–2.5	6	Local
	MIVIS	CNR, Italy	1993	VNIR–TIR	3–10	112	0.43–12.7	20-50-9-450	Local
	ROSIS	DLR, Germany	1993	VNIR	2	115	0.43–0.96	5	Local
	ProspecTIR	specTIR, USA	~2008	VNIR–SWIR	~0.5–5	653	0.40–2.5	3.3–12	Local
	SASI-600	ITRES, Canada	~1985	SWIR	~1–5	160	0.95–2.45	10	Local
	SEBASS	Aerospace, USA	~1985	TIR	~1–10	128	2.5–13.5	~5	Local
	TASI-600	ITRES, Canada	~1985	TIR	~1–5	32	8–11.4	125	Local
	HyperCam	Telops, Lux.	2014	TIR	tbd	256	3–12	tbd	Local
	TRWIS-III	TRW, Canada	1996	VNIR–SWIR	~0.5–11	384	0.4–2.45	5.2/6.2	Local

m, multispectral sensor; h, hyperspectral sensor; TBD, to be demonstrated.

[a] Full-width half-maximum (provided for hyperspectral sensors only).

[b] Modified from Staenz et al. (2013).

great relationship between spectra and soil properties, soil wet analysis cannot be completely substituted. This is because the analytical procedure to derive chemical information from soil spectroscopy is based on models between wet chemistry (as the independent variable) and reflectance data (as the dependent variables). Thus, the accuracy of the spectral data cannot exceed the reference wet chemistry information, or, in other words, these comparison papers assumed wet chemistry as the standard reference method for soil analysis. Nonetheless, the need for rapid, simultaneous, and accurate analysis of many soil properties in many soil samples favors the adoption of soil spectroscopy. In fact, Sousa Junior et al. (2011) determined that a measurement with a spectrometer in the laboratory takes 10 min including sample preparation (the sensor takes only 1 min to acquire 100 spectral readings), whereas a granulometric analysis takes 48 h considering the Bouhoucus method, for example. O'Rourke and Holden (2011) calculated the costs per sample, analytical accuracy, and time involved in SOC analysis, in order to identify the best method compared against Walkley–Black (Walkley and Black, 1934). The conclusion indicated that MIR spectroscopy and laboratory hyperspectral imaging were the cheapest techniques, with a cost of €0.45 and €1.26 per sample, respectively. In comparison, samples measured in a TOC analyzer costed €15.15 per sample.

An important point to use spectra is related with variances in wet analysis. Schwartz et al. (2012) observed a 20% difference in soil analysis inside the same laboratory and a 103% difference between laboratories, although for very specific analysis (petroleum contamination). Demattê et al. (2010) compared traditional soil analysis and its variations, and the relationship with spectra information. They concluded that there were 84%–89% agreement obtained for sand content and 74%–87% for clay between traditional laboratory variations and the estimated spectral data. Despite this, agriculture needs faster information. Looking toward this goal, Viscarra Rossel et al. (2006) showed the great advantages of SS analysis. In fact, Nanni and Demattê (2006a) already stated that the analysis of some soil attributes, such as clay and CEC, could already be optimized by spectral analysis, without complete substitution of traditional laboratory methods. Their data reached high R^2 values (>0.8), ratified by several other publications (e.g., Sheperd and Walsh, 2002; Araújo et al., 2014a). Recent studies have started using SS as a primary analytical method to measure soil properties. For example, Bradák et al. (2014) used NIR spectroscopy to quantify SOC in a paleoenvironmental investigation, as Gomez et al. (2008b) used MIR spectra to measure SOC in an investigation in Australia.

Another important point would be to define a protocol or a method for collecting spectra in the laboratory. For example, the system geometry, number of readings, and sample preparation should be the same to compare spectra among users. In reality, spectroscopy research has been done using different light sources, distances of sensors to soil samples and to the light source, sample preparation, and so on. Thus, data collection in SS has been done accordingly to individual preferences. To overcome this issue, in 2014, Ben-Dor et al. (personal communication; 2014, in press) proposed the first protocol for LSS data collection. As an analogy, SS is going on the same track of traditional soil analysis, with a good perspective to soon become commercially attractive and accurate enough for routine soil analysis in laboratories.

Recently, in view of the growing soil spectral community, Viscarra Rossel (2009) generated an initiative (Soil World Spectral Group, http://groups.google.com/group/soil-spectroscopy) in which all members of the soil spectral community were asked to join together and contribute their local SLs in order to generate a worldwide SL that would be accessible to all. This initiative, besides being the first attempt to gather spectral information on the world's soils, is an important step toward establishing a standard protocol and quality indicators that will be accepted by all members of the community. To that end, it is important to mention that special sessions dealing with soil spectroscopy have been organized in several leading conferences on earth and soil sciences (e.g., European Geosciences Union in 2007 and 2008; World Congress of Soil Science in 2010; WD 2014) and in specific workshops (e.g., European Facility For Airborne Research (EUFAR2) in 2014). These meetings expose many scientists from soil and related sciences to these new technologies.

In summary, the main purpose of SS is not to substitute traditional soil analyses, but to optimize them. Samples could be summarized in a primary evaluation by spectra and then followed to wet analysis. The good prediction results for many soil properties observed in the literature (see Section 24.5.2) show the potential of SS, attracting the attention of soil scientists and making SS an interesting and constantly evolving field. Soil spectroscopy is a relatively new science with much ongoing research, and the multitude of alternative tools of SS for soil characterization and attribute quantification make it an intriguing line of research with promising perspectives for soil and environmental analysis. Some important questions to be answered in the future include (1) should we need a databank to estimate a soil property by spectra?; (2) if yes, should it be local, regional, or global?; (3) how should data be processed and in what statistical packages?; and (4) should we test universal prediction models, or should they be regionalized depending on the types of soils? We hope that this chapter provides some guidance on how to achieve these answers.

24.9 Moisture Effects in Spectral Sensing

Soil moisture is a very important property in RS and PS spectroscopy not only because it is an important target soil property to detect, but also because it has a nonlinear interference on spectra (Lobel and Asner, 2002; Haubrock et al., 2008a). Water molecules alter spectra in shape and intensity. The wavebands at 1400 and 1900 nm are pronounced under higher moisture contents, while at 2200 nm, the opposite is true. Even though variable soil moisture affects our ability to accurately predict soil properties, it is the way in which water and soil minerals bond that provide spectral information that allows prediction of many soil constituents (Demattê et al., 2006).

For field soil spectra to be collected and properly used at its full potential, it is imperative to account for soil moisture. Aiming to reduce the effects of water in the soil spectra, without prior knowledge of the soil water and in a way that SLs of dried and ground samples can be used, Minasny et al. (2011) proposed a technique called external parameter orthogonalization (EPO), previously conceived by Roger et al. (2003) to eliminate the effect of temperature in SS data. The EPO technique requires a calibration using a set of soils scanned moist (intact or ground) so that a transformation can be applied to the SL in use and to any subsequent field scans of soils whose properties are to be predicted

(Ge et al., 2014a). Most recently, Ge et al. (2014a) tested the EPO concept on intact field moist soil cores and removed the effect of soil moisture and *intactness* so that a VIS–NIR SL of dried and ground could be used to predict clay content and SOC (Ge et al., 2014a). This promising technique has the goal of making in situ field prediction of soil constituents using a combination of field VIS–NIR spectroscopy and SLs of dried and ground samples.

From ground to space, Haubrock et al. (2008a,b) (Figure 24.6c) developed a solid technique for topsoil moisture retrieval at the field and space levels, based on the influence of soil-available water capacity (AWC), a proxy for soil moisture, on the edges of the spectral absorption band at 1900 nm, which is currently used in many SS applications. Their method, called normalized soil moisture index (NSMI), and the method from Whiting et al. (2004), based on the analysis of the water absorption feature at 2700 nm (soil moisture Gaussian model—SMGM), were both automatized and implemented in the HYSOMA toolbox (Chabrillat et al., 2011). SMGM seems to deliver slightly better estimates, although in general both methods deliver similar SM retrieval performance, for example, with an R^2 of 0.7 for airborne HyMap images with 4 m pixel size (Chabrillat et al., 2012).

Sobrino et al. (2012) estimated soil moisture at the aerial level using an airborne hyperspectral scanner (AHS) sensor and, at the orbital level using ASTER, by combining remotely sensed images with in situ measurements. Their methodology considered the correlation between surface temperature, NDVI, and emissivity, and allowed SM predictions with an RMSE of 0.05 and 0.06 $m^3\ m^{-3}$ from AHS and ASTER, respectively, compared with ground measurements.

24.10 Basic and Integrated Strategy: How to Make a Soil Map Integrating Spectral Sensing and Geotechnologies

Why do we need to use geotechnologies for soil mapping? During the fieldwork, the pedologist starts to create a mental picture of the soil boundaries envisioning a soil class map. In this task, several tools can be used, including RS. Aerial photographs have been extensively used in the past (and still today). Vink (1964) proved that the use of aerial photographs added efficiency to soil mapping, requiring less fieldwork compared to mapping procedures done without this product. Later on, Campos and Demattê (2004) highlighted the importance of using a colorimeter to quantify soil color in substitution to the visual comparison with Munsell soil color charts. They compared data from five pedologists that performed soil color for the same sample using the Munsell color chart approach. They observed a 17.5% and 8.7% agreement among pedologists for dried and moist samples, respectively. All pedologists superestimated the hue, with consequences for soil classification. Given that field light conditions are highly variable, and the eye sensitivity changes by person and with age, among other factors, we argue that automatic systems should be used for color determination. Bazaglia

Filho et al. (2013) compared soil maps of the same area produced by four experienced pedologists and observed important differences and inconsistencies among maps. These findings prove the necessity to aggregate other technologies in the soil mapping activity, not only to improve the accuracy of the information but also to minimize the subjectivity of pedologists. In this aspect, we can integrate SS and DSM methods for soil map production. Here, we will suggest a sequence that aims to assist users and guide future research on how to integrate SS and DSM. The sequence can be altered depending on user goals and tools available. Also, some references are cited in the sequence, but there are several other methods that could be applied presented in this paper or elsewhere.

The success of soil class mapping starts by understanding the classical soil survey technique (Legros, 2006). Afterward, the first step is to define the characteristics and objectives of the map (i.e., soil taxonomic level, map scale/spatial resolution, target users). Second, define the data, information, and tools you have (legacy soil data, spectrometers, images, equipments), and understand their advantages and limitations (e.g., looking at Table 24.5). One of the first used and most important SS tools employed in soil mapping was aerial photography, as described by Vink (1964). Since the stereoscope until the new aerial 3D visualization of the landscape, all instruments greatly improve the delineation of soil boundaries (Figure 24.16). Third, use orbital and aerial images with color compositions integrated with aerial photographs and elevation maps (e.g., DEMs) to achieve several goals, including (1) to define spots to collect soil samples (see Section 24.5.1); (2) to define toposequences to study and determine the soils distribution patterns in the area; (3) to define boundaries based on different physiographies (looking at the landscape), and colors (looking at the images), without attributing soil classes at this time; (4) to analyze quantitative information of soil surface by images; and (5) to relate landscape shapes and patterns with soil classes. Fourth, go to field and collect samples, bring them to the laboratory, and pass through spectral sensors (mainly VIS–NIR–SWIR–MIR); choose which samples should go to wet laboratory analysis. Another approach would be taking spectra directly in the field, although several issues such as moisture still remain. During field collection, measurements should be taken at different depths (Ben-Dor et al., 2008). Fifth, with the laboratory soil data and spectra of samples acquired in the laboratory (Nanni and Demattê, 2006a,b) or field (Waiser et al., 2007), the following activities can be pursued: (1) relate the spectra with landscape (Galvão et al., 2001) and weathering (Demattê and Garcia, 1999) patterns; (2) understand the soil alterations along toposequences (Demattê and Terra, 2014) based on spectra; (3) analyze spectra using quantitative (Brown et al., 2006) and qualitative (Demattê, 2002) methods to group samples and determine soil mapping units (Demattê et al., 2004a); (4) compare soil spectra with available minerals libraries (Clark et al., 2007; Baldridge et al., 2009) to estimate the presence and content of minerals; (5) relate all samples with an SL containing pedological data and analyze spectra for soil classification qualitatively (Bellinaso et al., 2010; Demattê et al., 2014)

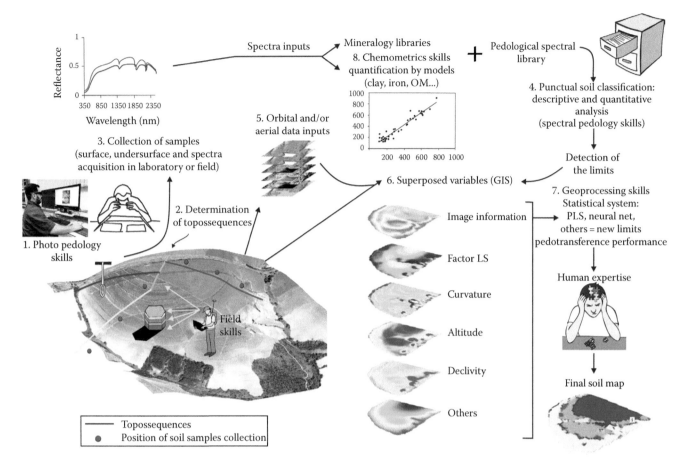

FIGURE 24.16 Suggested framework to produce a soil map by merging several techniques: photopedology, SS, spectral libraries, DSM, and statistical soil–landscape modeling. (Photograph with Caio Troula Fongaro.)

or quantitatively (Vasques et al., 2014), and/or for soil attribute estimation (Nocita et al., 2013); (6) use other equipments such as gamma ray sensors (Piikki et al., 2013) or electrical conductivity meters (Aliah et al., 2013) in the field; (7) incorporate laboratory soil data, SS data, and other available resources into DSM models (Behrens et al., 2010) and derive soil maps; (8) combine image radiometric data with elevation models; (9) derive models to map related geographical phenomena/properties influencing soil formation (e.g., geomorphic structures; Behrens et al., 2014); and finally (10) organize all available data and derived products in a GIS for publication. In all phases until the final results are achieved, including models and maps, human expertise and interpretation is required. Several skills are necessary, such as photopedology, spectral pedology, chemometrics, geoprocessing, pedometrics, and field experience. The methods described earlier are a suggestion and can be modified depending on the situation (objectives, scale, costs, equipments, and products available), or other strategies can always be proposed. The most important message is that today soil mapping is more complex for a pedologist since they have to integrate the classical experience with all these equipments, statistics, and modeling. In fact, there has been a great task to professionals to have all these skills, in addition to basic principles of pedology. On the other hand, technology is in constant evolution, and pedologists need to be able to use new available information, and thus, their background have to have these disciplines.

24.11 Potential of Spectral Sensing, Perspectives, and Final Considerations

The potential of SS in soil and environmental sciences is constantly growing and can be perceived based on the impressive developments of hardware and analytical tools going from ground to space sensors. In relation to GSS that started in the laboratory with, for example, Bowers and Hanks (1965), half a century later, we have advanced to extraterrestrial sensors exploring the soils from planet Mars (e.g., Murchie et al., 2009). GSS has reached early maturity with more than 60 years of papers and a strong theoretical background, but still new opportunities for GSS may arise with the development of better and less expensive sensors. Basic science in ground sensors has developed useful and fundamental relationships with several soil constituents such as water, granulometry, silicate clay type and content, mineralogy, Fe oxides, SOM, SOC, SIC, and CEC.

The applications toward soil inference are several, covering soil classification, mapping, quantification, conservation, and monitoring. The ground sensors are becoming more portable for the field, for example, VIS–NIR–MIR, gamma, radar, and apparent electrical conductivity, while computer technology and software are becoming cheaper and more powerful.

The perspectives to achieve undersurface information with sensor-mounted penetrometers have great importance. Indeed, this will not substitute the usual soil *pit* and the morphological description and sampling of soil horizons, but can certainly reduce the number of soil pits and samples necessary for fine-resolution (scale) mapping. Additionally, GSS has been adopted by PA management systems to optimize soil quality and crop production. It is still a difficult task to quantify attributes related to soil nutritional information (Ca, Mg, K, pH) because this information is primarily associated with soil water chemistry. However, correlations between soil water chemistry and soil physical (i.e., spectral or electromagnetic) properties, which can be sensed, are possible. Despite this weakness in detecting plant-available nutrients, GSS can assist PA with other information such as clay content, SOM, SOC, and CEC for indicating and optimizing soil sampling, assessing soil fertility, and quantifying soil water storage potential, among other properties. SS is not a panacea, but certainly can assist on the solution of some issues by providing continuous, quantitative, and accurate soil (spectral) data and mapping products, needed to assess and monitor soil status. Since 2000, an exponential increase in publications (e.g., Hartemink and McBratney, 2008; Ben-Dor et al., 2009; Chabrillat et al., 2013; Vasques et al., 2014) has promoted soil spectroscopy as a hot topic in soil science, with new interesting developing ideas. In fact, the question made by Janik et al. (1998) in their title "Can mid infrared diffuse reflectance analysis replace soil extractions?" was an indication that a revolution of new ideas on how to quantify soil properties was imminent. In fact, many papers in this research line have merged since then as observed in a review by Soriano-Disla et al. (2014). Today we can conclude that laboratory wet analysis cannot be substituted but certainly can be optimized by SS.

In 1993, Coleman et al. raised another question: "Spectral differentiation of surface soils and soil properties: is it possible from space platforms?" They answered this question with models that achieved 0.40 and 0.28 of R^2 for clay and iron quantification, respectively, using data from a sensor located 800 km from the target (Landsat). The authors admitted that the results were not very accurate, considering the use of a multispectral sensor, but that space-based spectroscopy had *potential*. This stimulated researchers. More recently, Nanni and Demattê (2006a) reached R^2 values of 0.67 and 0.95 also with Landsat and for the same elements, differing in the method to detect bare soil on the image, proving that SSS is no longer a *potential*, but a reality. The perspective is to improve these results with the future generation of optical satellite sensors, such as Copernicus sensors (Sentinel-2, to be launched in 2015) and hyperspectral EnMAP (launch planned for 2017) and other airborne or spaceborne hyperspectral sensors. Also, we expect that new methodologies on how to

isolate the soil information in image pixels emerge, thus making soil quantification and mapping more accurate.

Considering the actual development in aerial and orbital sensors, which now have a better spectral resolution, for example, hyperspectral sensors, we foresee improvements in soil property prediction using RS images, and thus, SLs will be important to understand and decompose the spectra in image pixels. Along the same lines, great opportunities for DSM applications are expected in the near future with the upcoming availability of satellite hyperspectral sensors that will routinely deliver high-spectral-resolution images for the entire globe, for example, EnMAP (Germany; Kaufmann et al., 2006), HISUI (Japan), HyspIRI (USA), HypXIM (France), PRISMA (Italy), and SHALOM (Israel–Italy). These hyperspectral satellite sensors are in development, with the earliest (EnMAP, HISUI) presently in phase D with a launch date around 2017 and the others (e.g., HyspIRI, HypXIM, PRISMA, SHALOM) planned for launching around 2020. They will have high SNR and pixel sizes from 8 (HypXIM) to 30 m (EnMAP) up to 60 m (HyspIRI), thus providing a range of options and image products for the soil science community. Table 24.5 summarizes these sensors and is described as follows.

With the development of the imaging spectroscopy concept in the early 1980s, the development of imaging systems was first focused on airborne instruments. Based on the expertise in the development of such systems, many terrestrial spaceborne missions have been under study, such as NASA's High-Resolution Imaging Spectrometer (Goetz and Davis, 1991), the Australian Resource Information and Environment Satellite (Roberts et al., 1997), and ESA's Ecosystem Changes through Response Analysis (Tobehn et al., 2002), to list just a few initiatives. Unfortunately, only a few, such as Hyperion (Middleton et al., 2013), made it into space the last decade. Table 24.5 provides an overview of current and upcoming civilian multispectral and imaging spectroscopy (hyperspectral) sensors currently operating for the imaging of the earth's soil surface. A survey of spaceborne missions currently operating or ready for launch is provided, completed with a survey of hyperspectral missions under development and a list of new initiatives currently in a planning stage. The latter is probably not a complete list of missions, but provides a good cross section of sensors, which might be in space around the 2020 time frame. With the launch in 2013 of Landsat 8 (former Landsat Data Continuity Mission) and the expected launch in 2015 of the first one of the Sentinel-2 satellites series, the new ESA flagship with a repeat rate of 10 days (5 days when the second sensor will be launched 2 years later), global earth coverage needs will be covered with additional spectral capabilities than previous multispectral sensors. In complement, with the expected launch of the HSE on the Resurs-P spacecraft in 2013, the next generation of imaging spectroscopy sensors is emerging. It will be followed by PRISMA, HISUI, and EnMAP, the three missions targeting a 30 m spatial resolution resulting in a swath width of 15–30 km and a 10 nm spectral resolution covering the VNIR and SWIR. These missions will replace the existing sensors currently in space in the 2016–2018 time frame, with increased data acquisition capacity and superior data quality compared to the technology demonstrators Hyperion and CHRIS. Although it is difficult to predict which of the hyperspectral sensors in the planning stage for a launch around 2020 or beyond will eventually be built and put in space, we can expect that the development of missions will continue and become more operational.

This overview of system shows the technical difficulties linked to the design and development of hyperspectral sensors. Due to the high requirements of hyperspectral systems toward higher spectral resolution and spectral coverage (up to 600 bands for airborne systems, up to 200 bands for spaceborne systems), then a compromise always has to be realized between the four pillars of space systems development: spectral resolution, spatial resolution, SNR, and temporal resolution. In hyperspectral missions, temporal resolution is sacrificed versus spectral resolution, so that all hyperspectral spaceborne systems have low revisit rate in comparison to multispectral missions with ~10 spectral bands. To avoid this difficulty, additional pointing capabilities such as in the EnMAP hyperspectral mission allow a higher revisit rate (4 days instead of 23 days in nadir mode) but then at the expense of adding the effect of different viewing angle. Then, the triangle (1) spectral resolution, (2) spatial resolution, and (3) SNR are the determining factors for system performances. The smaller the pixel, the lower the SNR. SNR can be improved only by increasing spectral bandwidth or increasing pixel size. As can be seen in the resulting design of the planned hyperspectral missions, this compromise leads to higher pixel size (Landsat-equivalent) for many missions (30 m EnMAP, HISUI, PRISMA, 60 m HypsIRI), and only two missions are considering smaller pixels (8 m HypXIM, 10 m SHALOM) at the expense of signal quality or lower spectral resolution. Data storage capacity/satellite downlink capabilities are the next limiting factors that prevent global coverage for hyperspectral missions.

Table 24.5 presents additionally a list of currently operating and new airborne hyperspectral sensors. The latter is probably not complete but shows the extensive development of airborne systems toward compacter and more portable systems (separated cameras for VIS–NIR and SWIR, e.g., Hyspex and AISA systems), with higher spectral resolution and more variable spectral coverage. One can note the recent development of TIR airborne hyperspectral sensors that will open new frontier of soil science as this spectral domain up to this point has mainly been available only from laboratory instrumentation.

It is difficult to understand the spectra of a soil type from space without knowing basic ground information. The link between field observations (e.g., from portable field sensors or from samples analyzed in the laboratory) and space sensors will be possible through soil SLs. After their start in the 1980s (Stoner and Baumgardner, 1981), soil SLs have evolved to a global SL soon to be published, with the participation of about 90 countries and coordinated by Dr. Raphael Viscarra Rossel (The Soil Spectroscopy Group, http://groups.google.com/group/soil-spectroscopy). The available EPO method to remove moisture effects from field samples will allow faster evaluation of soil properties (Ge et al., 2014a) in conjunction with soil SL. Irrespective of the acquisition level, from ground to space, hyperspectral ground information is still the basis for understanding and interpreting soil spectra generated by the new upcoming aerial or orbital sensors. An experienced "spectral" pedologist will be able to directly take measurements in situ and have a mental picture of the soil constituents, and thus soil class boundaries, in real-time. Removing or minimizing the need of laboratory analyses will create opportunities for the soil professional to cover larger areas with higher observation density and at a higher spatial resolution, reducing the time and cost of this activity.

Probably, the most important contribution of SS to soil science (since aerial photographs and now with many available ground and space sensors) relates to the identification, characterization, and delineation of soil bodies and mapping units across the landscape. Whereas ground sensors offer the opportunity to observe soils with great detail and fine sampling density, airborne and spaceborne sensors allow to extrapolate soil–spectra and soil–landscape relationships across large areas. Moreover, nowadays, other regions of the electromagnetic spectrum are being studied, such as TIR, microwaves, and gamma, with promising contributions to soil sensing. These sensors are already available at different acquisition levels, from ground to aerial and orbital.

In PA, the use of SS techniques has been part of the concept from the very beginning, and these techniques continue to develop, with diverse applications from soil status evaluation to crop yield monitoring. The demand for detailed information in farms stresses the need for fast and cheap methods, which includes both remote (airborne and spaceborne) and proximal (ground) sensing techniques. There is a growing interest in tractor-mounted sensors for real-time soil property quantification by SS in agriculture. Also, small UAVs that can carry several types of electromagnetic sensors, bringing the airborne techniques closer to the ground, will probably have great impact on soil sensing. Although most of the research so far using UAV has been on vegetation (Zhang and Kovacs, 2012), the possible applications are the same as those in traditional airborne or orbital-based SS. Both tractor-mounted and UAV-based sensing will require much research for future commercial applications.

Data integration and interpolation techniques that maximize information contained in multi-scale and multi-accuracy data sources need more development to achieve reliable and repeatable soil property predictions for DSM applications. For example, radar is still an important method mainly for soil moisture quantification, whereas gamma has been used for clay content estimation, among others. These sensors can be combined to assess both (and other) soil properties simultaneously.

These SS technologies are very attractive to soil scientists, especially younger ones, and care should be taken to focus the attention on the real "patient"—the soil—and not on the technology itself. The technology can help, but the real notion on how to use them comes from the human expertise and creativity, which the SS community has so far demonstrated. Thus, we have to avoid the paradox of the technology driving the science and the questions, because in reality, the opposite must be true.

We recommend multidisciplinary work that includes classical pedology (i.e., knowledge of soils), statistics, and understanding of the sensors (knowing their benefits and limitations) and their derived SS information. Effective and accurate SS must merge with terrain modeling and soil genesis knowledge to reach the highest levels of accuracy and detail for soil quantification, classification, and mapping. The inherent flexibility of SS allows merging micro- and macro-scale knowledge of soil properties, combining ground- to space-based data.

In conclusion, SS not only provides a general and flexible set of instruments and tools for users who need soil information, but also constitutes a real science domain under rapid development. Relevant SS applications that advance our knowledge of soils can contribute to understand soil-forming processes and soil patterns, both horizontally and vertically. Soil SS equipment is in constant evolution and, together with powerful computing and statistical tools, will support soil researchers reach the *next level* in soil assessment and mapping. It is a matter of time to coalesce technology and information to reach the main goal of soil science: to understand and preserve soils for the future through sustainable land use and management.

Acknowledgments

The authors thank Dr. Igo F. Lepsch for the revision of Section 24.1; the students Arnaldo Barros e Souza, Bruna Cristina Gallo, Caio Troula Fongaro, Danilo Jefferson Romero, Luis Gustavo Bedin, Marcus Vinicius Sato, Ana Paula Zanibão, João Paulo Brasiliano Camargo, Julia Antedomenico Cardoso De Morais, Lethícia Magno, Matheus Vinicius Rodrigues, and Veridiana Maria Sayão, and participants of the Geotechnologies in Soil Science Group (GEOSS, http://esalqgeocis.wix.com/english); and Gregory S. Rouze from Texas A&M University, for the assistance on part of gamma section.

References

Adamchuk, V. I., Viscarra Rossel, R.A., Marx, D.B., Samal, A.K., 2011. Using targeted sampling to process multivariate soil sensing data. *Geoderma* 163, 63–73.

Aliah, B.S.N., Kodaira, M., Shibusawa, S., 2013. Potential of visible-near infrared spectroscopy for mapping of multiple soil properties using real-time soil sensor. *Proceedings of SPIE* 8881, 8881071.

Alleoni, L.R.F., Mello, J.W.V., Rocha, W.S.D., 2009. Eletroquímica, adsorção e troca iônica no solo. In: Melo, V. F. and, Alleoni, L.R.F. (Eds.), Química e mineralogia do solo: Parte II–Aplicações. Sociedade Brasileira de Ciência do Solo, Viçosa, MG, pp. 69–129.

Anne, N.J.P., Abd-Elrahman, A.H., Lewis, D.B., Hewitta, N.A., 2014. Modeling soil parameters using hyperspectral image reflectance insubtropical coastal wetlands. *International Journal of Applied Earth Observation and Geoinformation* 33, 47–56.

Araújo, S.R., Demattê, J.A.M., Bellinaso, H., 2013. Analyzing the effects of applying agricultural lime to soils by VNIR spectral sensing: A quantitative and quick method. *International Journal of Remote Sensing* 34, 4750–4785.

Araújo, S.R., Wetterlind, J., Demattê, J.A.M., Stenberg, B., 2014a. Improving the prediction performance of a large national Vis-NIR spectroscopic library by clustering into smaller subsets or the use of data mining calibration techniques. *European Journal of Soil Science*, in press.

Araújo, S.R., Demattê, J.A.M., Vicente, S., 2014b. Soil contaminated with chromium by tannery sludge and identified by Vis-Nir-Mid spectroscopy techniques. *International Journal of Remote Sensing and Remote Sensing* 35, 3579–3593.

Asner, G.P., Borghi, C.E., Ojeda, R.A., 2003. Desertification in Central Argentina: Changes in ecosystem carbon and nitrogen from imaging spectroscopy. *Ecological Applications* 13, 629–648.

Ballabio, C., Fava, F., Rosenmund, A., 2012. A plant ecology approach to digital soil mapping, improving the prediction of soil organic carbon content in alpine grasslands. *Geoderma* 187, 102–116.

Ballari, D., de Bruin, S., Bregt, A.K., 2012. Value of information and mobility constraints for sampling with mobile sensors. *Computers & Geosciences* 49, 102–111.

Baptista, G.M.M., Corrêa, R.S., Santos, P.F., Madeira Netto, J.S. Meneses, P.R., 2011. Use of imaging spectroscopy for mapping and quantifying the weathering degree of tropical soils in central Brazil. *Applied and Environmental Soil Science*, doi:10.1155/2011/641328.

Barnes, R.J., Dhanoa, M.S., Lister, S.J., 1989. Standard normal variate transformation and de-trending of near-infrared diffuse reflectance spectra. *Applied Spectroscopy* 43, 772–777.

Barthes, B.G., Brunet, D., Ferrer, H., Chotte, J.-L., Feller, C., 2006. Determination of total carbon and nitrogen content in a range of tropical soils using near infrared spectroscopy: Influence of replication and sample grinding and drying. *Journal of Near Infrared Spectroscopy* 14(5), 341–348.

Bartholomeus, H., Epema, G., Schaepman, M., 2007. Determining iron content in Mediterranean soils in partly vegetated areas, using spectral reflectance and imaging spectroscopy. *International Journal of Applied Earth Observation and Geoinformation* 9, 194–203.

Bartholomeus, H., Kooistra, L., Stevens, A., van Leeuwen, M., van Wesemael, B., Ben-Dor, E., Tychon, B., 2011. Soil organic carbon mapping of partially vegetated agricultural fields with imaging spectroscopy. *International Journal of Applied Earth Observation and Geoinformation* 13, 81–88.

Batut, A., 1890. La Photographie Aerinne par Cerf-Volant. Gauthier-Villars. Gauthiers-Villars et Fils. Paris.

Baumgardner, M.F., Silva, L.F., Biehl, L.L., Stoner, E.R., 1985. Reflectance properties of soils. *Advances in Agronomy* 38, 1–43.

Bazaglia Filho, O., Rizzo, R., Lepsch, I.F., Prado, H., Gomes, F.H., Mazza, J.A., Demattê, J.A.M., 2013. Comparison between detailed digital and conventional soil maps of an area with complex geology. *Brazilian Journal of Soil Science* 37(5), 1136–1148.

Behrens, T., Schmidt, K., Ramirez-Lopez, L., Gallant, J., Zhu, A., Scholten, T., 2014. Hyper-scale digital soil mapping and soil formation analysis. *Geoderma* 213, 578–588.

Behrens, T., Zhu, A.X., Schmidt, K., Scholten, T., 2010. Multi-scale digital terrain analysis and feature selection for digital soil mapping. *Geoderma* 155, 175–185.

Bellinaso, H., Demattê, J.A.M., Araújo, S.R., 2010. Spectral library and its use in soil classification. *Brazilian Journal of Soil Science* 34, 861–870.

Bellon-Maurel, V., McBratney, A., 2011. Near-infrared (NIR) and mid-infrared (MIR) spectroscopic techniques for assessing the amount of carbon stock in soils–Critical review and research perspectives. *Soil Biology and Biochemistry* 43, 1398–1410.

Ben-Dor, E., 2011. Characterization of soil properties using reflectance spectroscopy. In: Thenkabail, P.S., Lyon, J.G., Huete, A. (Eds.), *Hyperspectral Remote Sensing of Vegetation*, CRC Press, Boca Raton, FL, pp. 513–557.

Ben-Dor, E., Banin, A., 1995a. Near-infrared analysis as a rapid method to simultaneously evaluate several soil properties. *Soil Science Society of America Journal* 59(2), 364–372.

Ben-Dor, E., Banin, A., 1995b. Near-infrared analysis (Nira) as a method to simultaneously evaluate spectral featureless constituents in soils. *Soil Science* 159(4), 259–270.

Ben-Dor, E., Chabrillat, S., Demattê, J.A.M., Taylor, G.R., Hill, J., Whiting, M.L., Sommer, S., 2009. Using imaging spectroscopy to study soil properties. *Remote Sensing of Environment* 113, S38–S55.

Ben-Dor, E., Heller, D., Chudnovsky, A., 2008. A novel method of classifying soil profiles in the field using optical means. *Soil Science Society of American Journal* 72, 1–13.

Ben-Dor, E., Irons, J.R., Epema, G.F., 1999. Soil reflectance. In: Rencz, A.R. (Ed.), *Remote Sensing for Earth Sciences: Manual of Remote Sensing*, Wiley, Hoboken, NJ, pp. 111–173.

Ben-Dor, E., Patkin, K., Banin, A., Karnieli, A., 2002. Mapping of several soil properties using DAIS-7915 hyperspectral scanner data–A case study over clayey soils in Israel. *International Journal of Remote Sensing*, 23, 1043–1062.

Bernath, P.F., 2005. *Spectra of Atoms and Molecules*. Oxford University, New York, NY.

Bishop, T.F.A., McBratney, A.B., 2001. A comparison of prediction methods for the creation of field-extent soil property maps. *Geoderma* 103(1–2), 149–160.

Boardman, J.W., Kruse, F.A., 2011. Analysis of imaging spectrometer data using N-dimensional geometry and a mixture-tuned matched filtering (MTMF) approach. *IEEE Transactions on Geoscience and Remote Sensing* 49(11), 4138–4152.

Boettinger, J.L., Ramsey, R.D., Bodily, J.M., Cole, N.J., Kienast-Brown, S., Nield, S.J., Saunders, A.M., Stum, A.K., 2008. Landsat spectral data for digital soil mapping. In: Hartemink, A.E., McBratney, A.B., Mendonça-Santos, M.L., (Eds.), *Digital Soil Mapping with Limited Data*, pp. 193–202.

Bornemann, L., Welp, G., Brodowski, S., Rodionov, A., Amelung, W., 2008. Rapid assessment of black carbon in soil organic matter using mid-infrared spectroscopy. *Organic Geochemistry* 39(11), 1537–1544.

Bowers, S.A., Hanks, J.R., 1965. Reflection of radiant energy from soil. *Soil Science* 100, 130–138.

Bradák, B., Kiss, K., Barta, G., Varga, G., Szeberényi, J., Józsa, S., Novothny, Á., Kovács, J., Markó, A., Mészáros, E., Szalai, Z., 2014. Different paleoenvironments of Late Pleistocene age identified in Verőce outcrop, Hungary: Preliminary results. *Quaternary International* 319, 119–136.

Breunig, F.M., Galvão, L.S., Formaggio, A.R., 2008. Detection of sandy soil surfaces using ASTER-derived reflectance, emissivity and elevation data: Potential for the identification of land degradation. *International Journal of Remote Sensing* 29, 1833–1840.

Bricklemyer, R.S., Brown, D.J., 2010. On-the-go VisNIR: Potential and limitations for mapping soil clay and organic carbon. *Computers and Electronics in Agriculture* 70(1), 209–216.

Brodský, L., Klement, A., Penizek, V., Kodesova, R., Boruvka, L., 2011. Building soil spectral library of the Czech soils for quantitative digital soil mapping. *Soil & Water Research* 6, 165–172.

Brown, D.J., 2007. Using a global VNIR soil-spectral library for local soil characterization and landscape modeling in a 2nd-order Uganda watershed. *Geoderma* 140, 444–453.

Brown, D.J., Shepherd, K.D., Walsh, M.G., Mays, M.D., Reinsch, T.G., 2006. Global soil characterization with VNIR diffuse reflectance spectroscopy. *Geoderma* 132(3–4), 273–290.

Browning, D.M., Duniway, M.C. 2011. Digital soil mapping in the absence of field training data: A case study using terrain attributes and semiautomated soil signature derivation to distinguish ecological potential. *Applied and Environmental Soil Science*, ID 421904, 12 p.

Bruinsma, J., 2009. *The resource outlook to 2050: By how much do land, water use and crop yields need to increase by 2050?* Expert Meeting on How to Feed the World in 2050. Rome: FAO/ESDD.

Brunet, D., Barthes, B.G., Chotte, J.-L., Feller, C., 2007. Determination of carbon and nitrogen contents in Alfisols, Oxisols and Ultisols from Africa and Brazil using NIRS analysis: Effects of sample grinding and set heterogeneity. *Geoderma* 139(1–2), 106–117.

Brunet, D., Woignier, T., Lesueur-Jannoyer, M., Achard, R., Rangon, L., Barthe's, B.G., 2009. Determination of soil content in chlordecone (organochlorine pesticide) using near infrared reflectance spectroscopy (NIRS). *Environmental Pollution* 157, 3120–3125.

Buol, S.W., Southard, R.J., Graham, R.C., McDaniel, P.A., 2011. *Soil Genesis and Classification*. Wiley-Blackwell, Hoboken, NJ.

Buringh, P., 1960. The application of aerial photography in soil surveys. In: Buringh, P.: *American Society of Photogrammetry and Remote Sensing*, Manual of Photographic Interpretation, Bethesda, MD, pp. 631–666.

Cambule, A.H., Rossiter, D.G., Stoorvogel, J.J., Smaling, E.M.A., 2012. Building a near infrared spectral library for soil organic carbon estimation in the Limpopo National Park, Mozambique. *Geoderma* 183, 41–48.

Campos, R.C., Demattê, J.A.M., 2004. Soil color: Approach to a conventional assessment method in comparison to an automatization process for soil classification. *Brazilian Journal of Soil Science* 28(5), 853–863.

Canadell, J.G., Quéré, C.L., Raupach, M.R., Field, C.B., Buitenhuis, E.T., Ciais, P., Conway, T. J., Gillett, N.P., Houghton, R.A., Marland, G., 2007. Contributions to accelerating atmospheric CO_2 growth from economic activity, carbon intensity and efficiency of natural sinks. *Proc. Natl. Acad. Sci. U.S.A.,* 104, 18866–18870.

Cantarella, H., Quaggio, J.A., Raij, B.van., Abreu, M.F., 2006. Variability of soil analysis in commercial laboratories: Implications for lime and fertilizer recommendations. *Communications in Soil Science and Plant Analysis* 37, 2213–2225.

Casa, R., Castaldi, F., Pascucci, S., Basso, B., Pignatti, S., 2013a. Geophysical and hyperspectral data fusion techniques for in-field estimation of soil properties. *Vadose Zone Journal*, 10p.

Casa, R., Castaldi, F., Pascucci, S., Palombo, A., Pignatti, S., 2013b. A comparison of sensor resolution and calibration strategies for soil texture estimation from hyperspectral remote sensing. *Geoderma* 197, 17–26.

Center for Food Safety. 2008. *Agricultural Pesticide Use in U.S. Agriculture.* Center for Food Safety. Washington, DC, pp. 2003.

Chabrillat, S., Goetz, A.F.H., Krosley, L., Olsen, H.W., 2002. Use of hyperspectral images in the identification and mapping of expansive clay soils and the role of spatial resolution. *Remote Sensing of Environment* 82, 431–445.

Chang, C.W., Laird, D.A., 2002. Near-infrared reflectance spectroscopic analysis of soil C and N. *Soil Science* 167(2), 110–116.

Chang, C.W., Laird, D.A., Mausbach, M.J., Hurburgh, C.R., 2001. Near-infrared reflectance spectroscopy-principal components regression analyses of soil properties. *Soil Science Society of America Journal* 65(2), 480–490.

Chang, G.W., Laird, D.A., Hurburgh, G.R., 2005. Influence of soil moisture on near-infrared reflectance spectroscopic measurement of soil properties. *Soil Science* 170(4), 244–255.

Chen, F., Kissel, D.E., West, L.T., Adkins, W., Rickman, D., Luvall, J.C., 2008. Mapping soil organic carbon concentration for multiple fields with image similarity analysis. *Soil Science Society of America Journal* 72, 186–193.

Chen, H., Pan, T., Chen, J., Lu, Q., 2011. Waveband selection for NIR spectroscopy analysis of soil organic matter based on SG smoothing and MWPLS methods. *Chemometrics and Intelligent Laboratory Systems* 107(1), 139–146.

Clark, R.N., 1999a. Project Corona. http://www.geog.ucsb.edu/~kclarke/Corona/Corona.html (Accessed January 14, 2014).

Clark, R.N., 1999b. Spectroscopy of rocks and minerals and principles of spectroscopy. In: Rencz, A.N. (Ed.), *Manual of Remote Sensing*, John Wiley and Sons, Toronto, ON, pp. 3–58.

Clark, R.N., Gallagher, A.J., Swayze, G.A., 1990. Material absorption band bepth mapping of imaging spectrometer data using a complete band shape least-squares fit with library reference spectra, *Proceedings of the Second Airborne Visible/Infrared Imaging Spectrometer (AVIRIS) Workshop.* JPL Publication, 90–54, 176–186.

Clark, R.N., Swayze, G.A., Wise, R., Livo, E., Hoefen, T., Kokaly, R., Sutley, S.J., 2007. USGS digital spectral library splib06a: U.S Geological Survey. Digital Data Series, Denver. 231p.

Christy, C.D., 2008. Real-time measurement of soil attributes using on-the-go near infrared reflectance spectroscopy. *Computers and Electronics in Agriculture* 61(1), 10–19.

Coleman, T.L., Agbu, P.A., Montgomery, O.L., 1993. Spectral differentiation of surface soils and soil properties: Is it possible from space platforms? *Soil Science*, 155, 283–293.

Coleman, T.L., Agbu, P.A., Montgomery, O.L., Gao, T., Prasad, S., 1991. Spectral band selection for quantifying selected properties in highly weathered soils. *Soil Science* 151, 355–361.

Colwell, R.N., 1997. History and place of photographic interpretation. In: Philipson, W.K. *Manual of Photographic Interpretation.* American Society for Photogrammetry & Remote Sensing, Bethesda, MD, pp. 33–48.

Condit, H.R., 1970. The spectral reflectance of American soils. *Photogrammetric Engineering* 36, 955–966.

Conforti, M., Buttafuoco, G., Leone, A.P., Aucelli, P.P.C., Robustelli, G., Scarciglia, F., 2013. Studying the relationship between water-induced soil erosion and soil organic matter using Vis-NIR spectroscopy and geomorphological analysis: A case study in southern Italy. *Catena* 110, 44–58.

Cook, S.E., Corner, R.J., Groves, P.R., Grealish, G.J., 1996. Use of airborne gamma radiometric data for soil mapping. *Australian Journal of Soil Research* 34, 183–194.

Cozzolino, D., Cynkar, W.U., Dambergs, R.G., Shah, N., Smith, P., 2013. In situ measurement of soil chemical composition by near-infrared spectroscopy: A tool toward sustainable vineyard management. *Communications in Soil Science and Plant Analysis* 44(10), 1610–1619.

Curcio, D., Ciraolo, G., D'Asaro, F., Minacapilli, M., 2013. Prediction of soil texture distributions using VNIR-SWIR reflectance spectroscopy. In: N. Romano, G. Durso, G. Severino, G.B. Chirico, M. Palladino (Eds.), *Four Decades of Progress in Monitoring and Modeling of Processes in the Soil-Plant-Atmosphere System: Applications and Challenges.* Procedia Environmental Sciences, pp. 494–503.

D'Acqui, L.P., Pucci, A., Janik, L.J., 2010. Soil properties prediction of western Mediterranean islands with similar climatic environments by means of mid-infrared diffuse reflectance spectroscopy. *European Journal of Soil Science* 61(6), 865–876.

Dalal, R.C., Henry, R.J., 1986. Simultaneous determination of moisture, organic-carbon, and total nitrogen by near-infrared reflectance spectrophotometry. *Soil Science Society of America Journal* 50(1), 120–123.

Daniel, K.W., Tripathi, X.K., Honda, K., 2003. Artificial neural network analysis of laboratory and in situ spectra for the estimation of macronutrients in soils of Lop Buri (Thailand). *Australian Journal of Soil Research* 41(1), 47–59.

Debaene, G., Niedzwiecki, J., Pecio, A., Zurek, A., 2014. Effect of the number of calibration samples on the prediction of several soil properties at the farm-scale. *Geoderma* 214, 114–125.

Demattê, J.A.M., 2002. Characterization and discrimination of soils by their reflected eletromagnetic energy. *Pesquisa Agropecuaria Brasileira,* 37, 1445–58.

Demattê, J.A.M., Bellinaso, H., Romero, D.J., Fongaro, C.T., 2014. Morphological Interpretation of Reflectance Spectrum (MIRS) using libraries looking towards soil classification. *Scientia Agricola* 71 (6), 509–520.

Demattê, J.A.M., Campos, R.C., Alves, M.C., Fiorio, P.R., Nanni, M.R., 2004a. Visible-NIR reflectance: A new approach on soil evaluation. *Geoderma* 121, 95–112.

Demattê, J.A.M., Demattê, J.L., Camargo, W., Fiorio, P., Nanni, M., 2001. Remote sensing in the recognition and mapping of tropical soils developed on topographic sequences. *Mapping Science and Remote Sensing* 38, 79–102.

Demattê, J.A.M., Epiphanio, J.C.N., Formaggio, A.R., 2003b. Influência da matéria orgânica e de formas de ferro na reflectância de solos tropicais [Organic matter and iron forms influence on the reflectance of tropical soils]. *Bragantia* 62, 451–464.

Demattê, J.A.M., Fiorio, P.R., Araújo, S.R., 2010. Variation of routine soil analysis when compared with hyperspectral narrow band sensing method. *Remote Sensing* 2, 1998–2016.

Demattê, J.A.M., Fiorio, P.R., Ben-Dor, E., 2009b. Estimation of soil properties by orbital and laboratory reflectance means and its relation with soil classification. *The Open Remote Sensing Journal* 2, 12–23.

Demattê, J.A.M., Galdos, M.V., Guimarães, R.V., Genú, A.M.; Nanni, M.R., Zulu, J., 2007a. Quantification of tropical soil attributes from ETM+/Landsat-7. *International Journal of Remote Sensing* 24, 257–275.

Demattê, J.A.M., Gama, M.P., Cooper, M., Araújo, J.C., Nanni, M.R., Fiorio, P.R., 2004b. Effect of fermentation residue on the spectral reflectance properties of soils. *Geoderma* 120, 187–200.

Demattê, J.A.M., Huete, A.R., Ferreira Jr., L., Alves, M C., Nanni, M.R., Fiorio, P.R., 2009a. Methodology for bare soil detection and discrimination by Landsat-TM image. *The Open Remote Sensing Journal* 2, 24–35.

Demattê, J.A.M., Marcondes, A., Simões, M.S., 2004c. Metodologia para reconhecimento de três solos por sensores: Laboratorial e orbital [Laboratory and orbital data: Methods on the recognation of three soils]. *Brazilian Journal of Soil Science* 28, 877–889.

Demattê, J.A.M., Moretti, D., Vasconcelos, A.C.F., Genú, A.M., 2005. Uso de imagens de satélite na discriminação de solos desenvolvidos de basalto e arenito na região de Paraguaçú Paulista [Satelite images on the discrimination of soils developed by basalt and arenit material from the region of Paraguaçu Paulista]. *Pesquisa Agropecuaria Brasileria* 40, 697–706.

Demattê, J.A.M., Nanni, M.R., 2003. Weathering sequence of soils developed from basalt as evaluated by laboratory (IRIS), airborne (AVIRIS) and orbital (TM) sensors. *International Journal of Remote Sensing* 24, 4715–4738.

Demattê, J.A.M., Nanni, M.R., 2006. Comportamento spectral da linha do solo obtida por espectrorradiometia laboratorial para diferentes classes de solo [Soil line behavior obtained by laboratorial spectroradiometry for different soil classes]. *Brazilian Journal of Soil Science* 30, 1031–1038.

Demattê, J.A.M., Nanni, M.R., Formaggio, A.R., Epiphanio, J.C.N., 2007b. Spectral Reflectance for the mineralogical evaluation of Brazilian low clay activity soils. *International Journal of Remote Sensing* 28, 4537–4559.

Demattê, J.A.M., Pereira, H.S., Nanni, M.R., Cooper, M., Fiorio, P.R., 2003a. Soil chemical alterations promoted by fertilizer application assessed by spectral reflectance. *Soil Science* 168, 730–747.

Demattê, J.A.M., Sousa, A.A., Alves, M., Nanni, M.R., Fiorio, P.R., Campos, R.C., 2006. Determing soil water status and other soil characteristics by spectral proximal sensing. *Geoderma* 135, 179–195.

Demattê, J.A.M., Terra, F.S., 2014. Spectral Pedology: A new perspective on evaluation of soils along pedogenetic alterations. *Geoderma* 218, 190–200.

Demattê, J.A.M., Terra, F.S., Quartaroli, C.F., 2012a. Spectral behavior of some modal soil profiles from São Paulo State, Brazil. *Bragantia* 71, 413–423.

Demattê, J.A.M., Vasques, G.M., Corrêa, E.A., Arruda, G.P., 2012b. Fotopedologia, espectroscopia e sistema de informação geográfica na caracterização de solos desenvolvidos do Grupo Barreiras no Amapá [Photopedology, spectroscopy and geographic information sistem in the characterization of soils developed by Barreiras Group from Amapa]. *Bragantia* 71, 438–446.

Deng, F., Minasny, B., Knadel, M., McBratney, A., Heckrath, G., Greve, M.H., 2013. Using vis-NIR spectroscopy for monitoring temporal changes in soil organic carbon. *Soil Science* 178(8), 389–399.

Denis, A., Stevens, A., Wesemael, B.V., Udelhoven, T., Tychon, B., 2014. Soil organic carbon assessment by field and airborne spectrometry in bare croplands: Accounting for soil surface roughness. *Geoderma* 226/227, 94–102.

Derenne, S., Rouzaud, J.N., Maquet, J., Bonhomme, C., Florian, P., Robert, F., 2003. Abundance, size and organization of aromatic moieties in insoluble organic matter of Orgueil and Murchison meteorites. In: 34th *Lunar and Planetary Science Conference.* Lunar and Planetary Institute, Houston, TX.

De Tar, W.R., Chesson, J.H., Penner, J.V., Ojala, J.C., 2008. Detection of soil properties with airborne hyperspectral measurements of bare fields. *Transactions of the ASABE* 51, 463–470.

Dierke, C., Werban, U., 2013. Relationships between gamma-ray data and soil properties at an agricultural test site. *Geoderma* 199, 90–98.

Dlugokencky, E., Pieter, T., *Trends in Atmospheric Carbon Dioxide.* Scripps Institution of Oceanography, La Jolla, CA. www.esrl.noaa.gov/gmd/ccgg/trends/ (Accessed June 25, 2014).

Doetterl, S., Stevens, A., Van Oost, K., van Wesemael, B., 2013. Soil organic carbon assessment at high vertical resolution using closed-tube sampling and vis-NIR spectroscopy. *Soil Science Society of America Journal* 77(4), 1430–1435.

Dogan, H.M., Kiliç, O.M., 2013. Modelling and mapping some soil surface properties of Central Kelkit Basin in Turkey by using Landsat-7 ETM+ images. *International Journal of Remote Sensing* 34, 5623–5640.

Dunn, B.W., Beecher, H.G., Batten, G.D., Ciavarella, S., 2002. The potential of near-infrared reflectance spectroscopy for soil analysis—A case study from the Riverine Plain of south- eastern Australia. *Australian Journal of Experimental Agriculture* 42(5), 607–614.

D'oleire-Oltmanns, S., Marzolff, I., Peter, K.D., Ries, J.B., 2012. Unmanned aerial vehicle (UAV) for monitoring soil erosion in Morocco. *Remote Sensing* 4, 3390–3416.

FAO. 2011. *The State of the World's Land and Water Resources for Food and Agriculture–Managing Systems at Risk.* Food & Agriculture Org, Rome. 47p.

FAO. 2013. *Statistical Yearbook 2013 World Food and Agriculture.* Food & Agriculture Org, Rome. 289p.

FAO. 2013b. *Global Soil Partnership Technical Report – State of the Art Report on Global and Regional Soil Information: Where are we? Where to go?* Food & Agriculture Org, Rome. 69p.

Fernandes, R.B.A., Barrón, V., Torrent, J., Fontes, M.P.F., 2004. Quantificação de óxidos de ferro de latossolos brasileiros por espectroscopia de reflectância difusa. *Brazilian Journal of Soil Science* 28, 245–257.

Fidêncio, P.H., Poppi, R.J., de Andrade, J.C., 2002. Determination of organic matter in soils using radial basis function networks and near infrared spectroscopy. *Analytica Chimica Acta* 453(1), 125–134.

Fiorio, P.R., Demattê, J.A.M., 2009. Orbital and laboratory spectral data to optimize soil analysis. *Scientia Agricola* 66, 250–257.

Fiorio, P.R., Demattê, J.A.M., Nanni, M.R., Formaggio, A.R., 2010. Soil spectral differentiation using laboratory and orbital sensor. *Bragantia* 69, 249–252.

Fiorio, P.R., Demattê, J.A.M., Nanni, M.R., Genú, A.M., Martins, J.A. 2014. *In situ* separation of soil types along transects employing Vis-NIR sensors: A new view of soil evaluation. *Revista Ciência Agronômica* 45, 433–442.

Fontan, J.M., Lopez-Bellido, L., Garcia-Olmo, J., Lopez-Bellido, R.J., 2011. Soil carbon determination in a Mediterranean vertisol by visible and near infrared reflectance spectroscopy. *Journal of Near Infrared Spectroscopy* 19(4), 253–263.

Franceschini, M.H.D., Demattê, J.A.M., Sato, M.V., Vicente, L.E., Grego, C.R., 2013. Abordagens semiquantitativa e quantitativa na avaliação da textura do solo por espectroscopia de reflectância bidirectional no VIS-NIR-SWIR. *Pesquisa Agropecuária Brasileira* 48, 1569–1582.

Fystro, G., 2002. The prediction of C and N content and their potential mineralisation in heterogeneous soil samples using Vis-NIR spectroscopy and comparative methods. *Plant and Soil* 246(2), 139–149.

Galvão, L.S., Formaggio, A.R., Couto, E.G., Roberts, D.A., 2008. Relationships between the mineralogical and chemical composition of tropical soils and topography from hyperspectral remote sensing data. *ISPRS Journal of Photogrammetry & Remote Sensing* 63, 259–271.

Galvão, L.S., Ícaro, V., Formaggio, A.R., 1997. Relationships of spectral reflectance and color among surface and subsurface horizons of tropical soil profiles. *Remote Sensing of Environment* 61, 24–33.

Galvão, L.S., Pizarro, M.A., Epiphanio, J.C.N., 2001. Variations in reflectance of tropical soils: Spectral-chemical composition relationships from AVIRIS data. *Remote Sensing of Environment* 75, 245–255.

Garfagnoli, F., Ciampalini, A., Moretti, S., Chiarantini, L., Vettori, S., 2013. Quantitative mapping of clay minerals using airborne imaging spectroscopy: New data on mugello (Italy) from SIM-GA prototypal sensor. *European Journal of Remote Sensing* 46, 1–17.

Ge, Y., Morgan, C.L.S., Ackerson, J.P., 2014a. VisNIR spectra of dried ground soils predict properties of soils scanned moist and intact. *Geoderma* 221, 61–69.

Ge Y., Morgan, C.L.S., Grunwald, S., Brown, D.J., Sarkhot, D.V., 2011. Comparison of soil reflectance spectra and calibration models obtained using multiple spectrometers. *Geoderma* 161, 202–211.

Ge, Y., Morgan, C.L.S., Thomasson, J.A., Waiser, T., 2007. A new perspective to near infrared reflectance spectroscopy: A wavelet approach. *Transactions of ASABE* 50, 303–311.

Ge, Y., Thomasson, J.A., Morgan, C.L.S., 2014b. Mid-infrared attenuated total reflectance spectroscopy for soil carbon and particle size determination. *Geoderma* 213, 57–63.

Geladi, P., MacDougall, D., Martens, H., 1985. Linerization and scatter-correction for near-infrared reflectance spectra of meat. *Applied Spectroscopy* 39, 491–500.

Genot, V., Colinet, G., Bock, L., Vanvyve, D., Reusen, Y., Dardenne, P., 2011. Near infrared reflectance spectroscopy for estimating soil characteristics valuable in the diagnosis of soil fertility. *Journal of Near Infrared Spectroscopy* 19(2), 117–138.

Genú, A.M., Demattê, J.A.M., Nanni, M.R., 2013b. Characterization and comparison of soil spectral response obtained from orbital (ASTER e TM) and terrestrial (IRIS). *Ambiência* 9, 279–288.

Genú, A.M., Roberts, D., Demattê, J.A.M., 2013a. The use of multiple endmember spectral mixture analysis for the mapping of soil attributes using Aster imagery. *Acta Scientiarum Agronomy* 35, 377–386.

Gerighausen, H., Menz, G., Kaufmann, H., 2012. Spatially explicit estimation of clay and organic carbon content in agricultural soils using multi-annual imaging spectroscopy data. *Applied and Environmental Soil Science,* Article ID 868090, 10 p.

Gernsheim, H., Gernsheim, A., 1952. Re-discovery of the world's first photograph. *Photograph Journal* 1, 118–121.

Ghaemi, M., Astaraei, A.R., Sanaeinejad, S.H., Zare, H., 2013. Using satellite data for soil cation exchange capacity studies. *International Agrophysics* 27, 409–417.

Gmur, S., Vogt, D., Zabowski, D., Moskal, L.M., 2012. Hyperspectral analysis of soil nitrogen, carbon, carbonate, and organic matter using regression trees. *Sensors* 12, 10639–10658.

Gobrecht, A., Roger, J.M., Bellon-Maurel, V., 2014. Major issues of diffuse reflectance nir spectroscopy in the specific context of soil carbon content estimation: A review. *Advances in Agronomy* 123, 145–175.

Godfray, H.C.J., Beddington, J.R., Crute, I.R., Haddad, L., Lawrence, D., Muir, J.F., Pretty, J., Robinson, S., Thomas, S.M., Toulmin, C., 2010. Food security: The Challenge of Feeding 9 Billion People. *Science* 327, 812–818.

Goetz, A.H.F., Davis, C.O., 1991. The high resolution imaging spectrometer (HIRIS): Science and instrument. *Journal of Imaging Systems and Technology* 3(2), 131–143.

Gogé, F., Gomez, C., Jolivet, C., Joffre, R., 2014. Which strategy is best to predict soil properties of a local site from a national Vis-NIR database? *Geoderma* 213, 1–9.

Gogé, F., Joffre, R., Jolivet, C., Ross, I., Ranjard, L., 2012. Optimization criteria in sample selection step of local regression for quantitative analysis of large soil NIRS database. *Chemometrics and Intelligent Laboratory Systems* 110(1), 168–176.

Gomez, C., Lagacherie, P., Coulouma, G., 2008a. Continuum removal versus PLSR method for clay and calcium carbonate content estimation from laboratory and airborne hyperspectral measurements. *Geoderma* 148(2), 141–148.

Gomez, C., Lagacherie, P., Coulouma, G., 2012. Regional predictions of eight common soil properties and their spatial structures from hyperspectral Vis–NIR data. *Geoderma* 189–190, 176–185.

Gomez, C., Viscara Rossel, R.A., McBratney, A.B., 2008b. Soil organic carbon prediction by hyperspectral remote sensing and field vis-NIR spectroscopy: An Australian case study. *Geoderma* 146(3–4), 403–411.

Goosen, D., 1967. *Aerial Photo Interpretation in Soil Survey.* Rome: Food & Agriculture Org.

Gras, J.-P., Barthes, B.G., Mahaut, B., Trupin, S., 2014. Best practices for obtaining and processing field visible and near infrared (VNIR) spectra of topsoils. *Geoderma* 214, 126–134.

Grasty, R.L., 1979. Gamma-ray spectrometric methods in uranium exploration: Theory and operational procedures. In: P.J. Hood. (Ed.), *Geophysics and Geochemistry in Search of Metallic Ores,* Geological Survey of Canada, Ottawa, ON, pp. 147–161.

Gregory, A.F., Horwood, J.L., 1961. A laboratory study of gamma-ray spectra at the surface of rocks (Mines Branch Research Report R.85). Department of Mines and Technical Surveys, Ottawa, ON. http://www.osti.gov/scitech/biblio/4833372 (Accessed June 15, 2014).

Grinand, C., Arrouays, D., Laroche, B., Martin, M.P., 2008. Extrapolating regional soil landscapes from an existing soil map: Sampling intensity, validation procedures, and integration of spatial context. *Geoderma* 143, 180–190.

Grinand, C., Barthes, B.G., Brunet, D., Kouakoua, E., Arrouays, D., Jolivet, C., Caria, G., Bernoux, M., 2012. Prediction of soil organic and inorganic carbon contents at a national scale (France) using mid-infrared reflectance spectroscopy (MIRS). *European Journal of Soil Science* 63(2), 141–151.

Guerrero, C., Stenberg, B., Wetterlind, J., Rossel, R.A.V., Maestre, F.T., Mouazen, A.M., Zornoza, R., Ruiz-Sinoga, J.D., Kuang, B., 2014. Assessment of soil organic carbon at local scale with spiked NIR calibrations: Effects of selection and extra-weighting on the spiking subset. *European Journal of Soil Science* 65(2), 248–263.

Guerschman, J.P., Hill, M.J., Renzullo, L.J., Barret, D.J., Marks, A.L., Botha, E.J., 2009. Estimating fractional cover of photosynthetic vegetation, non-photosynthetic vegetation and bare soil in the Australian tropical savanna region upscaling the EO-1 Hyperion and MODIS sensors. *Remote Sensing of Environment* 113(5), 928–945.

Hansena, M.K., Brown, D.J., Dennisona, P.E., Gravesa, S.A., Bricklemyerb, R.S., 2009. Inductively mapping expert-derived soil-landscape units within dambo wetland catenae using multispectral and topographic data. *Geoderma* 150, 72–84.

Hartemink, A.E., McBratney, A., 2008. A soil science renaissance. *Geoderma* 148(2), 123–129.

Hartemink, A.E., Minasny, B., 2014. Towards digital soil morphometrics. *Geoderma* 230-231, 305–317.

Haubrock, S.N., Chabrillat, S., Kuhnert, M., Hostert, P., Kaufmann, H., 2008a. Surface soil moisture quantification and validation based on hyperspectral data and field measurements. *Journal of Applied Remote Sensing* 2(1), 023552.

Haubrock, S.N., Chabrillat, S., Lemmnitz, C., Kaufmann, H., 2008b. Surface soil moisture quantification models from reflectance data under field conditions. *International Journal of Remote Sensing* 29(1), 3–29.

Hbirkou, C., Pätzold, S., Mahlein, A.K., Welp, G., 2012. Airborne hyperspectral imaging of spatial soil organic carbon heterogeneity at the field scale. *Geoderma* 175–176, 21–8.

He, Y., Huang, M., Garcia, A., Hernandez, A., Song, H., 2007. Prediction of soil macronutrients content using near-infrared spectroscopy. *Computers and Electronics in Agriculture* 58(2), 144–153.

Heinze, S., Vohland, M., Joergensen, R.G., Ludwig, B., 2013. Usefulness of near-infrared spectroscopy for the prediction of chemical and biological soil properties in different long-term experiments. *Journal of Plant Nutrition and Soil Science* 176(4), 520–528.

Hengl, T., Toomanianb, N., Reutera, H.I., Malakoutic, M.J., 2007. Methods to interpolate soil categorical variables from profile observations: Lessons from Iran. *Geoderma* 140, 417–427.

Hershel, W., 1800. Experiments on the refrangibility of the invisible rays of the sun. *Philos Trans R Soc* 90, 225–283.

Hewson, R.D., Cudahy, T.J., Jones, M., Thomas, M., 2012. Investigations into soil composition and texture using infrared spectroscopy (2–14 μm). *Applied and Environmental Soil Science*, doi:10.1155/2012/535646.

Hilwig, F.W., Goosen, D., Katsieris, D., 1974. Preliminary Results of the Interpretation of ERTS-1 Imagery for a Soil Survey. *ITC Journal* 3, 289–312.

Hively, W.D., McCarty, G.W., Reeves III, J.B., Lang, M.W., Oesterling, R.A., Delwiche, S.R., 2011. Use of airborne hyperspectral imagery to map soil properties in tilled agricultural fields. *Applied and Environmental Soil Science* 1–13.

Hollas, J.M., 2004. Electromagnetic radiation and its interaction with atoms and molecules. In: Hollas, J.M. (Ed.), *Modern Spectroscopy,* John Wiley, Chichester, WS, pp. 27–39.

Horwitz, H.M., Nalepka, R.F., Hyde, P.D., Morganstern, J.P., 1971. Estimating the proportion of objects within a single resolution element of a multispectral scanner. *Proceedings of the 7th International Symposium on RS of Environment.* Environmental Research Institute of Michigan, MI, pp. 1307–1320.

Hou, S., Wang, T., Tang, J., 2011. Soil Types extraction based on MODIS image. *Procedia Environmental Sciences* 10, 2207–2212.

Huang, X.W., Senthilkurnar, S., Kravchenko, A., Thelen, K., Qi, J.G., 2007. Total carbon mapping in glacial till soils using near-infrared spectroscopy, Landsat imagery and topographical information. *Geoderma* 141(1–2), 34–42.

Hunt, G.R., 1982. Spectroscopic properties of rocks and minerals. In: Carmichael, R.S., *Handbook of Physical Properties of Rocks.* CRC Press, Boca Raton, FL., pp. 295–385.

Hunt, G.R., Salisbury, J.W., 1970. Visible and near infrared spectra of minerals and rocks: I Silicate minerals. *Modern Geology* 1, 283–300.

IAEA. 2003. *Guidelines for Radioelement Mapping Using Gamma Ray Spectrometry Data* (IAEATECDOC-1363). International Atomic Energy Agency, Vienna.

Igne, B., Reeves, J.B., III, McCarty, G., Hively, W.D., Lund, E., Hurburgh, C.R., Jr., 2010. Evaluation of spectral pretreatments, partial least squares, least squares support vector machines and locally weighted regression for quantitative spectroscopic analysis of soils. *Journal of Near Infrared Spectroscopy* 18(3), 167–176.

Islam, K., Singh, B., McBratney, A., 2003. Simultaneous estimation of several soil properties by ultra-violet, visible, and near-infrared reflectance spectroscopy. *Australian Journal of Soil Research* 41(6), 1101–1114.

IUSS Working Group WRB. 2014. World Reference Base for Soil Resources 2014. International soil classification system for naming soils and creating legends for soil maps. World Soil Resources Reports No. 106. Rome: FAO.

Janik, L.J., Forrester, S.T., Rawson, A., 2009. The prediction of soil chemical and physical properties from mid-infrared spectroscopy and combined partial least-squares regression and neural networks (PLS-NN) analysis. *Chemometrics and Intelligent Laboratory Systems* 97(2), 179–188.

Janik, L.J., Merry, R.H., Skjemstad, J.O., 1998. Can mid infrared diffuse reflectance analysis replace soil extractions? *Australian Journal of Experimental Agriculture* 38, 681–696.

Janik, L.J., Skjemstad, J.O., 1995. Characterization and analysis of soils using midinfrared partial least-squares.2. Correlations with some laboratory data. *Australian Journal of Soil Research* 33(4), 637–650.

Jean-Philippe, S.R., Labbé, N., Franklin, J.A., Johnson, A., 2012. Detection of mercury and other metals in mercury contaminated soils using mid-infrared spectroscopy. *Proceedings of the International Academy of Ecology and Environmental Sciences* 2, 139–149.

Jenny, H., 1941. *Factors of Soil Formation: A System of Quantitative Pedology.* McGraw-Hill, New York, NY.

Jensen, J.R., 2005. *Introductory Digital Image Processing: A Remote Sensing Perspective (3rd edn).* Prentice-Hall, Upper Saddle River, NJ.

Jensen, J.R., 2006. Electromagnetic radiation principles. In: Jensen, J.R. (Ed.), *Remote Sensing of the Environment: An Earth Resource Perspective.* Prentice-Hall, Upper Saddle River, NJ, pp. 29–51.

Kamau-Rewe, M., Rasche, F., Cobo, J.G., Dercon, G., Shepherd, K.D., Cadisch, G., 2011. Generic prediction of soil organic carbon in Alfisols using diffuse reflectance Fourier-transform mid-infrared spectroscopy. *Soil Science Society of America Journal* 75(6), 2358–2360.

Kaufmann, H., Segl, K., Chabrillat, S., Hofer, S., Stuffler, T., Mueller, A., Richter, R., Schreier, G., Haydn, R., Bach, H., 2006. EnMAP–A hyperspectral sensor for environmental mapping and analysis. In: *Proceedings of the 2006 IEEE International Geoscience and Remote Sensing Symposium (IGARSS 2006) & 27th Canadian Symposium on Remote Sensing,* Denver.

Keen, B.A., Haines, W.B., 1925. Studies in soil cultivation. I. The evolution of a reliable dynamometer technique for use in soil cultivation experiments. *The Journal of Agricultural Science* 15, 375–386.

Kemper, T., Stefan, S., 2002. Estimate of heavy metal contamination in soils after a mining accident using reflectance spectroscopy. *Environment Science Technology* 36, 2742–2747.

Keshava, N., Mustard, J.F., 2002. Spectral unmixing. *IEEE Signal Processing* 19(1), 44–57.

King, C., Baghdadi, N., Lecomte, V., Cerdan, O., 2005. The application of remote-sensing data to monitoring and modelling of soil erosion. *Catena* 62, 79–93.

Kinoshita, R., Moebius-Clune, B.N., van Es, H.M., Hively, W.D., Bilgili, A.V., 2012. Strategies for soil quality assessment using visible and near-infrared reflectance spectroscopy in a Western Kenya chronosequence. *Soil Science Society of America Journal* 76(5), 1776–1788.

Knadel, M., Stenberg, B., Deng, F., Thomsen, A., Greve, M.H., 2013. Comparing predictive abilities of three visible-near infrared spectrophotometers for soil organic carbon and clay determination. *Journal of Near Infrared Spectroscopy* 21(1), 67–80.

Knadel, M., Thomsen, A., Greve, M.H., 2011. Multisensor on-the-go mapping of soil organic carbon content. *Soil Science Society of America Journal* 75(5), 1799–1806.

Kodaira, M., Shibusawa, S., 2013. Using a mobile real-time soil visible-near infrared sensor for high resolution soil property mapping. *Geoderma* 199, 64–79.

Kooistra, L., Leuven, R.S.E.W., Wehrens, R., Nienhuis, P.H., Buydens, L.M.C., 2003. A comparison of methods to relate grass reflectance to soil metal contamination. *International Journal of Remote Sensing* 24, 4995–5010.

Kuang, B., Mouazen, A.M., 2011. Calibration of visible and near infrared spectroscopy for soil analysis at the field scale on three European farms. *European Journal of Soil Science* 62(4), 629–636.

Kuang, B., Mouazen, A.M., 2011. Calibration of visible and near infrared spectroscopy for soil analysis at the field scale on three European farms. *European Journal of Soil Science* 62(4), 629–636.

Kuang, B., Mouazen, A.M., 2012. Influence of the number of samples on prediction error of visible and near infrared spectroscopy of selected soil properties at the farm scale. *European Journal of Soil Science* 63(3), 421–429.

Kuang, B., Mouazen, A.M., 2013a. Effect of spiking strategy and ratio on calibration of on-line visible and near infrared soil sensor for measurement in European farms. *Soil & Tillage Research* 128, 125–136.

Kuang, B., Mouazen, A.M., 2013b. Non-biased prediction of soil organic carbon and total nitrogen with vis-NIR spectroscopy, as affected by soil moisture content and texture. *Biosystems Engineering* 114(3), 249–258.

Kusumo, B.H., Hedley, C.B., Hedley, M.J., Hueni, A., Tuohy, M.P., Arnold, G.C., 2008. The use of diffuse reflectance spectroscopy for in situ carbon and nitrogen analysis of pastoral soils. *Australian Journal of Soil Research* 46(6–7), 623–635.

Kusumo, B.H., Hedley, M.J., Hedley, C.B., Tuohy, M.P., 2011. Measuring carbon dynamics in field soils using soil spectral reflectance: Prediction of maize root density, soil organic carbon and nitrogen content. *Plant and Soil* 338(1–2), 233–245.

Kweon, G., Maxton, C., 2013. Soil organic matter sensing with an on-the-go optical sensor. *Biosystems Engineering* 115(1), 66–81.

Lagacherie, P., Baret, F., Feret, J.B., Madeira Netto, J., Robbez-Masson, J.M., 2008. Estimation of soil clay and calcium carbonate using laboratory, field and airborne hyperspectral measurements. *Remote Sensing of Environment* 112, 825–835.

Lagacherie, P., Sneep, A.R., Gomez, C., Bacha, S., Coulouma, G., Hamrounid, M.H., Mekki, I., 2013. Combining Vis–NIR hyperspectral imagery and legacy measured soil profiles to map subsurface soil properties in a Mediterranean area (Cap-Bon, Tunisia). *Geoderma* 209–210, 168–176.

Lal, R., 1999. Soil management and restoration for C sequestration to mitigate the accelerated greenhouse effect. *Progress in Enviromental Science* 1, 307–326.

Lal, R., 2003. Soil erosion and the global carbon budget. *Environment International* 29, 437–450.

Lang, S., 2008. Object-based image analysis for remote sensing aplications: Modeling reality-dealing with complexity. In: Blaschke, T., Lang, S., and Hay, G. J. (Eds.), *Object-Based Image Analysis*. Springer, New York, NY, pp. 1–25.

Legros, J.P., 2006. *Mapping of the Soil*. Science Pub Inc., Enfiel, NH.

Liberti, M., Simoniello, T., Carone, M.T., Coppola, R., D'Emilio, M., Macchiato, M., 2009. Mapping badland areas using LANDSAT TM/ETM satellite imagery and morphological data. *Geomorphology* 106, 333–343.

Licciardi, G.A., Del Frate, F., 2011. Pixel unmixing in hyperspectral data by means of neural networks. *IEEE Transactions on Geoscience and Remote Sensing* 49(11), 4163–4172.

Liu, X., Liu, J., 2013. Measurement of soil properties using visible and short wave-near infrared spectroscopy and multivariate calibration. *Measurement* 46(10), 3808–3814.

Liu, Y.Y., Dorigo, W.A., Parinussa, R.M., de Jeu, R.A.M., Wagner, W., McCabe, M.F., Evans, J.P., van Dijk, A.I.J.M., 2012. Trend-preserving blending of passive and active microwave soil moisture retrievals. *Remote Sensing of Environmet* 123, 280–297.

Lobel, D.B., Asner, G.P., 2002. Moisture effects on soil reflectance. *Soil Science Society of America Journal* 66, 722–727.

Lu, P., Wang, L., Niu, Z., Li, L., Zhang, W., 2013. Prediction of soil properties using laboratory VIS-NIR spectroscopy and Hyperion imagery. *Journal of Geochemical Exploration* 132, 26–33.

Ludwig, B., Nitschke, R., Terhoeven-Urselmans, T., Michel, K., Flessa, H., 2008. Use of mid-infrared spectroscopy in the diffuse-reflectance mode for the prediction of the composition of organic matter in soil and litter. *Journal of Plant Nutrition and Soil Science-Zeitschrift Fur Pflanzenernahrung Und Bodenkunde* 171(3), 384–391.

Lugassi, R., Ben-Dor, E., Eshel, G., 2014. Reflectance spectroscopy of soils post-heating - Assessing thermal alterations in soil minerals. *Geoderma* 213, 268–279.

Löhrer, R., Bertrams, M., Eckmeier, E., Protze, J., Lehmkuhl, F., 2013. Mapping the distribution of weathered Pleistocene wadi deposits in Southern Jordan using ASTER, SPOT-5 data and laboratory spectroscopic analysis. *Catena* 107, 57–70.

Madeira Netto, J., 2001. Comportamento espectral dos solos. In: Meneses, P.R. and Madeira Netto, J.S. (Ed.). *Sensoriamento remoto: reflectância dos alvos naturais*. Brasília: UnB, Planaltina: Embrapa Cerrados, cap. 4, 127–154.

Madeira Netto, J.S.R., Robbez-Masson, J.M., Martins, E., 2007. Visible–NIR hyperspectral imagery for discriminating soil types in the La Peyne watershed (France). In: P. Lagacherie, A.B. McBratney, M. Voltz (Eds.), *Digital Soil Mapping: An Introductory Perspective*. Elsevier, Amsterdam, N.L., pp. 219–233.

Madejová, J., Komadel, P., 2001. Baseline studies of the clay minerals society source clays: Infrared methods. *Clays and Clay Minerals* 49, 410–432.

Mahmood, H.S., Hoogmoed, W.B., van Henten, E.J., 2012. Sensor data fusion to predict multiple soil properties. *Precision Agriculture* 13(6), 628–645.

Mahmood, H.S., Hoogmoed, W.B., van Henten, E.J., 2013. Proximal gamma-ray spectroscopy to predict soil properties using windows and full-spectrum analysis methods. *Sensors* 13(12), 16263–16280.

Mahmood, K., 1987. *Reservoir Sedimentation–Impact, Extent and Mitigation*. World Bank, Washington, DC.

Malley, D.F., Martin, P.D., McClintock, L.M., Yesmin, L., Eilers, R.G., Haluschak, P., 2000. Feasibility of analysing archived Canadian prairie agricultural soils by near infrared reflectance spectroscopy. In: A.M.C. Davies, R. Giangiacomo (Eds.), *Near Infrared Spectroscopy: Proceedings of the Ninth International Conference*. NIR Publications, Chichester, U.K., pp. 579–585.

Mann, K.K., Schumann, A.W., Obreza, T.A., 2011. Delineating productivity zones in a citrus grove using citrus production, tree growth and temporally stable soil data. *Precision Agriculture* 12, 457–472.

Martelet, G., Drufin, S., Tourliere, B., Saby, N.P.A., Perrin, J., Deparis, J., Prognon, F., Jolivet, C., Ratié, C., Arrouays, D., 2013. Regional regolith parameter prediction using the proxy of airborne gamma ray spectrometry. *Vadose Zone Journal* 12(4).

Martínez-Casasnovas, J.A., 2003. A spatial information technology approach for the mapping and quantification of gully erosion. *Catena* 50, 293–308.

Masoud, A.A., 2014. Predicting salt abundance in slightly saline soils from Landsat ETM+ imagery using Spectral Mixture Analysis and soil spectrometry. *Geoderma* 217–218, 45–56.

Masserschmidt, I., Cuelbas, C.J., Poppi, R.J., De Andrade, J.C., De Abreu, C.A., Davanzo, C.U., 1999. Determination of organic matter in soils by FTIR/diffuse reflectance and multivariate calibration. *Journal of Chemometrics* 13(3–4), 265–273.

McBratney, A. B., Mendonça Santos, M. L., & Minasny, B. 2003. On digital soil mapping. *Geoderma* 117, 3–52.

McCarty, G.W., Reeves, J.B., 2006. Comparison of near infrared and mid infrared diffuse reflectance spectroscopy for field-scale measurement of soil fertility parameters. *Soil Science* 171(2), 94–102.

McCarty, G.W., Reeves, J.B., Reeves, V.B., Follett, R.F., Kimble, J.M., 2002. Mid-infrared and near-infrared diffuse reflectance spectroscopy for soil carbon measurement. *Soil Science Society of America Journal* 66(2), 640–646.

McCarty, G., Hively, W.D., Reeves, J.B., III, Lang, M., Lund, E., Weatherbee, O., 2010. Infrared sensors to map soil carbon in agricultural ecosystems. *Proximal Soil Sensing*.

McDonald, R.A., 1997. *CORONA: Between the Sun and the Earth: The first NRO Reconnaissance Eye in Space*. ASP&RS, Bethesda, MD.

McDowell, M.L., Bruland, G.L., Deenik, J.L., Grunwald, S., Knox, N.M., 2012. Soil total carbon analysis in Hawaiian soils with visible, near-infrared and mid-infrared diffuse reflectance spectroscopy. *Geoderma* 189, 312–320.

Middleton, E.M., Campbell, P.E., Huemmrich, K.F., Ong, L., Mandl, D., Frye, S., Landis, D.R., 2013. The hyperion imaging spectrometer on the Earth Observing One (EO-1) satellite: Over a dozen years in Space. In: *Proceedings of the International Geoscience and Remote Sensing Symposium (IAGARSS'13)*, Melbourne, Australia.

Miltz, J., Don, A., 2012. Optimising sample preparation and near infrared spectra measurements of soil samples to calibrate organic carbon and total nitrogen content. *Journal of Near Infrared Spectroscopy* 20(6), 695–706.

Minasny, B., McBratney, A.B., 2006. A conditioned Latin hypercube method for sampling in the presence of ancillary information. *Computers & Geosciences* 32, 1378–1388.

Minasny, B., McBratney, A.B., 2008. Regression rules as a tool for predicting soil properties from infrared reflectance spectroscopy. *Chemometrics and Intelligent Laboratory Systems* 94(1), 72–79.

Minasny, B., McBratney, A.B., Bellon-Maurel, V., Roger, J.M., Gobrecht, A., Ferrand L., Joalland, S., 2011. Removing the effect of soil moisture from NIR diffuse reflectance spectra for the prediction of soil organic carbon. *Geoderma* 167–168, 118–124.

Minasny, B., McBratney, A.B., Tranter, G., Murphy, B.W., 2008. Using soil knowledge for the evaluation of mid-infrared diffuse reflectance spectroscopy for predicting soil physical and mechanical properties. *European Journal of Soil Science* 59(5), 960–971.

Minasny, B., McBratney, A.B., Walvoort, D.J.J., 2007. The variance quadtree algorithm: Use for spatial sampling design. *Computers & Geosciences* 33, 383–392.

Minasny, B., Tranter, G., McBratney, A.B., Brough, D.M., Murphy, B.W., 2009. Regional transferability of mid-infrared diffuse reflectance spectroscopic prediction for soil chemical properties. *Geoderma* 153(1–2), 155–162.

Mladenova, I.E., Jackson, T.J., Njoku, E., Bindlish, R., Chan, S., Cosh, M.H., Holmes, T.R.H., de Jeu, R.A.M., Jones, L., Kimball, J., Paloscia, S., Santi, E., 2014. Remote monitoring of soil moisture using passive microwave-based techniques –Theoretical basis and overview of selected algorithms for AMSR-E Original Research Article. *Remote Sensing of Enviroment* 144, 197–213.

Molan, Y.E., Refahi, D., Tarashti, A.H., 2014. Mineral mapping in the Maherabad area, eastern Iran, using the HyMap remote sensing data. *International Journal of Applied Earth Observation and Geoinformation* 27, 117–127.

Morgan, C.L.S., Waiser, T., Brown, D.J., Hallmark, C.T., 2009. Simulated in situ characterization of soil organic and inorganic carbon with visible near-infrared diffuse reflectance spectroscopy. *Geoderma* 151, 249–256.

Moron, A., Cozzolino, D., 2002. Application of near infrared reflectance spectroscopy for the analysis of organic C, total N and pH in soils of Uruguay. *Journal of Near Infrared Spectroscopy* 10(3), 215–221.

Mouazen, A.M., Kuang, B., De Baerdemaeker, J., Ramon, H., 2010. Comparison among principal component, partial least squares and back propagation neural network analyses for accuracy of measurement of selected soil properties with visible and near infrared spectroscopy. *Geoderma* 158(1–2), 23–31.

Mouazen, A.M., Maleki, M.R., De Baerdemaeker, J., Ramon, H., 2007. On-line measurement of some selected soil properties using a VIS-NIR sensor. *Soil & Tillage Research* 93(1), 13–27.

Mulder, V.L., de Bruin, S., Schaepman, M.E., 2013. Representing major soil variability at regional scale by constrained Latin Hypercube Sampling of remote sensing data. *International Journal of Applied Earth Observation and Geoinformation* 21, 301–310.

Mulder, V.L., De Bruin, S., Schaepman, M.E., Mayr, T.R., 2011. The use of RS in soil and terrain mapping–A review. *Geoderma* 162, 1–19.

Munoz, J.D., Kravchenko, A., 2011. Soil carbon mapping using on-the-go near infrared spectroscopy, topography and aerial photographs. *Geoderma* 166(1), 102–110.

Murchie, S.L., 2009. Compact Reconnaissance Imaging Spectrometer for Mars investigation and data set from the Mars Reconnaissance Orbiter's primary science phase. *Journal of Geophysical Research* 114(E2), 1–15.

Nadal-Romero, E., Vicente-Serrano, S. M., Jiménez, I., 2012. Assessment of badland dynamics using multi-temporal Landsat imagery: An example from the Spanish Pre-Pyrenees. *Catena* 96, 1–11.

Nanni, M.R., Demattê, J.A.M., 2006a. Spectral reflectance methodology in comparison to traditional soil analysis. *Soil Science Society of America Journal* 2, 393–407.

Nanni, M.R., Demattê, J.A.M., 2006b. Comportamento da linha do solo obtida por espectrorradiometria laboratorial para diferentes classes de solos [Soil line bevaior obtained by laboratory data for different classes of soils]. *Brazilian Journal of Soil Science* 30, 1031–1038.

Nanni, M.R., Demattê, J.A.M., Chicati, M.L., Fiorio, P.R., Cézar, E., de Oliveira, R.B., 2012. Soil surface spectral data from Landsat imagery for soil class discrimination. *Acta Scientiarum* 34, 103–112.

Nanni, M.R., Demattê, J.A.M., Chicati, M.L., Oliveira, R.B., Cezar, E., 2011. Spectroradiometric data as support to soil classification. *International Research Journal of Agricultural Science* 1, 100–115.

Nanni, M.R., Demattê, J.A.M., Silva Junior, C.A., Romagnoli, F., Silva, A.A., Cezar, E., Gasparotto, A.C., 2014. Soil mapping by laboratory and orbital spectral sensing compared with a traditional method in a detailed level. *Journal of Agronomy*, in press.

Nguyen, T.T., Janik, L.J., Raupach, M., 1991. Diffuse reflectance infrared fourier treansform (DRIFT) spectroscopy in soil studies. *Australian Journal of Soil Research* 29, 49–67.

Nocita, M., Kooistra, L., Bachmann, M., Mueller, A., Powell, M., Weel, S., 2011. Predictions of soil surface and topsoil organic carbon content through the use of laboratory and field spectroscopy in the Albany Thicket Biome of Eastern Cape Province of South Africa. *Geoderma* 167–68, 295–302.

Nocita, M., Stevens, A., Noon, C., van Wesemael, B., 2013. Prediction of soil organic carbon for different levels of soil moisture using Vis-NIR spectroscopy. *Geoderma* 199, 37–42.

Nocita, M., Stevens, A., Toth, G., Panagos, P., van Wesemael, B., Montanarella, L., 2014. Prediction of soil organic carbon content by diffuse reflectance spectroscopy using a local partial least square regression approach. *Soil Biology & Biochemistry* 68, 337–347.

Obukhov, A.I., Orlov, D.S., 1964. Spectral reflectivity of the major soils group and possibility of using diffuse reflection in soil investigation. *Society of Soil Science* 1, 174–184.

Oinuma, K., Hayashi, H., 1965. Infrared study of mixed-layer clay minerals. *American Mineralogy* 50, 1213–1227.

O'Rourke, S.M., Holden, N.M., 2011. Optical sensing and chemometric analysis of soil organic carbon–A cost effective alternative to conventional laboratory methods? *Soil Use and Management* 27(2), 143–155.

Peter, K.D., d'Oleire-Oltmanns, S., Ries, J. B., Marzolff, I., Hssaine, A.A., 2014. Soil erosion in gully catchments affected by land-levelling measures in the Souss Basin, Morocco, analysed by rainfall simulation and UAV remote sensing data. *Catena* 113, 24–40.

Petersen, H., Wunderlich, T., al Hagrey, S.A., Rabbel, W., 2012. Characterization of some Middle European soil textures by gamma-spectrometry. *Journal of Plant Nutrition and Soil Science* 175(5), 651–660.

Petersen, S.O., Frohne, P., Kennedy, A.C., 2002. Dynamics of a soil microbial community under spring wheat. *Soil Science Society of American Journal* 66, 826–833.

Piikki, K., Söderström, M., Stenberg, B., 2013. Sensor data fusion for topsoil clay mapping. *Geoderma* 199, 106–116.

Pirie, A., Singh, B., Islam, K., 2005. Ultra-violet, visible, near-infrared, and mid-infrared diffuse reflectance spectroscopic techniques to predict several soil properties. *Australian Journal of Soil Research* 43(6), 713–721.

Priori, S., Bianconi, N., Costantini, E.A.C., 2014. Can gamma-radiometrics predict soil textural data and stoniness in different parent materials? A comparison of two machine-learning methods. *Geoderma* 226/227, 354–364.

Rabenarivo, M., Chapuis-Lardy, L., Brunet, D., Chotte, J.-L., Rabeharisoa, L., Barthes, B.G., 2013. Comparing near and mid-infrared reflectance spectroscopy for determining properties of Malagasy soils, using global or LOCAL calibration. *Journal of near Infrared Spectroscopy* 21(6), 495–509.

Raij, B.V., Andrade, J.C., Cantarella, H., Quaggio, J.A., 2001. Análise química para avaliação da fertilidade de solos tropicais. *Campinas Instituto Agronômico*.

Ramirez-Lopez, L., Behrens, T., Schmidt, K., Stevens, A., Dematte, J.A.M., Scholten, T., 2013. The spectrum-based learner: A new local approach for modeling soil vis-NIR spectra of complex datasets. *Geoderma* 195, 268–279.

Ramirez-Lopez, L., Schmidt, K., Behrens, T., Wesemael, B.v., Dematte, J.A.M., Scholten, T., 2014. Sampling optimal calibration sets in soil infrared spectroscopy. *Geoderma* 226/227, 140–150.

Ratcliff, A.W., Busse, M.D., Shestak, C.J., 2006. Changes in microbial community structure following herbicide (glyphosate) additions to forest soils. *Applied Soil Ecology* 34, 114–124.

Reeves, J.B., Follett, R.F., McCarty, G.W., Kimble, J.M., 2006. Can near or mid-infrared diffuse reflectance spectroscopy be used to determine soil carbon pools? *Communications in Soil Science and Plant Analysis* 37(15–20), 2307–2325.

Reeves, J.B., McCarty, G.W., 2001. Quantitative analysis of agricultural soils using near infrared reflectance spectroscopy and a fibre-optic probe. *Journal of Near Infrared Spectroscopy* 9(1), 25–34.

Reeves, J.B., McCarty, G.W., Hively, W.D., 2010. Mid- versus near-infrared spectroscopy for on-site analysis of soil. *Proximal Soil Sensing.*

Reeves III, J.B., 2010. Near-versus mid-infrared diffuse reflectance spectroscopy for soil analysis emphasizing carbon and laboratory versus on-site analysis: Where are we and what needs to be done? *Geoderma* 158, 3–14.

Richards, J.A., 2013. *Remote Sensing Digital Image Analysis.* Springer, Berlin, 494p.

Richter, N., Chabrillat, S., Kaufmann, H., 2007. Enhanced quantification of soil variables linked with soil degradation using imaging spectroscopy. In Reusen, I., Cools, J. (Eds), *Proceedings of the 5th EARSeL Workshop "Imaging Spectroscopy: Innovation in Environmental Research"*, Bruges, Belgium, pp. 23–25.

Richter, N., Jarmer, T., Chabrillat, S., Oyonart, C., Hostert, P., Kaufmann, H., 2009. Free iron oxide determination in Mediterranean soils using diffuse reflectance spectroscopy. *Soil Science Society of American Journal* 73, 72–81.

Ries, J. B., Marzolff, I., 2003. Monitoring of gully erosion in the Central Ebro Basin by large-scale aerial photography taken from a remotely controlled blimp. *Catena* 50, 309–328.

Rivero, R.G., Grunwald, S., Bruland, G.L., 2007. Incorporation of spectral data into multivariate geostatistical models to map soil phosphorus variability in a Florida wetland. *Geoderma* 140(4), 428–443.

Roberts, D.A., Gardner, M., Church, R., Ustin, S., Scheer, G., Green, R.O., 1998. Mapping chaparral in the Santa Monica Mountains using multiple endmember spectral mixture models. *Remote Sensing of Environment* 65(3), 267–279.

Roberts, E., Huntington, J., Denize, R., 1997. The Australian Resource Information and Environment Satellite (ARIES), Phase A Study. In: *Proceedings of the 11th AIAA/USU Small Satellite Conference*, SSC97-III-2, Logan, Utah, pp. 1–10.

Roger, J.M., Chauchard, F., Bellon-Maurel, V., 2003. EPO–PLS external parameter orthogonalisation of PLS application to temperature-independent measurement of sugar content of intact fruits. *Chemometrics and Intelligent Laboratory Systems* 66(2), 191–204.

Rongjiang, Y., Jingsong, Y., Xiufang, Z., Xiaobing, C., Jianjun, H., Xiaoming, L., Meixian, L., Hongbo, S., 2012. *A new soil sampling design in coastal saline region using EM38 and VQT method.* CLEAN–Soil, Air, Water 40, 972–979.

Salisbury, J.W., D'Aria, D.M., 1992a. Infrared (8–14 µm) remote sensing of soil particle size. *Remote Sensing of Enviroment* 42, 157–165.

Salisbury, J.W., D'ária, D.M., 1994. Emissivity of terrestrial materials in the 3–5 µm atmospheric window. *Remote Sensing of Enviroment* 47, 345–361.

Salisbury, J.W., D'ária, D.M., 1992b. Emissivity of terrestrial materials in the 8–14 µm atmospheric window. *Remote Sensing of Enviroment* 42, 83–106.

Sankey, J.B., Brown, D.J., Bernard, M.L., Lawrence, R.L., 2008. Comparing local vs. global visible and near-infrared (VisNIR) diffuse reflectance spectroscopy (DRS) calibrations for the prediction of soil clay, organic C and inorganic C. *Geoderma* 148(2), 149–158.

Sannino, F., Gianfreda, L., 2001. Pesticide influence on soil enzymatic activities. *Chemosphere* 45, 417–425.

Savitzky, A., Golay, M.J.E., 1964. Smoothing and differentiation of data by simplified least squares procedures. *Analytical Chemistry* 36, 1627–1639.

Schirrmann, M., Gebbers, R., Kramer, E., 2012. Field scale mapping of soil fertility parameters by combination of proximal soil sensors, In: *International Conference of Agricultural Engineering—CIGR-Ageng 2012*, Valencia, Spain, July 8–12, 2012. Diazotec, S.L., Valencia, pp. 1–6 (at: http://cigr.ageng2012.org/images/fotosg/tabla_2137_C0411.pdf. Accessed: 2010/2010/2013).

Schmidtlein, S., 2004. Coarse-scale substrate mapping using plant functional response types. *Erdkunde* 58, 137–151.

Schreier, H., 1977. Quantitative predictions of chemical soil conditions from multispectral airborne, ground and laboratory measurements. In: *Proc. Of 4th Canadian Symposium on Remote Sensing*, CASI, Ottawa, ON, pp. 107–112.

Schuler, U., Erbe, P., Zarel, M., Rangubpit, W., Surinkum, A., Stahr, K., 2011. A gamma-ray spectrometry approach to field separation of illuviation-type WRB reference soil groups in northern Thailand. *Journal of Plant Nutrition and Soil Science* 174, 536–544.

Schulthess, C.P., 2011. Historical perspective on the tools that helped shape soil chemistry. *Soil Science Society of America Journal* 75, 2009–2036.

Seid, N.M., Yitaferu, B., Kibret, K., Ziadat, F., 2013. Soil-landscape modeling and RS to provide spatial representation of soil attributes for an Ethiopian watershed. *Applied and Environmental Soil Science* 2013.

Selige, T., Böhner, J., Schmidhalter, U., 2006. High resolution topsoil mapping using hyperspectral image and field data in multivariate regression modeling procedures. *Geoderma* 136, 235–244.

Sellitto, V.M., Fernandes, R.B.A., Barrón, V., Colombo, C., 2009. Comparing two different spectroscopic techniques for the characterization of soil iron oxides: Diffuse versus bi-directional reflectance. *Geoderma* 149, 2–9.

Serra, O., 1984. *Fundamentals of Well-Log Interpretation (Vol. 1): The Acquisition of Logging Data: Dev. Pet. Sci., 15A*. Elsevier, Amsterdam. 415p.

Shepherd, K.D., Walsh, M.G., 2002. Development of reflectance spectral libraries for characterization of soil properties. *Soil Science Society of America Journal* 66, 988–998.

Sherman, D.M. Waite, T.D., 1985. Electronic spectra of Fe^{+3} oxides and oxide hydroxides in the near IR to near UV. *American Mineralogy* 70, 1296–1309.

Shruthi, R.B.V., Kerle, N., Jetten, V., Abdellah, L., Machmach, I., 2014. Quantifying temporal changes in gully erosion areas with object oriented analysis. *Catena* 128, 262–277.

Simbahan, G. C., Dobermann, A., 2006. Sampling optimization based on secondary information and its utilization in soil carbon mapping. *Geoderma* 133, 345–362.

Singh, P., Ghoshal, N., 2010. Variation in total biological productivity and soil microbial biomass in rainfed agroecosystems: Impact of application of herbicide and soil amendments. *Agriculture, Ecosystem, Environment* 137, 241–250.

Sobrino, J.A., Franch, B., Mattar, C., Jiménez-Muñoz, J.C., Corbari, C.A., 2012. A method to estimate soil moisture from Airborne Hyperspectral Scanner (AHS) and ASTER data: Application to SEN2FLEX and SEN3EXP campaigns. *Remote Sensing of Environment* 117, 415–428.

Soil Survey Staff. 2013. Rapid Assessment of U.S. Soil Carbon (RaCA) project. USDA-Natural Resources Conservation Service. http://www.nrcs.usda.gov/wps/portal/nrcs/detail/soils/survey/?cid=nrcs142p2_054164 (Accessed December 2013).

Soil Survey Staff. 2014. *Keys to Soil Taxonomy*. USDA-Natural Resources Conservation Soil Survey Staff. 2014.

Soliman, A., Brown, R., Heck, R.J., 2011. Separating near surface thermal inertia signals from a thermal time series by standardized principal component analysis. *International Journal of Applied Earth Observation and Geoinformation* 13, 607–615.

Somers, B., Asner, G.P., Tits, L., Coppin, P., 2011. Endmember variability in spectral mixture analysis: A review. *Remote Sensing of Environment* 115(7), 1603–1616.

Sommer, M., Wehrhan, M., Zipprich, M., Weller, U., zu Castell, W., Ehrich, S., Tandler, B., Selige, T., 2003. Hierarchical data fusion for mapping soil units at field scale. *Geoderma* 112, 179–196.

Song, Y., Li, F., Yang, Z., Ayoko, G. A., Frost, R.L., Ji, J., 2012. Diffuse reflectance spectroscopy for monitoring potentially toxic elements in the agricultural soils of Changjiang River Delta, China. *Applied Clay Science* 64, 75–83.

Soriano-Disla, J.M., Janik, L.J., Viscarra Rossel, R.A., MacDonald, L.M., McLaughlin, M.J., 2014. The performance of visible, near-, and mid-infrared reflectance spectroscopy for prediction of soil physical, chemical and biological properties. *Applied Spectroscopy Reviews* 49, 139–186.

Sousa Junior, J.G., Demattê, J.A.M., Araújo, S.R., 2011. Terrestrial and orbital spectral models for the determination of soil attributes: Potential and costs. *Bragantia* 70(3), 610–621.

Spera, S.A., Cohn, A.S., VanWey, L.K., Mustard, J.F., Rudorff, B.F., Risso, J. Adami, M., 2014. Recent cropping frequency, expansion, and abandonment in Mato Grosso, Brazil had selective land characteristic. *Environmental Resserach Letters* 9, doi:10.1088/1748-9326/9/6/064010.

Srasra, E., Bergaya, F., Fripiat, J.J., 1994. Infrared spectroscopy study of tetrahedral and octahedral substitutions in an interstratified illite-smectite clay. *Clays Clay Min* 42, 237–241.

Staenz, K., Mueller, A., Heiden, U., 2013. Overview of terrestrial imaging spectroscopy missions. In: *Proceedings of the International Geoscience and Remote Sensing Symposium (IAGARSS'13)*, Melbourne, Australia, 4p.

Stark, E., Kes, K.L., Margoshes, M., 1986. Near-Infrared Analysis (NIRA): A technology for quantitative and qualitative analysis. *Appl Spec. Rev* 22, 335–399.

Stenberg, B., 2010. Effects of soil sample pretreatments and standardised rewetting as interacted with sand classes on Vis-NIR predictions of clay and soil organic carbon. *Geoderma* 158(1–2), 15–22.

Stenberg, B., Jonsson, A., Börjesson, T., 2002. Near infrared technology for soil analysis with implications for precision agriculture. In: A. Davies, R. Cho (Eds.), *Near Infrared Spectroscopy: Proceedings of the 10th International Conference*. Kyongju S. Korea, NIR Publications, Chichester, U.K., pp. 279–284.

Stenberg, B., Viscarra Rossel, R.A., 2010. Diffuse reflectance spectroscopy for high-resolution soil sensing. In: Viscarra Rossel, R.A., McBratney, A.B., Minasny, B. (Eds.). *Proximal Soil Sensing. Progress in Soil Science*, Springer, Dordrecht, 446p.

Stenberg, B., Viscarra Rossel, R.A., Mouazen, A.M., Wetterlind, J., 2010. Visible and near infrared spectroscopy in soil science. *Advances in Agronomy* 107, 164–206.

Stevens, A., Miralles, I., van Wesemael, B., 2012. Soil organic carbon predictions by airborne imaging spectroscopy: Comparing cross-validation and validation. *Soil Science Society of American Journal* 76, 2174–2183.

Stevens, A., Nocita, M., Toth, G., Montanarella, L., van Wesemael, B., 2013. Prediction of soil organic carbon at the European scale by visible and near infrared reflectance spectroscopy. *PLoS One* 8(6), e66409.

Stevens, A., Udelhoven, T., Denis, A., Tychon, B., Lioy, R., Hoffmann, L., Wesemael, B. van., 2010. Measuring soil organic carbon in croplands at regional scale using airborne imaging spectroscopy. *Geoderma* 158, 32–45.

Stevens, A., Wesemael, B. van, Vandenschrick, G., Touré, S., Tychon, B., 2006. Detection of carbon stock change in agricultural soils using spectroscopic techniques. *Soil Science Society in American Journal* 70, 844–850.

Stevens, A., Wesemael, B.V., Bartholomeus, H., Rosillon, D., Tychon, B., Ben-Dor, E., 2008. Laboratory, field and airborne spectroscopy for monitoring organic carbon content in agricultural soils. *Geoderma* 144, 395–404.

Stoner, E.R., Baumgardner, M.F., 1981. Characteristic variations in reflectance of surface soils. *Soil Science Society of American Journal* 45, 1161–1165.

Stoner, E.R., Baumgardner, M.F., Biehl, L.L., Robinson, B F., 1980. Atlas of soil reflectance properties. Agricultural Experiment Station, Indiana Research Purdue University, West Lafayette, IN, p. 75.

Sudduth, K.A., Hummel, J.W., 1993. Soil organic matter, CEC, and moisture sensing with a portable NIR spectrophotometer. *Transactions of the ASAE* 36(6), 1571–1582.

Sudduth, K.A., Kitchen, N.R., Sadler, E.J., Drummond, S.T., Myers, D.B., 2010. VNIR spectroscopy estimates of within-field variability in soil properties. *Proximal Soil Sensing*.

Sullivan, D.G., Shaw, J.N., Rickman, D., 2005. IKONOS imagery to estimate surface soil property variability in two Alabama physiographies. *Soil Science Society of America Journal* 69, 1789–1798.

Sumfleth, K., Duttmann, R., 2008. Prediction of soil property distribution in paddy soil landscapes using terrain data and satellite information as indicators. *Ecological Indicators* 8, 485–501.

Summers, D., Lewis, M., Ostendorf, B., Chittleborough, D., 2011. Visible near-infrared reflectance spectroscopy as a predictive indicator of soil properties. *Ecological Indicators* 11(1), 123–131.

Sørensen, L.K., Dalsgaard, S., 2005. Determination of clay and other soil properties by near infrared spectroscopy. *Soil Science Society of America Journal* 69(1), 159–167.

Taylor, J.A., Short, M., McBratney, A.B., Wilson, J., 2010. Comparing the ability of multiple soil sensors to predict soil properties in a scottish potato production system. *Proximal Soil Sensing*.

Taylor, M.J., Smettem, K., Pracilio, G., Verboom, W., 2002. Relationships between soil properties and high-resolution radiometrics, central eastern Wheatbelt, western Australia. *Exploration Geophysics* 33, 95–102.

Tekin, Y., Tumsavas, Z., Mouazen, A.M., 2012. Effect of moisture content on prediction of organic carbon and pH using visible and near-infrared spectroscopy. *Soil Science Society of America Journal* 76, 188–198.

Terhoeven-Urselmans, T., Schmidt, H., Joergensen, R.G., Ludwig, B., 2008. Usefulness of near-infrared spectroscopy to determine biological and chemical soil properties: Importance of sample pre-treatment. *Soil Biology & Biochemistry* 40(5), 1178–1188.

Thomsen, I.K., Bruun, S., Jensen, L.S., Christensen, B.T., 2009. Assessing soil carbon lability by near infrared spectroscopy and NaOCl oxidation. *Soil Biology & Biochemistry* 41(10), 2170–2177.

Tian, Y., Zhang, J., Yao, X., Cao, W., Zhu, Y., 2013. Laboratory assessment of three quantitative methods for estimating the organic matter content of soils in China based on visible/near-infrared reflectance spectra. *Geoderma* 202, 161–170.

Tobehn, C., Kassebom, M., Schmälter, E., Fuchs, J., Del Bello, U., Bianco, P., Battistelli, E., 2002. SPECTRA, Surface Process and Ecosystem Changes Through Response Analysis. In: *Proceedings of the German Aerospace Congress DGLR-2002-185*, Bonn, Germany.

Todorova, M., Atanassova, S., Ilieva, R., 2009. Determination of soil organic carbon using near-infrared spectroscopy. *Agricultural Science and Technology* 1(2), 45–50.

Tucker, C.J., 1979. Red and photographic infrared linear combinations for monitoring vegetation. *Remote Sensing of Environment* 8(2), 127–150.

Ucha, J.M., Botelho, M., Vilas Boas, G.S., Ribeiro, L.P., Santana, P.S., 2002. Uso do radar penetrante no solo (GPR) na investigação dos solos dos Tabuleiros Costeiros do litoral norte do Estado da Bahia [Experimental use of groud-penetrating radar (GPR) to investigate tablelands in the Northern Coast of Bahia, Brazil]. *Brazilian Journal of Soil Science* 26, 373–480.

Uno, Y., Prasher, S.O., Patel, R.M., Strachan, I.B., Pattey, E., Karimi, Y., 2005. Development of field-scale soil organic matter content estimation models in Eastern Canada using airborne hyperspectral imagery. *Canadian Biosystems Engineering* 47, 1.9–1.14.

USGS. 2014. *Earth Observing 1 (EO-1). http://eo1.usgs.gov/.* (Accessed March 9, 2014).

Ustin, S.L., Gamon, J.A., 2010. Remote sensing of plant functional types. *New Phytologist* 186, 795–816.

Vågen, T.G., Winowiecki, L.A., Abegaz, A., Hadgu, K.M., 2013. Landsat-based approaches for mapping of land degradation prevalence and soil functional properties in Ethiopia. *Remote Sensing of Environment* 134, 266–275.

van der Klooster, E., van Egmond, F.M., Sonneveld, M.P.W., 2011. Mapping soil clay contents in Dutch marine districts using gamma-ray spectrometry. *European Journal of Soil Science* 62(5), 743–753.

van der Marel, H.W., Beutelspacher, H., 1975. *Atlas of Infrared Spectroscopy of Clay Minerals and their Admixtures.* Elsevier Scientific Publishing Company, Amsterdam, NH, 396p.

van der Meer, F., de Jong, S., 2003. Spectral mapping methods: Many problems, some solutions. In: *3rd EARSeL Workshop on Imaging Spectroscopy*, Hersching, 13–16 May 2003, pp. 146–162.

van der Meer, F.D., van der Werff, H.M.A., van Ruitenbeek, F.J.A., Hecker, C.A., Bakker, W.H., Noomen, M.F., van der Meijde, M., Carranza, J.M., Smeth, J.B., Woldai, T., 2012.

Multi- and hyperspectral geologic remote sensing: A review. *International Journal of Applied Earth Observation and Geoinformation* 14, 112–128.

Vane, G., Green, R.O., Chrien, T.G., Enmark, H.T., Hansen, E.G., Porter, W.M., 1993. The airborne visible/infrared imaging spectrometer (AVIRIS). *Remote Sensing of Enviroment* 44, 127–143.

van Egmond, F.M., Loonstra, E.H., Limburg, J., 2010. Gamma ray sensor for topsoil mapping: The mole. *Proximal Soil Sensing*.

van Groenigen, J.W., Mutters, C.S., Horwath, W.R., van Kessel, C., 2003. NIR and DRIFT-MIR spectrometry of soils for predicting soil and crop parameters in a flooded field. *Plant and Soil* 250(1), 155–165.

Varmuza, K., Filzmoser, P., 2009. *Introduction to Multivariate Statistical Analysis in Chemometrics*. CRC Press, Boca Raton, FL, 321p.

Vasques, G.M., Demattê, J.A.M., Viscarra Rossel, R., Ramirez Lopez, L., Terra, F.S., 2014. Soil classification using visible/near-infrared diffuse reflectance spectra from multiple depths. *Geoderma* 223–225, 73–78.

Vendrame, P.R.S., Marchao, R.L., Brunet, D., Becquer, T., 2012. The potential of NIR spectroscopy to predict soil texture and mineralogy in Cerrado Latosols. *European Journal of Soil Science* 63(5), 743–753.

Vicente, L. E., & Souza Filho, C. R. 2011. Identification of mineral components in tropical soils using reflectance spectroscopy and advanced spaceborne thermal emission and reflection radiometer (ASTER) data. Remote Sensing of Environment, 115, 1824–1836.

Vink, A.P.A., 1964. Aerial Photographs and the Soil Sciences. In: *Aerial Surveys and Integrated Studies*. Proc Toulouse Conf., Paris, pp. 81–141.

Viscarra Rossel, R.A., 2009. The soil spectroscopy group and the development of a global spectral library. In: 3rd Global Workshop on Digital Soil Mapping, 30 September–3 October 2008. Utah State University, Logan, UT. http://groups.google.com/group/soil-spectroscopy. (Accessed March 3, 2014).

Viscarra Rossel, R.A., Adamchuk, V.I., Sudduth, K.A., Mckenzie, N.J., Lobsey, C., 2011a. Proximal soil sensing: an effective approach for soil measurements in space and time. *Advances in Agronomy* 113, 237–282.

Viscarra Rossel, R.A., Behrens, T., 2010. Using data mining to model and interpret soil diffuse reflectance spectra. *Geoderma* 158(1–2), 46–54.

Viscarra Rossel, R.A., Cattle, S.R., Ortega, A., Fouad, Y., 2009. In situ measurements of soil colour, mineral composition and clay content by vis-NIR spectroscopy. *Geoderma* 150(3/4), 253–266.

Viscarra Rossel, R.A., Chappell, A., De Caritat, P., Mckenzie, N.J., 2011b. On the soil information content of visible–near infrared reflectance spectra. *European of Journal Soil Science* 62(3), 442–453.

Viscarra Rossel, R.A., Jeon, Y.S., Odeh, I.O.A., Mcbratney, A.B., 2008. Using a legacy soil sample to develop a mid-IR spectral library. *Australian Journal of Soil Research* 46, 1–16.

Viscarra Rossel, R.A., Lark, R.M., 2009. Improved analysis and modelling of soil diffuse reflectance spectra using wavelets. *European Journal of Soil Science* 60(3), 453–464.

Viscarra Rossel, R.A., Taylor, H.J., McBratney, A.B., 2007. Multivariate calibration of hyperspectral γ-ray energy spectra for proximal soil sensing. *European Journal of Soil Science* 58(1), 343–353.

Viscarra Rossel, R.A., Walter, C., 2004. Rapid, quantitative and spatial field measurements of soil pH using an ion sensitive field effect transistor. *Geoderma* 119, 9–20.

Viscarra Rossel, R.A., Walvoort, D.J.J., McBratney, A.B., Janik, L.J., Skjemstad, J.O., 2006. Visible, near infrared, mid infrared or combined diffuse reflectance spectroscopy for simultaneous assessment of various soil properties. *Geoderma* 131(1–2), 59–75.

Viscarra Rossel, R.A., Webster, R., 2011. Discrimination of Australian soil horizons and classes from their visible-near infrared spectra. *European Journal of Soil Science* 62, 637–647.

Viscarra Rossel, R.A., Webster, R., 2012. Predicting soil properties from the Australian soil visible-near infrared spectroscopic database. *European Journal of Soil Science* 63(6), 848–860.

Vohland, M., Bossung, C., Fründ, H.C., 2009. A spectroscopic approach to assess trace–heavy metal contents in contaminated floodplain soils via spectrally active soil components. *Journal of Plant Nutrition and Soil Science* 172, 201–209.

Vohland, M., Emmerling, C., 2011. Determination of total soil organic C and hot water-extractable C from VIS-NIR soil reflectance with partial least squares regression and spectral feature selection techniques. *European Journal of Soil Science* 62(4), 598–606.

Vohland, M., Ludwig, M., Thiele-Bruhn, S., Ludwig, B., 2014. Determination of soil properties with visible to near- and mid-infrared spectroscopy: Effects of spectral variable selection. *Geoderma* 223, 88–96.

Vrieling, A., 2006. Satellite remote sensing for water erosion assessment: A review. *Catena* 65, 2–18.

Waiser, T.H., Morgan, C.L.S., Brown, D.J., Hallmark, C.T., 2007. In situ characterization of soil clay content with visible near-infrared diffuse reflectance spectroscopy. *Soil Science Society of America Journal* 71(2), 389–396.

Walkley, A., Black, I.A., 1934. An examination of the Degtjareff method for determining organic carbon in soils: Effect of variations in digestion conditions and of inorganic soil constituents. *Soil Science* 63, 251–263.

Wang, C.-K., Pan, X.-Z., Wang, M., Liu, Y., Li, Y.-L., Xie, X.-L., Zhou, R., Shi, R.-J., 2013a. Prediction of soil organic matter content under moist conditions using VIS-NIR diffuse reflectance spectroscopy. *Soil Science* 178(4), 189–193.

Wang, S.-Q., Li, W.-D., Li, J., Liu, X.-S., 2013b. Prediction of soil texture using FT-NIR spectroscopy and PXRF spectrometry with data fusion. *Soil Science* 178(11), 626–638.

Waruru, B.K., Shepherd, K.D., Ndegwa, G.M., Kamoni, P.T., Sila, A.M., 2014. Rapid estimation of soil engineering properties using diffuse reflectance near infrared spectroscopy. *Biosystems Engineering* 121, 177–185.

Wetterlind, J., Stenberg, B., 2010. Near-infrared spectroscopy for within-field soil characterization: Small local calibrations compared with national libraries spiked with local samples. *European Journal of Soil Science* 61(6), 823–843.

Wetterlind, J., Stenberg, B., Jonsson, A., 2008. Near infrared reflectance spectroscopy compared with soil clay and organic matter content for estimating within-field variation in N uptake in cereals. *Plant and Soil* 302, 317–327.

Wetterlind, J., Stenberg, B., Soderstrom, M., 2010. Increased sample point density in farm soil mapping by local calibration of visible and near infrared prediction models. *Geoderma* 156(3–4), 152–160.

White, J.L., Roth, C.B., 1986. Infrared spectrometry. In: Klute, A. (Ed.). *Methods of Soil Analysis: Part 1–Physical and Mineralogical Methods*. Soil Science Society of America, Madison, WI, pp. 291–330.

White, W.B., 1971. Infrared characterization of water and hidroxyl ion in the basic magnesium carbonate minerals. *American Mineralogy* 56, 46–53.

Whiting, M.L., Li, L., Ustin, S.L., 2004. Predicting water content using Gaussian model on soil spectra. *Remote Sensing of Environment* 89, 535–552.

Wijaya, I.A.S., Shibusawa, S., Sasao, A., Hirako, S., 2001. Soil parameters maps in paddy field using the real time soil spectrophotometer. *Journal of the Japanese Society of Agricultural Machinery* 63(3), 51–58.

Wong, M.T.F., Harper, R.J., 1999. Use of on-ground gamma-ray spectrometry to measure plant-available potassium and other topsoil attributes. *Australian Journal of Soil Research* 37(2), 267–278.

Wong, M.T.F., Oliver, Y.M., Robertson, M.J., 2009. Gamma-radiometric assessment of soil depth across a landscape not measurable using electromagnetic surveys. *Soil Science Society of America Journal* 73(40), 1261–1267.

Woodhouse, I.H., 2005. *Introduction to microwave remote sensing*. Michigan State University, East Lansing, MI, http://www.trfic.msu.edu/products/profcorner_products/Intro_Microwave.pdf (Accessed July 1, 2014).

Xie, H.T., Yang, X.M., Drury, C.F., Yang, J.Y., Zhang, X.D., 2011. Predicting soil organic carbon and total nitrogen using mid- and near-infrared spectra for Brookston clay loam soil in Southwestern Ontario, Canada. *Canadian Journal of Soil Science* 91(1), 53–63.

Yang, H., Kuang, B., Mouazen, A.M., 2012. Quantitative analysis of soil nitrogen and carbon at a farm scale using visible and near infrared spectroscopy coupled with wavelength reduction. *European Journal of Soil Science* 63(3), 410–420.

Zhan, W., Zhou, J., Ju, W., Li, M., Sandholt, I., Voogt, J., Yu, C., 2014. Remotely sensed soil temperatures beneath snow-free skin-surface using thermal observations from tandem polar-orbiting satellites: An analytical three-time-scale model. *Remote Sensing of Environment* 143, 1–14.

Zhang, C., Kovacs, J.M., 2012. The application of small unmanned aerial systems for precision agriculture: A review. *Precision Agriculture* 13(6), 693–712.

Zhang, T., Li, L., Zheng, B. 2013. Estimation of agricultural soil properties with imaging and laboratory spectroscopy. *Journal of Applied Remote Sensing* 7, 24.

Zhao, M.S., Rossiter, D.G., Li, D.C., Zhao, Y.G., Liu, F., Zhang, G.L.Z., 2014. Mapping soil organic matter in low-relief areas based on land surface diurnal temperature difference and a vegetation index. *Ecological Indicators* 39, 120–133.

Zhu, J., Morgan, C.L.S., Norman, J.M., Yue, W., Lowery, B., 2004. Combined mapping of soil properties using a multi-scale spatial model. *Geoderma* 118, 321–334.

Zhu, Y., Weindorf, D.C., Chakraborty, S., Haggard, B., Johnson, S., Bakr, N., 2010. Characterizing surface soil water with field portable diffuse reflectance spectroscopy. *Journal of Hydrology* 391, 133–140.

Zornoza, R., Guerrero, C., Mataix-Solera, J., Scow, K.M., Arcenegui, V., Mataix-Beneyto, J., 2008. Near infrared spectroscopy for determination of various physical, chemical and biochemical properties in Mediterranean soils. *Soil Biology & Biochemistry* 40(7), 1923–1930.

25

Remote Sensing of Soil in the Optical Domains

E. Ben-Dor
Tel Aviv University

José A.M. Demattê
University of São Paulo

Acronyms and Definitions

AHS	Airborne Hypersepctral Sensor
ASD	Analytic Spectral Devise
ASTER	Advanced Spaceborne Thermal Emission and Reflection
Atm	Atmosphere
AVIRIS	Airborne Visible InfraRed Imaging Spectrometer
BRDF	Bidirectional Reflectance Distribution Function
CASI	Compact Airborne Spectrographic Imager
CCD	Charge Couple Device
CD	Change Detection
CNES	Centre National d'Etudes Spatiales
DAIS	Digital Airborne Image Spectrometer
EART-1	LANDSAT-1
ESA	European Space Agency
FWHM	Full Width Half Max
FTIR-PAS	photo acoustic Furrier Transform Infra Red
ERSDAC	Earth Remote Sensing Data Analysis Center
GPS	Ground Position System
GSD	Ground Spatial Dimension
HRS	Hyperspectral Remote Sensing
ISRO	Indian Space Research Organisation
ITC	Faculty of Geo-InformationScience and Earth Observation
IR	Infra Red
LIDAR	Laser Imaging Detection and Ranging
LWIR	Long Wave Infra Red
MCT	Mercury Cadmium Telluride
NASA	National Aeronautics and Space Administration
MODIS	Moderate Resolution Imaging Spectroradiometer
MODTRAN	MODerate resolution atmospheric TRANsmission
MWIR	Mid Wave Infra Red
NIR	Near Infra Red
NOAA	National Oceanic and Atmospheric Administration
NSA	Normalized Spectral Area
OSACA	OSACA program
OM	Organic Matter
PLSR	Partial Least Square
RID	Reflectance Inflection Difference

RS	Remote Sensing
RMSE	Root Mean Square Error
SOC	Soil Organic Matter
SWIR	Short Wave Infra Red
SASI	Shortwave infrared Airborne Spectrographic Imager
SNR	Signal to Noise Ratio
SOC	Soil Organic Carbone
SPOT	Système Pour l'Observation de la Terre
TPH	Total Petroleum Hydrocarbon
TM	Thematic Mapper
UAV	Unmanned Airborne Vehicle
UV	Ultra Violet
VIS	Visible
2D	Two Dimension
3D	Three dimension

25.1 Introduction

In 1987, Mulder published a book entitled *Remote Sensing in Soil Science* (Mulder, 1987) that provided a comprehensive summary and background of all of the soil RS activities known at the time. Mulder's excellent review covered theory, sensors, and applications for soil using RS means. Since 1987, remarkable progress has been made in the soil RS arena, including electro-optic and space technologies, computing power, applied mathematics (for data manipulation), and soil spectral analysis and databases. After almost three decades, these significant advances in soil remote sensing have attracted many young as well as experienced users from the scientific community and from the industry. Many new users have entered the soil RS field and use the technology in different ways, making up specific scientific working groups that have created a unique subcommunity. With better accessibility to this infrastructure (in the laboratory, field, air and space domains), soil spectroscopy has become a very basic and powerful tool from both point and imaging spectral viewpoints. This chapter is thus aimed at covering some key historical stages of this promising technology and reviewing most of the advances in this arena to date. Based on past and present activities, this chapter also highlights the leading directions in the field and provides some thoughts on future possibilities for the remote sensing of soils.

25.2 Soil

25.2.1 Soil System

Soil has been defined as "The upper layer of the earth which may be dug, plowed, specifically, the loose surface material of the earth in which plants grow" (Thompson, 1957). Soil, as an anchoring medium for roots and a supplier of nutrient for crops, is a complex material that is extremely variable in its physical and chemical composition. It is formed from exposed masses of partially weathered rocks and minerals of the earth's crust. Soil formation, or genesis, is strongly dependent on the environmental conditions in both the atmosphere and lithosphere.

Soils are the product of five factors: climate, vegetation, organic matter (OM), topography, and parent materials. The great variability in soils is the result of myriad interactions among these factors and their influence on the formation of different soil profiles (Buol et al., 1973). The general equation describing the final soil body is

$$S = f(P, C, T, O, t) \tag{25.1}$$

where
  S represents the soil
  P is the parent material
  C is the climate
  T is the topography
  O is the OM
  t is related to time (relative age of the soil)

The high variation of soils makes it impossible to solve the aforementioned equation numerically or empirically. Soil serves as an important resource for food production for mankind and carries out other key environmental functions that are essential for human subsistence, such as water storage and redistribution, pollutant filtration, and carbon storage. The soil-forming factors segregate the weathered parent material into diagnostic horizons within the soil profile. In general, the profile, composed of several horizons, typically refers to the upper horizon A (termed the alluvial horizon), the intermediate horizon B (termed illuvial horizon), and the bottom horizon C (the transition to the parent material) (Figure 25.1). The number, nature, and development of

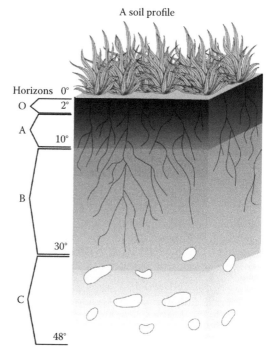

FIGURE 25.1 Illustration of the soil body along the soil profiles generated during the soil formation.

the horizons are products of aforementioned five soil-forming factors, and their relationships play a major role in soil-classification and mapping processes that require a description of the entire soil profile (Soil Survey Stuff, 1975). Pedology is one of the most important and ancient branches of soil science that is strongly related to soil genesis, formation, and mapping. Although soil mapping requires a description of the entire soil profile, observing the soil surface from close or far domains is the ultimate tool before and during any comprehensive field study aimed at generating a soil map (Simonson, 1987). In this respect, remote sensing, which sees mostly the upper part of the earth from afar, plays a major role in soil mapping, mostly for observations of the topmost (Ao) horizon where recently some effort is being taken toward enlarging this capability into the subsurface soil domains as will further be illustrated on in this chapter.

25.2.2 Soil Composition

The soil body is a mixture of three phases: solid, liquid, and gas. A typical soil volume may consist of about 50% pore space with temporally varying proportions of gas and liquid. The solid phase contains organic and inorganic matter in a complicated and generic mixture of primary and secondary minerals, organic components (fresh and decomposed), and salts. The solid phase consists of three main particle-size fractions: sand (2–0.2 mm), silt (0.2–0.002 mm), and clay (<0.002 mm), whose mixture is responsible for soil texture and structure. Soil texture is a function of the proportion of particle-size fractions (Figure 25.2), defining the soil as sandy, silty loam, or clayey. Soil structure is a function of the adhesive forces between the soil particles and describes the aggregation status of the solid particles (block, prism, grains, and others). These two properties play a major role in the soil's behavior and govern important soil characteristics such as drainage, porosity, fertility, and moisture that have importance on plant growth and erosion process. The inorganic portion of the solid phase consists of minerals, which are generally categorized as either primary or secondary. Primary minerals are derived directly from the weathering of parent materials and are formed under much higher temperatures and pressures than those found at the earth's surface. Secondary minerals are formed by geochemical weathering of the primary minerals. An extensive description of minerals in the soil environment is given by Dixon and Weed (1989), and readers who wish to expand their knowledge in this area are referred to this classic comprehensive text. In general, the dominant primary minerals are quartz, feldspar, orthoclase, and plagioclase. Some layer silicate minerals are mica and chlorite, and ferromagnesian silicate minerals include amphibole, peroxide, and olivine. The secondary minerals in the soil body (often termed clay minerals) are aluminosilicates with a layer structure, such as smectite, illite, vermiculite, sepiolite, kaolinite, and gibbsite. The type of clay mineral present in the solid

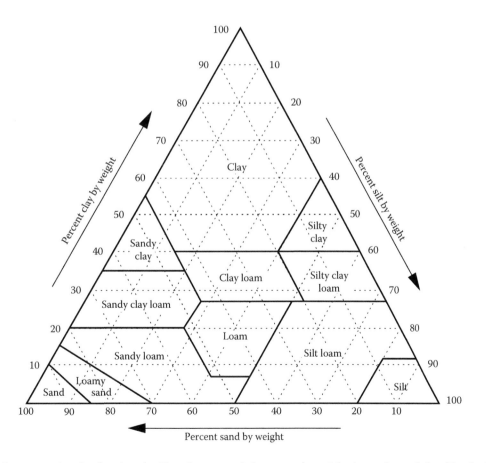

FIGURE 25.2 Soil texture triangle, showing the 12 major textural classes, and particle size scales as defined by the USDA.

phase of the soil is strongly dependent on the weathering stage of the parent material and can be a significant indicator of the environmental conditions under which the soil was formed (Singer, 2007). Other secondary minerals in soils are aluminum and iron (Fe) oxides and hydroxides, carbonates (calcite and dolomite), sulfates (gypsum), and phosphates (apatite). Most of these minerals are relatively insoluble in water and maintain an equilibrium with the water solution. Soluble salts such as halite may also be found in soils, but they are mobile in water and are sometimes transported to the soil matrix by external forces (e.g., wind or artificial irrigation). Clay minerals are most likely found in the fine-sized soil particles (<2 μm) and are characterized by relatively high specific surface areas (50–800 m^2/g). The primary minerals and other non-clay minerals are usually found in both the sand and silt fractions and consist of relatively low specific surface areas (<1 m^2/g). In addition to the inorganic components in the solid phase, organic components are also present. Although the OM content in mineral soils does not exceed 15% (and is usually less), it plays a major role in the soil's chemical and physical behavior (Schnitzer and Khan, 1978). OM is composed of decaying tissues from vegetation and the bodies of micro- and macrofauna. Soil organic matter (SOM) can be found in various stages of degradation, from coarse dead to complex fine components called humus (Stevenson, 1982). Its content is naturally higher in the upper soil horizon, making considerations of OM essential for remote sensing (RS) applications, where only the upper thin layer is detected.

The liquid and gas phases in soils are complementary to the solid phase and occupy about 50% of the soil's total volume. The liquid consists of components of water and dissolved anions and cations in various amounts and positions. The water molecules either fill the entire pore volume in the soil ("saturated"), occupy a portion of its pore volume ("wet"), or are adsorbed on the surfaces particles ("air dry"). Water status can be determined by the pressures needed to extract the water from the soil matrix (often call "matrix tension"). These range from 15 atm (the dry condition; wilting point for vegetation) to 0.3 atm (gravimetric water draining out) and 0 atm (the saturated stage). The composition of the soil's gaseous phase is normally very similar to that of the atmosphere, except for the concentrations of oxygen and carbon dioxide that vary according to the biochemical activity in the root zone due to biogenic respiration processes.

25.3 Remote Sensing

25.3.1 General

Remote sensing is the acquisition of information about an object or phenomenon without physical contact, using electromagnetic radiation (Elachi and Van Zyl, 2006). The term "remote sensing" was first introduced in 1960 by Evelyn L. Pruitt of the U.S. Office of Naval Research (Pruitt 1979). In general, remote sensing can be performed in two ways: passively, where the radiation is not controlled by the sensing system (e.g., the sun's radiation in photography), and actively, where the radiation is part of the sensing system (e.g., microwaves for radar). The RS discipline mostly uses air- and

spaceborne sensors but also portable instruments for close-range measurements in the field. As several types of electromagnetic radiation are available for both active and passive remote sensing (e.g., shortwave infrared [SWIR] and longwave infrared [LWIR], milli- and microwaves), we will limit our discussion to the passive radiation of the sun, which is actually the main source of radiation for remote sensing. This decision is backed up by Mulder et al. (2011) who reviewed RS means and demonstrated that across this region, all of the soil properties that can be remotely sensed use the solar spectral region, termed "optical" region (as the foreoptics are made of glass), see Table 25.1. This region can be separated into three parts: visible (VIS) 0.4–0.7 μm, near infrared (NIR) 0.7–1.0 μm, and SWIR 1.0–2.5 μm. Midwave infrared (MWIR) 2.5–5 nm and LWIR (8–14 μm) are other regions used in passive remote sensing of the earth and are products of the radiation emitted by the earth due to its black-body characteristics. In the MWIR, there is some conjunction between solar radiation and earth radiation due the characteristics of the Planck black-body function of the two bodies (sun and earth). Remote sensing can be performed with one or more instruments covering part or all of the aforementioned spectral regions, the choice being based mainly on the question at hand. RS sensors collect the radiation, disperse it into selected frequencies, measure the intensity at each frequency analogically, and then convert it to digital values for archiving and processing. The RS assembly consists of optics (e.g., lens and slit), a disperser element (e.g., prism or grating), and a detector (e.g., a charge-coupled device [CCD]). For every spectral region, there are specific materials that are sensitive to that radiation's frequencies, lenses, and dispersive elements. For solar radiation, the lens and prism are made of quartz and the detectors of Si (for the VIS–NIR region), InGaAs (for the NIR region), or HgCdTe (MCT) for the SWIR (Levinstein and Mudar, 1975). The final product of any RS sensor is governed by the quality of these components. There are several resolutions in the RS field related to the available technology as follows: spatial resolution, which refers to the size of the pixel (often termed ground sampling distance [GSD]), sensor swath and dimensionality (2D, flat, or 3D, elevation), spectral resolution (number of bands, bandwidths that measure by the full width at half maximum of the band [FWHM]) and sampling intervals, electronic sampling resolution (the number of bits for the stored digital data), radiometric resolution (the accuracy of conversion of the digital data into physical units), temporal resolution (time elapsed between acquisitions for the same area), and viewing resolution (the number of viewing angles capturing the same GSD). These resolutions are interrelated: for example, high spatial resolution requires low spectral resolution to ensure a high signal-to-noise ratio (SNR) (Figure 25.3). Optical RS means are categorized according to the sensor's spectral performance as follows: monospectral (the sensor carries a monochromatic band that is often termed panchromatic), multispectral (the sensor carries between 3 and around 7 semibroad spectral bands), superspectral (the sensor carries around 7–20 semibroad spectral bands), and hyperspectral (the sensor carries over 20 narrow spectral bands). The spectral data can be acquired in image or point domains. In the image domain, the CCD works to capture a response from every pixel on the ground that corresponds to every CCD pitch or divides

TABLE 25.1   A List of Soil Remote Sensing Applications that Can Be Remotely Sensed from Orbit and the Systems Requirements for Each Applications

Soil and Terrain Attributes	Radar		Lidar	Optical	
	Passive	Active		Multispectral	Spectroscopy
Terrain attributes					
Elevation	—	High	High	Medium	—
Slope	—	High	High	Medium	—
Aspect	—	High	High	Medium	—
Dissection	—	Medium-high	Medium-high	Low-medium	—
Landform unit	—	Medium-high	Medium-high	Medium-high	Low-medium
Digital soil mapping	—	High	Medium-high	Medium-high	Medium
Soil type	—	—	—	Medium	High
Soil attributes–proximal sensing					
Mineralogy	—	—	—	—	High
Soil texture	—	High	—	—	High
Iron content	—	—	—	—	Medium–high
Soil organic carbon	—	—	—	—	High
Soil moisture	High	High	—	—	High
Soil salinity	—	—	—	—	Medium
Carbonate content	—	—	—	—	Medium
Nitrogen content	—	—	—	—	High
Lichen	—	—	—	—	Medium–high
Photosynthetic vegetation	—	—	—	—	Medium–high
Nonphotosynthetic	—	—	—	—	Medium–high
Soil attributes-remote sensing					
Mineralogy	—	—	—	Medium	Medium–high
Soil texture	—	Medium	—	Medium	Medium
Iron content	—	—	—	Low	Medium
Soil organic carbon	—	—	—	Low	High
Soil moisture	Medium–high	Medium–high	—	Medium	Low-medium
Soil salinity	—	Medium–high	—	Low–medium	Medium
Carbonate content	—	—	—	Low–medium	Low–medium
Nitrogen content	—	—	—	—	Medium
Lichen	—	—	—	Low–medium	Medium
Photosynthetic vegetation	—	—	—	Medium	Medium–high
Nonphotosynthetic vegetation	—	—	—	Medium	Medium–high
Ellenberg indicator values	—	—	—	—	Low
Plant functional type	—	—	—	Low–medium	Low
Vegetation indices	—	—	—	High	Medium
Land cover	—	Low-medium	—	Medium-high	High
Land degradation	—	Low-medium	—	High	Low-medium

Source: Taken from Mulders, V.L., *Geoderma*, 162, 1, 2011.

[a] Feasibility (1–5) = weighted average scores for the number of studies reported, dataset quality, obtained result and applicability to field surveys. Low = 1; low–medium = 2; medium = 3; Mmdium–high = 4; high = 5.

onto one of the CCD axis into every ground pixel along one axis and its spectral information on the other axis. In the point domain, only one pixel is captured by the sensor, and the detector, which is formed in line-array architecture, uses it to measure the spectral information. Data acquisition in the image domain requires moving the sensor to form the image via line-by-line accumulation, known as "push broom" technology. The systems that collect pixel-by-pixel information and gather it into a final image are termed "whisk broom" sensors. The point (pixel) domain is mostly used

on the ground to measure an object's response by integrating the photons of the point GSD, which is based on the field of view of the foreoptic characteristics.

For soil applications, it is most important to retain high spectral resolution and high SNR in order to extract quantitative information on the soil object (see Section 25.7). High spatial resolution is also important, but this is strongly related to the question being asked. GSD resolutions of 1–30 m are very reasonable values for remote sensing of soils ranging from a selected plot to larger field

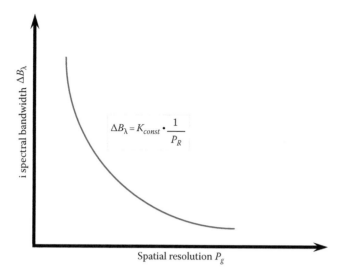

$$\Delta B_\lambda = K_{const} \cdot \frac{1}{P_R}$$

FIGURE 25.3 The relationship between geometric resolution (P_g = GSD) and spectral resolution as measured by the band width ($\Delta B\lambda$).

coverage, respectively. Fortunately, most of the RS means today, from either air or space domains, are characterized by these GSD characteristics. If the high spatial resolution is accompanied by high spectral resolution, then the quality of the RS capability increases, if the SNR is also kept high. This capability provides a better detection limit or interpretation of the data at hand and is sometimes crucial. Unfortunately, high spectral and spatial resolution is not yet possible from orbit; however, when it does become available, the exploitation of remote-sensing means will foster new applications. New technology that steers the sensor's foreoptic to the same pixel during its overpass and provides two- to fourfold better integration time will soon be available, enabling high spectral, spatial, and temporal resolution from orbit. This has already been achieved for the air domain, based on the relatively slow sensor speed that enables collecting a sufficient number of photons, even from small pixels. It should be remembered, however, that sometimes high resolution is simply overkill, and a comprehensive selection of sensor performance for the question at hand is critical.

25.3.1.1 Satellite Sensors

The orbital RS era started with the first Russian mission to space, Sputnik 1, on October 4th, 1957. A handheld manual camera enabled the first large view of the earth's surface, showing its curvature (which later became a major challenge for the RS operator who had to correct for it). However, at that time, space missions were restricted to the military/defense sectors, and the civilian community was not able to gain access to the data. There was no available space-observation technology for civilian applications, and several years passed before the first civilian sensor, ERTS 1 (also known as Landsat 1) from NASA was made available to all (1972). Today, many missions, sensors, platforms, and data are available to the public from several space agencies, as well as from the private sector. The distribution policy ranges from free of charge (NASA) to full-price data availability. The International Institute for Geo-Information Science and Earth Observation

(ITC) has published a list of 309 orbital sensors that shows how remote sensing has advanced since ERTS 1 (http://www.itc.nl/research/products/sensordb/allsensors.aspx). Table 25.2 provides a list of the main sensors in orbit covering the VIS–NIR–SWIR regions that were, are, and will be available to the public and can be used for soil applications. The sensors are divided into multi-, super-, and hyperspectral sensors consisting of different spatial, temporal, and spectral resolutions.

In contrast to the few, highly expensive satellites that were available 25 years ago, providing RS technology for scientists only at the demonstration and research levels, today, remote sensing from space (and the aerial domain) has become a commercial endeavor; it is less expensive, and the data are easier to process and can be used by all. Google Earth is a good example of how satellite data have become freely available, providing a new dimension in our understanding and exploration of the earth's surface from afar. A general list of all NASA satellites (along with available data archives) is provided at http://nssdc.gsfc.nasa.gov/nmc/spacecraftSearch.do. As most of the sensors are still characterized by broadbands and low spectral resolution, effort is being devoted to placing hyperspectral–high-spatial-resolution sensors in space, as they can provide quantitative rather than qualitative information of small areas (see more details in this chapter). An important list in this regard is given by Maksimenka for ESA, providing a summary of super- and hyperspectral orbital sensors in use today and planned for the near future, http://ubuntuone.com/2QsaOhOLPL7602cCOO3IZ5. The high spectral–spatial resolution data can serve as a significant management tool for precision farming activities covering, for example, soil cultivation and formation, contamination, degradation, and fertilization (see more further on in this chapter). Aside from spectral and spatial resolution, for soil applications, temporal resolution is also important. A high capability to provide information in a short time domain is important for properties such as soil moisture, SOM, and clay content variability. The temporal resolution may vary from daily to yearly coverage. In soils, the yearly temporal coverage is important for applications such as land use and coverage, change detection (CD) in large areas, and soil mapping. Figure 25.4 provides a scheme showing the temporal versus spatial resolutions required for soil applications (in comparison to climate and weather forecast applications) along with the currently operational satellites as adopted from Jensen (2011).

25.3.1.2 Airborne Sensors

In addition to the satellite sensors that provide large coverage and high temporal resolution, there are also sensors onboard air platforms that provide better spatial (and recently also better spectral) resolution but with small coverage of a given overpass. Airborne remote sensing began many years ago with aerial photography. The history of aerial photography and remote sensing is provided with some impressive facts and illustrations by Baumann (2010) at http://www.oneonta.edu/faculty/baumanpr/geosat2/RS%20History%20I/RS-History-Part-1.htm; the following text reparses

TABLE 25.2 A List of Past, Present, and Future Orbital Sensors for Soil Remote Sensing and Their Basic Characteristics

Sensor	Year	Mission	Sensor	Channels in VIS-NIR-SWIR	Spectral Range (mm)	Spatial Resolution (m)	Revisit (Day)	Swath (km)	Soil Application (Spectral Base)	Soil Applications (Spatial Base)	Sector
Multi spectral	1972, 1975, 1978	EART (Landsat 1-3)	MSS	4	0.5–1.1	60	18	170 × 185	General, color	Large area General view	NASA
	1995, 1997	IRS-1C-1D-	LISS-3	4	0.52–1.70	23.5/70.5	24	141/148	General/semi quantitative	Medium to large coverage Land use, general view	ISRO
	1988/1991	IRS-1A-1B	LISS-1,2	4	0.45–0.86	36/72.5	22	140	General/semi quantitative	Medium to large coverage Land use, general view	ISRO
	1986, 1990, 1993	SPOT 1-3	HRV	4	0.50–0.89	10/20	2–3	60	General/color	Medium to large coverage Land use (field to landscape)	CNES
	1998	SPOT-4	HRVIR	5	0.50–1.73	10/20	2–3	60	General/color + nonvisible color	Medium to large coverage Land use (field to landscape)	CNES
	1999, 2000	IKONOS 1-2		5	0.45–0.90	1/4	<3	11	General/color	Small coverage, field scale	Digital-Globe
	2000, 2001	QUICK BIRD 1-2		5	0.45–0.90	0.6/2.4	<1–5	20/40	General/color	Small coverage, field scale	Digital-Globe
	2002	SPOT-5		5	0.48–0.71	2.5/10	2–3	60	General/color	Small coverage, field scale	CNES
	2003	Orb View 3		5	0.45–0.90	1/4	<3	8	General/color	Small coverage, field scale	GeoEye
	2012, 2014	SPOT 6-7		5	0.45–0.89	1.5/8	2–3	60	General/color+	Medium to large coverage Land use (field to landscape)	CNES
	1982, 1984	Landsat 4-5	TM	6	0.45–2.35	30	18	170	General, semi quantitative	Medium to large coverage Land use	NASA
	1999	Landsat 7	ETM+	7	0.45–2.35	30/15	18	170	General/semi quantitative	Medium to large coverage Land use	NASA

(Continued)

TABLE 25.2 (*Continued*)   A List of Past, Present, and Future Orbital Sensors for Soil Remote Sensing and Their Basic Characteristics

Sensor	Year	Mission	Sensor	Channels in VIS-NIR-SWIR	Spectral Range (mm)	Spatial Resolution (m)	Revisit (Day)	Swath (km)	Soil Application (Spectral Base)	Soil Applications (Spatial Base)	Sector
Multi spectral	1997	Orb view 2		8	0.4–0.86	1100	1	2800	General	Large coverage	GeoEye
	2009	World View -2		8	0.40–900	0.46/2.08	1.1	16.4	Spectral base	Small coverage, field scale	Digital-Globe
	1997	Orb view 2		8	0.4–0.86	1100					GeoEye
	2001	EO-1	ALI	9	0.43–2.40	10/30	16	37			NASA
	2013	Landsat 8	OLI	9	0.43–2.29	30/15		170 × 183	Semi quantitative	Medium to large coverage Land use (field to landscape)	NASA
Super spectral	1999	Terra	ASTER	10	0.52–2.43	15/30	16	60	Semi quantitative / quantitative	Medium to large coverage Land use (field to landscape)	NASA/METI
	2017	Venus		12	0.4–1.0	5.3	2	27.5	Spectral based (limited)	Small coverage, field scale	ISA-CNES
	2018	Copernicus	Sentinel-2	13	0.4–2.4	10/60	2/3	250	General/Spectral Base (limited)	Medium to small coverage.	ESA
	2015	World view-3	CAVIS	16	0.45–2.365	0.34/4.1	1	13.1	Spectral base	Small coverage, field scale	Digital-Globe
	1999	Terra	MODIS	20	0.459–2.155	250/1000	1–2	2330	General, Semi quantitative	Large coverage General large view	NASA
Hyper spectral	2001	Chris-Proba		60	0.4–0.9	18	7	14	Spectral base (limited)	Medium coverage	ESA
	2019	ALOS-3	HISUI	185	0.4–2.5	30		90	Spectral base	Medium coverage	JAXA
	2017	En Map		228	0.42–2.45	30	4	30	Spectral base	Medium coverage	DLR
	2017	PRISMA		238	0.4–2.5	30	29	30	Spectral base	Medium coverage	ASI
	2001	EO-1	Hyperion	244	0.4–2.45	30	16	7.7	Spectral base	Medium coverage	NASA

Source: Modified from ITC list.

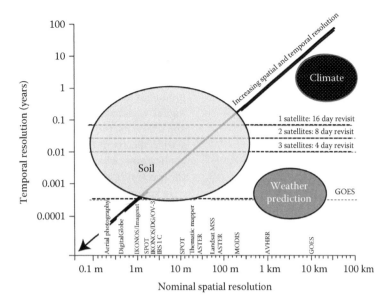

FIGURE 25.4 The spatial and temporal resolutions that are needed for soil applications and the orbital sensors available for that. Also given for comparison is the climate application. (The figure is modified from Jensen, R.J., *Remote Sensing of the Environment: An Earth Resources Perspective*, Pearson Prentice Hall Press, 2011, p. 592.)

the main points from his site. The first aerial photograph was taken in 1858, and its first practical use emerged 50 years later during World War I. The military on both sides of the conflict saw the value of using the airplane for reconnaissance work. Aerial observers, flying in two-seater airplanes with the pilots, performed aerial reconnaissance by sketching maps and verbally conveying the conditions on the ground. Toward the end of that war, the Germans and the British were monitoring the entire front at least twice a day with a total of half a million photographs from the English side and many more from the German side. The war brought major improvements in camera and product quality. As a result, in the late 1920s, the first books on aerial-photograph interpretation were published (a full list is given on Baumann's webpage). World War II brought tremendous growth and recognition to the field of aerial photography, which continues to this day. In 1938, the chief of the German General Staff, General Werner von Fritsch, stated "the nation with the best photoreconnaissance will win the war." Admiral J. F. Turner, commander of the American Amphibious Forces in the Pacific, stated that the importance of "photographic reconnaissance cannot be overemphasized." In parallel to its military development, aerial photography was used for soil applications. The first report on this use of aerial photography was published around 1927 for soil surveys of the United States (Bushnell, 1932), Australia (Prescott and Taylor, 1930), and the USSR (Levenhangst, 1930). Vink (1964) generated a checklist and bullet points on how to use aerial photographs for soil applications and highlighted the necessary methods to interpret the images and translate them into a soil map. Figure 25.5 provides an example of a final product conducted using stereo aerial photographs to interpret "possible soil boundaries" according to Vink's (1964) instructions. It can be seen that this map is still not informative and field work is

strongly needed. On the other hand, Vink's (1963a,b) work was the first to prove that the use of aerial photographs diminishing the number of observation on the development of a soil map, in comparison with a work for the same area and same scale, although without the use of this product. Its invaluable need for soil survey application was later well documented by Goosen (1967). This demonstrates the impact of aerial photography at that time for soil science, that is, in providing possible boundaries based only on the gray tones of the aerial photograph and indicating less need of field observations. Aerial photography is still a common way of mapping the surface, and many mapping agencies have instrumentation to interpret the analog data that today have mostly been converted to digital format. Other sensors, such as multispectral and hyperspectral sensors, are also available from the aerial domain, and over the last decade, they have been widely used for studying particular areas and zooming in on particular fields. Unmanned platforms (unmanned aerial vehicles [UAVs]) are also entering the field of aerial photography based on high-quality light cameras and GPS availability with an easy operational scheme. A recent example is given by d'Oleire et al. (2012), who monitored soil erosion in Morocco using a UAV. Ben-Dor et al. (2013) provide a comprehensive overview of hyperspectral RS (HSR) technology along with a full description of the main available airborne hyperspectral sensors worldwide; a detailed list can be found at http://www.tau.ac.il~rsl/rgomez. Those authors concluded that soon, HSR technology would move from the realm of scientific demonstration to become a practical commercial tool for remote sensing of the earth's surface, including soils. As it is possible to obtain high spectral and spatial resolutions using this technology onboard aircrafts, its contribution for soil applications is straightforward. Herein, we present the reasons why HSR technology is promising for soil monitoring.

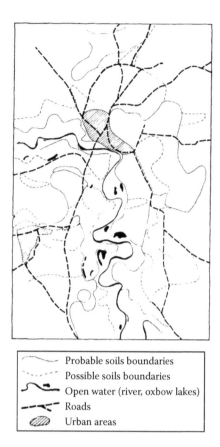

Probable soils boundaries	
Possible soils boundaries	
Open water (river, oxbow lakes)	
Roads	
Urban areas	

FIGURE 25.5 An example of the spatial map that was generated from aerial photography and then was used by soil surveyors to assess the soil entity by opening trenchers and describing the entire soil profiles. (Taken from Vink, A.P.A., *J. Soil Sci.*, 14, 88, 1963a; Vink, A.P.A., Planning of soil surveys in land development. International Institute for Land Reclamation and Improvement, publication 10, U.D.C. 631.47: 528.77, the Netherlands, 1963b, 53p.)

25.3.1.3 Sensor per Mission

Selection of a sensor that best suits a particular mission, and more specifically a particular soil application, must be tailored to the question at hand. Such questions might involve large or small areas, quantitative or qualitative information, CD or current-state position of the soil, as well as cost. Many of the satellite data available today are free of charge (mostly from governmental bodies such as NASA, NOAA, and ESA), whereas others can be costly as DigitalGlobe (https://www.digitalglobe.com/) and Imagesat (http://www.imagesatintl.com). Airborne data are also costly but may provide more information on a small scale based on its high-spatial-resolution domain and solve local problems by capturing high spectral/spatial information. Tactical considerations for a specific mission are important, and taking into account the processing time and spectral sensitivity of the sensor is crucial.

As across the VIS–NIR region, both radiation and detectors are at their maximum sensitivity, this region is well covered by most of the current sensors. In the SWIR regions, however, detector sensitivity and produced radiation intensity are poor, making this region more problematic and hence less available.

The evolution of remote sensing shows that the VIS sensors were the first to be used. Instead of a digital detector, a sensitive emulsion (film) for VIS radiation was used to capture visual information, allowing only limited interpretation of the RS data. Today, all sensors are equipped with a CCD assembly and the raw product is delivered in digital format. The computing power and digital visualization provide an innovative way of visualizing and analyzing the data using algorithms that can be shared with a wide spectrum of users. This is a real breakthrough for soil applications, where most of the information is hidden beneath the spectral responses of the solid and water phases.

In summary, there are different remote-sensor options related to the source of energy (passive or active sensors), the type of platform (ground, air, or space), the spectral region (optical, IR, microwave), the platform trajectory, the number and width of the spectral bands (e.g., panchromatic, multispectral, superspectral, hyperspectral), the spatial resolution (high, medium, low), the spatial coverage (point or image view), the temporal resolution (e.g., hourly, daily, or weekly revisiting frequencies), the radiometric resolution (e.g., 8, 12, 16 bits), and the collection system (push broom or whisk broom). The modern sensors provide information on spatial and spectral domains from aboveground elevations of a few meters (field sensors) to 800 km (orbital sensors), with spatial resolutions varying from a few centimeters to tens of meters and temporal coverage of minutes to days.

25.4 Remote Sensing of Soils

Remote sensing of soils refers to the product that can be obtained on soils from remote sensors. It should cover the information extracted and interpreted on a soil entity from afar. Although problems still exist in correctly remotely sensing the soil body, the advances made in both the technology and the know-how over the years are remarkable. In 1987, Mulder wrote that "it is still impossible to extrapolate the remote sensing information for the entire profile." Although for the most part, optical RS means cannot detect the entire soil body that extends from the surface down to the parent material, today, some innovative ideas can be used to probe below the soil surface. Most of the RS soil products in the optical domain characterize the surface, because the sun's radiation cannot penetrate more than 50 μm of the soil's surface (Ben-Dor et al., 1999). Accordingly, only the upper (mostly Ao) horizon can be sensed, with limited ability to extrapolate the information to the soil's deeper horizons. If the parent material is known (mostly from metadata such as geological maps), the Ao information can provide inferences about the soil body if the soil in question is a direct product of the lengthy rock-weathering process. However, as the soil surface is frequently characterized by a short time process, the upper soil horizon (Ao) can provide valuable information on the environmental (soil) processes such as soil degradation, dust accumulation, contamination, salinity, OM, crust formation, soil-surface moisture, soil runoff, and water infiltration (see further on). This information can be valuable for the farmer, and thus the RS product can assist decision makers in selecting appropriate action. For these applications, high-spectral-resolution sensors are needed. Obtaining spectral information from afar is not simple and requires high performance from both operational and infrastructural standpoints. Nonetheless, if the spectral-reflectance information can be captured from air and space domains in a precise manner, then information on the status of the soil surface becomes invaluable and promising. This makes reflectance spectroscopy a significant factor in the remote sensing of soils. Satellites can provide overviews of larger areas than airborne sensors and have a better temporal resolution. However, high spectral and spatial resolution from orbit is still problematic, whereas data storage and computing power are no longer limiting, thanks to cloud technology and computing power that can be shared by many computers. Despite some limitations, remote sensing of soils can assist the development of digital soil mapping by providing a good basis for soil surveyors and enabling the monitoring of surface processes in a unique way. New computing power, user-friendly software, and open-source algorithms combined with other sensors in the air and on the ground will enable (and in fact are enabling) new RS dimensions. It is obvious that soil, as one of the main covers of the earth's surface, is an ideal study target for RS technology, especially from satellites. Ge et al. (2011) provide a brief description of the history of remote sensing of soils from satellite sensors. They pointed out that in the late 1960s and early 1970s, soil scientists began to understand the capabilities of Landsat 1 data, from which differences in surface soils could be delineated (Kristof, 1971). Soon after, Kristof and Zachary (1974) reported partial success in delineating soil series in an Alfisol–Mollisol region through digital analysis of aerial multispectral data. It was only in 1972, when Landsat 1 data became available to the public, that the era of remote sensing of the soil began. This was mainly based on studies that realized that soil spectral information is a key factor in exploiting soil (optical and chemical) data obtained from afar. The first investigations of soil spectroscopy were published in around 1970 (e.g., Condit 1970, 1972). Those studies showed that the spectral information can be used as a tool to discriminate between soil families. Along these lines, it was shown that a dataset consisting of 160 different soil spectra could be categorized spectrally into three groups. This actually opened up the era of soil spectroscopy as the basis for soil RS disciplines. This topic is further elaborated upon later in this chapter.

25.5 Soil Reflectance Spectroscopy

25.5.1 Definition

The soil reflectance spectrum (ρ) is a collection of values obtained at every spectral band (λ) from the ratio of radiance (E) and irradiance (L) fluxes across most of the spectral region of the solar emittance function:

$$\rho(\lambda) = \frac{E(\lambda)}{L(\lambda)} \qquad (25.2)$$

The reflectance values are traditionally described, from a practical standpoint, by a relative ratio against a perfect reflector spectrum measured at the same geometry and position as the soils (Palmer, 1982; Baumgardner et al., 1985; Jackson et al., 1987).

To illuminate the value of the soil spectrum, this section will provide some historical notes and then a comprehensive theoretical background on the interaction of electromagnetic radiation with the soil matrix described by ρ.

25.5.2 Historical Notes

When spectrometers became available in around 1960, studies were conducted to elucidate the spectral responses of soils and pure minerals. From 1970 to 1982, Hunt and Salisbury conducted a comprehensive study on the spectral characteristics of pure minerals, which was published in (1970, 1971a–d, 1976, 1980). In 1965, a first attempt to demonstrate the quantitative capabilities of soil spectral information showed a correlation between moisture content and spectral response at several wavebands (Bowers and Hanks, 1965). Later, a systematic investigation of the relationship between soil spectral information and soil properties was conducted by Condit (1970) and then by Montgomery and Baumgardner (1974), who later systematically studied the soil reflectance of American soils (Stoner and Baumgardner, 1981). These latter authors also published the first "soil reflectance atlas" (Stoner et al., 1980); they demonstrated the importance of soil spectroscopy and for the first time, the spectral grouping of mostly U.S. soils into five major soil types (one of them, *type 5*, was from Brazil). Their soil spectral library was initially constructed with complete soil profiles (horizons) and its classification, which quickly became a classical tool for soil scientists and a fundamental reference source for future studies. The emerging activity in soil spectroscopy over the past decade, resulting in a vast accumulation of knowledge, led workers to complete Baumgardner's work on a global basis by establishing more libraries worldwide (e.g., Viscarra Rossel, 2009). In 1991, the first portable spectrometer hit the market (ASD, http://www.portableas.com/index.php/manufacturers/asd/), ringing in the era of portable and facile spectral sensing of soils (as well as other earth materials) in both field and laboratory domains (Goetz, 2009). This drove many scientists to the field of soil spectroscopy, and today, there is a strong scientific community in this field of interest from many aspects and contributing their know-how to the practical utilization of this promising tool. The significant contribution of soil reflectance lies in the possibility of extracting quantitative soil information from the spectrum by establishing a proximal-sensing approach, which historically started (in soils) around 1986; at that time, Dalal and Henry (1986) were the first to adopt the spectral data-mining technique based on Ben-Gera and Norris's (1968) approach developed a decade prior for wheat grains. From 1994 onward, after Ben-Dor and Banin (1994) showed the potential of the proxy technique to extract several soil properties (among them even "featureless"), soil spectral studies advanced rapidly. In 1983, the first hyperspectral airborne sensor arrived at NASA (Goetz, 2009), but it took years until this technology was adopted for a proximal-sensing approach in soils (Ben-Dor et al., 2002). Today, with some of the obstacles to extracting reflectance information from air and space domains having been partially overcome, this seems to be the direction that will enable exploitation of soil spectral information for the needs of mankind, such as soil monitoring, mapping, and cultivation. The following sections provide a theoretical background on soil reflectance to understand its capacity in the RS field, and then its applications in soil science will be demonstrated.

25.5.3 Radiation Interactions with a Volume of Soil

The process of radiation scattering by soils results from a multitude of quantum-mechanical interactions between the enormous number and variety of atoms, molecules, and crystals in a macroscopic volume of soil. In contrast to certain absorption features, most characteristics of the scattered radiation are not attributable to a specific quantum-mechanical interaction. The effects of a particular mechanism often become obscured by the composite effect of all of the interactions. The difficulty in accounting for the effects of a large number of complex quantum-mechanical interactions often leads to the use of non-quantum-mechanical models of electromagnetic radiation. Physicists frequently resort to the classical wave theory or even to geometrical optics to elucidate the effects of a macroscopic volume of matter on radiation.

25.5.3.1 Refractive Indices

When light passes through a medium, some part of it will always be absorbed. This can be conveniently taken into account by defining a complex index of refraction:

$$\tilde{n} = n + i\kappa \qquad (25.3)$$

Here, the real part of the refractive index n indicates the phase speed, while the imaginary part κ indicates the amount of absorption loss when the electromagnetic wave propagates through the material (i is the square root of -1). That κ corresponds to absorption can be seen by inserting this refractive index into the expression for the electrical field of a plane electromagnetic wave traveling in the z-direction. The wave number is related to the refractive index by

$$k = \frac{2\pi n}{\lambda_0} \qquad (25.4)$$

where λ_0 is the vacuum wavelength. With complex wave number and refractive index $n + i\kappa$, this can be inserted into the plane wave expression as

$$E(z,t) = \mathrm{Re}\left(E_0 e^{i(\tilde{k}z - \omega t)} \right) = \mathrm{Re}\left(E_0 e^{i(2\pi(n + i\kappa)z/\lambda_0 - \omega t)} \right)$$

$$= e^{-2\pi\kappa z/\lambda_0} \, \mathrm{Re}\left(E_0 e^{i(kz - \omega t)} \right) \qquad (25.5)$$

Here, we see that gives an exponential decay, as expected from the Beer–Lambert law. Since intensity is proportional to the square of the electrical field, the absorption coefficient becomes $4\pi\kappa/\lambda_0$.

κ is often called the extinction coefficient in physics, although this has a different definition in chemistry. Both n and κ are dependent on the frequency. Under most circumstances, $\kappa > 0$ (light is absorbed) or $\kappa = 0$ (light travels forever without loss). In special situations, especially in the gain medium of lasers, it is also possible for $\kappa < 0$, corresponding to an amplification of the light. In the soil matrix, the radiation travels through a thin layer of particles, is reflected back to the sensor, and provides a spectrum whose shape and nature are affected by the aforementioned process (consisting of both the real and imaginary part of the complex refractive index). Any substance in the soil matrix that affects the indices given earlier is termed a "chromophore." Knowing the chromophores' behavior can shed light on the physical and chemical constituents of the soil matrix under study. This information can either be derived by the naked eye's "color vision" (across the VIS region) or by careful analysis of the spectral responses (across the VIS–NIR–SWIR regions) according to the aforementioned theory. In general, due to its complexity, a given soil sample consists of a variety of chromophores, which vary with environmental conditions. In many cases, the spectral signals related to a given chromophore can overlap with other chromophores' signals, thereby hindering the assessment of the direct effect of the chromophore in question. Accordingly, it is important to understand the chromophores' physical processes as well as their origin and nature. Another point to mention is that in soil, there are many cases in which relationships between chromophoric and nonchromophoric properties exist.

We define the factors affecting soil spectra as "physical" if the real part of the refractive index is associated and "chemical" if the spectral changes are associated with the imaginary part of the refractive index. This terminology is adopted from the weathering processes in soil where "physical" weathering refers to "size" changes in the soil matrix with no chemical alteration and "chemical" weathering refers to chemical "alteration" of the soil materials. Figure 25.6 provides the possible light interactions within the thin layer of the soil surface.

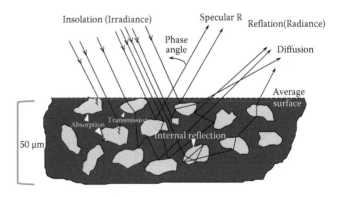

FIGURE 25.6 An illustration showing the interaction of light with the soil particles at the soil surface.

25.5.3.2 Chemical Chromophores

25.5.3.2.1 Physical Mechanism

Chemical chromophores are those materials that absorb incident radiation at discrete energy levels. The absorption process usually appears on a reflectance spectrum at positions attributed to specific chemical groups in various structural configurations. The interaction between radiation and matter occurs at the atomic and molecular levels. Electromagnetic radiation can be emitted or absorbed when an atom or molecule transitions between energy states. The energy of an emitted or absorbed photon equals the difference between the energy levels. Furthermore, the energy-level transitions must be accompanied by either a redistribution of the electric charge carried by electrons and nucleic protons or a reorientation of nuclear or electronic spins before a photon is emitted or absorbed (Hunt, 1982). A comprehensive description of the physical mechanisms describing the interactions of electromagnetic radiation with diverse minerals and rocks is provided by Clark (1999). This section focuses on the most common chromophores in the soil environment and their relationship with electromagnetic radiation across the VIS–NIR–SWIR spectral region.

25.5.3.2.2 Vibration Processes in the SWIR Region

The absorption or emission of shortwave radiation usually results from energy-level transitions accompanied by charge redistributions involving either the motion of atomic nuclei or the configuration of electrons in atomic and molecular structures. A molecule possesses several modes of vibration depending on the number and arrangement of its atoms. A molecule with N atoms may have $3N$-5 vibrational modes if the atoms are arranged linearly or $3N$-6 vibrational modes if the bonding is nonlinear (Castellan, 1983). The absorbed (or emitted) frequencies are called overtone bands when a vibrational mode transitions from one state to another that is more than one energy level above (or below) the original state. Combination bands refer to frequencies associated with transitions of more than one vibrational mode. These combined transitions occur when the energy of an absorbed photon is split between more than one mode (Castellan, 1983). Vibrational transitions corresponding to the fundamental bands are generally more likely to occur than transitions corresponding to combination and overtone bands (Castellan, 1983) and are usually stronger than the overtone transition. The fundamental transition of soil chromophores occurs mostly in the IR region (>2.5 μm), whereas the overtones occur in the SWIR region, mostly above 1 μm. The two basic vibrations that correspond to the atoms' motions in the chemical bonds' molecular processes in soil minerals are stretching (v) and bending (δ).

Aside from the overtone transitions, combination modes are also excited when the quantum mechanism enables combining different vibrational processes of the same bonds, such as stretching and bending. Combination and overtone bands associated with hydroxyl group (OH) vibrations are, for example, often apparent in soil reflectance spectra. OH groups are found on many soil minerals and the exact wavelength location of the associated bands depends upon which OH-bearing minerals are present in the soil. The one fundamental OH band due to oxygen–hydrogen stretching is found near 2.8 μm and the first overtone band due to this stretch is located near 1.4 μm. Absorption at this overtone band is the most common feature in the NIR spectra of terrestrial materials (Hunt, 1982). The hydrogen–oxygen stretch can also be coupled with other vibrations in the molecular structure of the soil minerals to create combination band features. Bending at magnesium–OH bonds coupled with stretching results in a combination band near 2.300 μm, and bending at aluminum-OH bonds coupled with stretching produces a combination band near 2.200 μm (Hunt, 1982).

25.5.3.2.3 Electron Processes in the VIS–NIR Region

At higher electromagnetic radiation energy (UV and VIS–NIR), the spectral response is associated with electron transitions. The locations of these bands are due to the relatively large gaps between electron energy states. The principles of quantum mechanics dictate that each electron of an atom, ion, or molecule can exist in only certain states corresponding to discrete energy levels. There are four possible electronic possibilities, termed crystal field, charge transfer, color center, and semiconductor. In soil and soil minerals, the first two are dominant. *A crystal field* can often be inferred from reflectance spectra of minerals containing transition metals. The allowable energy states for the unpaired transition-metal electrons and the gaps between states are determined primarily by the valence state of the ion and the coordination number and symmetry of the crystal site in which the ion occurs. The energy states are also influenced by the type of ligand surrounding the ion, by the interatomic metal–ligand distance, and by site distortion (Hunt, 1982). This latter author displayed the reflectance spectra of six minerals containing ferrous iron (Fe^{2+}) to demonstrate the effects of coordination number, site symmetry, and site distortion. *Charge transfer* is a mechanism involving the electrons of a specific ionic bond between adjacent ions. Charge transfer occurs when an electron migrates between the adjacent ions with a corresponding change in energy state. The usually prominent decrease in reflectance in the blue region of soil reflectance spectra is due to the charge transfer between iron and oxygen in Fe oxides (Hunt, 1982). The dark color of some minerals, such as magnetite (Fe_3O_4), which contain both ferrous (Fe^{3+}) iron, is due to charge transfer between these two ions (Nassau, 1980, 1983). *Color centers* refer to unpaired electrons and paired valence electrons that play a role in the interaction of shortwave radiation with soils. Molecular-orbital theory describes the distribution of paired-electron charges among the atoms of a molecule or crystal. In some cases, an electron pair may remain associated with a specific bond between two adjacent atoms or ions. In other cases, the charge may be distributed over several atoms or even throughout a crystal structure. *Semiconductors* interact like solids in which the allowable energy states of the bonding electrons are divided into two broadbands. The lower energy band is called the valence band and the higher-energy band, which contains all the excited states, is called the conduction band. Between these two bands is a region called the forbidden gap within which no electrons are allowed. To be absorbed by a semiconductor, a photon must carry at least enough energy to elevate an electron from the upper level of the valence band to the lower level of the conduction band (Nassau, 1980). The reflectance spectrum of a semiconductor is distinguished by an intense absorption edge, which marks the width of the forbidden gap (Hunt, 1980).

25.5.3.3 Physical Chromophores

In addition to chemical chromophores, the reflectance of light from the soil surface is dependent upon numerous physical processes, most of them related to the real part of the refractive index. Reflection, or scattering, is clearly described by Fresnel's equation and depends upon the angle of incidence radiation and upon the index of refraction of the material in question. In general, physical factors are those parameters that affect soil spectra in terms of Fresnel's equation (the real part of the refractive index), but which do not cause changes in the position of the specific chemical absorption. These parameters include particle size, particle's geometry, hydration stage, viewing angle, radiation intensity, incident angle, and azimuth angle of the source. Changes in these parameters are most likely to affect the shape of the spectral curve through changes in baseline height and absorption-feature intensities. In the laboratory, measurement conditions can be held constant; in the field, several of these parameters are unknown, which may complicate accurate assessments of their effect on soil spectra. Many studies, covering a wide range of materials, have shown that differences in particle size alter the shape of soil spectra (e.g., powdered material) (Hunt and Salisbury, 1970; Pieters, 1983; Baumgardner et al., 1985). Specifically, Hunt and Salisbury (1970) quantified particle-size difference effects of about 5% in absolute reflectance and noted that these changes occurred without altering the position of diagnostic spectral features. Under field conditions, aggregate-size rather than particle-size distribution may be more important in altering soil spectra (Orlov, 1966; Baumgardner et al., 1985). In the field, aggregate size may change over short periods due to tillage, soil erosion, aeolian accumulation, or physical crust formation (e.g., Jackson et al., 1990). Basically, the aggregate size, or more likely roughness, plays a major role in the shape of field and airborne soil spectra (e.g., Cierniewski, 1987, 1989). Escadafal and Hute (1991) showed strong anisotropic reflectance properties in five soils with rough surfaces. Cierniewski (1987) developed a model to account for soil roughness based on the soil-reflectance parameter, illumination properties, and viewing geometry for both forward and backward slopes. The model showed that the shading coefficient of the soil surface decreases with decreasing soil roughness. For soils on forward slopes of more than 20°, the shadowing coefficient also decreased with increasing solar altitude for the full interval of sun altitudes ranging from 0° to 90°. The model indicated that the relationship for soil slopes with surface roughness lower than 0.5 might be reversed for a specified range of solar altitudes. Using empirical observations of smoothed soil surfaces, Cierniewski (1987) showed that the model closely agrees with field observations. An excellent brief summary on multiple- and single-scattering models for soil particles with respect to the roughness effect is given by Irons et al. (1989).

25.5.3.4 Chromophores in Soils

This section focuses on the most common chromophores in the soil environment and their relationship with electromagnetic radiation across the VIS–NIR–SWIR spectral region. All features in the VIS–NIR–SWIR spectral regions have a clearly identifiable physical basis. In soils, three major chemical chromophores can be roughly categorized as follows: minerals (mostly clay and Fe oxides), OM (living and decomposing), and water (solid, liquid, and gas phases). Physical chromophores in soils are related to particle size and measurement geometry.

25.5.3.4.1 Soil Minerals

As already discussed previously, Hunt and Salisbury (1970–1980) have studied the details of the spectral behavior of many minerals on earth. Some minerals reported in their comprehensive review are encountered in the soil environment and will be discussed here, with additional information related to the soil medium.

Clay minerals: Of all clay mineral elements, only the OH group is spectrally active in the VIS–NIR–SWIR region. This group can be found as part of either the mineral structure (mostly in the octahedral position, which is termed lattice water) or the thin water molecule that is directly or indirectly attached to the mineral surface (termed adsorbed water). Three major spectral regions are active for clay minerals in general and for smectite minerals in particular: around 1.3–1.4 μm, 1.8–1.9 μm, and 2.2–2.5 μm. For Ca–montmorillonite (SCa-2)—a common clay mineral in the soil environment—the lattice OH features are found at 1.410 μm (assigned $2vOH$, where vOH symbolizes the stretching vibration around 3630 cm^{-1}) and at 2.206 μm (assigned $vOH + \delta OH$ where δOH symbolizes the bending vibration at around 915 cm^{-1}), whereas OH features of free water are found at 1.456 μm (assigned $vW + 2\delta W$, where vW symbolizes the stretching vibration at around 3420 cm^{-1} and δW the bending vibration at around 1635 cm^{-1}), 1.910 μm (assigned $v'W + \delta W$ where $v'W$ symbolizes the high-frequency stretching vibration at around 3630 cm^{-1}), and 1.978 μm (assigned for $vW + \delta W$). Note that these assigned positions can change slightly from one smectite to the next, depending upon their chemical composition and surface activity. The spectra of three smectite end members are given in Figure 25.7, as follows: montmorillonite (dioctahedral, aluminous), nontronite (dioctahedral, ferruginous), and hectorite (trioctahedral, manganese). The OH absorption feature of the $vOH + \delta OH$ in combination mode at around 2.2 μm is slightly but significantly shifted for each end member. In highly enriched Al smectite (montmorillonite), the Al–OH bond is spectrally active at 2.16–2.17 μm. In highly enriched iron smectite (nontronite), the Fe–OH bond is active at 2.21–2.24 μm, and in highly enriched magnesium smectite (hectorite), the Mg–OH bond is spectrally active at 2.3 μm. Based on these wavelengths, Ben-Dor and

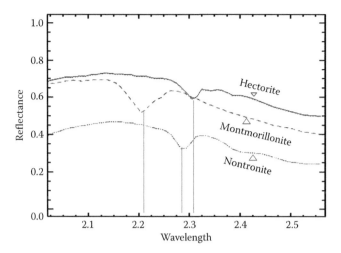

TABLE 25.3   Band Positions and Assignments for Calcite Mineral

	Band Position (µm)	Band Assignment
Hunt and Salisbury (1971)		
	2.55	$v_1 + v_3$
	2.35	$3v_3$
	2.16	$v_1 + 2v_3 + v_4$ or $3v_1 + 2v_4$
	2.00	$2v_1 + 2v_3$
	1.90	$v_1 + 3v_3$
Haxter (1958)	2.55	$2v_3 + 270 + 2 \times 416$
	2.37	$2v_3 + 270 + 3 \times 416$
Schroeder et al. (1962)	2.54	
Matossi (1928)	2.533	$2v_3 + v_1$
	2.500	$2v_3 + v_1$
	2.330	$3v_3$
	2.300	$3v_3$

Source: Gaffey 1986.

FIGURE 25.7   Reflectance spectra of three pure smectite endmembers in the SWIR region (nontronite = Fe-smectite; hectorite = Mg-smectite; montmorillonite Al-smectite). Note the different position of combination modes of ($v_{OH} + \delta_{OH}$) around 2.2 and 2.3 µm.

Banin (1990a) were able to find a significant correlation between the absorbance values derived from the reflectance spectra and the total content of Al_2O_3, MgO, and Fe_2O_3. Except for a significant lattice OH absorption feature at around 2.2 µm in smectite, invaluable information on OH in free water molecules can be measured at around 1.4 and 1.9 µm. Because smectite minerals contribute to the soil's relatively high specific surface area, which is covered by free and hydrated water molecules, these absorption features can be significant indicators of soil water content. Kaolinite and illite minerals are also spectrally active in the SWIR region as they both consist of octahedral OH sheets. In the case of kaolinite, a 1:1 mineral (one octahedral and one tetrahedral), the fraction of the OH group is higher than in 1:2 minerals (one octahedral and two tetrahedral), and therefore the lattice OH signals at around 1.4 and 2.2 µm are relatively strong, whereas the signal at 1.9 µm is very weak (because of relatively low surface area and adsorbed water molecules). In the case of gibbsite, an octahedral aluminum structure (1:0), the signal at 1.4 µm is even stronger, but that at 2.2 µm is shifted significantly to the IR region relative to kaolinite and presents an important diagnostic band at 2.265 µm. It should be noted that under relatively high SNR conditions, a second overtone feature of the structural OH ($3v_{OH}$) can be observed at around 0.95 µm in OH-layer-bearing minerals as well (Goetz et al., 1991). The affinity of water molecules to clay mineral surfaces under the same atmospheric conditions is correlated to the minerals' specific surface area. For the aforementioned minerals, the specific surface area follows the order smectite > vermiculite > illite > kaolinite > chlorite > gibbsite, and these usually provide a similar spectral sequence for the water-absorption feature near 1.9 µm (area and intensity). As smectite and kaolinite are often found in soils, they can also appear in a mixed-layer formation with spectral overlaps (Kruse et al. 1991).

Carbonates: Carbonates, particularly calcite and dolomite, are found in soils that are formed from carbonic parent materials or

in a chemical environment that permits calcite and dolomite precipitation. Carbonates, and especially those of fine particle size, play a major role in many of the soil chemical processes that are most likely to occur in the root zone. The C–O bond, part of the CO_3 radical in carbonate, is the spectrally active chromophore. Hunt and Salisbury (1970, 1971d) pointed out five major overtones and combination modes that describe the C–O bond in the SWIR region. Table 25.3 provides the band positions (calculated and observed from Gaffey, 1986) and their spectral assignments. In this table, $v1$ accounts for the symmetric C–O stretching mode, $v2$ for the out-of-plane bending mode, $v3$ for the antisymmetric stretching mode, and $v4$ for the in-plane bending mode in the IR region. Gaffey (1986) added two additional significant bands centered at 2.23–2.27 µm (moderate) and at 1.75–1.80 µm (very weak), and van der Meer (1995) summarized the seven possible calcite and dolomite absorption features with their spectral widths. It is evident that significant differences occur between the two minerals. This enabled Kruse et al. (1990), Ben-Dor and Kruse (1995), and others to differentiate between calcite and dolomite formations using airborne spectrometer data with 10 nm bandwidths. In addition to the seven major C–O bands, Gaffey and Reed (1987) were able to detect copper impurities in calcite minerals, as indicated by the broadband between 0.903 and 0.979 µm. However, such impurities are difficult to detect in soils because overlap with other strong chromophores may occur in this region. Gaffey (1985) showed that iron impurities in dolomite shift the carbonate's absorption bands toward longer wavelengths, whereas magnesium in calcite shifts the bands toward shorter wavelengths. As carbonates in soils are likely to be impure, it is only reasonable to expect that the carbonates' absorption-feature positions will differ slightly from one soil to the other.

Organic matter: The wide spectral range found by different workers to assess SOM content suggests that OM is an important chromophore across the entire spectral region. Figure 25.8 shows the reflectance spectra of coarse OM (in the NIR–SWIR region) isolated from an Alfisol and the humus compounds extracted

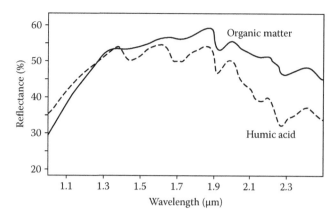

FIGURE 25.8 The reflectance spectrum of soil organic matter and the humic acid extracted from it. (Taken from Ben-Dor, E. et al., Soil spectroscopy, in *Manual of Remote Sensing*, 3rd edn., A. Rencz (ed.), John Wiley & Sons Inc., New York, 1999, pp. 111–189.)

from it. There are numerous absorption features that relate to the high number of functional groups in OM. These can all be spectrally explained by combination and vibration modes of organic functional groups (Chen and Inbar, 1994). Ben-Dor et al. (1997) referred the absorption peaks across the SWIR region to overtone and combination modes of nitrogen and carbon groups and across the VIS regions to charge transfer in the remaining chlorophyll, as well as to other smeared electronic processes that affect the entire VIS spectral region, flatting the reflectance spectrum accordingly. Dry litter and fresh OM show many absorption features across the SWIR region related to many chemical chromophores, such as starch, cellulose, lignin, and water.

Water: The various forms of water in soils are all active in the VIS–NIR–SWIR region (based on the vibration activity of the OH group) and can be classified into three major categories: (1) *hydration water*, which is incorporated into the lattice of the mineral (e.g., limonite [$Fe_2O_3.3H_2O$] and gypsum [$CaSO_4.4H_2O$]); (2) *hygroscopic water*, which is adsorbed on soil-surface areas as a thin layer; and (3) *free water*, which occupies the soil pores. Each of these categories influences the soil spectra differently, enabling identification of the water condition of the soil, and each is treated separately below. Three basic fundamentals in the IR regions exist for water molecules, particularly the OH group: $\upsilon w1$-asymmetric stretching, δw bending, and $\upsilon w3$-symmetric stretching vibrations. Theoretically, in a mixed system of water and minerals, combination modes of these vibrations can yield OH absorption features at around 0.95 μm (very weak), 1.2 μm (weak), 1.4 μm (strong), and 1.9 μm (very strong) related to $2\upsilon w1 + \upsilon w3$, $\upsilon w1 + \upsilon w3 + \delta w$, $\upsilon w3 + 2\delta w$, and $\upsilon w3 + \delta w$, respectively. The hydration water can be seen in minerals such as gypsum as strong OH absorption features at around 1.4 and 1.9 μm (Hunt et al., 1971b). However, free water changes the soil spectrum significantly as the real part of the refractive index is very dominant, causing a decrease in reflectance through the entire spectral range and consequently masking other possible features. This causes a problem in soil remote sensing when the soil is saturated or very wet.

Iron: This is the most abundant element on the earth's surface and the fourth most abundant element in the earth's crust (Dixon, 1989). Changes in its oxidation state, and consequently in its mobility, tend to occur under different soil conditions. The major Fe-bearing minerals in the earth's crust are the mafic silicates, Fe sulfides, carbonates, oxides, and smectite clay minerals. All Fe^{3+} oxides have striking colors, ranging from red and yellow to brown, due to selective light absorption in the VIS range caused by transitions in the electron shell. It is well known that even a small amount of Fe oxide can change the soil's color significantly. The red, brown, and yellow *hue* values, all caused by iron, are widely used in soil-classification systems in almost all countries. The iron feature assignments in the VIS–NIR region result from the electronic transition of iron cations (3+, 2+), either as the main constituent (as in Fe oxides) or as impurities (as in iron smectite). Hunt et al. (1971a) summarized the physical mechanisms responsible for Fe^{2+} (ferrous) and Fe^{3+} (ferric) spectral activity in the VIS–NIR region as follows: the ferrous ion typically produces a common band at around 1 μm due to the spin allowed during the transition between the E_g and T_{2g} quintet levels into which the D ground state splits into an octahedral crystal field. Other ferrous bands are produced by transitions from the $5T_{2g}$ to $3T_{1g}$ states at 0.55 μm, to $1A_{1g}$ at around 0.51 μm, to $3T_{2g}$ at 0.45 μm, and to $3T_{1g}$ at 0.43 μm. For the ferric ion, the major bands produced in the spectrum are the result of the transition from the $6A_{1g}$ ground state to $4T_{1g}$ at 0.87 μm, $4T_{2g}$ at 0.7 μm, and either $4A_{1g}$ or $4E_g$ at 0.4 μm.

Salts: Soil salts are reported to be Na_2CO_3, $NaHCO_3$, and NaCl, which are very soluble and mobile in the soil environment. In most cases, the spectra of these salts are featureless. However, indirect relationships with other chromophores can indicate their existence (e.g., OM, particle-size distribution). Hunt and Salisbury (1971c) reported an almost featureless spectrum of halite (NaCl 433B from Kansas), whereas later, Farifteh et al. (2007) reported some features of the salt, mostly related to the adsorbed water as it is a hygroscopic material, and confirmed Hick and Russell's (1990) hypothesis that there are certain wavelengths across the VIS–NIR–SWIR region that correlate to water features. Mougenot (1993) noted that in addition to an increase in reflectance with salt content, high salt content may mask ferric ion absorption in the VIS region. They concluded that salts are not easily identified in proportions below 10% or 15%. Salt is also visible in the VIS region due to its light tone, which reflects back radiation from the soil surface under dry conditions. This occurs mainly because the soluble salt migrates to the soil surface via capillary forces controlled by the evaporation process transporting water molecules from the soil body to the atmosphere.

To provide an overview of chemical chromophore activity in soils and to summarize this section, Figure 25.9 provides a spectrum from a Haploxeralf soil from Israel with the positions of all possible chromophores. Figure 25.10 provides six spectra of different soils from Israel consisting of different chromophores content as illustrates in Figure 25.9 across each spectral region segments. Figure 25.11 summarizes the chemical chromophores

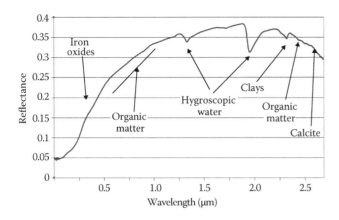

FIGURE 25.9   A representative spectrum of a semiarid soil from Israel (Haploxeralf) with all possible chromospheres associated with soils.

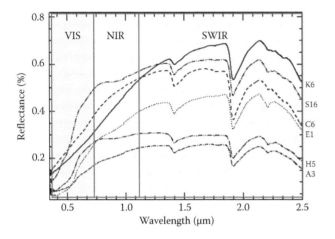

FIGURE 25.10   Reflectance spectra of six representative soils from arid and semiarid area from Israel (E1-Rhodoxeralf, A3-Rodoxeralf, H5-Xerert, C6-Xerothent, S16-Torriothent, K6-Caliorthid).

associated with soil and geological matter as collected from the literature and summarized by Ben-Dor et al. (1999). It also lists the intensities of each chromophore in the VIS–NIR–SWIR spectral regions as they appear in those studies. The current review demonstrates that high-resolution spectral data can provide additional, sometimes quantitative, information on soil properties that are strongly correlated with the chromophores, that is, primary and secondary minerals, OM, Fe oxides, water, and salt. It also demonstrates the importance of soil spectroscopy in designing a sensor for a soil mission and selecting the proper tools to interpret the results acquired by RS means using solar radiation.

25.6 Radiation Source and Atmospheric Windows

25.6.1 General

To acquire the chromophores by RS means, the radiation source and the medium through which it travels (the atmosphere) must be investigated. The Planck function is a physical expression

describing the energy emitted from a black body. The sun, as an ideal black body, is the main radiation source for remote sensing of the earth across the VIS–NIR–SWIR region. If a sensor is located far from the soil (air or space), radiation must travel from the source to the object and back to the sensor, thus crossing the atmosphere twice. The gasses and aerosols in the atmosphere interact with the radiation across this path and hinder soil reflectance at certain frequencies. Thus, the components in the atmosphere are spectrally active. This interaction has to be minimized as much as possible to obtain a signal from the soil. This can be done in two ways: (1) allocating the sensor bands across the high-transition spectral region of the atmosphere (known as atmospheric windows) or (2) determining the physical interaction of the radiation with the known atmospheric component using a physical calculation (known as "radiative transfer model"). Whereas in multi- and superspectral sensors, the spectral bands are usually located across atmospheric windows, in hyperspectral sensors, this is not possible, as the bands cover the entire spectral region, and thus the use of the radiative transfer model is called for to extract the soil reflectance. Masking the atmospheric attenuation from the sensor's radiance is termed "atmospheric correction." Figure 25.12 illustrates the spectral regions under which atmospheric attenuation can affect the soil spectrum. This figure shows the reflectance spectrum of an E-7 soil from Israel (Haploxeralf, taken from Tel Aviv University's spectral library) overlain on its simulated (soil) radiance as calculated by MODTRAN. The latter is normalized to the Planck sun function at the top of the atmosphere to illustrate only the atmosphere transmittance. As the atmospheric attenuation remains, the most affected spectral regions can be clearly seen. The VIS region is affected by aerosol scattering (monotonous decay from 0.4 to 0.8 μm) and absorption of ozone (around 0.6 μm), water vapor (0.73 and 0.82 μm), and oxygen (0.76 μm). The NIR–SWIR regions are affected by absorption of water vapor (0.94, 1.14, 1.38, 1.88 μm), oxygen (at around 1.3 μm), carbon dioxide (at around 1.56, 2.01, 2.08 μm), and methane (2.35 μm). Also seen are the absorption peaks of the soil chromophores at 2.33 μm (carbonates), 2.2 μm (clay), 1.9 and 1.4 μm (hygroscopic water), and 0.5, 0.6, and 0.9 μm (Fe oxides) that overlap with the aforementioned atmospheric chromophores. It can be seen that the most informative region for the soil has some overlap with the atmosphere (water vapor at 1.4 and 1.9 μm, oxygen at 0.76 μm), and if we do not allocate our spectral bands in this area, information about some soil attributes may be lost (moisture, Fe oxides, and OM).

Figure 25.12 provides also the radiation observed from a given pixel composed of the source (sun), the atmospheric transition, and the soil reflectance. The mixed radiation maintains a high response to the solar radiation source, followed by the atmospheric attenuation and, lastly, the soil response. Also seen is that the energy at the end of the SWIR region (2–2.5 μm) is low and may affect the quality of the information from this region based on the low SNR in this region due to the low radiation flux. It is hence demonstrated that the intensity of the soil target within the radiance observed by a sensor is quite small (>5%)

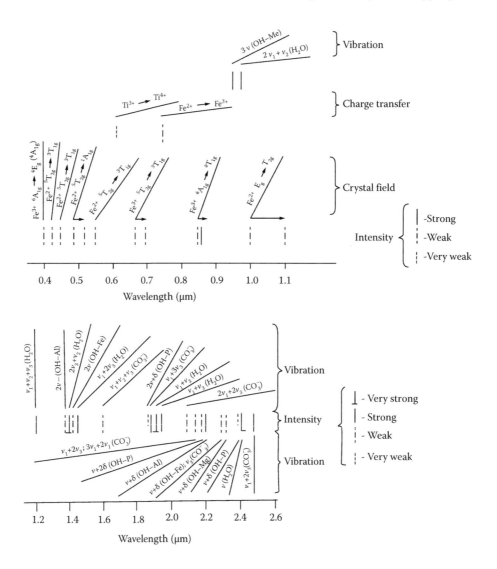

FIGURE 25.11 The active groups of the soil chromophores spectrum. For each possible mechanism, possible wavelength range and absorption feature intensity is given. The spectrum was generated using information presented in the literature. (Taken from Ben-Dor, E. et al., Soil spectroscopy, in *Manual of Remote Sensing*, 3rd edn., A. Rencz (ed.), John Wiley & Sons Inc., New York, 1999, pp. 111–189.)

and hence needs to be carefully isolated from the total radiance information at the sensor level. The atmospheric components are as follows: water vapor (at 0.68, 0.94, 1.12, 1.4, 1.9 μm), oxygen (0.76 and 1.3 μm), and CO_2 (2.105 and 2.015 μm). Ozone also sometimes plays a role in the VIS region at around 0.45–0.50 μm. Aerosol is also part of the atmosphere, and its scattering effect on the radiation is mostly in the VIS region (0.4–0.8 μm). The sun's radiation interacts with the atmospheric molecules (known as *Rayleigh* scattering) and with particles (known as Mie scattering). As previously discussed, the aforementioned components can be physically described by the radiative transfer equations that aim to remove the atmospheric effects, leaving the reflected information from the soil only. If the soil reflectance information can be extracted without artifacts from the sensor radiance, those soil chromophores may be useful for either qualitative or quantitative spectral utilization. A comprehensive description of methods to remove the atmospheric attenuations is given by

Ben-Dor et al. (2002) and Gao et al. (2006, 2009). As already mentioned, even weak spectral features in the soil spectrum can contain very useful information. Therefore, great caution must be taken before applying any quantitative models to soil reflectance spectra derived from air- or spaceborne hyperchannel sensors. Validation of the (atmospherically) corrected data is an essential step in ensuring that the reflectance spectrum contains reliable soil information. This section shows that atmospheric attenuation plays a major role in the final soil spectral products as the soil contributes less than 5% to the overall energy acquired by the sensor's detector.

25.6.2 Factors Affecting Soil Reflectance in Remote Sensing

Many factors can hinder the spectral information from a given soil sample or pixel. Residual atmospheric attenuation, mainly

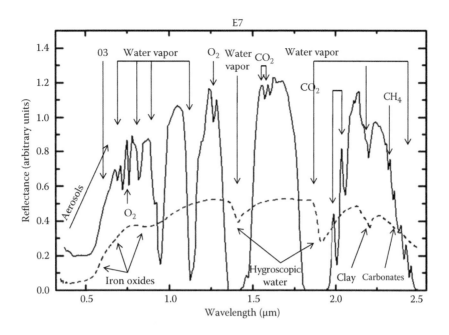

FIGURE 25.12 The reflectance spectrum of a typical soil from semiarid environment (dotted line) and its modeled radiance using MODTRAN (solid line). The atmospheric attenuation on the soil chromophores is specified.

across a highly active atmospheric spectral region, may be the first to strongly affect the soil spectrum. If not effectively removed, these residuals can contribute to spectral noise signals that are not related to soil. In many studies, some spectral ranges are ignored based on strong absorption of atmospheric constituents (e.g., 1.4 and 1.9 μm due to water vapor). In the multispectral sensor, this interference is minimal as most of the bands are allocated across atmospheric windows. In this case, however, a more continuous spectral effect, such as that obtained from aerosol scattering, may affect the soil reflectance, mostly in the VIS region. Lagacherie et al. (2008) presented a comparison of field, air, and space reflectance of soils, showing differences in the spectral information mostly due to different SNR values and the atmospheric residual in the corrected data. The soil surface is exposed to many natural effects, such as rain, erosion, and deposition, as well as fire and cultivation. These effects also play an important role in the accurate identification of a soil entity. This is mainly because such effects can hinder the real soil chromophores' interactions with solar radiation. In this case, even a thin layer on the soil surface can make a difference. In this respect moisture, fire, dust, crust, and roughness play a major role in the sensor's final response to the soil at any spectral resolution. The sensor quality also has an important effect on the soil spectrum. Airborne and spaceborne sensors vary in their SNR values, which are mostly lower than those used in the field and laboratory. A noisy spectral signature cannot be further analyzed as in many cases, it cannot be used for quantitative approaches. In the laboratory, soil-reflectance measurements are performed under controlled conditions and thus it is possible to standardize all spectral measurements and compare spectral libraries between users to allow a robust analysis (Piemstein et al., 2010, Ben Dor et al., 2015). In the field, however, this standardization is not yet achievable and spectra may

vary from one measurement to the next, mostly due to physical factors (e.g., viewing and illumination angles, particle-size and roughness distributions). Although correlations have been found between spectral responses of soils in the laboratory and field (e.g., Stevens et al., 2008), this is valid only for a select database and cannot be applied to sensor data from other areas. Some analytical manipulations (e.g., partial least squares [PLSR]) can decrease the variation between the two measurement domains, but upscaling the laboratory models to airborne data is still difficult. This problem calls for a common protocol and well-agreed-upon measurement scheme in the field, since it is the most relevant condition for remote sensors, and quantitative models are generated for that domain. In the field, reflectance measurements are fraught not only with variations in viewing angle and changes in illumination but also with variations in soil roughness, soil moisture, and soil sealing (Ben-Dor et al., 1999). Knadel et al. (2014) studied the spectral changes in four representative soils from Denmark with water retention ranging from wet to dry. Soil reflectance was found to decrease systematically, albeit not proportionally, with decreasing matric potential and increasing molecular layers. The changes in molecular layers were best captured by the soil reflectance of clay-rich soils, demonstrating the importance of an intercorrelative approach to the quantitative analysis of soil water content under dry and semidry conditions, as with other soil attributes. This study demonstrated the problem of the water's influence on soil spectra as discussed in Section 25.5.3.4.1.

Acquiring soil-reflectance data from air and space involves additional difficulties, such as homogeneity of the area being observed and problems on how to represent the sensor's pixel by ground-truth measurements, as the pixel size of the sensor is a major factor affecting the soil's spectral response. Hengl (2006) provided analytical and empirical rules for selecting the proper pixel size for a particular mission. He concluded that there is no ideal grid resolution; it has to be determined according to the property in question. For soil, this is a crucial element. If salinity is important, then the ground truth needs to be taken at high spatial distance using a high-spatial-resolution sensor. Using sensors that are not adequate for the mission is a waste of resources and time. In this regard, the grid selected for the question being asked must maximize the predictive capabilities or the informational content of the final processed map. Sometimes, the pixel size plays a more dominant role than the SNR values. Asner and Heidebrecht (2003) demonstrated lower accuracy in sensing bare soil areas from the Hyperion (30 m, low SNR) than from the AVIRIS (4.5 m, high SNR), but convolving the AVIRIS spectra into Hyperion's 30-m grid gave similar accuracy. Lagacherie et al. (2008) examined the performance of clay and calcium carbonate ($CaCO_3$) estimations using laboratory (ASD) versus airborne (HyMap) spectral scales. A significant decrease was observed going from laboratory to field and then to airborne domains. They indicated that the main factors inducing the uncertainties were the radiometric- and wavelength-calibration uncertainties of the HyMap sensor and possible residual atmospheric effects. In general, to represent a pixel area of a given sensor on the ground, it is necessary to set a ground truth of the given pixel at 4 px. This means that the reflectance property of this grid has to be measured in several locations along the 4 × 4 GDS area and then averaged to yield the "pixel spectral response." If the pixel size is large (such as in Landsat 8, 30 m), then a large (120 × 120 m) area is the minimal area required for the ground-validation mission. As soil-surface conditions may also affect the field spectrum based on the previously mentioned factors, a representative *pixel* area must be fully covered by both ground-reflectance measurements and sampling. Soil sampling must be planned according to a statistical (or other logical) framework (McKenzie and Ryan, 1999), and soil sampling should cover the 0–1 cm layer as much as possible (this is a compromise between what the sensor sees and what can be sampled under real field conditions). As this sampling is not easy, mis-sampling of soil can occur and hinder the classification results and accuracy based on laboratory analysis of the soil samples. As previously discussed, laboratory-based measurements provide an understanding of the chemical and physical properties of the soil reflectance. If the soil sample is not well represented, the spectral-based models or indices from the laboratory will not work for the field. This makes the models nonrobust and the transformation from one sensor to another almost impossible. As there is still no standardization process for field reflectance measurements, this problem continues to exist, even for high SNR data. Another problem is vegetation (or litter). Termed biospheric interference, these cover many soil surfaces worldwide and hinder direct sensing of the soil. Within the non-vegetated area, only a portion of the soils are characterized by an unaltered surface layer (e.g., as a consequence of soil tillage), and partial sensing of the natural soil surface can be interpolated into the vegetated area (Ben-Dor et al. 2002). A partial solution for this is "inferring soil properties through vegetation" (Huete, 2005) or the connection of vegetation to soil properties (e.g.,

Maestre and Cortina, 2002) where a recent study by Kopackova et al. (2014) demonstrated how this method can shed light on the information obtained from the root zone, mainly for heavy-metal content. However, it seems that this method cannot (yet) be used to obtain information on other important soil parameters, such as clay, silt and clay, OM, and $CaCO_3$, and thus it is still limited. Spectral unmixing is often used to account for the biogenic fraction (Asner and Heidebrecht 2002; Robichaud et al., 2007).

Radiation intensity on the soil pixel can also change the reflectance properties from a topographic view but can be corrected geometrically. A more difficult problem is the bidirectional reflectance distribution function (BRDF). This function assumes that the radiation source, the target, and the sensor are all points in the measurement space and that the ratio calculated between absolute values of radiance and irradiance is strongly dependent upon the geometry of their positions. Theories and models explaining the BRDF phenomenon in relation to soil components have been widely discussed and covered in the literature (Hapke 1981a,b, 1983, 1984, 1986, 1993; Pinty et al., 1989; Jacquemoud et al., 1992; Liang and Townshed, 1996). A number of models have been developed, which express soil bidirectional reflectance as a function of illumination and viewing direction. They cannot, however, be inverted to directly estimate soil properties on the basis of bidirectional reflectance observations nor can the equations be used to predict reflectance distributions on the basis of soil-property measurements in the field. Thus, the BRDF effects, although well studied, still play a major role in the final output of the soil spectral response (from all domains), and more research into this phenomenon is warranted. The soil spectral response can therefore be affected by the aforementioned factors in two ways: in its physical behavior, that is, spectral-baseline position (e.g., in the case of the BRDF effect), and in its chemical behavior, that is, new or absence of spectral features (e.g., in the case of atmospheric attenuation).

In summary, the spectral information from a given soil pixel (from either RS means or ground-truth measurements) may be affected by a variety of factors that can change the soil spectral response significantly. This calls for user caution in upscaling data from the laboratory to the field, in analyzing the data, and in exchanging data. Table 25.4 summarized the aforementioned factors in a qualitative scale. As clearly seen, the SWIR region is more sensitive to these factors as well as the hyperspectral systems. This illustrates that the hyperspectral sensors are the most sensitive systems for soil remote sensing. In other words, it can be said that if all of the aforementioned factors maintain small, then HRS can be the most sensitive way to remote sense the soil system.

25.7 Quantitative Aspects of Soil Spectroscopy

25.7.1 Proximal Sensing

Proximal sensing refers to the quantitative information on soil attributes that is mined from the soil-reflectance data. Today, quantitative soil spectroscopy is a mature discipline that has

TABLE 25.4 Factors Attenuate Soil Reflectance Quality from Remote Sensing Domains in Different Systems and Spectral Regions

Spectral Range and Sensing Systems	Factors Affecting Reflectance Tetrieval from RS Sensors						
	Atmosphere	SNR	Sensor Stability	Radiometry Calibration	BRDF	Spatial Resolution	Effect on Soil Surface Chromophores (e.g. Dust, Water, Crust)
VNIR-multi	+	++	++++	++	+++	++	+++
VNIR-super	++	+++	++++	+++	+++	+++	+++
VHIR-hyper	++++	++++	++++	+++++	++++	++++	++++++
SWIR-multi	++	+++	+++++	+++	++++	+++	++++
SWIR-super	+++	+++++	+++++	++++	++++	++++	++++
SWIR-hyper	++++	++++++	+++++++	++++++	++++++	++++++	+++++++

Notes: +, low; ++, moderate; +++, average; ++++, high; +++++, very high.

come quite a long way since the mid-1960s, when Bowers and Hanks (1965) published their paper on the correlation between soil reflectance and soil moisture content. That pioneering study, followed by a series of papers by Hunt and Salisbury (1970, 1980) and Hunt (1982), proved that water and minerals in the soil environment have unique spectral fingerprints that can be further used for specific recognition. Learning from several sectors' successes (e.g., food science, tobacco, textile), Dalal and Henry (1986) applied the proximal-sensing (proxy) approach to soils. In 1990, Ben-Dor and Banin demonstrated the power of reflectance spectroscopy in accounting for $CaCO_3$ content in the soil (Ben-Dor and Banin, 1990a) and, later, in monitoring the structural composition of smectite soil minerals in the laboratory (Ben-Dor and Banin, 1990b). Later still, when portable field spectrometers were introduced to the market (in around 1993), more scientists realized the potential of soil spectroscopy and consequently, more spectral libraries were assembled (e.g., Shepherd and Walsh, 2002; Bellinaso et al., 2010). A comprehensive summary of the quantitative applications of soil reflectance spectroscopy was provided by Ben-Dor et al. (2002) and Malley et al. (2004) and later by Viscarra Rossel et al. (2006). In April 2009, a world soil spectroscopy group was established by Viscarra Rossel (http://groups.google.com/group/soil-spectroscopy), who gathered soil spectra and corresponding attributes from more than 80 countries worldwide to generate a global soil spectral/attribute database, providing proxy capability to all. This initiative was based on the idea that since the quantitative approach in soil sciences had become well established and applicable, it should be more collaborative. This was the obvious step after understanding that only sharing information would help advance quantitative soil spectroscopy (e.g., Condit, 1970; Shepherd and Walsh., 2002; Brown et al., 2006). Comprehensive reviews on proxy applications for soils can be found in Malley et al. (2004) and Viscarra Rossel et al. (2006), and other important reviews focusing on soil-reflectance theory and applications can be found in Clark and Roush (1983), Irons et al. (1989), Ben-Dor et al. (1999), and Ben-Dor (2002). The number of national and international soil-spectral libraries is growing, constituting a database for spectral-based model generation for the proximal-sensing strategy. Ben-Dor et al. (2009) summarized the quantitative utilization of soil spectroscopy from airborne domains. It was pointed out that although

two communities are utilizing soil spectroscopy—soil scientists and the RS communities—there is a lack of consistency between the terminologies used by these two groups, leading to potential misunderstandings. The soil scientists define soil proximal sensing across the 400–2500 nm region as visNIR (as adopted from other disciplines such as the food sciences), whereas the RS community refers to this exact spectral region as VIS–NIR–SWIR (see Section 25.3.). As no common agreement has yet been reached, we suggest using the terms VIS/NIR/SWIR/TIR for any soil proxy analyses to emphasize the spectral regions (and accordingly, the chromophores) in which the analysis has been done for better performance (the TIR region is also divided into MWIR and LWIR).

25.7.2 Application Notes

In general, soil reflectance spectra are directly affected by chemical and physical chromophores, as already discussed. The spectral response is also a product of the interaction between these parameters, calling for a precise understanding of all chemical and physical reactions in soils. For example, even in a simple mixture of Fe oxides, clay, and OM, the spectral response cannot be judged simply by linear mixing models of the three end members. Strong chemical interactions between these components are, in most cases, nonlinear and rather complex. For instance, organic components, mostly humus, affect soil clay minerals in chemical and physical ways. Similarly, free Fe oxides may coat soil particles and mask photons that interact with the real mineral components or the Fe oxides themselves (and OM as well). In addition, the coating material may collate fine particles into coarse aggregates that may physically change the soil's spectral behavior from a physical standpoint. McBratney et al. (2003; 2006) also shed light on this technology through their pioneering work over the years. Brown et al. (2006) concluded that the spectral proxy technique in soil has the potential to replace or enhance standard soil-characterization techniques, basing their conclusion on 3768 soil samples from the United States. As noted earlier, in view of the growing soil spectral community, Viscarra Rossel (2009) generated an initiative (Soil World Spectral Group, http://groups.google.com/group/soil-spectroscopy) in which all members of the soil spectral community were asked to join together and contribute their local spectral library to generate a worldwide library that would be accessible to all. The world spectral library is composed (at the time of this writing, 2014) of about 20,000 soil spectra with their chemical attributes. This initiative, besides being the first attempt to gather spectral information on the world's soils, is an important step toward establishing standard protocols and quality indicators that will be accepted by all members of the growing soil-spectral community. To that end, it is important to mention that special sessions dealing with soil spectra have been organized at several leading conferences for both earth material and soil sciences (e.g., EGU 2007, 2008; WSC 2010, WD, 2014), along with specific workshops (e.g., EUFAR-2, 2014). As a consequence, many soil scientists who were unaware of this technology are now being exposed to it.

The quantitative option and the availability of field spectrometers enabled users to show that the soil-spectral-based technology can be used in the field. Some key works (among many that have been published) are mentioned here. Genu and Demattê (2006) evaluated 3300 samples using the multiple method with spectral proxy analysis and reached an R^2 of 0.74 and 0.53 for clay and OM, respectively. Demattê et al. (2009a) evaluated 1000 samples and determined $R^2 = 0.85$ for clay using a laboratory spectrometer. Nanni and Demattê (2006) suggested a reflectance inflection difference (RID) index, representing the difference between reflectance values at the highest and lowest points of inflection (or amplitude of spectral data in this range—demonstrating the height of the curve between the peak and the valley). This approach led them to use only some of the spectral information from the overall 400–2500 nm range. Using multiple stepwise statistics, they obtained $R^2 = 0.91$ and 0.89 for clay and SOM, respectively. Fiorio and Demattê (2009) obtained $R^2 = 0.83$ and 0.30 for clay and SOM, respectively, also with ground spectra analyzing 450 samples. Viscarra Rossel and Webster (2011) mapped the distribution of the spectral proxy approach's information from Australian soils. They concluded that the technique can provide integrative measures of soil properties and can act as an alternative to the conventional analytical method that can be effectively applied for both soil classification and environmental monitoring. As an example from Israel, Schwartz et al. (2012) evaluated the total petroleum hydrocarbon (TPH) content in the soil in a field. The technique has also reached the precision agriculture discipline, as demonstrated by several workers who used it to assess important soil attributes in the field (e.g., recent research by Debaene et al. (2014) on clay and carbon, Araújo et al. (2013) with cation-exchange capacity and pH, among others, and Barnes et al. (2003) with OM and electrical conductivity).

25.7.3 Constraints and Cautions in Using Proximal Remote Sensing for Soils

In reality, the number of chromophores in soils is quite limited relative to the soil's attributes. The factors affecting the soil spectrum (see Sections 7 and 8) also hinder the proxy analysis. In addition, there is no simple correlation between the spectroscopy and the chromophore content, as it requires a sophisticated data-mining approach with significant validation tests. For

example, using the spectral information for quantitative analysis, Karmanova (1981) selectively removed Fe oxides from soil samples and concluded that the effects of various iron compounds on the spectral reflectance and color of soils were not proportional to their relative contents. Another aspect is that the proxy models are not always robust and may be related to the soil population in the analysis. In fact, Araujo et al. (in press) clustered 7125 soil samples in relation with their mineralogy and gathered much stronger models than the global one provides. As pointed out by Bedidi et al. (1990, 1991), the normally accepted view of decreasing soil-baseline height with increasing moisture content (VIS region) does not hold for lateritic (highly leached, low-pH) soils. They concluded that the spectral behavior of such soils under various moisture conditions is more complex than originally thought. In this context, Galvao et al. (1995) showed spectra from laterite soils (VIS–NIR region) consisting of complex spectral features that appeared to deviate from those of other soils. Al-Abbas et al. (1972) found a correlation between clay content and reflectance data in the VIS–NIR–SWIR region and suggested that this is not a direct but rather an indirect relationship, strongly controlled by the OM chromophore. Another anomaly related to the interactions between soil chromophores was identified by Gerbermann and Neher (1979). They carefully measured the reflectance properties in the VIS region of a clay–sand mixture extracted from the upper horizon of a montmorillonite soil and found that "adding of sand to a clay soil decreases the percent of soil reflectance." This observation is in contrast with the traditional expectation from adding coarse (sand) to fine (clay) particles in a mixture (soil), that is, that this will tend to increase soil reflectance. Likewise, Ben-Dor and Banin (1994, 1995a–c) concluded that intercorrelations between feature and featureless properties play a major role in assessing unexpected information about soil solely from their reflectance spectra in either the VIS–NIR or SWIR regions. Ben-Dor and Banin (1995b) examined arid and semiarid soils from Israel and showed that "featureless" soil properties (i.e., properties without direct chromophores such as K_2O, total SiO_2, and Al_2O_3) can be predicted from the reflectance curves due to their strong correlation with "feature" soil properties (i.e., properties with direct chromophores). Csilag et al. (1993) best described the effect of multiple factors indirectly affecting soil spectra in their discussion on soil salinity, which can be considered a featureless property. They stated that "salinity is a complex phenomenon and therefore variation in the [soil] reflectance spectra cannot be attributed to a single [chromophoric] soil property." To get the most out of soil spectra, they examined the chromophoric properties of OM and clay content, among others, and ran a principal component analysis to fully account for the salinity status culled from the soil reflectance spectra.

The poor quality of the data acquired from orbit as compared to the laboratory may hinder the transfer of proxy models from the laboratory to orbit domains. This problem can be solved by generating models based on field conditions and measurements and using a better method to upscale laboratory data to orbital domains. This idea has been implemented by several users as will be discussed further on.

25.8 Soil Reflectance and Remote Sensing

25.8.1 General

A short review of remote sensing of soils from an optical perspective was published by GE and Sui (2011). A comprehensive description of soil spectral remote sensing can be found in Ben-Dor et al. (2008). Many studies have been conducted with the intention of classifying soils and their properties using optical sensors on board orbital satellites, such as Landsat MSS and TM, SPOT, and NOAA-AVHRR (e.g., Cipra, 1980; Frazier and Cheng, 1989; Kirein Young and Kruse, 1989; Agdu et al., 1990; Dobos et al., 2001). Qualitative classification approaches have traditionally been applied to multichannel data in cases of limited spectral information and in early 1990 also to hyperspectral data (Vane, 1993). Nevertheless, important qualitative, and sometimes even quantitative, information has been obtained on soil OM, soil degradation, and soil conditions (Price, 1990; Ben-Dor and Banin, 1995a, Metternicht et al. 2010). Huete (2004) has summarized some RS applications of soils (using different sensors and resolutions) and discussed on "properties controlled soil reflectance." His excellent overview of soil RS applications was strongly tied to the relationship between green vegetation, litter, and soils as based on the fact that most of the soils are altered by these substances. Later, Huete (2005) have provided more ideas on the how to use the spectral information from hyperspectral domains to measure bare soils and mixed soil–vegetation–litter and overlaying vegetation pixels in order to investigate the soil properties. As the technology progresses, soil spectra are becoming an important vehicle for the remote sensing of soils, and spectral libraries are being established to cover vast geographical areas worldwide (e.g., Latz et al., 1981; Price, 1995; Visscora Rossel, 2009; Montanarella et al., 2011). Although the use of these libraries has many limitations, it is understood that the spectral domain is very important for soil mapping and that effort has to be invested in super- and hyperspectral sensors. Over the past 30 years, HSR has developed to a point where it is now in high demand by many users (Ben-Dor et al., 2013). HSR technology provides high-spectral-resolution data with the aim of giving near-laboratory-quality reflectance or emittance information for each individual picture element (pixel) from far or near distances (Vane et al., 1984). This information enables the identification of objects based on the spectral absorption features of their chromophores and has found many uses in terrestrial and marine applications (Clark and Roush, 1984; Vane et al., 1984; Dekker et al., 2001). Figure 25.13 illustrates this concept, where the spectral information of a given pixel shows a new dimension that cannot be obtained by traditional point spectroscopy, air photography, or other multiband images. HSR can thus be described as an *expert* geographical information system (GIS) in which layers are built on a pixel-by-pixel basis, rather than with a selected group of points (McBratney et al., 2003). This enables spatial recognition of the phenomenon in question

with a precise spatial view and use of the traditional GIS-interpolation technique in precise thematic images. Since the spatial–spectral-based view may provide better information than viewing either the spatial or spectral views separately, imaging spectroscopy serves as a powerful and promising tool in the modern RS arena. Since 1983, when the first airborne imaging spectroscopy (AIS) sensor (Vane et al., 1984) ushered in the HSR era, this technique has been used mostly for geology, water, and vegetation applications. It appears that soil applications for HSR are quite limited because soils present a complex matrix and many of the previously discussed problems have not yet been resolved. Nonetheless, with the advent of better and lower cost HSR sensors, along with comprehensive studies by many scientists developing a wealth of innovative soil-spectral applications, the future of superspectral and HRS from airborne and spaceborne sensors is bright. Other previously discussed problems (see Sections 7 and 8) will hopefully be resolved in the near future, enabling this technology's use for other RS applications. A comprehensive overview of the pros and cons of hyperspectral technology for soils can be found in Ben-Dor et al. (2009).

25.8.2 Application of Soil Remote Sensing: Examples

25.8.2.1 Multi- to Hyperspectral Concept in Soil

Based on the theory presented earlier, this section provides case studies exemplifying the application of remote sensing to soils taking into account the pros and cons of this technology. We will focus on the multi- to hyperspectral domains, with special emphasis on the latter. In addition, we will demonstrate that low spectral resolution can be sufficient for some soil applications (e.g., Palacios-Orueta, and Ustin, 1998; Chen et al., 2000; Fox and Sabbagh, 2002; Dematte et al., 2007). A good example of this is the fact that some major soil properties are correlated with color, and soil surveyors describe soil colors that can be seen with the naked eye. From a RS perspective, Chen et al. (2000) were able to map SOM content from color aerial photographs by measuring the color tones and correlating them with the SOM content. Dematté et al. (2000, 2009) summarized the capabilities of the limited channel of the Landsat TM sensor from orbit to provide quantitative information on soils. Using the 6 TM bands in the VIS–NIR–SWIR region, they were able to allocate a reliable quantitative model for the Al_2O_3/Fe_2O_3 ratio using band 7. In another study, Nanni and Dematte (2006) showed that TM band 7 is mandatory for SiO_2 and TiO_2, whereas TM band 4 was not selected. The authors speculated that band 7 might be associated with the influence of younger, more clayey soils containing kaolinite minerals that are spectrally active around band 7. The quantitative potential of Landsat TM was shown by Ben-Dor and Banin (1995c), who convolved a soil spectral library to TM channels and were able to predict $CaCO_3$, SiO_2, loss on ignition (LOI), and specific surface area solely from the reflectance data. Landsat TM has also been used to establish a strong

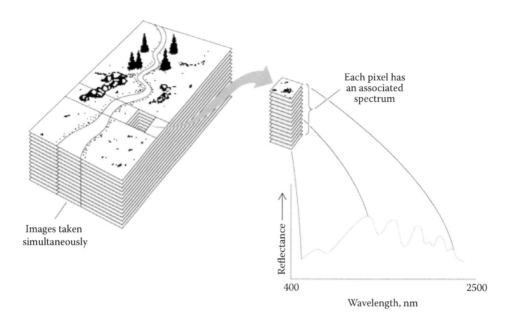

FIGURE 25.13 The spectral imaging concept where each pixel is described by a detailed spectrum consisting of many narrow continuous bands. (Taken from NASA website)

relationship between iron and reflectance (Frazier and Cheng, 1989; White et al., 1997; Tangestani and Moore, 2000; Jarmer, 2012), where recently, Kumar (2013) showed that from the classified image, they can point to the best fertilization areas using soil color. Better results were obtained later with the complete spectral information, as already discussed, using either point or image spectroscopy. The technology of spectral imaging has advanced soil remote sensing, with spectral know-how in the laboratory providing an innovative starting point. In general, imaging spectroscopy (HSR) is capable of generating qualitative and quantitative spatial indicators for ecologists, land managers, pedologists, and engineers. For soils, this technology has been used for the past 10 years to combine spatial information with the spectral one thus providing farmers with a spatially explicit quantitative overview of the soil properties and phenomena in question. This allows them to control their resources, such as irrigation, nutrients, and cultivation and obtain better yields per hectare. Since the HSR product is a geopositioned mosaic comprised many spectral points, traditional (quantitative) approaches that work successfully for point-spectrometry measurements in minerals and soils (e.g., Clark and Roush, 1984) may be suitable for the imaging domain. Despite its drawbacks, most of the applications developed for point spectrometry can be immediately adapted for the imaging spectroscopy domain. The following sections provide examples in which both multi- and hyperspectral technologies were used for various applications in soil, revealing this technology's potential for soil science.

25.8.2.2 Soil Organic Matter

SOM, or soil organic carbon (SOC) (SOM ≈ SOC × 1.72), plays a major role in many chemical and physical processes in the soil environment and therefore has a strong influence on soil-reflectance characteristics. Consequently, and as described

earlier, this can be seen in the tone of the soil's color (Chen et al., 2000). SOM is a mixture of decomposing tissues of plants, animals, and secreted substances. The sequence of OM decomposition in soils is strongly determined by the activity of soil microorganisms. The nature of SOM is responsible for many soil properties, such as compaction, fertility, water retention, and soil-structure stability, and it constitutes one of the major resources in the global carbon cycle (Stevens et al., 2008). The mature stage refers to the final stage of microorganism activity, when new, complex compounds, often called humus, are formed. The most important factors affecting the amount of SOM are those involved with soil formation, that is, topography, climate, time, type of vegetation, and oxidation state. OM, and especially humus, plays an important role in many of the soil's properties, such as aggregation, fertility, water retention, ion transformation, and color. SOM is part of the upper soil horizon that serves as the interface between the body of the soil, the biosphere, and the atmosphere. Since OM is mainly concentrated in the top Ao horizon that is exposed to the sun's radiation, it is a perfect property for RS assessments. This notion is strengthened by the fact that pure OM has unique spectral fingerprints (Ben-Dor et al., 1999) that can be correlated to content, composition, and maturity (Ben-Dor et al., 1997). Much attention has been devoted to OM from many perspectives. As OM has spectral activity throughout the entire VIS–NIR–SWIR region, especially the VIS portion, workers have extensively studied OM via RS (e.g., Kristof et al., 1971). Baumgardner et al. (1970) noted that if the SOM drops below 2%, it has only a minimal effect on soil reflectance. Montgomery (1976) indicated that OM content as high as 9% does not appear to mask the contribution of other soil parameters to soil reflectance. Galvao and Vitorello (1998) showed how OM affects the iron oxides' influence of the spectral reflectance and color of Brazilian tropical soils. In another study,

Schreier (1977(indicated that OM content is related to soil reflectance by a curvilinear exponential function. Mathews et al. (1973) found that OM correlates with reflectance values in the 0.5–1.2 µm range, whereas Beck et al. (1976) suggested that the 0.90–1.22 µm region is best suited for mapping OM in soils. Krishnan et al. (1980) used a slope parameter at around 0.8 µm to predict OM content, and Da-Costa (1979) found that simulated Landsat channels (# 4, 5, and 6) yield reflectance readings that are significantly correlated with SOC. The power of spectral information also led Ben-Dor et al. (1997) to exploit reflectance for the detection of SOM decomposition status and control of the soil biogenic activity aside from total SOM content. Vinogradov (1981) developed an exponential model to predict the humus content in the upper horizon of plowed forest soils by using reflectance parameters between 0.6 and 0.7 µm for two extreme end members (humus-free parent material and humus-enriched soil). Schreier (1977) found an exponential function to account for SOM content from reflectance spectra. Al-Abbas et al. (1972) used a multispectral scanner, with 12 spectral bands covering the 0.4–2.6 µm range, from an altitude of 1200 m and showed that a polynomial equation will predict the OM content from only five channels. They implemented the equation on a pixel-by-pixel basis to generate an organic content map of a 25 ha field. Dalal and Henry (1986) were able to predict the OM and total organic nitrogen content in Australian soils using wavelengths in the SWIR region (1.702–2.052 µm), combined with chemical parameters derived from the soils. Using similar methodology, Morra et al. (1991) showed that the SWIR region is suitable for identification of OM composition between 1.726 and 2.426 µm. Evidence that OM assessment from soil-reflectance properties is related to soil texture, and most likely to soil clay, was provided by Leger et al. (1979) and Al-Abbas (1972). Aber et al. (1990) noted that OM, including its stage of decomposition, affects the reflectance properties of mineral soil. Baumgardner et al. (1985) demonstrated that three organic soils with different decomposition levels yield different spectral patterns. Hill and Schütt (2000) successfully used the coefficients of a polynomial approximation of a spectral continuum between 0.4 and 1.6 µm to set up a statistical model to map organic carbon concentrations with multi- and hyperspectral imagery. As already discussed, based on the strong spectral relationship, SOM can also be estimated from soil color. Fox et al. (2002) presented a method in which the soil line Euclidean distance (SLED) could be used to estimate SOM from aerial color images. They reported coefficients of determination of 0.70 and 0.78 between observed and predicted SOM contents for two study sites. Nonetheless, these results were site dependent and did not work in another area, suggesting that the stage of SOM decomposition was not similar (see Ben-Dor et al., 1997). Ray et al. (2004) estimated SOC and nitrogen from an IKONOS multispectral image, but with only limited accuracy. This improved a little with Dematte et al.'s (2007) attempt to use Landsat TM images crossing a large geographical area (43,000 ha), which again showed "scene (local)-dependent" SOM determination from multispectral domains. In fact, research varies with OM

quantification, since some works demonstrates high quantitative values and others does not. In this aspect, it is interesting to note that several works performed in the tropics where OM is usually lower than in temperate areas show poor results. This can be probably due to the low values and variability in the tropics, which implies in the statistical models, as they are less detected. Despite this, it is likely that models for SOM are more related with local situations due to innumerous factors (climate, microorganisms, soil management). Jarmer et al. (2005) used a combination of CIE color coordinates (e.g., Escadafal, 1993) and specific spectral absorption features to parameterize statistical models to obtain maps of organic and inorganic carbon contents, as well as total iron content, on a regional scale. Chen et al. (2006, 2008) proposed using Euclidean distance for statistical clustering and the world neural network system to select fields with similar image properties, thereby ensuring the success of mapping SOC for a group of fields. Ben-Dor et al. (2002) were the first to use HSR technology from the air (DAIS 7915), adapting the spectral information modeling to quantitatively map SOM in Vertisol soils from Israel. Later, Stevens (Stevens et al., 2006, 2008, 2010) enlarged the HSR envelope and used CASI and SASI sensors over cultivated soils in Belgium, demonstrating the potential of HSR for mapping SOM, even with relatively low content. The assessed values of SOC ranged from a mean of 3.0% (5.8% max) to 1.7% (0.8% min). Assuming that the SOC-to-SOM ratio is about 0.58, the SOM content in these soils was rather low (6.8%–0.99%) but, in some cases, still higher than the 2% threshold set by Baumgardner et al. (1985) for spectral determination. Over these low SOM areas, Stevens et al. (2008) were able to map SOC on a pixel-by-pixel basis only if the VIS–NIR–SWIR regions were combined. Although the root mean square error (RMSE) of prediction was 0.17% SOC, which is double the value of the laboratory's accuracy, the processed SOC image was reliable and gave the first spatial overview of SOM distribution within a given field. Additional studies, such as those of Ben-Dor et al. (2002) and Toure and Tychon (2004), also showed the ability to derive SOM, but their accuracy was rather low. Zheng (2008) summarized all spectral attempts to obtain SOM and related components (such as total nitrogen and phosphorus) with point and imaging sensors from the air domain using several airborne HSR sensors. A comprehensive review of SOM estimation from reflectance data is given by Ladoni et al. (2010), showing varying results and means. In general, coefficients of determination from remote sensing of SOM vary from low (0.4 from Landsat TM; Dematte et al., 2007) to very high (0.96 from laboratory measurements; McCarty et al., 2002). The HSR sensors provided moderate (0.56, CASI, Uno et al. 2005) to high (0.89, HyMap, Selige, 2006) coefficients. This again demonstrates that finding a good model for predicting SOM is a challenging task with high potential. In mineral soils, the SOM content is rather low (less than 15% and about 0.1%–2% in arid soils) where, as stated earlier, less than 2% is almost undetectable spectrally. As the spectral fingerprints are strongly related to the decomposition stage of the OM, different locations may have different SOM spectral features. Other problems are soil

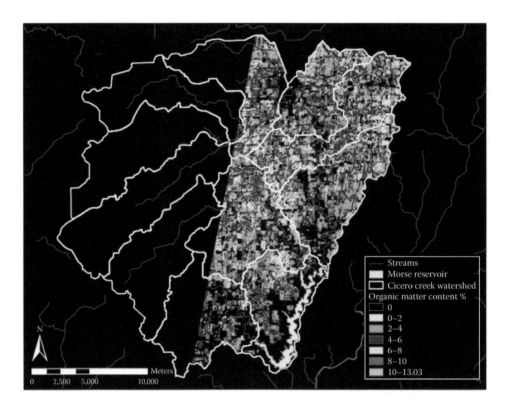

FIGURE 25.14 Prediction of soil organic matter distribution map in Cicero Creek Watershaed as generated from Hyperion data. (After Zheng, B., Using satellite hyperspectral imagery to map soil organic matter, total nitrogen and total phosphorus, A Master thesis submitted to the Department of Earth Science, University of Indiana, Indianapolis, IN, 2008, p. 81.)

moisture, which can hinder small SOM features, BRDF effects, and sensor SNR. It is concluded that SOM is an interesting chromophore that can be problematic and challenging. There are many studies on determining SOM by RS means, with varying results. As SOM is a very important issue in soil, more such studies are warranted, as the technology progresses. Figure 25.14 provides an example of a SOM map that was generated based on spectral information and HSR data from the Hyperion satellite sensor (taken from Zheng, 2008; Zhang et al., 2013). More SOM quantitative images that were generated from airborne HSR sensors can be seen in Stevens 2007 as well as with many others.

25.8.2.3 Change Detection in Soils

CD and multitemporal analysis of RS data are aimed at detecting various types of changes between two or more images taken at different times (Singh, 1989). The temporal information adds new data for the interpreter and "temporal resolution" (added to spectral and spatial resolution) to sensor-performance characterization. Adding a spectral dimension to the soil RS data provides better mapping capabilities than obtained from using limited spectral channels—even when the spatial resolution is high (e.g., Ben-Dor, 2002; Ben-Dor et al., 2005; Goidts and van Wesemael, 2007; Stevens et al., 2010; Hill et al., 2010; Hbirkou et al., 2012; Casa et al., 2013). Adding a temporal dimension to the spectral resolution provides even better capabilities. In practice, spectral resolution has the dominant role in the evolving HSR arena, and temporal resolution has been left behind. However,

temporal spectral analysis can cull information on interactions between the soil surface and the surrounding environment and accordingly can provide a better view of the factors affecting soil formation (Jenny, 1941). Rapid changes on the soil surface can occur from erosion, deposition, physical arrangement, self-segregation, and man-made activity (Lemos and Lutz, 1957). More specifically, the thin upper soil layer (which is ultimately sensed by optical sensors) may be altered by dust accumulation (Offer and Goossens, 2001), rust formation (Ona-Nguema et al., 2002), plowing activity (Fu et al., 2000), changes in particle-size distribution (Sertsu and Sánchez, 1978), vegetation coverage (Zhou et al., 2006), litter (Frey et al., 2003), and the formation of physical and biogenic crusts (Bresson and Boiffin, 1990; Karnieli et al., 1999; Valentin and Bresson, 1992). Until recently, applications for high spectral and temporal resolution data were scarce, mainly due to the high cost of data acquisition by airborne HSR. However, this situation is expected to change significantly as many satellite HSR sensors are in the pipeline with high spectral, spatial, and temporal resolution capabilities (Ben-Dor et al., 2013). A comprehensive overview of forthcoming HSR sensors is provided by Staenz et al., and Held (2012). As a result, the scientific community is starting to perform controlled experiments (Buddenbaum et al., 2012) to prepare the RS community for the multiresolution (spatial, spectral, temporal) approach. Methods to account for CD between areas that are well identified either spectrally or spatially (e.g., soil to vegetation) are well known and frequently used (Adar et al., 2014a). Methods to account for CD between the same land-cover categories (e.g., soil to soil) are more complex. In this case, the more information provided on the area in question, the better the discrimination capability will be. Spectral reflectance, as achieved from HSR sensors, can provide added information. Nonetheless, due to the aforementioned problems with factors affecting soil spectroscopy, this mission is very challenging, as a small spectral change might occur due to factors other than real change on the soil surface. Recently, Adar et al. (2014a) developed an approach in which the "factors affecting soil reflectance" can be estimated from HSR images. This method enables better CD analysis based on real spectral changes. Those authors demonstrated the approach using HyMap data acquired over an open-mining area in the Czech Republic at a 1-year interval. More recently, Adar et al. (2014b) conducted a controlled study to understand the capability of high spectral information for spatial discrimination between soil entities under optimal conditions, where "factors affecting soil reflectance" are minimized. An artificial soil matrix (made of 50 different soil samples, each in a 3 × 3 cm dish), which was measured by an image spectrometer (HySpex) under laboratory conditions, provided the database for this study. Several changes were made in the soil matrix between each data acquisition along with relocating some of the soils' original position. Using the VIS–NIR, SWIR, and VIS–NIR–SWIR spectral segments separately, with several known methods to detect possible (spectral) changes in a given pixel, it was found that the wider the spectral coverage, the better the discrimination capability. Figure 25.15 shows an example of the results obtained by this analysis using the VIS–NIR–SWIR regions (alone and together). As seen, only when the complete spectral region is used with a specific method to assess the spectral changes, the spatial changes could be obtained (compare image d–a). The authors

revealed limitations in identifying changes between different soils in three cases: (1) When the soils were within the same larger group of soil classes, very small changes could not be detected; (2) when there were opposing effects on the spectral signature, such as twice the Fe oxides and, at the same time, twice the OM, their effects might cancel each other out, resulting in very similar spectral signatures; and (3) when differences in some of the spectral absorption features are reduced as a result of similar and high average particle-size fraction, this reduces the albedo of the spectral signatures to a very similar level. Although Adar et al.'s (2014b) study demonstrated limitations for CD in soils and indicated that different algorithms can produce better results, they also concluded that CD in soils from both spectral and spatial domains is a difficult task and calls for caution in drawing any conclusions. They did, however, suggest overcoming these limitations by fusing chemometric capabilities with CD techniques and not relying purely on the spectral information of the image. To summarize, CD in soils can shed light on some quick processes on the soil surface but at the same time present difficulties in significantly distinguishing between the two (or more) soils.

25.8.2.4 Soil Salinity

Soil salinity is a dynamic property that emerges on the soil's surface mostly under arid and semiarid conditions and under secondary water utilization. It can therefore be effectively monitored by remote sensing as light spots of NaCl obtained under a high-salinity regime that can be monitored by aerial photography (Rao and Venkataratnam, 1991). At the lowest saline concentrations, the unaided eye cannot detect salinity effects and a better analytical tool is required. Airborne digital multispectral cameras and videography, usually with three to four

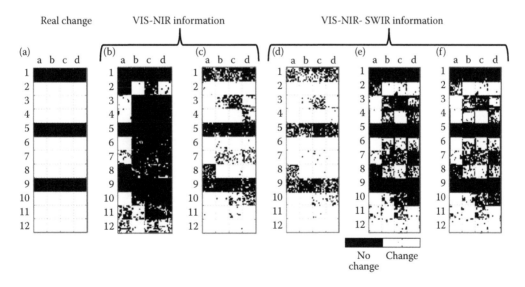

FIGURE 25.15 A demonstration of a spectral change detection of soils as done under laboratory (ideal) conditions. Every cell represents different soil that configured a soil matrix, which were then scanned by the HySpex image spectrometer at the laboratory. Black cells in image (a) represent soils without out any change. White cells in (a) represent areas where soil was replaced by other soils. Images (b–f) provide the results of different change detection analytical methods to point out possible changes within the cells. As seen image (d) provides the best results with minimum black pixels on the white cells and minimum white pixels on the black cells. (From Adar, S. et al., *Int. J. Remote Sens.*, 35, 1563, 2014a; Adar, S. et al., *Geoderma*, 216, 19, 2014b.)

channels in the VIS and IR regions, and color IR photographs were used as tools to identify and assess problem salinity areas in U.S. agriculture in the 1980s and 1990s. Everitt et al. (1988) used narrowband videography to detect and estimate the extent of salt-affected soils in Texas, United States, while Wiegand et al. (1991, 1992, 1994) analyzed and mapped the response of cotton to soil salinity using color IR photographs and videography with three bands (0.84–0.85 μm, 0.64–0.65 μm, and 0.54–0.55 μm) and a spatial resolution of 3.4 m. By relating video and field data such as soil electrical conductivity, plant height, and percent bare area, they determined the interrelations between plant, soil salinity, and spectral observations. These studies found that color IR composites and red narrowband images were better than green and NIR narrowband images (Escobar et al., 1998; Wiegand et al., 1992, 1994). Extensive research on the application of panchromatic and multispectral satellite imagery to map salt-affected areas has been conducted over the last four decades, mostly using panchromatic and multispectral (VIS–NIR–SWIR and/or thermal) sensor. Works by Csillag et al. (1993), Epema (1990), Metternicht and Zinck (1997), Rao et al. (1995), and Evans and Caccetta (2000) provide some examples of applications on different continents and in different environmental settings. All of these works were generally successful in mapping saline versus nonsaline surfaces. Some researchers have attempted to map salinity types (e.g., saline, alkaline) and degrees (e.g., low, moderate, high) (Metternicht and Zinck, 1996; Kalra and Joshi, 1996), with varying degrees of success. From the year 2000 onward, experimental hyperspectral satellite data from sensors like the CHRIS onboard the ESA mission PROBA-1, or Hyperion on EO-1, were assessed for their ability to identify and map salt-affected areas (Dutkiewicz, 2006; Schmid et al., 2007). A comprehensive description of all attempts to map soil salinity from multispectral satellite sensors, starting from old sensors such as Landsat MSS to newer ones such as IKONOS, is given in Metternicht and Zinck (2008). This book also covers

other RS means of detecting soil salinity, such as active and passive sensors in the microwave and thermal domains.

While the aforementioned means were being used mostly to locate saline soil areas, especially those that are visible to the naked eye, research was being directed to assessing soil salinity that is low in content or in its first stages of development, as such soil can be agrotechnically treated to obtain optimal cultivation under extreme conditions. In this case, full spectral information is needed and sophisticated analytical approaches to mining spectral information related to salt were developed (Farifteh et al., 2004, 2006; Huang and Foo, 2002). The added value was the identification of salt-affected areas before they become visible to the naked eye, as done by Ben-Dor et al. (2002) and later also by Howari et al. (2002) and Dehaan and Taylor (2002) using hyperspectral sensors. These studies were based on Taylor et al. (1994), who were the first to show that it is possible to use airborne superspectral data to map salinity by using the 24-band airborne Geoscan, and VIS–NIR/SWIR data, at Pyramid Hill, in Victoria, Australia.

Whereas the salinity is most important in the root zone, it is interesting to note that two recent innovative studies were able to correlate salinity level at 30 cm depth to surface reflectance as acquired by an airborne hyperspectral sensor (Figure 25.16) and form an indirect correlation between leaf reflectance of tomato and the electrical conductivity measured in the root zone. These data were then projected on a cartographic domain to generate soil salinity-affected areas for the farmer (Goldshleger et al., 2013).

Vegetation is an indirect factor that facilitates detection of salt in soils from reflectance measurements (Hardisky et al., 1983; Wiegand et al., 1994). Gausman et al. (1970), for example, pointed out that cotton leaves grown in saline soils have a higher chlorophyll content than leaves grown in low-salt soil. Hardisky et al. (1983) used the spectral reflectance of a *Spartina alterniflora* canopy to show a negative correlation between soil salinity and spectral vegetation indices. In the absence of vegetation,

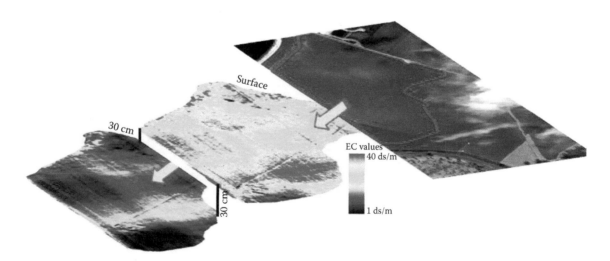

FIGURE 25.16 Soil salinity map as generated from an airborne AISA sensor (Eagle & Hawk sensors covering the VIS-NIR-SWIR region) used a spectral model generated from several samples taken at 0 and 30 cm depth. As seen, a favorable map was obtained for the salinity at 30 cm and not only on the soil surface.

salt's major influence is on the structure of the upper soil surface. Figure 25.17 shows saline and nonsaline spectra, taken from Everitt et al. (1988), in the VIS–NIR region. The saline soils had relatively higher albedo than the nonsaline ones. Furthermore, the saline soils had crusted surfaces that tended to be smoother than the generally rough surfaces of the nonsaline soils. Although Gausman et al. (1977) and Rao et al. (1995) reported similar trends in other soils, it should be noted that in soils with relatively high salt content, the opposite behavior can also be reasonably expected. This is because salt is a very hygroscopic material, which tends to decrease the soil albedo as water content rises. Because no direct significant spectral features are found in the VIS–NIR–SWIR region to identify sodic soil, indirect techniques are thought to be more appropriate for classifying salt-affected areas (Verma et al., 1994; Sharma and Bhargava, 1988). Salt in water is most likely to affect the hydrogen bond in water molecules, causing suitable spectral changes. Based on this, Hirschfeld (1985) suggested that high-spectral-resolution data are required. Support for this idea was given by Szilagyi and Baumgardner (1991), who reported that characterizing salinity status in soils is feasible with high-resolution laboratory spectra. A relatively high number of spectral channels are also important for identifying an indirect relationship between salinity and other soil properties that appear to consist of chromophores in the VIS–NIR–SWIR regions. Csillag et al. (1993) analyzed high-resolution spectra taken from about 90 soils in the United States and Hungary for chemical parameters, including clay and OM content, pH, and salt. They stated that because salinity is such a complex phenomenon, it cannot be attributed to a single soil property. While studying the capability of commercially available earth-observing optical sensors, they indicated that six broadbands in the VIS–NIR–SWIR region best discriminate soil salinity. These six channels were selected solely on the basis of their overall spectral distribution, which provided complete information about salinity status. Thus, it can be concluded that it is necessary to look at the entire spectral region to evaluate

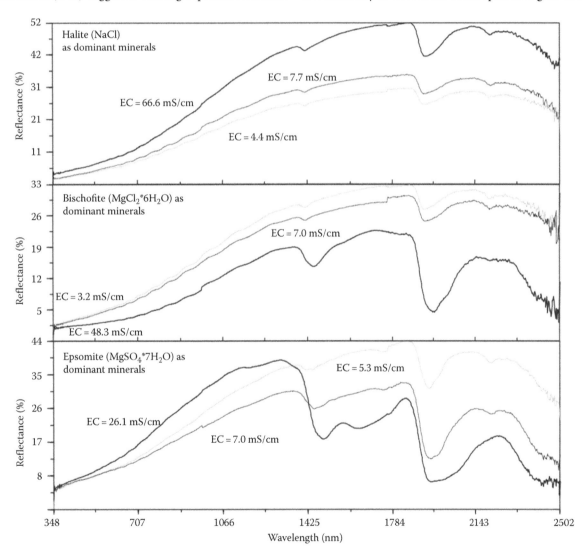

FIGURE 25.17 The reflectance spectra of a selected soil that was artificially contaminated with different evaporates in different content. (After Farifteh, J. et al., *Geoderma*, 130, 191, 2006.)

salinity levels in different environments and unknown soil systems. In summary, soil salinity is a property that can be monitored by RS means but requires high temporal, spectral, and spatial resolution.

25.8.2.5 Soil Moisture

Soil moisture is an important property, not only for assessing the available water content needed for plant utilization but also for assessing the direct exchange of soil water with the atmosphere (e.g., evaporation) and quantifying moisture effects on other chromophores. In fact, it is considered to be one of the most significant parameters in the soil system. It can vary from hygroscopic moisture (water left on the surface after equilibrium has been achieved with the atmosphere) to a saturated stage (water fills 50% of the soil pores). The effect of the water molecules on soil reflectance is strong and significant. Whereas hygroscopic water most likely shows the absorption features of OH molecules at 1.4 and 1.9 μm with a strong SWIR shoulder at 2.62 μm (demonstrating the imaginary part of the refractive index), in the saturated condition, the real part of the refractive index is dominant and hence the entire spectrum is affected such that the overall spectral-baseline height ("albedo") is lowered (visible to the naked eye as darker soil). In between these two moisture contents, the spectral signatures are affected by both mechanisms (real and imaginary parts), complicating the assessment of water content from soil reflectance. In most cases, the impact of soil moisture on the reflectance is unknown and therefore ignored. Muller and Decamps (2001) modeled reflectance changes due to soil moisture in a real field situation using multiband airborne Spot data. They showed that the impact of soil moisture on reflectance could be higher than the differences in reflectance due to the soil categories and hence calls for caution in applying soil remote sensing under wet conditions. On the other hand, several attempts have been made to map soil water content using soil reflectance information, some under laboratory conditions, and others in the field and in air and space domains. Nevertheless, assessing soil water content from reflectance measurements is still a challenge, in particular correcting for the water-masking effect that hinders the capture of other soil chromophores' activities. Under dry soil conditions (mostly represented by hygroscopic water), the absorption features at 1.4 and 1.9 μm, and others across the SWIR region, can be correlated with water, as shown by Bowers and Hanks (1965), Dalal and Henry (1986), and Ben-Dor et al. (1999). A study by Demattê et al. (2006) assessed the 1900 nm water combination band feature and others for practical use and found that the best interpretation of water content occurs when both dry and wet soil samples are spectrally measured.

A novel approach to reconstructing the soil's spectral signature was through the use of various water film depths related to moisture content. Thus, Bach and Mauser (1994) simulated the reflectance change in the soil spectra from dry to moist. They combined Lekner and Dorf's (1988) model for internal reflectance with the absorption coefficients from Palmer and Williams (1974) into Beer's law. Bach and Mauser (1994) simulated dry

through wet soil and applied the process for predicting water content to an AVIRIS image of a partially irrigated field and a field with dark organic soil at the Freiburg test site in Germany. Today, we have a better understanding of the causes to change the soil spectral and we are improving the methods for modeling water content in soils.

The challenge in determining soil moisture content across the VIS–NIR–SWIR region lies in the fact that the water molecules significantly affect all other spectral chromophores and thus may hinder the quantitative spectral approach to determining chromophores, such as OM, Fe oxide, clay, and carbonates. Accordingly, the water–radiation interaction is a very important issue in the soil proximal-sensing discipline, which is attracting more and more users and accumulating experience in the laboratory and field and, recently, from RS domains as well. Based on the strong effect of the real part of the refractive index, gray-level values in the VIS region enable estimating water content under certain amounts of water (Mouazen et al., 2005). Zhu et al. (2011) recently showed a good correlation between soil moisture and digital gray level under very moist soil conditions (25%–60%). Weidong et al. (2002) have shown that a better soil moisture prediction using soil spectroscopy can be determined by adjusting the soil types. Lobell and Asner (2002) demonstrated that the SWIR region is much more sensitive than the VIS region when assessing soil moisture and described an exponential relationship between the water content and soil reflectance values. Mouazen et al. (2006) showed that the soil moisture content can be estimated using the VIS–NIR and only part of the SWIR regions (306.5–1710.9 nm) where Whiting et al. (2004) suggested using the far SWIR region (2200–2500 nm) to estimate water content by fitting an inverted Gaussian function centered on the assigned fundamental water-absorption region at 2800 nm. As the far SWIR region is strongly affected by the left shoulder of the aforementioned fundamental absorption, a logarithmic soil spectrum continuum with convex hull boundary points was found to be correlated to water content. Based on the aforementioned method, they were also able to present a processed AVIRIS hyperspectral image that provides the soil-surface moisture content (Figure 25.18, Whiting et al., 2004). The spectral approach also attracted Haubrock et al. (2008), who successfully validated a new model for predicting gravimetric soil moisture. The method was termed normalized soil moisture index and combines reflectance values at 1800 and 2119 nm around the 1900 nm water combination bands. This index was applied to remotely sensed images and enabled the production of soil-surface moisture maps, generated from HyMap airborne images, which were found to be highly correlated with the field moisture content measured at the time of the overflight (Figure 25.19; Haubrock et al., 2008). Surprisingly, neither Whiting's nor Haubrock's methods have been implemented in practice, probably because they are relevant to certain conditions in which moisture is not high. As the real part of the refractive index is dominant at high moisture levels, modeling the spectral features of the water becomes difficult in certain soil moisture ranges and the

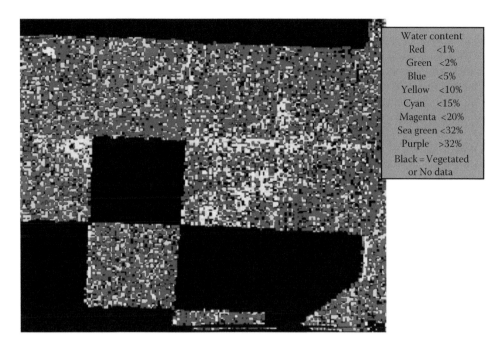

FIGURE 25.18 Surface water content (gravimetric) from AVIRIS data (May 3, 2003, near Lemoore, California) as estimated with the SMGM. (Generated by Whiting, M.L. et al., *Remote Sens. Environ.*, 89, 535, 2004.)

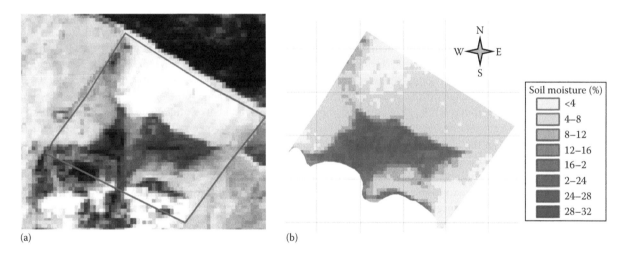

(a)   (b)

FIGURE 25.19 (a) Soil moisture content as estimated from HyMap images acquired on July 20, 2004, over Welzow, Germany, and (b) RGB image of the area encoded 0.619, 0.528, and 0.452 μm. (After Haubrock, S. et al., 2006.)

reflectance should be treated differently, as done by Zhu et al. (2011). This author used the brightness of the water film as an indicator to determine the soil moisture from the VIS region.

It should be mentioned here that in general, other remote methods exist to estimate soil water content, such as using thermal bands in the LWIR region (estimating the latent and sensible heat fluxes; Eltahir, 1998) or active microwave and millimetric wave (Eliran et al., 2013) spectral domains that are based on sensing the dielectric constant of the soil–water mixture. A comprehensive review of soil moisture assessment from orbital sensors is given by Serrano (2010). As the water reflectance properties are important not only for determining

the water content but also to sharpen and fine-tune other soil properties, it is essential that this field continue to be explored. Further work in reconstructing the spectra that will combine the spectral relationships of water content and soil components based on the physical nature of the materials and photon interactions is strongly needed. Other challenges include examining the models obtained under select conditions using satellite platforms and defining a robust way to assess water content at all levels, in all orders of soil. An excellent review on monitoring soil moisture content from all orbital RS means is given by Barrett and Petropoulos (2013). The optical means is partially covered by them mainly because the HRS is not yet widely operational from space, whereas other limitations are also reported, which are identical to what was discussed in Section 25.7. Nevertheless, they concluded that the high spectral and spatial resolutions of the HSR technology may open up another channel to better estimate soil moisture content from orbit.

In summary, it can be concluded that the soil-surface water content can be cautiously estimated using reflectance measurements, but due to the effect on other soil components, their spectral absorption requires proper attention. This approach has not yet been fully studied or developed in this innovative direction, that is, for use in HSR, although it seems to hold great promise. Other spectral regions, such as thermal, millimetric, and microwaves, may also be used for this mission.

25.8.2.6 Soil Carbonates

Carbonates, particularly calcite and dolomite, are found in soils that are formed from carbonic parent materials or in a chemical environment that permits calcite and dolomite precipitation. Carbonates, and especially those of fine particle size, play a major role in many of the soil chemical processes that are most likely to occur in the root zone. A relatively high concentration of fine carbonate particles may cause fixation of iron ions in the soil and consequently inhibition of chlorophyll production (chlorosis-driven carbonates). On the other hand, an absence of carbonate may affect the soil's buffering capacity and thus negatively affect the biochemical and physicochemical processes. Remote sensing allows distinguishing among the common carbonate minerals on the basis of unique spectral features found in the SWIR (as well as TIR) regions. The C–O bond, part of the $-CO_3$ radical in carbonate, is the spectrally active chromophore. Hunt and Salisbury (1970, 1971) pointed out that five major overtones and combination modes are available to describe the C–O bond in the SWIR region. Gaffey (1986) added two additional significant bands centered at 2.23–2.27 µm (moderate) and 1.75–1.80 µm (very weak), whereas van der Meer (1995) summarized the seven possible calcite and dolomite absorption features with their spectral widths. It is evident that significant differences occur between the two minerals. This enabled Kruse et al. (1990), Ben-Dor and Kruse (1995), and others to differentiate between calcite and dolomite formations using airborne spectrometer data with bandwidths of 10 nm. In addition to the seven major C–O bands, Gaffey and Reed (1987) were able to detect copper impurities in calcite minerals, as indicated by the broadband between 0.903 and 0.979 µm. However, such impurities are difficult to detect in soils, because overlap with other strong chromophores may occur in this region. Gaffey (1985) showed that iron impurities in dolomite shift the carbonate's absorption bands toward longer wavelengths, whereas magnesium in calcite shifts the band toward shorter wavelengths. As soil carbonates are most likely to be impure, it is only reasonable to expect that the carbonates' absorption-feature positions will differ slightly from one soil to the next. A correlation between reflectance spectra and soil carbonate concentration was first demonstrated by Ben-Dor and Banin (1990b). They used a calibration set of soil spectra

from Israel and $CaCO_3$ content to find three wavelengths that best predict the calcite content in the soil samples (1.8, 2.35, and 2.36 µm). They concluded that the strong and sharp absorption features of the C–O bands in the examined soils provide an ideal tool for studying the soil carbonate content solely from their reflectance spectra. The best obtained performance for quantifying soil carbonate content ranged between 10% and 60%. Since that pioneering work, several proxy models to assess soil carbonate content have been published (e.g., Balsam and Deaton, 1996; Thomasson, 2001). The use of the SWIR region to map carbonates from airborne HSR has been shown by many users in arable lands (e.g., Lagacherie et al., 2008; Gomez et al., 2008). However, in agricultural soil, where the plowing layer is mixed, estimating $CaCO_3$ content is still a significant challenge. This is especially true in heavily leached environments (low pH) where lime is required to improve soil function. Demattê et al. (2003) already determined alterations between controlled and limed soils. Along these lines, Viscarra Rossel et al. (2005) and later Araujo et al. (2013) developed a spectrally based concept to account, in the field, for the "lime requirement" content and demonstrated its applicability to precision agriculture. Mapping carbonate rocks from airborne domains using HSR technology is well documented (e.g., Kruse et al., 1990; Ben-Dor et al., 1995), as is mapping from satellites (e.g., Ninomiya et al., 2005; Gersman et al., 2008), but in soil, where the carbonate is mixed with other minerals, it is more difficult. Gmur et al. (2012) showed the efficiency of hyperspectral analysis associated with a regression tree to increase the prediction accuracy of carbonate in the soil ($R^2 = 0.95$). Lagacherie et al. (2008) examined how reflectance spectrometry can be used to estimate clay and $CaCO_3$ contents simultaneously in soil using both field and airborne measurements. They showed nine intermediate stages from the laboratory to HyMap sensor measurements crossing spatial and sensor characteristics such as radiometric quality, spectral resolution, spatial variability, illumination conditions, and surface status including roughness, soil moisture, and presence and nature of pebbles. They found significant relationships between clay and $CaCO_3$ contents from the spectral continuum removal values computed, respectively, at 2206 and 2341 nm, which persisted from an ASD spectrophotometer to the HyMap spectral imaging sensor. Decreasing performance was obtained going from the laboratory to the hyperspectral domains, indicating the factors affecting reflectance spectra as discussed in Section 7. In summary, it can be concluded that soil calcite content has significant potential for quantitative monitoring using spectral information across the end of the SWIR spectral region but, at the same time, has some constraints related to its mixture with other soil materials (minerals, SOM, and water), low solar energy, and accuracy deterioration in airborne relative to field and laboratory results.

25.8.2.7 Soil Contamination

Soil contamination refers to a process in which nonpedogenic constituents enter the soil volume with no relation to the soil's natural formation or generation. This refers mostly to short-term processes in the soil. Soils can be contaminated by various sources, either anthropogenic (e.g., hydrocarbon) or natural (e.g., dust accumulation). As various contaminants may change the soil's chemistry as well as its physical behavior, one would expect to be able to monitor such processes by spectral sensing means. Demattê et al. (2004a) examined the industrial by-product of sugarcane that was dumped into a nearby soil area and found it to significantly alter the soil's chemical properties. Accordingly, they found that this alteration was noticeable in the spectral reflectance of the soils via the magnitude of the signal, without much change in the general spectrum's shape. This is probably due to physical effects of the sugarcane by-product that may cause different aggregation stages in the natural and contaminated soils. Chemical contamination is also an important issue in the soil environment. Heavy-metal contamination of alluvial soils on river banks has been addressed in experimental studies that used the soil proxy approach (Kooistra et al., 2003, 2004; Wu et al., 2007; Xia et al., 2007). However, probably most operational applications of airborne HSR missions for monitoring soil contamination have been performed in the context of chronic or accidental pollution resulting from metal mining. For example, Chevrel et al. (2005) investigated six mining areas in the MINEO project, five in Europe (Portugal, United Kingdom, Germany, Austria, and Finland) and one in Greenland, using HyMap airborne-imaging-spectrometry data. HSR was used for mapping the extent and type of chronic contamination with heavy metals using primarily trace minerals of pyrite oxidation as an indirect indicator of potential contamination, forming an indispensable basis for environmental impact assessment, environmental monitoring of historical mining sites, and remediation planning. Within the framework of the EO-MINERS project (2010–2013), Chevrel (2013) used two HSR sensors (HyMap and AHS) to monitor the coal-ash contamination from open mines in the nearby urban and soil environments. Ren et al. (2009) have also estimated the soil contamination by As and Cu using reflectance spectroscopy of areas near mining activities. In April 1998, the dam of a mine tailings pond in Aznalcollar, Spain, collapsed and flooded a soil area of more than 4000 ha with pyrite-bearing sludge containing high concentrations of heavy metals. An emergency airborne RS mission, with the objective of assessing the extent of residual heavy-metal contamination after the first cleanup operations that lasted until 1999, was flown with HyMap covering the affected area in 1999 and 2000 (Kemper and Sommer, 2002, 2003, Garcia-Haro et al., 2005). As a first step, the possibility of adapting chemometric approaches to a quantitative estimation of heavy metals in the soils polluted by the mining accident was explored (Kemper and Sommer, 2002). Six months after the end of the first remediation campaign in early 1999, soil samples were collected for chemical analysis and VIS to SWIR reflectance (0.35–2.4 µm) was measured. Concentrations of As, Cd, Cu, Fe, Hg, Pb, S, Sb, and Zn in the samples were well above background values. Prediction of heavy metals was achieved by stepwise multiple linear regression analysis and by using an artificial neural network approach. This enabled the prediction of six out of the nine elements with high

accuracy. The best R^2 values between the predicted and chemically analyzed concentrations were As, 0.84; Fe, 0.72; Hg, 0.96; Pb, 0.95; S, 0.87; and Sb, 0.93. Results for Cd (0.51), Cu (0.43), and Zn (0.24) were not significant.

In the second step of the study, variable multiple end member spectral mixture analysis (VMESMA; Garcia-Haro et al., 2005) was used to analyze the HyMap data acquired in 1999 and 2000. A spectrally based zonal partition of the area was introduced to allow the application of different submodels to the selected areas. Based on an iterative feedback process, the unmixing performance could be improved in each stage until an optimum level was reached. The sludge quantities obtained by unmixing the hyperspectral data were confirmed by field observations and chemical measurements of samples taken in the area. Figure 25.20 shows the sludge-abundance map derived from the 1999 HyMap data using this iterative VMESMA approach. The semiquantitative estimate of sludge from residual pyrite-bearing material could be transformed into quantitative information to assess acidification risk and the distribution of residual heavymetal contaminants based on an artificial mixture experiment and the derivation of simple stoichiometric relationships. As a result, the sludge-abundance map could be rescaled to quantities of residual pyrite sludge, associated heavy metals, and acidification potential due to the need to counteract calcite buffering. Wu et al. (2005), who used reflectance spectroscopy to study the mercury contamination in suburban agricultural soils in the Nanjin region, China, revealed interesting results. They found correlations between mercury concentration and goethite and clay absorbance features at 496 and 2210 nm, respectively. They concluded that an intercorrelation between mercury and the aforementioned constituents is the key factor for obtaining a prediction of mercury, as it has no spectral fingerprints in the VIS–NIR–SWIR region. Although they have not yet been applied, the authors strongly recommended the use of operational RS techniques to fully implement this interesting finding for mapping of soil contamination with mercury. Another possible intercorrelation is with SOM. Malley and Williams (1997) showed that reflectance properties of sediments are associated with the content of OM, which acts as a chelating substance. As SOM has spectral fingerprints as previously discussed, the intercorrelation of the heavy metal bound to the SOM may enable extraction from the reflectance characteristics of the soil.

Several other studies have been published on the capacity of soil spectral information to detect heavy-metal content in soils. Wu et al. (2011) demonstrated that the intercorrelation of the nonspectral active constituents (Ni, Cr, Co, and Cd) with spectrally active soil components (Fe oxides, SOM, and clay) is the major predictive mechanism. They showed that a correlation with total iron (including active and residual iron) is the major mechanism by which cadmium can be spectrally active in the soil environment. Looking toward showing a mechanism of public control of the environment, Araujo et al. (2014) indicated that VIS–NIR, as combined with MWIR, was able to detect Cr variation on soils caused by chemical products discarded from leather industries with a 0.93 R^2. The authors observed strong alteration

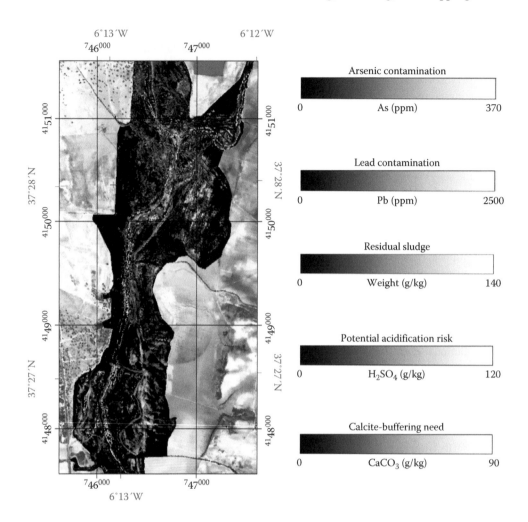

FIGURE 25.20 Contamination parameters derived from the sludge abundance image for June 1999: Arsenic and lead contamination in ppm, residual sludge in g/kg, maximum potential acidification after full oxidation of pyrite in g/kg, and neutralization requirements for full oxidation of pyrite in g/kg. (Taken from Kemper, T. and Sommer, S., *Environ. Sci. Technol.*, 36, 2742, 2002.)

on spectral features (i.e., in the 500 and 570 nm and 2400 cm^{-1} bands) and lower intensities in all spectra when high application of the product was performed. Other workers have demonstrated similar capability to assess heavy metals spectrally, such as Pandit et al. (2010) who modeled the concentrations of Pb and other heavy metals using soil reflectance and concluded that reflectance spectroscopy has a promising potential to map the spatial distribution of Pb abundance in soils. Another indirect way to study heavy-metal content in the root zone via remote sensing was demonstrated by Kooistra et al. (2004), Liu et al. (2010), and recently Kopackova et al. (2014), who studied the relationship between the leaves' spectral response to the heavy-metal content in the root zone. They found a good correlation between leaf spectra and both aluminum and basic cations in the soil solution of the 0–20 cm soil (organic) horizon. Associating the results with HyMap HSR data on a small scale could indirectly result in mapping the organic horizon status of the soil underneath the vegetation. This work shows that an innovative idea can play an important role in exploiting reflectance properties in all domains to understand the soil condition in general and pollution in particular.

Hydrocarbon contamination of soil is another important constituent that can be detected by reflectance spectroscopy. The spectral detection of hydrocarbons can be divided into two categories: direct sensing of hydrocarbons and indirect sensing of minerals altered by the hydrocarbons. Whereas sensing of the first category relies on the spectral fingerprints of the hydrocarbon materials, the second category examines the soil matrix (minerals) that might be affected by hydrocarbon contamination. Direct detection using spectral means was reported by Malley et al. (1999), who concluded that reflectance spectroscopy has good potential to work in the laboratory. Later, Hörig et al. (2001), using a HyMap sensor, showed that HSR technology can detect hydrocarbon signals in an artificially contaminated soil environment. Another comprehensive work in this area is Winkelmann's thesis dissertation (2005), in which she systematically studied the applicability of HSR for detecting contaminated sites (including soils). Souza Filho (2013) also demonstrated the ability of airborne HSR to quantitatively detect hydrocarbons from both the SWIR and TIR spectral ranges. Detection via indirect sensing has been reported by several authors, such as

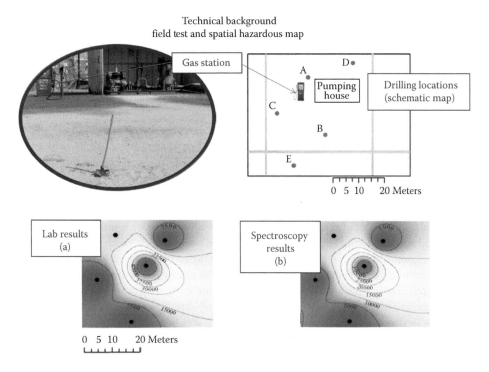

FIGURE 25.21 A contamination soil site with petroleum and the interpolation map generated from 5 samples using TPH analysis at the laboratory (a) and spectral-based model in the field (b).

Yang et al. (2000) and van der Meer et al. (2002) and also Bihong et al. (2007) who used ASTER data to detect seepage from hydrocarbon reservoirs. A recent attempt to map an oil spill from Landsat data (combined with radar data) was performed by Espinosa-Hernandez et al. (2013): Landsat images identified polluted areas over the bare soil. Several studies based on reflectance spectroscopy have demonstrated its ability to detect hydrocarbon species and their concentrations (Schwartz et al., 2011). Schwarz et al. (2012) also demonstrated that measurement of TPH from reflectance spectroscopy is as good as measurements performed by three certified laboratories using wet chemistry analyses and concluded that this approach has commercial applicability. Accordingly, a patent has been assigned for this application (US 20140012504 A1). Interpolating spectral measurements of TPH in the field on a GIS-based map enabled demonstrating the spatial distribution of the main source of the oil contamination from within a gas station area. Figure 25.21 shows a case study of a contaminated gas station where point spectral measurements and GIS interpolation provides similar results as laboratory measurements of the same points.

In summary, it can be concluded that the reflectance properties of soils enable the assessment of various contaminants in their environment and that HSR technology is proving to be a promising tool for this purpose. Many more ideas and research directions are still open and workers are encouraged to further explore and study this promising field.

25.8.2.8 Soil Aggregation and Roughness

Many studies have shown that particle-size differences alter the shape of soil spectra (in powdered shape) (Hunt and Salisbury,

1970; Pieters, 1983; Baumgardner et al., 1985). Specifically, Hunt and Salisbury (1970) quantified an effect of about 5% in absolute reflectance due to particle-size differences and showed that these changes occurred without altering the position of diagnostic spectral features. The physical process of the particle-size effect was related to changes in the real part of the refractive index resulting in attenuations in scattering and shading. Under field conditions, aggregate size rather than particle size may be more dominant in altering soil spectra (Orlov, 1966, Baumgardner et al., 1985). Soil aggregation is related to cementing agents in the soil such as clay, Fe oxides, and OM. Ben-Dor et al. (2008) demonstrated that rubification (the coating of quartz particles with Fe oxides) is responsible for sand-dune stabilization due to the cementing effect of Fe oxide. In the field, aggregate size may change over a short time frame due to tillage, soil erosion, artificial contamination (e.g., sugarcane by-product), aeolian accumulation, or physical crust formation (e.g., Dahms, 1993). Basically, the aggregate size, or more likely roughness, plays a major role in the shape of the soil spectra acquired in the field and air domains (e.g., Cierniewski 1987, 1989; Feingersh et al., 2007). Escadafal and Hute (1991) showed strong anisotropic reflectance properties of five soils with rough surfaces. Cierniewski (1987) developed a model to account for soil roughness based on the soil reflectance parameter, illumination properties, and viewing geometry for both forward and backward slopes. The model showed that the shading coefficient of the soil surface decreases with decreasing soil roughness. For soils on forward slopes of more than 20°, the shadowing coefficient also decreased when the solar altitude increased throughout the range of 0°–90°. The model indicated that the opposite relationship might hold for soil slopes with surface roughness lower than 0.5 for a specified range of solar altitudes. Using empirical observations of smoothed soil surfaces, Cierniewski (1987) showed that the model agrees closely with field observations. A brief and excellent summary on the multiple- and single-scattering models of soil particles with respect to the roughness effect is given by Irons et al. (1989). Soil aggregation is strongly correlated to the content of cementing agents in the soil such as OM, Fe oxides, carbonates, and clay. Selige et al. (2006) studied the variability of field topsoil texture and SOM using HyMap airborne hyperspectral imagery. They found a good correlation between reflectance and two related properties: OM and soil texture, suggesting that SOM is the direct chromophore for soil texture. Cierniewski et al. (2014) evaluated the fit between HSR bidirectional reflectance data of soil surfaces formed by a cultivator, a pulverizing harrow, and a smoothing harrow, collected under field conditions (as illuminated by direct and diffuse solar radiation), to their bidirectional reflectance equivalents measured in the laboratory with only a direct radiation component and soil roughnesses similar to those in the field. They found that the fit increased from 400 to 450 nm and decreased notably for wavelengths between 1950 and 2300 nm. A less significant decrease in fit was revealed at around of 700, 940, and 1140 nm. This again shows the constraint encountered when upscaling from the laboratory to field domains as a result of the soil's roughness characteristic. In another study, Piekarczyk (2013) performed a controlled study using broadband (VIS) albedo characteristics to study the optimal conditions for remote sensing of soils (in terms of radiation and general geometry). Using five different roughnesses of the same soil with the same solar zenith angle, the difference between the optimal time and the available times from current sun-synchronous satellites was examined (using NOAA-15 and MODIS). It was found that the MODIS, crossing the equator at 10:30 A.M. in an orbit that is far from optimal for the albedo approximation, is much less useful than the NOAA-15, crossing the equator at 7:30 A.M. Another conclusion drawn from that research was that the relationship between the broadband (blue sky) albedo of a bare cultivated soil surface and the angle of the solar zenith clearly depend on the soil roughness. In summary, it can be concluded that soil aggregation size may affect the reflectance properties of a soil surface, with a significant impact on the overall spectral signal. As the roughness may be associated with soil degradation, deposition, or other pedogenic effects, it can serve as an indicator for the RS of these processes but needs special attention to generate a quantitative model for that purpose.

25.8.2.9 Soil Sealing (Cover, Dust, and Crust)

Soil crusts and covers can be formed by different processes. The biogenic crust is one example of such interference, and mineral alteration by fire is another. Aeolian material and desert varnish are also good examples of surface crusts. Biogenic activity can be captured by RS means via what is called a biogenic crust. This crust is made up of cyanobacteria and strongly affects the general albedo of the soil reflectance as well as the SWIR

region absorption features (Karnieli et al., 1999). The biogenic crust shows photosynthetic activity when the soil water content is on the rise and is less active under dry conditions. Demattê et al. (1998) observed that the spectral behavior of the biological aggregates changes according to their chemical composition and is associated with the micro- and macrofauna activity in the soil pedon. They speculated that the animals bring soil particles from the undersurface to the surface, which can be distinguished using field spectroscopy. The crust is active in arable semiarid to arid soils, where cultivation is rare and vegetation even more so. Karnieli and Tsoar (1995) showed that overgrazing destroys the biogenic crust to the point that significant albedo differences can be obtained from satellite views. Similarly, they showed that these differences are significant across the Egypt–Israel border based on overgrazing activity on the Egyptian side and relatively low grazing on the Israeli side. Other studies have demonstrated similar effects across the globe (Lucht et al., 2000).

Another type of crust on the soil surface is termed "physical crust." This crust is formed by raindrops' energy (Morin et al., 1981), which causes segregation of fine particle sizes at the surface of the soil and hence affects its reflectance characteristics. This crust reduces water infiltration and accordingly, increases runoff, resulting in soil erosion. The crusting effect is more pronounced in saline soils and has been well studied in relation to the mineralogical and chemical changes in the soil surface (Agassi et al., 1981; Sheinberg, 1992) that also affect soil roughness (Cierniewski et al., 2013). The immediate observation after a rainstorm is enhanced "hue" and "value" of the soil color because of an increase in the fine fraction on the surface. One can assume that the reflectance spectrum of the "physical crust" will be totally different from that of the original soil, because it contains a greater clay fraction with a different textural component. In a study conducted by Goldshleger et al. (2019), using soils from Israel and the United States under artificial rain conditions, the spectral feature of clay minerals enabled detecting the mechanism of crust formation and distortion (under heavy rainstorm energy). The spectral changes observed at the soil surface were caused by changes in the soil's texture (fine-fraction enrichment), structure (from loose to compact), roughness, and mineralogy (clay minerals rather than primary minerals). Several innovative studies have shown a significant relationship between spectral information and the infiltration rate of water into the soil profile as measured in the laboratory (e.g., Ben-Dor et al., 2003; Goldshleger et al., 2005). The next requisite step in the rain-simulation studies was to test the use of an airborne HSR sensor to characterize a structural crust in the field (Ben-Dor et al., 2004). Using a spectral-based index (the normalized spectral area [NSA]), they were able to generate a possible erosion-hazard map of the soil area (Figure 25.22). An important question based on that finding was whether a generic spectral model could describe the crust's status, rather than the kinetics of the formation process. A study by Goldshleger et al. (2019) showed that the spectral model used to predict crust status might be more robust than originally thought. By using four soils from Israel and three soils from the United States subjected to rain events in a rain

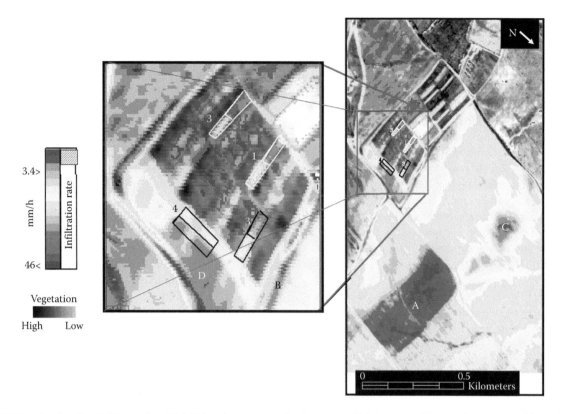

FIGURE 25.22 A colored scaled image (band 14, 576 nm) to account for the water infiltration rate of Loess soil as estimated from the spectral model that was based on the physical crust effect. Given also 4 controlled plots where the crust was broken by a gentle cultivation ("cold" colors) and those left with the crust ("warm" colors). Area A represents a favorable infiltration capability and area C a hazardous area with high heavy runoff potential. (After Ben-Dor, E. et al., 2006.)

simulator, promising results were obtained using a combined prediction equation for infiltration rate, with a cross-validation RMSE of 15.2% and a prediction-to-deviation ratio of 1.98.

Dust is another factor that can hinder remote sensing of the soil surface. The wind blowing the dust also has abrasive effects on the soil surface. Chappell et al. (2005) investigated the effect of soil-structure changes due to rain and wind-tunnel events. Their results showed that the spectral information can shed more light on the soil composition and structure generated by these two factors (rain crust and aeolian abrasion). Dust has two major effects: (1) It contaminates the atmosphere with a spectral signature that seems to interfere with soil-surface sensing and (2) it can accumulate on the soil surface as a thin layer that masks the true soil characteristics. One of the first observations of the effect of dust contamination on soil spectral information was reported by Montgomery and Baumgardner (1974). In a more recent study, Chudnovsky et al. (2009) demonstrated that an aeolian plume of Sahara dust has a significant effect on the clay mineral signals obtained from Hyperion data. They pointed out that the mineralogy of the dust plume can hinder the surface's spectral fingerprint. Dust accumulation on the soil surface has not yet been comprehensively studied. This is mainly because the spectral features of dust and the soil background may have similar features, and it is difficult to separate their contributions. This is unlike what can be seen over a vegetated background, especially if the dust is bright (Ben-Dor and Levin, 2000). Desert dust is most visible on a dark background (whereas anthropogenic dust is most visible on a light background) and hence is strongly affected by the underlying soil albedo. Li et al. (2013) demonstrated the impact of dust from a soil surface on nearby snow-covered areas as observed from MODIS and found significant indications of the dust-source area's characteristics. Musavi et al. (2013) demonstrated the influence of dust on four different soils using MODIS thermal data. They concluded that different soil types have different effects on surface-dust detection due to their spectral mixing process and demonstrated the connection between the real soil and the overlying accumulated dust. Okin and Painter (2004) studied the effect of soil-surface texture on spectral reflectance from the Mojave Desert. Sand plumes, eroded from the fields by wind, transported by saltation, and deposited downwind of the fields, were studied based on reflectance and on its correlation with grain-size distribution in the direction of wind transport. Analysis of AVIRIS-derived apparent surface reflectance demonstrated the expected negative correlation between effective grain size of the sand in the plume and reflectance, with the most significant correlations being in the SWIR region. The change in reflectance per mm change in particle diameter was -0.06 at $\lambda \sim 1.7$ µm and -0.08 at $\lambda \sim 2.2$ µm with $R^2 = 0.89$ and 0.93, respectively.

Another type of soil sealing is the crust formed on the soil surface during fire events. This crust is responsible for reducing infiltration and increasing erosion stages. Studies by Lugassi et al. (2010, 2014) showed that the high fire temperatures can alter the surface soil mineralogy, thereby hindering sensing of the real natural surface of soils that have been subjected to fire. They showed that Fe oxide species formed during the heating process can be good indicators of the temperature of the fire long after it has gone out. Other reflectance changes in burned soils have described by Kokaly et al. (2007). In summary, it can be concluded that the effects on top of the soil surface ("sealing") are important and can be clearly observed. Although they hinder correct sensing of the real soil body, the spectral effects can provide added information on processes in the "sealed" soils, such as water infiltration (in the case of a physical crust) and water runoff (in the case of burned soil).

25.8.2.10 Soil Iron

Just as OM acts as an important indicator for soils, Fe oxides provide significant information on the soil's formation and conditions (Schwertmann, 1988). Fe oxide content and species are strongly correlated with short- and long-term soil processes. An example of a short-term process is fire, whereas weathering exemplifies a long-term process. Fe oxide transformation often occurs under natural soil conditions. Hematite and goethite are common Fe oxides in soils and their relative content is strongly controlled by soil temperature, water regime, OM, and annual precipitation. Fe oxide is the major chromophore in the VIS–NIR region, contributing a *red-brown* color to the soil. It was thus obvious that RS imagery would prove successful in assessing Fe oxide coverage using soil color (Escadafal, 1993), and if spectral information was available, by modeling the absorption features of the active chromophore. Fe oxides (and Fe hydroxides) have specific absorption features that are located across the VIS–NIR region (based on electronic processes) and can be estimated from multispectral or imaging spectrometer images (Abrams and Hook, 1995). However, due to the occurrence of spectral oversampling, it is still problematic to quantitatively assess the soil's exact Fe oxide content (Deller, 2006). Different Fe oxide species have different colors. Hematitic soils are reddish and goethitic soils are yellowish-brown to yellowish. Hematite (α-Fe_2O_3) has Fe^{3+} ions in octahedral coordination with oxygen. Goethite (α-FeOOH) also has Fe^{3+} in octahedral coordination, but different site distortions along with the oxygen ligand (OH) provide the main absorption features that appear near 0.9 μm. Lepidocrocite (γ-FeOOH), which is associated with goethite but rarely with hematite, is another common unstable Fe oxide found in soils. It appears mostly in subtropical regions and is often found in the upper subsoil position (Schwertmann, 1988). Maghemite (γ-Fe_2O_3) is also found in soils, mostly in subtropical and tropical regions, but also occasionally in humid temperate areas. Ferrihydrite is a highly disordered Fe^{3+} oxide mineral found in soils in cool or temperate moist climates, characterized by young Fe oxide formations and soil environments that are relatively rich in other compounds (e.g., OM, silica). Fe oxides

are secondary minerals that are sensitive indicators of pH, Eh, relative humidity, and other environmental conditions. This enabled monitoring mine-waste remediation activities over contaminated soils (Crowley et al., 2003; Zabric et al., 2005). Iron-bearing minerals in the soil precipitate, ordered according to pH, are as follows: jarosite, pH < 3; schwertmannite, pH 2.8–4.5; mixtures of ferrihydrite and schwertmannite, pH 4.5–6.5; and ferrihydrite or a mixture of ferrihydrite and goethite, pH > 6.5. Based on the Fe oxides' spectral features, accordingly Zabcic et al. (2009), Kopackova (2014) and others were able to map surface pH, as shown in the Figure 25.23 example.

Iron associated with clay mineral structures is also an active chromophore in both the VIS–NIR and SWIR spectral regions. This can be seen in the nontronite-type mineral presented in Figure 25.24. Based on the structural OH–Fe features of smectite in the SWIR region, Ben-Dor and Banin (1990a) generated a predictive equation to account for the total iron content in a series of smectite minerals. The wavelengths that were selected automatically by their method were 2.2949, 2.2598, 2.2914, and 1.2661 μm. Stoner (1979) also observed a higher correlation between reflectance in the 1.55–2.32 μm region and iron content in soils, whereas Coyne et al. (1989) found a linear relationship between total iron content in montmorillonite and absorbance measured in the 0.6–1.1 μm spectral region. Ben-Dor and Banin (1995a) used spectra of 91 arid soils to show that their total iron content (both free and structural iron) can be predicted by multiple linear regression analysis and wavelengths 1.075, 1.025, and 0.425 μm. Obukhov and Orlov (1964) generated a linear relationship between reflectance values at 0.64 μm and the total percentage of Fe_2O_3 in other soils. Taranik and Kruse (1989) showed that a binary encoding technique for the spectral-slope values across the VIS–NIR spectral region is capable of differentiating a hematite mineral from a mixture of hematite–goethite–jarosite. It is important to mention that iron can often have an indirect influence on the overall spectral characteristics of soils. In the case of free Fe oxides, it is well known that soil particle size is strongly related to absolute Fe oxide content (Soileau and McCraken, 1967; Stoner and Baumgardner, 1981; Ben-Dor and Singer, 1987): as the Fe oxide content increases, the size fraction of the soil particles increases as well, because of the cementing effect of the free Fe oxides. As a result, problems resulting from different scattering effects are introduced into the soil analysis. Moreover, free Fe oxides, mostly in their amorphous state, can coat the soil particles with a film that prevents natural interaction between the soil particle (clay or nonclay minerals) and the sun's photons. Fe oxide minerals can be indicators for soil-stabilization processes (Ben-Dor et al., 2005). Karmanova (1981) found that well-crystallized iron compounds have the strongest effect on the spectral reflectance of soil and that removal of nonsilicate iron (mostly Fe oxides) helps enhance other chromophores in the soil. In this respect, Kosmas et al. (1984) demonstrated a second-derivative technique in the VIS region as a feasible approach for differentiating even small features of synthetic goethite from clays, and they suggested

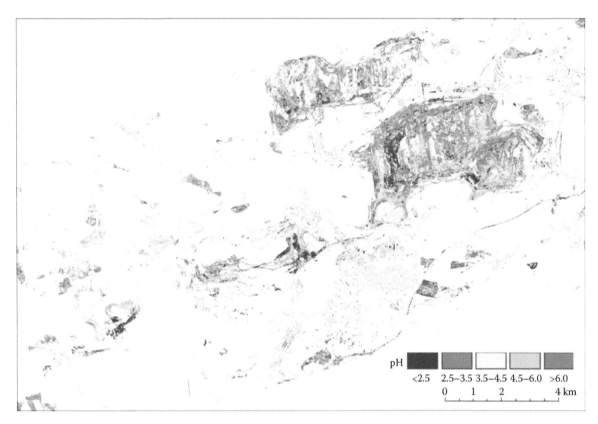

FIGURE 25.23 A pH map of Sokolov lignite basin Czech Republic as obtained from a spectral model based on iron oxides minerals. (After Kopačková, V., *Int. J. Appl. Earth Observ. Geoinform.*, 28, 28, 2014.)

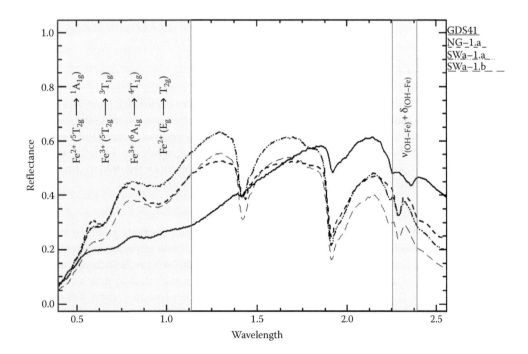

FIGURE 25.24 Reflectance spectra of three iron smectite minerals ("Nontronite") with emphasizing the SWIR spectral region where the structural iron combination mode of $\nu_{OH-Fe} + \delta_{OH-Fe}$ is active. Also given are possible electron transitions in the VIS-NIR region.

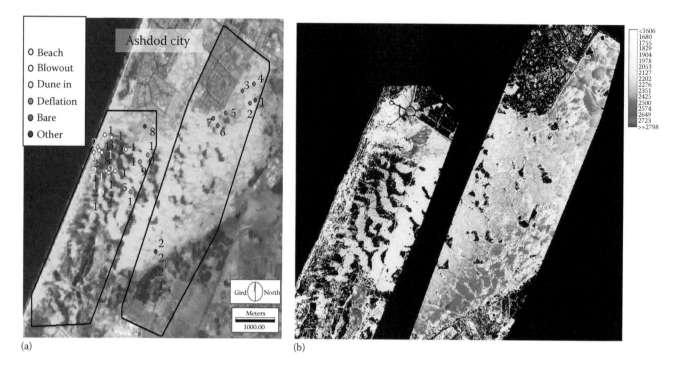

FIGURE 25.25 A color composite of Landsat 5 image (4,2,1) (a) and two CASI strips processed to map iron oxide content (b) of sand dune under stabilization process near Ashdod city Israel. (Taken from Ben-Dor, E. et al., *Geoderma*, 131, 1, 2005.)

that such a method be adopted to assess quantities of Fe oxide in mixtures. Based on these spectral characteristics, Scott and Pain (2008) showed the possibility of spectrally assessing the alteration of regolith Gerbermann and Neher (1979) showed that soil mixtures of clay and sand can be predicted from reflectance spectra. The first to map Fe oxides from the HSR domain were Ben-Dor et al. (2005), who modeled very low Fe oxide absorption features in a sand dune in Ashdod, Israel, from CASI data. Using this approach, they were able to account for the rubification process in the sand dune and for dune stabilization and soil formation over the dune. In Figure 25.25, the area as seen from Landsat in natural RGB band combination and from CASI data that were modeled to extract the free Fe oxide content is shown. Later, Bartholomeus et al. (2007) also applied HSR data from the ROSIS airborne image spectrometer and showed that quantification of soil iron content is possible over Mediterranean soils in Southern Spain. Fe_2O_3 was also well quantified and mapped by Landsat data by Dematte et al. (2009). The authors also observed a high correlation between laboratory Fe_2O_3 spectral readings, with R^2 0.82 and orbital R^2 0.67 in a 500 ha area with 500 samples. Lugassi et al. (2010) showed that alteration of goethite to hematite in soils subjected to fire can be depicted from the soil spectra and further used as an indicator to assess the temperature of the fire long after it has gone out. Another recent study by Lugassi et al. (2014) showed that minerals other than iron are altered during fire events and they can be used to assess the fire characteristics long after the actual fire event. While crystalline iron, such as hematite, occurs in high weathered tropical soils, amorphous iron occurs in low ones. Demattê and Garcia

(1999) observed that amorphous iron presents absence of spectral shape between 800 and 1000 nm, while crystalline iron is responsible for the convex shape. The same occurs between 400 and 600 nm where Sherman and Waite (1985) indicate a narrow convex shape for goethite, which were corroborated by Demattê and Garcia (1999), who observed the absence of this convexity when extracting goethite. These findings corroborate the importance of descriptive evaluation of spectra and not only quantitative information.

It can thus be concluded that iron is a very strong chromophore in soil and that its relationship with many of the soil's physical and chemical processes can be exploited by using its spectrum (or color) to track those processes. Based on the complexity of the iron component in the soil environment, as well as on the intercorrelation between iron and other soil components, sophisticated methods and data with relatively high spectral resolution are absolutely required to determine iron content from reflectance spectra.

25.8.2.11 Soil Classification and Taxonomy

As soil mapping requires grouping soil entities according to their chemical and physical characteristics, classification processes are a major issue. Soil classification is related to a taxonomic system that aims to give a common and agreed-upon definition and name to a given soil entity (Simonson, 1987). There are several such systems, the most common being the World Reference Base for Soil Resources classification protocol (FAO, 1998). These systems present the state and nature of several soils attributes, as well as their horizontal characteristics across the soil profile, which are then combined to define the soil name (class). Spectrally based soil classification was introduced by Stoner et al. (1980) as an atlas, which was published afterward by Stoner and Baumgardner (1981), who were able to group American soils into five spectral categories. This was the first method introduced to indicate soil entities based on soil spectra and was well studied. The limitation of this first soil's spectral atlas was that it used only surface soil samples that could not yet represent the soil profile as needed for soil classification. Other methods projected the spectral signatures into quantitative domain and determine several soil properties solely from the reflectance information (Dalal and Henry, 1986; Ben-Dor et al. 1995). Later, Dematté et al. (2002) proposed a complementary method to assist in descriptive spectral evaluations, with soil classification as the major goal. The authors indicated three major spectral aspects: complete shape, reflectance intensity, and absorption features. They also attempt for the necessity to evaluate spectra from each horizon, extract soil diagnostic information, compare spectra between them, and merge all data to reach soil classification. In fact, they observed that Ultisols have different spectra between A and B horizons mostly related to the clay gradient. An important descriptive difference was observed at about 2000 nm where A and B horizon spectra change because A has less clay (with more quartz) and B more clay (with less quartz). In fact, quartz raises reflectance intensity in SWIR as observed by White et al. (1997). The opposite observation was found in Oxisols. Usually, Oxisol description in the tropics indicates that there are very few differences between horizons, for example, in clay content and color. In fact, Dematté et al (2001, 2004b) conducted the first work to indicate a practical approach on how to perform soil-classification mission using reflectance spectra. Afterward, several works indicated the importance of spectra interpretation from all horizons on the same profile (Rizzo et al., 2014; Vasquez et al., 2014). Adopting a descriptive method, Bellinaso et al. (2010) created a database using spectra of all horizons from a given profile, indicating the importance of this characterization for soil classification. In this work, they revealed 75% efficiency in classifying 236 tropical soil profiles from five Brazilian states. Soil spectroscopy can also be used to describe variations in soil attributes along the profile and for soil classification (Ben-Dor et al., 2008). Odgers et al. (2011) developed a system for the continuous classification of soils of 262 profiles from Australia using the MWIR (2.5–5 μm) spectral region. Using the fuzzy K-means clustering algorithm, they showed a low level of confusion. Steffens and Buddenbaum (2013) studied diagnostic horizons of Luvisols through laboratory spectroscopic-imaging technique, indicating great potential for elucidating the processes and mechanisms of soil formation. Nanni et al. (2011) evaluated 18 types of soils and indicated that in many cases, there were differences between profiles that could only be depicted by a detailed spectral shape evaluation of samples taken from the soil-diagnostic horizons. In fact, the problem was how to evaluate spectra from different horizons at the same time using an automated system. To this end, Ben-Dor et al. (2008) were the first to introduce the 3HED assembly that is inserted into a small borehole and describes the

profile's chemical variations by spectral-based analysis, without having to sample the profile. This was the first goal on getting information from undersurface, since opening trenchers with 2 by 2 m is a difficult, high-cost, and time-consuming mission in traditional works. Elaborating on more tools for these descriptive evaluation, Dematté et al. (2014) described the multiple interpretations of reflectance spectra (MIRS) method, which relies on a detailed evaluation of spectral shapes, intensities, features, angles, and complete behavior across the VIS–NIR–SWIR region, crossing all of the diagnostic horizons along the soil's profile. Automated spectral-based systems are a more objective tool, as demonstrated by Vasques et al. (2014), which can practically avoid subjectivity of the interpreter. These authors proposed the insertion of spectra from each horizon from the same profile in the same system, observing a spectrum across the entire VIS–NIR–SWIR region from three sequent horizons (A, B1, B2), thus analyzing the spectra of all profiles as a continuum and reported 90% efficiency for soil-classification processes. Rizzo et al. (2014) ratified these findings and revealed a high correlation when classifying tropical soil profiles with the OSACA program (Carré and Jacobson, 2009). Other researchers also succeeded to classify soils based on spectroscopy, such as Nanni et al. (2004) who used stepwise multivariable analysis with 91% accuracy, Du et al. (2008) who used IR photoacoustic spectroscopy (FTIR-PAS) with 96% accuracy, and Linker (2008) who used MWIR with 95% accuracy. Using VIS–NIR–SWIR descriptive analysis, Dematté et al. (2012) distinguished 17 profiles of tropical soils derived from different parent materials. This was ratified by Cezar et al. (2013) who obtained 70% agreement with traditional field classification. Both stated that statistical methods are important, but cannot detect all of the particularities *inside* the spectral information, and thus human expertise and interpretation are still required. Whereas soil spectroscopy can generate soil attributes objectively, it cannot provide the morphological view of the soil that plays an important role in the soil-classification process. Another important property that the soil spectroscopy can provide is the objectively determination of the boarders between soil horizons. This allows an accurate profile spectral pattern description, which can then be projected into the *profile pattern* library accompanied with the soil-classification information as obtained by an expert. This library can be used then to classify unknown soil profile by applying multispectral classifier approaches. This technique was demonstrated by Dematté et al. (2004b), who determined soils classification by the descriptive correlation of known spectral patterns. To assist soil classification, image data can be used to support the traditional field work by indicating homogeneity and topography of the area in question (both are playing a major role in the soil-classification process). Dematté et al. (2009a) showed an error of 0.028% using discriminant analysis of Landsat images for soil classification. Each (nonvegetated) pixel along the study area was classified in the field using undersurface information and traditional method. The results indicated high performance between the traditional soil classification and the pixel (spectral) information. The spectrum of the surface of each classified soil was different in intensities and tendencies on shape. Accordingly, although image captures only the surface information, it shows a significant relationship with the under-surface soil properties that enabling discrimination between soil classes solely from the satellite imaga. Nonetheless, the readers need to realize that this is a very delicate procedure and must be taken and evaluated with care, because in some cases, different soil classes can have the same surface information. This needs an expert to evaluate and validate the RS results before transferring them to the end users for operation.

To summarize, classification processes are an important application that have progressed significantly in the last few years due to the use of soil spectroscopy with field spectrometers, airborne HSR sensors, and orbital sensors such as Landsat TM. Spectral information is particularly powerful in drawing the soil's boundaries more accurately than traditional aerial photography and in deriving information on the soil's horizon across the soil profile, by either measuring cores or using foreoptics that penetrate the soil profile. Nonetheless, for soil-mapping purposes, we strongly feel that merging RS technologies will advance soil-mapping missions. HSR with LIDAR and HSR with field profile spectroscopy are good examples of this.

The few success stories of soil classification using Landsat images strongly suggest that the HSR technology from orbit has a promising potential and needs further research. The problem to measure the spectral information of the soil horizons in a convenient way and the limitation to assess soil morphology from spectroscopy are issues that need to be further studied as well.

To summarize, soil-classification process as a punctual information process is an important application that have progressed significantly in the last few years due to the use of soil spectroscopy mostly in the field and laboratory. Favorite results from Landsat to classify soil pixels indicate that the emerging HSR technology from orbit has quite promising potential to that end. Soil classification from remote optical means still suffers from an inconvenient way to determine all of the horizons in a given profile and to a lesser extent to the fact that spectroscopy cannot provide the soil morphology.

25.8.2.12 Soil Mapping

Soil mapping involves gathering the soil's components into logical groups based on characteristics of the soil surface and profile according to taxonomical rules. Soil mapping is a basic stage in assessing a soil's agricultural potential for present and future activity. It is a difficult, time-consuming, and expensive task that requires difficult field and laboratory work along with professional manpower and a complex infrastructure. Soil mapping is carried out by expert soil surveyors who have a broad background in soil formation and genesis and are often termed pedologists. An important reference to understand soil mapping is Legros (2006) who provided an excellent perspective on how and why remote sensing is an important tool for soil mapping. Along with the spatial information required for soil-mapping processes, soil-field observations such as with auger (boreholes) are used to assist detection of soil boundaries. On a next step,

soil profile information is needed to classify the delineated polygon. Moreover, for the spatial domain, the soil surface's chemical and physical complexity also poses problems, significantly constraining the mapping process. Whereas the profile information is difficult to obtain with optical sensors, spatial variation can be obtained by remote sensing and, as has been demonstrated, with better accuracy if the spectral resolution is high. Airborne photographs have been intensively used in the past by soil surveyors to assist in soil-mapping procedures. The spatial variations of the soil surface were determined by grayscale analysis or color interpretation of *soil* boundaries. Vink (1963a) presented the information that can be culled from aerial photographs for soil pedological mapping and demonstrated the added value of this tool. Demattê and Garcia (1999) showed a significant relationship between spectral information and aerial photographs (drainage patterns) that could then be utilized in the field for soil pedological mapping. In another study, aerial photographs were used to map relief variations and were exploited to pinpoint areas for spectral sampling and measurement reaching a soil map (Demattê et al., 2001). This approach reached 90% accuracy with traditional field systems and demonstrated the strength of soil spectroscopy. The use of multispectral sensors from orbit enables locating pure soil pixels for soil sampling and false-alarm detection of nonpure (partially vegetated or covered with litter) soil pixels. Accordingly, Demattê et al. (2009a) developed a sequence-based technique to indicate an area of bare soil using RS means. This method used atmospheric-correction data and then evaluated pixels based on the soil-line concept and use of color compositions of 5,4,3 and 3,2,1 (red, green, blue of Landsat), calculating soil-adjusted and normalized-difference vegetation indices and, mostly, judging the spectral surface shape based on a soil databank from both ground and space (Landsat) domains. Based on this technique, several workers succeeded to quantify soil-surface properties of tropical Brazilian soils (Demattê and Nanni, 2003; Nanni and Demattê, 2006) using Landsat data, which were then projected to obtain better soil-classification results (Demattê et al., 2007, 2009b and Nanni et al., 2012). These latter authors observed that merging geological maps with spectral profiles increases mapping accuracy. Gaining experience with field, aerial, and orbital data, Demattê et al. (2001, 2004b) obtained 85% agreement between soil maps traditionally generated by soil surveyors and those using their RS method. As indicated by Legros (2006), one of the most important steps in soil mapping is determining soil delineation and afterward making their classification. But where is the limit of soils? Looking toward this task, Demattê et al. (2004b) and recently Demattê and Terra (2014) demonstrated that soils can be differentiated along toposequences by a *spectral pedology* analysis, which can detect their limits. In fact, these authors observed the close relationship between the pedogenetic alterations along the toposequence, which goes toward the detection of their limits. Thus, if we replicate several toposequences in a certain area, we will have several limits, which can be linked using relief information (e.g., by aerial photographs, as indicated by Vink, 1963), and we will reach the soil's polygons. This method does not implicate necessarily with soil classification instead only perform the *figure* of the map. In a subsequent step, inside the polygons, we would indicate the best spot to account for a profile; go inside with a sensor (see Viscarra Rossel and Webster 2011) or make a borehole with a fiber-optic method (see Ben-Dor et al. 2008). These spectral measurements can assist the classification of the soil entity and can then be further used with the image to generate the soil map.

In a different approach, Löhrer et al. (2013) identified and mapped the mineralogical composition of Pleistocene sediments using ASTER and SPOT-5 images associated with spectral laboratory work and also obtained good results. Although the orbital sensors see only the top 50 µm of the soil surface, the production of good soil-classification results indicates that the surface has, in some cases, a connection with below-surface dynamics of soil processes. Galvão et al. (2001) used AVIRIS data to show that surface-reflectance values and constituents (total iron, OM, TiO_2, Al_2O_3, and SiO_2) represent three important soil types from central Brazil (Terra Roxa Estruturada [S_{TE}], Latossolo Vermelho-Escuro [S_{LE}], and Areia Quartzosa [S_{AQ}]) i.e, Nitosol, Ferralsol, and Arenosol, respectively. Nonetheless, Demattê et al. (2009b) stated that some soils cannot be differentiated, mainly those that require morphological interpretation of the soil profiles. With the emergence of the new HSR technology (along with other multi- and superspectral data from orbit), the spectral approach was further extended to the notion of digital soil mapping (McBratney et al., 2003). Up until now, digital soil mappers have mainly used RS images as spatial data inputs representing the landscape variables that are related to soil, such as vegetation, topography, and parent material (soil covariates). Boettinger et al. (2008) reviewed the main indicators that could be retrieved from multispectral images for estimating these soil covariates. Along these lines, a modern RS approach can serve as a tool to assist soil scientists in obtaining up-to-date and accurate information on the soil surface in question. The recent implementation of HSR technology for the digital soil-mapping approach (Lagacherie et al., 2013) makes a significant contribution to soil-surface characterization, adding more information to both aerial photography and multispectral information. As most of the earth's soil areas are still unmapped, and those that have been mapped need updating, new methods (such as HSR technology) to map and classify soils are crucial. This issue is most relevant today with the rapidly growing world population, as it could lead to better utilization of soil resources, a critical issue for feeding mankind. Despite the constraints, reported in Section 7, to utilizing HSR technology, it has the proven capability to extract both quantitative and qualitative information on the soil surface, thereby providing a better way to account for the surface's complexity and variations.

To summarize, spectral information is particularly powerful for detecting soil limits and spots with punctual soil information with increasing observation density. Observation density is the most important rule we have to attempt on a soil mapping and is directly related with scale of the final product. This can assist on drawing the soil's boundaries and will mostly show better use if we achieve bare soil in continuous areas as indicated by several

methodologies stated before. If these observations are combined with traditional aerial photography, for deriving information on the soil's horizon across the soil profile, by either measuring cores or using foreoptics that penetrate the soil profile, we will certainly reach the best on soil mapping. Nonetheless, for soil-mapping purposes, we strongly feel that merging RS technologies will advance soil-mapping missions. HSR with LIDAR and HSR with field profile spectroscopy are good examples of this.

25.9 Summary and Conclusions

This chapter summarizes most of the key studies in soil remote sensing using the VIS–NIR–SWIR spectral region and provides the basic theory for optical remote sensing of soil. It demonstrates the significant progress that has been achieved in RS technology in the last 10 years. This progress has enabled the exposure and validation of many sensors from all domains for scientific and commercial use. Remote sensing of the soil is thus entering a new era, in which more data, information, infrastructures, and applications can be provided more efficiently to the end user. This calls for further development of both electro-optic technology and data-mining algorithms. Mulder et al. (2011) partially summarized the soil applications for remote sensing by all means, emphasizing mainly the optical region. Although their summary ignored some of the fundamental work done with optical remote sensing of soil at that time, their conclusion is important in terms of understanding the need for RS data with both high temporal and spectral resolution. The future will call for merging sensors, databases, and know-how and developing more accurate methods for data exchange and generating more robust models to be executed by many users. Another aspect that will need further attention is the establishment of standards and protocols for the acquisition of soil spectra in both the laboratory and field domains, as well as the development of quality indicators and assurance for the new RS data. New programs that place HSR sensors in orbit will enable better temporal resolution and wide coverage of all soils worldwide. This calls for more attention to this direction and adoption of the comprehensive know-how that has already been achieved in soil spectroscopy. Upscaling reflectance data from the ground to the air and space domains is still a bottleneck that needs to be overcome. Mulder (1987) stated that every work exploiting RS data requires field work, to either verify or expand the information into the soil profile. This has led us to realize that new ideas on how to extract soil-profile characteristics are important but still lacking. The idea of Ben-Dor et al. (2008) to sense the profile as a *catheterization* optical assembly provides a proof of concept for RS measurements of soil profiles but needs to be further developed. Combining such a method with the spatial view obtained by RS means (mainly HSR) may pave the way for future activity in the remote sensing of soils from afar. Other directions are also welcome, and more ideas on how to improve the remote sensing of soils are required. Nonetheless, the current achievements in remote sensing of soils are remarkable considering that not long ago, analog grayscale aerial photos were the basic and indeed only tool to remotely sense soils from afar. As new satellite missions covering all resolutions (spatial, spectral, and temporal) are being planned, and their availability tends to be for all, it is strongly anticipated that the remote sensing of soil will undergo further development, utilization, and exploitation to monitor soil status such that soil productivity will be benefited for all of mankind.

References

Aber, J., Wessman, C. A., Peterson, D. L., Mellilo, J. M., and Fownes, J. H. 1990. Remote sensing of litter and soil organic matter decomposition in forest ecosystems. *Remote Sensing of Biosphere Functioning* (Hobbs R. J and), 79, 87–101.

Abrams, M. and Hook, S. J. 1995. Simulated ASTER data for geologic studies. *IEEE Transactions on Geoscience and Remote Sensing*, 33, 692–699.

Adar, S., Shkolnisky, Y., and Ben-Dor, E. 2014a. A new approach for thresholding spectral change detection using multispectral and hyperspectral image data, a case study over Sokolov, Czech Republic. *International Journal of Remote Sensing*, 35, 1563–1584.

Adar, S., Shkolnisky, Y., and Ben-Dor, E. 2014b. Change detection of soils under small-scale laboratory conditions using imaging spectroscopy sensors. *Geoderma*, 216, 19–29.

Agassi, M., Shainberg, I., and Morrin, J. 1981. Effects of electrolyte concentration and soil sodicity on infiltration and crust formation. *Soil Science Society of America Journal*, 45, 848–851.

Agdu, P. A., Fehrenbacher, J. D., and Jansen, I. 1990. Soil property relationships with SPOT satellite digital data in the East central Illinois. *Soil Science Society of American Journal*, 54, 807–812.

Al-Abbas, H. H., Swain, H. H., and Baumgardner, M. F. 1972. Relating organic matter and clay content to multispectral radiance of soils. *Soils Science*, 114, 477–485.

Araújo, S. R., Demattê, J. A. M., and Bellinaso, H. 2013. Analyzing the effects of applying agricultural lime to soils by VNIR spectral sensing: A quantitative and quick method. *International Journal of Remote Sensing*, 34, 4570–4584.

Araújo, S. R., Demattê, J. A. M., and Vicente, S. 2014. Soil contaminated with chromium by tannery sludge and identified by Vis-Nir-Mid spectroscopy techniques. *International Journal of Remote Sensing*, 35, 3579–3593.

Asner, G. P. and Heidebrecht, K. B. 2002. Spectral unmixing of vegetation, soil and dry carbon cover in arid regions: Comparing multispectral and hyperspectral observations. *International Journal of Remote Sensing*, 23, 3939–3958.

Asner, G. P. and Heidebrecht, K. B. 2003. Imaging spectroscopy for desertification studies: Comparing AVIRIS and EO-1 Hyperion in Argentina drylands. *IEEE Transactions on Geoscience and Remote Sensing*, 41, 1283–1296.

Bach, H. and Mauser, W. 1994. Modelling and model verification of the spectral reflectance of soils under varying moisture conditions. *Geoscience and Remote Sensing Symposium, 1994. IGARSS'94. Surface and Atmospheric Remote Sensing: Technologies, Data Analysis and Interpretation International*, 4, 2354–2356.

Balsam, W. L. and Deaton, B. C. 1996. Determining the composition of late quaternary marine sediments from NUV, VIS, and NIR diffuse reflectance spectra. *Marine Geology*, 134, 31–55.

Barnes, E., M., Sudduth, K. A., Hummel, J. W., Lesch, S. M., Corwin, D. L., Yang, G., Doughtry, C. S. T., and Bausch, W. C. 2003. Remote-and ground-based sensor techniques to map soil properties. *Photogrammetric Engineering and Remote Sensing*, 69(6), 619–630.

Barrett, B. W. and Petropoulos, G. P. 2013. Satellite remote sensing of surface soil moisture. Remote Sensing of Energy Fluxes and Soil Moisture Content, 85.

Bartholomeus, H., Epema, G., and Schaepman, M. 2007. Determining iron content in Mediterranean soils in partly vegetated areas, using spectral reflectance and imaging spectroscopy. *International Journal of Applied Earth Observation and Geoinformation*, 9, 194–203.

Baumann, P.R. 2010. History of remote sensing, aerial photography. Available at: http://www.oneonta.edu/faculty/baumanpr/geosat2/RS%20History%20I/RS-History-Part-1.htm (Accessed: June 17, 2014).

Baumgardner, M. F., Kristof, S. J., Johannsen, C.J., and Zachary, A. L.1970. Effects of organic matter on multispectral properties of soils. *Proceedings of the Indian Academy of Science*, 79, 413–422.

Baumgardner, M. F., Silva, L. F., Biehl, L.L., and Stoner, E. R.1985. Reflectance properties of soils. *Advances in Agronomy*, 38, 1–44.

Beck, R. H., Robinson, B. F., McFee, W.H., and Peterson, J. B. 1976. *Information Note 081176*. Laboratory Application of Remote Sensing, Purdue University, West Lafayette, IN.

Bedidi, A., Cervelle, B., and Madeira, J. 1991. Moisture effects on spectral signatures and CIE-color of lateritic soils. *Proceedings of the 5th International Colloquium, Physical Measurements and Signatures in Remote Sensing*, Courchevel, France I, pp. 209–212.

Bedidi, A., Cervelle, B., Madeira, J., and Pouget, M. 1990. Moisture effects on spectral characteristics (visible) of lateritic soils. *Soil Science*, 153, 129–141.

Bellinaso, H., Demattê, J. A. M., and Araújo, S. R. 2010. Spectral library and its use in soil classification. *Revista Brasileira de Ciência do Solo*, 34, 861–870.

Ben-Dor, E. 2002. Quantitative remote sensing of soil properties. *Advances in Agronomy*, 75, 173–243.

Ben-Dor, E. and Banin, A. 1990a. Diffuse reflectance Spectra of smectite minerals in the near infrared and their relation to chemical composition. *Sciences Geologiques Bulletin*, 43, 24, 117–128.

Ben-Dor, E. and Banin, A. 1990b. Near infrared reflectance analysis of carbonate concentration in soils. *Applied Spectroscopy*, 44, 1064–1069.

Ben-Dor, E. and Banin, A. 1994. Visible and near infrared (0.4–1.1 mm) analysis of arid and semiarid soils. *Remote Sensing of Environment*, 48, 261–274.

Ben-Dor, E. and Banin, A. 1995a. Near infrared analysis (NIRA) as a rapid method to simultaneously evaluate, several soil properties. *Soil Science Society of American Journal*, 59, 364–372.

Ben-Dor, E. and Banin, A. 1995b. Near infrared analysis (NIRA) as a simultaneously method to evaluate spectral featureless constituents in soils. *Soil Science*, 159, 259–269.

Ben-Dor, E. and Banin, A. 1995c. Quantitative analysis of convolved TM spectra of soils in the visible, near infrared and short-wave infrared spectral regions (0.4–2.5 mm). *International Journal of Remote Sensing*, 18, 3509–3528.

Ben-Dor, E., Chabrillat, S., Demattê, J. A. M., Taylor, G. R., Hill, J., Whiting, M. L., & Sommer, S. 2009. Using imaging spectroscopy to study soil properties. *Remote Sensing of Environment*, 113, S38–S55.

Ben-Dor, E., Goldahlager, N., Benyamini, M., and Blumberg, D. G. 2003. The Spectral Reflectance properties of Soil's structural crust in the SWIR spectral region (1.2–2.5 µm). *Soil Science Society of American Journal*, 67, 289–299.

Ben-Dor, E., Goldshleger, N., Braun, O., Kindel, B., Goetz, A. F. H., Bonfil, D., Margalit, N., Binaymini, Y., Karnieli, A., and Agassi, M. 2004. Monitoring infiltration rates in semiarid soils using airborne hyperspectral technology. *International Journal of Remote Sensing*, 25(13), 2607–2624.

Ben-Dor, E., Inbar, Y., and Chen, Y. 1997. The reflectance spectra of organic matter in the visible near infrared and short wave infrared region (400–2,500 nm) during a control decomposition process. *Remote Sensing of Environment*, 61, 1–15.

Ben-Dor, E., Irons, J. A., and Epema, A. 1999. Soil spectroscopy. In: *Manual of Remote Sensing*, 3rd edn., A. Rencz (ed.), John Wiley & Sons Inc., New York, pp. 111–189.

Ben-Dor, E. and Kruse, F. A. 1995. Surface mineral mapping of Makhtesh Ramon Negev, Israel using GER 63 channel scanner data. *International Journal of Remote Sensing*, 18, 3529–3553.

Ben-Dor, E. and Levin, N. 2000 Determination of surface reflectance from raw hyperspectral data without simultaneous ground truth measurements. A case study of the GER 63-channel sensor data acquired over Naan Israel. *International Journal of Remote Sensing of Environment*, 21, 2053–2074.

Ben-Dor, E., Levin, N., Singer, A., Karnieli, A., Braun, O., and Kidron, G. J. 2005. Quantitative mapping of the soil rubification process on sand dunes using an airborne hyperspectral sensor. *Geoderma*, 131, 1–21.

Ben-Dor, E., Ong, C., and Lau, I. C. 2015. Reflectance measurements of soils in the laboratory: Standards and protocols. *Geoderma*, 245, 112–124.

Ben-Dor, E., Patkin, K., Banin, A., and Karnieli, A. 2002. Mapping of several soil properties using DAIS-7915 hyperspectral scanner data. A case study over clayey soils in Israel. *International Journal of Remote Sensing*, 23, 1043–1062.

Ben-Dor, E., Schläpfer, D., Plaza, A. J., and Malthus, T. 2013. Hyperspectral remote sensing. In: *Airborne Measurements for Environmental Research: Methods and Instruments*, M. Wndisch and J. L. Brenguier (eds.), Wiley-VCH Weinheim, Germany, pp. 413–456.

Ben-Dor, E. and Singer, A. 1987. Optical density of vertisol clays suspensions in relation to sediment volume and dithionite-citrate-bicarbonate extractable iron. *Clays and Clay Minerals*, 35, 311–317.

Ben-Dor, E., Taylor, R. G., Hill, J., Demattê, J. A. M., Whiting, M. L., Chabrillat, S., and Sommer, S. 2008. Imaging spectrometry for soil applications. *Advances in Agronomy*, 97, 321–392.

Ben-Gera, I. and Norris, K. H. 1968. Determination of moisture content in soybeans by direct spectrophotometry. *Israeli Journal of Agriculture Research*, 18, 124–132.

Bihong, F., Zheng, G., Nonomiya, Y., Wang, C., and Sum, G. 2007. Mapping hydrocarbon-induced mineralogical alteration in the northern Tian Shan using ASTER multispectral data. *Terra Nova*, 19, 225–231.

Boettinger, J. L., Ramsey, R. D., Bodily, J. M., Cole, N. J., Kienast-Brown, S., Nield, S. J., and Stum, A. K. 2008. Landsat spectral data for digital soil mapping. In: *Digital Soil Mapping with Limited Data*, A. E. Hantemink, A. B. McBratney, and M. D. L. Mendoncasntos (eds.), Springer, Dordrect, the Netherlands, pp. 193–202.

Bowers, S. and Hanks, R. J. 1965. Reflectance of radiant energy from soils. *Soil Science*, 100, 130–138.

Bresson, L. M. and Boiffin, J. 1990. Morphological characterization of soil crust development stages on an experimental field. *Geoderma*, 47, 301–325.

Brown, D. J., Shepherd, K. D., Walsh, M. G., Dewayne Mays, M., and Reinsch, T. G. 2006. Global soil characterization with VNIR diffuse reflectance spectroscopy. *Geoderma*, 132, 273–290.

Buddenbaum, H., Stern, O., Stellmes, M., Stoffels, J., Pueschel, P., Hill, J., and Werner, W. 2012. Field imaging spectroscopy of beech seedlings under dryness stress. *Remote Sensing*, 4, 3721–3740.

Buol, S. W., Hole, F. D., and McCracken, R. J. 1973. *Soil Genesis and Classification*. Iowa State University Press, Ames, IA.

Bushnell, T. M. 1932. A new technique in soil mapping. *American Soil Survey Association Bulletin*, 13, 74–81.

Carré, F. and Jacobson, V. 2009. Numerical classification of soil profile data using distance metrics. *Geoderma*, 148, 336–345.

Casa, R., Castaldi, F., Pascucci, S., Palombo, A., and Pignatti, S. 2013. A comparison of sensor resolution and calibration strategies for soil texture estimation from hyperspectral remote sensing. *Geoderma*, 197, 17–25. DOI: 10.1016/j.geoderma.2012.12.016.

Castellan, G. W. 1983. *Physical Chemistry*, 3rd edn., Addison-Wesley Publishing Co., Reading, MA, p. 943.

Cezar, E., Nanni, M. R., Chicati, M. L., Oliveira, R. B., and Demattê, J. A. M. 2013. Discriminação entre solos formados em região transicional por meio de resposta espectral. *Bioscience Uberlândia*, 29(3), 644–654.

Chappell, A., Zobeck, T., and Brunner, G. 2005. Induced soil surface change detected using on-nadir spectral reflectance to characteristics soil erodibility. *Earth Surface Processes and Landforms*, 30, 489–51.

Chen, F., Kissel, D. E., West, L. T., and Adkins, W. 2000. Field-scale mapping of surface soil organic carbon using remotely sensed imagery. *Soil Science Society of America Journal*, 64, 746.

Chen, F., Kissel, D. E., West, L. T., Adkins, W., Rickman, D., and Luvall, J. C. 2006. Feature selection and similarity analysis of crop fields for mapping organic carbon concentration in soil. *Computers and Electronics in Agriculture*, 54, 8–21.

Chen, F., Kissel, D. E., West, L.T., Adkins, W., Rickman, D., and Luvall, J. C. 2008. Mapping soil organic carbon concentration for multiple fields with image similarity analysis. *Soil Science Society of America Journal*, 72, 186–193.

Chen, Y. and Inbar, Y. 1994. Chemical and spectroscopical analysis of organic matter transformation during composting in relation to compost maturity. In: *Science and Engineering of Composting: Design, Environmental, Microbiology and Utilization Aspects*, H. A. J. Hoitink and H. M. Keener (eds.), Renaissance Publications, Worthington, OH, pp. 551–600.

Chevrel, S. 2005. MINEO two years later-did the project impulse a new era in imaging spectroscopy applied to mining environments. In: *Proceedings of the EARSEL SIG IS Fourth.* New quality in environmental studies, Warsaw, pp. 26–29.

Chevrel, S. 2013. Earth observation in support of mineral industry sustainable development-the EO-MINERS project. In: *23rd World Mining Congress 2013 Proceedings,* Queebec, Canda.

Chudnovsky, A., Ben-Dor, E., Kostinski, A. B., and Koren, I. 2009. Mineral content analysis of atmospheric dust using hyperspectral information from space. *Geophysical Research Letters*, 36, L15811.

Cierniewski, J. 1987. A model for soil surface roughness influence on the spectral response of bare soils in the visible and near infrared range. *Remote Sensing of Environment*, 23, 98–115.

Cierniewski, J. 1989. The influence of the viewing geometry of bare soil surfaces on their spectral response in the visible and near infrared. *Remote Sensing of Environment*, 27, 135–142.

Cierniewski, J., Karnieli, A., Kuśnierek, K., and Herrmann, I. 2013. Proximating the average daily surface albedo with respect to soil roughness and latitude. *International Journal of Remote Sensing*, 34, 9–10, 3416–3424.

Cierniewski, J., Kazmierowski, C., Krolewicz, S., Piekarczyk, J., Wróbel, M., and Zagajewski. 2014. Effects of different illumination and observation techniques of cultivated soils on their hyperspectral bidirectional measurements under field and laboratory conditions. *IEEE Journal of Selected Topics in Applied Earth Observations and Remote Sensing*, 7, 2525–2530.

Cipra, J. E., Franzmeir, D. P., Bauer, M. E., and Boyd, R. K. 1980. Comparison of multispectral measurements from some nonvegetated soils using Landsat digital data and a spectroradiometer. *Soil Science Society of American Journal*, 44, 80–84.

Clark, R. N. 1999. Spectroscopy of rocks and minerals, and principles of spectroscopy. *Manual of Remote Sensing*, 3, 3–58.

Clark, R. N. and Roush, T. L. 1984. Reflectance spectroscopy: Quantitative analysis techniques for remote sensing applications. *Journal of Geophysical Research*, 89, 6329–6340.

Condit, H. R. 1970. The spectral reflectance of American soils. *Photogrammetric Engineering*, 36, 955–966.

Condit, H. R. 1972. Application of characteristic vector analysis to the spectral energy distribution of daylight and the spectral reflectance of American soils. *Applied Optics*, 11, 74–86.

Coyne, L. M., Bishop, J. L., Sacttergood, T., Banin, A., Carle, G., and Orenberg, J. 1990. Near-infrared correlation spectroscopy: Quantifying iron and surface water in series of variably cation-exchanged montmorillonite clays. In: *Spectorscopic Characterization of Mineral and Their Surfaces*, L. M. Coyne, S. W. S. McKeever, and D. F. Blake (eds.), Food and Agriculture Organization (FAO) of the United Nations (UN), pp. 407–429.

Crowley, J. K., Williams, D. E., Hammarstrom, J. M., Piatak, N., Chou, I. M., and Mars, J. C. 2003. Spectral reflectance properties (0.4–2.5 μm) of secondary Fe-oxide, Fe-hydroxide, and Fe-sulphate-hydrate minerals associated with sulphide-bearing mine wastes. *Geochemistry: Exploration, Environment, Analysis*, 3, 219–228.

Csillag, F., Pasztor, L., and Biehl, L.L. 1993. Spectral band selection for the characterization of salinity status of soils. *Remote Sensing of Environment*, 43, 231–242.

Da-Costa, L. M. 1979. Surface soil color and reflectance as related to physicochemical and mineralogical soil properties. PhD dissertation, University of Missouri, Columbia, MO, p. 154.

Dahms, D. E. 1993. Mineralogical evidence for eolian contribution to soils of late Quaternary moraines, Wind River Mountains, Wyoming, USA. *Geoderma*, 59, 175–196.

Dalal, R. C. and Henry, R. J. 1986. Simultaneous determination of moisture, organic carbon and total nitrogen by near infrared reflectance spectroscopy. *Soil Science Society of America Journal*, 50, 120–123.

Debaene, G., Niedźwiecki, J., Pecio, A., and Żurek, A. 2014. Effect of the number of calibration samples on the prediction of several soil properties at the farm-scale. *Geoderma*, 214–215, 114–125.

Dehaan, R. L. and Taylor, G. R. 2002. Field-derived spectra of salinized soils and vegetation as indicators of irrigation-induced soil salinization. *Remote Sensing of Environment*, 80, 406–417.

Dekker, A. G., Brando, V. E., Anstee, J. M., Pinnel, N., Kutser, T., Hoogenboom, H. J., Pasterkamp, R.. 2001. Imaging spectrometry of water. In: *Imaging Spectrometry: Basic Principles and Prospective Applications: Remote Sensing and Digital Image Processing*, v. IV, Kluwer Academic Publishers, Dordrecht, the Netherlands, pp. 307–335, Chapter 11.

Deller, A. M. E. 2006. Facies discrimination in laterites using Landsat Thematic Mapper, ASTER and ALI data—Examples from Eritrea and Arabia. *International Journal of Remote Sensing*, 27, 2389–2240.

Demattê, J. A. M. 2002. Characterization and discrimination of soils by their reflected electromagnetic energy. *Pesquisa Agropecuaria Brasileira*, 37(10) 1445–1458.

Demattê, J. A. M., Bellinaso, H, Romero, D. J., and Fongaro, C. T. 2014. Morphological Interpretation of Reflectance Spectrum (MIRS) using libraries looking towards soil classification. *Scientia Agricola*, 71, 509–520.

Demattê, J. A. M., Campos, R. C., Alves, M. C., Fiorio, P. R., and Nanni, M. R. 2004b. Visible-NIR reflectance: A new approach on soil evaluation. *Geoderma*, 121, 95–112.

Demattê, J. A. M., Demattê, J. L. I., Camargo, W. P., Fiorio, P. R., and Nanni, M. R. 2001. Remote sensing in the recognition and mapping of tropical soils developed on topographic sequences. *Mapping Sciences and Remote Sensing*, 38(2), 79–102.

Demattê, J. A. M., Fiorio, P. R., and Ben-Dor, E. 2009b. Estimation of soil properties by orbital and laboratory reflectance means and its relation with soil classification. *Open Remote Sensing Journal*, 2, 12–23.

Demattê, J. A. M., Galdos, M. V., Guimarães, R. V., Genú, A. M., Nanni, M. R., and Zullo, Jr., J. 2007. Quantification of tropical soil attributes from ETM+/LANDSAT-7 data. *International Journal of Remote Sensing*, 28, 3813–3829.

Demattê, J. A. M., Gama, M. A. P., Cooper, M., Araújo, J. C., Nanni, M. R., and Fiorio, P. R. 2004a. Effect of fermentation residue on the spectral reflectance properties of soils. *Geoderma*, 120, 187–200.

Demattê, J. A. M. and Garcia, G. J. 1999. Alteration of soil properties through a weathering sequence as evaluated by spectral reflectance. *Soil Science Society of America Journal*, 63, 327–342.

Demattê, J. A. M., Huete, A. R., Ferreira Jr., L. G., Alves, M. C., Nanni, M. R., and Cerri, C. E. 2000. Evaluation of tropical soils through ground and orbital sensors. *International Conference Geospatial Information in Agriculture and Forestry* 2: 35–42.

Demattê, J. A. M., Huete, A. R., Ferreira Jr., L. G., Nanni, M. R., Alves, M.C., and Fiorio, P. R. 2009a. Methodology for bare soil detection and discrimination by Landsat TM image. *Open Remote Sensing Journal*, 2, 24–35.

Demattê, J. A. M., Mafra, L., and Bernardes, F. F. 1998. Comportamento espectral de materiais de solos e de estruturas biogênicas associadas [Soil spectral behavior associated with biogenic material]. *Revista Brasileira de Ciência do Solo [Brazilian Journal of Soil Science]*, 22, 621–630.

Demattê, J. A. M. and Nanni, M. R. 2003. Weathering sequence of soils developed from basalt as evaluated by laboratory (IRIS), airborne (AVIRIS) and orbital (TM) sensors. *International Journal of Remote Sensing*, 24, 4715–4738.

Demattê, J. A. M., Pereira, H. S., Nanni, M. R., Cooper, M., and Fiorio, P. R. 2003. Soil chemical alterations promoted by fertilizer application assessed by spectral reflectance. *Soil Science*, 168, 730–747.

Demattê, J. A. M., Sousa, A. A., Alves, M., Nanni, M. R., Fiorio, P. R., and Campos, R. C. 2006. Determining soil water status and other soil characteristics by spectral proximal sensing. *Geoderma*, 135, 179–195.

Demattê, J. A. M. and Terra, F. S. 2014. Spectral pedology: A new perspective on evaluation of soils along pedogenetic alterations. *Geoderma*, 217–218, 190–200.

Demattê, J. A. M., Terra, F. S., and Quartarolli, C. F. 2012. Spectral behavior of some modal soil profiles from São Paulo State, Brazil. *Bragantia*, 71, 413–423.

Dixon, J. B. and Weed, S. B. 1989. *Minerals in Soil Environments*. Soil Science Society of Soil Science Society of America Publishing, Madison, WI.

Dobos, E., Montanarella, L., Nègre, T., and Micheli, E. 2001. A regional scale soil mapping approach using integrated AVHRR and DEM data. *International Journal of Applied Earth Observation and Geoinformation*, 3, 30–42.

d'Oleire-Oltmanns, S., Marzolff, I., Peter, K. D., and Ries, J. B. 2012. Unmanned Aerial Vehicle (UAV) for monitoring soil erosion in Morocco. *Remote Sensing*, 4, 3390–3416.

Du, C., Linker, R., and Shaviv, A. 2008. Identification of agricultural Mediterranean soils using mid-infrared photoacoustic spectroscopy. *Geoderma*, 143, 85–90.

Dutkiewicz, A. 2006. Evaluating hyperspectral imagery for mapping the surface symptoms of dryland salinity. PhD dissertation, University of Adelaide, Adelaide, South Australia, Australia.

Elachi, C. and Van Zyl, J. J. 2006. *Introduction to the Physics and Techniques of Remote Sensing*, 28. John Wiley & Sons, Hoboken, NJ.

Eliran, A., Goldshleger, N., Yahalom, A., and Ben-Dor, E. 2013. Empirical model for backscattering at millimeter-wave frequency by bare soil subsurface with varied moisture content. *Geoscience and Remote Sensing*, 99, 1–5.

Eltahir, E. A. 1998. A soil moisture–rainfall feedback mechanism: 1. Theory and observations. *Water Resources Research*, 34, 765–776.

Epema, G. F. 1990. Effect of moisture content on spectral reflectance in a playa area in southern Tunisia. In: *Proceedings International Symposium, Remote Sensing and Water Resources*, Enschede, the Netherlands, pp. 301–308.

Escadafal, R. 1993. Remote sensing of soil color: Principles and applications. *Remote Sensing Reviews*, 7, 261–279.

Escadafal, R. and Hute, A. R. 1991. Influence of the viewing geometry on the spectral properties (high resolution visible and NIR) of selected soils from Arizona. In: *Proceedings of the fifth International Colloquium, Physical Measurements and Signatures in Remote Sensing*, Courchevel, France I, pp. 401–404.

Escobar, D. E., Everitt, J. H., Noriega, J. R., Cavazos, I., and Davis, M. R. 1998. A twelve-band airborne digital video imaging system (ADVIS). *Remote Sensing of Environment*, 66, 122–128.

Espinosa-Hernandez, A., Galvan-Pineda, J., Monsivais-Huertero, A., Jimenez-Escalona, J. C., and Ramos-Rodriguez, J. M. 2013. Delineation of hydrocarbon contaminated soils using optical and radar images in a costal region. *International Geoscience and Remote Sensing Symposium (IGARSS)*, art. no. 6721247, 676–679.

Evans, F. and Caccetta, P. 2000. Broad-scale spatial prediction of areas at risk from dryland salinity. *Cartography*, 29, 33–40.

Everitt, J. D., Escobar, D. E., Gerbermann, A. H., and Alaniz, M. A. 1988. Detecting saline soils with video imagery. *Photogrammetry Engineering and Remote Sensing*, 54, 1283–1287.

FAO. 1998. *World Reference Base for Soil Resources.* Food and Agriculture Organization of the United Nations, Rome, Italy.

Farifteh, J., Bouma, A., and van der Meijde, M. 2004. A new approach in the detection of salt affected soils: Integrating surface and subsurface measurements. In *Proceedings of 10th European Meeting of Environmental and Engineering Geophysics, P059,* Utrecht, the Netherlands.

Farifteh, J., Farshad, A., and George, R. 2006. Assessing salt-affected soils using remote sensing, solute modeling and geophysics. *Geoderma,* 130, 191–206.

Farifteh, J., van der Meer, F. M., C. Atzberger, and Carranza, E. J. M. 2007. Quantitative analysis of salt-affected soil reflectance spectra: A comparison of two adaptive methods (PLSR and ANN). *Remote Sensing of Environment,* 110, 59–78.

Feingersh, T., Ben-Dor, E., and Portugali, J. 2007. Construction of synthetic spectral reflectance imagery for monitoring of urban sprawl. *Environmental Modeling and Software,* 22, 335–348.

Fiorio, P. R. and Demattê, J. A. M. 2009. Orbital and laboratory spectral data to optimize soil analysis. *Scientia Agricola,* 66, 250–257.

Fox, G. A. and Sabbagh, G. J. 2002. Estimation of soil organic matter from red and near-infrared remotely sensed data using a soil line Euclidean distance technique. *Soil Science Society of America Journal,* 66, 1922–1929.

Frazier, B. E. and Cheng, Y. 1989. Remote sensing of soils in the Eastern Palouse region with Landsat Thematic Mapper. *Remote Sensing of Environment,* 28, 317–325.

Frey, S. D., Six, J., and Elliott, E. T. 2003. Reciprocal transfer of carbon and nitrogen by decomposer fungi at the soil–litter interface. *Soil Biology and Biochemistry,* 35, 1001–1004.

Fu, S., Cabrera, M. L., Coleman, D. C., Kisselle, K. W., Garrett, C. J., Hendrix, P. F., and Crossley, D. A. 2000. Soil carbon dynamics of conventional tillage and no-till agroecosystems at Georgia Piedmont—HSB-C models. *Ecological Modelling,* 131, 229–248.

Gaffey, S. J. 1985. Reflectance spectroscopy in the visible and near infrared (0.35–2.55 mm): Applications in carbonate petrology. *Geology,* 13, 270–273.

Gaffey, S. J. 1986. Spectral reflectance of carbonate minerals in the visible and near infrared (0.35–2.55 µm): Calcite, aragonite and dolomite. *American Mineralogist,* 71, 151–162.

Gaffey, S. J. and Reed, K. L. 1987. Copper in calcite: Detection by visible and near infra-red reflectance. *Economic Geology,* 82, 195–200.

Galvão, L. S., Pizarro, M. A., and Epiphanio, J. C. N. 2001. Variations in reflectance of tropical soils: Spectral-chemical composition relationships from AVIRIS data. *Remote Sensing of Environment,* 75, 245–255.

Galvão, L. S., Vitorello, I., and Paradella, W. L. 1995. Spectroradiometric discrimination of laterites with principle components analysis and additive modeling. *Remote Sensing of Environment,* 53, 70–75.

Galvão, L. S. and Vitorello, I. 1998. Role of organic matter in obliterating the effects of iron on spectral reflectance and colour of Brazilian tropical soils. *International Journal of Remote Sensing,* 19, 1969–1979.

Gao, Bo-C., Davis, C.O., and Goetz, A.F.H. 2006. A review of atmospheric correction techniques for hyperspectral remote sensing of land surfaces and ocean color. 4241659, pp. 1979–1981.

Gao, Bo-C., Montes, M.J., Davis, C.O., and Goetz, A.F.H. 2009. Atmospheric correction algorithms for hyperspectral remote sensing data of land and ocean. *Remote Sensing of Environment,* 113, 1, 17–24.

García-Haro, F. J., Sommer, S., and Kemper, T. 2005. A new tool for variable multiple endmember spectral mixture analysis (VMESMA). *International Journal of Remote Sensing,* 26, 2135–2162.

Gausman, H. W., Allen, W. A., Cardenas, R., and Bowen, R. L. 1970. Color photos, cotton leaves and soil salinity. *Photogrammetric Engineering,* 36: 454–459.

Gausman, H. W., Menges, R. M., Escobar, D. E., Everitt, J. H., and Bowen, R. L. 1977. Pubescence affects spectra and imagery of silverleaf sunflower (*Helianthus argophyllus*). *Weed Science,* 25, 437–440.

Ge, Y., Thomasson, J. A., and Sui, R. 2011. Remote sensing of soil properties in precision agriculture: A review. *Frontiers of Earth Science,* 5, 229–238.

Genú, A. M. and Demattê, J. A. M. 2006. Determination of soil attribute contents by means of reflected electromagnetic energy. *International Journal of Remote Sensing,* 27, 4807–4818.

Gerbermann, A. H. and Neher, D. D. 1979. Reflectance of varying mixtures of a clay soil and sand. *Photogrammetric Engineering and Remote Sensing,* 45, 1145–1151.

Gersman, R., Ben-Dor, E., Beyth, M., Avigad, D., Abraha, M., and Kibreab, A. 2008. Mapping of hydrothermally altered rocks by the EO-1 Hyperion sensor, Northern Danakil Depression, Eritrea. *International Journal of Remote Sensing,* 29, 3911–3936.

Gmur, S., Vogt, D., Zabowski, D., and Moskal, L. M. 2012. Hyperspectral analysis of soil nitrogen, carbon, carbonate, and organic matter using regression trees. *Sensors,* 12, 10639–10658.

Goetz, A. F. H.2009. Three decades of hyperspectral remote sensing of the Earth: A personal view. *Remote Sensing of Environment,* 113, S5–S16.

Goetz, F. A. H., Hauff, P., Shippert, M., and Maecher, G. A. 1991. Rapid detection and identification of OH-bearing minerals in the 0.9–1.0 mm region using new portable field spectrometer. In: *Proceeding of the Eighth Thematic Conference on Geologic Remote Sensing,* Denver, CO, Vol. I, pp. 1–11.

Goidts, E. and van Wesemael, B. 2007. Regional assessment of soil organic carbon changes under agriculture in Southern Belgium (1955–2005). *Geoderma,* 141, 341–354.

Goldshleger, N., Ben-Dor, E., Benyamini, Y., and Agassi, M. 2005. Soil reflectance as a tool for assessing physical crust arrangement of four typical soils in Israel. *Soil Science*, 169, 677–687.

Goldshleger, N., Ben-Dor, E., Chudnovsky, A., and Agassi, M. 2009. Soil reflectance as a generic tool for assessing infiltration rate induced by structural crust for heterogeneous soils. *European Journal of Soil Science*, 60, 1038–1951.

Goldshleger, N., Chudnovsky, A., and Ben-Binyamin, R. 2013. Predicting salinity in tomato using soil reflectance spectra. *International Journal of Remote Sensing*, 34, 6079–6093.

Gomez, C., Lagacherie, P., and Coulouma, G. 2008. Continuum removal versus PLSR method for clay and calcium carbonate content estimation from laboratory and airborne hyperspectral measurements. *Geoderma*, 148, 141–148.

Goosen, D. 1967. Aerial photo interpretation in soil survey. Food and Agricultural Organization, Rome, Italy.

Hapke, B. W. 1981a. Bidirectional reflectance spectroscopy I. Theory. *Journal of Geophysical Research*, 86, 3039–3054.

Hapke, B. W. 1981b. Bidirectional reflectance spectroscopy: 2 Experiments and observation. *Journal of Geophysical Research*, 86, 3055–3060.

Hapke, B. W. 1984. Bidirectional reflectance spectroscopy: Correction for macroscopic roughens. *Icarus*, 59, 41–59.

Hapke, B. W. 1986. Bidirectional reflectance spectroscopy 4: The extinction coefficient and the opposition effect. *Icarus*, 67, 264–280.

Hapke, B. W. 1993. *Theory of Reflectance and Emittance Spectroscopy*. Cambridge University Press, New York.

Hardisky, M. A., Klemas, V., and Smart, M. 1983. The influence of soil salinity, growth form, and leaf moisture on the spectral radiance of *Spartina alterniflora*. *Photogrammetric Engineering and Remote sensing*, 49, 77–83.

Haubrock, S., Chabrillat, S., Lemmnitz, C., and Kaufmann, H. 2008. Surface soil moisture quantification models from reflectance data under field conditions. *International Journal of Remote Sensing*, 29, 3–29.

Hbirkou, C., Pätzold, S., Mahlein, A. K., and Welp, G. 2012. Airborne hyperspectral imaging of spatial soil organic carbon heterogeneity at the field-scale. *Geoderma*, 175, 21–28.

Hengl, T. 2006. Finding the right pixel size. *Computers & Geosciences*, 32, 1283–1298.

Hick, R. T. and Russell, W. G. R. 1990. Some spectral considerations for remote sensing of soil salinity. *Australian Journal of Soil Research*, 28, 417–431.

Hill, J. and Schütt, B. 2000. Mapping complex patterns of erosion and stability in dry Mediterranean ecosystems. *Remote Sensing of Environment*, 74, 557–569.

Hill, J., Udelhoven, T., Vohland, M., and Stevens, A. 2010. The use of laboratory spectroscopy and optical remote sensing for estimating soil properties. In: *Precision Crop Protection—The Challenge and Use of Heterogeneity*, Springer, E. C. Oerke, R. Gerhards, G. Menz, and A. Sikora (eds.), Dordrect, the Netherlands, pp. 67–85.

Hirschfeld, T. 1985. Salinity determination using NIRA. *Applied Spectroscopy*, 39, 740–741.

Hörig, B., Kühn, F., Oschutz, F., and Lehaman, F. 2001. HyMap hyperspectral remote sensing to detect hydrocarbons. *International Journal of Remote Sensing*, 22, 1413–1422.

Howari, F. M., Goodell, P. C., and Miyamoto, S. 2002. Spectral properties of salt crusts formed on saline soils. *Journal of Environmental Quality*, 31, 1453–1461.

Huang, W. and Foo, S. 2002. Neural network modeling of salinity variation in Apalachicola River. *Water Research*, 36, 356–362.

Huete, A. R. 2004. Remote sensing of soils and soil processes. In: *Remote Sensing for Natural Resource Management and Environmental Monitoring*, S.L. Ustin (ed.), John Wiley & Sons, Hoboken, NJ, Vol. 4, pp. 3–52.

Huete, A. R. 2005. Estimation of soil properties using hyperspectral VIS/IR sensors. In: *Encyclopedia of Hydrological Sciences*, M. G. Anderson (ed.), J. Wiley and Sons.

Hunt, G. R. and Salisbury, J. W. 1970. Visible and near infrared spectra of minerals and rocks: I: Silicate minerals. *Modern Geology*, 1, 283–300.

Hunt, G. R. and Salisbury, J. W. 1971. Visible and near infrared spectra of minerals and rocks: II. Carbonates. *Modern Geology* 2, 23–30.

Hunt, G. R. and Salisbury, J. W. 1976. Visible and near infrared spectra of minerals and rocks: XI. Sedimentary rocks. *Modern Geology*, 5, 211–217.

Hunt, G. R., Salisbury, J. W., and Lenhoff, C. J. 1971a. Visible and near-infrared spectra of minerals and rocks: III Oxides and hydroxides. *Modern Geology*, 2, 195–205.

Hunt, G. R., Salisbury, J. W., and Lenhoff, C. J. 1971b. Visible and near-infrared spectra of minerals and rocks: IV Sulfides and sulfates. *Modern Geology*, 3, 1–14.

Hunt, G. R., Salisbury, J. W., and Lenhoff, C. J. 1971c. Visible and near-infrared spectra of minerals and rocks: Halides, phosphates, arsenates, vandates and borates. *Modern Geology*, 3, 121–132.

Hunt, G. R. 1982. Spectroscopic properties of rock and minerals. In: *Handbook of Physical Properties Rocks*, C.R. Stewart (ed.), CRC Press, Boca Raton, FL, p. 295.

Irons, J. R., Weismiller, R. A., and Petersen, G. W. 1989. Soil reflectance. In: *Theory and Application of Optical Remote Sensing*, G. Asrar (ed.), Willey Series in Remote Sensing, John Wiley & Sons, New York, 66–106.

Jackson, R. D., Moran, S., Slater, P. N., and Biggar, S. F. 1987. Field calibration of reflectance panels. *Remote Sensing of Environment*, 22, 145–158.

Jackson, R. D., Teillet, P. M., Slater, P. N., Fedosjsvs, G., Jasinski, M. F., Aase, J. K., and Moran, M. S. 1990. Bidirectional measurements of surface reflectance for view angle corrections of oblique imagery. *Remote Sensing of Environment*, 32, 189–202.

Jacquemoud, S., Baret, F., and Hanocq, J. F. 1992. Modeling spectral and bidirectional soil reflectance. *Remote Sensing of Environment*, 41, 123–132.

Jarmer, T. 2012. Using spectroscopy and satellite imagery to assess the total iron content of soils in the Judean Desert (Israel). In: *SPIE Remote Sensing, 853129–853129*, International Society for Optics and Photonics.

Jarmer, T., Lavée, H., Sarah, P., and Hill, J. 2005. The use of remote sensing for the assessment of soil inorganic carbon in the Judean Desert (Israel). In: *Proceedings of the First International Conference on Remote Sensing and Geoinformation Processing in the Assessment of Land Degradation and Desertification (RGLDD)*, University of Trier, Trier, Germany, pp. 68–75.

Jenny, H. 1941. Factors of soil formation, McGraw-Hill Book Company, New York, NY, pp. 281.

Jensen R. J. 2011. *Remote Sensing of the Environment: An Earth Resources Perspective*. Pearson Prentice Hall Press, p. 592.

Kalra, N. K. and Joshi, D. C. 1996. Potentiality of Landsat, SPOT and IRS satellite imagery for recognition of salt affected soils in Indian arid zone. *International Journal of Remote Sensing*, 17, 3001–3014.

Karmanova, L. A. 1981. Effect of various iron compounds on the spectral reflectance and color of soils. *Soviet Soil Science*, 13, 63–60.

Karnieli, A., Kidron, G., Ghassler, C., and Ben-Dor, E. 1999. Spectral characteristics of cyanobacteria soil crust in the visible near infrared and short wave infrared (400–2,500nm) in semiarid environment. *International Journal of Remote Sensing*, 69, 67–77.

Karnieli, A. and Tsoar, H. 1995. Spectral reflectance of biogenic crust developed on desert dune sand along the Israel-Egypt border. *Remote Sensing*, 16, 369–374.

Kemper, T. and Sommer, S. 2002. Estimate of heavy metal contamination in soils after a mining using reflectance spectroscopy. *Environmental Science and Technology*, 36, 2742–2747.

Kemper, T. and Sommer, S. 2003. Mapping and monitoring of residual heavy metal contamination and acidification risk after the Aznalcóllar mining accident (Andalusia, Spain) using field and airborne hyperspectral data. In: *Proceedings, Third EARSeL Workshop on Imaging Spectroscopy EARSeL Secretariat*, Paris, France, pp. 333–343.

Kierein-Young, K. and Kruse, F. A. 1989. Comparison of landsat thematic mapper images and geophysical and environmental reassert imaging spectrometer data for alteration mapping. In: *Proceedings of the Seventh Thematic Conference on Remote Sensing for Exploration Geology*, Calgary, Alberta, Canada, I, 349–359.

Knadel, M., Deng, F., Alinejadian, A., Wollesen de Jonge, L., Moldrup, P., and Greve, M. H. 2014. The effects of moisture conditions—From wet to hyper dry—On visible near-infrared spectra of danish reference soils. *Soil Science Society of America Journal*, 78, 422–433.

Kokaly, R. F., Rockwell, B. W., Haire, S. L., and King, T. V. 2007. Characterization of post-fire surface cover, soils, and burn severity at the Cerro Grande Fire, New Mexico, using hyperspectral and multispectral remote sensing. *Remote Sensing of Environment*, 106, 305–325.

Kooistra, L., Salas, E. A. L., Clevers, J. G. P. W., Wehrens, R., Leuven, R. S. E. W., Nienhuis, P. H., and Buydens, L. M. C. 2004. Exploring field vegetation reflectance as an indicator of soil contamination in river floodplains. *Environmental Pollution*, 127, 281–282.

Kooistra, L., Wanders, J., Epema, G. E., Leuven, R. S. E. W., Wehrens, R., and Buydens, L. M. C. 2003.The potential of field spectroscopy for the assessment of sediment properties in river floodplains. *Analytica Chimica Acta*, 484, 198–200.

Kopačková, V. 2014. Using multiple spectral feature analysis for quantitative pH mapping in a mining environment. *International Journal of Applied Earth Observation and Geoinformation*, 28, 28–42, 90.

Kopačková, V., Mišurec, J., Lhotáková, Z., Oulehle, F., and Albrechtová, J. 2014. Using multi-date high spectral resolution data to assess the physiological status of macroscopically undamaged foliage on a regional scale. *International Journal of Applied Earth Observation and Geoinformation*, 27, 169–186.

Kosmas, C. S., Curi, N., Bryant, R. B., and Franzmeier, D. P. 1984. Characterization of iron oxide minerals by second derivative visible spectroscopy. *Soil Science Society of American Journal*, 48, 401–405.

Krishnan, P., Alexander, J. D., Bulter, B. J., and Hummel, J. W. 1980. Reflectance technique for predicting soil organic matter. *Soil Science Society of American Journal*, 44, 1282.

Kristof, S. J. 1971. Preliminary multispectral studies of soils. *Journal of Soil and Water Conservation*, 26, 15–18.

Kristof, S. J., Baumgardner, M. F., and Johannsen, C. J. 1971. Spectral mapping of soil organic matter. Journal Paper No.5390, Agricultural Experiment Station, Purdue University, West Lafayette, IN.

Kristof, S. J. and Zachary, A. L. 1974. Mapping soil features from multispectral scanner data. *Photogrammetric Engineering and Remote Sensing*, 40, 1427–1434.

Kruse, F. A., Kierein-Young, and Boardman, J. W. 1990. Mineral mapping of Cuprite, Nevada with a 63-channel imaging spectrometer. *Photogrammetric Engineering and Remote Sensing*, 56, 83–92.

Kruse, F. A., Thiry, M., and Hauff, P. L. 1991. Spectral identification (1.2–2.5 mm) and characterization of Paris basin kaolinite/smectite clays using a field spectrometer. In *Proceedings of the Fifth International Colloquium, Physical Measurements and Signatures in Remote Sensing*, Courchevel, France I, pp. 181–184.

Kumar, N. S., Anouncia, S. M., and Prabu, M. 2013. Application of satellite remote sensing to find soil fertilization by using soil colour. *International Journal of Online Engineering*, 9, 44–49.

Ladoni, M., Bahrami, H. A., Alavipanah, S. K., and Norouzi, A. A. 2010. Estimating soil organic carbon from soil reflectance: A review. *Precision Agriculture*, 11, 82–99.

Lagacherie, P., Baret, F., Feret, J. B., Madeira Netto, J., and Robbez-Masson, J. M. 2008. Estimation of soil clay and calcium carbonate using laboratory, field and airborne hyperspectral measurements. *Remote Sensing of Environment*, 112, 825–835.

Lagacherie, P., Sneep, A. R., Gomez, C., Bacha, S., Coulouma, G., Hamrouni, M. H., and Mekki, I. 2013. Combining Vis–NIR hyperspectral imagery and legacy measured soil profiles to map subsurface soil properties in a Mediterranean area (Cap-Bon, Tunisia). *Geoderma*, 209, 168–176.

Latz, K., Weismiller, R. A., and Van Scoyoc, G. E. 1981. A study of the spectral reflectance of selected eroded soils of Indiana in relationship to their chemical and physical properties. LARS Technical Report 082181.

Leger, R. G., Millette, G. J. F., and Chomchan, S. 1979. The effects of organic matter, iron oxides and moisture on the color of' two agricultural soils of Quebec Can. *Journal of Soil Science*, 59, 191–202.

Legros, J. P. 2006. *Mapping of the Soil.* Science Publishers, Enfield, NH, p. 411.

Lekner, J. and Dorf, M. C. 1988. Why some things are darker when wet. *Applied Optics*, 27, 1278–1280.

Lemos, P. and Lutz, J. F. 1957. Soil crusting and some factors affecting it. *Soil Science Society of America Journal*, 21, 485–491.

Levenhangst, A. I. 1930. The use of aerial photographs in soil survey. *Ponhvoedenie*, 4, 116–122.

Levinstein, H. and Mudar, J. 1975. Infrared detectors in remote sensing. *Proceedings of the IEEE*, 63, 1.

Li, J., Okin, G. S., Skiles, S. M., and Painter, T. H. 2013. Relating variation of dust on snow to bare soil dynamics in the western United States. *Environmental Research Letters*, 8, 044054.

Liang, S. and Townshend, R. G. 1996. A modified Hapke model for soil bidirectional reflectance. *Remote Sensing of Environment*, 55, 1–10.

Linker, R. 2008. Soil classification mid-infrared spectroscopy. In: *IFIP International Federation for Information Processing*, Vol. 259. Computer and Haifa, Israel. Computing Technologies in Agriculture, Vol. 2, Daoliang Li, Springer, Boston, MS, pp. 1137–1146.

Liu, M., Liu, X., Li, M., Fang, M., and Chi, W. 2010. Neural-network model for estimating leaf chlorophyll concentration in rice under stress from heavy metals using four spectral indices. *Biosystems Engineering*, 106, 223–233.

Lobell, D.B. and Asner, G. P. 2002. Moisture effects on soil reflectance. *Soil Science Society America Journal*, 66, 722–727.

Löhrer, R., Bertrams, M., Eckmeier, E., Protze, J., and Lehmkuhl, F. 2013. Mapping the distribution of weathered Pleistocene wadi deposits in Southern Jordan using ASTER, SPOT-5 data and laboratory spectroscopic analysis. Catena, 107, 57–70.

Lucht, W., Hyman, A. H., Strahler, A. H., Barnsley, M. J., Hobson, P., and Muller, J. P. 2000. A comparison of satellite-derived spectral albedos to ground-based broadband albedo measurements modeled to satellite spatial scale for a semidesert landscape. *Remote Sensing of Environment*, 74, 85–98.

Lugassi, R., Ben-Dor, E., and Eshel, G. 2010. A spectral-based method for reconstructing spatial distributions of soil surface temperature during simulated fire events. *Remote Sensing of Environment*, 114, 322–331.

Lugassi, R., Ben-Dor, E., and Eshel, G. 2014. Reflectance spectroscopy of soils post-heating—Assessing thermal alterations in soil minerals. *Geoderma*, 213, 268–279.

Maestre, F. T. and Cortina, J. 2002. Spatial patterns of surface soil properties and vegetation in a Mediterranean semi-arid steppe. *Plant and Soil*, 241, 279–291.

Malley, D. F., Hunter, K. N., and Webste, G. R. 1999. Analysis of diesel fuel contamination in soils by near-infrared reflectance spectrometry and solid phase microextraction–gas chromatography. *Soil and Sediment Contamination*, 8, 481–489.

Malley, D. F. and Williams, P. C. 1997. Use of near-infrared reflectance spectroscopy in prediction of heavy metals in freshwater sediment by their association with organic matter. *Environmental Science and Technology*, 31, 3461–3467.

Malley, D. F., Martin P., and Ben-Dor, E. 2004. Application in analysis of soils. In: *Near Infrared Spectroscopy in Agriculture*, R. Craig, R. Windham, and J. Workman, (eds.), A three Societies Monograph (ASA, SSSA, CSSA), Madison Wisconsin, 44, 729–784, Chapter 26.

Mathews, H. L., Cunningham, R. L., and Peterson, G. W. 1973. Spectral reflectance of selected Pennsylvania soils. *Proceedings of the Soil Science Society of America Journal*, 37, 421–424.

McBratney, A. B., Minasny, B., and Rossel, R. V. 2006. Spectral soil analysis and inference systems: A powerful combination for solving the soil data crisis. *Geoderma*, 136, 272–278.

McBratney, A. B., Santos, M. D. L. M., and Minasny, B. 2003. On digital soil mapping. *Geoderma*, 117, 3–52.

McCarty, G. W., Reeves, J. B., Reeves, V. B., Follett, R. F., and Kimble, J. M. 2002. Mid-infrared and near-infrared diffuse reflectance spectroscopy for soil carbon measurement. *Soil Science Society of American Journal*, 66, 640–646.

McKenzie, N. J. and Ryan, P. J. 1999. Spatial prediction of soil properties using environmental correlation. *Geoderma*, 89, 67–94.

Metternicht, G., Zinck, J. A., Blanco, P. D., & Del Valle, H. F. 2010. Remote sensing of land degradation: Experiences from Latin America and the Caribbean. *Journal of Environmental Quality*, 39, 42–61.

Metternicht, G. I. and Zinck, J. A. 1996. Modelling salinity-alkalinity classes for mapping salt-affected topsoils in the semi-arid valleys of Cochabamba (Bolivia). *ITC-Journal*, 2, 125–135.

Metternicht, G. I. and Zinck, J. A. 1997. Spatial discrimination of salt- and sodium-affected soil surfaces. *International Journal of Remote Sensing*, 18, 2571–2586.

Metternicht, G. I. and Zinck, J. A. 2008. *Remote Sensing of Soil Salinization: Impact on Land Management.* CRC Press, Boca Raton, FL, p. 374.

Montanarella, L., Tóth, G., and Jones, A. 2011. Soil Component in the 2009 LUCAS Survey. p. 209–219. In: *Land Quality and Land Use Information in the European Union*. Publication Office of the European Union, Luxembourg. Tóth, G and Németh, T., Luxembourg.

Montgomery, O. L. 1976. An investigation of the relationship between spectral reflectance and the chemical, physical and genetic characteristics of soils. PhD thesis, Purdue University, West Lafayette, IN (Libr. Congr. no 79-32236).

Montgomery, O. L. and Baumgardner, M. F. 1974. The effects of the physical and chemical properties of soil and the spectral reflectance of soils. Information Note 1125 Laboratory for *Applications of Remote Sensing*, Purdue University, West Lafayette, IN.

Morin, Y., Benyamini, Y., and Michaeli, A. 1981. The dynamics of soil crusting by rainfall impact and the water movement in the soil profile. *Journal of Hydrology*, 52, 321–335.

Morra, M. J., Hall M. H., and Freeborn, L. L. 1991. Carbon and nitrogen analysis of soil fractions using near-infrared reflectance spectroscopy. *Soil Science Society of American Journal*, 55, 288–291.

Mouazen, A. M., De Baerdemaeker, J., and Ramon, H. 2005. Towards development of on-line soil moisture content sensor using a fibre-type NIR spectrophotometer. *Soil and Tillage Research*, 80, 171–183.

Mouazen, A. M., Karoui, R., De Baerdemaeker, J., and Ramon, H. 2006. Characterization of soil water content using measured visible and near infrared spectra. *Soil Science Society of America Journal*, 70, 1295–1302.

Mougenot, B. 1993. Effect of salts on reflectance on reflectance and remote sensing of salt affected soils. *Cahiers Orstom Serie Pedologie XXVIII*, 1, 45–54.

Mulder, M. A. 1987. *Remote Sensing of Soils*. Development in Soil Science 15, Elsevier Science Publishing, Amsterdam, the Netherland, p. 379.

Muller, E. and Decamps, H. 2001. Modeling soil moisture-reflectance. *Remote Sensing of Environment*, 76, 173–180.

Mulder, V. L., De Bruin, S., Schaepman, M. E., and Mayr, T. R. 2011. The use of remote sensing in soil and terrain mapping—A review. *Geoderma*, 162, 1–19.

Musavi, K. M. S., Bahrami, H. A., Effati, M., and Darvishi, B. A. 2013. Investigation of soil type effects on dust storms detection using day and night time multi-spectral MODIS images. *International Journal of Agriculture: Research and Review*, 3, 529–542.

Nanni, M. R. and Demattê, J. A. M. 2006. Spectral reflectance methodology in comparison to traditional soil analysis. *Soil Science Society of America Journal*, 70, 393–407.

Nanni, M. R., Demattê, J. A. M., Chicati, M. L., Fiorio, P. R., Cézar, E., and Oliveira, R. B. D. 2012. Soil surface spectral data from Landsat imagery for soil class discrimination. *Acta Scientiarum. Agronomy*, 34, 103–112.

Nanni, M. R., Demattê, J. A. M., Chicati, M. L., Oliveira, R. B., and Cezar, E. 2011. Spectroradiometric data as support to soil classification. *International Research Journal of Agricultural Science and Soil Science*, 1, 109–117,

Nanni, M. R., Demattê, J. A. M., and Fiorio, P. R. 2004. Soil discrimination analysis by spectral response in the ground level. *Pesquisa Agropecuaria Brasileira*, 39(10), 995–1006.

Nassau, K. 1980. The causes of color. *Scientific American*, 243, 106–124.

Nassau, K. 1983. *The Physics and Chemistry of Color*. John Wiley & Sons, New York, p. 454.

Ninomiya, Y., Fu, B., and Cudahy, T. J. 2005. Detecting lithology with Advanced Spaceborne Thermal Emission and Reflection Radiometer (ASTER) multispectral thermal infrared "radiance-at-sensor" data. *Remote Sensing of Environment*, 99, 127–139.

Obukhov, A. I. and Orlov, D. C. 1964. Spectral reflectance of the major soil groups and the possibility of using diffuse reflection in soil investigations. *Soviet Soil Science*, 2,174–184.

Odgers, N. P., McBratney, A. B., and Minasny, B. 2011. Bottom-up digital soil mapping. I. Soil layer classes. *Geoderma*, 163, 38–44.

Offer, Z. Y. and Goossens, D. 2001. Ten years of aeolian dust dynamics in a desert region (Negev desert, Israel): Analysis of airborne dust concentration, dust accumulation and the high-magnitude dust events. *Journal of Arid Environments*, 47, 211–249.

Okin, G. S. and Painter, T. H. 2004. Effect of grain size on remotely sensed spectral reflectance of sandy desert surfaces. *Remote Sensing of Environment*, 89, 272–280.

Ona-Nguema, G., Abdelmoula, M., Jorand, F., Benali, O., Gehin, A., Block, J. C., and Génin, J. M. R. 2002. Iron (II, III) hydroxycarbonate green rust formation and stabilization from lepidocrocite bioreduction. *Environmental Science and Technology*, 36, 16–20.

Orlov, D. C. 1966. Quantitative patterns of light reflectance on soils I: Influence of particles (aggregate) size on reflectivity. *Soviet Soil Science*, 13, 1495–1498.

Palacios-Orueta, A. and Ustin, S. L. 1998. Remote sensing of soil properties in the Santa Monica Mountains I. Spectral analysis. *Remote Sensing of Environment*, 65, 170–183.

Palmer, J. M. 1982. Field standards of reflectance. *Photogrammetric Engineering and Remote Sensing*, 48, 1623–1625.

Palmer, K. F. and Williams, D. 1974. Optical properties of water in the near infrared. *Journal of the Optical Society of America*, 64, 1107–1111.

Pandit, C. M., Filippelli, G. M., and Li, L. 2010. Estimation of heavy-metal contamination in soil using reflectance spectroscopy and partial least-squares regression. *International Journal of Remote Sensing*, 31, 4111–4123.

Piekarczyk, J. 2013. Effects of time of bare cultivated soils observation and their roughness on the average diurnal soil albedo approximation by satellite data. *Selected Topics in Applied Earth Observations and Remote Sensing*, 6(3), 1194–1198. DOI: 10.1109/JSTARS.2012.2234440.

Piemstein, A., Ben-Dor, E., and Notesko, G. 2010. Performance of three identical spectrometers in retrieving soil reflectance under laboratory conditions. *Soil Science Society of America Journal*, 75, 110–174.

Pieters, C. M. 1983. Strength of mineral absorption features in the transmitted component of near-infrared reflected light. First results from RELAB. *Journal of Geophysical Research*, 88, 9534–9544.

Pinty, B., Verstraete, M. M., and Dickson, R. E. 1989. A physical model for prediction bidirectional reflectance over bare soil. *Remote Sensing of Environment*, 27, 273–288.

Prescott, J. A. and Taylor, J. K. 1930. The value of aerial photography in relation to soil surveys and classification. *Journal CSIRO Australia*, 3, 229–230.

Price, J. C. 1990. On the information content of soil reflectance spectra. *Remote Sensing of Environment*, 33, 113–121.

Price, J. C. 1995. Examples of high resolution visible to near-infrared reflectance spectra and a standardized collection for remote sensing studies. *International Journal of Remote Sensing*, 16, 993–1000.

Pruitt, E. L. 1979. The office of naval research and geography. *Annals of the Association of American Geographers*, 69, 103–108.

Rao, B., Sankar, T., Dwivedi, R., Thammappa, S., Venkataratnam, L., Sharma, R., and Das, S. 1995. Spectral behaviour of salt-affected soils. *International Journal of Remote Sensing*, 16, 2125–2136.

Rao, B. R. M. and Venkataratnam, L. 1991. Monitoring of salt affected soils: A case study using aerial photographs, Salyut-7 space photographs, and Landsat TM data. *Geocarto International*, 6, 5–11.

Ray, S. S., Singh, J. P., Das, G., and Panigrahy, S. 2004. Use of high resolution remote sensing data for generating site-specific soil management plan. XX ISPRS Congress, Commission 7. Istanbul, Turkey The International Archives of the Photogrammetry. *Remote Sensing and Spatial Information Sciences*, 35, 127–131.

Ren, H. Y., Zhuang, D. F., Singh, A. N., Pan, J. J., Qiu, D. S., and Shi, R. H. 2009. Estimation of As and Cu contamination in agricultural soils around a mining area by reflectance spectroscopy: A case study. *Pedosphere*, 19, 719–725.

Rizzo, R., Dematté, J. A. M., and Terra, F. S. 2014. Using numerical classification of profiles on VIS-NIR spectra to distinguish soils from the Piracicaba region, Brazil. *Brazilian Journal of Soil Science*, 38, 372–385.

Robichaud, P. R., Lewis, S. A., Laes, D. Y., Hudak, A. T., Kokaly, R. F., and Zamudio, J. A. 2007. Postfire soil burn severity mapping with hyperspectral image unmixing. *Remote Sensing of Environment*, 108, 467–480.

Schmid, T., Gumuzzio, J., Koch, M., Mather, P., and Solana, J. 2007. Characterizing and monitoring semi-arid wetlands using multi-angle hyperspectral and multispectral data. In: *Proceedings Envisat Symposium, ESA SP-636*. Montreux, Switzerland, pp. 1–6.

Schnitzer, M. and Khan, S. U. 1978. *Soil Organic Matter*. Elsevier Publication, Amsterdam, the Netherlands, p. 320.

Schreier, H. May 17–18, 1977. Quantitative predictions of chemical soil conditions from multispectral airborne ground and laboratory measurements. In *Proceedings of the Fourth Canadian Symposium on Remote Sensing*, Canadian Aeronautics and Space Institute, Ottawa, Quebec, Canada, pp. 106–111.

Schwartz, G., Ben-Dor, E., and Eshel, G. 2012. Quantitative analysis of total petroleum hydrocarbons in soils: Comparison between reflectance spectroscopy and solvent extraction by 3 certified laboratories. *Applied and Environmental Soil Science*, 2012, 1–11.

Schwartz, G., Eshel, G., and Ben-Dor, E. 2011. Reflectance spectroscopy as a tool for monitoring contaminated soils. In *Soil Contamination*, S. Pascucci (ed.), InTECH Pub., Rijeka, Croetia, 67–90.

Schwertmann, U. 1988. Occurrence and formation of iron oxides in various pedoenvironment. In: *Iron in Soils and Clay Minerals*, J. W. Stucki, B. A. Goodman, and U. Schwertmann (eds.), NATO ASI series, Reidel Publishing Company, Dordrecht, the Netherlands, pp. 267–308.

Scott, K. M. and Pain, F. P. 2008. *Regolith Science*. CSIRO Publishing/Springer, Collingwood, Victoria, Australia, p. 461.

Selige, T., Bohner, J., and Schmidhalter, U. 2006. High resolution topsoil mapping using hyperspectral image and field data in multivariate regression modeling procedures. *Geoderma*, 136, 235–244.

Serrano, M. H. R. L. 2010. Satellite remote sensing of soil moisture. MSc Dissertation, University of Reading Department of Meteorology, p. 66.

Sertsu, S. M. and Sánchez, P. A. 1978. Effects of heating on some changes in soil properties in relation to an Ethiopian land management practice. *Soil Science Society of America Journal*, 42, 940–944.

Sharma, R. C. and Bhargava, G. P. 1988. Landsat imagery for mapping saline soils and wet lands in north-west India. *International Journal of Remote Sensing*, 9, 39–44.

Sheinberg I. 1992. Chemical and mineralogical components of crust. In: *Soil Crusting. Advances in Soil*, M. E. Sumner and B. A. Stewart (eds.), Science Lewis, London, U.K., pp. 39–53.

Shepherd, K. D. and Walsh, M. G. 2002. Development of reflectance spectral libraries for characterization of soil properties. *Soil Science Society of America Journal*, 66, 988–998.

Sherman, D.M. and Waite T.D. 1985. Electronic spectra of Fe3+ oxides and oxide hydroxides in the near-IR to near-UV. *American Mineralogist*, 70, 1262–1269.

Simonson, R. W. 1987. Historical aspects of soil survey and soil classification, 15–19.

Singer, A. 2007. *The Soils of Israel*. Springer-Verlag, Berlin, Germany, p. 306.

Singh, A. 1989. Review article digital change detection techniques using remotely-sensed data. *International Journal of Remote Sensing*, 10, 989–1003.

Soil Survey Staff. 1975. *Soil taxonomy, A Basic System of Soil Classification for Making and Interpreting Soil Survey*. Soil Conservation Service. U.S. Department of Agriculture Handbook No 436. Washington, DC

Soileau, J. M. and McCraken, R. J. 1967. Free iron and coloration in certain well-drained Costal Plain soils in relation to their other properties and classification. *Soil Science Society of American Proceedings*, 31, 248–255.

Souza Filho, C. 2013. Mapping of Geologic substrates impregnated with liquid hydrocarbons using proximal and airborne hyper spectral remote sensing: Potential applications for onshore exploration and leakage monitoring. Abstract un. IGARSS 2013 Melbourne July 25.

Staenz, K. and Held, A. 2012. Summary of current and future terrestrial civilian hyperspectral spaceborne systems. In *Geoscience and Remote Sensing Symposium (IGARSS), 2012 IEEE International*, pp. 123–125, IEEE.

Steffens, M. and Buddenbaum, H. 2013. Laboratory imaging spectroscopy of a stagnic Luvisol profile—High resolution soil characterisation, classification and mapping of elemental concentrations. *Geoderma*, 195–196, 122–132.

Stevens, A., Udelhoven, T., Denis, A., Tychon, B., Lioy, R., Hoffmann, L., and Van Wesemael, B. 2010. Measuring soil organic carbon in croplands at regional scale using airborne imaging spectroscopy. *Geoderma*, 158, 32–45.

Stevens, A., Van Wesemael, B., Bartholomeus, H., Rosillon, D., Tychon, B., and Ben-Dor, E. 2008. Laboratory, field and airborne spectroscopy for monitoring organic carbon content in agricultural soils. *Geoderma*, 144, 395–404.

Stevens, A., Van Wesemael, B., Vandenschrick, G., Touré, S., and Tychon, B. 2006. Detection of carbon stock change in agricultural soils using spectroscopic techniques. *Soil Science Society of America Journal*, 70, 844–850.

Stevenson, F. J. 1982. *Humus Chemistry*. John Wiley & Sons Inc., New York, p. 198.

Stoner, E. R. 1979. Physicochemical, site and bidirectional reflectance factor characteristics of uniformly-moist soils. PhD thesis, Purdue University, West Lafayette, IN.

Stoner, E. R. and Baumgardner, M. F. 1981. Characteristic variations in reflectance of surface soils. *Soil Science Society of American Journal*, 45, 1161–1165.

Stoner, E. R., Baumgardner, M. F., Weismiller, R. A., Biehl, L. L., and Robinson, B. F. 1980. Extension of laboratory-measured soil spectra to field conditions. In *Soil Science Society of America Proceedings*, Madison, WI, p. 4.

Szilagyi, A. and Baumgardner, M. F. 1991. Salinity and spectral reflectance of soils. In *Proceedings of ASPRS Annual Convention*, Baltimore, MD, pp. 430–438.

Tangestani, M. H. and Moore, F. 2000. Iron oxide and hydroxyl enhancement using the Crosta Method: A case study from the Zagros Belt, Fars Province, Iran. *International Journal of Applied Earth Observation and Geoinformation*, 2, 140–146.

Taranik, D. L. and Kruse, F. A. 1989. Iron minerals reflectance in geophysical and environmental research imaging spectrometer (GERIS) data. In *Proceedings of the Seventh Thematic Conference on Remote Sensing for Exploration Geology*, Calgary, Alberta I, pp. 445–458.

Taylor, G. R., Bennett, B. A., Mah, A. H., and Hewson, R. 1994. Spectral properties of salinised land and implications for interpretation of 24 channel imaging spectrometry. In *Proceedings of the First International Remote Sensing Conference and Exhibition*, Strasbourg, France, 3, 504–513.

Thomasson, J. A., Sui, R., Cox, M. S., and Al-Rajehy, A. 2001. Soil reflectance sensing for determining soil properties in precision agriculture. *ASAE* 44, 1445–1453.

Thompson, L. M. 1957. *Soils and Soil Fertility*. McGraw-Hill Book Company Inc., New York.

Touré, S. and Tychon, B. 2004. Airborne hyperspectral measurements and superficial soil organic matter. In *Proceedings of the Airborne Imaging Spectroscopy Workshop*.

Uno, Y., Prasher, S. O., Patel, R. M., Strachan, I. B., Pattey, E. and Karimi, Y. 2005. Development of field-scale soil organic matter content estimation models in Eastern Canada using airborne hyperspectral imagery. *Canadian Biosystems Engineering*, 45, 1.9–1.14.

Valentin, C. and Bresson, L. M. 1992. Morphology, genesis and classification of surface crusts in loamy and sandy soils. *Geoderma*, 55, 225–245.

Van der Meer, F. 1995. Spectral reflectance of carbonate mineral mixture and bidirectional reflectance theory: Quantitative analysis techniques for application in remote sensing. *Remote Sensing Reviews*, 13, 67–94.

Van der Meer, F.D., van Dijk, P.M., van der Werff H.M.A., and Yang Hong. 2002. Remote sensing and petroleum seepage: A review and case study. *Terra Nova*, 14, 1–17.

Vane, G. 1993. Airborne imaging spectrometry, special issue. *Remote Sensing of. Environment*, 44, 117–356.

Vane, G., Goetz, A. F., and Wellman, J. B. 1984. Airborne imaging spectrometer: A new tool for remote sensing. *IEEE Transactions on Geoscience and Remote Sensing*, vGE-22, 546–549.

Vasques, G. M., Demattê, J. A. M., Rossel, R. V., Lopez, L. R., and Terra, F. S. 2014. Soil classification using visible/near-infrared diffuse reflectance spectra from multiple depths. *Geoderma*, 223–225, 73–78.

Verma, K.S., Saxena, R.K., Barthwal, A.K., and Deshmukh, S. K. 1994. Remote sensing technique for mapping salt affected soils. *International Journal of Remote Sensing*, 15, 1901–1914.

Vink, A. P. A. 1963a. Soil survey as related to agricultural productivity. *Journal of Soil Science*, 14, 88–101.

Vink, A. P. A. 1963b. Planning of soil surveys in land development. International Institute for Land Reclamation and Improvement, publication 10, U.D.C. 631.47: 528.77, the Netherlands, 53.

Vink, A. P. A. 1964. Aerial photographs and the soil sciences. *Proceedings of the Toulouse Conference on Aerial Surveys and Integrated Studies*, pp. 81–141.

Vinogradov, B. V. 1981. Remote Sensing of the humus content of soils. *Soviet Soil Science*, 11, 114–123.

Viscarra Rossel, R. A. 2009. The soil spectroscopy group and the development of a global soil spectral library. *EGU General Assembly Conference Abstracts*, 11, 14021.

Viscarra Rossel, R. A., Gilbertson, M., Thylen, L., Hansen, O., McVey, S., McBratney, A. B., and Stafford, J. V. 2005. Field measurements of soil pH and lime requirement using an on-the-go soil pH and lime requirement measurement system.

In *Precision Agriculture'05. Papers Presented at the Fifth European Conference on Precision Agriculture*, Wageningen Academic Publishers, Uppsala, Sweden, pp. 511–520.

Viscarra Rossel, R. A., Walvoort, D. J. J., McBratney, A. B., Janik, L. J., and Skjemstad, J. O. 2006. Visible, near infrared, mid infrared or combined diffuse reflectance spectroscopy for simultaneous assessment of various soil properties. *Geoderma*, 131, 59–75.

Viscarra Rossel, R. A. and Webster, R. 2011. Discrimination of Australian soil horizons and classes from their visible–near infrared spectra. *European Journal of Soil Science*, 62, 637–647.

Weidong, L., Baret, F., Xingfa, Q., Qingxi, T., Lanfen, Z., and Bing, Z. 2002. Relating soil surface moisture to reflectance. *Remote Sensing of Environment*, 81, 238–246.

White, K., Walden, J., Drake, N., Eckardt, F., and Settlell, J. 1997. Mapping the iron oxide content of dune sands, Namib Sand Sea, Namibia, using Landsat Thematic Mapper data. *Remote Sensing of Environment*, 62, 30–39.

Whiting, M. L., Li, L., and Ustin, S. L. 2004, Predicting water content using Gaussian model on soil spectra. *Remote Sensing of Environment*, 89, 535–552.

Wiegand, C. L., Everitt, J. H., and Richardson, A. J. 1992. Comparison of multispectral video and SPOT-1 HRV observations for cotton affected by soil salinity. *International Journal of Remote Sensing*, 13, 1511–1525.

Wiegand, C., Richardson, A., Escobar, D., and Gerbermann, A. 1991. Vegetation indices in crop assessments. *Remote Sensing of Environment*, 35, 105–119.

Wiegand, C. L., Rhoades, J. D., Escobar, D. E., and Everitt, J. H. 1994. Photographic and videographic observations for determining and mapping the response of cotton to soil salinity. *Remote Sensing of Environment*, 49, 212–223.

Winkelmann, K. H. 2005. On the applicability of imaging spectrometry for the detection and investigation of contaminated sites with particular consideration given to the detection of fuel hydrocarbon contaminants in soil. Doctoral dissertation, Universitätsbibliothek.

Wu, Y., Chen, J., Ji, J., Gong, P., Liao, Q., Tian, Q., and Ma, H. 2007. A mechanism study of reflectance spectroscopy for investigating heavy metals in soils. *Soil Science Society of America Journal*, 71, 918–925.

Wu, Y., Zhang, X., Liao, Q., and Ji, J. 2011. Can contaminant elements in soils be assessed by remote sensing technology: A case study with simulated data. *Soil Science*, 176, 196–205.

Wu, Y. Z., Chen, J., Ji, J. F., Tian, Q. J., and Wu, X. M. 2005. Feasibility of reflectance spectroscopy for the assessment of soil mercury contamination. *Environmental Science and Technology*, 39, 873–878.

Xia, X. Q., Mao, Y. Q., Ji, J. F., Ma, H. R., Chen, J., and Liao, Q. L. 2007. Reflectance spectroscopy study of Cd contamination in the sediments of the Changjiang River, China. *Environmental Science & Technology*, 41, 3449–3454.

Yang, H., Zhang, J., van der Meer, F. D., and Kroonenberg, S. B. 2000. Imaging spectrometry data correlated to hydrocarbon microseepage. *International Journal of Remote Sensing*, 21, 197–202.

Zabcic, N., Rivard, B., and Ong, C. 2005. Mapping surface pH using airborne hyperspectral imagery at the Sotiel—Migollas Spain, 4th EARSeL and Warsaw University, Warsaw *Proceedings of Fourth EARSeL Workshop on Imaging Spectroscopy*.

Zabcic, N., Rivard, B., Ong, C., and Mueller, A. August 2009. Using airborne hyperspectral data to characterize the surface pH of pyrite mine tailings. *IEEE Workshop on Hyperspectral Image and Signal Processing: Evolution of Remote Sensing, Whispers'09*, Grenoble, France, pp. 26–28.

Zhang, T., Li, L., and Zheng, B. 2013. Estimation of agricultural soil properties with imaging and laboratory spectroscopy. *Journal of Applied Remote Sensing*, 7, 073587.

Zheng, B. 2008. Using satellite hyperspectral imagery to map soil organic matter, total nitrogen and total phosphorus. A Master thesis submitted to the Department of Earth Science University of Indiana, Indianapolis, IN, p. 81.

Zhou, Z. C., Shangguan, Z. P., and Zhao, D. 2006. Modeling vegetation coverage and soil erosion in the Loess Plateau Area of China. *Ecological Modeling*, 198, 263–268.

Zhu, Y., Wang, Y., Shao, M., and Horton, R. 2011. Estimating soil water content from surface digital image gray level measurements under visible spectrum. *Canadian Journal of Soil Science*, 91, 69–76.

Summary

26

Remote Sensing of Land Resources: Monitoring, Modeling, and Mapping Advances over the Last 50 Years and a Vision for the Future

Prasad S. Thenkabail
U.S. Geological Survey (USGS)

Acronyms and Definitions

ACD	Above ground carbon density
AGB	Above ground biomass
AgRISTARS	Agriculture and resources inventory surveys through aerospace remote sensing
ALS	Airborne laser scanning
APAR	Absorbed photosynthetically active radiation
ASAR	Advanced synthetic aperture radar onboard ENVISAT
ASS	Aerial spectral sensing
ATSR	Along-track scanning radiometer
AVHRR	Advanced very-high-resolution radiometer
CAI	Cellulose absorption indices
CDL	Cropland data layer
CHM	Canopy height model
COSMO-SkyMed	Constellation of Small Satellites for Mediterranean basin Observation (COSMO)-SkyMed
DTM	Digital terrain model
DSM	Digital surface model
EnMAP	Environmental Mapping and Analysis Program
ENVISAT	Environmental satellite
EO	Earth observing
EVI	Enhanced vegetation index
FAO	Food and Agriculture Organization of the United Nations
FLUXNET	A network of micrometeorological tower sites to measure carbon dioxide, water, and energy balance between terrestrial systems and the atmosphere
GEO	Group on Earth Observation
GIEWS	Global Information and Early Warning System
GIMMS	*Global Inventory Modeling and Mapping Studies data*
GIS	Geographic information systems
GLAI	Green leaf area index
GLAM	Global Agricultural Monitoring
GLAS	Geoscience Laser Altimeter System
GPP	Gross primary productivity
GPS	Global positioning systems
GSS	Ground spectral sensing
HNB	Hyperspectral narrow bands
HVI	Hyperspectral vegetation indices
HyspIRI	Hyperspectral Infrared Imager
ICESat	*Ice*, Cloud, and land Elevation Satellite
IRS	Indian Remote Sensing Satellites
JERS	Japanese Earth Resources Satellite
LACIE	Large Area Crop Inventory Experiment
LAI	Leaf area index
LiDAR	Light detection and ranging
LSP	Land surface phenology
LSS	Laboratory spectral sensing
LUC	Land use classes
LUE	Light use efficiency
LULC	Land use, land cover
LULCC	Land use, land cover change
MARS	Monitoring Agricultural Resources action of the European Commission European Union
MODIS	Moderate-resolution Imaging Spectroradiometer
MLS	Mobile laser scanning
NASA	National Aeronautics and Space Administration
NDTI	Normalized difference tillage index
NDVI	Normalized difference vegetation index
NOAA	National Aeronautics and Space Administration
NIR	nearinfrared
NP	Nonphotosynthetic vegetation
NPP	Net primary productivity
PAR	Photosynthetically active radiation
PolSAR	RADARSAT-2 polarimetric SAR ()
PRI	Photochemical reflectance index
PROSAIL	Combination of PROSPECT and SAIL, the two nondestructive physically based models to measure biophysical and biochemical properties
PROSPECT	Radiative transfer model to measure leaf optical properties spectra
PV	Photosynthetic vegetation
REDD	Reducing Emissions from Deforestation in Developing Countries
SAR	Synthetic aperture radar
SAIL	Scattering by arbitrary inclined leaves (SAIL)—a physically based model to measure and model canopy bidirectional reflectance
SAVI	Soil adjusted vegetation index
SPOT	Satellite Pour l'Observation de la Terre, Frech Earth Observing Satellites
STAARCH	Spatial temporal adaptive algorithm for mapping reflectance change
SWIR	Shortwave infrared
TLS	Terrestrial laser scanning
TM	Thematic mapper
UAV	Unmanned aerial vehicle
UNFCCC	United Nations Framework Convention on Climate Change
USAID	The United States Agency for International Development
USDA	United States Department of Agriculture
VHRI	Very-high-resolution imagery
VI	Vegetation index
VIS	Visible

26.1 Monitoring Photosynthesis from Space

Terrestrial biological activity is fundamental to the production of food, fiber, and fuel and is often considered the most important measure of global change (Running et al., 2000, 2004). Biological activity on Earth depends ultimately on solar radiation and its conversion into biochemical energy through photosynthesis. The fundamental paradigm measuring photosynthesis in terrestrial vegetation was first proposed by Monteith (1972) who showed us that stress-free annual crop productivity was linearly related to vegetation-absorbed photosynthetically active radiation (APAR) (e.g., Figure 26.1). Chapter 1 by Dr. Alfredo Huete et al. traces the development of various methods and approaches that have been applied in measuring, modeling, and mapping photosynthesis, accurately and routinely, using remote sensing data. In Chapter 1, they review the integration of remote sensing with traditional in situ methods and the more recent eddy covariance tower approach for estimating gross primary productivity (GPP) and net primary productivity (NPP) at global scales. Light Detection and Ranging (LiDAR) integration with field inventory plots now provides calibrated estimates of aboveground carbon stocks, which can be scaled up using satellite data of vegetation cover, topography, and rainfall to model carbon stocks. A series of six productivity models are presented and discussed, based on the light use efficiency (LUE) concept and primarily dependent on satellite data inputs. These include the following: (1) NPP derived from the integral of growing season normalized difference vegetation index (NDVI), as surrogate of vegetation APAR radiation and, more recently, integral of enhanced vegetation index (EVI); (2) biome-biogeochemical cycles (BGC) model that calculates daily GPP as a function of incoming solar radiation, light use conversion coefficients, and environmental stresses; (3) vegetation index–tower GPP relationships where spectral

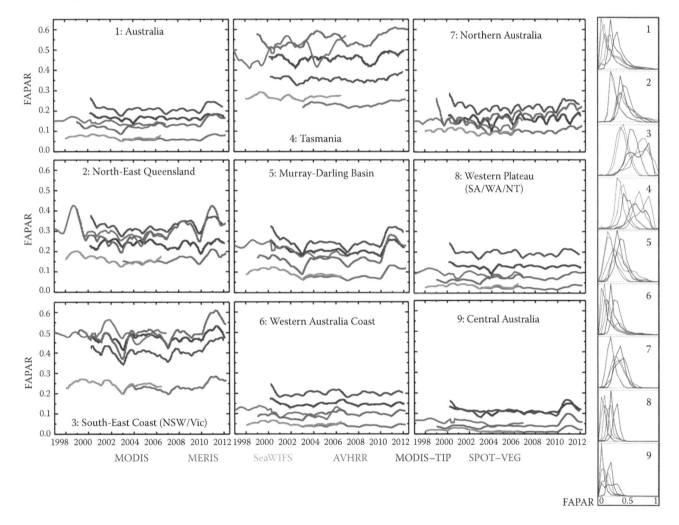

FIGURE 26.1 Time series of the different FAPAR satellite products (persistent vegetation component) for Australia and eight Australian drainage divisions. The plots on the right present the frequency histograms of each product (full FAPAR signal) for each region. (From Pickett-Heaps, C.A. et al., *Remote Sens. Environ.*, 140, 241, 2014, ISSN 0034-4257, http://dx.doi.org/10.1016/j.rse.2013.08.037.)

vegetation indices (VIs) are directly related to eddy covariance tower carbon flux measurements; (4) temperature and greenness (T-G) model that combines land surface temperature and EVI products from Moderate Resolution Imaging Spectrometer (MODIS); (5) greenness and radiation (G-R) model where chlorophyll-related spectral indices are coupled with measures of light energy, photosynthetically active radiation (PAR), to provide robust estimates of GPP; and (6) satellite-based vegetation photosynthesis model (VPM) that estimates GPP using satellite inputs of EVI and the land surface water index (LSWI), along with phenology and temperature scalars. Many of the limitations in productivity assessments concern the difficulty in deriving independent estimates of LUE, and the hyperspectral-based photochemical reflectance index (PRI), a scaled LUE measure based on light absorption processes by carotenoids, is discussed as a way to advance the accuracies of remote sensing retrievals of productivity. Significant and promising advances in direct estimates of GPP, even under stress conditions, have been demonstrated with new spaceborne measures of solar-induced chlorophyll fluorescence (SIF) based on near-infrared light reemitted from illuminated plants, as a by-product of photosynthesis, and thereby strongly correlated with GPP (Guanter et al., 2010, Frankenberg et al., 2011).

Remote sensing offers the ability to model and map productivity for the entire planet due to its synoptic and temporal coverage and thus complements other conventional approaches that are costly, time consuming, and very limited in coverage. Indeed, despite the rapid growth of FLUXNET (a network of micrometeorological tower sites that use eddy covariance methods to measure carbon dioxide, water, and energy balance between terrestrial systems and the atmosphere; Source: fluxnet.ornl.gov) there is not even a single tower present in many parts of the world. FLUXNET has been shown invaluable for validating remotely sensed measurements and process-based productivity models. However, the combination of in situ with satellite data offers much more than calibration opportunities. It will be through a better understanding of why satellite and in situ relationships hold, or don't hold, that will greatly advance and contribute to our comprehension of the carbon cycle mechanisms and scaling factors at play.

26.2 Vegetation Characterization Using Physically Based Models such as SAIL and PROSPECT

Canopy biophysical variables (e.g., leaf area index [LAI], fraction of absorbed photosynthetically active radiation [FAPAR]; see Table 26.1) are retrieved through methods such as (1) direct destructive measurements, (2) indirect nondestructive statistical modeling by relating biophysical quantities to spectral reflectivity in various wavebands or to indices derived from these wavebands, and (3) nondestructive physically based models such as the canopy bidirectional reflectance model called scattering by arbitrary inclined leaves (SAIL), leaf optical properties model based on the radiative transfer model called PROSPECT (Leaf Optical Properties Spectra), and a combination of these two called PROSAIL. Direct measurements of biophysical quantities (e.g., LAI, FAPAR, biomass, plant height, and canopy cover) are most accurate but require destructive sampling and are resource (e.g., time, money) intensive. Statistical approaches of estimating biophysical quantities by relating them to reflectivity of spectral wavebands or various VIs are nondestructive and less resource intensive compared to in situ methods and often explain over 80% variability in data (e.g., Figure 26.2). Physically based methods rely on inverting surface reflectance properties in various wavebands to determine biophysical quantities. Nondestructive methods are not as accurate as destructive method, but often explain over 80% variability in each of the biophysical quantity, are less resource intensive, and avoid destructive sampling. Darvishzadeh et al. (2008) showed a carefully selected spectral subset (Table 26.1) that contains sufficient information for a successful model inversion.

Chapter 2, written by Dr. Frédéric Baret, focuses on retrieving canopy biophysical quantities based on various physically based

TABLE 26.1   Spectral Subset Used for Successful Model Inversion in Physically Based Models like PROSAIL, SAIL, PROSPECT

Wavelength (nm)	Vegetation Parameters	References
466	Chlorophyll b	Curran (1989)
695	Total chlorophyll	Gitelson and Merzlyak (1997), Carter (1994)
725	Total chlorophyll, leaf mass	Horler et al. (1983)
740	Leaf mass, total	Horler et al. (1983)
786	Leaf mass	Guyot and Baret (1988)
845	Leaf mass, total, chlorophyll	Thenkabail et al. (2004)
895	Leaf mass, LAI	Schlerf et al. (2005), Thenkabail et al. (2004)
1114	Leaf mass, LAI	Thenkabail et al. (2004)
1215	Plant moisture, cellulose, starch	Curran (1989), Thenkabail et al. (2004)
1659	Lignin, leaf mass, starch	Thenkabail et al. (2004)
2173	Protein, nitrogen	Curran (1989)
2359	Cellulose, protein, nitrogen	Curran (1989)

Source: Darvishzadeh, R. et al., *Remote Sens. Environ.,* 112(5), 2592, 2008, ISSN 0034-4257, http://dx.doi.org/10.1016/j.rse.2007.12.003.

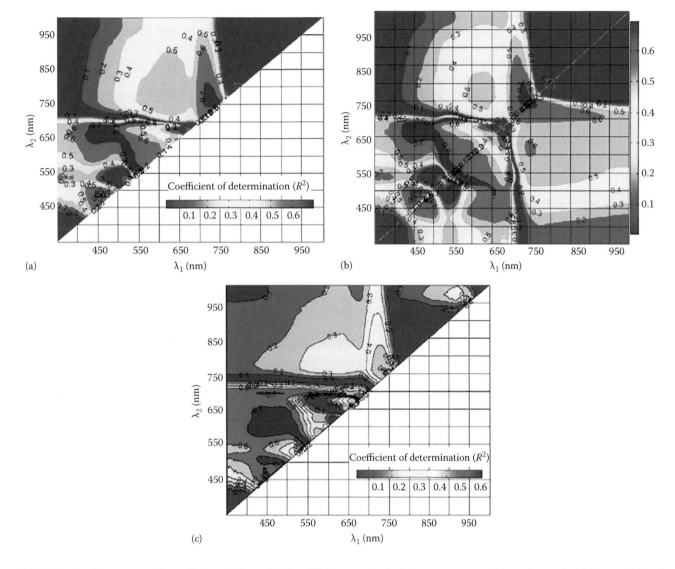

FIGURE 26.2 Contour maps for coefficients of determination (R^2) between rice leaf nitrogen concentrations and normalized (a), ratio (b), and difference (c) indices using two spectral bands (λ_1 and λ_2) ($n = 312$). (From Tian, Y.C. et al., *Field Crops Res.*, 120(2), 299, 2011, ISSN 0378-4290, http://dx.doi.org/10.1016/j.fcr.2010.11.002.)

models, through model inversion. They demonstrate how green leaf area index (GLAI; m²/m²), LAI, and FAPAR are retrieved through physically based radiative transfer model inversion methods. They address this through two approaches: (1) radiometric-driven approach that minimizes the distance between observed and simulated reflectance and emphasizes on outputs (radiometric data–driven approach) of the radiative transfer model and (2) canopy biophysical variables that are based on the VI approaches or are based on machine learning approaches and emphasize on inputs (the canopy biophysical variable–driven approach) of the radiative transfer model. The chapter then goes on to discuss the pros and cons of retrieval approaches. The chapter then provides limitations of these models and enumerate on the strategies of reducing their uncertainties.

Later in Chapter 9, some of the statistical approaches and methods of determining biophysical quantities from remote

sensing are discussed. Readers can also look into recent works of Marshall and Thenkabail (2015) on in situ and statistical methods as well as other novel work on statistical models for biophysical retrieval in Chapter 9 and by Thenkabail et al. (2000, 2004a, 2004b, 2004c, 2012, 2013, 2014).

26.3 Remote Sensing of Global Terrestrial Carbon

Methods for measuring carbon stocks and fluxes include satellite remote sensing, forest inventory, soil inventory, eddy flux, flask measurements, ecosystem modeling, and biome modeling (Bates et al., 2008). Micrometeorological eddy covariance studies and studies using forest inventory plots have yielded conflicting results regarding the sink strength of the mature tropical forests in the Amazon (Robinson et al., 2009). All these methods vary in complexity, precision, accuracy, and costs.

However, satellite remote sensing offers the most distinct advantages in consistency of data, synoptic coverage, global reach, and cost per unit area, repeatability, precision, and accuracy (Meng et al., 2009). Opportunities to significantly advance C storage and flux estimates through improved land use class (LUC) estimates and modeling exist with the evolution in spaceborne hyperspectral, hyperspatial, and advanced multispectral

sensors (e.g., Figure 26.3a), as a result of improvements in the spatial, spectral, radiometric, and temporal properties as well as in optics and signal-to-noise ratio of data. High spatial resolution allows location, while high spectral resolution allows identification of features. Hyperspectral remote sensing sensors allow direct measurement of canopy chemical content (e.g., chlorophyll, nitrogen), forest species, chemistry distribution, timber volumes, and water (Asner and Martin, 2008) and improved biophysical and yield characteristics (Thenkabail

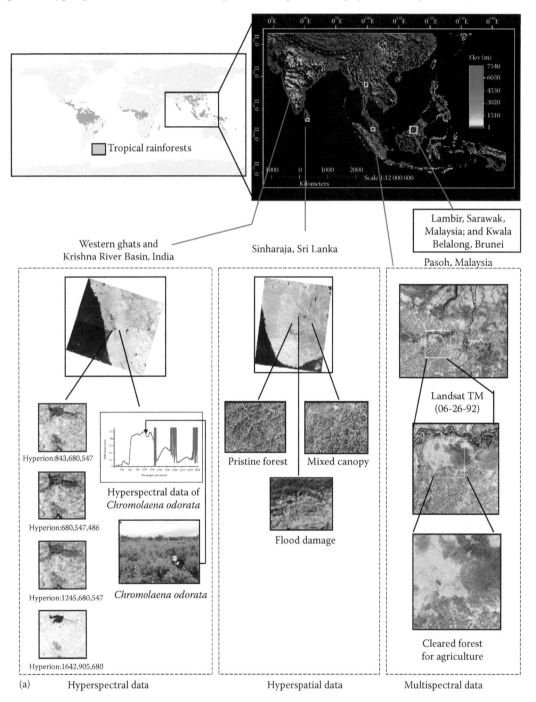

FIGURE 26.3 (a) Tropical forests. Snapshots of hyperspectral, hyperspatial, and advanced multispectral data are illustrated for certain random benchmark study areas in Wet Tropical Asian Bioregion (WTAB) in Sri Lanka, India, Malaysia, and Brunei. *(Continued)*

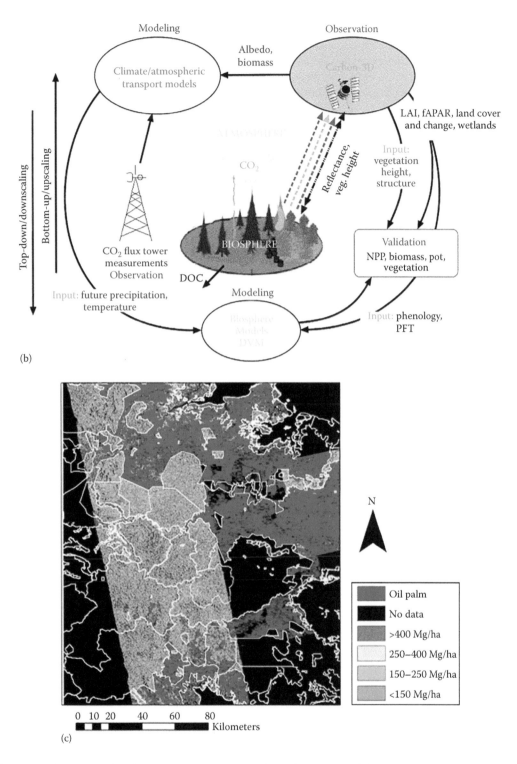

(b)

(c)

FIGURE 26.3 (*Continued*)   (b) Role of Carbon 3D in the observation and modeling strategy of the carbon cycle. (From Hese, S. et al., *Remote Sens. Environ.*, 94(1), 94, 2005, ISSN 0034-4257, http://dx.doi.org/10.1016/j.rse.2004.09.006.) (c) Biomass map of Sabah Malaysian Borneo. Biomass values were generated only from nonprecipitation-affected imagery. Oil palm areas were derived from all images and are pictured in red. Biomass values have been grouped for all high biomass areas (>250 Mg/ha), moderate biomass areas (two groups from 50 to 250 Mg/ha), and severely degraded areas (<50 Mg/ha). (From Morel, A.C. et al., *Forest Ecol. Manage.*, 262(9), 1786, 2011, ISSN 0378-1127, http://dx.doi.org/10.1016/j.foreco.2011.07.008.)

et al., 2004a,b). Thenkabail et al. (2004c) demonstrated an increased accuracy of about 30% in LUC and biomass when 30 hyperspectral wavebands are used relative to 6 nonthermal Landsat Thematic Mapper (TM) bands. Hyperspatial data have demonstrated the ability to extract individual tree crowns from 1 m panchromatic data. Agroforest successional stages have been mapped and their varying carbon sink strengths assessed using IKONOS (Thenkabail et al., 2004a). In contrast, forest structure variables (e.g., biomass, LAI) are poorly predicted by the older-generation sensors.

A model can be developed for characterizing forest structure, biomass yield, and carbon (C) storage based on hyperspectral, hyperspatial, and advanced multispectral data (e.g., Figure 26.24a) and the field-plot data by discerning a large number, say K, of LUCs. Letting τ_k, $k = 1...K$, represent the total C in the kth LUC, we propose that the model

$$\tau_k = \lambda_k A_k + \varepsilon_k \qquad (26.1)$$

is a practical model for regional estimates of C storage by LUC, which becomes possible as we can discern a sufficiently narrow number of LUC using advanced remote sensing. In this model, λ_k is average C per hectare in LUC_k, A_k is the total land area in LUC_k, and ε_k is the departure of τ_k from its expected value. We propose to use-field-plot data, supplemented with additional above- and belowground sampling, to evaluate λ_k for a suite of LUCs of importance in the wet tropical Asian bioregion (WTAB) region. Remotely sensed data will be used to evaluate A_k at various scales or pixel resolutions, radiometry, bandwidth, and time of acquisition.

The aforementioned model attempts to capitalize on having a sufficiently large number of LUC so that each is relatively homogeneous with respect to C and biomass variability. For example, Thenkabail et al. (2004a) have demonstrated the potential for such fine-resolution classification by LUC. The temporal change in τ_k is given by

$$\frac{d\tau_k}{dt} = \frac{d(\lambda_k A_k + \varepsilon_k)}{dt} = \frac{d\lambda_k}{dt} A_k + \lambda_k \frac{dA_k}{dt} + \frac{d\varepsilon_k}{dt} \qquad (26.2)$$

If the instantaneous rate of change in the aforementioned model is integrated into annual changes, the resulting discretized model of change in C storage and biomass is

$$\Delta\tau_k = \Delta\lambda_k + \lambda_k\Delta A_k + \Delta\varepsilon_k \qquad (26.3)$$

For many LUC, for example, mature upland forests, $\Delta\lambda_k$, are anticipated to be small, so that the model for $\Delta\tau_k$ will be dominated by the $\lambda_k\Delta A_k$ term. Moreover, ΔA_k is discernible from satellite data, whereas regional estimates of λ_k will be determined by field work. For LUCs that cannot be so finely discerned as to support an assertion that $\Delta\lambda \approx 0$, then a spatially averaged value, say $\overline{\Delta\lambda_k}$, will be used in (26.3).

Chapter 3 by Dr. Wenge Ni-Meister has exclusive focus on aboveground biomass (AGB) assessment and carbon stock estimations using multisensor remote sensing. Recent global

observation systems provide measurements of horizontal and vertical vegetation structure of ecosystems, which will be critical in estimating global carbon storage and assessing ecosystem response to climate change and natural and anthropogenic disturbances. Remote sensing overcomes the limitations associated with sparse field surveys; it has been used extensively as bases for inferring forest structure and AGB over large areas. They provide a systematic approach wherein they discuss AGB estimates based on

1. Optical remote sensing
2. Radar remote sensing
3. LiDAR remote sensing

Optical passive remote sensing data are sensitive to vegetation structure (LAI, crown size, and tree density), texture, and shadow, which are correlated with AGB. Radar data are directly related to AGB through measuring dielectric and geometrical properties of forests (Le Toan et al., 2011). LiDAR remote sensing is promising in characterizing vegetation vertical structure and height, which are then associated to ABG. Vegetation structure characteristics measured from satellite data are linked to field-based AGB estimates, and their relationships are used to map large-scale AGB from satellite data.

Many studies have demonstrated that LiDAR is significantly better at estimating biomass than passive optical or radar sensors used alone. LiDAR directly measures horizontal and vertical vegetation structure characteristics; AGB models developed from LiDAR structure metrics are significantly more accurate than those using radar or passive optical data. The use of LiDAR data, particularly spaceborne data, is limited by its sparse spatial sampling. Both radar and passive optical remote sensing provide large-scale imaging capability. However, both optical and synthetic aperture radar (SAR) estimates of AGB are limited by a loss of sensitivity with increasing biomass. Innovative approaches have been developed to fuse passive optical imagery or radar-imaging data with spatially extended point-based estimates of biophysical parameters derived from LiDAR to develop high-quality wall-to-wall AGB maps with unprecedented accuracy and spatial resolution. Particularly, the synergy of spaceborne large-footprint LiDAR (e.g., Ice, Cloud and land Elevation Satellite (ICESat) Geoscience Laser Altimeter System (GLAS)) and medium-resolution optical data, primarily from the MODIS, has been exploited to map AGB and carbon storage at regional to continental scales. A laser-based instrument being developed for the International Space Station will provide a unique 3D view of Earth's forests, helping to fill in missing information about their role in the carbon cycle (e.g., Figure 26.3b).

Such advances as remote sensing of carbon stock assessment in various landscapes such as forests, agroforests (e.g., Figure 26.3b,c), and other land use and land cover (LULC) categories will help in setting up an operational global carbon monitoring framework such as the one under the United Nations Framework Convention on Climate Change (UNFCCC) mechanism for Reducing Emissions from Deforestation in Developing Countries (REDD). An important advance presented and discussed in Chapter 3 is the use of multisensor data fusion approaches to increase accuracies of biomass quantification and carbon stock estimations. Readers should also look into Chapters 14 through 17 for other aspects of biomass and carbon estimations using remote sensing.

26.4 Remote Sensing of Agriculture

One of the main applications of remote sensing has always been agriculture. When the first Landsat was launched in 1972, one of the main uses of its data was for agriculture through such programs as the Large Area Crop Inventory Experiment (LACIE) and the agriculture and resources inventory surveys through aerospace remote sensing (AgRISTARS). Pioneering work of Compton Tucker (1979) showed the use of red and near-infrared bands for computing now widely used NDVI from any sensor that carries these wavebands. Remote sensing of agriculture is now common from any platform: ground based, airborne, spaceborne, or unmanned aerial vehicles (UAVs) using sensors gathering data in wavelengths including optical, thermal, radar, and LiDAR.

In Chapter 4 by Dr. Clement Atzberger et al., the importance of remote sensing of agriculture, type of its applications, and its evolution over the last 50 years is well documented. They identify remote sensing application of agriculture into the following main areas:

1. Qualitative crop monitoring involving changes that take place from within and across seasons. Changes such as deviation from normal conditions and changes from croplands to cropland fallows (also see Chapter 6)
2. Cropland classification and mapping including crop type identification leading to acreage estimation and phenological studies, identifying shifts in cultivation (also see Chapter 6)
3. Regression modeling involving spectral indices, FAPAR, and/or wavebands to predict crop growth and yield variables such as grain yield and biomass
4. Physical modeling of crop growth through remote sensing data assimilation in dynamic (process-driven) crop growth models
5. Data mining approaches in cropland studies

Chapter 4 makes an assessment of various remote sensing data used for different types of agricultural applications such as coarse-resolution high revisit data for crop yield and biomass estimation, hyperspectral data in quantifying biophysical and biochemical quantities (also see Chapter 9), and multispectral broadband data in cropland classification.

Over the years, a cropland study from remote sensing still provides a number of challenges. The issues involved include imagery resolution, time of acquisition, and number of images available during the season and their frequency, preprocessing and atmospheric correction, and classification methods and approaches (e.g., Figure 26.4). In Figure 26.4, multispectral QuickBird 2.44 m imagery is used for classification and mapping of crops and other land use using a number of different classification algorithms. When object-oriented segmentation approaches are used with different classification algorithms, they provide much better results. Similarly, crop biophysical and biochemical modeling

faces a number of challenges. Nevertheless, modern remote sensing is increasingly overcoming these challenges. For example, availability of imagery from multiple sensors (e.g., Landsat, Sentinel, Indian Remote Sensing Satellites or IRS, and Satellite Pour l'Observation de la Terre or SPOT) provides more frequent coverage of the same area that will overcome cloud cover issues and advance our ability to more precisely monitor phenology. Temporal images will also increase the classification accuracies.

Chapter 4 also provides an overview of existing global and regional remote sensing data–based agricultural monitoring systems such as the global food security support analysis data @ 30 m (GFSAD30) of United States Geological Survey (USGS) [http://geography.wr.usgs.gov/science/croplands/ and https://www.croplands.org/), Group on Earth Observation's (GEO) Global Agricultural Monitoring (GLAM), The United States Agency for International Development (USAID), Famine Early Warning Systems Network (FEWS-NET), United Nations' Food and Agriculture Organization (FAO) Global Information and Early Warning System (GIEWS), European Join Research Center's (JRC's), Monitoring Agricultural Resources action of the European Commission (MARS), European Union Global Monitoring for Food Security program (GMFS), and Crop Watch Program at the Institute of Remote Sensing Applications of the Chinese Academy of Sciences (CAS) (CropWatch). Evolution of these various crop-monitoring systems, from various agencies of the world, is the testament of the progress and maturity of remote sensing data in agricultural studies.

26.5 Agricultural Systems Studies through Remote Sensing

Humans started domesticating the agriculture some 10–12 thousand years ago. Early civilizations were primarily agrarian. Agricultural systems of the world have evolved over millennia and with these evolutions have also seen major changes in how agriculture is farmed and managed by humans. The present-day agriculture systems include both croplands and rangelands and occupy nearly one-third of the global terrestrial area. So an understanding and study of agricultural systems would involve every component of croplands and rangelands and their associations (e.g., Figure 26.5). Since croplands and rangelands are spread across the world and are dynamic by nature, remote sensing provides an ideal platform to characterize, model, and manage agricultural systems of the world routinely.

Chapter 5 by Dr. Agnès Bégué et al. provides approaches on applying remote sensing to study agricultural systems. They show us how remote sensing data are used to derive land cover, land use, and land systems, which are then tied to croplands, cropping systems, and cropland systems, respectively. Crop type is the first criterion used to characterize agricultural systems at the regional scale, followed by the cropping pattern, the water supply, and the cropland extent and fragmentation. Image resolutions (spatial, spectral, radiometric, and temporal) are all important in discerning particular features of agricultural systems. Especially, when the sensor spatial resolution is smaller than the objects of interest (fields, trees,

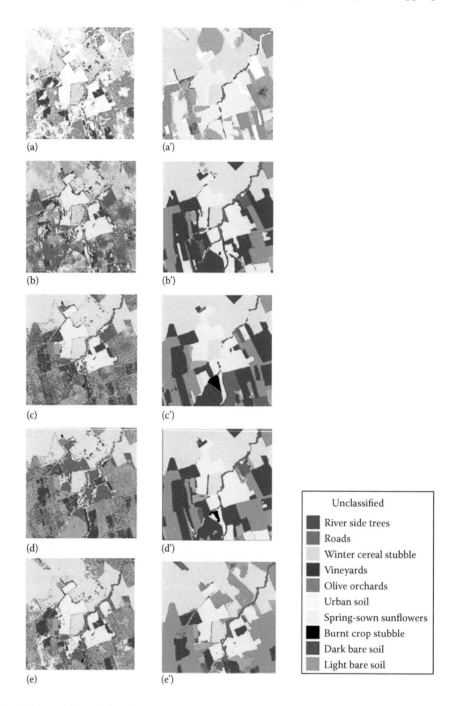

FIGURE 26.4   Result of the least (a, b, c, d, e) and most accurate (a', b', c', d', e') land use classifications based on QuickBird 2.44 m imagery with and without segmentation based on number of well-known algorithms: (a) P*, (b) MD*, (c) MC*, (d) SAM*, (e) ML* for pan-sharpened image and pixel as minimum information unit (MIU); (a') P, (b') MD, (c') MC, (d') SAM for multispectral image and pixel + object as MIU; (e') ML for pan-sharpened image and pixel + object as MIU. Note: P*, parallelepiped; MD*, minimum distance; MC*, Mahalanobis classifier distance; SAM*, spectral angle mapper; ML*, maximum likelihood. (From Castillejo-González, I.L. et al., *Comput. Electron. Agric.*, 68(2), 207, 2009, ISSN 0168-1699, http://dx.doi.org/10.1016/j.compag.2009.06.004.)

etc.), the agricultural landscape can be described as a mosaic of patches, and thus agricultural systems can be characterized and mapped directly using object-based analysis and landscape metrics (landscape agronomy). When the sensor spatial resolution is larger than the object of interest, the agricultural systems should be characterized indirectly by computing a large variety of satellite-derived indices, and environmental and socioeconomic variables are further processed with data mining techniques in order to stratify the agricultural lands. Dr. Agnès Bégué et al. in Chapter 5 provide concrete illustrations of mapping agroforestry in Bali using submeter to 2 m QuickBird imagery and double cropping in Brazil using high

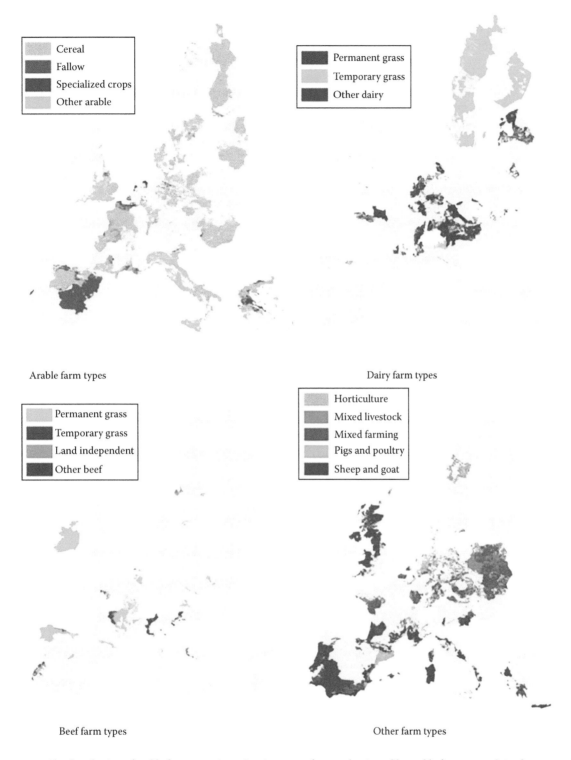

Arable farm types

Dairy farm types

Beef farm types

Other farm types

FIGURE 26.5   The distribution of arable farm types in agrienvironmental zones dominated by arable farm types, dairy farm types in agrienvironmental zones dominated by arable farm types, beef farm types in agrienvironmental zones dominated by beef farm types, and other farm types in agrienvironmental zones dominated by other farm types. (From Kempen, M. et al., *Agric. Ecosyst. Environ.*, 142(1–2), 51, 2011, ISSN 0167-8809, http://dx.doi.org/10.1016/j.agee.2010.08.001.)

temporal MODIS data (direct approach) and rainfed cropping in Mali also using MODIS time series data (indirect approach). Overall, they show us the potential of remote sensing to study agricultural systems at various levels such as region, landscape, field, and plant using appropriate imagery.

26.6 Global Food Security-support Analysis data @ Various Resolutions from Earth Observation Satellites

Globally, there is between 1.5 and 1.7 billion hectares of croplands of which around 400 million hectares are irrigated fully or partially when you consider cropping intensity (i.e., how many crops are grown over the same area in a calendar year; so one may account area twice when there are two crops a year over the

same land). An overwhelming proportion of rainfed croplands is cropped only once a year, during the rainy season. An overwhelming proportion of irrigated areas is cropped more than once a year. Cropland area estimates include-plantation crops. Understanding the importance of croplands is essential for managing its crop productivity (productivity per unit of land; kg/m^2) and water productivity (productivity per unit of water; m^3/m^2). All of this has huge implications on managing food security.

Chapter 6 by Dr. Pardhasaradhi Teluguntla et al. provides a comprehensive overview of the state of the art of mapping global agricultural croplands using multi-sensor, multi-resolutions (in terms of spatial, spectral, radiometric), and multi-temporal remotely sensed data from Earth Observing Satellites. A common application of remote sensing over the last five decades has been agricultural cropland mapping and separating them from other land cover (e.g., Figure 26.6a). However, most of these applications were limited to smaller areas. A few large area applications exist,

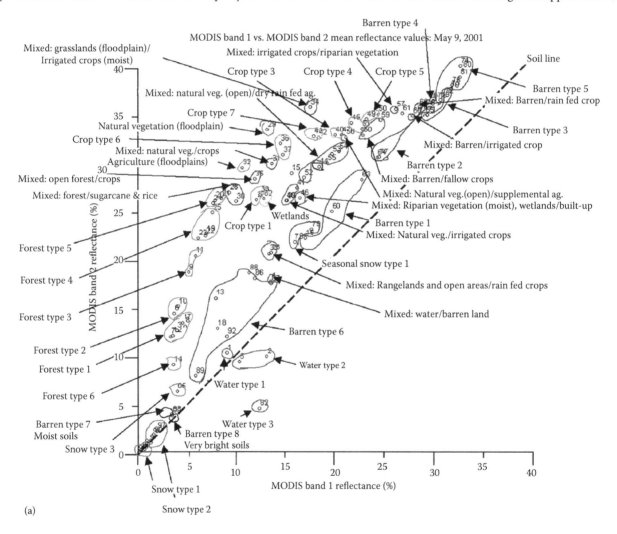

FIGURE 26.6 (a) RED-NIR single date (RN-SD) plot of 100 unsupervised classes. The 100 unsupervised classes are plotted taking mean class reflectance in MODIS band 1 (red) and band 2 (NIR). The classes are shown in brightness–greenness–wetness (BGW) feature space and their preliminary class names identified for further investigations during ground truthing. Similar to the figure shown earlier, RN-SDs were plotted for each of the 42 dates. (From Thenkabail, P.S. et al., *Remote Sens. Environ.*, 95(3), 317, 2005, ISSN 0034-4257, http://dx.doi.org/10.1016/j.rse.2004.12.018.)

(Continued)

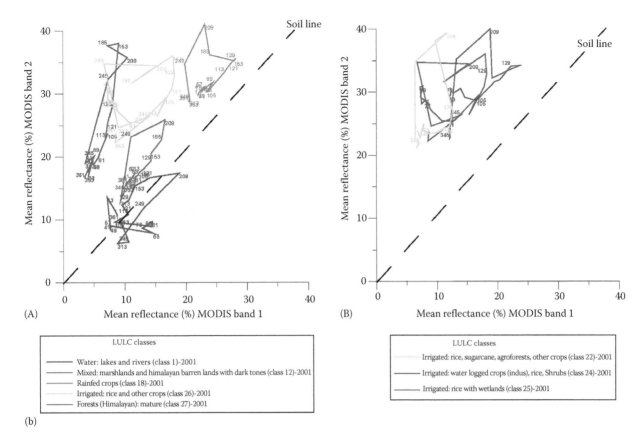

(b)

FIGURE 26.6 (*Continued*) (b) Space-time spiral curves (ST-SCs) to study subtle and not-so-subtle changes in LULC spectral separability. The ST-SCs are a unique and powerful representation of observing subtle and not-so-subtle changes over time mapped in 2D feature space. MODIS band reflectance in band 1 (red) and band 2 (NIR) is used to plot ST-SCs for (A) 5 spectrally distinct LULC classes and (B) 6 spectrally similar irrigated area classes. As the spectral properties of classes change over time, we can observe dates on which 2 or more spectral classes intersect (no spectral separability) or stay spectrally separate highlighting the near-continuous interval multitemporal data in LULC studies. (From Thenkabail, P.S. et al., *Remote Sens. Environ.*, 95(3), 317, 2005, ISSN 0034-4257, http://dx.doi.org/10.1016/j.rse.2004.12.018.)

such as the United State Department of Agriculture's (USDA's) cropland data layer (CDL). Global cropland products are limited to a few studies, but they provide large uncertainties due to factors such as coarse resolution of the imagery used and lack of field data and understanding needed to develop and test algorithms, so uncertainties in existing cropland products are substantial. Generally, it is agreed that crop products derived from remote sensing should include (also see: http://geography.wr.usgs.gov/science/croplands/ and https://www.croplands.org/),

- Cropland extent and areas
- Watering method: irrigated or rainfed (e.g., separating irrigated from rainfed; Figure 26.6b)
- Cropping intensity: single, double, triple, or continuous cropping
- Crop type: major global crops
- Change: how crops change over a given location over space and time

Accurate production of such products will also lead to far greater accuracies in cropland water use. Since 70%–80% of all human water use goes toward agriculture to produce food, it is of utmost importance. Accurate production of the aforementioned

products will lead to better assessment, planning, and management of crop productivity (productivity per unit of land; kg/m²) and water productivity (productivity per unit of water; m³/m²), which are very essential for increasing food production from population that is currently at little over 7.2 billion, but is expected to reach 9–10 billion by the year 2050. Chapter 6 shows us the remote sensing data, approaches, methods, and synthesis involved in use of remote sensing earth observation (EO) data for global cropland studies at various resolutions from 30 m (e.g., Landsat) and 250 m (e.g., MODIS) to 1000 m (e.g., Advanced Very High Resolution Radiometer [AVHRR]). Strengths and limitations of various remote sensing EO data, methods and algorithms required to routinely and repeatedly produce accurate cropland products, and approaches of using these data in food security analysis have been presented and highlighted.

26.7 Precision Farming and Remote Sensing

The concept and the idea of precision farming were established in the early quarter of the last century. Precision farming is variously referred to as site-specific farming, variable rate technology, or prescription farming. Concisely, precision farming can be understood as a customized subfield agricultural management decision system relying on information from advanced technologies such as remote sensing, geographic information systems (GIS), and global positioning systems (GPS). Usery et al. (1995) define precision farming as the application of a combination of advanced technologies to (1) improve agricultural crop productivity and (2) reduce environmental pollution through quantitative and qualitative information of within field variability (or site-specific variability) due to natural and human causes. Adoption of precision farming accelerated as sophisticated technologies such as remote sensing, GPS, GIS, and yield sensors improved and decreased in cost. An integrated approach to use these technologies has made precision farming realistic. The target of precision farming is to identify, map, quantify, and assess farm by farm (but spread across entire regions; e.g., Figure 26.7a) spatial and temporal variability for maximizing profits, sustainable production, and protecting the environment. Information requirements of precision farming (Moran et al., 1997) include

- Seasonally stable factors
- Soil properties, topography, prior management history, etc. (e.g., Figure 26.7a)

- Seasonally variable factors
- Crop biophysical parameters, crop growth, crop yield (e.g., Figure 26.7b), phenology, crop disease, weed infestation, nutrient deficiencies, soil characteristics, and evapotranspiration rates

Based on the earlier text, precision farming seeks to diagnose the cause of crop growth and/or yield variability and develop management strategies

Remote sensing technology is ideally suited to answer questions: How much are we producing (e.g., grain or biomass, or LAI) per unit of land (kg/m²; Figure 26.7b)? What is the variability in these production quantities within and between agricultural fields (e.g., Figure 26.7)? Providing such information at high spatial resolutions (typically <10 m but preferably submeter to a few meters) becomes invaluable for increasing crop productivity by targeting management to areas that produce less or are more responsive. Remote sensing is a generally well established, powerful, and accurate technology for quantifying crop biophysical quantities such as biomass, LAI, plant density, crop vigor, grain yield, and even plant height and canopy cover. As a result, remote sensing has been widely applied for better management of crops over the last 25 years. Especially

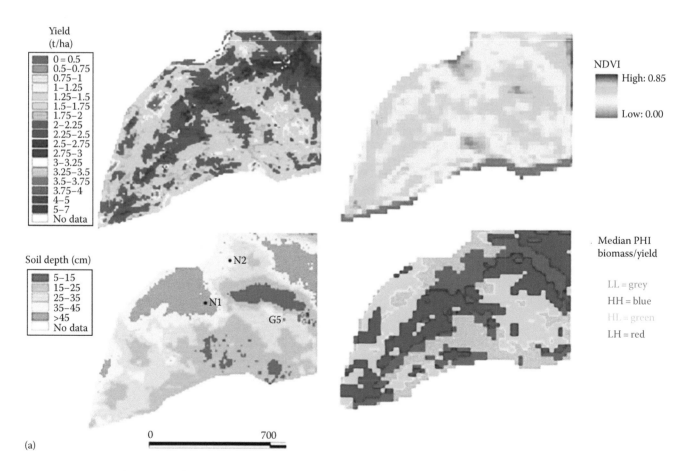

(a)

FIGURE 26.7   (a) Yield, NDVI, soil depth, and NDVI (biomass)/yield classification on a single field at Buntine, Western Australia. For biomass/yield classification LL, low biomass and low yield; HH, high biomass and high yield; HL, high biomass and low yield; LH, low biomass and high yield. (From Robertson, M. et al., *Field Crops Res.*, 104(1–3), 60, 2007, ISSN 0378-4290, http://dx.doi.org/10.1016/j.fcr.2006.12.013.)   (*Continued*)

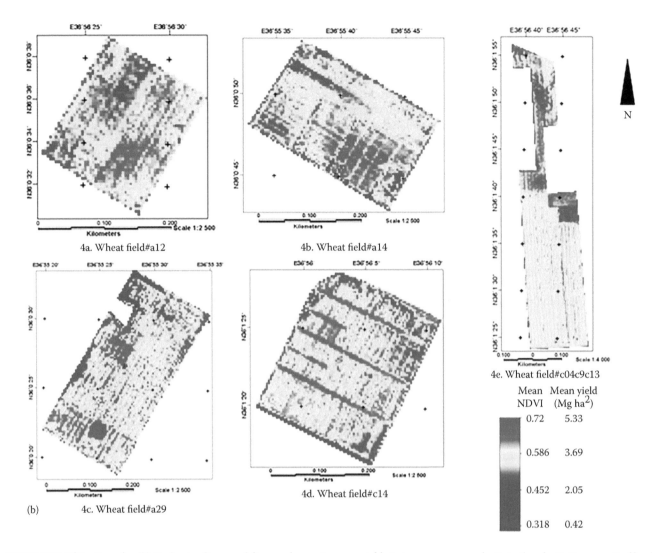

4a. Wheat field#a12

4b. Wheat field#a14

4c. Wheat field#a29

4d. Wheat field#c14

4e. Wheat field#c04c9c13

Mean NDVI	Mean yield (Mg ha^2)
0.72	5.33
0.586	3.69
0.452	2.05
0.318	0.42

(b)

FIGURE 26.7 (*Continued*)   (b) Understanding, modeling, and mapping crop yields to improve crop productivity (productivity per unit of land; kg/m^2) and water productivity (productivity per unit of water; m^3/m^2). Spatial distribution of wheat yield based on Landsat TM data. These are the typical maps used in precision farming to understand and improve crop and water productivity of crops.

when remote sensing is combined with other spatial technologies such as GIS, GPS, and spatial modeling, they become a powerful tool to understand and manage agricultural crops leading to improved productivity. However, precision farming requires an intimate understanding of crop stresses and their causes (e.g., nutrient, pest, disease, water deficiency, weeds, and even macro- and micronutrients). Whereas remote sensing can detect stress accurately, its ability to detect the nature of stress (e.g., whether from pests or nutrients) is still uncertain. Hyperspectral data acquired at very high spatial resolution and often combined with other types of data (e.g., thermal) have been helpful in making advances. For example, hyperspectral narrowbands (HNBs) and hyperspectral vegetation indices (HVIs) from specific wavelengths have enabled advances.

Dr. David Mulla, in Chapter 7, identifies specific HVIs for detecting N, P, or K, disease, insects, and weeds. For example, Dr. Mulla shows us that normalized difference vegetation index (NDVI), soil adjusted vegetation index (SAVI), and red edge are

good for detecting N, P, or K; fluorescence and PRI are good for detecting disease; aphid index, damage index, and leaf hopper index are good for insects; and ratio VI is good for weeds. Of course, there are indices like NDVI or SAVI that can detect multiple types of crop stress. Yet the uncertainty in understanding and modeling specific stress causes using remote sensing is very high. Some of these uncertainties can be reduced substantially if we have sufficient spatial resolution (e.g., centimeters) along with hyperspectral and other supplementary data (e.g., thermal). Chapter 7 by Dr. Mulla provides an overview of remote sensing technology use over the last 25 years and establishes the current state of the art. There is great scope for further research and development of precision farming applications using remote sensing to reduce uncertainties and increase accuracies of prediction. UAVs offer a new platform with much higher spatial resolution (e.g., few centimeters) as well as the ability to fly multiple sensors (e.g., hyperspectral, thermal) and offer new opportunities for precision farming applications of remote sensing. But these need to go hand in hand with methodological improvements. UAVs will, however, come with their own limitations such as the ability to get permission to fly, inability to carry heavier sensors, errors due to stability issues, and inability to cover large areas. Nevertheless, a combination of multisensor approach, with greater efforts in ideal sensor design (e.g., hyperspectral sensor that is of a few centimeter resolution), and improved methodological efforts will lead to wider application of remote sensing technology in precision farming leading to improved productivity in crop grain and biomass.

Remote sensing can be used to collect quantitative crop information on subareas within a farm, including

- Crop biophysical quantities—LAI, biomass
- Within field variations over time (temporal changes)
- Spatial variations in growth stage and phenology
- Weed and pest infestations
- Crop stress and disease
- Quantitative and qualitative changes within and between seasons

The aforementioned information can be linked to decision support systems such as pest or drought early warning systems, leading to county- or regional-scale agricultural decision support systems.

26.8 Mapping Tillage versus Nontilled Lands and Establishing Crop Residue Status of Agricultural Croplands Using Remote Sensing

The importance of mapping tilled versus no-till agricultural lands is discussed in Chapter 8 by Dr. Baojuan Zheng et al. Conservation tillage practices improve soil and cropland management. Conventional tillage practices often have detrimental impacts, including soil erosion, loss of organic matter, and leaching of nutrients. Conservation tillage practices leave crop residues (e.g., Figure 26.8) on the soil surface to retain moisture and resist soil and wind erosion. So the ability to reliably monitor tilled versus no-till lands on a field-by-field basis becomes an important component of soil and landscape management. Dr. Baojuan Zheng et al. survey state-of-the-art strategies to map tillage practices by estimating crop residue cover. Further, they survey applications of remote sensing tillage assessment using

1. Optical remote sensing
2. SAR remote sensing

For optical remote sensing, shortwave infrared (SWIR) bands in range of 2000–2300 nm are critical to distinguish live crops

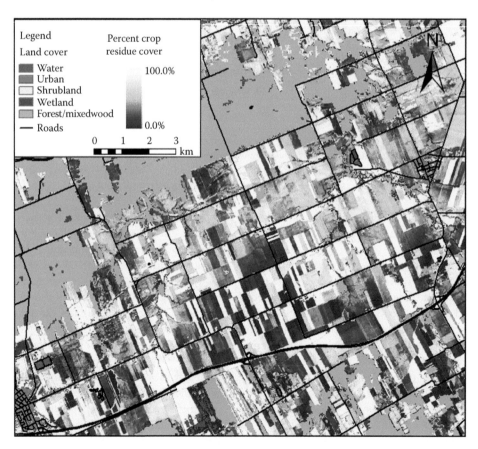

FIGURE 26.8 Percent crop residue cover map over the Casselman/St. Isidore study site derived from spectral mixture analysis on a SPOT image acquired on November 9, 2007. A land cover mask (nonagricultural areas) is overlaid on the residue cover map. (From Pacheco, A. and McNairn, H., *Remote Sens. Environ.*, 114(10), 2219, 2010, ISSN 0034-4257, http://dx.doi.org/10.1016/j.rse.2010.04.024.)

versus crop residue versus soil. Cellulose absorption bands centered at 2100 nm are especially valuable. Landsat bands 5 and 7 (for calculation of the Normalized Difference Tillage Index [NDTI]) are widely used in detecting crop residue, thereby distinguishing tilled and no-till conditions. Classification methods such as maximum likelihood, spectral angle mapping, and data mining (e.g., the random forest classifier and support vector machines [SVMs]) are widely used. However, narrowbands centered at 2030, 2100, and 2210 nm with bandwidths of ~10 nm or less are likely to provide the best results. When hyperspectral data are available (e.g., Hyperion, Environmental Mapping and Analysis Program [EnMAP], and Hyperspectral Infrared Imager [HyspIRI]), they can be used to compute cellulose absorption indices (CAIs). Accurate tillage assessment depends upon observing crop residue in fields within a short interval at the start of the growing season, just as farmers begin to plant crops. So timeliness of imagery is key to successful mapping of tillage status using optical images, which are subject to effects of cloud cover. Further, Zheng et al. explore SAR images acquired from multiple platforms at X, C, and L bands, showing that SAR backscatter coefficient thresholds are key to distinguishing tilled from no-till conditions. SAR images perform best when more than one frequency and polarization are used. Cloud penetration of SAR images offers important advantages. Research to investigate the role of polarimetric images will likely form an important topic supporting SAR applications for tillage assessment.

26.9 Hyperspectral Remote Sensing for Terrestrial Applications

Traditional remote sensing from sensors such as Landsat, SPOT, and IRS gathered data in broad spectral wavebands across the electromagnetic spectrum, typically in spatial resolution of greater than 20 m. The new-generation hyperspatial sensors (e.g., IKONOS, QuickBird) acquired data in very high spatial resolution (e.g., submeter to 5 m) but also in broad spectral wavebands across the electromagnetic spectrum. However, there is great scientific interest and need to gather data near continuously across the electromagnetic spectrum. This need is fulfilled by hyperspectral or imaging spectroscopy data (Goetz, 2010).

Chapter 9 by Dr. Prasad S. Thenkabail defined hyperspectral data as follows:

> "Remote sensing data is called hyperspectral when the data is collected contiguously over a spectral range, preferably in narrow bandwidths and in reasonably high number of bands."

So typical hyperspectral (imaging spectroscopy) data are gathered in very narrow bands (~1–10 nm bandwidths), contiguously across electromagnetic spectrum (e.g., 400–2500 nm), resulting in several 10s, or 100s, or 1000s of narrowbands of data.

Chapter 9 begins by enumerating key characteristics of ground-based, airborne, and spaceborne hyperspectral sensors. The spatial, spectral, radiometric, and temporal characteristics of the key spaceborne hyperspectral sensors are presented and

discussed. One of the first steps in hyperspectral data analysis is data mining. Data mining is extremely important from the point of view of reducing the data volumes to eliminate redundant bands, especially in the age of "big data." The value, importance, and approach of overcoming Hughes' phenomenon are discussed. A greater number of HNBs are very important to increase classification accuracies as well as to develop unique and more powerful HVIs. Chapter 9 shows that it is feasible to achieve increased accuracies of 30% or higher when ~20 HNBs are involved relative to 7 Landsat broadbands. The chapter also shows that there are several unique two-band HVIs (e.g., Table 26.2) targeted to study specific biophysical and biochemical quantities. Further multiband HVIs involving 3–10 HNBs often provide much higher R-square values relative to commonly known two-band indices like NDVI.

The key highlights of Chapter 9 include a summary of

1. 28 optimal HNBs for studying agricultural crops and vegetation
2. 6 categories of important HVIs for modeling crop and vegetation biophysical and biochemical quantities
3. 4, 6, 8, 10, 12, 16, and 20 HNB combinations to best classify croplands and vegetation

Further, Chapter 9 presents and illustrates various hyperspectral data analysis methods that are broadly grouped under two categories: feature extraction methods and information extraction methods. The various hyperspectral classification methods discussed and illustrated include spectral matching techniques, spectral mixture analysis (SMA), SVMs, and tree-based ensemble classifiers (e.g., Random Forest and Adaboost). Readers are also referred to detailed studies on hyperspectral remote sensing of vegetation and agricultural crops in Thenkabail et al. (2012).

26.10 Rangelands: A Global View

Rangelands represent relatively arid sites where potential vegetation is predominantly comprised of grasses, forbs, and shrubs (e.g., Figure 26.9). Examples include savannas, shrub- and grasslands, tundras, open woodlands, and chaparral. Rangelands are globally important as sources of forage for both domesticated and wild animals. Additionally, rangelands support unique flora and fauna and provide numerous ecosystem services. Roughly 24% of the terrestrial area or about 3 billion hectares (double the area of croplands) can be considered rangeland. Omitting deserts from the global terrestrial area, however, increases rangeland proportion to as much as 52% as mentioned in Chapter 10. So definition is key to how rangelands are assessed and accounted for.

Given the vastness and inaccessibility of rangelands, remote sensing offers the best opportunity to study rangelands. The key factors of rangeland studies using remote sensing are presented in detail in Chapter 10 by Dr. Matt Reeves et al. These factors are

1. Rangeland degradation studies that involve quantifying and modeling degradation of land and soil and ensuing desertification
2. Biomass, NPP, and forage quantification and modeling

TABLE 26.2   Some of the Best Two-Narrowband HVIs as per Early Research by Thenkabail et al. (2000)

Crop (Sample Size)	Crop Variable	Band Center and Width (nm)	Band Centers (λ_1 and λ_2) and Band Widths ($\Delta\lambda_1$ and $\Delta\lambda_2$) for Two-Band VI						
			Index 1 (nm)	Index 2 (nm)	Index 3 (nm)	Index 4 (nm)	Index 5 (nm)	Index 6 (nm)	Index 7 (nm)
1. Cotton (73) except for yield that has a sample size of 50	WBM (kg/m²) (see Figure 26.4)	λ_1	682	682	568	555	615	525	982
		$\Delta\lambda_1$	28	28	10	20	175	4	10
		λ_2	918	845	918	666	925	540	940
		$\Delta\lambda_2$	20	250	10	5	20	7	10
	LAI (m²/m²)	λ_1	682	550	678	550	568	525	940
		$\Delta\lambda_1$	15	30	15	50	4	60	10
		λ_2	940	682	865	675	915	540	980
		$\Delta\lambda_2$	60	20	275	50	15	60	10
	Yield (lmt/ha)	λ_1	540	696	678	540	690	678	670
		$\Delta\lambda_1$	30	4	30	40	10	30	50
		λ_2	678	940	940	684	720	860	970
		$\Delta\lambda_2$	20	20	50	20	20	290	20
2. Potato (25)	WBM (kg/m²)	λ_1	550	550	678	682	720	682	615
		$\Delta\lambda_1$	20	30	10	20	30	10	70
		λ_2	682	682	920	940	790	710	935
		$\Delta\lambda_2$	4	26	20	40	20	20	50
	LAI (m²m²) (see Figure 26.5)	λ_1	682	472	682	678	550	682	625
		$\Delta\lambda_1$	28	15	28	28	20	7	50
		λ_2	982	790	738	860	688	940	940
		$\Delta\lambda_2$	36	20	45	100	28	30	8
3. Soybean (27)	WBM (kg/m²) (see Figure 26.4)	λ_1	725	732	696	565	550	635	490
		$\Delta\lambda_1$	25	10	28	50	20	30	75
		λ_2	845	758	791	875	755	682	682
		$\Delta\lambda_2$	70	10	10	130	10	10	28
	LAI (m²/m²)	λ_1	625	495	495	730	682	690	418
		$\Delta\lambda_1$	30	30	30	20	10	40	510
		λ_2	688	685	670	840	790	860	855
		$\Delta\lambda_2$	15	20	40	60	20	100	90
4. Corn (17)	WBM (kg/m²)	λ_1	720	505	715	620	620	620	490
		$\Delta\lambda_1$	18	10	4	30	30	30	50
		λ_2	820	645	990	830	1.000	940	825
		$\Delta\lambda_2$	160	6	20	160	30	70	110
	LAI (m²/m²) (see Figure 26.5)	λ_1	620	550	635	470	495	495	650
		$\Delta\lambda_1$	20	30	20	8	50	50	4
		λ_2	590	684	720	740	760	1.000	800
		$\Delta\lambda_2$	10	28	10	2	40	25	165
5. All crops (151)[a]	WBM (kg/m²)	λ_1	682	655	525	660	525	640	510
		$\Delta\lambda_1$	30	90	20	30	50	120	100
		λ_2	910	920	682	875	675	880	670
		$\Delta\lambda_2$	20	20	25	250	30	280	60
	LAI (m²/m²)	λ_1	540	682	550	682	682	670	490
		$\Delta\lambda_1$	20	10	40	28	35	40	30
		λ_2	682	756	682	910	754	910	965
		$\Delta\lambda_2$	10	20	30	200	40	200	20

Source: Thenkabail, P.S. et al.,. *Remote Sens. Environ.*, 71, 158, 2000.

Note: These were linear models. However, often, two-band nonlinear and/or multiband linear models involving more than two narrowbands provide significantly improved results.

Band centers (λ_1 and λ_2) and bandwidths ($\Delta\lambda_1$ and $\Delta\lambda_2$) in nanometers. For example, an HVI involving two bands, one centered at 682 nm (bandwidth = 28 nm) and another centered at 918 nm (bandwidth 20 nm), provides the best index for modeling wet biomass of cotton crop. Ranking of indices: Index 1 always has the highest R-square value and hence ranked higher and index 7 has the lowest R-square of the seven indices listed for each variable of each crop.

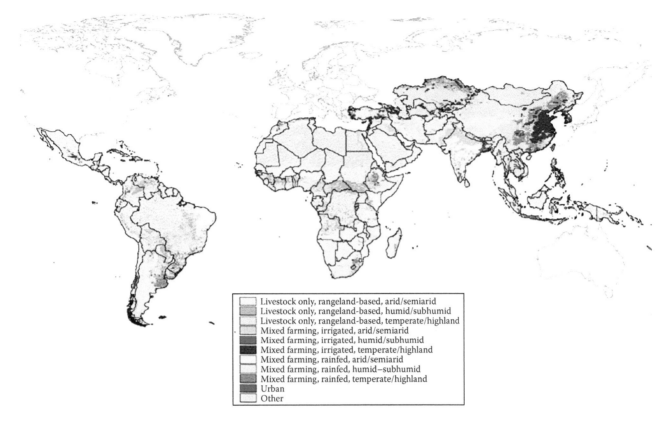

FIGURE 26.9   Global livestock production systems. (From Kruska, R.L. et al., *Agric. Syst.*, 77(1), 39, 2003, ISSN 0308-521X, http://dx.doi. org/10.1016/S0308-521X(02)00085-9.)

3. Assessment of rangeland fires that include (a) burned area determination, (b) burned area impact on land and atmosphere, and (c) fire progression and intensity monitoring
4. Land cover studies in rangelands
5. Rangeland extent mapping and monitoring

Chapter 10 also provides a history of rangeland studies and its evolution over time, starting from early days of Landsat in the 1970s to the current period. A wide array of satellites and sensors are routinely used in rangeland studies. For example, wide area fire studies are best conducted using thermal imagery from sensors such as Along-Track Scanning Radiometer (ATSR-2), Advanced ATSR, and MODIS. Degradation studies typically use indices such as NDVI, Tasseled Cap, or classification approaches. Given the large areal extent of rangelands, routine monitoring of large landscapes can be done using coarse spatial but high temporal resolution imagery such as AVHRR, MODIS, Visible Infrared Imaging Radiometer Suite, and Sentinel. But more detailed assessments of rangelands are often conducted using either hyperspatial or hyperspectral imagery.

26.11 Rangeland Biodiversity Studies

Rangelands are often the ideal landscapes for study using remote sensing (e.g., Figure 26.10). They are vast and contiguous, are relatively uninhabited, have a clear phenology based on precipitation, and have characteristic vegetation comprising of grasslands, shrublands, or some mixture of these. Maintenance of biodiversity is the goal for managing rangelands sustainably; rangeland health is monitored to prevent an irreversible loss of biodiversity. Dr. Raymond Hunt et al. in Chapter 11 identify 19 indicators of rangeland health of which 3 are the most important: (1) bare ground, (2) vegetation composition by plant functional type, and (3) presence of invasive plants. Spatial, spectral, temporal, and radiometric properties of sensors play key roles in determining what rangeland indicators can be studied and at what accuracy. Spectral unmixing of either medium-resolution or hyperspectral data is important for estimating the cover of bare ground and the amount of soil erosion. Land characterizations need to have sufficient temporal resolution; for example, functional types of the rangeland grasses (e.g., tallgrass C_3, tallgrass C_4, short-grass C_3, and short-grass C_3 study reported in Chapter 11) are best distinguished using temporal data. Better characterization of NPP, biomass, and rangeland plant functional types can be made using new satellite sensors that have high temporal coverage with medium spatial resolution (between 30 and 100 m pixels). Accurate characterization of rangeland species types and invasive species within these rangelands and study of their health will require either hyperspectral or hyperspatial data (<10 cm pixels). Invasive species are often a significant problem in most rangelands, and hence Hunt et al. provide substantial

FIGURE 26.10   False color images draped over the surface height model derived from LiDAR data for selected sites. Data recorded by Carnegie Airborne Observatory: (a) dense woodland in private reserve, L3 granite Sabi Sand; (b) highly impacted rangeland in communal rangeland with very low woody vegetation cover, L6 gabbro rangeland; and (c) cultivated and fallow fields in communal areas with large trees, L6 granite fields. (From Wessels, K.J. et al., *Forest Ecol. Manage.*, 261(1), 19, 2011, ISSN 0378-1127, http://dx.doi.org/10.1016/j.foreco.2010.09.012.)

discussion on remote sensing of invasive plants in Chapter 11. They found that spectral matching algorithms like spectral angle mapper and spectral diversity of classified objects have potential to estimate rangeland biodiversity directly, but the spectral similarity of green leaves will be a limit to the number of species individually characterized. For managing biodiversity, hyperspatial imagery acquired from low-flying manned or unmanned aircraft can be used to estimate most of the rangeland health indicators, especially bare ground, vegetation composition, and invasive species that are not spectrally unique. New ways of analyzing hyperspatial data include statistical analysis of transects over the landscape and methods based on computer vision research.

26.12   Methods of Characterizing, Mapping, and Monitoring Rangelands

Remote sensing methods, approaches, and techniques of rangeland characterization, mapping, and monitoring are many and depend on the parameter studied. Broadly, these are grouped as follows.

26.12.1 Rangeland Phenology and Productivity

Natural as well as human-induced (e.g., grazing) changes in rangeland dynamics are studied using VIs such as the NDVI, EVI, and a host of other VIs. Rangeland productivity parameters modeled using NDVI, EVI, and other variables include biomass, LAI, percent cover, fractional cover, species dominance, and species type. Phenology is characterized by taking, for example, a wide range of NDVI characteristics: cumulative NDVI over a season, NDVI at peak, NDVI at the start of season and/or end of season, NDVI amplitude, and so on (e.g., Figure 26.11).

26.12.2 Rangeland Ecological Characteristics

Distinct ecologies of rangelands are distinguished using differences in NDVI magnitude and timing.

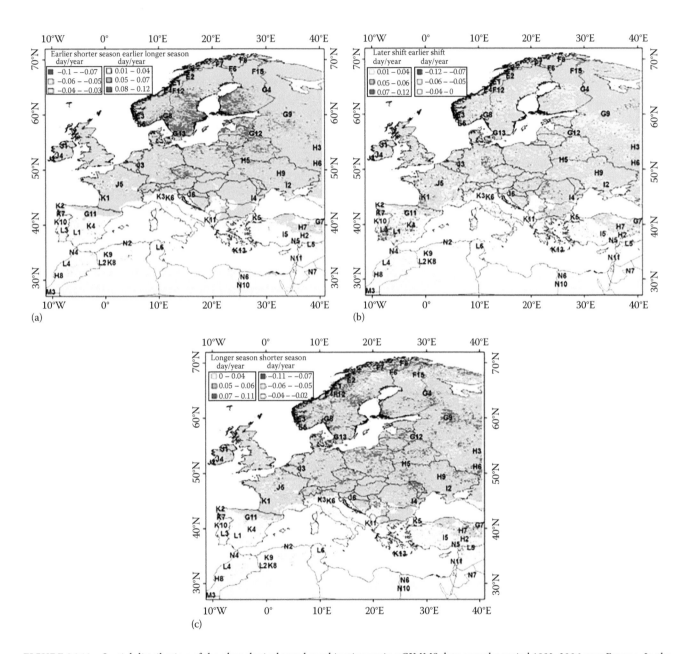

FIGURE 26.11 Spatial distribution of the phenological trend combinations using GIMMS data over the period 1982–2006 over Europe. In the background the primarily precipitation-driven areas (Mediterranean, light grey) and the primarily temperature-driven areas (central–northern Europe, dark grey) for phenological development are shown delineated after the Köppen–Geiger climate classification. (From Ivits, E. et al., *Global Planetary Change*, 88–89, 85, 2012, ISSN 0921-8181, http://dx.doi.org/10.1016/j.gloplacha.2012.03.010.) (a) Earlier longer seasons and earlier shorter seasons. Grey shaded areas, not significant trends and/or not significant spatial agglomerations. (b) Earlier shift of season and later shift of season. Grey shaded areas, not significant trends and/or not significant spatial agglomerations. (c) Longer seasons and shorter seasons. Grey shaded areas, not significant trends and/or not significant spatial agglomerations.

26.12.3 Rangeland Biological Diversity, Fuel Analysis, and Change Detection

The biological diversity, fuel loadings, and change detection are determined using VIs, well-known methods of classification such as decision trees, various supervised and unsupervised classification methods, as well as newer methods such as neural networks, random forest, and object-oriented classifications.

26.12.4 Rangeland Change Detection

Apart from classification approaches, image-differencing methods are of great importance for rangeland change detection.

26.12.5 Vegetation Continuous Fields

Vegetation continuous fields are mapped through the use of digital remote sensing data coupled with field-based measurements and advanced statistical modeling techniques.

26.13 Land Surface Phenology in Food Security Analysis

Dr. Molly Brown et al. in Chapter 13 identify the two basic approaches to measure food insecurity directly: anthropometrics measuring body weight, height, and age of the population and determining the individual consumption of food per day compared to average requirements. Both approaches require extensive, time-consuming, and costly data to be collected in households and communities. Early warning organizations require much more rapid and timely information about the probability of food insecurity in response to droughts, floods, and other environmental shocks. Modern-day remote sensing can provide a proxy for possible food insecurity through the measurement of land surface phenology (LSP). The NDVI time series provides a good proxy for LSP. Sensors such as the MODIS 250 m (6.25 ha per pixel) are ideal in time series coverage of LSP. The idea here is to look at LSP for normal, food secure year and compare it with food insecure years where food production has been affected. The ability to make these observations for every pixel enables us to study very small areas in regions with food insecure communities (e.g., Figure 26.12). Chapter 14 provides illustration of this in highly food insecure countries like Niger. Developing countries that have food insecure populations are also characterized by small, irregularly shaped fields where a variety of crops are cultivated. Ideally, LSP studied using pixel resolutions that can capture crop types would be ideal. Remote sensing data from high-resolution sensors like Landsat have sufficient resolution for identifying crop types in small fields but often do not have sufficient cloud-free images during the growing season to use the sensor for agricultural monitoring. MODIS 250 m data are ideal in terms of temporal coverage, but often the pixel will have more than one crop due to the small field size. Even then, LSP studies from a pixel with multiple crops will suffice in food security analysis. However, the change from 1 year to the next may just

be due to the change in the distribution of crop types within a pixel area and not a change in agricultural productivity. The use of LSP in global food security studies can have a powerful impact in understanding and monitoring food security. The ideal remote sensing platforms for LSP studies will be to acquire frequent data of good quality and sufficient resolution (e.g., 30 m or less) throughout the growing season. However, the current MODIS 250 m data that have excellent daily coverage are of great value.

26.14 Tropical Forest Characterization Using Multispectral Imagery

Forest carbon (C) estimates vary widely (Chapters 14 through 17) as a result of knowledge gaps, data and methods used, and rapid changes in tropical land use that may account for the "missing sink" of carbon in the global C budget. Depending on changes in land use and global climate, tropical forests can alternate between sources and sinks of atmospheric C, leading to uncertainty in future trends in forest C fluxes. The long-term net flux of carbon between terrestrial ecosystems and the atmosphere has been dominated by two factors: (1) changes in the total area of forests and (2) per hectare changes in forest biomass resulting from management and regrowth (Chapters 14 through 17). Apart from regional-level uncertainties in tropical forest carbon fluxes, uncertainties also exist in the regenerative capacity of forests and in harvest and management policies (Chapters 14 through 17).

The need to remove uncertainties and errors in estimates of tropical forest C storage and fluxes is more urgent than ever before. Under the UNFCCC, countries regularly report the state of their forest resources and emerging mechanisms, such as REDD+, and are likely to require temporally and spatially fine-grained assessments of carbon stocks.

In Chapter 14, Dr. E.H. Helmer et al. focus on the characteristics of tropical forests that are relevant to REDD+ and studied with various types of multispectral imagery. These parameters include aboveground live tree biomass (AGLB) or height, age (to estimate rates of C accumulation in regrowth), degradation, and forest type. In the chapter, they use coarse-, medium-, and high-resolution imagery to quantify and model these parameters. They highlight the use of remote sensing methods such as the following:

1. Forest type or tree species community mapping, which can be critical to REDD+ and are achieved over a range of resolutions with various classification methods. The detail at which these classes are mapped and accuracies at which they are mapped depend on the resolution of the imagery, methods used, and richness of data (e.g., how frequent the temporal images are or whether they include climate phenological extremes).

2. Forest degradation studies, which are conducted using multispectral imagery with pixel sizes less than about 10 m or with SMA of multispectral imagery with coarser spatial resolution. SMA decomposes green vegetation, nonphotosynthetic vegetation (NPV), soil, and shade. Normalized Difference Fraction Index (NDFI) is used for forest

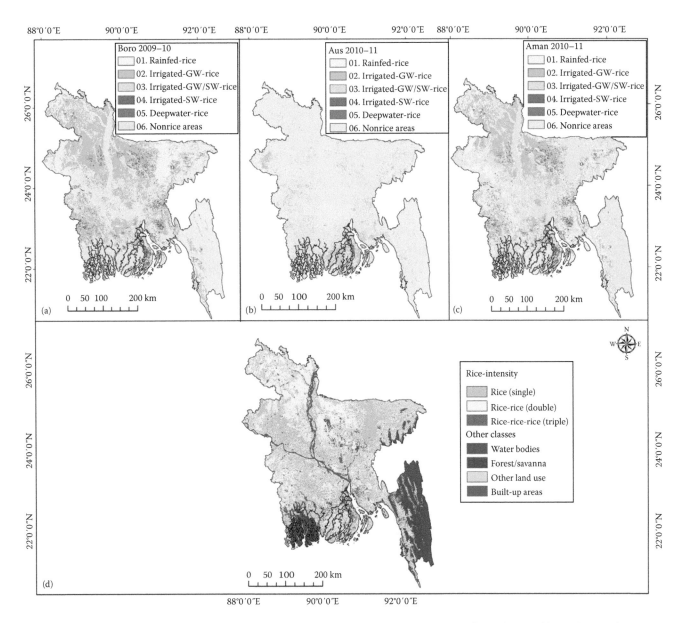

FIGURE 26.12 Spatial distribution of rice cultivation with season and irrigation source. (a) Spatial distribution of boro, (b) aus, (c) aman rice, and (d) net rice with other land use/land cover. (From Gumma et al., 2010, 2014.) Note: boro, aus, and aman are three distinct seasons.

degradation studies. NDFI values of 1 are for intact forests and −1 for bare soil. Forest degradation is of many types that include roads and trails for selective logging, slash, and burn agriculture that are detected using fine-resolution imagery.

3. AGLB has been modeled using spectral indices or bands of various multispectral sensors. C stored in tree biomass is relatively accurate when summed over large areas even though pixelwise estimates are somewhat uncertain because models underestimate AGLB at high biomass and overestimate it at low biomass.

4. When image time series span the age range of regrowth forests, spectral data can precisely estimate forest height or AGLB; stand age can also be determined, which is required for estimating rates of C removal from the atmosphere in tree biomass.

5. Fine spatial resolution imagery is used to characterize the distribution of canopy crown sizes, which is then used to estimate AGLB (also see Chapter 3) (Figure 26.13).

26.15 LiDAR and Radar for Forest Informatics

LiDAR and radar are both active sensors and as a result provide their own light source and hence have the ability to acquire data during day or night as well as offer better cloud penetration. Prof. Juha Hyyppa et al., in Chapter 15, provide an exhaustive state of the art on forest informatics assessed, modeled, and mapped by collecting 3D information using LiDAR and radar.

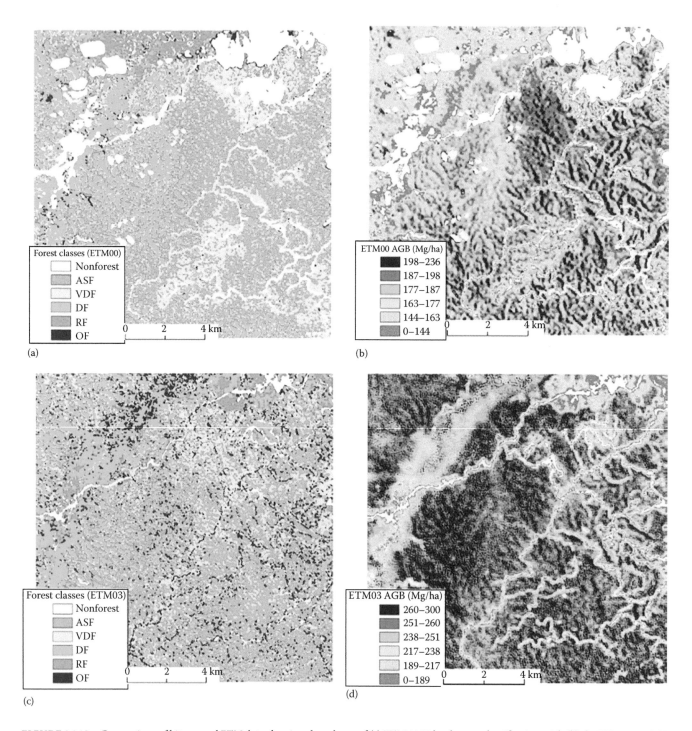

FIGURE 26.13 Comparison of bitemporal ETM data showing the subsets of (a) ETM 2000 land cover classification with (b) the ETM 2000 AGB estimates and (c) the ETM 2003 land cover map and the AGB predictions of (d) the radiometrically calibrated ETM 2003 image. (From Wijaya, A. et al., *Forest Ecol. Manage.*, 259(12), 2315, 2010, ISSN 0378-1127, http://dx.doi.org/10.1016/j.foreco.2010.03.004.)

LiDAR acquires data by illuminating the surface with laser in (1) 600–1000 nm, inexpensive but unsafe for eyes; (2) 1550 nm, eye safe but less accurate (note: this wavelength is widely applied by Riegl and Toposys; with these companies, there are no accuracy problems); (3) 1064 nm, safe for eye but greater attenuation in water; and (4) 532 nm, bathymetric

applications. The most common types of LiDAR data acquisition platforms are

1. Terrestrial (ground)-based laser scanning (TLS) or terrestrial LiDAR.
2. Mobile laser scanning (MLS) or mobile LiDAR.

3. Airborne laser scanning (ALS) or airborne LiDAR.
4. Spaceborne LiDAR missions. NASA's GLAS onboard NASA's ICESat is the first spaceborne LiDAR. ICESat GLAS acquired data from 2003 to 2009.

Radar data are collected in various bands by a number of spaceborne SAR over many years as outlined in the following text:

1. X band (frequency, 12.5–8 GHz; wavelength, 2.4–3.75 cm). Satellites include TerraSAR-X, TanDEM-x, and Constellation of Small Satellites for Mediterranean basin Observation (COSMO)-SkyMed. Data heavily used in military reconnaissance and surveillance
2. C band (frequency, 8–4 GHz; wavelength, 3.75–7.5 cm). RADARSAT and European Remote Sensing Satellite (ERS). Sea-ice surveillance. Penetration of vegetation limited to top layers.
3. S band (frequency, 4–2 GHz; wavelength, 7.5–15 cm). Meteorological applications (e.g., rainfall measurement)
4. L band (frequency, 2–1 GHz; wavelength, 15–30 cm). Satellites include Advanced Land Observing Satellite (ALOS) and Phased Array type L-band Synthetic Aperture Radar (PALSAR). Penetrates vegetation, ice, and glaciers studies
5. P band (frequency, 1–0.3 GHz; wavelength, 30–100 cm). Satellites include European Space Agency's (ESA) Explorer 7 (435 MGz). High-biomass penetration

An important advance is the ability to assess forest biomass and in turn carbon stocks and fluxes using data from ESA's Explorer 7 that acquires data in P-band synthetic aperture polarimetric radar operating at 435 MHz. Early days of radar were limited to 2D application with the real advantage of cloud penetration unlike optical sensors.

But as outlined by Juha Hyyppa et al., in Chapter 15, both LiDAR and radar provide point clouds (echo) that allow for creating 3D clouds of trees and other vegetation. The forest informatics derived by the LiDAR and radar 3D cloud of points (e.g., Figure 27.14) include parameters such as tree location, tree height, diameter at breast height, species, age, basal area, crown area, volume, biomass, and LAI. These forest variables from 3D point clouds are obtained through two approaches: (1) area-based approaches (ABAs) and (2) individual/single-tree detection approaches (ITDs). Regression models, neural networks, and random forest are some of the methods used in ABAs and ITDs. The 3D data further help derive digital terrain model (DTM), digital surface model (DSM), and canopy height model (CHM) and normalized DSM (CHM/nDSM).

Extensive discussion on how to derive forest informatics using LiDAR and radar 3D point clouds including methods and approaches used, strengths, and limitations is presented and discussed in Chapter 15.

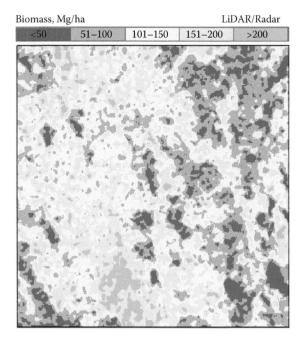

FIGURE 26.14 Forest biomass mapping from LiDAR and radar synergies. Biomass map from SRTM phase center height and PALSAR data developed from regression model using random biomass samples from LVIS-derived reference map. The image was smoothed using a 5 by 5 window (pixel size of 15 m). (From Sun, G. et al., *Remote Sens. Environ.*, 115(11), 2906, 2011, ISSN 0034-4257, http://dx.doi.org/10.1016/j.rse.2011.03.021.)

26.16 Hyperspectral Imager and LiDAR Data in Study of Forest Biophysical, Biochemical, and Structural Properties

When remote sensing data are acquired in narrowbands and contiguously over a wavelength range representing a spectrum, it is called hyperspectral data. The number of bands itself is not as critical a factor, so 20 or 30 or 100 or 200 narrowbands (typically, ≤10 nm bandwidth) over 400–2500 nm are quite commonly used as hyperspectral data. When hyperspectral data are collected in an image format using ground-based, airborne, or spaceborne sensors, such data are called hyperspectral imager (HSI). In contrast, LiDAR is an active sensor based on emitted laser pulses, which provides 3D information in the form of laser point clouds (see further details on LiDAR in Chapter 15). Dr. Gregory Asner et al. in Chapter 16 assess the HSI and LiDAR data uses in the study of forest:

1. Biophysical properties
2. Biochemical properties
3. Canopy physiological properties
4. Canopy structural and carbon properties

HSI is ideal for the study of biophysical, biochemical, and physiological properties of vegetation (e.g., Figure 26.15). LiDAR is

FIGURE 26.15 Demonstration of a virtual active hyperspectral LiDAR in automated point cloud classification. Hyperspectrally classified point cloud visualized from two viewing directions. The background points were left out, and only the needle (green) and trunk (brown) points are plotted. The red arrow points toward the point of measurement. (From Suomalainen, 2010, 2011.)

ideal for characterizing the structural and architectural properties of vegetation and for advancing biomass assessments. Dr. Gregory Asner et al. in Chapter 16 discuss all of these and show us the advances one can make in studying these features. Biophysical variables include LAI (m^2/m^2), biomass (kg/m^2), equivalent water thickness (mm), and leaf mass area (g/m^2). Biochemical properties include nitrogen, cellulose, lignin, pigments (e.g., chlorophyll a, b, total; anthocyanins, carotenoids), PAR, APAR, and LUE. These are studied using HSI-derived HVIs (see Thenkabail et al., 2014, 2013), radiative transfer models such as PROSPECT (Jacquemoud et al., 2009), and empirical spectroscopic algorithms (Asner et al., 2011). Specific HVIs can be applied to study specific properties, for example, biomass or LAI using narrow spectral bands centered at 680 and 910 nm or LUE using PRI (531 and 570 nm). In contrast, LiDAR is often used to study tree heights, AGB of vegetation, forest structure, and aboveground carbon density (ACD). Research is still in progress on establishing accuracies of LiDAR-estimated ACD with that of plot-based measurements. An important part of Chapter 16 is the illustration and enumeration of advances one can make in improved understanding of structural, architectural, biophysical, biochemical, and ACD characteristics of forests by integrating HSI and LiDAR data.

26.17 Tree and Stand Heights from Optical Remote Sensing

Over the years, remote sensing has been used to map and monitor forests in terms of their cover, type, distribution, species dominance, deforestation, tree crown, biomass, LAI, stand area, and a host of other parameters including forest health and change over time. But almost all these remote sensing measurements have been 2D. NASA created the first 3D global map of

forest heights using GLAS on ICESat (GLAS/ICESat), MODIS, and Tropical Rainfall Measuring Mission data (Lefsky, 2010). Greater accuracies and lesser uncertainties in forest biomass and carbon estimates are feasible through improved accuracies in measurement of tree heights. Traditionally, tree heights are measured by plot sampling in the field. This is extremely tedious, difficult in complex forests due to difficulty in accessibility, and resource prohibitory to cover forests of the world, which cover about 30% of the terrestrial area. Further, since rapid changes occur in forests particularly due to anthropogenic activities, repeated measurement of these changes is required.

Chapter 17 by Dr. Sylvie Durrieu and Dr. Cédric Véga et al. shows us the approaches and methods of estimating tree heights through 3D vertical measurement using digital photogrammetry and more recently (and increasingly) through LiDAR remote sensing. The main advantage of LiDAR is that it sees all the vegetation within the plot, whereas field-plot data may describe in a plot only a sample of trees for the sake of cost-effectiveness, and digital photogrammetry points see only dominant vegetation. Dr. Sylvie Durrieu and Dr. Cédric Véga et al. present and discuss in details the following:

1. The principles of height measurement using LiDAR and photogrammetry.
2. Stand level height assessments made through ABAs that include (i) field inventory at plot level of forest parameters, (ii) extracting LiDAR point clouds of the forest parameters for inventoried plots, (iii) establishing empirical models linking LiDAR data with field-plot data for each forest parameter, and (iv) extrapolating the empirical models same over the entire forests leading to large area inventories of forest parameters. The process involves developing models relating LiDAR or photogrammetric 3D data versus field-plot data of forest characteristics (e.g., tree height, basal area) and applying the same over larger forest areas. For greater accuracies, large area inventories require segmenting forest types into distinct categories and developing the stand level models for each of these forest categories separately.
3. Individual tree height assessment through (a) raster-based approaches (e.g., canopy height modeling, detecting tree apices, measuring tree crowns), (b) point-based approaches, and (c) hybrid approaches. For tree height modeling using raster-based approach, the tree crown data may or may not be required, but CHM and detecting tree apices are required. The point-based approach detects not only the dominant trees (apices) but also overtopped trees taking advantage of the ALS 3D point cloud (e.g., Figure 26.16). Hybrid approaches, combining raster-based and point-based approaches, have shown improved individual tree extraction.

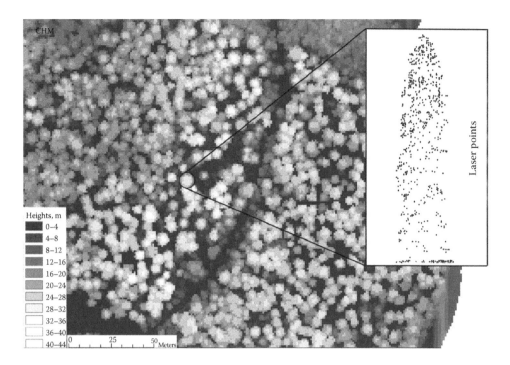

FIGURE 26.16 Single-tree biomass modeling using airborne laser scanning. Example of ALS points inside one tree canopy segment and of CHM with 0.5 m grid size. (From Kankare, V. et al., *J. Photogramm. Remote Sens.*, 85, 66, 2013, ISSN 0924-2716, http://dx.doi.org/10.1016/j.isprsjprs.2013.08.008.)

26.18 Study of Biodiversity from Space

Chapter 18 by Dr. Thomas Gillespie et al. provides a lucid outline on how to use remote sensing data from space in biodiversity studies. They approach this by looking at three categories of biodiversity assessment using optical, radar, LiDAR, and thermal remote sensing data from space. The three categories are

1. Mapping
2. Modeling
3. Monitoring

Mapping presentation and discussions include vegetation categories and invasive species. Mapping of vegetation categories and broad habitats can be, typically, performed using 30 m Landsat or better resolution. However, accurate and detailed mapping of species or individual trees requires hyperspectral data (5 m or better) from sensors such as QuickBird, GeoEye, and IKONOS and/or hyperspectral data from sensors such as Hyperion. The ability to map animals from space is limited due to the lack of coverage of frequent images at sufficiently high spatial resolution and also data from specific spectral bands such as thermal that can detect body heat from animals if data are within a meter or so.

Spatial modeling for biodiversity studies such as understanding species richness, ecosystems richness for habitats, and carrying capacity will be a powerful tool. This would involve using a wide array of spatial data such as precipitation, land use/cover, soils, and elevation and then performing spatial models

for planning and decision-making process (e.g., Which lands to conserve? Where are the richest habitats for biodiversity?).

Monitoring biodiversity is important for understanding factors such as habitat loss or degradation and productivity changes and assessing development and conservation.

Many biodiversity studies like identifying species or trees require hyperspatial data that are also hyperspectral. Other biodiversity studies like habitat mapping require more temporal data at moderate resolutions like Landsat 30 m or MODIS time series (e.g., Figure 26.17). Monitoring animals will require hyperspatial data that are possibly thermal as well.

In Chapter 18, Dr. Thomas Gillespie et al. provide the remote sensing data characteristics needed for biodiversity studies and present the state of the art in mapping, modeling, and monitoring biodiversity.

26.19 Multiscale Habitat Mapping and Monitoring Using Satellite Data and Advanced Image Analysis Techniques

There are wide-ranging habitats on the planet that house plants and animals, such as the forest, savanna, desert, and wetland. The ability of EO data to map and study habitats varies widely depending on the detail at which a habitat needs to be mapped and the basic characteristics of remote sensing data like their

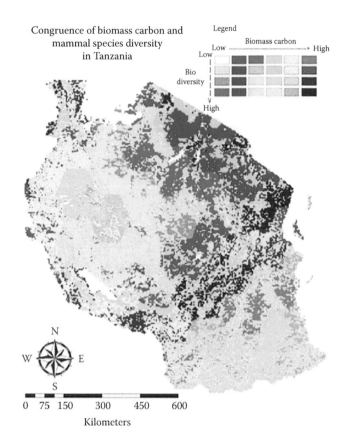

Congruence of biomass carbon and mammal species diversity in Tanzania

FIGURE 27.17 A framework for integrating biodiversity concerns into national REDD+ programs. Example national scale map for Tanzania displaying congruence values between carbon and biodiversity at the scale of a 5 km grid and across all vegetation types. Map generated using freely available land cover data from MODIS, mammal data from the freely available African mammal databank (African Mammals Databank and African carbon data provided by UNEP-WCMC, based on multiple sources (Khan, 2011)). This kind of simple overlay map can help in identifying those areas of both high opportunity (strong positive correlation in carbon and biodiversity values) and risk (low in carbon but high in biodiversity) in the REDD+ planning process. (From Gardner, T.A. et al., *Biolog. Conserv.*, 154, 61, 2012, ISSN 0006-3207, http://dx.doi.org/10.1016/j.biocon.2011.11.018.)

spatial, spectral, and temporal resolutions. If the need of mapping is to discern a single species of tree or shrub or grass, the requirement of spatial resolution could be submeter to few meters. If the need of habitat mapping is to get a broad understanding of density of forest cover, then coarse resolution like 30 or 250 m may suffice. However, if the goal of the habitat mapping is to get a broad understanding of habitat land cover over vast areas, even 1 km data that are more temporally rich may be needed.

In Chapter 9, Dr. Stefan Lang provides habitat mapping protocols using a wide array of EO data. The biodiversity of each habitat defines the richness of plant and animal species contained in these habitats. They begin with providing the importance of habitat studies by referring to GEO, identifying biodiversity as one of the nine societal beneficial areas. Central to their chapter is the strategy, approaches, and methods adopted in two

major European Union projects such as Multiscale Service for Monitoring NATURA 2000 Habitats of European Community Interest (MS.MONINA) and Biodiversity Multi-SOurce Monitoring System (BIO_SOS). For example, as highlighted in Chapter 9, these projects use

1. Very high spatial resolution imagery (VHRI) from sensors like WorldView-2 for fine-scale habitat mapping
2. Imaging spectroscopy (hyperspectral) data from Compact High Resolution Imaging Spectrometer/Project for On-Board Autonomy or Hyperion for plant biophysical and biochemical properties
3. LiDAR data from ICESat/GLAS, to determine tree height and 3D biomass
4. X-band radar for from fine-resolution sensors like TerraSAR-X to differentiate plant species
5. Thermal VHRI can be used to even count number of cows in rangelands

However, in order to map fine details of habitats such as individual species, one may require a combination of hyperspatial, hyperspectral, and other data (e.g., bathymetry) analyzed with an ensemble of algorithms (e.g., Figure 26.18).

Chapter 19 provides a sensor suitability table showing what forest, grassland, heathland, and wetland habitat variables are mapped and at what detail by various low-, medium-, very high-, hyperspectral-, laser-, and microwave sensors.

26.20 Ecological Characterization of Vegetation Using Optical Sensors

The launch of optical sensors, starting with Soviet's Sputnik and NOAA AVHRR, changed our view of the world on how we study planet Earth. In Chapter 20, Dr. Conghe Song et al. provided an exhaustive series of optical satellites, their brief history and characteristics, and their value in studying vegetation. These satellites provide data in distinct wavebands and have unique spectral, spatial, radiometric, and temporal coverage of the entire planet Earth. Hence, quantifying, modeling, and mapping of vegetation from remote sensing became widespread, especially with the launch of the first Landsat in 1972. Vegetation characterization has been grouped into two broad categories by Dr. Conghe Song et al.:

Vegetation structure that include
    Vegetation cover
    Forest successional stages
    LAI
    Biomass and NPP
Vegetation functions that include
    LSP
    FAPAR
    Chlorophyll
    LUE
    GPP/NPP

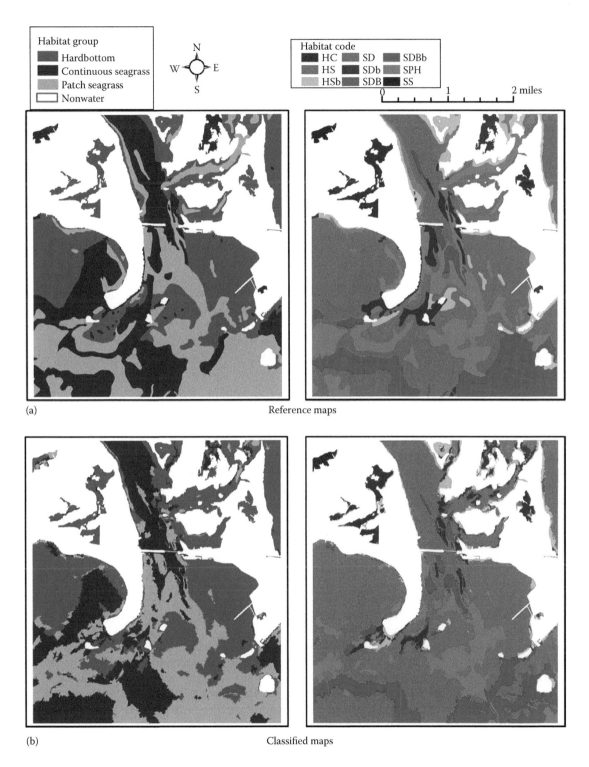

FIGURE 26.18 Habitat maps from (a) reference data and (b) classification results of the fused dataset (hyperspectral imagery, aerial photography, and bathymetry data) and ensemble analysis of random forest, SVMs, and k-nearest neighbor. Six code-level habitats were observed: HC (soft coral, hard coral, sponge, and algae hardbottom), HS (hardbottom with perceptible sea grass [<50%]), SD (moderate to dense, continuous beds of sea grass), SDB (moderate to dense nearly continuous beds [sea grass > 50%], with blowouts and/or sand or mud patches), SPH (dense patches of sea grass [>50%] in a matrix of hardbottom), and SS (sparse continuous beds of sea grass). (From Zhang, C., *J. Photogramm. Remote Sens.*, Available online 27 June 2014, ISSN 0924-2716, http://dx.doi.org/10.1016/j.isprsjprs.2014.06.005.)

Chapter 22 shows us

1. Vegetation cover modeling using various techniques that include statistical regression with VIs, classifications, and SMA
2. Forest successional stages characterized by physically based models, empirical models, change detection approaches
3. LAI algorithms based on VIs, radiative transfer models
4. Biomass and NPP through regression models, k-nearest neighbor algorithms, machine learning algorithms, and biophysical approaches
5. LSP through VIs
6. FAPAR through empirical models involving VIs, biophysical models
7. Chlorophyll assessment through VIs and radiative transfer models
8. LUE through PRI
9. GPP and NPP based on LUE and other process-based models using remotely sensed data as inputs

As we can clearly see, this is an exhaustive list of vegetation parameters that are widely used in a wide array of global and local studies such as the primary productivity of GPP and NPP, understanding ecosystems, and assessing degradation and changes over space and time (e.g., Figure 26.19). These products are often the most accurate data on vegetation and their characteristics that feed into global change models and climate models.

26.21 Land Cover Change Detection

Remote sensing is an ideal way to observe, quantify, and monitor land cover and land cover changes (LCLCC) over space and time. The advantage remote sensing offers is repeated coverage, synoptic views over large areas, global coverage, and the ability to study LCLCC in different resolutions or scales and using consistent data over time. Remote sensing does not directly provide information on land use but is inferred from land cover. So LULC studies are widespread through the use of a plethora of remotely sensed data. Many types of LULC applications are possible through remote sensing. These applications involve forests, grasslands, croplands, and so forth. The degree of detail one can study in LULC and LULC change (LULCC) will depend on the characteristics of remotely sensed data and their spatial, spectral, radiometric, and temporal resolution. Also, the degree of detail depends on methods, techniques, and approaches used in classifying and synthesizing. All these factors influence the details at which LULC and land use and land use change (LULUC) are mapped and their accuracies achieved (e.g., Figure 26.20).

Chapter 21 by John Rogan and Nathan Mietkiewicz provides a background on the importance of land cover change detection studies, outlines the theory and practice, enumerates on trends of change detection studies, outlines and discusses methods and approaches, and provides an assessment of map accuracy strategies. They show us how the 40+ years of spaceborne remote sensing archives, from various sensors such as 1–10 km AVHRR,

250–1000 m MODIS, and 30 m Landsat, is helping us study and understand land cover trends in any part of the world. For example, there is now a monthly continuous record of AVHRR Global Inventory Modeling and Mapping Studies (GIMMS) NDVI data for 30 years (1982–2012); Landsat 4–7 band data epochs of the 1970s, 1980s, 1990s, 2000s, 2005s, and 2010s; also Landsat 4–7 band record of 40 years (even through number of coverage for any location on the earth may differ widely); and 15-year record of MODIS (2–36 bands). Such a record has enabled land cover studies in various spatial, spectral, temporal, and radiometric resolutions.

Chapter 21 reviews the three types of change detection approaches: (1) monotemporal change detection where only a single image of an area is involved; (2) bitemporal change detection where two images, of two distinct dates, of an area are involved; and (3) temporal trend analysis where a continuous series of images (e.g., monthly maximum value composites over 1 or more years) of an area are involved. They compare several automated change detection approaches that include (1) disturbance index, (2) MODIS global disturbance index, (3) Carnegie Landsat Analysis System–Lite (CLASlite) (a forest cover automated change detection algorithm), (4) vegetation change tracker, and (5) Spatial Temporal Adaptive Algorithm for mapping Reflectance Change (STAARCH). They demonstrate the CLASlite method of change detection for three study areas (rural, urban, coastal) using Landsat images for eight eras (1985–1993, 1993–1995, 1995–1999, 1999–2002, 2002–2009, 2009–2010, 2010–2011, and 2011–2013) in Central Massachusetts. CLASlite method partitions each scene into proportional fractional cover types of bare ground (B), photosynthetic vegetation (PV), and NPV for every pixel. They use very high spatial resolution data (submeter to 5 m) obtained from Google Earth as reference data to establish forest versus nonforest class accuracies using classic error matrices.

26.22 Radar Remote Sensing in Land Use and Land Cover Mapping and Change Detection

Radar remote sensing has some unique features such as cloud penetration and all-day imaging ability (since it is an active sensor) when compared with optical remote sensing. Radar data are acquired over 0.3–100 cm wavelength, in one or more of the four polarization (HH, HV, VH, VV), three modes (SpotLight, StripMap, and ScanSAR), various incident angles, repeat frequency, and resolutions (range and azimuth). Radar data are also processed at various levels: slant range data, ground range data, and geocoded and orthorectified data. How the radar data are acquired and processed is important in determining what applications are these data used. An application such as change detection and interferometry requires radar data acquisition to have identical parameters (e.g., orbit, incidence angle, and polarization). In Chapter 22, Dr. Zhixin Qi and Dr. Anthony Gar-On Yeh discuss a number of applications of radar data that

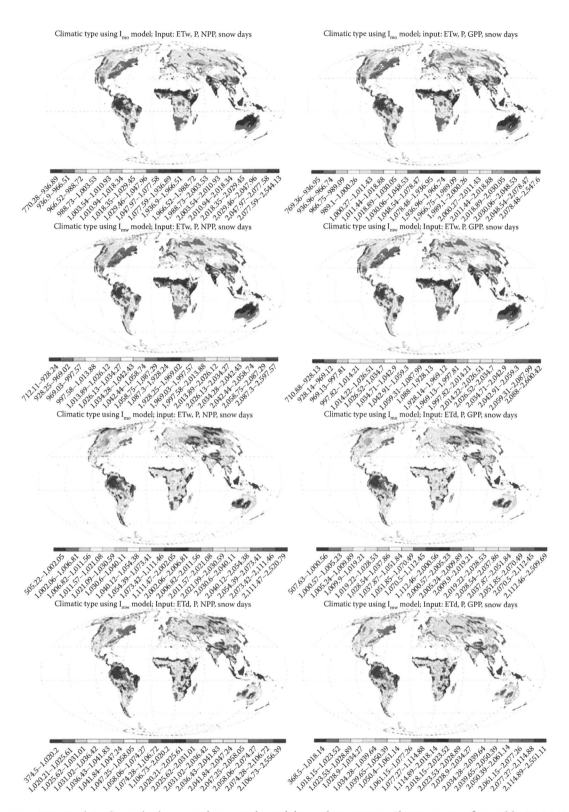

FIGURE 26.19 Terrestrial Earth couple climate–carbon spatial variability and uncertainty. Climatic types influenced by NPP MODIS (left hand) and GPP MODIS (right hand). (From Alves, M.C. et al., *Global Planetary Change*, 111, 9, 2013, ISSN 0921-8181, http://dx.doi.org/10.1016/j.gloplacha.2013.08.009.)

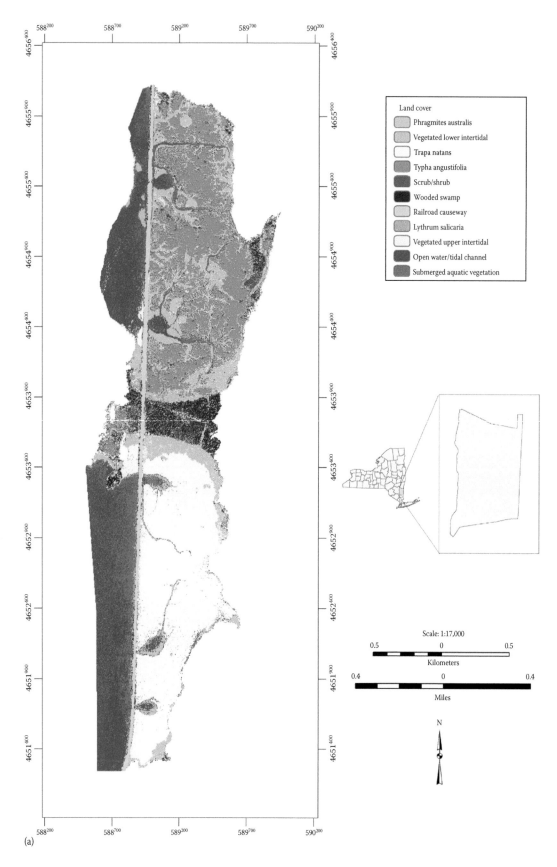

Land cover
- Phragmites australis
- Vegetated lower intertidal
- Trapa natans
- Typha angustifolia
- Scrub/shrub
- Wooded swamp
- Railroad causeway
- Lythrum salicaria
- Vegetated upper intertidal
- Open water/tidal channel
- Submerged aquatic vegetation

Scale: 1:17,000

Kilometers

Miles

N

(a)

FIGURE 26.20 Tivoli Bays land cover map produced using IKONOS data with two methods: (a) Method 1 relied solely on the four spectral bands (blue, green, red, near infrared) of the IKONOS image.   *(Continued)*

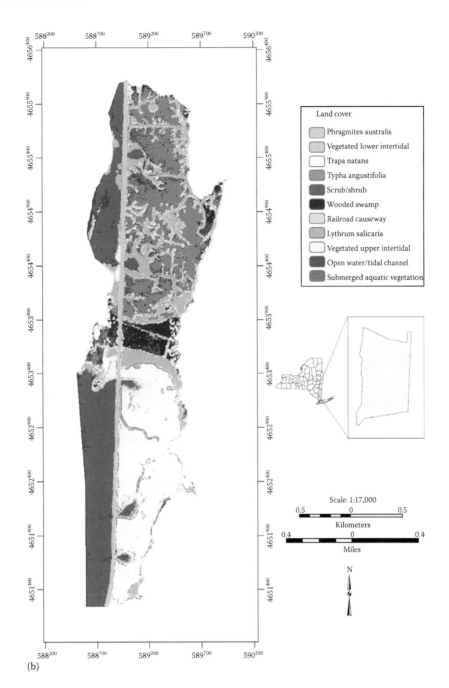

FIGURE 26.20 (*Continued*) Tivoli Bays land cover map produced using IKONOS data with two methods: (b) Method 2 used a maximum-likelihood classification of the four spectral bands, supplemented by local texture information (variance) calculated in a moving 3 by 3 pixel (Method 2) or 5 by 5 pixel (Method 3) window, superposed separately on each band of the IKONOS image. Ultimately, eight bands were used in the classifications for Method 2. (From Laba, M. et al., *Remote Sens. Environ.*, 114(4), 876, 2010, ISSN 0034-4257, http://dx.doi.org/10.1016/j.rse.2009.12.002.)

include LULC classification, forest inventory and mapping, crop and vegetation identification, urban environments, snow and ice, and a number of others. These studies are reported based on data acquired in various frequencies and wavelengths by a wide array of spaceborne SAR sensors such as Environmental Satellite Advanced Synthetic Aperture Radar (Envisat ASAR), TanDEM-X, TerraSAR-X, RADARSAT constellation, ERS, Japanese Earth Resource Satellite (JERS), ALOS PALSAR, and

COSMO-SkyMed. Chapter 22 highlights the following strengths of SAR data in various applications:

1. LULC classification: A wide array of SAR data have been used in classifying forest types (e.g., primary, secondary, slash and burn agriculture, regrowth or regenerative forests, plantations) or LULC classes (e.g., cropped areas, bare soil areas, forestry, forest clear cut, forest burnt areas,

water bodies). A large number of LULC classes as mentioned earlier are, for example, mapped when single date ALOS PALSAR fine/dual beam is combined with multitemporal Envisat ASAR. Some studies have shown overall classification accuracies as high as 87% for several land cover classes such as built-up areas, water, barren land, crop/natural vegetation, lawn, banana fields, and forests by combining the SAR textural, polarimetric, and interferometric information extracted from RADARSAT-2 polarimetric SAR (PolSAR) images. Combining multiple-frequency SAR scenes (e.g., L band, P band, PolSAR, polarimetric interferometry SAR) and fusing them provided greater accuracies in LULC classification than using single-frequency SAR images. Numerous studies have reported significant improvement in classification accuracies when radar data are combined with optical data.

2. Forest species identification has been successfully performed using SAR data with multiple incident angles and variation of backscatter coefficient in various incident angle.

3. Crop classification: Crops such as corn, soybeans, cereals, and hay pasture are classified with 70%–89% accuracy with SAR data, with accuracies increasing with an increasing number of temporal coverage and multiple frequencies.

4. Biophysical characterization of forests have been conducted using interferometric coherence maps derived from ERS-1 and ERS-2 SAR images and from JERS-1 SAR images.

5. Urban applications: Increased accuracies in urban mapping was possible when very high resolution optical imagery (e.g., ,QuickBird) is combined with SAR data.

Radar data have also been used extensively for AGB estimations and for carbon stock assessments (e.g., Figure 26.21). However, radar data have high noise and large geometric distortion relative to optical imagery, hence requiring its own specialized algorithms to process data. This is nowhere as developed as for optical sensors (Nolte et al., 2001).

Radar data can provide complementary/supplementary information to optical remote sensing to advance our understanding and better map, model, and monitor land themes. So wherever feasible, it is better to use optical and radar data to complement/supplement information of each sensor type.

26.23 Global Carbon Budgets and Remote Sensing

The global carbon budget consists of four terms, atmosphere, land, oceans, and fossil fuels, as presented by Dr. Richard Houghton in Chapter 23. The long-term net flux of carbon between terrestrial ecosystems and the atmosphere has been dominated by two factors (Chapters 15 through 19 and 25): (1) changes in the total area of forests and (2) per hectare changes in forest biomass resulting from management and regrowth.

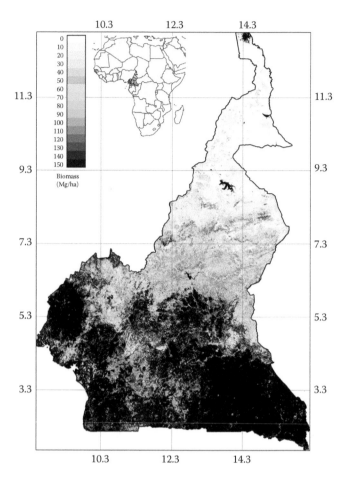

FIGURE 26.21 Biomass assessment in the Cameroon rainforests and savannas using ALOS PALSAR data. Pixels saturate at 150 Mg/ha. Dense forest classes were masked out using the GlobCover 2009 land cover map (Bontemps et al., 2011). Figure shows the north–south AGB gradient. (From Mermoz, S., *Remote Sens. Environ.*, Available online 15 May 2014, ISSN 0034-4257, http://dx.doi.org/10.1016/j.rse.2014.01.029.)

Apart from regional level uncertainties, the carbon flux of tropical forests is greatly influenced by uncertainty in the regenerative capacity of forests and in harvest and management policies (Chapters 23 and 24).

Chapter 23 highlights the need to keep track of global carbon budget annually in order to determine how much carbon is emitted to the atmosphere, how much is absorbed by land and oceans, and how much stays in the atmosphere. Currently, of the 32 billion tons of carbon (C) emitted to the atmosphere each year due to anthropogenic activity, tropical forests sequester about 4.25 billion tons, soils and other vegetation another 4.25 billion tons, and oceans 8.5 billion tons, leaving the residual 15 billion tons in the atmosphere (Lewis et al., 2009). Also, land use change, mainly from deforestation in the tropics, is responsible for estimated net emissions of about 6 billion tons of greenhouse gases—greater than the emissions from all the world's planes, ships, trucks, and cars (Lewis et al., 2009, see also Chapters 14 through 17). Dr. Richard Houghton points out that the fraction of carbon that remains in the atmosphere has been remarkably constant over the last 50 years, with the increase in emissions compensated by an increase in sinks on land and in the oceans. Estimates of C storage in terrestrial ecosystems are still very approximate. For example, Lewis et al. (2009) report a wide range (0.29–0.66 Mg C/ha/year) in tropical forest C storage. The Wet Tropical Asian Bioregion forests, for example, contain high C density of up to 500 Mg/ha (Lasco, 2004) but are changing rapidly due to selective logging, forest conversion to agriculture resulting in C density of less than 40 Mg/ha, and conversion to plantations (agroforests), which are responsible for at least a 50% decline in forest C density.

In order to track carbon sources and sinks from land, first, Chapter 23 discusses the *bookkeeping model* of early days, which uses annual rates of land cover change and standard growth and decomposition rates per hectare to calculate annual changes in carbon pools as a result of management. This highly aggregated approach did not use remote sensing as input but was based on statistics available from national and international sources, which were subjective and approximate.

The current approach to global carbon budgets enumerated by Dr. Houghton in Chapter 23 is "based on two broad types of explanatory mechanisms [that] account for the loss and accumulation of carbon on land: (1) disturbances and recovery (structural mechanisms) and (2) the differential effects of environmental change (e.g. CO_2, N deposition, climate) on photosynthesis and respiration (metabolic mechanisms)." This approach uses significant remote sensing. The flux of carbon from LULCC is based on disturbances and recovery, especially from medium resolution (30–100 m). Analyses include

1. Rates of change in forest area
2. Biomass density
3. Measurement of changes in carbon density

Houghton presents an example of estimating flux of carbon from LULCC taking the UNFCCC REDD+. Under this mechanism, when a country reduces emissions, it is eligible for carbon credits. I recommend readers to pay attention to the nine issues inherent in estimating the flux of carbon from LULCC outlined by Dr. Richard Houghton in Chapter 23. These nine issues are

1. Definitions
2. Assigning a carbon density to the areas deforested
3. Committed versus actual emissions
4. Gross and net emissions of carbon from LULCC
5. Initial conditions
6. Full carbon accounting
7. Accuracy and precision
8. Attrition
9. Uncertainties

Opportunities to significantly improve estimates of C storage and flux through improved estimates of LUC and modeling are possible with the evolution in spaceborne hyperspectral, hyperspatial, and advanced multispectral sensors, as a result of improvements in the spatial, spectral, radiometric, and temporal properties as well as in optics and signal-to-noise ratio of

data (e.g., Figure 26.22). High spatial resolution allows location, while high spectral resolution allows identification of features. Hyperspectral remote sensing sensors allow direct measurement of canopy chemical content (e.g., chlorophyll, nitrogen), forest species, chemistry distribution, timber volumes, and water and improved biophysical and yield characteristics (Chapter 18). Thenkabail et al. (2004a) demonstrated an increased accuracy of about 30% in LUC and biomass when 30 hyperspectral wavebands are used relative to six nonthermal Landsat TM bands. Hyperspatial data have demonstrated the ability to extract individual tree crowns from 1 m panchromatic data. Agroforest successional stages have been mapped and their varying carbon sink strengths assessed using IKONOS (Thenkabail et al., 2004b). In contrast, forest structure variables (e.g., biomass, LAI) are poorly predicted by the older-generation sensors. One also has to look at the new Orbiting Carbon Observatory-2 launched in 2014 to study CO_2 in the column of air over the Earth's surface, which will further advance our understanding of CO_2 sources and sinks.

26.24 Spectral Sensing of Soils

Spectral sensing implies gathering near-continuous or noncontinuous spectral data of targets as images or spectral behaviors from different platforms (i.e., remote and proximal sensing). Depending on the level of acquisition, we classified spectral sensing into laboratory spectral sensing (LSS), field spectral sensing (FSS), ground spectral sensing (GSS), aerial spectral sensing (ASS), and space spectral sensing (SSS). Chapter 24 by José A.M. Demattê dwells deep into how soils can be evaluated by spectral sensing using platforms from ground to space. The chapter shows a new way to *see* soils and study their characteristics by spectral sensing point of view. In this way, the chapter indicates the study of soil properties of all types such as soil organic matter and carbon, pH, plant nutrients (e.g., N, Ca, Mg, K, P, and Na), soil particle size (clay, sand, and silt) content, moisture, and color. A large portion of the chapter includes summaries on how the spectral bands and indices are correlated with soil properties using linear and nonlinear modeling. Linear modeling includes statistical methods involving linear and multilinear regressions, principal component analyses, and partial least squares regression. Nonlinear modeling methods include SVM, boosted regression trees, and artificial neural networks. An extensive discussion of literature shows how various wavebands (absorption features) in spectral sensing help decipher soil information. What is important to note is that one to multiple wavebands or indices can be used to obtain important correlations with soil properties (typically, R^2 of 0.80 or above). For example, the most important wavebands for predicting soil water are 1350–1450, 1890–1990, and 2220–2280 nm. Minerals like goethite and hematite are predicted with R^2 values as high as 0.8 using a simple spectral band depth calculation in the visible wavelengths.

As also pointed out in the chapter, no single problem has plagued soil scientists more than the identification of the spatial boundaries of an individual soil body on the landscape (e.g., Figure 26.23). Chapter 24 by José A.M. Demattê demonstrates several strategies on how to use spectral data to assessment soils

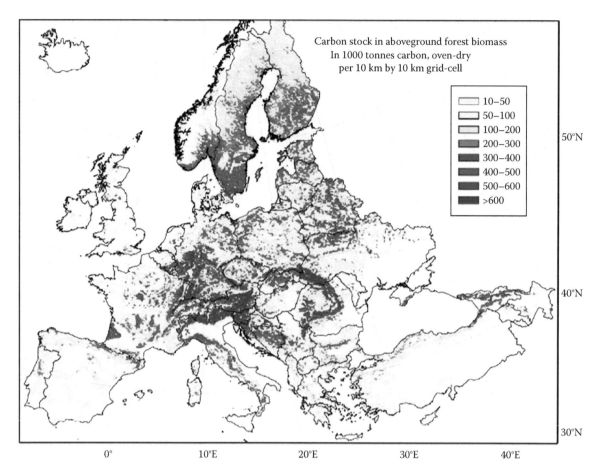

FIGURE 26.22 Total carbon stock of aboveground forest biomass for the European Union countries calculated separately for broadleaves and conifers with a spatial resolution of 500 m MODIS data. Aggregated biomass conversion and expansion factors were used to convert the remote sensing–based growing stock classification results to carbon stock of the aboveground forest biomass. (From Gallaun, H. et al., *Forest Ecol. Manage.*, 260(3), 252, 2010, ISSN 0378-1127, http://dx.doi.org/10.1016/j.foreco.2009.10.011.)

indicating advantages and limitations of each platform. It further demonstrates the ability of spectral sensing to assist and produce accurate (approximately 80%) pedological maps comparable to traditional approaches. The aspect of studying soils from vegetated areas involves their use and coverage, and PV and NPV and has been discussed in this chapter. Use of soil mapping for precision farming needs detailed spatial information of soil physical and chemical properties like nutrient status and water-holding capacity where field spectral sensing has been applied. Soil conservation requires a large-scale understanding of relief and slope, erosion susceptibility, drainage systems, and vegetation cover, which aerial and space spectral sensing can be very useful. Study of soil profiles at depths up to 2 m can employ new ground-penetrating spectral sensing equipments that will help establish soil information. Gamma ray spectroscopy helps produce fast and cost-effective soil maps for soil properties associated with parent material. Radar helps penetrate soils to few centimeters to study soil moisture, salinity, and other properties. Further, this chapter deals with variations about the study of soil properties from different platforms as well as building spectral libraries from these data. The chapter also indicates on how to use spectral sensing associated with pedotransference system to build pedological maps.

An exciting option, going forward, is the expected launch of various hyperspectral sensors in coming years. These sensors include EnMAP (Germany), Hyperspectral Imager Suite (Japan), HyspIRI (United States), Hyperspectral-X Imager (France), PRecursore IperSpettrale della Missione Applicativa, Hyperspectral PRecursor of the Application Mission (PRISMA) (Italy), and Spaceborne Hyperspectral Applicative Land and Ocean Mission (Israel–Italy). The idea of spectral sensing from space and building spectral libraries for soil applications is an exciting one, given the uniformity of such a data collection on a routine temporal basis.

26.25 Soil Studies from Remote Sensing

Soils are foundation of agriculture and all vegetation on the planet and have any number of other uses in preserving our environments and sustaining our livelihoods. Ideal soils for agriculture are balanced in contributions from mineral components (sand, 0.05–2 mm; silt, 0.002–0.05 mm; clay, <0.002 mm), soil organic matter, air, and water (Parikh and James, 2012). Soils are also places that house many living beings such as microbes, fungi, earthworms, and mites. So the study of basic soil properties and understanding their fertility and soil degradation are

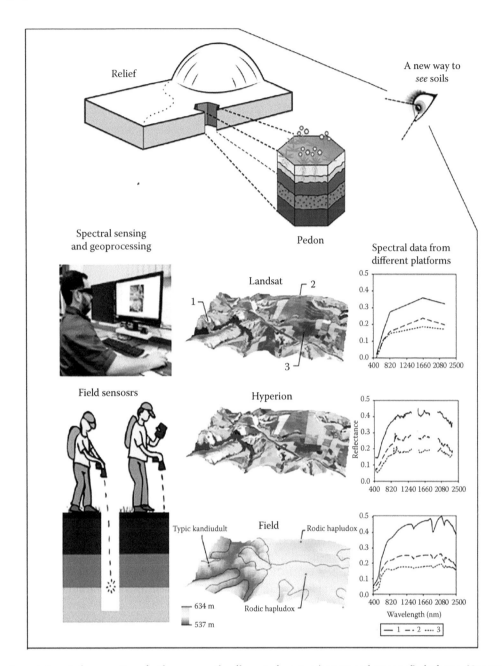

FIGURE 26.23 A new integrated perspective of soil assessment by all spectral sensing (remote and proximal) platforms. (Courtesy: Dr. Alexandre Dematte, lead author of Chapter 24.)

of utmost importance for agriculture, carbon storage, biomass sustainability, and livelihood of plants, animals, and humans. Soil formation is a result of five factors (climate, parent material, time, organic matter, and topography) that leads the world soils to vary widely, from location to location, even within a small area. Soil formations have occurred over thousands of years and are heavily influenced by climate as enumerated by the International Soil Reference and Information Centre (ISRIC). The ISRIC defines the soils of the world into the following broad climate-driven themes: (1) tropical soils, strongly weathered and leached with low nutrient with only lush vegetation to replenish soils; (2) arid soils, low precipitation and high evapotranspiration

leading easily soluble components like calcium carbonate and gypsum left behind after evaporation of water; (3) temperate climate soils, soil formation restricted to the warmer part of the season and hence less deep, but less weathered; (4) subarctic and northern temperate soils, melting of large glaciers from last ice removed most of the soils and hence new soils have formed after the ice retreat and hence are relatively young and immature; and (5) arctic climate soils, soil formation is highly restricted and is permanently frozen (permafrost). So a global or a local study of soils using a wide array of satellite sensors is considered both cost effective and powerful (World Reference Base for Soil Resources; Figure 26.24, FAO-GIS, 1998).

Dominant soils of the world

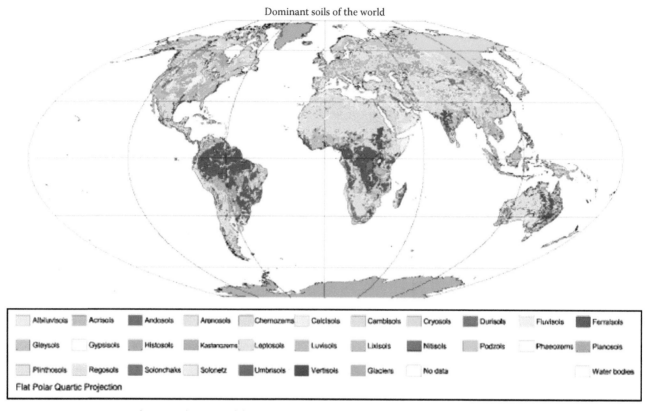

FIGURE 26.24　Dominant reference soils groups of the world. (From FAO 1995, 2003.)

In Chapter 25, Dr. Eyal Ben Dor and Dr. Jose Demattê provide a comprehensive assessment of soil studies using optical remote sensing. They show us how remote sensing is used widely and successfully, to map soil's properties: (1) organic matter, (2) salinity, (3) degradation and change, (4) moisture, (5) carbonates, (6) contamination, (7) aggregation, and roughness, (8) sealing, (9) classification and taxonomy, and (10) pedomapping. It is clear from their synthesis that much of success is achieved in characterizing, and/or quantifying, and/or mapping surface soil moisture, organic matter, texture, and color. Organic matter has spectral activity throughout the entire visible (VIS)–near-infrared (NIR)–SWIR region. Researchers have shown wavelengths such as 425–695, 500–1200, 900–1220, 1926–2032, and 1726–2426 nm as effective in soil organic matter studies. Saline versus nonsaline soils as well as salinity types (e.g., saline, alkaline) and salinity degrees (e.g., low, moderate, high) are successfully delineated using optical remote sensing using data from the VIS–NIR–SWIR spectrum. However, often the uses of multiple bands across 400–2500 nm when used to classify and determine soil properties provide far better results. This may involve, for example, the use of 10 or 20 HNBs used to classify an area and determine soil characteristics like organic matter, salinity, and moisture. However, optical data can only penetrate soils to some degree, resulting in measuring, to most extent, only surface properties. In a summary of all soil applications obtained from remote sensing, they concluded that the optical region is the most widely used. Nonetheless, a soil study conducted using nonoptical remote sensing has great value but accordingly is not part of Chapter 25. For example, thermal data are, often, used in determining salt effects and soil moisture. Microwave (both active and passive) data are widely used to quantify and map surface soil moisture as well as near-surface (<20 cm) soil moisture.

Acknowledgments

I thank the lead authors and coauthors of each of the chapters for providing their insights and edits of my chapter summaries.

References

Alves, M.C., Carvalho, L.G., and Oliveira, M.S. 2013. Terrestrial Earth couple climate–carbon spatial variability and uncertainty. *Global and Planetary Change*, 111, 9–30, ISSN 0921-8181, http://dx.doi.org/10.1016/j.gloplacha.2013.08.009.

Asner, G.P. and Martin, R.E. 2008. Spectral and chemical analysis of tropical forests: Scaling from leaf to canopy levels. *Remote Sensing of Environment*. 112(10): 3958–3970.

Asner, G.P., Martin, R.E., Knapp, D.E., Tupayachi, R., Anderson, C., Carranza, L., Martinez, P., Houcheime, M., Sinca, F., and Weiss, P. 2011. Spectroscopy of canopy chemicals in humid tropical forests. *Remote Sensing of Environment* 115, 3587–3598.

Bates, B.C., Kundzewicz, Z.W., Wu, S., and Palutikof, J.P., Eds. 2008. *Climate Change and Water.* Technical Paper of the Intergovernmental Panel on Climate Change, IPCC Secretariat, Geneva, Switzerland, 210pp.

Bontemps, S., Defourny, P., Van Bogaert, E., Arino, O., Kalogirou, V., and Ramos Perez, J. 2009. Globcover 2009: Products description and validation report URL: http://www.ionia1.esrin.esa.int/docs/GLOBCOVER2009_Validation_Report_2 2 (2011).

Carter, G.A. 1994. Ratios of leaf reflectances in narrow wavebands as indicators of plant stress. *International Journal of Remote Sensing*, 15, 697–703.

Castillejo-González, I.L., López-Granados, F., García-Ferrer, A., Peña-Barragán, J.M., Jurado-Expósito, M., Sánchez de la Orden, M., and González-Audicana, M. 2009. Object- and pixel-based analysis for mapping crops and their agro-environmental associated measures using QuickBird imagery. *Computers and Electronics in Agriculture*, 68(2), 207–215, ISSN 0168-1699, http://dx.doi.org/10.1016/j.compag.2009.06.004.

Curran, P.J. 1989. Remote sensing of foliar chemistry. *Remote Sensing of Environment*, 30(3), 271–278.

Darvishzadeh, R., Skidmore, A., Schlerf, M., and Atzberger, C. 2008. Inversion of a radiative transfer model for estimating vegetation LAI and chlorophyll in a heterogeneous grassland. *Remote Sensing of Environment*, 112(5), 2592–2604, ISSN 0034-4257, http://dx.doi.org/10.1016/j.rse.2007.12.003.

FAO 1995, 2003. *The Digitized Soil Map of the World and Derived Soil Properties*. (version 3.5) FAO Land and Water Digital Media Series 1. FAO, Rome, Italy.

Frankenberg, C., Fisher, J.B., Worden, J., Badgley, G., Saatchi, S.S., Lee, J.-E. et al. 2011. New global observations of the terrestrial carbon cycle from GOSAT: Patterns of plant fluorescence with gross primary productivity. *Geophysical Research Letters*, 38, L17706, doi:10.1029/2011GL048738.

Gallaun, H., Zanchi, G., Nabuurs, G.-J., Hengeveld, G., Schardt, M., and Verkerk, P.J. 2010. EU-wide maps of growing stock and above-ground biomass in forests based on remote sensing and field measurements. *Forest Ecology and Management*, 260(3), 252–261, ISSN 0378-1127, http://dx.doi.org/10.1016/j.foreco.2009.10.011.

Gallaun, H., Zanchi, G., Nabuurs, G.-J., Hengeveld, G., Schardt, M., and Verkerk, P.J. 2010. EU-wide maps of growing stock and above-ground biomass in forests based on remote sensing and field measurements. *Forest Ecology and Management*, 260, 252–261.

Gardner, T.A., Burgess, N.D., Aguilar-Amuchastegui, N., Barlow, J., Berenguer, E., Clements, T., Danielsen, F. et al. 2012. A framework for integrating biodiversity concerns into national REDD+ programmes. *Biological Conservation*, 154, 61–71, ISSN 0006-3207, http://dx.doi.org/10.1016/j.biocon.2011.11.018.

Gitelson, A.A. and Merzlyak, M.N. 1997. Remote estimation of chlorophyll content in higher plant leaves. *International Journal of Remote Sensing*, V18(12), 2691–2697.

Goetz, A.F.H. 2010. Three decades of hyperspectral remote sensing of the Earth: A personal view. *Remote Sensing of Environment*, 113(Suppl. 1), S5–S16, ISSN 0034-4257, http://dx.doi.org/10.1016/j.rse.2007.12.014.

Guanter, L., Frankenberg, C., Dudhia, A., Lewis, P.E., José G.-D., Akihiko, K., Hiroshi, S., and Grainger, R.G. 2012. Retrieval and global assessment of terrestrial chlorophyll fluorescence from GOSAT space measurements. *Remote Sensing of Environment*, 121, 236–251, ISSN 0034-4257, http://dx.doi.org/10.1016/j.rse.2012.02.006.

Gumma, M.K., Thenkabail, P.S., Maunahan, A., Islam, S., and Nelson, A. 2010. Mapping seasonal rice cropland extent and area in the high cropping intensity environment of Bangladesh using MODIS 500 m data for the year 2010. *Journal of Photogrammetry and Remote Sensing*, 91, 98–113, ISSN 0924-2716, http://dx.doi.org/10.1016/j.isprsjprs.2014.02.007.

Gumma, M.K., Thenkabail, P.S., Maunahan, A., Islam, S., and Nelson, A. May 2014. Mapping seasonal rice cropland extent and area in the high cropping intensity environment of Bangladesh using MODIS 500 m data for the year 2010. *ISPRS Journal of Photogrammetry and Remote Sensing*, 91, 98–113, http://dx.doi.org/10.1016/j.isprsjprs.2014.02.007.

Guyot, G. and Baret, F. 1988. Utilisation de la haute resolution spectrale pour suivre l'état des couverts végétaux. *Proceedings of the Fourth International Colloquium on Spectral Signatures of Objects in Remote Sensing*. ESA SP-287, Aussois, France, pp. 279–286.

Hese, S., Lucht, W., Schmullius, C., Barnsley, M., Dubayah, R., Knorr, D., Neumann, K., Riedel, T., and Schröter, K. 2005. Global biomass mapping for an improved understanding of the CO2 balance—The Earth observation mission Carbon-3D. *Remote Sensing of Environment*, 94(1), 94–104, ISSN 0034-4257, http://dx.doi.org/10.1016/j.rse.2004.09.006.

Horler, D.N.H., Dockray, M., and Barber, J. 1983. The red edge of plant leaf reflectance. *International Journal of Remote Sensing*, 4(2), 273–288.

Ivits, E., Cherlet, M., Tóth, G., Sommer, S., Mehl, W., Vogt, J., and Micale, F. 2012. Combining satellite derived phenology with climate data for climate change impact assessment. *Global and Planetary Change*, 88–89, 85–97, ISSN 0921-8181, http://dx.doi.org/10.1016/j.gloplacha.2012.03.010.

Jacquemoud, S., Verhoef, W., Baret, F., Bacour, C., Zarco-Tejada, P.J., Asner, G.P., François, C., and Ustin, S.L. 2009. Prospect + sail: A review of use for vegetation characterization. *Remote Sensing of Environment*, 113, S56–S66.

Kankare, V., Räty, M., Yu, X., Holopainen, M., Vastaranta, M., Kantola, T., Hyyppä, J., Alho, P., and Viitala, R. 2013. Single tree biomass modelling using airborne laser scanning. *Journal of Photogrammetry and Remote Sensing*, 85, 66–73, ISSN 0924-2716, http://dx.doi.org/10.1016/j.isprsjprs.2013.08.008.

Kempen, M., Elbersen, B.S., Staritsky, I., Andersen, E., and Heckelei, T. 2011. Spatial allocation of farming systems and farming indicators in Europe. *Agriculture, Ecosystems & Environment*, 142(1–2), 51–62, ISSN 0167-8809, http://dx.doi.org/10.1016/j.agee.2010.08.001.

Khan, S.I., Hong, Y., Wang, J., Yilmaz, K.K., Gourley, J.J., Adler, R.F., Brakenridge, G.R., Policelli, F., Habib, S., and Irwin, D. 2011. Satellite remote sensing and hydrologic modeling for flood inundation mapping in Lake Victoria Basin: Implications for hydrologic prediction in ungauged basins. *IEEE Transactions on Geoscience and Remote Sensing*, 49(1), 85–95, http://sadiqibrahimkhan.files.wordpress.com/2013/04/sadiq-2011_ieee.pdf.

Kruska, R.L., Reid, R.S., Thornton, P.K., Henninger, N., and Kristjanson, P.M. 2003. Mapping livestock-oriented agricultural production systems for the developing world. *Agricultural Systems*, 77(1), 39–63, ISSN 0308-521X, http://dx.doi.org/10.1016/S0308-521X(02)00085-9.

Laba, M., Blair, B., Downs, R., Monger, B., Philpot, W., Smith, S., Sullivan, P., and Baveye, P.C. 2010. Use of textural measurements to map invasive wetland plants in the Hudson River National Estuarine Research Reserve with IKONOS satellite imagery. *Remote Sensing of Environment*, 114(4), 876–886, ISSN 0034-4257, http://dx.doi.org/10.1016/j.rse.2009.12.002.

Lasco, R.D. 2004. Forest carbon budgets in Southeast Asia following harvesting and land cover change. *Science in China. Series 3*, 45(Suppl.), 55–64.

Le Toan, T., Quegan, S., Davidson, M.W.J., Balzter, H., Paillou, P., Papathanassiou, K., Plummer, S. et al. 2011. The BIOMASS mission: mapping global forest biomass to better understand the terrestrial carbon cycle. *Remote Sensing of Environment*, 115, 2850–2860.

Lefsky, M.A. 2010. A global forest canopy height map from the moderate resolution spectroradiometer and the geoscience laser altimeter system. *Geophysical Research Letters*, 37(15), 1–5. doi:10.1029/2010GL043622.

Lewis, S.L., Gabriela, L.-G., Bonaventure, S., Kofi, A.-B., Baker, T.R., Ojo, L.O., Phillips, O.L. et al. 2009. *Nature*, 457, 1003–1006. doi:10.1038/nature07771.

Marshall, M.T. and Thenkabail, P.S. 2014. Biomass modeling of four leading World crops using hyperspectral narrowbands in support of HyspIRI mission. *Photogrammetric Engineering and Remote Sensing*, 80(4), 757–772.

Meng, Q., Cieszewski, C., and Madden, M. 2009. Large area forest inventory using Landsat ETM+: A geostatistical approach. *Journal of Photogrammetry and Remote Sensing*, 64 (1), 27–36.

Mermoz, S., Toan, T.L., Villard, L., Réjou-Méchain, M., and Seifert-Granzin, J. 2014. Biomass assessment in the Cameroon savanna using ALOS PALSAR data. *Remote Sensing of Environment*, Available online 15 May 2014, ISSN 0034-4257, http://dx.doi.org/10.1016/j.rse.2014.01.029.

Monteith, J.L. 1972. Solar radiation and productivity in tropical ecosystems. *Journal of Applied Ecology*, 9(3), 747.

Moran, M.S., Inoue, Y., and Barnes, E.M. 1997. Opportunities and limitations for image-based remote sensing in precision crop management. *Remote Sensing of Environment*, 61, 319–346.

Morel, A.C., Saatchi, S.S., Malhi, Y., Berry, N.J., Banin, L., Burslem, D., Nilus, R., Ong, R.C. 2011. Estimating aboveground biomass in forest and oil palm plantation in Sabah, Malaysian Borneo using ALOS PALSAR data. *Forest Ecology and Management*, 262(9), 1786–1798, ISSN 0378-1127, http://dx.doi.org/10.1016/j.foreco.2011.07.008.

Nolte, C., Kotto-Same, J., Moukam, A., Thenkabail, P.S., Weise, S.F., Woomer, P.L., and Zapfack, L. 2001. *Land-Use Characterization and Estimation of Carbon Stocks in the Alternatives to Slash-and-Burn Benchmark Area in Cameroon*. Resource and Crop Management Division (RCMD) Monograph No. 28, RCMD, IITA, Ibadan, Nigeria. 27pp.

Pacheco, A. and McNairn, H. 2010. Evaluating multispectral remote sensing and spectral unmixing analysis for crop residue mapping. *Remote Sensing of Environment*, 114(10), 2219–2228, ISSN 0034-4257, http://dx.doi.org/10.1016/j.rse.2010.04.024.

Parikh, S.J. and James, B.R. 2012. Soil: The foundation of agriculture. *Nature Education Knowledge*, 3(10), 2.

Pickett-Heaps, C.A., Canadell, J.G., Briggs, P.R., Gobron, N., Haverd, V., Paget, M.J., Pinty, B., Raupach, M.R. 2014. Evaluation of six satellite-derived Fraction of Absorbed Photosynthetic Active Radiation (FAPAR) products across the Australian continent. *Remote Sensing of Environment*, 140, 241–256, ISSN 0034-4257, http://dx.doi.org/10.1016/j.rse.2013.08.037.

Robertson, M., Isbister, B., Maling, I., Oliver, Y., Wong, M., Adams, M., Bowden, B., Tozer, P. 2007. Opportunities and constraints for managing within-field spatial variability in Western Australian grain production. *Field Crops Research*, 104(1–3), 60–67, ISSN 0378-4290, http://dx.doi.org/10.1016/j.fcr.2006.12.013.

Robinson, D.T., Brown, D.G., and Currie, W.S. 2009. Modelling carbon storage in highly fragmented and human-dominated landscapes: Linking land-cover patterns and ecosystem models. *Ecological Modelling*, 220(9–10), 1325–1338.Running, S.W., Heinsch, F.A., Zhao, M., Reeves, M., Hashimoto, H., and Nemani, R.R. 2004. A continuous satellite-derived measure of global terrestrial primary production. *BioScience*, 54(6), 547–560.

Running, S.W., Thornton, P.E., Nemani, R.R., and Glassy, J.M. 2000. Global terrestrial gross and net primary productivity from the earth observing system. In O. Sala, R. Jackson, and H. Mooney (Eds.), *Methods in Ecosystem Science*, (pp. 44–57). New York: Springer-Verlag.

Schlerf, M., Atzberger, C., and Hill, J. 2005. Remote sensing of forest biophysical variables using HyMap imaging spectrometer data. *Remote Sensing of Environment*, 95(2), 177–194.

Sun, G., Ranson, K.J., Guo, Z., Zhang, Z., Montesano, P., and Kimes, D. 2011. Forest biomass mapping from lidar and radar synergies. *Remote Sensing of Environment*, 115(11), 2906–2916, ISSN 0034-4257, http://dx.doi.org/10.1016/j.rse.2011.03.021.

Suomalainen, J., Hakala, T., Kaartinen, H., Räikkönen, E., and Kaasalainen, S. 2010. Demonstration of a virtual active hyperspectral LiDAR in automated point cloud classification. *Journal of Photogrammetry and Remote Sensing*, 66(5), 637–641, ISSN 0924-2716, http://dx.doi.org/10.1016/j.isprsjprs.2011.04.002.

Suomalainen, J., Hakala, T., Kaartinen, H., Räikkönen, E., and Kaasalainen, S. 2011. Demonstration of an active hyperspectral LiDAR in automated point cloud classification. *ISPRS Journal of Photogrammetry and Remote Sensing*, 66(5), 637–641; doi:10.1016/j.isprsjprs.2011.04.002. http://www.sciencedirect.com/science/article/B6VF4-52T0C29-1/2/1275bc180f8685cfced6c705099c4469.

Thenkabail, P.S., Enclona, E.A., Ashton, M.S., Legg, C., and De Dieu, M.J. 2004b. Hyperion, IKONOS, ALI, and ETM+ sensors in the study of African rainforests. *Remote Sensing of Environment*, 90(1), 23–43.

Thenkabail, P.S., Enclona, E.A., Ashton, M.S., and Van Der Meer, V. 2004c. Accuracy assessments of hyperspectral waveband performance for vegetation analysis applications. *Remote Sensing of Environment*, 91(2–3): 354–376.

Thenkabail, P.S., Gumma, M.K., Teluguntla, P., and Mohammed, I.A., 2014. Hyperspectral Remote sensing of vegetation and agricultural crops. Highlight article. *Photogrammetric Engineering and Remote Sensing*, 80(4), 697–709.

Thenkabail, P.S., Lyon, G.J., and Huete, A. 2012. *Hyperspectral Remote Sensing of Vegetation.* CRC Press- Taylor and Francis group, Boca Raton, London, New York, p. 781 (80+ pages in color). Reviews of this book: http://www.crcpress.com/product/isbn/9781439845370.

Thenkabail, P.S., Mariotto, I., Gumma, M.K., Middleton, E.M., Landis, D.R., and Huemmrich, F.K., 2013. Selection of hyperspectral narrowbands (HNBs) and composition of hyperspectral twoband vegetation indices (HVIs) for biophysical characterization and discrimination of crop types using field reflectance and Hyperion/EO-1 data. *IEEE Journal of Selected Topics in Applied Earth Observations and Remote Sensing*, 6(2), 427–439, doi:10.1109/JSTARS.2013.2252601.

Thenkabail, P.S., Schull, M., and Turral, H. 2005. Ganges and Indus river basin land use/land cover (LULC) and irrigated area mapping using continuous streams of MODIS data. *Remote Sensing of Environment*, 95(3), 317–341, ISSN 0034-4257, http://dx.doi.org/10.1016/j.rse.2004.12.018.

Thenkabail, P.S., Smith, R.B., and De-Pauw, E. 2000. Hyperspectral vegetation indices for determining agricultural crop characteristics. *Remote Sensing of Environment*, 71, 158–182.

Thenkabail, P.S., Stucky, N., Griscom, B.W., Ashton, M.S., Diels, J., Van Der Meer, B., and Enclona, E. 2004a. Biomass estimations and carbon Stock calculations in the oil palm plantations of African derived savannas using IKONOS data. *International Journal of Remote Sensing*, 25(23), 5447–5472.

Tian, Y.C., Yao, X., Yang, J., Cao, W.X., Hannaway, D.B., and Zhu, Y. 2011. Assessing newly developed and published vegetation indices for estimating rice leaf nitrogen concentration with ground- and space-based hyperspectral reflectance. *Field Crops Research*, 120(2), 299–310, ISSN 0378-4290, http://dx.doi.org/10.1016/j.fcr.2010.11.002.

Tucker, C.J. 1979. Red and photographic infrared linear combinations for monitoring vegetation. *Remote Sensing of Environment*, 8, 127–150.

Usery, E.L., Pocknee, S., and Boydell, B. 1995. Precision farming data management using geographic information systems. *Photogrammetric Engineering and Remote Sensing*, 61(11), 1383–1391.

Wessels, K.J., Mathieu, R., Erasmus, B.F.N., Asner, G.P., Smit, I.P.J., van Aardt, J.A.N., Main, R. et al. 2011. Impact of communal land use and conservation on woody vegetation structure in the Lowveld savannas of South Africa. *Forest Ecology and Management*, 261(1), 19–29, ISSN 0378-1127, http://dx.doi.org/10.1016/j.foreco.2010.09.012.

Wijaya, A., Liesenberg, V., and Gloaguen, R. 2010. Retrieval of forest attributes in complex successional forests of Central Indonesia: Modeling and estimation of bitemporal data. *Forest Ecology and Management*, 259(12), 2315–2326, ISSN 0378-1127, http://dx.doi.org/10.1016/j.foreco.2010.03.004.

Zhang, C. 2014. Applying data fusion techniques for benthic habitat mapping and monitoring in a coral reef ecosystem. *Journal of Photogrammetry and Remote Sensing*, Available online 27 June 2014, ISSN 0924-2716, http://dx.doi.org/10.1016/j.isprsjprs.2014.06.005.

Index

T - #0536 - 071024 - C888 - 279/216/39 - PB - 9780367868970 - Gloss Lamination